Encyclopedia of Nonlinear Science

Alwyn Scott
Editor

LONDON AND NEW YORK

Published in 2005 by
Routledge
Taylor & Francis Group
711 Third Avenue, New
York, NY 10017

Published in Great Britain by
Routledge
Taylor & Francis Group
2 Park Square,
Milton Park, Abingdon,
Oxfordshire OX14 4RN

Routledge is an imprint of the Taylor & Francis Group, an informa business

Library of Congress Cataloging-in-Publication Data

Encyclopedia of nonlinear science/Alwyn Scott, Editor
p. cm.
Includes bibliographical references and index.
ISBN 978-1-57958-385-9 (hbk)
ISBN 978-1-138-01214-1 (pbk)
1. Nonlinear theories-Encyclopedias. 1. Scott, Alwyn, 1931–
QA427, E53 2005
003:75—dc22 2004011708

Contents

Introduction

Among the several advances of the 20th century, nonlinear science is exceptional for its generality. Although the invention of radio was important for communications, the discovery of DNA structure for biology, the development of quantum theory for theoretical physics and chemistry, and the invention of the transistor for computer engineering, nonlinear science is significant in all these areas and many more. Indeed, it plays a key role in almost every branch of modern research, as this *Encyclopedia of Nonlinear Science* shows.

In simple terms, nonlinear science recognizes that the ``whole is more than a sum of its parts,'' providing a context for consideration of phenomena like tsunamis (tidal waves), biological evolution, atmospheric dynamics, and the electrochemical activity of a human brain, among many others. For a research scientist, nonlinear science offers novel phenomena, including the emergence of coherent structures (an optical soliton, e.g., or a nerve impulse) and chaos (characterized by the difficulties in making accurate predictions for surprisingly simple systems over extended periods of time). Both these phenomena can be studied using mathematical methods described in this *Encyclopedia*. From a more fundamental perspective, a wide spectrum of applications arises because nonlinear science introduces a paradigm shift in our collective attitude about causality. What is the nature of this shift?

Consider the difference between linear and nonlinear analyses. Linear analyses are characterized by the assumption that individual effects can be unambiguously traced back to particular causes. In other words, a compound cause is viewed as the linear (or algebraic) sum of a collection of simple causes, each of which can be uniquely linked to a particular effect. The total effect responding to the total cause is then considered to be just the linear sum of the constituent effects.

A fundamental tenet of nonlinear science is to reject this convenient, but often unwarranted, assumption. Of course, the notion that components of complex causes can interact among themselves is not surprising to any thoughtful person who manages to get through an ordinary day of normal life, and it is not at all new. Twenty-four centuries ago, Aristotle described four types of cause (material, efficient, formal, and final), which overlap and intermingle in ways that were often overlooked in 20th-century thought but are now under scrutiny. Consider some examples of linear scientific thinking that are presently being reevaluated in the context of nonlinear science.

— Around the middle of the 20th century, behavioral psychologists adopted the theoretical position that human mental activity can be reduced to a sum of individual responses to specific stimuli that have been learned at earlier stages of development. Current research in neuroscience shows this perspective to be unwarranted.

— Some evolutionary psychologists believe that particular genes, located in the structure of DNA, can always be related in a one-to-one manner to individual features of an adult organism, leading to hunts for a ``crime gene'' that seem abhorrent to moralists. Nonlinear science suggests that the relation between genes and features of an adult organism is more intricate than the linear perspective assumes.

— The sad disintegration of space shuttle Columbia on the morning of February 1, 2003, set off a search for ``the cause of the accident,'' ignoring Aristotelian insights into the difficulties of defining such a concept, never mind sorting out the pieces. Did the mishap occur because the heat-resistant tiles were timeworn (a material cause)? Or because 1.67 pounds of debris hit the left wing at 775 ft/s during takeoff (an efficient cause)? Perhaps a management culture that discounted the importance of safety measures (a formal cause) should shoulder some of the blame.

— Cultural phenomena, in turn, are often viewed as the mere sum of individual psychologies, ignoring the grim realities of war hysteria and lynch mobs, not to mention the "tulip craze" of 17th-century Holland, the more recent "dot-com bubble," and the outbreak of communal mourning over the death of Princess Diana.

Evolution of the Science

As the practice of nonlinear science involves such abstruse issues, one might expect its history to be checkered, and indeed it is. Mathematical physics began with the 17th-century work of Isaac Newton, whose formulation of the laws of mechanical motion and gravitation explained how the Earth moves about the Sun, replacing a final cause (God's plan) with an efficient cause (the force of gravity). Because it assumed that the net gravitational force acting on any celestial body is the linear (vector) sum of individual forces, Newton's theory provides support for the linear perspective in science, as has often been emphasized. Nonetheless, the mathematical system Newton developed (calculus) is the natural language for nonlinear science, and he used this language to solve the two-body problem (collective motion of Earth and Moon)—the first nonlinear system to be mathematically studied. Also in the 17th century, Christiaan Huygens noted that two pendulum clocks (which he had recently invented) kept exactly the same time when hanging from a common support. (Confined to his room by an indisposition, Huygens observed the clocks over a period of several days, during which the swinging pendula remained in step.) If the clocks were separated to opposite sides of the room, one lost several seconds a day with respect to the other. From small vibrations transmitted through the common support, he concluded, the two clocks became synchronized—a typical nonlinear phenomenon.

In the 18th century, Leonhard Euler used Newton's laws of motion to derive nonlinear field equations for fluid flow, which were augmented a century later by Louis Navier and George Stokes to include the dissipative effects of viscosity that are present in real fluids. In their generality, these equations defied solution until the middle of the 20th century when, together with the digital computer, elaborations of the Navier–Stokes equations provided a basis for general models of the Earth's atmosphere and oceans, with implications for the vexing question of global warming. During the latter half of the 19th century, however, special analytic solutions were obtained by Joseph Boussinesq and related to experimental observations of hydrodynamic solitary waves by John Scott Russell. These studies—which involved a decade of careful observations of uniformly propagating "heaps of water" on canals and in wave tanks—were among the earliest research programs in the area now recognized as nonlinear science. At about the same time, Pierre François Verhulst formulated and solved a nonlinear differential equation—sometimes called the logistic equation—to model the population growth of his native Belgium.

Toward the end of the 19th century, Henri Poincaré returned to Newton's original theme, presenting a solution of the three-body problem of celestial motion (e.g., a planet with two moons) in a mathematical competition sponsored by the King of Sweden. Interestingly, a serious error in this work was discovered prior to its publication, and he (Poincaré, not the Swedish king) eventually concluded that the three-body problem cannot be exactly solved. Now regarded by many as the birth of the "science of complexity," this negative result had implications that were not widely appreciated until the 1960s, when numerical studies of simplified atmospheric models by Edward Lorenz showed that nonlinear systems with as few as three degrees of freedom can readily exhibit the nonlinear phenomenon of chaos. (A key observation here was of an unanticipated sensitivity to initial conditions, popularly known as the "butterfly effect" from Lorenz's speculation that "the flap of a butterfly's wings in Brazil [might] set off a tornado in Texas.")

During the first half of the 20th century, the tempo of research picked up. Although still carried on as unrelated activities, there appeared a notable number of experimental and theoretical studies now recognized as precursors of modern nonlinear science. Among others, these include Albert Einstein's nonlinear theory of gravitation; nonlinear field theories of elementary particles (like the recently discovered electron) developed by Gustav Mie and Max Born; experimental observations of local modes in molecules by physical chemists (for which a nonlinear theory was developed by Reinhard Mecke in the 1930s, forgotten, and then redeveloped in the 1970s); biological models of predator-prey population dynamics formulated by Vito Volterra (to describe year-to-year variations in fish catches from the Adriatic Sea); observations of a profusion of localized nonlinear entities in solid-state physics (including ferromagnetic domain walls, crystal dislocations, polarons, and magnetic flux vortices in superconductors, among others); a definitive experimental and theoretical study of nerve impulse propagation on the giant axon of the squid by Alan Hodgkin and Andrew Huxley; Alan Turing's theory of pattern formation in the development of biological organisms; and Boris Belousov's observations of pattern formation in a chemical solution, which were at first ignored (under the mistaken assumption that they violated the second law of thermodynamics) and later confirmed and extended by Anatol

Zhabotinsky and Art Winfree. Just as the invention of the laser in the early 1960s led to numerous experimental and theoretical studies in the new field of nonlinear optics, the steady increases in computing power throughout the second half of the 20th century enabled ever more detailed numerical studies of hydrodynamic turbulence and chaos, whittling away at the long-established Navier–Stokes equations and confirming the importance of Poincaré's negative result on the three-body problem.

Thus, it was evident by 1970 that nonlinearity manifests itself in several remarkable properties of dynamical systems, including the following. (There are others, some no doubt waiting to be discovered.)

— Many nonlinear partial differential equations (wave equations, diffusion equations, and more complicated field equations) are often observed to exhibit localized or lump-like solutions, similar to Russell's hydrodynamic solitary wave. These ``coherent structures'' of energy or activity emerge from initial conditions as distinct dynamic entities, each having its own trajectory in space-time and characteristic ways of interacting with others. Thus, they are ``things'' in the normal sense of the word. Interestingly, it is sometimes possible to compute the velocity of emergent entities (their speeds and shapes) from initial conditions and express them as tabulated functions (theta functions or elliptic functions), thereby extending the analytic reach of nonlinear analysis. Examples of emergent entities include tornadoes, nerve impulses, magnetic domain walls, tsunamis, optical solitons, Jupiter's Great Red Spot, black holes, schools of fish, and cities, to name but a few. A related phenomenon, exemplified by meandering rivers, bolts of lightning, and woodland paths, is called filamentation, which also causes spotty output beams in poorly designed lasers.

— Surprisingly simple nonlinear systems (Poincaré's three-body problem is the classic example) are found to have chaotic solutions, which remain within a bounded region, while the difference between neighboring solution trajectories grows exponentially with time. Thus, the course of a solution trajectory is strongly sensitive to its initial conditions (the ``butterfly effect''). Chaotic solutions arise in both energy-conserving (Hamiltonian) systems and dissipative systems, and they are fated to wander unpredictably as trajectories that cannot be accurately extended into the future for unlimited periods of time. As Lorenz pointed out, the chaotic behavior the Earth's atmosphere makes detailed meteorological predictions problematic, to the delight of the mathematician and the despair of the weatherman. Chaotic systems also exhibit ``strange attractors'' in the solution space, which are characterized by fractal (non-integer) dimensions.

— Nonlinear problems often display threshold phenomena, meaning that there is a relatively sharp boundary across which the qualitative nature of a solution changes abruptly. This is the basic property of an electric wall switch, the trigger of a pistol, and the flip-flop circuit that a computer engineer uses to store a bit of information. (Indeed, a computer can be viewed as a large, interconnected collection of threshold devices.) Sometimes called ``tipping points'' in the context of social phenomena, thresholds are an important part of our daily experience, where they complicate the relationship of causality to legal responsibility. Was it the last straw that broke the camel's back? Or did all of the straws contribute to some degree? Should each be blamed according to its weight? How does one assign culpability for the *Murder on the Orient Express*?

— Nonlinear systems with several spatial coordinates often exhibit spontaneous pattern formation, examples of which include fairy rings of mushrooms, oscillatory patterns of heart muscle activity under fibrillation (leading to sudden cardiac arrest), weather fronts, the growth of form in a biological embryo, and the Gulf Stream. Such patterns can be chaotic in time and regular in space, regular in time and chaotic in space, or chaotic in both space and in time, which in turn is a feature of hydrodyamic turbulence.

— If the input to (or stimulation of) a nonlinear system is a single frequency sinusoid, the output (or response) is nonsinusoidal, comprising a spectrum of sinusoidal frequencies. For lossless nonlinear systems, this can be an efficient means for producing energy at integer multiples of the driving frequency, through the process of harmonic generation. In electronics, this process is widely used for digital tuning of radio receivers. Taking advantage of the nonlinear properties of certain transparent crystals, harmonic generation is also employed in laser optics to create light beams of higher frequency, for example, conversion of red light to blue.

— Another nonlinear phenomenon is the synchronization of weakly coupled oscillators, first observed by the ailing Huygens in the winter of 1665. Now recognized in a variety of contexts, this effect crops up in the frequency locking of electric power generators tied to the same grid and the coupling of biological rhythms (circadian rhythms in humans, hibernation of bears, and the synchronized flashing of Indonesian fireflies), in addition to many applications in electronics. Some suggest that neuronal firings in the neocortex may be mutually synchronized.

— Shock waves are familiar to most of us as the boom of a jet airplane that has broken the sound barrier or the report of a cannon. Closely related from a mathematical perspective are the bow wave of a speedboat, the breaking of onshore surf, and the sudden automobile pileups that can occur on a highway that is carrying traffic close to its maximum capacity.

— More complicated nonlinear systems can be hierarchical in nature. This comes about when the emergence of coherent states at one level provides a basis for new nonlinear dynamics at a higher level of description. Thus, in the course of biological evolution, chemical molecules emerged from interactions among the atomic elements, and biological molecules then emerged from simpler molecules to provide a basis for the dynamics of a living cell. From collections of cells, multicellular organisms emerged, and so on up the evolutionary ladder to creatures like ourselves, who comprise several distinct levels of biological dynamics. Similar structures are observed in the organization of coinage and of military units, not to mention the hierarchical arrangement of information in the human brain.

Often, qualitatively related behaviors—involving one or more of such nonlinear manifestations—are found in models that arise from different areas of application, suggesting the need for interdisciplinary communications. By the early 1970s, therefore, research in nonlinear science was in a state that the physical chemists might describe as "supersaturated." Dozens of people across the globe were working on one facet or another of nonlinear science, often unaware of related studies in traditionally unrelated fields. During the mid-1970s, this activity experienced a "phase change," which can be viewed as a collective nonlinear effect in the sociology of science. Unexpectedly, a number of conferences devoted entirely to nonlinear science were organized, with participants from a variety of professional backgrounds, nationalities, and research interests eagerly contributing. Solid-state physicists began to talk seriously with biologists, neuroscientists with chemical engineers, and meteorologists with psychologists. As interdisciplinary barriers crumbled, these unanticipated interactions led to the founding of centers for nonlinear science and the launching of several important research journals amid an explosion of research activity. By the early 1980s, nonlinear science had gained recognition as a key component of modern inquiry, playing a central role in a wide spectrum of activities. In the terminology introduced by Thomas Kuhn, a new paradigm had been established.

About this Book

The primary aim of this *Encyclopedia* is to provide a source from which undergraduate and graduate students in the physical and biological sciences can study how concepts of nonlinear science are presently understood and applied. In addition, it is anticipated that teachers of science and research scientists who are unfamiliar with nonlinear concepts will use the work to expand their intellectual horizons and improve their lectures. Finally, it is hoped that this book will help members of the literate public—philosophers, social scientists, and physicians, for example—to appreciate the wealth of natural phenomena described by a science that does not discount the notion of complex causality.

An early step in writing the *Encyclopedia* was to choose the entry subjects—a difficult task that was accomplished through the efforts of a distinguished Board of Advisers (see page xiii), with members from Australia, Germany, Italy, Japan, Russia, the United Kingdom, and the United States. After much sifting and winnowing, an initial list of about a thousand suggestions was reduced to the 438 items given on pages 1–1010. Depending on the subject matter, the entries are of several types. Some are historical or descriptive, while others present concepts and ideas that require notations from physics, engineering, or mathematics. Although most of the entries were planned to be about a thousand words in length, some—covering subjects of greater generality or importance—are two or four times as long.

Of the many enjoyable aspects in editing this *Encyclopedia,* the most rewarding has been working with those who wrote it—the contributors. The willing way in which these busy people responded to entry invitations and their enthusiastic preparation of assignments underscores the degree to which nonlinear science has become a community with a healthy sense of professional responsibility. In every case, the contributors have tried to present their ideas as simply as possible, with a minimum of technical jargon. For a list of the contributors and their affiliations, see pages xv–xxxi from which it is evident that they come from about 30 different countries, emphasizing the international character of nonlinear science.

A proper presentation of the diverse professional perspectives that make up nonlinear science requires careful organization of the *Encyclopedia,* which we attempt to provide. Although each entry is self-contained, the links among them can be explored in several ways. First, the **Thematic List** on pages xxxix–xliii groups entries within several categories, providing a useful summary of related entries through which the reader can surf. Second, the entries have "**See also**" notes, both within the text and at the end of the entry, encouraging the reader to browse outwards from a starting node. Finally, the **Index** contains a detailed list of topics that do not have their own entries but are discussed within the context of broader entries. If you cannot find an entry on a topic you expected to find, use the **Thematic List or Index** to locate the title of the entry that contains the item you seek. Additionally, all entries have selected bibliographies or suggestions for further reading, leading to original research and textbooks that augment the overview approach to which an encyclopedia is necessarily limited. Although much of nonlinear science evolved from applied mathematics, many of the entries contain no equations or mathematical symbols and can be absorbed by the general reader. Some entries are necessarily technical, but efforts have been made to explain all terms in simple English. Also, many entries have either line diagrams expanding on explanations given in the text or photographs illustrating typical examples. Typographical errors will be posted on the encyclopedia web site at http://www.routledge-ny.com/ref/nonlinearsci/.

The editing of this *Encyclopedia of Nonlinear Science* culminates a lifetime of study in the area, leaving me indebted to many. First is the Acquisitions Editor, Gillian Lindsey, who conceived of the project, organized it, and carried it from its beginnings in London across the ocean to publication in New York. Without her dedication, quite simply, the *Encyclopedia* would not exist. Equally important to reaching the finished work were the efforts of the advisers, contributors, and referees, who, respectively, planned, wrote, and vetted the work, and to whom I am deeply grateful. On a broader time-span are colleagues and students from the University of Wisconsin, Los Alamos National Laboratory, the University of Arizona, and the Technical University of Denmark, with whom I have interacted over four decades. Although far too many to list, these collaborations are fondly remembered, and they provide the basis for much of my editorial judgment. Finally, I express my gratitude for the generous financial support of research in nonlinear science that has been provided to me since the early 1960s by the National Science Foundation (USA), the National Institutes of Health (USA), the Consiglio Nazionale delle Ricerche (Italy), the European Molecular Biology Organization, the Department of Energy (USA), the Technical Research Council (Denmark), the Natural Science Research Council (Denmark), the Thomas B. Thriges Foundation (Denmark), and the Fetzer Foundation (USA).

Alwyn Scott
Tucson, Arizona 2004

Editorial Advisory Board

List of Contributors

Ablowitz, Mark J.
Professor, Department of Applied Mathematics, University of Colorado, Boulder, USA
Ablowitz–Kaup–Newell–Segur system

Aigner, Andreas A.
Research Associate, Department of Mathematical Sciences, University of Exeter, UK
Atmospheric and ocean sciences
General Circulation models of the atmosphere
Navier–Stokes equation
Partial differential equations, nonlinear

Albano, Ezequiel V.
Instituto de Investigaciones Fisicoquímicas Teóricas y Aplicadas (INIFTA) University of La Plata, Argentina
Forest fires

Aratyn, Henrik
Professor, Physics Department, University of Illinois at Chicago, USA
Dressing method

Aref, Hassan
Dean of Engineering and Reynolds Metals Professor Virginia Polytechnic Institute & State University, USA
Bernoulli's equation
Chaos vs. turbulence
Chaotic advection
Cluster coagulation
Hele-Shaw cell
Newton's laws of motion

Arrowsmith, David
Professor, School of Mathematical Sciences, Queen Mary University of London, UK
Symbolic dynamics
Topology

Athorne, Christopher
Senior Lecturer, Department of Mathematics, University of Glasgow, UK
Darboux transformation

Bahr, David
Assistant Professor, Department of Computer Science, Regis University, Colorado, USA
Glacial flow

Ball, Rowena
Department of Theoretical Physics, Australian National University, Australia
Fairy rings of mushrooms
Kolmogorov cascade
Singularity theory

Barnes, Howard
Unilever Research Professor of Industrial Rheology, Department of Mathematics, University of Wales Aberystwyth, Wales
Rheology

Barthes, Mariette
Groupe de Dynamique des Phases Condensées UMR CNRS 5581, Université Montpellier 2, France
Rayleigh and Raman scattering and IR absorption

Beck, Christian
Professor, School of Mathematical Sciences, Queen Mary University of London, UK
Free energy
Multifractal analysis
String theory

Beeckman, Jeroen
Department of Electronics and Information Systems Ghent University, Belgium
Liquid crystals

Benedict, Keith
Senior Lecturer, School of Physics and Astronomy, University of Nottingham, UK
Anderson localization
Frustration

Bergé, Luc
Commissariat à l'Energie Atomique, Bruyères-le-Châtel, France
Development of singularities
Filamentation
Kerr effect

Berland, Nicole
Chimie Général et Organique Lycée Faidherbe de Lille, France
Belousov–Zhabotinsky reaction

Bernevig, Bogdan A.
Physics Department, Massachusetts Institute of Technology, USA
Holons

Biktashev, Vadim N.
Lecturer in Applied Maths, Mathematical Sciences, University of Liverpool, UK
Vortex dynamics in excitable media

Binczak, Stephane
Laboratoire d'Electronique, Informatique et Image, Université de Bourgogné, France
Ephaptic coupling
Myelinated nerves

Biondini, Gino
Assistant Professor, Department of Mathematics, Ohio State University, USA
Einstein equations
Harmonic generation

Blair, David
Professor, School of Physics, The University of Western Australia, Australia
Gravitational waves

Boardman, Alan D.
Professor of Applied Physics, Institute for Materials Research, University of Salford, UK
Polaritons

Bollt, Erik M.
Associate Professor, Departments of Mathematics & Computer Science and Physics, Clarkson University, Potsdam, N.Y., USA
Markov partitions
Order from chaos

Boon, J.-P.
Professor, Faculté des Sciences, Université Libre de Bruxelles, Belgium
Lattice gas methods

Borckmans, Pierre
Center for Nonlinear Phenomena & Complex Systems, Université Libre de Bruxelles, Belgium
Turing patterns

Boumenir, Amin
Department of Mathematics, State University of West Georgia, USA
Gel'fand–Levitan theory

Bountis, Tassos
Professor, Department of Mathematics and Center for Research and Application of Nonlinear Systems, University of Patras, Greece
Painlevé analysis

Boyd, Robert W.
Professor, The Institute of Optics, University of Rochester, USA
Frequency doubling

Bradley, Elizabeth
Associate Professor, Department of Computer Science, University of Colorado, USA
Kirchhoff's laws

Bullough, Robin
Professor, Mathematical Physics, University of Manchester Institute of Science and Technology, UK
Maxwell–Bloch equations
Sine-Gordon equation

Bunimovich, Leonid
Regents Professor, Department of Mathematics, Georgia Institute of Technology, USA
Billiards
Deterministic walks in random environments
Lorentz gas

Busse, Friedrich (Adviser)
Professor, Theoretical Physics, University of Bayreuth, Germany
Dynamos, homogeneous
Fluid dynamics
Magnetohydrodynamics

Calini, Annalisa M.
Associate Professor, Department of Mathematics, College of Charleston, USA
Elliptic functions
Mel'nikov method

Caputo, Jean Guy
Laboratoire de Mathématiques, Institut National des Sciences Appliquées de Rouen, France
Jump phenomena

Censor, Dan
Professor, Department of Electrical and Computer Engineering, Ben-Gurion University of the Negev, Israel
Volterra series and operators

Chen, Wei-Yin
Professor, Department of Chemical Engineering, University of Mississippi, USA
Stochastic processes

Chernitskii, Alexander A.
Department of Physical Electronics, St. Petersburg Electrotechnical University, Russia
Born–Infeld equations

Chiffaudel, Arnaud
CEA-Saclay (Commissariat à l'Énergie Atomique) & CNRS (Centre National de la Recherche Scientifique), France
Hydrothermal waves

Choudhury, S. Roy
Professor, Department of Mathematics, University of Central Florida, USA
Kelvin–Helmholtz instability
Lorenz equations

Christiansen, Peter L.
Professor, Informatics and Mathematical Modelling and Department of Physics, Technical University of Denmark, Denmark
Separation of variables

Christodoulides, Demetrios
Professor, CREOL/School of Optics, University of Central Florida, USA
Incoherent solitons

Coskun, Tamer
Assistant Professor, Department of Electrical Engineering, Pamukkale University, Turkey
Incoherent solitons

Cruzeiro, Leonor
CCMAR and FCT, University of Algarve, Campus de Gambelas, Faro, Portugal
Davydov soliton

Cushing, J.M.
Professor, Department of Mathematics, University of Arizona, USA
Population dynamics

Dafilis, Mathew
School of Biophysical Sciences and Electrical Engineering, Swinbume University of Technology, Australia
Electroencephalogram at mesoscopic scales

Davies, Brian
Department of Mathematics, Australian National University, Australia
Integral transforms
Period doubling

Davis, William C.
Formerly, Los Alamos National Laboratory USA
Explosions

deBruyn, John
Professor, Department of Physics and Physical Oceanography, Memorial University of Newfoundland, Canada
Phase transitions
Thermal convection

Deconinck, Bernard
Assistant Professor, Department of Applied Mathematics University of Washington, USA
Kadomtsev–Petviashvili equation
Periodic spectral theory
Poisson brackets

Degallaix, Jerome
School of Physics, The University of Western Australia, Australia
Gravitational waves

Deift, Percy
Professor, Department of Mathematics, Courant Institute of Mathematical Sciences, New York University, USA
Random matrix theory IV: Analytic methods
Riemann–Hilbert problem

Deryabin, Mikhail V.
Department of Mathematics, Technical University of Denmark, Denmark
Kolmogorov–Arnol'd–Moser theorem

Dewel, Guy (deceased)
Formely Professor, Faculté des Sciences Université Libre de Bruxelles, Belgium
Turing patterns

Diacu, Florin
Professor, Department of Mathematics and Statistics, University of Victoria, Canada
Celestial mechanics
N-body problem

Ding, Mingzhou
Professor, Department of Biomedical Engineering Univeristy of Florida, USA
Intermittency

Dmitriev, S.V.
Researcher, Institute of Industrial Science, University of Tokyo, Japan
Collisions

Dolgaleva, Ksenia
Department of Physics, M.V. Lomonosov Moscow State University, Moscow and The Institute of Optics, University of Rochester, USA
Frequency doubling

Donoso, José M.
E.T.S.I. Aeronauticos, Universidad Politecnica, Madrid, Spain
Ball lightning

Doucet, Arnaud
Signal Processing Group, Department of Engineering, Cambridge University, UK
Monte Carlo methods

Dritschel, David
Professor, Department of Applied Mathematics, The University of St. Andrews, UK
Contour dynamics

Dupuis, Gérard
Chimie générale et organique, Lycée Faidherbe de Lille, France
Belousov–Zhabotinsky reaction

Easton, Robert W.
Professor, Department of Applied Mathematics, University of Colorado, Boulder, USA
Conley index

Eckhardt, Bruno
Professor, Fachbereich Physik, Philipps Universität, Marburg, Germany
Chaotic Advection
Maps in the complex plane
Periodic orbit theory
Quantum chaos
Random matrix theory I: Origins and physical applications
Shear flow
Solar system
Universality

Efimo, I.
Associate Professor of Biomedical Engineering, Stanley and Lucy Lopata Endowment, Washington University, Missouri, USA
Cardiac muscle models

Eilbeck, Chris (Adviser)
Professor, Department of Mathematics, Heriot-Watt University, UK
Discrete self-trapping system

Elgin, John
Professor, Maths Department, Imperial College of Science, Technology and Medicine, London, UK
Kuramoto–Sivashinsky equation

Emmeche, Claus
Associate Professor and Head of Center for the Philosophy of Nature and Science Studies, University of Copenhagen, Denmark
Causality

Enolskii, Victor
Professor, Heriot-Watt University, UK
Theta functions

Falkovich, Gregory
Professor, Department of Physics of Complex Systems, Weizmann Institute of Science, Israel
Mixing
Turbulence

Falqui, Gregorio
Professor, Mathematical Physics Sector, International School for Advanced Studies, Trieste, Italy
Hodograph transform
N-soliton formulas

Faris, William G.
Professor, Department of Mathematics, University of Arizona, USA
Martingales

Feddersen, Henrik
Research Scientist, Climate Research Division, Danish Meteorological Institute, Denmark
Forecasting

Fedorenko, Vladimir V.
Senior Scientific Researcher, Institute of Mathematics, National Academy of Science of Ukraine, Ukraine
One-dimensional maps

Fenimore, Paul W.
Theoretical Biology and Biophysics Group, Los Alamos National Laboratory, USA
Protein dynamics

Flach, Sergej (Adviser)
Max Planck Institut für Physik komplexer Systeme, Germany
Discrete breathers
Symmetry: equations vs. solutions

Flaschka, Hermann (Adviser)
Professor, Department of Mathematics, The University of Arizona, USA
Toda lattice

Fletcher, Neville
Professor, Department of Electronic Materials Engineering, Australian National University, Australia
Overtones

Floría, Luis Mario
Department of Theory and Simulation of Complex Systems, Instituto de Ciencia de Materiales de Aragon, Spain
Aubry–Mather theory
Commensurate-incommensurate transition
Frenkel–Kontorova model

Forrester, Peter
Department of Mathematics and Statistics, University of Melbourne, Australia
Random matrix theory II: Algebraic developments

Fowler, W. Beall
Emeritus Professor, Physics Department, Lehigh University, USA
Color centers

Fraedrich, Klaus
Professor, Meteorologisches Institut, Universität Hamburg, Germany
Atmospheric and ocean sciences
General circulation models of the atmosphere

Freites, Juan Alfredo
Department of Physics and Astronomy, University of California, Irvine, USA
Molecular dynamics

Frieden, Roy
Optical Sciences Center, University of Arizona in Tucson, USA
Information theory

Friedrich, Joseph
Professor, Lehrstuhl für Physik Weihenstephan Technische Universität München, Germany
Hole burning

Fuchikami, Nobuko
Department of Physics, Tokyo Metropolitan University, Japan
Dripping faucet

Gallagher, Marcus
School of Information Technology & Electrical Engineering, The University of Queensland, Australia
McCulloch–Pitts network
Perceptron

Garnier, Nicolas
Laboratoire de Physique, Ecole Normale Supérieure de Lyon, France
Hydrothermal waves

Gaspard, Pierre P.
Center for Nonlinear Phenomena & Complex Systems Université Libre de Bruxelles, Belgium
Entropy
Maps
Quantum theory
Rössler systems

Gendelman, Oleg
Faculty of Mechanical Engineering, Israel Institute of Technology, Israel
Heat conduction

Giuliani, Alessandro
Environment and Health Departmant, Istituto Superiore di Sanitá, Rome, Italy
Algorithmic complexity

Glass, Leon
Isadore Rosenfeld Chair and Professor of Physiology, McGill University, Canada
Cardiac arrhythmias and the electrocardiogram

Glendinning, Paul
Professor, Department of Mathematics, University of Manchester Institute of Science and Technology, UK
Hénon map
Invariant manifolds and sets
Routes to chaos

Goriely, Alain
Professor, Department of Mathematics, University of Arizona, USA
Normal forms theory

Grand, Steve
Director, Cyberlife Research Ltd., Shipham, UK
Artificial life

Gratrix, Sam
Maths Department, Imperial College of Science, Technology and Medicine, UK
Kuramoto–Sivashinsky equation

Grava, Tamara
Mathematical Physics Sector, International school for Advanced Studies, Trieste, Italy
Hodograph transform
N-soliton formulas
Zero-dispersion limits

Grimshaw, Roger
Professor, Department of Mathematical Sciences, Loughborough University, UK
Group velocity
Korteweg–de Vries equation
Water waves

Haken, Hermann (Adviser)
Professor Emeritus, Fakultät für Physik, University of Stuttgart, Germany
Gestalt phenomena
Synergetics

Halburd, Rodney G.
Lecturer, Department of Mathematical Sciences, Loughborough University, UK
Einstein equations

Hallinan, Jennifer
Institute for Molecular Bioscience, The University of Queensland, Australia
Game of life
Game theory

Hamilton, Mark
Professor, Department of Mechanical Engineering, University of Texas at Austin, USA
Nonlinear acoustics

Hamm, Peter
Professor, Physikalisch-Chemisches Institut, Universität Zürich, Switzerland
Franck–Condon factor
Hydrogen bond
Pump-probe measurements

Hasselblatt, Boris
Professor, Department of Mathematics, Tufts University, USA
Anosov and Axiom-A systems
Measures
Phase space

Hastings, Alan
Professor, Department of Environmental Science and Policy, University of California, USA
Epidemiology

Hawkins, Jane
Professor, Department of Mathematics, University of North Carolina at Chapel Hill, USA
Ergodic theory

Helbing, Dirk
Institute for Economics and Traffic, Dresden University of Technology, Germany
Traffic flow

Henry, Bruce
Department of Applied Mathematics, University of New South Wales, Australia
Equipartition of energy
Hénon–Heiles system

Henry, Bryan
Department of Chemistry and Biochemistry, University of Guelph, Canada
Local modes in molecules

Hensler, Gerhard
Professor, Institut für Astronomie, Universitäts-Sternwarte Wien, Austria
Galaxies

Herrmann, Hans
Institute for Computational Physics, University of Stuttgart, Germany
Dune formation

Hertz, John
Professor, Nordic Institute for Theoretical Physics, Denmark
Attractor neural networks

Hietarinta, Jarmo
Professor, Department of Physics, University of Turku, Finland
Hirota's method

Hill, Larry
Technical Staff Member, Detonation Science & Technology, Los Alamos National Laboratory, USA
Evaporation wave

Hjorth, Poul G.
Associate Professor, Department of Mathematics, Technical University of Denmark, Denmark
Kolmogorov–Arnol'd–Moser theorem

Holden, Arun
Professor of Computational Biology, School of Biomedical Sciences, University of Leeds, UK
Excitability
Hodgkin–Huxley equations
Integrate and fire neuron
Markin–Chizmadzhev model
Periodic bursting
Spiral waves

Holstein-Rathlou, N.-H.
Professor, Department of Medical Physiology, University of Copenhagen, Denmark
Nephron dynamics

Hommes, Cars
Professor, Center for Nonlinear Dynamics in Economics and Finance, Department of Quantitative Economics, University of Amsterdam, The Netherlands
Economic dynamics

Hone, Andrew
Lecturer in Applied Mathematics, Institute of Mathematics & Actuarial Science, University of Kent at Canterbury, UK
Extremum principles
Ordinary differential equations, nonlinear
Riccati equations

Hood, Alan
Professor, School of Mathematics and Statistics, University of St Andrews, UK
Characteristics

Houghton, Conor
Department of Pure and Applied Mathematics, Trinity College Dublin, Ireland
Instantons
Yang–Mills theory

Howard, James E.
Research Associate, Department of Physics, University of Colorado at Boulder, USA
Nontwist maps
Regular and chaotic dynamics in atomic physics

Ivey, Thomas A.
Department of Mathematics, College of Charleston, USA
Differential geometry
Framed space curves

Jiménez, Salvador
Professor, Departamento de Matemáticas, Universidad Alfonso X El Sabio, Madrid, Spain
Charge density waves
Dispersion relations

Joannopoulos, John D.
Professor, Department of Physics, Massachusetts Institute of Technology, USA
Photonic crystals

Johansson, Magnus
Department of Physics and Measurement Technology, Linköping University, Sweden
Discrete nonlinear Schrödinger equations

Johnson, Steven G.
Assistant Professor, Department of Mathematics, Massachussetts Institute of Technology, USA
Photonic crystals

Joshi, Nalini
Professor, School of Mathematics and Statistics, University of Sydney, Australia
Solitons

Kaneko, Kunihiko
Department of Pure and Applied Sciences, University of Tokyo, Japan
Coupled map lattice

Kantz, Holger
Professor of Theoretical Physics, Max Planck Institut für komplexer Systeme, Germany
Time series analysis

Kennedy, Michael Peter
Professor of Microelectronic Engineering, University College, Cork, Ireland
Chua's circuit

Kevrekidis, I.G.
Professor, Department of Chemical Engineering, Princeton University, USA
Wave of translation

Kevrekidis, Panayotis G.
Assistant Professor, Department of Mathematics and Statistics, University of Massachusetts, Amherst, USA
Binding energy
Collisions
Wave of translation

Khanin, Konstantin
Professor, Department of Mathematics, Heriot-Watt University, UK
Denjoy theory

Khovanov, Igor A.
Department of Physics, Saratov State University, Russia
Quasiperiodicity

Khovanova, Natalya A.
Department of Physics, Saratov State University, Russia
Quasiperiodicity

King, Aaron
Assistant Professor, Department of Ecology and Evolutionary Biology, University of Tennessee, Knoxville, USA
Phase plane

Kirby, Michael J.
Professor, Department of Mathematics, Colorado State University, USA
Nonlinear signal processing

Kirk, Edilbert
Meteorologisches Institut, Universität Hamburg, Germany
General circulation models of the atmosphere

Kivshar, Yuri (Adviser)
Nonlinear Physics Center, Australian National University, Australia
Optical fiber communications

Kiyono, Ken
Research Fellow of the Japan Society for the Promotion of Science, Educational Physiology Laboratory, University of Tokyo, Japan
Dripping faucet

Knott, Ron
Department of Mathematics, University of Surrey, UK
Fibonacci series

Kocarev, Liupco
Associate Research Scientist, Institute for Nonlinear Science, University of California, San Diego, USA
Damped-driven anharmonic oscillator

Konopelchenko, Boris G.
Professor, Dipartimento di Fisica, University of Lecce, Italy
Multidimensional solitons

Konotop, Vladimir V.
Centro de Física Teórica e Computacional Complexo Interdisciplinar da Universidade de Lisboa, Portugal
Wave propagation in disordered media

Kosevich, Arnold
B. Verkin Institute for Low Temperature Physics and Engineering, National Academy of Sciences of Ukraine, Kharkov, Ukraine
Breathers
Dislocations in crystals
Effective mass
Landau–Lifshitz equation
Superfluidity
Superlattices

Kovalev, Alexander S.
Institute for Low Temperature Physics and Engineering, National Academy of Sciences of Ukraine, Ukraine
Continuum approximations
Topological defects

Kramer, Peter R.
Assistant Professor, Department of Mathematical Sciences, Rensselaer Polytechnic Institute, USA
Brownian motion
Fokker–Planck equation

Krinsky, Valentin
Professor, Institut Non-Lineaire de Nice, France
Cardiac muscle models

Kuramoto, Yoshiki (Adviser)
Department of Physics, Kyoto University, Japan
Phase dynamics

Kurin, V.
Institute for Physics of Microstructures, Russian Academy of Science, Russia
Cherenkov radiation

Kuvshinov, Viatcheslav I.
Professor, Institute of Physics, Belarus Academy of Sciences, Belarus
Black holes
Cosmological models
Fractals
General relativity

Kuzmin, Andrei
Professor, Institute of Physics, Belarus Academy of Sciences, Belarus
Fractals

Kuznetsov, Vadim
Advanced Research Fellow, Department of Applied Mathematics, University of Leeds, UK
Rotating rigid bodies

LaBute, Montiago X.
Theoretical Biology and Biophysics Group, Los Alamos National Laboratory, USA
Protein structure

Lakshmanan, Muthusamy
Professor, Department of Physics, Bharathidasan University, Tiruchirapalli, India
Equations, nonlinear
Nonlinear electronics
Spin systems

Landa, Polina S.
Professor, Department of Physics, Moscow State University, Russia
Feedback
Pendulum
Quasilinear analysis
Relaxation oscillators

Landsberg, Peter
Professor, Faculty of Mathematical Studies, University of Southampton, UK
Detailed balance

Lansner, Anders
Department of Numerical Analysis and Computer Science (NADA), Royal Institute of Technology (KTH), Sweden
Cell assemblies
Neural network models

Lee, John
Professor, Department of Mechanical Engineering, McGill University, Canada
Flame front

Lega, Joceline
Associate Professor, Department of Mathematics, University of Arizona, USA
Equilibrium
Fredholm theorem

Lepeshkin, Nick
The Institute of Optics, University of Rochester, USA
Frequency doubling

Levi, Decio
Professor, Dipartimento di Ingegneria Electronica, Università degli Studi Roma tre, Italy
Delay-differential equations

Lichtenberg, Allan J.
Professor, Department of Electrical Engineering and Computer Science, University of California at Berkeley, USA
Arnol'd diffusion
Averaging methods
Electron beam microwave devices
Fermi acceleration and Fermi map
Fermi–Pasta–Ulam oscillator chain
Particle accelerators
Phase-space diffusion and correlations

Liley, David
School of Biophysical Sciences and Electrical Engineering, Swinburne University of Technology, Australia
Electroencephalogram at mesoscopic scales

Lonngren, Karl E.
Professor, Department of Electrical and Computer Engineering, University of Iowa, USA
Plasma soliton experiments

Losert, Wolfgang
Assistant Professor, Department of Physics, IPST and IREAP, University of Maryland, USA
Granular materials
Pattern formation

Lotrič, Maja-Bračič
Faculty of Electrical Engineering, University of Liubljana, Slovenia
Wavelets

Luchinsky, Dmitry G.
Department of Physics, Lancaster University, UK
Nonlinearity, definition of

Lücke, Manfred
Institut für Theoretische Physik, Universität des Saarlandes, Saarbrücken, Germany
Thermo-diffusion effects

Lunkeit, Frank
Meteorologisches Institut, Universität Hamburg, Germany
General circulation models of the atmosphere

Ma, Wen-Xiu
Department of Mathematics, University of South Florida, USA
Integrability

Macaskill, Charles
Associate Professor, School of Mathematics and Statistics, University of Sydney, Australia
Jupiter's Great Red Spot

MacClune, Karen Lewis
Hydrologist, SS Papadopulos & Associates, Boulder, Colorado, USA
Glacial flow

Maggio, Gian Mario
ST Microelectronics and Center for Wireless Communications (CWC), University of California at San Diego, USA
Damped-driven anharmonic oscillator

Maini, Philip K.
Professor, Centre for Mathematical Biology, Mathematical Institute, University of Oxford, UK
Morphogenesis, biological

Mainzer, Klaus
Professor, Director of the Institute of Interdisciplinary Informatics, Department of Philosophy of Science, University of Augsburg, Germany
Artificial intelligence
Cellular nonlinear networks
Dynamical systems

Malomed, Boris A.
Professor, Department of Interdisciplinary Studies, Faculty of Engineering, Tel Aviv University, Israel

Complex Ginzburg–Landau equation
Constants of motion and conservation laws
Multisoliton perturbation theory
Nonlinear Schrödinger equations
Power balance

Manevitch, Leonid
Professor, Institute of Chemical Physics, Russia
Heat conduction
Mechanics of solids
Peierls barrier

Manneville, Paul
Laboratoire d'Hydrodynamique (LadHyX), École Polytechnique, Palaiseau, France
Spatiotemporal chaos

Marklof, Jens
School of Mathematics, University of Bristol, UK
Cat map

Marsden, Jerrold E.
Professor of Control and Dynamical Systems California Institute of Technology, Pasadena, USA
Berry's phase

Martínez, Pedro Jesús
Department of Theory and Simulation of Complex Systems, Instituto de Ciencia de Materiales de Aragon, Spain
Frenkel–Kontorova model

Masmoudi, Nader
Associate Professor, Department of Mathematics, Courant Institute of Mathematical Sciences, New York University, USA
Boundary layers
Rayleigh–Taylor instability

Mason, Lionel
Mathematical Institute, Oxford University, UK
Twistor theory

Mayer, Andreas
Institute for Theoretical Physics, University of Regensburg, Germany
Surface waves

McKenna, Joe
Professor, Department of Mathematics, University of Connecticut, USA
Tacoma Narrows Bridge collapse

McLaughlin, Kenneth
Associate Professor, Department of Mathematics, University of North Carolina at Chapel Hill, USA
Random matrix theory III: Combinatorics

McLaughlin, Richard
Associate Professor, Department of Mathematics, University of North, Carolina, Chapel Hill, USA
Plume dynamics

McMahon, Ben
Theoretical Biology and Biophysics Group, Los Alamos National Laboratory, USA
Protein dynamics
Protein structure

Meiss, James
Professor, Department of Applied Mathematics, University of Colorado at Boulder, USA
Hamiltonian systems
Standard map
Symplectic maps

Minkevich, Albert
Professor of Theoretical Physics, Belorussian State University, Minsk, Belarus
Cosmological models
General relativity

Miura, Robert
Professor, Department of Mathematical Sciences, New Jersey Institute of Technology, USA
Nonlinear toys

Moloney, Jerome V.
Professor, Department of Mathematics, University of Arizona, USA
Nonlinear optics

Moore, Richard O.
Assistant Professor, Departmant of Mathematical Sciences, New Jersey, Institute of Technology, USA
Harmonic generation

Mørk, Jesper
Professor, Optoelectronics, Research Center COM, Technical University of Denmark, Denmark
Semiconductor laser

Mornev, Oleg
Senior Researcher, Institute of Theoretical and Experimental Biophysics, Russia
Geometrical optics, nonlinear
Gradient system
Zeldovich–Frank-Kamenetsky equation

Mosekilde, E.
Professor, Department of Physics, Technical University of Denmark, Denmark
Nephron dynamics

Mueller, Stefan C.
Department of Biophysics, Otto-von-Guericke-Universität Magdeburg, Germany
Scroll waves

Mullin, Tom
Professor of Physics and Director of Manchester Centre for Nonlinear Dynamics, University of Manchester, UK
Bifurcations
Catastrophe theory
Taylor–Couette flow

Mygind, Jesper
Professor, Department of Physics, Technical University of Denmark, Denmark
Josephson junctions
Superconducting quantum interference device

Nakamura, Yoshiharu
Associate Professor, Institute of Space and Astronautical Science, Kanagawa, Japan
Plasma soliton experiments

Natiello, Mario
Centre for Mathematical Sciences, Lund University, Sweden
Lasers
Winding numbers

Newell, Alan
Professor, Department of Mathematics, University of Arizona, USA
Inverse scattering method or transform

Newton, Paul K.
Professor, Department of Aerospace and Mechanical Engineering, University of Southern California, USA
Berry's phase
Chaos vs. turbulence

Neyts, Kristiaan
Professor, Department of Electronics and Information Systems, Ghent University, Belgium
Liquid crystals

Nicolis, G.
Professor, Faculté des Sciences, Université Libre de Bruxelles, Belgium
Brusselator
Chemical kinetics
Nonequilibrium statistical mechanics
Recurrence

Nuñez, Paul
Professor, Brain Physics Group, Department of Biomedical Engineering, Tulane University, USA
Electroencephalogram at large scales

Olsder, Geert Jan
Faculty of Technical Mathematics and Informatics, Delft University of Technology, The Netherlands
Idempotent analysis

Olver, Peter J.
Professor, School of Mathematics, University of Minnesota, USA
Lie algebras and Lie groups

Ostrovsky, Lev (Adviser)
Professor, Zel Technologies/Univeristy of Colorado, Boulder, Colorado, USA, and Institute of Applied Physics, Nizhny Novgorod, Russia
Hurricanes and tornadoes
Modulated waves
Nonlinear acoustics
Shock waves

Ottova-Leitmannova, Angelica
Department of Physiology, Michigan State University, USA
Bilayer lipid membrance

Palmer, John
Professor, Department of Mathematics, University of Arizona, USA
Monodromy preserving deformations

Pascual, Pedro J.
Associate Professor, Departamento de Ingenieria Informática, Universidad Autonoma de Madrid, Spain
Charge density waves

Pedersen, Niels Falsig
Professor, Department of Power Engineering, Technical University of Denmark, Denmark
Long Josephson junctions
Superconductivity

Pelinovsky, Dmitry
Associate Professor, Department of Mathematics, McMaster University, Canada
Coupled systems of partial differential equations
Energy analysis
Generalized functions
Linearization
Manley–Rowe relations
Numerical methods
N-wave interactions
Spectral analysis

Pelletier, Jon D.
Assistant Professor, Department of Geosciences, University of Arizona, USA
Geomorphology and tectonics

Pelloni, Beatrice
Mathematics Department, University of Reading, UK
Boundary value problems
Burgers equation

Petty, Michael
Professor, Centre for Molecular and Nanoscale Electronics, University of Durham, UK
Langmuir–Blodgett films

Peyrard, Michel
Professor of Physics, Laboratoire de Physique, Ecole Normale Supérieure de Lyon, France
Biomolecular solitons

Pikovsky, Arkady
Department of Physics Universität Potsdam, Germany
Synchronization
Van der Pol equation

Pitchford, Jon
Lecturer, Department of Biology, University of York, UK
Random walks

Pojman, John A.
Professor, Department of Chemistry and Biochemistry, The University of Southern Mississippi, USA
Polymerization

Pumiri, A.
Directeur de Recherche, Institut Non-Lineaire de Nice, France
Cardiac muscle models

Pushkin Dmitri O.
Department of Theoretical and Applied Mechanics, University of Illinois, Urbana–Champaign, USA
Cluster coagulation

Rabinovich, Mikhail
Research Physicist, Institute for Nonlinear Science, University of California at San Diego, USA and Institute of Applied Physics, Russian Academy of Sciences
Chaotic dynamics

Rañada, Antonio F.
Facultad de Fisica, Universidad Complutense, Madrid, Spain
Ball lightning

Recami, Erasmo
Professor of Physics, Faculty of Engineering, Bergamo State University, Bergamo, Italy
Tachyons and superluminal motion

Reucroft, Stephen
Professor of Physics, Northeastern University, Boston, USA
Higgs boson

Ricca, Renzo L.
Professor, Dipartimento di Matematica e Applicazioni, Università di Milano-Bicocca, Milan, Italy
Knot theory
Structural complexity

Robinson, James C.
Mathematics Institute, University of Warwick, UK
Attractors
Dimensions
Function spaces
Functional analysis

Robnik, Marko
Professor, Center for Applied Mathematics and Theoretical Physics, University of Maribor, Slovenia
Adiabatic invariants
Determinism

Rogers, Colin
Professor, Australian Research Council Centre of Excellence for Mathematics and Statistics of Complex Systems, School of Mathematics, University of New South Wales, Australia
Bäcklund transformations

Romanenko, Elena
Senior Scientific Researcher, Institute of Mathematics, National Academy of Science of Ukraine, Ukraine
Turbulence, ideal

Rosenblum, Michael
Department of Physics, University of Potsdam, Germany
Synchronization
Van der Pol equation

Rouvas-Nicolis, C.
Climatologie Dynamique, Institut Royal Météorologique de Belgique, Belgium
Recurrence

Ruijsenaars, Simon
Center for Mathematics and Computer Science, The Netherlands
Derrick–Hobart theorem
Particles and antiparticles

Rulkov, Nikolai
Institute for Nonlinear Science, University of California at San Diego, USA
Chaotic dynamics

Sabatier, Pierre
Professor, Physique Mathématique, Université Montpellier II, France
Inverse problems

Sakaguchi, Hidetsugu
Department of Applied Science for Electronics and Materials, Kyushu University, Japan
Coupled oscillators

Salerno, Mario
Professor, Departimento di Fisica "E.R. Caianiello", Università degli Studi, Salerno, Italy
Bethe ansatz
Salerno equation

Sandstede, Bjorn
Associate Professor, Department of Mathematics, Ohio State University, USA
Evans function

Satnoianu, Razvan
Centre for Mathematics, School of Engineering and Mathematical Sciences, City University, UK
Diffusion
Reaction-diffusion systems

Sauer, Tim
Professor, Department of Mathematics, George Mason University, USA
Embedding methods

Savin, Alexander
Professor, Moscow Institute of Physics and Technology, Russia
Peierls barrier

Schaerf, Timothy
School of Mathematics and Statistics, University of Sydney, Australia
Jupiter's Great Red Spot

Schattschneider, Doris
Professor, Department of Mathematics, Moravian College, Pennsylvania, USA
Tessellation

Schirmer, Jochen
Professor, Institute for Physical Chemistry, Heidelberg, Germany
Hartee approximation

Schmelcher, Peter
Institute for Physical Chemistry, University of Heidelberg, Germany
Hartree approximation

Schöll, Eckehard
Professor, Institut für Theoretische Physik, Technische Universität Berlin, Germany
Avalanche breakdown
Diodes
Drude model
Semiconductor oscillators

Schuster, Peter
Institut für Theoretische Chemie und Molekulare Strukturbiologie, Austria
Biological evolution
Catalytic hypercycle
Fitness landscape

Scott, Alwyn (Editor)
Emeritus Professor of Mathematics, University of Arizona, USA
Candle
Discrete self-trapping system
Distributed oscillators
Emergence
Euler–Lagrange equations
Hierarchies of nonlinear systems
Laboratory models of nonlinear waves
Lifetime
Matter, nonlinear theories of
Multiplex neuron
Nerve impulses
Neuristor
Quantum nonlinearity
Rotating-wave approximation
Solitons, a brief history
State diagrams
Symmetry groups
Tachyons and superluminal phenomena
Threshold phenomena
Wave packets, linear and nonlinear

Segev, Mordechai
Professor, Technion-Israel Institute of Technology, Haifa, Israel
Incoherent solitons

Shalfeev, Vladimir
Head of Department of Oscillation Theory, Nizhni Novgorod State University, Russia
Parametric amplification

Sharkovsky, Alexander N.
Institute of Mathematics, National Academy of Sciences of Ukraine, Ukraine
One-dimensional maps
Turbulence, ideal

Sharman, Robert
National Center for Atmospheric Research, Boulder, Colorado, USA
Clear air turbulence

Shinbrot, Troy
Associate Professor, Department of Chemical and Biochemical Engineering, Rutgers University, USA
Controlling chaos

Shohet, J. Leon
Professor, Department of Electrical and Computer Engineering, University of Wisconsin-Madison, USA
Nonlinear plasma waves

Siwak, Pawel
Department of Electrical Engineering, Poznan University of Technology, Poland
Integrable cellular automata

Skufca, Joe D.
Department of Mathematics, US Naval Academy, USA
Markov partition

Skufca, Joseph
Center for Computational Science and Mathematical Modelling, University of Maryland, USA
Markov partitions

Smil, Vaclav
Professor, Department of Environment, University of Manitoba, Canada
Global warming

Sobell, Henry M.
Independent scholar, New York, USA
DNA premelting

Solari, Hernán Gustavo
Departamento Física, University of Buenos Aires, Argentina
Lasers
Winding numbers

Soljačić, Marin
Principal Research Scientist, Research Laboratory of Electronics, Massachusetts Institute of Technology, USA
Photonic crystals

Sørensen, Mads Peter
Associate Professor, Department of Mathematics, Technical University of Denmark, Denmark
Collective coordinates
Multiple scale analysis
Perturbation theory

Sornette, Didier
Professor, Laboratoire de Physique de la Matiere Condensee, Université de Nice - Sophia Antipolis, France
Sandpile model

Sosnovtseva, O.
Lecturer, Department of Physics, Technical University of Denmark, Denmark
Nephron dynamics

Spatschek, Karl
Professor, Institut für Theoretische Physics 1, Heinrich-Heine-Universität Düsseldorf, Germany
Center manifold reduction
Dispersion management

Stadler, Michael A.
Professor, Institut für Physchologie and Kognitionsforschung, Bremen, Germany
Gestalt phenomena

Stauffer, Dietrich
Institute for Theoretical Physics, University of Cologne, Germany
Percolation theory

Stefanovska, Aneta
Head, Nonlinear Dynamics and Synergetics Group Faculty of Electrical Engineering, University of Ljubljana, Slovenia
Flip-flop circuit
Inhibition
Nonlinearity, definition of
Quasiperiodicity
Wavelets

Storb, Ulrich
Institut für Experimentelle Physik, Otto-von-Guericke-Universität, Magdeburg, Germany
Scroll waves

Strelcyn, Jean-Marie
Professeur, Département de Mathématiques, Université de Rouen, Mont Saint Aignan Cedex, France
Poincaré theorems

Suris, Yuri B.
Department of Mathematics, Technische Universität Berlin, Germany
Integrable lattices

Sutcliffe, Paul
Professor of Mathematical Physics, Institute of Mathematics & Acturial Science, University of Kent at Canterbury, UK
Skyrmions

Sverdlov, Masha
TEC High School, Newton, Massachusetts, USA
Hurricanes and Tornadoes

Swain, John David
Professor, Department of Physics, Northeastern University, Boston, USA
Doppler shift
Quantum field theory
Tensors

Tabor, Michael
Professor, Department of Mathematics, University of Arizona, USA
Growth patterns

Tajiri, Masayoshi
Emeritus Professor, Department of Mathematical Sciences, Osaka Prefecture University, Japan
Solitons, types of
Wave stability and instability

Tass, Peter
Professor, Institut für Medizin, Forschungszentrum Jülich, Germany
Stochastic analysis of neural systems

Taylor, Richard
Associate Professor, Materials Science Institute, University of Oregon, USA
Lévy flights

Teman, Roger
Laboratoire d'Analyse Numerique, Université de Paris Sud, France
Inertial manifolds

Thompson, Michael
Emeritus Professor (UCL) and Honorary Fellow, Department of Applied Mathematics and Theoretical Physics, University of Cambridge, UK
Duffing equation
Stability

Tien, H. Ti (deceased)
Formerly Professor, Membrane Biophysics Laboratory, Michigan State University, USA
Bilayer lipid membranes

Tobias, Douglas J.
Associate Professor, Department of Chemistry, University of California at Irvine, USA
Molecular dynamics

Toda, Morikazu
Emeritus Professor, Tokyo University of Education, Japan
Nonlinear toys

Trueba, José L.
Departmento di Mathemáticas, y Fisica Aplicadas y Ciencias de la Natura, Universidad Rey Juan Carlos, Móstoles, Spain
Ball lightning

Tsimring, Lev S.
Research Physicist, Institute for Nonlinear Science, University of California, San Diego USA
Avalanches

Tsinober, Arkady
Professor, Iby and Aladar Fleischman Faculty of Engineering, Tel Aviv University, Israel
Helicity

Tsironis, Giorgos P.
Department of Physics, University of Crete, Greece
Bjerrum defects
Excitons
Ising model
Local modes in molecular crystals

Tsygvintsev, Alexei
Maitre de Conférences, Unité de Mathématiques Pures et Appliquées, École Normale Superieure de Lyon, France
Poincaré theorems

Tuszynski, Jack
Department of Physics, University of Alberta, Canada
Critical phenomena
Domain walls
Ferromagnetism and ferroelectricity
Fröhlich theory
Hysteresis
Order parameters
Renormalization groups
Scheibe aggregates

Ustinov, Alexey V.
Physikalisches Institut III, University of Erlangen-Nürnberg, Germany
Josephson junction arrays

van der Heijden, Gert
Centre for Nonlinear Dynamics, University College London, UK
Butterfly effect
Hopf bifurcation

Vázquez, Luis
Professor, Faculted de Informática, Universidad Complutense de Madrid, Spain. Senior Researcher and Cofounder of the Centro de Astrobiología, Instituo Nacional de Técnica Aeroespacial, Madrid, Spain

Charge density waves
Dispersion relations
FitzHugh–Nagumo equation
Virial theorem
Wave propagation in disordered media

Verboncoeur, John P.
Associate Professor, Nuclear Engineering Department, University of California, Berkeley, USA
Electron beam microwave devices

Veselov, Alexander
Professor, Department of Mathematical Sciences, Loughborough University, UK
Huygens principle

Vo, Ba-Ngu
Electrical and Electronic Engineering Department, The Univeristy of Melbourne, Victoria, Australia
Monte Carlo methods

Voiculescu, Dan-Virgil
Professor, Department of Mathematics, University of California at Berkeley, USA
Free probability theory

Voorhees, Burton H.
Professor, Department of Mathematics, Athabasca University, Canada
Cellular automata

Wadati, M.
Professor, Department of Physics, University of Tokyo, Japan
Quantum inverse scattering method

Walter, Gilbert G.
Professor Emeritus, Department of Mathematical Sciences, University of Wisconsin-Milwaukee, USA
Compartmental models

Waymire, Edward C.
Professor, Department of Mathematics, Oregon State University, USA
Multiplicative processes

West, Bruce J.
Chief Scientist, Mathematics, US Army Research Office, North Carolina, USA
Branching laws
Fluctuation-dissipation theorem
Kicked rotor

Wilhelmsson, Hans
Professor Emeritus of Physics, Chalmers University of Technology, Sweden
Alfvén waves

Wilson, Hugh R.
Centre for Vision Research, York University, Canada
Neurons
Stereoscopic vision and binocular rivalry

Winfree, A.T. (Adviser) (deceased)
Formerly, Department of Ecology and Evolutionary Biology, University of Arizona, USA
Dimensional analysis

Wojtkowski, Maciej P.
Professor, Department of Mathematics, University of Arizona, USA
Lyapunov exponents

Yakushevich, Ludmilla (Adviser)
Researcher, Institute of Cell Biophysics, Russian Academy of Sciences, Russia
DNA solitons

Young, Lai-Sang (Adviser)
Professor, Courant Institute of Mathematical Sciences, New York University, USA
Anosov and Axiom-A systems
Horseshoes and hyperbolicity in dynamical systems
Sinai–Ruelle–Bowen measures

Yiguang, Ju
Assistant Professor, Department of Mechanical and Aerospace Engineering, Princeton University, USA
Flame front

Yukalov, V.I.
Professor, Bogolubov Laboratory of Theoretical Physics, Joint Institute for Nuclear Research, Russia
Bose–Einstein condensation
Coherence phenomena

Zabusky, Norman J.
Professor, Department of Mechanical and Aerospace Engineering, Rutgers University, USA
Visiometrics
Vortex dynamics of fluids

Zbilut, Joseph P.
Professor, Department of Molecular Biophysics and Physiology, Rush University, USA
Algorithmic complexity

Zhou, Xin
Professor, Department of Mathematics, Duke University, USA
Random matrix theory IV: Analytic methods
Riemann–Hilbert problem

Zolotaryuk, Alexander V.
Bogolyubov Institute for Theoretical Physics, Ukraine
Polarons
Ratchets

Zorzano, María-Paz
Young Researcher, Centro de Astrobiología, Instituto Nacional de Técnica Aeroespacial, Madrid, Spain
FitzHugh–Nagumo equations
Virial Theorem

List of Entries

Thematic List of Entries

General

HISTORY OF NONLINEAR SCIENCE

Bernoulli's equation, Butterfly effect, Candle, Celestial mechanics, Davydov soliton, Determinism, Feedback, Fermi–Pasta–Ulam oscillator chain, Fibonacci series, Hodgkin–Huxley equations, Introduction, Integrability, Lorenz equations, Manley–Rowe relations, Markin–Chizmadzhev model, Martingales, Matter, nonlinear theory of, Poincaré theorems, Solar system, Solitons, a brief history, Tacoma Narrows Bridge collapse, Van der Pol equation, Zeldovich–Frank-Kamenetsky equation

COMMON EXAMPLES OF NONLINEAR PHENOMENA

Avalanches, Ball lightning, Brownian motion, Butterfly effect, Candle, Clear air turbulence, Diffusion, Dripping faucet, Dune formation, Explosions, Fairy rings of mushrooms, Filamentation, Flame front, Fluid dynamics, Forest fires, Glacial flow, Global warming, Hurricanes and tornadoes, Jupiter's Great Red Spot, Nonlinear toys, Order from chaos, Pendulum, Phase transitions, Plume dynamics, Solar system, Tacoma Narrows Bridge collapse, Traffic flow, Water waves

Methods and Models

ANALYTICAL METHODS

Bäcklund transformations, Bethe ansatz, Center-manifold reduction, Characteristics, Collective coordinates, Continuum approximations, Dimensional analysis, Dispersion relations, Dressing method, Elliptic functions, Energy analysis, Evans function, Fredholm theorem, Gel'fand–Levitan theory, Generalized functions, Hamiltonian systems, Hirota's method, Hodograph transform, Idempotent analysis, Integral transforms, Inverse scattering method or transform, Kirchhoff's laws, Multiple scale analysis, Multisoliton perturbation theory, Nonequilibrium statistical mechanics, Normal forms theory, N-soliton formulas, Painlevé analysis, Periodic spectral theory, Perturbation theory, Phase dynamics, Phase plane, Poisson brackets, Power balance, Quantum inverse scattering method, Quasilinear analysis, Riccati equations, Rotating-wave approximation, Separation of variables, Spectral analysis, Stability, State diagrams, Synergetics, Tensors, Theta functions, Time series analysis, Volterra series, Wavelets, Zero-dispersion limits

COMPUTATIONAL METHODS

Averaging methods, Cellular automata, Cellular nonlinear networks, Characteristics, Compartmental models, Contour dynamics, Embedding methods, Extremum principles, Fitness landscape, Forecasting, Framed space curves, Hartree approximation, Integrability, Inverse problems, Lattice gas methods, Linearization, Maps, Martingales, Monte-Carlo methods, Numerical methods, Recurrence, Theta functions, Time series analysis, Visiometrics, Volterra series and operators, Wavelets

TOPOLOGICAL METHODS

Bäcklund transformations, Cat map, Conley index, Darboux transformation, Denjoy theory, Derrick–Hobart theorem, Differential geometry, Extremum principles, Functional analysis, Horseshoes and hyperbolicity in dynamical systems, Huygens principle, Inertial manifolds, Invariant manifolds and sets, Knot theory, Kolmogorov–Arnol'd–Moser theorem, Lie algebras and Lie groups, Maps, Measures, Monodromy-preserving deformations, Multifractal analysis, Nontwist maps, One-dimensional maps, Periodic orbit theory, Phase plane, Phase space, Renormalization groups, Riemann–Hilbert problem, Singularity theory, Symbolic dynamics, Symmetry groups, Topology, Virial theorem, Winding numbers

CHAOS, NOISE AND TURBULENCE

Attractors, Aubry–Mather theory, Butterfly effect, Chaos vs. turbulence, Chaotic advection, Chaotic dynamics, Clear air turbulence, Dimensions, Entropy, Ergodic theory, Fluctuation-dissipation theorem, Fokker–Planck equation, Free probability theory, Frustration, Hele-Shaw cell, Horseshoes and hyperbolicity in dynamical systems, Lévy flights, Lyapunov exponents, Martingales, Mel'nikov method, Order from chaos, Percolation theory, Phase space, Quantum chaos, Random matrix theory, Random walks, Routes to chaos, Spatiotemporal chaos, Stochastic processes, Turbulence, Turbulence, ideal

COHERENT STRUCTURES

Biomolecular solitons, Black holes, Breathers, Cell assemblies, Davydov soliton, Discrete breathers, Dislocations in crystals, DNA solitons, Domain walls, Dune formation, Emergence, Fairy rings of mushrooms, Flame front, Higgs boson, Holons, Hurricanes and tornadoes, Instantons, Jupiter's Great Red Spot, Local modes in molecular crystals, Local modes in molecules, Multidimensional solitons, Nerve impulses, Polaritons, Polarons, Shock waves, Skyrmions, Solitons, types of, Spiral waves, Tachyons and superluminal motion, Turbulence, Turing patterns, Wave of translation

DYNAMICAL SYSTEMS

Anosov and axiom-A systems, Arnol'd diffusion, Attractors, Aubry–Mather theory, Bifurcations, Billiards, Butterfly effect, Cat map, Catastrophe theory, Center manifold reduction, Chaotic dynamics, Coupled map lattice, Deterministic walks in random environments, Development of singularities, Dynamical systems, Equilibrium, Ergodic theory, Fitness landscape, Framed space curves, Function spaces, Gradient system, Hamiltonian systems, Hénon map, Hopf bifurcation, Horseshoes and hyperbolicity in dynamical systems, Inertial manifolds, Intermittency, Kicked rotor, Kolmogorov–Arnol'd–Moser theorem, Lyapunov exponents, Maps, Measures, Mel'nikov method, One-dimensional maps, Pattern formation, Periodic orbit theory, Phase plane, Phase space, Phase-space diffusion and correlations, Poincaré theorems, Reaction-diffusion systems, Rössler systems, Rotating rigid bodies, Routes to chaos, Sinai–Ruelle–Bowen measures, Standard map, Stochastic processes, Symbolic dynamics, Synergetics, Universality, Visiometrics, Winding numbers

GENERAL PHENOMENA

Adiabatic invariants, Algorithmic complexity, Anderson localization, Arnol'd diffusion, Attractors, Berry's phase, Bifurcations, Binding energy, Boundary layers, Branching laws, Breathers, Brownian motion, Butterfly effect, Causality, Chaotic dynamics, Characteristics, Cluster coagulation, Coherence phenomena, Collisions, Critical phenomena, Detailed balance, Determinism, Diffusion, Domain walls, Doppler shift, Effective mass, Emergence, Entropy, Equilibrium, Equipartition of energy, Excitability, Explosions, Feedback,

Filamentation, Fractals, Free energy, Frequency doubling, Frustration, Gestalt phenomena, Group velocity, Harmonic generation, Helicity, Hopf bifurcation, Huygens' principle, Hysteresis, Incoherent solitons, Inhibition, Integrability, Intermittency, Jump phenomena, Kolmogorov cascade, Lévy flights, Lifetime, Mixing, Modulated waves, Multiplicative processes, Nonlinearity, definition of, N-wave interactions, Order from chaos, Order parameters, Overtones, Pattern formation, Period doubling, Periodic bursting, Power balance, Quantum chaos, Quantum nonlinearity, Quasiperiodicity, Recurrence, Routes to chaos, Scroll waves, Shear flow, Solitons, Spiral waves, Structural complexity, Symmetry: equations vs. solutions, Synergetics, Tachyons and superluminal motion, Tessellation, Thermal convection, Threshold phenomena, Turbulence, Universality, Wave packets, linear and nonlinear, Wave propagation in disordered media, Wave stability and instability

MAPS

Aubry–Mather theory, Bäcklund transformations, Cat map, Coupled map lattice, Darboux transformation, Denjoy theory, Embedding methods, Fermi acceleration and Fermi map, Hénon map, Maps, Maps in the complex plane, Monodromy preserving deformations, Nontwist maps, One-dimensional maps, Periodic orbit theory, Recurrence, Renormalization groups, Singularity theory, Standard map, Symplectic maps

MATHEMATICAL MODELS

Ablowitz–Kaup–Newell–Segur system, Attractor neural network, Billiards, Boundary value problems, Brusselator, Burger's equation, Cat map, Cellular automata, Compartmental models, Complex Ginzburg–Landau equation, Continuum approximations, Coupled map lattice, Coupled systems of partial differential equations, Delay-differential equations, Discrete nonlinear Schrödinger equations, Discrete self-trapping system, Duffing equation, Equations, nonlinear, Euler–Lagrange equations, Fitzhugh–Nagumo equation, Fokker–Planck equation, Frenkel–Kontorova model, Game of life, General circulation models of the atmosphere, Hénon–Heiles system, Integrable cellular automata, Integrable lattices, Ising model, Kadomtsev–Petviashvili equation, Knot theory, Korteweg–de Vries equation, Kuramoto–Sivashinsky equation, Landau–Lifshitz equation, Lattice gas methods, Lie algebras and Lie groups, Lorenz equations, Markov partitions, Martingales, Maxwell–Bloch equation, McCulloch–Pitts network, Navier–Stokes equation, Neural network models, Newton's laws of motion, Nonlinear Schrödinger equations, One-dimensional maps, Ordinary differential equations, nonlinear, Partial differential equations, nonlinear, Random walks, Riccati equations, Salerno equation, Sandpile model, Sine-Gordon equation, Spin systems, Stochastic processes, Structural complexity, Symbolic dynamics, Synergetics, Toda lattice, Van der Pol equation, Zeldovich–Frank-Kamenetsky equation

STABILITY

Attractors, Bifurcations, Butterfly effect, Catastrophe theory, Controlling chaos, Development of singularities, Dispersion management, Dispersion relations, Emergence, Equilibrium, Excitability, Feedback, Growth patterns, Hopf bifurcation, Lyapunov exponents, Nonequilibrium statistical mechanics, Stability

Disciplines

ASTRONOMY AND ASTROPHYSICS

Alfvèn waves, Black holes, Celestial mechanics, Cosmological models, Einstein equations, Galaxies, Gravitational waves, Hénon–Heiles system, Jupiter's Great Red Spot, N-body problem, Solar system

BIOLOGY

Artificial life, Bilayer lipid membranes, Biological evolution, Biomolecular solitons, Cardiac arrhythmias and electro cardiogram, Cardiac muscle models, Catalytic hypercycle, Compartmental models, Davydov soliton, DNA premelting, DNA solitons,

CHEMISTRY

CONDENSED MATTER AND SOLID-STATE PHYSICS

EARTH SCIENCE

ENGINEERING

FLUIDS

NEUROSCIENCE

phenomena, Hodgkin–Huxley equations, Inhibition, Integrate and fire neuron, Multiplex neuron, Myelinated nerves, Nerve impulses, Neural network models, Neurons, Pattern formation, Perceptron, Stereoscopic vision and binocular rivalry, Stochastic analyses of neural systems, Synergetics

NONLINEAR OPTICS

Cherenkov radiation, Color centers, Damped-driven anharmonic oscillator, Dispersion management, Distributed oscillators, Excitons, Filamentation, Geometrical optics, nonlinear, Harmonic generation, Hole burning, Kerr effect, Lasers, Liquid crystals, Maxwell–Bloch equations, Nonlinear optics, Optical fiber communications, Photonic crystals, Polaritons, Polarons, Pump-probe measurements, Rayleigh and Raman scattering and IR absorption, Semiconductor laser, Tachyons and superluminal motion

PLASMA PHYSICS

Alfvén waves, Ball lightning, Charge density waves, Drude model, Dynamos, homogeneous, Electron beam microwave devices, Magnetohydrodynamics, Nonlinear plasma waves, Particle accelerators, Plasma soliton experiments

SOCIAL SCIENCE

Economic system dynamics, Epidemiology, Game theory, Hierarchies of nonlinear systems, Population dynamics, Synergetics, Traffic flow

SOLID MECHANICS AND NONLINEAR VIBRATIONS

Avalanche breakdown, Bilayer lipid membranes, Bjerrum defects, Charge density waves, Cluster coagulation, Color centers, Detailed balance, Dislocations in crystals, Domain walls, Frustration, Glacial flow, Granular materials, Growth patterns, Heat conduction, Hydrogen bond, Ising model, Kerr effect, Langmuir–Blodgett films, Liquid crystals, Local modes in molecular crystals, Mechanics of solids, Molecular dynamics, Nonlinear acoustics, Protein dynamics, Ratchets, Rheology, Sandpile model, Scheibe aggregates, Shock waves, Spin systems, Superlattices, Surface waves, Tessellation, Topological defects

THEORETICAL PHYSICS

Berry's phase, Black holes, Born–Infeld equations, Celestial mechanics, Cherenkov radiation, Cluster coagulation, Constants of motion and conservation laws, Cosmological models, Critical phenomena, Derrick–Hobart theorem, Detailed balance, Einstein equations, Entropy, Equipartition of energy, Fluctuation-dissipation theorem, Fokker–Planck equation, Free energy, Galaxies, General relativity, Gravitational waves, Hamiltonian systems, Higgs boson, Holons, Instantons, Matter, nonlinear theory of, N-body problem, Newton's laws of motion, Particles and antiparticles, Quantum field theory, Quantum theory, Regular and chaotic dynamics in atomic physics, Rotating rigid bodies, Skyrmions, String theory, Tachyons and superluminal motion, Twistor theory, Virial theorem, Yang–Mills theory

AB INITIO CALCULATIONS

See **Molecular dynamics**

ABLOWITZ–KAUP–NEWELL–SEGUR SYSTEM

In 1967, Gardner, Greene, Kruskal, and Miura (or GGKM) (Gardner et al., 1967) showed that the Kortegweg–de Vries (KdV) equation

$$q_t + 6qq_x + q_{xxx} = 0, \qquad (1)$$

with rapidly decaying initial data on $-\infty < x < \infty$, can be linearized using direct and inverse scattering methods associated with the linear Schrödinger equation

$$v_{xx} + [k^2 + q(x,t)]v = 0. \qquad (2)$$

The KdV equation is of practical interest, having been first derived in the study of long water waves (Korteweg & de Vries, 1895) and subsequently in several other areas of applied science.

In the method proposed by Gardner et al., the solitary wave (soliton) solution to the KdV equation (1)

$$q = 2\kappa^2 \operatorname{sech}^2 \kappa(x - 4\kappa^2 t - x_0)$$

and multisoliton solutions are associated with the discrete spectrum of Equation (2). The discrete eigenvalues were shown to be invariants of the KdV motion; for example, the above soliton solution is associated with the discrete eigenvalue of Equation (2) at $k = \mathrm{i}\kappa$.

At that time, it was not clear whether the method could be applied to other physically significant equations. In 1972, however, Zakharov and Shabat (1972) used an operator formalism developed by Lax (1968) to show that the nonlinear Schrödinger (NLS) equation

$$\mathrm{i}q_t + q_{xx} + \sigma|q|^2 q = 0, \qquad (3)$$

with rapidly decaying initial data on $-\infty < x < \infty$, could also be linearized by direct and inverse scattering methods.

The NLS equation was known to arise in many physical contexts (Benney & Newell, 1967) and in 1973 Hasegawa and Tappert showed that the NLS equation describes the long-distance dynamics of nonlinear pulses in optical fibers (Hasegawa & Tappert, 1973). Motivated by these developments and indications that other equations fit into this category, David Kaup, Alan Newell, Harvey Segur, and the present author (Ablowitz et al., 1973, 1974) studied the following modification of the Zakharov–Shabat system:

$$\begin{aligned} v_{1x} &= -\mathrm{i}\zeta v_1 + q v_2, \\ v_{2x} &= \mathrm{i}\zeta v_2 + r v_1, \end{aligned} \qquad (4)$$

$$\begin{aligned} v_{1t} &= A v_1 + B v_2, \\ v_{2t} &= C v_1 + D v_2. \end{aligned} \qquad (5)$$

In Equations (4) and (5), v_1 and v_2 are auxiliary functions obeying the postulated linear systems; Equation (4) play the same role as Equation (2), whereas Equation (5) determine the temporal evolution of the functions v_1 and v_2. (The evolution equation associated with the auxiliary function v for the KdV equation was not given above.) The method establishes that the functions $q = q(x,t)$ and $r = r(x,t)$ satisfy nonlinear equations when the (yet to be determined) functions A, B, C, and D are properly chosen.

The key to this approach is to make Equations (4) and (5) compatible, that is, set the x-derivative of v_{it} equal to the t-derivative of v_{ix}. In other words, we set the x-derivative of the right-hand side of Equations (5) equal to the t-derivative of the right-hand side of Equations (4). The result of this calculation yields the following equations for A, B, C, and D:

$$\begin{aligned} A_x &= qC - rB, \\ B_x + 2\mathrm{i}\zeta B &= q_t - 2Aq, \\ C_x - 2\mathrm{i}\zeta C &= r_t + 2Ar, \\ D &= -A. \end{aligned} \qquad (6)$$

In Ablowitz et al. (1973, 1974; see also Ablowitz & Segur, 1981), methods to solve these equations are described. The simplest procedure is to look for finite power series expansions such as $A = \sum_{i=0}^{N} \zeta^i A_i$ and similarly for B and C.

For example, with $N = 2$, we find with $r = \mp q^*$ that the nonlinear Schrödinger equation (3) with $\sigma = \pm 1$ is a necessary condition. In this case there are 11 equations for the nine unknowns $\{A_i, B_i, C_i\}$, $i = 0, 1, 2$, and the remaining two equations determine the nonlinear evolution equations for q and r (in this case NLS when $q = \mp r^*$).

With $N = 3$ and $r = -1$, we find that q must satisfy the KdV equation. Also, with $r = \mp q$, the modifed KdV equation

$$q_t \pm 6q^2 q_x + q_{xxx} = 0 \tag{7}$$

results.

If we look for expansions containing inverse powers of ζ, additional interesting equations can be obtained. For example, postulating $A = a/\zeta$, $B = b/\zeta$, $C = c/\zeta$ results in the sine-Gordon and sinh-Gordon equations

$$u_{xt} = \sin u, \tag{8}$$

$$u_{xt} = \sinh u, \tag{9}$$

where $q = -r = -u_x/2$ in Equation (8) and $q = r = u_x/2$ in Equation (9). The sine-Gordon equation has been known to be an important equation in the study of differential geometry since the 19th century (cf. Bianchi, 1902), and it has found applications in the 20th century as models for dislocation propagation in crystals, domain walls in ferromagnetic and ferroelectric materials, short-pulse propagation in resonant optical media, and magnetic flux propagation in long Josephson junctions, among others.

Thus, a number of physically interesting nonlinear wave equations are obtained from the above formalism. In Ablowitz et al. (1973, 1974; see also Ablowitz & Segur, 1981), it was further shown as to how this approach could be generalized to a class of nonlinear equations described in terms of certain nonlinear evolution operators that were subsequently referred to in the literature as recursion operators.

Further, the whole class of nonlinear equations with rapidly decaying initial data on $-\infty < x < \infty$ was shown to be linearized via direct and inverse scattering methods. Special soliton solutions are associated with the discrete spectrum of the linear operator (4), and via (5) the discrete eigenvalues were shown to be invariants of the motion. In subsequent years, asymptotic analysis of the integral equations yielded the long-time behavior of the continuous spectrum, which in turn showed the ubiquitous role that the Painlevé equations play in integrable systems (cf. Ablowitz & Segur, 1981).

Because this formulation is analogous to the method of Fourier transforms, the method was termed the inverse scattering transform or simply the IST.

Mark J. Ablowitz

See also **Integrability; Inverse scattering method or transform; Korteweg–de Vries equation; Nonlinear Schrödinger equations; Solitons**

Further Reading

Ablowitz, M.J., Kaup, D.J., Newell, A.C. & Segur, H. 1973. Nonlinear equations of physical significance. *Physical Review Letters*, 31: 125–127

Ablowitz, M.J., Kaup, D.J., Newell, A.C. & Segur, H. 1974. The inverse scattering transform–Fourier analysis for nonlinear problems. *Studies in Applied Mathematics*, 53: 249–315

Ablowitz, M.J. & Segur, H. 1981. *Solitons and the Inverse Scattering Transform*, Philadelphia, PA: Society for Industrial and Applied Mathematics

Benney, D.J. & Newell, A.C. 1967. The propagation of nonlinear envelopes. *Journal of Mathematics and Physics* (Name changed to: *Studies in Applied Mathematics*), 46: 133–139

Bianchi, L. 1902. *Lezioni de Geometria Differenziale*, 3 vols, Pisa: Spoerri

Gardner, C.S, Greene, J.M., Kruskal, M.D. & Miura, R.M. 1967. Method for solving the Korteweg–deVries equation. *Physical Review Letters*, 19: 1095–1097

Hasegawa, A. & Tappert, F. 1973. Transmission of stationary nonlinear optical pulses in dispersive dielectrical fibers. I. Anamolous dispersion. *Applied Physics Letters*, 23: 142–144

Korteweg, D.J. & de Vries, F. 1895. On the change of form of long waves advancing in a rectangular canal, and on a new type of long stationary waves. *Philosophical Magazine*, 39: 422–443

Lax, P.D. 1968. Integrals of nonlinear equations of evolution and solitary waves. *Communications in Pure and Applied Mathematics*, 21: 467–490

Zakharov, V.E. & Shabat A.B. 1972. Exact theory of two-dimensional self-focusing and one-dimensional self-modulation of waves in nonlinear media. *Soviet Physics, JETP*, 34: 62–69

ABLOWITZ–LADIK EQUATION

See **Discrete nonlinear Schrödinger equations**

ACOUSTIC SOLITONS

See **Nonlinear acoustics**

ACTION POTENTIAL

See **Nerve impulses**

ACTION-ANGLE VARIABLES

See **Hamiltonian systems**

ACTIVATOR-INHIBITOR SYSTEM

See **Reaction-diffusion systems**

ADIABATIC APPROXIMATION

See **Davydov soliton**

ADIABATIC INVARIANTS

Adiabatic invariants, denoted by I, are approximate constants of motion of a given dynamical system (not necessarily Hamiltonian), which are approximately preserved during a process of slow change of the system's parameters (denoted by λ). This change is on a time scale T, which is supposed to be much larger than any typical dynamical time scale such as traversal time or the period of the shortest periodic orbits.

This is an asymptotic statement, in the sense that the adiabatic invariants are better preserved, the slower the driving of the system. In other words, the switching function $\lambda = \lambda(t)$ varies more slowly on the typical evolutionary time scale T, and the preservation is perfect in the limit $T \to \infty$.

The important point is that while the system's parameters $\lambda(t)$ and their dynamical quantities such as the total energy and angular momentum can change by arbitrarily large amounts, their combination involved in the adiabatic invariant I is preserved to a very high degree of accuracy, and this allows us to calculate changes of important quantities in dynamical systems. Examples arise in celestial mechanics, in other Hamiltonian systems, and in the motion of charged particles in magnetic and electric fields.

The accuracy of preservation can be calculated in systems with one degree of freedom and is exponentially good with T if the switching function $\lambda(t)$ is analytic (of class C^∞); that is to say, the change of the adiabatic invariant ΔI is of the form

$$\Delta I = \alpha \exp(-\beta T), \tag{1}$$

where α and β are known constants. If, however, the switching function $\lambda(t)$ is only of class C^m (m-times continuously differentiable), then the change of the adiabatic invariant ΔI during an adiabatic change over a time period of length T is algebraic only, namely

$$\Delta I = \alpha T^{-(m+1)}. \tag{2}$$

In both cases, $\Delta I \to 0$ as $T \to \infty$.

The fact that the evolutionary time scale T is large compared to the typical shortest dynamical time scales (average return time, etc.) suggests the averaging method or the so-called averaging principle. Here the long-term evolution (adiabatic evolution) of the system can be calculated by replacing the actual dynamical system with its averaged correspondent, obtained by averaging over the shortest dynamical time scales (the fast variables). Such a procedure is well known, for example, in celestial mechanics where the secular effects of the third-body perturbations of a planet are obtained by averaging the perturbations over one revolutionary period of the perturbers. This was done by Carl Friedrich Gauss in 1801 in the context of studying the dynamics of planets.

The adiabatic invariants can be easily calculated in one-dimensional systems and in completely integrable systems with N degrees of freedom. Something is known about the ergodic Hamiltonian systems, while little is known about adiabatic invariants in mixed-type Hamiltonian systems (with divided phase space), where for some initial conditions in the classical phase space, we have regular motion on invariant tori and irregular (chaotic) motion for other (complementary) initial conditions.

One elementary example is the simple (mathematical) pendulum, of point mass m and of length l with the declination angle φ, described by the Hamiltonian

$$H = \frac{p_\varphi^2}{2ml^2} - mgl \cos\varphi, \tag{3}$$

where $p_\varphi = ml^2\dot{\varphi}$ is the angular momentum. For small oscillations $\varphi \ll 1$, around the stable equilibrium $\varphi = 0$. It is described by the harmonic Hamiltonian

$$H' = \frac{p_\varphi^2}{2ml^2} + \frac{mgl}{2}\varphi^2. \tag{4}$$

Here the angular oscillation frequency is $\omega = 2\pi\nu = \sqrt{g/l}$, where ν is the frequency and g is the gravitational acceleration. We denote the total energy of the Hamiltonian H' by E. Paul Ehrenfest discovered that the quantity $I = E/\omega$ is the adiabatic invariant of the system, so the change of $E(t)$ on large time scales $T \gg 1/\nu$ is such that $I = E(t)/\omega(t)$ remains constant. Therefore, if for example the length of the pendulum $l = l(t)$ is slowly, adiabatically changing, then the energy of the system will change according to the law

$$E = E_0\sqrt{\frac{l_0}{l}}, \tag{5}$$

where E_0 and l_0 are the initial values and E and l the final values of the two variables. One can easily show that the oscillation amplitude φ_0 changes as $l^{-3/4}$ as the length l changes. This is an elementary example of a dynamically driven system in which the change of energy E can be very large, as is the change of ω, but $I = E/\omega$ is a well-preserved adiabatic invariant; in fact, it is exponentially well preserved if the switching function $\lambda(t)$ is analytic.

More generally, for Hamiltonian systems $H(q, p, \lambda)$ with one degree of freedom, whose state is described by the coordinate q and canonically conjugate momentum p in the phase space (q, p), and $\lambda = \lambda(t)$ is the system's parameter (slowly changing on time scale T), one can show that the action integral

$$I(E, \lambda) = I(E(t), \lambda(t)) = \frac{1}{2\pi}\oint p\, dq \tag{6}$$

is the adiabatic invariant of the system, where the contour integral is taken at a fixed total energy E and a fixed value of λ. In this case, $2\pi I$ is interpreted as the area inside the curve $E = \text{const.}$ in the phase plane (q, p). The accuracy is exponentially good if $\lambda(t)$ is an analytic function and algebraic if it is of class C^m. Moreover, the theorem holds true only if the frequency ω is nonzero.

This implies that a passage through a separatrix (in the phase space of a one-dimensional system) is excluded because $\omega = 0$; thus a different approach is necessary with a highly nontrivial result. When crossing a separatrix of a one-dimensional double potential well from outside in an adiabatic way going inside, a bifurcation takes place, and the capture of the trajectory in either of the two wells is possible with some probabilities. These probabilities can be calculated quite easily, and the spread of the adiabatic invariant ΔI after such a passage can also be calculated, but this is more difficult. Important applications are found in celestial mechanics, where an adiabatic capture of a small body near a resonance with a planet can take place; in plasma physics; and in quantum mechanics of states close to the separatrix (in the semiclassical limit).

This is an interesting result, because I is precisely that quantity which according to the "old quantum mechanics" of Bohr and Sommerfeld has to be quantized, that is, made equal to an integer multiple of Planck's constant $\hbar$. Of course, the old quantum mechanics is generally wrong, but it can be a good approximation to the solution of the Schrödinger equation. Even then, strictly speaking, the quantization condition in the sense of EBK or Maslov quantization, must be written in the form

$$I = \frac{1}{2\pi} \oint p \, \mathrm{d}q = \left(n + \frac{\alpha}{4}\right) \hbar, \tag{7}$$

where $n = 0, 1, 2, \ldots$ is the quantum number and α is the Maslov index, that is, the number of caustics (projection singularities) round the cycle $E = \text{const.}$ in the phase plane. For smooth systems with quadratic kinetic energy, it is typically $\alpha = 2$. Thus, at this semiclassical level, we have the semiclassical adiabatic invariant, stating that in one-dimensional systems under an adiabatic change, the quantum number (and thus the eigenstate) is preserved. This agrees with the exact result in the theory of the Schrödinger equation in quantum mechanics. Round a closed loop in a parameter space, a quantum system returns to its original state, except for the phase. (This closed-loop phase change is essentially the so-called Berry's phase.)

The method of averaging can also be used in N-dimensional Hamiltonians $H = H(\boldsymbol{q}, \boldsymbol{p})$, where $\boldsymbol{q}$ and $\boldsymbol{p}$ are N-dimensional vectors, but it works only in two extreme cases: the integrable case and the ergodic case.

In a classical integrable Hamiltonian system we have N analytic, global, and functionally independent constants of motion $A_i = A_i(\boldsymbol{q}, \boldsymbol{p})$, $i = 1, 2, \ldots, N$, pairwise in involution; that is, all Poisson brackets $\{A_i, A_j\}$ vanish identically everywhere in phase space. The orbits in phase space are then confined to an invariant N-dimensional surface, and according to the Liouville–Arnol'd theorem the topology of these surfaces must be the topology of an N-dimensional torus. Then an action integral $I = (1/2\pi) \oint \boldsymbol{p} \cdot \mathrm{d}\boldsymbol{q}$ along a closed loop on a torus will be zero if the loop can be continuously shrunk to a point on the torus. But there are loops that cannot be shrunk to a point due to the topology of the torus. Then the integral I is different from zero, otherwise its value does not depend on the particular loop, so in a sense it is a topological invariant of the torus. On an N-dimensional torus, there are N such independent elementary closed loops C_i, $i = 1, 2, \ldots, N$. The integrals that we call simply actions or action variables

$$I_i = \frac{1}{2\pi} \oint_{C_i} \boldsymbol{p} \cdot \mathrm{d}\boldsymbol{q} \tag{8}$$

are then the most natural momentum variables on the torus, whilst angle variables $\boldsymbol{\Theta}$ specifying the position on the torus labeled by $\boldsymbol{I}$ can be generated from the transformation

$$\boldsymbol{\Theta} = \frac{\partial S(\boldsymbol{I}, \boldsymbol{q})}{\partial \boldsymbol{I}}, \tag{9}$$

where $S = \int \mathbf{p} \cdot \mathrm{d}\boldsymbol{q}$ is an action integral on the torus.

Applying the averaging principle (the method of averaging), one readily shows that for an integrable system the actions $\boldsymbol{I}$ are N adiabatic invariants, provided the system is nondegenerate, which means that the frequencies

$$\boldsymbol{\omega} = \frac{\partial H}{\partial \boldsymbol{I}} \tag{10}$$

on the given torus are not rationally connected; that is to say, there is no integer vector $\boldsymbol{k}$ such that $\boldsymbol{\omega} \cdot \boldsymbol{k} = 0$. The problem is that during an adiabatic process the frequencies $\boldsymbol{\omega}$ will change, and therefore, strictly speaking, there will be infinitely many points of $\lambda = \lambda(t)$, where $\boldsymbol{\omega} \cdot \boldsymbol{k} = 0$, which will, strictly speaking, invalidate the theorem. However, it is thought that if the degree of resonances or rationality conditions $\boldsymbol{\omega} \cdot \boldsymbol{k} = 0$ is of a very high order, meaning that all components of $\boldsymbol{k}$ are very large, then the adiabatic invariants I_i will be quite well preserved. But low-order resonances (rationality conditions) must be excluded. The details of such a process call for further investigation.

When the N actions I_i of an integrable system are quantized in the sense of Maslov, as explained above in the one-dimensional case, we again find agreement, at this semiclassical level, with quantum mechanics: in a family of integrable systems, all N quantum numbers and the corresponding eigenstates are preserved under an adiabatic change.

Another extreme of classically ergodic and thus fully chaotic systems has been considered already by Hertz. He found that in such ergodic Hamiltonian systems the phase space volume enclosed by the energy surface $H(\boldsymbol{q},\boldsymbol{p})=E=\text{constant}$ is the adiabatic invariant, denoted by

$$\Omega(E)=\int_{H(\boldsymbol{q},\boldsymbol{p})\leq E}\mathrm{d}^N\boldsymbol{q}\,\mathrm{d}^N\boldsymbol{p}. \qquad (11)$$

Of course, here it is required that while the system's parameter $\lambda(t)$ is slowly changing, the system itself must be ergodic for all $\lambda(t)$. Sometimes, this condition is difficult to satisfy, but sometimes it is easily fulfilled. Examples are the stadium of Bunimovich with varying length of the straight line between the two semicircles, or the Sinai billiard with varying radius of the circle inside a square. For an ergodic two-dimensional billiard of area $\mathcal{A}$ and point mass m, we have

$$\Omega(E)=2\pi m\mathcal{A}E. \qquad (12)$$

Therefore, when $\mathcal{A}$ is adiabatically changing, the energy E of the billiard particle is changing reciprocally with $\mathcal{A}$. Diminishing $\mathcal{A}$ implies increasing E, and this can be interpreted as work being done against the "pressure" of only one particle, if we define the pressure as the time average of the momentum transfer at collisions with the boundary of our ergodic billiard. There is a formalism to proceed with this analysis close to the thermodynamic formalism, as derived from statistical mechanics, except that here we are talking about time averages rather than phase averages of classical variables.

Again, this general result for ergodic systems is interesting from the quantum point of view because $\mathcal{N}=\Omega(E)/(2\pi\hbar)^N$ is precisely the number of energy levels below the energy E in the semiclassical limit of very large $\mathcal{N}$, which is known as the Thomas–Fermi rule. It is the number of elementary quantum Planck cells inside the volume element $H(\boldsymbol{q},\boldsymbol{p})\leq E$. Indeed, quantum mechanically, the eigenstate and the (energy counting sequential) main quantum number $\mathcal{N}$ are preserved under an adiabatic change.

In case of a mixed-type Hamiltonian system, which is a typical case in nature, adiabatic theory is in its infancy. Moreover, in three or higher degrees of freedom, we have universal diffusion on the Arnol'd web, which is dense on the energy surface, even for KAM-type Hamiltonian systems that are very close to integrability, like our solar planetary system. On the Arnol'd web we have diffusional chaotic motion, and there is a rigorous theory by Nekhoroshev giving a rigorous upper bound to the diffusion rate in such a case. However, when compared with numerical calculations, it is found that the diffusion rate is many orders of magnitude smaller than the Nekhoroshev limit. In other words, the actual diffusion time is much longer than estimated by Nekhoroshev, implying that there we have some approximate adiabatic invariant for long times, but not very long times.

MARKO ROBNIK

See also **Averaging methods; Berry's phase; Billiards; Phase space; Quantum theory; Quasilinear analysis**

Further Reading

Arnol'd, V.I. 1989. *Mathematical Methods of Classical Mechanics*, 2nd edition, New York and Heidelberg: Springer

Cary, J.R. & Rusu, P. 1992. Separatrix eigenfunctions. *Physical Review* A, 45: 8501–8512

Cary, J.R. & Rusu, P. 1993. Quantum dynamics near a classical separatrix. *Physical Review A*, 47: 2496–2505

Landau, L.D. & Lifshitz, E.M. 1996. *Mechanics—Course of Theoretical Physics*, vol. 1, 3rd edition, Oxford: Butterworth-Heinemann

Landau, L.D. & Lifshitz, E.M. 1997. *Quantum Mechanics: Non-Relativistic Theory—Course of Theoretical Physics*, vol. 3, 3rd edition, Oxford: Butterworth-Heinemann

Lichtenberg, A.J. & Lieberman, M.A. 1992. *Regular and Chaotic Dynamics*, 2nd edition, New York and Heidelberg: Springer

Lochak, P. & Meunier, C. 1988. *Multiphase Averaging for Classical Systems*, New York and Heidelberg: Springer

Reinhardt, W.P. 1994. Regular and irregular correspondences. *Progress of Theoretical Physics Supplement*, 116: 179–205

ALFVÉN WAVES

The essence of Hannes Alfvén's contributions to cosmic and laboratory plasmas is his idea of combining electromagnetics and hydrodynamics (Alfvén, 1942), thus introducing the new concept of magnetohydrodynamics (MHD). Electromagnetic waves associated with the motion of conducting liquids in magnetic fields, now known as Alfvén waves, were first observed experimentally (Lundquist, 1949; Lehnert, 1954). Later on, waves of this nature have turned out to be fundamental constituents of numerous phenomena in all parts of the universe (Fälthammar, 1995; Wilhelmsson, 2000). In the pioneering experiments, liquid mercury was used by Lundquist, and liquid sodium by Lehnert, who achieved higher electrical conductivity and lower density, leading to a higher Lundquist number (lower damping). Alfvén used his early results to give a possible explanation for sunspots and the solar cycle (periodicity in the Sun's activity) (Alfvén, 1942). Alfvén noticed that the Sun has a general magnetic field and that solar matter is a good conductor, thus fulfilling idealized requirements for the notion of an electromagnetic wave in a gaseous conductor or plasma.

At a very early age, Alfvén was given a copy of a popular astronomy book by Camille Flammarion, which greatly stimulated his lifelong interest in astronomy and astrophysics. His early experiences building radio receivers at the school radio club were also important for his later activities. Interestingly, another great scientist, Albert Einstein, received a small compass as a present when he was five years old, which entirely absorbed his interest. He asked everybody around him what a magnetic field was and what gravity was, and later on in his life he admitted that this early experience might have influenced his lifelong scientific activities. Other similarities between the two scientists were that in their professional work both Einstein and Alfvén were very creative individualists, striving for simplicity of their solutions, and being skilled in many areas, they often looked at problems with fresh eyes. Both received Nobel Prizes in physics: Einstein in 1922, Alfvén in 1970.

The simplest form of an Alfvén wave, a propagation of an electromagnetic wave in a highly conducting plasma, was first rejected by critics on the grounds that it could not be correct, otherwise it would already have been discovered by Maxwell. Furthermore, experiments had been performed with magnetic fields and conductive media by Ampère and others long ago. Nevertheless, "The Alfvén wave, in fact, is the very foundation on which the entire structure of magnetohydrodynamics (MHD) is erected. Beginning from a majestic original simplicity, it has acquired a rich and variegated character, and has ended up dictating most of the low-frequency dynamics of magnetized plasmas" (Mahajan, 1995).

To visualize the interaction between the magnetic field and the motion of the conductive fluid, one may use an analogy with the theory of stretched strings to obtain a wave along the magnetic lines of force with a velocity v_A, where

$$v_A{}^2 = \frac{B^2}{\mu_0 \rho} \tag{1}$$

and ρ is the mass density of the fluid, μ_0 is the permittivity, and B is the magnetic field.

The variations in velocity and current are mutually perpendicular and the magnetic field variations are in the direction of the fluid velocity variations, all variations being perpendicular to the direction of propagation. One may say that the variations of the magnetic field lines are frozen to those of the fluid motion, as can be deduced from electromagnetic equations, together with the hydrodynamic equation for the case of an incompressible fluid of infinite conductivity. The Alfvén wave is a low-frequency wave ($\omega < \omega_{ci}$, ω_{ci} being the ion cyclotron frequency) for which the displacement current is negligible. In fact, there are two types of Alfvén waves, for which

$$\omega/k_\parallel = v_A \text{ (torsional or shearwave)} \tag{2}$$

and

$$\omega/k = v_A \text{ (compressional wave)} \tag{3}$$

where ω is the frequency, $k^2 = k_\parallel^2 + k_\perp^2$, with $k_\parallel$ and $k_\perp$ being the wave numbers along and perpendicular to the magnetic field. For the shear wave, the frequency depends only on $k_\parallel$ and not on $k_\perp$, which has profound consequences and leads to a continuous spectrum (Mahajan, 1995).

For determining plasma stability and in selecting schemes for plasma heating and current drive in fusion plasma devices, the understanding of Alfvén wave dynamics is of great importance and has led to a vast literature. Nonlinear effects are of relevance to large-amplitude disturbances frequently observed in laboratory and space plasmas (Wilhelmsson, 1976). The formation and propagation of Alfvén vortices with geocosmophysical and pulsar (electron-positron plasma) applications are just two examples. Alfvén waves have also found interesting applications in solid-state plasmas in semiconductors as well as in metals and semimetals. Such studies have resulted in refined methods of measuring magnetic fields.

It is often said that the universe consists 99% of plasma. Alfvén used to say that it seems as if only the crust of the Earth is not plasma. In the mid-1960s, this author gave a talk at the Royal Institute of Technology in Stockholm about plasmas in solids (electrons and holes), which Hannes Alfvén himself attended. Among other things, the talk described recent observations of Alfvén waves in such plasmas, and Alfvén said: "Ah, they are here also. How interesting, I did not know that."

It was not until the middle of the 20th century that more intensive investigations on Alfvén waves in space and laboratory plasmas began. The slow development of the field of space plasmas was possibly because many physicists were not acquainted with the fact that electric currents can be distributed in large volumes and magnetic fields in such volumes can be present. Since then, the gigantic laboratory of the universe from the aurora originating in the Earth's magnetosphere to quasars at the rim of the universe has attracted immense interest with regard to Alfvén waves.

When propagating in inhomogeneous plasmas, for example, in the magnetosphere, the Alfvén wave experiences many interesting phenomena, including mode coupling, resonant mode conversion, and resonant absorption. We now know that shear Alfvén waves lie behind the phenomena of micropulsations in the geomagnetic field and also acceleration of particles. Micropulsations were detected a hundred years ago with simple magnetometers on the ground. It took more than 50 years before it was understood that they were related to the magnetosphere. Solar physics is another fascinating field where Alfvén waves occur, giving rise to sunspots (Alfvén, 1942). The vast amount of energy exhibited in eruptions of particles on the solar surface, originating in the interior of the Sun, is probably transported by Alfvén waves. These also play a role in heating the solar corona. Alfvén waves were first identified in the solar wind by means of spacecraft measurements by the end of the 1960s. They also occur in the exosphere of comets. A new and promising area of research is laboratory astrophysics using high-intensity particle and photon beams that may shed light on superstrong fields in plasmas.

For applications to confinement and heating of fusion plasmas, for example, in Tokamak devices, shear Alfvén waves have been studied in toroidal plasmas, accounting for nonuniform plasmas in axisymmetric situations. It is believed that the remaining exciting challenges lie in the area of nonlinear physics of shear Alfvén waves and associated particle dynamics and anomalous losses of α particles in a deuterium-tritium plasma. Collective modes in inhomogeneous plasmas as well as energy and particle transport in plasmas with transport barriers are of paramount importance for the design of a future Tokamak power plant (Parail, 2002).

Nonlinear transport processes in laboratory and cosmic plasmas have much in common (Wilhelmsson, 2000; Wilhelmsson & Lazzaro, 2001). Similarities (and discrepancies) could be highly indicative and beneficial for an improved understanding of specific phenomena as well as for plasma dynamics in general—possibly even for describing the evolution of the universe (Wilhelmsson, 2002).

HANS WILHELMSSON

See also **Magnetohydrodynamics; Nonlinear plasma waves; Plasma soliton experiments**

Further Reading

Alfvén, H. 1942. Existence of electromagnetic-hydrodynamic waves. *Nature*, 150: 405–406

Fälthammar, C.-G. 1995. Hannes Alfvén. In *Alfvén Waves in Cosmic and Laboratory Plasmas*, edited by A.C.-L. Chian, A.S. de Assis, C.A. de Azevedo, P.K. Shukla & L. Stenflo, Proceedings of the International Workshop on Alfvén Waves, *Physica Scripta*, T60: 7

Lehnert, B. 1954. Magnetohydrodynamic waves in liquid sodium. *Physical Review*, 94: 815

Lundquist, S. 1949. Experimental demonstration of magnetohydrodynamic waves. *Nature*, 164: 145

Mahajan, S.M. 1995. Spectrum of Alfvén waves, a brief review. *Physica Scripta*, T60: 160–170

Marston, E.H. & Kao Y.H. 1969. Damped Alfvén waves in bismuth. A determination of charge-carrier relaxation times. *Physical Review*, 182: 504

Parail, V.V. 2002. Energy and particle transport in plasmas with transport barriers. *Plasma Physics and Controlled Fusion*, 44: A63–85

Wilhelmsson, H. (editor). 1982. The physics of hot plasmas. Proceedings of the International Conference on Plasma Physics, Göteborg, May 1982, *Physica Scripta*, T2 (1 and 2)

Wilhelmsson, H. (editor). 1976. *Plasma Physics: Nonlinear Theory and Experiments*, New York and London: Plenum Press

Wilhelmsson, H. 2000. *Fusion: A Voyage through the Plasma Universe*, Bristol and Philadelphia: Institute of Physics Publishing

Wilhelmsson, H. 2002. Gravitational contraction and plasma fusion burn; universal expansion and the Hubble Law. *Physica Scripta*, 66: 395

Wilhelmsson, H. & Lazzaro, E. 2001. *Reaction-Diffusion Problems in the Physics of Hot Plasmas*, Bristol and Philadelphia: Institute of Physics Publishing

ALGORITHMIC COMPLEXITY

The notion of complexity as an object of scientific interest is relatively new. Prior to the 20th century, the main concern was that of simplicity, with complexity being the denigrated opposite. This idea of simplicity has had a long history enshrined in the dictum of the 14th-century Franciscan philosopher, William of Occam, that "pluritas non est ponenda sine necessitate" [being is not multiplied without necessity], and passed on simply as "Occam's razor," or more prosaically, "keep it simple" (Thorburn, 1918).

Indeed, the razor has been invoked by such notables as Isaac Newton and, in modern times, Albert Einstein and Stephen Hawking to justify parsimony in the adoption of physical principles. Although the dictum has proved its usefulness as a support for many scientific theories, the last century witnessed a gradual concern for simplicity's complement. Implicit was a recognition that beyond logical partitions was a need to quantify the simple/complex continuum. Perhaps the first milestone on the road to quantifying complexity came with Claude Shannon's famous information entropy in the late 1940s. Although it was not specifically developed as a complexity measure, the information connection made by Warren Weaver soon provided an impetus for sustained interest in information as a unifying concept for complexity (Weaver, 1948).

Shannon approached information as a statistical measure of receiving a message (Shannon, 1948): if $p_1, p_2, \ldots, p_N$ are the probabilities of receiving messages $m_1, m_2, \ldots, m_N$, then the information carried is defined by

$$I = -\sum_{i}^{N} p_i \log_2 p_i. \qquad (1)$$

Information is typically referred to as a measure of surprise; that is, the more unlikely a message, the more information it contains. To some degree, information is related to the notion of randomness in that the more regular (less complex, less random) something is, the less surprise is available. A simple calculation of this entropy demonstrates that the maximum of the function is achieved when all probabilities are equal. Shannon had a measure of capacities of a communication channel as his goal and did not concern himself with individual objects of the messages. Nonetheless, the quantification in terms of probabilities provided a basis for viewing complexity.

This theme was soon independently taken up by Ray Solomonoff (1964), Andrey Kolmogorov (1965), and Gregory Chaitin (1966). In a sense, Somolonoff was looking for a way to measure the effect of Occam's razor; that is, how can one measure objectively the simplicity of a scientific theory? Kolmogorov and Chaitin, on the other hand, were interested in a measure of complexity of individual objects, as opposed to Shannon's average. This Kolmogorov–Chaitin complexity has come to be known variously as algorithmic complexity, algorithmic entropy, and algorithmic randomness, among other designations. Both Kolmogorov and Chaitin were interested in binary number strings as objects and the ability to define the complexity of a string in terms of the shortest algorithm that prints out the string. Again, regularity and randomness is involved (Gammarman & Vovk, 1999). Consider, for example, the simple bit string, 101010101010, . . .; the minimal program to write the string requires only the pattern 10, the length of the string, and the "repeat, write" instructions, or

$$K(s) = \min\{|p| : s = C_T(p)\}, \tag{2}$$

where $K(s)$ is the Kolmogorov complexity of the string, $|p|$ is the program length in bits, and $C_T(p)$ is the result of running program p on a universal Turing machine T.

Clearly, the recognition that patterns play an important role in defining complexity re-emphasized their importance in terms of data compression. In the early 1950s, David Huffman recognized their importance, and algorithmic complexity reaffirmed their utility with the ascendancy of computers and their demand for storage space (Huffman, 1952). Thus, numerous coding schemes were developed to take advantage of the fact that a simple algorithm can compress long data streams based upon the idea that recurrent patterns exist.

The efforts of Somonoloff, Kolmogorov, and Chaitin spawned numerous alternative measures of complexity, often seeking to address identified deficiencies in the definitions (Shalizi & Crutchfield, 2001). Among the deficiencies pointed out were the following: (i) Complexity is defined in terms of randomness—it is maximized by random strings. Is this what is really sought? (ii) Complexity is uncomputable, since there is no algorithm to compute it on a universal computing machine. (iii) Complexity does not provide information regarding structural patterns or organizations that have the same amount of disorder.

These questions were compounded by the expanding field of nonlinear dynamics. Kolmogorov's earlier entropy (1958)—developed to determine the rate of information creation—were among the invariants used to distinguish chaotic systems and an inferred complexity. Moreover, the description of physical dynamical systems became an additional issue (Zurek, 1989). Physical processes were typically described along the continuum of two extremes: periodic or random. However, both such systems are simply described, one by a recurrent pattern and the other by a statistical description. While information is high in a random system, it is low in a periodic process. Amalgams of two such processes might require considerable computational effort, yet no metric sufficiently expressed this. Certainly, such combined processes (random and periodic) may exhibit moderate information but the most concise description may be quite complex.

Some of these difficulties have been addressed with varying degrees of acceptance (Wakerberger et al., 1994). Increasingly, however, it appears that the question is evolving along two different lines: a formal approach (rules) with its main ramifications redounding to mathematics and computer science and a physical approach (equations) dealing with the characterization of systems.

Both approaches have in common the emphasis on the reconstruction (prediction) of an observed system and on the need to give the most parsimonious recipe for generating the studied entity. The metrics of complexity is thus expressed in terms of program lines for the mathematics-oriented option and in dimensionality for the physics-oriented definition, but there is the same basic notion of complexity as the inverse of compressibility of a given object (Boffetta et al., 2002).

This notion of compressibility has an immediate translation in terms of both multidimensional statistics and technology. In multidimensional statistics, the compressibility of a given data set corresponds to the percentage of explained variance by the optimal (generally in a least-squares sense) model of the data. More generally, something can be compressed if there exists some sort of correlation structure linking the different portions of a system, the existence of such correlations implying that the information about one part of the system is implicit in another part. Thus, all the information is not needed to reconstruct the entire system. It is evident how this concept corresponds to the cognate concept of redundancy, bringing us back to the notion of entropy (Giuliani et al., 2001).

Clearly, the diverse algorithms designed to measure complexity suggest a commonality. The question remains as to whether one metric is sufficient for its characterization.

Joseph P. Zbilut and Alessandro Giuliani

See also **Entropy; Information theory; Structural complexity**

Further Reading

Boffetta, G., Cencini, M., Falcioni, M. & Vulpiani, A. 2002. Predictability: a way to characterize complexity. *Physics Reports*, 356: 367–474

Chaitin, G.J. 1966. On the length of programs for computing finite binary sequences. *Journal of the Association for Computing Machinery*, 13: 547–569
Gammerman, A. & Vovk, V. 1999. Kolmogorov complexity: sources, theory and applications. *The Computer Journal*, 42: 252–255
Gell-Mann, M. & Lloyd, S. 1999. Information measures, effective complexity, and total information. *Complexity*, 2: 44–52
Giuliani, A., Colafranceschi, M., Webber Jr., C.L. & Zbilut, J.P. 2001. A complexity score derived from principal components analysis of nonlinear order measures. *Physica* A, 301: 567–588
Huffman, D.A. 1952. A method for the construction of minimum redundancy codes. *Proceedings IRE*, 40: 1098–1101
Kolmogorov, A.N. 1958. A new metric invariant of transitive dynamical systems and automorphism in Lebesgue spaces. *Doklady Akademii Nauk SSSR [Proceedings of the Academy of Sciences of the USSR]*, 119: 861–864
Kolmogorov, A.N. 1965. Tri podkhoda k opredeleniiu poniatiia "kolichestvo informatsii". [Three approaches to the quantitative definition of information.] *Problemy Peredachy Informatsii [Problems Information Transmission]*, 1: 3–11
Lempel, A. & Ziv, J. 1976. On the complexity of finite sequences. *IEEE Transactions on Information Theory*, 22: 75–81
Li, M. & Vitányi, P. 1997. *An Introduction to Kolmogorov Complexity and Its Applications*, 2nd edition, New York: Springer
Salomon, D. 1998. *Data Compression, the Complete Reference*, New York: Springer
Shalizi, C.R. & Chrutchfield, J.P. 2001. Computational mechanics: pattern and prediction, structure and simplicity. *Journal of Statistical Physics* 104: 819–881
Shannon, C.E. 1948. A mathematical theory of communication. *Bell System Technical Journal*, 27: 379–423
Solomonoff, R.J. 1964. The formal theory of inductive inference, parts 1 and 2. *Information and Control*, 7: 1–22, 224–254
Thorburn, W.M. 1918. The myth of Occam's razor. *Mind*, 27: 345–353
Wackerberger, R., Witt, A., Atmanspacher, H., Kurths, J. & Scheingraber, H. 1994. A comparative classification of complexity measures. *Chaos, Solitons & Fractals*, 4: 133–173
Weaver, W. 1948. Science and complexity. *American Scientist*, 36: 536–544
Zurek, W. 1989. Thermodynamic cost of computation, algorithmic complexity, and the information metric. *Nature*, 341: 119–124

ALL-OR-NOTHING RESPONSE

See **Nerve impulses**

ALMOST PERIODIC FUNCTIONS

See **Quasiperiodicity**

AMBIGUOUS FIGURES

See **Cell assemblies**

ANDERSON LOCALIZATION

Anderson localization is a phenomenon associated with the interference of waves in random media. Although Philip Anderson's original publication (Anderson, 1958) was actually motivated by experiments on the propagation of spin-waves in random magnets, the greatest application of the concept has been the study of electrical transport phenomena in metals and semiconductors. Over the past 10 years, more attention has been focused on other wave phenomena in random media, particularly optical phenomena.

Our current understanding of electronic transport in metals and semiconductors is based on the Schrödinger equation for the wave function of conduction electrons of the form

$$-\frac{\hbar^2}{2m^*}\nabla^2\psi(\boldsymbol{r}) + [U(\boldsymbol{r}) + V(\boldsymbol{r})]\,\psi(\boldsymbol{r}) = E\psi(\boldsymbol{r}),$$

where $U(\boldsymbol{r})$ is a periodic potential representing the regular lattice in the solid and $V(\boldsymbol{r})$ is a random function of position, which represents the presence of impurities in the system. In the absence of the random potential, the allowed energies of such an electron fall within a series of bands separated by energy gaps. The eigenfunctions in the absence of the random potential are all of the form $\mathrm{e}^{\mathrm{i}\boldsymbol{k}\cdot\boldsymbol{r}}u_{j,\boldsymbol{k}}(\boldsymbol{r})$ where the wavevector $\boldsymbol{k}$ lies within the first Brillouin zone, j labels the energy bands and the Bloch function, $u_{j,\boldsymbol{k}}(\boldsymbol{r})$, has the periodicity of the regular potential, U. Such states are *extended* in the sense that their support covers the entire system and hence they can contribute to electrical conduction.

The eigenstates of an electron subject to a random potential may be of two types. Some are extended, although there may be strong local modulations in the amplitude. These states can contribute to electrical conduction through the material, even at zero temperature, as they connect the two ends of a sample. Other states, however, are localized in that their amplitude vanishes exponentially outside a specific finite region. These states are referred to as *Anderson localized* and can only contribute to conduction via thermal activation.

One can understand the existence of localized states by considering the low-energy states of an electron moving in a very rough, random potential, $V(\boldsymbol{r})$ (Lee & Ramakrishnan, 1985). The lowest energy states will be those bound to very deep troughs in the potential function. The mixing between states localized in different wells will be very weak because states with significant spatial overlap will have very different energies, while states with similar energies are spatially well separated so that the wave function overlaps are exponentially small.

The scale on which the wave functions of localized states decay to zero defines the localization length ξ, which depends on the energy of the state and the strength of the disorder. The balance between extended and localized states depends on the strength of the disorder and the spatial dimensionality of the

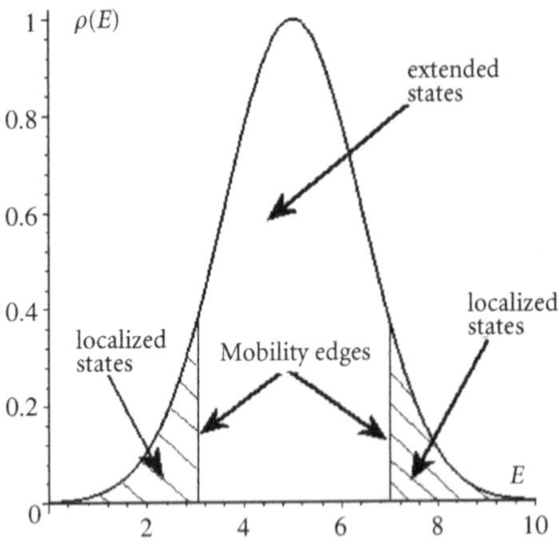

Figure 1. A schematic plot of the density of states showing a single disorder broadened band for a 3-d system. The states in the band center are extended while those in the tails are localized (shaded regions); the mobility edges between the two types of state are marked.

system. In a one-dimensional (1-d) system, all of the electronic eigenstates are strongly localized by any amount of disorder. In two dimensions, it is believed that all states are actually localized, but that the localization length can be very long in the center of a band. The application of a strong magnetic field to a disordered 2-d electron system, such as may be formed at a semiconductor heterojunction at low temperature, causes the conduction band to break up into a sequence of disorder broadened Landau bands, each with an extended state at its center—an essential feature of the quantum Hall effect. In three dimensions, the eigenstates at the center of a band are truly extended while those in the low- and high-energy tails are localized. It is believed that there are two well-defined critical energies within the band at which the nature of the states changes so that localized and extended states do not co-exist at a given energy.

The critical energies are usually referred to as mobility edges because the zero-temperature conductance of the system will be zero when the Fermi energy lies in the regime of localized states but nonzero in the extended regime (see Figure 1). The location of the mobility edges depends on the strength of the disorder —in very clean systems, only the states in the tails of a band will be localized while in a very dirty system the mobility edges may meet in the band center so that all states are localized. The behavior at the mobility edge has been studied by performing experiments on a series of devices with increasing amounts of disorder. The transition between metallic (conducting) and insulating behavior is closely analogous to other phase transitions. Mott (1973) supposed that this metal-insulator transition was first order, with the conductivity jumping from a fixed finite value, $\sigma_{\min}$, to zero. In 1979, a renormalization group analysis was carried out by the so-called "gang of four," based on earlier scaling arguments by Thouless and co-workers (Thouless, 1974), which favored a continuous metal-insulator transition for 3-d systems (Abrahams et al., 1979). The discussion of the change in nature of the states from extended to localized in terms of a zero-temperature quantum phase transition has been very fruitful. In this description, the localization length, ξ, plays the same role as the correlation length for fluctuations in a thermal transition. It is supposed that ξ diverges at the mobility edges according to a universal power law

$$\xi \sim |E - E_c|^{-\nu} .$$

Numerical evidence indicates that the value of the exponent is indeed universal and has the value $\nu \sim 1.6$.

Although the underlying physics of Anderson localization is that of linear waves in random media, the discussion can be recast in terms of nonlinear models without disorder, which are in the same family of statistical field theories used to describe thermal phase transitions, specifically nonlinear sigma models (Efetov, 1997). This has led to the notion that the spatial variation of the wave functions of states at the mobility edge displays a multifractal character.

The application of these ideas to other wave phenomena in random media has been slower. It is much harder to observe strong localization in bosonic and classical wave systems, but recently much experimental work has been carried out on optical and acoustic localization (see John (1990) for a good introduction). This work shows that Anderson localization is not an essentially quantum mechanical effect but is ubiquitous for wave propagation in random media. Similarly, the interplay between the physics of randomly disordered systems and quantum chaos is also proving very rich and fruitful (Altshuler & Simons, 1994).

There are a number of other mechanisms whereby wave excitations may become spatially localized. The propagation of excitations within macromolecules, for example, may become localized both because of interference effects associated with "random" changes in structure and also because of self-trapping effects associated with nonlinearity in the wave equation for these modes. In the case of electrons in solids, electronic excitations may become localized both by random variations in potential and by interaction effects that give rise to the so-called Mott transition. Such interaction effects are responsible for the phenomenon of Coulomb blockade observed in semiconductor nanostructures.

Keith Benedict

See also **Discrete self-trapping system; Local modes in molecular crystals**

Further Reading

Abrahams, E., Anderson, P.W., Licciardello, D.C. & Ramakrishnan, T.V. 1979. Scaling theory of localization: Absence

of quantum diffusion in two dimensions. *Physical Review Letters*, 42: 673
Altshuler, B. & Simons, B.D. 1994. Universalities: from Anderson localization to quantum chaos. In *Mesoscopic Quantum Physics, Proceedings of the 61st Les Houches Summer School,* edited by E. Akkermans, G. Montambaux, J.-L. Pichard & J. Zinn-Justin, Amsterdam: North-Holland
Anderson, P.W. 1958. The absence of diffusion in certain random lattices. *Physical Review*, 109: 1492
Efetov, K. 1997. *Supersymmetry in Disorder and Chaos*. Cambridge and New York: Cambridge University Press
John, S. 1990. The localization of waves in disordered media. In *Scattering and Localization of Classical Waves in Random Media,* edited by P. Sheng. Singapore: World Scientific, pp. 1–96
Lee, P.A. & Ramakrishnan, T.V. 1985. Disordered electronic systems. *Reviews of Modern Physics*, 57: 287
Mott, N.F. 1973. In *Electronic and Structural Properties of Amorphous Semiconductors*, edited by P.G. LeComber & J. Mort, London: Academic Press, p. 1
Thouless, D.J. 1974. Electrons in disordered systems and the theory of localization. *Physics Reports*, 13: 93

ANNIHILATION (KINK-ANTIKINK)

See **Sine-Gordon equation**

ANOSOV AND AXIOM-A SYSTEMS

Two classes of dynamical systems exhibiting chaotic behavior were axiomatically defined and systematically studied for the first time in the 1960s. Previous studies had concentrated on more specific situations. Axiom-A systems were introduced by Stephen Smale in his seminal paper (Smale, 1967). Anosov systems, which are a special case of Axiom-A systems, were studied independently in Moscow around the same period. Today, Anosov and Axiom-A systems are valued as idealized models of chaos: while the conditions defining Axiom A are too stringent to include many real-life examples, it is recognized that they have features shared in various forms by most chaotic systems.

Definitions

First, we give the definitions in the discrete-time case.

Let f be a smooth invertible map (for basic notions, *See* **Phase space**). A compact invariant set of f is said to be *hyperbolic* if at every point in this set, the tangent space splits into a direct sum of two subspaces E^{u} and E^{s} with the property that these subspaces are invariant under the differential $\mathrm{d}f$, that is, $\mathrm{d}f(x)E^{\mathrm{u}}(x) = E^{\mathrm{u}}(f(x))$, $\mathrm{d}f(x)E^{\mathrm{s}}(x) = E^{\mathrm{u}}(f(x))$, and that $\mathrm{d}f$ expands vectors in E^{u} and contracts vectors in E^{s}.

If $E^{\mathrm{u}} = \{0\}$ in the definition above, then the invariant set is made up of attracting fixed points or periodic orbits. Similarly, if $E^{\mathrm{s}} = \{0\}$, then the orbits are repelling. If neither subspace is trivial, then the behavior is locally "saddle-like," that is to say, relative to the orbit of a point x, most nearby orbits diverge exponentially fast both in forward and in backward time. This is why hyperbolicity is a mathematical notion of chaos.

An *Anosov diffeomorphism* is a smooth invertible map of a compact manifold with the property that the entire space is a hyperbolic set.

Axiom A, which is a larger class, focuses on the part of the system that is not transient. More precisely, a point x in the phase space is said to be nonwandering if every neighborhood U of x contains an orbit that returns to U. A map is said to satisfy Axiom A if its nonwandering set is hyperbolic and contains a dense set of periodic points.

Definitions in the continuous-time case are analogous: f above is replaced by the time-t-maps of the flow, and the tangent spaces now decompose into $E^{\mathrm{u}} \oplus E^0 \oplus E^{\mathrm{s}}$ where E^0, which is 1-d, represents the direction of the flow lines.

Phase Space Structures and Properties

Anosov and Axiom-A systems are defined by the behavior of the differential. Corresponding to the linear structures left invariant by $\mathrm{d}f$ are nonlinear structures, namely stable manifolds tangent to E^{s} and unstable manifolds tangent to E^{u}.

Thus, two families of invariant manifolds are associated with an Anosov map and each one of these fills up the entire phase space; they are sometimes called the *stable* and *unstable foliations*. The leaves of these foliations are transverse at each point, forming a kind of (topological) coordinate system. The map f expands distances along the leaves of one of these foliations and contracts distances along the leaves of the other. For Axiom-A systems, one has a similar local product structure or "coordinate system" at each point in the nonwandering set, but the picture is local, and there are gaps: the stable and unstable leaves do not necessarily fill out open sets in the phase space.

In addition to these local structures, Axiom-A systems have a global structure theorem known as *spectral decomposition.* It says that the nonwandering set of every Axiom-A map can be written as $X_1 \cup \cdots \cup X_r$ where the X_i are disjoint closed invariant sets on which f is topologically transitive. The X_i are called basic sets. Each X_i can be decomposed further into a finite union $\bigcup X_{i,j}$, where each $X_{i,j}$ is invariant and topologically mixing under some iterate of f. (Topological transitivity and mixing are irreducibility conditions; *See* **Phase space**.) This decomposition is reminiscent of the corresponding result for finite-state Markov chains.

One of the reasons why hyperbolic sets are important is their robustness: they cannot be perturbed away. More precisely, let f be a map with a hyperbolic set Λ that is locally maximal, that is, it is the largest invariant set in some neighborhood U. Then for every map g that is C^1 near f, the largest invariant set Λ' of g in

U is again hyperbolic; moreover, f restricted to Λ is *topologically conjugate* to g restricted to Λ'. This is mathematical shorthand for saying that not only are the two sets Λ and Λ' topologically indistinguishable, but the orbit structure of f on Λ is indistinguishable from that of g on Λ'.

The above phenomenon brings us to the idea of *structural stability*. A map f is said to be *structurally stable* if every map g, that is C^1 near f is topologically conjugate to f (on the entire phase space). It turns out that a map is structurally stable if and only if it satisfies Axiom A and an additional condition called strong transversality.

Next, we discuss the idea of pseudo-orbits versus real orbits. Letting $d(\cdot,\cdot)$ be the metric, a sequence of points $x_0, x_1, x_2, \ldots$ in the phase space is called an *ε-pseudo-orbit* of f if $d(f(x_i), x_{i+1}) < \varepsilon$ for every i. Computer-generated orbits, for example, are pseudo-orbits due to round-off errors. A fact of consequence to people performing numerical experiments is that in hyperbolic systems, small errors at each step get magnified exponentially fast. For example, if the expansion rate is ≥ 3, then an ε-error made at one step is tripled at each subsequent step, that is, after only $O(|\log \varepsilon|)$ iterates, the error is $O(1)$, and the pseudo-orbit bears no relation to the real one. There is, however, a theorem that states that every pseudo-orbit is shadowed by a real one. More precisely, given a hyperbolic set, there is a constant C such that if $x_0, x_1, x_2, \ldots$ is an ε-pseudo-orbit, then there is a phase point z such that $d(x_i, f^i(z)) < C\varepsilon$ for all i. Thus, paradoxical as it may first seem, this result asserts that on hyperbolic sets, each pseudo-orbit approximates a real orbit, even though it may deviate considerably from the one with the same initial condition.

The shadowing orbit corresponding to a bi-infinite pseudo-orbit is, in fact, unique. From this, one deduces the following Closing Lemma: for any hyperbolic set, there is a constant C such that the following holds: every finite orbit segment $x, f(x), \ldots, f^{n-1}(x)$ that nearly closes up, that is, $d(x, f^{n-1}(x)) < \varepsilon$ for some small ε, lies within $< C\varepsilon$ of a genuine periodic orbit of period n. Thus, hyperbolic sets contain many periodic points.

Examples

A large class of Anosov diffeomorphisms comes from linear toral automorphisms, that is, maps of the n-dimensional torus induced by $n \times n$ matrices with integer entries, det $= \pm 1$, and no eigenvalues of modulus one. (*See* **Cat map** for a detailed example of this). We remark that due to their structural stability (nonlinear), perturbations of linear toral automorphisms continue to have the Anosov property. This remark also applies to all of the examples below. In fact, all known Anosov diffeomorphisms are

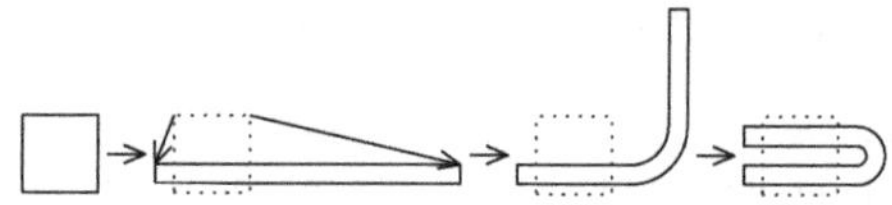

Figure 1. The horseshoe.

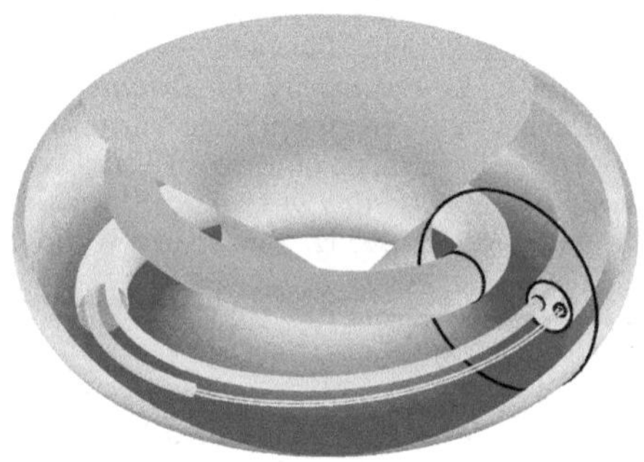

Figure 2. The solenoid.

topologically identical to a linear toral automorphism (or a slight generalization of these).

Geodesic flows describe free motions of points on manifolds. Let M be a manifold. Given $x \in M$ and a unit vector v at x, there is a unique geodesic starting from x in the direction v. The geodesic flow φ^t is given by $\varphi^t(x, v) = (x', v')$, where x' is the point t units down the geodesic and v' is the direction at x'. Geodesic flows on manifolds of strictly negative curvature are the main examples of Anosov flows. They were studied by Jacques Hadamard (ca. 1900) and Gustav Hedlund and Eberhard Hopf (in the 1930s) considerably before Anosov theory was developed.

Smale's *horseshoe* is the prototypical example of a hyperbolic invariant set. This map, so called because it bends a rectangle B into the shape of a horseshoe and puts it back on top of B, is shown in Figure 1. The set $\{x\colon f^n(x) \in B \text{ for all } n = 0, \pm 1, \pm 2, \ldots\}$ is hyperbolic (*See* **Horseshoes and hyperbolicity in dynamical systems; Phase space**).

Finally, we mention the *solenoid* (see Figure 2, and also in the color plate section as the Smale solenoid), which is an example of an Axiom-A attractor. Here, the map f is defined on a solid torus $M = S^1 \times D_2$, where D_2 is a 2-d disk. It is easiest to describe it in two steps: first it maps M into a long thin solid torus, which is then placed inside M winding around the S^1 direction twice. The attractor is given by $\Lambda = \bigcap_{n \geq 0} f^n(M)$.

Symbolic Coding of Orbits and Ergodic Theory

An important tool for studying the orbit structure of Axiom-A systems is the *Markov partition*, constructed for Anosov systems by Sinai and extended to Axiom-A basic sets by Bowen. Given a partition $\{R_1, \ldots, R_k\}$ of the phase space, there is a natural way to attach to each point x in the phase space a sequence of symbols, namely $(\ldots, a_{-1}, a_0, a_1, a_2, \ldots)$ where $a_i \in \{1, 2, \ldots, k\}$ is the name of the partition element containing $f^i(x)$, that is, $f^i(x) \in R_{a_i}$ for each i. In general, not all sequences are realized by orbits of f. Markov partitions are designed so that the set of symbol sequences that correspond to real orbits has Markovian properties; it is called a shift of finite type (*See* **Symbolic dynamics**).

The ergodic theory of Axiom-A systems has its origins in statistical mechanics. In a 1-d lattice model in statistical mechanics, one has an infinite array of sites indexed by the integers; at each site, the system can be in any one of a finite number of states. Thus, the configuration space for a 1-d lattice model is the set of bi-infinite sequences on a finite alphabet. Identifying this symbol space with the one from Markov partitions, Sinai and Ruelle were able to transport some of the basic ideas from statistical mechanics, including the notions of Gibbs states and equilibrium states, to the ergodic theory of Axiom-A systems.

The notion of *equilibrium states*, which is equivalent to Gibbs states for Axiom-A systems, has the following meaning in dynamical systems in general: given a potential function φ, an invariant measure is said to be an equilibrium state if it maximizes the quantity

$$h_\mu(f) - \int \varphi \, d\mu,$$

where $h_\mu(f)$ denotes the Kolmogorov–Sinai entropy of f and the supremum is taken over all f-invariant probability measures μ. In particular, when $\varphi = 0$, this measure is the measure that maximizes entropy; and when $\varphi = \log|\det(df|_{E^u})|$, it is the Sinai–Ruelle–Bowen (SRB) measure. From a physical or observational point of view, SRB measures are the most important invariant measures for dissipative dynamical systems (*See* **Sinai–Ruelle–Bowen measures**).

Periodic Points and Their Growth Properties

We discuss briefly some further results related to the abundance of periodic points in Axiom-A systems.

For an Axiom-A diffeomorphism f, if $P(n)$ is the number of periodic points of period $\leq n$, then $P(n) \sim e^{hn}$ where h is the topological entropy of f. That is to say, the dynamical complexity of f is reflected in its periodic behavior. An analogous result holds for Axiom-A flows.

Finally, we mention the dynamical zeta function, which sums up the periodic information of a system. In the discrete-time case, $\zeta(z) := \exp \sum_{n=1}^{\infty} P(n) z^n / n$ has been shown to be a rational function analytic on $|z| < e^{-h}$. In the continuous-time case, the zeta function is given by $\zeta(z) := \prod_\gamma (1 - \exp(-z\, l(\gamma)))^{-1}$, where the product is taken over all (nonstationary) periodic orbits γ and $l(\gamma)$ is the smallest positive period of γ. This function is known to be meromorphic on a certain domain, but the locations of its poles, which are intimately related to correlation decay properties of the system, remain one of the yet unresolved issues in Axiom-A theory.

Boris Hasselblatt and Lai-Sang Young

See also **Cat map; Horseshoes and hyperbolicity in dynamical systems; Phase space; Sinai–Ruelle–Bowen measures; Symbolic dynamics**

Further Reading

Bowen, R. 1975. *Equilibrium States and the Ergodic Theory of Anosov Diffeomorphisms*, Berlin and New York: Springer

Branner, B. & Hjorth, P. (editors). 1995. *Real and Complex Dynamical Systems. Proceedings of the NATO Advanced Study Institute held in Hillerød, June 20–July 2, 1993*, Dordrecht and Boston: Kluwer

Fielder, B. (editor). 2002. *Handbook of Dynamical Systems*, Vol. 2, Amsterdam and New York: Elsevier

Hasselblatt, B. & Katok, A. (editors). 2002. *Handbook of Dynamical Systems*, Vol. 1A, Amsterdam and New York: Elsevier

Hasselblatt, B. & Katok, A. 2003. *Dynamics: A First Course*, Cambridge and New York: Cambridge University Press

Katok, A. & Hasselblatt, B. 1995. *Introduction to the Modern Theory of Dynamical Systems*, Cambridge and New York: Cambridge University Press

Smale, S. 1967. Differentiable dynamical systems. *Bulletin of the American Mathematical Society*, 73: 747–817

ANTISOLITONS

See **Solitons, types of**

ANTI-STOKES SCATTERING

See **Rayleigh and Raman scattering and IR absorption**

ARNOL'D CAT MAP

See **Cat map**

ARNOL'D DIFFUSION

For near-integrable Hamiltonian systems with more than two degrees of freedom, stochastic and regular trajectories are intimately co-mingled in the $2N$-dimensional phase space. Stochastic layers in phase space exist near resonances of the motion. The thickness of the layers expands with increasing perturbation, leading to primary resonance overlap, motion across the layers, and the appearance of strong stochasticity in the motion. In the limit of weak perturbation, however, primary resonance overlap does not occur. A new physical behavior of the motion then makes its appearance: motion along the resonance layers called Arnol'd diffusion (AD). For two degrees of freedom, with a weak perturbation, two-dimensional

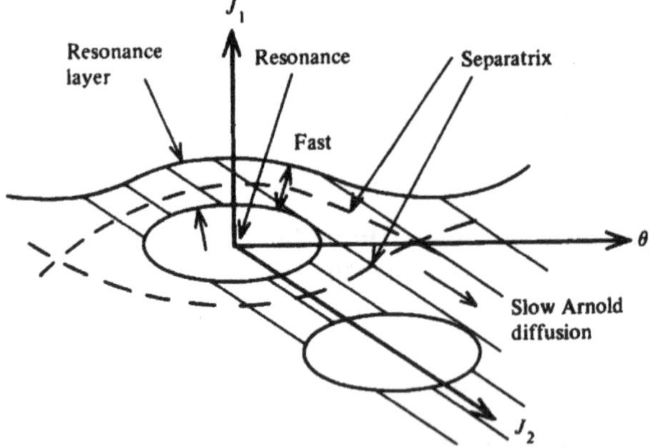

Figure 1. Illustration of the directions of the fast diffusion across a resonance layer and the slow diffusion along the resonance layer.

Kolmogorov–Arnol'd–Moser (KAM) surfaces divide the three-dimensional energy "volume" in phase space into a set of closed volumes each bounded by KAM surfaces, much as lines isolate regions of a plane. For $N > 2$ degrees of freedom, the N-dimensional KAM surfaces do not divide the $(2N-1)$-dimensional energy volume into distinct regions. Thus, for $N > 2$, in the generic case, all stochastic layers of the energy surface in phase space are connected into a single complex network—the Arnol'd web. The web permeates the entire phase space, intersecting or lying infinitesimally close to every point. For an initial condition within the web, the subsequent stochastic motion will eventually intersect every finite region of the energy surface in phase space, even in the limit as the perturbation strength approaches zero.

The merging of stochastic trajectories into a single web was proved (Arnol'd, 1964) for a specific nonlinear Hamiltonian. A general proof of the existence of a single web has not been given, but many computational examples support the conjecture. From a practical point of view, there are two major questions with respect to AD in a particular system: what is the relative measure of stochastic trajectories (fraction of the phase space that is stochastic) in the region of interest? And for a given initial condition, how fast will system points diffuse along the thin threads of the Arnol'd web?

We illustrate the motion along the resonance layer in Figure 1. A projection of the motion onto the J_1, θ_1 plane is shown, illustrating a resonance with a stochastic layer. At right angles to this plane, the action of the other coordinate J_2 is shown. If there are only two degrees of freedom in a conservative system, the fact that the motion is constrained to lie on a constant energy surface restricts the change in J_2 for J_1 constrained to the stochastic layer. However, if there is another degree of freedom, or if the Hamiltonian is time dependent, then this restriction is lifted, and motion along the stochastic layer in the J_2 direction can occur.

The diffusion rate (D) along a layer has been calculated by Chirikov (1979) for the important case of three resonances, and by Tennyson et al. (1979) for an equivalent mapping model, which they called a stochastic pump. These models predict, for a single action I corresponding to J_2 in Figure 1,

$$D = (\Delta I)^2/t \propto \mathrm{e}^{-A/\varepsilon^{1/2}}, \tag{1}$$

where t is the time, ε is a perturbation parameter, and $A \approx 1$. For coupling among many resonances, a rigorous upper bound on the diffusion rate (Nekhoroshev, 1977) generally overestimates the rate by orders of magnitude. Using a similar formalism with a somewhat more restrictive class of Hamiltonians, but still encompassing most physical problems, the upper bound can be improved (Benettin et al., 1985; Lochak & Neistadt, 1992) to give what they considered to be an optimal upper bound:

$$D \propto \mathrm{e}^{-A/\varepsilon^{\gamma}}, \ \gamma \approx N^{-1}. \tag{2}$$

If N is large, such an exponentially small diffusion could only hold for very small ε (specified within the theory), otherwise the exponential factor could be essentially unity. Also, an upper bound must be related to the fastest local diffusion. This may be much more rapid than an average global diffusion, which would be controlled by the portions of the phase space where the diffusion is slowest. For upper bound calculations, consult the original papers of Nekhoroshev (1977), Benettin et al. (1985), and Lochak & Neistadt (1992).

The simplest way to calculate local AD is to couple two standard maps together with a weak coupling term $\mu \sin(\theta_n + \phi_n)$, where θ_n and ϕ_n are the map phases and $\mu \ll 1$. Using the stochastic pump model, with a regular orbit (in the absence of coupling) in the (I, θ) map being driven by stochasticity in the (J, ϕ) map, the Hamiltonian of the mapping is approximated as $H \approx H_i + H_j$, with

$$\begin{aligned} H_i &= I^2/2 + K_i \cos\theta + 2\mu\cos(\theta+\phi), \\ H_j &= J^2/2 + K_j \cos\phi + 2K_j \cos\phi \cos 2\pi n, \end{aligned} \tag{3}$$

where n is the time normalized to mapping periods. We have retained only the lowest Fourier term from the mapping frequency in the H_j equation of (3), and considered that the stochasticity in H_i is driven by the coupling. To calculate the changes in H_i per iteration due to kicks delivered by (J, ϕ), we take the derivative

$$\begin{aligned} \frac{\partial H_i}{\partial n} &= \frac{\mathrm{d}H_i}{\mathrm{d}n} \\ &= \frac{\mathrm{d}}{\mathrm{d}n}[2\mu \cos(\theta+\phi)] \\ &\quad + 2\mu\frac{\mathrm{d}\theta}{\mathrm{d}n}\sin[\theta+\phi(n)]. \end{aligned} \tag{4}$$

For rotational orbits $\theta = w_i n + \theta_0$, scaling the time variable to revolutions of the map ($s = \omega_j n$), and defining the ratio of frequencies ($Q_0 = \omega_i/\omega_j = \omega_i/K_j^{1/2}$), Equation (4) is integrated to obtain

$$\Delta H_i = 2\mu Q_0 \left(\cos\theta_0 \int_{-\infty}^{\infty} \sin[Q_0 s + \phi(s)]\mathrm{d}s + \sin\theta_0 \int_{-\infty}^{\infty} \cos[Q_0 s + \phi(s)]\mathrm{d}s \right). \quad (5)$$

The first of the integrals in (5) integrates to zero; the second is a Mel'nikov–Arnol'd integral (Chirikov, 1979, Appendix A), which can be evaluated to give the change in ΔH_i over one characteristic half-period of the (J, ϕ) map. Squaring ΔH_i and averaging over θ_0 gives

$$\langle(\Delta H_i)^2\rangle = 32\pi^2 Q_0^4 \mu^2 \frac{\sinh^2(\pi Q_0/2)}{\sinh^2(\pi Q_0)}. \quad (6)$$

To determine the diffusion constant D, divide $\langle(\Delta H_i)^2\rangle$ by twice the average number of iterations in this half-period

$$\overline{T}_j = \frac{1}{\omega_j} \ln \left| \frac{32\mathrm{e}}{w_1} \right|, \quad (7)$$

where $w_1 = \Delta H / H_{\mathrm{separatrix}}$ is the relative energy of the edge of the stochastic region, $w_1 = 8\pi(2\pi/K_j^{1/2})^3 \times \mathrm{e}^{-\pi^2/K_j^{1/2}}$, and e is the base of natural logarithms. Combining (6) and (7), and using $\Delta H_i = I\Delta I$, the diffusion constant in action space can be approximated in a form that exhibits the main Q_0 scaling:

$$D \approx 16\mu^2 n Q_0^2 \mathrm{e}^{-\pi Q_0}, \quad (8)$$

where we have assumed that $I \approx \omega_i$. Comparing (8) to (1) with $Q_0 = \omega_i/K_j^{1/2}$, we observe that $K_j \propto \varepsilon$, the perturbation parameter. The numerical results agreed well with (8) (see Lichtenberg & Aswani (1998) and references therein).

Chirikov et al. (1979) found, numerically, that one could distinguish the diffusion in a range where ε was sufficiently large and a single resonance was dominant, such that a three-resonance model scaling as in (1) holds, from a range of smaller values of ε with many overlapping weak resonances, where the scaling in (2) applies. The results of their numerical investigation demonstrated the transition between the two regimes. In another approach, the diffusion through a large number of weakly coupled standard mappings was determined numerically, with the strength of the coupling controlled in a manner such that the three-resonance model could be applied in a statistical manner to determine the diffusion rate (Lichtenberg & Aswani, 1998).

These studies indicate that the basic models can be used to determine Arnol'd diffusion in multidimensional systems if the system parameters can be sufficiently controlled. For more information on these and related topics, the reader is referred to Chirikov (1979) and to Lichtenberg & Lieberman (1991, Chapter 6).

ALLAN J. LIICHTENBERG

See also **Kolmogorov–Arnol'd–Moser theorem; Phase space diffusion and correlations; Standard map**

Further Reading

Arnol'd, V.I. 1964. Instability of dynamical systems with several degrees of freedom. *Russian Mathematical Surveys*, 18: 85

Benettin, G., Galgani, L. & Giorgilli, A. 1985. A proof of Nekoroshev's theorem for the stability times of nearly integrable Hamiltonian systems. *Celestial Mechanics*, 37: 1–25

Chirikov, B.V. 1979. A universal instability of many-dimensional oscillator systems. *Physics Reports*, 52: 265–379

Chirikov, B.V., Ford, J. & Vivaldi, F. 1979. Some numerical studies of AD in a simple model. In *Nonlinear Dynamics and the Beam-Beam Interaction*, edited by M. Month & J.C. Herrera, New York: American Institute of Physics

Lichtenberg, A.J. & Aswani, A.M. 1998. Arnold diffusion in many weakly coupled mappings. *Physical Review* E, 57: 5325–5331

Lichtenberg, A.J. & Lieberman, M.A. 1991. *Regular and Chaotic Dynamics*, 2nd edition, New York: Springer

Lochak, P. & Neistadt, A.I. 1992. Estimates in the theorem of N.N. Nekhoroshev for systems with quasi-convex Hamiltonian. *Chaos*, 2: 495–499

Nekhoroshev, N.N. 1977. An exponential estimate of the time of stability of nearly itegrable Hamiltonian systems. *Russian Mathematical Surveys*, 32: 1–65

Tennyson, J.L., Lieberman, M.A. & Lichtenberg, A.J. 1979. Diffusion in near-integrable Hamiltonian systems with three degrees of freedom. In *Nonlinear Dynamics and the Beam-Beam Interaction*, edited by M. Month & J.C. Herrera, New York: American Institute of Physics

ARNOL'D TONGUES

See **Coupled oscillators**

ARTIFICIAL INTELLIGENCE

Artificial intelligence (AI) is a field of research in computer science reproducing intelligent reasoning. AI programs are mainly based on logic-oriented symbolic languages such as, for example, Prolog (Programming in Logic) or LISP (List Programming). Historically, AI was inspired by Alan Turing's question: "Can machines think?" According to the Turing test for AI, a machine is intelligent if a human user cannot distinguish whether he or she is interacting and communicating with a machine or a human being. Thus, before starting with AI, a general concept of computer and computabilty must be defined in computer science.

In 1936, Turing and Emil Post independently suggested the following definition of computability.

A "Turing machine" consists of: (a) a control box in which a finite program is placed, (b) a potentially infinite tape, divided lengthwise into squares, and (c) a device for scanning, or printing on one square of the tape at a time, and for moving along the tape or stopping, all under the command of the control box.

If the symbols used by a Turing machine are restricted to a stroke / and a blank *, then every natural number x can be represented by a sequence of x strokes (e.g., 3 by ///), each stroke on a square of the Turing tape. The blank is used to denote that the square is empty (or the corresponding number is zero). In particular, a blank is necessary to separate sequences of strokes representing numbers. Thus, a Turing machine computes a numerical function f with arguments $x_1, \ldots, x_n$ if the machine program starts with the input tape $\ldots * x_1 * x_2 * \ldots * x_n * \ldots$ and stops after finite steps with an output $\ldots * x_1 * x_2 * \ldots * x_n * f(x_1, \ldots, x_n) * \ldots$ on the tape. From a logical point of view, John von Neumann's general-purpose computer is a technical realization of a universal Turing machine that can simulate any kind of Turing program.

Besides Turing machines, there are many other mathematically equivalent procedures for defining computability (e.g., register machines, recursive functions) that are mathematically equivalent. According to Alonzo Church's thesis, the informal intuitive notion of an algorithm is identical to one of these equivalent mathematical concepts, for example, the program of a Turing machine.

With respect to AI, the paradigm of effective computabilty implies that the mind is represented by program-controlled machines, and mental structures refer to symbolic data structures, while mental processes implement algorithms. Historically, the hard core of AI was established during the Dartmouth Conference in 1956 when leading researchers, such as John McCarthy, Alan Newell, Herbert Simon, and others from different disciplines, formed the new scientific community of AI. If human thinking can be represented by an algorithm, then according to Church's thesis, it can be represented by a Turing program that can be computed by a universal Turing machine. Thus, human thinking could be simulated by a general-purpose computer and, in this sense, Turing's question ("Can machines think?") must be answered with a "yes." The premise that human thinking can be codified and represented by recursive procedures is, of course, doubtful. Even processes of mathematical thinking can be more complex than recursive functions.

The first period of AI was dominated by questions of heuristic programming, which means the automated search for human problem solutions in trees of possible derivations, controlled and evaluated by heuristics. In 1962, these simulative procedures were generalized and enlarged for the so-called General Problem Solver (GPS), which was assumed to be the heuristic framework of human problem solving. But GPS could only solve some insignificant problems in a formalized microworld. Thus, AI researchers tried to construct specialized systems of problem solving that use the specialized knowledge of human experts.

The architecture of an "expert system" consists of the following components: knowledge base, problem-solving component (interference system), explanation component, knowledge acquisition, and dialogue component. Knowledge is the key factor in the performance of an expert system. The knowledge is of two types. The first type is the facts of the domain that are written in textbooks and journals in the field. Equally important to the practice of a field is the second type of knowledge, called heuristic knowledge, which is the knowledge of good practice and judgment in a field. It is experimental knowledge, that art of good guessing, that a human expert acquires over years of work.

Expert systems are computational models of problem-solving procedures that need symbolic representation of knowledge. Unlike program-controlled serial computers, the human brain and mind are characterized by learning processes without symbolic representations. With respect to the architecture of von Neumann computers, an essential limitation derives from the sequential and centralized control, but complex dynamical systems like the brain are intrinsically parallel and self-organized.

In their famous paper "A Logical Calculus of the Ideas Immanent in Nervous Activity" in 1943, Warren McCulloch and Walter Pitts offered a complex model of neurons as threshold logic units with excitatory and inhibitory synapses. Their "McCulloch–Pitts neuron" fires an impulse along its axon at time $t+1$ if the weighted sum of its inputs and weights at time t exceeds the threshold of the neuron. The weights are numbers corresponding to the neurochemical interactions of the neuron with other neurons. But, in a McCulloch–Pitts network, the function of an artificial neuron is fixed for all time. McCulloch and Pitts succeeded in demonstrating that a network of formal neurons of their type could compute any finite logical expression.

In order to make a neural computer capable of complex tasks, it is necessary to find procedures of learning. A learning procedure is nothing else than an adjustment of the many weights so that the desired output vector (e.g., a perception) is achieved. The first learning neural computer was Frank Rosenblatt's "Perceptron" (1957). Rosenblatt's neural computer is a feedforward network with binary threshold units and three layers. The first layer is a sensory surface called a "retina" that consists of stimulus cells (S-units). The S-units are connected with the intermediate layer by fixed weights that do not change during the learning process. The elements of the intermediate layer are called associator cells (A-units). Each A-unit has a fixed weighted input of some S-units. In other words, some S-units project their output onto an A-unit. An S-unit may also project its output onto several A-units. The intermediate layer is completely connected with the output layer, the elements of which are called response cells (R-units). The weights between the intermediate layer and the output layer are variable and thus able to learn.

The Perceptron was viewed as a neural computer that can classify a perceived pattern in one of several possible groups. In 1969, Marvin Minsky and Seymour Papert proved that Perceptrons cannot recognize and distinguish the connectivity of patterns, in general. The Perceptron's failure is overcome by more flexible networks with supervised and unsupervised learning algorithms (e.g., Hopfield systems, Chua's cellular neural networks, Kohonen's self-organizing maps).

In the age of globalization, communication networks such as the Internet are a tremendous challenge to AI. From a technical point of view, we need intelligent programs distributed in the nets. There are already more or less intelligent virtual organisms (agents), learning, self-organizing, and adapting to our individual preferences of information, to select our e-mails, to prepare economic transactions, or to defend against attacks of hostile computer viruses, like the immune system of our body. Although the capability to manage the complexity of modern societies depends decisively on progress in AI, we need computational ecologies with distributed AI to support human life and not human-like robots to replace it.

KLAUS MAINZER

See also **Artificial life; Cell assemblies; Game of life; McCulloch–Pitts network; Neural network models; Perceptron**

Further Reading

Lenat, D.B. & Guha, R.V. 1990. *Building Large Knowledge-Based Systems*, Reading, MA: Addison-Wesley

Mainzer, K. 2003. *Thinking in Complexity. The Computational Dynamics of Matter, Mind, and Mankind*, 4th edition, Berlin and New York: Springer

Minsky, M. & Papert, S.A. 1969. *Perceptrons*, Cambridge, MA MIT Press

Minski, M. 1985. *The Society of Mind*, New York: Simon & Schuster

Nilson, N.J. 1982. *Principles of Artificial Intelligence*, Berlin and New York: Springer

Palm, G. (editor). 1984. *Neural Assemblies. An Alternative Approach to Artificial Intelligence*, Berlin and New York Springer

ARTIFICIAL LIFE

The term *artificial life* (AL) was coined in 1987 by Christopher Langton, who organized a workshop by that name in frustration with the lack of a forum for discussing work on the computer simulation of biological systems. In Langton's characterization, AL seeks to "contribute to theoretical biology by locating life-as-we-know-it within the larger picture of life-as-it-could-be" (Langton, 1989). In other words, AL aims to use computer simulation to synthesize alternative life-like systems and, thus, find out which characteristics and principles are essential and which are merely contingent on how life happened to evolve on this planet. While other branches of biology may use simulation to understand specific mechanisms, AL is broader, more abstract, and highly interdisciplinary, in addition to implying certain ideological convictions. Chief among these is the assumption that life is a process, rather than a metaphysical substance or an atomic property of matter, which emerges in a bottom-up fashion from local interactions among suitably arranged populations of individually lifeless components.

Opinions differ on whether such artificial systems may be logically equivalent to their natural counterparts and therefore really alive, or whether they are simply life-like simulacra. The former view is called the strong AL hypothesis, to associate it with a similarly functionalist standpoint known as Strong Artificial intelligence. However, the strong position in AL is considerably more tenable than its AI analog, which fails to distinguish between emergent and explicitly predetermined sources of behavior. Related to this "strong versus weak" argument is the unresolved question of whether life is an absolute category in nature at all or simply a useful way of grouping certain phenomena.

Early History

Attempts to construct living or life-like artifacts from mechanical parts date back at least to the ancient Greeks, and we can presume that many of these experiments were motivated by questions similar to those posed today. Nevertheless, these early systems tended to employ the "if it quacks like a duck it is a duck" principle, and so were only superficially lifelike, rather than in the deeper sense presently hoped for. One of the most ingenious of these early automata was indeed a duck (or at least something that moved, ate, defecated, and quacked like one), built by Jacques de Vaucanson around 1730.

Mary Shelley's *Frankenstein* explores similar issues in a fictional context. Contrary to popular belief, Shelley's monster was apparently not made from human body parts but from raw materials (cadavers are only mentioned with regard to Frankenstein's research). These components were then imbued with the "spark of life" (which Shelley associates with electricity) in order to animate them. Her viewpoint was still partially vitalistic, but there is a link between Shelley and her contemporary Charles Babbage, whose interpretation of intelligence (if not life itself) was more formalized, abstract, and mechanical. The mechanization of the mind continued with George Boole's logical algebra and then the work of Alan Turing and John von Neumann on automating thought processes, which led directly to the invention of the digital computer and the beginnings of artificial intelligence. It was a similar inquiry into the abstract nature of life, as distinct from mind, that prompted von Neumann's investigations into self-replicating machinery and Turing's work on embryogenesis and "unorganized machines" (related to neural networks).

Formal Methods

While complexity theory is concerned with the manner in which complex behavior arises from simple systems, AL is interested in how systems generate continually increasing levels of complexity. The most striking feature of living systems is their ability to self-organize and self-maintain—a property that Humberto Maturana and Francisco Varela have termed "autopoiesis" (Maturana & Varela, 1980). Evolution, embryogenesis, learning, and the development of social organizations are therefore the mechanisms of primary interest to AL researchers.

The key features of AL models are the use of populations of semi-autonomous entities, the coupling of these through simple local interactions (no centralized control and little or no globally accessible information), and the consequent emergence of collective, persistent phenomena that require a higher level of description than that used to describe their substrate. Conventional mathematical notation is not usually appropriate for such distributed and labile systems, and the individual computer programs are often their own best description. There are, however, a number of frequently used abstract structures and formal grammars, including the following:

Cellular automata, in which the populations are arrays of finite state machines and interactions occur between neighboring cells according to simple rules. Under the right conditions, emergent entities (such as the glider in John Conway's Game of Life) arise and persist on the surface of the matrix, interacting with other entities in computationally interesting ways.

Genetic algorithms, in which the populations are genomes in a gene pool and interactions occur between their phenotypes and some form of stressful environment. Natural selection (or sometimes human choice) drives the population to adapt and grow ever fitter, perhaps solving real practical problems in the process.

L-systems, or *Lindenmayer systems*, which provide a grammar for defining the growth of branching (often plant-like) physical structures, as insights into morphology and embryology.

Autonomous agents, which are composite code and data objects, representing mobile physical entities (robots, ants, stock market traders) embedded in a real or simulated environment. They interact locally by sensing their environment and receiving messages from other agents, giving rise to emergent phenomena of many kinds including cooperative social structures, nest-building, and collective problem-solving.

Autocatalytic networks, in which the populations are of simulated enzymes and the interactions are equivalent to catalysis. Such networks are capable of self-generation and a growth in complexity, mimicking the bootstrapping process that presumably gave rise to life on Earth.

Current Status

Like most new fields, AL has undergone cycles of hubris and doubt, innovation and stasis, and differentiation and consolidation. The listing of topics for the latest in the series of workshops started by Langton in 1987 is as broad as ever, although probably the bulk of AL work today (2004) is focused on artificial evolution. Most research concentrates on fine details, while the basic philosophical questions remain largely unanswered. Nevertheless, AL remains one of relatively few fields where one can ask direct questions about one's own existence in a practical way.

STEVE GRAND

See also **Catalytic hypercycle; Cellular automata; Emergence; Game of life; Hierarchies of nonlinear systems; Turing patterns**

Further Reading

Adami, C. 1998. *Introduction to Artificial Life*, New York: Springer
Boden, M.A. (editor). 1996. *The Philosophy of Artificial Life*, Oxford and New York: Oxford University Press
Langton, C.G. (editor). 1989. *Artificial Life*, Redwood City, CA: Addison-Wesley
Levy, S. 1992. *Artificial Life: The Quest for a New Creation*, New York: Pantheon
Maturana, H.R. & Varela, F.J. 1980. *Autopoiesis and Cognition: The Realization of the Living*, Dordrecht and Boston: Reidel

ASSEMBLY OF NEURONS

See **Cell assemblies**

ATMOSPHERIC AND OCEAN SCIENCES

Earliest works on the study of the atmosphere and ocean date back to Aristotle and his student Theophrastus in 350 BC and further progressed through Torricelli's invention of the barometer in 1643, Boyle's law in 1657, and Celsius's invention of the thermometer in 1742 (due to Galileo in 1607). The first rigorous theoretical model for the study of the atmosphere was proposed by

Vilhelm Bjerknes in 1904, following which many scientists began to apply fundamental physics to the atmosphere and ocean. The advent of these theoretical approaches and the invention of efficient communication technologies in the mid-20th century made numerical weather prediction feasible and was in particular encouraged by Lewis Fry Richardson and John von Neumann in 1946, using the differential equations proposed by Bjerknes. Today, advanced numerical modeling and observational techniques exist, which are constantly being developed further in order to understand and study the complex nonlinear dynamics of the atmosphere and ocean.

This overview article summarizes the governing equations used in atmospheric and ocean sciences, features of atmosphere-ocean interaction, and processes for an idealized geometry and structure with reference to a one-dimensional vertical scale (Figure 1), a two-dimensional vertically averaged scale (Figures 3(a) and 4(a)), a two-dimensional zonally averaged meridional scale (Figures 3(b) and 4(b)), and a three-dimensional scale (Figure 2), and regimes of interacting systems (such as El Niño and Southern Oscillation and North Atlantic Oscillation) (Figures 5–7). The entry serves as an introduction to the many nonlinear processes taking place (for example, chaos, turbulence) and provides a few illustrative examples of self-organizing coherent structures of the nonlinear dynamics of the atmosphere and ocean.

Governing Equations

The combined atmosphere and ocean system can be regarded as a huge volume of fluid resting on a rotating oblate spheroid with varying surface topography moving through space, with an interface (which in general is discontinuous) between two fluid masses of differing densities. This coupled atmosphere-ocean system is driven by energy input through solar radiation (see Figure 1), gravity (for example, through interaction with other stellar bodies such as the Sun and Moon, i.e., tides), and inertia. The entire fluid is described by equations for conserved quantities such as momentum, mass (of air, water vapor, water, salt), and energy together with equations of state for air and water (*See* **Fluid dynamics; Navier–Stokes equation**). The movement of large water or air masses in a rotating reference frame adds to the complexity of motions, due to the presence of Coriolis forces, introduced by Coriolis in 1835.

Atmosphere-ocean interactions can be defined as an exchange of momentum, heat, and water (vapor and its partial masses: salts, carbon, oxygen, nitrogen, etc.) between air and water masses. The governing equations in the Euler formulation and a cartesian coordinate system are given by:

(i) The conservation of momentum

$$\frac{\mathrm{d}\boldsymbol{u}}{\mathrm{d}t} + 2\Omega \times \boldsymbol{u} = -\frac{1}{\rho}\nabla p - \boldsymbol{g} + \boldsymbol{F}_{\text{ext}} + \boldsymbol{F}_{\text{fric}}, \quad (1)$$

where the second term $2\Omega \times \boldsymbol{u}$ is the term due to the Coriolis force (Ω is the angular velocity of the Earth; $|\Omega| = 7.29 \times 10^{-5}\ \mathrm{s}^{-1}$), and forces due to a pressure gradient ∇p, gravity ($|\boldsymbol{g}| = 9.81\ \mathrm{m\,s^{-2}}$) and external ($\boldsymbol{F}_{\text{ext}}$) as well as frictional ($\boldsymbol{F}_{\text{fric}}$) forces are included. Note that the operator $\mathrm{d}/\mathrm{d}t$ is defined by

$$\frac{\mathrm{d}\boldsymbol{v}}{\mathrm{d}t} = \left(\frac{\partial}{\partial t} + \boldsymbol{v} \cdot \nabla\right)\boldsymbol{v}.$$

(ii) The conservation of mass (or continuity equation)

$$\frac{1}{\rho}\frac{\mathrm{d}\rho}{\mathrm{d}t} + \nabla \cdot \boldsymbol{u} = 0. \quad (2)$$

Note that there are alternative formulations such as the Lagrangian and impulse-flux form for these equations, and cartesian coordinate systems can be mapped to different geometries such as spherical coordinates by appropriate transformations.

(iii) The conservation of energy (First Law of Thermodynamics) and Gibbs's equation (Second Law of Thermodynamics)

$$\frac{\mathrm{d}Q}{\mathrm{d}t} = \frac{\mathrm{d}\varepsilon}{\mathrm{d}t} + p\frac{\mathrm{d}\alpha}{\mathrm{d}t},$$

$$\frac{\mathrm{d}\eta}{\mathrm{d}t} = \frac{1}{T}\frac{\mathrm{d}\varepsilon}{\mathrm{d}t} + \frac{p}{T}\frac{\mathrm{d}\alpha}{\mathrm{d}t} - \sum \frac{\mu_i}{T}\frac{\mathrm{d}\gamma_i}{\mathrm{d}t}, \quad (3)$$

where Q is the heat supply (sensible, latent, and radiative heat fluxes; see Figure 1), T is the temperature, ε the internal energy and α the specific volume ($\alpha = 1/\rho$), η the entropy, μ_i the chemical potentials, and γ_i the partial masses. The conservation of energy states in brief that the change in heat is balanced by a change in internal energy and mechanical work performed, and Gibbs's equation determines the direction of an irreversible process, relating entropy to a change in internal energy, volume, and partial masses.

(iv) The conservation of partial masses of water and air, that is, salinity for water, where all constituents are represented as salts and water vapor for air, yield equations similar to (2)

$$\frac{1}{\rho_{\mathrm{v}}}\frac{\mathrm{d}\rho_{\mathrm{v}}}{\mathrm{d}t} + \nabla \cdot \boldsymbol{u} = W_v$$

and

$$\frac{\mathrm{d}\rho s}{\mathrm{d}t} + \rho s\nabla \cdot \boldsymbol{u} = W_{\mathrm{s}}, \quad (4)$$

where ρ_{v} is the density of water vapor, s the specific salinity (gram salts per gram water), and W_v, W_s

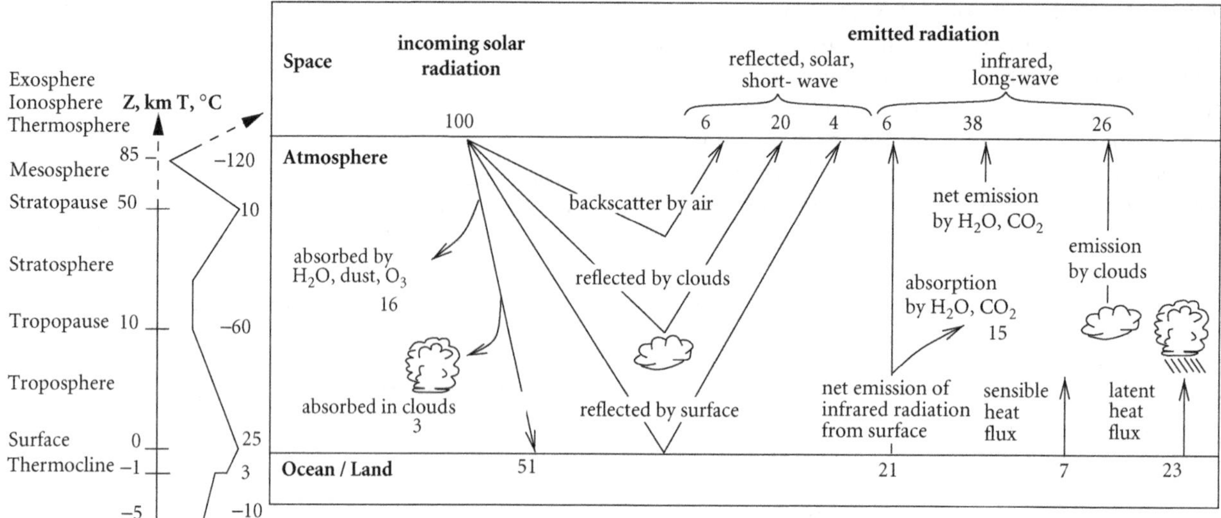

Figure 1. Sketch of the vertical structure of the atmosphere–ocean system and radiation balance and processes in the global climate system. Adapted from National Academy of Sciences (1975). Note that lengths are not to scale and and temperatures indicate only global averages.

contain possible source and sink terms as well as the effect of molecular diffusion in terms of the concentration flux density $\boldsymbol{S}\,(-\nabla\cdot\boldsymbol{S})$ and possible phase changes.

(v) The equation of state for a mixture of salts and gases for air and water, whose constituent concentrations are virtually constant in the atmosphere and ocean is

$$p \approx \rho R T\,(1 + 0.6078\, q)\,, \tag{5}$$

where R is the gas constant for dry air ($R = 287.04\,\mathrm{J\,kg^{-1}\,K^{-1}}$) and $q = \rho_v/\rho$ is the specific humidity. Similarly, the equation of state for near incompressible water is

$$\rho \approx \rho_0[1 - \alpha(T - T_0) + \beta(S - S_0)], \tag{6}$$

where ρ_0, T_0, and S_0 are reference values for density, temperature, and salinity ($\rho_0 = 1028\,\mathrm{kg\,m^{-3}}$, $T_0 = 283\,\mathrm{K}(= 10°\mathrm{C})$, $S_0 = 35‰$), and α and β are the coefficients of thermal expansion and saline contraction ($\alpha = 1.7 \times 10^{-4}\,\mathrm{K^{-1}}$, $\beta = 7.6 \times 10^{-4}$), see Krauss (1973); Cushman-Roisin (1993).

Equations detailed in (i)–(v) form a set of hydrothermodynamic equations for the atmosphere-ocean system to which various approximations and scaling limits can be applied. Among them are the shallow-water equations, primitive equations, the Boussinesq and anelastic approximation, quasigeostrophic, and semigeostrophic equations and variants or mixtures of these. These equations have to be solved with appropriate boundary conditions and conditions at the air-sea interface; for details refer to Krauss (1973), Gill (1982) and Kraus & Businger (1994). For studies of the upper atmosphere, further equations for the geomagnetic field can also be taken into account (Maxwell's equations).

Atmospheric Structure and Circulation

In the vertical dimension, several atmospheric layers can be differentiated (see Figure 1). Figure 2 gives the length and time scales of typical atmospheric processes.

From sea level up to about 2 km is the atmospheric boundary layer, characterized by momentum, heat, moisture, and water transfer between the atmosphere and its underlying surface. Above the boundary layer is the troposphere (Greek, *tropos* meaning *turn, change*) that constitutes most of the total mass of the atmosphere (about 10 km height) and is largely in hydrostatic balance characterized by a decrease in temperature. Above the troposphere and stratosphere, which contains the ozone layer, temperatures rise throughout. The mesosphere, which is bounded by the stratopause (about 50 km height) below and mesopause (about 85 km height) above, is a layer of very thin air where temperatures drop to extreme lows. Above the mesopause, temperatures increase again throughout the thermosphere (from about 85 km to 700 km), the largest layer of the atmosphere, where the ionosphere is located (between about 100 km and 300 km). The ionosphere contains ionized atoms and free electrons and permits the reflection of electromagnetic waves. Above the thermosphere is the exosphere, which is the outermost layer of the atmosphere and the transition region between the atmosphere and outer space, the magnetosphere in particular, where atoms can escape into space beyond the so-called escape velocity and where the Van Allen belt is situated.

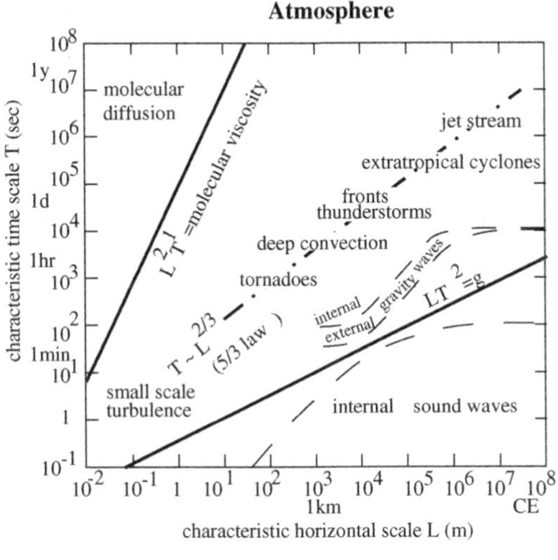

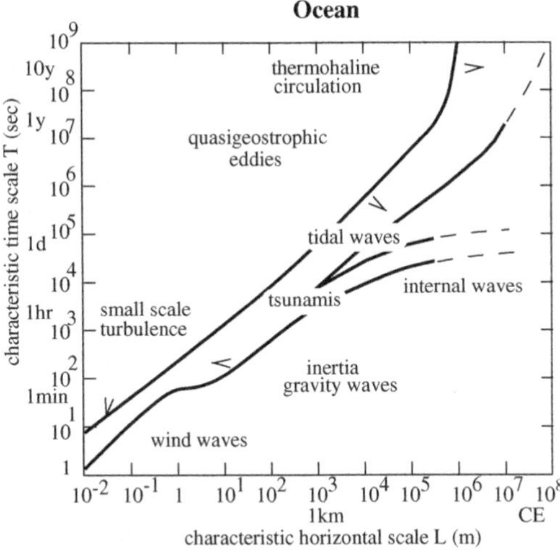

Figure 2. Schematic logarithmic time and horizontal length scales of typical atmospheric and oceanic phenomena. Note that Richardson's $L \propto T^{3/2}$ relation and CE stands for circumference of the Earth. Modified from Lettau (1952), Smagorinsky (1974), and World Meteorological Organization (1975).

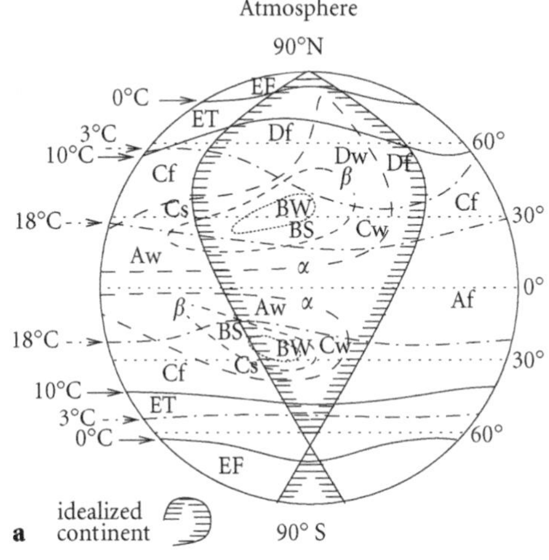

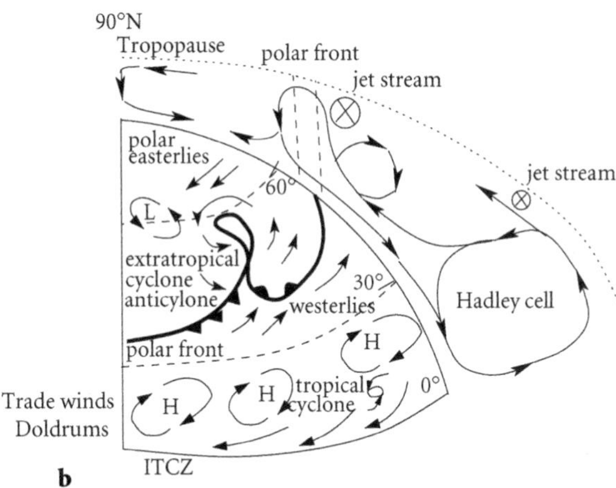

Figure 3. Sketch of the near-surface climate and atmospheric circulation of the Earth with an idealized continent. (a) Averaged isothermals of the coldest month (dashed-dot, −3°C, 18°C) and the warmest month (solid, 0°C, 10°C) and periodical dry season boundaries α and dry climate β. The following climate regions are indicated: wet equatorial climate (Af), tropical wet/dry climate (Aw), desert climate (BW), steppe climate (BS), sinic climate (Cw), Mediterranean climate (Cs), humid subtropical climate (Cf), humid continental climate (Df), continental subarctic climate (Dw), tundra climate (ET) and snow and ice climate (EF). Modified from Köppen (1923). (b) The zonal mean jet streams (primary circulation) and mass overturning (secondary circulation) in a meridional height section, the subtropical highs (H) and subpolar lows (L), polar easterlies, westerlies, polar front, trade winds, and intertropical convergence zone (ITCZ). ▲ denotes a cold front and ● a warm front. Adapted from Palmen (1951), Defant and Defant (1958), and Hantel in Bergmann & Schäfer (2001).

A low (high) in meteorology refers to a system of low (high) pressure, a closed area of minimum (maximum) atmospheric pressure (closed isobars, or contours of constant pressure) on a constant height chart. A low (high) is always associated with (anti)cyclonic circulation, thus also called a cyclone (anticyclone). Anticyclonic means clockwise in the Northern Hemisphere (and counterclockwise in the Southern Hemisphere). Cyclonic means counterclockwise in the Northern Hemisphere (and clockwise in the Southern Hemisphere). At zeroth order, a balance of pressure gradient forces and Coriolis forces, that is, geostrophic balance, occurs, leading to the flow of air along isobars instead of across (in the direction of the pressure gradient). A front is a discontinuous interface or a region of strong gradients between two air masses of differing densities or temperatures, thus encouraging conversion of potential into kinetic energy (examples are polar front, arctic front, cold front, and warm front).

Hurricanes and typhoons (local names for tropical cyclones) transport large amounts of heat from low to mid and high latitudes and develop over oceans. Little is known about the initial stages of their formation, although they are triggered by small low-pressure systems in the Intertropical Convergence Zone (*See* **Hurricanes and tornadoes**). Because of their strong winds, cyclones are particularly active in inducing

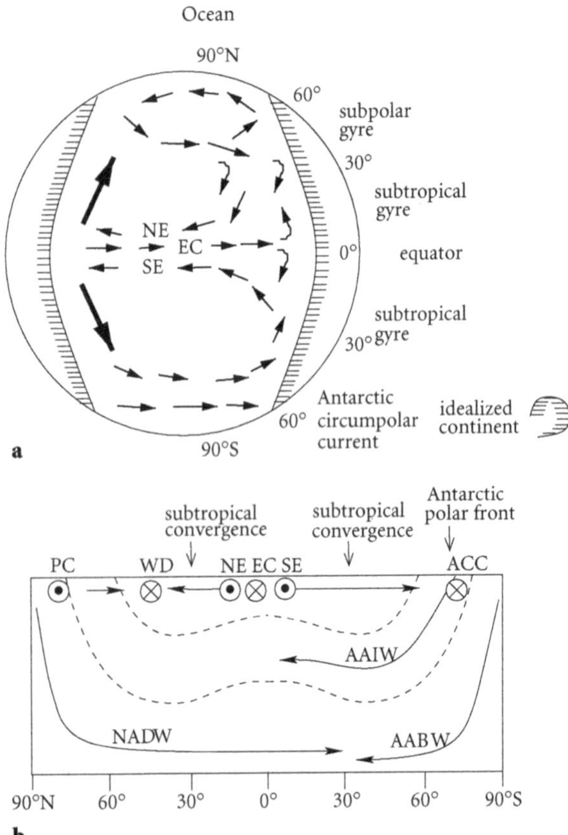

Figure 4. Sketch of the oceanic circulation of an idealized basin (a) the global wind-induced distribution of ocean currents (primary circulation) and (b) the zonal mean thermohaline circulation (secondary circulation) in a meridional depth section showing upper, intermediate, deep and bottom water masses. NE denotes the north equatorial, EC the equatorial and SE the south equatorial current. PC stands for polar current, ACC for Antarctic circumpolar current, WD for west drift, AAIW for Antarctic intermediate waters, NADW for North Atlantic deep water, and AABW for Antarctic bottom water. Adapted from Hasse and Dobson (1986).

upwelling and wind-driven surface water transport. Extratropical cyclones are frontal cyclones of mid to high latitudes (see Figures 2 and 3).

Meteorology and oceanography are concerned with understanding, predicting, and modeling the weather, climate, and oceans due to their fundamental socio-economic and environmental impact. In meteorology, one distinguishes between short (1–3 days) and medium-range (4–10 days) numerical weather prediction models (NWPs) for the atmosphere and general circulation models (GCMs, *See* **General circulation models of the atmosphere**). While NWPs for local and regional weather prediction are usually not coupled to ocean models, GCMs are global three-dimensional complex coupled atmosphere-ocean models (which even include the influence of land masses), used to study global climate change, modeling radiation, photochemistry, transfer of heat, water vapor, momentum, greenhouse gases, clouds, ocean temperatures, and ice boundaries. The atmosphere-ocean interface couples the "fast" processes of the atmosphere with the comparably "slow" processes of the ocean through evaporation, precipitation, and momentum interaction. GCMs are validated using statistical techniques and correlated to the actual climate evolution. Additionally, the application of GCMs to different planetary atmospheres, for example, on Mars and Jupiter, leads to a greater understanding of the planet's history and environment.

The complexity of the dynamics of the atmosphere and ocean is largely due to the intrinsic coupling between these two large masses at the air-sea interface.

Ocean

Ocean circulation is forced by tidal forces (also known to force atmospheric tides), due to gravitational attraction, wind stress, applied shear forces acting on the interface, and external, mainly solar, radiation, penetrating into the sea surface and affecting the heat budget and water mass due to evaporation. Primary sources of tidal forcing, earliest work on which was undertaken by Pierre-Simon Laplace in 1778, are the Moon and the Sun. One discerns between diurnal, semidiurnal, and mixed-type tides.

In the ocean, one distinguishes between two types of ocean currents: surface (wind-driven) and deep circulation (thermohaline circulation). Separating the surface and deep circulation is the thermocline, a small layer of strong gradient of temperature, salinity, and density, acting as an interface between the two types of circulations.

Surface circulation ranging up to 400 m in depth is forced by the prominent westerly winds in the mid-latitudes and trade winds in the tropical regions (see Figures 3 and 4), which are both forced by solar heating and Coriolis forces leading to expansion of water near the equator and decreased density, but increased salinity due to evaporation. An example of the latter is the Gulf Stream in the North Atlantic. The surface wind stress, solar heating, Coriolis forces, and gravity lead to the creation of large gyres in all ocean basins with clockwise (anticyclonic) circulation in the northern hemisphere and counterclockwise circulation in the southern hemisphere. The North Atlantic Gyre, for example, consists of four currents: the north equatorial current, the Gulf Stream, the North Atlantic current, and the Canary current.

Ekman transport, the combination of wind stress and Coriolis forces, leads to a convergence of water masses in the center of such gyres, which increases the sea surface elevation. The layer of Ekman transport can be 100–150 m in depth and also leads to upwelling due to conservation of mass on the western (eastern) coasts for winds from the north (south) in the Northern (Southern) Hemisphere. As a consequence, nutrient-rich

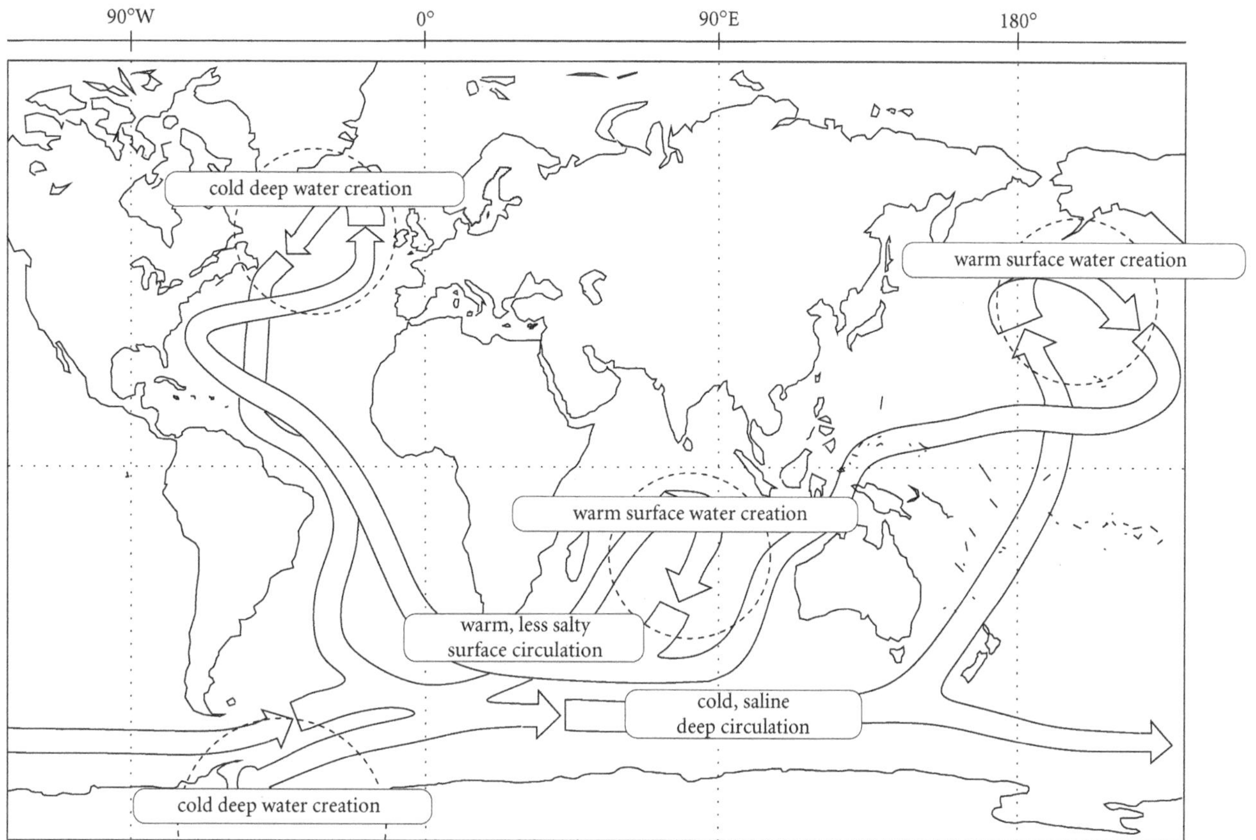

Figure 5. Sketch of the global conveyor belt through all oceans, showing the cold saline deep circulation, the warm, less salty surface circulation, and the primary regions of their creation. Note that this circulation is only characteristic of the actual global circulation. Adapted from Broecker (1987).

deep water is brought to the surface. With the opposite wind direction, Ekman transport acts to induce downwelling.

Another important combination of forces is the balance of Coriolis forces and gravity (pressure gradient forces), which is called geostrophic balance, leading to the movement of mass along isobars instead of across (geostrophic current), similar to the atmosphere. The boundary currents along the eastern and western coastlines are the major geostrophic currents in a gyre. The western side of the gyre is stronger than the eastern due to the Earth's rotation, called western intensification.

Deep circulation makes up 90% of the total water mass and is driven by density forces and gravity, which in turn is a function of temperature and salinity. High-density deep water originates in the case of extreme cooling of the sea surface in the polar regions, sinking to large depths as a density current, a strongly nonlinear phenomenon. When the warm Gulf Stream waters, which have increased salinity due to excessive evaporation in the tropics, move north due to the North Atlantic Gyre, they are cooled by Arctic winds from the north and sink to great depths forming the high-density Atlantic deep waters (see Figure 5). The downward transport of water is balanced by upward transport in low- and mid-latitude regions.

The most prominent example of the interaction between atmospheric and ocean dynamics is the global conveyor belt, which links the surface (wind-driven) and deep (thermohaline) circulation to the atmospheric circulation. The global conveyor belt is a global circulatory system of distinguishable and recognizable water masses traversing all oceans (see Figure 5). The water masses of this global conveyor belt transport heat and moisture, contributing to the climate globally. In Earth's history, the global conveyor belt has experienced flow reversals and perturbations leading to changes in the global circulatory system. The rather recent anthropogenic impact on climate and oceans through greenhouse gas emissions has the potential to create instability in this large-scale dynamical system, which could alter Earth's climate and have devastating environmental and agricultural effects.

ENSO and NAO

Another example of atmosphere-ocean coupling is the combination of the El Niño and Southern Oscillation (ENSO). The El Niño ocean current (and associated

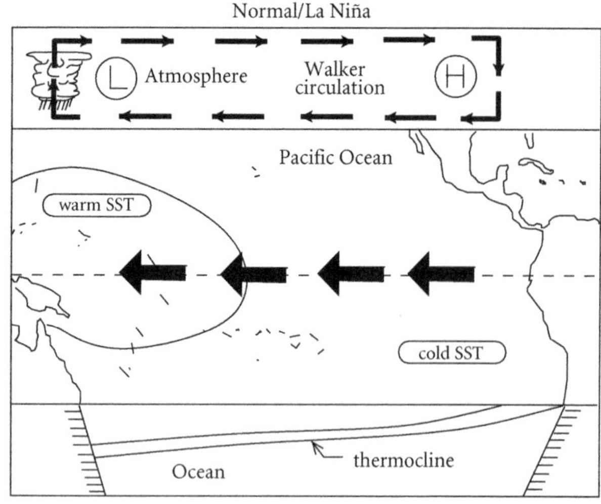

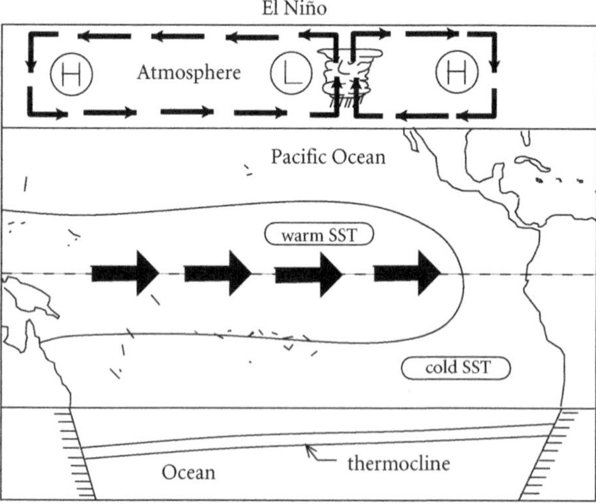

Figure 6. Sketch of the El Niño in the tropical Pacific, showing a reversal in (trade) wind direction from easterlies to westerlies during an El Niño period bringing warmer water (warm corresponds to a positive sea-surface temperature [SST]) close to the South American coast, displacing the equatorial thermocline downwards. Note the change in atmospheric tropical convection and associated heavy rainfall. After McPhaden, NOAA/TAO (2002) and Holton (1992).

wind and rain change) is named from the Spanish for Christ Child, due to its annual occurrence off the South American coast around Christmas, and may also be sensitive to anthropogenic influence (see Figure 6). The Southern Oscillation occurs as a 2–5-year periodic reversal in the east-west pressure gradient associated with the present equatorial wind circulation, called Walker circulation, across the Pacific leading to a reversal in wind direction and changes in temperature and precipitation. The easterly wind in the West Pacific becomes a westerly. As a consequence, the strong trade winds are weakened, affecting climate globally (e.g., crop failures in Australia, flooding in the USA, and the monsoon in India). The Southern Oscillation in turn leads to large-scale oceanic fluctuations in the circulation of the Pacific Ocean and sea-surface temperatures, which is called El Niño. The interannual variability, though, is not yet fully understood; consideration of a wider range of tropical and extratropical influences is needed. A counterpart to the ENSO in the Pacific is the North Atlantic Oscillation (NAO), which is essentially an oscillation in the pressure difference across the North Atlantic and is described further in Figure 7.

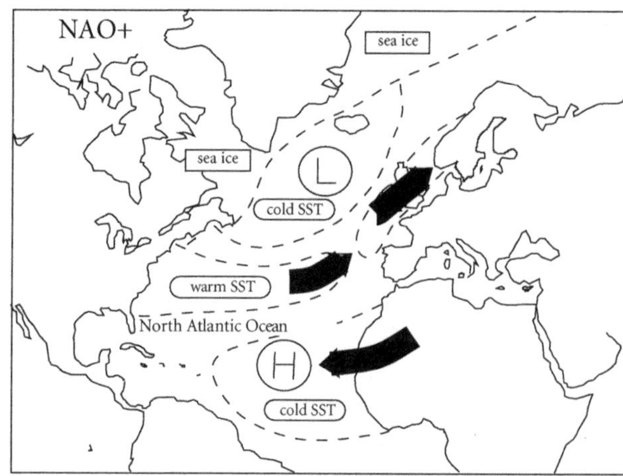

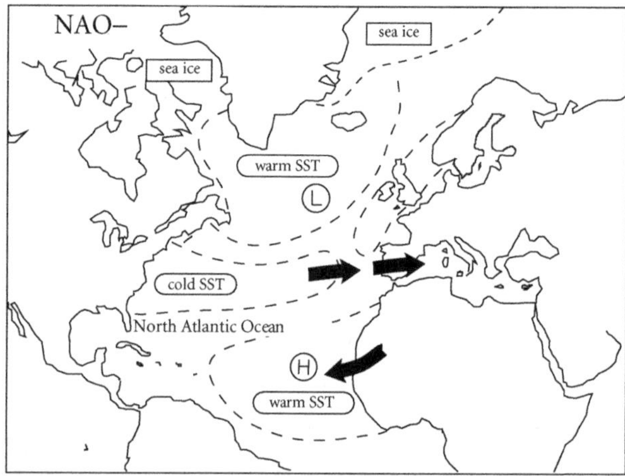

Figure 7. Sketch of the North Atlantic Oscillation (NAO) during the northern hemisphere winter season. Positive NAO (NAO+) showing an above-usual strong subtropical high-pressure center and subpolar low, resulting in increased wind strengths and storms crossing the Atlantic towards northern Europe. NAO+ is associated with a warm wet winter in Europe and cold dry winter in North America. Central America experiences mild wet winter conditions. Negative NAO (NAO–) shows a weaker subtropical high and subpolar low, resulting in lower wind speeds and weaker storms crossing the Atlantic toward southern Europe and receded sea ice masses around Greenland. NAO– is associated with cold weather in northern Europe and moist air in the Mediterranean. Central America experiences colder climates and more snow. Adapted from Wanner (2000).

Monsoons

The monsoons (derived from Arabic, *mauism*, meaning season or shift in wind) are seasonally reversing winds and one of the most pertinent features of the global atmospheric circulation. The best-known examples are the monsoons over the Indian Ocean and, to some extent, the western Pacific Ocean (tropical region of Australia), the western coast of Africa, and the Carribbean. Monsoons are characteristic for wet summer and dry winter seasons, associated with strong winds and cyclone formation. They occur due to differing thermal characteristics of the land and sea surfaces. Land, having a much smaller heat capacity than the ocean, emits heat from solar radiation more easily, leading to upward heat (cumulus) convection. In the summer season, this leads to a pressure gradient and thus wind from the land to the ocean in the upper layers of the atmosphere and subsequent conserving flow of moisture-rich air from the sea back inland at lower levels. This leads to monsoonal rains, increased latent heat release, and intensified monsoon circulation. During the monsoons of the winter season, the opposite of the summer season monsoon takes place, although less pronounced, since the thermal gradient between the land and sea is reversed. The winter monsoons thus lead to precipitation over the sea and cool dry land surfaces.

ANDREAS A. AIGNER AND KLAUS FRAEDRICH

See also **Fluid dynamics; General circulation models of the atmosphere; Hurricanes and tornadoes; Lorenz equations; Navier–Stokes equation**

Further Reading

Apel, J. 1989. *Principles of Ocean Physics*, London: Academic Press

Barry, R.G., Chorley, R.J. & Chase, T. 2003. *Atmosphere, Weather and Climate*, 8th edition, London and New York: Routledge

Bergmann, K., Schaefer C. & von Raith, W. 2001. *Lehrbuch der Experimentalphysik, Band 7, Erde und Planeten*, Berlin: de Gruyter

Cushman-Roisin, B. 1993. *Introduction to Geophysical Fluid Dynamics*, Englewood Cliffs, NJ: Prentice–Hall

Defant, A. & Defant, Fr. 1958. *Physikalische Dynamik der Atmosphäre*, Frankfurt: Akademische Verlagsgesellschaft

Gill, A. 1982. *Atmosphere–Ocean Dynamics*, New York: Academic Press

Hasse, L. & Dobson, F. 1986. *Introductory Physics of the Atmosphere and Ocean*, Dordrecht and Boston: Reidel

Holton, J.R. 1992. *An Introduction to Dynamic Meteorology*, 3rd edition, New York: Academic Press

Kraus, E.B. & Businger, J.A. 1994. *Atmosphere–Ocean Interaction*, New York: Oxford University Press, and Oxford: Clarendon Press

Krauss, W. 1973. *Dynamics of the Homogeneous and Quasi-homogeneous Ocean*, vol I, Berlin: Bornträger

LeBlond, P.H. & Mysak, L.A. 1978. *Waves in the Ocean*, Amsterdan: Elsevier

Lindzen, R.S. 1990. *Dynamics in Atmospheric Physics*, Cambridge and New York: Cambridge University Press

Pedlosky, J. 1986. *Geophysical Fluid Dynamics*, New York: Springer

Philander, S.G. 1990. *El Niño, La Niña, and the Southern Oscillation*, New York: Academic Press

ATTRACTOR NEURAL NETWORKS

Neural networks with feedback can have complex dynamics; their outputs are not related in a simple way to their inputs. Nevertheless, they can perform computations by converging to attractors of their dynamics. Here, we analyze how this is done for a simple example problem: associative memory, following the treatment by Hopfield (1984) (see also Hertz, et al., 1991, Chapters 2 and 3).

Let us assume that input data are fed into the network by setting the initial values of the units that make it up (or a subset of them). The network dynamics then lead to successive changes in these values. Eventually, the network will settle down into an attractor, after which the values of the units (or some subset of them) give the output of the computation. The associative memory problem can be described in the following way: there is a set of p patterns to be stored. Given, as input, a pattern that is a corrupted version of one of these, the attractor should be a fixed point as close as possible to the corresponding uncorrupted pattern.

We focus on networks described by systems of differential equations such as

$$\tau_i \frac{\mathrm{d}u_i}{\mathrm{d}t} + u_i(t) = \sum_{j \neq i} w_{ij} g[u_j(t)]. \tag{1}$$

Here, $u_i(t)$ is the net input to unit i at time t and $g(\ )$ is a sigmoidal activation function ($g' > 0$), so that $V_i = g(u_i)$ is the value (output) of unit i. The connection weight to unit i from unit j is denoted w_{ij}, and τ_i is the relaxation time. We can also consider discrete-time systems governed by

$$V_i(t+1) = g\left[\sum_j w_{ij} V_j(t)\right]. \tag{2}$$

Here, it is understood that all units are updated simultaneously. In either case, the "program" of such a network is its connection weights w_{ij}.

In general, three kinds of attractors are possible: fixed point, limit cycle, and strange attractor. There are conditions under which the attractors will always be fixed points. For nets described by the continuous dynamics of Equation (1), a sufficient (but not necessary) condition is that the connection weights be symmetric: $w_{ij} = w_{ji}$. General results about the stability of recurrent nets were proved by Cohen & Grossberg (1983). They showed, for dynamics (1), that there is a Lyapunov function, that is, a function of the state variables u_i, which always decreases under the dynamics, except for special values of the u_i at which it does not change. These values are fixed points. For values of the u_i close to such a point, the system will evolve either toward it (an attractor) or away from it (a repellor). For almost all starting states, the dynamics will end at one of the attractor's fixed points. Furthermore, these are the only attractors.

We treat the case $g(u) = \tanh(\beta u)$ and consider the *ansatz*

$$w_{ij} = \frac{1}{N} \sum_{\mu=1}^{p} \xi_i^\mu \xi_j^\mu. \tag{3}$$

That is, for each pattern, there is a contribution to the connection weight proportional to the product of sending (ξ_j^μ) and receiving (ξ_i^μ) unit activities when the network is in a stationary state $V_i = \xi_i^\mu$. This is just the form of synaptic strength proposed by Hebb (1949) as the basis of animal memory, so this *ansatz* is sometimes called a Hebbian storage prescription.

To see how well the network performs this computation, we examine the fixed points of (1) or (2), which solve

$$V_i = \tanh\left(\beta \sum_j w_{ij} V_j\right). \tag{4}$$

The quality of retrieval of a particular stored pattern ξ_i^μ is measured by the quantity $m_\mu = N^{-1} \sum_i \xi_i^\mu V_i$. Using (4), with the weight formula (3), we look for solutions in which the configuration of the network is correlated with only one of the stored patterns, that is, just one of the m_μ's is not zero. If the number of stored patterns $p \ll N$, we find a simple equation for m_μ:

$$m_\mu = \tanh(\beta m_\mu). \tag{5}$$

This equation has nontrivial solutions whenever the gain $\beta > 1$ and for β large, $m_\mu \to 1$, indicating perfect retrieval. If the gain is high enough, there are other attractors in addition to the ones we have tried to program into the network with the choice (3), but by keeping the gain between 1 and 2.17 we can limit the attractor set to the desired states.

When p is of the same order as N, the analysis is more involved. We define a parameter $\alpha = p/N$. For small α, the overlaps m_μ between the stored patterns and the fixed points are less than, but still close to, 1. However, there is a critical value of α, $\alpha_c(\beta)$, above which there are no longer fixed points close to the patterns to be stored and the memory breaks down catastrophically. One finds $\alpha_c(1) = 0$ and, in the limit $\beta \to \infty$, $\alpha_c(\beta) \to 0.14$.

Thus, attractor computation works in this system over a wide range of the model parameters α and β. It can be shown to be robust with respect to many other variations, including dilution (random removal of connections), asymmetry (making some of the $w_{ij} \neq w_{ji}$), and quantization or clipping of the weight values. Its breakdown at the boundary $\alpha_c(\beta)$ is a collective effect like a phase transition in a physical system.

The weight formula (3) was only an educated guess. It is possible to obtain better weights, which reduce the crosstalk and increase α_c, by employing systematic learning algorithms.

It is also possible to extend the above-described model to store pattern sequences by including suitable delays in the discrete-time dynamics (2).

It appears that attractor networks play a role in computations in the brain. One example of current interest is working memory: some neurons in the prefrontal cortex that are selectively sensitive to a particular visual stimulus exhibit continuing activity after the stimulus is turned off, even though the animal sees other stimuli. Thus, they seem to be involved in the temporary storage of visual patterns. Computational network models based on the simple concepts described above (Renart et al., 2001) are able to reproduce the main features seen in recordings from these neurons.

JOHN HERTZ

See also **Cellular nonlinear networks; McCulloch–Pitts network; Neural network models**

Further Reading

Amit, D.J. 1989. *Modeling Brain Function*, Cambridge and New York: Cambridge University Press

Cohen, M. & Grossberg, S. 1983. Absolute stability of global pattern formation and parallel memory storage by competitive neural networks. *IEEE Transactions on Systems, Man and Cybernetics*, 13: 815–826 etc.

Hebb, D.O. 1949. *The Organization of Behavior*, New York: Wiley

Hertz, J.A., Krogh, A.S., & Palmer, R.G. 1991. *Introduction to the Theory of Neural Computation*, Redwood City, CA: Addison-Wesley

Hopfield, J.J. 1984. Neurons with graded responses have collective computational properties like those of two-state neurons. *Proceedings of the National Academy of Sciences USA*, 79: 3088–3092

Renart, A., Moreno, R., de la Rocha, J., Parga, N. & Rolls, E.T. 2001. A model of the IT-PF network in object working memory which includes balanced persistent activity and tuned inhibition. *Neurocomputing*, 38–40: 1525–1531

ATTRACTORS

A wide variety of problems arising in physics, chemistry, and biology can be recast within the framework of dynamical systems. A dynamical system is made up of two parts: the phase space, which consists of all possible configurations of the physical system, and the "dynamics," a rule describing how the state of the system changes over time. The fundamental insight of the theory is that some problems, which initially appear extremely complicated, can be greatly simplified if we are prepared to concentrate on their long-term behavior, that is, what happens eventually.

This idea finds mathematical expression in the concept of an attractor.

The simplest possibility is that the system settles down to a constant state (e.g., a pendulum damped by air resistance will end up hanging vertically downward). In the phase space, this corresponds to an attractor that is a single "fixed point" for the dynamics. If the system settles down to a repeated oscillation, then this corresponds to a "periodic orbit," a closed curve in the phase space. For two coupled ordinary differential equations (ODEs), it is a consequence of the Poincaré–Bendixson Theorem that these fixed points and periodic orbits are essentially the only two kinds of attractors that are possible (see Hirsch & Smale, 1974 for a more exact statement). In higher dimensions, it is possible for the limiting behavior to be quasi-periodic with k different frequencies, corresponding to a k-torus in the phase space (cf. Landau's picture of turbulence as in Landau & Lifschitz, 1987).

However, with three or more coupled ODEs (or in one-dimensional maps), the attractor can be an extremely complicated object. The famous "Lorenz attractor" was perhaps the first explicit example of an attractor that is not just a fixed point or (quasi) periodic orbit. Edward Lorenz highlighted this in the title of his 1963 paper, "Deterministic Nonperiodic Flow." The phrase "strange attractor" was coined by Ruelle & Takens (1971) for such complicated attracting sets. These attractors, and the chaotic dynamics associated with them, have been the focus of much attention, particularly in relation with the theory of turbulence (the subject of Ruelle & Takens' paper; see also Ruelle, 1989). There is no fixed definition of a "strange attractor"; some authors use the phrase as a signature of chaotic dynamics, while others use it to denote a fractal attractor (e.g., Grebogi et al. (1984) discuss "strange nonchaotic attractors").

Over the years, various authors have given precise (but different) definitions of an attractor: Milnor (1985) discusses many of these (and proposes a new one of his own). Most definitions require that an attractor attract a "large set of initial conditions and satisfy some kind of minimality property" (without this, the whole phase space could be called an attractor).

We refer to the set of all those points in the phase space whose trajectories are attracted to some set A as the basin of attraction of A and write this $B(A)$. There are essentially two choices of what it means to attract a large set of initial conditions: the more common one is that $B(A)$ contains an open neighborhood of A, while Milnor (1985) suggested that a more realistic requirement is that $B(A)$ has positive Lebesgue measure.

Exactly what type of minimality assumption we require depends on what we want our attractor to say about the dynamics. At the very least, there should be no smaller (closed) set with the same basin of attraction: this excludes any "unnecessary" points from the attractor. A consequence of this minimality property is that the attractor is invariant: if A is the attractor of a map f, this means that $f(A) = A$ (there is, of course, a similar property for the attractor of a flow). In particular, this means that it is possible to talk about the "dynamics on the attractor."

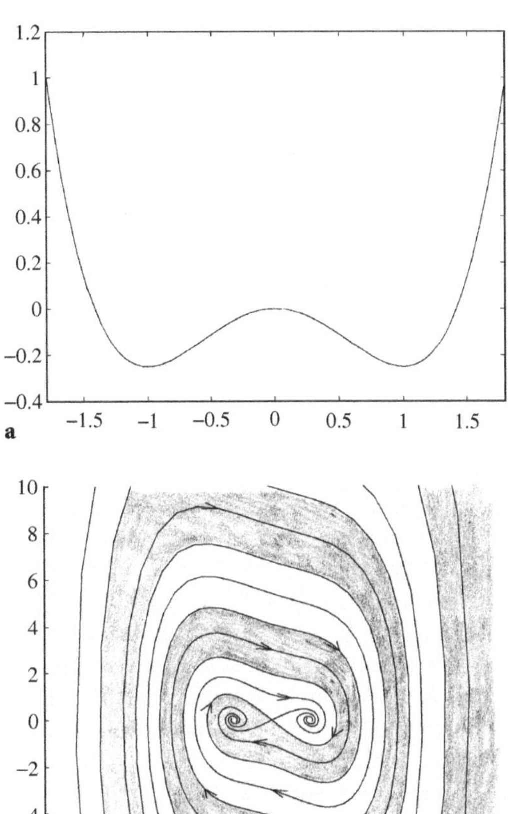

Figure 1. (a) A symmetric double-well potential. (b) Phase portrait of a particle moving in the potential of (a) with friction.

If we want one attractor to describe the possible asymptotic dynamics of *every* initial condition, then there is no need to impose any further minimality assumption. Figure 1(b) shows the phase portrait for a particle moving with friction in the symmetric double-well potential of Figure 1(a); the basin of attraction of the fixed point corresponding to the bottom of the left-hand well is shaded. (The equations of motion are $\dot{x} = y$ and $\dot{y} = -\frac{y}{2} + x - x^3$.) We could say that the attractor consists of the three points $\{(-1, 0), (1, 0), (0, 0)\}$, but this discards much of the information contained in the phase portrait. This motivates the further requirement that an attractor be "indecomposable": it should not be possible to split it into two disjoint invariant subsets. (Some definitions require there to be a dense orbit in the attractor: essentially, this means that one trajectory "covers" the entire attractor, so in particular the attractor cannot be split into two pieces.) This gives us two possible attractors: $(-1, 0)$ and $(1, 0)$ (the origin does not attract a neighborhood of itself, nor any set of positive measure).

In this example, the boundary between the basins of attraction of the two competing attractors is a smooth curve. However, in many examples this boundary is a fractal set. This was first noticed by McDonald et al. (1985), who observed that near a fractal boundary, it is harder to predict the asymptotic behavior of imprecisely known initial conditions. An extreme version of this occurs with the phenomenon of "riddled basins," first observed by Alexander et al. (1992): arbitrarily close to a point attracted to one attractor; there can be a point attracted to another. In this case, an arbitrarily small change in the initial condition can lead to completely different asymptotic behavior. (In addition to treating some analytically tractable examples, Alexander et al. (1992) give an impressive array of pictures from their numerical simulations.)

Attractors can also be meaningfully defined for the infinite-dimensional dynamical systems arising from partial and functional differential equations (e.g., Hale, 1988; Robinson, 2001; Temam 1988/1996), and for random and nonautonomous systems (Crauel et al. (1997) adopt an approach that includes both these cases).

JAMES C. ROBINSON

See also **Chaos vs. turbulence; Dynamical systems; Fractals; Phase space; Turbulence**

Further Reading

Alexander, J.C., Yorke, J.A., You, Z. & Kan, I. 1992. Riddled basins. *International Journal of Bifurcation and Chaos*, 2: 795–813

Crauel, H., Debussche, A. & Flandoli, F. 1997. Random attractors. *Journal of Dynamics and Differential Equations*, 9: 307–341

Grebogi, C., Ott, E., Pelikan, S. & Yorke, J.A. 1984. Strange attractors that are not chaotic. *Physica* D, 13: 261–268

Hale, J.K. 1988. *Asymptotic Behavior of Dissipative Systems*, Providence, RI: American Mathematical Society

Hirsch, M.W. & Smale, S. 1974. *Differential Equations, Dynamical Systems and Linear Algebra*. New York: Academic Press

Landau, L.D. & Lifschitz, E.M. 1987. *Fluid Mechanics*, 2nd edition, Oxford: Pergamon Press

Lorenz, E.N. 1963. Determininstic non-periodic flow. *Journal of Atmospheric Science*, 20: 130–141

Milnor, J. 1985. On the concept of attractor. *Communications in Mathematical Physics*, 99: 177–195

Robinson, J.C. 2001. *Infinite-Dimensional Dynamical Systems*, Cambridge and New York: Cambridge University Press

Ruelle, D. 1989. *Chaotic Evolution and Strange Attractors*, Cambridge and New York: Cambridge University Press

Ruelle, D. & Takens, F. 1971. On the nature of turbulence. *Communications in Mathematical Physics*, 20: 167–192

Temam, R. 1996. *Infinite-dimensional Dynamical Systems in Mechanics and Physics*, 2nd edition, Berlin and New York: Springer

AUBRY–MATHER THEORY

Named after Serge Aubry and John Mather, who independently shaped the seminal ideas, the Aubry–Mather theory addresses one of the central problems of modern dynamics: the characterization, of nonintegrable Hamiltonian time evolution beyond the realm of perturbation theory. In general terms, when a Hamiltonian system is near-integrable, perturbation theory provides a rigorous generic description of the invariant sets of motion (closed sets containing trajectories) as smooth surfaces (KAM tori), each one parametrized by the rotation number ω of the angle variable (angle-action coordinates): all the trajectories born and living inside the invariant torus share this common value of ω. An invariant set has an associated natural invariant measure, which describes the measure-theoretical (or statistical) properties of the trajectories inside the invariant set. The invariant measure on a torus is a continuous measure, so that the distribution function of the angle variable is continuous. In this near-integrable regime of the dynamics, perturbative schemes converge adequately and future evolution is—to a desired arbitrary degree—predictable for arbitrary initial conditions on each torus.

Far from integrable Hamiltonian dynamics, what is the fate of these invariant natural measures, or invariant sets of motion, beyond the borders of validity of perturbation theory? The answer is that each torus breaks down and its remaining pieces form an invariant fractal set, called by Percival a cantorus (or Aubry–Mather set) characterized by the rotation number value common to all trajectories in the cantorus. The statistical properties of the trajectories on the invariant cantorus are now described by a purely discrete measure or Cantor distribution function (see Figure 1).

Basic Theorems

The formal setting of Aubry–Mather theory for the transition from regular motion on invariant tori to orbits on hierarchically structured nowhere dense cantori is the class of maps of a cylindrical surface $\mathcal{C} = S^1 \times \mathcal{R}$ (cylindrical coordinates (u, p)) (see Figure 2) onto itself,

$$f : \mathcal{C} \to \mathcal{C}, \qquad (1)$$

characterized by preservation of areas (symplectic) and the "twist" property, meaning that the torsion produced by an iteration of the map on a vertical segment of the cylindrical surface converts it into a part of the graph

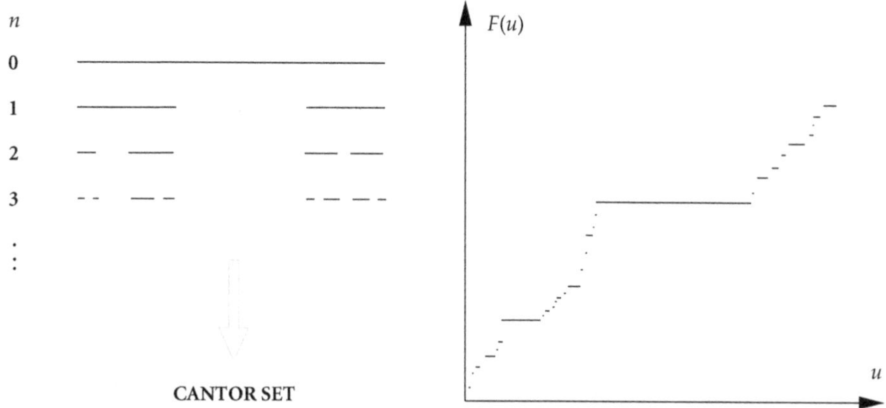

Figure 1. Left: Construction of a Cantor set from the unit real interval (or circle S^1) as a limiting process. At each step n, a whole piece is cut out from each remaining full segment. Right: The distribution function $F(u)$ of the projection onto the angular component u_n of a cantorus orbit. $F(u)$ is the limiting proportion of the values of $u_n < u$, $-\infty < n < +\infty$.

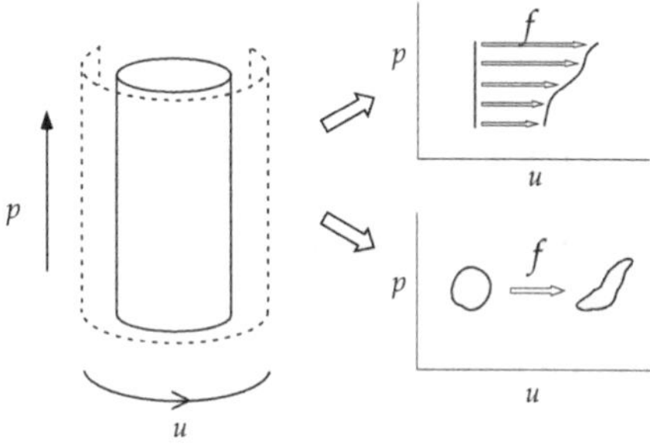

Figure 2. Schematic illustration on the unfolded cylindrical surface $\mathcal{C} = S^1 \times \mathcal{R}$ of the twist (upper right) and area-preserving (lower right) properties of the map $F : \mathcal{C} \to \mathcal{C}$. In the upper right area the curve is a single-valued function $\varphi(u)$ of the angular variable.

of some function $\varphi(u)$ of the angular coordinate. More explicitly, if we denote by $(u', p') = f(u = 0, p)$ the image of the vertical segment, then u' is a monotone function of p, so that (u', p') is the graph of a single-valued function.

An area-preserving twist map has associated an action-generating function related to the map via a variational (extremal action) principle:

- An action-generating function, $H(x, x')$, of a twist map is a two-variable function that is strictly convex:

$$\frac{\partial^2 H}{\partial u \partial u'} \leq K < 0. \tag{2}$$

- If u_0 is a critical point of $\mathcal{L}(x) = H(u_{-1}, x) + H(x, u_1)$, then a certain sequence (u_{-1}, p_{-1}), (u_0, p_0), (u_1, p_1) is a segment of a cylinder orbit of f.

(Given a sequence $\{u_j\}_{j=0}^n$ with fixed ends ($u_0 = a$, $u_n = b$), the associated action functional $\mathcal{L}$ is the sum $\sum_{j=0}^{n-1} H(u_j, u_{j+1})$.

A cylinder orbit is called ordered when it projects onto an angular sequence ordered in the same way as a uniform rotation of angle ω. An invariant set is a minimal invariant set if it does not include proper invariant subsets and is called ordered if it contains only ordered orbits. The proper definition of an Aubry–Mather set is a minimal invariant ordered set that projects one-to-one on a nowhere dense Cantor set of the circle S^1.

The following points comprise the main core (Golé, 2001; Katok & Hasselblatt, 1995) of the Aubry–Mather theory:

- For each rational value $\omega = p/q$ of the rotation number, there exist (Poincaré–Birkhoff theorem) at least two ordered periodic orbits (Birkhoff periodic orbits of type (p, q)), which are obtained by, respectively, minimizing and maximizing the action over the appropriate set of angular sequences.

  In general, periodic orbits of rational rotation number $\omega = p/q$ do not form an invariant circle, in which case there are nonperiodic orbits approaching two different periodic orbits as $n \to -\infty$ and as $n \to +\infty$, called heteroclinic orbits (or homoclinic in some contexts). These orbits connect two Birkhoff periodic orbits through a minimal action path. Usually, the number of map iterations needed for such action-minimizing orbits to pass over the action barriers is exponentially small.
- For each irrational value of ω, there exists either an invariant torus or an Aubry–Mather set. There are also homoclinic trajectories connecting orbits on the Aubry–Mather set.

The hierarchical structure of gaps that break up the torus has its origins in the path-dependent action barriers. Note also that heteroclinics to nearby periodic orbits (of rational rotation number) pass over nearly the same action barriers, leading to a somewhat metaphorical view of nearby resonances biting the tori and leaving gaps. Certainly, the action barriers fractalize the invariant measure according to the proximity of the irrational rotation number ω to rationals.

The explanatory power of the extremal action principle in the fractalization of invariant sets of motion suggests one of the immediate physical applications of this theory. Indeed, Aubry's work was originally motivated by the equilibrium problem of a discrete field (u_n) $n \in \mathcal{Z}$, under some energy functional, whose extremalization defines equilibrium field configurations. Under some conditions on the energy-generating function $H(u, u')$, both are equivalent mathematical physics problems, and a one-to-one correspondence between orbits (u_n, p_n) and equilibrium field configurations (u_n) does exist (Aubry, 1985).

Application to the Generalized Frenkel–Kontorova Model

From this perspective, the Aubry–Mather theory gives rigorous variational answers in the description of equilibrium discrete nonlinear fields such as the generalized Frenkel–Kontorova (FK) model with convex interactions. Although the terminology changes, every aspect of cylinder dynamics has a counterpart in the equilibrium problem of this interacting nonlinear many-body model.

- Commensurate (periodic) field configurations correspond to Birkhoff periodic orbits, and as such they are connected by the field configurations associated to heteroclinics, which are here called discrete (sine-Gordon) solitons or elementary discommensurations.
- Incommensurate (quasiperiodic) field configurations can correspond either to tori trajectories or to Aubry–Mather trajectories. The macroscopic physical properties (formally represented by averages on the invariant measure) of the field configuration experience drastic changes when passing from one case (tori) to the other (Aubry–Mather sets). This transition (called breaking of analiticity by Aubry) has been characterized as a critical phenomenon using renormalization group methods by (MacKay, 1993).

The Aubry–Mather theory puts on a firm basis what is known as discommensuration theory, which is the description of a generic incommensurate or (higher-order) commensurate field configuration as an array of discrete field solitons, strongly interacting when tori subsist, but almost noninteracting and deeply pinned when only Cantor invariant measures remain. Aubry's work provided a satisfactory understanding of the complexity of the phase diagrams and the singular character of the equations of state of the generalized FK model (Griffiths, 1990).

Luis Mario Floría

See also **Commensurate–incommensurate transition; Frenkel–Kontorova model; Hamiltonian systems; Kolmogorov–Arnol'd–Moser theorem; Phase transitions; Standard map; Symplectic maps**

Further Reading

Aubry, S. 1985. *Structures incommensurables et brisure de la symmetrie de translation I* [Incommensurate structures and the breaking of traslation symmetry I]. In *Structures et Instabilités*, edited by C. Godréche, Les Ulis: Editions de Physique, pp. 73–194

Golé, C. 2001. *Symplectic Twist Maps. Global Variational Techniques*, Singapore: World Scientific

Griffiths, R.B. 1990. Frenkel–Kontorova models of commensurate-incommensurate phase transitions. In *Fundamental Problems in Statistical Mechanics VII*, edited by H. van Beijeren, Amsterdam: North-Holland, pp. 69–110

Katok, A. & Hasselblatt, B. 1995. *Introduction to the Modern Theory of Dynamical Systems*. Cambridge and New York: Cambridge University Press

MacKay, R.S. 1993. *Renormalisation in Area-preserving Maps*, Singapore: World Scientific

AUTO-BÄCKLUND TRANSFORMATION

See **Bäcklund transformations**

AUTOCATALYTIC SYSTEM

See **Reaction-diffusion systems**

AUTOCORRELATION FUNCTION

See **Coherence phenomena**

AUTONOMOUS SYSTEM

See **Phase space**

AUTO-OSCILLATIONS

See **Phase plane**

AVALANCHE BREAKDOWN

Charge transport in condensed matter is simply described by the current density j as a function of the local electric field E. For bulk materials, the current density per unit area is given by $j(E) = -env$, where $e > 0$ is the electron charge, n is the conduction electron density per unit volume, and v is the drift velocity. In the simplest case, v is a linear function of the field: $v = -\mu E$, with mobility μ.

Thus, the conductivity $\sigma = j/E = en\mu$ is proportional to the number of conduction electrons. In metals, n is given by the number of valence electrons, which is temperature independent. In semiconductors, however, the concentration of electrons in the conduction band varies greatly and is determined by generation-recombination (GR) processes that induce transitions between valence band, conduction band, and impurity levels (donors and acceptors). Charge carrier concentration depends not only upon temperature but also upon the electric field, which explains why the conductivity can change over many orders of magnitude. A GR process that depends particularly strongly on the field is impact ionization, the inverse of the Auger effect. It is a process in which a charge carrier with high kinetic energy collides with a second charge carrier, transferring its kinetic energy to the latter, which is thereby lifted to a higher energy level. The kinetic energy is increased by the local electric field, which heats up the carriers. As a certain minimum energy is necessary to overcome the difference in the energy levels of the second carrier, the impact ionization probability depends in a threshold-like manner on the applied voltage.

Impact ionization processes may be classified as band-band processes or band-trap processes depending on whether the second carrier is initially in the valence band and makes a transition from the valence band to the conduction band, or whether it is initially at a localized level (impurity, donor, acceptor), and makes a transition to a band state. Further, impact ionization processes are classified as electron or hole processes according to whether the ionizing hot carrier is a conduction band electron or a hole in the valence band.

Schematically, impact ionization may be written as one of the following reaction equations, in analogy with chemical kinetics:

$$\mathrm{e} \longrightarrow 2\mathrm{e} + \mathrm{h}, \tag{1}$$

$$\mathrm{e} + \mathrm{e_t} \longrightarrow 2\mathrm{e} + \mathrm{h_t}, \tag{2}$$

$$\mathrm{h} \longrightarrow 2\mathrm{h} + \mathrm{e}, \tag{3}$$

$$\mathrm{h} + \mathrm{h_t} \longrightarrow 2\mathrm{h} + \mathrm{e_t}, \tag{4}$$

where e and h denote band electrons and holes, respectively, and $\mathrm{e_t}$ and $\mathrm{h_t}$ stand for electrons and holes trapped at impurities (donors, acceptors, or deep levels). The result of the process is carrier multiplication (avalanching), which may induce electrical instabilities at sufficiently high electric fields. Impact ionization from shallow donors or acceptors is responsible for impurity breakdown at low temperatures. Being an autocatalytic process (i.e., each carrier ionizes secondary carriers that might, in turn, impact ionize other carriers), it induces a nonequilibrium phase transition between a low- and high-conductivity state and may lead to a variety of spatiotemporal instabilities, including current filamentation, self-sustained oscillations, and chaos. The conductivity saturates when all impurities are ionized. Band-to-band impact ionization eventually induces avalanche breakdown, limiting the bias voltage that can be safely applied to a device. The conductivity increases much more strongly than during impurity breakdown because of the large number of valence band electrons available for ionization.

Impurity impact-ionization breakdown at helium temperatures (ca. 5 K) in p-Ge, n-GaAs, and other semiconductor materials has been thoroughly studied both experimentally and theoretically as a model system for nonlinear dynamics in semiconductors. It displays S-shaped current–voltage characteristics because the GR kinetics incorporating impact ionization from at least two impurity levels (ground state and excited state) allows for three different values of the carrier density $n(E)$ in a certain range of fields E. As a result of the negative differential conductivity, a variety of temporal and spatiotemporal instabilities occur, ranging from stationary and breathing current filaments and traveling charge density waves to various chaotic scenarios.

Band-to-band impact ionization of a reverse biased p–n junction is the basis of a variety of electronic devices. A number of these devices depend on a combination of impact ionization of hot electrons and transit time effects. The IMPATT (impact ionization avalanche transit time) diodes can generate the highest continuous power output at frequencies > 30 GHz. The originally proposed device (Read diode) involves a reverse biased $\mathrm{n^+}$–p–i–$\mathrm{p^+}$ multilayer structure, where $\mathrm{n^+}$ and $\mathrm{p^+}$ denote strongly n- or p-doped semiconductor regions, and i denotes an intrinsic (undoped) region. In the $\mathrm{n^+}$–p region (avalanche region), carriers are generated by impact ionization across the bandgap; the generated holes are swept through the i region (drift region) and collected at the $\mathrm{p^+}$ contact. When a periodic (ac) voltage is superimposed on the time-independent (dc) reverse bias, a π phase lag of the ac current behind the voltage can arise. This phase lag is due to the finite buildup time of the avalanche current and the finite time carriers take to cross the drift region (transit-time delay). If the sum of these delay times is approximately one-half cycle of the operating frequency, negative conductance is observed; in other words, the carrier flow drifts opposite to the ac electric field. This can be achieved by properly matching the length of the drift region with the drift velocity and the frequency. Other devices using the avalanche breakdown effect are the TRAPATT (trapped plasma avalanche triggered transit) diode and the avalanche transistor. The Zener diode is a p–n junction that exhibits a sharp increase in the magnitude of the current at a certain reverse voltage where avalanche breakdown sets in. It is used to stabilize and limit the dc voltage in circuits (overload and transient suppressor) since the current can vary over a large range at the avalanche breakdown threshold without a noticeable change in the voltage. The original Zener effect, on the other hand, is due to quantum mechanical tunneling across the bandgap at high fields, and is effective in highly doped (resulting in narrow depletion layers) Zener diodes at lower breakdown voltages.

ECKEHARD SCHÖLL

See also **Diodes; Drude model; Nonlinear electronics; Semiconductor oscillators**

Further Reading

Landsberg, P.T. 1991. *Recombination in Semiconductors*, Cambridge and New York: Cambridge University Press

Schöll, E. 1987. *Nonequilibrium Phase Transitions in Semiconductors*, Berlin: Springer

Schöll, E. 2001. *Nonlinear Spatio-temporal Dynamics and Chaos in Semiconductors*, Cambridge and New York: Cambridge University Press

Schöll, E., Niedernostheide, F.-J., Parisi, J., Prettl, W. & Purwins, H. 1998. Formation of spatio-temporal structures in semiconductors. In *Evolution of Spontaneous Structures in Dissipative Continuous Systems*, edited by F.H. Busse & S.C. Müller, Berlin: Springer, pp. 446–494

Shaw, M.P., Mitin, V.V., Schöll, E. & Grubin, H.L. 1992. *The Physics of Instabilities in Solid State Electron Devices*, New York: Plenum Press

AVALANCHES

An avalanche is a downhill slide of a large mass, usually of snow, ice, or rock debris prompted by a small initial disturbance. Avalanches, along with landslides, are one of the major natural disasters that still present significant danger for people in the mountains. On average, 25 people die in avalanches every winter in Switzerland alone. Dozens of people were killed on September 23, 2002, in a gigantic avalanche in Northern Ossetia, Russia, when a 150 m thick chunk of the Kolka Glacier broke off and triggered an avalanche of ice and debris that slid some 25 km along Karmadon gorge. In 1999, some 3000 avalanches occurred in the Swiss Alps.

Avalanches vary widely in size, from minor slides to large movements of snow reaching a volume of 10^5 m^3 and a weight of 30,000 tons. The speed of the downhill snow movement can reach 100 m/s. There are two main types of avalanches-loose avalanche and slab avalanche: depending on the physical properties of snow. Soft dry snow typically produces loose avalanches that form a wedge downward from the starting point, mainly determined by the physical properties of the granular material. In wet or icy conditions, on the other hand, a whole slab of solid dense snow may slide down. The initiation of the second type occurs as a fracture line at the top of the slab. The study of real avalanches and landslides is mostly an empirical science that is traditionally a part of geophysics and draws from the physics of snow, ice, and soil. Semi-empirical computer codes have been developed for prediction of avalanches dependent on the weather conditions (snowfall, wind, temperature profiles) and topography.

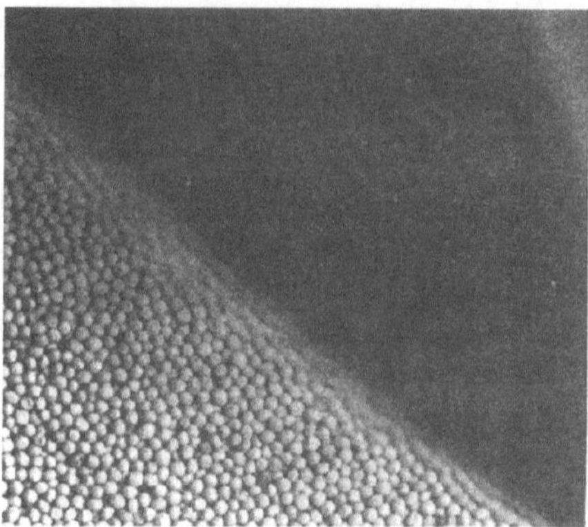

Figure 1. Only several layers of mustard seeds are involved in the rolling motion inside the avalanche: moving grains are smeared out in this long-exposure photograph. Reproduced with permission from Jaeger et al. (1998).

More fundamental aspects of avalanche dynamics have been studied in controlled laboratory experiments with dry or wet granular piles, or sandpiles. Granular slope can be characterized by two *angles of repose*—the *static* angle of repose θ_s which is the maximum angle at which the granular slope can remain static, and the *dynamic* angle of repose θ_d, or a minimum angle at which the granular flow down the slope can persist. Typically, in dry granular media, the difference between static and dynamic angles of repose is about 2–5°, for smooth glass beads $\theta_s \approx 25°$, $\theta_d \approx 23°$. Avalanches may occur in the bistable regime when the slope angle satisfies $\theta_d < \theta < \theta_s$. The bistability is explained by the need to *dilate* the granular material for it to enter flowing regime (*Bagnold's dilatancy*).

An avalanche can be initiated by a small localized fluctuation from which the fluidized region expands downhill and sometimes also uphill, while the sand always slides downhill. An avalanche in a deep sandpile usually involves a narrow layer near the surface (see Figure 1). Avalanches have also been studied in finite-depth granular layers on inclined planes. The two-dimensional structure of a developing avalanche depends on the thickness of the granular layer and the slope angle. For thin layers and small angles, wedge-shaped avalanches are formed similar to the loose snow avalanches (Figure 2a). In thicker layers and at higher inclination angles, avalanches have a balloon-type shape that expands both down- and uphill (Figure 2b).

The kinematics of the fluidized layer in one dimension can be described by a set of hydraulic equations for the local thickness $R(x,t)$ of the layer of rolling particles flowing over a sandpile of immobile particles with variable profile $h(x,t)$ (BCRE model, after Bouchaud et al., (1994)),

$$\partial_t R = -v\partial_x R + \Gamma(R,h) + (\text{diffusive terms}), \quad (1)$$

$$\partial_t h = -\Gamma(R,h) + (\text{diffusive terms}), \quad (2)$$

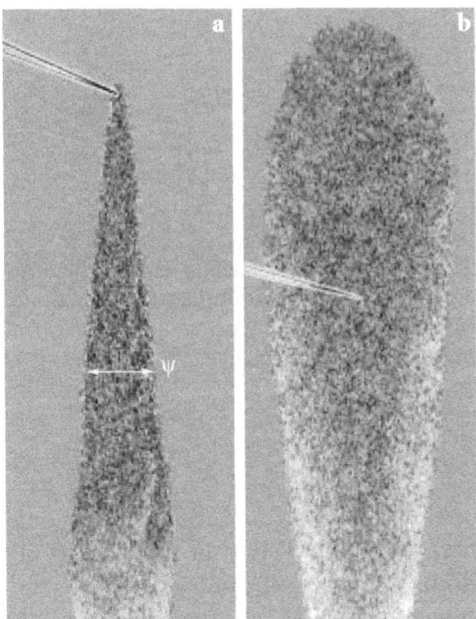

Figure 2. Structure of the avalanche in a thin (4 grain diameters) layer of glass beads: (a) wedge-shaped avalanche for $\theta = 31.5°$; (b) balloon-shaped avalanche propagating both up- and downhill for $\theta = 32.5°$. Reprinted by permission from *Nature* (Daerr & Douady, 1999).Copyright (1999) Macmillan Publishers Ltd.

where Γ is the entrainment flux of immobile particles into the rolling layer and the downhill transport velocity v is assumed constant. Γ becomes positive when the local slope becomes steeper than the static repose angle θ_s, and in the simplest case, $\Gamma = \gamma R(\partial_x h - \tan\theta_s)$. This model allows for a complete analytical treatment.

A more sophisticated continuum theory of granular avalanches is based on the fluid dynamics (Navier–Stokes) equations coupled with a phenomenological description of the first-order phase transition from a static to a fluidized state driven by the local shear stress (Aranson & Tsimring, 2001). The local phase state is described by the local order parameter ρ that is controlled by a Ginzburg–Landau-type equation with bistable free energy $F(\rho, \delta)$:

$$\partial_t \rho = D\nabla^2 \rho - \partial_\rho F(\rho, \delta) \tag{3}$$

The control parameter δ in this equation depends on the ratio of shear to normal stress. This theory can describe a variety of "partially fluidized" granular flows, including avalanches in sandpiles. In a "shallow-water" approximation, it yields the BCRE-type equations for the local slope and the thickness of the rolling layer.

The wide distribution of scales in real avalanches led Bak et al. (1988) to propose a "sandpile cellular automaton" (*See* **Sandpile model**) as a paradigm model for *self-organized criticality* (SOC), the phenomenon that occurs in slowly driven nonequilibrium spatially extended systems when they asymptotically reach a critical state characterized by a power-law distribution of event sizes. The BTW model is remarkably simple, yet it exhibits a highly nontrivial behavior. The sandpile is formed on a lattice by dropping "grains" on a random site from above, one at a time. "Grains" form stacks of integer height at each lattice site. After each grain dropping the sandpile is allowed to relax. Relaxation occurs when the slope (a difference in heights of two adjacent stacks) reaches a critical value ("angle of repose") and the grain hops to a lower stack. This may prompt a series of subsequent hops and so trigger an avalanche. The size of the avalanche is determined by the number of grains set into motion by adding a single grain to a sandpile. In the asymptotic regime in a large system, the avalanche size distribution becomes scale-invariant, $P(s) \propto s^{-\alpha}$ with $\alpha \approx 1.5$.

The relevance of this model and its generalizations to real avalanches is still a matter of debate. The sandpile model is defined via a single repose angle, and so its asymptotic behavior has the properties of the critical state for a second-order phase transition. Real sandpiles are characterized by two angles of repose and thus exhibit features of the first-order phase transition. Experiments with avalanches in slowly rotating drums do not confirm the scale-invariant distribution of avalanches. However, in such experiments, the internal structures of the sandpile (the force chains) are constantly changing in the process of rotation. In other experiments with large monodispersed glass beads dropped on a conical sandpile, SOC with $\alpha \approx 1.5$ was observed. The characteristics of the size distribution depend on the geometry of the sandpile and the physical and geometrical properties of grains. SOC was also observed in the avalanche statistics in a three-dimensional pile of long rice; however, a smaller scaling exponent $\alpha \approx 1.2$ was measured for the avalanche size distribution.

An avalanche in a pile of sand has been used as a metaphor in many other physical phenomena including the avalanche diodes, vortices in type-II superconductors, Barkhausen effect in ferro-magnetics, $1/f$ noise, and.

LEV TSIMRING

See also **Granular materials; Sandpile model**

Further Reading

Aranson, I.S. & Tsimring, L.S. 2001. Continuum description of avalanches in granular media. *Physical Review E*, 64: 020301

Bak, P., Tang, C. & Wiesenfeld, K. 1988. Self-organized criticality. *Physical Review A*, 38: 364–374

Bouchaud, J.-P., Cates, M.E., Ravi Prakash, J. & Edwards, S.F. 1994. A model for the dynamics of sandpile surfaces. *Journal de Physique I*, 4: 1383–1410

Daerr, A. & Douady, S. 1999. Two types of avalanche behaviour in granular media. *Nature*, 399: 241–243

Duran, J. 1999. *Sands, Powders, and Grains: An Introduction to the Physics of Granular Materials*, Berlin and New York: Springer

Jaeger, H.M., Nagel, S.R. & Behringer, R.P. 1996. Granular solids, liquids, and gases. *Reviews of Modern Physics*, 68: 1259–1273

Jensen, H.J. 1998. *Self-Organized Criticality*, Cambridge: Cambridge University Press

Nagel, S.R. 1992. Instabilities in a sandpile. *Reviews of Modern Physics*, 64(1): 321–325

Rajchenbach, J. 2000. Granular flows. *Advances in Physics*, 49(2): 229–256

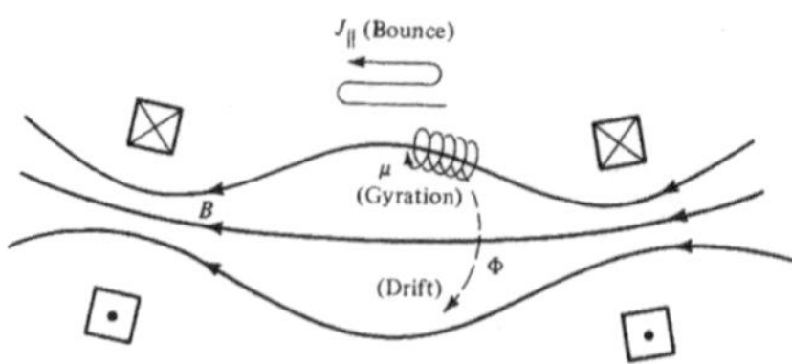

Figure 1. A hierarchy of adiabatic invariants for the charged particle gyrating in a nonaxisymmetric, magnetic mirror field. The three adiabatic invariants are the magnetic moment μ, the longitudinal invariant $J_{||}$, and the guiding center flux invariant Φ.

AVERAGING METHODS

Averaging methods are generally used for dynamical systems of two or more degrees of freedom when time scales or space scales are well separated. An average over the rapidly varying coordinates of one degree of freedom, considering the coordinates of the second degree of freedom to be constant during the average, can, with appropriate variables, retain a time-invariant quantity that enters into the solution of the slower motion. This solution, in turn, supplies a parameter to the rapid motion, which can then be solved in a lower-dimensional space. The averaging method is closely related to the calculation of *adiabatic invariants*, which are the approximately constant integrals of the motion that are obtained by averaging over the fast angle variables.

The lowest-order calculation is generally straightforwardly performed in canonical coordinates. A transformation from momentum and position coordinates (p, q), for the fast oscillation, to action-angle form (J, θ) gives a constant of the motion J, if all other variables are held constant. The action J is then the constant parameter in the equation for the slower motion. It is not always convenient to transform to action-angle form directly, but the underlying constants are related to the action variables.

The formal expansion procedure that is employed is to develop the solution in an asymptotic series. The mathematical method applied to ordinary differential equations was developed by Nikolai Bogoliubov (Bogoliubov & Mitropolsky, 1961) and in a somewhat different form by Martin Kruskal (1962). The expansion techniques can be formally extended to all orders in the perturbation parameter but are actually divergent. For multiple periodic systems, higher-order local nonlinear resonances between the degrees of freedom may destroy the ordering in their neighborhood. We will return to this problem below.

Averaging over the fastest oscillation of an N-degree-of-freedom system reduces the number of freedoms to $N-1$. A second average over the next fastest motion then produces a second adiabatic invariant to reduce the freedoms to $N-2$. This process may be continued to obtain a hierarchy of adiabatic invariants, until the system is reduced to one degree of freedom, which can be integrated to obtain a final integrable equation. The process is well known in plasma physics where, for a charged particle gyrating in a magnetic mirror field, we first find the magnetic moment invariant μ associated with the fast gyration, then find the longitudinal invariant $J_{||}$ associated with the slower bounce motion, and finally find the flux invariant Φ associated with the drift motion. The three degrees of freedom are shown in Figure 1. The small parameters in this case are ε_1, the ratio of bounce frequency to gyration frequency; ε_2, the ratio of guiding center drift frequency to bounce frequency; and ε_3, the ratio of the frequency of the time-varying magnetic field to the drift frequency. This example motivated the development of averaging methods. The derivations of these invariants are given in detail in Northrop (1963) or in other plasma physics texts.

Although the asymptotic expansions are formally good to all orders in a small dimensionless parameter of the form $\varepsilon = |\dot{\omega}/\omega^2|$, where ω is the frequency of the fast oscillation that is slowly changing in time and $\dot{\omega} \equiv \mathrm{d}\omega/\mathrm{d}t$, the series generally diverge. The physical reason is that resonances or near-resonances between degrees of freedom lead to small denominators in the coefficients of terms. For nonlinear coupled oscillatory systems, exact resonances for certain values of the action locally change the structure of the phase-space orbits so that they do not follow the values obtained by averaging. This led to the development of the secular perturbation theory (Born, 1927), in which a local transformation of the coordinates around the resonance can be made. The frequency of the oscillatory motion in the neighborhood of the exact resonance is then slow compared with the other frequencies in the transformed coordinates, and averaging can then be applied locally. A review of the various methods, their limitations, practical examples, and reference to original sources can be found in Lichtenberg (1969) and Lichtenberg & Lieberman (1991).

The above discussion is related to the study of finite-dimensional systems governed by ordinary differential equations. The methods are usually applied to relatively low-dimensional systems, for example, the motion of a magnetically confined charged particle as described above. However, averaging methods are also applied to systems governed by partial differential equations, such as nonlinear wave propagation problems and wave instabilities. For example, waves on discrete oscillator chains can be obtained by first averaging over the discreteness using a Taylor expansion.

ALLAN J. LICHTENBERG

See also **Adiabatic invariants; Breathers; Collective coordinates; Modulated waves; Solitons**

Further Reading

Bogoliubov, N.N. & Mitropolsky, Y.A. 1961. *Assymptotic Methods in the Theory of Nonlinear Oscillators*, New York: Gordon & Beach

Born, M. 1927. *The Mechanics of the Atom*, London: Bell

Kruskal, M.D. 1962. Asymptotic theory of Hamiltonian systems with all solutions nearly periodic. *Journal of Mathematical Physics*, 3: 806–828

Lichtenberg, A.J. 1969. *Phase Space Dynamics of Particles*, New York: Wiley

Lichtenberg, A.J. & Lieberman M.A. 1991. *Regular and Chaotic Dynamics*, 2nd edition, New York: Springer

Northrop, T.G. 1963. *The Adiabatic Motion of Charged Particles*, New York: Wiley

B

BÄCKLUND TRANSFORMATIONS

Bäcklund transformations (BTs) originated in investigations conducted in the late 19th century into invariance properties of pseudospherical surfaces, namely surfaces of constant negative Gaussian curvature. In 1862, it was Edmond Bour who derived the well-known sine-Gordon equation

$$\omega_{uv} = \frac{1}{\rho^2} \sin \omega \tag{1}$$

via the Gauss–Mainardi–Codazzi system for pseudospherical surfaces with total curvature $\mathcal{K} = -1/\rho^2$, parametrized in asymptotic coordinates. In 1883, Albert Bäcklund published his now classical result whereby pseudo-spherical surfaces may be generated in an iterative manner. Thus, if $\boldsymbol{r}$ is the position vector of a pseudospherical surface Σ corresponding to a seed solution ω of Equation (1) and ω' denotes the Bäcklund transformation of ω via the BT

$$\left.\begin{aligned} \omega'_u - \omega_u &= \frac{2\beta}{\rho} \sin\left(\frac{\omega' + \omega}{2}\right), \\ \omega'_v + \omega_v &= \frac{2}{\beta\rho} \sin\left(\frac{\omega' - \omega}{2}\right), \end{aligned}\right\} \mathbb{B}_\beta \tag{2}$$

then the position vector $\boldsymbol{r}'$ of the one-parameter class of surfaces Σ' corresponding to ω' is given by (Bäcklund, 1883)

$$\boldsymbol{r}' = \boldsymbol{r} + \frac{\mathrm{L}}{\sin \omega} \times \left[\sin\left(\frac{\omega - \omega'}{2}\right) \boldsymbol{r}_u + \sin\left(\frac{\omega + \omega'}{2}\right) \boldsymbol{r}_v\right], \tag{3}$$

where $L = \rho \sin \zeta$ and $\beta = \tan(\zeta/2)$, ζ being the constant angle between the normals to Σ and Σ' and β being termed the Bäcklund parameter. Sophus Lie subsequently observed that $\mathbb{B}_\beta$ may be decomposed according to $\mathbb{B}_\beta = \mathbb{L}_\beta^{-1} \mathbb{B}_{\beta=1} \mathbb{L}_\beta$, where $\mathbb{L}_\beta$ and $\mathbb{L}_\beta^{-1}$ are Lie invariances. Thus, Lie transformations play a crucial role in intruding the Bäcklund parameter β into the parameter-independent "Bianchi" transformation $\mathbb{B}_{\beta=1}$ to produce $\mathbb{B}_\beta$.

It was in 1892 that Luigi Bianchi in his masterly paper *Sulla Trasformazione di Bäcklund per le Superficie Pseudosferiche* established that the BT $\mathbb{B}_\beta$ admits a commutation property $\mathbb{B}_{\beta_2}\mathbb{B}_{\beta_1} = \mathbb{B}_{\beta_1}\mathbb{B}_{\beta_2}$, a consequence of which is a nonlinear superposition principle embodied in what is termed a "permutability theorem."

Bianchi's Permutability Theorem

If ω is a seed solution of the sine-Gordon equation (1), let ω_1, ω_2 denote the BT of ω via $\mathbb{B}_{\beta_1}$ and $\mathbb{B}_{\beta_2}$, that is, $\omega_1 = \mathbb{B}_{\beta_1}(\omega)$ and $\omega_2 = \mathbb{B}_{\beta_2}(\omega)$. Let $\omega_{12} = \mathbb{B}_{\beta_2}(\omega_1)$ and $\omega_{21} = \mathbb{B}_{\beta_1}(\omega_2)$. Then, imposition of the commutativity requirement $\omega_{12} = \omega_{21}$ yields a new solution of (1), namely

$$\begin{aligned} \Omega &= \omega_{12} = \omega_{21} \\ &= \omega + 4 \tan^{-1}\left[\frac{\beta_2 + \beta_1}{\beta_2 - \beta_1} \tan\left(\frac{\omega_2 - \omega_1}{4}\right)\right]. \end{aligned} \tag{4}$$

This result is commonly encapsulated in what is termed a "Lamb diagram" as shown in Figure 1. This solution-generation procedure may be iterated via what is sometimes termed a Bianchi lattice. At each iteration, a new Bäcklund parameter β_i is introduced.

The discovery of the BT for the iterative construction of pseudospherical surfaces along with its concomitant permutability theorem led to an intensive search by geometers at the turn of the 20th century for other classes of privileged surfaces that possess Bäcklund-type transformations. In this connection, Luther Eisenhart, in the preface to his monograph *Transformations of Surfaces* published in 1922, asserted that: "During the past twenty-five years many of the advances in differential geometry of surfaces in Euclidean space have had to do with transformations of surfaces of a given type into surfaces of the same type." Thus,

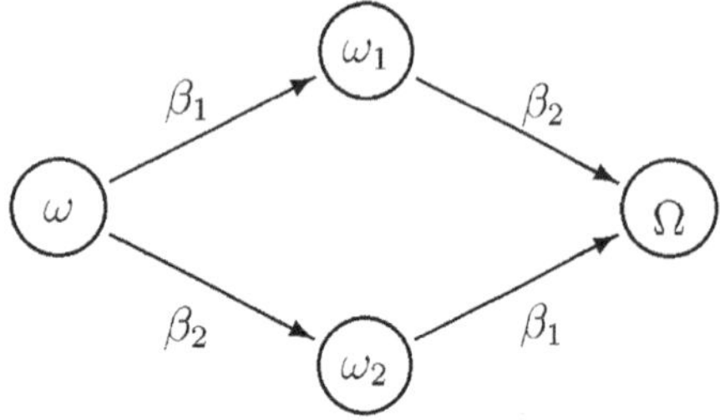

Figure 1. The Lamb diagram.

distinguished geometers such as Bianchi, Calapso, Darboux, Demoulin, Guichard, Jonas, Tzitzeica, and Weingarten all conducted extensive investigations into various classes of surfaces that admit BTs.

The particular Lamé system descriptive of triply orthogonal systems in the case when one of the constituent coordinate surfaces is pseudospherical was shown by Bianchi (1885) to admit an auto-BT, that is, a BT that renders the system invariant. Bianchi followed this in 1890 with the construction of a BT for the Gauss–Mainardi–Codazzi system associated with the class of hyperbolic surfaces with Gaussian curvature $\mathcal{K} = -1/\rho^2$ subject to the constraint

$$\rho_{uv} = 0, \tag{5}$$

where u, v are asymptotic coordinates. In 1899, Gaston Darboux constructed a BT for the nonlinear system

$$\begin{aligned} \theta_{xx} + \theta_{yy} + \kappa_1\kappa_2 e^{2\theta} &= 0, \\ \kappa_{1,y} + (\kappa_1 - \kappa_2)\theta_y &= 0, \\ \kappa_{2,x} + (\kappa_2 - \kappa_1)\theta_x &= 0 \end{aligned} \tag{6}$$

descriptive of isothermic surfaces with fundamental forms

$$\mathrm{I} = e^{2\theta}(dx^2 + dy^2), \quad \mathrm{II} = e^{2\theta}(\kappa_1 dx^2 + \kappa_2 dy^2), \tag{7}$$

where κ_1, κ_2 are principal curvatures and x, y are conjugate coordinates. The classical BT for system (6) has been set in a modern solitonic context by Cieśliński (1997).

In the first decade of the 20th century, the Romanian geometer Gheorghe Tzitzeica embarked upon an investigation of an important class of surfaces for which, in asymptotic coordinates, the Gauss–Mainardi–Codazzi system reduces to the nonlinear hyperbolic equation

$$(\ln h)_{uv} = h - h^{-2} \tag{8}$$

to be rediscovered some 70 years later in a soliton context. Tzitzeica (1910) not only constructed a BT for (8) but also set down what, in modern terms, is a linear representation containing a spectral parameter. Tzitzeica surfaces may be subsumed in the more general class of projective minimal surfaces for which BTs can be established (Rogers & Schief, 2002). In particular, this class contains the Demoulin system (1933)

$$(\ln h)_{uv} = h - \frac{1}{hk}, \quad (\ln k)_{uv} = k - \frac{1}{hk}. \tag{9}$$

The application of BTs in physics began with the work of Seeger et al. (1953) on crystal dislocation theory. Therein, within the context of Frenkel and Kontorova's dislocation theory, the superposition of so-called "eigenmotions" was obtained via the permutability relation (4). The interaction of what are today called breathers with kink-type dislocations was both described analytically and displayed graphically. The typical solitonic features to be subsequently discovered for the Korteweg–de Vries (KdV) equation

$$u_t + 6uu_x + u_{xxx} = 0, \tag{10}$$

(namely, preservation of velocity and shape following interaction as well as the concomitant phase shift) were all recorded. Bianchi's permutability theorem was subsequently employed in an investigation of the propagation of ultrashort optical pulses in a resonant medium by Lamb (1971).

A BT for the Korteweg–de Vries equation (10), namely

$$\begin{aligned} (\Lambda + \Lambda')_x &= \beta - \tfrac{1}{2}(\Lambda - \Lambda')^2, \\ (\Lambda' + \Lambda')_t &= (u - u')(u_{xx} - u'_{xx}) \\ &\quad -2(u_x^2 + u_x u'_x + u'^2_x), \end{aligned} \tag{11}$$

where

$$\Lambda = \int_\infty^x u(\sigma, t)d\sigma \tag{12}$$

was established by Wahlquist and Estabrook (1973). The spatial part of the BT was used to construct a permutability theorem, whereby multi-soliton solutions may be generated. This permutability theorem makes a remarkable appearance in numerical analysis as the so-called ε-algorithm.

A BT for the celebrated nonlinear Schrödinger (NLS) equation

$$iq_t + q_{xx} + 2|q|^2 q = 0 \tag{13}$$

was established by Lamb (1974) employing a direct method due to Clairin (1902) and by Chen (1974) via the inverse scattering transform (IST) formalism. The BT adopts the form

$$\begin{aligned} q_x + q'_x &= (q - q')(4\beta^2 - |q + q'|^2)^{1/2}, \\ q_t + q'_t &= i(q_x - q'_x)(4\beta^2 - |q + q'|^2)^{1/2} \\ &\quad + \frac{i}{2}(q + q')(|q + q'|^2 + |q - q'|^2), \end{aligned} \tag{14}$$

the spatial part of which may be used to construct a permutability theorem (Rogers & Shadwick, 1982).

Crum's theorem may be adduced to show that, at the level of the linear representation of soliton equations, the action of the BT is to add a discrete eigenvalue to the spectrum. The role of BTs in the context of the IST and their action on reflection coefficients is treated in detail by Calogero and Degasperis (1982).

That the Toda lattice equation

$$\ddot{y}_n = \exp[-(y_n - y_{n-1})] - \exp[-(y_{n+1} - y_n] \quad (15)$$

admits a BT, namely

$$\begin{aligned} \dot{y}_n - \dot{y}'_{n-1} &= \beta\,[\exp\{-(y'_n - y_n)\} \\ &\quad - \exp\{(y'_{n-1} - y_{n-1})\}], \\ \dot{y}'_n - \dot{y}_n &= \beta^{-1}[\exp\{-(y_{n+1} - y'_n)\} \\ &\quad - \exp\{-(y_n - y'_{n-1})\}] \end{aligned} \quad (16)$$

was established by Wadati and Toda (1975). BTs for a range of integrable differential-difference as well as integro-differential equations may be conveniently derived via Hirota's bilinear operator approach (see Rogers & Shadwick, 1982).

BTs have by now been constructed for the gamut of known solitonic equations as well as their Painlevé reductions (Gromak, 1999). The importance of BTs in soliton theory with regard to such aspects as multi-soliton generation, geometric connections, and integrable discretization is well established. Moreover, BTs also have extensive applications in continuum mechanics (Rogers & Shadwick, 1982). Important connections between infinitesimal BTs as originally introduced in a gas dynamics context (Loewner, 1952) and the construction of 2 + 1 dimensional solitonic systems have also been uncovered.

Colin Rogers

See also **Hirota's method; Inverse scattering method or transform; N-soliton formulas; Sine-Gordon equation; Solitons**

Further Reading

Bäcklund, A.V. 1883. Om ytor med konstant negativ krökning. *Lunds Universitets Årsskrift*, 19: 1–48

Bianchi, L. 1885. Sopra i sistemi tripli ortogonali di Weingarten. *Annali di Matematica*, 13: 177–234

Bianchi, L. 1890. Sopra alcone nuove classi di superficie e di sistemi tripli ortogonali. *Annali di Matematica*, 18:301–358

Bianchi, L. 1892. Sulla traformazione di Bäcklund per le superficie pseudosferiche. *Rendiconti Lincei*, 5: 3–12

Calogero, F. & Degasperis, A. 1982. *Spectral Transform and Solitons*, Amsterdam and New York: North-Holland

Chen, H.H. 1974. General derivation of Bäcklund transformations from inverse scattering problems. *Physical Review Letters*, 33: 925–928

Cieśliński, J. 1997. The Darboux–Bianchi transformation for isothermic surfaces. Classical results versus the soliton approach. *Differential Geometry and Its Applications*, 7: 1–28

Clairin, J. 1902. Sur les transformations de Bäcklund. *Annales de l'Ecole Normale Supérieure*, 27: 451–489

Darboux, G. 1899. Sur les surfaces isothermiques. *Comptes Rendus*, 128: 1299–1305

Demoulin, A. 1933. Sur deux transformations des surfaces dont les quadriques de Lie n'ont que deux ou trois points charactéristiques. *Bulletin de l'Académie Belgique*, 19: 479–501, 579–592, 1352–1363

Gromak, V. 1999. Bäcklund transformations of Painlevé equations and their applications. In *The Painlevé Property: One Century Later*, edited by R. Conte, New York: Springer

Konopelchenko, B.G & Rogers, C. 1993. On generalised Loewner systems: novel integrable equations in 2+1-dimensions. *Journal of Mathematical Physics*, 34: 214–242

Lamb, G.L. Jr. 1971. Analytical descriptions of ultra short optical pulse propagation in a resonant medium. *Reviews of Modern Physics*, 43: 99–124

Lamb, G.L. Jr. 1974. Bäcklund transformations for certain nonlinear evolution equations. *Journal of Mathematical Physics*, 15: 2157–2165

Loewner, C. 1952. Generation of solutions of systems of partial differential equations by composition of infinitesimal Bäcklund transformations. *Journal d'Analyse Mathématique*, 2: 219–242

Rogers, C. & Schief, W.K. 2002. *Bäcklund and Darboux Transformations: Geometry and Modern Applications in Soliton Theory*, Cambridge and New York: Cambridge University Press

Rogers, C. & Shadwick, W.F. 1982. *Bäcklund Transformations and Their Applications*, New York: Academic Press

Seeger, A., Donth, H. & Kochendörfer, A. 1953. Theorie der Versetzungen in Eindimensionalen Atomreihen III. Versetzungen, Eigenbewegungen und ihre Wechselwirkung, *Zeitschrift für Physik*, 134: 173–193

Tzitzeica, G. 1910. Sur une nouvelle classe de surfaces. *Comptes Rendus*, 150: 955–956

Wadati, M. & Toda, M. 1975. Bäcklund transformation for the exponential lattice. *Journal of the Physical Society of Japan*, 39: 1196–1203

Wahlquist, H.D. & Estabrook, F.B. 1973, Bäcklund transformations for solutions of the Korteweg–de Vries equation. *Physical Review Letters*, 31: 1386–1390

BAKER MAP

See **Maps**

BAKER–AKHIEZER FUNCTION

See **Integrable lattices**

BALL LIGHTNING

Properties

Ball lightning is an impressive natural phenomenon for which there is yet no accepted scientific explanation. It consists of flaming balls or fireballs, usually bright white, red, orange, or yellow, which appear unexpectedly sometimes near the ground, following the discharge of a lightning flash, or in midair coming from a cloud.

Most observations of ball lightning are associated with thunderstorms, and they exhibit the following more detailed properties: (i) Their shape is usually spherical or spheroidal with diameters between 10 and

50 cm. (ii) They tend to move horizontally. (iii) The observed distribution of lifetimes has a most probable value between 2 and 5 s and an average value of about 10 s or higher (some cases of more than 1 min having been reported). (iv) Ball lightning is bright enough to be clearly seen in daylight, the visible output being in the range 10–150 W (similar to that of a home electric light bulb). (v) Some balls have appeared within aircraft, traveling inside the fuselage along the aisle from front to rear. (vi) There are reports of odors, similar to those of ozone, burning sulfur, or nitric oxide, and of sounds, mainly hisses, buzzes, or flutters. (vii) Most balls decay silently, but some expire with an explosion. (viii) Ball lightning has killed or injured people and animals and damaged trees, buildings, cars, and electric equipment. (ix) Fires have been started showing that there is something hot inside. In such events, the released energy has been estimated to be between 10 kJ and more than 1 MJ. (x) Ball lightning has never been produced in laboratories, in spite of many attempts and some interesting results, including anode spots and luminous objects that decay very quickly. Consequently, the properties of ball lightning are derived from reports by witnesses, who are often excited by the phenomenon and have no scientific training.

A possibly related phenomenon has been observed in submarines, after a short circuit of the batteries. Balls of plasma that float in air for several seconds have appeared at the electrodes. On these occasions, the current was about 150 kA and the energy was estimated to be between 200 and 400 kJ.

Classification of the Models

Three main characteristics must be accounted for by a successful model but seem very difficult to explain: the tendency toward horizontal motion (hot air or plasma in air tends to rise), relatively long lifetimes, and contradictions among witnesses. For example, some report that balls are cold since they did not feel any warmth when one passed nearby, while others were burned and needed medical care.

The many different models proposed to explain the phenomenon can be classified into two groups, according to whether the energy source is internal or external. In the first group, some are based on plasmoids (equilibrium configurations of plasmas), high-density plasmas with quantum mechanical properties, closed loops of currents confined by their own magnetic field (in some cases, the linking of the currents playing an important role), vortex structures as whirlwinds or rotating spheres, bubbles containing microwave radiation, chemical reactions or combustion, fractal structures, aerosols, filaments of silicon, carbon nanotubes, nuclear processes, or new physics, and even primordial mini black holes. In the second group, some assume that the balls are powered by electrical discharges or by high-frequency microwave focused from thunderclouds. None of them is generally accepted.

Chemical and Electromagnetic Models

The association of ball lightning with electrical discharges suggests strongly that they have an electromagnetic nature. However, Michael Faraday argued that ball lightning cannot be an electric phenomenon as it would decay almost instantaneously, in contrast to its observed lifetime of at least several seconds. Finkelstein & Rubinstein (1964) used the time-independent magnetic virial theorem to place a stringent upper limit to the energy of a fireball. This limit has been viewed as a compelling argument against electromagnetic models, stimulating non-electromagnetic chemical approaches.

Recently, aerosol models have received considerable attention. In one model (Abrahamson & Dinniss, 2000), a lightning discharge vaporizes silicon dioxide in the soil that—after interacting with carbon compounds—is transformed into pure silicon droplets of nanometer scale. These droplets become coated with an insulating coat of oxides and are polarized, after which they become aligned with electric fields and form networks of filaments, in loose structures called "fluff balls." In another model (Bychkov, 2002), the discharges pick up organic material from the soil and transform it into a kind of "spongy ball" that can hold electric charges. Models of this type fail to explain that some balls appear in mid air, where there is neither silicon nor organic nor any other similar material.

Electromagnetic models that include chemical effects are promising candidates for an explanation of ball lightning. Indeed, there are now counterarguments to the three main objections that express Faraday's argument in modern language. These are based on the radiated output, the pinch effect, and magnetic pressure.

The first objection is that the power emitted by a plasma of the size of a ball lightning is too high (one liter of air plasma at 15,000 K emits about 5 MW, several orders of magnitude too much). This may be, however, an indication that most of the ball is at ambient temperature, only a small fraction being hot, concentrated in filamentary structures (as hot current streamers). If this fraction is of the order of 1 ppm, the radiated output would be of the order of 10–100 W, in agreement with reports.

But the solution to the first problem raises another one. Any plasma current channel inside the ball would be necked and cut in a very short time by the pinch effect (the Lorentz force); thus, a ball structured by such currents could not last long enough. However, in 1958, Chandrasekhar and Woltjer showed in an astrophysical context that plasmas relax to minimum energy states, verifying the condition that the electric current and the magnetic field are parallel so $\nabla \times \boldsymbol{B} = \lambda \boldsymbol{B}$, in which there can be no pinch effect since the Lorentz force vanishes. They concluded that such states, known as force-free fields, can confine large amounts of magnetic energy. Although the minimum energy of an uncontained plasma (as in ball lightning) is zero and corresponds to an infinitely expanded magnetic field, an almost force-free condition could be attained first in a very short time at a finite radius, a slow expansion continuing afterward with negligible pinch effect. This could take several seconds.

The third objection is based on the magnetic virial theorem, which states that a system of charges in electromagnetic interactions has no equilibrium state in the absence of external forces, because the large magnetic pressure must produce an explosion with no other force to compensate it. But it is not certain that the fireballs are in equilibrium; they could be just in metastable states with slow evolution, the streamers, moreover, clearly not being in equilibrium themselves. Still more important, the force-free condition annihilates the magnetic pressure or at least reduces it to a much smaller value if the field is almost force-free. Furthermore, the problem needs a much more complex analysis than has been offered up to now. For instance, one must include the thermochemical and quantum effects on the transport processes in the plasma as well as other nonlinear effects.

Faddeev and Niemi (2000) have proposed compelling arguments that challenge certain widely held views on plasmas, showing that the virial theorem does allow nontrivial equilibrium states of streamers and electromagnetic fields inside a background of plasma, which are "topologically stable solitons that describe knotted and linked flux tubes of helical magnetic fields." This kind of configuration was proposed in 1998 by Rañada et al. (2000) in the context of ball lightning. Therefore, it seems that the virial theorem does not necessarily support Faraday's view.

That ball lightning may contain force-free magnetic configurations of plasmas seems plausible. Because electric conduction in air proceeds through thin channels called streamers—as happens in ordinary lightning—it can be imagined that plasma inside the fireball consists of a self-organized set of metastable, highly conductive, wire-like or filamentary currents. Furthermore, unusually long-lived filaments (even closed loops) in high-density structures have been theoretically predicted and experimentally observed in many plasma systems within a great range of length scales, for instance, in astrophysics, tokamaks, and ordinary discharges in air. Thus, filamentary structures are currently receiving considerable attention.

The strongly nonlinear behavior of a plasma is enhanced when filamentary structures appear, leading to a complex non-isotropic system, which should be studied within a more general theory than ideal magnetohydrodynamics (MHD). Still, the main features of such systems can be described in the framework of resistive MHD. The important dissipative effects depend on the transport coefficients, such as thermal and electrical conductivities, which are highly nonlinear functions of the electromagnetic fields and the temperature, as well as of the chemical and quantum properties.

From the point of view of the MHD approximation, the dimensionless parameters of the plasma inside ball lightning may be quite similar to those found in other plasma scenarios, implying stable or metastable currents along a set of closed loops in filamentary structures. An interesting and unexplored possibility is the establishment, inside the streamers, of a quasicollision-free highly conductive regime in the direction of the magnetic field, which is strong and parallel to the streamers axis. In such a regime, both the electric and the thermal conductivities would become highly anisotropic. The first would be considerably enhanced along the axis of the streamer. On the other hand, both conductivities would be greatly reduced in the transverse directions, behaving as $1/B^2$ according to classical predictions. In this way, the dissipation and the spreading in the streamers would be much smaller than in ordinary regimes and should produce a long-lived strongly magnetized global plasma structure within an intricate stabilizing topology of filamentary currents.

In summary, even though the phenomenon of ball lightning has been known for many years, there is still no accepted theory to explain it—the alternatives currently most favored being the chemical and the electromagnetic models. The latter seem promising now, after recent results showed that some classical objections are not always applicable. As an example, a number of filamentary plasma structures have generated considerable interest, which are similar to stable plasma scenarios observed in nature, for instance, in astrophysics. These kinds of models could possibly embody chemical and electromagnetic elements.

Antonio F. Rañada, José L. Trueba, and José M. Donoso

See also **Helicity; Magnetohydrodynamics; Nonlinear plasma waves**

Further reading

Abrahamson, J. & Dinniss, J. 2000. Ball lightning caused by oxidation of nanoparticle networks from normal lightning strikes on soil. *Nature*, 403: 519–521

Barry, J.D. 1980. *Ball Lightning and Bead Lightning. Extreme Forms of Atmospheric Electricity*, New York: Plenum Press

Bychkov, V.L. 2002. Polymer-composite ball lightning. *Philosophical Transactions of the Royal Society* A, 360: 37–60

Faddeev, L. & Niemi, A.J. 2000. Magnetic geometry and the confinement of electrically conducting plasmas. *Physical Review Letters*, 85: 3416–3419

Finkelstein, D. & Rubinstein, J. 1964. Ball lightning. *Physical Review* A, 135: 390
Ohtsuki, Y.-H. (editor). 1988. *Science of Ball Lightning (Fire Ball)*, Singapore: World Scientific
Rañada, A.F., Soler, M. & Trueba, J.L. 2000. Ball lightning as a force-free magnetic knot. *Physical Review* E, 62: 7181–7190
Singer, S. 1971. *The Nature of Ball Lightning*, New York and London: Plenum Press
Smirnov, B.M. 1993. Physics of ball lightning. *Physics Reports*, 224: 151–236
Stenhoff, M. 1999. *Ball Lightning: An Unsolved Problem in Atmospheric Physics*, Dordrecht and New York: Kluwer
Trubnikov, B.A. 2002. Current filaments in plasmas. *Plasma Physics Reports*, 28: 312–326

BANACH SPACE

See **Function spaces**

BASIN OF ATTRACTION

See **Phase space**

BAXTER'S Q-OPERATOR

See **Bethe ansatz**

BELOUSOV–ZHABOTINSKY REACTION

In 1950, Boris Pavlovich Belousov worked at the Institute of Biophysics of the Ministry of Public Health of the USSR when he observed that the reaction between citric acid, bromate ions, and ceric ions (as catalyst) produced a regular periodic and reproducible change of color between an oxidized state and a reduced state. A temporal oscillating reaction appeared like a chemical clock. His 1951 paper on this study was largely rejected by the science community because it seemed to violate the Second Law of thermodynamics.

In 1961, Anatol Zhabotinsky, a Russian postgraduate student guided by his professor, Simon Shnoll, modified the previous reaction by replacing citric acid with malonic acid and adding ferroin sulfate as an indicator. As the oxidized state of ferroin is blue and the reduced one is red, he was able to observe an oscillating temporal reaction with larger oscillating amplitudes. This reaction

$$\begin{aligned} &3CH_2(CO_2H)_2 + 4BrO_3^- \\ &\quad = 4Br^- + 9CO_2 + 6H_2O \end{aligned} \tag{1}$$

was named the Belousov–Zhabotinsky (BZ) reaction. The first publications in English, recognizing the works of Belousov and Zhabotinsky, were done in 1967 by the Danish scientist Hans Degn. However, Zhabotinsky was unable to propose a complete mechanism for the system. Experimentally, the periodic evolution of the potential of the reacting solution can be followed by a potentiometric method, as shown in Figure 1.

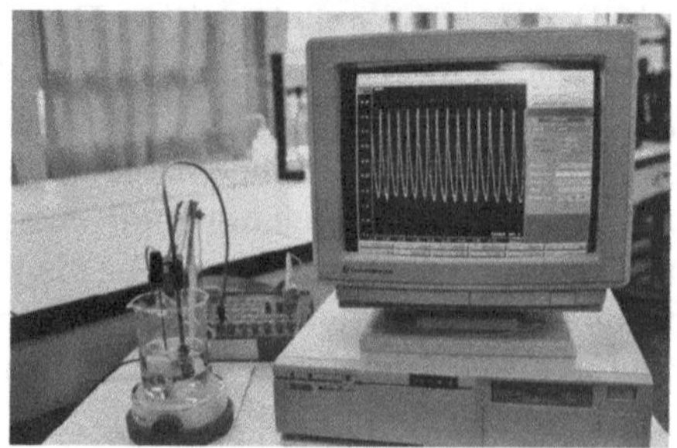

Figure 1. A photograph showing the periodic potential obtained between a platinum electrode and a reference electrode immersed in a BZ solution.

In 1972, Richard Field, Endre Körös, and Richard Noyes, at the University of Oregon, studied this mechanism using a bromide-selective electrode to follow the reaction. They proposed 18 steps involving 21 chemical species, using the same principles of chemical kinetics and thermodynamics that govern ordinary chemical reactions—this was the FKN mechanism. In 1974, the same scientists proposed a simplified mechanism with penetrating chemical insight: the "Oregonator" (in honor of the University of Oregon).

Compound:	BrO_3^-	Organic species	HBrO
Notation:	A	B	P

Compound:	$HBrO_2$	Br^-	Ce^{4+}
Notation:	X	Y	Z

With these notations, the FKN mechanism is

- $A + Y \to X + P$,
- $A + X \to 2X + 2Z$,
- $X + Y \to 2P$,
- $2X \to A + P$,
- $B + Z \to (f/2)Y +$ other products.

The second step is fundamental for the observation of oscillations; it is an autocatalytic reaction or retroaction loop. In the fifth step, B represents all oxidizable organic species present, and f is a stoichiometric factor. The kinetic differential equations are:

$$\begin{aligned} \frac{d[X]}{dt} &= k_1[A]\cdot[Y] + k_2[A]\cdot[X] \\ &\quad - k_3[X]\cdot[Y] - 2k_4[X]^2, \end{aligned} \tag{2}$$

$$\frac{d[Y]}{dt} = -k_1[A]\cdot[Y] - k_3[X]\cdot[Y] + 2k_5\frac{f}{2}[B]\cdot[X], \tag{3}$$

$$\frac{d[Z]}{dt} = 2k_2[A]\cdot[X] - k_5[B][Z]. \tag{4}$$

The rate constants are:

Rate constants:	k_1	k_2	k_3	k_4	k_5
Value (l/mol s):	1.28	2.4×10^6	33.6	3×10^3	1

This system is clearly nonlinear. Solutions can be obtained by numerical methods, and for $0.5 \leq f \leq 2.4$, some oscillating temporal solutions are observed, whose solutions depend on the initial conditions. According to the Second Law of thermodynamics, all spontaneous chemical changes in a homogeneous and closed system involve a decrease in free enthalpy of this system. If a fluctuation disrupts the system close to equilibrium, the system will return irrevocably to this stable state, making oscillations impossible. Nonetheless, it is possible to observe oscillations when the system is far from equilibrium. One of the striking properties of nonlinear systems is the effect of fluctuations (of the concentrations of intermediates), which can transform an unstable system into new states that are more organized than the initial state. Ilya Prigogine (who was awarded the 1977 Nobel Prize in chemistry for his work on thermodynamics) gave the name "dissipative structures" to such systems to emphasize the importance of irreversible phenomena far from equilibrium.

Continuing to work on oscillating reactions, in 1970, Zhabotinsky published, with Zaikin, a paper that announced the existence of two-dimensional waves in the BZ reaction; also in 1972, Arthur Winfree observed spiral wave patterns in a BZ reaction (see Figure 2). This reaction took place without stirring in a thin (approximately 2 mm thick) layer of reactants poured into a Petri dish. Blue concentric circles (called targets) radiated across the dish on a red background and self-generating spirals appeared. Soon afterward, scientists reported that a blue target center can produce waves of oxidation propagating through the reduced medium, and as the waves advance toward the interior of the center, they transform from red to blue. The period of oscillation is variable, but the speed of propagation is roughly constant. Thus, the BZ reaction proves to be a stationary spatiotemporal oscillating reaction.

With the aid of the FKN mechanism, Field and Noyes showed how to understand the development of such target waves. The diffusion is the transport of species from the areas of high concentrations to those of low concentrations. When there is a coupling between a chemical reaction with an autocatalytic step (or feedback-retroaction loop) as in the BZ reaction with the diffusion of species, spatial organization phenomena can occur; thus these are called "reaction-diffusion systems." In such systems, molecules react chemically with each other when they collide, and as

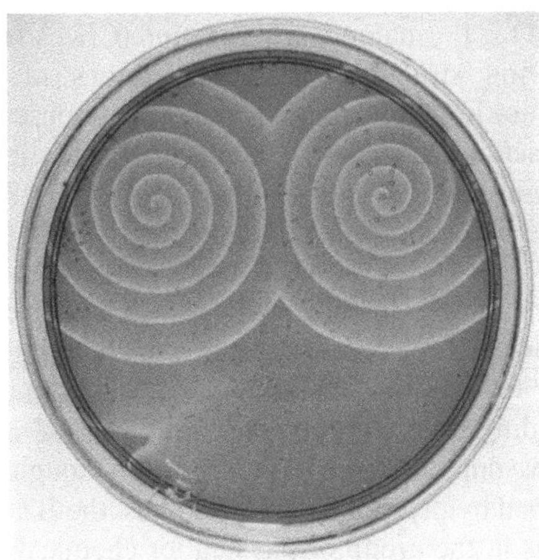

Figure 2. Spiral waves in a BZ reaction (Courtesy of A.T. Winfree). See text for details.

the concentrations of components change, a chemical wave propagates.

In 1984, Oleg Mornev (at the Institute of Theoretical and Experimental Biophysics of the Russian Academy of Sciences) showed that in an infinite plane stationary system of reaction-diffusion oscillations, Snell's sine law of refraction was verified. Thus the simple rule

$$\frac{\sin \psi_1}{\sin \psi_2} = \frac{v_1}{v_2} \tag{5}$$

dictates the angles ψ_1 and ψ_2 when waves hit an interface separating two regions with different speeds (v_1 and v_2) of wave propagation. This result was surprising because reaction-diffusion systems are nonlinear; thus, the medium is an active and an integral part of the wave. In fact, ψ_1 cannot be set arbitrarily but is slaved with ψ_2 due to the nature of the two regions.

In 1993, Zhabotinsky (at the Department of Chemistry, Brandeis University) demonstrated Snell's law experimentally. In 1998, Rui Dilaö and Joaquim Sainhas (at the Instituto Superio Tecnico, Lisbon Portugal) showed the following constraint in a reaction-diffusion system: after the reaction, the medium must be chemically identical to its initial form. In other words, while the waves are propagating and reactions are taking place, the medium has different properties, but these waves transform the medium back to the original species. By solving reaction-diffusion equations under this constraint, Dilaö and Sainhas showed (using computer simulations) that their formulation agrees with experiments, and mathematically chemical waves obey Snell's sine law. Continuing to work on the phenomenon of refraction, Mornev is developing formulations that hold both in infinite and finite media.

There are other examples of reaction-diffusion systems. In 1983, Patrick de Kepper (a French chemist in Toulouse) highlighted Turing structures in a ClO_2^-, I^-, malonic acid reaction (CIMA reaction). In Turing structures, stationary zones of varying concentrations appear in space. For these observations, the reaction must have steps with retroaction loops containing activators and other steps with inhibitors, and the activators must diffuse more slowly than the inhibitors.

Although the intense study of oscillating chemical reactions and nonlinear dynamics in chemistry is only about 30 years old, its progress has been impressive. Depending on the initial conditions, the BZ reactions can have unpredictable behaviors, even though they are described by deterministic laws. Thus, the BZ reaction belongs to the group of physical or chemical systems that exhibits deterministic chaos.

Gérard Dupuis and Nicole Berland

See also **Brusselator; Chemical kinetics; Fairy rings of mushrooms; Reaction-diffusion systems; Turing patterns; Vortex dynamics in excitable media**

Further Reading

Dilao, R. & Sainhas, J. 1998. Wave optics in reaction-diffusion systems. *Physical Review Letters*, 80: 5216

Epsein, I.R. & Pojman, J.A. 1998. *An Introduction to Nonlinear Chemical Dynamics: Oscillations, Waves, Patterns and Chaos*, Oxford and New York: Oxford University Press

Field, R.J., Körös, E. & Noyes, R.M. 1972. *Journal of American Chemical Society*, 94: 8649

Gray, P. & Scott, S.K. 1990. *Chemical Oscillations and Instabilities, Nonlinear Chemical Kinetics*, Oxford: Clarendon Press and New York: Oxford University Press

Mornev, O.A. 1984. Elements of the "optics" of autowaves. In *Self-Organization of Autowaves and Structures Far from Equilibrium*, edited by V.I. Krinsky, Berlin: Springer, pp. 111–118

Zaikin, A.N. & Zhabotinsky, A.M. 1970. Concentration wave propagation in two dimensional liquid phase self oscillating system. *Nature*, 225: 535–537

Zhabotinsky, A.M. 1964. *Biofizika*, 9, 306.

Zhabotinsky, A.M., Eager, M.D. & Epstein, I.R. 1993. Refraction and reflection of chemical waves. *Physical Review Letters*, 71: 1526–1529

BENJAMIN–BONA–MAHONY EQUATION

See **Water waves**

BENJAMIN–FEIR INSTABILITY

See **Wave stability and instability**

BENJAMIN–ONO EQUATION

See **Solitons, types of**

BERNOULLI SHIFT

See **Maps**

BERNOULLI'S EQUATION

Bernoulli's equation is possibly the best-known result in fluid mechanics—and the most frequently abused. Bernoulli's equation may be viewed as an energy-conservation budget for a fluid particle as it travels up and down the "hills" of potential energy, due to the fields of gravity and the pressure within the fluid, acquiring and relinquishing kinetic energy. In its simplest form, it states that

$$V^2/2 + p/\rho + gz = C, \qquad (1)$$

where p is the pressure in the fluid of density ρ, V is the flow speed, and z is the vertical coordinate. (The flow takes place in a uniform gravitational field of acceleration g.) The sum on the left-hand side of Equation (1) is a constant, C. The first term is the kinetic energy of the fluid per unit mass, the second and third terms are the potential energy (again per unit mass) in the combined energy "landscape" of pressure and gravity.

The result is credited to Daniel Bernoulli (1700–1782), son of Johann Bernoulli (1667–1748), and to his monograph *Hydrodynamica*, initiated in 1729 and ultimately published in 1738. The history of the equation is, however, much richer than this simple sequence of events would suggest. Hunter Rouse puts the issues this way in a book containing English translations of the writings of both the son and the father:

> Why [these] works should have been singled out for translation seems at first sight rather obvious, if only because of the frequency with which the name Bernoulli is on the hydraulician's lips. But it is only Daniel to whom one is making reference, and the word is gradually spreading that the theorem bearing his name is nowhere to be found in his habitually cited *Hydrodynamica*. Not until the last few years has mention of either the work *Hydraulica* or its author Johann Bernoulli appeared in the fluids literature with any frequency, and this almost exclusively in the writings of C. Truesdell. It is Truesdell's thesis that, whereas Daniel has received too much credit for the formulation of the Bernoulli theorem, Johann has received too little. (Carmody et al., 1968).

A complicated set of circumstances ensued in which both father (Johann) and son (Daniel) sent their manuscripts to Leonhard Euler for comment and this led to Johann's manuscript, which appears to have been composed later than Daniel's, being published in the Memoirs of the Imperial Academy of Science in St. Petersburg in 1737 and 1738 (although these were not printed until a decade later). Indeed, Johann's treatise first appeared in his collected works published in Switzerland in 1743. While Daniel's treatise has the gist of what we today call Bernoulli's equation, Johann's treatment is more mature and complete.

In the form stated in Equation (1), Bernoulli's equation applies only to steady, constant-density, irrotational flow, that is, to a flow pattern that is unchanging in time and that has no vorticity.

More refined versions may be derived. Thus, in a steady, constant-density flow with vorticity, Equation (1) still holds along each streamline, but the "constant" on the right-hand side may vary from streamline to streamline. Indeed, the gradient of this changing "Bernoulli constant," ∇C, equals the Lamb vector, the vector product of flow velocity and vorticity,

$$\boldsymbol{V} \times \boldsymbol{\omega} = \nabla C.$$

If the flow is irrotational but unsteady, a version of Bernoulli's equation again holds, but the constant on the right-hand side of (1) is replaced by (minus) the time derivative of the velocity potential. (In an irrotational flow, the velocity field is the gradient of a scalar known as the velocity potential.) With $V = -\nabla\phi$, where ϕ is the velocity potential, we obtain Bernoulli's equation in the form

$$(\nabla\phi)^2/2 + p/\rho + gz = -\frac{\partial\phi}{\partial t}, \qquad (2)$$

which, coupled with the condition of irrotational flow,

$$\Delta\phi = 0, \qquad (3)$$

gives a system of two partial differential equations for the fields p and ϕ.

Bernoulli's equation in the simplistic form "high flow speed implies low pressure and vice versa" is often applied as a first, crude explanation of many flow phenomena from the ability to balance a ball atop a plume of air to the lift on an airfoil in flight. Some of these explanations are too simplistic, not to say incorrect. Nevertheless, Bernoulli's equation, when properly applied under the assumptions that ensure its validity, can be an extremely useful and powerful tool of fluid flow analysis.

It is remarkable—and important to note—that Bernoulli's equation (1) is not invariant to a Galilean transformation, ordinarily a prerequisite for a physical law to be useful. Thus, if one wants to use Bernoulli's equation (1) to calculate the pressure distribution for flow around an object, assuming the velocity field is known, it is essential to do so in a frame of reference in which the flow satisfies the necessary assumptions, in particular, that the flow is steady. The correct result is obtained by carrying out such a calculation in a frame of reference moving with the body. If the calculation is attempted in the "laboratory frame" through which the object is moving, one has to tackle the much more complex version of Bernoulli's equation given in (2). If the version in Equation (1) is applied, one obtains an incorrect result.

Hassan Aref

See also **Fluid dynamics**

Further Reading

Batchelor, G.K. 1967. *An Introduction to Fluid Dynamics*, Cambridge: Cambridge University Press

Carmody, T., Kobus, H. & Rouse, H. 1968. *Hydrodynamica by Daniel Bernoulli and Hydraulica by Johann Bernoulli*, translated from the Latin, with a preface by Hunter Rouse, New York: Dover

Lamb, H. 1932. *Hydrodynamics*, 6th edition, Cambridge: Cambridge University Press

BERRY'S PHASE

Consider the parallel transport of an orthonormal frame along a line of constant latitude on the surface of a sphere. In going once around the sphere, the frame undergoes a rotation through an angle $\Delta\theta = 2\pi \cos\alpha$, where α is the colatitude. This may be shown using the geometry of Figure 1. As is also evident from the figure, this phase shift is purely geometric in character—it is independent of the time it takes to traverse the closed loop.

This construction underlies the well-known phase shift exhibited by the Foucault pendulum as the Earth rotates through one full period. Although arising through a dynamical process involving two widely separated time scales (the period of the Earth's rotation and the oscillation period of the pendulum), the phase shift in this and other examples is now understood in a more unified way. *Holonomic* effects such as these arise in a host of applications ranging from problems in superconductivity theory, fiber optic design, magnetic resonance imaging (MRI), amoeba propulsion and robotic locomotion and control, micromoter design, molecular dynamics, rigid-body motion, vortex dynamics in incompressible fluid flows (Newton, 2001), and satellite orientation control. For a survey and further references on the use of phases in locomotion problems, see Marsden & Ostrowski (1998).

That the falling cat learns quickly to re-orient itself optimally in mid-flight while maintaining zero angular momentum is a manifestation of the fact that controlling and manipulating a system's internal or shape variables can lead to phase changes in the external, or group variables, a process that can be exploited and has deeper connections to problems related to the dynamics of Yang–Mills particles moving in their associated gauge field, a link that is the *falling cat theorem* of Montgomery (1991a) (see further discussion and references in Marsden (1992) and Marsden & Ratiu (1999)). One can read many of the original articles leading to our current understanding of the geometric phase in the collection edited by Shapere & Wilczek (1989).

Problems of this type have a long and complex history dating back to work on the circular polarization of light in an inhomogeneous medium by Vladimirskii and Rytov in the 1930s and by Pancharatnam in the 1950s, who studied interference patterns produced by plates of

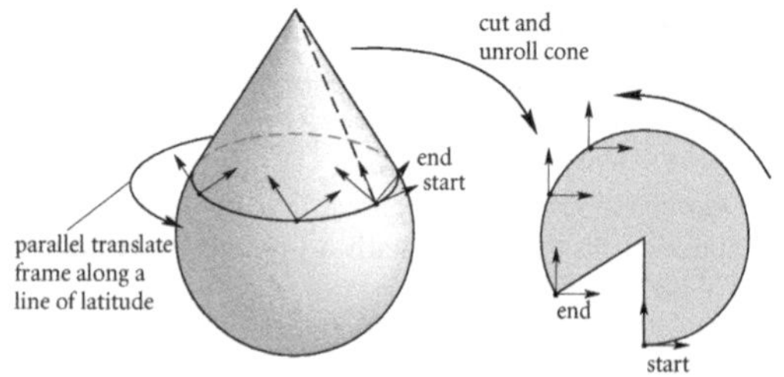

Figure 1. Parallel transport of a frame around a line of latitude.

an anisotropic crystal. Much of this early history is described in the articles by Michael Berry (Berry, 1988, 1990). The more recent literature was initiated by his earlier articles (Berry, 1984, 1985), which investigated the evolution of quantum systems whose Hamiltonian depends on external parameters that are slowly varied around a closed loop. The adiabatic theorem of quantum mechanics states that for infinitely slow changes of the parameters, the evolution of the complex wave function, governed by the time-dependent Schrödinger equation, is instantaneously in an eigenstate of the *frozen* Hamiltonian. At the end of one cycle, when the parameters recur, the wave function returns to its original eigenstate, but with a phase change that is related to the geometric properties of the closed loop. This phase change now goes by the name *Berry's phase*. Geometric developments started with the work of Simon (1983), and Marsden et al. (1989). One can introduce a bundle of eigenstates of the slowly varying Hamiltonian, as well as a natural *connection* on it; the Berry phase is then the bundle holonomy associated with this connection, while the curvature of the connection, when integrated over a closed two-dimensional (2-d) surface in parameter space gives rise to the first Chern class characterizing the topological twisting of this bundle.

The classical counterpart to Berry's phase was originally developed by Hannay (1985) (hence the terminology *Hannay's angle*) and is most naturally described by considering slowly varying integrable Hamiltonian systems in action-angle form. If we let $(I_1, \ldots, I_n; \theta_1, \ldots \theta_n)$ represent the action-angle variables of a given integrable system, then the governing Hamiltonian can be expressed as $\mathcal{H}(I_1, \ldots, I_n; R(t))$, where $R(t)$ is a slowly varying parameter that cycles through a closed loop in time period T, that is, $R(t+T) = R(t)$, $\dot{R}(t) \sim \varepsilon R$, $\varepsilon \ll 1$. The configuration space for the system is an n-dimensional torus $\mathbb{T}^n$ and we seek a formula for the angle variables as the parameter or parameters slowly evolve around the closed loop $\mathcal{C}$ in parameter space. The time-dependent system is governed by

$$\dot{\mathbf{I}} = \dot{R}(t) \cdot \frac{\partial \mathbf{I}}{\partial R}, \tag{1}$$

$$\dot{\boldsymbol{\theta}} = \boldsymbol{\omega}(\mathbf{I}) + \dot{R}(t) \cdot \frac{\partial \boldsymbol{\theta}}{\partial R}, \tag{2}$$

where

$$\omega(\mathbf{I}) \equiv \frac{\partial \mathcal{H}}{\partial \mathbf{I}}.$$

Since R is slowly varying, we can average the system around level curves of the frozen (i.e., $\varepsilon = 0$) Hamiltonian. If we let $\langle \quad \rangle$ denote this phase-space average, then the averaged canonical system becomes

$$\dot{\mathbf{I}} = \dot{R}(t) \cdot \left\langle \frac{\partial \mathbf{I}}{\partial R} \right\rangle \tag{3}$$

$$\dot{\boldsymbol{\theta}} = \boldsymbol{\omega}(\mathbf{I}) + \dot{R}(t) \cdot \left\langle \frac{\partial \boldsymbol{\theta}}{\partial R} \right\rangle. \tag{4}$$

The well-known adiabatic theorem of quantum mechanics guarantees that the action variable is nearly constant due to its adiabatic invariance, whereas the angle variables can be integrated over period T

$$\theta_T^i = \int_0^T \omega^i(\mathbf{I})\mathrm{d}t + \int_0^T \dot{R}(t) \cdot \left\langle \frac{\partial \theta^i}{\partial R} \right\rangle \mathrm{d}t \tag{5}$$

$$= \theta_\mathrm{d} + \theta_\mathrm{g}. \tag{6}$$

The first term, θ_d, called the dynamic phase is due to the frozen system, while the second term, θ_g, arises from the time variation. This geometric phase can be rewritten in a revealing manner as

$$\theta_\mathrm{g} = \int_0^T \dot{R}(t) \cdot \left\langle \frac{\partial \theta^i}{\partial R} \right\rangle \mathrm{d}t \tag{7}$$

$$= \oint \left\langle \frac{\partial \theta^i}{\partial R} \right\rangle \mathrm{d}R. \tag{8}$$

The contour integral is taken over the closed loop $\mathcal{C}$ in parameter space. Although arising through a dynamical process, it is ultimately a purely geometric quantity that results from a delicate balance of two compensating effects in the limit $\varepsilon \to 0$. On the one hand, $T \to \infty$ in (7), while on the other, $\dot{R}(t) \to 0$. Their rates exactly balance so that the integral leaves a residual term in the limit $\varepsilon = 0$, as given in (8).

A nice example developed in Hannay (1985) is the *bead-on-hoop* problem in which a frictionless bead is constrained to slide along a closed planar wire hoop that encloses area $\mathcal{A}$ and has perimeter length $\mathcal{L}$. As the bead slides around the hoop, the hoop is slowly rotated about its vertical axis (which is aligned with the gravitational vector) through one full revolution. We are interested in the angular position of the bead with respect to a fixed point on the hoop after one full revolution of the hoop. When compared with its angular position had the hoop been held fixed (the frozen problem), this angle difference would represent the geometric phase and is given by

$$\Delta\theta = -8\pi^2 \mathcal{A}/\mathcal{L}^2. \tag{9}$$

Montgomery (1991b) shows that modulo 2π, we have the following *rigid-body phase formula*:

$$\Delta\theta = -\Lambda + 2ET/R. \tag{10}$$

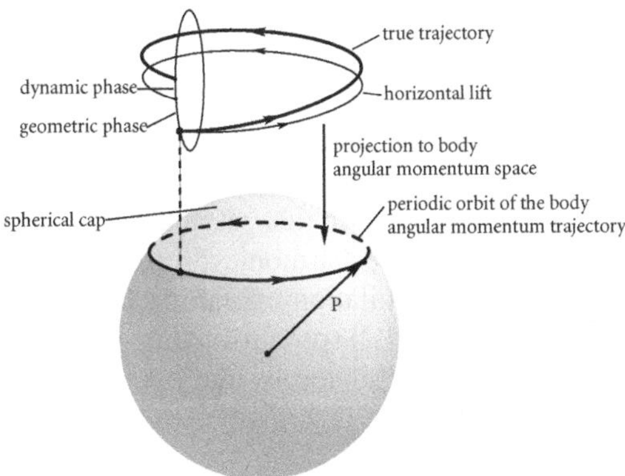

Figure 2. The geometry of the rigid-body phase formula.

Let us explain the notation in this remarkable formula. When a rigid body is freely spinning about its center of mass, one learns in mechanics that this dynamics can be described by the Euler equations, which are equations for the body angular momentum $\boldsymbol{\Pi}$. This vector in $\mathbb{R}^3$ moves on a sphere (of radius $R = \|\boldsymbol{\Pi}\|$) and describes periodic orbits (or exceptionally, heteroclinic orbits). This orbit is schematically depicted by the closed curve on the sphere shown in Figure 2. However, the full dynamics includes the dynamics of the rotation matrix for describing the attitude of the rigid body as well as its conjugate momentum. There is a projection from the full dynamic phase space (which is 6-d) to the body angular momentum space (which is 3-d). After one period of the motion on the sphere, the actual rigid-body motion was not periodic, but it had rotated about the spatial angular momentum vector by an angle $\Delta\theta$, the left-hand side of the above formula. The quantity Λ is the spherical angle subtended by the cap shown in the figure, E is the energy of the trajectory, and T is the period of the closed orbit on the sphere. A detailed history of this formula is given in Marsden & Ratiu (1999).

PAUL K. NEWTON AND JERROLD E. MARSDEN

See also **Adiabatic invariants; Averaging methods; Hamiltonian systems; Integrability; Phase space**

Further Reading

Berry, M.V. 1984. Quantal phase factors accompanying adiabatic changes. *Proceedings of the Royal Society, London* A, 392: 45–57

Berry, M.V. 1985. Classical adiabatic angles and quantal adiabatic phase. *Journal of Physics* A, 18: 15–27

Berry, M.V. 1988. The geometric phase. *Scientific American*, December, 46–52

Berry, M.V. 1990. Anticipations of the geometric phase. *Physics Today*, 43(12), 34–40

Hannay, J.H. 1985. Angle variable holonomy in adiabatic excursion of an integrable Hamiltonian. *Journal of Physics* A, 18: 221–230

Marsden, J.E. 1992. *Lectures on Mechanics*, Cambridge and New York: Cambridge University Press

Marsden, J.E., Montgomery, R. & Ratiu, T. 1989. Cartan–Hannay–Berry phases and symmetry. *Contemporary Mathematics*, 97: 279–295

Marsden, J.E. & Ostrowski, J. 1998. Symmetries in motion: geometric foundations of motion control, *Nonlinear Science Today*. (http://link.springer-ny.com)

Marsden, J.E. & Ratiu, T. 1999. *Introduction to Mechanics and Symmetry*, 2nd edition, New York: Springer

Montgomery, R. 1991a. Optimal control of deformable bodies and its relation to gauge theory. In *The Geometry of Hamiltonian Systems*, edited by T. Ratiu, New York: Springer, pp. 403–438

Montgomery, R. 1991b. How much does a rigid body rotate? A Berry's phase from the 18th century, *American Journal of Physics*, 59: 394–398

Newton, P.K. 2001. *The N-Vortex Problem: Analytical Techniques*, New York: Springer, Chapter 5

Shapere, A. & Wilczek, F. (editors). 1989. *Geometric Phases in Physics*, Singapore: World Scientific

Simon, B. 1983. Holonomy, the quantum adiabatic theorem, and Berry's phase. *Physical Review Letters*, 51(24): 2167–2170

BETHE ANSATZ

The Bethe ansatz is the name given to a method for exactly solving quantum many-body systems in one spatial dimension (1-d) or classical statistical lattice models (vertex models) in two spatial dimensions (Baxter, 1982; Korepin et al., 1993). The method was developed by Hans Bethe in 1931 (Bethe, 1931) in order to diagonalize the Hamiltonian of a chain of N spins with isotropic exchange interactions, introduced by Werner Heisenberg some years before as the simplest model for a 1-d magnet. This result was achieved by assuming the wave function to be of the form

$$f(x_1, x_2, ..., x_M) = \sum_P A_P \, \mathrm{e}^{\mathrm{i}\sum_{j=1}^{M} k_{P_j} x_j} \qquad (1)$$

with the sum performed on all possible permutations P of M distinct wave numbers $\{k_1, ..., k_M\}$, corresponding to down spins in the system (Bethe ansatz). By imposing invariance under the physical symmetries of the system (discrete translations and total spin rotations), Bethe obtained conditions on the coefficients A_P, which were satisfied if a set of M nonlinear equations (Bethe equations) in N complex parameters (Bethe numbers) were fulfilled. Surprisingly, the wave functions thus constructed were simultaneous eigenfunctions not only of the translation operator, the total spin $\boldsymbol{S}$, and its projection S_z along the z-direction but also of the isotropic Heisenberg Hamiltonian

$$H = \sum_{i=1}^{N} \left(\boldsymbol{S}_i \cdot \boldsymbol{S}_{i+1} - \frac{1}{4} \right). \qquad (2)$$

The energy and the crystal momentum were expressed as symmetric functions of the Bethe numbers; thus, the eigenvalue problem for H was reduced to the solution of an algebraic problem—solution of the Bethe equations.

This remarkable result was possible because of the existence of additional symmetries of the Heisenberg Hamiltonian, which emerged thanks to the ansatz made by Bethe on the wave function. In this original formulation, the method is known as the coordinate Bethe ansatz.

Progress in clarifying the role of symmetries in the Bethe ansatz, the link with integrable systems, and the algebraic aspect of the method was achieved by the Saint Petersburg (formerly Leningrad) School in the course of developing the quantum inverse scattering method (QISM) (Faddeev, 1984; Sklyanin & Faddeev, 1978; Korepin et al., 1993), after the work of Baxter on the integrability of vertex models (Baxter, 1982). In this approach, a key role is played by the monodromy operator defined as $\tau(\lambda)=L_N(\lambda)L_{N-1},\ldots,L_1(\lambda)$, with λ being a complex number (spectral parameter) and $L_n(\lambda)$ being the quantum local Lax operator defined for the isotropic Heisenberg model as

$$L_n(\lambda)=\begin{pmatrix} \lambda+\mathrm{i}S_n^z & \mathrm{i}S_n^- \\ \mathrm{i}S_n^+ & \lambda-\mathrm{i}S_n^z \end{pmatrix} \quad (3)$$

with S^+, S^- raising and lowering spin-$\frac{1}{2}$ operators. Note that L_n can be viewed as an operator acting on the space $h_n \otimes V$, where h_n ($\equiv C^2$) is the physical Hilbert at site n (the space of couples of complex numbers), and V is an auxiliary space related to the matrix representation of L_n (for the present case, V is also identified with C^2). The product of Lax operators, taken in the auxiliary space, coincides with the usual matrix multiplication, so that the monodromy matrix can be re-written as

$$\tau(\lambda)=\begin{pmatrix} A(\lambda) & B(\lambda) \\ C(\lambda) & D(\lambda) \end{pmatrix}, \quad (4)$$

with A, B, C, D operators acting on the full physical Hilbert space: $H=\prod \otimes h_i$. As is known from QISM (Korepin et al., 1993; Sklyanin & Faddeev, 1978; Faddeev, 1984), the commutation relations between elements of the monodromy matrix can be written in a compact form as

$$R(\lambda-\mu)\,(\tau(\lambda)\otimes\tau(\mu))=(\tau(\mu)\otimes\tau(\lambda))\,R(\lambda-\mu), \quad (5)$$

where R is a 4×4 matrix (quantum R-matrix) satisfying the Yang–Baxter equation. From Equation (5), it follows that the trace of the monodromy operator (also known as the transfer matrix) $T(\lambda)\equiv \mathrm{tr}(\tau(\lambda))=A(\lambda)+D(\lambda)$ gives rise, for different values of the spectral parameter, to an abelian algebra of operators

$$[T(\lambda), T(\mu)]=0. \quad (6)$$

One can prove that the Hamiltonian is also an element of this algebra so that, for a system of N sites, there are N quantum integrals of motion in involution, corresponding to Liouville integrability in the classical limit. The diagonalization of the Hamiltonian and the other integrals of motion is thus reduced to the solution of the eigenvalue problem for the transfer matrix T. This problem can be solved by the so-called algebraic Bethe ansatz, a procedure that resembles the algebraic diagonalization of the harmonic oscillator by means of creation and annihilation operators. It relies on the existence of a vector $|\Omega\rangle$ (pseudovacuum) in the Hilbert space, which is annihilated by the operator C of the monodromy matrix

$$C(\lambda)|\Omega\rangle=0. \quad (7)$$

For the Heisenberg chain, $|\Omega\rangle$ can be chosen as $|\Omega\rangle=\prod_{i=1}^N \otimes|\uparrow\rangle_i$ with $|\uparrow\rangle_i$ denoting the spin up state at site i. From Equation (3) it is clear that L_n acts on the state $|\uparrow\rangle_n$ as a triangular matrix, and the same is true for $\tau(\lambda)$ acting on $|\Omega\rangle$; thus, Equation (7) is automatically satisfied (C plays the role of an annihilation operator). From Equations (3) and (4), it is also evident that $A(\lambda)|\Omega\rangle=(\lambda+\mathrm{i}/2)^N|\Omega\rangle$, $D(\lambda)|\Omega\rangle=(\lambda-\mathrm{i}/2)^N|\Omega\rangle$. Moreover, one can show that the operator B in Equation (4) can be used as a creation operator.

By taking N different values of the spectral parameter $\lambda_1, \lambda_2, \ldots, \lambda_N$, one constructs a trial wave function as

$$|\Psi(\lambda_1,\ldots,\lambda_N)\rangle=\prod_{i=1}^N B(\lambda_i)|\Omega\rangle. \quad (8)$$

A direct calculation shows that

$$T(\lambda)|\Psi(\lambda_1,\ldots,\lambda_N)\rangle = \Lambda(\lambda)|\Psi(\lambda_1,\ldots,\lambda_N)\rangle + \text{unwanted terms}, \quad (9)$$

where the unwanted terms can be calculated using the commutation relations of A and D with B, obtained from Equation (5). The unwanted terms, however, are eliminated if the λ_i are taken as solutions of the Bethe equations that, for the isotropic Heisenberg chain, are of the form

$$\left(\frac{\lambda_\alpha-\mathrm{i}/2}{\lambda_\alpha+\mathrm{i}/2}\right)^N=-\prod_{\beta=1}^M \frac{\lambda_\alpha-\lambda_\beta-\mathrm{i}}{\lambda_\alpha-\lambda_\beta+\mathrm{i}}, \quad \alpha=1,2,\ldots,M. \quad (10)$$

The set of states obtained from Equation (8) in correspondence of the solutions of this system of nonlinear equations can be shown to be complete. The diagonalization of $T(\lambda)$, and hence of the Hamiltonian and all the quantum integrals of motion, is thus reduced to the problem of solving the Bethe equations. For finite size systems, this is a difficult problem to solve due to the nonlinearity of the equations, and one usually resorts to numerical tools. In the thermodynamical limit, however, it is possible to obtain exact solutions of the energy spectrum by deriving linear integral equation for the density distribution of the Bethe solutions in a complex plane (however, this requires an assumption on the nature of the solution known as the string hypothesis).

The algebraic Bethe ansatz has been successfully applied to a large class of many-body problems, including anisotropic generalizations of the Heisenberg chain, the Hubbard model, and the Kondo model, and has stimulated a variety of related approaches including Baxter's q-operator method (Baxter, 1982) and the notion of quantum groups. Recent progress in the computation of correlation functions of quantum-integrable many-body problems have also been made using the Bethe ansatz (Korepin et al., 1993).

MARIO SALERNO

See also **Quantum inverse scattering method; Salerno equation**

Further Reading

Baxter, R.J. 1982. *Exactly Solved Model of Statistical Mechanics*, New York: Academic Press

Bethe, H. 1931. Zur Theorie der Metalle I. Eigenwerte and Eigenfunktionen der Linearen Atomkette. *Zeitschrift für Physik*, 71: 205–226

Faddeev, L.D. 1984. Integrable models in 1+1 dimensional quantum field theory. In *Recent Advances in Field Theory and Statistical Mechanics*, Les Houches 1982, edited by J.B. Zuber & R. Stora, Amsterdam: North-Holland

Korepin, V.E., Bogoliubov, N.M. & Izergin, A.G. 1993. *Quantum Inverse Scattering Method and Correlation Functions*, Cambridge and New York: Cambridge University Press, and references therein

Sklyanin, E.K. & Faddeev, L.D. 1978. Quantum mechanical approach to completely integrable field theory models. *Soviet Physics Doklady*, 23: 902

BIFURCATIONS

Bifurcations are critical events that arise in systems when an external control parameter is varied (Arnol'd et al., 1994). For small values of the parameter the system will be linear and a unique fixed point will exist. As the parameter is changed to ranges, where nonlinearity becomes important, instabilities in the form of new fixed points or solutions with qualitatively different dynamical behavior may arise at bifurcations. These critical events are of mathematical and practical interest since their analysis can be performed, and they form organizing centers for observed dynamics (Guckenheimer & Holmes, 1986).

As an example, consider the simple physical system of a plastic ruler that is compressed lengthwise between your hands. This was first considered by Leonhard Euler in 1744 and is often referred to as the Euler strut problem (Acheson, 1997). At small forces, the ruler is approximately straight and supports the applied load. This is called the trivial state of the system. However, as the load increases, buckling takes place so that the ruler is deflected up or down. The straight trivial state becomes unstable and is replaced by a pair of solutions where each corresponds to one of the buckled states. If the ruler and the application of the load were both perfect, this would provide a physical example of a symmetry-breaking supercritical pitchfork bifurcation of the type shown in the bifurcation diagram of Figure 1(a).

The symbol X represents the deflection of the center of the ruler, which is used as the measure of the state of the system shown plotted as a function of the applied load λ. The symmetry that is broken is the mirror-plane symmetry of the straight ruler. The bifurcation is called supercritical because the nontrivial branches have the same stability as the trivial state, and it is termed pitchfork due to its shape.

When the bifurcating solutions have a stability opposite to the trivial, the bifurcation is called subcritical. A sketch of such a bifurcation is given in Figure 1(b) where an increase in the parameter λ would involve a jump to a large X state when λ is increased beyond the critical value λ_c. In order to regain the trivial state, λ would then have to be reduced to reach the folded part of the solution branches, and a sudden change back to the trivial state would occur. Hence, hysteresis takes place between the two transitions, and such a path is labeled C in Figure 1(b). The pair of folds in Figure 1(b) are called saddle-node bifurcations (Iooss & Joseph, 1990). A physical example of this is provided by the buckling of an elastic wire such as the outer portion of a bicycle brake cable. When a short length is held vertically, it will stand upright. If you push the remaining length upward through your hand, it will eventually become long enough so that gravity will cause it to fall over through a large angle of deflection. Now pull it back downward through your hand and you will find that the deflected state remains over a range of lengths before flipping back to the vertical. This is an example of a hysteresis loop.

The two models in Figures 1(a) and (b) contain a reflection symmetry. If this is not present, the bifurcation is transcritical and an example of such a bifurcation is given in Figure 1(c). In the physical example of the buckling of the ruler, this type of bifurcation would be observed if a constant side load

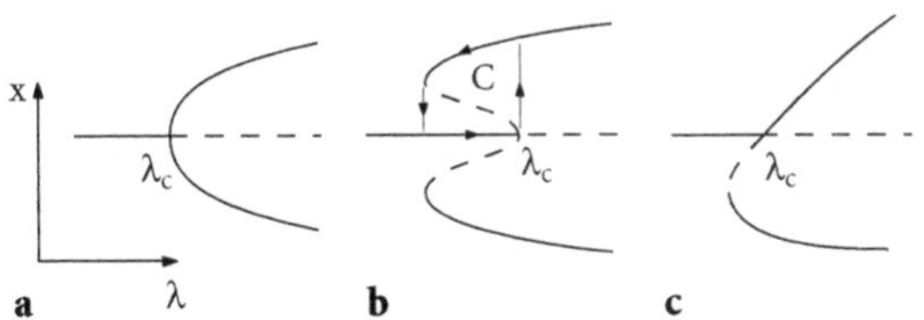

Figure 1. Sketches of (a) supercritical pitchfork, (b) subcritical pitchfork and (c) transcritical bifurcations. Solid lines indicate stable solutions and dashed lines indicate unstable.

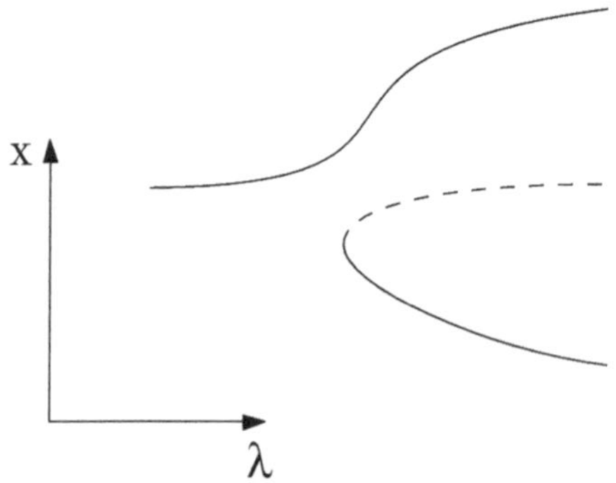

Figure 2. Imperfect pitchfork bifurcation.

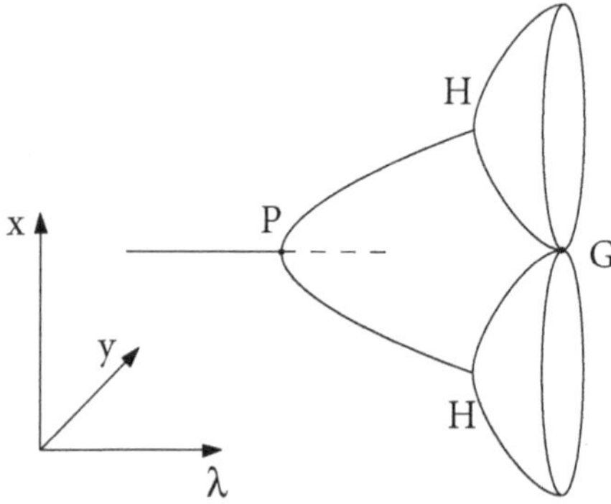

Figure 3. Schematic of a gluing bifurcation sequence.

was applied in addition to the end load in the example of the ruler. In this case, the mid-plane symmetry is automatically broken.

Of course, in any real system, physical imperfections will be present. These can be taken into account using the imperfect bifurcation theory (Golubitsky & Schaeffer, 1985). The effect of an imperfection is to disconnect the supercritical pitchfork bifurcation as shown in Figure 2. It can be seen that there is one state that evolves smoothly with an increase in parameter λ and another disconnected branch. In the example of the ruler, imperfections could arise from irregularities in the shape of the ruler or a slight imbalance in the applied load. The lower limit of the disconnected branch is defined by another type of bifurcation, a saddle node. The disconnected state can be attained, either by variation of two parameters (e.g., variation of side load for the Euler strut), or by a discontinuous or sudden jump in the parameter λ. In the latter case, there is a finite chance that the system will land on the disconnected solution. Examples of observations of such behavior in fluid flows are provided by Taylor–Couette flow (Pfister et al., 1988), a flow through a sudden expansion (Fearn et al., 1990), and convection (Arroya & Saviron, 1992).

Another very important bifurcation is a Hopf where a simply periodic cycle arises from a fixed point as a parameter is changed. As in the case of pitchforks, Hopf bifurcations may also be super- or subcritical, with hysteresis present in the latter case. An interesting feature of supercritical Hopf bifurcations is that the system takes more and more time to reach the equilibrium state as the bifurcation point nears. The observed long-term dynamics are analogous to critical slowing down (Landau & Lifshitz, 1980) in phase transitions and have been found in a fluid flow (Pfister & Gerdts, 1981), for example.

Further interesting global bifurcations (Glendinning, 1994) may occur when pitchfork and Hopf bifurcations occur sequentially. An example of this is shown schematically in Figure 3, where the pair of asymmetric states that arise at the pitchfork P then undergo a pair of Hopf bifurcations at the points labeled H. The cycles that arise at H then join together at a gluing bifurcation (Coullet et al., 1984) when λ is increased. This point is marked G in Figure 3, and it is an example of a homoclinic bifurcation. In this case, a single large orbit is formed from the pair of cycles as λ is increased beyond G with the period going to infinity exactly at G. Interesting dynamical behavior including chaos can be observed near such points in experiments when physical imperfections are taken into account (Glendinning et al., 2001).

TOM MULLIN

See also **Catastrophe theory; Critical phenomena; Equilibrium; Hopf bifurcation; Phase transitions**

Further Reading

Acheson, D. 1997. *From Calculus to Chaos*, Oxford and New York: Oxford University Press

Arnol'd, V.I., Afrajmovich, V.S., Il'yashenko, Yu.S. & Shil'nikov, L.P. 1994. *Bifurcation Theory and Catastrophe Theory*, Berlin and New York: Springer

Arroyo, M.P. & Saviron, J.M. 1992. Rayleigh Bénard convection in a small box-spatial features and thermal-dependence of the velocity field. *Journal of Fluid Mechanics*, 235: 325–348

Coullet, P., Gambaudo, J.M. & Tresser, C. 1984. Une nouvelle bifurcation de codimension 2: le collage de cycles. *Comptes Rendu de l'Academie des Sciences de Paris, Ser. I–Mathematische* 299: 253–256

Fearn, R.M., Mullin, T. & Cliffe, K.A. 1990. Nonlinear flow phenomena in a symmetric sudden-expansion. *Journal of Fluid Mechanics*, 211: 595–608

Glendinning, P. 1994. *Stability, Instability and Chaos: An Introduction to the Theory of Nonlinear Differential Equations*, Cambridge and New York: Cambridge University Press

Glendinning, P., Abshagen, J. & Mullin, T. 2001. Imperfect homoclinic bifurcations. *Physical Review* E, 64: 036208

Golubitsky, M. & Schaeffer, D.G. 1985. *Singularities and Groups in Bifurcation Theory I*, Berlin and New York: Springer

Guckenheimer, J. & Holmes, P.J. 1986. *Nonlinear Oscillations, Dynamical Systems, and Bifurcations of Vector Fields*, 2nd edition, Berlin and New York: Springer

Iooss, G. & Joseph, D.D. 1990. *Elementary Stability and Bifurcation Theory*, 2nd edition, Berlin and New York: Springer

Landau, L.D. & Lifshitz, E.M. 1980. *Statistical Physics*, part 1, 3rd edition, London: Pergamon

Pfister, G. & Gerdts, U. 1981. The dynamics of Taylor wavy vortex flow. *Physics Letters A*, 83: 23–27

Pfister, G., Schmidt, H. Cliffe, K.A. & Mullin, T. 1988. Bifurcation phenomena in Taylor–Couette flow in a very short annulus. *Journal of Fluid Mechanics*, 191: 1–18

BI-HAMILTONIAN STRUCTURE

See **Integrable lattices**

BILAYER LIPID MEMBRANES

When a group of unknown researchers reported the artificial assembly of a bimolecular lipid membrane in vitro (at a 1961 symposium on the plasma membrane sponsored by the American and New York Heart Association), it was initially met with skepticism. The research group led by Donald O. Rudin began their report with a description of mundane soap bubbles, followed by "black holes" in soap films, ending with an invisible "black" lipid membrane made from extracts of cows' brains. The reconstituted structure (7.5 nm thick) was created just like a cell membrane separating two aqueous solutions. The speaker then said:

> upon adding one, as yet unidentified, heat-stable compound. . .from fermented egg white. . . to one side of the bathing solutions. . .lowers the resistance. . .by 5 orders of magnitude to a new steady state. . .which changes with applied potential. . .Recovery is prompt. . . the phenomenon is indistinguishable. . . from the excitable alga *Valonia*. . ., and similar to the frog nerve action potential (Ottova & Tien, 2002).

The first report was published a year later (Mueller et al., 1962). In reaction to that report, the subsequent inventor of liposomes (artificial spherical bilayer lipid membranes) wrote recently in an article entitled "Surrogate cells or Trojan horses" (Bangham, 1995):

> . . .a preprint of a paper was lent to me by Richard Keynes, then Head of the Department of Physiology [Cambridge University], and my boss. This paper was a bombshell . . .They [Mueller, Rudin, Tien, and Wescott] described methods for preparing a membrane . . . not too dissimilar to that of a node of Ranvier. . .The physiologists went mad over the model, referred to as a "BLM", an acronym for Bilayer or by some for Black Lipid Membrane. They were as irresistible to play with as soap bubbles.

Indeed, the Rudin group was playing with soap bubbles using equipment purchased from a local toy shop. But scientific experimentation with soap bubbles began with the observations of Robert Hooke (who coined the word "cell" in 1665 to describe the structure of a thin slice of cork tissue observed through a microscope he had constructed), with his observation of "black spots" in soap bubbles and films. Years later, Isaac Newton estimated the blackest soap film (when light waves reflecting from one layer of soap molecules destructively interfere with light waves reflecting from the second layer of soap molecule) to be about $3–8\times10^{-6}$ in. thick. (Modern measurements give thicknesses between 5 and 9 nm, depending on the soap solution used.)

Origins of the Lipid Bilayer Concept

The recognition of the lipid bilayer as a model for biomembranes dates back to the work of Hugo Fricke, in the 1920s and 1930s, who calculated the thickness of red blood cell (RBC) membranes to be between 3.3 and 11 nm, based on frequency-dependent measurements of the impedance of cell suspensions. Modern measurements on experimental bilayer lipid membranes (BLMs) and biomembranes confirm Fricke's estimation of the thickness of the plasma membrane (Tien & Ottova, 2000).

In his 1917 studies of the molecular organization of fatty acids at the air-water interface, Irving Langmuir had demonstrated that a simple trough apparatus could provide the data to estimate the dimensions of a molecule. Evert Gorter and F. Grendel (respectively, a pediatrician and a chemist) used Langmuir's trough to determine the area occupied by lipids extracted from red blood cell (from human, pig, or rat sources) "ghosts" (empty membrane sacs) and found that there was enough lipid to form a layer two molecules thick over the whole cell surface. In other words

$$\frac{\text{surface area occupied (from monolayer experiment)}}{\text{surface area of red blood cell}} \cong 2. \tag{1}$$

Thus, Gorter and Grendel suggested that the plasma membrane of red blood cells may be thought of as a lipid bilayer, with the polar (hydrophilic) head groups oriented outward.

Experimental Realization

The structure of black soap films led to the realization by Rudin and his co-workers in 1960 that a soap film in its final stages of thinning has a structure composed of two fatty acid monolayers sandwiching an aqueous solution as follows:

air | monolayer | soap solution | monolayer | air.

With the above background in mind, Rudin et al. simply proceeded to make a BLM under an aqueous solution, which may be represented as follows:

aqueous solution | BLM | aqueous solution.

Their effort was successful (Tien & Ottova, 2001, p. 86). Rudin and his colleagues showed that a

BLM formed from brain extracts and separating two aqueous solutions was self-sealing to punctures, with many physical and chemical properties similar to those of biomembranes. Upon modification with a certain compound called excitability-inducing molecule (EIM), this otherwise electrically inert structure became excitable, displaying characteristic features similar to those of action potentials of the nerve membrane.

By the end of the early 1970s, it had been determined that an unmodified bilayer lipid membrane separating two similar aqueous solutions is about 5 nm thick and is in a liquid-crystalline state with the following electrical properties: membrane potential ($E_m \simeq 0$), membrane resistivity ($R_m \simeq 10^9\ \Omega\,\mathrm{cm}$), membrane capacitance ($C_m \simeq$ 0.5–1 $\mu\mathrm{F\,cm^{-2}}$), and dielectric breakdown ($V_b > 250{,}000$ V/cm). In spite of its very low dielectric constant ($\varepsilon \simeq$ 2–7), this liquid-crystallline BLM is surprisingly permeable to water (8–24 μm/s) (Tien & Ottova, 2000).

The Lipid Bilayer Principle

In spite of their variable compositions, the fundamental structural element of all biomembranes is a liquid-crystalline phospholipid bilayer. Thus, the lipid bilayer principle of cell or biological membranes may be summarily stated as follows: all living organisms are made of cells, and every cell is enclosed by a plasma membrane, the indispensable component of which is a lipid bilayer. The key property of lipid bilayer-based cells is that they are separated from the environment by a permeability barrier that allows them to preserve their identity, take up nutrients, and remove waste. This 5 nm thick liquid-crystalline lipid bilayer serves not only as a physical barrier but also as a conduit for transport, a reactor for energy conversion, a transducer for signal processing, a bipolar electrode for redox reactions, or a site for molecular recognition.

The liquid-crystalline lipid bilayer of biomembranes not only provides the physical barrier separating the cytoplasm from its extracellular surroundings, it also separates organelles inside the cell to protect important processes and events. More specifically, the lipid bilayer of cell membrane must prevent its molecules of life (genetic materials and many proteins) from diffusing away. At the same time, the lipid bilayer must keep out foreign molecules that are harmful to the cells. To be viable, the cell must also communicate with the environment to continuously monitor the external conditions and adapt to them. Further, the cell needs to pump in nutrients and release toxic products of its metabolism. How does the cell carry out all of these multi-faceted activities?

A brief answer is that the cell depends on its lipid-proteins-carbohydrate complexes (i.e., glycoproteins, proteolipids, glycolipids, etc.) embedded in the lipid bilayer to gather information about the environment in various ways. Examples include communication with hundreds of other cells about a variety of vital tasks such as growth, differentiation, and death (apoptosis). Glycoproteins are responsible for regulating the traffic of material to and from the cytoplasmic space. Paradoxically, the intrinsic structure of cell membranes creates a bumpy obstacle to these vital processes of intercellular communication. The cell shields itself behind its lipid bilayer, which is virtually impermeable to all ions (e.g., Na^+, K^+, Cl^-) and most polar molecules (except H_2O). This barrier must be overcome, however, for a cell to inform itself of what is happening in the world outside, as well as to carry out vital functions. Thus, over millions and millions of years of evolution, the liquid-crystalline lipid bilayer—besides acting as a physical restraint—has been modified to serve as a conduit for material transport, as a reactor for energy conversion, as a bipolar electrode for redox reactions, as a site for molecular recognition, and other diverse functions such as apoptosis and signal transduction.

Insofar as membrane transport is concerned, cells make use of three approaches: simple diffusion, facilitated diffusion, and active transport. Although simple diffusion is an effective transport mechanism for some substances such as water, the cell must make use of other mechanisms for moving substances in and out of the cell. Facilitated diffusion utilizes membrane channels to allow charged molecules, which otherwise could not diffuse across the lipid bilayer. These channels are especially useful with small ions such as K^+, Na^+, and Cl^-. The number of protein channels available limits the rate of facilitated transport, whereas the speed of simple diffusion is controlled by the concentration gradient. Under active transport, the expenditure of energy is necessary to translocate the molecule from one side of the lipid bilayer to the other, in contrast to the concentration gradient. Similar to facilitated diffusion, active transport is limited by either the capacity of membrane channels or the number of carriers present.

Today, ion channels are found ubiquitously. To name a few, they are in the plasma membrane of sperm, bacteria, and higher plants; the sarcoplasmic retculum of skeletal muscle, nerve membrane, synaptic vesicle membranes of rat cerebral cortex, and the skin of carps. As a weapon of attack, many toxins released by living organisms such as dermonecrotic toxin, hemolysin, brevetoxin, and bee venom are polypeptide-based ion-channel formers. For example, functioning of membrane proteins, in particular, ionic channels, can be modulated by alteration of their arrangement in membranes (e.g., electroporation, Tien & Ottova, 2003).

At the membrane level, most cellular activities involve some kind of lipid bilayer-based receptor-ligand contact interactions. Outstanding examples among these are ion-sensing, molecular recognition (e.g., antigen-antibody binding and enzyme-substrate interaction), light conversion and detection, gated channels, and active transport. The development of self-assembled bilayer lipid membranes (BLMs and liposomes) has made it possible to investigate directly the electrical properties and transport phenomena across a 5 nm thick biomembrane element separating two aqueous phases. A modified or reconstituted BLM is viewed as a dynamic structure that changes in response to environmental stimuli as a function of time, as described by the so-called dynamic membrane hypothesis. Under this hypothesis, each type of receptor interacts specifically with its own ligand. That is, the so-called G-receptor is usually coupled to a guanosine nucleotide-binding protein that in turn stimulates or inhibits an intracellular, lipid bilayer-bound enzyme. G-protein-linked receptors mediate the cellular responses to a vast variety of signaling molecules, including local mediators, hormones, and neurotransmitters, which are as varied in structure as they are in function. G-protein-linked receptors usually consist of a single polypeptide chain, which threads back and forth across the lipid bilayer up to seven times. The members of this receptor family have a similar amino acid sequence and functional relationship. The binding sites for G-proteins have been reported to be the second and third intracellular loops and the carboxy-terminal tail. The endogenous ligands, such as hormones, neurotransmitters, and exogenous stimulants such as odorants, belonging to this class are important target analytes for biosensor technology Tien & Ottova (2003).

H. Ti. Tien and Angelica Ottova-Lueitmannova

See also **Langmuir–Blodgett films; Nerve impulses; Neurons**

Further Reading

Bangham, A.D. 1995. Surrogate cells or Trojan horses. *BioEssays*, 17: 1081–1088

Mueller, P., Rudin, D.O., Tien H.T. & Wescott, W.C. 1962. Reconstitution of cell membrane structure in vitro and its transformation into an excitable system. *Nature*, 194: 979–980

Ottova A. & Tien, H.T. 2002. The 40th anniversary of bilayer lipid membrane research. *Bioelectrochemistry*, 56: 171–173

Ottova, A., Tvarozek, V. & Tien, H.T. 2003. Supported BLMs. In *Planar Lipid Bilayers (BLMs) and Their Applications*, edited by H.T. Tien & A. Ottova-Leitmannova, Amsterdam: Elsevier

Tien, H.T. 1974. *Bilayer Lipid Membranes (BLM): Theory and Practice*, New York: Marcel Dekker

Tien, H.T. & Ottova, A.L. 2000. *Membrane Biophysics: As Viewed from Experimental Bilayer Lipid Membranes (Planar Lipid Bilayers and Spherical Liposomes)*, Amsterdam and New York: Elsevier Science

Tien, H.T. & Ottova, A. 2001. The lipid bilayer concept and its experimental realization: from soap bubbles, the kitchen sink, to bilayer lipid membranes. *Journal of Membrane Science*, 189: 83–117

Tien, H.T. & Ottova, A. 2003. The bilayer lipid membrane (BLM) under electrical fields. *IEEE Transactions on Dielectrics and Electrical Insulation*, 10(5): 717–727

BILLIARDS

In mathematical physics, the singular noun "billiards" denotes a dynamical system corresponding to the inertial motion of a point mass within a region that has a piecewise smooth boundary. The reflections from the boundary are taken to be elastic; that is, the angle of reflection equals the angle of incidence. This model arises naturally in optics, acoustics, and classical and statistical mechanics. In fact, two fundamental models in statistical mechanics, gas of hard spheres (Boltzmann gas) and the Lorentz gas, are billiards. The billiards concept occupies a central position in nonlinear physics because it provides ideal visible models for analysis of dynamical properties leading to classical chaos and an ideal testing ground for the semiclassical analysis of quantum systems.

Billiards models are Hamiltonian systems. Hence, the phase volume is preserved under the dynamics, and the system can be studied in the framework of ergodic theory. In particular, the boundary of the billiard region is supposed to be only piecewise smooth; that is, it consists of smooth components. Therefore, the dynamics of billiards is not defined for orbits that hit singular points of the boundary. However, the phase volume of such orbits equals zero. The dynamics of billiards is completely defined by the shape of its boundary. A smooth component of the boundary is called dispersing, focusing, or neutral if it is convex inward, outward the billiard region, or if it is flat (has zero curvature), respectively. Any billiard orbit is a broken line in its configuration space.

The classical examples of integrable billiards are provided by circular and elliptical boundaries. Configuration spaces of these billiards are foliated by caustics, which are smooth curves (surfaces γ) such that if one link of the billiard orbit is tangent to γ, then every other link of this orbit is tangent to γ. Billiards in a circle has one family of caustics formed by (smaller) concentric circles, while billiards in an ellipse has two families of caustics (confocal ellipses and confocal hyperbolas), which are separated by orbits such that each link intersects a focus of the ellipse. Birkhoff's conjecture (Birkhoff, 1927) claims that among all billiards inside smooth convex curves, only billiards in ellipses are integrable. Berger (1990) has shown that in three dimensions (*d*), only billiards in ellipsoids produce foliations of a billiard region by smooth convex caustics. However, it does not imply that only billiards in ellipsoids are integrable because if a billiard in $d > 2$ has an invariant hypersurface then this hypersurface does not necessarily consist of rays tangent to some hypersurface in the configuration space. Using KAM theory, Lazutkin has shown that if a billiards boundary is strictly convex, with a sufficiently smooth curve and its curvature never vanishes; then there exists an uncountable number of smooth caustics in the vicinity of the boundary, and moreover, the phase volume of the orbits tangent to these caustics is positive (Lazutkin, 1991).

An opposite situation occurs when a boundary is everywhere dispersing. Such models were introduced by Sinai (1970), in his seminal paper and they are called Sinai (or dispersing) billiards (Figure 1(a)). Sinai billiards have the strongest chaotic properties; that is, they are ergodic, mixing, have a positive metric entropy, and are Bernoulli systems. If a (narrow) parallel beam of rays is made to fall onto a dispersing boundary, then after reflection it becomes divergent and, therefore, the distance between the rays in this beam increases with time. It is the mechanism of dispersing that generates sensitive dependence on initial conditions (hyperbolicity) and is responsible for strong chaotic properties of dispersing billiards.

On the other hand, focusing boundaries produce the opposite effect. Indeed, a narrow parallel beam of rays after reflection from the focusing boundary becomes convergent; that is, the distance between rays in such a beam decreases with time. Therefore, it has been the general understanding that a dispersing boundary always produces chaotization of the dynamics, while a focusing boundary produces stabilization of the dynamics. However, there exists another mechanism of chaos in billiards (and, in general, in Hamiltonian systems), which is called defocusing (Bunimovich, 1974, 1979). The point is that a narrow parallel beam of rays, after focusing because of reflection from a focusing boundary, may become divergent provided that a free path between two consecutive reflections from the boundary is long enough. Assuming that the time of divergence exceeds (averaged over all orbits) the time of convergence, one obtains chaotic billiards. One of the first, and the most famous, example of such billiards is called a stadium (Figure 1(b)). One obtains a stadium by cutting a circle into two semi-circles and connecting them by two common tangent segments. The length of these segments could be arbitrarily small, which demonstrates that the mechanism of defocusing can work under small deformations of even the integrable (a circle) billiards. Focusing billiards can have as strong chaotic properties as Sinai's billiards do (Bunimovich, 2000).

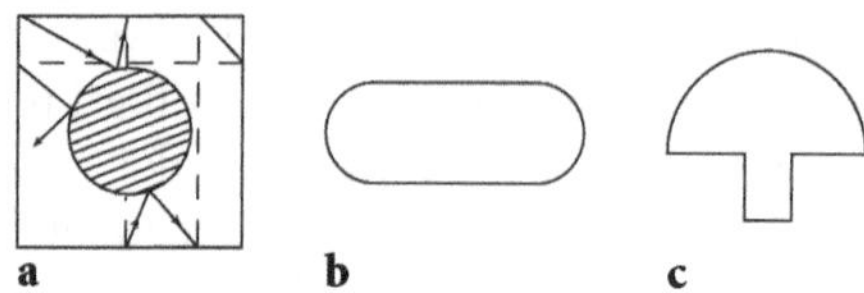

Figure 1. (a) Sinai billiard. (b) Stadium. (c) Mushroom.

There are no other mechanisms of chaos in billiards. Indeed, billiards in polygons and polyhedrons have zero metric entropy (Boldrighini et al., 1978). Nevertheless, a typical billiard in a polygon is ergodic (Kerckhoff et al., 1986). Because focusing components can form parts of the boundary of integrable as well as chaotic billiards, a natural question is whether there are some restrictions. Two classes of focusing components admissible in chaotic billiards were found (Wojtkowski, 1986; Markarian, 1988). The most general class of such focusing components is formed by absolutely focusing mirrors (AFM) (Bunimovich, 1992). AFMs form a new notion in geometric optics. A mirror γ (or a smooth component of a billiards' boundary) is called absolutely focusing if any narrow parallel beam of rays that falls on γ becomes focused after its last reflection in a series of consecutive reflections from γ. Observe that a mirror is focusing if any parallel beam of rays becomes focused just after the first reflection from this mirror. AFMs can also be characterized in terms of their local properties (Donnay, 1991; Bunimovich, 1992).

Generic Hamiltonian systems are neither integrable nor chaotic. Instead, their phase spaces get divided into KAM-islands and chaotic sea(s). The only clear and clean example of this phenomenon is a billiard in a mushroom (Bunimovich, 2001). The mushroom consists of a semicircular hat sitting on a foot (Figure 1(c)). A mushroom becomes a stadium when the width of the foot equals the width of the hat.

Clearly, the mechanism of dispersing works is higher than two dimensions as well (Sinai, 1970). It is not obvious at all for the mechanism of defocusing because of astigmatism. However, chaotic focusing billiards also do exist in dimension $d \geq 3$ (Bunimovich & Rehacek, 1998). But one pays a price of astigmatism by not allowing the focusing component to be as large as it can be in $d = 2$.

Many properties of classical dynamics of billiards are closely related to the properties of the corresponding quantum problem. Consider the Schrödinger equation with a potential equal to zero inside the billiard region and equal to infinity on the boundary. The eigenfunctions become uniformly distributed over the regions of ergodic billiards for high wave numbers (Shnirelman, 1991). On the contrary, there exist infinite series of eigenfunctions localized in the vicinity of convex caustics of billiards (Lazutkin, 1991).

Leonid Bunimovich

See also **Ergodic theory; Horseshoes and hyperbolicity in dynamical systems; Lorentz gas**

Further Reading

Berger, M. 1990. Sur les caustiques de surfaces en dimension 3. *Comptes Rendu de l'Academie de Sciences*, 311: 333–336

Birkhoff, G. 1927. *Dynamical Systems*. New York, American Mathematical Society

Boldrighini, C., Keane, M. & Marchetti, F. 1978. Billiards in polygons. *Annals of Probability*, 6: 532–540
Bunimovich, L.A. 1974. On billiards close to dispersing. *Mathematical USSR Sbornik*, 95: 49–73 (originally published in Russian)
Bunimovich, L.A. 1979. On the ergodic properties of nowhere dispersing billiards. *Communications in Mathematical Physics*, 65: 295–312
Bunimovich, L.A. 1992. On absolutely focusing mirrors. In *Ergodic Theory and Related Topics*, edited by U. Krengel, et al., Berlin and New York: Springer, pp. 62–82
Bunimovich, L.A. 2000. Billiards and other hyperbolic systems with singularities. In *Dynamical Systems, Ergodic Theory and Applications*, edited by Ya. G. Sinai, Berlin: Springer
Bunimovich, L.A. 2001. Mushrooms and other billiards with divided phase space. *Chaos*, 11: 802–808
Bunimovich, L.A. & Rehacek, J. 1998. How many dimensional stadia look like. *Communications in Mathematical Physics*, 197: 277–301
Donnay, V. 1991. Using integrability to produce chaos: billiards with positive entropy. *Communications in Mathematical Physics*, 141: 225–257
Kerckhoff, S., Mazur, H. & Smillie, J. 1986. Ergodicity of billiard flows and quadratic differentials. *Annals of Mathematics*, 124: 293–311
Lazutkin, V. F. 1991. *The KAM Theory and Asymptotics of Spectrum of Elliptic Operators*. Berlin and New York: Springer
Markarian, R. 1988. Billiards with Pesin region of measure one. *Communications in Mathematical Physics*, 118: 87–97
Shnirelman, A. I. 1991. On the asymptotic properties of eigenfunctions in the regions of chaotic motion. Addendum in *The KAM Theory and Asymptotics of Spectrum of Elliptic Operators* by V. F. Lazutkin. Berlin and New York: Springer
Sinai, Ya. G. 1970. Dynamical systems with elastic reflections. Ergodic properties of dispersing billiards. *Russian Mathematical Surveys*, 25: 137–189 (originally published in Russian 1970)
Wojtkowski, M. 1986. Principles for the design of billiards with nonvanishing Lyapunov exponents. *Communications in Mathematical Physics*, 105: 391–414

BINDING ENERGY

When two particles form a bound state under a certain kind of physical interaction, the resulting state has an energy smaller than the sum of the rest energies of the constituent elements of such a bound state. That is why, by definition (or one could say by construction), bound states are ones in which work has to be done to separate the constituents.

The energy that one has to provide (equivalently the work) in order to separate a bound state into its elements is called the binding energy, E_b, and from the above, it can be directly inferred that $E_b > 0$. The equivalent mass to this energy (under the Einstein relation) also bears a name and is called the "mass defect," $\Delta m = E_b/c^2$, where c is the speed of light.

Examples of binding energy can be easily found among the fundamental forces in nature, such as the gravitational force, the electromagnetic force, and the nuclear force.

Considering an approximately circular (in reality, elliptical) motion of the Earth around the Sun, equating the gravitational force $F_g = GM_sM_e/R^2$ (where G is the gravitational constant, the subscripts s and e denote Sun and Earth, respectively, and R is their relative distance) with the centripetal force $F_c = M_e v^2/R$, one obtains $v = \sqrt{GM_s/R}$, leading to a kinetic energy $E_k = GM_eM_s/(2R)$, which combined with the potential energy of $E_p = -GM_eM_s/R$, results in a binding energy for the solar system of the form

$$E_b = G\frac{M_e M_s}{2R}. \qquad (1)$$

Using the relevant masses for the Earth and Sun and their separation, this quantity can be approximately calculated as $E_b \approx 2.6 \times 10^{33}$ J ($\Delta m \approx 2.9 \times 10^{16}$ kg). However, what actually matters in terms of physical "observability" is the ratio of mass defect to the bound state mass (the closer this ratio is to 1, the greater the possibility of observing the mass defect). In the case of the gravitational system $\Delta M/M_b \approx 1.5 \times 10^{-14}$; hence, the mass defect for the gravitational force will not be observable.

Similar calculations can be performed classically for the hydrogen atom (following the same path, but substituting $G \to 1/(4\pi\varepsilon_0)$, $M_s \to |q_e|$, and $M_e \to q_e$, where ε_0 is the dielectric constant in a vacuum and q_e is the charge of the electron). In this case, for the electrostatic force,

$$E_b = \frac{1}{2}\frac{q_e^2}{4\pi\varepsilon_0 R}. \qquad (2)$$

In this case, however, $R \approx 0.53 \times 10^{-10}$ m (while in the previous example, it was $\approx 1.5 \times 10^{11}$ m!). In the case of the hydrogen atom, $E_b \approx 13.6$ eV and $\Delta m \approx 2.5 \times 10^{-35}$ kg. The ratio $\Delta M/M_b \approx 1.5 \times 10^{-8}$, indicating that in this case also it is not possible to observe the mass defect.

In the case of the nuclear force, however, the ratio of $\Delta M/M_b$ is of order 10^{-3}, and hence it is possible to observe the mass defect. For example, the mass of an α particle consisting of two protons and two neutrons is 6.6447×10^{-27} kg, while the individual masses of these particles add up to 6.6951×10^{-27} kg, resulting in a binding energy of 28.3 MeV and $\Delta M/M_b \approx 0.0075$. In fact, a very common diagram in nuclear physics is the so-called nuclear binding energy curve (see, e.g., http://hyperphysics.phy-astr.gsu.edu/hbase/nucene/nucbin.html), which shows the binding energy of various elements as a function of their mass number. In this graph, the larger the E_b, the more stable the element; iron (with atomic number $A = 56$ and binding energy 8.8 MeV/nucleon) is the most stable element. Lighter elements can yield energy by fission, while heavier elements can yield energy by means of fusion, emitting energies in the MeV range.

Binding Energy in Nonlinear Systems

Naturally, bound states of multiple waves can be formed in nonlinear systems. To fix ideas, we will examine such bound states and their corresponding binding energies in the specific context of the well-known sine-Gordon equation. For a detailed exposition of the features and applications of this equation, see Dodd et al. (1982). The sine-Gordon equation in (1 + 1) dimensions is

$$u_{tt} = u_{xx} - \sin(u). \tag{3}$$

Perhaps, the best-known nonlinear wave solution of this equation consists of the topological soliton (kink), which is of the form (in the static case)

$$u(x) = 4\tan^{-1}(e^{sx}), \tag{4}$$

$s \in \{-1, 1\}$, where the case of $s = 1$ corresponds to a kink, while $s = -1$ corresponds to an antikink. The energy of such a static kink solution

$$E = \int_{-\infty}^{\infty} dx \left[\frac{1}{2}(u_t^2 + u_x^2) + 1 - \cos(u)\right] \tag{5}$$

can be calculated as $E = 8$.

Another elemental solution of the equation is the breather-like solution of the form

$$u(x,t) = 4\tan^{-1}\left[\frac{(1-\omega^2)^{1/2}}{\omega}\frac{\sin(\omega t)}{\cosh\left((1-\omega^2)^{1/2}x\right)}\right]. \tag{6}$$

This exponentially localized in space, periodic in time solution can be considered as a result of a merger of a kink and an antikink. Hence, this is perhaps the simplest example of a bound state in this nonlinear system.

The bound state character of this solution can also be revealed by the expression for its energy. Using expression (6) in Equation (5), we obtain

$$E_{\text{breather}} = 16(1-\omega^2)^{1/2}. \tag{7}$$

Hence, this energy, for any $\omega \in (0, 1)$ (ω is the frequency of the internal breathing oscillation), is less than the sum of the kink and antikink energies, verifying that the binding energy of such a state is

$$E_b = 16\left[1 - (1-\omega^2)^{1/2}\right]. \tag{8}$$

It is also worthwhile to note that the energy of such a breather excitation varies in the interval (0, 16) depending on its frequency. Hence, there is no threshold for the excitation of such a wave, but even for small amounts of energy, such waveforms will be excited (large frequency/small period ones for small excitation energy).

One can generalize the solution of the form (6) in a periodic breather lattice solution of the sine-Gordon equation in the form (see, e.g., McLachlan, 1994)

$$u(x,t) = 4\tan^{-1}[a\,\mathrm{sn}(bt, k^2)\mathrm{dn}(cx, 1-m^2)], \tag{9}$$

where $\mathrm{sn}(x,k)$ and $\mathrm{dn}(x,k)$ are the Jacobi elliptic functions with modulus k. Here

$$a = \sqrt{\frac{k}{m}},\ b = \sqrt{\frac{m}{(m+k)(1+mk)}} \text{ and } c = \sqrt{\frac{k}{(m+k)(1+mk)}}.$$

One can then evaluate the energy (per breather) of this infinite periodic breather lattice configuration (for details of the calculation, the interested reader is directed to Kevrekidis et al., 2001) to be

$$E = 16\sqrt{\frac{k}{(k+m)(1+km)}}E(1-m^2), \tag{10}$$

where $E(1-m^2)$ is the complete elliptic integral of the second kind.

Depending on the values of the elliptic moduli, k and m, the expression of Equation (9) represents a lattice of different entities. For $m, k \to 0$, it corresponds to genuine sine-Gordon breathers. For $k, m \to 1$, the limit gives the "pseudosphere" solution $u = 4\tan^{-1}(\tanh\frac{t}{2})$, which resembles a π-kink but in time rather than space (see McLachlan, 1994). On the other hand, the $k \to$ finite, $m \to 0$ limit gives the kink-antikink pair solution $4\tan^{-1}(t\ \mathrm{sech}\ x)$. This solution has the character of a kink-antikink pair "breathing" in time.

The above different limits illustrate why Equation (10) is an important result, since it can be used (see below) to obtain the asymptotic interaction between entities such as breathers, pseudospheres, or kink-antikink pairs. When taking the appropriate above limits of expression (10), the leading-order term will be the energy of a single such entity. However, the correction to that will be the (per particle) binding energy in a configuration of multiple such entities (or, as it is often referred to, the energy of interaction between two such entities).

To calculate the breather-breather interaction (their binding energy), we take the limit $m, k \to 0$, with $k/m = (1-\omega^2)/\omega^2$, where ω is the breather frequency (see McLachlan, 1994), to obtain

$$E = \left[16\sqrt{1-\omega^2} - 8m^2\frac{(1-\omega^2)^{3/2}}{\omega^2} + 6m^4\frac{(1-\omega^2)^{5/2}}{\omega^4}\right]E(1-m^2). \tag{11}$$

Hence, the corrections to the (single) breather energy in Equation (11) correspond to the binding energy of the formed breather bound states. Similar expressions can be found for the pseudosphere:

$$E = 4\pi - \frac{\pi}{2}(k-1)^2 + \frac{\pi}{4}(m-1)^2 + \frac{3\pi}{32}(m-1)^2(k-1)^2 \tag{12}$$

(again, 4π is the energy of a single pseudosphere) and for kink-antikink pairs

$$E = \left[16 - 8\left(k + \frac{1}{k}\right) + 2m^2\left(2 + 3k^2 + \frac{3}{k^2}\right)\right] \times E(1 - m^2). \tag{13}$$

Similar examples of breather lattices also exist for other equations, such as the well-known Korteweg–de Vries (KdV) equation, and one can again infer breather-breather state binding energies in a similar manner.

In general, one can say that the concept of a bound state for nonlinear evolutionary partial differential equations supporting soliton (or solitary wave) solutions persists in a form very similar to the way it manifests itself for fundamental physical forces and their particle carriers. In the present case, the elements of the bound states are the nonlinear waves proper (a feature reminiscent of the particle-like character of such waves, manifest evidently also in their interactions). In a number of (most often integrable) cases, where the form of the bound state solutions is analytically tractable, the calculation of the bound state energy and of the energy of its constituent elements again provides information, through the difference between the two, for the binding energy (or energy of interaction) of such waves.

P.G. KEVREKIDIS

See also **Breathers; Partial differential equations, nonlinear; Sine-Gordon equation; Solitons**

Further Reading

Dodd, R.K., Eilbeck, J.C., Gibbon, J.D. & Morris, H.C. 1982. *Solitons and Nonlinear Wave Equations*, London: Academic Press

Kevrekidis, P.G., Saxena, A. & Bishop, A.R. 2001. *Physical Review* E, 64: 026613

McLachlan, R. 1994. *Math. Intelligencer*, 16: 31

BIOLOGICAL EVOLUTION

The term *evolution* defines a process that is driven by internal and/or external forces. In quantum mechanics, an evolution operator conducts change in time. Cosmic evolution in the standard model aims at a consistent description of the process from the "big bang" to the present universe. "Prebiotic evolution" deals with chemical precursors of present-day life and is determined by the conditions at the early Earth, be it in the primordial atmosphere, in the surrounding of volcanic hot springs at the sea floor, or at some other location. "Biological evolution" follows the prebiotic scenario, and it shaped and is still shaping the biosphere on Earth. A temporal change in the biosphere manifests itself as the appearance, alteration, and extinction of biological species. This view was not generally accepted before Charles Darwin. Influenced by the geologist Charles Lyell and his concept of uniformitarianism, Darwin and the proponents of the theory of evolution suggested that changes in the biosphere occur gradually, continuously, or at least, in small steps. In this aspect, which is not essential for the mechanism of evolution, Darwin's theory contrasted the view held by the majority of his contemporaries, who assumed constancy of biological species and change exclusively through catastrophic events leading to mass extinction (Ruse, 1979). The opponents of evolutionary thinking, Louis Agassiz, Georges Cuvier, and others, considered species as invariant entities. The remnants of extinct species in the fossil record were interpreted by them as witnesses from earlier worlds destroyed by punctual events, the great deluge, and other catastrophes that wiped out major parts of the organismic world. In society, the concept of evolution was heavily attacked by representatives of the Christian Churches because it was seen to be in conflict with the *Genesis* report in the Bible (Ruse, 2001). During the 20th century, European religious thought has reconciled religious belief and the idea of an evolving biosphere. In North America, the strong opposition of some groups of religious fanatics led to the peculiar development of Creationism, whose claim of being an alternative to the theory of evolution is rejected by the established scientific community (NAS, 1999).

The current theory of biological evolution originated from two epochal contributions by Charles Darwin and Gregor Mendel. Darwin conceived a mechanism for evolutionary change of the biosphere based on variation and selection, and he gathered empirical data providing evidence for the action of natural and artificial selection, the latter exercised in animal breeding and nursery gardens. Darwin's principle (published in *On The Origin of Species by Natural Selection* in 1859) has two consequences: species adapt to their environments and are related to their ancestors in terms of phylogenies, or branches of an ancestral tree of species. In 1866, Gregor Mendel introduced quantitative statistics into the evaluation of data in biology and performed the first precisely controlled fertilization experiments with plants. He discovered and interpreted correctly the action of genes in determining the properties of organisms. Mendel's work was considered irrelevant by the evolutionists of the second half of the 19th century and was "rediscovered" around 1900. Only in 1930 were the Darwinian concept of selection and Mendel's rules of inheritance combined to a common mathematical formalism by the population geneticists Ronald Fisher, John Haldane, and Sewall Wright (for a recent text in population genetics, see Hartl & Clark, 1997). In the 1940s, finally, Darwinian evolution and Mendelian genetics were united in the *Synthetic* or *Neo-Darwinian Theory of Evolution* by the works of the experimental biologists Theodosius

Dobzhansky, Julian Huxley, Ernst Mayr, and others (Mayr, 1997). In the second half of the 20th century, molecular biology put evolutionary theory on firm fundamentals, chemistry and physics. Comparison of genes and, more recently of whole genomes, allows for reconstruction of phylogenies on the basis of nucleotide sequence divergence through mutation (Judson, 1979); the exploration of molecular structures provides insights into the chemistry of present day life; and knowledge of biomolecular properties eventually led to the construction of laboratory systems that allow for observation of evolution of molecules in the test tube (Spiegelman, 1971; Watts & Schwarz, 1997).

Darwinian evolution results from the interplay of variation and selection, both being consequences of reproduction in populations. Variation operates on genomes or genotypes, which are polynucleotide sequences carrying the genetic information, and occurs in two fundamentally different ways: (i) mutation causes local changes in genomic sequences, whereas (ii) recombination exchanges corresponding segments between two genotypes. Selection is based on differences in *fitness* being a property of the phenotype. The phenotype is defined as the union of all, structural as well as dynamic, properties of an individual organism. Unfolding of the phenotype is programmed by the genome; but, at the same time, requires a highly specific environment. In addition, it is influenced by epigenetic factors (epigenetic refers to every nonenvironmental factor that interferes with the development of the organism, except those encoded in the nucleotide sequence of DNA; many epigenetic factors are already understood at the molecular level, and involve specific modifications of genomic DNA). Fitness, in essence, counts the number of fertile descendants reaching the reproductive age. It has two major components: (i) the probability of survival to reproduction, and (ii) the number of viable and fertile offspring.

To illustrate selection in a population of n asexually reproducing phenotypes, we consider a continuous-time model that describes change by a differential equation

$$\frac{\mathrm{d}x_i}{\mathrm{d}t} = x_i(f_i - \Phi), \quad i = 1, \ldots, n$$

$$\text{with } \Phi(t) = \sum_{j=1}^{n} f_j x_j = \overline{f}. \qquad (1)$$

The variables denote the frequencies of reproducing variants: $x_i(t) = N_i(t)/\sum_{j=1}^{n} N_j(t)$, with $N_i(t)$ counting the number of individuals with phenotype S_i or genotype I_i at time t. (For several genotypes giving rise to the same phenotype, see *neutrality* below.) Fitness values f_i when averaged over the entire population yield the mean fitness expressed by a time-dependent flux $\Phi(t)$. Frequencies of variants with fitness values above average, $f_i > \Phi$, increase with time, those of below-average variants, $f_i < \Phi$, decrease and as a consequence, the mean fitness increases. The flux $\Phi(t)$ is a nondecreasing function of time and selection continues until all variants, except the fittest, have died out (*See also* **Fitness landscape**). For two variants, I_0 and I_1, the solution boils down to

$$\left[\frac{x_1}{x_0}\right](t) = \left[\frac{x_1}{x_0}\right](0) \cdot \exp(mt)$$

or

$$\left[\frac{X_1}{X_0}\right]_t = \left[\frac{X_1}{X_0}\right]_0 w^t,$$

where the upper equation refers to continuously varying $x(t)$ and the lower equation refers to population to discrete time variables X_t with synchronized reproduction. The *Malthusian* fitness difference $m = f_1 - f_0$ is related to the *Darwinian* relative fitness $w = (1 + f_1)/(1 + f_2)$ by $m \approx \ln w$ (see Hartl & Clark (1997)). The conditions for selection are $m > 0$ or $w > 1$, respectively. An example is shown in Figure 1.

Sexual reproduction of diploid organisms involves Mendelian genetics (see Figure 2). Every gene (A) comes in two copies, identical or different, which are chosen from a reservoir of variants A_i called alleles. Recombination occurs in the process of reproduction when the two copies are separated and reassembled in pieces through random combination. The differential equation (1) is extended to describe selection in the diploid case in the form of Fisher's selection equation:

$$\frac{\mathrm{d}x_i}{\mathrm{d}t} = x_i \cdot \left(\sum_{j=1}^{n} a_{ij}x_j - \Phi\right)$$

$$= x_i\,(\overline{a_i} - \Phi), \quad i = 1, \ldots, n \qquad (2)$$

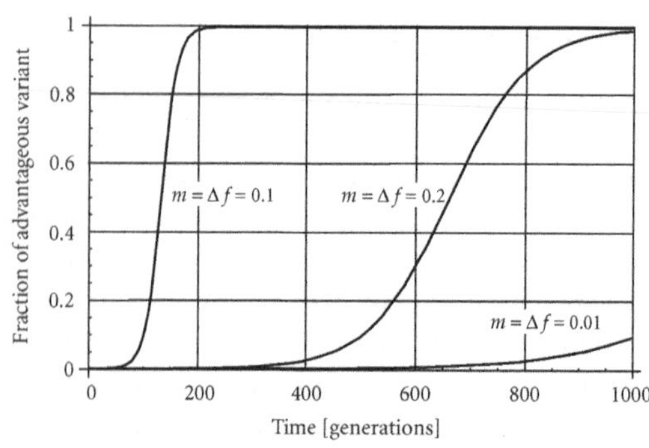

Figure 1. Illustration of selection in populations. The plotted curves represent the frequencies of advantageous mutants I_1 in a population of individuals I_0 with a Malthusian fitness difference of $m = \Delta f = f_1 - f_0 = 0.1, 0.02$, and 0.01. The population size is $N = 10{,}000$, and the mutants were initially present in a single copy: $N_1(0) = 1$ or $x_1(0) = 0.0001$.

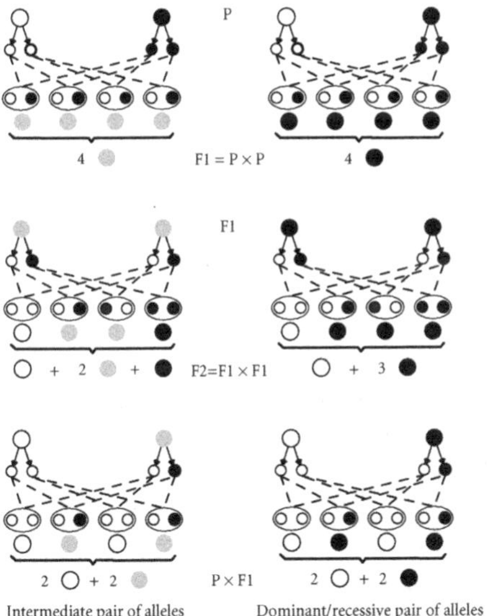

Figure 2. Mendelian genetics. In sexual reproduction, the two parental genomes are split into pieces and recombined randomly, which means each of the four alleles has a 50% chance to be incorporated in the genome of an offspring. Mendel's laws are of a statistical nature and hold as mean values in the limit of large numbers of observations. Two cases are shown: (i) the heterozygote unfolds into a phenotype with intermediate properties (gray through blending of black and white), and (ii) the property of one allele (black) is dominant. In the latter case, the other allele (white) is called recessive. Interbreeding of two homozygous individuals (parent generation P) leads to a first offspring generation (F1) of identical heterozygous individuals; the phenotypes in the next (F2) generation show a distribution of 1:2:1 in the intermediate and 1:3 in the dominant/recessive case. Crossing of the (recessive) parent genotype with an F1-individual yields a 1:1 ratio of phenotypes.

with

$$\overline{a_i} = \sum_{j=1}^{n} a_{ij} x_j,$$

$$\Phi(t) = \overline{a} = \sum_{i=1}^{n} \overline{a_i} x_i = \sum_{i=1}^{n} \sum_{j=1}^{n} a_{ij} x_i x_j.$$

The variables refer to alleles A_j rather than to whole genomes, and the rate coefficient a_{ij} represents the individual fitness values for the combination $A_i A_j$. Fitness is assumed to be independent of the positioning of alleles, $A_i A_j$ or $A_j A_i$, and hence, $a_{ij} = a_{ji}$ holds. The term $\overline{a_i}$ is the population-averaged mean fitness of the allele combinations carrying A_i at least once: $A_i A_j$, $j = 1, \ldots, n$. Fisher's fundamental theorem states that the flux $\Phi(t) = \sum_{i=1}^{n} \overline{a_i} x_i = \overline{a}$ is a nondecreasing function of time, but the outcome of selection need not be unique as optimization might end in a local optimum of Φ (*See also* **Fitness landscape**). For example, in the two-allele case, inferiority of the *heterozygote* A_1A_2, $a_{12} < \min\{a_{11}, a_{22}\}$, results in bistability since homogenous populations of either *homozygote*, A_1A_1 or A_2A_2, represent stable equilibrium points. Then, the initial conditions determine the outcome of selection.

The optimization principle is not universally valid: when mutation is included or when more complex cases of recombination are considered, optimization of mean fitness is restricted to certain ranges of initial conditions, whereas different behavior is observed for other starting values. Still, optimization remains an important heuristic in evolution as it is frequently observed.

Innovation is introduced into genes by mutation consisting of a local change in the sequence of nucleotides resulting from an imperfect replication of genetic information or externally caused damage. Two scenarios are distinguished: (i) rare mutation treated by conventional population genetics and typically occurring with multicellular organisms and most bacteria, and (ii) frequent mutation handled by quasispecies theory (Eigen, 1971; Eigen & Schuster, 1977) and determining evolution of viruses. Higher mutation rates are often advantageous because they allow for adaptation, but there exists an error threshold of replication beyond which inheritance breaks down because too many mutations destroy the genetic message. RNA viruses are under a strong selection constraint by the host and their mutation rates are close to the error threshold.

The idea that genotypes and phenotypes are related one-to-one turned out to be wrong. Molecular genetics revealed a high degree of *neutrality* (Kimura, 1983): many different genotypes give rise to the same phenotype. Advantageous mutations are rare; deleterious mutations are eliminated by selection thus leaving a majority of observed changes in the genomes to result from neutral mutations. Neutrality gives rise to random drift of populations in genotype space, which was also found to be important for the mechanism of evolution since it allows populations to escape from minor local fitness optima or evolutionary traps (Schuster, 1996) and Schuster in Crutchfield & Schuster, 2003). Random drift leads to an almost constant mutation rate per year and nucleotide independent of the species being tantamount to a *molecular clock of evolution*. This clock is used for dating in the reconstruction of phylogenies from comparison of present-day genome sequences. Molecular clock dates yield substantially longer time spans compared with those from the fossil record. The discrepancy seems to be reconcilable because paleontological datings are too young and molecular clock datings are too old by systematic errors (Benton & Ayala, 2003).

The Darwinian mechanism is powerful because it makes no reference to the specific nature of the reproducing entities. Therefore, it is likewise valid for molecules, viruses, bacteria, or higher organisms. Selection based on the Darwinian principle is observed in many disciplines outside biology, for example, in physics and chemistry, in economics, and in the social sciences.

Since its introduction, the theory of evolution has undergone changes and modifications. The rejection of catastrophic events as an important source of change in the history of life on Earth was a political issue rather than one based on scientific data. Geological evidence for fallings of large meteorites as well as major floods is now available, and such events wiped out substantial parts of the biosphere. The paleontological record reflects the interplay between continuous evolution and external influences, which resulted in epochs of gradual development interrupted by punctuated events. Interestingly, evolution of bacteria or molecules under constant conditions also showed punctuation without external triggers: populations "wait" during quasistationary periods for rare mutations that initiate fast periods of change.

Still, there are open problems in current evolutionary theory. Recent sequence data challenge the idea of a tree of life. Although animal phylogeny appears to be on a firm basis, there are problems with the reconstruction of a tree-like history of plant species. Prokaryote evolution cannot be cast into a tree: archebacteria and eubacteria exchange genetic information across species and kingdoms. Such horizontal gene transfer occurs frequently and obscures the descendance of species. Darwinian evolution, although successful in describing the mechanisms of optimization and adaptations of species, is unable to provide explanations for the major evolutionary transitions that lead from one hierarchical level of life to the next higher forms (Maynard Smith & Szathmáry, 1995; Schuster , 1996). Examples of such transitions are the origin of the genetic code; the transition from the prokaryotic to the eukaryotic cell; the transition from unicellular organisms to multicellular plants, fungi, and animals; the transition from solitary animals to animal societies; and eventually the transition to man and human societies. Common to all these transitions is the integration of individual competitors as cooperating elements into a novel functional unit. Simple model mechanisms have been proposed that can explain cooperation of competitors (see, e.g., the hypercycle Eigen & Schuster, 1978), but no real solution to the problem has been found yet.

PETER SCHUSTER

See also **Catalytic hypercycle; Fitness landscape**

Further Reading

Benton, M.J. & Ayala, F.J. 2003. Dating the tree of life. *Science*, 300: 1698–1700

Crutchfield, J.P. & Schuster, P. (editors). 2003. *Evolutionary Dynamics: Exploring the Interplay of Selection, Accident, Neutrality, and Function*, Oxford and New York: Oxford University Press

Eigen, M. 1971. Selforganization of matter and the evolution of biological macromolecules. *Naturwissenschaften*, 58: 465–523

Eigen, M. & Schuster, P. 1977. The hypercycle. A principle of natural self-organization. Part A: Emergence of the hypercycle. *Naturwissenschaften*, 64: 541–565

Eigen, M. & Schuster, P. 1978. The hypercycle. A principle of natural self-organization. Part B: The abstract hypercycle. *Naturwissenschaften*, 65: 7–41

Hartl, D.L. & Clark, A.G. 1997. *Principles of Population Genetics*, 3rd edition, Sunderland, MA: Sinauer Associates

Judson, H.F. 1979. *The Eighth Day of Creation. The Makers of the Revolution in Biology*, London: Jonathan Cape and New York: Simon and Schuster

Kimura, M. 1983. *The Neutral Theory of Molecular Evolution*, Cambridge and New York: Cambridge University Press.

Maynard Smith, J. & Szathmáry, E. 1995. *The Major Transitions in Evolution*, Oxford and New York: Freeman

Mayr, E. 1997. The establishment of evolutionary biology as a discrete biological discipline. *BioEssays*, 19: 263–266

National Academy of Sciences (NAS). 1999. *Science and Creationism. A View from the National Academy of Sciences*, 2nd edition, Washington, DC: National Academy Press

Ruse, M. 1979. *The Darwinian Revolution*, Chicago, IL: University of Chicago Press

Ruse, M. 2001. *Can a Darwinian Be a Christian? The Relationship Between Science and Religion*, Cambridge and New York: Cambridge University Press

Schuster, P. 1996. How does complexity arise in evolution? *Complexity*, 2(1): 22–30

Spiegelman, S. 1971. An approach to the experimental analysis of precellular evolution. *Quarterly Reviews of Biophysics*, 4: 213–253

Watts, A. & Schwarz, G. (editors). 1997. *Evolutionary Biotechnology—From Theory to Experiment. Biophyscial Chemistry*, vol. 66, nos. 2–3, Amsterdam: Elsevier, pp. 67–284

BIOMOLECULAR SOLITONS

Biological molecules are complex systems that evolve in an ever-changing environment and nevertheless exhibit a remarkable stability of their functions. This feature, which is reminiscent of the exceptional stability of solitons in the presence of perturbations, is perhaps what led to suggestions that solitons could have a role in some biological functions. Beyond this analogy, there are more solid arguments to consider the role of nonlinearity in biological molecules. They are very large atomic assemblies performing their function through large conformational changes, which have a cooperative character because they involve many atoms moving in a coherent manner, and are highly nonlinear due to their amplitude of motion, which is much larger than the standard thermal motions observed in small molecules. Additional nonlinearities can originate from the coupling of different degrees of freedom, as proposed for proteins. Dispersion, necessary to balance the effect of nonlinearity in order to obtain soliton-like excitations, is introduced by the discreteness of the molecular lattice, which behaves in a manner different from a continuous medium.

Besides conformational changes involved in many biological functions, issues important for biological molecules are energy transport and storage, and charge transport. Nonlinear excitations have been proposed as possible contributors to these phenomena, in the two main classes of biological molecules: nucleic acids and proteins (Peyrard, 1995).

Following Erwin Schrödinger in his prophetic book *What Is Life?* (Schrödinger, 1944), the nucleic acid DNA can be viewed as an "aperiodic crystal." The static structure of DNA is a fairly regular pile of flat base pairs, linked by hydrogen bonds and connected by sugar–phosphate strands that form a double helix (Calladine & Drew, 1997). The lack of periodicity occurs because the base pairs can be either $A-T$ (adenine–thymine) or $G-C$ (guanine–cytosine), their sequence defining the genetic code. The static picture that emerges from crystallographic data has little to do with actual DNA in a living cell. The genetic code, buried in the double helix, would not be accessible if DNA were not a highly dynamical structure. Biologists have observed the "breathing of DNA," which is a fluctuational opening in which one or a few base pairs open temporarily. These motions are probed experimentally by monitoring deuterium-hydrogen exchange, based on the assumption that the imino-protons that bind the bases can only be exchanged for open base pairs. DNA double helix is also opened by enzymes during the transcription of a gene, that is, the reading of the genetic code.

This phenomenon is complex, but there are related experimental observations that are more amenable to physical analysis, on the thermal denaturation of DNA. When the double helix is heated, one first observes local openings over a few to a few tens of base pairs. These grow and invade the whole molecule, leading to a thermal separation of the two strands, which can be monitored by measuring the UV absorbance of the molecule, which is highly sensitive to the disturbance of the base stacking. This "DNA melting"—which appears as a phase transition in one dimension—poses challenging questions because, in order to cause the local openings, one has to break many hydrogen bonds between the bases, which requires the localization of a large amount of thermal energy in a small region of the molecule. Nonlinear effects could be at the origin of this phenomenon.

All these observations led to many investigations and models of the nonlinear dynamics of the DNA molecule. A description at the scale of the individual atoms is not necessary to analyze base-pair openings, so the bases are generally described as rigid objects in these models. The earliest attempt to describe DNA opening in terms of solitons is due to Englander et al. (1980), who viewed it as a cooperative motion involving 10 or more base pairs and propagating as a localized defect along the molecule. This idea was further formalized by Yomosa (1984), and Takeno & Homma (1983), who introduced a coupled base rotator model for the structure and dynamics of DNA. The general ideas behind this approach are schematized in Figure 1(a).

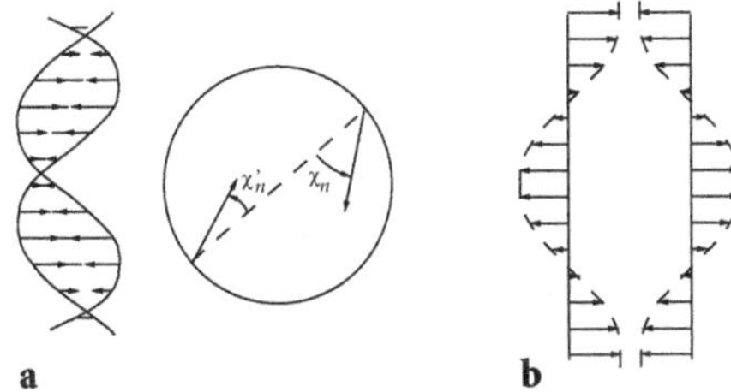

Figure 1. (a) Schematic view of the plane base rotator model for DNA. (b) A symmetric open state of the model.

Only rotational degrees of freedom of the bases are introduced (denoted by angles χ_n and χ_n' in Figure 1). The pairing of the bases is described by an on-site potential $V(\chi_n, \chi_n')$, which, in its simplest form is $V(\chi_n, \chi_n') = A(1 - \cos\chi_n) + A(1 - \cos\chi_n') + B(1 - \cos\chi_n \cos\chi_n')$, and the stacking along the molecule is represented by a potential $W(\chi_n, \chi_n', \chi_{n-1}, \chi_{n-1}' = S[1 - \cos(\chi_n - \chi_{n-1})] + S[1 - \cos(\chi_n' - \chi_{n-1}')]$, where A, B, S are constants. Adding the kinetic energy of the bases $\frac{1}{2}I(\dot\chi_n^2 + \dot\chi_n'^2)$, where I is the moment of inertia of the bases around their rotation axis, and summing over n, one obtains the Hamiltonian of the model. Various nonlinear excitations are possible depending on the symmetries of the motion (such as $\chi_n = \chi_n'$, $\chi_n = -\chi_n'$) and the values of the constants. If the stacking interaction is strong enough, a continuum approximation can be made. This approximation replaces the discrete variables $\chi_n(t)$ by the function $\chi(x, t)$ and finite differences such as $\chi_n - \chi_{n-1}$ by derivatives $a(\partial\chi/\partial x)$, where a is the spacing between the bases and x denotes the continuous coordinate along the helix axis. When $A = 0$, in its simplest form, the model leads to a sine-Gordon equation

$$I\frac{\partial^2\chi}{\partial t^2} - Sa^2\frac{\partial^2\chi}{\partial x^2} + B\sin\chi = 0, \qquad (1)$$

which has topological solutions such as the one schematized in Figure 1(b), where the bases undergo a 2π rotation, generating an open state that may slide along the chain.

Models for the rotation of the base pairs have been further refined by Yakushevich (1998) and are discussed in the entry on **DNA solitons**.

Another point of view was chosen later by Dauxois et al. (1993), who were interested in the statistical physics of DNA thermal denaturation. This problem had been studied by Ising models, which simply use a two-state variable equal to 0 or 1 to specify whether a base pair is closed (0) or open (1). Such models cannot describe the intermediate states, but they can be generalized by introducing a real variable $y_n(t)$ that measures the stretching of the hydrogen bonds in a base pair that is equal to 0 in the equilibrium structure and grows to infinity when the two strands are fully separated. With such a variable, a natural shape of the on-site potential is the Morse potential $V(y_n) = D[\exp(-\alpha y_n) - 1]^2$ (D and α are constants), which has a minimum corresponding to the binding of the two bases in their equilibrium state by the hydrogen bonds, and a plateau at large y_n, which is associated to the vanishing of the pairing force $\partial V/\partial y_n$ when the bases are far apart.

Such a model does not have topological solitons, but its nonlinear dynamics leads to localized oscillatory modes, called breathers, which are approximately described by solitons of the nonlinear Schrödinger equation in the continuum limit and turn into permanently open states at a high temperature (*See* **Breathers**). These studies have focused attention on the importance of discreteness for nonlinear energy localization. In DNA, the stacking interactions are not very strong, and this is why imino-proton exchange experiments can detect the exchange on one base pair while the neighboring base pairs are not affected. As a result a continuum approximation is very crude. What could appear as a problem because it complicates analytical studies of the nonlinear dynamics turns out to have a far-reaching consequence because it has been shown that discreteness is *crucial* for the existence and formation of nonlinear localized modes (Sievers & Takeno, 1988; MacKay & Aubry, 1994), which correspond to the "breathing" of DNA observed by biologists. It is important to notice that the existence of these nonlinear solutions is not linked to a particular mathematical expression of the potentials. Instead, it is a generic feature of nonlinear lattices having interactions qualitatively similar to those that connect the bases in DNA. Moreover, it has also been shown that thermal fluctuations can self-localize in such lattices so it is likely that related nonlinear excitations could exist in DNA. But discreteness has another consequence. Large-amplitude modes are strongly localized due to their high nonlinearity. Their width becomes of the order of the spacing between the bases and they lose the translational invariance of solitons in continuum media. The image of freely moving solitons has to be corrected by the pinning effect of discreteness, and the translation of the nonlinear excitations in DNA, if it occurs, has to be activated, for instance, by thermal fluctuations.

Proteins are much more complex than DNA because they do not have a quasi-periodic structure, but some of their substructures are nevertheless fairly regular. They are biological polymers composed of amino acids of the general formula

```
          H   H
          |   |
H−O−C−C−N−H
          ||  |
          O   R
```

where R is an organic radical that determines the amino acid. These building blocks are linked by a peptide bond that can be viewed as a result of the elimination of a water molecule between consecutive amino acids, leading to the generic formula

```
     H   H       H   H       H   H
     |   |       |   |       |   |
−C−C−N−C−C−N−C−C−N−
 ||  |       ||  |       ||  |
 O   R1      O   R2      O   R3
```

A given protein is defined by its sequence of amino acids chosen by 20 possible types and the length of the chain (typically 150–180 residues), but this so-called primary structure does not determine the function that depends on the spatial organization of the residues. Segments of the chain tend to fold into secondary structures having the shape of helices (called α-helices) or sheets (called β-sheets) stabilized by hydrogen bonds formed mainly between the negatively charged C=O groups and the positively charged protons linked to the nitrogen atom of a peptide bond. The different components of the secondary structure assemble in the tertiary structure, which is the functional form of the protein.

Proteins perform numerous functions and one of them is the storage and transport of the energy released by the hydrolysis of adenosine-triphosphate (ATP), which plays the role of the fuel necessary for many biological processes, such as muscle contraction. The hydrolysis of a single ATP molecule releases approximately 0.4 eV, which is transmitted to a protein for later use. This raises a puzzling question because, if this energy were distributed among all the degrees of freedom of a protein, each atom would carry such a small amount that the energy would be useless. There must be a mechanism that maintains this energy sufficiently localized, and moreover, as it will not be used at the site where it has been released, it must be transported efficiently within the molecule. Recent experiments at the molecular scale have shown that the hydrolysis of a single ATP molecule can be used for *several* steps of a molecular motor involved in muscle contraction (Kitamura et al., 1999), providing evidence of the temporary storage of the energy.

Attempting to understand these phenomena in 1973, Alexander Davydov noticed that the energy released by ATP hydrolysis almost coincides with 2 quanta of the vibrational energy of the C=O bond, which led him to the conclusion that this energy was stored as vibrational energy in the peptide bond (Scott, 1992). He conjectured that it could stay localized through an extrinsic nonlinearity associated with a distortion of the chain of hydrogen bonds that spans the α-helix. The underlying mechanism is similar to the one leading to the formation of a polaron in solid-state physics (Ashcroft & Mermin, 1976). The vibration of the $C = O$ bond strains the lattice in its vicinity, resulting in slight displacements of the neighboring amino acids. But, as the frequency of the $C = O$ vibration is affected by its interactions with the neighboring atoms, the frequency of the excited $C = O$ bond becomes slightly shifted and no longer coincides with the resonating frequencies of the neighboring $C = O$ bonds, preventing an efficient transfer of energy to the neighboring sites. As a consequence, the energy released by the ATP hydrolysis does not spread along the protein.

Therefore, the basic idea behind the mechanism proposed by Davydov is nonlinear energy localization due to the shift of the frequency of an oscillator when it is excited. For the protein, it is not due to an intrinsic nonlinearity of the $C = O$ bond (as was the case for the Morse potential linking the bases in a pair for DNA), but due to a coupling with another degree of freedom, which is an acoustic mode of the lattice of amino acids connected by hydrogen bonds.

As only a few quanta of the $C = O$ vibrational motion are excited, the theory cannot ignore quantum effects, and in order to go beyond the qualitative picture discussed above, one has to solve the time-dependent Schrödinger equation. Davydov proposed a simple *ansatz* to describe the quantum state of the system. In this simple approximation, the motion of the self-trapped energy packet is described by a discrete form of the nonlinear Schrödinger equation. When one introduces proper parameters for the α-helix, the calculation of the soliton width shows that it is much broader than the lattice spacing, which should allow its motion without pinning by discreteness. As a result, energy transfer by solitons in the α-helix is plausible, but a definitive conclusion about the existence of such solitons is still pending. This is because the role of thermal fluctuations, which could destroy the coherence of the lattice distortion around the excited $C = O$ site and hence the self-trapping, and the extent of quantum effects are hard to evaluate quantitatively (Peyrard, 1995). A direct experimental observation on a protein has not been possible up to now.

These uncertainties prompted physicists and physical chemists to experimentally investigate model systems that are simpler than proteins but, nevertheless, show chemical bonds comparable to the peptide bonds in proteins. Crystalline acetanilide consists of quasi-one-dimensional chains of hydrogen-bonded peptide groups. In the early 1980s, it was recognized by spectroscopic studies that the $C = O$ stretching and $N-H$ stretching bands of crystalline acetanilide exhibit anomalies, and tentative explanations involve self-trapped states similar to the Davydov solitons. These ideas have been confirmed by recent pump–probe experiments (Edler et al., 2002). A direct observation of self-trapping could be achieved, and it appears that the crystal structure is essential to stabilize the excitation that decays 20 times faster for isolated molecules than for molecules linked by hydrogen bonds in the crystal. Although the lifetime of the self-trapped state (20 ps) is shorter than expected by Davydov, this study supports the original idea of the importance of the coupling with the lattice degrees of freedom. A possible translational motion of the self-trapped state and its possible role for biological functions are still open questions.

The *News and Views* section of the journal *Nature* attests that nonlinear excitations in biomolecules have been the object of strong controversy, ranging from enthusiastic approval (Maddox, 1986, 1989) to harsh criticisms (Frank-Kamenetskii, 1987), which were justified by some of the overstatements by theoreticians. Today, passionate opinions have subsided and experiments at the scale of a single molecule have become feasible, showing us how biomolecules work or take their shape. Thus, it appears likely that while freely moving solitons along DNA or protein α-helices may not exist, nonlinear excitations leading to energy localization or storage, and perhaps transport, could well provide useful clues to understand some of the phenomena occurring in biomolecules.

MICHEL PEYRARD

See also **Davydov soliton; DNA premelting; DNA solitons; Pump-probe measurements**

Further Reading

Ashcroft, N.W. & Mermin, D.A. 1976. *Solid State Physics*, Philadelphia: Saunders Company

Calladine, C.R. & Drew, H.R. 1997. *Understanding DNA: The Molecule and How It Works*, 2nd edition, San Diego and London: Academic Press

Dauxois, T., Peyrard, M. & Bishop, A.R. 1993. Dynamics and thermodynamics of a nonlinear model for DNA denaturation. *Physical Review E*, 47: 684–695 and R44–R47

Edler, J., Hamm, P. & Scott, A.C. 2002. Femtosecond study of self-trapped vibrational excitons in crystalline acetanilide. *Physical Review Letters*, 88 (1–4): 067403

Englander, S.W., Kallenbach, N.R., Heeger, A.J., Krumhansl, J.A. & Litwin, S. 1980. Nature of the open state in long polynucleotide double helices: Possibility of soliton excitations. *Proceedings of the National Academy of Sciences USA*, 777: 7222–7227

Frank-Kamenetskii, M. 1987. Physicists retreat again. *Nature*, 328: 108

Kitamura, K., Tokunaga, M., Iwane A.H. & Yanagida, T. 1999. A single myosin head moves along an actin filament with regular steps of 5.3 nanometres. *Nature*, 397: 129–134

MacKay, R.S. & Aubry, S. 1994. Proof of existence of breathers for time-reversible or Hamiltonian networks of weakly coupled oscillators. *Nonlinearity*, 7: 1623–1643

Maddox, J. 1986. Physicists about to hi-jack DNA? *Nature*, 324: 11

Maddox, J. 1989. Towards the calculation of DNA. *Nature*, 339: 577

Peyrard, M. (editor). 1995. *Nonlinear Excitations in Biomolecules*, Berlin and New York: Springer

Schrödinger, E. 1944. *What is Life? The Physical Aspect of the Living Cell*, Cambridge: Cambridge University Press

Scott, A.C. 1992. Davydov's soliton. *Physics Reports*, 217: 1–67

Sievers, A.J. & Takeno, S. 1988. Intrinsic localized modes in anharmonic crystals. *Physical Review Letters*, 61: 970–973

Takeno, S. & Homma, S. 1983. Topological solitons and modulated structures of bases in DNA double helices. *Progress of Theoretical Physics*, 70: 308–311

Yakushevich, L.V. 1998. *Nonlinear Physics of DNA*, Chichester and New York: Wiley

Yomosa, S. 1984. Solitary excitations in deoxyribonucleic acid (DNA) double helices. *Physical Review* A, 30: 474–480

BIONS

See **Breathers**

BIRGE–SPONER RELATION

See **Local modes in molecules**

BIRKOFF–SMALE THEOREM

See **Phase space**

BISTABILITY

See **Equilibrium**

BISTABLE EQUATION

See **Zeldovich–Frank-Kamenetsky equation**

BJERRUM DEFECTS

Ice is the most common and important member in a class of solids in which the conduction of electricity is carried almost exclusively by protons. Its dc electrical conductivity reaches the level of $10^{-7}\,\Omega^{-1}\,\mathrm{m}^{-1}$, placing ice among the semiconductors. Known protonic semiconductors include disordered forms of solid water and several salt hydrates and gas hydrates. In the case of lithium hydrazinium sulfate ($LiN_2H_5SO_4$), one finds quasi-one-dimensional hydrogen-bonded chains (HBCs) along the c-crystallographic axis. Electrical conductivity is three orders of magnitude larger in the c-direction (compared with the perpendicular directions), demonstrating that proton conductivity is directly related to the presence of hydrogen bonds. In addition to inorganic crystals, protonic conductivity plays a significant role in biological systems, where it participates in energy transduction and formation of proton pumps. Of particular significance is proton transport across cellular membranes through the use of hydrogen-bonded side-chains of proteins embedded in membrane pores.

As is shown in Figures 1 and 2, protonic conductivity in HBCs takes place through ionic defects and bonding or "Bjerrum" defects (named after Danish physical chemist Niels Bjerrum). Ionic defects are formed by an excess proton (H_3O^+) or a proton vacancy (HO^-), while bonding defects are misfits in the orientations of neighboring atoms resulting in either vacant bonds (L defect) or placing two protons in the same bond (D defect). Bonding defects do not obey the Bernal–Fowler rule of one proton per hydrogen bond for the ideal ice crystal.

When a proton is transported along an HBC through an ionic defect, after the passage of the proton, the chain remains blocked to further proton movement since all chain protons have been moved, say, from the left-hand to the right-hand side of each hydrogen bond. The chain gets unblocked through cooperative rotations, that is, through the passage of a bonding defect. Thus, protons move in an HBC through coordinated ionic and Bjerrum defects, using a mechanism that is also found in hydrogen-bonded protein side chains. Coordinated proton transport in biological macromolecules leads to the formation of proton pumps that channel protons across membranes and, through reversals, produce cyclic motor actions of mechanical nature. Defects in HBCs are topological in nature, with the rotational activation energy being smaller than that of ionic defect energy. The total charge of the topological ionic and bonding defects is not the same; in ice, the ionic defect charge is $e_I = 0.64e$ (e is the proton charge), while the bonding defect charge is $e_B = 0.36e$. Only after a coordinated passage of an ionic and bonding defect is one entire proton charge transferred across the HBC.

A simple one-dimensional cooperative model of an HBC is similar to the Frenkel–Kontorova model but with two alternating barriers modeling bonding and ionic activation energies. The minima separating the barriers correspond to equilibrium positions of protons that interact mutually through dipole-dipole interactions. In equilibrium under this model, there is initially one proton per hydrogen bond; transitions of protons over the large barriers correspond to ionic defects, while bonding defects result from transitions over the smaller barrier. Both classes of defects are modeled through topological solitons. There are two kink solutions corresponding to HO^- and L-bonding defects, while the corresponding antikinks are the H_3O^+ and D-bonding defects, respectively.

This simple model can be made quantitative through the introduction of the one-dimensional Hamiltonian

$$H = \sum_n \left[\frac{p_n^2}{2} + \frac{1}{2}(u_{n+1} - u_n)^2 + \omega V(u_n) \right], \quad (1)$$

where u_n and p_n are the dimensionless displacement from an equilibrium position and momentum, respectively, of the nth hydrogen that is coupled to its nearest

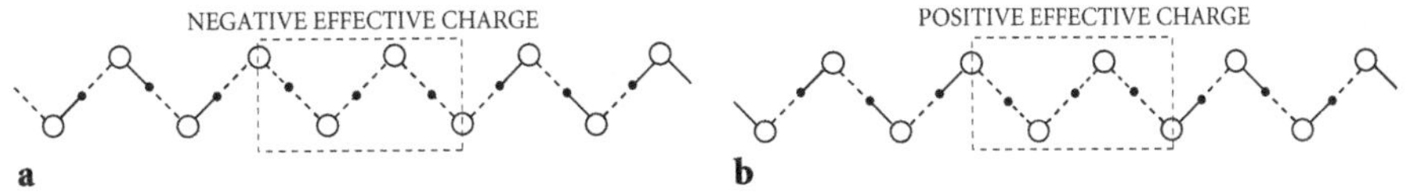

Figure 1. Ionic defects present in a hydrogen bonded chain: (a) negative ionic defect HO^- and (b) positive ionic defect H_3O^+. Large open circles denote ions, for example, oxygen ions in ice, while small black dots are protons. A hydrogen bond that links two ions contains a covalent part (solid line) that places in equilibrium the proton closer to one of the two oxygens. In the ionic defect region, there is a gradual transition in the equilibrium locations of protons within the hydrogen bonds; this transitional region is modeled through a topological soliton.

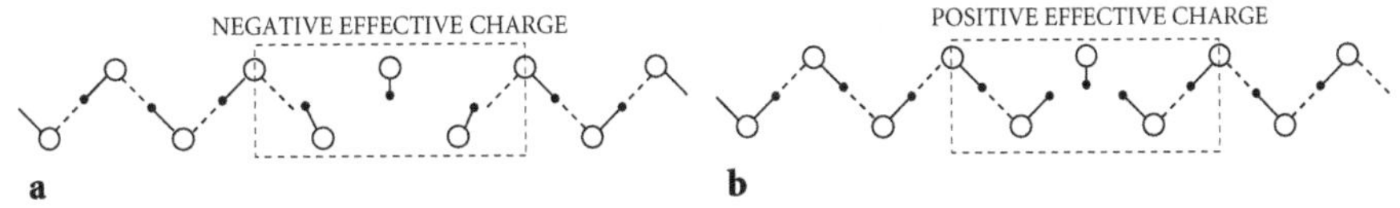

Figure 2. Bonding defects in a hydrogen bonded chain: (a) negative bonding defect (L) and (b) positive bonding defect (D). Molecular rotations introduce additional protons or remove protons from the quasi-one-dimensional HBC and produce bonding defects.

neighboring protons through harmonic spring interaction, while ω sets the energy scale. A typical choice for $V(u_n)$, the nonlinear substrate potential that models the ionic and bonding barriers, is

$$V(u_n) = \frac{2}{1-\alpha^2}\left[\cos\left(\frac{u_n}{2}\right) - \alpha\right]^2. \tag{2}$$

The substrate potential (2) is periodic and (for appropriate values of the parameter α) is doubly periodic with two distinct alternating maxima that separate degenerate minima. In this model, one assumes one proton per unit cell, the latter consisting of the larger ionic barrier with its adjacent minima, one on each side. In the strongly cooperative limit, where neighboring hydrogen displacements do not differ substantially, one obtains for the proton displacement $u(x,t)$ that becomes a function of the continuous space variable x as well as time t the double sine-Gordon partial differential equation:

$$u_{tt} - c^2 u_{xx} + \frac{\omega^2}{1-\alpha^2}\left[-\sin u + 2\alpha \sin\frac{u}{2}\right] = 0, \tag{3}$$

where c is the speed of sound of the linearized lattice oscillations. This sine-Gordon model has as solutions two sets of soliton kinks as well as their corresponding antikinks representing L-Bjerrum (kink I), D-Bjerrum (antikink I), HO^- ionic (kink II) and H_3O^+ (antikink II). More complex nonlinear models can be constructed that also include an acoustic interaction between neighboring ions as well as coupling of protons with ions. In these cases, one obtains two component solitons where the defects in the proton sublattice are topological solitons that induce a polaronic-like deformation in the ionic lattice. This more complex defect can travel along an HBC when an external electric field is applied in the system. Numerical simulations demonstrate that these nonlinear defects do indeed encompass some of the basic dynamical properties of the ionic and bonding defects found in hydrogen-bonded networks.

G.P. Tsironis

See also **Frenkel–Kontorova model; Hydrogen bond; Sine-Gordon equation; Topological defects**

Further Reading

Hobbs, P.V. 1974. *Ice Physics*, Oxford: Clarendon Press

Pnevmatikos, St. 1988. Soliton dynamics of hydrogen-bonded networks: a mechanism for proton conductivity. *Physical Review Letters*, 60: 1534–1537

Pnevmatikos, St., Tsironis, G.P. & Zolotaryuk, A.V. 1989. Nonlinear quasiparticles in hydrogen-bonded systems. *Journal of Molecular Liquids*, 41: 85–103

Savin, A.V, Tsironis, G.P. & Zolotaryuk, A.V. 1997. Reversal effects in stochastic kink dynamics. *Physical Review* E, 56: 2457–2466

Zolotaryuk, A.V., Pnevmatikos, St. & Savin, A.V. 1991. Charge transport by solitons in hydrogen-bonded materials. *Physical Review Letters*, 67: 707–710

BLACK HOLES

A massive body like the Earth is characterized by an "escape velocity," which is the speed that a moving particle must have on leaving the surface of the body to leave the attraction of gravity. Consider a bullet at the surface of the Earth that is moving upward with a speed of 11.2 km/s and neglect atmospheric friction. Such a bullet will just escape Earth's gravitational field by exchanging its kinetic energy for the potential energy of the gravitational field. If the mass of the Earth were compressed into a smaller radius, this escape velocity would be larger, because the gravitational energy to be overcome by the kinetic energy of the bullet would be larger.

In 1916, Karl Schwarzschild used Einstein's gravitational field equations to show that if a body of mass m is compressed to a radius

$$r_s = \frac{2Gm}{c^2}, \qquad (1)$$

where G is the gravitational constant, then an object traveling at the speed of light c will be unable to escape the influence of gravity (Schwarzschild, 1916). For a body having the mass of the Earth, this "Schwarzschild radius" is about 1 cm, and for the Sun, it is about 3 km. Interestingly, Schwarzschild's idea was first suggested in the 18th century (Mitchell, 1783; Laplace, 1796).

The term "black hole" was coined by John Archibald Wheeler in 1967 to denote a cosmic object with its mass concentrated within the Schwarzschild radius. Neither particles nor light can overcome gravitational attraction and travel outside the sphere of radius r_s. Interestingly, Stephen Hawking (1974) has shown that—due to quantum fluctuations—a black hole should radiate as a black body with the temperature

$$T = \frac{hc}{4\pi k r_s}, \qquad (2)$$

where k and h are, respectively, the Boltzmann and Planck constants. Indirect evidence for black holes is provided by the fact that there do not exist stable cold stars with masses larger than about three Sun masses. Under its own gravitational field, according to theory, such a star should collapse into a black hole. In 1931, Subrahmanyan Chandrasekhar was the first to conclude that above some critical mass of white dwarfs, the equation of state (of a quantum relativistic gas of degenerate Fermi particles) is too weak to counter the gravitational forces, leading to the formation of black holes. Both Lev Landau (1932) and Arthur Eddington (1924) rejected this implication of relativistic quantum mechanics rather than accept the possibility of black holes. Albert Einstein also concluded that Schwarzschild singularities do not exist in the real world (Einstein, 1939).

In 1939, however, J. Robert Oppenheimer and his colleagues used general relativity (rather than Newtonian gravity) to show that when all thermonuclear sources of energy are exhausted (with no further outward pressure due to radiation), a sufficiently heavy star will continue to contract indefinitely, never reaching equilibrium (Oppenheimer & Volkoff, 1939; Oppenheimer & Snyder, 1939). Oppenheimer et al. further noted that if one considers stellar collapse from the inside, a stationary observer sees the stellar surface moving inward to the Schwarzschild sphere and finally sees the surface freeze as it nears the Schwarzschild sphere. Moreover, they showed that observers who move inward with the collapsing matter do not observe such freezing; these observers could cross the critical surface ("event horizon") after a finite time on their own clocks, after which they have no possibility of sending a signal that could be detected by an observer located outside the collapsing matter.

Recently, the scientific history of black holes has been characterized by a rapid growth of observational, theoretical, and mathematical studies, in which the discovery of such compact objects becomes the main purpose (Thorne et al., 1986). Currently, the most important classes are black holes of stellar masses (about 3–10 solar masses) and super-massive black holes. The most convincing candidates for stellar black holes are binary X-ray sources, one component of which is an ordinary star and the other component is a black hole or neutron star (Novikov & Zeldovich, 1966). Estimates of the masses of compact objects in these systems are essentially greater than three solar masses, and one example of such a system is Cygnus X-1 (V 1357 Cyg). The present number of systems mentioned as possible candidates for black holes with stellar masses is about 20, all of which are X-ray sources in binary systems (Novikov & Frolov, 2001). In the case of super-massive black holes and nuclei of Seyfert galaxies, interpretations of the observable effects using the black hole theory seem the most simple and natural; for example, Galaxy M 31 is a black hole candidate having a mass of about 3×10^7 the Sun's mass (Novikov & Frolov, 2001).

Presently, the concept of black holes continues to be confirmed by direct observations and is used to explain observable astronomical effects related to exceptionally strong emission of energy. Thus, it is expected that in the future new astronomical objects will be detected near black holes, and new physical phenomena will be discovered that can be interpreted using the black hole concept. Along these lines, it is interesting to note recent work in which concepts from thermodynamics and information theory (such as temperature and entropy) are connected with black holes based on Hawking's ideas (Markov, 1965; Hawking, 1977). Also of interest are "artificial black holes," which do not compress a large amount of mass into a small volume, but reduce the speed of light in a moving medium to less than the speed of the medium, thereby creating an event horizon (Leonhardt, 2001).

VIATCHESLAV KUVSHINOV

See also **Binding energy; Einstein equations; General relativity**

Further Reading

Chandrasekhar, S. 1931. The maximum mass of ideal white dwarfs. *Astrophysical Journal*, 74: 81

Eddington, A. 1924. A comparison of Whitehead's and Einstein's formulae. *Nature*, 113: 192

Einstein, A. 1939. On a stationary system with spherical symmetry consisting of many gravitating masses. *Annals of Mathematics*, 40: 922

Hawking, S.W. 1974. Black hole explosions. *Nature*, 248: 30–31

Hawking, S.W. 1977. Gravitational instantons. *Physics Letters* A, 60: 81
Landau, L.D. 1932. On the theory of stars. *Physikalische Zeitschrift der Sowjetunion*, 1: 285
Laplace, P.-S. 1796. *Exposition du système du monde*, [Description of the World System], Paris
Leonhardt, U. 2001. A laboratory analogue of the event horizon using slow light in an atomic medium. *Nature*, 415: 406–409
Markov, M.A. 1965. Can the gravitational field prove essential for the theory of elementaryparticles? *Progress of Theoretical Physics*, (Suppl): 85–95
Mitchell, J. 1783. *Transactions of the Royal Society of London*, 74: 35
Novikov, I.D. & Frolov, V.P. 2001. Black holes in the Universe. *Physics-Uspekhi*, 44(3): 291
Novikov, I.D. & Zeldovich, Ya.B. 1966. *Nuovo Cimento*, 4 (Suppl.): 810
Oppenheimer, J.R. & Snyder, M. 1939. On continued gravitational contraction. *Physical Review*, 56: 455
Oppenheimer J.R. & Volkoff, G.M. 1939. On massive neutron cores. *Physical Review*, 55: 374–381
Schwarzschild, K. 1916. Über das Gravitational eines Massenpunktes nach der Einsteineschen Theory. *Sitzungsberichte der Preusischen Akademie der Wissenschaften zu Berlin, Physik-Mathematik, Kl.*: 189–196
Thorne, K.S., Price, R.H. & MacDonald, D.A. (editors). 1986. *Black Holes: The Membrane Paradigm*, New Haven: Yale University Press

BLOCH DOMAIN WALL

See **Domain walls**

BLOCH FUNCTIONS

See **Periodic spectral theroy**

BLOWOUT BIFURCATION

See **Intermittency**

BLOW-UP (COLLAPSE)

See **Development of singularities**

BOHR–SOMMERFELD QUANTIZATION

See **Quantum theory**

BOOMERONS

See **Solitons, types of**

BORN–INFELD EQUATIONS

Classical linear vacuum electrodynamics with point massive charged particles has two limiting properties: the electromagnetic energy of a point particle field is infinity, and a Lorentz force must be postulated to describe interactions between point particles and an electromagnetic field. Nonlinear vacuum electrodynamics can be free of these imperfections.

Gustav Mie (1912–1913) considered a nonlinear electrodynamics model in the framework of his "Fundamental unified theory of matter." In this theory the electron is represented by a nonsingular solution with a finite electromagnetic energy, but Mie's field equations are noninvariant under the gauge transformation for an electromagnetic four-potential (addition of the four-gradient of an arbitrary scalar function).

Max Born (1934) considered a nonlinear electrodynamics model that is invariant under the gauge transformation. A stationary electron in this model is represented by an electrostatic field configuration that is everywhere finite, in contrast to the case of linear electrodynamics when the electron's field is infinite at the singular point (see Figure 1). The central point in Born's electron is also singular because there is a discontinuity of electrical field at this point (hedgehog singularity). The full electromagnetic energy of this electron's field configuration is finite.

Born and Leopold Infeld (1934) then considered a more general nonlinear electrodynamics model, which has the same solution associated with the electron. Called Born–Infeld electrodynamics, this model is based on the Born–Infeld equations, which have the form of Maxwell's equations, including electrical and magnetic field strengths $\boldsymbol{E}$, $\boldsymbol{H}$, and inductions $\boldsymbol{D}$, $\boldsymbol{B}$ with nonlinear constitutive relations $\boldsymbol{D}=\boldsymbol{D}(\boldsymbol{E},\boldsymbol{B})$, $\boldsymbol{H}=\boldsymbol{H}(\boldsymbol{E},\boldsymbol{B})$ of a special kind. For inertial reference frames and in the region outside of field singularities, these equations are

$$\left\{\begin{aligned} \operatorname{div}\boldsymbol{B} &= 0, \\ \frac{1}{c}\frac{\partial \boldsymbol{B}}{\partial t} + \operatorname{curl}\boldsymbol{E} &= 0, \\ \operatorname{div}\boldsymbol{D} &= 0, \\ \frac{1}{c}\frac{\partial \boldsymbol{D}}{\partial t} - \operatorname{curl}\boldsymbol{H} &= 0, \end{aligned}\right. \tag{1}$$

where

$$\begin{aligned} \boldsymbol{D} &= \frac{1}{\mathcal{L}}(\boldsymbol{E}+\chi^2\mathcal{J}\boldsymbol{B}), \\ \boldsymbol{H} &= \frac{1}{\mathcal{L}}(\boldsymbol{B}-\chi^2\mathcal{J}\boldsymbol{E}), \end{aligned} \tag{2}$$

$$\begin{aligned} &\mathcal{L} \equiv \sqrt{|1-\chi^2\mathcal{I}-\chi^4\mathcal{J}^2|}, \\ &\mathcal{I} = \boldsymbol{E}\cdot\boldsymbol{E}-\boldsymbol{B}\cdot\boldsymbol{B}, \quad \mathcal{J} = \boldsymbol{E}\cdot\boldsymbol{B}. \end{aligned} \tag{3}$$

Relations (2) can be resolved for $\boldsymbol{E}$ and $\boldsymbol{H}$:

$$\begin{aligned} \boldsymbol{E} &= \frac{1}{\mathcal{H}}\left(\boldsymbol{D}-\chi^2\boldsymbol{\mathcal{P}}\times\boldsymbol{B}\right), \\ \boldsymbol{H} &= \frac{1}{\mathcal{H}}\left(\boldsymbol{B}+\chi^2\boldsymbol{\mathcal{P}}\times\boldsymbol{D}\right), \end{aligned} \tag{4}$$

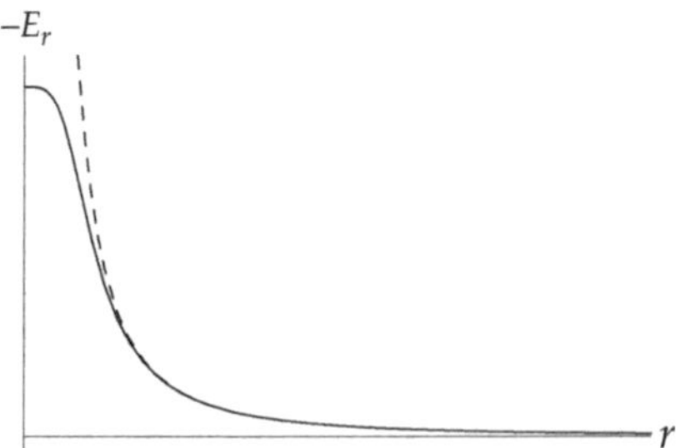

Figure 1. Radial components of electrical field for Born's electron and purely Coulomb field (dashed).

where $\mathcal{H}=\sqrt{1+\chi^2\left(\boldsymbol{D}^2+\boldsymbol{B}^2\right)+\chi^4\mathcal{P}^2}$, $\mathcal{P}\equiv\boldsymbol{D}\times\boldsymbol{B}$. Using relations (4) for Eq. (1), the fields $\boldsymbol{D}$ and $\boldsymbol{B}$ are unknown.

The symmetrical energy-momentum tensor for Born–Infeld equations has the following components:

$$T^{00}=\frac{1}{4\pi\,\chi^2}\left(\mathcal{H}-1\right),\quad T^{0i}=\frac{1}{4\pi}\mathcal{P}^i,$$
$$T^{ij}=\frac{1}{4\pi}\left\{\delta^{ij}\left[\boldsymbol{D}\cdot\boldsymbol{E}+\boldsymbol{B}\cdot\boldsymbol{H}-\chi^{-2}\left(\mathcal{H}-1\right)\right]\right.$$
$$\left.-\left(D^i\,E^j+B^i\,H^j\right)\right\}. \qquad (5)$$

In spherical coordinates, the field of Born's static electron solution may have only radial components

$$D_r=\frac{e}{r^2},\quad E_r=\frac{e}{\sqrt{\bar{r}^4+r^4}}, \qquad (6)$$

where e is the electron's charge and $\bar{r}\equiv\sqrt{|\chi\,e|}$. At the point $r=0$, the electrical field has the maximum absolute value

$$|E_r(0)|=\frac{|e|}{\bar{r}^2}=\frac{1}{\chi}, \qquad (7)$$

which Born and Infeld called the absolute field constant.

The energy of field configuration (6) is

$$m=\int T^{00}\,\mathrm{d}V=\frac{2}{3}\beta\frac{\bar{r}^3}{\chi^2}, \qquad (8)$$

where the volume integral is calculated over the whole space, and

$$\beta\equiv\int_0^\infty\frac{\mathrm{d}r}{\sqrt{1+r^4}}=\frac{\left[\Gamma(\frac{1}{4})\right]^2}{4\sqrt{\pi}}\approx 1.8541. \qquad (9)$$

In view of the definition for $\bar{r}$ below (6), Equation (8) yields

$$\bar{r}=\frac{2}{3}\beta\frac{e^2}{m}. \qquad (10)$$

Considering m as the mass of electron and using (7), Born & Infeld (1934) estimated the absolute field constant $\chi^{-1}\approx 3\times 10^{20}$ V/m. Later, Born & Schrödinger (1935) gave a new estimate (two orders of magnitude less) based on some considerations taking into account the spin of the electron. (Of course, such estimates may be corrected with more detailed models.)

An electrically charged solution of the Born–Infeld equations can be generalized to a solution with the singularity having both electrical and magnetic charges (Chernitskii, 1999). A corresponding hypothetical particle is called a dyon (Schwinger, 1969). Nonzero (radial) components of fields for this solution have the form

$$D_r=\frac{\mathfrak{E}}{r^2},\quad E_r=\frac{\mathfrak{E}}{\sqrt{\bar{r}^4+r^4}},$$
$$B_r=\frac{\mathfrak{M}}{r^2},\quad H_r=\frac{\mathfrak{M}}{\sqrt{\bar{r}^4+r^4}}, \qquad (11)$$

where $\mathfrak{E}$ is the electric charge and $\mathfrak{M}$ is the magnetic one; $\bar{r}\equiv\left[\chi^2\left(\mathfrak{E}^2+\mathfrak{M}^2\right)\right]^{1/4}$. The energy of this solution is given by Equation (8) with this definition for $\bar{r}$. It should be noted that space components of electromagnetic potential for the static dyon solution have a line singularity.

A generalized Lorentz force appears when a small, almost constant field $\tilde{\boldsymbol{D}}$, $\tilde{\boldsymbol{B}}$ is considered in addition to the moving dyon solution. The sum of the field $\tilde{\boldsymbol{D}}$, $\tilde{\boldsymbol{B}}$ and the field of the dyon with varying velocity is taken as an initial approximation to some exact solution. Conservation of total momentum gives the following trajectory equation (Chernitskii, 1999):

$$m\frac{\mathrm{d}}{\mathrm{d}t}\frac{\boldsymbol{v}}{\sqrt{1-\boldsymbol{v}^2}}=\mathfrak{E}\left(\tilde{\boldsymbol{D}}+\boldsymbol{v}\times\tilde{\boldsymbol{B}}\right)$$
$$+\mathfrak{M}\left(\tilde{\boldsymbol{B}}-\boldsymbol{v}\times\tilde{\boldsymbol{D}}\right), \qquad (12)$$

where $\boldsymbol{v}$ is the velocity of the dyon, and m is the energy for static dyon defined by (8).

A solution with two dyon singularities (called a bidyon) having equal electric ($\mathfrak{E}=e/2$) and opposite magnetic charges can be considered as a model for a charged particle with spin (Chernitskii, 1999). Such a solution has both angular momentum and magnetic moment.

A plane electromagnetic wave with arbitrary polarization and form in the direction of propagation (without coordinate dependence in a perpendicular plane) is an exact solution to the Born–Infeld equations. The simplest case assumes one nonzero component of the vector potential ($A_y\equiv\phi(t,x)$), whereupon Equations (1) reduce to the linearly polarized plane wave equation

$$\left(1+\chi^2\,\phi_x^2\right)\phi_{tt}-\chi^2\,2\,\phi_x\,\phi_t\,\phi_{xt}$$
$$-\left(c^2-\chi^2\,\phi_t^2\right)\phi_{xx}=0 \qquad (13)$$

with indices indicating partial derivatives. Sometimes called the Born–Infeld equation, Equation (13) has solutions $\phi=\zeta(x^1-x^0)$ and $\phi=\zeta(x^1+x^0)$, where $\zeta(x)$ is an arbitrary function (Whitham, 1974). Solutions comprising two interacting waves propagating in opposite directions are obtained via a hodograph transform (Whitham, 1974). Brunelli & Ashok (1998) have found a Lax representation for solutions of this equation.

A solution to the Born–Infeld equations, which is the sum of two circularly polarized waves propagating in different directions, was obtained by Erwin Schrödinger (1943).

Equations (1) with relations (2) have an interesting characteristic equation (Chernitskii, 1998)

$$g^{\mu\nu}\frac{\partial\Phi}{\partial x^{\mu}}\frac{\partial\Phi}{\partial x^{\nu}}=0, \quad g^{\mu\nu}\equiv g^{\mu\nu}-4\pi\,\chi^{2}\,T^{\mu\nu}, \quad (14)$$

where $\Phi(x^{\mu})=0$ is an equation of the characteristic surface and $T^{\mu\nu}$ are defined by (5). This form for $g^{\mu\nu}$, including in addition the energy-momentum tensor, is special for Born–Infeld equations.

The Born–Infeld model also appears in the quantized string theory (Fradkin & Tseytlin, 1985) and in Einstein's unified field theory with a nonsymmetrical metric (Chernikov & Shavokhina, 1986). In general, this nonlinear electrodynamics model is connected with ideas of space-time geometrization and general relativity (see Eddington, 1924; Chernitskii, 2002).

ALEXANDER A. CHERNITSKII

See also **Einstein equations; Hodograph transform; Matter, nonlinear theory of; String theory**

Further Reading

Born, M. 1934. On the quantum theory of the electromagnetic field. *Proceedings of the Royal Society of London* A, 143: 410–437

Born, M. & Infeld, L. 1934. Foundation of the new field theory. *Proceedings of the Royal Society of London* A, 144: 425–451

Born, M. & Schrödinger, E. 1935. The absolute field constant in the new field theory. *Nature*, 135: 342

Brunelli, J.C. & Ashok, D. 1998. A Lax representation for Born–Infeld equation. *Physics Letters* B, 426: 57–63

Chernikov, N.A. & Shavokhina, N.S. 1986. The Born–Infeld theory as part of Einstein's unified field theory. *Soviet Mathematics*, (Izvestiya Vgsikish Uchebnykh Zaverdenii), 30(4): 81–83

Chernitskii, A.A. 1998. Light beams distortion in nonlinear electrodynamics. *Journal of High Energy Physics*, 11, 15: 1–5

Chernitskii, A.A. 1999. Dyons and interactions in nonlinear (Born–Infeld) electrodynamics. *Journal of High Energy Physics*, 12, 10: 1–34

Chernitskii, A.A. 2002. Induced gravitation as nonlinear electrodynamics effect. *Gravitation & Cosmology*, 8 (Suppl.), 123–130

Eddington, A.S. 1924. *The Mathematical Theory of Relativity*, Cambridge: Cambridge University Press

Fradkin, R.S. & Tseytlin, A.A. 1985. Nonlinear electrodynamics from quantized strings. *Physics Letters* B, 163: 123–130

Mie, G. 1912–13. Grundlagen einer theorie der materie. *Annalen der Physik*, 37: 511–534; 39: 1–40: 40: 1–66

Schrödinger, E. 1942. Dynamics and scattering-power of Born's electron. *Proceedings of the Royal Irish Academy* A, 48: 91–122

Schrödinger, E. 1943. A new exact solution in non-linear optics (two-wave system). *Proceedings of the Royal Irish Academy* A, 49: 59–66

Schwinger, J. 1969. A magnetic model of matter. *Science*, 165: 757–761

Whitham, G.B. 1974. *Linear and Nonlinear Waves*, New York: Wiley

BOSE–EINSTEIN CONDENSATION

Bose–Einstein condensation (BEC) is the occupation of a single quantum state by a large number of identical particles, which implies that the particles are bosons, satisfying Bose–Einstein statistics and allowing for many particles to pile up in the same quantum state. This is in contrast to fermions, satisfying Fermi–Dirac statistics, for which the Pauli exclusion principle forbids the occupation of any single quantum state by more than one particle.

The role of quantum correlations caused by Bose–Einstein statistics is crucial for the occurrence of BEC. These statistics were advanced by Satyendranath Bose (1924) for photons, having zero mass, and generalized by Albert Einstein (1924) to particles with nonzero masses. Einstein (1925) also described the phenomenon of condensation in ideal gases. The possibility of BEC in weakly nonideal gases was theoretically demonstrated by Nikolai Bogolubov (1947). The wave function of Bose-condensed particles in dilute gases satisfies the Gross–Pitaevskii equation, suggested by Gross (1961) and Pitaevskii (1961). Its mathematical structure is that of the nonlinear Schrödinger equation. Experimental evidence of BEC in weakly interacting confined gases was achieved 70 years after Einstein's prediction, almost simultaneously, by three experimental groups (Anderson et al., 1995; Bradley et al., 1995; Davis et al., 1995).

To say that many particles are in the same quantum state implies that these particles display state coherence, a particular example of coherence phenomena requiring the particles to be strongly correlated with each other. The necessary conditions may be qualitatively understood by applying the de Broglie duality arguments to an ensemble of atoms in thermodynamic equilibrium at temperature T. Then the thermal energy of an atom is given by k_BT, where k_B is the Boltzmann constant. This energy defines the thermal wavelength

$$\lambda_T \equiv \sqrt{2\pi\hbar^2/m_0k_BT} \quad (1)$$

for an atom of mass m_0, with $\hbar$ being the Planck constant. Thus, an atom can be associated with a matter wave characterized by the wavelength (λ_T). Atoms become correlated with each other when their related waves overlap, which requires that the wavelength be larger than the mean interatomic distance, $\lambda_T > a$. The average atomic density $\rho \equiv N/V$ for N atoms in volume V is related to the mean distance a through the equality $\rho a^3 = 1$. Hence, condition $\lambda_T > a$ may be rewritten as $\rho\lambda_T^3 > 1$. With the thermal wavelength (1), this yields the inequality

$$T < \frac{2\pi\hbar^2}{m_0 k_B}\rho^{2/3}, \tag{2}$$

which implies that state coherence may develop if the temperature is sufficiently low or the density of particles is sufficiently high.

An accurate description of BEC for an ideal gas is based on the Bose–Einstein distribution

$$n(\boldsymbol{p}) = \left[\exp\left(\frac{\varepsilon_p - \mu}{k_B T}\right) - 1\right]^{-1}, \tag{3}$$

describing the density of particles with a single-particle energy $\varepsilon_p = p^2/2m_0$ for a momentum $\boldsymbol{p}$ and with a chemical potential μ. The latter is defined from the condition $N = \sum_p n(\boldsymbol{p})$ for the total number of particles. Assuming the thermodynamic limit

$$N \to \infty, \qquad V \to \infty, \qquad \frac{N}{V} \to \text{const}$$

allows the replacement of summation over $\boldsymbol{p}$ by integration. Then, the fraction of particles, condensing to the state with $\boldsymbol{p} = 0$ is

$$n_0 \equiv \frac{N_0}{N} = 1 - \left(\frac{T}{T_c}\right)^{3/2} \tag{4}$$

below the condensation temperature

$$T_c = \frac{2\pi\hbar^2\rho^{2/3}}{m_0 k_B \zeta^{2/3}}, \tag{5}$$

where $\zeta \approx 2.612$. Above the critical temperature (5), $n_0 = 0$. The latter is about half of the right-hand side of inequality (2).

The condensate fraction (4) is derived for an ideal (noninteracting) Bose gas. A weakly nonideal (weakly interacting) Bose gas also displays Bose–Einstein condensation, although particle interactions deplete the condensate so that at zero temperature the condensate fraction is smaller than unity ($n_0 < 1$). A system is called weakly interacting if the characteristic interaction radius r_{int} is much shorter than the mean interparticle distance ($r_{\text{int}} \ll a$). This inequality can be rewritten as $\rho r_{\text{int}}^3 \ll 1$, and such a system is termed dilute.

Superfluid liquids, such as liquid ^{4}He, are far from being dilute, but it is commonly believed that the phenomenon of superfluidity is somehow connected with BEC. Although an explicit relation between the superfluid and condensate fractions is not known, theoretical calculations and experimental observations for superfluid helium estimate the condensate fraction at $T = 0$ as $n_0 \approx 0.1$.

A strongly correlated pair of fermions can be treated approximately as a boson, allowing superfluidity in liquid ^{3}He to be interpreted as the condensation of coupled fermions. Similarly, superconductivity is often compared with the condensation of the Cooper pairs that are formed by correlated electrons or holes. One should understand, however, that the superconductivity of fermions is analogous to but not identical to BEC of bosons.

An ideal object for the experimental observation of BEC is a dilute atomic Bose gas confined in a trap and cooled down to temperatures satisfying condition (2). Such experiments with different atomic gases have been recently realized, BEC has been explicitly observed, and a variety of its features have been carefully investigated. It has been demonstrated that the system of Bose-condensed atoms displays a high level of state coherence.

There exist different types of traps (single- and double-well), magnetic, optical, and their combinations, which make it possible to confine atoms for sufficiently long times of up to 100s. Using a standing wave of laser light, multi-well periodic effective potentials called optical lattices have been obtained, which have allowed the demonstration of a number of interesting effects, including Bloch oscillations, Landau–Zener tunneling, Josephson current, Wannier–Stark ladders, Bragg diffraction, and so on.

Displaying a high level of state coherence, an ensemble of Bose-condensed atoms forms a matter wave that is analogous to a coherent electromagnetic wave from a laser. Therefore, a device emitting a coherent beam of Bose atoms is called an atom laser.

The realization of BEC of dilute trapped gases is important for several reasons. First, this demonstrated the phenomenon predicted by Einstein in the 1920s. Note that a direct observation of BEC in superfluid helium—despite enormous experimental efforts—has never been achieved. Second, dilute atomic gases are simple statistical systems that can serve as a touchstone for testing different theoretical approaches. Finally, Bose-condensed trapped gases display a variety of interesting properties that promise diverse practical applications.

V.I. YUKALOV

See also **Coherence phenomena; Critical phenomena; Lasers; Nonequilibrium statistical mechanics; Nonlinear optics; Nonlinear Schrödinger equations; Order parameters; Phase transitions; Quantum nonlinearity; Quantum theory; Superconductivity; Superfluidity**

Further Reading

Anderson, M.H., Ensher, J.R., Matthews, M.R., Wieman, C.E. & Cornell, E.A. 1995. Observation of Bose–Einstein condensation in a dilute atomic vapor. *Science*, 269: 198–201

Bogolubov, N.N. 1947. On the theory of superfluidity. *Journal of Physics*, 11: 23–32

Bose, S.N. 1924. Plancks gesetz und lichtquantenhypothese. *Zeitschrift für Physik*, 26: 178–181

Bradley, C.C., Sackett, C.A., Tollett, J.J. & Hulet, R.G. 1995. Evidence of Bose–Einstein condensation in an atomic gas with attractive interactions. *Physical Review Letters*, 75: 1687–1690
Coleman, A.J. & Yukalov, V.I. 2000. *Reduced Density Matrices*, Berlin: Springer
Courteille, P.W., Bagnato, V.S. & Yukalov, V.I. 2001. Bose–Einstein condensation of trapped atomic gases. *Laser Physics*, 11: 659–800
Dalfovo, F., Giorgini, S., Pitaevskii, L.P. & Stringari, S. 1999. Theory of Bose–Einstein condensation in trapped gases. *Reviews of Modern Physics*, 71: 463–512
Davis, K.B., Mewes, M.O., Andrews, M.R., van Drutten, N.J., Durfee, D.S., Kurn, D.M. & Ketterle, W. 1995. Bose–Einstein condensation in a gas of sodium atoms. *Physical Review Letters*, 75: 3969–3973
Einstein, A. 1924. Quantentheorie des einatomigen idealen gases. *Sitzungsberichte der Preussischen Akademie der Wissenschaften, Physik-Mathematik*, 261–267
Einstein, A. 1925. Quantentheorie des einatomigen idealen gases. Zweite abhandlung. *Sitzungsberichte der Preussischen Akademie der Wissenschaften, Physik-Mathematik*, 3–14
Gross, E.P. 1961. Structure of a quantized vortex in boson systems. *Nuovo Cimento*, 20: 454–477
Huang, K. 1963. *Statistical Mechanics*, New York: Wiley
Klauder, J.R. & Skagerstam, B.S. 1985. *Coherent States*, Singapore: World Scientific
Lifshitz, E.M. & Pitaevskii, L.P. 1980. *Statistical Physics: Theory of Condensed State*, Oxford: Pergamon
Nozières, P. Pines, D. 1990. *Theory of Quantum Liquids: Superfluid Bose Liquids*, Redwood, CA: Addison-Wesley
Parkins, A.S. & Walls, D.F. 1998. The physics of trapped dilute-gas Bose–Einstein condensates. *Physics Reports*, 303: 1–80
Pitaevskii, L.P. 1961. Vortex lines in an imperfect Bose gas. *Journal of Experimental and Theoretical Physics*, 13: 451–455
Ter Haar, D. 1977. *Lectures on Selected Topics in Statistical Mechanics*, Oxford: Pergamon
Yukalov, V.I. & Shumovsky, A.S. 1990. *Lectures on Phase Transitions*, Singapore: World Scientific
Ziff, R.M., Uhlenbeck, G.E. & Kac, M. 1977. The ideal Bose-Einstein gas revisited. *Physics Reports*, 32: 169–248

BOSONS

See **Quantum nonlinearity**

BOUNDARY LAYERS

The Navier–Stokes system is the basic mathematical model for viscous incompressible flows. It reads

$$(\mathrm{NS}_\nu)\qquad \begin{cases} \partial_t \boldsymbol{u} + \boldsymbol{u}\cdot\nabla\boldsymbol{u} - \nu\Delta\boldsymbol{u} + \nabla p = 0, \\ \mathrm{div}(\boldsymbol{u}) = 0, \\ \boldsymbol{u} = 0 \quad \text{on} \quad \partial\Omega, \end{cases} \tag{1}$$

where $\boldsymbol{u}$ is the velocity, p is the pressure, and ν is the viscosity. We can define a typical length scale L and a typical velocity U. The dimensionless parameter or Reynolds number, $Re = UL/\nu$, is very important to compare the properties of different flows. Indeed, two flows having the same Re have the same properties. When Re is very large (ν very small), the Navier–Stokes system (NS_ν) behaves like the Euler system

$$(\mathrm{Euler})\qquad \begin{cases} \partial_t \boldsymbol{U} + \boldsymbol{U}\cdot\nabla\boldsymbol{U} + \nabla p = 0, \\ \mathrm{div}(\boldsymbol{U}) = 0, \\ \boldsymbol{U}.n = 0 \quad \text{on} \quad \partial\Omega. \end{cases} \tag{2}$$

In the region close to the boundary, the length scale becomes very small and we cannot neglect viscous effects. In 1905, Ludwig Prandtl suggested that there exists a thin layer called the boundary layer, where the solution $\boldsymbol{u}$ undergoes a sharp transition from a solution to the Euler system to the no-slip boundary condition $\boldsymbol{u}=0$ on $\partial\Omega$ of the Navier–Stokes system. In other words, $\boldsymbol{u} = \boldsymbol{U} + \boldsymbol{u}_{\mathrm{BL}}$ where $\boldsymbol{u}_{\mathrm{BL}}$ is small except near the boundary.

To illustrate this, we consider a two-dimensional (planar) flow $\boldsymbol{u} = (u, v)$ in the half-space $\{(x, y) \mid y > 0\}$ subject to the following initial condition $\boldsymbol{u}(t=0, x, y) = \boldsymbol{u}_0(x, y)$, boundary condition $\boldsymbol{u}(t, x, y=0) = 0$, and $\boldsymbol{u} \to (U_0, 0)$ when $y \to \infty$.

Taking the typical length and velocity of order one, the Reynolds number reduces to $Re = \nu^{-1}$. Let $\varepsilon = Re^{-1/2} = \sqrt{\nu}$.

Near the boundary, the Euler system is not a good approximation. We introduce new independent variables and new unknowns

$$\tilde{t} = t, \quad \tilde{x} = x, \quad \tilde{y} = \frac{y}{\varepsilon},$$

$$(\tilde{u}, \tilde{v})(\tilde{t}, \tilde{x}, \tilde{y}) = \left(u, \frac{v}{\varepsilon}\right)(\tilde{t}, \tilde{x}, \varepsilon\tilde{y}).$$

Notice that when $\tilde{y}$ is of order one, $y = \varepsilon\tilde{y}$ is of order ε. Rewriting the Navier–Stokes system in terms of the new variables and unknowns yields

$$\begin{cases} \tilde{u}_{\tilde{t}} + \tilde{u}\tilde{u}_{\tilde{x}} + \tilde{v}\tilde{u}_{\tilde{y}} - \tilde{u}_{\tilde{y}\tilde{y}} - \varepsilon^2\tilde{u}_{\tilde{x}\tilde{x}} + p_{\tilde{x}} = 0, \\ \varepsilon^2(\tilde{v}_{\tilde{t}} + \tilde{u}\tilde{v}_{\tilde{x}} + \tilde{v}\tilde{v}_{\tilde{y}} - \tilde{v}_{\tilde{y}\tilde{y}}) - \varepsilon^4\tilde{v}_{\tilde{x}\tilde{x}} + p_{\tilde{y}} = 0, \\ \tilde{u}_{\tilde{x}} + \tilde{v}_{\tilde{y}} = 0. \end{cases} \tag{3}$$

Neglecting the terms of order ε^2 and ε^4 yields

$$\begin{cases} \tilde{u}_{\tilde{t}} + \tilde{u}\tilde{u}_{\tilde{x}} + \tilde{v}\tilde{u}_{\tilde{y}} - \tilde{u}_{\tilde{y}\tilde{y}} + p_{\tilde{x}} = 0, \\ p_{\tilde{y}} = 0, \qquad \tilde{u}_{\tilde{x}} + \tilde{v}_{\tilde{y}} = 0. \end{cases} \tag{4}$$

Since p does not depend on $\tilde{y}$, we deduce that the pressure does not vary within the boundary layer and can be recovered from the Euler system (2) when $y=0$, namely $p_x(t, x) = -(U_t + UU_x)(t, x, y=0)$, since $V(t, x, y=0) = 0$.

Going back to the old variables, we obtain

$$\begin{cases} u_t + uu_x + vu_y - \nu u_{yy} + p_x = 0, \\ u_x + v_y = 0 \end{cases} \tag{5}$$

which is the so-called Prandtl system. It should be supplemented with the following boundary conditions:

$$\begin{cases} u(t, x, y=0) = v(t, x, y=0) = 0, \\ (u, v)(t, x, y) \to (U(t, x, 0), 0) \quad \text{as} \quad y \to \infty. \end{cases} \tag{6}$$

Formally, a good approximation of $\boldsymbol{u}$ should be $\boldsymbol{U}+\boldsymbol{u}_{\mathrm{BL}}$, where U is the solution of the Euler system (2) and $\boldsymbol{u}_{\mathrm{BL}}+\boldsymbol{U}(t,x,0)$ is the solution of the Prandtl system (5), (6).

Replacing the Navier–Stokes system by the Euler system in the interior and the Prandtl system near the boundary requires a justification. Mathematically, this can be formulated as a convergence theorem when ν goes to 0; namely, $\boldsymbol{u}-(\boldsymbol{U}+\boldsymbol{u}_{\mathrm{BL}})$ goes to 0 when ν goes to 0 in L^{∞} or in some energy space (see Masmoudi, 1998 for a special case). In its whole generality, this is still a major open problem in fluid mechanics. This is due to problems related to the well-posedness of the Prandtl system as well as problems related to the instability of some solutions to the Prandtl system, which may prevent the convergence.

Let us explain the first problem for the steady Prandtl system

$$\begin{cases} uu_x+vu_y-\nu u_{yy}+p_x=0, \\ u_x+v_y=0 \end{cases} \tag{7}$$

in $\Omega=\{(x,y)\,|\,0<x<X\,,\,y>0\}$ subject to the following extra boundary condition $u(x=0,y)=u_0(y)$. Here, x should be thought of as a time-like variable. If we assume that $U,u_0\geq 0$ and $u>0$ if $y>0$, we can introduce the von Mises transformation

$$(x,y)\to(x,\psi)\quad\text{and}\quad w=u^2,$$

where $\psi_y=u$, $\psi_x=-v$ and $\psi(x,0)=0$. In (x,ψ), the steady Prandtl system reads

$$w_x=\nu\sqrt{w}\,w_{\psi\psi}-2p_x,$$

which is a degenerate parabolic equation, with the boundary conditions

$$w(x,0)=0,\quad w(0,\psi)=w_1(\psi),$$
$$w(x,\psi)\to U^2(x)\quad\text{as }\psi\to\infty,$$

where $w_1(\int_0^y u_1(s)\mathrm{d}s)=u_1^2(y)$. Using this new equation, one can prove existence for the steady Prandtl system (see Oleinik and Samokhin, 1999). In the case of favorable pressure gradient, namely $p_x\leq 0$, the solution is global ($X=+\infty$). If $p_x>0$, then a separation of the boundary layer may occur. x_0 is said to be a point of separation if $u_y(x_0,0)=0$ and $u_y(x,0)>0$ for $0<x<x_0$. Qualitatively, the separation of the boundary layer is caused by a downward flow that drives the boundary layer away from the boundary. In that case, the assumption that the tangential velocity is large compared with the normal one is not valid, and the derivation of the Prandtl system is not justified.

A second obstacle to the convergence can come from the instability of the solution to the Prandtl system itself, if we consider a two-dimensional shear flow $\boldsymbol{u}^{\mathrm{s}}=(u^{\mathrm{s}}(y),0)$, which is a steady solution of the Euler system. It is well known that the linear stability of such a flow is linked to the presence of inflection points in the profile u^{s}. A necessary condition of instability is that the profile has an inflection point. The solution to the Prandtl system with initial data $\boldsymbol{u}^{\mathrm{s}}$ and $U=0$ is just the solution of a heat equation $u_t=\nu u_{yy}$. If the profile $\boldsymbol{u}^{\mathrm{s}}$ is linearly unstable for the Euler system, then u is not a good approximation of the Navier–Stokes system when ν goes to 0 (see Grenier, 2000).

The boundary layer theory is a very powerful tool in asymptotic analysis and is present in many different fields of partial differential equations, including the magnetohydrodynamic flow boundary layer. In fluid mechanics and atmospheric dynamics, we can also mention the Ekman layer, which is due to the balance between the viscosity and the rapid rotation of a fluid (Coriolis forces). In kinetic theory, systems of conservation laws, passage to the limit from a parabolic to a hyperbolic system, different types of boundary layers arise.

NADER MASMOUDI

See also **Fluid dynamics; Navier–Stokes equation**

Further Reading

Grenier, E. 2000. On the nonlinear instability of Euler and Prandtl equations. *Communications on Pure and Applied Mathematics*, (53): 1067–1091

Grenier, E. & Masmoudi, N. 1997. Ekman layers of rotating fluids, the case of well prepared initial data. *Communications in Partial Differential Equations*, 22: 953–975

Masmoudi, N. 1998. The Euler limit of the Navier-Stokes equations, and rotating fluids with boundary. *Archive Rational Mechanics and Analysis*, 142(4): 375–394

Oleinik, O.A. & V.N. Samokhin. 1999. *Mathematical Models in Boundary Layer Theory*, Boca Raton, FL: Chapman & Hall

Prandtl, L. 1905. Mathematiker-Kongresses. *Boundary Layer*, Heidelberg: *Verhandlung Internationalen*, pp. 484–494

Weinan, W.E. 2000. Boundary layer theory and the zero-viscosity limit of the Navier–Stokes system. *Acta Mathematica Sinica*, 16(2): 207–218

BOUNDARY VALUE PROBLEMS

For a given ordinary or partial differential equation, a boundary value problem (BVP) requires finding a solution of the equation valid in a bounded domain and satisfying a set of given conditions on the boundary of a domain. To define a boundary value problem, therefore, one needs to give an equation, a domain, and an appropriate number of functions supported on the boundary of the given domain, defining the boundary conditions. For example,

$q_t+q_x+q_{xxx}=0$, the PDE,
$x\in[0,\infty)$, $t\in[0,T]$, the domain,
$q(x,0)=f_0(x)$, $q(0,t)=g_0(t)$, the boundary conditions.

Finding a solution of a given BVP is more difficult than finding a function that satisfies the PDE, because of the constraint imposed on the solution at the boundary of the domain. Indeed, for nonlinear equations, there exists no general method to find the solution of a given BVP. The question of the solvability of such a problem must also be addressed, and in general, it does not have an easy answer (in the above example, if we prescribe two rather than just one condition at the boundary $x=0$, there exists, in general, no solution). The existence and uniqueness of the solution of a given BVP can be guaranteed only when a specific, well-defined number of boundary conditions are prescribed, and this number depends on the highest-order derivatives appearing in the equation with respect to each variable (Coddington & Levinson, 1955).

For linear ordinary differential equations (ODEs), there is a general methodology for solving a BVP, based on defining the particular solution of a related problem. This particular solution is called the Green function associated with the BVP, and it depends on the boundary conditions. The Green function is used to define an integral operator, and if this operator is sufficiently regular, one can use it to express the solution of the original problem (Stackgold, 1979). This approach is powerful and fairly general, but it is not always successful, and it cannot be used for nonlinear equations.

No general methods are available to construct solutions for nonlinear ODEs or even to assert their existence. Most techniques rely on perturbing in some way the solution of an associated linearized problem or an integrable nonlinear problem. If one hopes to extract information about the nonlinear problem by studying a corresponding linearized one, the correct way to linearize must also be evaluated. Examples of such techniques are branching theory, eigenvalue perturbation, and boundary conditions or domain perturbation.

For linear PDEs in two variables, the classical approach for solving a BVP (going back to Jean D'Alembert's work in the 1740s) is separation of variables. The aim of this technique is to reduce the problem to two distinct linear problems for two ODEs. However, the separability of the problem depends on the specific domain and boundary conditions. For example, depending on the specific boundary conditions prescribed, the ODE one obtains may lead, via the associated Green function, to a non-self-adjoint transform problem, for which few general results are available.

An important theoretical result for the solvability of BVP for linear PDEs is the fundamental principle of Ehrenpreis (1970), which states that there always exists an appropriate generalization of the Fourier transform capable of representing the general solution of a BVP for a linear PDE in the variables (x_1, x_2), posed in a smooth convex domain. This result assumes the well-posedness of the problem. It then ensures that there exists a measure $\rho(k)$ and a contour Γ in the complex plane such that the solution of the problem can be expressed as an integral in the form

$$\int_\Gamma e^{f_1(k)x_1+f_2(k)x_2}\,d\rho(k),$$

where k is a complex parameter. The functions $f_1(k)$ and $f_2(k)$ are given explicitly. For example, in the case of an evolution equation $u_t = Lu$, where $u=u(x,t)$ and L is a linear differential operator, the representation takes the form

$$u(x,t) = \int_\Gamma e^{ikx-i\omega(k)t}\,d\rho(k), \qquad (1)$$

where $\omega(k)$ is the dispersion relation of the equation. In representation (1), the dependence on the solution variables (x,t) is explicit; the integration variable k is called the spectral parameter, and such a representation is said to be spectrally decomposed. However, this result is, in general, not constructive, as $\rho(k)$ and Γ are not known. In some cases, it is possible, to obtain this representation via separation of variables, but this is not always the case. Consider, for example, the second-order BVP

$$\begin{aligned} iq_t + q_{xx} &= 0, 0<x<\infty,\ 0<t<\infty,\\ q(x,0) &= q_0(x), q(0,t) = f(t), \end{aligned} \qquad (2)$$

where $q=q(x,t)$ and it is assumed that all functions are infinitely differentiable and vanish as $x\to\infty$. By separating variables, one obtains an ODE in x, which can be solved using the sine transform. Assuming that a unique solution exists, this procedure yields the representation

$$q(x,t) = \frac{2}{\pi}\int_0^\infty \sin(kx)e^{-ik^2t} \times\left(\hat{q}_0(k) + ik\int_0^t e^{ik^2s} f(s)ds\right)dk,$$

where

$$\hat{q}_0(k) = \int_0^\infty \sin(kx)q_0(x)dx.$$

This representation is not in form (1) as the variable t also appears as a limit of integration. This fact not only makes this representation less convenient for extracting information about the t dependence of the solution but also makes the rigorous proof of existence and uniqueness of a solution more cumbersome, as the relevant integral is not uniformly convergent as $x\to 0$.

For nonlinear PDEs, no general method is available (Logan, 1994). Perturbation techniques can be of some use in the study of evolutionary PDEs of the form $u_t + P(u) = 0$, where $u=u(t,x)$ and Pu is a nonlinear ODE containing the x-derivatives. Solutions of this problem such that $u_t=0$ are called steady-state solutions: these are the solutions that are independent of time. In this context, one studies the linearized stability of the steady state by using the same perturbative techniques discussed for ODEs, as this yields information about the qualitative behavior in time of the solution of the nonlinear problem. The results available in this area are, in general, of limited applicability and practical use for finding explicit solutions.

The special class of nonlinear PDEs known as integrable deserves separate consideration. For these equations, there exists a particular linearizing technique, the inverse scattering transform, which yields the solution of the Cauchy problem. Some of these equations, such as the Korteweg–de Vries and sine-Gordon equations, have been considered on simple domains, and specific BVPs have been solved by ad hoc PDE techniques. The first such result was obtained already 40 years ago (Cattabriga, 1959), but recently this field has witnessed a new surge of interest. To obtain such results, the nonlinear problem is often considered as a linear problem, with the nonlinear term considered as a nonhomogeneous (or forcing) term; thus, the analysis is based on the analysis of the linearized equations by classical PDE techniques (Bona et al., 2001). A different approach involves the attempt to extend the inverse scattering linearizing technique to BVPs, as done, for example, in Leon (2002) for the sine-Gordon equation.

Recently, a general approach to solving BVPs for two-dimensional linear PDEs has been proposed and successfully used to solve many different types of such problems (Fokas, 2000). Its relevance is enhanced by the fact that this approach can be generalized to treat integrable nonlinear PDEs. This methodology yields a spectral transform associated directly to the PDE rather than to transforms associated to the two ODEs obtained by separating variables. For Example (2), this yields, for the solution, the representation

$$q(x,t) = \frac{1}{2\pi}\int_0^\infty e^{ikx-ik^2t}\hat{q}_0(k)dk + \frac{1}{2\pi}\int_\Gamma e^{ikx-ik^2t}\hat{q}(k)dk,$$

where Γ is the boundary of the first quadrant of the complex k-plane, and

$$\hat{q}(k) = \frac{\hat{q}_0(k)+\hat{q}_0(-k)}{2} - k\int_0^\infty e^{ik^2t} f(t)dt.$$

This representation is in Ehrenpreis form, and in addition, measure and contour are explicitly constructed.

The above-mentioned approach provides a unification of the integral representation of the solution of a linear PDE in terms of the Ehrenpreis fundamental principle with the inverse scattering transform for integrable nonlinear PDEs. Indeed, when the problem reduces to an initial value problem for decaying solutions (i.e., the domain for the spatial variable is the whole real line, and the solution is assumed to vanish at $\pm\infty$), the transform obtained is precisely the inverse scattering transform.

The essential ingredients of this approach are the reformulation of the PDE as the closure condition of a certain differential form, and the definition in the complex plane of a Riemann–Hilbert problem depending on both the PDE and the domain. The differential form can be found algorithmically for linear PDEs and is equivalent to the Lax pair formulation for integrable nonlinear PDEs (Lax, 1968). The solution of this Riemann–Hilbert problem (which can be found in closed form in many cases) takes the role of the classical Green formula, and yields an integral representation for the solution, which is independent of the particular boundary conditions and indeed contains all the boundary values of the solution. What this approach crucially provides (when the definition domain is connected) is a global relation among these boundary values, which is the tool necessary to express the solution only in terms of the given boundary conditions and to prove rigorously problems with well-posedness, as well as existence and uniqueness results.

BEATRICE PELLONI

See also **Integrability; Inverse scattering method or transform; Riemann–Hilbert problem; Separation of variables**

Further Reading

Bona, J., et al. 2001. A non-homogeneous boundary value problem for the Korteweg–de Vries equation. *Transactions of the American Mathematical Society*, 354: 427–490

Cattabriga, L. 1959. Un problema al contorno per una equazione parabolica di ordine dispari. *Annali della Scuola Normale Superiore di Pisa*, 13: 163–203.

Coddington, E.A. & Levinson, N. 1955. *Theory of Ordinary Differential Equations*, New York: McGraw-Hill

Ehrenpreis, L. 1970. *Fourier Analysis in Several Complex Variables*, New York: Wiley-Interscience

Fokas, A.S. 2000. On the integrability of linear and nonlinear PDEs. *Journal of Mathematical Physics*, 41: 4188

Ghidaglia, J.L. & Colin, T. 2001. An initial-boundary value problems for the Korteweg–de Vries equation posed on a finite interval. *Advances in Differential Equations*, 6(12): 1463–1492

Lax, P.D. 1968. Integrals of nonlinear equations of evolution and solitary waves. *Communications in Pure and Applied Mathematics*, 21: 467–490

Leon, J. 2002. Solution of the Dirichlet boundary value problem for the sine-Gordon equation. *Physics Letters* A, 298, 343–252

Logan, J.D. 1994. *An Introduction to Nonlinear Partial Differential Equations*, New York: Wiley-Interscience

Stackgold, I. 1979. *Green's Functions and Boundary Value Problems*, New York: Wiley-Interscience

BOUSSINESQ EQUATIONS

See **Water waves**

BOX COUNTING

See **Dimensions**

BRAIN WAVES

See **Electroencephalogram at large scales**

BRANCHING LAWS

In this entry, we briefly trace the history of a familiar phenomenon, branching, in physical and biological systems and the laws governing them. The simplest type of branching tree is one in which a single conduit enters a vertex and two conduits emerge. This dichotomous process is clearly seen in the patterns of biological systems, such as botanical trees, neuronal dendrites, lungs, and arteries, as well as in the patterns of physical systems, such as lightning, river networks, and fluvial landscapes. The quantification of branching through the construction of the mathematical laws that govern them can be traced back to Leonardo da Vinci (1452–1519). In his *Notebooks*, he writes (Richter, 1970):

> All the branches of a tree at every stage of its height when put together are equal in thickness to the trunk [below them]. All the branches of a water [course] at every stage of its course, if they are of equal rapidity, are equal to the body of the main stream.

He also admonished his readers with: "Let no man who is not a Mathematician read the elements of my work." This statement increases in significance when we consider that da Vinci wrote nearly two centuries before Galileo (Galilei, 1638), who is generally given the credit for establishing the importance of mathematics in modern science.

The first sentence in the da Vinci quote is further clarified in subsequent paragraphs of the *Notebooks*. With the aid of da Vinci's sketch reproduced in Figure 1, this sentence has been interpreted as follows: if a tree has a trunk of diameter d_0 that bifurcates into two limbs of diameters d_1 and d_2, the three diameters are related by

$$d_0^a = d_1^a + d_2^a \qquad (1)$$

Simple geometrical scaling yields the diameter exponent $\alpha = 2$, which corresponds to rigid pipes carrying fluid from one level of the tree to the next, while retaining a fixed cross-sectional area through successive generations of bifurcation. Although the pipe model has a number of proponents from hydrology, the diameter exponent for botanical trees was determined empirically by Cecil D. Murray in 1927 to be insensitive to the kind of botanical tree and to have a value 2.59 rather than 2 (Murray, 1927). Equation (1) is referred to as Murray's law in the physiology literature.

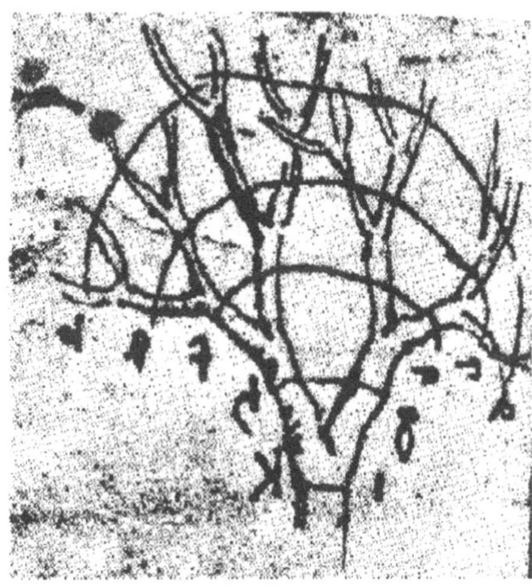

Figure 1. Sketch of tree from Leonardo da Vinci's Notebooks, PL XXVII (Richter, 1970).

The significance of Murray's law was not lost on D'Arcy Thompson (1942). In the second edition of his classic *On Growth and Form*, first published in 1917, Thompson argues that the geometrical properties of biological systems can often be the limiting factor in the development and final function of an organism. This is stated in his principle of similitude, which is a generalization of certain observations made by Galileo regarding the trade-off between the weight and strength of bone (Galilei, 1638). Thompson goes on to argue that the design principle for biological systems is that of energy minimization.

The second sentence in the da Vinci quotation is as suggestive as the first. In modern language, we would interpret it to mean that the flow of a river remains constant as tributaries emerge along the river's course. This equality must be true in order for the water to continue flowing in one direction and not stop and reverse course at the mouth of a tributary. Using the pipe model introduced above, and minimizing the energy with respect to the pipe radius, yields $\alpha = 3$, in Equation (1). Thus, the value of the diameter exponent obtained empirically by Murray falls between the theoretical limits of geometric self-similarity and hydrodynamic conservation, $2 \leq \alpha \leq 3$.

In da Vinci's tree, it is easy to assign a generation number to each of the limbs, but the counting procedure can become complicated in more complex systems like the generations of the bronchial tubes in the mammalian lung. One form taken by the branching laws is that the ratio of the radii of the tubes (from one generation to the next) is constant, that is, by the scaling relation

$$\frac{r_j}{r_{j+1}} = R. \qquad (2)$$

Equation (2) is analogous to Horton's law for river trees and fluvial landscapes, which involves the ratio of the number of branches in successive generations of branches, rather than radii (Mandelbrot, 1977). In either case, the parameter R determines the branching law and Equation (2) implies a geometrical self-similarity in the tree, as anticipated by Thompson. In the branching of bronchial airways, $d_1 = d_2$ at each stage of the tree, so that from Equation (1) we deduce the relationship between the radii of the pipes between successive generations as

$$r_j = 2^{1/\alpha} r_{j+1}. \tag{3}$$

In the case of the lung, the diameter of an airway is reduced by a factor $2^{-1/3}$ at each generation, since $\alpha = 3$ for the bronchial tree. Therefore, after j generations, $r_j = r_0 \exp(-j/\lambda)$, where the exponential rate of reduction, $\lambda = \ln(2)/3$, is the same for each generation beyond the trachea r_0, as argued by Weibel (2000).

A less space-filling value of the scaling index is obtained for the arterial system, where it is empirically determined that $\alpha = 2.7$. For a general non-integer branching index, the scaling relation Equation (3) defines a fractal tree. Such trees have no characteristic scale length and were first organized and discussed as a class by Benoit Mandelbrot (1977)—the father of fractals. The classical approach relied on the assumption that biological processes, like their physical counterparts, are continuous, homogeneous, and regular. However, most biological systems and many physical ones are discontinuous, inhomogeneous, and irregular and are necessarily that way in order to perform a particular function, such as gas exchange in lungs and arteries.

An entirely different kind of fractal tree is that of neuronal dendrites. The branching trees of neurons interleave the brain and form the communication system within the body. In the neurophysiology literature, Equation (1) is known as Rall's law with $\alpha = 1.5$ (Rall, 1959). More recent measurements of the scaling index, at each generation of dendritic branching, show a change with generation number; that is, the single parameter R is replaced with R_j. This non-constant scaling coefficient implies that Thompson's principle of similitude is violated.

A fractal model of the bronchial tree assumes that the ratio of successive generations of radii is dependent on the generation number, giving rise to a renormalization group relation, with the solution given by

$$r_j = \frac{a(j)}{j^u}, \quad j > 0. \tag{4}$$

Here, the average radius is an inverse power law in the generation number j, modulated by a slowly oscillating function $a(j)$ as observed in the human, dog, rat, and hamster data shown in Figure 2 (West & Deering, 1994). In this way, the fractal concept is used as a design principle in biology (Weibel, 2000; West, 1999; Mandelbrot, 1977) and in the development of branching laws.

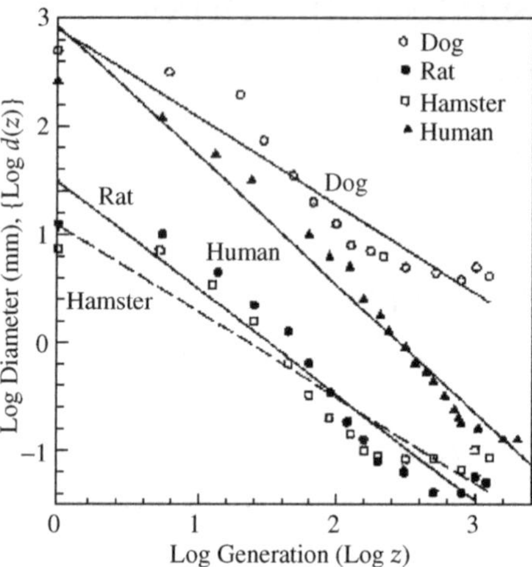

Figure 2. The variation in diameter d of the bronchial airways is depicted as a function of generation number j for rats, hamsters, humans, and dogs. The modulated inverse power law from the fractal model of the bronchial airway is observed in each case (West & Deering, 1994).

BRUCE J. WEST

See also **Fibonacci series; Geomorphology and tectonics; Martingales; Multiplex neuron**

Further Reading

Galilei, G. 1638. *Dialogue Concerning Two New Sciences*, translated by H. Crew & A. deSalvio in 1914, New York: Dover, 1954

Mandelbrot, B.B. 1977. *The Fractal Geometry of Nature*, San Francisco: W.H. Freeman

Murray, C.D. 1927. A relationship between circumference and weight and its bearing on branching angles. *Journal of General Physiology*, 10: 725–729

Rall, W. 1959. Theory of physiological properties of dendrites. *Annals of New York Academy of Science*, 96: 1071–1091

Richter, J.P. 1970. *The Notebooks of Leonardo da Vinci*, vol. 1, New York: Dover, unabridged edition of the work first published in London in 1883

Thompson, D.W. 1942. *On Growth and Form*, 2nd edition, Cambridge: Cambridge University Press, republished New York: Dover, 1992

Weibel, E.R. 2000. *Symmorphosis: On Form and Function in Shaping Life*, Cambridge, MA: Harvard University Press

West, B.J. 1999. *Physiology, Promiscuity and Prophecy at the Millennium: A Tale of Tails*, Singapore: World Scientific

West, B.J. & Deering, W. 1994. Fractal physiology for physicists: Levy statistics. *Physics Reports*, 246: 1–100

BREATHERS

The term *breather* (also called a "bion") arose from studies of the sine-Gordon (SG) equation

$$u_{tt} - u_{xx} + \sin u = 0, \tag{1}$$

which has localized solutions that oscillate periodically with time and decay exponentially in space. Such a solution of Equation (1) is given by

$$u(x,t) = 4\tan^{-1}\left[\frac{\lambda \sin \omega t}{\omega \cosh \lambda x}\right], \quad \lambda = \sqrt{1-\omega^2}, \tag{2}$$

which is shown in Figure 1.

Although the breather of Equation (2) is a non-topological soliton of Equation (1), it can be considered as a bound state of two topological solitons of the SG equation (kink and antikink), one of which is shown in Figure 2(a). The kink and antikink oscillate with respect to each other with the period $T = 2\pi/\omega$. Thus, such a soliton is also called a "doublet." A sketch of the bion at small frequencies ($\omega \ll 1$) and large enough t is presented in Figure 2(b). At some initial time, the kink and antikink move outward in opposite directions and separate in space with increasing time up to some finite distance (at $t = T/4$). The kink and antikink components of the breather never become fully free of distortions in their shapes due to interactions between each other, and finally oscillate in a kind of bound state.

At $1-\omega^2 \ll 1$, Equation (2) reduces to a small-amplitude breather $u(x,t) = 4\,Re\,\psi(x,t)$, where

$$\psi(x,t) = -\mathrm{i}\,\frac{\lambda \exp(-\mathrm{i}\omega t)}{\cosh \lambda x} \tag{3}$$

and $\lambda = \sqrt{2(1-\omega)}$. Equation (3) is a soliton solution of the nonlinear Schrödinger (NLS) equation

$$\mathrm{i}\psi_t + \tfrac{1}{2}\psi_{xx} - \psi + |\psi|^2\psi = 0, \tag{4}$$

which is regarded as a breather and can be written as $\psi(x,t) = \phi(x)\exp(\mathrm{i}\omega t)$. In this form, the spatial dependence of the soliton amplitude and the time dependence of the phase (of the complex function ψ) are separated. As a result, the nonlinearity appears only in the amplitude, but not in the phase of the NLS soliton. Although such a separation of the spatial and time dependencies in a soliton expression does not take place in the general form of the SG breather, the limiting case of the SG breather coincides with the amplitude of the NLS soliton.

At present, it is not known whether other nonlinear Klein–Gordon equations similar to Equation (1), but differing from it only by the nonlinear term, possess exact breather solutions (Segur & Kruskal, 1987). However, if certain nonlinear terms in a Klein–Gordon-type equation differ only slightly from $\sin u$ (slightly perturbed SG equation), a breather-like solution may persist in the first order with respect to the perturbation (Birnir et al., 1994).

The breather of the SG equation can move along the space coordinate axis with a stationary velocity V. As Equation (1) is a relativistic invariant equation (invariant under a Lorentz transformation

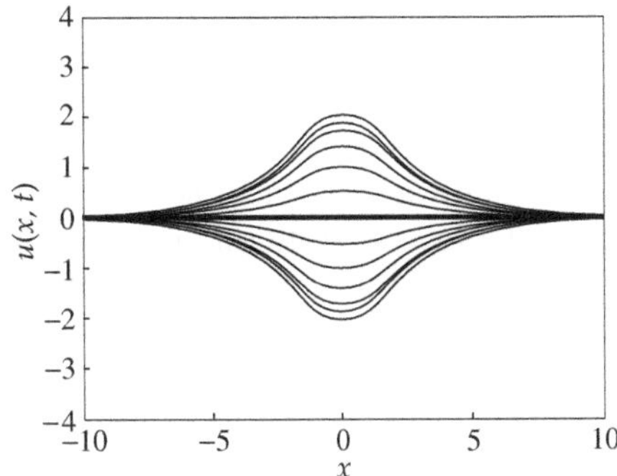

Figure 1. $u(x,t)$ from Equation (2) versus x for 26 different times equally spaced and covering one period, with $\lambda = 0.5$.

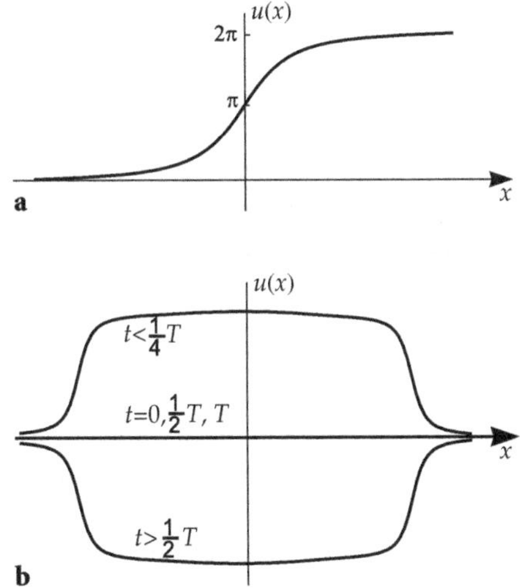

Figure 2. (a) A sketch of the sine-Gordon kink. (b) Three profiles of the kink-antikink oscillations.

of the independent variables), one can obtain a moving breather from Equation (2) substituting $x \to (x - Vt)/(1-V^2)^{1/2}$ and $t \to (t - Vx)/(1-V^2)^{1/2}$. Consequently, the moving breathers form a two-parametric (ω and V) family of solutions of the SG equation.

Possible values of the breather parameters ω and V can be compared with the dispersion relation ($\omega^2 = 1 + k^2$) for small vibrations (phonons) described by the linearized version of Equation (1). These phonons have frequencies $\omega > 1$ and phase velocities $\omega/k > 1$. A breather frequency, on the other hand, is smaller than the minimum frequency of the phonons ($\omega < 1$), and the breather velocity is smaller than the minimum phonon phase velocity ($V < 1$). Therefore, the dynamical breather parameters lie outside of the spectrum of the linear vibrations. Although the time dependence of the breather includes the higher temporal harmonics of the oscillations, the phonons cannot be resonantly excited by the breather. Thus, breathers and phonons are asymptotically independent vibrational modes of the system described by the SG equation. This asymptotic independence of nonlinear excitations (breathers and kinks) and phonons follows from the integrability of Equation (1).

An important way to study nonlinear integrable equations is the inverse scattering method. According to this method, breathers are characterized by poles in the complex phase plane of scattering parameters for the equation under consideration.

It is known that several nonlinear differential equations possess breather solutions. The Landau–Lifshitz (LL) equation provides an example of a nonlinear equation generalizing the results that are described by the SG and NLS equations. The breather-like solution of the LL equation has a more complicated form than the one presented above; however, it is also a two-parameter soliton called a dynamic magnetic soliton (Kosevich et al., 1977). Its oscillatory behavior is characterized by a frequency ω, and its center can move with a velocity V. In the general case, the magnetic soliton can be described by some complex function of x and t, but the time and spatial dependencies are not separated in the analytical expression for such a soliton.

An important class of breathers the so-called discrete breathers (also known as intrinsic localized modes, self-localized anharmonic modes, or nonlinear localized excitations). These are solutions of a nonlinear equation on a lattice, and they are periodic in time and localized in space. Although most such investigations are performed by numerical calculations, there exist nonlinear dynamic equations on a lattice possessing exact analytical breather solutions. One of them is the following discrete version of the NLS equation proposed by Ablowitz and Ladik (AL) in 1976 (Ablowitz & Ladik, 1976):

$$\mathrm{i}\partial_t\psi_n - (\psi_{n+1} + \psi_{n-1})(1 + |\psi_n|^2) = 0. \qquad (5)$$

The AL lattice is integrable and it allows for breather-like solutions, the simplest of which has a form close to that of Equation (2):

$$\psi = \frac{\sinh\beta\exp(-\mathrm{i}\omega t)}{\cosh\beta(n - x_0)}, \qquad (6)$$

where n is the integer number of a lattice site, $x_0 =$ constant, and $\omega = -2\cosh\beta$.

Arnold Kosevich

See also **Discrete breathers; Discrete nonlinear Schrödinger equations; Integrability; Inverse scattering method or transform; Sine-Gordon equation; Solitons**

Further Reading

Ablowitz, M.J. & Ladik, J.F. 1976. Nonlinear differential-difference equations and Fourier analysis. *Journal of Mathematical Physics*, 17: 1011

Birnir, B., McKean, H.P. & Weinstein, A. 1994. The rigidity of sine-Gordon breathers. *Communications on Pure and Applied Mathematics*, 47: 1043

Flach, S. & Willis, C.R. 1998. Discrete breathers. *Physics Reports*, 295: 181

Kosevich, A.M., Ivanov, B.A. & Kovalev, A.S. 1977. Nonlinear localized magnetization wave of a ferromagnet as a bound-state of a large number of magnons. *Pis'ma Zhurnal Eksperimental'noy i Teoreticheskoy Fiziki*, 25: 516 (in Russian); *JETP Letters*, 25: 486

Kosevich, A.M., Ivanov, B.A. & Kovalev, A.S. 1990. Magnetic solitons. *Physics Reports*, 194: 117

Segur, H. & Kruskal, M.D. 1987. Nonexistence of small-amplitude breather solutions in ϕ^4 theory. *Physical Review Letters*, 58: 747

BROUWER'S FIXED POINT THEOREM

See **Winding numbers**

BROWNIAN MOTION

In 1828, Robert Brown, a leading botanist, observed that a wide variety of particles suspended in liquid exhibit an intrinsic, irregular motion when viewed under a microscope. While not the first to witness such motion, his experimental focus on this phenomenon, which would bear his name, established its universality and intrinsic nature, thereby raising it as an issue for fundamental scientific inquiry (Nelson, 1967). Deutsch has recently raised the question of whether Brown actually witnessed Brownian motion or fluctuations due to some external contaminating influence (Peterson, 1991). Indeed, while Brownian motion is a ubiquitous phenomenon, not all irregular motions can be ascribed to Brownian motion. The dancing of dust particles in sunlight is dominated by imperceptible turbulent currents, not Brownian motion. True Brownian motion is generally only visible on scales of microns and below, but has important macroscopic ramifications because all microscopic particles manifest it. For example, Brownian motion makes possible both the fine-scale mixing of initially segregated substances in nature and industry, as well as the passive transport of ions, nutrients, and fuel, which allow biological cells to support life.

The origin of Brownian motion remained under debate throughout the 19th century, with Cantoni, Delsaux, Gouy, and C. Weiner proposing that thermal motions in the suspending liquid were responsible, as discussed in Einstein (1956, pp. 86–88), Gallavotti (1999, Chapter 8), and Russel et al. (1989, pp. 65–66). Attempts to examine this hypothesis quantitatively were hampered by the inability to measure accurately the velocity of particles undergoing Brownian motion, since such motion loses coherence over time scales (microseconds) that are shorter than those which experimental observations were able to resolve. In one of three ground-breaking papers that Einstein published in 1905, he offered a statistical mechanical means for theoretical calculations involving Brownian motion (Einstein, 1956). Einstein realized that the quantity involving Brownian motion that can be best observed under a microscope in an experiment is the "diffusivity":

$$D = \lim_{t \to \infty} \frac{|X(t) - X(0)|^2}{2t}, \tag{1}$$

where $X(t)$ denotes the observed displacement of the Brownian particle along a fixed direction at time t. In practice, t is simply taken as some satisfactorily long time of observation, and there is no need for fine temporal resolution as there would be if the velocity were to be measured. Einstein employed a random walk model for his analysis and showed that the diffusivity defined in (1) is identical to the diffusion constant that describes the macroscopic evolution of the concentration density $n(\boldsymbol{x}, t)$ of a large number of Brownian particles:

$$\frac{\partial n(\boldsymbol{x}, t)}{\partial t} = D \sum_{j=1}^{3} \frac{\partial^2 n(\boldsymbol{x}, t)}{\partial x_j^2}. \tag{2}$$

Through an elegant argument based on *equilibrium* statistical mechanical arguments, Einstein showed that the diffusivity D of a Brownian particle must be related to its friction coefficient ξ in the following way:

$$D = \frac{k_B T}{m\xi}, \tag{3}$$

where k_B is Boltzmann's constant and T is the absolute temperature (measured in Kelvin scale), and m is the particle's mass. The friction coefficient ξ appears in the relation between the drag force F_{drag} and velocity v of the particle in steady-state motion (assuming a low Reynolds number):

$$F_{\text{drag}} = m\xi v. \tag{4}$$

For a sphere of radius a moving through a fluid with dynamic viscosity μ, the friction coefficient is given by $\xi = 6\pi\mu a/m$. The remarkable property of the "Einstein relation" in (3) is that it links a quantity D pertaining to statistically unpredictable dynamical fluctuations to a quantity ξ, which involves deterministic, steady-state properties. Later work generalized the Einstein relation (3) to "fluctuation-dissipation theorems," which express the structure of the spontaneous statistical fluctuations in a wide class of physical systems to the structure of the dissipative (frictional) dynamics (Kubo et al., 1991, Chapter 1).

The basic theory of Brownian motion was developed by Einstein in 1905, a time when the premises of the atomic theory of matter were still not yet fully agreed upon (Gallavotti, 1999; Nelson, 1967). Einstein realized that a careful observation of Brownian motion and his relation between the diffusivity of a Brownian particle and its mobility could be used to calculate the number of particles making up a given mass of fluid if the atomic theory were valid. Under a microscope sufficient to resolve the Brownian motion of a particle, all quantities in (3) are directly observable except for Boltzmann's constant k_B. Therefore, the Einstein relation (3) can be used to compute a value for k_B based on Brownian motion data. Now, k_B is in turn related to Avogadro's number N_A, which is the number of molecules in a "mole" (a certain well-defined macroscopic amount) of a substance. The Brownian motion data and the Einstein relation, therefore, furnish an independent prediction for Avogadro's number N_A and, thereby, the number of molecules per unit mass of the fluid. In other words, the number (and therefore mass) of the individual fluid particles could be calculated without having to observe them at an individual level, an experimental feat that has become possible only in recent years. Instead, their individual mass and number could be assessed through their collective influence on a much larger and, therefore, observable immersed particle. In 1908, Jean Perrin experimentally confirmed that the value of N_A computed in this way agreed with those obtained from other techniques (Gallavotti, 1999), providing strong support for the atomic theory of matter. Since the 1970s, Brownian motion has been investigated in the laboratory through dynamic light scattering techniques (Russel et al., 1989, Chapter 3).

The most idealized mathematical representation of Brownian motion with diffusivity D is defined as $(2D)^{1/2}W(t)$, where $W(t)$ is a canonical continuous random process with Gaussian statistics such that $W(0) = 0$, $\langle W(t) \rangle = 0$, and

$$\langle (W(t) - W(t'))^2 \rangle = |t - t'|. \tag{5}$$

This mathematical Brownian motion is often referred to as the Wiener process (Borodin & Salminen, 2002; Gallavotti, 1999; Nelson, 1967). This idealized Brownian motion has independent increments (no inertia). Physical Brownian motion, of course, has some small inertia as well as several other complicating influences from the fluid environment and from the presence of other nearby Brownian particles (Russel et al., 1989). These extra features can be built into a dynamical description using the mathematical Brownian motion as the basic noise input with influence mediated by the other physical parameters. The mathematical Brownian motion has a similar role in modeling noise input in a wide variety of stochastic models in physics, biology, finance, and other fields. More precisely, the Levy–Khinchine theorem indicates that in any system affected by noise in a continuous way such that the noise on disjoint time intervals is independent can be modeled in terms of mathematical Brownian motion (Reichl, 1998, Chapters 4, 5).

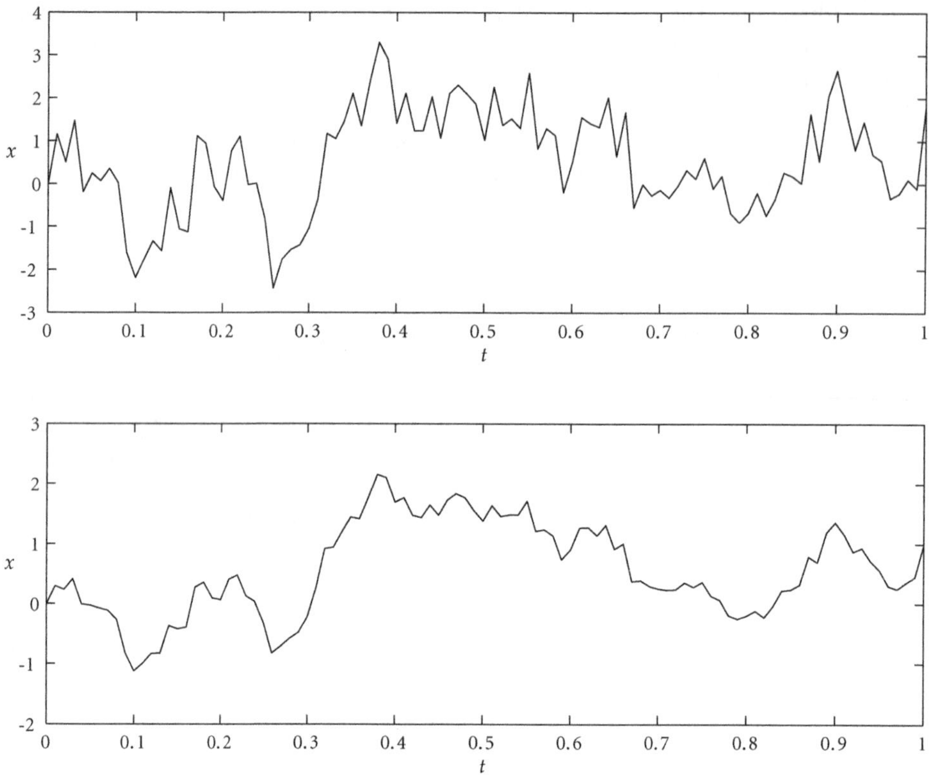

Figure 1. Sample trajectories of fractional Brownian motion using Fourier-wavelet method (Elliott et al., 1997). Top panel: $H = \frac{1}{3}$, lower panel: $H = \frac{2}{3}$. Both simulations used the same random numbers.

Discontinuous noise-induced jumps, in contrast, are modeled in terms of Poisson processes or more generally Lévy processes (Reichl, 1998, Chapters 4, 5). Continuous noise with long-range correlations (so that the independent increment property is not satisfied), on the other hand, can often be usefully modeled in terms of "fractional Brownian motion" (FBM) (Mandelbrot, 2002). This is an idealized Gaussian random process $Z(t)$ with $Z(0) = 0$, $\langle Z(t) \rangle = 0$, and

$$\langle (Z(t) - Z(t'))^2 \rangle = |t - t'|^{2H} \qquad (6)$$

where the Hurst exponent H is chosen from the interval $0 < H < 1$. The FBM with $H = \frac{1}{2}$ corresponds to ordinary Brownian motion with independent increments. FBMs with $\frac{1}{2} < H < 1$ have positive, long-ranged correlations with less rough trajectories and large excursions, while FBMs with $0 < H < \frac{1}{2}$ have negative, long-ranged correlations with rougher trajectories and a more oscillatory character (Figure 1). All FBMs have a statistical self-similarity property; the statistics of the rescaled FBM $\lambda^{-H} Z(\lambda t)$ are identical to those of the original $Z(t)$. That is, these processes have no finite length or time scale associated to them and can be thought of as random fractals. Fractional Brownian motions are therefore particularly appropriate for modeling systems with fluctuations occurring over a wide range of scales; cutoff lengths can be introduced by filtering an input FBM. Models built from FBMs have been developed in turbulence theory, natural landscape and cloud structures, surface adsorption processes, neural signals in biology, and self-organized critical systems such as earthquakes, forest fires, and sandpiles.

PETER R. KRAMER

See also **Fluctuation-dissipation theorem; Fluid dynamics; Fokker–Planck equation; Lévy flights; Random walks**

Further Reading

Borodin, A.N. & Salminen, P. 2002. *Handbook of Brownian Motion: Facts and Formulae*, 2nd edition, Basel: Birkhäuser

Einstein, A. 1956. *Investigations on the Theory of the Brownian Movement*, edited with notes by R. Fürth, translated by A.D. Cowper, New York: Dover

Elliott, Jr, F.W., Horntrop, D.J. & Majda, A.J. 1997. A Fourier-wavelet Monte Carlo method for fractal random fields. *Journal of Computational Physics*, 132(2): 384–408

Gallavotti, G. 1999. *Statistical Mechanics*, Berlin: Springer

Kubo, R., Toda, M. & Hashitsume, N. 1991. *Statistical Physics*, vol. 2, 2nd edition, Berlin: Springer

Mandelbrot, B.B. 2002. *Gaussian Self-affinity and Fractals*, Chapter IV. New York: Springer

Mazo, R.M. *Brownian Motion. Fluctuations, Dynamics and Applications*, Oxford and New York: Oxford University Press

Nelson, E. 1967. *Dynamical Theories of Brownian Motion*, Princeton, NJ: Princeton University Press

Peterson, I. 1991. Did Brown see Brownian motion? *Science News*, 139: 287

Reichl, L.E. 1998. *A Modern Course in Statistical Physics*. 2nd edition, New York: Wiley

Russel, W.B., Saville, D.A. & Schowalter, W.R. 1989. *Colloidal Dispersions*. Cambridge and New York: Cambridge University Press

BRUSSELATOR

The Brusselator is an autocatalytic model involving two intermediates. It illustrates how the fundamental laws of thermodynamics and chemical kinetics as applied to open systems far from equilibrium can give rise to self-organizing behavior and to dissipative structures in the form of temporal oscillations and spatial pattern formation.

Chemical kinetics imposes stringent conditions on the concentrations of the species involved in a reaction scheme and on the associated parameters. In a scheme consisting entirely of elementary steps, the overall rates are given (in an ideal system) by mass action kinetics, featuring particular combinations of products of concentrations preceded by stoichiometric coefficients (integer numbers specifying how the relevant constituents are produced or consumed). This guarantees the positivity of the solutions of the mass balance equations. A second condition is detailed balance, which requires that in chemical equilibrium, each individual reaction step is balanced by its inverse (*See* **Detailed balance**). This gives rise to relations linking the concentrations of initial reactants and final products to the rate constants, independent of the concentrations of the intermediates.

An additional set of requirements stems from the fact that self-organization in chemical kinetics must arise through an instability. One reason for this is that equilibrium and the states in its vicinity, obtained as the constraints are gradually increased (called the "thermodynamic branch"), are stable. To overcome this property and evolve to states that are qualitatively different from equilibrium, new branches of solutions must be generated, which can only take place through the mechanisms of instability and bifurcation. This, in turn, requires that the non-equilibrium constraints exceed a critical value (Glansdorff & Prigogine, 1971). Because the evolution laws generated by chemical kinetics at the macroscopic level of description are dissipative, the bifurcating states are attractors, attained by families of initial conditions belonging to an appropriate part of phase space. This guarantees structural stability, that is to say, the robustness of the solution toward small perturbations—a condition to be fulfilled by a model aiming to describe a physical phenomenon. Note that non-equilibrium instabilities and self-organization collapse when the system becomes closed to the external environment or when the kinetics involves only first-order steps, in which case one obtains a monotonic decay to a unique steady state (Denbigh et al., 1948).

Following the pioneering work of Alfred Lotka, several authors in the late 1940s, proposed models of open nonlinear systems deriving from mass action kinetics and giving rise to sustained oscillations (Lotka, 1956; Moore, 1949). These models do not have structural stability (as they give rise to a continuum of initial condition-dependent solutions), and they do not exhibit the role of the constraints in a transparent manner (as they are usually formulated in the limit of irreversible reactions). When the non-equilibrium constraints are explicitly accounted for, it is found (Lefever et al., 1967) that there is no instability threshold in these models. As the first known chemical model that is both fully compatible with the laws of thermodynamics and chemical kinetics and generates dissipative structures through non-equilibrium instabilities, the Brusselator is free from such deficiencies.

Model Presentation

In the interest of transparency, one desires a minimal model, and if oscillatory behavior is one of the required properties, this necessitates two coupled variables representing the concentrations of intermediate products. As in the models of the Lotka family, one seeks steps that are not only nonlinear but also include feedback processes, the simplest chemical version of which is autocatalysis. But contrary to these models, one now needs to scan the whole range of near to far from equilibrium situations and to undergo an instability somewhere in this range. As the Lotka-type models contain only second-order steps, a natural solution is to amend them by replacing these steps by a third-order one. This leads to the scheme (Prigogine & Lefever, 1968)

$$\begin{aligned} &A \underset{k_{-1}}{\overset{k_1}{\rightleftharpoons}} X, \quad B + X \underset{k_{-2}}{\overset{k_2}{\rightleftharpoons}} Y + D, \\ &2X + Y \underset{k_{-3}}{\overset{k_3}{\rightleftharpoons}} 3X, \quad X \underset{k_{-4}}{\overset{k_4}{\rightleftharpoons}} E \end{aligned} \tag{1}$$

Hanusse, Tyson, and Light have shown that a two-variable system compatible with the above requirements necessarily comprises a third-order step. Here A, B are the initial reactants, D, E the final products, and X, Y the intermediates: X can be thought of as an activator generating Y at its own expense, which acts as an inhibitor if the B concentration is large.

From the standpoint of irreversible thermodynamics, the Brusselator can be driven out of equilibrium through two independent constraints (affinities) related to the overall reactions

$$A \underset{k_{-1}k_{-4}}{\overset{k_1k_4}{\rightleftharpoons}} E, \; B \underset{k_{-3}k_{-2}}{\overset{k_3k_2}{\rightleftharpoons}} D.$$

This offers sufficient flexibility to allow one to take the limit of purely irreversible steps and of fixed reactant and product concentrations (also referred to as pool chemical approximation, ensuring that the (X, Y) subsystem becomes open to the external environment), while satisfying the positivity of concentrations, detailed balance, and mass conservation (Lefever et al., 1988). When diffusion is also included, and upon performing a suitable scaling transformation, this leads to the Brusselator equations

$$\frac{\partial X}{\partial t} = A - (B+1)X + X^2Y + D_x\nabla^2 X,$$
$$\frac{\partial Y}{\partial t} = BX - X^2Y + D_y\nabla^2 Y \qquad (2)$$

in which B, D_x/D_y, and the system size usually play the role of the parameters controlling the instabilities.

A number of variants of this canonical form have also been developed, including Brusselator in an open well-stirred reactor, Brusselator in a non-ideal system, including coupling between non-equilibrium instabilities and phase transitions, and coupling with external fields or advection.

Behavior of the Solutions

Since the first bifurcation analysis of the Brusselator equations (Nicolis & Auchmuty, 1974), several studies have been devoted to the various modes of spatiotemporal organization generated by Equations (2): limit cycles, Turing patterns, and traveling waves in one-dimensional systems (Nicolis & Prigogine, 1977); spatiotemporal chaos arising from the diffusive coupling of local limit cycle oscillators (Kuramoto, 1984); patterns in two- and three-dimensional systems including patterns arising from the interference of different instability mechanisms such as Turing, Hopf (De Wit, 1999); and the effect of confinement (Herschkowitz-Kaufman & Nicolis, 1972). Many phenomena now known to be generic have first been discovered on these Brusselator-based analyses, which have also helped to test the limits of traditional theoretical approaches and to explore new methodologies such as normal forms and phase dynamics.

The Brusselator has also been used to explore possible thermodynamic signatures of dissipative structures. No clearcut tendencies seem to exist, suggesting that global thermodynamic quantities like entropy and entropy production do not provide adequate measures of dynamic complexity. Finally, attention has been focused on the new insights afforded when the mean-field equations (2) are augmented to account for fluctuations, a study for which the Brusselator is well suited thanks to its mechanistic basis. The interest here is to provide a fundamental understanding of how large-scale order can be sustained despite the locally prevailing thermal disorder. Early accounts of the results, with emphasis on critical behavior in the vicinity of bifurcations, can be found in Nicolis & Prigogine (1977) and in Walgraef et al. (1982).

G. Nicolis

See also **Chemical kinetics; Detailed balance; Emergence; Turing patterns**

Further reading

Denbigh, K.G., Hicks, M. & Page, F.M. 1948. Kinetics of open systems. *Transactions of the Faraday Society*, 44: 479–494

De Wit, A. 1999. Spatial patterns and spatio-temporal dynamics in chemical systems. *Advances in Chemical Physics*, 109: 453–513

Glansdorff, P. & Prigogine, I. 1971. *Thermodynamic Theory of Structure, Stability and Fluctuations*, London and New York: Wiley

Herschkowitz-Kaufman, M. & Nicolis, G. 1972. Localized spatial structures and nonlinear chemical waves in dissipative systems. *Journal of Chemical Physics*, 56: 1890–1895

Kuramoto, Y. 1984. *Chemical Oscillations, Waves and Turbulence*, Berlin and New York: Springer

Lefever, R., Nicolis, G. & Prigogine, I. 1967. On the occurrence of oscillations around the steady state in systems of chemical reactions far from equilibrium. *Journal of Chemical Physics*, 47: 1045–1047

Lefever, R., Nicolis, G. & Borckmans, P. 1988. The Brusselator: it does oscillate all the same. *Journal of the Chemical Society, Faraday Transactions*, 1. 84: 1013–1023

Lotka, A. 1956. *Elements of Mathematical Biology*, New York: Dover

Moore, M.J. 1949. Kinetics of open reaction systems: chains of simple autocatalytic reactions. *Transactions of the Faraday Society*, 45: 1098–1109

Nicolis, G. & Auchmuty, J.F.G. 1974. Dissipative structures, catastrophes, and pattern formation: a bifurcation analysis. *Proceedings National Academy of Sciences USA*, 71: 2748–2751

Nicolis, G. & Prigogine, I. 1977. *Self-organization in Nonequilibrium Systems*, New York: Wiley

Prigogine, I. & Lefever, R. 1968. Symmetry-breaking instabilities in dissipative systems. II. *Journal of Chemical Physics*, 48: 1695–1700

Walgraef, D., Dewel, G. & Borckmans, P. 1982. Nonequilibrium phase transitions and chemical instabilities. *Advances in Chemical Physics*, 491: 311–355

BULLETS

See **Solitons, types of**

BURGERS EQUATION

In 1915, Harry Bateman considered a nonlinear equation whose steady solutions were thought to describe certain viscous flows (Bateman, 1915). This equation, modeling a diffusive nonlinear wave, is now widely known as the Burgers equation, and is given by

$$u_t + uu_x = \frac{\mu}{2} u_{xx}, \qquad (1)$$

where μ is a constant measuring the viscosity of the fluid. It is a nonlinear parabolic equation, simply describing a temporal evolution where nonlinear convection and linear diffusion are combined, and it can be derived as a weakly nonlinear approximation to the equations of gas dynamics.

Although nonlinear, Equation (1) is very simple, and interest in it was revived in the 1940s, when Dutch physicist Jan Burgers proposed it to describe a mathematical model of turbulence in gas (Burgers, 1940). As a model for gas dynamics, it was then studied extensively by Burgers (1948), Eberhard Hopf (1950), Julian Cole (1951), and others, in particular; after the discovery of a coordinate transformation that maps it to the heat equation. While as a model for gas turbulence the equation was soon rivaled by more complicated models, the linearizing transformation just mentioned added importance to the equation as a mathematical model, which has since been extensively studied. The limit $\mu \to 0$ is a hyperbolic equation, called the inviscid Burgers equation:

$$u_t + uu_x = 0. \qquad (2)$$

This limiting equation is important because it provides a simple example of a conservation law, capturing the crucial phenomenon of shock formation. Indeed, it was originally introduced as a model to describe the formation of shock waves in gas dynamics.

A first-order partial differential equation for $u(x,t)$ is called a conservation law if it can be written in the form

$$u_t + (f(u))_x = 0\,.$$

For Equation (2), $f(u) = u^2/2$. Such conservation laws may exhibit the formation of shocks, which are discontinuities appearing in the solution after a finite time and then propagating in a regular manner. When this phenomenon arises, an initially smooth wave becomes steeper and steeper as time progresses, until it forms a jump discontinuity—the shock.

Once a discontinuity forms, the solution is no longer a globally differentiable function; thus, the sense in which it can be considered as a solution of the PDE must be clarified. A discontinuous function $(u(x,t))$ can still be considered as a solution in the weak sense if it satisfies

$$\int\int_D \left(u\varphi_t + \frac{1}{2}u^2\varphi_x \right) \mathrm{d}x\mathrm{d}t = 0, \qquad (3)$$

where D is any rectangle in the (x,t) plane, and $\varphi(x,t)$ is any smooth function vanishing on the boundary ∂D. Any regular solution is a weak solution, as is seen by multiplying the equation by $\varphi(x,t)$, integrating by parts along ∂D, and using Green's theorem.

In physical applications, one often considers the discontinuous solution as the limit, as $\mu \to 0$, of smooth solutions of the viscous Equation (1). This idea is correct from a physical point of view, as it takes into account the significance of these solutions as a physical description of gas dynamics. From the form of the equation (or its weak formulation (3)), one can derive the velocity v_s of a shock separating two regimes, u_r to the right and u_l to the left of a discontinuity. The result is the Rankine–Hugoniot formula, valid in general for conservation laws, which for the case of the Burgers equation yields

$$s = \tfrac{1}{2}(u_r + u_l). \qquad (4)$$

Even this, however, is not enough to guarantee uniqueness of the solution, because there are several ways of writing the equation in the form of a conservation law. Often, the way to select the physically relevant solution is to consider the vanishing viscosity limit. To obtain this solution mathematically, an additional entropy condition, that $u_l > s > u_r$, must be imposed.

Besides its significance as a model for shocks, the Burgers equation is prominent among PDEs because it is completely integrable. Indeed, the nonlinear change of variable

$$u = -\mu(\log\psi)_x \qquad (5)$$

transforms Equation (1) into the heat equation $\psi_t = \psi_{xx}$, with initial conditions transforming simply into initial conditions for this latter equation: if $u(x,0) = f(x)$ is the given condition, then the corresponding initial condition for the heat equation is given by

$$\psi(x,0) = \exp\left[-\frac{1}{\mu}\int_0^x f(\xi)\mathrm{d}\xi \right].$$

The relation between the Burgers and the heat equation was already mentioned in an earlier book (Forsyth, 1906), but the former had not been recognized as physically relevant; hence, the importance of this connection was seemingly not noticed at the time. Using the transformation of Equation (5), known as the Cole–Hopf transformation, it is easy to solve the initial value problem for this equation. Recently, a generalization of the Cole–Hopf transformation has been successfully used to linearize the boundary value problem for the Burgers equation posed on the semiline $x > 0$ (Calogero & De Lillo, 1989).

The existence of this linearizing transformation, which is a transformation of Bäcklund type (Rogers & Shadwick, 1982) relating the solutions of two different PDEs, stimulated work to extend this approach to a generalized version of the Burgers equation, such as the Korteweg–de Vries–Burgers equation, given by

$$u_t + uu_x = \frac{\mu}{2}u_{xx} - \varepsilon u_{xxx}, \quad \varepsilon > 0. \qquad (6)$$

Although it was found out that such a directly linearizing transformation did not exist, efforts in this direction were rewarded by several discoveries. Indeed, the importance of a linearizing transformation became evident when the inverse scattering transform (IST) was discovered, leading to the full analytical understanding of the solution of the Cauchy problem for the Korteweg–de Vries (KdV) equation, and later all integrable evolution equations in one spatial dimension, such as KdV, the nonlinear Schrödinger, and the sine-Gordon equations.

A crucial step in the discovery of the IST was an observation made by Robert Miura. In analogy to gas dynamics, he noted that one needs conservation laws to compute jump conditions across the region where the solution is small and essentially dispersive, isolating the solitonic part, which is thought of as a kind of reversible shock. This led to the connection between the KdV and modified KdV equation via the Miura transformation and eventually to the IST through which these nonlinear equations are solved through a series of linear problems (Gardner et al., 1967).

Nowadays, the Burgers equation is used as a simplified model of a kind of hydrodynamic turbulence (Case & Chiu, 1969), called Burgers turbulence. Burgers himself wrote a treatise on the equation now known by his name (Burgers, 1974), where several variants are proposed to describe this particular kind of turbulence. Generalizations such as the KdV–Burgers equation (6) arose from the need to model more complicated physical situations and introduce more factors than those that the Burgers equation takes into account. Lower-order friction terms may be considered that reduce the amplitude of the wave, although in a different scale and manner than the reduction due to the higher-order diffusion term u_{xx}. For example, the KdV–Burgers equation is an appropriate model when a different higher-order amplitude-reducing effect, namely dispersion, is introduced. Depending on the relative sizes of μ and ε, this equation may exhibit either an essentially shock-like structure, with the presence of dispersive tail, or mainly dispersive phenomena; thus, Equation (6) has been proposed as a natural model for hydrodynamic turbulence. In the context of the study of gas dynamics (particularly turbulent and vorticity phenomena), the Burgers equation has also been used to model phase diffusion along vortex filaments.

Beatrice Pelloni

See also **Constants of motion and conservation laws; Inverse scattering method or transform; Shock waves; Turbulence**

Further Reading

Bateman, H. 1915. Some recent research on the motion of fluids. *Monthly Weather Review*, 43: 163–170

Burgers, J. 1940. Application of a model system to illustrate some points of the statistical theory of free turbulence. *Proceedings of the Nederlandse Akademie van Wetenschappen*, 43: 2–12

Burgers, J. 1948. A mathematical model illustrating the theory of turbulence, *Advances in Applied Mechanics*, 1: 171–199

Burgers, J. 1974. *The Nonlinear Diffusion Equation: Asymptotic Solutions and Statistical Problems*, Dordrecht and Boston: Reidel

Calogero, F. & De Lillo, S. 1989. The Burgers equation on the semiline. *Inverse Problems*, 5: L37

Case, K.M. & Chiu, S.C. 1969. Burgers turbulence models. *Physics of Fluids*, 12: 1799–1808

Cole, J. 1951. On a quasilinear parabolic equation occuring in aerodynamics. *Quarterly Journal of Applied Mathematics*, 9:225–236

Forsyth, A.R. 1906. *Theory of Differential Equations*, Cambridge: Cambridge University Press

Gardner, C.S., Greene, J.M., Kruskal, M.D. & Miura, R.M. 1967. Method for solving the Korteweg-de Vries equation. *Physical Review Letters*, 19: 1095–1097

Hopf, E. 1950. The partial differential equation $u_t + uu_x = \mu u_{xx}$. *Communications in Pure and Applied Mathematics*, 3: 201–230

Lax, P.D. 1973. *Hyperbolic Systems of Conservation Laws and the Mathematical Theory of Shock Waves*, Philadelphia: Society for Industrial and Applied Mathematics

Newell, A.C. (editor). 1974. *Nonlinear Wave Motion*, Providence, RI: American Mathematical Society

Rogers, C. & Shadwick, W.F. 1982. *Bäcklund Transformations and Their Applications*, New York: Academic Press

Sachdev, P.L. 1987. *Nonlinear Diffusive Waves*, Cambridge and New York: Cambridge University Press

Smoller, J. 1983. *Shock Waves and Reaction-Diffusion Equations*, Berlin and New York: Springer

Whitham, G.B. 1974. *Linear and Nonlinear Waves*, New York: Wiley-Interscience

BUTTERFLY EFFECT

The Butterfly Effect serves as a metaphor for what in technical language is called "sensitive dependence on initial conditions" or "deterministic chaos," the fact that small causes can have large effects.

As recounted by Gleick (1987, Chapter 1), in the early 1960s, Edward Lorenz was carrying out computer simulations on a 12-dimensional weather model. One day, he decided to run a particular time series for longer. In order to save time, he restarted his code from data from a previous printout. After returning from a coffee break, he found that the weather simulation had diverged sharply from that of his earlier run. After some checks, he could only conclude that the difference was caused by the difference in initial conditions: he had typed in only the first three of the six decimal digits that the computer worked with internally. Apparently, his assumption that the fourth digit would be unimportant was false.

Lorenz realized the importance of his observation: "If, then, there is any error whatever in observing the present state—and in any real system such errors seem inevitable—an acceptable prediction of an instantaneous state in the distant future may well be impossible" (Lorenz, 1963, p. 133). Indeed, the error made by discarding the fourth and higher digits is so small that it can be imagined to represent the effect of the flap of the wings of a butterfly. Lorenz originally used the image of a seagull. The more lasting name seems to have come from his address at the annual meeting of the American Association for the Advancement of Science in Washington, December 29, 1972, which was entitled "Predictability: does the flap of a butterfly's wings in Brazil set off a tornado in Texas?" The text of this talk was never published but is presented in its original form as an appendix in Lorenz (1993).

Sensitive dependence on initial conditions forces us to distinguish between determinism and predictability, two concepts often confused by scientists and popular writers alike. Determinism has to do with how Nature (or, less ambitiously, any system under consideration) behaves, while predictability has to do with what we, human beings, are able to observe, analyze, and compute. We have determinism if we have a law or a formula describing exactly, and fully, how the system behaves given its present state. To have predictability we need, in addition, to be able to measure the present state of the system with sufficient precision and to compute with the given formula (to solve the equations) in a sufficiently accurate computational scheme.

Determinism is most famously expressed by Pierre-Simon Laplace (1814, p.2):

> An intelligence that, at a given instant, could comprehend all the forces by which nature is animated and the respective situation of the beings that make it up, if moreover it were vast enough to submit these data to analysis, would encompass in the same formula the movements of the greatest bodies of the universe and those of the lightest atoms. For such an intelligence nothing would be uncertain, and the future, like the past, would be open to its eyes.

Laplace's dramatic statement is often erroneously interpreted as a belief in perfect predictability now rendered untenable by the chaos theory. But he was describing determinism: given the state of the system (the universe) at some time, we have a formula (a set of differential equations) that gives, in principle, the state of the system at any later time. Nowhere will one find a claim about the computability, by us humans, of all the consequences of the laws of mechanics. Indeed, the quote appears in the introduction of a book on probability. Laplace is, in fact, assuming incomplete knowledge from the start and uses probabilities to make rational inferences. If it were not for quantum mechanics, Laplace's statement would still stand, unaffected by deterministic chaos.

To illustrate the problems with computability, consider the simple but important example of the (deterministic) Bernoulli shift map defined by

$$f : [0, 1] \to [0, 1] : \; x_{n+1} = 2x_n \pmod 1.$$

On numbers in binary representation, this map has a particularly simple effect: shift the binary point one place to the right and discard the first digit. For example, if $x_0 = 0.10110$ (which corresponds to the decimal 0.6875), then $x_1 = 0.01100$ (decimal 0.375). Now, any rational starting number x_0 is represented by a repeating sequence of 0s and 1s and hence leads to a periodic orbit of f, while any irrational x_0 is represented by a nonrepeating sequence of 0s and 1s and hence leads to a nonperiodic orbit. This latter sequence would appear as unpredictable as the sequence of heads and tails generated by flipping a coin, the quintessentially random process. Since there is an irrational number arbitrarily close to every rational number and vice versa, the map exhibits a sensitive dependence on initial conditions. In practice, on a computer, numbers are always represented with finite precision; hence, the computations become completely meaningless once—after a finite number of iterations—all significant digits have been removed. In the standard 32-bit (4-byte) floating point arithmetic with 23-bit mantissa, this will be after roughly 23 iterations.

The significance of the Bernoulli shift map is that dynamical systems theory tells us that its dynamics lies at the heart of the so-called "horseshoe dynamics," which in turn is commonly found in (the wide class of) systems with homoclinic (i.e., expanding and reinjecting) orbits (Wiggins, 1988). It means that in many situations, all we can say about a system's dynamics is of a statistical nature. A quantitative measure of the sensitivity on initial conditions, and therefore a measure of the predictability horizon, is provided by the leading Lyapunov exponent.

The possibility of small causes having large effects (in a perfectly deterministic universe) was anticipated by many scientists before Lorenz, and even before the birth of dynamical systems theory, which is generally accepted to have its origins in Poincaré's work on differential equations toward the end of the 19th century. Maxwell (1876, p. 20) wrote: "There is a maxim which is often quoted, that 'The same causes will always produce the same effects.'" After discussing the meaning of this principle, he adds: "There is another maxim which must not be confounded with [this], which asserts that 'Like causes produce like effects.' This is only true when small variations in the initial circumstances produce only small variations in the final state of the system." He then gives the example of how a small displacement of railway points sends a train on different courses.

Others have often used the image of the weather:

Wiener (1954/55):

> It is quite conceivable that the general outlines of the weather give us a good, large picture of its course for hours or possibly even for days. However, I am profoundly skeptical of the unimportance of the unobserved part of the weather for longer periods.

> To assume that these factors which determine the infinitely complicated pattern of the winds and the temperature will not in the long run play their share in determining major features of weather, is to ignore the very real possibility of the self-amplification of small details in the weather map. A tornado is a highly local phenomenon, and apparent trifles of no great extent may determine its exact track. Even a hurricane is probably fairly local where it starts, and phenomena of no great importance there may change its ultimate track by hundreds of miles.

Poincaré (1908, p. 67):

> Why have meteorologists such difficulty in predicting the weather with any certainty? Why is it that showers and even storms seem to come by chance, so that many people think it quite natural to pray for rain or fine weather, though they would consider it ridiculous to ask for an eclipse by prayer? We see that great disturbances are generally produced in regions where the atmosphere is in unstable equilibrium. The meteorologists see very well that the equilibrium is unstable, that a cyclone will be formed somewhere, but exactly where they are not in a position to say; a tenth of a degree more or less at any given point, and the cyclone will burst here and not there, and extend its ravages over districts it would otherwise have spared. If they had been aware of this tenth of a degree, they could have known it beforehand, but the observations were neither sufficiently comprehensive nor sufficiently precise, and that is the reason why it all seems due to the intervention of chance.

Even earlier, Franklin (1898, p. 173) had used an analogy surprisingly similar to Lorenz's:

> ... an infinitesimal cause may produce a finite effect. Long range detailed weather prediction is therefore impossible, and the only detailed prediction which is possible is the inference of the ultimate trend and character of a storm from observations of its early stages; and the accuracy of this prediction is subject to the condition that the flight of a grasshopper in Montana may turn a storm aside from Philadelphia to New York!

Duhem (1954, p. 141) used Hadamard's theorem of 1898 on the complicated geodesic motion on surfaces of negative curvature to "expose fully the absolutely irremediable physical uselessness of certain mathematical deductions." If such incomputable behavior is possible in mechanics, "the least complex of physical theories," Duhem goes on to ask rhetorically, "Should we not meet that ensnaring conclusion in a host of other, more complicated problems, if it were possible to analyse the solutions closely enough?"

Many had contemplated the possibility of sensitive dependence on initial conditions, but Lorenz was the first to see it actually happening quantitatively in the numbers spit out by his Royal McBee computing machine, and to be sufficiently intrigued by it to study it more closely in the delightfully simple system of equations now bearing his name (Lorenz, 1963). Indeed, while most scientists, with Duhem, had looked to complicated systems for unpredictable behavior, Lorenz found it in simple ones and thereby made it amenable to analysis.

GERT VAN DER HEIJDEN

See also **Chaotic dynamics; Determinism; General circulation models of the atmosphere; Horseshoes and hyperbolicity in dynamical systems; Lorenz equations**

Further Reading

Bricmont, J. 1995. Science of chaos or chaos in science? *Physicalia Magazine*, 17: 159–208

Duhem, P. 1954. *The Aim and Structure of Physical Theory*, translated by Ph.P. Wiener, Princeton: Princeton University Press (original French edition, *La Théorie Physique: Son Objet, Sa Structure, 2ème éd.*, Paris: Marcel Rivière & Cie, 1914; 1st edition, 1906)

Franklin, W.S. 1898. A book review of P. Duhem, *Traité Elémentaire de Méchanique Chimique fondée sur la Thermodynamique*, vols. I and II, Paris, 1897, *The Physical Review*, 6: 170–175

Gleick, J. 1987. *Chaos: Making a New Science*, London: Heinemann, and New York: Viking

Laplace, P.-S. 1814. *Philosophical Essay on Probabilities*, translated from the fifth French edition of 1825 by A.I. Dale, Berlin and New York: Springer, 1995 (first edition published in French, 1814)

Lorenz, E.N. 1963. Deterministic nonperiodic flow. *Journal of the Atmospheric Sciences*, 20: 130–141

Lorenz, E.N. 1993. *The Essence of Chaos*, Seattle: University of Washington Press and London: University College London Press

Maxwell, J.C. 1876. *Matter and Motion*, New York: Van Nostrand

Poincaré, H. 1908. *Chance*. In *Science and Method*, pp. 64–90, translated by F. Maitland, London: Thomas Nelson and Sons, 1914 (original French edition, *Science et Méthode*, Paris: E. Flammarion, 1908)

Wiener, N. 1954/55. Nonlinear prediction and dynamics. *Proceedings of the Third Berkeley Symposium on Mathematical Statistics and Probability*, Berkeley: University of California Press, vol. 3, pp. 247–252; *Mathematical Reviews*, 18: 949 (1957); reprinted in 1981. *Norbert Wiener: Collected Works with Commentaries*, vol. III, edited by P. Masani, Cambridge, Massachusetts: MIT Press, pp. 371–376

Wiggins, S. 1988. *Global Bifurcations and Chaos*, Berlin and New York: Springer

CALOGERO–MOSER MODEL

See **Particles and antiparticles**

CANDLE

At about the time that John Scott Russell was systematically studying hydrodynamic solitons on a Scottish canal, Michael Faraday, the brilliant English experimental physicist and physical chemist, organized his annual Christmas Lectures on facets of natural philosophy for young people. These included a series on the candle that began with the claim (Faraday, 1861):

> There is no better, there is no more open door by which you can enter into the study of natural philosophy than by considering the physical phenomena of a candle.

Although this assertion may have startled some of his listeners, Faraday went on to support it with a sequence of simple yet elegant experiments that clearly expose the structure and composition of a candle flame, demonstrating a stream of energy-laden vapor feeding into the flame and suggesting an analogy with the process of respiration in living organisms (Day & Catlow, 1994). (An engraving showing Faraday presenting one of these lectures can be found on a recent British 20-pound note.)

While the details are intricate (Fife, 1988), the flame of a candle can be regarded globally as a dynamic process balancing two flows of energy: the rate at which energy is dissipated by the flame (through emission of heat and light) and the rate at which energy is released from the wax as the flame eats its way down the candle. Let us define variables as follows.

- P is the power dissipation by the flame, in units of (say) joules per hour.
- E is the chemical energy stored in the wax of the candle, in units of (say) joules per centimeter.
- v is the speed at which the flame moves down the candle.

If the rates of dissipation and energy input are equal, then the velocity of the flame is determined by the power balance condition

$$P = vE, \tag{1}$$

implying $v = P/E$ (cm/h). Metaphorically, the flame must digest energy at the same rate at which it is eaten.

Consider a family of cylindrical candles with various diameters (d), and assume the dissipation rates of their flames to be independent of the sizes of the candles. Since stored chemical energy is proportional to the area of cross section ($E \propto d^2$), power balance implies

$$v \propto 1/d^2. \tag{2}$$

Some measured flame speeds for typical candles are plotted on a log–log scale in Figure 1, where the dashed line of slope -2 indicates a $1/d^2$ dependence. From this figure, it is evident that the inverse square dependence of Equation (2) is obeyed for larger candles. For smaller candles, v is somewhat less than expected, because the flames are not so large.

Although Equation (2) was derived for candles, Equation (1) is quite general, expressing a global constraint that governs the dynamics of many kinds of nonlinear diffusion, including the propagation of nerve impulses (Scott, 2002). For a smooth axon described by the Hodgkin–Huxley equations, power balance is established between electrostatic energy released from the fiber membrane and ohmic dissipation by circulating ionic currents, implying $v \propto \sqrt{d}$. (Plotted in Figure 1, this dependence would have a slope of $+\frac{1}{2}$.) For myelinated nerves, on the other hand, evolutionary pressures require that $v \propto d$, corresponding to a slope of unity in Figure 1.

In the language of nonlinear dynamics, a candle flame provides a physical example of an attractor, which is evidently stable because moderate disturbances (small gusts of air) do not extinguish the flame by forcing it out of its basin of attraction. As the air becomes still, the flame returns to its original shape and size. The task of lighting a candle, on the other hand, requires getting the wick hot enough—above an

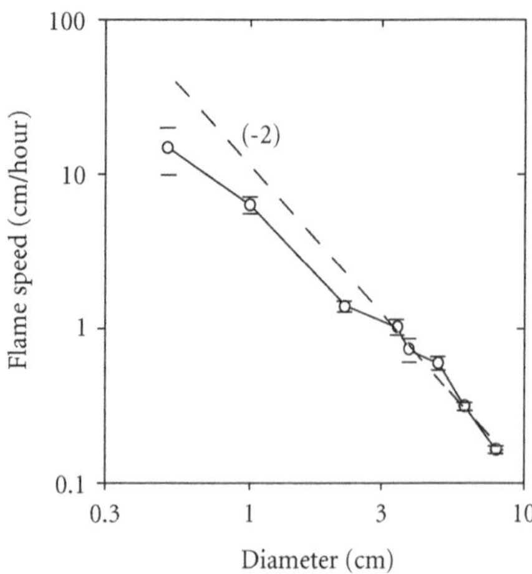

Figure 1. Measurements of flame speeds (v) for candles of different diameters (d). The error bars indicate rms deviations of about six individual measurements. (Data courtesy of Lela Scott MacNeil Scott (2003).)

ignition threshold and into the basin of attraction—so that a viable flame is established. Qualitatively similar conditions govern the firing of a nerve axon, leading to an all-or-nothing response.

From a more intuitive perspective, an ignition threshold implies the power balance indicated in Equation (1), where the corresponding flame is unstable. Above the threshold, instability arises from the establishment of a positive feedback loop, in which the flame releases more than enough chemical energy than is needed to maintain its temperature. Such a positive feedback loop is represented by the diagram

Release of energy (vE)
$\downarrow \quad \uparrow$
Dissipation of energy (P),

with the gain about the loop being greater than unity, implying an increase in the flame size with time (Scott, 2003). Eventually, this temporal increase is limited by nonlinear effects in the release and dissipation of energy, reducing the loop gain to unity as the fully developed flame is established.

The candle flame is an example of a reaction-diffusion (or autocatalytic) process, going back to an early suggestion by Robert Luther, a German physical chemist (Luther, 1906). Following a lecture demonstration of a chemical wave, Luther claimed that such systems should support traveling waves at a speed proportional to $\sqrt{D/\tau}$ where D is the diffusion constant for the reacting components (in units of distance squared per unit of time) and τ is a delay time for the onset of the reaction. During the 1930s, autocatalytic systems were studied in the context of genetic diffusion through spatially dispersed biological species, and in 1938 the equation

$$D\frac{\partial^2 u}{\partial x^2} - \frac{\partial u}{\partial t} = \frac{1}{\tau}u(u-a)(u-1), \tag{3}$$

where a is a threshold parameter lying in the range $(0, \frac{1}{2}]$, was used by Soviet scientists Yakov Zeldovich and David Frank-Kamenetsky to represent a flame front. These authors showed that uniform traveling waves solutions of Equation (3) propagate at a fixed speed given by the expression

$$v = (1-2a)\sqrt{D/2\tau}, \tag{4}$$

which includes both Luther's factor $\left(\sqrt{D/\tau}\right)$ and the power balance condition. Long overlooked by the neuroscience community, this early work on flame propagation offers a convenient model for the leading edge of a nerve impulse, confirming Faraday's intuition (Scott, 2002).

Alwyn Scott

See also **Attractors; Flame front; Hodgkin–Huxley equations; Power balance; Zeldovich–Frank-Kamenetsky equation**

Further Reading

Day, P. & Catlow, C.R.A. 1994. *The Candle Revisited: Essays on Science and Technology*, Oxford and New York: Oxford University Press

Faraday, M. 1861. *A Course of Six Lectures on the Chemical History of a Candle*. Reprinted as *Faraday's Chemical History of a Candle*, Chicago: Chicago Review Press, 1988

Fife, P.C. 1988. *Dynamics of Internal Layers and Diffusive Interfaces*, Philadelphia: Society for Industrial and Applied Mathematics

Luther, R. 1906. Räumliche Fortpflanzung chemischer Reaktionen. *Zeitschrift für Elektrochemie* 12(32): 596–600 (English translation in *Journal of Chemical Education* 64 (1987): 740–742)

Scott, A.C. 2002. *Neuroscience: A Mathematical Primer*, Berlin and New York: Springer

Scott, A.C. 2003. *Nonlinear Science: Emergence and Dynamics of Coherent Structures*. 2nd edition, Oxford and New York: Oxford University Press

CANONICAL VARIABLES

See **Hamiltonian systems**

CANTOR SETS

See **Fractals**

CAPACITY DIMENSION

See **Dimensions**

CAPILLARY WAVES

See **Water waves**

CARDIAC ARRHYTHMIAS AND THE ELECTROCARDIOGRAM

In the early 1900s, Willem Einthoven developed the string galvanometer to measure the potential differences in the body surface associated with the heartbeat and introduced a nomenclature for the deflections of the electrocardiogram that are still used today. For this work, Einthoven was awarded a Nobel prize in 1924 (Katz & Hellerstein, 1964).

The electrocardiogram (ECG) is a measurement of the potential difference between two points on the surface of the body. Because the heart generates waves of electrical activation that propagate through the heart during the cardiac cycle, ECG measurements reflect cardiac activity. Over the past century, physicians have learned how to interpret the electrocardiogram to diagnose a variety of different cardiac abnormalities. Although interpreting an ECG is difficult, this entry introduces the basic principles.

In order to appreciate the ECG, it is first necessary to have a rudimentary knowledge about the spread of the cardiac impulse in the heart. The heart is composed of four chambers, the right and left atria, and the right and left ventricles (see Figure 1). The atria are electrically connected to each other, but are insulated from the ventricles everywhere except in a small region called the atrioventricular (AV) node. The ventricles are also electrically connected to each other. The rhythm of the heart is set by the sinoatrial node located in the right atrium, which acts as the pacemaker of the heart. From a mathematical perspective, this pacemaker is an example of a nonlinear oscillator. Thus, if the rhythm is perturbed, for example, by delivering a shock to the atria, then in general the timing of subsequent firings of the sinus node may be reset (i.e., they occur at different times than they would have if the shock had not been delivered), but the frequency and amplitude of the oscillation will remain the same. A wave of excitation initiated in the sinus node travels through the atria, then through the atrioventricular node, and then through specialized Purkinje fibers to the ventricles. The wave of electrical excitation is associated with a wave of mechanical contraction so that the cardiac cycle is associated with contraction and pumping of the blood through the body. The right and left atria are comparatively small chambers and act as collection points for blood. The right atrium collects blood from the body and the left atrium collects blood from the lungs. The right ventricle pumps blood to the lungs to be oxygenated, whereas the left ventricle pumps blood that has returned to the heart from the lungs to the rest of the body. The right atrium and right ventricle are separated by the tricuspid valve that prevents backflow of blood during the ventricular contraction. Similarly, the left atrium and left ventricle are separated by the mitral valve. In order to pump the blood, the ventricles are comparatively large and muscular.

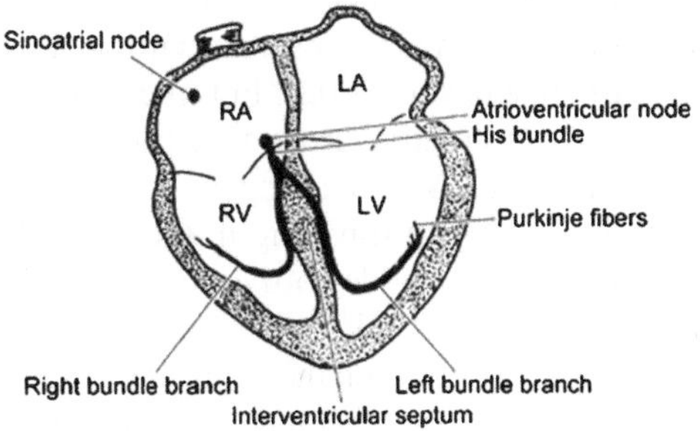

Figure 1. A schematic diagram of the heart. Adapted from Goldberger & Goldberger (1994) with permission.

In the normal ECG, there are several main deflections labeled the P wave, the QRS complex, and the T wave, Figure 2a (Goldberger & Goldberger, 1994). The P wave is associated with the electrical activation of the atria, the QRS complex is associated with the electrical activation of the ventricles, and the T wave is associated with the repolarization of the ventricles. The duration of the PR interval reflects the conduction time from the atria to ventricles, which is typically 120–200 ms. The duration of the QRS complex reflects the time that it takes for the wave of excitation to activate the ventricles. Because of the specialized Purkinje fibers, the wave of activation spreads rapidly through the ventricles so that the normal duration of the QRS complex is less than 100 ms. The time interval from the beginning of the QRS complex to the end of the T wave, called the QT interval, reflects the time that the ventricles are in the contraction phase. The

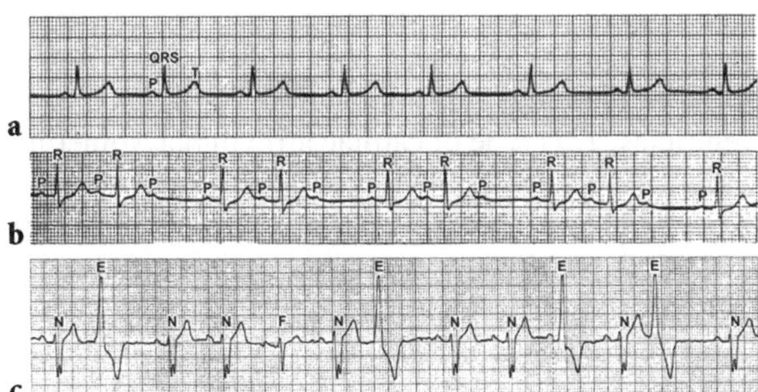

Figure 2. Sample electrocardiograms. In all traces, one large box represents 0.2 s. (a) The normal electrocardiogram. The P wave, QRS complex, and T wave are labeled. (b) 3:2 Wenckebach rhythm, an example of a second-degree heart block. There are 3 P waves for each R wave in a repeating pattern. (c) Parasystole. The normal beats, labeled N, occur with a period of about 790 ms, and the abnormal ectopic beats, labeled E, occur with a regular period of 1300 ms s. However, when ectopic beats fall too soon after the normal beats, they are blocked. Normal beats that occur after an ectopic beat are also blocked. If a normal and ectopic beat occur at the same time, the complex has a different geometry, labelled F for fusion. In this record, the number of normal beats occurring between ectopic beats is either 4, 2, or 1, satisfying the rules given in the text. Panels (a) and (b) are adopted from Goldberger & Goldberger (1994), with permission. Panel (c) is adapted from Courtemanche et al. (1989) with permission.

duration of QT interval depends somewhat on the basic heart rate. It is shorter when the heart is beating faster. For heart beats in the normal range, the QT interval is typically of the order of 300–450 ms.

On examining an ECG, one first looks for P waves. The presence of the P wave indicates that the heart beat is being generated by the normal pacemaker. In the normal heart, each P wave is followed by a QRS complex and then a T wave. The heart rate is often measured by time intervals between two consecutive R waves. Abnormally fast heart rates, faster than about 90 beats per minute, are called tachycardia, and abnormally slow heart rates, slower than about 50 beats per minute, are called bradycardia.

Reduced to the basics, all cardiac arrhythmias (i.e., abnormal cardiac rhythms) are associated with abnormal initiation of a wave of cardiac excitation, abnormal propagation of a wave of cardiac excitation, or some combination of the two. Given such a simple underlying concept, it is not surprising that mathematicians have been attracted to the study of cardiac arrhythmias, or that many cardiologists are mathematically inclined. However, despite the apparent simplicity, cardiac arrhythmias can manifest themselves in many different ways, and it is still not always possible to figure out the mechanism of an arrhythmia in any given individual. The following is focused on some arrhythmias that are well understood and that have interesting mathematical analyses.

One class of cardiac arrhythmias is associated with conduction defects through the AV node. In first-degree heart block, the PR interval is elevated above its normal value, but each P wave is followed by a QRS complex and T wave. However, in second-degree heart block, there are more P waves than QRS complexes, as some of the atrial activations do not propagate to the ventricles. This type of cardiac arrhythmia, sometimes called Wenckebach rhythms (after Karel Frederik Wenckebach, a Dutch-born Austrian physician, who studied these rhythms at the beginning of the 20th century), has repeatedly attracted theoretical interest (Katz & Hellerstein, 1964). It is common to classify Wenckebach rhythms by a ratio giving the number of P waves to the number of QRS complexes. For example, Figure 2b shows a 3:2 heart block.

In the 1920s, Balthasar van der Pol and J. van der Mark developed a mathematical model of the heart as coupled nonlinear oscillators that displayed striking similarities to the Wenckebach rhythms. We now understand that in a number of different models, as the frequency of atrial activation is increased, different types of $N : M$ heart block can be observed (van der Pol & van der Mark, 1928). In fact, theoretical models have demonstrated that if there is $N : M$ heart block at one stimulation frequency and an $N' : M'$ heart block at a higher frequency, then the $N + N' : M + M'$ heart block is expected at some intermediate stimulation frequency. This result provides a mathematical classification complementary to the cardiological classification, and can be confirmed in clinical settings (Guevara, 1991). Finally, in third-degree heart block, there is a regular atrial rhythm and a regular ventricular rhythm (at a slower frequency), but there is no coupling between the two rhythms. Such rhythms in mathematics are called quasi-periodic.

A different type of rhythm that appeals to mathematicians is called parasystole. In the "pure" case, the normal sinus rhythm beats at a constant frequency, and an abnormal (ectopic) pacemaker in the ventricles beats at a second slower frequency (Glass et al., 1986; Courtemanche et al., 1989). Figure 2c shows the normal (N) beats and the ectopic (E) beats. If the ectopic pacemaker fires at a time outside the refractory period of the ventricles, then there is an abnormal ectopic beat, identifiable on the ECG by a morphology distinct from the normal beat, and the following normal sinus beat is blocked. If the normal and abnormal beats occur at the same time, this leads to a fusion (F) beat.

Surprisingly, this simple mechanism has amazing consequences that can be appreciated by forming a sequence of integers that counts the number of sinus beats between two ectopic beats. In general, for fixed sinus and ectopic frequencies and a fixed refractory period, in this sequence, there are at most three integers, where the sum of the two smaller integers is one less than the largest integer. Moreover, given the values of the parameters, it is possible to predict the three integers. The mathematics for this problem is related to the "gaps and steps" problem in number theory. Both AV heart block and parasystole lead to mathematical predictions of cardiac arrhythmias in humans, which have been tested in experimental models and in humans. Such arrhythmias are diagnosed and treated when necessary, by physicians who have no knowledge of the underlying mathematics. Thus, to date, the mathematical analysis of these arrhythmias is of little medical interest.

From a medical perspective, the most important class of arrhythmias is called re-entrant arrhythmias. In these arrhythmias, the period of the oscillation is set by the time an excitation takes to travel in a circuitous path, rather than the period of oscillation of a pacemaker. The re-entrant circuit can be found in a single chamber of the heart or can involve several anatomical features of the heart (Josephson, 2002). In typical atrial flutter, there is a wave circulating around the tricuspid valve in the right atrium; in Wolf–Parkinson–White syndrome, there can be excitation traveling in the normal circuit from atria to the AV node to the ventricles, but then traveling retrogradely back to the atria via an abnormal accessory pathway between the ventricles and the atria. Also in some patients who have had a heart attack, there is a re-entrant circuit contained entirely in the ventricles. In all these three re-entrant arrhythmias, a part of the circuit is believed to be a comparatively thin strand of tissue.

Considering these re-entrant arrhythmias from a mathematical perspective, the wave often appears to be circulating on a one-dimensional ring. This conceptualization developed by cardiologists has an important implication for therapy: "if you cut the ring, you can cure the rhythm." By inserting catheters directly into a patient's heart and delivering radio frequency radiation to precisely identified loci, the cardiologist destroys heart tissue and can often cure these serious arrhythmias. In these cases, the cardiologist is thinking like a topologist since changing the topology of the heart cures the arrhythmia. Moreover, there is a body of mathematics that has studied the properties of excitation traveling on one-dimensional rings (Glass et al., 2002).

Other reentrant arrhythmias are not as well understood and are not as easily treated. Many theoretical and experimental studies (*See* **Cardiac muscle models**) have documented spiral waves circulating stably in two dimensions and scroll waves circulating in three dimensions (Winfree, 2001). Since real hearts are three-dimensional, and there is still no good technology to image excitation in the depth (as opposed to the surface) of the cardiac tissue, the actual geometry of excitation waves in cardiac tissue associated with some arrhythmias is not as well understood and is now the subject of intense study.

From an operational point of view, it is suggested that any arrhythmia that cannot be cured by a small localized lesion in the heart is a candidate for a circulating wave in two or three dimensions. Such rhythms include atrial and ventricular fibrillation. In these rhythms, there is evidence of strong fractionation (breakup) of excitation waves giving rise to multiple small spiral waves and patterns of shifting blocks. Tachycardias can also arise in the ventricles in patients other than those who have experienced a heart attack, or perhaps occasionally in hearts with completely normal anatomy, and in these individuals it is likely that spiral and scroll waves are the underlying geometries of the excitation. A particularly dangerous arrhythmia, polymorphic ventricular tachycardia (in which there is a continually changing morphology of the ECG complexes), is probably associated with spiral and scroll waves that undergo a meander. New technologies are presenting unique opportunities to image cardiac arrhythmias in model systems and the clinic, and nonlinear dynamics is suggesting new strategies for controlling cardiac arrhythmias. For a summary of advances up until 2002 see the collection of papers in Christini & Glass (2002).

Despite the great advances in research and medicine over the past 100 years, there is still a huge gap between what is understood and what actually happens in the human heart. The only way to appreciate this gap is to toss out the models and start looking at real data from patients who are experiencing complex arrhythmia as measured on the ECG. All who plan to model cardiac arrhythmias are encouraged to take this step.

LEON GLASS

See also **Cardiac muscle models; Scroll waves; Spiral waves; Van der Pol equation**

Further Reading

Christini, D. & Glass, L. (editors). 2002. Focus issue: mapping and control of complex cardiac arrhythmias. *Chaos*, 12(3)

Courtemanche, M., Glass, L., Bélair, J., Scagliotti, D. & Gordon, D. 1989. A circle map in a human heart. *Physica* D, 40: 299–310

Glass, L., Goldberger, A., & Bélair, J. 1986. Dynamics of pure parasystole. *American Journal of Physiology*, 251: H841–H847

Glass, L., Nagai, Y., Hall, K., et al. 2002. Predicting the entrainment of reentrant cardiac waves using phase resetting curves. *Physical Review* E, 65: 021908

Goldberger, A.L. & Goldberger, E. 1994. *Clinical Electrocardiography: A Simplified Approach*, 5th edition, St Louis: Mosby

Guevara, M.R. 1991. Iteration of the human atrioventricular (AV) nodal recovery curve predicts many rhythms of AV block. In *Biomechanics, Biophysics, and Nonlinear Dynamics of Cardiac Function*, edited by L. Glass, P. Hunter, P. & A., McCulloch Theory of Heart: New York: Springer, pp. 313–358

Josephson, M.E. 2002. *Clinical Cardiac Electrophysiology: Techniques and Interpretations*, Philadelphia and London: Lippincott Williams & Wilkins

Katz, L.N. & Hellerstein, H.K. 1964. Electrocardiography. In *Circulation of the Blood: Men and Ideas*, edited by A.P. Fishman & D.W. Richards, Oxford and New York: Oxford University Press, pp. 265–354

van der Pol, B. & van der Mark, J. 1928. The heartbeat considered as a relaxation oscillation, and an electrical model of the heart. *Philosophical Magazine*, 6: 763–765

Winfree, A.T. 2001. *The Geometry of Biological Time*, 2nd edition, Berlin and New York: Springer

CARDIAC MUSCLE MODELS

Cardiac muscle was created by evolution to pump the blood, and contractions of cardiac muscle, as of any muscle, are governed by calcium (Ca) ions. Increased concentration of calcium ions inside a cardiac cell ($[\mathrm{Ca}]_i$) induces a contraction, and diminished concentration induces relaxation (diastole). Calcium ion concentration in a cardiac cell is governed by many mechanisms. Importantly, a signal to increase $[\mathrm{Ca}]_i$ is given by an abrupt increase in membrane potential E, which is called an action potential (AP). The membrane potential E in cardiac cells is described by reaction diffusion equations.

Cardiac models have the mathematical structure of the well-known Hodgkin–Huxley (HH) equations:

$$\partial E/\partial t = -I(E, g_1, \ldots, g_N))/C + \nabla\cdot(D\nabla E), \quad (1)$$

$$\partial g_i/\partial t = (\tilde{g}_i(E) - g_i)/\tau_i(E),\ i = 1, \ldots, N, \quad (2)$$

where g_i are the gating variables (describing opening and closing of the gates of ionic channels), $\tilde{g}_i(E)$ are the steady values of those variables, $\tau_i(E)$ are associated time constants, $I(\cdots)$ is the transmembrane ionic current, and the diffusive term $\nabla \cdot (D\nabla E)$ describes the current flowing from the neighboring cells. In this description of the coupling, the anisotropy of the heart is properly described with an anisotropic diffusivity tensor. Equations (1) and (2) are similar to standard HH equations (with $N=3$), and this formulation was used for the first cardiac model introduced by Denis Noble in 1962. As more and more ionic channels were discovered, they were incorporated into more detailed models: $N=6$ for a modified Beeler–Reuter (BR) model, and $N>10$ for the Luo–Rudy model and for recent Noble models. Although the original HH formulation was based on analytic functions for the $\tilde{g}_i(E)$ and $\tau_i(E)$, experimentally measured functions can be used in the equations, as shown in Figure 1. This results in both physical transparency and faster numerical calculations.

The functions $g_i(E)$ and $\tau_i(E)$ have a simple physical interpretation—the dynamics of each gating variable g_i are governed by the membrane potential E only. For fixed E, Equations (2) are linear and independent; each of them describes relaxation to the steady value $\tilde{g}_i(E)$ with the time constant $\tau_i(E)$. The characteristic times $\tau_i(E)$ scan four orders of magnitude (see Figure 1(a)), leading to qualitative understanding using time-scale separation.

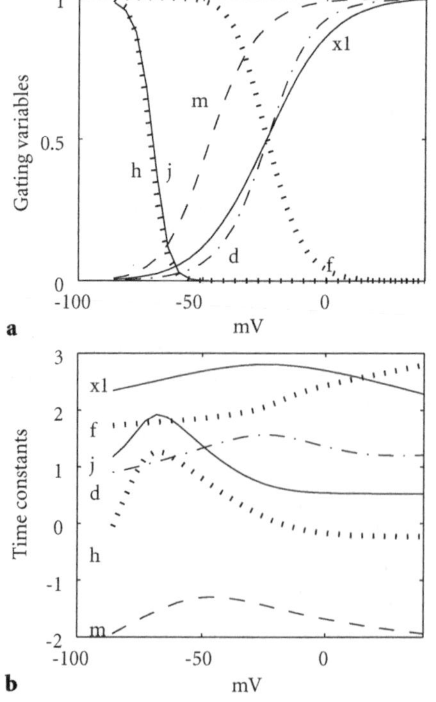

Figure 1. BR model with eight variables. (a) Gating variables $\tilde{g}_i(E)$. (b) Time constants $\tau_i(E)$ (log scale).

Noble's 1962 Model

This original cardiac model (N4 model), which is the key for understanding all subsequent models, consists of the following four equations (Noble, 1962).

$$\partial E/\partial t = -(I_{\text{Na}}(E,m,h) + I_{\text{K}}(E,n))/C + D\nabla^2 E, \qquad (3)$$

$$\partial n/\partial t = (\tilde{n}(E) - n)/\tau_n, \qquad (4)$$

$$\partial m/\partial t = (\tilde{m}(E) - m)/\tau_m, \qquad (5)$$

$$\partial h/\partial t = (\tilde{h}(E) - h)/\tau_h, \qquad (6)$$

where

$$I_{\text{Na}}(E,m,h) = (g_{\text{Na}}m^3h + \varepsilon_1)(E - E_{\text{Na}}), \qquad (7)$$

$$I_{\text{K}}(E,n) = (\alpha g_{\text{K}}n^4 + \varepsilon_2)(E - E_{\text{K}}), \qquad (8)$$

A cardiac action potential has a duration of about 200 ms, which is about two orders of magnitude longer than that of a typical nerve impulse. To describe cardiac action potential, just one time constant in the HH equations was increased by a factor of 100, two small terms (ε_1 and ε_2) were added, and other time constants were adjusted to incorporate cardiac experimental results.

To observe how the N4 model works, note that it has three gating variables: m, n, and h. The characteristic time τ_n is about 100 ms, while τ_m and τ_h are almost two orders of magnitude faster (shorter). This permits adiabatic elimination of the fast equations (5) and (6), thereby replacing Equations (3)–(6) with a system of equations only (system N2) (Krinsky & Kokoz, 1973):

$$\partial E/\partial t = -(I_{\text{Na}}(E,\tilde{m},\tilde{h}) + I_{\text{K}}(E,n))/C + D\nabla^2 E, \qquad (9)$$

$$\partial n/\partial t = (\tilde{n}(E) - n)/\tau_n. \qquad (10)$$

Note that the current I_{Na} becomes

$$I_{\text{Na}} = (g_{\text{Na}}\tilde{m}^3(E)\tilde{h}(E) + \varepsilon_1)(E - E_{\text{Na}}), \qquad (11)$$

which contains only the known functions $\tilde{m} \equiv \tilde{m}(E)$ and $\tilde{h} \equiv \tilde{h}(E)$. The behavior of this model is illustrated in Figure 2.

Nullclines $dE/dt = 0$, and $dn/dt = 0$ for a Purkiknje fiber are shown in Figure 2(a). A Purkinje fiber has a pacemaker activity, which the nullclines (in the phase plane) show directly. Note that there is only one fixed point S, which is unstable, thus giving rise to limit cycle oscillations.

The nullclines for a myocyte are shown in Figure 2(b). There is no pacemaker activity in myocytes; thus, the nullclines show the absence of a limit cycle. Although there are two unstable fixed points (S′ and S″), they do not induce a limit cycle because there is a third fixed point (S) that is stable and determines the resting potential.

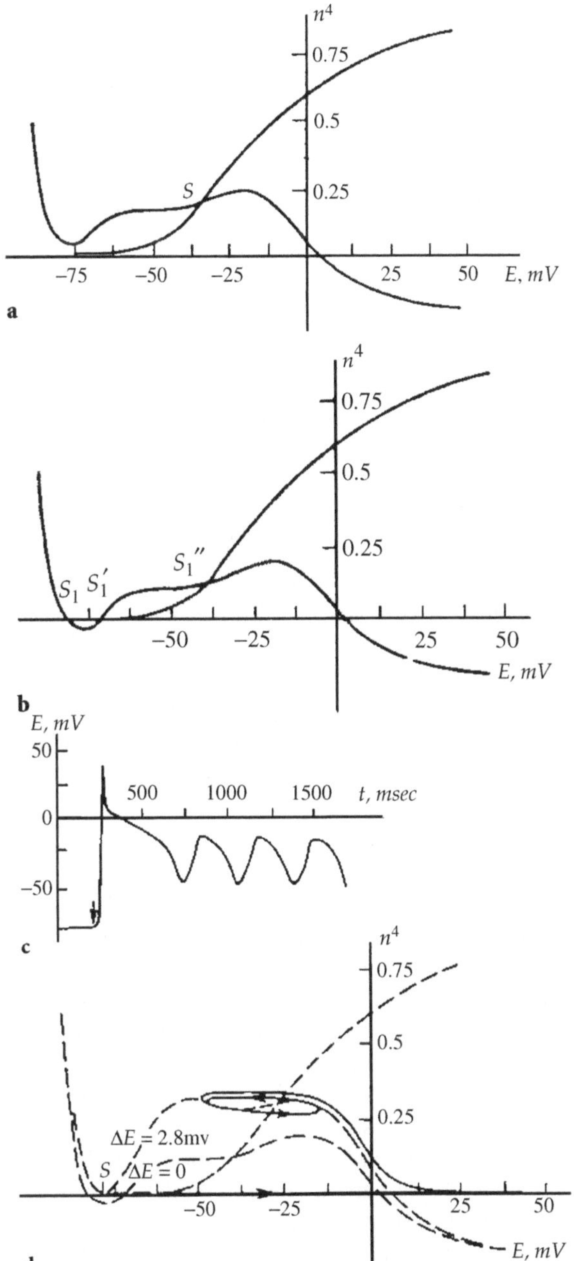

Figure 2. Analysis of the N4 model based on adiabatic elimination of two fast variables. (a) A Purkiknje fiber with a pacemaker activity. The nullclines of the reduced system N2 contain only one fixed point S that is unstable, thus giving rise to the limit cycle oscillations. (b) An excitable myocyte. There are three fixed points: S′ and S″ are unstable, and point S determines the resting potential. (c) Analysis of the effect of an arrhythmogenic drug aconitine, showing oscillations of the plateau of action potential (AP). (d) As in (c), where the nullclines show a small-amplitude limit cycle on the plateau of AP. (Parameters: (a) $\alpha = 1.2$; (b) $\alpha = 1.3$; (c) and (d) $\alpha =$ 1.3; h $= \tilde{h}(E - \delta E)$; $\delta E = 2.8$ mV.)

Analysis of the effect of the arrhythmogenic drug aconitine (inducing oscillations on the plateau of the action potential) is shown in Figures 2(c) and (d). Aconitine induces dangerous oscillations because of a shift of the voltage dependence of Na

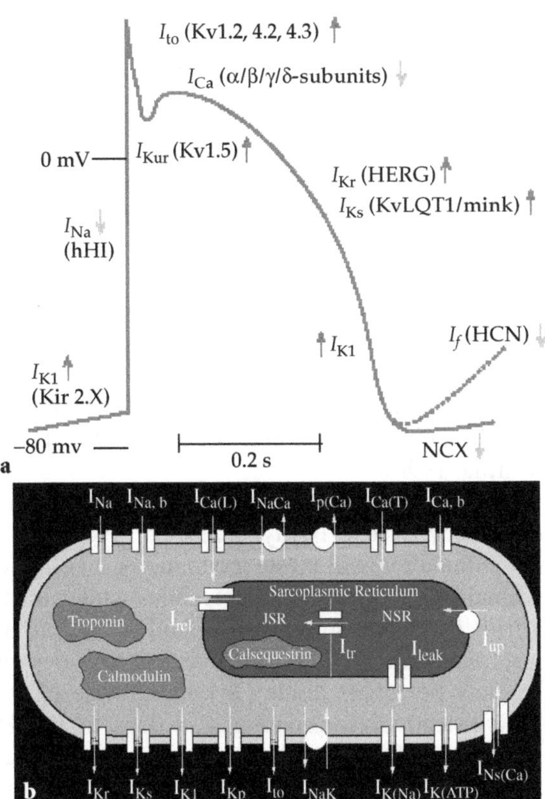

Figure 3. (a) Action potential showing the principal currents that flow in each phase, with the corresponding subunit clones shown in parentheses. (Courtesy of S. Nattel) (b) Main ionic currents (Courtesy of Y. Rudy).

inactivation variable h. Oscillations on the plateau of action potential are shown in Figure 2(c), and the corresponding nullclines are shown in Figure 2(d). The shift of $h(E)$ dependence results in the disappearance of the stable resting point and appearance of a small-amplitude limit cycle on the plateau of the AP. The electrophysiological characteristics in the full (N4) and reduced (N2) models are the same with an accuracy of 0.1–0.2 mV (Krinsky & Kokoz, 1973).

Contemporary Models

Recent models include more ionic currents (see Figures 1 and 3), and also incorporate a change of intracellular ionic concentrations. For example, the BR model contains an additional equation for the concentration $[\mathrm{Ca}]_i$ of intracellular Ca:

$$\partial Ca_i / \partial t = I_{\mathrm{Ca}}. \tag{12}$$

The Luo–Rudy (LR) model, and the Noble model are widely known. The LR model (Rudy, 2001) can be downloaded from http://www.cwru.edu/med/CBRTC/LRdOnline. Noble models are described at http://www.cellml.org/ and http://cor.physiol.ox.ac.uk/. The model of human ventricular myocyte (Ten Tusscher et al.

2003) is also available at http://www-binf.bio.uu.nl/khwjtuss/HVM.

In contemporary models the gap between the 100 and 1 ms time scales has been filled by many ionic currents, and (contrary to what was seen in the first cardiac model) the graphs even intersect each other (see Figure 1(b)). This makes the separation for fast and slow variables dependent upon the phase of AP, so results as in Figure 2 cannot be obtained directly. Instead, events on every time scale must be analyzed separately, eliminating adiabatically equations with a faster time scale and considering variables with a larger time scale fixed, or even postulating a model with only two to four equations (Keener & Sneyd, 1998; Fenton & Karma, 1998).

Models that do not follow the HH formalism are needed because HH-type models predict that a point stimulation will create a circular or an elliptical distribution of membrane potential, while in the experiments a quadrupolar distribution was found. Thus, *bidomain models* were created that describe separately potentials inside (E_i) and outside (E_o) of a cell instead of considering $E_o = 0$ as in the HH formulation. These models correctly reproduce many important electrophysiological effects, and turn out to be useful for understanding the mechanisms of cardiac defibrillation (Trayanova et al., 2002). The integration of these models, however, is computationally more expensive than the integration of HH models. *Markov chain models* are used to describe transitions between states of single ionic channels linking genetical defects with cardiac arrhythmias. They also depart from the HH formalism. *Anatomical models* incorporate cardiac geometry and tissue structure, but they require months of laborious measurements on anatomical slices cut from the heart. A new approach is being developed at INRIA in France, which aims to create models for every cardiac patient and is intended to be used in clinics. Cardiac contractions are measured and incorporated into the model. NMR tomography methods were used that permit obtaining anatomical data in a few minutes, not months, see http://www-sop.inria.fr/epidaure/. More cardiac models and authors can be found at http://www.cardiacsimulation.org/.

Dynamics of Myocardial Tissue

In the past, breakthroughs were due to very simple models beginning with the pioneering work of Norbert Wiener and Arturo Rosenbluth (Wiener & Rosenbluth, 1946). They led the way to understanding rotating waves using a cellular automata model, where a cardiac cell can be in three states only: rest, excitation, and refractory. This model explained anatomical reentry (a wave rotating around an obstacle (Wiener & Rosenbluth, 1946)), and predicted a free rotating wave. Partial differential equations yield more refined refined results (Keener & Sneyd, 1998).

As two- and three-dimensional studies of rotating waves are time consuming, it is often convenient to use a two-variable model, permitting increased speed of calculations by two orders of magnitude. Numerical simulation of ionic models can be accelerated either by adiabatic elimination of fast variable m (Na) activation or by slowing down its dynamics and increasing the diffusion coefficient to keep the propagation velocity unchanged.

Propagation Failure

As cardiac cells are connected via gap junctions, an excitation can be blocked when propagating from one cardiac cell to another, similar to the propagation failure in myelinated nerves. When an excitation propagates from auricles (A) to ventricles (V) via the AV node, a periodic pattern can be observed: for example, from every three pulses, only two pulses propagate, and the third pulse is blocked (3:2 Wenckebach periodicity). Other periodicities $N : (N - 1)$ can also be observed. Propagation block can be observed on any cardiac heterogeneity when the period T of stimulation is shorter than the restoration time R (refractory period). Usually, Wenckebach rhythms with only small periods N are observed because

$$T_N \sim N^{-2}, \tag{13}$$

where T_N is an interval of T that can yield. Wenckebach rhythms with period N. When an excitation block (Wenckebach rhythm) occurs in a two- or three-dimensional excitable medium, it generically gives rise to a wave break that evolves into a rotating wave. For cardiac muscle, initiation of a rotating wave is a dangerous event, often leading to life-threatening cardiac arrhythmias.

A new approach is being developed at INRIA in France, to create patient-specific models to be used in clinics. A 3-d electro-mechanical model of the heart is automatically adpted to a time series of volumetric cardiac images gated on the ECG (Ayache et al., 2001) providing useful quantitative parameters on the heart function in a few minutes, not months, see http://www-sop.inria.fr/epidaure/.

V. Krinsky, A. Pumir, and I. Efimov

See also **Hodgkin–Huxley equations; Myelinated nerves; Neurons; Scroll waves; Synchronization; Van der Pol equation**

Further Reading

Ayache, N., Chapelle, D., Clément, F., Coudiére, Y., Delingette, H., Désidéri, J.A., Sermesant, M. Sorine, M. & Urquiza, J. 2001. Towards model-based estimation of the cardiac electro-mechanical activity from ECG signals and ultrasound images. In *Functional Imaging and Modeling of the Heart (FIMH'01), Helsinki, Finland*, Lecture Notes in Computer Sciences, vol. 2230, Berlin, Springer, pp. 120–127

Chaos, topical issue: Ventricular fibrillation, 8(1), 1998
Fenton, F. & Karma, A. 1998. Vortex dynamics in three-dimensional continuous myocardium with fiber rotation: Filament instability and fibrillation. *Chaos*, 8: 20–47
Journal of Theoretical Biology, topical issue devoted to the work of Winfree, 2004
Keener, J. & Sneyd, J. 1998. *Mathematical Physiology*, New York: Springer
Krinsky, V. & Kokoz, Ju. 1973. Membrane of the Purkinje fiber. reduction of the noble equations to a second order system. *Biophysics*, 18: 1133–1139
Noble, D. 1962. A modification of the Hodgkin–Huxley equations applicable to Purkinje fiber action and pacemaker potential. *Journal of Physiology*, 160: 317–352
Noble, D. & Rudy, Y. 2001. Models of cardiac ventricular action potentials: iterative interaction between experiment and simulation. *Philosophical Transactions of the Royal Society A*, 359: 1127–1142
Rudy, Y. 2001. The cardiac ventricular action potential. In *Handbook of Physiology: A Critical, Comprehensive Presentation of Physiological Knowledge and Concepts. Section 2, The cardiovascular system. vol. 1, the heart*, edited by E. Page, H.A. Fozzard & R.J. Solaro, Oxford: Oxford University Press, pp. 531–547
Pumir, A., Romey, G. & Krinsky, V. 1998. De-excitation of cardiac cells. *Biophysical Journal*, 74: 2850–2861
Sambelashvili, A. & Efimov, I.R. 2002. Pinwheel experiment re-revisited. *Journal of Theoretical Biology*, 214: 147–153
TenTusscher, K.H.W.J., Noble, D., Noble, P.J. & Panfilov, A.V. 2003. A model of the human ventricular myocyte. *American Journal of Physiology*, 10: 1152
Trayanova, N., Eason, J. & Aguel, F. 2002. Computer simulations of cardiac defibrillation: a look inside the heart. *Computing and Visualization in Science*, 4: 259–270
Wiener, N. & Rosenbluth, N. 1946. The mathematical formulation of the problem of conduction of impulses in a network of connected excitable elements, specifically in cardiac muscle. *Archivos del Instituto de cardiologia de Mexico*, 16: 205–265

CASIMIRS

See **Poisson brackets**

CAT MAP

The cat map is perhaps the simplest area-preserving transformation that exhibits a high degree of chaos. In the development of the theory of dynamical systems, it served as a guiding example to illustrate new concepts such as entropy (Sinai, 1959) and Markov partitions (Adler & Weiss, 1967). The cat map owes its name to an illustration by V.I. Arnol'd showing the image of a cat before and after the application of the map. In the mathematical literature it is also referred to as "hyperbolic toral automorphism."

The torus M, which topologically has the shape of a doughnut, may be described by the points in the unit square (see Figure 1) where opposite sides are identified. Alternatively, one may think of a point on M as a point in the plane $\mathbb{R}^2$ modulo integer translations in $\mathbb{Z}^2$. This yields a representation of M as the coset space $\mathbb{R}^2/\mathbb{Z} := \{x + \mathbb{Z}^2 : x \in \mathbb{R}^2\}$. The cat map is now a mapping $\phi : M \to M$ defined by $x \mapsto Ax \pmod{\mathbb{Z}^2}$, where

$$x = \begin{pmatrix} x_1 \\ x_2 \end{pmatrix}, \qquad A = \begin{pmatrix} a & b \\ c & d \end{pmatrix}, \tag{1}$$

provided $a, b, c, d \in \mathbb{Z}$ are chosen such that $|\det A| = 1$ and A has eigenvalues $\lambda_\pm$ with modulus not equal to one. (This implies that both eigenvalues are real and distinct.) The matrix A used in our illustration is

$$A = \begin{pmatrix} 2 & 1 \\ 1 & 1 \end{pmatrix}. \tag{2}$$

Let us now explore some of the dynamical properties that show that the cat map is indeed completely "chaotic" (*See* **Chaotic dynamics**). Sensitivity on initial conditions is measured by the rate of divergence of two nearby points, x and $x + \delta x$, under iterations of ϕ. For any smooth map ϕ, the Taylor expansion for small δx yields $\phi(x + \delta x) = \phi(x) + D\phi_x \delta x + O(\delta x^2)$ where $D\phi_x$ is the differential of ϕ at x; it may be viewed as a linear map from TM_x to $TM_{\phi(x)}$, the tangent spaces at x and $\phi(x)$, respectively.

Because ϕ is linear in the case of the cat map, the above Taylor expansion is in fact exact, that is, $\phi(x + \delta x) = \phi(x) + D\phi_x\, \delta x$ with $D\phi_x = A$. If $v_\pm$ are the eigenvectors of A corresponding to the eigenvalues $\lambda_\pm$, let us denote by E_x^+ and E_x^- the subspaces of TM_x spanned by v_+ and v_-, respectively. As $\lambda_+ \neq \lambda_-$, we have

$$TM_x = E_x^+ \oplus E_x^-. \tag{3}$$

Furthermore, since $|\lambda_+| = 1/|\lambda_-| > 1$, we find

$$\|D\phi_x(\xi)\| \geq e^{\lambda}\|\xi\| \quad \text{if } \xi \in E_x^+, \tag{4}$$

$$\|D\phi_x(\xi)\| \leq e^{-\lambda}\|\xi\| \quad \text{if } \xi \in E_x^-, \tag{5}$$

where $\lambda = \ln|\lambda_+| > 0$ and $\|\cdot\|$ is the Euclidean norm. Hence, ϕ is expanding in the direction of v_+, and contracting in the direction of v_-, which will therefore be referred to as the unstable and stable directions, respectively. [Here, inequalities (4) and (5) are in fact equalities; inequalities become necessary in the case of more general Anosov maps, if one seeks uniform bounds with λ independent of x.] For the nth forward or backward iterates of the map (n a positive integer), we have, by the above arguments with ϕ replaced by $\phi^{\pm n}$,

$$\|D\phi_x^n(\xi)\| \geq e^{n\lambda}\|\xi\|, \quad \|D\phi_x^{-n}(\xi)\| \leq e^{-n\lambda}\|\xi\|, \quad \text{if } \xi \in E_x^+, \tag{6}$$

$$\|D\phi_x^n(\xi)\| \leq e^{-n\lambda}\|\xi\|, \quad \|D\phi_x^{-n}(\xi)\| \geq e^{n\lambda}\|\xi\|, \quad \text{if } \xi \in E_x^-. \tag{7}$$

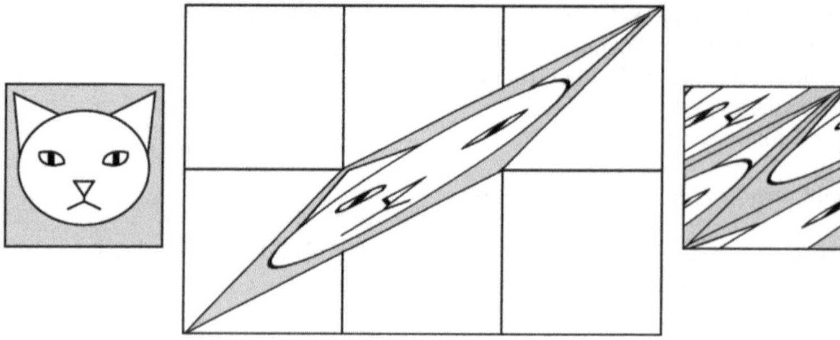

Figure 1. The cat map: the image of a cat in the unit square (left) is stretched by the matrix A (middle) and then re-assembled by cutting and translating (without rotation) the different parts of the cat's face back into the unit square (right). (Illustration by Federica Vasetti.)

The expansion/contraction is thus exponentially fast in time. Relations (3), (6), and (7) are equivalent to the statement that the cat map is an Anosov system. Special features of Anosov systems are ergodicity, mixing, structural stability, exponential proliferation of periodic orbits, and positive entropy h.

There is a particularly simple formula for the entropy h due to Sinai (1959), which states that $h = \lambda = \ln|\lambda_+|$. The number of fixed points of the nth iterate ϕ^n is equal to $|\det(1 - A^n)| = |(1 - \lambda_+^n)(1 - \lambda_-^n)|$ and is therefore asymptotically given by $\exp(nh)$, for large n. The notion of ergodicity implies that for any $f \in \mathrm{L}^1(M, \mathrm{d}x)$, for almost every $x \in M$ (that is, for all x up to a set of Lebesgue measure zero), we have

$$\lim_{N\to\infty} \frac{1}{N} \sum_{n=0}^{N-1} f(\phi^n x) = \langle f \rangle, \quad \langle f \rangle := \int_M f(x)\,\mathrm{d}x. \tag{8}$$

"Mixing" means that, if $f, g \in \mathrm{L}^2(M, \mathrm{d}x)$, then

$$\lim_{n\to\pm\infty} \int_M f(\phi^n x)\, g(x)\,\mathrm{d}x = \langle f \rangle \langle g \rangle. \tag{9}$$

Although the mixing property follows from general arguments for Anosov systems, it can be proved for ϕ directly by means of Fourier analysis. What is more, the rate of convergence in (9) is in fact exponentially fast in n for suitably smooth test functions f, g ("exponential decay of correlations") and super-exponentially fast for analytic ones.

Markov partitions are a powerful tool in the analysis of dynamical systems. In the case of the cat map the torus is divided into a finite collection of non-overlapping parallelograms $P_1, \ldots, P_N$ whose sides point in the directions of the eigenvectors v_+ and v_-, such that if $\phi(P_i)$ (or $\phi^{-1}(P_i)$) intersects with P_j, then it extends all the way across P_j. Let us construct an N by N matrix B whose coefficients are $B_{ij} = 1$ if the intersection of $\phi(P_i)$ with P_j is non-empty, and $B_{ij} = 0$ otherwise. A symbolic description of the dynamics of the cat map can now be obtained as follows. Consider doubly infinite sequences of the form $\cdots b_{-2}b_{-1}\underline{b_0}b_1b_2\ldots$, where b_n is an integer $1, \ldots, N$ with the condition that the number b_n can be followed by b_{n+1} only if $B_{b_n b_{n+1}} = 1$. To each such sequence we can associate a point x on M by requiring that for every $n \in \mathbb{Z}$, the parallelogram R_{b_n} contain the iterate $\phi^n(x)$. The symbolic dynamics of ϕ is now given by shifting the mark by one step to the right: the new word $\cdots b_{-1}b_0\underline{b_1}b_2b_3\cdots$ indeed represents the point $\phi(x)$. The dynamical properties of ϕ are thus encoded in the matrix B. In particular, the higher the rate of mixing of ϕ, the smaller the number of coefficients B_{ij} that are zero. This in turn means that we have fewer restrictions on $b_n \mapsto b_{n+1}$ and a typical orbit will be represented by a more "random" sequence of symbols.

Cat maps, as well as higher-dimensional toral automorphisms, are featured in most textbooks on dynamical systems and ergodic theory. An introduction to the basic concepts may be found in Arnol'd & Avez (1968), and more advanced topics, such as entropy, Markov partitions, and structural stability, are discussed, for example, in Adler & Weiss (1970), Katok & Hasselblatt (1995), Pollicott & Yuri (1998), and Shub (1987).

JENS MARKLOF

See also **Anosov and Axiom-A systems; Chaotic dynamics; Horseshoes and hyperbolicity in dynamical systems; Maps; Markov partitions; Measures**

Further Reading

Adler, R.L. & Weiss, B. 1967. Entropy, a complete metric invariant for automorphisms of the torus. *Proceedings of the National Academy of Sciences of the United States of America*, 57: 1573–1576

Adler, R.L. & Weiss, B. 1970. *Similarity of Automorphisms of the Torus*, Providence, RI: American Mathematical Society

Arnol'd, V.I. & Avez, A. 1968. *Ergodic Problems of Classical Mechanics*, New York and Amsterdam: Benjamin

Katok, A. & Hasselblatt, B. 1995. *Introduction to the Modern Theory of Dynamical Systems*, Cambridge and New York: Cambridge University Press

Pollicott, M. & Yuri, M. 1998. *Dynamical Systems and Ergodic Theory*, Cambridge and New York: Cambridge University Press
Sinai, Ya.G. 1959. On the concept of entropy for a dynamic system. *Doklady Akademii Nauk SSSR*, 124: 768–771
Shub, M. 1987. *Global Stability of Dynamical Systems*, Berlin and New York: Springer

CATALYTIC HYPERCYCLE

The concept of the "hypercycle" was invented in the 1970s in order to characterize a functional entity that integrates several autocatalytic elements into an organized unit (Eigen, 1971; Eigen & Schuster, 1977, 1978a,b). This concept is a key to understanding the dynamics of living organisms.

A catalytic hypercycle is defined as a cyclic network of autocatalytic reactions (Figure 1). Autocatalysts, in general, compete when they are supported by the same source of energy or material. Hypercyclic coupling introduces mutual dependence of elements and suppresses competition. Consequently, the fate of all members of a hypercycle is identical to that of the entire system and, in other words, no element of a hypercycle dies out provided the hypercycle as such survives.

The current view of biological evolution distinguishes periods of dominating Darwinian evolution based on variation, competition, and selection interrupted by rather short epochs of radical innovations often called *major transitions* (Maynard Smith & Szathmáry, 1995; Schuster, 1996). In the course of biological evolution major transitions introduce higher hierarchical levels. Examples are (i) the origin of translation from nucleic acid sequences into proteins including the invention of the genetic code, (ii) the transition from independent replicating molecules to chromosomes and genomes, (iii) the transition from the prokaryotic to the eukaryotic cell, (iv) the transition from independent unicellular individuals to differentiated multicellular organisms, (v) the transition from solitary animals to animal societies, and (vi) presumably a series of successive transitions from animal societies to humans. All major transitions introduce a previously unknown kind of cooperation into biology. The hypercycle is one of very few mechanisms that can deal with cooperation of otherwise competing individuals. It is used as a model system in prebiotic chemistry, evolutionary biology, theoretical economics, as well as in cultural sciences.

The simplest example of a catalytic hypercycle is the *elementary hypercycle*. It is described by the dynamical system

$$\frac{\mathrm{d}x_i}{\mathrm{d}t} = x_i \left(f_i x_{i-1} - \sum_{j=1}^{n} f_j x_{j-1} x_j \right); \quad i, j = 1, 2, \ldots, n; \; i, j = \bmod n. \tag{1}$$

The catalytic interactions within a hypercycle form a directed closed loop comprising all elements, often called Hamiltonian arc: $1 \to 2 \to 3 \to \cdots \to n \to 1$ (Figure 1). Hypercycles are special cases of replicator equations of the class

$$\frac{\mathrm{d}x_i}{\mathrm{d}t} = x_i \left(\sum_{j=1}^{n} a_{ij} x_j - \sum_{j=1}^{n} \sum_{k=1}^{n} a_{jk} x_j x_k \right); \quad i, j, k = 1, 2, \ldots, n \tag{2}$$

with $a_{ij} = f_i \cdot \delta_{i-1,j}$; $i, j = \bmod n$. (The 'mod n' function implies a cyclic progression of integers, $1, \ldots, n-1, n, 1, \ldots$. The symbol $\delta_{i,j}$ represents Kronecker's symbol: $\delta = 1$ for $i = j$ and $\delta = 0$ for $i \neq j$.)

For positive rate parameters and initial conditions inside the positive orthant (the notion of an *orthant* refers to the entire section of a Cartesian coordinate system in which the signs of variables do not change. In n-dimensions, the positive orthant is defined by $\{x_i > 0 \; \forall i = 1, 2, \ldots, n\}$.). The trajectory of a hypercycle remains within the orthant:

$$f_i > 0, \; x_i(0) > 0 \; \forall i = 1, 2, \ldots, n \Longrightarrow x_i(t) > 0 \; \forall t \geq 0.$$

In other words, none of the variables is going to vanish and hence, the system is *permanent* in the sense that no member of a hypercycle dies out in the limit of long times, $\lim_{t \to \infty} x_i(t) \neq 0 \; \forall i = 1, \ldots, n$. The existence of a Hamiltonian arc, that is, a closed loop of directed edges visiting all nodes once, is a sufficient condition for permanence (Hofbauer & Sigmund, 1998). It is also a necessary condition for low-dimensional systems with $n \leq 5$, but there exist permanent

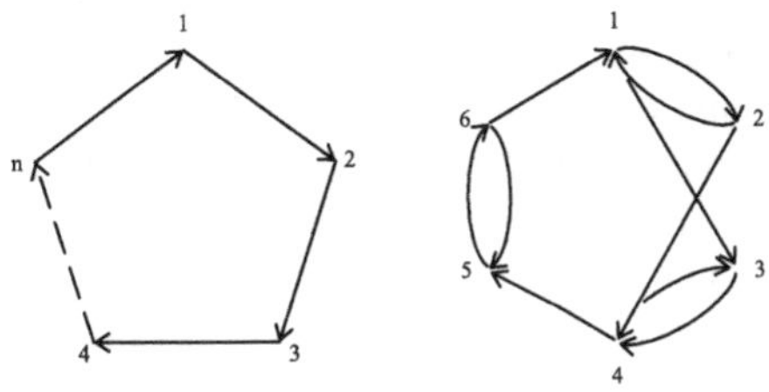

Figure 1. Definition of hypercycles. Replicator equations as described by the differential equation (2) can be symbolized by directed graphs: the individual species are denoted by nodes and two nodes are connected by an edge, $j \cdot \longrightarrow \cdot i$, if and only if $a_{ij} > 0$. The graphs of hypercycles consist of single Hamiltonian arcs as sketched on the left-hand side of the figure. These dynamical systems are permanent independent of the choice of rate parameters f_i. For $n \leq 5$ they represent the only permanent systems, but for $n \geq 6$ the existence of a single Hamiltonian arc is only a sufficient but not a necessary condition for permanence. The graph on the right-hand side, for example, does not contain a Hamiltonian arc but the corresponding replicator equation is permanent for certain choices of rate parameters (Hofbauer & Sigmund, 1998).

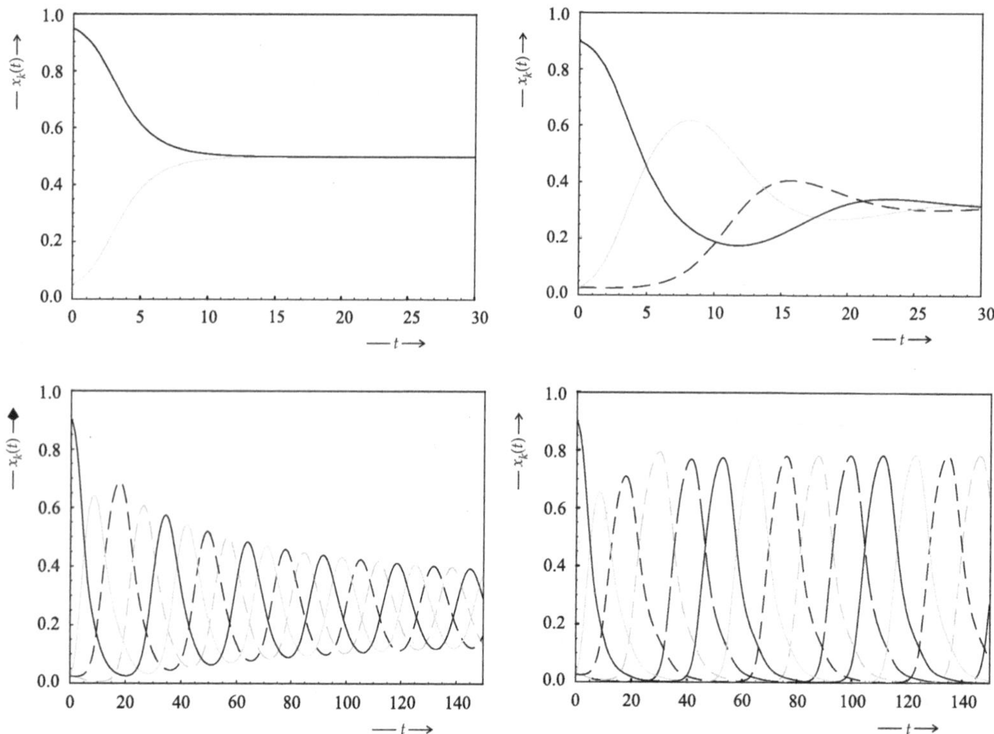

Figure 2. Solution curves of small elementary hypercycles. The figure shows the solution curves of Equation (1) with $f_1 = f_2 = \ldots = f_n = 1$ for $n = 2$ (upper left picture), $n = 3$ (upper right picture), $n = 4$ (lower left picture), and $n = 5$ (lower right picture). The initial conditions were $x_1(0) = 1 - (n - 1) \cdot 0.025$ and $x_k(0) = 0.025\ \forall\, k = 2, 3, \ldots, n$. The sequence of the curves $x_k(t)$ is $k = 1$ full black line, $k = 2$ full gray line, $k = 3$ hatched black line, $k = 4$ hatched grey line, and $k = 5$ black line with long hatches. The cases $n = 2$, 3, and 4 have stable equilibrium points in the middle of the concentration space $c = (1/n, 1/n, \ldots, 1/n)$; Equation (1) with equal rate parameters, $n = 4$, and linearized around the midpoint c exhibits a marginally stable "center" and very slow convergence is caused by the nonlinear term, which becomes smaller as the system approaches c. For $n = 5$, the midpoint c is unstable and the trajectory converges toward a limit cycle (Hofbauer et al., 1991).

dynamical systems for $n \geq 6$ without a Hamiltonian arc; one example is shown in Figure 1.

The dynamics of Equation (1) remains qualitatively unchanged when all rate parameters are set equal: $f_1 = f_2 = \ldots = f_n = f$, which is tantamount to a *barycentric* transformation of the differential equation (Hofbauer, 1981). The hypercycle is invariant with respect to a rotational change of variables, $x_i \Longrightarrow x_{i+1}$ with $i = 1, 2, \ldots, n$; $i \bmod n$, it has one equilibrium point in the center, and its dynamics depends exclusively on n. Some examples with small n are shown in Figure 2. The systems with $n \leq 4$ converge toward stable equilibrium points, whereas the trajectories of Equation (1) with $n \geq 5$ approach limit cycles. Independent of n, elementary hypercycles do not sustain chaotic dynamics.

Hypercycles have two inherent instabilities, which are easily illustrated for molecular species: (i) The members of the cycle may also catalyze the formation of nonmembers that do not contribute to the growth of the hypercycle, and thus hypercycles are vulnerable to parasitic exploitation (Eigen & Schuster, 1978a,b), and (ii) concentrations of individual species in oscillating hypercycles ($n \geq 5$) go through very small values, and these species might become extinct through random fluctuations. More elaborate kinetic mechanisms can stabilize the system in Case (ii). Exploitation by parasites, Case (i), can be avoided by compartmentalization. Competition between different hypercycles is characterized by a strong nonlinearity in selection (Hofbauer, 2002): once a hypercycle has been formed and established, it is very hard to replace it by another hypercycle. Epochs with hypercyclic dynamics provide explanations for "once for ever" decisions or "frozen accidents."

PETER SCHUSTER

See also **Artificial life; Biological evolution**

Further Reading

Eigen, M. 1971. Self-organization of matter and the evolution of biological macromolecules. *Naturwissenschaften*, 58: 465–523

Eigen, M. & Schuster, P. 1977. The hypercycle. A principle of natural self-organization. Part A: emergence of the hypercycle. *Naturwissenschaften*, 64: 541–565

Eigen, M. & Schuster, P. 1978a. The hypercycle. A principle of natural self-organization. Part B: The abstract hypercycle. *Naturwissenschaften*, 65: 7–41

Eigen, M. & Schuster, P. 1978b. The hypercycle. A principle of natural self-organization. Part C: The realistic hypercycle. *Naturwissenschaften*, 65: 341–369

Hofbauer, J. 1981. On the occurrence of limit cycles in the Volterra–Lotka equation. *Nonlinear Analysis, 5: 1003–1007*
Hofbauer, J. 2002. Competitive exclusion of disjoint hypercycles. Zeitschrift für Physikalische Chemie, 216: 35–39
Hofbauer, J. & Sigmund, K. 1998. *Evolutionary Games and Replicator Dynamics, Cambridge and New York: Cambridge University Press*
Hofbauer, J., Mallet-Paret, J. & Smith, H.L. 1991. Stable periodic solutions for the hypercycle system. Journal of Dynamics and Differential Equations, 3: 423–436
Maynard Smith, J. & Szathmáry, E. 1995. *The Major Transitions in Evolution, Oxford and New York: Freeman*
Schuster, P. 1996. How does complexity arise in evolution? Complexity, 2(1): 22–30

CATASTROPHE THEORY

Many natural phenomena (cell division, the bursting of bubbles, the collapse of buildings, and so on) involve discontinuous changes, whereas the majority of applied mathematics is directed toward modeling continuous processes. On the other hand, catastrophe theory is primarily concerned directly with singular behavior and as such deals with properties of discontinuities directly. This approach has found use in many and diverse fields and at one time was heralded as a new direction in mathematics, uniting singularity and bifurcation theories and their applications (Zeeman, 1976).

A simple mechanical system that illustrates the important ideas of discontinuous changes and hysteresis is provided by Zeeman's catastrophe machine (Zeeman, 1972, Poston & Stewart, 1996). A sketch showing its construction is given in Figure 1(a). It is recommended that readers make such devices for themselves and experiment with them.

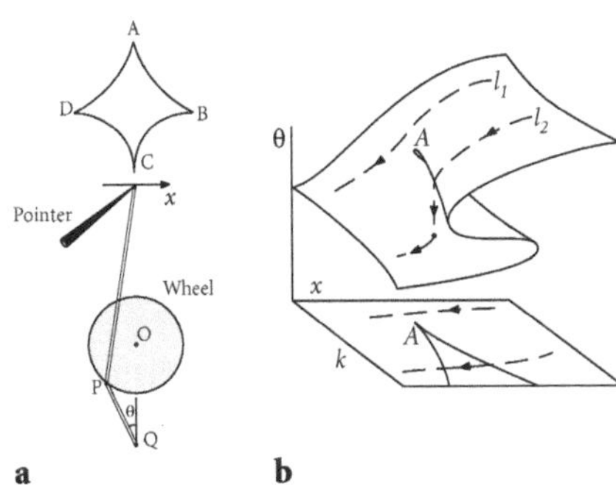

Figure 1. (a) Sketch of Zeeman's catastrophe machine, (b) A sketch of the three-dimensional solution surface and its projection onto two dimensions near the cusp A.

The lines between Q, P, and the pointer represent rubber bands attached to a disk that rotates around O. Movement of the pointer such that it remains outside of the region ABCD results in a smooth motion of the wheel from one equilibrium state to another. This is illustrated by the path ℓ_1 in Figure 1(b), where x and θ are as defined in Figure 1(a) and k is a measure of the stiffness of the bands. However, starting with the pointer below AB and moving the pointer horizontally to cross AD will cause an anticlockwise jump in the wheel. This is equivalent to following the path ℓ_2 in Figure 1(b). This equilibrium will remain until AB is crossed when the pointer is moved backwards. The loci DAB form a cusp in the parameter space of the system where AD and AB are lines of folds that meet at the cusp point A. The cusp is a projection down onto the plane of a three-dimensional folded surface. The region labeled ABCD comprises four such cusps, each of which can be described by a simple cubic equation. Indeed, the set can be obtained from an approximate model for the machine (Poston & Stewart, 1996).

The term *catastrophe* was introduced by René Thom in 1972 to describe such discontinuous changes in a system where a parameter is changed smoothly. Zeeman (1976) then coined the phrase *catastrophe theory*, and an explosion of applications arose ranging from the physical to the social sciences. An important idea put forward is that there are seven elementary catastrophes that classify most types of observed discontinuous behavior. They are the fold, cusp, swallowtail, and butterfly catastrophes and the elliptic, hyperbolic, and parabolic umbilic catastrophes. These describe all possible discontinuities for potential systems that are controlled by up to four variables. They are ordered according to their typicality of occurrence with the fold being the most common. A path of folds will be represented by a line of singular behavior in parameter space and a cusp will be formed when two such lines meet. Indeed, these two singularities are sufficient to cover most of the observed macroscopic critical behavior in practical applications.

One area of application where catastrophe theory has been used with considerable success is in optical caustics (Nye, 1999). Common experience of this phenomenon is observation of a distant light source through a drop of water on a window pane where a web of bright lines separated by dark regions can often be seen. The bright lines on the bottom of swimming pools are also examples of caustics where the bright sunlight is focused by the surface of the water. In this case, the line caustics are examples of paths of folds in the ray surfaces. An example of an optical cusp is provided by strong sunlight focused on the surface of a cup of coffee where two principal fold lines are made to meet by the curvature of the cup. In an outstanding series of experiments (Berry, 1976), a laser beam was shone through a water drop and a whole sequence of catastrophes was uncovered including swallowtails. All of the observed patterns can be reproduced in detail using the equations of ray optics (Nye, 1999).

Catastrophe theory has also been used to explore in some detail the state selection process in Taylor–Couette flow between concentric cylinders. In this case, even numbers of vortices are generated in the flow field, and the number that is realized depends on the length of the cylinders. For a given length of the cylinder, one state develops smoothly, with control parameter and neighboring states delimited by fold lines in parameter space. The fold lines meet in a cusp that has been observed experimentally (Benjamin, 1978) and calculated numerically (Cliffe, 1988) from the Navier–Stokes equations.

There has been considerable criticism of catastrophe theory on both technical and practical grounds (Arnol'd, 1986; Arnol'd et al., 1994). For example, it is known that critical behavior or bifurcations in some multidimensional gradient systems do not reduce to critical points of potentials (Guckenheimer, 1973). Also, the ideas have been applied to a wide range of social, financial, and biological applications where the governing rules are not known or are very primitive. Very often, it is a case of re-interpretation of common experience in terms of technical mathematical language, which is most often qualitative rather than quantitative (Arnol'd, 1986). Hence, it is often the case that disparate systems superficially appear the same, but closer examination reveals that they are quite different in terms of important details.

Tom Mullin

See also **Bifurcations; Critical phenomena; Development of singularities; Equilibrium; Taylor–Couette flow**

Further Reading

Arnol'd, V.I. 1986. *Catastrophe Theory*, Berlin and New York: Springer

Arnol'd, V.I., Afrajmovich, V.S. Il'yashenko, Yu.S. & Shil'nikov, L.P. 1994. *Bifurcation Theory and Catastrophe Theory*, Berlin and New York: Springer

Benjamin, T.B. 1978. Bifurcation phenomena in steady flows of a viscous fluid. *Proceedings of the Royal Society of London*, Series A, 359: 1–43

Berry, M.V. 1976. Waves and Thomas theorem. J *Advances in Physics*, 25(1): 1–26

Cliffe, K.A. 1988. Numerical calculations of the primary-flow exchange process in the Taylor problem. *Journal of Fluid Mechanics*, 197: 57–79

Guckenheimer, J. 1973. *Bifurcation and catastrophe*. In *Dynamical Systems*: Proceedings of the Symposium of University of Babia Salvador, 1971, pp. 95–109

Nye, J.F. 1999. *Natural Focusing and Fine Structure of Light: Caustics and Wave Dislocation*, Bristol: Institute of Physics Publishing

Poston, T. & Stewart, I. 1996. *Catastrophe Theory and Its Applications*, New York: Dover

Saunders, P.T. 1980. *An Introduction to Catastrophe Theory*, Cambridge and New York: Cambridge University Press

Thom, R. 1972. *Structural Stability and Morphogenesis*. Reading, MA: Benjamin

Zeeman, E.C. 1976. Catastrophe theory: a reply to Thom. In *Dynamical Systems Warwick 1974*, edited by A. Manning, Berlin and New York: Springer, pp. 405–419

Zeeman, E.C. 1972. A catastrophe machine. In *Towards a Theoretical Biology*, vol. 4, edited by C.H. Waddington, Edinburgh: Edinburgh University Press, pp. 276–282

CAUSALITY

Basic to science as well to common sense is the root notion of causality, that things and processes in the world we experience are not totally random, but ordered in specific ways that allow for rational understanding through explanations of various types. In the transition from a mythological worldview to a rational one, the notion of guilt (in Greek *aitia*), as in a criminal being guilty of a crime, was metaphorically used to describe nonpersonal natural processes whenever one phenomenon would necessarily follow another. As one aim of modern science is to uncover the deep structure of the world beyond our immediate experience, its explanations deal with the different determinants (or causes) of processual order. Three classical conceptions of inquiry, associated with the traditions of Plato, Aristotle, and Archimedes, have provided influential ideas about the role of causal explanations in science.

In the Platonic tradition, certain properties of nature could be derived from a priori given mathematical structures. As discovered by the Pythagorean philosophers of nature, a specific mathematical structure (relations between small whole numbers) could be the key to a part of nature (such as acoustics), so why not see whether that same structure could also describe other areas (such as astronomy)? Although the latter attempt failed, the general idea of using the power of demonstrative or formal necessity in mathematics as a descriptive tool of natural causality is still vital in many areas of science, including cosmology and high-energy physics. Also, the use of analog mathematical structures to describe phenomena has become standard.

The Aristotelian tradition did not refuse mathematical description but saw it only as a tool in a search for the real causes of things. For Aristotle, there were four kinds of causes: the material cause (*hyle* or *causa materialis*) that describes the stuff of which something is made, the formal cause (*eidos* or *causa formalis*) that describes the organization of something, the efficient cause (*to kineti'kon* or *causa efficiens*) that describes the active forces by which the phenomenon comes into being, and the final cause (*to 'telos* or *causa finalis*) that describes the purpose that it serves. Thus, for a house, bricks and mortar are its material cause, the plan of the house is its formal cause, the mason building it is its efficient cause, and the purpose of sheltering a family is its final cause.

In that ancient world of Aristotle, each phenomenon generally served a purpose. Aristotle did not consider the four causes as necessarily separate aspects of nature, but more like principles of explanation that may sometimes merge, as in the sprouting acorn becoming an oak tree where the formal, efficient, and final causes work together to actualize the characteristics of an adult oak. The popular renaissance critique of the final cause as implying the paradox of a future state (a goal) influencing a present state led to a dismissal of any pluralist conception of causes. In the subsequent mechanical world picture, only efficient causes were left as explanatory. The life sciences could not live up to this reduction but continued as a descriptive natural history with an essentially Aristotelian outlook, at least until Darwin would explain the final cause of adaptations by the efficient causes of natural selection and heredity. Yet, even the Darwinian paradigm could not account for the nonlinear mechanisms of self-organization in the organism's embryonic development. Such goal-like (teleological) properties of development and self-reproduction remained necessary yet unexplained preconditions for the mechanics of natural selection.

The Archimedean tradition was founded by disciplines more physical than mathematical, although combining the two, such as optics, astronomy, mechanics, and music theory. The mathematical relations discovered by Archimedes (ca. 287–212 BC) in his books on mechanics were not a priori, as in the Platonic tradition, but derived from experience. However, the Aristotelian pursuit after the causes of the phenomena, especially the final ones, was regarded as metaphysical and so ignored. The Archimedean tradition includes such names as Ptolemaios, Johannes Kepler, Galileo Galilei, and Isaac Newton. Kepler started out as a Platonic, aiming to explain the Copernican system (which placed the Sun at the center of the solar system, in opposition to the Ptolemaic system) by regular polyhedrons, but failed and found the right laws for planetary movements through a mathematical analysis of Tycho Brahe's empirical observations. Galileo found his laws of falling bodies by eschewing the search for a hypothetical cause of gravitational force and instead using measures proportional to the velocity of a moving body for the effect of this force.

Although the mechanical worldview emphasizes only the role of efficient causes as principles of explanation in physics, the very idea of cause gave way for a long period to skepticism about proving any real existence of causes (the positivism of David Hume), eventually seeing the concept of cause as a feature of the observing subject (the transcendental idealism of Immanuel Kant). Yet, in physics, the laws of nature as expressed in terms of mathematics came to play the explanatory role of the causes of a system's movement. It was assumed that any natural system could be encoded into some formalism (e.g., a set of differential equations representing the basic laws governing the system) and that the entailment structure of that formalism perfectly mirrored the (efficient) causal structure of that part of nature. This view was compatible with a micro-determinism where a system's macroscopic properties and processes are seen as completely determined by the behavior of the system's constituent particles, governed by deterministic laws. This view was deeply questioned by quantum physics, and by Rosen's work on fundamental limits on dynamic models of causal systems.

The complexity of causality, especially in goal-directed systems, was presaged by cybernetic research in the 1940s, dealing with negative feedback control (in animals and artifacts such as self-guiding missiles) and the role of information processing for the regulation of dynamic systems. A paradigmatic example is the closed causal loops connecting various physiological levels of hormones in the body, essential for maintaining a constant internal environment (homeostasis)—a modern version of the ancient symbol of uroboros, the snake biting its own tail.

The emergence of nonlinear science in the late 20th century increased interest in the old idea that causal explanations may not all reduce to simple one-to-one correspondences between cause and effect. The realization that complex systems may occupy different areas in phase space characterized by qualitatively distinct attractors, eventually separated by fractal borders, has questioned micro-determinism even more than the fact that many such nonlinear systems have a high sensitivity to the initial conditions (the butterfly effect).

Another insight is that complex things often self-organize as high-level patterns via processes of local interactions between simple entities. This emergence of wholes (or collective behavior of units) may be mimicked in causal explanations. Instead of top-down reductive explanations, nonlinear science provides additional bottom-up explanations of emergent phenomena. Although these explanations are still reductive (in the methodological sense that one can show exactly what is going on from step to step in a simulation of the system), the complexity makes prediction impossible; thus, computational shortcuts to predict a future state can rarely be found.

As an emergent whole is formed bottom-up, its organization constrains its components in a top-down manner, that has been called downward causation (DC). There are three interpretations of DC: in strong DC, the emergent whole (a human mind) effectuates a change in the very laws that govern the lower-level (like free will might suspend what normally determines the action of the brain's neurons). This interpretation is often related to vitalist and dualist conceptions of life and mind and is hard to reconcile with science. In medium DC, lower-level laws remain unaffected; yet, their boundary conditions are constrained by the emergent pattern (a mental representation), which is considered just as real as the components of the system (neuronal signaling). Here, the state of the higher level works as a factor selecting which of the many possible next states of the high level may emerge from the low level. In weak DC, the emergent higher levels are seen as regulated by stable (cyclic or chaotic) attractors for the dynamics of the lower level. The fact that a biological species consists of stable organisms is not solely a product of natural selection, but is a result of such internal, formal properties in the system's organization—the job of natural selection being to sort out the possible stable organisms and find those most fit for the given milieu (Kauffman, 1993; Goodwin, 1994). It should be emphasized that DC is not a form of efficient causation (involving a temporal sequence from cause to effect), rather it is a modern version of the Aristotelian formal and final cause.

Nonlinear science may be said to integrate a Platonic appreciation of universality (as found in the equations governing the passage to chaos in systems of quite distinct material nature), an Aristotelian acceptance of several types of causes, and an Archimedean pragmatism regarding the deeper status of determinism and causality. The latter is reflected in the fact that although deterministic chaos characterizes a large class of systems, this does not imply that these systems (or nature) are fully deterministic. The determinism refers to the mathematical tools used rather than an ontological notion of causality.

CLAUS EMMECHE

See also **Biological evolution; Butterfly effect; Determinism; Feedback**

Further Reading

Depew, D.J. & Weber, B.H. 1995. *Darwinism Evolving: System Dynamics and the Genealogy of Natural Selection*, Cambridge, MA: MIT Press

Emmeche, C., Stjernfelt, F. & Køppe, S. 2000. Levels, Emergence, and three versions of downward causation. In *Downward Causation. Minds, Bodies and Matter*, edited by P.B. Andersen, C. Emmeche, N.O. Finnemann & P.V. Christiansen, aarhus: Aarhus University Press, pp. 13–34

Fox, R.F. 1982. *Biological Energy Transduction: The Uroboros*, New York: Wiley

Goodwin, B. 1994. *How the Leopard Changed Its Spots: The Evolution of Complexity*, New York: Scribner's

Kauffman, S.A. 1993. *The Origins of Order. Self-organization and Selection in Evolution*, Oxford and New York: Oxford University Press

Pedersen, O. 1993. *Early Physics and Astronomy: A Historical Introduction*, Cambridge and New York: Cambridge University Press

Rosen, R. 2000. *Essays on Life Itself*, New York: Columbia University Press

Weinert, F. (editor). 1995. *Laws of Nature: Essays on the Philosophical, Scientific and Historical Dimensions*, Berlin and New York: Walter de Gruyter

CAUSTICS

See **Catastrophe theory**

CAVITY SOLITONS

See **Solitons, types of**

CELESTIAL MECHANICS

Although its origins can be traced back in antiquity to the first attempts of explaining the apparently irregular wandering of the planets, celestial mechanics was born in 1687 with the release of Isaac Newton's *Principia*. In 1799, Pierre-Simon Laplace introduced the term *mécanique céleste* (Laplace, 1799), which was adopted to describe the branch of astronomy that studies the motion of celestial bodies under the influence of gravity. Celestial mechanics is researched and developed by astronomers and mathematicians; the methods used to investigate it including numerical analysis, the theory of dynamical systems, perturbation theory, the quantitative and qualitative theory of differential equations, topology, the theory of probabilities, differential and algebraic geometry, and combinatorics.

Ptolemy's idea of the epicycles—according to which planets are orbiting on small circles, whose centers move on larger circles, whose centers move on even larger circles around the Earth—dominated astronomy in antiquity and the Middle Ages. In 1543, after working for more than 30 years on a new theory, Copernicus finished writing *De Revolutionibus*, a book in which he expressed the motion of the planets with respect to a heliocentric reference system, that is, one with the Sun at its origin. This allowed Kepler to use existing observations and formulate three laws of planetary motion, published in 1609 in *Astronomia Nova*:

(i) *The law of motion*: every planet moves on an ellipse having the sun at one of its foci.
(ii) *The law of areas*: every planet moves such that the segment planet-sun sweeps equal areas in equal intervals of time.
(iii) *The harmonic law*: the squares of the periods of any two planets are to each other as the cubes of their mean distances from the sun.

But all these achievements were empirical, based on observations, not on deductions obtained from a more general physical law.

In 1666, Newton came up with the idea that the attractive force responsible for the free fall of objects might be the same as the one keeping the Moon in its orbit. He conjectured that the expression of this force is directly proportional to the product of the masses and inversely proportional to the square of the distance between bodies. The tools of calculus, which he had invented independent of—and at about the same time as—Gottfried Wilhelm von Leibniz, allowed him to proceed with the computations. Two decades later, in *Principia*, Newton proved the correctness of his theory. Kepler's laws follow as consequences. They are obtained from the differential equations of the Newtonian two-body problem (also called the Kepler problem) given by a potential energy of the form $U(r) = -Gm_1m_2/r$, where G is the gravitational constant and r is the distance between the bodies of masses m_1 and m_2.

After Newton, mathematicians, such as Johann Bernoulli, Alexis Clairaut, Leonhard Euler, such as Jean d'Alembert, Laplace, Joseph-Louis Lagrange, Siméon Poisson, Carl Jacobi, Karl Weierstrass, and Spiru Haretu, attacked various theoretical questions of celestial mechanics (e.g., the 2- and 3-body problem, the lunar problem, the motion of Jupiter's satellites, and the stability of the solar system) mostly with the quantitative tools of analysis, algebra, and the theory of differential equations. On the practical side, the first resounding success in the field was the prediction of the return of Halley's comet, which occurred in 1758—as the calculations had shown. An even more spectacular achievement came in 1846 with the discovery of the planet Neptune on the basis of the perturbation theory through computations independently performed by John Couch Adams and Urbain Jean-Joseph Le Verrier. Having its origin in one of Euler's papers, which applied the calculus of trigonometric functions to the 3-body problem, perturbation theory is now an independent branch of mathematics (see, e.g., Verhulst, 1990; Guckenheimer & Holmes, 1983) that is often used in celestial mechanics.

An important theoretical advance was achieved by Henri Poincaré toward the end of the 19th century, when the questions of celestial mechanics—especially those concerning the Newtonian 3-body problem—received substantial attention. While working on this problem, Poincaré understood that the quantitative methods of obtaining explicit solutions for differential equations are not strong enough to help him make significant progress; thus, he tried to describe the qualitative behavior of orbits (e.g., stability, the motion in the neighborhood of collisions and at infinity, existence of periodic solutions) even when their expressions were too complicated or impossible to derive, which is the case in general. His ideas led to the birth of several branches of mathematics, including the theory of dynamical systems, nonlinear analysis, chaos, stability, and algebraic topology (Barrow-Green, 1997; Diacu & Holmes, 1996).

Today's astronomers working in celestial mechanics are primarily interested in questions directly related to the solar system, such as the accurate prediction of eclipses, orbits of comets and asteroids, the motion of Jovian moons, Saturn's rings, and artificial satellites. The invention of the electronic computer had a significant impact on the practical aspects of the field. The development of numerical methods allowed researchers to obtain good approximations of the planet's motion for long intervals of time. These types of results are also used in astronautics. No space mission, from the Sputnik, Apollo, and Pioneer programs to the space shuttle, the Hubble telescope launch, and the recent international space collaboration projects, could have been possible without the contributions of celestial mechanics.

Contemporary mathematicians active in the field are mostly dealing with theoretical issues, as, for example, the study of the general N-body problem and its particular cases (Wintner, 1947) ($N = 2, 3, 4$, the collinear, isosceles, rhomboidal, Sitnikov, and planetary problems, central configurations, etc.), attempting to answer questions regarding motion near singularities and at infinity, periodic orbits, stability and chaos, oscillatory behavior, Arnol'd diffusion, etc. Some researchers also study alternative gravitational forces like that suggested by Manev (Diacu et al., 2000; Hagihara, 1975; Moulton, 1970), which offers a good relativistic approximation at the level of the solar system.

Celestial mechanics and mathematics have always influenced each other's development, a trend that is far from slowing down today. The contemporary needs of space science bring a new wave of interest in the theoretical and practical aspects of celestial mechanics, making its connections with mathematics stronger than ever before.

FLORIN DIACU

See also ***N*-body problem; Perturbation theory; Solar system**

Further Reading

Barrow-Green, J. 1997. *Poincaré and the Three-Body Problem*, Providence, RI: American Mathematical Society

Diacu, F. & Holmes, P. 1996. *Celestial Encounters—The Origins of Chaos and Stability*, Princeton, NJ: Princeton University Press

Diacu, F., Mioc, V. & Stoica, C. 2000. Phase-space structure and regularization of Manev-type problems, *Nonlinear Analysis*, 41: 1029–1055

Guckenheimer, J. & Holmes, P. 1983. *Nonlinear Oscillations, Dynamical Systems, and Bifurcations of Vector Fields*, Berlin and New York: Springer

Hagihara, Y. 1975. *Celestial Mechanics*, vol. 2, part 1, Cambridge, MA: MIT Press

Laplace, P.-S. 1799. *Traité de mécanique céleste*. 5 vols. Paris, 1799–1825

Moulton, J.R. 1970. *An Introduction to Celestial Mechanics*, Dover

Verhulst, F. 1990. *Nonlinear Differential Equations and Dynamical Systems*, Berlin and New York: Springer

Wintner, A. 1947. *The Analytical Foundations of Celestial Mechanics*, Princeton, NJ: Princeton University Press

CELL ASSEMBLIES

The term *cell assembly* became well established as a neuropsychological concept with the publication in 1949 of Donald Hebb's book *The Organization of Behavior*. A cell assembly forms when a group of cortical neurons gets wired together through experience-dependent synaptic potentiation induced by nearly synchronous activity in pairs of connected neurons. Once formed, it serves as a cooperative and holistic mental representation. The mutual excitation within the active cell assembly influences network dynamics such that (i) the entire group of cells can be activated by stimulating only a part of it and (ii) once active, the ensemble displays afteractivity outlasting the stimulus triggering it. An important extension of the concept was suggested by Peter Milner, who proposed that negative feedback from lateral inhibition is essential to prevent mass activation of the entire network and the emergence of epileptic-like activity (Milner, 1957). This also introduces an element of competitive interaction between the cell assemblies allowing one assembly at a time to dominate the network. Hebb further proposed that cell assemblies activated in succession would wire together to form "phase sequences" that might be the physiological substrate of chains of association and the flow of thought. The concept of cell assemblies has been extensively elaborated in the context of cortical associative memory (see, e.g., Fuster, 1995).

Notably, abstract attractor network models (Hopfield, 1982; Rolls and Treves, 1997) may be regarded as mathematical instantiations of Hebb's cell assembly theory. They have been useful, for example, to estimate how many assemblies can be stored in a network of a given size, the existence of spurious attractors, and the effect of sparse activity and diluted network connectivity.

The extensive recurrent neuronal circuitry required for the formation of cell assemblies is abundant in the cerebral cortex in the form of horizontal intracortical and cortico-cortical connections. The latter are myelinated and support fast communication over large distances. The existence of experience-dependent synaptic changes in the form of "Hebbian" long-term synaptic potentiation is very well experimentally established today. In light of what we now know about brain circuitry and neuronal response properties, it is reasonable to assume that a cell assembly may extend across large areas of sensory, motor, and association cortices, perhaps even involving subcortical structures like the thalamus and basal ganglia. Rather than individual neurons, the actual nodes of the cell assembly are likely to be cortical modules, that is, minicolumns, comprising a couple of hundred neurons and with a diameter of about 30 µm (Mountcastle, 1998). Feature detectors in primary sensory areas like the visual orientation columns described by David Hubel and Torsten Wiesel are typical examples. Although the modular organization of higher-order cortical areas is less obvious, it has been proposed that the cortical sheet actually comprises a mosaic of such minicolumns. A single-cell assembly would then engage a small fraction of them distributed over a large part of the brain.

The reverberatory afteractivity in an active cell assembly was proposed by Hebb (1949) to correspond to "a single content in perception" and last some hundred milliseconds to half a second. Notably, this is of the same order as the duration of visual fixations and the period of cognitively relevant neurodynamical phenomena evident in EEG (e.g., the theta rhythm) and evoked potentials (Pulvermueller et al., 1994). Persistent activity over seconds but otherwise of the same origin has recently been proposed as a mechanism underlying working memory (Compte et al., 2000).

The cell assembly theory connects the cellular and synaptic levels of the brain with psychological phenomenology. It suggests explanations for the close interaction between memory and perception, including the holistic and reconstructive nature of Gestalt perception, perceptual fusion of polymodal stimuli, and perceptual illusions. Typical examples are perceptual completion and filling in as when looking at the Kaniza triangle and perceptual rivalry as demonstrated by the slowly alternating perception of an ambiguous three-dimensional stimulus like the Necker cube (Figure 1). An analogy to Gestalt perception in the motor domain would be motor synergies, and their existence can also be understood based on the cell assembly theory. In the motor control domain, however, the

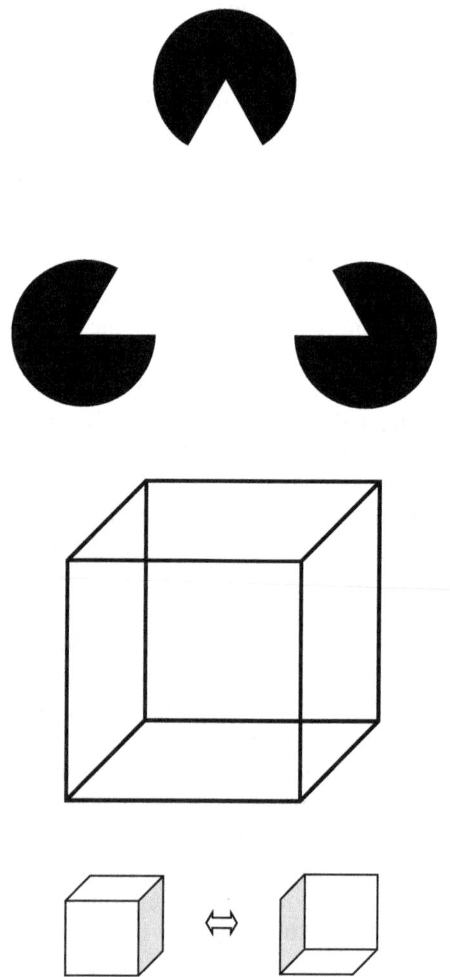

Figure 1. Two perceptual illusions that may be understood in terms of cell assembly dynamics.

temporal component of the underlying neurodynamics is critical, and for instance, finely tuned temporal sensory-motor coordination cannot be explained within this paradigm alone.

The neurodynamics of cell assemblies has been studied in a network with biophysically detailed model neurons (see, e.g., Lansner and Fransén, 1992; Fransén and Lansner, 1998), showing that the cellular properties of cortical pyramidal cells promote sustained activity. Moreover, the experimentally observed level of mutual excitation is sufficient to support such activity, and the measured magnitude of cortical lateral inhibition is effective in preventing coactivation of assemblies. The time to activate an entire assembly from a part of it was found to be within 50–100 ms in accordance with psychological experimental results. Modeling has further shown that neuronal adaptation due to accumulation of slow afterhyperpolarization (presumably together with synaptic depression) is a likely cause of termination of activity in an active assembly. Afteractivity may typically last some 300 ms, but the network dynamics is quite sensitive to endogenous monoamines such as serotonin, which acts by modulating the neuronal conductances underlying such adaptation. Further, simulations demonstrate that a cell assembly can survive even if the average conduction delay between the participating neurons increases to 10 ms, corresponding to a spatial extent of about 50 mm at axonal conduction velocities of 5 m/s. Despite quite powerful mutual excitation, reasonably low firing rates of cortical cells can be obtained in models with saturating synapses with slow kinetics (e.g., NMDA receptor gated channels) together with cortical feedback inhibition.

Although biologically highly plausible and supported by computational models, solid experimental evidence for the existence of cell assemblies is still lacking. Detection of their transient and highly distributed and diluted activity requires simultaneous noninvasive measurement in awake animals of the activity in a large number of neurons with high spatial and temporal resolution. This is still beyond the reach of current experimental techniques. Nevertheless, Hebb's original proposal has remained a vital hypothesis for more than half a century and it continues to inspire much experimental and computational research aimed at understanding how the brain works.

ANDERS LANSNER

See also **Attractor neural network; Electroencephalogram at large scales; Gestalt phenomena; Neural network models; Neurons**

Further Reading

Compte, A., Brunel, N., Goldman-Rakic, P.S. & Wang, X.-J. 2000. Synaptic mechanisms and network dynamics underlying visuospatial working memory in a cortical network model. *Cerebral Cortex*, 10: 910–923

Fransén, E. & Lansner, A. 1998. A model of cortical associative memory based on a horizontal network of connected columns. *Network: Computation in Neural Systems*, 9: 235–264

Fuster, J.M. 1995. *Memory in the Cerebreal Cortex*. Cambridge, MA: MIT Press

Hebb, D.O. 1949. *The Organization of Behavior*, New York: Wiley

Hopfield, J.J. 1982. Neural networks and physical systems with emergent collective computational properties, *Proceedings of the National Academy of Sciences, USA*, 81: 3088–3092

Lansner, A. & Fransén, E. 1992. Modeling Hebbian cell assemblies comprised of cortical neurons. *Network: Computation in Neural Systems*, 3: 105–119

Milner, P.M. 1957. The cell assembly: Mark II. *Psychological Review*, 64: 242–252

Mountcastle, V.B. 1998. *Perceptual Neuroscience: The Cerebral Cortex*. Cambridge, MA: Harvard University Press

Pulvermueller, F., Preissl, H., Eulitz, C., Pantev, C., Lutzenberger, W., Elbert, T. & Birbaumer, N. 1994. Brain rhythms, cell assemblies and cognition: evidence from the processing of words and pseudowords. *Psycoloquy*, 5(48)

Rolls, E. & Treves, A. 1997. *Neural Networks and Brain Function*. Oxford and New York: Oxford University Press

Scott, A.C. 2002. *Neuroscience: A Mathematical Primer*. Berlin and New York: Springer

CELLULAR AUTOMATA

Following a suggestion by Stanislaw Ulam, John von Neumann developed the concept of cellular automata (CA) in 1948. Von Neuman wanted to formalize a set of primitive logical operations that were sufficient to evolve the complex forms of organization necessary for life. In doing this, he constructed a two-dimensional self-replicating automaton and initiated not only the study of CA but also the idea, now popular among students of complexity theory, that highly complex global behavior can be generated from local interaction rules. Much of the interest in CA has been motivated by their ability to generate complex spatial and temporal patterns from simple rules. Because of this, they provide a rich modeling environment, having the twin virtues of mathematical tractability and representational robustness (see Burk, 1970, for discussion of much of the early work on CA).

The non-obvious connection between simple local rules of interaction and emergent complex global patterns offers a possible approach to an explanation of complexity through determination of its generating local interactions. Some researchers have gone so far as to suggest that the universe itself is a CA, or CA-like object, and that sets of local generative interaction rules can replace differential equations as the standard mathematical expression for physical models.

CA are spatially and temporally discrete symbolic dynamical systems defined in terms of a lattice of sites (or cells), an alphabet of symbols that can be assigned to lattice sites, and a local update rule that determines what symbol is assigned to each site at time $t+1$ on the basis of the site values in a local neighborhood of that site at time t. Given the local neighborhood structure, a neighborhood state is defined by an assignment of symbols from the alphabet to each site in the neighborhood. The local update rule can be specified in terms of a look-up table that lists the set of neighborhood states together with the symbol that each one assigns to the designated site at the next time step.

For example, if the lattice is isomorphic to the integers, the alphabet is 0, 1, and the neighborhood of any site s is the set of sites $s-1, s, s+1$, then there are 256 possible update rules, defined by the look-up table:

000	001	010	011	100	101	110	111
x_0	x_1	x_2	x_3	x_4	x_5	x_6	x_7

(The updated symbol is entered in the central cell. This defines what are called nearest-neighbor rules.) Here, each of the x_i is either 0 or 1 depending on the specific rule. Thus, they can be thought of as components of the rule.

A cellular automaton can be defined in any number of dimensions. Von Neuman's original automaton was two dimensional, as is what is perhaps the best-known CA—Conway's Game of Life (Gardner, 1970). This is one of the simplest two-dimensional CA known to be equivalent to a universal Turing machine.

If Σ is the full state space for a CA, then the local update rule defines a global mapping $\psi : \Sigma \to \Sigma$. Much of the analytical work on CA has been directed at determining the mathematical properties of the map ψ. A fundamental paper published by Hedlund in 1969 showed that CA are just the shift-commuting endomorphisms of the shift dynamical system. It is also known that surjectivity (a function is surjective, or onto, if every state has at least one predecessor) of the map ψ is decidable only in dimension one and that for one-dimensional additive rules (i.e., those satisfying the condition $\psi(\mu+\mu')=\psi(\mu)+(\mu')$), injectivity is equivalent to a certain rule-dependent complex polynomial having no roots that are nth roots of unity for any n. (A function is injective, or one to one, if every state has at most one predecessor. A function that is both surjective and injective is called reversible.)

From the early 1960s until the early 1980s, much of the work on CA was either simple applications or mathematical analysis. The terminology was not settled, and work can be found under the names *cellular structures, cellular spaces, homogeneous structures, iterative arrays, tessellation structures*, and *tessellation arrays*.

As computers became powerful enough to support the intense calculations required, however, an experimental mathematics approach became possible. In addition, solution of systems of differential equations by computer makes use of numerical combination rules on a discrete lattice, and CA are the simplest examples of such rules, adding impetus to interest in their study.

Concurrent with the appearance of powerful computers, work was initiated on the physics of computation, and the construction of reversible automata that, it was supposed, would be a discrete equivalent to the time-invariant differential equations of physics. More or less simultaneously, Stephen Wolfram began to publish a series of papers that popularized the study of elementary automata (see Wolfram, 1994), and by the mid-1980s, CA had emerged as a major field of interest among research in the field of complex systems theory.

Because of their generality as a modeling platform, CA have found wide application in many areas of science. In chemistry and physics, they have provided models of pattern formation in reaction-diffusion systems, the evolution of spiral galaxies, spin exchange systems and Ising models, fluid and chemical turbulence (especially as lattice gas automata), dendritic crystal growth, and solitons, among other applications.

Spatially recurring patterns that propagate in the space-time evolution of certain CA have been likened to particles moving in physical space-time. The interactions of these "particles" are important in attempts to use CA for computational tasks (e.g., Crutchfield & Mitchell, 1995). It has also been pointed out that these particles are analogous to the defects, or coherent structures found in pattern formation processes in condensed matter physics, and to solitons in hydrodynamics. The best-known examples of such particles are the so-called "gliders" that occur in Conway's game of life.

Numerous connections have also been shown between fractals and cellular automata. Rescaling the space-time output of a CA often generates a fractal, as, for example, the two-site rule defined by 00, 11 $\to$ 0, 01, 10 $\to$ 1 generates the well-known Sierpinski gasket (Peitgen, Jürgen & Saupe, 1992).

In biology and medicine, CAs have been applied in models of heart fibrillation, developmental processes, evolution, propagation of diseases infectious, plant growth, and ecological simulations.

In computation, CAs have been used as parallel computers, sorters, prime number sieves, and tools for encryption and for image processing and pattern recognition. Some automata have the capacity for universal computation, although how to implement this capacity remains problematic.

Cellular automata have also been used to model social dynamics (Axtell & Epstein, 1996), the spread of forest fires, neural networks, and military combat situations. Extensive references to these applications and others can be found in Voorhees (1995) and Ilachinski (2001).

Work on CA has also stimulated work on other systems that generate complex patterns based on local rules. There are close connections to the fields of artificial life, random Boolean networks, genetic programming and evolutionary computation (Mitchell, 1996), and the general theory of computational mechanics.

A web search under the key word "cellular automata" will turn up literally hundreds of sites devoted to various aspects of their study. A particularly useful program for the study of CA, Boolean networks, and other discrete iterated systems is *Discrete Dynamics Lab* (Wuensche & Lesser, 1992), available for downloading from http://www.ddlab.com.

BURTON H. VOORHEES

See also **Artificial life; Chaotic dynamics; Emergence; Fractals; Game of life; Integrable cellular automata; Lattice gas methods; Neural network models; Solitons**

Further Reading

Axtell, R. & Epstein, J.M. 1996. *Growing Artificial Societies: Social Science from the Bottom Up*, Cambridge, MA: MIT Press

Burk, A.W. (editor). 1970. *Essays on Cellular Automata*, Champaign, IL: University of Illinois Press

Crutchfield, J.P. & Mitchell, M. 1995. The evolution of emergent computation. *Proceedings of the National Academy of Sciences*, 92(10): 10,742–10,746

Doolen, G.D. (editor). 1991. *Lattice Gas Methods: Theory, Applications, and Hardware*, New York: Elsevier

Gardner, M. 1970. The fantastic combinations of John Conway's new solitaire game of life. *Scientific American*, 223: 120–123

Ilachinski, A. 2001. *Cellular Automata*, Singapore: World Scientific

Mitchell, M. 1996. *An Introduction to Genetic Algorithms*, Cambridge, MA: MIT Press

Peitgen, H.-O, Jürgens, H. & Saupe, D. 1992. *Chaos and Fractals: New Frontiers in Science*, New York: Springer

Toffoli, T. & Margolis, N. 1987. *Cellular Automata Machines: A New Environment for Modeling*, Cambridge, MA: MIT Press

Voorhees, B.H. 1995. *Computational Analysis of One-Dimensional Cellular Automata*. Singapore: World Scientific

Wolfram, S. 1994. *Cellular Automata and Complexity*, Reading, MA: Addison-Wesley

Wuensche, A. & Lesser, M. 1992. *The Global Dynamics of Cellular Automata*, Reading, MA: Addison-Wesley

CELLULAR NONLINEAR NETWORKS

The development of cellular nonlinear networks (CNN) is embedded in the history of the electronic and computer industry, which is characterized by three revolutions: cheap computing power via microprocessors (since the 1970s), cheap bandwidth (since the end of the 1980s), and cheap sensors and MEMS (micro-electromechanical system) arrays (since the end of the 1990s). These research and technology breakthroughs led the way for several important economic enterprises such as the PC industry of the 1980s, the Internet industry of the 1990s, and the future analog computing industry, which is growing, together with optical and nanoscale implementations on the atomic and molecular level. Analog cellular computers have been the technical response to the sensors revolution, mimicking the autonomy and physiology of sensory and processing organs.

The CNN was invented by Leon O. Chua and Lin Yang in Berkeley in 1988. The main idea behind the CNN paradigm is Chua's so-called "local activity principle," which asserts that no complex phenomena can arise in any homogeneous media without local activity. Obviously, local activity is a fundamental property in microelectronics, where, for example, vacuum tubes and, later on, transistors are locally active devices in the electronic circuits of radios, televisions, and computers. The demand for local activity in neural networks was motivated by practical technological reasons. In 1985, John Hopfield theoretically suggested a neural network, which, in principle, could overcome the failures of pattern recognition in Frank Rosenblatt's perceptron (*See* **Perceptron**). However, its globally connected architecture was highly impractical for technical realizations in VLSI (very-large-scale-integrated) circuits of microelectronics: the number of wires in a fully connected Hopfield network grows exponentially with the size of the array. A CNN only needs electrical interconnections in a prescribed sphere of influence.

In general, a CNN is a nonlinear analog circuit that processes signals in real time. It is a multicomponent system of regularly spaced identical ("cloned") units, called cells, which communicate with each other directly only through their nearest neighbors. However, the locality of direct connections also permits global information processing. Communications between nondirectly (remote) connected units are obtained on passing through other units. The idea that complex and global phenomena can emerge from local activities in a network dates back to John von Neumann's first paradigm of cellular automata (CA). In this sense, the CNN paradigm is a higher development of the CA paradigm under the new conditions of information processing and chip technology. Unlike conventional cellular automata, CNN host processors accept and generate analog signals in continuous time with real numbers as interaction values. Furthermore, the CNN paradigm allows deep insights into the dynamic complexity of computational processes. While the classification of complexity by CA was more or less inspired by empirical observations of pattern formation in computer experiments, the CNN approach delivers a mathematically precise measure of dynamic complexity. The basic idea is to understand cellular automata as a special case of CNNs that can be characterized by a precise code for attractors of nonlinear dynamical systems and by a unique complexity index.

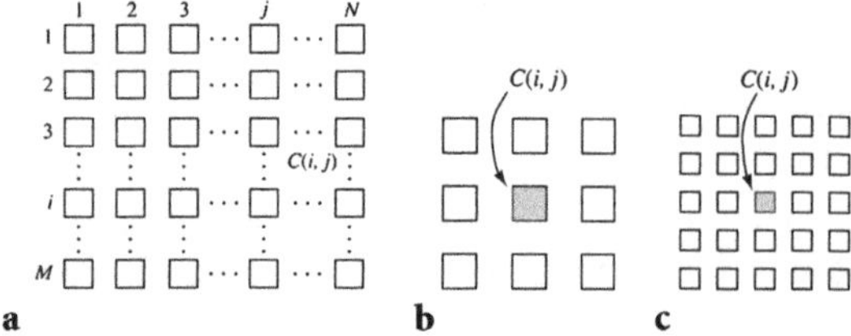

Figure 1. Standard CNN with array (a), 3×3 and 5×5 neighborhood (b,c).

Mathematical Definition

A CNN is defined by (1) a spatially discrete set of continuous nonlinear dynamical systems (cells or neurons) where information is processed into each cell via three independent variables (input, threshold, initial state) and (2) a coupling law relating relevant variables of each cell to all neighboring cells within a predescribed sphere of influence. A standard CNN architecture consists of an $M \times N$ rectangular array of cells $C(i, j)$ with cartesian coordinates (i, j) with $i = 1, 2, ..., M$ and $j = 1, 2, ..., N$ (Figure 1a). Figures 1b–c show examples of cellular spheres of influence as 3×3 and 5×5 neighborhoods. The dynamics of a cell's state is defined by a nonlinear differential equation (CNN state equation) with scalars for state x_{ij}, output y_{ij}, input u_{ij}, and threshold z_{ij}, and coefficients, called "synaptic weights", modeling the intensity of synaptic connections of the cell $C(i, j)$ with the inputs (feedforward signals) and outputs (feedback signals) of the neighboring cells $C(k, l)$. The CNN output equation connects the states of a cell with the outputs.

The majority of CNN applications use space-invariant standard CNNs with a cellular neighborhood of 3×3 cells and no variation of synaptic weights and cellular thresholds in the cellular space. A 3×3 sphere of influence at each node of the grid contains nine cells with eight neighboring cells and the cell in its center. In this case, the contributions of the output (feedback) and input (feedforward) weights can be reduced to two fixed 3×3 matrices, which are called feedback (output) cloning template **A** and feedforward (input) cloning template **B**. Thus, each CNN is uniquely defined by the two cloning templates **A**, **B**, and a threshold z, which consist of $3 \times 3 + 3 \times 3 + 1 = 19$ real numbers. They can be ordered as a string of 19 scalars with a uniform threshold, nine feedforward, and nine feedback synaptic weights. This string is called a "CNN gene" because it completely determines the dynamics of the CNN. Consequently, the universe of all CNN genes is called the "CNN genome." In analogy to the human genome project, steady progress can be made by isolating and analyzing various classes of CNN genes and their influences on CNN genomes.

Applications

In visual computing, the triple **A**, **B**, z, and its 19 real numbers can be considered as a CNN macroinstruction on how to transform an input image into an output image. Simple examples are a subclasses of CNNs with practical relevance such as the class $C(\mathbf{A}, \mathbf{B}, z)$ of space-invariant CNNs with excitatory and inhibitory synaptic weights, the zero-feedback (feedforward) class $C(0, \mathbf{B}, z)$ of CNNs without cellular feedback, the zero-input (autonomous) class $C(\mathbf{A}, 0, z)$ of CNNs without cellular input, and the uncoupled class $C(\mathbf{A}^0, \mathbf{B}, z)$ of CNNs without cellular coupling. In $\mathbf{A}^0$, all weights are zero except for the weight of the cell in the center of the matrix. Their signal flow and system structure can be illustrated in diagrams that can easily be applied to electronic circuits as well as to typical living neurons.

CNN templates are extremely useful for standards in visual computing. Simple examples are CNNs detecting edges either in binary (black-and-white) input images or in gray-scale input images. An image consists of pixels corresponding to the cells of a CNN with binary or gray scale. Logic operators can also be realized by simple CNN templates in order to combine CNN templates for visual computing. The logic NOT CNN operation inverts intensities of all binary image pixels, the foreground pixels becoming the background, and vice versa. The logic AND (logic OR, respectively) CNN operation performs a pixel-wise logic AND (logic OR operation, respectively) on corresponding elements of two binary images. These operations can be used as elements of some Boolean logic algorithms that operate in parallel on data arranged in the form of images.

The simplest form of a CNN can be characterized via Boolean functions. We consider a space-invariant binary CNN belonging to the uncoupled class $C(\mathbf{A}^0, \mathbf{B}, z)$ with a 3×3 neighborhood that maps any static 3×3 input pattern into a static binary 3×3 output pattern. It can be uniquely defined by a Boolean function of nine binary input variables, where each variable denotes one of the nine pixels within the sphere of influence of a cell. Although there are infinitely many distinct templates of the class $C(\mathbf{A}^0, \mathbf{B}, z)$, there are only a finite number of distinct combinations of 3×3 pattern of black and white cells, namely, $2^9 = 512$. As each binary nine input pattern can map to either 0 (white) or 1 (black), there are 2^{512} distinct Boolean maps of nine binary variables. Thus, every binary standard CNN can be uniquely characterized by a CNN truth table, consisting of 512 rows with one for each distinct 3×3 black-and-white pattern, nine input columns with one for each binary input variable, and one output column with binary values of the output variable.

$2^{512} \approx 1.3408 \times 10^{154} > 10^{154}$ is an "immense" number (in the sense proposed by Walter Elsasser), although the uncoupled $C(\mathbf{A}^0, \mathbf{B}, z)$ CNNs are only a small subclass of all CNNs. So, the question arises as to which subclass of Boolean functions exactly characterizes the uncoupled CNNs. In their critique of the perceptron (1969), M. Minsky and S. Papert introduced the concept of linearly separable and nonseparable Boolean functions. It can be proven that the class $C(\mathbf{A}^0, \mathbf{B}, z)$ of all uncoupled CNNs with binary inputs and binary outputs is identical to the linearly separable class of Boolean functions. Thus, linearly nonseparable Boolean functions such as, for example, the XOR function cannot be realized by an uncoupled CNN. But the uncoupled CNNs can be used as elementary building blocks that are connected by CNNs of logical operations. It can be proved that every Boolean function of nine variables can be realized by using uncoupled CNNs with nine inputs and either one logic OR CNN, or one logic AND CNN, in addition to one logic NOT CNN.

Every uncoupled CNN $C(\mathbf{A}^0, \mathbf{B}, z)$ with static binary inputs is completely stable in the sense that any solution converges to an equilibrium point. The waveform of the CNN state increases or decreases monotonically to the equilibrium point if the state at this point is positive or negative. Moreover, except for some degenerate cases, the steady-state output solution can be explicitly calculated by an algebraic formula without solving the associated nonlinear differential equations. Obviously, this is an important result to characterize a CNN class of nonlinear dynamics with robust CNN templates. Completely stable CNNs are the workhouses of the most current CNN applications. But there are also even simpler CNNs with oscillatory or chaotic behavior. Future applications will exploit the immense potentials of the unexplored terrains of oscillatory and chaotic operating regions. Then, Cellular Neural Networks will actually be transformed to Cellular Nonlinear Networks with all kinds of phase transitions and attractors of nonlinear dynamics.

Complexity Paradigm

From the perspective of nonlinear dynamics, it is convenient to think of standard CNN state equations as a set of ordinary differential equations with the components of the CNN gene as bifurcation parameters. Then, the dynamical behavior of standard CNNs can be studied in detail. Numerical examples deliver CNNs with limit cycles and chaotic attractors. The emergence of complex structures in nature can be explained by the nonlinear dynamics and attractors of complex systems. They result from the collective behavior of interacting elements in a complex system. The different paradigms of complexity research promise to explain pattern formation and pattern recognition in nature by specific mechanisms (e.g., Prigogine's chemical dissipation, Haken's work on lasers). From the CNN point of view, it is convenient to study the subclass of autonomous CNNs where the cells have no inputs. In these systems, it can be explained how patterns can arise, evolve, and sometimes converge to an equilibrium by diffusion-reaction processes. Pattern formation starts with an initial uniform pattern in an unstable equilibrium that is perturbed by small, random displacements. Thus, in the initial state, the symmetry of the unstable equilibrium is disturbed, leading to rather complex patterns. Obviously, in these applications, cellular networks do not refer only to neural activities in nerve systems, but to pattern formation in general.

A CNN is defined by the state equations of isolated cells and the cell coupling laws. For simulating reaction-diffusion processes, the coupling law describes a discrete version of diffusion (with a discrete Laplacian operator). CNN state equations and CNN coupling laws can be combined in a CNN reaction-diffusion equation, determining the dynamics of autonomous CNNs. If we replace their discrete functions and operators by their limiting continuum version, then we obtain the well-known continuous partial differential equations of reaction-diffusion processes that have been studied in different complexity approaches. Chua's version of the CNN reaction-diffusion equation delivers computer simulations of these pattern formations in chemistry and biology (e.g., concentric, auto, and spiral waves). On the other hand, for any nonlinear partial differential equation, many appropriate CNN equations can be associated with it. In many cases, it is sufficient to study the computer simulations of associated CNN equations, in order to understand the nonlinear dynamics of these complex systems.

CNN Universal Machine and Programming

There are practical and theoretical reasons for introducing a CNN Universal Machine (CNN-UM). From an engineering point of view, it is totally impractical to implement different CNN components or templates with different hardwired CNNs. Historically, John von Neumann's general-purpose computer was inspired by Alan Turing's universal machine in order to overcome all the different hardware machines of the 1930s and 1940s for different applications. From a theoretical point of view, CNN-UM opens new avenues of analog neural computers. In the CNN-UM, analog (continuous) and logic operations are mixed and embedded in an array computer. It is a complex nonlinear system, which combines two different types of operations, namely continuous nonlinear array dynamics and continuous time with local and global logic. Obviously, the mixture of analog and digital components considerably resembles to neural information processing in living organisms. The stored program, as a sequence of templates, could be considered as a genetic code for the CNN-UM. The elementary genes are the templates.

After the introduction of the architecture with standard CNN universal cells and the global analog programming unit (GAPU), the complete sequence of an analog CNN program can be executed on a CNN-UM. The description of such a program contains the global task, the flow diagram of the algorithm, the description of the algorithm in a high level α (analog) programming language, and the sequence of macro-instructions by a compiler in the form of an analog machine code (AMC). At the lowest level, the chips are embedded in their physical environment of circuits. The AMC code will be translated into hardware circuits and electrical signals. At the highest level, the α compiler generates a macro-level code called analog macro-code (AMC). The input of the α compiler is the description of the flow diagram of the algorithm using the language. In Figure 2, the levels of the software and the core engines are described. The analog macro code is used for software simulations running on a Pentium chip in a PC and for applications in a CNN-UM Chip with a CNN Chip Prototyping System (CCPS).

The CNN-UM is technically realized by analog and digital VLSI implementation. It is well known that any complex system of digital technology can be built from a few implemented building blocks by wiring and programming. In the same way, the CNN-UM, also containing analog building blocks, can be constructed. A core cell needs only three building blocks of a capacitor, resistor, and a VCCS (voltage-controlled current source). If a switch, a logic register, and a logic gate are added to the three building blocks, the extended CNN cell of the CNN-UM can be implemented. In principle, six building blocks plus wiring are sufficient to build the CNN-UM: resistor, capacitor, switch, VCCS, logic register, logic gate. As in a digital computer, stored programmability can also be introduced for analog neural computers, enabling the fabrication of visual microprocessors. Similar to classical microprocessors, stored programmability needs a complex computational infrastructure with high-level language, compiler, macro-code, interpreter, operating system, and physical code, in order to make it understandable for the human user. Using this computational infrastructure, a visual microprocessor can be programmed by downloading the programs onto the chips, as in the case of classical digital microprocessors. Writing a program for an analog CNN algorithm is as easy as writing a BASIC program.

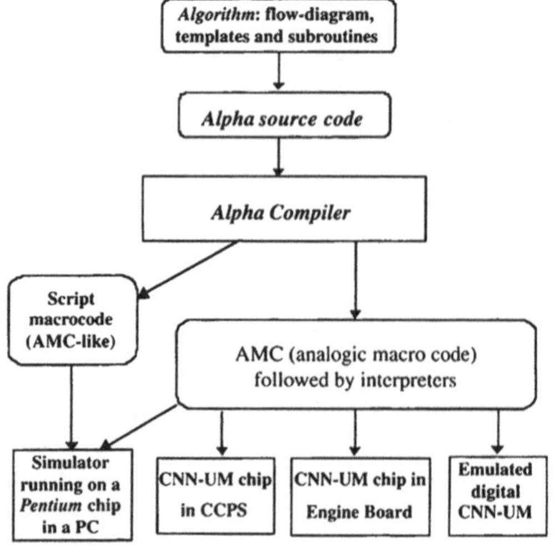

Figure 2. Levels of the software and the core engines in the CNN-UM.

With respect to computing power, CNN computers offer an orders-of-magnitude speed advantage over conventional technology when the task is complex. There are also advantages in size, complexity, and power consumption. A complete CNN-UM on a chip consists of an array of 64×64 0.5 μm micron CMOS cell processors. Each cell is endowed not only with a sensor for direct optical input of images and video but also with communication and control circuitries, as well as local analog and logic memories. CNN cells are interfaced with their nearest neighbors, as well as with the outside world. A CNN chip with 4096 cell processors on a chip means more than 3.0 Tera-OPS (operations per second) equivalent of computing power, which is about a 1000 times faster than the computing power of an advanced Pentium processor. By exploiting the state-of-the-art vertical packaging technologies, close to 10^{15} OPS CNN-UM architectures can be constructed on chips with 200×200 arrays. Thus, CNN universal chips will realize Tera-OPS or even Penta-(10^{15}) OPS, which are required for high-speed target recognition and tracking, real-time visual inspection of manufacturing processes, and intelligent machine vision capable of recognizing context-sensitive and moving scenes.

KLAUS MAINZER

See also **Attractor neural network; Cellular automata; Integrable cellular automata**

Further Reading

Chua, L.O. 1998. *A Paradigm for Complexity*, Singapore: World Scientific

Chua, L.O., Gulak, G., Pierzchala, E. & Rodriguez-Vázquez (editors). 1998. Cellular neural networks and analog VLSI. *Analog Integrated Circuits and Signal Processing. An International Journal*, 15(3)

Chua, L.O. & Roska, T. 2002. *Cellular Neural Networks and Visual Computing: Foundations and Applications*, Cambridge and New York: Cambridge University Press

Chua, L.O., Sbitnev, V.I. & Yoon, S. 2003. A nonlinear dynamics perspective of Wolfram's new kind of science. Part II: universal neuron. *International Journal of Bifurcation and Chaos in Applied Sciences and Engineering*, 13: 2377–2491

Chua, L.O. & Yang, L. 1988. Cellular neural networks: theory. *IEEE Transactions on Circuits and Systems*, 35: 1257–1272; Cellular neural networks: applications. *IEEE Transactions on Circuits and Systems*, 35: 1273–1290

Chua, L.O., Yoon, S. & Dogaru, R. 2002. A nonlinear dynamics perspective of Wolfram's new kind of science. Part I: threshold of complexity. *International Journal of Bifurcation and Chaos in Applied Sciences and Engineering*, 12: 2655–2766

Elsassee, W.M. 1987. *Reflections on a Theory of Organisms: Holism in Biology*, Frelighsburg, Quebec: Editions Orbis

Huertas, J.L., Che, W.K. & Madan, R.N. (editors). 1999. *Visions of Nonlinear Science in the 21st Century. Festshrift Dedicated to Leon O. Chua on the Occasion of his 60th Birthday*, Singapore: World Scientific

Mainzer, K. 2003. *Thinking in Complexity: The Computational Dynamics of Matter, Mind, and Mankind*, 4th edition, Berlin and New York: Springer

Tetzlaff, R. (editor). 2002. Cellular Neural Networks and Their Applications. *Proceedings of the 7th IEEE International CNN Workshop*, Singapore: World Scientific

CENTER MANIFOLD REDUCTION

A dynamical system might be difficult to solve, even numerically. To better understand its behavior in the neighborhood of an equilibrium, a reduction can be performed. For this, one starts with the eigenvalue spectrum λ of the linearized system. In the linear evolution $\sim \exp(\lambda t)$, eigenvalues with $\Re\lambda < 0$ ($\Re\lambda > 0$) are called stable (unstable), whereas those with $\Re\lambda = 0$ are called central.

In the neighborhood of an equilibrium point, P, of a dynamical system, in general, three different types of invariant manifolds exist: the trajectories belonging to the stable manifold M^{s} are being attracted by P, whereas those of the unstable manifold M^{u} are being repelled. The dynamics on the center manifold M^{c} depend on the nonlinearities. For the linearized problem, $E^{\mathrm{s}} \equiv M^{\mathrm{s}}$, $E^{\mathrm{u}} \equiv M^{\mathrm{u}}$, and $E^{\mathrm{c}} \equiv M^{\mathrm{c}}$ are uniquely determined linear subspaces that span the whole space. The transition to the nonlinear system causes (only) deformations of the linearly determined manifolds M_{s}, M_{u}, and M_{c}. However, the form of the latter critically depends on the nonlinear terms.

Let us elucidate that the behavior for the very simple system (Grosche et al., 1995)

$$\dot{x} = -xy, \quad \dot{y} = -y + x^2. \tag{1}$$

Here, the dot means differentiation with respect to time t. The equilibrium point P is $(0, 0)$, and the linearized system is

$$\dot{x} = 0, \quad \dot{y} = -y. \tag{2}$$

Here, we have the stable manifold E^{s} being identical to the y-axis and the center manifold E^{c} being identical to the x-axis. The linearized problem can be visualized by the graph shown in Figure 1.

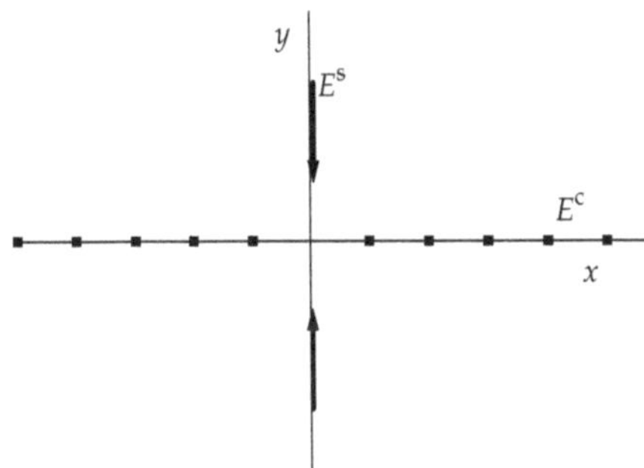

Figure 1. The linear stable (E^{s}) and center manifold (E^{c}) for example (1).

Both manifolds will be deformed in the transition to the nonlinear system. The (nonlinear, perturbed) center manifold can be described in the present example by

$$M^{\mathrm{c}} : y = h(x) \tag{3}$$

with $h(0) = h'(0) = 0$. Using that ansatz for the center manifold in (1), we obtain

$$\dot{x} = -x\, h(x). \tag{4}$$

Differentiating (3) with respect to t, leads to

$$\dot{y} = h'(x)\, \dot{x} \tag{5}$$

or

$$-h(x) + x^2 = h'(x)[-xh(x)], \tag{6}$$

that is, a differential equation for $h = h(x)$. Performing a power series ansatz $h(x) = cx^2 + dx^3 + \cdots$, we find that $c = 1$. The (nonlinear) center manifold is thus given by

$$y = x^2 + \cdots, \tag{7}$$

and the dynamics on it follows from

$$\dot{x} = -x^3; \tag{8}$$

that is, the trajectories are being attracted by P. For an illustration, see Figure 2.

More General Theoretical Background

Let us now generalize the idea and consider a system of ordinary differential equations (ODEs)

$$\dot{a} = Aa + N(a, b), \tag{9}$$

$$\dot{b} = Bb + M(a, b), \tag{10}$$

describing the dynamics of amplitudes $a_1, \ldots, a_n$ and $b_1, \ldots, b_m$ of n linear marginal stable modes and m linear stable modes, respectively $[(a_1, \ldots, a_n) := a,$

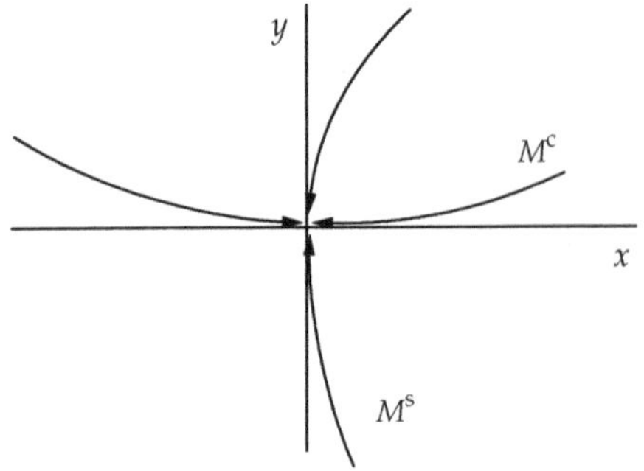

Figure 2. The stable (M^s) and center manifold (M^c) for example (1).

$(b_1, \ldots, b_m) := b]$. This implies that the real parts of eigenvalues of the matrix A vanish, and the real parts of the eigenvalues of the matrix B are negative. The functions $N(a, b), M(a, b) \in C^r$ on the right-hand sides of Equations (9) and (10) represent the nonlinear terms. Let E^c be the n-dimensional (generalized) eigenspace of A and E^s be the m-dimensional (generalized) eigenspace of B. Under these assumptions, the center manifold theorem provides the following statement (Guckenheimer & Holmes, 1983):

> There exists an invariant C^r manifold M^s and an invariant C^{r-1} manifold M^c that are tangent at $(a, b) = (0, 0)$ to the eigenspaces E^s and E^c, respectively. The stable manifold M^s is unique but the center manifold M^c is not necessarily unique.

Locally, the center manifold M^c can be represented as a graph,

$$M^c = \{(a, b) | b = h(a)\}, \; h(0) = 0, \; Dh(0) = 0, \quad (11)$$

where the C^{r-1} function h is defined in a neighborhood of the origin, and Dh denotes the Jacobi matrix. Introducing (11) in Equations (9) and (10), we obtain

$$\dot{a} = Aa + N(a, h(a)), \quad (12)$$

$$Dh(a)\,[Aa + N(a, h(a))] = Bh(a) + M(a, h(a)). \quad (13)$$

The solution h of Equation (13) can be approximated by a power series. The ambiguity of the center manifold is manifested by the fact that h is determined only modulo C^∞, a non-analytic function; thus, the power series approximation of the function h is unique. The importance of the center manifold theory is reflected by the following theorem (Marsden & McCracken, 1976; Carr, 1981):

> If there exists a neighborhood U^c of $(a, b) = (0, 0)$ on M^c, so that every trajectory starting in U^c never leaves it, then there exists a neighborhood U of $(a, b) = (0, 0)$ in $\Re^n \times \Re^m$, so that every trajectory starting in U converges to a trajectory on the center manifold.

Therefore, it is sufficient to discuss the dynamics on the center manifold, described by Equation (12). If all solutions are bounded to some neighborhood of the origin, then we have described all features of the asymptotic behavior of Equations (9) and (10). In order to fulfill the condition, the function $N(a, h(a))$ has to be expanded up to a sufficiently high order. We end up with normal forms, for example, the third order may be adequate.

Very often, the problems contain parameters and, in addition, the systems may be infinite dimensional. In both cases, one can generalize the theory presented so far. Parameters can be taken into account by expanding Equations (9) and (10) to

$$\dot{a} = A(\Lambda)a + N(a, b, \Lambda), \quad (14)$$

$$\dot{b} = B(\Lambda)b + M(a, b, \Lambda), \quad (15)$$

$$\dot{\Lambda} = 0, \quad (16)$$

where $\Lambda = (a_{n+1}, \ldots, a_{n+l})$ contains l parameters. The center manifold now has dimension $n + l$.

PDE Reduction and Symmetry Considerations

The theory is also valid in the infinite-dimensional case, if the spectrum of the linear operator can be split into two parts. The first part contains a finite number of eigenvalues whose real parts are zero, and the second part contains (an infinite number of) eigenvalues with negative real parts that are bounded away from zero.

To elucidate the power of center manifold reduction, let us consider the partial differential equation (PDE)

$$\frac{\partial \phi}{\partial t} + \phi \frac{\partial \phi}{\partial y} + \alpha \frac{\partial^2 \phi}{\partial y^2} + \beta \frac{\partial^3 \phi}{\partial y^3} + \frac{\partial^4 \phi}{\partial y^4} + \nu\phi = 0. \quad (17)$$

All coefficients α, β, and ν are nonnegative. In the following, we treat β as a fixed parameter, and consider the dynamics in dependence in α and ν. The linearization with $\phi \equiv 0$ as the equilibrium solution leads to

$$\omega = -\beta k^3 + \mathrm{i}\left(-k^4 + \alpha k^2 - \nu\right) \quad (18)$$

when we assume a unit cell of length 2π with periodic boundary conditions. A typical dependence of the linear growth (or damping) rate $\gamma := \Im\omega$ is shown in Figure 3 for $\alpha = 5.25$ and $\nu = 3.8$.

The case of two unstable modes ($k = 1, 2$) is already highly nontrivial. Let us choose $\alpha = \alpha_c = 5$ and $\nu = \nu_c = 4$. Then, the modes $\phi^{(1)} = \sin y$, $\phi^{(2)} = \cos y$, $\phi^{(3)} = \sin 2y$, and $\phi^{(4)} = \cos 2y$ belonging to $k = 1$ and 2, respectively, are marginally stable. We introduce

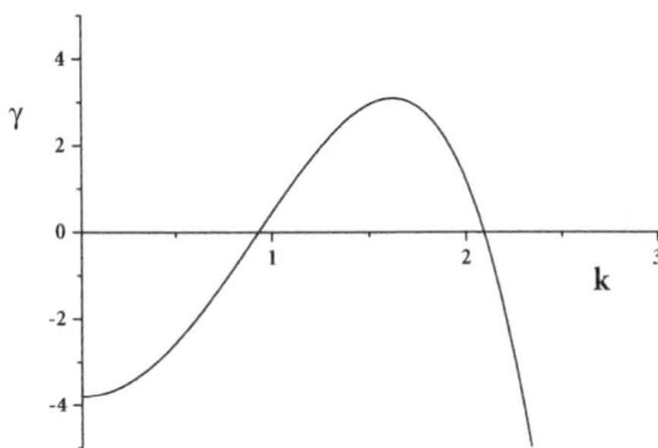

Figure 3. Growth rate curve, with two unstable modes at $k=1$ and $k=2$ for the PDE (17).

the four (real) amplitudes a_1, a_2, a_3, and a_4, as well as $\alpha_5=\alpha-\alpha_c$ and $\alpha_6=\nu-\nu_c$. The center manifold theory will allow us to derive a closed set of nonlinear amplitude equations

$$\dot{a}_n = f_n(a_1,\ldots,a_6),\ n=1,\ldots,6, \tag{19}$$

which are valid in the neighborhood of the critical point α_c, ν_c. One has $f_5\equiv f_6\equiv 0$. The other functions f_n are written as a power series in a_n,

$$f_n = \sum_{1\le m\le 6} A_n^m\, a_m + \sum_{1\le m\le p\le 6} A_n^{mp}\, a_m\, a_p + \cdots . \tag{20}$$

The dynamics on the center manifold is characterized by $a_1,\ldots,a_6$. Thus, we can make the ansatz (Carr, 1981)

$$\phi(y,t) = \sum_{1\le n\le 4} a_n(t)\,\phi^{(n)}(y) + \sum_{1\le n\le m\le 6} a_n(t)\,a_m(t)\phi^{(nm)}(y), \tag{21}$$

where the $\binom{7}{2}=21$ new functions $\phi^{(nm)}$ and, of course, the next $\binom{8}{3}=56$ functions $\phi^{(nmp)}$, and so on, can be chosen orthogonal to $\phi^{(n)}, n=1,\ldots,4$.

The technical procedure is now as follows. One inserts ansatz (21) into the basic equation (17) and compares equal orders in the amplitudes. For example, in the second-order, one collects equal powers $a_r a_s$; the "coefficients" (being equated to zero) will determine the unknown functions $\phi^{(nm)}$ via ODEs. Taking into account the (periodic) boundary conditions, we have to satisfy the solvability conditions. Collecting equal powers of the amplitudes a_n, we find the solutions for the coefficients A_r^{np}. With these values, we can solve for $\phi^{(mn)}$. This procedure should be continued to higher orders.

Actually, when written explicitly, one faces considerable work (in second order, we have to solve for 84, and in third order for 224 coefficients $A_r^{\cdots}$, and so on). One can simplify the calculations by making use of symmetries.

Translational invariance implies the following. If $\phi(y)$ is a solution,

$$T_{y_0}\phi(y) := \phi(y+y_0) \tag{22}$$

will also satisfy the dynamical equation (17), where y_0 is a real shift parameter. (In the case $\beta\equiv 0$, we also have the mirror symmetry $\phi(y)=\phi(-y)$.)

Remember the structure of the center manifold reduction: the modes $\phi^{(nm)}$, $\phi^{(nmp)}, \ldots$ have to be determined from inhomogeneous differential equations. The inhomogeneities contain (in nonlinear forms) the marginal modes $\phi^{(r)}$, $r=1,\ldots,4$. Thus, the so-called slaved modes can be written in symbolic form as

$$\phi^{(m\ldots)} = h^{(m\ldots)}\left[\{\phi^{(r)}\}\right]. \tag{23}$$

Thus, the following symmetry should hold:

$$T_{y_0}h^{(m\ldots)}\left[\{\phi^{(r)}\}\right] = h^{(m\ldots)}\left[T_{y_0}\{\phi^{(r)}\}\right]. \tag{24}$$

The consequences of the translational symmetry are most easily seen when combing the marginal modes to

$$\varphi := \sum_{r=1}^{4} a_r\phi^{(r)} \equiv \Im\left(c_1 e^{iy} + c_2 e^{2iy}\right) \tag{25}$$

with the complex amplitudes $c_1 := a_1 + ia_2$, $c_2 := a_3 + ia_4$. The (complex) amplitude equations are

$$\dot{c} = g_n(c_1,c_2,a_5,a_6),\ n=1,2, \tag{26}$$

$$\dot{a}_m = 0,\ m=5,6. \tag{27}$$

The translational symmetry (22) requires

$$e^{iny_0} g_n(c_1,c_2,a_5,a_6) = g_n\left(e^{iy_0}c_1, e^{i2y_0}c_2, a_5, a_6\right) \tag{28}$$

for $n=1,2$. The vector field (g_1,g_2) is called equivariant with respect to the operation

$$(c_1,c_2)\to\left(e^{iy_0}c_1, e^{i2y_0}c_2\right). \tag{29}$$

The most general form of vector fields being equivariant under operation (28) is

$$(g_1,g_2) = \left(c_1P_1 + \bar{c}_1c_2Q_1,\ c_2P_2 + c_1^2Q_2\right), \tag{30}$$

where P_1, P_2, Q_1, and Q_2 are polynomials in $|c_1|^2, |c_2|^2$, and $\Re(c_1^2\bar{c}_2)$; of course, they can also depend on a_5 and a_6.

Keeping in mind the symmetry properties, the general form of the amplitude equations reduces to

$$\dot{c}_1 = \lambda c_1 + \mathcal{A}\bar{c}_1c_2 + \mathcal{C}c_1|c_1|^2 + \mathcal{E}c_1|c_2|^2 + \mathcal{O}(|c|^4), \tag{31}$$

$$\dot{c}_2 = \mu c_2 + \mathcal{B}c_1^2 + \mathcal{D}c_2|c_1|^2 + \mathcal{F}c_2|c_2|^2 + \mathcal{O}(|c|^4), \tag{32}$$

$$\dot{a}_5 = \dot{a}_6 = 0. \tag{33}$$

A straightforward analysis leads to

$$\begin{aligned} &\lambda = a_5 - a_6 - \mathrm{i}\beta, \quad \mu = 4a_5 - a_6 + \mathrm{i}8\beta, \\ &\mathcal{A} = \tfrac{1}{2}, \quad \mathcal{B} = -\tfrac{1}{2}, \\ &\mathcal{C} = 0, \quad \mathcal{D} = -\tfrac{3}{4(20-\mathrm{i}9\beta)}, \\ &\mathcal{E} = \tfrac{1}{2}D, \quad \mathcal{F} = -\tfrac{1}{12(15-\mathrm{i}4\beta)}. \end{aligned} \tag{34}$$

This completes the center manifold reduction.

Very interesting conclusions result, for example, with respect to the number of modes and their interplay in time, from the systematic treatment with the center manifold theory. For example, one interesting aspect is that the present codimension-two analysis can describe successive bifurcations of one unstable mode, which, in some cases can lead to chaos in time.

Karl Spatschek

See also **Inertial manifolds; Invariant manifolds and sets; Synergetics**

Further Reading

Carr, J. 1981. *Applications of Center Manifold Theory*, New York: Springer

Grosche, G., Ziegler, V., Ziegler, D. & Zeidler, E. (editors). 1995. *Teubner-Taschenbuch der Mathematik, Teil II*, Stuttgart: Teubner

Guckenheimer, J. & Holmes, P. 1983. *Nonlinear Oscillations, Dynamical Systems, and Bifurcations of Vector Fields*, Berlin and New York: Springer

Marsden, J.E. & McCracken, M. 1976. *The Hopf Bifurcation and Its Applications*, New York: Springer

CENTRAL LIMIT THEOREM

See **Martingales**

CHAOS VS. TURBULENCE

The notion of chaos has its genesis in the work of Henri Poincaré (*See* **Poincaré theorems**) on the three-body problem of celestial mechanics. Poincaré realized that this problem cannot be reduced to quadratures and solved in the manner of the two-body problem. A precise definition of chaos or non-integrability can be given in terms of the absence of conserved quantities necessary to yield a solution. It took several decades for the full significance of non-integrable dynamical systems to be appreciated and for the term "chaos" to be introduced (*See* **Chaotic dynamics**). An important step was the 1963 paper by Edward N. Lorenz, entitled "Deterministic Nonperiodic Flow" (Lorenz, 1963), on a model describing thermal convection in a layer of fluid heated from below. The Lorenz model truncates the basic fluid dynamical equations, written in terms of Fourier amplitudes, to just three modes (*See* **Lorenz equations**):

$$\begin{aligned} \dot{x} &= -\sigma x + \sigma y, \\ \dot{y} &= -xz + rx - y \\ \dot{z} &= xy - bz. \end{aligned} \tag{1}$$

In this system, x is the time-dependent amplitude of a stream-function mode, while y and z are mode amplitudes of the temperature field. The parameters σ, r, and b depend on the geometry, the boundary conditions, and the physical parameters of the fluid. Equations (1) are a subset of the full, infinite system of mode amplitude equations, chosen such that it exactly captures the initial instability of the thermally conducting state to convecting rolls when the parameter r, known as the Rayleigh number, is increased.

What Lorenz observed in numerical solutions of (1), and verified by analysis, was that very complicated, erratic solutions would arise when r was increased well beyond the conduction-to-convection transition. In fact, Lorenz had found the first example of what is today called a strange attractor (*See* Figure 1 and **Attractors**). System (1) is clearly deterministic, yet it can produce non-periodic solutions. There were other intriguing aspects of the solutions to (1) in the chaotic regime. Solutions arising from close initial conditions would separate exponentially in time, leading to an apparently random dependence on initial conditions of the solution after a finite time (*See* **Butterfly effect**). Today, this would be associated with the existence of a positive characteristic Lyapunov exponent. A list of "symptoms" can be established that are shared by systems having the property of chaos, including: complex temporal evolution, exponential separation from close initial conditions, a strange attractor in phase space (if the system is dissipative), and positive Lyapunov exponents. An important difference from Poincaré's work was that Lorenz's system described a dissipative system in which energy is not conserved.

From the start, the potential connection between chaos and other concepts in statistical physics, such as ergodicity and turbulence, was of central interest. For example, chaos was thought to imply ergodic

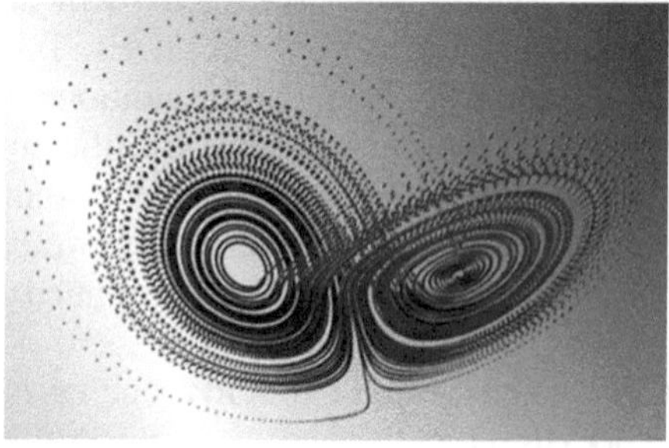

Figure 1. Strange attractor associated with the Lorenz equations. Reproduced with permission from Images by Paul Bourke, http://astronomy.swin.edu.au/ pbourke/fractals/lorenz/.

behavior in the sense of the "ergodic hypothesis" underlying equilibrium statistical mechanics (*See* **Ergodic theory**). Similarly, the connection between chaos and turbulence was sought, particularly appropriate given that Lorenz's model was of a fluid flow. Experiments on other fluid systems by Gollub, Swinney, Libchaber, and later many others established that the transition from laminar to turbulent flow typically takes place through a regime of chaotic fluid motion. The well-known route to chaos via period-doubling bifurcations of Mitchell J. Feigenbaum belongs here as well (Feigenbaum, 1980; Eckmann, 1981). In view of this, it is natural to think that turbulent flow itself is simply some kind of chaotic flow state.

Turbulence is a common state of fluid flow that shares several "symptoms" with chaotic dynamical systems, but also has distinct features not easily duplicated by chaos. The word "turbulence" was apparently first used by Leonardo da Vinci to describe a complex flow. In mathematical terms, turbulent flows should be solutions of the Navier–Stokes equation, usually written in the dimensionless form (*See* **Navier–Stokes equation**)

$$\frac{\partial \boldsymbol{u}}{\partial t} + \boldsymbol{u} \cdot \nabla \boldsymbol{u} = -\nabla p + R^{-1} \Delta \boldsymbol{u}, \qquad (2)$$

$$\nabla \cdot \boldsymbol{u} = 0. \qquad (3)$$

We have restricted attention to incompressible flows by insisting in (3) that the velocity field $\boldsymbol{u}(\boldsymbol{x}, t)$ be divergence free. In (2) the field p represents the pressure—the constant density has been absorbed in the nondimensionalization. The sole dimensionless parameter R is Reynolds number. In terms of physical variables $R = UL/\nu$, where U is a typical scale of velocity, L a typical length scale of the flow, and n is the kinematic viscosity of the fluid. For small values of R, say $0 < R \leq 1$, the flow is laminar. For moderate R, say $1 < R \leq 100$, various periodic flow phenomena may arise, such as the shedding of vortices from blunt bodies. For large R, the flow eventually breaks down into many interacting eddies—this is turbulent flow. Since most flowing fluid is, in fact, flowing at large R, turbulence is the prevailing flow state of fluids in our surroundings (oceans and atmosphere), in the universe in general, in many industrial processes, and to some extent, within our bodies.

The characterization of what makes a flow turbulent is not nearly so clear as what makes a dynamical system chaotic. First, the issue of whether the particular set of nonlinear partial differential equations (2) and (3) even has a smooth solution for all time, given smooth initial conditions, is still unsettled and is one of the prize challenges set by the Clay Mathematics Institute (http://www.claymath.org). In spite of several attempts, a convincing example of a flow with smooth initial conditions, evolving under (2) and (3), that develops a singularity in a finite time has not been found. Conversely, there is no proof that solutions with the requisite number of derivatives will exist for all time.

Turbulent flows are also recognized by a variety of "symptoms." The flow velocity as a function of time at any given point in a turbulent flow is a random function (roughly a Gaussian). However, the overall nature of the velocity field viewed as a random vector field is not Gaussian. The random nature of turbulent velocity fields is today thoroughly familiar to the flying public. The randomness is not just temporal at a fixed point in space; the spatial variation of the flow field at a given time constitutes a multitude of interacting eddies of different sizes. Because of their random character, turbulent flows stir vigorously, leading to rapid dispersal of a passively advected substance or a field, such as temperature, and to a rapid exchange of momentum with contiguous fluid. In the classic pipe flow experiment of Osborne Reynolds, for example, in which the transition from laminar to turbulent flow was first demonstrated to depend only on the dimensionless number R, a streak of dye introduced at the inlet would remain a thin streak (except for a bit of molecular diffusion) when the flow in the pipe was laminar. When the flow rate was increased and the flow became turbulent, the dye rapidly dispersed across the pipe.

In a turbulent flow, the large scales of motion, which are typically in contact with some kind of forcing from the outside, will generate smaller scales through interactions and instabilities. This process continues through a broad range of length scales, ultimately reaching small scales where molecular dissipation is effective and quells the motion altogether. The repeated process of "handing down" energy from larger scales to smaller scales is a key process in turbulence. It is usually referred to as the Kolmogorov cascade (*See* **Kolmogorov cascade**). The qualitative nature of this process was already envisaged by Lewis Fry Richardson and was described by him in an adaptation of a verse by Jonathan Swift:

Big whorls have little whorls,
Which feed on their velocity;
And little whorls have lesser whorls,
And so on to viscosity (in the molecular sense).

Because of its broad range of length scales, the energy in a turbulent flow may be considered partitioned among modes of different wavenumbers k. The energy spectrum $E(k)$ is defined such that $E(k)\,\mathrm{d}k$ is the amount of kinetic energy of the turbulent flow associated with motions with wavenumbers between k and $k + \mathrm{d}k$. The cascade implies a transfer of energy from scale to scale with a characteristic energy flux per unit mass, ε, which must also be equal to the rate at which energy is fed to the flow from the largest scales, and to the rate at which energy is dissipated by viscosity at the smallest scales. A simple dimensional argument then (*See* **Dimensional**

analysis) gives the dependence of $E(k)$ on ε and k to be

$$E(k) = C\varepsilon^{2/3}k^{-5/3}. \tag{4}$$

This is the well-known Kolmogorov spectrum, predicted by Andrei N. Kolmogorov in 1941 (Hunt et al., 1991; Frisch, 1995) and only subsequently verified by experiments in a tidal channel (see Figure 2).

Turbulence has many further intriguing statistical properties, which remain subjects of active research. A major shift in our thinking on turbulence occurred in the late 1960s and in the 1970s when experiments by Kline and Brown & Roshko demonstrated that even in turbulent shear flows at very large Reynolds number, one can identify coherent structures that organize the flow to some extent (Figure 3). Later investigations have shown that even in homogeneous, isotropic turbulence, the flow is often organized into strong filamentary vortices. The persistence of these organized structures, which can dominate the flow for long times and interact dynamically, forces a strong coupling among the spectral modes, reducing the effective number of degrees of freedom of the problem.

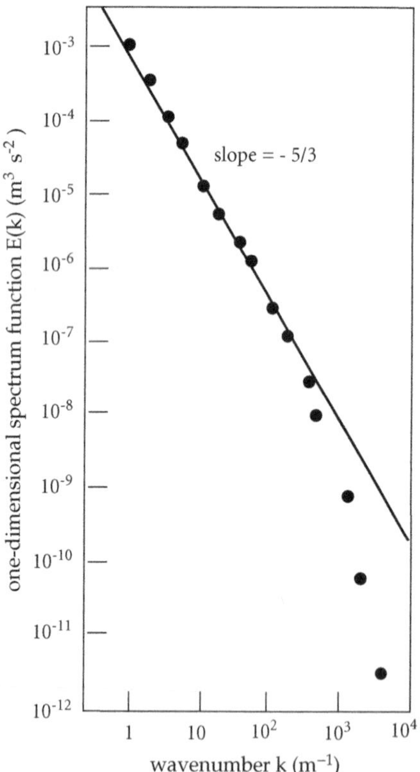

Figure 2. One-dimensional spectrum in a tidal channel from data in Grant et al. (1962).

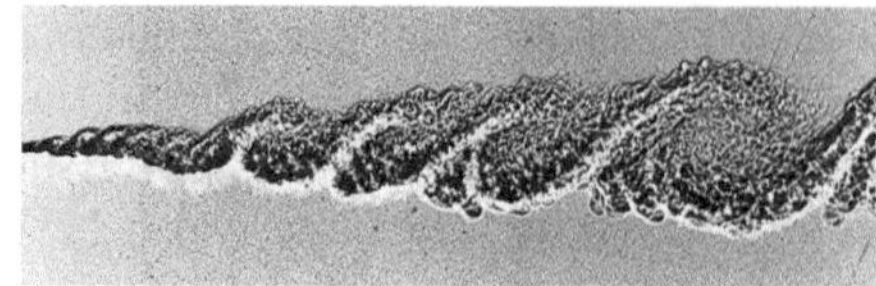

Figure 3. Coherent structures in a turbulent mixing layer. From Brown & Roshko (1974), reprinted from *An Album of Fluid Motion*, M. Van Dyke, Parabolic Press, 1982.

Chaos and turbulence both describe states of a deterministic dynamical system in which the solutions appear random. Our current understanding of chaos is largely restricted to few-degree-of-freedom systems. Turbulence, on the other hand, is a many-degree-of-freedom phenomenon. It seems somewhat unique to fluid flows—related phenomena such as plasma turbulence or wave turbulence appear to be intrinsically different. The emergence of collective modes in the form of coherent structures in turbulence amidst the randomness is an intriguing feature, somewhat reminiscent of the mix between regular "islands" and the "chaotic sea" observed in chaotic, low-dimensional dynamical systems. The coherent structures themselves approximately form a deterministic, low-dimensional dynamical system. However, it seems impossible to fully eliminate all but a finite number of degrees of freedom in a turbulent flow—the modes not included explicitly form an essential, dissipative background, often referred to as an eddy viscosity, that must be included in the description.

Turbulence is intrinsically spatiotemporal, whereas chaotic behavior in a fluid system can be merely temporal with a simple spatial structure. It is possible for the flow field to be perfectly regular in space and time, yet the trajectories of fluid particles moving within the flow will be chaotic. This is the phenomenon of chaotic advection (*See* **Choatic advection**), which points out the hugely increased complexity of a turbulent flow relative to chaos in a dynamical system.

PAUL K.A.NEWTON AND HASSAN AREF

See also **Attractors; Butterfly effect; Celestial mechanics; Chaotic advection; Chaotic dynamics; Diffusion; Ergodic theory; Kolmogorov cascade; Lorenz equations; Lyapunov exponents; Navier–Stokes equation; N-body problem; Partial differential equations, nonlinear; Period doubling; Phase space; Poincaré theorems; Routes to chaos; Shear flow; Thermal convection; Turbulence**

Further Reading

Aref, H. & Gollub, J.P. 1996. Application of dynamical systems theory to fluid mechanics. *Research Trends in Fluid Dynamics, Report of the US National Committee on Theoretical and Applied Mechanics*, edited by J.L. Lumley et al., New York: AIP Press, pp. 15–30

Eckmann, J.P. 1981. Roads to turbulence in dissipative dynamical systems. *Reviews of Modern Physics*, 53: 643–654

Feigenbaum, M.J. 1980. Transition to aperiodic behavior in turbulent systems. *Communications in Mathematical Physics*, 77: 65–86

Frisch, U. 1995. *Turbulence—The Legacy of A. N. Kolmogorov*, Cambridge and New York: Cambridge University Press
Grant, H.L., Stewart, R.W. & Moilliet, A. 1962. Turbulent spectra from a tidal channel. *J. Fluid. Mech.*, 12: 241–268
Hunt, J.C.R., Phillips, O.M., & Williams, D. (editors). 1991. Turbulence and stochastic processes: Kolmogorov's ideas 50 years on. *Proceedings of the Royal Society, London* A, 434: 1–240
Lorenz, E.N. 1963. Deterministic nonperiodic flow. *Journal of Atmospheric Sciences*, 20: 130–141
Ruelle, D. 1991. *Chance and Chaos*, Princeton, NJ: Princeton University Press

CHAOTIC ADVECTION

In fluid mechanics, *advection* means the transport of material particles by a fluid flow, as when smoke from a chimney is blown by the wind. The term *passive advection* is sometimes used to emphasize that the substance being carried by the flow is sufficiently inert that it follows the flow entirely, the velocity of the advected substance at every point and every instant adjusting to that of the prevailing flow.

To describe the kinematics of a fluid, two points of view may be adopted: the Eulerian representation focuses on the velocity field $\mathbf{u}$ as a function of position and time, $\mathbf{u}(\mathbf{x}, t)$; the Lagrangian representation emphasizes the trajectories $\mathbf{x}_P(t)$ of a fluid particle as it is advected by the flow. The two points of view are linked by stating that the value of the velocity field at a given point in space and instant in time equals the velocity of the fluid element passing through that same point at that instant, that is,

$$\dot{\mathbf{x}}_P(t) = \mathbf{u}(\mathbf{x}_P(t), t). \tag{1}$$

The Eulerian representation is used extensively for measurements and numerical simulations of fluid flow since it allows one to fix the points in space and time where the field is to be determined. The Lagrangian representation, on the other hand, is often more natural for theoretical analysis, as it explicitly addresses the nonlinearity of the Navier–Stokes equation.

For a given flow, the equations of motion (1), sometimes called the *advection equations*, are a system of ordinary differential equations that define a dynamical system. These equations can be integrable or non-integrable. Chaotic advection appears when the equations are non-integrable and the trajectories of fluid elements become chaotic. The dynamical system defined by (1) has two or more degrees of freedom. For a two-dimensional time-independent or steady flow, there are just two degrees of freedom and no chaotic motion is possible. However, already for a 2-d time-dependent or a 3-d steady flow, there are enough degrees of freedom to allow for chaotic trajectories. In other words, chaotic advection can appear even for flows that would otherwise be considered laminar.

The phenomenon of chaotic advection is also known as Lagrangian chaos, or sometimes Lagrangian turbulence. Usually, the word turbulence refers to the Eulerian representation and to flows in which the velocity field fluctuates across a wide range of spatial and temporal scales with limited correlations. In such flows, the trajectories of fluid elements are always chaotic. By contrast, chaotic advection or Lagrangian chaos can arise in situations where the velocity field is spatially coherent and the time dependence is no more complicated than a simple periodic modulation.

Many examples have now been given to illustrate the point that the complexity of the spatial structure of material advected by a flow can be much greater than one might surmise from a picture of the instantaneous streamlines of the flow. Thus, in the paper that introduced the notion of chaotic advection (Aref (1984) and Figure 1), the case of two stirrers that act alternately on fluid confined to a disk was considered. Each stirrer was modeled as a point vortex that could be switched on and off. There are several parameters in the system, such as the strengths and positions of the vortex stirrer and the time interval over which each acts. For a wide range of parameter values, the dynamics is as shown in Figure 1; after just a few periods, the 10,000 particles being advected are spread out over a large fraction of the disk.

Chaotic advection gives rise to very efficient stirring of a fluid. Material lines are stretched at a rate given by the Lyapunov exponent. In bounded flows, these exponentially growing material lines have to be folded back over and over again, giving rise to ever finer and denser striations. They are familiar from the mixing of paint or from marbelized paper. On the smallest scales diffusion, takes over and smoothes the steep gradients, giving rise to mixing on the molecular scale. The interplay between stirring and diffusion is the

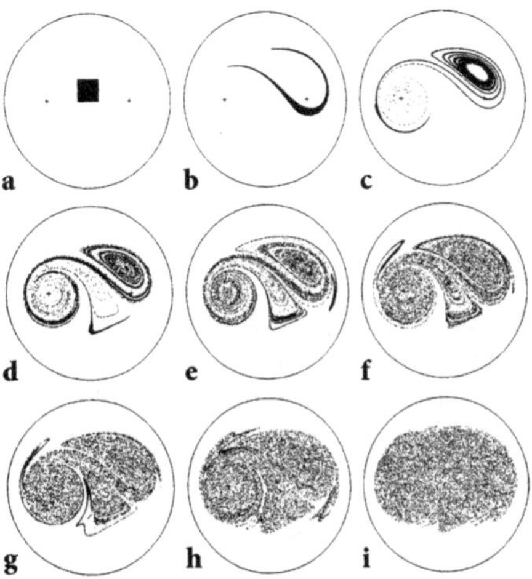

Figure 1. Spreading of 10,000 particles in a cylindrical container (disk) under the alternating action of two stirrers. The positions of the stirrers are marked by crosses. (a) initial distribution; (b)–(g) positions of the particles after 1, 2, ..., 6 periods; (h) after 9 periods; (i) after 12 periods. From (Aref, 1984).

source of the efficient mixing in the presence of chaotic advection. This phenomenon is being exploited in various procedures for mixing highly viscous fluids, including applications to materials processing, in micro-fluidics, and even in large-scale atmospheric, oceanographic, and geological flows. It may play a role in the feeding of microorganisms.

In the case of 2-d incompressible flows the equations of motion allow for an interesting connection to Hamiltonian dynamics. The velocity field can be represented through a stream function $\psi(x, y, t)$, so that $\mathbf{u} = \nabla \times \psi \mathbf{e}_z$ and Equations (1) for the trajectories of fluid elements become

$$\dot{x} = \frac{\partial \psi}{\partial y}, \qquad \dot{y} = -\frac{\partial \psi}{\partial x}. \qquad (2)$$

The relation to Hamilton's canonical equations is established through the identification x=position, y=momentum, and ψ=Hamilton function. Thus, what is the phase space in Hamiltonian systems can be visualized as the position space in the hydrodynamic situation. The structures that appear in 2-d periodically driven flows are, therefore, similar to the phase space structures in a Poincaré surface of section for a chaotic Hamiltonian system, and the same techniques can be used to analyze the transport of particles and the stretching and folding of material lines.

The phenomena that arise in chaotic advection by simple flows may be relevant to turbulent flows when a separation of length and time scales is possible. Consider, for example, the small-scale structures that appear in the density of a tracer substance when the molecular diffusivity κ of the tracer is much smaller than the kinematic viscosity ν of the liquid, that is, in a situation where the Schmidt number $\mathrm{Sc} = \nu/\kappa$ is much larger than one. Then, the velocity field is smooth below the Kolmogorov scale, $\lambda_K = (\nu^3/\varepsilon)^{1/4}$, where ε is the kinetic energy dissipation, but the scalar field has structures on even smaller scales, down to $\lambda_s = (\mathrm{Sc})^{-1/2}\lambda_K$. These arise from Lagrangian chaos with a randomly fluctuating velocity field. The patterns produced in this so-called Batchelor regime are strikingly similar to the ones observed in laminar flows.

On larger scales, ideas from chaotic advection are relevant when there are large-scale coherent structures with slow spatial and temporal evolution. Typical examples are 2-d or quasi-2-d flows, for example, in the atmosphere or in the oceans. Fluid volumes can be trapped in regions bounded by separatrices or by stable and unstable manifolds of stagnation points and may have very little exchange with their surroundings. Such a reduction in stirring appears to occur in the Wadden sea (Ridderinkhof & Zimmerman, 1992).

Equations (1) apply in this form to fluid elements and ideal particles only. For realistic particles with finite volume and inertia, further terms must be added. A significant change in the qualitative side is that the effective velocity field for inertial particles can have a nonvanishing divergence even for incompressible flows (Maxey & Riley, 1983).

The book by Ottino, (1989) and the two conference proceedings (Aref, 1994; IUTAM, 1991) provide good starting points for entering the many aspects of chaotic advection and Lagrangian chaos in engineering applications, geophysical flows, turbulent flows, and theoretical modeling. Historical remarks may be found in the Introduction to Aref, (1994) and in Aref, (2002). Today, the term chaotic advection designates an established subtopic of fluid mechanics that is used as a classification keyword by leading journals and conferences in the field.

HASSAN AREF AND BRUNO ECKHARDT

See also **Chaotic dynamics; Chaos vs. turbulence; Dynamical systems; Hamiltonian systems; Lyapunov exponents; Turbulence**

Further Reading

Aref, H. 1984. Stirring by chaotic advection. *Journal of Fluid Mechanics*, 143: 1–21

Aref, H. (editor). 1994. Chaos applied to fluid mixing. *Chaos, Solitons and Fractals*, 4: 1–372

Aref, H. 2002. The development of chaotic advection. *Physics of Fluids*, 14: 1315–1325

IUTAM Symposium on fluid mechanics of stirring and mixing. 1991. *Physics of Fluids*, 3: 1009–1496

Maxey, M. & Riley, J. 1983, Equation of motion for a small rigid sphere in a nonuniform flow. *Physics of Fluids*, 26: 883–889

Ottino, J.M. 1989. *The Kinematics of Mixing: Stretching, Chaos and Transport*, Cambridge: Cambridge University Press

Ridderinkhof, H. & Zimmermann, J.T.F. 1992. Chaotic stirring in a tidal system. *Science*, 258: 1107–1111

CHAOTIC BILLIARDS

See **Billiards**

CHAOTIC DYNAMICS

When we say "chaos", we usually imagine a very complex scene with many different elements that move in different directions, collide with each other, and appear and disappear randomly. Thus, according to everyday intuition, the system's complexity (e.g., many degrees of freedom) is an important attribute of chaos. It seems reasonable to think that in the opposite case, for example, a system with only a few degrees of freedom, the dynamical behavior must be simple and predictable. In fact, this point of view is Laplacian determinism.

The discovery of dynamical chaos has destroyed this traditional view. Dynamical chaos is a phenomenon that can be described by mathematical models for many natural systems, for example, physical, chemical, biological, and social, which evolve in time according to a deterministic rule and demonstrate capricious and seemingly unpredictable behavior. To illustrate such behavior, consider a few examples.

Examples

Hyperion: Using Newton's laws, one can compute relatively easily all future solar eclipses not only for the next few hundred years but also for thousands and millions of years into the future. This is indicative of a real predictability of the system's dynamical behavior. But even in the solar system, there exists an object with unpredictable behavior: a small irregularly shaped moon of Saturn, Hyperion. Its orbit is regular and elliptic, but its altitude in the orbit is not. Hyperion is tumbling in a complex and irregular pattern while obeying the laws of gravitational dynamics. Hyperion may not be the only example of chaotic motion in the solar system. Recent studies indicate that chaotic behavior possibly exists in Jovian planets (Murray & Holman, 1999), resulting from the overlap of components of the mean motion resonance among Jupiter, Saturn, and Uranus. Chaos in Hamiltonian systems, which represent the dynamics of the planets, arises when one resonance is perturbed by another one (*See* **Standard map**).

Chaotic mixing is an example of the complex irregular motion of particles in a regular periodic velocity field, like drops of cream in a cup of coffee; see Figure 1. Such mixing, caused by sequential stretching and folding of a region of the flow, illustrates the general mechanism of the origin of chaos in the phase space of simple dynamical systems (*See* **Chaotic advection; Mixing**).

Billiards: For its conceptual simplicity, nothing could be more deterministic and completely predictable than the motion of a single ball on a billiard table. However, in the case of a table bounded by four quarters of a circle curved inward (Sinai billiard), the future fate of a rolling billiard ball is unpredictable beyond a surprisingly small number of bounces. As indicated by Figure 2, a typical trajectory of the Sinai billiard is irregular and a statistical approach is required for a quantitative description of this simple mechanical system. Such an irregularity is the result of having a finite space and an exponential instability of individual trajectories resulting in a sensitive dependence on initial conditions. Due to the curved shape of the boundary, two trajectories emanating from the same point but in slightly different directions with angle δ between them, hit the boundary $\partial\Omega$ (see Figure 2) at different points that are $c\delta$ apart where $c > 0$. After a bounce, the direction of the trajectories will differ by angle $(1+2c)\delta$, and because an actual difference between the directions is multiplied by a factor $\mu = (1+2c) > 1$, the small perturbation δ will grow more or less exponentially (Sinai, 2000). Such sensitive dependence on initial conditions is the main feature of every chaotic system.

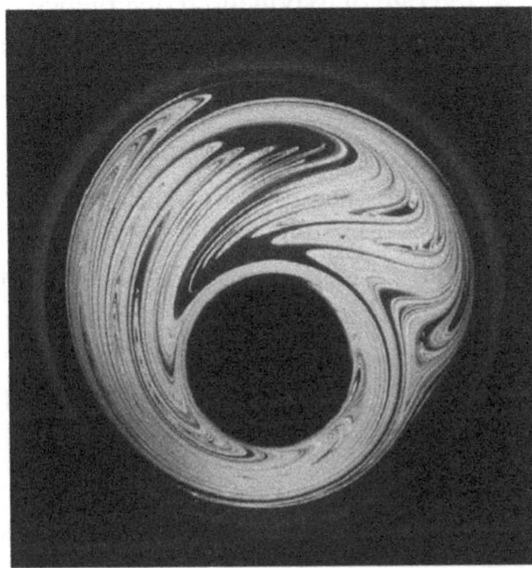

Figure 1. Mixing of a passive tracer in a Newtonian flow between two rotating cylinders with different rotation axes. The rotation speed of the inner cylinder is modulated with constant frequency. The flow is stretched and folded in a region of the flow. The repetition of these operations leads to a layered structure—folds within folds, producing a fractal structure (Ottino, 1989).

A Markov map: To understand in more detail how randomness appears in a nonrandom system, consider a simple dynamical system in the form of a one-dimensional map

$$x_{n+1} = 2x_n \bmod 1. \qquad (1)$$

Since the distance between any two nearby trajectories ($|x_n - x'_n| \ll 1$) after each iteration increases at least two times ($|\mathrm{d}x_{n+1}/\mathrm{d}x_n| = 2$), any trajectory of the map is unstable. The map has a countable infinity of unstable periodic trajectories, which can be seen as fixed points when one considers the shape of the map $x_{n+k} = F^{(k)}(x_n)$; see Figure 3(b). Since all fixed points and periodic trajectories are repelling, the only possibility left for the most arbitrarily selected initial condition is that the map will produce a chaotic motion that never exactly repeats itself. The irregularity of such dynamics can be illustrated using a binary symbolic description ($s_n = 0$ if $x_n < \frac{1}{2}$ and $s_n = 1$ if $x_n \geq \frac{1}{2}$). In this case, any value of x_n can be represented as a binary

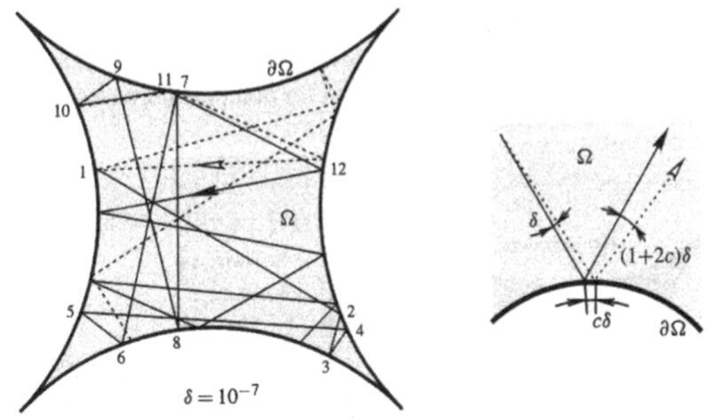

Figure 2. Illustration of the trajectory sensitivity to the initial conditions in a billiard model with convex borders.

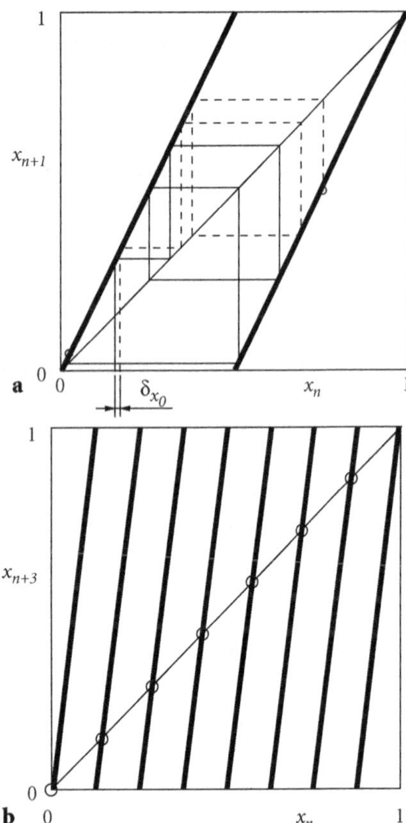

Figure 3. Simple map diagram: (a) two initially close trajectories diverge exponentially; (b) illustration of the increasing of the number of unstable periodic trajectories with the number of iterations.

decimal

$$x_n = 0.s_{n+1}s_{n+2}s_{n+3}\ldots \equiv \sum_{j=n+1}^{\infty} 2^{-j}s_j.$$

If the initial state happens to be a rational number, it can be written as a periodic sequence of 0's and 1's. For instance, 0.10111011101110111... is the rational number $\frac{11}{15}$. Each iteration $x_n \to x_{n+1}$ of map (1) corresponds to setting the symbol s_{n+1} to zero and then moving the decimal point one space to the right (this is known as a *Bernoulli shift*). For example, the iterations of the number $\frac{11}{15}$ yield

$$\begin{aligned}&0.10111011101110111\ldots,\\&0.01110111011101110\ldots,\\&0.11101110111011101\ldots,\\&0.11011101110111011\ldots,\\&0.10111011101110111\ldots,\end{aligned}$$

which illustrates a periodic motion of period 4. Selecting an irrational number as the initial condition, one chooses a binary sequence that cannot be split into groups of 0's and 1's periodically repeated an infinite number of times. As a result, each iteration of the irrational number generates a new irrational number. Since the irrational numbers appear in the interval $x_n \in [0, 1]$ with probability one, one can observe only the aperiodic (chaotic) motions. Random-like behavior of the chaotic motions is illustrated in a separate figure in the color plate section (See the color plate section for a comparison of chaos generated by Equation (1) and a truly random process).

The degree of such chaoticity is characterized by *Lyapunov exponents* that can be defined for one-dimensional maps ($x_{n+1} = f(x_n)$). The stability or instability of a trajectory with the initial state x_0 is determined by the evolution of neighboring trajectories starting at $\tilde{x}_0 = x_0 + \delta x_0$ with $|\delta x_0| \ll 1$. After one iteration

$$\tilde{x}_1 = x_1 + \delta x_1 = f(x_0 + \delta x_0) \approx f(x_0) + \left.\frac{\mathrm{d}f}{\mathrm{d}x}\right|_{x=x_0} \delta x_0.$$

Now, the deviation is $\delta x_1 \approx f'(x_0)\delta x_0$. After the nth iteration it becomes $\delta x_n = (\prod_{m=0}^{n-1} f'(x_m))\delta x_0$. The evolution of the distance between the two trajectories is calculated by taking the absolute value of this product. For infinitesimally small perturbations and large enough n, it is expected that $|\delta x_n| = \alpha^n |\delta x_0|$, where

$$\alpha \approx \lim_{n\to\infty} \left(\left| \frac{\delta x_n}{\delta x_0} \right| \right)^{1/n} = \left(\prod_{m=0}^{n-1} |f'(x_m)| \right)^{1/n}$$

or

$$\ln \alpha \approx \lambda = \lim_{n\to\infty} \frac{1}{n} \sum_{m=0}^{n-1} \ln\left(|f'(x_m)|\right). \qquad (2)$$

Limit (2) exists for a typical trajectory x_m and defines the Lyapunov exponent, λ, which is the time average of the rate of exponential divergence of nearby trajectories. For map (1) $f' = 2$ for all values of x and, therefore, $\lambda = \ln 2$ (*See* **Lyapunov exponents**).

Assuming that the initial state cannot be defined with absolute accuracy, the prediction of the state of the map after a sufficiently large number of iterations becomes impossible. The only description that one can use for defining that state is a statistical one. The statistical ensemble in this case is the ensemble of initial conditions. The equation of evolution for the initial state probability density $\rho_{n+1}(F(x))$ can be written as (Ott, 1993, p. 33)

$$\rho_{n+1}(F(x)) = \sum_{j=1,2} \rho_n(x) \Big/ \left| \frac{\mathrm{d}F(x)}{\mathrm{d}x} \right|_j , \qquad (3)$$

where the summation is taken over both branches of $F(x)$. Considering the evolution of a sharp initial distribution $\rho_0(x)$, one can see that at each step this distribution becomes smoother. As n approaches infinity, the distribution asymptotically approaches the steady state $\rho(x) = 1$.

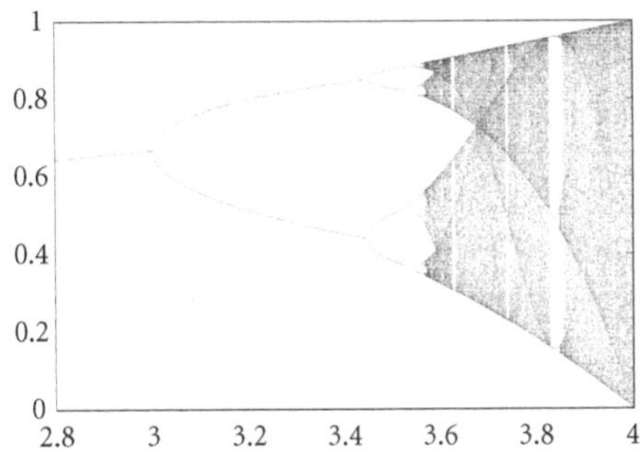

Figure 4. Bifurcation diagram for the logistic map.

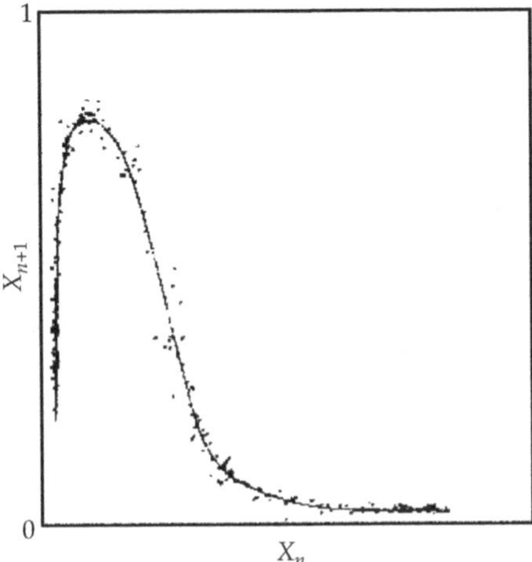

Figure 5. Return map measured in the Belousov–Zhabotinsky autocatalytic reaction.

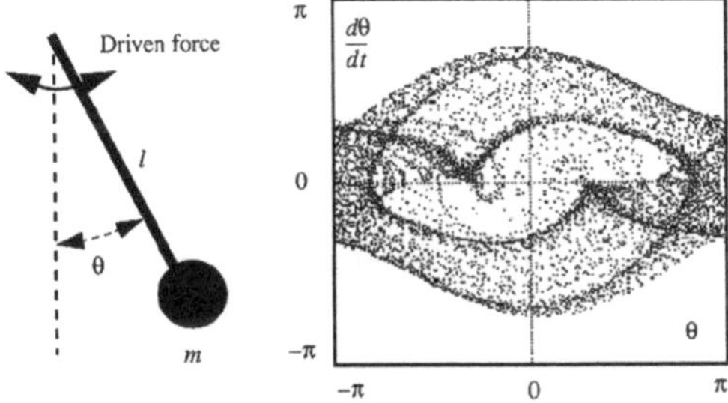

Figure 6. Chaotic oscillation of a periodically driven pendulum, in phase-space plot of angular velocity versus angular position (Deco & Schürmann, 2000).

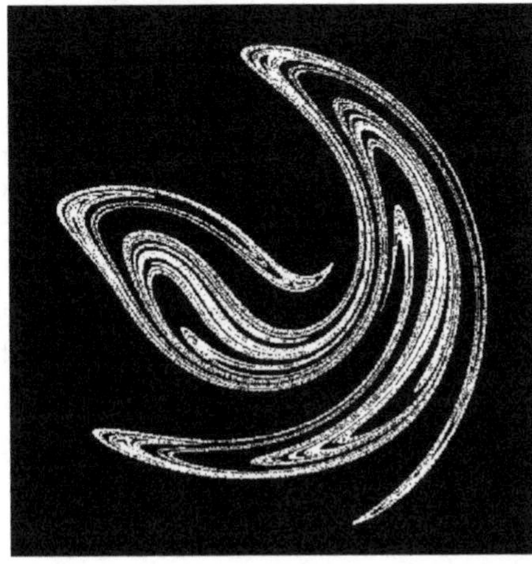

Figure 7. Ueda attractor. The fractal structure of the attractor is typical for all chaotic sets (compare this picture with Figure 1) (Ueda, 1992).

Population dynamics: A popular model of population growth is the logistic map $x_{n+1} = \alpha x_n(1 - x_n)$, $0 \le \alpha \le 4$ (*See* **Population dynamics**). The formation of chaos in this map is illustrated in the bifurcation diagram shown in Figure 4. This diagram presents the evolution of the attracting set as the value of α grows. Below the Feigenbaum point $\alpha_\infty = 3.569\ldots$, the attractor of the map is periodic. Its period increases through a sequence of period-doubling bifurcations as the value of α approaches α_∞ (*See* **Period doubling**). For $\alpha > \alpha_\infty$, the behavior is chaotic but some windows of periodic attractors exist (*See* **Order from chaos**).

Belousov–Zhabotinsky (BZ) autocatalytic reaction: In the BZ reaction (*See* **Belousov–Zhabotinsky reaction**), an acid bromate solution oxidizes malonic acid in the presence of a metalion catalyst and other important chemical components in a well-stirred reactor (Roux et al., 1983). The concentration of the bromide ions is measured and parameterized by the return map (plotting a variable against its next value in time) $x_{n+1} = \alpha x_n \exp[-bx_n]$ (see Figure 5). This map exhibits chaotic behavior for a very broad range of parameter values.

Simple chaotic oscillators: The dynamics of the periodically driven pendulum shown in Figure 6 is described by

$$\frac{d^2\Theta}{dt^2} + \nu\frac{d\Theta}{dt} + \frac{g}{l}\sin\Theta = B\cos 2\pi ft, \qquad (4)$$

where the term on the right-hand side is the forcing (sinusoidal torque) applied to the pivot and f is the forcing frequency. Chaotic motions of the pendulum computed for $\nu = 0.5$, $g/l = 1$, $B = 1.15$, $f = 0.098$, and visualized with stroboscopic points at moments of time $t = i/f$ are shown in Figure 6.

A similar example of chaotic behavior was intensively studied in an oscillator where the restoring force is proportional to the cube of the displacement (Ueda, 1992, p. 158)

$$\frac{d^2\Theta}{dt^2} + \nu\frac{d\Theta}{dt} + \Theta^3 = B\cos t. \qquad (5)$$

The stroboscopic image (with $t = i$) of the strange attractor in this forced Duffing-type oscillator computed with $\nu = 0.05$ and $B = 7.5$ is shown in Figure 7.

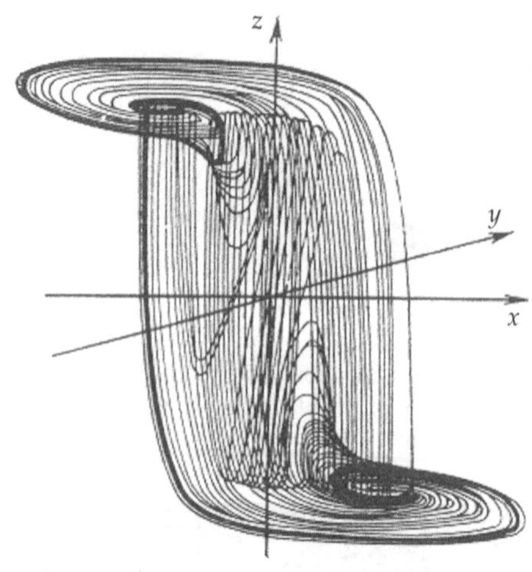

Figure 8. Chaotic attractor generated by electric circuit, which is a modification of van der Pol oscillator: $\dot{x}=hx+y-gz$; $\dot{y}=-x$; $\mu\dot{z}=x-f(x)$; where $f(x)=x^3-x$ (Pikovsky & Rabinovich, 1978).

Figure 8 presents a chaotic attractor generated by an electronic circuit. Such circuits are a popular topic in engineering studies today.

Characteristics of Chaos

Lyapunov exponents: Consider the Lyapunov exponents for a trajectory $\tilde{\boldsymbol{x}}(t)$ generated by a d-dimensional autonomous system

$$\frac{\mathrm{d}\boldsymbol{x}}{\mathrm{d}t}=\boldsymbol{F}(\boldsymbol{x}), \tag{6}$$

with initial condition $\boldsymbol{x}_0$, $\boldsymbol{x}\in\Re^d$. Linearizing Equation (6) about this solution, one obtains a linear system which describes the evolution of infinitesimally small perturbations $\boldsymbol{w}=\boldsymbol{x}(t)-\tilde{\boldsymbol{x}}(t)$ of the trajectory, in the form

$$\frac{\mathrm{d}\boldsymbol{w}}{\mathrm{d}t}=M(\tilde{\boldsymbol{x}})\boldsymbol{w}, \tag{7}$$

where $M(\boldsymbol{x})=\partial\boldsymbol{F}(\boldsymbol{x})/\partial\boldsymbol{x}$ is the Jacobian of $\boldsymbol{F}(\boldsymbol{x})$ that changes in time in accordance with $\tilde{\boldsymbol{x}}(t)$. In d-dimensional phase space of (7), consider a sphere of initial conditions for perturbations $\boldsymbol{w}(t)$ of diameter l, that is, $|\boldsymbol{w}(0)|\leq l$. The evolution of this ball in time is governed by linear system (7) and depends on trajectory $\tilde{\boldsymbol{x}}(t)$. As the system evolves in time, the ball transforms into an ellipsoid. Let the ellipsoid have d principal axes of different length l_j, $j=1,d$. Then, the values of Lyapunov exponents of the trajectory $\tilde{\boldsymbol{x}}(t)$ are defined as

$$\lambda_j(\tilde{\boldsymbol{x}})=\lim_{t\to\infty}\left[\left(\frac{1}{t}\right)\ln\left(\frac{l_j(\tilde{\boldsymbol{x}},t)}{l(\boldsymbol{x}_0,0)}\right)\right]. \tag{8}$$

Although limit (8) depends on $\tilde{\boldsymbol{x}}(t)$, the *spectrum of the Lyapunov exponents* λ_j for the selected regime of chaotic oscillations generated by (6) is independent of the initial conditions for the typical trajectories and characterizes the chaotic behavior.

The Lyapunov exponents, λ_j, can be ordered in size: $\lambda_1\geq\lambda_2\geq\cdots\geq\lambda_d$. Self-sustained oscillations in autonomous time-continuous systems always have at least one Lyapunov exponent that is equal to zero. This is the exponent characterizing the stretching of phase volume along the trajectory. The spectrum of λ_j for chaotic trajectories contains one or more Lyapunov exponents with positive values.

Kolmogorov–Sinai entropy is a measure of the degree of predictability of further states visited by a chaotic trajectory started within a small region. Due to the divergence, a long-term observation of such a trajectory gives more and more information about the actual initial condition of the trajectory. In this sense, one may say that a chaotic trajectory creates information. Consider a partitioning of the d-dimensional phase space into small cubes of volume ε^d. Observing a continuous trajectory during T instances of time, one obtains a sequence $\{i_0,i_1,\ldots,i_T\}$, where $i_0,i_1,\ldots$ are the indexes of the cubes consequently visited by the trajectory. As a result, the type of the trajectory observed during the time interval from 0 to T is specified by the sequence $\{i_0,i_1,\ldots,i_T\}$. As Kolmogorov and Sinai showed, in dynamical systems whose behavior is characterized by exponential instability, the number of different types of trajectories, K_T, grows exponentially with T:

$$0<H=\lim_{T\to\infty}\frac{1}{T}\log K_T.$$

The quantity H is the Kolmogorov–Sinai (KS) entropy.

The number of unique random sequences $\{i_0, i_1,\ldots,i_T\}$ that can be obtained without any rules applied increases exponentially with T. In the case of nonrandom sequences where there is a strict law for the generation of future symbols, like the periodic motion, the number of possible sequences grows in time slower than the exponent. Since the exponential growth takes place for the segments of trajectories in the unstable dynamical system producing chaos, such a dynamical system is capable of generating "random" sequences. The Kolmogorov–Sinai entropy is a measure of such "randomness" in a "nonrandom" system, for example, a dynamical system.

Since both KS entropy and Lyapunov exponents reflect the properties of the divergence of the nearby trajectories, these characteristics are related to each other. The formula describing this relation is given by Ruelle's Inequality

$$H\leq K=\sum_{j=1}^{m}\lambda_j\geq 0 \tag{9}$$

where m is the number of positive λ_i ($K=0$, when $m=0$). The equality $H=K$ holds when the system has a physical measure (Sinai–Ruelle–Bowen measure) (Young, 1998). The invariant set of trajectories characterized by a positive Kolmogorov–Sinai entropy is a chaotic set.

Forecasting

If a sufficiently long experimental time series capturing the chaotic process of an unknown dynamical system is available in the form of scalar data $\{x_n\}_{n=0}^{N}$, it is possible, in principle, to predict x_{N+m} with finite accuracy for some $m \geq 1$. Such predictions are based on the assumption that the unknown generating mechanism is time independent. As a result, "what happened in the past may happen again—even stronger: that what is happening now has happened in the past" (Takens, 1991). In classical mechanics (no dissipation), this idea of "what happens now has happened in the past" is related to the Poincaré Recurrence Theorem.

Usually, the prediction procedure consists of two steps: first, it is necessary to consider all values of n in the "past," that is, with $n < N$, such that $\sum_{k=0}^{K} |x_{n-k} - x_{N-k}| < \varepsilon$, where ε is a small constant. If there are only a few of such n, then one can try again with a smaller value of K or a larger value of ε. In the second step, it is necessary to consider the corresponding elements x_{n+l} for all the values of n found in the first step. Finally, taking a union of the ε-neighborhoods of all these elements, one can predict that x_{N+l} will be in this union.

To understand when and why forecasting is possible and when it is not, it is reasonable to use characteristics such as *dimension* and *entropy* that can be computed directly from time series (Takens, 1991). If we want to make a catalog of essentially different segments of length $k+1$ in $\{x_n\}_{n=0}^{N}$, this can be done with $C(k, \varepsilon, N)$ elements. $C(k, \varepsilon, N)$ is a function of N that has a limit $C(k, \varepsilon) = \lim_{N \to \infty} C(k, \varepsilon, N)$, and for prediction, we need $C(k, \varepsilon) \ll N$.

The quantitative measure for the way in which $C(k, \varepsilon)$ increases as ε goes to zero is

$$D = \lim_{k \to \infty} \left(\overline{\lim_{\varepsilon \to 0}} \frac{C(k, \varepsilon, N)}{\ln(1/\varepsilon)} \right). \tag{10}$$

If D is large, the prediction is problematic. The quantity D defined by (10) is the dimension of the time series.

The quantitative measure for the way in which $C(k, \varepsilon)$ increases with k is

$$H = \overline{\lim_{\varepsilon \to 0}} \left(\lim_{k \to \infty} \frac{C(k, \varepsilon, N)}{k} \right), \tag{11}$$

This is the entropy of the time series. For the time series generated by a differentiable dynamical system, both the dimension and entropy are finite, but for a random time series they are infinite. Suppose each x_n is taken at random in the interval [0,1] (with respect to the uniform distribution) and for each $n_1, \ldots, n_k$ (different), the choices of $x_{n_1}, \ldots, x_{n_k}$ are independent. For such time series, one can find: $C(k, \varepsilon) = (1 + \lfloor 1/2\varepsilon \rfloor)^{k+1}$, where $\lfloor 1/2\varepsilon \rfloor$ is the integer part of $1/2\varepsilon$. From this formula, it immediately follows that both dimension and entropy in such random time series are infinite.

Models of the Earth's atmosphere are generally considered as chaotic dynamical systems. Due to the unstability, even infinitesimally small uncertainties in the initial conditions grow exponentially fast and make a forecast useless after a finite time interval. This is known as the butterfly effect. However, in the tropics, there are certain regions where wind patterns and rainfall are so strongly determined by the temperature of the underlying sea surface, that they do not show such sensitive dependence on the atmosphere. Therefore, it should be possible to predict large-scale tropical circulation and rainfall for as long as the ocean temperature can be predicted (Shukla, 1998).

History

The complex behavior of nonlinear oscillatory systems was observed long before dynamical chaos was understood. In fact, the possibility of complex behavior in dynamical systems was discovered by Henri Poincaré in the 1890s in his unsuccessful efforts to prove the regularity and stability of planetary orbits. Later on, experiments with an electrical circuit by van der Pol and van der Mark (1927) and the double-disk model experiments of the magnetic dynamo (Rikitake, 1958) also indicated the paradoxically complex behavior of a simple system. At that time, several mathematical tools were available to aid the description of the nontrivial behavior of dynamical systems in phase space, such as homoclinic Poincaré structures (homoclinic tangles). However, at the time, neither physicists nor mathematicians realized that deterministic systems may behave chaotically. It was only in the 1960s that the understanding of randomness was revolutionized as a result of discoveries in mathematics and in computer modeling (Lorenz, 1963) of real systems. An elementary model of chaotic dynamics was suggested by Boris Chirikov in 1959. During the last few decades, chaotic dynamics has moved from mystery to familiarity.

Standard map and homoclinic tangle: The standard map (Chirikov, 1979) is an area-preserving map

$$\begin{aligned} I_{n+1} &= I_n + K \sin \Theta_n, \\ \Theta_{n+1} &= I_n + \Theta_n + K \sin \Theta_n, \end{aligned} \tag{12}$$

where Θ is an angle variable (computed modulo 2π) and k is a positive constant. This map was proposed as a model for the motion of a charged particle in a magnetic field. For K larger than K_{cr}, map

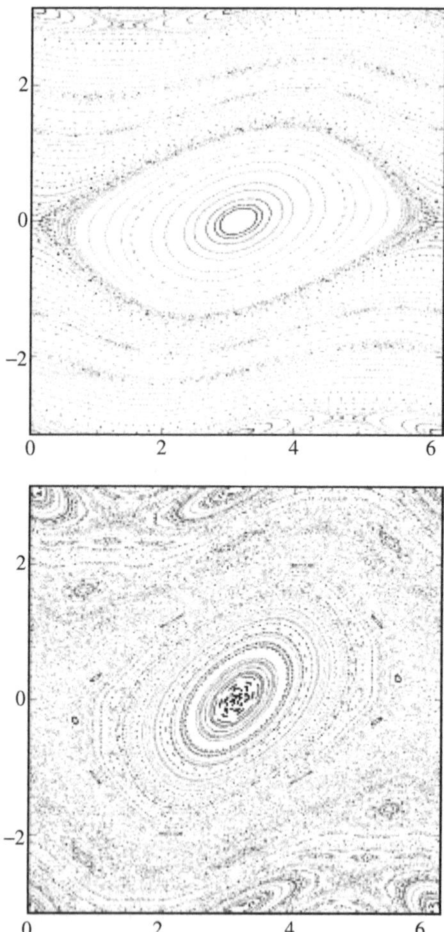

Figure 9. Examples of chaos in the standard map for two different values of K. The coexistence of the "chaotic sea" and "regular islands" that one can see in the panel on the right is typical for Hamiltonian systems with chaotic regimes (Lichtenberg & Lieberman, 1992).

(12) demonstrates an irregular (chaotic) motion; see Figure 9. The complexity of the phase portrait of this map is related to the existence of *homoclinic tangles* formed by stable and unstable manifolds of a saddle point or saddle periodic orbits when the manifolds intersect transversally. The complexity of the manifold's geometry stems from the fact that, if stable and unstable manifolds intersect once, then they must intersect an infinite number of times. Such a complex structure results in the generation of a horseshoe mapping, which persistently stretches and then folds the area around the manifolds generating a chaotic motion. The layers of the chaotic motion are clearly seen in Figure 9.

Lorenz system: The first clear numerical manifestation of chaotic dynamics was obtained in the Lorenz model. This model is a three-dimensional dynamical system derived from a reasonable simplification of the fluid dynamics equations for thermal convection in a liquid layer heated from below. The differential equations $\dot{x}=\sigma(y-x)$, $\dot{y}=rx-y-xz$, $\dot{z}=-bz+xy$ are written for the amplitude of the first horizontal harmonic of the vertical velocity (x), the amplitude of the corresponding temperature fluctuation (y), and a uniform correction of the temperature field (z) (Lorenz, 1963). σ is the Prandtl number, r is the reduced Rayleigh number, and b is a geometric factor. The phase portrait of the Lorenz attractor, time series, and the return mapping generated on the Poincaré cross section computed for $r=28$, $\sigma=10$, and $b=\frac{8}{3}$ are presented in Figure 10. A simple mechanical model illustrating the dynamical origin of oscillations in the Lorenz system is shown in Figure 11 (*See* **Lorenz equations**).

Definition of Chaos

As was shown above, dynamical chaos is related to unpredictability. For quantitative measurment of the unpredictability, it is reasonable to use the familiar characteristics *dimension* and *entropy*. These characteristics are independent: it is possible to generate a time series that has a high dimension and at the same time entropy equal to zero. This is a quasiperiodic motion. It is also simple to imagine a low-dimensional dynamical system with high entropy (see, e.g., the map in Figure 3).

Various definitions of chaos exist, but the common feature of these definitions is the sensitive dependence on initial conditions that was formalized above as positive entropy. Thus, dynamical chaos is the behavior of a dynamical system that is characterized by finite positive entropy.

Chaotic Attractors and Strange Attractors

A region in the phase space of a dissipative system that attracts all neighboring trajectories is called an attractor. An attractor is the phase space image of the behavior established in the dissipative system, for example, a stable limit cycle is the image of periodic oscillations. Therefore, the image of chaotic oscillations is a chaotic attractor.

A *chaotic attractor* (CA) possesses the following two properties that define any attractor of the dynamical system:

- There exists a bounded open region U containing a chaotic attractor ($\mathrm{CA} \in U$) in the phase space such that all points from this neighborhood converge to a chaotic attractor when time goes to infinity.
- A chaotic attractor is invariant under the evolution of the system,

In addition, the motion on a chaotic attractor has to be chaotic, for example:

- each trajectory of a chaotic attractor has at least one positive Lyapunov exponent.

Such types of attractors represent some regimes of chaotic oscillations generated by a Lorenz system and

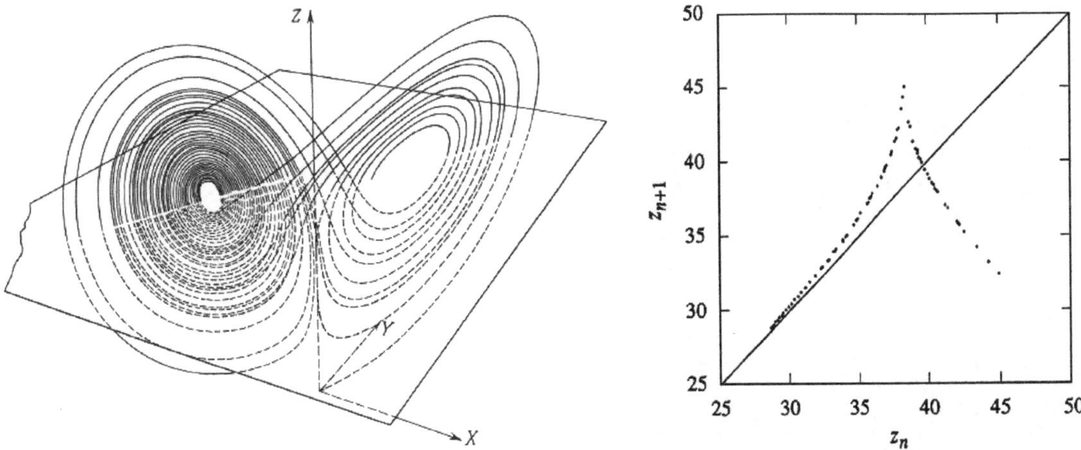

Figure 10. Lorenz attractor (left) and the return map $z_n + 1 = F(z_n)$ plotted for maximum values of variable z for the attractor trajectory (right).

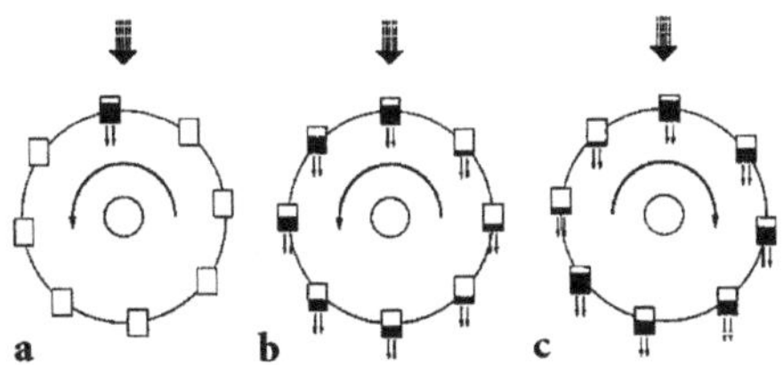

Figure 11. A toy model invented by Willem Malkus and Lou Howard illustrates dynamical mechanisms analogous to oscillations and chaos in the Lorenz system. Water steadily flowing into the top (leaky) bucket makes it heavy enough to start the wheel turning. When the flow is large enough, the wheel can start generating chaotic rotations characterized by unpredictable switching of the rotation direction; see Strogatz (1994, p. 302) for details.

the piece-wise linear maps. However, most of the chaotic oscillations observed in dynamical systems correspond to attractors that do not precisely satisfy the latter property. Although almost all trajectories in such attractors are unstable, some stable periodic orbits may exist within the complex structure of unstable trajectories. Chaos in such systems is persistent both in physical experiments and in numerical simulations because all of these stable orbits have extremely narrow basins of attraction. Due to natural small perturbations of the system, the trajectory of the system never settles down on one of the stable orbits and wanders within the complex set of unstable orbits.

The definition of a *strange attractor* is related to the complicated geometrical structure of an attractor. A strange attractor is defined as an attractor that cannot be presented by a union of the finite number of smooth manifolds. For example, an attractor whose topology can be locally represented by the direct product of a Cantor set to a manifold is a strange attractor. In many cases, the geometry of a chaotic attractor satisfies the definition of a strange attractor. At the same time, the definition of a strange attractor can be satisfied in the case of a nonchaotic strange attractor. This is an attractor that has fractal structure, but does not have positive Lyapunov exponents.

The origin of chaotic dynamics in dissipative systems and Hamiltonian systems in many cases is the same and is related to coexistence in the phase space of infinitely many unstable periodic trajectories as a part of homoclinic or heteroclinic tangles.

The Lorenz attractor, as for many other attractors in systems with a small number of degrees of freedom, can appear through a finite number of easily observable bifurcations. The bifurcation of a sudden birth and death of a strange attractor is called a crisis. Usually, it is related to the collision of the attractor with an unstable periodic orbit or its stable manifold (Arnol'd et al., 1993; Ott, 1993).

Order in Chaos

How does the dynamical origin imprint in chaos? Or in other words, how can the rules or order of the dynamical system be found inside a chaotic behavior? Consider the images (portraits) of the dynamical chaos shown in Figures 7, 8, and 10. The elegance of these images reflects the existence of order in dynamical chaos.

The dynamical origin of such elegance is very similar: different trajectories with close initial conditions have to be close in time $t_l \approx 1/\lambda$, where λ is the maximally positive Lyapunov exponent. The domain occupied by the strange attractor in phase space is finite; thus, the divergence of the phase space flow changes to convergence, and as a result of sequential action of divergence and convergence of the phase flow in the finite domain, the mixing of trajectories occurs. Such mixing can be illustrated with the motions of liquids in the physical space experimentally observed by Ottino (1989; see Figure 1).

Another way to recognize the existence of order in chaos is to analyze its dependence on a control parameter. The macroscopic features of real stochastic

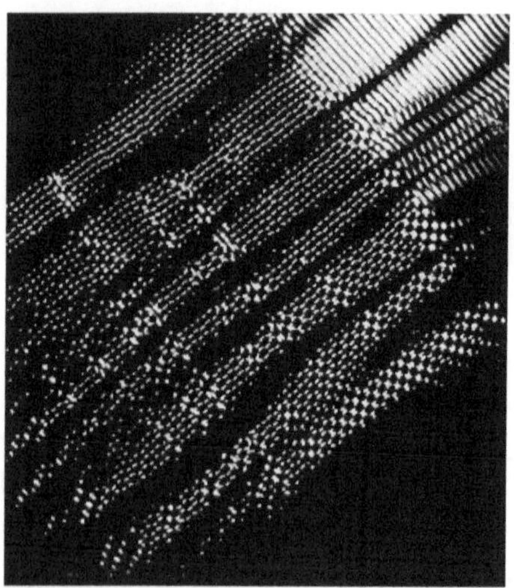

Figure 12. Appearance of spatiotemporal chaos in the extended Faraday experiment: chaotic patterns on the surface of the liquid layer in the oscillating gravitational field. The irregular chain of the localized structures—dark solitons—can be seen beneath a background of the square capillary lattice (Gaponov-Grekhov & Rabinovich, 1988).

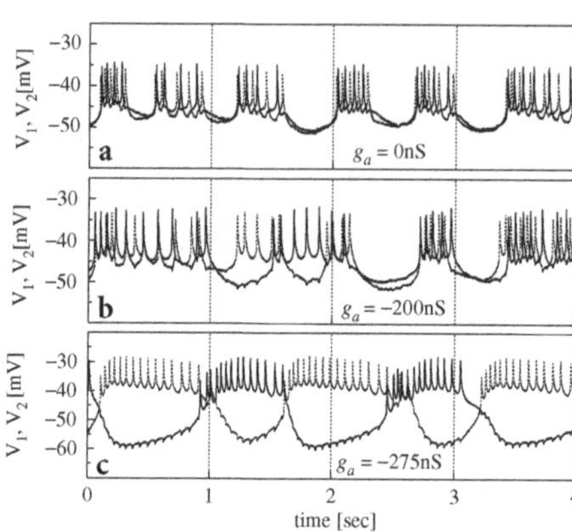

Figure 13. Dynamics of chaotic bursts of spikes generated by two living neurons coupled with an electrical synapse—a gap junction (Elson et al., 1998). Chaotic busts in naturally coupled neurons synchronize (a). When natural coupling is compensated by additional artificial coupling g_a, the chaotic oscillations are independent oscillations (b). The neurons coupled with negative conductivity fire in the regimes of antiphase synchronization (c).

processes, for example, Brownian motion or developed turbulence, depend on this parameter and change without any revolutionary events such as bifurcations. But for dynamical chaos, the picture is different. A continuous increase of control parameters of the logistic map does not necessarily gradually increase the degree of chaos: within chaos, there are windows—intervals of control parameter values in which the chaotic behavior of the system changes to stable periodic behavior, see Figure 4.

In a spatially extended system, for example, in convection or Faraday flow, order within chaos is related to the existence of coherent structures inside the chaotic sea (Rabinovich et al., 2001); see Figure 12.

Spatiotemporal Chaos

Similar to regular (e.g., periodic) motions, low-dimensional chaotic behavior is observed not only in simple (e.g., low-dimensional) systems but also in systems with many, and even with infinite number of degrees of freedom. The dynamical mechanisms behind the formation of low-dimensional chaotic spatiotemporal patterns in dissipative and nondissipative systems are different. In conservative systems, such patterns are related to the chaotic motion of particle-like localized structures. For example, a soliton that is described by a nonlinear Schrödinger equation with the harmonic potential

$$\mathrm{i}\frac{\partial a}{\partial t} + \beta\frac{\partial^2 a}{\partial^2 x} + \left(|a|^2 + \alpha \sin qx\right) a = 0 \qquad (13)$$

moves chaotically in physical space x and reminds us of the chaotic motion occurring in the phase space of a parametrically excited conservative oscillator (the equations of such an oscillator can be derived from (13) for slow variables characterizing the motion of the soliton center mass).

The interaction of the localized structures (particles) in a finite area, large in comparison with the size of the structure, can also lead to the appearance of spatiotemporal chaos. It was observed that collisions of solitons moving in two-dimensional space result in chaotic scattering similar to the chaotic motion observed in billiards (Gorshkov et al., 1992).

In dissipative nonlinear media and high-dimensional discrete systems, the role of coherent structures is also very important (such as defects in convection, clusters of excitations in neural networks, and vortices in the wake behind a cylinder; see Rabinovich et al., 2001). However, the origin of low-dimensional chaotic motions in such systems is determined by dissipation. There are two important mechanisms of finite dynamics (including chaos) that are due to dissipation: (1) the truncation of the number of excited modes (in hydrodynamic flows) due to high viscosity of the small-scale perturbations and (2) the synchronization of the modes or individual oscillators. Dissipation makes synchronization possible not only among periodic modes or oscillators but even in the case when the interacting subsystems are chaotic (Afraimovich et al., 1986). Figure 13 illustrates the synchronization of chaotic bursts of spikes observed experimentally in two coupled living neurons. In the case of a dissipative lattice of chaotic elements (e.g., neural lattices or models of an extended autocatalytic chemical reaction), complete synchronization leads to the onset

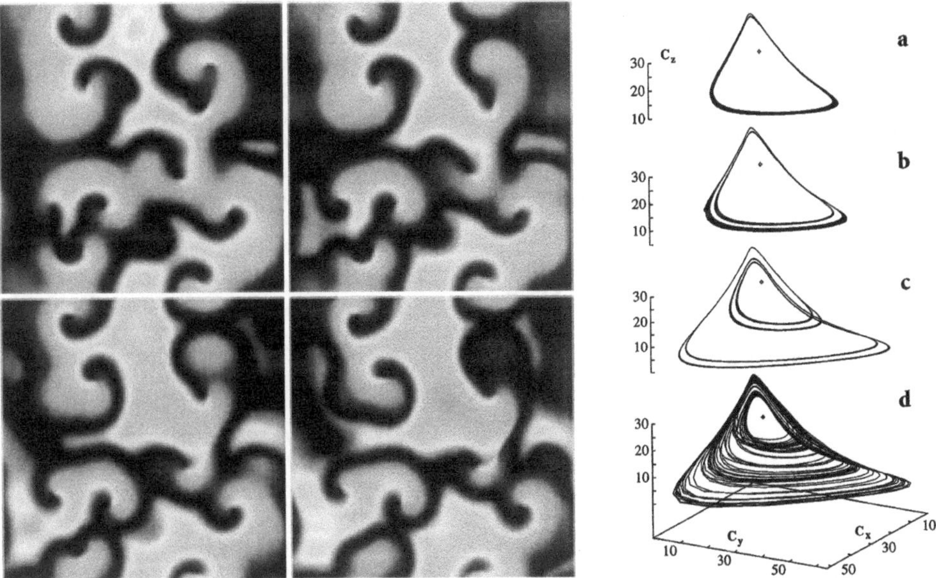

Figure 14. Coherent patterns generated in the chaotic medium with Rössler-type dynamics of medium elements. Left: an example of coherent patterns with defects. Right: evolution of the attractor with increasing distance r from a defect. The attractor changed from the limit cycle of period T at $r = r_1$ to the period $2T$ limit cycle at $r = r_2 > r_1$, then to the period $4T$ limit cycle at $r = r_3 > r_2$, and finally to the chaotic attractor for $r = r_4 > r_3$ (Goryachev & Kapral, 1996).

of a spatially homogeneous chaotic state. When this state becomes unstable against spatial perturbations, the system moves to the spatiotemporal chaotic state. A snapshot of such spatiotemporal chaos, which is observed in the model of chaotic media consisting of diffusively coupled Rössler-type chaotic oscillators, is presented in Figure 14. Figure 14 also illustrates the sequence of period-doubling bifurcations that are observed in the neighborhood of the defect in such a medium.

Edge of Chaos

In dynamical systems with many elements and interconnections (e.g., complex systems), the transition between ordered dynamics and chaos is similar to phase transitions between states of matter (crystal, liquid, gas, etc.). Based on this analogy, an attractive hypothesis named "edge of chaos"(EOC) appeared at the end of the 1980s. EOC suggests a fundamental equivalence between the dynamics of phase transitions and the dynamics of information processing (computation). One of the simplest frameworks in which to formulate relations between complex system dynamics and computation at the EOC is a cellular automaton. There is currently some controversy over the validity of this idea (Langton 1990; Mitchel et al., 1993).

Chaos and Turbulence

The discovery of dynamical chaos has fundamentally changed the accepted concept of the origin of hydrodynamic turbulence. When dealing with turbulence at finite Reynolds number, the main point of interest is the established irregular motion. The image of such irregularity in the phase space could be a chaotic attractor. Experiments in closed systems, for example, one in which fluid particles continuously recirculate through points previously visited, have shown the most common scenarios for the transition to chaos. These are (i) transition through the destruction of quasiperiodic motion that was observed in Taylor–Couette flow (Gollub & Swinney, 1975); (ii) period-doubling sequence observed in Rayleigh–Bénard convection (Libchaber & Maurer, 1980); and (iii) transition through intermittency (Gollub & Benson, 1980). Observation of these canonical scenarios for particular flows proved the validity of the concept of dynamical origin of the transition to turbulence in closed systems. It is possible to reconstruct a chaotic set in the phase space of the flow directly from observed data; see Brandstäter et al. (1982). At present it is difficult to say how dynamical chaos theory can be useful for the understanding and description of the developed turbulence.

The discovery and understanding of chaotic dynamics have important applications in all branches of science and engineering and, in general, to our evolving culture. An understanding of the origins of chaos in the last decades has produced many clear and useful models for the description of systems with complex behavior, such as the global economy (Barkly Russel, 2000), the human immune system (Gupta et al., 1998), animal behavior (Varona et al., 2002), and more. Thus, chaos theory provides a new tool for the unification of the sciences.

M.I. Rabinovich and N.F. Rulkov

See also **Attractors; Billiards; Butterfly effect; Chaos vs. turbulence; Controlling chaos; Dripping faucet; Duffing equation; Entropy; Fractals; Hénon map; Horseshoes and hyperbolicity in dynamical systems; Intermittency; Kicked rotor; Lorenz equations; Lyapunov exponents; Maps; Maps in the complex plane; Markov partitions; Multifractal analysis; One-dimensional maps; Order from chaos; Period doubling; Phase space; Quasiperiodicity; Rössler systems; Routes to chaos; Sinai–Ruelle–Bowen measures; Spatiotemporal chaos; Synchronization; Time series analysis**

Further Reading

Abarbanel, H.D.I. 1996. *Analysis of Chaotic Time Series*, New York: Springer

Afraimovich, V.S., Verichev, N.N. & Rabinovich, M.I. 1986. Stochastic synchronization of oscillations in dissipative systems. *Izvestiya Vysshikh Vchebnykh Zavedenii Radiofizika. RPQAEC*, 29: 795–803

Arnol'd, V.I., Afraimovich, V.S., Ilyashenko, Yu.S. & Shilnikov, L.P. 1993. Bifurcation theory and catastrophe theory. In *Dynamical Systems*, vol. 5, Berlin and New York: Springer

Barkly Russel, J., Jr. 2000. *From Catastrophe to Chaos: A General Theory of Economic Discontinuities*, 2nd edition, Boston: Kluwer

Brandstäter, A., Swift, J., Swinney, H.L., Wolf, A., Doyne Farmer, J., Jen, E. & Crutchfield, P.J. 1982. Low-dimensional chaos in a hydrodynamic system. *Physical Review Letters*, 51: 1442–1445

Chirikov, V.A. 1979. A universal instability of many-dimensional oscillator systems. *Physics Reports*, 52: 264–379

Deco, G. & Schürmann, B. 2000. *Information Dynamics: Foundations and Applications*, Berlin and New York: Springer

Elson, R.C., Selverston, A.I., Huerta, R., Rulkov, N.F., Rabinovich, M.I. & Abarbanel H.D.I. 1998. Synchronous behavior of two coupled biological neurons. *Physical Review Letters*, 81: 5692–5695

Gaponov-Grekhov, A.V. & Rabinovich, M.I. 1988. *Nonlinearity in Action: Oscillations, Chaos, Order, Fractals*, Berlin and New York: Springer

Gollub, J.P. & Benson, S.V. 1980. Many routes to turbulent convection. *Journal of Fluid Mechanics*, 100: 449–470

Gollub, J.P. & Swinney, H.L. 1975. Onset of turbulence in rotating fluid. *Physical Review Letters*, 35: 927–930

Gorshkov, K.A., Lomov, A.S. & Rabinovich, M.I. 1992. Chaotic scattering of two-dimensional solitons. *Nonlinearity*, 5: 1343–1353

Goryachev, A. & Kapral, R. 1996. Spiral waves in chaotic systems. *Physical Review Letters*, 76: 1619–1622

Gupta, S., Ferguson, N. & Anderson, R. 1998. Chaos, persistence, and evolution of strain structure in antigenically diverse infectious agent. *Science*, 280: 912–915

Langton, C.C. 1990. Computation at the edge of chaos—phase transitions and emergent computation. *Physica* D, 42: 12–37

Libchaber, A. & Maurer, J. 1980. Une expérience de Rayleigh-Bénard en géométrie réduite; multiplication, accrochage et démultiplication de fréquences. *Journal de Physique Colloques*, 41: 51–56

Lichtenberg, A.J. & Lieberman, M.A. 1992. *Regular and Chaotic Dynamics*, Berlin and New York: Springer

Lorenz, E.N. 1963. Deterministic nonperiodic flow. *Journal of Atmospheric Science*, 20: 130–136

Mitchel, M., Hraber, P. & Crutchfield, J. 1993. Revisiting the edge of chaos: evolving cellular automata to perform computations. *Complex Systems*, 7: 89–130

Murray, N. & Holman, M. 1999. The origin of chaos in the outer solar system. *Science*, 283: 1877–1881

Ott, E. 1993. *Chaos in Dynamical Systems*, Cambridge and New York: Cambridge University Press

Ottino, J.M. 1989. *The Kinetics of Mixing: Stretching, Chaos, and Transport*, Cambridge and New York: Cambridge University Press

Pikovsky, A.S. & Rabinovich, M.I. 1978. A simple generator with chaotic behavior. *Soviet Physics Doklady*, 23: 183–185 (see also Rabinovich, M.I. 1978. Stochastic self-oscillations and turbulence. *Soviet Physics Uspekhi*, 21: 443–469)

Rabinovich, M.I., Ezersky, A.B. & Weidman, P.D. 2001. *The Dynamics of Patterns*, Singapore: World Scientific

Rikitake, T. 1958. Oscillations of a system of disk dynamos. *Proceedings of the Cambridge Philosophical Society*, 54: 89–105

Roux, J.C., Simoyi, R.H. & Swinney, H.L. 1983. Observation of a strange attractor. *Physica* D, 8: 257–266

Shukla, J. 1998. Predictability in the midst of chaos: a scientific basis for climate forecasting. *Science*, 282: 728–731

Sinai, Ya.G. 2000. *Dynamical Systems, Ergodic Theory and Applications*, Berlin and New York: Springer

Strogatz, S.H. 1994. *Nonlinear Dynamics and Chaos: With Applications to Physics, Biology, Chemistry, and Engineering* Reading, MA: Addison-Wesley

Takens, F. 1991. Chaos, In *Structures in Dynamics: Finite Dimensional Deterministic Studies*, edited by H.W. Broer, F. Dumortier, S.J. van Strien & F. Takens, Amsterdam: North-Holland Elsevier Science

Ueda, Y. 1992. *The Road to Chaos*, Santa Cruz, CA: Aeirial Press

van der Pol, B. & van der Mark, B. 1927. Frequency demultiplication. *Nature*, 120: 363–364

Varona, P., Rabinovich, M.I., Selverston, A.I. & Arshavsky, Yu.I. 2002. Winnerless competition between sensory neurons generates chaos: A possible mechanism for molluscan hunting behavior. *CHAOS* 12: 672–677

Young, L.S. 1998. Developments in chaotic dynamics. *Notices of the AMS*, 17: 483–504

CHARACTERISTICS

Systems of first-order partial differential equations describe many different physical phenomena from the behavior of fluids, gases, and plasmas. To introduce the Method of Characteristics, consider the simple scalar conservation law of the form

$$\frac{\partial U}{\partial t} + A(U)\frac{\partial U}{\partial x} = 0. \qquad (1)$$

Here, $U = U(x, t)$, where x is a spatial coordinate and t is the time coordinate. The function $A(U)$ defines the speed of propagation of a disturbance and either may be independent of U, in which case equation (1) is a linear partial differential equation, or it may depend explicitly on the dependent variable U, in which case the equation is a nonlinear partial differential equation.

It is important to specify the initial or boundary conditions that the solution $U(x, t)$ must satisfy.

Consider the simple case where the function is known at the initial time $t = 0$. Thus,

$$U(x, 0) = F(x). \tag{2}$$

The idea is to simplify Equation (1) by choosing a suitable curve in the x-t plane. This curve can be written in parametric form as

$$x = x(s), \quad t = t(s), \tag{3}$$

where s is the parameter that can be thought of as measuring the distance along the curve. To understand how to select the particular form of the curve, note that, using (3),

$$U(x, t) = U(x(s), t(s)), \tag{4}$$

implies that U is a function of s. Hence, the chain rule gives the derivatives of U along the curve as

$$\frac{\mathrm{d}U}{\mathrm{d}s} = \frac{\partial U}{\partial t}\frac{\mathrm{d}t}{\mathrm{d}s} + \frac{\partial U}{\partial x}\frac{\mathrm{d}x}{\mathrm{d}s}. \tag{5}$$

Comparing the right-hand side of (5) with the left-hand side of (1), it is clear that they are identical, provided the parametric form of the curve is chosen as

$$\frac{\mathrm{d}t}{\mathrm{d}s} = 1, \tag{6}$$

$$\frac{\mathrm{d}x}{\mathrm{d}s} = A(U), \tag{7}$$

and (1) reduces to

$$\frac{\mathrm{d}U}{\mathrm{d}s} = 0, \Rightarrow U = \text{constant along the curve.} \tag{8}$$

The curve satisfied by (6) and (7) is called the *characteristic curve*. Along this curve, U is constant. However, the value of the constant may be different on each characteristic curve. The solution of the characteristic equations requires some initial conditions for x and t. These are taken as

$$x = x_0, \quad t = 0, \quad \text{at } s = 0. \tag{9}$$

Note that x_0 covers the same domain as x. Solving (6)–(8) yields

$$t = s, \quad x = A(U)s + x_0, \tag{10}$$

on using the initial conditions (9). x_0 can be thought of as a constant of integration and so it has a fixed value along the characteristic curve. This implies, using (2), that

$$U(x, 0) = F(x_0). \tag{11}$$

Note that the particular characteristic curve is determined by the value of the parameter x_0. Hence, eliminating the parameter s and solving for x_0 in terms of x, U, and t, the solution given by (11) is

$$U(x, t) = F(x_0) = F(x - A(U)t). \tag{12}$$

If $A(U)$ is a constant, say c, then the solution is simply

$$U(x, t) = F(x - ct), \tag{13}$$

but if $A(U)$ depends explicitly on U, then the solution is an *implicit solution*.

The characteristic curves in this example are straight lines in the x-t plane with a gradient given by $A(U)$. When $A(U) = c$ is a constant, the characteristic curves are parallel straight lines. This means that the shape of the initial disturbance propagates unchanged, to the right if c is a positive constant. If $A(U)$ depends on U, then the characteristic curves are straight lines but with different gradients. There exists the possibility that the characteristic curves may cross, and this corresponds to the formation of a shock. When the characteristic curves diverge, the solution exhibits an expansion fan that can be expressed in terms of a similarity variable.

Note that the method of characteristics can be used when $A = A(U, x, t)$ depends explicitly on the space and time coordinates. In this case, the coupled equations, (6)–(8), may be solved numerically. A detailed description of the method of characteristics for general first-order partial differential equations is given in Rubenstein & Rubenstein (1993).

Example: Burgers Equation

Consider the case when $A(U) = U$. Then, (1) becomes the inviscid Burgers equation

$$\frac{\partial U}{\partial t} + U\frac{\partial U}{\partial x} = 0, \tag{14}$$

and the solution satisfying the initial condition

$$U(x, 0) = \begin{cases} 0, & x < -1, \\ 1 + x, & -1 \le x < 0, \\ 1 - x, & 0 \le x \le 1, \\ 0, & x > 1 \end{cases} \tag{15}$$

is

$$U = \begin{cases} 0, & x < -1 + Ut, \\ 1 + x - Ut, & -1 + Ut \le x < Ut, \\ 1 - x + Ut, & Ut \le x \le 1 + Ut, \\ 0, & x > 1 + Ut. \end{cases} \tag{16}$$

Thus, solving for U in each region gives the solution as

$$U = \begin{cases} 0, & x < -1, \\ (1 + x)/(1 + t), & -1 \le x < t, \\ (1 - x)/(1 - t), & t \le x \le 1, \\ 0, & x > 1. \end{cases} \tag{17}$$

Note that the solution becomes multi-valued for $t > 1$. This can be understood by considering the characteristic curves defined by (9). In the x-t plane, they are straight lines of the form

$$x = Ut + x_0, \tag{18}$$

so that the gradient depends on the value of U at $t = 0$ and $x = x_0$. Thus, using the initial conditions, the characteristic curves, valid in the region $t \le x \le 1$, can be expressed as

$$x = (1 - x_0)t + x_0, \tag{19}$$

for $0 \le x_0 \le 1$. Considering two different values of x_0, say x_a and x_b, the straight lines cross when

$$x = (1 - x_a)t + x_a = (1 - x_b)t + x_b,$$

$$\Rightarrow \qquad t = 1 \quad \text{and} \quad x = 1. \tag{20}$$

Hyperbolic Systems of Several Dependent Variables

Systems of first-order hyperbolic equations can be expressed in vector and matrix form as

$$\frac{\partial \boldsymbol{U}}{\partial t} + \boldsymbol{A}(\boldsymbol{U}, x, t)\frac{\partial \boldsymbol{U}}{\partial x} = 0, \tag{21}$$

where $\boldsymbol{U}$ is a column vector of n elements containing the dependent variables and $\boldsymbol{A}$ is an $n \times n$ matrix whose coefficients may depend on the dependent variables. The problem is linear if the matrix $\boldsymbol{A}$ has elements independent of $\boldsymbol{U}$ and nonlinear otherwise. The characteristic curves in this case are given by the equations

$$\frac{\mathrm{d}t}{\mathrm{d}s} = 1, \tag{22}$$

$$\frac{\mathrm{d}x}{\mathrm{d}s} = \lambda_i(\boldsymbol{U}, x, t) \tag{23}$$

for $i = 1, 2, ..., n$ and where λ_i is an eigenvalue of the matrix $\boldsymbol{A}$. Here, it is assumed that the matrix $\boldsymbol{A}$ has n distinct eigenvalues. For the linear problem, and in particular, for the case where the matrix $\boldsymbol{A}$ has constant coefficients, the full solution can be obtained by using a suitable linear combination of dependent variables so that the equations reduce to a set of simple advection equations. Hence, the first step is to determine the eigenvalues, λ_i, of the matrix $\boldsymbol{A}$ and the corresponding eigenvectors $\boldsymbol{z}_i$, where

$$\boldsymbol{A}\boldsymbol{z}_i = \lambda_i \boldsymbol{z}_i, \tag{24}$$

Next, use the change of variable

$$\boldsymbol{U} = \boldsymbol{Q}\boldsymbol{V}, \tag{25}$$

where $\boldsymbol{Q}$ is an $n \times n$ matrix whose jth column is the jth eigenvector $\boldsymbol{z}_j$. Substituting into (21) yields

$$\boldsymbol{Q}\frac{\partial \boldsymbol{V}}{\partial t} + \boldsymbol{A}\boldsymbol{Q}\frac{\partial \boldsymbol{V}}{\partial x} = 0. \tag{26}$$

Finally, pre-multiplying by $\boldsymbol{Q}^{-1}$, the inverse of $\boldsymbol{Q}$, results in a decoupled system of equation, since $\boldsymbol{Q}^{-1}\boldsymbol{A}\boldsymbol{Q}$ is a diagonal matrix whose elements are the eigenvalues λ_i. Thus, the final set of n equations are

$$\frac{\partial V_i}{\partial t} + \lambda_i \frac{\partial V_i}{\partial x} = 0, \tag{27}$$

for $i = 1, 2, ..., n$. The solutions to (27) are simply

$$V_i = F_i(x - \lambda_i t), \tag{28}$$

where F_i is an arbitrary function determined by the initial conditions. Once all the solutions for V_i are determined from the initial conditions, the solution in terms of the original variables is obtained using (25). Note that while the original variables may depend on all the characteristic variables, the V_i solution is *constant* along the ith characteristic curve.

Example: The Second-Order Wave Equation
The second-order wave equation

$$\frac{\partial^2 U}{\partial t^2} = c^2 \frac{\partial^2 U}{\partial x^2} \tag{29}$$

can be expressed as a pair of first-order equations as

$$\frac{\partial U}{\partial t} = -c\frac{\partial p}{\partial x},$$
$$\frac{\partial p}{\partial t} = -c\frac{\partial U}{\partial x}.$$

Thus,

$$\boldsymbol{A} = \begin{pmatrix} 0 & c \\ c & 0 \end{pmatrix}. \tag{30}$$

The eigenvalues are simply $\lambda_1 = c$ and $\lambda_2 = -c$ and the corresponding eigenvectors are

$$\boldsymbol{z}_1 = \begin{pmatrix} 1 \\ -1 \end{pmatrix}, \qquad \boldsymbol{z}_2 = \begin{pmatrix} 1 \\ 1 \end{pmatrix}. \tag{31}$$

Thus,

$$\boldsymbol{Q} = \begin{pmatrix} 1 & 1 \\ -1 & 1 \end{pmatrix}, \qquad \boldsymbol{Q}^{-1} = \frac{1}{2}\begin{pmatrix} 1 & -1 \\ 1 & 1 \end{pmatrix}. \tag{32}$$

Equation (27) reduces to the pair of equations

$$\frac{\partial V_1}{\partial t} - c\frac{\partial V_1}{\partial x} = 0, \qquad \frac{\partial V_2}{\partial t} + c\frac{\partial V_2}{\partial x} = 0. \tag{33}$$

The solutions are $V_1 = F_1(x+ct)$ and $V_2 = F_2(x-ct)$ and, in terms of the original variables, the solution is

$$\begin{aligned} U &= F_1(x+ct) + F_2(x-ct), \\ p &= -F_1(x+ct) + F_2(x-ct). \end{aligned} \qquad (34)$$

Riemann Invariants

A Riemann invariant may be thought of as a function of the dependent variables that is constant along a characteristic curve. In the previous example, it is clear that $U + p = 2F_2(x-ct)$, so $U + p$ is a Riemann invariant along the characteristic curve defined by $x - ct = \text{constant}$. Similarly, $U - p$ is constant along $x + ct = \text{constant}$.

Example: Isentropic Flow

The dimensionless equations for *isentropic* fluid flow with $p/\rho^\gamma = 1$ can be expressed in terms of the fluid velocity, u, and the sound speed $c = (\gamma p/\rho)^{1/2}$ as

$$\frac{\partial u}{\partial t} + u\frac{\partial u}{\partial x} + \frac{2}{\gamma - 1}c\frac{\partial c}{\partial x} = 0, \qquad (35)$$

$$\frac{\partial c}{\partial t} + \frac{\gamma - 1}{2}c\frac{\partial u}{\partial x} + u\frac{\partial c}{\partial x} = 0. \qquad (36)$$

A detailed derivation of these equations is given in Kevorkian (1989). The matrix $\boldsymbol{A}$ is

$$\boldsymbol{A} = \begin{pmatrix} u & \frac{2}{\gamma-1}c \\ \frac{\gamma-1}{2}c & u \end{pmatrix} \qquad (37)$$

having eigenvalues $\lambda_1 = u + c$ and $\lambda_2 = u - c$. There-are two characteristic curves given by the solution of the coupled differential equations

$$\frac{\mathrm{d}t}{\mathrm{d}s} = 1,$$

$$\frac{\mathrm{d}x}{\mathrm{d}s} = \lambda_i,$$

where $i = 1$ or $i = 2$. For $i = 1$, the initial conditions are $t = 0$ and $x = \xi$ at $s = 0$, which implies that the characteristic curve is defined by $\xi = x - \int_0^t (u+c)\,\mathrm{d}s = \text{constant}$. For $i = 2$, $t = 0$, and $x = \eta$ at $s = 0$, the second curve is defined by $\eta = x - \int_0^t (u-c)\,\mathrm{d}s = \text{constant}$. Multiplying (35) by $(\gamma - 1)/2$ and then adding and subtracting (36) gives two equations

$$\frac{\partial R}{\partial t} + (u+c)\frac{\partial R}{\partial x} = 0,$$

$$\frac{\partial S}{\partial t} + (u-c)\frac{\partial S}{\partial x} = 0,$$

where $R = c + (\gamma - 1)u/2$ and $S = c - (\gamma - 1)u/2$. R and S are Riemann invariants since R is constant along the characteristic curve defined by $\xi = \text{constant}$) and S is constant along $\eta = \text{constant}$). A more detailed derivation of Riemann invariants is described in Kevorkian (1989).

ALAN HOOD

See also **Burgers equation; Coupled systems of partial differential equations; Hodograph transform; Shock waves**

Further Reading

Kevorkian, J. 1989. *Partial Differential Equations: Analytical Solution Techniques*, New York: Chapman & Hall

Rubenstein, I. & Rubenstein, L. 1993 *Partial Differential Equations in Classical Mathematical Physics*, Cambridge and New York: Cambridge University Press

CHARGE DENSITY WAVES

A charge density wave (CDW) is a collective transport phenomenon, whose origin lies in the interaction between electrons and phonons in a solid (Grüner & Zettl 1985; Grüner, 1988). As envisioned by Rudolph Peierls in 1930 in some quasi-one-dimensional metals (where the influence of one electron to each other electron is much stronger than in higher dimensions), the elastic energy needed to displace the position of the atoms may be balanced by a lowering of conduction electron energy.

In such cases, the more stable configuration may have a periodic distortion of the lattice; thus, there is a modulation of the electronic charge density, which gives rise to a CDW. The wave vector turns out to be $Q = 2k_\mathrm{F}$, where k_F is the Fermi wave vector, and the electronic density becomes

$$\delta\rho = \rho_0 \cos(2k_\mathrm{F}x + \phi).$$

Due to this periodic lattice distortion, a gap at the Fermi level appears, and the conduction electrons lower their kinetic energy. At high temperatures, thermal excitation of electrons across the band gap makes the normal metallic state stable. When the temperature is sufficiently low, a second-order phase transition (known as the Peierls transition) takes place, and a CDW is formed.

In 1954, Herbert Fröhlich suggested that if Q was not commensurate with the lattice constant, the CDW energy would be independent of the phase ϕ, and thus, an electrical current would appear under any electric field, independent of its intensity. For a while, this phenomenon was speculated to be a possible origin of superconductivity. Interestingly, the interplay and relationship among CDWs, superconductivity, and spin density waves is still a field of study (Gabovich et al., 2002).

If the translational invariance of ϕ is disrupted, there is a phase for which the CDW energy is the lowest, and there is also a minimum threshold field to overcome this energy reduction and to initiate the conduction. A possible cause of the invariance break could be that the CDW is commensurate with the lattice. Although this case is unusual and mostly of theoretical interest, such a CDW (with a period quasimultiple of the lattice constant) may contain solitons in the form of constant phase zones, separated by abrupt change areas. This soliton behavior is modeled by sine-Gordon-like equations. However, empirical evidence suggests that the origin of the pinning of the CDW to the lattice and the appearance of a threshold field stem from impurities.

Experimental evidence of CDW behavior became available in the 1970s, and nowadays several materials show CDW behavior, both inorganic, like $NbSe_3$, NbS_3, or $K_{0.3}MoO_3$ ("blue bronze"), and organic, like (fluoranthene)$_2PF_6$. Evidence of this kind of transition is detected through magnitudes affected by the gap at the Fermi level, including magnetic susceptibility, resistivity, thermoelectric power, scattering experiments where the CDW wave vector manifests itself, and more recently, by means of scanning tunneling microscope images.

Conductivity is among the more interesting properties of CDWs. The dielectric constants for these materials are high, and conductivity suffers an abrupt change from insulating to metallic values of orders of magnitude. The Hall and thermoelectric effects suggest that their conductivity consists of an ohmic linear term and a CDW nonlinear term.

The response of the CDW to a field higher than the threshold value is twofold. First, there appears a high-frequency coherent current, or narrow band noise, which seems to be due to the displacement of the CDW over the pinning potential. Second, a low-frequency broad band noise, incoherent response, is also detected. It is also found that the conductivity saturates for high values of the external field, and it seems that this is due to electrons leaving the CDW region or due to the elimination of $2k_F$ phonons.

When an a.c. field is present, the CDW exhibits a strong dependence on the field frequency, and its conductivity, σ_{ac}, also saturates for high frequencies. There also appear an induced conductivity, σ_{dc}, which increases when V_{ac} increases, and some interference phenomena between the narrow band noise and the a.c. The external field $E_{ac}\cos\omega_e t$ causes oscillations of the current at frequency ω_e, and if there is also a d.c. field, E_{dc}, there are oscillations at frequency ω_i corresponding to the narrow band noise. These two frequencies may interact to produce mode-locking phenomena when they are commensurate ($n\omega_i = m\omega_e$).

In CDW systems, this locking shows up in the step structure of the differential resistance as a function of the d.c. field. As the external d.c. changes, the nonlinearity of the system keeps the relation between both frequencies constant, ω_i/ω_e, over a finite interval of the external parameter, corresponding to the intervals where dV/dI is constant. When the external parameter moves far from the locking region, the system undergoes a transition to an unlocked state, which is quasi-periodic, with two incommensurate frequencies. The interference between the internal frequency ω_i and the external one ω_e is the origin of the coherent and incoherent responses of the system. Usually, the low-frequency region of the power spectra consists of a broad band noise, while the narrow components show up at high frequencies as narrow band noise. A systematic elimination of the broad band noise when the CDW entered mode locking (Sherwin & Zettl, 1985) and a reinforcement of this noise in the unlocked regime have been observed. The interplay between the internal frequency and the external one may give rise to chaotic behavior, with a period-doubling route to chaos.

Studied in the context of self-organized criticality, CDWs are an example of systems that reorganize themselves near the edge of stability, and any small change in the external electrical field gives rise to a drastic change in the response of the CDW (high increase of conductivity).

Although several models have been proposed to explain CDW behavior, none is completely satisfactory. The classical model considers the CDW as a rigid carrier, without any internal degree of freedom, using the forced oscillator equations with some analogy to the Josephson junctions. The tunneling model focuses on the gap in the excitation spectrum of the CDW, explaining the nonlinear conductivity and the scale relationship between $\sigma_{ac}(\omega)$ and $\sigma_{dc}(E)$. However, these models do not explain the interference phenomena between the narrow band noise and the external field frequency ω.

There are other models that consider the internal degrees of freedom of the CDW (segmenting the CDW either through a hydrodynamical description or the Kelmm–Schrieffer model), but none of them completely explains the phenomenology observed in a CDW. Another interesting model is the Fukuyama–Lee–Rice model, which treats the CDW as a classical extended elastic medium, interacting with impurities and an electric field. Discrete versions of these models have also been used. For the commensurate case, Frenkel–Kontorova and soliton models (such as the sine-Gordon equation) have been used.

Several applications have been suggested for these materials, including tunable condensers, optical detectors, memory devices, and switches, among others.

Luis Vázquez, P. Pascual, and S. Jiménez

See also **Coupled oscillators; Frenkel–Kontorova model; Polarons; Sine-Gordon equation; Superconductivity**

Further Reading

Brown, S. & Grüner, G. 1994. Charge and spin density waves. *Scientific American*, April 1994: 50–56

Gabovich A.M., Voitenko, A.I. & Ausloos, M. 2002. Charge-and spin-density waves in existing superconductors: competition between Cooper pairing and Peierls or excitonic instabilities. *Physics Reports*, 367: 583–709

Grüner, G. & Zettl, A. 1985. Charge density wave conduction: a novel collective transport phenomenon in solids. *Physics Reports*, 119: 117–232

Grüner, G. 1988. The dynamics of charge-density waves. *Reviews of Modern Physics*, 60: 1129–1181

Sherwin, M. & Zettl A. 1985. Complete charge density-wave mode locking and freeze-out of fluctuations in $NbSe_3$. *Physical Review B*, 32: 5536–5539

Thorne, R.E. 1996. Charge-density-wave conductors. *Physics Today*, May 1996: 42–47

CHEMICAL KINETICS

Chemical kinetics is a well-defined field of physical chemistry that arose in the 1850s as a complement to the investigation of chemical equilibria. The question of how fast a reactive mixture in a closed vessel reaches equilibrium gave rise to the concept of reaction velocity. The mass action law, enunciated by Cato Guldberg and Peter Waage in 1863, provided a quantitative expression of the velocity of an elementary reaction step in a homogeneous medium in terms of the concentrations or the mole fractions of the reactants involved, and a parameter known as the rate constant. Chemical kinetics is intrinsically nonlinear, since the law of mass action features products of concentrations of the species involved.

Early Developments

Evidence that chemical reactions can generate complex behavior was reported in the early days of chemical kinetics (Pacault & Perraud, 1997). In 1899, Wilhelm Ostwald discovered that in a reaction involving chromium in concentrated acid solution, the release of hydrogen gas was periodic. In 1906, Robert Luther observed propagating chemical reaction fronts in connection with the catalytic hydrolysis of alkyl sulfates. These studies remained isolated for a long time. Possible origins, including the systematic study of reaction mechanisms, were hardly touched upon and there was little or no modeling effort. Not surprisingly, therefore, they came to be regarded by the scientific community as curiosities or even as artifices. On the theoretical side in the 1920s, Alfred Lotka devised a model formally deriving from chemical kinetics and giving rise to sustained oscillations. As the model did not apply to any known chemical system, it was discarded by chemists but was far better received in population dynamics where it played a seminal role. This connection was further enforced in 1926 when Vito Volterra advanced an explanation of ecological cycles in connection with predator-prey systems, using ideas similar to those of Lotka.

The Phenomenology of Nonlinear Chemical Kinetics

Nonlinear chemical kinetics in its modern form owes much to the Belousov–Zhabotinsky reaction (Zhabotinsky, 1964; Field et al., 1972) dealing with the oxidation of a weak acid by bromate in the presence of a metal ion redox catalyst. In addition to the possibility of displaying long records of oscillatory behavior in batch (closed reactor), this reaction gave rise for the first time to a thorough mechanistic study which highlighted the important role of feedback in the onset of complex behavior, in addition to nonlinearity.

Nonlinear phenomena in chemical kinetics have been observed on whole classes of systems giving rise to a large variety of complex behaviors, as reviewed in a Faraday discussion held in 2001. Quantitative phase diagrams have been constructed separating different behavioral modes as some key parameters are varied (Gray & Scott, 1990; Epstein & Pojman, 1998).

Open Well-Stirred Reactors

Simple periodic, multi-periodic, and chaotic oscillations are observed as the residence time of reactants (inversely proportional to their pumping rate into the reactor) is varied. A second type of phenomenon is multistability, the possibility of exhibiting more than one simultaneously stable steady-state concentration level. A third type of phenomenon is excitability whereby, once perturbed, a system performs an extended excursion resembling a single pulse of an aborted oscillation, before settling back to its stable steady state. Finally, an interesting phenomenology pertains to combustion reactions, where the dependence of the rate constant on temperature is the source of a universal (reaction mechanism-independent) positive feedback.

Open Unstirred Reactors

In the absence of stirring, chemical dynamics coexists with transport phenomena. This can give rise to the generic phenomenon of propagating wave fronts. In a bistable system, a typical front may join the two stable states, with one of them progressing at the expense of the other. In two- and three-dimensional reactors undergoing excitable or oscillatory dynamics, the fronts can take the more spectacular form of circles (target patterns), rotating spirals, and scrolls. An exciting form of spatial self-organization is spontaneous symmetry-breaking leading to sustained steady-state patterns, anticipated by Alan Turing in 1952 and realized experimentally by Patrick De Kepper and coworkers; see Figure 1 (Turing, 1952; De Kepper et al., 2000).

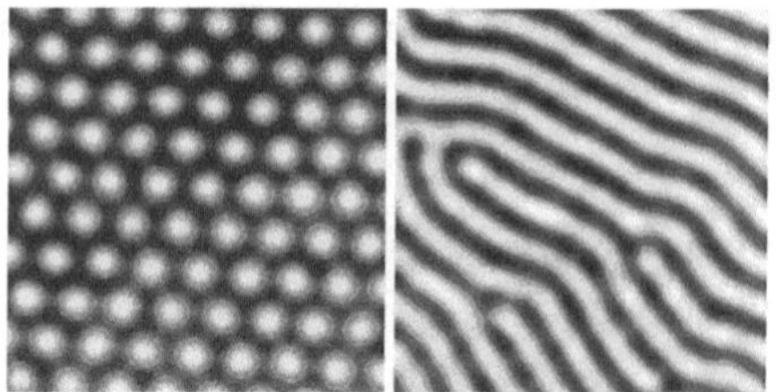

Figure 1. Stationary concentration patterns arising in the chlorite–iodide–malonic acid reaction beyond a symmetry-breaking instability (courtesy of P. De Kepper).

Heterogeneous Systems

Since the late 1980s, a series of novel developments has been initiated following the encounter of nonlinear chemical kinetics with surface science as it is manifested, for instance, in heterogeneous catalysis. Complex behavior in all the above-mentioned forms is observed. Furthermore, the development of sophisticated monitoring techniques, such as field ion microscopy, opens the perspective of monitoring chemical dynamics at the nanoscale level (Hildebrand et al., 1999).

Theoretical Developments: Dynamical Systems and Nonlinear Chemical Kinetics

The essence of nonlinear chemical kinetics is captured by the reaction-diffusion equations (Nicolis & Prigogine, 1977)

$$\frac{\partial c_i}{\partial t} = v_i(\{c_j\}, k_\alpha, \Delta H_\alpha, \cdots) + D_i \nabla^2 c_i, \qquad (1)$$

where $c_i (i = 1, \ldots, n)$ denotes the concentrations or the temperature, k_α and ΔH_α the rate constants and heats of reaction of the steps involved, and D_i the mass or heat diffusivity coefficients. The rate function v_i accounts for the nonlinearities and feedbacks, whereas the contribution of transport processes is linear.

The reaction-diffusion equations (1) exhibit nonlinearity in its simplest expression, as a property arising from intrinsic and local cooperative events. Because of this, complex behavior may arise in the absence of spatial degrees of freedom and persist even when few variables are present. In thermodynamic language, reactions are purely dissipative processes, whereas in nonlinear mechanics, inertia plays a very important role in the onset of complex behavior. Understanding how purely dissipative systems can come to terms with the restrictions imposed by the laws of thermodynamics and statistical mechanics has stimulated several fundamental developments (Glansdorff & Prigogine, 1971; Nicolis & Prigogine, 1977). It has also led to the design of canonical models, such as the Brusselator, that are being used with success to test ideas and to assess the limits of validity of approximations.

The intrinsic parameters k and D in Equations (1) has dimensions of $[\text{time}]^{-1}$ and $[(\text{length})^2/\text{time}]$, respectively. It follows that a reaction-diffusion system possesses intrinsic time (k^{-1}) and space ($(D/k)^{1/2}$) scales. This places nonlinear kinetics at the forefront for understanding the origin of endogenous rhythmic and patterning phenomena as observed, in particular, in biology and in materials science. In thermodynamic equilibrium, these intrinsic time and length scales remain dormant, owing to detailed balance. Nonequilibrium allows for the excitation and eventual stabilization of finite amplitude disturbances bearing these characteristic scales.

Equations (1) form the basis of interpretation of the experiments surveyed above. They also constitute some of the earliest and most widely used models of bifurcation and chaos theories. The classical tools used in their analysis are stability theory and the reduction to normal form (amplitude) equations using perturbation techniques and/or symmetry arguments, complemented by interactive numerical simulations (Nicolis, 1995). A most interesting development is the prediction of an impressive variety of intrinsically generated spatial and spatiotemporal patterns, including spatiotemporal chaos, when two or more mechanisms of instability are interfering.

Nonlinear Chemical Kinetics in the Life Sciences

Research in nonlinear chemical kinetics has led to a semi-quantitative interpretation of a wide spectrum of dynamical behaviors in biochemistry (Goldbeter, 1996). This has been possible thanks to the development of models in which the involvement of cooperative enzymes in some key steps provides the principal source of nonlinearity and feedback. Glycolytic oscillations, calcium oscillations and waves, the cell division cycle, cAMP-induced aggregation in amoebae, and synchronization in cell populations are among the main achievements of this effort that helped to identify the principal mechanisms behind the observed behavior.

Nonlinear kinetics has also been a source of inspiration for approaching dynamical phenomena of crucial importance in biology, in which modeling involving a few variables and/or well-established molecular mechanisms is still not available. Immune response, the electrical activity of the brain, embryonic development, cooperative processes such as food recruitment or building activity in social insects, and, last but not least, chemical and biochemical evolution itself (Eigen & Schuster, 1979; *See* **Biological evolution**) have been explored in one way or the other in the light of the concepts and techniques of nonlinear kinetics.

As a closing remark, Equations (1) anticipate a decoupling between the evolution laws of the macroscopic observables and dynamics at the microscopic level, which may actually break down when reactive systems are embedded on low-dimensional supports owing to the generation of anomalous inhomogeneous fluctuations. This leads to interesting synergies among nonlinear chemical kinetics, statistical mechanics, and computational science (Nicolis, 2001).

G. Nicolis

See also **Belousov–Zhabotinsky reaction; Brusselator; Population dynamics; Reaction-diffusion systems; Turing patterns**

Further Reading

De Kepper, P., Dulos, E., Boissonade, J., De Wit, A., Dewel, G. & Borckmans, P. 2000. Reaction–diffusion patterns in confined chemical systems, *Journal of Statistical Physics*, 101: 495–508

Eigen, M. & Schuster, P. 1979. *The Hypercycle: A Principle of Natural Self-organization*, Berlin and New York: Springer

Epstein, I.R. & Pojman, J.A. 1998. *An Introduction to Nonlinear Chemical Dynamics*, Oxford and New York: Oxford University Press

Field, R.J., Körös, E. & Noyes, R. 1972. Oscillations in chemical systems. II. Thorough analysis of temporal oscillation in the bromate–cerium–malonic acid system. *Journal of the American Chemical Society*, 94: 8649–8664

Glansdorff, P. & Prigogine, I. 1971. *Thermodynamic Theory of Structure, Stability and Fluctuations*, London and New York: Wiley

Goldbeter, A. 1996. *Biochemical Oscillations and Cellular Rhythms*, Cambridge and New York: Cambridge University Press

Gray, P. & Scott, S.K. 1990. *Chemical Oscillations and Instabilities*, Oxford: Clarendon Press and New York: Oxford University Press

Hildebrand, M. Kuperman, M., Wio, H., Mikhailov, A.S. & Ertl, G. 1999. Self-organized chemical nanoscale microreactors. *Physical Review Letters*, 83: 1475–1478

Nicolis, G. 1995. *Introduction to Nonlinear Science*, Cambridge and New York: Cambridge University Press

Nicolis, G. 2001. Nonlinear kinetics: at the crossroads of chemistry, physics and life sciences. *Faraday Discussions*, 120: 1–10

Nicolis, G. & Prigogine, I. 1977. *Self-organization in Nonequilibrium Systems*, New York: Wiley

Pacault, A. & Perraud J.-J. 1997. *Rythmes et Formes en Chimie*, Paris: Presses Universitaires de France

Royal Society of Chemistry. 2002. *Nonlinear Chemical Kinetics: Complex Dynamics and Spatio-temporal Patterns*. Faraday Discussion no. 120, London: Royal Society of Chemistry; pp. 1–431

Turing, A.M. 1952. The chemical basis of morphogenesis. *Philosophical Transactions of the Royal Society of London, Series B*, 237: 37–72

Zhabotinsky, A.M. 1964. Periodic liquid phase oxidation reactions, *Doklady Akademie Nauk* SSSR, 157: 392–395

CHERENKOV RADIATION

Cherenkov radiation is the electromagnetic radiation of a charged particle moving uniformly in a medium with a velocity exceeding the velocity of light (c) in that medium. It was discovered experimentally by Pavel Cherenkov in 1934 and was theoretically explained by Igor Tamm and Il'ja Frank in 1937. (In 1958, Cherenkov, Tamm, and Frank were awarded a Nobel Prize in physics for this work.) Earlier, Cherenkov radiation was theoretically predicted by Arnol'd Sommerfeld, who, in 1904, solved a formal problem on the radiation of a charged particle moving in vacuum with a velocity $v > c$, and by Oliver Heaviside (Heaviside, 1950) at the end of the 19th century.

At present, all radiation phenomena of waves of any origin, created by a source that moves with a velocity exceeding the phase velocity of the waves, are regarded as Cherenkov radiation. Common examples that can be observed in ordinary life include waves created on a water surface by moving objects, the so-called bow waves according to a theory developed by Lord Kelvin (William Thomson) in the middle of the 19th century, and acoustic shock waves brought about in the atmosphere by a supersonic jet or a rocket, first described by Ernst Mach in 1877.

Essentially, Cherenkov radiation can be understood from the following simple considerations. A source moving steadily with a velocity $\boldsymbol{v}$ and depending on coordinates and time as $f(\boldsymbol{r}-\boldsymbol{v}t)$ can be presented in the form of a Fourier integral $f(\boldsymbol{r},t)=\int f(\boldsymbol{k})\mathrm{e}^{\mathrm{i}\boldsymbol{k}\boldsymbol{r}-\mathrm{i}\boldsymbol{k}\boldsymbol{v}t}\mathrm{d}^3k$, which suggests that each Fourier harmonic has frequency $\boldsymbol{k}\boldsymbol{v}$. In electrodynamics, the role of the source is played by the distribution of charge density $\rho(\boldsymbol{r}-\boldsymbol{v}t)$ and current $\boldsymbol{j}=\rho\boldsymbol{v}(\boldsymbol{r}-\boldsymbol{v}t)$, and in hydrodynamics by external forces and moving particles.

If a source is to excite the waves in a medium whose wave vector $\boldsymbol{k}$ and frequency ω are related by the dispersion equation $\omega=\omega(\boldsymbol{k})$, a resonance condition must be satisfied, under which the wave frequency $\omega(\boldsymbol{k})$ coincides with the external force frequency $\boldsymbol{k}\boldsymbol{v}$. This equality yields the Cherenkov condition

$$\omega(\boldsymbol{k})=\boldsymbol{k}\cdot\boldsymbol{v}, \tag{1}$$

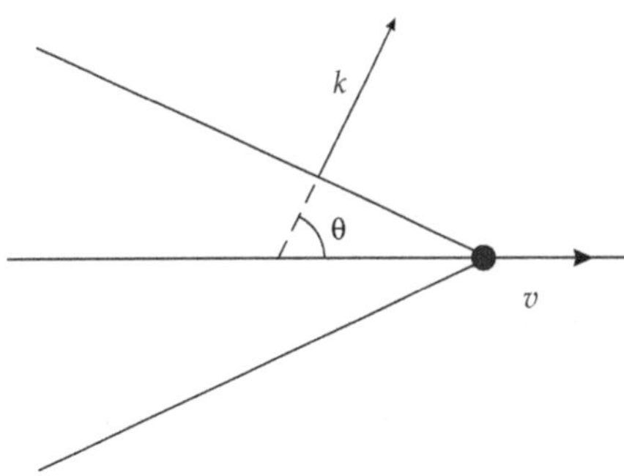

Figure 1. Wave front configuration by Cherenkov radiation, $\boldsymbol{v}$ is the particle velocity, and $\boldsymbol{k}$ is the wave vector of the emitted wave.

which can be fulfilled only if the moving source velocity exceeds the phase velocity of the waves $v > v_{\text{ph}} = \omega(\boldsymbol{k})/k$.

The radiated wave vectors form a characteristic Cherenkov cone, which is similar to the Mach cone in hydrodynamics. Thus, the angle θ between the wave vector of the radiated wave and the direction of particle velocity is determined by the Cherenkov condition of Equation (1), which requires $\cos\theta = v_{\text{ph}}/v$, showing that the frequency distributions of radiation are related to each other. Figure 1 shows the wave front configuration of a wave of frequency ω, emitted by a moving source.

Nowadays, Cherenkov radiation is used in high-energy physics as an important experimental tool (the Cherenkov counter), enabling identification and velocity measurements of fast charged particles, and in electronics for Cherenkov electron oscillators and amplifiers of electromagnetic waves, such as traveling-wave tubes and backward-wave oscillators. Many diverse phenomena can be related to manifestations of Cherenkov radiation, including Landau damping and instabilities in plasma physics, and the Kelvin–Helmholtz instability in hydrodynamics (excitation of waves on a water surface by wind).

VLADISLAV V. KURIN

See also **Dispersion relations; Shock waves**

Further Reading

Heaviside, O. 1950. *Electromagnetic Theory*, 3 vols, London: The Electrician; reprinted New York: Dover

Jackson, J.D. 1998. *Classical Electrodynamics*, New York: Wiley

Landau, L.D. & Lifshitz, E.M. 1963. *Theoretical Physics*, vol. 6: *Fluid Mechanics*, Oxford: Pergamon Press

Landau, L.D. & Lifshitz, E.M. 1963. *Theoretical Physics*, vol. 8: *Electrodynamics of Continious Media*, Oxford: Pergamon Press

Landau, L.D. & Lifshitz, E.M. 1963. Theoretical Physics, vol. 10: *Physical Kinetics*, Oxford: Pergamon Press

CHI(2) MATERIALS AND SOLITONS

See **Nonlinear optics**

CHIRIKOV MAP

See **Maps**

CHUA'S CIRCUIT

Having witnessed futile attempts at producing chaos in an electrical analog of the Lorenz equations while on a visit to Japan in 1983, Leon Chua was prompted to develop a chaotic electronic circuit. He realized that chaos could be produced in a piecewise-linear circuit if it possessed at least two unstable equilibrium points—one to provide stretching and the other to fold the trajectories. With this insight and using nonlinear circuit theory (Chua et al., 1987), he systematically identified those third-order piecewise-linear circuits containing a single voltage-controlled nonlinear resistor that could produce chaos. Specifying that the V–I characteristic of the nonlinear resistor N_R should be chosen to yield at least two unstable equilibrium points, he invented the circuit shown in Figure 1.

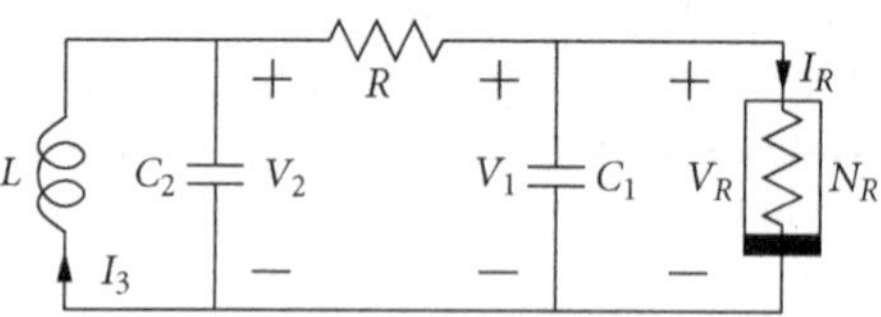

Figure 1. Chua's circuit consists of a linear inductor L, two linear capacitors (C_1 and C_2), a linear resistor R, and a voltage-controlled nonlinear resistor N_R (called a Chua diode).

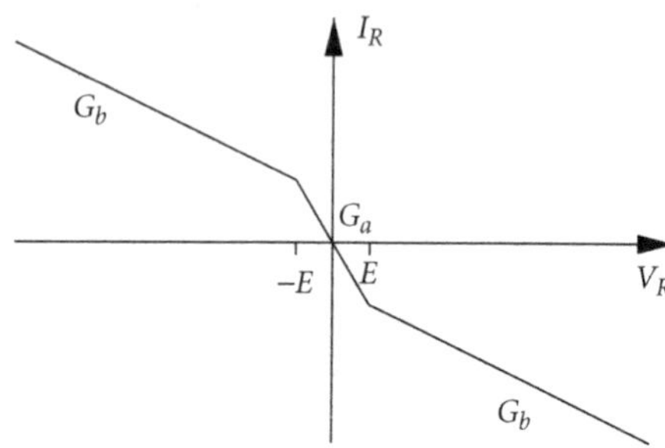

Figure 2. The V–I characteristic of the Chua diode N_R has breakpoints at $\pm E$ and slopes G_a and G_b in the inner and outer regions, respectively.

This circuit is described by three ordinary differential equations

$$\begin{aligned} C_1\frac{\mathrm{d}V_1}{\mathrm{d}t} &= -GV_1 - f(V_1) + GV_2, \\ C_2\frac{\mathrm{d}V_2}{\mathrm{d}t} &= GV_1 - GV_2 + I_3, \\ L\frac{\mathrm{d}I_3}{\mathrm{d}t} &= -V_2, \end{aligned} \qquad (1)$$

where $G = 1/R$. Also, $f(\cdot)$ is the V–I characteristic of the nonlinear resistor N_R (known as a Chua diode), which has a piecewise-linear V–I characteristic defined by

$$f(V_R) = G_bV_R + \tfrac{1}{2}(G_a - G_b)(|V_R + E| - |V_R - E|), \qquad (2)$$

where $\pm E$ are the breakpoints in the characteristic, and G_a and G_b are the slopes in the inner and outer regions, respectively, as shown in Figure 2.

If the values of the circuit parameters are chosen such that the circuit contains three equilibrium points (two in the outer regions and one at the origin), all of which are unstable with saddle-focus stability, then a homoclinic trajectory can be formed, potentially producing chaos.

Soon after its conception, the rich dynamical behavior of Chua's circuit was confirmed by computer simulation and experiment and in 1986 was proven to exhibit chaos in the sense of Shilnikov (Chua et al., 1986). An intensive effort since then to understand every aspect of the dynamics of this circuit has resulted in its widespread acceptance as a powerful paradigm for learning, understanding, and teaching about nonlinear dynamics and chaos (Madan, 1993; Chua, 1992).

By adding a linear resistor R_0 in series with the inductor, Chua's circuit has been generalized to the Chua oscillator (Chua, 1993), with the last of Equations (1) changing to

$$L\frac{dI_3}{dt} = -V_2 - R_0 I_3. \qquad (3)$$

Chua's oscillator is canonical in the sense that it is equivalent (topologically conjugate) to a 13-parameter family of three-dimensional ordinary differential equations with odd-symmetric piecewise-linear vector fields. The circuit can exhibit every dynamical behavior known to be possible in a system described by a continuous odd-symmetric three-region piecewise-linear vector field. Unlike the Lorenz or Rössler equations, which have more complex multiplicative nonlinearities, the only nonlinearity in Chua's circuit is a scalar function of one variable. With an appropriate choice of parameters, the circuit can be made to follow the classic period-doubling, intermittency, and torus breakdown routes to chaos; in addition, over 60 different types of strange attractors have been reported in Chua's oscillator.

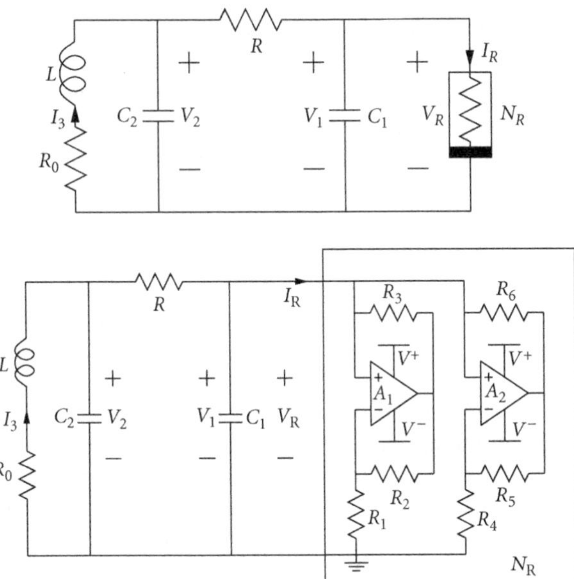

Figure 3. Robust practical implementation of Chua's circuit/oscillator using two op amps and six resistors to realize the Chua diode. In the case of Chua's circuit, R_0 is zero; in Chua's oscillator, R_0 may assume negative or positive values. Component values for Chua's circuit are listed in Table 1.

Chua's circuit oscillator can be realized in a variety of ways by using standard or custom-made electronic components. Since all of the linear elements (capacitors, resistors, and inductor) are readily available as two-terminal devices, only the nonlinear diode must be synthesized using a combination of standard electronic components. The most robust practical realization of Chua's circuit/oscillator, shown in Figure 3, uses two operational amplifiers (op-amps) and six resistors to implement the nonlinear diode (Kennedy, 1992).

The op-amp subcircuit consisting of A_1, A_2, and R_1 through R_6 functions as a Chua diode with V–I

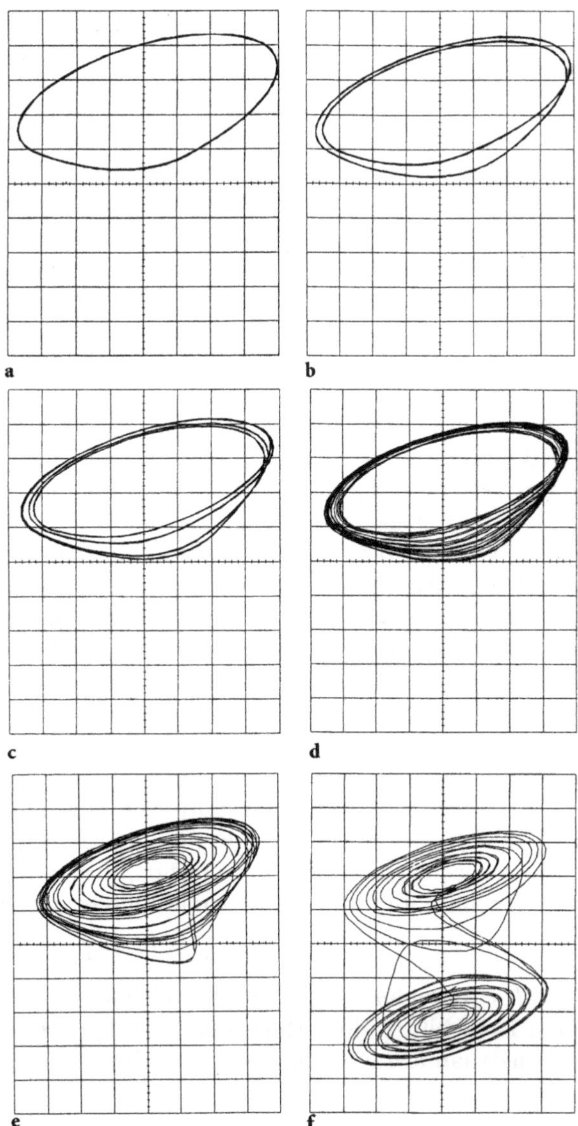

Figure 4. Typical experimental bifurcation sequence in Chua's circuit (component values as in Table 1) recorded using a Hitachi VC-6025 Digital Storage Oscilloscope. Horizontal axis V_2 200 mV/div; vertical axis V_1 1 V/div. (a) $R = 1.83$ kΩ, period 1; (b) $R = 1.82$ kΩ, period 2; (c) $R = 1.81$ kΩ, period 4; (d) $R = 1.80$ kΩ, Spiral attractor; (e) $R = 1.76$ kΩ, Spiral attractor; (f) $R = 1.73$ kΩ, double-scroll attractor [reproduced from Kennedy (1993)].

Element	Description	Value	Tolerance (%)
A_1	Op-amp ($\frac{1}{2}$ AD712 or TL082)		
A_2	Op-amp ($\frac{1}{2}$ AD712 or TL082)		
C_1	Capacitor	10 nF	± 1
C_2	Capacitor	100 nF	± 1
R	Potentiometer	2 kΩ	
R_1	$\frac{1}{4}$W Resistor	3.3 kΩ	± 1
R_2	$\frac{1}{4}$W Resistor	22 kΩ	± 1
R_3	$\frac{1}{4}$W Resistor	22 kΩ	± 1
R_4	$\frac{1}{4}$W Resistor	2.2 kΩ	± 1
R_5	$\frac{1}{4}$W Resistor	220 Ω	± 1
R_6	$\frac{1}{4}$W Resistor	220 Ω	± 1
L	Inductor (TOKO type 10RB)	18 mH	± 5

Table 1. Component list for Chua's circuit.

characteristic as shown in Figure 2. Using two 9 V power supplies for the analog devices AD712, op-amps set their saturation voltages at approximately $\pm$ 8.3 V, yielding breakpoints $E \approx$ 1V. With $R_2 = R_3$ and $R_5 = R_6$, the V–I characteristic of the Chua diode is defined by $G_a = -1/R_1 - 1/R_4 = -25/33$ mS and $G_b = 1/R_3 - 1/R_4 = -9/22$ mS. Note that the value of the resistance R_0 is ideally zero in Chua's circuit. In practice, the real inductor L has a small parasitic resistance that can be modeled by R_0; the TOKO-type 10RB inductor is preferred because it has a sufficiently low parasitic resistance R_0 for this application.

By reducing the value of the variable resistor R from 2000 Ω to zero, with all other components as in Table 1, the circuit exhibits a Hopf bifurcation from dc equilibrium, a sequence of period-doubling bifurcations to a spiral attractor, periodic windows, a double-scroll strange attractor, and a boundary crisis (Kennedy, 1995).

Although the diode characteristic in Equation (2) is piecewise-linear, qualitatively similar behavior is observed when a smooth nonlinearity such as a cubic is used instead. The piecewise-linear nonlinearity is more convenient for circuit realization, while the smooth nonlinearity is more appropriate for bifurcation analysis.

MICHAEL PETER KENNEDY

See also **Attractors; Bifurcations; Chaotic dynamics; Hopf bifurcation; Horseshoes and hyperbolicity in dynamical systems; Period doubling; Routes to chaos**

Further Reading

Chua, L.O. 1992. The genesis of Chua's circuit. *Archiv für Elektronik und Übertragungstechnik*, 46(4): 250–257

Chua, L.O. 1993. Global unfolding of Chua's circuit. *IEICE Transactions Fundamentals*, E76A(5): 704–734

Chua, L.O., Desoer, C.A. & Kuh, E.S. 1987. *Linear and Nonlinear Circuits*, New York: McGraw-Hill

Chua, L.O., Komuro, M. & Matsumoto, T. 1986. The double scroll family—Parts I and II. *IEEE Transactions on Circuits and Systems*, 33(11): 1073–1118

Kennedy, M.P. 1992. Robust op amp realization of Chua's circuit. *Frequenz*, 46(3–4): 66–80

Kennedy, M.P. 1993. Three steps to chaos—Parts I and II. *IEEE Transactions on Circuits and Systems—Part I*, 40(10): 640–674

Kennedy, M.P. 1995. Experimental chaos from autonomous electronic circuits. *Philosophical Transactions of the Royal Society London* A, 353(1701): 13–32

Madan, R.N. 1993. *Chua's Circuit: A Paradigm for Chaos*, Singapore: World Scientific

CIRCLE MAP

See **Denjoy theory**

CLEAR AIR TURBULENCE

Description

In 1966, the National Committee for Clear Air Turbulence officially defined *clear air turbulence* (CAT) as "all turbulence in the free atmosphere of interest in aerospace operations that is not in or adjacent to visible convective activity (this includes turbulence found in cirrus clouds not in or adjacent to visible convective activity)." FAA Advisory Circular AC 00-30B (1997) has simplified this somewhat to "turbulence encountered outside of convective clouds." Thus, CAT is considered to mean turbulence in the clear air, not in or near convective clouds, usually at upper levels of the atmosphere (above 6 km).

CAT was first observed by high-flying fighter aircraft in the mid-to-late 1940s. It was expected that turbulence encounters would be rare at high levels due to the lack of clouds at upper levels. However, it was soon discovered that turbulence encounters in clear air were not only frequent but sometimes quite severe. Since then, CAT encounters have become a significant problem for commercial aircraft flying at cruising altitudes (18,000–45,000 ft above the mean sea level). In fact, various reviews of National Transportation Safety Board (NTSB) reports indicate that in the U.S., turbulence encounters account for approximately 65% of all weather-related accidents or incidents for commercial aircraft; probably more than half of these are due to CAT. Although most turbulence encounters are generally just an annoyance to passengers and crew, on average there are about eight commercial aircraft turbulence-related incidents per year that are significant enough to be reported to the NTSB, accounting for 10 serious and 32 minor injuries. Fortunately, fatalities and substantial damage to the aircraft structure are rare but can occur, as shown in Figure 1.

It should be noted that only a certain range of frequencies or wavelengths of turbulent eddies is felt by aircraft as bumpiness. For most commercial aircraft, this wavelength is anywhere from about 10 m to 1 km.

Figure 1. Damage sustained to a DC-8 cargo aircraft in an encounter with CAT on 9 December 1992 at 31,000 ft over the Rocky Mountains near Evergreen, Colorado. Note the loss of left outboard engine and approximately 6 m of the wing. (Photo by Kent Meiries.)

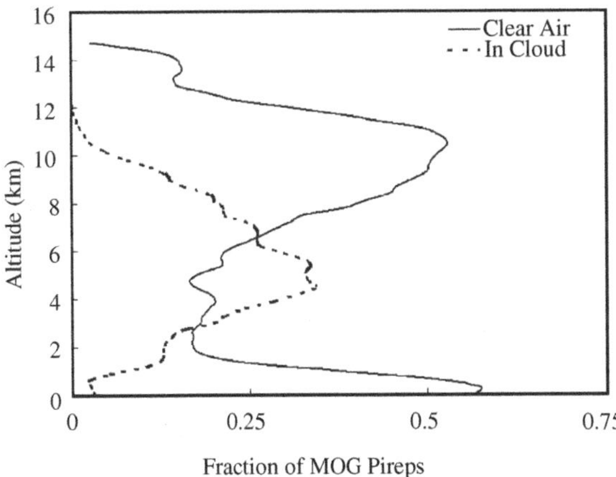

Figure 2. Vertical distribution of the fraction of moderate or greater-intensity (MOG) turbulence pilot reports taken over a two-year period. Solid lines indicate reports in clear air, and dashed lines indicate reports in cloud.

Shorter wavelengths are integrated out over the aircraft structure and longer wavelengths are felt as "waves" and do not generally have vertical accelerations large enough to be felt as "bumps."

CAT can be measured quantitatively with instrumented research aircraft or remotely by instruments such as clear air radar and lidar, but by far the majority of measurements are through pilot reports (PIREPs) of turbulent encounters. PIREPs usually report turbulence on intensity scales of smooth, light, moderate, severe, or extreme. Although definitions of these categories are provided in terms of normal accelerations or airspeed fluctuations, there is still an amount of subjectivity associated with these reports. The pilot reporting system is fairly successful in warning other aircraft of turbulence regions encountered, but to use PIREPs to deduce CAT climatology is difficult, since they are biased by air traffic patterns, non-uniform reporting practices, and the tendency to avoid known turbulence areas. One way to reduce these biases is to examine the ratio of moderate or severe PIREPs to total reports in the three-dimensional airspace averaged over many years of reports. This has been done over the continental U.S. (Sharman et al., 2002). The distribution of this ratio of moderate or greater (MOG) severity PIREPs to total PIREPs by altitude for both CAT and in-cloud encounters is shown in Figure 2. Note that above about 8 km, the majority of reports are in clear air. Similar analyses show the occurrence of CAT to be about twice as frequent in winter as in summer. CAT encounters also tend to be more frequent and more severe over mountainous regions, for example, the Colorado Rockies.

One characteristic of CAT is its patchiness in both time and space. These patches tend to be relatively thin compared with their length (the median thickness is about 500 m, whereas the median horizontal dimension is about 50 km); the median duration is about 6 hours (Vinnichenko et al., 1980). Within a patch, the turbulence may be continuous or may occur in discrete bursts that may be very severe but very narrow (1–2 km).

Relation to Kelvin–Helmholtz Instability

From Figure 2, it can be seen that the altitude of maximum occurrence of CAT is at upper levels near the tropopause and jet stream levels. This relation has been known since the 1950s. For example, Bannon (1952) noted that severe CAT tended to occur above and below the jet stream core on the low-pressure side. These areas tend to have large values of the vertical shear of the horizontal wind, and this led to the hypothesis (e.g., Dutton & Panofsky, 1970) that, at least in some cases, CAT may be related to Kelvin–Helmholtz (KH) instability (KHI, *See* **Kelvin–Helmholtz instability**). The KHI process occurs in stably stratified shear flows when dynamic instabilities due to wind shear exceed the restoring forces due to stability. KH waves (also known as "billows") are, in fact, commonly observed in the atmosphere near the top of clouds, where the KH distortions become visible (Figure 3).

Further, the KHI connection to CAT has been verified on occasion by simultaneous measurements of KH billows by high-powered radar and aircraft measurements of turbulence (Browning et al., 1970). Although the figure shows a KH wave train at an instant in time, the KHI process is an evolutionary one, where waves develop, amplify, roll up, and break down into turbulent patches. The resultant turbulent mixing will eventually destroy the wave structure and mix out the shear, and density distributions that created it, but if larger scale processes continue to reinforce the shears, the entire process may reinitiate.

The names associated with KHI derive from the early works of Hermann von Helmholtz (1868), who realized the destabilizing effects of shear, and later of Lord Kelvin (William Thomson) (1871), who posed and solved the instability problem mathematically. Richardson (1920), using simple energy considerations, deduced that a sufficient condition for

Figure 3. An example of Kelvin–Helmholtz billows observed in the presence of clouds.

stability of disturbances in shear flow occurs when the restoring force of stability (as measured by the buoyancy frequency N) is greater than the destabilizing force of the mean horizontal velocity shear ($\mathrm{d}U/\mathrm{d}z$) in the vertical (z) direction. Thus, when the ratio $N^2/(\mathrm{d}U/\mathrm{d}z)^2$ is greater than unity, the flow should be stable. In honor of Richardson's insight, it has become common to refer to this ratio as the Richardson number (Ri). The linear problem for various stratified shear flow configurations is well reviewed in the texts by Chandrasekhar (1961, Chapter 11) and by Drazin and Reid (1981). The sufficient condition for stability to linear two-dimensional disturbances is that $Ri > 0.25$; $Ri < 0.25$ is necessary but not sufficient for instability. More recently, Abarbanel et al. (1984), using the method of Arnol'd, were able to show that the necessary and sufficient condition for Liapunov stability to three-dimensional nonlinear disturbances is $Ri >$ unity, in agreement with Richardson's deduction.

Through these theoretical studies, laboratory experiments (e.g., Thorpe, 1987), and more recently, very high-resolution numerical simulations (e.g., Werne & Fritts, 1999), considerable progress has been made in understanding the intricacies of KHI. One (probably common) method in which KHI is initiated in the atmosphere is through longer wavelength gravity-wave-induced perturbations that lead to local reductions (e.g., in the crest of the wave) in Ri to a value small enough to initiate instability. Gravity waves are ubiquitous in the atmosphere and can be generated in a variety of ways, for example, by flow over mountains or by strong updrafts and downdrafts in convective storms. However, the processes by which KHI may lead to turbulent breakdowns within three-dimensional transient gravity waves is not yet completely understood. It should be mentioned that gravity wave breakdown into turbulent patches may also occur through other mechanisms besides KHI. Examples include convective overturning in large-amplitude waves or nonlinear wave–wave interactions (see the reviews by Wurtele et al. (1996) and Staquet & Sommeria (2002) for discussions of some of these effects). Further, other instability mechanisms besides KHI may lead to CAT, for example, inertial instability or critical-level instability. Thus, the processes by which CAT may be generated at any given time and place are complex, involving many different sources, making its forecasting quite difficult.

One new promising avenue of research is the use of high-resolution numerical simulations of the atmosphere to reconstruct the atmospheric processes that led to particularly severe encounters of CAT (e.g., Clark et al., 2000; Lane et al., 2003). These types of studies have only recently become possible with advances in computing capabilities that allow model runs to contain both the large-scale processes that create conditions conducive to turbulence and the smaller scales that may affect aircraft. Further studies such as these, along with continued theoretical and numerical studies and field measurement campaigns, should lead to a better understanding of CAT genesis and evolution processes.

Until this understanding is available, forecasting of CAT must be accomplished by empirical means. This is done by forecasting various large-scale atmospheric conditions that are known through experience to be related to CAT. Until recently, these diagnostics for likely regions of turbulence had to be performed by laborious weather map analyses of jet streams and upper-level fronts. Nowadays, these diagnostics can be computed from the output of routine numerical weather prediction forecast models. However, the reliability of these turbulence diagnostics is highly variable, and at the moment it seems that better success may be achieved by combining the various diagnostics within an artificial intelligence framework (Tebaldi et al., 2002).

Robert Sharman

See also **Kelvin–Helmholtz instability; Turbulence**

Further Reading

Abarbanel, H.D.I., Holm, D.D., Marsden, J.E. & Ratiu, T. 1984. Richardson number criterion for the nonlinear stability of three-dimensional stratified flow. *Physical Review Letters*, 52: 2352–2355

Bannon, J.K. 1952. Weather systems associated with some occasions of severe turbulence at high altitude. *Meteorological Magazine*, 81: 97–101

Browning, K.A., Watkins, C.D., Starr, J.R. & McPherson, A. 1970. Simultaneous measurements of clear air turbulence at the tropopause by high-power radar and instrumented aircraft. *Nature*, 228: 1065–1067

Chandrasekhar, S. 1961. *Hydrodynamic and Hydromagnetic Stability*, Oxford: Clarendon Press

Clark, T.L., Hall, W.D., Kerr, R.M., Middleton, D., Radke, L., Ralph, F.M., Nieman, P.J. & Levinson, D. 2000. Origins of aircraft-damaging clear-air turbulence during the 9 December 1992 Colorado downslope windstorm: numerical simulations and comparison with observations. *Journal of the Atmospheric Sciences*, 57: 1105–1131

Drazin, P.G. & Reid, W.H. 1981. *Hydrodynamic Stability*, Cambridge: Cambridge University Press

Dutton, J.A. & Panofsky, H.A. 1970. Clear air turbulence: a mystery may be unfolding. *Science*, 167: 937–944
von Helmholtz, H. 1868. On discontinuous movements of fluids. *Philosophical Magazine*, 36: 337–346 (originally published in German, 1862)
Lane, T.P., Sharman R.D., Clark T.L. & Hsu, H.-M. 2003. An investigation of turbulence generation mechanisms above deep convection. *Journal of the Atmospheric Sciences*, 60: 1297–1321
Kelvin, Lord. 1871. Hydrokinetic solutions and observations. *Philosophical Magazine*, 42: 362–377
Richardson, L.F. 1920. The supply of energy from and to atmospheric eddies. *Proceedings of the Royal Society, London*, A97: 354–373
Sharman, R., Fowler, T.L., Brown, B.G. & Wolff, J. 2002. Climatologies of upper-level turbulence over the continental U. S. and oceans. Preprints, 10th Conference on Aviation, Range, and Aerospace Meteorology, Portland OR: American Meteorological Society, J29–J32
Staquet, C. & Sommeria, J. 2002. Internal gravity waves: from instabilities to turbulence. *Annual Reviews of Fluid Mechanics*, 34: 559–593
Tebaldi, C., Nychka D., Brown B.G., and Sharman, R. 2002. Flexible discriminant techniques for forecasting clear-air turbulence. *Environmetrics*, 13(8): 859–878
Thorpe, S.A. 1987. Transitional phenomena and the development of turbulence in stratified fluids: a review. *Journal of Geophysical Research*, 92: 5321–5248
Vinnechenko, N.K., Pinus, N.Z., Shmeter, S.M. & Shur, G.N. 1980. *Turbulence in the Free Atmosphere*, 2nd edition, New York: Consultants Bureau
Werne, J. & Fritts, D.C. 1999. Stratified shear turbulence: evolution and statistics. *Geophysical Research Letters*, 26: 439–442
Wurtele, M.G., Sharman, R.D. & Datta, A. 1996. Atmospheric lee waves. *Annual Reviews of Fluid Mechanics*, 28: 429–476

CLUSTER COAGULATION

In 1916, nine years before he was awarded the Nobel prize for his studies of colloidal solutions, the Austro-Hungarian chemist Richard Zsigmondy (1865–1929) brought forth the first model for cluster coagulation. Interpreting the behavior of aqueous solutions of gold colloidal particles, he posited that each cluster of particles is surrounded by a sphere of influence. According to this model, clusters execute independent Brownian motions when their spheres of influence do not overlap. Whenever the spheres of influence of a pair of clusters touch, the clusters instantaneously stick together to form a new cluster. This kind of non-equilibrium kinetics (*See* **Nonequilibrium statistical mechanics**) has proven to be truly ubiquitous: bond formation between polymerization sites; the coalescence of rain drops, smog, smoke, and dust; the aggregation of bacteria into colonies; the formation of planetesimals from sub-micron dust grains; the coalescence arising in genetic trees; and even the merging of banks to form ever-larger financial institutions are all examples of cluster coagulation.

Cluster coagulation results in a broad distribution of cluster sizes described by $\{n_i(t)\}_{i=1,\ldots,\infty}$, where $n_i(t)$ is the number of clusters of size i present in the system at time t. The size of a cluster is defined as the number of unit clusters that it comprises. The primary goal of coagulation theory is to determine the evolution of $n_i(t)$ for all i.

The most important theory of coagulation was given by the Polish physicist Marian Smoluchowski (1872–1917) (Smoluchowski, 1916, 1917). In 1916, prompted by a request from Zsigmondy to provide a mathematical description of coagulation, Smoluchowski postulated that (1) clusters are randomly distributed in space and this feature persists throughout the coagulation process, (2) only collisions between pairs of clusters are significant, and (3) the number of new clusters of size $i+j$, formed per unit time and unit volume due to collisions of clusters of sizes i and j, is proportional to the product of the concentrations $c_i = n_i/V$ and $c_j = n_j/V$:

$$\frac{\text{number of new clusters}}{V\Delta t} = K_{i,j}c_i c_j. \tag{1}$$

Here, V is the volume of the coagulating system and $K_{i,j}$ is the collision frequency factor, also called the *coagulation kernel*.

The rate equation describing the evolution of $c_i(t)$ follows from the balance between the total number of clusters of size i created and annihilated as a result of coagulation:

$$\dot{c}_i(t) = \frac{1}{2}\sum_{j=1}^{i-1} K_{i-j,j}c_{i-j}(t)c_j(t) - \sum_{j=1}^{\infty} K_{i,j}c_i(t)c_j(t), \quad i = 1, \ldots, \infty. \tag{2}$$

Here, $\dot{c}_i(t)$ is the time derivative of the concentration $c_i(t)$.

This equation—in fact, the chain of nonlinear ordinary differential equations—is called the Smoluchowski coagulation equation (SCE). It describes the evolution of homogeneous aggregating systems with the distribution $c_i(t)$, provided knowledge of $K_{i,j}$ and an initial distribution $c_i(0)$. The SCE does not depend on the spatial dimension in which the coagulation process is taking place. According to modern terminology, Smoluchowski theory is a mean field (*See* **Phase transitions**) theory of nonequilibrium growth. It neglects fluctuations of the concentrations c_k; that is, it presumes the existence of the thermodynamic limit: $V \to \infty$, $n_k \to \infty$, $n_k/V \to c_k$. For a broad variety of aggregating systems, this proves to be a reasonable assumption. However, the first assumption of Smoluchowski, that correlations in the distribution of the cluster may be disregarded, is not always satisfied. Aggregating systems fulfilling this assumption are called well-mixed. In low-dimensional systems with no "external" mixing mechanism, the cluster collisions are often not able to provide sufficient mixing. This may result in correlation build-up and, therefore, a breakdown of Equation (2).

The last, essential, albeit obvious, condition for the applicability of the SCE is that interactions between aggregating clusters be treated as instantaneous collisions, that is, $\tau_{col} \ll \tau_{coag}$, where τ_{col} is the characteristic time scale of collisions and τ_{coag} is the characteristic time scale of coagulation. Thus, similar to the Boltzmann kinetic equation (*See* **Nonequilibrium statistical mechanics**), the SCE describes a slow evolution of the distribution due to fast collisions.

For a continuous distribution $c(t, v)$, SCE takes the form of an integro-differential equation (Müller, 1928):

$$\partial_t c(t, v) = \frac{1}{2}\int_0^v K(u, v-u)c(t,u)c(t,v-u) \\ -c(t,v)\mathrm{d}u \int_0^\infty K(u,v)c(t,u)\,\mathrm{d}u. \quad (3)$$

Here u, v are the physical sizes of clusters.

Although the SCE establishes a firm foundation for our understanding of a great variety of cluster coagulation processes, other models have also been devised. They include the Oort–van de Hulst–Safronov equation, which describes cluster coagulation as a continuous growth process, and various stochastic models, such as Kingman's coalescent and the Marcus–Lushnikov process. In this article, we limit ourselves to the Smoluchowski coagulation theory.

The mathematical structure of SCE can be traced to the master equation for stochastic processes:

$$\dot{P}_i(t) = \sum_j \left(w_{i,j}P_j(t) - w_{j,i}P_i(t)\right). \quad (4)$$

Here, $P_i(t)$ is the probability of finding the system in a state i at time t, and $w_{i,j}$ is the probability of transition from state j to state i per unit time. Under the aforementioned assumptions of Smoluchowski theory, $P_i = c_i$, $w_{i,j} = K_{i,j}P_i$. Thus, we arrive at the probabilistic interpretation of $K_{i,j}$ as the probability of coagulation of a pair of clusters i and j in unit volume per unit time.

Calculating the coagulation kernel for a particular coagulation mechanism is a separate problem that is beyond the scope of this article. However, a few remarks are appropriate. First, due to the aforementioned time-scale separation, calculation of K can be treated as a stationary problem. Second, the probability of cluster collisions will depend on the cluster geometry. Very often, the aggregates prove to be fractals (*See* **Fractals; Pattern formation**) having no characteristic size. The coagulation kernel K will then depend on their fractal dimension D_f. Third, the probability of coagulation of a pair of clusters is a product of the probability of their collision and the sticking efficiency E. The latter is defined as the probability of clusters merging once they have collided. Two practically important limiting cases are distinguished for coagulation of diffusing clusters: diffusion-limited cluster-cluster aggregation (DLCA), when $E \approx 1$, and reaction-limited cluster-cluster aggregation (RLCA), when $E \ll 1$. DLCA and RLCA produce aggregates of different fractal dimensions and have kinetics of different speeds. Fourth, when the coagulation mechanism is scale-free and the aggregates are fractals, the coagulation kernel K should also be scale-free. In mathematical terms this amounts to the following requirement on the function $K(u, v)$:

$$K(\lambda u, \lambda v) = \lambda^\alpha K(u, v) \quad (5)$$

for any real $\lambda > 0$. Such kernels are called homogeneous, and α is called the homogeneity index. Several examples of widely used kernels are listed in Table 1.

Our present knowledge of when solutions of (2) exist and are unique is limited, as the nonlinearity of SCE presents challenging problems for rigorous mathematical analysis. Existence and uniqueness of solutions for all times have been proven for the kernels $K(u, v) \leq C(u + v)$, where C is a constant. This result has recently been extended to the kernels $K(u, v) \leq r(u)r(v)$, where $r(v) = o(v)$, as $v \to \infty$ (Norris, 1999; Leyvraz, 2003).

The distribution function is significant because any macroscopic property characterizing a given coagulating system can be calculated in the continuous (discrete) case as an integral (sum) over the distribution. From a mathematical point of view, the distribution function is simply a time-dependent measure on the set of all cluster sizes. Therefore, it is natural to look for *weak solutions* to SCE. Weak solutions can be conveniently defined in the continuous case by means of a Laplace transformed SCE, which describes the time evolution of the Laplace transformation of the concentration.

Weak solutions are inverse Laplace transforms of the solutions to this equation. The discrete case can be treated analogously with the help of generating functions. In fact, most of the presently known exact solutions of SCE were obtained by this approach.

Since the total mass of clusters is, apparently, conserved during collisions, the SCE is expected to conserve the first moment of the distribution

$$M_1(t) = \int_0^\infty vc(t, v)\,\mathrm{d}v. \quad (6)$$

It appears as a deceptively simple exercise to prove this by substituting (6) into SCE. The proof, however, hinges on the condition that the infinite sums involved are convergent. Violation of this condition gives rise to the important phenomenon of *gelation*, also known as

Coagulation mechanism	Kernel	Originator (year)
"Mating"	2	Smoluchowski (1916)
Brownian motion	$\frac{(r(u)+r(v))^2}{r(u)r(v)}$; $r(v) \propto v^{1/D_f}$	Smoluchowski (1916)
Isotropic turbulent shear	$(r(u)+r(v))^3$; $r(v) \propto v^{1/D_f}$	Saffman and Turner (1956)
Gravitational coalescence	$(r(u)+r(v))^2 \, \lvert r(u)^2 - r(v)^2 \rvert$; $r(v) \propto v^{1/D_f}$	Findheisen (1939)
Polymerization (RA_∞ model)	uv	Flory (1953)
Polymerization (ARB_∞ model)	$u+v$	

Table 1. Examples of kernels. All kernels are given up to a non-dimensional prefactor. D_f is the fractal dimension of the coagulates.

the gel-sol transition or runaway growth. It was first predicted for the kernel $K(u,v)=uv$, which serves as a model of polymerization (*See* **Polymerization**) in which new links are formed randomly between polymerization sites. In the mean field approximation, this model also describes random graph growth and bond percolation (*See* **Percolation theory**). An exact solution of this problem shows that starting with monodisperse initial conditions, the first moment $M_1(t)$ begins to decay after a finite time, t_c, whereas the second moment $M_2(t)$, measuring the average cluster size, diverges. This behavior corresponds to the formation of an infinite cluster, or gel, in a finite time due to the coagulation kinetics "accelerating" with growing cluster sizes. The sol mass M_1 decreases, as part of it is being lost to the gel. This kind of kinetics has also proved to be a key to the explanation of the rapid growth of Jupiter and planetesimal growth in the terrestrial planets.

It has been shown that $M_1(t)=M_1(0)$ for all times; that is, the system is nongelling, when $K(u,v) \le C(u+v)$. A wealth of data suggests that this is, in fact, the exact bound separating gelation at finite times from nongelling behavior, although a rigorous proof has not yet been given.

The nonlinear character of SCE complicates mathematical analysis for arbitrary kernels, and the set of exactly solvable kernels is limited. Considerable progress in understanding coagulation kinetics has been achieved for the wide class of homogeneous (or asymptotically homogeneous) kernels by looking for similarity solutions. This can be done in two different ways, which we shall refer to as the self-preservation theory and the self-similarity theory.

The first approach embodies the notion of a single characteristic size in the system, which can be chosen to be equal to an average cluster size, $s(t)$. The asymptotic solution of SCE is then expected to have a self-preserving (scaling) form:

$$c(t,v) = s(t)^{-2}\Phi\left(v/s(t)\right). \quad (7)$$

The self-preserving form has been further studied with the objective of identifying when it will lead to a power-law distribution asymptotically (Leyvraz, 2003). This approach allows one to obtain sensible results for a large variety of problems, including some extensions of SCE, in an almost automatic manner. However, since the scaling hypothesis is postulated, one cannot estimate a priori the accuracy of these solutions. Experimental data on coagulation often display a power-law distribution over some range of cluster sizes.

The second approach deals with a coagulating system maintained at steady state by an external source of monomers. By analogy with the scaling theories for turbulent flows and the theory of critical phenomena, it may be expected that the steady-state distribution at large sizes will "forget" the forcing scale v_0 and, therefore, will evolve to a scale-free form,

$$c(v) = \text{const} \times v^{-\tau}. \quad (8)$$

A careful mathematical analysis has shown that this is indeed the case for a wide range of homogeneous kernels, and that the asymptotic distribution equals

$$c(v) = \left(\frac{E}{\kappa}\right)^{1/2} v^{-(3+\alpha)/2}. \quad (9)$$

Here, E is the total influx of mass into the system due to the forcing and κ is a kernel-dependent constant.

Dmitri O. Pushkin and Hassan Aref

See also **Brownian motion; Dimensional analysis; Fractals; Nonequilibrium statistical mechanics; Pattern formation; Percolation theory; Phase transitions; Polymerization**

Further Reading

Aldous, D.J. 1999. Deterministic and stochastic methods for coalescence (aggregation, coagulation): review of the mean-field theory for probabilists. *Bernoulli*, 5: 3

Drake, R.L. 1972. A general mathematical survey of the coagulation equation. In *Topics in Current Aerosol Research*, edited by G.M. Hidy and J.R.Brock, part 2, Oxford: Pergamon Press

Family, F. & Landau, D.P. (editors). 1984. *Kinetics of Aggregation and Gelation*, Amsterdam and New York: Elsevier Science

Findheisen, W. 1939. Zur Frage der Regentropfenbildung in reinen Wasserwolken. *Meteorologische Zeitschrift*, 56: 365–368

Flory, P. 1953. Principles of Polymer Chemistry. Ithica: Cornell University Press

Friedlander, S.K. 1960. Similarity consideration for the particle-size spectrum of a coagulating, sedimenting aerosol. 17(5): 479

Friedlander, S.K. 2000. *Smoke, Dust, and Haze: Fundamentals of Aerosol Dynamics*, 2nd edition, Oxford and New York: Oxford University Press

Friedlander, S.K. & Wang, C.S. 1966. The self-preserving particle size distribution for coagulation by Brownian motion. *Journal of Colloid and Interface Science*, 22: 126

Galina, H. & Lechowicz, J.B. 1998. Mean-field kinetic modeling of polymerization: the Smoluchowski coagulation equation. *Advances in Polymer Science*, 137: 135–172

Hunt, J.R. 1982. Self-similar particle-size distributions during coagulation: theory and experimental verification. *Journal of Fluid Mechanics*, 122: 169

Leyvraz, F. 2003. Scaling theory and exactly solved models in the kinetics of irreversible aggregation. *Physics Reports*, 383: 95–212

Müller, H. 1928. Zur algemeinen Theorie der raschen Koagulation. *Kolloid-chemische Beihefte*, 27: 223

Norris, J.R. 1999. Uniqueness, non-uniqueness, and a hydrodynamic limit for the stochastic coalescent. *Annals of Applied Probability*, 9: 78

Pushkin, D.O. & Aref, H. 2002. Self-similarity theory of stationary coagulation. *Physics of Fluids*, 14(2): 694

Saffman, P.G. & Turner, J.S. 1956. On the collision of drops in turbulent clouds. *Journal of Fluid Mechanics*, 1: 16–30

Smoluchowski, M.V. 1916. Drei vorträge über diffusion, *Physikalische Zeitschrift*, 17: 557

Smoluchowski, M.V. 1917. Versuch einer mathematischen theorie der Koagulationskinetik kolloider lösungen. *Zeitschrift für Physikalische Chemie*, 92: 129

van Dongen, P.G.J. & Ernst, M.H. 1985. Dynamic scaling in the kinetics of clustering. *Physical Review Letters*, 54: 1396

van Dongen, P.G.J. & Ernst, M.H. 1988. Scaling solution of Smoluchowski coagulation equation. *Journal of Statistical Physics*, 50: 295

CNOIDAL WAVE

See **Elliptic functions**

COHERENCE PHENOMENA

The word *coherence* comes from the Latin *cohaerens*, meaning "being in relation." Thus, coherence phenomena are those displaying a high level of correlation between several objects.

From the physical point of view, it is necessary to distinguish between two types of coherence: state coherence, which characterizes correlations between static properties of the considered objects, and transition coherence, which describes correlated dynamical processes. These types of coherence are two sides of the same coin, and one obtains a better insight from considering them together.

To gain an intuitive idea of these two types of coherence, imagine a group of soldiers all standing at attention, without moving. This corresponds to state coherence. If the soldiers were all in different positions (some standing, some sitting, some lying down), there would be no state coherence between them. Now, imagine well-aligned rows of soldiers in a parade, moving synchronously with respect to each other. This corresponds to transition coherence. Also, if they were to march with different speeds and in different directions, transition coherence would be absent.

Coherence is related to the existence of a kind of order—be it a static order defining the same positions or an ordered motion of a group. Then, the antonym of *coherence* is *chaos*. Thus, state chaos means the absence of any static order among several objects, and transition chaos implies an absolutely disorganized motion of an ensemble of constituents.

The notion of coherence is implicit in the existence of correlation among several objects (enumerated with the index $i = 1, 2, \ldots, N$). Each object, placed in the spatial point $\boldsymbol{r}_i$, at time t, can be associated with a set $\{Q_\alpha(\boldsymbol{r}_i, t\}$ of observable quantities labeled by α. To formalize the definition of state and transition coherence, one may write $Q_i^\alpha = Q_\alpha(\boldsymbol{r}_i, t)$, where Q_i^z corresponds to a state property of an object, while Q_i^x and Q_i^y describe its motion. As an illustration, assume that Q_i^α are spin components. Another example assumes that Q_i^z is the population difference of a resonant atom, while Q_i^x and Q_i^y are its transition dipoles. Instead of considering the latter separately, it is convenient to introduce the complex combinations $Q_i^\pm \equiv Q_i^x \pm \mathrm{i} Q_i^y$. (In general, Q_i^α are not restricted to classical quantities but may be operators.)

If the system is associated with a statistical operator $\hat{\rho}$, then the observable quantities are the statistical averages

$$\langle Q_i^\alpha \rangle \equiv \mathrm{Tr} \hat{\rho}\, Q_i^\alpha, \tag{1}$$

expressed by means of the trace operation. A convenient way of describing the system features is by introducing dimensionless quantities, normalized to the number of objects N and to the maximal value $Q \equiv \max\langle Q_i^z \rangle$. Then, one may define the state variable

$$s \equiv \frac{1}{QN} \sum_{i=1}^{N} \langle Q_i^z \rangle \tag{2}$$

and the transition variable

$$u \equiv \frac{1}{QN} \sum_{i=1}^{N} \langle Q_i^- \rangle. \tag{3}$$

One may distinguish two opposite cases, when the individual states of all objects are the same and when they are randomly distributed. These two limiting cases give

$$|s| = \begin{cases} 1, & \text{state coherence}, \\ 0, & \text{state chaos}. \end{cases} \tag{4}$$

Next consider the transition characteristic (3) and collective motion of an ensemble of oscillators. Again, there can be two opposite situations, when the oscillation frequencies of all oscillators, as well as their initial phases, are identical and when these take randomly different values. For the corresponding limiting cases of completely synchronized oscillations and of an absolutely random motion, respectively, one has

$$|u| = \begin{cases} 1, & \text{transition coherence}, \\ 0, & \text{transition chaos}. \end{cases} \tag{5}$$

In the intermediate situation, one may say that there is partial state coherence if $0 < |s| < 1$ and partial transition coherence when $0 < |u| < 1$.

Accepting that coherence is not necessarily total, it is convenient to define qualitative characteristics for partial coherence by introducing correlation functions. Let $Q_\alpha^+(\boldsymbol{r}, t)$ denote the Hermitian conjugation for an operator $Q_\alpha(\boldsymbol{r}, t)$. When $Q_\alpha(\boldsymbol{r}, t)$ is a classical function, Hermitian conjugation means complex conjugation. For any two operators from the set $\{Q_\alpha(\boldsymbol{r}, t)\}$, one may define the correlation function

$$C_{\alpha\beta}(\boldsymbol{r}_1, t_1, \boldsymbol{r}_2, t_2) \equiv \langle Q_\alpha^+(\boldsymbol{r}_1, t_1) Q_\beta(\boldsymbol{r}_2, t_2) \rangle. \tag{6}$$

The function $C_{\alpha\alpha}(\ldots)$ for coinciding operators is called the autocorrelation function. There is also a shifted correlation function

$$B_{\alpha\beta} \equiv \langle Q_\alpha^+ Q_\beta \rangle - \langle Q_\alpha^+ \rangle \langle Q_\beta \rangle,$$

where, for brevity, the spatiotemporal variables are not written explicitly. For describing coherent processes, it is convenient to use the normalized correlation function

$$K_{\alpha\beta} \equiv \frac{\langle Q_\alpha^+ Q_\beta \rangle}{(\langle Q_\alpha^+ Q_\alpha \rangle \langle Q_\beta^+ Q_\beta \rangle)^{1/2}}, \tag{7}$$

which is sometimes termed a "coherence function." Functions (6) and (7) can be specified as second-order correlation functions since, in general, it is possible to define higher-order correlation functions, such as the $2p$-order function

$$C_{\alpha_1 \ldots \alpha_{2p}} = \langle Q_{\alpha_1}^+ \cdots Q_{\alpha_p}^+ Q_{\alpha_{p+1}} \cdots Q_{\alpha_{2p}} \rangle,$$

which are closely related to reduced density matrices.

Correlations are usually strongest among nearby spatiotemporal points. Thus, function (7) varies in the interval $0 \leq |K_{\alpha\beta}| \leq 1$, being maximal for the autocorrelation function $|K_{\alpha\alpha}| = 1$ at the coinciding points $\boldsymbol{r}_1 = \boldsymbol{r}_2$, $t_1 = t_2$. When either the spatial or temporal distance between two points increases, correlations diminish; this is named *correlation decay*. At an asymptotically large distance, the correlation function (6) for two local observables displays the property of correlation weakening (correlation decoupling)

$$\langle Q_\alpha^+(\boldsymbol{r}_1, t_1) Q_\beta(\boldsymbol{r}_2, t_2) \rangle \simeq \langle Q_\alpha^+(\boldsymbol{r}_1, t_1) \rangle \langle Q_\beta(\boldsymbol{r}_2, t_2) \rangle, \tag{8}$$

where either $|\boldsymbol{r}_1 - \boldsymbol{r}_2| \to \infty$ or $|t_1 - t_2| \to \infty$. It is important to stress that property (8) holds only for local observables; thus, for operators representing no observable quantities, correlation decoupling generally has no meaning.

Coherence characteristically implies correlations between similar objects, which require the use of autocorrelation functions. To describe coherence decay, it is also necessary to fix a point from which this decay is measured (usually at $\boldsymbol{r} = 0$ and $t = 0$), whereupon coherence decay is studied by considering an autocorrelation function

$$C_\alpha(\boldsymbol{r}, t) \equiv \langle Q_\alpha^+(\boldsymbol{r}, t)\, Q_\alpha(0, 0) \rangle. \tag{9}$$

In many cases, there exists a spatial direction of particular importance, for example, the direction of field propagation. It is natural to associate this special direction with the longitudinal z-axis and the transverse direction with the radial variable $r_\perp$. The characteristic scale of coherence decay in the longitudinal direction is called coherence length l_{coh}, where

$$l_{\text{coh}}^2 \equiv \frac{\int z^2 |C_\alpha(\boldsymbol{r}, t)|^2 \, d\boldsymbol{r}}{\int |C_\alpha(\boldsymbol{r}, t)|^2 \, d\boldsymbol{r}}, \tag{10}$$

and the integration is over the entire space volume. Coherence decay in the transverse direction is classified as transverse coherence radius r_{coh}, where

$$r_{\text{coh}}^2 \equiv \frac{\int r_\perp^2 |C_\alpha(\boldsymbol{r}, t)|^2 \, d\boldsymbol{r}}{\int |C_\alpha(\boldsymbol{r}, t)|^2 \, d\boldsymbol{r}}. \tag{11}$$

For isotropic systems, one replaces $r_\perp$ by the spherical radius r and obtains a coherence radius from equation (11). It is natural to call $A_{\text{coh}} \equiv \pi r_{\text{coh}}^2$ the coherence area and $V_{\text{coh}} \equiv A_{\text{coh}} l_{\text{coh}}$ the coherence volume. The typical scale of temporal correlation decay is termed the coherence time t_{coh}, where

$$t_{\text{coh}}^2 \equiv \frac{\int_0^\infty t^2 |C_\alpha(\boldsymbol{r}, t)|^2 \, dt}{\int_0^\infty |C_\alpha(\boldsymbol{r}, t)|^2 \, dt}. \tag{12}$$

As seen, the coherence length (10) and coherence radius (11) are related to a fixed moment of time, while the coherence time (12) defines the temporal coherence decay at a given spatial point. Equations (10)–(12) all have to do with a particular coherence phenomenon characterized by the correlation function (9).

Phase transitions in equilibrium statistical systems are collective phenomena demonstrating different types of state coherence arising under adiabatically slow variation of thermodynamic or system parameters (temperature, pressure, external fields, and so on). Phase transitions are conventionally specified by means of order parameters, which are defined as statistical averages of operators corresponding to some local observables. The order parameter is assumed to be zero in a disordered phase and nonzero in an ordered phase. For example, the order parameter at Bose–Einstein condensation is the fraction or density of particles in the single-particle ground state. The order parameter for superconducting phase transition is the density of Cooper pairs or the related gap in the excitation spectrum. Superfluidity is characterized by the fraction or density of the superfluid component. For magnetic phase transitions, the order parameter is the average magnetization. Thermodynamic phases can also be classified by order indices.

Let the autocorrelation function (9) be defined for the operator related to an order parameter. Then, for a disordered phase, the coherence length is close to the interparticle distance and the coherence time is roughly the interaction time. But for an ordered phase, the coherence length is comparable with the system size and the coherence time becomes infinite.

Taking account of heterophase fluctuations in the quasiequilibrium picture of phase transitions, there appear mesoscopic coherent structures, with the coherence length being much larger than interparticle distance, but much smaller than the system size. The coherence time of these mesoscopic coherent structures (their lifetime) is much longer than the local equilibrium time, although it may be shorter than the observation time. Such coherent structures are similar to those arising in turbulence.

Electromagnetic coherent radiation by lasers and masers presents a good example of transition coherence. Such radiation processes are accompanied by interference patterns, a phenomenon that is typical of coherent radiation and can be produced by atoms, molecules, nuclei, or other radiating objects. Interference effects caused by light beams are studied in nonlinear optics. But coherent radiation and related interference effects also exist in other ranges of electromagnetic radiation frequencies, including infrared, radio, or gamma regions. Moreover, there exist other types of field radiation, such as acoustic radiation or emission of matter waves formed by Bose-condensed atoms. Registration of interference between a reference beam and that reflected by an object is the basis for holography, which is the method of recording and reproducing wave fields.

The description of interference involves correlation functions. Let $Q_i(t)$ represent a field at time t, produced by a radiator at a spatial point $\boldsymbol{r}_i$. The radiation intensity of a single emitter may be defined as

$$I_i(t) \equiv \langle Q_i^+(t)\, Q_i(t)\rangle, \tag{13}$$

whereupon the radiation intensity for an ensemble of N emitters is

$$I(t) = \sum_{i,j=1}^{N} \langle Q_i^+(t)\, Q_j(t)\rangle. \tag{14}$$

Separating the sums with $i = j$ and with $i \neq j$ yields

$$I(t) = \sum_{i=1}^{N} I_i(t) + \sum_{i\neq j}^{N} \langle Q_i^+(t)\, Q_j(t)\rangle, \tag{15}$$

which shows that intensity (14) is not simply a sum of the intensities (13) of individual emitters but also includes the interference part, expressed through the autocorrelation functions of type (9). The first term in equation (15) is the intensity of incoherent radiation, while the second term corresponds to the intensity of coherent radiation.

V.I. YUKALOV

See also **Bose–Einstein condensation; Chaotic dynamics; Critical phenomena; Ferromagnetism and ferroelectricity; Lasers; Nonequilibrium statistical mechanics; Nonlinear optics; Order parameters; Phase transitions; Spatiotemporal chaos; Spin systems; Structural complexity; Superconductivity; Superfluidity; Turbulence**

Further Reading

Andreev, A.V., Emelyanov, V.I. & Ilinski, Y.A. 1993. *Cooperative Effects in Optics*, Bristol: Institute of Physics Publishing

Benedict, M.G., Ermolaev, A.M., Malyshev, V.A., Sokolov, I.V. & Trifonov, E.D. 1996. *Superradiance: Multiatomic Coherent Emission*, Bristol: Institute of Physics Publishing

Bogolubov, N.N. 1967. *Lectures on Quantum Statistics*, Vol. 1, New York: Gordon and Breach

Bogolubov, N.N. 1970. *Lectures on Quantum Statistics*, Vol. 2, New York: Gordon and Breach

Coleman, A.J. & Yukalov, V.I. 2000. *Reduced Density Matrices*, Berlin: Springer

Klauder, J.R. & Skagerstam, B.S. 1985. *Coherent States*, Singapore: World Scientific

Klauder, J.R. & Sudarshan, E.C.G. 1968. *Fundamentals of Quantum Optics*, New York: Benjamin

Lifshitz, E.M. & Pitaevskii, L.P. 1980. *Statistical Physics: Theory of Condensed State*, Oxford: Pergamon Press

Mandel, L. & Wolf, E. 1995. *Optical Coherence and Quantum Optics*, Cambridge and New York: Cambridge University Press

Nozières, P. & Pines, D. 1990. *Theory of Quantum Liquids: Superfluid Bose Liquids*, Redwood, CA: Addison-Wesley
Perina, J. 1985. *Coherence of Light*, Dordrecht: Reidel
Scott, A.C. 1999. *Nonlinear Science: Emergence and Dynamics of Coherent Structures*, Oxford and New York: Oxford University Press
Ter Haar, D. 1977. *Lectures on Selected Topics in Statistical Mechanics*, Oxford: Pergamon Press
Yukalov, V.I. 1991. Phase transitions anad heterophase fluctuations. *Physics Reports*, 208: 395–492
Yukalov, V.I. & Yukalova, E.P. 2000. Cooperative electromagnetic effects. *Physics of Particles and Nuclei*, 31: 561–602

COHERENT EXCITON

See **Excitons**

COHERENT STRUCTURES

See **Emergence**

COLE–HOPF TRANSFORM

See **Burgers equation**

COLLAPSE

See **Development of singularities**

COLLECTIVE COORDINATES

The soliton equations describe a number of important nonlinear physical phenomena. However, in real life, these phenomena are not precisely modeled by say the sine-Gordon (SG) equation, the nonlinear Schrödinger (NLS) equation, or some of the other pure soliton equations. Corrective and often small terms are added to include, for example, inhomogeneities, dissipation, or energy input. The resulting wave phenomena possess modified solitonic features that can be treated approximately starting with a pure soliton solution and then allow the parameters of the soliton solution to vary slowly with time under the influence of the perturbations instead of being constant. Solution parameters that are chosen to vary with time are called *collective coordinates*. They encompass the influence of the perturbations in the pure soliton equations. The advantage of introducing collective coordinates is to reduce a problem with infinitely many degrees of freedom to a problem with a few degrees of freedom (Kivshar & Malomed, 1989; Sánchez & Bishop, 1998).

To illustrate the use of a collective coordinate approach, we shall investigate the NLS equation with the perturbative term εR

$$\mathrm{i}u_t + u_{xx} + 2|u|^2 u = \varepsilon R. \tag{1}$$

Here, $u = u(x, t)$, where x is the spatial coordinate and t is time. Subscripts x and t denote partial derivatives with respect to these variables. The simple single soliton solution of the pure NLS equation ($\varepsilon = 0$) reads

$$u(x, t) = a\,\mathrm{sech}(a\theta)\exp(\mathrm{i}\xi\theta + \mathrm{i}\sigma), \tag{2}$$

where $\theta = x - x_0$. This solution possesses four parameters $(a, \xi, x_0, \sigma) = (y_1, y_2, y_3, y_4)$, which we shall choose as collective coordinates. For weak perturbations, $\varepsilon \ll 1$, we shall assume that the collective coordinates depend slowly on time t due to the influence of εR. In addition, the perturbation leads to a radiation field of small amplitude, which is neglected. A variational approach is employed to determine the time evolution of the collective coordinates, but this is not the only method available. In the framework of a variational approach, the collective coordinates are the generalized coordinates. Although perturbations may destroy the Hamiltonian property of the pure soliton equations, dissipative effects and external nonconservative forces can be accounted for in the variation of a Lagrangian function by introducing generalized forces associated with the generalized coordinates. Below, this is done by introducing a generalized force for each collective coordinate as in classical mechanics.

The unperturbed NLS equation (and its complex conjugate) can be derived from the Lagrangian density (Caputo et al., 1995; Scott, 2003);

$$\mathcal{L}(u, u^*, u_t, u_t^*, u_x, u_x^*) = \frac{\mathrm{i}}{2}(u^* u_t - u_t^* u) - |u_x|^2 + |u|^4. \tag{3}$$

The total Lagrangian function is

$$L(y_i, \dot{y}_i) = \int_{-\infty}^{\infty} \mathcal{L}\,\mathrm{d}x, \tag{4}$$

where we denote the collective coordinates $y_i(t)$, $i = 1, 2, 3, 4$, and $\dot{y}_i$ is the time derivative of y_i. Together with Equation (1), the variation of total Lagrangian leads to the Euler–Lagrange equations (Caputo et al., 1995; Scott, 2003)

$$\frac{\partial L}{\partial y_i} - \frac{\mathrm{d}}{\mathrm{d}t}\left(\frac{\partial L}{\partial \dot{y}_i}\right) = \varepsilon \int_{-\infty}^{\infty} R \frac{\partial u^*}{\partial y_i}\mathrm{d}x + \mathrm{c.c.}, \tag{5}$$

where c.c. stands for complex conjugate of the preceding term on the right-hand side. The inhomogeneous term on the right-hand side is interpreted as the generalized force associated with the collective coordinate $y_i(t)$, which is a key result as we do not rely on a perturbation with a strict Hamiltonian nature. Another important feature is that the above approach provides as many dynamical equations as we choose collective coordinates; thus it is straightforward to determine the generalized forces associated with each collective coordinate.

To illustrate the method, we calculate the total Lagrangian for the NLS equation using the simple single soliton solution in (2)

$$L = \frac{2}{3}a^3 - 2a\xi^2 + 2a\xi\frac{\mathrm{d}x_0}{\mathrm{d}t} - 2a\frac{\mathrm{d}\sigma}{\mathrm{d}t}. \quad (6)$$

Consider the perturbation

$$\varepsilon R = -\mathrm{i}\Gamma u + \frac{\mathrm{i}g_0 u}{1 + p/p_\mathrm{s}}, \quad (7)$$

describing light propagation through an optical fiber amplifier with a loss factor Γ, gain g_0, and power $p = \int_{-\infty}^{\infty} |u|^2 \mathrm{d}x$. The constant p_s is a saturation power level that is characteristic for a given fiber amplifier. From Equation (5), the resulting dynamical equations for the collective coordinates are

$$\frac{\mathrm{d}a}{\mathrm{d}t} = -2\Gamma a + \frac{2g_0 a}{1 + 2a^2/p_\mathrm{s}}, \quad (8)$$

$$\frac{\mathrm{d}\xi}{\mathrm{d}t} = 0, \quad (9)$$

$$\frac{\mathrm{d}x_0}{\mathrm{d}t} = 2\xi, \quad (10)$$

$$\frac{\mathrm{d}\sigma}{\mathrm{d}t} = a^2 + \xi^2, \quad (11)$$

which is a drastic simplification compared to the original perturbed NLS problem. As for the sine-Gordon system, one can define a power balance condition by requiring $\dot{a}(t) = 0$, implying an equilibrium amplitude a_∞ given as $a_\infty = \sqrt{(g_0 - \Gamma)p_\mathrm{s}/2\Gamma}$.

A number of other strategies have been designed to determine slow time variation of collective coordinates in perturbed soliton solutions. These include slow variation of scattering data of the inverse scattering theory for pure solitons (Kivshar & Malomed, 1989; Lamb, 1980), more direct perturbation approaches (McLaughlin & Scott, 1978), and utilizing Hamiltonian structures in cases where the perturbation leads to Hamiltonian systems (Caputo & Flytzanis, 1991). An important result of perturbations is the formation of trailing radiation fields, which are low-amplitude linear waves created as perturbed solitons propagates (McLaughlin & Scott, 1978; Kivshar & Malomed, 1989; Willis, 1997). The solitons lose energy to the radiation field, and the variational approach used here can be extended to include such radiation.

MADS PETER SØRENSEN

See also **Constants of motion and conservation laws; Energy analysis; Euler–Lagrange equations; Hamiltonian systems; Inverse scattering method or transform; Nonlinear optics; Nonlinear Schrödinger equations; Perturbation theory; Solitons**

Further Reading

Caputo, J.G. & Flytzanis, N. 1991. Kink-antikink collisions in sine-Gordon and ϕ^4 models: problems in the variational approach. *Physical Review* A, 44(10): 6219–6225

Caputo, J.G., Flytzanis, N. & Sørensen, M.P. 1995. The ring laser configuration studied by collective coordinates. *Journal of the Optical Society of America* B, 12(1): 139–145

Kivshar, Y.S. & Malomed, B.A. 1989. Dynamics of solitons in nearly integrable systems. *Reviews of Moderen Physics*, 61: 763–915

Lamb, G.L. 1980. *Elements of Soliton Theory*, New York: Wiley

McLaughlin, D.W. & Scott, A.C. 1978. Perturbation analysis of fluxon dynamics. *Physical Review A*, 18: 1652–1680

Sánchez, A. & Bishop, A.R. 1998. Collective coordinates and length-scale competition in spatially inhomogeneous soliton-bearing equations. *SIAM Review*, 40(3): 579–615

Scott, A.C. 2003. *Nonlinear Science: Emergence and Dynamics of Coherent Structures*, 2nd edition, Oxford and New York: Oxford University Press

Willis, C.R. 1997. Spontaneous emission of a continuum sine-Gordon kink in the presence of a spatially periodic potential. *Physical Review E*, 55(5): 6097–6100

COLLISIONS

An interesting consequence of Hamiltonian structures is that there typically exist symmetries and invariances that allow one to generate mobile localized states from standing ones. For instance, the sine-Gordon (SG) equation (Dodd et al., 1982)

$$u_{tt} = u_{xx} - \sin(u), \quad (1)$$

where the subscripts denote partial derivatives, is invariant under the Lorentz transformation $x \to x' = \gamma(x - vt)$ and $t \to t' = \gamma(t - vx)$. $\gamma = \sqrt{1 - v^2}$. As a result, static kinks and antikinks, corresponding to the two signs of the solution

$$u(x,t) = \pm 4\tan^{-1}(x - x_0), \quad (2)$$

can be boosted to any subluminal speed $v < 1$, as

$$u(x,t) = \pm 4\tan^{-1}\left[\gamma(x - x_0 - vt)\right]. \quad (3)$$

Similarly, standing waves of the nonlinear Schrödinger (NLS) equation

$$\mathrm{i}u_t = -u_{xx} - |u|^2 u \quad (4)$$

can be boosted by Galilean invariance into traveling ones with any speed v of the form (Sulem & Sulem, 1999)

$$u(x,t) = (2\alpha)^{1/2}\mathrm{e}^{\mathrm{i}(v/2)x + \mathrm{i}(\alpha - (v^2/4))t} \times \mathrm{sech}\,(x - x_0 - vt). \quad (5)$$

Note that localized solutions of dissipative partial differential equations do not typically share such features, because their traveling wave speeds are determined by the dynamics rather than initial conditions. Hereafter, we will focus on Hamiltonian models.

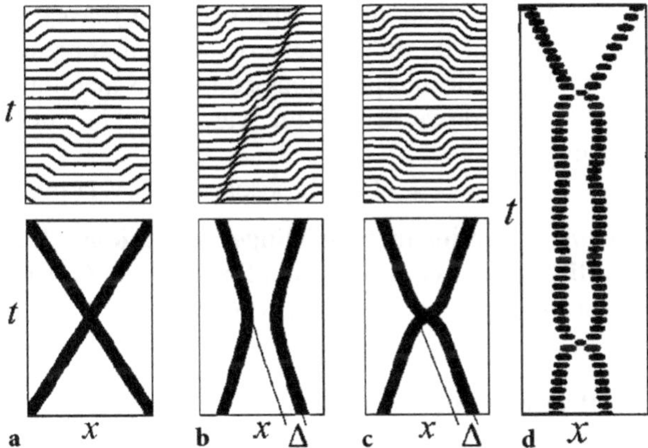

Figure 1. Elastic collisions: (a) linear-shaped antikink and kink in the linear wave equation; (b) kink-kink repulsive collision in SG; (c) antikink-kink attractive collision in SG. Top panels show $u(x,t)$, while bottom panels show the trajectories of soliton cores in the (x,t)-plane. (d) Inelastic collisions between two NLS solitons in the presence of weak perturbation.

Given the mobility of the localized coherent structures, it is natural to consider the outcome of their collisions. In fact, it was the "solitary" nature of such interactions (Zabusky & Kruskal, 1965) that inspired the term *soliton* in the case of integrable systems.

Linear versus Nonlinear Collisions

Consider the collision of two wavepackets that are governed by a linear wave equation. In this case, the superposition principle inherent in linearity guarantees that the two packets do not "feel" each other and survive the collision without change of shape, speed, or trajectory (see Figure 1(a)).

On the other hand, nonlinear dynamics offers more interesting collisions. Here, the result of the interaction of two excitations does not resemble their sum and (at least) a phase shift Δ is present (compare Figures 1(b) and 1(c) with 1(a)). In 1(b) and 1(c), the kink-kink and the antikink-kink collisions in the nonlinear SG equation are shown. The soliton cores do not merge in mutually repulsive collisions 1(b) while they do so in attractive collisions 1(c).

Elastic versus Inelastic Collisions

In fully integrable systems, solitons have the remarkable property (often used to define them) of colliding elastically (Ablowitz & Segur, 1981). In these special systems, the dynamics are severely restricted by the existence of an infinite set of conservation laws. Although realistic systems are typically non-integrable, the non-integrability in many applications is weak and can be treated by including small perturbative terms in integrable equations. Such perturbations are called Hamiltonian if the total energy of the perturbed system remains a dynamical invariant. While collisions in linear systems are much simpler than those in integrable nonlinear systems (such as the SG and NLS equations), the latter can, in turn, also be very different from the much more complex picture of inelastic collisions in near-integrable or more generally non-integrable systems.

In general non-integrable systems, the inelasticity of collisions may be manifested through emission of radiation, excitation of soliton internal modes, and energy exchange between solitons. Internal modes are the long-lived, spatially localized, oscillatory excitations (corresponding to the point spectrum of the linearization around the wave). These can typically be excited only for a particular sign of the perturbation (Campbell et al., 1983).

Small-amplitude radiation waves correspond to an irreversible chunk of energy lost in the collision (radiated toward the boundaries). Such modes are extended, plane waves of the continuous spectrum. Notice that the energy of internal modes can be partly restored to solitons if they collide again, while this is not possible for radiation waves (in the first-order approximation).

Strong energy exchange between nonlinear waves in near-integrable systems can also occur in the absence of the above two mechanisms (and for different types and signs of the original perturbation). There are two necessary conditions for this recently manifested mechanism (Dmitriev et al., 2001, 2002). It can be observed only if the energy exchange is not forbidden by the conservation laws existing in the perturbed system and if the collision is of attractive type, as in Figure 1(c). For example, in the SG equation perturbed by the term εu_{xxxx}, where energy and momentum are conserved for the one (free)-parameter kink-solitons, energy exchange is possible only when more than two solitons participate in the collision. The energy exchange between only kinks or only antikinks is also not possible because SG solitons with the same parity repel each other. Similarly, the effect is not possible in the Korteweg–de Vries (KdV) equation, where soliton interactions are always mutually repulsive.

In the NLS equation, in-phase solitons attract each other, while out-of-phase solitons repel. Each soliton has two parameters (amplitude and phase), and for many practically important perturbations, there are two conserved quantities: the Hamiltonian and L^2 norm of the solution. Thus, energy exchange between two nearly in-phase NLS solitons is possible in the presence of a weak perturbation. The effect of radiationless energy exchange between solitons survives even for a very weak perturbation, as it decreases linearly with a decrease in perturbation amplitude, while other effects of the perturbation decay faster. If the perturbation is not small, the energy exchange effect mingles with radiation emission and possibly with the excitation of internal modes.

Probabilistic Nature of Soliton Collisions

In many examples of perturbed, non-integrable models related to applications in optics, fluid mechanics, or condensed-matter physics (Kivshar & Malomed, 1989), the result of soliton collisions can be predicted only in a probabilistic sense. The following sources of stochastic behavior can be identified. First, chaotic soliton scattering can arise from resonant interaction with the soliton internal modes (Campbell et al., 1986; Gorshkov et al., 1992). Second, in discrete systems, the result of inelastic collisions can be sensitive to the coordinate of collision point, x_c, with respect to the lattice site (Dmitriev et al., 2003; Papacharalampous et al., 2003). Because the coordinate of the collision point usually cannot be controlled, it is natural to describe the result of the collision as a function of the random variable x_c. Finally, the result of the collision can be extremely sensitive to some other uncontrolled characteristics, such as the relative phase of the colliding solitons, as has been demonstrated to dramatically affect the collisions between NLS solitons or between kinks and breathers in SG (Dmitriev et al., 2001, 2002, 2003; Papacharalampous et al., 2003).

This last source of randomness is important when energy exchange between solitons is possible. In Figure 1(d), an example of inelastic interaction between two NLS solitons in the presence of quintic perturbation, $\varepsilon|u|^4u$ in Equation (4), with a small $\varepsilon > 0$ is presented (Dmitriev & Shigenari, 2002). (The regions of the (x,t)-plane are shown, where the real part of the solution exceeds a certain value.) After each collision, the properties of the solitons such as the frequency and amplitude are different depending on the collision phase. With a certain probability, the solitons can attain (after a number of collisions) a velocity sufficient to overcome their weak mutual attraction. In the example presented, the solitons split after the fourth collision. Emission of extended wave radiation is monitored and found to be very weak in this case; as a result, the solitons may continue to collide for a very long time, forming a two-soliton bound state. However, the probability P to obtain a bound state with the lifetime T decays algebraically as $P \sim T^{-\alpha}$ (Dmitriev & Shigenari, 2002), which is a manifestation of the chaotic character of their interaction.

In conclusion, linear waves do not "feel" each other, and nonlinear waves of integrable equations emerge unscathed from collisions, retaining their solitary character. However, solitary wave collisions in more realistic, non-integrable models remain a fascinating topic, where a number of basic mechanisms (such as internal mode resonances, extended wave radiation, and radiationless energy exchange) have been elucidated, but the full picture is far from complete. The probabilistic interpretation of collisions mentioned above may prove a fruitful viewpoint in future studies.

P.G. Kevrekidis and S.V. Dmitriev

See also **Nonlinear Schrödinger equations; Partial differential equations, nonlinear; Sine-Gordon equation; Solitons; Solitons, types of**

Further Reading

Ablowitz, M.J. & Segur, H. 1981. *Solitons and the Inverse Scattering Transform*, Philadelphia: SIAM

Campbell, D.K., Schonfeld, J.F. & Wingate, C.A. 1983. Resonance structure in kink–antikink interactions in $\phi/4$ field theory. *Physica* D, 9: 1–32

Campbell, D.K., Peyrard, M. & Sodano, P. 1986. Kink-antikink interactions in the double sine-Gordon equation. *Physica* D, 19: 165–205

Dmitriev, S.V., Kivshar, Yu.S. & Shigenari, T. 2001. Fractal structures and multiparticle effects in soliton scattering. *Physical Review* E, 64: 056613

Dmitriev, S.V., Semagin, D.A., Sukhorukov, A.A. & Shigenari, T. 2002. Chaotic character of two-soliton collisions in the weakly perturbed nonlinear Schrödinger equation. *Physical Review* E, 66: 046609

Dmitriev, S.V. & Shigenari, T. 2002. Short-lived two-soliton bound states in weakly perturbed nonlinear Schrödinger equation. *Chaos*, 12: 324

Dmitriev, S.V., Kevrekidis, P.G., Malomed, B.A. & Frantzeskakis, D.J. 2003. Two-soliton collisions in a near-integrable lattice system. *Physical Review* E, 68: 056603

Dodd, R.K., Eilbeck, J.C., Gibbon, J.D. & Morris, H.C. 1982. *Solitons and Nonlinear Wave Equations*, London: Academic Press

Gorshkov, K.A., Lomov, A.S. & Gorshkov, M.I. 1992. Chaotic scattering of two-dimensional solitons. *Nonlinearity*, 5: 1343–1353

Kivshar, Yu.S. & Malomed, B.A. 1989. Dynamics of solitons in nearly integrable systems. *Reviews of Modern Physics*, 61: 763–915

Papacharalampous, I.E., Kevrekidis, P.G., Malomed, B.A. & Frantzeskakis, D.J. 2003. Soliton collisions in the discrete nonlinear Schrödinger equation. *Physical Review* E, 68: 046604

Sulem, C. & Sulem, P.L. 1999. *The Nonlinear Schrödinger Equation*, New York: Springer

Zabusky, N.J. & Kruskal, M.D. 1965. Interactions of solitons in a collisionless plasma and the recurrence of initial states. *Physical Review Letters*, 15: 240–243

COLOR CENTERS

Gemstones are brightly colored because of the presence of *color centers*, atomic-scale imperfections that absorb light in otherwise transparent crystals. Historically, the term *color centers* has been associated with such imperfections in a special class of crystals called alkali halides, because it was in these relatively simple transparent hosts that the scientific study of color centers flourished. A German research program that began in the 1930s soon led to the recognition that alkali halides are excellent hosts for scientific studies of defects in nonmetallic solids, and the understanding of fundamental properties in these materials has been of great significance in the studies of similar phenomena in more complicated (and sometimes more practical) situations and materials.

Sodium chloride—common table salt—is the most familiar alkali halide. It consists of positively charged sodium and negatively charged chlorine ions, alternating positions in a simple cubic array. In the perfect crystal, each ion has six nearest neighbors of opposite charge. Simple defects may involve chemical impurities, such as a positively charged silver or thallium ion on an alkali site, or the removal of one or more ions from normal positions. The most fundamental of the latter class is the F center (from the German *Farbzentren*), a halogen ion vacancy that has trapped an electron. Defects involving more than one F center, or chemical impurities next to one or more F centers, may also occur.

Understanding the properties of color centers in detail requires some knowledge of quantum concepts. For example, the trapped electron of an F center can exist only in certain quantum states. In order for light to be absorbed, the energy of an incident photon must match the energy difference between the lowest quantum state and a higher one. This energy difference depends on the host crystal, and so F centers induce different colors in different alkali halides.

The situation becomes more complicated when one considers that the electron is not trapped in a static host, but rather that the neighboring ions are vibrating. The electron's interaction with these vibrations, whose time scale is long with respect to that of the (light) electron's motion, means that the energy difference between ground and excited electronic states does not have a single value, but rather a distribution of values about some mean. Thus, for example, whereas the mean photon absorption energy for F centers at low temperatures in NaCl is 2.77 eV, the mean width of the distribution of absorption energies is 0.26 eV.

This *electron-lattice interaction* has other consequences. In most cases, an F center at low temperature that has been excited by light will re-emit light. However, the mean energy of the emitted photon is found to be considerably smaller than that of the absorbed photon. This *Stokes shift* is exemplified in NaCl, where the mean photon emission energy at low temperatures is 0.98 eV.

Why is there a Stokes shift? In simple terms, the energy-level structure of the F center is determined by the mean position of the neighboring ions. However, the mean position of the neighboring ions is in turn determined by the quantum state of the F center electron. After the electron is excited from the ground state to another quantum state, the neighboring ions relax in response to the change in the average force exerted on them by the electron. This relaxation then leads to a change in the energy-level separations, as well as other fundamental aspects of their properties. The emitted photon has an energy smaller than the absorbed photon had, and energy is conserved in the total cycle as the relaxation processes create quanta of lattice vibrations, or phonons, which remove the excess energy.

This cyclic process may be visualized by means of a configuration coordinate diagram. In the simplest case, this diagram consists of two equal parabolas, displaced vertically (in energy, E) and horizontally (in the displacement of neighboring ions, R), as shown in Figure 1. Each parabola represents the vibration of neighboring ions and leads to quantized vibrational states. According to the Franck–Condon principle, electronic transitions take place vertically on the diagram—the massive ions do not respond instantly to the excitation by the photon—so absorption corresponds to a transition from A to B, emission C to D. This picture is highly oversimplified, but it does represent a physical and visual way to understand phenomena associated with the optical properties of F centers (and by extension, many other types of defects in both insulators and semiconductors).

We now consider a perfect alkali halide crystal in which one electron has been removed by light (ionizing radiation) from the array of negative ions. At low temperatures, this is found to lead to the phenomenon of self-trapping, whereas in semiconductors, the removal of an electron from a valence band of occupied states leads to motion of the empty state, or hole, and resulting electrical conduction; in most alkali halides the missing charge becomes localized in space, or self-trapped. What happens in detail is that the halogen ion that has lost one electron (and become a neutral atom) can form a chemical bond with a nearby halogen ion. In the process, both of these move toward each other, the lattice around them relaxes, and the missing charge is equally shared by this halogen molecule-ion.

Self-trapping is not a universal process; it "costs" energy to localize a quantum particle, and the energy

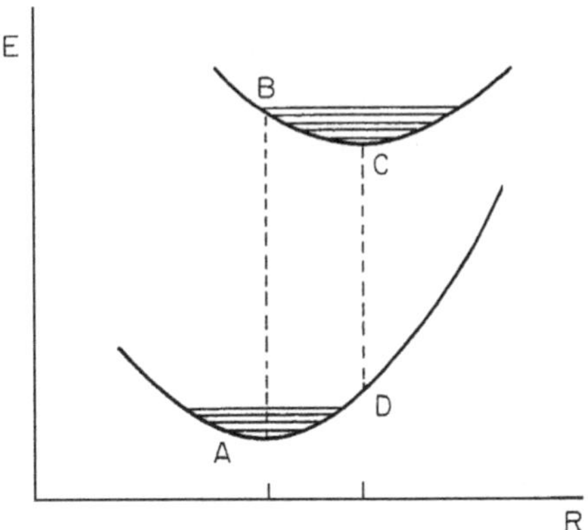

Figure 1. Schematic configuration coordinate diagram.

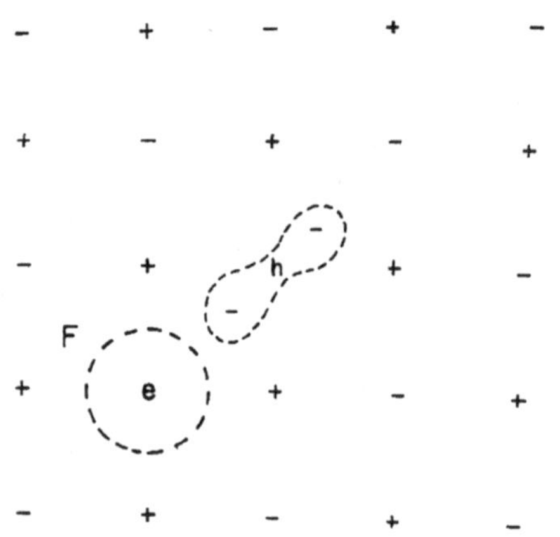

Figure 2. Adjacent F center and self-trapped hole center in an alkali halide. "+" and "−" denote alkali and halogen ions, respectively.

gained by chemical bonding and relaxation must overcome this. Self-trapping of holes occurs in most alkali halides, but not in semiconductors and not in many other insulators. Also of interest is the creation of defects by the self-trapping of excitation energy in alkali halides, leading to a *self-trapped exciton*. To approach this, we consider the trapping of an electron by a self-trapped hole. Since the hole is effectively positive (it resulted from the removal of an electron), we might imagine that the trapped electron would find itself in loosely bound quantum states about the self-trapped hole. However, there is another possibility. If it does not cost too much energy for the halogen molecule-ion to move into one halogen site, rather than be shared by two sites, then the electron may be trapped in the other, empty halogen site. One then has an F center next to a halogen molecule-ion; hence, two defects have been formed. This situation is shown in Figure 2. Since the two defects are adjacent, this may be an unstable arrangement, and indeed there is a finite probability that the system will revert back to the perfect crystal, either before or after the emission of a photon. But, in many cases, it is found that this nearest-neighbor arrangement is the precursor for the creation of a stable defect pair: the halogen molecule-ion may migrate through the halogen sublattice, not by long-range atomic motion, but rather by short-range halogen motion accompanied by motion of the hole. This results in sequential sharing of the hole by the halogens as the hole migrates away from the F center.

W. BEALL FOWLER

See also **Excitons; Franck–Condon factor; Quantum theory**

Further Reading

Crawford, J.H. & Slifkin, L.M. (editors). 1972 (vol. 1), 1975 (vol. 2). *Point Defects in Solids*, New York, Plenum

Fowler, W.B. (editor). 1968. *Physics of Color Centers*, New York: Academic

Hayes, W. & Stoneham, A.M. 1985. *Defects and Defect Processes in Nonmetallic Solids*, New York: Wiley

Schulman, J.H. & Compton, W.D. 1962. *Color Centers in Solids*, New York: Macmillan

Song, K.S. & Williams, R.T. 1993, 1996. *Self-Trapped Excitons*, Berlin and New York: Springer

COMMENSURATE-INCOMMENSURATE TRANSITION

When some local property in a crystalline solid (atomic positions or orientation of local magnetic moments) develops a spatial modulation with a wavelength b that differs from the underlying lattice spacing a, one speaks of a "modulated phase." A modulated phase is said to be "commensurate" when the ratio b/a is rational and "incommensurate" when b/a is irrational.

Modulated phases are experimentally observed by the appearance of "satellite spots" in X-ray, neutron, or electron diffraction patterns (Janssen & Janner, 1987; Cummins, 1990). Observations of spatially modulated structures are abundant in condensed matter physical systems, such as the ferrimagnetic phases of the rare earths and their compounds, long-period structures of binary alloys, graphite intercalation compounds, or the polytypic phases of spinelloids, perovskites, and micas, among other minerals.

The wavelength b characterizing the modulation varies with external parameters, such as temperature, pressure, or magnetic field. Sometimes, this variation is smooth, but often it remains constant at a rational locking value through some range of values of the external parameter before changing to another rational locking value, and so on.

The ubiquity of modulated phases shows that the physical origin of these behaviors cannot be tied to particularities of specific types of systems, but must be of a general character. It is widely recognized that modulated phases appear whenever different terms in the free energy of the system compete, each one trying to impose its own characteristic length scale. The external parameters control the relative strength of the competing interactions and new compromises are reached as they vary.

An insight into the physics of modulated phases was obtained through detailed analyses of simple model systems of competing interactions. Although these simple models are unlikely to fit experiments on specific materials, they help to understand the complexity of behaviors that emanate from length-scale competition and to discern essential features.

One of the best-studied model systems with competing interactions is the axial next-nearest-neighbors Ising (ANNNI) model (motivated by the magnetic structures of erbium), which was introduced by R.J. Elliot in 1961 (Yeomans, 1988). This is an Ising model with a two-state spin, $S = \pm 1$, on each site of a cubic lattice. Interactions between spins on nearest-neighbor sites are ferromagnetic, but there are second (next-nearest) neighbor antiferromagnetic interactions along the axial direction, z. The Hamiltonian is

$$H = -\frac{1}{2}J_0 \sum_{i,j,j'} S_{i,j}S_{i,j'} - J_1 \sum_{i,j} S_{i,j}S_{i+1,j} - J_2 \sum_{i,j} S_{i,j}S_{i+2,j}, \qquad (1)$$

where i indicates the two-dimensional layers perpendicular to the axial direction, and j, j' are nearest-neighbor sites within a layer. Both J_0 and J_1 are positive (ferromagnetic interactions), thus favoring the same value of neighboring spins, but $J_2 < 0$ (antiferromagnetic) so that it favors opposite values of second neighbor spins along the z-direction.

In the absence of thermal effects ($T = 0$), the favored spin configurations along the z-direction are the ferromagnetic alignment $(\cdots ++++++ \cdots)$ if $\kappa = -J_2/J_1 < \frac{1}{2}$, and the antiphase $(\cdots ++--++ \cdots)$ configuration if $\kappa > 1/2$. Exactly at $\kappa = \frac{1}{2}$, any configuration containing a stripe of two or more spins "+" followed by a stripe of two or more spins "−", etc., has the same energy; thus there is a multiphase point.

A convenient notation is to use $\langle n_1, \ldots, n_m \rangle$ to represent a state in which a set of stripes of width n_i of alternating spins repeat. For example, $(\cdots ++--+++--++--- \cdots)$ is denoted by $\langle 223 \rangle$. Ferromagnetic and antiphase configurations are consistently denoted by $\langle \infty \rangle$ and $\langle 2 \rangle$, respectively. When temperature increases from zero, entropic effects regulate the competition among the ferro and antiferro interactions, and a flower (called a "devil's flower" by Per Bak) of petal-like phases of modulated structures opens up from the multiphase point.

Figure 1 shows how new commensurate phases appear as temperature increases, demonstrating qualitatively what is commonly observed in experiments—locking to a few short-wavelength commensurate phases separated either by first-order phase transitions (as in the low-temperature regime of Figure 1) or by regions where the wave vector appears to vary smoothly (as at higher temperatures).

The following question naturally arises: what determines the behavior of the modulation wave vector as parameters vary? The answer to this question is closely tied to a central paradigm of nonlinear science, the soliton concept in the context of its discrete counterpart called discommensuration.

To clarify this issue, consider the simplest model of competing interactions (Griffiths, 1990), the Frenkel–Kontorova model, which can be visualized as an array of atoms at positions (u_n), $-\infty < n < +\infty$, experiencing

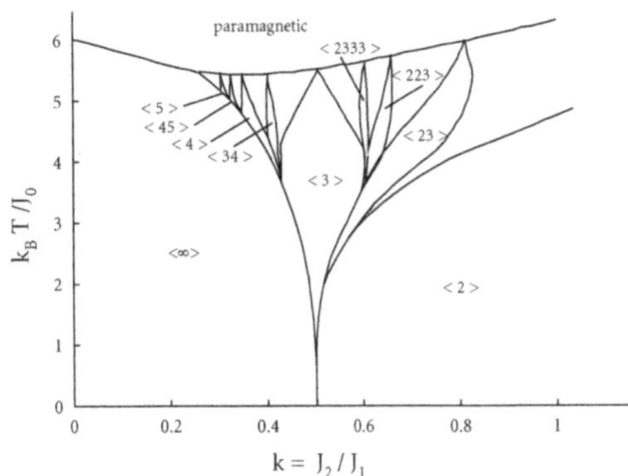

Figure 1. Mean-field phase diagram of the ANNNI model showing the main commensurate phases.

a periodic substrate potential $V(u) = V(u+1)$ and nearest-neighbor interaction $W(\Delta u)$. The Hamiltonian is

$$H = \sum_n [KV(u_n) + W(\Delta u_n)], \qquad (2)$$

where the parameter K controls the relative strength of the interactions. The standard Frenkel–Kontorova model corresponds to

$$V(u) = \frac{1}{(2\pi)^2}[1 - 2\cos(2\pi u)],$$

$$W(\Delta u) = \frac{1}{2}(\Delta u)^2 - \sigma \Delta u. \qquad (3)$$

Note that V favors an integer value of the interspacing Δu, while W favors a uniform value σ, so both interactions compete to determine the configuration (u_n), characterized here by the average interspacing $\omega = \langle \Delta u \rangle$.

If (as in the standard model) the interaction $W(\Delta u)$ is a convex function, a complete rigorous characterization of the model phase diagram was given by Aubry (1985). Thus, we restrict the analysis to convex $W(\Delta u)$.

In the thermodynamic limit of the system, care must be taken when defining what is meant by configurations of minimum energy or even by the energy of a configuration. Thus, the mean energy ε per particle of a configuration (u_n) is defined as

$$\varepsilon = \lim_{(N-M)\to\infty} \frac{1}{N-M} \times \sum_{j=M}^{N-1} [KV(u_j) + W(u_{j+1} - u_j)]. \qquad (4)$$

A minimum energy configuration (MEC) (u_j) is that for which the arbitrary displacement of any finite segment $(u_j + \delta_j)$, $M < j < N$, of the configuration always produces a nonnegative energy change. For an MEC, both limits (ω and ε) exist. Moreover, for every real number ω, there is at least an MEC with average spacing ω, and its mean energy per particle $\varepsilon(\omega)$ is a well-defined function. A ground state is an MEC that is "recurrent," meaning that any arbitrary finite segment of the configuration approximately reappears somewhere else, within arbitrary precision. One often speaks of the ground state that corresponds to fixed values of the parameters (K and σ for the standard model), meaning the ground-state configuration for which $\varepsilon = \min_\omega \varepsilon(\omega)$ occurs at the fixed values of the parameters. The minimum ε, which depends on the parameter values, is the ground state energy, and the value of ω where that minimum occurs is the ground-state average spacing.

The ground-state phase diagram of the standard Frenkel–Kontorova model is shown in Figure 2. Some low-order commensurate phases (rational ω) are shown there, but in between any two of them, there are an infinite number of others. The ground-state average spacing ω as a function of the tension σ, for a fixed value of $K = 1$, is shown in Figure 3. This is a continuous function with plateaus of (rational) constant value, which correspond to commensurate phases. This function is called "devil's staircase" or Cantor function. No first-order phase transitions between commensurate phases of different commensurabilities are allowed in convex models.

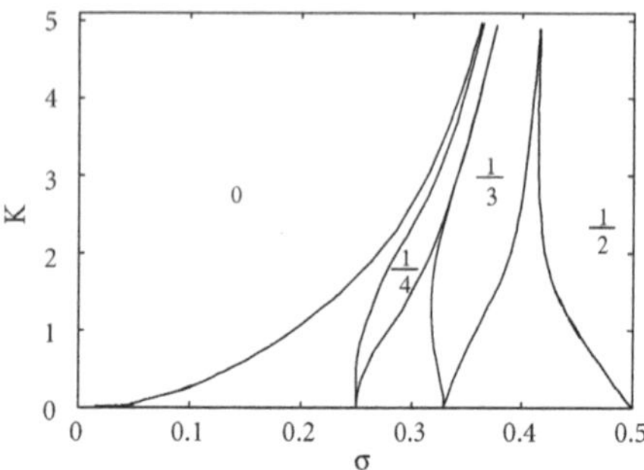

Figure 2. Ground-state phase diagram of the standard Frenkel–Kontorova model. Each point in the diagram is characterized by the value of ω, average spacing of the groundstate configurations for the parameters (K, σ) of the model. The numbers in the figure are values of ω. Only a few commensurate phases are plotted, as there are infinitely many commensurate and incommensurate phases in between any two of them.

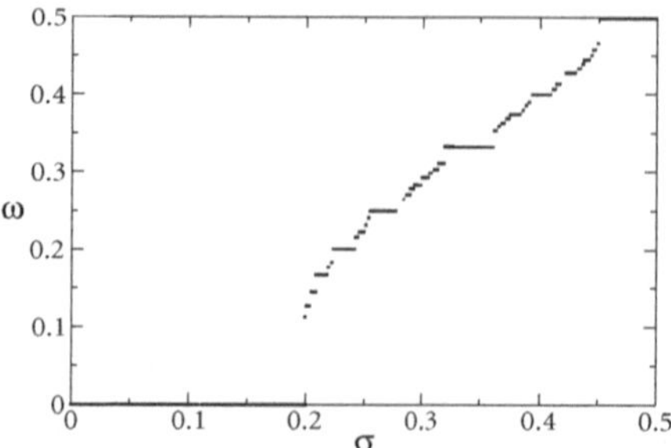

Figure 3. ω as a function of σ at K = 1, a devil's staircase, there being a "plateau" for each rational value of ω.

For rational values of $\omega_c = p/q$, commensurate ground states are periodic configurations: $\sigma_{qp}(u_n) := (u_{n+q} - p) = (u_n)$. But there are also two MECs with average spacing (ω_c) that are nonrecurrent and called elementary "discommensurations" (DC). These can be described as a localized compression (retarded DC) or a localized expansion (advanced DC) over a commensurate ground state. The distortion being localized, they have the same mean energy per particle $\varepsilon(\omega_c)$ of the ground state. However, the difference of the non-averaged energy of a DC and the ground state (the extra energy of the localized compression or expansion) is generally nonzero; this is called the "creation energy" of the DC.

If a configuration contains two elementary DCs superimposed on a commensurate ground state, the energy difference with respect to the ground state can be understood as the sum of the creation energies of the DCs plus the interaction energy. If both DCs are of the same type, the energy decreases with the interdistance (repulsive interaction). The interaction is attractive for DCs of opposite type. Elementary DCs have an *excess length* Δ with respect to the ground state, which is exactly $\Delta = 1/q$ for an advanced DC and $\Delta = -1/q$ for a retarded one.

The creation energy of a DC varies linearly with the tension σ, the slope being the excess length Δ of the DC. Inside the corresponding commensurate plateau, the creation energy of an advanced DC increases with σ and disappears at the left edge σ^-_{C-IC} of the plateau, while the creation energy of the retarded DC decreases with σ, disappears at the right edge σ^+_{C-IC} (see Figure 4). Inside the commensurate plateau, the creation of an elementary DC is energetically costly, but right after either edge (σ^-_{C-IC} or σ^+_{C-IC}) is crossed, it becomes energetically advantageous to create as many DCs (of the correct type) as possible, were it not for the repulsive interaction between them that prevents condensation of an infinite number of DCs.

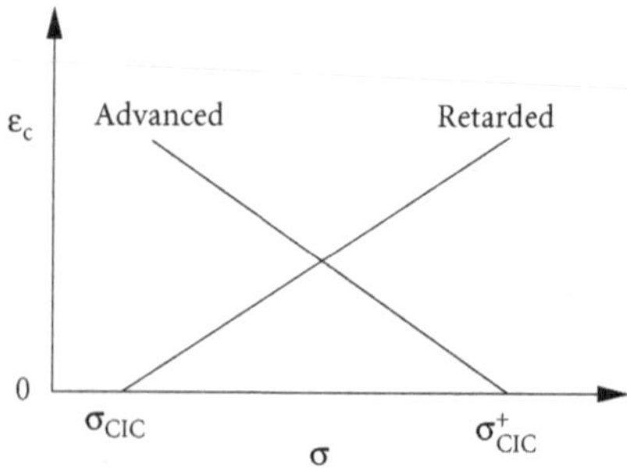

Figure 4. Creation energy ε_c of both types of elementary discommensurations for values of σ inside the tongue associated with a rational ω. At a certain critical value of σ this energy becomes negative and a commensurate–incommensurate transition occurs, changing the average spacing of the ground-state configuration.

The compromise between these two effects determines the density c of DCs, and, as each DC carries an excess length Δ, the average spacing of the new structure is $\omega = \omega_c + c\Delta$. For very small values of $|\sigma - \sigma_{C-IC}|$, where the interdistance (c^{-1}) is so large as to approximate a continuous variable, the scaling behavior of the ground-state average spacing ω in the commensurate-incommensurate transition is then

$$|\omega - \omega_c| \sim -(\ln|\sigma - \sigma_{C-IC}|)^{-1}. \quad (5)$$

For nonconvex interactions, the basis for an interpretation of the transitions between modulated phases in terms of interacting DCs is less rigorously supported. However, the central idea that the creation of localized defects or discommensurations drives these transitions and that the character (repulsive, attracting, or oscillatory) of their interaction determines the type of transition has been successfully applied, and undoubtedly it provides the most physical way of understanding why and how transitions between modulated phases occur. This fact adds to the many merits of the soliton paradigm in nonlinear science.

LUIS MARIO FLORÍA

See also **Aubry–Mather theory; Frenkel–Kontorova model; Ising model; Phase transitions**

Further Reading

Aubry, S. 1985. Structures incommensurables et brisure de la symmetrie de translation I [Incommensurate structures and the breaking of traslation simmetry I]. In *Structures et Instabilités*, edited by C. Godrèche, Les Ulis: Editions de Physique, pp. 73–194

Cummins, H.Z. 1990. Experimental studies of structurally incommensurate crystal phases. *Physics Reports*, 185: 211–409

Griffiths, R.B. 1990. Frenkel–Kontorova models of commensurate–incommensurate phase transitions. In *Fundamental Problems in Statistical Mechanics VII*, edited by H. van Beijeren, Amsterdam: North-Holland, pp. 69–110

Janssen, T. & Janner, A. 1987. Incommensurability in crystals. *Advances in Physics*, 36: 519–624

Yeomans, J. 1988. The theory and application of axial Ising models. *Solid State Physics*, 41: 151–200

COMPACT SET

See **Topology**

COMPACTON

See **Solitons, types of**

COMPARTMENTAL MODELS

Compartmental models, which model the flow of material between components of a system, have been applied to the study of nutrient or energy flow in an ecosystem, to analyses of tracer kinetics in an organism, and to the theory of epidemics, among other applications.

The construction of a compartmental model is straightforward and intuitive, but the estimation of the associated parameters is not. The system is divided into homogeneous compartments, and the flow of "material" (meaning nutrients, energy, money, goods, electrons, radioactive tracers, information, etc.) between them is traced, keeping account of the level of material in each compartment. The compartments are represented by boxes and the flows are represented by arrows (see Figure 1).

This representation is a directed graph in which the compartments are vertices and the flows are arcs. To analyze a system, however, the model must be quantized by studying the rate of change of the level x_i of the ith compartment in time as

$$x_i = x_i(t). \quad (1)$$

The flow between the ith and jth compartment, designated f_{ij}, is a rate, measured in terms of quantity of material per unit time.

A differential equation describing the behavior of x_i may be obtained by equating the time rate of change of x_i to the difference between the flows coming in and those going out of the ith compartment. Thus,

$$\frac{dx_i}{dt} = \sum_k f_{ki} - \sum_j f_{ij}.$$

$$\text{rate of change} = \text{flows in} - \text{flows out} \quad (2)$$

The f_{ij} may be constant or variable in time and are usually functionally dependent on the x_i's.

A widely used assumption (particularly in physiology and medicine) is that

$$f_{ij} = a_{ij}x_i, \quad (3)$$

which corresponds to a certain fraction a_{ij} of the level of material x_i in compartment i passing to compartment j per unit time. In the ecological literature, this is often

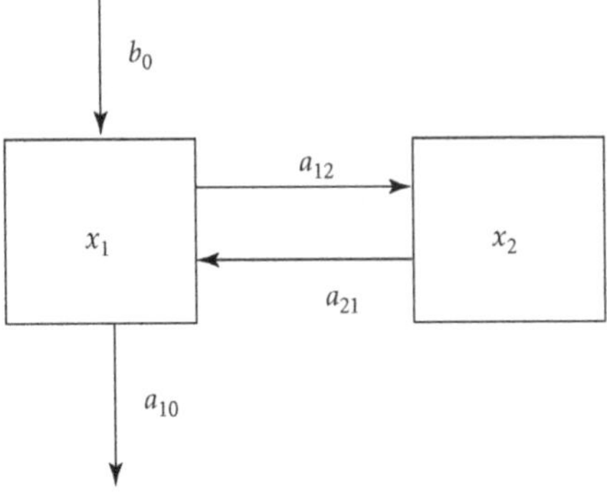

Figure 1. A compartmental model of tracer kinetics.

called donor-controlled flow. (The a_{ij} may be functions of time but are usually assumed to be independent of the x_is.)

Tracer Kinetics

Tracer kinetics refers to a technique for studying the flow of a drug or other material through a biological system (often a human body) near steady state. A small quantity of radioactive isotope or other recognizable form of the material (the tracer) is introduced, so that the tracer flows may be modeled by linear equations.

As a simple example, consider lipoprotein metabolism. One compartment corresponds to the blood plasma and the other corresponds to the extravascular space. A unit dose of tracer is injected into the blood and the flow follows the diagram in Figure 1. The quantity of interest is the flow out of compartment 1 (excretion), which cannot be measured directly, although the level of compartment 1 over time is known.

The differential equations are those obtained from (2) by taking the flow rates to be donor controlled; thus,

$$\begin{aligned} \mathrm{d}x_1/\mathrm{d}t &= -a_{10}x_1 - a_{12}x_1 + a_{21}x_2, \\ \mathrm{d}x_2/\mathrm{d}t &= a_{12}x_1 - a_{21}x_2, \end{aligned} \tag{4}$$

with initial conditions

$$x_1(0) = 1, \quad x_2(0) = 0. \tag{5}$$

These equations can be solved for $x_1(t)$ in terms of the three parameters a_{10}, a_{12}, a_{21}; since x_1 is known for all t, this result can be used to estimate these parameters by fitting a number of nonlinear equations.

General Differential Equations and Solutions

The differential equations for a system of n compartments can be written in vector form as

$$\mathrm{d}\boldsymbol{x}/\mathrm{d}t = A\boldsymbol{x} + \boldsymbol{f}, \tag{6}$$

where A is the matrix given by

$$A = \begin{bmatrix} a_{11} & a_{21} & \cdots & a_{n1} \\ a_{12} & a_{22} & \cdots & a_{n2} \\ \cdots & \cdots & \cdots & \cdots \\ a_{1n} & a_{2n} & \cdots & a_{nn} \end{bmatrix} \tag{7}$$

and $\mathbf{x} = [x_1, x_2, ..., x_n]^{\mathrm{T}}$ is the vector of levels of material in the n compartments and $\boldsymbol{f}$ is the input vector. The a_{ij} for $i \neq j$ are the flow rates, while the diagonal elements of A are

$$a_{ii} = -\sum_{j \pm i} a_{ij} - a_{i0}. \tag{8}$$

Although the input to a compartmental system is often to a single compartment, it may be distributed to input compartments. In general, we assume that the input to the system is $\boldsymbol{f} = B\boldsymbol{u}$, where $\boldsymbol{u}$ is a vector function distributed to the various compartments by B. Typically, B has one column but as many rows as there are compartments.

The outputs (measurements) of the system may be similarly combined. If the system is assumed donor to be controlled, the output will have the form

$$\boldsymbol{y} = C\boldsymbol{x}, \tag{9}$$

where the output of the compartments is directed to one or more measuring devices with measurements $\boldsymbol{y}(t)$. (Here, C is a general sampling matrix that is usually composed of 0s and 1s.)

The problem is then translated to the differential equation form

$$\begin{aligned} \frac{\mathrm{d}\mathbf{x}}{\mathrm{d}t} &= A\mathbf{x} + B\mathbf{u}, \\ \mathbf{y} &= C\mathbf{x}, \\ \mathbf{x}(0) &= \mathbf{0}. \end{aligned} \tag{10}$$

where we have assumed that the initial levels are $\mathbf{0}$. (This is no restriction as an initial value can be incorporated into $B\boldsymbol{u}$.) Solutions of the system are determined by matrices A, B, and C as

$$\mathbf{y}(t) = C\mathbf{x}(t) = C\int_0^t \mathrm{e}^{A(t-s)}B\mathbf{u}(s)\mathrm{d}s \tag{11}$$

where $\boldsymbol{x}(0) = 0$. Typically, however, the flow rates in the matrix A are unknown and must be determined from knowledge of $\mathbf{u}(t)$ and $\boldsymbol{y}(t)$, which is called the identification problem. A system is said to be identifiable if there is a unique solution to this problem, which is not always the case, and even if a solution does exist, it may be difficult to find.

Parameter Estimation

The problem of actually estimating the parameters of an identifiable problem is often an ill-posed problem, meaning that the estimated output may be close to the true output, while the parameters are still quite different from the true parameters. Moreover, such parameters are usually sensitive to slight perturbations of the data.

Suppose we are given a discrete set of observations/measurements made at times $t_i, i = 1, \ldots, m$ such that

$$y_i = y(t_i) + \varepsilon_i, \tag{12}$$

where ε_i are assumed to be independent normal random variables with mean 0 and constant variance. The standard approach to estimation is to solve for A in the least-squares sense

$$\min_{a_{ij}} \sum_i |y_i - y(t_i)|^2 \tag{13}$$

subject to the condition that A is a compartmental matrix of the appropriate model. (This is a nonlinear least-squares problem, since $y(t)$ is a linear combination of exponential functions.)

Another approach to estimation involves two stages; the data are first fitted to a sum of exponentials (Anderson, 1983)

$$g(t) = \sum_{k=1}^{n} c_k e^{\lambda_k t}. \tag{14}$$

In most nonpathological cases, the number of exponentials to be considered is the same as the size of the model. The estimates $\hat{c}_k$ and $\hat{\lambda}_k$ are obtained by least squares subject to $\lambda_k \leq 0$. Then, the Laplace transform of the fitted function $g(t)$ is compared with the Laplace transform of $y(t)$. As both are rational functions, the numerator and denominator are polynomials whose coefficients must be approximately equal.

There is also the method of peeling, which avoids least squares. This involves fitting an exponential function to the later data values. This corresponds to the eigenvalue of smallest magnitude. The difference between the data and this function is then used to approximate the second eigenvalue, and so on.

GILBERT G. WALTER

See also **Continuum approximations; Neural network models**

Further Reading

Anderson, D.H. 1983. *Compartmental Models and Tracer Kinetics,* Berlin and New York: Springer

Bellman, R.E. & Aström, K.J. 1970. On structural identifiability. *Mathematical Biosciences*, 7: 329–339

Dennis, J.E. & Schnabel, R. 1983. *Numerical Methods for Unconstrained Optimization and Nonlinear Equations*, Englewood Cliffs, NJ: Prentice-Hall

Godfrey, K. 1983. *Compartmental Models and Their Applications*, New York: Academic Press

Jacquez, S.A. 1996. *Compartmental Analysis in Biology and Medicine*, Ann Arbor: University of Michigan Press

Walter, E. 1987. *Identifiability of Parametric Models*, Oxford and New York: Pergamon

Walter, G.G. & Contreras, M. 1999. *Compartmental Modelling with Networks*, Boston: Birkhäuser

COMPLEX GINZBURG–LANDAU EQUATION

Ginzburg–Landau (GL) is a generic name for a class of basic models of nonlinear science, with a wide range of applications. They combine most fundamental features that underlie pattern formation in nonlinear media: dissipation (linear and nonlinear) and gain (provided by an intrinsic instability), conservative nonlinearity (accounting for wave mixing), and diffusion and/or dispersion (dependence of the dissipation rate and wave frequency on the wave number). Usually, the diffusion and dispersion are linear, but in more sophisticated models they may be both linear and nonlinear. In most cases, GL equations are local, but nonlocal (integral) generalizations are sometimes used. Both single GL equations and systems thereof play important roles as fundamental models.

The name of this class of equations first appeared in a seminal paper by Vitaly Ginzburg and Lev Landau (1950), which presented a phenomenological theory of superconductivity (V.L. Ginzburg was awarded the Nobel Prize in Physics in 2003 for this work.) In a standard form (Tinkham, 1996), these equations are

$$\eta\left(\partial_t + \mathrm{i}\kappa\phi\right)\psi = -\left(\mathrm{i}\kappa^{-1}\nabla + \boldsymbol{A}\right)^2\psi + \left(1 - |\psi|^2\right)\psi, \tag{1}$$

$$\boldsymbol{A}_t + \nabla\phi = -\nabla\times(\nabla\times\boldsymbol{A}) + (2\mathrm{i}\kappa)^{-1} \times\left(\psi^*\nabla\psi - \psi\nabla\psi^*\right) - |\psi|^2\boldsymbol{A} + \nabla\times\boldsymbol{H}, \tag{2}$$

where ψ is the wave function of superconducting electrons, also known as the order parameter of the superconductivity, $\boldsymbol{A}$ and $\boldsymbol{H}$ are the vector potential of the intrinsic magnetic field and an externally applied field, ϕ is the electrostatic potential (related to $\boldsymbol{A}$ by an arbitrary gauge condition, e.g., $\phi_t + \nabla\cdot\boldsymbol{A} = 0$), η and κ are the dissipative and so-called GL parameters, ∇ is the gradient, and subscripts stand for the partial derivatives. A solution with $\psi = 0$ represents a normal state with no superconductivity. In their original work, Ginzburg and Landau (1950) introduced a time-independent version of the equations, while the full time-dependent system was proposed later.

In the modern theory of nonlinear pattern formation (Cross & Hohenberg, 1993), the name GL is applied to equations resembling Equation (1) [while there is usually no counterpart of Equation (2)], which govern the evolution of a physically or phenomenologically defined order parameter in a dynamical medium. For instance, in the case of chemical-reaction waves, the order parameter determines local concentrations of reactants (Ipsen et al., 2000); in applications to fluid mechanics, the velocity field is expressed in terms of the order parameter, and in numerous applications to optics the order parameter is a local amplitude of the electromagnetic field (Arecchi et al., 1999).

The first generic example of the GL equation of this type is the complex cubic GL (CCGL) equation ("cubic" refers to the nonlinearity), which is closest in its form to Equation (1), but the name "complex" stresses that its coefficients are complex:

$$\psi_t = g\psi + (a + \mathrm{i}b)\nabla^2\psi - (d + \mathrm{i}c)|\psi|^2\psi. \tag{3}$$

The first term on the right-hand side of Equation (3) accounts for the gain, provided that $g > 0$ (its imaginary part can be trivially removed), a and b are real coefficients accounting for the diffusion and dispersion, while d and c are parameters controlling the nonlinear dissipation and frequency shift. Note that a and d must be positive, otherwise (3) is an ill-posed equation. In the case $|a| \ll |b|$, $|d| \ll |c|$, Equation (3) may be treated as a perturbed version of the nonlinear Schrödinger (NLS) equation.

In the case $b=c=0$, Equation (3) is called a real GL equation, to stress that all its coefficients are real, although ψ remains complex. The real GL equation may be represented in the gradient form, $\psi_t = -\delta L\{\psi,\psi^*\}/\delta\psi^*$, where $\delta/\delta\psi^*$ stands for the functional derivative, and $\int(-|\psi|^2+|\nabla\psi|^2+(1/2)|\psi|^4)\,dV$ is a real Lyapunov functional. A consequence of the gradient representation is that L may only decrease, $dL/dt \le 0$. This fact simplifies the dynamics of the real GL equation.

A fundamental feature of Equation (3) is that its zero solution, $\psi=0$, becomes unstable as the gain g changes its sign from negative to positive. In this case, a transition from the stable trivial solution to a nontrivial state is called *supercritical*. In particular, the supercritical transition in the one-dimensional (1-d) case yields a solitary-pulse (SP) solution that can be found in an exact analytical form,

$$u = A\,[\cosh(\kappa x)]^{-(1+i\mu)}\exp(-i\omega t), \qquad (4)$$

where A, κ, μ, and ω are uniquely determined by parameters of Equation (3). If the CCGL equation reduces to a perturbed NLS equation, the SP (4) can be obtained from the NLS soliton by means of the perturbation theory, provided that $bc<0$ (otherwise, the NLS equation does not have bright-soliton solutions). However, the SP solution (4) is always unstable because the zero background around it is unstable.

In many cases (for instance, in the case of thermal convection in a binary fluid), a nontrivial state may be excited by a finite-amplitude perturbation, while the trivial solution is stable against small perturbations. The simplest model that describes the corresponding *subcritical* transition to a nontrivial state is the cubic-quintic complex GL (CQCGL) equation, first proposed by Petviashvili and Sergeev (1984) as

$$\psi_t = -g\psi + (a+ib)\nabla^2\psi + (d-ic)|\psi|^2\psi - (f+ih)|\psi|^4\psi. \qquad (5)$$

Here, $g>0$ implies stability of the zero solution, the last term with $f>0$ guarantees overall stabilization of the system, the coefficient h accounts for a quintic nonlinear correction to the wave frequency, while $d>0$ provides for a possibility of *nonlinear* gain. The CQCGL equation may give rise to nontrivial states, coexisting with the stable zero solution, if the nonlinear gain coefficient exceeds a value $d_{\min}=2\sqrt{gf}$. An important result, obtained by means of analytical and numerical methods, is that the 1-d and 2-d versions of the CQCGL equation support SP solutions that may be *stable* (in the 2-d case, the localized pulse may carry vorticity, having a spiral structure). If all the parameters g, a, d, f, and h are small, the 1-d pulse can be constructed on the basis of the NLS soliton by means of the perturbation theory. However, the CQCGL equation does not make it possible to find stable SP solutions in an exact analytical form (one exact solution for a 1-d SP is known, but it is always unstable).

Patterns in nonlinear dissipative media may be supported not only by intrinsic gain; another possibility is to apply an external field, which is, for instance, the case for the pattern formation in laser cavities (Arecchi et al., 1999). In this case, the appropriate CCGL equation is

$$\psi_t = -g\psi + (a+ib)\nabla^2\psi - (d+ic)|\psi|^2\psi + P, \qquad (6)$$

where the driving term, induced by the external field, may be of two different types: direct drive, $P=\varepsilon$, or parametric drive, $P=-i\omega_0\psi+\varepsilon\psi^*$, where the asterisk stands for complex conjugation, ω_0 fixes the frequency, and ε is the drive's amplitude. Equation (6) with either type of drive can support stable SPs in 1-d and 2-d cases (in the case of the direct drive, SP settles on a nonzero background).

An important generalization is to consider systems of coupled GL equations. These may describe counter-propagating waves (for instance, in thermal convection in a binary-fluid layer), or second-harmonic generation in a lossy medium. In the latter case, the nonlinearity is not cubic, but *quadratic*, viz., $\psi_1^*\psi_2$ in the equation for the fundamental-frequency field ψ_1, and ψ_1^2 in the equation for the second-harmonic field ψ_2.

An alternative to the CQCGL equation is a system originating in nonlinear optics, in which the stability of the zero solution is provided by an extra linear equation,

$$\begin{aligned}\psi_t &= g\psi + (a+ib)\nabla^2\psi - (d+ic)|\psi|^2\psi - i\kappa\chi,\\ \chi_t &= -G\chi - i\kappa\psi,\end{aligned} \qquad (7)$$

where κ is a real coupling constant, and $G>0$ is the loss coefficient in the additional equation. The zero solution in this system is stable if the loss in the second equation and coupling are strong enough, $G>g$ and $\kappa^2>Gg$. System (7) has exact SP solutions, in which both fields ψ and χ take the form (4) (with different amplitudes), but, in contrast to the CCGL equation proper, in this case the pulse may be *stable* (Atai & Malomed, 1998).

Yet another type of a system occurs if, due to a specific nature of the underlying physical problem, the complex order parameter ψ is coupled to an extra real-order parameter ϕ, which accounts for the existence of a conserved quantity in the medium (Matthews & Cox, 2000): $\psi_t=\psi+\nabla^2\psi-|\psi|^2\psi-\phi\psi$, $\phi_t=\nabla^2[\sigma\phi+\mu(|\psi|^2)]$. In this simplest version of the system, $\sigma>0$ and μ are real constants.

A common feature of the various GL equations displayed above is their universality. Each is a generic representative of a class of models with given qualitative properties (for instance, super- or subcritical character of the excitation of nontrivial states, and the absence or presence of a conserved quantity). In more specific situations, there arise generic equations of other types. In particular, the complex Swift–Hohenberg (SH) equation describes a situation (for instance, in Rayleigh–Bénard convection) when the instability of the zero solutions appears at a finite wave number k_0 of small perturbations; thus,

$$\psi_t = g\psi - (a+ib)\left(\nabla^2+k_0^2\right)^2\psi - (d+ic)\,|\psi|^2\psi \quad (8)$$

with $g, a, d > 0$. Quasi-1-d solutions of Equation (8) can be looked for as $\psi\,(x, y, t) = \Psi(x, y, t)\,\exp\,(ik_0 x)$, where Ψ is a slowly varying function, whose x- and y-dependences are characterized by large scales $X, Y \gg k_0^{-1}$. An asymptotic consideration is then consistent in the case $X \sim Y^2$, reducing the SH equation (8) to an anisotropic complex Newell–Whitehead–Segel equation, $\Psi_t = g\Psi + (a+ib)\left(2i\partial_x + \partial_y^2\right)^2\Psi - (d+ic)\,|\Psi|^2\Psi$.

GL equations generate rich dynamics. The simplest exact solutions are plane waves (PWs). In the case of the CCGL equation (3) (here, it is set by rescaling $g = a = d \equiv 1$), a family of PWs is

$$\psi = A\,\exp\,(iQx - i\omega t),\ A = \sqrt{1 - Q^2},$$
$$\omega = c + (b - c)Q^2, \quad (9)$$

where Q is a wave number (a parameter of the solution family) taking values $-1 < Q < +1$. Note that the group velocity of this PW is $d\omega/dQ = 2(b-c)Q$.

The CQCGL equation gives rise to two different PWs for given Q, with amplitudes $A_\pm = (2f)^{-1} \times \left[1 \pm \sqrt{1 - 4f\,(1+Q^2)}\right]$ (under the same normalization $g = a = d \equiv 1$). This solution exists in the interval $Q^2 < (4f)^{-1}(1-4f)$, and only A_+ may correspond to a stable PW.

The stability of PWs is an important issue. In the CCGL equation, a condition for stability against long-wave perturbations with a large wavelength takes a simple form, $Q^2 < (1+bc)\,/\,\left(3+2c^2+bc\right)$, which is called a generalized Eckhaus stability criterion (the original Eckhaus criterion, which applies to the real GL equation with $b = c = 0$, is $Q^2 < 1/3$). Obviously, this criterion cannot hold unless $1 + bc > 0$ (the Benjamin–Feir–Newell (BFN) condition). The consideration of finite-wavelength perturbations gives rise to more complex stability conditions. Therefore, following Aranson and Kramer (2002), the structure of a full stability region for the PWs can be shown by means of its cross sections in the space (b, c, Q); see Figure 1. The figure makes a distinction between convective and absolute instabilities, when, respectively, the growing perturbation is traveling away or staying put. The transition from the 1-d CCGL equation to its multidimensional counterpart does not import extra instabilities to the PWs.

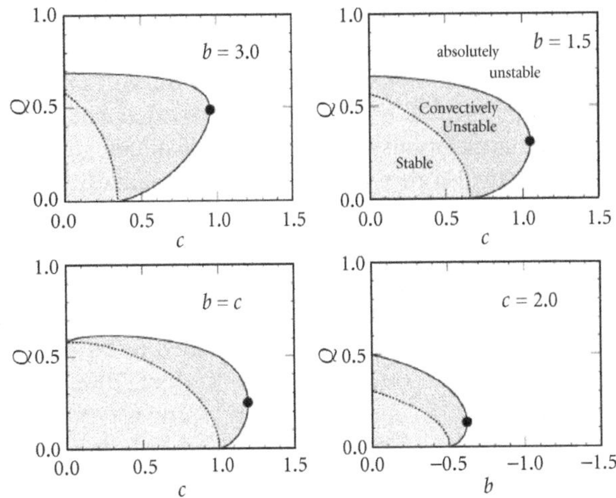

Figure 1. A set of cross sections of the stability region for the plane-wave solutions (9) of the cubic complex GL equation (3) in the space $(b, c, |Q|)$. The two top panels, the left bottom panel, and the right bottom panel show, respectively, the cross sections $b =$ const, $b = -c$, and $c =$ const. The filled circles mark turning points on the border between absolute and convective instabilities.

If the BFN combination $1 + bc$ becomes negative, the PW develops phase turbulence, which means that $|\psi|$ remains roughly constant, while the phase of the complex phase ψ demonstrates spatiotemporal chaos. In the 1-d case, close to the BFN instability threshold, the chaotic evolution of the phase gradient $p \equiv \phi_x$ obeys the Kuramoto–Sivashinsky equation, which, in a rescaled form, is $p_t + p_{xx} + p_{xxxx} + pp_x = 0$. Deeper into the instability region, phase-slip points (PSPs) arise, at which the amplitude $|\psi|$ disappears. Multiple creation of PSPs leads to a transition from the phase turbulence to defect turbulence, which is distinguished by random dynamics of the PSP ensemble. Mixed turbulence at the border between these two types also occurs (Aranson & Kramer, 2002).

In the case when PWs are stable, shock waves can be generated by collision between two PWs (9) with different wave numbers Q_1 and Q_2. Although exact solutions for shocks are not available, they can be obtained in an approximate form, provided that $Q_1 - Q_2$ is small, or the coefficients b and c are small. In particular, a transient layer between the two PWs moves at the velocity $v = (b-c)(Q_1 + Q_2)$, which is exactly the mean of the group velocities of the colliding PWs. In the 2-d case, PWs can collide obliquely; in this case, shocks take the form of a domain wall. Generally, the shocks are stable.

Besides the shocks (which are sources that emit PWs), the 1-d CCGL equation also gives rise to sinks, that is, localized hole-type structures that absorb PWs. If the CCGL equation (3) is a perturbed version of the NLS equation, sinks can be constructed as perturbed counterparts of NLS dark solitons, provided that $bc > 0$ (particular solutions for the sinks are available in an exact analytical form). A standing sink is actually a PSP, as $|\psi|$ vanishes at its center, and it may be dynamically stable. A moving sink is a finite dip in the profile of $|\psi|$; it is structurally unstable, as it is either decelerated (turning into a standing sink) or accelerated (eventually vanishing) by a quintic term added to the CCGL equation (Lega, 2001; Aranson & Kramer, 2002).

The 2-d CCGL equation displays spiral waves (SWs) in the form $\psi = A(r)\exp(\mathrm{i}N\theta - \mathrm{i}\omega t)$, where r and θ are the polar coordinates, N is an integer vorticity, and $A(r)$ is a complex function. An asymptotic form of the SW is $A(r) \approx \sqrt{1-Q^2}\exp(\mathrm{i}Qr)$ at $r \to \infty$, where Q is related to ω as in Equation (9), and $A(r) \sim r^N$ at $r \to 0$. (In the case of the real GL equation, the SW has $Q = 0$, which corresponds to a *vortex* solution. Similar solutions are generated by Equations (1) and (2), which represent *Abrikosov vortices* in superconductivity. For the prediction of these vortices, A.A. Abrikosov shared the Nobel Prize for Physics in 2003.) The asymptotic wave number Q is an eigenvalue of the 2-d CCGL equation, as it is uniquely selected by parameters of the equation. All the SWs with $N > 1$ are unstable. The SW with $N = 1$ may be subject to specific instabilities localized near its core (Aranson & Kramer, 2002).

An extension of the SW is a vortex line in three dimensions, which, in particular, may be closed into a ring. A vortex line with an additional wave number directed along its axis (twisted vortex) is also possible. The dynamics of 3-d vortex lines are quite complicated (Aranson & Kramer, 2002).

Boris Malomed

See also **Nonlinear Schrödinger equations; Partial differential equations, nonlinear; Pattern formation; Spatiotemporal chaos; Superconductivity**

Further Reading

Aranson, I.S. & Kramer, L. 2002. The world of the complex Ginzburg–Landau equation. *Reviews of Modern Physics*, 74: 99–143

Arecchi, F.T., Boccaletti, S. & Ramazza, P. 1999. Pattern formation and competition in nonlinear optics. *Physics Reports*, 318: 1–83

Atai, J. & Malomed, B.A. 1998. Exact stable pulses in asymmetric linearly coupled Ginzburg–Landau equations. *Physics Letters* A, 246: 412–422

Cross, M.C. & Hohenberg, P.C. 1993. Pattern-formation outside of equilibrium. *Reviews of Modern Physics*, 65: 851–1112

Ginzburg V.L. & Landau, L.D. 1950. On the theory of superconductivity. *Zhurnal Eksperimentalnoy i Teoreticheskoy Fiziki (USSR)*, 20: 1064–1082 (in Russian) [English translation: in *Men of Physics*, vol. 1. 1965. Oxford: Pergamon Press, pp. 138–167]

Ipsen, M., Kramer, L. & Sorensen, P.G. 2000. Amplitude equations for description of chemical reaction-diffusion systems. *Physics Reports*, 337: 193–235

Lega, J. 2001. Traveling hole solutions of the complex Ginzburg–Landau equation: a review. *Physica* D, 152: 269–287

Matthews, P.C. & Cox, S.M. 2000. Pattern formation with a conservation law. *Nonlinearity*, 13: 1293–1320

Petviashvili, V.I. & Sergeev, A.M. 1984. Spiral solitons in active media with an excitation threshold. *Doklady Akademii Nauk SSSR*, 276: 1380–1384 (in Russian)

Tinkham, M. 1996. *Introduction to Superconductivity*. New York: McGraw-Hill

COMPLEXITY, MEASURES OF

See **Algorithmic complexity**

CONDENSATES

See **Bose–Einstein condensation**

CONLEY INDEX

In a five-page paper published in the Proceedings of the 1970 International Congress of Mathematicians (Conley, 1971), Charles Conley gave the first definition of his index. For context, he chose the phase space to be a compact connected metric space X and F to be the space of flows on X with the compact open topology. (A flow is a continuous function such that $f(x,0) = x$, $f(x,t+s) = f(f(x,t),s)$ for all choices of x, t, s.)

For a compact subset Y of X define the set $\mathrm{Inv}(Y,f) = \{y \in Y : f(y,t) \in Y \text{ for all } t\}$. This set is the maximal invariant set contained in Y. An isolating neighborhood for a flow f is a compact subset N of X such that $\mathrm{Inv}(N,f)$ is contained in the interior of N. The set, $\mathrm{Inv}(N,f)$, is the isolated invariant set associated with N.

It is easy to show that isolating neighborhoods persist. If N is an isolating neighborhood for f, then there exists a neighborhood U of f in F such that N is an isolating neighborhood for every flow g in U. Suppose that M is an isolating neighborhood for g. If $\mathrm{Inv}(M,g) = \mathrm{Inv}(N,g)$ then the isolated invariant set $\mathrm{Inv}(M,g)$ is said to be a local continuation of the set $\mathrm{Inv}(N,f)$. Two invariant sets $\mathrm{Inv}(N,f)$ and $\mathrm{Inv}(M,g)$ are related by continuation if there is a finite sequence of local continuations linking one to the other.

For a flow f and an arbitrary compact subset W of X, one defines forward and backward exit time functions from W into the extended real line $[-\infty,\infty]$ as follows:

$$t^+(x) = \sup\{t \ge 0 : f(x,[0,t]) \subset W\},$$
$$t^-(x) = \inf\{t \le 0 : f(x,[t,0]) \subset W\}.$$

Certain subsets of W are associated with these functions. The forward asymptotic set is the set $A^+ = \{x : t^+(x) = \infty\}$, and the backward asymptotic set is the set $A^- = \{x : t^-(x) = -\infty\}$. Note that the maximal invariant set $\mathrm{Inv}(W, f)$ is the set $A^+ \cap A^-$. Forward and backward exit sets are the sets $W^{\pm} = \{x : t^{\pm}(x) = 0\}$.

An isolating block B for a flow f is a special type of isolating neighborhood with the following property. The boundary of B is the union of the exit sets B^+ and B^-. The intersection of these sets is the "tangency" set τ of boundary points that immediately exit in both time directions. Thus, a block has no internal tangencies where an orbit comes to the boundary from inside the block and does not exit.

If B is an isolating block for a flow f, one can show that the exit time functions are continuous. Using these functions and the flow, one may define deformation retractions of $B - A^+$ onto B^- and $B - A^-$ onto B^+. For example, a retraction r of $B - A^-$ onto B^+ is defined by the formula $r(x) = f(x, t^+(x))$ (This property was first used by Wazewski (1954).)

Suppose that f is a flow on R^3 and also that T is a solid torus that is an isolating block for the flow. Suppose that the exit set is a disk D on the boundary of T. Then, the invariant set $\mathrm{Inv}(T, f)$ must not be empty. If it were empty, one could define a deformation retraction of T onto D, which is impossible.

Under these definitions, the following theorem is fundamental (Conley & Easton, 1971). Given an isolating neighborhood N for a flow f, there exists an isolating block B contained in N such that $\mathrm{Inv}(N, f) = \mathrm{Inv}(B, f)$.

We use this theorem to define the Conley index of the set $\mathrm{Inv}(N, f)$ to be the homotopy types of the pair of quotient spaces $[B/B^+]$ and $[B/B^-]$. These spaces are obtained by collapsing the exit sets to points and using the quotient topology on the resulting spaces.

Consider, for example, the flow f defined on R^2 by $f((x, y), t) = (\mathrm{e}^{-t}x, \mathrm{e}^{t}y)$. This flow has the origin as a saddle point. Let B be a square centered at the origin. Then B is an isolating block for the flow. The exit set consists of the top and bottom sides of B. The quotient space $[B/B^-]$ has the homotopy type of the pair of spaces consisting of a circle and a point on this circle.

The Conley Index has the following properties (Conley, 1978; Herman et al., 1988; Smoller, 1983; Easton, 1998; Hofer & Zehnder, 1995; Mischaikow, 2002):

(i) The Conley index is well defined. Thus it is independent of the choice of block B.
(ii) If $\mathrm{Inv}(N, f)$ and $\mathrm{Inv}(M, g)$ are two isolated invariant sets that are related by continuation, then they have the same Conley index.
(iii) The index $[B/B^-]$ of a saddle point for a smooth flow on a manifold is a sphere together with a point on the sphere. The sphere has the same dimension as that of the unstable manifold of the saddle point. Thus, the Conley index is a generalization of the Morse index of a saddle point.
(iv) The index of two disjoint isolated invariant sets is the "sum" or "join" of their indices.

Traveling Waves

One of the early applications of the Conley index was to find traveling waves for reaction-diffusion equations (Smoller, 1983). We will use as an example the FitzHugh–Nagumo (FN) equations $u_t = \varepsilon v$, $v_t = v_{xx} + f(v) - u$, which are a simplification of the Hodgkin–Huxley equations used to model nerve impulses. The parameter ε is assumed to be small, and one seeks a solution of the form $U(x, t) = u(s)$, $V(x, t) = v(s)$, where $s = x + \theta t$ and θ is the wave velocity.

Substituting the trial solutions into the FN equations, one obtains the following system of ordinary differential equations: $\mathrm{d}u/\mathrm{d}s = (\varepsilon/\theta)v$, $\theta\, \mathrm{d}v/\mathrm{d}s = \mathrm{d}^2v/\mathrm{d}s^2 + f(v) - u$, where $f(v)$ is assumed have the general shape of a cubic equation that is decreasing, increasing, and then decreasing. To be specific, we take $f(v) = -v(v-1)(v-2)$. The corresponding first-order system of ordinary differential equations is $u' = \sigma v$, $v' = w$, $w' = \theta w + u - f(v)$, where $\sigma = \varepsilon/\theta$. Our goal is to find a periodic solution of this system for small values of the parameters sigma and theta, and thereby to find a periodic solution of the FN equations.

One can completely understand the phase portrait of system when the parameters σ and θ are set to zero. In this case, $u(t)$ is constant and the equations for v, w form a Hamiltonian system with Hamiltonian $H(v, w) = w^2/2 + F(v) - uv$ with $F(v) = \int_0^u f(r)\mathrm{d}r = -v^2(v-2)^2/4$. The phase portrait in the $u = 0$ plane has saddle points at $(v, w) = (0, 0)$, $(v, w) = (2, 0)$ and a center at $(v, w) = (1, 0)$. The saddle points are connected by heteroclinic orbits implicitly defined by the equation $H(v, w) = 0, 0 < v < 2$. Note that the set of equilibrium points for the system is the set $\{(u, v, w) : u = f(v), w = 0\}$.

Next consider the system with $\sigma = 0, \theta > 0$. For this system, we have $(d/dt)H(u(t), v(t), w(t)) = \theta w^2(t)$. Let $a(u) < b(u) < c(u)$ denote the three solutions of the cubic equation $u - f(u) = 0$. One can show that there are values $0 < u_1 < u_2$ such that for $j = 1, 2$, there is a heteroclinic solution joining the equilibrium points $(u_j, a(u_j), 0)$ and $(u_j, c(u_j), 0)$ in the plane $u = u_j$. We now have a cycle consisting of the two heteroclinic orbits together with arcs of equilibrium points $\{(u, a(u), 0) : u_1 < u < u_2\}$ and $\{(u, c(u), 0) : u_1 < u < u_2\}$. This cycle is an invariant set for the system. However, the cycle is not isolated since it intersects the set of equilibrium points in two arcs. The two arcs of equilibrium points may be viewed as normally hyperbolic invariant manifolds, whose stable and unstable manifolds intersect transversally along the two heteroclinic orbits.

Finally, consider the system for small positive values of sigma. In this case, u is increasing when u is positive and decreasing when negative. The hard part is to construct an isolating block, which topologically is a solid torus containing the cycle. The transversal intersection noted above is essential to this construction. The cycle is no longer invariant when sigma is positive. However, one shows that the isolating block must contain a periodic solution of the full system of equations. The periodic solution thus constructed may be viewed as a periodic traveling wave solution of the FN equations.

Applications to Discrete Dynamical Systems

Consider the discrete dynamics generated by iterating a homeomorphism f of a compact metric space X. It is natural to study orbits with "errors" such as truncation or round-off errors in numerical algorithms. Thus, an ε-chain for f is a finite sequence $(y_0, y_1, y_2, \ldots)$ such that $d(f(y_n), y_{n+1}) \leq \varepsilon$ where $d(x, y)$ is the distance function on X. Conley (1978) and Bowen (1975) both understood the importance of studying orbits with errors. Bowen asked when such an orbit could be shadowed by a true orbit of the system.

Conley defined the ε-chain recurrent set $\mathrm{CR}(f, \varepsilon)$ to be the set of points that are contained in periodic ε-chains of length at least 2. The chain recurrent set is the set $\mathrm{CR}(f) = \cap \{\mathrm{CR}(f, \varepsilon){:}\varepsilon > 0\}$. Points in $\mathrm{CR}(f)$ are chain equivalent if for any positive epsilon there is a periodic ε-chain containing both points. He showed that every orbit uniformly approaches a unique chain equivalence class in $\mathrm{CR}(f)$. This result is known as the Conley decomposition theorem, and it generalizes Smale's decomposition of an Axiom A system into basic sets (Bowen, 1975).

An isolating block for f is a compact subset N of X such that whenever x, $f(x)$, $f^2(x) \in N$, then $f(x)$ is contained in the interior of N. This is something like having no internal tangencies. The exit set of N is the set $E = \{x \in N : f(x) \ni \mathrm{int}(N)\}$. Because N is a block, the exit set is compact. Define an equivalence relation on N making all points of E equivalent and all points not in E equivalent only to themselves. Let $N^{\#}$ denote the space of equivalence classes with the quotient topology obtained by projecting N onto $N^{\#}$ by sending a point to the equivalence class to which it belongs. Let $E^{\#}$ denote the image of E. The index space of N is the pair $(N^{\#}, E^{\#})$. Define an index map $f^{\#} : N^{\#} \to N^{\#}$ by $f^{\#}(x) = E^{\#}$ if $x = E^{\#}$ or $f(x) \in E^{\#}$. Otherwise, define $f^{\#}(x) = f(x)$. Note that the index map is continuous. The pair $(N^{\#}, f^{\#})$ plays the role of the Conley index in this context. If the index map is not homotopic to a constant, then one can prove that the set $\mathrm{Inv}(N, f)$ is non-empty.

Smale's horseshoe map in the plane may be used as an example. Suppose that a rectangle B is mapped across itself so that the image crosses the rectangle in a horizontal strip, then curves back and crosses the rectangle again in another horizontal strip above the first. In this case, the exit set consists of three vertical strips, one in the center of the rectangle and the other two on the left and right sides containing the vertical edges of B. The index space $[B/B^-]$ has the homotopy type of a figure of eight and the index map is non-trivial.

Sequences of compact sets (called "windows") may sometimes be constructed to contain an orbit with errors. If each window in the sequence is "correctly" mapped across the next one, then a true orbit runs through the sequence of windows and shadows the orbit with errors (Easton, 1998).

Robert W. Easton

See also **Anosov and Axiom-A systems; FitzHugh–Nagumo equation; Horseshoes and hyperbolicity in dynamical systems**

Further Reading

Bowen, R. 1975. *Equilibrium States and the Ergodic Theory of Anosov Diffeomorphisms*, New York: Springer

Conley, C.C. 1971. On the continuation of invariant sets of a flow. In *Proceedings of the International Congress of Mathematicians 1970*, Paris: Gauthiers-Villars, pp. 909–913

Conley, C.C. & Easton, R.W. 1971. Isolated invariant sets and isolating blocks. *Transactions of the American Mathematical Society* 158: 35–61

Conley, C.C. 1978. *Isolated Invariant Sets and the Morse Index*, Providence, RI: American Mathematical Society.

Easton, R. 1998. *Geometric Methods for Discrete Dynamical Systems*, Oxford and New York: Oxford University Press.

Herman, M., McGehee, R., Moser, J. & Zehnder, E. (editors). 1988. Charles Conley Memorial Volume, Special Issue of *Ergodic Theory and Dynamical Systems*, 8

Hofer, H. & Zehnder, E. 1995. Symplectic invariants and Hamiltonian dynamics. *The Floer Memorial Volume Progress in Mathematics*, vol. 133, Basel: Birkhäuser.

Mischaikow, K. 2002. Topological techniques for efficient rigorous computations in dynamics, *Acta Numerica*,vol. 11, 435–477

Smoller, J. 1983. *Shock Waves and Reaction-diffusion Equations*, New York: Springer.

Wazewski, T. 1954. Sur un principe topologique de l'examen de l'allure asymptotique des integrales des equations differentielles, *Proceedings of the International Congress of Mathematicians*, vol. 3, 132–139

CONSTANTS OF MOTION AND CONSERVATION LAWS

Although nonlinear spatiotemporal processes may be very complicated, they frequently obey simple constraints in the form of conservation laws. It is sometimes possible to construct one or several constants of motion (also called dynamical invariants, DIs), in the form of spatial integrals of local densities expressed in terms of the physical fields and their derivatives, which are conserved in time, as a consequence of the underlying dynamics. Such commonly known conserved quantities as energy, momentum, and angular momentum belong to this class.

Typically, the existence of conservation laws can be established if the underlying dynamics is dissipation-free; however, a specific DI may sometimes also exist in dissipative systems. Examples of the latter are provided by the diffusion equation, $u_t = u_{xx}$, and its important nonlinear counterparts in the form of the Burgers equation,

$$u_t = u_{xx} + uu_x, \tag{1}$$

and Cahn–Hilliard equation,

$$u_t + \left(u - u^3 + u_{xx}\right)_{xx} = 0 \tag{2}$$

(the subscripts stand for the partial derivatives). They all conserve $\int_{-\infty}^{+\infty} u(x,t)\mathrm{d}x$, which is simply the total mass of the substance in the case of diffusion.

A more sophisticated example of the "dissipative conservation" occurs in physically important models based on the nonlinear Schrödinger (NLS) equation with special additional terms:

$$\mathrm{i}u_t + u_{xx} + F'(|u|^2)u = \varepsilon Q, \tag{3}$$

where the function F' describes conservative nonlinearity of the medium (the prime stands for the derivative with respect to the argument of F; in particular, $F = \left(|u|^2\right)^2$ corresponds to the most generic case of the cubic NLS equation), ε is a real parameter, and the "special perturbation" is, for instance, the nonlinear Landau-damping term in the NLS equation for Langmuir waves in plasmas,

$$Q = -u \int_{-\infty}^{+\infty} \mathrm{d}x' \left(x - x'\right)^{-1} |u(x')|^2, \tag{4}$$

or the stimulated Raman scattering term in the equation for electromagnetic waves in nonlinear optical fibers, $Q = \left(|u|^2\right)_x u$. While these terms are dissipative ones, the corresponding perturbed NLS equation conserves the single DI, namely, the total *wave action* (alias "number of quanta"),

$$N = \int_{-\infty}^{+\infty} |u(x,t)|^2 \mathrm{d}x. \tag{5}$$

In the general case, equations that govern the dissipation-free spatiotemporal dynamics can be derived from the underlying action functional $S\left\{u^{(n)}\right\}$: $\delta S/\delta u^{(n)} = 0$, where $u^{(n)}(\boldsymbol{r},t)$ is the nth field variable, $\boldsymbol{r}$ is the set of the spatial coordinates, $\delta/\delta u^{(n)}$ stands for the variational (functional) derivative, and the action is expressed in terms of the Lagrangian density $\mathcal{L}$, so that

$$S = \int\int \mathcal{L}\left(u^{(n)}, \nabla u^{(n)}, u_t^{(n)}\right) \mathrm{d}\boldsymbol{r}\mathrm{d}t. \tag{6}$$

For instance, the density

$$\mathcal{L} = (1/2)\left[u_t^2 - (\nabla u)^2\right] - F(u) \tag{7}$$

yields a nonlinear Klein–Gordon (NKG) equation for a single real field u,

$$u_{tt} - \nabla^2 u + F'(u) = 0. \tag{8}$$

The fundamental nature of DIs in Lagrangian systems is established by a theorem that was published by Emmy Noether in 1918: any *continuous symmetry* of the system, that is a family of transformations of the field variables, spatial coordinates, and time, which depend on an arbitrary continuous parameter ξ and leave the action invariant, generates a constant of motion. If the infinitesimal symmetry transformation is written in the form

$$u^{(n)} \to u^{(n)} + U_n\,\mathrm{d}\xi,\ \boldsymbol{r} \to \boldsymbol{r} + \boldsymbol{R}\,\mathrm{d}\xi,$$
$$t \to t + T\,\mathrm{d}\xi, \tag{9}$$

then the main result following from the Noether theorem is the continuity equation $\mathcal{I}_t + \nabla \cdot \boldsymbol{J} = 0$, with the following density and current:

$$\mathcal{I} = \sum_n \frac{\partial \mathcal{L}}{\partial\left(u_t^{(n)}\right)}\left(\boldsymbol{R}\cdot\nabla u^{(n)} + T u_t^{(n)} - U_n\right)$$
$$+ T\mathcal{L}, \tag{10}$$

$$\boldsymbol{J} = \sum_n \frac{\partial \mathcal{L}}{\partial\left(\nabla u^{(n)}\right)}\left(\boldsymbol{R}\cdot\nabla u^{(n)} + T u_t^{(n)} - U_n\right)$$
$$+ \boldsymbol{R}\mathcal{L}, \tag{11}$$

($\partial/\partial\left(\nabla u^{(n)}\right)$ is realized as a vector with the components $\partial/\partial\left(u_x^{(n)}\right)$, $\partial/\partial\left(u_y^{(n)}\right)$, $\partial/\partial\left(u_z^{(n)}\right)$). Then, assuming, as usual, that the fields disappear at $|\boldsymbol{r}|\to\infty$, the continuity equation immediately yields the conservation law in the form of $\mathrm{d}I/\mathrm{d}t = 0$, with $I \equiv \int \mathcal{I}\,\mathrm{d}\boldsymbol{r}$. A detailed derivation of this fundamental result can be found in the book by Bogoliubov & Shirkov (1973); for discussion of the Noether theorem in various contexts, see also Sulem & Sulem (1999), and Whitham (1974) If the underlying equations are complex, the Lagrangian density and all the DIs are nevertheless real.

The obvious invariance of the action against arbitrary temporal and spatial shifts, which are described by Equation (9) with $U_n = 0$ and, respectively, $\boldsymbol{R} = 0$, $T = 1$, or $\boldsymbol{R}_j = \boldsymbol{e}_j$, $T = 0$ ($\boldsymbol{e}_j$ is the unit vector corresponding to the jth spatial coordinate) gives rise to the conservation of the energy (Hamiltonian) H and momentum $\boldsymbol{P}$. For the important classes of the NKG and multidimensional NLS, Equations (8) and (3), Equation (10) yields

$$H_{\rm NKG} = \int \left[\frac{1}{2} \left(u_t^2 + (\nabla u)^2 \right) + F(u) \right] d\boldsymbol{r},$$

$$\boldsymbol{P}_{\rm NKG} = - \int u_t \nabla u \; d\boldsymbol{r}, \tag{12}$$

$$H_{\rm NLS} = \int \left[|\nabla u|^2 - F(|u|^2) \right] d\boldsymbol{r},$$

$$\boldsymbol{P}_{\rm NLS} = {\rm i} \int \left(u \nabla u^* - u^* \nabla u \right) d\boldsymbol{r}, \tag{13}$$

where * stands for the complex conjugation (the transition from the Lagrangian to Hamiltonian density as per Equation (10) in the case of the temporal-shift invariance is called the Legendre transformation). The invariance against rotations in the three-dimensional space leads to the conservation of the angular momentum,

$$\boldsymbol{M} = \int (\boldsymbol{r} \times \mathcal{P}) \; d\boldsymbol{r}, \tag{14}$$

where $\mathcal{P}$ is the density in the expressions for the momentum in Equations (12)–(14); in the two-dimensional case, there is only one component of the conserved angular momentum. Additionally, in the NLS-type equations, the invariance against the phase shift (alias *gauge invariance*), $u \to u \exp({\rm i}\xi)$ with an arbitrary constant ξ, generates the conservation of the above-mentioned wave action (5), which is $\int |u|^2 \, d\boldsymbol{r}$ in the multidimensional case.

Another important class of models in one dimension is based on equations of the Korteweg–de Vries (KdV) type for a real function $u(x,t)$,

$$u_t + u_{xxx} + F''(u) u_x = 0 \tag{15}$$

(the most important cases of the KdV equation proper and modified KdV equation correspond to $F = u^3$ and $F = u^4$, respectively). The Lagrangian representation of Equation (15) is possible in terms of the potential field v, defined so that $v_x \equiv u$, but the Hamiltonian and momentum are expressed solely in terms of the original field u,

$$H_{\rm KdV} = \int_{-\infty}^{+\infty} \left[\frac{1}{2} u_x^2 - F(u) \right] dx,$$

$$P_{\rm KdV} = \int_{-\infty}^{+\infty} u^2 \, dx. \tag{16}$$

The invariance of the action, written in terms of the potential v, against the arbitrary shift $v \to v + \xi$ additionally generates the conservation of the "mass," $\int_{-\infty}^{+\infty} u \, dx$.

Besides being a dynamical invariant, the Hamiltonian gives rise to a canonical representation of the equation(s) in the Hamiltonian form, which is dual to the Lagrangian representation. In particular, for the complex and real equations of the NLS and KdV types, respectively, this representation is

$$u_t = -{\rm i} \frac{\delta H}{\delta u^*}, \quad u_t = \frac{\partial}{\partial x} \frac{\delta H}{\delta u}. \tag{17}$$

The conservation of H itself and the conservation of the mass in the KdV-type equations are immediate consequences of the general form of Equations (17). The conservation of the wave action in the NLS-type equation is also a consequence of its representation in the form of Equations (17).

If a multicomponent Lagrangian system possesses an additional ("isotopic") symmetry against linear transformations of the components, this also gives rise to a specific DI. An important example is a system of coupled NLS equations of Manakov's type (Manakov, 1973)

$${\rm i} \begin{pmatrix} u_t \\ v_t \end{pmatrix} + \nabla^2 \begin{pmatrix} u \\ v \end{pmatrix} + F' \left(|u|^2 + |v|^2 \right) \begin{pmatrix} u \\ v \end{pmatrix} = 0, \tag{18}$$

which are invariant against rotation in the plane of (u, v). In this case, Equation (10) gives rise to the DI ("isotopic spin") in the form

$$S = {\rm i} \int \left(u v^* - u^* v \right) d\boldsymbol{r}. \tag{19}$$

A very special situation arises for the DIs in the case of integrable equations, that is, those that are amenable to the application of the inverse scattering transform (IST) (Ablowitz & Segur, 1981; Newell, 1984; Zakharov et al., 1984). Integrable equations have an infinite set of hidden dynamical symmetries, which, unlike the above-mentioned elementary invariances against temporal and spatial shifts, spatial rotations, phase shift, etc., do not have a straightforward meaning. In compliance with the Noether theorem, each hidden symmetry generates the corresponding DI, which is an integral expression with a density that, unlike those corresponding to the elementary DIs (see Equations (12), (4), and (16), involves higher-order derivatives. For instance, in the integrable KdV equation (15) with $F = u^3$, the first higher-order DI is

$$I = \int_{-\infty}^{+\infty} \left(u_{xx}^2 + 5u^2 u_{xx} + 5u^4 \right) dx. \tag{20}$$

In fact, it was an empirical discovery of several higher-order DIs in the KdV equation that was a major incentive for the study that had resulted in the discovery of the IST technique.

The IST provides a systematic method to derive the infinite set of the DIs in terms of the corresponding scattering data, into which the original wave field is mapped to make the hidden integrability explicit (Ablowitz & Segur, 1981; Newell, 1984; Zakharov et al., 1984). The use of the scattering data makes it possible to explicitly introduce a full system of the action-angle variables for the integrable equations, and demonstrate that the infinite set of the action variables is in one-to-one correspondence with the set of the DIs. It is also possible to prove; that all the DIs are *in involution* among themselves; that is, the Poisson bracket between any two DIs, defined as per the corresponding symplectic (Hamiltonian) structure, is zero. Thus, integrable equations are direct counterparts, for the case of the infinite number of degrees of freedom, of finite-dimensional Hamiltonian systems that are Liouville-integrable; that is, with a set of DIs that are in involution, their number being equal to the number of the degrees of freedom.

The presence of the infinite set of the DIs in the integrable equations helps to understand such a well-known property as the completely elastic character of collisions between solitons (Ablowitz & Segur, 1981; Newell, 1984; Zakharov et al., 1984): roughly speaking, the necessity to satisfy the infinite set of the conservation laws leaves no room for changes of the solitons, except for phase shifts. On the other hand, some equations amenable to the application of the IST technique, such as, for instance, the standard three-wave system,

$$\left(u_{1,2}\right)_t + c_{1,2}\left(u_{1,2}\right)_x = \mathrm{i}u_{2,1}^* u_3, \quad (u_3)_t = \mathrm{i}u_1 u_2, \tag{21}$$

where $u_{1,2}$ and u_3 are, respectively, the "daughter" and pump waves, and $c_{1,2}$ are group velocities, feature nontrivial "soliton reactions"—for instance, a spontaneous split of a pump-wave soliton into separating daughter ones. This possibility is explained by the fact that the above-mentioned one-to-one correspondence between the infinite sets of the degrees of freedom and DIs does not really hold for these equations: the set of the DIs is not "infinite enough" (Fokas & Zakharov, 1992). Such equations are sometimes called "solvable," to stress their difference from the genuinely integrable ones (integrable in the sense of Liouville, as generalized to systems with infinitely many degrees of freedom).

Integrable lattice (discrete) models feature another important property: due to the lack of the continuous translational invariance, lattice systems lack the momentum conservation. Nevertheless, integrable lattice models do possess a conserved momentum, due to their hidden symmetry. For example, the Ablowitz–Ladik equation,

$$2\mathrm{i}\frac{\mathrm{d}u_n}{\mathrm{d}t}+(u_{n+1}+u_{n-1}-2u_n)+|u_n|^2(u_{n+1}+u_{n-1})=0, \tag{22}$$

which is an integrable discretization of the cubic NLS equation, conserves the real momentum in the form of

$$P=\sum_{n=-\infty}^{+\infty}(\psi_n\psi_{n+1}^*-\psi_n^*\psi_{n+1}). \tag{23}$$

In fact, conservation of the momentum is a specific integrability feature of discrete models, in contrast to continuum ones.

Elementary DIs find specific applications in systems perturbed by small nonconservative terms to order ε. In that case, the conservation laws no longer hold; however, using evolution (balance) equations for the former DI(s) is a convenient way to derive effective equations of motion for solitons (or other collective nonlinear excitations) in the weakly perturbed model. For instance, in the cubic NLS equation (3) with the above-mentioned terms (Kerr and stimulated Raman scattering ones), $F=\left(|u|^2\right)^2$ and $Q=\left(|u|^2\right)_x u$, an exact soliton solution with arbitrary amplitude η and velocity c, in the case of $\varepsilon=0$, is

$$\begin{aligned} u_{\text{sol}} &= \eta\,\operatorname{sech}\left(\eta\,(x-ct)\right) \\ &\quad\times\exp\left[\mathrm{i}(c/2)x+\mathrm{i}\left(\eta^2-c^2\right)t\right]. \end{aligned} \tag{24}$$

In the presence of small $\varepsilon>0$, the wave action (4) remains a DI (see above), and the balance equation for the formerly conserved momentum P_{NLS}, see Equation (13), is

$$\frac{\mathrm{d}P}{\mathrm{d}t}=2\varepsilon\int_{-\infty}^{+\infty}\left[\left(|u|^2\right)_x\right]^2\mathrm{d}x. \tag{25}$$

Substitution of the unperturbed soliton (24) into Equation (25), and into the conservation of the wave action, yields evolution equations for the amplitude and velocity:

$$\frac{\mathrm{d}\eta}{\mathrm{d}t}=0,\quad \frac{\mathrm{d}c}{\mathrm{d}t}=\frac{16}{15}\varepsilon\eta^4. \tag{26}$$

For further details and references, see the review by Kivshar & Malomed (1989).

BORIS MALOMED

See also **Hamiltonian systems; Integrability; Integrable lattices; Inverse scattering method or transform; Korteweg–de Vries equation; N-wave interactions; Symmetry groups**

Further Reading

Ablowitz, M. & Segur, H. 1981. *Solitons and the Inverse Scattering Transform*, Philadelphia: SIAM

Bogoliubov, N.N. & Shirkov, D.V. 1973. *Introduction to the Theory of Quantized Fields*, Moscow: Nauka (in Russian); English translation, 2nd edition: New York: Wiley, 1980

Fokas, A.S. & Zakharov, V.E. 1992. The dressing method and nonlocal Riemann-Hilbert problems. *Journal of Nonlinear Science*, 2: 109–134

Kivshar, Y.S. & Malomed, B.A. 1989. Dynamics of solitons in nearly integrable systems. *Reviews of Modern Physics*, 61: 763–915

Manakov, S.V. 1973. Theory of two-dimensional stationary self-focusing of electromagnetic waves. *Zhurnal Eksperimentalnoy i Teoreticheskoy Fiziki*, 65: 505–516 (in Russian); translated in *Soviet Physics—Journal of Experimental and Theoretical Physics*, 38: 248 (1974)

Newell, A.C. 1984. *Solitons in Mathematics and Physics*, Philadelphia: SIAM

Sulem, C. & Sulem, P.-L. 1999. *The Nonlinear Schrödinger Equation*, New York: Springer

Whitham, G.B. 1974. *Linear and Nonlinear Waves*, New York: Wiley

Zakharov, V.E., Manakov, S.P., Novikov, S.P. & Pitaevskii, L.P. 1984. *Theory of Solitons*, New York: Consultants Bureau

CONTINUITY EQUATION

See **Constants of motion and conservation laws**

CONTINUOUS SPECTRUM

See **Inverse scattering method or transform**

CONTINUUM APPROXIMATIONS

In general, a physical system belongs to one of three broad classes: (i) media with distributed parameters (electromagnetic fields, fluids, or liquid), (ii) discrete media (crystal lattices, polymers, or macromolecules), and (iii) artificial periodic systems (layered structures, lattices of nano-dots, or Josephson arrays). In the first class, the dynamics of a system is described by partial differential equations (PDE) for the field variable $u(x,t)$ and in the other two cases by discrete differential equations (DDE) for the field variable at the lattice sites $u(n,t)=u_n(t)$. For simplicity, only one-dimensional (1-d) models are discussed here.

Well-known examples of discrete nonlinear dynamical systems include the following:

- The discrete 1-dimensional elastic chain with a nonlinear interaction between the nearest neighbors (generalized Fermi–Pasta–Ulam (FPU) model), whose equation of motion reads

$$m\frac{\mathrm{d}^2u_n}{\mathrm{d}t^2}=\varphi'(u_{n+1}-u_n)-\varphi'(u_n-u_{n-1}), \quad (1)$$

  where u_n is the displacement of the nth atom in a chain; a prime indicates the derivative with respect to the argument, and $\varphi(u_n-u_{n-1})$ is the energy of the interatomic interaction (Fermi et al., 1955). The particular choice $\varphi(\xi)=A\xi^2/2+\alpha\xi^3/3+\beta\xi^4/4$ and $\varphi(\xi)=c\exp(-p\xi)+q\xi$ represents α-FPU (for $\beta=0$), β-FPU (for $\alpha=0$), and Toda models.

- The discrete 1-d chain with linear interatomic interaction exposed to a nonlinear external field (discrete Frenkel–Kontorova (FK) model) with the following equation of motion:

$$m\frac{\mathrm{d}^2u_n}{\mathrm{d}t^2}=A(u_{n+1}+u_{n-1}-2u_n)-w'(u_n), \quad (2)$$

  where $w(u_n)$ is the nonlinear on-site external potential. The particular choice $w=U(1-\cos(2\pi u_n/a))$, where a is the interatomic distance, corresponds to the traditional FK-model (Frenkel & Kontorova, 1939).

- 1-d photonic crystals (periodic arrays of optical waveguides) or the discrete spin lattice, which may be described in the context of the discrete nonlinear Schrödinger equation (DNLS)

$$\mathrm{i}\frac{\mathrm{d}\psi_n}{\mathrm{d}t}+B(\psi_{n+1}+\psi_{n-1}-2\psi_n) +F(|\psi_n|^2)\psi_n=0, \quad (3)$$

  where in the simplest case F is a linear function of the argument. ψ_n denotes the value of the effective field at the nth element of the discrete system, which can be assigned different physical meanings for various applications.

Even in the 1-d case, the solution of the nonlinear DDE poses a fairly complicated mathematical problem and only a few of them can be solved exactly (the Toda and Ablowitz–Ladik equations). Thus, it is often easier to study discrete problems in the "continuum approximation" (CA) within the framework of PDEs. Clearly, some information about nonlinear dynamics of discrete systems is lost, and some phenomena cannot be described in this continuum limit; but in the long wave limit, this approach provides a good qualitative and even quantitative agreement with the results for a discrete system investigation.

In the CA, the discrete number of the atom site n is replaced with the continuous coordinate $x: na\to x$, with a being the interatomic equilibrium distance or the period of mesoscopic periodical structure, and $u_n(t)=u(na,t)$ is replaced with $u(x,t)$. The finite differences $u_{n\pm1}-u_n$ have to be expanded in Taylor series

$$u_{n\pm1}-u_n=\pm a\frac{\partial u}{\partial x}+\frac{a^2}{2}\frac{\partial^2u}{\partial x^2}\pm\frac{a^3}{6}\frac{\partial^3u}{\partial x^3}+\frac{a^4}{24}\frac{\partial^4u}{\partial x^4}\cdots. \quad (4)$$

This expansion is valid under the condition

$$|(u_{n+1}-u_n)/u_n|\ll 1. \quad (5)$$

For linear waves of the form $u_n=u_0\sin(kna-\omega t)$ this expansion agrees with the long wavelength approximation $ak\ll 1$, where k is a wave number.

Substitution of expansion (4) in DDE (1) in the leading approximation yields for the α-FPU and Toda models the Boussinesq equation

$$m\frac{\partial^2 u}{\partial t^2} - Aa^2\frac{\partial^2 u}{\partial x^2} - \frac{Aa^4}{12}\frac{\partial^4 u}{\partial x^4} - 2\alpha a^3\frac{\partial u}{\partial x}\frac{\partial^2 u}{\partial x^2} = 0. \quad (6)$$

(In the case of the β-FPU model, the modification of this equation with the nonlinear term $3\beta a^4(\partial u/\partial x)^2(\partial^2 u/\partial x^2)$ is obtained.)

Equations (2) for discrete FK model within continuum limit are transformed in the leading approximation into the nonlinear Klein–Gordon (NKG) equation

$$m\frac{\partial^2 u}{\partial t^2} - A\,a^2\frac{\partial^2 u}{\partial x^2} + w'(u) = 0. \quad (7)$$

In the particular case of the periodic external potential $w = U(1 - \cos u_n)$, one obtains the sine-Gordon (SG) equation. Finally, the CA for DNLS equation (3) reduces to the usual partial differential nonlinear Schrödinger equation

$$\mathrm{i}\frac{\partial \psi}{\partial t} + Ba^2\frac{\partial^2 \psi}{\partial t^2} + F(|\psi|^2)\psi = 0. \quad (8)$$

Examples (6)–(8) demonstrate that a different number of terms in expansion (4) are to be taken into account in different situations. In the case of Equation (1), the dispersion relation is

$$\omega = 2\sqrt{(A/m)}\,\sin ak/2, \quad (9)$$

which is of acoustic type with a weak dispersion in the limit $k \to 0$. Within the CA, this weak dispersion is governed by the fourth spatial derivative alone in expansion (4), and it is necessary to retain the dispersion term in (6). The dispersion relation for Equation (6) is consistent with the exact formula (9) only in the long wave limit $ak \ll 1$.

Unfortunately, the dispersion term $-(Aa^4/12)\,\partial^4 u/\partial x^4$ in (6) necessitates an additional boundary conditions for this equation and results in the appearance of additional nonphysical solutions with small frequencies and $ak \sim 1$. To avoid these side effects, a regularization of the expansion over the discreteness parameter a can be performed (Rosenau, 1987). This corresponds to substitution of the relation $\partial^2/\partial x^2 \simeq (m/Aa^2)\partial^2/\partial t^2$ into the dispersion term in (6) and leads to yet another version of the CA for Equation (1) containing the term with mixed derivatives:

$$m\frac{\partial^2 u}{\partial t^2} - Aa^2\frac{\partial^2 u}{\partial x^2} - \frac{ma^2}{12}\frac{\partial^4 u}{\partial x^2 \partial t^2} - 2\alpha a^3\frac{\partial u}{\partial x}\frac{\partial^2 u}{\partial x^2} = 0. \quad (10)$$

The above estimation of CA application area ($ak \ll 1$) holds for long wave small amplitude envelope solitons in the nonlinear case as well. But in general, this condition can differ for solitons of different types. For example, CA descriptions of Boussinesq solitons of the type $u(x,t) = u(x - vt)$ describe the solutions of FPU model (1) only under the condition $v/v_c - 1 \ll 1$, where $v_c = \sqrt{Aa^2/m}$.

Only the lowest-order terms of expansion (4) have so far been taken into account for discrete systems within the CA. In general, retaining the next terms with higher powers of spatial derivatives exceeds the accuracy of CA. But in some way, such extended versions of the CA also take into account the discreteness of the systems and can lead to interesting and important physical results. Retaining the fourth order derivative in (4) transforms the nonlinear KGE (7) into

$$m\frac{\partial^2 u}{\partial t^2} - Aa^2\frac{\partial^2 u}{\partial x^2} - \frac{Aa^4}{12}\frac{\partial^4 u}{\partial x^4} + w'(u) = 0. \quad (11)$$

In the case of a sinusoidal external force, the corresponding generalized equation (dispersive SG equation) has steady-state bounded kinks solutions (4π-kink solitons) for some particular values of their velocities (Bogdan et al., 2001). This result obtained within the CA is in agreement with the numerical result for the corresponding discrete system (2) (Alfimov et al., 1993). The inclusion of yet higher terms of expansion (4) in the nonlinear parts of discrete equations in the CA gives rise to the nonlinear dispersion and leads to an existence of exotic solitons such as "compactons" and "peakons."

The CA is not restricted to the long-wave limit. For high-frequency short waves with wave numbers $k \simeq \pi/a$ and $\omega - \omega_{\max} \ll \omega_{\max}$ (where $\omega_{\max} = 2\sqrt{A/m}$ for (1) and (2)), the CA for the slowly varying envelope of antiphase oscillations ($u_n = (-1)^n v_n$) in the β-FPU-model results in a PDE with the Euclidean differential part

$$m\frac{\partial^2 v}{\partial t^2} + Aa^2\frac{\partial^2 v}{\partial x^2} + 4Av + 16\beta v^3 = 0. \quad (12)$$

The breather solution of this equation (Kosevich & Kovalev, 1975) within the CA describes the "intrinsic modes," which are currently being widely discussed since the pioneering paper of Sievers and Takeno (1988). In more complicated diatomic chains with the gap in the linear waves spectrum at $k = \pi/2a$, the so-called "gap solitons" (breathers with frequencies lying in the gap) can be described in CA for the envelopes of the antiphase oscillations of atoms from two sublattices.

To this point, applications of the CA for DDE models of discrete systems were discussed. Often, the opposite approach is used where the corresponding PDEs are investigated numerically in some discrete schemes as a system of DDE (Dodd et al., 1982). The finite-differences method is one of the most popular in this case: the initial function $u(x,t)$ is defined on the rectangular net of the (x,t) plane at points $x=hn, t=h't$. The partial derivatives are replaced by the finite differences

$$\frac{\partial u(x,t)}{\partial x} = \frac{u_{n+1}-u_n}{h}, \quad \frac{\partial^2 u(x,t)}{\partial x^2} = \frac{u_{n+1}+u_{n-1}-2u_n}{h^2}, \ldots \tag{13}$$

Generally, sampling over all variables is performed, but in some hybrid methods space sampling alone is carried out and the resulting system of ODEs is solved by using standard computer codes.

Space sampling is commonly used for complicated biological systems. In order to simulate the behavior of a single neuron, for example, its continuous structure may be sliced into a large number of small segments (compartments). This procedure is called the "compartmental" approach, and within it the continuous PDEs are replaced by sets of ODEs. The advantage of this modeling approach is that it imposes no restrictions on the properties of each compartment and permits great flexibility at the level of resolution. Compartmental methods make it possible to develop the realistic models that have a close relationship with the relevant experimental data.

ALEXANDER S. KOVALEV

See also **Compartmental models; Delay-differential equations; Discrete nonlinear Schrödinger equations; Dispersion relations; Fermi–Pasta–Ulam oscillator chain; Frenkel–Kontorova model; Partial differential equations, nonlinear; Peierls barrier; Sine-Gordon equation**

Further Reading

Alfimov, G., Eleonskii, V., Kulagin, N. & Mitskevich, N. 1993. Dynamics of topological solitons in models with nonlocal interaction. *Chaos*, 3: 405–414

Bogdan, M., Kosevich, A. & Maugin, G. 2001. Soliton complex dynamics in strongly dispersive medium. *Wave Motion*, 34: 1–26

Dodd, R., Eilbeck, J., Gibbon, J. & Morris, H. 1982. *Solitons and Nonlinear Wave Equations*, London: Academic Press

Eisenberg, H., Silberberg, Y., Marandotti, R., Boyd, A. & Aitchison, J. 1998. Discrete spatial optical solitons in waveguide arrays. *Physical Review Letters*, 81: 3383–3386

Fermi, E., Pasta, J. & Ulam, S. 1955. Studies of nonlinear problems. 1965. *Collected Works of E. Fermi*, vol. II, Chicago: University Chicago Press

Frenkel, J. & Kontorova, T. 1939. On the theory of plastic deformation and twinning. *Physical Journal USSR*, I: 137–149 (originally published in Russia, 1938)

Kosevich, A. & Kovalev, A. 1975. Self-localization of vibrations in 1D anharmonic chain. *Soviet Physics, Journal of Experimental and Theoretical Physics*, 40: 891–896

Rosenau, P. 1987. Dynamics of dense lattices. *Physical Review*, B36: 5868–5876

Sievers, A. & Takeno, S. 1988. Intrinsic localized modes in anharmonic crystals. *Physical Review Letters*, 61: 970–973

CONTOUR DYNAMICS

A wide variety of fluid dynamical problems involve the material advection of a tracer field $q(\boldsymbol{x},t)$, expressed by

$$\frac{\mathrm{D}q}{\mathrm{D}t} \equiv \frac{\partial q}{\partial t} + \boldsymbol{u}\cdot\nabla q = 0, \tag{1}$$

where $\boldsymbol{u}(\boldsymbol{x},t)$ is the fluid velocity. The value of q thus does not change following an infinitesimal element or particle. That is, for a particle at $\boldsymbol{x}=\boldsymbol{X}(\boldsymbol{a},t)$, where $\boldsymbol{a}$ is a vector label (e.g., the initial position of the particle), Equation (1) implies that $q=q(\boldsymbol{a})$, a constant, and $\partial\boldsymbol{X}/\partial t=\boldsymbol{u}(\boldsymbol{X},t)$, which is just the statement that the particle moves with the local fluid velocity. The collective effect of this transport is a rearrangement of the tracer field q by the velocity field $\boldsymbol{u}$.

Depending on the nature of $\boldsymbol{u}$, this may lead to highly intricate distributions of q, even starting from simple initial conditions. Moreover, there are important applications in which $\boldsymbol{u}$ depends on q itself, often in a nonlocal manner, that is, in which the entire field of q contributes to $\boldsymbol{u}$ at any given point $\boldsymbol{x}$. A specific example relevant to the present topic is provided by the two-dimensional Euler equations governing the behavior of an inviscid, incompressible fluid:

$$\frac{\mathrm{D}\boldsymbol{u}}{\mathrm{D}t} = -\frac{\nabla p}{\rho}, \tag{2}$$

$$\nabla\cdot\boldsymbol{u} = 0, \tag{3}$$

where p is the pressure and ρ is the density (here constant), and where now the velocity field is two dimensional: $\boldsymbol{u}=(u,v)$. Taking the curl of Equation (2) gives an equation for the scalar vorticity $\zeta \equiv \partial v/\partial x - \partial u/\partial y$:

$$\frac{\mathrm{D}\zeta}{\mathrm{D}t} = 0, \tag{4}$$

which is identical to Equation (1) if we take $q=\zeta$. Thus, the vorticity is materially conserved in this system. But it also induces the velocity field $\boldsymbol{u}$, that transports it. Equation (3) is satisfied generally by considering

$$u = -\partial\psi/\partial y \quad \text{and} \quad v = \partial\psi/\partial x, \tag{5}$$

where $\psi(\boldsymbol{x},t)$ is called the streamfunction. Substituting these components into the definition of ζ leads to a Poisson equation for ψ:

$$\nabla^2\psi = \zeta. \tag{6}$$

Given the distribution of ζ, this equation (with suitable boundary conditions) may be inverted to find ψ, whose spatial derivatives provide u and v. The inversion of this equation can be formally accomplished by using the Green function $G(\boldsymbol{x};\, \boldsymbol{x}')$ of Laplace's operator ∇^2; in two dimensions,

$$G(\boldsymbol{x};\, \boldsymbol{x}') = (2\pi)^{-1} \log |\boldsymbol{x}' - \boldsymbol{x}|. \tag{7}$$

(G is the solution to Equation (6) for $\zeta = \delta(\boldsymbol{x}' - \boldsymbol{x})$, a singular delta distribution of vorticity having a unit spatial integral.)

Consider henceforth an unbounded two-dimensional fluid. Then, the formal solution to the inversion problem is

$$\psi(\boldsymbol{x}, t) = \iint G(\boldsymbol{x};\, \boldsymbol{x}')\zeta(\boldsymbol{x}', t)\, \mathrm{d}x'\mathrm{d}y', \tag{8}$$

which shows explicitly that the flow field at any point depends on the vorticity field at all points. Moreover, the integration over space implies that the field of ψ is generally smoother than that of ζ.

The evolution of the flow in this case consists of two basic steps:

- *inversion*—the recovery of the velocity field $\boldsymbol{u}$ from the distribution of ζ and
- *advection*—the transport of fluid particles to the next instant of time.

Now, a two-dimensional plane is carpeted by an infinite number of such particles, and therefore, this view of the evolution may appear to be quite complex. However, the material conservation of q (or ζ) affords an enormous simplification. First, note that, if one exchanges two particles labeled $\boldsymbol{a}$ and $\boldsymbol{a}'$ having the same value of q, this does not alter the distribution of q, and as a result the velocity field $\boldsymbol{u}$ remains unchanged. This is a "particle relabeling" symmetry, and in general, it gives rise to an infinite number of globally conserved quantities (the spatial integrals of any functional of q). This symmetry implies that contours of fixed q consist of the same set of fluid particles for all time. They are called material contours.

Contour dynamics arises from representing the distribution of q by a finite set of contours C_k, $k = 1, 2, \ldots, n$, between which q is spatially uniform, and across which q jumps by Δq_k (defined to be the value of q to the left of C_k minus that to the right of C_k). The contours here are still material ones—the particles just on either side of a contour retain their distinct values of q. Between two contours, any two fluid particles may be exchanged without altering q or $\boldsymbol{u}$. This implies that only the contours matter for determining the velocity field. Also, since the contours are material, their advection suffices to evolve the entire distribution of q. This is the basis of contour dynamics.

To see how this works for the two-dimensional Euler equations, it remains to be shown as to how one can calculate the velocity field directly from the contours C_k. The starting point is Equation (8), in which we consider $\zeta = q$ to be a piecewise-uniform field. For the moment, we need only use the property $G(\boldsymbol{x};\, \boldsymbol{x}') = g(\boldsymbol{x}' - \boldsymbol{x})$ satisfied by the Green function of the Laplace operator. Also, we need only consider one contour at a time and afterwards linearly superpose the results, since the relation between q and $\boldsymbol{u}$ is linear. We may take $q = 0$ outside of the (closed) contour, so that $q = \Delta q$ inside it (denoted by the region R below). Nonzero exterior q simply gives rise to solid-body rotation, and may be superposed afterwards. Then, from Equation (5), we have

$$\begin{aligned}(u(\boldsymbol{x}, t), v(\boldsymbol{x}, t)) &= \left(-\frac{\partial}{\partial y}, \frac{\partial}{\partial x}\right) \Delta q \iint_R g(\boldsymbol{x}' - \boldsymbol{x})\, \mathrm{d}x'\mathrm{d}y' \quad (9)\\ &= -\Delta q \iint_R \left(-\frac{\partial}{\partial y'}, \frac{\partial}{\partial x'}\right) g(\boldsymbol{x}' - \boldsymbol{x})\, \mathrm{d}x'\mathrm{d}y' \quad (10)\\ &= -\Delta q \oint_C g(\boldsymbol{X}' - \boldsymbol{x})\, (\mathrm{d}X', \mathrm{d}Y'), \quad (11)\end{aligned}$$

where we have used the symmetry of g with respect to $\boldsymbol{x}$ and $\boldsymbol{x}'$ in the second line and Stokes' theorem in the third, and where $\boldsymbol{X}'$ denotes a point on the contour C. The velocity field anywhere thus depends only on the shape of C.

For a set of contours, the velocity field is required only on the contours themselves to evolve q. The contours thus form a closed dynamical system, contour dynamics, governed by

$$\frac{\mathrm{d}\boldsymbol{X}_j}{\mathrm{d}t} = \boldsymbol{u}(\boldsymbol{X}_j) = -\sum_{k=1}^{n} \Delta q_k \oint_{C_k} g(\boldsymbol{X}'_k - \boldsymbol{X}_j)\mathrm{d}\boldsymbol{X}'_k \tag{12}$$

for all points $\boldsymbol{X}_j$ on contours C_j, $j = 1, 2, \ldots, n$.

These equations were first derived for the two-dimensional Euler equations by Zabusky et al. (1979), following earlier work by Berk & Roberts (1967), who derived a similar contour-based model for the two-dimensional Vlaslov equations in plasma physics. These authors also developed numerical methods for contour dynamics in which the contours were discretized into a finite set of points or nodes, originally connected by straight line segments. A wide variety of numerical methods have since been developed, many of which are summarized in the review articles of Dritschel (1989) and Pullin (1992). They principally differ in terms of the choice of the interpolation between nodes (linear, quadratic, cubic; local and global splines); the method of numerical quadrature used to evaluate the contour integral over the segment connecting adjacent nodes (trapezoidal, Gaussian, explicit); the method of redistributing, inserting, and removing nodes to maintain an accurate representation of the contour shape; and the procedure used, if any, to remove fine-scale structure (e.g., filaments and thin bridges connecting two separating regions)—a procedure coined "contour surgery" (Dritschel, 1988).

Contour dynamics has been used to study a wide variety of problems, from the interaction of two vortex patches (having just one contour each), to the filamentation and stripping of nested vortices (having many contours to represent a continuum) (Dritschel, 1989; Pullin, 1992). The numerical method illustrated next is described in Dritschel (1988, 1989). This method uses local cubic splines between contour nodes, explicit quadrature to first order in the departure of the contour from a straight line between nodes, node redistribution based on maintaining a local node density proportional to the square root of contour curvature, and automatic surgery whenever contours or contour parts get closer than a prescribed cutoff scale δ. This scale and the precise formula for the node density are chosen to balance the errors arising from surgery and node redistribution. A fourth-order Runge–Kutta scheme is used for the time integration.

An example is presented next of the collapse of three vortex patches (see also (Rogberg & Dritschel, 2000)). The centers of the vortices are initially chosen at points where equivalent delta-distributed point vortices of the same circulation (spatial integral of q) are known to collide in finite time. Two of the vortices have $q = +2\pi$ and radii 1 and $2/\sqrt{5}$, while the third has $q = -2\pi$ and radius $\frac{2}{3}$. The two positive vortices are initially separated by a distance $d = 5$, and the negative vortex is placed at a distance $d\sqrt{17/27}$ and at an angle 225° relative to the joint center of the two positive vortices. The vortices are then all shifted so that the joint center of all three vortices lies at the origin. Starting from this configuration, the collapse time for the point vortices is 7.70059886...

Figure 1 illustrates the evolution of the vortices—in the upper left-hand frame, the initial conditions are shown, while the remaining frames (to the right and then downwards) are spaced at unit increments in time starting from $t = 5$ and ending at $t = 15$. By $t = 6$, the two positive vortices begin to merge (they are separated by only a thin channel of irrotational fluid). Thereafter, the flow grows rapidly in complexity, as many filaments are generated and small vortices roll up at the tips of some of the filaments. Notably, the negative vortex does not distort significantly, but merely acts to bring the two positive vortices together.

The complexity just illustrated is typical of many vortex interactions. An accurate, robust numerical method must be able to capture this generic behavior, at least over time scales when the flow is reasonably predictable. To see how well the current method performs, we next examine how the results vary with spatial resolution. Two additional simulations were performed at half and double the average point spacing used in Figure 1. The results are compared in Figure 2 at the final time, $t = 15$, when the numbers of nodes in the three simulations are, from low to high resolution, 2740, 10,738 and 27,297 (at $t = 0$, the numbers of nodes are 183, 349, and 682). Note the cutoff scale $\delta = 0.000225$, 0.0009, and 0.0036 in the three simulations—there is a factor of 4 difference in δ between resolutions. The agreement is striking even in the detailed structure. The most visible differences show up in the lengths of the filaments, which are removed more readily at low resolution. These filaments, however, contribute negligibly to the velocity field, and retaining them makes little difference to the evolution of the flow.

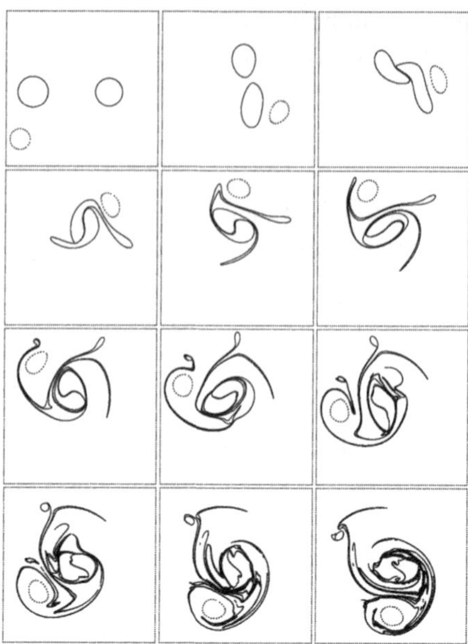

Figure 1. The collapse of three vortex patches. The initial condition is shown in the upper left frame. Time proceeds to the right and downwards in increments of one unit, from $t = 5$ to $t = 15$. The window of view is $-5.0 < x < 5.0$ and $-5.8 < y < 4.2$. The negative vortex is rendered with a short-dashed line (with a dash between each node), while the positive vortices are rendered with a continuous solid line.

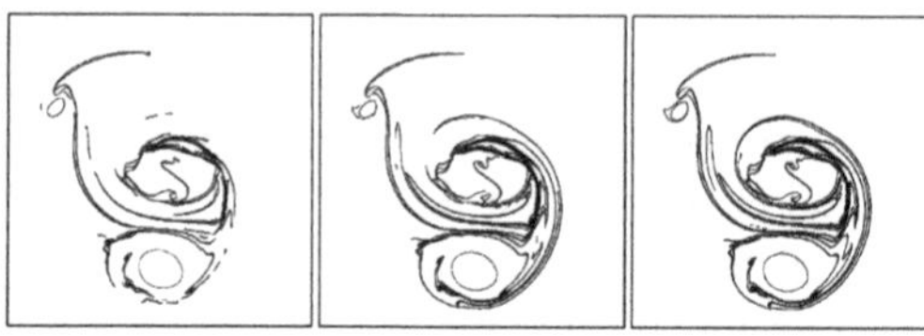

Figure 2. Comparison, at $t = 15$, of three contour dynamics simulations of vortex collapse. Resolution increases from left to right, doubling between each frame (the node spacing parameter is $\mu = 0.12, 0.06$, and 0.03, and the large-scale length $L = 1$ in all cases; consult Dritschel (1989) or Dritschel & Ambaum (1997) for further details). The domain of view is the same as used in the previous figure.

Contour dynamics has since been applied in a variety of diverse fields. Its largest growth has occurred in the field of atmospheric and oceanic dynamics, where the potential vorticity plays the role of the materially conserved tracer q often to a very good approximation (Hoskins et al., 1985). Indeed, its application to this field is on a much sounder footing than it is to the fields it was originally developed for: plasma physics and aeronautics. The two-dimensional approximation is a particularly severe one in aeronautics, since real flows do not preserve two-dimensional symmetry, unless constrained in some manner. In the atmosphere and oceans, rotation and stratification serve to constrain the flow to be two dimensional, or more appropriately, layerwise two dimensional, and furthermore one may extend contour dynamics to study such flows (indeed, the equations are formally no different than given those by (12); see (Dritschel, 2002)).

Finally, the use of contours to carry out tracer advection—the fast and accurate part of contour dynamics—has been combined with more traditional approaches of computing the velocity field (the inversion step) to produce a particularly fast, accurate, and versatile numerical method called the contour-advective semi-Lagrangian (CASL) algorithm (Dritschel & Ambaum, 1997; Dritschel et al., 1999; Dritschel & Viúdez, 2003). This latest development allows the extension of the contour approach to much more realistic sets of equations, and has significantly widened the applicability of the original contour dynamics method.

DAVID DRITSCHEL

See also **Chaotic advection; Euler–Lagrange equations; Vortex dynamics of fluids**

Further Reading

Berk, H.L. & Roberts. K.V. 1967. The water-bag model. *Methods in Computational Physics*, 9: 87–134

Dritschel, D.G. 1988. Contour surgery: a topological reconnection scheme for extended integrations using contour dynamics. *Journal of Computational Physics*, 77: 240–266

Dritschel, D.G. 1989. Contour dynamics and contour surgery: numerical algorithms for extended, high-resolution modelling of vortex dynamics in two-dimensional, inviscid, incompressible flows. *Computer Physics Reports*, 10: 77–146

Dritschel, D.G. 2002. Vortex merger in rotating stratified flows. *Journal of Fluid Mechanics*, 455: 83–101

Dritschel, D.G. & Ambaum, M.H.P. 1997. A contour-advective semi-Lagrangian numerical algorithm for simulating fine-scale conservative dynamical fields. *Quarterly Journal of the Royal Meteorological Society*, 123: 1097–1130

Dritschel, D.G., Polvani, L.M. & Mohebalhojeh, A.R. 1999. The contour-advective semi-Lagrangian algorithm for the shallow water equations. *Monthly Weather Review*, 127(7): 1551–1565

Dritschel, D.G. & Viúdez, A. 2003. A balanced approach to modelling rotating stably-stratified geophysical flows. *Journal of Fluid Mechanics*, 488: 123–150. See also: www-vortex.mcs.st-and.ac.uk.

Hoskins, B.J., McIntyre, M.E. & Robertson, A.W. 1985. On the use and significance of isentropic potential-vorticity maps. *Quarterly Journal of the Royal Meteorological Society*, 111: 877–946

Pullin, D.I. 1992. Contour dynamics methods. *Annual Review of Fluid Mechanics*, 24: 89–115

Rogberg, P. & Dritschel, D.G. 2000. Mixing and transport in two-dimensional vortex interactions. *Physics of Fluids*, 12(12): 3285–3288

Zabusky, N.J., Hughes, M.H. & Roberts, K.V. 1979. Contour dynamics for the Euler equations in two dimensions. *Journal of Computational Physics*, 30: 96–106

CONTROL PARAMETERS

See **Bifurcations**

CONTROLLING CHAOS

It may seem paradoxical that chaotic systems—which are extremely sensitive to the tiniest fluctuations—can be controlled; yet, the earliest reference to this idea appears around 1950, when John von Neumann presaged just that. Nowadays, laboratory demonstrations of the control of chaos have been realized in chemical, fluid, and biological systems, and the intrinsic instability of chaotic celestial orbits is routinely used to advantage by international space agencies who divert spacecraft to travel vast distances using only modest fuel expenditures.

A variety of techniques for chaos control has been implemented since around 1990 when the first concrete analyses appeared, including traditional feedback and open-loop methods, neural network applications, shooting methods, Lyapunov function approaches, and synchronization to both simple and complex external signals. These techniques resolve the paradox implied by chaos control in different ways, but they all make use of the fact that chaotic systems can be productively controlled if disturbances are countered by small and intelligently applied impulses. Just as an acrobat balances about an unstable position on a tightrope by the application of small correcting movements, a chaotic system can be stabilized about any of an infinite number of unstable states by continuous application of small corrections.

Two characteristics of chaos make the application of control techniques even more fruitful. First, chaotic systems alternately visit small neighborhoods of an infinite number of periodic orbits. The presence of an infinite number of periodic orbits embedded within a chaotic trajectory implies the existence of an enormous variety of different behaviors within a single system. Thus, the control of chaos opens up the potential for tremendous flexibility in operating performance within a single system.

As an example, Figure 1 depicts the Lorenz attractor, used to model fluid convection. Embedded within the gray attractor are innumerable periodic orbits, such as the solid figure-8 orbit and the more complicated dashed one. For practical systems such as chemical reactors or fluidized beds, the presence of multiple

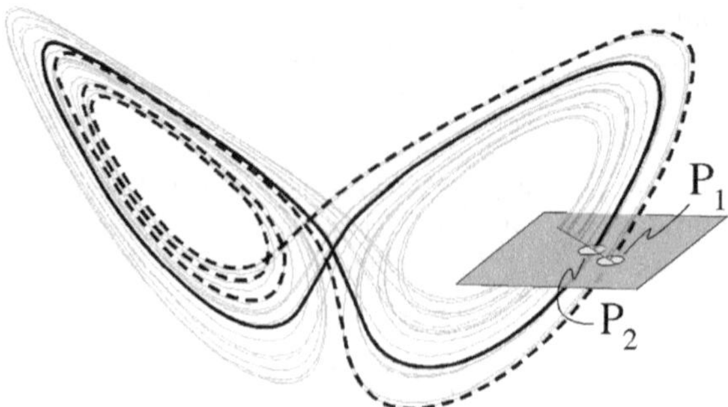

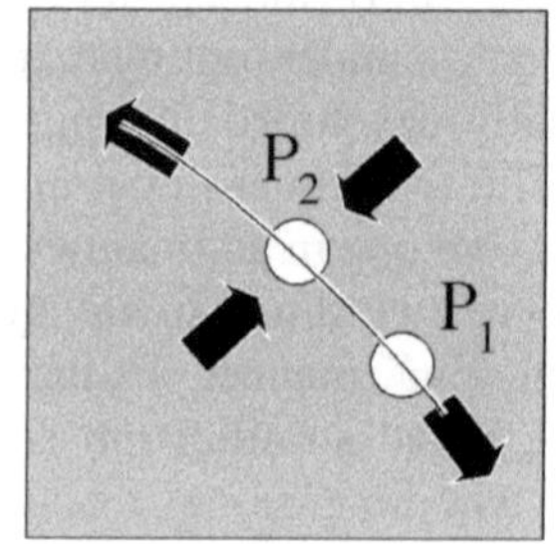

Figure 1. Left: Lorenz attractor with two of its embedded unstable periodic orbits highlighted. Right: unstable points P_1 and P_2 in the surface of section indicated. The unstable direction is denoted by outgoing arrows, and stable direction is denoted by ingoing arrows.

co-existing states implies that one chaotic system could be operated in multiple different states, thus potentially performing the function of several separate units.

A second characteristic of chaos that is important for control applications is the exponential sensitivity of the phenomenon. That is, the fact that the state of a chaotic system can be drastically altered by the application of small perturbations means two things: such a system if uncontrolled can be expected to fluctuate wildly, and if controlled can be directed from one state to a very different one using only very small controls.

Traditional feedback control remains among the most widely used methods of control for chaotic systems. To implement feedback control, one waits until a chaotic trajectory by chance lands near a desired periodic point and then applies small variations to an accessible system parameter in order to repeatedly nudge the trajectory closer to that point. As an example, consider the plot to the right in Figure 1, where we depict two periodic points as they appear on a "surface of section" formed in this case by recording every intersection between the Lorenz chaotic attractor and the half plane, $Z=0$, $X>0$. To control the state to remain near point P_2 (so the trajectory stays near the figure-8 trajectory shown to the left), one needs to apply variations in a parameter, p, that directs the state toward P_2 along the unstable direction (or directions in more complicated problems) indicated by outgoing arrows in Figure 1. One can establish the direction in which a parametric control moves the chaotic state either experimentally, by varying the parameter and recording the future variation of the system state, or analytically, by determining the Jacobian of the flow or mapping where available. Nudging the state closer to P_2 amounts to what is termed "pole placement" in traditional control literature, and numerous reports of alternative strategies for selecting parameter variations appear in the literature. Strategies include simple pole placement, optimal control, neural network approaches, simple proportional control, periodic forcing, and control dependent on the distance from P_2. Most of these strategies have proven successful under appropriate conditions, and the choice of strategy depends principally on details of the control goal required and the computational resources available to meet that goal.

All of these strategies require that the system state must lie close to the desired state in order to achieve control. In such a case, the system dynamics can be linearized, making control calculations rapid and effective. Fortunately, in chaotic systems, one can rely on ergodicity to ensure that the system state will eventually wander arbitrarily close to the desired state. By the same token, if it is desired to switch the system between one accessible state (say P_1) and a second (say P_2), one can merely release control from P_1 and re-apply a new control algorithm once the system strays close to P_2, which it is certain to do by ergodicity.

In higher-dimensional or slowly varying systems, the time taken for the state to move on its own from one state to another can be prohibitive, and for this reason fully nonlinear control strategies have been devised that use chaotic sensitivity to steer the system state from any given initial point to a desired state. Since chaotic systems amplify control impulses exponentially, the time needed to steer such a system can be quite short. These strategies have been demonstrated both in systems in which a large effect is desired using very modest parameter expenditures (energy or fuel) and in systems in which rapid switching between states is needed (computational or communications applications).

On the other hand, in both linear and nonlinear control approaches, one needs to repeatedly re-apply control over a time that is short compared with the inverse of the fastest growing growth rate of the system in order to counter the potential amplification of ubiquitous noises. Computational and experimental analyses have demonstrated that this is readily done in typical chaotic systems and that control can be robustly achieved. Because large but rare noise events can occur, however, controlled states occasionally break free when the system encounters an anomalous large noise. In this case, bounds have been established for the frequency and duration of these noise-induced excursions.

Numerous biological control applications have been proposed since the first introduction of the notion of chaotic control. Among the first applications were studies of intrinsic nonlinear control mechanisms involved in autonomic and involuntary functions such as the regulation of internal rhythms and the control of gait and balance. These studies confirm that nontrivial control algorithms are involved in the maintenance of normal physiological function and that provocative insights into pathological conditions can be gained (such as cardiac and breathing arrythmias and motor tremor). Further work has shown that networks of chaotic devices, under prescribed conditions, can be brought into synchronization, and strong indications have been presented that neuronal signaling may rely on nonlinear synchronization. Additional experimental studies are promising for the control of unwanted fluctuations (e.g., during fibrillation of the heart) or for the so-called "anticontrol" of synchronized periodic signals in focal epilepsy. In both studies, the goal is to use feedback control methods to steer a diseased organ using small electrical stimulation: in the former state, toward a stabilized state, and in the latter, away from a synchronized state.

TROY SHINBROT

See also **Chaotic dynamics; Feedback; Lorenz equations**

Further reading

Alekseev, V.V. & Loskutov, A.Y. 1987. Control of a system with a strange attractor through periodic parametric action. *Soviet Physics Doklady*, 32: 1346–1348

Ditto, W.L. & Pecora, L.M. 1993. Mastering chaos. *Scientific American*, 78–84

Garfinkel, A., Spano, M.L., Ditto, W.L. & Weiss, J.N. 1992. Controlling cardiac chaos. *Science*, 257:1230–1235

Glass, L. & Zeng, W. 1994. Bifurcations in flat-topped maps and the control of cardiac chaos. *International Journal of Bifurcation & Chaos*, 4: 1061–1067

Hayes, S., Grebogi, C. & Ott, E. Communicating with Chaos. *Physical Review Letters*, 70: 3031–3014

Hübler, A., Georgii, R., Kuckler, M., Stelzl, W. & Lscher, E. 1988. Resonant Stimulation of nonlinear damped oscillators by Poincaré maps. *Helvetica Physica Acta*, 61: 897–900

Lima, R. & Pettini, M. 1990. Suppression of chaos by resonant parametric perturbations. *Physics Review A*, 41: 726–733

Ott, E., Grebogi, C. & Yorke, J.A. 1990. Controlling chaos. *Physical Review Letters*, 64: 1196–1199

Pecora, L.M. & Carroll, T.J. 1990. Synchronization in chaotic systems. *Physical Review Letters*, 64: 821–824

Schiff, S.J., Jerger, K., Duong, D.H., Chang, T., Spano, M.L. & Ditto, W.L. 1994. Controlling chaos in the brain. *Nature*, 370: 615–620

Shinbrot, T., Ott, E., Grebogi, C. & Yorke, J.A. 1993. Using small perturbations to control chaos. *Nature*, 363: 411–417

COSMOLOGICAL MODELS

Relativistic cosmology—the science of the structure and evolution of the universe—is based on the building and investigation of cosmological models (CMs), which describe geometrical properties of physical space-time, the matter, composition and structure of the universe, and physical processes at different stages of the universe's evolution. Prominent in cosmology is the hot big bang CM, which is based on solutions of Alexander Friedmann's cosmological equations for homogeneous isotropic models deduced in the framework of Einstein's general relativity theory (GR). Because of its large-scale structure (galaxies, clusters of galaxies, etc.), the universe is homogeneous and isotropic only on the largest scales from 100 Mpc. (The pc or parsec is an astronomical unit of distance equal to 3.2616 light years; thus, an Mpc is 3.2616 million light years.)

The most important feature of Friedmann's CM is its nonstationary character, which was confirmed by Edwin Hubble's discovery of cosmological expansion in 1929. In this formulation, the geometrical properties of physical space depend on the value of energy density ρ relative to a critical density

$$\rho_{\text{crit}} = \frac{3H_0^2}{8\pi G}, \tag{0}$$

where H_0 is the expansion rate (Hubble parameter) at the present epoch and G is Newton's gravitational constant. If $\Omega = \rho/\rho_{\text{crit}} = 1$, the 3-space is flat; if $\Omega > 1$, 3-space possesses positive curvature and if $\Omega < 1$, 3-space possesses negative curvature. Corresponding CMs are flat, closed, and open CM, respectively.

All Friedmann CMs have a beginning in time (or cosmological singularity), where energy density and curvature invariants diverge. Their evolutions depend on properties of matter. In the case of ordinary matter with a positive energy density and a nonnegative pressure, the evolution of flat and open models has the character of expansion, and closed models recollapse after an expansion stage. The assumption that the temperature was very high at the initial stage of cosmological explanation (hot CM) was confirmed by the discovery of the cosmic microwave background (CMB) radiation in 1965, with a present epoch temperature of about $T = 2.7$ K. The theory of nucleosynthesis of light elements (hydrogen, helium, deuterium, lithium, etc.) into the first few minutes of cosmological expansion based on the framework of the hot big bang CM is in accord with empirical data.

Advances in both theory and technology during the last 20 years have launched cosmology into a most exciting period of discovery. By using precise instruments (telescopes, satellites, spectroscopes), several cosmological research programs are being carried out, including investigations of the anisotropy of CMB and supernovae observations. Cosmological observations have not only strengthened and expanded the hot big bang CM but they have also revealed surprises.

Recent measurements of the anisotropy of the CMB have provided convincing evidence that the spatial geometry is very close to being uncurved (flat) with $\Omega = 1.0 \pm 0.03$. The currently known components of the universe include ordinary baryonic matter, cold dark matter (CDM), massive neutrinos, the CMB and other forms of radiation, and dark energy. The sum of the values for these densities derived empirically is equal to the critical density (to within their margins of error). The largest contributions to energy density are from two components—CDM and dark energy. About 30% of the total mass-energy is dark matter, composed of particles probably formed early in the universe. Two thirds is in smooth dark energy whose gravitational effects began causing the expansion of the universe to speed up just a few billion years ago. The remarkable fact that the expansion is accelerating can be accounted for within GR, as the source of gravity is proportional to $(\rho + 3p)$, where the pressure p and energy density ρ describe the bulk properties of "the substance." A substance with pressure more negative than one-third its energy has repulsive gravity in GR. Such a situation occurs, for example, for gravitating vacuum (positive cosmological constant), for which $p = -\rho$.

In addition to breakthrough empirical observations, creative theoretical ideas are also driving progress in cosmology. The development of cosmology during the last 20 years shows that profound connections exist between the elementary particles on the smallest scales and the universe on the largest. Using unified gauge theories of elementary particles, an inflationary scenario was formulated, which resolves a number of problems of standard Friedmann cosmology: flatness and the problem of horizon, among others. According to inflation, small bits of the universe underwent a burst of expansion when the universe was extremely young, explaining the homogeneity and isotropy of the universe at initial stages of cosmological expansion. Based on the framework of an inflationary CM, the appearance of quantum fluctuations with a nearly scale-invariant distribution by transition to radiation-dominated era was predicted, explaining the large scale structure of the universe.

The inflationary CM as well as others discussed above are singular, which is an outstanding problem of GR. Assuming that the Planck era (when the universe was sufficiently dense to require a quantum mechanical treatment) existed, some quantum gravitation theory is necessary to construct a regular CM. At present, the superstring theory is a candidate for such a theory.

Some regular CMs have been constructed in a "brane world," under which our universe is thought to exist as a slice (or membrane) through a higher-dimensional space. By using scalar fields with a negative potential, a solution for an oscillating CM was obtained; thus, the Big Crunch takes place in such models before Big Bang. Resolving the problem of cosmological singularity requires that the gravitation theory not only admits regular solutions for CM but also excludes singular solutions. This suggests gauge theories of gravitation (Poincaré gauge theory or metric-affine gauge theory), leading to regular bouncing solutions for CMs.

The building of more realistic CMs requires the resolution of fundamental cosmological problems. According to present knowledge, our universe is flat and 13 Gyr old, and it is expanding at the current rate of $H_0 = 72 \pm 8\,\text{km sec}^{-1}\,\text{Mpc}^{-1}$. Measurements of the past rate reveal that the universe is presently in a period of cosmic acceleration. The contribution of ordinary matter to the overall mass energy is small, with more than 95% of the Universe existing in new and unidentified forms of matter and energy. What is the composition of dark matter (axions, neutralinos, or other exotic particles)? What is the nature of dark energy (quantum vacuum energy or scalar fields)? What is the field that drives inflation? Answers to these and other questions will change the picture presented above.

Viachaslav Kuvshinov and Albert Minkevich

See also **Black holes; Einstein equations; Galaxies; General relativity**

Further Reading

Gasperini, M. & Veneziano, G. 2003. The pre-big bang scenario in string cosmology. *Physics Reports*, 373: 1–212

Khoury, J., Ovrut, B.A, Seiberg, N., Steinhardt, P.J. & Turok, N. 2002. From big crunch to big bang. *Physical Review* D, 65: 086007

Kolb, E.W. & Turner, M.S. 1990. *The Early Universe*, Reading, MA: Addison-Wesley

Linde, A.D. 1990. *Particle Physics and Inflationary Cosmology*, Chur: Harwood Academic

Steinhardt, P.J. & Turok, N. 2002. A Cyclic Model of the Universe. hep-th/0111030

CONVECTION

See **Fluid dynamics**

CONVECTIVE INSTABLITY

See **Wave stability and instability**

CORRELATION DIMENSION

See **Dimensions**

CORRESPONDENCE PRINCIPLE

See **Quantum nonlinearity**

COUPLED MAP LATTICE

Originally introduced in the study of spatiotemporal chaos, the coupled map lattice (CML) can be presented as a dynamical model for the evolution of a spatially extended system in time (Kaneko, 1983). CMLs have been widely used, not only as a tool for the study of spatiotemporal chaos but also for pattern dynamics in physics, chemistry, ecology, biology, brain theory, and information processing.

A CML is a dynamical system with discrete time (map), discrete space (lattice), and a continuous state. It consists of dynamical elements on a lattice, which interact (are coupled) with suitably chosen sets of other elements.

The construction of a CML is carried out as follows. First, choose a (set of) field variable(s) on a lattice. This (set of) variable(s) is on a macroscopic, not a microscopic level. Second, decompose the phenomenon of interest into independent units (e.g., convection, reaction, diffusion, and so on). Third, replace each unit by simple parallel dynamics (procedure) on a lattice, where the dynamics consists of a nonlinear transformation of the field variable at each lattice point and/or a coupling term among suitably chosen neighbors. Finally, carry out each unit dynamics (procedure) successively.

As a simple and widely used example, consider a phenomenon that is created by a locally chaotic process and by diffusion, and choose a suitable lattice model on a coarsegrained level for each process. As the simplest choice, we can adopt some one-dimensional map for chaos, and a discrete Laplacian operator for the diffusion. The former process is given by $x_n'(i) = f(x_n(i))$, where $x_n(i)$ is a variable at time n and lattice site i, $(i = 1, 2, \ldots, N)$, whereas $x_n'(i)$ is introduced as the intermediate value. The discrete Laplacian operator for diffusion is given by

$$x_{n+1}(i) = (1-\varepsilon)x_n'(i) + \frac{\varepsilon}{2}\{x_n'(i+1) + x_n'(i-1)\}. \qquad (0)$$

Combining the above two processes, the CML is given by

$$x_{n+1}(i) = (1-\varepsilon)f(x_n(i)) + \frac{\varepsilon}{2}\{f(x_n(i+1)) + f(x_n(i-1))\}. \qquad (1)$$

The mapping function $f(x)$ is chosen to depend on the type of local chaos. For example, one can choose the logistic map, $f(x) = rx(1-x)$, as a typical model for chaos. As the map dynamics are well studied, dynamical systems theory can be applied to understand behaviors of the CML.

By adopting different procedures, one can construct models for different types of spatially extended dynamical systems. For problems of phase transition dynamics, it is useful to adopt a map with bistable fixed points (e.g., $f(x) = \tanh x$) as a local dynamics. The choice of a different type of coupling, as well as the extension to a higher-dimensional space is straightforward. By changing the procedures in the CML, one can easily construct a model for dynamical phenomena in space-time. Examples include spinodal decomposition, crystal growth, boiling, convection, and cloud dynamics, among others.

Universality Classes of the Phenomena

Phenomena found in one CML are often observed in a wide variety of systems, and they form a universality class common to such systems. CMLs thus work as a tool to predict novel phenomenology forming such qualitative universality classes.

In the model of Equation (1), the following phenomena have been discovered: (i) spatial bifurcation and frozen chaos, (ii) spatiotemporal intermittency (STI), (iii) Brownian motion of chaotic defects, and (iv) global traveling wave by local phase slips. These phenomena are observed in a wide variety of systems, including experiments. In particular, STI is now regarded as a universal route to fully developed spatiotemporal chaos. In fully developed spatiotemporal chaos, statistical mechanics theory is developed by taking advantage of the discreteness in spacetime.

If one adopts a two-dimensional lattice system, spiral pattern dynamics are often observed. For example, by taking a local map with an excitable state, the formation of spiral waves is studied, including turbulence due to the break-up of a spiral wave pair. Such a model is studied in relation to the pattern dynamics in reaction-diffusion systems as well as wave propagation in cardiac tissue.

Another straightforward extension is a spatially asymmetric coupling. In an open fluid flow, for example, there is coupling from up-flow to down-flow, instead of the diffusion. The CML $x_{n+1}(i) = (1-\varepsilon) f(x_n(i)) + \varepsilon f(x_n(i))$ gives a prototype model for such a case. In this open flow system, it is important to distinguish absolute instability from convective instability. If a small perturbation against a reference state grows in a stationary frame, it is called "absolute instability," while if the perturbation grows only in a frame moving with a specific velocity, it is called "convective instability." This convective instability leads to spatial bifurcation from a homogeneous state to down-flow convective chaos.

Globally Coupled Maps with Applications to Biology

An extension of CML to global coupling is interesting, and often important for biological problems. Thus a globally coupled map (GCM) was introduced as a mean-field-type extension of a CML, written as

$$x_{n+1}(i) = (1-\varepsilon)f(x_n(i)) + (\varepsilon/N)\sum_{j=1}^{N} f(x_n(j)). \quad (2)$$

One important notion here is clustering. The elements split into several clusters, within which all the elements oscillate in synchronization. Depending on the numbers of clusters in the GCM, there are phase transitions among a coherent phase, an ordered phase, a partially ordered phase, and a desynchronized phase, as the parameter describing the nonlinearity in $f(x)$ is increased. In the partially ordered phase, there are many attractors with different numbers of clusterings and with a variety of partitions. Dynamically, the system spontaneously switches between ordered states through disordered states, known as chaotic itinerancy. In the desynchronized phase, nontrivial collective motion is observed with some hidden coherence among elements. This demonstrates the existence of macroscopic chaos different from microscopic chaos represented by each map $x_{n+1} = f(x_n)$. This observation may shed new light on the origin of collective behavior by an ensemble of cells, such as an electroencephalogram (EEG) in the brain.

Often, a biological system has both internal dynamics and interactions. Chemical dynamics in a cell includes both intra-cellular reactions associated with gene expressions and cell-cell interactions. Since a CML or GCM is a model for such intra-inter dynamics, the concepts developed in this area will be relevant to biological problems. For example, clustering leads to differentiation of the states of elements. The theory for cell differentiation and robust developmental process may be based on this dynamic differentiation.

KUNIHIKO KANEKO

See also **Cellular automata; Cluster coagulation; Maps**

Further Reading

Chaté, H. & Courbage, M. (editiors). 1997. Special issue on lattice dynamics. *Physica* D, 103: 1–612

Kaneko, K. 1986. *Collapse of Tori and Genesis of Chaos in Dissipative Systems* Singapore: World Scientific 1986 (PhD thesis originally published 1983)

Kaneko, K. (editior). 1992. Chaos focus issue on coupled map lattices. *Chaos*, 2(3): 279–408

Kaneko, K. (editor). 1993 *Theory and Applications of Coupled Map Lattices*, Chichester and New York: Wiley

Kaneko, K. & Tsuda, I. 2000. *Complex Systems: Chaos and Beyond—A Constructive Approach with Applications in Life Sciences*, Berlin and New York: Springer

Kaneko, K. & Tsuda, I. 2003. Chaos focus issue on chaotic itinerancy. *Chaos*, 13(3): 926–1164

COUPLED OSCILLATORS

The simplest coupled oscillator is a pair of linearly coupled harmonic oscillators, which is used as a model for a wide variety of physical systems—including the interactions of musical instruments and tuning forks, lattice vibrations, electrical resonances, and so on—in which energy tunnels back and forth between two sites at a difference (beat) frequency. If there are many elementary oscillators that are nonlinear, coupled systems exhibit more varied nonlinear phenomena. There are two types of coupled nonlinear oscillators: those described by Hamiltonian (energy-conserving) dynamics, and systems in which energy is not conserved. In addition to coupled pendula, examples of the first kind include the Fermi–Pasta–Ulam model and the Toda lattice.

Coupled nonlinear oscillators that do not conserve energy can be viewed as coupled limit cycle oscillators. A limit cycle oscillator (also called a self-sustained oscillator) is described as an attractor in a dissipative dynamical system. A typical dissipative dynamical system that exhibits a limit cycle oscillation is van der Pol's equation

$$\frac{\mathrm{d}^2x}{\mathrm{d}t^2} - \varepsilon\left(1-x^2\right)\frac{\mathrm{d}x}{\mathrm{d}t} + \omega^2 x = 0 \quad (1)$$

in which the character of the oscillation varies from sinusoidal and energy-conserving to a strongly dissipative (blocking or relaxation) oscillation through the variation of a parameter (ε) from zero to large values (van der Pol, 1934).

Among the varieties of limit cycle oscillators, the behavior of a quasilinear oscillator (small ε) can be expressed by a sinusoidal wave, $x(t) = A\sin(\omega t + \phi_0)$. The wave shape of a relaxation oscillator (large ε), on the other hand, is composed of alternating fast and slow motions, similar to the spikes and slow recovery motions in a firing neuron, and stick-slip oscillations in frictional motions.

Although the limit cycle oscillation has a certain natural amplitude and frequency, the phase variable, for example, $\phi = \omega t + \phi_0$ for a quasilinear oscillator, is a neutral mode, sensitively perturbed by an external force. If the external force is periodic with a frequency close to the natural frequency of the limit cycle oscillator, the phase of the limit cycle oscillator tends to approach the phase of the external periodic force. If the external force is sufficiently strong, the phase difference $\Delta\phi(t) = \phi(t) - \phi_e(t)$ between the limit cycle oscillator and the external force is fixed. This phenomenon—termed *phase* or *frequency locking*—occurs more easily when ε is large, the frequency of the limit cycle oscillator is close to that of the external force, and the coupling (K) is large.

Regions in the (ω, ε, K) parameter space where frequency locking is observed are termed "Arnol'd

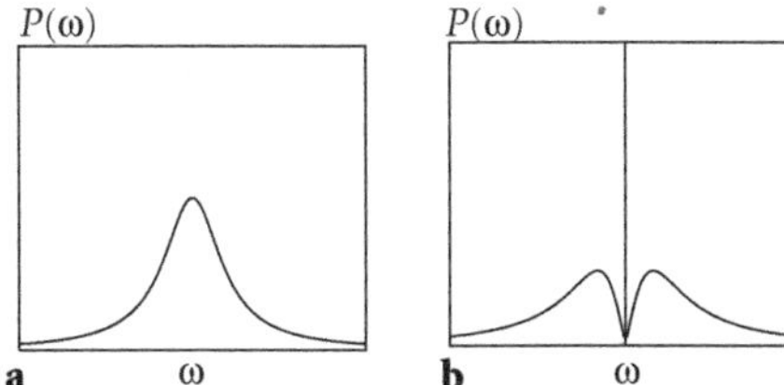

Figure 1. Frequency distribution $P(\omega)$ (a) in an asynchronous state for $K < K_c$ and (b) in a mutually entrained state for $K > K_c$.

tongues" owing to their peculiar shape. The frequency ratio between the limit cycle and the external force is 1:1 in the above frequency locking. In general, n:m frequency lockings are possible, where n and m are small integers.

For a collection of coupled limit cycle oscillators with slightly different natural frequencies, frequency locking (called mutual entrainment) also occurs, as was first observed by Christiaan Huygens in the 17th century. He found that the motions of pendulum clocks suspended from the same wooden beam come to coincide with each other perfectly. Nobert Wiener analyzed such systems in the 1950s, showing that the power spectrum of the waves should have a peak close to 10 Hz, and he inferred that a similar shape of the power spectra of electroencephalogram (EEG) is due to mutual entrainment in coupled neural oscillators (Wiener, 1958). Buck and Buck reported that rhythmical flashes of South Asian fireflies were mutually synchronized (Buck & Buck, 1976).

Mutual entrainment of coupled limit cycle oscillators has been studied by Winfree (2000) and also by Kuramoto, who considered a coupled phase oscillator model, noting the neutrality of phase variables (Kuramoto, 1984). The simplest model with global coupling has the form

$$\dot{\phi}_i = \omega_i + \frac{K}{N}\sum_{j=1}^{N}\sin(\phi_j - \phi_i), \qquad (2)$$

where ϕ_i and ω_i represent the phase and the natural frequency of the ith oscillator, N is the total number of oscillators, and K is a coupling constant.

For $K < K_c$, the motion of each oscillator is independent and the frequency of the ith oscillator is the same as ω_i. However, for $K > K_c$, collective oscillation appears and a number of oscillators are entrained to the collective oscillation. Figure 1 displays a typical frequency distribution for $K < K_c$ and $K > K_c$. The δ-function peak in the frequency distribution implies mutual entrainment and a depression is seen around the deserved frequency for $K > K_c$.

The Josephson junction is a quantum device composed of two weakly coupled superconductors. With the current bias current below a critical value, the superconducting current flows without a voltage drop. If the bias is above the critical current, the phase difference (ϕ) between the Josephson junction is not constant in time, and the voltage drop (V) between the Josephson junction equals $\hbar\dot{\phi} = 2\,\mathrm{eV}$. This is called the AC Josephson effect. Thus the Josephson junction behaves as a kind of limit cycle oscillator above the critical current. If microwaves with frequency ω_0 are applied to the Josephson junction, $n:1$ frequency locking occurs, and the voltage becomes $V = n\hbar\omega_0/2e$. With N Josephson junctions coupled in series, the total voltage across the array is given by $V = Nn\hbar\omega_0/2e$. Such series arrays are currently used to establish the international standard of voltage (*See* **Josephson junction arrays**).

Hidetsugu Sakaguchi

See also **Chaotic dynamics; Phase dynamics; Synchronization; Van der Pol equation**

Further Reading

Buck, J. & Buck, E. 1976. Synchronous fireflies. *Scientific American*, 234: 74–85

Kuramoto, Y. 1984. *Chemical Oscillations, Waves, and Turbulence*, Berlin: Springer

van der Pol, B. 1934. The nonlinear theory of electric oscillations. *Proceedings of the IRE*, 22: 1051–1086

Wiener, N. 1958. *Nonlinear Problems in Random Theory*, Cambridge, MA: MIT Press

Winfree, A.T. 2000. *When Time Breaks Down*, Berlin and New York: Springer

COUPLED SYSTEMS OF PARTIAL DIFFERENTIAL EQUATIONS

Coupled systems of nonlinear partial differential equations (PDEs) are often derived to simplify complicated systems of governing equations in theoretical and applied sciences (Engelbrecht et al., 1988). Nonlinear electromagnetic theory, fluid dynamics, and systems in general relativity are difficult computational problems even with the help of numerical algorithms and the latest computer technologies. Using additional assumptions on properties of nonlinear wave processes in physical systems, however, one can derive coupled systems of nonlinear PDEs from the original governing equations, which simplify the analysis.

The main effects of nonlinear waves (such as nonlinearity, dispersion, diffraction, diffusion, damping and driven forces, and resonances) can be described with coupled nonlinear PDEs. Such systems may exhibit simple solutions such as traveling solitary waves and periodic waves, and some can be solved with the inverse scattering transform methods. Coupled systems comprise various combinations of nonlinear evolution equations that describe long solitary waves (Korteweg–de Vries and Boussinesq equations), envelope waves (nonlinear Schrödinger equations), kinks and breathers (sine-Gordon equations), and traveling fronts and pulses (reaction-diffusion systems). Here, we present a few examples.

Long surface water waves occur in oceans, seas, and lakes. The tsunami wave (Bryant, 2001) is an example of a nonlinear surface wave that arises following underwater earthquakes or underwater volcano eruptions and may reach heights of 20–30 m as it comes ashore. Because tsunamis are as long as tens and hundreds of kilometers, the ocean can be considered as shallow for such waves. This shallow-water approximation reduces the Euler equations for water waves to the Boussinesq system of coupled PDEs (Whitham, 1974):

$$\begin{aligned} u_t + uu_x + g\eta_x - (h^3/3)u_{txx} &= 0, \\ \eta_t + hu_x + \eta u_x + u\eta_x &= 0, \end{aligned}$$

where $\eta = \eta(x,t)$ is the wave surface elevation, $u = u(x,t)$ is the horizontal velocity, h is the water depth, and g is the gravitational acceleration.

The linear Boussinesq equation takes the form of the wave equation: $\eta_{tt} - c^2\eta_{xx} = 0$, which exhibits a two-wave solution $\eta = f(x-ct) + g(x+ct)$, where $f(x)$, $g(x)$ are arbitrary functions and $c = \sqrt{gh}$ is the wave speed. When the two waves are separated in space, small nonlinearity and dispersion are captured in the unidirectional Korteweg–de Vries (KdV) equation (Johnson, 1997):

$$\eta_t + c\left(1 + \frac{3\eta}{2h}\right)\eta_x + \frac{ch^2}{6}\eta_{xxx} = 0.$$

Different modes of long weakly nonlinear waves may travel with the same speed, exchanging energy by means of wave resonances. Because ocean water is stratified in density and shear flow, gravity waves can propagate along internal interfaces of the ocean stratification. Resonant interaction of internal wave modes in stratified shear flows is described by the system of coupled KdV equations (Grimshaw, 2001):

$$\boldsymbol{u}_t + A\boldsymbol{u}_x + B(\boldsymbol{u})\boldsymbol{u}_x + C\boldsymbol{u}_{xxx} = 0,$$

where A, $B(\boldsymbol{u})$, C are matrices and $\boldsymbol{u} = \boldsymbol{u}(x,t)$ is the vector for amplitudes of different internal wave modes.

Optical pulses may consist of electromagnetic waves in optical fibers, waveguides, and transmission lines. The propagation of optical pulses due to a balance between nonlinearity and dispersion is based on the paraxial approximation of the Maxwell equations with nonlinear refractive indices. This perturbation technique results in the nonlinear Schrödinger (NLS) equation (Newell & Moloney, 1992):

$$\mathrm{i}\psi_t + \tfrac{1}{2}\omega''(k_0)\psi_{xx} + \gamma(k_0)|\psi|^2\psi = 0,$$

where $\psi = \psi(x,t)$ is the envelope amplitude of a wave packet with the carrier wave number k_0. Depending on the relative signs of the dispersion coefficient $\omega''(k_0)$ and the nonlinearity coefficient $\gamma(k_0)$, wave perturbations are focused or defocused in the time evolution of the NLS equation. Interactions between waves with two orientations of polarization (ψ_y and ψ_z with propagation in the x-direction) can be represented in a normalized form as

$$\begin{aligned} \mathrm{i}\psi_{y,t} + \psi_{y,xx} + 2(|\psi_y|^2 + |\psi_z|^2)\psi_y = 0 \\ \mathrm{i}\psi_{z,t} + \psi_{z,xx} + 2(|\psi_y|^2 + |\psi_z|^2)\psi_z = 0. \end{aligned}$$

These are a pair of coupled NLS equations that are integrable by the inverse scattering transform method and display vector solitons (Manakov, 1974). Under collisions, the polarization vectors of two vector solitons change.

In wavelength-division-multiplexing optical systems, optical signals are transmitted through parallel channels at different carrier wave numbers (up to 40 channels in latest communication lines). Incoherent interaction of optical pulses at nonresonant frequencies is described by the system of coupled NLS equations (Akhmediev & Ankiewicz, 1997):

$$\mathrm{i}\boldsymbol{\psi}_t + D\boldsymbol{\psi}_{xx} + E(|\boldsymbol{\psi}|^2)\boldsymbol{\psi} = 0,$$

where D and $E(|\boldsymbol{\psi}|^2)$ are matrices and $\boldsymbol{\psi} = \boldsymbol{\psi}(x,t)$ is the vector for optical pulses in different channels. If the coupling between optical pulses is coherent (as in birefringent fibers, waveguide couplers, phase mixers, and resonant optical materials), the system of coupled NLS equations takes a general form:

$$\mathrm{i}\boldsymbol{\psi}_t + A\boldsymbol{\psi} + \mathrm{i}B\boldsymbol{\psi}_x + \mathrm{i}C(\boldsymbol{\psi})\boldsymbol{\psi}_x + D\boldsymbol{\psi}_{xx} + E(\boldsymbol{\psi})\boldsymbol{\psi} = \mathbf{0}.$$

The coupled NLS equations describe phase-matching resonance in quadratic χ^2 materials, gap solitons in periodic photonic gratings under Bragg resonance, Alfvén waves in plasmas, and other applications (Newell & Moloney, 1992).

In conservative nonlinear systems, wave dynamics of small amplitudes occur typically in the neighborhood of local minima of potential energy. Wave oscillations in the system of nonlinear massive pendulums are described by the Frenkel–Kontorova dislocation model. In a continuous approximation, the Frenkel–Kontorova lattice model reduces to the sine-Gordon (SG) equation (Braun & Kivshar, 1998):

$$\varphi_{tt} - c^2\varphi_{xx} + \sin\varphi = 0,$$

where φ is the angle between a pendulum and the vertical axis in a mechanical model. The nonlinear pendulums swing on a rigid rod under the gravity force and couple to each other with elastic springs. More complicated models of molecular crystals and ferromagnetics in solid state mechanics, stacked Josephson contacts in superconductivity, and strings in the general relativity theory are formulated as coupled systems of sine-Gordon equations (Maugin, 1999). The coupled Klein–Gordon equations take the form

$$\boldsymbol{\varphi}_{tt} - C\boldsymbol{\varphi}_{xx} + \boldsymbol{f}(\boldsymbol{\varphi}) = 0,$$

where C is a positive-definite matrix and $\boldsymbol{f}(\boldsymbol{\varphi})$ is the nonlinear vector function of components of the vector $\boldsymbol{\varphi} = \boldsymbol{\varphi}(x,t)$.

In more general systems, the energy of a nonlinear wave changes in time due to active and dissipative forces. The simplest system of this type is the nonlinear heat equation, which models the flame propagation (Zeldovich et al., 1985):

$$u_t = Du_{xx} + f(u),$$

where $u = u(x,t)$ is the temperature and D is the diffusivity constant. A complex form of the nonlinear heat equation (known as the Ginzburg–Landau equation) is derived for the amplitude of the most unstable wave mode (Newell & Moloney, 1992).

Active and dissipative systems include typically pairs of coupled activators and inhibitors. Coupled activator-inhibitor equations, known as the reaction-diffusion systems, are derived from the governing equations of thermodynamics in the form (Remoissenet, 1999):

$$\boldsymbol{u}_t = C\boldsymbol{u}_x + D\boldsymbol{u}_{xx} + \boldsymbol{f}(\boldsymbol{u}),$$

where C and D are matrices, and $\boldsymbol{f}(\boldsymbol{u})$ is a nonlinear vector function of components of the vector $\boldsymbol{u} = \boldsymbol{u}(x,t)$. Reaction-diffusion systems exhibit static, traveling, and pulsating nonlinear wave structures such as fronts and impulses. Coupled reaction-diffusion systems include the FitzHugh–Nagumo and Hodgkin–Huxley equations for nerve impulses, ephaptic coupling among nerve impulses, and models of the global dynamics of heart.

DMITRY PELINOVSKY

See also **Ephaptic coupling; Nonlinear optics; Reaction-diffusion systems; Sine-Gordon equation; Water waves**

Further Reading

Akhmediev, N. & Ankiewicz, A. 1997. *Solitons, Nonlinear Pulses, and Beams*, London: Chapman & Hall

Braun, O.M. & Kivshar, Yu.S. 1998. Nonlinear dynamics of the Frenkel–Kontorova model. *Physics Reports*, 306: 1–109

Bryant, T., 2001. *Tsunami: The Underrated Hazard*, Cambridge and New York: Cambridge University Press

Engelbrecht, J.K., Fridman V.E. & Pelinovski E.N., 1988. *Nonlinear Evolution Equations*, London: Longman and New York: Wiley

Grimshaw, R. (editor). 2001. *Environmental Stratified Flows*, Boston: Kluwer

Johnson, R.S. 1997. *A Modern Introduction to the Mathematical Theory of Water Waves*, Cambridge and New York: Cambridge University Press

Manakov, S.V. 1974. On the theory of two-dimensional stationary self-focusing of electromagnetic waves. *Soviet Physics, JETP*, 38: 248–253

Maugin, G.A. 1999. *Nonlinear Waves in Elastic Crystals*, Oxford and New York: Oxford University Press

Newell, A.C. & Moloney, J.V. 1992. *Nonlinear Optics*, Redwood City, CA: Addison-Wesley

Remoissenet, M. 1999. *Waves Called Solitons*. Berlin and New York: Springer

Whitham, G. 1974. *Linear and Nonlinear Waves*. New York: Wiley

Zeldovich, Ya.B., Barenblatt, G.I., Librovich, V.B. & Makhviladze, G.M. 1985. *The Mathematical Theory of Combustion and Explosions*. New York: Consultants Bureau

CRITICAL PHENOMENA

The term *critical phenomenon* is used synonymously with "phase transition," which involves a change of one system phase to another and occurs at a characteristic temperature (called a transition temperature or a critical temperature: T_c). There are several different kinds of phase transitions such as melting, vaporization, and sublimation, as well as solid-solid, conducting-superconducting, and fluid-superfluid transitions. In systems undergoing phase transitions, an emergence of long-range order is seen in which the value of a physical quantity at one arbitrary point in the system is correlated with its value at another point a long distance away.

A classification scheme of phase transitions which remains the most popular was originally proposed by Paul Ehrenfest. According to this scheme, a transition for which the first derivative of the free energy with respect to temperature is discontinuous is called a first-order phase transition; thus, the heat capacity, C_p, at a first-order transition is infinite. A second-order phase transition is one in which the first derivative of the thermodynamic potential with respect to temperature is continuous, but its second derivative is discontinuous, so the heat capacity is discontinuous but not infinite at the transition. Near a second-order phase transition (due to the reduction of rigidity of the system), critical fluctuations dominate as their amplitudes diverge.

A useful concept in analyzing phase transitions is that of a critical exponent. In general, if a physical quantity $Q(T)$ either diverges or tends to a constant value (see Figure 1) as T tends to T_c, it can be characterized by defining the reduced temperature ε as

$$\varepsilon \equiv \frac{T - T_c}{T_c}. \tag{1}$$

The associated critical exponent is

$$\mu = \lim_{\varepsilon \to 0} \frac{\ln Q(\varepsilon)}{\ln \varepsilon}. \tag{2}$$

The most important critical exponents are denoted as $\alpha, \beta, \gamma, \delta, \upsilon$, and η and describe the specific heat, order parameter, isothermal susceptibility, response to an external field, the correlation length, and the pair correlation function, respectively. (See Table 1 where

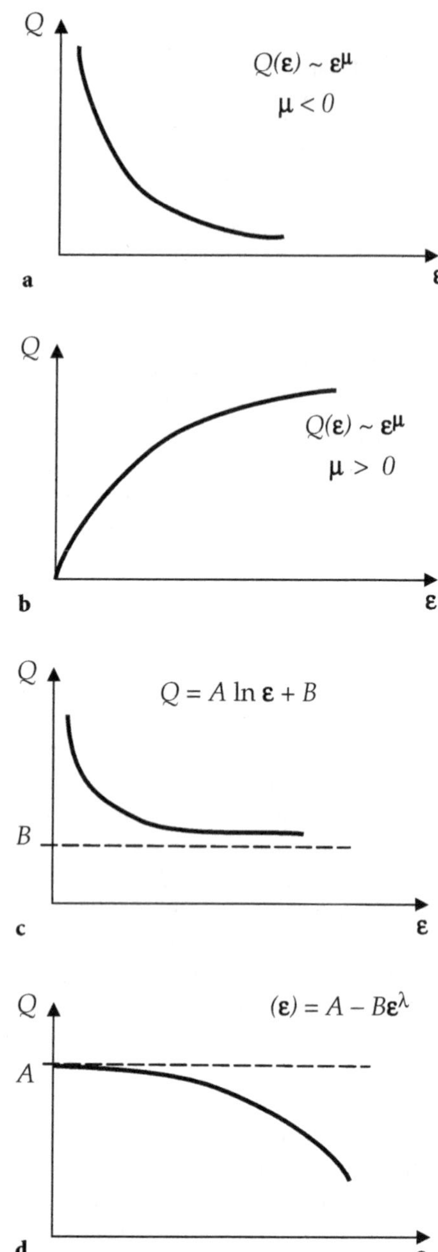

Figure 1. The four generic behaviors near criticality.

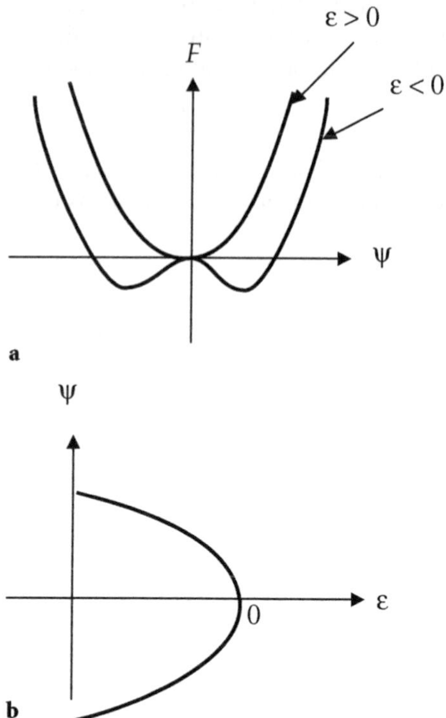

Figure 2. A prototype of second-order phase transitions according to Landau.

the primed exponents are introduced for temperatures below the critical temperature while the unprimed exponents are valid above the critical temperature.)

The mean field approximation (Landau theory) describes the physics of phase transitions well except in the immediate vicinity of the critical point where order parameter fluctuations are large. It is assumed that close to T_c, the free energy F can be expanded in a Taylor series of the order parameter ψ. Introducing the reduced temperature ε as a control parameter, the simplest such expansion is (see Figure 2)

$$F(T, V, \psi) = F_0 + a\varepsilon\psi^2 + A_4\psi^4, \tag{3}$$

where $a > 0$ and $A_4 > 0$. Solving the equilibrium conditions for ψ yields $\psi = 0$ for $\varepsilon > 0$ and

$$\psi = \pm\left(-\frac{a\varepsilon}{2A_4}\right)^{1/2} \quad \text{for } \varepsilon < 0, \tag{4}$$

thus $\beta = 0.5$ is obtained. Calculating the entropy

$$S = \frac{\partial F}{\partial T} = S_0 + \frac{a^2}{2A_4}\frac{\varepsilon}{T_c} \quad \varepsilon \leq 0, \tag{5}$$

where for $\varepsilon > 0$, $S = S_0$ is the entropy of the disordered phase, which gives the specific heat as

$$C_v = T\frac{\partial S}{\partial T} = C_0 + \frac{a^2}{2A_4T_c}T \quad \varepsilon \leq 0, \tag{6}$$

where for $\varepsilon > 0$, $C_v = C_0$ is the specific heat of the disordered phase. Hence, a discontinuity occurs at T_c (see Figure 3).

$$\Delta C = \frac{a^2}{2A_4T_c}. \tag{7}$$

Thus, $\alpha = 0$.

Including in F an external field h coupled to ψ

$$F = F_0 + a\varepsilon\psi^2 + A_4\psi^4 - h\psi \tag{8}$$

Exponent	Definition (liquid-vapor) Specific heat at constant volume	Definition (magnetic) Specific heat at constant H
α'	$C_v \sim (-\varepsilon)^{-\alpha'}$	$C_H \sim (-\varepsilon)^{-\alpha'}$
α	$C_v \sim \varepsilon^{-\alpha}$	$C_H \sim \varepsilon^{-\alpha}$
	Density difference	Magnetization
β	$\rho_L - \rho_G \sim (-\varepsilon)^{\beta}$	$M \sim (-\varepsilon)^{\beta}$
	Isothermal compressibility	Isothermal susceptibility
γ'	$\kappa_T \sim (-\varepsilon)^{-\gamma'}$	$\chi_T \sim -\varepsilon^{-\gamma'}$
γ	$\kappa_T \sim \varepsilon^{-\gamma}$	$\chi_T \sim \varepsilon^{-\gamma}$
	Pressure-density critical isotherm	Magnetic field-magnetization
δ	$P - Pc \sim \lvert\rho_L - \rho_G\rvert^{\delta}$ $(T = T_c)$	$H \sim \lvert M\rvert^{\delta}$ $(T = T_c)$
	Correlation length	Correlation length
ν'	$\xi \sim (-\varepsilon)^{-\nu'}$	$\xi \sim -\varepsilon^{-\nu'}$
ν	$\xi \sim \varepsilon^{-\nu}$	$\xi \sim \varepsilon^{-\nu}$
	Density–density pair correlation	Spin-spin pair correlation
η	$\Gamma(r) \sim \lvert r\rvert^{-(d-2+\eta)}$	$\Gamma(r) \sim \lvert r\rvert^{-(d-2+\eta)}$

Table 1. The definitions of critical exponents for liquid-vapor and magnetic systems. [The primed (unprimed) exponents are for temperatures below (above) T_c.]

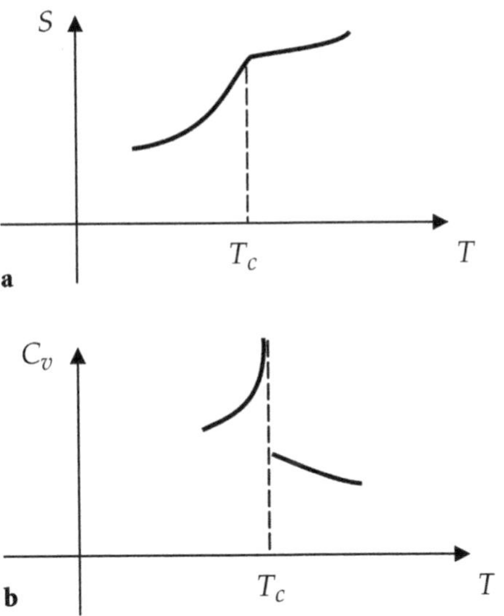

Figure 3. Plots of $S(T)$ and $C_v(T)$ in the Landau model of a second-order phase transition.

and minimizing F with respect to ψ yields an equation of state of the form

$$h = 2\psi\lfloor a\varepsilon + 2A_4\psi^2\rfloor. \tag{9}$$

Because the susceptibility $\chi \equiv \partial\psi/\partial h$, we find

$$\chi = [2a\varepsilon + 12A_4\psi^2]^{-1}. \tag{10}$$

As $\psi \to 0$ as $T \to T_c$, the third exponent is $\gamma = 1$. At $T = T_c$, the equation of state (9) simplifies to:

$$h \cong 4A_4\psi^3, \tag{11}$$

and hence $\psi \sim h^{1/3}$ giving $\delta = 3$. The quartic expansion in the Landau model invariably leads to the classical critical exponents: $\alpha = 0$, $\beta = 0.5$, $\gamma = 1$, and $\delta = 3$.

While the Landau theory cannot describe spatial fluctuations, following Ginzburg and Landau's proposal, it can be extended to consider the free energy to be a functional:

$$F(\psi(\vec{r}), T) = \int d^3\mathbf{r}[A_2\psi^2 + A_4\psi^4 - h\psi + D(\nabla\psi)^2] \tag{12}$$

where D describes the energy due to spatial inhomogeneities. Applying a variational principle to F results in a nonlinear Klein–Gordon equation for the order parameter

$$h = 2A_2\psi + 4A_4\psi^3 - 2D\nabla^2\psi. \tag{13}$$

A linearized solution of Equation (13) in spherical coordinates is

$$\psi = \frac{h_0}{4\pi D}\frac{e^{-r/\xi}}{r}, \tag{14}$$

where $\xi \sim A_2^{-1/2}$ is the correlation length that diverges as $T \to T_c$ so that the critical exponent $\upsilon = 0.5$. Fourier transforming the order parameter according to

$$\psi(\vec{r}) \equiv L^{-d/2} \sum_{k<k_0} \psi_k e^{i\vec{k}\cdot\vec{r}}, \tag{15}$$

where d is the spatial dimensionality and the cutoff wavelength $k_0 = \Lambda^{-1}$ corresponds to the smallest periodicity, F becomes (for $h = 0$)

$$F = \sum_{k<k_0} |\psi_k|^2(A_2 + Dk^2) + L^{-d} \sum_{k,k',k''<k_0} A_4\psi_k^*\psi_{k'}^*\psi_{k''}\psi_{k+k'-k''} \tag{16}$$

Ignoring mode-mode coupling provides the basis for a Gaussian approximation, where

$$F \cong \sum_{k<k_0} |\psi_k|^2 (A_2 + Dk^2). \tag{17}$$

The Fourier transform of the correlation function is found as

$$\Gamma(k) = L^{-d} \int e^{-i\vec{k}(\vec{r}-\vec{r}')} < \psi(\vec{r})\psi(\vec{r}') > d\vec{r}\, d\vec{r}'$$
$$= \langle |\psi_k|^2 \rangle \cong \frac{n}{2}(A_2 + Dk^2)^{-1} \tag{18}$$

Therefore, as $T \to T_c$, we find that $\Gamma(k) \sim k^{\eta-2}$ with $\eta = 0$ in the Gaussian approximation and also $\Gamma(r) \sim r^{-(d-2)}$. Furthermore, the Gaussian approximation introduces a new value of the specific heat critical exponent $\alpha = 2 - d/2$.

JACK A. TUSZYŃSKI

See also **Bifurcations; Ferromagnetism and ferroelectricity; Order parameters; Phase transitions**

Further Reading

Anderson, P.W. 1984. *Basic Notions of Condensed Matter Physics*, Menlo Park, CA: Benjamin Cummings

Landau, L.D. & Lifshitz, E.M. 1959. *Statistical Physics*, London: Pergamon

Ma, S.-K. 1976. *Modern Theory of Critical Phenomena*, New York: Benjamin

Reichl, L.E. 1979. *A Modern Course in Statistical Physics*, Austin: University of Texas Press

Stanley, H.E. 1971. *Introduction to Phase Transitions and Critical Phenomena*, Oxford: Clarendon Press and New York: Oxford University Press

White, R.H. & Geballe, T. 1979. *Long Range Order in Solids*, New York: Academic Press

Yeomans, J.M. 1992. *Statistical Mechanics of Phase Transitions*, Oxford: Clarendon Press and New York: Oxford University Press

CRITICAL POINTS

See **Critical phenomena**

CRYSTAL DISLOCATIONS

See **Dislocations in crystals**

CRYSTALS, MOLECULAR

See **Local modes in molecular crystals**

CUBIC MAP

See **Maps**

DAMPED-DRIVEN ANHARMONIC OSCILLATOR

Many problems in engineering and the applied sciences are modeled by low-dimensional forced oscillators. Examples include vibrating mechanical and structural engineering systems (beams, bridges), marine systems (ships, oil platforms), electronic circuits and devices (Josephson junctions), as well as biological oscillators. While the linear behavior of such systems has long been well understood, the potential dynamics of nonlinear oscillators is far more complex and many open problems remain.

Damped Nonlinear Pendulum

Let us consider a mathematical pendulum and denote with θ the angular displacement of the pendulum from the equilibrium (hanging down) position. The equation of motion for the oscillations of a damped nonlinear pendulum of unitary mass is

$$\frac{\mathrm{d}^2\theta}{\mathrm{d}t^2} + \eta\frac{\mathrm{d}\theta}{\mathrm{d}t} + \frac{g}{l}\sin(\theta) = 0, \qquad (1)$$

where l is the length of the pendulum and g is the gravitational acceleration. This equation expresses Newton's second law with the terms on the left-hand side, respectively, representing acceleration, damping, and gravitation. Equation (1) may be rewritten as two first-order equations:

$$\begin{cases} \dot{\theta} &= \omega, \\ \dot{\omega} &= -\eta\omega - \frac{g}{l}\sin(\theta), \end{cases} \qquad (2)$$

where ω denotes the angular velocity. Then, one can use (θ, ω) as phase space coordinates.

Fixed points: The dynamics of the nonlinear pendulum can be analyzed as trajectories in the phase space. The rest position $\theta = \omega = 0$ is called a *fixed point*. It is easy to see that almost all phase space trajectories spiral into the fixed point at $(0, 0)$. This is a linearly stable fixed point, since if a small perturbation is made from the fixed point, the perturbation decays in time (exponentially for small enough perturbations). There is a second fixed point at $(\pi, 0)$ corresponding to the pendulum pointing vertically up. The fixed point at $(\pi, 0)$ is linearly unstable because a small perturbation from this fixed point grows exponentially. Only very carefully tuned initial conditions will lead to a trajectory ending on the unstable fixed point, and almost all perturbations to the initial condition will lead to a trajectory that may approach close to the unstable fixed point, but eventually spirals into the stable fixed point. The $(0, 0)$ fixed point is attracting, and in this case the basin of attraction, that is, the set of initial conditions leading to trajectories that approach the fixed point, is the whole phase space except for a set of zero measure.

An example of a trajectory of system (2) converging towards the stable fixed point at $(0, 0)$ is shown in Figure 1.

The Periodically Driven Damped Pendulum

Let us now consider the nonlinear damped pendulum of Equation (1) when a harmonic driving force is applied. The equation of motion for the oscillations of a damped, sinusoidally driven nonlinear pendulum is

$$\frac{\mathrm{d}^2\theta}{\mathrm{d}t^2} + \gamma\frac{\mathrm{d}\theta}{\mathrm{d}t} + \sin(\theta) = A_{\mathrm{D}}\cos(\omega_{\mathrm{D}}t), \qquad (3)$$

where A_{D} is the forcing amplitude and the angular velocity of the forcing, ω_{D}, may be different from the natural frequency of the pendulum. Note, with respect to Equation (1), that the time variable has been rescaled so that the period of small oscillations of the undamped and undriven pendulum is unity, and the (scaled) dissipation coefficient is now denoted by γ. Physically, the situation corresponds to driving the pendulum, feeding in energy to resupply the energy dissipated.

Small-Amplitude Oscillations

For small driving amplitudes, A_{D}, and assuming a small initial condition, one can approximate $\sin\theta \approx \theta$ and

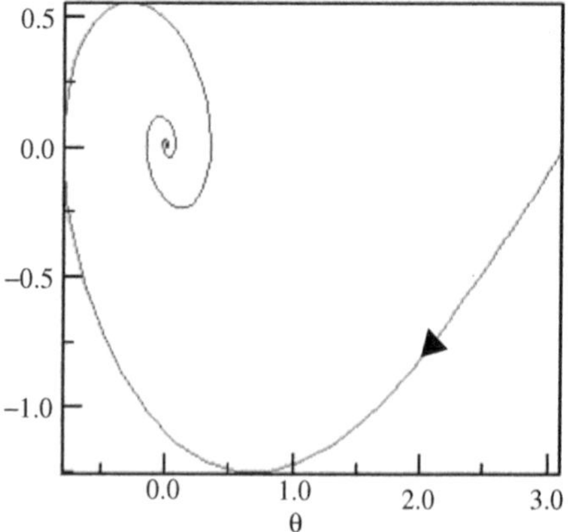

Figure 1. Typical trajectory for the nonlinear pendulum of Equation (2). All initial conditions lead to an orbit spiraling into the fixed point at (0, 0).

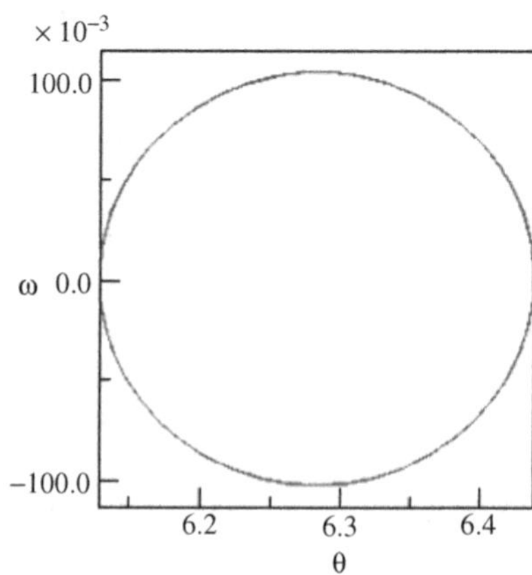

Figure 2. If small driving is turned on ($A_D = 0.1$), the steady-state solution is a stable limit cycle. Note that, for simplicity, the transient dynamics are not shown.

solve Equation (3) analytically:

$$\theta = \frac{A_D}{\sqrt{(1-\omega_D^2)^2 + \gamma^2\omega_D^2}}\cos(\omega_D t + \Phi) + A_0 e^{-\gamma t/2}\cos(\omega t + \varphi_0), \tag{4}$$

where

$$\tan\Phi = -\frac{\gamma\omega_D}{(1-\omega_D^2)}, \quad \omega = \sqrt{1 - \frac{\gamma^2}{4}}, \quad \gamma < 2 \tag{5}$$

while the constants A_0 and φ_0 depend upon the initial conditions (θ_0, ω_0), that is, for $t = 0$.

Equation (4) describes the resonant response (the first term) oscillating at the applied frequency, together with decaying free oscillations (the second term), depending on the initial conditions. This kind of solution is called an attracting *limit cycle.*

A typical limit cycle at this small amplitude, for $A_D = 0.1$, is shown in Figure 2 and is seen to be close to circular. From now on, although the transient may be quite complicated, only the asymptotic behavior, that is, the behavior of the attractors will be considered.

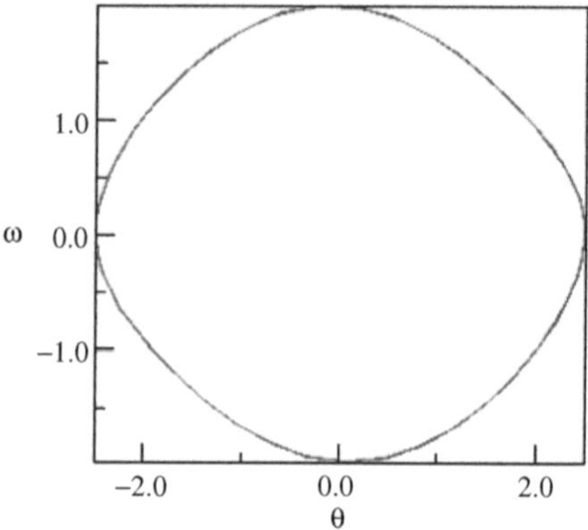

Figure 3. Large-amplitude limit cycle, for $A_D = 0.9$.

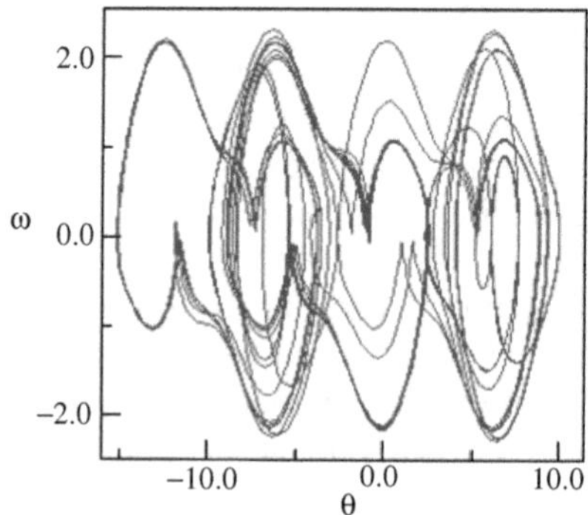

Figure 4. Chaotic attractor, for driving amplitude $A_D = 1.15$.

Large-Amplitude Solutions

What happens for large driving amplitudes? Here, there are no analytic solutions, and one must proceed numerically. To gain some intuition it is more convenient to view the dynamics in the phase space. To this end, Equation (3) may be converted to autonomous form by using three variables, as follows:

$$\begin{cases} \dot\theta &= \omega, \\ \dot\omega &= -\gamma\omega - \sin(\theta) + A_D\cos(\theta_D), \\ \dot{\theta_D} &= \omega_D, \end{cases} \tag{6}$$

where the phase of the driving, θ_D, has been introduced.

If one increases the driving strength to $A_D = 0.9$, a large-amplitude limit cycle is observed, as shown in Figure 3. This limit cycle bifurcates for increasing values of the driving strength according to the well-known period-doubling route to chaos, leading to a chaotic attractor for $A_D = 1.09$.

Then, increasing the driving to $A_D = 1.15$ results in an even more irregular chaotic behavior, as illustrated in Figure 4. But, perhaps at first sight surprisingly, increasing the driving further, for example, to $A_D = 1.35$ gives periodic motion again. In this case, though, there is a drift along with the oscillation—the pendulum is oscillating but also consistently rotating all the way around.

In summary, numerical simulations show that both periodic and chaotic solutions of the driven damped pendulum are possible, depending on the particular choice of the system parameters. Chaos may indeed occur since the driven pendulum equation is equivalent to a three-dimensional nonlinear autonomous system. Note the contrast with linear theory, where a damped system driven periodically *must* eventually respond periodically.

GIAN MARIO MAGGIO AND LJUPCO KOCAREV

See also **Chaotic dynamics; Coupled oscillators; Distributed oscillators; Pendulum; Relaxation oscillators; Semiconductor oscillators**

Further Reading

Andronov, A.A., Vitt, A.A. & Khaikin, S.E. 1966. *Theory of Oscillators*, New York: Dover

Baker, G.L. & Gollub, J.P. 1996. *Chaotic Dynamics*, Cambridge and New York: Cambridge University Press

Guckenheimer, J. & Holmes, P. 1983. *Nonlinear Oscillations, Dynamical Systems, and Bifurcations of Vector Fields*, New York: Springer

Hagedorn, P. 1988. *Non-Linear Oscillations*, Oxford: Clarendon Press

Mitropolskii, Y.A. & Van Dao, N. 1997. *Applied Asymptotic Methods in Nonlinear Oscillations*, Dordrecht: Kluwer

Ott, E. 1993. *Chaos in Dynamical Systems*, Cambridge: Cambridge University Press

Parker, T.S. & Chua, L.O. 1989. *Practical Numerical Algorithms for Chaotic Systems*, New York: Springer

Thompson, J.M.T. & Stewart, H.B. 2002. *Nonlinear Dynamics and Chaos*, 2nd edition, New York: Wiley

DARBOUX TRANSFORMATION

From a historical point of view, the Darboux transformation is the specialization of a more general result due to Théodore Moutard. The prior result is of application in the classical theory of surfaces (Darboux, 1887–96) and takes the following form.

Let $\phi_0(x, y)$ be any particular integral of the hyperbolic partial differential equation

$$\phi_{xy} + q(x, y)\phi = 0, \tag{1}$$

$q(x, y)$ being a given function. We may rewrite this equation in the form

$$(D_x^+ D_y^- + D_y^+ D_x^-)\phi = 0, \tag{2}$$

where the Ds are differential operators, $D_u^{\pm} = \partial_u \pm \phi_0^{-1}\phi_{0u}$. Now let $\psi(x, y)$ and $\phi(x, y)$ be functions satisfying the pair of first-order partial differential equations

$$\begin{aligned} D_x^- \phi &= -D_x^+ \psi, \\ D_y^- \phi &= D_y^+ \psi, \end{aligned} \tag{3}$$

then (1) is equivalent to the identity $[D_x^+, D_y^+]\psi = 0$. Alternatively, the identity $[D_x^-, D_y^-]\phi = 0$ yields an equation for ψ, namely,

$$\psi_{xy} + p(x, y)\psi = 0, \tag{4}$$

where $p(x, y) = q(x, y) + 2(\ln \phi_0)_{xy}$. The class of equations of form (1) is thus covariant under maps $\mu_{\phi_0} : (\phi, q) \mapsto (\psi, p)$.

The map μ_{ϕ_0} is the *Moutard* transformation and its value lies in its relating, and generating, hyperbolic equations integrable in finite terms (Darboux, 1887–96).

Darboux (1882) specialized this result to the case where (1) carries a symmetry, that is, where $q(x, y) = Q(x + y)$ is a function of $x + y$ only. Putting $z = x + y$, $\bar{z} = x - y$ and applying the standard technique of separation of variables leads to $\phi(x, y) = \exp(\lambda\bar{z})\Phi(z, \lambda)$ and the equation for Φ,

$$\Phi_{zz} + Q(z)\Phi = \lambda^2\Phi. \tag{5}$$

Given a solution Φ_0 to this equation with $\lambda = \lambda_0$, there is a corresponding $\phi_0(x, y) = \exp(\lambda_0\bar{z})\Phi_0(z, \lambda_0)$ and after a simple, constant rescaling of Φ, Equations (3) become

$$\begin{aligned} \Psi &= (\partial_z - \Phi_0^{-1}\Phi_{0z})\Phi, \\ (\lambda^2 - \lambda_0^2)\Phi &= (\partial_z + \Phi_0^{-1}\Phi_{0z})\Psi \end{aligned} \tag{6}$$

and Equation (5) becomes

$$\Psi_{zz} + P(z)\Psi = \lambda^2\Psi. \tag{7}$$

Here $P(z) = Q(z) + 2(\ln \Phi_0)_{zz}$.

The map $\delta_{\lambda_0} : (\Phi, Q) \mapsto (\Psi, P)$ is the *Darboux* transformation (DT). Note that it creates from the Φ equation one member of a family of equations depending on the free parameter λ_0.

Starting with $Q = 0$ and a specific choice λ_0 for λ, we may take $\Phi_0 = c_1 \cosh \lambda_0(z + c_2)$. Then $P(z) = 2\lambda_0^2 \operatorname{sech}^2 \lambda_0(z + c_2)$ and so, from the trivial ("vacuum") potential, the DT has generated a family of potential barriers involving the chosen λ_0 and a parameter of translation, c_2, in the z-direction.

The linear PDE (5) plays a crucial role in the Inverse Scattering Transform (IST). We suppose the potential Q to depend upon an additional parameter t ("time"): $Q = Q(z, t)$. The pair of linear PDE (Lax pair),

$$L_Q\Phi = \lambda^2\Phi, \tag{8}$$

$$M_Q\Phi = \partial_t\Phi, \tag{9}$$

where $L_Q = \partial_z^2 + Q$ and $M_Q = 4\partial_z^3 + 6Q\partial_z + 3Q_z$, is integrable in the Cauchy sense provided Q satisfies the nonlinear PDE

$$Q_t = Q_{zzz} + 6QQ_z, \tag{10}$$

namely, the Korteweg–de Vries (KdV) equation. This follows from the operator relation (Frobenius condition)

$$L_Q t = [M_Q, L_Q] \equiv M_Q L_Q - L_Q M_Q. \tag{11}$$

The DT, $\Psi = \mathcal{D}\Phi$ for $\mathcal{D} = \partial_z + \alpha$, $\alpha = -\Phi_0^{-1}\Phi_{0z}$, is seen to play the role of an intertwining operator:

$$\mathcal{D}L_Q = L_P\mathcal{D}, \tag{12}$$

and

$$\partial_t\Psi = M_P\Psi + \{\alpha_t + \mathcal{D}M_Q - M_P\mathcal{D}\}\Phi. \tag{13}$$

Some calculation shows that the bracketed term is vanishing and we have the Lax pair

$$L_P\Psi = \lambda^2\Psi, \tag{14}$$

$$M_P\Psi = \partial_t\Psi, \tag{15}$$

in the DT transformed variables Ψ and P. Consequently, the transformed potential P will also satisfy the KdV equation (10).

Returning now to the simple example above, Φ_0 will satisfy Equation (9) with $Q = 0$, namely $\Phi_{0t} = 4\Phi_{0zzz}$, so that $\Phi_0 = c_1\cosh(\lambda_0 z + 4\lambda_0^3 t + c_3)$ and the DT generates the traveling wave (1-soliton) solution to the KdV,

$$P = 2\lambda_0^2\,\mathrm{sech}^2(\lambda_0 z + 4\lambda_0^3 t + c_3). \tag{16}$$

Imagine now that we wish to form a sequence of DTs. Starting with (5), we make a choice Φ_0, λ_0 to form (7) via (6). Now for (7) we choose Ψ_1, λ_1 and use

$$\Theta = (\partial_z - \Psi_1^{-1}\Psi_{1z})\Psi$$

to generate Θ, satisfying

$$\Theta_{zz} + R(z)\Theta = \lambda^2\Theta,$$

where $R = P + 2(\ln\Psi_1)_{zz}$. But since Ψ_1 satisfies (6) with $\lambda = \lambda_1$, it can itself be constructed as

$$\Psi_1 = \Phi_{1z} - \Phi_0^{-1}\Phi_{0z}\Phi_1,$$

where Φ_1 satisfies (5) with $\lambda = \lambda_1$. Then simple algebra yields formulae for Θ and R:

$$\Theta = \begin{vmatrix} \Phi_0 & \Phi_1 \\ \Phi_{0z} & \Phi_{1z} \end{vmatrix}^{-1} \begin{vmatrix} \Phi_0 & \Phi_1 & \Phi \\ \Phi_{0z} & \Phi_{1z} & \Phi_z \\ \Phi_{0zz} & \Phi_{1zz} & \Phi_{zz} \end{vmatrix}, \tag{17}$$

$$R = Q + 2\ln\left(\begin{vmatrix} \Phi_0 & \Phi_1 \\ \Phi_{0z} & \Phi_{1z} \end{vmatrix}\right)_{zz}. \tag{18}$$

From the symmetry of these formulae under interchange of λ_0 and λ_1, it follows that the order of composition of DTs is immaterial. The fundamental result of Crum (1955) is that, starting with $n+1$ pairs (Φ_i, λ_i), $i = 0, \ldots, n$, associated with (5) one obtains the n-fold DT:

$$\begin{aligned}\delta_0 \ldots \delta_n(\Phi, Q) = {} & (W^{-1}(\Phi_0, \ldots, \Phi_n) \\ & \times W(\Phi_0, \ldots, \Phi_n, \Phi), Q \\ & + 2\ln W(\Phi_0, \ldots, \Phi_n)_{zz}),\end{aligned}$$

where W denotes the Wronskian function of its arguments.

This result, when applied to a set of n single soliton solutions of the KdV, yields an explicit n-soliton solution.

Generally speaking, a transformation of dependent variables that carries solutions of one DE to solutions of another is termed a *Bäcklund* transformation (BT). The term *Darboux* transformation refers to the special case of linear equations. We have described above a DT for a Lax pair that gives rise to a BT for the KdV equation.

For DTs of higher order, ordinary linear differential equations have been developed. For such results the intertwining property can be related to the factorization of differential operators, an approach to solvable Schrödinger potentials pioneered in quantum mechanics. This naturally leads to their algebraic formulation in terms of $\mathcal{D}$-modules. Moutard (sometimes called the Binary DT) and Darboux transformations have also been generalized to higher dimensions and utilized for solving associated analogues of the KdV with three independent variables, the Kadomtsev–Petviashvili, Novikov–Veselov, and Davey–Stewartson equations, to name but three.

The double logarithmic derivative occurring in the expression of the DT transformed potentials is of fundamental importance in the bilinear theory of soliton equations originally due to Hirota.

In the Hamiltonian theory of soliton equations the DT is seen to play the role of a canonical transformation.

A comparatively recent application of DTs has been in the theory of integrable systems with discrete variables, that is, systems defined on lattices (Antoniou & Lambert, 2000).

Finally, it should be remarked that the geometrical insights of Gaston Darboux, Luigi Bianchi, and their contemporaries are being rediscovered and developed within the modern theory of integrable systems and that Darboux and Moutard transformations are playing a key role in this resurgence (Rogers & Schief, 2002).

There are a number of monographs where further details about DTs and BTs may be found, for example, Calogero & Degasperis (1982); Konopelchenko (1987); Matveev & Salle (1991); Rogers & Shadwick (1982). For more recent reviews of work in this area see Coley (2001).

CHRIS ATHORNE

See also **Bäcklund transformations; Inverse scattering method or transform**

Further Reading

Antoniou, I. & Lambert, F. (editors). 2000. Integrability and chaos in discrete systems. Proceedings Brussels Meeting 1997. *Chaos, Solitons and Fractals*, 11(1–3)

Calogero, F. & Degasperis, A. 1982. *The Spectral Transform and Solitons*, Amsterdam: North-Holland

Coley, A. (editor). 2001. Bäcklund and Darboux transformations. The geometry of solitons, *Proceedings AARMS-CRM Workshop, Halifax, Nova Scotia, 1999*, Providence, RI: American Mathematical Society

Crum, M. 1955. Associated Sturm–Liouville systems. *Quarterly Journal of Mathematics*, 6: 121–128

Darboux, G. 1882. Sur une proposition relative aux équations linéaires. *Comptes Rendus de l'Academie des Sciences, Paris*, 94: 1343; 1456–1459

Darboux, G. 1887–96. *Leçons sur la Théorie des Surfaces*, Parts I–IV, Paris: Gauthier-Villars

Konopelchenko, B.G. 1987. *Nonlinear Integrable Equations*, Berlin and New York: Springer

Matveev, V.B. & Salle, M.A. 1991. *Darboux Transformations and Solitons*, Berlin and New York: Springer

Rogers, C. & Shadwick, W. 1982. *Bäcklund Transformations and their Applications*, New York: Academic Press

Rogers, C. & Schief, W.K. 2002. *Bäcklund and Darboux Transformations*, Cambridge and New York: Cambridge University Press

DAVEY–STEWARTSON EQUATION

See **Multidimensional solitons**

DAVYDOV SOLITON

In 1972, a meeting was convened in order to search for the general principles that underlie biological processes such as muscle contraction, active transport, enzyme catalysis, and oxidative phosphorylation (Green, 1974). One of the main questions asked was whether there was a crisis in bioenergetics. While some participants disagreed that a crisis existed, others proposed solutions to the crisis. One solution was the "conformon" of Green and Ji, defined as the free energy associated with a localized conformational strain, which was to store and transport energy (Green, 1974, pp. 419–437). McClare's paper, on the other hand, suggested that vibrational excited states and resonant energy transfer constitute a fundamental step in energy transduction in biological systems (Green, 1974, pp. 74–97). The main objection to McClare's proposal was that the lifetime of vibrational excited states, thought to be in the subpicosecond time range, was considered too short for these states to be useful. Alexander Davydov's model, proposed in connection with a mechanism for muscle contraction (Davydov, 1982), aimed to answer this objection.

The Davydov Model

The Davydov model describes energy transfer in the hydrogen-bonded spines that stabilize protein α-helices:

$$\cdots \underset{n-1}{\text{H-N-C=O}} \cdots \underset{n}{\text{H-N-C=O}} \cdots \underset{n+1}{\text{H-N-C=O}} \cdots$$

The idea is that the energy liberated in the hydrolysis of adenosine triphosphate (ATP) creates up to two quanta of amide-I, an excited vibrational state in the peptide group that is essentially a stretching vibration in the C=O bond (Davydov, 1982). This vibration excitation propagates from one group to the next because of the dipole-dipole interaction between the groups. But it also interacts with the neighboring hydrogen bonds, leading to a deformation of the lattice and a lower energy state. This new state, which is constituted by an amide-I excitation and its associated hydrogen bond distortion, is the Davydov soliton. Structurally, the Davydov soliton is, in fact, a subtle local conformational change of the α-helix, similar to the conformon.

Davydov's Hamiltonian has three parts:

$$\hat{H} = \hat{H}_{\text{ex}} + \hat{H}_{\text{int}} + \hat{H}_{\text{ph}}, \tag{1}$$

where $\hat{H}_{\text{ex}}$, the exciton Hamiltonian, describes the transfer of an amide-I excitation between adjacent sites

$$\hat{H}_{\text{ex}} = -V \sum_{n=1}^{N} \left[\left(\hat{a}_n^{\dagger} \hat{a}_{n-1} + \hat{a}_n^{\dagger} \hat{a}_{n+1} \right) \right], \tag{2}$$

$-V$ being the dipole-dipole interaction energy between neighboring sites, $\hat{a}_n^{\dagger}(\hat{a}_n)$ the creation (annihilation) operator for a quantum quasiparticle (the amide-I excitation) at site n, and N being the number of sites (the peptide groups) in the lattice; $\hat{H}_{\text{ph}}$, the phonon Hamiltonian, describes the vibrations of the peptide groups n in the one-dimensional chain:

$$\hat{H}_{\text{ph}} = \frac{1}{2} \sum_{n=1}^{N} \left[\kappa \left(\hat{u}_n - \hat{u}_{n-1} \right)^2 + \frac{\hat{P}_n^2}{M} \right], \tag{3}$$

$\hat{u}_n$ being the displacement operator from the equilibrium position of site n, $\hat{P}_n$ the momentum operator of site n, M the mass of each site, and κ the elasticity constant of the lattice; $\hat{H}_{\text{int}}$, the interaction Hamiltonian, describes the interaction of amide-I excitation with the motions of the lattice sites:

$$\hat{H}_{\text{int}} = \chi \sum_{n=1}^{N} \left[\left(\hat{u}_{n+1} - \hat{u}_{n-1} \right) \hat{a}_n^{\dagger} \hat{a}_n \right], \tag{4}$$

χ being an anharmonic parameter arising from the coupling between the vibrational excitation and the lattice displacements.

Davydov's Hamiltonian (Equations (1–4)) is formally similar to the Holstein Hamiltonian for the interaction of electrons with a polarizable lattice. The difference is that while Davydov's Hamiltonian couples quantum particles to acoustic phonons, the Holstein Hamiltonian couples quantum particles to optical phonons.

In Equations (1–4) both the excitation and the lattice are treated quantum mechanically. Exact wavefunctions for the system are yet to be determined in this approximation, and different solutions have been proposed in the literature (see references in Scott, 1992; Cruzeiro-Hansson & Takeno, 1997). On the other hand, when the lattice is treated classically, an approximation that has been called mixed quantum-classical (Cruzeiro-Hansson & Takeno, 1997), it is possible to determine exact wave functions. They are

$$|\psi(t)\rangle = \sum_{i=1}^{N} \varphi_i(\{u_m\}, \{p_m\}, t)\, \hat{a}_i^\dagger\, |0\rangle, \qquad (5)$$

where the dependence of the probability amplitude for an excitation at site i on the displacements and momenta of all the sites is not specified *a priori*. The equations of motion for the quantum variables are derived from Schrödinger's equation, while the equations of motion for the classical variables are derived from Hamilton's equations. For a one-quantum state they are

$$\begin{aligned} \mathrm{i}\hbar \frac{\mathrm{d}\varphi_n}{\mathrm{d}t} &= -V\,(\varphi_{n-1} + \varphi_{n+1}) \\ &\quad + \chi\,(u_{n+1} - u_{n-1})\,\varphi_n, \end{aligned} \qquad (6)$$

$$\begin{aligned} M \frac{\mathrm{d}^2 u_n}{\mathrm{d}t} &= \chi\,(|\,\varphi_{n+1}\,|^2 - |\,\varphi_{n-1}\,|^2) \\ &\quad + \kappa\,(u_{n+1} + u_{n-1} - 2\,u_n) \end{aligned} \qquad (7)$$

The numerical simulation of these equations (and of extensions of these equations to the three interacting chains that are present in an α-helix (Scott, 1992)) confirmed Davydov's analytical studies in the continuum limit by showing that solitons can form and travel along the chains. It was also found that in the discrete chains there is a threshold in the nonlinearity parameter χ above which localized excitations can form. This threshold tends to zero as the length of the system increases.

Thermal Stability of Localized States

While the zero temperature results summarized above allow for the existence of soliton states in proteins, an important question is whether these states are stable at biological temperatures and, if not, how long they last. Coupling to a thermal bath was modeled by extending the equations of motion into Langevin equations, that is, adding stochastic forces $F(n)$ and damping terms $-\Gamma \mathrm{d}u_n/\mathrm{d}t$ to the right-hand side of Equation (7), so that these terms obey the fluctuation-dissipation relations (Lomdahl & Kerr, 1985): $\langle F_n(t)\, F_m(t')\rangle \geq 2M\Gamma k_B T\, \delta_{nm}\delta(t-t')$, k_B being the Boltzmann constant and T the temperature. The result is that soliton solutions disperse in a few picoseconds at biological temperatures. On the other hand, exact quantum Monte Carlo (QMC) simulations (Wang et al., 1989) showed that the distortion induced by the excitation increased with temperature. This conflict was resolved by showing that the coupling of a classical bath to a mixed quantum-classical system leads to a classical behavior of the quantum part (Cruzeiro-Hansson & Takeno, 1997). While the states of a classical excitation at finite temperature are predominantly delocalized, the states of a quantum excitation are predominantly localized. A set of dynamic equations, that leads to the same equilibrium averages as the QMC, leads to the following dynamics (Cruzeiro-Hansson & Takeno, 1997): instead of coherent propagation along the chain, as is expected from solitons, localized amide-I excitations jump stochastically from site to site, changing their shape and their velocity as they travel. It has been suggested that this stochastic propagation is a very robust way of transferring energy in proteins because it can survive mutations more than coherent propagation. Thus these simulations indicate that the Davydov model is indeed a possible mechanism for energy transfer in proteins (Cruzeiro-Hansson & Takeno, 1997).

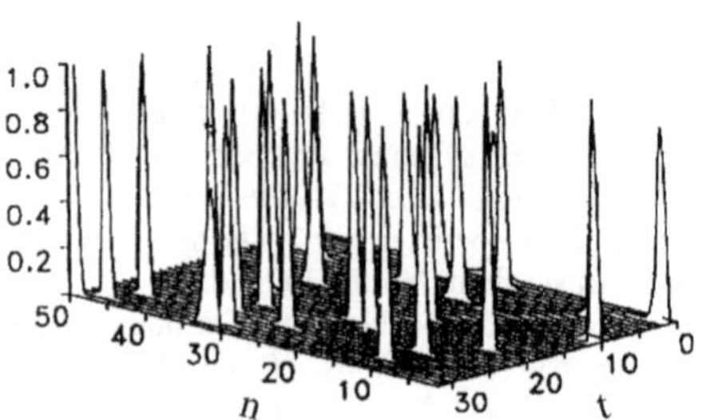

Figure 1. Dynamics of an amide-I excitation at biological temperatures.

Experimental Evidence

The first evidence for a Davydov-like state was obtained in a crystal of acetanilide (ACN), which includes hydrogen-bonded chains identical to those found in proteins. Careri and co-workers found an anomalous line, red-shifted by 15 cm^{-1} with respect to the amide-I vibration in that crystal (Scott, 1992). A series of studies established that this spectral line could not be attributed to a Fermi resonance, a Davydov splitting, or a phase transition. Its interpretation in terms of an interaction of an amide excitation with optical phonons provided good fits of the variation of its intensity with temperature as well as of its overtone frequencies. Careri and Wyman (see Scott, 1992) integrated the Davydov model in the so-called "turning wheel" model of enzyme action in which it is assumed that the binding of ligands leads to the formation of solitons that are used to promote the enzymatic cycle.

An early objection to the role of vibrational excited states in proteins was their short lifetime, which was supposed to be in the subpicosecond range. Recently, Austin and co-workers have measured the lifetime of amide-I vibrations in myoglobin and found it to be 15 ps (Xie et al., 2000). Hamm and co-workers have also recently applied nonlinear spectroscopy to the study of ACN, not only to the amide-I band (Edler & Hamm, 2002) but also to the NH vibrations (Edler et al., 2002) that have approximately twice the energy of amide-I and a binding energy 20 times larger. In a pump-probe femtosecond spectroscopy experiment, they found that the ground state recovery for NH at room temperature is 20 ps and for amide-I at 90 K, it is 35 ps, and that the lifetime of NH stretch in the crystal is 20 times greater than in isolated ACN. Computer simulations show that vibrational excitations can travel tens of nanometers within the lifetime of these excitations (Cruzeiro-Hansson & Takeno, 1997). Edler & Hamm (2002) also find that while the low-temperature amide-I excitations are self-trapped, the higher temperature ones are Anderson-localized, in agreement with the predictions of the computer simulations (Cruzeiro-Hansson & Takeno, 1997).

Future Work

The theoretical work on the Davydov model shows that localized vibrational excited states can form in proteins and that they can be used to transfer energy from the active site to other regions of the proteins where this energy is used for work. The experimental evidence available corroborates this possibility. Further experiments should test the role of excited vibrational states in protein function in a more direct way. A theoretical challenge is to integrate the Davydov model in the full protein cycle and find out how vibrational energy transfer may eventually lead to a conformational change (Cruzeiro–Hansson & Silva, 2001), the well-known way in which proteins fold or transform into their permanent three-dimensional shapes.

LEONOR CRUZEIRO

See also **Biomolecular solitons; Excitons**

Further Reading

Cruzeiro-Hansson, L. & Takeno, S. 1997. Davydov model: the quantum, mixed quantum-classical and full classical systems. *Physical Review*, E 56(1): 894–906

Cruzeiro-Hansson, L. & Silva, P.A.S. 2001. Protein folding: thermodynamic versus kinetic control. *Journal of Biological Physics*, 27: S6–S9

Davydov, A.S. 1982. *Biology and Quantum Mechanics*, Oxford–Frankfurt: Pergamon

Edler, J. & Hamm, P. 2002. Self-trapping of the amide I band in a peptide model crystal. *Journal of Chemical Physics*, 117: 2415–2424

Edler, J., Hamm, P. & Scott, A.C. 2002. Femtosecond study of self-trapped vibrational excitons in crystalline acetanilide. *Physical Review Letters*, 88(6): 067403

Green, D.E. (editor). 1974. The mechanism of energy transduction in biological systems. *Annals of the New York Academy of Sciences*, 227: 1–668

Lomdahl, P.S. & Kerr, W.C. 1985. Do Davydov solitons exist at 300K? *Physical Review Letters*, 55(11): 1235–1238

Scott, A. 1992. Davydov's soliton. *Physics Reports*, 217(1): 1–67

Wang, X., Brown, D.W. & Lindenberg, K. 1989. Quantum Monte Carlo simulation of the Davydov model. *Physical Review Letters*, 62(15): 1796–1799

Xie, A.H., van der Meer, L., Hoff, W. & Austin, R.H. 2000. Long-lived amide I vibrational modes in myoglobin. *Physical Review Letters*, 84: 5435–5438

DELAY COORDINATES

See **Embedding methods**

DELAY-DIFFERENTIAL EQUATIONS

A delay-differential equation (DDE) comprises an unknown function and certain of its derivatives, evaluated at arguments that differ by fixed numerical values. For example, $\dot{x}(t) = F(t, x(t), x(t-r))$ is a retarded DDE for $r > 0$.

DDEs can be characterized by the number of variables, the differential order, the order of the highest derivative appearing, and the difference order (one less than the number of different arguments appearing). DDEs (also called functional differential equations or difference-differential equations) generalize the concept of differential equations by allowing the state of the system to depend on states different from the present one, via, for example, a feedback mechanism.

DDEs appear in many fields, from economics to biology and from chemistry to mathematics, as the following examples suggest.

Bacterial Infection

In a book on the study of malaria published in 1911 (Ross, 1911), Ross considered a differential equation model of malarial epidemics based on the interaction of two populations, that of infected humans, described by $h(t)$, and that of infected mosquitoes, by $m(t)$. The relevant nonlinear DDEs are

$$\dot{h}(t) = m(t-u)\frac{p - h(t-u)}{p} - (M+r)h(t),$$

$$\dot{m}(t) = h(t-v)\frac{q - m(t-v)}{p} - (N+s)m(t), \quad (1)$$

with p and q being the total populations, M and N the death rates, and r and s the recovery rates for human and mosquito populations, respectively. The delay from the time of a bite to the time at which the human or the mosquito is infective is $u = 0.5$ months for humans and $v = 0.6$ months for mosquitoes.

Enzyme Catalysis

Roussel (1996) presents a model of enzyme catalysis, proposed by Adrian Brown in 1902, in which the enzyme-substrate complex has a fixed lifetime τ before giving rise to its reaction products

$$E(t) + S(t) \rightarrow P(t+\tau) + E(t+\tau). \qquad (2)$$

The DDEs generated by this model are

$$\begin{aligned} \frac{dS(t)}{dt} &= -kE(t)S(t), \\ \frac{dE(t)}{dt} &= -kE(t)S(t) + kE(t-\tau)S(t-\tau), \qquad (3) \\ \frac{dP(t)}{dt} &= kE(t-\tau)S(t-\tau), \end{aligned}$$

where k is the rate constant and by $E(t)$, $P(t)$, and $S(t)$, we mean the enzyme, product, and substratum concentrations.

Economic Dynamics

Delays can emerge in the dynamics of economic processes in two ways. First, there is a lag between the time economic decisions are taken and the time the decisions bear fruit. Also, the expected future values of a variable are functions of its current and past values. Let us consider a world in which infinitely many houses are inhabited, and let us consider the problem of finding the best time for constructing a new house (Asea & Zak, 1999).

The individuals' preferences are represented by a continuous, strictly increasing, and concave utility function $U(c(t))$. Subjective discount rates are characterized by $\rho > 0$. In this model, it takes $r > 0$ periods to obtain new capital. The solution of this planning problem is given by

$$\max_{\{c(t)\}} \int_0^\infty U(c(t))e^{-\rho t}\,dt, \qquad (4)$$

subject to the DDE

$$\frac{dk(t)}{dt} = f(k(t-r)) - \delta k(t-r) - c(t), \qquad (5)$$

with initial condition $k(t) = \phi(t)$, for all $t\varepsilon[-r, 0]$. $f(\cdot)$ is the production function; $c(t)$, the rate at which capital depreciates, such that $0 < c(t) \le f(k(t-r))$, $\delta \in [0, 1]$; and $k(t)$, the productive capital stock at time t.

Mathematics

If one considers the symmetry reduction of a nonlinear differential-difference equation with respect to a combination of continuous and discrete symmetries, then the initial equation reduces to a DDE. As an example, consider the Toda lattice (Levi & Winternitz, 1993)

$$\ddot{u}_n(t) = e^{u_{n-1}(t)-u_n(t)} - e^{u_n(t)-u_{n+1}(t)}, \qquad (6)$$

and assume a reduction with respect to $\partial_n + a\partial_t$ where a is an arbitrary real parameter. Equation (6) then reduces to

$$\frac{d^2u(\eta)}{d\eta^2} = e^{u(\eta+a)-u(\eta)} - e^{u(\eta)-u(\eta-a)}, \qquad (7)$$

where $\eta = t - an$.

The equations considered in these examples are all instances of a general DDE, which, in the simple case of a linear first-order equation for just one field, can be written as

$$a_0\frac{du(t)}{dt} + a_1\frac{du(t-\sigma)}{dt} + b_0u(t) + b_1u(t-\sigma) = f(t). \qquad (8)$$

An equation of the form (8) is said to be a DDE of *retarded* type if $a_0 \neq 0$ and $a_1 = 0$; it is said to be of *neutral* type if $a_0 \neq 0$ and $a_1 \neq 0$; and of *advanced* type if $a_0 = 0$ and $a_1 \neq 0$.

In applications, an equation of *retarded* type may represent the behavior of a system in which the rate of change of $u(t)$ depends on its past and present values. A *neutral* equation represents a system in which the present rate of change depends on past rates of changes as well as its present and past values. An *advanced* type equation may represent a system in which its rate of change depends on its present and future values.

If $a_0 = a_1 = 0$, Equation (8) is a pure difference equation, while if $a_0 = b_0 = 0$ or $a_1 = b_1 = 0$, it is a pure differential equation. In either case, $f(t)$ is a forcing function.

Let us compare the solution techniques for DDEs with those of ordinary differential equations (ODEs) and note some of their peculiar features. For simplicity we limit ourselves to retarded DDEs. For more details, see Bellman & Cooke (1963), Hale (1977), Hale & Verduyn Lunel (1993), Driver (1977), Bainov & Mishev (1991), Kuang (1993), and Gyori & Lada (1991).

Because retarded DDEs depend on previous history, the initial condition at one point is not sufficient to obtain the present time behavior. What one needs depends on the discrete order of the equation. If the equation is a DDE of first order, then the initial solution on a whole delay interval is needed.

For constant coefficient DDEs, an algebraic method of solution is provided by the "method of steps," which also provides a constructive proof of the existence of the solution. To illustrate this method, consider a DDE generalization of the logistic equation

$$\frac{dx(t)}{dt} = -cx(t-1)[1+x(t)], \qquad t > 0, \qquad (9)$$

with the initial condition $x(t)=\phi(t)$ for $t\in[-1,0]$. To solve Equation (9), we divide the interval $[0,\infty)$ into steps of the size of the delay and solve recursively in each interval. We use the solution obtained in one interval to solve Equation (9) in the next one. For example, the solution in the interval [0, 1] is given by

$$x(t)=[\phi(0)+1]\mathrm{e}^{-c\int_0^t\phi(s-1)\mathrm{d}s}-1,$$

which is obtained as a solution of the ODE

$$\frac{\mathrm{d}x(t)}{\mathrm{d}t}=-c\phi(t-1)[1+x(t)]. \tag{10}$$

For linear DDEs we can construct, as in the case of linear ODEs, the characteristic equation, by looking at exponential solutions. In this case, however, the characteristic equation is given by a nonlinear algebraic equation. For example in the case of Equation (8), with $a_1=0$, we have

$$h(\lambda)=a_0\lambda+b_0+b_1e^{-\lambda\sigma}=0. \tag{11}$$

Once the characteristic equation is solved, a particular solution of the DDE is obtained by applying the Laplace transform (Bellman & Cooke, 1963).

As we have seen, the nature of the method of solution of a DDE is similar to that of an ODE. Nevertheless, DDEs exhibit more complicated behaviors, even in the linear case. For example, scalar linear first-order homogeneous DDEs with real coefficients can have nontrivial oscillating solutions unlike ODEs (Kalecki, 1935). Moreover, solutions to DDEs may be discontinuous and, depending on the initial conditions, a solution may also not exist (Winston & Yorke, 1969). As in the case of ODEs, series solutions can be used to approximate solutions to nonlinear DDEs (Bellman & Cooke, 1963); however, the solutions obtained are often complicated and obscure.

We can gain a better insight into the solution using qualitative theory and stability analysis to obtain properties of the dynamics of a nonlinear DDE by looking at its linearization.

The stability of a fixed point of a DDE is defined by examining the roots of the characteristic equation $h(\lambda)$. Thus, a fixed point of a DDE is stable if all roots of $h(\lambda)$ have negative real parts. As the characteristic equation (11) is transcendental, it has an infinity of roots, and it is not guaranteed that all roots will have real parts, strictly negative or positive. So fixed points of DDEs will often be saddle points. Moreover, stability may depend crucially on the initial data (Driver, 1977).

The stability of homogeneous scalar DDEs of the first order has been studied by Hayes (Bellman & Cooke, 1963). These results can be extended to nonlinear systems by linearizing the DDE around a stable solution and then using a generalization of the Poincaré–Lyapunov theorem. In such a way, one can show that DDEs often admit periodic solutions after a sequence of Hopf bifurcations. Chaotic orbits may also exist, with the structure of the orbits depending critically on the smoothness of the feedback mechanism.

DECIO LEVI

See also **Bifurcations; Equations, nonlinear; Feedback; Hopf bifurcation; Integral transforms; Ordinary differential equations, nonlinear; Poincaré theorems; Quasilinear analysis; Stability; Symmetry: equations vs. solutions**

Further Reading

Asea, P.K. & Zak, P.J. 1999. Time-to-build and cycles. *Journal of Economic Dynamics & Control*, 23: 1155–1175

Bainov, D.D. & Mishev, D.P. 1991. *Oscillation Theory for Neutral Differential Equations with Delay*, Bristol: Adam Hilger

Bellman, R. & Cooke, K.L. 1963. *Differential-Difference Equations*, New York: Academic Press

Driver, R.D. 1977. *Ordinary and Delay Differential Equations*, New York: Springer

Gyori, I. & Ladas, P. 1991. *Oscillation Theory of Delay Differential Equations: with Applications*, Oxford: Clarendon Press

Hale, J.K. 1977. *Theory of Functional Differential Equations*, New York: Springer

Hale, J.K. & Verduyn Lunel, S.M. 1993. *Introduction to Functional Differential Equations*, New York: Springer

Kalecki, M. 1935. A macroeconomic theory of business cycles. *Econometrica*, 3: 327–344

Kuang, Y. 1993. *Delay Differential Equations with Applications in Population Dynamics*, Boston: Academic Press

Levi, D. & Winternitz, P. 1993. Symmetries and conditional symmetries of differential-difference equations. *Journal of Mathematical Physics*, 34: 3713–3730

Ross, R. 1911. *The Prevention of Malaria*, 2nd edition, London: John Murray

Roussel, M.R. 1996. The use of delay differential equations in chemical kinetics. *Journal of Physical Chemistry*, 100: 8323–8330

Winston, E. & Yorke, J.A. 1969. Linear delay differential equations whose solutions become identically zero. *Académie de la République Popolaire Roumaine*, 14: 885–887

DENJOY THEORY

The theory developed by Arnaud Denjoy (1884–1974) showed that any sufficiently smooth orientation-preserving diffeomorphism T of the unit circle S^1 with an irrational rotation number ρ is topologically equivalent to a linear rotation by the angle $2\pi\rho$ (Denjoy, 1932). Informally, diffeomorphism is a smooth invertible map such that its inverse is also smooth.

Circle diffeomorphisms arise naturally in many physical problems. For instance, in the case of Hamiltonian systems with two degrees of freedom, such diffeomorphisms appear as Poincaré first return maps for the two-dimensional invariant tori. When the rotation number is irrational, circle diffeomorphisms represent an important model for quasi-periodic dynamics (*See* **Quasiperiodicity**). The Denjoy theory implies the following important fact: if two smooth circle maps have the same irrational rotation number then the topological structure of their trajectories is exactly the same.

The topological equivalence means that circle diffeomorphisms are conjugated to a linear rotation with the help of a homeomorphic change of variables. Namely, there exists a homeomorphism, ϕ, which is an invertible map that is continuous together with its inverse, such that $T \circ \phi = \phi \circ T_\rho$, where T_ρ is the linear rotation by the angle $2\pi\rho$ and $\circ$ stands for a composition of two maps. Denjoy's theorem holds if T is absolutely continuous and $\log T'(x)$ has bounded total variation: $V = \mathrm{Var}_{S^1} \log T'(x) < \infty$. The last condition is satisfied if T is C^2-smooth and $T'(x) > 0$. The conjugacy ϕ is defined uniquely up to an arbitrary rotation T_α. In fact, a mapping ϕ of the unit circle S^1 that satisfies condition $T \circ \phi = \phi \circ T_\rho$ can be constructed for any quasi-periodic homeomorphisms T. This means that any homeomorphism T with irrational rotation number ρ is semiconjugate to T_ρ. However, if T is not regular enough, ϕ may not be a homeomorphism. To construct ϕ it is enough to take two arbitrary points x_0 and y_0 and define their forward trajectories by T and T_ρ, respectively: $x_i = T^i x_0$, $y_i = T_\rho^i y_0$, $i \geq 1$. Now one can define ϕ on $\{y_i\}$ by letting $\phi(y_i) = x_i$, $i \geq 0$ and extending ϕ by continuity to the whole unit circle. This can be done since any trajectory of a linear rotation by an irrational angle is everywhere dense. It is easy to see that a conjugacy ϕ is a homeomorphism if and only if T is transitive; that is, all its trajectories are dense on S^1.

When the total variation V is bounded, the transitivity of T follows from the Denjoy inequality:

$$\exp(-V) \leq \prod_{i=0}^{q_n-1} T'(x_i) \leq \exp(V),$$

where q_n are the denominators of the convergents $p_n/q_n = [k_1, k_2, \ldots, k_n]$, and $\rho = [k_1, k_2, \ldots, k_n, \ldots]$ is the continued fraction expansion for ρ. The condition $T \in C^2(S^1)$ that implies topological equivalence is almost sharp. Indeed, Denjoy constructed counterexamples where $T \in C^1(S^1)$ and the derivative $T'(x)$ is a Hölder continuous function with an arbitrary Höder exponent $0 < \alpha < 1$. In these examples T is not transient and, hence, is not conjugate to T_ρ.

An important extension of the Denjoy theory is connected with the problem of smoothness of the conjugacy ϕ. It is natural to ask when the homeomorphism ϕ is at least C^1-smooth, which implies not only topological but also asymptotic metrical equivalence between T and T_ρ. In this case, the unique probability invariant measure for T is absolutely continuous with respect to the Lebesgue measure. The first progress in this direction was made by Arnol'd (1961), who proved that for analytic diffeomorphisms, T that are close enough to the linear rotation T_ρ, a conjugacy ϕ is analytic provided the rotation number ρ is Diophantine, that is, $\|\rho - p/q\| \geq 1/q^{2+\delta}$ for some $\delta > 0$ and all integers p, q. Diophantine numbers form a set of positive Lebesgue measure and, hence, are typical in the Lebesgue sense. Arnol'd has also constructed counterexamples in the case of nontypical rotation numbers, which show that the smooth theory cannot be constructed for all irrational rotation numbers. In these counterexamples, ϕ is not differentiable, and the invariant measure for T is essentially singular with respect to Lebesgue measure. Arnol'd's results are of the KAM-type (Kolmogorov–Arnol'd–Moser) and, hence, have a local character. However, as it was conjectured by Arnol'd, in the one-dimensional case the local condition of T being close to T_ρ should not be necessary, and the global result should hold for all T smooth enough. Such a global result has been proven by Herman (1979) in the case when ρ satisfies certain Diophantine condition and $T \in C^3(S^1)$. Later Herman's results were extended to a wider class of rotation numbers (Yoccoz, 1984) and to diffeomorphisms $T \in C^{2+\varepsilon}(S^1)$ (Khanin & Sinai, 1987; Sinai & Khanin, 1989; Katznelson & Ornstein, 1989).

Finally, we mention another extension of the Denjoy theory to the case of diffeomorphisms with singularities. Such mappings appear, for example, in the case of critical invariant tori in Hamiltonian systems with two degrees of freedom. The extension of the Denjoy theory to this case is a subject of the so-called rigidity theory. The main aim is to find conditions which imply that two topologically equivalent homeomorphisms that have the same local structure of their singular points are, in fact, C^1-smoothly conjugate to each other. Significant progress in this direction has been made in the last 5 years in the case of mappings with one singular point (de Faria & de Melo, 1999, 2000; Yampolsky, 2001; Khanin & Khmelev, 2003). Note that the presence of singularities makes rigidity stronger than in the case of smooth diffeomorphisms. The arithmetical properties of the rotation numbers are less important, and one should expect C^1-rigidity for all irrational rotation numbers.

Konstantin Khanin

See also **Kolmogorov–Arnol'd–Moser theorem; Maps; Quasiperiodicity**

Further Reading

Arnol'd, V.I. 1961. Small denominators. I. Mapping the circle onto itself. *Izvestiya Akademii Nauk SSSR Seriya Mathematicheskaya*, 25: 21–86

Cornfeld, I. P., Fomin, S. V. & Sinai, Ya. G. 1982. *Ergodic Theory*, New York: Springer

Denjoy, A. 1932. Sur les courbes définies par les équations différentielles à la surface du tore. *Journal des Mathematiques Pures et Appliques*, ser. 9, 11: 333–375

de Faria, E. & de Melo, W. 1999. Rigidity of critical circle mappings. I. *Journal of the European Mathematical Society (JEMS)*, 1: 339–392
de Faria, E. & de Melo, W. 2000. Rigidity of critical circle mappings. II. *Journal of the European Mathematical Society (JEMS)*, 13: 343–370
Herman, M. 1979. Sur la conjugaison différentiable des difféomorphismes du cercle à des rotations. *Publications Mathématiques de l'Institut des Hautes Études Scientifiques*, 49: 5–233
Katznelson, Y. & Ornstein, D. 1989. The differentiability of the conjugation of certain diffeomorphisms of the circle. *Ergodic Theory & Dynamical Systems*, 9: 643–680
Khanin, K.M. & Sinai, Ya.G. 1987. A new proof of M. Herman's theorem. *Communications in Mathematical Physics*, 112: 89–101
Khanin, K. & Khmelev D. 2003. Renormalizations and rigidity theory for circle homeomorphisms with singularities of the break type. *Communications in Mathematical Physics*, 235: 69–124
Sinai, Ya.G. & Khanin, K.M. 1989. Smoothness of conjugacies of diffeomorphisms of the circle with rotations. *Russian Mathematical Surveys*, 44: 69–99
Yampolsky, M. 2001. The attractor of renormalization and rigidity of towers of critical circle maps. *Communications in Mathematical Physics*, 218: 537–568
Yoccoz, J.-C. 1984. Conjugaison différentiable des difféomorphismes du cercle dont le nombre de rotation vérifie une condition diophantienne. *Annales Scientifique de l'École Normale Superierieure*, 4(17): 333–359

DERIVATIVE NLS EQUATION

See **Nonlinear Schrödinger equations**

DERRICK–HOBART THEOREM

The Derrick–Hobart scaling argument concerns certain solutions of nonlinear partial differential equations that arise as models for elementary particles; thus they are mostly of the relativistic variety. To appreciate the context in which the argument arose and the way it is used, some introductory remarks on relativistic quantum field theory are in order.

There are only a few interacting relativistic quantum field theories that have been solved explicitly, in the sense that physically relevant quantities (particle spectrum, scattering, form factors, and so on) are known in closed form. For all of these models the dimension of space-time equals two. To gain more insight into higher-dimensional models, it has become standard practice to study the field theory first as a classical field theory. The underlying idea is that (via the Feynman path integral) one can use classical findings to obtain nonperturbative information on the quantum version. In particular, the presence of nonconstant, smooth, stable, time-independent, finite-energy, classical solutions is believed to signal the presence of an associated stable quantum particle.

The notion of "stability" refers to small fluctuations around such a classical finite-energy solution. To first order, such variations do not change the energy, as expressed by the Euler–Lagrange equation. To second order, however, the energy may become smaller, in which case the corresponding quantum particle is considered to be unstable. (Think of a ball resting on top of a hill. A little push makes it roll down.)

In order to study the existence of nonconstant finite-energy solutions (either stable or unstable), an argument due independently to Derrick (1964) and (in a somewhat different form) to Hobart (1963) is often useful. These authors were concerned with three-dimensional space (four-dimensional space-time), but the argument can be extended without difficulty to an arbitrary space dimension N. Briefly, the argument is as follows.

Assume $\phi(x,t)$, $x \in R^N$, $t \in R$, is a scalar field on $(N+1)$-dimensional space-time, whose dynamics is given by the Lagrangian

$$\mathcal{L} = \tfrac{1}{2}(\partial_t\phi^2 - \nabla\phi\cdot\nabla\phi) - V(\phi), \tag{1}$$

with $V(y)$ being a potential function. Now let $\phi(x)$ be a (smooth) time-independent nonconstant solution to the Euler–Lagrange equation, with finite energy

$$E = E_{\text{kin}} + E_{\text{pot}}, \quad E_{\text{kin}} = \frac{1}{2}\int_{R^N} \nabla\phi\cdot\nabla\phi \mathrm{d}x,$$

$$E_{\text{pot}} = \int_{R^N} V(\phi)\mathrm{d}x. \tag{2}$$

Starting from the above data, Derrick's key idea is to consider the family of scaled functions

$$\phi_\lambda(x) = \phi(\lambda x). \tag{3}$$

Clearly, the energy associated with ϕ_λ is given by

$$E_\lambda = \lambda^{(2-N)}E_{\text{kin}} + \lambda^{-N}E_{\text{pot}}, \tag{4}$$

so that

$$(\mathrm{d}E_\lambda/\mathrm{d}\lambda)_{\lambda=1} = (2-N)E_{\text{kin}} - NE_{\text{pot}}, \tag{5}$$

$$(\mathrm{d}^2E_\lambda/\mathrm{d}\lambda^2)_{\lambda=1} = (2-N)(1-N)E_{\text{kin}} + N(N+1)E_{\text{pot}}. \tag{6}$$

Since ϕ_λ makes the energy stationary for $\lambda = 1$, we have

$$(\mathrm{d}E_\lambda/\mathrm{d}\lambda)_{\lambda=1} = 0. \tag{7}$$

Hence (5) yields

$$E_{\text{pot}} = \frac{2-N}{N}E_{\text{kin}}, \tag{8}$$

which entails

$$(\mathrm{d}^2E_\lambda/\mathrm{d}\lambda^2)_{\lambda=1} = 2(2-N)E_{\text{kin}}. \tag{9}$$

Let us now draw the relevant conclusions from this simple calculation. Since $\phi(x)$ is nonconstant, its kinetic energy E_{kin} is positive. For $N > 2$, then, (9) says that the finite-energy solution cannot be stable. This is the first consequence, an instability result for $N > 2$. It does not involve restrictions on the potential $V(y)$.

Assuming from now on that $V(y) \geq 0$, far stronger conclusions can be drawn. Indeed, since ϕ_λ is a solution for $\lambda = 1, \phi_1 = \phi$ makes the energy stationary. But since ϕ is nonconstant, we have $E_{\text{kin}} > 0$, and since $V \geq 0$, we also have $E_{\text{pot}} \geq 0$. Therefore, the right-hand side of (5) is negative for $N > 2$, a contradiction. A second consequence, therefore, is the absence of finite-energy nonconstant solutions for $V \geq 0$ and $N > 2$.

Retaining the assumption $V \geq 0$, one can draw a conclusion for $N = 2$, too. Indeed, it then follows that $E_{\text{pot}} = 0$, so that ϕ must satisfy $V(\phi) = 0$; moreover, the second variation (6) vanishes.

For $N = 1$ the variation formulas (5), (6) have no useful consequences. Indeed, in two-dimensional space-time there do exist stable time-independent finite-energy solutions, as exemplified by the one-soliton and one-antisoliton solutions of the sine-Gordon theory.

In applications of Derrick's argument, one usually encounters positive potentials and invokes the latter consequences sketched above. Thus, it is used to the effect that for $N \geq 2$, time-independent finite-energy solutions must be constant (the so-called vacuum solutions). Some caveats should be heeded, however.

First, it is important to keep track of the above steps in models that are not of the above form, since the reasoning may need to be suitably modified. Second, even when this can be done at face value, it should be observed that the above argument, although convincing at first sight, is not a rigorous proof. Indeed, the scaling variation that is involved has a global character, whereas the Euler–Lagrange equation is derived by considering local variations. More in detail, one needs to control boundary terms that can *a priori* spoil the above derivation. (This was already realized in Hobart (1963).)

We exemplify these related issues with two models described by Lagrangians that are different from the above, namely a (special) Yang–Mills/Higgs model in physical space ($N = 3$) and a class of nonlinear σ-models for $N \geq 2$. In the first setting, explicit static finite-energy solutions were obtained in Prasad & Sommerfield (1975) and Bogomolnyi (1976). (These are nowadays called BPS monopoles.) The energy of these solutions is manifestly not scale-invariant, contradicting (7) for the case at hand. Inspection of the solution shows that this is due to poor decay at spatial infinity; it entails that the pertinent boundary term cannot be ignored.

Turning to $O(3)$ σ-models, one can once more study the issue of finite-energy solutions by adapting Derrick's scaling argument. For $N = 2$ (now viewed as Euclidean space-time) this yields no conclusion, since the energy is scale-invariant. In this case, the so-called instanton and anti-instanton solutions do exist, and they are stable for topological reasons.

For $N > 2$, the scaling argument leads to the absence of finite-energy solutions. In this particular setting, the heuristic reasoning can be corroborated. More specifically, the boundary term can be rigorously controlled. The pertinent result (Garber et al. (1979), Theorem 5.1) has later been used by differential geometers to prove the nonexistence of harmonic maps, which are closely related to the above type of solution.

SIMON RUIJSENAARS

See also **Matter, nonlinear theory of; Skyrmions; Virial theorem; Yang–Mills theory**

Further Reading

Bogomolnyi, E.B. 1976. The stability of classical solutions. *Soviet Journal of Nuclear Physics*, 24: 449–454

Derrick, G.H. 1964. Comments on nonlinear wave equations as models for elementary particles. *Journal of Mathematical Physics*, 5: 1252–1254

Garber, W.-D., Ruijsenaars, S.N.M., Seiler, E. & Burns, D. 1979. On finite-action solutions of the nonlinear σ-model. *Annals of Physics*, 119: 305–325

Hobart, R.H. 1963. On the instability of a class of unitary field models. *Proceedings of the Physical Society, London*, 82: 201–203

Prasad, M.K. & Sommerfield, C.M. 1975. Exact classical solution for the 't Hooft monopole and the Julia-Zee dyon. *Physical Review Letters*, 35: 760–762

DETAILED BALANCE

This entry provides a qualitative discussion of equilibrium, a more quantitative discourse of principles such as detailed balance (which are needed in the description of equilibrium phenomenon), and a brief presentation of the Einstein relation between mobility and diffusion, which can be related to the above topics.

The Problem of Time

One often says that a system has reached an equilibrium state if its physical variables are constant in time. Because of fluctuations that cannot be removed, however, it is better to regard the system as in equilibrium when there are no systematic trends in the time averages of its physical parameters. Here, averages are considered over all the microscopic constituents of the system, whether they are elementary particles, atoms, molecules, or larger objects. Equilibrium can be established among these constituents.

Thus, a system that is in equilibrium cannot reveal the time variable among its broad characteristics. In other words, there is no way of telling which way time is running if one's observations are confined to an equilibrium system. Formulated as a philosophical puzzle about the nature of time, this subject has spawned a library of books and papers, with little agreement among the authors (see Landsberg, 1982; Smith, 1993; Price, 1996; Davies, 1995).

Some Relevant Principles of Statistical Mechanics

Here and below, we shall deal with a number of important principles that may or may not hold in any given case and are related to each other. To make these matters quantitative, denote by P_i the probability of finding a system of interest in any one of the ith group of states, G_i in number. The probability per unit time that a transition occurs from a state of group i to a state of group j is denoted by A_{ij}. The transition rate $i \rightarrow j$ can be written as

$$R_{ij} = P_i A_{ij} G_j. \tag{1}$$

If there are W available groups of states, the time rate of change of P_i is

$$\dot{P}_i = \sum_{l=1}^{W} (R_{li} - R_{il}) \quad (i = 1, 2 \ldots W). \tag{2}$$

To be tractable the A_{ij} have to be independent of time. The first sum gives the transitions into states i and the second sum gives the transitions out of the states i. To simplify the picture one can replace a typical group of states i by a single state, i.e., one can put $G_i = 1$.

Now some additional general principles can be defined. The existence of the A_{ij} can be deduced from quantum mechanical perturbation theory, but it is then valid only for a restricted time interval. One often finds the symmetry relation

$$A_{ij} = A_{ji} \quad (\text{all } i, j) \tag{3}$$

as a result of the Hermitian character of the perturbation operator. In statistical mechanics, one also uses the principle of Equation (3). It can then be independent of perturbation theory and is regarded instead as resulting from adequate statistical assumptions. It is then called the *principle of microscopic reversibility*.

Next we have the principle of detailed balance which asserts that at a certain time t the forward and reverse transition rates between two groups of states are equal at a certain time; thus,

$$R_{ij} = R_{ji} \quad (\text{all } i, j). \tag{4}$$

If Equation (4) holds, one sees that $\dot{P}_i$ vanishes for all i. In fact, we can define a steady state by

$$\dot{P}_i = 0 \quad (\text{all } i). \tag{5}$$

Such a state need not be an equilibrium state since the system may, for example, be continuously raised to a high energy state by some external influence and then drop back continuously, for example, by the emission of radiation. Thus, one sees that Equation (4) implies Equation (5), but not conversely. For more details, see Lifschitz & Pitaewski (1981) and Landsberg (1991).

A Simple Example from the Solid State

One can use detailed balance arguments to infer the form of an unknown emission rate from a known absorption rate, as will now be shown by an example (Landsberg, 1991, p. 391). The idea is to obtain an expression for the equilibrium absorption rate per unit volume of photons of frequency ν_0 in a semiconductor of refractive index μ and, hence, to infer spontaneous emission rates per unit volume.

The probability of a single photon of vacuum wavelength λ_0 being absorbed per unit time per unit volume is

$$P(\lambda_0) = c\alpha(\lambda_0)/V\mu(\lambda_0). \tag{6}$$

The dimensions $(LT^{-1} \cdot L^{-1} \cdot L^{-3})$ are easily verified to be correct. To find the volume rate of excitation in the solid by photons in the vacuum wavelength range $d\lambda_0$, $P(\lambda_0)$ has to be multiplied by the number of relevant photon modes $(8\pi\mu^3\lambda_0^{-4} V d\lambda_0)$, and also by their equilibrium occupation probability at temperature T:

$$1/[\exp(ch/\lambda_0 kT) - 1]. \tag{7}$$

But not all photons of wavelength λ_0 will, when absorbed, produce one electron-hole pair. We denote by $\alpha'(\lambda_0)/\alpha(\lambda_0) (\leq 1)$ the probability of this happening per absorbed photon. Hence, the equilibrium absorption rate (per unit volume) of photons in the wavelength range $d\lambda_0$ with production of an electron-hole pair is

$$\alpha'/\alpha \frac{8\pi\mu^2\alpha c\lambda^{-4}}{\exp(ch/\lambda_0 kT) - 1} d\lambda_0$$

or

$$\frac{8\pi\alpha'}{h^3} \frac{\mu}{c^2} (kT)^3 \frac{x^2 dx}{\exp x - 1}. \tag{8}$$

Here $x = h\nu_0/kT$, and the second of these expressions is like the first, except that it is in terms of frequencies. According to detailed balance, the new inference is that these expressions give the rate per unit volume of spontaneous radiated recombination of electron-hole pairs with the emission of photons in the range $d\lambda_0$ or $d\nu_0$. Note that we have passed from absorption to emission data. This widely used result was first given by W. van Roosbroeck and W. Shockley in 1954. For other examples of the use of the principle of detailed balance in solid state physics, see Landsberg (1991).

The Einstein Relation

The Einstein Relation is basic to solid states physics and rests on the assumption that in a steady state the flux of charged particles due to an electric field must be balanced by diffusion of these particles induced by their density gradients. These two effects are due to well-known and simple forces. The first is a particle flux due to diffusion (with diffusion coefficient D, say). It can be written $-D\mathrm{d}n/\mathrm{d}x$ for one-dimensional motion, where n is the density of particles and $\mathrm{d}n/\mathrm{d}x$ the gradient ("grad n" in three dimensions). The minus sign shows that the force acts to the left if the concentration n increases to the right. The second force on the charged particles is due to a built-in or externally applied electric field E, which is a vector in three dimensions. Here we deal merely with the one-dimensional problem, and note that E can be replaced by $-\mathrm{d}V/\mathrm{d}x$, where V is the electrostatic potential at the point considered. The flux of particles can be written as $n\nu E = -n\mu \mathrm{d}V/\mathrm{d}x$, where ν is the so-called mobility of the particles.

In order to obtain the Einstein relation in its simplest form, one has to equate the two forces

$$-n\mu\frac{\mathrm{d}V}{\mathrm{d}x} = D\frac{\mathrm{d}n}{\mathrm{d}x}, \tag{9}$$

which implies that

$$\frac{\mathrm{d}(\ln n)}{\mathrm{d}x} = -\frac{\mu}{D}\frac{\mathrm{d}V}{\mathrm{d}x} \tag{10}$$

giving the simple result

$$n = n_0 \exp(-\mu V/D). \tag{11}$$

As we also know that the stationary state in an electric field at a temperature T is governed by the Boltzmann distribution

$$n = n_0 \exp(-eV/kT), \tag{12}$$

where n_0 is a constant and k is Boltzmann's constant. Comparison yields the Einstein relation

$$\mu = eD/kT. \tag{13}$$

This result connects the mobility of charged particles in a field with their diffusion coefficient. At first sight this seems unexpected because one side of the equation deals with the mechanical characteristic of diffusion.

The extension to three dimensions is not the only generalization that can be made. For example, a similar Einstein relation holds for thermal current density, and generalizations have also been made for large departures from equilibrium (Landsberg, 1991). A further variety of special cases arises for different assumptions about the shape of the semiconductor bands that can occur; for example, they can be degenerate or nondegenerate, parabolic or nonparabolic, etc., and the results can be given in a table of formulae. (Einstein's paper was published in *Annalen der Physik und Chemie* in 1905, the first of three important papers published by him in that year.)

The principle of detailed balance emerged somewhat hesitantly in the 1920s, based on Einstein's 1917 paper on transition possibilities. It was named by Fowler and Milne following other authors and other names. A brief historical survey is given by ter Haar (1955).

PETER LANDSBERG

See also **Diffusion; Stochastic processes**

Further Reading

Coveney, P. & Highfield, R. 1991. *The Arrow of Time*, London: Allen, 1990 and New York: Fawcett Columbine

Davies, P. 1995. *About Time: Einstein's Unfinished Revolution*, New York: Simon and Schuster

Einstein, A. 1905. Die von der molekularkinetischen Theorie der Wärme geforderte Bewegung. *Annalen der Physik und Chemie*, 17: 549–560

Landsberg, P.T. (editor). 1982. *The Enigma of Time*, Bristol: Adam Hilger

Landsberg, P.T. 1991. *Thermodynamics and Statistical Mechanics*, Oxford and New York: Oxford University Press

Landsberg, P.T. 1991. *Recombination in Semiconductors*, Cambridge and New York: Cambridge University Press

Lifschitz, E.M. & Pitaewski, L.P. 1981. *Physical Kinetics*, Oxford and New York: Pergamon Press

Price, H. 1996. *Time's Arrow and Archimedes' Point*, Oxford and New York: Oxford University Press

Smith, Q. 1993. *Language and Time*, Oxford and New York: Oxford University Press

ter Haar, D. 1955. Foundations of statistical mechanics. *Reviews of Modern Physics*, 27: 289

DETERMINISM

Determinism is a philosophical and scientific notion, and discussions about it are as old as philosophy and science themselves. Richard Taylor writes "Determinism is the general philosophical thesis which states that for everything that ever happens there are conditions such that, given them, nothing else could happen" (Taylor, 1996). This seems to be the most general formulation of determinism. In philosophy, he continues, "There are five theories of determinism to be considered, which can for convenience be called ethical determinism, logical determinism, theological determinism, physical determinism, and psychological determinism." Here we shall confine ourselves only to physical determinism in the natural sciences, except in the concluding section.

In physics, the deterministic view developed along with the experimental approach to research, in the sense that phenomena are reproducible under the same unchanged external conditions, implying that the same cause leads to the same consequences under the same conditions. The quantitative description of physical reality began with Galileo Galilei; although some early developments are due to Pythagoras. However, Isaac Newton was the first to lay down the complete basis of classical mechanics, which at the time was considered to be the origin of all physical phenomena. His laws of mechanics plus the law of gravitation enabled him to reproduce and mathematically derive the motion of the planets, observations of which were empirically well known by the beginning of the 16th century and formulated in Johannes Kepler's laws of celestial mechanics. With the rise and development of classical mechanics the view of determinism developed, with the opinion that all natural laws can be described by dynamical equations, either ordinary differential equations (as, for example, in celestial mechanics) or partial differential equations (as, for example, in the dynamics of fluids). In each case precise knowledge of the initial conditions (all positions and all velocities) completely determines the entire future and entire past of the system. When pushed to its extremum, this view implies complete deterministic evolution of the entire universe, including all its smallest and largest details. The French mathematician Pierre Simon de Laplace, about one century after Newton, wrote (in an often quoted passage):

> We ought then to regard the present state of the universe as the effect of its antecedent state and the cause of the state that is to follow. An intelligence knowing, in any instant of time, all forces acting in nature, as well as the momentary positions of all things of which the universe consists, would be able to comprehend the motions of the largest bodies in the world and those of the smallest atoms in one single formula, provided it were sufficiently powerful to subject all data to analysis: to it, nothing would be uncertain, both future and past would be present before its eyes. (Laplace, 1814)

We can comment on Laplace's statement from our modern perspective. First, to store and process data of *infinite precision* about the state of the entire universe is problematic, as it would require a computer that would be of comparable size and complexity to the entire universe. Thus, its presence has to be taken into account, since—obeying the same mechanical laws as the rest of the universe—it would itself disturb the universe. Therefore, we can conclude that Laplace's "intelligence" (sometimes known as Laplace's daemon) cannot exist, and consequently his idea is fiction. Second, infinite precision of all the initial conditions (positions and momenta) can never be achieved in practice. And when the precision is finite, the existence of chaos (positive Lyapunov exponents) implies sensitive dependence on initial conditions and exponential divergence of nearby trajectories. In other words, there is a finite time horizon exists in general chaotic mechanic systems, beyond which nothing at all can be predicted (Lyapunov time). Therefore, the modern notion of omnipresent chaotic behavior makes Laplace's idea impossible to implement, even in principle. Third, the universe is not described by classical mechanics, but by quantum mechanics, classical mechanics being just a useful or even excellent approximation in observing and describing the motions of sufficiently large bodies. Quantum mechanics tells us, through Heisenberg's principle of uncertainty, that momenta and positions cannot be measured simultaneously with infinite precision, but we have instead the inequality $\Delta x \Delta p_x \geq \hbar/2$, where Δx and Δp_x are the uncertainties of position x and the conjugated momentum p_x. So, Laplace's initial conditions can never be known to arbitrary precision, even in principle.

Quantum mechanics is the correct description of physical reality, with the Schrödinger equation as the starting tool, for nonrelativistic systems. The quantum theory has been further developed by Paul Dirac for relativistic quantum systems and by the quantum field theory up to the unifying field theories, which capture three fundamental interactions (electromagnetic, weak, and strong interactions), but not yet gravity. The Schrödinger equation is a deterministic equation of motion of the wave function ψ, which contains the complete description of the quantum state of a given system. Importantly, ψ itself is a statistical quantity and thus not deterministic: it gives merely probabilities for the given system to be found (by measurement) in a given state.

This is the so-called Copenhagen interpretation of quantum mechanics, initiated by Max Born in 1926 and further developed by Niels Bohr and his colleagues, according to whom there is no determinism in physical reality. This view was strongly opposed by Albert Einstein and colleagues, who accepted the quantum theory as correct but thought that it was an incomplete theory, to be supplemented (through future research) by a more general deterministic theory, uncovering further "hidden variables," which seem to be ignored in present-day quantum mechanics. Many attempts have been made to find such a classical theory of fields to deduce the quantum theory but without success. There are also certain predictions such as Bell's inequalities that are the testing ground of whether quantum theory can in principle be an extended classical deterministic field theory. So far the answer is no, at least for a large class of "local hidden variables theories," and today we do have experimental confirmations where Bell's inequalities are experimentally violated, meaning that the quantum theory and its prediction for the outcome of such experiments is correct. Therefore, the statistical interpretation of quantum mechanics of Bohr's Copenhagen school, together with the strongly counter-intuitive notion of nonlocality, is proven to be correct, and these nondeterministic properties of quantum mechanics are being used in technological applications (such as quantum information theory, quantum teleportation, and quantum computing).

It is, of course, a philosophical shock to learn that the world is not deterministic, but there seems to be no way out. One of the main causes is the process of quantum measurement, which as a process is *not* described by the Schrödinger equation and seems to be the primary source of quantum indeterminism. Quantum measurement is the main source for the generally accepted statistical interpretation of quantum mechanics. Still, the potential of a classical nonlinear field theory (including its turbulent solutions) seems largely unexplored as a description of physical reality. Even classical nonlinear dynamics is not deterministic (even in principle) due to the existence of chaos. A nonlinear classical field theory is even richer, for example the complex turbulent solutions of the Navier–Stokes equations.

In a deterministic world, there would be no place for free will in the lives of human beings or other living creatures. Everything would be predetermined by the initial state before our life, even if we do not have information about that, which implies that we cannot be aware of our predestination. Since the world is not deterministic, there is room for free will and free choice. However, it might be that the world is deterministic, if we do not observe it, and is not deterministic as soon as we "touch" it. Therefore, determinism can never be proved (in analogy with Kurt Gödel's famous incompleteness theorem). Thus, our free will may materialize as soon as we interact with the world, otherwise we would be completely predestined, but isolated from the rest of the world, which is of course not possible. The issue of classical and quantum measurement lies at the bottom of such discussions. It leads to the general conclusion that the world ultimately is not deterministic, but determinism might be a good approximation under certain conditions imposed on the measurement process.

MARKO ROBNIK

See also **Butterfly effect; Causality; Chaotic dynamics; Lyapunov exponents; Quantum theory; Turbulence**

Further Reading

Belavkin, V.P. 2002. Quantum causality, stochastics, trajectories and information. *Reports on Progress in Physics*, 65: 353–420

Bell, J.S. 1987. *Speakable and Unspeakable in Quantum Mechanics*, Cambridge and New York: Cambridge University Press

Edward U. Condon, author. Mechanics, Quantum. 1980. *The New Encyclopaedia Britannica: Macropaedia*, 15th edition, vol. 11, Chicago: Encyclopaedia Britannica: 793

Laplace, P.S. 1814. *Essai philosophique sur le probabilités*, Paris: Courier, 1814; as *Philosophical Essay on Probabilities*, Berlin and New York: Springer, 1995

Peres, A. 1995. *Quantum Theory: Concepts and Methods*, Dordrecht: Kluwer

Philip W. Goetz. (editor). 1980. Determinism. *The New Encyclopaedia Britannica: Micropaedia*, 15th edition, vol. III, Chicago: Encyclopaedia Britannica: 494

Taylor, R. 1996. Determinism. In *The Encyclopedia of Philosophy*, vol. 2, edited by Paul Edwards, New York: Macmillan and London: Simon & Schuster, 359

Wheeler, J.A. & Zurek, W.H. (editors). 1983. *Quantum Theory and Measurement*, Princeton, NJ: Princeton University Press

DETERMINISTIC WALKS IN RANDOM ENVIRONMENTS

A "deterministic walk in a random environment" (DWRE) is the name given to a system generated by the motion of some object (such as, a particle, signal, wave, ant, read/write head of the Turing machine) on a graph. At each time step, the object hops from a vertex to one of its neighboring vertices. The choice of neighbor is completely determined by the type of deterministic scattering rule or scatterer, located at the vertex. A random environment is formed by the scatterers that are assumed to be initially randomly (usually independently) distributed among the vertices. DWREs (in their simplest form and under different names) were introduced in various branches of science (Gunn & Ortuño, 1985; Langton, 1986; Ruijgrok & Cohen, 1988) as paradigms, for example, for propagation of a signal in a random media, evolutionary dynamics, growth processes, and the computational environment.

In the early numerical studies, graphs were regular lattices and usually two types of scatterers were considered in each model. The most studied case was that of the regular quadratic lattice with left and right rotators, which rotate the particle to the left or to the right by an angle $\pi/2$, or left and right mirrors aligned along the two diagonals of the lattice. Two classes of such models have been extensively studied numerically (Cohen, 1992). The first class corresponds to the case when there is no feedback of the moving particle to the environment; that is, a particular type of scatterer is fixed at each site of the lattice forever. Another class is formed by models with flipping scatterers, when a scatterer at a site changes (deterministically) after every visit of a particle to this site.

In statistical physics, these models naturally appear as deterministic Lorentz lattice gases (but with a random distribution of scatterers). The scatterers are not spheres (disks) as in the classical Lorentz gas. Instead, say in the d-dimensional cubic lattice, there are $(2d)^{2d}$ different types of scatterers because each vertex in this case has $2d$ incoming and $2d$ outcoming edges. In theoretical computer science, these models are referred to as many-dimensional Turing machines because the changes of scatterer type at each vertex occur deterministically—according to some program written on an infinite tape divided into commands, for example, to change a given scatterer to some other type (Bunimovich & Khlabystova, 2002a).

Although a DWRE reminds one of random walks, these systems are essentially different. The major difference with random walks is that instead of carrying out a random experiment (like flipping a coin), the particle chooses each step deterministically. Formally, DWREs are deterministic cellular automata, but their behavior reflects a mixture of deterministic dynamics and a random environment. Their dynamics is often counterintuitive (Cohen, 1992) because one's intuition is essentially based on exactly understood (completely solved) systems and models. There are many such models among purely deterministic and purely stochastic systems; however, there were basically no completely understood systems with a mixture of deterministic and stochastic features. Some subclasses of DWREs provide such exactly solvable models (Bunimovich, 2000).

Although closest to stochastic systems, DWREs have fixed environments. This seems counterintuitive, but the evolution of scattering types makes the entire dynamics more deterministic than in the case where an (initially random) distribution of scatterers is frozen.

In many cases, DWRE systems are equivalent to various models from percolation theory (Bunimovich & Troubetzkoy, 1992). Not only the structure of the graph (lattice) but also the types of scatterer in the model determine the corresponding percolation problem. For instance, the mirror's model in the square lattice is reduced to the percolation problem on the square lattice, while the rotator's model is reduced to the percolation problem on some nonplanar graph (Bunimovich & Troubetzkoy, 1992).

Perhaps the most widely known DWRE models are Langton's Ant (Langton, 1986) or the flipping rotators model on the square lattice (Ruijgrok & Cohen, 1988), which are solvable again with rather counterintuitive results (Bunimovich & Troubetzkoy, 1993). If all vertices are occupied with rotators, then all orbits (particle's paths) are unbounded. If, on the other hand, one allows vertices to be empty with positive probability (i.e., the third, straight-ahead scatterer is allowed), then the particle's path becomes bounded with probability one.

The results, both numerical and mathematical, continued to surprise until "Walks in Rigid Environments" (WRE) were introduced and analyzed (Bunimovich, 2000). WREs employ a new integer parameter r, $1 \leq r \leq \infty$, which is called the rigidity of the environment. The rigidity determines how many times the particle must collide with the given scatterer in order to change its type. In other words, the scatterer at a given vertex changes its type immediately after the rth visit of the particle to this site. Therefore, WREs interpolate between DWREs with fixed environments (where $r = \infty$) and DWREs with flipping environments (where $r = 1$).

WREs on a one-dimensional lattice $\mathbb{Z}$ are completely solved (Bunimovich, 2000; Bunimovich & Khlabystova, 2002b). In this case, there are only four types of scatterers. Two of them (forward scatterer and backscatterer) are symmetric with respect to the reflection of $\mathbb{Z}$, which is the only nontrivial symmetry of the one-dimensional lattice. The other two scatterers, which always send the particle to the right (or to the left), do not respect this symmetry. Therefore, the WRE with the last two types of scatterers has the same behavior for all values of the rigidity r. This model demonstrates a diffusive type of behavior, in which the particle eventually visits all vertices and the mean square displacement of the particle is proportional to t. On the contrary, WREs with forward and back scatterers demonstrate totally different behavior depending on the parity of the rigidity r. For even rigidities, the particle eventually visits all vertices again but its motion is subdiffusive. The most interesting behavior occurs for odd values of the rigidity. In this case the particle—after a short initial period of seemingly irregular motion near the origin—starts to propagate in one direction with random velocity. This phenomenon of (eventual) propagation reminds one of "gliders" in Conway's Game of Life. However, in a WRE this propagation occurs for all initial configurations of environment, while in the Game of Life, gliders appear as only very special solutions. The phenomenon of eventual propagation in one direction is not restricted to one-dimensional WREs. For instance, the same behavior is demonstrated by the model with right and left rotators on the triangular lattice (Grosfils et al., 1999).

If the rigidity $r < \infty$, then one can also investigate the dynamics of the environment. Observe that if $r < \infty$, then it makes sense to consider not one but many moving particles as well. Indeed, even though the particles do not interact directly, they do effectively interact by changing environments with each other. The evolution of the environment in such models can be chaotic having a positive and even infinite metric entropy (Bunimovich & Troubetzkoy, 1993). This means, in particular, that the moving particle sees itself surrounded by almost any possible environment of scatterers at different moments of time.

An important difference between DWREs and random walks is that in the continuous limit DWREs become completely deterministic (Bunimovich & Khlabystova, 2002b), whereas biased random walks under proper scaling of probabilities become in the continuous limit stochastic diffusion processes.

LEONID BUNIMOVICH

See also **Game of life; Lorentz gas; Random walks**

Further Reading

Bunimovich, L.A. 2000. Walks in rigid environments. *Physica* A, 279:169–179

Bunimovich, L.A. & Khlabystova, M.A. 2002a. Lorentz lattice gases and many-dimensional Turing machines. In *Collision-Based Computing*, edited by A. Adamatzky, London: Springer

Bunimovich, L. A. & Khlabystova, M.A. 2002b. Walks in rigid environments: continuous limits. *Journal of Statistical Physics*, 108: 905–925

Bunimovich, L. A. & Troubetzkoy, S.E. 1992. Recurrence properties of the Lorentz lattice gases. *Journal of Statistical Physics*, 67: 289–302

Bunimovich, L. A. & Troubetzkoy, S.E. 1993. Topological properties of flipping Lorentz lattice gas models. *Journal of Statistical Physics*, 72: 297–307

Cohen, E.G.D. 1992. New types of diffusion in lattice gas cellular automata. In *Microscopic Simulations of Complex Hydrodynamic Phenomena*, edited by M. Mareschal & B.L. Holian, New York: Plenum

Grosfils, P., Boon, J.P., Cohen, E.G.D. & Bunimovich, L.A. 1999. Propagation and self-organization in lattice random media. *Journal of Statistical Physics*, 97: 575–608

Gunn, J.M.F. & Ortuño, M. 1985. Percolation and motion in a simple random environment. *Journal of Physics* A, 18: 1095–1099

Langton, C.G. 1986. Studying artificial life with cellular automata. *Physica* D, 22: 120–149

Ruijgrok, T.W. & Cohen, E.G.D. 1988. Deterministic lattice gas models. *Physics Letters* A, 133: 415–419

DEVELOPMENT OF SINGULARITIES

For some nonlinear PDEs, solutions of the Cauchy (initial-value) problem may exist only until a finite moment $t = t_c$, and they cannot be continued afterwards. At $t = t_c$, the solution loses its initial smoothness, and a singularity appears, which causes the blow-up of the solution. For conservative systems where the wavefield develops sharply diverging gradients, this blow-up phenomenon is often called collapse.

Historically, the first example of collapse follows from the so-called Hopf equation: $u_t + uu_x = 0$, which describes the one-dimensional velocity of dust with zero pressure. This equation admits the generic implicit solution $u = F(x - ut)$, where F is determined from the initial velocity profile. For any compact distribution F, there exists a couple of coordinates (x_c, t_c), at which the spatial derivative of u reaches infinity.

Collapse also occurs in the high-dimensional solutions to the nonlinear Schrödinger (NLS) equation, consisting of a point-singularity that achieves the fate of nonlinear waves undergoing self-focusing (Kelley, 1965). To describe this process, let us consider the NLS equation

$$i\partial_t \psi + \nabla^2 \psi + |\psi|^2 \psi = 0, \tag{1}$$

where t is a time variable and the Laplacian $\nabla^2 = \partial_x^2 + \partial_y^2 + \partial_z^2 + \cdots$ accounts for the dispersion of a wave-packet along D orthogonal spatial axes $[\boldsymbol{r} = (x, y, z, ...)]$. The wave function ψ evolves from the spatially localized initial datum $\psi(\boldsymbol{r}, 0) \equiv \psi_0(\boldsymbol{r})$, assumed to belong to the Hilbert space H^1 with finite norm $\|\psi\|_{H^1} = (\|\psi\|_2^2 + \|\nabla\psi\|_2^2)^{1/2}$, where $\|f\|_p \equiv (\int |f|^p \mathrm{d}\boldsymbol{r})^{1/p}$. Two invariants are associated with ψ, namely, the L^2 norm N (sometimes called mass, power, or number of particles) and the Hamiltonian H:

$$N = \|\psi\|_2^2, \quad H = \|\nabla\psi\|_2^2 - \tfrac{1}{2}\|\psi\|_4^4. \tag{2}$$

From Equation (1), the following virial equality can be established (Glassey, 1977):

$$N d_t^2 \langle r^2 \rangle = 4\{2H + (1 - D/2)\|\psi\|_4^4\}, \tag{3}$$

where $\langle r^2 \rangle = \int r^2 |\psi|^2 \mathrm{d}\boldsymbol{r}/N$ denotes the mean-squared radius of the solution ψ. By a double integration in time, Equation (3) shows that, whenever $D \geq 2$, there exist initial conditions for which $\langle r^2 \rangle$ vanishes at finite time, which is the signature of a wave collapse. For finite norms N, the inequality $N \leq (2/D)^2 \langle r^2 \rangle \times \|\nabla\psi\|_2^2$ implies that the gradient norm diverges in collapse regimes. As H is finite, the collapse dynamics makes the L^4 norm $\|\psi\|_4^4$ blow up in turn and $\max|\psi|$ diverges, by virtue of the mean-value theorem $\int |\psi|^4 \mathrm{d}\boldsymbol{r} \leq \max_r |\psi|^2 \times N$. This leads to a finite-time blow-up, at which the solution ψ ceases to exist in H^1. Blow-up generally occurs before $\langle r^2 \rangle$ reaches zero (Rasmussen & Rypdal, 1986).

While $H < 0$ arises from Equation (3) as a sufficient condition for collapse, sharper requirements can be derived by means of the so-called Sobolev inequality

$$\|\psi\|_4^4 \leq C \|\nabla\psi\|_2^D \times \|\psi\|_2^{4-D}. \tag{4}$$

- In the critical case $D = 2$, this inequality can be used to bound H from below, so that the gradient norm blows up only if N fulfills the constraint $N > N_c$. The best constant in Equation (4) is exactly $C_{\text{best}} = 2/N_c$, and it involves the quantity $N_c = \int \phi_0^2 \,\mathrm{d}\boldsymbol{r} = 11.68$, where ϕ_0 is the radially symmetric soliton solution of $-\phi_0 + \nabla^2\phi_0 + \phi_0^3 = 0$ (Weinstein, 1983). N_c provides the minimum power that ψ must necessarily contain at $t = 0$ for producing a collapse, justifying the existence of a critical power for the 2-dimensional self-focusing of optical beams in nonlinear Kerr media.
- In the supercritical case $D = 3$, a criterion for collapse, sharper than $H < 0$, can be established from a combination of Equations (3) and (4) as $H < N_c^2/N$ for gradient norms initially above $3N_c^2/N$ (Kuznetsov et al., 1995). N_c again corresponds to the mass of the three-dimensional soliton solution ϕ_0 satisfying $-\phi_0 + \nabla^2\phi_0 + \phi_0^3 = 0$.

Once the collapse is triggered, the solution self-focuses and shrinks isotropically in a self-similar way near the singularity point t_c. It is, thus, convenient to introduce the self-similar substitution:

$$\psi(\boldsymbol{r}, t) = a^{-\alpha}(t)\phi(\boldsymbol{\xi}, \tau)\mathrm{e}^{\mathrm{i}\lambda\tau - \mathrm{i}\beta\xi^2/4}, \tag{5}$$

where $\boldsymbol{\xi} = \boldsymbol{r}/a(t)$, $\tau(t) \equiv \int_0^t \mathrm{d}u/a^2(u)$, and $\beta = -a\dot{a}$ (dot indicates differentiation with respect to time). Here, the parameter λ is positive for localization of ϕ. The function $a(t)$ represents the scale length that vanishes as collapse develops, and ϕ converges to an exactly self-similar form $\phi(\boldsymbol{\xi})$, which no longer depends explicitly on time $[\partial_\tau \phi \to 0]$. Inserting (3) into H shows that the right balance between the gradient and L^4 norms requires $\alpha = 1$, in order to preserve the finiteness of the Hamiltonian. With $\alpha = 1$, $N = a^{D-2}(t) \int |\phi|^2 \mathrm{d}\boldsymbol{\xi}$ and for radial solutions, Equation (1) transforms to

$$\mathrm{i}\partial_\tau \phi + \xi^{1-D}\partial_\xi \xi^{D-1}\partial_\xi \phi + |\phi|^2\phi + \varepsilon[\xi^2 - \xi_T^2]\phi = 0, \tag{6}$$

where $\xi_T^2 \equiv \varepsilon^{-1}[\lambda + \mathrm{i}\beta(D/2 - 1)]$ is viewed as a turning point, with $\varepsilon \equiv \frac{1}{4}(\beta^2 + \partial_\tau \beta)$. As $a(t) \to 0$, ϕ can be treated by means of quasi-self-similar techniques (Bergé, 1998). In the limit $\partial_\tau \phi \to 0$, ε converges to $\beta^2/4$ and the solution ϕ is split into a nonlinear core (ϕ_c) extending in the range $\xi < \xi_T$ and a linear tail (ϕ_T) defined in the complementary spatial domain $\xi > \xi_T$, where the nonlinearity vanishes. As a result, the wave function ψ reads near the collapse point:

$$\psi(\boldsymbol{r}, t) = \frac{\mathrm{e}^{\mathrm{i}\lambda \int_0^t \mathrm{d}u/a^2(u)}}{a(t)} \times \left\{ \phi_c\left(\frac{r}{a}, \varepsilon\right) \mathrm{e}^{-\mathrm{i}\beta r^2/4a^2} \Big|_{0 \le r < r_T} + \frac{C(\beta)}{(r/a)^{1+\mathrm{i}\lambda/\beta}} \Big|_{r_T < r < r_{\max}} \right\}, \tag{7}$$

where $|C(\beta)|^2$ evolves like $(2/\beta|\xi_T|^{D-2})\mathrm{e}^{-\lambda\pi/\beta}$, $r_T = a(t)|\xi_T|$ and $r_{\max}$ bounds the self-similarity domain. The length $a(t)$ is identified from the continuity equation describing the mass exchanges between the core and tail parts of ψ. The dynamics of self-similar collapses then vary with the space dimension number as follows.

- *Strong collapse*: For $D = 2$, the size $a(t)$ behaves with a double-logarithmic correction: $a(t) \simeq a_0 \sqrt{t_c - t}/\sqrt{\ln\ln[1/(t_c - t)]}$, coming from $\beta \simeq \lambda\pi/\ln[\tau(t)]$ and $\tau(t) \simeq \ln[1/(t_c - t)]$. As $t \to t_c$, the exponential contribution in tail (5) thus decreases to zero, while the core boundary ξ_T increases slowly to infinity. Collapse thus takes place with a core solution providing the principal contribution in the wavefunction ψ. N is mainly given by $\int |\phi_c|^2 \mathrm{d}\boldsymbol{\xi}$, which relaxes to the critical value $N_c = 11.68$. The mass stays mostly located around the center, which meets the definition of *strong collapse* (Zakharov & Kuznetsov, 1986).
- *Weak collapse*: For $D = 3$, β attains a fixed point $\beta_0 \neq 0$, leading to the scaling law $a(t) \simeq a_0(t_c - t)^{1/2}$. The power is no longer preserved self-similarly in space, since $N = a(t) \int |\phi|^2 \mathrm{d}\boldsymbol{\xi}$. This integral behaves as $N \simeq N_{\mathrm{core}}(t) + N_{\mathrm{tail}}(t)$, where $N_{\mathrm{core}}(t) \sim a(t)$ vanishes, while $N_{\mathrm{tail}}(t) \simeq 4\pi|C(\beta)|^2 r_{\max}$ contains almost all the initial mass as $a(t) \to 0$. A 3-d collapse is thus accompanied by an expulsion of particles towards the large-distance domain $r \gg r_T(t)$, which characterizes a *weak collapse*. The solution ψ blows up at the center, where N_{core} becomes zero. Accordingly, ψ extends in the outer domain with the stationary density $r^2|\psi|^2 \to |C(\beta_0)|^2 = \mathrm{const}$.

These different collapses can be tested numerically by adding a small nonlinear dissipation term in Equation (1), i.e., by changing ψ_t into $\psi_t + \beta|\psi|^m\psi$ with $\beta \ll 1$ and $m > 1$. For $D = 2$, dissipation removes a substantial amount of energy per collapse event, because N remains mostly confined at the center. This leads to a step-wise decrease of N (Kosmatov et al., 1991). In contrast, for $D = 3$, nonlinear dissipation cannot remove much energy, the major part of $N \simeq N_{\mathrm{tail}}$ being transferred to large distances. This causes a monotonical decrease of N.

- *Superstrong collapse*: For weak and strong collapses, only the energy captured by the singularity dissipates. A different situation is realized if collapsing solutions can sustain dissipation through power loss into a persistent region of high intensity. Such solutions approach a quasi stationary state, described by $\psi_{rr} + (D - 1)\psi_r/r + |\psi|^2\psi = 0$, in the limit of small $\beta \to 0$. For instance, when $D = 4$, Equation (1) admits the stationary radial state $\psi(r) = B\mathrm{e}^{\mathrm{i}\ln r\sqrt{B^2-1}}/r$ $(B > 1)$, for which the power density $|\psi|^2$ flows into the singularity with a constant energy flux equal to $B^2\sqrt{B^2 - 1}$. This collapse, which received the adjective of *superstrong*, has been numerically detected at high dimension numbers $D > 3$ (Kosmatov et al., 1991).

So far, the discussion has remained within the realm of the one-wave component NLS equation with a cubic nonlinearity. It is thus worth underlining the following.

- The previous results can be generalized to a power-law nonlinearity, when the cubic term $|\psi|^2\psi$ of Equation (1) is replaced by $|\psi|^{2n}\psi$ with $n > 1$ (Rasmussen & Rypdal, 1986; Bergé, 1998). Solutions with $D = 2/n$ follow the route of a strong collapse, while solutions defined for $D > 2/n$ collapse weakly. Superstrong collapses apply to the dimensional configurations $D > 2 + 1/n$.
- Several NLS equations coupled through their cubic nonlinearities often serve to model the self- and cross-interactions of multiple wave components (or different polarizations) in vector systems. Such systems promote blow-up phenomena, which can be examined by means of the above analytical tools (Bergé, 2001).
- Blow-up may take place in solutions of PDEs other than the NLS equation. For example, investigations of the solutions to the generalized D-dimensional Korteweg–de Vries (KdV) equation

$$q_t + q^n q_x + (\nabla^2 q)_x = 0 \tag{8}$$

suggest that, whereas no collapse occurs for values of the product $nD < 4$, collapsing states can arise and adopt a self-similar shape provided that $nD \geq 4$ (Blaha et al., 1989). The mathematical proof for this statement is presently incomplete.

LUC BERGÉ

See also **Filamentation; Kerr effect; Nonlinear Schrödinger equations; Virial theorem**

Further Reading

Bergé, L. 1998. Wave collapse in physics: Principles and applications to light and plasma waves. *Physics Reports*, 303: 259–370

Bergé, L. 2001. Nonlinear wave collapse. In *Spatial Solitons*, edited by S. Trillo & W. Torruellas, Berlin: Springer, pp. 247–267

Blaha, R., Laedke, E.W. & Spatschek, K.H. 1989. Collapsing states of generalized Korteweg–de Vries equations. *Physica* D, 40: 249–264

Glassey, R.T. 1977. On the blowing-up of solutions to the Cauchy problem for nonlinear Schrödinger equations. *Journal of Mathematical Physics*, 18: 1794–1797

Kelley, P.L. 1965. Self-focusing of optical beams. *Physical Review Letters*, 15: 1005–1008

Kosmatov, N.E., Shvets, V.F. & Zakharov, V.E. 1991. Computer simulation of wave collapses in the nonlinear Schrödinger equation. *Physica* D, 52: 16–35

Kuznetsov, E.A., Rasmussen, J., Rypdal, K. & Turitsyn, S.K. 1995. Sharper criteria for the wave collapse. *Physica* D, 87: 273–284

Rasmussen, J. & Rypdal, K. 1986. Blow-up in nonlinear Schrödinger equations-I: a general review. *Physica Scripta*, 33: 481–497

Weinstein, M.I. 1983. Nonlinear Schrödinger equations and sharp interpolation estimates. *Communications in Mathematical Physics*, 87: 567–576

Zakharov, V.E. & Kuznetsov, E.A. 1986. Quasiclassical theory of three-dimensional wave collapse. *Zhurnal Eksperimental'noi i Teoreticheskoi Fiziki* (USSR JETP), 91: 1310–1324 [Trans. in *Soviet Physics JETP*, 64: 773–780]

DEVIL'S STAIRCASE

See **Fractals**

DIFFEOMORPHISM

See **Maps**

DIFFERENTIAL GEOMETRY

A *topological manifold of dimension n* is a topological space M that can locally be identified with an open set in $\mathbb{R}^n$; a superscript on the symbol for M is often used to indicate the dimension. For example, a circle is a one-dimensional manifold, denoted S^1, while a figure-eight is not a manifold. In more detail, M must have a set of *coordinate charts* ϕ_α, each of which is a homeomorphism from an open set $U_\alpha \subset M$ to an open set in $\mathbb{R}^n$, such that every point of M is in the domain of a chart. If the intersection $U_{\alpha\beta}$ of two domains U_α and U_β is nonempty, then the change of coordinates $\phi_\alpha \circ \phi_\beta^{-1}$ is a continuous map from $\phi_\beta(U_{\alpha\beta})$ to $\phi_\alpha(U_{\alpha\beta})$ with a continuous inverse.

A *differentiable manifold* is a topological manifold M^n equipped with charts such that, on the overlaps, $\phi_\alpha \circ \phi_\beta^{-1}$ is differentiable with a differentiable inverse. (Here, "differentiable" or "smooth" functions are those whose partial derivatives exist and are continuous to all orders; however, for C^k *manifolds*, the changes of coordinates are only required to be differentiable up to order k.) If $x^1, \ldots, x^n$ and $\bar{x}^1, \ldots, \bar{x}^n$ are the coordinate functions for two overlapping charts, then the Jacobian determinant $||\partial x^i/\partial \bar{x}^j||$ must be nonzero at each point.

For example, the subset M of $\mathbb{R}^{n+1}$ defined by an equation form $f(x^0, x^1, \ldots, x^n) = 0$ is a differentiable manifold if f is a smooth function and at each point $p \in M$, at least one partial derivative (say, $\partial f/\partial x^0$) is nonzero. Then, by the Implicit Function Theorem, the projection from a neighborhood of p (i.e., an open subset of M containing p) onto the $x^1 \cdots x^n$ coordinate hyperplane is differentiable with a differentiable inverse. So, the n-dimensional spheres S^n are compact differentiable manifolds. Similarly, the group SL_n of $n \times n$ matrices with determinant one is a noncompact manifold of dimension $n^2 - 1$.

A function $f : M \to \mathbb{R}$ is smooth if $f \circ \phi$ is smooth for any chart ϕ. More generally, if M^m and N^n are two differentiable manifolds, we say a mapping $F : M \to N$ is smooth if it is smooth with respect to coordinate charts on both ends, that is, $\psi \circ F \circ \phi^{-1}$ is smooth from $\mathbb{R}^m$ to $\mathbb{R}^n$ for any charts ϕ on M and ψ on N. When $F : M \to N$ also has a smooth inverse, it is a *diffeomorphism*, and M and N are *diffeomorphic*. For example, the hyperboloid $x^2 + y^2 - z^2 = 1$ is diffeomorphic to the cylinder $x^2 + y^2 = 1$, and any open interval on the real line $\mathbb{R}$ is diffeomorphic to all of $\mathbb{R}$.

Vector Fields and 1-Forms

A *tangent vector* at a point $p \in M^n$ is a linear operator v on smooth functions f defined near p, such that (i) $\mathsf{v}(f_1 f_2) = f_1 \mathsf{v}(f_2) + f_2 \mathsf{v}(f_1)$, and (ii) $\mathsf{v}(f_1) = \mathsf{v}(f_2)$ if $f_1 = f_2$ on a neighborhood of p (i.e., f_1, f_2 have the same *germ* at p). For example, if $c : \mathbb{R} \to M$ defines a curve in M such that $c(t) = p$, then we define the tangent vector $c'(t)$ to the curve by $c'(t)(f) := (\mathrm{d}/\mathrm{d}t) f(c(t))$, where the symbol ":=" indicates a definition. If $\mathsf{v} = c'(t)$, then we say that the curve is *tangent to* v at p.

The set of tangent vectors at p form a vector space of dimension n, denoted T_pM. If $x^1, \ldots, x^n$ are local coordinates near p, then the partial derivative operators $\partial/\partial x^i$ are a basis for T_pM. The set of all tangent vectors at all points of M is itself a differentiable manifold of dimension $2n$, since we can adjoin coordinates

$y^1, \dots, y^n$ and locally write all tangent vectors as

$$\mathsf{v} = \sum_{i=1}^{n} y^i \frac{\partial}{\partial x^i}.$$

This manifold is the *tangent bundle* of M, and is denoted TM.

A *vector field* on M smoothly assigns a tangent vector at each point; in local coordinates, we specify a vector field by giving the y^i as smooth functions of x^i. If f is a smooth function on M and $\boldsymbol{V}$ is a vector field, then $\boldsymbol{V}(f)$ is another smooth function.

Given a vector field $\boldsymbol{V}$ and a point $p \in M$ at which $\boldsymbol{V} \neq 0$, existence theorems for systems of ODE (e.g., Picard's Theorem) imply that there exist a neighborhood U of p and a one-parameter family of smooth, one-to-one mappings $F_t : U \to M$, such that for any fixed $q \in U$, the curve $F_t(q)$ is tangent to $\boldsymbol{V}$ for every t. The F_t's are called the *flow by vector field* $\boldsymbol{V}$.

The *Lie bracket* $[\boldsymbol{V}_1, \boldsymbol{V}_2]$ of two vector fields $\boldsymbol{V}_1, \boldsymbol{V}_2$ is a third vector field defined by

$$[\boldsymbol{V}_1, \boldsymbol{V}_2](f) := \boldsymbol{V}_1(\boldsymbol{V}_2(f)) - \boldsymbol{V}_2(\boldsymbol{V}_1(f)).$$

If y_1^i and y_2^i are the components of $\boldsymbol{V}_1, \boldsymbol{V}_2$ in local coordinates, then

$$[\boldsymbol{V}_1, \boldsymbol{V}_2] = \sum_{i,j=1}^{n} \left(y_1^i \frac{\partial y_2^j}{\partial x^i} - y_2^i \frac{\partial y_1^j}{\partial x^i} \right) \frac{\partial}{\partial x^j}. \tag{1}$$

The vector fields are said to *commute* if $[\boldsymbol{V}_1, \boldsymbol{V}_2]$ is identically zero; then, flow by $\boldsymbol{V}_1$ commutes with flow by $\boldsymbol{V}_2$.

A *differential 1-form at p* is a linear mapping function from T_pM to $\mathbb{R}$. For example, given a differentiable function f defined near p, we define the 1-form $\mathrm{d}f$ by $\mathrm{d}f(\mathsf{v}) := \mathsf{v}(f)$ for all $\mathsf{v} \in T_pM$. The vector space of 1-forms at p is denoted T_p^*M. Given local coordinates, the differentials $\mathrm{d}x^1, \dots, \mathrm{d}x^n$, which satisfy $\mathrm{d}x^i(\partial/\partial x^j) = \delta_j^i$, are a basis for T_p^*M. While tangent vectors generalize directional derivatives in $\mathbb{R}^n$, 1-forms generalize gradients; however, 1-forms cannot be identified with tangent vectors without using some nondegenerate bilinear form on tangent vectors, for example, a Riemannian metric or symplectic form on M.

A *differential 1-form* on M assigns a 1-form at each point; in local coordinates, these appear as

$$\omega = \sum_{i=1}^{n} z_i \mathrm{d}x^i,$$

where the z_i are some smooth functions of the x^i. If $\boldsymbol{V}$ is a vector field on M, then $\omega(\boldsymbol{V})$ is a function on M.

Given a smooth mapping $F : M \to N$, we can transform a tangent vector $\mathsf{v} \in T_pM$ to a tangent vector $F_*\mathsf{v}$ at $F(p)$ by defining

$$F_*\mathsf{v}(f) := \mathsf{v}(f \circ F)$$

for any function $f : N \to \mathbb{R}$. This is called the *pushforward* of v. We similarly define the pushforward $F_*\boldsymbol{V}$ of a vector field $\boldsymbol{V}$ on M. We can also transform a 1-form ω on N to a 1-form $F^*\omega$ on M by defining

$$F^*\omega(\mathsf{v}) := \omega(F_*\mathsf{v})$$

for any tangent vector v to M. This is called the *pullback* of ω.

If $\boldsymbol{V}$ is a vector field and F_t is the flow by $\boldsymbol{V}$, we define the *Lie derivative* of a 1-form ω with respect to $\boldsymbol{V}$ by

$$\mathcal{L}_{\boldsymbol{V}}\omega := \left.\frac{\mathrm{d}}{\mathrm{d}t}\right|_{t=0} F_t^*(\omega).$$

If the components of $\boldsymbol{V}$ and ω in local coordinates are y^i and z_i, respectively, then

$$\mathcal{L}_{\boldsymbol{V}}\omega = \sum_{i,j=1}^{n} \left(y^i \frac{\partial z_j}{\partial x^i} - z_i \frac{\partial y^j}{\partial x^i} \right) \mathrm{d}x_j. \tag{2}$$

Some useful properties of the Lie derivative are that it obeys the product rule and commutes with d, that is,

$$\begin{aligned} \mathcal{L}_{\boldsymbol{V}}(f\boldsymbol{W}) &= (\mathcal{L}_{\boldsymbol{V}} f)\boldsymbol{W} + f\,\mathcal{L}_{\boldsymbol{V}}\boldsymbol{W}, \\ \mathcal{L}_{\boldsymbol{V}}(f\omega) &= f\,\mathcal{L}_{\boldsymbol{V}}\omega + (\mathcal{L}_{\boldsymbol{V}} f)\omega, \\ \mathcal{L}_{\boldsymbol{V}}(\mathrm{d}f) &= \mathrm{d}(\mathcal{L}_{\boldsymbol{V}} f), \end{aligned}$$

where we define $\mathcal{L}_{\boldsymbol{V}} f := \boldsymbol{V}(f)$ and $\mathcal{L}_{\boldsymbol{V}}\boldsymbol{W} := [\boldsymbol{V}, \boldsymbol{W}]$. (Note than one can derive (2) using these properties.)

Higher Degree Differential Forms and Topology

A *differential k-form at p* is a multilinear function on k-tuples of vectors in T_pM that is skew-symmetric; that is, its value is multiplied by -1 whenever two adjacent vectors in the k-tuple are exchanged. For example, a 2-form may be constructed from two 1-forms ω^1, ω^2 using the *wedge product*:

$$\omega^1 \wedge \omega^2(\mathsf{v}_1, \mathsf{v}_2) := \omega^1(\mathsf{v}_1)\omega^2(\mathsf{v}_2) - \omega^1(\mathsf{v}_2)\omega^2(\mathsf{v}_1).$$

(Note that this is zero if ω^1 and ω^2 are linearly dependent.) More generally, the wedge product of 1-forms $\omega^1, \dots, \omega^k$ is defined by

$$\begin{aligned} &\omega^1 \wedge \ldots \wedge \omega^k(\mathsf{v}_1, \ldots \mathsf{v}_k) \\ &\quad := \sum_{\sigma} (-1)^{\sigma} \omega^1(\mathsf{v}_{\sigma(1)})\omega^2(\mathsf{v}_{\sigma(2)}) \cdots \omega^k(\mathsf{v}_{\sigma(k)}), \end{aligned}$$

where the sum is over all permutations σ of $1, 2, \dots, k$ and $(-1)^{\sigma}$ is the sign of the permutation. On an n-dimensional manifold, the vector space of k-forms at a point has dimension $\binom{n}{k}$ and is spanned by wedge products of 1-forms; in particular, there are no forms of degree higher than n.

A 2-form may also be constructed from a single 1-form ω by taking the *exterior derivative* $\mathrm{d}\omega$, which satisfies

$$\mathrm{d}\omega(\boldsymbol{V}_1, \boldsymbol{V}_2) = \boldsymbol{V}_1(\omega(\boldsymbol{V}_2)) - \boldsymbol{V}_2(\omega(\boldsymbol{V}_1)) - \omega([\boldsymbol{V}_1, \boldsymbol{V}_2]). \tag{3}$$

(Although the right-hand side is defined using vector fields, its value at a point p depends only on the values of $\boldsymbol{V}_1, \boldsymbol{V}_2$ at p.) The exterior derivative of a k-form is a $(k+1)$-form; it can be calculated inductively using linearity and the product rule

$$d(\alpha \wedge \beta) = \mathrm{d}\alpha \wedge \beta + (-1)^{\mathrm{degree}(\alpha)} \alpha \wedge \mathrm{d}\beta.$$

If the exterior derivative of a k-form is identically zero on M, the form is *closed*. If a k-form is an exterior derivative of a $(k-1)$-form, it is *exact*. An important property of the exterior derivative is that $\mathrm{d}(\mathrm{d}\alpha) = 0$, i.e., exact forms are closed. The *Poincaré Lemma* asserts that a closed k-form α is *locally* exact; that is, in the vicinity of any given point a $(k-1)$-form β is defined such that $\alpha = \mathrm{d}\beta$. However, not every closed form is *globally* exact (e.g., the 1-form $\mathrm{d}\theta$ on S^1). Moreover, the *de Rham Theorem* asserts that the dimension of the vector space of closed k-forms modulo exact k-forms is a topological invariant of M; that is, two manifolds cannot be homeomorphic (or diffeomorphic) unless these dimensions, known as the *Betti numbers*, match up. (For example, the Euler characteristic is determined by the Betti numbers.)

THOMAS A. IVEY

See also **Invariant manifolds and sets; Lie algebras and Lie groups; Topology**

Further Reading

Isham, C.J. 1999. *Modern Differential Geometry for Physicists*, Singapore: World Scientific

Ivey, T. & Landsberg, J.M. 2003. *Cartan for Beginners: Differential Geometry via Moving Frames and Exterior Differential Systems*, Providence, RI: American Mathematical Society

Warner, F.W. 1983. *Foundations of Differentiable Manifolds and Lie Groups*, Berlin and New York: Springer

DIFFUSION

When a small drop of ink is poured onto a soft gel, the ink molecules disperse through by diffusion. Similarly, the spread of heat through a medium can also be a diffusive process (called heat conduction). Many other applications of diffusion arise in biology, combustion, economics, chemical engineering, and geophysics, among other fields.

Mathematically, the diffusion equation takes the form (Crank, 1975)

$$u_t = \Delta u, \tag{1}$$

where $u = u(\boldsymbol{x}, t)$ is the state variable, representing, for example, the density of concentration of some substance, at time $t \geq 0$ and position $\boldsymbol{x}$ in R^n, where Δu denotes the Laplacian of u with respect to the space variable $\boldsymbol{x}$. Equation (1) is an example of a parabolic equation of evolution. If it holds for all $\boldsymbol{x}$ in R^n, then the problem is fully specified once appropriate initial conditions

$$u(\boldsymbol{x}, 0) = u_0(\boldsymbol{x}) \tag{2}$$

are known. If Equation (1) holds in a limited domain $\Omega \subset R^n$, then some boundary conditions must be imposed on u at $\partial\Omega$ that are compatible with the physical situation.

The diffusion equation can be viewed as a balance law (Grindrod, 1996). Let $Q(\boldsymbol{x}, t)$ be the net creation rate of particles at $\boldsymbol{x} \in \Gamma \subset \Omega$ and time t, and let $\boldsymbol{J}(\boldsymbol{x}, t)$ be the flux density. For any unit vector $\boldsymbol{n} \in R^n$, the scalar product $\boldsymbol{J} \cdot \boldsymbol{n}$ is the net rate at which particles cross a unit area in a plane perpendicular to $\boldsymbol{n}$ (take the plus sign in the direction of $\boldsymbol{n}$). Assuming that the rate of change of mass in Ω is due to particle creation or degradation inside Γ and the inflow and the outflow of particles through the boundary $\partial\Gamma$, we have

$$\frac{\mathrm{d}}{\mathrm{d}t} \int_\Gamma u \, \mathrm{d}\boldsymbol{x} = -\int_{\partial\Gamma} \boldsymbol{J} \cdot \boldsymbol{n} \, \mathrm{d}S + \int_\Gamma Q \, \mathrm{d}\boldsymbol{x}, \tag{3}$$

where $\int_\Gamma u \, \mathrm{d}\boldsymbol{x}$ denotes the population mass in Γ. If the solution is smooth enough, then applying the divergence theorem to the right-hand side in (3) gives

$$\frac{\mathrm{d}}{\mathrm{d}t} \int_\Gamma u \, \mathrm{d}\boldsymbol{x} = \int_\Gamma \nabla \cdot \boldsymbol{J} \, \mathrm{d}S + \int_\Gamma Q \, \mathrm{d}\boldsymbol{x}. \tag{4}$$

As Γ was arbitrary in Ω, Equation (4) implies that

$$u_t = -\nabla \cdot \boldsymbol{J} + Q \tag{5}$$

at every point in Ω, which is the required balance law.

In practice, depending on the process studied, one must specify the flux $\boldsymbol{J}$ and the source term Q. For example, in combustion or in chemistry, one uses Fick's law:

$$\boldsymbol{J} = -D\nabla u. \tag{6}$$

Here $D > 0$ is a constant called diffusivity with physical units of $\mathrm{m}^2\mathrm{s}^{-1}$. The minus sign in (6) accounts for the fact that the particles are transported from high to low densities. Using (6) in (5) gives the usual reaction-diffusion equation

$$u_t = D\Delta u + Q(\boldsymbol{x}, t, u, \ldots). \tag{7}$$

A heuristic derivation of the diffusion equation (1) involves the notion of a random walk. Consider a one-dimensional (1-d) random walker that at each time step t_i hops from its position x_n one unit to the left or right, $x_{n\pm1}$. The change in probability $p(x_n, t_i)$ to find the walker at x_n at t_i is equal to the sum of probabilities for it to hop into the point minus the sum of probabilities to hop off the point:

$$p(x_n, t_i) - p(x_n, t_{i-1}) = \frac{1}{2}(p(x_{n+1}, t_{i-1}) + p(x_{n-1}, t_{i-1})) - p(x_n, t_{i-1}). \quad (8)$$

Introducing the space and time scales of the motion, this equation can be rearranged as

$$\frac{p(x_n, t_i) - p(x_n, t_{i-1})}{\delta_t} = \frac{\delta_x^2}{2\delta_t}\left[\frac{p(x_{n-1}, t_{i-1}) - 2p(x_n, t_{i-1}) + p(x_{n+1}, t_{i-1})}{\delta_x^2}\right]. \quad (9)$$

Denoting $D = \delta_x^2/2\delta_t$, Equation (9) becomes a discrete approximation of Equation (1), and by taking the limit as $\delta_t, \delta_x \to 0$, keeping D finite, we recover Equation (1).

A rigorous derivation of the diffusion equation is obtained via stochastic calculus. The motion of individual particles is described by stochastic differential equations with the positions of the particles being modeled as random variables in R^n. The global behavior depends on the type of stochastic process governing the motion of the particles. Typically, one describes this in terms of the probability distribution of the random variable. In many cases, one finds that all its moments of order higher than 2 vanish. Consequently, the distribution of the population density, u, satisfies the second-order Fokker–Planck (or Kolmogorov's forward equation) (Øksendal, 2000)

$$u_t = \nabla(D(\boldsymbol{x}, t)u) - \nabla(C(\boldsymbol{x}, t)u). \quad (10)$$

Equation (10) is a diffusion equation with a nonconstant, inhomogeneous diffusivity and convective term. It often arises for particles whose individual speeds are random deviations from some externally applied convection velocity. Examples arise in fluid flow and in biology (dispersals of population dispersals, e.g. chemotaxis), among other fields.

Particularly important cases are when the diffusivity depends on the population density such as in biology or ecology (Aronson & Weinberger, 1975). For example, insect dispersal models use the fact that the rate of spread of the population is increased at higher insect density. This is usually modeled with an equation of the form

$$u_t = \Delta(D(u)\nabla u) + f(u), \quad (11)$$

where typically $D(u) = D_0 u^m$, $m > 0$ (Okubo & Levin, 2001; Murray, 2002). Density-dependent diffusion equations appear in many other fields. In physics, impurities are diffused into semiconductor materials (in the processing of silicon-based electronic devices)

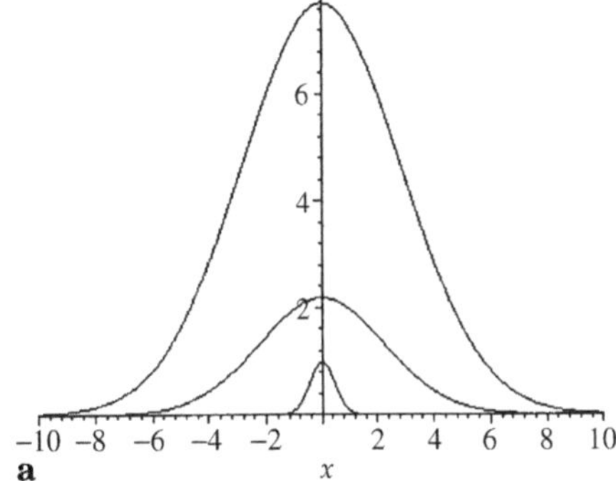

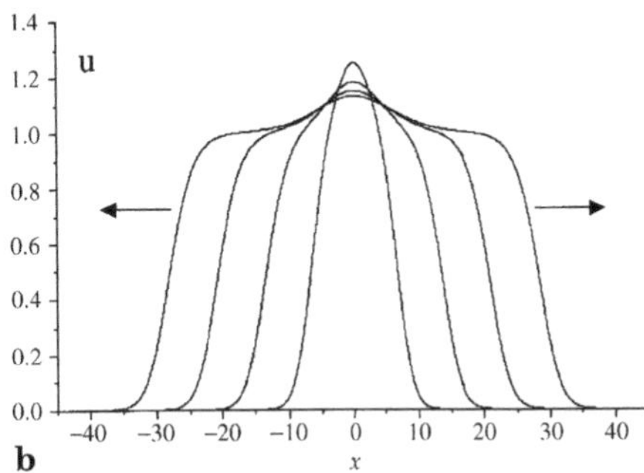

Figure 1. (a) Linear diffusion solutions of Equation (13). (b) Nonlinear diffusion solutions of Equation (14).

so the diffusivity is a function of the density of the semiconductor. Other examples include models of crystal growth, porous media, magnetic flux vortices in superconductors, surface reactions, and so on.

The behavior of the solution to Equation (1) is well understood. If Q is linear in u, then the solution is found by Fourier transform or eigenfunction-expansion techniques. For example, the linear diffusion equation

$$u_t = \Delta u + u \quad (12)$$

with the initial condition $u(x, 0) = \delta(x)$ (Dirac's delta function) has, in (1-d), the fundamental solution

$$u(x, t) = \frac{1}{2\sqrt{\pi t}} \exp\left(t - \frac{x^2}{4t}\right), t > 0. \quad (13)$$

Figure 1a illustrates the behavior of Equation (13) as a function of x for various times. Due to the linear source term, the solution grows exponentially (is unbounded). Another feature of solution (13) is the "paradox of infinite speed propagation." For all $x \neq 0$, $u(x, 0) = 0$, but $u(x, t) > 0$ for all $t > 0$. However, the diffusion equation describes well the global behavior of mass, as it can be easily verified that the center of mass does propagate with a finite speed.

The behavior of solutions of Equation (7) changes dramatically if Q is no longer linear in u. Consider the typical nonlinear autonomous form now called the Fisher–KPP equation and first investigated by Kolmogorov, Petrovsky, and Piscounoff (1937) and, separately, by Fisher (1937) to model the process of genetic diffusion:

$$u_t = \Delta u + u(1 - u). \qquad (14)$$

The only solution evolving from a positive compactly supported initial data is always bounded and propagates in the form of a traveling wave with constant speed $v = 2$, see Figure 1b for an illustration. This is due to the combined action of diffusion and local nonlinearity and has been used in many models applied in biology (genetics, ecology, population dynamics, etc.), chemistry, combustion, economics, physics, etc. The equation similar to (14) but with a cubic nonlinearity (bistable model) instead of the quadratic was proposed in 1938 as a model for a flame front by Zeldovich and Frank-Kamenetsky (1938).

RAZVAN A. SATNOIANU

See also **Fokker–Planck equation; Heat conduction; Reaction-diffusion systems; Zeldovich–Frank-Kamenetsky equation**

Further Reading

Aronson, D.G. & Weinberger H.F. 1975. Nonlinear diffusion in population genetics, combustion and nerve pulse propagation. In *Partial Differential Equations and Related Topics*, edited by J.A. Goldstein, New York: Springer

Crank, J. 1975. *The Mathematics of Diffusion*, Oxford: Clarendon Press

Fisher, R.A. 1937. The wave of advance of advantageous genes. *Annuals of Eugenics*, 7, 353–369

Grindrod, P. 1996. *The Theory and Applications of Reaction–Diffusion Equations*, 2nd edition, Oxford: Clarendon Press and New York: Oxford University Press

Kolmogorov, A., Petrovsky, I. & Piscounoff, N. 1937. Etude de l'equation de la diffusion avec croissance de la quantité de matière et son application á un problème biologique. *Moscow Univ. Bull. Math.*, 1: 1–25

Murray, J.D. 2002. *Mathematical Biology*, vol. I, 3rd edition: *An Introduction*, Berlin and New York: Springer

Øksendal, B. 2000. *Stochastic Differential Equations*, Berlin and New York: Springer

Okubo, A. & Levin, S.A. 2001. *Diffusion and Ecological Problems: Modern Perspectives*, Berlin and New York: Springer

Zeldovich, Ya.B. & Frank-Kamenetsky, D.A. 1938. K teorii ravnomernogo rasprostranenia plameni [Toward a theory of uniformly propagating flames]. *Doklady Akademii Nauk SSSR*, 19: 693–697

DIMENSIONAL ANALYSIS

The dimensions of quantities in any equation (in physics these are mass, length, time, charge, temperature, angle, and so on) must be the same on both sides; otherwise an equality would be violated by changing units. From such reasoning, it is often possible to derive valuable insights without delving into mechanisms. This "dimensional analysis" is an old idea; Lord Rayleigh (John William Strutt), an early vigorous exponent of the method, called it the "principle of similitude,"

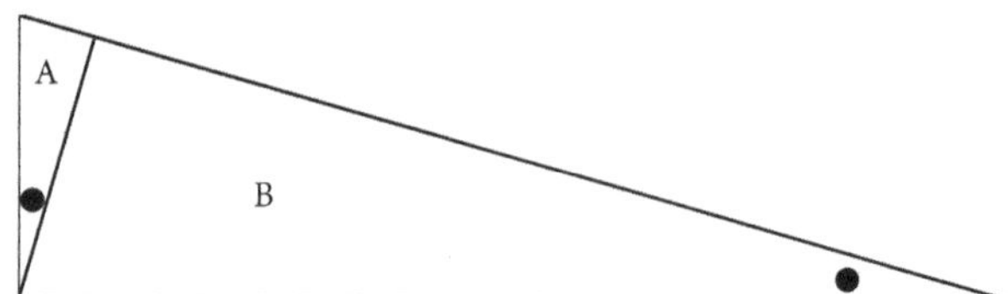

Figure 1. A right triangle is broken into two smaller right triangles of the same proportions.

which he extolled in the following terms (Rayleigh, 1915): "It happens not infrequently that results in the form of 'laws' are put forward as novelties on the basis of elaborate experiments, which might have been predicted a priori after a few minutes' consideration."

The principle was familiar to Galileo, Newton, Fourier, Reynolds, and Maxwell, and was widely used in engineering around 1900. Edgar Buckingham formalized it in what is now called the Pi (for "product") Theorem, that any functional relation among N quantities represented by real numbers and collectively involving $U < N$ basic units can be rewritten as a dimensionless, constant function of $N - U$ dimensionless, multiplicative combinations of those variables (Buckingham, 1914). Unless N and U are trivially small, some systematic procedure is helpful for finding all possible ways of combining variables into dimensionless constants (Birkhoff, 1950; Coyle & Ballico-Lay, 1984).

Interestingly, there have been attempts to discover the laws of economics and finance by similar procedures, starting of course from a different list of fundamental units (DeJong, 1967).

As most articles on dimensional analysis expound abstract principles, two examples are presented here, one drawn from the ancient roots of mathematics and one from the nonlinear physics of shock waves.

Pythagorean Theorem

The area of a right triangle is uniquely determined by the length of the "long" side and one of the other ("wrong") angles. Area is some universal function of Long and Angle, let us say the smaller Angle, dotted in Figure 1. Do we have to figure out exactly what function? Maybe not. We know Area has to be proportional to Long^2 to make the dimensions come out right, and the other factor must be some dimensionless function of the dimensionless Angle (or equivalently of the ratio between the two shorter sides). Break this triangle into two smaller ones of exactly the same shape by constructing a perpendicular from the right angle to the Long side. Call the two areas Area A and Area B. So Area = Area A + Area B.

Each of these similar triangles has for its own long side one of the big triangle's shorter sides, and all the angles are the same as in the big one. Thus $\text{Long}^2 f(\text{Angle}) = \text{ShortA}^2 f(\text{Angle}) + \text{ShortB}^2 f(\text{Angle})$, where Angle means the dotted smaller angle, which is the same in all three cases. So unless $f(\text{Angle}) = 0$ (i.e., unless the areas are all 0 anyhow), then

$$\text{Long}^2 = \text{ShortA}^2 + \text{ShortB}^2. \tag{1}$$

This is the theorem attributed to Pythagoras. We obtained it by merely insisting on dimensional consistency (Goldenfeld, 1992).

Atomic Explosions

From any big explosion in air, shock waves propagate outwards, slowing as the hemisphere of destruction expands. How fast does the hemisphere expand? You might think that an answer to this question necessarily involves a complex variety of considerations about sound, chemistry, thermal physics, and so on. Indeed it does, but let us consider an atomic explosion from the perspective of dimensional analysis.

Suppose the results are pretty much the same for any blast of the same total energy. If that were so, how could the expansion of the shock—its distance from ground zero (R) as a function of time (t)—depend on the energy (E) of the blast?

Let us denote the unit of length by the symbol L and the unit of time by the symbol T. (Thus L might be meters and T seconds.) Then energy (which is mass times velocity squared) has units (ML^2T^{-2}), where M denotes the unit of mass.

Since no combination of the three factors R, T, and E is unitless, we need to include another factor that involves mass. Such a factor is the air density, ρ. This factor is clearly important because without air there would be no shock wave but only bomb parts flying at fixed speeds through the vacuum of space. So throw ρ into the stew with units ML^{-3}. How to relate these four quantities (R, t, E, and ρ) to obtain a unitless result?

From the Pi Theorem, there is only one unitless combination of our four quantities (R, t, E, and ρ) that are expressed in terms of three basic units (L, t, and M). Thus,

$$\frac{R^5 \rho}{t^2 E} = a, \tag{2}$$

where a is a dimensionless number that Geoffrey Taylor estimated (in 1941) to be 0.926 (Taylor, 1950a).

In 1949, Taylor verified his analysis by plotting the log of R (in meters) against the log of t (in seconds) for the first atomic explosion—the Trinity blast of 16 July 1945 in Alamogordo, New Mexico. According to Equation (2), log R versus log$t$ should be a straight line with a dimensionless slope $\frac{2}{5}$. Except for the first point at 0.1 ms, the log-log plot in Figure 2 is indeed close to

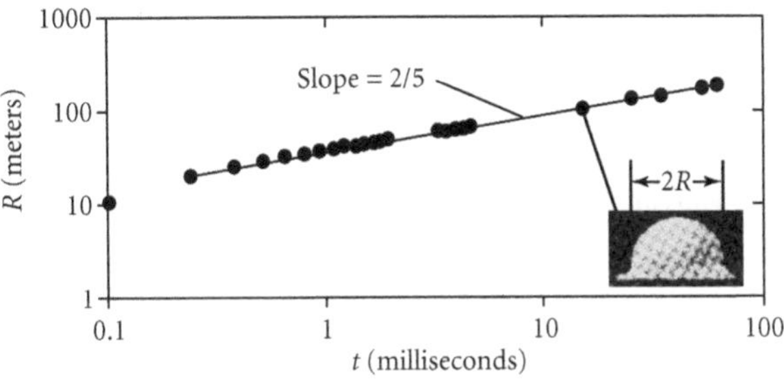

Figure 2. A log-log plot of the expansion of the Trinity fireball. From the origin of the blast, a hemispherical shock of radius R expands as a function of t during the first 62 ms. (The inset shows is a photograph of the blast at 15 ms, and the data are from Taylor, 1950b.)

a straight line with slope $\frac{2}{5}$. In dimensional terms,

$$\frac{R^5}{t^2} = 6 \times 10^{13} \text{m}^5\text{s}^{-2} \tag{3}$$

to an experimental uncertainty of about 10%. Because a and ρ are known, the data in Figure 2 reveal the blast energy, which Taylor computed to be the equivalent of about 20 ktons of TNT (Taylor, 1950b).

In 1947, several of the photographs upon which Figure 2 is based were published by *Life*, a popular magazine of the time. Using these data, assuming $a = 1$, and taking $\rho = 1.25\ \text{kg m}^{-3}$, Equations (2) and (3) imply a blast energy of $8.4 \times 10^{10}\ \text{kg m}^2\ \text{s}^{-2}$ (or joules), which is equivalent to about 20 ktons of TNT. It may be presumed that interested parties made this simple estimate without delay.

Although these two examples may seem like magic, dimensional analysis has some limitations (Rayleigh, 1915). For example, this approach is useless in the absence of clear functional relations among mathematical quantities. Furthermore, one may doubt that the list of dimensioned variables presumed to be relevant is complete and does not include superfluous items. Finally, the desired relations may be unknown functions of dimensionless combinations of the pertinent variables.

A.T. WINFREE

See also **Explosions; Nerve impulses; Shock waves**

Further Reading

Birkhoff, G. 1950. *Hydrodynamics: A Study in Logic, Fact, and Similitude*, Princeton, NJ: Princeton University Press (Chapter 3 especially is a primary source and particularly lucid)

Bridgman, P.W. 1922. *Dimensional Analysis*, New Haven, CT: Yale University Press; often reprinted

Buckingham, E. 1914. On physically similar systems: illustrations of the use of dimensional equations. *Physical Review*, 4: 345–376

Buckingham, E. 1915. Model experiments and the form of empirical equations. *Transactions of the Americal Society of Mechanical Engineers*, 37: 263

Coyle, R.G. & Ballico-Lay, B. 1984. Concepts and software for dimensional analysis in modeling. *IEEE Transactions on*

Systems, Man, and Cybernetics SMC-14(3): 478–482 (Other than MatLab, the only source I know for software in this area)

DeJong, F.J. 1967. *Dimensional Analysis for Economists*, Amsterdam: North-Holland (The only effort I know to adapt dimensional analysis to economics and finance)

Goldenfeld, N. 1992. *Lectures on Phase Transitions and the Renormalization Group*, Reading, MA: Addison-Wesley (Contains full analysis of shock wave from explosions, esp. nuclear, following Taylor, 1952)

Rayleigh, Lord, 1915. The principle of similitude. *Nature*, 95: 66–68

Taylor, G.I. 1950a. The formation of a blast wave by a very intense explosion. I. Theoretical discussion. *Proceedings of the Royal Society of London*, 201A: 159–174

Taylor, G.I. 1950b. The formation of a blast wave by a very intense explosion. II. The atomic explosion of 1945. *Proceedings of the Royal Society of London*, 201A: 175–186

DIMENSIONS

The classical, integer-valued definition of dimension (see Hurewicz & Wallman, 1941) is defined inductively: the empty set has dimension -1, and a set has dimension n if n is the smallest integer such that every point has arbitrarily small neighborhoods whose boundaries have dimension less than n. This gives the "right" answer for smooth curves and surfaces, whose dimension we know intuitively.

In order to describe more accurately the complicated fractal sets that arise in nonlinear dynamics, we need to introduce more subtle definitions. Surprisingly, there are several generalizations of dimensions that still assign the intuitively correct dimensions to the above-noted well-behaved sets, and we recall two of them here.

The first of these is the "box-counting" dimension, also known as the Minkowski dimension, the fractal dimension, the entropy dimension, the capacity dimension, and the limit capacity: a litany of names that testifies to its popularity. For a subset X of $\mathbb{R}^n$, take a fixed array of boxes of side δ, and count the number $N_\delta(X)$ of these boxes that intersect with X. If $N_\delta(X) \sim \delta^{-d}$ as $\delta \to 0$, then X has box-counting dimension d. This can be made mathematically precise by defining

$$d_{\text{box}}(X) = \limsup_{\delta \to 0} \frac{\log N_\delta(X)}{-\log \delta}. \qquad (1)$$

(For alternative definitions that give the same quantity see Falconer (1990).) While the box-counting dimension is simple to define, it is not without problems. For example, the set $S = \{0\} \cup \{1/n : n = 1, 2, \ldots\}$, an unlikely candidate for a fractal, has box-counting dimension $\frac{1}{2}$. We now introduce another widely used definition of dimension that does not suffer from this anomaly.

We could try to define a notion of the "d-dimensional volume" of X as the limit of $N_\delta(X)\delta^d$ as $\delta \to 0$, but such a definition does not even agree with the standard definition of volume (Lebesgue measure) when d is an integer. Instead, the proper generalization of Lebesgue measure to non-integer dimensions is d-dimensional Hausdorff measure. Essentially, we cover a set $X \subset \mathbb{R}^n$ by a collection of balls of radii $r_i \le \delta$, and then let $\mathcal{H}^d(X)$ be the limit of $\sum r_i^d$ as δ tends to zero. More precisely, we define

$$\mathcal{H}^d(X) = \lim_{\delta \to 0} \inf \left\{ \sum_i r_i^d \; : \; r_i \le \delta \text{ and } X \subseteq \cup_i B_{r_i}(x_i) \right\}$$

(the notation $B_r(x)$ denotes an open ball centered at x of radius r). The resulting measure is proportional to Lebesgue measure when d is an integer. The "Hausdorff dimension" of X is the smallest value of d for which $\mathcal{H}^d(X)$ is finite,

$$d_{\text{H}}(X) = \inf\{d > 0 : \mathcal{H}^d(X) < \infty\}.$$

Since $\mu_d(X, \delta) \le N_\delta(X)\delta^d$, we always have $d_{\text{H}}(X) \le d_{\text{box}}(X)$ (and this inequality can be strict: the set S defined above has zero Hausdorff dimension). While harder to estimate in practice, the Hausdorff dimension is easier to deal with theoretically.

If we want to estimate the dimension of the attractor $\mathcal{A}$ of a dynamical system, it is useful to have a method based on dynamical quantities. In 1980, Douady & Oesterlé showed how to obtain a bound on $d_{\text{H}}(\mathcal{A})$, the dimension of the attractor of an iterated C^1 map f on $\mathbb{R}^n$. Denote by $\mathrm{D}f(x)$ the matrix of partial derivatives of f, i.e., $[\mathrm{D}f]_{ij} = \partial f_i / \partial x_j$, and let $\lambda_1(x) \ge \lambda_2(x) \ge \lambda_n(x)$ be the logarithms of the eigenvalues of $[\mathrm{D}f(x)^T \mathrm{D}f(x)]^{1/2}$. Now set

$$d(x) = j + \frac{\lambda_1(x) + \cdots + \lambda_j(x)}{|\lambda_{j+1}(x)|}, \qquad (2)$$

where j is the largest integer for which $\lambda_1(x) + \cdots + \lambda_j(x) \ge 0$ (note that $j \le d(x) < j+1$). If $d > d(x)$, then any infinitesimal d-volume near x is contracted under the application of f, so

$$d_{\text{H}}(\mathcal{A}) \le \sup_{x \in \mathcal{A}} d(x). \qquad (3)$$

Hunt (1996) showed that the right-hand side of (3) also bounds $d_{\text{box}}(\mathcal{A})$. (A similar approach also works for the attractors of flows by taking f to be the time T map, for some suitable T. Constantin & Foias (1985) have proved a version of (3) for the attractors of infinite-dimensional dynamical systems.)

However, the box-counting and Hausdorff dimensions give equal weighting to all points in the attractor, while it is possible to have regions of the attractor that are visited very rarely. In such a situation, it can be more natural to consider invariant measures rather than attractors. As a (canonical) example of such a measure, suppose that $f : \mathbb{R}^n \to \mathbb{R}^n$ generates a dynamical system on $\mathbb{R}^n$. Then for any set X, we can define

$$\mu(X) = \lim_{m \to \infty} \frac{1}{m} \operatorname{card}\{k : f^k(x) \in X, \; 1 \le k \le m\},$$

where x is a point in the basin of attraction of $\mathcal{A}$. The quantity $\mu(X)$ is the proportion of time spent in X by a "typical trajectory" on the attractor.

There are various ways of defining the dimension of a measure μ. We could define the Hausdorff/box-counting dimension of μ to be the dimension of its support,

$$d_{\mathrm{box/H}}(\mu) = \inf\{d_{\mathrm{box/H}}(E) : \mu(E) = 1\},$$

but this still discounts the dynamical information contained in μ. Kaplan & Yorke (1979) defined the *Lyapunov dimension* of μ, $d_{\mathrm{L}}(\mu)$, precisely as in (2), but replacing $\lambda_j(x)$ by the Lyapunov exponents associated with μ (the asymptotic growth rates of infinitesimal displacements about trajectories through μ-almost every choice of initial condition). In 1981, Ledrappier showed that for a very general class of dynamical systems, $d_{\mathrm{H}}(\mu) \leq d_{\mathrm{L}}(\mu)$ (the inequality can be strict), while

$$d_{\mathrm{box}}(\mathcal{A}) \leq \sup_{\text{all invariant ergodic } \mu} d_{\mathrm{L}}(\mu).$$

(Kaplan & Yorke had originally conjectured that $d_{\mathrm{box}}(\mathcal{A}) = d_{\mathrm{L}}(\mu)$.)

We now give two definitions of dimension that take into account the spatial structure of μ. The correlation at scale δ is defined by

$$C(\delta) = \int_{X\times X:\, |x-y|\leq\delta} \mathrm{d}\mu(x)\,\mathrm{d}\mu(y),$$

which gives the probability that two points chosen according to the probability measure μ lie within δ of each other. If $C(\delta) \sim \delta^d$ as $\delta \to 0$, then d is the *correlation dimension* $d_{\mathrm{corr}}(\mu)$. This was introduced by Grassberger & Procaccia (1983), who demonstrated that this quantity is particularly suited to numerical calculation.

Alternatively, define the "δ-entropy" $K_\delta(\mu) = -\sum_i \mu(B_i) \ln \mu(B_i)$, where $\{B_i\}$ is an array of boxes of side δ; the *information dimension* is given by

$$d_{\mathrm{inf}}(\mu) = \lim_{\delta\to 0} \frac{K_\delta(\mu)}{-\log\delta}.$$

(Ruelle (1989) refers to $d_{\mathrm{H}}(\mu)$ as the "information dimension" of μ: this should serve to emphasize how important it is when discussing dimensions to be explicit about the definition.)

Three of these dimensions occur as part of a scale of dimension-like quantities (see Grassberger, 1983). If B_i is an array of boxes of side δ, set

$$K_\delta(q) = \frac{1}{1-q} \log \sum_i \mu(B_i)^q$$

("the Renyi q entropy"). Note that $K_\delta(0) = \log N_\delta$ (supp μ),

$$\lim_{q\to 1} K_\delta(q) = -\sum_i \mu(B_i) \ln \mu(B_i) = K_\delta(\mu)$$

and that since $C(\delta) \leq \sum_i \mu(B_i)^2 \leq C(\delta\sqrt{n})$, we have $\log C(\delta) \leq K_\delta(2) \leq \log C(\delta\sqrt{n})$. Now define the *Renyi dimensions* $D_q(\mu)$ by

$$D_q(\mu) = \lim_{\delta\to 0} \frac{K_\delta(q)}{-\log\delta}.$$

Then $D_0(\mu) = d_{\mathrm{box}}(\mu)$, $D_1(\mu) = d_{\mathrm{inf}}(\mu)$, and $D_2(\mu) = d_{\mathrm{corr}}(\mu)$. Since D_q is non-increasing in q, we have in particular $d_{\mathrm{corr}}(\mu) \leq d_{\mathrm{inf}}(\mu) \leq d_{\mathrm{box}}(\mu)$.

The Renyi dimensions are similar to quantities used to define the "multi-fractal spectrum." The theory relates the numbers $\tau(q) = (1-q)D_q$ to the frequency of various scaling behaviors about points on a fractal set: roughly, if for some small ε the number of δ-mesh cubes B_i with

$$\delta^{\alpha+\varepsilon} \leq \mu(B_i) < \delta^\alpha$$

scales like $\delta^{-f(\alpha)}$, then

$$f(\alpha(q)) = \tau(q) + q\alpha(q),$$

where $q = f'(\alpha(q))$. The curve $f(\alpha)$ is the multi-fractal spectrum of the measure μ. (As remarked by Falconer (1990), the tempting interpretation of the "fractal spectrum" as the dimension of sets of points x where $\mu(B_\delta(x)) \sim \delta^\alpha$ is incorrect: typically the dimension of such sets will be zero or the same as that of the whole space, see Genyuk (1997/98).) Although these ideas have proved useful in the theory of turbulence (e.g. Frisch, 1995), their mathematical foundations have still to be fully resolved.

JAMES C. ROBINSON

See also **Attractors; Fractals; Lyapunov exponents; Measures**

Further Reading

Constantin, P. & Foias, C. 1985. Global Lyapunov exponents, Kaplan–Yorke formulas and the dimension of the attractor for 2D Navier–Stokes equation. *Communications in Pure and Applied Mathematics*, 38: 1–27

Douady, A. & Oesterlé, J.D. 1980. Dimension de Hausdorff des attracteurs. *Compes Rendus de l'Academies des Sciences, Paris Sèries A–B*, 290: A1135–A1138

Falconer, K. 1990. *Fractal Geometry*, Chichester and New York: Wiley

Frisch, U. 1995. *Turbulence*, Cambridge and New York: Cambridge University Press

Genyuk, J. 1997/98. A typical measure typically has no local dimension. *Real Analysis Exchange*, 23: 525–537

Grassberger, P. 1983. Generalized dimensions of strange attractors. *Physics Letters A*, 97: 227–230

Grassberger, P. & Procaccia, I. 1983. Measuring the strangeness of strange attractors. *Physica* D, 9: 189–208

Hunt, B. 1996. Maximal local Lyapunov dimension bounds the box dimension of chaotic attractors. *Nonlinearity*, 9: 845–852

Hurewicz, W. & Wallman, H. 1941. *Dimension Theory*, Princeton, NJ: Princeton University Press

Kaplan, J.L. & Yorke, J.A. 1979. Chaotic behavior of multidimensional difference equations. In *Functional Differential Equations and Approximation of Fixed Points*, Berlin: Springer, 204–227
Ledrappier, F. 1981. Some relations between dimension and Lyapounov exponents. *Communications in Mathematical Physics*, 81: 229–238
Ruelle, D. 1989. *Chaotic Evolution and Strange Attractors*, Cambridge and New York: Cambridge University Press

DIODES

Many of the desirable features of a wide variety of solid-state electronic devices are based on nonlinear current-voltage characteristics. These involve situations where a sufficiently high bias is applied to the device so that it either switches from one conductive state to another or oscillates between different conductive states.

A two-terminal electronic device is called a diode. In general, the current I through a diode depends upon the polarity of the applied voltage U and exhibits a nonlinear voltage dependence. This includes the important special case of a rectifier diode that conducts the current for one polarity of voltage and blocks it for the other. A rectifier can be realized by a p–n junction (a semiconductor that is p-doped on one side and n-doped on the other side), and it can be described approximately by the current-voltage characteristic

$$I = I_s\left(e^{eU/k_BT} - 1\right), \quad (1)$$

where I_s is the saturation current for reverse bias $U < 0$, k_B is Boltzmann's constant, and T is the temperature (Figure 1a). For forward bias (positive voltage applied to the p side), the majority carriers (holes from the p side and electrons from the n side) flow towards the junction where they recombine, while for reverse bias they are pulled away from the junction.

The Schottky diode is a rectifier diode, which consists of a metal-semiconductor contact. Interface states and to a small degree the difference in work functions between the metal and the semiconductor give rise to a potential barrier (Schottky barrier) and depletion of majority carriers in the barrier region for reverse bias. The current-voltage characteristic is similar to Equation (1).

Depending upon bias conditions, doping profiles, and device geometry, various other terminal functions of diodes are possible, which involve nonlinear behavior of the conductivity, the capacitance, and the inductance of the device.

The Zener diode is a p–n junction that exhibits a sharp increase in the magnitude of the current at a certain well-controlled reverse bias (Figure 1b). Depending upon the specific device structure, it is either due to avalanche breakdown (multiplication of carriers by impact ionization occurring at high voltage) or due to tunneling across the bandgap (occurring at lower voltage). It is used to stabilize and limit the dc voltage in circuits, utilizing the property that the current can vary over a large range at the breakdown threshold without noticeable change in voltage.

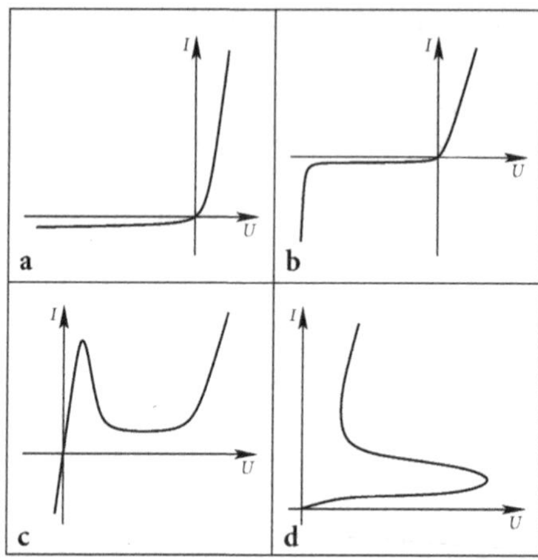

Figure 1. Typical nonlinear current-voltage characeristics of diodes. (a) p–n diode, (b) Zener diode, (c) Tunnel diode, (d) p–i–n diode (schematic).

A class of diodes exhibits a nonmonotonic dependence of the current I upon voltage U. Negative differential conductance $dI/dU < 0$ can arise due to various mechanisms and may result in self-generated oscillations and complex self-organized spatiotemporal patterns. A famous example is the Esaki tunnel diode (Figure 1c) that, in its original version, consists of a heavily doped p–n junction. In thermal equilibrium, the Fermi level lies within the conduction band on the n side and within the valence band on the p side. When a small forward bias is applied, electrons can tunnel from the n- to the p-side where they find empty states in the valence band. With increasing bias, the filled conduction band states on the n side move up, and the overlap with empty states in the valence band decreases; consequently, the tunneling current decreases, and $dI/dU < 0$. With a further increase of the bias, diffusion of the electrons over the barrier sets in as in a normal p–n junction, and the current increases again.

Modern variants of the tunnel diode are double-barrier resonant tunneling structures and superlattices that consist of alternating layers of different semiconductor materials (heterostructures) forming potential barriers and quantum wells on length scales of a few nanometers. The current density across the barrier between two wells is maximum if there is maximum overlap between the occupied states in one well and the available unoccupied states in the other, that is, if the energies are in resonance. For low bias, equivalent levels in adjacent wells are approximately in resonance, while for higher bias the ground energy level in one quantum well becomes aligned with the second level in the neigboring well. Thus, resonant tunneling produces an $I(U)$ characteristic similar to Figure 1c. High-frequency oscillations up to 150 GHz can be generated in resonant tunneling structures and superlattices.

Gunn diodes are used to generate and amplify microwaves at frequencies typically beyond 1 GHz (these devices are called "diodes" because they are two-terminal devices, but no p–n junction is involved). The mechanism is based upon field-induced intervalley transfer of electrons from a high- to a low-mobility valley in the conduction band, and the manifestation of the current instability (oscillations or switching) is determined primarily by the cathode contact boundary condition. Traveling high-field domains (Gunn domains), which show up as transit time oscillations, represent an important mode of operation.

Multilayered structures with alternating p- and n-doping represent another class of diodes that exhibit negative differential conductance, high-power microwave oscillations well above 30 GHz, and complex spatiotemporal current density patterns. Starting from the basic n^+–p–i–p^+ structure, where "+" denotes high doping and i is an intrinsic (undoped) layer, various modifications like IMPATT (impact ionization avalanche transit time) diodes, TRAPATT (trapped plasma avalanche-triggered transit) diodes, p–i–n diodes (with double injection of electrons and holes), or $n^+p^+np^-p^+$ devices, where "−" denotes low doping, have been studied. These structures typically display bistability and switching from a low- to a high-conductivity state where carrier multiplication and avalanche breakdown set in (Figure 1d). Negative differential conductance is also exhibited by multilayer systems composed of layers of different semiconductor materials (heterojunctions) like the heterostructure hot electron diode or real-space transfer devices.

A p–n diode may also be operated as a nonlinear capacitor. As the depletion-layer width increases with increasing reverse bias, its capacitance decreases in a controlled way depending upon the doping profile. This effect is used in varactor (*var*iable re*actor*) diodes to tune the capacitance and for parametric amplification. State-of-the-art developments include studies on p–n diodes with embedded quantum dot nanostructures, which strongly exhibit nonlinear capacitance-voltage characteristics as signatures of the charging of these quantum dots.

Nonlinear inductors represent another class of two-terminal devices. A Josephson junction consists of two superconductors separated by a thin nonsuperconducting region. The tunneling of superconducting electron pairs produces a current at zero voltage called the Josephson current. Switching can be obtained from this state to a voltage state by either current overdrive or a magnetic field. The dynamic response of the Josephson junction can be represented by an equivalent circuit with an intrinsic field-dependent inductance.

ECKEHARD SCHÖLL

See also **Avalanche breakdown; Josephson junctions; Nonlinear electronics; Semiconductor oscillators**

Further Reading

Böer, K.W. 2002. *Survey of Semiconductor Physics*, 2nd edition, New York: Plenum

Ibach, H. & Lüth, H. 2003, *Solid-State Physics*, Berlin: Springer

Schöll, E. 2001. *Nonlinear Spatio-Temporal Dynamics and Chaos in Semiconductors*, Cambridge and New York: Cambridge University Press

Shaw, M.P., Mitin, V.V., Schöll, E. & Grubin, H.L. 1992. *The Physics of Instabilities in Solid State Electron Devices*, New York: Plenum Press

Sze, S.M. 1981. *Physics of Semiconductor Devices*, New York: Wiley

Sze, S.M. 1998. *Modern Semiconductor Device Physics*, New York: Wiley

DIRAC'S DELTA FUNCTION

See **Generalized functions**

DISCRETE BREATHERS

The study of dynamical nontopological localization in translationary invariant nonlinear Hamiltonian lattices has experienced considerable development during the late 1990s (Sievers & Page, 1995; Aubry, 1997; Flach & Willis, 1998). The discreteness of space—that is, the use of a spatial lattice—is crucial in order to provide structural stability for spatially localized excitations. Spatial discreteness is a very common situation for various applications from, for example, solid-state physics.

To make things precise, consider a d-dimensional hypercubic spatial lattice with discrete translational invariance. Each lattice site is labeled by a d-dimensional vector l with integer components. To each lattice site we associate one pair of canonically conjugated coordinates and momenta X_l, P_l that are real functions of time t. Let us then define some Hamiltonian H being a function of all coordinates and momenta and further require that H has the same symmetries as the lattice. The dynamical evolution of the system is given by the usual Hamiltonian equations of motion. Without loss of generality, let us consider that H is a nonnegative function and that $H=0$ for $X_l = P_l = 0$ (for all l's). We call this state the classical ground state. Generalizations to other lattices and larger numbers of degrees of freedom per lattice site are straightforward.

When linearizing the equations of motion around $H=0$, we obtain an eigenvalue problem. Due to translational invariance the eigenvectors will be spatially extended plane waves, and the eigenvalues Ω_q (frequencies) form a phonon spectrum, that is, Ω_q is a function of the wave vector q. Due to the translation symmetry of the Hamiltonian, Ω_q will be periodic in

q. Moreover, the phonon spectrum will be bounded, that is, $|\Omega_q| \le \Omega_{\max}$. Depending on the presence or absence of Goldstone modes Ω_q might be gapless (zero belongs to the spectrum, spectrum is acoustic) or exhibit a gap ($|\Omega_q| \ge \Omega_{\min}$, spectrum is optical). Increasing the number of degrees of freedom per lattice site induces several branches in Ω_q with possible gaps between them.

Let us search for spatially localized time periodic solutions of the full nonlinear equations of motion, that is, $X_{|l| \to \infty} \to 0$, and

$$X_l(t) = X_l(t + T_b) + \lambda k_l,\ P_l(t) = P_l(t + T_b), \quad (1)$$

with k_l being integers and λ a spatial period (the equations of motion should be invariant under shifts of X_l by multiples of λ if applicable.) These solutions are called discrete breathers. If $k_l \neq 0$ for a finite subset of lattice sites, the solutions are sometimes called "rotobreathers."

If a solution exists, we can expand it into a Fourier series in time, that is, $X_l(t) = \sum_k A_{k_l} e^{ik\omega_b t}$ ($\omega_b = 2\pi / T_b$). Spatial localization implies $A_{k_l \to \infty} \to 0$. Inserting these series into the equations of motion results in a set of coupled algebraic equations for the Fourier amplitudes (Flach & Willis, 1998). Consider the spatial tail of the solution where all Fourier amplitudes are small and should further decay to zero with growing distance from the excitation center. Since all amplitudes are small, the equations of motion can be linearized. This procedure decouples the interaction in k-space, and we obtain for each k a linear equation for A_{k_l} with coupling over l. This equation will contain $k\omega_b$ as a parameter. It will, in fact, be identical to the above-discussed equation linearized around $H = 0$, and it will contain $k\omega_b$ instead of Ω_q (Flach & Willis, 1998). If $k\omega_b = \Omega_q$, the corresponding amplitude A_{k_l} will not decay in space, instead it will oscillate. To obtain localization, we arrive at the nonresonance condition (Flach & Willis, 1998)

$$k\omega_b \neq \Omega_q. \quad (2)$$

This condition has to be fulfilled for all integer k. For an optical spectrum Ω_q, frequency ranges for ω_b exist, which satisfy this condition. For acoustic spectra, $k = 0$ has to be considered separately (Flach et al., 1997).

The nonresonance condition is only a necessary condition for generic occurrence of discrete breathers. More detailed analysis shows that breathers being periodic orbits bifurcate from band edge plane waves (Flach, 1996). The condition for this bifurcation is an inequality involving parameters of expansion of H around $H = 0$ (Flach, 1996). Rigorous existence proofs for weakly coupled anharmonic oscillators use, for example, the implicit function theorem (MacKay & Aubry, 1994).

Discrete breathers (periodic orbits) appear generically as one-parameter families of periodic orbits. The parameter of the family can be, for example, the frequency (or energy, action, etc.). Note that we do not need any topological requirement on H (no energy barriers). Indeed, breather families possess limits where the breather delocalizes and its amplitude becomes zero.

With the help of the nonresonance conditions, we can exclude the generic existence of spatially localized solutions that are quasi-periodic in time. Indeed, in the simplest case, we would have to satisfy a nonresonance condition $k_1\omega_1 + k_2\omega_2 \neq \Omega_q$ for ω_1/ω_2 being irrational and all possible pairs of integers k_1, k_2. This is impossible (Flach, 1994).

The nonresonance condition also explains why breather solutions are nongeneric for nonlinear Hamiltonian field equations, since Ω_q becomes unbounded for a spatially continuous system. Consequently, generically an infinite number of unavoidable resonances destroys the breather existence there.

Note that in many cases breathers can be easily excited by choosing some localized perturbation of the lattice system. Integrating numerically the equations of motion, we find that the energy distribution is not delocalizing but stays essentially localized over several orders of magnitude of the characteristic phonon periods. These numerical results clearly show that breathers are not only interesting solutions but can be rather typical and robust depending on the system's parameters.

Note that breathers can also exist for autonomous forced damped systems (MacKay & Sepulchre, 1998). In these systems, contrary to the Hamiltonian ones, breather periodic orbits do not come in one-parameter families of the frequency ω_b, but correspond to limit cycle attractors that are isolated in the system's phase space.

Sergej Flach

See also **Breathers; Discrete nonlinear Schrödinger equations; Discrete self-trapping system; Fermi–Pasta–Ulam oscillator chain; Josephson junction arrays; Local modes in molecular crystals; Multidimensional solitons; Nonlinear Schrödinger equations; Solitons**

Further Reading

Aubry, S. 1997. Breathers in nonlinear lattices: existence, linear stability and quantization. *Physica* D, 103: 201

Flach, S. 1994. Conditions on the existence of localized excitations in nonlinear discrete systems. *Physical Review* E, 50: 3134

Flach, S. 1996. Tangent bifurcation of band edge plane waves, dynamical symmetry breaking and vibrational localization. *Physica* D, 91: 223

Flach, S., Kladko, K. & Takeno. S. 1997. Acoustic breathers in two-dimensional lattices. *Physical Review Letters*, 79: 4838

Flach, S. & Willis, C.R. 1998. Discrete breathers. *Physics Reports*, 295: 182

MacKay, R.S. &. Aubry, S. 1994. Proof of existence of breathers for time-reversible or Hamiltonian networks of weakly coupled oscillators. *Nonlinearity*, 7: 1623

MacKay, R.S. & Sepulchre, J.A. 1998. Stability of discrete breathers. *Physica* D, 119: 148

Sievers, A.J. & Page, J.B. 1995. Unusual Anharmonic Local Mode Systems. In *Dynamical Properties of Solids VII Phonon Physics The Cutting Edge*, edited by G.K. Horton and A.A. Maradudin, Amsterdam: Elsevier

DISCRETE NONLINEAR SCHRÖDINGER EQUATIONS

With a fairly generous definition, a discrete nonlinear Schrödinger (DNLS) equation is any equation that can be obtained from a nonlinear Schrödinger (NLS) equation of general form

$$\mathrm{i}\frac{\partial \phi}{\partial t} + \Delta\phi + f(|\phi|^2)\phi = 0, \qquad (1)$$

by employing some finite-difference approximation to the operators acting on the space-time-dependent continuous field $\phi(\boldsymbol{r};t)$. In (1), $\Delta = \nabla^2$ is the Laplace operator acting in one, two, or three spatial dimensions, and f is a quite general function that, for most purposes, is taken to be differentiable and with $f(0)=0$. In the most well-known case of cubic nonlinearity, $f(|\phi|^2)=\gamma|\phi|^2$, Equation (1) is often referred to as *the* NLS equation, and is integrable with the inverse scattering method if the number of spatial dimensions is one. Here we use the term DNLS equation to denote the set of coupled ordinary differential equations resulting from discretizing *all* spatial variables in (1), while keeping the time-variable t continuous. However, one may also consider equations with discrete time ("fully discrete NLS equations"), as well as equations with only some of the spatial dimensions discretized ("discrete-continuum NLS equations"). The former are of interest as algorithms for numerical solution of (1), while the latter may describe pulse propagation in *arrays* of coupled nonlinear optical fibers (Aceves et al., 1995).

The simplest example of a DNLS equation can be formally obtained by just replacing the Laplacian operator in (1) with the corresponding discrete Laplacian. Thus, for the one-dimensional (1-d) case, we let $\phi_n(t) \equiv \phi(x=na;t)$ where a is the lattice parameter, so that for the particular case of cubic nonlinearity, the following equation is obtained:

$$\mathrm{i}\frac{\mathrm{d}\phi_n}{\mathrm{d}t} + C(\phi_{n+1} - 2\phi_n + \phi_{n-1}) + \gamma|\phi_n|^2\phi_n = 0, \qquad (2)$$

where $C=1/a^2$. This set of differential–difference equations with purely diagonal ("on-site") nonlinearity is sometimes called the diagonal DNLS (DDNLS) equation, but since it is by far the most studied example of a DNLS equation, it is most commonly referred to as simply *the* DNLS equation. Extensions to higher dimensions are straightforward, so that, for example, for a 2-d lattice with $x=ma$, $y=na$, the DNLS equation reads

$$\mathrm{i}\frac{\mathrm{d}\phi_{m,n}}{\mathrm{d}t} + C(\phi_{m+1,n} + \phi_{m-1,n} + \phi_{m,n+1} + \phi_{m,n-1} - 4\phi_{m,n}) + \gamma|\phi_{m,n}|^2\phi_{m,n} = 0. \qquad (3)$$

The study of DDNLS equations has a long and fascinating history, beginning in the 1950s within solid-state physics with Holstein's model for polaron motion in molecular crystals (Holstein, 1959); reappearing in the 1970s within biophysics with Davydov's model for energy transport in biomolecules (see, e.g., Scott, 1999, Chapter 5.6), in the 1980s within physical chemistry in the theory of local modes of small molecules (see, e.g., Scott, 1999, Chapter 5.4), and within nonlinear optics modeling coupled nonlinear waveguides (see, e.g., Hennig & Tsironis, 1999, Chapter 1.4); and most recently around the turn of the century within matter wave physics in the description of a dilute Bose–Einstein condensate trapped in a periodic potential (Trombettoni & Smerzi, 2001). A brief account of experimental verifications of the validity of the DNLS description in the two latter contexts available at the time of writing was given in Eilbeck & Johansson (2003, Chapter 10). In addition, the DDNLS equation has played a central role in the development of the general theory for intrinsic localized modes ("discrete breathers") in systems of coupled anharmonic oscillators during the 1990s (Flach & Willis, 1998). The reader should also note that the DDNLS equation is a particular example of the more general "discrete self-trapping" (DST) systems (described under a separate entry), where the general DST dispersion matrix describing interactions between lattice sites is restricted to nearest-neighbor couplings. Thus, the general theory described for DST systems is also applicable for the DDNLS equation.

The reason for the ubiquity of the DDNLS equation in nonlinear lattice systems is analogous to that of the NLS equation for continuum systems: it takes into account dispersion (through the nearest-neighbor interaction terms) as well as nonlinearity (the term $\gamma|\phi_n|^2\phi_n$) at the lowest order of approximation. It can be derived, for example, from a general system of coupled anharmonic oscillators using a "rotating wave approximation" (RWA), where it is assumed that each oscillator approximately can be described by a complex rotating-wave amplitude as $u_n(t)=\mathrm{Re}(\phi_n \mathrm{e}^{-\mathrm{i}\omega t})$ (see, e.g., Flach & Willis, 1998, Chapter 2.3; Scott, 1999, Chapter 5.3.1). Thus, the RWA assumes time-periodic solutions to have a purely harmonic time dependence, neglecting the generation of all higher harmonics. This approximation can be justified for small-amplitude oscillations in weakly coupled oscillator chains, using perturbational techniques with expansions on multiple time scales (see, e.g., Flach & Willis, 1998, Chapter 2.2, for a general outline, and Morgante et al., 2002, Chapter 2.2, for details).

As for general DST systems, the DDNLS equation has, in addition to the energy (Hamiltonian) $H=\sum_n \left(C|\phi_{n+1}-\phi_n|^2-\frac{\gamma}{2}|\phi_n|^4\right)$ (where $\mathrm{i}\phi_n$ and ϕ_n^* are canonical conjugated variables), a second conserved quantity, which is the excitation number $N=\sum_n |\phi_n|^2$. The conservation of excitation number results (through Noether's theorem) from the invariance of the equation under infinitesimal transformations of the overall phase ($\phi_n \to \phi_n \mathrm{e}^{\mathrm{i}\varepsilon}$). As a consequence, the DDNLS equation is integrable for two degrees of freedom but nonintegrable for larger systems. Still, the existence of a second conserved quantity has some notable consequences, which makes the DDNLS equation nongeneric among general Hamiltonian lattice systems, such as:

- It has purely harmonic time-periodic solutions $\phi_n(t)=A_n\mathrm{e}^{-\mathrm{i}\omega t}$ with time-independent A_n ("stationary solutions").
- It has continuous families of time-quasi-periodic solutions, with two incommensurate frequencies, which may be spatially exponentially localized also in infinite systems ("quasi-periodic breathers") (see, e.g., Eilbeck & Johansson, 2003, Chapter 7; Kevrekidis et al., 2001, Chapter 2.4).

From a mathematical point of view, it is highly interesting that there also exist discretizations of the integrable 1-d cubic NLS equation that conserve its integrability. The most famous integrable DNLS equation is the so-called Ablowitz–Ladik (AL) DNLS equation, the integrability of which was first proven by Ablowitz & Ladik (1976). It is obtained by replacing the nonlinear term $\gamma|\phi|^2\phi$ with an "off-diagonal" discretization $\frac{\gamma}{2}|\phi_n|^2(\phi_{n+1}+\phi_{n-1})$, yielding

$$\mathrm{i}\frac{\mathrm{d}\phi_n}{\mathrm{d}t}+C(\phi_{n+1}-2\phi_n+\phi_{n-1}) +\frac{\gamma}{2}|\phi_n|^2(\phi_{n+1}+\phi_{n-1})=0. \qquad (4)$$

Due to its integrability, it is possible to obtain exact analytical solutions to (4) describing, for example, traveling waves (in terms of elliptic functions, see, e.g., Scott, 1999, Chapter 5.3.2), solitons, and multisolitons. In particular, for $\gamma/C>0$, there is a (bright) one-soliton solution given by (with rescalings such that $C=1, \gamma=2$):

$$\phi_n(t)=\sinh\beta\,\mathrm{sech}[\beta(n-vt)]\mathrm{e}^{\mathrm{i}(kn+\omega t+\alpha)}, \qquad (5)$$

where β, k, and α are free parameters,

$$v=(2/\beta)\sinh\beta\sin k,$$

and $\omega=2(\cosh\beta\cos k-1)$. On the other hand, when $\gamma/C<0$, there are dark-soliton solutions with nonvanishing amplitude as $|n|\to\infty$ (Vekslerchik & Konotop, 1992).

The ALDNLS equation (4) also has a Hamiltonian structure with

$$H=\sum_n\Big[-C\phi_n^*(\phi_{n+1}+\phi_{n-1}) +\frac{4C^2}{\gamma}\log\left(1+\frac{\gamma}{2C}|\phi_n|^2\right)\Big],$$

although the conjugated variables $\mathrm{i}\phi_n$ and ϕ_n^* are noncanonical and the corresponding Poisson bracket deformed (see, e.g., Faddeev & Takhtajan, 1987, p. 303; Scott, 1999, Chapter 5.3.2). There is, at this writing, no known direct physical application of the ALDNLS equation; however, it is commonly used as a starting point for perturbational studies of physically more relevant equations such as (2). A particularly interesting model allowing interpolations between Equations (2) and (4) is the so-called "Salerno equation," which is described under a separate entry.

As a final example, we mention a rather complicated DNLS equation, which was introduced and proven to be integrable by Izergin and Korepin in 1981. This Izergin–Korepin equation reads

$$\mathrm{i}\frac{\mathrm{d}\phi_n}{\mathrm{d}t}=4\phi_n+\frac{P_{n,n+1}}{Q_{n,n+1}}+\frac{P_{n,n-1}}{Q_{n,n-1}}, \qquad (6)$$

where

$$P_{n,n\pm1}=-\phi_n-\phi_{n\pm1}\sqrt{\left(1-\frac{\gamma}{8}|\phi_n|^2\right)\left(1-\frac{\gamma}{8}|\phi_{n\pm1}|^2\right)} +\frac{\gamma}{4}\phi_n|\phi_{n\pm1}|^2+\frac{\gamma}{16}\left(|\phi_n|^2\phi_{n\pm1}+\phi_n^2\phi_{n\pm1}^*\right) \times\sqrt{\frac{1-\frac{\gamma}{8}|\phi_{n\pm1}|^2}{1-\frac{\gamma}{8}|\phi_n|^2}}$$

and

$$Q_{n,n\pm1}=1-\frac{\gamma}{8}\left(|\phi_n|^2+|\phi_{n\pm1}|^2\right) -\frac{\gamma}{8}\left(\phi_n\phi_{n\pm1}^*+\phi_n^*\phi_{n\pm1}\right) \times\sqrt{\left(1-\frac{\gamma}{8}|\phi_n|^2\right)\left(1-\frac{\gamma}{8}|\phi_{n\pm1}|^2\right)} +\frac{\gamma^2}{32}|\phi_n|^2|\phi_{n\pm1}|^2.$$

It is associated with a lattice Heisenberg magnet model, and turns into the cubic NLS equation in the continuum limit (see Faddeev & Takhtajan, 1987, pp. 299, for details).

MAGNUS JOHANSSON

See also **Discrete self-trapping system; Nonlinear Schrödinger equations; Rotating wave approximation; Salerno equation**

Further Reading

Ablowitz, M.J. & Ladik, J.F. 1976. Nonlinear differential-difference equations and Fourier analysis. *Journal of Mathematical Physics*, 17(6): 1011–1018

Aceves, A.B., De Angelis, C., Luther, G.G., Rubenchik, A.M. & Turitsyn, S.K. 1995. All-optical-switching and pulse amplification and steering in nonlinear fiber arrays. *Physica* D, 87(1–4): 262–272

Eilbeck, J.C. & Johansson, M. 2003. The discrete nonlinear Schrödinger equation—20 years on. In *Localization and Energy Transfer in Nonlinear Systems: Proceedings of the Third Conference, San Lorenzo de El Escorial, Spain, 17–21 June 2002*, edited by L. Vázquez, M.P. Zorzano and R. MacKay. Singapore: World Scientific

Faddeev, L.D. & Takhtajan, L.A. 1987. *Hamiltonian Methods in the Theory of Solitons*. Berlin and Heidelberg: Springer

Flach, S. & Willis, C.R. 1998. Discrete breathers. *Physics Reports*, 295(5): 181–264

Hennig, D. & Tsironis, G.P. 1999. Wave transmission in nonlinear lattices. *Physics Reports*, 307(5–6), 333–432

Holstein, T. 1959. Studies of polaron motion. *Annals of Physics*, 8: 325–389

Kevrekidis, P.G., Rasmussen, K.Ø. & Bishop, A.R. 2001. The discrete nonlinear Schrödinger equation: a survey of recent results. *International Journal of Modern Physics* B, 15(21): 2833–2900

Morgante, A.M., Johansson, M., Kopidakis, G. & Aubry, S. 2002. Standing wave instabilities in a chain of nonlinear coupled oscillators. *Physica* D, 162(1–2): 53–94

Scott, A. 1999. *Nonlinear Science: Emergence & Dynamics of Coherent Structures*. Oxford and New York: Oxford University Press

Trombettoni, A. & Smerzi, A. 2001. Discrete solitons and breathers with dilute Bose-Einstein condensates. *Physical Review Letters*, 86(11): 2353–2356

Vekslerchik, V.E. & Konotop, V.V. 1992. Discrete nonlinear Schrödinger equation under non-vanishing boundary conditions. *Inverse Problems*, 8(6): 889–909

DISCRETE SELF-TRAPPING SYSTEM

In the early 1980s, experimental evidence suggested that vibrational energy in natural proteins (specifically the CO stretch oscillation of the peptide unit) might become self-localized, with implications for the storage and transport of energy in biological organisms (Careri et al., 1984). Because the structures of natural proteins take a wide variety of shapes, the following equation—called the discrete self-trapping (DST) equation (Eilbeck et al., 1985)—was proposed to capture the essential features of self-localization:

$$\left(\mathrm{i}\frac{\mathrm{d}}{\mathrm{d}t} - \omega_0\right) \boldsymbol{U} + \gamma D(|\boldsymbol{U}|^2)\boldsymbol{U} + \varepsilon M\boldsymbol{U} = 0. \quad (1)$$

Here, $\boldsymbol{U}(t) \equiv \mathrm{col}(u_1, u_2, \ldots, u_f)$ is a column vector representing the amplitudes of f oscillatory modes, each of which is described in the rotating wave approximation by the complex amplitudes: $u_1(t), u_2(t), \ldots, u_f(t)$. With $\gamma = 0$ and $\varepsilon = 0$, these modes (sites) oscillate independently and sinusoidally at the site frequency ω_0.

In the last term of Equation (1), $M = [m_{jk}]$ is an $f \times f$ dispersion matrix, expressing energetic interactions among the f modes stemming from electromagnetic couplings. This matrix is real and symmetric ($m_{ij} = m_{ji}$), and its diagonal elements can be chosen to represent small variations in the site frequencies from ω_0. Thus, with $\gamma = 0$ but with the dispersion parameter ε not zero, there will be f modes of oscillation given by eigenvectors of the matrix $(M - \omega_0 I)$.

In the second term of Equation (1), the parameter γ introduces nonlinearity into the formulation, where $D(|\boldsymbol{U}|^2) \equiv \mathrm{diag}|(|u_1|^2, |u_2|^2, \ldots, |u_f|^2)$ is a diagonal matrix. Thus, with $\varepsilon = 0$ but with the nonlinear parameter γ not equal to zero, each site is an independent (uncoupled) anharmonic oscillator.

From a physical perspective, there are two types of nonlinearity: intrinsic (stemming from weakening of electronic bonding with mode amplitude) and extrinsic (arising from interactions between localized vibrations and the lattice) (Scott, 2003). Intrinsic nonlinearity governs energy localization in small molecules (such as "local modes" of the CH stretch oscillations in benzene), which has been known to physical chemists since the 1920s (Ellis, 1929).

With both γ and ε not equal to zero, the DST equation displays an interesting variety of regular and chaotic motions for various values of the energy

$$H = \sum_j \left(\omega_0 |u_j|^2 - \frac{\gamma}{2}|u_j|^4\right) - \varepsilon \sum_{j,k} m_{jk} u_j^* u_k \quad (2)$$

and the "mass" $N = |u_1|^2 + |u_2|^2 + \cdots + |u_f|^2$. Numerical studies of these motions provide insights into the dynamics of vibrational energy in proteins and in small molecules. (For descriptions at small levels of oscillatory energy, Equation (1) is conveniently quantized, Scott et al., 1994.)

The scaling $\boldsymbol{U} \to \boldsymbol{U}\exp(-\mathrm{i}\omega_0 t)$ reduces (1) to the standard form of the DST equation

$$\mathrm{i}\frac{\mathrm{d}}{\mathrm{d}t}\boldsymbol{U} + \gamma D(|\boldsymbol{U}|^2)\boldsymbol{U} + \varepsilon M\boldsymbol{U} = 0. \quad (3)$$

An important class of solutions of the DST equation are the so-called stationary solutions ("stationary" because amplitudes are time-independent), which satisfy the ansatz

$$\boldsymbol{U}(t) = \boldsymbol{y}\exp(\mathrm{i}\omega t).$$

Inserting this into (3), we see that the constant vector $\boldsymbol{y} = \mathrm{col}(y_1, y_2, \ldots, y_f)$ satisfies the nonlinear eigenvalue problem

$$-\omega\boldsymbol{y} + \gamma D(|\boldsymbol{y}|^2)\boldsymbol{y} + \varepsilon M\boldsymbol{y} = 0, \quad (4)$$

where ω plays the role of the eigenvalue. In many cases, the vector $\boldsymbol{y}$ can be chosen to be real. In the limit $\varepsilon \to 0$ (alternatively $\gamma \to \infty$), called the "anti-integrable" limit, we have $y_i = 0$ or $|y_i|^2 = \omega/\gamma$, and in the real case, we have $y_i = \pm\sqrt{N/K}$, where K is the number of nonzero y_i on the chain. Starting from these solutions, branches of solutions for nonzero ε can then be generated numerically by path-following methods (Eilbeck et al., 1984). In the limit $\gamma \to 0$, (4) becomes a linear eigenvalue problem.

The values of y_i in the anti-integrable limit serve as a useful classification scheme for the branch, where we denote the three different (real) limits by the symbols $\cdot$, $\uparrow$, and $\downarrow$, respectively. The simplest stationary localized solution has only one nonzero mode or site amplitude in the anti-integrable limit, with $y_k = \sqrt{N}$, $y_i = 0, i \neq k$. For small but nonzero ε the amplitudes on the sites $\neq k$ are small and tend to zero exponentially as $|k - i| \to \infty$.

Such a localized solution was called a soliton in the early work on the DST equation but is now more often referred to as a breather due to its internal degree of freedom ω. A single breather in the center of a 1-dimensional lattice would be written as $(\ldots\cdots\uparrow\cdots\ldots)$, and would appear as in Figure 1. More complicated breather structures are possible, such as $(\ldots\cdots\uparrow\uparrow\cdots\ldots)$ and $(\ldots\cdots\uparrow\downarrow\cdots\ldots)$.

Breathers satisfying (4) are referred to as stationary breathers; more general mobile solutions are of interest. Feddersen (1991b) carried out a numerical study of solutions of the form

$$U_n(t) = \psi(n - ct)\mathrm{e}^{\mathrm{i}(kn-\omega t)}$$

in the case where M is a tridiagonal matrix with constant coefficients (nearest-neighbor interactions). This case is now known as the discrete nonlinear Schrödinger (DNLS) equation. Note that the "carrier wave" speed ω/k is different from the "envelope" speed c. He found branches of localized solutions to high accuracy, but the existence of such solutions is still an open question.

The stability criteria for stationary solutions of the DST equation were studied by Carr & Eilbeck (1985). An important step is to consider perturbations in a frame rotating with the stationary solutions

$$U_n(t) = [y_n + \varepsilon_n(t)]\,\mathrm{e}^{\mathrm{i}\omega t}.$$

Once this trick is carried out, the study of linear stability reduces to an algebraic eigenvalue problem, at least when f is finite, which greatly simplifies the problem. One noteworthy point is that the resulting stability matrix is not self-adjoint, so branches can change stability at points away from bifurcation points.

On a finite lattice, modeling vibrational modes on small molecules, the stationary solutions of the DST equation can often be found exactly. For $f = 3$ and $M_{ij} = 1$ if $i \neq j$ (modeling NH_3 stretching oscillations in ammonia), we obtain the bifurcation diagram shown in Figure 2. Details are discussed in Eilbeck et al. (1985). Note the change of stability of the $\uparrow\downarrow 0$ branch at two places, and the local-mode branch, where the

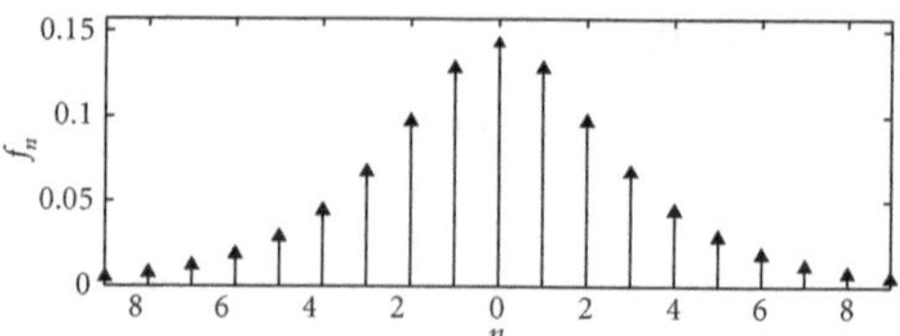

Figure 1. Stationary Breather on a DST (DNLS) lattice.

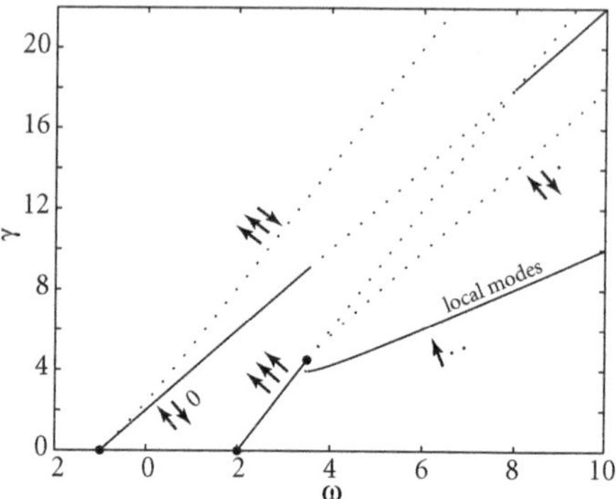

Figure 2. Stationary solutions on a DST lattice with $f = 3$. Solid (dotted) lines indicate stable (unstable) solutions.

vibrational energy is localized on a single degree of freedom.

In the large f case, in an application of the DST formulation to the dynamics of a globular protein, Feddersen considered interactions among CO stretch oscillations in adenylate kinase, which comprises 194 amino acids ($f = 194$) (Feddersen, 1991a). Since the structure of this enzyme has been determined by X-ray analysis, the $f(f-1)/2 = 18721$ off-diagonal elements of the dispersion matrix (M) were calculated from Maxwell's equations. Also, diagonal elements were selected from a random distribution, and the degree of localization of a particular solution was defined by evaluating the quotient $\sum |u_j|^4 / \sum |u_j|^2$.

In this study at experimentally reasonable levels of nonlinearity (γ), stable localized solutions were observed near some but not all of the amino acids. Interestingly, this anharmonic localization was observed to be distinctly different from Anderson localization, a property of randomly interacting linear systems. Thus, none of the stationary states, that were observed to be highly localized at large γ, remained so as γ was made small. Also, none of the states, that were localized at $\gamma = 0$ (i.e., Anderson localized), remained so as γ was increased to a physically reasonable level.

Other studies and applications of the DST equation include models for Bose–Einstein condensates and coupled optical fiber arrays. On the theoretical side, work has been carried out in the study of nonstationary and chaotic solutions on both small and large lattices. A list of citations to Eilbeck et al. (1985) is presently maintained at http://www.ma.hw.ac.uk/~chris/dst/.

ALWYN SCOTT AND CHRIS EILBECK

See also **Anderson localization; Discrete breathers; Discrete nonlinear Schrödinger equations; Local modes in molecular crystals; Local modes in molecules; Rotating wave approximation**

Further Reading

Careri, G., Buontempo, U., Galluzzi, F., Scott, A.C., Gratton, E. & Shyamsunder, E. 1984. Spectroscopic evidence for Davydov-like solitons in acetanilide. *Physical Review* B 30: 4689–4702

Carr, J. & Eilbeck, J.C. 1985. Stability of stationary solutions of the discrete self-trapping equation. *Physics Letters* A 109: 201–204

Eilbeck, J.C., Lomdahl, P.S. & Scott, A.C. 1984. Soliton structure in crystalline acetanilide. *Physical Review* B, 30: 4703–4712

Eilbeck, J.C., Lomdahl, P.S. & Scott, A.C. 1985. The discrete self-trapping equation. *Physica* D 16: 318–338

Ellis, J.W. 1929. Molecular absorption spectra of liquids below 3 μ. *Transactions of the Faraday Society*, 25: 888–897

Feddersen, H. 1991a. Localization of vibrational energy in globular protein. *Physics Letters* A, 154: 391–395

Feddersen, H. 1991b. Solitary wave solutions to the discrete nonlinear Schrödinger equation. In *Nonlinear Coherent Structures in Physics and Biology*, edited by M. Remoissenet & M. Peyrard, Berlin and New York: Springer, 159–167

Scott, A.C., Eilbeck, J.C. & Gilhøj, H. 1994. Quantum lattice solitons. *Physica* D, 78: 194–213

Scott, A.C. 2003. *Nonlinear Science: Emergence and Dynamics of Coherent Structures*. 2nd edition, Oxford and New York: Oxford University Press

DISLOCATIONS IN CRYSTALS

Dislocations in crystals are linear topological defects near which the regular crystalline atomic arrangement is broken. The dislocation as a structural defect is characterized by the following peculiarity: the regular crystal lattice structure is greatly distorted only in the near vicinity of an isolated line (the dislocation axis), and the region of irregular atomic arrangement has transverse dimensions of the order of the interatomic distance. Nevertheless, deformation occurs even far from the dislocation core, and this is a consequence of a topological property of the dislocation. The deformation at a distance from the dislocation axis may be seen by passing along a closed contour around the dislocation core and considering the displacement vector of each crystalline site from its position in an ideal lattice. Upon calculating the total increment of this displacement vector at the end of the pass, one sees that the total atomic displacement is nonzero and equals one atomic translation period.

Among the many microscopic models of dislocations, the simplest takes the dislocation to be the edge of an extra half-plane present in the crystal lattice. In the conventional atomic scheme of this model (Figure 1), the trace of the half-plane coincides with the upper semiaxis y, and its edge coincides with the z-axis; thus, one has an "edge" dislocation. If one surrounds

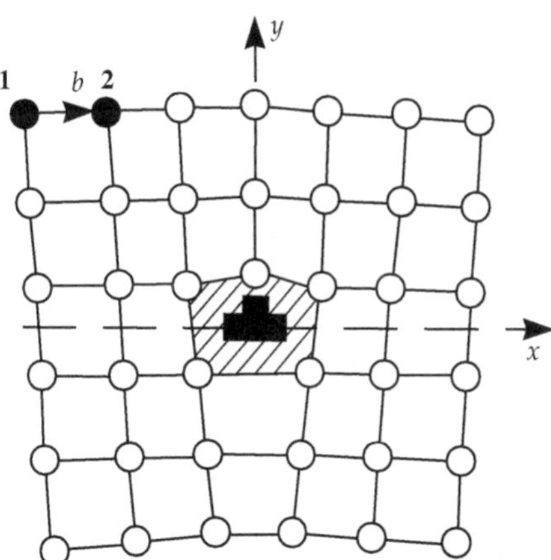

Figure 1. Schematic atomic arrangement in the vicinity of an edge dislocation.

the dislocation axis with a tube with radius of the order of several interatomic distances, the crystal outside this tube may be regarded as ideal and subject only to elastic deformation (crystal planes are connected to one another almost regularly), and inside the tube the atoms are considerably displaced relative to their equilibrium positions and form the "dislocation core." The atoms of the dislocation core are distributed over the contour of the shaded pentagon in Figure 1. The closed pass is the external contour starting from point 1 and finishing at point 2. The vector connecting atoms 1 and 2 is denoted by $\boldsymbol{b}$ and called the "Burgers vector." Possible values of the Burgers vectors in a crystal are determined by its crystallographic structure and correspond, as a rule, to a small number of certain directions in a crystal. The dislocation lines are arranged arbitrarily, although their arrangement is limited by a set of definite crystallographic planes. The dislocation line cannot end inside a crystal. It must either leave the crystal with each end on the crystal surface or form a dislocation loop.

The main topological property of the dislocation implies that a dislocation in a crystal is a specific line D having the following general property. After a circuit around the closed contour L enclosing the line D (see Figure 2), the elastic displacement vector $\boldsymbol{u}$ changes by a certain finite increment $\boldsymbol{b}$ equal to one of the lattice periods. This property can be written as

$$\oint_L \mathrm{d}u_i = \oint_L \frac{\partial u_i}{\partial x_k}\mathrm{d}x_k = -b_i, \qquad (1)$$

assuming that the direction of the circuit is related to a chosen direction of the tangent vector $\boldsymbol{\tau}$ to the dislocation line. The dislocation line is in this case a line of singular points of the elastic fields.

Macroscopic considerations (based on elasticity theory) allow one to find a concentration of strains and stresses near the dislocation. The elastic field created by a dislocation leads to its interaction with

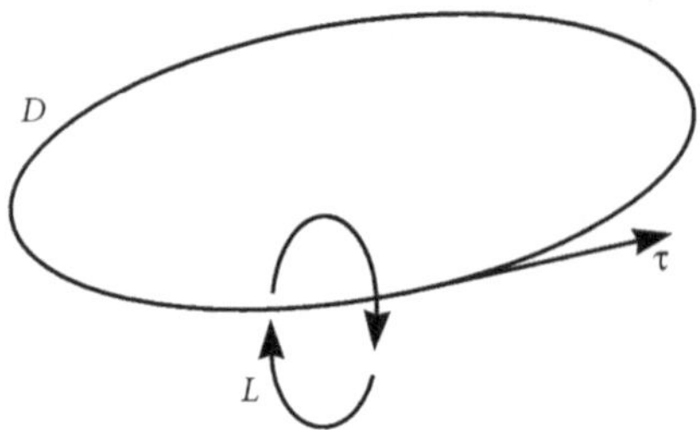

Figure 2. Mutual coordination of the vector $\boldsymbol{\tau}$ and circuit around the contour L.

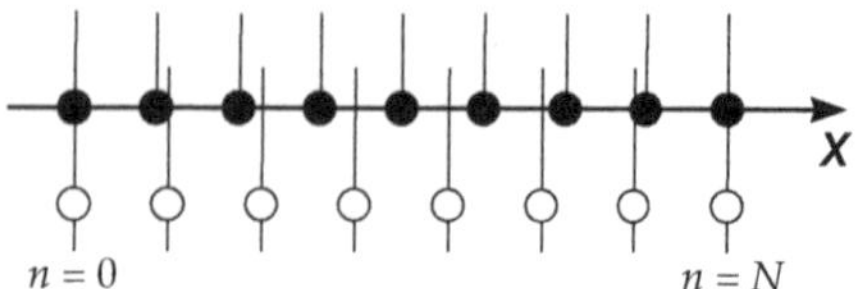

Figure 3. Edge dislocation in the Frenkel–Kontorova model.

other dislocations and other types of crystal defects. In particular, the dislocation can accumulate point defects near its core. If such defects are color centers, the dislocation line can be decorated (*See* **Color centers**).

Displacements of the dislocation line in the plane determined by the vectors $\boldsymbol{b}$ and $\boldsymbol{\tau}$ (called a "slip plane") have no effect on the crystal continuity. Therefore, a comparatively easy mechanical motion of the dislocation is possible in principle in this plane. But any displacement of the dislocation produces some plastic deformation, so moving dislocations are carriers of crystal plasticity.

Macroscopic dislocation theory can describe a contribution of the dislocations to mechanical properties of real crystals and proposes a physical explanation of the crystal plasticity. However, the macroscopic theory cannot propose a structure of the dislocation core and explain possible mechanisms of the dislocation motion. The Soviet scientists Yakob Frenkel and Tatiana Kontorova were the first to present a simple 1-dimensional microscopic dislocation model (Frenkel & Kontorova, 1938). To formulate their model, it is convenient to analyze the following picture. The atomic chain (a set of black circles in Figure 3) is an edge series of one half of a plane crystal ($y > 0$, Figure 1) displaced in a certain way with respect to another half of a crystal (a substrate at $y < 0$). The influence of the nondisplaced half of the crystal on the atoms distributed along the x-axis can be qualitatively described assuming that those atoms are in a given external periodic field whose period coincides with the interatomic distance on the substrate a. The chain energy is then determined not only by a relative displacement of neighboring atoms but also by an absolute displacement of separate atoms u_n (n is the atomic number) in the external field. If the external periodical field has a simple sine-shaped dependence, the equation of chain motion has the form

$$m\frac{d^2u_n}{dt^2} = \alpha(u_{n+1} + u_{n-1} - 2u_n) - U_0\sin(2\pi u_n/a), \qquad (2)$$

where m is the atom mass and α and U_0 are constant parameters. The displacements of the atoms in the chain are performed in such a way that $N+1$ atoms are situated over N crystal sites on the substrate. This atom distribution corresponds to the following boundary conditions for the displacements:

$$u_n = a \quad \text{at} \quad n = -\infty, \qquad u_n = 0 \quad \text{at} \quad n = \infty. \qquad (3)$$

Equation (2) and the boundary conditions (3) are basic in the Frenkel–Kontorova model.

In a long-wavelength approximation, the function of the discrete number u_n is replaced by a continuous function of the coordinate $u(x)$ where $x = na$, and Equation (2) transforms into the sine-Gordon equation (SG) equation

$$\frac{\partial^2 w}{\partial t^2} = s^2\frac{\partial^2 w}{\partial x^2} - \omega_0^2 \sin w, \qquad w = 2\pi u/a, \qquad (4)$$

where $s^2 = a^2\alpha/m$ and $\omega_0^2 = U_0/m$.

The SG equation (4) possesses the following (kink-soliton) solution satisfying the boundary conditions (3): $u(x) = \tan^{-1}\exp[-(x - x_0/l)]$, where $l = s/\omega_0$ and $x_0 =$ constant. This kink describes the distribution of atoms in the core of the Frenkel–Kontorova dislocation, the parameter l determines the dislocation semiwidth, and x_0 is the coordinate of the dislocation center. Later Peierls (1940) proposed a semimicroscopic model of a straight dislocation line (directed along the z-axis in Figure 1) in a 3-d isotropic elastic medium. In this model, it was assumed that strains around the dislocation are described by the theory of elasticity; however, the interaction energy of both semi-spaces above (+) and under (−) the slide plane depends periodically on the relative displacements $v(x) = w(x)^+ - w(x)^-$ (the x-axis is a trace of the slide plane) and is described by a sine-shaped periodic function of $v(x)$. The main equation of Peierls's theory has the following form:

$$a\int_{-\infty}^{\infty}\left(\frac{dv}{dx}\right)\frac{d\xi}{x-\xi} + M\ \sin v = 0, \qquad (5)$$

where the integral means its principal value and M is some dimensionless combination of elastic modules. Equation (5) has a soliton solution very close to the Frenkel–Kontorova dislocation $w(x) = \pi + 2\tan^{-1}[(x - x_0)/l]$ where $l = \pi a/M$.

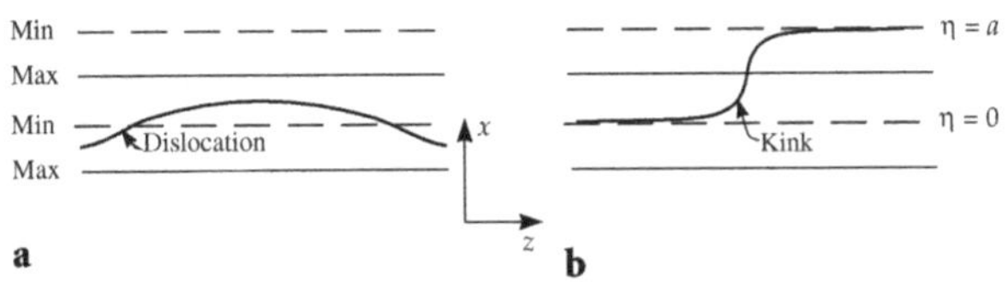

Figure 4. Two types of dislocation motion in the Peierls potential field. (a) Dislocation oscillates in its own trough. (b) The dislocation forms a kink moving along the z-axis.

Both the chain energy in the 1-d case and crystal energy in the 3-d case (calculated for the Frenkel–Kontorova and Peierls models, respectively) do not depend on the coordinate of the dislocation center x_0. Thus, the dislocation can glide, changing its location on the slide plane without action of any external field. This fact is a result of using the continuous approximation. Taking into account discreteness of the crystal and following the Peierls theory, Nabarro (1947) showed that a moving dislocation is influenced by some crystal field periodically depending on the dislocation center x_0 (later called the "Peierls relief"). Therefore, the dislocation can start moving only if an external stress acting on it exceeds a certain value called the "Peierls barrier."

Many dynamic problems in dislocation theory can be analyzed using the so-called "string model." This model treats the dislocation line as a heavy string under tension lying on a corrugated surface. An effective mass per unit length of the dislocation m and its line tension T_D are of a field origin and are associated with the inertia and energy of dislocation dynamic elastic field (Kosevich, 1964). The corrugated surface describes the Peierls relief. Troughs of this surface correspond to the potential minima of the slide plane occupied by a straight-line dislocation in equilibrium (see Figure 4). If the z-axis is directed along the equilibrium position of the dislocation and its transverse displacements η go along the x-axis, then the equation of motion of the dislocation has the form (Seeger, 1956):

$$m\frac{\partial^2\eta}{\partial t^2} - T_D\frac{\partial^2\eta}{\partial x^2} + b\sigma_p \sin(2\pi\eta/a) = b\sigma , \qquad (6)$$

where σ_p is the Peierls stress and σ is the applied stress. In the case $\sigma = 0$, this equation is equivalent to Equation (4), proposed by Frenkel and Kontorova. Equation (6) describes both small amplitude vibrations of the dislocation effected by an external oscillation force and the so-called kink-mechanism of the transverse displacements of the dislocation along the x-axis effected by a strong stationary driving force.

ARNOLD KOSEVICH

See also **Color centers; Frenkel–Kontorova model; Peierls barrier; Sine-Gordon equation; Topological defects**

Further Reading

Frenkel, J. & Kontorova, T. 1938. On the theory of plastic deformation and twinning. *Physikalische Zeitschrift der Sowjetunion*, 13: 1

Kosevich, A.M. 1964. Dynamical theory of dislocations. *Uspekhi Fizicheskikh Nauk*, 84: 579 (in Russian); translated in 1965. *Soviet Physics, Uspekhi*, 7: 837

Kosevich, A.M. 1979. Crystal dislocations and the theory of elasticity, in *Dislocations in Solids*, edited by F.R.N. Nabarro Amsterdam: North Holland, p. 33

Nabarro, F.R.N. 1947. Dislocations in a simple cubic lattice *Proceedings of Physical Society, London*, A59: 256

Nabarro, F.R.N. 1967. *Theory of Crystal Dislocations*, Oxford: Clarendon Press

Peierls, R.E. 1940. The size of a dislocation. *Proceedings of Physical Society, London*, A52: 34

Seeger, A. 1956. On the theory of the low-temperature internal friction peak observed in metals. *Philosophical Magazine*, 1: 651

DISPERSION MANAGEMENT

The propagation of optical pulses in fiber lines is usually limited by dispersion and nonlinearity. Dispersion causes a broadening of the pulses whereas nonlinearity is the reason for four-wave mixing (FWM). A balance of dispersion and nonlinearity may produce the so-called optical soliton. Dispersion management (manipulations of the chromatic dispersion along the line) is an attractive technique that allows enhancement of the performance of fiber communication links both for soliton and nonsoliton transmission. Dispersion-managed solitons can be viewed as novel kinds of information carriers with many attractive features that will lead to further improvement of the transmission capacity of fiber links. The starting point of all considerations in this area is the cubic nonlinear Schrödinger (NLS) equation for an optical mode in a cylindrical fiber with (constant) dispersion and Kerr nonlinearity (Newell & Maloney, 1992):

$$\mathrm{i}\,\frac{\partial u}{\partial z} + \frac{1}{2}\,\frac{\partial^2 u}{\partial t^2} + |u|^2 u = 0. \qquad (1)$$

This equation is written in dimensionless form (u is the normalized intensity, z the normalized propagation distance, and t is the normalized time); for more details, see, for example, Hasegawa & Kodama (1995).

From FWM calculations (Agrawal, 1995), one knows that the FWM efficiency depends on the phase mismatch and thereby on the dispersion coefficient, D. The larger the phase mismatch (proportional to dispersion), the smaller the efficiency of FWM. On the other hand, small D values are necessary for high bit-rates B. One reason is the so-called Gordon–Haus effect (Gordon & Haus, 1986). Solitons adjust themselves to external perturbations, and thereby they can be considered as quite robust. However, this adjustment also has disadvantages. Any re-shaping, for example, in the presence of amplifier noise, also has consequences

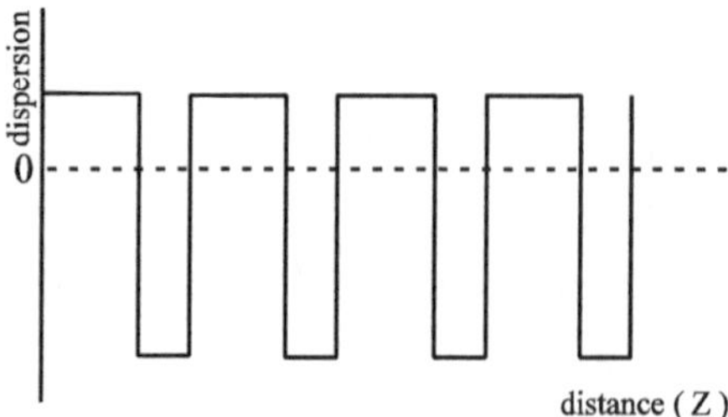

Figure 1. Schematic plot of the dispersion coefficient that varies in space when dispersion management is applied.

for the velocities. Changing velocities cause a timing jitter, and the latter limits the maximal possible bit-rate. Note that the general soliton solution of Equation (1)

$$u_s = A\ \mathrm{sech}\,[A(t+\Omega z)-q_0]\,\mathrm{e}^{-\mathrm{i}\Omega t-(\mathrm{i}/2)(A^2-\Omega^2)z+\mathrm{i}\phi_0} \quad (2)$$

is moving with velocity $v = -1/\Omega$. The detailed calculation (Gordon & Haus, 1986) gives an upper limit for the bit-rate B and the length L of the system in the form

$$(BL)^3 \leq \text{upper limit} \sim \frac{1}{D}. \quad (3)$$

One clearly recognizes that low dispersion favors high bit-rates over large distances. On the other hand, as was mentioned already, low dispersion is dangerous with respect to FWM. Dispersion management is an elegant way out of this dilemma.

Principle of Dispersion Management

The idea of using a dispersion-compensating fiber (DCF) to overcome the dispersion of the standard mono-mode fiber (SMF) was proposed in 1980 (Lin et al., 1980). The simplest optical-pulse equalizing system consists of a transmission fiber (SMF or dispersion-shifted fiber (DSF)) and an equalizer fiber with the opposite dispersion (DCF to compensate SMF or, for instance, SMF to compensate DSF with normal dispersion). The incorporation of a fiber with normal dispersion reduces (or, in the ideal case, eliminates) the total dispersion of the fiber span between two amplifiers. Low average dispersion is good for reducing the Gordon–Haus effect while high local dispersion reduces the FWM efficiency. In Figure 1, we indicate the principle of dispersion management by a schematic drawing.

Dispersion management means that the coefficient in front of the second time derivative becomes Z-dependent (we change notation in order to allow for different normalization parameters), that is, the simplest form of the dispersion-managed nonlinear Schrödinger equation becomes (Kodama, 1998)

$$\mathrm{i}\,\frac{\partial q}{\partial Z}+\frac{1}{2}\,\mathrm{d}(Z)\,\frac{\partial^2 q}{\partial T^2}+|q|^2 q=0. \quad (4)$$

A short remark: if $\mathrm{d}(Z)$ would be $\neq 0$ everywhere in the fiber, the variation of $\mathrm{d}(Z)$ can be eliminated by using a new coordinate Z' via

$$\frac{\mathrm{d}Z'}{\mathrm{d}Z}=d(Z). \quad (5)$$

Then a Z-dependent coefficient would appear in front of the nonlinear term. Again, we would end up with an equation with a rapidly varying coefficient. However, characteristic for nearly all practical dispersion management arrangements is the vanishing of $d(z)$ at certain positions z.

Weak Dispersion Management

For weak dispersion management of small-amplitude pulses, one can start with the equation

$$\mathrm{i}\frac{\partial u}{\partial z}+d(z)\frac{\partial^2 u}{\partial t^2}+\varepsilon|u|^2 u=0. \quad (6)$$

Here,

$$\begin{aligned} d(z) &= \langle d\rangle+\tilde{d} \ \text{ for } \ 0\leq z\leq l,\\ d(z) &= \langle d\rangle-\tilde{d} \ \text{ for } \ l\leq z\leq 2l, \end{aligned} \quad (7)$$

and so on, is the z-dependent dispersion. The z-coordinate is made dimensionless with the dispersion length of the local dispersion, that is, $\tilde{d}\sim O(1)$. The averaged dispersion $\langle d\rangle$ is assumed to be small, and we introduce the smallness parameter $\varepsilon \ll 1$ via $\langle d\rangle \approx \varepsilon$. Nonlinearity will also be scaled by ε. The strength of the dispersion management is characterized by

$$R(z)=\int_0^z (d-\langle d\rangle)\,\mathrm{d}s. \quad (8)$$

Weak dispersion management means $R\sim O(\varepsilon^{1/2})$; that is, R is assumed to be small. From the definition $R\sim O(\tilde{d}\cdot l)$ holds. Therefore, the length l (of the parts with constant dispersion) is also small compared with the dispersion length, $l\sim O(\varepsilon^{1/2})$.

One can use (at least three different) averaging methods to calculate the average behavior of pulses over large distances. The first one is a direct method, starting with an expansion. The second one uses the Lie-transform technique; it is, therefore, much more systematic than the direct method. The third method makes use of a Bogolyubov transformation, which is very elegant in practice, but the ansatz is not as straightforward as the Lie transformation. The ansatz has to be specifically chosen for each problem. In the direct method (Kivshar & Spatschek, 1995; Yang et al., 1997), one expands u in orders of $\varepsilon^{1/2}$ and applies a multiple scale technique,

$$\begin{aligned} u &= U+u_1+u_2+\cdots,\\ \frac{\partial}{\partial z} &= \frac{\partial}{\partial \xi}+\frac{\partial}{\partial z_0}+\frac{\partial}{\partial z_1}+\cdots \end{aligned} \quad (9)$$

with $U = U(z_0, z_1, z_2, \ldots)$ and $u_k = u_k(\xi, U)$. One may use the scaling $\xi \sim O(\varepsilon^{-1/2})$, $z_k \sim O(\varepsilon^{k/2})$, $u_k \sim O(\varepsilon^{k/2})$, where $\langle u \rangle = U$, $\langle u_k \rangle = 0$.

The Lie transformation technique was originally developed for ordinary differential equations but can also be used for a partial differential equation (Neyfeh, 1973). To apply the Lie transformation one rearranges Equation (6) into the form

$$\begin{aligned}\frac{\partial u}{\partial z} &= \mathrm{i}\langle d\rangle u_{tt} + \mathrm{i}\varepsilon|u|^2 u + \mathrm{i}\tilde{d}u_{tt} \\ &= X_0 + \tilde{d}X_{0D} = X[u, u^*]. \qquad (10)\end{aligned}$$

X depends on the infinite set of variables $(u, u_t, u_{tt}, \ldots, u^*, u_t^*, u_{tt}^*, \ldots)$, indicated by $X[u, u^*]$. Obviously $X_0 \sim O(\varepsilon)$, $X_{0D} \sim O(1)$. The transformation

$$\begin{aligned}u &= \mathrm{e}^{\vec{\phi}\nabla} v = v + \phi + \frac{1}{2!}\left(\vec{\phi}\cdot\nabla\right)\phi \\ &+ \frac{1}{3!}\left((\vec{\phi}\cdot\nabla)\phi\cdot\nabla\right)\phi + \cdots \qquad (11)\end{aligned}$$

transforms Equation (10) into

$$\frac{\partial v}{\partial z} = Y[v, v^*]. \qquad (12)$$

The right-hand side can, in principle, be derived appropriately in order to identify v with the averaged intensity. Here, $\vec{\phi} = (\phi, \phi_t, \ldots)$ and $\nabla = (\frac{\partial}{\partial v}, \frac{\partial}{\partial v_t}, \ldots)$ are infinite-dimensional vectors.

In the third method (Arnol'd, 1988), one uses a Floquet–Lyapunov transformation

$$u(z,t) = \mathrm{e}^{\mathrm{i}R(z)\partial_t^2} B(z,t) \qquad (13)$$

as well as the Bogolyubov–Krylov transformation

$$\begin{aligned}B &= v + \mathrm{i}f|v|^2 v + (g+p)N_1(v) \\ &+ \mathrm{i}hN_2(v) + q|v|^4 v + O(\varepsilon^{5/2}), \qquad (14)\end{aligned}$$

with coefficients to be determined appropriately during the calculation. All three methods lead to the same result. Over fairly large distances, the averaged solution is still a fundamental soliton, that is, robust. This follows from the fact that *to lowest order*, the basic equation for the averaged pulse intensity is the *integrable* cubic nonlinear Schrödinger equation

$$\mathrm{i}v_z + \langle d\rangle v_{tt} + \varepsilon|v|^2 v = \frac{1}{2}\varepsilon\left\langle R^2\right\rangle N_2(v). \qquad (15)$$

The correction term on the right-hand side contains a polynomial N_2 in v, v_t, $v_{tt}, \cdots$. One should bear in mind, on the other hand, that on the short scale (length $2l$ of one of the periodic elements), the exact (non-averaged) solution is oscillating (breathing) with the frequency following from the periodicity in the dispersion map. Besides the scaling $\tilde{d} \sim O(1)$ and $R \sim O(\varepsilon^{1/2})$, other scalings are possible and tractable by the methods mentioned above, as long as R is small (weak dispersion management).

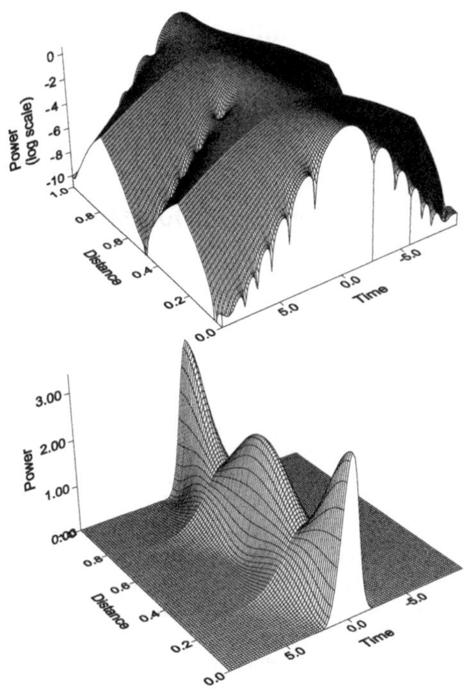

Figure 2. Plot of a DM soliton. The lower part shows the breathing over one element of the dispersion map. The upper graph shows the fine structure on a logarithmic axis, indicating the characteristic deviations from the fundamental soliton.

Strongly Dispersion-Managed Solitons

Recently (Nijhof et al., 1998), the strong dispersion management (R is no longer small) became of increasing interest. The chirp (pulse phase has a quadratic time-dependence) is a new and one of the most important features of the (strongly dispersion managed) DM soliton (recall that the conventional soliton is unchirped). In addition, the DM soliton is not anymore of the sech-type. The DM soliton consists of a self-similar energy-containing core surrounded by oscillating tails. Such tails manifest themselves as non-self-similar modulations of the soliton profile during the compensation period, although their amplitudes are rather small compared with the main peak (see Figure 2).

The most surprising feature of the DM soliton is that it can propagate stably along the line with zero or even with normal average dispersion (in contrast to the fundamental soliton that propagates stably only in the anomalous dispersion region). These observations indicate that an average model describing the evolution of the breathing pulse differs from the nonlinear Schrödinger equation. In other words, strong dispersion management imposes such a strong perturbation that a carrier pulse in this case is no longer the NLS soliton. For more details, see Nijhof et al. (1998).

Experimental Results

One can estimate the potential of dispersion management from successful experiments, for example, Carter et al. (1999).

KARL SPATSCHEK

See also **Nonlinear optics; Optical fiber communications**

Further Reading

Agrawal, G.P. 1995. *Nonlinear Fiber Optics*, San Diego: Academic Press

Arnol'd, V.I. 1988. *Geometrical Methods in the Theory of Ordinary Differential Equations*, New York: Springer

Carter, G.M., Grigoryan, V.S., Mu, R.-M., Menyuk, C.R., Sinha, P., Carruthers, T.F., Dennis, M.L. & Duling, I.N. 1999. Transmission of dispersion-managed solitons at 20 Gbit/s over 20 000 km. *Electronics Letters*, 35(3): 233

Gordon, J.P. & Haus, H.A. 1986. Random walk of coherently amplified solitons in optical fiber transmission. *Optics Letters*, 11: 665

Hasegawa, A. & Kodama, Y. 1995. *Solitons in Optical Communications*, Oxford: Clarendon Press

Kivshar, Yu.S. & Spatschek, K.H. 1995. Nonlinear dynamics and solitons in the presence of rapidly varying periodic perturbations. *Chaos, Solitons, and Fractals*, 5: 2551–2569

Kodama, Y. 1998. Nonlinear pulse propagation in dispersion managed system. *Physica* D, 123: 255–266

Lin, C., Kogelnik, H. & Cohen, L.G. 1980. Optical pulse equalization and low dispersion transmission in single-mode fibers in the 1.3–1.7 mm spectral region. *Optics Letters*, 5: 476–478

Newell, A.C. & Moloney, J.V. 1992. *Nonlinear Optics*, Redwood City, CA: Addison-Wesley

Neyfeh, A.H. 1973. *Perturbation Methods*, New York: Wiley

Nijhof, J., Doran, N., Forysiak, W. & Bernston, A. 1998. Energy enhancement of dispersion-managed solitons and WDM. *Electronics Letters*, 34: 481–482

Yang, T.-S., Golovchenko, A., Pilipetskii, A.N. & Menyuk, C.R. 1997. Dispersion-managed soliton interactions in optical fibers. *Optics Letters*, 22: 793–795

DISPERSION RELATIONS

Dispersion relations (DRs) are associated with wave equations to characterize the nature of their temporal and spatial evolution. For a linear wave equation in one dimension, the DR expresses the wavenumber ($k = 2\pi/\lambda$) as a function of the frequency ($\omega = 2\pi/T$), or vice-versa, where λ and T are, respectively, the wavelength and temporal period of an elementary periodic solution. Through the superposition principle, the behaviors of all solutions of linear equations are determined by the DR. In nonlinear cases, DRs also depend on wave amplitude and can be used in perturbative analyses of quasiharmonic systems. For more than one spatial dimension, the wave number is a vector.

Dispersion Relations for Linear Equations

Fourier analysis establishes a one-to-one correspondence between DRs and equations in such a way that we can obtain one from the other. Consider as an example, the Klein–Gordon equation,

$$u_{tt} - u_{xx} + m^2 u = 0, \tag{1}$$

where m is a constant and the subscripts indicate partial derivatives. Substituting a plane-wave solution of the form $u(t,x) = a e^{i(\omega t - kx)}$ into (1), we obtain the DR

$$\omega^2 = m^2 + k^2. \tag{2}$$

For linear equations, the phase velocity $v_\varphi(k)$ and the group velocity $v_g(k)$ are defined as

$$v_\varphi(k) = \frac{\omega(k)}{k}, \quad v_g(k) = \frac{d\omega(k)}{dk}. \tag{3}$$

The phase velocity of a wave number k is the speed of propagation of the corresponding harmonic wave component or mode. The group velocity corresponds to the propagation speed of a wave packet, and thus is the speed of energy transmission for the system (Whitham, 1974). This distinction is especially important in the propagation of wave trains, where signals travel with the group velocity.

An equation (or the wave that is represented by its solution) is said to be *dispersive* if $v_\varphi(k)$ is not a constant (or equivalently if $v_\varphi \neq v_g$ for all k). In such cases, the initial profile of the wave is not conserved. For the Klein–Gordon equation we have

$$v_\varphi(k) = \pm\frac{\sqrt{m^2+k^2}}{k}, \quad v_g(k) = \pm\frac{k}{\sqrt{m^2+k^2}}, \tag{4}$$

and the equation is dispersive except in the special case $m = 0$ (which corresponds to the wave equation).

The DR also plays an important role in simulations of Equation (1) by finite differences. The usual (von Neumann) stability conditions are equivalent to requiring that the phase velocity in the mesh be greater or equal to the phase velocity of the equation (Ritchmyer & Morton, 1967). Other aspects that mark deviations of numerical solutions from continuous solutions are governed by discrete DRs (Trefethen, 1982).

Amplitude-Dependent Dispersion Relations

For nonlinear equations, solutions cannot, in general, be expressed as a superposition of plane waves and thus, in principle, there is not dispersion relation. There are nevertheless two possible applications of the concept. The first is in the small-amplitude regime where an approximate DR can be established by neglecting the higher order terms, and the second is when plane waves are particular solutions of the nonlinear equation. In both cases, the DRs are amplitude dependent. We illustrate both cases with a nonlinear Klein–Gordon equation and the cubic Schrödinger equation.

The nonlinear Klein–Gordon equation with cubic term

$$u_{tt} - u_{xx} + m^2 u + g u^3 = 0 \qquad (5)$$

does not have exact plane wave solutions. (The case $m = 1$, $g = -1/3!$ corresponds for instance to the first two terms expansion of the sine-Gordon equation.) Nevertheless, we may assume a term of the form $a\cos(\omega t - kx)$ in the solution. Substituting and retaining only the lower-order terms, we obtain the approximate DR

$$-\omega^2 + k^2 + m^2 + \frac{3}{4} g a^2 + \cdots = 0 \qquad (6)$$

provided $m^2 \gg g a^2$. The presence of coupled higher-order terms together with the approximate DR gives a picture of how such modes get excited and evolve.

The previous analysis can also be applied, for instance, to the cubic nonlinear Schrödinger equation

$$\mathrm{i}\phi_t - \phi_{xx} + |\phi|^2 \phi = 0, \qquad (7)$$

but this equation is special in the sense that it has particular solutions in the form of plane waves $\phi = a e^{\mathrm{i}(\omega t - kx)}$. Although the general solution cannot be expressed as a superposition of such plane waves, we have an amplitude-dependent DR of the form

$$\omega = k^2 + a^2. \qquad (8)$$

In this case, wave dispersion can be compensated by nonlinearity and solitons may appear.

L. Vázquez and S. Jiménez

See also **Modulated waves; Wave packets, linear and nonlinear**

Further Reading

Ritchmyer, R.D. & Morton, K.W. 1967. *Difference Methods for Initial-Value Problems*, New York: Wiley-Interscience

Trefethen, L.N. 1982. Group velocity in finite difference schemes. *SIAM Review*, 24: 113–136

Whitham, G.B. 1974. *Linear and Nonlinear Waves*, New York: Wiley-Interscience

DISTRIBUTED OSCILLATORS

Two modes of oscillation sharing a common source of energy may have a stationary solution of equal mode amplitudes, but this solution may be unstable for the following reason. One of the modes will inevitably gain a bit of amplitude (due to noise), and if it then consumes more than its share of the available power, it will grow further, causing the other mode to decrease. Interestingly, this effect depends upon the spatial dimension of the oscillating system. Related dynamics arise in diverse areas of engineering and science, including flip-flop circuits, multimode lasers, interacting biological species, and the neocortex of a human brain.

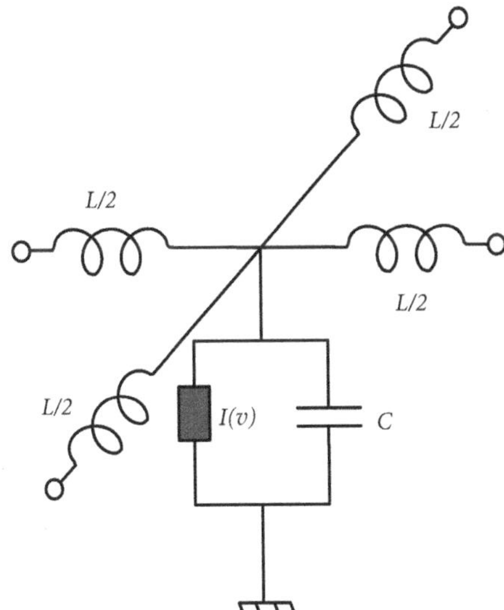

Figure 1. A unit cell of a two-dimensional oscillator array.

As a model for mode interactions, consider a two-dimensional array of identical oscillators, each interacting only with its nearest neighbors. Power flow among the modes of such an array can be studied using the method of harmonic balance, introduced by Balthasar van der Pol in 1934 (van der Pol, 1934) and developed by applied mathematicians and engineers in the Soviet Union during the 1930s and 1940s (Andronov et al., 1966; Kryloff & Bogoliuboff, 1947).

As shown in Figure 1, a unit cell of the array consists of a nonlinear conductance $I(v)$ in parallel with a capacitance C, connecting a lattice point to a ground plane. Each lattice point is attached to its four nearest neighbors through an inductance L. The nonlinear conductance is represented as

$$I(v) = -Gv\left(1 - 4v^2/3V_0^2\right), \qquad (1)$$

so for $-V_0/2 < v < +V_0/2$ the differential conductance $(\mathrm{d}I/\mathrm{d}v)$ is negative.

Assume a large, square $(N \times N)$ lattice of these unit cells with zero-voltage (short circuit) boundary conditions on the edges, as in Figure 2. In a zero-order approximation, the nonlinear conductance can be neglected, and in a continuum approximation, this linear, lossless system supports an arbitrary number (say n) modes. Thus, the total voltage is

$$v(x,y,t) = \mathcal{V}_1(x,y)\cos\theta_1 + \mathcal{V}_2(x,y) \times \cos\theta_2 + \cdots + \mathcal{V}_n(x,y)\cos\theta_n, \qquad (2)$$

which depends on space (x and y) and time (t) where

$$\mathcal{V}_i(x,y) \equiv V_i \cos k_{xi}x \cos k_{yi}y \quad \text{and} \quad \theta_i \equiv \omega_i t + \phi_i.$$

Noting that the energy in the ith mode is related to its amplitude by $U_i = (N^2C/8)V_i^2$ and averaging the

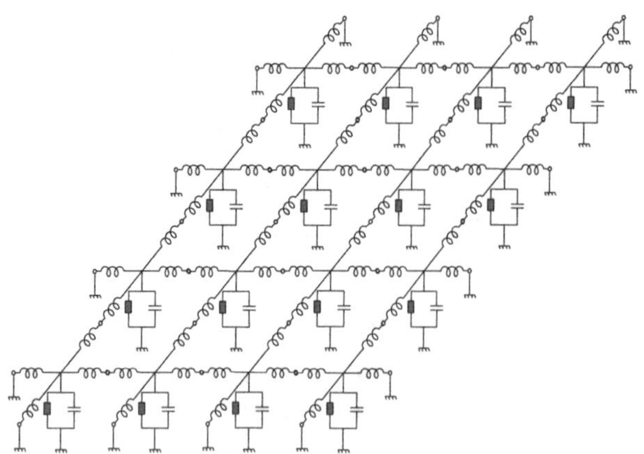

Figure 2. A 4×4 array ($N^2 = 16$) of the unit cells in Figure 1.

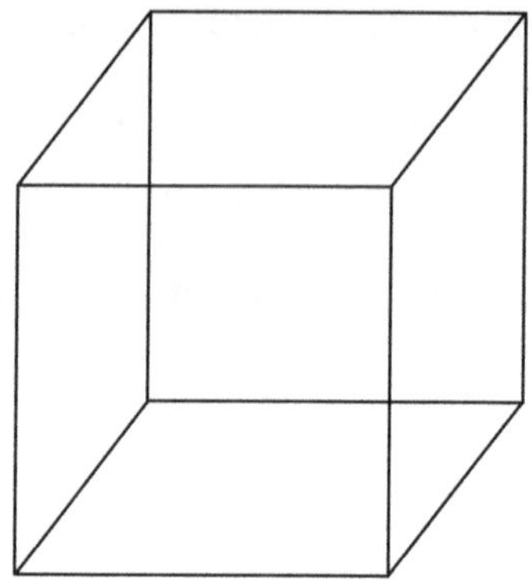

Figure 3. A Necker cube.

product of current (given by Equation (1)) and voltage (given by Equation (2)) over space and time gives the rate of change in mode energies as functions of mode energies (Scott, 2003).

The rate of change of energy (or power) into the first mode is

$$\frac{dU_1}{d\tau} = U_1\left[1 - \alpha\left(\frac{9}{8}U_1 + U_2 + \cdots + U_n\right)\right], \quad (3)$$

with $\tau \equiv Gt/C$ and $\alpha \equiv 4/(N^2CV_0^2)$. For n excited modes, there is a set of n equations—each similar to Equation (3) but with the indices appropriately altered—for the rates of change of the mode energies as functions of those energies. These nonlinear, autonomous equations have the same form as those introduced by Vito Volterra to describe the interaction of biological species competing for the same food supply (Volterra, 1937), and they are similar to a formulation suggested by Peter Greene (Greene, 1962) to describe interactions between assemblies of neurons in the brain.

Generalizing to a system of d-space dimensions, equations of interacting mode energies become

$$\begin{aligned} \frac{dU_1}{d\tau} &= U_1\left[1 - \alpha\left(KU_1 + U_2 + \cdots + U_n\right)\right], \\ &\vdots \\ \frac{dU_n}{d\tau} &= U_n\left[1 - \alpha\left(U_1 + U_2 + \cdots + KU_n\right)\right], \end{aligned} \quad (4)$$

where

$$K = 3^d/2^{d+1}.$$

For $K > 1$ (implying $d \geq 2$), analysis of Equations (4) indicates multimode stability. In other words, several modes can stably exist for two or more spatial dimensions. For one spatial dimension, on the other hand, two modes of equal amplitude are unstable, and only a single mode can be stably established. This observation is important in the design of semiconductor lasers, where large transverse dimensions are introduced to increase the power output. Unfortunately, such a design allows several modes to oscillate together, thereby decreasing the coherence and spectral purity of the output beam.

Multimode oscillator arrays have been realized using semiconductor tunnel diodes (Scott, 1971), superconductor tunnel diodes (Hoel, et al., 1972), and integrated circuits (Aumann, 1974). In these experiments, a variety of stable multimode oscillations have been observed, some quasiperiodic and others periodic, indicating mode locking. In a manner qualitatively similar to that proposed by Greene in his model of the brain's neocortex (Greene, 1962), the oscillator array can be induced to switch from one stable multi-mode configuration to another.

At the level of subjective perception, similar switchings in the brain are observed as one stares at Figure 3. Constructed by Louis Albert Necker (a Swiss geologist) in the mid-1800s, this image seems to jump from one metastable orientation to another, like the flip-flop circuit of a computer engineer. Defining order parameters (ξ_1 and ξ_2) to represent neural activities corresponding to the two perceptions of the Necker cube, Hermann Haken has recently suggested the equations

$$\begin{aligned} \frac{d\xi_1}{dt} &= \xi_1[1 - C\xi_1^2 - (B + C)\xi_2^2], \\ \frac{d\xi_2}{dt} &= \xi_2[1 - C\xi_2^2 - (B + C)\xi_1^2] \end{aligned} \quad (5)$$

as an appropriate dynamic description (Haken, 1996). With $U_j = \xi_j^2$ ($j = 1, 2$), Equations (5) are identical to Equations (4), showing only single-mode stability for $B > 0$. Study of this system in the (ξ_1, ξ_2) phase plane reveals two stable states: $(1/\sqrt{C}, 0)$ and $(0, 1/\sqrt{C})$, and jumping back and forth between these two states models one's subjective experience of Figure 3.

Alwyn Scott

See also **Cell assemblies; Lasers; Population dynamics; Quasilinear analysis; Synergetics; Tacoma Narrows Bridge collapse**

Further Reading

Andronov, A.A., Vitt, A.A. & Khaikin, S.E. 1966. *Theory of Oscillators*, Oxford and New York: Pergamon

Aumann, H.M. 1974. Standing waves on a multimode ladder oscillator. *IEEE Transactions on Circuits and Systems*, CAS-21: 461–462

Greene, P.H. 1962. On looking for neural networks and "cell assemblies" that underlie behavior. *Bulletin Mathematical Biophysics* 24: 247–275 and 395–411

Haken, H. 1996. *Principles of Brain Functioning: A Synergetic Approach to Brain Activity, Behavior and Cognition*, Berlin: Springer

Hoel, L.S., Keller, W.H., Nordman, J.E. & Scott, A.C. 1972. Niobium superconductive tunnel diode integrated circuit arrays. *Solid State Electronics*, 15: 1167–1173

Kryloff, N. & Bogoliuboff, N. 1947. *Introduction to Nonlinear Mechanics*, Princeton, NJ: Princeton University Press

Scott, A.C. 1971. Tunnel diode arrays for information processing and storage. *IEEE Transactions on Systems, Man, and Cybernetics*, SMC-1: 267–275

Scott, A.C. 2003. *Nonlinear Science: Emergence and Dynmaics of Coherent Structures*, 2nd edition. Oxford and New York: Oxford University Press

van der Pol, B. 1934. The nonlinear theory of electric oscillations. *Proceedings of the Institute of Radio Engineers*, 22: 1051–1086

Volterra, V. 1937. Principes de biologie mathématique. *Acta Biotheoretica*, 3: 1–36

DNA PREMELTING

DNA premelting phenomena are spontaneous dynamical processes that occur well below the melting temperature of DNA. They reflect DNA "breathing," a process that combines the transient unstacking of base-pairs (allowing planar drugs and dyes to intercalate into DNA), with the transient breaking and rejoining of hydrogen-bonds connecting base-pairs in limited DNA regions (allowing tritium nuclei in hydrogen bonds connecting base-pairs to exchange with protons in water). The formation and propagation of chain-slippage structures either within or between homologous DNA molecules, along with a variety of structural phase transitions in DNA are also included in discussions of premelting. This entry describes how concepts of kink-antikink-bound states (breather solitons) have been used to assist in understanding a wide range of DNA premelting phenomena.

Conceptually, the lowest energy kink-antikink-bound state in DNA (termed a "premelton") contains a central hyperflexible beta-DNA core region modulated into B-DNA on either side through kink and antikink boundaries. As the kink and antikink move together, for example, the energy density in the central beta-DNA core region rises. This energy is used first to enhance (alternate) base-pair unstacking, and next, to stretch and to eventually break hydrogen-bonds connecting base-pairs. As kink and antikink move apart, the reverse occurs. Energy within this central core region falls, allowing hydrogen-bonds to reform, and (alternate) base-pairs to partially restack. Isoenergetic breather-motions such as these can facilitate the intercalation of drugs and dyes into DNA and allow tritium exchange to occur at temperatures well below the melting temperature.

Beta-DNA is a structural form that differs from either B- or A-DNA. Evidence for its existence comes from studies of intercalation by drugs and dyes into DNA and the binding of certain proteins to DNA. Although double-stranded, beta-DNA is unique in being both hyperflexible and metastable. Its hyperflexibility suggests it be a liquid-like phase, lying intermediate between the more rigid B- and A-forms and the melted single-stranded form. Both properties necessitate beta-DNA to be pinned by an intercalator or held by a protein in order to be studied in detail. The structure is composed of repeating units called beta-structural elements. These are a family of base-paired, dinucleotide structures, each possessing the same mixed sugar-puckering pattern (i.e., C3′ endo (3′–5′) C2′ endo) and having similar backbone conformational angles, but varying in the degree of base-unstacking. Lower energy forms contain base-pairs that are partially unstacked, while higher energy forms contain base-pairs completely unstacked.

Beta-DNA is an intermediate in DNA melting, lying on the minimal energy pathway connecting B-DNA with single-stranded melted DNA. Three distinctly different sources of nonlinearity appear as DNA chains unwind, and these determine the sequence of conformational changes that occur along this pathway.

The first two sources of nonlinearity stem from changes in the sugar-pucker conformations and base-pair stacking. These changes require small energies (ca. kT) and appear as part of the initial structural distortions accompanying DNA unwinding. Starting with B-DNA, the effect of unwinding DNA is to counterbalance this with an equal but opposite right-handed superhelical writhing to keep the linking invariant. This is achieved through a modulated beta-alternation in sugar-puckering along the chains, accompanied by the gradual partial unstacking of alternate base-pairs. The lowest energy beta-DNA structure emerges as an end result. Its metastability reflects the presence of additional energies in its structure that cause the partial unstacking of alternate base-pairs.

The third source of nonlinearity arises from the stretching and ultimate rupture of hydrogen bonds connecting base-pairs. At first, beta-DNA accommodates further unwinding through the gradual loss of superhelical writhing. This reflects the appearance of beta-structural elements having increasingly higher energy (these have base-pairs further unstacked and unwound). Eventually, however, a limit is reached and further unwinding begins to stretch hydrogen-bonds that connect the base-pairs. Continued unwinding results in the disruption of these hydrogen-bonds, and the appearance

Figure 1. Molecular structure of a B–B premelton.

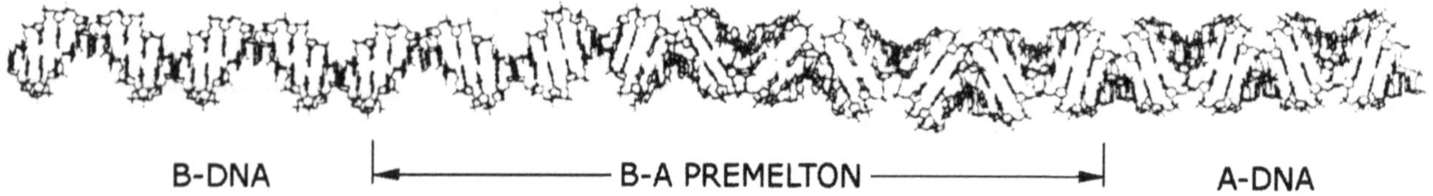

Figure 2. Molecular structure of a B–A premelton.

of single-stranded DNA. This final sequence of conformational change defines the boundary that connects beta-DNA to single-stranded melted DNA.

Using nonlinear least-squares methods, it has been possible to form premelton structures within B-DNA and within A-DNA (i.e., B–B premeltons, and A–A premeltons), as well as hybrid structures that connect the two (i.e., B–A premeltons and A–B premeltons) (see Figures 1 and 2). Such hybrid premeltons are constructed by connecting the central beta-DNA core with either type of kink-antikink boundary. Importantly, B–B and A–A premeltons are nontopological, whereas B–A and A–B premeltons are topological.

The B to A structural phase transition can be understood in the following way. In the presence of suitable thermodynamic conditions, kink and antikink within premeltons in B-DNA structure begin to move apart to form larger and larger core regions, whose centers modulate into A–DNA structure. Eventually, B–A premeltons and A–B premeltons form, and these continue to move apart, leaving a growing A-DNA region within. Such a mechanism is reversible and illustrates how a bifurcation within the central core region of this low-energy kink-antikink bound state structure can give rise to the B- to A-DNA structural phase transition.

Bifurcations arising within premeltons having higher energy (containing longer beta-DNA core regions that undergo more vigorous breather-motions) can lead to the formation of two additional types of higher energy kink-antikink bound states in DNA. These are called "branch-migratons" and "dislocatons." Each gives rise to different types of chain-slippage events, called double- and single-strand branch-migration.

Branch-migratons arise from breathing events that cause DNA chains to come apart transiently, and to then snap back, at nucleotide base sequences having 2-fold symmetry. Weakly hydrogen-bonded hairpin-like structures initially form. These are lengthened by a series of kinetic steps in which hydrogen-bonds connecting base-pairs within dinucleotide elements in vertical stems are broken and simultaneously reformed in horizontal stems (or vice versa) in a concerted 2-fold symmetric process. This phenomenon is called "cruciform-extrusion." A branch-migraton contains four stems; each stem contains kink (or antikink) boundaries connecting beta-DNA core regions with surrounding B–DNA.

Dislocatons arise at repetitive base sequences, for example, in poly-d(G-A): poly-d(C-T). Again, these structures form as a result of particularly energetic breathing events that cause DNA chains to come apart, and to then "snap back," forming small single-stranded bubble-like protrusions on opposite chains. These protrusions can then move apart, leaving growing regions of beta-DNA in between. The centers of these regions modulate into kink and antikink boundaries, and these, in turn, continue to move apart leaving B-DNA. The net result is the appearance of pairs of dislocatons, each moving in opposite directions along DNA. Movement involves single-chain slippage and is remarkably similar to the mechanism underlying moving crystal lattice dislocations, hence, the term dislocaton.

The formation of these two different kinds of higher energy kink-antikink bound state structures is another example of bifurcations emanating within the centers of premeltons. The underlying source of nonlinearity that determines the path of the bifurcation is the breaking and reforming (after chain-slippage) of hydrogen bonds connecting dinucleotide base-pairs. The decision as to whether branch-migratons or dislocaton pairs form is determined by information coded in the nucleotide base-sequence. This information constitutes the bias.

The combined presence of torsional and writhing strain energies found in negatively superhelical circular

DNA increases the probability that branch-migratons and dislocatons form at the appropriate sequences. These energies are first used to form premeltons. They are next used to form dislocatons or branch-migratons, and to propagate chain-slippage events.

Although B- and A-DNA are right-handed double-helical structures, DNA molecules containing the alternating poly-d(G–C): poly-d(G–C) sequence can, under certain conditions, assume a left-handed double-helical conformation (i.e., in the presence of high salt and/or negative superhelicity). This structure, called Z-DNA, contains the dinucleotide (G–C) as the asymmetric unit, held together by base-pairs. Being a left-handed double-helical structure, Z-DNA contains sugar-phosphate backbone conformations radically different from either B- or A-DNA.

The B to Z transition is proposed to occur as a result of a bifurcation that takes place during the formation of the dislocaton. As before, DNA breathing within the premelton takes place, followed by chains snapping back to form pairs of single-stranded, bubble-like protrusions on opposite chains. As pairs of protrusions move apart, Z-DNA forms within. The molecular boundaries that allow the helix to "swing left," capitalize on the additional flexibility and length provided by the single-stranded DNA regions on opposite DNA chains. B to Z boundaries form simultaneously on both ends in a concerted two-fold symmetric process. This is a direct consequence of the nonlinearity that ties the process together. A prediction of the model is the appearance of single-stranded DNA regions juxtaposed to beta-DNA regions within each B–Z junction, which is supported by experimental evidence.

HENRY M. SOBELL

See also **Biomolecular solitons; DNA solitons; Domain walls; Sine-Gordon equation**

Further Reading

Banerjee, A. & Sobell, H.M. 1983. Presence of nonlinear excitations in DNA structure and their relationship to DNA premelting and to drug intercalation. *Journal of Biomolecular Structure and Dynamics*, 1: 253–262

Bond, P.J., Langridge, R., Jennette, K.W. & Lippard, S.J. 1975. X-ray fiber diffraction evidence for neighbor exclusion binding of a platinum metallointercalation reagent to DNA. *Proceedings of the National Academy of Sciences*, 72: 4825–4829

Crothers, D.M. 1968. Calculation of binding isotherms for heterogeneous polymers. *Biopolymers*, 6: 575–583

Jessee, B., Gargiulo, G., Razvi, F. & Worcel, A. 1982. Analogous cleavage of DNA by micrococcal nuclease and a 1,10-phenanthroline-cuprous complex. *Nucleic Acids Research*, 10: 5823–5834

Lerman, L.S. 1961. Structural considerations in the interaction of DNA and acridines. *Journal of Molecular Biology*, 3, 18–30

Pohl, F.M., Jovin, T.M., Baehr, W., & Holbrook, J.J. 1972. Ethidium bromide as a cooperative effector of a DNA structure. *Proceedings of the National Academy of Sciences*, 69: 3805–3809

Printz, M.P. & von Hippel, P. 1965. Hydrogen exchange studies of DNA structure. *Proceedings of the National Academy of Sciences*, 53: 363–367

Sobell, H.M. 1985. Kink–antikink bound states in DNA Structure. In *Biological Macromolecules and Assemblies*, vol. 2: *Nucleic Acids and Interactive Proteins*, edited by F.A. Jurnack & A. McPherson, New York: Wiley, pp. 172–232

Sobell, H.M. 1985. Actinomycin and DNA transcription. *Proceedings of the National Academy of Sciences*, 82: 5328–5331

Stasiak, A., DiCapua, E. & Koller, T. 1983. Unwinding of duplex DNA in complexes with recA protein. *Cold Spring Harbor Symposia on Quantitative Biology*, 47: 811–820

DNA SOLITONS

In the early 1970s, Alexander Davydov and his colleagues began studies of nonlinear conformational waves in α-helix proteins. Englander et al. (1980) extended these ideas to DNA, interpreting open states (also called bubbles or local unwound regions, evidenced in hydrogen exchange experiments) as solitary waves, which can be viewed as kink solutions of the sine-Gordon (SG) equation. To understand the relation between the SG equation and DNA, we begin with models of the structure and dynamics of DNA.

Tightly packed in the nuclei of living cells, DNA consists of two polynucleotide chains (Figure 1). The chemical structure of each chain consists of periodically repeating phosphate groups and sugar rings (shown by dark ribbons), and irregular nitrous bases (shown in gray). Sequences of four types of bases (adenine, thymine, guanine, and cytosine) occur in DNA, forming the genetic code. These two polynucleotide chains interact weakly through hydrogen bonds and wind around each other to produce the double helix.

The internal structure of DNA is rather flexible. Thermal collisions with molecules of the surrounding solution, action of radiation, and local interactions with proteins, drugs, or other ligands induce several types of internal motions. Among these are

- small-amplitude oscillations of individual atoms;
- rotational, transverse, and longitudinal displacements of the atomic groups (phosphate groups, sugars, and bases);
- motions of the double chain fragments, having several base pairs lengths;

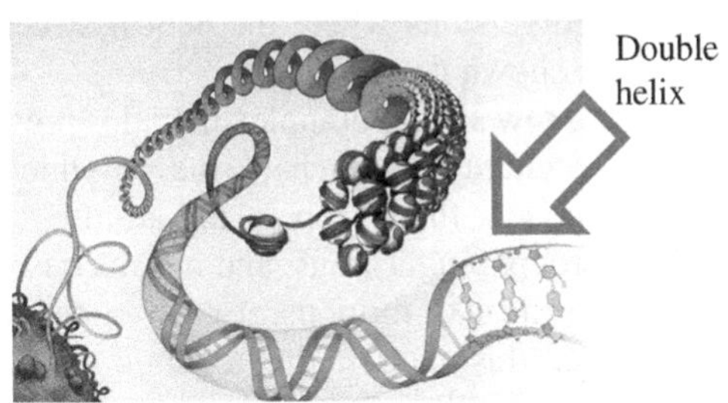

Figure 1. DNA molecule packing into the cell nucleus (courtesy of Nicolas Bouvier).

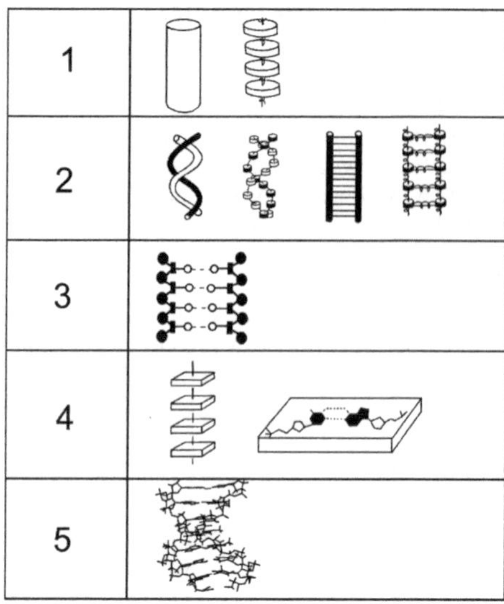

Figure 2. Hierarchy of models of DNA structure and dynamics.

- local unwinding of the double helix; and
- transitions of DNA fragments from one conformational form to another (e.g., from A- to B-form).

Thus, the DNA molecule is a complex dynamical system with many types of internal motions having different energies, velocities, amplitudes, frequencies, and lifetimes (McCammon & Harvey, 1987).

To describe the motions of DNA, different approximate models may be used, as indicated in Figure 2. The simplest model resembles a fragment of elastic thread (or rod), which ignores the internal structure. This model emerges naturally from microphotos where the DNA molecule does indeed resemble a thin elastic thread. Under this formulation only three coupled differential equations are needed: one for longitudinal displacements, another for angular displacements, and a third for transverse displacements. In Figure 2 a discrete analog of the model is placed nearby in the same row.

In the second row of Figure 2, a more complex model recognizes that DNA consists of two polynucleotide chains while ignoring the internal structure. Thus, the model has two weakly interacting elastic threads that wind around each other in a double helix. A mathematical formulation of this model requires six coupled differential equations—three for each of the two helices. The discrete version of the model and two simplified versions where the helical structure is neglected are shown nearby.

In the third row a more complex model is presented, taking into account that each of the polynucleotide chains comprises three types of atomic groups (phosphate groups, sugar rings, and nitrous bases) and representing solid-like motions of each of the atomic groups. Under this model, the required number of equations dramatically increases.

The model shown in the fourth row takes into account the internal motions of the atomic groups while neglecting variations of the base pairs (genetic code). Finally, the most accurate model is shown in the fifth line, which represents positions and motions of all atoms of the DNA molecule. The number of differential equations required to describe the motions in both of these models is inconveniently large at present levels of computer development.

To see the relation between DNA and the SG equation, apply the first model of the second row to study the opening of base pairs. The mathematical formulation of this model comprises six coupled differential equations that are nonlinear because base pair opening is a motion of large amplitude. Following Englander et al. (1980) we assume that rotational motions of bases around the sugar-phosphate chains make the greatest contribution to the process of opening, reducing the number of equations to two. Considering the rotational motions of bases in only one of the chains (accounting for the other chain as some averaged external field) reduces the number of equations from two to one.

To obtain the form of this single equation, an analogy between rotational motions of DNA bases and rotational motions of pendula in a mechanical model of the SG equation is convenient (*See* **Laboratory models of nonlinear waves**). Thus, the pendula of the mechanical model can be considered as analogs of DNA bases, the horizontal spring as an analog of one of the two sugar-phosphate chains, and the gravitational field as an averaged field formed by the second chain interacting with the first chain through hydrogen bonds. The analogy between these two dynamical systems suggests that the equation describing rotational motions bases in DNA should be an SG equation, with parameters defined as follows:

- I is the moment of inertia of bases.
- K is the coefficient of rigidity of the sugar-phosphate chain.
- V_0 is a potential function that represents interactions of bases through hydrogen bonds.
- R_0 is the radius of DNA.
- a is the distance between bases along the chains.

The variable of the equation is Ψ (the angle of rotations of bases around one of the two sugar-phosphate chains), and the equation itself is

$$I\frac{\partial^2}{\partial t^2}\Psi - KR_0^2a^2\frac{\partial^2}{\partial z^2}\Psi + V_0\sin\Psi = 0, \qquad (1)$$

where z indicates distance along the molecule. Kink solutions of this equation describe the opening of base pairs.

Another type of DNA soliton has been obtained by Peyrard & Bishop (1989), who studied melting and denaturation of DNA. Supposing that transverse (rather than rotational) motions of bases make the most contribution to the process, they reduced the number of equations from six to two, and after a simple linear transformation, they obtained two independent equations describing transverse displacements in phase and out of phase. The first of the equations was nonlinear and its solitary wave solutions were interpreted as bubbles appearing in the DNA structure as temperature was increased.

Solitary wave solutions obtained by Englander et al. (1980) and by Peyrard and Bishop (1989) were improved later by investigators who considered other types of internal motions and their interactions, effects of dissipation and inhomogeneity, and interactions with surroundings. These refinements led to additional solitary wave solutions that have been used to interpret experimental data on hydrogen-tritium exchange, resonant microwave absorption, and neutron scattering by DNA. Such studies help us to understand transitions between different DNA forms, long-range effects, regulation of transcription, DNA denaturation, protein synthesis (e.g., insulin production), and carcinogenesis (Scott, 1985; Gaeta et al., 1994; Yakushevich, 1998).

LUDMILA YAKUSHEVICH

See also **Biomolecular solitons; Davydov soliton; DNA premelting; Laboratory models of nonlinear waves; Sine-Gordon equation**

Further Reading

Davydov, A.S. & Kislukha, N.I. 1973. Solitary excitations in one-dimensional chains. *Physica Status Solidi* B, 59: 465–470

Davydov, A.S. & Suprun, A.D. 1974. Configurational changes and optical properties of alpha-helical proteins. *Ukrainian Physical Journal*, 19: 44–50

Englander, S.W., Kallenbach, N.R., Heeger, A.J., Krumhansl, J.A. & Litwin, A. 1980. Nature of the open state in long polynucleotide double helices: possibility of soliton excitations. *Proceedings of the National Academy of Sciences (USA)*, 77: 7222–7226

Gaeta, G., Reiss, C., Peyrard, M. & Dauxois, T. 1994. Simple models of nonlinear DNA dynamics. *Rivista del Nuovo Cimento*, 17: 1–48

McCammon, J.A. & Harvey, S.C. 1987. *Dynamics of Proteins and Nucleic Acids*. Cambridge: Cambridge University Press

Peyrard, M. (editor). 1995. *Nonlinear Excitations in Biomolecules*, Berlin and New York: Springer

Peyrard, M. & Bishop, A.R. 1989. Statistical mechanics of a nonlinear model for DNA denaturation. *Physical Review Letters*, 62: 2755–2758

Scott, A.C. 1969. A nonlinear Klein–Gordon equation. *American Journal of Phyics*, 37: 52–61

Scott, A.C. 1985. Solitons in biological molecules. *Comments on Molecular and Cellular. Biophysics*, 3: 5–57

Yakushevich, L.V. 1998. *Nonlinear Physics of DNA*, Chichester and New York: Wiley

DOMAIN WALLS

In ferromagnetic materials, small regions of correlated magnetic moments formed below the critical temperature are called domains. Domain walls are two-dimensional structures that separate distinct domains of order and form spontaneously when a discrete symmetry (such as time-reversal symmetry in magnets) is broken at a phase transition. With each subdivision of a substance into distinct domains, there is a decrease in the bulk energy because the order parameter value inside each domain minimizes its free energy. However, there is a simultaneous increase in the energy of interaction between differently aligned domains giving rise to an extra surface energy at the boundaries between neighboring domains. Consequently, this competition leads to an average domain size that gives the lowest overall free energy in a material sample, which is quantified below.

The energy of a ferromagnetic domain wall is calculated as arising from the exchange interactions between spins augmented by the anisotropy energy, while the exchange energy for N spins of magnitude S comprising a domain wall varies as

$$E_{\text{exch}} = \frac{\pi^2 J S^2}{N}, \tag{1}$$

where J is the exchange constant and the anisotropy energy is

$$E_{\text{anis}} = K N a^3, \tag{2}$$

where a is the lattice constant and K the anisotropy constant. Minimizing the sum with respect to N yields

$$N = \sqrt{\frac{\pi^2 J S^2}{K a^3}} \tag{3}$$

giving the domain width as $\Delta = Na$.

A Bloch domain wall is a region separating two (magnetic) domains within which magnetization changes gradually by rotating in the plane perpendicular to the line along the direction from one domain to the next. This way the magnetization direction experiences a reversal by 180° without changing its magnitude. The energy associated with a domain wall decreases with the width of the wall, and domain wall thickness is found as a minimization problem involving the anisotropy energy. A Neel domain wall, on the other hand, involves magnetization reversal in the plane perpendicular to the boundary between two domains.

Domains undergo reorganization under the effects of applied fields and can move in space. This occurs especially in the initial phase of remagnetization favoring those domains that are aligned with it and thus setting their boundary in motion to occupy more space, and it is followed by reorientation of the magnetization within each domain that is not aligned with the field.

Domain walls also exist in other systems, including crystals, ferroelectrics, metals, alloys, liquid crystals, and so on. In annealing metals, for example, domain walls appear as the grain boundaries between two sharply different compositions. Generically, the underlying physical quantity is called the order parameter and is specific for a given substance. (For the annealing metal it is a real field, while in superfluid helium it is a complex-valued field.) Over most of the sample, the order parameter has a constant magnitude, but the sign (when it is real) or the phase (when complex) is not fixed and can change from place to place. A real order parameter field may be positive in one region of space and negative in its neighborhood, the continuity of the field implies that it must cross the zero value on a surface between them. This transition region is a domain wall.

In all types of critical systems, domain walls arise due to the competition between the bulk part of the free energy, which in the Landau theory of phase transitions is a quartic polynomial in the order parameter ϕ, and the surface energy term, which is due to inhomogeneities and varies as the square of the order parameter gradient following Ginzburg. Minimizing this type of free energy functional

$$F = \int \left[A_2\phi^2 + A_4\phi^4 + D\left(\frac{d\phi}{dx}\right)^2 \right] dx \quad (4)$$

leads to a stationary nonlinear Klein–Gordon (NLKG) equation for the order parameter as a function of the spatial variable

$$D\phi'' = A\phi + B\phi^3, \quad (5)$$

where $A = A_2$ and $B = 2A_4$. One of its stable solutions is proportional to $\phi_0 \tanh(x/\xi)$ where $\xi = \sqrt{-8D/A}$, which describes a smooth function that interpolates between the two homogeneous phases $\phi = \pm\sqrt{-A/B}$. For magnetization as an order parameter, this solution represents a magnetic domain wall (in one space dimension), and for ferroelectrics, where the order parameter is a polarization vector, this represents a ferroelectric domain wall. For crystals undergoing structural phase transitions, there can also be a kinetic energy term in the free energy functional leading to a standard form of the NLKG equation

$$-m\ddot{\phi} + D\phi'' = A\phi + B\phi^3. \quad (6)$$

This solution is a moving domain wall (or kink)

$$\phi = \phi_0 \tanh[(x - vt)/\xi] \quad (7)$$

as shown in Figure 1 (Krumhansl & Schrieffer, 1975).

Nonlinear traveling solitary waves have also been investigated in ferroelectrics where kinks representing domain walls were shown to carry an electric dipole flip (Benedek et al., 1987). Domain walls in ferroelectrics are typically several unit cells wide, while in ferromagnets their thickness is several hundred or even thousands of unit cells. This difference is due to the exchange interactions between spins that are much stronger than the dipole-dipole interactions in ferroelectric crystals.

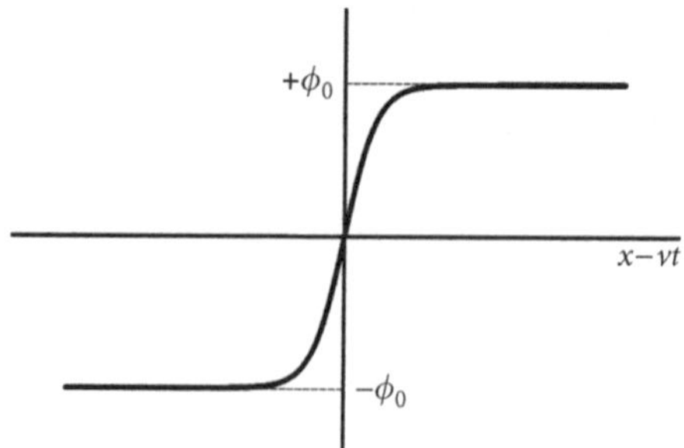

Figure 1. A typical form of a domain wall (or kink).

Geometry	Defect
Sheet-like	domain walls, membranes
Line-like	vortices, strings
Point-like	monopoles, hedgehogs

Table 1. Geometry of space and the corresponding topological defects.

There also exist cylindrical domains in magnets. As can be seen from Table 1, domain walls are examples of topological defects, and as such, they are very common in all broken-symmetry phenomena that take place slowly enough to allow for the generation of defects. Modern particle physics predicts that phase transitions occurred in the early universe following the Big Bang. Of particular interest to cosmology is the production of topological defects, which may be sheet-like, line-like, or point-like concentrations of energy.

Jack A. Tuszyński

See also **Critical phenomena; Ferromagnetism and ferroelectricity; Order parameters; Topological defects**

Further Reading

Anderson, P.W. 1984. *Basic Notions of Condensed Matter Physics*, Menlo Park, CA: Benjamin Cummings

Benedek, G., Bussmann-Holder, A. & Bilz, H. 1987. Nonlinear travelling waves in ferroelectrics. *Physical Review*, B36: 630–638

Kittel, C. 1996. *Introduction to Solid State Physics*, 7th edition, New York: Wiley

Krumhansl, J.A. & Schrieffer, J.R. 1975. Dynamics and statistical mechanics of a one-dimensional model Hamiltonian for structural phase transitions. *Physical Review*, B11: 3535–3545

White, R.H. & Geballe, T. 1979. *Long Range Order in Solids*, New York: Academic Press

DOPPLER SHIFT

The Doppler shift (or Doppler effect) is named after the Austrian scientist Christian Doppler, who in 1842 noticed that the observed frequency of waves from a source depends on how it is moving relative to the observer. In its simplest form, applicable to motions that are slow compared with the wave speed, it is quite easy to visualize and understand. If a source of waves at some frequency f is approaching an observer, the frequency f' that the observer detects is higher than would be the case were the two at rest with respect to each other. From the point of the observer, he is "running into" the waves, while from the point of view of the source, the waves are being "bunched up." The opposite effect occurs when a source is moving away from an observer, and the general situation can be expressed as

$$\frac{f'}{f} = 1 + (v/c)\cos\theta, \quad (1)$$

where v is the relative speed, c the speed of the waves, and θ the angle between the wave propagation and the relative velocity. The $\cos\theta$ factor is simply a statement that the only important component of the relative velocity which can contribute to the shift is that along the direction of wave propagation. As long as v is small compared with the speed of light, this expression works well for most applications, and is quite appropriate to discuss the frequency shift of reflected radar signals used to detect speeding motorists (where c would be the speed of light) or the rise and fall in the pitch of a train whistle or an ambulance siren as it approaches and passes (in which case c would be the speed of sound).

One of the most famous experiments ever done to demonstrate the Doppler shift was done by the Dutch meteorologist Buys-Ballot who put a group of musicians on a train and then had them hold a constant note while racing past him as he stood on the platform.

The Doppler shift often appears more than once in a single application. For example, in police radar, electrons in the metal of an approaching car see a higher frequency radar signal than the police officer emits. They then re-emit radio waves at a higher frequency that is in turn seen at a still higher frequency by the officer being approached.

When speeds involved are very large, the effects of special relativity come into play. While it is possible to outrun a sound wave, the speed of light is the same no matter how the source moves. If we take c to be the speed of light and consider the Doppler shift for light (or any other electromagnetic wave), we have

$$\frac{f'}{f} = \frac{1 + (v/c)\cos\theta}{\sqrt{1 - v^2/c^2}}. \quad (2)$$

A good discussion of special relativity including the Doppler shift is that of Rindler (1991), and a particularly elegant geometric derivation of the formula under very general conditions can be found in Burke (1980). The additional term in the denominator of Equation (2) is due to time dilation whereby a moving clock appears to go more slowly than one at rest. In this case there is a Doppler shift even when $\cos\theta = 0$, which is referred to as the "transverse Doppler effect." One might imagine that it would be very difficult to detect, but in the special case where the sources are in random thermal motion, the leading $(v/c)\cos\theta$ term averages to zero and merely broadens spectral lines of atoms without actually shifting them. This "thermal Doppler effect" due to the transverse Doppler effect was measured by Pound and Rebka in 1960 using the Mössbauer effect. Interestingly, the nontransverse Doppler shift itself played an important role in that measurement.

The Doppler effect has numerous practical applications. In addition to its use already mentioned in measuring automobile velocities, these include Doppler radar for studying weather based on measuring the speeds of raindrops blown by wind (see, e.g., Doviak & Zrnic, 1993) and imaging moving tissues such as the heart in echocardiography, or measuring the speed of blood flow through an artery (see, e.g., Evans & McDicken, 2000).

The Doppler effect is also a valuable tool for pure science and has played a key role in many experiments upon which our present view of the world is based. Edwin Hubble used it to determine the velocities of distant objects in space from shifts in their spectral lines (shifted toward the red end of the spectrum, hence redshift), finding that distant objects seem to be moving away from us with speeds that increase with their distance (Christianson, 1995). This discovery is the basis of modern cosmology and is one of the strongest pieces of evidence that the universe had its origins in a Big Bang. The Doppler shift has also been used in delicate and beautiful experiments that demonstrate the time dilation effects of gravity. Both these gravitational effects on time and the relativistic effect in the transverse Doppler shift must be taken into account in order for the Global Positioning System (GPS) to function properly.

JOHN DAVID SWAIN

See also **Einstein equations; Gravitational waves**

Further Reading

Burke, W.L. 1980. *Spacetime, Geometry, Cosmology*, Mill Valley, CA: University Science Books

Christianson, G.E. 1995. *Edwin Hubble: Mariner of the Nebulae*, New York: Farrar Straus & Giroux

Doviak, R.J. & Zrnic, D.S. 1993. *Doppler Radar and Weather Observations*, New York: Academic Press

Evans, D.H. & McDicken, W.N. 2000. *Doppler Ultrasound: Physics, Instrumental, and Clinical Applications*, 2nd edition, New York: Wiley

Rindler, W. 1991. *Introduction to Special Relativity*, 2nd edition, Oxford and New York: Oxford University Press

DOUBLE-WELL POTENTIAL

See **Equilibrium**

DRESSING METHOD

The dressing method is a technique of constructing and solving nonlinear partial differential equations of integrable models. It is based on dressing transformations that are symmetries of nonlinear partial differential equations and act on the Lax operators as gauge transformations on the connection (Zakharov & Shabat, 1974). Accordingly, the form of the linear spectral problem and the zero-curvature conditions are preserved.

The basic concept of the dressing method (Zakharov & Shabat, 1974) is that starting from known solutions of the underlying linear problem, one obtains new solutions of the transformed ("dressed") linear problem. Suppose Ψ_0 is a solution of the following linear problems:

$$\left(\frac{\partial}{\partial t_i} - E_i^{(0)}\right)\Psi_0\,(t_1, \ldots, t_n, \lambda) = 0, \quad i = 1, \ldots, n, \tag{1}$$

where $t_1, t_2, \ldots, t_n$ are independent flow variables and $E_i^{(0)}$ are complex $m \times m$ mutually commuting matrices that depend on the spectral parameter λ. The matrix Ψ_0 is referred to as a "bare" wave function and commuting operators $L_i^{(0)} = \partial/\partial t_i - E_i^{(0)}$ as "bare" Lax operators.

For the soliton equations in $1+1$ (one spatial and one time) dimensions, the dressing transformations are generated by the dressing matrices $\Theta_\pm$ which can be defined by the Riemann–Hilbert factorization problem (Zakharov & Shabat, 1974; Faddeev & Takhtajan, 1987). Given a Lie group of functions $g(\lambda)$ on the contour $\mathcal{C}$ in the complex plane, one finds a new (dressed) wave function through the factorization problem:

$$\Psi_0(t_1, \ldots, t_n, \lambda)g(\lambda)\Psi_0^{-1}(t_1, \ldots, t_n, \lambda) = \Theta_-^{-1}(t_1, \ldots, t_n, \lambda)\Theta_+(t_1, \ldots, t_n, \lambda), \tag{2}$$

where $\Theta_-^{-1} = (\Psi_0 g \Psi_0^{-1})_-$ and $\Theta_+ = (\Psi_0 g \Psi_0^{-1})_+$ have analytic continuation inside or outside the contour $\mathcal{C}$, respectively. The dressing transformation defines a new wave function $\Psi^{(g)} = \Theta_-\Psi_0 g = \Theta_+\Psi_0$, which is a solution of new linear problems:

$$\left(\frac{\partial}{\partial t_i} - E_i\right)\Psi^{(g)}(t_1, \ldots, t_n, \lambda) = 0, \tag{3}$$

where

$$E_i = \frac{\partial\Theta_-}{\partial t_i}\Theta_-^{-1} + \Theta_- E_i^{(0)}\Theta_-^{-1} = \frac{\partial\Psi^{(g)}}{\partial t_i}(\Psi^{(g)})^{-1} \tag{4}$$

satisfy the compatibility conditions also known as zero-curvature conditions:

$$\frac{\partial E_j}{\partial t_i} - \frac{\partial E_i}{\partial t_j} - \left[E_i\,,\,E_j\right] = 0. \tag{5}$$

Comparing the expression for E_i from Equation (4) with $E_i^{(0)} = (\partial\Psi_0/\partial t_i)\Psi_0^{-1}$, following from Equation (1), one sees that the dressing transformations preserve the form of the Lax connections $L_i = \partial/\partial t_i - E_i$. A related approach also exists, which yields the dressing matrix in terms of the Fredholm type of integral operator entering the Gel'fand–Levitan–Marchenko equation (Zakharov & Shabat, 1974). Both dressing approaches are equivalent.

By considering two successive dressing transformations associated with two group elements g_1 and g_2, one naturally arrives at the concept of the group of the dressing transformations with $\Psi^{(g_1 g_2)} = (\Psi^{(g_1)})^{(g_2)}$. The general theory of dressing transformations and its group was developed by Semenov–Tian–Shansky (1985), who also introduced a Poisson bracket covariant under the dressing group action on the phase space of functions E_i. With such a bracket, the group of the dressing transformations induces on a phase space a Lie–Poisson action and turns out to be a symmetry of the phase space. Furthermore, it was observed that the group of the dressing transformations appears as a semi-classical limit of the quantum group symmetry of the two-dimensional integrable quantum field theories. Hence the group of the dressing transformations appears to be a classical precursor of the quantum group structure of an integrable field system in two dimensions (Babelon & Bernard, 1992). In many integrable models, the N-soliton solutions can be thought of as elements of the dressing group orbit of the vacuum state. Accordingly, successive dressing transformations can be used to build N-soliton solutions from the vacuum solution. In Babelon & Bernard (1993), the authors presented the construction of N-soliton solutions by dressing transformations in the sine-Gordon model. The dressing group also admits an elegant interpretation within the Kyoto school (Date et al., 1983) approach to the integrable models appearing in this context as a transformation group of the τ-function.

For equations in $2+1$ (two spatial and one time), dimensions, one applies the dressing technique based on a nonlocal Riemann–Hilbert problem for Kadomtsev–Petviashvili I and Davy–Stewartson I equations as well as the N-wave equations. The nonlocal (D-bar) $\bar{\partial}$ problem is required for the Kadomtsev-Petviashvili II and Davey–Stewartson II equations (Zakharov & Manakov, 1985).

The nonlocal (D-bar) $\bar{\partial}$-problem gives rise to the $\bar{\partial}$-dressing method in which an $N \times N$ quasi-analytical matrix function of $\lambda, \bar{\lambda}$ $\chi(\lambda, \bar{\lambda}, x)$, $x \in \mathbb{C}^n$ satisfies the following nonlocal $\bar{\partial}$-problem:

$$\frac{\partial\chi(\lambda, \bar{\lambda})}{\bar{\lambda}} = \hat{R}\chi = \iint_C d\lambda' d\bar{\lambda}' \chi(\lambda', \bar{\lambda}') R(\lambda', \bar{\lambda}', \lambda, \bar{\lambda}), \tag{6}$$

with R being a kernel of a linear integral operator $\hat{R}$. The set of commuting differential operators defined in terms of commuting rational matrix functions I_i:

$$D_i\chi = \frac{\partial\chi}{\partial x_i} + \chi I_i(\lambda), \qquad i = 1, \ldots, n, \tag{7}$$

defines an integrable system. Let the integral linear operator $\hat{R}$ commute with all differential operators D_i, $[D_i, \hat{R}]=0$. Then, the choice $I_1=\lambda$, $I_2=\lambda^2$, $I_3=\lambda^3$ leads for $N=1$ to the KP II equation: $(u_t+u_{xxx}++6uu_x)_x+3u_{yy}=0$ with $u=2\partial\chi_1/\partial x_1$ where χ_1 is the term in an asymptotic expansion ($\lambda\to\infty$) of $\chi=1+\chi_1/\lambda+\chi_2/\lambda^2+\ldots$.

HENRIK ARATYN

See also **Bäcklund transformations; Darboux transformation; Hirota's method; Inverse scattering method or transform; Multidimensional solitons; N-soliton formulas; Riemann–Hilbert problem**

Further Reading

Babelon, O. & Bernard, D. 1992. Dressing symmetries. *Communications in Mathematical Physics*, 149: 279–306

Babelon, O. & Bernard, D. 1993. Affine solitons: a relation between tau functions, dressing and Bäcklund transformations. *International Journal of Modern Physics A*, 8: 507–543

Date, E., Jimbo, M., Kashiwara, M. & Miwa, T. 1983. Transformation groups for soliton equations. In *Nonlinear Integrable Systems-Classical and Quantum Theory*, edited by M. Jimbo & T. Miwa. Singapore: World Scientific

Faddeev, L.D. & Takhtajan, L.A. 1987. *Hamiltonian Methods in the Theory of Solitons*, Berlin and Heidelberg: Springer

Semenov–Tian–Shansky, M.A. 1985. Dressing transformations and Poisson group actions, *Publications of Research Institute of Mathematical Sciences, Kyoto University*, 21: 1237–1260

Zakharov, V.E. & Manakov, S.V. 1985. Construction of multidimensional integrable non-linear systems and their solutions, *Functional Analysis and Applications*, 19 (N2): 11–25

Zakharov, V.E. & Shabat, A.B. 1974. A scheme for integrating the nonlinear equations of mathematical physics by the method of the inverse scattering problem. I. *Functional Analysis and Applications*, 8: 226–235

Zakharov, V.E. & Shabat, A.B. 1979. Integration of the nonlinear equations of mathematical physics by the method of the inverse scattering problem. II. *Functional Analysis and Applications*, 13: 166–174

DRIPPING FAUCET

A dripping faucet may easily be seen in everyday life. Its rhythm is sometimes regular but sometimes not, which sensitively depends on the flow of water. If the rhythm is irregular, one might blame it on noise due to unseen influences such as small air vibrations. However, it is nowadays well known that the irregularity arises from deterministic chaos instead of stochastic noise. In fact, this system is a good example showing that chaos is not only a mathematical product but also a phenomenon ubiquitous in the real world.

Chaotic dripping was originally suggested by Otto Rössler (Rössler, 1977) and the first experimental study was performed by Robert Shaw and his colleagues (Shaw, 1984; Martien et al., 1985). They measured the time interval (T_n) between the nth drip and its successor, and obtained a time series $\{T_1, T_2, \cdots\}$. To detect possible determinism from a nonperiodic time series, they made a return map (plot of (T_n, T_{n+1}) for each n). A return map for nonperiodic dripping is shown in the inset of Figure 1(a). If irregular numbers $T_1, T_2, \cdots$ were generated stochastically by throwing the dice, the plots would look like a set of random points with no particular structure. The observed map actually exhibited a clear structure, which implies a deterministic rule existing in the seemingly random outcomes.

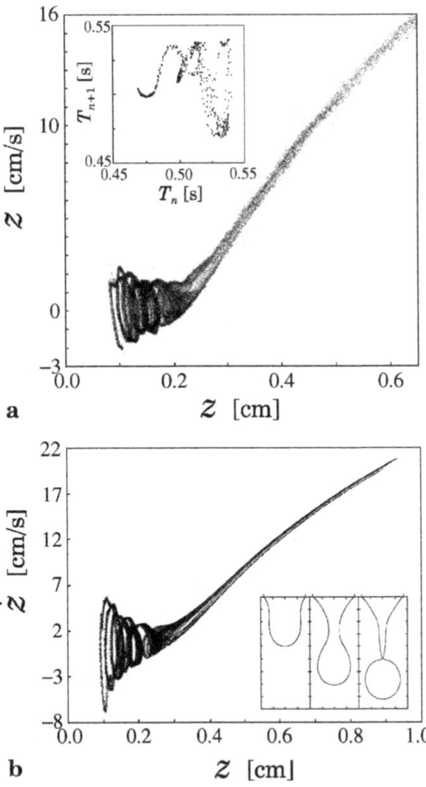

Figure 1. (a) An experimental strange attractor reconstructed from the observation of the oscillation, deformation, and breakup of drops hanging from a nozzle (7 mm inner diameter and 10 mm outer diameter). The projection of the orbit onto the plane $(z, \dot{z})$ is presented, where z is the center of mass and $\dot{z}$ is its velocity. The flow rate $Q=0.24$ g/s ($\sim$ 2 drips/s on average). Inset of (a): return map of the dripping time interval (T_n). (b) Fluid dynamic simulation for 7 mm nozzle diameter. $Q=0.32$ g/s. Inset of (b): drop deformation.

As suggested by Rössler, the deterministic randomness, chaos is expected if two oscillating variables couple: (i) damped oscillation of the drop position (z, the center of mass) due to the surface tension of the water and (ii) relaxation oscillation of the mass of the drop (m) due to the filling-and-discharging process. A minimal model including these variables is the so-called mass-spring model described as

$$\frac{\mathrm{d}}{\mathrm{d}t}\left(m\frac{\mathrm{d}z}{\mathrm{d}t}\right)=-kz-\gamma\frac{\mathrm{d}z}{\mathrm{d}t}+mg, \qquad (1)$$

where g is the acceleration of gravity, k is the spring constant, and γ is the damping parameter. The mass increases at a constant flow rate Q as

$$\frac{\mathrm{d}m}{\mathrm{d}t} = Q. \tag{2}$$

The model assumes that when the position z reaches a critical point, a part of the total mass (Δm) breaks away. Shaw used this model to explain the dripping dynamics as low-dimensional chaos.

Since then, dripping faucets have attracted many physicists, and a wide range of nonlinear behavior has been reported, such as strange attractor, period-doubling bifurcation, Hopf bifurcation, intermittency, hysteresis, crisis, and satellite drop formation. On the other hand, theoretical studies mainly rested on the mass-spring model, and any direct link between this simple model and the physics of drop formation was not known. The basic dynamics inherent in the complex behavior of the dripping faucet system was revealed quite recently, owing to detailed analyses of (i) experiments for a wide range of flow rates (Katsuyama & Nagata, 1999), (ii) fluid dynamic simulations using a new algorithm (Fuchikami et al., 1999), and (iii) the improved mass-spring model based on the fluid dynamic simulations (Kiyono & Fuchikami, 1999).

The fluid dynamic simulations clearly visualized how drops are formed and pinched off repeatedly (see figures in color plate section):

- The water under the faucet, increasing at a constant flow rate, forms a drop, which bulges until m reaches $\sim m_{\mathrm{crit}}$ (the maximum mass for the static stable state) and the surface tension is overwhelmed by the gravitational force.
- Then its sides begin to shrink, forming a rapidly narrowing neck (necking process) and the drop is soon pinched off (breakup). In the necking process, the drop undergoes almost free-fall.
- Because the water is stretched downward at the breakup moment, the surface tension works as a restoring force just after the breakup, which causes oscillation of the water. Thus the position z, the center of mass of the water, moves downward with up-and-down oscillation.
- After m reaches m_{crit} again, the necking followed by the breakup is repeated.
- The phase of oscillation at the onset of the necking process affects the breakup moment, the mass at that moment, and so the remnant mass ($m_{\mathrm{r}} = m - \Delta m$).
- If the phase changes periodically, the motion is periodic (for example, in the period-two motion T_1, T_2, T_1, T_2, $\cdots$, the phase changes as θ_1, θ_2, θ_1, θ_2, $\cdots$), while the phase is random in chaotic motion.

The essential information obtained from the fluid dynamic simulations is as follows:

(i) The state point of the water is well described in the limited phase space $(z, \mathrm{d}z/\mathrm{d}t, m)$, even if the water, an infinite-dimensional system, deforms its shape in a complex way. (Remember z denotes the center of mass.)

(ii) The instability of the shape of water induces the instability of the chaotic orbit. In other words, stretching of the chaotic attractor mainly occurs in the beginning of the necking process.

(iii) After the breakup, the renewed (i.e., remnant) mass realizes various values, while the renewed position and velocity are confined in a small region, well approximated as constants. Thus the attractor is compressed and becomes low dimensional.

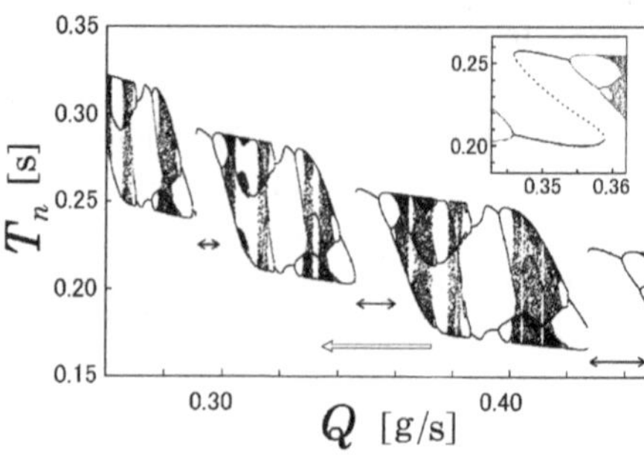

Figure 2. A section of the bifurcation diagram of the improved mass-spring model obtained by decreasing the control parameter Q, the flow rate. Hysteresis is observed in certain ranges of Q (indicated with $\leftrightarrow$). Inset: a hysteresis curve.

These features have also been confirmed experimentally. Figure 1 presents a recent experiment (a) and the corresponding fluid dynamic simulation result (b). The low-dimensional (almost one-dimensional) pattern of the return map (inset of (a)) suggests that the system can be described by a low-dimensional dynamical system. The experimental trajectory in the phase space $(z, dz/dt, m)$ was reconstructed from the continuous change of the drop shape observed. The spiral orbit indicates that the drop oscillates several times and then makes free-fall in the necking process before breakup. The drop shape obtained from the simulation (inset of (b)) is very close to the experimental result.

The observation of the drop formation process made it possible to improve the traditional mass-spring model by taking account of several points ignored so far, which include

- the mass dependence of the spring constant: $k = k(m)$;
- the necking process by setting $k = 0$ for $m > m_{\mathrm{crit}}$; and
- the mass dependence of the remnant mass: $m_{\mathrm{r}} = m_{\mathrm{r}}(m)$, where m is the mass just before the breakup.

Bifurcation diagrams (plot of T_n versus Q) obtained from the improved mass-spring model reproduce the global structure of experimental bifurcation diagrams for a wide range of flow rates, Q. As seen in Figure 2, one distinct feature is a repetition of similar "units." The neighboring units are very similar but gradually become complex as Q is increased. Each unit is characterized by an integer that is the number of oscillations of each drop before breakup. This number decreases with increasing Q because the drop mass reaches the critical value sooner. In Figure 2, for example, there are three units and the corresponding numbers are 6, 5, 4 from left to right. Units in a range of relatively large Q include various types of bifurcations, such as period-doubling cascade to chaos, intermittency, and hysteresis (inset of Figure 2), while units in a range of sufficiently small Q include just period one motion. The bifurcation diagrams for small faucet diameters also exhibit a relatively simple structure.

The improved mass-spring model systematically explains the characteristic complexities of low-dimensional chaos (Ott, 1993) inherent in dripping dynamics. However, the model can be applied only when Q is small enough so that drops are clearly separated from each other. For larger flow rates, experimental results are so complex that approximations used in the fluid dynamic simulations do not work. New theoretical approaches are required, especially to interpret Hopf bifurcation and statistical features of satellite drops.

NOBUKO FUCHIKAMI AND KEN KIYONO

See also **Bifurcations; Chaotic dynamics**

Further Reading

Fuchikami, N., Ishioka, S. & Kiyono, K. 1999. Simulation of a dripping faucet. *Journal of the Physical Society of Japan*, 68: 1185–1196

Katsuyama, T. & Nagata, K. 1999. Behavior of the dripping faucet over a wide range of the flow rate. *Journal of the Physical Society of Japan*, 68: 396–400

Kiyono, K. & Fuchikami, N. 1999. Dripping faucet dynamics by an improved mass-spring model. *Journal of the Physical Society of Japan*, 68: 3259–3270

Kiyono, K., Katsuyama, T., Masunaga, T. & Fuchikami, N. 2003. Picture of the low-dimensional structure in chaotic dripping faucets. *Physics Letters* A, 320: 47–52

Martien, P., Pope, S.C., Scott, P.L. & Shaw, R.S. 1985. The chaotic behavior of the leaky faucet. *Physics Letters* A, 110: 399–404

Ott, E. 1993. *Chaos in Dynamical Systems*, Cambridge and New York: Cambridge University Press

Rössler, O.E. 1977. Chemical turbulence. In *Synergetics: Proceedings of the International Workshop on Synergetics at Schloss Elmau, Bavaria*, edited by Hermann Haken, Berlin: Springer, 174–183

Shaw, R. 1984. *Dripping Faucet as a Model Chaotic System*, Santa Cruz: Aerial Press

DRUDE MODEL

In 1900, Paul Drude developed his theory of metallic conduction of electricity, following the discovery of the electron by Joseph John Thomson in 1897. Applying the kinetic theory of gases to a metal, considered as a gas of electrons, he made the following assumptions:

(i) Each electron moves according to Newton's law of motion in the presence of external fields until it collides with other electrons or ions. Between collisions, interactions with both the other electrons and with the ions are neglected.
(ii) Collisions are instantaneous events that abruptly alter the velocity of an electron. The probability per unit time that an electron experiences a collision is given by $1/\tau$, where τ is called the collision time, mean free time, or relaxation time.
(iii) Electrons are assumed to achieve thermal equilibrium with their surroundings only through collisions. After each collision, the electron emerges with a randomly directed velocity whose magnitude is given by the temperature.

Note that in contrast to a classical gas of neutral molecules, the electrons move against a background of positively charged immobile ions.

The Drude model considers an average electron representative for the whole ensemble. Newton's equation of motion is

$$m\frac{\mathrm{d}^2}{\mathrm{d}t^2}\boldsymbol{r} + \frac{m}{\tau}\frac{\mathrm{d}}{\mathrm{d}t}\boldsymbol{r} = -e\boldsymbol{E}, \tag{1}$$

where $\boldsymbol{r}$ is the spatial drift coordinate (excluding the random thermal motion), m is the effective mass of the electron, $e > 0$ is the electron charge, and $\boldsymbol{E}$ is the applied electric field. The second term is a friction term arising from collisions. This equation can be rewritten in terms of the electron momentum $\boldsymbol{p} = m\boldsymbol{v} = m\mathrm{d}\boldsymbol{r}/\mathrm{d}t$, where $\boldsymbol{v}$ is the drift velocity,

$$\frac{\mathrm{d}}{\mathrm{d}t}\boldsymbol{p} + \frac{\boldsymbol{p}}{\tau} = -e\boldsymbol{E}. \tag{2}$$

In the overdamped case the momentum induced by the electric field is given by

$$\boldsymbol{p} = -e\tau\boldsymbol{E}. \tag{3}$$

The current density is obtained by multiplying the drift velocity of the mean electron by the electron charge and by the density per unit volume n of all electrons

$$\boldsymbol{j} = -en\boldsymbol{v} = en\frac{e}{m}\tau\boldsymbol{E} = en\mu\boldsymbol{E}, \tag{4}$$

where the drift mobility $\mu = (e/m)\tau$ has been introduced. This is Ohm's law, with constant conductivity $\sigma = en\mu$.

The classical Drude model neglects electron-electron interactions, heating of the electrons, non-equilibrium dynamics, and quantum transport effects, and is thus restricted to low fields, moderate carrier densities and temperature ranges. If any of these conditions are violated, nonlinear and non-Ohmic conduction arises. Such effects are abundant in metals and semiconductors and may lead to instabilities and bifurcations of self-sustained oscillations or self-organized spatiotemporal patterns (Shaw et al., 1992, Schöll, 2001).

In the simplest case, a local, instantaneous conductivity $\sigma = en\mu$ still exists, but n or μ depends upon the field E (generation-recombination or drift instability, respectively), leading to a nonlinear or even nonmonotonic current density-field relation $j = en(E)\mu(E)E$, possibly with a range of negative differential conductivity (NDC), where $\mathrm{d}j/\mathrm{d}E = en\mu + e(\mathrm{d}n/\mathrm{d}E)\mu(E)E + en(\mathrm{d}\mu/\mathrm{d}E)E < 0$.

In an extension of the simple Drude picture, those nonlinearities may be due to a dependence of the momentum relaxation time upon the field (drift instability) or upon the electron temperature that, in turn, depends upon E (electron overheating instability, as a result of changes in the dissipation of energy and momentum). They may also be due to dependence of the carrier density upon field (generation-recombination instability, as in avalanche breakdown). More complex situations arise if the (semi)conductor consists of several layers of different materials, and intrinsic inhomogeneities render the notion of a local conductivity inappropriate. In particular, this applies to low-dimensional nanoscale structures (Ferry & Goodnick, 1997) or mesoscopic conductors (Datta, 1995), where the characteristic length scales may be such that the device dimension becomes smaller than the mean free path, the quantum mechanical phase relaxation length, and the de Broglie wavelength of the electron. Then various nonlinear transport regimes occur, where ballistic or non-equilibrium or coherent quantum effects dominate.

ECKEHARD SCHÖLL

See also **Avalanche breakdown; Diodes; Semiconductor oscillators**

Further Reading

Ashcroft, N.W. & Mermin, N.D. 1976. *Solid State Physics*, Philadelphia: Saunders College

Datta, S. 1995. *Electronic Transport in Mesoscopic Systems*, Cambridge and New York: Cambridge University Press

Drude, P. 1900. Zur Elektronentheorie der Metalle. *Annalen der Physik*, 1: 566–613

Ferry, D.K. & Goodnick, S.M. 1997. *Transport in Nanostructures*, Cambridge and New York: Cambridge University Press

Schöll, E. 2001. *Nonlinear Spatio-temporal Dynamics and Chaos in Semiconductors*, Cambridge: Cambridge University Press

Schöll, E. (editor). 1998. *Theory of Transport Properties of Semiconductor Nanostructures*, London: Chapman & Hall

Shaw, M.P., Mitin, V.V., Schöll, E. & Grubin, H.L. 1992. *The Physics of Instabilities in Solid State Electron Devices*, New York: Plenum Press

DUFFING EQUATION

Serious studies of forced nonlinear oscillators appeared early in the 20th century when Georg Duffing (1918) examined mechanical systems with nonlinear restoring forces and Balthasar van der Pol studied electrical systems with nonlinear damping. Subsequently, any equation of the form

$$a\ddot{x} + b\dot{x} + f(x) = F \sin \omega t \qquad (1)$$

was called Duffing's equation, the nonlinearity often being polynomial, usually cubic (Hayashi, 1964). Here, $\dot{x}$ represents the time derivative $\mathrm{d}x/\mathrm{d}t$.

Linear resonance has $f(x) = cx$ $(c > 0)$, and Duffing's extension models many mechanical and electrical phenomena. With symmetry, f is odd, giving the first nonlinear approximation

$$f(x) = cx + dx^3. \qquad (2)$$

If $d > 0$, the system is *hardening* and the resonant peak tilts to the right. A *softening* system $(d < 0)$ has a peak tilted to the left (as in Figure 2). Duffing's method of successive approximation and a variety of averaging and perturbation techniques can estimate these tilts for conditions of weak nonlinearity.

Twin-Well Duffing Oscillator

Taking Equations (1) and (2) with $c < 0$, $d > 0$ gives the "twin-well" Duffing oscillator governed by the potential energy $V(x) = \frac{1}{2}cx^2 + \frac{1}{4}dx^4$. Trajectories lie in and across the two symmetric potential wells separated by the hill-top at $x = 0$. This is a useful archetypal model for studies of chaos (Guckenheimer & Holmes, 1983). The undamped, unforced system $(b = F = 0)$ has two symmetrically placed orbits that (in infinite time) leave and return to the phase-space saddle at $x = 0$, $\dot{x} = 0$: we say they are "homoclinic to the saddle." For small b and F the corresponding invariant manifolds exhibit a homoclinic tangency on an arc in the (b, F) control space. Beyond this global bifurcation a homoclinic tangle generates "horseshoe dynamics" giving chaos and fractal basin boundaries (see Figure 3). The (b, F) arc can be estimated by Mel'nikov perturbation analysis.

Ueda's Chaos

For hardening with $c = 0$, Ueda (1980, 1992) mapped intricate regimes of subharmonics and chaos. Ueda's equation is

$$\ddot{x} + k\dot{x} + x^3 = B \cos t. \qquad (3)$$

With $k = 0.05$, $B = 7.5$, all solutions settle onto a unique chaotic attractor, irrespective of $x(0), \dot{x}(0)$: its basin of attraction is the whole starting plane.

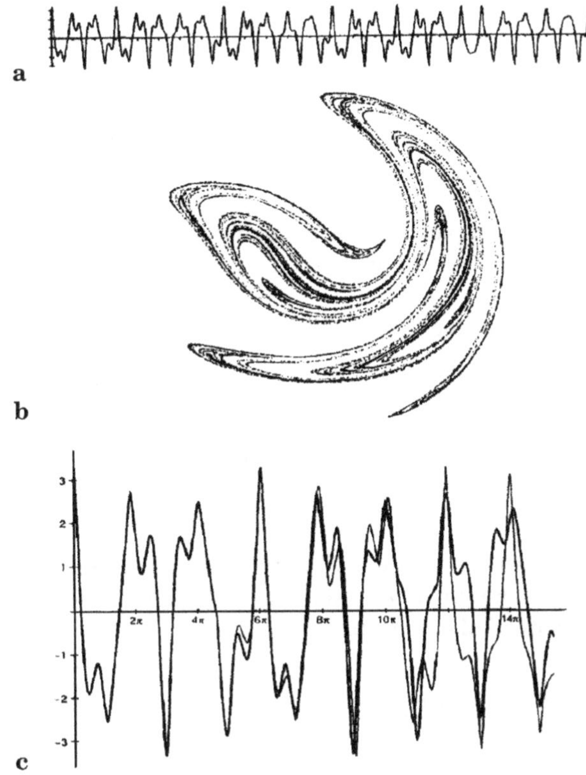

Figure 1. Waveform (a) shows a plot of $x(t)$ at the end of a long computer time integration during which any start-up transient is effectively dissipated, while the large fractal dot diagram (b) shows this chaotic attractor in a Poincaré section sampling $x(t)$ and $\dot{x}(t)$ stroboscopically at the period of the forcing. Two waveforms in (c) show the exponential divergence from two adjacent starts on the attractor where $(x(0), \dot{x}(0))$ are $(3, 4)$ and $(3.01, 4.01)$, respectively. Reproduced from Thompson & Stewart (2002) with the permission of John Wiley.

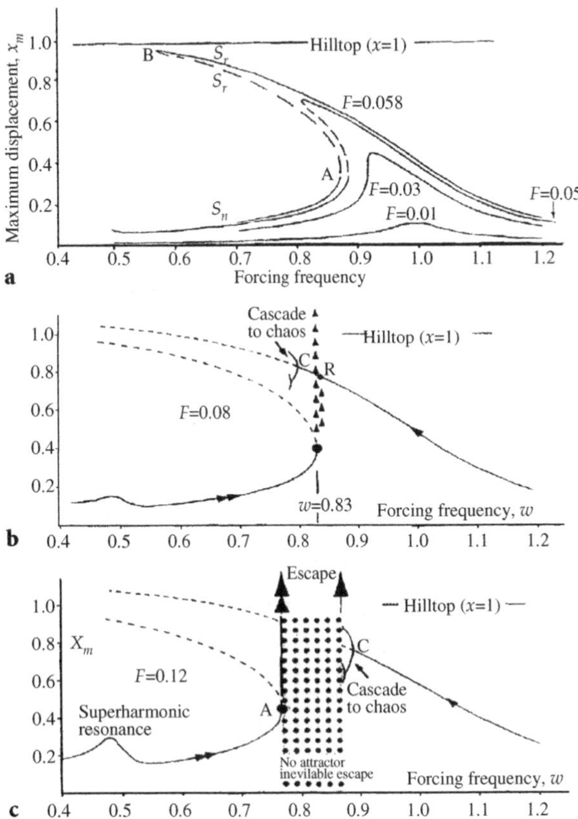

Figure 2. Three resonance response diagrams for Equation (4), where the maximum of the steady-state response, x_m, is plotted against the forcing frequency, ω, for fixed values of F (all with $\beta = 0.1$). At higher F, in (c), there is a regime with no attractor, implying inevitable escape from either fold A or the chaotic crisis ending the cascade from C. Reproduced from Thompson & Stewart (2002) with the permission of John Wiley.

The steady chaotic response is shown in Figure 1. In each cycle, sheets of trajectories are folded and compressed, as in making flaky pastry. This mixing produces divergence (Figure 1(c)) that quickly makes adjacent motions totally uncorrelated, although both remain on the fractal attractor. This divergence serves to identify chaos: it is quantified by Lyapunov exponent techniques (Guckenheimer & Holmes, 1983).

Escape from a Potential Well

Asymmetric models were used by Hermann Helmholtz to model vibrations in the human ear. An archetypal example introduced by Thompson (1989) is

$$\ddot{x} + \beta\dot{x} + x - x^2 = F \sin \omega t. \qquad (4)$$

This Helmholtz–Thompson equation governs escape from the well,

$$V(x) = \tfrac{1}{2}x^2 - \tfrac{1}{3}x^3. \qquad (5)$$

Such escape is a universal problem in science, from activation energies of molecular dynamics to the gravitational collapse of massive stars. Failures in electrical systems are triggered when the underlying system escapes from a well: if power generators slip out of synchronization, an entire city can be blacked out. Naval research (Thompson, 1997) is directed toward the capsizing of vessels under sinusoidal forcing by ocean waves.

Equation (4) is used by Thompson & Stewart (2002) to illustrate a variety of complex phenomena including chaos, fractal boundaries, and indeterminacies. These arise in a wide class of systems involving nonlinear damping, different well shapes, different direct and parametric forcing, and hardening characteristics. They are not detected by perturbation and averaging methods.

Chaos in Nonlinear Resonance

Three response diagrams for Equation (4) are shown in Figure 2. The quadratic form of the restoring force is technically *softening* for $x > 0$ and *hardening* for $x < 0$, but its overall effect is *softening*. Thus in 2(a), the peaks tilt to the left, as for a cubic Duffing oscillator with

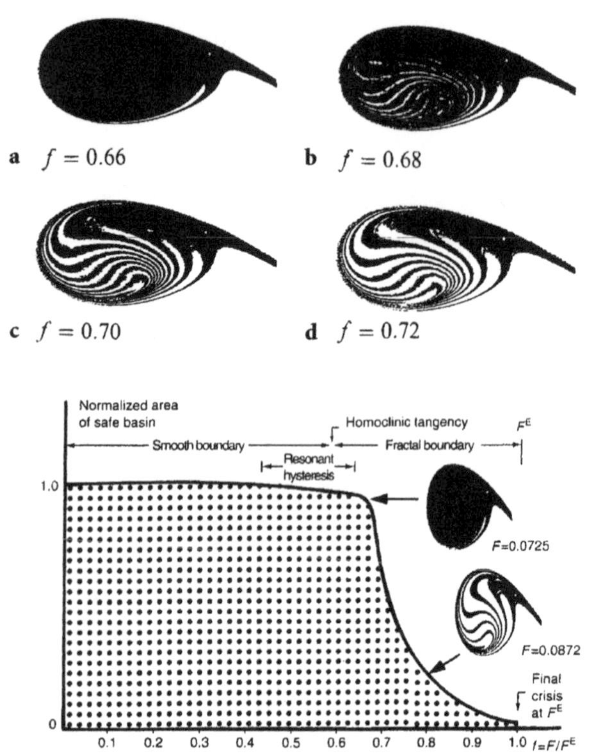

Figure 3. In the space of the coordinates $x(0)$, $\dot{x}(0)$, the four safe basins of attraction, (a)–(d), comprise all those starts that do not give escape: we see how they vary with f, the basin being dramatically eroded by fractal fingers at $f \approx 0.68$. This basin erosion is quantified in (e) by plotting the area of the safe basin against f to give an integrity diagram. Reproduced from Thompson & Stewart (2002) with the permission of John Wiley.

$d < 0$. At the low value of $F = 0.01$, where the small response is almost linear, the untilted peak lies over $\omega = 1$ (the natural frequency). At $F = 0.056$ we see the hysteresis response of a softening oscillator, with a jump to resonance at fold A as ω increases, and a jump from resonance at B as ω decreases. Between these *cyclic folds*, there are three steady periodic solutions with frequency ω: nonresonant attractor, S_n; resonant attractor, S_r; unstable saddle, D_r.

In 2(b), the resonant branch loses stability in a period-doubling cascade to chaos, and the jump at A is indeterminate.

Fractal Boundaries, Indeterminate Bifurcations

Transient motions arise from different starting values of $x(0)$, $\dot{x}(0)$, and Figure 3 shows how the safe basin of attraction varies with $f = F/F^E$. Here F^E is the steady-state escape magnitude, at which a slowly evolving system would jump out of the well. The observed fractal structure is generated at a homoclinic tangency. The sudden drop in the integrity diagram at the "Dover cliff" can be used as a design criterion. An engineering system should not be operated at values of $f > 0.68$, even though there are still stable motions within the well up to $f = 1$ (Thompson & Stewart, 2002).

Fractal basin boundaries and their chaotic transients generate great complexities in all forms of Duffing's equation once the response is significantly nonlinear: they are a topic of active research (Stewart et al., 1995). An example is the indeterminate jump from fold A in Figure 2(b). Locally, this is a normal saddle-node fold, but it is located precisely on a fractal boundary. The outcome from such a "tangled saddle-node" is unpredictable (Thompson & Soliman, 1991). Depending sensitively on how A is approached, the jump may settle onto attractor R or escape out of the well; it may also settle onto a co-existing subharmonic motion.

MICHAEL THOMPSON

See also **Attractors; Chaotic dynamics; Damped-driven anharmonic oscillator; Mel'nikov method; Van der Pol equation**

Further Reading

Duffing, G. 1918. *Erzwungene Schwingungen bei Veränderlicher Eigenfrequenz*, Braunschweig: Vieweg

Guckenheimer, J. & Holmes, P. 1983. *Nonlinear Oscillations, Dynamical Systems and Bifurcations of Vector Fields*, New York: Springer

Hayashi, C. 1964. *Nonlinear Oscillations in Physical Systems*, Princeton, NJ: Princeton University Press

Stewart, H.B., Thompson, J.M.T., Ueda, Y. & Lansbury, A.N. 1995. Optimal escape from potential wells: patterns of regular and chaotic bifurcation. *Physica* D, 85: 259–295

Thompson, J.M.T. 1989. Chaotic phenomena triggering the escape from a potential well. *Proceedings of the Royal Society of London*, A, 421: 195–225

Thompson, J.M.T. 1997. Designing against capsize in beam seas: recent advances and new insights. *Applied Mechanics Reviews*, 50: 307–325

Thompson, J.M.T. & Soliman, M.S. 1991. Indeterminate jumps to resonance from a tangled saddle-node bifurcation. *Proceedings of the Royal Society of London*, A, 432: 101–111

Thompson, J.M.T. & Stewart, H.B. 2002. *Nonlinear Dynamics and Chaos*, 2nd edition, Chichester and New York: Wiley

Ueda, Y. 1980. Steady motions exhibited by Duffing's equation: a picture book of regular and chaotic motions. In *New Approaches to Nonlinear Problems in Dynamics*, edited by P.J. Holmes, Philadelphia: SIAM, pp. 311–322

Ueda, Y. 1992. *The Road to Chaos*, Santa Cruz: Aerial Press

DUNE FORMATION

Dunes are sand formations, found on land, that have heights ranging from 1 to 500 m and have been shaped by the wind. These topographical structures are found typically where large masses of sand have accumulated, which can be in the desert or along the beach; thus, one distinguishes desert dunes and coastal dunes. Dunes can be mobile or fixed. Fixed dunes are older and are either "fossilized," that is, transformed into a cohesive material, a precursor to sandstone, or fixed because of

Figure 1. Barchan dunes near Laâyoune, Morocco.

the vegetation or because the average wind over some period at their location is zero. Otherwise the sand moves if the winds are strong enough, which means typically stronger than $4\,\mathrm{ms}^{-1}$.

The beautiful landscapes (Figure 1; see also color plate section) formed by dunes are characterized by very gentle hills interrupted by sharp edges called brink lines, delimiting regions of steeper slope, called slip-faces, lying in the wind shadow. Depending on the amount of available sand and the variation of the wind direction, one distinguishes different typical dune morphologies that have been classified by geographers into over a hundred categories. The most well known are longitudinal, transverse, and barchan dunes. Barchans (from an Arabic word) are crescent-shaped mobile dunes that appear when the wind always comes from the same direction and there is not much sand present. Their movement ranges from 5 to 50 m per year and is inversely proportional to their height. If sufficient sand is available to cover all the surface, then transverse dunes appear (from the merger of many barchans). Longitudinal dunes, that is, along the direction of the wind, are observed when the wind periodically changes its direction over about 30°. Other famous dune types are star dunes, ergs, parabolic dunes, and draas.

The driving force for sand motion is the drag imposed by the wind on the grains at the surface. A given dune morphology can therefore be understood as an aerodynamic instability close to a mobile surface. A complete mathematical description of the problem therefore needs the equation of motion of the wind velocity field coupled with an equation of motion of the granular surface. The right formulation of these equations requires a good understanding of the transport mechanism of sand.

Three types of transport can be distinguished according to the size of a sand grain: creep, saltation (bouncing), and suspension. The only mechanism relevant for dune formation is saltation, which drags grains typically of 100–300 μm diameter. The mechanism of saltation, first described by Ralph Bagnold in his pioneering work (Bagnold, 1941), consists of grains, which once lifted out from their granular bed are accelerated by the wind and then impact against the surface, ejecting new grains. These grains are again accelerated and eject a further splash of grains, so that the number of grains flying above the surface increases exponentially until the total momentum transferred from the air to the grains saturates to its maximum capacity, and the wind can no longer pick up more sand from the dune. On dunes in the field, these saltating grains form a sheet of grains floating typically 5 cm above the surface. The wind typically is turbulent and has a logarithmic profile as a function of height, which (due to the presence of the grains) is strongly modified close to the surface. The height of the boundary layer is less than 1 cm. Using the techniques of Jackson & Hunt (1975), one can calculate the shear force of the wind at the surface in an approximate form and obtain reasonable agreement with measurements on dunes. The saturated flux of sand at the surface is a function of this shear stress and has been described by various phenomenological expressions, the first one given by Bagnold (1941) and subsequent ones by Lettau & Lettau (1969), and by Sørensen (1981). A full description also requires taking into account the transient length before (or after) reaching saturation. Together with mass conservation one can then close the system of equations, giving at the end a full set of equations of motions (Sauermann et al., 2001).

When the local slope exceeds a value of typically 35°, the angle of repose, the sand begins to slide in the form of avalanches giving a second mechanism of sand transport driven by gravity. The slip-faces all have this slope. The edges separating them from the purely wind-driven regions are just given by the brink lines. Over these regions the wind field develops recirculation eddies of velocities typically below the minimum threshold for grain motion. When the saturation length is less than the size of these low-velocity regions, the sand grains get trapped. This is the principal instability underlying dune morphology: the dunes become traps of sand for more sand. The typical saturation length of about 10 m also means that no dune below 1.5 m in height is stable under Earth's meteorological conditions; the dune is too short for the air to reach its saturation point, and the dune suffers erosion.

Dunes have been studied on all continents and their shape, sand flux, velocity, and granulometry have been presented in many publications. Several books review the subject (Pye & Tsoar, 1990). For different morphologies one finds specific shapes and scaling laws, but systematic studies exist only for barchans. Sand fluxes are typically measured with traps, but a more sophisticated metrology (e.g., acoustic, optic) is evolving. The limitation factors are the fluctuations of the wind fields and the climate. In the arid regions of the world, in particular the poor countries in the Sahara, dune motion poses an important threat to housing, roads, and fields, and sand removal constitutes a significant economic factor in these countries. Many empirical techniques of dune fixing and dune destruction have been developed, mostly applied to coastal dunes, which are in fact disappearing in many places, sometimes damaging the fragile dune ecosystem.

H.J. Herrmann

See also **Avalanches; Geomorphology and tectonics; Sandpile model**

Further Reading

Bagnold, R.A. 1941. *The Physics of Blown Sand and Desert Dunes*, London: Methuen

Hersen, P., Douady, S. & Andreotti, B. 2002. Relevant length scale of barchan dunes. *Physical Review Letters*. 89: 264301

Jackson, P. S. & Hunt, J.C.R. 1975. Turbulent wind flow over a low hill. *Quarterly Journal of the Royal Meteorological Society*, 101–929

Lettau, K. & Lettau, H. 1969. Bulk transport of sand by the barchans of the Pampa de La Joya in Southern Peru. Zeitschrift für Geomorphologie, N.F. 13(2): 182–195

Pye, K. & Tsoar, H. 1990. *Aeolian Sand and Sand Dunes*, London: Unwin Hyman

Sauermann, G., Kroy, K. & Herrmann, H.J. 2001. A continuum saltation model for sand dunes. *Physical Review* E, 64: 31305

Sørensen, M. 1991. An analytic model of wind-blown sand transport. *Acta Mechanica*, (Suppl.) 1: 67–81

DYM EQUATION

See **Solitons, types of**

DYNAMIC PATTERN FORMATION

See **Synergetics**

DYNAMIC SCALING FUNCTION

See **Routes to chaos**

DYNAMICAL SYSTEMS

A dynamical system is a time-dependent, multicomponent system of elements with local states determining a global state of the whole system. In a planetary system, for example, the state of the system at a certain time is the set of values that completely describe the system (the position and momentum of each planet). The states can also refer to moving molecules in a gas, excitation of neurons in a neural network, nutrition of organisms in an ecological system, supply and demand of economic markets, or behavior of social groups in human societies.

The dynamics of a system (the change of system states with time) is given by linear or nonlinear differential equations. In the case of nonlinearity, several feedback activities take place between the elements of the system: in the solar system, the movement of the Earth is determined by the gravitation not only of the Sun, but of all the other celestial bodies of the system, which attract each other gravitationally.

For deterministic processes (for example, movements in a planetary system), each future state is uniquely determined by the present state. A conservative (Hamiltonian) system such as an ideal pendulum is characterized by the reversibility of time direction and conservation of energy. Dissipative systems, for example, a real pendulum with friction, are irreversible. The time-dependent development of a system's state is geometrically represented by orbits (trajectories) in a state space or phase space, which is defined by the multidimensional vectors of the nonlinear system (*See* **Phase space**).

In addition to continuous processes, we can also consider discrete processes of changing states at certain points of time. Difference equations are important for modeling measured data at discrete points of time, which are chosen equidistant or defined by other measurement devices.

Random events (Brownian motion in a fluid, mutation in evolution) are represented by additional fluctuation terms. Classical stochastic processes, such as the billions of unknown molecular states in a fluid, are defined by time-dependent differential equations with distribution functions of probabilistic states. Stochastic nonlinear differential equations (such as Fokker–Planck equation, Master equation) are also used to model phase transitions of complex systems, including migration dynamics of populations, traffic dynamics, data dynamics in the Internet, among others.

In quantum systems, the dynamics of quantum states are determined by Schrödinger's equation. Although it is a deterministic differential equation of a wave function, its observables (position and momentum of a particle) depend on Heisenberg's uncertainty principle, which only allows probabilistic forecasts.

During the 17th–19th centuries, classical physics viewed the universe as a deterministic and conservative system. The astronomer and mathematician Pierre-Simon Laplace assumed that all future states of the universe could be computed or determined if all forces acting in nature and the initial states of all celestial bodies are known at one instant of time (Laplacian determinism). Laplace's assumption was correct for linear and conservative dynamical systems such as a harmonic oscillator. However, at the end of the 19th century, Henri Poincaré discovered that celestial mechanics is not a completely computable system even if it is considered as a deterministic and conservative system. The mutual gravitational interactions of more than two celestial bodies (the many-body problem) correspond to nonlinear and non-integrable equations with instabilities and sometimes chaos (*See* ***N*-body problem**).

According to Laplacian determinism, similar causes effectively determine similar effects. Thus, in the phase space, trajectories starting close to each other also remain close to each other during time evolution. Dynamical systems with deterministic chaos exhibit an exponential dependence on initial conditions for bounded orbits: the separation of trajectories with close initial states increases exponentially. (The rate at which nearby orbits diverge from each other after small perturbations is measured by Lyapunov exponents.) Consequently, tiny deviations of initial states lead to exponentially increasing computational efforts for future data, limiting long-term prediction, although the dynamics is, in principle, uniquely determined.

The sensitivity of chaotic dynamics to small changes in initial conditions is known as the "butterfly effect": small and local causes (local perturbations of the weather, for example) can lead to unpredictable large and global effects in unstable states (*See* **Butterfly effect**). According to the famous KAM theorem of Andrei Kolmogorov, Vladimir Arnol'd, and Jürgen Moser, trajectories in the phase space of classical mechanics are neither completely regular nor completely irregular, but depend sensitively on the chosen initial conditions. Changes of states in a dynamical system that change the stability of solutions to its nonlinear equations are associated with bifurcations of orbits in the corresponding phase space (*See* **Bifurcations**).

Dynamical systems can be classified by the effects of their dynamics on a region of the phase space. A conservative system is defined by the fact that, during time evolution, the volume of a region remains constant, although its shape may be transformed. An attractor is a region of a phase space into which all trajectories departing from an adjacent region, the so-called basin of attraction, converge. There are different kinds of more or less complex attractors. Chaotic attractors are highly complex structures of nonperiodic orbits in a bounded region of the phase space between regular behavior and stochastic behavior or noise. Although high-dimensional dynamical systems (such as the stock market, the economy, and society) cannot be formulated in all detail, approximate models may provide qualitative insights into their complex behavior.

KLAUS MAINZER

See also **Chaotic dynamics; Determinism; Equations, nonlinear**

Further Reading

Abarbanel, H.D.I. 1996. *Analysis of Observed Chaotic Data*, New York: Springer

Abraham, R.H. & Shaw, C.D. 1992. *Dynamics: The Geometry of Behavior*, 2nd edition, Redwood City, CA: Addison-Wesley

Arnol'd, V.I. 1978. *Mathematical Methods of Classical Mechanics*, Berlin and New York: Springer

Guckenheimer, J. & Holmes, P. 1983. *Nonlinear Oscillations, Dynamical Systems, and Bifurcations of Vector Fields*, New York: Springer

Hirsch, M.W. & Smale, S. 1974. *Differential Equations, Dynamical Systems, and Linear Algebra*. New York: Academic Press

Kaplan, D. & Glass, L. 1995. *Understanding Nonlinear Dynamics*, New York: Springer

Mainzer, K. 2003. *Thinking in Complexity. The Computational Dynamics of Matter, Mind, and Mankind*, 4th edition, New York: Springer

Rand, D.A. & Young, L.S. (editors). 1981. *Dynamical Systems and Turbulence*, Berlin and New York: Springer

Shilnikov, L.P., Shilnikov, A.L., Turaev, D.V. & Chua, L.O. 2001. *Methods of Qualitative Theory of Nonlinear Dynamics*, Singapore: World Scientific

DYNAMICAL ZETA FUNCTIONS

See **Periodic orbit theory**

DYNAMOS, HOMOGENEOUS

Dynamos convert mechanical energy into electromagnetic energy, the most familiar example being the bicycle dynamo. Nearly all electricity consumed by mankind is generated by dynamos in power plants—electrochemical processes also generate electric power, for example, in batteries, but these play only a minor role. The simplest dynamo is the disk dynamo originally invented by Faraday and shown in Figure 1. A metal disk rotates about its axis with an angular velocity $\boldsymbol{\Omega} = \boldsymbol{k}\Omega$. When the disk is permeated by an initial magnetic field $\boldsymbol{B}_0$ parallel to the axis, an electromotive force is generated between the axis and rim of the disk,

$$U = \boldsymbol{\Omega} \cdot \int_{r_1}^{r_0} \boldsymbol{B}_0 \mathrm{d}r, \qquad (1)$$

where r_1 and r_0 denote the radii of axis and disk. The electromotive force U can be used to drive an electric current J through the circuit indicated in Figure 1. Denoting by L and R the inductivity and the ohmic resistance of the circuit, we obtain for the current J

$$L\frac{\mathrm{d}J}{\mathrm{d}t} + RJ = U. \qquad (2)$$

The current J flowing through the circuit is associated with a magnetic field $\boldsymbol{B}_1$, which may replace the initial field $\boldsymbol{B}_0$. The integral $2\pi \int_{r_1}^{r_0} \boldsymbol{k} \cdot \boldsymbol{B}_1 \mathrm{d}r/J$ describes

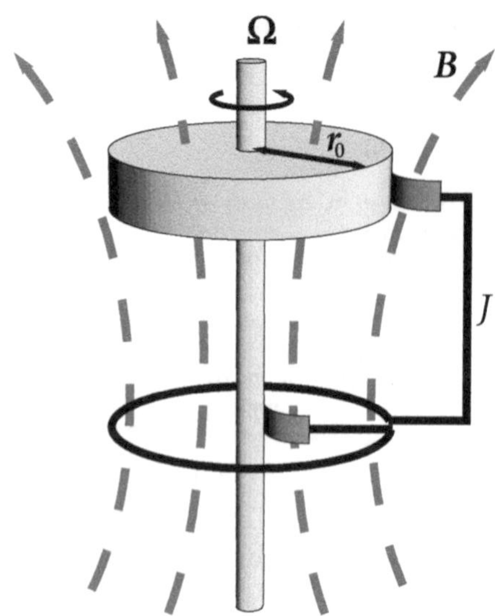

Figure 1. The disk dynamo.

the mutual inductivity M between circuit and disk. Equation (2) for the self-exited disk dynamo can thus be written in the form

$$L\frac{\mathrm{d}J}{\mathrm{d}t} = (\Omega M/2\pi - R)J, \tag{3}$$

which allows for exponentially growing solutions once the dynamo condition

$$\Omega > 2\pi R/M \tag{4}$$

is satisfied.

The disk dynamo is an inhomogeneous dynamo since it depends on an inhomogeneous distribution of electrical conductivity as given by the wiring of the circuit. The dynamo would not work if the sense of wiring around the axis would be opposite to that shown in the figure or if it would be short-circuited by immersion in a highly conducting fluid. While it is generally believed that planetary and stellar magnetic fields are generated by dynamos in the electrically conducting interiors of these celestial bodies, these dynamos must be homogeneous ones because they operate in singly connected finite domains of essentially homogeneous electrical conductivity. In 1919, Larmor first proposed this idea as an explanation for the magnetic field of sunspots. It was doubtful for a long time whether homogeneous dynamos were possible. Cowling proved in 1934 that axisymmetric fields could not be generated by a homogeneous dynamo. But, in 1958, Backus and Herzenberg independently demonstrated in a mathematically convincing way that homogeneous dynamos are indeed possible.

The velocity fields that are required to drive a homogeneous dynamo are necessarily more complex than the simple rotation velocity of the disk dynamo. A new mathematical discipline called dynamo theory has evolved and continues to be an active field of research. One distinguishes between the kinematic dynamo problem based on a linear equation such as (3) with a prescribed velocity field and the magnetohydrodynamic dynamo problem in which the influence of the growing magnetic field on the velocity field is taken into account. Obviously, a magnetic field cannot grow exponentially forever. Just as the Lorentz force produced by the magnetic field together with the current density in the disk opposes the rotation of the disk in Figure 1, it also changes the velocity field of the magnetohydrodynamic dynamo. The external torque T_e applied to the disk in the figure must be increased in the presence of dynamo action in order to sustain the rotation rate Ω. The equilibrium strength of the magnetic field will thus be determined by the available torque T_e.

In the case of the geodynamo driven by convection flows in the liquid iron core of the Earth, as well as in the case of other planetary or stellar dynamos, the equations of motions together with the equation of magnetic induction (*See* **Magnetohydrodynamics**) must be solved to determine the strength of the generated magnetic field as a function of the relevant parameters. Extensive computer simulations have been performed in recent years. Some examples can be found in Jones et al. (2003). The complex numerical simulations of magnetohydrodynamic dynamos share the following properties with the simple disk dynamo:

(i) For given external parameters, there always exists a solution without magnetic field besides the dynamo solution, just as the disk of Figure 1 can rotate with $\Omega > \Omega_c$ in the absence of any initial field $\boldsymbol{B}_0$.

(ii) The existence of a dynamo solution requires that the magnetic Reynolds number Rm exceeds a critical value Rm_c. Rm is defined by

$$Rm = V d\sigma\mu, \tag{5}$$

where V is a typical velocity, d is a characteristic length such as the radius of the iron core in the case of the Earth, and σ and μ are the electrical conductivity and the magnetic permeability of the fluid. The inverse product $\lambda = 1/\sigma\mu$ is called the magnetic diffusivity. The dimensionless parameter Rm corresponds to the quantity $\Omega M/R$ in the case of the disk dynamo. The nonmagnetic solution exists, but it is unstable for $Rm > Rm_c$.

(iii) Magnetohydrodynamic dynamos exist in two forms that are identical except for the sign of the magnetic field $\boldsymbol{B}$. This property reflects the fact that the Lorentz force is given by an expression quadratic in $\boldsymbol{B}$.

Property (iii) is the basic reason that the geomagnetic field has often switched its polarity in the geologic past. These "reversals" have occurred nearly randomly about every 200,000 years on average. In contrast, the solar magnetic field reverses every 11 years in a surprisingly periodic fashion.

For the description of magnetic fields associated with a spherical system, one often uses the general representation for a solenoidal vector field,

$$\boldsymbol{B} = \nabla \times (\nabla h \times \boldsymbol{r}) + \nabla g \times \boldsymbol{r}, \quad (6)$$

in terms of a poloidal and a toroidal component each of which is described by a scalar function, the poloidal function h, and the toroidal function g. Without loss of generality, the averages of h and g over surfaces $|\boldsymbol{r}| = \text{constant}$ can be assumed to vanish. A homogeneous dynamo usually requires the interaction of both components of the magnetic field. It can be shown (Kaiser et al., 1994) that a magnetic field with vanishing toroidal part cannot be generated. It is also generally believed that a purely poloidal field cannot be generated either. But a proof of this hypothesis has not yet been given. The functions h and g can be separated into their axisymmetric parts $\bar{h}$ and $\bar{g}$ and non-axisymmetric parts, $\check{h} = h - \bar{h}$ and $\check{g} = g - \bar{g}$. The component $\bar{g}$ can easily be generated in a spherical dynamo through a stretching of the axisymmetric poloidal field by a differential rotation. This process is known as the ω-effect. The amplification of $\bar{h}$ requires the interaction of the non-axisymmetric components of magnetic fields and velocity fields. This is often called the α-effect. This latter effect can, of course, also be used for the generation of $\bar{g}$ in the absence of a differential rotation. Accordingly, one distinguishes between $\alpha\omega$- and α^2-dynamos. These concepts were originally introduced within the framework of mean-field magnetohydrodynamics for which the reader is referred to the book by Krause & Raedler (1980).

F.H. Busse

See also **Alfvén waves; Fluid dynamics; Magnetohydrodynamics; Nonlinear plasma waves**

Further Reading

Childress, S. & Gilbert, A.P. 1995. *Stretch, Twist, Fold: The Fast Dynamo*, Berlin and New York: Springer

Davidson, P.A. 2001. *An Introduction to Magnetohydrodynamics*, Cambridge and New York: Cambridge University Press

Jones, C.A., Soward, A.M. & Zhang, K. 2003. *Earth's Core and Lower Mantle*, London and New York: Taylor & Francis

Kaiser, R., Schmitt, B.J. & Busse, F.H. 1994. On the invisible dynamo. *Geophysical and Astrophysical Fluid Dynamics*, 77: 91–109

Krause, F. & Raedler, K.-H. 1980. *Mean-field Magnetohydrodynamics and Dynamo Theory*, Oxford and New York: Pergamon Press

Moffatt, H.K. 1978. *Magnetic Field Generation in Electrically Conducting Fluids*, Cambridge and New York: Cambridge University Press

EARTHQUAKES

See **Geomorphology and tectonics**

ECKHAUS INSTABILITY

See **Wave stability and instability**

ECONOMIC SYSTEM DYNAMICS

Economic dynamics is concerned with fluctuations in the economy. Most economic variables, such as gross domestic product (GDP), production, unemployment, interest rates, exchange rates, and stock prices, exhibit perpetual fluctuations over time. These fluctuations are characterized by sustained growth of production and employment as well as large oscillations in relative changes or growth rates. The fluctuations vary from fairly regular business cycles in macroeconomic variables to very irregular fluctuations, for example, in stock prices and exchange rates, in financial markets. In this note, we discuss some approaches to the theory of economic fluctuations, emphasizing the role of nonlinear dynamic models.

In contrast to many dynamic phenomena in natural sciences, uncertainty always plays a role in an economy, at least to some extent. Therefore, a purely deterministic model seems inappropriate to describe fluctuations in the economy, and a stochastic dynamic model is needed. Nevertheless, a key question in economic dynamics is whether a simple, nonlinear dynamic model can explain a *significant part* of observed economic fluctuations.

Brief History

There are two contrasting viewpoints concerning the explanation of observed economic fluctuations. According to the first (New Classical) viewpoint, the main source of fluctuations is to be found in exogenous, random shocks (news about economic fundamentals) to an inherently stable, often linear economic system. Without any external shocks to economic fundamentals (preferences, endowments, technology, etc.), the economy would be stable and converge to the unique steady-state (growth) path. According to the second, opposing (Keynesian) viewpoint, economic fluctuations are not caused by chance or random impulses, but should be explained by nonlinear economic laws of motion. Even without any external shocks to the fundamentals of the economy, fluctuations in prices or other economic variables may arise. It is an old Keynesian theme that fluctuations are not determined by economic fundamentals only, but are also driven by volatile, self-fulfilling expectations ("animal spirits," market psychology).

The view that business cycles are driven by external random shocks was propagated in the 1930s, for example, by Ragnar Frisch and Jan Tinbergen (sharing the first Nobel Prize in Economic Sciences in 1969 "for having developed and applied dynamic models for the analysis of economic processes"). They observed that simple, linear systems buffeted with noise can mimic time series similar to those observed in real business cycle data. To several economists this approach was unsatisfactory, however, because it does not provide an *economic* explanation of business cycles, but rather attributes them to external, random events. In the 1940s and 1950s Nicholas Kaldor, John Hicks, and Richard Goodwin developed nonlinear dynamic models with locally unstable steady states and stable limit cycles as an explanation for business cycles. These early nonlinear business cycle models, however, suffered from a number of serious shortcomings. First of all, the laws of motion were too "ad hoc," and in particular they were *not* derived from rational behavior, that is, from utility and profit maximizing principles. Secondly, the simulated time series from the models were too regular compared with observed business cycles, even when small dynamic noise was added to the models. Finally, expectation rules were "ad hoc," and along the regular cycles, agents made "systematic" forecasting errors.

The Role of Expectations

The most important difference between economics and natural sciences is perhaps the fact that an economic system is an *expectations feedback* system. Decisions of economic agents are based upon their expectations and beliefs about the future state of the economy. Through these decisions, expectations feed back into the economy and affect actual realization of economic variables. These realizations lead to new expectations, in turn affecting new realizations, implying an infinite sequence of expectational feedback. For example, in the stock market, optimistic expectations that stock prices will rise will lead to a larger demand for stocks, which will cause stock prices to rise. This process may lead to a self-fulfilling speculative bubble in the stock market. A theory of expectation formation is, therefore, a crucial part of economics, in particular for modeling dynamic asset markets.

In the early business cycle models, simple, ad hoc expectations rules were employed, such as naive expectations (where the forecast of the economic variable is simply the latest observation of that variable) or adaptive expectations (where the forecast is a weighted average of the previous forecast and the latest observation). An important problem with simple forecasting rules is that typically agents make systematic forecasting errors, especially when there are regular cycles. A smart agent would learn from her mistakes and adapt her expectations rule accordingly. Another problem is that if an agent is to use a simple forecasting rule, it is far from clear which simple rule to choose in a particular model. With the development of empirical, econometric analysis of business cycles, it became clear that unrestricted models of expectations preclude a systematic inquiry into business fluctuations. These considerations led to the development of *rational expectations*, a solution to the expectations feedback system proposed by John Muth (1961) and applied to macroeconomics, for example, by Robert Lucas and Thomas Sargent. Rational expectations means that agents use all available information, including economic theory, to form optimal forecasts and that, on average, expectations coincide with realizations. In a deterministic model, without noise and randomness, rational expectations implies perfect foresight (no mistakes at all); in a stochastic model, rational expectations coincides with the conditional mathematical expectations given all available information (no mistakes on average, no systematic bias).

In the 1970s and 1980s, the rational expectations critique culminated in the development of New Classical economics and real business cycle models, based upon rational expectations, intertemporal utility and profit maximization, and perfectly competitive markets. This approach outdated the early Keynesian nonlinear business cycle models of the 1950s. Due to the discovery of deterministic chaos and other developments in nonlinear dynamics, however, the last two decades have witnessed a strong revival of interest in nonlinear endogenous business cycle models.

Nonlinear Dynamics

In mathematics and physics, things changed dramatically in the 1970s due to the discovery of deterministic chaos, the phenomenon that simple, deterministic laws of motion can generate unpredictable time series. This discovery shattered the Laplacian deterministic view of perfect predictability and made scientists realize that long-run prediction may be fundamentally impossible, even when laws of motion are known exactly. Inspired by "chaos theory," economists (e.g., Richard Day and Jean-Michel Grandmont) started looking for nonlinear, deterministic models generating erratic time series similar to the patterns observed in real business cycles. This search led to new, simple nonlinear business cycle models within the paradigm of rational expectations, optimizing behavior and perfectly competitive markets, generating chaotic business fluctuations.

In the 1980s, several economists (e.g., William Brock, Davis Dechert, Jose Scheinkman, and Blake LeBaron) also employed nonlinear methods, such as correlation dimension tests, from the natural sciences to look for evidence of nonlinearity and low deterministic chaos in economic and financial data. This turned out to be a difficult task because the methods employed require very long time series and the methods are very sensitive to noise. One can say that evidence for low-dimensional deterministic chaos in economic and financial data is weak (but it seems fair to add that because of the sensitivity to noise, the hypothesis of chaos buffeted with dynamic noise has not been rejected) but evidence for nonlinearity is strong. In particular, Brock, Dechert, and Scheinkman have developed a general test (the BDS-test), based upon ideas from U-statistics theory and correlation integrals, to test for nonlinearity in a given time series; see Brock et al. (1996) and Brock, Hsieh, & LeBaron (1991) for the basic theory, references, applications, and extensions. The BDS test has become widely used in economics and also in physics.

Bounded Rationality

Already in the 1950s, Herbert Simon pointed out that rationality requires unrealistically strong assumptions about the computing abilities of agents and proposed that *bounded rationality*, with limited computing capabilities and with agents using habitual rules of thumb instead of perfectly optimal decision rules, would be a more accurate description of human behavior. Nevertheless, as noted above, rational expectations became the dominating paradigm in dynamic economics in the 1970s and 1980s. Nonlinear dynamics, the possibility of chaos, and its implications for limited predictability shed important new light on the expectations hypothesis, however. In a simple (linear) stable economy with a unique steady-state path, it seems natural that agents can learn to have rational expectations, at least in the long run. A representative, perfectly rational agent model nicely fits into a linear view of a globally stable and predictable economy. But how could agents have rational expectations or perfect foresight in a complex, nonlinear world, with prices and quantities moving irregularly on a strange attractor and sensitivity to initial conditions? A boundedly rational world view with agents using simple forecasting strategies, perhaps not perfect but at least approximately right, seems more appropriate for a complex nonlinear world. These developments contributed to a rapidly growing interest in bounded rationality in the 1990s (see, e.g., the survey in Sargent (1993)). A boundedly rational agent forms expectations based upon observable quantities and adapts her forecasting rule as additional observations become available. Adaptive learning may converge to a rational expectations equilibrium, or it may converge to an "approximate rational expectations equilibrium," where there is at least some degree of consistency between expectations and realizations (see, e.g., Evans & Honkapohja (2001) for an extensive and modern treatment of adaptive learning in macroeconomics).

Interacting Agents and Evolutionary Models

The representative agent model has played a key role in economics for a long time. An important motivation for the dominance of the rational agent model dates back to the 1950s, to Milton Friedman who argued that nonrational agents will be driven out of the market by rational agents, who will trade against them and simply earn higher profits. In recent years, however, this view has been challenged, and heterogeneous agent models are becoming increasingly popular, especially in financial market modeling (see, e.g., Kirman (1992) for a critique on representative agent modeling).

Many heterogeneous agent models are artificial, computer simulated markets. This work views the economy as a complex evolving system composed of many different, boundedly rational, interacting traders, with strategies, expectations and realizations co-evolving over time (see, e.g., work at the Santa Fe Institute collected in Anderson et al. (1988)). Two typical trader types arising in many heterogeneous agent financial market models are fundamentalists and chartist or technical traders. Fundamentalists base their investment decisions upon market fundamentals such as dividends, earnings, interest rates, or growth indicators. In contrast, technical traders pay no attention to economic fundamentals but look for regular patterns in past prices and base their investment decision upon simple trend following trading rules. An evolutionary competition between these different trader types, where traders tend to follow strategies that have performed well in the recent past, may lead to irregular switching between the different strategies and result in complicated, irregular asset price fluctuations. It has been shown, for example, by Brock and Hommes (1997, 1998), that in these evolutionary systems, rational agents and/or fundamental traders do not necessarily drive out all other trader types, but that the market may be characterized by perpetual evolutionary switching between competing trading strategies. Nonrational traders may survive evolutionary competition in the market (see, e.g., Hommes (2001) for a survey and many relevant references). Lux & Marchesi (1999) show that these types of interacting agent models are able to generate many of the stylized facts, such as unpredictable returns, clustered volatility, fat tails, and long memory, observed in real financial markets.

Future Perspective

A good feature of the rationality hypothesis is that it puts natural discipline on agents' forecasting rules and minimizes the number of free parameters in dynamic economic models. In contrast, the "wilderness of bounded rationality" leaves too many degrees of freedom in modeling, and it is far from clear which out of a large class of habitual rules of thumb is most reasonable. Stated differently in a popular phrase: "there is only one way (or perhaps a few ways) one can be right, but there are many ways one can be wrong." The philosophy underlying the evolutionary approach is to use simple forecasting rules based upon their performance in the recent past. In this type of modeling, "evolution decides who is right." Bounded rationality, heterogeneity, adaptive learning, and evolutionary competition all create natural nonlinearities. Nonlinearity is, therefore, likely to play an increasingly important role in the future of economic dynamics.

CARS HOMMES

See also **Dynamical systems; Forecasting; Game theory; Time series analysis**

Further Reading

Anderson, P.W., Arrow, K.J. & Pines, D. (editors). 1988. *The Economy as a Complex Evolving System*, Redwood City, CA: Addison-Wesley

Brock, W.A. Dechert, W.D., Scheinkman, J.A. & LeBaron, B. 1996. A test for independence based upon the correlation dimension. *Econometric Review*, 15: 197–235

Brock, W.A. & Hommes, C.H. 1997. A rational route to randomness. *Econometrica*, 65: 1059–1095

Brock, W.A. & Hommes, C.H. 1998. Heterogeneous beliefs and routes to chaos in a simple asset pricing model. *Journal of Economic Dynamics and Control*, 22: 1235–1274

Brock, W.A., Hsieh, D.A. & LeBaron, B. 1991. *Nonlinear Dynamics, Chaos and Instability: Statistical Theory and Economic Evidence*, Cambridge, MA: MIT Press

Day, R.H. 1996. *Complex Economic Systems*, Cambridge, MA: MIT Press

DeGrauwe, P., DeWachter, H. & Embrechts, M. 1993. *Exchange Rate Theory. Chaotic Models of Foreign Exchange Markets*, Oxford: Blackwell

Evans, G.W. & Honkapohja, S. 2001. *Learning in Macroeconomics*, Princeton, NJ: Princeton University Press
Friedman, M. 1953. The case of flexible exchange rates. In *Essays in Positive Economics*, Chicago: University of Chicago Press
Frisch, R. 1933. Propagation problems and impulse problems in dynamic economics. In *Economic essays in honor of Gustav Cassel*, London: George Allen and Unwin, Ltd; reprinted in *Readings in Business Cycles*, edited by R.A. Gordon and L.R. Klein, Homewook, IL: Richard D. Irwin, Inc, 1965
Gabisch, G. & Lorenz, H.W. 1987. *Business Cycle Theory. A Survey of Methods and Concepts*. Berlin: Springer
Goodwin, R.M. 1951. The nonlinear accelerator and the persistence of business cycles. *Econometrica*, 19: 1–17
Grandmont, J.M. 1985. On endogenous competitive business cycles. *Econometrica*, 53: 995–1045
Hicks, J.R. 1950. *A contribution to the theory of the trade cycle*. Oxford: Clarendon Press
Hommes, C.H. 2001. Financial markets as nonlinear adaptive evolutionary systems. *Quantitative Finance*, 1: 149–167
Kaldor, N. 1940. A model of the trade cycle. *Economic Journal*, 50: 78–92
Kirman, A. 1992. Whom or what does the representative individual represent? *Journal of Economic Perspectives*, 6: 117–136
Lorenz, H.W. 1993. *Nonlinear Dynamical Economics and Chaotic Motion*, 2nd edition, Berlin: Springer
Lucas, R.E. 1971. Econometric testing of the natural rate hypothesis. In *The Econometrics of Price Determination*, edited by O. Eckstein, Washington, DC: Board of Governors of the Federal Reserve System and Social Science Research Council
Lux, T. & Marchesi, M. 1999. Scaling and criticality in a stochastic multi-agent model of a financial market. *Nature*, 397: 498–500
Muth, J.F. 1961. Rational expectations and the theory of price movements. *Econometrica*, 29: 315–335
Sargent, T.J. 1993. *Bounded Rationality in Macroeconomics*, Oxford: Clarendon Press
Simon, H.A. 1957. *Models of Man*, New York: Wiley

EFFECTIVE MASS

Effective mass is a physical quantity characterizing the dynamics of a particle or quasiparticle with an energy ($\mathcal{E}$) that is quadratic in the components of the momentum ($\boldsymbol{p}$). With a dispersion relations of the form

$$\mathcal{E} = \mathcal{E}_0 + \tfrac{1}{2}\,\mu_{ik}\,p_i p_k\,, \quad \mathcal{E}_0 = \text{const.}, \tag{1}$$

the tensor $\mu_{ik} = m_{ik}^{-1}$ is a tensor of reciprocal effective masses, and m_{ik} is an effective mass tensor. Considering Equation (1) as the Hamiltonian function of a free particle, one can determine a particle velocity using a canonical equation of motion

$$v_i = \partial\mathcal{E}/\partial p_i, \quad i = 1, 2, 3. \tag{2}$$

From the Hamiltonian function (1) and relations (2) a Lagrangian function of the particle is

$$L = L_0 + \tfrac{1}{2}\,m_{ik}\,v_i v_k, \quad L_0 = \text{const.} \tag{3}$$

As the Lagrangian function of a free particle coincides with its kinetic energy, the effective mass tensor can be associated with kinetic energy (3), which is quadratic in components of the velocity.

Having its origin in the mechanics of particles, the definition of the effective mass can be connected with the dynamics of wave packets. According to the de Broglie principle of wave-corpuscular dualism, the energy $\mathcal{E}$ and momentum $\boldsymbol{p}$ of a particle correspond to the frequency ω and wave vector $\boldsymbol{k}$ of some wave packet, as $\mathcal{E} = \hbar\omega$ and $\boldsymbol{p} = \hbar\boldsymbol{k}$ where $\hbar$ is Planck's constant. From this point of view, any quasiparticle excitation in condensed matter behaves as a particle-like wave packet, and Equation (2) coincides with the definition of a group velocity of the wave packet.

The effective mass tensor for a free Newtonian particle or quasiparticle in an isotropic media (for example, excitations in the superfluid liquid He) has the simple form: $m_{ik} = m\,\delta_{ik}$, $i, k = 1, 2, 3$.

The dispersion relation of type (1) for the quasiparticles described by band theory in a periodic structure (electrons in metals and phonons in crystals) takes place at vicinities of singular points in the **p**-space. These are at the points where energy $\mathcal{E}(\boldsymbol{p})$ has a minimum (then μ_{ik} is positively definite), at the points near the maximum of $\mathcal{E}(\boldsymbol{p})$ (then μ_{ik} is negatively definite), and at the so-called conical points, when the main values of the tensor μ_{ik} have different signs.

In the general band theory of electrons and semiconductors, energy is a more complicated (arbitrary) function of the momentum $\mathcal{E} = \mathcal{E}(\boldsymbol{p})$, and the tensor of the reciprocal effective mass is defined as

$$\mu_{ik} = \frac{\partial^2\mathcal{E}}{\partial p_i \partial p_k}, \quad i, k = 1, 2, 3, \tag{4}$$

which can be a function of the momentum. The effective mass tensor allows one to calculate the acceleration of a particle under the action of the external force $\boldsymbol{f}$. As an equation of the particle motion is

$$\frac{\mathrm{d}\boldsymbol{p}}{\mathrm{d}t} = \boldsymbol{f}, \tag{5}$$

Equations (2) and (5) lead to the following equation for the acceleration:

$$m_{ik}\,\frac{\mathrm{d}v_k}{\mathrm{d}t} = f_i, \quad i = 1, 2, 3. \tag{6}$$

The force $\boldsymbol{f}$ is determined, for example, by the electric field effect on a charged particle.

The effective mass has another definition for an electron (or a charged particle) moving in a static magnetic field $\boldsymbol{B}$. In such a case the force $\boldsymbol{f}$ is a Lorentz force, and Equation (5) implies that the electron moves under the following conditions: $\mathcal{E}(\boldsymbol{p}) =$ constant and $p_B =$ constant, where p_B is a projection of the momentum on the magnetic field direction. Thus, an electron trajectory in $\boldsymbol{p}$-space is a section of the isoenergy surface $\mathcal{E}(\boldsymbol{p}) = \mathcal{E}$ with the plane $p_B = p$. If this section is a closed curve and has a sectional area $S(\mathcal{E}, p)$, the electron motion is periodic in time and characterized by the "cyclotron frequency" $\omega_c = eB/(m^*c)$, where e is the electron charge, and c is the velocity of light. Thus, the effective mass m^* is equal to

$$m^* = \frac{1}{2\pi}\frac{\partial S}{\partial \mathcal{E}}. \qquad (7)$$

Cyclotron resonance is the most convenient experimental method for measuring the effective mass defined in Equation (6).

ARNOLD KOSEVICH

See also **Dispersion relations; Group velocity; Wave packets, linear and nonlinear**

Further Reading

Haken, H. 1976. *Quantum Field Theory of Solids*, Amsterdam: North-Holland

Kittel, C. 1987. *Quantum Theory of Solids*, New York: Wiley

Slater, J.C. 1951. *Quantum Theory of Matter*, New York: McGraw-Hill

EIFFEL JUNCTION

See **Long Josephson junction**

EIGENVALUES AND EIGENVECTORS (BOUND STATE)

See **Inverse scattering method or transform**

EIKONAL CURVATURE EQUATION

See **Geometrical optics, nonlinear**

EINSTEIN EQUATIONS

After Albert Einstein (1905) published his special theory of relativity, Hermann Minkowski (1908) delivered a seminar in which he said

> Henceforth space by itself, and time by itself, are doomed to fade away into mere shadows, and only a kind of union of the two will preserve an independent reality.

This "kind of union" is called space-time. In special relativity, two observers in inertial frames moving at constant velocity with respect to each other will not, in general, agree on the distance between two objects or the duration of some interval of time they both observe. However, suppose an observer sees a particle at time t and at (cartesian) coordinates (x, y, z) and later observes the particle at time $t + \mathrm{d}t$ to be at the point $(x + \mathrm{d}x, y + \mathrm{d}y, z + \mathrm{d}z)$. If both observers record these measurements, they will both agree on the value of the quantity

$$\mathrm{d}s^2 = c^2\mathrm{d}t^2 - \mathrm{d}x^2 - \mathrm{d}y^2 - \mathrm{d}z^2 \qquad (1)$$

(where c is the speed of light in vacuum), which is called the metric on Minkowski space-time. In tensor notation, $\mathrm{d}s^2 = \sum \eta_{\mu\nu}\,\mathrm{d}x^\mu \mathrm{d}x^\nu$, where the four values of the subscripts μ, ν correspond to the coordinates on the four-dimensional space-time, and $\eta_{\mu\nu} = \mathrm{diag}(1, -1, -1, -1)$ is the metric tensor. Minkowski space-time is flat; that is, it can be identified with its tangent space.

In general relativity, described by the Einstein equations, the constants $\eta_{\mu\nu}$ are replaced by more general, coordinate-dependent metric coefficients $g_{\mu\nu}$, and space-time is allowed to have nonzero curvature. This curvature manifests itself as gravitation and is caused by the presence of matter. In any space-time, test particles follow geodesics (paths that minimize distance locally, i.e., generalizations of straight lines) of the space-time. In the now-famous words of the astrophysicist John Archibald Wheeler (Misner, Thorne & Wheeler, 1973),

> Matter tells space-time how to curve and curved space tells matter how to move.

The latter part of the sentence corresponds to the geodesics; the first part is the content of the Einstein equations, which in tensor notation are

$$G_{\mu\nu} = 8\pi\, T_{\mu\nu}, \qquad \mu, \nu = 0, 1, 2, 3, \qquad (2)$$

where we are using units in which the speed of light and Newton's gravitational constant both have the value 1. The tensor $T_{\mu\nu}$ is called the stress-energy or energy-momentum tensor, and its components measure several physical properties of continuous matter. In vacuum, $T_{\mu\nu} = 0$. The tensor $G_{\mu\nu}$ is called the Einstein tensor. It is constructed from $g_{\mu\nu}$ and its first two derivatives, and it is a measure of the curvature of space-time. In short, the left-hand side of the Einstein equations (2) measures curvature, which encodes the geometry of the space-time, while the right-hand side of the equations measures energy, momentum, and stress, and so encodes the physical properties of the matter. The Einstein equations can be derived from an action principle and are constructed in such a way as to satisfy a generalization of the law of conservation of energy.

Apart from the flat space-time of Minkowski, the most important solutions of the Einstein equations are those with spherical symmetry. In Schwarzschild coordinates, the metric of any vacuum, asymptotically flat, spherically symmetric space-time has the form

$$ds^2 = \left(1 - \frac{2M}{r}\right) dt^2 - \left(1 - \frac{2M}{r}\right)^{-1} dr^2 \\ -r^2(d\theta^2 + \sin^2\theta\, d\phi^2), \qquad (3)$$

where M is a constant (Schwarzschild, 1916). The Schwarzschild metric (3) is used to describe the external field of a nonrotating spherically symmetric matter distribution (such as a star) of mass M. Expression (3) becomes singular at the so-called Schwarzschild radius $r = 2M$. In ordinary stellar models, the Schwarzschild radius lies deep in the interior of the star where the vacuum metric (3) does not apply. However, if a star undergoes gravitational collapse, then its radius shrinks to zero and the Schwarzschild radius is in the vacuum region.

By computing certain curvature invariants, it can be shown that there is no physical singularity at $r = 2M$. In other words, there is nothing wrong with the space-time manifold at $r = 2M$; it is simply that the coordinates r and t are bad here. This phenomenon is called a coordinate singularity. Use of the so-called Kruskal–Szekeres coordinates allow us to explicitly extend the space-time past this singularity. The surface $r = 2M$ is the event horizon of a black hole. Nothing can escape from the interior of the event horizon, including light, and will eventually fall into the (physical) singularity at $r = 0$ in finite proper time, where the theory breaks down.

Although a great number of exact solutions of the Einstein equations are known (Stephani et al., 2003), they describe very special and often unphysical situations. In order to study more general situations, extensive numerical studies have been undertaken. There are many ways in which the numerical evolution of the Einstein equations is especially difficult. The dependent variables of the Einstein equations are metric coefficients which describe the space-time manifold on which the independent variables (the space-time coordinates) live. Many issues stem from the fact that there is no preferred frame in general relativity meaning that one has to deal with the choice of an appropriate gauge and to recognize and avoid coordinate singularities. The Einstein equations are a system of ten strongly coupled nonlinear PDEs.

The Einstein equations as written in (2) are not in evolution form. For numerical purposes the Einstein equations are usually written in terms of an evolution variable λ. A Cauchy or characteristic approach corresponds to the normal of the $\lambda = \text{const}$ hypersurfaces being time-like or null, respectively. Einstein's equations then project to a set of constraint equations on the hypersurfaces and a set of evolution equations. If the constraint equations are satisfied on an "initial" hypersurface, then they are preserved under evolution to other hypersurfaces.

Numerical studies using the Cauchy approach have been used to model the collision of axisymmetric black holes (Anninos et al., 1995), and they have led to the discovery of critical phenomena in spherical collapse (Choptuik, 1993). One drawback of this method is that boundary conditions are usually artificially imposed on the hypersurfaces to avoid integrating out to infinity (although sometimes conformal methods can be used to include space-like infinity). The characteristic approach has led to the first unlimited evolution of a single black hole space-time (Gomez et al., 1998). While this approach allows for long-time simulations, it is only valid in the absence of crossing characteristics or caustics.

General relativity makes a number of predictions that have been tested. These include the excess advance in the perihelion of Mercury, the bending of light rays near the sun, time delays in radar signals passing near the sun, and gravitational lensing of distant galaxies by nearer ones. The most important prediction still awaiting confirmation is the existence of gravitational waves. A number of detectors have now been built around the world and observations are expected soon.

ROD HALBURD AND GINO BIONDINI

See also **Black holes; Cosmological models; General relativity; Gravitational waves**

Further Reading

Anninos, P., Hobill, D., Seidel, E., Smarr, L. & Suen, W.-M. 1995. Head-on collision of two equal mass black holes, *Physical Review* D, 52: 2044–2058

Choptuik, M.W. 1993. Universality and scaling in gravitational collapse of a massless scalar field. *Physical Review Letters*, 70: 9–12

Einstein, A. 1905. Zur elecktrodynamik bewegter Körper. *Annalen der Physik*, 17: 891–921

Gomez, R., Lehner, L., Marsa, R.L. & Winicour, J. 1998. Moving black holes in 3D, *Physical Review*, D57: 4778–4788

Minkowski, H. 1908. Space and time. Translated in *The Principle of Relativity*, New York: Dover, 1923

Misner, C.W., Thorne, K.S. & Wheeler, J.A. 1973. *Gravitation*, San Francisco: Freeman

Schwarzschild, K. 1916. Über das Gravitationsfeld einer Kugel aus inkompressibler Flussigkeit nach der Einsteinschen Theorie, *Sitzungsberichte der Preussischen Akademie der Wissenschaften, Sitzung der Physikalisch-mathematischen Klasse*, 424–434

Stephani, H., Kramer, D., MacCallum, M., Hoenselaers, C. & Herlt, E. 2003. *Exact Solutions of Einstein's Field Equations*, Cambridge and New York: Cambridge University Press

Wald, R.M. 1984. *General Relativity*, Chicago: University of Chicago Press

ELASTIC AND INELASTIC COLLISIONS

See **Collisions**

ELECTROENCEPHALOGRAM AT LARGE SCALES

Since the first human scalp recordings of the tiny electric currents generated by the brain (electroencephalogram or EEG) were obtained in the mid-1920s, EEG has been recognized as a genuine (if often opaque) window on the mind, allowing observations of brain processes to be correlated with behavior and cognition. EEG has important applications in medicine, including epilepsy, head trauma, drug overdose, brain infection, sleep disorder, coma, stroke, tumor, monitoring anesthesia depth, and fundamental cognitive studies. Most EEG power occurs at frequencies below about 15 Hz in scalp (not cortical) recordings. EEG and magnetoencephalography (MEG) are the only technologies with sufficient temporal resolution to follow the fast dynamic changes associated with cognition; however, EEG spatial resolution is poor relative to modern brain structural imaging methods such as positron emission tomography (PET), computed tomography (CT), and magnetic resonance imaging (MRI). Each scalp electrode records electrical activity at very large scales, involving cortical tissue containing perhaps 10^7–10^9 neurons.

Mesoscopic and Microscopic Sources

Scalp potentials are generated by micro-current sources at cell membranes. Sorting out the complex relations between micro-sources and macroscopic scalp potentials is facilitated by assuming an intermediate (mesoscopic) descriptive scale that recognizes the columnar structure of the neocortex. From this perspective, the mesoscopic source strength of a volume of tissue is its electric current dipole moment per unit volume (microamps/mm^2), designated as $\boldsymbol{P}(\boldsymbol{r}', t)$. This function represents the weighted average of micro-source activity in a volume of the neocortex (near $\boldsymbol{r}'$) that is large compared with the scale of individual neurons yet small compared with the size of a typical EEG electrode. For the idealized case of micro-sources of one sign confined to a superficial cortical layer and micro-sources of opposite sign confined to a deep layer, $\boldsymbol{P}(\boldsymbol{r}', t)$ is roughly the mesoscopic current density across a cortical column (≈ 1 mm^2). The contributing electrical activities are primarily dendritic post-synaptic currents, although currents associated with axonal action potentials (spikes) may also contribute (see Figure 1).

Mesoscopic Sources and Scalp Recordings

Human neocortical sources form dipole layers over which the function $\boldsymbol{P}(\boldsymbol{r}', t)$ varies with cortical location ($\boldsymbol{r}'$), measured in and out of cortical folds. In a few special cases, $\boldsymbol{P}(\boldsymbol{r}', t)$ may be approximated by a few cm-scale active regions, consisting of focal sources as in focal epilepsy or mid-latency components of evoked potentials (before tangential spread to other cortical locations). In general, however, $\boldsymbol{P}(\boldsymbol{r}', t)$ is widely distributed, perhaps over the entire cortical surface. Most EEGs are believed to be generated as a linear sum of contributions from cortical sources, in which case cortical or scalp potential may be approximated by the following integral of dipole moment over the cortical surface:

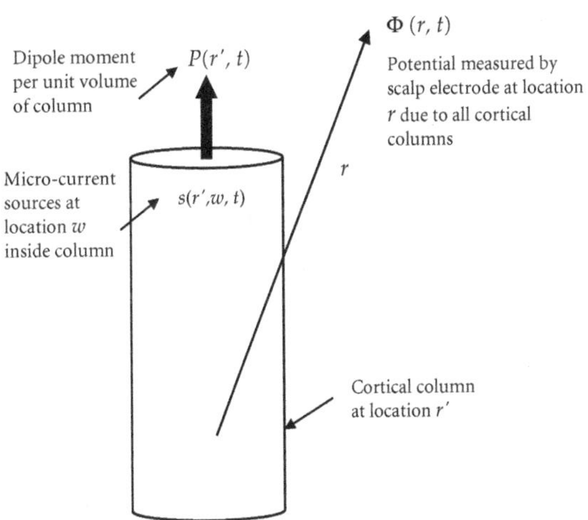

Figure 1. A mesoscopic tissue mass (for example, a mm-scale cortical column containing millions of volume micro-current sources $s(\boldsymbol{r}', \boldsymbol{w}, t)$) produces a current dipole moment per unit volume $\boldsymbol{P}(\boldsymbol{r}', t)$, or meso-source strength. Cortical or scalp potential due to brain sources is the weighted integral of $\boldsymbol{P}(\boldsymbol{r}', t)$ over the brain volume or, in the case of exclusively cortical sources, the integral over the cortical surface.

$$\Phi(\boldsymbol{r}, t) = \int_S \boldsymbol{G}(\boldsymbol{r}, \boldsymbol{r}') \cdot \boldsymbol{P}(\boldsymbol{r}', t)\, \mathrm{d}S(\boldsymbol{r}'). \qquad (1)$$

The Green function $\boldsymbol{G}(\boldsymbol{r}, \boldsymbol{r}')$ contains all geometric and conductive information about the head volume conductor. Cortical dipole moment may in turn be expanded in a series of basis functions $\boldsymbol{p}_n(\boldsymbol{r})$:

$$\boldsymbol{P}(\boldsymbol{r}, t) = \sum_{n=0}^{\infty} \xi_n(t) \boldsymbol{p}_n(\boldsymbol{r}). \qquad (2)$$

An idealized model neocortex consists of a thin spherical shell with $\boldsymbol{P}(\boldsymbol{r}', t)$ everywhere normal to the surface, reflecting the columnar structure of the closed neocortical surface. An appropriate choice of basis functions for this idealized cortex is the set of spherical harmonics $Y_{lm}(\theta, \phi)\boldsymbol{a}_r$, where $\boldsymbol{a}_r$ is a unit vector in the radial direction and (θ, ϕ) are the usual spherical coordinates. The single sum over n in Equation (2) may then be expressed as a double sum $l = 0, \infty; m = -l, +l$, associated with the spherical geometry. Combining Equations (2) and (1) yields the following series expansion for the cortical $\boldsymbol{r}_C$ or scalp $\boldsymbol{r}_S$ surface potential $\Phi(\boldsymbol{r}, t)$ in terms of a new set of basis functions $\phi_n(\boldsymbol{r})$ that are surface integrals of the dot product of the Green's function $\boldsymbol{G}(\boldsymbol{r}, \boldsymbol{r}')$ with the basis functions $p_n(\boldsymbol{r})$. Thus,

$$\Phi(\boldsymbol{r},t)=\sum_{n=0}^{\infty}\xi_n(t)\phi_n(\boldsymbol{r}). \tag{3}$$

General Dynamic Properties of Cortical and Scalp Potentials

As summarized below, EEG exhibits many dynamic behaviors, depending on recording location, physiologic state, and subject.

(i) Often complex physical or biological systems can be adequately characterized by only a few terms in Equation (3). The time-dependent coefficients $\xi_n(t)$ are called order parameters in the field of synergetics and may be governed by nonlinear differential or integral equations. The basis functions $\phi_n(\boldsymbol{r})$ may be chosen (bottom up) by physiological theory or (top down) by experimental data, for example, by constructing Karhunen–Loeve expansions (also called principal components analysis) in which the $\phi_n(\boldsymbol{r})$ are chosen as the most efficient set of orthogonal functions representing a data record.

(ii) The basis functions $\phi_n(\boldsymbol{r})$ are typically ordered in terms of progressively higher spatial frequencies as in Fourier series. The index n is then a measure of the two-dimensional spatial frequency of the corresponding basis function. Many systems exhibit a correspondence between spatial and temporal frequencies such that Fourier transforms of the order parameters $\xi_n(\omega)$ peak at higher frequencies ω for higher spatial frequencies n. For linear wave phenomena, such a correspondence is called the dispersion relation.

(iii) The large-scale spatiotemporal dynamics of cortical potential $\Phi(\boldsymbol{r}_\mathrm{c},t)$ are believed to be very similar to the dynamics of the mesoscopic cortical sources $\boldsymbol{P}(\boldsymbol{r},t)$. The head volume conductor acts as a low-pass spatial filter, resulting mainly from the low electrical conductivity of skull and the physical separation (1–2 cm) between sources and electrodes. Scalp potential $\Phi(\boldsymbol{r}_\mathrm{s},t)$ is then a spatial low-pass representation of cortical potential $\Phi(\boldsymbol{r}_\mathrm{c},t)$ or mesoscopic source function $\boldsymbol{P}(\boldsymbol{r},t)$. Comparisons of cortical potential (ECoG) with scalp potential (EEG) show that this spatial filtering results in temporal filtering. That is, the Fourier transform of cortical potential $\Phi(\boldsymbol{r}_\mathrm{c},\omega)$ typically contains much more relative power at higher frequencies (say 15–40 Hz) than the scalp potential transform $\Phi(\boldsymbol{r}_\mathrm{s},\omega)$, recorded in the same brain state, an observation qualitatively consistent with normal wave dispersion relations.

(iv) EEG phenomena typically exhibit larger amplitudes at lower frequencies. In deep sleep and moderate to deep anesthesia, $\Phi(\boldsymbol{r}_\mathrm{S},t)$ is typically a few 100 μV with nearly all power in the delta range (0–4 Hz) at all scalp locations. The eyes closed, waking alpha rhythm (ca. 40 μV) is normally dominated by one or two spectral peaks in the 8–13 Hz range at widespread scalp locations. Low-amplitude beta activity (ca. 13–20 Hz) superimposed on alpha rhythms is more evident in frontal cortex. Alcohol and hyperventilation typically lower alpha frequencies and increase amplitudes; barbiturates increase beta activity. Scalp EEG activity is more consistent with limit cycle modes $\xi_n(t)$ than low-dimensional chaos.

(v) Scalp EEG is a weighted space average of many cortical rhythms that can look different in different cortical regions. Alpha rhythms have been recorded from nearly the entire upper cortical surface, including frontal and prefrontal areas. Differences in ECoG waveforms between cortical areas are largely eliminated with anesthesia, suggesting shifts from more functional localization to more globally dominated brain states.

(vi) Both globally coherent and locally dominated behavior can occur within the alpha band, depending on narrow band frequency, measurement scale and brain state. Upper alpha (ca. 10 Hz) and theta (ca. 4–6 Hz) phase locking between cortical regions during mental calculations often occurs, consistent with neural network formation. At the same time, quasi-stable alpha phase structures consistent with global standing waves have been observed.

Neocortical Dynamic Theory

The apparent balance among locally and globally dominated dynamic processes has been estimated by phase synchronization among other measures. Cortical or thalamic interactions with time delays due to rise and decay times of post-synaptic potentials (local theory) have been modeled. Network frequencies are then determined only by local tissue properties. In global theories, characteristic frequencies depend on the entire neocortex/cortico-cortical fiber system. Excitatory synaptic action density $F(\boldsymbol{r},t)$ may be expressed in terms of action potential density $\Gamma(\boldsymbol{r},t)$ by

$$F(\boldsymbol{r},t)=\int_0^{\infty}\mathrm{d}\nu\int_S R(\boldsymbol{r},\boldsymbol{r}_l,\nu)\times\Gamma\left(\boldsymbol{r}_l,t-\frac{|\boldsymbol{r}-\boldsymbol{r}_l|}{v}\right)\mathrm{d}S(\boldsymbol{r}_l) \tag{4}$$

The outer integral is over distributed cortico-cortical propagation speeds ν, and the inner integral is over the neocortical surface S. The dependence of cortico-cortical fiber density with distance $|\boldsymbol{r}-\boldsymbol{r}_l|$ is expressed by the function $R(\boldsymbol{r},\boldsymbol{r}_l,\nu)$. This linear relation between the dependent variables $F(\boldsymbol{r},t)$ and $\Gamma(\boldsymbol{r},t)$ may be combined with a variety of nonlinear local equations in the same dependent variables. Such local/global theories include both network-to-global and global-to-network interactions.

PAUL NUÑEZ

See also **Cell assemblies; Electroencephalogram at mesoscopic scales; Gestalt phenomena; Synergetics**

Further Reading

Edelman, G.M. & Tononi, G. 2000. *A Universe of Consciousness*, New York: Basic Books

Freeman, W.J. 1975. *Mass Action in the Nervous System*, New York: Academic Press

Haken, H. 1983. *Synergetics. An Introduction*, 3rd edition, Berlin: Springer

Malmuvino, J. & Plonsey R. 1995. *Bioelectromagetism*, New York: Oxford University Press

Niedermeyer, E. & Lopes da Silva, F.H. (editors). *electroencephalography. Basic Principals, Clinical Applications, and Related Fields*, 4th edition. London: Williams and Wilkins

Nuñez, P.L. 1981. *Electric Fields of the Brain: The Neurophysics of EEG*, New York: Oxford University Press

Nuñez, P.L. 1995. *Neocortical Dynamics and Human EEG Rhythms*, New York: Oxford University Press

Nuñez, P.L., Wingeier, B.M. & Silberstein, S.B. 2001. Spatial-temporal structure of human alpha rhythms: theory, microcurrent sources, multiscale measurements, and global binding of local networks. *Human Brain Mapping*, 13: 125–164

Nuñez, P.L. 2000. Toward a quantitative description of large scale neocortical dynamic function and EEG. *Behavioral and Brain Sciences*, 23: 371–437

Penfield, W. & Jasper, H.D. 1954. *Epilepsy and the Functional Anatomy of the Human Brain*, London: Little, Brown and Co

Scott, A.C. *Stairway to the Mind*, Berlin and New York: Springer, 1995

Uhl, C. (editor). 1999. *Analysis of Neurophysiological Brain Functioning*, Berlin: Springer 1999

Wilson, H.R. & Cowan, J.D. 1973. A mathematical theory of the functional dynamics of cortical and thalamic nervous tissue. *Kybernetik*, 13: 55–80

ELECTROENCEPHALOGRAM AT MESOSCOPIC SCALES

The two most dominant theoretical positions that have influenced the development of explanations of behavior are functional modularity, where specific behaviors are believed to reside in distinct cortical locations, and mass action, in which behavior is posited to arise out of the cooperative activity of distributed neural structures comprising the brain. The first of these positions predominates in the medical and cognitive sciences, based upon observations collected over many years in which selective behavioral deficits were observed to occur in response to the specific destruction or stimulation of various areas of the brain. This has led to considering the brain as a collection of interconnected functionally specialized modules. This view is generally known as modularism and in the context of higher functions is also referred to as cortical localizationism. Modules are thought to embody algorithms, acting mechanistically on input to produce output, in a computer-like manner. Mass action, however, considers behavior as best understood as arising out of cooperative neural activity occurring over a number of interacting spatial and temporal scales. The consequence of this view is that behavior can now be characterized by the observation of dynamical patterns of brain activity.

The dynamics of the brain can be observed at a number of different spatial and temporal scales. These range from the level of the ion channel, synapse, and neuron (microscopic), to the level of neuronal population (mesoscopic), up to the level of large aggregates of brain tissue (macroscopic). The associated methods can generally be divided into those that measure some form of electromagnetic activity and those that reflect the metabolic correlates of neural activity. At the mesoscopic scale electromagnetic measures typically have millisecond temporal resolution whereas metabolic measures (e.g., fMRI, PET, near-infrared spectroscopy, diffuse optical tomography) have comparatively poor temporal resolution (s), while both mesoscopic and metabolic measures have comparable spatial resolution (mm). Electromagnetic measurements of brain activity better reflect the time scales of neuronal activity associated with the dynamics of behavior. For these reasons theories of mesoscopic brain dynamics are built around state variables that characterize the electromagnetic activity associated with neural populations. The most important mesoscopic electromagnetic measures are those of the local field potential (LFP) and the electrocorticogram (ECoG) which predominately reflect the quasi-static electromagnetic fields produced in response to the ionic currents generated by synchronized synaptic activity. It is generally thought that in the cortex this synchronized activity is linearly related to the spatially averaged soma membrane potential of populations of excitatory (pyramidal) neurons.

For both excitatory and inhibitory neuronal populations the mean soma membrane potential determines the mean rate of neuronal action potential generation (or firing rate). In the simplest case, the mean population firing rate is a sigmoidal function of the mean soma membrane potential for functionally equivalent members of the same neuronal population. More generally, such a relationship will be time-variant. Because the mean soma membrane potential is a function of neuronal population synaptic input which itself is a function of the mean neuronal population firing rates, the mean soma membrane potential can be used as a canonical state variable to both characterize and develop theories of the dynamics of neuronal populations.

At the level of the neuronal population, the spatially averaged soma membrane potential is typically defined over the characteristic scales of the short-range

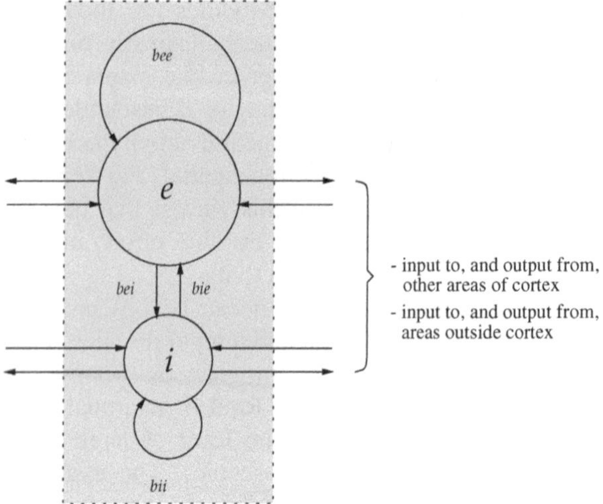

Figure 1. The neurons of mammalian cortex interact with each other locally (by short range connections) and globally (by long-range connections). Short-range connections serve to diffusely interconnect the intermixed excitatory (e) and inhibitory (i) neuronal populations of cortex. The characteristic scales of these connections loosely organize cortex into a sheet of overlapping "modules" or cylinders that span the entire thickness of cortex. The resulting pattern is often referred to as columnar organization with a single one of these cylinders known as a cortical macrocolumn. One way of quantifying the strength of interaction between these local neuronal populations is to specify the mean number of connections (or synapses) neurons of one population receive from neurons of the same or another population (b_{ij}).

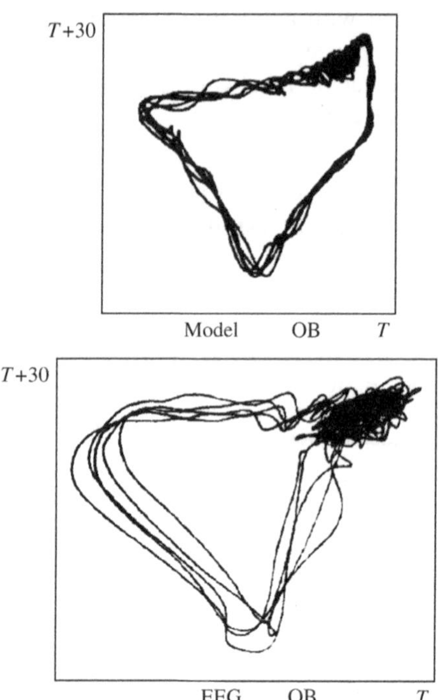

Figure 2. Delay embedded phase plane plots comparing EEG seizure activity recorded from rat olfactory bulb (a phylogenetically primitive form of cortex present in all mammals), top panel, and with a mesoscopic theory of olfactory bulb dynamics (bottom panel). The delay embedding is 30 ms for experimental and simulated time series of 1s duration, with time increasing counterclockwise. Figure reproduced with permission from Freeman (1987), Copyright Springer-Verlag.

intracortical fibers. These intracortical fibers comprise the recurrent axonal branchings of the pyramidal neurons as well as all the axonal fibres of the inhibitory interneurons. Detailed morphometric studies of cortex have established that the typical characteristic spatial scale of these fibers ranges anywhere between 30 μm to 1 mm (Braitenberg & Schüz, 1998). The advantage of using this spatially averaged state variable is that it has a commensurate spatial scale to LFP and ECoG recordings.

Most mesoscopic theories of neuronal dynamics have considered cerebral cortex to consist of two functionally distinct neural populations—excitatory (e) and inhibitory (i)—reflecting the respective pyramidal and interneuron populations. The earliest models of cortical mass action attempted to describe neuronal dynamics exclusively in terms of short- and long-range excitatory interactions between excitatory populations. While a number of interesting analytical solutions were obtained using this approach, they are no longer of any significant physiological relevance because they did not incorporate the activity of inhibitory neurons. Subsequent models, based on anatomical considerations, have addressed this deficiency by the incorporation of all forms of feedforward and feedback connectivity between excitatory and inhibitory neuronal populations. Starting with Wilson & Cowan (1972), most modern theories functionally incorporate local $e \to e$, $e \to i$, $i \to e$, and $i \to i$ connections, as illustrated in Figure 1. However, a notable exception is the influential body of work by Lopes da Silva et al. in which all the significant population neurodynamics arise out of the reciprocal interactions between excitatory and inhibitory neurons.

In the majority of theories the dynamical response of the mean soma membrane potential to synaptic activity induced by population neuronal firings is generally described using a differential formalism.

$$\dot{h}_n = -a_n h_n + \sum_{r=1}^{N} g_{nr}(h_n) I_{nr}, \tag{1}$$

$$\mathcal{D}_{nr} I_{nr} = S_{nr}(h_r) \quad n, r = 1 \ldots N, \tag{2}$$

where N is the number of locally interacting neuronal populations and the a_n are real constants which correspond approximately to the reciprocal of the mean neuronal membrane time constant. In most theories $N = 2$. h_n is the mean soma membrane potential of the nth neuronal population, and S_{nr} defines functions that take into account the topology of neuronal population connectivity as well as the sigmoidal relationships between mean soma membrane potential and mean firing rate. $g_{nr}(h_n) I_{nr}$ represents the postsynaptic current induced in local neuronal population n by neuronal population r, whose temporal evolution is determined by the temporal differential operator $\mathcal{D}_{nr}$. This operator takes into account bulk neurotransmitter kinetics and neuronal cable properties.

Theories of mesoscopic dynamics can generally be distinguished by the form and order of $\mathcal{D}_{nr}$, g_{nr} and S_{nr}, and N. For instance, in the original work of Wilson and Cowan, $\mathcal{D}_{nr}$ was of zero order in time, whereas in the theories of Lopes da Silva, $\mathcal{D}_{nr}$ is of first order and $g_{nr}(h_n) = \text{constant}$. More recent theories have considered $g_{nr}(h_n)$ linear in h_n and $\mathcal{D}_{nr}$ second order in time due to more detailed physiological considerations, (Liley et al., 2002). Differing forms and parametrizations of $\mathcal{D}_{nr}$, g_{nr}, and S_{nr} have been shown to give rise to a wide range of dynamics (e.g., limit cycle, chaos, and filtered noise) some of which bear strong similarities to experimental recordings (see Figure 2; see also figures in color plate section).

DAVID LILEY AND MATHEW DAFILIS

See also **Cell assemblies; Electroencephalogram at large scales; Nerve impulses; Neurons; Synergetics**

Further Reading

Braitenberg, V. & Schüz, A. 1998. *Cortex: Statistics and Geometry of Neuronal Connectivity*. Berlin and New York: Springer

Freeman, W.J. 1975. *Mass Action in the Nervous System*. New York: Academic Press

Freeman, W.J. 1987. Simulation of chaotic EEG patterns with a dynamic model of the olfactory system. *Biological Cybernetics*, 56(2–3): 139–150

Freeman, W.J. 2000. *Neurodynamics: An Exploration in Mesoscopic Brain Dynamics*. London and New York: Springer

Liley, D.T.J., Cadusch, P.J. & Dafilis, M.P. 2002. A spatially continuous mean field theory of electrocortical activity. *Network: Computation in Neural Systems*, 13(1): 67–113

van Gelder, T. 1999. Dynamic approaches to cognition. In *The MIT Encyclopedia of the Cognitive Sciences*, edited by R.A. Wilson & F.C. Keil, Cambridge, MA: MIT Press

van Rotterdam, A., Lopes da Silva, F.H., van den Ende, J., Viergever, M.A. & Hermans, A.J. 1982. A model of the spatial-temporal characteristics of the alpha rhythm. *Bulletin of Mathematical Biology*, 44(2): 283–305

Wilson, H.R. & Cowan, J.D. 1972. Excitatory and inhibitory interactions in localized populations of model neurons. *Biophysical Journal*, 12: 1–24

ELECTRON BEAM MICROWAVE DEVICES

Electron beam devices have been used in a wide variety of applications since the beginning of vacuum technology early in the 20th century. These include simple diodes and triodes, used in early radio; cathode-ray tubes, used for display in oscilloscopes, televisions, and computers; lithography; high-voltage devices used to produce X-rays; sources for electron accelerators; and the energy source for microwave devices, with applications in communications, microwave ovens, industrial processes, and electronic warfare. For the purposes of this entry, we will limit our discussion to just one application, the production of electromagnetic energy at microwave frequencies, where nonlinear space charge effects tend to play a major role. For a wider range of applications, including the basic theory, texts are available such as Harmon (1953) and Gilmour (1994). It should be noted that solid state devices, not involving electron beams, are now often used at low power, particularly as oscillators to drive microwave electron-beam amplifiers.

Generation of microwaves with electron beams involves the interaction of a beam with electromagnetic fields across gaps of discrete cavities or synchronously with electromagnetic fields propagating on a slow wave structure. The basic physical principle is that a bunched beam interacts with the decelerating phase of an electric field to transfer energy from the beam to the field. This mechanism is the converse of the interaction in linear accelerators and synchrotrons that operate by interaction of a field in an accelerating phase with a bunched beam to transfer energy from the fields to the beam. In a device such as a traveling-wave tube (TWT) or magnetron, in which the beam interacts with a traveling wave, there is a natural bunching action as the wave develops, such that most of the beam particles are decelerated, producing the wave amplification. In a klystron, with a bunching cavity and an energy extraction cavity, the proper phase relation must be established, which is, in fact, equivalent to the synchronous condition in the TWT.

The earliest beam-type device to be extensively used for microwave generation is the magnetron, a coaxial cylindrical device with a potential between inside cathode and outside anode cylinders, and a magnetic field along the cylindrical axis. The electrons accelerated from the cathode toward the anode perform cycloidal motion due to the magnetic field with an average azimuthal drift velocity in the combined electric and magnetic fields given by $\boldsymbol{v}_{\text{d}} = \boldsymbol{E} \times \boldsymbol{B}/B^2$. The anode consists of a series of microwave cavities with the operating frequency, chosen such that the cavities operate in the π-mode, that is, 180° phase shift between cavities. The fields and dimensions are chosen such that the electrons are in synchronism with the fundamental Fourier component of the fields that appear across the cavity openings. Through a complicated nonlinear process, the electrons bunch at a phase with respect to the traveling microwave field to give energy to the field, finally being collected on the anode structure with an energy considerably less than the energy eV, where V is the cathode-anode accelerating voltage. Simplified models describe some of the processes (Hutter, 1965), but most development has been experimental, assisted more recently by detailed numerical calculations (Lemke et al., 2000). The magnetron configuration described above operates naturally as an oscillator. It has high efficiency and usually operates at high power. It was an essential element in the development of radar during World War II, which was important in winning the air war over Great Britain. The magnetron has become ubiquitous in the consumer market, as the power source in the household microwave oven. Magnetrons, despite being inexpensive and robust, have a number of unfavorable characteristics, including excess noise, narrow bandwidth, and difficulty in tuning, which limit their applications.

The klystron, which was developed roughly during the same period of time as the magnetron, is more flexible because it can operate, with small modifications, as either an amplifier or an oscillator, and versions operate over very large ranges of power and frequency. The basic configuration is two cylindrical microwave cavities with an electron beam passing through their center. The cavities have a reentrant shape such that the central hole consists of a narrow gap. The beam is velocity modulated by an alternating electric field across the first cavity gap which becomes density modulated in the second gap. At very low beam density, the electron trajectories can be taken to be kinematic, with the bunching distance related to the beam velocity and the perturbed velocity created by the gap fields. However, the usual operation is in a regime where space charge effects are fundamental to the operation. The excitation produces two waves, fast and slow space charge waves, which give a beat modulation distance $\lambda_b/4 = \pi v_0/2\omega_{pb}$, where v_0 is the beam velocity and ω_{pb} is the plasma frequency of the beam. The trajectories are shown schematically in Figure 1. The beam plasma frequency is reduced from the natural electron oscillation $\omega_p = ne^2/m\epsilon_0$ due to the transverse beam dimensions. A second cavity at a position $z = \lambda_b/4$ from the first cavity is excited by a maximum of the radio-frequency current in the beam. The amplifier is converted into an oscillator by either feeding back some of the signal externally from the second cavity to the first cavity or by reflecting the modulated beam, after traversing only a single cavity, to self-excite it as a reflex klystron. See a basic text such as Harmon (1953), or more detail in Hutter (1965), for mathematical analysis.

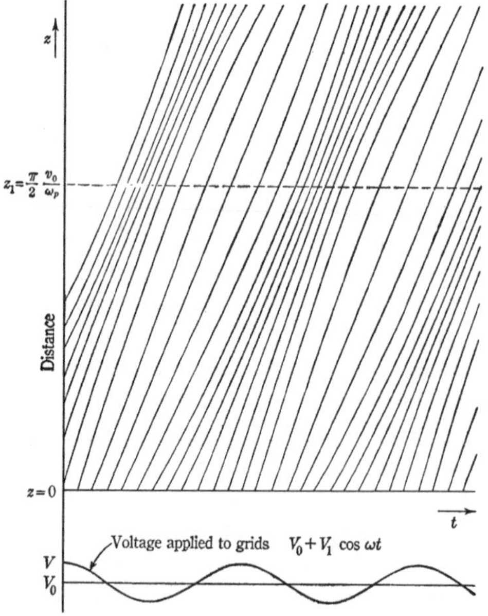

Figure 1. Distance-time diagram of a klystron indicating how velocity variations at the input grids result in density variations farther down the electron stream (after Harmon, 1953).

The traveling-wave tube (TWT), developed somewhat later, competes for applications with the klystron at all but the highest powers. The device transfers energy from a beam to a slow wave structure, usually a helix, by a resonant interaction in which the beam becomes naturally modulated in the presence of the wave. A linear, or small signal, analysis is sufficient to obtain the amplifying properties of the device for many low-power applications. However, as the microwave power becomes comparable to the beam power, nonlinear considerations become very important. As in the klystron, space charge is a fundamental source of nonlinearity. If significant power is extracted from the beam, the wave velocity on the circuit must be reduced along its length to maintain the coherent synchronous interaction. Due to the periodicity of slow wave structures, the propagation characteristics have operating regions for which the group velocity $v_g = d\omega/dk$ is opposite in direction to the phase velocity $v_{ph} = \omega/k$ (Brillouin, 1953). If the TWT is operated with such parameters, then wave energy is propagated backward from the direction in which the beam is moving and in which the wave amplitude, by interaction with the beam, is growing. This feedback allows the beam to self-excite the wave and therefore become an oscillator, which is called a backward wave oscillator (BWO). The operating conditions of a BWO are inherently nonlinear. In addition to the general books already mentioned, a detailed account of TWTs by one of the scientists associated with their development, is Pierce (1950). Sketches of the three basic types of devices are shown in Figures 2, 3, 4. There are many variants of these basic configurations.

Because electron beams are the power source and the active interacting medium for both klystron and TWT amplifiers and their associated oscillators, beam design is of great importance in obtaining good operating

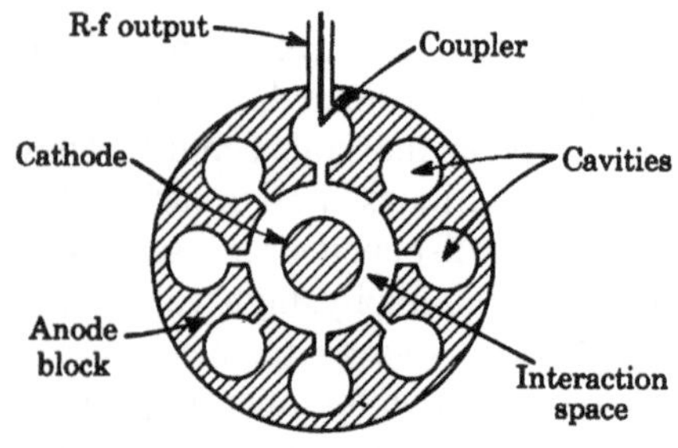

Figure 2. Multicavity magnetron (after Hutter, 1965).

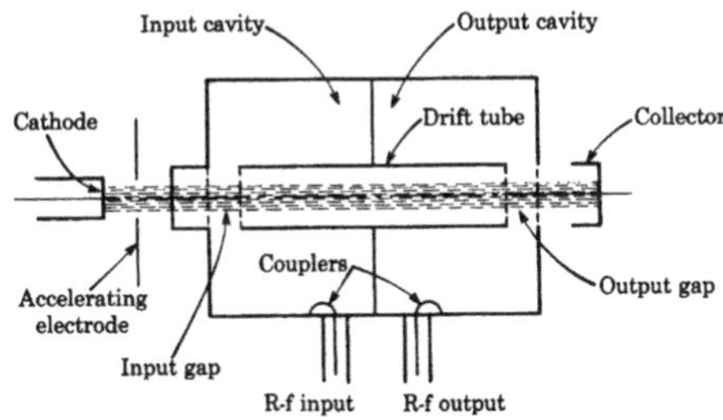

Figure 3. Two-cavity klystron (after Hutter, 1965).

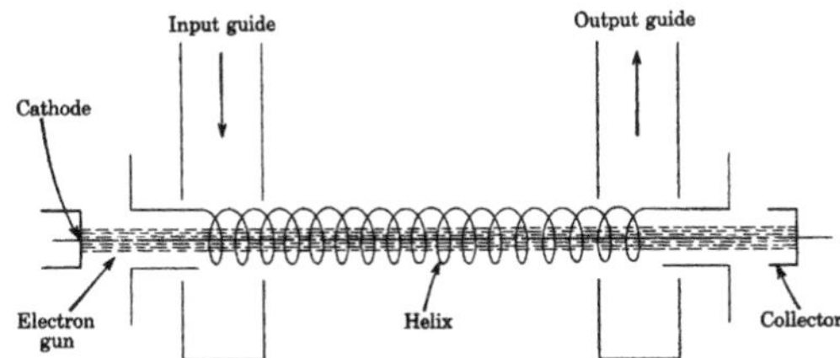

Figure 4. Helix-type traveling-wave tube (after Hutter, 1965).

characteristics. Some of the issues involved are collimation (minimizing transverse emittance), energy spread (minimizing longitudinal emittance), and high current and/or current density. The collimation is often accomplished with a uniform magnetic field along the beam axis, but radial space charge forces also lead to beam rotation. Alternatively, electric or magnetic lenses can be employed. An important consideration is to launch the beam from an accelerating region in a smooth manner; this led to the electrode shape known as the Pierce diode. An important method of confining a beam with significant space charge is to inject the beam into a uniform magnetic field from a magnetic field-free region. The proper choice of magnetic field leads to azimuthal rotation that just takes up the space charge potential, lending to smooth, uniform axial flow, called Brillouin flow. Many of the basic topics of beam dynamics and beam design are covered in Pierce (1954).

Some of the important considerations that have motivated the development of various microwave beam devices have been noise characteristics, particularly achieving low noise to amplify very low signals; frequency ranges, particularly pushing devices to increasingly high frequencies for broadband applications; tunability; and power and efficiency, particularly to obtain high power at high efficiency. For certain applications, there are also other types of constraints such as ruggedness, reliability, and weight limits, which we will not consider here. Low noise requirements have tended to favor TWTs over klystrons on the front end of receivers. Higher frequencies, used to obtain more bandwidth in communication applications, have led to miniaturization of klystrons, but other types of devices such as free electron lasers can obtain even higher frequencies for some applications (Freund & Antonsen, 1996). Some of the key issues motivating nonlinear analysis have arisen from the requirement of obtaining high efficiency at high power. Currently, the state of the art in high power is to produce 1GW in a pulse of 0.1kJ. For studies of more recent analysis and developments the reader is referred to Benford & Swegle (1992) and Barker & Schamiloglu (2001).

To obtain high efficiency the majority of the electrons should be trapped in the wave field, such that they can be decelerated. This requires matching the phase space of the beam emittance to the acceptance of the wave field, usually by some additional bunching mechanisms. Some of the basic ideas are treated in Lichtenberg (1969), but each device needs detailed numerical trajectory calculations to optimize trapping. In addition to trapping, the exiting spent beam must be decelerated to retrieve the excess energy (a depressed collector). Reducing the collector voltage is limited by the requirement that the average beam energy of the exiting beam must be greater than its energy spread so as not to turn electrons around. This condition requires knowledge of the nonlinear characteristics of the longitudinal phase space emittance.

ALLAN J. LICHTENBERG AND JOHN P. VERBONCOEUR

See also **Particle accelerators**

Further Reading

Barker, R.J. & Schamiloglu, E. (editors). 2001. *High Power Microwave Sources and Technologies*. New York: IEEE Press

Benford, J. & Swegle, J. 1992. *High Power Microwaves*, Norwood, MA: Artech House

Brillouin, L. 1953. *Wave Propagation in Periodic Structures*. New York: Dover

Freund, H.P. & Antonsen, Jr., T.M. 1996. *Principles of Free Electron Lasers*, London: Chapman & Hall

Gilmour, A.S. 1994. *Principles of Traveling Wave Tubes*, Boston: Artech House

Harmon, W.W. 1953. *Fundamentals of Electronic Motion*, New York: McGraw-Hill

Hutter, R.G.E. 1965. *Beam and Wave Electronics in Microwave Tubes*, Cambridge, MA: Boston Technical Publishers

Lemke, R.W., Genoni, T.C. & Spencer, T.A. 2000. Effects that limit efficiency in relativistic magnetrons. *IEEE Transactions on Plasma Science*, 28: 887–897

Lichtenberg, A.J. 1969. *Phase-Space Dynamics of Particles*, New York: Wiley

Pierce, J.R. 1950. *Traveling-Wave Tubes*, New York: Van Nostrand

Pierce, J.R. 1954. *Theory and Design of Electron Beams*, 2nd edition, New York: Van Nostrand

ELLIPTIC FUNCTIONS

Elliptic functions were first introduced as inverses of elliptic integrals, so called because the integral for the arclength of the ellipse studied by John Wallis in the 17th century (Stillwell, 1989) was the first such example. By the end of the 18th century, in particular after the work of Leonhard Euler, Joseph-Louis Lagrange, and Adrien-Marie Legendre, mathematicians had realized that integrals of the form $\int dx/\sqrt{P(x)}$, where $P(x)$ is a cubic or quartic polynomial, could not be expressed in terms of elementary functions (or their inverses). Many elliptic integrals arise naturally in the solutions of problems in mechanics (Lawden, 1989). For example, the solution of the simple pendulum equation

$$\ddot{\theta} + \omega^2 \sin\theta = 0$$

can be obtained using conservation of energy $\frac{1}{2}(\dot{\theta})^2 - \omega^2\cos\theta = E, E > -\omega^2$, to arrive at the inversion problem for the integral

$$t - t_0 = \int_0^{\theta} \frac{d\theta}{\sqrt{2E + \omega^2 - \omega^2 \sin^2(\theta/2)}} d\theta.$$

Substituting $z = \sin(\theta/2)$ and defining $k^2 = \omega^2/(E + \omega^2) \in (0, 1)$, one arrives at the expression

$$K(z; k) = \int_0^z \frac{dz}{\sqrt{1 - z^2}\sqrt{1 - k^2 z^2}} = \frac{\omega}{k}(t - t_0). \quad (1)$$

In the late 19th century, Niels Henrik Abel, Carl Jacobi, and Carl Friedrich Gauss observed the similarity between $K(z; k)$ and the integral defining the transcendental function arcsin z as

$$\text{arcsine}\, z = \int_0^z \frac{dz}{\sqrt{1 - z^2}}$$

(which is indeed the limiting case of (1) as $k \to 0$, giving the solution for the small amplitude oscillations of the pendulum), and introduced the *Jacobi elliptic sine function* $\text{sn}(z; k)$ as the inverse of $K(z; k)$

$$z = K(\text{sn}(z; k); k).$$

While the circular (trigonometric) functions $\sin z$, $\cos z$, $\tan z$, etc., are singly periodic functions of the complex variable z, satisfying $f(z + 2\pi) = f(z)$, elliptic functions can be characterized as the doubly periodic functions, which only possess pole singularities. Elliptic functions satisfy

$$f(z + 2\omega_1) = f(z), \quad f(z + 2\omega_2) = f(z),$$

where ω_1 and ω_2 (the *periods* of f) are complex numbers such that ω_1/ω_2 is not purely real. The region of the complex plane with vertices $\{0, 2\omega_1, 2\omega_2, 2\omega_1 + 2\omega_2\}$ is called the *fundamental period parallelogram* (see Figure 1).

The simplest elliptic functions are those with two poles: the *Jacobi elliptic functions*, with two simple poles of opposite residues, and the *Weierstrass elliptic functions*, with a single double pole with zero residue, all others can be constructed from these.

Jacobi Elliptic Functions

The three most common Jacobi elliptic functions are $\text{sn}(z; k)$, introduced earlier, and the associated

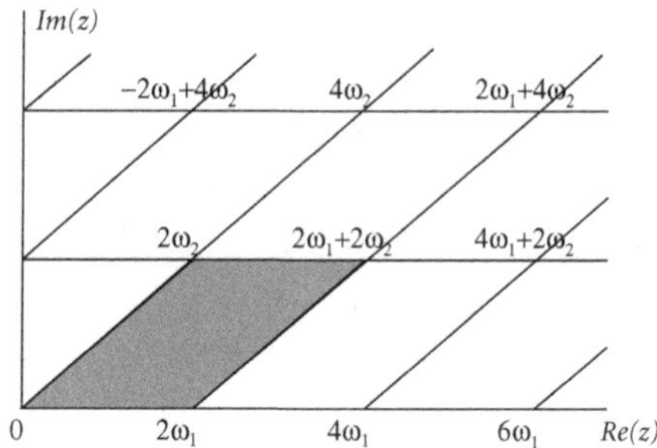

Figure 1. The period lattice of an elliptic function.

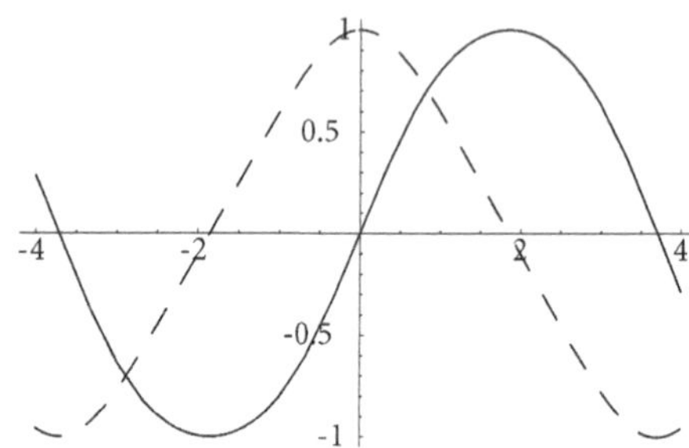

Figure 2. Graphs of $\text{sn}(x; k)$ (solid line) and $\text{cn}(x; k)$ (dashed line), for real x, $k = 0.5$.

$\text{cn}(z; k) = \sqrt{1 - \text{sn}^2(z; k)}$ (a generalization of the cosine function) and $\text{dn}(z; k) = \sqrt{1 - k^2\text{sn}^2(z; k)}$. Some of their main properties, analogues of the properties of trigonometric functions, are listed below; their proofs are elementary (Whittaker & Watson, 1943) (see Figure 2):

Symmetry:

$$\text{sn}(-u) = -\text{sn}\, u, \quad \text{cn}(-u) = \text{cn}\, u, \quad \text{dn}(-u) = \text{dn}\, u.$$

Derivatives:

$$\frac{d}{du}\text{sn}\, u = \text{cn}\, u\, \text{dn}\, u, \quad \frac{d}{du}\text{cn}\, u = -\text{sn}\, u\, \text{dn}\, u,$$

$$\frac{d}{du}\text{dn}\, u = -k^2 \text{sn}\, u\, \text{cn}\, u.$$

Addition formulas: For example,

$$\text{sn}(u + v) = \frac{\text{sn}\, u\, \text{cn}\, v\, \text{dn}\, v + \text{sn}\, v\, \text{cn}\, u\, \text{dn}\, u}{1 - k^2 \text{sn}^2 u\, \text{sn}^2 v.}$$

Periods: Define the *complete elliptic integral*

$$K = \int_0^1 \frac{dz}{\sqrt{(1 - z^2)(1 - k^2 z^2)}},$$

the *complementary modulus* k' such that $k^2 + k'^2 = 1$, and the constant $K' = K(1; k')$. Then, for example,

$$\text{sn}(u + 4K) = \text{sn}\, u, \quad \text{sn}(u + 2iK') = \text{sn}\, u.$$

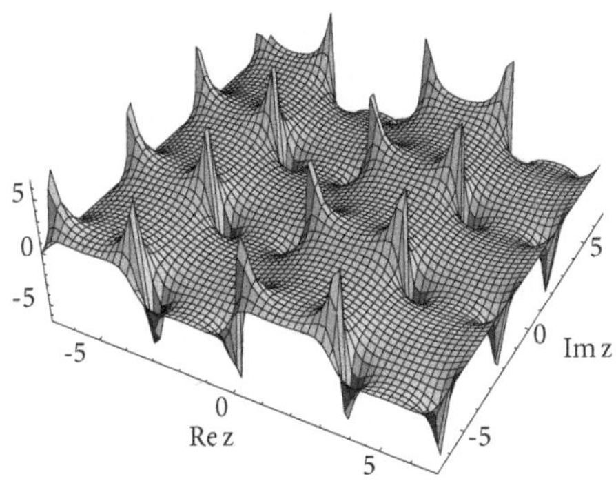

Figure 3. The doubly periodic function $\mathrm{sn}(z; k)$: graph of $Re[\mathrm{sn}(z; 0.3)]$.

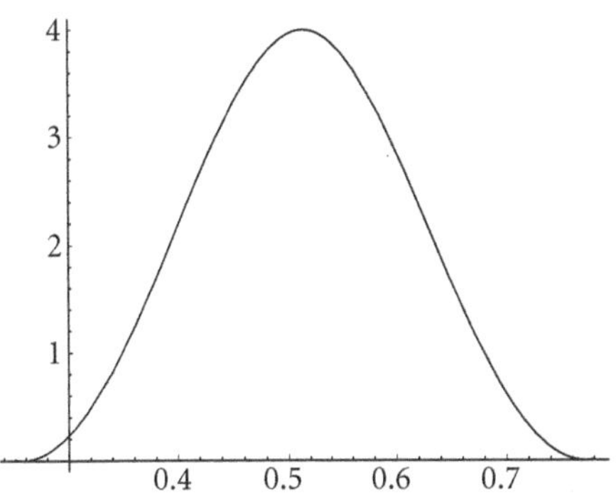

Figure 4. The profile of the cnoidal wave solution of the KdV equation, $A = 4$, $k = 0.3$.

Poles: For example, $\mathrm{sn}(z; k)$ is analytic except at points congruent to $\mathrm{i}K'$ and $2K + \mathrm{i}K'$ (mod $2K$, $2\mathrm{i}K'$), which are simple poles.

Weierstrass Elliptic Functions

This class of elliptic functions was introduced by Weierstrass by means of infinite partial-fraction and product expansions, rather than as inversions of elliptic integrals. Indeed, such construction exists for all elliptic functions. Given $\omega_1, \omega_3 \in \mathbb{C}$, $\omega_2 = \omega_1 + \omega_2$, we define the most famous of Weierstrass' elliptic functions, the $\wp$-function, on the period lattice $\{m2\omega_1 + n2\omega_3 \mid m, n \in \mathbb{Z}\}$:

$$\wp(z) = \frac{1}{z^2} + \sum_{(m,n)\neq(0,0)} \left\{ \frac{1}{(z - (m2\omega_1 + n2\omega_3))^2} - \frac{1}{(m2\omega_1 + n2\omega_3)^2} \right\}. \qquad (2)$$

$\wp(z)$ is a meromorphic function with double poles at each point of the period lattice and satisfies the following differential equation:

$$[\wp'(z)]^2 = 4[\wp(z)]^3 - g_2[\wp(z)] - g_3 = 4(\wp(z) - e_1)(\wp(z) - e_2)(\wp(z) - e_3), \qquad (3)$$

where $e_i = \wp(\omega_i)$.

Among applications of the Weierstrass $\wp$-function is the profile of a wave traveling in water (see, e.g., Walker, 1996), as described by the Korteweg–de Vries (KdV) equation

$$u_t + u_x + 12uu_x + u_{xxx} = 0. \qquad (4)$$

Seek a traveling-wave solution by setting $u(x,t) = f(x - ct)$, for some constant speed c. Then f satisfies the ordinary differential equation $(1 - c)f' + 12ff' - + f''' = 0$ which, after twice integrating, leads to

$$(f')^2 = 4f^3 - (1 - c)f^2 + Af + B,$$

A, B constants of integration. By comparison with (3), the form of f must be $f(s) = \alpha\wp(\beta s) + \gamma$, for suitable constants α, β, γ. Since the Weierstrass $\wp$-function can be shown to be related to the square of Jacobi elliptic functions (Whittaker & Watson, 1943), this leads to the notable *cnoidal wave* solution of KdV (see Figure 4):

$$u(x,t) = A\,\mathrm{cn}^2\left(\frac{\sqrt{A}}{k}(x - ct); k\right),$$

$$c = 1 + 4A\,\frac{2k^2 - 1}{k^2}.$$

Finally, we briefly mention hyperelliptic functions (or abelian functions), as inverses of integrals of rational functions $R(z, \sqrt{P(z)})$, with $P(z)$ a quintic or higher-degree polynomial (Baker, 1995). For example, the equations for the Kovalevsky top were shown by Sophia Kovalevsky to be integrable in terms of hyperelliptic functions (Whittaker, 1960).

ANNALISA M. CALINI

Further Reading

Baker, H.F. 1995. *Abelian Functions: Abel's Theorem and the Allied Theory of Theta Functions*. Reissue of the 1897 edition, Cambridge and New York: Cambridge University Press

Byrd, P.F. & Friedman, M.D. 1954. *Handbook of Elliptic Integrals for Engineers and Physicists*, New York: Wiley

Lawden, D.F. 1989. *Elliptic Functions and Applications*, Berlin and New York: Springer

Rauch, H.E. & Lebowitz, A. 1973. *Elliptic Functions, Theta Functions and Riemann Surfaces*, Baltimore: Williams and Wilkins

Stillwell, J. 1989. *Mathematics and Its History*, 2nd edition, Berlin and New York: Springer, Chapter 12

Walker, P.L. 1996. *Elliptic Functions: A Constructive Approach*, Berlin and New York: Springer

Whittaker, E.T. 1960. *A Treatise on the Analytical Dynamics of Particles and Rigid Bodies: With an Introduction to the Problem of Three Bodies*, Cambridge: Cambridge University Press (originally published 1904)

Whittaker, E.T. & Watson, G.N. 1943. *A Course of Modern Analysis*, 4th edition, Cambridge: Cambridge University Press

EMBEDDING METHODS

The modeling of a deterministic dynamical system relies on the concept of a phase space, which is a theoretical representation of the totality of possible system states. In general, a system state consists of information of positions and velocities needed to specify all future system states. For a system that has a mathematical model, the phase space is known from the equations of motion.

For experimental and natural chaotic dynamical systems, the full state space is unknown. Embedding methods have been developed as a means to reconstruct the phase space and develop new predictive models. One or more signals from the system must be observed as a function of time. The time series are then used to build a proxy of the observed states.

Mathematical theorems show how the observed time series can be used. A famous theorem of Hassler Whitney from the 1930s holds that a generic map from an n-manifold to $2n+1$-dimensional Euclidean space is an embedding. In particular, the image of the n-manifold is completely unfolded in the larger space because $2n+1$ signal traces measured from a system can be considered as a map from the set of states to $2n+1$-dimensional space, Whitney's theorem implies that each state can be identified uniquely by a vector of $2n+1$ measurements, thereby reconstructing the phase space.

The contribution of Floris Takens (1981) was to show that the same goal could be reached with a single measured quantity. He proved that instead of $2n+1$ generic signals, the time-delayed versions $[y(t), y(t-\tau), y(t-2\tau), \ldots, y(t-2n\tau)]$ of one generic signal would suffice to embed the n-dimensional manifold. There are some technical assumptions that must be satisfied, restricting the number of low-period orbits with respect to the time-delay τ and repeated eigenvalues of the periodic orbits. This result was roughly contemporaneous with similar theoretical results by D. Aeyels and a more empirical account by Packard et al. (1980).

The idea of using time delayed coordinates to represent a system state is reminiscent of the theory of ordinary differential equations, where existence theorems say that a unique solution exists for each $[y(t), \dot{y}(t), \ddot{y}(t), \ldots]$. For example, in Newtonian many-body dynamics, current knowledge of the position and momentum of each body suffices to uniquely determine the future dynamics. The time derivatives can be approximated by delay-coordinate terms as

$$\left[y(t), \frac{y(t)-y(t-\tau)}{\tau}, \frac{y(t)-2y(t-\tau)+y(t-2\tau)}{\tau^2}, \ldots\right].$$

The emergence of chaos and fractal geometry in physical systems motivated a reassessment of the original theory, which applies to smooth manifold attractors. It was shown (Sauer et al., 1991) that an attractor of (possibly fractional) box-counting dimension d can always be reconstructed with m generic observations, or with m time-delayed versions of one generic observation, where m is any integer greater than $2d$.

Embedding ideas were later extended beyond autonomous systems with continuously measured time series. A version was designed for excitable media, where information may be transmitted by spiking events, extending usage to possible neuroscience applications. An embedding theorem for skew systems by J. Stark explores extensions of the methodology when one part of a system is driving another.

Although the theory implies that an arbitrary time delay is sufficient to reconstruct the attractor, efficiency with a limited amount of data is enhanced by particular choices of the time delay τ. Methods for choosing an appropriate time delay have centered on measures of linear autocorrelation and mutual information (Fraser & Swinney, 1986). Further, in the absence of knowledge of the phase space dimension n, a choice of the number of embedding dimensions m must also be made. A number of ad hoc methods have been proposed that try to estimate whether the image has been fully unfolded by a given m-dimensional map.

The success of embedding in practice depends heavily on the specifics of the application. In particular, the hypothesis of a generic observation function creating the time series is often problematic. A mathematically generic observation monitors by definition all degrees of freedom of the system. The extent to which this is true affects the faithfulness of the reconstruction. If there is only a weak connection from some degrees of freedom to the observation function, the data requirements for a satisfactory reconstruction may be prohibitive in practice. Other factors that limit success are difference in time scales between different parts of the system, as well as system and observational noise.

Applications of embedding time-series data (Ott et al., 1994; Kantz & Schreiber, 1997) have been extensive since Takens's theorem was published. Many techniques of system characterization and identification were made possible, including determination of periodic orbits and symbolic dynamics, as well as approximation of attractor dimensions and Lyapunov exponents of chaotic dynamics. In addition, researchers have focused on methods of time series prediction and nonlinear filtering for noise reduction, the use of chaotic signals for communication, and the control of chaotic systems.

TIM SAUER

See also **Chaos vs. turbulence; Controlling chaos; Fractals; Phase space; Time series analysis**

Further Reading

Eckmann, J.-P. & Ruelle, D. 1985. Ergodic theory of chaos and strange attractors. *Reviews of Modern Physics*, 57: 617–652

Fraser, A.M. & Swinney, H.L. 1986. Independent coordinates for strange attractors from mutual information. *Physical Review* A, 33: 1134–1140

Kantz, H. & Schreiber, T. 1997. *Nonlinear Time Series Analysis*, Cambridge and New York: Cambridge University Press

Ott, E., Sauer, T. & Yorke, J.A. 1994. *Coping with Chaos: Analysis of Chaotic Data and the Exploitation of Chaotic Systems*, New York: Wiley Interscience

Packard, N., Crutchfield, J., Farmer, D. & Shaw, R. 1980. Geometry from a time series. *Physical Review Letters*, 45: 712–715

Sauer, T., Yorke, J.A. & Casdagli, M. 1991. Embedology. *Journal of Statistical Physics*, 65: 579–616

Takens, F. 1981. *Detecting Strange Attractors in Turbulence*, Berlin: Springer

EMERGENCE

Like many commonly used words, the term *emergence* has several meanings. In its weakest sense, a metaphor is provided by Michelangelo Buonarroti's famous sculptures entitled *The Prisoners*, which show human figures struggling to free themselves from their lithic confines, suggesting the artist's view of the creative process.

In simplest terms, there is little mystery here; the sculptor merely removes the unnecessary marble to expose the finished work, as Michelangelo himself is said to have pointed out. Thus, his prisoners are emerging only in an elementary sense which philosopher Robert Van Gulick calls "specific value emergence" and defines as follows (Van Gulick, 2001):

> Specific Value Emergence: The whole and its parts have features of the *same kind*, but have different *specific subtypes* or *values* of that kind. For example, a bronze statue has a given mass as does each of the molecular parts of which it is composed, but the mass of the whole is different in value from that of any of its material parts.

Moving beyond this limited sense, Van Gulick defines various degrees of "modest emergence" in these terms.

> Modest Emergence: The whole has features that are *different in kind* from those of its parts (or alternatively that *could* be had by its parts). For example, a piece of cloth might be purple in hue even though none of the molecules that make up its surface could be said to be purple. Or a mouse might be alive even if none of its parts (or at least none of its subcellular parts) were alive.

Modest emergence thus arises in a spectrum of different ways depending upon the degree of difference between a phenomenon and the base out of which it emerges, with the coherent structures of nonlinear science providing many examples (Scott, 2003).

Among the more modest types of emergence, one would include solitons of the Korteweg–deVries (KdV) equation, which emerge out of a nonlinear partial differential equation (PDE) in response to certain initial conditions. Although KdV solitons are independent dynamic entities, their speeds and shapes are determined via the inverse scattering transform (IST) method from the initial conditions applied to the system. Somewhat less modest would be the various solitary wave solutions of Hamiltonian (energy conserving) systems for which IST formulations are not currently known and may not exist, precluding the prediction of solitary wave speeds from initial conditions.

Further decreasing modesty (increasing robustness) of emergence leads to the nerve impulse, which has several model PDEs (Hodgkin–Huxley, FitzHugh–Nagumo, etc.) in addition to those many physiological manifestations (the action potentials of the brain) which compose our mental activity (Scott, 2002). While propagating on a uniform system with constant speed and shape, a nerve impulse differs fundamentally from solitary waves of Hamiltonian systems for the following reason: a nerve impulse (like the flame of a candle) does not conserve energy. The nonlinear dynamics of a nerve impulse involves a balance between the release and dissipation of energy, so the process is open and thus does not have a Hamiltonian formulation. This, in turn, implies that the dynamic behavior of a nerve impulse changes greatly upon reversal of the direction of time, whereas the qualitative behavior of a Hamiltonian system is insensitive to time reversal.

Under this distinction, we can gauge the relative modesty of several other types of emergence that arise in the realms of nonlinear science. Vortex solutions of viscosity-free fluids (superfluids, for example) would be more modestly emergent than those of (more or less) viscous fluids, in which dissipative processes cause the dynamics to (more or less) rapidly forget the information received from the initial conditions. As residents of Tornado Alley in the U.S. midlands know well, tornados are famously ill-behaved, detached from their initial conditions and moving quite wildly in response to local variations of pressure, humidity, temperature, and so on.

A deeper meaning of Michelangelo's metaphor suggests the emergence of living organisms from the oily brine of the Hadean oceans some three thousand million years ago. Life is even less modestly (more robustly) emergent, with the "arrow of time" clearly constraining us all and playing a key role in the unpredictable drama of biological evolution (Gould, 1989). In other words, the emergence of life is far more intricate than the emergence of John Scott Russell's soliton from the prow of his test vessel on the Union Canal.

Yet more robust (less immodest) than the emergence of life is the phenomenon of human consciousness, which philosophers have struggled for centuries to understand and many find qualitatively different from its material substrate. To include the qualitative aspects of emergence at this far end of the scale, Van Gulick introduces the following definition.

> RADICAL EMERGENCE: The whole has features that are both different in kind from those had by its parts, and of a kind whose nature and existence is not necessitated by the features of its parts, their mode of combination and the law-like regularities governing the features of its parts.

Whether human consciousness offers an example of radical emergence is currently controversial among cognitive scientists, neuroscientists, psychologists, cultural anthropologists, philosophers, and others interested in the nature of mind. On the one hand, reductive materialists assert that all of reality must "in principle" reduce to a physical basis (Kim, 1999), whereas substance dualists claim that the human mind differs ontologically (in its nature) from physical reality (Chalmers, 1996). Situated between these two positions, property dualists suggest that the human mind may radically emerge from intricate interactions among the various nonlinear dynamic levels of body, brain, and culture (Scott, 1995; Van Gulick, 2001).

ALWYN SCOTT

See also **Biological evolution; Game of life; Morphogenesis, biological**

Further Reading

Chalmers, D. 1996. *The Conscious Mind: In Search of a Fundamental Theory*, Oxford and New York: Oxford University Press

Gould, S.J. 1989. *Wonderful Life: The Burgess Shale and the Nature of History*, New York and London: Norton

Kim, J. 1999. *Mind in a Physical World*, Cambridge, MA: MIT Press

Scott, A.C. 1995. *Stairway to the Mind: The Controversial New Science of Consciousness*, Berlin and New York: Springer

Scott, A.C. 2002. *Neuroscience: A Mathematical Primer*, Berlin and New York: Springer

Scott, A.C. 2003. *Nonlinear Science: Emergence and Dynamics of Coherent Structures*, 2nd edition, Oxford and New York: Oxford University Press

Van Gulick, R. 2001. Reduction, emergence and other recent options on the mind/body problem. *Journal of Consciousness Studies*, 8(9–10): 1–34

ENDOMORPHISM

See **Maps**

ENERGY ANALYSIS

In time-reversible Hamiltonian systems, the total energy of the system is constant in time. A local change of the energy density W at a point x in time t is balanced by the energy flux S from/to the point x according to the energy balance equation:

$$\frac{\partial W}{\partial t} + \frac{\partial S}{\partial x} = 0.$$

In irreversible (active and/or dissipative) systems, the total energy changes in time due to the energy sinks/sources of density ρ are given by the extended the extended energy balance equation:

$$\frac{\partial W}{\partial t} + \frac{\partial S}{\partial x} = -\rho.$$

In mathematical physics, energy balance is used for analysis of well-posedness of partial differential equations. Solutions of well-posed differential equations remain bounded in a suitable function space, starting with a bounded initial data (Strauss, 1992). For illustration, we consider the energy balance for the heat equation $u_t = u_{xx}$ that takes the form

$$\frac{\partial}{\partial t}\left(u^2\right) - 2\frac{\partial}{\partial x}\left(uu_x\right) = -2\left(u_x\right)^2.$$

Suppose the initial data $u(x,0)$ belong to space of real-valued $L^2(\mathcal{R})$ functions such that the initial energy is bounded: $E(0) = \int_{-\infty}^{\infty} u^2(x,0)\,\mathrm{d}x < \infty$. The energy $E(t)$ decreases at later times such that $0 \le E(t) = \int_{-\infty}^{\infty} u^2(x,t)\,\mathrm{d}x \le E(0)$. These simple estimates of energy analysis immediately imply the following properties of the heat equation:

(i) The zero solution $u(x,t) = 0$ is unique.
(ii) A general solution $u(x,t)$ is asymptotically stable in space of $L^2(\mathcal{R})$ functions.
(iii) A general solution $u(x,t)$ decays uniformly to zero in the L^2-norm sense: $\lim_{t\to\infty} ||u(\cdot,t)||_{L^2(\mathcal{R})} = 0$.

A real-valued energy can be introduced for complex functions, for example, in the Schrödinger equation $\mathrm{i}u_t = u_{xx}$, where the energy balance takes the form

$$\frac{\partial}{\partial t}|u|^2 + \mathrm{i}\frac{\partial}{\partial x}\left(\bar{u}u_x - \bar{u}_x u\right) = 0.$$

The Schrödinger equation is reversible, and the total energy is constant in time such that $0 \le E(t) = \int_{-\infty}^{\infty} |u|^2(x,t)\,\mathrm{d}x = E(0)$. As a result, the solution is well-posed in the space of complex-valued $L^2(\mathcal{R})$ functions with the following properties:

(i) The zero solution $u(x,t) = 0$ is unique.
(ii) A general solution $u(x,t)$ is neutrally stable in the L^2-norm sense.

Partial differential equations can be well-posed in energy space, which is different from the space of square integrable functions. For the wave equation $u_{tt} = c^2 u_{xx}$, two balance equations take the form

$$\frac{\partial}{\partial t}\left(u_t^2 + c^2 u_x^2\right) - 2c^2\frac{\partial}{\partial x}\left(u_t u_x\right) = 0$$

and

$$\frac{\partial}{\partial t}(u_t u_x) - \frac{1}{2}\frac{\partial}{\partial x}\left(u_t^2 + c^2 u_x^2\right) = 0.$$

The first equation prescribes the positive-definite quantity, $E(t) = \int_{-\infty}^{\infty}\left(u_t^2 + c^2 u_x^2\right) \mathrm{d}x$, which defines the energy space $H^1(\mathcal{R})$ for solutions of the wave equation, such that $0 \leq E(t) = E(0)$. The second equation prescribes the sign-indefinite quantity, $P(t) = \int_{-\infty}^{\infty} u_t u_x \, \mathrm{d}x$, referred to as the momentum, such that $P(t) = P(0)$.

In modeling of various physical phenomena, momentum balance equations are useful for analysis of integral properties of solutions of underlying equations at infinite or finite intervals. For instance, adiabatic dynamics of localized pulses, envelope wave packets, and radiative wave trains can be studied with the momentum balance equations, when a solitary wave changes under the action of external perturbations, variations of physical parameters, internal instabilities, various resonances, and interactions with other wave structures (Kivshar & Malomed, 1989). In the simplest version of the soliton perturbation theory, effects of external perturbations to a physical system are captured by slow variations of soliton parameters. The dynamical rate of change of soliton parameters is found by substituting the soliton solutions in the momentum balance equations. For illustration, we consider the perturbed sine-Gordon equation with dissipative and external harmonic forces:

$$\begin{aligned} u_{tt} - u_{xx} + \sin u &= \varepsilon\, R(u) \\ &= \varepsilon\left[-\alpha u_t + \Gamma \sin(\omega t)(1 - \cos u)\right], \end{aligned}$$

where $\varepsilon \ll 1$, α is a damping parameter and Γ is the amplitude of the external force such that $\omega < 1$ (no resonance occurs). With the account of the perturbation, the balance equation for momentum is

$$\frac{\partial}{\partial t}(u_t u_x) - \frac{1}{2}\frac{\partial}{\partial x}\left(u_t^2 + u_x^2 + 2\cos u\right) = \varepsilon\, R(u) u_x.$$

As a result, the rate of change of momentum $P(t) = \int_{-\infty}^{\infty} u_t u_x \, dx$ is given by

$$\frac{\mathrm{d}P}{\mathrm{d}t} = \varepsilon \int_{-\infty}^{\infty} u_x R[u] \mathrm{d}x.$$

provided that $\lim_{|x|\to\infty}\left(u_t^2 + u_x^2\right) = 0$ and $\lim_{x\to\infty} \cos u = \lim_{x\to-\infty} \cos u = 1$. The unperturbed sine-Gordon equation at $\varepsilon = 0$ has the kink solution:

$$u_k(x, t) = 4 \arctan\left(\exp\left(\frac{x - vt - x_0}{\sqrt{1 - v^2}}\right)\right),$$

where $|v| < 1$ is the kink's velocity and x_0 is its position. The momentum of the unperturbed kink is a function of its velocity: $P_k(v) = -8v/\sqrt{1 - v^2}$. The kink solution satisfies the aforementioned vanishing conditions at infinity. Assuming that the velocity of the kink $v = v(t)$ changes due to the external perturbation, we use the momentum balance equation and find a particle-like equation of motion for the kink's adiabatic dynamics:

$$\frac{\mathrm{d}P_k}{\mathrm{d}t} + \varepsilon\, \alpha\, P_k = 2\pi\, \varepsilon\, \Gamma \sin(\omega t).$$

The equation has a simple solution:

$$\begin{aligned} P_k(t) &= C\mathrm{e}^{-\varepsilon \alpha t} \\ &\quad - \frac{2\pi \varepsilon \omega \Gamma}{\omega^2 + \varepsilon^2 \alpha^2} \cos \omega t \\ &\quad + \frac{2\pi \varepsilon^2 \alpha \Gamma}{\omega^2 + \varepsilon^2 \alpha^2} \sin \omega t, \end{aligned}$$

where C is found from initial condition: $P_k(0) = P_0$.

Adiabatic dynamics of kinks and solitary waves often generate a strong radiation field that takes away part of momentum of the localized solution. The radiation field can be taken into account from other balance equations, such as the mass balance. Radiative effects usually occur in the second order of the perturbation theory, leading to radiative decay of solitary waves or their perturbations (Pelinovsky & Grimshaw, 1996). For illustration, we consider dynamics of solitary waves in the critical KdV equation:

$$u_t + 15u^4 u_x + u_{xxx} = 0.$$

The balance equations for mass and momentum of a nonlinear field are:

$$\frac{\partial}{\partial t}(u) + \frac{\partial}{\partial x}\left(3u^5 + u_{xx}\right) = 0$$

and

$$\frac{\partial}{\partial t}(u^2) + \frac{\partial}{\partial x}\left(5u^6 + 2uu_{xx} - u_x^2\right) = 0.$$

Solitary waves are given by special solutions of the critical KdV equation:

$$u_s(x, t) = \left[\sqrt{v}\, \mathrm{sech}(2\sqrt{v}(x - vt - x_0))\right]^{1/2}.$$

Solitary waves may change adiabatically due to internal perturbation of the initial data, such that v becomes a function of t. The radiation field is generated behind the solitary wave due to the uni-directional property of the dispersion relation for the linear KdV equation: $u_t + u_{xxx} = 0$. The radiation field can be found from the mass balance equation

$$\lim_{x\to-\infty} u(x, t) = -\frac{1}{v}\frac{\mathrm{d}M_s}{\mathrm{d}t},$$

where $M_s(v) = \int_{-\infty}^{\infty} u_s(x, t)\, \mathrm{d}x = M_0/v^{1/4}$ and M_0 is constant. The momentum balance equation leads to a closed particle-like equation of motion for the solitary wave's adiabatic dynamics:

$$\frac{\mathrm{d}}{\mathrm{d}t}\left[\left(\frac{\mathrm{d}M_s}{\mathrm{d}v}\right)^2\frac{\mathrm{d}v}{\mathrm{d}t}\right] = -v\left[\lim_{x\to-\infty} u(x,t)\right]^2 = -\frac{1}{v}\left(\frac{\mathrm{d}M_s}{\mathrm{d}v}\right)^2\left(\frac{\mathrm{d}v}{\mathrm{d}t}\right)^2.$$

The dynamical equation has a unique solution:

$$v(t) = v_0\left(\frac{t_0}{t_0 - t}\right)^2,$$

where v_0 and t_0 are constants. The exact solution defines the scaling law for self-similar blowup in the critical KdV equation. The blowup rate is modified due to generation of the radiation field, which makes the balance between the left- and right-hand sides of the momentum balance equation. To summarize, the balance equations for energy, momentum, and mass can be used for both qualitative and quantitative estimates of the interaction between localized and radiative components of the nonlinear wave in solutions of nonlinear wave equations.

DMITRY PELINOVSKY

See also **Constants of motion and conservation laws; Multisoliton perturbation theory; Power balance**

Further Reading

Kivshar, Yu.S. & Malomed, B.A. 1989. Dynamics of solitons in nearly integrable systems. *Reviews of Modern Physics*, 61: 763–915

Pelinovsky, D.E. & Grimshaw, R.H.J. 1996. An asymptotic approach to solitary wave instability and critical collapse in long-wave KdV-type evolution equations. *Physica* D, 98: 139–155

Strauss, W.A. 1992. *Partial Differential Equations: An Introduction*, New York: Wiley

ENERGY CASCADE

See **Turbulence**

ENERGY OPERATORS

See **Quantum nonlinearity**

ENTRAINMENT

See **Coupled oscillators**

ENTROPY

Entropy is a quantity characterizing disorder, or randomness. This conclusion is the result of successive advances since the pioneering works by Sadi Carnot on the fundamentals of steam engines in 1824 and by Rudolf Clausius, who, between 1851 and 1865, developed the concept of entropy (from $\tau\rho o\pi\eta$ meaning "transformation" in Greek) as a thermodynamic state variable. Called S, this quantity varies as $\mathrm{d}S = \mathrm{d}Q/T$ in an equilibrium system at temperature T exchanging a quantity $\mathrm{d}Q$ of heat and no matter reversibly with its environment.

Thereafter, Ludwig Boltzmann (1896/1898), Max Planck (1901), Josiah Willard Gibbs (1902), and others discovered the statistical meaning of entropy. Boltzmann and Gibbs introduced the concepts of probability and statistical ensemble in the context of mechanics. If P_α is the probability (satisfying $0 \le P_\alpha \le 1$ and $\sum_\alpha P_\alpha = 1$) that the system is found in the microstate α specified by a set of observables such as energy and linear or angular momenta, the thermodynamic entropy is given by

$$S = -k_\mathrm{B}\sum_\alpha P_\alpha \ln P_\alpha, \tag{1}$$

where $k_\mathrm{B} = 1.38065\ 10^{-23}\mathrm{J\,K}^{-1}$ is the so-called Boltzmann constant, although it was originally introduced by Planck (Sommerfeld, 1956).

The entropy S measures the disorder in a statistical ensemble composed of a very large number, $\mathcal{N}$, of copies of the system. Each copy is observed in a microstate α occurring with the probability P_α. The microstates of all the $\mathcal{N}$ copies form a random list $\{\alpha_1, \alpha_2, \ldots, \alpha_\mathcal{N}\}$. Typically, $\mathcal{N}_\alpha = \mathcal{N}P_\alpha$ copies are found in the microstate α, and $\sum_\alpha \mathcal{N}_\alpha = \mathcal{N}$. The copies being statistically independent, the total number of possible lists of the microstates of the ensemble is thus equal to

$$W = \frac{\mathcal{N}!}{\prod_\alpha \mathcal{N}_\alpha!}, \tag{2}$$

where $\mathcal{N}! = \mathcal{N}\times(\mathcal{N}-1)\times\cdots\times 3\times 2$ denotes the factorial of the integer $\mathcal{N}$. In the limit $\mathcal{N}\to\infty$, according to Stirling's formula $\mathcal{N}! \simeq (\mathcal{N}/\mathrm{e})^\mathcal{N}$, where $\mathrm{e} = 2.718\ldots$ denotes the Naperian base, entropy (1) is given in terms of the logarithm of the number W of possible lists (originally called complexions by Boltzmann):

$$S \simeq \frac{k_\mathrm{B}}{\mathcal{N}}\ln W. \tag{3}$$

If the system is perfectly ordered, all the copies in the ensemble would be in the same microstate, and there would be a single possible complexion $W = 1$ so that the entropy would vanish ($S = 0$). In contrast, if the system were completely disordered with A equiprobable microstates α, the entropy would take the maximum value $S = k_\mathrm{B}\ln A$. For partial disorder, the entropy takes an intermediate value.

In spatially extended homogeneous systems, the thermodynamic entropy is an extensive quantity. If the system is covered by $\mathcal{N}$ disjoint windows of observations of volume V, the microstates $\{\alpha_1, \alpha_2, \ldots, \alpha_{\mathcal{N}}\}$ in the $\mathcal{N}$ windows form one among W possible lists (2). If the volume V of the observation window is large enough, the thermodynamic entropy is again given by Equation (3). In this case, the entropy per unit volume obtained by dividing the entropy S by the volume V is a measure of spatial disorder.

Entropy can also be interpreted as a quantity of information required to specify the microstate of the system. Indeed, the recording of the random microstates $\{\alpha_1, \alpha_2, \ldots, \alpha_{\mathcal{N}}\}$ of the statistical ensemble requires one to allocate at least $I = \log_2 W$ bits of information in the memory of a computer. This number of bits is related to the entropy (1) by $I \simeq \mathcal{N} S/(k_B \ln 2)$. Such connections between entropy and information have been developed since the works by Leo Szilard in 1929 and Léon Brillouin around 1951.

Equation (1) for the entropy is very general. It applies not only to equilibrium but also to out-of-equilibrium systems provided the states α are understood as coarse-grained states. In a classical system of N particles, the coarse-grained states α should correspond to cells of volume h^{3N} in the phase space of the positions and momenta of the particles, where $h = 6.626\ 10^{-34}$ Js is Planck's constant of quantum mechanics.

In 1902, Gibbs suggested that the second law of thermodynamics is a consequence of a dynamics having the mixing property according to which the statistical averages of observables or the coarse-grained probabilities P_α converge to their equilibrium values. Not all systems are mixing, but for those that are, entropy (1) converges toward its equilibrium value. In mixing systems, the statistical correlations in the initial probability distribution tend to disappear on finer and finer scales in phase space and are shared among more and more particles during a causal time evolution. The approach to the thermodynamic equilibrium may thus be described in terms of asymptotic expansions of the probability distributions in the long-time limits $t \to \pm\infty$. Both limits are not equivalent because, in the limit $t \to +\infty$, the probability distributions remain smooth in the unstable phase-space directions but become singular in the stable ones and vice versa in the other limit $t \to -\infty$, which may appear as an irreversibility or time arrow in the long-time description. The irreversibility in the increase of the entropy is thus closely related to the problems of identifying all the degrees of freedom guaranteeing the causality of the time evolution and of reconstructing the initial conditions, which is of great importance for the understanding of historical processes such as biological and cosmological evolution.

During recent decades, it has been shown that the increase of entropy does not preclude the formation of structures in equilibrium or non-equilibrium systems, nor in self-gravitating systems. At equilibrium, the homogeneity of pressure, temperature, and chemical potentials does not prevent inhomogeneities in the densities as is the case in crystals, in vortex states of quantum superfluids, and in mesomorphic phases of colloidal systems where equilibrium self-assembly occurs. Besides, open non-equilibrium systems can remove entropy to their environment, leading to far-from-equilibrium self-organization into spatial structures such as Turing patterns, self-sustained oscillations in such systems as chemical clocks, or complex processes such as biological morphogenesis.

While the entropy per unit volume characterizes spatial disorder at a given time, a concept of entropy per unit time was introduced in 1949 by Claude Shannon in development of his information theory, as a measure of temporal disorder in random or stochastic processes. It is defined in the same way as standard entropy but replacing space by time. In 1959, Andrei N. Kolmogorov and Yakov G. Sinai applied Shannon's idea to deterministic dynamical systems with an invariant probability measure, and they defined a metric entropy per unit time in analogy with Equation (1), considering the states α as the sequences $\omega_1\omega_2 \ldots \omega_n$ of the phase space cells ω_j successively visited at time intervals Δt by the trajectories during the time evolution of the system. In order to get rid of the arbitrariness of the coarse-grained cells ω_j, Kolmogorov and Sinai considered the supremum (least upper bound) of the entropy per unit time with respect to all possible partitions $\mathcal{P}$ of the phase space into cells ω_j, defining

$$\begin{aligned} h_{\mathrm{KS}} &= \mathrm{Sup}_{\mathcal{P}} \lim_{n\to\infty} -\frac{1}{n\Delta t} \\ &\quad \times \sum_{\omega_1\omega_2\ldots\omega_n} P_{\omega_1\omega_2\ldots\omega_n} \ln P_{\omega_1\omega_2\ldots\omega_n}, \end{aligned} \tag{4}$$

where the probability P is evaluated with the given invariant measure (Cornfeld et al., 1982).

In isolated chaotic dynamical systems, the temporal disorder of the trajectories finds its origin in the sensitivity to initial conditions because the Kolmogorov–Sinai entropy per unit time is equal to the sum of positive Lyapunov exponents λ_i, $h_{\mathrm{KS}} = \sum_{\lambda_i>0} \lambda_i$, as proved by Yakov B. Pesin in 1977 (Eckmann & Ruelle, 1985). We notice that the entropy per unit time h_{KS} differs from the irreversible entropy production defined by the time derivative of the standard thermodynamic entropy S in an isolated system. Indeed, the entropy per unit time may take a positive value for a system of particles already at thermodynamic equilibrium where entropy production vanishes. In spatially extended chaotic systems, the spatiotemporal disorder can be characterized by a further concept of entropy per unit time and unit volume.

A so-called topological entropy has also been introduced as the rate of proliferation of cells in successive partitions iteratively refined by the dynamics (Eckmann & Ruelle, 1985). In finite chaotic systems, the topological entropy is the rate of proliferation of periodic orbits as a function of their period. The topological entropy is not smaller than the Kolmogorov–Sinai entropy: $h_{\mathrm{top}} \geq h_{\mathrm{KS}}$.

PIERRE GASPARD

See also **Algorithmic complexity; Biological evolution; Cosmological models; Information theory; Lyapunov exponents; Measures; Mixing; Morphogenesis, biological; Nonequilibrium statistical mechanics; Pattern formation; Phase space; Stochastic processes; Turing patterns**

Further Reading

Boltzmann, L. 1896/1898. *Vorlesungen über Gastheorie*, 2 vols, Leipzig: Barth; as *Lectures on Gas Theory*, translated by S.G. Brush, Berkeley, University of California Press, 1964

Cornfeld, I.P., Fomin, S.V. & Sinai, Ya.G. 1982. *Ergodic Theory*, Berlin: Springer

Eckmann, J.-P. & Ruelle, D. 1985. Ergodic theory of chaos and strange attractors. *Reviews of Modern Physics*, 57: 617–656

Gibbs, J.W. 1902. *Elementary Principles in Statistical Mechanics*, New Haven, CT: Yale University Press

Planck, M. 1901. Uber das Gesetz der Energieverteilung im Normalspektrum. *Annalen der Physik*, 4: 553–563 (First historical publication of $S = k_{\mathrm{B}} \ln W$)

Sommerfeld, A. 1956. *Thermodynamics and Statistical Mechanics*, New York: Academic Press

ENVELOPE EQUATIONS

See **Nonlinear Schrödinger equations**

ENVELOPE SOLITONS

See **Solitons, types of**

EPHAPTIC COUPLING

Neurons communicate with each other by different means. The most investigated of these is synaptic transmission, which can be either chemical or electrical. Chemical synapses transmit neural signals from presynaptic to postsynaptic membranes via neurotransmitters and are widely found in vertebrate neurons, while electrical synapses correspond to an electronic coupling through specialized gap junctions and are more common in invertebrates (Jefferys, 1995). Therefore, both transmission processes need a specialized anatomical structure to create points of contact between neurons but there are other possibilities.

Since the work of Elwald Hering in 1882, it has been known that electrical communication can also occur between neurons when neuronal membranes are closely apposed but not contiguous (Scott, 2002). Called ephaptic coupling by Angelique Arvanitaki (who defined the term *ephapse* as "... the locus of contact or close vicinity of the active functional surfaces, whether this contact be experimental or brought by natural means"), this process relies on current spread through the extracellular space, which may influence the membrane dynamics of a second fiber (Arvanitaki, 1942). Early experimental evidence of ephaptic interactions was also provided by Bernhard Katz and Otto Schmitt on a pair of naturally adjacent unmyelinated fibers from the limb nerve of a crab (Katz & Schmitt, 1940). They showed that an impulse traveling on one fiber changes the excitability of the other fiber. Furthermore, impulses on adjacent fibers with similar speed and launched at about the same time become synchronized.

In the 1970s, Markin proposed a theoretical description of ephaptic coupling between two parallel unmyelinated fibers based on the assumption that they share an external series resistance per unit length proportional to the ionic resistivity of the extracellular medium (Markin, 1970). Using a piecewise constant function for the transmembrane ionic current, yielding to a leading-edge analysis, Markin concluded that under normal physiological conditions, ephaptic coupling does not allow transmission of an impulse from a fiber to an adjacent one, as verified in numerous experiments. Nevertheless, he also suggested that a synchronization of two impulses is to be expected with a longitudinal distance δ equal to zero.

A recent leading-edge analysis (Scott, 2002) considered a more appropriate representation of the ionic current (sodium ions), the resulting system corresponding to the Zeldovich–Frank-Kamenetsky (ZF) equation. In the case of a small ephaptic coupling, a perturbation theory showed that $\delta = 0$ corresponds to a stable locking of pairs of impulses if the threshold parameter a of the membrane is above a critical value, otherwise stability occurs with δ increasing when a decreases.

Other possibilities of synchronization have been found when studying the influence of complete impulses including a leading edge and a recovery part. Based on the FitzHugh–Nagumo model, studies have shown that two other locking distances δ are also stable and separated by two unstable ones (Scott, 2002; Eilbeck et al., 1981). Recently, it has also been shown that ephaptic coupling influences the speed of conduction of synchronized impulses, the speed decreasing when the coupling increases (Binczak et al., 2001).

In mammals, motor or sensory nerves are often organized in fiber bundles. These fibers are myelinated, and the active nodes of the membrane are separated

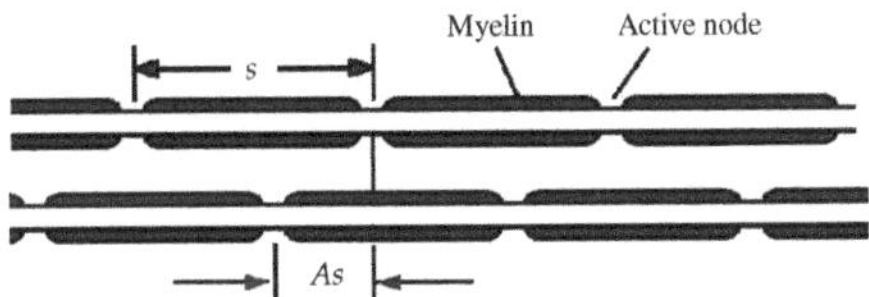

Figure 1. Sketch of two adjacent and parallel myelinated nerve fibers.

by a myelin sheath, which acts as an insulator. Thus, ephaptic coupling may be influenced by the alignment of the active nodes between two adjacent fibers (Binczak et al., 2001), as illustrated in Figure 1 where A is an alignment parameter.

Using a ZF description for the ionic current, a leading-edge analysis suggests that synchronization of impulses occurs whatever the alignment A. Nevertheless, an alignment of the nodes ($A = 1$) leads to a stronger synchronization while staggered nodes ($A = \frac{1}{2}$) allow a broader impulse coupling. Furthermore, when impulses are synchronized, an alignment of adjacent nodes reduces the critical longitudinal internodal distance at which propagation fails. Staggered nodes, on the other hand, lead to a more robust medium in order to prevent propagation failure.

From a functional perspective, synchronization of impulses on parallel and adjacent fibers might provide a means to adjust and organize the timing of impulses necessary for coordination and computation of neuronal information. Furthermore, ephaptic interactions may be important in neurological pathophysiology. Indeed, it has been observed that the transmission of impulses occurs from a fiber to an adjacent one when nerves are damaged, as when the nerves end in a neurisma after a nerve crush injury or a nerve compression (Seltzer & Devor, 1979) or when the ionic composition of the extracellular medium is changed (Ramon & Moore, 1978). Demyelination diseases, such as multiple sclerosis, are also a cause of ephaptic connections leading to pathological synchronization (Jefferys, 1995). Finally, ephaptic phenomena have been reported between muscles and motor nerves causing possible cramps and spasms and between cardiac cells, implying a possible involvement of ephaptic coupling in cardiac arrhythmias (Suenson, 1984).

Stephane Binczak

See also **FitzHugh–Nagumo equation; Myelinated nerves; Neurons; Zeldovich–Frank-Kamenetsky equation**

Further Reading

Arvanitaki, A. 1942. Effects evoked in an axon by the activity of a contiguous one. *Journal of Neurophysiology*, 5: 89–108

Binczak, S., Eilbeck, J.C. & Scott, A.C. 2001. Ephaptic coupling of myelinated nerve fiber. *Physica D*, 148: 159–174

Eilbeck, J.C., Luzader, S.D. & Scott, A.C. 1981. Pulse evolution on coupled nerve fibers. *Bulletin of Mathematical Biology*, 43(3): 389–400

Jefferys, J.G.R. 1995. Nonsynaptic modulation of neuronal activity in the brain: electric currents and extracellular ions. *Physiological Reviews*, 75 (4): 689–723

Katz, B. & Schmitt, O.H. 1940. Electric interaction between two adjacent nerve fibers. *Journal of Physiology*, 97: 471–488

Markin, V.S. 1970. Electrical interactions of parallel non-myelinated fibers. I and II. *Biophysics*, 15: 122–133 and 713–721

Ramon, F. & Moore, J.W. 1978. Ephaptic transmission in squid giant axons. *American Journal of Physiology: Cell Physiology*, 234: C162–C169

Seltzer, Z. & Devor, M. 1979. Ephaptic transmission in chronically damaged peripherical nerves. *Neurology*, 29: 1061–1064

Scott, A.C. 2002. *Neuroscience: A Mathematical Primer*, Berlin and New York: Springer

Suenson, M. 1984. Ephaptic impulse transmission between ventricular myocardial cells in vitro. *Acta Physiologica Scandinavica*, 120: 445–455

EPIDEMIOLOGY

Throughout human history diseases have played an important role. The black death in Europe in the middle ages is one example, and more recent examples include the flu pandemic of 1918 and 1919 and diseases such as AIDS and various childhood diseases. Additionally, the study of population dynamics in ecology has been greatly advanced by the study of diseases for several reasons. First, the data on the incidence of childhood diseases are both extensive and accurate. Second, the processes involved in the dynamics of childhood diseases (essentially, infection and either recovery or death) are relatively simple and straightforward and well understood.

In particular, many questions of scientific or practical interest can be answered using relatively simple models. A fundamental question is why does a disease die out before everyone has the disease? And, what fraction of the population needs to be vaccinated so a disease can be controlled or eliminated? Answering this question can explain why smallpox was more easily eradicated than other so-called childhood diseases.

Perhaps the simplest epidemic model, which also introduces many of the ideas, is the case of a single epidemic of a disease, first studied in detail by Kermack and McKendrick (1927). We focus here on diseases that are caused by microparasites, so individuals either have the disease or do not, as opposed to diseases caused by macroparasites (such as tapeworms) where the number of infectious agents in individuals needs to be considered explicitly in order to understand the disease dynamics. Different models are needed to understand epidemics and diseases which are endemic (Kermack & McKendrick, 1932).

We assume a constant population size and divide the population into three classes: susceptible, infective, and removed. Since the time scale of an epidemic is much shorter than the time scale of changes in human population sizes, we can simplify the system by ignoring any demographic influences in the simplest model for an epidemic, which corresponds to the assumption of a constant population size. In other words, the population size is assumed constant, and births and deaths are ignored (although there are models that do take this into account) because the time scale of an epidemic is short relative to the time scale of human population dynamics.

We assume that the rate at which susceptibles become infected is simply proportional to the product of the number of susceptible and infective individuals, corresponding to an assumption of random encounters. The rate at which infective individuals recover is assumed to be a constant. Then, with S the number of susceptibles, I the number of infectives, and R the number of removed individuals, the dynamics are given by

$$\frac{\mathrm{d}S}{\mathrm{d}t} = -\beta SI, \qquad (1)$$

$$\frac{\mathrm{d}I}{\mathrm{d}t} = \beta SI - \gamma I, \qquad (2)$$

$$\frac{\mathrm{d}R}{\mathrm{d}t} = \gamma I. \qquad (3)$$

Under the assumption that the total population size N remains constant, we can use the relationship $N = S + I + R$, and reduce the system (1)–(3) to one based on just the first two equations. The phase plane analysis of the resulting system is facilitated by the fact that the formula

$$\frac{\mathrm{d}I}{\mathrm{d}S} = \frac{\gamma}{\beta S} - 1 \qquad (4)$$

is so simple. As first discussed by Kermack and McKendrick (1927), one can solve this system explicitly by integration and using approximations. From this one can see that the solution curves which start with I arbitrarily small return to $I = 0$ before all individuals in the population become infected, or in other words the number of susceptibles at $t = \infty$ is positive.

The qualitative behavior of the system (1)–(3) essentially depends on a single nondimensional parameter, the reproductive number for the disease. The reproductive number is defined as the mean number of infective individuals produced by a single infective individual, which can clearly be calculated by multiplying the rate at which a single infective individual produces new infections by the mean period of infectivity for a single individual. Under our assumptions, the mean rate of infection is simply βS and the mean infective period is $1/\gamma$. Thus the reproductive number is simply

$$R_0 = \frac{\beta S}{\gamma}. \qquad (5)$$

The dynamics are governed by the observation that the number of infectives will increase if $R_0 > 1$ and will decrease if $R_0 < 1$. For the case of a single infective introduced into a population of susceptibles, we can use the total population N instead of S in the formula for R_0.

The import of the observation about the importance of R_0 (and results from integrating (4)) is typically summarized in the *threshold theorem* which states that there will be an epidemic if the population initially satisfies $R_0 > 1$, which may be a reflection of the population size. From integrating (4) one finds that the total number of individuals who get the disease is dramatically larger if the population is initially above the threshold.

Kermack & McKendrick (1932) further studied the cases of endemic diseases, where it is necessary to take into account the demography of the population, since the time scale is long enough that births and deaths need to be explicitly included. Other more complex systems were also studied (Kermack & McKendrick, 1933).

Modifications of the basic equations have been used to study the dynamics of sexually transmitted diseases, including AIDS. Here, an important modification has been to break up the population into classes based on different encounter rates.

More recent work on disease dynamics has played a central role in population biology, as the data sets for the incidence of childhood diseases in the United Kingdom and many large U.S. cities are among the longest, most accurate, and most detailed records of populations available (Bjornstad & Grenfell, 2001). In particular, the prevaccination dynamics of measles has been carefully analyzed using a variety of approaches, based essentially on extensions of the basic model (Equations (1)–(3)), modified to include a seasonally varying contact rate and, in some cases, stochasticity. Studies of this data has led to important substantial advances in the analysis of the kinds of time series available: relatively short with substantial stochastic influences. Analysis of these time series using nonlinear methods have demonstrated that at least some of the dynamics may be chaotic. Further and more recent efforts have focussed on spatiotemporal dynamics (Rohani et al., 1999) and on applied questions (Keeling et al., 2002) like the recent hoof and mouth epidemic in the UK.

ALAN HASTINGS

See also **Chaotic dynamics; Phase plane; Population dynamics**

Further Reading

Bjornstad, O.N. & Grenfell, B.T. 2001. Noisy clockwork: time series analysis of population fluctuations in animals. *Science*, 293: 638–643

Diekmann, O. & Heesterbeek, J.A.P. 2000. *Mathematical Epidemiology of Infectious Diseases*, Chichester and New York: Wiley

Keeling, M.J., Woolhouse, M.E.J., Shaw, D.J., Matthews, L., Chase-Topping, M., Haydon, D.T., Cornell, S.J., Kappey, J., Wilesmith, J. & Grenfell, B.T. 2002. Dynamics of the 2001 UK foot and mouth epidemic: stochastic dispersal in a heterogeneous landscape. *Science*, 294: 813–817

Kermack, W.O. & McKendrick, A.G. 1927. Contributions to the mathematical theory of epidemics. *Proceedings of the Royal Society of London A*, 115: 700–721

Kermack, W.O. & McKendrick, A.G. 1932. Contributions to the mathematical theory of epidemics. *Proceedings of the Royal Society of London A*, 138: 55–83

Kermack, W.O. & McKendrick, A.G. 1933. Contributions to the mathematical theory of epidemics. *Proceedings of the Royal Society of London A*, 141: 94–122

Rohani, P., Earn, D.J.D. & Grenfell, B.T. 1999. Opposite patterns of synchrony in sympatric disease metapopulations. *Science*, 286: 968–971

EQUATIONS, NONLINEAR

Nonlinear equations arise in a wide variety of forms in all branches of science, engineering, and technology as well as in other fields such as economics and social dynamics. These include algebraic, differential (ordinary/partial/delay), difference, differential-difference, integro-differential, and integral equations.

Nonlinear algebraic equations such as polynomial (quadratic, cubic, quartic, etc.) and transcendental equations in one or more variables and simultaneous equations have a chequered history. A large number of root-searching algorithms and methods of solutions are available in the literature.

With the advent of differential calculus in the 17th century, differential equations started to play a pivotal role in scientific investigations. This is particularly so for evolutionary problems, as differential equations are obvious candidates for the dynamical description of natural phenomena. Since most natural processes are nonlinear, it is no wonder that nonlinear differential equations arise frequently in theoretical descriptions (Ablowitz & Clarkson, 1991; Murray, 2002; Lakshmanan & Rajasekar, 2003; Scott, 2003).

Nonlinear ordinary differential equations (ODEs) occur in a wide variety of situations depending upon the nature of the problem. Their solution properties, such as integrability, non-integrability, and chaos, depend upon the order of the highest derivative, number of dependent variables, degree of nonlinearity, whether homogeneous or inhomogeneous, and on the values of the parameters. Some ubiquitous nonlinear ODEs and their significance are indicated in Table 1.

Nonlinear partial differential equations (PDEs) also have a long history dating back to the early days of differential calculus. For example, the basic equations of fluid dynamics, such as the Euler equation and the Navier–Stokes equation, are highly nonlinear. Many of the equations describing nontrivial surfaces in geometry are essentially nonlinear. The properties of solutions of nonlinear PDEs depend heavily on the number of dependent and independent variables, the order, the nature of nonlinearity, and the boundary conditions. However, it is also useful to classify the nonlinear PDEs into nonlinear dispersive and nonlinear diffusive/dissipative types. Further, they can

	Name	Equation	Significance
1.	Logistic equation	$\dot{x} = ax - bx^2$	Population growth model
2.	Bernoulli equation	$\dot{x} + P(t)x = Q(t)x^n$	Linearizable
3.	Riccati equation	$\dot{x} + P(t) + Q(t)x + R(t)x^2 = 0$	Admits nonlinear superposition principle
4.	Lotka–Volterra equation	$\dot{x} = ax - xy$, $\dot{y} = xy - by$	Population dynamics; exhibits limit cycle
5.	Anharmonic oscillator	$\ddot{x} + \omega_0^2 x + \beta x^3 = 0$	Integrable by Jacobian elliptic function(s)
6.	Pendulum equation	$\ddot{\theta} + \frac{g}{L}\sin\theta = 0$	Integrable by Jacobian elliptic function(s)
7.	Painlevé II equation	$\ddot{x} = 2x^3 + tx + \alpha$	Satisfies Painlevé property
8.	Duffing oscillator	$\ddot{x} + p\dot{x} + \omega_0^2 x + \beta x^3 = f\cos\omega t$	Exhibits chaotic dynamics
9.	Damped driven pendulum	$\ddot{\theta} + \alpha\dot{\theta} + \omega_0^2\sin\theta = \gamma\cos\omega t$	Exhibits chaotic dynamics
10.	Lorenz equation	$\dot{x} = \sigma(y - x)$, $\dot{y} = -xz + rx - y$, $\dot{z} = xy - bz$	Prototypical example of chaotic motion
11.	Hénon–Heiles system	$\ddot{x} + x + 2xy = 0$, $\ddot{y} + y + x^2 - y^2 = 0$	Hamiltonian chaos
12.	Mackey–Glass equation	$\dot{x} + \frac{ax_\tau}{(1+x_\tau^{10})} + bx = 0$, $x_\tau = x(t-\tau)$, τ : constant	Delay-differential system

Table 1. Some important nonlinear ordinary differential equations ($\cdot = \mathrm{d}/\mathrm{d}t$)

	Name	Equation	Significance
I	*Dispersive Equations*		
1.	Korteweg–de Vries equation	$u_t + 6uu_x + u_{xxx} = 0$	Integrable soliton eqn. in (1+1) dimensions
2.	Sine-Gordon equation	$u_{xt} = \sin u$	Integrable soliton eqn. in (1+1) dimensions
3.	Nonlinear Schrödinger equation	$iq_t + q_{xx} + 2\|q\|^2 q = 0$	Integrable soliton eqn. in (1+1) dimensions
4.	Heisenberg ferromagnetic spin equation	$\mathbf{S}_t = \mathbf{S} \times \mathbf{S}_{xx},\ \ \mathbf{S} = (S_1, S_2, S_3)$	Integrable soliton eqn. in (1+1) dimensions
5.	Kadomtsev–Petviashvili equation	$(u_t + 6uu_x + u_{xxx})_x + 3\sigma^2 u_{yy} = 0$, $\sigma^2 = \pm 1$	Integrable soliton eqn. in (2+1) dimensions
6.	Davey–Stewartson equation	$iq_t + \frac{1}{2}(q_{xx} + q_{yy}) + \alpha\|q\|^2 q + q\phi = 0$, $\phi_{xx} - \phi_{yy} + 2\alpha(\|q\|^2)_{xx} = 0, \alpha = \pm 1$	Integrable soliton eqn. in (2+1) dimensions
7.	ϕ^4 equation	$u_{tt} - u_{xx} + u - u^3 = 0$	Nonintegrable equation
II	*Diffusive equations*		
8.	Burgers equation	$u_t + uu_x - u_{xx} = 0$	Linearizable through Cole–Hopf transformation
9.	FitzHugh–Nagumo equation	$u_t = u_{xx} + u - u^3 - v$, $v_t = a(u - b)$	Nerve impulse propagation model
10.	Kuramoto–Sivashinsky equation	$u_t = -u - u_{xx} - u_{xxxx} - uu_x$	Spatiotemporal patterns and chaos
11.	Ginzburg–Landau equation	$u_t = (a + ib)\nabla^2 u + (c + id)(\|u\|^2)u$	Spatiotemporal patterns and chaos

Table 2. Some important nonlinear partial differential equations (suffix denotes partial derivative with respect to that variable)

be classified into integrable and non-integrable PDEs. A select set of such equations along with their significance is given in Table 2.

Another class of interesting nonlinear equations is the so-called difference equations/recurrence equations/iterated maps of the form $\boldsymbol{x}_{n+1} = \boldsymbol{F}(\boldsymbol{x}_n, n)$, where the independent/time variable n takes discrete values $n = 0, 1, 2, \ldots$, and $\boldsymbol{x}_n$ stands for the m dependent variables, while $\boldsymbol{F}$ is an m-dimensional nonlinear function. These equations often correspond to various finite-difference schemes of nonlinear ODEs/PDEs, such as the explicit Euler or implicit midpoint. For example, the logistic differential equation $\mathrm{d}x/\mathrm{d}t = ax - bx^2$ can be approximated by Euler's forward difference scheme as $x_{n+1} - x_n = h(ax_n - bx_n^2)$, where $h > 0$. But nonlinear difference equations can arise as dynamical systems on their own merit as in population dynamics or radioactive decay. The most famous example is the logistic map,

$$x_{n+1} = \lambda x_n(1 - x_n),\quad 0 \le x \le 1,\quad 0 \le \lambda \le 4,$$

as a prototypical model exhibiting a period-doubling bifurcation route to chaos. A two-dimensional example is the Hénon map, $x_{n+1} = 1 - ax_n^2 + y_n$, $y_{n+1} = bx_n$, $a, b > 0$ showing self-similar and fractal nature. Spatiotemporal patterns have been identified in coupled logistic map lattices. There also exist several integrable families of maps (Kosmann-Schwarzbach et al., 1997): for example, the McMillan map,

$$x_{n+1} + x_{n-1} = \frac{2\mu x_n}{1 - x_n^2}. \tag{1}$$

Other examples include integrable discrete Painlevé equations; Quispel, Robert, and Thompson (QRT) map; and so on. Several integrable nonlinear differential-difference and difference-difference equations also exist (Ablowitz & Clarkson, 1991). The Toda lattice equation

$$\ddot{x}_n = \exp(-(x_n - x_{n-1})) - \exp(-(x_{n+1} - x_n)),$$
$$n = 0, 1, 2, \ldots, N \tag{2}$$

is an integrable soliton system. Examples of integrable nonlinear difference-difference equations can be constructed from integrable PDEs.

Nonlinear integro-differential equations arose (Davis, 1962) with the work of Vito Volterra when he introduced a hereditary component composed of the sum of individual factors encountered in the past as a modification to the logistic equation as

$$\frac{\mathrm{d}y}{\mathrm{d}t} = ay - by^2 + y\int_c^t K(t - s)y(s)\,\mathrm{d}s, \tag{3}$$

where a, b, c are parameters and $K(t)$ is a given function. A special case of such an equation is the nonlinear integral equation

$$y(x) = f(x) + \lambda \int_a^b K(x, s, y(s))\,\mathrm{d}s, \qquad (4)$$

where λ is a parameter. Some interesting special cases correspond to $f = 0$, $K = \bar{k}(x, t)y^n$ and $f = 0$, $K = G(x, t)\exp y$ (Bratu equation), where $\bar{k}$ and G are specified functions. One of the most important nonlinear integro-differential equations in physics is the Boltzmann transport equation for the distribution function $f(\boldsymbol{r}, \boldsymbol{v}, t)$ of a dilute gas

$$\left(\frac{\partial}{\partial t} + \boldsymbol{v}_1 \cdot \nabla_r + \frac{\boldsymbol{F}}{m} \cdot \nabla_{v_1}\right) f(\boldsymbol{r}, \boldsymbol{v}_1, t) = \int d^3 v_2 \int d\Omega\, \sigma(\Omega) |\boldsymbol{v}_2 - \boldsymbol{v}_1| (f_1' f_2' - f_1 f_2), \qquad (5)$$

where $f_i' = f(\boldsymbol{r}, \boldsymbol{v}_i', t)$, $f_i = f(\boldsymbol{r}, \boldsymbol{v}_i, t)$, $i = 1, 2$. In the integrable soliton case, there exist several interesting integro-differential equations, for example, the Benjamin–Ono equation,

$$u_t + 2uu_x + Hu_{xx} = 0, \qquad (6)$$

where Hu is the Hilbert transform.

MUTHUSAMY LAKSHMANAN

See also **Maps; Ordinary differential equations, nonlinear; Partial differential equations, nonlinear**

Further Reading

Ablowitz, M.J & Clarkson, P.A. 1991. *Solitons, Nonlinear Evolution Equations and Inverse Scattering*, Cambridge and New York: Cambridge University Press

Davis, H.T. 1962. *Introduction to Nonlinear Differential and Integral Equations*, New York: Dover

Ince, E.L. 1956. *Ordinary Differential Equations*, New York: Dover

Kosmann-Schwarzbach, Y., Grammaticos, B. & Tamizhmani, K.M., (editors). 1997. *Integrability of Nonlinear Systems*, Berlin and New York: Springer

Lakshmanan, M. & Rajasekar, S. 2003. *Nonlinear Dynamics: Integrability, Chaos and Patterns*, Berlin: Springer

Murray, J.D. 2002. *Mathematical Biology*, Berlin and New York: Springer

Scott, A.C. 2003. *Nonlinear Science: Emergence and Dynamics of Coherent Structures*, 2nd edition, Oxford: Oxford University Press

EQUILIBRIUM

The word *equilibrium* suggests a balance, typically between antagonistic forces, which in turn often implies absence of motion. Consider a rigid rod, which can freely (i.e., with almost no friction) rotate in a vertical plane about a horizontal axis located near one of its ends. Because of gravity, the rod naturally assumes a position where it is suspended from the horizontal axis, and points downward (see Figure 1a). In this situation, the gravity force acting on the rod is balanced by the reaction force at the suspension point, and no motion occurs. The rod is in equilibrium. If displaced a little bit from this position, the rod oscillates about the horizontal axis and, because of friction, eventually returns to its initial position, which is therefore stable. The other equilibrium position of the rod, which is unstable, is that shown in Figure 1b, where the rod points upward. A slight perturbation will take the rod away from this position, toward the stable equilibrium. Consider now a marble rolling on an uneven floor. Typically, it will keep rolling until it stops in a depression of the floor or reaches a wall or a corner of the room. It is clear from this example that depending on the configuration of the floor, there may be more than one stable equilibrium position. As in the case of the rod, there also exist unstable equilibria, which correspond to local maxima of the floor surface. In both of these examples, motion takes place as to decrease the potential energy of the system, which increases with elevation.

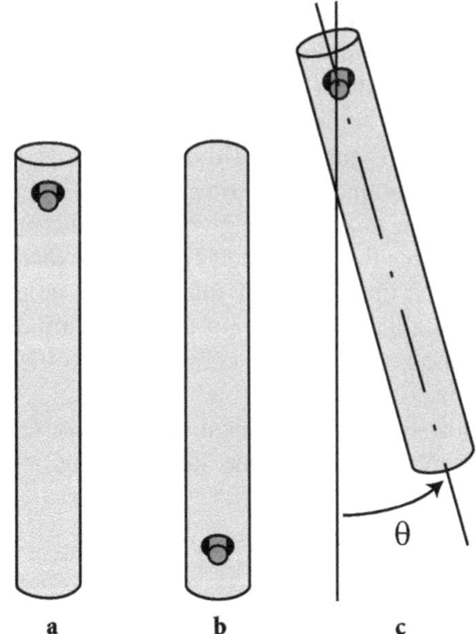

Figure 1. Stable (a) and unstable (b) equilibrium positions of a simple pendulum. (c) The angle θ is measured from the vertical.

From a mathematical point of view, an equilibrium corresponds to a stationary solution of a differential system or a map (see, for instance, Arnol'd, 1992). Quite often, these equations are mathematical models of physical, chemical, or biological systems. Consider the differential equation

$$\frac{\mathrm{d}^2\theta}{\mathrm{d}t^2} = -\sin(\theta) \qquad (1)$$

describing the motion (in dimensionless form) of a simple pendulum, similar to the rigid rod discussed above but without friction. Here, the variable θ measures the angle between the pendulum and the vertical (see Figure 1c). Time-independent solutions of (1) are obtained by setting the right-hand side of this equation to zero. This gives an infinite number of critical points $\theta = 2p\pi$ and $(2p+1)\pi$, where p is an integer. The first family of solutions can be identified with the stable equilibrium of the pendulum discussed above; the second family corresponds to its unstable equilibria.

Nondissipative mechanical systems conserve their total energy. Typically, the latter can be written in dimensionless form as

$$E(u) \equiv \tfrac{1}{2}(\dot{u})^2 + V(u) = \text{constant}, \qquad (2)$$

where u is a time-dependent variable which describes the position of the system and $\dot{u}$ is its derivative. Taking the derivative of (2) with respect to time and dividing by $\dot{u}$ gives

$$\frac{\mathrm{d}^2 u}{\mathrm{d}t^2} = -\frac{\mathrm{d}V}{\mathrm{d}u}. \qquad (3)$$

In the case of the pendulum, $u = \theta$ and E is obtained by multiplying (1) by $\dot{\theta} = \mathrm{d}\theta/\mathrm{d}t$ and integrating over time. It reads

$$E = \frac{1}{2}\left(\frac{\mathrm{d}\theta}{\mathrm{d}t}\right)^2 - \cos(\theta). \qquad (4)$$

The first term on the right-hand side of E corresponds to the (dimensionless) kinetic energy of the pendulum and the second term to its potential energy V. The graph of V as a function of θ is plotted in Figure 2a. We see that solutions $\theta = 2p\pi$ are minima of V, and solutions of the form $\theta = (2p+1)\pi$ are maxima of V. This can be understood in general as follows. Since the right-hand side of (3) vanishes at an equilibrium point, the latter corresponds to an extremum of the potential energy $V(u)$. Because $-\mathrm{d}V/\mathrm{d}u$ can be interpreted as a force, a minimum of V corresponds to a stable equilibrium, for which the force is restoring. Similarly, maxima of V identify unstable equilibria, as illustrated in Figure 2b. In dynamical systems terminology (see, for instance, Guckenheimer & Holmes, 1990), solutions $\theta = 2p\pi$ of (1) are centers and are *neutrally* stable; solutions $\theta = (2p+1)\pi$ are saddles and are unstable. If we now add a friction term to Equation (1), we get

$$\frac{\mathrm{d}^2\theta}{\mathrm{d}t^2} = -\sin(\theta) - c\frac{\mathrm{d}\theta}{\mathrm{d}t}. \qquad (5)$$

This equation has the same equilibria as (1) but now the total energy E decreases as a function of time, $\mathrm{d}E/\mathrm{d}t = -c\,(\mathrm{d}\theta/\mathrm{d}t)^2$, and as a consequence,

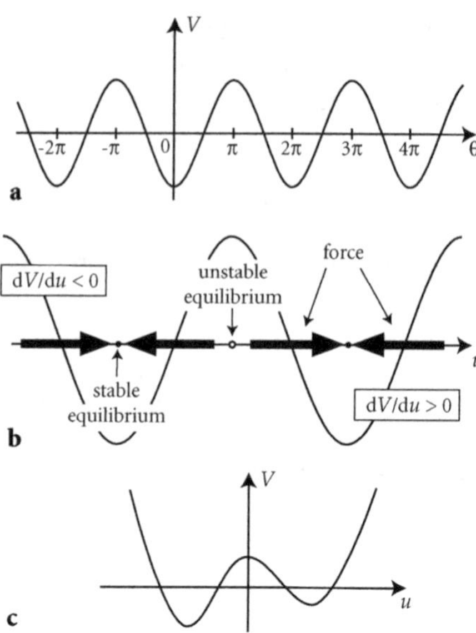

Figure 2. (a) Potential energy $V(\theta) = -\cos(\theta)$ of the simple pendulum. (b) Minima of the potential energy correspond to stable equilibria, whereas maxima describe unstable equilibria. (c) Double-well potential.

the equilibrium corresponding to solutions of the form $\theta = 2p\pi$, where p is an integer, is *globally asymptotically* stable.

In the situation depicted in Figure 2c, V is a double-well potential and the system is *bistable*. Generically, the two minima correspond to different values of the potential energy; the absolute minimum is then the most stable equilibrium of the system; the other minimum corresponds to a metastable equilibrium position.

In the case of maps, equilibria (i.e., stationary solutions) are often called fixed points. For instance, the map

$$u_{n+1} = 2u_n - |u_n|^2 u_n \qquad (6)$$

with $u_n \in \mathbb{R}$ has three fixed points, given by $u_n = u_\mathrm{e}$ for all values of n, where $u_\mathrm{e} = 0$ or $u_\mathrm{e} = \pm 1$. They are obtained by substituting the unknown equilibrium value u_e for u_n in Equation (6) and solving for u_e. If we now consider the same map but let u_n be complex, we see that we have, together with the solution $u_\mathrm{e} = 0$, a whole family of equilibria, given by $u_\mathrm{e} = \exp(\mathrm{i}\varphi)$, where φ is an arbitrary real number. This happens because Equation (6) with u_n complex has the following gauge symmetry: if u_n is a sequence of iterates of the map, then so is $u_n \exp(\mathrm{i}\varphi)$, but with, of course, a different initial condition, $u_0 \exp(\mathrm{i}\varphi)$. More generally, by applying a symmetry of the equation to one of its solutions, one typically obtains another solution. If one starts this iterative process with an equilibrium, a collection of other equilibria is generated. Moreover, if the symmetry is continuous, as it is the case for the gauge invariance of Equation (6) with u_n complex, a continuous family of relative equilibria is created. Similarly, if the symmetry is discrete, a discrete family of equilibria is obtained. For instance, Equation (6) with u_n real has the symmetry $u_n \to -u_n$; as a consequence if $u_e = 1$ is an equilibrium, so is $u_e = -1$.

Similar ideas apply to differential equations. Equations (1) and (5) have the discrete symmetry $\theta \to \theta + 2\pi$. The Landau equation

$$\frac{du}{dt} = u - |u|^2 u, \quad u \in \mathbb{C}, \tag{7}$$

which is the continuous version of Equation (6) with $u_n \in \mathbb{C}$, has the continuous symmetry $u \to u \exp(i\varphi)$, and its equilibria are given by $u = 0$ and $u = \exp(i\varphi)$, where φ is an arbitrary real number.

JOCELINE LEGA

See also **Pendulum; Stability; Symmetry: equations vs. solutions**

Further Reading

Arnol'd, V.I. 1992. *Ordinary Differential Equations*, 3rd edition, Berlin and New York: Springer

Guckenheimer, J. & Holmes, P. 1990. *Nonlinear Oscillations, Dynamical Systems, and Bifurcations of Vector Fields*, New York and London: Springer

EQUIPARTITION OF ENERGY

In 1845, the Scottish scientist John James Waterston submitted a large manuscript outlining a kinetic theory of gases to the Royal Society of London for publication in *Philosophical Transactions*. The central idea of this theory was that heat should be viewed as nothing but the energy associated with the motions of the enormous numbers of molecules, which themselves constitute the gas. Among his results, Waterston noted that "in mixed media the mean square molecular velocity is inversely proportional to the specific weight of the molecules," where the constant of proportionality is to be identified as a measure of thermodynamic temperature. This result is the essence of the Principle of Equipartition of Energy. The manuscript was rejected for publication at the time (as "nonsense, unfit even for reading before the Society") and then published (Waterston, 1893) almost 50 years later when it was (re)discovered by Lord Rayleigh who prefaced the paper with the suggestion that Waterston should be ranked "among the very foremost theorists of all ages."

In the years while Waterston's manuscript lay dormant, the kinetic theory of gases became established, mainly through the collected works of Rudolf Clausius, James Clerk Maxwell, and Ludwig Boltzmann (Brush, 1965, 1966) utilizing statistical methods to describe equilibrium properties. In 1868 Boltzmann obtained one of the most important results of this period, the (Maxwell–Boltzmann) distribution of energy among the molecules for a gas in thermal equilibrium. Let ε denote the energy of a molecule with r degrees of freedom, and position and momentum coordinates in the range $d\omega = \delta q_1 \ldots \delta q_r \delta p_1 \ldots \delta p_r$; the number of such molecules is proportional to the product of the extension $d\omega$ and the Boltzmann factor

$$\exp(-\varepsilon/kT).$$

In the Boltzmann factor, $k = 1.38 \times 10^{-16}$ erg/deg is a constant and T is the absolute temperature. The statistical justification for the Maxwell–Boltzmann distribution is that it is the most probable distribution for the distribution of energy among the possible states for a large number of molecules making up a system at constant energy.

Boltzmann derived the Principle of Equipartition of Energy from the Maxwell–Boltzmann distribution in 1871. In Boltzmann's formulation the mean energy associated with each variable that contributes a quadratic term to the total energy of the molecule has the same value $kT/2$. Thus, the mean energy of a molecule in a system is independent of mass—lighter molecules travel faster. From the mean energy per molecule, it is a simple matter to multiply by the total number of molecules to find the total energy and then to differentiate the total energy with respect to the temperature to find the specific heat at constant volume.

For example, for a perfect monatomic gas of N point atoms where $E = \sum_{i=1}^{N} mv_i^2/2$, there are $3N$ quadratic terms. Thus the internal energy of the gas is $U = 3NkT/2$, and the specific heat at constant volume is $C_V = 3Nk/2$. A simple thermodynamic calculation for the ratio of the specific heat at constant pressure to the specific heat at constant volume yields $\gamma = \frac{5}{3}$ in good agreement with experimental values for the inert gases. In the case of a crystal of N point atoms vibrating about their equilibrium positions according to Hooke's law (harmonic oscillators), there are three quadratic terms for the kinetic energy and three for the potential energy for each particle. Thus, the total internal energy is $U = 3NkT$ and the specific heat is $C_V = 3Nk$, in agreement with the empirical result obtained by Pierre-Louis Dulong and Alexis-Thérèse Petit in 1819.

In 1918 Richard Tolman derived the general Equipartition Principle (Tolman, 1918),

$$\left\langle q_i \frac{\partial \varepsilon}{\partial q_i} \right\rangle = \left\langle p_j \frac{\partial \varepsilon}{\partial p_j} \right\rangle = kT$$

which agrees with Boltzmann's result in the case where ε is a quadratic function of the q_i and p_j but can be applied more generally when ε is not a quadratic function.

Toward the end of the 19th century, some of the predictions from the classical Equipartition Principle were found to be at odds with experimental results.

Examples include the energy distribution from black body radiation, the specific heats of low temperature solids, and the specific heats of diatomic gases. The eventual resolution of these problems played a pivotal role in the upheaval of classical mechanics and the subsequent revolution of quantum mechanics but paradoxically strengthened the atomistic and probabilistic basis on which this principle was founded. The essential new idea of the quantum mechanics is that atoms have sets of discrete energy levels and radiation is absorbed and emitted in discrete quanta. Boltzmann's method of obtaining the Maxwell–Boltzmann distribution as the most probable distribution according to the way that small units of energy could be partitioned was employed by Max Planck in his famous derivation of the radiation law (Planck, 1972) in 1900. The Boltzmann factor is immediately adaptable to the quantum situation. In thermal equilibrium the ratio of the number of particles n_i in one given energy level E_i to the number of particles n_j in another energy level E_j is given by

$$\frac{n_i}{n_j} = \exp\left(-\frac{E_i - E_j}{kT}\right).$$

The word *particle* in this context embraces molecules, electrons, sound waves, light waves, and other dynamical quantities with well-defined energies. As an example, the Boltzmann ratio leads to the average energy of a harmonic oscillator whose quantum energy is an integer multiple of $h\nu$ as

$$\bar{\varepsilon} = \frac{1}{2}h\nu + \frac{h\nu}{e^{h\nu/kT} - 1}.$$

In the low-temperature regime this provides reasonable agreement with the experimental result, but it only agrees with the classical Equipartition Principle $\bar{\varepsilon} = kT$ at high temperatures, $T \gg h\nu/k$.

The Maxwell–Boltzmann distribution is a stationary or steady-state distribution. It can be derived (Tolman, 1948) from the steady-state micro-canonical ensemble distribution for which the density of ensemble copies in phase space is uniform in a narrow energy range $E, E + \delta E$ but zero elsewhere. The dynamical justification for the Maxwell–Boltzmann distribution then rests upon the (quasi-)ergodic hypothesis that the time spent by the phase space trajectory for the dynamical system in a given region of the energy surface will be proportional to the micro-canonical density of ensemble copies in that region. Thus, from a dynamical perspective, ergodicity is a necessary condition for Equipartition of Energy. It follows that in isolated integrable systems Equipartition of Energy has no dynamical justification.

The Equipartition principle emerged again as a catalyst for change in the mid-20th century when its failure observed in computer experiments (Fermi et al., 1955) triggered studies into solitons on the one hand and deterministic Hamiltonian chaos on the other (Ford, 1992). The systems investigated by Fermi, Pasta, and Ulam (FPU) could be considered as collections of harmonic oscillators (harmonic modes of vibration) weakly coupled by nonlinear interactions. The expectation of FPU was that energy initially supplied to one of the harmonic oscillators would become uniformly distributed among all harmonic oscillators. Following the FPU experiments, there has been an enormous literature, attempting to recover the equipartition result, particularly for larger nonlinearities and larger numbers of oscillators. Note however that Boltzmann's principle of equipartition of energy predicts uniform energy sharing among the harmonic modes of an isolated linear chain (a result that cannot be justified dynamically), and it does not make predictions for an isolated nonlinear chain. For energy sharing in an isolated nonlinear chain Tolman's general principle of equipartition can be employed, but the result of uniform energy sharing among the harmonic modes should not generally be expected (Henry & Szeredi, 1995). Moreover, nonlinear chains can sustain intrinsic local nonlinear modes that may persist for very long times, and integrable nonlinear chains support nonlinear normal modes that are more fundamental than the harmonic modes.

The principle of equipartition of energy is one of the most fundamental results in the history of the atomistic description of matter. Inconsistencies between this principle and experimental results in different settings have provided a catalyst for some of the most profound theoretical developments in science in the 20th century. A contemporary problem is the failure of the Equipartition Principle in granular materials (Feitosa & Menon, 2002), the resolution of which may play a fundamental role in the development of a comprehensive kinetic theory of granular systems.

BRUCE HENRY

See also **Ergodic theory; Fermi–Pasta–Ulam oscillator chain; Local modes in molecules; Quantum theory**

Further Reading

Boltzmann, L. 1896. *Lectures on Gas Theory*, New York: Dover (English Translation by S.G. Brush. Originally published in German in two parts by J.A. Barth, Leipzig, 1896 (Part I) and 1898 (Part II), under the title *Vorlesungen über Gastheorie*)

Brush, S.G. 1965, 1966. *Kinetic Theory*, vols. 1(2), Oxford: Pergamon Press

Ehrenfest, P. & Ehrenfest, T. 1959. *The Conceptual Foundations of the Statistical Approach in Mechanics*, Ithaca: Cornell University Press (English Translation of No. 6, vol. IV 2II, *Encyklopädie der mathematischen Wissenschaften*, 1912)

Feitosa, K. & Menon, N. 2002. Breakdown of energy equipartition in a 2D binary vibrated granular gas. *Physical Review Letters*, 88(19): 198301

Fermi, E., Pasta, J.R. & Ulam, S.M. 1955. *Studies of Nonlinear Problems*, Los Alamos Scientific Laboratory Report N. LA-1940 (Electronic Access: http://www.osti.gov/accomplishments/pdf/A80037041/A80037041.pdf)

Ford, J. 1992. The Fermi–Pasta–Ulam problem: paradox turns discovery. *Physics Reports*, 213(5): 271–310

Henry, B.I. & Szeredi, T. 1995. New equipartition results for normal mode energies of anharmonic chains. *Journal of Statistical Physics*, 78: 1039–1053

Planck, M. 1972. On the theory of the energy distribution law of the normal spectrum. In *Planck's Original Papers in Quantum Physics*, edited by H. Kangro with D. ter Haar, & S. Brush (transl.). London: Taylor & Francis (English translation of Planck, M. 1900. Zur Theorie des Gesetzes der Energieverteilung im Normalspectrum. *Verh deutsch phys ges*, 2: 202–237)

Tolman, R.C. 1918. A general theory of energy partition with applications to quantum theory. *Physical Review*, 11: 261–275

Tolman, R.C. 1948. *The Principles of Statistical Mechanics*, Oxford: Oxford University Press, Chapter IV.

Waterston, J.J. 1893. On the physics of media that are composed of free and perfectly elastic molecules in a state of motion. *Philosophical Transactions of the Royal Society London*, 183A: 5–79

ERGODIC THEORY

Ergodic theory is the statistical study of groups of motions of a space, either physical or mathematical, with a measurable structure on it. The origins of ergodic theory can be traced back to the mid-19th century when containers of gas particles were first viewed as sets of randomly moving objects rather than as a collection of individual particles moving under known forces. The word *ergodic* was introduced by Ludwig Boltzmann in the context of the statistical mechanics of gas particles, and it comes from two Greek words *ergon* (work) and *odos* (path).

The mathematical setting in which ergodic theory is studied is as follows. Starting with a space X that represents all possible states of some system which changes over time under known forces, a point $x \in X$ corresponds to one state in the space. The measurable structure consists of a collection of measurable sets $\mathcal{B}$ on X along with a probability measure μ. The measure μ is a function that associates to each set B in $\mathcal{B}$ a number between 0 and 1; this number is the measure of B and we write its measure as $\mu(B)$. A probability measure has the property that $\mu(X) = 1$. Instead of tracking the path of each object in the system, one studies the statistical properties of the motion. The subject evolved from statistical mechanics applied to the study of systems of gas particles moving according to classical laws of physics. In principle, the path of each gas particle can be tracked and its entire history and future can be known; in practice, the complete determination of the paths of the gas molecules is not feasible.

After t units of time, $F_t(x)$ is the point in X corresponding to where x ends up. If $t \in \mathbb{R}$, then the *orbit* of x is the set $\{F_t(x) | t \in \mathbb{R}\}$; F_t is called a flow and defines an action of the group $\mathbb{R}$ on X. Frequently, one uses discrete time intervals and writes $T^n(x) \equiv F_n(x)$ for each integer n, so the orbit of x is a discrete set $\{T^n(x) | n \in \mathbb{Z}\}$ and the group acting on X is $\mathbb{Z}$. In the discrete setting the transformation T is the generating map T^1, and T^n is T composed with itself n times. In classical ergodic theory, the measure μ is preserved under the action; i.e., for any set $A \in \mathcal{B}$, $\mu(A) = \mu(F_t A)$ for all $t \in \mathbb{R}$ or $\mu(A) = \mu(T^n A)$ for all $n \in \mathbb{Z}$. One of the main advantages of the ergodic theoretic point of view is that one can ignore some orbits if they only form a set of measure 0 in the space X; therefore, one uses the terminology *μ-almost everywhere* (or *for μ-a.e. x*) to refer to a property that holds on a set of points of measure 1 in X but perhaps fails to hold on some set of measure 0.

Boltzmann's original ergodic hypothesis has come to be known as the statement that *time average equals space average*; he conjectured (in a certain classical setting) that for any integrable function f, for μ-a.e. x,

$$\lim_{n \to \infty} \frac{1}{n} \sum_{k=0}^{n-1} f(T^k x) = \int_X f \mathrm{d}\mu. \tag{8}$$

In this expression, the time average of f for n time steps along the orbit of x is represented by averaging $f(x)$ with $f(Tx)$, $f(T^2 x)$, ..., and $f(T^{n-1}x)$, since each application of T represents the passage of one unit of time (the left-hand side of the expression). The space average of f is obtained by integrating f over the entire space X, giving the right-hand side of the expression. For a flow the ergodic hypothesis is

$$\lim_{t \to \infty} \frac{1}{t} \int_0^t f[F_s(x)] \mathrm{d}s = \int_X f \mathrm{d}\mu. \tag{9}$$

The conjecture is false as stated; it only holds when the action is μ-ergodic, and this realization led to the definition of an ergodic action.

The setting of ergodic theory has been greatly enlarged to include the actions of many other groups (Zimmer, 1984). A nonsingular action Φ of a group G on a space $(X, \mathcal{B}, \mu)$ consists of an action of G on X such that the map $\Phi : G \times X \to X$ is measurable and for each $g \in G$ the map $\phi_g(x) = \Phi(g, x)$ is a nonsingular automorphism of X; that is, ϕ_g is an invertible measurable transformation and for any $B \in \mathcal{B}$, $\mu(B) = 0$ if and only if $\mu(\phi_g^{-1} B) = \mu(\phi_{g^{-1}} B) = 0$. The group operation on G is reflected in the action since $\phi_{g_1 g_2}(x) = \phi_{g_2}(\phi_{g_1} x)$ for all $g_1, g_2 \in G$ and almost every (μ-a.e.) x.

An action is *ergodic* if whenever $\phi_g(A)=A$ for all $g\in G$, then $\mu(A)=0$ or 1. The study has been extended beyond group actions to the study of ergodic equivalence relations, and the assumption that the measure μ be a probability measure is frequently dropped. One also considers the actions of semigroups when the action being studied is not invertible.

The Birkhoff ergodic theorem for a measure-preserving transformation T states that the limit of the left-hand side in Equation (8) exists for μ-a.e. x, is an integrable function f^*, and $f^*(Tx)=f^*(x)$ for μ-a.e. x. Furthermore, f^* is the constant $\int_X f\,\mathrm{d}\mu$, the space average of f, precisely when the action is ergodic (Birkhoff, 1931). Any point x satisfying the theorem is called a generic point and generates a *typical orbit*. There are many ergodic theorems; perhaps the simplest is the von Neumann ergodic theorem (von Neumann, 1932). By definition, a function $f\in L^2$ if $\int|f|^2\mathrm{d}\mu<\infty$; moreover defining $f\circ T^k(x)=f(T^kx)$, for measure-preserving T, $f\circ T^k\in L^2$ as well. Von Neumann's theorem is also called the Mean or L^2 ergodic theorem because while no individual orbit is tracked (no x appears in the statement), the average difference (L^2 integral) between the time average and the limit function f^* must be small.

Theorem 1. (von Neumann or L^2 Ergodic Theorem) *If T is a measure-preserving transformation and $f\in L^2$, then there is a function $f^*\in L^2$ such that*

$$\int_X \left|\frac{1}{n}\sum_{k=0}^{n-1} f\circ T^k - f^*\right|^2 \mathrm{d}\mu \to 0 \quad as \quad n\to\infty.$$

If $(X,\mathcal{B},\mu)$ is a probability space and the transformation T (or group action) on X is non-ergodic, there is a disintegration or decomposition of μ into measures μ_y, indexed by points y in another probability space Y, with its own measure ρ, such that $\mu(A)=\int_Y \mu_y(A)\,\mathrm{d}\rho$ for every $A\in\mathcal{B}$. Furthermore, the limit function f^* in the ergodic theorems is constant with respect to each measure μ_y; that is, T is μ_y ergodic for ρ–a.e. y. The decomposition is independent of the function f; this is referred to as the *ergodic decomposition* of the measure μ with respect to T (Rohlin, 1949). One of the central problems in ergodic theory is to determine when two measure-preserving group actions are conjugate via a measure-preserving isomorphism. To this end, invariants such as entropy and spectral properties have been the subject of much study.

There is a hierarchy of statistical properties associated to ergodic actions, including weak mixing, mixing, K-automorphisms, and Bernoulli automorphisms. As a transformation moves up the hierarchy (in the order listed), the more chaotic the behavior of the system is expected to be.

Applications to Dynamical Systems and Chaos

The simplest example of an ergodic transformation is irrational rotation on the circle with respect to Lebesgue measure. The most random transformation is a Bernoulli shift with an independent identically distributed measure on it; this includes coin tosses.

Overlaps of the topological setting of dynamical systems with ergodic theory exist, and much of ergodic theory highlights the interface between the topological and measurable structure of a group G acting on $(X,\mathcal{B},\mu)$. In 1935, Hedlund proved the ergodicity of the geodesic flow on the unit tangent bundle of a surface of constant negative curvature; in 1940, Hopf extended the result to establish ergodicity of the geodesic flow on arbitrary manifolds with negative sectional curvature. In this setting the invariant measure is the Liouville measure.

Modern ergodic theory was started by Andrei Kolmogorov with the formal development of Boltzmann's notion of entropy, and developed in the 1960's and 1970's to include many differentiable actions. Applications include fluid dynamics, coding theory, number theory, complex dynamics, and cellular automata.

JANE HAWKINS

See also **Chaotic dynamics; Dynamical systems; Entropy; Symbolic dynamics**

Further Reading

Birkhoff, G. 1931. Proof of the ergodic theorem. *Proceedings of the National Academy of Sciences USA*, 17: 656–660

Cornfeld, I., Fomin, S. & Sinai, Y. 1982. *Ergodic Theory*, New York: Springer

Furstenberg, H. 1981. *Recurrence in Ergodic Theory and Combinatorial Number Theory*, Princeton: Princeton University Press

Halmos, P. 1956. *Ergodic Theory*, New York: Chelsea Publishing

Katok, A. & Hasselblatt, B. 1995. *Introduction to the Modern Theory of Dynamical Systems*, Cambridge and New York: Cambridge University Press

Petersen, K. 1983. *Ergodic Theory*, Cambridge and New York: Cambridge University Press

Rohlin, V. A. 1949. On the fundamental ideas of measure theory. *American Mathematical Society Translations*, 10 (1): 1–54 (original Russian article 1949)

von Neumann, J. 1932. Proof of the quasiergodic hypothesis. *Proceedings of the National Academy of Sciences USA*, 18: 70–82

Walters, P. 1982. *An Introduction to Ergodic Theory*, New York: Springer

Zimmer, R. 1984. *Ergodic Theory and Semisimple Groups*, Boston: Birkhäuser

ESAKI TUNNEL DIODE

See **Diodes**

EULER'S METHOD

See **Numerical methods**

EULERIAN DESCRIPTION

See **Fluid dynamics**

EULER–LAGRANGE EQUATIONS

In the dynamics of energy conserving systems there are two equivalent formulations: one based on differential equations and the other based on the principle of least action (Goldstein, 1951). At first glance, these two approaches seem to have little in common as one stems from iterating infinitesimal changes along a solution trajectory, whereas the other depends on evaluating all possible paths between two end points and selecting the path with the minimum action.

To see how this goes, consider a simple classical system: the nonrelativistic point particle of unit mass. In three dimensions this particle has three degrees of freedom each labeled by coordinates $q_i(t)$ for $i=1,2,3$. The motion of the particle is determined by a Lagrangian functional $L(q_i,\dot{q}_i)$ which depends on both the position components, $q_i(t)$, and the velocity components, $\dot{q}_i(t)$, of the particle. This functional is

$$L=\sum_i\left[\dot{q}_i^2/2-V(q_i)\right], \tag{1}$$

where $V(q_i)$ is some potential in which the particle moves; thus, the Lagrangian is defined as the difference between the kinetic and the potential energy of the system. The motion of the particle is determining by minimizing the action, which is the time integral of the Lagrangian

$$S=\int_{t_1}^{t_2} L(q_i,\dot{q}_i)\,\mathrm{d}t, \tag{2}$$

and the path taken by the particle is the one for which the action is a minimum. In other words, $\delta S=0$, where δ implies a differential change in S under a small change in the path.

In calculating the equations of motion, a small variation in the path of the particle $\delta q_i(t)$ is assumed with the endpoints fixed, that is, $\delta q_i(t_1)=\delta q_i(t_2)=0$. To calculate δS, the Lagrangian is varied with respect to both changes in the position and the velocity

$$\delta S=\int_{t_1}^{t_2}\sum_i\left(\frac{\partial L}{\delta q_i}\delta q_i+\frac{\partial L}{\partial \dot{q}_i}\delta\dot{q}_i\right)\mathrm{d}t. \tag{3}$$

Integration by parts of the last term gives

$$\delta S=\int_{t_1}^{t_2}\sum_i\left[\frac{\partial L}{\partial q_i}\delta q_i-\frac{\mathrm{d}}{\mathrm{d}t}\left(\frac{\partial L}{\partial \dot{q}_i}\right)\delta q_i\right]\mathrm{d}t, \tag{4}$$

so the condition $\delta S=0$ implies that

$$\frac{\partial L}{\partial q_i}-\frac{\mathrm{d}}{\mathrm{d}t}\left(\frac{\partial L}{\partial \dot{q}_i}\right)=0 \tag{5}$$

for all i. These are the Euler–Lagrange (EL) equations of motion. For the system of Equation (1), they are $\mathrm{d}^2q_i/\mathrm{d}t^2+\partial V/\partial q_i=0$, which is Newton's second law.

In some cases the motion is constrained, for example to curves or surfaces. Such a constraint can be expressed as a function $f(q_i)=0$. Augmenting the Lagrangian to $\tilde{L}=L+\lambda f$ (where λ is called a Lagrange multiplier) and minimizing S with respect to $\tilde{L}$ then yields equations for motion under the constraint (Gel'fand & Fomin, 2000). As a simple example, consider the system of Equation (1) with $V=(q_1^2+q_2^2+q_3^2)/2$ moving under the constraint $q_1=q_2$ or $q_1-q_2=0$. Then the EL equations for the augmented Lagrangian $\tilde{L}=L+\lambda(q_1-q_2)$ are $\mathrm{d}^2(q_1+q_2)/\mathrm{d}t^2+(q_1+q_2)=0$ and $\mathrm{d}^2q_3/\mathrm{d}t^2+q_3=0$.

An alternative to the Lagrangian formulation (where positions and velocities are the fundamental variables) is the Hamiltonian formulation. Here the positions and momenta (p_i) are independent variables, where

$$p_i\equiv\partial L/\partial\dot{q}_i \tag{6}$$

and a Hamiltonian is defined through the Legendre transformation

$$H(q_i,p_i)\equiv\sum_i p_i\dot{q}_i-L(q_i,\dot{q}_i). \tag{7}$$

Because this Hamiltonian is conserved under the dynamics, it is often interpreted as the energy of a dynamical system. For the example of Equation (1),

$$H=\sum_i\left[\dot{q}_i^2/2+V(q_i)\right], \tag{8}$$

the sum of kinetic and potential energies.

The generalization from a point particle to a three-dimensional field may be viewed as the replacement of $q_i(t)$ by a field amplitude: $\phi(x,y,z,,t)$ (Goldstein, 1951). In this case, EL equations can be derived from a variational principle applied to an action integral of the form

$$S=\int\mathcal{L}\left(\phi,\phi_t,\phi_x,\phi_y,\phi_z\right)\,\mathrm{d}x\,\mathrm{d}y\,\mathrm{d}z\,\mathrm{d}t, \tag{9}$$

where $\mathcal{L}$ is a Lagrangian density and the subscripts indicate partial derivatives. The condition for minimum action then gives

$$\frac{\partial \mathcal{L}}{\partial \phi} - \frac{\partial}{\partial t}\frac{\partial \mathcal{L}}{\partial \phi_t} - \frac{\partial}{\partial x}\frac{\partial \mathcal{L}}{\partial \phi_x} - \frac{\partial}{\partial y}\frac{\partial \mathcal{L}}{\partial \phi_y} - \frac{\partial}{\partial z}\frac{\partial \mathcal{L}}{\partial \phi_z} = 0, \tag{10}$$

which is the EL equation in field theory.

A simple example of this field formulation is provided by the sine-Gordon equation in three-dimensional space. The Lagrangian density is given by

$$\mathcal{L} = \tfrac{1}{2}\left(\phi_t^2 - \phi_x^2 - \phi_y^2 - \phi_z^2\right) + (\cos\phi - 1), \tag{11}$$

for which Equation (10) implies

$$\phi_{xx} + \phi_{yy} + \phi_{zz} - \phi_{tt} = \sin\phi. \tag{12}$$

Because not all dynamics conserve energy, the Lagrangian formulation is not applicable to all physically motivated systems. Reaction-diffusion equations, for example, are excluded from this class (Scott, 2003).

ALWYN SCOTT

See also **Extremum principles; Hamiltonian systems; Newton's laws of motion; Poisson brackets**

Further Reading

Gel'fand, I.M. & Fomin S.V. 2000. *Calculus of Variations*, New York: Dover

Goldstein, H. 1951. *Classical Mechanics*, Reading, MA: Addison–Wesley

Scott, A.C. 2003. *Nonlinear Science: Emergence and Dynamics of Coherent Structures*, 2nd edition, Oxford and New York: Oxford University Press

EVANS FUNCTION

Traveling waves are one-dimensional patterns that arise in numerous applications, for instance, as defects, flame fronts, nerve impulses, shock waves, and solitons. Typically, traveling waves are only observed in experiments or in numerical simulations if they are stable.

In the early 1970s, Evans developed a number of tools designed to determine the stability of pulses, that is of traveling waves that are localized in space as solutions to a system of partial differential equations (PDEs) (Evans, 1975). The class of PDEs considered by Evans serve as models for the propagation of impulses in nerve axons that include the FitzHugh–Nagumo equation and the Hodgkin–Huxley equations. The strategy to determine the stability of waves consists of two steps: (i) linearize the nonlinear PDE about the wave and calculate the spectrum of the resulting linear operator, and (ii) show that the wave is stable for the nonlinear PDE if the spectrum contains no unstable elements. This entry focuses on the first step, see Sandstede (2002) for references concerning (ii).

The spectrum of the linearization $\mathcal{L}$ about a pulse or a front is the union of two sets, namely the point spectrum (which consists of all isolated eigenvalues) and the essential spectrum (which is also often referred to as the continuous or radiation spectrum). The essential spectrum can be computed easily as it involves only the homogeneous background states of the pulse or front.

The key problem is therefore to find all isolated eigenvalues in the complement Ω of the essential spectrum in the complex plane. To address this issue for his class of nerve-axon models, Evans constructed an analytic function $D(\lambda)$ (now referred to as the Evans function) that maps the region Ω into the complex plane with the property that $D(\lambda)$ vanishes precisely at isolated eigenvalues of the linear operator $\mathcal{L}$. In fact, the order of a root λ_* of the Evans function $D(\lambda)$ coincides with the algebraic multiplicity of λ_* as an eigenvalue of $\mathcal{L}$. Here, ℓ is the order of a root λ_* of $D(\lambda)$ if the first $\ell - 1$ derivatives of $D(\lambda)$ vanish at λ_* but the ℓth derivative does not.

Note that, if the underlying PDE is invariant under space translations, then $\lambda = 0$ is an eigenvalue. The corresponding eigenfunction is given by the spatial derivative of the wave. Hence, if $\lambda = 0$ is not in the essential spectrum, the Evans function $D(\lambda)$ vanishes at $\lambda = 0$. Proving linear stability of a pulse, therefore, amounts to calculating the Evans function $D(\lambda)$ and showing that it does not have any zero in the closed right half-plane except possibly a simple zero at $\lambda = 0$.

The derivative $D'(0)$ is actually given by the Mel'nikov integral, computed with respect to the wave speed, c, of the pulse. This integral measures how the stable and unstable manifolds of the background state intersect as the wave speed c of the pulse is varied. In other words, if the existence problem of the wave is understood well enough, then $D'(0)$ is computable.

This can be exploited as follows. The Evans function $D(\lambda)$ is, in fact, real whenever λ is real, and $D(\lambda)$ can be normalized so that $D(\lambda) > 0$ for all sufficiently large $\lambda \gg 1$. Here, λ is the temporal growth rate so that unstable eigenvalues have positive real part. Thus, if $D'(0) < 0$, then the pulse is necessarily unstable with an odd (and therefore nonzero) number of unstable real eigenvalues. Consequently, the derivative $D'(0)$ of the Evans function provides a parity instability index.

The function $D(\lambda)$ introduced by Evans was then later used by Jones and by Yanagida to prove that the fast pulse arising in the FitzHugh–Nagumo equation is stable. Both proofs used the parity index introduced above: after establishing that there is precisely one eigenvalue besides $\lambda = 0$ that lies close to the imaginary axis or in the right half-plane, a computation of the parity index at $\lambda = 0$ showed that it is positive, so that the other eigenvalue must lie in the left half-plane.

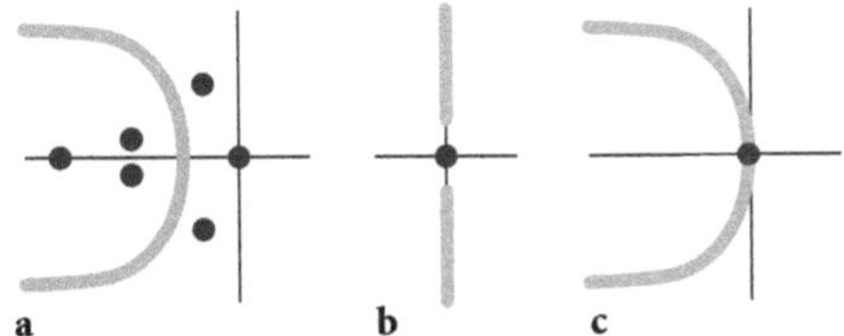

Figure 1. The three illustrations show typical spectra of the linearization about (a) a stable pulse in a dissipative PDE, (b) a soliton in an integrable system, and (c) a viscous shock wave. The horizontal and vertical axes are (Re λ, Im λ). The point and essential spectrum are shown as discs and lines, respectively.

Alexander et al. (1990) developed the Evans function further and also related it to certain topological invariants. Since then, Evans-function calculations have been utilized in various applications, for instance, in nonlinear optics, plasma physics, and thin films (Sandstede, 2002; Zumbrun, 2001).

An important class of PDEs to which the Evans function has been applied is singularly perturbed equations that exhibit two different spatial scales. In this situation, the Evans function can often be calculated from the Evans functions for the slow and the fast subsystem (Doelman et al., 2002).

The Evans function is also a useful tool for investigating the stability of multi-hump pulses, or n-pulses, which are composed of n widely separated copies of a primary pulse. If the primary pulse is stable, so that $\lambda = 0$ is a simple eigenvalue, then such an n-pulse has exactly n critical eigenvalues near the origin which encode the interaction properties of the n individual pulses. These eigenvalues can be calculated using the Evans function (Sandstede, 2002).

The original construction of the Evans function works only outside of the essential spectrum. For Hamiltonian PDEs, the essential spectrum of stable pulses lies on the imaginary axis, see Figure 1(b). Thus, it is possible that eigenvalues move out of the essential spectrum and into the right half-plane once a small perturbation is added to the underlying PDE. The points in the essential spectrum from which such eigenvalues can potentially emerge and the exact location of the bifurcating eigenvalues for concrete perturbations can be found by extending the Evans function across the essential spectrum. For integrable PDEs such as the Korteweg–de Vries or the nonlinear Schrödinger equation, it is possible to compute the extended Evans function explicitly using inverse scattering theory (see Kapitula & Sandstede (2002) for references).

Similar issues occur for the linearization about viscous shock waves where the essential spectrum always touches the origin, see Figure 1(c). An extension of the Evans function across the essential spectrum to a full neighborhood of $\lambda = 0$ allows one to generalize the parity instability index, given by an appropriate sign condition on the derivative $D'(0)$, to viscous shocks. In addition, the extended Evans function plays an important role in the nonlinear stability analysis of viscous shocks (Zumbrun, 2001).

BJÖRN SANDSTEDE

See also **Shock waves; Stability; Wave stability and instability**

Further Reading

Alexander, J.C., Gardner, R.A. & Jones, C.K.R.T. 1990. A topological invariant arising in the stability analysis of travelling waves. *Journal für die Reine und Angewandte Mathematik*, 410: 167–212

Doelman, A., Gardner, R.A. & Kaper, T.J. 2002. A stability index analysis of 1-D patterns of the Gray–Scott model. *Memoirs of the American Mathematical Society*, 155(737): xii+64 pp.

Evans, J. 1975. Nerve axon equations (iv): The stable and unstable impulse. *Indiana University Mathematics Journal*, 24: 1169–1190

Kapitula, T. & Sandstede, B. 2002. Edge bifurcations for near integrable systems via Evans-function techniques. *SIAM Journal on Mathematical Analysis*, 33: 1117–1143

Sandstede, B. 2002. Stability of travelling waves. In *Handbook of Dynamical Systems,* vol. 2, edited by B. Fiedler, Amsterdam: North-Holland

Zumbrun, K. 2001. Multidimensional stability of planar viscous shock waves. In *Advances in the Theory of Shock Waves*, edited by H. Freistühler, A. Szepessy, T.-P. Liu & G. Métiver, Boston: Birkhäuser

EVAPORATION WAVE

Under special conditions a liquid can be superheated to a few hundred degrees Celsius above its nominal boiling point. The thermal energy associated with this excess temperature can be released by rapid evaporation, also called a *vapor explosion*. The largest explosions on Earth are vapor explosions in the form of volcanic eruptions. For example, the Mount St. Helens blast was estimated at 400 Mtons TNT equivalent (Decker & Decker, 1981), which is nearly the sum of all worldwide nuclear explosions to date (about 500 Mtons).

Vapor explosions also pose a hazard in industry. In one incident, 100 lb of molten steel fell into an open trough containing 78 gallons of water. The resulting explosion killed one person and injured many, cracked the 20-in. thick concrete floor, and shattered 6000 panes of glass (Reid, 1976). If a vapor explosion involves a flammable liquid, then the physical explosion can transition to a more powerful chemical explosion.

An *evaporation wave* is a vapor explosion that proceeds across the system with an identifiable propagation velocity. Evaporation waves are an interesting example of self-organized dynamical complexity and comprise a controlled form of vapor explosion that facilitates the detailed study of liquid fragmentation and dispersal mechanisms.

Liquid superheating is caused by cohesion between molecules. Hence, bubbles—even very small

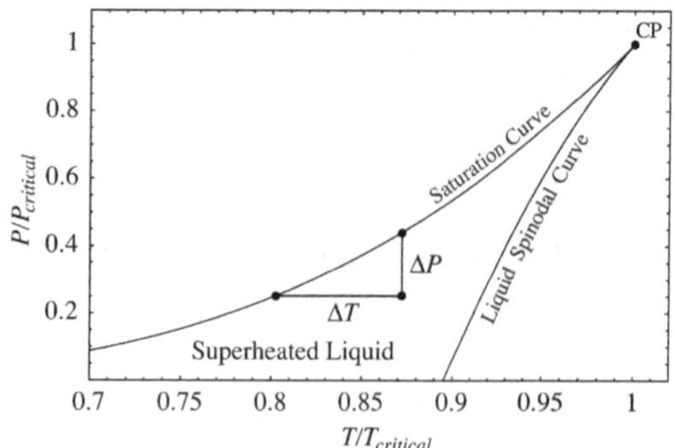

Figure 1. Existence region for superheated liquid from the Redlich–Kwong (R–K) equation of state. R–K predicts universal behavior in a reduced coordinate system.

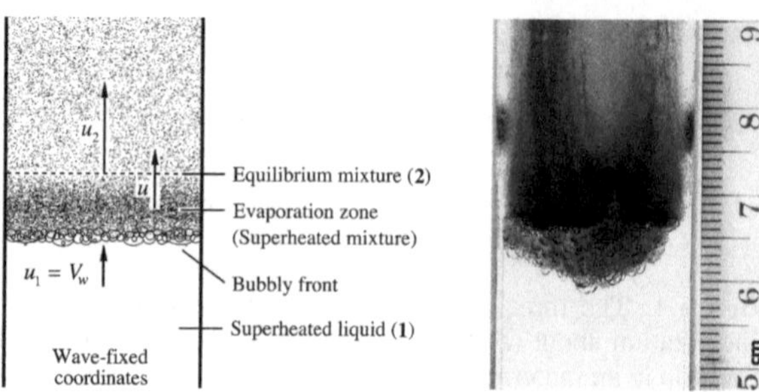

Figure 2. Schematic diagram of an evaporation wave (a), and a photograph of an evaporation wave in freon-12 at 20°C (b). The front is moving down at $0.6\,\mathrm{m\,s^{-1}}$.

ones that reside in most liquids under normal conditions—comprise weak spots that inhibit superheating. Even a completely pure liquid can only be superheated to a point, called the superheat limit, where it becomes mechanically unstable. The theoretical superheat limit occurs at a locus of states called the *spinodal curve* (Debenedetti, 1996). Superheated liquid can exist between the saturation and spinodal curves as shown in Figure 1. The amount of superheat is quantified by the excursion of the liquid state from the saturation curve, for example, the degree ΔT by which the liquid temperature exceeds the equilibrium temperature at its pressure.

The specific enthalpy required to evaporate a liquid is the latent heat of evaporation L. A superheated liquid stores an excess specific enthalpy $C_p \Delta T$, where C_p is the liquid specific heat at constant pressure. The Jakob number, $\mathrm{Ja} \equiv (C_p \Delta T)/L$, is equal to the liquid mass fraction χ that can evaporate adiabatically at constant pressure. If $\mathrm{Ja} > 1$ then complete evaporation can occur. This condition is possible in *retrograde* liquids, which are those possessing a sufficiently large specific heat (Thompson et al., 1986). Nowhere near complete evaporation is required to produce quasisteady evaporation waves. However, sufficient evaporation must occur so that the liquid phase is swept along by the vapor; otherwise, the upstream liquid will cool and the process will quench.

The first mention of an evaporation wave appears to have been by Terner (1962), who performed vertical shock tube experiments on heated water. Terner believed that an evaporation wave followed depressurization, though he did not definitively show it.

The first direct observation appears to have been by Friz (1965), who studied the expulsion of superheated water from a vertical glass tube into the atmosphere. The tube was sealed on top by a diaphragm, and the water was heated to temperatures from 105°C to 125°C. The gas space above the water was pressurized to 3.5 bar, whereupon a cutter was activated to puncture the seal. Friz describes that upon depressurization "a great number of bubbles" formed throughout the water column. An "acceleration front" several centimeters thick then propagated into the bubbly mixture at $1\text{–}2\,\mathrm{m\,s^{-1}}$, transforming it into a spray that was ejected upward. The many bubbles were probably caused by pre-pressurization, which increased the amplitude of the initial expansion wave so that it reflected from the test cell base in tension, causing cavitation.

Grolmes and Fauske (1970) performed similar experiments in heated water, freon-11, and methanol. Their configuration was like Friz's, except that (i) the initial pressure was the vapor pressure, and (ii) the fluid was expelled into a nearly evacuated reservoir. By paying careful attention to the liquid purity and test cell cleanliness, they were able to suppress *all* nucleation within the liquid column. Upon depressurization boiling erupted on the liquid surface and proceeded with sufficient vigor to expel the generated spray from the container as depicted in Figure 2a. The boiling surface receded at a constant average rate of $0.3\text{–}0.5\,\mathrm{m\,s^{-1}}$, depending on the liquid and conditions.

Thompson et al. (1987) performed similar experiments on the retrograde fluorocarbon perfluoro-*n*-hexane (C_6F_{14}). At the highest (nearly critical) temperatures, the lead expansion wave attempted to lower the pressure past the superheat limit, whereupon homogeneous nucleation erupted and blocked the drop. An approximate plateau was maintained near the spinodal, into which a slower evaporation wave propagated in a manner similar to Friz's. Such *wave splitting* occurs when material properties abruptly change within the wave, in a way that lowers the sound speed. At lower temperatures, homogeneous nucleation likely erupted upon reflection of the initial expansion wave from the test cell base, and an evaporation wave initiated after the reflected wave reached the free surface. The waves became progressively less distinct as the initial temperature was lowered, presumably because heterogeneous nucleation from the metal side-walls and the transducer ports became increasingly important.

Shepherd and Sturtevant (1982) performed experiments on butane drops that were heated slowly in an ethylene glycol host fluid to the superheat limit

($\Delta T = 105°C$), whereupon they exploded with a sharp "crack." High-speed photographs showed that evaporation initiated at a single nucleation site at or near the drop surface and spread into pure liquid as a miniature evaporation wave resembling Grolmes and Fauske's. Thus, heating to the superheat limit has produced waves in pure liquid, whereas depressurizing to the superheat limit has produced waves in bubbly liquid.

Shepherd and Sturtevant suggested that the *Landau instability* of premixed flames was the essential fragmentation mechanism. Frost and Sturtevant (Frost, 1988) repeated the exploding droplet experiments in pentane, iso-pentane, and ethyl ether. They obtained several photographs in which the evaporating interface had a mottled texture resembling an orange peel. This suggested that they had caught the interface at the early stages of instability, which they believed grew to the point of pinching off droplets to produce the observed rough-looking interfaces and spray flow. About the same time Mesler (1988) proposed that the *secondary nucleation* mechanism of surface boiling—whereby receding film-caps from ruptured bubbles strike the liquid surface and entrain vapor to create new nucleation sites—was responsible.

Hill and Sturtevant (Hill, 1991) performed Grolmes and Fauske-type experiments in freons 12 and 114, with the goal of observing detailed propagation mechanisms. Figure 2b is a photograph from that study. Several observations suggested that a three-step cycle occurs: (i) bubbles grow at the leading edge in accordance with classical theory; (ii) these are consumed en masse by fragmentation waves that sweep around the bubbly layer transversely to the main propagation at speeds of order the axial flow, and which shatter the superheated interstitial liquid into droplets; (iii) a small fraction of the droplets are propelled forward to strike the leading edge and nucleate new bubbles à la Mesler. Whether such behavior extends toward the superheat limit, or whether a nucleation-free evaporative instability dominates at some point, is unknown.

Simões-Moreira and Shepherd (1999) performed similar experiments in heated dodecane ($C_{12}H_{26}$)—a highly retrograde liquid—with the intention of observing complete evaporation waves. For $Ja > 1$ a simple plane evaporation wave, or a convoluted but simply connected surface like Frost's "orange peel," would be energetically admissible. Instead they observed a scenario qualitatively like previous experiments, except that the droplets evaporated in the downstream flow.

From a gas-dynamic viewpoint an evaporation wave is analogous to a flame in a combustible mixture. Both are classified as *deflagrations*, or *exothermic discontinuities*. The ideal theory (e.g., Hayes, 1958) considers a plane, steady wave. The conservation equations for mass, momentum, and energy are applied between the upstream state **1**, and an arbitrary state in the reaction or evaporation zone. The flow is assumed to be one-dimensional (no turbulence), and viscosity and heat conduction are neglected. Eliminating the fluid velocities between the three conditions yields the Hugoniot curve $\mathcal{H}$, which specifies the locus of possible downstream states given **1**:

$$h_1 - h(v, P) = \tfrac{1}{2}(P_1 - P)(v_1 + v), \tag{1}$$

where h is the specific enthalpy, P is the pressure, and v is the specific volume. Combining the mass and momentum equations gives the Rayleigh line $\mathcal{R}$:

$$\frac{P - P_1}{v - v_1} = -(\rho_1 V_w)^2 = -\dot{m}^2, \tag{2}$$

a line in P–v space with negative slope equal to the square of the mass flux $\dot{m}$. Since the same value of $\dot{m}$ applies throughout the structure of the steady wave, $\mathcal{R}$ traces the locus of states in the reaction or evaporation zone. Steady wave solutions are given by the intersection of $\mathcal{R}$, with $\mathcal{H}$ evaluated at the end state **2**.

We now specialize the analysis to an evaporation wave with pure (or nearly pure) upstream liquid, with a two-phase downstream mixture described by

$$h = h_l(1 - \chi) + h_v\chi \quad \text{and} \quad v = v_l(1 - \chi) + v_v\chi, \tag{3}$$

where χ is the mass fraction of vapor, and l and v denote the liquid and vapor phases. We neglect slip between the two phases keep to and invoke four additional assumptions that are each usually accurate to a percent or better: (i) the liquid enthalpy is conserved upon depressurization from the initial state **0** to the established upstream state **1** ($h_0 = h_1$); (ii) the liquid density is constant throughout ($\rho_0 = \rho_1 = \rho_l = \rho_{l2}$); (iii) the vapor density is small compared with the liquid density ($\rho_v \ll \rho_l$)—true except near the critical point; (iv) the kinetic energies are small compared with the thermal energy ($\chi_2 = Ja_2$).

In the evaporation zone the liquid droplets are still somewhat superheated, with bulk enthalpy (averaged over a cross-section) h_l. Motivated by the reaction progress variable used in combustion, we define an *evaporation progress variable*, $0 \le \lambda \le 1$:

$$\lambda \equiv \frac{h_0 - h_l}{h_0 - h_{l\phi}(P)}, \tag{4}$$

where $h_{l\phi}(P)$ is the equilibrium liquid enthalpy. The Jakob number is then written as

$$Ja(P) \equiv \frac{h_0 - h_{l\phi}(P)}{h_v(P) - h_{l\phi}(P)}. \tag{5}$$

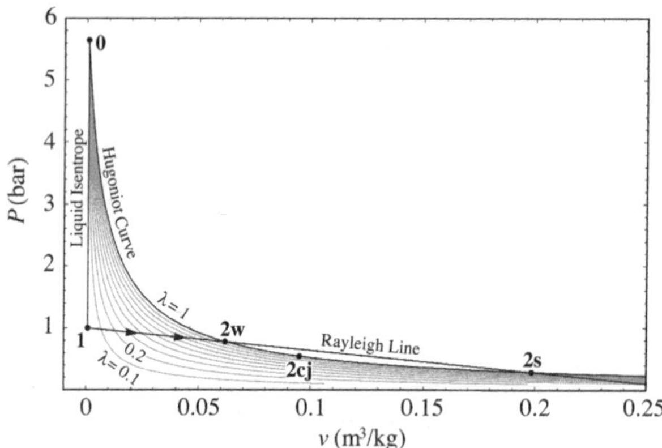

Figure 3. Thermodynamic construction for an evaporation wave—the liquid expansion isentrope, Hugoniot curve, Rayleigh line, and Chapman–Jouguet point.

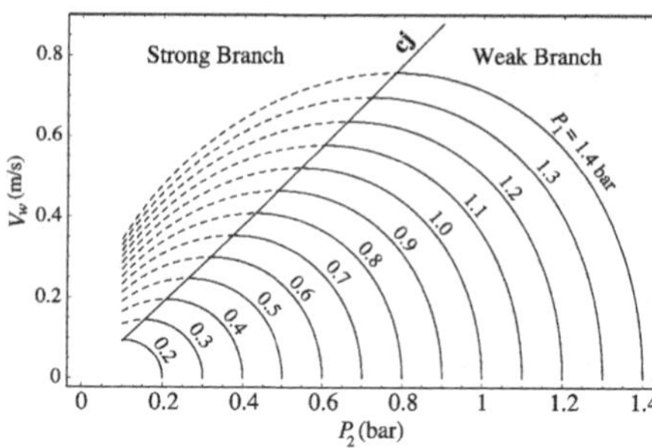

Figure 4. Plot of the evaporation wave speed relation, Equation (7), for freon-12.

Combining Equations (3b)–(5) with the energy equation gives an implicit formula for $\mathcal{H}$:

$$v = v_0 + \frac{v_v(P)\lambda\,\mathrm{Ja}(P)}{1 - \mathrm{Ja}(P)(1-\lambda)}. \tag{6}$$

Equation (6) is plotted in Figure 3 for several values of λ. A particular $\mathcal{R}$ is overlaid, which intersects $\mathcal{H}$ in two places. The high-pressure intersection at **2w** is the *weak solution*; the low-pressure intersection at **2s** is the *strong solution*. One can show that the end state **2** is subsonic for the weak branch and supersonic for the strong branch. In progressing along $\mathcal{R}$ from **1** to **2w**, the local state crosses increasing values of λ. This is a physically realizable process. But there is no physical way for evaporation to proceed beyond **2w** to **2s**—which is why strong deflagrations are not observed.

In the special case where $\mathcal{R}$ intersects $\mathcal{H}$ at the tangency point the emerging flow is exactly sonic; this is the Chapman–Jouguet (**cj**) point. It is clear from Equation (2) and Figure 3 that for a given **1**, **2cj** defines the maximum allowable wave speed. In some experiments, the test cell is *choked*, which corresponds to sonic outflow in the laboratory frame. In an evaporation wave, the two-phase flow speed is much faster than the wave speed, such that the difference between the laboratory and wave frames is negligible. Therefore, if an unobstructed test cell of constant cross-sectional area is choked, the wave can be considered **cj**.

Combining $\mathcal{R}$ with $\mathcal{H}$ evaluated at **2** gives an explicit equation for the wave speed:

$$V_w = \sqrt{\frac{v_0\,\delta P_2}{\mathrm{Jk}(P_2)}},\quad \text{where } \mathrm{Jk}(P) \equiv \mathrm{Ja}(P)\left(\frac{v_v(P)}{v_0}\right)$$

and

$$\delta P \equiv P_1 - P. \tag{7}$$

δP_2 is the overall wave pressure-amplitude, one measure of its strength. Jk is a useful alternative definition of the Jakob number that arises about as frequently as Ja, and which unfortunately goes by the same name (and usually the same symbol).

The liquid initial conditions v_0 and h_0, and the external reservoir pressure P_r, are set a priori and remain essentially unchanged after depressurization. Assuming that the test cell is long enough for equilibrium to be achieved at or before the exit: if the flow is unchoked (weak solution), then $P_2 = P_r$; if the flow is choked (**cj** solution), then $P_2 > P_r$ and a sonic condition applies that specifies P_2. But P_1, and therefore δP_2, is unknown.

Figure 4 plots Equation (7) for various values of P_1. Each curve has a velocity maximum, representing its own **cj** solution, the locus of which is given by

$$V_{w\mathbf{cj}}(P_2) = \sqrt{\frac{v_0}{-\mathrm{d}Jk/\mathrm{d}P_2}}. \tag{8}$$

Predicted and measured wave speeds can deviate by up to 50%. Two issues likely dominate. First, the theory assumes the flow to be strictly one dimensional, but Hill's experiments show that the transient lateral flow velocity in the evaporation zone is of order the mean axial velocity. This results in a higher-than-calculated pressure drop. Second, evaporation is very rapid where fragmentation occurs, but slows downstream where droplets are convecting. Thus, the end state **2** may not exist within the tube.

To close the theory—to calculate the wave speed without measuring P_1—we must add a statement about the physical mechanisms peculiar to evaporation waves. This task is actually simpler for waves in bubbly liquids than for waves in pure ones. For Thompson et al.'s highest-temperature experiments, the upstream state was essentially at the superheat limit, which can be computed from initial conditions. In Friz's experiments, there was time for multiple acoustic reflections in the upstream liquid. He assumed P_1 to be the equilibrium value at the initial temperature, which gave reasonable results.

For waves in pure liquids there is no upstream nucleation to modulate the pressure, and one must address the boiling dynamics in the evaporation zone. The problem is analogous to the calculation of flame speed, which is controlled by the rate of heat conduction (and often convection) from the reaction zone to the reactants. This is where transport properties such as thermal and mass diffusivity enter. No such physically based model exists for evaporation waves, although Reinke and Yadigaroglu (2001) have proposed an empirical correlation based on existing data.

LARRY HILL

See also **Dimensional analysis; Explosions; Flame front; Fluid dynamics**

Further Reading

Debeneditti, P.G. 1996. *Metastable Liquids: Concepts & Principles.* Princeton, NJ: Princeton University Press

Decker, R. & Decker, B. 1981. The eruptions of Mount St Helens. *Scientific American*, 244(3): 68–80

Friz, G. 1965. Coolant ejection studies with analogy experiments. *Proceedings of the Conference on Safety Fuels & Core Design in Large Fast Power Reactors*, US Atomic Energy Commission Report ANL-7120: 890–894

Frost, D.L. 1988. Dynamics of explosive boiling of a droplet. *Physics of Fluids*, 31(9): 2554–2561

Grolmes, M.A. & Fauske, H.K. 1970. Modeling of sodium expulsion with freon-11. *ASME Paper* 70-HT-24

Hayes, W.D. 1958. The basic theory of gasdynamic discontinuities. In *Fundamentals of Gas Dynamics*, edited by H.W. Emmons, Princeton, NJ: Princeton University Press

Hill, L. 1991. An experimental study of evaporation waves in a superheated liquid. Ph.D. Thesis, California Institute of Technology, Pasadena, CA, USA

Mesler, R. 1988. Explosive boiling: a chain reaction involving secondary nucleation. *Proceedings of the ASME National Heat Transfer Conference*, 96: 487–491

Reid, R.C. 1976. Superheated liquids. *American Scientist*, 64: 146–156

Reinke, P. & Yadigaroglu, G. 2001. Explosive vaporization of superheated liquids by boiling fronts. *International Journal of Multiphase Flow*, 27: 1487–1516

Shepherd, J.E. & Sturtevant, B. 1982. Rapid evaporation at the superheat limit. *Journal of Fluid Mechanics*, 121: 379–402

Simões-Moreira, J.R. & Shepherd, J.E. 1999. Evaporation waves in superheated dodecane. *Journal of Fluid Mechanics*, 382: 63–86

Terner, E. 1962. Shock tube experiments involving phase changes. *Industrial and Engineering Chemistry Process Design and Development*, 1(2): 84–86

Thompson, P.A., Carofano, G.C. & Kim, Y.-G. 1986. Shock waves and phase changes in a large-heat-capacity fluid emerging from a tube. *Journal of Fluid Mechanics*, 166: 57–92

Thompson, P.A., Chaves, H., Meier, G.E.A., Kim, Y.-G. & Speckmann, H.-D. 1987. Wave splitting in a fluid of large heat capacity. *Journal of Fluid Mechanics*, 185: 385–414

EXCITABILITY

Nineteen century natural history identified irritability—the active response to stimulation—as a characteristic of living systems; this vague concept has developed into excitability, a characteristic of some nonlinear systems.

An excitable system has a stable resting state, and the response to a small brief perturbation is small, with an amplitude that varies smoothly with the perturbation amplitude. The response to a sufficiently large perturbation, above a threshold, is qualitatively different, undergoing a stereotyped, large amplitude excursion in at least one of the state variables before return to the resting state. For a spatially extended excitable system, the subthreshold response is localized while the suprathreshold response is a traveling wave, or traveling wave train. This dynamics—a stable steady state (a mathematical, but not thermodynamic equilibrium), a threshold, and a return to the steady state—requires at least two processes, a fast nonlinear excitation and a slower recovery process.

The excitation process can be exemplified by the bistability of a candle. The unlit candle is stable, and the threshold for igniting the candle arises from a positive feedback, with heat melting and vaporizing the wax, and the vapor providing the fuel sustaining the flame. This positive feedback loop gives the all-or-none nature of the flame—either it is self-sustaining or not, and the threshold separates these two states. The burnt candle is stable; there is no recovery process.

An example of excitation and recovery is provided by the nerve impulse, where the resting state potential is stable and the energy required for the action potential is obtained from the electrochemical gradients across the membrane. The voltage dependence of the Na^+ conductance provides the positive feedback. Recovery is by the slower, voltage-dependent K^+ conductance, and propagation by spread of depolarization along the fiber (Cole, 1968; Aidley, 1998). In both these examples, the spatial spread of activity can be considered as traveling-wave solution of a reaction-diffusion equation in which there is a balance between the active nonlinear term and the diffusive term. In typical chemical excitable systems, all the variables diffuse, with similar diffusion coefficients; in nerve and muscle excitation only the fast excitation variable (transmembrane voltage) spreads and there is only one nonzero diffusion coefficient. The velocity of the traveling-wave solution varies as the root of the diffusion coefficient.

The simplest excitable dynamical system has two kinetic variables, a fast excitation u and a slower recovery v

$$\varepsilon\,\mathrm{d}u/\mathrm{d}t = f(v,u), \quad \mathrm{d}v/\mathrm{d}t = g(v,u), \tag{1}$$

and the nullcline for the fast excitation system has the characteristics of a cubic that is, for a range of u values the solution of $f(v,u)=0$ has three branches. If, as in Figure 1, the nullclines have a single intersection

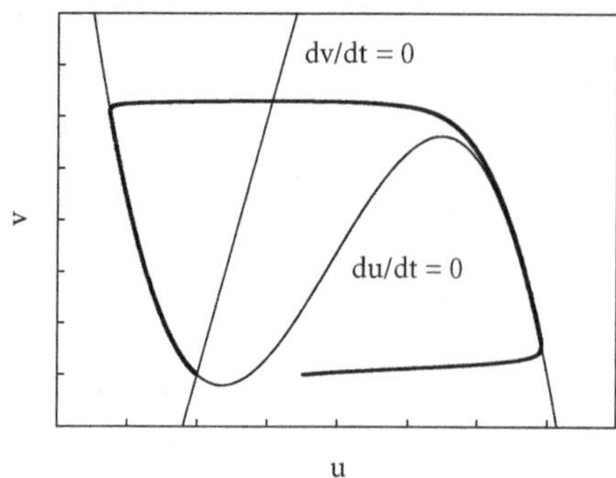

Figure 1. State space of two-variable excitable system, with u the fast excitation variable and v the slower recovery variable. There is a stable equilibrium and a single suprathreshold response.

on the left branch, there is a single globally stable, but excitable, steady state. A large enough perturbation gives a trajectory that jumps to the right branch, moves slowly close to the right branch, and then jumps back to the left branch, returning slowly to rest. If the nullclines intersect on the central branch the resultant equilibrium can be unstable, and the system is no longer excitable but oscillatory.

The key concepts of excitability—threshold and all-or-none response—came from the physiology of nerve and muscle tissue. Nerve and muscle are electrically excitable tissues; their excitability is electrochemical, due to nonlinear membrane current-voltage relations produced by voltage-dependent conductances. The suprathreshold response is the action potential that acts as a trigger, where the intensity of the response is determined by the rate of action potentials. Repetitive activity—that can be idealized as a periodic response—is a biologically significant behavior and may be forced, by inputs to an excitable system, or endogenous, as in a pacemaking system, due to stable limit cycle surrounding an unstable equilibrium when the equilibrium state loses its stability at a Hopf bifurcation as a parameter (say, an applied current or maximal membrane conductance) is changed. Also rate coding phase synchronization effects can be significant—synchronization of bursting discharges in the mammalian central nervous systems provides a candidate mechanism for cognitive effects (Eckhorn, 1999). In spatially extended cells, there can be repetitive traveling-wave trains—effectively one-dimensional in a nerve fiber, and three-dimensional in heart muscle. In the heart three-dimensional scroll waves provide the mechanism for re-entrant arrhythmias that can lead to sudden cardiac death. Some intracellular systems show chemical excitability: a localized increase in intracellular [Ca^{2+}] can trigger calcium-induced calcium release from intracellular organelles, producing intracellular calcium oscillations, or traveling waves. An example is a [Ca^{2+}] wave triggered by fertilization of an oocyte (Dupont & Goldbeter, 1997). Chemical excitability, via cyclic AMP triggering cellular release of cyclic AMP, underlies the wave phenomena seen in the aggregation of the slime mold *Dictyostelium*, a simple example of morphogenesis (Tyson & Murray, 1989). In all these biological examples, high-order, mechanistically detailed stiff systems of equations have been developed, validated, and investigated numerically, but much of the essential phenomenology can be captured by low-order caricatures. Simple models of biological populations can also show excitable behavior, for example, the initiation of an epidemic.

The Belousov–Zhabotinsky reaction, initially developed as an oscillatory organic reaction as an analogue of biochemical oscillations, was the first of several autocatalytic chemical reactions that, in a flow reactor with concentrations that allow bistability, addition of a feedback species can lead to oscillations (Sagues & Epstein, 2003). These chemical oscillatory systems, in thin films and bulk media, also show traveling waves—target patterns, spirals, and scrolls—and with some parameter values are excitable. Models of autocatalytic processes with cubic excitable kinetics are widely used to represent oscillatory and wave phenomena in chemistry, especially combustion problems (Gray & Scott, 1990).

ARUN V. HOLDEN

See also **FitzHugh–Nagumo equation; Hodgkin–Huxley equations; Hopf bifurcation; Nerve impulses; Neurons; Periodic bursting; Threshold phenomena**

Further Reading

Aidley, D.J. 1998. *The Physiology of Excitable Cells*, Cambridge and New York: Cambridge University Press

Cole, K.S. 1969. *Membranes, Ions and Impulses*, Berkley: University of California Press

Dupont, G. & Goldbeter, A. 1997. Modelling oscillations and waves of cytosolic calcium. *Nonlinear Analysis—Theory, Methods and Applications*, 30: 1781–1792

Eckhorn, R. 1999. Neural mechanisms of visual feature binding investigated with microelectrodes and models. *Visual Cognition* 6: 231–265

Gray, P. & Scott, S.K. 1990. *Chemical Oscillations and Instabilities: Non-linear Chemical Kinetics*, Oxford: Clarendon Press and New York: Oxford University Press

Holden, A.V., Markus. M. & Othmer, H.G. 1991. *Nonlinear Wave Processes in Excitable Media*, New York: Plenum

Keener, J.P & Sneyd, J. 1998. *Mathematical Physiology*, New York: Springer

Sagues F. & Epstein, I.R. 2003. Nonlinear chemical dynamics. *Dalton Transactions*, (7): 1201–1217

Tyson, J.J. & Murray, J.D. 1989. Cyclic-AMP waves and aggregation of *Dictyostelium amebas*. *Development*, 106: 421–426

EXCITABLE MEDIUM

See **Reaction-diffusion systems**

EXCITONS

Traditionally, an exciton is defined as a quantum of electronic excitation energy traveling in a periodic structure, whose motion is characterized by a wave vector. It is generated when an electron from a filled electronic orbital of a molecule is transferred through an optical or electrical excitation to a high-energy unoccupied electronic orbital, leaving behind it a hole. Coulomb interaction binds the excited negative electron and the positively charged hole producing an electrically neutral bound state, and this bound electron-hole pair is the exciton. Once created the exciton can be destroyed through emission of light (radiative recombination) or heat (nonradiative recombination), electron-hole separation (exciton dissociation), and subsequent absorption of the electron by an acceptor site (electron transfer). It can change its spin state through intersystem crossing, while at high excitation densities, excitons can mutually collide and dissociate (exciton-exciton annihilation). Furthermore, the whole excitation can move to other sites leading to energy transfer, one of its most important properties.

Excitons occur in molecular crystals, inorganic semiconductors, and conjugated polymers. Three different types of excitons are known. The Frenkel exciton is an electronic excitation of a single molecular unit with both electron and hole located on the same molecule. As any molecule is equally likely to be excited and if there is nonzero coupling between adjacent molecules, this exciton is transferred from one molecular unit to another. Frenkel excitons appear in ionic, molecular, and noble gas crystals having a typical binding energy of 1 eV. In the Wannier exciton the electron-hole distance is much larger than the lattice spacing, and the Coulomb interaction gets screened by the crystal dielectric constant forming a hydrogen-like bound system. They appear mostly in inorganic semiconductors and have a small binding energy of about 0.1 eV and a large radius of typically 50 Å. Finally, the charge-transfer exciton is intermediate between Frenkel and Wannier excitons with an electron-hole distance of a few lattice spacings.

Although excitons are neutral they can be found in definite spin states depending on the relative orientation of electron and hole spins. For antiparallel electron-hole spins a singlet exciton with total spin zero is formed with short lifetime of the order of picoseconds due to fast decay through an optically allowed transition leading to fluorescent emission from this state. There are also three degenerate spin one states leading to a triplet exciton that produces emission termed "phosphorescence." Optical transitions between triplet excited and single ground state are not allowed due to the forbidden spin flip, resulting in conjugated polymers in triplet exciton lifetime of the order of milliseconds at low temperatures. Thermal energy assists triplet motion and at higher temperatures exciton motion process in an incoherent diffusive mode, although some quantum mechanical coherence may be retained. Coherent exciton dynamics at finite temperatures is investigated theoretically through generalized Langevin equations.

Self-trapped excitons (STEs) form a special class of excitations that involve excitons coupled strongly to vibrational degrees of freedom of molecules. Low-dimensional and, in particular, quasi-one-dimensional materials are ideal systems for studying STEs because reduced dimensionality results in strong electron-phonon interaction. Systems under experimental study include halogen-bridged MX materials, where X is a halogen (chlorine, bromine, etc.) and the metal is, for instance, platinum or crystalline hydrogen-bonded acetanilide (ACN).

The STE can be of electronic or vibrational nature if, in the latter case, specific vibrational modes couple strongly to other phonon system modes leading to vibron excitons. In ACN, C=O as well as NH stretching modes have been linked with STEs through the temperature dependence of specific peaks in their absorption spectra, while in the metal-halogen material PtCl, resonant Raman spectroscopy experiments have indicated the existence of self-localized modes.

From a more formal point of view, STEs can be classified as polarons, solitons, or (discrete) breathers, and depending on their specific nature, they may have some differences in their precise physical properties. A generic mathematical model for exciton self-trapping is provided by the discrete nonlinear Schrödinger equation (DNLS) or discrete self-trapping equation

$$\mathrm{i}\frac{\mathrm{d}\psi_n}{\mathrm{d}t} = \varepsilon_n \psi_n + V(\psi_{n+1} + \psi_{n-1}) - \gamma |\psi_n|^2 \psi_n, \quad (1)$$

where ψ_n is the probability amplitude for an exciton to be located in a molecular unit at a crystal site n with local energy ε_n, V is the nearest neighbor overlap, and γ is proportional to exciton-phonon coupling. In this approximate, semiclassical description, the nonlinear term in the equation arises from strong electron-phonon coupling in the original exact problem that can be introduced, for instance, by a model such as the Holstein model for molecular crystals. The conserved number $\sum_n |\psi_n|^2 = N$ of the DNLS equation corresponds to the number of excitation quanta present in the system. In the weak coupling limit where $V \to 0$, the STEs of DNLS are completely localized in a given site with energy equal to

$$E_N = N\varepsilon - \frac{\gamma}{2}N^2, \quad (2)$$

where all local site energies are taken to be identical, that is, $\varepsilon_n = \varepsilon$ for all n. The energy of the STE is thus lower than $N\varepsilon$, the energy of the delocalized exciton and also red-shifted by an amount that grows quadratically in the number of quanta N. These spectral features have been observed experimentally in spectra of both PtCl and ACN as well as in other systems. Precise theoretical analysis of resonant Raman spectra of PtCl as well as femtosecond infrared pump-probe spectroscopy experiments in ACN that go beyond the weak coupling DNLS energy in Equation (2) have demonstrated the existence of STEs in these systems. In the case of ACN, an amide-I exciton lifetime of the order of 2 ps was found. If, as originally conjectured by Davydov, similar modes with larger lifetimes exist in biological macromolecules such as proteins, STEs may participate in energy transfer processes of biological significance.

G.P. Tsironis

See also **Davydov soliton; Discrete breathers; Discrete nonlinear Schrödinger equations; Discrete self-trapping system; Polaritons; Polarons**

Further Reading

Dexter, D.L. & Knox, R.S. 1965. *Excitons*, New York: Wiley

Edler, J, Hamm, P. & Scott, A. C. 2002. Femtosecond study of self-trapped vibrational excitons in crystaline acetanilide. *Physical Review Letters*, 88, 067403: 1–4

Eilbeck, J.C, Lomdahl, P.S. & Scott, A.C. 1985. The discrete self-trapping equation. *Physica D*, 16: 318–338

Hennig, D. & Tsironis, G.P. 1999. Wave transmission in nonlinear lattices. *Physics Reports*, 307: 333–432

Kenkre, V.M. & Reineker P. 1982. *Exciton Dynamics in Molecular Crystals and Aggregates*, Berlin: Springer

Pope, M. & Swenberg, C.E. 1982. *Electronic Processes in Organic Crystals*, Oxford: Clarendon Press and New York: Oxford University Press; 2nd edition, 1999

Scott, A.C. 1992. Davydov's soliton. *Physics Reports*, 217: 1–67

Voulgarakis, N.K., Kalosakas, G., Bishop, A.R. & Tsironis, G.P. 2001. Multiquanta breather model for PtCl. *Physical Review* B, 64 (020301): 1–4

EXPLOSIONS

The English word *explode* comes from the Latin word *explodere*, which means to drive an actor from the stage by rhythmic clapping. In applied science, a sudden deposition of energy leads to an explosion. Since on a streamline of a flow we have $dE/dt = -p dv/dt$ (where E is internal energy, t is time, p is pressure, and v is volume), rapid deposition of energy will give high pressure, or rapid motion as volume expands, or both. How large and how rapid the energy deposition must be to have an explosion depends on external circumstances, so the concept of explosion is investigated here by looking at some examples.

Low Explosives and Propellants

The words *energetic materials* are used to describe substances that have fuel and oxidizer in the same molecule or in separate molecules intimately mixed. Production and purification of potassium nitrate (KNO_3) were the keys to making materials that burned rapidly and could be used for fireworks, frightening enemies, setting enemy ships on fire, and entertainment at fairs. After 1300 AD, these materials were used in guns, and the mixtures were improved for better performance. High-strength black powder is about six parts of potassium nitrate and one part each of charcoal and sulfur. Black powder has been replaced by smokeless powders made from nitrocellulose or nitrocellulose with nitroglycerin, with various additives. Modern rocket motors usually use a coarse mixture of ammonium perchlorate (NH_4ClO_4) as oxidizer, and rubber, aluminum powder, and high explosive molecules as fuels. Black powder was used in mining from 1650 to 1900; it has since been replaced by high explosives.

All of these materials deliver their energy by burning, which is also called deflagration. That is, energy is transferred from the reacting region forward to ignite the unreacted material mainly by heat conduction. The wave speed of the flame is slow compared with the speed of sound, so that the pressure is nearly constant in the flame. The distance from the reacting region to the wave front varies with the pressure, so the speed of the flame and its rate of release of energy depend on the external confinement. In a mine, the apparent power of black powder varies with the strength of the rock and the existing cracks in the rock. The powder in a gun is usually in the form of separate pellets or grains, and the rate of burning depends on the exposed surface area that supports the flame. As the projectile moves along the barrel, the volume to be occupied by the gases increases at an accelerating rate. It is advantageous to have the rate of burning also accelerate, and this is accomplished by shaping the grains. For example, the grains may be in the form of cylindrical pellets with a number of holes parallel to the axis. As the pellet burns, the holes are enlarged, exposing more area and accelerating the rate of gas production.

High Explosives

High explosives are also energetic materials with the fuel and oxidizer in intimate mixture or in the same molecule. They explode by a process called detonation, and the speed of the reaction is 100,000–10,000,000 times faster than burning in the same material. The rapid reaction produces extremely high pressures and speeds. It also makes the detonation process almost independent of confinement or boundary effects. High explosives detonate the same way every time, without being influenced by exterior changes.

Detonation proceeds as a wave traveling at high speed, from 4000 to 9000 ms^{-1} in various explosives, with a shock wave at the front of the wave structure. The compression in the shock heats the explosive enough to make it react rapidly. The energy released in the reaction supports the shock. The shock is supersonic, going as fast as the available energy can drive it, so nothing occurs ahead of the shock. In the reaction zone the flow is subsonic so energy can be transferred, but at the end of the reaction zone the flow becomes supersonic, and anything that happens in that supersonic flow cannot come forward to affect the reaction zone. It is for these reasons that the detonation is almost independent of the external conditions. Only at the edges can it be perturbed.

High explosives were discovered about the middle of the 19th century by chemists developing new dyes. Ascanio Sobrero, an Italian chemist, first prepared nitroglycerin and several other high explosives in 1846 and 1847. Alfred Nobel in 1864 developed a method of initiating nitroglycerin reliably by a fast compression wave, or shock wave, but it was dangerous to handle and ill-suited to mining operations. In 1867 Nobel was granted a patent for an explosive prepared by mixing nitroglycerin with a porous absorbent such as charcoal, diatomaceous earth, or fine sawdust. This explosive, called dynamite, proved to be just what was needed for the developing industrial revolution. In a few years it replaced black powder throughout the world. Many compositions competed with dynamite: a dictionary of explosives published in 1891 listed more than a thousand explosives.

After the middle of the 20th century, dynamite was replaced by explosives prepared from ammonium nitrate and fuel oil as granular mixtures, as slurries, or as emulsions. In the United States alone the annual use of high explosives is more than five billion pounds per year, or nearly 20 pounds for each person. A pound of explosive will break about 3000 pounds of rock, so about 60,000 pounds of rock is broken for each person in the United States each year. A large fraction of the explosive is employed to break the rock overburden of coal seams, other uses are in other mining, road construction, lumbering, and farming. Almost any product we purchase has had explosive used for it at some stage.

The first high-precision application of high explosives was to nuclear weapons, developed during World War II, where explosive was used for the precise dynamic assembly of the fissionable material. The nuclear weapons laboratories around the world have contributed to the detailed theory of detonation, to numerical modeling of detonations, to the development of powerful and safe explosives, and to the experimental techniques for studying explosives.

Nonlinear relations are especially important to detonation. The shock wave forms because the stress-strain behavior of the explosive is nonlinear. Materials become stiffer as they are compressed, so the sound speed increases with pressure, and a compression wave steepens to become a shock. The extremely strong nonlinear dependence of the reaction rate on temperature is also very important. It is the on-off switch for the release of energy. For example, a typical rate law for an explosive might be $d\lambda/dt = k(1-\lambda)e^{(-T^*/T)}$, where k is the limiting rate; λ is a progress variable for the reaction, going from zero to one as the explosive goes to products; t is the time; T is the temperature; and T^* is the activation temperature. If $k = 10^{20}\ s^{-1}$ and $T^* = 25{,}000$ K, the time required for half of the explosive to react when kept at constant temperature is 341 million years at 300 K (room temperature), 8.64 ms at 600 K, and 8.03 ns at 900 K.

Nuclear Weapons

In nuclear fission weapons, U^{235} or Pu^{239} is bombarded by neutrons that cause it to split into two lighter nuclei, releasing a large amount of energy and also emitting more neutrons. If more neutrons are generated than escape through the surface, the exponential increase in the number of neutrons leads to an explosion. A fusion weapon is initiated by a fission weapon and releases energy as hydrogen, deuterium, tritium, and lithium fuse to make heavier nuclei. These weapons release a million or more times as much energy as the same weight of conventional chemical explosive. The energy is often quoted in kilotons or megatons. These units are based on an assumed yield for TNT of 1000 $calg^{-1}$ or 4187 Jg^{-1}. The metric tonne is 10^6 g and the kiloton is a thousand times that, so a kiloton is 10^{12} cal or 4.187×10^{12} J, and a megaton is 4.187×10^{15} J.

Asteroid Collision with Earth

About 65 million years ago an asteroid hit the earth leaving the crater Chicxulub in Yucatan, Mexico, and perturbing the conditions on earth to the extent that dinosaurs and many other animals and plants became extinct. The energy deposited was the kinetic energy of the asteroid. The exact details cannot be known, but if the asteroid was 12 km in diameter with density 2500 kg m^{-3} and relative speed 20 km s^{-1}, its kinetic energy was 4.5×10^{23} J. An obvious time constant for the energy release is the diameter divided by the speed, 0.6 s. The kinetic energy divided by the time gives the power as 7.5×10^{23} W; the actual value is a few times smaller.

An asteroid of the Chicxulub size is not slowed much by the Earth's atmosphere and reaches the Earth's surface. A smaller asteroid is slowed more, and the compression wave from the air travels through the asteroid. If the asteroid is small enough, the compression wave reaches the rear surface before the asteroid reaches the ground. The high pressure inside the asteroid causes it to expand and break up into small pieces that are then slowed more by the air. The kinetic energy is transferred suddenly to the air, and there is an explosion above the surface of the earth. The Tunguska Event in Central Siberia in 1908 flattened trees over an area of about 2000 km^2. It was caused by an asteroid about 60 m diameter that exploded about 8 km above the surface. Its kinetic energy was about 4×10^{16} J.

Volcanoes

The 1883 eruption of the Indonesian volcano Krakatau, which blew ash to a height of 80 km and ejected 22 km^3 of rock, was heard in Australia 4600 km away. Tsunamis caused by the eruption reached heights of 40 m and killed 34,000 people.

Mount St. Helens, in the state of Washington, erupted in 1980 and has been studied extensively. It ejected 2.7 km^3 of rock, devastated an area of 550 km^2, and blew down an estimated 10^7 trees.

A huge volcanic explosion occurred in Yellowstone, United States, about 2,000,000 years ago. About 3000 km^3 of rock were ejected. No explosion of such a magnitude has been experienced during the period of human civilization.

Novae and Supernovae

A nova occurs in a close binary star system made up of a red giant and a white dwarf. Hydrogen-rich matter from the red giant is pulled onto the surface of the white dwarf. When enough matter is accumulated, a nuclear fusion detonation occurs, causing the ejection of hot surface gases and resulting in an extraordinary increase in luminosity. A supernova occurs when a massive star exhausts its nuclear fuel and its core collapses to become a neutron star. The outer layers of the star are attracted by the gravitational pull of the core, and then rebound from it. A shock wave generated in the collision propagates outward and blows off the surface gases.

W.C. Davis

See also **Dimensional analysis; Flame front; Shock waves**

Further Reading

Baker, W.E., 1973. *Explosions in Air*, Austin: University of Texas Press

Baker, W.E., Cox, P.A., Westine, P.S., Kulesz, J.J. & Strehlow, R.A. 1983. *Explosion Hazards and Evaluation*, Amsterdam and New York: Elsevier

Federoff. B.T. et al. 1960. *Encyclopedia of Explosives and Related Items*, 10 vols, Dover, NJ: US Army Armament Research and Development Command (Picatinny Arsenal), (PATR-2700)

Glasstone, S. & Dolan, P.J. (editors). 1977. *The Effects of Nuclear Weapons*, 3rd edition, Washington: United States Department of Defences

International Astronomical Union, 1991. *Physics of Classical Novae*, Berlin and New York: Springer

Kinney, G.F. & Graham, K.J. 1985. *Explosive Shocks in Air*, 2nd edition, Berlin and New York: Springer

Sigurdsson, H. et al. (editors). 2000. *Encyclopedia of Volcanoes*, San Diego and London: Academic Press

Wheeler, J. Craig 2000. *Cosmic Catastrophes: Supernovae, Gamma Ray Bursts, and Adventures in Hyperspace*, Cambridge and New York: Cambridge University Press

EXTREMUM PRINCIPLES

Extremum principles are ubiquitous in mathematics, with wide applications in areas ranging from genetic theory (Narain, 1993) to thermodynamics (Velasco & Fernandez-Pineda, 2002). The simplest sort of extremum principle occurs in differential geometry, whenever one considers a differentiable function $f(x_1, \ldots, x_N)$ of N variables and wishes to determine the critical points of f, namely those $\boldsymbol{x}$ that satisfy

$$\nabla f(\boldsymbol{x}) = 0.$$

The Hessian matrix H has entries given by the second derivatives

$$H_{jk} = \frac{\partial^2 f}{\partial x_j \partial x_k}(\boldsymbol{x})$$

(assuming these exist). If all the eigenvalues of the Hessian are positive (negative), then this ensures that $\boldsymbol{x}$ is a local minimum (maximum); eigenvalues of mixed sign correspond to a saddle point. The general characterization of functions in terms of their behavior near critical points is the subject of Morse theory (Morse & Cairns, 1969).

Probably the most widespread type of extremum principles are those arising from the application of the calculus of variations (Gel'fand & Fomin, 1963). One of the first historical examples of this sort is the brachistochrone problem posed by Johann Bernoulli the elder in 1696: given a particle sliding from rest along a path between two fixed points under the influence of gravity, find the plane curve such that the time taken is minimized. The solution curve (a cycloid) was obtained by both Johann and Jakob Bernoulli, and also by Newton, Leibniz, and l'Hôpital. Nowadays, the problem is usually solved directly through the use of the total time functional

$$T = \int_{x_0}^{x_1} \sqrt{\frac{1+(y')^2}{2g(y_0-y)}}\,\mathrm{d}x \equiv \int_{x_0}^{x_1} F(y, y')\,\mathrm{d}x,$$
$$y' = \frac{\mathrm{d}y}{\mathrm{d}x}, \qquad (1)$$

where $y(x)$ is the plane curve joining the points (x_0, y_0) and (x_1, y_1) and g is the acceleration due to gravity. To minimize the time, it is necessary that the first variation of the functional T should vanish,

$$\delta T = 0,$$

which leads to the Euler–Lagrange equations

$$\frac{\partial F}{\partial y} - \frac{\mathrm{d}}{\mathrm{d}x}\left(\frac{\partial F}{\partial y'}\right) = 0 \qquad (2)$$

The solution of (2) is given parametrically in terms of (x_0, y_0) and another constant h by

$$x(\theta) = x_0 + (\theta - \sin\theta)/(2h^2),$$

$$y(\theta) = y_0 - (1 - \cos\theta)/(2h^2).$$

In complete analogy to the Hessian condition in finite dimensions, the positivity of the second variation $\delta^2 T$ ensures that the cycloid yields a minimum of the functional $T[y]$.

In physics perhaps the most important extremum principle of all is the Principle of Least Action. An early proponent of this notion was Pierre Louis de Maupertuis, who employed an extremality argument to solve a problem in statics (Beeson, 1992). Maupertuis, who was tutored by Johann Bernoulli, was in regular correspondence with Leonhard Euler, and the work of the latter laid the foundation for the subsequent development of the subject. The enormous generality of the least-action principle became apparent from the Lagrangian formulation of classical mechanics (Goldstein, 1950), as derived from the action functional

$$S[\boldsymbol{q}] = \int_{t_0}^{t_1} L(\boldsymbol{q}, \dot{\boldsymbol{q}}, t)\,\mathrm{d}t. \qquad (3)$$

The variable t is the time, and the vector $\boldsymbol{q}$ and its time derivative $\dot{\boldsymbol{q}}$ denote the generalized coordinates and velocities, respectively. Requiring that the action should be stationary with respect to variations that vanish at the initial and final times, so

$$\delta S = 0, \quad \text{with } \delta\boldsymbol{q}(t_0) = 0 = \delta\boldsymbol{q}(t_1), \qquad (4)$$

it follows (after an integration by parts) that L must satisfy the Euler–Lagrange equations

$$\frac{\partial L}{\partial \boldsymbol{q}} - \frac{\mathrm{d}}{\mathrm{d}t}\left(\frac{\partial L}{\partial \dot{\boldsymbol{q}}}\right) = 0.$$

The extremality condition (4) is truly fundamental, since once the Lagrangian L has been specified appropriately, then all the classical equations of motion for a mechanical system (in particular, Newton's Second Law) are a direct consequence. Even the geodesic equations for a Riemannian manifold arise in this way (Choquet-Bruhat, 1968), by taking the purely kinetic (free particle) Lagrangian

$$L(\boldsymbol{q}, \dot{\boldsymbol{q}}) = g_{jk}(\boldsymbol{q})\dot{q}^j\dot{q}^k,$$

where g_{jk} is the metric tensor (the Einstein summation convention is assumed).

In parallel to the development of mechanics, a corresponding extremum principle was employed in optics, namely Fermat's Principle of Least Time. This is more correctly formulated within the theory of geometrical optics (Kline & Kay, 1965), as the requirement that the path taken by light traveling between two points P_1 and P_2 is such that the optical distance is *extremized*, that is,

$$\delta \int_{P_1}^{P_2} n(x, y, z)\,\mathrm{d}s = 0, \qquad (5)$$

where $n(x, y, z)$ is the refractive index at the point (x, y, z) in the medium and s denotes arc length along the path. Contrary to Fermat's original formulation, if the optical distance is extremal, then in general, the time taken for the light to travel between P_1 and P_2 need not be minimized—this only holds in certain special cases.

The precise analogy between geometrical optics and mechanics was fully appreciated and exploited by Hamilton. Choosing a parameter τ along the light path such that $\boldsymbol{q} = (x, y, z)^T$ satisfies

$$|\dot{\boldsymbol{q}}|^2 = n^2, \quad \mathrm{d}s = |\dot{\boldsymbol{q}}|\,\mathrm{d}\tau, \qquad (6)$$

the equations arising from the variational principle (5) are just

$$\ddot{\boldsymbol{q}} = \nabla\left(\tfrac{1}{2}n^2\right), \quad n = n(\boldsymbol{q}),$$

which are equivalent to Newton's equations for a particle of unit mass moving in a potential $-\frac{1}{2}n^2$. In this mechanical analogy, τ should be reinterpreted as the time, while the first equation (6) implies that the total energy of the particle is zero.

The Principle of Least Action is also essential in field theory. A scalar field theory in N dimensions has the action

$$S[\phi] = \int_V \mathcal{L}(\phi, \phi_\mu, \ldots)\,\mathrm{d}^N x$$

with the Lagrangian density $\mathcal{L}$ being a function of the field ϕ, its first derivatives ϕ_μ ($\mu = 1, \ldots, N$), and possibly higher derivatives. The appropriate variational principle for the classical field equations is

$$\delta S = 0 \quad \text{with } \delta\phi(\boldsymbol{x}) = 0 \quad \text{for } \boldsymbol{x} \in \partial V,$$

so that the variations vanish on the boundary of the space-time volume V. In the path integral formalism of quantum field theory, as pioneered by Richard Feynman (Feynman & Hibbs, 1965), the central object is the partition function, defined in terms of the action by

$$Z = \int e^{iS[\phi]}\,\mathcal{D}\phi,$$

where $\mathcal{D}\phi$ = path integral measure. When the action is coupled to an external source, $S \to S + \int J(\boldsymbol{x})\phi(\boldsymbol{x})\,d^N x$, then the vacuum expectation values of the field operators are obtained by taking successive functional derivatives $\delta/\delta J$ of Z. This is the key technique for calculating Feynman rules and perturbative scattering amplitudes in gauge field theory (Bailin & Love, 1986).

Another sort of extremum principle appears in game theory (Owen, 1982). In the simplest case of two-person zero-sum games, with two players denoted "row" and "column" and strategies labelled by an index j ($j = 1, \ldots, M$) for the row player and an index k ($k = 1, \ldots, N$) for the column player, the payoffs to the row player form an $M \times N$ matrix $\mathcal{A}$. The matrix element $\mathcal{A}_{jk}$ is just the payoff to the row player arising from the pure strategy combination (j, k); since the game is zero-sum, the corresponding payoff to the column player is $-\mathcal{A}_{jk}$. If the maximum of the row minima in the payoff matrix is equal to the minimum of the column maxima, then this is a saddle point (or minimax point), corresponding to an optimum solution of the game. However, a saddle point does not exist in general, and so it is necessary to resort to mixed strategies, taking the set of probability distributions $\mathcal{S}_X \times \mathcal{S}_Y$ on the strategy space, where

$$\mathcal{S}_X = \left\{ \boldsymbol{x} \in \mathbb{R}^M \mid x_j \in [0,1],\ \sum_{j=1}^{M} x_j = 1 \right\},$$

$$\mathcal{S}_Y = \left\{ \boldsymbol{y} \in \mathbb{R}^N \mid y_k \in [0,1],\ \sum_{k=1}^{N} y_k = 1 \right\}.$$

Then the Minimax Theorem, due to John von Neumann (von Neumann & Morgenstern, 1947), states that

$$\max_{\boldsymbol{x}\in\mathcal{S}_X} \min_{\boldsymbol{y}\in\mathcal{S}_Y} \boldsymbol{x}^{\mathrm{T}}\mathcal{A}\boldsymbol{y} = v = \min_{\boldsymbol{y}\in\mathcal{S}_Y} \max_{\boldsymbol{x}\in\mathcal{S}_X} \boldsymbol{x}^{\mathrm{T}}\mathcal{A}\boldsymbol{y}. \quad (7)$$

The quantity v is the expected payoff to the row player corresponding to some optimal mixed strategy combination $(\boldsymbol{x}, \boldsymbol{y})$ and is called the *value* of the game. The existence of a more general sort of equilibrium in n-person, noncooperative games has been proved by John Nash (1951). Nash's proof relies on properties of convex functions (as does the standard proof of the Minimax Theorem). Convex analysis has also been used recently in thermodynamics, for a general proof of the minimum energy principle, starting from the maximum entropy principle and using the concavity of the entropy (Prestipino & Giaquinta, 2003).

ANDREW HONE

See also **Euler–Lagrange equations; Game theory; Geometrical optics, nonlinear; Quantum field theory**

Further Reading

Bailin, D. & Love, A. 1986. *Introduction to Gauge Field Theory*, Bristol: Adam Hilger

Beeson, D. 1992. *Maupertuis: An Intellectual Biography*, Oxford: Voltaire Foundation

Choquet-Bruhat, Y. 1968. *Géométrie Différentielle et Systèmes Extérieurs*, Paris: Dunod

Feynman, R.P. & Hibbs, A.R. 1965. *Quantum Mechanics and Path Integrals*, New York: McGraw-Hill

Gel'fand, I.M. & Fomin, S.V. 1963. *Calculus of Variations*, Englewood Cliffs, NJ: Prentice-Hall

Goldstein, H. 1950. *Classical Mechanics*, Reading, MA: Addison-Wesley

Kline, M. & Kay, I.W. 1965. *Electromagnetic Theory and Geometrical Optics*, New York: Interscience Publishers, Wiley

Morse, M. & Cairns, S.S. 1969. *Critical Point Theory in Global Analysis and Differential Topology*, New York: Academic Press

Narain, P. 1993. On an extremum principle in the genetical theory of natural selection. *Journal of Genetics*, 72(2–3): 59–71

Nash, J. 1951. Non-cooperative games. *Annals of Mathematics*, 54: 286–295

Owen, G. 1982. *Game Theory*, 2nd edition, London: Academic Press

Prestipino, S. & Giaquinta, P.V. 2003. The concavity of entropy and extremum principles in thermodynamics. *Journal of Statistical Physics*, 111(1-2): 479–493

Velasco, S. & Fernandez-Pineda, C. 2002. A simple example illustrating the application of thermodynamic extremum principles. *European Journal of Physics*, 23: 501–511

von Neumann, J. & Morgenstern, O. 1947. *Theory of Games and Economic Behaviour*, Princeton, NJ: Princeton University Press

FABRY–PEROT RESONANCES

See **Laser**

FAIRY RINGS OF MUSHROOMS

Much of the mythology and mysticism formerly associated with fairy rings is summed up by the words of Prospero in Shakespeare's *The Tempest*, V.i: "Ye elves of hills. . .that/ By moonshine do the green sour ringlets make,/ Whereof the ewe not bites, and you whose pastime/ Is to make midnight mushrooms. . .." However, in what could be interpreted as a minor triumph of reason over superstition, Prospero's invocation is not to the sorcery that was supposedly imbued in fairy rings, but a declaration that he is giving up magic forever.

Fairy rings are not caused by supernatural beings, witches, moles, snails mating, or lightning—these being some of the early explanations, all the more hilarious *now* for being propounded so seriously back *then*, for the annular rings of dead grass, fringed on both edges by concentric rings of over-lush grass, that are often evident in grassy fields. They are caused by fungi. One of the first scientific investigations to establish this fact was reported by Wollaston (1807), whose explanation of fairy rings is still broadly accepted today. Despite this demystification, their appeal to the human imagination remained strong. Kipling (1906) wrote *Puck of Pooks Hill*, a story in which magical manifestations occur when children perform *A Midsummer Night's Dream* in a fairy ring, and Conan Doyle in *The Coming of the Fairies* (1923) only reluctantly admitted that fairy rings were due to fungal growth.

The mode of life involving progressive radial increase from a central point is not unusual among fungi. As pointed out by Ramsbottom (1953), hundreds of fungi grow in circular patterns, including microscopic ones such as *Penicillium* molds as well as those with macroscopic fruiting bodies such as mushrooms and puff-balls. Although the conditions governing initiation of a ring are still obscure, many scientific studies have documented the kinetics of expansion, the species involved, and the biology and ecology of fairy rings. Two excellent and complementary studies are reported in Shantz & Piemeisel (1917) and Dowson et al. (1989).

The underground body of a fairy ring fungus, consisting of a network of filaments called the mycelium, grows radially outwards as it consumes organic matter in the soil. Behind the fungus front, the mycelial mass dies, so that the advancing live fungus front is actually a strange sort of disconnected organism. The dead filaments form a dense and water-repellant mat. Grass in this advancing annular region dies, due simply to physiological drought. Inside this bare region, the grass can grow luxuriantly because the dead filaments eventually decay, providing nitrogen-rich fertilizer. In advance of the fungal front, the grass also grows dark and luxuriant because of the peculiar (to us) eating habits of fungi: they exude digestive enzymes into the medium, then absorb this pre-digested food. Left-over digested food stimulates the grass forward of the fungus.

The radial growth rate of grassland fairy rings has been measured between 99 and 350 mm/year (Dickinson, 1979) and diameters of tens and hundreds of meters have been recorded. The obvious but rather awesome conclusion is that fairy rings may be among the world's oldest living organisms, since many rings are estimated to be several centuries old and some are believed to be 600 or 700 years old. It would be a fascinating exercise to obtain supporting evidence, such as historical records or results from scientific dating methods, for the ages of large fairy rings.

At favorable times of the year, fruiting bodies—mushrooms or toadstools—may be put forth around the circumference of a fairy ring. In woodlands these are often the only obvious manifestation of a fairy ring, since leaf-litter usually covers the ground (Figure 1). Because they depend on the roots of a tree for nutrient supply, woodland fairy rings are referred to as "tethered" rings, whereas those in grassland, for which the nutrient source is spread though the ground, are called "free" rings. The growth of tethered rings is coupled to the radial growth of the host tree roots, and they tend to reach an equilibrium diameter, determined

Figure 1. A typical fairy ring in grass.

by the mature size of the tree, rather than increase indefinitely (Gregory, 1982).

What happens when two or more fairy rings meet? Usually, the putative intersecting portions of same-species rings are extinguished, because of direct competition for resources or as each reaches the other's annular dead zone where nutrients and moisture are depleted in any case. Rings of different species either continue growing through each other or only the dominant species may survive. Rings of different species may also form inside one another.

A simple mathematical model for the ecology of fairy ring systems was developed by Parker-Rhodes (1955), from which was derived estimates for the proportion of ground covered by rings at a given time, the age distribution of rings, and geometric factors affecting inter- and intra-specific competition, such as rate of birth of new rings per unit area and distance between their centers. This model was extended by Stevenson & Thompson (1976) to include boundary effects and a more realistic treatment of the rings as annuli rather than discs. Their conclusions had some interesting implications for the management and control of fairy rings.

Why would anyone want to control them? Some people, sadly, consider fairy rings to be a pathogenic nuisance, a disease to be eradicated; and it is true that they can spoil the appearance and functionality of lawns, golf courses, playing fields, and pastures. The modeling results of Stevenson & Thompson (1976) indicated that if one's aim is to control a population of harmful rings by using an antagonistic but innocuous species of rings, it is more effective in the long term to choose the *smallest* growth rate available. In their work, the rings' growth rate was assumed constant.

If we admit, however, that the growth rate is variable, then fairy rings must be modeled as a dynamical system, and since they spread in two spatial dimensions, the appropriate dynamical description is a system of partial differential equations. Davidson et al. (1997) modeled the spatiotemporal dynamics of fungal mycelia by nonlinear reaction-diffusion equations describing the coupled evolution of the mycelial biomass and nutrient substrate concentration. They found that qualitative features of the development of fairy rings, such as the annular advancing front, degeneration of colony centers, and extinction of the interface between two colliding fronts, were reflected in the structure of solutions to the equations.

This type of predictive computational modeling and simulation of radial growth patterns is likely to become more important in future—and not only because it gives fairy rings the ultimate modern imprimatur and authority of the computer. Other organisms are known to exhibit the fairy ring habit of growth. In the semi-arid rangelands of Australia, for example, certain species of saltbush grow radially outwards from a central origin forming a slowly increasing ring of foliage, the interior of which is bare ground. The patterns on the landscape formed by these blue-gray saltbush rings on the red soil are very striking from the air. The saltbush is a nutritious food source for sheep; thus one might be interested in management strategies that *increase* the covering fraction of saltbush and in factors that affect its growth. In this and other similar problems, the results from mathematical modeling of fungus fairy rings will provide valuable insights.

ROWENA BALL

See also **Growth patterns; Pattern formation; Reaction-diffusion systems**

Further Reading

Davidson, F.A., Sleeman, B.D., Rayner, A.D.M., Crawford, J.W. & Ritz, K. 1997. Travelling waves and pattern formation in a model for fungal development. *Journal of Mathematical Biology*, 35: 589–608

Dickinson, C.H. 1979. Fairy rings in Norfolk. *Bulletin of the British Mycological Society*, 13: 91–94

Dowson, C.G., Rayner, A.D.M. & Boddy, L. 1989. Spatial dynamics and interactions of the woodland fairy ring fungus, *Clitocybe nebularis*. *New Phytologist*, 111: 699–705

Gregory, P.H. 1982. Fairy rings: free and tethered. *Bulletin of the British Mycological Society*, 16: 161–163

Parker-Rhodes, A.F. 1955. Fairy ring kinetics. *Transactions of the British Mycologcal Society*, 38 (1): 59–72

Ramsbottom, J. 1953. *The New Naturalist Mushrooms & Toadstools*, London: Collins

Shantz, H.L. & Piemeisel, R.L. 1917. Fungus fairy rings in eastern Colorado and their effect on vegetation. *Journal of Agricultural Research*, 11: 191–247

Stevenson, D.R. & Thompson, C.J. 1976. Fairy ring kinetics. *Journal of Theoretical Biology*, 58: 143–163

Wollaston, W.H. 1807. On fairy-rings. *Philosophical Transactions of the Royal Society of London*, 2: 133–138

FARADAY WAVES

See **Surface waves**

FEEDBACK

Long before the invention of electrical amplifiers and generators, feedback was used widely in mechanical control devices, including the governor that was devised by James Watt in 1784 to stabilize the speed of a steam engine. The theory of such a governor was developed in the latter half of the 19th century by James Clerk Maxwell and by Ivan Alekseevich Vyshnegradskiy. Although the results obtained by Maxwell and Vyshnegradskiy seem contradictory, they present complementary aspects of the analysis. We consider briefly the detailed discussion of these results as given by Neimark et al. (1985).

The Lagrange equations for a steam engine with a governor are

$$I(\vartheta)\ddot{\varphi} + I'_{\vartheta}(\vartheta)\dot{\vartheta}\dot{\varphi} = M_{\mathrm{r}}(\dot{\varphi}, \vartheta) - M_{\mathrm{l}},$$
$$I_{\mathrm{g}}\ddot{\vartheta} - \tfrac{1}{2} I'_{\vartheta}(\vartheta)\dot{\varphi}^2 + U'_{\vartheta}(\vartheta) + h\dot{\vartheta}, \qquad (1)$$

where $I(\vartheta)$ is the moment of inertia of the engine and governor about the engine shaft, φ is the rotation angle of the engine shaft, ϑ and I_{g} are, respectively, the angular deviation and the moment of inertia of the governor balls, $M_{\mathrm{r}}(\dot{\varphi}, \vartheta)$ is the driving torque, M_{l} is the moment of load, and $-U'_{\vartheta}(\vartheta)$ and $\dot{\vartheta}$ are, respectively, the moments of gravity and viscous friction.

For ideal functioning, Maxwell assumed that the friction factor h must be as small as possible and the governor sensitivity $\mathrm{d}\vartheta_0/\mathrm{d}\Omega$ (where $\Omega = \dot{\varphi}_0$, φ_0 and ϑ_0 are the equilibrium values of φ and ϑ) must be as large as possible, and he gave recommendations how these conditions could be fulfilled. Vyshnegradskiy, on the other hand, understood that the friction plays an important role by promoting the stability of control. Therefore, he proposed adding a new element to the governor, called the cataract. In addition, Vysnegradskiy showed that for stability of control $\mathrm{d}\Omega/\mathrm{d}M_{\mathrm{l}}$ must be negative and constructed his famous diagram in the plane $\mathrm{d}\Omega/\mathrm{d}M_{\mathrm{l}}$, Ih/I_{g}, now known as the Vysnegradskiy diagram. Maxwell's ideas have also been realized in practice by a device called an "isodromic" in which feedback is effected by a rod that changes its length due to stretching or compressive force.

In the context of electronic amplifier theory, a detailed mathematical treatment of linear feedback was given by Henrik Bode (1950). Here the basic formula is

$$G = \frac{A}{1 - \mu A}, \qquad (2)$$

which was famously discovered by Harold Black in 1927 (Brittain & Black, 1997). In this equation, A is the (large) forward gain of an amplifier, μ is the loss of a passive feedback circuit, μA is the open-loop gain, and G is the closed-loop gain. Assuming the amplifier does not oscillate and making $\mu A \gg 1$ implies $G \approx 1/\mu$. Thus, the closed-loop gain is essentially determined by the properties of a passive circuit (μ), which is far less sensitive to variations in temperature and aging than the

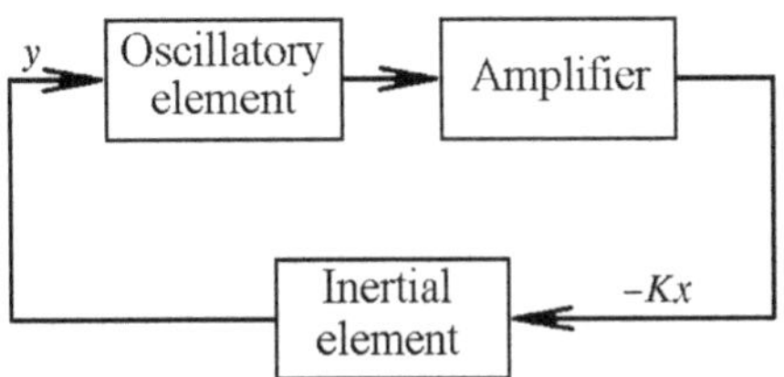

Figure 1. The block diagram of the simplest self-oscillatory system with inertial excitation.

forward gain (A). Long-distance telephone service—which requires a large number of highly stable repeater amplifiers—was made possible by Black's invention.

To avoid low-frequency oscillations, the amplifier system of Equation (2) is designed with the open-loop gain having a phase shift of 180° at low frequencies, which corresponds to negative feedback. To avoid oscillations resulting from positive feedback at higher frequencies (where the phase shift approaches 360°), it is necessary and sufficient that the complex mapping of the frequency axis by the open-loop gain function does not surround the point +1 in the μA-plane—the Nyquist criterion (Bode, 1950).

Positive feedback reveals itself in the appearance of negative friction or negative resistance, whereas negative feedback causes the stabilization of device parameters. In Watt's governor, for example, negative feedback stabilizes the speed of rotation, and in other devices it can stabilize frequency variations, the gain factor, and so on. In certain systems, however, negative feedback can result in self-excitation of oscillations. Such systems were called "systems with inertial excitation" by Babitzky & Landa (1982, 1984), and they often arise in physical and engineering studies. The excitation of oscillations in such systems is conditioned by the so-called inertial interaction between dynamical variables occurring as a result of inertia in the feedback loop.

A block diagram of the simplest self-oscillatory system with inertial excitation is shown in Figure 1. The inertial interaction between dynamical variables, like negative friction, can be both linear and nonlinear. A linear interaction can, under certain conditions, lead to spontaneous oscillations, whereas a nonlinear interaction can result in the hard excitation of oscillations (requiring that a threshold be overcome).

A simple system with linear inertial interaction is

$$\ddot{x} + 2\delta\dot{x} + \omega_0^2 x = -ky + f(x, \dot{x}, y),$$
$$\dot{y} + \gamma y = ax + \varphi(x, \dot{x}, y), \qquad (3)$$

where $f(x, \dot{x}, y)$ and $\varphi(x, \dot{x}, y)$ are nonlinear functions free from linear terms, and the parameter a is proportional to the gain of the amplifier. The inertia of the feedback loop is characterized by the parameter γ. The condition of self-excitation of oscillations can be shown to be $\gamma \leq \gamma_{cr}$, where

$$\gamma_{cr} = -\delta + \sqrt{\delta^2 + ak/2\delta - \omega_0^2}.$$

Because γ_{cr} should be positive, self-excitation can occur only for $ak \geq 2\delta\omega_0^2$. The condition $\gamma \leq \gamma_{cr}$ signifies that the feedback loop must be sufficiently inertial (an inertia-less feedback loop corresponds to $\gamma \to \infty$).

Systems with inertial feedback need not be self-excited. A child on a swing, for example, is an inertial system that is not self-excited. Interestingly, the oscillation of a swing is often inaccurately presented in textbooks as an example of parametric excitation. If oscillations are excited by a child who stands on the swing and lifts her center of gravity up and down at the proper moments, the system is not parametrically excited, because the oscillation frequency of the center of gravity is not constant but tuned according to variations of the frequency of the oscillations of the swing. Thus, a child on a swing is a control system with feedback that is self-oscillatory but not self-excited. For excitation of self-oscillations some finite initial perturbation is necessary (i.e., the excitation of oscillations is hard).

The simplest equations describing oscillations of a swing are

$$\ddot{x} + 2\delta\dot{x} + \omega_0^2 x = -bxy, \quad \dot{y} + \gamma y = ax^2, \tag{4}$$

where the variable y describes the position of the child's center of gravity and γ is the inertial factor of the control circuit. If the feedback is slightly inertial (i.e., γ is too large to put $y \approx (a/\gamma)x^2$), then the variable x obeys the Duffing equation, which has no self-oscillatory solution. Self-oscillations exist only for modest values of γ, when the feedback loop is sufficiently inertial.

Examples of self-oscillatory systems with inertial excitation include the Lorenz system, the Helmholtz resonator with non-uniformly heated walls, a heated wire with a weight at its center, the Vallis model for nonlinear interaction between ocean and atmosphere, a modified Brusselator, an air-cushioned body, and a lumped model of the "singing" flame (Landa, 1996, 2001). Note that the mechanism of self-excitation of oscillations in some continuous system is similar to the examples considered above. Among these are the Rijke tube and a distributed model of the "singing" flame (Landa, 1996).

It must be emphasized that feedback can be nonlinear as well as linear. Although nonlinear feedback cannot cause self-excitation of oscillations, it is of primary importance in their development, and many examples are known. In addition to the swing, noted above, nonlinear feedback plays a role in population dynamics, and it is related to the development of turbulence in a subsonic submerged jet, where the feedback is realized via an acoustic wave. In general, positive feedback plays a key role in the emergence of coherent structures (Landa, 1996).

POLINA LANDA

See also **Candle; Flame front; Nerve impulses; Population dynamics; Tacoma Narrows Bridge collapse**

Further Reading

Babitzky, V.I. & Landa, P.S. 1982. Self-excited vibrations in systems with inertial excitation. *Soviet Physics, Doklady*, 27: 826–827

Babitzky, V.I. & Landa, P.S. 1984. Auto-oscillation systems with inertial self-excitation. *Zeitschrift fuer Angewandte Mathematik und Mechanik*, 64: 329–339

Bode, H.W. 1950. *Network Analysis and Feedback Amplifier Design*, New York: Nostrand

Brittain, J.E. & Black, 1997. Black and the negative feedback amplifier. *Proceedings of the IEEE*, 85: 1335–1336

Landa, P.S. 1996. *Nonlinear Oscillations and Waves in Dynamical Systems*, Dordrecht and Boston: Kluwer

Landa, P.S. 2001. *Regular and Chaotic Oscillations*, Berlin and New York: Springer

Neimark, Yu.I., Kogan, N.Ya. & Savel'ev, V.P. 1985. *Dinamicheskie Modeli Teorii Upravleniya* [*Dynamical Models of the Control Theory*], Moscow: Science

FEIGENBAUM THEORY

See **Routes to chaos**

FERMI ACCELERATION AND FERMI MAP

In a seminal paper, Enrico Fermi (1949) proposed two related methods of producing cosmic rays by accelerating charged particles by repeated collisions with moving magnetic fields. The particles could either be trapped by magnetic mirroring (*See* **Averaging methods**) or be deflected without trapping. In either case, the charged particles gain energy if the magnetic field is moving opposite to the particle's motion and lose energy if the motions are in the same direction. For random motions of the magnetic fields the net effect is stochastic acceleration; that is, the probability is higher for an accelerating collision, due to the relative velocities. The mechanism could explain the power law of proton energies and was consistent with the magnitude of the cosmic ray energies. However, because of the competition between energy loss due to ionization and energy gain from the stochastic heating, the process could not reasonably explain the existence of high mass cosmic rays. More recently, after the discovery of supernovas, and observing magnetic fields and motions of supernova remnants (SNRs), the trapping version of Fermi's acceleration mechanism has been closely examined. A general understanding, developed over the last few years, is that shock waves in SNRs can repeatedly accelerate trapped charged particles to produce both proton and higher mass cosmic rays (see a recent account by Malkov et al. (2002), and references therein).

Well before the recent interest in cosmic ray production by SNRs, the original basic idea of Fermi acceleration was being investigated with simple models. In particular, the question was asked whether the nonlinear dynamics of the particle motion would lead to stochastic acceleration even if the particles were acted on by periodic forces. A simple mapping model to detect this effect is that of a ball bouncing between a fixed and an oscillating wall. The model, developed by Ulam in 1961, was very straightforward to implement on a computer. The results were explained analytically and confirmed numerically in work during the 1960s and early 1970s by, among others, Zaslavsky and Chirikov, Brahic, and Lieberman and Lichtenberg (see Lichtenberg & Lieberman, 1991). They showed that for smooth forcing functions, the phase plane divides up into three distinct regions with increasing ball velocity: (1) a low-velocity region in which all period 1 fixed points are unstable, leading to stochastic motion over almost the entire region; (2) an intermediate velocity region in which islands of stability, surrounding elliptic fixed points, are imbedded in a stochastic sea; and (3) a higher-velocity region in which bands of stochastic motion, near the separatrices of the island trajectories joining the hyperbolic fixed points, are isolated from each other by regular orbits. The existence of region (3), in which invariant curves span the entire range of phase, bounds the energy gain of the particle. If the forcing function is not sufficiently smooth, then region (3) does not exist, in agreement with the Kolmogorov–Arnol'd–Moser (KAM) theory.

Because the Fermi particle acceleration mechanism was one of the first considered for determining the regions of parameter space where KAM surfaces exist and is easily approximated by simple mappings for which numerical solutions are attainable for "long times," it has become a bellweather problem in understanding the dynamics of nonlinear Hamiltonian systems with the equivalent of two degrees of freedom. Various versions both with analytic and non-analytic wall oscillation functions, and with physically moving walls or walls that just impart momentum, have been analyzed. The basic Ulam model, together with an interesting variant of it, is shown in Figure 1. The model in (a) is homologous to many physical problems (see below) and has therefore been of considerable importance. A simplified version of the model in (b) leads to the *standard map*.

A simplified form of the Ulam map can be constructed if the oscillating wall imparts momentum to the ball, according to the wall velocity, without the wall changing its position in space. The problem defined in this manner has many of the features of the more physical problem and can be analytically treated with

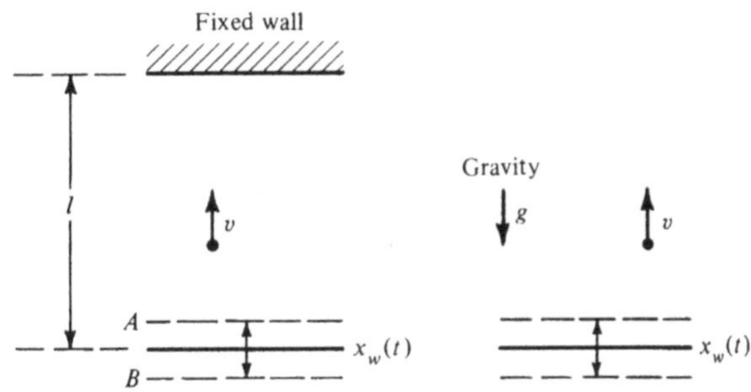

Figure 1. Fermi acceleration models. (a) Version in which a particle bounces between an oscillating wall and a fixed wall. (b) Version in which the particle returns to an oscillating wall under the action of a constant gravitational acceleration.

various wall-forcing functions. In this simplified form, the mapping is

$$u_{n+1} = |u_n + f(\psi_n)|,$$
$$\psi_{n+1} = \psi_n + \frac{2\pi M}{u_{n+1}}, \qquad (1)$$

where u is the velocity, normalized to the maximum wall velocity, and ψ is the phase of the oscillating wall at impact, with $2\pi M/u$ a conventionally used form for phase advance. Equation (1) is the product of two involutions and is, therefore area, preserving; that is, it satisfies the Jacobian condition

$$\frac{\partial(u_{n+1}, \psi_{n+1})}{\partial(u_n, \psi_n)} = 1. \qquad (2)$$

The mapping in (1) serves as an approximation (with suitably defined variables) to many physical systems in which the transit time between kicks is inversely proportional to a velocity. The absolute-value signs correspond to the velocity reversal, at low velocities, $u < 1$ but have no effect on the region $u > 1$, which is the primary region of interest. The forcing function $f(\psi)$ is often a sinusoid in physical problems, but it may have other forms.

The basic method to analyze mapping models is to expand about a fixed point $u_{n+1} = u_n$, $\psi_{n+1} = \psi_n + 2\pi n$, and examine the linear stability. Alternatively, a local linearization of the phase advance equation about the phase-stable value of u linearizes the phase-advance equation but retains the nonlinearity in the forcing term. This expansion leads to the standard map, whose nonlinear stability properties have been extensively analyzed. These procedures have determined the basic phase-plane motion, as described above. A difference between the exact and simplified mappings, which should be noted, is that the canonical variables are different, leading to different variables in which an invariant distribution is uniform. For the simplified problem, a proper canonical set of variables is the ball velocity and phase just before the nth impact with the moving wall. The normalized velocity u then has a uniform invariant distribution. A detailed theoretical and numerical study of various forms of the Fermi map can be found in Lichtenberg & Lieberman (1991, Chapters 3 and 4), where many of the original references can be found. The Fermi map together with the standard map has also led to extensive analysis of phase space diffusion and correlations, which can also be found in the above reference, Chapter 5.

Various elaborated forms of the Fermi map have been applied to a variety of problems. Two such mapping models are electron cyclotron resonance heating (ECRH) in magnetic mirrors (Lieberman & Lichtenberg, 1973) and radio-frequency heating in capacitively driven discharges (Goedde et al., 1988). Although these physical models are generally more complicated than that given in (1), much of the basic analysis is similar. In particular, the expansion around fixed points can be made to investigate linear stability, and nonlinear expansions can obtain the standard mapping from which the KAM barriers to heating can be obtained. Weak dissipation can also be included in models, but this results in a contracting phase space not discussed here (*See* **Chaotic Dynamics**).

Another variant of the basic mapping is one for which the sign in the phase advance equation changes at some value of the action. Such mappings naturally occur in the dynamics of particles in circular accelerators where the sign transition occurs with increasingly relativistic motion in two frequency cyclotron heating and is an underlying cause of the period 3 "catastrophe" of the standard map. When the phase advance equation changes sign, the stable and unstable fixed points exchange their ψ-values, and the phase space structure between these fixed points undergoes a change in topology called reconnection. For further discussion of this phenomenon and references to the original literature, see Lichtenberg & Lieberman (1991, Sections 3.3 and 5.5).

ALLAN J. LICHTENBERG

See also **Averaging methods; Chaotic dynamics; Kolmogorov–Arnol'd–Moser theorem; Maps; Phase space diffusion and correlations; Standard map**

Further Reading

Fermi, E. 1949. On the origin of cosmic radiation. *Physical Review*, 75: 1169–1174

Goedde, C.G., Lichtenberg, A.J. & Lieberman, M.A. 1988. Self-consistent stochastic electron heating in radio frequency discharges. *Journal of Applied Physics*, 64: 4375–4383

Lichtenberg, A.J. & Lieberman, M.A. 1991. *Regular and Chaotic Dynamics*, 2nd edition, New York: Springer

Lieberman, M.A. & Lichtenberg, A.J. 1973. Theory of electron cyclotron resonance heating—II. Long-time and stochastic effects. *Plasma Physics*, 15: 125–150

Malkov, M.A., Diamond, P. & Jones, T.W. 2002. On the possible reason for non-detection of TEV protons in SNRs. In *Bifurcation Phenomena in Plasmas*, edited by S.I. Itoh & Y. Kawai, Fukuoka: Kyushu University

FERMI RESONANCE

See **Harmonic generation**

FERMI–PASTA–ULAM OSCILLATOR CHAIN

A closely examined model of a high-dimensional system comprises a number of masses constrained to move along a line, with a specified force law governing their interaction. In the early 1950s, Enrico Fermi, John Pasta, and Stan Ulam (FPU) (Fermi et al., 1955) numerically examined such a chain of coupled oscillators, with a Hamiltonian of the general form

$$H(p,q) = \sum_{k=1}^{N}\left[\frac{1}{2}p_k^2 + \frac{1}{2}(q_{k+1}-q_k)^2 + \frac{1}{3}\alpha(q_{k+1}-q_k)^3 + \frac{1}{4}\beta(q_{k+1}-qk)^4\right], \quad (1)$$

where each of N particles has unit mass and unit harmonic coupling constant and the end points are either fixed (vertical position $q_1 = q_{N+1} = 0$) or periodic ($q_{N+1} = q_1$). The parameters α, β represent the nonlinearity in the forces between the particles in the chain. In the original FPU numerical computations either α or β was set equal to zero, and fixed end points were used to correspond to a discretized nonlinear spring. Most subsequent work was done with $\alpha = 0$ and is known as the FPU-β model. The FPU-β model can be normalized so that the significant parameter is βE, where E is the total energy, $H(p,q) = E$.

The original idea of Fermi was that the nonlinearity of the springs would lead to mode mixing such that thermodynamic properties in the lattice could be studied. The original numerical work placed the initial energy in the lowest mode of the linearized system, which at low energy becomes a quasimode of the nonlinear system. The N independent normal modes Q_k of the harmonic system with fixed end point boundary conditions give

$$q_i = \sqrt{\frac{2}{N+1}}\sum_{k=1}^{N} Q_k \sin\left(\frac{ik\pi}{N+1}\right), \quad (2)$$

for which the amplitudes q_i of the mass points form a half-sine for the lowest mode and pick up an additional zero for each higher mode. The harmonic mode frequencies are

$$\omega_k = 2\sin\left(\frac{\pi k}{2N+2}\right). \quad (3)$$

The original numerical study surprisingly showed recurrences, rather than a thermodynamic spreading of energy among the modes. Subsequent studies explained these recurrences as resulting from near resonances among various combinations of modes (Bivins et al., 1973). However, the question of whether the nonlinearity led to long-term mixing of the energy remained an open question.

Although the original question of whether the nonlinearity led to equipartition among the degrees of freedom went unanswered during the 1960s, the FPU-β system became rightly famous for having inspired the early development of soliton theory. Zabusky & Kruskal (1965) showed that a Taylor expansion to transform the FPU chain into a continuous differential equation resulted in the Korteweg–de Vries (KdV) or the modified Korteweg–de Vries (mKdV) equation, depending on whether the α or β term in (1) was retained. For the FPU-β chain (mKdV equation), we have

$$u_\tau + 12u^2 u_\xi + u_{\xi\xi\xi} = 0, \qquad (4)$$

where time and length variables have been rescaled by $\tau = h^3 t/24$ and $\xi = x - ht$; subscripts τ, ξ denote differentiation with respect to that variable.

The periodic solutions of (4), stationary in the frame $\xi - c\tau$, may be obtained by integrating to give

$$\tfrac{1}{2}u_\xi^2 + u^4 - \tfrac{1}{2}cu^2 + A \equiv \tfrac{1}{2}u_\xi^2 + P(u) = 0, \qquad (5)$$

where A is a constant of integration. Equation (5) is in the form of a one-degree-of-freedom Hamiltonian, which is therefore integrable. Solutions have been obtained in terms of the Jacobi elliptic functions (or cnoidal waves) $\mathrm{cn}(\xi, q)$, with q (the modulus) taken as a parameter with $0 \le q^2 < 1$. This formalism led to the observations of the stability properties of solitons, and more generally motivated the development of the inverse scattering technique (Gardner et al., 1967). However, the explanation that long-time recurrences were due to initial conditions that were superpositions of stable solitons was not a complete picture and was ultimately misleading.

An explanation for the mechanism leading to stochastic diffusion of energy over the $2N$-dimensional phase space, which also predicted a transition from regular to stochastic motion with increasing energy, was put forward by Izrailev & Chirikov (1966). Using the transformation to modes as in (2), they postulated that if the interaction of pairs of neighboring modes was sufficient to create overlapping beat modes, then the overlap would create global stochasticity. The concept that had been developed by Chirikov for low degree-of-freedom Hamiltonian systems, known as the Chirikov overlap criterion, was applied to the high-dimensional system by isolating a few modes containing the energy. The criterion had been confirmed numerically and later studied in great detail using the *standard map*. The criterion can be shown to be roughly equivalent to requiring that the nonlinear frequency shift with the energy in a single mode be equal to the mode separation (Lichtenberg & Lieberman, 1991, Section 6.5b). Substituting (2) into (1), the nonlinear shift in mode frequency of mode k can be calculated approximately as $\Delta\omega_k \simeq 3\beta E_k \omega_k/4N$. From (3), the frequency separation between low-frequency modes with $\omega_k \simeq \pi k/N$, $k \ll N$, is $\delta\omega_k \simeq \pi/N$, such that for overlap ($\Delta\omega_k/\delta\omega_k \ge 1$)

$$\beta E_k/N \ge 4/3k. \qquad (6)$$

Stochastic energy transfer among modes, leading to approximate equipartition among modes, has been numerically found at much lower values of energy density than given by (6). Numerical studies of equipartition (e.g., Livi et al. (1985); Pettini & Landolphi, 1990) found that (6) roughly corresponds to a transition between an inverse time to obtain some measure of equipartition that scales as $\tau^{-1} \propto (E/N)^2$ at low energy, to a time scaling as $\tau^{-1} \propto (E/N)^{2/3}$ at high energy. The latter scaling models a random process that is very strongly stochastic. Furthermore, the prediction that mode overlap determines the transition between regular motion and stochastic diffusion leading to equipartition does not hold if the energy is initially placed in one or a few high-frequency modes. There is local mode mixing, but the high-frequency modes interact only weakly with the low frequencies, as theoretically predicted by Benettin et al. (1987), such that equipartition is not observed above the mode overlap transition on a fast time scale.

The mKdV soliton was found to become unstable from a low-frequency mode, leading to exponential growth of higher frequencies, which correspond quite closely to similar growth of higher frequencies in an oscillator chain (Driscoll & O'Neil, 1976). However, this low-frequency instability is neither a necessary nor sufficient condition for a transition to equipartition, as the soliton theory does not describe the high-frequency modes, but it does present a physical picture of the process that can hold a low-frequency mode together in the absence of coupling to high-frequency modes and how it can break down.

A more coherent picture of the underlying processes leading to equipartition has been developed recently. Basic phase space arguments indicate that, for relatively high-dimensional systems with not too small a perturbation strength, the probability will be high that a generic initial condition will lie in a stochastic portion of the phase space. Arnol'd diffusion will then transport energy over most of the degrees of freedom, essentially leading to equipartition, on some time scale. However, the time to equipartition can be exponentially slow.

Focusing on the region of weaker stochasticity, a transition was numerically found between power law and exponentially long time scales as the energy density $\varepsilon = E/N$ of the system is decreased. This latter transition is of prime importance for the observation of equipartition, as it essentially separates observable times from those that are not observable. For the FPU chain, the main mechanisms leading to equipartition in this lower energy regime are that resonant interaction of some set of low-frequency modes, in which a significant portion of the energy resides, can lead to local superperiod (very low frequency) beat oscillations that are stochastic. The beat oscillations, which increase with energy, can become comparable to frequency differences between high-frequency modes. This results in Arnol'd diffusion transferring energy to high-frequency modes, on a power law time scale (DeLuca et al., 1995). Furthermore, the appearance of stochasticity corresponds to the onset of the mKdV instability. The driving frequency for diffusion is associated with the libration frequency of the resonance, Ω_{B}. Using resonant normal form perturbation theory to isolate the most important coupling to the high-frequency modes, the energy transfer to high-frequency modes by Arnol'd diffusion, depends exponentially on the frequency ratio as

$$\frac{\mathrm{d}E}{\mathrm{d}t} \propto \exp(-\pi\delta\omega_h/2\Omega_{\mathrm{B}}), \quad (7)$$

where $\delta\omega_h$ is the difference frequency between two high-frequency modes. When $\Omega_{\mathrm{B}} \sim \delta\omega_h$, the exponential factor is of order unity, allowing strong diffusion of energy to high-frequency modes, and equipartition on computationally observable time scales. In a separate work, an estimate of the scaling, with energy density, of the equipartition time for $E \gg E_c$ was found theoretically to be $T_{\mathrm{eq}} \propto (N/E)^3$, which agrees with numerical computations (De Luca et al., 1999). The somewhat weaker quadratic scaling found in earlier work, as quoted above, was probably due to the use of a less accurate measure of equipartition.

If the energy is initially placed in high-frequency modes, the equipartition process is significantly different from that starting from low-frequency initial conditions. In this case, the dynamics is transiently mediated by the formation of unstable nonlinear structures. First, there is an initial fast stage in which the mode breaks up into a number of breather-like structures. Second, on a slower time scale, these structures coalesce into one large unstable structure. These structures have been called chaotic breathers (CB). Because a single large CB closely approximates a stable breather, the final decay stage, toward equipartition, can be very slow. This behavior has been observed in oscillator chains approximating the Klein–Gordon equation with various force laws and the FPU-β model (Cretegny et al., 1998; Mirnov et al., 2001). In Cretegny et al. (1998), the energy was placed in the highest frequency mode with strict alternation of the amplitudes from one oscillator to the next (periodic boundary conditions). This configuration is stable up to a particular energy, for which there exist exact discrete solutions. Beyond this energy a parametric instability occurs, leading to the events described above. However, the nonlinear evolution does not depend on special initial conditions but will generically evolve from any high-frequency mode initial condition that has predominantly the alternating amplitude symmetry. One does not know, in this generic situation, whether there exists any true energy threshold. As discussed with respect to low-frequency mode initial conditions, the practical thresholds refer to observable time scales. From a phase space perspective, it is intuitively reasonable that for a large number of oscillators and not too low an initial energy, the generic set of initial conditions will lie in a chaotic layer, but the chaotic motion can remain close to a regular orbit for very long times (Lichtenberg & Lieberman, 1991).

Considerable insight into the behavior of a nonlinear oscillator chain, starting from high-frequency mode initial conditions, can be obtained by introducing an envelope function for the displacements of the oscillators. Low-order expansions produce PDEs that have integrable solutions in the form of envelope solitons, analogous to the solitons produced from low-frequency initial conditions (Kosevich, 1993). Higher-order terms usually destroy the integrability.

Substituting the envelope function $\psi_i(t) = (-1)^i q_i(t)$ in (1) (with $\alpha = 0$), using the continuous variable $x = ai$, a Taylor's expansion in a, the lattice period, yields

$$\begin{aligned}&\psi_{tt} + 4\psi + 16\beta\psi^3\\&\quad + a^2\{\psi_{xx} + \beta(12\psi\psi_x^2 + 12\psi^2\psi_{xx})\}\\&\quad + a^4\{(1/12)\psi_{xxxx} + \beta(3\psi_x^2\psi_{xx} + 3\psi\psi_{xx}^2\\&\quad + 4\psi\psi_x\psi_{xxx} + \psi^2\psi_{xxxx})\} + \cdots = 0. \quad (8)\end{aligned}$$

Comparing (8) with (4) qualitatively explains why relaxation is accompanied by the formation of sharply localized states if energy is initially deposited in the high-frequency part of the spectrum where the effect of dispersion is small, while only broad nonlinear structures are formed if the energy is initially in the low-frequency modes where the dispersion is large. Using a dimensionless variable, introducing the rotating wave approximation (RWA) $\cos^3 \omega t \simeq (\frac{3}{4})\cos \omega t$ and neglecting terms proportional to a^4 and higher, (8) can be integrated to yield

$$(-\omega^2 + 4)\psi^2 + \psi_x^2 + \beta(6\psi^4 + 9\psi^2\psi_x^2) = C_1, \quad (9)$$

which has integrable trajectories in phase space similar to (5) but includes a high-frequency drive ω. Depending on the boundary conditions, these phase trajectories represent multiple or single breathers.

As with low-frequency initial conditions, the soliton solution of the envelope, obtained from (9), becomes unstable with increasing energy. For most numerical studies of oscillator chains, the initial state imposed on the system is mainly that of a single linear mode. This state is generally not close to an equilibrium. The initial state rapidly relaxes, governed by the nonlinear equations. Numerical studies indicate that the chaotic breathers that are formed in the nonlinear processes are probably marginally stable, which accounts for their long-lived existence.

After a set of chaotic breathers has been formed, on a short time scale by a modulational instability or breakup relaxation, the breathers coalesce, on a longer time scale into a single chaotic breather. This process has been well documented numerically (Cretegny et al., 1998; Mirnov et al., 2001) and an analytic estimate of the process made (Kosevich & Lepri, 2000; Mirnov et al., 2001) with the breather coalescence time found to scale as $\tau_B^{-1} \propto E_B/N$, the breather energy density. The background mode spectrum beats with the breather to transfer energy to low-frequency modes, resulting in equipartition on a slower time-scale as $T_{eq} \propto (E/N)^{-2}$ (Mirnov et al., 2001).

Although we have concentrated on the seminal problem of the FPU-β lattice, the results are qualitatively connected to other types of oscillator chains. For example, if the cubic, rather than the quartic, potential is retained in (1), then the Hamiltonian is equivalent to the lowest-order nonlinear expansion of the integrable Toda lattice. Consequently, the chain is considerably more stable. Other interesting types of lattices are composed of discretizations of Klein–Gordon equations. In particular, with a quartic potential, the dynamics of the Klein–Gordon Hamiltonian

$$H = \sum_{i=1}^{N} \left[\tfrac{1}{2}p_i^2 + \tfrac{1}{2}(q_{i+1} - q_i)^2 + \tfrac{1}{2}m^2 q_i^2 + \tfrac{1}{4}\beta q_i^4 \right], \tag{10}$$

has been compared with the FPU-β chain, showing both similarities and differences (Pettini & Cerruti-Sola, 1991). The stability of "discrete breathers" has also been examined for various chains (Cretegny et al., 1998).

ALLAN J. LICHTENBERG

See also **Arnol'd diffusion; Breathers; Discrete breathers; Frenkel–Kontorova model; Korteweg–de Vries equation; Phase space diffusion and correlations; Solitons; Solitons, a brief history; Standard map; Toda lattice**

Further Reading

Benettin, G., Galgani, L. & Giorgilli, A. 1987. Realization of holonomic constraints and freezing of high frequency degrees of freedom in the light of classical perturbation theory. *Communications in Mathematical Physics*, 113: 87–103

Bivins, R.L., Metropolis, N. & Pasta, J.R. 1973. Nonlinear coupled oscillators: model equation approach. *Journal of Computational Physics*, 12: 65–87

Cretegny, T., Dauxois, T., Ruffo, S. & Torcini, A. 1998. Localization and equipartition of energy in the β-FPU chain: chaotic breathers. *Physica* D, 121: 109–126

De Luca, J., Lichtenberg, A.J. & Lieberman, M.A. 1995. Time scale to ergodicity in the Fermi–Pasta–Ulam system. *Chaos*, 5: 283–297

De Luca, J., Lichtenberg, A.J. & Ruffo, S. 1999. Finite times to equipartition in the thermodynamic limit. *Physical Review* E, 60: 3781–3786

Driscoll, C.F. & O'Neil, T.M. 1976. Explanation of instabilities observed on the Fermi–Pasta–Ulam lattice. *Physical Review Letters*, 37: 69–72

Fermi, E., Pasta, J.R. & Ulam, S.M. 1955. Studies of nonlinear problems. *Los Alamos Scientific Report*, LA-1940; also in *Collected Works of Enrico Fermi* 1965. Chicago: University of Chicago Press, 2: 978–988

Gardner, C.S., Greene, J.M., Kruskal, M.D. & Mura, R.M. 1967. Method for solving the Korteweg–de Vries equation. *Physical Review Letters*, 19: 1095–1097

Izrailev, F.M. & Chirikov, B.V. 1966. Statistical properties of the nonlinear string. *Soviet Physics-Doklady*, 11: 30–32

Kosevich, Y.A. 1993. Nonlinear envelope-function equation and strongly localized vibrational modes in anharmonic lattices. *Physical Review* B, 47: 3138–3151

Kosevich, Y.A. & Lepri, S. 2000. On modulation instability and energy localization in harmonic lattices at finite energy density. *Physical Review* B, 61: 299–314

Lichtenberg, A.J. & Lieberman, M.A. 1991. *Regular and Chaotic Dynamics*, 2nd edition, New York: Springer

Livi, R., Pettini, M., Ruffo, S., Sparglioni, M. & Vulpiani, A. 1985. Equipartition threshold in nonlinear large Hamiltonian systems: the Fermi–Pasta–Ulam model. *Physical Review* A, 31: 1039–1045

Mirnov, V.V., Lichtenberg, A.J. & Guclu, H. 2001. Chaotic breather formation, coalescence, and evolution to energy equipartition in an oscillator chain. *Physica* D, 157: 251–282

Pettini, M. & Cerruti-Sola, M. 1991. Strong stochasticity threshold in nonlinear large Hamiltonian systems: effect on mixing times. *Physical Review* A, 44: 975–987

Pettini, M. & Landolphi, M. 1990. Relaxation and ergodicity breaking in nonlinear Hamiltonian dynamics. *Physical Review* A, 41: 768–783

Zabusky, N.J. & Kruskal, M.D. 1965. Interactions of solitons in a collisionless plasma, and the recurrence of initial states. *Physical Review Letters*, 15: 240–243

FERROMAGNETISM AND FERROELECTRICITY

Iron, nickel, cobalt, and some rare earths (e.g., gadolinium) are characterized by long-range ferromagnetic order. This originates at the atomic level and causes the unpaired electron spins to line up parallel to each other within a region of space called a magnetic domain. Domains range from 0.1 mm to a few mm in size. Within each domain, the net magnetization is large and homogenous, but over the entire sample it averages out to zero due to random orientations of spins. An externally applied magnetic field can cause the material to become macroscopically magnetized, as the magnetic domains already aligned in the direction of this field grow at the expense of their neighbors and those neighbors reorient their magnetizations towards the field direction.

Ferromagnets have very high magnetic susceptibilities (χ), ranging from 1000 up to 100,000 (meaning the ferromagnet is that much easier to magnetize than free space), and they tend to stay magnetized following the application of an external magnetic field. This tendency to remember magnetic history is called "hysteresis." All ferromagnets have a maximum temperature, called the Curie temperature (T_c), which for iron is about 1043 K. Above T_c the ferromagnetic phase changes into a paramagnetic phase, in which induced magnetism is proportional to the applied field. Ferromagnets also respond mechanically to an applied magnetic field, changing length slightly in the direction of the applied field, a property that is called magnetostriction.

In addition to ferromagnets, there exist other magnetically ordered compounds with parallel-oriented sublattices. The simplest such example is antiferromagnetism with two antiparallel sublattice magnetizations. To measure the degree of order in a complex magnetic phase, as many order-parameter components as there are distinct sublattices may be needed. For ferromagnets, the order parameter is the net magnetization. In antiferromagnets, the order parameter is the staggered magnetization: $\boldsymbol{M}_1 - \boldsymbol{M}_2$, where $\boldsymbol{M}_1$ and $\boldsymbol{M}_2$ are the magnetization vectors for the two sublattices.

In 1907, Pierre Weiss proposed a phenomenological theory of ferromagnetism, building on the model of paramagnetism that Paul Langevin introduced in 1905. The Langevin function: $L = \coth(x) - 1/x$ (where $x = \mu H/kT$) describes the paramagnetic susceptibility of N non-interacting classical spins in a magnetic field of intensity H. Weiss assumed that spins interact with each other through a molecular field proportional to the average magnetization in the sample. So that

$$H_{\text{eff}} = H + \lambda M. \tag{1}$$

By replacing spin-spin interactions with the interaction of a single spin (S) and all its neighbors taken as an average field, the nonlinear problem was approximately solved giving rise to ferromagnetism for T below $T_c = \lambda C$, where

$$C = \frac{N\mu^2 S(S+1)}{3k} \tag{2}$$

with k denoting the Boltzmann constant.

As a consequence, the Langevin expression for susceptibility in paramagnetism

$$\chi = \frac{M}{H} = \frac{C}{T} \tag{3}$$

becomes the Curie–Weiss relation

$$\chi = \frac{C}{T - T_c} \tag{4}$$

above T_c, with spontaneous magnetization below T_c for ferromagnets. This is characteristic of a second-order phase transition. Spin alignment stems from the exchange interactions between spins whose energy can be expressed via the Ising Hamiltonian $\mathcal{H}$ in the case of strong uniaxial anisotropy favoring alignment parallel or antiparallel to the z-axis of quantization depending on the sign of the exchange constant J:

$$\mathcal{H} = -\sum_{i,j} J S_i^z S_j^z, \tag{5}$$

where S_i^z and S_j^z are the z-components of the spin vectors at the lattice sites i and j, respectively. In the absence of anisotropy, one uses the Heisenberg Hamiltonian that couples the spin vectors in a scalar product. Free-energy expansion of the Landau type can be obtained within the Curie–Weiss approximation for a Hamiltonian that includes the Zeeman interaction with an external magnetic field and an Ising-type spin-spin interaction.

Ferroelectricity was discovered in the beginning of the 20th century as a property of ionic, covalent, molecular crystals, and even polymers that possess electrical polarization either spontaneously (e.g., Rochelle salt) or under mechanical stress (piezoelectricity) or temperature changes (pyroelectricty). The net polarization of a ferroelectric crystal can be reoriented by applying an electric field. In ferroelectric phase transitions, a change in the crystal structure is accompanied by the appearance of spontaneous polarization. Ferroelectric phase transitions can be either displacive or order-disorder type. In displacive transitions (e.g., $BaTiO_3$), atoms or molecules exhibit small (compared with the unit cell) positional shifts with long-range correlations. These transitions are caused by phonons and the order parameter is the amplitude of the related lattice distortion giving rise to a change of the lattice structure. Displacive transitions are described using a continuous Landau–Ginzburg model with ensuing solitary waves. In order-disorder transitions (e.g., $NaNO_2$), atoms or molecules order themselves on distances comparable to the unit cell. A transformation between randomly distributed atomic positions of their local double-well potential bottoms ($T > T_c$) and an ordered arrangement ($T < T_c$) takes place. Models of order-disorder transitions use the Ising Hamiltonian with an effective (not real) spin variable.

Ferroelectric phase transitions involve symmetry changes in the crystal structure that are manifested by the emergence of an order parameter: spontaneous polarization vector $\boldsymbol{P}$. For second-order transitions, the symmetry group of the ferroelectric phase is a subgroup of that in the paraelectric phase (in which $\boldsymbol{P}$ is proportional to the applied electric field $\boldsymbol{E}$). In some cases, such as the onset of ferroelectricity with a transverse optical branch, a so-called soft mode is responsible for the transition. The soft-mode's frequency ω_k for the wave vector k tends to 0 as $T \to T_c$. A special type of ferroelectric phase transition involves incommensurate phases where spontaneous polarization develops a spatial modulation with a wavelength that is incommensurate with lattice periodicity. The occurrence of incommensurate phases is usually explained by competition between long- and short-range forces, for example, in the Frenkel–Kontorova model. As in ferromagnets, ferroelectrics develop domains in which a particular orientation of polarization is selected. These domains can range in sizes from submicroscopic to macroscopic, and the region between two neighboring domains is called a domain wall.

Jack A. Tuszyński

See also **Critical phenomena; Domain walls; Frenkel–Kontorova model; Hysteresis; Ising model; Order parameters**

Further Reading

Bruce, A.D. & Cowley, R.A. 1981. *Structural Phase Transitions*, London: Taylor & Francis

Kittel, C. 1996. *Introduction to Solid State Physics*, 7th edition, New York: Wiley

Landau, L.D. & Lifshitz, E.M. 1959. *Statistical Physics*, London: Pergamon Press

Lines, M.E. & Glass, A.M. 1977. *Principles and Applications of Ferroelectric and Related Materials*, Oxford: Clarendon Press

Stanley, H.E. 1971. *Introduction to Phase Transitions and Critical Phenomena*, Oxford: Clarendon Press and New York: Oxford University Press

White, R.H. & Geballe, T. 1979. *Long Range Order in Solids*, New York: Academic Press

FIBERS, OPTICAL

See **Optical fiber communications**

FIBONACCI SERIES

Leonardo of Pisa (approx. 1175 to around 1250), also known as Leonardo Pisano, referred to himself as Fibonacci (or son of Bonacci), the name by which he is usually called today. As the son of a customs officer, he had opportunities to travel around the Mediterranean coast and observe many commercial practices. He saw that the Hindu-Arabic system of ten digits and its algorithms for arithmetic had many advantages over Roman numerals. His book of 1202 (revised 1228), *Liber Abaci*, meaning *The Book of the Abacus* or *The Book of Reckoning*, described the system in the Italian vernacular, and so the decimal system became common in Europe.

In his book are some problems and puzzles that were meant as arithmetic practice using the new system. One referred to a problem about rabbits in a field: "A certain man put a pair of rabbits in a place surrounded on all sides by a wall. How many pairs of rabbits can be produced from that pair in a year if it is supposed that every month each pair begets a new pair, which from the second month on becomes productive?" It is based on the assumption that the pairs mate according to the same conditions. Each month, the number of rabbit pairs in the field is 1, 1, 2, 3, 5, 8, 13, 21, 34, 55, 89, 144. So the answer is 144 pairs after 12 months. The number of rabbits in any month is found by adding the number of rabbits alive in the last month (since we assume none die or are eaten) and adding to it the number of new rabbits born that month, which is one for each pair that was alive in the month before that. The rule is therefore "add the last two numbers to get the next." The next number in the series is 89 plus 144, or 255. The series was probably known before this and Fibonacci merely copied it from another source. The sequence is now called the Fibonacci sequence in his honor, but this name was not given until the late 1800s when Edouard Lucas rediscovered the series and wrote about some of its many properties.

Earlier than this, it had been noticed that the Fibonacci numbers (though not called that) appear in the number of petals of flowers of many plant species, in the arrangements of leaves round a branch, or seed whorls on a seed head. The study of such features of plants is termed phyllotaxis. There are flowers with 3 petals (clover) and many with 5, but none with 7 or 9 leaves or petals. The numbers 4, 6, and 10 also occur but the arrangement is of two sets of 2, 3, or 5. No scientific justification for this was given until Douady and Couder wrote about it in 1993 where they showed that if the growing tip of a plant (the meristem) produces a new primordial cell that becomes a leaf or petal or branch, then the optimal arrangement is for the new cell to be produced at the rate of 1.618 per turn of the growing point. This produces the least overlapping for leaves and the maximum exposure to collect sunlight, or for seeds, it provides the most compact packing (given the simple method of growth of the meristem) with even placing of seeds no matter how large the seed head becomes.

This number, precisely computed as $g \equiv (\sqrt{5}+1)/2 = 1.6180339887\ldots$, is a famous number in geometry called the golden section (also golden ratio, golden mean, divine proportion). It appears as the ratio of diagonals to sides in a pentagon. The golden section as a decimal number never recurs, and so cannot be expressed exactly as a fraction (i.e., it is irrational). It has the property that its reciprocal (1 divided by the golden section) is 0.6180339887.... This is exactly 1 less than the golden section itself, giving rise to the definition of the golden section as "the positive number that is 1 less than its reciprocal." In other words, $1/g = g - 1$ or $1 = g^2 - g$, a quadratic equation that is solved by the value above together with $-1/g = -0.6180339887\ldots$.

The fractions that best approximate g are the ratios of two successive Fibonacci numbers: 3/2, 5/3, 8/5, 13/8, 21/13 in the sense that no fraction with numbers smaller than the numerator of one of these terms will give a better approximation. It is this that explains why the Fibonacci numbers appear in phyllotaxis since the optimal arrangements are best approximated using Fibonacci numbers (or, indeed any series where the last two numbers are added to produce the next).

Since the Fibonacci rule is so simple, it is also found in other mathematical situations. For instance, suppose that houses can be made in two sizes, single (separate) and double (two attached), the latter taking twice the frontage along a road as the single ones. Given a road that is long enough for n houses on one side, how many arrangements are there that the architect can choose from? For instance, a road that can have 3 houses on it could have 3 singles, or else a double followed by a single or a single first and then a double—3 ways. For 4 houses there are 5 arrangements; for 5 houses, 8 possibilities. The number of arrangements is always a Fibonacci number.

The rule also applies to the family tree of a male honeybee. Male bees are produced from unfertilized eggs and female bees from fertilized eggs. So males have one parent and females have two. A male bee thus has one parent (F) and two grandparents (M and F), three great-grandparents and so on with 5, 8, 13, 21 as we go back each generation.

In two dimensions, a rectangle with sides in the golden ratio is called a golden rectangle. It has been observed in the shape of many parts of the Parthenon on the Acropolis in Athens, although none of the original plans of the building are extant. The Greeks knew of the ratio, and Euclid's *Elements* shows how to find and construct the golden section point on any line.

RON KNOTT

See also **Branching laws**

Further Reading

For further information, references, applications, and related mathematics, see www.mcs.surrey.ac.uk/Personal/R.Knott/Fibonacci/

Boncompagni, B. 1852. *Della vita e delle opere di Leonardo Pisano*, Rome: Tipografia delle Belle Arti (The only complete printed version of Fibonacci's 1228 edition of *Liber Abbaci.*)

Gies, J. & Gies, F. 1969. *Leonard of Pisa and the New Mathematics of the Middle Ages*, New York: Crowell (Another book with much on the background to Fibonacci's life and work.)

Grimm, R.E. 1973. The autobiography of Leonardo Pisano. *Fibonacci Quarterly*, 11: 99–104

Horadam, A.F. 1985. Eight hundred years young. *The Australian Mathematics Teacher*, 31: 123–134 (An interesting and readable article on Fibonacci, his names and origins as well as his mathematical works. It refers to and expands upon the article by Grimm.)

Parshall, Karen Hunger. *The Art of Algebra from al-Khwarizmi to Viète: A Study in the Natural Selection of Ideas*, http://www.lib.virginia.edu/science/parshall/algebra.html (Contains a brief biography of Fibonacci if you want to read more about the history of mathematics.)

Smith, D.E. 1923. *History of Mathematics*, vol. 1, Boston and London: Ginn; reprinted, New York: Dover, 1958 (Gives a complete list of other books that he wrote and is a fuller reference on Fibonacci's life and works.)

FIBRILLATION

See **Cardiac arrhythmias and electrocardiogram**

FILAMENTATION

When we look at a window on rainy days, we may observe a curious phenomenon: the rain forms a sheet of water on the glass. A single drop—from time to time—introduces a perturbation on this surface that is sufficient to trigger small-scale channels breaking the layer of liquid and escaping from it. Filamentary structures of water emerge from the unstable flow, destroying the initial expanse. The water sheet has thus decayed into filaments through an instability that is called filamentation.

Filamentation also occurs in plasma physics and nonlinear optics. In this context, an intense optical beam propagating in a focusing (Kerr) medium may break up into several spots due to the small inhomogeneities affecting its initial distribution, as evidenced by self-focusing experiments in solids, liquids, and gases (see, e.g., Bespalov & Talanov, 1966; Campillo et al., 1973). This phenomenon can be understood from the equation for the paraxial self-focusing of optical beams that describes the propagation of the slowly varying complex envelope ψ of a scalar electric field with central frequency ω_0 through a nonlinear medium. The envelope ψ, expressed in the frame moving with the group velocity, obeys the nonlinear Schrödinger (NLS) equation

$$i\partial_z\psi + \nabla_\perp^2\psi + f(|\psi|^2)\psi = 0, \qquad (1)$$

where $\nabla_\perp^2 \equiv \partial_x^2 + \partial_y^2$ represents wave diffraction and $f(|\psi|^2)$ models the nonlinear response of the medium. Here, the radial (x, y) and longitudinal (z) variables are normalized with respect to the beam waist w_0 and to the Rayleigh length $z_0 = \pi n_0 w_0^2/\lambda_0$, depending on the linear refractive index n_0 and central wavelength λ_0.

For an unsaturated Kerr medium, $f(s) = s$ and solutions of Equation (1) collapse at finite distance whenever their power $P = \int |\psi|^2 d\boldsymbol{r}$ exceeds the threshold $P_c \simeq 11.7$. Nonlinear responses of optical materials are, however, generally saturating, for example, like $f(s)=s/(1+\beta s)$ with $\beta=|\psi|_{\max}^{-2}\ll 1$. Such saturations occur in two-level systems, such as sodium and rubidium atomic vapors (Soto-Crespo et al., 1992).

Modeling filament formation requires a perturbation theory, following which small fluctuations break up steady-state solutions, expressed as $\psi_s(\boldsymbol{r},z)=\phi(\boldsymbol{r})e^{i\lambda z}$, where ϕ satisfies the differential equation

$$-\lambda\phi+\nabla_\perp^2\phi+f(\phi^2)\phi=0 \quad (2)$$

and $\lambda=$ const. Stability of ϕ can thus be investigated from the perturbed solution

$$\psi(x,y,z)=[\phi(x,y,\lambda)+\varepsilon\{v(x,y,z)+iw(x,y,z)\}]e^{i\lambda z}, \quad (3)$$

where v and w are real-valued functions with amplitude parameter $\varepsilon\ll 1$.

Linearizing Equation (1) with respect to these functions yields the eigenvalue problem

$$\partial_z v=L_0 w,\quad -\partial_z w=L_1 v, \quad (4)$$

where L_0 and L_1 represent the self-adjoint operators

$$L_0=\lambda-\nabla_\perp^2-f(\phi^2),$$
$$L_1=\lambda-\nabla_\perp^2-[f(\phi^2)+2f'(\phi^2)\phi^2], \quad (5)$$

with $f'(\phi^2)=\partial f(u)/\partial u|_{u=\phi^2}$. Combining Equations (4), we then obtain $\partial_z^2 v=-L_0L_1 v$, and different filamentation-like instabilities may be investigated from this general formalism.

- *Modulational instability (MI):* Originally, Bespalov & Talanov (1966) proposed a modulational instability theory, following which oscillatory perturbations with an exponential growth rate split the beam envelope taken as a background uniform solution into small-scale cells. Perturbative modes are chosen as $v,w\sim\cos(k_x x)\cos(k_y y)e^{\gamma z}$, and they apply to a plane wave ϕ which, by definition, satisfies $\nabla_\perp^2\phi=0$, so that $\lambda=f(\phi^2)$. The growth rate γ is then given by Equation (4):

$$\gamma^2=k^2[2A-k^2],\quad A\equiv uf'(u)|_{u=\phi^2},$$
$$k^2=k_x^2+k_y^2. \quad (6)$$

  Plane waves are unstable with $\gamma^2>0$ in the range $0<k<\sqrt{2A}$ and the maximum growth rate $\gamma_{\max}=A$ is attained for $k=k_{\max}=\sqrt{A}$. This instability promotes the beam breakup into a wave-train of small-scale filaments regularly distributed in the diffraction plane with the transversal spacing $\lambda_{\text{mod}}\simeq 2\pi/k_{\max}$ and longitudinal length $\sim\gamma^{-1}$. In practical uses, the number of filaments is close to the ratio $P_{\text{in}}/P_{\text{fil}}$, where P_{in} is the power of the input beam and P_{fil} the power enclosed in one filament. Considering each filament with radial symmetry, the evaluation $P_{\text{fil}}\simeq 2\pi\int_0^{\lambda_{\text{mod}}/2} r|\phi|^2\,dr\simeq 2.65P_c$ holds for unsaturated Kerr media [$f(s)=s$].

- *Filamentation on a ring*: For broad beams, self-focusing often takes place as a regular distribution of dots superimposed upon ring-like diffraction patterns, so that filamentation may not develop over the entire surface of the input beam. To model this instability, the Laplacian in Equation (1) can be rewritten as $\nabla_\perp^2=r^{-1}\partial_r r\partial_r+r^{-2}\partial_\theta^2$, where θ denotes the azimuthal angle. Unstable modes $v,w\sim\cos(m\theta)e^{\gamma_m z}$ with azimuthal index number m break up a spatial ring, which is modeled by a uniform background solution ϕ lying on a circular path with length $s=\bar{r}\theta$, where $\bar{r}$ is the mean radius of the ring. Equation (4) then yields (Soto-Crespo et al., 1992)

$$\gamma_m^2=\left(\frac{m}{\bar{r}}\right)^2\left(2A-\frac{m^2}{\bar{r}^2}\right) \quad (7)$$

  and the maximum number of modulations on the ring is provided by the integer part number $m_{\max}=\text{int}\{\bar{r}^2A\}^{1/2}$. An example of filamentation on a ring-diffraction pattern emerging from a tenth-order super-Gaussian beam is shown in Figure 1 for $f(s)=s$ in the presence of a random-phase noise. About seven to eight filaments emerge and rapidly collapse.

- *Transverse instability*: Although elementary, MI theory has been considered as the starting point for understanding laser filamentation, until plane waves are replaced by bounded solutions. These can be the soliton modes of the NLS equation, and the resulting instability is termed transverse instability. It appears when a soliton ϕ is perturbed by oscillatory modulations developing along one axis. For instance, when the Laplacian reduces to the 1-dimensional operator $\nabla_\perp^2=\partial_x^2$ in Equation (2), a soliton solution $\phi(x,\lambda)=\sqrt{2\lambda}\,\text{sech}(\sqrt{\lambda}x)$ can undergo perturbative modes v,w decomposed as $V(x),W(x)\cos(k_y y)e^{\gamma z}$. These perturbations are local in x and they promote the formation of bunches, periodically distributed over the y-axis. The operators L_0 and L_1 in Equation (5) are transformed as $L_0=\lambda-\partial_x^2-f(\phi^2)+k_y^2$ and $L_1=L_0-2f'(\phi^2)\phi^2$. For the cubic nonlinearity ($f(s)=s$), the instability growth rate is close to the theoretical curve $\gamma^2=1.08k_y^2(3\lambda-k_y^2)$ (Rypdal & Rasmussen, 1989). The same reasoning holds when the Laplacian of Equation (1) includes a third dimension, for example, a time variable, along which 2-dimensional spatial solitons are subject to periodic fluctuations. Because bounded solutions ϕ are not always accessible analytically, numerical computations are often required for determining γ^2 (Akhmediev et al., 1992).

Besides filament formation, Equations (4) and (5) provide information on the inner stability of solitons, known as orbital stability, which refers to the ability

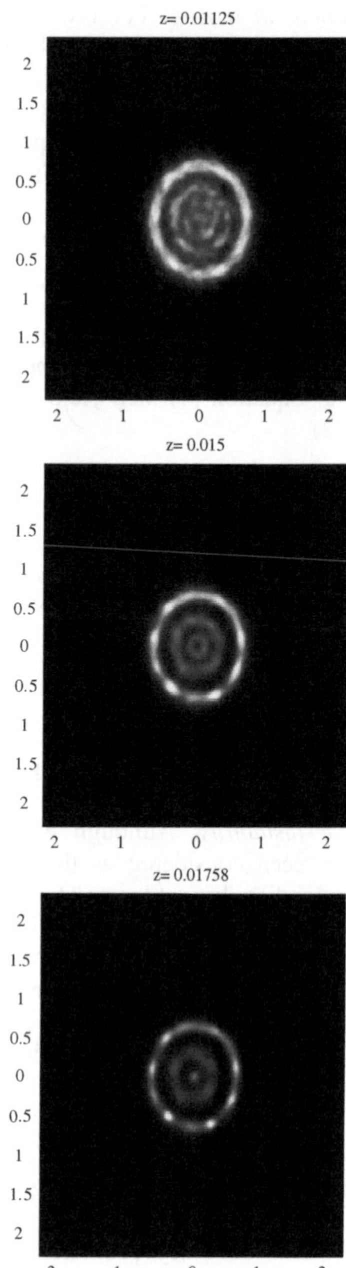

Figure 1. Filamentation pattern numerically computed for a tenth-order super-Gaussian beam with $P_{in} = 30P_c$ in a cubic medium.

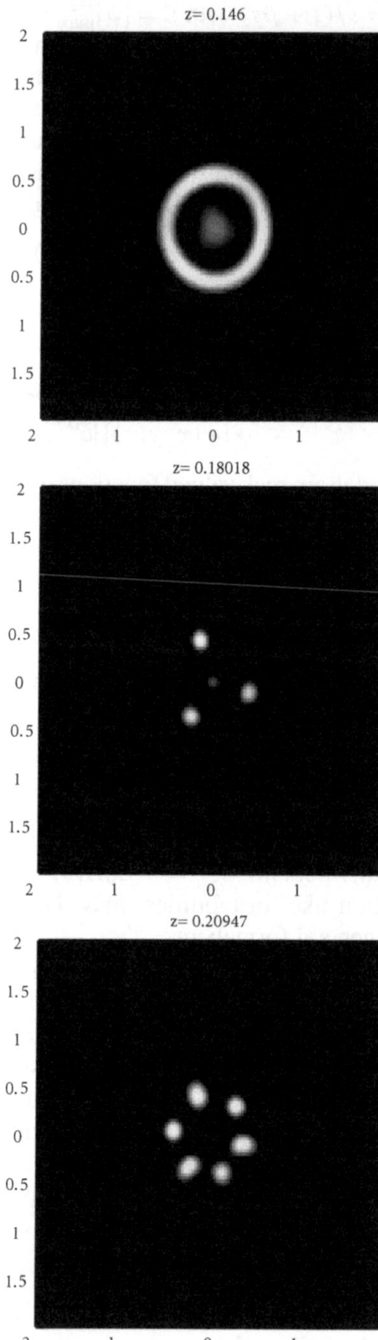

Figure 2. Filamentation pattern numerically computed for a Gaussian beam with $P_{in} = 20P_c$ in a saturable medium.

of initial solutions near equilibrium states to converge to robust soliton shapes. Standard procedures allow the stability criterion $dP(\phi)/d\lambda > 0$ to be established, where $P(\phi) \equiv \int \phi^2 d\mathbf{r}$ is the soliton power (Kuznetsov et al., 1986). Orbital stability applies to filaments formed in saturable media with $f(s) = \alpha s - \beta s^2$, $f(s) = \alpha s/(1 + \beta s)$, or $f(s) = \alpha(1 - e^{-s})$, where α and β are positive constants. Relaxation of filaments to stable solitons promoted by the nonlinearity $f(s) = s/(1 + 2 \times 10^{-3}s)$ is shown in Figure 2.

On the whole, filamentation follows from linear stability analyses, which are valid as long as $\varepsilon(v + iw)$ remains smaller than ϕ, thus applying to early stages in the beam propagation. At later stages, filaments develop into a fully nonlinear regime, and they may interact. From the interplay between diffraction and nonlinearity, two filaments with radius ρ and separated by the distance $\Delta > 2\rho$ can fuse whenever Δ is below a critical value, Δ_c. This critical distance can be evaluated by the balance between the free and interaction contributions in the Hamiltonian of Equation (1) (Bergé et al., 1997). With no saturation, each filament whose

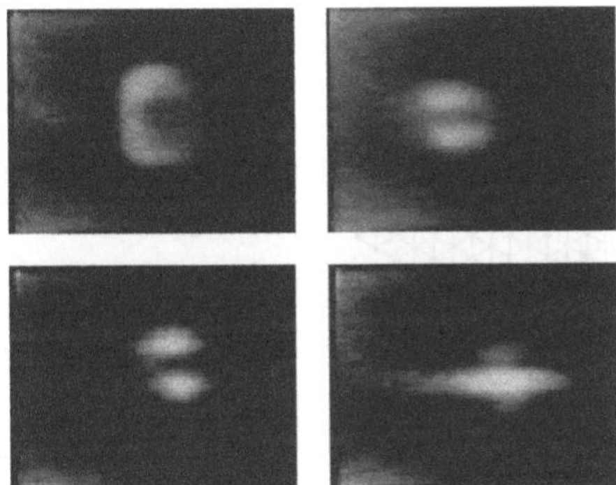

Figure 3. Filamentation and coalescence event in a two-spot pattern produced by a 50 fs pulse with 5 mJ energy propagating in air at the increasing distances (from left to right and from top to bottom): $z = 2.5$, 4.5, 6.5, and 8.5 m (Tzortzakis et al., 2001).

power is above critical creates its own attractor, at which it mostly freely collapses. This constraint is softened by including saturation, so that filaments with powers above critical are able to coalesce into an intense central lobe. As an example, the coalescence of two spots resulting from the propagation of modulationally unstable femtosecond pulses in air is shown in Figure 3 (Tzortzakis et al., 2001).

Modulational instability and multisoliton-like generation take place in various nonlinear media, such as biased photorefractive crystals or quadratically nonlinear optical materials favoring the coupling of fundamental and second harmonic fields (Fuerst et al., 1997). Filamentation moreover occurs in inertial fusion confinement (IFC) experiments as a harmful instability destroying the homogeneity in the beam energy distributed in the focal spot. For IFC, it has detrimental consequences to the hydrodynamics of the plasma created by a laser beam and contributes to the growth of parametric instabilities, which dissipate part of the laser energy. To limit its influence, optical smoothing techniques can be employed. For instance, random phase plates are used to generate a diffraction pattern composed of speckles, whose size λ_{sp} is smaller than the optimal wavelength λ_{mod} that maximizes the filamentation growth rate. This contributes to suppress laser filamentation (Labaune et al., 1992).

LUC BERGÉ

See also **Development of singularities; Kerr effect; Nonlinear optics**

Further Reading

Akhmediev, N.N., Korneev, V.I. & Nabiev, R.F. 1992. Modulational instability of the ground state of the nonlinear wave equation: optical machine gun. *Optics Letters*, 17: 393–395

Bergé, L., Schmidt, M.R., Rasmussen, J. Juul, Christiansen, P.L. & Rasmussen K.Ø. 1997. Amalgamation of interacting light beamlets in Kerr-type media. *Journal of the Optical Society of America* B, 14: 2550–2562

Bespalov, V.I. & Talanov, V.I. 1966. Filamentary structure of light beams in nonlinear media. *Zhurnal Eksperimental'noi i Teoreticheskoi Fiziki, Pis'ma v Redaktsiyu* (USSR JETP), 3: 471–476 [Translated in *JETP Letters*, 3: 307–310]

Campillo, S.L., Shapiro, S.L. & Suydam, B.R. 1973. Periodic breakup of optical beams due to self-focusing. *Applied Physics Letters*, 23: 628–630

Fuerst, R.A., Baboiu, D.-M., Lawrence B., Torruellas, W.E., Stegeman, G.I., Trillo, S. & Wabnitz, S. 1997. Spatial modulational instability and multisolitonlike generation in a quadratically nonlinear optical medium. *Physical Review Letters*, 78: 2756–2759

Kuznetsov, E.A., Rubenchik, A.M. & Zakharov, V.E. 1986. Soliton stability in plasmas and hydrodynamics. *Physics Reports*, 142: 103–165

Labaune, C., Baton, S., Jalinaud, T., Baldis, H.A. & Pesme, D. 1992. Filamentation in long scale length plasmas: experimental evidence and effects of laser spatial incoherence. *Physics of Fluids* B, 4: 2224–2231

Rypdal, K. & Rasmussen, J. Juul. 1989. Stability of solitary structures in the nonlinear Schrödinger equation. *Physica Scripta*, 40: 192–201

Soto-Crespo, J.M., Wright, E.M. & Akhmediev, N.N. 1992. Recurrence and azimuthal-symmetry breaking of a cylindrical Gaussian beam in a saturable self-focusing medium. *Physical Review* A, 45: 3168–3175

Tzortzakis, S., Bergé, L., Couairon, A., Franco, M., Prade, B. & Mysyrowicz, A. 2001. Break-up and fusion of self-guided femtosecond light pulses in air. *Physical Review Letters*, 86: 5470–5473

FINGERING

See **Hele-Shaw cell**

FINITE ELEMENT METHODS

See **Numerical methods**

FINITE-DIFFERENCE METHODS

See **Numerical methods**

FISHER'S EQUATION

See **Zeldovich–Frank-Kamenetsky equation**

FISKE STEPS

See **Josephson junctions**

FITNESS LANDSCAPE

The notion of landscape is chosen in analogy to terrestrial landscapes, which are functions over a two-dimensional space, $f(x, y)$. Commonly, landscape is altitude h as a function of latitude θ and longitude φ: $h(\theta, \varphi)$. In the history of physics, the landscape concept was first applied to motion in the gravitational field resulting in the concept of a potential energy surface. The gravitational potential V depends on altitude, latitude, the mass of the particle m, and the gravitational acceleration on the surface of the Earth as a function of latitude, $g(\theta)$: $V(\theta,\varphi)=m\,g(\theta)\,h(\theta,\varphi)$. The metaphor of a landscape embedded in the gravitational field suggests an exploration of optimal downhill paths following the negative gradient of the potential, $-\mathbf{grad}\ V(\boldsymbol{x}) = -\,(\partial V/\partial x_1, \partial V/\partial x_2, \ldots, \partial V/\partial x_n)$, with infinitesimal step width and vanishing kinetic energy corresponding to infinitesimally slow motion.

The landscape metaphor is used today in many different disciplines where the optimization of a nonsimple cost function is the primary goal. Such a cost function may depend on any number of variables, for example, spatial coordinates and strengths of external fields, and then the space upon which a landscape is built will be high-dimensional. Landscapes describing complex systems exhibit a large number of local minima and maxima. Models of disordered systems, spin glasses in particular, were the first examples studied in detail. Spin glasses are traditionally studied by statistical mechanics and still represent the best-studied cases of statistics on complex landscapes (Binder, 1986; Dotsenko et al., 1990).

The notion of fitness landscape was introduced into evolutionary biology by Sewall Wright in 1932 as a metaphor for the visualization of Darwinian evolution as an optimization process (see Wright, 1967). The Darwinian mechanism operates on populations over the course of many generations. It is based on genetic variation of individuals through mutation and recombination and selection of the variants with largest reproductive success. Reproductive success of a genotype I is measured in terms of its fitness value f, which counts the number of (fertile) descendants in the next generation. The concept of a fitness landscape or mapping ϕ assigns a fitness value f_k to every genotype I_k: $\phi(I_k)=f_k$. Population genetics describes the evolution of a population by means of the time dependence in the distribution of genotypes: At time t the genotype I_k is assumed to be present with frequency $x_k(t)$ in a population of N individuals distributed over n types or variants. The frequencies fulfill $\sum_{k=1}^{n} x_k = 1$, and for asexual reproduction, we have $\mathrm{d}x_k/\mathrm{d}t = x_k(f_k-\bar{f})$ with $\bar{f}=\sum_{k=1}^{n} x_k f_k$ being the mean fitness of the population. The time derivative of the mean fitness,

$$\frac{\mathrm{d}\bar{f}}{\mathrm{d}t} = \overline{f^2} - \left(\bar{f}\right)^2 = \mathrm{var}\{f\} \geq 0, \qquad (1)$$

is equal to the variance of the fitness. The mean fitness of a population is a nondecreasing function of time and, hence, subjected to optimization. The frequencies of all variants with fitness values larger than average, $f_k > \bar{f}$, increase whereas those of genotypes that are less productive than average, $f_k < \bar{f}$ decrease until the genotype becomes extinct. This process continues until the mean fitness $\bar{f}$ has reached its maximum because all variants except the fittest have

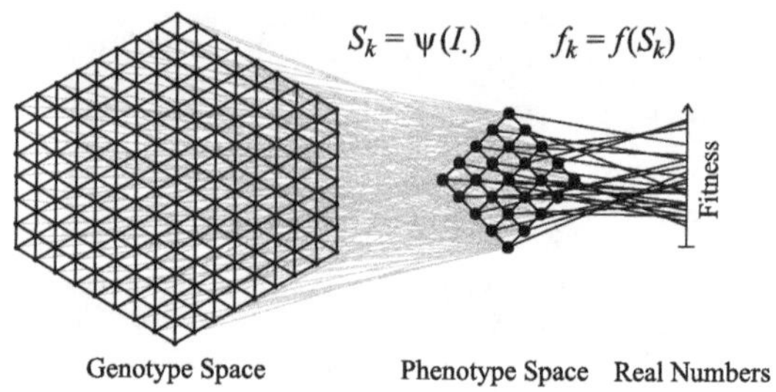

Figure 1. Fitness landscapes illustrated by two consecutive mappings, one from genotype space into phenotype space and the other from phenotype space into the positive real numbers, $\mathbb{R}_+$, representing the fitness values. In agreement with the available data on biopolymer landscapes, we are dealing with the phenomenon of neutrality: There are many more genotypes than phenotypes, and different phenotypes may have fitness values that cannot be distinguished by selection.

disappeared. The optimization principle also holds for sexual reproduction with recombination, as long as variants at a single gene locus are considered, and is called Fisher's fundamental theorem in this context. More general models of population dynamics describe variation explicitly. When mutation is included and several genetic loci are considered, the optimization principle is no longer valid. However, it works in the mild form as an optimization heuristic in the sense that optimization is observed in almost all cases.

The dichotomy of biological evolution—genetic variation in the form of mutation and recombination changes genotypes, whereas selection operates on phenotypes—suggests a splitting of the conventional fitness landscapes into two successive mappings: Genotypes are mapped onto a space of phenotypes, and the phenotypes are evaluated by a second map to yield fitness values (Figure 1):

Biological evolution:
genotype $\Longrightarrow$ phenotype $\Longrightarrow$ fitness,
RNA evolution:
sequences $\Longrightarrow$ structure $\Longrightarrow$ replication rate.

Evolution experiments with RNA molecules in the test tube (Biebricher & Gardiner, 1997) are sufficiently simple to allow for a description of the fitness landscape in molecular terms. The RNA sequence forms a molecular structure that determines the replication rate parameters that are the molecular counterparts of fitness values.

In biological landscape metaphors, the genotypes are materialized as polynucleotide sequences, DNA or RNA. These are strings built from four classes of symbols, {**A**, **T**(**U**), **G**, **C**}. In RNA **T** is replaced by **U**. All sequences of given chain length are subsumed in a discrete sequence space, $\mathcal{I}$, with the Hamming distance between two sequences I_i and I_j, $d_{ij}^{(h)}$ serving as metric. The fitness landscape can be expressed by

$$\phi\ :\ \{\mathcal{I}; d_{ij}^{(h)}\} \Longrightarrow \mathbb{R}_{(+)}, \qquad (2)$$

where the plus sign implies a restriction to strictly positive fitness values. The structures are points in another discrete metric space $\mathcal{S}$ with some distance measure between structures $d_{ij}^{(s)}$. The fitness landscape is properly split into two mappings

$$\psi : \{\mathcal{I}; d_{ij}^{(h)}\} \Longrightarrow \{\mathcal{S}; d_{ij}^{(s)}\}$$

and

$$f : \{\mathcal{S}; d_{ij}^{(s)}\} \Longrightarrow \mathbb{R}_{(+)}. \tag{3}$$

Sequence-structure maps of RNA have been studied by means of computer models (Schuster, 2001, 2003) on the simplified level of RNA secondary structure. As sketched in Figure 1, the number of sequences is much larger than the number of secondary structures. Hence, many sequences may give rise to the same structure. In population biology this phenomenon is known as neutrality in sequence space, and it was discussed already in the 1980s by Kimura (1983) when the first biopolymer sequences became accessible. Neutrality was also found to be highly relevant for the efficiency of evolutionary processes because it allows for escape from local optima that would otherwise trap populations for long times (Fontana & Schuster, 1998).

The nature of landscapes can be analyzed by means of an expansion of the landscape in eigenfunctions of the Laplacian on the underlying discrete space (Reidys & Stadler, 2002). The expansion coefficients of this expansion allow for estimates of the hardness of optimization. Whenever we have only a single dominating expansion coefficient, the optimization problem is much simpler than in the case of equally important blending of two or more eigenfunctions. Biopolymer landscapes of free energies of conformations or other relevant properties turned out to be rather complex as several eigenfunctions of the Laplacian were found to be important.

PETER SCHUSTER

See also **Biological evolution; Spin systems**

Further Reading

Biebricher, C.K. & Gardiner, W.C. 1997. Molecular evolution of RNA in vitro. *Biophysical Chemistry*, 66: 179–192

Binder, K. 1986. Spin glasses: experimental facts, theoretical concepts and open questions. *Reviews of Modern Physics*, 58: 801–976

Dotesenko, V.S., Feigel'man, M.V. & Ioffe, L.B. 1990. Spin glasses and related problems. *Soviet Science Reviews A. Physics*, 15: 1–250

Fontana, W. & Schuster, P. 1998. Continuity in evolution. On the nature of transitions. *Science*, 280: 1451–1455

Kimura, M. 1983. *The Neutral Theory of Molecular Evolution*, Cambridge and New York: Cambridge University Press

Reidys, C.M. & Stadler, P.F. 2002. Combinatorial landscapes. *SIAM Review*, 44: 3–54

Schuster, P. 2001. Evolution *in silico* and in vitro: the RNA model. *Biological Chemistry*, 382: 1301–1314

Schuster, P. 2003. Molecular insights into the evolution of phenotypes. In *Evolutionary Dynamics—Exploring the Interface of Accident, Selection, Neutrality, and Function*, edited by J.P. Crutchfield & P. Schuster, Oxford and New York: Oxford University Press, pp.163–215

Wright, S. 1967. "Surfaces" of selective value. *Proceedings of the National Academy of Sciences USA*, 58: 165–172

FITZHUGH–NAGUMO EQUATION

Around 1960, Jin-ichi Nagumo (from the University of Tokyo) visited with Richard FitzHugh (at the U.S. National Institutes of Health), who was adapting the relaxation-oscillator models of Karl Bonhoeffer and Balthasar van der Pol to provide a more simple formulation of nerve membrane switching than the four-variable Hodgkin–Huxley system. From this visit emerged the FitzHugh–Nagumo (FN) system as a two-variable oscillator model (FitzHugh, 1961), which led to a simplified formulation of neural action potentials (Nagumo et al., 1962).

Neurons display relaxation oscillations par excellence. When the voltage across the cell membrane (in the trigger zone of the soma) exceeds a threshold, they fire an action potential and then gradually decay to the resting state. If the exciting input is constantly above the threshold, the neuron fires repeatedly. The frequency of such firing depends on the intensity of the input (total input coming from neighboring neurons). FitzHugh sought to reduce the Hodgkin–Huxley model of this dynamics to a two-variable model for which phase-plane analysis applies. His general observations were that the gating variables of the Hodgkin–Huxley model, n (potassium activation) and h (sodium inactivation), have slow kinetics relative to m (or sodium activation), and that for typical nerves $n+h \approx 0.8$. Furthermore, FitzHugh noticed that the voltage nullcline had the shape of a cubic function and the n-nullcline could be approximated by a straight line, both within the physiological range of the variables. This led to the following two-variable model that provides a phase space qualitative explanation of the formation and decay of the action potential.

$$\begin{aligned} \dot{v} &= +v(v-\theta)(1-v) - y + I, \\ \dot{y} &= \varepsilon(v - \gamma y), \end{aligned} \tag{1}$$

where the dots indicate time derivatives.

Like all relaxation oscillators, this oscillator has a slow accrual phase and a fast release phase. Here v is the scaled voltage or membrane potential (namely, the output of the neuron), and y the single recovery variable accounting for the slow dynamics (sodium inactivation and potassium activation variables). The difference in time scales between sodium activation (m) and sodium inactivation and potassium activation (n and h) is represented by $\varepsilon \ll 1$. As ε increases, so does the frequency of oscillation. The constant γ

is a shunting parameter, $0 < \theta < 1$ is a thresholding parameter, and I is an externally applied current.

Such oscillators exhibit the characteristic dynamics of a real neuron. When fed by a constant amount of low current, the oscillator gradually increases its voltage (or membrane charge) until it reaches the threshold. Upon reaching threshold, the neuron fires and quickly releases the accumulated charge. If the current source is sufficient and is constantly applied, this results in stable limit-cycle oscillations.

The FN oscillator is applied in neuroscience to study the synchronization properties of neurons, which is particularly easy for relaxation oscillators such as the FN model. Following the Hebbian rule, neurons that fire closely in time strengthen their synaptic connections, and therefore synchronization in neural activity is important to the way the states of a brain evolve in time. Furthermore, information in neural tissues seems to be coded in the spatiotemporal firing of sets of neurons that fire synchronously, and tissues that are anatomically separated may operate together when they are synchronized.

An ensemble of coupled excitable oscillators of the FN type makes up an excitable medium, through which electrical signals can propagate. This is the case for a nerve and for other excitable media such as muscles. To model the spatiotemporal evolution of membrane potential and its generated current flows in such structure, one treats the axon or dendrite branch of the nerve as a membrane cylinder, with x denoting the distance along the cylinder. If ρ is the intracellular resistivity (in ohm-cm) and d the diameter, then the axial current is

$$I_i = -(\pi d^2/4\rho)\frac{\partial V}{\partial x}, \qquad (2)$$

which follows directly from Ohm's law. Conservation of current then requires

$$C_m\frac{\partial V}{\partial t} = \frac{d}{4\rho}\frac{\partial^2 V}{\partial x^2} - I_{\text{ion}} + I(x,t), \qquad (3)$$

where I_{ion} is the ionic current flowing through a unit area of the membrane surface and $I(x,t)$ and C are an external current and membrane capacitance per unit area.

Applying Equation (3) to the FN case, suggests that the electrical propagation in nerves can be represented as a reaction-diffusion system of the form

$$\frac{\partial^2 V}{\partial x^2} - \frac{\partial V}{\partial t} = +F(V) - Y + I(x,t),$$
$$\frac{\partial Y}{\partial t} = \varepsilon(V - \gamma Y). \qquad (4)$$

Here $F(V)$ is a function with cubic shape such as $F(V)=V(V-\alpha)(1-V)$ or some similar function. [Such diffusion models can also be extended to more dimensions with $\partial^2 V/\partial x^2 + \partial^2 V/\partial y^2$ and $V(x,y;t)$.]

Under a traveling-wave analysis, let $I(x,t) = 0$ and $z = x + ct$ and define $\mathrm{d}V/\mathrm{d}z \equiv W$. Then Equations (4) become a system of ordinary differential equations

$$\frac{\mathrm{d}V}{\mathrm{d}z} = W,$$
$$\frac{\mathrm{d}W}{\mathrm{d}z} = F(V) - Y + cW,$$
$$\frac{\mathrm{d}Y}{\mathrm{d}z} = \frac{\varepsilon}{c}(V - \gamma Y), \qquad (5)$$

where one seeks traveling-waves solutions as trajectories $(V(z), W(z), Y(z))$ that approach the origin $(0,0,0)$ as $z \to \pm\infty$ (Scott, 2003).

With $\varepsilon = 0$, Y is constant. In this case, the only stable traveling wave solution is a level change from one of the outer zeros of $F(V) - Y$ to the other; in other words, a moving impulse-like solution does not exist. For $0 < \varepsilon \ll 1$, however, there is an impulse-like traveling-wave solution (or solitary wave). At arbitrarily small values of ε, this solitary wave continues to exist but with the front and back edges of the wave becoming far apart. These front and back edges interpolate between the slowly varying regions and are called boundary layers (as in hydrodynamics). As $\varepsilon \to 0$ on a scale where the impulse length is unity, the boundary layers reduce to step functions.

One of the main research areas related to these reaction-diffusion models in electrophysiology focuses on pattern formation and cardiac rhythm disturbances. FN equations have been used to investigate a variety of unusual front-bifurcation and pattern-formation processes. The precise conditions for wave front formation and subsequent wave propagation in an excitable medium are critical for understanding the genesis and possible control of re-entrant (spiral-wave) cardiac arrhythmias.

L. Vazquez and M.-P. Zorzano

See also **Boundary layers; Hodgkin–Huxley equations; Neurons; Reaction-diffusion systems; Relaxation oscillators**

Further Reading

FitzHugh, R. 1961. Impulses and physiological states in theoretical models of nerve membranes. *Biophysiscal Journal*, 1: 445–466

Nagumo, J., Arimoto, S. & Yoshizawa, S. 1962. An active pulse transmission line simulating nerve axons. *Proceedings of the Institute of Radio Engineering and Electronics*, 50: 2061–2070

Scott, A. 2002. *Neuroscience: A Mathematical Primer*, Berlin and New York: Springer

Scott, A. 2003. *Nonlinear Science: Emergence and Dynamics of Coherent Structures*, 2nd edition, Oxford and New York: Oxford University Press

FIXED POINTS

See **Equilibrium**

FLAME FRONT

Combustion waves can propagate over a wide range of burning velocities that differ by more than three orders of magnitude for the same mixture. At one end of the velocity spectrum, we have a laminar flame (deflagration) that propagates at a typical velocity of about half a meter per second for common fuel-air mixtures at normal conditions. At the other end, the combustion wave propagates as a detonation whose speed is of the order of a couple of thousand meters per second in the same mixture. In between these limits, we have an almost continuous range of turbulent burning velocities. Both laminar flames and detonation wave are intrinsically unstable and have the morphology of a transient cellular structure (Figure 1).

A laminar flame is essentially an isobaric diffusion-reaction wave. Its propagation speed is determined by the rate of diffusion transport of heat from the reaction zone to the cold unburned mixture and the characteristic time of heat release of the chemical reactions. A laminar flame speed (S_L) is proportional to the square root of the product of thermal diffusivity (D_{th}) and the reaction rate (w_r) (i.e., $S_L \sim \sqrt{D_{th} w_r}$). The reaction rate is given by $w_r \sim \exp(-E/RT_f)$ in which E is the activation energy and T_f is the flame temperature. Because the activation energy is very large in general, the reaction rate is extremely temperature sensitive. Thus, any fluctuation in the flame temperature will result in a large variation in the reaction rate leading to the development of instability of the flame front. Due to the large density and temperature changes across the flame, a strong thermal expansion of the burned gas results from the conservation of mass. The flame as a strong density interface as well as an expansion wave is subject to a number of dynamic instability mechanisms in the presence of an acceleration field. Furthermore, competition between heat and mass diffusion across the flame results in thermal diffusion instability. Rapid density changes across the flame also give rise to acoustic wave generation that can couple with increase in burning rate to induce acoustic driven instability. Thus, in practice, there is a wealth of instability mechanisms that render laminar flames unstable. Various instability mechanisms can be at work simultaneously and can influence each other. However, historically, each instability mechanism was isolated and studied individually and was, thus, named after the original researchers.

The flame as a density interface is unstable when subjected to an acceleration field (for example, gravity). If the flame propagates upward, the light burned gas (ρ_b) is at the bottom and the heavy unburned gas (ρ_u) is on the top. The lighter fluid will be driven upward (i.e., buoyancy) whereas the heavier fluid is driven downward by gravity (g). This motion will destabilize the interface, and hence, small perturbations on the flame surface will grow with time (t). If the perturbed flame surface (F) is defined as $A\exp(\sigma t + ikx)$, where k is the wave number and x the space coordinate, the disturbance growth rate σ can be determined from normal mode stability analysis as

$$\sigma = \sqrt{gk\gamma/(2-\gamma)}, \quad \gamma = 1 - \rho_b/\rho_u. \qquad (1)$$

Therefore, instability occurs at all wavelengths but growth is faster at shorter wavelengths. This phenomenon was first discovered by Lord Rayleigh (1883) and later by Geoffrey Ingram Taylor (1950), and thus, it is referred to as the Rayleigh–Taylor instability. This instability is common to all density gradient fields in the presence of an acceleration field normal to it.

Due to the density change across the flame, the flow velocity increases as the density ratio because the flame is approximately isobaric. The expansion across

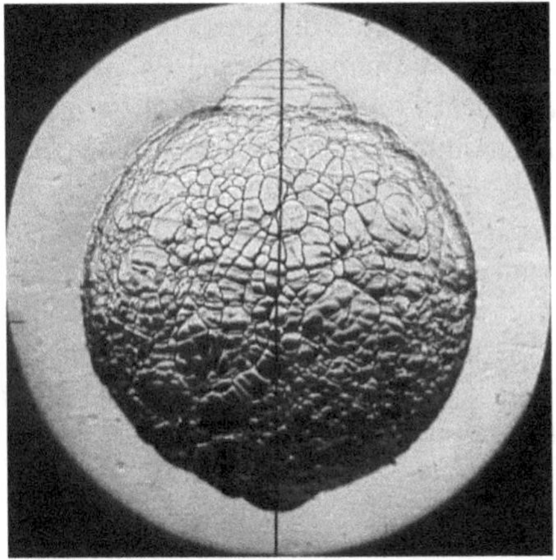

a

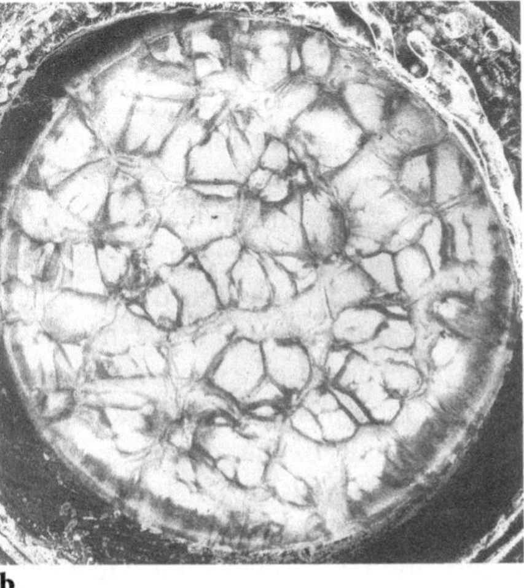

b

Figure 1. Cellular structures of laminar flame and detonation front. (a) Laminar outward propagating spherical flame (rich dimethyl ether–air at equivalence ratio of 1.2 and 10 atm). (b) Detonation: unstable cellular detonation front as recorder upon reflection from a soot-coated glass plate (C_2H_2–O_2 mixture at 10 mmHg).

the front will induce a divergent flow field ahead of a curved flame (Williams, 1985). This divergence slows down the local flow speed ahead of the curved flame front, and assuming the flame speed to be constant, this will result in a growth of the curvature of the flame. This instability was first discovered by G. Darrieus (1938) and Lev Landau (1944). The Darrieus–Landau instability was obtained by treating the flame as a surface of discontinuity moving at a constant speed. In the limit of small density change, stability analysis gives the growth rate as

$$\sigma = \gamma k S_{\mathrm{L}}/2. \tag{2}$$

Therefore, it was concluded that a flame front is unstable to perturbations at all wavelengths, with growth rate proportional to wave number (i.e., perturbations grow faster for small wavelengths). Unfortunately, this conclusion was contrary to later laboratory observations of small-scale stable flames.

The deficiency of this model was the result of neglecting the finite thicknesses of the flame front, which influences the flame speed when the flame is curved. To include the effect of flame thickness, George Markstein (1951) proposed a phenomenological model by adding a modification of flame speed due to curvature and showed that curvature decreases the flame speed and tends to stabilize the flame at short wavelengths inhibiting its growth. A more rigorous derivation of the dispersion equation including diffusion effect on the flame speed (in the limit of small density jump) was given later by Gregory Sivashinsky (1983) as

$$\sigma = \frac{\gamma k S_{\mathrm{L}}}{2} - \left[\frac{\beta}{2}(\mathrm{Le}-1)+1\right] \times D_{\mathrm{th}}k^2 - 4\frac{D_{\mathrm{th}}^3}{S_{\mathrm{L}}^2}k^4 + \sqrt{gk\gamma/(2-\gamma)}, \tag{3}$$

where $D_{\mathrm{th}}k^2$ in the second term of Equation (3) represents the thermal relaxation via transverse thermal diffusion that stabilizes the flame. Le is the Lewis number (i.e., the ratio of thermal diffusivity to mass diffusivity) and (Le − 1) in the second term denotes the competition between heat loss via thermal diffusion and the enthalpy gain via mass diffusion. As a result, flame temperature will increase or decrease if Le is less or larger than unity. Equation (3) shows that if Le − 1 is less than $-2/\beta$ (β is the reduced activation energy), diffusion transport will destabilize the flame. This is the mechanism of the cellular instability. The third term in Equation (3) is the thermal relaxation to the modification of flame temperature caused by the heat and mass diffusion (Clavin, 1985). Therefore, for long wavelength disturbance, the hydrodynamic instability dominates (Figure 1a). At short wavelengths, diffusion relaxation stabilizes the flame. At moderate wavelengths, the competition between heat and mass transfer induces cellular instability at small Le. At large Le, flame temperature is very sensitive to the mass diffusion of the deficient reactant. Coupling between the diffusion and temperature sensitive reaction yields pulsating and spinning waves. This traveling wave is often seen in lean propane-air flames.

By further considering the effect of flame curvature on flow field, Equation (3) can be normalized as an evolution equation of flame front

$$F_t + \tfrac{1}{2}(\nabla F)^2 + \nabla^2 F + 4\nabla^4 F = 0 \tag{4}$$

This is the so-called Kuramoto–Sivashinsky equation which was independently developed by Yoshiki Kuramoto (Kuramoto & Tsuzuki, 1976) for the study of phase turbulence in the Belousov–Zhabotinsky reaction and by Gregory Sivashinsky (1983) for thermal diffusive instabilities of flame fronts. This equation has also been used to model directional solidification and weak fluid turbulence. However, the assumption of small density jump used in Equations (3) and (4) is not rigorous in practical flames. A unified model considering large density jump and Le was obtained by Class Andreas et al. (2003).

Heat release by combustion results in an increase in the specific volume of the product gases and thus generates acoustic waves (Chu, 1956). The acoustic waves play two roles in affecting combustion: (1) inducing pressure-heat release coupling via pressure-dependent reactions, and (2) increasing flame surface area via the baroclinic torque (Meshkov instability). If the changes of pressure and chemical heat release are in phase, the acoustic instability occurs (Rayleigh, 1877; Markstein, 1953; Clavin, 2002). Lord Rayleigh first used this criterion (Rayleigh criterion) to explain the singing flame and Rijke's tone (where heating the bottom of a tube causes it to produce sound). The acoustic instability causes the major problems of noise and vibration in combustors (Putnam & Dennis, 1953). On the other hand, volumetric heat loss reduces the flame speed and changes the resident time of the emitting gases. This coupling triggers the radiation induced instability for weak flames (Ju et al., 2000).

At the upper limit of propagation of combustion waves, the propagation mechanism is not due to diffusion. The flame instability mechanisms discussed above are too slow to be relevant in detonation wave instability. A detonation wave is a supersonic compression wave where mixture is ignited by the adiabatic compression of the leading shock. The classical structure of a detonation wave was formulated by Yakov Zeldovich, John von Neumann, and W. Döring (ZND) independently in the early 1940s and consists of a leading shock followed by the reaction zone after a short induction length (Zeldovich, 1940; von Neumann, 1942; Döring, 1943). Gas dynamic theory gives the detonation wave speed as proportional to the square root of the chemical heat release and does not involve any non-equilibrium rate processes. Again due to the high-temperature sensitivity of the reaction rates, small temperature fluctuations due to variation of the leading shock speed will result in large variations in the induction length and reaction rates, hence the coupling between the energy release zone and the leading shock. The instability yields a transient three-dimensional cellular detonation front (Lee, 1984). The unstable cellular detonation front consists of an ensemble of interacting transverse shock waves with the leading shock front. The cell boundaries (Figure 1b) are formed by the intersections of the transverse shocks. Shock interactions (Mach reflections) also give rise to the formation of shear layers which lead to turbulence generation due to Kelvin–Helmholtz instability. Chemical reactions in cellular detonations occur in disjointed piecemeal zones embedded within the complex of interacting shocks and shear layers.

The instability of the laminar ZND detonation structure was demonstrated theoretically by standard normal mode stability analysis using the one-dimensional Euler equation (e.g., Erpenbeck, 1964; Lee & Stewart, 1990). In one dimension, unstable detonations are referred to as pulsating detonations that go from harmonic oscillations near the stability limit to highly nonlinear and eventually to chaotic oscillations with the increase of the activation energy. By examining the bifurcation diagram, it is interesting to find that the path to higher instability mode follows closely the Feigenbaum route (Feigenbaum, 1983) of a period-doubling cascade observed in many nonlinear systems. One-dimensional pulsating detonation as well as two- and three-dimensional cellular detonations have been reproduced qualitatively via numerical simulation using the reactive Euler equations (Bourlioux et al., 1991; Short & Stewart, 1999). However, the detailed description of the turbulent structure and chemical reactions requires resolutions far beyond current computing capabilities.

In between the two limits of laminar flames and detonations, there is a continuous range of flame speeds that depend on turbulence. The morphology of a turbulent flame is a time-dependent cellular or wrinkled surface. Turbulent flame is, in fact, an unstable flame, and the effect of turbulence is to increase the burning rate via faster transport and increase in burning surface area. In the limit of very intense turbulence where mixing and reaction rates are comparable, auto-ignition may result, and thus, the mechanism becomes similar to that of a detonation. It differs only in the manner in which auto-ignition is achieved by turbulent mixing of fresh mixture with hot products or by adiabatic heating of the leading shock. Thus, nature tends to maximize the burning rate of a mixture, and instability is a route to optimize the burning rate for given initial and boundary conditions.

YIGUANG JU AND JOHN LEE

See also **Candle; Explosions; Forest fires; Kuramoto–Sivashinsky equation; Reaction-diffusion systems; Zeldovich–Frank-Kamenetsky equation**

Further Reading

Bourlioux, A., Majda, A.J. & Roytburd, V. 1991. Theoretical and numerical structure for unstable one-dimensional detonations. *SIAM Journal of Applied Maths*, 51: 303–343

Class Andreas G., Matkowsky, B.J. & Klimenko, A.Y. 2003. Stability of planar flames as gas dynamic discontinuities. *Journal of Fluid Mechanics*, 491: 51–63

Clavin, P. 1985. Dynamic behaviour of premixed flame fronts in laminar and turbulent flows. *Progress in Energy and Combustion Science*, 11: 1–39

Clavin, P. 2002. Dynamics of combustion fronts in premixed gases: from flames to detonation. *Proceedings of the Combustion Institute*, 29: 569

Chu, B.T. 1956. Stability of systems containing a heat source—the Rayleigh criterion, National Advisory Committee for Aeronautics research memorandum, RM56D27

Darrieus, G. 1938. Propagation d'un front de flame, unpublished work presented at La Technique Moderne and le Congrès de Méchanique Appliquée, Paris

Döring, W. 1943. On detonation processes in gases. *Annals of Physics, Leipzig*, 43: 421–436

Erpenbeck, J. 1964. Stability of idealized one-reaction detonations. *Physics of Fluids*, 7: 684–696

Feigenbaum, M. 1983. Universal behaviour in nonlinear systems. *Physica* D, 7: 16–39

Ju, Y., Law, Chung K., Maruta, K. & Niioka, T. 2000. Radiation induced instability of stretched premixed flames. *Proceedings of the Combustion Institute*, 28: 1891–1900

Kuramoto, Y. & Tsuzuki, T. 1976. Persistent propagation of concentration waves in dissipative media far from thermal equilibrium. *Progress of Theoretical Physics*, 55: 356–369

Landau, L.D. 1944. On the theory of slow combustion. *Acta Physiocochimica URSS*, 19: 77–85

Lee, J.H.S. 1984. Dynamic parameters of gaseous detonation. *Annual Reviews of Fluid Mechanics*, 16: 311–316

Lee, H. & Stewart, D. 1990. Calculation of linear detonation instability: one-dimensional instability of plane detonation. *Journal of Fluid Mechanics*, 216: 103–132

Markstein, G.H. 1951. Experimental and theoretical studies of flame front instability. *Journal of the Aeronautical Sciences*, 18: 199–209

Markstein, G.H. 1953. Instability phenomena in combustion waves. *Proceedings of the Combustion Institute*, 14: 44–59

Putnam, A.A. & Dennis, W.R. 1953. A study of burner oscillations of the organ-pipe type. *Transaction of the ASME*, 75: 15–28

Rayleigh, L. (John William Strutt). 1877. *The Theory of Sound*, vol.2, London: Macmillan; reprinted New York: Dover, 1945, p. 226

Rayleigh, L. (John William Strutt). 1883. Investigation of the character of the equilibrium of an incompressible heavy fluid of variable density. *Proceedings of the London Mathematical Society*, 14: 170–177

Short, M. & Stewart, D. 1999. Multi-dimensional stability of weak-heat-release detonation. *Journal of Fluid Mechanics*, 382: 103–135

Sivashinsky, G.I. 1983. Instabilities, pattern formation, and turbulence in flames. *Annual Reviews of Fluid Mechanics*, 15: 179–199

Taylor, G.I. 1950. The instability of liquid surfaces when accelerated in a direction perpendicular to their planes.

Proceedings of the Royal Society of London A, 201: 192–196
von Neumann, J. 1942. Theory of detonation waves, Proj. Report No. 238, OSRD report No.549. In *Von Neumann, Collected Works*, vol. 6, edited by A.J. Taub, Oxford: Pergamon Press, 1963
Williams, F.A. 1985. *Combustion Theory*, 2nd edition, New York: Benjamin Cumming
Zeldovich, Y.B. 1940. On the theory of the propagation of detonations in gaseous systems. *Experimental and Theoretical Physics SSSR*, 10: 542

FLIP-FLOP CIRCUIT

A bistable circuit is one that can exist indefinitely in either one of two stable states and that can be induced to make an abrupt transition from one stable state to the other by means of an external excitation. Bistable circuits are known by a variety of names, such as bistable multivibrator, Eccles–Jordan circuit (after the inventors), trigger circuit, binary, and flip-flop, the latter being the term that we adopt. Flip-flops are used for the performance of many digital operations such as counting and storing of binary information.

In general, digital (switching) circuits may be either combinational or sequential. In combinational circuits the Boolean relation that describes the output function is at any moment uniquely determined by the inputs. The output function is independent of all prior input or output conditions and the circuit is without memory. In contrast, the sequential circuit contains feedback and its outputs depend not only on the present inputs but, in general, also on the entire past history of inputs. The flip-flop is a basic sequential circuit that functions as a basic logic memory element. It has two distinct states of equilibrium and may, therefore, be used as a single binary-digit (bit) storage device. The two stable states are referred to by various names such as TRUE and FALSE, HIGH and LOW, or 1 and 0.

Sequential circuits can be presented using a graphical tool known as a state diagram. In a state diagram, nodes are used to represent the different states of a circuit, and connections between nodes are used to show transitions between states with different inputs that act as conditioning signals. Figure 1 represents a general bistable circuit. Let us denote the present state of the circuit as $Q(t)$ and the state that follows after a time interval Δt as $Q(t + \Delta t)$. Then the circuit stays in state 1 as long as condition 1 is applied, but moves to state 0 when condition 2 is applied. Similarly, it stays in state 0 as long as condition 3 is applied, and moves to state 1 when condition 4 occurs.

Four types of flip-flops are presently in use, RS (conditional **S**et and **R**eset), JK (unconditional set and reset), D (**D**elay), and T (**T**rigger), and they have different sets of conditions for transition between the two states. For example, the characteristic equation for

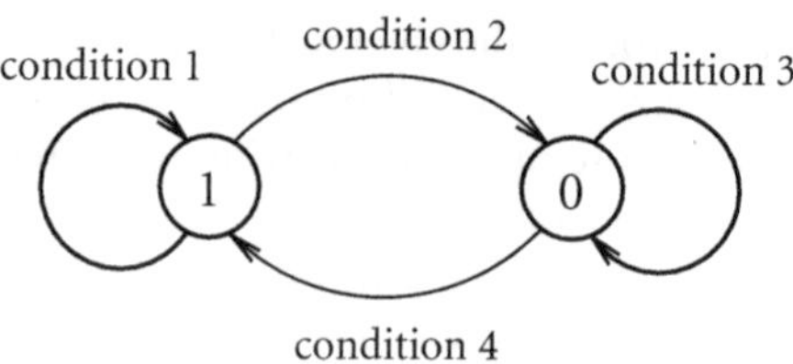

Figure 1. State diagram of a general bistable circuit.

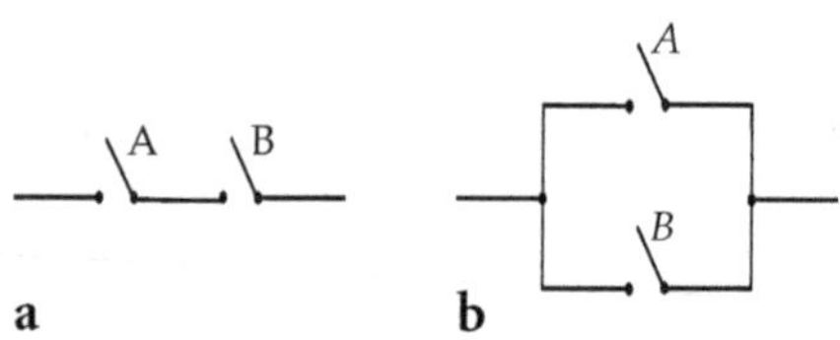

Figure 2. Basic switching elements, (a) AND and (b) OR. An ideal switch consists of a pair of contacts that have zero internal (or switch closed) resistance and infinite leakage (or switch open) resistance. Transitions from one state to the other should be instantaneous. It should attain the open and closed states with equal probability. Although these characteristics have been approached most closely with every new technological generation, we deal in reality with approximations to them.

the RS flip-flop is given as

$$Q(t + \Delta t) = S + Q(t) \cdot \overline{R} \qquad (1)$$

under the constraint

$$S \cdot R = 0. \qquad (2)$$

Here $+$ (OR), $\cdot$ (AND), and $-$ (NOT, also known as inversion and complement) are operations from switching algebra. Switching circuits which represent the AND, $A \cdot B$, and the OR, $A + B$, functions are presented in Figures 2a and b, respectively.

The mathematical theory of switching circuits, first postulated by Shannon (1938, 1949), is based on Boolean algebra. It is defined on a set U that consists of two values, 0 and 1, two basic binary operations, $\cdot$ (which is also called AND, product or conjunction), and $+$ (which is also called OR, sum or disjunction), and a set of basic postulates that were derived and proven by Edward Huntington in 1904, based on the work of George Boole in 1847.

Switching circuits have been designed from various technologies; for example, vacuum tubes were used in their early development. They have become dramatically faster and dramatically smaller in size with every new technology. The advent of integrated circuits (ICs), in which many discrete components (diodes, transistors, and resistors) are fabricated at the same time on one chip of silicon, has led to many different types of switching circuits in IC form. Recent research in nanoelectronics has introduced the concepts of resonant tunneling diodes, electronic quantum cellular automata, single electron transistors, and molecular electronic devices.

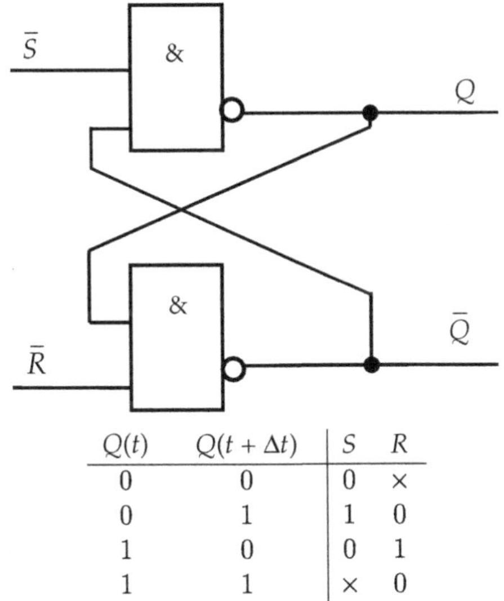

$Q(t)$	$Q(t+\Delta t)$	S	R
0	0	0	×
0	1	1	0
1	0	0	1
1	1	×	0

Figure 3. The RS flip-flop with NAND gates (top) and its truth-table (bottom). Its characteristic equation is $Q(t+\Delta t)=\overline{\overline{S}\cdot\overline{Q}(t)}=\overline{\overline{S}\cdot\overline{\overline{Q(t)}\cdot\overline{R}}}=S+Q(t)\cdot\overline{R}$. Here we used de Morgan's laws: for any two subsets A and B of the U, we have $\overline{A\cdot B}=\overline{A}+\overline{B}$ and $\overline{A+B}=\overline{A}\cdot\overline{B}$.

The basic switching circuits are called gates. Often NAND and NOR gates are used, as they are easier to implement. The RS flip-flop can be designed simply by interconnecting a pair of two-input NAND gates, with appropriate feedback, as shown in Figure 3. The truth-table in Figure 3 gives input values, R and S, that condition changes from a present state $Q(t)$ to a next state $Q(t+\Delta t)$. It is constructed by considering the physical action of the circuit shown in the same figure. For example, to stay in state 1, the equivalent to input condition 1 in Figure 1 is $R=0$, and S need not be defined; its value is ×. The entries marked with × correspond to "not allowed" or "don't care" inputs because, under these conditions, when both R and S are present simultaneously, the operation of the circuit becomes uncertain.

The Δt is a delay that occurs between the present state and the next state. This delay is essential in the operation of sequential circuits. With respect to the delay, two types of sequential circuits can be distinguished: synchronous and asynchronous. The synchronous sequential circuits trigger on receiving a certain clock pulse. The maximum frequency of the clock is defined by the operational time of the slowest element in the circuit. State transitions in a synchronous sequential circuit thus occur with a constant frequency. The delay in asynchronous sequential circuits changes from transition to transition, i.e. $\Delta t \neq$ const.

The RS flip-flop, introduced above, is a bistable multivibrator. It stays in one of two states that can be changed only when an external input is applied. If there are no external inputs, but one state causes a transition to the other state and this repeats continuously, the multivibrator is said to be astable. It is not stable in either state and alternates between them at a specific frequency. The astable multivibrator is, in fact, a free-running oscillator. The frequency is usually determined by a capacitor placed between the two gates and a resistor in parallel with each gate. Thus the delay, Δt, and consequently the frequency, are determined by the values of the capacitor and resistors.

In general, every oscillator contains a positive and negative feedback loop at the same time. For example, a unit with one excitatory and one inhibitory neuron that are mutually connected is a neuronal oscillator. The model proposed by Wilson and Cowan (*See* **Inhibition**), who introduced an inhibitory neuron and by this a positive along with the negative feedback loop, is today known as the Wilson–Cowan oscillator. As the exchange of energy and matter is continuous in every biological process that is based on existence of both positive and negative feedback loops, an oscillator can be seen as a basic biological unit.

Flip-flop circuits are today used as a general paradigm for bistability in a very wide range of nonlinear systems, one of many examples being the phenomenon of stochastic resonance (Fauve & Heslot, 1983).

ANETA STEFANOVSKA

See also **Coupled oscillators; Feedback; Inhibition; State diagrams**

Further Reading

Fauve, S. & Heslot, F. 1983. Stochastic resonance in a bistable system. *Physics Letters* A, 97: 5–7

Shannon, C.E. 1938. A symbolic analysis of relay and switching circuits. *Transactions of the AIEE*, 57: 713–723

Shannon, C.E. 1949. The synthesys of two-terminal switching circuits. *Bell Systems Technical Journal*, 28: 59–98

FLOQUET THEORY

See **Periodic spectral theory**

FLUCTUATION-DISSIPATION THEOREM

In the 19th century, there were two schools of thought on the existence of atoms, those that believed in them and those that did not. Remarkably, it was not until the early 20th century that the experiments of Jean Perrin—summarized in his 1913 book *Atoms* (Perrin, 1913)—established once and for all the existence of atoms. One of the laws Perrin experimentally verified was the fluctuation-dissipation relation, which interrelates the physical notions of randomness (through fluctuations) and determinism (through dissipation) (Montroll & West, 1979).

No matter how carefully experiments are done, they never yield the same value of a physical observable from one measurement to the next. The collection of values from such measurements is called an ensemble, and the number of times a measurement falls in an assigned interval divided by the total number of measurements in the ensemble yields a probability, which when all the intervals are taken together yields a probability distribution function. The best representation of this ensemble of measurements is the mode of the distribution (or ensemble average) as first noted by Carl Friedrich Gauss (1809). The deviations of a physical observable from its average value are called fluctuations, which are typically small and random. In physical systems, the source of these fluctuations is thermal agitation of the atoms. Thus, the thermodynamic properties of an equilibrium physical system are determined by the probability density, through the averages, and not by the instantaneous values of the positions and velocities of the atoms.

The changes in macroscopic physical systems over time comprise a combination of deterministic dynamics and microscopically induced macroscopic fluctuations. Albert Einstein, in 1905, was the first to fully appreciate the influence of these fluctuations on macroscopic transport phenomena in his investigations of the phenomenon of equilibrium diffusion (Furth, 1956). Einstein showed that the strength of the fluctuations, as measured by their mean-square level through the diffusion coefficient, D, is directly proportional to the temperature of the ambient fluid, $kT/2$, which is the average kinetic energy per degree of freedom of the ambient fluid particles. The constant of proportionality between the temperature and the diffusion coefficient is the dissipation time per unit mass, λ/m, thereby inter-relating the fluctuations and dissipation of the medium through the particle's motion

$$D = \frac{2\lambda kT}{m}. \tag{1}$$

Equation (1) was the first fluctuation-dissipation relation and demonstrates that macroscopic fluctuations and dissipation have the same microscopic origin.

Two decades before Einstein's analysis of diffusion, Walther Nernst (1884) investigated the combined process of mobility and diffusion to determine the size of the charge on an individual particle. He established that the mobility is proportional to the ratio of the diffusion coefficient to the temperature, and the charge of the particle is the proportionality constant. This ratio is usually called the Einstein equation even though Nernst discovered it. One can more generally define the mobility of a particle as the terminal velocity per unit force, through a generalized Nernst relation

$$D = kT \times \text{mobility} \tag{2}$$

for any sort of particle that is free to move but is subject to frictional drag. The linear transport coefficient (here the mobility) was only partially understood until Harry Nyquist (in considering the fluctuations in the current flowing through an electrical resistance) showed that the circuit impedance can be used to compute the fluctuations arising from the thermal agitation of the electrons (Nyquist, 1928). Nyquist's form of the fluctuation-dissipation theorem gives the mean-square voltage fluctuations as proportional to the product of the temperature and resistance, with the proportionality constant being the bandwidth. Alternatively, the temperature can be written as the ratio of the spectral density of the random fluctuations of the electromotive force $S_e(\omega)$ to the real part of the impedance $Z(\omega)$:

$$S_e(\omega) = 2kT \operatorname{Re} Z(\omega) \tag{3}$$

at the frequency ω. When the Nyquist relation (3) was first derived, the applicability of the underlying reasoning to other linear transport processes involving thermal noise was not appreciated.

Let us consider the dynamical equation constructed by Paul Langevin concerning the forces acting on a particle in a fluid (Langevin, 1908). The equation of motion for a particle of mass m is

$$m\frac{du}{dt} + \frac{u}{\mu} = F + f, \tag{4}$$

where u is the particle velocity, μ is the mobility (the inverse of the dissipation), F is an external driving force, and f is a fluctuating force with zero mean produced by the thermal agitation of the ambient fluid particles. The Fourier transform of Equation (4) yields the impedance for the particle $Z(\omega) \equiv \bar{F}(\omega)/\bar{u}(\omega) = im\omega + \mu^{-1}$. The average of the transformed equation yields the Nernst relation for a constant external force: $\langle \bar{u}(0) \rangle = \mu F$. The diffusion coefficient is given by the integral over the velocity autocorrelation function, which using Parseval's theorem can be written in terms of the velocity spectral density, $S_u(\omega)$, as

$$D = 2S_u(0). \tag{5}$$

This expression for the diffusion coefficient is valid for any form of the velocity autocorrelation function and velocity spectral density, under the condition that the underlying process is stationary in time.

In general, one can conclude that the autocorrelation function of the random force in a physical system is proportional to the dissipation in that system, with the proportionality constant given by the temperature kT. This relation can be summarized as

$$\langle f(t)f(t') \rangle = 2D\delta(t - t'), \tag{6}$$

so that the diffusion coefficient determines the strength of the fluctuations. Further, there are time-dependent generalizations of this form of the fluctuation–dissipation relation, where the δ function is replaced with a memory kernel, as well as extensions from the classical to the quantum domain, see, for example, Lindenberg & West (1990) for a review. However, each elaboration contains essentially the same information, namely, that microscopic dynamics are amplified to macroscopic fluctuations and dissipation. A complete understanding of the macroscopic phenomena of fluctuations and dissipation therefore requires an understanding of microscopic dynamics.

BRUCE J. WEST

See also **Brownian motion; Diffusion; Fokker–Planck equation**

Further Reading

Furth, R. (editor). 1956. *Albert Einstein, Investigations on the Theory of Brownian Motion*, New York: Dover; originally published in German, 1926

Gauss, F. 1809. *Theoria Motus Corporum Coelestrium*, Hamburg:

Langevin, P. 1908. Comptes Rendus Acad. Sci. Paris, 530

Lindenberg, K. & West, B.J. 1990. *The Nonequilibrium Statistical Mechanics of Open and Closed Systems*, New York: VCH

Montroll, E.W. & West, B.J. 1979. An enriched collection of stochastic processes. In *Fluctuation Phenomena*, edited by E.W. Montroll & Lebowitz, J. Amsterdam: North-Holland

Nernst, W. 1884. *Zeitschrift für Physikalische Chemie*, 9: 613

Nyquist, H. 1928. Thermal agitation of electric charge in conductors. *Physical Review*, 32: 110

Perrin, J. 1990. *Atoms*, Woodbridge: Ox Bow Press; originally published as *Les Atomes* in French, 1913

FLUID DYNAMICS

The field of fluid dynamics is devoted to the study of flows of matter in the liquid or gaseous state or in the form of multiple phases including suspensions of solid particles in liquids or gases. While it is in principle possible (and sometimes appropriate) to apply Newton's laws to individual molecules of a fluid and to describe the flow as an average over trajectories of the particles (*See* **Molecular dynamics**), it is usually far more efficient to consider liquids and gases as a continuum and to apply the equations for the dynamics of continuous media.

Conservation Laws of Continuous Media

The laws of the conservation of mass, momentum, and energy can be written in the form

$$\frac{\mathrm{d}}{\mathrm{d}t}\int_V \rho\,\mathrm{d}^3V = \int_V \frac{\partial\rho}{\partial t}\,\mathrm{d}^3V = -\oint_{\partial V} \rho n_j v_j\,\mathrm{d}^2A, \tag{1a}$$

$$\frac{\mathrm{d}}{\mathrm{d}t}\int_V \rho v_i\,\mathrm{d}^3V = -\oint_{\partial V} \rho v_i n_j v_j\,\mathrm{d}^2A, + \int_V \rho F_i\,\mathrm{d}^3V + \oint_{\partial V} S_{ij}n_j\,\mathrm{d}^2A, \tag{1b}$$

$$\begin{aligned}\frac{\mathrm{d}}{\mathrm{d}t}\int_V &\rho(e+v_jv_j/2)\,\mathrm{d}^3V \\ &= -\oint_{\partial V} v_jn_j\rho(e+v_iv_i/2)\mathrm{d}^2A \\ &+ \int_V \rho F_iv_i\mathrm{d}^3V + \int_V \rho q\,\mathrm{d}^3V \\ &+ \oint_{\partial V} v_iS_{ij}n_j\,\mathrm{d}^2A - \oint_{\partial V} h_jn_j\,\mathrm{d}^2A,\end{aligned} \tag{1c}$$

where V denotes an arbitrary volume fixed in space and ∂V denotes its surface. ρ is the density of the continuous medium, v_i is the velocity vector, and ρF_i is the body force density acting on the medium. A typical example for F_i is the acceleration of gravity. e is the internal energy per unit mass, ρq is the heat source density, and h_i denotes the heat flux. Contributions like absorption and emission of radiation have not been included explicitly in the energy balance (1c). But in optically thick media they can be subsumed under q and h_i. The index notation has been used where the index i refers to the coordinates 1, 2, 3 of a Cartesian system of coordinates and where the summation over indices occurring twice in any term is implied. Relativistic effects have been neglected in writing Equations (1) that are thus valid only for velocities small compared with the velocity of light.

The fundamental quantity of the dynamics of a continuous medium is the stress tensor S_{ij} which describes the surface force $S_{ij}n_j$ exerted on a surface element with the normal unit vector n_i. More exactly, $S_{ij}n_j$ is the force per unit area exerted by the material pierced by n_i on the material pierced by $-n_i$. The concept of the surface force was advanced long before the atomistic nature of materials became generally accepted. Surface forces in the ideal sense do not exist in nature. But since the forces between molecules in materials act only over atomic distances, the concept of surface forces has turned out to be very useful. The limitation should be kept in mind, however, in the fluid dynamical treatment of the dynamics of galaxies, for instance.

For applications of the conservation laws (1) it is convenient to formulate them in differential form by taking the limit $V \to 0$ and using Gauss' theorem,

$$\left(\frac{\partial}{\partial t} + v_j\partial_j\right)\rho + \rho\partial_jv_j = 0, \tag{2a}$$

$$\rho\frac{\partial}{\partial t}v_i + \rho v_j\partial_jv_i = \rho F_i + \partial_jS_{ij}, \tag{2b}$$

$$\frac{\partial}{\partial t}\rho e + \partial_j(\rho v_je) = S_{ij}\partial_jv_i + \rho q - \partial_jh_j. \tag{2c}$$

In deriving (2b) from (1b) and (2c) from (1c), we have used relationships (2a) and (2b), respectively. While Equations (2) apply to all continua, including solid materials, for applications to fluids the stress tensor is separated into two parts,

$$S_{ij} = -p\delta_{ij} + S'_{ij}, \tag{3}$$

where p is the thermodynamic pressure and S'_{ij} is the part of the stress tensor which depends on viscous friction and other dissipative effects. Then $S'_{ij}\partial_j v_i \equiv \rho\Phi$ is the irreversible conversion of work into heat. After using (2b), we may write (2c) in the form

$$\frac{\mathrm{D}e}{\mathrm{D}t} + p\frac{\mathrm{D}\rho^{-1}}{\mathrm{D}t} = \Phi + q - \rho^{-1}\partial_j h_j, \tag{4}$$

where the material derivative

$$\frac{\mathrm{D}}{\mathrm{D}t} \equiv \frac{\partial}{\partial t} + v_j\partial_j \tag{5}$$

has been introduced that describes the change in time with respect to the frame of reference moving with the fluid. This is also referred to as the Lagrangian description in contrast to the Eulerian description where the time derivative at a fixed point in space is taken.

In applications, the pressure p and the temperature T are usually the most readily known thermodynamic variables. We thus use the relationship $e = h - p/\rho$ for simple materials where h is the specific enthalpy and obtain

$$\mathrm{d}h = c_p\,\mathrm{d}T + \left.\frac{\partial h}{\partial p}\right|_T \mathrm{d}p = c_p\,\mathrm{d}T + \frac{1}{\rho}(1 - \alpha T)\,\mathrm{d}p, \tag{6}$$

where some simple thermodynamic relationships have been used and $\alpha \equiv \rho(\partial\rho^{-1}/\partial T)_p$ is the coefficient of thermal expansion. As a final result, we thus obtain the energy equation in its most useful form

$$T\frac{\mathrm{D}s}{\mathrm{D}t} \equiv c_p\frac{\mathrm{D}T}{\mathrm{D}t} - \frac{\alpha T}{\rho}\frac{\mathrm{D}p}{\mathrm{D}t} = \Phi + q - \frac{1}{\rho}\partial_j h_j. \tag{7}$$

Fourier's law, $h_i = -\lambda\partial_i T$, is commonly used for the heat flux, and for laboratory applications, it is a good approximation to neglect $\alpha T\rho^{-1}\mathrm{D}p/\mathrm{D}t + \Phi$ in comparison with $c_p\,\mathrm{D}T/\mathrm{D}t$. In this case, the familiar heat equation

$$\left(\frac{\partial}{\partial t} + v_j\partial_j\right)T = q + \kappa\partial_j\partial_j T \tag{8}$$

is obtained where $\lambda =$ const. has been assumed and where $\kappa = \lambda/\rho c_p$ is the thermal diffusivity.

Dynamics of Inviscid Fluids

The simplest equation of fluid dynamics is obtained when viscous, elastic, and other effects are neglected and $S_{ij} = -p\delta_{ij}$ is assumed. The resulting Euler equation,

$$\left(\frac{\partial}{\partial t} + v_j\partial_j\right)v_i = -\rho^{-1}\partial_i p + F_i, \tag{9}$$

is most easily solved when the fluid is incompressible and $\rho =$ const. can be assumed. In that case

$$\partial_j v_j = 0 \tag{10}$$

holds and an equation of state connecting ρ and p is not needed.

A more general form of the equations is obtained in the barotropic case when the density is prescribed as a function of the pressure alone as, for instance, in the form of a polytropic relationship

$$p = p_0\left(\frac{\rho}{\rho_0}\right)^{\gamma}. \tag{11}$$

This expression is valid, for example, for adiabatic changes of an ideal gas in which case γ equals the ratio of specific heats, $\gamma = c_p/c_v$. The equation for sound waves can be obtained when Equations (2a) and (9) are linearized around the static equilibrium state with $\rho = \rho_0$, $p = p_0$,

$$\frac{\partial^2}{\partial t^2}\rho = \gamma\frac{p_0}{\rho_0}\partial_j\partial_j\rho, \tag{12}$$

where $\gamma p_0/\rho_0 = c_s^2$ is the square of the speed of sound. For general fluids, the relationship $c_s^2 = \partial p/\partial\rho$ holds. The ratio of a characteristic velocity divided by the sound speed is called the Mach number Ma. The domains of fluid flow with Ma > 1 are usually separated from regions with Ma < 1 by shock fronts in which energy is dissipated and frictional processes must be taken into account.

A fundamental consequence of Equation (9) is Kelvin's theorem,

$$\frac{\mathrm{D}}{\mathrm{D}t}\oint_C \boldsymbol{v}\cdot\mathrm{d}\boldsymbol{l} = 0, \tag{13}$$

where the integral is called the circulation and must be taken over a closed curve C moving with the fluid. Theorem (13) holds for barotropic fluids when the force field F_i is conservative; that is, it can be written as the gradient of a single-valued scalar function. The property that the circulation is an invariant of fluid motion represents an elegant formulation of Helmholtz's vortex theorem, which states that the strength of a vortex tube moving with the fluid remains unchanged,

$$\frac{\mathrm{D}}{\mathrm{D}t}\int_S \nabla\times\boldsymbol{v}\cdot\mathrm{d}^2\boldsymbol{A} = 0, \tag{14}$$

where the surface S is bounded by a closed material curve, C. $\nabla\times\boldsymbol{v}$ is called the vorticity field of the velocity field $\boldsymbol{v}$. The manifold of vorticity vectors intersecting the curve C generates as tangential vectors the surface of the vortex tube. Kelvin's theorem and Helmhotz's theorem are connected, of course, by Stokes' theorem.

An important consequence of theorem (13) or (14) is that motions starting in a fluid from rest have vanishing vorticity if the force field is conservative and viscous effects can be neglected. An incompressible fluid with vanishing vorticity obeys the equation

$$\boldsymbol{v} = \nabla\phi \quad \text{with} \quad \nabla^2\phi = 0. \tag{15}$$

Such flows are called potential flows.

Dynamics of Viscous Fluids

In the general case, the stress tensor S'_{ij} may be a rather complex function of the properties of the fluid and its motions. The various problems arising in this connection are the subject of the field of rheology. Fortunately, the most important fluids such as air and water and many others can be described as Newtonian fluids for which a linear homogeneous relationship holds between S'_{ij} and the velocity gradient tensor $\partial_i v_j$,

$$S'_{ij} = \mu(\partial_i v_j + \partial_j v_i - \tfrac{2}{3}\delta_{ij}\partial_k v_k) + \tilde{\mu}\delta_{ij}\partial_k v_k, \tag{16}$$

where μ is the dynamic viscosity and $\tilde{\mu}$ is the bulk viscosity. It is difficult to measure the latter property. Because $\tilde{\mu} = 0$ can be shown to hold for mono-atomic gases, this relationship is often generally applied in which case the Navier–Stokes equations are obtained,

$$\rho\left(\frac{\partial}{\partial t} + v_j\partial_j\right)v_i = -\partial_i p + \rho F_i + \partial_j\left(\mu(\partial_i v_j + \partial_j v_i - \tfrac{2}{3}\delta_{ij}\partial_k v_k)\right). \tag{17}$$

For a given force field F_i, this equation together with the equation of continuity (2a) and a barotropic equation of state, $\rho = \rho(p)$, provides a complete set for the determination of the variables v_i, ρ, p (*See* **Navier–Stokes equations**).

An even simpler set of equations can be obtained for incompressible fluids with $\rho =$ const., $\mu =$ const.,

$$\left(\frac{\partial}{\partial t} + v_j\partial_j\right)v_i = -\partial_i\frac{p}{\rho} + F_i + \nu\partial_j\partial_j v_i, \tag{18a}$$

$$\partial_j v_j = 0, \tag{18b}$$

where the kinematic viscosity $\nu = \mu/\rho$ has been introduced. Using U as typical velocity and d as a typical length scale, we find for the ratio of the last terms on the left- and right-hand sides of Equation (18a)

$$\frac{Ud}{\nu} = \mathrm{Re} \tag{19}$$

which is called the Reynolds number. In the low Reynolds number limit, the Stokes equation is obtained in which the term $v_j\partial_j v_i$ in Equation (18a) is dropped. In the high Reynolds number limit, the Euler equation does not become applicable, in general, because of the existence of a turbulent cascade where energy is transferred to smaller and smaller scales such that the term $\nu\partial_j\partial_j v_i$ cannot be neglected in the limit $\nu \to 0$ (*See* **Turbulence**).

Buoyancy-driven Flows

Among the force fields that are not conservative and therefore tend to generate motions with vorticity, buoyancy forces are the most common ones. Especially in atmospheric, oceanic, and astrophysical applications, buoyancy-driven flows are of fundamental importance. In order to keep as much as possible of the convenient properties of the Navier–Stokes equations for incompressible fluids, the Boussinesq approximation is usually introduced in which the temperature dependence of the density, say $\rho = \rho_0(1 - \gamma(T - T_0))$, is taken into account in the gravity force term only, while ρ is replaced by ρ_0 elsewhere in the equations. Together with the heat equation, the following set of equations is obtained:

$$\left(\frac{\partial}{\partial t} + \boldsymbol{v}\cdot\nabla\right)\boldsymbol{v} = -\nabla\pi - \gamma(T - T_0)\boldsymbol{g} + \nu\nabla^2\boldsymbol{v}, \tag{20a}$$

$$\nabla\cdot\boldsymbol{v} = 0, \tag{20b}$$

$$\left(\frac{\partial}{\partial t} + \boldsymbol{v}\cdot\nabla\right)T = \kappa\nabla^2 T, \tag{20c}$$

where $\boldsymbol{g}$ is the gravity vector and where $\nabla\pi$ includes all terms that can be written as a gradient. Because the pressure dependence of the density is neglected in the Boussinesq approximation, the contribution Φ must also be neglected (Busse, 1989).

A simple solution described by these equations is the flow between two parallel plates with separation d that are kept at different temperatures and that are inclined at an angle χ with respect to the horizontal plane (see Figure 1).

$$\boldsymbol{v} = z(z^2 - d^2/4)\gamma g \sin\chi(T_2 - T_1)\boldsymbol{i}, \tag{21a}$$

$$T = T_0 + (T_2 - T_1)z/d. \tag{21b}$$

A Cartesian system of coordinates has been assumed with the z-coordinate normal to the plates and the origin in the middle between the plates. $\boldsymbol{i}$ is the unit vector in the direction of inclination parallel to the x-axis as shown in Figure 1. T_1 and T_2 are the fixed temperatures at the boundaries $z = -0.5$ and $+0.5$, respectively, and $T_0 = (T_2 + T_1)/2$. Since in laboratory realizations of this configuration the space between the plates will be enclosed by side walls, the condition that the total mass transport between the plates vanishes has been imposed. Of special interest is the case $\gamma(T_2 - T_1) < 0$

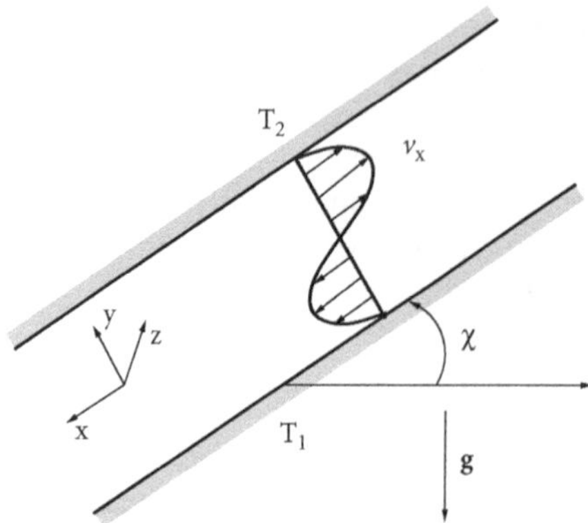

Figure 1. Inclined fluid layer heated from above.

when the density increases opposite to the direction of gravity such that an unstable stratification results. In a horizontal layer where solution (21a) vanishes, the instability occurs in the form of Rayleigh–Bénard convection. The dimensionless control parameter for this case, $\text{Ra} \equiv \gamma g(T_2 - T_1)d^3/\kappa\nu$, is called the Rayleigh number. Convection rolls set in when Ra exceeds a critical value Ra_c depending on the boundary conditions. Besides buoyancy driven rolls aligned with the x-axis for $\chi \neq 0$ hydrodynamic, instabilities may set in in the form of transverse vortices. Since these forms of secondary states of motion are themselves subject to secondary instabilities, sequences of bifurcations are observed leading to increasingly complex patterns of fluid flow as indicated in Figure 2 for a few examples. The variety of patterns that can be realized continue to be a subject of intense experimental and theoretical research (*See* **Taylor–Couette flow**).

A role similar to that of buoyancy is played by the temperature dependence of surface tension that may give rise to instabilities much like Rayleigh–Bénard convection. Instead of the parameter Ra, the Marangoni number $M = \xi(T_2 - T_1)d/\kappa\eta$ is used as a control parameter, in this case where $\xi = \partial\sigma/\partial T$ denotes the derivative with respect to temperature of the surface tension σ.

Dynamics of Rotating Fluids

The Navier–Stokes equations (17) are invariant with respect to a Galileo transformation; that is, they retain their form in the transition from one inertial frame of reference to another one. When the transformation is made from an inertial to a frame of reference rotating with the constant angular velocity $\boldsymbol{\Omega}$, it must be taken into account that the transformation of a vector field $\boldsymbol{a}$ is different from that of a scalar field ϕ,

$$\frac{\partial}{\partial t'}\boldsymbol{a} = \left(\frac{\partial}{\partial t} + \boldsymbol{\Omega}\times\boldsymbol{r}\cdot\nabla - \boldsymbol{\Omega}\times\right)\boldsymbol{a}, \qquad (22a)$$

$$\frac{\partial}{\partial t'}\phi = \left(\frac{\partial}{\partial t} + \boldsymbol{\Omega}\times\boldsymbol{r}\cdot\nabla\right)\phi; \qquad (22b)$$

that is, a constant vector field in an inertial frame becomes time dependent as seen from the rotating frame. Using $\boldsymbol{v}' = \boldsymbol{v} - \boldsymbol{\Omega}\times\boldsymbol{r}$ where $\boldsymbol{r}$ is the position vector, we find as Navier–Stokes equation in the rotating frame

$$\left(\frac{\partial}{\partial t'} + \boldsymbol{v}'\cdot\nabla\right)\boldsymbol{v}' + 2\boldsymbol{\Omega}\times\boldsymbol{v}' = -\nabla\frac{p}{\rho} - \nabla\Psi + (\boldsymbol{\Omega}\times\boldsymbol{r})\times\boldsymbol{\Omega} + \nu\nabla^2\boldsymbol{v}' \qquad (23a)$$

$$\nabla\cdot\boldsymbol{v}' = 0. \qquad (23b)$$

We have restricted our attention to the incompressible case with $\rho =$ const. and have assumed that the force field $\boldsymbol{F}$ is conservative, that is, $\boldsymbol{F} = -\nabla\Psi$. Because the centrifugal force is also conservative it can be combined with $\boldsymbol{F}$; that is, $\Psi' = \Psi - |\boldsymbol{\Omega}\times\boldsymbol{r}|^2/2$ can be used as the potential. The other new term in Equations (23) is the Coriolis force, $-2\boldsymbol{\Omega}\times\boldsymbol{v}'$, which is responsible for the fact that the dynamics of rotating systems is quite different from that of nonrotating systems. Since the Coriolis force is typically the largest term on the left-hand side of Equation (23a) and since viscous friction is usually negligible in the interior of the fluid, the approximate relationship

$$2\boldsymbol{\Omega}\times\boldsymbol{v} = -\nabla\pi \qquad (24)$$

is obtained which is called the geostrophic balance because of its importance in meteorology and oceanography. Here, $\rho\pi$ is the dynamic pressure, $\pi = p/\rho + \Psi'$, and for simplicity, the prime of $\boldsymbol{v}$ has been dropped. When the curl of balance (24) is taken and Equation (23b) is used, the famous Proudman–Taylor theorem is obtained,

$$2\boldsymbol{\Omega}\cdot\nabla\boldsymbol{v} = 0. \qquad (25)$$

This theorem states that in a rapidly rotating system, a steady velocity field of a nearly inviscid fluid must be independent of the coordinate in the direction of the axis of rotation. As a consequence, $\boldsymbol{v}$ will depend only on the x-, y-coordinates of a cartesian system with the z-coordinate pointing in the direction of $\boldsymbol{\Omega}$. Since $\boldsymbol{v}$ must also satisfy $\nabla\cdot\boldsymbol{v} = 0$, a single scalar function, the stream function ψ, is sufficient to describe $\boldsymbol{v}$. Assuming boundaries perpendicular to $\boldsymbol{\Omega}$, we find

$$\boldsymbol{v} = \nabla\psi\times\boldsymbol{\Omega}/\Omega \quad \text{with} \quad \psi = -\pi(x,y)/2\Omega$$

and

$$\Omega = |\boldsymbol{\Omega}|. \qquad (26)$$

The form of the velocity field indicates that the flow parallels the isobars in contrast to the situation in a nonrotating system where the flow is usually directed perpendicular to the pressure gradient. The

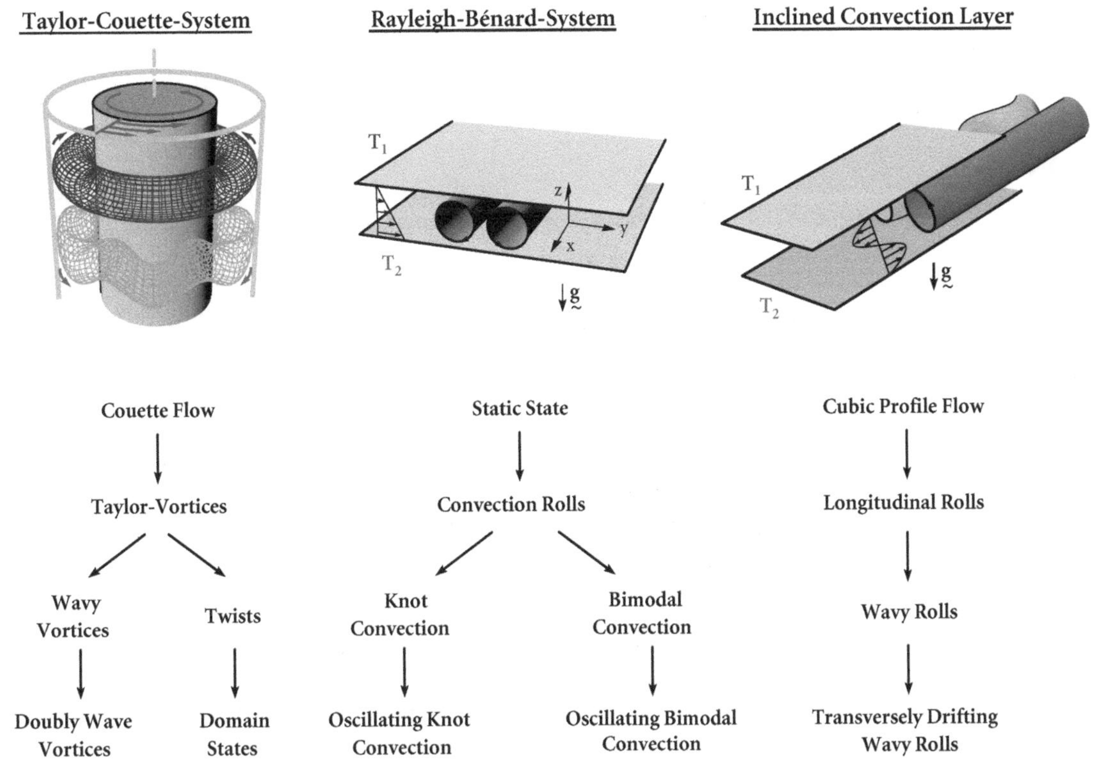

Figure 2. Examples of bifurcation sequences.

geostrophic wind (26) flows cyclonically (i.e., in the sense of rotation) around low-pressure regions and anticyclonically around high-pressure regions, as can be seen on weather maps.

The fact that the function $\pi(x, y)$ has remained undetermined so far is known as the geostrophic degeneracy. Terms of higher order in the equations of motion are needed to remove this degeneracy and to introduce time dependences on scales much longer than the rotation period $2\pi/\Omega$ of the system. A three-dimensional nature of the velocity field enforced, for instance, by boundary conditions can thus be introduced as a perturbation of the geostrophic solution. Rossby waves are a typical example for the quasigeostrophic dynamical response of rotating fluid systems.

The friction term in Equation (23a) is characterized by the highest spatial derivatives and thus can become large wherever strong variations of the velocity field occur, as, for example, near solid boundaries where the tangential velocity component must vanish. Thin Ekman boundary layers are formed here with the typical thickness $\sqrt{\nu/\Omega}$. But internal shear layers in the interior of the fluid can also be realized. The Ekman number defined by

$$E = \nu/\Omega d^2, \tag{27}$$

where d is a typical length scale of the system in the direction of the axis of rotation, is often used as a dimensionless parameter in the dynamics of rotating fluids.

F.H. Busse

See also **Boundary layers; Molecular dynamics; Navier–Stokes equation; Taylor–Couette flow; Thermal convection; Turbulence; Vortex dynamics of fluids**

Further Reading

Acheson, D.J. 1990. *Elementary Fluid Dynamics*, Oxford: Clarendon Press and New York: Oxford University Press

Batchelor, G.K. 1967. *An Introduction to Fluid Dynamics*, Cambridge: Cambridge University Press

Busse, F.H. 1989. Fundamentals of Thermal Convection. In *Mantle Convection*, edited by W. Peltier, London: Gordon and Breach

Greenspan, H.P. 1968. *The Theory of Rotating Fluids*, Cambridge and New York: Cambridge University Press

Swinney, H.L. & Gollub, J.P. (editors). 1985. *Hydrodynamic Instabilities and Transition to Turbulence*, 2nd edition, Berlin and New York: Springer

Tritton, D.J. 1988. *Physical Fluid Dynamics*, 2nd edition, Oxford: Clarendon Press and New York: Oxford University Press

FLUX FLOW OSCILLATOR

See **Long Josephson junction**

FLUX QUANTIZATION

See **Josephson junction**

FLUXON

See **Long Josephson junction**

FOCUSING SYSTEM

See **Nonlinear optics**

FOKKER–PLANCK EQUATION

The Fokker–Planck equation (FPE) plays a role in stochastic systems analogous to that of the Liouville equation in deterministic mechanical systems. Namely, the FPE describes in a statistical sense how a *collection* of initial data evolves in time. To be precise, let the phase space describing the system be parameterized by the state vector $\boldsymbol{y}$, and let $\boldsymbol{Y}(t)$ denote the value of the state vector assumed by the system at time t. Suppose the dynamics of the system is governed by a stochastic differential equation (SDE) of the form

$$\mathrm{d}\boldsymbol{Y}(t) = \boldsymbol{U}(\boldsymbol{Y}(t), t)\,\mathrm{d}t + \sigma(\boldsymbol{Y}(t), t) \cdot \mathrm{d}\boldsymbol{W}(t). \quad (1)$$

The deterministic component of the dynamics is described by the drift vector $\boldsymbol{U}(\boldsymbol{y}, t)$, while the random part is driven by a vector of independent Brownian motions $\boldsymbol{W}(t)$ that are coupled to the system through the matrix $\sigma(\boldsymbol{y}, t)$.

Alternatively, we can define the system dynamics as a "diffusion process" in the relevant phase space such that

$$\begin{cases} \lim_{\tau \downarrow 0} \tau^{-1}\langle \boldsymbol{Y}(t+\tau) - \boldsymbol{Y}(t)\rangle = \boldsymbol{U}(\boldsymbol{Y}(t), t), \\ \lim_{\tau \downarrow 0} \tau^{-1}\langle (\boldsymbol{Y}(t+\tau) - \boldsymbol{Y}(t)) \\ \quad \otimes (\boldsymbol{Y}(t+\tau) - \boldsymbol{Y}(t))\rangle = 2\mathsf{D}(\boldsymbol{Y}(t), t), \\ \lim_{\tau \downarrow 0} \tau^{-1}\langle |\boldsymbol{Y}(t+\tau) - \boldsymbol{Y}(t)|^{\gamma}\rangle = 0 \text{ for } \gamma > 2, \end{cases} \quad (2)$$

where $\langle\cdot\rangle$ denotes an average over the random component of the dynamics. The SDE and diffusion process descriptions are equivalent, with $2\mathsf{D} = \sigma\sigma^{\dagger}$.

One useful way to describe the evolution of stochastic systems over a finite (rather than infinitesimal) time interval is through the probability transition density $p(t, \boldsymbol{y}|t', \boldsymbol{y}')$, which describes the likelihood that if the state variable assumes the value $\boldsymbol{y}'$ at time t', then it will assume a value near $\boldsymbol{y}$ at the later time t. More precisely, p is defined in terms of conditional probability as

$$\mathrm{Prob}(\boldsymbol{Y}(t) \in B | \boldsymbol{Y}(t') = \boldsymbol{y}') = \int_B p(t, \boldsymbol{y}|t', \boldsymbol{y}')\,\mathrm{d}\boldsymbol{y} \quad (3)$$

for all nice (Borel) sets B in the phase space. The FPE, also known as the Kolmogorov forward equation, is a partial differential equation that describes how the probability transition density evolves when the system can be described as an SDE (1) or diffusion process (2):

$$\begin{cases} \dfrac{\partial p(t, \boldsymbol{y}|t', \boldsymbol{y}')}{\partial t} \\ \quad = \dfrac{\partial}{\partial \boldsymbol{y}}\left(-\boldsymbol{U}(\boldsymbol{y}, t)p + \dfrac{\partial}{\partial \boldsymbol{y}}(\mathsf{D}(\boldsymbol{y}, t)p)\right), \\ p(t = t', \boldsymbol{y}|t', \boldsymbol{y}') = \delta(\boldsymbol{y} - \boldsymbol{y}'). \end{cases} \quad (4)$$

This equation is to be solved for $t > t'$ with fixed data for the source variables $(t', \boldsymbol{y}')$ and appropriate boundary conditions in $\boldsymbol{y}$ (Risken, 1989, Chapter 4).

The probability transition density can be used to describe the probability density for the state vector at any moment of time:

$$\mathrm{Prob}(\boldsymbol{Y}(t) \in B) = \int_B \phi(t, \boldsymbol{y})\,\mathrm{d}\boldsymbol{y} \quad (5)$$

by simply integrating against the prescribed probability distribution of the states at the initial time t':

$$\phi(t, \boldsymbol{y}) = \int p(t, \boldsymbol{y}|t', \boldsymbol{y}')\phi(t', \boldsymbol{y}')\,\mathrm{d}\boldsymbol{y}'. \quad (6)$$

Alternatively, $\phi(t, \boldsymbol{y})$ can be shown to satisfy the FPE (4) but with initial data $\phi(t', \boldsymbol{y}')$ prescribed more generally as a nonnegative function with integral one rather than as a delta function.

The FPE was first applied to describe Brownian motion. In the most idealized case, where inertia is completely neglected, Einstein showed that the statistics of the position $\boldsymbol{x} = \boldsymbol{X}(t)$ of a Brownian particle obeys an FPE that coincides with the ordinary diffusion equation:

$$\frac{\partial \phi(t, \boldsymbol{x}|t', \boldsymbol{x}')}{\partial t} = \kappa \Delta_{\boldsymbol{x}} \phi, \quad (7)$$

where $\Delta_{\boldsymbol{x}}$ is the Laplace operator and the diffusion coefficient $\mathsf{D} = \kappa\mathsf{I}$ is a constant scalar multiple of the identity matrix. The stochastic differential description of this model is simply

$$\mathrm{d}\boldsymbol{X}(t) = (2\kappa)^{1/2}\,\mathrm{d}\boldsymbol{W}(t). \quad (8)$$

The effects of inertia and external forces can be incorporated by passing to a phase space description including both the position $\boldsymbol{x} = \boldsymbol{X}(t)$ and the velocity $\boldsymbol{v} = \boldsymbol{V}(t)$ of the Brownian particle. The equations of motion can then be written in terms of Newton's law with a random forcing component proportional to $\mathrm{d}\boldsymbol{W}$:

$$\begin{cases} \mathrm{d}\boldsymbol{X}(t) = \boldsymbol{V}(t)\,\mathrm{d}t, \\ m\,\mathrm{d}\boldsymbol{V}(t) = -m\xi\boldsymbol{V}(t)\,\mathrm{d}t + \boldsymbol{f}(\boldsymbol{X}(t), t)\,\mathrm{d}t \\ \qquad +(2k_{\mathrm{B}}Tm\xi)^{1/2}\,\mathrm{d}\boldsymbol{W}(t), \end{cases} \tag{9}$$

where ξ is a friction coefficient, k_{B} is Boltzmann's constant, T is the absolute temperature, and $\boldsymbol{f}(\boldsymbol{x}, t)$ represents the deterministic part of the external applied force. The equivalent Fokker–Planck description for the phase space probability transition density in this system reads

$$\begin{aligned} &\frac{\partial p(t, \boldsymbol{x}, \boldsymbol{v}|t', \boldsymbol{x}', \boldsymbol{v}')}{\partial t} \\ &= \mathbf{grad}_{\boldsymbol{v}} \cdot \left[\left(\xi\boldsymbol{v} - \frac{\boldsymbol{f}(\boldsymbol{x}, t)}{m} + \frac{k_{\mathrm{B}}T\xi}{m}\mathbf{grad}_{\boldsymbol{v}}\right) p\right] \\ &\quad -\mathbf{grad}_{\boldsymbol{x}} \cdot (\boldsymbol{v}p) \end{aligned} \tag{10}$$

Observe how this equation generalizes the Liouville equation for deterministic mechanics to include Brownian motion. The simplified diffusion equation (7) can be obtained through an asymptotic limit of the full FPE (10) with $\kappa = k_{\mathrm{B}}T/(m\xi)$, when $\boldsymbol{f}(\boldsymbol{x}, t) = 0$ and $(k_{\mathrm{B}}T/(m\xi^2\ell^2))^{1/2} \ll 1$ with ℓ a characteristic length of the system (Bocquet, 1997; Risken, 1989). A similar reduction to a partial differential equation in coordinate space, called the Smoluchowski equation, is also possible for collections of interacting particles where the friction tensor depends on the particle configuration (Titulaer, 1980). From the FPE for Brownian motion, one can compute various statistical properties such as its mean-square displacement, the spectrum of its fluctuations, the rate of relaxation of initial velocity to thermal equilibrium values, and the probability distribution for the time at which the particle first achieves a certain location or surmounts a potential barrier (Wax, 1954).

The FPE has found useful application in computing similar statistical properties in numerous other systems, such as lasers, polymers, particle suspensions, quantum electronic systems, molecular motors, and finance. In some instances, the system is not at first formulated in one of the senses (1) or (2) which immediately imply an FPE. Rather, the systems are more naturally represented in terms of a master equation with a complete description of the rates at which the system (randomly) jumps from one state to another. The FPE is an appropriate approximation to this system when certain asymptotic conditions, such as small jumps, a large system size, or a separation of time scales between resolved and unresolved variables, are met (Grabert, 1982; Risken, 1989; van Kampen, 1981).

A variety of techniques for practical analysis of FPEs can be found in Risken (1989). The main alternative to analyzing stochastic systems through the FPE is through the stochastic differential (or Langevin) formulation (1). The FPE has the merit of being a deterministic partial differential equation that can be solved once through either analytical or numerical methods. The stochastic differential description (1) by contrast consists of ordinary differential equations for which individual realizations (samples) are faster to simulate through Monte Carlo methods (Kloeden & Platen, 1992). But obtaining a statistical description of the general behavior of the system requires the generation of a large number of realizations.

Certain statistical quantities described above can be easily calculated using either the solution of the FPE or by operating directly on the FPE itself. On the other hand, the stochastic differential system (1) must be used if one wishes to characterize particular realizations of the stochastic system. Moreover, the stochastic differential description can be readily generalized to cases in which the random driving has temporal correlations (Kubo et al., 1991; Moss & McClintock, 1989). The FPE formulation can be modified to a "fractional" formalism for certain self-similar temporal correlation structures (Metzler & Klafter, 2000), but further generalizations are much more complicated and usually require some sort of approximation (Moss & McClintock, 1989).

PETER R. KRAMER

See also **Brownian motion; Fluctuation-dissipation theorem; Nonequilibrium statistical mechanics; Phase-space diffusion and correlations; Stochastic processes**

Further Reading

Bocquet, L. 1997. High friction limit of the Kramers equation: The multiple time-scale approach. *American Journal of Physics*, 65(2): 140–144

Grabert, H. 1982. *Projection Operator Techniques in Nonequilibrium Statistical Mechanics*, Berlin and New York: Springer [FPE as approximation to more detailed models when separation of time scales exists]

Kloeden, P.E. & Platen, E. 1992. *Numerical Solution of Stochastic Differential Equations*, Berlin and New York: Springer [Sections 1.7, 2.4, 4.1, 4.7]

Kubo, R., Toda, M. & Hashitsume, N. 1991. *Statistical Physics*, 2 vols, 2nd edition, Berlin and New York: Springer [Chapter 2 has a derivation of hierarchy of statistical descriptions for physical systems]

Metzler, R. & Klafter, J. 2000. The random walk's guide to anomalous diffusion: a fractional dynamics approach. *Physics Reports*, 339(1): 1–77

Moss, F. & McClintock, P.V.E. (editors). 1989. *Noise in Nonlinear Dynamical Systems*, vol. 1. Cambridge and New York: Cambridge University Press [A collection of expository articles on FPE methods with extensions to temporal correlations in noise]

Risken, H. 1989. *The Fokker–Planck Equation*, 2nd edition, Berlin: Springer

Titulaer, U.M. 1980. Corrections to the Smoluchowski equation in the presence of hydrodynamic interactions. *Physica* A, 100(2): 251–265

van Kampen, N.G. 1981. *Stochastic Processes in Physics and Chemistry*, Amsterdam: North-Holland [Chapter 8 has a careful discussion of physical applicability of FPE]
Wax, N. (editor). 1954. *Selected Papers on Noise and Stochastic Processes*, New York: Dover

FORCED SYSTEMS

See **Damped-driven anharmonic oscillator**

FORECASTING

The success of Newtonian classical mechanics led in the 18th and 19th centuries to a strong belief in determinism; that is, the notion that the past determines the present and the future. As a consequence, a superior intellect ("Laplace's Daemon") would be able to forecast the future given enough information about the present and the past. However, the discovery of chaos in a large class of deterministic nonlinear dynamical systems sets a limit as to what Laplace's Daemon can forecast.

One of the characteristics of deterministic chaos is that the future state of a dynamical system depends sensitively on the initial conditions—infinitesimal perturbations will grow exponentially. This effect is also known more poetically as the "butterfly effect" (from a lecture by Edward Lorenz in 1972: "Does the flap of a butterfly's wings in Brazil set off a tornado in Texas?"). Since there will always be some uncertainty—however small—about the observed state of any physical system, a consequence of extreme sensitivity on the initial conditions is that the future state of such a system can only be forecast to lie within a given error tolerance a limited time ahead. As even Laplace's Daemon would not be able to determine the exact present state of the system, she would not be able to forecast the state of the future, but as she would know the uncertainty about the present, she would be able to work out all possible future states.

Uncertainty in the initial conditions is appropriately expressed in terms of a multivariate probability density function (pdf). A forecast is then given by the time evolution of the initial pdf, subject to the underlying dynamical system. Typically, the initial pdf is sharp, centered around the best estimate of the observed state of the physical system, but with time the pdf is smeared out in a complicated fashion until, eventually, all information about the initial condition is lost. For the short-range evolution, when the pdf remains "sufficiently" sharp, a forecast can be given in deterministic terms. For the intermediate-range evolution, when the pdf still has sufficient structure, a forecast can (and should) be given in probabilistic terms. Beyond that, a forecast will be no better than a guess based on the observed long-term behavior of the system.

When the governing dynamical equations are known, the time evolution of the pdf can be expressed in terms of the Liouville equation, which is a special case of the stochastic-differential Fokker–Planck equation. In theory, the Liouville equation could be applied to weather forecasting where dynamical models have been developed of the atmospheric circulation, but in practice, numerical problems and computational limits need to be overcome. An alternative approach is to make a Monte Carlo-type estimation of the time evolution of the pdf, that is, numerical integration of an ensemble of a large number of initial conditions that are generated by random sampling of the initial pdf. Application to weather forecasting of this Monte–Carlo approach is, however, also problematic, primarily due to the very high dimension ($\sim 10^7$) of the phase space of atmospheric circulation models that requires a very large number of Monte Carlo integrations to be performed (beyond the capacity of present-day supercomputers) in order to sufficiently reduce sampling uncertainty.

The practical approach that is applied to today's weather forecasting is to generate a relatively small ensemble (comprising typically 50–100 members) of initial perturbations to the observed initial state of the atmosphere. The perturbations are calculated in a way that ensures that the corresponding trajectories of the dynamical model equations initially diverge rapidly from the observed initial state. Thus, the time evolution of the individual ensemble members is not representative of the time evolution of the "true" pdf, but the ensemble is useful for the weather forecaster who is often most interested in the most likely development of the weather and possible major deviations from this development.

Forecasting becomes increasingly problematic when we also take into account that the forecast model is not perfect; that is, there are errors in the model (perhaps due to incorrect assumptions or poor spatial resolution), the effect of which is that the model does not behave like the real world. In the area of weather forecasting, model errors cause severe problems for long-range forecasts. A long-range weather forecast, or seasonal forecast, valid out to typically six months ahead, aims at predicting slowly varying surface conditions, such as sea surface temperature, sea ice cover, and soil moisture, and the effect they have on the preferred state of the atmospheric circulation in certain regions. A prominent example is prediction of El Niño conditions (increased sea surface temperature in the central and eastern tropical Pacific), which have an influence on worldwide weather.

The effect of model errors is in practice taken into account by basing seasonal weather forecasts on ensembles made up of integrations of more than one model. In this way, a pseudo-stochastic element is introduced into the forecast through the different formulations of the models, particularly for processes that take place on spatial scales that are smaller than the resolution of the model grid.

Forecasting is also of interest in areas in which models are not readily expressed in terms of known physical laws, most notably, in economics. When a physical model is not available, forecasting has to be based on more empirical methods. For example, a statistical forecast model would be based on the forecaster's notion of cause and effect, while an empirically constructed dynamical model could be derived from analysis of the combined behavior of many observed variables.

Due to the nondeterministic nature of economic evolution, economic forecasts will often be associated with substantial uncertainty. However, it is worth noting that a forecast need not be very accurate to be useful. If, in the long run, an economic forecast is just slightly better than a guess, then money can be made from the forecast.

In practical forecasting, no matter in what field, there is almost always an element of human expert judgment involved, and overall the expert judgment probably remains the most widespread forecasting method.

HENRIK FEDDERSEN

See also **Butterfly effect; Chaotic dynamics; Determinism; Economic system dynamics; Fokker–Planck equation; General circulation models of the atmosphere**

Further Reading

Ehrendorfer, M. 1994. The Liouville equation and its potential usefulness for the prediction of forecast skill. Part I: theory. *Monthly Weather Review*, 122: 703–713

Houghton, J. 1991. The Bakerian lecture, 1991. The predictability of weather and climate. *Philosophical Transactions of the Royal Society of London* A, 337: 521–572

Lorenz, E. 1963. Deterministic non-periodic flow. *Journal of Atmospheric Sciences*, 20: 130–141

Sivillo, J.K., Ahlquist, J.E. & Toth, Z. 1997. An ensemble forecasting primer. *Weather and Forecasting*, 12: 809–818

Smith, L.A. 2000. Disentangling uncertainty and error: On the predictability of nonlinear systems. In *Nonlinear Dynamics and Statistics* (Chapter 2), edited by A.I. Mees, Boston: Birkhauser

FOREST FIRES

The modeling of forest fires (FF) is based upon the concept of interacting particle systems (Liggett, 1985) for the description of complex systems. Within this framework, attention is on the interactions among particles (i.e., the trees), disregarding any attempt to understand the behavior of each particular tree. Using local interaction rules among neighboring trees, interesting collective behavior can be observed on a global scale.

FF models are defined so that trees may grow on the nodes of a surface network with periodic boundary conditions (Bak et al., 1990; Drossel & Schwabl, 1992, 1993; Christensen et al., 1993; Grassberger, 1993; Albano, 1995). Each node can assume three different states: (s1) occupied by a growing tree, (s2) occupied by a burning tree, and (s3) empty. The last corresponds to the absence of a tree that occurs after the completion of the tree burning process. Local interaction rules govern the dynamics of the system, which is updated as a cellular automation. For a given configuration, the time evolution is obtained by applying the following rules: (r1) a tree grows on an empty node with probability P; (r2) a burning tree evolves into an empty node; and (r3) a tree catches fire from a neighbor burning tree with probability $1 - G$. Thus, P is the growing probability of trees and G is the immunity of each tree to catch fire (Drossel & Schwabl, 1993).

Qualitatively, one expects that if the immunity is nonzero, the fire fronts present for $G = 0$ will become more and more fuzzy when G is increased. Of course, the fire cannot propagate for $G = 1$. For large values of P ($P \to 1$) and if the immunity is low ($G \to 0$), one observes coexistence of fire, trees, and empty nodes. Within this active state, fire fronts exhibit a rich variety of spatiotemporal patterns, including the development of spirals, and the collision between fronts. However, by keeping P constant and increasing G, the fire will eventually cease and the system will become trapped in an inactive state where the network is completely occupied by trees (see Figure 1). Since the spontaneous ignition of trees is not considered, the system will remain in the inactive state forever. So, for some critical values of the parameters $P \equiv P_c$ and $G \equiv G_c$, this simple kind of FF model exhibits irreversible phase transitions (Marro & Dickman, 1999) between active states with fire propagation and an inactive state where the fire becomes irreversibly extinguished. Figure 1 shows the critical curve at the border between these two states. In the case of standard reversible phase transitions and in marked contrast to forest fire models, the state of the system can be reversibly selected by tuning the control parameter (Binder & Heermann, 2002), that is, the temperature or pressure.

In spite of the irreversible nature of the transitions, a statistical treatment of this kind of FF model reveals that they share the relevant features of many phase transitions and critical phenomena occurring under equilibrium conditions. Therefore, FF models can be described by using a well-established framework in the field of statistical physics. In fact, close to the transition point, it is well known that the correlation length among particles is quite large—diverging to infinite at the critical point (Binder & Heermann, 2002). Therefore, at the critical state $P \to P_c$, $G \to G_c$, the burning of a single tree would affect the whole forest. The onset of this collective behavior, characteristic of critical phenomena, washes out the microscopic details (i.e., the mechanisms keeping every tree alive)

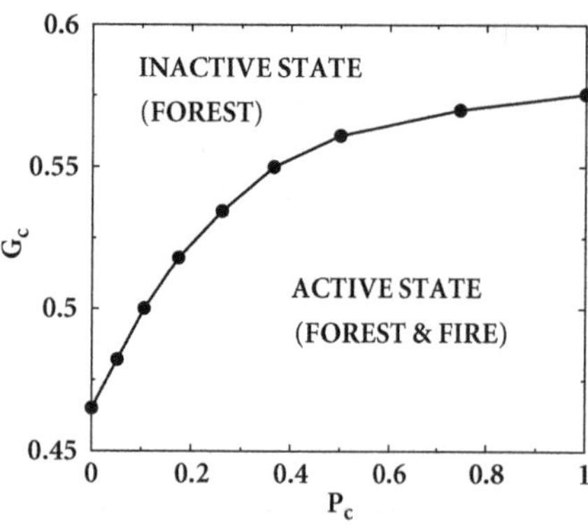

Figure 1. Plot of the critical threshold G_C versus P_C. Filled circles are critical values determined quite accurately by means of epidemic simulations. The critical line (solid curve) is drawn to guide the eye. The upper part of the figure corresponds to the inactive state of the system where the fire becomes irreversibly extinguished, while the lower part corresponds to the stationary state with fire fronts.

allowing the relevant behavior of the whole system to be caught by means of a small set of simple local rules, in agreement with the concept of interacting particle systems (Liggett, 1985).

A powerful method used to characterize and study an irreversible transition consists of performing the so-called epidemic (or spreading) simulations (ES) (Grassberger, 1989). The idea behind ES is to start the simulations from a configuration very close to the inactive state and follow the temporal evolution of the system under consideration. For this purpose, a network fully occupied by trees, except for a small patch of burning trees, is employed. Depending on the values of the parameters P and G, such a small fire would either propagate or become extinguished. During the propagation the following quantities are measured:

(i) The average number of burning trees $N(t)$;
(ii) The survival probability of the fire $P(t)$, that is, the probability that the fire is still ignited at time t;
(iii) The average mean-square distance $R^2(t)$ over which the fire has spread.

Due to the existence of a diverging correlation length, it is usually assumed that the measured quantities obey simple power laws given by

$$N(t) \propto t^{\eta}, \quad P(t) \propto t^{-\delta} \quad \text{and} \quad R^2 \propto t^{z}, \qquad (1)$$

where η, δ, and z are critical exponents. These exponents have been evaluated by means of extensive numerical simulations, yielding $\eta = 0.212 \pm 0.005$, $\delta = 0.459 \pm 0.005$, and $z = 1.115 \pm 0.010$ (Albano, 1995). This result allows us to place the FF model with immune trees within the universality class of directed percolation (DP) (Grassberger, 1989). It should be stressed that DP was previously proposed in the field of physics as a model for the systematic dripping of a fluid through a lattice with randomly occupied bonds (Kinzel, 1983), and similar irreversible behavior is observed in many disciplines, such as catalysis (the interruption of chemical reactions due to the poisoning of the catalysts (Loscar & Albano, 2003)), biology (the epidemic propagation of a disease (Marro & Dickman, 1999)), ecology (the extinction of species when resources become exhausted or the environment is changed (Rozenfeld & Albano, 2001)), and sociology (models for a society of interacting individuals (Monetti & Albano, 1997)).

By introducing an additional parameter given by the lightning probability F, which can be thought of as the probability of spontaneous ignition of a tree, to the above-formulated FF model, an interesting behavior with two characteristic time scales is observed. In the limit $F \to 0$, the onset of fire is quite sporadic and occurs on large time scales. After ignition, however, fire propagates during relatively short periods of time (Drossel & Schwabl, 1992; Christensen et al., 1993; Grassberger, 1993). By analogy to sandpile models, these rapid events are called avalanches. A statistical analysis of the area of the forest reached by fire shows that it lacks a characteristic size. So, in contrast to the standard Gaussian distribution, which describes many natural phenomena exhibiting a typical (average) size, avalanches of all magnitude are expected to be observed. Under these circumstances, it is believed that the time-scale separation effectively tunes the system into a scale-free state. This behavior is the typical sign of criticality observed in studies of the phase transitions of many substances and is due to the existence of a diverging correlation length.

Finally, several attempts to reproduce forest fires in the laboratory should be mentioned. Nahmias et al. (1989, 1996) have built a square network sample containing combustible and noncombustible blocks randomly distributed with a variable concentration. The effect of randomness on the propagation of the fire is studied by considering both the presence and absence of wind. It is found that the results are consistent with critical models of percolation. Also, Zhang et al. (1992) have studied forest-fire propagation by means of paper-burning experiments. It is found that fire fronts exhibit self-affine scaling behavior that can be described by a nonlinear stochastic differential equation of the Kardar–Parisi–Zhang type with Gaussian noise.

EZEQUIEL ALBANO

See also **Cellular automata; Critical phenomena; Flame front; Reaction-diffusion systems; Sandpile model**

Further Reading

Albano, E.V. 1995. Spreading and finite-size scaling study of the critical behaviour of a forest-fire model with immune trees. *Physica* A, 216: 213–226

Bak, P., Chen, K. & Tang, C. 1990. A forest-fire model and some thoughts on turbulence. *Physics Letters*, 147A: 297–300
Binder, K. & Heermann, D. 2002. *Monte Carlo Simulations in Statistical Physics: An Introduction*, 4th edition, Berlin and New York: Springer
Christensen, K., Flyvbjerg, H. & Olami, Z. 1993. Self-organized critical forest-fire model: mean-field theory and simulation results in 1 to 6 dimensions. *Physical Review Letters*, 71: 2737–2740
Drossel, B. & Schwabl, F. 1992. Self-organized critical forest-fire model. *Physical Review Letters*, 69: 1629–1632
Drossel, B. & Schwabl, F. 1993. Forest-fire model with immune trees. *Physica* A, 199: 183–97
Grassberger, P. 1989. Directed percolation in 2 + 1 dimensions. *Journal of Physics A (Mathematical & General)*, 22: 3673–3680
Grassberger, P. 1993. On a self-organized critical forest-fire model. *Journal of Physics A (Mathematical & General)*, 26: 2081–2089
Kinzel, W. 1983. Directed percolation. In *Percolation Structures and Processes.* Annals of the Israel Physical Society. vol. 5, edited by G. Deutscher, R. Zallen & J. Adler Jerusalem: Israel Physical Society, pp.425–445
Liggett, T.M. 1985. *Interacting Particle Systems*, New York: Springer
Loscar, E.S. & Albano, E.V. 2003. Critical behaviour of irreversible reaction systems. *Reports on Progress in Physics*, 66: 1343–1382
Marro, J. & Dickman, R. 1999. *Nonequilibrium Phase Transitions and Critical Phenomena*, Cambridge and New York: Cambridge University Press
Monetti, R.A. & Albano, E.V. 1997. On the emergence of large-scale complex behaviour in the dynamics of a society of living individuals: the stochastic game of life. *Journal of Theoretical Biology*, 187: 183–194
Nahmias, J., Téphany, H. & Duarte, J. 1996. Percolation de sites en combustion. *Comptes Rendues de l´Academie des Sciences Paris*, 332: 113–119
Nahmias, J., Téphany, H. & Guyon, E. 1989. Propagation de la combustion sur un réseau hétérogène bidimensionnel. *Revue Physique Applique*, 24: 773–777
Rozenfeld, A.F. & Albano, E.V. 2001. Critical and oscillatory behaviour of a system of smart preys and predators. *Physical Review* E, 63: 061907-1–061907-6
Zhang, J., Zhang, Y.-C., Alstrom, P. & Levinsen, M.T. 1992. Modeling forest fire by paper-burning experiment, a realization of the interface growth mechanism. *Physica* A, 189: 383–389

FOURIER TRANSFORM

See **Integral transforms**

FOUR-WAVE MIXING

See ***N*-wave interactions**

FRACTAL BASIC BOUNDARIES

See **Attractors**

FRACTAL DIMENSION

See **Dimensions**

FRACTALS

Many spatial patterns and objects in nature are either irregular or fragmented to such an extreme degree that it is very hard to describe their form. For example, the coastline of a typical oceanic island or of a continent is neither straight, nor circular, nor elliptic, and no other classical curve can serve to describe it. Similarly, there is no Euclidean surface to capture adequately the boundaries of clouds or of rough turbulent wakes. To fill this gap, Benoit Mandelbrot proposed to use a family of shapes called *fractals* for geometric representation of the shapes mentioned above and many other irregular patterns. The word *fractal* derives from the Latin *fractus* (an adjective from the verb *frangere*, meaning "to break"), capturing the irregular characteristics of geometrical objects it describes. One can think of a fractal as an irregular set consisting of parts similar to the whole. It means that a fractal is a rough or fragmented geometrical shape that can be subdivided into parts, each of which is (at least approximately) similar at reduced size to the whole. Roughness present at any resolution of a fractal object distinguishes it from Euclidean shapes. The transition from a Euclidean geometry to a fractal one reflects the conceptual jump from translational invariance of traditional Euclidean shapes to continuous scale invariance of fractal objects. The concept of fractals is useful for describing various natural objects, such as clouds, coasts, and river or road networks. However, the fractal structure of natural objects is only observed over a limited range of scales, beyond which the fractal description breaks down.

During the study of geometry of irregular patterns, it became clear that a proper understanding of irregularity or fragmentation must, among other things, involve an analysis of the intuitive notion of dimension. Topological dimension cannot properly describe strongly irregular fractal structures. Let us consider the simplest example of a fractal, namely the Cantor set, see Figure 1. For the construction of the Cantor set an iteration procedure is used. At the zeroth level, the construction of the triadic Cantor set begins with the unit interval, that is, all points on the line between 0 and 1. The first level is obtained from the zeroth one by deleting all points between $\frac{1}{3}$ and $\frac{2}{3}$, that is the middle third of the initial interval. The second level is obtained from the first level by deleting the middle third of each remaining interval at the first level. In general, the next level is obtained from the previous one by deleting the middle third of all intervals obtained from the previous level. This process continues ad infinitum. The result is a collection of points with topological dimension $d_t = 0$ since its total measure (length) is zero.

It is seen that the notion of dimension from Euclidean geometry is not very useful for the Cantor set and similar objects since it does not distinguish between the rather complex set of elements and a single point that also has a vanishing topological dimension. To cope

Figure 1. The construction of the Cantor set.

with this degeneracy, concepts of fractal dimensions were introduced for quantifying such fractal sets. Many paradoxical aspects of the Cantor set can be traced to the fact that it is dimensionally discordant. Furthermore, the discordant character of basic fractals is not at all a minor nuisance. Rather, it is such a basic feature that we shall use it to give a tentative definition of the concept of fractal. The simplest nontrivial dimension that generalizes the topological dimension is the so-called fractal dimension, which in this context can be defined as follows:

$$D_c = \lim_{n\to+\infty} \frac{\ln N_n}{\ln 1/l_n}. \qquad (1)$$

Here, N_n is the number of observable elements at the nth level, l_n is the measure (length) of the smallest element at the nth level, and $1/l_n$ is the resolution. For the triangle Cantor set at the nth level, the set consists of $N_n = 2^n$ segments, each of which has length $l_n = \frac{1}{3}^n$. Therefore, the capacity dimension of the triangle Cantor set equals

$$D_c^{\text{cantor}} = \frac{\ln 2}{\ln 3} \simeq 0.63. \qquad (2)$$

We see that fractal dimension is noninteger. Mathematically, we can say that a fractal is a set of points whose fractal dimension exceeds its topological dimension.

Felix Hausdorff developed another way to quantify a fractal set. He suggested examining the number $N(\varepsilon)$ of small intervals needed to cover the set at a scale ε. The Hausdorff dimension is defined by considering the quantity

$$M \equiv \lim_{\varepsilon\to 0} \inf_{\varepsilon_m<\varepsilon} \sum_m \varepsilon_m^d, \qquad (3)$$

which is constructed by summing d-dimensional volumes ε_m^d of balls of radii ε_m not exceeding ε that cover the fractal set. The "inf" means that all partitions of balls of radius less than ε that cover the set are considered and the one that gives the smallest possible value for the sum is kept. Hausdorff demonstrated that there is a special value D_H for the exponent d, called the *Hausdorff dimension*, such that for $d < D_H$ measure $M = \infty$ as $\varepsilon \to 0$ and for $d > D_H$ measure $M = 0$ in the limit $\varepsilon \to 0$.

Now the reader may ask, why do we need more than one fractal dimension? Actually, as is seen below, there is an infinite set of fractal dimensions. The reason for this number of quantities describing a single fractal object is the *fractal form*. The form of Euclidean shapes is well described by topology. However, as was noted by Mandelbrot the concept of form possesses mathematical aspects other than topological ones. Topology is the branch of mathematics that teaches us, for example, that all pots with two handles are of the same form because, if they were made of an infinitely flexible clay, each could be molded into any new opening, closing up any old one. Obviously, this particular aspect of the notion of form is not useful in the study of individual coastlines, since it simply indicates that they are all topologically identical to each other (and to a circle). This identity is underlined by the fact that the topological dimension in each case equals 1. By way of contrast, it will be seen that coastlines of different "degrees of irregularity" tend to have different fractal dimensions. Differences in fractal dimensions express differences in nontopological aspects of form, which can be called the *fractal form*.

Now let us consider the cumulative distribution function of mass along the Cantor set. Let the initial line's length and mass both be equal to 1, and define the cumulative distribution function for the abscissa x as being the mass contained between 0 and x. It is assumed that the total mass of the pattern is not changed while building the Cantor set. It means that we do not take away one-third of each line, but divide it in to two parts and compress them until the length of each part equals one-third of the initial line length at the previous step. Iterating this procedure up to infinity, we obtain the massive Cantor set. Because there is no mass in the intermissions, the distribution function remains constant along almost the whole length of the unit interval. However, since hammering does not affect the total mass, the distribution must manage to increase somewhere from the point of coordinates $(0, 0)$ to the point of coordinates $(1, 1)$. It increases over infinitely many, infinitely small, highly clustered jumps corresponding to the points of the Cantor set. The plot of the cumulative distribution function shown in Figure 2 is called the *Devil's staircase*. It is the plot of a continuous function on the unit interval, whose derivative is 0 almost everywhere, but it rises from 0 to 1 (see Figure 2). The cumulative sums of the widths and of the heights of the steps are both equal to 1, and one finds in addition that this curve has a well-defined length equal to 2. A curve of finite length is called *rectifiable* and is of dimension $D = 1$. This example has the virtue of demonstrating that sharp irregularities do not necessarily prevent a curve from being of dimension $D = 1$, as long as they remain sufficiently few and scattered.

You can see from Figure 2 that the Devil's staircase is a function which not differentiable at the Cantor set points, but has zero derivative everywhere else. Because a Cantor set has measure zero, this function has

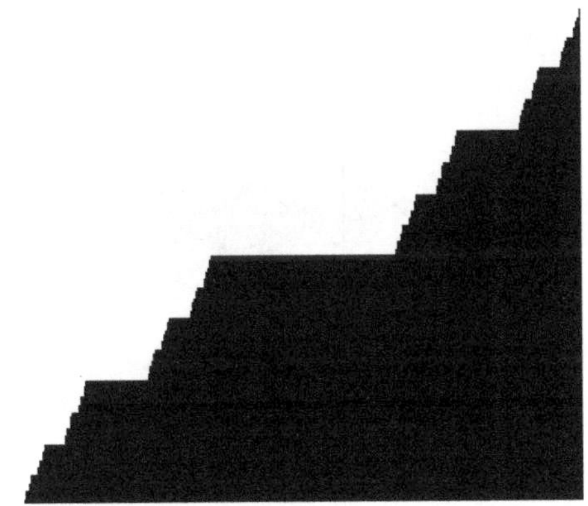

Figure 2. The Devil's staircase.

Figure 3. The Koch snowflake.

zero derivative practically everywhere and rises only on Cantor set points.

Now we consider the fractal whose dimension, in contrast to the Cantor set, exceeds unity. As an example we consider the *Koch snowflake*; see Figure 3. This is a fractal, also known as the Koch island, which was first described by Helge von Koch in 1904. It is built by starting with an equilateral triangle, removing the inner third of each side, building another equilateral triangle at the location where the side was removed, and then repeating the process indefinitely (see Figure 4). The border of the snowflake is a curve. Indeed, its area vanishes, but on the other hand each stage of its construction increases its total length by the ratio 4/3. Thus, the limit border is of infinite length. It is also continuous, but it has no definite tangent in almost all of its points because it has, so to speak, a corner almost everywhere. Its nature borders on that of a continuous function without a derivative. This object has a fractal nature. The capacity dimension of the Koch snowflake equals $D_c^{\text{Koch}} = \ln 4/\ln 3 \simeq 1.26$. Though the total length of the border is infinite, the area

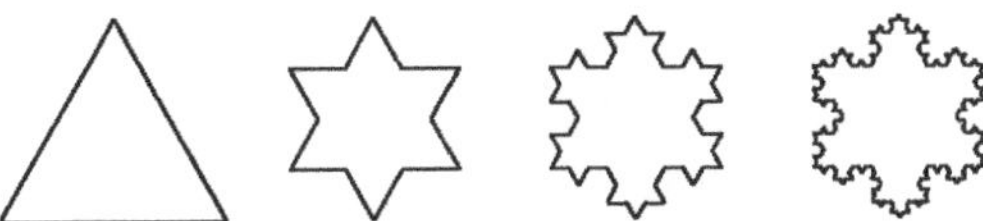

Figure 4. The construction of the Koch snowflake.

of the Koch island originated from the initial triangle with unit side is finite and equals $S^{\text{Koch}} = 2\sqrt{5}/5$.

Fractals arising in nature usually have a more complex scaling relation than simple fractals (like the Cantor set) described above. The single fractal dimension is not enough to describe their structure, because these fractal sets usually involve a range of scales that depend on their location within the set (the simplest example is the Cantor set with some measure distributed on it). Such fractals are called *multifractals*.

The multifractal formalism is a statistical description that provides global information on the self-similarity properties of fractal objects. For practical realization of the method, one primarily covers the fractal set under study by a regular array of cubic boxes of some given mesh size l. The measure of weight p_n of a given box n is then defined as the sum of the measure of interest within the box. A simple fractal dimension α is defined by the relation:

$$p_n \sim l^{\alpha}. \quad (4)$$

Multifractal analysis is a generalization in which α may change from point to point and is a local quantity. The standard method to test for multifractal properties consists in calculating the moments of order q of the measure p_n defined by

$$M_q(l) = \sum_{n=1}^{n(l)} p_n^q. \quad (5)$$

Here $n(l)$ is the total number of non-empty boxes. Varying q, one can characterize the inhomogeneity of the pattern, for instance, the values of moments with large q are controlled by the densest boxes. If scaling holds, then the generalized dimension D_q is defined by the relation:

$$M_q(l) \sim l^{(q-1)D_q}. \quad (6)$$

For instance, D_0, D_1, and D_2 are called *capacity* (defined above), *information*, and *correlation* dimensions, respectively.

In multifractal analysis, one also determines the number $N_\alpha(l)$ of boxes having similar local scaling characterized by the same exponent α. Using it, one can introduce the multifractal singularity spectrum $f(\alpha)$ as the fractal dimension of the set of singularities of strength α according to the following relation:

$$N_\alpha(l) \sim (1/l)^{f(\alpha)}. \quad (7)$$

There is a general relationship between a moment of order q and a singularity strength α, expressed

Figure 5. Examples of Julia sets.

mathematically as a Legendre transformation:

$$f(\alpha) = q\alpha - (q-1)D_q. \tag{8}$$

Fractal sets can be impressive. For example, see Figure 5 where a number of so-called *Julia sets* is shown. (See also color plate section.)

Mathematically, the Julia set can be introduced as follows. Take some function $f(z)$ and consider the sequence obtained when $f(z)$ is iterated starting from the point $z = a$:

$$a, f(a), f(f(a)), f(f(f(a))), \text{etc.} \tag{9}$$

Depending on the initial condition a and form of the function f, it may happen that these values stay small or they do not, that is, repeatedly applying f to yield arbitrary large values. So the set of all numbers (initial conditions) is partitioned; into two parts, and the Julia set associated with the function f is the boundary between these sets. The "filled" Julia set includes those numbers $z = a$ for which the iterates of f applied to remain bounded. If one considers complex numbers rather than real ones, it is the complex plane that is partitioned into two sets, and the resulting picture can be quite striking; see Figure 5. That is an example of iterating a quadratic function defined in the complex plane. Linear functions do not yield interesting partitions of the complex plane, but quadratic and higher-order polynomials do.

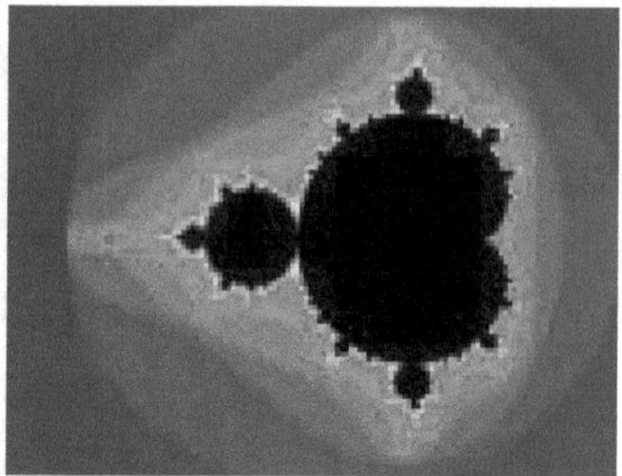

Figure 6. The Mandelbrot set.

Consider the most studied family of quadratic polynomials $f(z) = z^2 + \mu$ parametrized by a complex variable μ. As μ varies, the Julia set will vary on the complex plane. Some of these Julia sets will be connected, and some will be disconnected. Those values of μ for which the Julia set is connected are called the *Mandelbrot set* in the parameter plane. The boundary between the Mandelbrot set and its complement is often called *Mandelbrot separator curve*. The Mandelbrot set is the black shape in Figure 6 (see also color plate section). The Mandelbrot set is the set of points in the complex μ-plane that do not go to infinity when iterating the map $z_{n+1} = z_n^2 + \mu$ starting with $z = 0$. One can avoid the use of the complex numbers by substitution $z = x + iy$ and $\mu = a + ib$, and computing the orbits in the ab-plane for two-dimensional mapping:

$$\begin{aligned} x_{n+1} &= x_n^2 - y_n^2 + a, \\ y_{n+1} &= 2x_n y_n + b, \end{aligned} \tag{10}$$

with initial conditions $x = y = 0$ or equivalently $x = a$, $y = b$.

Now let us consider a natural phenomenon exhibiting fractal properties, namely, Brownian motion. Consider a fluid mass in equilibrium, for example, water in a glass; all the parts appear completely motionless. However, if one puts a small enough particle into the water, it is observed that instead of rising or descending regularly (depending on the density of the particle),

Figure 7. The example of Brownian motion of a colloidal particle.

it moves with a perfectly irregular movement. The segment of motion of a colloidal particle is presented in Figure 7 as it is seen under the microscope. The successive positions of a particle in equal time intervals are marked by points, then joined by straight lines having no physical reality whatsoever. If this particle's position were marked down 100 times more frequently, each segment would be replaced by a polygon relatively just as complicated as the whole drawing, and so on. Here, we have the natural example of the *scaling property* of the colloidal particle's motion. It is easy to see that in practice the notion of tangent is meaningless for such curves. This property of the trajectory reminds one of the Koch's snowflake described above, because when a Brownian trajectory is examined increasingly closely, its length increases without bound. Thus, we see that the trajectory of a colloidal particle has features peculiar to fractal objects, and we can qualify Brownian motion as being *fractal* and giving us a natural example of a fractal pattern.

VITCHESLAV KUVSHINOV AND ANDREI KUZMIN

See also **Brownian motion; Dimensions; Multifractal analysis**

Further Reading

Barnsley, M. 1988. *Fractals Everywhere*, Boston: Academic Press

Devaney, R. & Keen, L. (editors). 1989. *Chaos and Fractals: The Mathematics Behind the Computer Graphics*, Providence, RI: American Mathematical Society

Falcone, K.J. 1985. *The Geometry of Fractal Sets*, Cambridge and New York: Cambridge University Press

Feder, J. 1988. *Fractals*, New York and London: Plenum Press

Mandelbrot, B.B. 1988. *Fractal Geometry of Nature*, New York: Freeman

Sornette, D. 2000. *Critical Phenomena in Natural Sciences. Chaos, Fractals, Self-organization, and Disorder: Concepts and Tools*, Berlin and New York: Springer

FRAMED SPACE CURVES

The Frenet–Serret Frame

Let $\gamma : I \to \mathbb{R}^3$ be a C^3 regular parametrized space curve, with I an open interval on the real line; by regular we mean that $\gamma'(t)$ is never zero. We define the unit tangent vector $\boldsymbol{t}$ by $\boldsymbol{t} = \gamma'/|\gamma'|$ and an arclength coordinate s by $s = \int |\gamma'|\,\mathrm{d}t$. An inflection point of the curve is one where $\boldsymbol{t}'$ is zero; away from such points, we may define the unit normal by $\boldsymbol{n} = \boldsymbol{t}'/|\boldsymbol{t}'|$. Then the binormal vector is $\boldsymbol{b} = \boldsymbol{t} \times \boldsymbol{n}$, where $\times$ denotes the cross product of vectors in $\mathbb{R}^3$.

Expressing the derivatives of these mutually orthogonal unit vectors in terms of themselves produces the Frenet equations

$$\begin{aligned} \dot{\boldsymbol{t}} &= \kappa \boldsymbol{n}, \\ \dot{\boldsymbol{n}} &= -\kappa \boldsymbol{t} + \tau \boldsymbol{b}, \\ \dot{\boldsymbol{b}} &= -\tau \boldsymbol{n}, \end{aligned}$$

where the dot denotes derivative with respect to arclength s, and κ and τ are the *curvature* and *torsion* functions of the curve, respectively. Curvature may be defined in terms of $\boldsymbol{t}$ by $\kappa = |\dot{\boldsymbol{t}}|$, so that if κ is identically zero then $\gamma(t)$ is a straight line; more generally, inflection points, where κ vanishes, may be defined as points where the curve has at least second-order contact with a line.

We will say that a regular curve is *nondegenerate* if it has no inflection points. For such curves, we may define τ in terms of the scalar triple product as $\tau = (\boldsymbol{t}, \dot{\boldsymbol{t}}, \ddot{\boldsymbol{t}})/\kappa^2$. Thus, τ depends on the third derivative of γ while κ is a second-order quantity. Moreover, τ is identically zero if and only if $\gamma(t)$ lies in a fixed plane parallel to $\boldsymbol{t}$ and $\dot{\boldsymbol{t}}$.

Curvature and torsion are unchanged if we reverse the orientation of the curve (i.e., the direction of $\boldsymbol{t}$), or if the curve is rotated or translated within $\mathbb{R}^3$. (However, τ is changed by a minus sign if the curve is reflected, and both functions are multiplied by ρ^{-1} when the curve is scaled by a factor $\rho > 0$.) Conversely, uniqueness theorems for linear systems of differential equations, as applied to the Frenet equations, imply that two nondegenerate space curves with the same curvature and torsion, as functions of arclength, are congruent under an orientation-preserving isometry of $\mathbb{R}^3$ (congruence theorem).

Natural Frames

Although the vector $\kappa\boldsymbol{n}$ is well defined for any regular curve, by itself $\boldsymbol{n}$ is undefined at inflection points, and even at simple inflection points (where $\boldsymbol{t} \times \dot{\boldsymbol{t}}$ vanishes only to first order), κ is nondifferentiable and $\boldsymbol{n}$ is discontinuous. One can get around this problem and obtain a C^1 orthonormal framing along the curve, by using *natural frames*, also known as relatively parallel adapted frames (Bishop, 1975). Such frames satisfy equations of the form

$$\begin{aligned}\dot{t} &= k_1 m_1 + k_2 m_2,\\ \dot{m}_1 &= -k_1 t,\\ \dot{m}_2 &= -k_2 t,\end{aligned} \tag{1}$$

where t is the usual unit tangent vector, m_1 is a unit normal to t, $m_2 = t \times m_1$, and k_1, k_2 are the *natural curvatures*. The natural frame is not unique, but any two such frames for the same oriented curve differ only by rotating m_1 and m_2 around t through an angle which is constant along the curve.

Any regular curve that is C^k for $k \geq 2$ has a C^{k-1} natural frame. When the curve also has a Frenet frame, and we suppose $n = \cos\theta\, m_1 + \sin\theta\, m_2$, then $k_1 = \kappa\cos\theta, k_2 = \kappa\sin\theta$, and $\theta = \int \tau\, \mathrm{d}s$. Thus, a space curve is planar when its natural curvatures (k_1, k_2), when graphed in $\mathbb{R}^2$, lie along a fixed line through the origin. More generally, a space curve lies on a fixed sphere of radius ρ when (k_1, k_2) lie along a line distance $1/\rho$ from the origin.

The above congruence theorem is true for regular curves with degeneracies when suitably rephrased in terms of natural frames and natural curvatures.

The Hasimoto Correspondence

The vortex filament flow

$$\frac{\partial \gamma}{\partial t} = \frac{\partial \gamma}{\partial x} \times \frac{\partial^2 \gamma}{\partial x^2} \tag{VFF}$$

is an evolution equation for space curves that models the self-induced motion of a vortex filament in an ideal fluid taking only local effects into account. This model, originally formulated by L. Da Rios in 1906 (Ricca, 1996), is also known as the *localized induction approximation*. Note that if x is an arclength parameter at a given time, then it remains so for all t. We will take x as such, and then the right-hand side of (VFF) is κB when expressed in terms of the Frenet frame.

Hasimoto (1972) showed that solutions to this equation induce solutions to the focusing cubic nonlinear Schrödinger equation

$$\mathrm{i}q_t + q_{xx} + \tfrac{1}{2}|q|^2 q = 0, \tag{NLS}$$

by virtue of letting

$$q = \kappa \mathrm{e}^{\mathrm{i}\theta} \quad \text{with} \quad \theta = \int \tau\, \mathrm{d}s. \tag{H}$$

However, the "constant of integration" cannot be taken to be an arbitrary function of t; in fact, substituting (H) into (NLS) shows that we must have $\theta_t = \kappa^{-1}\ddot{\kappa} - \tau^2 + \frac{1}{2}\kappa^2$. Thus, $q(x, t)$ is unique up to multiplication by a unit modulus constant. Unfortunately, definition (H) for the Hasimoto map does not work at points of inflection (unless $\ddot{\kappa}$ vanishes to the same order as κ). However, it may be defined for arbitrary regular curves by using the natural curvatures of an evolving natural frame (see Calini, 2000, Section 1.4), whereupon $q = k_1 + \mathrm{i}k_2$.

DNA and White's Formula

At its lowest level of structure, DNA consists of two right-handed helices connected by base pairs, resembling the rungs of a twisted ladder. At lesser magnification the double helix, seen as a single strand, usually coils around itself in a left-handed fashion called *negative supercoiling*. Moreover, as the base pairs are separated during DNA replication and the ladder is untwisted, the supercoiling relaxes (becomes less negative) or may even reverse itself to become positive supercoiling. The dynamic interplay of these two levels of structure in DNA is governed by White's formula (White, 1970; Pohl, 1980). Although the formula, as stated below, applies only to closed loops of DNA (which occur in bacteria and in the mitochondria of some cells), it also governs the behavior of long strands of DNA, since in living cells the DNA is clamped to a protein scaffold at intervals of roughly 100,000 base pairs (Cozzarelli, 1992).

Let $\mathcal{C} \subset \mathbb{R}^3$ be a closed oriented regular curve and let v be a C^1 unit normal vector field along $\mathcal{C}$. (In DNA, we should regard $\mathcal{C}$ as one of the helices and v as pointing along the base pairs.) Let $\tilde{\mathcal{C}} = \mathcal{C} + \varepsilon v$ have the same orientation. For $\varepsilon > 0$ sufficiently small, the linking number $\mathrm{Lk}(\mathcal{C}, \tilde{\mathcal{C}})$ is independent of ε and will be denoted by Lk. Define the total twist of v by the integral

$$\mathrm{Tw} = \frac{1}{2\pi}\int (t, v, \dot{v})\, \mathrm{d}s.$$

The twisting of v about γ obviously contributes to $\mathrm{Lk}(\mathcal{C}, \tilde{\mathcal{C}})$, but there is also a contribution independent of v, due to the "writhing" of $\mathcal{C}$ about itself.

White's formula says that

$$\mathrm{Lk} = \mathrm{Tw} + \mathrm{Wr},$$

where Wr is the *writhe* of $\mathcal{C}$, defined as follows. On $D = \{(x, y) \in \mathcal{C} \times \mathcal{C} | x \neq y\}$, define the vector-valued function $e = (x - y)/|x - y|$. Extend e continuously to the two components of ∂D, as the forward- and backward-pointing unit tangent vector. Then

$$\mathrm{Wr} = \frac{1}{4\pi}\int_{\overline{D}} e^*\, \mathrm{d}S,$$

where $\mathrm{d}S$ is the standard element of area on the unit sphere S^2 and $\overline{D}$ is the closure of D. If s_1 and s_2 are arclength parameters on D, then we may also write

$$\mathrm{Wr} = \frac{1}{4\pi}\int\int \left(e, \frac{\partial e}{\partial s_1}, \frac{\partial e}{\partial s_2}\right) \mathrm{d}s_1\, \mathrm{d}s_2.$$

When $\mathcal{C}$ is nondegenerate and $\boldsymbol{v} = \boldsymbol{n}$, the Frenet normal, Lk is called the *self-linking* or *helicity* of $\mathcal{C}$, and White's formula specializes to

$$\mathrm{SL}(\mathcal{C}) = \mathrm{Wr} + \int \tau \, \mathrm{d}s,$$

a formula which was discovered much earlier by Calagareanu and later extended to regular closed curves (Pohl, 1980).

In DNA, with each base pair twisting the ladder by about 34°, Tw is positive and on the order of 10^5. The fact that DNA supercoiling is left-handed is reflected by a negative value of Wr, leading to a relatively small overall linking number. However, this number is rarely zero; the surgery that unlinks each of the two strands, a necessary step in cell division, is accomplished by enzymes known as *topoisomerases* (Sumners, 1995).

Thomas A. Ivey

See also **DNA premelting; DNA solitons**

Further Reading

Bishop, R. 1975. There is more than one way to frame a curve. *American Mathematical Monthly*, 82: 246–251

Calini, A. 2000. Recent developments in integrable curve dynamics. In *Geometric Approaches to Differential Equations*, edited by P. Vassiliou & I. Lisle, Cambridge and New York: Cambridge University Press

Cozzarelli, N. 1992. Evolution of DNA topology: implications for its biological roles. In *New Scientific Applications of Geometry and Topology*, edited by D. Sumners, Providence, RI: American Mathematical Society

Hasimoto, H. 1972. A soliton on a vortex filament. *Journal of Fluid Mechanics*, 51: 477–485

Pohl, W. 1980. DNA and differential geometry. *Mathematical Intelligencer*, 3: 20–27

Ricca, R. 1996. The contributions of Da Rios and Levi-Civita to asymptotic potential theory and vortex filament dynamics. *Fluid Dynamical Research*, 18: 245–268

Sumners, D. 1995. Lifting the curtain: using topology to probe the hidden action of enzymes. *Notices of the American Mathematical Society*, 42: 528–537

White, J. 1970. Some differential invariants of submanifolds of Euclidean space. *Journal of Differential Geometry*, 4: 207–223

FRANCK–CONDON FACTOR

The concept of Franck–Condon transitions between electronic energy levels originates from the Born–Oppenheimer approximation, which adiabatically separates the fast electronic degrees of freedom of a molecule from the much slower nuclear degrees of freedom (since the nuclei are much heavier than the electrons, electronic motions occur as if the nuclei were fixed in place). Within the Born–Oppenheimer approximation, the total wave function $\Psi(q, Q)$ can be written as a product of an electronic and a nuclear (vibrational) wave function:

$$\Psi(q, Q) = \Psi^{(\mathrm{el})}(q; Q)\varphi^{(\mathrm{vib})}(Q), \tag{1}$$

where the electronic wave function $\Psi^{(\mathrm{el})}(q; Q)$ depends only parametrically on the nuclear coordinates Q. One first calculates the electronic eigenstates for a particular nuclear configuration Q, which is assumed to be fixed in space, and assembles electronic potential energy surfaces by varying the nuclear coordinates Q step-by-step. The nuclear wave function is subsequently calculated by solving the Schrödinger equation on these potential energy surfaces, assuming that the electrons can adapt adiabatically to a changing nuclear configuration.

As usual, the intensity of a transition between two electronic states k and l with vibrational quanta n and m is given by the transition dipole matrix element

$$\mu_{kn,lm}^2 = \langle \Psi_{k,n} | \hat{\mu} | \Psi_{l,m} \rangle^2 = \left(\frac{\partial \mu}{\partial q} \right)^2 \langle \Psi_{k,n} | \hat{q} | \Psi_{l,m} \rangle^2. \tag{2}$$

When incorporating the product wave function of Equation (1), we obtain

$$\mu_{kn,lm}^2 = \left(\frac{\partial \mu}{\partial q} \right)^2 \langle \Psi_k^{(\mathrm{el})} | \hat{q} | \Psi_l^{(\mathrm{el})} \rangle^2 \langle \varphi_{k,n}^{(\mathrm{vib})} | \varphi_{l,m}^{(\mathrm{vib})} \rangle^2 . \tag{3}$$

The transition dipole matrix element separates into a product because (a) the transition dipole operator $\partial\mu/\partial q \cdot \hat{q}$ acts on the electronic coordinates q only and (b) we have assumed that the transition dipole operator does not depend on the nuclear coordinate Q (*Franck–Condon approximation*). The second term $\langle \varphi_{k,n}^{(\mathrm{vib})} | \varphi_{l,m}^{(\mathrm{vib})} \rangle^2$ is the *Franck–Condon factor*.

Figure 1a shows possible transitions between two electronic potential energy surfaces (S_0 and S_1). At not too high temperatures, only the vibrational ground state $\nu = 0$ in the electronic ground state S_0 will be populated. When both potential energy surfaces have the same shape and their lowest points are not displaced, the Franck–Condon factors $\langle \varphi_n^{(\mathrm{vib})} | \varphi_m^{(\mathrm{vib})} \rangle^2$ reduce to the condition of orthonormality of the vibrational eigenstates (i.e., the vibrational wave functions are the same on both potential energy surfaces). Hence, in that case, only the $\nu = 0 \to \nu' = 0$ transition (the *zero-phonon transition*) carries oscillator strength, while all others are dark. However, when both potential energy surfaces are displaced (i.e., the excited state has a different nuclear coordinate to the ground state), the absorption spectrum will consist of a series of absorption lines (a Franck–Condon progression) as depicted in Figure 1b. The intensity of these lines scales with the Frank–Condon factors $\langle \varphi_{k,n}^{(\mathrm{vib})} | \varphi_{l,m}^{(\mathrm{vib})} \rangle^2$. In the spectra of Figure 1b, the energy origin relates to the frequency of the zero-phonon transition, and

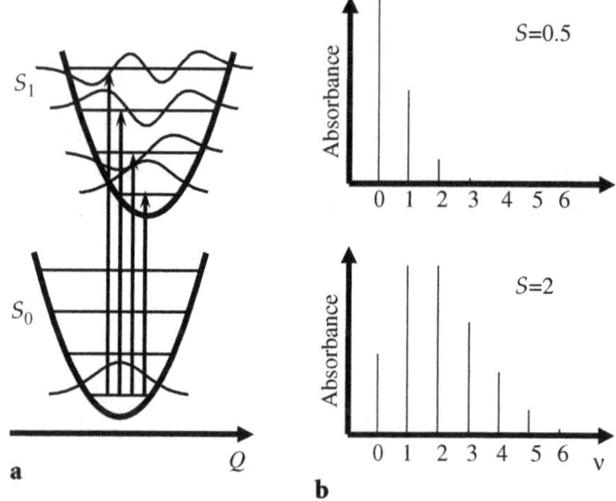

Figure 1. (a) Displaced potential energy surfaces of the electronic ground state S_0 and the first electronic excited state S_1 with the possible transitions depicted at a temperature of 0 K. The overlap integrals $\langle\varphi_i^{(\text{vib})}|\varphi_j^{\prime(\text{vib})}\rangle^2$ of the vibrational wave functions in both electronic states determine the strength of the individual absorption lines. (b) Absorption spectra for Huang–Rhys parameters $S = 0.5$ and 2, respectively.

the energy spacing between the various lines is given by the frequency ω of the oscillator. We call normal modes with such a displacement of the potential energy surfaces Franck–Condon active modes.

An analytic expression for the Franck–Condon factors can be given when the potential energy surfaces are assumed to be harmonic (as in Figure 1a):

$$V_{S_0} = \frac{1}{2}\hbar\omega Q^2,$$

$$V_{S_1} = E_{01} + \frac{1}{2}\hbar\omega Q^2 - \bar{x}\hbar\omega Q, \qquad (4)$$

where Q is the dimensionless oscillator coordinate, E_{01} the energy separation between the electronic ground and first excited states at $Q = 0$, and ω the vibrational frequency. The dimensionless displacement $\bar{x}$ between potential energy surfaces is a measure of the strength of the coupling between electronic and vibrational transition. It is common to introduce an alternative coupling parameter, the Huang–Rhys parameter:

$$S \equiv \frac{x^2}{2} \qquad (5)$$

with which the Franck–Condon factors can be expressed as (Fitchen, 1968)

$$\langle\varphi_{S_1,n}^{(\text{vib})}|\varphi_{S_0,m}^{(\text{vib})}\rangle^2 = \mathrm{e}^{-S}\frac{m!}{n!}S^{n-m}\left[L_m^{n-m}(S)\right]^2, \qquad (6)$$

where $L_j^{i-j}(S)$ are the Laguerre polynominals. For the important case where $T = 0$ K, the only occupied initial level is $m = 0$ and $L_0^n(S) = 1$, and we obtain

$$\langle\varphi_{S_1,n}^{(\text{vib})}|\varphi_{S_0,0}^{(\text{vib})}\rangle^2 = \frac{S^n}{n!}\mathrm{e}^{-S}. \qquad (7)$$

For $S < 1$, the zero phonon line is the most intense transition, while for larger displacements, $S > 1$, the intensity is shifted toward transitions with higher vibrational quanta on the electronically excited state S_1 (Figure 1b). This finding can be explained in simple words with the Franck–Condon principle, which states that in a classical picture, an electronic transition occurs very quickly and is most likely to occur without changes in the positions of the nuclei. Hence, the transition is vertical from the bottom of the S_0 potential energy surface to the S_1 surface. The quantum mechanical formulation of this principle is that the intensity of a vibronic transition is proportional to the square of the overlap integral between the vibrational wave functions of the two states that are involved in the transition.

The concept of a Franck–Condon transition was originally formulated for electronic states. However, it can be used in a more general sense and appears whenever an adiabatic separation of timescales can be made between two molecular degrees of freedom. For example, this is the case when a high-frequency vibrational mode is coupled to a low-frequency vibrational mode. In that case, $\Psi^{(\text{el})}$ in the expressions above is replaced by the wave function of the high-frequency mode, but the rest of the formalism remains unchanged. Most prominent examples are hydrogen bonds XH$\cdots$Y, where the high-frequency X$-$H stretching vibration is coupled to the low-frequency XH$\cdots$Y hydrogen bond vibration. One of the striking features of hydrogen bonds is an extraordinarily large anharmonic coupling between both modes that leads to a Franck–Condon-like progression in the absorption spectrum of the X$-$H stretching band.

PETER HAMM

See also **Color centers; Hydrogen bond**

Further Reading

Fitchen, D.B. 1968. Zero-phonon transitions. In *Physics of Color Centers*, edited by W.B. Fowler, New York: Academic Press, pp. 293–350

FREDHOLM THEOREM

The Fredholm theorem gives a necessary and sufficient condition for the existence of a solution to a linear equation of the form $A\boldsymbol{x} = \boldsymbol{y}$, where A is a linear operator and $\boldsymbol{x}$ and $\boldsymbol{y}$ are vectors. Consider the matrix

$$A = \begin{pmatrix} 1 & 4 & 5 \\ 2 & 5 & 7 \\ 3 & 6 & 9 \end{pmatrix}, \qquad (1)$$

and the linear map $\mathcal{F}$, which associates the vector $A\boldsymbol{x}$ to each vector $\boldsymbol{x}$ in $\mathbb{R}^3$. If x_i, $i = 1\ldots3$, are the components of $\boldsymbol{x}$, we have

$$\mathcal{F}\boldsymbol{x} = A\boldsymbol{x} = x_1 \begin{pmatrix} 1 \\ 2 \\ 3 \end{pmatrix} + x_2 \begin{pmatrix} 4 \\ 5 \\ 6 \end{pmatrix} + x_3 \begin{pmatrix} 5 \\ 7 \\ 9 \end{pmatrix} = (x_1 + x_3) \begin{pmatrix} 1 \\ 2 \\ 3 \end{pmatrix} + (x_2 + x_3) \begin{pmatrix} 4 \\ 5 \\ 6 \end{pmatrix}. \tag{2}$$

The third equality comes from the fact that the last column of A can be written as the sum of its first two columns. It is then clear that the equation $\mathcal{F}\boldsymbol{x} = \boldsymbol{y}$ has a solution if and only if $\boldsymbol{y}$ is a linear combination of the first two columns of A. In this case, there is more than one solution because the kernel or null space of A is not trivial.

If we equip $\mathbb{R}^3$ with the inner product $\langle \boldsymbol{x}, \boldsymbol{y} \rangle = \sum_{i=1}^{3} x_i\, y_i$, the transpose A^{T} of A, defined by $\{A^{\mathrm{T}}\}_{ij} = \{A\}_{ji}$, is such that

$$\langle A\boldsymbol{x}, \boldsymbol{v} \rangle = \langle \boldsymbol{x}, A^{\mathrm{T}}\boldsymbol{v} \rangle, \tag{3}$$

for any two vectors $\boldsymbol{x}$ and $\boldsymbol{v}$ in $\mathbb{R}^3$. This equality shows that $\boldsymbol{v}$ is orthogonal to the range of A (i.e., $\langle A\boldsymbol{x}, \boldsymbol{v} \rangle = \boldsymbol{0}$ for every $\boldsymbol{x}$ in $\mathbb{R}^3$) if and only if $\boldsymbol{v}$ is in the kernel of A^{T}; that is, $A^{\mathrm{T}}\boldsymbol{v} = \boldsymbol{0}$. As a consequence, the existence of a solution to the equation $\mathcal{F}\boldsymbol{x} = \boldsymbol{y}$ (i.e., $A\boldsymbol{x} = \boldsymbol{y}$), which is equivalent to $\boldsymbol{y}$ being in the range of A, can be restated as $\boldsymbol{y}$ being orthogonal to the kernel of A^{T}. This is the contents of the Fredholm theorem, which is often presented as the following alternative theorem.

Fredholm alternative: For a matrix A, either the equation $A\boldsymbol{x} = \boldsymbol{y}$ always has a solution or the kernel of A^* is nontrivial. Here A^* is the adjoint of A, which is its transpose (resp. Hermitian transpose) if A has real (resp. complex) entries. The Fredholm alternative can be stated in a similar fashion for a special class of infinite-dimensional linear operators. In this case, $A = \lambda I - \mathcal{L}$, where $\mathcal{L}$ is a compact operator, λ is a nonzero complex number, and I is the identity operator (see for instance Reed & Simon, 1980; Kreyszig, 1989; Dunford & Schwartz, 1988).

Applications of this theorem involve multiple scales analysis and perturbation theory. Assume for instance that we want to solve the eigenvalue problem $\mathcal{L}\boldsymbol{x} = \lambda\boldsymbol{x}$, where the linear operator $\mathcal{L}$ can be written as $\mathcal{L} = \mathcal{L}_0 + \varepsilon\mathcal{L}_1$ and ε is a small parameter. If we look for expansions of the eigenvector $\boldsymbol{x} = \boldsymbol{x}_0 + \varepsilon\boldsymbol{x}_1 + \cdots$ and eigenvalue $\lambda = \lambda_0 + \varepsilon\lambda_1 + \cdots$ in powers of ε, substitute in the equation, and equate the left-and right-hand sides at each power of ε, we obtain a hierarchy of linear equations of the form

$$(\mathcal{L}_0 - \lambda_0 I)\,\boldsymbol{x}_i = \mathcal{J}_i, \quad i = 0, 1, \ldots, \tag{4}$$

where I is the identity operator and the right-hand side, $\mathcal{J}_i$, of (4) only depends on $\boldsymbol{x}_k$ for $k < i$. In particular, $\mathcal{J}_0 = \boldsymbol{0}$ and $\mathcal{J}_1 = \lambda_1\boldsymbol{x}_0 - \mathcal{L}_1\boldsymbol{x}_0$. At each step of this iterative process, the linear equation $(\mathcal{L}_0 - \lambda_0 I)\,\boldsymbol{x}_i = \mathcal{J}_i$ has a solution if and only if $\mathcal{J}_i$ is orthogonal to the kernel of the adjoint of $\mathcal{L}_0 - \lambda_0 I$. This typically gives the value of the coefficient λ_i of the expansion of the eigenvalue λ in powers of ε. Once λ_i is known, one can then solve the equation $(\mathcal{L}_0 - \lambda_0 I)\,\boldsymbol{x}_i = \mathcal{J}_i$ for $\boldsymbol{x}_i$. A similar method can be applied to solve equations of the form

$$\mathcal{L}\boldsymbol{x} = \mathcal{G}(\boldsymbol{x}), \tag{5}$$

where $\mathcal{L} = \mathcal{L}_0 + \varepsilon\mathcal{L}_1$ and $\mathcal{G}$ is a nonlinear function of $\boldsymbol{x}$. Such equations are often encountered in bifurcation theory. Assume for instance that $\mathcal{G}$ is polynomial, say $\mathcal{G} = \langle \boldsymbol{x}, \boldsymbol{x} \rangle\boldsymbol{x}$, and that one knows a solution $\boldsymbol{x}_0$ to the linear equation $\mathcal{L}_0\boldsymbol{x}_0 = \boldsymbol{0}$. Small-amplitude solutions to the nonlinear problem can be approximated by introducing a small parameter that measures the size of $\boldsymbol{x}$ and rescaling. For instance, the scaling $\boldsymbol{x} = \sqrt{\varepsilon}\,\boldsymbol{y}$ turns (5) into $(\mathcal{L}_0 + \varepsilon\mathcal{L}_1)\,\boldsymbol{y} = \varepsilon\langle \boldsymbol{y}, \boldsymbol{y} \rangle\boldsymbol{y}$. A solution of the form $\boldsymbol{y} = \boldsymbol{y}_0 + \varepsilon\boldsymbol{y}_1 + \cdots$ can then be sought, using an iterative process similar to the one discussed above.

JOCELINE LEGA

See also **Perturbation theory**

Further Reading

Dunford, D. & Schwartz, J.T. 1988. *Linear Operators. Part I: General Theory*, New York and Chichester: Wiley

Kreyszig, E. 1989. *Introductory Functional Analysis with Applications*, New York and Chichester: Wiley

Reed, M. & Simon, B. 1980. *Methods of Modern Mathematical Physics. I. Functional Analysis*, Boston and London: Academic Press

FREE ENERGY

The concept of "free energy" is a key concept to characterize physically relevant states in statistical mechanics (see, e.g., Mandl, 1988; Reichl, 1998). Given an equilibrium system of statistical mechanics with energy levels E_i of the microstates i, the (Helmholtz) free energy F is defined as

$$F(\beta) = -\frac{1}{\beta}\log Z(\beta), \tag{1}$$

where

$$Z(\beta) = \sum_i \mathrm{e}^{-\beta E_i} \tag{2}$$

is the partition function and β is the inverse temperature. Apparently, the free energy is different from the internal energy U given by

$$U = -\frac{\partial}{\partial\beta}\log Z(\beta) = \langle E_i \rangle. \tag{3}$$

The difference is given by the entropy S times the absolute temperature T:

$$F = U - TS. \tag{4}$$

This equation can also be regarded as describing a Legendre transformation from U to F. Equilibrium states minimize the free energy (rather than the internal energy)—in this sense F is more relevant than U. The minimum of F can be achieved in two contrasting ways, either by making the internal energy U small, or by making the entropy S large (or both). Changes of the free energy are related to the maximum amount of work a system can do at constant pressure and temperature.

The basic principle underlying statistical mechanics, the maximum entropy principle, can also be formulated as a "principle of minimum free energy" (Beck & Schlögl, 1993). In this formulation, one starts from fixed intensive quantities (e.g., the inverse temperature β) and considers a free energy function $F[p]$ as a function of an a priori arbitrary set of probabilities $[p]=\{p_1, p_2, \ldots\}$. The probabilities p_i can describe any state, for example, also non-equilibrium and nonthermal states. One then requires

$$F[p]=\sum_i p_i E_i+\beta^{-1} p_i \log p_i=U-\beta^{-1} S \quad (5)$$

to have a minimum in the space of all possible probability distributions. This minimum is achieved for the canonical distribution

$$p_i=\frac{1}{Z} \mathrm{e}^{-\beta E_i}. \quad (6)$$

There are straightforward generalizations of this principle for systems with several intensive quantities, for example, grand canonical ensembles.

For nonlinear dynamical systems, the concept of free energy is often seen in a much broader sense. Various generalized types of free energies can be defined. A key ingredient for this more general approach is the so-called thermodynamic formalism of dynamical systems (Beck & Schlögl, 1993). One defines partition functions that contain information on certain fluctuating quantities associated with the dynamical system, for example, local singularities of the invariant density or local Lyapunov exponents. To proceed to a (formal) statistical mechanics description, one then defines a free energy function for these partition functions quite analogous to Equation (1).

Examples are "static," "dynamic," and "expansion" free energies of dynamical systems, as well as the so-called "topological pressure." Let us here consider the static free energy in somewhat more detail. This free energy function contains information on the spectrum of singularities of the invariant density of the dynamical system and is closely related to very important quantities that characterize multifractals, the so-called Rényi dimensions. One covers the attractor (or repeller) under consideration with small d-dimensional cubes ("boxes") of volume ε^d, where d is an integer dimension large enough to embed the attractor. The static partition function is then defined as

$$Z(\beta)=\sum_i p_i^\beta=\sum_i \mathrm{e}^{-\beta \alpha_i V}, \quad (7)$$

where p_i is the invariant probability measure associated with box i and β is a formal "inverse temperature" parameter. To illustrate the similarities with statistical mechanics, we have written in the above equation $p_i=: \varepsilon^{\alpha_i}=e^{-\alpha_i \cdot V}$, where α_i is like a fluctuating energy associated with each box i and $V=-\log \varepsilon$ plays the role of a formal volume variable. One can then study the free energy density associated with this partition function,

$$f(\beta):=\lim_{V \rightarrow \infty}-\frac{1}{\beta V} \log Z(\beta). \quad (8)$$

This static free energy, up to a trivial factor, is indeed identical with the Rényi dimensions D_β:

$$f(\beta)=\frac{\beta-1}{\beta} D_\beta. \quad (9)$$

In this way, methods borrowed from statistical mechanics play an important role for the statistical description of dynamical systems.

While the above generalized free energies, of relevance for nonlinear dynamical systems, have only formal analogies with conventional free energies, a more fundamental generalization of statistical mechanics suitable for complex systems has been suggested by Tsallis (1988). In this so-called non-extensive statistical mechanics approach (Abe & Okamoto, 2001; Kaniadakis et al., 2002), the entire formalism of statistical mechanics is generalized by starting from more general entropy measures, the Tsallis entropies. These are defined by

$$S_q=\frac{1}{q-1}\left(1-\sum_i p_i^q\right). \quad (10)$$

Here the p_i are probabilities associated with the microstates i of a physical system. Note that the Tsallis entropies look somewhat similar to the Rényi information measures but are indeed different (there is no logarithm). The parameter q, called "entropic index," can take on any real value but is in practice often close to 1. For $q \rightarrow 1$, one can easily check that the Tsallis entropies reduce to the ordinary Shannon entropy $S_1=-\sum_i p_i \log p_i$. The principal idea of non-extensive statistical mechanics is to do everything we know from ordinary statistical mechanics, but start from the more general entropy measures. Naturally, ordinary statistical mechanics is contained as a special case $(q=1)$ in this more general formalism.

The maximum entropy principle yields power-law generalizations of the canonical ensemble,

$$p_i = \frac{1}{Z_q(\beta)}(1 + (q-1)\beta E_i)^{-1/(q-1)}, \qquad (11)$$

which reduce to Equation (6) for $q \to 1$. As a consequence, there is then also a more general free energy $F_q(\beta)$, which is parametrized by the entropic index q. One can basically show that the entire formalism of thermodynamics has simple q-generalizations, for example, for the generalized free energy there is a relation of the form

$$F_q = U_q - T S_q, \qquad (12)$$

where the index q indicates that the q-generalized canonical ensemble is chosen. This type of formalism has interesting physical applications, for example, for fully developed turbulence and for scattering processes in high-energy physics (Beck, 2002).

CHRISTIAN BECK

See also **Chaotic dynamics; Dimensions; Entropy; Fractals; Multifractal analysis**

Further Reading

Abe, S. & Okamoto, Y. (editors). 2001. *Nonextensive Statistical Mechanics and Its Applications*, Berlin and New York: Springer

Beck, C. & Schlögl, F. 1993. *Thermodynamics of Chaotic Systems*, Cambridge and New York: Cambridge University Press

Beck, C. 2002. Nonextensive methods in turbulence and particle physics. *Physica A*, 305: 209-217

Kaniadakis, G., Lissia, M. & Rapisarda, A. (editors). 2002. *Nonextensive Thermodynamics and Physical Applications*, Physica A, 305 (special volume)

Mandl, F. 1988. *Statistical Physics*, 2nd edition, Chichester and New York: Wiley

Reichl, L.E. 1998. *A Modern Course in Statistical Physics*, 2nd edition, New York: Wiley

Tsallis, C. 1988. Possible generalization of Boltzmann–Gibbs statistics. *Journal of Statistical Physics*, 52: 479-487

FREE PROBABILITY THEORY

A new probabilistic context emerged in the 1980s (Voiculescu, 1985) from studies in operator algebras, a mathematical subject with ties to quantum physics. Free probability theory (FPT) is developing along a parallel to the basics of usual probability theory, drawn from assumptions designed for quantities with the highest degree of noncommutativity. FPT has turned out to have important connections to random matrix theory (Voiculescu, 1991) as well as to some combinatorics (Speicher, 1998) and to certain models in physics (Voiculescu, 1997).

To introduce FPT, one takes the view that "a probability theory" deals with expectation values $\varphi(a)$ for the elements a, called random variables, of a set A endowed with operations of addition, multiplication, and multiplication by scalars. We may describe FPT as noncommutative probability theory (i.e., the products ab and ba can be different for $a, b \in A$) with independence defined by a new relation called free independence. In usual probability theory, A consists of numerical functions on some space of events and clearly $ab = ba$ in this case. By contrast, noncommutative probability occurs typically when A consists of quantum mechanical observables Q. These are linear operators on a vector space $\mathcal{H}$ with an inner product $\langle\cdot,\cdot\rangle$ and $\varphi(Q) = \langle Q\xi, \xi\rangle$ where $\xi \in \mathcal{H}$ is a fixed vector with $\langle\xi, \xi\rangle = 1$. Under quantum mechanics, the usual definition of independence of Q_1 and Q_2 requires commutation $Q_1Q_2 = Q_2Q_1$ and $\varphi(f(Q_1)g(Q_2)) = \varphi(f(Q_1))\varphi(g(Q_2))$ for polynomials f and g. Free independence of Q_1, Q_2, by contrast requires that

$$\varphi(\dots f(Q_1)g(Q_2)h(Q_1)\dots) = 0 \qquad (1)$$

whenever

$$\dots, \varphi(f(Q_1)) = 0, \varphi(g(Q_2)) = 0,$$
$$\varphi(h(Q_1)) = 0, \dots, \qquad (2)$$

where $\dots, f, g, h, \dots$ are polynomials and Q_1, Q_2 alternate as arguments. The definition of free independence extends to sets of variables by replacing the one-variable polynomials $f, g, h, \dots$ by corresponding polynomials in noncommuting variables from those sets. Note that if Q_1, Q_2 satisfy free independence, then in most cases $Q_1Q_2 \neq Q_2Q_1$.

The FPT analogue of the central limit theorem (Voiculescu, 1985; Voiculescu et al., 1992) states $\varphi(Q_j) = 0$, $\varphi(Q_j^2) = 1$ and $|\varphi(Q_j^k)| \le C_k$, then

$$S_n = n^{-1/2}(Q_1 + \dots + Q_n) \qquad (3)$$

will have in the limit, the moments of a semicircle distribution, that is,

$$\lim_{n\to\infty} \varphi(S_n^k) = (2\pi)^{-1}\int_{-2}^{2} t^k (4 - t^2)^{1/2}\,dt. \qquad (4)$$

Many other topics in classical probability theory have FPT analogs. The list includes the following (Voiculescu et al., 1992; Biane & Speicher, 1998; Voiculescu, 2002):

- Addition and multiplication of freely independent random variables corresponds to highly nonlinear free convolution operations on the distributions.
- The FPT infinitely divisible laws and stable laws have been classified, and there are corresponding processes with free increments.
- Semicircular processes are the FPT analog of Gaussian processes, and in particular there is a concept of free Brownian motion.
- There are stochastic integrals with respect to free Brownian motion.

- There is a free entropy theory that deals with FPT analogues of Shannon's differential entropy and of the Fisher information.

The semicircle distribution, which plays the role of the Gauss law in FPT, was well known from Wigner's work as the limit-distribution of eigenvalues of large symmetric random matrices with independent identically distributed Gaussian entries (the Gaussian orthogonal ensemble). The explanation of this surprising coincidence was found to be that free independence often occurs asymptotically among large random matrices (Voiculescu, 1991). For instance, if for each n we are given two sets $\{A_1^{(n)}, \dots, A_m^{(n)}\}$ and $\{B_1^{(n)}, \dots, B_p^{(n)}\}$ of $n \times n$ Hermitian random matrices, then the requirement that their joint distribution (as matrix-valued random variables) be the same as that of $\{UA_1^{(n)}U^{-1}, \dots, UA_m^{(n)}U^{-1}, B_1^{(n)}, \dots, B_p^{(n)}\}$ for all unitary matrices U is the key to asymptotic free independence of the two sets (Voiculescu, 1998). For $n \times n$ random matrices T, the expectation functional is $\varphi_n(T) = n^{-1}E(TrT)$, where E denotes the expectation of the numerical random variable TrT. Asymptotic free independence means the equalities of free independence hold after taking the limit $n \to \infty$.

A random matrix T is thus a classical matrix-valued random variable and at the same time a noncommutative random variable. Passing to the noncommutative context discards information about the entries, since the expectation values $\varphi_n(T^k)$ remember only the eigenvalue distribution of T.

Asymptotic free independence results for random matrices imply that their limit behavior is governed by FPT. For instance, for the sum or product of random matrices satisfying free independence asymptotically, the limit distribution of eigenvalues is computed by the free convolution operations.

Among the simplest cases of asymptotic free independence are a pair of independent matrices from the Gaussian ensembles or a constant matrix and one from a Gaussian ensemble. Also, if X_n is a rectangular $[\alpha n] \times n$ Gaussian random matrix with independent and identically distributed (i.i.d.) entries, then $X_n^* X_n$ and a constant $n \times n$ random matrix will exhibit asymptotic free independence. Besides these simplest instances of asymptotic free independence, there are many others involving unitary matrices, random permutation matrices or matrices with non-Gaussian i.i.d. entries (Voiculescu, 2000). Also, Gaussian random band-matrices can be handled with more involved FPT techniques (Shlyakhtenko, 1996). Most asymptotic free independence results for complex Hermitian or unitary matrices also hold for real symmetric or orthogonal matrices. The differences between distributions of eigenvalue spacings between the real and complex cases occur at another scale, which is wiped out by taking the limit of the φ_n as $n \to \infty$.

FPT methods have also been used for physics models related to asymptotics of random matrices, like the large N Yang–Mills two-dimensional quantum chromodynamics in papers of I.M. Singer, R. Gopakumar and D. Gross, and M. Douglas (Voiculescu, 1997). Also occurring in physics are multimatrix-models, corresponding to ensembles of k-tuples of $N \times N$-matrices with the densities $c_N \exp(-NTrP(T_1, \dots, T_k))$ that have connections to the analog of entropy in FPT (Voiculescu, 2002). An indication about where FPT and quantum field theory connect is given by the combinatorial approach to FPT (Speicher, 1998) that deals with noncrossing partitions of $\{1, \dots, n\}$, that is, with planar Feynman diagrams.

DAN VIRGIL VOICULESCU

See also **Quantum field theory; Random matrix theory I, II, III, IV**

Further Reading

Biane, P. & Speicher, R. 1998. Stochastic calculus with respect to free Brownian motion and analysis on Wigner space. *Probability Theory and Related Fields*, 112: 373–409

Shlyakhtenko, D. 1996. Random Gaussian band matrices and freeness with amalgamation. *International Mathematics Research Notices*, 20: 1013–1025

Speicher, R. 1998. Combinatorial theory of the free product with amalgamation and operator-valued free probability theory. *Memoirs of the American Mathematical Society*, 627

Voiculescu, D. 1985. Symmetries of some reduced free product C^*-algebras. In *Operator Algebras and Their Connections with Topology and Ergodic Theory*, edited by H. Araki et al., Berlin and New York: Springer

Voiculescu, D. 1991. Limit laws for random matrices and free products. *Inventiones Mathematicae*, 104: 201–220

Voiculescu, D.V. (editor). 1997. *Free Probability Theory*, Providence, RI: American Mathematical Society

Voiculescu, D. 1998. A strengthened asymptotic freeness result for random matrices with applications to free entropy. *International Mathematics Research Notices*: 41–63

Voiculescu, D. 2000. Lectures on free probability theory. In *Lectures on Probability Theory and Statistics, École d'Été de Probabilities de Saint-Flour XXVIII –1998*, edited by M. Emery, A. Nemirovski & D. Voiculescu, New York: Springer

Voiculescu, D. 2002. Free entropy. *Bulletin of the London Mathematical Society*, 34: 257–278

Voiculescu, D.V., Dykema, K.J. & Nica, A.M. 1992. *Free Random Variables*, Providence, RI: American Mathematical Society

FRENKEL–KONTOROVA MODEL

In 1938, Yakob Frenkel and Tatiana Kontorova developed in some detail a model originally proposed by Ulrich Dehlinger (see Nabarro, 1967) for studying stationary dislocations in a crystal, and its motion. A perfect crystalline (regular lattice) solid subjected to small deformations returns to a perfect lattice arrangement of atoms after the forces causing the deformations are released. This regime of small deformations is called "elastic." Larger deformations can lead to rearrangements of the crystal lattice. In this other regime, called "plastic," the perfect lattice is not restored after tension is released. The attempt to describe the energetics involved in those irreversible rearrangements (or lattice dislocations, see Figure 1(b)) led to the posing of a theoretical physics problem of a discrete elastic one-dimensional field of displacements (atomic positions) experiencing a spatially periodic potential (Dehlinger, 1929; Frenkel & Kontorova, 1939; Frank & Van der Merwe, 1949). Figure 1(c) shows a schematic representation of Dehlinger's *Verhakung* (interlocking), as he named his idea.

The Frenkel–Kontorova (FK) model has been used for modeling nanostructural properties of adsorbed monolayers of atoms on crystalline surfaces, nonlinear magnetic excitations in magnets, ionic conductors, and many other solid-state physics problems. Today it has become a textbook mathematical model with direct applications in condensed matter, statistical physics, nonlinear fields, and material science, often under different names, notably the "discrete sine-Gordon" or "periodic Klein–Gordon" equation.

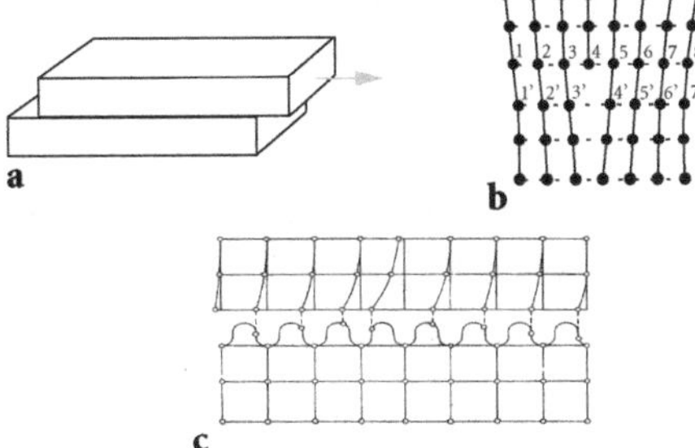

Figure 1. (a) Schematic net outcome of a plastic deformation. Relative shift between layers does not occur uniformly: The slip process can take place sequentially by the mechanism of dislocation motion shown in (b). (c) Dehlinger's picture of the imperfection, that he named *Verhakung*, where the atoms of the upper row, in this displacement, experience a periodic potential on account of their interaction with atoms in the lower row.

The Generalized Frenkel–Kontorova Model

In the generalized FK model, a discrete one-dimensional field (u_n) $(-\infty < n < +\infty)$, has a potential energy density

$$H(u_{n+1}, u_n) = K\,V(u_n) + W(\Delta u_n), \qquad (1)$$

where $W(\Delta u)$ is the interaction between neighbor components of the field, $V(u)$ is a periodic function $V(u+1) = V(u)$, and K is a parameter controlling the relative strength of both (V and W) contributions to the energy density.

The standard version of the model corresponds to the simplest choices of sinusoidal on-site potential V and harmonic spring interaction W:

$$V = \frac{1}{(2\pi)^2}\left[1 - \cos(2\pi u)\right],$$
$$W(\Delta u) = \frac{1}{2}(\Delta u)^2 - \sigma\,\Delta u, \qquad (2)$$

where σ is the unstretched length of the spring, the value at which W is a minimum.

Keeping V as above, the following choice for W

$$W = \frac{1}{(2\pi)^2}\left[1 - \cos(2\pi(\Delta u - \gamma))\right] \qquad (3)$$

(known as a "chiral XY model" in a magnetic field) models a system of magnetic moments with strong planar (XY) anisotropy (here described by the angles $2\pi u_n$), with chiral (preferred relative angle γ) interactions W, and K in Equation (1) measuring the intensity of the external magnetic field. Other choices are motivated by different physical situations of interest.

To complete the formal definition of the generalized FK model, one must define its time evolution and this choice depends on the specific application one has in mind. The standard model with Hamiltonian dynamics

$$\frac{d^2u_n}{dt^2} = u_{n+1} - 2u_n + u_{n-1} - \frac{K}{2\pi}\sin(2\pi u_n) \qquad (4)$$

corresponds to the simplest discretized version of the integrable partial differential equation known as sine-Gordon (SG):

$$\frac{d^2u}{dt^2} = \frac{d^2u}{dx^2} - \sin(2\pi u). \qquad (5)$$

This justifies the name of "discrete SG model" often used for the FK model.

Some physical applications, for example, in Josephson junction devices, suggest introducing dissipative terms in Equation (4) (involving first-time derivatives $\dot{u}_n$) and more complicated functional forms of the right-hand-side dependence on u_n and u_{n+1}. The richness of dynamical behaviors shown by the different dynamical FK models is a continuing source of interest in both theory development and applications (Braun & Kivshar, 1998, 2004; Floría & Mazo, 1996).

The Equilibrium Problem for Convex FK Models

A simpler problem, but quite complex, is the characterization of equilibrium properties of the generalized FK model. The physical motivation for studying this problem comes mainly from the observed abundance of modulated phases in minerals and man-made materials and compounds, and the need to understand the peculiar multiphase diagrams shown by experimental studies.

For the case in which W is not a convex function, like the chiral XY example above, the difficulties of a complete rigorous characterization of equilibria are much greater than in the case of convex W interaction. For the latter, Aubry's work allowed a satisfactory theory in the infinite system size (thermodynamic) limit, by showing a one-to-one correspondence between equilibrium configurations (u_n), which satisfy the (force-equilibrium) equations

$$\frac{\partial(H(u_n, u_{n-1}) + H(u_{n+1}, u_n))}{\partial u_n} = 0 \quad (6)$$

and orbits of a symplectic twist map of the cylinder.

When an equilibrium configuration is such that any segment of length $L = M - N + 1$, $(u_N, \ldots, u_M)$, can only increase its energy as result of arbitrary fluctuations of $u_{N+1}, \ldots, u_{M-1}$, the field configuration is called a minimum energy configuration (MEC).

There are two types of MEC, which are both recurrent—meaning that given any integer p and any number $\varepsilon > 0$, there are integers m and $q > 0$ such that both inequalities: $|u_{p+q} - u_p - m| < \varepsilon$ and $|u_{p+q+1} - u_{p+1} - m| < \varepsilon$ are satisfied. These two types of recurrent MEC are called commensurate (or periodic) phases and incommensurate (or quasi-periodic) phases, respectively. Remarkably, there are nonrecurrent MECs, which are rigorous discrete versions of the soliton solutions of continuous partial differential equations (like the integrable sine-Gordon model). These are called "discommensurations" or "homoclinic field configurations," and they correspond to the original motivation (Dehlinger's *Verhakung*) of the FK model: localized dislocations over the recurrent (periodic or quasi-periodic) substrate phase.

Commensurate Phases

These are periodic configurations with rational p/q average spacing ω (defined as $\lim_{L\to\infty}(u_M - u_N)/L$), where $u_{n+p} - q = u_n$. Such field configurations are generically pinned, meaning that one has to put some finite energy E_{PN} on the system to displace the configuration over the path-dependent barriers of energy (called "Peierls–Nabarro barriers") separating configurations that are equivalent by translations σ_{mn}.

A pinned configuration has a finite coherence length (or decay range) ξ of fluctuations, meaning that if a field component (the center u_{n_0}) is kept displaced by an external force, the rest of components u_n are displaced with exponentially decaying amplitude $\exp(-|n - n_0|/\xi)$. Also, the linear spectrum of small oscillations (phonon spectrum) of a pinned configuration has a finite gap $\Delta > 0$.

Incommensurate or Quasi-periodic Phases

Recurrent configurations with irrational average spacing ω can be viewed as limits of sequences of periodic phases of mean spacing $\omega_n \to \omega$ as $n \to \infty$. The physical properties of these structures depend on the parameter K of the model: For each irrational ω there is a critical value $K_c(\omega)$ such that:

- For $K < K_c$, the quasi-periodic MEC is sliding (unpinned), so that $E_{\text{PN}} = 0$. Consequently, localized induced fluctuations extend through the system so that the decay range diverge, $\xi \to \infty$. The spectrum of phonons of a sliding configuration is gapless, $\Delta = 0$. The distribution function of the fractional parts $\text{Frac}(u_n)$ in this phase is a continuous function.
- For $K > K_c$, the incommensurate phase is pinned, $E_{\text{PN}} > 0$, the decay range is finite, and a gap Δ appears in the phonon spectrum. The distribution function of the values $\text{Frac}(u_n)$ is now a Cantor function; there are infinitely many forbidden intervals (gaps) of values of the field (modulo 1).

MacKay (1993) has characterized this drastic change in the physical properties of quasi-periodic fields as a critical phenomenon (already called by Aubry as a "transition by breaking of analiticity") using renormalization group methods for the case of noble irrational ω and the standard model (2).

Discrete Solitons (DSs) or Discommensurations (DCs)

Field configurations can connect asymptotically ($n \to \pm\infty$) two commensurate configurations (usually one is a translation of the other, though they can be different in general) by an energy minimizing path of exponentially localized length (the width of the DS) around a lattice site (the center of the DS). The width is of the order of the decay range ξ of the recurrent (periodic) substrate.

These structural defects are the elementary excitations of the modulated phase, or localized compressions (advanced DCs) or expansions (retarded DCs) superimposed on the regular substrate modulation. If ω is close to a rational ω_0, the modulated phase corresponding to a mean spacing ω can be correctly described as an array of localized DCs (advanced if $\omega > \omega_0$, retarded if $\omega < \omega_0$) over a periodic substrate of mean spacing ω_0. The interaction energy between neighboring DCs decays exponentially with the quotient interdistance/width, $\exp(-1/(\xi|\omega - \omega_0|))$, so that for ω very close to ω_0 and/or high values of the pinning parameter K (ξ small), the elementary excitations are almost non-interacting (and pinned). However, at lower values of K the tail overlapping of the DCs makes the interactions strong enough, and the array of strongly interacting DCs becomes sliding.

Though with less rigorous formal basis, the DC theory of modulated phases is a well-established formal description of equilibrium for nonconvex FK models. In these cases, interactions between DCs can be of either (attractive or repulsive) sign and even alternating with interdistance. Also, the inner structure of DC can be more complex than in the convex case.

Luis Mario Floría and Pedro Jesús Martínez

See also **Aubry–Mather theory; Commensurate-incommensurate transition; Dislocations in crystals; Josephson junction arrays; Peierls barrier; Sine-Gordon equation**

Further Reading

Braun, O.M. & Kivshar, Yu. S. 1998. Nonlinear dynamics of the Frenkel–Kontorova model. *Physics Reports*, 306: 1–108

Braun, O.M. & Kivshar, Yu. S. 2004. *The Frenkel–Kontorova Model*, Berlin and New York: Springer

Dehlinger, U. 1929. Zur theorie der rekristallisation reiner metalle [On the theory of recrystallization in pure metals]. *Annalen der Physik*, (5)2: 749–793

Floría, L.M. & Mazo, J.J. 1996. Dissipative dynamics of the Frenkel–Kontorova model. *Advances in Physics*, 45: 505–598

Frank, F.C. & van der Merwe, J.H. 1949. One dimensional dislocations I & II. *Proccedings of the Royal Society (London)* A, 198: 205–225

Frenkel, Y. & Kontorova, T. 1939. On the theory of plastic deformation and twinning. *Journal of Physics (Moscow)*, 1: 137–149 (original Russian publication 1938)

MacKay, R.S. 1993. *Renormalisation in Area-preserving Maps*, Singapore: World Scientific

Nabarro, F.R.N. 1967. *Theory of Crystal Dislocations*, Oxford: Clarendon Press

FREQUENCY DOUBLING

Frequency doubling, or second-harmonic generation, is the process in which an intense laser beam propagating through a nonlinear optical medium produces light at double the frequency of the input beam (Bloembergen, 1969; Boyd, 2002; Shen, 1984; Zernike & Midwinter, 1973). It was the first laser-induced nonlinear optical effect to be reported, and its observation by Franken et al. (1961) marked the beginning of the field of nonlinear optics. This work involved the use of ruby laser radiation at 694.3 nm propagating though a quartz crystal to produce second-harmonic radiation at 347.2 nm (Franken et al., 1961).

Our laboratory demonstration of second-harmonic generation is shown in Figure 1 (also in color plate section). The 800 nm radiation of a titanium-doped sapphire laser is converted into the second harmonic at 400 nm in passing through a lithium niobate crystal. A prism is used to separate the second harmonic from the fundamental radiation exiting of the crystal. A side view showing the ray trajectories is represented in Figure 1c. We can see the red light of the laser beam illuminating the crystal and the blue light of the second harmonic exiting the crystal, together with some portion of unconverted red light. Typically, not all of the incident radiation is converted into the second harmonic. The quantity describing the effectiveness of converting the incident radiation into a harmonic is called the conversion efficiency.

The physical origin of the process of second-harmonic generation can be understood as follows. The electric field of a monochromatic optical wave incident upon a nonlinear medium forces the electrons bound in atoms and molecules to oscillate about their equilibrium positions. Thus, it induces a polarization (dipole moment per unit volume) in the medium that depends on the strength of the electric field. This polarization plays the role of a secondary source of electromagnetic radiation.

The Lorentz model, which treats the atom as a harmonic oscillator, provides a good description of the linear atomic response. The electric force F_E exerted on the electron by the electromagnetic field displaces the electron with respect to the nucleus. The restoring force F_{rest} has the opposite action, tending to return the electron to its equilibrium position. If the applied field is weak, the restoring force is linear with respect to the electron displacement x from equilibrium, that is,

$$F_{rest} = -m\omega_0^2 x.$$

Here m is the mass of the electron and ω_0 is the resonance frequency of the oscillator. Therefore, the potential energy function of the medium is quadratic with respect to x (see Figure 2a, the solid line):

$$U = -\int F_{rest}\,dx = \tfrac{1}{2}m\omega_0^2 x^2.$$

The parabolic shape of the potential energy function describes harmonic oscillations of the electron at the applied optical field frequency. In such a way, the induced polarization gives rise to the radiation at the applied field frequency. Figure 2b shows the waveform of the incident monochromatic electromagnetic wave of frequency ω (the top graph). For the case of a medium with linear response, there is no distortion of the waveform associated with the polarization of the medium (the middle graph).

In order to describe the nonlinear response, the model has to be extended by allowing the possibility of a nonlinearity in the restoring force (Boyd, 2002). The lowest-order nonlinear response is modelled by adding a contribution quadratic in the displacement so that

$$F_{rest} = -m\omega_0^2 x - max^2.$$

Here a characterizes the strength of the nonlinearity. The potential energy function of such a system has an additional term that is cubic with respect to x (Figure 2a, the dashed line):

$$U = -\tfrac{1}{2}m\omega_0^2 x^2 + \tfrac{1}{3}max^3.$$

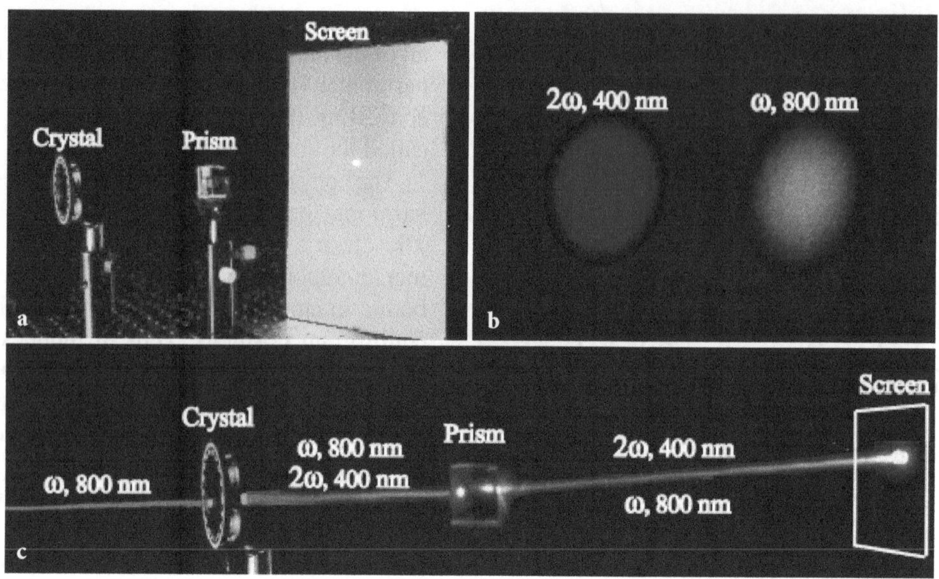

Figure 1. Second-harmonic generation in lithium niobate: (a) Experimental setup; (b) close-up view of the observation screen; (c) side view showing ray trajectories.

Because of the nonlinearity of the restoring force, the atomic response can show a harmonic distortion of the waveform associated with polarization oscillation (Figure 2b the bottom graph). In this case, the induced polarization will give rise to radiation not only at the same frequency as the applied field, but also at higher frequencies (harmonics) as well. A nonlinear effect can occur only if the applied field is sufficiently strong to produce a large displacement x.

More formally, the nonlinear response of a medium to a strong incident field can be described by a nonlinear susceptibility (Bloembergen, 1969; Boyd, 2002; Shen, 1984). In the case of a linear response to the applied field, the induced polarization P depends linearly upon the electric field strength E according to the relationship

$$P = \chi^{(1)}E, \qquad (1)$$

where the constant of proportionality $\chi^{(1)}$ is the linear susceptibility. In the case of a nonlinear response, optical response can be described by expanding polarization P as a power series in the field strength E:

$$P = \chi^{(1)}E + \chi^{(2)}E^2 + \chi^{(3)}E^3 + \cdots . \qquad (2)$$

The quantities $\chi^{(2)}$ and $\chi^{(3)}$ represent the second- and third-order nonlinear optical susceptibilities, respectively. Here, we will consider only the second-order nonlinear response.

Second-order nonlinear interactions can occur only in noncentrosymmetric media. A typical value of the second-order nonlinear susceptibility in crystals is $\chi^{(2)} \simeq 5 \times 10^{-8}$ esu (6.3×10^{-7} SI). We refer to

$$P^{(2)} = \chi^{(2)}E^2 \qquad (3)$$

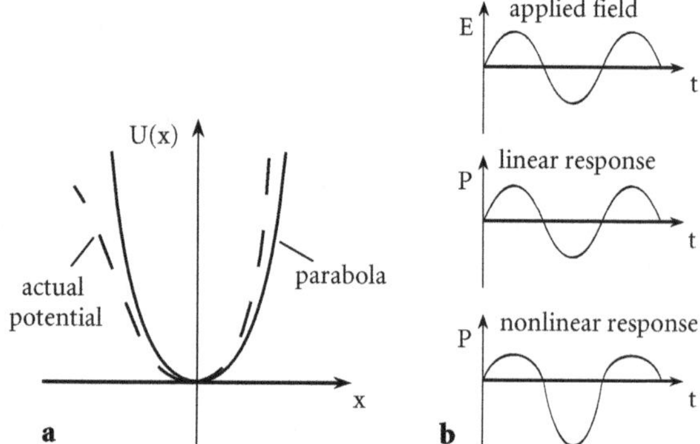

Figure 2. (a) Potential energy function in case of linear response (solid line) and nonlinear noncentrosymmetric response (dashed line); (b) polarization waveforms associated with the atomic response.

as the second-order nonlinear polarization. The strength of the applied optical field can be represented by

$$E = E_0 \exp^{-i\omega t} + \text{c.c.} \qquad (4)$$

Substituting (4) into the equation for the second-order nonlinear polarization (3), we can obtain

$$P^{(2)} = 2\chi^{(2)}EE^* + (\chi^{(2)}E^2 \exp^{-i2\omega t} + \text{c.c.}). \qquad (5)$$

We can see from (5) that the second-order polarization consists of a constant contribution (the first term) and a contribution at frequency 2ω (the second term). This doubled frequency contribution leads to the generation of radiation at the second-harmonic frequency 2ω.

Frequency doubling can also be understood by means of a virtual energy level description (Figure 3) (Boyd, 2002). The diagram shows that for a lossless

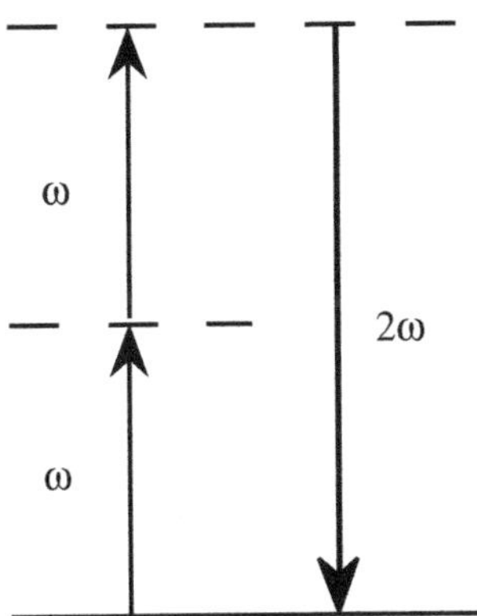

Figure 3. Energy-level diagram describing second-harmonic generation.

medium, the creation of a photon at a doubled frequency 2ω must be accompanied by the destruction of two photons at frequency ω, or according to Manley–Rowe relations: the rate at which a photon at frequency 2ω is created is equal to the rate at which two photons at frequency ω are destroyed.

Second-harmonic generation is a process in which the initial and final quantum mechanical states (energy levels) are identical. Therefore, the energy conservation and momentum conservation laws need to be satisfied. According to the momentum conservation law, the momentum of the two incident photons of frequency ω, $2\hbar\boldsymbol{k}_\omega$, must be equal to the momentum of the radiated photon of frequency 2ω, $\hbar\boldsymbol{k}_{2\omega}$:

$$2\hbar\boldsymbol{k}_\omega = \hbar\boldsymbol{k}_{2\omega}. \tag{6}$$

Here $\boldsymbol{k}_\omega$ and $\boldsymbol{k}_{2\omega}$ are the wave vectors of the incident wave and the second harmonic, respectively. According to (6), in order for the second-harmonic generation to effectively occur, it is necessary for the following phase matching condition to be satisfied:

$$2\boldsymbol{k}_\omega = \boldsymbol{k}_{2\omega}. \tag{7}$$

The dispersion relation for light propagating in a medium is

$$k_\omega = n(\omega)\frac{\omega}{c}, \tag{8}$$

where $n(\omega)$ is the index of refraction for the wave of frequency ω. Thus, condition for perfect phase matching with collinear beams (7) requires that

$$n(\omega) = n(2\omega). \tag{9}$$

In practice, the condition of perfect phase matching is often difficult to achieve because the refractive index of materials displays an effect of normal dispersion: $n(\omega)$ increases monotonically with ω. Thus, condition (9) cannot be satisfied.

There are several ways to achieve phase matching. One of the most common ways is to use the birefringence displayed by many crystals. Birefringence is the dependence of the refractive index on the direction of polarization of the optical radiation. In order to achieve phase matching through the use of birefringent crystals, the highest-frequency wave (2ω) should be polarized in the direction that gives it the lower of two possible refractive indices. There is another technique, known as quasi-phase matching, that can be used when normal phase matching cannot be implemented (Armstrong et al., 1962; Boyd, 2002). This technique involves the use of periodically poled materials, that are fabricated so that the orientation of one of the crystalline axis is inverted periodically as a function of position within the material. It results in periodic alternation of the sign of $\chi^{(2)}$, that can compensate for a nonzero wave vector mismatch $\Delta\boldsymbol{k}$.

Second-harmonic generation can usefully convert the coherent output of a fixed-frequency laser to a different spectral region (Zernike & Midwinter, 1973). For example, the infrared radiation of an Nd:YAG laser operating at 1064 nm can be converted into 532 nm visible radiation with a conversion efficiency of more than 50%.

KSENIA DOLGALEVA, NICK LEPESHKIN, AND ROBERT W. BOYD

See also **Harmonic generation; Manley–Rowe relations; Nonlinear optics**

Further Reading

Armstrong, J.A., Bloembergen, N., Ducuing, J. & Pershan, P.S. 1962. Interaction between light waves in a nonlinear dielectric. *Physical Review*, 127: 1918–1939

Bloembergen, N. 1969. *Nonlinear Optics*, New York: Plenum Press

Boyd, R.W. 2002. *Nonlinear Optics*, San Diego: Academic Press

Franken, P.A., Hill, A.E., Peters, C.W. & Weinreich, G. 1961. Generation of optical harmonics. *Physical Review Letters*, 7: 118–119

Shen, Y.R. 1984. *The Principles of Nonlinear Optics*, New York: Wiley

Zernike, F. & Midwinter, J.E. 1973. *Applied Nonlinear Optics*, New York: Wiley

FREQUENCY LOCKING

See **Coupled oscillators**

FRÖHLICH THEORY

Starting in the late 1960s and continuing until his death in 1991, Herbert Fröhlich developed a theory of biological coherence that was based on quantum interactions between dipolar constituents of biomolecules. Fröhlich advocated momentum-space correlations within a living system such as a membrane, a cell, or an organism. This dynamic order would be a characteristic feature that distinguishes living systems from inanimate matter. The key assumptions

DISORDERED STATE (Unpolarized)

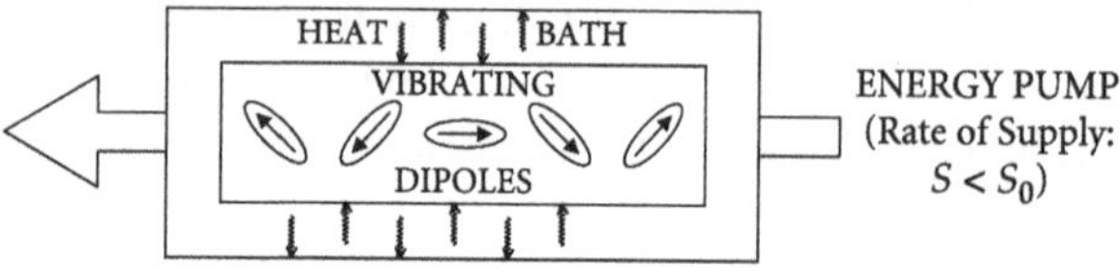

ORDERED STATE (Polarized)

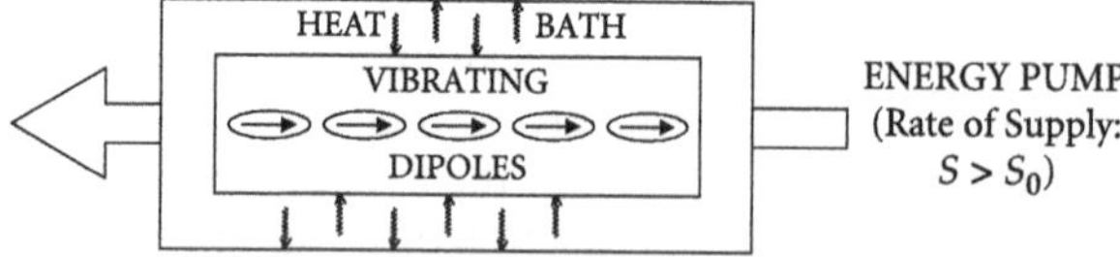

Figure 1. Illustration of the Fröhlich model.

of Fröhlich's theory can be listed as follows:

- (i) a continuous supply of metabolic energy (also referred to as energy pumping) above a threshold level,
- (ii) the presence of thermal noise due to physiological temperature,
- (iii) internal organization of the biosystem that promotes functional features,
- (iv) the existence of a large transmembrane potential difference, and
- (v) a nonlinear interaction between two or more types of degrees of freedom.

As a result of these nonlinear interactions, in addition to the global minimum that characterizes a biological system in its nonliving state, a metastable energy minimum is achieved in the living state.

The Fröhlich model of biological coherence is based on a condensate of quanta of collective polar vibrations. It is a non-equilibrium property due to the interactions of the system with both the surrounding heat bath and an energy supply (see Figure 1). The energy is channeled into a single collective mode that becomes strongly excited. Most importantly, the Fröhlich model relies on the nonlinearity of internal vibrational mode interactions and in this respect is somewhat reminiscent of the laser action principle. Associated with this dynamically ordered macroscopic quantum state is the emergence of polarization due to the ordering of dipoles in biomolecules such as membrane head groups. Furthermore, Fröhlich predicted the generation of coherent modes of excitation such as dipole oscillations in the microwave frequency range. Nonlinear interactions between dynamic degrees of freedom were predicted to result in the local stability of the polarized state and in the long-range frequency-selective interactions between two systems with these properties.

In the search for empirical support of this model, Fröhlich placed an emphasis on the presence of dipole moments in many biomolecular systems that would then oscillate in synchrony in the frequency range of $10^{11}-10^{12}$ Hz due to their nonlinear interactions. Consequently, due to the resonant dipole-dipole coupling in a narrow frequency range, the entire biological system can be seen as a giant oscillating dipole. An alternative picture developed within the Fröhlich theory was that of a Bose–Einstein condensation in the space of dipole oscillations. The Hamiltonian postulated by Wu and Austin (1977) takes the form

$$H = \sum_i \omega_i a_i^\dagger a_i + \sum_i \Omega b_i^\dagger b_i + \sum_i \Theta_i P_i^\dagger P_i + \frac{1}{2}\sum_{i,j,k}(\chi a_i^\dagger a_j b_k^\dagger + \chi^* a_j a_i^\dagger b_k) + \sum_{i,j}(\lambda b_i a_j^\dagger + \lambda^* b_i^\dagger a_j) + \sum_{i,j}(\xi P_i a_j^\dagger + \xi^* P_i^\dagger a_j), \quad (1)$$

where $(a_i^\dagger, a_i)$, $(b_i^\dagger, b_i)$, and $(P_i^\dagger, P_i)$ are the cell, heat-bath, and energy-pump creation and annihilation (boson-type) operators, respectively. A kinetic rate equation was derived for this model that indicates Bose-type condensate in the frequency domain with a stationary occupation number dependence of the dipole modes given by

$$N_i = \left[e^{\beta(\omega_i - \mu)} - 1\right]^{-1}. \quad (2)$$

The nonlinear coupling comes from the dipole-phonon interaction proportional to χ. Provided that the oscillating dipoles are within a narrow band of resonance frequencies ($\omega_{\min} \leq \omega_i \leq \omega_{\max}$) and the coupling constants (χ, λ and ξ) are large enough, long-range ($\sim 1\ \mu$m) attractive forces are expected to act between the dipoles.

The effective potential between two interacting dipoles which initially vibrate with frequencies ω_1 and ω_2 is given by

$$U(r) = \frac{E}{r^3} + \frac{F}{r^6}. \quad (3)$$

The van der Waals coefficient F is negative (attractive) unless the average distance between the interacting particles is very short. However, the van der Waals attraction between particles rapidly decreases with distance making it an unlikely candidate mechanism for biological coherence. The coefficient E can be both positive (repulsion) and negative (attraction) depending on the occupation of the new eigenstates of the system. Population inversion leads to attraction while thermal equilibrium for occupation numbers leads to repulsion between particles. In particular, E is given by

$$E = \frac{\hbar e^2 Z \mid \bar{A} \mid \gamma}{4\bar{\omega} M}\left(\frac{1}{\varepsilon'(\omega_+)} - \frac{1}{\varepsilon'(\omega_-)}\right), \quad (4)$$

where M is the mass of a dipole, e is the electron charge, Z is the number of elementary charges on each dipole, $\bar{A}$ is an angle constant, and $\varepsilon'(\omega)$ is the real part of the frequency-dependent dielectric constant. The quantities $\omega_{\pm}$ are the new dipole vibration frequencies

$$\omega_{\pm} = \left[\frac{1}{2}(\omega_1{}^2+\omega_2{}^2) \pm \left(\frac{1}{4}\left(\omega_1{}^2-\omega_2{}^2\right)+\frac{\beta_0{}^4}{(\varepsilon'_{\pm})^2}\right)^{1/2}\right]^{1/2}, \quad (5)$$

where $\beta_0 = \gamma^2 e^2 \sqrt{Z_1 Z_2}/Mr^3$. In the resonant frequency case, the effective interaction energy between two oscillating dipoles was found to be long-range, depending on distance as r^{-3}.

Most of the expected condensation of dipolar vibrations was envisaged by Fröhlich to take place in the cell membrane due to its strong potential on the order of 10–100 mV across the thickness of 5–10 nm giving an electric field intensity of $1–20 \times 10^6$ V/m. The resultant dipole-dipole interactions were calculated to show a resonant long-range order at a high-frequency range of $10^{11}-10^{12}$ Hz with a propagation velocity of about 10^3m/s. In addition to membrane dipoles, several other candidates for Fröhlich coherence were considered, including double ionic layers, dipoles of DNA and RNA molecules, and plasmon oscillations of free ions in the cytoplasm.

Applications of the Fröhlich theory were made to explain cancer proliferation, where a shift in the resonant frequency was seen to affect the cell-cell signaling, the brain waves, and enzymatic chemical reactions.

Various experiments have been reported that appeared to demonstrate the sensitivity of metabolic processes to certain frequencies of electromagnetic radiation above the expected Boltzmann probability level. Raman scattering experiments by Webb (1980) found non-thermal effects in *E. coli* but these could not be reproduced by other labs. Irradiation of yeast cells by millimeter waves showed increased growth at specific frequencies (Grundler, et al., 1983). Rouleaux formation of human erythrocytes was explained in terms of Fröhlich's resonant dipole-dipole attraction (Paul et al., 1983) but did not rule out standard coagulation processes. While some of these experiments illustrate nonthermal effects in living matter that would require nonlinear and non-equilibrium interactions for explanation, no unambiguous experimental proof has been furnished to date to support Fröhlich's hypothesis.

It is interesting to note that while Fröhlich sought evidence for frequency selection in biological systems, another famous physicist, Alexander S. Davydov tried to find spatial localization of vibrational energy in biological systems such as DNA and peptides. There is a conceptual link between these two approaches as shown by Tuszyński et al. (1983) that involves self-focusing in reciprocal (Fröhlich) or real space (Davydov).

JACK A. TUSZYŃSKI

See also **Bose–Einstein condensation; Cluster–coagulation; Davydov soliton; Synergetics**

Further Reading

Davydov, A.S. 1982. *Biology and Quantum Mechanics*, Oxford: Pergamon Press

Fröhlich, H. 1968. Long range coherence and energy storage in biological systems. *International Journal of Quantum Chemistry*, 2: 641–649

Fröhlich, H. 1972. Selective long range dispersion forces between large systems. *Physics Letters*, 39A: 153–155

Fröhlich, H. 1980. The biological effects of microwaves and related questions. In *Advances in Electronics and Electron Physics*, vol. 53, edited by L. Marton and C. Marton, New York: Academic Press, pp. 85–152

Grundler, W., Eisenberg, C.P. & Sewchand, L.S. 1983. *Journal of Biological Physics*, 11: 1

Paul, R., Chatterjee, R., Tuszyński, J.A. & Fritz, O.G. 1983. Theory of long-range coherence in biological systems. I: the anomalous behaviour of human erythrocytes. *Journal of Theoretical Biology*, 104: 169

Tuszyński, J.A., Paul, R., Chatterjee, R. & Sreenivasan, S.R. 1983. Relationship between Fröhlich and Davydov models of biological order. *Physical Review* A, 30: 2666

Webb, S.J. 1980. Laser-Raman spectroscopy of living cells. *Physics Reports*, 60: 201–224

Wu, T.M. & Austin, S. 1977. Bose condensation in biosystems. *Physics Letters* A, 64: 151

FRONT PROPAGATION

See **Reaction-diffusion systems**

FRUSTRATION

Frustration is the inevitable consequence of trying to simultaneously satisfy a number of conflicting constraints. In general, when faced with such conflicting requirements one's only way out is to compromise. Inevitably, there are usually many widely different ways to compromise, some better than others. In physical systems, frustration arises from competition between different contributions to the total energy of a system. Generally, each contribution can be separately minimized by a unique choice of state but the different contributions have different minimum-energy states. The presence of frustration then typically leads to a situation in which there are many, very different, states with similar low energies separated from one another by "barriers" of much higher energy states. This leads to a proliferation of low-energy dynamical modes in the system and, especially in the presence of disorder, slow (glassy) dynamics.

A simple physical example that shows the way in which frustration leads to complexity is afforded by the Ising model (*See* **Spin systems**). Each site of a regular lattice bears a simple binary variable—a spin than can

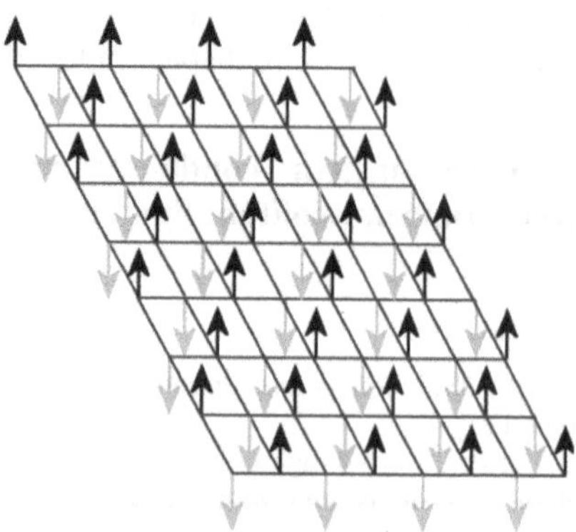

Figure 1. The ground state configuration of the 2-d Ising model on a square lattice.

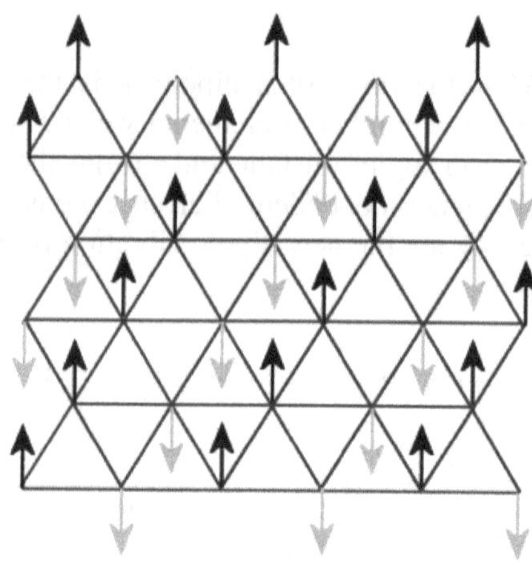

Figure 2. One of the degenerate ground state configurations of the 2-d Isig model on a triangular lattice.

point either up or down. The energy of the whole system has a contribution from each bond on the lattice. A given bond contributes an energy $+J$ if the spins at either end of the bond point in the same direction and an energy $-J$ if they point in opposite directions (this corresponds to the anti-ferromagnetic Ising model). First, consider the model on a two-dimensional square lattice with N sites (this model was solved exactly by Lars Onsager in 1940) as shown in Figure 1.

We can divide the lattice into two sub-lattices such that each site on one sub-lattice is surrounded entirely by sites on the other sub-lattice. The lowest energy state of the system is trivially found—we make all the spins on one sub-lattice point in one direction and all the spins on the other sub-lattice point in the opposite direction (clearly there are two such states with the same energy related by global inversion of all the spins). Each bond of the lattice connects two oppositely oriented spins so that the total energy is $-2NJ$. Any other configuration of spins must result in at least four bonds connecting like spins and so has energy $\geq -2JN + 4J$. This situation is simple and unfrustrated—all bonds can be simultaneously "happy."

Now consider the same simple model, with the same rules for assigning energy, but on a triangular lattice as shown in Figure 2. It is fairly simple to see that there is no way of simultaneously making all the bonds happy. Each elementary plaquette of the lattice has three spins, each of which wants to be in a different state to its neighbors—at least one bond must be unsatisfied (i.e., have energy $+J$). The minimum energy states of this system have alternate "stripes" of aligned spins so that each plaquette has only one unsatisfied bond giving a minimum energy of $-NJ$. A little thought will tell one that there are six equivalent configurations with this minimum energy: three choices for the direction of the stripes as well as the possibility of reversing all spins.

A very simple situation in which there is geometrical frustration arises when one considers the adsorption of one material onto a surface of a different material. Suppose that we have a substrate of material A, which is flat, and we deposit atoms of material B onto it. Further suppose that a bulk sample of pure A atoms has a lattice parameter a, while a bulk sample of pure B atoms has a lattice parameter b. When only a few atoms of material B have been deposited, they will simply bond with A atoms at the surface. Once a large number of B atoms have been deposited, they will begin to come into contact and be able to bond with each other. In the case that $b \approx a$, there is no frustration—the B atoms simply attach themselves to surface A atoms and quite happily bond to other B atoms when they are adjacent. This happy situation is referred to as epitaxial growth and happens for the technologically important situation in which material A is the semiconductor gallium arsenide and material B is the alloy aluminium gallium arsenide. However if $b \not\approx a$, then the growth of B is frustrated—there is a competition between the desire of the B atoms to attach themselves to A atoms and their desire to be a distance b from the nearest B atom. If the energetics are such that one set of bond energies is much greater than the other, the system will either form an adsorbed, strained layer, with B–B neighbor distance set by the substrate (a) or it will detach itself entirely. If the energetics are more closely balanced, then the system will find a compromise, often involving the formation of dislocations in the surface layer. There are, of course, many ways of introducing such dislocations to relieve the strain. A simple model for this effect was proposed by Frenkel and Kontorova (*See* **Frenkel–Kontorova model**).

An interesting system in which the frustration can be continuously tuned experimentally is the Josephson junction array (*See* **Josephson junction arrays**). This is a system composed of a lattice of superconducting islands separated by weak links. Each island is characterized by a phase variable and the total energy of the array can be written

$$E = -\varepsilon_J \sum_{\langle i,j \rangle} \cos\left(\theta_i - \theta_j - A_{ij}\right),$$

where the sum is over nearest-neighbor sites, θ_i is the phase on the ith island, ε_J is the Josephson energy, and A_{ij} is a twist variable along a nearest-neighbor bond. The twist variables must satisfy

$$\sum A_{ij} = 2\pi \frac{\Phi}{\phi_0},$$

where the sum is around all the bonds on an elementary plaquette of the lattice, Φ is the magnetic flux through that plaquette, and ϕ_0 is the flux quantum $h/2e$. When the flux per plaquette is an integer number of flux quanta, the system is unfrustrated and its ground state simply has all phases equal. When Φ/ϕ_0 is a rational fraction, the system is uniformly frustrated and its ground states exhibit periodic spatial ordering. If Φ/ϕ_0 is irrational (or, in finite-sized systems if the commensuration length exceeds the system size), then the system exhibits complex, aperiodic ordering. The introduction of frustration introduces a greater range of elementary excitations and a greater degeneracy in the lowest energy states.

Spin glasses are magnetic alloys in which there are competing interactions between magnetic ions that can be of either sign. The canonical model of a spin glass was devised by S.F. Edwards and P.W. Anderson (Edwards & Anderson, 1975) in which each site of a regular lattice bears a classical vector spin. Nearest-neighbor pairs of spins interact via an exchange coupling drawn from a quenched (i.e., nondynamical) random distribution that includes both ferromagnetic and antiferromagnetic possibilities. The energy of a given configuration of spins is then given by

$$E = -\sum_{\langle i,j \rangle} J_{ij} \mathbf{S}_i \cdot \mathbf{S}_j.$$

The combination of frustration and quenched randomness leads to very rich behavior. A mean field version of this model in which all pairs of spins interact weakly was proposed by Sherrington and Kirkpatrick (1975) and solved by Parisi (1979) by introducing the notion of replica symmetry breaking. An alternative approach based on an equation (the so-called TAP equation) for the number of metastable states was presented by Thouless, Anderson, and Palmer (Mezard et al., 1987; Fischer & Hertz, 1991). The relevance of the notions developed for this mean field model to real finite range spin glasses (the EA model) has been questioned and the so-called droplet picture for these was developed (Fischer & Hertz, 1991). The debate over the relative merits of the droplet picture and the broken replica symmetry picture has resurfaced recently and is still highly contentious. Irrespective of which detailed picture is correct, it is clear that the combination of frustration and quenched randomness leads to spin glass systems having many, closely competing low-energy configurations. This leads to a hierarchy of free energy valleys and barriers causing the dynamics of these systems to show very complex and slow dynamical behavior such as aging in which the dynamical response of the system depends on its thermal history. Much current attention is focussed on quantum mechanical frustrated systems both regular (spin liquid states in quantum antiferromagnets) and disordered (quantum spin glasses, etc.).

KEITH BENEDICT

See also **Ferromagnetism and ferroelectricity; Frenkel–Kontorova model; Ising model; Josephson junction arrays; Spin systems**

Further Reading

Binder, K. & Young, A.P. 1996. Spin glasses: experimental facts, theoretical concepts and open questions. *Reviews of Modern Physics*, 58: 801–976

Edwards, S.F. & Anderson, P.W. 1975. Theory of spin glasses. *Journal of Physics* F, 5: 965–974

Fischer, K.H. & Hertz, J.A. 1991. *Spin Glasses*, Cambridge and New York: Cambridge University Press

Mezard, M., Parisi, G. & Virasoro, M.A. 1987. *Spin Glass Theory and Beyond*, Singapore: World Scientific

Parisi, G. 1979. Infinite number of order parameters for spin-glasses. *Physical Review Letters*, 43: 1754–1756

Sherrington, D. & Kirkpatrick, S. 1975. Solvable model of a spin glass. *Physical Review Letters*, 35: 1792–1796

Young, A.P. 1997. *Spin Glasses and Random Fields*, Singapore: World Scientific

FUNCTION SPACES

There are classes of spaces that, while maintaining some essential features of the n-dimensional Euclidean space ($\mathbb{R}^n$), are still general enough to include spaces of functions as particular examples. This entry covers both these abstract classes and some particular function spaces.

A *metric space* retains only the notion of distance. To be a metric on a space X, $d(x, y)$ should be nonnegative and satisfy

(i) $d(x, x) = 0$ for every $x \in X$;
(ii) $d(x, y) = d(y, x)$ for every $x, y \in X$; and
(iii) the triangle inequality: $d(x, y) \le d(x, z) + d(z, y)$ for every x, y, and $z \in X$.

More precisely, it is the pair (X, d), rather than X alone, that is the metric space. For $x, y \in \mathbb{R}^n$, the standard metric is just given by $d(x, y) = |x - y|$.

A metric leads to a notion of convergence: $x_j \to x$ in X as $j \to \infty$ if for any $\varepsilon > 0$, there is an N such that $d(x_j, x) < \varepsilon$ whenever $j \ge N$. In Euclidean spaces, a sequence converges if and only if it is a Cauchy sequence. In the context of a metric space, $\{x_j\}_{j=1}^{\infty}$ is Cauchy if for any $\varepsilon > 0$, there exists an N such that $d(x_i, x_j) < \varepsilon$ whenever $i, j \ge N$. A metric space in which every Cauchy sequence converges is called *complete*.

The general version of a length is a norm. Consider a real (or complex) vector space X [which satisfies the property: if x and y are elements of X, then so is $x + \lambda y$ for any $\lambda \in \mathbb{R}$ (or $\mathbb{C}$)]. Then a norm on X is a nonnegative function $\|\cdot\|$ such that

(iv) $\|x\| \geq 0$ with equality if and only if $x = 0$;
(v) $\|\lambda x\| = |\lambda| \|x\|$ for every $x \in X$ and $\lambda \in \mathbb{R}$ (or $\mathbb{C}$); and
(vi) $\|x + y\| \leq \|x\| + \|y\|$ for every $x, y \in X$ (the triangle inequality again).

Note that X needs to be a vector space so that if x and y are in X, so are λx (as in (v)) and $x + y$ (as in (vi)). The pair $(X, \|\cdot\|)$ is called a *normed space*. A complete normed space is called a *Banach space*.

Given a norm, the distance between x and y, $d(x, y) = \|x - y\|$, defines a metric; thus, normed spaces are less general than metric spaces.

The standard notion of the length of a vector in $\mathbb{R}^n$ is not just an ad hoc definition satisfying (iv)–(vi). Vectors are specified relative to n orthogonal coordinate axes, and the expression for the length follows from Pythagoras's theorem. Since the mathematical formalization of orthogonality involves the dot product ($\mathbf{x} \cdot \mathbf{y} = \sum_{j=1}^{n} x_j y_j$), it is natural to consider generalizing this concept.

An *inner product space* is a real (or complex) vector space X equipped with an inner product: that is, a function (x, y) that associates a real (or complex) number with any two elements $x, y \in X$ and satisfies

(vii) $(x, x) \geq 0$, with equality if and only if $x = 0$;
(viii) $(x, y) = \overline{(y, x)}$ for all $x, y \in X$; and
(ix) $(\lambda x + \mu y, z) = \lambda(x, z) + \mu(y, z)$ for every $x, y, z \in X$ and every $\lambda, \mu \in \mathbb{R}$ (or $\mathbb{C}$).

A complete inner product space is called a *Hilbert space*.

Due to (vii), it is possible to set $\|x\| = (x, x)^{1/2}$; the Cauchy–Schwarz inequality

$$|(x, y)| \leq \|x\| \, \|y\|, \tag{1}$$

a consequence of (vii)–(ix), can then be used to show that $\|\cdot\|$ is a norm on X. Thus, every inner product space can also be viewed as a normed space, implying that inner product spaces are more restrictive than normed spaces.

There is one abstract setting that includes all the above: the only concept retained is the notion of an open set. Such *topological spaces* (Sutherland, 1975) consist of a space X and a topology $\mathcal{T}$—the collection of all the open sets in X. This collection must satisfy:

(x) $\emptyset$ and X are elements of $\mathcal{T}$;
(xi) if $O_1, O_2 \in \mathcal{T}$, then $O_1 \cap O_2 \in \mathcal{T}$; and
(xii) the union of any collection of sets in $\mathcal{T}$ is also in $\mathcal{T}$.

Any metric space gives rise to a topological space by taking $\mathcal{T}$ to be the collection of all of its open sets, but there are some notions of convergence (e.g.,weak convergence, *See* **Functional analysis**) that do not correspond to any choice of metric. Because of this, topological vector spaces (vector spaces equipped with a topology) form the basis of advanced functional analysis (Rudin, 1991).

We now give some simple examples of function spaces, in which I denotes any interval (finite or infinite) in $\mathbb{R}$.

The space $C^0(I)$ consists of all real-valued continuous functions on I. The standard norm on this space is the supremum or uniform norm $\|\cdot\|_\infty$, defined as

$$\|f\|_\infty = \sup_{x \in I} |f(x)| \tag{2}$$

(essentially the maximum value of f on the interval I, provided that this is attained). Since the uniform limit of continuous functions is itself continuous, equipped with the uniform norm $C^0(I)$ is complete, and so a Banach space. For functions defined on the whole real line, such uniform convergence is often too strong. More realistic is uniform convergence on compact intervals, that is, equivalent to convergence in the metric

$$d_K(f, g) = \sum_{j=1}^{\infty} 2^{-j} \times \min\left(\sup_{x \in [-j, j]} |f(x) - g(x)|, 1 \right) \tag{3}$$

which cannot be derived from a norm.

The space $C^k(I)$ consists of all continuous functions whose first k derivatives are also continuous,

$$C^k(I) = \{f \in C^0(I) : \mathrm{d}^j f/\mathrm{d}x^j \in C^0(I),\ j = 1, \ldots, k\}. \tag{4}$$

Its standard norm (which makes C^k a Banach space) is formed by adding the maximum value of f and its first k derivatives,

$$\|f\|_{C^k} = \sum_{j=0}^{k} \|\mathrm{d}^j f/\mathrm{d}x^j\|_\infty. \tag{5}$$

The theory of generalized functions uses the space $\mathcal{D}(\mathbb{R})$ of infinitely differentiable functions with compact support in $\mathbb{R}$ ("test functions"). A sequence $\{\phi_n\}_{n=1}^{\infty} \in \mathcal{D}(\mathbb{R})$ converges to ϕ in $\mathcal{D}$ if there is a fixed compact set K containing the support of each ϕ_n, and $\mathrm{d}^j\phi_n/\mathrm{d}x^j$ converges uniformly to $\mathrm{d}^j\phi/\mathrm{d}x^j$ for every

$j=0, 1, 2, \cdots$. This form of convergence gives rise to a topology, but there is no corresponding metric.

Another family of Banach spaces are the "Lebesgue spaces" $L^p(I)$, consisting of all Lebesgue integrable functions on I for which the L^p norm defined by

$$\|f\|_{L^p} = \begin{cases} \left(\int_I |f(x)|^p \, \mathrm{d}x\right)^{1/p}, 1 \leq p < \infty, \\ \inf\{M : |f(x)| \leq M \text{ almost everywhere in } I\}, \\ \qquad p = \infty \end{cases} \tag{6}$$

is finite. (See Priestley (1997) for a readable introduction to Lebesgue integration.) The space $L^2(I)$ of square integrable functions is a Hilbert space: for two functions $f, g \in L^2(I)$ one can define an inner product by setting

$$(f, g) = \int_I f(x)\overline{g(x)} \, \mathrm{d}x. \tag{7}$$

Note that this is a very natural space of functions to consider physically: if $u(x)$ denotes a velocity then the L^2 norm of $u(x)$ is proportional to the kinetic energy.

The modern theory of partial differential equations relies heavily on *Sobolev spaces* (Adams, 1975; Evans, 1998; Gilbarg & Trudinger, 1983). These allow discussion of the degree of differentiability of functions that are only weakly differentiable: a function f defined on an open interval I has weak derivative $Df = g$ if there is a function $g \in L^1(I)$ such that for every infinitely differentiable function ϕ with compact support in I

$$\int_I g(x)\phi(x) \, \mathrm{d}x = -\int_I f(x)\phi'(x) \, \mathrm{d}x. \tag{8}$$

(The right-hand side would be the result of integrating $\int f'(x)\phi(x) \, \mathrm{d}x$ by parts if f were a C^1 function. Although this is similar to the definition of the derivative of a generalized function, the weak derivative must be an element of L^1.) By requiring successive weak derivatives $D^j f$ to be in $L^p(I)$, we obtain the Sobolev space $W^{k,p}(I)$:

$$W^{k,p}(I) = \{f \in L^p(I) : D^j f \in L^p(I), \\ j = 1, \ldots, k\}. \tag{9}$$

Given the norm

$$\|f\|_{W^{k,p}} = \left(\sum_{j=0}^{k} \|D_j f\|_{L^p}^p\right)^{1/p}, \tag{10}$$

these are Banach spaces. The spaces $H^k(I) \equiv W^{k,2}(I)$ are Hilbert spaces when equipped with the inner product

$$(f, g)_{H^k} = \sum_{j=0}^{k} (D^j f, D^j g)_{L^2}. \tag{11}$$

James C. Robinson

See also **Functional analysis; Generalized functions; Topology**

Further Reading

Adams, R.A. 1975. *Sobolev Spaces*, New York: Academic Press

Evans, L.C. 1998. *Partial Differential Equations*, Providence, RI: American Mathematical Society

Gilbarg, D. & Trudinger, N.S. 1983. *Elliptic Partial Differential Equations of Second Order*, Berlin: Springer

Priestley, H.A. 1997. *Introduction to Integration*, Oxford: Clarendon Press and New York: Oxford University Press

Rudin, W. 1991. *Functional Analysis*, New York: McGraw-Hill

Sutherland, W.A. 1975. *Introduction to Metric and Topological Spaces*, Oxford: Clarendon Press

FUNCTIONAL ANALYSIS

Introduction

Systems that are extended in space are usually modeled by partial differential equations (PDEs). The solution of such an equation will be a function of both space and time, for example, $u(x, t)$, and so the state of the system is specified by the *function* $f(\cdot) = u(\cdot, t)$. Since the resolution of such a function as a Fourier series

$$f(x) = \sum_{k \in \mathbb{Z}} c_k \mathrm{e}^{\mathrm{i}kx}, \qquad c_{-k} = \overline{c_k},$$

requires an infinite number of coefficients, the appropriate phase space for a PDE is generally infinite-dimensional. For example, the stability of a solution (be it stationary, periodic, or more general still) will depend on the eigenvalues of some linear operator that acts on this infinite-dimensional phase space.

Broadly speaking, functional analysis can be viewed as the study of infinite-dimensional spaces and the properties of maps (often linear) defined on them. Functional analysis is required for any rigorous treatment of PDEs (e.g., Evans, 1998, or Robinson, 2001) and many problems in the theory of ordinary differential equations.

In this entry we discuss two topics, both central to the subject. First, we give a very brief outline of spectral theory, which generalizes ideas familiar from linear algebra; and then we discuss the notion of "weak convergence" that goes some way to circumventing the problems arising from a fundamental difference between finite- and infinite-dimensional spaces.

For the sake of simplicity, we will discuss these two topics only in the context of a Hilbert space rather than for a general Banach space (*See* **Function spaces**). We denote this Hilbert space by H, its norm by $\|\cdot\|$, and its inner product by $(\cdot, \cdot)$. Note that $\mathbb{R}^n$ is a finite-dimensional Hilbert space. For more details and a more general treatment, see the suggestions for further reading.

Spectral Theory

Initially motivated by Sturm–Liouville boundary value problems and the related theory of integral equations, spectral theory has become an important part of functional analysis. The theory generalizes ideas from finite-dimensional linear algebra to linear operators on infinite-dimensional spaces.

A map $A : H \to H$ is *linear* if

$$A(x + \lambda y) = Ax + \lambda Ay \quad \text{for every } x, y \in H, \ \lambda \in \mathbb{R},$$

and is *bounded* if, for some $M > 0$,

$$\|Ax\| \leq M\|x\| \qquad \text{for every } x \in H. \qquad (1)$$

[If $H = \mathbb{R}^n$, then any linear map is bounded, but this is not true when H is infinite-dimensional.]

When $H = \mathbb{R}^n$, the eigenvalues of a matrix A are all those complex numbers λ for which $A - \lambda I$ is not invertible (so that $Ax = \lambda x$ for some nonzero x). When A is a linear operator on an infinite-dimensional space, the *spectrum* of A consists of all the values of λ for which $R_\lambda(A) = (A - \lambda I)^{-1}$ lacks one or more of the following "nice" properties:

(i) $R_\lambda(A)$ exists,
(ii) $R_\lambda(A)$ is bounded, and
(iii) $R_\lambda(A)$ can be defined on a dense subset of H (another "nice" property whose exact meaning is unimportant here).

In general, this spectrum can be divided into three distinct pieces:

- the point spectrum $\sigma_\mathrm{p}(A)$ (eigenvalues): (i) does not hold, so that there is a nonzero $x \in H$ (the "eigenfunction") with $Ax = \lambda x$;
- the continuous spectrum $\sigma_\mathrm{c}(A)$: (i) and (iii) hold, but not (ii), and
- the residual spectrum $\sigma_\mathrm{r}(A)$: (i) holds but (iii) does not.

If $H = \mathbb{R}^n$, then whenever (i) holds so do (ii) and (iii), and the spectrum consists only of the eigenvalues of A.

When $H = \mathbb{R}^n$ and A is a real symmetric matrix, results from linear algebra guarantee that (a) all its eigenvalues are real, (b) the eigenvectors corresponding to distinct eigenvalues are orthogonal, and (c) the eigenvectors form a basis for $\mathbb{R}^n$. To obtain a similar result when H is infinite-dimensional we have to impose certain restrictions on A. The original applications to boundary value problems and integral equations motivated the following two definitions that are useful here: an operator $A : H \to H$ is *compact* if the image $\{Ax_n\}_{n=1}^\infty$ of any bounded sequence $\{x_n\}_{n=1}^\infty$ has a convergent subsequence [if $H = \mathbb{R}^n$, then any linear map is compact, cf. below]; and a bounded map from a Hilbert space H into itself is *self-adjoint* if

$$(Ax, y) = (x, Ay) \quad \text{for every} \quad x, y \in H$$

[when $H = \mathbb{R}^n$, this means that A is a real symmetric matrix].

If A is a compact, self-adjoint operator that is also invertible, then it behaves much like a real symmetric matrix: (a) all of its eigenvalues are real; (a′) the residual spectrum is empty and there are at most a countable number of eigenvalues, which are bounded and can only have zero as an accumulation point; (b) eigenfunctions corresponding to distinct eigenvalues are orthogonal; and (c) the eigenfunctions form a basis for H. This is the celebrated Hilbert–Schmidt theorem, the rigorous result that justifies the approach of Sturm–Liouville theory, that is, using eigenfunctions as a basis in which to expand solutions as a "generalized Fourier series" (see Kreyszig, 1978; Renardy & Rogers, 1992, for example).

Weak Convergence

A space is finite-dimensional if and only if its unit ball is compact. Put another way, any bounded sequence in a finite-dimensional space has a convergent subsequence (in $\mathbb{R}$ this is the Bolzano–Weierstrass theorem): this result is extremely useful, but it is not true in an infinite-dimensional space.

However, it is possible to define a weaker notion of convergence ("weak convergence") and so recover a form of this compactness property in certain infinite-dimensional spaces. To motivate the definition we make two observations. First, if x and y are two distinct elements of a Hilbert space H, then, it is possible to find a $z \in H$ such that

$$(z, x) \neq (z, y),$$

"inner products can distinguish elements of H." Second, it is also possible to show that if $(z, x) = (z, y)$ for every $z \in H$, then we must have $x = y$: "inner products can determine elements of H."

Because of these two results, it is reasonable to define a notion of convergence based on inner products. In a Hilbert space, a sequence x_n converges weakly to x, written $x_n \rightharpoonup x$, if

$$(z, x_n) \to (z, x) \quad \text{for every element} \quad z \in H.$$

Although a bounded sequence in an infinite-dimensional Hilbert space need not have a convergent subsequence, it will have a subsequence that converges weakly. Such convergence is often sufficient for applications; in particular, it is fundamental to many results in the theory of existence and uniqueness for solutions of PDEs.

[The Riesz Representation theorem guarantees that for any bounded linear map $\rho : H \to \mathbb{R}$, there is a $z \in H$ such that $\rho(x) = (z, x)$ for all $x \in H$. In a general Banach space B, the "inner product with" z used above is replaced by "any bounded linear map from B into $\mathbb{R}$". These maps are known as "linear functionals," and gave rise to the name of the subject to which they are central.]

JAMES C. ROBINSON

See also **Function spaces; Generalized functions; Topology**

Further Reading

Evans, L.C. 1998. *Partial Differential Equations*, Providence, RI: American Mathematical Society

Kreyszig, E. 1978. *Introductory Functional Analysis with Applications*, New York: Wiley

Meise, R. & Vogt, D. 1997. *Introduction to Functional Analysis*, Oxford: Clarendon Press and New York: Oxford University Press

Renardy, M. & Rogers, R.C. 1992. *An Introduction to Partial Differential Equations*, New York: Springer

Robinson, J.C. 2001. *Infinite-dimensional Dynamical Systems*, Cambridge and New York: Cambridge University Press

Rudin, W. 1991. *Functional Analysis*, New York: McGraw-Hill

Yosida, K. 1980. *Functional Analysis*, Berlin: Springer

G

GALAXIES

Galaxies are dense agglomerations of matter in the Universe. They consist of gas, dust, and stars as a major fraction. In addition, for kinematical reasons a hypothetical "dark matter" component is required. Their formation dates back to the early Universe, almost 14 billion years ago. Although some galaxies were already listed as nebulous objects in Charles Messier's catalogue of nebulae and star clusters from 1784, their discovery as detached "island universes" only dates from 1924, when Edwin Hubble resolved the outer parts of Messier Object 31 (M31, the Andromeda Nebula) into stars and measured their distances.

Within 2.2 million light years ($1\,\text{ly} = 9.46 \times 10^{12}$ km), M31 is the closest massive galaxy and is similar to our Milky Way galaxy, both belonging to a galaxy type that is characterized by a rotating disk of stars and gas showing spiral patterns. In addition, such spiral galaxies possess a central spheroid of old stars called a bulge and an extended spheroidal halo of old single stars and bound star clusters, called globular clusters.

All spiral galaxy labels today begin with the prefix "*S*". For decreasing bulge-to-disk ratios, Hubble classified the sequence of spirals from Sa to Sc, some of which also show an innermost bar structure where the spiral arms emerge at the ends. Hubble distinguished these as SBs as distinct from normal Ss.

Another morphological galaxy type exists with an elliptical shape, almost no particular substructure, relatively old stars, and depleted in gas: ellipticals. According to their minor-to-major axes ratio b/a, Hubble denoted them by E10$(1 - b/a)$ reaching from E0 (circular) to E7 ($b/a = 0.3$) at most.

Lenticular galaxies (SOs) form the link between spirals and ellipticals in Hubble's famous "tuning-fork diagram" (see Figure 1). Hubble galaxies have masses between several 10^{10} to 10^{12} solar masses ($m_s = 1.9891 \times 10^{30}$ kg) and diameters of 100,000 light years and more. More refined classification schemes are possible such as those by de Vaucouleurs (1959) and Sandage (1961). Today an extension exists to Sds, and even further to irregularly shaped galaxies (Is) with lower mass and brightness and no central bulge.

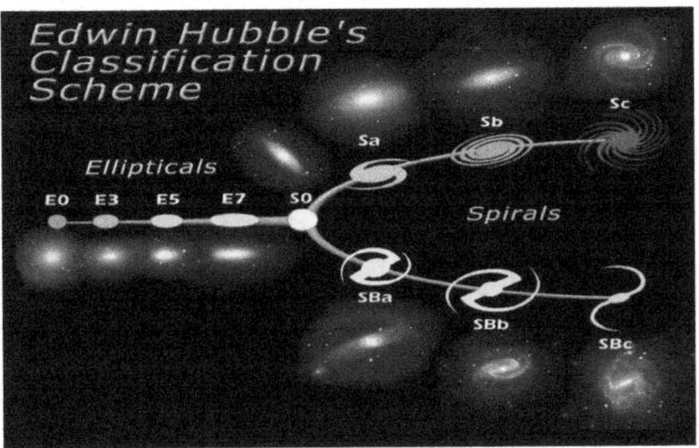

Figure 1. Hubble's "tuning-fork diagram" of galaxies. With courtesy from STScI.

Internal and Dynamical Structure of Galaxies

Ellipticals

According to the stellar mass distribution, any galaxy becomes dimmer with distance *R* from the geometrical center. The radial dependence however, differs between morphological types and their components. Ellipticals follow in all directions the empirical de Vaucouleurs law with brightness falling off with $\exp(-R^{1/4})$. Although it seems plausible that rotation causes ellipticity, since the 1970s, it has been known that in most ellipticals their regular rotation velocity is smaller than the irregular proper motion of stars with speeds of 200–250 km/s. They are therefore denoted as "hot stellar systems." The elliptical shape is thus formed by the anisotropic velocity dispersion of stars. Moreover, triaxiallity of the figure axes, as for a flattened cigar, can be derived for some elliptical galaxies if the elliptical contours of equal brightness are somehow twisted. Most elliptical galaxies can be divided additionally into boxy- or disky-shaped contours considering the fourth-order cosine in a trigonometric expansion.

Spirals

Spiral galaxies are the most complex systems. While their bulges' brightness profiles resemble ellipticals, they are rotationally supported. Although the halos are not dense enough to show a continuous stellar distribution, one can take, for example, globular clusters as representative and find a power law for the density of $R^{-2} \dots R^{-3.5}$. While in the halo, the large irregular velocities of stars (determined in our Milky Way) and globular clusters, their age, and the lack of significant amounts of gas are similar to those of ellipticals (without knowing the halo's ellipticity), disks consist of 10–20% gas and of stars, both rotating with velocities of 200–250 km/s. Because the velocity dispersion of stars is much less, in the range of only 10–60 km/s, disks are therefore "cold systems" and rotationally flattened. Their radial face-on brightness drops with $\exp(-R/\alpha)$ where α is the so-called scale length of around 10,000 light years.

Although spiral arms are exceptionally trailing, they are not the result of structures wound by differential rotation because their pitch angles, that is, the angle between an arm and a circle, should be much smaller than observed. Because the arm structure is pronounced in the visual by the brightness of young stars that form out of cold gas condensed within the arms, among different possible processes of arm formation, density waves are the most favored.

Since the characteristic velocities v, for example, the rotation of spiral disks and velocity dispersion in elliptical galaxies, are almost constant with R, although the visible mass decreases, this invokes the existence of dark matter. Because in equilibrium the centrifugal force $F_c = mv^2/R$ acting on the mass element m is balanced by the gravitational force of a mass M_R included within R, $F_g \propto mM_R/R^2$, for observed v=const. M_R has to increase with R. This means that the dark matter contribution dominates in the outermost regime. In contrast, less bright ellipticals have been found recently where a decline of velocities at large radii is measured, so the dark matter content remains controversial.

Formation and Evolution of Galaxies

Galaxies assemble into larger units by means of gravitation, thus forming galaxy groups like our Local Group and galaxy clusters with numerous members, as for example, the Virgo Cluster at a distance of 50 million light years. In the center of the clusters, ellipticals dominate while spirals permeate the whole cluster. This leads to the suggestion that ellipticals are formed by merging events in the densest cluster regions and in the early universe.

While the Hubble-type galaxies are massive, dwarf galaxies with less mass exist. In our vicinity the large and small magellanic clouds (LMC, SMC) as satellites of the Milky Way represent a dwarf irregular type, consisting of gas and stars as spirals but with less regular structures. With masses of $5 \times 10^9 m_s$, the LMC lies at the upper-mass range of dwarf galaxies, while other Milky Way galaxy satellites, dwarf spheroidals, form the low-mass end of the dwarf ellipticals. With 10^6–$10^7 m_s$ their brightness is so low that observations are still incomplete. Dwarf ellipticals are the most frequent galaxy type in the present Universe resembling Hubble's Es.

The class of dwarf galaxies is of substantial importance for our understanding of galaxy formation and the evolution of the Universe, because they serve as the building blocks of massive galaxies in the cosmological picture of hierarchical clustering. Although their accretion by mature galaxies by means of tidal friction is observable, their destruction rate during the course of the universe is yet unclear because it is also possible that the dwarf galaxy types are replenished by other processes such as tidal tails of merging galaxies.

GERHARD HENSLER

See also **Cosmological models; Gravitational waves; Hénon–Heiles system; Spiral waves**

Further Reading

Binney, J. & Merrifield, M. 1998. *Galactic Astronomy*, Princeton Series in Astrophysics, Princeton, NJ: Princeton University Press

Binney, J. & Tremaine, S. 1987. *Galactic Dynamics*, Princeton Series in Astrophysics, Princeton, NJ: Princeton University Press

Combes, F., Mazure, A. & Boisse, F. 2001. *Galaxies and Cosmology*, Berlin and Heidelberg: Springer

Sandage, A. 1961. *Hubble Atlas of Galaxies*, Washington, DC: Carnegie Institution of Washington

Sparke, L.S. & Gallagher, J.S. 2000. *Galaxies in the Universe*, Cambridge and New York: Cambridge University Press

de Vaucouleurs, G. 1959. Classification and morphology of external galaxies, vol. 53, *Handbuch der Physik*, Berlin: Springer, p. 275

GAME OF LIFE

The rules underlying Life are simple, according to computer scientists. Biologists are inclined to be skeptical, but they do agree that the cellular automaton known as the Game of Life provides fascinating insights into the phenomena of self-organization and emergence in systems of interacting agents.

Biological life has been around for at least 3.8 billion years, but the Game of Life was invented by mathematician John Conway in 1970 and publicized by Martin Gardner in his "Mathematical Games" column in *Scientific American*. It is probably the best-known example of the class of algorithms known as cellular automata (CA). A CA is a one- or two-dimensional array of cells, each of which can exist in a number of states. Time in a CA is discrete; at each time step, every cell updates itself on the basis of its current state and those of its neighbors. Cellular automata can exhibit surprisingly complex global temporal

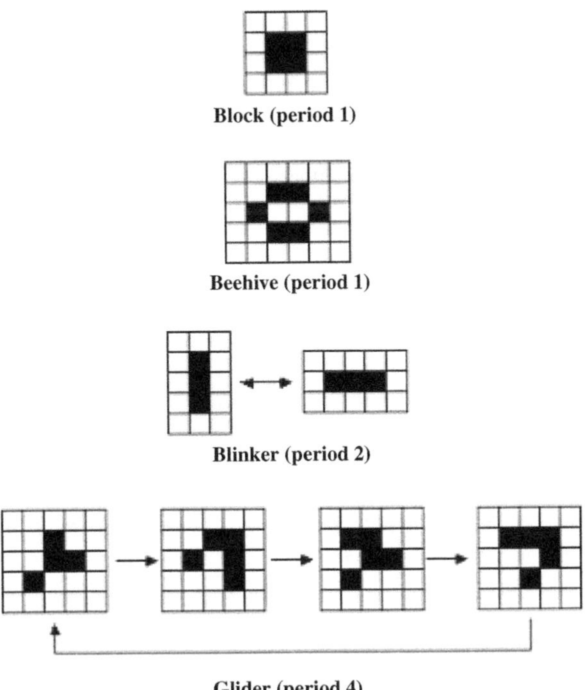

Figure 1. Some commonly seen patterns in the Game of Life.

dynamics, arising from extremely simple rules applied on a local scale.

The standard Game of Life uses a two-dimensional grid. Cells can be either on (alive) or off (dead). The neighborhood of a cell is the eight cells surrounding it. The rules of Life are, indeed, simple; if a live cell has two or three live neighbors, it stays alive. A dead cell with exactly three live neighbors comes alive (is born). In all other cases the cell dies, either of overcrowding (with more than three live neighbors) or loneliness (with fewer than two). At each time step, all cells update their states simultaneously. That's all there is to Life! No wonder the biologists are dubious. But from this simple foundation, complex, consistent patterns of activity emerge. If the grid is seeded at random with live cells, the first few time steps are a turmoil of apparently random activity. However, identifiable patterns of live cells quickly emerge. Interesting patterns exhibit periodic behavior, cycling between a number of different states in a deterministic manner. Many patterns settle into a limit cycle of length one—a stable point attractor, or "still life" in the Life terminology. Others, known to Life practitioners as "oscillators," have longer limit cycles.

So consistent are these patterns that hundreds have been identified, named, and studied by Life enthusiasts all over the world (see e.g., Figure 1). There are several catalogues of Life patterns available on the Web; a particularly nice site is http://hensel.lifepatterns.net/.

Although most initial configurations eventually settle to a stable state or cyclic set of states, this is not always the case. Life aficionados have identified initial states that generate new states indefinitely.

So is Life more than just a generator of interesting patterns? The answer, of course, is "yes." Cellular automata in general, and Life in particular, have interesting theoretical properties. Stephen Wolfram identified four broad classes of dynamic behavior common to one-dimensional (Wolfram, 1984) and two-dimensional (Wolfram, 1985) CAs.

Class 1: Evolution leads to a homogeneous state (analogous to a point attractor in a nonlinear dynamic system),

Class 2: Evolution leads to a set of separated simple stable or periodic structures (analogous to limit cycles),

Class 3: Evolution leads to a chaotic pattern (analogous to the chaotic attractors found in continual dynamic systems),

Class 4: Evolution leads to complex localized structures, sometimes long-lived.

Although the universal applicability and hence usefulness of Wolfram's classification system has been questioned (e.g., Eppstein, 2000), it is still widely accepted. The Game of Life falls, serendipitously, into class 4, the class for which Wolfram hypothesizes that "class-4 cellular automata are generically capable of universal computation, so that they can implement arbitrary information-processing procedures" (Wolfram, 1985). It has, in fact, been proven that Life is a universal cellular automaton—one which can emulate a Turing machine, capable of performing universal

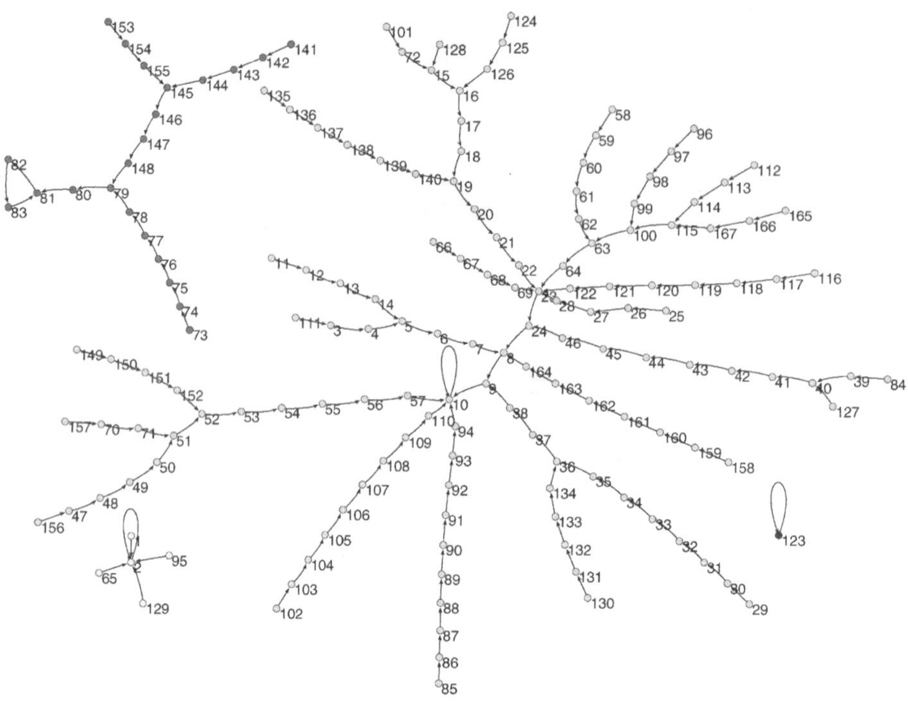

Figure 2. State transition graph for a cellular automaton. States are arbitrarily numbered. The state space contains four basins of attraction, three of which contain a point attractor, while the fourth leads to a limit cycle of length three (states 81, 82 and 83).

computation. The universality of Life was proved in the early 1980s (Berlekamp et al., 1982), and a Turing machine that can be extended to a universal Turing machine was implemented by Paul Rendell in 2000 (Adamatzky, 2002).

An overview of the temporal dynamics of a two-dimensional CA such as the Game of Life can be provided by a state transition graph, in which each possible pattern of on and off cells in the CA makes up a unique state. If the number of cells in the lattice is s and the number of cell states is k, then there are k^s possible states for that CA. Each state forms a node in the graph. Since the rules are deterministic, each state will map to a single new state in the next timestep, and the transition between the two states is represented by a directed link (an arc) between the two nodes. A state transition graph depicts the course of evolution of the CA from any given starting point (Figure 2).

Inspection of a state transition diagram leads to the conclusion that evolution in a CA is not deterministically reversible; some states have two or more antecedent states. Some states, in contrast, have no preceding states. The latter are known as Garden of Eden states. The existence of Garden of Eden states in the Game of Life was predicted by CA theory, but actually identifying such a state is a nontrivial task. Several Garden of Eden states have been identified by trial and error, but the search for a reliable algorithm for their identification continues.

The rules of Life are simple, but from them arise complex, emergent behaviors—coherent forms arising at different levels of organization and interacting to produce new forms and patterns. This complexity arising from underlying simplicity makes the Game of Life an ideal toy world for the study of many of the fascinating phenomena of complex systems.

JENNIFER HALLINAN

See also **Artificial life; Attractors; Biological evolution; Cellular automata; Order from chaos; State diagrams**

Further Reading

Adamatzky, A. (editor). 2002. *Collision-based Computing*, Heidelberg: Springer

Berlekamp, E.R., Conway, J.H. & Guy, R.K. 1982. *Winning Ways for Your Mathematical Plays*, vol. II. *Games in Particular*, London: Academic Press

Eppstein, D. 2000. Searching for spaceships. In *More Games of No Chance*, edited by R.J. Nowakowski, Cambridge and New York: Cambridge University Press, pp. 433–453

Gardner, M. 1970. Mathematical games: the fantastic combinations of John Conway's new solitaire game "life." *Scientific American*, 223: 120–123

Wolfram, S. 1984. Universality and complexity in cellular automata. *Physica* D, 10: 1–35

Wolfram, S. 1985. Two-dimensional cellular automata. *Journal of Statistical Physics*, 38: 901–946

The Game of Life can be played online at:
http://www.math.com/students/wonders/life/life.html
http://www.bitstorm.org/gameoflife/
http://hensel.lifepatterns.net/

GAME THEORY

Humans play games. From the formalized warfare of chess to the Machiavellian machinations of politics to the subtleties of sexual pursuit, interactions between individuals with desires and priorities that are often conflicting and contradictory lies at the heart of human society. The formal study of games, however, is a relatively recent phenomenon and has moved rapidly from its beginnings as a mathematical tool to aid gamblers to its current status as an essential paradigm in fields as diverse as economics and evolutionary biology.

Today, *game theory* can be defined as "the analysis of rational behavior under circumstances of strategic interdependence, when an individual's best strategy depends upon what his opponents are likely to do" (Varoufakis, 2001). Individuals may be people, corporations, nations, animals, species, or any other entity that can be said to exhibit strategic behavior.

Formal game theory began in 1713, when Pierre-Rémond de Montmort first proposed the concept of a minimax solution to a card game called Le Her. A minimax solution to a two-player game is one in which an individual chooses his strategy so as to minimize the maximum loss or risk that he/she will incur. It was James Waldegrave, the originator of Le Her, who actually produced a minimax solution to the game.

Other concepts of fundamental importance to modern game theory also emerged in the 18th century, a result of work by Daniel Bernoulli in his 1738 analysis of the St. Petersburg paradox (Dimand & Dimand, 1992). These concepts included utility (a measure of the desirability to the player of each possible outcome of the game), the maximization of expected utility, diminishing marginal utility (the decrease in the amount of benefit derived from consuming each additional unit of a product or service), and risk aversion as a parameter of a utility function.

The work on minimax game solutions remained an isolated curiosity until the 1920s, when Émile Borel published a series of short papers on strategic games in the Proceedings of the French Academy of Sciences between 1921 and 1927. In these papers, he defined the normal form of a game: a matrix representation of the game in which each player tries to work out the best strategy independent of the sequence of moves. Borel later claimed to have proven the minimax theorem, but this does not appear to be the case. In fact, he may not even have stated the theorem.

The first formal proof of the minimax theorem for two-person games with any finite number of pure strategies was given by John von Neumann in a paper presented to the Gottingen Mathematical Society on December 7, 1926 (Dimand & Dimand, 1992). He proved that in a zero-sum game (one in which one player's gain is the other's loss), there exists a unique set of mixed strategies, one per player, which equalizes the payoffs that each player can gain regardless of the strategy adopted by the other player.

At about the same time, the Princeton economist Oskar Morgenstern was pondering mixed strategy game theoretic issues, as exemplified by that master of bluff, Sherlock Holmes, and the strategies he should adopt to avoid his arch enemy, Professor Moriarty. In 1944, von Neumann and Morgenstern collaborated to produce *The Theory of Games and Economic Behavior*, the seminal publication in this area. In 1947, they revised the book to include expected utility theory, under which games are expressed in terms of the players' perceptions of the inherent desirability and likelihood of their outcomes, and players never expect other players to hold mistaken beliefs—the assumption of complete rationality, which has been fundamental to much ensuing work. Economists were initially reluctant to accept game theory, but since the publication of von Neumann and Morgenstern's book, the theory has undergone extensive development, and has been applied to an enormous variety of problems in economics, to the point where Leonard (1992) could assert that "game theory plays a central role in economic theory."

Minimax theory assumes that each player has perfect knowledge about the game and that the game is zero-sum. In such a case, the best strategy for each player is independent of the strategy adopted by the other player. In most real-world situations, this is not the case; more often, the best strategy for one player depends on what the other players choose to do. The extension of minimax theory to n-player, noncooperative games was achieved by the Princeton mathematician John Nash in a paper published in 1950. In this paper, he defined the Nash equilibrium. A Nash equilibrium is a set of strategies such that no player could improve his/her payoff, given the strategies of all other players in the game, by changing his/her strategy. Nash proved that all noncooperative games have a Nash equilibrium and, thereby, established an analytical structure within which all situations of conflict and cooperation could be studied. For this work, he received the 1994 Nobel Prize for Economics, together with John Harsanyi and Reinhard Selten.

Game theory was also applied with considerable success to other fields of research, perhaps most notably to evolutionary biology. The concepts of game theory transfer readily to evolutionary biology—the values of different outcomes, which in economic theory are measured as utility, are readily interpreted in terms of Darwinian fitness. Moreover, the somewhat sweeping assumption of complete rationality in the behavior of the agents is replaced by the concept of evolutionary stability.

Game-theoretic concepts were first explicitly applied to the study of evolution by Lewontin (1961), who saw the agents in the game as a species on the one hand, against nature on the other. The utility of this approach was quickly recognized, and the focus shifted to modeling interactions between individuals. In this context, the concept of the Nash equilibrium was rediscovered independently by John Maynard Smith and G.R. Price in 1973 as the Evolutionarily Stable Strategy (ESS—"a strategy such that, if all the members of a population adopt it, no other strategy can invade" (Maynard Smith, 1982, p. 204). An ESS represents the solution to a game.

In the last three decades, game theory has been used to provide a framework for the analysis of a wide range of biological phenomena, including the evolution of sex ratios, parental investment in offspring, patterns of animal dispersal, competition for resources (see, e.g., Maynard Smith, 1982) and the evolution of cooperation (Axelrod, 1984). In biology, as in economics, game theory has become an essential tool for the theorist, providing a structured framework for the analysis of a wide range of phenomena.

JENNIFER HALLINAN

See also **Artificial life; Biological evolution; Economic system dynamics**

Further Reading

Axelrod, R. 1984. *The Evolution of Cooperation*, New York: Basic Books

Dimand, R.W. & Dimand, M. 1992. The early history of the theory of strategic games from Waldegrave to Borel. In *Toward a History of Game Theory*, edited by E.R. Weintraub, Durham: Duke University Press

Leonard, R.J. 1992. Creating a context for game theory. In *Toward a History of Game Theory*, edited by E.R. Weintraub, Durham: Duke University Press

Lewontin, R.C. 1961. Evolution and the theory of games. *Journal of Theoretical Biology*, 1: 382–403

Maynard Smith, J. 1982. *Evolution and the Theory of Games*, Cambridge: Cambridge University Press

Nash, J. 1950. Equilibrium points in N-person games. *Proceedings of the National Academy of Sciences*, 36: 48–49

Varoufakis, Y. 2001. General introduction: game theory's quest for a single, unifying framework for the social sciences. In *Game Theory: Critical Concepts in the Social Sciences*, edited by Y. Varoufakis, vol. 1. London and New York: Routledge

von Neumann, J. & Morganstern, O. 1953. *Theory of Games and Economic Behavior*, 3rd edition, New York: Wiley (1st edition, 1944)

GAP SOLITONS

See **Solitons, types of**

GARDNER (GGKM) EQUATION

See **Inverse scattering method**

GAUSSIAN BEAM

See **Nonlinear optics**

GAUSSIAN ENSEMBLES

See **Free probability theory**

GEL'FAND–LEVITAN THEORY

In inverse problems, sometimes also called backward problems, one is given the solution and required to find the underlying equation. Inverse spectral problems deal with the recovery of unknown coefficients of a differential operator from the knowledge of its spectral data. Their importance stems from the fact that coefficients in a differential equation usually model the physical structure or composition of a certain material. Nondestructive testing, geophysical prospecting, radar, and medical imaging are just some of the important applications of inverse problems. By analyzing reflected waves, radar can locate, track, and sometimes identify a target. In geophysical prospecting, the travel time of reflected underground acoustic waves can reveal deposits of petroleum and natural gas. The field of medical imaging has witnessed the development of many non-invasive and safe procedures using the above ideas. For example, ultrasonography using the three-dimensional Doppler effect can help visualize the fetal vascular system for prenatal diagnosis. High-resolution ultrasound cardiovascular imaging not only reconstructs a real-time picture of the heart but also measures the rate of blood flow, thus giving early warnings for clogged arteries, heart attacks, and strokes.

A simple rule of the thumb is that data used should be equivalent to the information to be reconstructed. For example, if we are looking for a square integrable function, the data should be equivalent to its sequence of Fourier coefficients, while an analytic function is equivalent to its Taylor coefficients. Can we recover a matrix of size $n \times n$ from its n eigenvalues? Although in general to define a matrix we would need n^2 entries, possible exceptions would be matrices that are fully determined by their first rows—symmetric Toeplitz or Hankel matrices, for example. In all kinds of inverse problems, one faces two major issues: uniqueness of the solution and the algorithm for its reconstruction. One of the best understood inverse problems, which deals with the Schrödinger operator, was solved by Gel'fand and Levitan in 1951 (see Gel'fand & Levitan, 1955).

The One-dimensional Schrödinger Equation

Consider the one-dimensional Schrödinger operator

$$\begin{cases} L(y) = -y''(x,\lambda) + q(x)y(x,\lambda) = \lambda y(x,\lambda), \\ \quad x \in [0, \infty), \\ y'(0,\lambda) - hy(0,\lambda) = 0, \end{cases} \tag{1}$$

where q is a real continuous function and h is a real constant. The continuity of q ensures the existence of a unique solution of the initial value problem at $x=0$ for every $\lambda \in \mathbb{C}$. The boundary value problem (1) is regular at $x=0$ and singular at $x=\infty$. The operator L, which acts in the Hilbert space of square integrable functions on the positive half line, is self-adjoint so its eigenvalues are real. We usually normalize solutions of (1) by the condition $y(0,\lambda)=1$, which leads to $y'(0,\lambda)=h$. To determine the spectrum of L (the set of values λ such that the inverse operator of $L-\lambda I_d$ either does not exist or is unbounded), we need to examine the behavior of the solutions $y(x,\lambda)$ as $x \to \infty$. If the solution decays fast enough and is square integrable, that is, has a finite energy $\alpha_n = \int_0^\infty |y(x,\lambda_n)|^2 \, dx < \infty$, then λ_n is an eigenvalue and belongs to the discrete part of the spectrum. For other solutions that could be approximated by continuous functions or do not grow faster than a polynomial in x, λ belongs to the continuous spectrum; otherwise λ is not in the spectrum. A more precise result using the theory of distributions (Gel'fand & Shilov, 1967) shows that λ is in the spectrum if and only if the solution $y(x,\lambda)$ is the derivative of a function that cannot grow faster than $x^{3/2+\varepsilon}$ as $x \to \infty$. This characterization threw a new light on the mysterious behavior of the continuous spectrum.

For example, when $q=h=0$, the spectrum of L is continuous on $[0,\infty]$. Indeed its "eigenfunctions" are $\cos\left(x\sqrt{\lambda}\right)$ which are not square integrable and so there are no eigenvalues, while for $\lambda>0$ the solutions $\cos\left(x\sqrt{\lambda}\right)$ are bounded and so grow less than $x^{3/2+\varepsilon}$ as $x\to\infty$, which means that $\lambda>0$ is in the continuous spectrum. On the other hand, when $\lambda<0$, the solutions $\cos\left(x\sqrt{\lambda}\right) = \cosh\left(x\sqrt{-\lambda}\right)$ have an exponential growth, and so $\lambda<0$ cannot be not in the spectrum.

Once the operator L is known to be self-adjoint, the solution $y(x,\lambda)$ forms the kernel of a transform, which is similar to the Fourier cosine transform,

$$F(\lambda) = \int_0^\infty f(x)y(x,\lambda)\,dx,$$

and its inverse transform is defined explicitly by

$$f(x) = \int_\sigma F(\lambda)\overline{y(x,\lambda)}\,d\rho(\lambda).$$

The function ρ, which is called the spectral function, is nondecreasing, is right-continuous, has jumps at the eigenvalues $\rho(\lambda_n) - \rho(\lambda_n - 0) = 1/\alpha_n$, and is continuous and increasing only on the continuous part of the spectrum. For example, when $q=h=0$, the spectral function is simply

$$\rho_0(\lambda) = \begin{cases} (2/\pi)\sqrt{\lambda} & \text{if } \lambda \geq 0, \\ 0 & \text{if } \lambda < 0. \end{cases}$$

Obviously when $q \neq 0$, the new spectral function ρ would record all new spectral changes, which leads to the inverse spectral problem: by comparing ρ to ρ_0 can we recover the perturbation q?

Gel'fand–Levitan Theory

If we are given a spectral function ρ, can we find its associated Sturm–Liouville differential operator given in (1)? This is in essence the Gel'fand–Levitan theory, which produced an elegant algorithm for the recovery of the potential q and the boundary condition at $x=0$.

As explained earlier, the spectrum depends on the growth of solutions and a key point is their representations by the Fourier transform,

$$y(x,\lambda) = \cos\left(x\sqrt{\lambda}\right) + \int_0^x K(x,t)\cos\left(t\sqrt{\lambda}\right)dt. \quad (2)$$

The kernel $K(x,t)$ is continuous and contains all information about q, namely, that

$$\frac{d}{dx}K(x,x) = \frac{1}{2}q(x) \text{ and } K(0,0) = h. \quad (3)$$

Thus, it is a matter of finding $K(x,t)$ in order to recover q. To do so, we first form the function

$$F(x,t) = \int_{-\infty}^\infty \frac{\sin\left(x\sqrt{\lambda}\right)}{\sqrt{\lambda}} \frac{\sin\left(t\sqrt{\lambda}\right)}{\sqrt{\lambda}}\,d\sigma(\lambda), \quad (4)$$

where the given spectral function ρ is used in

$$\sigma(\lambda) = \begin{cases} \rho(\lambda) - \frac{2}{\pi}\sqrt{\lambda} & \text{if } \lambda \geq 0, \\ \rho(\lambda) & \text{if } \lambda < 0 \end{cases} \quad (5)$$

and set

$$f(x,t) = \frac{\partial^2 F(x,t)}{\partial x \partial t}.$$

The crux of the theory is that $f(x,t)$ and $K(x,t)$ satisfy a linear integral equation, for each fixed x

$$K(x,t) + \int_0^x K(x,s)f(s,t)\,ds = -f(x,t)$$
$$\text{for } 0 \leq t \leq x. \quad (6)$$

Thus given the spectral function ρ, we can form $F(x,t)$ by (4), yielding the kernel $f(x,t)$ and then for each fixed x solve the Fredholm integral equation (6) for $K(x,t)$. As for matrices, a Fredholm integral equation has a solution if the null space contains only the trivial solution. Once the existence of a solution K is guaranteed, its smoothness is examined. It is shown that $f(x,t)$ and $K(x,t)$ have the same degree of smoothness which implies the smoothness of q by (3). The original

Gel'fand–Levitan result in (1951) is based on the linear equation (6) and can be summarized as follows:

Theorem 1 (Gel'fand and Levitan Theory). *If a nondecreasing function $\rho(\lambda)$ satisfies*

(A) *for arbitrary real x the integral*

$$\int_{-\infty}^{0} \exp\left(x\sqrt{|\lambda|}\right) \mathrm{d}\rho(\lambda) \text{ exists,}$$

(B) *the integral*

$$a(x) = \int_{1}^{\infty} \frac{\cos\left(x\sqrt{\lambda}\right)}{\lambda} \mathrm{d}\sigma(\lambda)$$

exists for all $0 \leq x < \infty$, *while* $a(x)$ *has continuous derivatives up to the fourth order for all* $0 \leq x < \infty$ *and if the set of points of increase of* ρ *has at least one finite accumulation point, then there exists just one differential operator of the second order defined by* (1) *which has* $\rho(\lambda)$ *as its spectral function. The function* $q(x)$ *and the number* h *are defined by* (3), *where* $K(x,t)$ *is a solution of* (6).

However, the original theory had a gap between the necessary and sufficient conditions. Ten years later, Levitan and Gasymov proved a stricter version, which we state as the Gel'fand–Levitan–Gasymov theory found in Gasymov & Levitan (1964) that contains necessary and sufficient conditions.

Theorem 2 (Gel'fand–Levitan–Gasymov). *The monotonically increasing function ρ is the spectral function of a Sturm–Liouville of type* (1) *with a function q having m integrable derivatives and a number h if and only if the following conditions are satisfied:*

(A) *If* $E(\lambda) = \int_0^\infty f(x)\cos\left(x\sqrt{\lambda}\right)\mathrm{d}x$, *where* $f \in L^2(0,\infty)$ *and of compact support, then*

$$\int |E(\lambda)|^2 \mathrm{d}\rho(\lambda) = 0 \Longrightarrow$$

$$f = 0 \text{ almost everywhere.}$$

(B) *The limit* $\Phi(x) = \lim_{N\to\infty} \int_{-\infty}^{N} \cos\left(x\sqrt{\lambda}\right)\mathrm{d}\sigma(\lambda)$, *where* $\sigma(\lambda)$ *is defined by* (5), *exists boundedly in every finite range of values of x and* Φ *has* $m+1$ *locally integrable derivatives with* $\Phi(0) = -h$.

There are also interesting results for the regular case that is defined by

$$\begin{cases} L(y) = -y''(x,\lambda) + q(x)y(x,\lambda) = \lambda y(x,\lambda), \\ \quad 0 \leq x \leq \pi, \\ y'(0,\lambda) - hy(0,\lambda) = 0, \\ y'(\pi,\lambda) - Hy(\pi,\lambda) = 0, \end{cases} \tag{7}$$

where q is continuous and $q, h, H \in \mathbb{R}$. Let the norming constants be defined by $\alpha_n = \int_0^\pi |y(x,\lambda_n)|^2 \mathrm{d}x$.

Note that since q is continuous, its knowledge is equivalent to its sequence of Fourier coefficients. Thus, the data ought to be at least an infinite sequence of numbers in order to recover the potential q.

Theorem 3. *If all the* $\alpha_n > 0$,

$$\sqrt{\lambda_n} = n + \frac{a_0}{n} + \frac{a_1}{n^3} + O\left(\frac{1}{n^4}\right)$$

and

$$\alpha_n = \frac{\pi}{2} + \frac{b_0}{n^2} + O\left(\frac{1}{n^3}\right),$$

where a_0, a_1, *and* b_0 *are constants, then there exists an absolutely continuous function* $q(x)$ *corresponding to the given* λ_n *and* α_n.

Theorem 4. *The numbers* $\{\lambda_n\}_{n=0}^\infty$ *and* $\{\alpha_n\}_{n=0}^\infty$ *are the spectral data of some boundary value problem* (7) *with* q'' *being square integrable if and only if the following asymptotic estimates hold:*

$$\sqrt{\lambda_n} = n + \frac{a_0}{n} + \frac{a_1}{n^3} + \frac{\gamma_n}{n^3}$$

and

$$\alpha_n = \frac{\pi}{2} + \frac{b_0}{n^2} + \frac{\tau_n}{n^3},$$

where $\lambda_n \neq \lambda_m$ *and* $\alpha_n > 0$ *and the series* $\sum_{n=1}^\infty \gamma_n^2$ *and* $\sum_{n=1}^\infty \alpha_n^2$ *are convergent.*

The problem of generalizing the theory to higher dimensional operators remains open and depends heavily on the idea of transformation operators that map generalized functions. In Boumenir (1991), it is also proved that the Gel'fand–Levitan theory is based on the factorization of operators whose symbol is given by the spectral functions.

Discrete Case

Consider now a discrete version of the Sturm–Liouville problem

$$Bv = \lambda Av, \tag{8}$$

where B is a Jacobi matrix and A is a diagonal matrix with positive entries (positive definite). Expressing (8) in a vectorial form, we end up with recurrence relation defined by

$$\begin{cases} c_n y_{n+1} = (a_n\lambda + b_n)\, y_n - c_{n-1}y_{n-1}, \\ \quad n = 0, \ldots, m-1, \\ y_{-1} = 0, \quad \text{and } y_m + hy_{m-1} = 0, \end{cases} \tag{9}$$

where $b_n \in \mathbb{R}$, $a_n > 0$, and $c_n > 0$. We look for a nontrivial finite sequence $y_{-1}, \ldots, y_m$ which satisfies (9). To this end, we normalize the solution vectors by $y_0 = 1/c_{-1}$, then knowing the first terms y_{-1} and

y_0 allows us to compute recursively the remaining entries that are now functions of λ. It is easy to see that eigenvalues λ_r are nothing other than the zeros of the last condition $y_m(\lambda) + hy_{m-1}(\lambda) = 0$. For a given eigenvector, define its squared norm by $\alpha_r = \sum_{k=0}^{m-1} a_k |y_k(\lambda_r)|^2$. The following theorem is found in Atkinson (1964) or in Teschl (2000).

Theorem 5. *Assume that we are given* $h \in \mathbb{R}$, $\{a_k > 0\}$, *eigenvalues* $\{\lambda_r\}_{0 \le r \le m-1}$, *norming constants* $\{\rho_r\}_{0 \le r \le m-1}$, *then there exists* $\{c_k\}_{-1 \le k \le m-1}$ *which are positive and constants* $\{b_k\}_{0 \le k \le m-1}$ *such that the boundary value problem has the set* $\{\lambda_r\}_{0 \le r \le m-1}$ *as its eigenvalues.*

AMIN BOUMENIR

See also **Inverse problems; Inverse scattering method or transform; Quantum inverse scattering method**

Further Reading

Atkinson, F.V. 1964. *Discrete and Continuous Boundary Problems*, New York and London: Academic Press

Boumenir, A. 1991. A comparison theorem for self-adjoint operators. *Proceedings of the American Mathematical Society*, 111(1): 161–175

Gasymov, M.G. & Levitan, B.M. 1964. Determination of a differential equation by two of its spectra. *Russian Mathematical Surveys*, 19(2): 1–63

Gel'fand, I.M. & Levitan, B.M. 1955. On the determination of a differential equation from its spectral function. *American Mathematical Society Translations*, (2)1: 253–304

Gel'fand, I.M. & Shilov, G.E. 1967. *Generalized Functions*, vol. 3: *Theory of Differential Equations*, New York and London: Academic Press

Levitan, B.M. 1987. *Inverse Sturm–Liouville Problems*, Utrecht: VNU Science Press

Marchenko, V.A. 1986. *Sturm–Liouville Operators and Applications*, Basel: Birkhäuser

Teschl, G. 2000. *Jacobi Operators and Completely Integrable Nonlinear Lattices*, Providence, RI: American Mathematical Society

GENERAL CIRCULATION MODELS OF THE ATMOSPHERE

Atmospheric general circulation models (AGCMs) simulate the dynamical, physical, and chemical processes of planetary atmospheres. For the Earth's atmosphere (*See* **Atmospheric and ocean sciences**), they are based on the thermo-hydrodynamic equations, which consist of the conservation of momentum, mass, and energy with the ideal gas law in coordinates suitable for the rotating planet.

In the presently used form, they were first derived by Vilhelm Bjerknes in 1904; subsequently, Lewis Fry Richardson (1922) proposed numerical weather prediction (NWP) as a practical application which, in 1950, was successfully performed by Jule Charney, R. Fjørtoft, and John von Neumann on an electronic computer based on a simplified set of these equations. While numerical weather prediction models utilize the atmospheric short-term memory for forecasting, AGCMs developed since the 1960s extend applications to longer time scales simulating seasonal and climate variability (for a personal recollection, see Smagorinsky, 1983).

Since then, numerical weather prediction and atmospheric general circulation modeling have enjoyed continuous advances, which are attributed to the following gains and improvements: (1) observational data accuracy, analysis, and assimilation; (2) insight into dynamical and physical processes, numerical algorithms, and computer power; and (3) the use of model hierarchies to study individual atmospheric phenomena. With simulations of the Earth system and that of other planets envisaged, a broad field of science has been established, which is of vital importance socio-economically, agriculturally, politically, and strategically.

Observations

Since the foundation of the World Meteorological Organization (WMO) and international treaties to monitor and record meteorological and oceanographical data, the collection of global data through local weather and oceanic stations has been systematically organized and has become a truly globalized system through the deployment of satellites and introduction of remote sensing facilities. The availability of extensive global data has enabled the extension of NWP models to complex global GCMs, thus facilitating simulations on larger time scales (months and years instead of hours and days) and validation of NWPs in return. Furthermore, it has become possible to study climate history and make estimates of future climates, which is particularly important due to likely anthropogenic impacts.

General Circulation Models

Atmospheric general circulation models have two basic components. First, the dynamical core consists of the primitive equations (the conservation equations with vertical momentum equation approximated by hydrostatic equilibrium, that is, balancing the vertical pressure gradient and the apparent gravitational forces) under adiabatic conditions. Second, physical processes contribute to the diabatic sources and sinks interacting with the dynamical core. They are incorporated as parameterizations, mostly in a modular format: solar and terrestrial radiation, the hydrological cycle (with phase transitions manifested in evaporation and transpiration, cloud, and precipitation processes), the planetary boundary layer communicating between the free atmosphere and the ground (soil with vegetation, snow and ice cover; ocean with sea ice), and atmospheric chemistry. Most of these parameterizations enter the thermodynamic energy equation as heat sources or sinks.

Dynamical core

State-of-the-art atmospheric general circulation models commonly utilize the so-called primitive equation approximation of the Navier–Stokes equations. In addition to the dry dynamics, equations describing the transport of other constituents such as water vapor, cloud liquid water and ice, trace gases, and particles (aerosols) can be an integral part of the dynamical core. To integrate the equations, they are discretized in space and time where finite differences and spectral methods are the most dominant. For more details on the governing equations; *See* the entries for **Atmospheric and ocean sciences; Fluid dynamics; Navier–Stokes equation**.

(i) Horizontal discretization: In the horizontal, grid point, or spectral representations of the dependent model variables are used. Different grid structures and finite difference schemes have been designed to reduce the error (Messinger & Arakawa, 1976). An alternative approach is the spectral method. The dependent variables are represented in terms of orthogonal functions where appropriate basis functions are the spherical harmonics. The maximum wave number of the expansion defines the resolution of the model. Since the computation of products is expensive, only linear terms are evaluated in the spectral domain. To compute the nonlinear contributions, the variables are transformed into grid point space every time step, where the respective products are computed and transformed back to spectral space. Necessary derivatives are computed during the transformation. This spectral-transform procedure (Eliassen et al.,1970; Orszag, 1970) makes the spectral approach computationally competitive with finite difference schemes. For low resolutions, the spectral method is, in general, more accurate than the grid point method. However, spectral methods are less suitable for the treatment of scalar fields, which exhibit sharp gradients and, for physical reasons, must maintain a positive-definite value (e.g., water vapor, cloud water, chemical tracers). Therefore, selected fields are often treated separately in the grid point domain using, for example, semi-Lagrangian techniques. Recently, with increasing model resolutions and the need for transporting more species (e.g., for chemical submodels), grid point models are attracting more attention again, while novel grid structures are introduced, for instance, the spherical icosahedral grid of the German Weather Service model GME (Majewski et al., 2000).

(ii) Vertical discretization: In general, finite differences and numerical integration techniques are used for the derivatives and integrals in the vertical. The vertical coordinate can be defined in different ways. The isobaric coordinate eliminates the density from the equations and simplifies the continuity equation compared with a z-coordinate system. However, the intersection of low-level pressure surfaces with the orography enforces time-dependent lower boundary conditions which are difficult to treat numerically. This problem can be avoided if terrain following sigma (σ) coordinates are used, where sigma is defined by the pressure divided by the surface pressure $\sigma = p/p_0$. Unfortunately, the sigma coordinate leads to a formulation of the pressure gradient force which, in the presence of steep orography, is difficult to treat. The advantages of both sigma and pressure coordinates are combined by introducing a hybrid coordinate system with a smoothed transition from σ to p with height.

Physical Processes and Parameterizations

Many processes that are important for large-scale atmospheric flow cannot be explicitly resolved by the model due to its given spatial and temporal resolution. These processes need to be parameterized; that is, their effect on the large-scale circulation needs to be formulated in terms of the resolved grid-scale variables. The most prominent processes in building a parameterization package of an atmospheric general circulation model are long- and short-wave radiation, cumulus convection, large-scale condensation, cloud formation and the vertical transport due to turbulent fluxes in the planetary boundary layer, and the effect of different surface characteristics such as vegetation on the surface fluxes. Additional processes such as the excitation of gravity waves and their impact on the atmospheric momentum budget or the effect of vertical eddy fluxes above the boundary layer are often considered. Because the land surface provides a time-dependent boundary condition that acts on time scales comparable to the atmosphere, great effort has been made to include land surface and soil processes in the atmospheric parameterization package. More recently, the effect of the interaction of various chemical species and their reactions with atmospheric circulation are also being considered. In addition to the direct relation between the resolved atmospheric flow and the effect of the parameterized processes, there are various other interactions among the individual processes that have to be taken into account. For typical comprehensive atmospheric general circulation models, Figure 1 displays the interrelations between the adiabatic dynamics, providing the spatial and temporal distribution of the dependent model variables and the various processes being parameterized.

Model Hierarchy

General circulation models of reduced complexity are continuously developed to supplement comprehensive

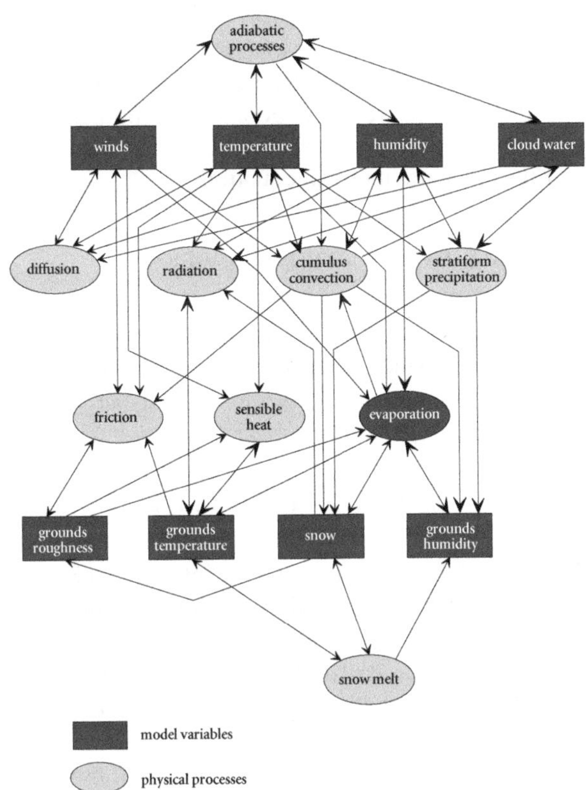

Figure 1. Interactions in comprehensive GCMs (schematic).

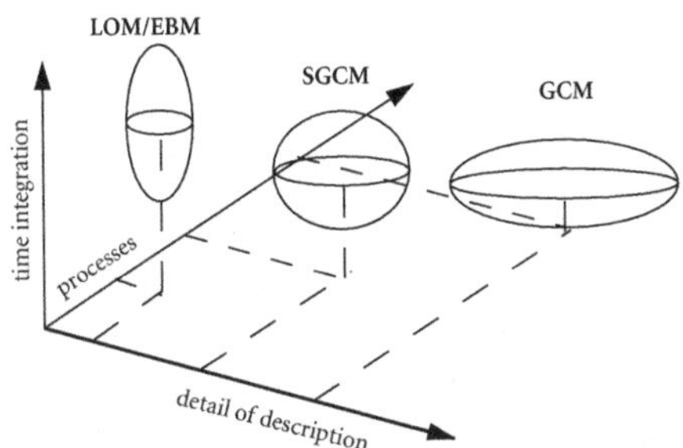

Figure 2. A model hierarchy of general circulation models.

GCMs, to gain insight into atmospheric phenomena (see Figure 2) and for educational purposes: when utilizing the full set of equations, the model spectrum ranges from simple GCMs (SGCMs) with analytic forms of heating and friction to low-order models (LOMs). A prominent LOM is the Lorenz model (*See* **Lorenz equations**), which approximately describes the nonlinear convection dynamics in the vicinity of a critical point for the stream function and temperature in a set of ordinary differential equations. It can be regarded as including first-order nonlinear effects (temperature advection) to a linear model, which leads to chaotic behavior. The Lorenz model is used to study predictability and serves as a paradigm for phase-space behavior of atmospheric GCMs. Utilizing thermal energy conservation only, another spectrum of models (energy balance models, or EBMs) is obtained by averaging in certain spatial directions. This leads to the horizontally averaged one-dimensional radiative-convective models; the one-dimensional energy balance model when averaged vertically and longitudinally for studying climate feedback and stability; and two-dimensional statistical–dynamical models when averaged longitudinally where dynamical processes are being parameterized. A prominent EBM example is the globally averaged or zero-dimensional energy balance model. With ice-albedo and water vapor-emissivity feedbacks included, climate catastrophes leading to a snowball earth and runaway greenhouse can be demonstrated. With random forcing and periodic solar radiation input (e.g., Milankovich cycles), stochastic resonance emerges.

KLAUS FRAEDRICH, ANDREAS A AIGNER, EDILBERT KIRK, AND FRANK LUNKEIT

See also **Atmospheric and ocean sciences; Fluid dynamics; Forecasting; Lorenz equations; Navier–Stokes equation**

Further Reading

Eliassen, E., Machenhauer, B. & Rasmussen, E. 1970. *On a Numerical Method for Integration of the Hydrodynamical Equations with a Spectral Representation of the Horizontal Fields.* Report No. 2, Institut for Teoretisk Meteorologi, University of Copenhagen

Majewski, D., Liermann, D., Prohl, P., Ritter, B., Buchhold, M., Hanisch, T., Paul, G., Wergen, W. & Baumgardner, J. 2000. *The global icosahedral–hexagonal grid point model GME-operational version and high resolution tests*, ECMWF, Workshop Proceedings, Numerical methods for high resolution global models, pp. 47–91

McGuffie, K. & Henderson-Sellers, A. 1997. *A Climate Modelling Primer*, New York: Wiley

Messinger, R. & Arakawa, A. (1976). *Numerical Methods Used in Atmospheric Models*, Geneva: WMO

Orszag, S.A. 1970. Transform method for calculation of vector coupled sums: application to the spectral form of the vorticity equation. *Journal of Atmospheric Sciences*, 27: 890–895

Richardson, L.F. 1922. *Weather Prediction by Numerical Process*, Cambridge: Cambridge University Press

Smagorinsky, J. 1983. The beginnings of numerical weather prediction and general circulation modeling: early recollections. In *Theory of Climate*, edited by B. Saltzman, New York: Academic Press, pp. 3–38

Trenberth, K. 1992. *Climate System Modelling*, Cambridge and New York: Cambridge University Press

GENERAL RELATIVITY

Called *general relativity*, Albert Einstein's theory of gravitation was created as a generalization of his special relativity theory. As special relativity is a theory of physical space-time (neglecting gravitational effects), general relativity is a theory of physical space-time in the presence of gravitation.

While Maxwell's theory of electromagnetism is a relativistic theory that is covariant with respect to Lorentz transformations, Newton's theory of gravitation is incompatible with special relativity. In 1907, two years after proposing special relativity, Einstein was preparing a review of special relativity when he suddenly wondered how Newtonian gravitation would have to be modified to fit in with special relativity. Einstein described this as "the happiest thought of my life." He proposed the *equivalence principle* as "the complete physical equivalence of a gravitational field and the corresponding acceleration of the reference frame" that "extends the principle of relativity to the case of uniformly accelerated motion of the reference frame."

In fact, Einstein's equivalence principle is a generalization of the so-called *weak* equivalence principle, which dates from Galileo and Newton and states that the inertial mass and gravitational mass of any object are equal. Thus (neglecting friction), the acceleration of different bodies in a gravitational field is independent of their masses and other physical characteristics, and hence, with given initial conditions, their motions will be the same.

The next important step was made by Einstein in his 1912 papers, where he concluded that "if all accelerated systems are equivalent, then Euclidean geometry cannot hold in all of them." Further investigations by Einstein to find the correct form of equations for a gravitational field were connected with applications of Riemannian geometry and tensor analysis. The final form of equations of general relativity was given by Einstein in his paper "The Field Equations of gravitation," submitted on 25 November 1915. At about the same time, David Hilbert submitted a paper entitled "Foundations of Physics," which also contains the correct field equations for gravitation, introduced by applying a variation principle.

Physical Space-time and Gravitating Matter in General Relativity

According to general relativity, physical space-time in a gravitational field is non-Euclidean; so to describe the properties of space-time, Einstein applied Riemannian geometry. Without gravitation, physical space-time is a flat pseudo-Euclidean Minkowski 4-continuum, where free particles move uniformly and linearly along geodesic worldlines of Minkowski space-time. Einstein's key idea is that in a gravitational field, particles move along geodesic lines of curved space-time, and in accordance with the Equivalence principle, their movement does not depend on the particles' characteristics. Thus, motion for an observer is motion along curves in 3-space with variable velocity. Curvature of space-time in general relativity is created by sources of gravitational field. In general relativity, the role of sources of gravitational field is played by the energy-momentum tensor describing the distribution and motion of gravitating matter. Energy (or mass) density (the source of gravitation in Newton's theory) is only a component of the energy-momentum tensor. Besides energy (mass), a gravitational field in general relativity is also created by momentum and other components of an energy-momentum tensor. The dependence between geometrical properties of physical space-time and gravitating matter is described by Einstein's gravitational equations.

Einstein Gravitational Equations

The principal geometrical characteristics of a gravitational field in general relativity are given by the metric tensor g_{ik}, which determines the square of distance between two infinitesimally close points of pseudo-Riemannian 4-space-time

$$\mathrm{d}s^2 = g_{ik}\,\mathrm{d}x^i\,\mathrm{d}x^k \quad (i, k = 0, 1, 2, 3). \tag{1}$$

By means of the metric tensor, time intervals and spatial distances can be defined; thus the formula for proper time fixed by a clock at rest in some reference frame is $\mathrm{d}\tau = (1/c)\sqrt{g_{00}\,\mathrm{d}x^0\,\mathrm{d}x^0}$. Because the value of g_{00} in a gravitational field depends on location, time flow depends on gravitational field.

Einstein's gravitational equations are nonlinear second-order differential equations with respect to the metric tensor and have the form

$$R_{ik} - 1/2 g_{ik} R = 8\pi G/c^4 T_{ik}, \tag{2}$$

where R_{ik} is the so-called Ricci tensor (a contraction of the curvature tensor), R is the scalar curvature, T_{ik} is the energy-momentum tensor, G is Newton's gravitational constant, and c is the velocity of light in a vacuum. Einstein's equations are covariant with respect to arbitrary coordinate transformations.

Equation (2) can be changed by adding to the right-hand side, the so-called cosmological term Λg_{ik}, where Λ is a cosmological constant introduced by Einstein. This term describes energy density and pressure of the vacuum, and it plays an essential role in cosmology, leading to the effect of gravitational repulsion if $\Lambda > 0$. In the case of weak gravitational fields, when the variation of metric with respect to the Minkowski metric is small, the Einstein equations lead to Newton's law of gravitational attraction and allow one to find first relativistic corrections. In the case of strong gravitational fields, Einstein equations can give new physical results, including black holes and gravitational waves.

Experimental Verification and Bounds of Applicability

Several classical effects of general relativity have been verified observationally, including the bending of light in a gravitational field, gravitational redshift, the advance of the perihelion of the planet Mercury (43 s of arc per century), and retarding of the propagation of light in a gravitational field. The first three effects were discussed by Einstein even before the creation of general relativity.

The *weak* equivalence principle has been verified to high precision (10^{-12}), and general relativity provides a basis for relativistic astrophysics and cosmology. The Hot Big Bang model was built within the framework of general relativity. Over the past two decades, the role of general relativity has grown in connection with discoveries in cosmology—in particular, acceleration of cosmological expansion, dark matter and dark energy, and other problems that need to be resolved.

General relativity is a classical theory, and a consistent quantum theory of gravitation has not yet been developed. In fact, the formulation of a quantum-gravitation theory requires a unified theory of all fundamental physical interactions. At present, the most popular candidate for such a unified theory is the superstring theory. This theory is in higher-dimensional space, and it leads to a generalization of Einstein's gravitation theory. A second problem with general relativity is the presence of gravitational singularities (cosmological singularities, collapsing systems, etc.). According to theorems by Stephen Hawking and Roger Penrose, this problem is connected under certain conditions with internal properties of the gravitational equations of general relativity. As with classical theory, general relativity is inapplicable near singular states.

The creation and development of Einstein's gravitation theory was a triumph of 20th-century science, providing the basis of gravitation theory, relativistic astrophysics, and cosmology. Within the frame of its applicability, general relativity will remain a great achievement of human culture.

VIACHASLAV KUVSHINOV AND ALBERT MINKEVICH

See also **Black holes; Cosmological models; Einstein equations; Gravitational waves; String theory; Tensors; Twistor theory**

Further Reading

Einstein, A. 1989–1996. *The Collected Papers of Albert Einstein*, vol. 2: The Swiss Years: Writings, 1900–1909, vol. 3: The Swiss Years: Writings, 1909–1911; vol. 4: The Swiss years: Writings, 1912–1914, Princeton, NJ: Princeton University Press

Fok, V.A. 1961. *The Theory of Space, Time and Gravitation*, New York: Pergamon Press

Landau, L.D. & Lifshitz, E.M. 1973. *Therie of field*, Moscow: Nauka

Landau, L.D. & Lifshitz, E.M. 1984. *The Classical Theory of Fields*, 4th edition, Oxford and New York: Pergamon Press

Misner, C., Thorne, K. & Wheeler, J. 1973. *Gravitation*, San Francisco: Freeman

Weinberg, S. 1972. *Gravitation and Cosmology: Principles and Applications of the General Theory of Relativity*, New York: Wiley

Zel'dovich, Ya.B. & Novikov, I.D. 1971. *Relativistic Astrophysics*, 2 vols. Chicago: University of Chicago Press

GENERALIZED FUNCTIONS

Generalized functions were introduced into quantum mechanics by Paul Dirac, who defined the delta function $\delta(x)$ as follows (Dirac, 1958):

$$\delta(x) = 0 \text{ for } x \neq 0$$

and

$$\int_{-\varepsilon}^{+\varepsilon} \delta(x) = 1 \text{ for } \varepsilon > 0. \tag{1}$$

Although $\delta(x)$ is not a true function (from a mathematical perspective) because it is undefined at $x = 0$, the delta function is widely used in physics to approximate functions that are localized in space or time. Examples of such idealizations include the concepts of point mass or point charge and the spatially localized action of a pick on a guitar string.

In an engineering context, the delta function is called the unit impulse function and is used to approximate functions of large amplitude and short duration such as the striking of a golf ball or the instantaneous charging of an electrical capacitor (Guillemin, 1953). Since such point sources are only idealizations, they can be represented by the limiting procedure, $\delta(x) = \lim_{a\to\infty} \delta_a(x)$ for a family of piecewise continuous functions $\delta_a(x)$, where a is a continuous parameter. Among others, the following four functions, $\delta_a(x)$, are popular approximations of the delta function $\delta(x)$ (Pauli, 1973):

- *Unit impulse (box)*: $\delta_a(x) = a/2$ for $|x| < 1/a$, and $\delta_a(x) = 0$ for $|x| > 1/a$.
- *Finite impulse response filter*: $\delta_a(x) = \sin ax/\pi x$.
- *Gaussian pulse*: $\delta_a(x) = a\mathrm{e}^{-\pi a^2 x^2}$.
- *Lorentzian pulse*: $\delta_a(x) = a/(1 + \pi^2 a^2 x^2)$.

In the limit $a \to \infty$, all representations $\delta_a(x)$ have zero width, infinite peak amplitude, and unit area, that is, all the approximations converge to the delta function $\delta(x)$ (see Figure 1).

Closely related to Dirac's delta function is Oliver Heaviside's step function, which was introduced in the late 19th century and has been widely used in electronics and communications research since the 1920s (Heaviside, 1950; Guillemin, 1953). The Heaviside step $H(x)$ is defined as follows:

$$H(x) = 1 \text{ for } x > 0 \text{ and } H(x) = 0 \text{ for } x < 0. \tag{2}$$

The derivative of the Heaviside step function is recognized as Dirac's delta function because $H'(x) = 0$ for $x \neq 0$ and for $\varepsilon > 0$

$$\int_{-\varepsilon}^{+\varepsilon} H'(x)\,\mathrm{d}x = H(+\varepsilon) - H(-\varepsilon) = 1. \tag{3}$$

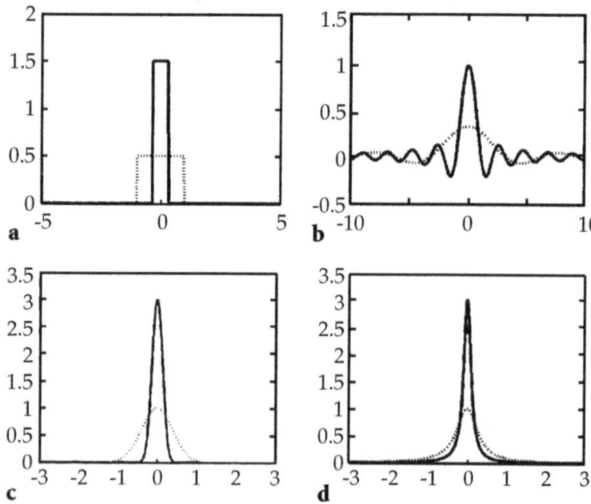

Figure 1. Approximations of the Dirac delta function $\delta(x)$ for $a=1$ (dotted curves) and $a=3$ (solid curves): (a) the box, (b) the impulse response filter, (c) the Gaussian pulse, and (d) the Lorentzian pulse.

In engineering terms, the step function represents an instantaneous jump from zero to unit value of some physical quantity such as the signal voltage at the input terminals of an amplifier.

Because the delta and step are not defined at $x=0$, they are not true functions; thus mathematicians call them distributions, implying rules that assign numbers to integral expressions (Strauss, 1992). To understand this perspective, consider a real function, $f(x)$ which has derivatives for all values of x that approach zero faster than any power of x. In other words, $\lim_{|x|\to\infty}|x|^p f^{(m)}(x)=0$ for any $m\geq 0$ and $p\geq 0$. Such functions $f(x)$ are called test functions for integral distributions. As is seen from the box approximation of $\delta_a(x)$ in the limit $a\to\infty$, the delta function $\delta(x)$ assigns the number $f(0)$ to the integral distribution associated with a test function $f(x)$:

$$\int_{-\infty}^{\infty} f(x)\delta(x)\,\mathrm{d}x = f(0). \tag{4}$$

More generally, the delta function $\delta(x)$ and its derivatives $\delta^{(m)}(x)$ satisfy the following fundamental property:

$$\begin{aligned}&\int_{-\infty}^{\infty} f(x)\delta^{(m)}(x-\xi)\,\mathrm{d}x\\ &= (-1)^m \int_{-\infty}^{\infty} f^{(m)}(x)\delta(x-\xi)\,\mathrm{d}x\\ &= (-1)^m f^{(m)}(\xi),\end{aligned} \tag{5}$$

where $f(x)$ is a test function. A brief list of useful delta function properties follows from the fundamental property in Equation (5) (Gel'fand & Shilov, 1964):

- *Even function*: $\delta(-x)=\delta(x)$.
- *Scaling transformation*: $|\xi|\delta(\xi x)=\delta(x)$.
- *Factorization*: $2|\xi|\delta(x^2-\xi^2)=\delta(x-\xi)+\delta(x+\xi)$.
- *Projection formula*: $\frac{1}{(x-\xi\mp \mathrm{i}0)}=\mathrm{PV}\left[\frac{1}{(x-\xi)}\right]\pm\pi\mathrm{i}\delta(x-\xi)$ (where PV stands for principal value).
- *Fourier transform*: $\hat{\delta}(k)=\int_{-\infty}^{\infty}\delta(x)\mathrm{e}^{-\mathrm{i}kx}\,\mathrm{d}x=1$.
- *Spectral representation*: $2\pi\delta(x)=\int_{-\infty}^{\infty}\mathrm{e}^{\mathrm{i}kx}\,\mathrm{d}x$.

Dirac studied orthogonality relations for the wave function $\psi_\lambda(x)$ of the stationary Schrödinger equation with a potential

$$-\frac{h}{2m}\psi_\lambda''(x)+U(x)\psi_\lambda(x)=\lambda\psi_\lambda(x). \tag{6}$$

In this context, square integrable wave functions for different levels of energy λ are orthogonal with respect to the inner product

$$\langle\psi_{\lambda'}|\psi_\lambda\rangle = \int_{-\infty}^{\infty}\bar{\psi}_{\lambda'}(x)\psi_\lambda(x)\,\mathrm{d}x = \delta_{\lambda',\lambda}, \tag{7}$$

where $\delta_{\lambda',\lambda}=0$ for $\lambda'\neq\lambda$ and $\delta_{\lambda,\lambda}=1$. Similar orthogonality relations for wave functions of continuous spectra diverge. For example, linear waves in free space (when $U(x)=0$) take the form $\psi_\lambda(x)=\Psi(x;k)=\mathrm{e}^{\mathrm{i}kx}$ for $\lambda=hk^2/2m$. They are periodic in x with period $L=2\pi/k$. If $k=k_n=2\pi n/L$, the periodic function satisfies the following orthogonality relations on the finite interval $x\in[-L/2,L/2]$:

$$\langle\Psi(k_{n'})|\Psi(k_n)\rangle = \int_{-L/2}^{L/2}\mathrm{e}^{\mathrm{i}(k_n-k_{n'})x}\,\mathrm{d}x = L\delta_{n',n}. \tag{8}$$

When linear waves in free space are taken over the whole real axis (i.e., $L\to\infty$), the inner product of the wave function $\Psi(x;k)$ with itself diverges. The delta function $\delta(k)$ replaces this divergence in the closed form

$$\begin{aligned}\langle\Psi(k')|\Psi(k)\rangle &= \int_{-\infty}^{\infty}\bar{\Psi}(x;k')\Psi(x;k)\,\mathrm{d}x\\ &= 2\pi\delta(k'-k),\end{aligned} \tag{9}$$

where $\delta(k)=0$ for $k\neq 0$ and $\delta(0)=\infty$. The singularity of the delta function $\delta(k)$ is uniquely specified in the distribution (integral) sense by requiring that

$$\int_{-\infty}^{\infty}\delta(k)\,dk = 1. \tag{10}$$

The delta function $\delta(x)$ represents not only the orthogonality of the wave functions $\psi_\lambda(x)$ of the stationary Schrödinger equation, but also their completeness with the completeness relation:

$$\frac{1}{2\pi}\int_{-\infty}^{\infty}\bar{\Psi}(x';k)\Psi(x;k)\,dk+\sum_{n}\bar{\Psi}_n(x')\Psi_n(x) = \delta(x'-x), \qquad (11)$$

where $\Psi(x;k)$ are wave functions of the continuous spectrum and $\Psi_n(x)$ are wave functions of the discrete spectrum. With the use of the delta function $\delta(x)$, a test function $f(x)$ can be expanded into a complete and orthogonal set of wave functions:

$$f(x)=\frac{1}{2\pi}\int_{-\infty}^{\infty}\hat{f}(k)\Psi(x;k)\,dk+\sum_{n}\hat{f}_n\Psi_n(x), \qquad (12)$$

where coefficients of the expansion are

$$\hat{f}(k)=\langle\Psi(k)|f\rangle, \qquad \hat{f}_n=\langle\Psi_n|f\rangle. \qquad (13)$$

Generalized functions are widely used in electric circuit theory, communications, spectral analysis, integral transforms, and Green function solutions of equations of mathematical physics (Guillemin, 1953; Pauli, 1973; Strauss, 1992). For example, the Fourier transform is based on the spectral decomposition above for $\Psi(x;k)=e^{ikx}$ and $\Psi_n(x)=0$.

DMITRY PELINOVSKY

See also **Function spaces; Functional analysis; Integral transforms; Quantum theory; Spectral analysis**

Further Reading

Dirac, P.A.M. 1958. *The Principles of Quantum Mechanics*, 4th edition, Oxford: Clarendon Press

Gel'fand, I.M. & Shilov, G.E. 1964. *Generalized Functions*, vol. 1, *Properties and Operations*, New York and London: Academic Press

Guillemin, E.A. 1953. *Introductory Circuit Theory*, New York: Wiley

Heaviside, O. 1950. *Electromagnetic Theory*, New York: Dover

Pauli, W. 1973. *Pauli Lectures on Physics:* Vol. 5, *Wave Mechanics*, Cambridge, MIT Press

Strauss, W.A. 1992. *Partial Differential Equations, An Introduction*, New York: Wiley

GEOMETRICAL OPTICS, NONLINEAR

Nonlinear geometric optics arose in the second half of the 20th century in several areas of research, including nonlinear optics and reaction-diffusion systems. To understand the origin of such problems and the underlying physical phenomena, it is helpful to consider briefly the historical path that has led to them.

Classical Results

Traditionally, geometric optics comprised all of optics. It developed from empirical laws of propagation for light beams, which are bent in a non-uniform medium and refract and reflect at the interfaces of media possessing different optical properties. Remarkable successes were achieved, including inventions of the microscope and the telescope around the beginning of the 17th century. At the same time, some important scientific problems were studied; thus in 1620, Willebrord Snell established the law of refraction at the interface of two transparent media. Assuming angles are measured with respect to a line normal to the boundary, Snell's law states that at any angle of incidence (ψ_1), the ratio

$$\frac{\sin\psi_1}{\sin\psi_2}=\text{constant}\equiv n_{12}, \qquad (1)$$

where ψ_2 is the angle of refraction and the constant n_{12} is the reciprocal refractive index. Interestingly, this important law was independently discovered by René Descartes about a decade later. Pierre Fermat's subsequent formulation of the principle of least time—which governs the propagation of light rays—completed the phenomenological theory of geometric optics.

In the 17th century, two competitive hypotheses were advanced for the physical nature of light: the corpuscular hypothesis of Isaac Newton and the wave theory of Christian Huygens. Two centuries later, after the theoretical studies and experimental works of Thomas Young, Augustin-Jean Fresnel, and François Arago, the wave theory of light triumphed. The true nature of light was revealed in the 19th century to be based on James Clerk Maxwell's theory of electromagnetism, which predicted the existence of electromagnetic waves that propagate in vacuum with constant velocity $c=1/\sqrt{\varepsilon_0\mu_0}\approx 3\times 10^8$ m/s, where ε_0 and μ_0 are the dielectric permittivity and magnetic permeability of vacuum, respectively.

This value—obtained entirely from independent measurements of electric and magnetic fields—is precisely equal to the light velocity in vacuum, which was first measured by Ole Roemer two centuries before (using astronomical data on time intervals between eclipses of Jupiter's satellites). Together with the subsequent experimental discovery and investigation of properties of electromagnetic waves inspired by Hermann Helmholtz and performed by Heinrich Hertz, this result has convinced physicists of the electromagnetic nature of light.

This electromagnetic theory of light explains most optical phenomena. For example, it yields Snell's law of refraction, giving the constant in Equation (1) as

$$n_{12}=v_1/v_2=\sqrt{\varepsilon_2/\varepsilon_1}=n_2/n_1. \qquad (2)$$

Here v_i is the velocity of light in the ith medium $(i=1,2)$, and ε_i and $n_i=c/v_i=\sqrt{\varepsilon_i/\varepsilon_0}$ are the dielectric permittivity and the absolute refractive index of this medium, respectively.

Physical mechanisms underlying the refractive index were studied by Helmholtz, Paul Drude, and Hendrik Lorentz at the close of the 19th century. They merged the electromagnetic theory with the idea of electrons as charged particles bound in atoms and molecules which are dislocated under the influence of an electromagnetic field of a light wave. (From the modern point of view, such electrons occupy exterior atomic shells; thus, they are referred to as optical electrons.) Further developments of such ideas connect optical phenomena with the dynamic response of an optical medium to the electromagnetic waves propagating through it.

When a light wave propagates through a dielectric medium, its electric field ($\boldsymbol{E}$) displaces the optical electrons, inducing a wave of electric polarization ($\boldsymbol{P}$). The latter generates a secondary electromagnetic wave which is added to the primary wave, modifying the polarization wave, and so on, ad infinitum. In other words, the electromagnetic wave and the response of an optical medium determine each other. As the relations

$$\begin{aligned} \varepsilon \boldsymbol{E} &= \varepsilon_0 \boldsymbol{E} + \boldsymbol{P}, \quad \boldsymbol{P} = \varepsilon_0 \chi \boldsymbol{E}, \\ n &= c/v = \sqrt{\varepsilon/\varepsilon_0} \end{aligned} \tag{3}$$

are always fulfilled (where χ is the dielectric susceptibility of the medium), the absolute refraction index of the medium can be expressed via χ by the formula

$$n = \sqrt{\varepsilon/\varepsilon_0} = \sqrt{1+\chi}. \tag{4}$$

If the light wave is sinusoidal and sufficiently weak, the response of the medium is readily calculated; thus the susceptibility χ is expressed via the oscillatory parameters of optical electrons which, in turn, depend on the frequency (ω), but not on the wave amplitude. Weakness of a light wave means that intensity $\boldsymbol{E}$ of an electric field in the wave must be much smaller than characteristic intensities of intra-atomic electric fields. The greatest intra-atomic intensities (reached in the hydrogen atom) are about 5×10^9 V/cm, whereas the intensities of light fields generated by ordinary (nonlaser) sources of light are about 1 V/cm. Thus, the constitutive equation relating the electric polarization $\boldsymbol{P}$ with the electric field intensity $\boldsymbol{E}$ is linear for ordinary light.

Nonlinear Optics

In the latter half of the 20th century, lasers were invented, which could generate fields with intensities of about 10^7 V/cm, and in 1989, beams of light were produced with electric fields of more than ten times the intra-atomic values. In such strong fields, dielectric susceptibility χ becomes dependent on the electric field intensity, making the constitutive equation nonlinear. Both the dielectric permittivity ε and the absolute refraction index n of the medium also depend on the electric field intensity $\boldsymbol{E}$, and these dependencies are modified by additional influences, such as heating of the medium.

Studies of optical effects in strong light fields use the basic ideas of geometric optics, for example, the concept of refractive index. Consider a uniform isotropic dielectric medium. Adverting to Equation (3), let us write the expansion

$$\boldsymbol{P} = \varepsilon_0\chi^{(1)}\boldsymbol{E} + \varepsilon_0\chi^{(2)}|\boldsymbol{E}|^2 + \varepsilon_0\chi^{(3)}|\boldsymbol{E}|^2\boldsymbol{E} + \cdots. \tag{5}$$

Here, even powers of $\boldsymbol{E}$ are excluded by symmetry consideration, so $\chi = \boldsymbol{P}/\varepsilon_0\boldsymbol{E} = \chi^{(1)} + \chi^{(3)}|\boldsymbol{E}|^2 + \cdots$, which implies (see Equation (4))

$$\begin{aligned} n &= \sqrt{1+\chi^{(1)}+\chi^{(3)}\boldsymbol{E}|^2} \approx n_0 + 2n_3|\boldsymbol{E}|^2, \\ n_0 &= \sqrt{1+\chi^{(1)}}, \quad n_3 = \frac{\chi^{(3)}}{4\sqrt{1+\chi^{(1)}}}, \end{aligned} \tag{6}$$

because the nonlinear term is small. Averaging n on the time period $T=2\pi/\omega$ of oscillations of electric field $\boldsymbol{E}=\boldsymbol{E}_0(x,y,z)\sin\omega t$ yields

$$n = n_0 + n_3|\boldsymbol{E}_0|^2, \tag{7}$$

where n_0 is the linear refractive index and $n_3|\boldsymbol{E}_0|^2$ describes the nonlinear correction to this index.

Equations (3) and (7) indicate that a bounded cylindrical beam of light can create an optical heterogeneity in the medium through which it propagates. Suppose the beam is Gaussian so the amplitude of electric oscillations varies transversely as $|\boldsymbol{E}_0|(r) = A_0\exp[r^2/(2r_0^2)]$, where A_0 is the value of amplitude on the axis of beam, r is distance from the beam axis, and r_0 is a phenomenological constant determining the typical width of the beam. Then,

$$\begin{aligned} v &= c/(n_0 + n_3A_0^2) \text{ on the axis of the beam;} \\ v &= c/n_0 \text{ far from the axis of the beam.} \end{aligned}$$

If the inequality $n_3>0$ ($n_3<0$) is fulfilled, the axial velocity of light is smaller (greater) than the peripheral velocity. In first case, the plane wave front will become concave in the direction of light propagation, and self-focusing of the beam will occur; while in the second case, self-defocusing of the beam will occur. Both these phenomena have been observed; in particular, self-focusing was predicted by Gurgen Askarian in 1962 and then experimentally observed by N.P. Pilipetsky and A.R. Rustamov in 1965.

Nonlinear Reaction-Diffusion Systems

One can image an excitable medium (EM) as a spatial region G occupied by a medium, in which the processes of autocatalytic production, destruction, and diffusion of some substances occur. In the simplest case, when only one substance of concentration $u = u(\mathbf{r}, t)$ is involved (here $\mathbf{r}$ is a point in three-dimensional space and t is time), the dynamics of u are described by two relationships: the continuity equation

$$u_t + \operatorname{div} \boldsymbol{J} = f(u) \tag{8}$$

(where $u_t = \partial u/\partial t$), and the Fick diffusion law

$$\boldsymbol{J} = -D \operatorname{grad} u. \tag{9}$$

For constant diffusivity D, these two equations reduce to the single parabolic reaction-diffusion equation

$$u_t = D\Delta u + f(u). \tag{10}$$

Here, the symbols div, grad, and Δ designate the spatial divergence, gradient operator, and Laplace operator, respectively; $\mathbf{J} = \mathbf{J}(\mathbf{r}, t)$ is the vector of diffusion flux density of the substance; and the function $f(u)$ (called the kinetic function of the active medium) determines the dependence of the production/destruction rate of substance per unit volume of the concentration u.

In general, an EM is able to support the propagation of traveling-wave fronts and impulses. In the steady regime, plane excitation waves propagate with constant speed preserving their spatial profile. Both the speed and the wave profile are determined by physical parameters of the medium in which the waves propagate and are independent of initial conditions; thus such waves are called *autowaves* (AW). This term was coined by Rem Khokhlov (an early Russian specialist on nonlinear physics and president of Moscow State University) in a 1974 presentation as an official opponent during Anatol Zhabotinsky's defense of his doctoral thesis on periodic chemical reactions. (An enthusiastic mountain climber, Khokhlov sadly perished in the Caucasian mountains in 1977.) The significance of AWs stems from the fact that they frequently occur in physical, chemical and biological applications, including muscles, the nervous system, and the heart.

To reveal the qualitative properties of AW propagation, one needs to solve these equations; but even in the spatially homogeneous case—when the medium is described by Equation (10) with constant D—this problem is challenging. Thus, it is of interest to consider an approach in which the general properties of EM are specialized by means of some simple axioms. Interestingly, this approach leads to an analog of traditional geometric optics, as was first shown by Israel Gel'fand and Sergei Fomin in 1961 (Gel'fand & Fomin, 1961).

When either a traveling-wave front or the leading edge of an impulse moves through an EM, the latter switches from the resting state to the excited state (in the case of impulse, the life time of excitable state is finite and equal to the pulse duration). Neglecting the front width, one may imagine the leading edge as a surface that separates the resting and excited zones, and a motion of this surface (which models the motion of an AW) is interpreted as propagation of excitation. Thus, Gel'fand and Fomin introduced two axioms: (i) each spatial point of the EM can be in one of two states: either in the resting state or in the excited one; (ii) if at some time moment t the excitation has reached some spatial point P, then P immediately becomes a source of propagating excitation. In the case of a nonhomogeneous anisotropic EM, the time period $\mathrm{d}t$ of motion of excitation along the infinitesimal path connecting the points $x(s)$ and $x(s + \mathrm{d}s) = x(s) + (\mathrm{d}x/\mathrm{d}s)\,\mathrm{d}s$ (here s is the parameter) depends both on the point $x(s)$ and on the vector $[\mathrm{d}x/\mathrm{d}s\ :\ \mathrm{d}t = f(x, \mathrm{d}x/\mathrm{d}s)\,\mathrm{d}s]$. The function $f(x, \mathrm{d}x/\mathrm{d}s)$ (which is assumed to be a strict convex function of the second argument) is a key element of the theory; thus, if the point $x_1 = x(s_1)$ is excited, then the time period after which the point $x_2 = x(s_2)$ will be excited is given by the expression

$$\min \int_{x_1}^{x_2} f(x, \mathrm{d}x/\mathrm{d}s)\,\mathrm{d}s,$$

where the minimum is taken over all curves connecting the points x_1 and x_2. (Indeed, if the excitation propagating from the point x_1 along all possible paths has already reached the point x_2 along some path, then all paths that connect x_1 and x_2, but take more time, are insignificant.) This is equivalent to Fermat's principle of least time, adapted to dynamic processes in EM. Thus, Gel'fand and Fomin developed a variational theory of propagation of excitation, endowing it with proper Lagrange–Hamiltonian–Jacobian equations that are formally equivalent to the variational formalism of geometric optics based on Fermat's principle.

The Gel'fand–Fomin formulation follows earlier studies of cardiology by Norbert Wiener and Arturo Rosenblueth, whose system of axioms involves an additional (refractory) state (Wiener & Rosenblueth, 1946). Any point goes to the refractory state immediately upon being excited, after which it cannot be excited during the finite time period R. Supposing the velocity v of propagation of excitation to be constant, Wiener and Rosenblueth developed the theory of circulation of an excitation around non-excitable obstacles and determined the least critical length λ of the obstacle, around which the excitation can stationary circulate: $\lambda = Rv$. Further development of these geometric approaches was undertaken by those involved in a seminar on mathematical biology (organized by Gel'fand at Moscow State University in the early 1960s) and led to a prediction of the existence of spiral waves (Balakhovsky, 1965).

Current geometric theories of spiral waves are concerned with properties that were long ignored, including the dependence of an AW leading-edge velocity on the curvature of the edge. Around 1980, Werner Ebeling, Yakov Zeldovich, Yoshiki Kuramoto, and Vladimir Zykov independently showed that the velocity of a curved leading edge must differ from the velocity of a plane edge—if the edge is concave (convex) in the direction of its propagation then it must accelerate (decelerate). This fact is readily understood in the case of a combustion wave where ignition occurs more rapidly (slowly) ahead of concave (convex) parts of the flame front than ahead of the plane parts due to focusing (diverging) the lines of heat flux. The formula for the velocity v of a weakly bent wave front that propagates in two-dimensional EM and obeys Equation (10) is

$$v = v_0 - DK, \tag{11}$$

where v_0 is the velocity of plane front and K is the curvature of the wave front. The domain of applicability of this equation is given by the inequality $Kl_f \ll 1$, where l_f is a characteristic width of the front.

Equation (11) implies that excitation of sufficiently small circular patches of EM will die out. The radii of such patches are bounded above by the critical value $\rho_c = |K_c|^{-1} = Dv_0^{-1}$ which corresponds to the zero value of v in Equation (11). In a more precise theory of curved AW dynamics (Kuramoto, 1980), Equation (11) emerges as an eikonal equation in the "nonlinear geometric optics" of an AW; thus, it is deeply involved in the modern geometric theory of spiral waves (Mikhailov et al., 1994; Elkin et al., 1998).

Interestingly, a version of Snell's law for AWs emerges from such studies which differs from that of classical optics in several important ways:

- Assuming the underlying reaction-diffusion equation to be Equation (10), Equation (1) is replaced by

$$\frac{\tan\psi_1}{\tan\psi_2} = \text{constant} = \frac{D_1}{D_2}, \tag{12}$$

  where D_1 and D_2 are the diffusion constants in the incident and refractive regions.
- This is a *local* law which determines the structure of refracted concentration fronts only near the interface between homogeneous regions rather than far from it.
- Equation (12) is obeyed for nonstationary as well as stationary refractions.
- In the regime of stationary refraction, when the incident wave and the refracted one shape a planar form and move with constant velocities, the angles of incidence and refraction are determined by the conditions (Mornev, 1984)

$$\sin\psi_1 = \sqrt{D_1/(D_1 + D_2)}$$

  and

$$\sin\psi_2 = \sqrt{D_2/(D_1 + D_2)}.$$

If a plane AW is normally incident on the boundary ($\psi_1 = 0$) and D_2 is sufficiently larger than D_1, the wave will be forced to stop. Assuming $f(u)$ to be a cubic-shaped function possessing three zeroes $u_0 < u_b < u_1$ and satisfying the conditions $f(u) > 0$ at $(u < u_0)$ and $(u_b < u < u_1)$, $f(u) < 0$ at $(u_0 < u < u_b)$ and $(u > u_1)$, the condition for the AW to die at the boundary is

$$\frac{D_2}{D_1} > \frac{\sigma_+}{\sigma_-},$$

where σ_- (σ_+) denotes the least (greatest) of two values $|\int_{u_0}^{u_b} f(u)\,du|$, $\int_{u_b}^{u_1} f(u)\,du$ (Mornev, 1984).

O. A. Mornev

See also **Diffusion; Fairy rings of mushrooms; Nonlinear optics; Reaction-diffusion systems; Spiral waves; Zeldovich–Frank-Kamenetsky equation**

Further Reading

Balakhovsky, I.S. 1965. Nekotorye rezhimy dvizheniya vozbuzhdeniya v ideal'noi vozbudimoi tkani. [Some regimes of motion of excitations in ideal excitable tissue.] *Biofizika*, 10(6): 1063–1067

Born, M. & Wolf, E. 2002. *Principles of Optics*, 7th edition, Cambridge and New York: Cambridge University Press

Elkin, Yu.E., Biktashev, V.N. & Holden, A.V. 1998. On the movement of excitation wave breaks. *Chaos, Solitons, and Fractals*, 9: 1597–1610

Gel'fand, I.M. & Fomin, S.V. 1961. *Variatsionnoe ischislenie* [*Variational Calculus*], Moscow: Gosudarstvennoe Izdatel'stvo Fiziko-Matematicheskoii Literatury

Kuramoto, Y. 1980. Instability and turbulence of wavefronts in reaction–diffusion systems. *Progress of Theoretical Physics*, 63(6): 1885–1903

Mikhailov, A.S., Davydov, V.A. & Zykpv, V.S. 1994. Complex dynamics of spiral waves and motion of curves. *Physics* D, 70: 1–39

Mornev, O.A. 1984. Elements of the "optics" of autowaves. In *Self-Organization of Autowaves and Structurees Far from Equilibrium*, edited by V.I. Krinsky, Berlin: Springer

Vinogradova, M.B., Rudenko, O.V. & Sukhorukov, A.P. 1990. *Teoriya Voln* (*Wave Theory*), 2nd edition, Moscow: Nauka

Wiener, N. & Rosenblueth, A. 1946. The mathematical formulation of the problem of conduction of impulses in a network of connected excitable elements, specially in cardiac muscle. *Archivos del Instituto de Cardiologia de Mexico* XVI (3–4): 205–265

GEOMORPHOLOGY AND TECTONICS

Geomorphology deals with the evolution of Earth's surface by gravity (e.g., landslides) and the sculpting action of wind, river flow, and ice. Tectonics deals with the deformation and uplift of rocks, including the behavior of earthquakes. Nonlinearity arises in many geomorphic systems in the feedbacks between the shape of the surface and the fluid flow above the surface. The shapes of sand ripples and dunes, for example, control the flow of wind above the surface, which, in turn, controls the spatial distribution of erosion and deposition. Over time, erosion and deposition modify the shape of the surface and the flow of wind in a positive feedback loop. The flow of wind over a dune and the sediment transport caused by the wind are both nonlinear processes. The relationship between the sediment flux moved by the wind over a dune, for example, and the shear stress exerted by the wind is strongly nonlinear, including both a threshold shear stress for particle entrainment and a nonlinear power-law relationship between flux and shear stress above that threshold.

Some of the nonlinear feedback relationships in geomorphic systems affect practical issues. The feedback between vegetation density and soil erosion, for example, can lead to dust bowl conditions if low vegetation density promotes wind erosion, stripping the soil to further inhibit vegetation growth in a positive feedback loop. In tectonic systems, nonlinear feedbacks also govern much of the interesting behavior. Deformations in rock or along a fault plane, for example, can become localized by the feedback between shear strength and strain rate. Rocks often become weaker as they are strained, in turn focusing more deformation to the areas of highest strain. The rheology of rocks, including their nonlinear dependence on strain rate and time, thus plays a very important role in our understanding of mountain belts.

Self-organized, periodic landforms have received considerable attention in geomorphology. Sand ripples, dunes, and yardangs (all examples of eolian landforms), flutes (elongated ridges of subglacial sediment), drumlins (sculpted large mounds of subglacial debris), finger lakes, cirques, sorted stone stripes and circles (all glacial and periglacial landforms), discontinuous ephemeral streams and step-pool sequences (fluvial landforms), beach cusps and spits (coastal landforms) all have a characteristic width or spacing (Figure 1) that is controlled in part by the fluid movement on or above the surface and its relationship to erosion and deposition. The dynamic processes governing these systems are disparate and complex, but feedback between the surface and the fluid flow on or above the surface plays a key role in all of these examples. Finger lakes, for example, are elongated glacial troughs that developed on the margins of former ice sheets as glacier flow was focused into incipient bedrock depressions, concentrating glacial erosion in troughs in a positive feedback that forms deeper basins when the ice retreats. The Finger Lakes of central New York are the typical examples, but similar lakes occur along other former ice margins. Step-pool sequences form in mountain river channels by feedback between the roughness of the channel bed, the flow velocity in the channel, and the selective entrainment of particles on the bed. A channel reach with an initially rougher bed than nearby reaches will have slower flow velocity, promoting the deposition of coarse particles along the channel reach and a further decrease in flow velocity. This feedback produces channel reaches characterized by shallow gradients and fine particles alternating with steep gradients and coarse particles. In some cases, numerical models have been developed that enable the wavelengths of these periodic landforms to be predicted.

Figure 1. Sand ripples with a characteristic width or spacing. White Sands, New Mexico, courtesy of National Park Service.

In many if not most cases, landform evolution does not lead to periodic topography at all. River networks, coastal erosion of a rugged shoreline, and the dissolution of limestone to form cave networks (or karst topography above ground) are all examples of processes that create chaotic landforms with no apparent characteristic scale. Many of these landforms have an underlying order, however, by virtue of their self similarity. Self-similar landforms are those that have a similar appearance and statistical structure at a wide range of spatial scales. Rugged coastlines are the classic example of fractals popularized by Benoit Mandelbrot (1982). Long before the fractal structure of coastlines had been described, however, geomorphologists were interested in the self-similarity of river networks. To examine this self-similarity, streams are first ordered according to their position in a drainage network. The Strahler order defines all channels with no upstream tributaries as first-order channels. Whenever any two streams of like order join at a tributary, they form a stream of the next highest order. Horton's law states the principle of self-similarity of drainage networks mathematically: the ratio of the number of streams of order n to the number of streams of order $n+1$ is equal to approximately 4 and is independent of n (Rodriguez-Iturbe and Rinaldo (2001) provide an excellent review). Similar relationships exist between channel lengths, areas, and the Strahler order.

Horton's laws are satisfied in many different kinds of networks; however, so the modern view is that Horton's laws are not unique properties of drainage networks (Kirchner, 1993). Other relationships, such as the angles of tributary junctions and the relationships between channel slope and the Stahler order, contain important information about the self-organization of channel networks.

The Earth's crust exhibits nonlinear, critical behavior in several ways. First, earthquakes occur over a wide range of sizes with a frequency-size distribution characterized by a power law. This observation, known as the Gutenberg–Richter law, is the most fundamental rule of seismology. Seismicity also exhibits temporal correlations that have self-similar properties. Omori's law, for example, states that the frequency of foreshocks or aftershocks is inversely proportional to the time before or since the mainshock. Theoretical models for fault behavior have been devised based upon a simple model of blocks (representing one fault plane) frictionally coupled to a table (the other fault plane) and elastically coupled to one another and to a driver plate (representing the regional tectonic stress) (e.g., Turcotte, 1997). The model builds up stress until all the elements are near the threshold for slipping. At that point, the slippage of one block can transfer stress to other blocks in a cascade that produces earthquakes that follow the Gutenberg–Richter law. This phenomenon is called stick-slip friction. The discrete nature of the slider-block model appears to have an analog in real faults; the behavior of faults can be characterized according to the size and strength of asperities (sticky spots) on the fault plane, which are like the individual blocks in the slider-block system. The aftershock behavior of earthquakes does not appear to arise naturally in the simplest slider-block models; instead, some viscous coupling between blocks is necessary to reproduce Omori's law (Pelletier, 2000), suggesting that the punctuated process of seismic events is linked to a steady, creeping motion of the fault motion over longer time scales.

The crust also exhibits nonlinear critical behavior in the way that magma makes its way through the crust and is released as volcanic eruptions. The frequency–size distribution of volcanic eruptions appears to follow a power-law distribution analogous to the Gutenberg–Richter distribution in terms of the volume of material released. Magmatic and volcanic activity is also clustered in time in a way that is generally analogous to Omori's law. The processes of fluid movement through the crust are very different compared with the stresses on an earthquake fault, but a model for the spatial interaction of many fluid conduits all near the threshold for eruption produces a model very similar to the slider-block model (Pelletier, 1999). This kind of model appears to be most consistent with the Rayleigh distillation model of geochemical mixing (Turcotte, 1997), which describes the abundances of certain geochemical elements in the crust.

Jon D. Pelletier

See also **Avalanches; Branching laws; Dune formation; Evaporation wave; Feedback; Fractals; Glacial flow; Rheology; Sandpile model**

Further Reading

Kirchner, J.W. 1993. Statistical inevitability of Horton's laws and the apparent randomness of stream channel networks. *Geology*, 21: 591–594

Mandelbrot, B.B. 1982. *Fractal Geometry of Nature*, New York: W.H. Freeman

Pelletier, J.D. 1999. Statistical self-similarity of magmatism and volcanism. *Journal of Geophysical Research*, 104: 15,425–15,438

Pelletier, J.D. 2000. Spring-block models of seismicity: review and analysis of a structurally heterogeneous model coupled to a viscous asthenosphere. In *GeoComplexity and the Physics of Earthquakes*, edited by J.B. Rundle, D.L. Turcotte, & W. Klein, Washington, DC.: American Geophysical Union

Rodriguez-Iturbe, I. & Rinaldo, A. 2001. *Fractal River Basins: Chance and Self-Organization*, Cambridge and New York: Cambridge University Press

Turcotte, D.L. 1997. *Fractals and Chaos in Geology and Geophysics*, 2nd edition, Cambridge and New York: Cambridge University Press

GESTALT PHENOMENA

The Gestalt idea was introduced to science by Ernst Mach (1868) and Christian von Ehrenfels (1890). Mach stated that the spontaneous creation of order, that is, order arising without any external control, can be shown in inanimate nature. Von Ehrenfels characterized Gestalt qualities, that is, higher order qualities emerging from basic elements, by two criteria: (1) supersummativity, which means that the elements of a pattern presented individually to a person give, in the totality of the experience, a poorer impression than the total experience of a person to whom all the elements are presented; and (2) transposition, which means the characteristics of a Gestalt quality are retained even if all the elements which exhibit the Gestalt quality are changed in a certain way (for example, the transposition of a melody).

Wolfgang Köhler (1920) delivered the earliest formulation of a concept of self-organization of perception. The idea that perception must necessarily be understood as a process of autonomous creation of order runs through all his works. Starting from the observation of spontaneous Gestalt tendencies in experience, Köhler made it his primary task to design and test a model of brain function in which the phenomenal organization of the perceptual world is explained as not only stimulus-dependent, but as strongly dependent upon the perceptual system's own inner dynamics. In the development of Gestalt theory, Köhler was mainly oriented to the observations and theoretical concepts of physics. In accordance with the assumption of linear thermodynamics, the general systemic tendency towards final equilibrium was considered in Köhler's time to be the only basic principle of self-organization. This principle can easily be demonstrated in cognition by recursive experiments of serial reproduction of complex patterns. These patterns follow the "principle of prägnanz" towards very simple and stable configurations (Stadler & Kruse, 1990; Kanizsa & Luccio, 1990). In perception, this principle states that people will perceive the most orderly or regular thing they can out of the stimuli that are presented to them. Köhler applied the model of physical fields striving independently to balance forces directly to the way in which the visual system functions. At the time, this almost provocatively contradicted the findings of neuroanatomy and neurophysiology. On the basis of his theoretical model of perception, the brain is not seen as a complex network of many different interacting neurons but as a homogeneous conductor of bioelectric forces.

Köhler's argument was not primarily the postulation of electromagnetic field forces acting in the brain independently of neuroanatomical structures (which has been refuted by most contemporary brain scientists) but the idea of self-organization in the brain. He was fully aware of the fact that the general principle of development of linear thermodynamics has been rather unsuitable for application to biological systems as long as it was oriented exclusively towards the final equilibrium. This was criticized by Köhler himself in 1955: "Although this is a perfectly good principle, it cannot, in its present formulation, be applied to the organism. For the organism is obviously not a closed system; moreover, while the direction indicated by the principle may be called 'downward', the direction of events in healthy organisms is on the whole clearly not 'downward' but, in a good sense, 'upward'."

Modern Developments

The brain is conceived as a self-organizing system that can be treated by means of synergetics (Haken, 1983, 1990). Pattern recognition is understood as pattern formation (of activities of the neural net). Incomplete (visual) data are complemented by a dynamic associative memory. In both cases (pattern formation and recognition), incomplete data generate order parameters that compete with each other. In general, one order parameter wins and generates, according to the slaving principle of synergetics, the complete pattern. In this processes, idealized (Gestalt) patterns may be incorporated. Of particular interest are ambiguous figures, such as Figure 1 (young woman vs. old woman). Here, two or more interpretations (percepts) are possible, and two (or more) order parameters may win the competition. The final outcome is determined by an order parameter dynamics in which the dynamics of attention parameters is included.

Figure 1. Young woman vs. old woman.

The mathematical approach (algorithm of the "synergetic computer") is as follows: The images of different objects are decomposed into their pixels that are lumped together as pixel vectors

$$\boldsymbol{v}_\mu = (v_{\mu 1}, v_{\mu 2}, ..., v_{\mu N}), \quad \mu = 1, ..., M. \tag{1}$$

The label μ is associated with an interpretation (the name of a person, say), and the adjoint vectors $\boldsymbol{v}_\mu^+$ are defined by

$$(\boldsymbol{v}_\mu^+ \boldsymbol{v}_\nu) = \delta_{\mu\nu}. \tag{2}$$

The activity pattern of the neural net is written as

$$\boldsymbol{q}(t) = \sum_{\mu=1}^{M} \xi_\mu(t) \boldsymbol{v}_\mu + \boldsymbol{r}(t), \tag{3}$$

where $\boldsymbol{r}$ is a residual term that vanishes in the course of time. The order parameters are defined by

$$\xi_\mu(t), \tag{4}$$

where the initial value at the observation time t_0 is given by

$$\xi_\mu(0) = (\boldsymbol{v}_\mu^+ \boldsymbol{q}(0)). \tag{5}$$

The order parameters obey competition equations that can be derived from a potential function V

$$\mathrm{d}\xi_\mu(t)/\mathrm{d}t = -\partial V/\partial \xi_\mu, \tag{6}$$

where

$$V = V(\xi_1, ..., \xi_M; \lambda_1, ..., \lambda_M) \tag{7}$$

does not only depend on the order parameters but also on the attention parameters λ_j. The competition equations read explicitly

$$\mathrm{d}\xi_\mu(t)/\mathrm{d}t = \left(\lambda_\mu - B\sum_{\mu'\neq\mu}^{M} \xi_{\mu'}^2 - D\sum_{\mu'=1}^{M} \xi_{\mu'}^2\right)\xi_\mu \tag{8}$$

with positive constants λ_μ, B, C. The winning order parameter fixes the activity pattern (3). In the case of ambiguous patterns (such as that of Figure 1), equations for the order parameters ξ_1 and ξ_2 and the attention parameters read

$$\mathrm{d}\xi_1/\mathrm{d}t = (\lambda_1 - C\xi_1^2 - (B+C)\xi_2^2)\xi_1 - \partial V_\mathrm{b}/\partial\xi_1, \tag{9}$$

$$\mathrm{d}\xi_2/\mathrm{d}t = (\lambda_2 - C\xi_2^2 - (B+C)\xi_1^2)\xi_2 - \partial V_\mathrm{b}/\partial\xi_2, \tag{10}$$

$$\mathrm{d}\lambda_j/\mathrm{d}t = \gamma(1 - \lambda_j - \xi_j^2), \quad j = 1, 2. \tag{11}$$

The bias potential V_b is defined by

$$V_\mathrm{b} = 2B\xi_1^2\xi_2^2\left(1 - 4\alpha\frac{\xi_1^2 - \xi_2^2}{\xi_1^2 + \xi_2^2}\right). \tag{12}$$

The parameter α is determined by the percentage of that perception that is seen first. It also determines the relative length of the perception times that occur in the oscillatory motion of the order parameters ξ_1, ξ_2 that represent the switch from one percept to the other one and back again.

HERMANN HAKEN AND MICHAEL A. STADLER

See also **Cell assemblies; Emergence; Synergetics**

Further Reading

von Ehrenfels, C. 1890. Über Gestaltqualitäten [On Gestalt qualities]. In *Foundations of Gestalt Theory*, edited by B. Smith, München: Philosophia-Verlag, pp. 82–117

Haken, H. 1983. Synopsis and introduction. In *Synergetics of the Brain*, edited by E. Başar, H. Flohr, H. Haken & A.J. Mandell, Berlin and New York: Springer, pp. 3–25

Haken, H. 1990. *Synergetic Computers and Cognition: A Top-down Approach to Neural Nets*, Berlin and New York: Springer

Kanizsa, G. & Luccio, R. 1990. The phenomenology of autonomous order formation in perception. In *Synergetics of Cognition*, edited by H. Haken & M. Stadler, Berlin and New York: Springer, pp. 186–200

Köhler, W. 1920. *Die physischen Gestalten in Ruhe und im stationären Zustand*, Braunschweig: Vieweg

Köhler, W. 1955. Direction of processes in living systems. *Scientific Monthly*, 8: 29–32

Mach, E. 1868. Die Gestalten der Flüssigkeit [The Gestalts of fluids]. In *Populärwissenschaftliche Vorlesungen*, 4th edition, 1910, Leipzig: Barth

Stadler, M. & Kruse, P. 1990. The self-organization perspective in cognition research: historical remarks and new experimental approaches. In *Synergetics of Cognition*, edited by H. Haken & M. Stadler, Berlin and New York: Springer, pp. 32–52

GINZBURG–LANDAU EQUATION

See **Complex Ginzburg–Landau equation**

GLACIAL FLOW

Glaciers are defined as multi-year features, consisting of snow and ice, which flow down-slope under the force of gravity. The broadness of this definition means that there exists a continuum of glaciers that ranges from small, multi-year snow patches with surface areas of the order of 100 square meters, to the Columbia Glacier in Alaska, with a surface area of over 1100 square kilometers (the District of Columbia would fit within its terminus). Yet, in spite of the broad range of features encapsulated in this definition, the same basic physical processes are common to all of the world's roughly 160,000 different glaciers, and most of these processes are nonlinear.

Glaciers deform under their own weight, behaving like highly viscous fluids. Mass input is greatest at the glacier's upper elevations where colder temperatures result in a greater percentage of precipitation falling as snow. Mass loss is greatest at the glacier terminus, the lowest point on the glacier, where temperatures and melting are highest and where ablation (mass loss from melting, evaporation, sublimation, and in special cases, iceberg formation) equals flow. This imbalance results in a continuous mass transfer from the upper reaches to the lower reaches. The basis of the equations governing glacial flow is therefore mass conservation. However, unlike liquid water, where an applied stress, τ, (i.e., a squeeze) causes a linear, proportional deformation or strain, glaciers have a nonlinear stress-strain response. Although measurements in remote mountain locations are difficult and limited, field studies and laboratory experiments have shown that

$$\dot{\varepsilon} = A\tau^n, \tag{1}$$

where n is constant and generally assigned a value of 3 (though values ranging from 1.5 to 4.2 can be found in the literature), $\dot{\varepsilon}$ is the rate of deformation, and $1/A$ is a nonlinear measure of the viscosity. This glacier "flow law" is a version of pseudoplastic flow and is similar to dry sand dune flows (which use a value of 2 for n). Viscoplastic flows, such as clay-water mixtures, and Bagnold macro-viscous flows, such as mud-flows, are also similar, differing primarily in the value of the exponent n.

The nonlinear flow law is the source of many fractal, self-similar, and nonlinear scaling properties. The basic scaling relationships are simple: discharge of ice through a given cross section of the glacier is proportional to glacier depth raised to the power $(n+2)$, and glacier flow velocity is proportional to glacier depth raised to the power $(n+1)$ (Patterson, 1994). However, Bahr (1997) has shown that in conjunction with mass and momentum conservation, the nonlinear flow law implies nonlinear scaling relationships between surface area (a parameter easily measured with satellites) and many difficult to measure but fundamental properties. Glacier thickness, volume, mass balance, velocity, flux, and other parameters relate to the surface area by exponents of 3/8, 11/8, etc. Using these unusual scaling exponents, the volume of ice in the world's glaciers can be predicted based solely on observed surface areas.

Equation (1) applies to basic glacier flow under constant stress. In reality, glaciers are rarely under consistent stress throughout. In areas where glaciers are under tensional stress, if the stress becomes too high, the ice becomes brittle and fractures, resulting in crevassing. Crevassing can be mathematically described using fracture mechanics (Smith, 1976; Sassolas et al., 1995). Crevassing can occur as the glacier flows over large drops in the bed as a result of varying flow speeds (Harper et al., 1998). Particularly dramatic crevassing occurs in the lower reaches of retreating tidewater glaciers as a result of faster flow at the glacier terminus than in the ablation zone (the area of the glacier that is annually losing more mass than it is gaining). Because the lower sections of tidewater glaciers are near or at floatation, unlike land-based glaciers, tidewater glacier crevassing can result in calving (the formation of icebergs from pieces that are broken off the terminus). Calving from tidewater glaciers can be modeled using both fracture mechanics and percolation theory (Bahr, 1995).

Tidewater glacier calving contributes to an unexpected nonlinearity in the tidewater glacier terminus position. Most glaciers move back and forth with changes in climate as the balance between melt and accumulation shifts. For tidewater glaciers, however, there is an additional loss of mass through calving. The tidewater glacier calving rate increases with water depth, but water depth is typically minimized by a pile of debris deposited at the end of the glacier. If the glacier terminus retreats backwards off the debris pile, then the water depth increases and the calving rate increases, further increasing glacier retreat (Meier, 1993; van der Veen, 2002). This is a classic positive feedback loop scenario, and is the cause of the dramatic and rapid retreats recently seen in many glaciers that terminate in water, such as the Columbia Glacier in Alaska and many of the New Zealand glaciers that terminate in lakes. While many of the world's glaciers are slowly retreating due to changes in climate, these tidewater glaciers retreat nonlinearly in response to even the smallest climatic perturbations.

In addition to surface and internal processes, such as flow and crevassing, glaciers exhibit nonlinear behavior in their basal processes. For temperate glaciers (those not frozen to their beds), glacial flow is a combination of ice deformation and sliding at the glacier bed. Motion tends to be stick-slip, very similar to the nonlinear slider-block models of earthquakes (Bahr & Rundle, 1996; Fischer & Clarke, 1997). This gives rise to the grinding of the underlying rock, plucking of rocks out of the bed, and deformation of the bed in places where it is a fine-grained matrix. Lubrication appears to increase flow rates, as it does in sub-surface faults (Patterson, 1994). At the extreme end of lubricated basal flow, we find surging glaciers. These glaciers appear to build up water and water pressure at the glacier bed. Some mechanism or pressure trigger allows this water to be periodically catastrophically released (the Variegated Glacier in Alaska, for example, surged in 1906, 1947, 1964–65, and 1982–83), resulting in rapid flow and over-extension of the glacier (Patterson, 1994).

Finally, the overall structure of large glaciers is fractal. Large glaciers, such as the Talkeetna and Columbia Glaciers in Alaska have multiple upper branches that coalesce into one outlet tongue, similar to a river system or branching tree. Measurements have shown that the structure is statistically self-similar with fractal dimensions ranging from roughly 1.6 for small- to mid-sized mountain glaciers, to 2.0 for large glaciers and space-filling ice sheets such as Greenland (Bahr & Peckham, 1996).

KAREN LEWIS MACCLUNE AND DAVID BAHR

See also **Avalanches; Dune formation; Geomorphology and tectonics; Sandpile model**

Further Reading

Bahr, D.B. 1995. Simulating iceberg calving with a percolation model. *Journal of Geophysical Research*, 100(B4): 6225–6232

Bahr, D.B. 1997. Global distributions of glacier properties: a stochastic scaling paradigm. *Water Resources Research*, 33(7): 1669–1679

Bahr, D.B. & Peckham, S. 1996. Observations of self-similar branching topology in glacier networks. *Journal of Geophysical Research*, 101(B11): 25511–25521

Bahr, D.B. & Rundle, J.B. 1996. Stick-slip statistical mechanics of motion at the bed of a glacier. *Geophysical Research Letters*, 23(16): 2073–2076

Fischer, U.H. & Clarke, G.K.C. 1997. Stick-slip sliding behavior at the base of a glacier. *Annals of Glaciology*, 24: 390–396

Harper, J.T., Humphrey N.F. & Pfeffer, W.T. 1998. Crevasse patterns and the strain rate tensor: a high-resolution comparison. *Journal of Glaciology*, 44(146): 68–76

Hooke, R. 1998. *Principles of Glacier Mechanics*, Saddle River, NJ: Prentice-Hall

Meier, M.F. 1993. Columbia Glacier during rapid retreat: interactions between glacier flow and iceberg calving dynamics. *Workshop on the Calving Rate of West Greenland Glaciers in Response to Climate Change*, Copenhagen

Patterson, W.S.B. 1994. *The Physics of Glaciers*, 3rd edition, New York: Elsevier

Sassolas, C., Pfeffer, T. & Amadei, B. 1995. Stress interaction between multiple crevasses in glacier ice. *Cold Regions Science and Technology*, 24: 107–116

Sharp, R.P. 1988. *Living Ice*, Cambridge and New York: Cambridge University Press

Smith, R.A. 1976. The application of fracture mechanics to the problem of crevasse penetration. *Journal of Glaciology*, 17(76): 223–228

van der Veen, C.J. 2002. Calving glaciers. *Progress in Physical Geography*, 26(1): 96–122

GLOBAL WARMING

Few modern scientific concerns have achieved such notoriety as the possibility of relatively rapid anthropogenic global warming through increased CO_2 emissions. This complex problem became a matter of considerable public attention during the 1980s, and during the 1990s, the first attempt was made at international management of the challenge (the Kyoto Protocol under the United Nations Framework Convention on Climatic Change). However, scientific awareness of CO_2-induced climatic change is not new, and the underlying physical processes were understood from the beginning of studies concerning the absorption of radiation by the atmosphere. Later research resulted in a deeper understanding of the dynamics of the biospheric carbon cycle, and global atmospheric circulation models have been adopted, and adapted, for assessing the future course of tropospheric CO_2 levels. In spite of all of these advances, much remains unclear and uncertain.

Early Studies

Several years before his death in 1830, the French mathematician Joseph Fourier concluded that the atmosphere acts like the glass of a greenhouse, letting light through and retaining the invisible rays emanating from the ground (Fourier, 1822). In modern scientific terms, the atmosphere is highly (though not perfectly) transparent to incoming (shortwave) solar radiation, but it is a strong absorber of certain wavelengths in the outgoing (longwave) infrared spectrum produced by the reradiation of absorbed sunlight.

John Tyndall was the first scientist to study this process in detail by measuring the absorptive properties of air and its key constituent molecules (water vapor and about a dozen different compounds). He used a sensitive galvanometer to measure the electric current passing through gases irradiated by heat. In 1861, Tyndall concluded that water vapor accounts for most of the atmospheric absorption and hence "every variation of this constituent must produce a change in climate. Similar remarks would apply to the carbonic acid diffused through the air. . ." (Tyndall, 1861). The next major contribution to the field came just before the end of the 19th century when Svanté Arrhenius offered the first calculations of the global surface temperature rise resulting from naturally changing atmospheric CO_2.

Arrhenius's conclusions contained all of the key qualitative modern results. He found that geometric increases of CO_2 will produce a nearly arithmetic rise in surface temperatures, that the warming will be smallest near the equator and highest in polar regions, that the Southern hemisphere will be less affected, and that the warming will reduce temperature differences between night and day (Arrhenius, 1896). His quantitative results also resembled those of today's best global climatic models: he predicted that the increase in average annual temperature will be about 50°C in the tropics and just over 6°C in the Arctic. All of these findings applied to natural fluctuations of atmospheric CO_2: Arrhenius concluded (correctly) that future anthropogenic carbon emissions would be largely absorbed by the ocean and (incorrectly, as he grossly underestimated future fossil fuel combustion) that the accumulation would amount to only about 3 ppm in half a century.

The link between CO_2 and climate change was resurrected in 1938 by George Callendar who calculated a more realistic temperature rise with doubling of CO_2 concentrations (1.5°C rise) and documented a slight global warming trend of 0.25°C for the preceding half a century (Callendar, 1938). In his later writings, he also recognized the importance of carbon emissions from land-use changes. In 1956, Gilbert Plass performed the first computerized calculation of the radiation flux in the main infrared region of CO_2 absorption (Plass, 1956). His results (average surface temperature rise of 3.6°C with the doubled atmospheric CO_2) were published a year before Roger Revelle and Hans Suess summarized the problem with continuing large-scale fossil fuel combustion in such a way that the key sentence has become a citation classic:

> Thus human beings are now carrying out a large scale geophysical experiment of a kind that could not have happened in the past nor be reproduced in the future. Within a few centuries we are returning to the atmosphere and oceans the concentrated organic carbon stored in sedimentary rocks over hundreds of millions of years. (Revelle & Suess, 1957)

An almost instant response to this concern was the setting up of the first two permanent stations for the measurement of background CO_2 concentrations, at Mauna Loa in Hawai'i and at the South Pole. Accumulating measurements began showing a steady rise of atmospheric CO_2 at these two remote locations, but, once again, attention to the problem of potential global warming eased during the 1960s and began to grow only in the aftermath of OPEC's two sudden oil price hikes during the 1970s.

Numerical Models of Anthropogenic Global Warming

By the late 1960s, improvements in computer capabilities made it possible to run the first three-dimensional models of global atmospheric circulation and use them to simulate the effects of higher CO_2 levels. Most of these simulations looked at possible effects arising from the doubling of preindustrial CO_2, that is, after reaching levels around 600 ppm. Initial simulations indicated a 2.93°C rise with the doubling of the CO_2 level to 600 ppm (Manabe & Wetherald, 1967). Increases in computing power (subject to Moore's famous law) and better understanding of interactions between the atmosphere, oceans, and the terrestrial biosphere has led to increasingly more realistic models of global climate. Another important refinement was the inexplicably delayed consideration of other greenhouse gases (above all, of CH_4, N_2O, and chlorofluorocarbons) whose combined radiative forcing is now slightly higher than that of carbon dioxide (about 1.5 and 1.4 $W\,m^{-2}$).

By the late 1990s, the best models coupled the atmosphere's physical behavior with changes on land and in the ocean, and with simulations of some key features of carbon and sulfur cycles and of atmospheric chemistry (Houghton et al., 2001). At the same time, even our best numerical models still represent the atmosphere with a relatively coarse grid and are incapable of reproducing the intricacies and multiple feedbacks that determine the course and the rate of climate change.

One of the most important sources of potential error in the climate models is the treatment of clouds. The best general circulation models represent fairly well some essential gross features of global atmospheric physics but their iterative calculations are done at such widely spaced points of three-dimensional grids that it is impossible to treat cloudiness in a realistic manner. And yet clouds are key determinants of the planetary radiation balance because they have, on balance, a pronounced net cooling effect. Because clouds account for about half of the Earth's albedo (the fraction of incident radiation that is reflected), even relatively small changes in their properties could have an appreciable effect on the course of global warming.

Other unresolved matters include the response of terrestrial biota (Will carbon sequestration take place mostly in short- or long-lived tissues or in soil?), marine algae (especially their role in forming clouds), sudden releases of methane (rising temperatures may lead to catastrophic emissions from methane hydrates), and effects of orbital and solar influences (particularly a very high correlation between the solar cycles shorter than the 11-year mean and higher average land temperature of the Northern Hemisphere, and a link between mid-atmospheric temperature and changing intensity of radiation over the sunspot cycle). If the global forecasts are uncertain, regional predictions are particularly questionable. The most complex coupled models now provide reasonably reliable simulations of climate down to the sub-continental level but their results still have unacceptably large variations on regional scales.

Geological Evidence for Global Warming and Cooling

Indirect or proxy markers (such as isotopic and trace chemical analysis on tree rings, ice, or sediment cores) make it clear that a substantial decline of atmospheric CO_2 preceded the most extensive and longest lasting (some 70 million years, or Ma) glaciation of the entire Phanerozoic era that began about 330 Ma ago. Approximate reconstruction of CO_2 levels for the past 300 Ma—since the formation of the Pangea whose eventual break-up led to the current distribution of oceans and land masses—indicates, first, a pronounced rise (about five times the current level during the Triassic period), followed by a steep decline (Berner, 1998; Figure 1, top). Boron-isotope ratios of planktonic foraminifer shells point to CO_2 levels above 2000 ppm 60–50 Ma ago (with peaks above 4000 ppm), followed by an erratic decline to less than 1000 ppm by 40 Ma ago, and relatively stable and low (below 500 ppm) concentrations ever since the early Miocene 24 Ma ago (Pearson & Palmer, 2000; Figure 1, bottom).

Reliable record of atmospheric CO_2 is available only for the past 420,000 years thanks to the analyses of air bubbles from ice cores retrieved from Antarctica and Greenland. Preindustrial CO_2 levels never dipped below 180 ppm and never rose above 300 ppm (Raynaud et al., 1993; Petit et al., 1999; Figure 2) and their oscillations are highly positively correlated with changing temperatures. But these correlations are not a proof of a clear cause-and-effect relationship as there are no obvious lead-lag sequences. Other recent paleoclimatic studies actually found signs of decoupling of atmospheric CO_2 and global climate during the Phanerozoic eon and particularly during the early to middle Miocene, when a warm period coexisted with low CO_2 levels (Veizer et al., 2000; Pagani et al., 1999). These findings confirm the complexity of climate change where cause and effect are difficult to assign: atmospheric CO_2 may have

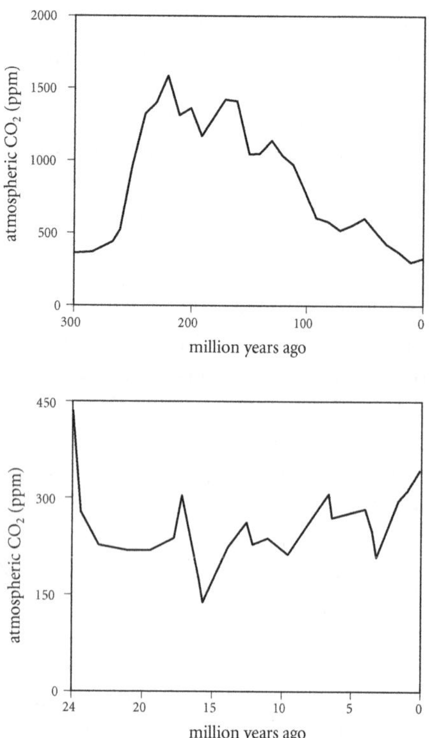

Figure 1. Atmospheric CO_2 concentrations during the past 300 and 24 million years. Based on Berner (1998) and Pearson and Palmer (2000).

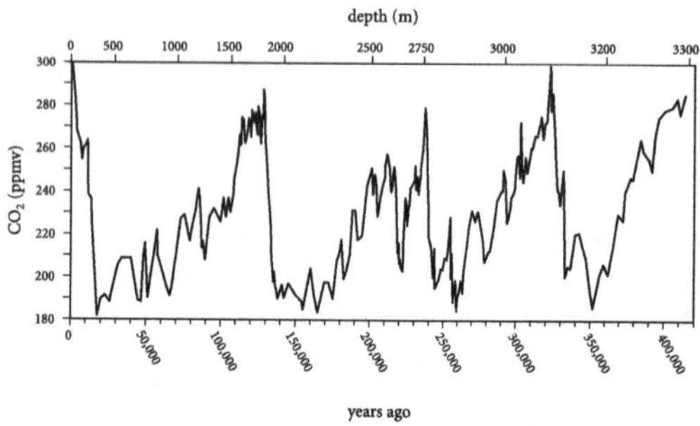

Figure 2. Atmospheric CO_2 concentrations during the past 420,000 years derived from air bubbles in Antarctica's Vostok ice core. Based on Petit et al. (1999).

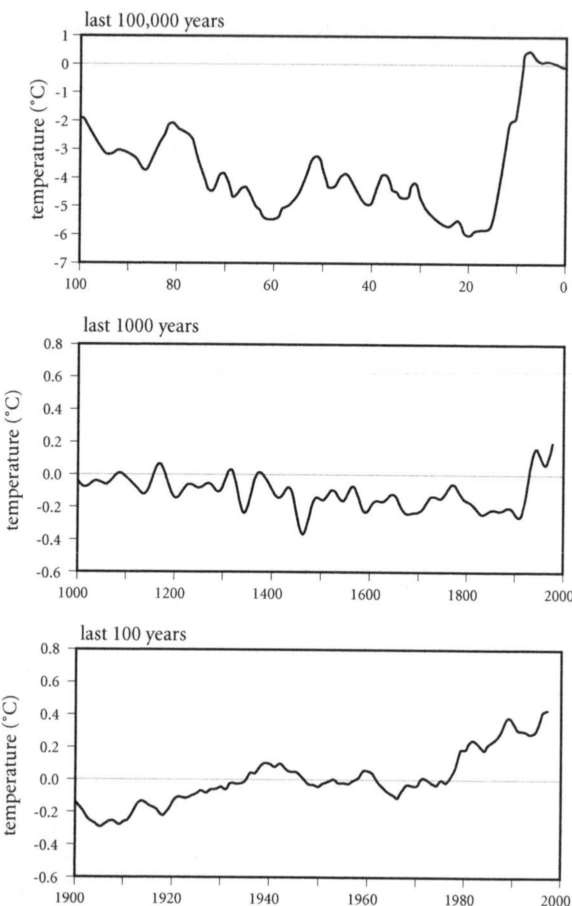

Figure 3. Reconstructed temperature trends during the past 100,000 years (from the Vostok ice core), 1000 years (for the Northern Hemisphere), and a 5-year running mean from instrumental temperature measurements for the past 100 years. Reproduced from Smil (2002).

been a primary climate driver but the evidence is not conclusive (Kump, 2002). The most likely pacemaker during the Pleistocene period was small changes in the Earth's orbit around the Sun; massive methane releases from gas hydrates and volcanic activity must be also considered.

Recent Evidence for Global Warming

During the time between the rise of the first high civilizations (5000–6000 years ago) and the beginning of the fossil fuel era, atmospheric CO_2 levels had fluctuated within an even narrower range of 250–290 ppm. Subsequent anthropogenic emissions pushed atmospheric concentrations of CO_2 to a high of 370 ppm by the year 2000. Paleoclimatic studies of the Northern Hemisphere during the last millennium show warming periods during the 12th and 18th centuries and pronounced cooling during the 15th century (the Little Ice Age). A demonstrable cooling trend between the late 18th and the early 20th centuries was followed by an unprecedented rate of warming that has brought the average planetary temperature to levels higher than at any time during the past 1000 years (Figure 3).

The most extensive studies of the existing global record of surface temperatures have detected long-term planetary warming of, respectively, 0.5°C and 0.78°C since the middle of the 19th century (Jones et al., 1986; Hansen & Lebedeff, 1988). Changes in measurement techniques (different thermometers), station locations (from downtowns to suburbs) and station environment (increasing urban heat island effect), and until very recently, highly inadequate coverage of large areas of the Southern Hemisphere complicate the interpretation of this shift, which has distinct spatial patterns with areas of more pronounced warming and regions of slight cooling. However, the most recent (post-1976) spell of warming has been almost global and the 1990s were the warmest decade since the beginning of the

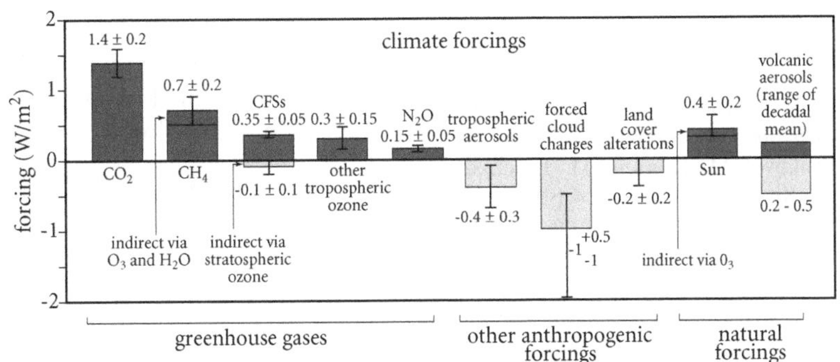

Figure 4. Estimates of cumulative radiation forcings by greenhouse gases and aerosols between 1850 and 2000 according to Hansen et al. (2000).

instrumental record in the 1850s and the warmest ten years of the millennium in the Northern Hemisphere.

According to the general circulation models, the warming should have been more pronounced. The best explanation of the discrepancy between the models and the actual temperature record is that the warming was partially counteracted by sulfate aerosols. The combined direct and indirect effect of all greenhouse gases resulted in a total anthropogenic forcing of about $2.8\,W\,m^2$ by the late 1990s (Hansen et al., 2000; Figure 4). This is equal to a little more than 1% of solar radiation reaching the ground.

Future Climate

If the atmospheric warming was primarily the function of radiative forcing, then the level of greenhouse gas emissions would be the key variable. Recent emission scenarios for CO_2 alone offer a very large range of concentrations, 540–970 ppm, by the year 2100. CH_4 levels may range even wider, from just above 1500 ppb to more than 3600 ppb. The aggregate radiative forcing may thus be anywhere between 4 and $9\,W\,m^{-2}$ and the climate sensitivity would then range between 1.5°C and 4°C with 2.2–3°C considered to be the most likely by the latest IPCC assessment. Broad consensus from the latest generation of models foresees that this climatic change would cool the stratosphere while raising the tropospheric temperatures in a distinct spatial pattern, with the warming more pronounced on the land (and during nights) and with increases of about two to three times the global mean in higher latitudes in winter than in the tropics, and greater in the Arctic than in the Antarctic.

There are many effective ways to slow down the greenhouse gas emissions and reduce their environmental impact. Most significantly, the affluent countries could largely retain their quality of life while reducing their energy and material consumption by at least a third. While the means are available, the will to act, nationally and internationally, is mostly absent. Global warming is a complex natural process but its anthropogenic enhancement calls for a fundamentally moral solution that runs against some basic human propensities: consume less and do so more efficiently.

VACLAV SMIL

See also **Atmospheric and ocean sciences; Forecasting; General circulation models of the atmosphere**

Further Reading

Alverson, K.D., Bradley, R.S. & Pedersen, T.F. (editors). 2003. *Paleoclimate, Global Change, and the Future*, Berlin and New York: Springer

Arrhenius, S. 1896. On the influence of carbonic acid in the air upon the temperature of the ground. *Philosophical Magazine, Series 5*, 41: 237–276

Berner, R.A. 1998. The carbon cycle and CO_2 over Phanerozoic time: the role of land plants. *Philosophical Transactions of the Royal Society of London* B, 353: 75–82

Callendar, G.S. 1938. The artificial production of carbon dioxide and its influence on temperature. *Quarterly Journal of the Royal Meteorological Society*, 64: 223–237

Fourier, J.B.J. 1822. *Théorie Analytique de la Chaleur*, Paris: Firmin Didot

Hansen, J. & Lebedeff, S. 1988. Global surface air temperatures: update through 1987. *Geophysical Research Letters*, 15: 323–326

Hansen J., Sato, M., Ruedy, R., Lacis, A. & Oinas, V. 2000. Global warming in the twenty-first century: an alternative scenario. *Proceedings of the National Academy of Sciences USA*, 97: 9875–9880

Houghton, J.T., Ding, Y., Griggs, D.J., Noguer, M., van der Linden, P.J. & Xiaosu, D. (editors). 2001. *Climate Change 2001: The Scientific Basis*, Cambridge and New York: Cambridge University Press

Jones P.D., Wigley, T.M.L. & Wright, P.B. 1986. Global temperature variations between 1861 and 1984. *Nature*, 322: 430–434

Kump, L.R. 2002. Reducing uncertainty about carbon dioxide as a climate driver. *Nature*, 419: 188–190

Lozán, J.L., Grassl, H. & Hupfer, P. (editors). 2001. *Climate of the 21st Century: Changes and Risks, Scientific Facts*, Hamburg: Wissenschaftliche Auswertungen

Manabe, S. & Wetherald, R.T. 1967. The effects of doubling CO2 concentration on the climate of a general circulation model. *Journal of the Atmospheric Sciences*, 32: 3–15

Pagani, M., Arthur, M.A. & Freeman, K.H. 1999. Miocene evolution of atmospheric carbon dioxide. *Paleoceanograohy*, 14: 273–292

Pearson, P.N. & Palmer, M.R.. 2000. Atmospheric carbon dioxide concentrations over the past 60 million years. *Nature*, 406: 695–699

Petit, J.R., Jouzel, J., Raynaud, D., Barkov, N.I., Barnola, J.-M., Basile, I., Bender, M., Chappellaz, J., Davis, M., Delaygue, G., Delmotte, M., Kotlyakov, V.M., Legrand, M., Lipenkov, V.Y., Lorius, C., Pepin, L., Ritz, C., Saltzman, E. & Stievenard, M. 1999. Climate and atmospheric history of the past 420,000 years from the Vostok ice core, Antarctica. *Nature*, 399: 429–426

Plass, G.N. 1956. The carbon dioxide theory of climatic change. *Tellus*, 8: 140–154

Raynaud, D., Jouzel, J., Barnola, J.M., Chappellaz, J., Delmas, R.J. & Lorius C. 1993. The ice record of greenhouse gases. *Science*, 259: 926–934

Revelle, R. & Suess, H.E. 1957. Carbon dioxide exchange between atmosphere and ocean and the question of an increase of atmospheric CO_2 during the past decades. *Tellus*, 9: 18–27

Smil, V. 2002. *The Earth's Biosphere: Evolution, Dynamics, and Change*, Cambridge, MA: MIT Press

Tyndall, J. 1861. On the absorption and radiation of heat by gases and vapours, and on the physical connection of radiation, absorption, and conduction. *Philosophical Magazine and Journal of Science*, 22: 169–194, 273–285

Veizer, J., Godderis, Y. & François, L.M. 2000. Evidence for decoupling of atmospheric CO_2 and global climate during the Phanerozoic eon. *Nature*, 408: 698–701

Woodwell, G.M. & Mackenzie, F.T., eds. 1995. *Biotic Feedbacks in the Global Climatic System*, Oxford and New York: Oxford University Press

GOLDEN MEAN

See **Fibonacci series**

GRADIENT SYSTEM

In the study of dynamic systems, it is often observed that the rate of evolution of some system in its phase space is proportional to the gradient of a state function. A system of this type is called a *gradient system*, and the state function—which governs the course of its evolution—is its *potential*.

If the state s of a system is given by n state variables $s_1, \ldots, s_n$ and $G(s) = G(s_a, \ldots s_{,n})$ is the potential of the system, then one can write (in matrix form)

$$\dot{s} = -\hat{k} \cdot \left(\frac{\partial G}{\partial s}\right)^{\mathrm{T}}. \qquad (1)$$

Here the rate of change of state s is given by the column vector $\dot{s} = |\dot{s_1}, \ldots, \dot{s_n}|^{\mathrm{T}}$, where the superscript T indicates transposition; $\partial G(s)/\partial s = |\partial G/\partial s_1, \ldots, \partial G/\partial s_n|$ is the row vector of gradient of potential $G(s)$; and $\hat{k}$ is a coefficient of proportionality, represented by the nonsingular matrix $\hat{k} = |k_{ij}|$, $\det |k_{ij}| \neq 0$. The minus sign in Equation (1) is chosen for convenience. The coordinate representation of this equation is

$$\dot{s_i} = -\sum_{j=1}^{n} k_{ij} \frac{\partial G}{\partial s_j} \quad (i = 1, \ldots, n). \qquad (2)$$

Equations (1) and (2) are quite general. For example, if the order n is an even number ($n = 2m, m \geq 1$), and $\hat{k}$ is a $2m \times 2m$ block-diagonal matrix composed of the skew-symmetric blocks,

$$\hat{k} = \mathrm{diag}|\hat{k}_1, \ldots, \hat{k}_m|, \hat{k}_1 = \hat{k}_2 = \ldots = \hat{k}_m = \begin{pmatrix} 0 & -1 \\ 1 & 0 \end{pmatrix}, \qquad (3)$$

then Equations (1) and (2) are the usual Hamiltonian system (Fomenko, 1995).

However, gradient systems differ radically from Hamiltonian systems. Essential restrictions that specialize the definition of gradient systems and predetermine their qualitative properties are on the structure of matrix $\hat{k}$. For a gradient system: (i) The nonsingular matrix $\hat{k}$ is assumed to be symmetric: ($\hat{k} = \hat{k}^{\mathrm{T}}$ or $k_{ij} = k_{ji}$); hence, it possesses exactly n nonzero eigenvalues that the real numbers. (ii) All eigenvalues of $\hat{k}$ are assumed to have the same signs, either positive or negative.

Note that requiring that the rates $\dot{s}$ be proportional to the gradient of potential $G(s)$ does not oblige the coefficient of proportionality $\hat{k}$ to be a constant matrix. In general, the elements of $\hat{k}$ can (smoothly) depend on the current state $s : k_{ij} = k_{ji}(s)$, suggesting a third requirement of a gradient system: (iii) The properties (i) and (ii) must be fulfilled everywhere in the state space of system. Thus, a gradient system is defined as a system, whose evolution follows Equations (1) and (2), with a matrix satisfying conditions (i)–(iii).

The key dynamic difference between Hamiltonian and gradient systems is that Hamiltonian systems preserve the values of system potential, the Hamiltonian function ($\dot{G}(s) = 0$). The behavior of a gradient system is quite different because the matrix $\hat{k}$ in (3) is symmetric rather then skew-symmetric; thus $\dot{G}(s) \neq 0$.

From conditions (i)–(iii), one can demonstrate that the evolution of a gradient system preserves the sign of $\dot{G}(s)$; hence, the potential $G(s)$ either decreases or increases monotonically. To see this, consider the dissipative function of a gradient system, which is introduced by the quadratic form

$$\Gamma = \frac{1}{2}\dot{s}^{\mathrm{T}}\hat{\gamma}\dot{s} = \frac{1}{2}\sum_{i,j=1}^{n} \gamma_{ij}\dot{s}_i\dot{s}_j \quad \text{where} \quad \hat{\gamma} = \hat{k}^{-1}. \qquad (4)$$

The matrix $\hat{\gamma}$ inherits the properties (i)–(iii) of matrix $\hat{k}$, and the constancy of sign $\dot{G}(s)$ under the evolution of a gradient system follows directly from the relation

$$\dot{G}(s) = -2\Gamma. \qquad (5)$$

Relation (5) indicates that the sign of $\dot{G}(s)$ is opposite to the sign of dissipative function; thus, as the latter is positively (negatively) definite, the potential of gradient system strictly decreases (increases).

As an elementary example, the ordinary equation $\dot{u} = f(u)$ is a gradient system because it admits the representation $\dot{u} = -k[\mathrm{d}G(u)/\mathrm{d}u]$, $k = 1$, $G(u) = -\int f(u)\mathrm{d}u$. The corresponding dissipative function is $\Gamma = \dot{u}^2/2$, and under time-dependent solutions of this equation, the potential decreases monotonically.

As a second example, consider a Newtonian particle of mass m that is changing its position $\boldsymbol{r}$ with the speed $\dot{\boldsymbol{r}}$ and acceleration $\ddot{\boldsymbol{r}}$ by the action of forces of two kinds: a potential force $\boldsymbol{Q}^{(\mathrm{P})} = -\,\mathrm{grad}G(\boldsymbol{r})$ (where $G(\boldsymbol{r})$ is the potential energy), and a friction force $\boldsymbol{Q}^{(\mathrm{D})} = -\gamma\dot{\boldsymbol{r}}$, $\gamma = \mathrm{const} > 0$. The Newtonian vector equation of motion of this particle is $m\ddot{\boldsymbol{r}} = \boldsymbol{Q}^{(\mathrm{D})} + \boldsymbol{Q}^{(\mathrm{P})} \equiv -\gamma\dot{\boldsymbol{r}} - \mathrm{grad}G(\boldsymbol{r})$, or, equivalently,

$$\dot{\boldsymbol{r}} = \boldsymbol{v}, \quad m\dot{\boldsymbol{v}} + \gamma\boldsymbol{v} = -\mathrm{grad}G(\boldsymbol{r}). \tag{6}$$

This system has two different limits—"Galilean" and "Aristotelian." The Galilean limit corresponds to the situation when dissipative force $\boldsymbol{Q}^{(\mathrm{D})}$ is negligible in comparison with the d'Alembert inertia force $\boldsymbol{Q}^{(\mathrm{I})} = -\,m\dot{\boldsymbol{v}}$. At this limit—which is realized under motion through a vacuum—the term $\gamma\boldsymbol{v}$ in Equations (6) vanishes, and it reduces to the form

$$\dot{\boldsymbol{r}} = \boldsymbol{v}, \quad m\dot{\boldsymbol{v}} = -\mathrm{grad}G(\boldsymbol{r}). \tag{7}$$

We refer to this limit as Galilean because Galileo proposed and experimentally demonstrated that gravitational force determines the acceleration of a falling body rather then its speed.

The Aristotelian limit, on the other hand, describes so-called creeping motions, in which the d'Alembert inertia force $\boldsymbol{Q}^{(I)}$ is much weaker than the dissipative force $\boldsymbol{Q}^{(D)}: |\boldsymbol{Q}^{(D)}| = m|\dot{\boldsymbol{v}}| \ll |\boldsymbol{Q}^{(I)}| = \gamma|\boldsymbol{v}|$. This limit is realized in the viscosity limited motion of particle, which occurs in a strongly viscous medium in the presence of a potential force field. In this limit, Equations (6) asymptotically reduce to

$$\dot{\boldsymbol{r}} = -k\mathrm{grad}G(\boldsymbol{r}), \quad k = \gamma^{-1}. \tag{8}$$

The order of Equation (8) with respect to time is less by half then corresponding order of (6). In this case, the current state s of particle is fully determined by current value of radius-vector $\boldsymbol{r}$. Now, the rate of change of state is proportional to the gradient of potential energy; therefore, the Newtonian particle undergoes creeping motion in a gradient system. The dissipative function of this system is

$$\Gamma = \gamma\boldsymbol{v}^2/2 \quad (\gamma > 0), \tag{9}$$

and the potential $G(\boldsymbol{r})$ always decreases monotonically as $\dot{G} = -\gamma|\mathrm{grad}G(\boldsymbol{r})|^2$. This limit is called Aristotelian because it manifests the Aristotelian principle that "velocity of a body is proportional to a force acting upon the body." (This principle is not wrong; it merely corresponds to one of two possible limits of classical macroscopic dynamics which is realized in the presence of strong friction.)

The Newtonian example can be extended to the more general case of constrained system of n degrees of freedom, which move under stationary holonomic constraints and generalized forces of three types: the potential forces $Q_i^{(\mathrm{P})}$, the dissipative forces $Q_i^{(\mathrm{D})}$ of viscous friction, and the d'Alembert inertia forces ($Q_i^{(\mathrm{I})} = 1, \ldots, n$). A related example is provided by electrical networks that are constructed with linear lumped elements: resistances (R), inductances (L), and capacitances (C). From an electromechanical analogy (Gantmacher, 1975), the considerations discussed above are extended to this case almost automatically. The analogs of Galilean and Aristotelian limits also exist here. The first limit corresponds to the situation when all resistances are negligible; it displays the LC subclass of general RLC networks. The networks belonging to this subclass are described by systems of ordinary differential equations that involve second-order time derivatives; as a rule, they are not gradient systems. The electrical analog of the Aristotelian limit corresponds to the networks with negligible inductances (RC subclass of general RLC networks). The networks making up this subclass are gradient systems.

The gradient systems considered in these examples are characterized by the monotonic diminution of potential G in their motions. Formally, this property follows from positiveness of dissipative functions related to these systems. In the mechanical examples, where the potential G is given by the potential energy of a mechanical system, its diminution can be explained physically by the action of viscous friction, which dissipates the energy of system and fully converts it to heat Q at a rate $\mathrm{d}Q/\mathrm{d}t = 2\Gamma$.

The opposite case, which corresponds to a monotonic *increase* of G, can be viewed as the action of "negative friction" transferring energy from exterior sources to the moving system; for example, processes during evolution of the genetic structure of biological populations which are described by classical Fisher–Haldane–Wright equations. The gradient representations of these equations were developed by Svirezhev in 1972 and by Shahshahani in 1979. (Svirezhev's results now can be found in the comprehensive monograph of Svirezhev & Pasekov, 1990.) In the context of their results, the famous Fisher's Fundamental Theorem of Natural Selection—which asserts that the mean fitness of population always increases with the rate proportional to the genotypical diversity of the population—is a simple consequence of the gradient properties of the equations. In this case, the role of potential of the corresponding gradient system is played by the mean fitness of population, whereas the dissipative function is connected with a certain measure of genotypical diversity.

Owing to the presence of the monotonically varying quantity G, any gradient system is forbidden to return to states it once already left. In particular, such a system cannot perform nontrivial periodic motions (different from the states of rest). These properties are widely used in applied mathematics, including numerical gradient methods of searching for minima and maxima of multivariable functions.

In the above examples, gradient systems with finite degrees of freedom were considered. However, in various problems one deals with nonlocal spatially distributed systems in continuous media, occupying some d-dimensional region X of physical space, whose dynamics inherits the main features of finite-dimensional gradient systems. These objects form a class of gradient systems of an infinite number of degrees of freedom, which are referred to as *continuum gradient systems*. A nontrivial example is the system

$$\dot{u} = -k(u, \boldsymbol{x})\frac{\delta G[u]}{\delta u},$$

$$G[u] = \int_X g(u, \nabla u, \boldsymbol{x})\mathrm{d}X, \tag{10}$$

where $u = u(\boldsymbol{x}, t)$ is a function describing the state of the system at the spatial point $\boldsymbol{x} \equiv (x_i, \ldots, x_d) \in X$ and time t, and the integral functional $G[u]$ is a potential of the continual gradient system. Also ∇u denotes the spatial gradient, $\nabla u = |\partial u/\partial x_1, \ldots, \partial u/\partial x_d|$; $\mathrm{d}X = \mathrm{d}x^1 \ldots \mathrm{d}x^d$ is the space volume element of the region X occupied by the continuous medium; and $\delta/\delta u$ is a functional derivative, which acts upon the functional $G[u]$ according to the rule

$$\frac{\delta G}{\delta u} = \frac{\partial g}{\partial u} - \operatorname{div}\left[\frac{\partial g}{\partial(\nabla u)}\right]$$

$$= \frac{\partial g}{\partial u} - \sum_{\mu=1}^{d} \frac{\partial}{\partial x^\mu}\frac{\partial g}{\partial(\partial u/\partial x^\mu)}. \tag{11}$$

Substituting (11) into the second part of (10), one can show that equations (10) are equivalent to the equation

$$\dot{u} + \operatorname{div}\boldsymbol{J} = Q \tag{12}$$

where $\boldsymbol{J}$ and Q are given by the expressions

$$\boldsymbol{J} = -k\frac{\partial g}{\partial(\nabla u)}, \tag{13}$$

$$Q = -k\frac{\partial g}{\partial u} - \nabla k \cdot \frac{\partial g}{\partial(\nabla u)}. \tag{14}$$

Equation (12) has the form of standard continuity equations for some substance filling the spatial region X; therefore, the quantities u and $\boldsymbol{J}$ can be interpreted as the density of this substance and as a vector of density of the corresponding spatial transport flux, respectively. From this standpoint, the function in the second part of continuity equation (12), defined by relationship (14), describes the rate of production of these substances per unit space volume, whereas expression (13) relating the flux $\boldsymbol{J}$ to the density u and its spatial gradient ∇u can be interpreted as a peculiar nonlinear generalization of well-known linear phenomenological laws such as Fick's law of diffusion or Ohm's law of electrical conduction. Hence, the continuum physical systems obeying the functional gradient equation (10) constitute a specific subclass in the class of reaction diffusion systems. This subclass comprises many physically important systems, including the parabolic equation

$$\gamma(\boldsymbol{x})\dot{u} = \operatorname{div}[D(\boldsymbol{x})\nabla u] + f(u, \boldsymbol{x})$$

$$= D(\boldsymbol{x})\Delta u + \nabla D(\boldsymbol{x}) \cdot \nabla u + f(u, \boldsymbol{x}) \tag{15}$$

defined in the space region X. Here $\gamma(\boldsymbol{x}) > 0$, $D(\boldsymbol{x}) > 0$, and $f(u, \boldsymbol{x})$ are the given functions of their arguments. Special cases of (15) include the linear diffusion equation $\dot{u} = \Delta u$, as well as the nonlinear reaction diffusion equation

$$\dot{u} = \Delta u + f(u), \tag{16}$$

which arises in several branches of natural science.

In mathematical genetics, Equation (16) describes the gene exchange waves traveling in the populations of biological organisms (Fisher's equation). In addition, this equation describes flame propagation as well as the switching processes in some physical, chemical, and biological nonlinear media (Zeldovich–Frank-Kamenetsky equation). Also, the time-dependent Ginzburg–Landau equation, which arises in physical kinetics and in synergetics for the phase transitions in spatially distributed self-organizing systems, has the form of Equation (16).

Remarkably, all these examples can be interpreted as processes of time evolution of certain gradient continuum systems. The functional gradient representation (10) is important because it suggests their general qualitative properties by the analogy with finite-dimensional case. For example, the potential $G[u]$ in (10) decreases monotonously in all time-dependent solutions if the latter is endowed with impermeability boundary conditions [see (18)]

$$J_n|_{\partial X} = 0 \leftrightarrow \boldsymbol{n}\left.\frac{\partial g}{\partial(\nabla u)}\right|_{\partial X} = 0, \tag{17}$$

where the subscript ∂X indicates that corresponding expressions are considered at the boundary ∂X of space region X enclosing the continual system, $J_n|_{\partial X}$ is the normal component of $\boldsymbol{J}$ at the boundary, and $\boldsymbol{n}$ is unit normal vector at ∂X. Indeed, the full derivative of the energy functional $G[u]$ with respect to time t is given by the elegant relation

$$\frac{\mathrm{d}G}{\mathrm{d}t} = -\int_X k\left(\frac{\delta G}{\delta u}\right)^2 \mathrm{d}X \leq 0. \qquad (18)$$

Defining the *dissipative functional* by the formula

$$\Gamma[u, \dot{u}] = \int_X \frac{1}{2k}\dot{u}^2 \mathrm{d}X = \frac{1}{2}\int_X \gamma \dot{u}^2 \mathrm{d}X,$$
$$\gamma = k^{-1},$$

puts (18) into the simple form of Equation (5), which appeared in the finite-dimensional case.

As in the finite-dimensional case, the monotone decrease of potential means that time evolution is *unidirectional*—these systems cannot return to the states once left. In particular, they can have neither solutions that are periodic in space and time nor traveling-impulse solutions with complete recovery of the initial state. If they possess moving wave-front solutions, contra directional wave fronts cannot be reflected after collisions with each other. On the contrary, if some spatially distributed system isolated from the outer world by the impermeable boundaries can modify its state supporting the propagation of solitary/periodic traveling pulses, then it cannot be described by continuum gradient equations.

O.A. MORNEV

See also **Diffusion; Flame front; Nerve impulses; Reaction-diffusion systems; Synergetics; Zeldovich–Frank-Kamenetsky equation**

Further Reading

Fomenko, A.T. 1995. *Symplectic Geometry: Methods and Applications*, 2nd edition, New York: Gordon and Breach (original Russian edition 1988)

Gantmacher, F. 1975. *Lectures in Analytical Mechanics*, Moscow: Mir Publications (original Russian edition 1966)

Loskutov, A.Yu. & Mikhailov, A.S. 1996. *Foundations of Synergetics*, 2nd edition, Berlin and New York: Springer (original Russian edition 1990)

Mornev, O.A. 1997. Dynamical principle of minimum energy dissipation for systems with ideal constraints and viscous friction. *Russian Journal of Physical Chemistry*, 71(12): 2077–2081 [Translated from *Zhurnal Fizicheskoi Khimii*, 1997, 71(12): 2293–2298]

Mornev, O.A. 1998. Modification of the biot method on the basis of the principle of minimum dissipation (with an application to the problem of propagation of nonlinear concentration waves in an autocatalytic medium). *Russian Journal of Physical Chemistry*, 72(1): 112–118 [Translated from *Zhurnal Fizicheskoi Khimii*, 1998, 72(1): 124–131]

Mornev, O.A. & Aliev, R.R. 1995. Local variational principle of minimum dissipation in the dynamics of reaction-diffusion systems. *Russian Journal of Physical Chemistry*, 69(8): 1325–1328 [Translated from *Zhurnal Fizicheskoi Khimii*, 1995, 69(8): 1466–1469]

Scott, A.C. 2002. *Neuroscience: A Mathematical Primer*, Berlin and New York: Springer

Scott, A.C. 2003. *Nonlinear Science: Emergence and Dynamics of Coherent Structures*, 2nd edition, Oxford and New York: Oxford University Press

Svirezchev, Yu.M. & Pasekov, V.P. 1990. *Fundamentals of Mathematical Evolutionary Genetics*, Dordrecht: Kluwer (original Russian edition 1982)

GRANULAR MATERIALS

What do coffee powder, wheat, mustard seeds, granulated sugar, cement, sand, and rocks have in common? They are granular materials, assemblies of solid objects, from tens of micrometers to meter sized, that are generally not bound together by significant attractive forces. Such assemblies of objects possess unique physical and dynamic characteristics. Even though individual grains are solid particles, the assembly of grains behaves distinctly from ordinary solids, fluids, or gases (Jeager et al., 1996; Duran, 1997). For example, a sand dune is reliably solid-like and can sustain the weight of a person. However, if the sand from the dune is placed in an hourglass, it will flow rapidly and at a predictable rate.

The curious behavior of granular matter has long attracted the attention of scientists. In 1885, Osborne Reynolds observed that in order for grains to flow past each other, they have to move out of each other's way, leading to a dilation of granular matter under shear (Reynolds, 1885).

Some of the first modern work on granular matter was inspired by Per Bak, C. Tang, and K. Wiesenfeld (1987), who developed the concept that many systems may self-organize into a critical state. Granular avalanching was thought to be one of the prominent experimental realizations of their theory of self-organized criticality (SOC). The idea was that on a sandpile, grains might self-organize through avalanching to form a heap with a critical angle, which is called the angle of repose of the sandpile.

While experiments showed that SOC does not describe the behavior of real sandpiles (Nagel, 1992), the initial experiments showed some of the puzzling properties of granular matter and inspired a significant resurgence of research into granular matter. Here we describe three of the main questions of recent work:

When and How Does Granular Matter Flow?

In most situations when granular matter flows, such as during emptying of silos or during natural rock avalanches, only part of the material behaves like a fluid. This is illustrated in Figure 1 in a long exposure image of the side of an avalanche flowing down the

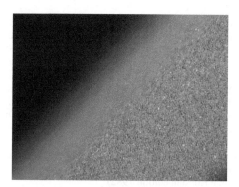

Figure 1. Long exposure image of the side of an avalanche flowing down the side of a granular pile. A clear separation into fluid-like and solid-like regions is visible. (Courtesy of N. Taberlet and P. Richard, University of Rennes).

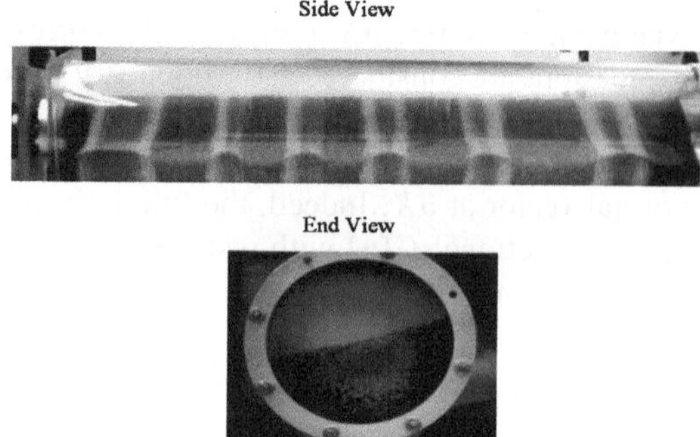

Figure 2. Unmixing of grains of three different sizes in a rotating horizontal cylinder. After 5 min of rotation at 15 rpm, material is separated into axial and radial bands. Both a side view and end view of the cylinder are shown.

side of a granular pile. The flow leaves behind a pile with a known surface angle, the angle of repose (Nagel, 1992). Predicting the timing and extent of such partial fluidization is part of the current challenge of modeling dense granular flows.

Once flowing, hydrodynamic equations may describe the flow behavior (Losert et al., 2000), though velocity gradients on length scales of a few particle diameters can call the validity of continuum models with local equilibria into question.

More dilute flows can be modeled as a system of hard spheres in which some energy is lost during each collision. Based on such a picture, kinetic theories of driven granular media have been developed, for example by Goldhirsch & Zanetti (1993). Since energy is continuously lost in collisions and must be added through gravity or shaking, granular flows are non-equilibrium driven, dissipative systems that exhibit interesting instabilities. One example is the clustering instability, in which a dilute system of particles can spontaneously develop localized, dense clusters (Goldhirsch & Zanetti, 1993). The basic mechanism for and clustering is that a local increase in particle density increases the number of collisions in that region, and, thus, slowing particles down and trapping them in the dense region.

How Can Different Kinds of Particles be Mixed or Separated?

During flow, mixtures of grains tend to spontaneously unmix (Shinbrot & Muzzio, 2000), as illustrated in Figure 2. The figure shows unmixing in a mixture of glass particles of three different sizes after several rotations in a half-filled horizontal cylinder. The phenomenology of granular unmixing is complex. In a horizontally rotating drum, for example, materials can segregate both radially and axially by size or by density. There is, to date, no simple model for unmixing, but a wealth of empirical experimental data and several simple physical mechanisms, such as percolation through voids, shear flows, and convection, all play some role in the unmixing process.

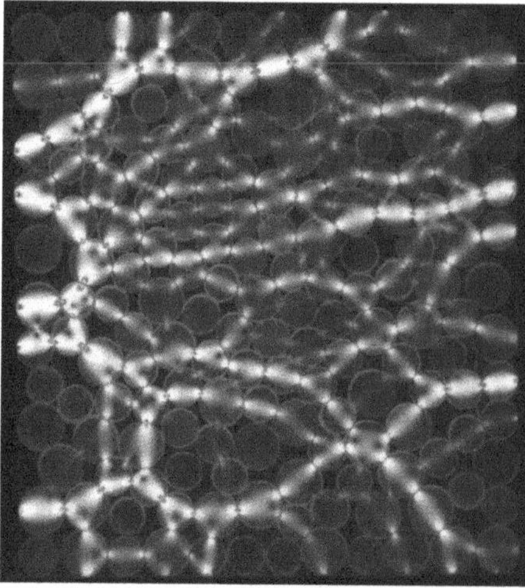

Figure 3. Forces in a 2-dimensional assembly of disks sheared between concentric cylinders. The disks are birefringent and placed between crossed polarizers, so that only points subject to large forces are visible as bright regions. (Courtesy of B. Utter and R.P. Behringer, Duke University).

How Are Forces Transmitted Through Granular Matter?

In civil engineering, the mechanical behavior of granular matter is modeled by solid-like equations (Nedderman, 1992). Zooming in to the scale of grains, forces can only be transmitted at particle contacts, and only repulsive forces are permitted at each contact. This can lead to very inhomogeneous force distributions and locally very large forces as is indeed seen, for example, in grain silos. Experiments with birefringent disks that highlight the location and magnitude of contact forces (Howell et al., 1999) have helped in our microscopic understanding of the way forces are transmitted through granular matter (Figure 3). Strong inhomogeneities in the force magnitude, anisotropies in the direction of stress transmission, and large rearrangements in the force distribution, even for small changes in structure are observed.

Two emerging concepts to explain the properties of such dense granular matter are jamming, the property of a system to get trapped in some intermediate state which may be generic to granular matter as well as thermal systems such as glasses, and the idea of stress chains, lines of particles that carry disproportionately large stress, as can be seen in Figure 3.

To conclude, basic questions about granular matter remain. Granular matter encompasses different kinds of particles with many variables, such as the frictional properties of grain-grain or grain-boundary contacts, the deformability, surface roughness, polydispersity, and grain shape. It remains difficult to predict whether a particular granular parameter, such as the shape of individual grains will qualitatively alter flow, forces, or mixing behavior, and thus, has to be taken into account in modeling. While models for flow, forces, and segregation have been developed that agree with experimental data under various conditions, a broadly valid model of granular flow based on simple physical insights similar to the Navier–Stokes equation for fluids has so far proven elusive. Similarly, no consensus about the most suitable equation for granular solids has yet emerged, nor has a clear separation between the solid-like and liquid-like regime emerged.

WOLFGANG LOSERT

See also **Avalanches; Cluster coagulation; Dune formation; Sandpile model**

Further Reading

Bak, P., Tang, C. & Wiesenfeld, K. 1987. Self-organized criticality: an explanation of the 1/f noise. *Physical Review* A, 59: 381–384

Duran, J. 1997. *Sands, Powders, and Grains: An Introduction to the Physics of Granular Materials*, Berlin and New York: Springer

Goldhirsch, I. & Zanetti, I. 1993. Clustering instability in dissipative gases. *Physical Review Letters*, 70: 1619–1622

Howell, D.W., Veje, C.T. & Behringer, R.P. 1999. Stress fluctuations in a 2D granular couette experiment: a critical transition. *Physical Review Letters*, 82: 5241–5244

Jaeger, H.M., Nagel, S.R. & Behringer, R.P. 1996. *Reviews of Modern Physics*, 68: 1259–1273

Knight, J.B., Jaeger, H.M. & Nagel, S.R. 1993. *Physical Review Letters*, 70: 3728–3731

Losert, W., Bocquet, L., Lubensky, T.C. & Gollub, J.P. 2000. Particle dynamics in sheared granular matter. *Physical Review Letters*, 85: 1428–1431

Nagel, S.R. 1992. Instabilities in a sandpile. *Reviews of Modern Physics*, 64: 321—325

Nedderman, R.M. 1992. *Statics and Kinematics of Granular Materials*, Cambridge and New York: Cambridge University Press

Reynolds, O. 1885. On the dilatancy of media composed of rigid particles in contact. *Philosophical Magazine*, 20: 469

Shinbrot, T. & Muzzio, F.J. 2000. *Physics Today*, March 25

GRAVITATIONAL WAVES

In 1687, Newton published *The Philosophiae Naturalis Principia Mathematica* in which he first proposed his law of gravitation. According to this law, the gravitational force of attraction between two bodies is always proportional to their masses and inversely proportional to the square of their distance apart, and it acts instantaneously through infinite distance. Less than two and half centuries later, Newton's theory was radically revised.

According to Einstein's general theory of relativity (published in 1913), gravitation is not a force of attraction but rather the force required to prevent the natural motion of matter, which is to follow a geodesic in space time. The geodesic or shortest path in four-dimensional space time describes a free-fall trajectory such as the motion of a planet around the sun. The space time is curved by the presence of matter or energy. The predictions of general relativity are very close to those of Newtonian theory as long as gravity is weak and velocities are slow, but they diverge in strong gravity due to two aspects of the theory. The first is the change in geometry due to the non-Euclidean properties of space time. The second stems from the nonlinear aspects of general relativity, which arise because gravitational energy is itself a source of curvature.

Eight years prior to publishing his general relativity theory, Einstein published the special theory of relativity, which predicted that neither matter nor information could ever travel faster than the speed of light. This means that the curvature at a point generated by a mass at another point only achieves its value at a retarded time, a time set by the travel time for light between the two points.

The general theory of relativity changed our vision of a Euclidean flat space that had been assumed since the days of Newton. In non-Euclidean space, the sum of the angles of a triangle does not equal 180° and the area of a circle is not always πr^2. Space time can be considered as an elastic membrane. The deformations are described by the Einstein curvature tensor G, while the sources of curvature (i.e., the mass-energy distribution) are described by the stress-energy tensor T. Einstein's field equations are then expressed by the equation (Misner et al., 1973)

$$\mathsf{T} = \frac{c^4}{8\pi G}\mathsf{G}. \quad (1)$$

The constant $c^4/8\pi G$ (where c is the speed of light and G is Newton's gravitational constant) is a very large number which can be considered as the "spring constant" of space time. Because it is so large, only very small curvatures are generated even by very large values of mass-energy distribution. A natural consequence of the membrane analogy is the existence of waves; ripples in the curvature of the membrane which propagate through the membrane. This concept can also be deduced from Einstein's field equations.

Equation (1) is a set of ten nonlinear equations. Except in simplified situations, these equations are difficult to solve directly. Matter creates curvature and curvature influences the motion of matter. This mutual influence gives rise to nonlinear phenomena in gravitational wave propagation. Unlike the theory of electromagnetism, the theory of general relativity is intrinsically nonlinear. Since gravity is itself a source of curvature, there can be a gravitational interaction between gravitational waves. While gravitational wave signals are normally expected to be very small (and hence amenable to a linearized theory), at their sources, the nonlinear aspect of the theory makes prediction extremely difficult. The following list offers some examples of expected nonlinear phenomena in gravitational wave propagation:

- For gravitational waves emerging from the birth of a black hole, the gravitational redshift of the waves reduces the total emitted energy. The mass-energy of the gravitational waves themselves constitute the redshift.
- Like electromagnetic radiation, gravitational waves can undergo gravitational lensing. A mass in the path of the waves creates a background curvature. This curvature focuses the waves by modifying the wave front. This effect can enhance the intensity of dim sources. The gravitational lensing effect has successfully been observed for electromagnetic radiation.
- A time-varying curvature can amplify waves. If a wave modulates the space curvature, a second incident wave will be amplified or its frequency shifted. An optical parametric oscillator represents the equivalent phenomena for optical radiation. Similar to the optical Kerr effect, a gravitational wave can interact with its own self-generated background curvature. In the early universe, it is predicted that gravitational waves from the Big Bang may have been parametrically amplified by the action of inflation.
- A strong background curvature (e.g., from a black hole) can scatter a gravitational wave passing in its vicinity.

Under gravitational wave astronomy, it may be possible to find evidence of these predicted phenomena. The first indirect experimental proof of gravitational waves was provided in 1984 by Weisberg and Taylor. By studying the pulsar PSR 1913+16, they showed that the period of the pulsar around its companion star decreased exactly as predicted by the Einstein equations (Weisberg et al., 1981). A part of the pulsar orbital energy is converted to gravitational radiation.

The orbital parameters of binary stars are usually deduced from measurements of the Doppler shift of the radiated electromagnetic waves. According to Newtonian gravitational theory, the mass m_1 of the pulsar and the mass m_2 of its companion cannot be determined with confidence. All results obtained are proportional to an unknown parameter: the sine (normally denoted $\sin i$) of the angle between the orbital plane and the line of sight. This parameter can be evaluated in the framework of general relativity. From this theory, five independent parameters (among them, the advance of the periastron (the point of closet approach of the two stars), the evolution of the period, or the Einstein parameter) which are functions of m_1, m_2, and $\sin i$ can be measured. The overdetermined system of five equations determines with an unprecedented accuracy (error less than 0.5%) the mass of the pulsar. Moreover, the compatible results from the five independent equations confirm the general theory of relativity in the strong field and radiative regime. This high-precision validation of general relativity indirectly implies that the velocity of gravitational waves is equal to the speed of light.

Direct detection of gravitational waves on Earth is one of the most exciting challenges of today's science. In the 1960s, Joseph Weber invented and developed the first gravitational wave detectors, consisting of large (about a ton) vibration-isolated cylinders of aluminium or niobium (Weber, 1960). The gravitational waves excite the longitudinal resonance in the cylinder, which behaves as an extremely low-loss mechanical oscillator. The motion of the cylinder is monitored to very high precision. Most of its motion is thermal vibration but the small effect of the gravitational waves appear as very small perturbations in the amplitude or phase of the vibration. The mechanical oscillations, converted into an electrical signal, are recorded and analyzed by sophisticated predictive filter algorithms. A worldwide network of five resonant bars still uses Weber's technique (see Figure 1).

A new generation of gravity-wave detectors is based on laser interferometry (Blair, 1991). Gravitational waves passing through a Michelson interferometer change the relative length difference between the two perpendicular arms of the interferometer. Thus the variation of the gravitational field can be converted to an optical phase variation. Due to the weakness of gravitational wave interaction with matter, the detectors must be able to measure a length variation of less than 10^{-18} m, which is close to the quantum limit. Compared with the resonant bar, the interferometric devices exhibit a larger bandwidth and higher sensitivity. The sensitivity is limited by seismic and thermal noise at low frequency and photon-counting noise at higher frequency.

Five international projects, similar to the example shown in Figure 2, are using long base interferometry

Figure 1. Photograph of the interior of Niobe, the resonant bar at the University of Western Australia. The bar is a one and half ton cylinder of niobium cooled to 5 K to limit the thermal noise.

Figure 2. Aerial view of the French–Italian gravitational waves detector VIRGO near Pisa in Italy. On this photo, the two perpendicular 3 km vacuum pipes containing the arms of the interferometer are clearly distinct (Bradaschia et al., 1990) (Reproduced with permission from EGO).

to detect gravitational waves. The network of, interferometers (situated in the United States, Europe, Japan, and Australia) will enable the localization of sources of gravitational radiation in space as well as determination of their polarization. By measuring the arrival time of the waves at different detectors, the speed of the gravity waves can be calculated. This speed could be influenced by two independent factors: a coupling of the wave with a strong curvature background (difficult to detect on Earth) and the possibility that the graviton, the particle associated with the gravitational waves, is not massless (Will, 1998).

The direct detection of gravitational waves is expected to occur before 2012. This gravitational spectrum will open a new window to studying the universe, comparable to the revolution fathered by the invention of radio astronomy in the 1950s. The direct mapping of the gravity spectrum will broaden our knowledge about the universe, from the earliest moment of the Big Bang to the end of stars.

JEROME DEGALLAIX AND DAVID BLAIR

See also **Einstein equations; General relativity**

Further Reading

Blair, D.G. 1991. *The Detection of Gravitational Waves*, Cambridge and New York: Cambridge University Press

Bradaschia, C. et al. 1990. The VIRGO project: a wide band antenna for gravitational wave detections. *Nuclear Instruments and Methods in Physics Research* A, 289: 518–528

Misner, C.W., Thorne, K.S. & Wheeler, J.A. 1973. *Gravitation*, San Francisco: W.H. Freemann

Weber, J. 1960. Detection and generation of gravitational waves. *Physical Review*, 117: 306–313

Weisberg, J.M., Taylor, J.H. & Fowler, L.A. 1981. *Gravitational Waves from an Orbiting Pulsar*, Scientific American, p. 74

Will, C.M. 1998. Bounding the mass of the graviton using gravitational waves observations of inspiralling compact binaries. *Physical Review* D, 57: 2061–2068

GRAVITY WAVES

See **Water waves**

GREEN'S FUNCTION

See **Boundary layer problems**

GREY SOLITON

See **Solitons, types of**

GROSS-PITAEVSKII EQUATION

See **Nonlinear Schrödinger equations**

GROUP VELOCITY

Waves often propagate as a packet, or group, within which there are several wave crests. A common illustration is given by the wave pattern formed when a stone is thrown into a pond. An axially symmetric ring of waves on the water surface propagates outwards as a wave group; within the group, however, one can see that the wave crests propagate through the group, apparently from the rear to the front. The speed of the wave crests, called the *phase velocity*, is different from that of the group as a whole, this speed being called the *group velocity*. For water waves, the phase velocity is greater than the group velocity, except for the very short waves dominated by surface tension.

This important distinction between phase and group velocity arises in all physical systems which support waves. The concept of group velocity appears to have been first formulated by William Hamilton in 1839 (quoted by Havelock, 1914). The first recorded observation of the group velocity of a (water) wave is due to John Scott Russell in 1844 (Russell, 1844) (note also his remark that "the sound of a cannon travels faster than the command to fire it" (Russell, 1885)). However, our present understanding of group velocity is usually attributed to George Stokes, who used it as the topic of a Smith's Prize examination paper in 1876.

For simplicity, consider a linearized system, with a single spatial variable x and time represented by t. Then a sinusoidal wave has the representation,

$$u(x,t) = a\cos(kx - \omega t + \phi). \tag{1}$$

Here k is the wave number, ω is the wave frequency, a is the constant amplitude, and ϕ is a constant phase. Thus, this sinusoidal wave has a wavelength $\lambda = 2\pi/k$ and a wave period $T = 2\pi/\omega$. Equation (1) is a kinematic expression, valid for all physical systems which support waves. The dynamics of the system are governed by the dispersion relation

$$\omega = \omega(k), \tag{2}$$

defining the frequency as a function of wave number. The phase velocity $c = \omega/k$ is likewise a function of wavenumber.

To obtain the group velocity, consider an argument first advanced by Stokes (1876) and later in more general form by Lord Rayleigh (1881). Form the linear superposition of two sinusoidal waves, each of the form (1) with frequencies $\omega \pm \Omega$, wave numbers $k \pm K$, and with equal amplitudes a and phases ϕ which can be written as

$$u(x,t) = 2a\cos(kx - \omega t\phi)\cos(Kx - \Omega t). \tag{3}$$

This expression can be viewed as a wave packet, or group, with a dominant wave number k and frequency ω, and with a group velocity of Ω/K. Taking the limit $K \to 0$, one sees that the group velocity is given by

$$c_g = \frac{d\omega}{dk}, \tag{4}$$

which can be obtained by differentiation of the dispersion relation (2).

A more general argument that links the origin of a wave packet to initial conditions uses Fourier superposition to represent the solution of an initial-value problem in the form

$$u(x,t) = \int_{-\infty}^{\infty} F(k)\exp(i(kx - \omega t))\,dk + \text{c.c.}, \tag{5}$$

where $F(k)$ is the Fourier transform of $u(x,0)$ and c.c. stands for complex conjugate. Thus, the initial conditions determine the Fourier transform, each component of which evolves independently with frequency ω related to the wave number k through the dispersion relation (2). To obtain a wave packet, we suppose that the initial conditions are such that $F(k)$ has a dominant component centered at $k = k_0$. The dispersion relation is then approximated by

$$\omega = \omega_0 + b_1(k - k_0) + b_2(k - k_0)^2, \tag{6}$$

where

$$b_1 = \frac{d\omega}{dk} = c_g, \quad \text{and} \quad b_2 = \frac{1}{2}\frac{d^2\omega}{dk^2}. \tag{7}$$

Here, both b_1, b_2 are evaluated at $k = k_0$. Expression (5) then becomes

$$u(x,t) \approx A(x,t)\exp(i(k_0 x - \omega_0 t)) + \text{c.c.}, \tag{8}$$

where

$$A(x,t) = \int_{-\infty}^{\infty} F(k_0 + \kappa)\exp[i(\kappa(x - c_g t)) - ib_2\kappa^2 t]\,d\kappa, \tag{9}$$

where the variable of integration has been changed from k to $\kappa = k - k_0$. Here, the sinusoidal factor $\exp(i(k_0 x - \omega_0 t))$ is a carrier wave with a phase velocity ω_0/k_0, while the (complex) amplitude $A(x,t)$ describes the wave packet. Since the term proportional to κ^2 in the exponent in (9) is a small correction term, it can be seen that to the leading order, the amplitude A propagates with the group velocity c_g, while the aforementioned small correction term gives a dispersive correction term proportional to $t^{-1/2}$. Indeed, it can be shown that as $t \to \infty$

$$A(x,t)|_{x=c_g t} \sim F(k_0)\left(\frac{2\pi}{|b_2|t}\right)^{1/2}\exp\left(\frac{i\pi}{4}\text{sign}\, b_2\right). \tag{10}$$

This same result can be obtained directly from (5) by using the method of stationary phase, valid here in the limit when $t \to \infty$ (see, for instance, Lighthill, 1978); In this case, it is not necessary to also assume that $F(k)$ is centered at k_0, and so wave packets are the generic long-time outcome of initial-value problems.

It is useful to note that the group velocity appears naturally in the kinematic theory of waves (Lighthill, 1978; Whitham, 1974). Thus, let the wave field be defined asymptotically by

$$u(x,t) \sim A(x,t)\exp(i\theta(x,t)) + \text{c.c.}, \tag{11}$$

where $A(x,t)$ is the (complex) wave amplitude, and $\theta(x,t)$ is the phase, which is assumed to be rapidly varying compared with the amplitude. Then it is natural to define the local wave frequency and wave number by

$$\omega = -\frac{\partial \theta}{\partial t}, \qquad k = \frac{\partial \theta}{\partial x}. \qquad (12)$$

Note that expression (10) has the required form (11). Then cross-differentiation leads to the kinematic equation for the conservation of waves,

$$\frac{\partial k}{\partial t} + \frac{\partial \omega}{\partial x} = 0. \qquad (13)$$

But now, if we suppose the dispersion relation (2) holds for the frequency and wave number defined by (12), then we readily obtain

$$\frac{\partial k}{\partial t} + c_g \frac{\partial k}{\partial x} = 0 \qquad (14)$$

with a similar equation for the frequency. Thus, both the wave number and frequency propagate with the group velocity, a fact that can also be seen in (10). Equation (14) is itself a simple wave equation, which can be readily integrated by the method of characteristics, or rays. It is important to note that the group velocity c_g is a function of k, so that (14) is a nonlinear equation.

Next, suppose that the physical system contains several spatial variables, represented by the vector $\boldsymbol{x}$. Then the phase variable $(kx - \omega t)$ in (1, 5) is replaced by $(\boldsymbol{k} \cdot \boldsymbol{x} - \omega t)$, where $\boldsymbol{k}$ is the vector wave number. The dispersion relation (2) is replaced by

$$\omega = \omega(\boldsymbol{k}). \qquad (15)$$

The phase velocity is now a vector ($\boldsymbol{c}$), has a magnitude $\omega/|\boldsymbol{k}|$, and is in the direction of $\boldsymbol{k}$, and the definition of group velocity, replacing (4), is

$$\boldsymbol{c}_g = \nabla_{\boldsymbol{k}} \omega. \qquad (16)$$

Thus, the group velocity can differ from the phase velocity in both magnitude and direction. A striking example of the latter arises for internal waves, whose group velocity is perpendicular to the phase velocity (see Lighthill, 1978).

Since the group velocity is the velocity of the wave packet as a whole, it is no surprise to find that it can also be identified with energy propagation. Indeed, it can be shown that in most linearized physical systems, an equation of the following form can be derived:

$$\frac{\partial \boldsymbol{E}}{\partial t} + \nabla \cdot (\boldsymbol{c}_g \boldsymbol{E}) = 0, \qquad (17)$$

where $\boldsymbol{E}$ is the wave action density, and is proportional to the square of the wave amplitude $|A|^2$, with the factor being a function of the wavenumber $\boldsymbol{k}$. Typically, the wave action is just the wave energy density divided by the frequency, at least in inertial frames of reference.

So far the discussion has remained within the realm of linearized theory, and it remains to mention the consequences of nonlinearity. Analogous definitions and concepts can be developed, and at least for weakly nonlinear waves, the main outcome is that a dependence on wave amplitude (more strictly, on $|A|^2$) appears in the dispersion relation. This has the consequence that the phase and group velocities both inherit a weak dependence on the wave amplitude. It can be shown that at least within the confines of weakly nonlinear theory, in the case of just a single active spatial dimension, the complex wave amplitude A is governed by the nonlinear Schrödinger (NLS) equation,

$$\mathrm{i}\left(\frac{\partial A}{\partial t} + c_g \frac{\partial A}{\partial x}\right) + b_2 \frac{\partial^2 A}{\partial x^2} + \mu |A|^2 A = 0. \qquad (18)$$

Here recall that b_2 is defined in (7), and μ is a nonlinear coefficient which is system-dependent.

If A is assumed to depend only on t then Equation (18) has the plane wave solution

$$A = A_0 \exp(\mathrm{i}\mu |A_0|^2 t), \qquad (19)$$

which shows that the nonlinear coefficient μ has the physical interpretation that $-\mu|A_0|^2$ is the nonlinear correction to the wave frequency. It can be shown that this plane wave is stable (unstable) according as $\mu b_2 < (>) 0$ (see Whitham, 1974). In the unstable case, $\mu b_2 > 0$, a perturbed plane wave will evolve into one or more solitons, where the soliton family is given by (see Zakharov and Shabat, 1972),

$$A(x,t) = \alpha \exp(-\mathrm{i}\mu\alpha^2 t)\operatorname{sech}(\gamma\alpha(x - c_g t - x_0)),$$
$$\text{where} \quad \gamma^2 = \frac{\mu}{b_2}. \qquad (20)$$

Here, x_0 is a phase constant determining the location of the soliton at $t = 0$, and the amplitude α is a free parameter. In linear theory, the wave packet profile is determined by the initial conditions, whereas the influence of even weak nonlinearity, when it is balanced by weak dispersion, results in the generic sech profile.

ROGER GRIMSHAW

See also **Modulated waves; Nonlinear Schrödinger equations; Wave packets, linear and nonlinear**

Further Reading

Havelock, T.H. 1914. *The Propagation of Disturbances in Dispersive Media*, Cambridge: Cambridge University Press

Lighthill, M.J. 1978. *Waves in Fluids*, Cambridge: Cambridge University Press

Rayleigh, Lord, 1881. On the velocity of light. *Nature*, XXIV: 382

Russell, J.S. 1844. *Report on Waves*, 14th meeting of the British Association for the Advancement of Science, London: BAAS, pp. 311–339

Russell, J.S. 1885. *The Wave of Translation in the Oceans of Water, Air and Ether*, London: Trübner

Stokes, G.G. 1876. Problem 11 of the Smith's Prize examination papers (February 2, 1876). In *Mathematical and Physical Papers of George Gabriel Stokes*, vol. 5, New York: Johnson Reprint Co., 1966, p. 362

Whitham, G.B. 1974. *Linear and Nonlinear Waves*, New York: Wiley

Zakharov, V.E. & Shabat, A.B. 1972. Exact theory of two-dimensional self-focusing and one-dimensional self-modulation of waves in nonlinear media. *Soviet Physics, JETP*, 34: 62–69

GROWTH PATTERNS

Patterns are ubiquitous in nature. They can range from the broad stripes on a zebra, to the intricate design of a snowflake, to the rib patterns on a saguaro cactus. The idea that naturally occurring patterns might have some underlying and universal governing mathematical structure has long been a source of fascination to researchers. The study of biological patterns was pioneered by D'Arcy Thompson in his landmark treatise *On Growth and Form* (Thompson, 1942). This was followed by a notable contribution by the mathematician (and Enigma code breaker) Alan Turing in a now famous paper "The Chemical Basis of Morphogenesis" (Turing, 1952). A more contemporary account of many biological patterns and their mathematical modeling is given in James D. Murray's textbook *Mathematical Biology* (Murray, 2003). But to what extent are patterns universal? Are there any connections among the three examples cited above? Inevitably, the underlying physical and biological processes governing a particular pattern-forming process can be very different, even if the topology of the resultant patterns and certain "defects" in those patterns look very similar. In some cases, different governing physical processes end up having essentially the same mathematical formulation, and in these cases, commonality of pattern structure is less surprising.

Here, we will concentrate on two physical pattern-forming systems that are popular topics of contemporary research in condensed-matter physics, namely, the fingering patterns at the interface of fluid mixtures and formation of dendritic crystals in solidification. Some of the patterns seen in these two systems are also seen in growing bacterial colonies.

The Hele-Shaw Cell: Fluid Fingering

This classic problem was first studied at the end of the 19th century, but even to this day there are still unresolved issues relating to the stability of the patterns exhibited. The experimental setup is relatively simple, consisting merely of a thin film of fluid sandwiched between two glass plates into which air (or another fluid) is pumped. As the air is pumped into the fluid film, one might imagine that it simply forms an expanding circular planar bubble. Initially this appears to be the case, but the circular perimeter rapidly becomes wavy and these "waves" then grow into long finger-like structures which, in turn can split or side branch forming highly complicated patterns. A typical pattern is shown in Figure 1a. The physical principles governing the growing air-fluid interface are well understood. The key idea is that the normal velocity of the air-fluid boundary (i.e., the velocity in the direction normal to each point of the interface) is governed by the gradient of the pressure across the interface. Because the fluid layer is so thin, the equations of fluid mechanics are greatly simplified, and one ends up with governing equations of the form

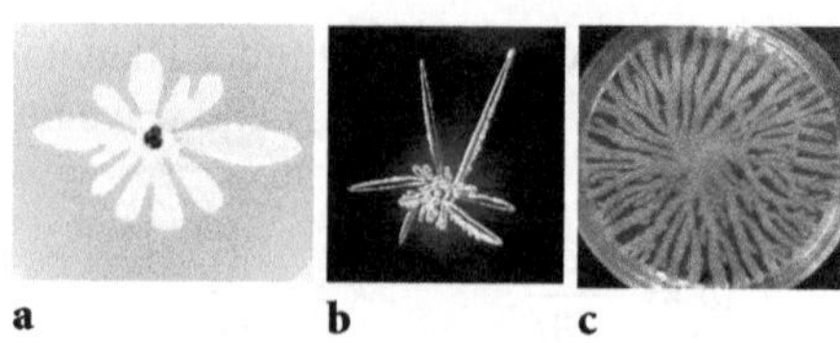

Figure 1. Three different types of growth patterns: (a) Hele-Shaw fingers; (b) dendritic crystals; (c) *Bacillus substilis* bacterial colonies.

$$\nabla^2 p = 0, \tag{1}$$

$$u_{\rm n} = -\frac{b^2}{12\mu}(\boldsymbol{\nabla} p)\cdot \boldsymbol{n}, \tag{2}$$

$$\Delta p|_{\partial\Omega} = \sigma\kappa. \tag{3}$$

The first equation, in which $p = p(x, y)$ denotes the pressure, is a statement of the incompressibility of the fluid. In the second equation, $u_{\rm n}$ denotes the normal velocity (in the normal direction $\boldsymbol{n}$) of the interface. This is determined by the pressure gradient, the thickness of the fluid film b, and the fluid viscosity μ. The third equation describes the pressure jump Δp across the air-fluid interface $\partial\Omega$, and this depends on the surface tension σ and the interface curvature κ (the Young–Laplace law). Unlike more standard systems of partial differential equations in which the equation and boundary conditions are specified, the solution to this problem involves finding the boundary, $\partial\Omega$, itself. This is an example of a Stefan problem. At first sight the equations might look linear, but the coupling to the boundary dynamics makes them effectively highly nonlinear. Finding an exact solution of these equations is very difficult. A standard "linear stability" analysis is relatively straightforward and can explain why the initial circular air-fluid interface becomes wavy. But beyond this, finding solutions for the finger-like structures, let alone fingers whose tips split, is (still) a difficult problem. A pioneering early study of the Hele-Shaw cell is due to Philip Saffman and Geoffrey Taylor (1958), and many hundreds of papers have been published on the problem in subsequent years.

Solidification and Dendritic Crystal Growth

At first sight this seems like a very different process. Cooling a drop of a molten substance (or a supersaturated salt solution) can produce very striking needle crystals that grow with approximately constant velocity and develop side branches. A typical pattern is shown in Figure 1b. Many questions come to mind: why do finger-like structures form; what determines their propagation speed; why do they form side branches; and is there any connection with the finger-like structures seen in the Hele-Shaw cell? Before addressing these questions it is worth noting that historically the study of dendritic growth had its origins in metallurgical studies associated with the casting of canons. It was discovered that during the casting process the solidifying metal developed a crystalline structure that could weaken the canon—to the point of its shattering when it was fired!

The physical process determining the growth of the solid phase is that of cooling, with the rate of cooling being determined by the rate at which heat can diffuse away from the solid-melt interface. The speed at which the the solid front advances is determined by the temperature gradient across the interface. Using these basic principles, the governing equations are found to be

$$\frac{\partial T}{\partial t} = \nabla^2 T, \tag{4}$$

$$u_{\mathrm{n}} = -\alpha(\boldsymbol{\nabla} T).\boldsymbol{n}, \tag{5}$$

$$T|_{\partial\Omega} = T_{\mathrm{m}} - \beta\kappa. \tag{6}$$

Here, the first equation is just the diffusion equation for the temperature field T. The second equation for the normal velocity, u_{n}, of the solidification front is exactly analogous to the one for the Hele-Shaw cell. The third equation expresses the temperature drop at the solid-melt interface $\partial\Omega$ in which the normal melting temperature T_{m} is corrected by subtle effects related to the curvature of the interface (the Gibbs–Thompson effect). In more sophisticated theories, there is an additional correction proportional to the velocity of the moving front itself. Here, α and β are certain parameters reflecting the physical properties of the crystalizing substance. Thus, despite the very different physics, the mathematical description of the Hele-Shaw cell and a solidifying melt are almost identical Stefan problems. Again, solving the equations is difficult. Apart from a linear stability analysis (leading to the so-called Mullins–Sekerka instability criterion (Langer 1980)), solutions for the dendritic structures and a description of the side-branching structure in particular are very difficult. A classic review of dendritic growth is Langer (1980) and a more recent textbook is Davis (2001).

Other Pattern-Forming Systems

The formulation of the above problems indicates that two quantities appear to play a key role: the gradient of a field variable (pressure or temperature in the above examples) that drives the evolution of the pattern, and curvature of the interface (Pelcé, 1988; Ben-Jacob & Garik, 1990). In fact, these two quantities also play important roles in other pattern-forming systems. In almost every problem whose pattern is in the form of a propagating front or interface (another classic example is flame propagation), the curvature, and the parameter that multiplies it (e.g., surface tension in the case of the Hele-Shaw cell), plays an important role in determining the basic pattern wavelength.

In the introduction, we cited bacterial colony formation as another example of a pattern-forming system. A typical pattern, of a colony of *Bacillus subtilis* on an agar plate, is shown in Figure 1c. What drives the formation of these patterns? A fundamental concept is that of chemotaxis, namely, the response of the organisms to a concentration gradient. Thus, flux of cells in a medium is typically determined by the nutrient gradient. This principle enables one to write down systems of coupled diffusion and hydrodynamic equations for the concentration of cells responding to the nutrient gradient, and for the nutrient concentration itself, which is clearly going to be depleted in regions of high cell concentration. The ensuing nonlinear feedback between these quantities can lead to a rich, and much-studied, pattern structure (Lega & Passot, 2003).

MICHAEL TABOR

See also **Branching laws; Cluster coagulation; Flame front; Hele-Shaw cell; Navier–Stokes equation; Pattern formation; Turing patterns**

Further Reading

Ben-Jacob, E. & Garik, P. 1990. The formation of patterns in non-equilibrium growth. *Nature*, 343: 523–530

Bensimon, D., Kadanoff, L.P., Liang, S., Shraiman, B.I. & Tang, C. 1986. Viscous flows in two dimensions. *Reviews of Modern Physics*, 58: 977–999

Davis, S.H. 2001. *Theory of Solidification*, Cambridge and New York: Cambridge University Press

Langer, J.S. 1980. Instabilities and pattern formation in crystal growth. *Reviews of Modern Physics*, 52: 1–28

Lega, J. & Passot, T. 2003. Hydrodynamics of bacterial colonies: a model. *Physical Review* E, 67: 031960

Murray, J.D. 2003. *Mathematical Biology*, 3rd edition, Berlin and New York: Springer

Pelcé, P. 1988. *Dynamics of Curved Fronts*, Boston: Academic Press

Saffman, P.G. & Taylor, G.I. 1958. The penetration of a fluid into a porous medium or Hele-Shaw cell containing a more viscous fluid. *Proceedings of the Royal Society of London* B, 245: 312–329

Thompson, D.W. 1942. *On Growth and Form*, 2nd edition, Cambridge: Cambridge University Press

Turing, A.M. 1952. The chemical basis of morphogenesis. *Philosophical Transactions of the Royal Society of London* B, 237: 37–72

GUTZWILLER TRACE FORMULA

See **Quantum theory**

H

HAMILTONIAN SYSTEMS

A system of $2n$, first order, ordinary differential equations

$$\dot{z} = J\nabla H(z,t), \quad J = \begin{pmatrix} 0 & I \\ -I & 0 \end{pmatrix} \qquad (1)$$

is a Hamiltonian system with n degrees of freedom. (When this system is non-autonomous, it has $n+1/2$ degrees of freedom.) Here H is the Hamiltonian, a smooth scalar function of the extended phase space variables z and time t; the $2n \times 2n$ matrix J is called the "Poisson matrix"; and I is the $n \times n$ identity matrix. The equations naturally split into two sets of n equations for *canonically conjugate* variables, $z=(q,p)$, as follows.

$$\dot{q} = \partial H/\partial p, \quad \dot{p} = -\partial H/\partial q.$$

Here the n coordinates q represent the configuration variables of the system (positions of the component parts), and their canonically conjugate momenta p represent the impetus associated with movement. These equations generalize Newton's second law: $F = ma = \mathrm{d}p/\mathrm{d}t$, to systems (like particles in magnetic fields, or motion in non-inertial reference frames) where the momentum is not simply mass times velocity. The Hamiltonian usually represents the total energy of the system; thus, if $H(q, p)$ does not depend explicitly upon t, then its value is invariant, and Equations (1) are a conservative system. More generally, however, Hamiltonian systems need not be conservative.

William Rowan Hamilton first gave this reformulation of Lagrangian dynamics in 1834 (Hamilton, 1835). However, Hamiltonian dynamics is much more than just a reformulation. It leads, for example, to Henri Poincaré's geometrical insight that gave rise to symplectic geometry, and it provides a compact notation in which the concept of integrability is most naturally expressed and in which perturbation theory can be efficiently carried out. Moreover, nearly integrable Hamiltonian systems exhibit a remarkable stability expressed by the famous results of Kolmogorov–Arnol'd–Moser (KAM) theory and also Nekhoroshev theory. The Hamiltonian formulation also provides the foundation of both statistical and quantum mechanics.

Importantly, virtually all of the dynamical laws of physics—from the motion of a point particle to the interaction of complex quantum fields—have a formulation based on Equations (1). For example, frictionless mechanical systems are described by a Hamiltonian $H(q,p) = K(p) + V(q)$, where K is the kinetic energy (which is often quadratic in p) and V is the potential energy. For example, an ideal planar pendulum consists of a point particle of mass m attached to a massless rigid rod of length L whose other end is attached to a frictionless pivot. The most convenient configuration variable for the pendulum is $q=\theta$, the angle of the rod from the vertical. The gravitational potential energy of the system is then $V = -mgL\cos\theta$, and its kinetic energy is $K = p^2/(2mL^2)$, where p is the angular momentum about the pivot. For this case Equations (1) become

$$\dot{\theta} = \frac{\partial H}{\partial p} = \frac{p}{mL^2}, \quad \dot{p} = -\frac{\partial H}{\partial q} = -mgL\sin\theta. \quad (2)$$

The point $(0,0)$ is a stable (elliptic) equilibrium corresponding to the pendulum hanging down, at rest. The point $(\pm\pi, 0)$ is also an equilibrium but is an unstable (saddle) point. The unstable eigenvector of the saddle is the beginning of the unstable manifold, W^{u}, a trajectory that is backwards asymptotic to the saddle. By energy conservation, the unstable manifold corresponds to a branch of the energy contour $E = mgL$, which again joins the saddle point (after the pendulum has undergone one full rotation). Thus, this trajectory is forward asymptotic to the saddle as well; that is, it lies on the stable manifold, W^{s}, of the saddle. Orbits of this kind are called *homoclinic*. In this case the homoclinic orbit separates the trajectories oscillating about the elliptic equilibrium from those in which the pendulum undergoes complete rotations, thus it is called a *separatrix*. Orbits near the elliptic equilibrium oscillate with the frequency of the linearized, harmonic oscillator, $\omega = \sqrt{g/L}$. The frequency of oscillation decreases monotonically as the amplitude increases, approaching zero logarithmically at the separatrix. The frequency of rotation of the solution grows again from zero as the energy is further increased.

Canonical Structure

The geometrical structure of Hamiltonian systems arises from the preservation of the loop action, defined by

$$A[\gamma] = \oint_\gamma p\,\mathrm{d}q - H\,\mathrm{d}t, \qquad (3)$$

where γ is any closed loop in (q, p, t)-space. A consequence of Equations (1) is that if each point on a loop γ_0 is evolved with the flow to obtain a new loop γ_t, then $A[\gamma_0] = A[\gamma_t]$: the loop action is known as the *Poincaré invariant.*

The Hamiltonian form of Equations (1) is not preserved under an arbitrary coordinate transformation (unlike the Euler–Lagrange equations for a Lagrangian system). A canonical transformation $(q, p) \to (q', p')$ preserves the form of the Equations (1). Canonical transformations can be obtained by requiring that the Poincaré invariant of Equation (3) be the same in the new coordinate system, or locally that

$$\left(p'\,\mathrm{d}q' - H'\,\mathrm{d}t\right) - (p\,\mathrm{d}q - H\,\mathrm{d}t) = \mathrm{d}F$$

is the total differential of a function F. If F is represented as a function of a selection of half of the variables (q, p) and the complementary half of (q', p'), it is a "generating" function for the canonical transformation. For example, a function $F(q, q', t)$ implicitly generates a canonical transformation through

$$p = -\frac{\partial F}{\partial q}, \quad p' = \frac{\partial F}{\partial q'},$$
$$H'(q', p', t) = H(q, p, t) - \frac{\partial F}{\partial t}. \qquad (4)$$

In order that this transformation be well defined, the first equation must be inverted to find $q'(q, p)$; this requires that the matrix $\partial^2 F/\partial q \partial q'$ be nonsingular. An autonomous canonical transformation is also called a *symplectic map.* Canonical transformations are often employed to simplify the equations of motion. For example, Hamilton's equations are especially simple if the new Hamiltonian is a function of only the momentum variables, $H'(p')$. If we can find a transformation to such a coordinate system then the system is said to be integrable. In general, such transformations do not exist; one of the consequences is chaotic motion.

Integrability

Loosely speaking, a set of differential equations is integrable if it can be explicitly solved for arbitrary initial conditions (Zakharov, 1991). The explicit solution, when inverted, yields the initial conditions as invariant functions along the orbits of a system—the initial conditions are constants of motion. A Hamiltonian $H(q, p)$ is said to be Liouville integrable if it can be transformed to a canonical coordinate system in which it depends only on new momenta. When the energy surfaces are compact and the new momenta are everywhere independent, Arnol'd showed that it is always possible to choose the momentum variables so that their conjugate configuration variables are periodic angles ranging from 0 to 2π (Arnol'd, 1989). These coordinates are called *action-angle variables*, denoted $(\theta, \mathcal{J})$.

As the Hamiltonian is a function only of the actions, $H(\mathcal{J})$, Equation (1) becomes $\dot{\mathcal{J}} = 0$, and $\dot{\theta} = \Omega(\mathcal{J}) = \partial H/\partial \mathcal{J}$. A system is anharmonic when the frequency vector Ω has a nontrivial dependence on $\mathcal{J}$. Thus, for an integrable system, motion occurs on the n-dimensional tori $\mathcal{J}$ = constant. Orbits helically wind around the torus with frequencies Ω that depend upon the torus chosen. When the frequency is nonresonant (there is no integer vector m for which $m \cdot \Omega = 0$), then the motion is dense on the torus.

Any one degree-of-freedom, autonomous Hamiltonian system is locally integrable. A Hamiltonian with more than one degree of freedom, such as pendulum with an oscillating support, is typically not integrable. Systems that are separable into non-interacting parts are integrable, and there are also a number of classical integrable systems with arbitrarily many degrees of freedom. These include the elliptical billiard, the rigid body in free space, the Neumann problem of the motion of a particle on a sphere in a quadratic potential, the Toda lattice, and the Calogero–Moser lattice (Arnol'd, 1988).

Hamiltonian Chaos

The problem of understanding the motion of a slightly perturbed integrable system originated with the desire to understand the motion of the planets. The Kepler problem corresponding to the gravitational interaction of two spherical bodies is integrable; however, once other effects (such as the mutual forces among planets) are included, there appear to be no general, explicit solutions. Poincaré in particular addressed the question of the stability of the solar system, finally realizing that the convergence of perturbation series for the solutions could not be guaranteed and discovering the phenomenon of transverse homoclinic intersections that is a harbinger of chaos (Poincaré, 1892).

Consider the problem

$$H(\theta, \mathcal{J}) = H_0(\mathcal{J}) + \varepsilon H_1(\theta, \mathcal{J}) + \cdots,$$

where the perturbation H_1 depends periodically on the angles and can be expanded in a Fourier series. When

$\Omega(\mathcal{J})$ is nonresonant, a formal sequence of canonical transformations can be constructed to find a set of coordinates in which H is independent of the angle. The problem is the occurrence of denominators in the coefficients proportional to resonance conditions $m \cdot \Omega(\mathcal{J})$ for integer vectors m. Even for actions where the frequencies are incommensurate, it is always possible to make $m \cdot \Omega(\mathcal{J})$ arbitrarily small by choosing large enough integers m. Thus, a priori bounds on the convergence fail. This is called the *problem of small denominators*.

Chirikov realized that small denominators signal the creation of topologically distinct regions of motion (Chirikov, 1979; MacKay & Meiss, 1987). Near a typical resonance, one can use averaging methods to approximate the motion by an integrable pendulum-like Hamiltonian, effectively discarding all of the terms in H_1 except for those that are commensurate with the resonance, that is, the Fourier modes $H_m(\mathcal{J})$ with $m \cdot \Omega = 0$. Thus, orbits near to a resonance are trapped in an effective potential well. The domain of the trapped motion has the width in action of the corresponding pendulum separatrix; it is typically proportional to the square root of the mth Fourier amplitude of H_1. If we can treat the resonances independently, then each gives rise to a corresponding separatrix. However, as the perturbation amplitude grows, this approximation must break down as it predicts the overlap of neighboring separatrices. In 1959, Chirikov proposed this resonance overlap condition as an estimate of the onset of global chaos. Renormalization theory gives a more precise criterion (*See* **Standard map**).

This picture, together with the fact that rational numbers are dense, leads to the expectation that none of the invariant tori of an integrable system persist when it is perturbed with an arbitrarily small perturbation. Surprisingly, the Fermi–Pasta–Ulam computational experiment in 1955 (Fermi et al., 1965; Weissert, 1997) failed to find this behavior. Indeed, KAM theory, initiated by Andrei Kolmogorov in the 1950s and developed in the 1960s by Vladimir Arnol'd and Jürgen Moser, proves persistence of most of the invariant tori (de la Llave, 2001; Pöschel, 2001). This holds when the perturbation is small enough, provided that the system satisfies an anharmonicity or nondegeneracy condition, it is sufficiently differentiable, and the frequency of the torus is sufficiently irrational. The irrationality condition is that $|m \cdot \Omega| > c/|m|^{\tau}$ for all nonzero integer vectors m and some $c > 0$ and $\tau \geq 1$; this is a *Diophantine condition*. Each of these conditions is essential, though some systems (such as the solar system for which the frequencies are degenerate) can be reformulated so that KAM theory applies.

Resonant tori and tori whose frequencies are nearly commensurate lie between the Diophantine tori. Generally, these tori are destroyed by a small perturbation and either form new, secondary tori trapped in a resonance or are replaced by a zone of chaotic motion that is found in the neighborhood of the stable and unstable manifolds of the resonance. These generically intersect and give rise to a homoclinic tangle or trellis that contains a Smale horseshoe. In the case that the Hamiltonian is analytic, the size of the chaotic region is exponentially small in ε and, thus, can be difficult to detect (Gelfreich & Lazutkin, 2001).

For small perturbations of an integrable Hamiltonian, it remains an open problem to show that a nonzero volume of initial conditions behaves chaotically, in the sense that they have positive Lyapunov exponents. Numerical investigations indicate that orbits in the chaotic zones do have positive Lyapunov exponents, and that these domains form a "fat fractal" (a fractal with positive measure). There are also many examples of uniformly hyperbolic dynamics (especially for the case of billiards (Bunimovich, 1989)) which can also have properties such as mixing and ergodicity.

The problem of the nonlinear stability of a typical system is also open (*See* **Symplectic maps**). However, N.N. Nekhoroshev showed in 1977 that for an analytic system, the actions drift very little for very long times (at most by an amount that is proportional to a power of ε for times that are exponentially long in ε (Lochak, 1993; Pöschel, 1993)). Thus, while it is possible that a KAM torus is unstable, for most practical purposes, they appear to be stable.

Generalizations

Many partial differential equations (PDEs) also have a Hamiltonian structure. For a PDE with independent variables (x, t), the canonical variables are replaced by fields $(q(x,t), p(x,t))$ and the partial derivatives in (1) by functional or Frechêt derivatives, so that

$$\frac{\partial q}{\partial t} = \frac{\delta H}{\delta p}, \quad \frac{\partial p}{\partial t} = -\frac{\delta H}{\delta q}.$$

The Hamiltonian functional H is the integral of an energy density. For example, the wave equation has the Hamiltonian $H[q,p] = \int \mathrm{d}x \frac{1}{2}\left(p^2 + c^2(\partial_x q)^2\right)$. Other nonlinear wave equations such as the integrable nonlinear Schrödinger, Korteweg–de Vries, and sine-Gordon equations also have Hamiltonian formulations.

James D. Meiss

See also **Adiabatic invariants; Chaotic dynamics; Constants of motion and conservation laws; Ergodic theory; Euler–Lagrange equations; Fermi–Pasta–Ulam oscillator chain; Hénon–Heiles system; Horseshoes and hyperbolicity in dynamical systems; Lyapunov exponents; Mel'nikov method; Pendulum; Phase space; Poisson brackets; Standard map; Symplectic maps; Toda lattice**

Further Reading

Arnol'd, V.I. (editor). 1988. *Dynamical Systems III*, New York: Springer

Arnol'd, V.I. 1989. *Mathematical Methods of Classical Mechanics*, New York: Springer

Bunimovich, L.A. 1989. Dynamical systems of hyperbolic type with singularities. In *Dynamical Systems*, edited by Ya. Sinai, Berlin: Springer, p. 278

Chirikov, B.V. 1979. A universal instability of many-dimensional oscillator systems. *Physics Reports*, 52: 265–379

de la Llave, R. 2001. A tutorial on KAM theory. In *Smooth Ergodic Theory and Its Applications (Seattle, WA, 1999)*, Providence, RI: American Mathematical Society, pp. 175–292

Fermi, E., Pasta, J. & Ulam, S. 1965. Studies of nonlinear problems. In *Collected Papers of Enrico Fermi*, vol. 2, edited by E.Segré, Chicago: University of Chicago Press, pp. 977–988

Gelfreich, V.G. & Lazutkin, V.F. 2001. Separatrix splitting: perturbation theory and exponential smallness. *Russian Mathematical Surveys*, 56(3): 499–558

Hamilton, W.R. 1835. On the application to dynamics of a general mathematical method previously applied to optics. *British Association Report, 1834*, pp. 513–518

Lochak, P. 1993. Hamiltonian perturbation theory: periodic orbits, resonances and intermittancy. *Nonlinearity*, 6: 885–904

MacKay, R.S. & Meiss, J.D. (editors). 1987. *Hamiltonian Dynamical Systems: A Reprint Selection*, London: Adam Hilger

Poincaré, H. 1892. *Les Methodes Nouvelles de la Mechanique Celeste*, Paris: Gauthier–Villars

Pöschel, J. 1993. Nekhoroshev estimates for quasi–convex Hamiltonian systems. *Mathematische Zeitschrift*, 213:187–216

Pöschel, J. 2001. A lecture on the classical KAM theorem. In *Smooth Ergodic Theory and Its Applications (Seattle, WA, 1999)*, Providence, RI: American Mathematical Society, pp. 707–732

Weissert, T.P. 1997. *The Genesis of Simulation in Dynamics: Pursuing the Fermi–Pasta–Ulam Problem*, New York: Springer

Zakharov, V.E. (editor). 1991. *What Is Integrability?* Berlin: Springer

HARMONIC GENERATION

Harmonic generation is the phenomenon whereby new frequency components of a wave are created upon interaction with a nonlinear medium. The concept of a harmonic frequency is familiar to anyone with a basic knowledge of music: two notes whose frequencies form a ratio of small whole numbers (e.g., 2:1, 3:2, 5:4) produce a sound that is pleasing or "harmonious" compared with two notes with incommensurate frequencies. More generally, any periodic function $f(t+T)=f(t)$ can be written as an infinite Fourier series containing the first harmonic (fundamental frequency) $\omega=2\pi/T$, the second harmonic, and higher harmonic frequencies $\omega_m=m\omega$:

$$f(t)=\sum_{m=-\infty}^{\infty} c_m\,\mathrm{e}^{-\mathrm{i}\omega_m t}. \qquad (1)$$

The function $f(t)$ could represent the displacement of a guitar string, the amplitude of a wave in the ocean, or the electric field strength of a radio wave or a beam of light. If f satisfies a linear dynamic equation, then the principle of superposition holds, and no energy can be exchanged among the different Fourier coefficients c_m. Under nonlinear dynamics, however, the coefficients c_m are coupled to each other, and energy can be transferred back and forth among them.

Harmonic generation refers to the case in which energy originally at frequency ω generates a new wave component at a harmonic frequency ω_m. A closely related phenomenon is that of parametric amplification, in which a frequency ω_m provides the energy source to amplify a signal at a related frequency $\omega_{m'}$.

In optics, harmonic generation is based upon the nonlinear response of some medium to electromagnetic radiation at optical frequencies. The origin of harmonic generation is related to Theodore Maiman's invention of the ruby laser in 1960. Exploiting the high intensities and the coherency properties of the newly available laser output, in 1961 Peter Franken and his colleagues used the nonlinear polarization response of crystal quartz to produce second-harmonic light at a free-space wavelength of 347 nm from the ruby laser output at 694 nm, marking the birth of nonlinear optics.

By the late 1960s, lithium niobate ($LiNbO_3$) had emerged as the nonlinear crystal of choice for parametric amplification and harmonic generation. Much of the work taking place in the following decades was dedicated to improving the efficiency of the generation of harmonics, either by finding materials with better properties (higher transparency, stronger nonlinearity, etc.) or by aiding the energy transfer through phase matching the various waves involved in the process. More recently, improvements in material processing have made possible the technique of quasi-phase-matching (QPM), in which better phase matching is achieved through the engineering of a spatial variation of the nonlinear coefficient of the medium. In contrast to previous phase-matching techniques, QPM allows a wide, continuous range of frequencies to be coupled to their harmonics without imposing constraints on the orientation of the crystal; the orientation can then be chosen to maximize the nonlinear coefficient seen by the fundamental frequency and its harmonic as they pass through the crystal.

In optics, harmonic generation is often used to convert energy from one frequency band to another. The generated frequency may be more suitable for telecommunications or spectroscopy, for example, and may be in a region of the electromagnetic spectrum that is not directly available from high-powered lasers. The generation of a harmonic might also be a way of obtaining a coherent copy of an optical pulse, where the copy can be compressed, broadened, chirped, squeezed, or simply measured to determine the properties of the original pulse. Depending on the type of physical process involved and the particular application considered, one refers to the relevant phenomena as second-harmonic generation, sum- or difference-frequency generation, optical parametric generation, or optical parametric oscillation.

The canonical example for studying harmonic generation is an anharmonic oscillator, which can be taken as a simplified model of coherent light propagating through a nonlinear medium. In optics, the nonlinearity can arise due to the material polarization (electronic response) of the medium, or due to a coupling between the light and lattice vibrations in the medium (optical phonons). Referring to the former phenomenon, the polarization response can be expanded in a power series of the applied field, where the coefficients, the nonlinear susceptibility tensors of the material, can be obtained through quantum-mechanical calculations. In general, the quadratic polarization of a medium will be its leading order nonlinear response unless symmetries of the medium rule it out. In this case, one can then reduce Maxwell's equations to the nonlinear wave equation

$$\frac{\partial^2 E}{\partial z^2} - \frac{n^2}{c^2}\frac{\partial^2 E}{\partial t^2} = \frac{\chi^{(2)}}{c^2}\frac{\partial^2 (E^2)}{\partial t^2}, \tag{2}$$

where n is the refractive index of light in the medium, c is the speed of light, and $E(z,t)$ is the electromagnetic field strength which depends on a longitudinal dimension z and on time t.

The nonlinear susceptibility $\chi^{(2)}$ characterizes the quadratic response of the medium to an incident field, where the linear susceptibility has been included in the definition of the refractive index n. To observe the exchange of energy between two frequencies ω_1 and ω_2 mediated by this nonlinearity, the Fourier expansion of Equation (1) can be augmented as

$$E(z,t) = c_1(z)\,\mathrm{e}^{\mathrm{i}\theta_1(z,t)} + c_2(z)\,\mathrm{e}^{\mathrm{i}\theta_2(z,t)} + \text{c.c.}, \tag{3}$$

where "c.c." denotes the complex conjugate, $\theta_j(z,t) = k_j z - \omega_j t$ is a rapidly varying optical phase, and $k_j = n\omega_j/c$ is the linear wave number. Under the assumption that the coefficients $c_j(z)$ vary much more slowly in z than the optical phase (slowly varying envelope approximation), substituting Equation (3) into Equation (2) gives

$$2\mathrm{i}k_1 c_1'(z)\mathrm{e}^{\mathrm{i}\theta_1} + 2\mathrm{i}k_2 c_2'(z)e^{\mathrm{i}\theta_2} + \text{c.c.} = \frac{\chi^{(2)}}{c^2}\frac{\partial^2}{\partial t^2}(E^2), \tag{4}$$

where

$$E^2 = c_1^2\mathrm{e}^{\mathrm{i}(2\theta_1)} + c_2^2\mathrm{e}^{\mathrm{i}(2\theta_2)} + 2c_1^* c_2\mathrm{e}^{\mathrm{i}(-\theta_1+\theta_2)} + 2c_1c_2\mathrm{e}^{\mathrm{i}(\theta_1+\theta_2)} + |c_1|^2 + |c_2|^2 + \text{c.c.} \tag{5}$$

In order for the terms on the right of Equation (4) to contribute significantly to the growth of $c_1(z)$ or $c_2(z)$ on the left, the rapid phase rotations given by the exponentials must resonate, which occurs if $\theta_2 = 2\theta_1$, or $\omega_2 = 2\omega_1$. The resulting equations will then be

$$c_1'(z) = \frac{\mathrm{i}\omega_1\chi^{(2)}}{n_1 c}\, c_1^* c_2 \mathrm{e}^{\mathrm{i}\Delta kz}, \tag{6a}$$

$$c_2'(z) = \frac{\mathrm{i}\omega_2\chi^{(2)}}{n_2 c}\, c_1^2 \mathrm{e}^{-\mathrm{i}\Delta kz}. \tag{6b}$$

The phase mismatch term $\Delta k = k_1 - 2k_2$ arises from a more careful calculation that takes into account that the medium's linear susceptibility (and therefore the refractive index n) is a function of frequency ω; thus, it is not generally true that $\omega_2 = 2\omega_1$ implies $k_2 = 2k_1$.

Equations (6) can be solved exactly, but it is more instructive to consider the quantities that are invariant in this system of equations. These invariants reflect physical quantities that are conserved during the mixing process (called Manley–Rowe invariants):

$$I_0 = \frac{1}{2}\varepsilon_0 c\left(n_1|c_1|^2 + n_2|c_2|^2\right), \tag{7a}$$

$$I_1 = \frac{1}{2}\varepsilon_0 c\left(\frac{n_1|c_1|^2}{\hbar\omega_1} + 2\frac{n_2|c_2|^2}{\hbar\omega_2}\right). \tag{7b}$$

The first of these, I_0, is the total irradiance, or incident optical power per unit of surface area. That I_0 is invariant follows from conservation of energy, which is to be expected given that no absorption has been accounted for here in the linear susceptibility. The quantity I_1 sums the irradiance from the two frequencies, dividing each by the energy of a single photon at the respective frequency ($E_j = \hbar\omega_j$) before summing. The invariance of I_2, therefore, reflects a law of conservation of photon flux, as the destruction of two photons at the fundamental frequency ω_1 must bring about the creation of a single photon at the second harmonic, $\omega_2 = 2\omega_1$.

In the spectroscopy of molecular crystals, the fundamental frequency of a vibrating mode sometimes lies close to an overtone or combination of other mode frequencies, whereupon both frequencies are pushed away from each another. In such a case, the overtone or combination band borrows intensity from the fundamental in a phenomenon called Fermi resonance.

Richard O. Moore and Gino Biondini

See also **Frequency doubling; Manley–Rowe relations; Nonlinear optics; Parametric amplification**

Further reading

Armstrong, J.A., Bloembergen, N., Ducuing, J. & Pershan, P.S. 1962. Interactions between light waves in a nonlinear dielectric. *Physical Review*, 127: 1918

Bloembergen, N. 1965. *Nonlinear Optics*, Singapore: World Scientific

Boyd, R.W. 2003. *Nonlinear Optics*, 2nd edition, San Diego: Academic Press

Dunn, M.H. & Ebrahimzadeh, M. 1999. Parametric generation of tunable light from continuous-wave to femtosecond pulses. *Science*, 286: 1513

Franken, P.A., Hill, A.E., Peters, C.W. & Weinreich, G. 1961. Generation of optical harmonics. *Physical Review Letters*, 7: 118

Maiman, T.H. 1960. Stimulated optical radiation in ruby. *Nature*, 187: 493

Schubert, M. & Wilhelmi, B. 1986. *Nonlinear Optics and Quantum Electronics*, New York: Wiley

HARMONIC OSCILLATORS

See **Damped-driven anharmonic oscillator**

HARTREE APPROXIMATION

In the quantum-mechanical treatment of a system with many degrees of freedom, the Hartree approximation writes the multidimensional correlated wave function as a simple product of one-dimensional or low-dimensional (orbital) functions. This ansatz reduces the full Schrödinger equation to a set of simpler equations for the orbital functions. The Hartree product approximation, first proposed by Douglas R. Hartree (1928) for many-electron atoms, can in principle be applied to any many-particle system, for example, electrons in an atom, molecule, or crystal. However, the Hartree ansatz is often physically inadequate as it disregards the fundamental permutation symmetry requirements for wave functions of nondistinguishable particles (fermions or bosons). In the case of electron systems, the Hartree approximation soon gave way to the more appropriate Hartree–Fock (HF) approximation that is based on a fully antisymmetrized product representation (Slater determinant) of the many-electron wave function proposed by Vladimir A. Fock (Fock, 1930a). For systems with distinct degrees of freedom, on the other hand, such as the multimode nuclear dynamics in molecules, the Hartree product representation still represents a basic approximation. (see, e.g., Beck et al., 2000).

For the many-electron system of an atom or molecule, the (nonrelativistic) stationary Schrödinger equation can be written in the form (see, e.g., Landau & Lifshitz, 1977)

$$\hat{H}\Phi_n = E_n\Phi_n, \tag{1}$$

where the Hamiltonian $\hat{H} = \hat{H}_o + \hat{V}$ consists of a one-particle part,

$$\hat{H}_o = \sum_i \hat{h}_i, \quad \hat{h}_i = -\frac{\hbar^2}{2m}\nabla^2 - \sum_\alpha \frac{Z_\alpha e^2}{|\boldsymbol{R}_\alpha - \boldsymbol{r}_i|} \tag{2}$$

i.e., $\hat{H}_o$ is the sum of the Hamiltonians $\hat{h}_i$ associated with the motion of the ith electron in the field of the nuclei with charges eZ_α at positions $\boldsymbol{R}_\alpha$, and a two-particle part,

$$\hat{V} = \frac{1}{2}\sum_{i\neq j} \frac{e^2}{|\boldsymbol{r}_i - \boldsymbol{r}_j|} \tag{3}$$

corresponding to the Coulomb repulsion of the electrons. The complexity of the many-electron problem is due to the latter part of the Hamiltonian, as it prevents a separation of the N-electron Schrödinger equation into N independent single-particle equations. The idea underlying the Hartree and the Hartree–Fock approximation is to replace the full electron-electron interaction by a new approximately separable effective Coulomb repulsion (or "mean field"). In the Hartree case, the optimal mean-field formulation is obtained by making use of a variational principle (Fock, 1930b),

$$\delta\langle\Phi|\hat{H}|\Phi\rangle = 0 \tag{4}$$

with the trial wave function given by a Hartree product

$$\Phi(\boldsymbol{x}_1, \boldsymbol{x}_2, \ldots, \boldsymbol{x}_N) = \phi_1(\boldsymbol{x}_1)\phi_2(\boldsymbol{x}_2)\ldots\phi_N(\boldsymbol{x}_N) \tag{5}$$

of normalized single-electron functions (orbitals) $\phi_i(\boldsymbol{x}_i)$. Here $\boldsymbol{x}_i = (\boldsymbol{r}_i, s_i)$ comprises in a short-hand notation the set of spatial and spin variables of the ith electron. The resulting equations for the orbitals read

$$(\hat{h} + \hat{v}_k^H)\phi_k(\boldsymbol{x}) = \varepsilon_k\phi_k(\boldsymbol{x}), \quad k = 1, 2, \ldots, N, \tag{6}$$

where for each orbital $\phi_k(\boldsymbol{x})$ a distinct mean field

$$\hat{v}_k^H = \sum_{l\neq k} \hat{J}_l, \qquad \hat{J}_l = e^2 \int \frac{|\psi_l(\boldsymbol{r}')|^2}{|\boldsymbol{r}' - \boldsymbol{r}|}\, \mathrm{d}\boldsymbol{r}' \tag{7}$$

augments the one-particle Hamiltonian $\hat{h}$. The physical meaning of $\hat{v}_k^H$ is the Coulomb energy arising from the interaction of the electron in orbital $\phi_k(\boldsymbol{x})$ and the charge distribution associated with the other orbitals (the operators $\hat{J}_l$ are referred to as Coulomb operators).

Equations (6) are a set of nonlinear integro-differential equations in the orbitals $\phi_1(\boldsymbol{x}_1)$, $\phi_2(\boldsymbol{x}_2), \ldots, \phi_N(\boldsymbol{x}_N)$. Starting with an initial guess for the orbitals $\phi_k(\boldsymbol{x})$, these equations have to be solved in an iterative way until self-consistency is reached. Here it is essential to select appropriate orbital solutions of the individual one-particle eigenvalue problems (6) so that in the N-electron product wave function the Pauli exclusion principle is at least approximately fulfilled. The resulting potential energy operators (7) constitute the Hartree self-consistent field (SCF).

In the many-electron problem, the Hartree approximation is only of historical or pedagogical interest. It was soon succeeded by the more rigorous and practical Hartree–Fock (HF) approximation (Fock, 1930a) based on the proper antisymmetrized product (Slater determinant) representation

$$\Phi(\boldsymbol{x}_1, \boldsymbol{x}_2, \ldots, \boldsymbol{x}_N) = \begin{vmatrix} \phi_1(\boldsymbol{x}_1) & \phi_1(\boldsymbol{x}_2)\ldots & \phi_1(\boldsymbol{x}_N) \\ \phi_2(\boldsymbol{x}_1) & \phi_2(\boldsymbol{x}_2)\ldots & \phi_2(\boldsymbol{x}_N) \\ \vdots & \vdots & \vdots \\ \phi_N(\boldsymbol{x}_1) & \phi_N(\boldsymbol{x}_2)\ldots & \phi_N(\boldsymbol{x}_N) \end{vmatrix} \quad (8)$$

of the N-electron wave function. This leads to one common eigenvalue problem for the HF orbitals,

$$(\hat{h} + \hat{v}^{\mathrm{HF}})\phi_k(\boldsymbol{x}) = \varepsilon_k \phi_k(\boldsymbol{x}), \quad k = 1, 2, \ldots, N, \quad (9)$$

where the HF mean field operator

$$\hat{v}^{\mathrm{HF}} = \sum_{l=1}^{N} (\hat{J}_l - \hat{K}_l) \quad (10)$$

now contains in addition to the Coulomb contributions a so-called exchange part $-\sum_{l=1}^{N} \hat{K}_l$, and the nonlocal exchange operators $\hat{K}_l$ are defined by

$$\hat{K}_l \phi(\boldsymbol{x}) = e^2 \int \frac{\phi_l^*(\boldsymbol{x}')\phi(\boldsymbol{x}')\,\mathrm{d}\boldsymbol{x}'}{|\boldsymbol{r}' - \boldsymbol{r}|} \phi_l(\boldsymbol{x}). \quad (11)$$

Here the integration sums over the spin variable. The HF eigenvalue equation (9) represents a nonlinear problem that again has to be solved in a self-consistent way. It should be noted that because of $(\hat{J}_l - \hat{K}_l)\phi_l = 0$, there arise no self-interaction contributions in the self-consistent HF mean field. Usually, the HF or "molecular" orbitals (MO) are written as linear combinations (LC) of basis functions or "atomic" orbitals (AO). This LCAO representation transforms the integro-differential HF equation (9) into an algebraic eigenvalue problem, and the self-consistency solution is achieved by successive matrix diagonalization. Both with respect to practical applicability and theoretical significance, the HF (or SCF) method is fundamental for electronic structure computations in atoms, molecules, and the solid state. In the LCAO form it is a main ingredient in all existing quantum chemistry program packages. Most of the methods aiming at the description of electron correlation, that is, the effects beyond the HF one-particle picture, are based on the SCF orbital energies and integrals as input data.

Although in electronic structure calculations the Hartree approximation is no longer used, it is still a viable means in the description of nuclear dynamics, that is, the nuclear motion accompanying electronic states and processes. Because one deals here with distinguishable "particles" and not with fermions, the Hartree and not the Hartree–Fock approximation is the appropriate choice. Here one assumes a specific kind of separation of the nuclear and electronic degrees of freedom that has its physical foundation on the large mass differences of electrons and nuclei and correspondingly large differences in the respective velocities (Born & Oppenheimer, 1927). Setting molecular rotations aside, one may write the total wave function as a product of an electronic wave function $\Phi(\boldsymbol{x}; \boldsymbol{R})$ and a wave function $\chi(\boldsymbol{R})$ associated with the vibrational motion of the nuclei:

$$\Psi(\boldsymbol{x}, \boldsymbol{R}) = \Phi(\boldsymbol{x}; \boldsymbol{R})\chi(\boldsymbol{R}) \quad (12)$$

Here $\boldsymbol{x}$ and $\boldsymbol{R}$ denote collectively the electronic and nuclear variables. The electronic function is an eigenfunction of the electronic Hamiltonian at spatially fixed nuclei and, thus, depends parametrically on the nuclear coordinates $\boldsymbol{R}_\alpha$. The product approximation (12) is referred to as Born–Oppenheimer approximation. Having determined the electronic wave function and energy $E(\boldsymbol{R})$, the Schrödinger equation for the vibrational wave function can be formulated as follows:

$$\left(\sum_\alpha -\frac{\hbar^2}{2M_\alpha}\nabla_\alpha^2 + \hat{U}(\boldsymbol{R})\right)\chi(\boldsymbol{R}) = E_{\mathrm{tot}}\,\chi(\boldsymbol{R}) \quad (13)$$

Here the potential energy term $\hat{U}(\boldsymbol{R})$ is the sum of the nuclear repulsion energy and the electronic energy $E(\boldsymbol{R})$. Since we are here not interested in translational or rotational motion of the molecule as a whole, it is advantageous to replace the Cartesian coordinates used so far by a set of $M = 3N - 6$ nuclear coordinates $Q_1, Q_2, \ldots, Q_M$ describing, for example, the displacements from a reference nuclear configuration. Assuming a separable form of the kinetic energy in the new variables Q_i, that is,

$$\hat{T} = \sum_{i=1}^{M} \hat{t}_i \quad (14)$$

the Hartree ansatz

$$\chi(Q_1, Q_2, \ldots, Q_M) = \chi_1(Q_1)\chi_2(Q_2)\ldots\chi_M(Q_M) \quad (15)$$

leads to the following M nonlinear coupled one-dimensional eigenvalue equations,

$$(\hat{t}_i + \hat{u}_i)\chi_i(Q_i) = e_i\chi_i(Q_i), \quad i = 1, \ldots, M, \quad (16)$$

where the Hartree mean fields $\hat{u}_i$ are given by

$$\hat{u}_i = \int U(Q_1, Q_1, \ldots, Q_M)\prod_{j\neq i}(|\chi(Q_j)|^2\,\mathrm{d}Q_j). \quad (17)$$

The Hartree product representation is particularly useful in the context of time-dependent methods which solve the time-dependent Schrödinger equation by propagating an initial wave fuction numerically over a sufficiently long period of time. A very efficient and accurate example of such methods is the multi-configuration time-dependent Hartree (MCTDH) method (Beck et al., 2000). Here the representation of the wave function is given by a linear combination of several Hartree products.

PETER SCHMELCHER AND JOCHEN SCHIRMER

See also **Quantum theory**

Further Reading

Beck, M.H., Jäckle, A., Worth, G.A. & Meyer, H.-D. 2000. The multiconfiguration time-dependent Hartree method: a highly efficient algorithm for propagating wavepackets. *Physics Reports*, 324: 1–105

Born, M. & Oppenheimer, R. 1927. On the quantum theory of molecules. *Annalen der Physik*, 84: 457

Fock, V. 1930a. Self-consistent field mit Austausch für Natrium. [Self-consistent field with exchange for Sodium]. *Zeitschrift für Physik*, 62: 795–805

Fock, V. 1930b. Näherungsmethode zur Lösung des quantenmechanischen Mehrkorperproblems. [Approximate method for solution of quantum many body problem]. *Zeitschrift für Physik*, 61: 126–148

Hartree, D.R. 1928. *Proceedings of the Cambridge Philosophical Society*, 24: 111

Landau, L.D. & Lifshitz, E.M. 1977. *Course of Theoretical Physics*, vol. 3: *Quantum Mechanics*, Oxford: Pergamon Press

HAUSDORFF DIMENSION

See **Dimensions**

HEAT CONDUCTION

There are two main aspects of the theory of heat conduction that are intensively treated in nonlinear science. The first is related to the possibility of nonlinear terms in the phenomenological equations of heat conduction. The second aspect is related to the microscopic nature of heat conduction and especially to deriving Fourier's law (proportionality of heat flux to the temperature gradient) from the fundamental laws of motion at an atomic level.

Equations of Heat Conduction

In the phenomenological theory of heat conduction, the nonlinearity is normally manifested in temperature dependence of the thermal conductivity coefficient (or concentration dependence of the diffusion coefficient in the related diffusion problem). Such dependence leads to a nonlinear modification of Fourier's law. In the one-dimensional case, this equation may be written as

$$\frac{\partial T}{\partial t} = \frac{\partial}{\partial x}\left[\kappa(T)\frac{\partial T}{\partial x}\right], \qquad (1)$$

where $T(x,t)$ is the temperature field, $\kappa(T) = \vartheta/\rho C$, ϑ is the heat conductivity coefficient, ρ is the density, and C is the heat capacity of the specimen. All parameters introduced above may be significantly temperature-dependent.

As an example of nonlinear heat conduction, consider heat transfer in a hot plasma driven primarily by radiation. The internal energy of the photon gas is $(4\sigma/c)T^4$, where c is the velocity of light and σ is the Stefan–Boltzmann constant. The heat capacity $C = 16\sigma T^3/3c$. Using results of kinetic theory and plasma physics,

$$\lambda \sim T^{7/2}\rho^{-2}, \qquad (2)$$

where λ is the mean free path one can derive the final temperature dependence of the thermal conductivity coefficient: $\kappa(T) \sim T^{13/2}$. In the more common case of molecular heat conduction, the transfer is realized by gas molecules, and therefore, the average velocity of the molecules $v \sim T^{1/2}$ and λ is independent of the temperature ($\lambda \sim 1/\rho$); accordingly, the heat conductivity $\kappa \sim T^{1/2}$. More general models deal with the general form of the temperature dependence, like $\kappa = \kappa_0 T^{\nu}$, $\nu > 0$.

Probably, the most impressive consequence of the temperature dependence of the thermal conductivity coefficient is the phenomenon of so-called "heat inertia" when the heat energy remains localized in the system and does not dissipate away. If $T(x,t)$ is the solution of the Cauchy problem for Equation (1) with initial temperature

$$T(x,t_0) = \begin{cases} T_m(1-|x|/x_0)^{2/\nu}, & |x| \le x_0, \\ 0, & |x| \ge 0, \end{cases}$$

then $T(x,t) = 0$ for $|x| \ge x_0$, $t_0 < t < t_0 + t^*$, $t^* = x_0^2\nu/2\kappa_0(\nu+2)T_m^{\nu}$; that is, the heat energy is localized in the initial region for a certain time interval.

Microscopic Nature of Heat Conduction

The problem of heat conduction in dielectric crystals goes back to Rudolph Peierls (1926). A linear model of the crystalline lattice, being useful in many other physical problems, turns out to be insufficient when explaining the finiteness of the coefficient of thermal conductivity. The normal process of the heat conduction implies formation of a linear temperature gradient as the temperatures of the boundaries of the specimen are different. In an ideal linear crystal, however, any excitation is presented as a sum of independent vibrational modes that transfer energy from one boundary to the other without any loss. Therefore, a temperature gradient cannot be formed. This is why an account of the nonlinearity of interactions is necessary to describe the finite heat conductivity. Peierls has suggested that this condition is also a sufficient one because the phonon collision in a weakly nonlinear crystal is governed by the following relationships:

$$\omega_1 + \omega_2 = \omega_1' + \omega_2',$$
$$\boldsymbol{k}_1 + \boldsymbol{k}_2 = \boldsymbol{k}_1' + \boldsymbol{k}_2' + m\boldsymbol{p}. \tag{3}$$

The values with a prime correspond to the state after collision. The relationship for frequencies ω_i expresses the energy conservation. However the conservation of the quasimomenta $\boldsymbol{k}_i$ can be violated by a multiple of Bragg vector $\boldsymbol{p}$. The processes with $m = 0$ are referred to as normal phonon scattering. The case with $m \neq 0$ is referred to as Umklapp process. The latter case means that a certain quantity of the quasimomentum is transferred to the lattice. According to Peierls, the fact that the heat flux is scattered due to the Umklapp processes leads to normal heat conductivity. This picture was accepted until Fermi, Pasta, and Ulam examined it in one of the first computer simulations ever accomplished, finding that a one-dimensional chain with weak anharmonicity of the interaction potential did not demonstrate normal heat conduction (*See* **Fermi–Pasta–Ulam oscillator chain**). Instead, the energy was preserved in a few low-frequency vibrational modes. This famous numerical experiment led to great progress in nonlinear dynamics, including the discovery of lattice solitons, but the problem of the finite heat conductivity of a dielectric crystal again turned out to be unsolved.

It is easy to understand the essence of the problem of finding a sufficient condition for normal heat conductivity. Let us consider a set of equivalent particles arranged along a straight line and interacting only via purely elastic collisions. In every collision the interacting particles mutually exchange their velocities. Any pulse propagates through the system without loss, and therefore, a temperature profile is not formed. Similar properties were proved for the chain with exponential potential of interaction between the nearest neighbors (the so-called Toda lattice).

This problem remains unsolved to date despite numerous efforts. The lack of reliable analytical methods has resulted in many numerical simulations dealing with various lattice models. For one-dimensional systems the results available may be summarized as follows.

- A narrow class of completely integrable systems demonstrates no formation of linear temperature profile at all. Consequently, the coefficient of thermal conductivity cannot be defined. Examples are the linear chain and the Toda lattice.
- For many non-integrable chains, a linear temperature profile is formed and the coefficient of thermal conductivity may be defined for any given size N of the specimen. However, in the thermodynamic limit $N \to \infty$, the coefficient diverges: $\kappa \to \infty$. Therefore, such systems cannot be treated as having normal heat conductivity. Examples are the chain with parabolic and quartic potential among many others.
- A few one-dimensional chains were recently demonstrated to have normal heat conductivity. These examples are the chain of rotators, the Frenkel–Kontorova chain, and some other models.

It should be mentioned that some chains seem to "switch" between different types of behavior with respect to normal heat conductivity with changes of temperature and/or parameters of the potential of interaction.

It seems that different regimes of heat conduction in one-dimensional systems are related to the properties of nonlinear excitations in every concrete model. For instance, the normal heat conduction in the chain of rotators was found to be related to the breakdown of nonlinear acoustic waves and formation of localized breathers, which scatter the heat flux. Similar mechanisms were discovered in other models exhibiting normal heat conduction.

For two-dimensional systems very little information is available, but some investigations suggest that the heat conductivity for different types of potentials is finite or logarithmically divergent. No classification similar to the one-dimensional case can yet be provided.

For the three-dimensional case the situation is as unknown, but the field of molecular dynamics is developing very rapidly and new results are expected to arrive soon.

Oleg Gendelman and Leonid Manevitch

See also **Fermi–Pasta–Ulam oscillator chain; Frenkel–Kontorova model; Integrable lattices; Solitons; Toda lattice**

Further Reading

Bonetto, F., Lebowitz, J. & Rey-Bellet, L. 2000. In *Mathematical Physics 2000*, edited by A. Fokas, A. Grigoryan, T. Kibble, & B. Zegarlinski, River Edge, NJ : Imperial College Press

Fermi, E., Pasta, J.R. & Ulam, S.M. 1955. *Studies of Nonlinear Problems*, Los Alamos Scientific Laboratory Report No. LA–1940

Gendelman, O.V. & Savin, A.V. 2000. Normal heat conductivity in one-dimensional chain of rotators. *Physical Review Letters*, 84: 2381–2384

Lepri, S., Livi, R. & Politi, A. 2001. Thermal conduction in classical low-dimensional lattices, preprint: www.lanl.gov, arXiv:cond-mat/0112193

Peierls, R.E. 1955. *Quantum Theory of Solids*, London: Oxford University Press

Samarskii, A.A., Galaktionov, V.A., Kurdiumov, S.P. & Mikailov, A.P. 1987. *Regimes with Sharpening in the Problems for Quasilinear Parabolic Equations*, Moscow, Nauka (in Russian)

Toda, M. 1967. Vibration of a chain with nonlinear interactions. *Journal of the Physical Society of Japan*, 22: 431–36; Wave propagation in anharmonic lattices. *Journal of the Physical Society of Japan*, 23: 501–06

HEAVISIDE STEP FUNCTION

See **Generalized functions**

HELE-SHAW CELL

A very simple device invented by Henry Selby Hele-Shaw in the late 1890s, the Hele-Shaw cell consists of two closely spaced parallel plates, held in place by a frame, between which is a layer of a viscous liquid. Hele-Shaw invented this simple instrument in order to be able to exhibit the streamlines of a flow to students at University College of Liverpool by projecting the flow in the thin liquid layer onto a large screen. He found that minute air bubbles provided excellent flow tracers, and developed the technique to be able to display in a spectacular manner the flow patterns past a variety of objects.

Hele-Shaw discovered that when the distance between the two plates was very small (he used water as the fluid, and glass plates mounted within 0.02 in. of each other), the flow pattern around a circular cylinder perpendicular to the plates closely matched the potential flow solution that was known from the theory of two-dimensional flow of an ideal fluid. Apparently, the viscous liquid between the plates was flowing in this narrow space as though it were an inviscid fluid. George Stokes explained these observations by writing the equation of motion for a viscous fluid—today known as the Navier–Stokes equation—and averaging the flow (today known as plane Poiseuille flow) across the narrow gap between the two bounding plates in the Hele-Shaw cell. One then finds that the averaged, essentially two-dimensional flow velocity, v, is proportional to the gradient of the pressure, p, in the fluid:

$$v = -(d^2/12\mu)\nabla p, \qquad (1)$$

where d is the separation between the plates and μ is the viscosity of the fluid. Thus, flow in a Hele-Shaw cell is potential flow, with a velocity potential proportional to the pressure, and this explains the coincidence between Hele-Shaw's observations and the theory of two-dimensional, ideal flow. As Stokes wrote, "[Hele-Shaw's experiments] afford a complete graphical solution, experimentally obtained, of a problem which from its complexity baffles mathematicians except in a few simple cases."

Hele-Shaw flows illustrate two-dimensional potential flow around objects with sharp edges and thus show no separation. This is quite remarkable, because even the slightest viscosity would immediately alter the flow pattern were the fluid not constrained by the two plates of the Hele-Shaw cell.

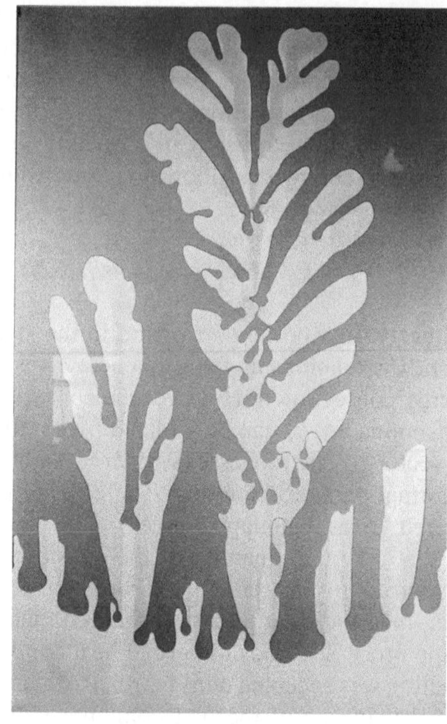

Figure 1. Viscous fingering patterns in a Hele-Shaw cell at a gravitationally unstable interface between two immiscible fluids—viscous silicone oil and air. (Reprinted with permission from Gallery of Fluid Motion, *Physics of Fluids*, 28(9). Copyright 1985, American Institute of Physics. Courtesy of T. Maxwell.)

A controversy with Osborne Reynolds ensued about the validity of the experiments since they appeared to invalidate Reynolds' criterion of a transition to turbulence based solely on the value of "Reynolds number." This led to several acrimonious exchanges of notes and letters in *Nature* in 1898 and 1899 between Reynolds and Hele-Shaw.

The Hele-Shaw cell has found wide application in later years, after it was realized that Equation (1) has the form of the velocity-pressure relationship one would expect for fluid flow in a porous material obeying Darcy's law. Thus, the Hele-Shaw cell became a paradigmatic instrument for studying flows in porous media, a subject of great interest to petroleum engineering.

When two immiscible fluids of different viscosities are present in a Hele-Shaw cell, a number of interesting phenomena occur. In the late 1950s, Geoffrey Taylor and Philip Saffman explored the evolution of "fingers" when a less viscous fluid (air) was pushed into a more viscous fluid (water, glycerin, or oil) in a Hele-Shaw cell. Similar "fingering" occurs when water is forced into oil-carrying porous rock in secondary oil recovery.

The subject of viscous fingering is similar to other pattern-forming instabilities, dendritic crystal growth in particular, and has become a much studied experimental system. The experimental data tend to be very visually appealing (Figure 1). Many variations on the basic setup have been explored and a voluminous literature exists. The degree of structure of the two-fluid interface depends sensitively on the ratio of viscosities of the two fluids in the Hele-Shaw cell, and the pattern is sensitive to small perturbations, such as imperfections on either plate or tiny air bubbles resident in the cell.

The scale of the smallest fingers observed in a two-fluid Hele-Shaw experiment is set by the surface tension at the interface between the two immiscible fluids. When the surface tension is very small, apparently fractal patterns emerge.

HASSAN AREF

See also **Fractals; Navier–Stokes equation; Pattern formation; Rayleigh–Taylor instability**

Further reading

Hele-Shaw, J.H.S. 1898. The flow of water. *Nature*, 58: 34–36

Lamb, H. 1932. *Hydrodynamics*, 6th edition, Cambridge: Cambridge University Press

Saffman, P.G. & Taylor, G.I. 1958. The penetration of a fluid into a porous medium or Hele-Shaw cell containing a more viscous liquid. *Proceedings of the Royal Society of London* A, 245: 312–329

Wooding, R.A. & Morel-Seytoux, H.J. 1976. Multiphase fluid flow through porous media. *Annual Reviews of Fluid Mechanics*, 8: 233–274

HELICAL WAVES

See **Nonlinear plasma waves**

HELICITY

Intuitively, it is clear that a vortex having a component of velocity along its axis is a helical structure. Many structures fall under this category, including such diverse structures as Taylor–Gortler vortices, leading edge and trailing vortices shed from wings and slender bodies, streamwise vortices in boundary layers and free shear flows, Langmuir circulations in the ocean and analogous structures in the atmosphere, tornados, and rotating storms. Similar structures are observed in space and in laboratory plasmas. Such structures lack reflection symmetry. One of the key quantities characterizing reflection symmetry (or its lack) of a fluid flow is the so-called helicity, which roughly measures how parallel velocity and vorticity are in a fluid flow.

Definition

Helicity $\mathcal{H}_B$ of a solenoidal vector field $\boldsymbol{B}$ ($\mathrm{div}\boldsymbol{B}=0$) in a domain D within any surface S on which $\boldsymbol{n}\cdot\boldsymbol{B}=0$ is defined as the integral

$$\mathcal{H}_B = \int_D \boldsymbol{A}\cdot\boldsymbol{B}\,\mathrm{d}V \qquad (1)$$

with $\boldsymbol{A}$ being the vector potential of $\boldsymbol{B}$. Here $\boldsymbol{B}=\mathrm{curl}\boldsymbol{A}$ and the quantity $\mathfrak{h}_B=\boldsymbol{A}\cdot\boldsymbol{B}$ is the helicity density of the field $\boldsymbol{B}$. Both $\mathcal{H}_B$ and $\mathfrak{h}_B$ are pseudoscalars: they change sign under the parity transformation, that is, under change from a right-handed to a left-handed frame of reference. Helicity as defined in Equation (1) is gauge invariant, that is, invariant under replacement of $\boldsymbol{A}$ by $\boldsymbol{A}+\mathrm{grad}\phi$, where ϕ is any single-valued scalar field. This type of invariance should not be confused with time invariance of $\mathcal{H}_B$ in a nondissipative medium (see below). It is important that $\mathcal{H}_B$ is gauge invariant only as defined in Equation (1) for volumes bounded by the field surfaces, that is, $\boldsymbol{n}\cdot\boldsymbol{B}|_s=0$. Generalizations of Equation (1) for open field structures (i.e., such that $\boldsymbol{n}\cdot\boldsymbol{B}|_s\neq 0$) are not gauge invariant. As a rule these generalizations are also not well-defined topologically. Another generalization is the cross-helicity for two solenoidal fields $\boldsymbol{B}_1$ and $\boldsymbol{B}_2$ defined as $\mathcal{H}_{B_1,B_2}=\int_D \boldsymbol{A}_1\cdot\boldsymbol{B}_2\,\mathrm{d}V$. For closed structures, that is, $\boldsymbol{n}\cdot\boldsymbol{B}_{1,2}|_s=0$, cross-helicity is gauge invariant and is symmetric, $\mathcal{H}_{B_1,B_2}=\mathcal{H}_{B_2,B_1}$. Of special interest in physics is a magnetic field, $\boldsymbol{B}$, in electrically conducting fluids or plasmas (with $\mathcal{H}_B$ termed magnetic helicity) and vorticity $\boldsymbol{\omega}$ in fluid flows with $\mathcal{H}_\omega=\int_D \boldsymbol{u}\cdot\boldsymbol{\omega}\,\mathrm{d}V$ (kinetic helicity), where $\boldsymbol{u}$ is the velocity and $\boldsymbol{\omega}=\mathrm{curl}\,\boldsymbol{u}$.

Geometrical and Topological Meaning

Helicity is important at a fundamental level in relation with the linkage(s) of field lines, such as magnetic lines or vortex lines of fluid flow (see Figure 1). The simplest example of two linked vortex tubes is shown in Figure 1 with fluxes $\Phi_{1,2}$ associated with each tube, that are equal to their circulations $\kappa_{1,2}$. Hence, $\mathcal{H}_B$ is directly related to the most basic topological invariant of two

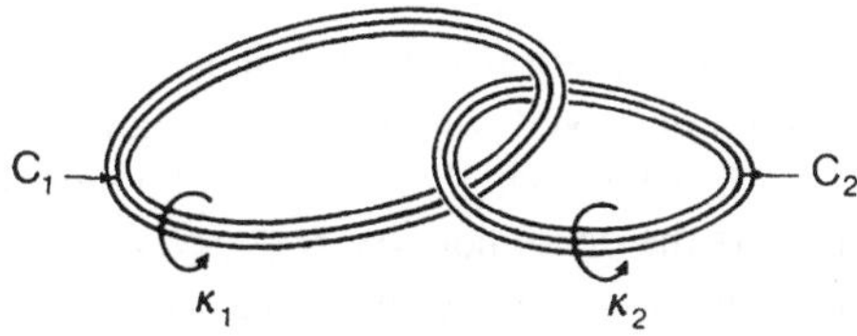

Figure 1. Prototype configuration of linked field tubes with nonzero helicity (Moffatt & Tsinober, 1992). Here the field is assumed to be identical to zero except in two closed tubes with axes $C_{1,2}$ and vanishingly small cross section. The field lines are untwisted within each tube, i.e. each line is a closed curve passing once round the tube, and unlinked with its neighbors in the same tube. In such a case $H_B = \pm 2L_{1,2}F_1F_2$, where $L_{1,2}$ is the linking (or winding) number of the two tubes (in the figure $L_{1,2}=1$), $F_{1,2}$ are the fluxes (equal to their circulations $\kappa_{1,2}$) associated with each tube, and the + or − sign is chosen depending on whether the linkage is right- or left-handed.

linked curves—their linking (winding) number $L_{1,2}$, which was defined by Gauss in 1833 as

$$L_{12} = L_{21} = \frac{1}{4\pi} \oint_{C_1} \oint_{C_2} \frac{\boldsymbol{R} \cdot (\mathrm{d}\boldsymbol{x}_1 \times \mathrm{d}\boldsymbol{x}_2)}{R^3}, \quad (2)$$

where $\boldsymbol{R} = \boldsymbol{x}_1 - \boldsymbol{x}_2$, $\boldsymbol{x}_1 \varepsilon C_1$, and $\boldsymbol{x}_2 \varepsilon C_2$. In the general case of an infinite number of field lines, one can approximate the field via a finite number of N of small flux tubes each with flux Φ_i ($i = 1, \ldots, N$). Then $\mathcal{H}_B = L_{ij}\Phi_i\Phi_j$, where L_{ij} is the linking number of tubes i and j, and summation is assumed over the repeated indices. In other words, the helicity $\mathcal{H}_B$ can be interpreted as the sum of linking numbers of all pairs of the field lines weighted by the flux. This interpretation remains valid also in the case when the field lines are not closed upon themselves and even may wander chaotically, and the integral of Equation (1) is interpreted as the asymptotic Hopf invariant, that is, asymptotic linking number. The cross-helicity $\mathcal{H}_{B,\omega}$ is interpreted as a measure of mutual linkage of the two fields $\boldsymbol{B}$ and $\boldsymbol{\omega}$.

Invariance in Nondissipative Media

The magnetic field, $\boldsymbol{B}$, in a perfectly conducting fluid and vorticity, $\boldsymbol{\omega}$, in an inviscid fluid is governed by the equations

$$\frac{\partial \boldsymbol{B}}{\partial t} = \mathrm{curl}(\boldsymbol{u} \times \boldsymbol{B}), \quad (3)$$

$$\frac{\partial \boldsymbol{\omega}}{\partial t} = \mathrm{curl}(\boldsymbol{u} \times \omega), \quad (4)$$

describing frozen-field transport of the vector fields $\boldsymbol{B}$ and $\boldsymbol{\omega}$. Under this nondissipative evolution both the magnetic helicity $\mathcal{H}_B$ and the helicity of vorticity $\mathcal{H}_\omega$ are conserved. Invariance of helicity is directly associated with invariance of the topology of $\boldsymbol{B}$ and $\boldsymbol{\omega}$.

In cases when the fluid flow is influenced by the presence of the magnetic field, that is, by the Lorenz force (the term $\mathrm{curl}(\boldsymbol{j} \times \boldsymbol{B})$ added to the left-hand side of Equation (4)), the helicity of vorticity $\mathcal{H}_\omega$ is not conserved whereas the magnetic helicity remains conserved. Another conserved quantity in this latter case is the cross-helicity $\mathcal{H}_{B,\omega} = \int \boldsymbol{u} \cdot \boldsymbol{B}\, \mathrm{d}V$. In principle, one can choose the gauge ϕ in such a way that the helicity density $\mathfrak{h}_B = \boldsymbol{A} \cdot \boldsymbol{B}$ is a Lagrangian invariant, that is, it is pointwise conserved along the paths of fluid particles and therefore for any fluid volume. Such a choice is possible both for magnetic field and for nonconducting fluid flows.

Dissipative effects (finite electrical resistivity and viscosity) are responsible for the reconnection of field lines and thus for the evolution of helicity. In such a case, helicity is either created or destroyed just as if $\boldsymbol{n} \cdot \boldsymbol{B} \neq 0$ on the boundary of the domain.

Kinetic Helicity and Turbulent Dynamo

Helicity plays an important role in the process of growth of magnetic fields in electrically conducting fluids due to fluid motion, called the dynamo process. In turbulent flows with nonzero kinetic helicity, there exists the so-called α-effect, the main ingredient of which is that the large-scale (mean) magnetic field results from currents induced by an electromotive force that is (roughly) proportional to the kinetic helicity $\mathcal{H}_\omega$. An important aspect is that this electromotive force is (generally) nonvanishing for nonvanishing electrical resistivity. There is a buildup of magnetic helicity of the generated large-scale magnetic field, and the importance of finite electrical resistivity is in the dissipation of the small scale (on the energy containing scales of fluid turbulence) magnetic helicity of the opposite sign that allows the build up of the large magnetic helicity (Brandenburg, 2001). Recent laboratory experiments with liquid sodium confirm the possibility of generating a magnetic field by helical fluid flows. The growing magnetic field reacts back on the fluid flow via the Lorenz force that results in a decrease of the electromotive force driving the electrical currents (α-quenching), that is, in a nonlinear saturation in the growth of the mean magnetic field.

Kinetic Helicity and Fluid Turbulence

Present knowledge of the effects of helicity on fluid turbulence is controversial. A difficulty in the assessment of the role of helicity in the dynamics of turbulence stems from the fact that—unlike kinetic energy—helicity is not a positively defined quantity. At a speculative level it is expected that nonvanishing helicity comprises a constraint reducing in some sense the nonlinear processes, and thereby turbulence. The simplest argument is that since $(\boldsymbol{u} \cdot \boldsymbol{\omega})^2 + (\boldsymbol{u} \times \boldsymbol{\omega})^2 = \boldsymbol{u}^2\boldsymbol{\omega}^2$, larger $\langle |\boldsymbol{u} \cdot \boldsymbol{\omega}| \rangle$ implies reducing of nonlinear processes associated with $\langle |\boldsymbol{u} \times \boldsymbol{\omega}| \rangle$. Indeed, the so-called Beltrami flows with $\boldsymbol{\omega} = \boldsymbol{\lambda} u$, possessing in some sense maximal helicity, are linear. However, existing evidence weakly supports the above argument (Moffatt & Tsinober, 1992; Tsinober, 2001). On the other hand, if $\langle \boldsymbol{u} \cdot \boldsymbol{\omega} \rangle \neq 0$, this is a clear indication of direct and bidirectional coupling of large and small scales. Therefore, in flows with nonzero mean helicity, the coupling between small and large scales is stronger than otherwise. This stronger coupling may aid creation of large scale structures out of small scale turbulence—a process that is frequently called inverse energy cascade. There is little understanding of these processes, although there are suggestions that in the presence of kinetic helicity in compressible fluids there exists an effect (Hα-effect) analogous to the above-mentioned α-effect for a magnetic field.

More subtle issues include spontaneous breaking of relexional symmetry and emergence of helicity in initially helicity-free flows. The evidence is based on indications from stability analysis, laboratory observations and quantitative measurements of helicity, and direct numerical simulations of Navier–Stokes equations (Kholmyansky et al., 2001). Finally, kinetic helicity has a considerable effect on the effective diffusivity of a passive scalar that may be enhanced by the presence of helicity by more than a factor of two. In addition, helicity leads to a new "skew-diffusion" effect, that is, appearance of a component of turbulent flux of passive scalar perpendicular to the local mean gradient of a passive scalar.

Applications

In atmospheric physics, the so-called storm-relative environmental (kinetic) helicity (SREH) is routinely used in the U.S. with moderate success as a parameter for characterizing, interpreting, and forecasting tornado and supercell rotating storms (Weisman & Rotunno, 2000). The origin of these strucures is not understood, though there are suggestions that the Hα-effect (due to the Coriolis force) and intensive heat transport in turbulent convection are among the important physical mechanisms contributing to their formation (Levina et al., 1999). One has to add to this list at least one more factor, the vertical wind shear.

In plasma physics, magnetic helicity plays a central role in magnetic confinement and in the relaxation of plasmas to a minimum-energy state, for example, in toroidal laboratory plasmas and spheromaks. Interestingly, during this process involving reconnection due to small but finite resistivity and turbulence, the total magnetic helicity is approximately conserved and constrains the energy of the magnetized plasma in equilibrium. This minimum energy state is generally a force-free (Beltrami) field, curl $\boldsymbol{B}=\lambda\boldsymbol{B}$ with λ constant.

In astrophysics, magnetic helicity is important in a number of aspects in solar physics (coronal loops, solar wind), the astrophysical dynamo problem (Brown et al., 1999), and cosmology (primordial magnetic field).

ARKADY TSINOBER

See also **Alfvén waves; Magnetohydrodynamics; Nonlinear plasma waves; Topology; Vortex dynamics of fluids; Winding numbers**

Further Reading

Brandenburg, A. 2001. The inverse cascade and nonlinear alpha-effect in simulations of isotropic helical hydromagnetic turbulence. *Astrophysical Journal*, 550: 824–840

Brown, M.R., Canfield, R.C. & Pevtsov, A.A. 1999. *Magnetic Helicity in Space and Laboratory Plasmas*, Washington, DC: American Geophysical Union

Kholmyansky, M., Shapiro-Orot, M. & Tsinober, A. 2001. Experimental observations of spontaneous breaking of reflexional symmetry in turbulent flows, *Proceedings of the Royal Society of London*, 457: 2699–2717

Levina, G.V., Moiseev, S.S. & Rutkevich, P.B. 1999. Hydrodynamic alpha-effect in a convective system. In *Nonlinear Instability, Chaos and Turbulence*, edited by L. Debnath & D.N. Riahi, Southampton and Boston: WIT Press

Moffatt, H.K. & Tsinober, A. 1992. Helicity in laminar and turbulent flows. *Annual Reviews of Fluid Mechanics*, 24: 281–312

Ricca, R.L. (editor). 2001. *An Introduction to the Geometry and Topology of Fluid Flows*, Dordrecht and Boston: Kluwer

Tsinober, A. 1995. Variability of anomalous transport exponents versus different physical situations in geophysical and laboratory turbulence. In *Levy Flights and Related Topics in Physics*, edited by M. Schlesinger, G. Zaslavsky & U. Frisch, Berlin and New York: Springer

Tsinober, A. 2001. *An Informal Introduction to Turbulence*, Dordrecht and Boston: Kluwer

Weisman, M. & Rotunno, R. 2000. The use of vertical wind shear versus helicity in interpreting supercell dynamics. *Journal of Atmospheric Sciences*, 57: 1452–1472

HÉNON MAP

By the mid-1970s, examples such as the Lorenz equations had convinced researchers that strange attractors could arise in differential equations modeling physical systems. Unfortunately, the length of time needed to compute solutions coupled with strong contraction rates made it very difficult to observe fractal structures numerically in these examples with the computers then available. The Hénon map provided the first simple equation in which this fractal structure is easily observed.

Michel Hénon's approach was based on the idea of return maps. The dynamics of differential equations can be modelled by invertible maps, and so evidence of fractal structure in the attractor of an invertible map shows that these objects can exist in differential equations. The Hénon map is a very simple nonlinear difference equation

$$\begin{aligned} x_{n+1} &= y_n + 1 - ax_n^2, \\ y_{n+1} &= bx_n, \end{aligned} \qquad (1)$$

where the parameters a and b were chosen as $a=1.4$ and $b=0.3$ by Hénon (1976), although other values are also interesting. The attractor, together with some blowups of parts of the attractor, are shown in Figure 1. These pictures are very easy to generate: successive points on an orbit are obtained by evaluating the algebraic expressions on the right-hand side of (1) as the following rough MATLAB program shows:

```
x(1)=0.3; y(1)=0.2;
  %% Initial conditions
N=5000;
  %% N is the number of iterates
a=1.4;b=0.3;
  %% Sets the parameters
for i=1:N
```

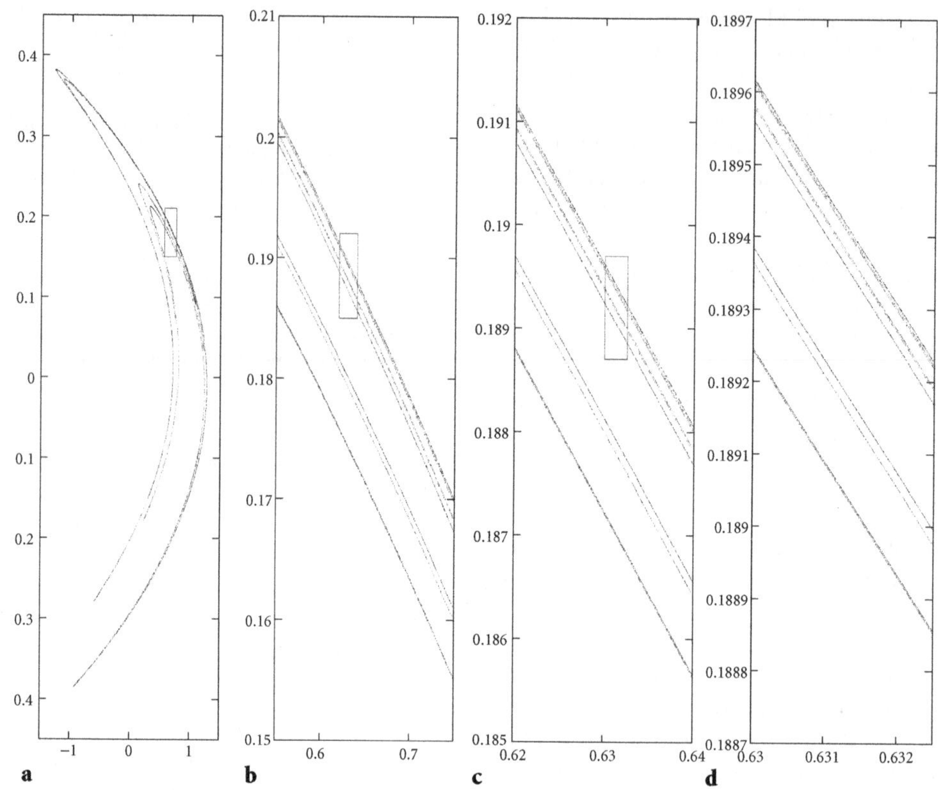

Figure 1. (a) Numerically computed attractor of the Hénon map, (1) with $a = 1.4$ and $b = 0.3$; (b), (c), and (d) are blowups of the boxed regions of (a), (b), and (c), respectively, showing the fractal structure of the attractor. Each figure contains the first 5000 points to land in the displayed region.

```
    %% Begin the iteration loop here
  x(i+1)=y(i)+1-a*x(i)^2;
  y(i+1)=b*x(i);
    %% That calculated the next
    point
  end
  plot(x,y,'k.')
    %% Plot the iterates in the
    %% (x,y) plane
```

Despite the simplicity of this program, Hénon used a mainframe (IBM 7040) to perform the 5 million iterates he needed to get a reasonable number of points in the equivalent of Figure 1(d). Figure 1 uses a slightly more sophisticated program to generate the attractor and zoom in on the rectangular regions indicated so that 5000 points can be plotted in each of the blowup regions. This involved 35,724,657 iterations of the map order to get 5000 points in the smallest blowup region of Figure 1(d). This level of computation would have been almost unthinkable when Hénon wrote his paper.

The Hénon attractor pictured in Figure 1 is computationally cheap, requiring no more than the most simple algebraic operations. Also, the numerical evidence for fractal structure in the attractor is sufficiently convincing that most researchers have come to accept that it is a strange attractor or, at least, to suspend their disbelief. For this reason it has become a canonical example of chaotic motion. Almost every new technique or relevant theoretical result is applied to the Hénon map as part of the evaluation of the method. Early papers on phase space reconstruction, dimension calculations, chaotic prediction, chaotic control and synchronization, periodic orbit expansions, invariant measure algorithms, etc., have all used the Hénon map as an important test example. Given this general level of acceptance, it may come as a surprise to learn that it is still not known whether there really is a strange attractor for the Hénon map at the standard parameter values ($a = 1.4$, $b = 0.3$).

Hénon (1976) gave a number of reasons for looking at orbits of (1):

> ... we try to find a model problem which is as simple as possible, yet exhibits the same essential properties as the Lorenz system. Our aim is (i) to make the numerical exploration faster and more accurate...; (ii) to provide a model which might lend itself more easily to mathematical analysis.

As we have seen, Hénon's aim of making the numerical exploration of apparently chaotic attractors more straightforward succeeded spectacularly. However, he could not have imagined how hard it would be to answer the theoretical questions posed by this deceptively simple map.

Hénon's intuitive explanation for his map in terms of folding, stretching, and contraction is much closer to the formation of horseshoes rather than to the Lorenz model, which has discontinuities in the natural return map. As such, the Hénon map has become a paradigm for the formation of horseshoes as parameters vary (see Devaney & Nitecki, 1979), and it is the more general question of how the attractors of the Hénon map change as the parameters vary that has occupied most theoretical approaches.

By defining a new y variable $y_{\text{new}} = b^{-1} y_{\text{old}}$, Equation (1) can be written in the form of a more general, Hénon-like map:

$$\begin{aligned} x_{n+1} &= -\varepsilon y_n + f_a(x_n), \\ y_{n+1} &= x_n, \end{aligned} \qquad (2)$$

where $\varepsilon = -b$ and $f_a(x) = 1 - ax^2$ gives the Hénon map in the new coordinates. This formulation emphasizes the relationship between Hénon-like maps and one-dimensional maps: if $\varepsilon = 0$ then the x equation decouples and x evolves according to the one-dimensional difference equation $x_{n+1} = f_a(x_n)$, which in the original case, (1), is just the standard quadratic family. The Jacobian of the map is ε, so positive ε corresponds to orientation-preserving maps, which is more natural in the context of return maps, although this means that $b < 0$ in Hénon's original formulation, (1). Early efforts toward proving that strange attractors exist in the Hénon map concentrated on extending results for one-dimensional maps to the two-dimensional case with $\varepsilon > 0$ small. On the negative side, Holmes & Whitley (1984) showed that however small ε is, some periodic orbits of the Hénon map appear in a order different from the order in which they appear in the quadratic family. On the positive side, Gambaudo, van Strien, & Tresser (1989) showed that for sufficiently small $\varepsilon > 0$, the first complete period-doubling cascade is associated with the original period two orbit.

The major breakthrough on the existence of strange attractors was made by Benedicks & Carleson (1991). Using delicate mathematical analysis, they were able to show that if $\varepsilon > 0$ is small enough and a is close to $a = -2$ (the equivalent of $\mu = 4$ for the standard formulation of the quadratic map, $\mu x(1 - x)$), then there is a positive measure of parameter values for which the Hénon map has a strange attractor. This result was generalized by Mora & Viana (1993) who showed that Hénon-like maps arise naturally near homoclinic bifurcations of maps. It had long been recognized that these bifurcations occur in the Hénon map (see, for example, Holmes & Whitley (1984)), so this made it possible to deduce the existence of strange attractors at values of ε that are not small. Indeed, this important paper provides a method of proving the existence of strange attractors for a set of parameter values with positive measure in a wide variety of model systems. Despite all these advances, these results only prove that there exist such parameter values and they do not give methods for proving that a strange attractor exists at a *given* parameter value.

Two other avenues of research suggested by the one-dimensional, $\varepsilon = 0$, limit in (2) have led to interesting developments. We have already noted that the one-dimensional order of periodic orbits is not preserved if $\varepsilon > 0$. However, a beautiful theory of partial orders based on period and knot type has emerged, which shows that the existence of some periodic orbits implies the existence of some others in two-dimensional maps (see Boyland (1994) for more details). The second adaptation of one-dimensional approaches is based on the idea of the symbolic dynamics (or *kneading theory*) of unimodal maps. The main idea, introduced by Cvitanovic et al. (1988) and developed by de Carvalho (1999), is to produce symbolic models of the dynamics of the Hénon map by relating the dynamics to modifications of the full horseshoe. This is done by *pruning* the horseshoe, that is, identifying regions of the horseshoe with dynamics that are not present in the Hénon map under consideration and judiciously removing these regions together with their images and preimages, leaving a pruned horseshoe that can still be accurately described.

Much of the current interest in the Hénon map involves the existence and construction of invariant measures for the attractors. This work should lead to a good statistical description of properties of orbits and averages along orbits. So, even now, this simple two-dimensional map with a single nonlinear term is motivating important questions in dynamical systems.

PAUL GLENDINNING

See also **Attractors; Bifurcations; Chaotic dynamics; Horseshoes and hyperbolicity in dynamical systems; Markov partitions; One-dimensional maps; Routes to chaos; Sinai–Ruelle–Bowen measures; Symbolic dynamics**

Further Reading

Benedicks, M. & Carleson, L. 1991. The dynamics of the Hénon map. *Annals of Mathematics*, 133: 73–169

Boyland, P. 1994. Topological methods in surface dynamics. *Topology and Its Applications*, 58: 223–298

de Carvalho, A. 1999. Pruning fronts and the formation of horseshoes. *Ergodic Theory and Dynamical Systems*, 19: 851–894

Cvitanovic, P., Gunaratne, G. & Procaccia, I. 1988. Topological and metric properties of Hénon-type strange attractors. *Physical Review* A, 38: 1503–1520

Devaney, R. & Nitecki, Z. 1979. Shift automorphisms in the Hénon mapping. *Communications in Mathematical Physics*, 67: 137–146

Gambaudo, J.-M., van Strien, S. & Tresser, C. 1989. Hénon-like maps with strange attractors: there exist C^∞

Kupka–Smale diffeomorphisms on $\mathbf{S}^2$ with neither sinks nor sources. *Nonlinearity*, 2: 287–304

Hénon, M. 1976. A two-dimensional mapping with a strange attractor. *Communications in Mathematical Physics*, 50: 69–77

Holmes, P. & Whitley, D. 1984. Bifurcations of one- and two-dimensional maps. *Philosophical Transactions of the Royal Society (London)* A, 311: 43–102

Mora, L. & Viana, M. 1993. Abundance of strange attractors. *Acta Mathematica*, 171: 1–71

HÉNON–HEILES SYSTEM

Although the molecules of a gas move freely and collide frequently, the stars in a galaxy move under the constraint of long-range interactions and do not collide. The motion of any one star in a galaxy can be modeled as if it moves alone under the constraint of a fixed gravitational galactic potential with cylindrical symmetry. The energy of the star is conserved as is the angular momentum with respect to an axis through the center of the galactic plane and at right angles to it. These two conserved quantities are isolating integrals of the motion. In the absence of a third isolating integral, the distribution of stellar velocities in a galaxy with an axisymmetric potential should be circularly symmetric in the meridian plane, a result contrary to observation (Ollongren, 1965).

The astronomer Michel Hénon and a graduate student Carl Heiles introduced a mathematical model to numerically investigate the third integral problem when Hénon visited Princeton University in 1962. In "order to have more freedom of experimentation" in their numerical studies, they considered the simplified model system, with two degrees of freedom, described by the Hamiltonian (Hénon & Heiles, 1964)

$$H = \tfrac{1}{2}\left(\dot{x}^2 + \dot{y}^2\right) + \tfrac{1}{2}\left(x^2 + y^2\right) + x^2y - \tfrac{1}{3}y^3,$$

where the last three terms are called V. The choice of coefficients in the cubic potential confers a ternary symmetry on the equipotential lines. The equipotential line for $V = \frac{1}{6}$ is an equilateral triangle, and this is the limiting energy at which the motion is bounded. In the Hénon–Heiles model, the original problem of the third isolating integral is transformed into whether there exists an isolating integral other than the constant energy itself. The Hénon–Heiles Hamiltonian also describes the center of mass reduction of a three-particle periodic chain with linear and quadratic nearest neighbor elastic forces and is, thus, a member of the class of systems investigated in the numerical studies of Fermi, Pasta, and Ulam (Ford, 1992). More generally, the Hénon–Heiles Hamiltonian is one of the simplest non-integrable Hamiltonian systems known. It has become a paradigm for studies of chaos and non-integrability in Hamiltonian systems with two degrees of freedom.

The solutions of the equations of motion corresponding to the Hénon–Heiles Hamiltonian can be represented by trajectories in the four-dimensional phase space spanned by $(x, \dot{x}, y, \dot{y})$. If the energy is fixed in the range $0 \le E \le \frac{1}{6}$ then these trajectories are bounded and recurrent on a three-dimensional energy surface. No techniques are known to construct exact algebraic formulae for the trajectories for arbitrary initial conditions. Approximate solutions can be obtained using perturbation methods or numerical methods. In their numerical studies, Hénon and Heiles plotted the points of intersection between phase-space trajectories and a two-dimensional surface of section in the phase-space. For recurrent trajectories, successive intersection points on the surface of section define an area-preserving two-dimensional discrete time dynamical system or map that shares the important dynamical features of the original system (Poincaré, 1993).

A convenient Poincaré map for the Hénon–Heiles Hamiltonian at fixed energy, in the range $0 \le E \le \frac{1}{6}$, is obtained with the surface of section $x = 0$ and $\dot{x} \ge 0$. The coordinate pair $(y, \dot{y})$ in this case uniquely determines an initial point for a trajectory since $\dot{x} = \sqrt{2E - \dot{y}^2 - y^2 + 2y^3/3}$. Thus, the dynamical description is reduced to a study of the Poincaré map from one intersection point $(y_n, \dot{y}_n)$ to the next intersection point $(y_{n+1}, \dot{y}_{n+1})$. The existence of an additional isolating integral in the Hénon–Heiles Hamiltonian would restrict a trajectory to intersect an invariant curve in the surface of section. In the absence of an additional isolating integral, the pattern of intersection points should fill an area on the surface of section. In numerical studies, where the number of intersection points is finite, discrimination between these two signatures is subjective since any finite set of points can be aligned to a smooth curve. Despite this, the interpretation of the intersection points as invariant curves or filled area in the Hénon–Heiles study has been regarded as unambiguous.

Figure 1 shows the intersection points between numerical phase space trajectories and a Poincaré surface of section in the Hénon–Heiles system at four different values of the energy, $E = \frac{1}{24}, \frac{1}{12}, \frac{1}{8}$, and $\frac{1}{6}$.

At $E = \frac{1}{24}$ and $E = \frac{1}{12}$, the intersection points appear to lie on invariant curves. At $E = \frac{1}{8}$, intersection points for some trajectories appear to lie on invariant curves whereas intersection points for other trajectories explore areas on the surface of section. At $E = \frac{1}{6}$, intersection points for most trajectories do not appear to lie on invariant curves (and typically the separation distance between trajectories started from nearby initial conditions diverges exponentially in time, a signature of chaos). A feature that is not revealed in the surface of section portraits at $E = \frac{1}{6}$, but is clear in three-dimensional phase space portraits (Henry & Grindlay, 1994), is intermittency whereby a single trajectory switches in time between regular behavior (close to invariant curves) and irregular behavior (filling areas).

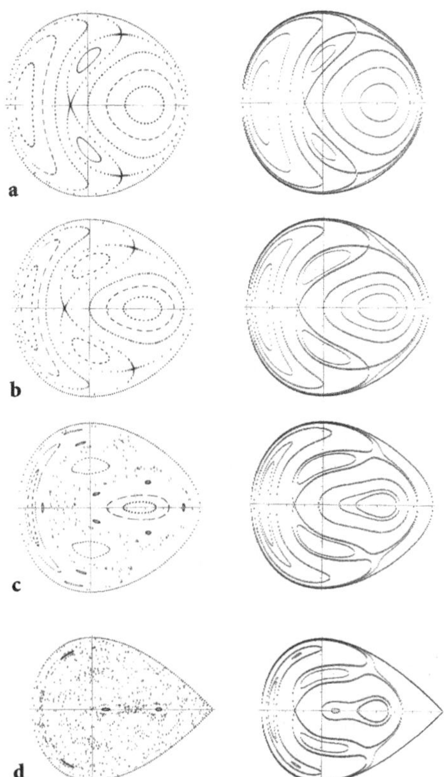

Figure 1. Surface of section portraits for the Hénon–Heiles Hamiltonian at four different values of the energy; (a) $E = 1/24$, (b) $E = 1/12$, (c) $E = 1/8$, (d) $E = 1/6$. The figures on the left are from numerical integrations and the figures on the right are from Birkhoff–Gustavson normal form theory. The figures have been reproduced from Gustavson (1966).

The transition from regular behavior to irregular behavior as the system energy is increased in the Hénon–Heiles model is characteristic of a more general transition from regular behavior to irregular behavior in Hamiltonian systems with two degrees of freedom and in two-dimensional area-preserving maps. A detailed theoretical understanding of this phenomena encompasses collective results from the Birkhoff–Gustavson normal form theory, the Kolmogorov–Arnol'd–Moser (KAM) theory, the Poincaré–Birkhoff theory, the Aubry–Mather theory, the theory of heteroclinic tangles (Poincaré, 1993), and the theory of overlapping resonances (Ford, 1992).

The Birkhoff–Gustavson normal form (Gustavson, 1966) represents the Hamiltonian as an infinite series of harmonic oscillator Hamiltonians through an infinite series of successive canonical transformations. The Hénon–Heiles Hamiltonian is strictly non-integrable at all values of the system energy (Ito, 1985); thus the normal form series for this Hamiltonian does not converge. On the other hand, any finite truncation of the normal form series constitutes an integrable Hamiltonian that can be considered to be "close" in some sense to the Hénon–Heiles Hamiltonian. The additional conserved quantity for an integrable normal form Hamiltonian can be readily calculated and used to plot the associated invariant curves on the Poincaré surface of section. The KAM theory proves that for a non-integrable Hamiltonian sufficiently close to an integrable Hamiltonian, most of the phase space trajectories are confined to invariant tori, similar to the invariant tori of the integrable system. Thus the surface of section portraits for the Hénon–Heiles Hamiltonian should map out the invariant curves from the truncated normal form Hamiltonian whenever this Hamiltonian is sufficiently close to the Hénon–Heiles Hamiltonian. The difference between the truncated normal form series and the Hénon–Heiles Hamiltonian increases with system energy. This can be seen in Figure 1 where the invariant curves from an eighth order truncation (Gustavson, 1966) are compared side by side with the numerical results for the Hénon–Heiles Hamiltonian.

In Hamiltonian systems with two degrees of freedom, the tori are characterized by a winding number ω_1/ω_2. The KAM theory proves the existence of invariant tori whose winding numbers are sufficiently irrational that

$$\left|\frac{\omega_1}{\omega_2} - \frac{r}{s}\right| > \frac{k(\varepsilon)}{s^{5/2}}$$

where r and s are co-prime integers and $k(\varepsilon)$ approaches zero as ε approaches zero. Here ε measures the strength of the non-integrable perturbation on the integrable Hamiltonian. The volume of phase-space not filled by invariant tori approaches zero as ε approaches zero. Tori whose winding numbers do not satisfy the stability condition under perturbation breakup according to the Poincaré–Birkhoff theory (rational winding numbers) or the Aubry–Mather theory (irrational winding numbers). The numerical results for the Hénon–Heiles Hamiltonian are consistent with an increasing breakup of tori with increasing energy as less and less tori satisfy the stability criteria.

The motion of an asteroid around the Sun can also be described approximately by a simplified model Hamiltonian with two degrees of freedom. The simplest non-trivial model is a plane circular restricted three-body problem involving the asteroid, the Sun, and Jupiter (Berry, 1978). The two-body motion of the asteroid around the Sun is a Kepler ellipse with frequency ω_A, and the two-body motion of Jupiter about its center of mass with the Sun is taken to be a circle with frequency ω_J. In the restricted three-body problem the effect of the asteroid on the motion of the Sun and Jupiter is neglected, the motion of the asteroid is dominated by the Sun, and the motion of Jupiter is considered to be a perturbation on the motion of the asteroid. The Hamiltonian for this restricted three-body problem is non-integrable, and an application of KAM theory to the problem identifies the frequency ratio ω_A/ω_J with the winding number for invariant tori. Thus, asteroids should be expected to be found in stable orbits at distances from the Sun where ω_A/ω_J is sufficiently irrational. Indeed, as had been noted by Daniel Kirkwood in 1866, there is an abundance of asteroids at these locations but gaps in the asteroid belt at locations where the frequency ratio is 3:1, 5:2, 7:3, or 2:1; however a detailed understanding of the Kirkwood gaps is still an area of active research (Ferraz-Mello, 1999).

BRUCE HENRY

See also **Kolmogorov–Arnol'd–Moser theorem; N-body problem; Normal forms theory; Poincaré theorems; Solar system**

Further Reading

Berry, M.V. 1978. Regular and irregular motion. In *Topics in Nonlinear Dynamics*, edited by S. Jorna, New York: American Institute of Physics, pp.16–120

Contopoulos, G. 1963. On the existence of a third integral of motion. *The Astronomical Journal*, 68(1): 1–14

Ferraz-Mello, S. 1999. Slow and fast diffusion in asteroid-belt resonances: a review. *Celestial Mechanics & Dynamical Astronomy*, 73(1–4): 25–37

Ford, J. 1992. The Fermi–Pasta–Ulam problem: paradox turns discovery. *Physics Reports*, 213(5): 271–310

Gustavson, F.G. 1966. On constructing formal integrals of a Hamiltonian system near an equilibrium point. *The Astronomical Journal*, 71(8): 670–686

Gutzwiller, M.C. 1990. *Chaos in Classical and Quantum Mechanics*, New York: Springer

Hénon, M. & Heiles, C. 1964. The applicability of the third integral of motion: some numerical experiments, *The Astronomical Journal*, 69(1): 73–99

Henry, B.I. & Grindlay, J. 1994. From dynamics to statistical mechanics in the Hénon–Heiles model: dynamics. *Physical Review* E, 49(4): 2549–2558

Ito, H. 1985. Non-integrability of Hénon-Heiles system and a theorem of Ziglin. *Kodai Mathematical Journal*, 8: 120–138

Ollongren, A. 1965. Theory of stellar orbits in the galaxy. *Annual Review of Astronomy and Astrophysics*, 3: 113–134

Poincaré, H. 1993. *New Methods of Celestial Mechanics, vol. 3. Integral Invariants and Asymptotic Properties of Certain Solutions*. Woodbury, New York: American Institute of Physics. (English translation of *Les Méthodes nouvelles de la Mécanique Céleste, vol. 3: Invariants integraux. Solutions periodiques du deuxieme genre Solutions doublement asymptotiques*, 1899)

HETEROCLINIC INTERSECTION

See **Phase space**

HETEROCLINIC TRAJECTORY

See **Phase space**

HIERARCHIES OF NONLINEAR SYSTEMS

Although we must deal with many hierarchical structures in the course of life—social, military, and monetary, among others—two are of particular interest to practitioners of nonlinear science: the biological and cognitive hierarchies.

Biological Hierarchy

It is an empirical fact that living creatures are hierarchical in structure, organized according to a scheme that is roughly as follows.

Biosphere
Species
Organisms
Organs
Cells
Processes of replication
Genetic transcription
Biochemical cycles
Biomolecules
Molecules
Atoms

Several comments are in order. First, it is only the general nature of this hierarchy that is of interest here, not the details. Second, the nonlinear dynamics at each level of description generate emergent structures, and nonlinear interactions among these structures provide a basis for the dynamics at the next higher level (Scott, 2003). Thus, molecules emerge from the nonlinear forces among their constituent atoms, biomolecules emerge from molecules, biochemical cycles (such as the Krebs cycle, which processes energy in a living cell) emerge from interactions among biomolecules, and so on. Third, the emergence of a new dynamic entity usually involves presence of a closed causal loop, which leads to positive feedback and exponential growth that is ultimately limited by nonlinear effects.

Finally, it should be noted that philosophers disagree about the ontological status of an emergent entity. Are the levels mere designations, convenient for academic organization, or do they mark qualitatively different realms of reality?

In attempting to answer this question, it is important to know whether the upper levels of the biological hierarchy can be derived from lower levels, which leads to the concept of reductionism. In other words, can the upper levels of the hierarchy (organisms, species, biosphere) be logically derived from the properties of the lower levels? Or are they in some sense independent? In response to such questions, condensed matter physicist Philip Anderson has commented (Anderson, 1972): "The ability to reduce everything to simple fundamental laws does not imply the ability to start from those laws and reconstruct the universe."

Why not? Like the flame of a candle, the Belousov–Zhabotinsky reaction, or a nerve impulse, the dynamics of biological systems arise from nonlinear reaction-diffusion processes, in which the particular atoms participating in the reactions change with time. Thus, exact knowledge of the positions and velocities of the constituent atoms of an organism at one time tells little about the behavior at a later time (Bickhard & Campbell, 2000; Scott, 2002). Of course, a dedicated reductionist would respond that the explanatory net has not been cast far enough—all the atoms and radiation that wander into and out of the organism must be followed—but few seriously propose to do this as it would require computing in detail the dynamics of the entire universe.

Cognitive Hierarchy

Just as the phenomenon of life arises from the hierarchical nature of the biological realm, it has been suggested that the mysteries of mind may stem from an analogous hierarchy, describing levels of activity of the brain. Again ignoring details, this cognitive hierarchy may be sketched as follows.

Human cultures
Phase sequences
Complex assemblies
...
Assemblies of assemblies of assemblies
Assemblies of assemblies
Assemblies of neurons
Neurons
Nerve impulses
Nerve membranes
Membrane proteins
Molecules
Atoms

To the previous comments on the biological hierarchy, the following can be added. Although the biological levels are observed—through x-ray diffraction, via electron and light microscopy, or more directly—several levels of the cognitive hierarchy are theoretical constructs. This is particularly so for the cell assemblies and their constituent subassemblies, which Donald Hebb proposed to underlie the dynamic nature of human thought (Hebb, 1949). Thus a particular complex cell assembly that embodies a particular idea (your memory of your grandmother, for example) might involve interactions among several (visual, auditory, conceptual) subassemblies as suggested by Figure 1.

In this figure, the shaded areas do not suggest that all neurons in that region are firing, but only those (perhaps ten thousand or so) that have become specific to this particular memory through many related experiences and interactions. The neurons of this assembly are interconnected in such a manner that they are "capable of interacting briefly as a closed system, delivering facilitation to other such systems and usually having a specific motor facilitation" (Hebb, 1949). As a complex assembly comprises subassemblies that in turn are composed of yet more basic constituent assemblies, the concept is inherently hierarchical.

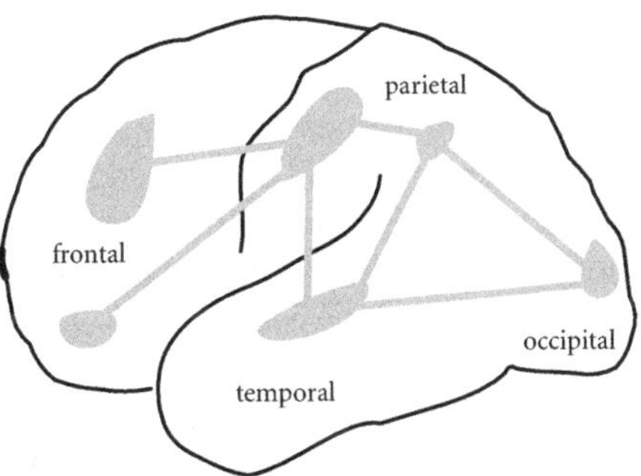

Figure 1. A sketch of the left side of the human brain, suggesting how the active subassemblies (shaded areas) of a complex cell assembly might be distributed over various lobes of the neocortex.

Although Hebb's theory is in accord with a substantial amount of psychological data (Hebb, 1980; Scott, 2002), it is difficult to imagine placing electrodes in the neocortex so that the simultaneous firings of a substantial number of these assembly neurons could be recorded. Nontheless, a substantial amount of experimental and numerical data is accumulating in support of this theory.

At higher levels of the cognitive hierarchy, we recognize the phase sequence, in which the focus of thought moves from one complex assembly to the next in a train of thought or a dream. At yet higher levels, important aspects of cultural dynamics—ingrained social prejudices, for example, or an aversion to eating certain foods—may remain unrecognized as they play important roles in human behavior.

ALWYN SCOTT

See also **Cell assemblies; Electroencephalogram at large scales; Emergence; Neural network models**

Further Reading

Anderson, P.W. 1972. More is different: broken symmetry and the nature of the hierarchical structure of science. *Science*, 177: 393–396

Bickhard, M.H. & Campbell, D.T. 2000. Emergence. In *Downward Causation: Minds, Bodies and Matter*, edited by P.B., Andersen, C. Emmeche, N.O. Finnemann & P.V. Christiansen, Aarhus, Denmark: Aarhus University Press

Hebb, D.O. 1949. *Organization of Behavior: A Neuropsychological Theory*, New York: Wiley

Hebb, D.O. 1980. The structure of thought. In *The Nature of Thought*, edited by P.W. Jusczyk & R.M. Klein, Hillside, NJ: Lawrence Erlbaum Associates, pp. 19–35

Scott, A.C. 1995. *Stairway to the Mind*, New York: Springer

Scott, A.C. 2002. *Neuroscience: A Mathematical Primer*, Berlin and New York: Springer

Scott, A.C. 2003. *Nonlinear Science: Emergence and Dynamics of Coherent Structures*, 2nd edition, Oxford and New York: Oxford University Press

HIGGS BOSON

Particle physics is devoted to discovering and investigating the fundamental constituents of all the matter in the universe (at all times), and the forces of nature by which these constituents interact with each other. The theoretical framework that currently aids our understanding is called the Standard Model. Although there is no experimental result that significantly disagrees with the Standard Model, it is not a complete theory; indeed, there are several reasons why the Standard Model is not a theory at all in the normal sense of the word. Perhaps the most important of these is that it contains a large number of parameters (such as the fundamental particle masses) that are not predicted but determined experimentally and then incorporated into the theoretical framework.

A principle mission of particle physics is the development of a complete, fundamental theory that unifies all of the known forces into one. Historically, this process seems to be taking place one pair of forces at a time (for example, James Clerk Maxwell famously unified electricity and magnetism in the 19th century), and an important feature of the Standard Model is that it contains the unification of the short-range weak nuclear force and the long-range electromagnetic force. The strong nuclear force is not yet unified and gravity is completely excluded from the standard model.

Although the successful implementation of electroweak unification encourages particle physicists to consider the Standard Model as a low energy approximation of a more fundamental theory, electroweak unification comes at a price. The quantum of the electromagnetic field is the photon, which has zero rest mass. The weak field has a chargeless quantum, known as the Z particle, and a charged quantum, known as the W particle. The rest masses of the Z and the W are both almost 100 times the proton rest mass. This huge difference in the masses of the particles associated with the electromagnetic and weak force fields is known as electroweak symmetry breaking, and in order to interpret it, physicists have introduced into the Standard Model an arbitrary mathematical trick.

In a series of papers some 40 years ago, Englert and Brout (and independently, Higgs) pointed out that a symmetry-breaking idea used in the theory of superconductivity could also be used to break the electroweak symmetry. This mathematical trick is known in particle physics as the Higgs mechanism. In the theory of superconductivity, the idea explains how photons appear to acquire mass inside a superconductor. Introducing it into the Standard Model provides an explanation of where the Z and W masses come from.

The Higgs mechanism is most readily viewed in terms of a constant, ubiquitous energy field that has no preferred direction in space. This field is known as the Higgs field. It is this Higgs field that breaks the electroweak symmetry, giving mass to both Z and W while retaining a massless photon. Just as there are well-defined quanta, known as bosons, associated with each of the known force fields, so it is with the hypothetical Higgs field. In the simplest version of the Higgs mechanism, there is just one quantum and it is known as the Higgs particle or Higgs boson.

So, to recap, the Higgs particle is intimately associated with the Higgs field, and the Higgs field is assumed to be responsible for the mass difference between the quanta of the electromagnetic and weak fields. It is not too much of a stretch to expect that the masses of all matter particles (quarks and leptons) somehow get their masses from the Higgs mechanism. Indeed, the coupling of every particle to the Higgs field has strength proportional to its mass, and so this interpretation seems natural. Unfortunately, the Higgs mechanism does not reduce the number of unknown parameters of the Standard Model—there is still one per particle, plus a few more.

Curiously, the assumed Higgs particle has quite well-determined properties. Its mass, like all other particle masses, is not predicted but must be measured. As a function of an assumed mass, however, one can use the Standard Model to calculate the rate at which it will be produced in a high-energy collision of two elementary particles. The Higgs particle is expected to be highly unstable, and it is also straightforward to calculate, using the Standard Model, how it will decay. In other words, what it will decay into and with what relative rates. These Standard Model predictions provide a basis for experimental searches for the Higgs particle.

The particle experimentalists perform experiments in which they collide, at the highest available energy, protons with protons, or protons with antiprotons, or electrons with positrons. They then examine the debris of the collision for evidence of a Higgs particle whose properties match those suggested by the Standard Model.

In this manner, the experimental techniques of particle physics have been able, over the past decade or so, to eliminate the possibility of a Higgs particle with mass below about 120 proton masses or above about 200 proton masses. When the Higgs mechanism was first introduced to particle physics, the mass of the Higgs particle could have been anywhere between zero and about 1000 times the proton mass. One of the remarkable achievements of particle physics over the past few decades has been the elimination of most of this mass region. Of course, this elimination is of a statistical nature and the best thing that physicists can do is to assign a measure of the statistical confidence to these limits. In late 2002, most particle physicists were confident that if the Standard Model Higgs particle exists, its mass is probably in the mass range of approximately 120–200 proton masses (somewhere between the atomic masses of tin and gold). The Tevatron (proton-antiproton collider accelerator at Fermilab with energy close to 2 TeV) has a chance to investigate the lower 10% of this mass range; the Large Hadron Collider (LHC; proton-proton collider accelerator under construction at CERN with energy 14 TeV, scheduled to start operation in or after 2007) will finish the job. Thus, by 2020 at the latest, we will know whether there is a Higgs particle, and therefore whether Nature uses the Higgs mechanism to generate the masses of fundamental particles.

STEPHEN REUCROFT

See also **Born–Infeld equations; Matter, nonlinear theory of; Particles and antiparticles; Skyrmions; String theory**

Further Reading

Englert, F. & Brout, R. 1964. Broken symmetry and the mass of the gauge vector mesons. *Physical Review Letters*, 13(9): 321

Gunion, J., Haber, H., Kane, G. & Dawson, S. 2000. *The Higgs Hunter's Guide*, Reading, MA: Perseus

Halzen, F. & Martin, A. 1984. *Quarks and Leptons*, New York: Wiley

Higgs, P. 1964. Broken symmetries and the masses of gauge bosons. *Physical Review Letters*, 13(16): 508

Higgs, P. 1964. Broken symmetries, massless particles and gauge fields. *Physics Letters*, 12(2): 132

Kane, G. 1993. *Perspectives on Higgs Physics*, River Edge, NJ: World Scientific

HILL'S EQUATION

See **Periodic spectral theory**

HIROTA'S METHOD

Ryogo Hirota originally proposed his bilinear method as a "direct method" for constructing multisoliton solutions to integrable equations. The method is "direct" in the sense that one can apply it without resorting to deeper mathematical properties, for example, to those that are needed in order to apply the inverse scattering transform (IST). In practice, this means that Hirota's method is applicable to a wider class of equations. Nevertheless, the bilinear method is related to deep mathematical foundations of soliton theory (Sato theory), as has been shown by the members of the Kyoto school (Sato, Date, Jimbo, Miwa, Kashiwara).

Prototypical Result: The KdV Equation

Let us see how Hirota's bilinear method works for the Korteweg–de Vries (KdV) equation

$$u_{xxx} + 6uu_x + u_t = 0. \tag{1}$$

It is known from IST that multisoliton solutions can be written as $u = 2\partial_x^2 \log(\det M)$, where the entries in M are of the form $c_0 + c_1 e^{c_2 x + c_3 t}$ for some constants c_i. Thus, $\det M$ is a sum of exponentials and, therefore, an entire function. This inspired Hirota to write the KdV equation in terms of a new dependent variable F defined by (Hirota, 1971)

$$u = 2\partial_x^2 \log F. \tag{2}$$

Then the KdV equation (1) becomes

$$\partial_x[F_{xxxx}F - 4F_{xxx}F_x + 3F_{xx}^2 + F_{xt}F - F_xF_t] = 0. \tag{3}$$

Note the derivative outside the brackets, which means that the bilinearization actually works most naturally for the potential KdV equation.

In order to shorten the notation, let us introduce the bilinear derivative operator D as follows:

$$\begin{aligned} D_x^n f \cdot g &= (\partial_{x_1} - \partial_{x_2})^n f(x_1)g(x_2)\big|_{x_2=x_1=x} \\ &\equiv \partial_\varepsilon^n f(x+\varepsilon)g(x-\varepsilon)\big|_{\varepsilon=0}. \end{aligned} \tag{4}$$

Then the KdV equation may be written in the simple form

$$(D_x^4 + D_xD_t)\, F \cdot F = 0. \tag{5}$$

Since F was assumed to be a sum of exponentials, let us try to construct a one-soliton solution (1SS) to (5) using the ansatz

$$F = 1 + e^\eta, \quad \eta = px + \omega t + \eta^0. \tag{6}$$

A direct computation shows that this is indeed a solution, provided that $\omega = -p^3$; this condition on the parameters is called the *dispersion relation*. From (2) we get the usual soliton solution

$$u = \frac{2p^2 e^\eta}{(1+e^\eta)^2} = \frac{p^2}{2\cosh^2(\eta/2)}, \tag{7}$$

where $\eta = px - p^3 t + \eta^0$.

A particularly nice feature of the bilinear method is that two-soliton solutions (2SSs) are also very easy to obtain. We assume that a 2SS is constructed from two 1SSs perturbatively as follows:

$$F = 1 + e^{\eta_1} + e^{\eta_2} + a_{12}e^{\eta_1+\eta_2}, \tag{8}$$

where $\eta_i = p_i x - p_i^3 t + \eta_i^0$.

A direct computation then shows that this is indeed a solution of (5), provided that the *phase factor* is given by

$$a_{12} = \left(\frac{p_1 - p_2}{p_1 + p_2}\right)^2. \tag{9}$$

Using (2) one gets the nonlinear form of the 2SS, which is already rather complicated. The above 1SS and 2SS ansätze actually work (with suitable a_{12}) for *any* equation having a bilinear form of type $P(D)F \cdot F = 0$, with $P(0) = 0$.

The above can be extended to a 3SS. By considering various asymptotic limits (where one of the three solitons is far away), one finds that the proper ansatz cannot have new free coefficients, but must have the form

$$F = 1 + e^{\eta_1} + e^{\eta_2} + e^{\eta_3} + a_{12}e^{\eta_1+\eta_2} + a_{13}e^{\eta_1+\eta_3} + a_{23}e^{\eta_2+\eta_3} + a_{12}a_{13}a_{23}e^{\eta_1+\eta_2+\eta_3}. \quad (10)$$

In contrast to the 2SS case, this ansatz only works for *integrable* equations of type $P(D)F \cdot F = 0$, for example, the KdV, Sawada–Kotera, and Kadomtsev–Petviashvili equations. In fact, the existence of multisoliton solutions can be used as a criterion of integrability (Hietarinta, 1990).

Examples of Bilinear Equations

Let us next consider the nonlinear Schrödinger equation (NLS)

$$i\phi_t + \phi_{xx} + 2|\phi|^2\phi = 0. \quad (11)$$

It is bilinearized with the substitution

$$\phi = G/F, \quad G \text{ complex, } F \text{ real,} \quad (12)$$

which yields

$$F\,[(iD_t + D_x^2)G \cdot F] - G\,[D_x^2 F \cdot F - 2|G|^2] = 0. \quad (13)$$

For normal (bright) solitons we split this into two equations for two functions:

$$\begin{cases} (iD_t + D_x^2)G \cdot F = 0, \\ D_x^2 F \cdot F = 2|G|^2. \end{cases} \quad (14)$$

The 1SS ansatz is given by

$$G = e^{\eta}, \quad F = 1 + a\,e^{\eta+\eta^*}, \quad \eta = px + \omega t, \quad p \text{ and } \omega \text{ complex} \quad (15)$$

and from the equations one finds a dispersion relation for the complex parameters, $i\omega + p^2 = 0$, and the value of the phase factor, $a = 1/(p + p^*)^2$.

Hirota's method works also for discrete systems. Consider the (semi-)discrete Toda-lattice equation

$$\ddot{y}_n = e^{-(y_n - y_{n-1})} - e^{-(y_{n+1} - y_n)}. \quad (16)$$

First one goes over to the dependent variable $r_n = y_n - y_{n-1}$, and then the substitution $e^{r_n} = 1 + \partial_t^2 \log f_n$ yields

$$(D_t^2 + 2) f_n \cdot f_n = 2 f_{n-1} f_{n+1}. \quad (17)$$

Since $e^{\alpha D_x} f(x) \cdot g(x) = f(x + \alpha)g(x - \alpha)$, we can write (17) also in the form

$$\left[D_t^2 - 4\sinh^2\left(\tfrac{1}{2}D_n\right)\right] f_n \cdot f_n = 0. \quad (18)$$

The 1SS is given by

$$f_n(t) = 1 + e^{2(pn+\omega t)}, \quad (19)$$

where $\omega = \pm \sinh(p)$.

The above types of bilinearizations work, *mutatis mutandis*, for several classes of equations, but unfortunately there is no algorithmic method of finding the bilinearizing substitution for a given nonlinear equation. In fact, it is not even clear a priori how many dependent or even independent variables one should use. There are some general guidelines; for example, in order to bilinearize higher members in a hierarchy, one usually needs extra independent variables. If the 1SS and 2SS are known, one may use them in order to make an educated guess about the bilinearizing substitution.

The ultimate bilinear equation, from which many other bilinear equations can be derived as particular limits, is the Hirota–Miwa equation

$$(z_1 e^{D_1} + z_2 e^{D_2} + z_3 e^{D_3})F \cdot F = 0, \quad (20)$$

where z_i are constants and D_i are arbitrary linear combinations of $D_x, D_y, \ldots$.

The General Perturbative Approach for Constructing Soliton Solutions

Once a reasonable bilinear form has been found, one can try to find soliton solutions using a perturbative expansion in a fictitious parameter ε (Hirota, 1976):

$$F = f_0 + \varepsilon f_1 + \varepsilon^2 f_2 + \cdots, \quad (21)$$

and similarly for the other dependent variables appearing in the bilinear equations. One first chooses a suitable "vacuum" solution (no solitons), that is, constant values for the ε^0 terms, with all other terms vanishing. Next for a 1SS, one uses a minimal nontrivial entry (typically $e^{px+qy+\omega t}$) at order ε, that is, for f_1, and normally one should then be able to truncate the expansion at this or the next level (cf. (6) or (15)). For a 2SS, the previous 1SS solution is generalized at the ε level by taking the sum of two terms (cf. (8)), and as a result the expansion will truncate later (for KdV at ε^2, but for NLS we need even powers up to 4 for F and odd powers up to 3 for G). If the expansion does not truncate, it is probable that the equation is non-integrable. What helps here is that since $P(\boldsymbol{D})e^{\boldsymbol{a}\cdot x} \cdot e^{\boldsymbol{b}\cdot x} = P(\boldsymbol{a} - \boldsymbol{b})e^{(\boldsymbol{a}+\boldsymbol{b})\cdot x}$, the highest order terms in $P(\boldsymbol{D})F \cdot F$ vanish automatically.

In a systematic construction of multisoliton solutions, one uses determinants of Wronskian or Grammian type, or Pfaffians (Nimmo, 1990). Their applicability follows from the fact that determinants often satisfy quadratic identities, of which the Laplace expansion (a Plücker relation) is a typical example. The type of the matrix used is intimately related to the above-mentioned phase factor of the 2SS.

Bäcklund Transformations

In addition to providing a simple method for constructing multisoliton solutions, Hirota's method is also natural for studying Bäcklund transformations (BTs). In general terms, the starting set of equations and variables is doubled (while adding a new free parameter), and their combination is manipulated in order to get a pair of bilinear equations that are *linear* in each set of variables. Furthermore, if variables of one set are eliminated, one should get back the original equations. For example, the BT of the KdV equation is (Hirota, 1980)

$$\begin{cases} (D_x^3 + 3\lambda D_x + D_t)F' \cdot F = 0, \\ (D_x^2 - \mu D_x - \lambda)F' \cdot F = 0, \end{cases} \tag{22}$$

and if one eliminates F' from this pair, one obtains (a derivative of) (5).

Sato Theory

The reason Hirota's bilinear method works so well is that it is a reflection of an elegant fundamental theory called Sato theory, where bilinear identities (expressed using Young or Maya diagrams) play a major role. The theory naturally yields several infinite hierarchies of equations in an infinite number of variables, and their finite-dimensional reductions are the usual soliton equations. For an introduction, see Ohta et al. (1988) and Miwa et al. (2000).

Because there is so much mathematical structure underlying the bilinear approach, it is not surprising that bilinear forms of various equations have appeared before in the literature (although only in passing). For example, in 1902, Painlevé found bilinear forms for the first three of the six Painlevé equations using a substitution like (2); his idea was to express the equations in terms of entire functions. Indeed, the functions that arise in Sato theory are also entire functions, called τ-functions, and so one often refers to the bilinearizing functions (F and G above, for example) as τ-functions.

Summary

Hirota's bilinear method is a very effective technique in constructing multisoliton solutions. For a given equation, one should first find a one-soliton solution, and using it, one can often guess a method for bilinearizing the nonlinear equation. If that is successful, one can search for multisoliton solutions using the perturbative approach.

It is believed that all integrable evolution equations are obtained as reductions from a limited number of hierarchies, as described in Miwa et al. (2000), although the reductions can be highly nontrivial. Conversely, for a given new integrable equation, one should try to understand its origin as a reduction of one of the known hierarchies, and the bilinear form and the phase factor are helpful in this process.

In practical applications, one must nevertheless be careful, first to avoid bilinearizations that trivialize the system. The usual bilinearizing substitution (2) usually yields a multilinear system. One can forcibly bilinearize it by separating it into smaller bilinear parts and requiring each one of them to vanish independently. But if at the end there are more equations than unknown functions, the set of bilinear equations so obtained may include conditions that did not exist in the original equations.

Also, the existence of a bilinear form does not imply integrability. Bilinear forms have also been written for many non-integrable models, and the existence of a bilinear form does not even imply the existence of N-soliton solutions, nor can one always assume that the bilinearizing functions are τ-functions. Indeed, for a large class of bilinear equations (containing non-integrable equations), it is possible to find two-soliton solutions exhibiting elastic scattering, but that does not imply the existence of a *general* N-soliton solution for any $N > 2$.

JARMO HIETARINTA

See also **Bäcklund transformations; Inverse scattering method or transform; N-soliton formulas; Solitons**

Further Reading

Hietarinta, J. 1990. Hirota's bilinear method and integrability. In *Partially Integrable Evolution Equations in Physics*, edited by R. Conte & N. Boccara, Dordrecht: Kluwer, pp. 459–478

Hirota, R. 1971. Exact solution of the Korteweg–de Vries equation for multiple collision of solitons. *Physical Review Letters*, 27: 1192–1194

Hirota, R. 1976. Direct method of finding exact solutions of nonlinear evolution equations. In *Backlund Transformations, the Inverse Scattering Method, and Their Applications*, edited by R. Miura, New York: Springer, pp. 40–68

Hirota, R. 1980. Direct methods in soliton theory. In *Solitons*, edited by R. Bullough & P.J. Caudrey, New York: Springer, pp. 157–176

Miwa, J., Jimbo, M. & Date, E. 2000. *Solitons: Differential Equations, Symmetries and Infinite Dimensional Algebras*, Cambridge and New York: Cambridge University Press

Nimmo, J.J.C. 1990. Hirota's method. In *Soliton Theory: A Survey of Results*, edited by A.P. Fordy, Manchester: Manchester University Press, pp. 75–96

Ohta, Y., Satsuma, J., Takahashi, D. & Tokihiro, T. 1988. An elementary introduction to Sato theory. *Progress in Theoretical physics, Supplementum*, 94: 210–241

HODGKIN–HUXLEY EQUATIONS

Classical experiments on the compound action potential produced by vertebrate nerve trunks provided all the components for a reaction-diffusion model for the propagation of an action potential, with the nerve impulse as a nonlinear traveling wave produced by the nonlinear "reaction" of excitation and the diffusive spread of voltage. Even as late as the 1930s, however, it was not clear if nerve excitation was a membrane phenomenon, or if the the extracellular potential changes were a consequence of some chemical excitation propagating through the axoplasm.

The identification of the 100–500 μm diameter giant axons of the squid as single nerve fibers provided a preparation large enough for excitation and propagation in a single nerve fiber to be analyzed in detail. Wires can be inserted down the axoplasm, one to pass current, one to record potential, and one to short circuit the resistance of the intracellular axoplasm, providing a space clamp under which the membrane potential is spatially uniform. Current can be passed between the internal electrode and external electrodes, exciting a membrane action potential that is recorded as the difference in potential V across the membrane.

The membrane current I_m is the sum of a capacitative and an ionic current:

$$I_{\mathrm{m}} = C_{\mathrm{m}}\,\mathrm{d}V/\mathrm{d}t + I_{\mathrm{ion}} \tag{1}$$

with the membrane capacitance $C_{\mathrm{m}} = 1\,\mu\mathrm{F\,cm}^{-2}$. Direct recordings of the membrane action potential show that the potential changes sign, from a resting value of -60 mV to a peak value of $+40$ mV. The peak of the action potential varies logarithmically with the extracellular Na^+ concentration, showing the role of sodium ion (Na^+) current in the generation of the action potential.

The voltage clamp technique uses a feedback circuit to control the potential across the membrane. Using rectangular command pulses, the membrane potential can be changed from its resting value to an arbitrary value within a fraction of a millisecond. During a voltage clamp, the only current flowing across the membrane is ionic and is equal and opposite to the current that is injected to maintain the clamp. Net ionic currents can be measured, and dissected into their components by ion substitution experiments. From the estimated ionic currents and their electrochemical gradients, the ionic conductances changes can be calculated. Under a depolarizing voltage clamp, the sodium(Na^+)-conductance change is fast and transient, while the potassium (K^+)-conductance change is slower, delayed, and maintained.

Hodgkin & Huxley (1952) fitted their extensive series of experimental measurements of ionic currents obtained under voltage clamp and synthesized their empirical equations into a quantitative model for the propagating action potential.

For axoplasmic resistance R and membrane capacitance C, both per unit length of axon, the spread of membrane potential V with distance x (cm) and time t (ms) is described by a reaction–diffusion equation:

$$C\frac{\partial V}{\partial t} = \frac{1}{R}\frac{\partial^2 V}{\partial x^2} - I_{\mathrm{ion}}, \tag{2}$$

where I_{ion} is a nonlinear function of V and t .

The membrane ionic current I_{ion} is assumed to be composed of three independent components, a Na^+-current I_{Na}, a K^+-current I_{K}, and a leakage current density I_{L} ($\mu\mathrm{A\,cm}^{-2}$) that flow through separate conductance pathways that have different ion-selectivities, maximal conductances, and voltage-dependent kinetics. Thus,

$$I_{\mathrm{ion}} = I_{\mathrm{Na}} + I_{\mathrm{K}} + I_{\mathrm{L}}. \tag{3}$$

The instantaneous current-voltage relation for each pathway is linear, so the current is the product of a conductance and a driving force—the difference between the potential V and the equilibrium (Nernst) potential for the pathway. The Nernst potential is determined by the intracellular (axoplasm) and extracellular (artificial sea water) ionic concentrations, and $V_{\mathrm{Na}} = -115$, $V_{\mathrm{K}} = +12$, and $V_{\mathrm{L}} = -10.613$ mV (the unreasonable precision for V_{L} is to fix $V = 0$ as a stable equilibrium solution). The conductance is the product of the maximal conductance $\bar{g}$ and gating variables—activation m and inactivation h for Na^+-channels activation n for K^+-channels. The activation variables are raised to a certain power (3 for m, 4 for n) that empirically reproduces the delayed, sigmoid increase in current and may be interpreted as the number of gating structures per channel. Thus

$$\begin{aligned} I_{\mathrm{Na}} &= \bar{g}_{\mathrm{Na}} m^3 h (V - V_{\mathrm{Na}}), \\ I_{\mathrm{K}} &= \bar{g}_{\mathrm{K}} n^4 (V - V_{\mathrm{K}}), \\ I_{\mathrm{L}} &= g_{\mathrm{L}} (V - V_{\mathrm{L}}). \end{aligned} \tag{4}$$

Each gating variable obeys first-order kinetics

$$\begin{aligned} \mathrm{d}m/\mathrm{d}t &= \alpha_m(1-m) - \beta_m m, \\ \mathrm{d}h/\mathrm{d}t &= \alpha_h(1-h) - \beta_h h, \\ \mathrm{d}n/\mathrm{d}t &= \alpha_n(1-n) - \beta_n n \end{aligned} \tag{5}$$

with rate coefficients α and β for opening or closing that are empirical functions of voltage:

$$\begin{aligned} \alpha_n &= \frac{0.01(10-V)}{\exp(10-V)/10-1}, \\ \beta_n &= 0.125\exp(-V/80), \\ \alpha_m &= \frac{0.1(25-V)}{\exp(25-V)/10-1}, \\ \beta_m &= 4\exp(-V/18), \\ \alpha_h &= 0.07\exp(-V/20), \\ \beta_h &= \frac{1}{\exp(30-V)/10+1}. \end{aligned} \tag{6}$$

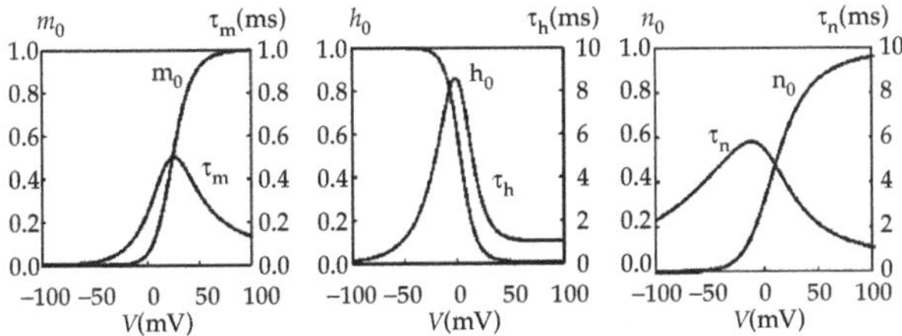

Figure 1. Voltage dependence of time coefficients and steady state values for gating variables m, h, and n of HH equations.

The rate coefficients α and β are derived from the experimentally estimated time constants τ and steady-state values illustrated in Figure 1, for example,

$$\tau_m = 1/(\alpha_m + \beta_m), \quad m_0 = \alpha_m/(\alpha_m + \beta_m). \quad (7)$$

Numerical solution (on a mechanical hand calculator) of the membrane equations reproduced both the subthreshold and action potential responses to initial depolarizations and a decrease in latency seen with increasing amplitude above threshold.

Computations of the total conductance change during the membrane action potential; total ion fluxes, time course of refractory period, and damped oscillatory responses matched experimental observations. The equations were based on experiments performed at 6.3°C; a Q_{10} of 3 (increasing temperature by 10° caused the rates to increase by a factor of 3) allowed simulation of the temperature dependence of the action potential. A solitary propagating solution of Equations (1)–(6) was obtained by assuming a wave solution traveling with a constant velocity of θ and making a coordinate transformation $\xi = x - \theta t$ to reduce Equation (1) to an autonomous system of ordinary differential equations. A value of $\theta = 18.8\,\mathrm{ms}^{-1}$ was found by trial and error to give an appropriately bounded solution in this traveling-coordinate system, which is close to the experimental estimate of $20\,\mathrm{ms}^{-1}$. For this work Alan Hodgkin and Andrew Huxley shared the Nobel Prize in Physiology or Medicine in 1963.

Early digital computations showed the continuous HH membrane equations to have a quasi-threshold, with the size of the action potential increasing smoothly with stimulus intensity. This contradiction of the "all-or-none" nature of the nerve impulse required control of variables to a higher resolution than thermal noise would allow in practice. However, similar behavior was found both numerically and experimentally at high temperatures (Cole et al., 1970). The membrane equations respond to steady injected currents by a repetitive discharge with a rate from 50 to 125 s^{-1}, with a logarithmic relation between rate and injected current density, as found in sensory coding. The numerical and experimental responses to periodic stimulation include phase-locked, entrained, and even chaotic responses. The close agreement between numerical

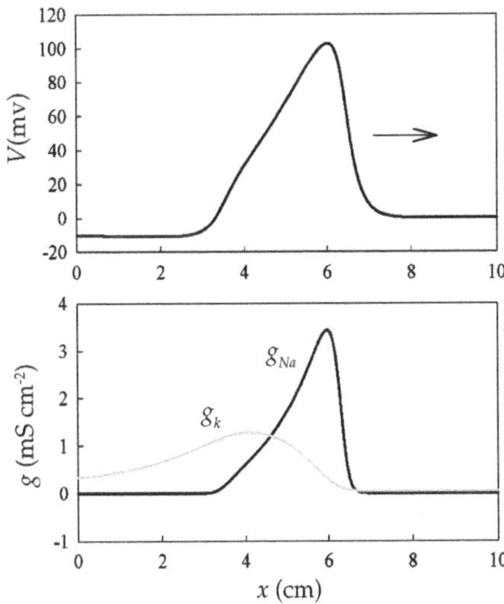

Figure 2. Constant velocity travelling wave solution of HH equations. Transmembrane voltage (upper) and membrane permeability (lower).

and experimental results has allowed chaos in the periodically forced squid axon to be a test signal for the validation of methods of quantifying chaos.

The equations successfully reproduced experimental data on membrane potential, ionic currents, and ion fluxes and provided a quantitative theory for the excitation and propagation of the nerve impulse. The equations were an empirical description of the currents seen under voltage clamp that also gave a quantitatively accurate reconstruction of the action potential. The form of the equations suggested a simple model of gated pores, with each pore gated by a system of four independent gates, and each gate opening or closing independently, with voltage-dependent, first-order kinetics. This physical model, of gated ionic conductance channels, has been substantiated by the electrophysiological and molecular biological characterization of single ionic channels, their kinetics and gating (Hille, 2001). The Hodgkin–Huxley formalism has also been successfully applied to the analysis of excitable membranes of nerve, smooth, skeletal, and cardiac muscle.

The introduction of digital computing into bioscience laboratories allowed numerical solution of the full partial differential equation system (Cooley & Dodge, 1966). Both stable and unstable traveling- wave solutions are found (Rinzel & Keller, 1973), and the effects of changes in diameter and axonal branching have been described.

As the first quantitatively accurate description of excitation, the HH equations have been used as a prototype for studying the behavior of excitation cells in general, as well as in circumstances with parameter values that may have no biological relevance for the squid axon. The development of autorhythmicity, by changes in parameters (injected steady current, maximal conductances, extracellular ionic concentrations, temperature), has been shown to be by simple Hopf bifurcation; the bifurcation curves in parameter space have been mapped (Holden & Winlow, 1984) and explained using singularity theory (Golubitsky & Schaeffer, 1985). The HH equations were in fact a case study for the first numerical package for using continuation algorithms to track Hopf bifurcation curves. Phase-resetting behavior and annihilation of repetitive activity by appropriately timed perturbations has been computed and found in experiments (Guttman et al., 1984).

The importance of the HH equations in physiology has led to interest from mathematicians, either by generalizing the equations to characterize the mathematical aspects of excitation (Carpenter, 1977) or to allow exotic behavior such as bursting or by simplifying the equations to allow analysis or to facilitate numerical computations.

ARUN V. HOLDEN

See also **Excitability; FitzHugh–Nagumo equation; Hopf bifurcation; Markin–Chizmadzhev model; Nerve impulses; Neurons; Periodic bursting**

Further Reading

Carpenter, G. 1977. A geometric approach to singular perturbation problems with applications to nerve impulse equations. *Journal of Differential Equations*, 23: 335–367

Cole, K.S. 1969. *Membranes, Ions and Impulses*, Berkley: University of California Press

Cole, K.S., Guttman, R. & Bezanilla, F. 1970. Nerve membrane excitation without threshold. *Proceedings of the National Academy of Science of the USA*, 65: 884–891

Cooley, J.W. & Dodge, F.A., Jr. 1966. Digital computer solutions for excitation and propagation of the nerve impulse. *Biophysical Journal*, 6: 583–599

Golubitsky, M. & Schaeffer, D.G. 1985. *Singularities and Groups in Bifurcation Theory: vol. 1*, Berlin and New York: Springer, pp. 382–396

Guttman, R., Lewis, S. & Rinzel, J. 1984. Control of repetitive firing in squid axonal membrane as a model of a neurone oscillator. *Journal of Physiology*, 305: 377–395

Hille, B. 2001. *Ion Channels of Excitable Membranes*, 3rd edition, Sunderland, MA: Sinauer

Hodgkin, A.L. & Huxley, A.F. 1952. A quantitative description of membrane current and its application to conduction and excitation in nerve. *Journal of Physiology*, 117: 500–544

Holden, A.V. & Winslow, W. 1984. Neuronal activity as the behaviour of a differential system. *IEEE Transactions SMC*, 13: 711–719

Rinzel, J. & Keller, J.B. 1973. Traveling wave solutions of a nerve conduction equation. *Biophysical Journal*, 13: 1313–1337

HODOGRAPH TRANSFORM

Consider an N-component system with dependent variables

$$u(x,t) = [u_1(x,t), u_2(x,t), \ldots u_N(x,t)]$$

that satisfy the partial differential equations

$$\frac{\partial u_i}{\partial t} = \sum_{j=1}^{N} v_{ij}(u)\frac{\partial u_j}{\partial x}, \quad v_{ij}(u) \in \mathbb{R}, \quad i = 1, \ldots, N. \tag{1}$$

Under invertible smooth changes of variables of the form

$$u_i = u_i(w_1, w_2, \ldots, w_N), \quad i = 1, 2, \ldots, N, \tag{2}$$

the coefficients v_{ij} transform as a tensor,

$$v_{ij}(u_1, u_2, \ldots, u_N) \to \tilde{v}_{rs}(w_1, w_2, \ldots, w_N) = \sum_{i,j=1}^{N} \frac{\partial w_r}{\partial u_i} v_{ij}(u(w))\frac{\partial u_j}{\partial w_s}.$$

Now let us assume that the eigenvalues $\lambda_1, \lambda_2, \ldots, \lambda_N$ of the matrix v_{ij} are real and distinct, so system (1) is hyperbolic. Then it is possible to reduce (1), using transformation (2) to a diagonal form

$$\frac{\partial w_i}{\partial t} = \lambda_i(w)\frac{\partial w_i}{\partial x}, \quad i = 1, 2, \ldots, N, \tag{3}$$

where $w(x,t) = (w_1(x,t), w_2(x,t), \ldots, w_N(x,t))$. The variables $w_1, w_2, \ldots w_N$ are called "Riemann invariants," and the coefficients $\lambda_1(w), \lambda_2(w), \ldots, \lambda_N(w)$ are the corresponding characteristic velocities.

For $N = 2$ it is always possible locally to reduce (1) to Riemann invariant form, while for $N \geq 3$ this is not true in general. When system (1) is reducible to the diagonal form (3), it is sometimes possible to integrate it through the so-called hodograph transform. For $N \leq 2$, this fact is well known (see Courant & Hilbert, 1962); while for $N \geq 3$, the result was proved by Tsarev (1985). Let us consider the two-component system

$$\frac{\partial w_1}{\partial t} = \lambda_1(w_1, w_2)\frac{\partial w_1}{\partial x}, \quad \frac{\partial w_2}{\partial t} = \lambda_2(w_1, w_2)\frac{\partial w_2}{\partial x}. \tag{4}$$

To integrate (4), we map it into a linear set of equations, by interchanging the role of the dependent and independent variables, the so-called hodograph transform $(x, t) \to (w_1, w_2)$. In this transformation

$$\partial_2 x = -\frac{\partial_t w_1}{J}, \quad \partial_1 x = \frac{\partial_t w_2}{J}, \quad \partial_2 t = \frac{\partial_x w_1}{J}, \quad \partial_1 t = -\frac{\partial_x w_2}{J},$$

where $x = x(w_1, w_2)$, $t = t(w_1, w_2)$, $\partial_i = \partial/\partial w_i$, $i = 1, 2$ and $J = \partial_x w_1 \partial_t w_2 - \partial_t w_1 \partial_x w_2$ is the Jacobian. Then system (4) is mapped into the linear equations

$$\partial_2 x + \lambda_1 \partial_2 t = 0, \quad \partial_1 x + \lambda_2 \partial_1 t = 0. \tag{5}$$

The hodograph transform guarantees a linear equation, but it may not be useful in practice. Let us write (5) in the equivalent form

$$\partial_2 \underbrace{(x+\lambda_1 t)}_{\chi_1} = t\,\partial_2\lambda_1,$$
$$\partial_1 \underbrace{(x+\lambda_2 t)}_{\chi_2} = t\,\partial_1\lambda_2. \quad (6)$$

Then the functions $w_1(x,t)$ and $w_2(x,t)$, defined implicitly by

$$\chi_1(w_1,w_2) = \lambda_1(w_1,w_2)t + x,$$
$$\chi_2(w_1,w_2) = \lambda_2(w_1,w_2)t + x, \quad (7)$$

solve (4) if

$$\frac{\partial_2\chi_1}{\chi_1-\chi_2} = \frac{\partial_2\lambda_1}{\lambda_1-\lambda_2}, \quad \frac{\partial_1\chi_2}{\chi_2-\chi_1} = \frac{\partial_1\lambda_2}{\lambda_2-\lambda_1}. \quad (8)$$

Indeed from (7)

$$t = \frac{\chi_1-\chi_2}{\lambda_1-\lambda_2},$$

and substituting the above into (6), we obtain the linear overdetermined system (8).

As an example, consider the Born–Infeld equation

$$(1-\psi_t^2)\psi_{xx} + 2\psi_x\psi_t\psi_{xt} - (1+\psi_x^2)\psi_{tt} = 0, \quad (9)$$

which was formulated by Max Born in the early 1930s as a nonlinear field model for elementary particles. It is a simple matter to check that either

$$\psi = \Psi(x-t) \quad \text{or} \quad \psi = \Psi(x+t)$$

are solutions of (9) for any function Ψ. In particular, the solution can be chosen in the form of a single hump to give the solitary wave appearance. The equation is hyperbolic for solution with $1+\psi_x^2-\psi_t^2>0$. We notice that the solitary waves have constant characteristic velocity ± 1 and avoid the usual breaking expected for nonlinear hyperbolic waves. Interacting waves for system (9) were obtained through the hodograph method by Barbishov & Chernikov (1966). First, if new variables

$$\xi = x-t, \quad \eta = x+t, \quad u_1 = \psi_\xi, \quad u_2 = \psi_\eta$$

are introduced, we obtain the equivalent system

$$\partial_\eta u_1 - \partial_\xi u_2 = 0,$$
$$u_2^2\partial_\xi u_1 - (1+2u_1u_2)\partial_\eta u_1 + u_1^2\partial_\eta u_2 = 0. \quad (10)$$

The Riemann invariants are

$$w_1 = \frac{\sqrt{1+4u_1u_2}-1}{2u_2}, \quad w_2 = \frac{\sqrt{1+4u_1u_2}-1}{2u_1},$$

so system (10) reduces to the form

$$\partial_\xi w_1 = w_2^2\partial_\eta w_1, \quad \partial_\xi w_2 = \frac{1}{w_1^2}\partial_\eta w_2. \quad (11)$$

Performing the hodograph transform $(\eta,\xi)\to(w_1,w_2)$, system (11) is mapped to

$$w_1^2\partial_1\xi + \partial_1\eta = 0, \qquad \partial_2\xi + w_2^2\partial_2\eta = 0,$$

where $\partial_1 \equiv \partial_{w_1}$ and $\partial_2 \equiv \partial_{w_2}$. Eliminating η leads to the simple form

$$\partial_1\partial_2\xi = 0.$$

The above equation is readily integrated as

$$x-t = \xi = F(w_1) - \int w_2^2 G'(w_2)\,\mathrm{d}w_2,$$
$$x+t = \eta = G(w_2) - \int w_1^2 F'(w_1)\,\mathrm{d}w_1,$$

where F and G are arbitrary functions, and it follows that

$$\partial_1\psi = w_1F'(w_1), \quad \partial_2\psi = w_2G'(w_2).$$

It is convenient to introduce

$$F(w_1) = \rho, \quad G(w_2) = \sigma,$$
$$w_1 = \Phi_1'(\rho), \quad w_2 = \Phi_2'(\sigma),$$

so that the corresponding expression for ψ reads

$$\psi = \Phi_1(\rho) + \Phi_2(\sigma),$$
$$x-t = \rho - \int_{-\infty}^{\sigma} \Phi_2'^2(\sigma)\,\mathrm{d}\sigma,$$
$$x+t = \sigma + \int_{\rho}^{+\infty} \Phi_1'^2(\rho)\,\mathrm{d}\rho.$$

If $\Phi_1(\rho)$ and $\Phi_2(\sigma)$ are localized, say, they are nonzero in $-1<\rho<0$ and $0<\sigma<1$, then

$$\psi = \Phi_1(x-t) + \Phi_2(x+t), \quad t<0,$$

while for $t\to+\infty$, the solution approaches

$$\psi = \Phi_1\left(x-t+\int_{-\infty}^{\infty}\Phi_2'^2(\sigma)\,\mathrm{d}\sigma\right)$$
$$+\Phi_2\left(x+t-\int_{-\infty}^{+\infty}\Phi_1'^2(\rho)\,\mathrm{d}\rho\right).$$

Each wave receives a displacement in the direction opposite to its direction of propagation equal to

$$\int_{-\infty}^{+\infty}\Phi_i'^2(\tau)\,\mathrm{d}\tau, \quad i=1,2.$$

The interaction is remarkably similar to the interaction of solitary waves even though the Born–Infeld equation belongs to a different class of systems. We remark that the above analysis is valid provided that the mapping from the (x,t) plane to the (ρ,σ) plane is nonsingular.

Application to Multiphase Averaging

The generalized hodograph method is used to integrate (Dubrovin & Novikov, 1989) the multiphase averaged equations or Whitham equations (Whitham, 1974; Flaschka et al., 1980). These equations are obtained by averaging conservation laws over the family of multiphase solutions of one-dimensional integrable nonlinear evolution equations like, for example, the Korteweg–de Vries equation. The Whitham equations describe the modulation of the wave parameters and wave numbers of the multiphase solutions over time and space scales of order εt and εx, where t and x are the independent variables of the original nonlinear evolution equations and $0 < \varepsilon \ll 1$. When the Whitham equations are hyperbolic, the technique of integration follows from a result of Tsarev (1985) that generalizes the hodograph transform to many dependent variables. The following condition is sufficient for integrability

$$\partial_i \left(\frac{\partial_k \lambda_j}{\lambda_k - \lambda_j} \right) = \partial_k \left(\frac{\partial_i \lambda_j}{\lambda_i - \lambda_j} \right), \quad i \neq j \neq k, \ i, j, k = 1, \ldots, N.$$

The integration of (3) is obtained as follows. If $\chi_1(w), \ldots, \chi_N(w)$ solves the linear over-determined system

$$\partial_i \chi_j = \frac{\chi_i - \chi_j}{\lambda_i - \lambda_j} \partial_i \lambda_j, \quad i \neq j, \ i, j = 1, \ldots, N, \quad (12)$$

then the solution $w_1(x,t), \ldots, w_N(x,t)$ of the generalized hodograph transform

$$x + \lambda_i(w)t = \chi_i(w), \quad i = 1, \ldots, N,$$

satisfies (3). Conversely, every solution of (3) can be obtained in this way in the neighborhood of a point (x_0, t_0) where $\partial_x w_i$, $i = 1, 2, \ldots, N$, is not vanishing. The solution of the linear overdetermined system (12) corresponding to the Whitham equations can be obtained by algebro-geometric integration.

GREGORIO FALQUI AND TAMARA GRAVA

See also **Born–Infeld equations; Characteristics; Horseshoes and hyperbolicity in dynamical systems; Modulated waves; Periodic spectral theory; Shock waves; Solitons**

Further Reading

Barbishov, B.M. & Chenikov, N.A. 1966. Solution of the two plane wave scattering problem in a nonlinear scalar field theory of the Born–Infeld type, *JETP.*, 23: 1025–1033

Courant, R. & Hilbert, D. 1962. *Methods of Mathematical Physics*, New York: Interscience

Dubrovin, B. & Novikov, S.P. 1989. Hydrodynamic of weakly deformed soliton lattices. Differential geometry and Hamiltonian theory, *Russian Mathematical Surveys*, 44(6): 35–124

Flaschka, H., Forest, M. & McLaughlin, D.W. 1980. Multiphase averaging and the inverse spectral solution of the Korteweg–de Vries equation, *Communications on Pure and Applied Mathematics*, 33: 739–784

Tsarev, S.P. 1985. Poisson brackets and one-dimensional Hamiltonian systems of hydrodynamic type. *Soviet Mathematics Doklady*, 31: 488–491

Whitham, G.B. 1974. *Linear and Nonlinear Waves*, New York: Wiley

HOLE BURNING

Hole burning is a nonlinear technique for high-resolution spectroscopy, which was first demonstrated in magnetic resonance spectroscopy (Bloembergen et al., 1948). Hole burning employs the spectral purity of a laser to remove from an inhomogeneously broadened spectral line a narrow homogeneous line, causing a dip (or hole) in the spectrum. Thus, the technique is aimed at revealing the homogeneous line shape in the presence of strong inhomogeneous broadening.

The homogenous line shape is usually a Lorentzian whose full width $\Delta\omega_h$ is determined by the phase relaxation time T_2 of the system being considered

$$\Delta\omega_h = 1/T_2 . \quad (1)$$

T_2 is itself composed of two different times T_1 and T_2^*. T_1 is the energy relaxation time, that is, the time by which a non-equilibrium population reaches equilibrium. T_2^* is the pure dephasing time, the time in which an ensemble of molecules randomizes its initial coherent motion. From the Bloch equation, it can be readily shown that

$$1/T_2 = 1/T_1 + 2/T_2^* . \quad (2)$$

Hence, the homogeneous line shape function contains all the information on energy relaxation and dephasing processes (Abragam, 1961).

The inhomogeneous width $\Delta\omega_i$ is often much larger than $\Delta\omega_h$, so that the homogeneous line shape is completely masked. Inhomogeneous line broadening occurs since the microenvironments of the molecules, atoms, etc. under consideration are different, so the respective resonance frequencies experience a dispersion. Accordingly, the associated line shape is, as a rule, a Gaussian. In magnetic resonance this dispersion has its origin in the field inhomogeneities, in optical spectroscopy it comes from structural imperfections of the matrix in which the probe molecule is embedded.

Hole burning is one possible method for revealing the homogenous line shape. It is a special variant of saturation spectroscopy for the case that $\Delta\omega_h \ll \Delta\omega_i$. There are two main branches of this technique, namely saturation via the power and saturation via the radiation dose of the applied field. A two-state system, for example, an electronic two-level system of a probe molecule in an imperfect lattice, can only be saturated

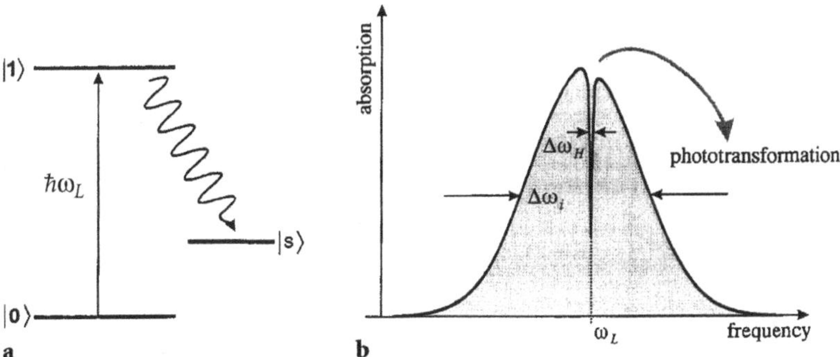

Figure 1. (a) Saturation of a three-level system. The storage level $|s\rangle$ is accessible from the excited state $|1\rangle$ by nonradiative processes. ω_L: laser frequency. (b) Schematic representation of the spectrum associated with saturation of the three-level system. $\Delta\omega_H$: holewidth. In the limit of vanishing saturation $\Delta\omega_H \rightarrow 2\Delta\omega_h$.

via the power of the radiation field: power is absorbed as long as there is a population difference between the lower and the upper state. As the power is increased, the population of the two states may equalize. Absorption and stimulated emission are balanced and the sample becomes transparent at the frequency of the radiation field. This phenomenon is called saturation. Note that power saturation of a two-level system is nonpersistent; that is, it decays with the energy relaxation time T_1. If the radiation field is tuned through the saturated transition, a dip or "hole" appears in the spectrum. The width of the hole in a power-saturated transition, $\Delta\omega_p$, however, deviates from the width of the homogenous line-shape function because it depends via the so-called Rabi frequency ω_1 on the power of the saturating radiation:

$$\Delta\omega_p = \Delta\omega_h[1 + \omega_1^2 T_1 T_2]^{1/2}. \tag{3}$$

ω_1 is proportional to the amplitude of the radiation field. Note that although the width becomes power dependent, the shape of the power-saturated transition retains its Lorentzian shape.

A very different saturation process is associated with hole burning involving three levels (Gorokhovskii et al., 1974; Kharlamov et al., 1974). Although power-saturation of an electronic two-level system is possible, it is of much less importance compared with saturation processes involving three levels, say the groundstate of a molecule $|0\rangle$, its first excited singlet state $|1\rangle$, and a long-lived intermediate $|s\rangle$ that acts as a storage state (Figure 1a). This intermediate, for instance, could be a photochemical state, a structural (e.g., a conformational) state or a long-lived spin state. It is not directly accessible to the radiation field, but can only be populated from state $|1\rangle$ via some radiative or radiationless process. Upon irradiating the $|0\rangle \rightarrow |1\rangle$ transition with a laser, population from a small range around the laser frequency ω_L is transferred to $|s\rangle$. Accordingly, a hole appears around ω_L in the absorption spectrum (Figure 1b). This hole may be persistent if the storage state $|s\rangle$ lives sufficiently long, as is the case for many photochemical states. In this case, the technique is called "photochemical hole burning." At sufficiently low temperatures, say, a few K, T_2^* becomes very large so that $\Delta\omega_h$ becomes lifetime-limited (Equations (1) and (2)). For a dye molecule with a typical lifetime of 10 ns dissolved in an organic glass, which gives rise to a typical inhomogeneous width of 300 wave numbers, this means that the burnt-in hole can be 4 to 5 orders of magnitude narrower than the inhomogeneous width. Because of this improvement in resolution, spectral hole burning is essentially a low-temperature technique (Friedrich & Haarer, 1984).

Persistent spectral hole burning may be characterized as a zero-power technique because holes can be burnt with vanishingly small power just by increasing the irradiation time. Yet, despite the fact that power saturation can be avoided, there is saturation broadening that, in contrast to the two-level case, depends on the irradiation dose: as population is transferred to the level $|s\rangle$, the number of absorbers in the center of the hole decreases steadily. Hence, the population transfer slows down in the center but still continues in the margins of the hole. This gives rise to an additional broadening in a similar way as has been discussed above for power broadening. Accordingly, $\Delta\omega_h$ is obtained from an extrapolation of the hole width to zero radiation dose.

Since the power of the radiation field can be made vanishingly small, the optics itself is perfectly linear. The nonlinear behavior of optical hole burning comes in via the persistent changes of the complex index of refraction through the irradiated energy. The advantage of the technique lies in the persistency of these changes. This makes it easy to exploit the high resolution, for instance, in measuring the influence of small external perturbations in the presence of strong inhomogeneous broadening, such as the influence of small electric, magnetic, or pressure fields. Persistent hole burning has been one of the major techniques to investigate the dynamics of low-temperature glasses via so-called spectral diffusion experiments

(Friedrich & Haarer, 1986), and it has gained much attraction in high-resolution spectroscopy of biological molecules (Friedrich, 1995). There are also quite a few technical applications, mostly in optical data storage (Moerner, 1988).

JOSEF FRIEDRICH

See also **Nonlinear optics**

Further Reading

Abragam, A. 1961. *Principles of Nuclear Magnetism*, Oxford and New York: Oxford University Press

Bloembergen, N., Purcell, E.M. & Pound, R.V. 1948. Relaxation effects in nuclear magnetic resonance absorption. *Physical Review*, 73: 679–712

Friedrich, J. 1995. Hole burning spectroscopy and physics of proteins. In *Biochemical Spectroscopy*, edited by K. Sauer, San Diego: Academic Press

Friedrich, J. & Haarer, D. 1984. Photochemical hole burning: a spectroscopic study of relaxation processes in polymers and glasses. *Angewandte Chemie*, International Edition, 23: 113–140

Friedrich, J. & Haarer, D. 1986. Structural relaxation processes in polymers and glasses as studied by high resolution optical spectrocopy. In *Optical Spectrocopy of Glasses*, edited by I. Zschokke, Dordrecht: Reidel

Gorokhovskii, A.A., Kaarli, R.K. & Rebane, L.A. 1974. Hole burning in the contour of a purely electronic line in a Shpol'skii system. *JETP Letters*, 20: 216–218

Kharlamov, B.M., Personov, R.I. & Bykovskaya, L.A. 1974. Stable gap in absorption spectra of solid solutions of organic molecules by laser irradiation. *Optics Communications*, 12: 191–193

Moerner, W.E. (editor). 1988. *Persistent Spectral Hole Burning: Science and Application*, Berlin and New York: Springer

HOLONS

In the literature on nonlinear science, the term *holon* has arisen in two quite different senses: as a type of elementary quantum excitation and in the context of emergent biological and social phenomena.

Holons in Physics

In the physics community, holons are particles with zero spin and charge ± 1 (in units of electron charge e) obtained in the fractionalization of an electron or a hole. They are related to the phenomena of spin-charge separation and fractionalization and are the supersymmetric partner of spinons—particles with zero charge and spin $\frac{1}{2}$. Holons and spinons are neither bosons (statistics 0) nor fermions (statistics -1) (statistics α represents the phase that the two-particle wavefunction accumulates when one particle is taken and moved around another), but obey fractional statistics $\frac{1}{2}$ and are, therefore, called semions.

Recently, it has become possible to experimentally realize one-dimensional samples by growing crystals, such as $KCuF_3$, $SrCuO_2$, or Sr_2CuO_3, with strong anisotropy due to antiferromagnetic spin-spin interaction. Photoemission spectroscopy experiments on these crystals show two distinct bands in the energy dispersion, one belonging to the spinon, and the other to the holon. In theory and in experiments, holons have been observed only in one spatial dimension (Kim et al., 1996, 1997; Fujisawa et al., 1998, 1999), with credible efforts to discover them in higher dimensions having remained so far unsuccessful.

The simplest model where holons show up is the supersymmetric t-J model with $1/r^2$ interaction (Kuramoto–Yokohama (KY)-model) (Kuramoto & Yokohama, 1991). This is a system of electrons on a lattice with periodic boundary conditions, where it is forbidden that two electrons be on the same lattice site. Electrons can hop from site to site, there is a Coulomb interaction term, and there are also antiferromagnetic spin-spin interactions. The Hamiltonian is supersymmetric, in the sense that it costs zero energy for a charge to be introduced in the system. The elementary excitations of the system are spinons and holons. The holons have a dispersion relation:

$$E = -\frac{J}{2}\left[\left(\frac{\pi}{2}\right)^2 - q^2\right], \tag{1}$$

where E is the holon energy, q is the holon momentum, and J is an energy scale in the system. The configuration space of positive energy holons is halved with respect to that of an integral particle. Although the equations governing the dynamics of spinons and holons are different, they yield the same form for the interaction between a holon and a spinon and between two spinons. The interaction between the holon and the spinon in one spatial dimension is inversely proportional to the distance between these particles. When the holon and the spinon are at the same point in space, the interaction between them is divergent and together they form the electron. However, the interaction does not diverge fast enough to prevent an instability of the electron toward decay into the two particles, and indeed there is a small but finite matrix element for electron to holon and spinon decay.

An experimentally accessible quantity is the spectral function for a holon-spinon pair:

$$\begin{aligned} A_{\text{holon-spinon}}(\omega, q) = {} & \frac{1}{\pi^2 J q}\sqrt{\frac{J[q+\frac{\pi}{2}]^2-\omega}{\omega - J[q-\frac{\pi}{2}]^2}} \\ & \times\Theta\left[\omega - J[q-\frac{\pi}{2}]^2\right] \\ & \Theta\left[J[\frac{\pi^2}{4}+q(\pi-q)]-\omega\right], \end{aligned} \tag{2}$$

where ω and q are the energy and momentum at which the measurements are being conducted and J is the relevant energy scale in the system. The physical consequence of these measurements is the proof of instability of the electron, which is no longer a legitimate excitation of the system but breaks up into a holon and a spinon (Bernevig et al., 2001, 2002).

Holons in Biological and Social Systems

Biological and social systems are typically organized as hierarchies, in which nonlinear dynamics at each level of the hierarchy leads to the emergence of new dynamic entities that provide a basis for the nonlinear dynamics of the next higher level (Scott, 1995).

A well-known example of such a hierarchy is the structure of a military unit. Thus, an infantry division comprises four regiments, each of which comprises four battalions, which consist of four companies, and so on, up to 15,000 individual riflemen. Similar hierarchical structure is displayed by living organisms—for example, a human being is composed of a collection of organs, and the organs are composed of cells, which are made up of biochemicals (proteins, DNA, etc.), down to our constituent atoms.

In the 1960s, Arthur Koestler proposed the term *holon* to imply a general feature of a biological or social system that has a unique identity, yet is made up of lower-level features that comprise parts of the whole (Koestler, 1968). In this sense, holons are components of complex dynamic systems that are efficient and adaptable to disturbances, both internal and external. For example, you are a holon, as is your liver and each of your skin cells.

Interestingly, the human neocortex seems also to be hierarchically organized into closely interconnected assemblies of neurons that embody the holonic components of complex thought (Scott, 1995, 2002).

BOGDAN A. BERNEVIG

See also **Cell assemblies; Emergence; Hierarchies of nonlinear systems; Quantum field theory**

Further Reading

Bernevig, B.A., Giuliano, D. & Laughlin, R.B. 2001. Spinon attraction in spin-$\frac{1}{2}$ antiferromagnetic spin chains. *Physical Review Letters*, 86: 3392–3395

Bernevig, B.A., Giuliano, D. & Laughlin, R.B. 2002. Coordinate representation of the one-spinon one-holon wave function and spinon-holon interaction. *Physical Review* B, 65: 195112

Fujisawa, H. et al. 1998. Spin-charge separation in single chain compound Sr_2CuO_3 studied by angle resolved photo emission. *Solid State Communications*, 106: 543

Fujisawa, H. et al. 1999. Angle-resolved photoemission study of Sr_2CuO_3. *Physical Review* B, 59: 7358–7361

Kim, C. et al. 1996. Observation of spin-charge separation in one-dimensional $SrCuO_2$. *Physical Review Letters*, 77: 4054

Kim, C. et al. 1997. Separation of spin and charge excitations in one-dimensional $SrCuO_2$. *Physical Review* B, 56: 15,589–15,595

Koestler, A. 1968. *Ghost in the Machine*, New York: Macmillan

Kuramoto, Y. & Yokohama, M. 1991. Exactly soluble supersymmetric t–J-type model with long-range exchange and transfer. *Physical Review Letters*, 67: 1338–1341

Scott, A. 1995. *Stairway to the Mind*, Berlin and New York: Springer

Scott, A. 2002. *Neuroscience: A Mathematical Primer*, Berlin and New York: Springer

HOLSTEIN MODEL

See **Davydov soliton**

HOLSTEIN–PRIMAKOV TRANSFORMATION

See **Spin systems**

HOMEOMORPHISM

See **Maps**

HOMOCLINIC INTERSECTION

See **Phase space**

HOMOCLINIC TRAJECTORY

See **Phase space**

HOPF BIFURCATION

The term *Hopf bifurcation* (also called Poincaré–Andronov–Hopf bifurcation) refers to the local birth or death of a periodic solution (self-excited oscillation) from an equilibrium as a parameter crosses a critical value. It is the simplest bifurcation not just involving equilibria and, therefore, belongs to what is sometimes called dynamic (as opposed to static) bifurcation theory. In a differential equation, a Hopf bifurcation typically occurs when a complex conjugate pair of eigenvalues of the linearized flow at a fixed point becomes purely imaginary. This implies that a Hopf bifurcation can only occur in systems of dimension two or higher.

That a periodic solution should be generated in this event is intuitively clear from Figure 1. When the real parts of the eigenvalues are negative the fixed point is a stable focus (Figure 1(a)); when they cross zero and become positive the fixed point becomes an unstable focus, with orbits spiralling out. But this change of stability is a local change, and the phase portrait sufficiently far from the fixed point will be qualitatively unaffected. If the nonlinearity makes the far flow contracting, then orbits will still be coming in and we expect a periodic orbit to appear

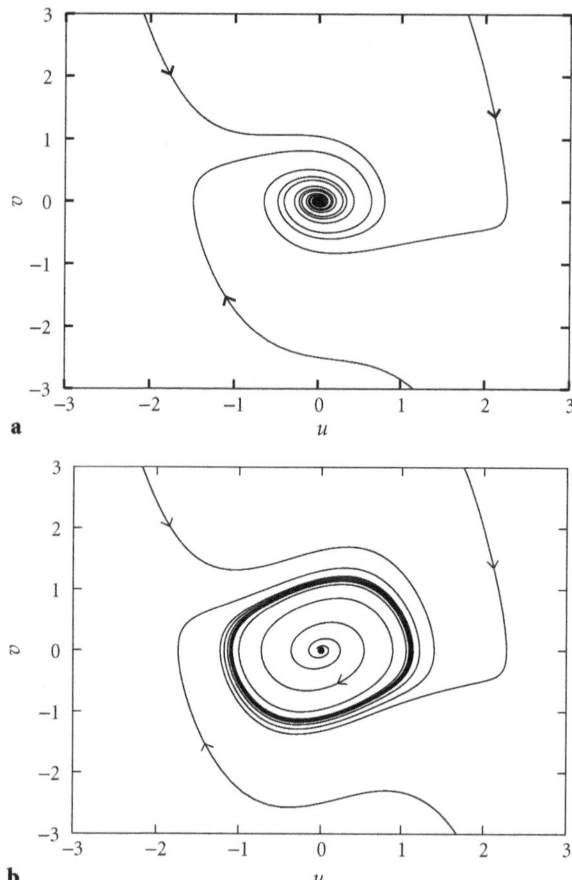

Figure 1. Phase portraits of (2) for (a) $\mu = -0.2$, (b) $\mu = 0.3$. There is a supercritical Hopf bifurcation at $\mu = 0$.

where the near and far flow find a balance (as in Figure 1(b)).

The Hopf bifurcation theorem makes the above precise. Consider the planar system

$$\begin{aligned} \dot{x} &= f_\mu(x, y), \\ \dot{y} &= g_\mu(x, y), \end{aligned} \tag{1}$$

where μ is a parameter. Suppose it has a fixed point that without loss of generality, we may assume to be located at the origin of the (x, y) plane. Let the eigenvalues of the linearized system about this fixed point be given by $\lambda(\mu), \bar{\lambda}(\mu) = \alpha(\mu) \pm \mathrm{i}\beta(\mu)$. Suppose further that for a certain value of μ (which we may assume to be zero), the following conditions are satisfied:

(i) $\alpha(0) = 0, \quad \beta(0) = \omega \neq 0$
(nonhyperbolicity condition: conjugate pair of imaginary eigenvalues);

(ii) $\left.\dfrac{\mathrm{d}\alpha(\mu)}{\mathrm{d}\mu}\right|_{\mu=0} = d \neq 0$
(transversality condition: eigenvalues cross imaginary axis with nonzero speed);

(iii) $a \neq 0$, where

$$\begin{aligned} a = {} & \frac{1}{16}\left(f_{xxx} + f_{xyy} + g_{xxy} + g_{yyy}\right) \\ & + \frac{1}{16\omega}\left(f_{xy}(f_{xx} + f_{yy}) - g_{xy}(g_{xx} + g_{yy})\right. \\ & \left. - f_{xx}g_{xx} + f_{yy}g_{yy}\right), \end{aligned}$$

with $f_{xy} = (\partial^2 f_\mu/\partial x \partial y)\big|_{\mu=0} (0, 0)$, etc. (genericity condition).

Then a unique curve of periodic solutions bifurcates from the origin into the region $\mu > 0$ if $ad < 0$ or $\mu < 0$ if $ad > 0$. The origin is a stable fixed point for $\mu > 0$ (resp. $\mu < 0$) and an unstable fixed point for $\mu < 0$ (resp. $\mu > 0$) if $d < 0$ (resp. $d > 0$), while the periodic solutions are stable (resp. unstable) if the origin is unstable (resp. stable) on the side of $\mu = 0$, where the periodic solutions exist. The amplitude of the periodic orbits grows like $\sqrt{|\mu|}$ while their periods tend to $2\pi/|\omega|$ as $|\mu|$ tends to zero. The bifurcation is called "supercritical" if the bifurcating periodic solutions are stable, and "subcritical" if they are unstable.

This two-dimensional version of the Hopf bifurcation theorem was known to Alexandr A. Andronov and his co-workers from around 1930 (Andronov et al., 1966) and had been suggested by Henri Poincaré (1892). Eberhard Hopf (1942) proved the result for arbitrary (finite) dimensions. Through center-manifold reduction, the higher-dimensional version essentially reduces to the planar one provided that apart from the two purely imaginary eigenvalues no other eigenvalues have zero real part. In his proof (which predates the center-manifold theorem), Hopf assumes the functions f_μ and g_μ to be analytic, but C^5 differentiability is sufficient (a proof can be found in Marsden & McCracken (1976)). Extensions exist to infinite-dimensional problems such as differential delay equations and certain classes of partial differential equations (including the Navier–Stokes equations) (e.g., Marsden & McCracken, 1976, Sections 8 and 9).

Example. Consider the oscillator $\ddot{x} - (\mu - x^2)\dot{x} + x = 0$ (an example of a so-called Liénard system), which, with $u = x$, $v = \dot{x}$, we can write as the first-order system

$$\begin{aligned} \dot{u} &= v, \\ \dot{v} &= -u + (\mu - u^2)v. \end{aligned} \tag{2}$$

The origin is a fixed point for each μ, with eigenvalues $\lambda(\mu), \bar{\lambda}(\mu) = \frac{1}{2}\left(\mu \pm \mathrm{i}\sqrt{4 - \mu^2}\right)$. The system has a Hopf bifurcation at $\mu = 0$. We have $d = \frac{1}{2}$ and $a = -\frac{1}{8}$, so the bifurcation is supercritical and there is a stable isolated periodic orbit (limit cycle) if $\mu > 0$ for each sufficiently small μ (see Figure 1).

Hopf Bifurcation for Maps

There is a discrete-time counterpart of the Hopf bifurcation that occurs when a pair of complex conjugate eigenvalues of a map crosses the unit circle. It is slightly more complicated than the version for flows. The corresponding theorem was first proved independently by Naimark (1959) and Sacker (1965), and the bifurcation is, therefore, sometimes called the Naimark–Sacker bifurcation. A proof can again be found in Marsden & McCracken (1976).

Consider the planar map $f_\mu : \mathbb{R}^2 \to \mathbb{R}^2$, with parameter μ, and suppose it has a fixed point that without loss of generality, we may assume to be located at $(x, y) = (0, 0)$. Suppose further that at this fixed point, Df_μ has a complex conjugate pair of eigenvalues $\lambda(\mu), \bar{\lambda}(\mu) = |\lambda(\mu)|e^{\pm i\omega(\mu)}$ and that for a certain value of μ (which we may assume to be 0), the following conditions are satisfied:

(i) $|\lambda(0)| = 1$ (nonhyperbolicity condition: eigenvalues on the unit circle);

(ii) $\lambda^k(0) \neq 1$ for $k = 1, 2, 3, 4$ (nonstrong-resonance condition);

(iii) $\left.\dfrac{d|\lambda(\mu)|}{d\mu}\right|_{\mu=0} = d \neq 0$ (transversality condition);

(iv) $a \neq 0$, where

$$a = -\mathrm{Re}\left[\frac{(1-2e^{ic})e^{-2ic}}{1-e^{ic}} c_{11}c_{20}\right] - \tfrac{1}{2}|c_{11}|^2 - |c_{02}|^2 + \mathrm{Re}(e^{-ic}c_{21}), \qquad c = \omega(0),$$

with

$$c_{20} = \tfrac{1}{8}[(f_{xx} - f_{yy} + 2g_{xy}) + i(g_{xx} - g_{yy} - 2f_{xy})],$$

$$c_{11} = \tfrac{1}{4}[(f_{xx} + f_{yy}) + i(g_{xx} + g_{yy})],$$

$$c_{02} = \tfrac{1}{8}[(f_{xx} - f_{yy} - 2g_{xy}) + i(g_{xx} - g_{yy} + 2f_{xy})],$$

$$c_{21} = \tfrac{1}{16}[(f_{xxx} + f_{xyy} + g_{xxy} + g_{yyy}) + i(g_{xxx} + g_{xyy} - f_{xxy} - f_{yyy})]$$

(genericity condition).

Then an invariant simple closed curve bifurcates into either $\mu > 0$ or $\mu < 0$, depending on the signs of d and a. This invariant circle is attracting if it bifurcates into the region of μ where the origin is unstable (a supercritical bifurcation) and repelling if it bifurcates into the region where the origin is stable (a subcritical bifurcation).

Note that this result says nothing about the dynamics *on* the invariant circle. In fact, the dynamics on the circle has the full complexity of so-called "circle maps" (including the possibility of having attracting periodic orbits on the invariant circle) and depends sensitively on any perturbation (see the example below). Consequently, unlike the Hopf bifurcation for flows, the Hopf bifurcation for maps is not structurally stable.

Example. Consider the following family of maps:

$$f_\mu\begin{pmatrix} x \\ y \end{pmatrix} = (1 + d\mu + a(x^2 + y^2)) \times \begin{pmatrix} \cos(c + b(x^2 + y^2)) & -\sin(c + b(x^2 + y^2)) \\ \sin(c + b(x^2 + y^2)) & \cos(c + b(x^2 + y^2)) \end{pmatrix} \times \begin{pmatrix} x \\ y \end{pmatrix}. \tag{3}$$

The origin $(x, y) = (0, 0)$is a fixed point for each μ. The Jacobian matrix of f_μ at this fixed point is

$$Df_\mu(0, 0) = (1 + d\mu)\begin{pmatrix} \cos c & -\sin c \\ \sin c & \cos c \end{pmatrix} \tag{4}$$

and the eigenvalues are $\lambda(\mu), \bar{\lambda}(\mu) = (1 + d\mu)e^{\pm ic}$. The map takes a simpler, semi-decoupled form in polar coordinates $r = \sqrt{x^2 + y^2}$, $\theta = \arctan(y/x)$:

$$\begin{pmatrix} r \\ \theta \end{pmatrix} \mapsto \begin{pmatrix} r(1 + d\mu + ar^2) \\ \theta + c + br^2 \end{pmatrix}. \tag{5}$$

This five-parameter map is in fact the *normal form* for the Hopf bifurcation up to cubic terms (i.e., by a smooth change of coordinates, we can bring any f_μ into this form (plus higher-order terms)). The parameters a, c, and d in (3) and (5) are precisely those defined in the conditions above. We choose $a = -0.02$, $b = c = 0.1$, $d = 0.2$. The map then undergoes a supercritical Hopf bifurcation at $\mu = 0$, as can be confirmed by a simple graphical analysis of the decoupled r map (for $a > 0$, it would be subcritical). For sufficiently small $\mu > 0$, we have an attracting invariant circle given by $r = \sqrt{-d\mu/a}$ (see Figure 2). On the circle the map is given by $\theta \mapsto \theta + c - bd\mu/a$. This is simply a rotation through a fixed angle $\phi = c - bd\mu/a$, giving periodic orbits if $2\pi/\phi \in \mathbb{Q}$, or dense (irrational) orbits if $2\pi/\phi \in \mathbb{R} \setminus \mathbb{Q}$.

If the Hopf bifurcation occurs in a map associated with the return map (Poincaré map) near a periodic orbit of an autonomous flow, then the bifurcation is often called a secondary Hopf bifurcation. In this case, the invariant curve corresponds to an invariant torus because the flow and attracting periodic orbits on the circle correspond to mode-locked periodic motion on the torus, while dense orbits correspond to quasi-periodic motion.

Degenerate Hopf Bifurcations

If one or more of the listed conditions for a Hopf bifurcation are not satisfied (for instance, because of symmetry), one may still have the emergence of a

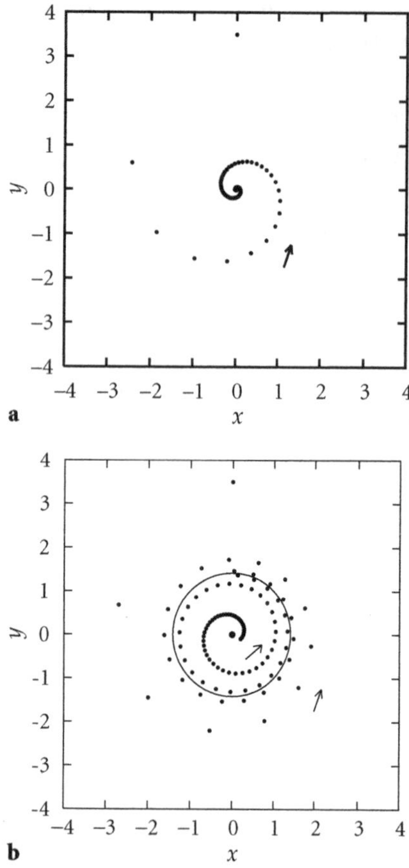

Figure 2. Phase portraits of (3) for (a) $\mu = -0.2$, (b) $\mu = 0.2$. There is a supercritical Hopf bifurcation at $\mu = 0$. ($a = -0.02$, $b = c = 0.1$, $d = 0.2$.)

periodic orbit, but some of the conclusions of the theorem may cease to hold true. The bifurcation is then called a "degenerate Hopf bifurcation." For instance, if the transversality condition is not fulfilled, the fixed point may not change stability, or multiple periodic solutions may bifurcate. An important case is provided by a Hamiltonian system for which complex eigenvalues come in symmetric quadruples, and therefore, the transversality condition cannot be satisfied. This is why the analogous bifurcation in Hamiltonian systems (the so-called "Hamiltonian–Hopf bifurcation" (van der Meer, 1985)) is much more complicated requiring, for one thing, a four-dimensional phase space.

Applications

The balance between local excitation and global damping mentioned above occurs commonly in physical systems. Thus, the Hopf bifurcation underlies many spontaneous oscillations, including airfoil flutter and other wind-induced oscillations (for example, the Tacoma–Narrows bridge collapse) in structural engineering systems, vortex shedding in fluid flow around a solid body at sufficiently high stream velocity, LCR oscillations in electrical circuits, relaxation oscillations (the van der Pol oscillator), the periodic firing of neurons in nervous systems (the FitzHugh–Nagumo equation), oscillations in autocatalytic chemical reactions (the Belousov–Zhabotinsky reaction) as described by the Brusselator and similar models, oscillations in fish populations (as described by Volterra's predator-prey model), and periodic fluctuations in the number of individuals suffering from an infectious disease (as described by epidemic models), among others.

GERT VAN DER HEIJDEN

See also **Bifurcations; Center manifold reduction; Normal forms theory; Phase space; Tacoma Narrows Bridge collapse**

Further Reading

Andronov, A.A., Vitt, A.A. & Khaikin, S.E. 1966. *Theory of Oscillators*, translated from the Russian by F. Immirzi, edited and abridged by W. Fishwick, Oxford: Pergamon Press

Hopf, E. 1942. Abzweigung einer periodischen Lösung von einer stationären Lösung eines Differentialsystems. *Berichten der Mathematisch-Physischen Klasse der Sächsischen Akademie der Wissenschaften zu Leipzig*, XCIV: 1–22. An English translation, with comments, is included as Section 5 in Marsden & McCracken (1976)

Marsden, J.E. & McCracken, M. 1976. *The Hopf Bifurcation and Its Applications*, Berlin and New York: Springer

Naimark, J. 1959. On some cases of periodic motions depending on parameters. *Doklady Akademii Nauk SSSR*, 129: 736–739

Poincaré, H. 1892. *Les Méthodes Nouvelles de la Mécanique Céleste, vol. 1*, Paris: Gauthier-Villars

Sacker, R.S. 1965. On invariant surfaces and bifurcations of periodic solutions of ordinary differential equations. *Communications on Pure and Applied Mathematics*, 18: 717–732

van der Meer, J.C. 1985. *The Hamiltonian Hopf Bifurcation*, Berlin and New York: Springer

HOPFIELD MODEL

See **Cellular nonlinear networks**

HORSESHOES AND HYPERBOLICITY IN DYNAMICAL SYSTEMS

In dynamical systems, *hyperbolicity* refers to the phenomenon in which nearby orbits diverge exponentially fast. It implies instability and sensitive dependence on initial conditions; when occurring on a wide enough scale, it implies *chaos*. This entry surveys and puts into perspective the core ideas in hyperbolic theory, one of the most developed branches in the mathematical theory of dynamical systems today.

For definiteness, we discuss only discrete-time systems, that is, systems generated by the iteration of a map f of a space (usually Euclidean space or a manifold) to itself, leaving analogous results in the continuous-time case to the reader.

Hyperbolic Fixed Points

A linear map $T : \mathbb{R}^n \to \mathbb{R}^n$ is called hyperbolic if none of its eigenvalues lies on the unit circle. A nonlinear map f is said to have a hyperbolic fixed point at p if $f(p) = p$ and $Df(p)$ is a hyperbolic linear map. Thus, there are three kinds of hyperbolic fixed points: attracting (when the moduli of all the eigenvalues of $Df(p)$ are < 1), repelling (when they are all > 1), and saddle type (when some are > 1 and some are < 1). In the saddle case, p has a stable manifold $W^{\mathrm{s}}(p)$ and an unstable manifold $W^{\mathrm{u}}(p)$ consisting of points the orbits of which tend to p in forward and backward time, respectively.

Smale's Horseshoe

Stephen Smale's horseshoe (Smale, 1967) can be seen as a globalization of the idea of a saddle-type fixed point to an invariant set with complicated dynamics. Geometrically, the presence of a horseshoe implies stretching and folding. In terms of orbit types, it implies the existence of random motion as unpredictable as the repeated flipping of a coin (see below).

A version of the horseshoe map is shown in Figure 1: f stretches the square B in the horizontal direction, compresses it in the vertical direction, bends the resulting rectangle into the shape of a horseshoe, and puts it back on top of B as shown. The two shaded vertical strips are mapped onto the two shaded horizontal strips, the union of which is equal to $B \cap f(B)$. Reasoning inductively, we see that after n iterates, $\cap_{i=0}^{n} f^i(B)$ is the union of 2^n disjoint horizontal strips. Iterating backwards as well as forwards, we see that Λ, the set of points that remain in B in all forward and backward times, is the product of two Cantor sets.

If we label the left vertical strip in B "L" and the right one "R" (or "head" and "tail"), then every point x in Λ can be coded into a bi-infinite sequence of L and R where the ith coordinate is L if and only if $f^i x$ is in the left strip. It is easy to see that this defines a one-to-one correspondence between the points in Λ and the set of all possible bi-infinite strings in L and R. An immediate consequence of this coding is that Λ contains many periodic points, one corresponding to each finite block of L and R.

By *horseshoes*, one generally refers to a much larger class of objects than that depicted in Figure 1. The map f is assumed to be invertible, but it does not have to be linear anywhere. There is a box B (or a region B that can be deformed into a box) that is stretched and compressed by f and mapped to a set, which crosses over B finitely many times. Moreover, for points that remain in B, f has well-defined expanding and contracting directions, that is, it is hyperbolic. For an n-dimensional analog of the linear model in Figure 1, imagine $B = D_k \times D_{n-k}$, where the "horizontal" direction represents a k-dimensional disk and the "vertical" direction an $(n-k)$-dimensional disk (see also the Smale solenoid picture in the color plate section).

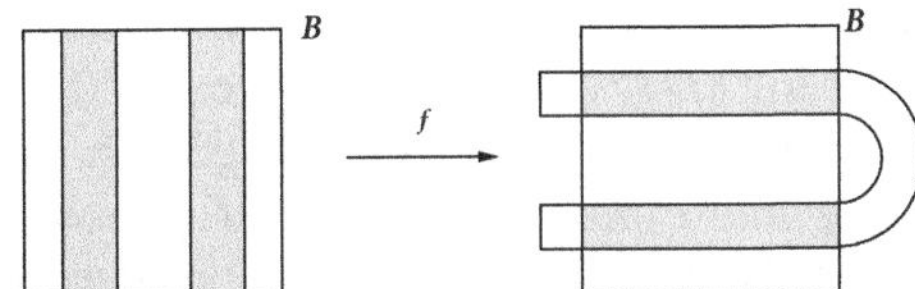

Figure 1. The horseshoe map: f sends the square B to the horseshoe on the right.

Smale's idea for the horseshoe was influenced by the work of Norman Levinson, who in the late 1940s studied a simplified version of the periodically forced van der Pol equation and proved that the resulting oscillator contains infinitely many periodic orbits with distinct periods. This map was shown to have a horseshoe by M. Levi many years later.

Dynamical complexity near homoclinic orbits was noted by Henri Poincaré. (A homoclinic point is a point in $W^s(p) \cap W^u(p)$ where p is a fixed point of saddle type; *See* **Phase space**.) An important result due to Smale says that transverse homoclinic orbits are always accompanied by horseshoes. Thus, locating transverse, homoclinic points is a means of detecting chaos.

Finally, while the presence of horseshoes implies the existence of chaotic behavior, it should be pointed out that from the probabilistic or observational point of view, this may be transient chaos. The reasons are as follows. Horseshoes have Lebesgue measure zero. It is possible to have a horseshoe Λ and at the same time for the orbit of almost every point to go to a stable equilibrium. In such a scenario, a typical orbit may come near Λ, spend some time near Λ mimicking its orbits, before it heads for its eventual destination. An experimenter tracking this orbit will observe chaotic behavior but only for a finite time period.

Uniform and Non-uniform Hyperbolicity

More general than the idea of a horseshoe is that of a uniformly hyperbolic invariant set. Let f be a smooth invertible map. A compact f-invariant set Λ is called uniformly hyperbolic if everywhere on Λ, the tangent space splits into Df-invariant subspaces $E^{\mathrm{u}} \oplus E^{\mathrm{s}}$, with $\|Df(v)\| > \|v\|$ for $v \in E^{\mathrm{u}}$, $\|Df(v)\| < \|v\|$ for $v \in E^{\mathrm{s}}$, $v \neq 0$. From the 1960s to 1970s, a detailed theory was developed for a class of dynamical systems that are uniformly hyperbolic either on their entire phase spaces or on certain important invariant sets. This theory is called Axiom A theory or uniform hyperbolic theory.

A weaker form of hyperbolicity was introduced in the 1970s. The setting here consists of a pair (f, μ) where f is a map and μ is an f-invariant Borel probability measure. Oseledec's multiplicative ergodic theorem says that at μ-almost every x, the limit $\lim_{n\to\infty} \frac{1}{n} \log \|Df^n(x)v\| = \lambda(x, v)$ exists for every tangent vector v. These asymptotic growth rates are called *Lyapunov exponents*. The pair (f, μ) is said to be non-uniformly hyperbolic if there is a positive measure set of x such that $\lambda(x, v) > 0$ for some v and < 0 for other v; that is, there is some expansion and some contraction. Pesin's paper (Pesin, 1977) helped launch a systematic study of these systems. The hyperbolicity here is non-uniform in the sense that there may be points that take arbitrarily long for the expanding and contracting behaviors to manifest themselves. Indeed, on a set of μ-measure zero, the limit above may not exist.

We put into perspective the ideas introduced: Horseshoes are examples of uniformly hyperbolic sets. They occur widely, and their occurrence does not preclude other types of dynamical behaviors. Axiom A is a more stringent condition; it requires that all important parts of the phase space be uniformly hyperbolic. This idealized picture excludes many real-life examples; at the same time it has permitted the development of a rich and extensive theory, one that is useful beyond Axiom A. As for the relation between uniform and non-uniform settings, the latter is clearly more flexible and therefore larger in scope. But the contexts are not identical: the properties of (f, μ) depend on μ, and most maps have many invariant measures. Not all invariant measures are equally important, however (*See* **Sinai–Ruelle–Bowen measures**). To complete this circle of ideas, we mention the following result of Katok: if a non-uniformly hyperbolic system has positive entropy and no zero Lyapunov exponents, then nearly all of its entropy is carried by horseshoes.

For more on uniform hyperbolic theory, *See* **Anosov and Axiom-A systems**.

Highlights from General Non-uniform Theory

The results below are very general. They hold for *all* invertible maps f (for which both f and f^{-1} are twice continuously differentiable) acting on compact domains in finite dimensions. For more detailed expositions, see Eckmann and Ruelle (1985) and Young (1993).

(1) *Local nonlinear theory* (Pesin, 1977): It is shown that corresponding to negative and positive Lyapunov exponents are measurable families of local stable and unstable manifolds.

(2) *Structure of conservative systems*, that is, when μ has a density (Pesin, 1977): Assume there are no zero Lyapunov exponents. Then the phase space is made up of at most countably many ergodic components, each one of which is, up to a permutation of sets, mixing.

(3) *Relation among entropy, Lyapunov exponents, and dimension*: For conservative systems, there is the following entropy formula (Pesin, 1997):

$$h_\mu(f) = \int \sum_{\lambda_i > 0} \lambda_i m_i \, d\mu.$$

Here $h_\mu(f)$ is Kolmogorov–Sinai entropy, and the λ_i are distinct Lyapunov exponents with multiplicity m_i. In general, i.e. for arbitrary invariant measures, "=" in the formula above is replaced by "$\leq$" (Ruelle, 1978), and dimension enters to give the following equalities (Ledrappier-Young, 1985):

$$h_\mu(f) = \int \sum_{\lambda_i > 0} \lambda_i \delta_i \, d\mu = -\int \sum_{\lambda_i < 0} \lambda_i \delta_i \, d\mu.$$

Here δ_i is a notion of *partial dimension*; it gives the dimension of μ in the direction of the subspace of vectors whose growth rates are equal to λ_i; in particular $0 \leq \delta_i \leq m_i$. Together the results in (3) express the following two basic principles: randomness is created entirely from expansion; and dissipation can be measured in "wasted" expansion, that is, by the gap in Ruelle's inequality or, equivalently, by the dimensions of the invariant measure.

(4) *Sinai–Ruelle–Bowen (SRB) measures*: A special invariant measure (to play the role of volume) has been identified for dissipative systems with strange attractors (*See* **Sinai–Ruelle–Bowen measures**).

Examples

A large number of dynamical systems are believed to fit into the framework of non-uniform hyperbolic theory, but relatively few (that are not uniformly hyperbolic) have been rigorously studied. This is due in part to technical difficulties in proving the positivity of Lyapunov exponents in the absence of invariant cones (*See* **Lyapunov exponents**). Among the systems successfully analyzed, the most prominent are (1) Billiards flows, both scattering billiards (including the periodic Lorentz gas) and some focusing billiards (including the stadium), and geodesic flows on manifolds of nonpositive curvature. These systems are conservative and have natural invariant measures. (2) One-dimensional maps (such as the logistic family) and a class of attractors with one direction of instability (including the Hénon maps). For these maps, μ either has a density or is an SRB measure.

LAI-SANG YOUNG

See also **Anosov and Axiom-A systems; Chaotic dynamics; Lyapunov exponents; Sinai–Ruelle–Bowen measures**

Further Reading

Eckmann, J.-P. & Ruelle, D. 1985. Ergodic theory of chaos and strange attractors. *Reviews of Modern Physics*, 57: 617–656

Guckenheimer, J. & Holmes, P. 1983. *Nonlinear Oscillations, Dynamical Systems, and Bifurcations of Vector Fields*, New York: Springer

Pesin, Ya.B. 1977. Characteristic Lyapunov exponents and smooth ergodic theory. *Russian Mathematics Surveys*, 32: 55–114

Smale, S. 1967. Differentiable dynamical systems. *Bulletin of the American Mathematical Society*, 73: 747–817

Young, L.-S. 1993. Ergodic theory of differentiable dynamical systems. In *Proceedings of the NATO Advanced Study Institute, Hillerφd, June 20–July 2, 1993*, edited by B. Branner & P. Hjorth, Dordrecht: Kluwer

Young, L.-S. 1999. Ergodic theory of chaotic dynamical systems. In *Proceedings of the 12th International Congress of Mathematics and physics, Brisbane, Australia, 1997*, International Press, pp. 131–143

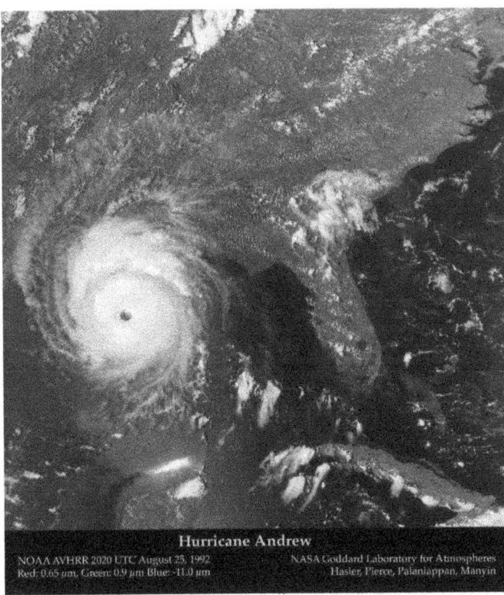

Figure 1. Satellite view of Hurricane Andrew (1992), 25 August 1992 at 20:20 UT. Courtesy of National Oceanic and Atmospheric Administration.

HURRICANES AND TORNADOES

As examples of nonlinear emergent structures, hurricanes and tornadoes are both violent and destructive atmospheric storms, albeit of different scales: a hurricane is much larger than a tornado and lives longer. The word *hurricane* derives from the Spanish word *huracan* which, in turn, stems from the Caribbean name for the god of evil; *tornado* is a variant of the Spanish *tronada* for thunderstorm. Both are associated with very strong winds circling around a center, called the eye, which is the calmest region of the storm with the lowest pressure.

Due to the Coriolis force (which stems from the Earth's rotation), hurricane winds spiral around the eye in a counter-clockwise direction in the Northern Hemisphere, whereas in the Southern Hemisphere the opposite is true. The term *hurricane* is often restricted to storms occurring in the North Atlantic Ocean; the same phenomenon over the West Pacific Ocean is called a typhoon, and in the South Pacific, a cyclone. For simplicity, in what follows, the term *hurricane* is used for the entire phenomenon.

A hurricane forms as a tropical cyclone in areas over warm ocean waters (at temperatures over 26.5°C) under a number of other preconditions such as a moist atmosphere with a sufficiently strong vertical temperature gradient, and latitudes exceeding approximately 8° from the equator (the latter condition provides sufficient Coriolis force to create a wind rotation). Its energy source is associated with latent heat release. The low-pressure center takes in thermal energy and moisture from the ocean surface, the warm air ascends because of convection, and the higher pressure in the cold upper levels of the atmosphere pushes it outward. When the warm air ascends it cools and the temperature of the air may fall below its dew point, releasing water vapor that is able to condense into water droplets. The

Figure 2. Photo of a tornado in Seymour, Texas (1979). Courtesy of National Oceanic and Atmospheric Administration.

condensation releases heat that in turn warms the air, increasing its ascent; at this stage the storm works as a self-sustaining heat engine. When the wind speed exceeds 74 mph (119 km h^{-1}), the tropical cyclone is classified as a hurricane. In some hurricanes wind speeds reach over 280 km h^{-1}. A mature hurricane is nearly circularly symmetrical (see Figure 1) and is moved by the airstreams in which it is embedded.

A tornado is a dark, funnel-shaped cloud containing violently rotating air that extends downward to the Earth (Figure 2). It is often accompanied by lightning and a roaring sound. Although it is not yet completely known how tornadoes form, the mechanism of tornado formation is presently attributed to mechanical and thermodynamical processes (the role of atmospheric electricity is disputable). Tornadoes are formed within massive and powerful storms called supercells. The supercells develop inside cumulonimbus clouds, which owe their name to the Latin *cumulus*, meaning mound or heap, and *nimbus*, which refers to rain. Cumulonimbus clouds may reach up to 18 km into the stratosphere, where the distinctive anvil-shaped top of a cloud consisting of ice crystals is formed. Cumulonimbus clouds often form along a squall line (a wall of clouds and harsh weather often associated with a cold front) where, as a result of convective instability, a supercell is generated.

Once a supercell storm is formed, wind shear creates a mesocyclone, a region of rotating updrafts within a mature thunderstorm. Eventually, the mesocyclone reaches below the cumulonimbus cloud, whereupon it is considered a tornado cyclone. As air is pulled into the tornado cyclone, the water vapor condenses into a visible funnel cloud. When the latter touches the surface of the Earth, it is finally considered a full-fledged tornado. As time goes on, the tornado takes on a brownish color because of the dirt and debris it has drawn in. This whole process is termed the organizational stage, which leads to the mature stage when the tornado is at its largest and strongest. Eventually, when its source of warm, humid air is gone, the tornado begins to degenerate. This occurs when the atmosphere is finally stabilized due to convective processes. The tornado enters a shrinking stage and begins gradually weakening as it loses fuel. The funnel decreases to a thin column and becomes fragmented and disorganized in the decaying stage, although it may still remain destructive.

Hurricanes and tornadoes are destructive atmospheric phenomena, and not only because of strong winds. Since both hurricanes and tornadoes are storms, they are typically associated with excessive rainfall that can cause flooding. Another side effect of a hurricane is the so-called storm surge, which occurs when the ocean rises above its normal tide level. In fact, a hurricane can cause even more damage through storm surge than through strong winds. Still another potential side effect is the formation of tornadoes by thunderstorms embedded in the rainbands of a hurricane.

Both hurricanes and tornadoes are categorized by their wind speeds, from weak to devastating or violent, under the Saffir–Simpson scale for hurricanes and the Fujita–Pearson scale for tornadoes. While a hurricane can be hundreds of kilometers in diameter, the diameter of an average tornado is approximately 400 m. Tornadoes generally last from several minutes to several hours, and they travel up to several hundred kilometers before dying, whereas hurricanes last for several days and travel much greater distances. In the United States, about six hurricanes and 600 tornadoes are reported in an average year.

Some typical parameters of hurricanes and tornadoes are given in Table 1.

Since 1950, hurricanes have been given human names, such as Betty or George. Each year a new list of names is made, with male and female names alternating alphabetically, and the lists are reused every six years. Examples of exceptionally deadly hurricanes are the one that occurred in the Bay of Bengal in 1737, which claimed about 300,000 lives, many lost to the huge sea waves; and an 1881 typhoon in China with a similar death toll. More recent examples include the typhoon Vera (1959) that struck central Japan, causing over 5000 deaths and destroying about 40,000 homes; the hurricane Flora (1963), which presumably killed more than 6000 people in the Caribbean area, and Camille (1969), with 256 casualties and over 5 billion dollars of damage in the southern United States. The most destructive United States hurricane of record was hurricane Andrew. It blasted its way across south Florida on August 24, 1992; Louisiana and the Bahamas were also impacted. The hurricane caused 26.5 billion dollars in damage in the United States, mainly due to the winds.

	Hurricanes	Tornadoes
Wind speed (kph)	119–300	60–over 500
Diameter	200–1300 km	few m–1.5 km
Lifetime	1–30 days	1–180 min

Table 1. Some parameters of hurricanes and tornadoes (by order of magnitude).

Regarding tornadoes in the United States, extremely destructive was the Tri-State Tornado of 1925, which killed 689 people and injured nearly 2000. In the United States, tornadoes are most prevalent in "Tornado Alley," a group of central plains states. Since the terrain there is relatively flat, warm air coming up from the Gulf of Mexico and cold air coming down from Canada often clash, to form some of the most fertile tornado-producing storms in the world. One such tornado hit central Oklahoma in May 1999. Actually, it was a tornado outbreak—a collection of tornadoes all striking at the same time in the same general area. More than 70 tornadoes were observed in the region of Kansas, Oklahoma, and Texas. They caused 40 deaths and injured 675 people, with total damages estimated at 1.2 billion dollars.

From the physical viewpoint, hurricanes and tornadoes are examples of environmental vorticity, requiring interdisciplinary, nonlinear studies for their comprehension. Appropriate mathematical models comprise a set of nonlinear hydrodynamic and thermodynamic equations, and for hurricanes it is crucial to account for the Earth's rotation through the Coriolis force. Although a wide network of meteorological stations exists and available models provide some useful insights, the detailed modeling and prediction of hurricanes and tornadoes remains an exciting challenge for future practitioners of nonlinear science.

LEV OSTROVSKY AND MASHA SVERDLOV

See also **Atmospheric and ocean sciences; General circulation models of the atmosphere; Vortex dynamics of fluids**

Further Reading

Elsner, J. & Birol Kara, A. 1999. *Hurricanes of the North Atlantic*, Oxford and New York: Oxford University Press
Grazulis, T. 2001. *The Tornado: Nature's Ultimate Windstorm*, Norman: University of Oklahoma Press
Longshore, D. 1998. *Encyclopedia of Hurricanes, Typhoons, and Cyclones*, New York: Facts on File
Pielke, R., Jr. & Pielke, R., Sr. 1997. *Hurricanes: Their Nature and Impacts on Society*, New York: Wiley
Tufty, B. 1987. *1001 Questions Answered about Hurricanes, Tornadoes, and Other Natural Air Disasters*, New York: Dover

HUYGENS PRINCIPLE

Huygens principle (HP) is a notion that goes back to the classical *Traité de la Lumière* (Treatise on Light) by Christiaan Huygens, which was published in 1690. In its original meaning, HP gives the following geometric description of the wave fronts. Consider a wave front at the moment t as the source of the new (secondary) waves emanating from points of this front. HP states that the new wave front at a later time is the envelope of the fronts of these secondary waves (see Figure 1). This principle was further elaborated by Augustin Fresnel who added the superposition principle for the amplitudes of the secondary waves to explain the phenomenon of diffraction.

In the second half of the 19th century, various mathematical aspects of the HP were discussed in the pioneering works of Gustav Kirchhoff, Eugenio Beltrami, and Vito Volterra. At the same time it, became evident that different authors were using the term *Huygens principle* with different meanings. It was Jacques Hadamard who brought clarity in this area. In his fundamental "Lectures on Cauchy's Problem" published in 1923, Hadamard introduced the notion of the "Huygens principle in the narrow sense" (minor premise) (Hadamard, 1923). In physical language one can formulate this principle in the following way: *an instantaneous signal remains instantaneous for every observer at each later time*. This property implies that a localized disturbance will have an effect localized

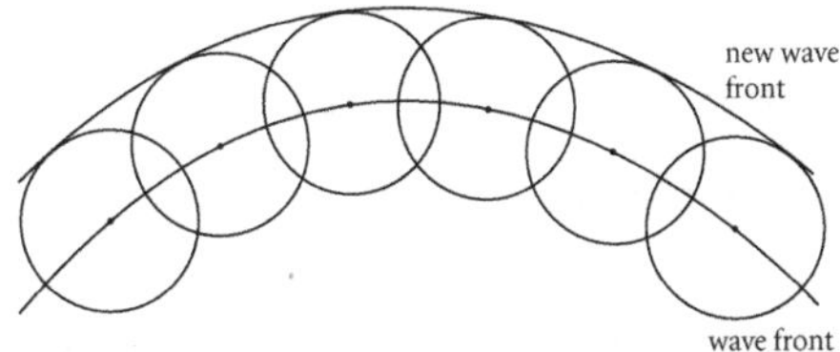

Figure 1. Diagram related to Huygens' principle.

in time at any point (for example, for sound waves after some time there will be complete silence).

Mathematically, HP in the narrow sense means that the fundamental solution of the corresponding hyperbolic equation vanishes not only outside but also inside the characteristic conoid and, thus, must be located on it. Equivalently, a hyperbolic equation satisfies the HP in the narrow sense if the solution of the Cauchy initial-value problem at a point P depends not on all the Cauchy data, but only on its part on the intersection of the characteristic conoid with vertex P with the Cauchy surface.

In contrast to the original principle ("major premise" in Hadamard's terminology), which holds for a general class of wave propagations, HP in the narrow sense is actually a remarkable property valid only for very special equations. In particular, this never happens in two dimensions, where one has wave diffusion. One can see this when a pebble falls in water, where the front wave is followed by the so-called "residual waves," and any object on the water surface is hit by the subsequent waves many times.

To explain this phenomenon, consider the fundamental solution of the wave equation

$$\left(\frac{\partial^2}{\partial t^2} - \frac{\partial^2}{\partial x_1^2} - \cdots - \frac{\partial^2}{\partial x_n^2}\right)\Phi = 0,$$

$$\Phi(x,0) = 0, \qquad \frac{\partial \Phi}{\partial t}(x,0) = \delta(x).$$

This solution has the form

$$\Phi = \frac{1}{2\pi}\frac{\theta(t^2 - |x|^2)}{\sqrt{t^2 - |x|^2}}$$

when $n = 2$, and

$$\Phi = \frac{1}{2\pi}\delta(t^2 - |x|^2)$$

when $n = 3$. Here θ is the Heaviside step-like function and δ is the Dirac delta-function. Thus, we see that indeed for $n = 3$, Φ is located on the characteristic cone $t^2 - |x|^2 = 0$, which explains the sharpness of signal transmission in our three-dimensional world, while in two dimensions this is not the case.

Corresponding formulas for the solutions of the Cauchy problem for the wave equation in two and three spatial dimensions were found by Simeon Poisson and Kirchhoff, and for general n by Orazio Tedone. From these formulas it follows that for the Euclidean spaces, the HP in the narrow sense holds in all odd dimensions starting from 3 and never holds in even dimensions.

Hadamard raised the question of how to describe all second-order hyperbolic equations that satisfy the HP in the narrow sense (*Hadamard's problem*). Hadamard found a criterion for this, but it was not effective enough to answer this question. There was a common belief that the HP in the narrow sense holds only for the wave equations in the Euclidean spaces of odd dimensions until 1953 when Karl Stellmacher found other examples (Stellmacher, 1953). He showed that the hyperbolic equation

$$\left(\frac{\partial^2}{\partial t^2} - \frac{\partial^2}{\partial x_1^2} - \cdots - \frac{\partial^2}{\partial x_n^2} + \frac{m(m+1)}{x_1^2}\right)\phi = 0$$

with integer nonnegative m satisfies the HP in the narrow sense if the dimension n is odd and large enough: $n \geq 2m+3$.

Later in 1967–68, Stellmacher and Lagnese solved the Hadamard problem in the class of the hyperbolic equations of the special form

$$\left(\frac{\partial^2}{\partial t^2} - \frac{\partial^2}{\partial x_1^2} - \cdots - \frac{\partial^2}{\partial x_n^2} + u(x_1)\right)\phi = 0,$$

when the potential depends only on one of the coordinates. They showed that the corresponding potentials are rational and (in modern terminology) can be described as the results of the Darboux transformations applied to $u = 0$ (Lagnese & Stellmacher, 1967).

The first examples of the Huygensian potentials $u(x)$ depending on more than one coordinate were found in 1993 by Berest and Veselov, who discovered a close relation of the Hadamard problem with the theory of quantum integrable systems of Calogero–Moser type (Berest & Veselov, 1994; Chalykh et al., 1999).

For the hyperbolic equations on the manifolds with nontrivial metrics, the Hadamard problem is still open even when the number of spatial variables is three. There are known only several particular cases when HP in the narrow sense is satisfied, including the modified wave equation on the symmetric spaces with even multiplicities, in particular on the spheres (Lax & Phillips, 1978) and simple compact Lie groups (Helgason, 1984), and wave equations on the spaces with the plane wave metrics (Günther, 1965). For a review of the results on the HP for other relativistic wave equations (e.g., Maxwell's equations), see Günther (1988).

Because the HP in the narrow sense is a very rare phenomenon, it is natural to ask about a weaker version. Such a version called the "generalized Huygens' principle" was introduced by Lax and Courant and is valid for a general hyperbolic equation. It states that the singularities of the solution at the point P depend only on the singularities of the initial data and only so far as these data are presented on the characteristic conoid of P. This implies that "in an approximate, and for that matter, usually satisfactory, sense any hyperbolic system does preserve the sharp signals, though in general slightly blurred" (Courant & Hilbert, 1962).

Finally, one should mention that the theory of HP is closely related to the deep theory of lacunae in the domains of dependence of hyperbolic equations (Petrovskii, 1945; Atiyah et al., 1970, 1973).

ALEXANDER P. VESELOV

See also **Characteristics; Darboux transformation; Generalized functions; Gravitational waves; Quantum nonlinearity**

Further Reading

Atiyah, M., Bott, R. & Gårding L. 1970 and 1973. Lacunae for hyperbolic differential equations with constant coefficients. *Acta Mathematica,* 124: 109–89 and 131: 145–206

Berest, Yu.Yu. & Veselov, A.P. 1994. Huygens' principle and integrability. *Russian Mathematical Surveys*, 49(6): 5–77

Chalykh, O.A., Feigin, M.V. & Veselov, A.P. 1999. Multidimensional Baker–Akhiezer functions and Huygens' principle. *Communications in Mathematical Physics*, 206: 533–566

Courant, R. & Hilbert, D. 1962. *Methods of Mathematical Physics*, New York: Interscience Publishers

Günther, P. 1965. Ein Beispiel einer nichttrivialen Huygensschen Differentialgleichung mit vier unabhängigen Variablen. *Archive for Rational Mechanics and Analysis*, 18: 103–106

Günther, P. 1988. *Huygens' Principle and Hyperbolic Equations*, Boston: Academic Press

Hadamard, J. 1923. *Lectures on Cauchy's Problem in Linear Partial Differential Equations*, New Haven: Yale University Press

Helgason, S. 1984. Wave equations on homogeneous spaces. *Lecture Notes in Mathematics*, 1077: 252–287

Huygens, Ch. 1690. *Traité de la Lumière*, Hague: Pierre vander Aa; as *Treatise on Light*, New York: Dover, 1962, and London: Dawson, 1966

Lagnese, J.E. & Stellmacher, K.L. 1967. A method of generating classes of Huygens' operators. *Journal of Mathematics and Mechanics*, 17: 461–472

Lax, P.D. & Phillips, R.S. 1978. An example of Huygens' principle. *Communications in Pure and Applied Mathematics*, 31: 415–421

Petrovskii, I.G. 1945. On the diffusion of waves and lacunas for hyperbolic equations. *Matematicheskii Sbornik,* 17: 289-370 (Russian) (Also in Petrowsky, I.G. *Selected works.* Part I. Gordon and Breach, London, 1996)

Stellmacher, K.L. 1953. Ein Beispiel einer Huygensschen Differentialgleichung. *Nachrichten der Akademie der Wissenschaften in Goettingen. Mathematisch-physikalisches Klasse,* IIa, 10: 133–138

HYDRAULIC JUMPS

See **Jump phenomena**

HYDRODYNAMIC LIMIT

See **Fluid dynamics**

HYDRODYNAMIC SOLITARY WAVES

See **Water waves**

HYDROGEN BOND

The hydrogen bond is a weak chemical bond, which nevertheless is fundamentally important in stabilizing secondary structure motifs in biological macromolecules such as proteins and DNA. Also, the peculiar properties of liquid water are attributed to a complex hydrogen bond network between neighboring water molecules. A hydrogen bond is mediated by a proton sitting in between two negatively charged atoms, the more negative of which is called the hydrogen bond donor and the other the hydrogen bond acceptor. Figure 1a shows a prototype hydrogen bond, using the N–H and the C=O group of two peptide units in an α-helix as an example (one of the most important secondary structure motif of proteins).

Hydrogen bonding leads to a dramatic distortion of the potential energy surfaces that determine the nuclei positions, giving rise to strongly anharmonic or even double-well potentials. Much of our knowledge about hydrogen bond potentials stems from vibrational spectroscopy, see, for example, Pimentel & McClellan (1976); Hadzi & Bratos (1976); Henri-Rousseau & Blaise (1998). The absorption band of the high-frequency vibration of the donor-proton bond (i.e., the NH bond in the example considered here) changes considerably upon formation of a hydrogen bond. The most evident effects are a strong red-shift with respect to the free group, an intensity increase and band broadening, often accompanied by a peculiar band shape with rich substructure. These effects could be reasonably explained using a model of the electronic (ground state) potential energy surface as a function of two coordinates, the high-frequency N–H stretching coordinate q, and the low-frequency hydrogen bond coordinate Q (see Figure 1a). A Taylor expansion of that potential energy surface yields (Hadzi & Bratos, 1976; Henri-Rousseau & Blaise, 1998)

$$V(Q,q) = \tfrac{1}{2}m\omega^2 q^2 + \tfrac{1}{2}M\Omega^2 Q^2 + \chi' Q q^2 + \cdots, \quad (1)$$

where the first two terms represent the harmonic normal modes of the N–H stretching and the hydrogen bond vibration with oscillation frequencies ω and Ω and effective masses m and M, respectively. Typical vibrational frequencies of N–H stretching modes lie in the range of $\omega = 3000\text{–}3600\,\text{cm}^{-1}$, while hydrogen bond vibrations are found around $\Omega = 50\text{–}300\,\text{cm}^{-1}$, depending on the masses of the molecular groups involved. This translates into typical spring constants of $k_{\text{N–H}} \approx 650\,\text{N/m}$ for the N–H bond and $K_{\text{N–O}} \approx 10-20\,\text{N/m}$ for the hydrogen bond. In the case of most molecular systems, the harmonic approximation is extremely good, and higher order terms in Equation (1) can be treated as weak perturbation (being responsible, for example, for energy dissipation). Harmonic normal modes have the property to decouple completely. Linear

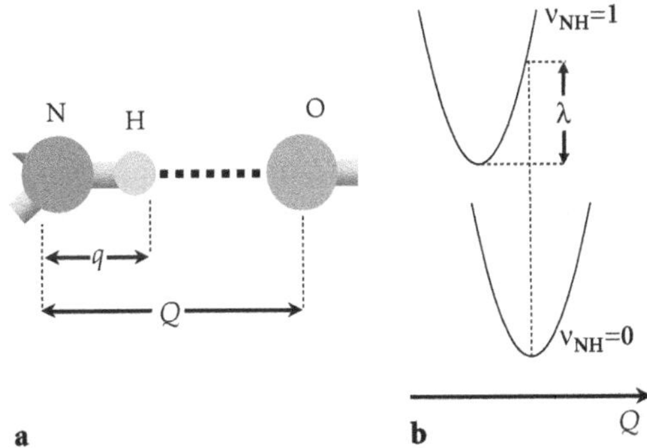

Figure 1. (a) A prototype hydrogen bond. The discussion presented here applies for other hydrogen bonds in essentially the same way. (b) Potential energy surfaces of the $\nu_{\text{NH}} = 0$ and the $\nu_{\text{NH}} = 1$ level.

phenomena, such as propagation of linear phonon wave packets, can be fully understood in the harmonic approximation.

However, one of the striking features of hydrogen bonds is an extraordinarily large anharmonicity, described by third, and higher, order terms in the expansion Equation (1). This anharmonicity leads to appreciable nonlinear behavior, a property that makes hydrogen-bonded crystals interesting objects to study in nonlinear science.

The problem can be simplified significantly when taking into account the more than one order of magnitude difference in the N–H and the hydrogen bond vibration frequency, allowing an adiabatic separation of timescales to be introduced. This is done in exactly the same way as the Born–Oppenheimer approximation separates off the fast motion of the electrons from the slow motion of the nuclei. In the case of a hydrogen bond, the motion of the fast N–H vibration adiabatically adapts to the position of the coordinate Q of the hydrogen bond. When the coordinate Q is held fixed, one can recast Equation (1) in the form

$$V(Q,q) = \tfrac{1}{2}m\omega_{\text{eff}}^2 q^2 + \tfrac{1}{2}M\Omega^2 Q^2 \quad (2)$$

with

$$\omega_{\text{eff}} \approx \omega + \frac{\chi'}{m\omega} Q \equiv \omega + \chi Q. \quad (3)$$

In this approximation, the vibration frequency of the N–H stretching mode, and hence its excitation energy, varies linearly with the hydrogen bond distance Q. The other third-order term not considered in Equation (1) ($\sim Q^2 q$) leads to a Q-dependent shift of the N–H-bond-length without change of its excitation energy. Such a linear dependence is observed experimentally for a large variety of hydrogen-bonded crystals (Pimentel & McClellan, 1976), which allows direct measurement of the nonlinear coupling constant: $\chi_{\text{N–H}} \approx 500-1500\,\text{cm}^{-1}/\text{Å}$ or

$\chi_{N\text{–}H} \approx 1-3 \times 10^{-10}$ N. The nonlinear coupling constant varies significantly with hydrogen bond type.

Starting from Equations (2) and (3), one can construct potential energy surfaces (see Figure 1b), leading to the displaced oscillator picture that is well known from electronic Franck–Condon transitions. However, keep in mind that all this is happening on the *electronic ground state* potential surface! Each potential energy surface corresponds to a vibrational excitation level $\nu_{N\text{–}H} = 0, 1, \ldots$ of the N–H stretching mode, and describes the total energy of the system as a function of the hydrogen bond coordinate Q. It is the nonlinear coupling term χ that gives rise to the displacements of the potential energy surfaces. The reorganization energy λ, that is, the energy the system gains by relaxation towards the bottom of excited state potential energy surface after a vertical "Franck–Condon"-like excitation (see Figure 1b), is given by

$$\lambda = \frac{\chi^2}{2M\Omega^2} = \frac{\chi^2}{2K_{N\cdots O}}. \qquad (4)$$

Biological macromolecules, such as proteins and DNA, often form one-dimensional quasicrystals with approximate translational symmetry, whose structures are stabilized by hydrogen bonds. This observation motivated Alexander Davydov to speculate about nonlinear collective phenomena in bio-macromolecules (Davydov, 1979). Davydov discussed only the C=O vibration in a N–H···OC-hydrogen bond, which shows the same phenomenon with a smaller, yet still appreciable, nonlinearity $\chi_{CO} \approx 300$ cm^{-1}/Å or $\chi_{CO} \approx 6 \times 10^{-11}$ N (Careri et al., 1984). Dipole–dipole interaction between adjacent C=O and N–H vibrations tends to delocalize the excitation along the crystal, forming a vibrational exciton (vibron). Coupling of the vibron to lattice-deformation modes, mediated through the nonlinearity χ of the hydrogen bonds, subsequently self-localizes the excitation. The hydrogen bond is getting stronger after excitation of the high frequency vibration (see Figure 1b), leading to a contraction of the macromolecular backbone. As a result, solitons and/or polarons may be formed.

PETER HAMM

See also **Color centers; Davydov soliton; Frank–Condon factor**

Further Reading

Careri, G., Buontempo, U., Galluzzi, F., Scott, A.C., Gratton, E. & Shyamsunder, E. 1984. Spectroscopic evidence for Davydov-like solitons in acetanilide *Physical Review* A, 30(8): 4689–4702

Davydov, A.S. 1979. Solitons in molecular systems. *Physica Scripta*, 20: 387–394

Hadzi, D. & Bratos, S. 1976. Vibrational spectroscopy of the hydrogen bond. In *The Hydrogen Bond: Recent Developments in Theory and Experiment*, edited by P. Schuster, G. Zundel & C. Sandorfy, Amsterdam: Elsevier, pp. 565–611

Henri-Rousseau O. & Blaise P. 1978. The infrared spectral density of weak hydrogen bonds within the linear response theory. *Advances in Chemical Physics*, 103: 1–186

Pimentel, G.C. & McClellan, A. 1960. *The Hydrogen Bond*, San Francisco: Freemann

HYDROTHERMAL WAVES

Hydrothermal waves are traveling waves produced by a supercritical instability of the thermocapillary flow of a thin liquid layer with a free surface subjected to a horizontal temperature gradient.

This article presents experiments in one spatial dimension, then gives some indications about the physical basis for wave effects. Extended systems are then considered and applications are described.

Experimental Observation in One Spatial Dimension

Hydrothermal waves were first experimentally observed in hot-wire convection experiments (Vince & Dubois, 1992), and a geometrically simpler system was then imagined by Daviaud & Vince (1993) that led to the first clear experimental observation of hydrothermal waves.

The classical experiment is as follows: a channel whose long sides are thermoregulated copper blocks contains a fluid with free surface. Its typical section is depicted in Figure 1. The horizontal gradient of temperature on the free surface leads to a gradient of surface tension. This results in a stationary flow of the surface from the hot side (low surface tension) toward the cold side (high surface tension): this is the Marangoni effect. Due to mass conservation, there is a return flow in the bottom of the layer. This is the basic thermocapillary flow (Kuhlnan, 1999; Schatz & Neitzel, 2001), the instability of which leads to hydrothermal waves (Smith & Davis, 1983; Davis, 1987). A similar flow is present in the melted wax of a candle (the cold side being the external boundary and the hot side being the wick). For visualization purposes, silicon oil is used because it is transparent and allows shadowgraphy.

For fluid depth h larger than the capillary length λ_c, one observes first when increasing the temperature gradient corotative rolls with their axis parallel to the cold and hot side of the container. Then hydrothermal waves of type 1 (HW1, plane waves) appear on top of the rolls (Figure 2a) (Riley & Neitzel, 1998). HW1 are emitted by "line"-sources that extend over the whole extension between the hot and the cold sides. For smaller depth $h < \lambda_c$, waves of type 2 (HW2) are observed (Figure 2b,c); they are circular and emitted by point sources located on the cold side of the container (Burguete et al., 2001; Garnier & Chiffaudel, 2001).

When the fluid depth is even larger (typically larger than the depth λ_{th} for which thermocapillary

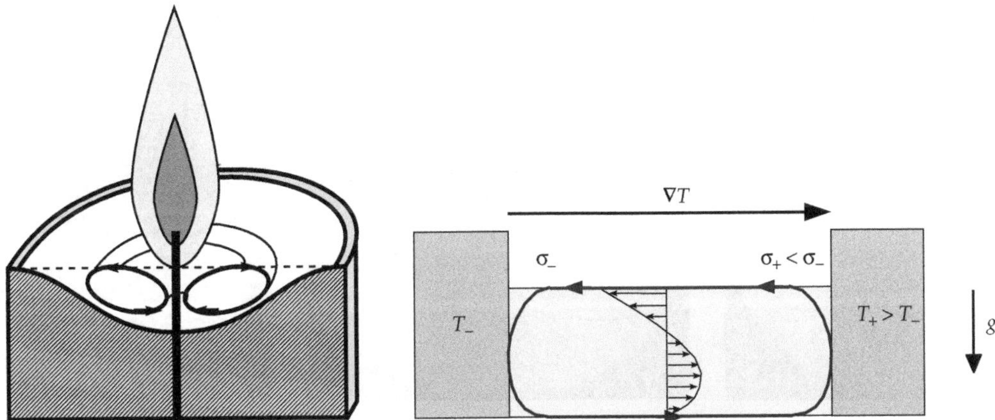

Figure 1. Thermocapillary basic flow created by Marangoni effect and susceptible to instability into hydrothermal waves. Left: the melted wax below the flame of a candle is an example of thermocapillary flow. Right: typical section of a laboratory setup. A horizontal temperature gradient is applied to a thin liquid layer with a free surface. Hydrothermal waves (not drawn) will propagate in the horizontal direction orthogonal to the temperature gradient ∇T, and with a small component along the gradient.

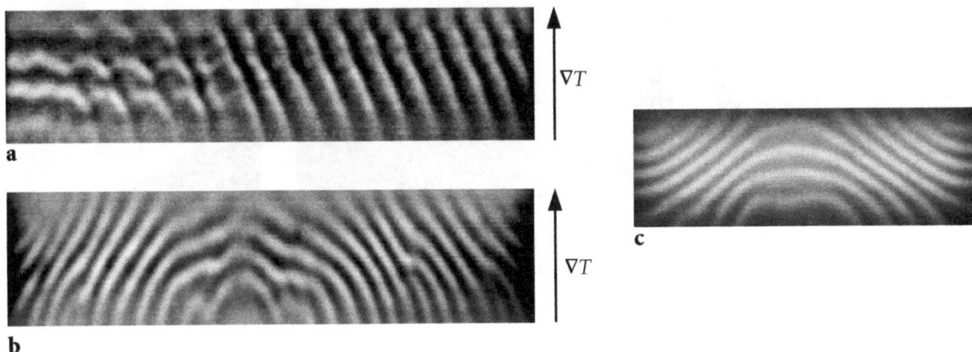

Figure 2. Shadowgraphic images of hydrothermal waves in rectangular cells. (a): $L_x = 30$ mm, $L_y = 180$ mm, $h = 2.75$ mm, $\Delta T = 7$ K. A single right-going wave of type 1 is present. The left side is a source, the amplitude of the wave is weak, and stationary corotative rolls orthogonal to the gradient are present. (b): $L_x = 30$ mm, $L_y = 180$ mm, $h = 1$ mm, $\Delta T = 7.3$ K. (c): $L_x = 30$ mm, $L_y = 90$ mm, $h = 1$ mm, $\Delta T = 5$ K. Hydrothermal waves of type 2, with a source located at the center of the cold side of the channel.

forces balances thermogravity forces), hydrothermal wave instability is replaced by a stationary instability: parallel rolls with axis aligned with the temperature gradient.

Physical Mechanisms

Vertical Temperature Gradient: Bénard–Marangoni Instability

Pearson (1958) gave a simple mechanism to explain hexagon formation in Bénard–Marangoni convection. In that case, the temperature gradient is purely vertical (cold at the surface, and hot at the bottom) and the velocity of the fluid is zero before the instability. Let us consider a positive temperature perturbation at the surface of the fluid: a point has a temperature larger than its surroundings. Because surface tension decreases with temperature for simple fluids, the surface tension is smaller at that point. This implies a differential stress on the free surface, and according to the equations of motion, the fluid has to flow away from that point (fluids flow from regions of small surface tension toward regions of large surface tension). But due to mass conservation, some fluid must rise up from the bulk of the fluid to the point on the surface (in the same way that wax climbs around the wick and flows away on the surface towards cooler regions). This ascending fluid is at a higher temperature because of the vertical temperature gradient. In conclusion, any positive temperature perturbation at the free surface is amplified; this means that there is an instability, namely, the Bénard–Marangoni instability into stationary hexagons, as can be shown by a linear stability analysis.

Horizontal Temperature Gradient: Hydrothermal Waves

Basically, the mechanism is the same, but one has to take into account, first, that the basic flow has

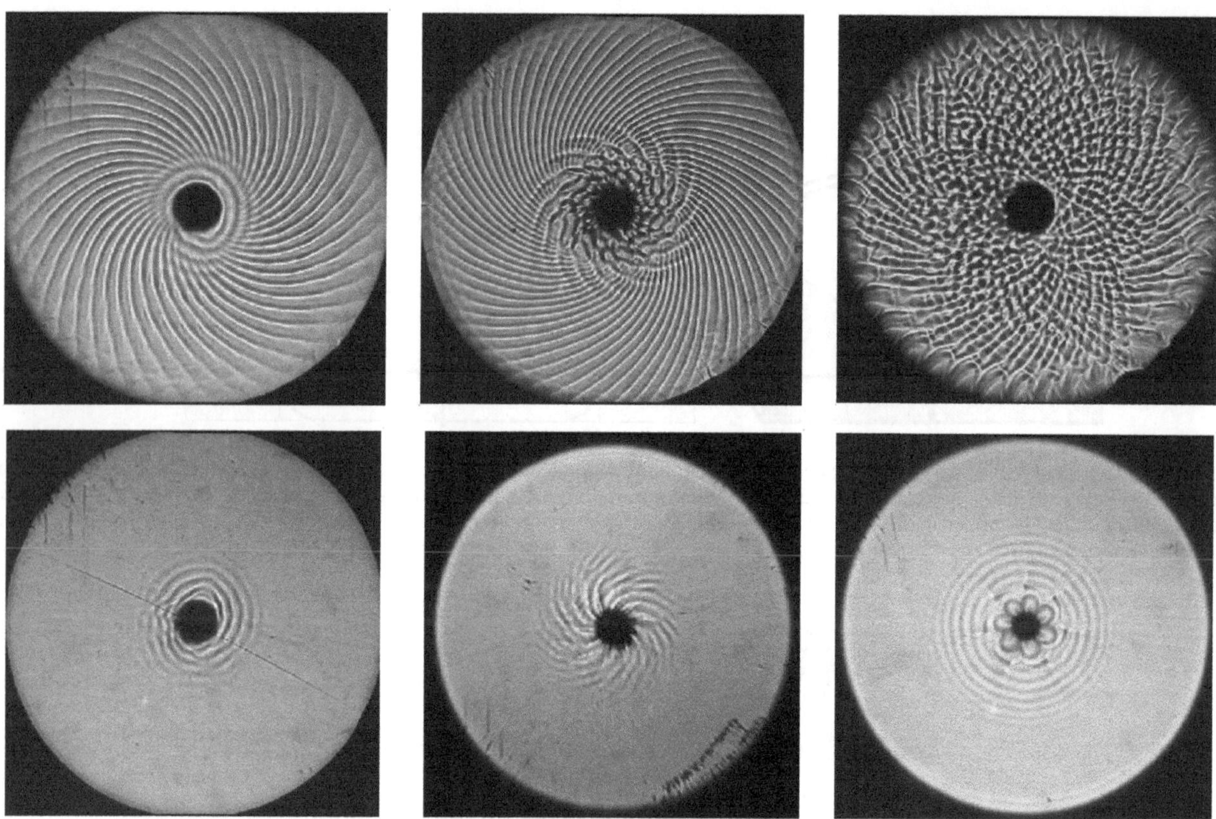

Figure 3. Top three pictures: extended patterns: single HW1 ($h = 1.9$ mm, $\Delta T = 14.25$ K), HW1 and HW2 together ($h = 1.2$ mm, $\Delta T = 20$ K), turbulent HW1 ($h = 1.9$ mm, $\Delta T = 25$ K). Bottom three pictures: localized patterns: HW2 ($h = 1.2$ mm, $\Delta T = 10$ K), inverted spirals ($h = 1.2$ mm, $\Delta T = -9$ K), and flowers ($h = 1.9$ mm, $\Delta T = -7$ K).

a finite velocity (see Figure 1), and second, that the vertical temperature profile is nonmonotonous. This intricacy results in a time-dependent pattern, in contrast to the stationary hexagons described above. Time-dependence is due to the existence of two different timescales: one for the relaxation of thermal perturbations (depending on the thermal diffusion coefficient) and the other for the relaxation of velocity perturbations (depending on kinematic viscosity). The ratio of those two timescales is called the Prandtl number, Pr.

Smith (1988) expressed two different tentative mechanisms depending on Pr in the limit cases of a flow dominated by inertial effects ($\mathrm{Pr} \to 0$) or by viscous effects ($\mathrm{Pr} \to \infty$). The relaxation of temperature and velocity perturbations then occur on very different timescales. Depending on the signs of the temperature and velocity gradients, an oscillatory behavior is shown to be unstable and to propagate along the horizontal temperature gradient (small Pr), or perpendicularly to it (large Pr).

Finally, note that hydrothermal waves are not necessarily associated with surface deflections (in contrast to gravity waves for example): they are an instability mode of the temperature field in the bulk (Pelacho & Burguete, 1999).

Experiments in 2-d

Recent experiments in extended geometries (the two horizontal dimensions are large compared to the fluid height) revealed a large variety of hydrothermal waves instability modes. Several of these obtained in cylindrical geometry (the candle geometry) are reproduced in Figure 3. The control parameter is defined as $\Delta T = T_{\mathrm{ext}} - T_{\mathrm{int}}$. This quantity can be positive or negative, and both cases are not equivalent. This is due to the presence of curvature (Garnier & Normand, 2001) that may also localize patterns near the center.

Applications

Many technological applications in which the Marangoni effect is present will involve hydrothermal waves—even when manufacturing is carried out in low gravity, where gravity-dependent, buoyancy-driven flow is reduced, for example, floating zone purification of silicon crystals, photographic films production, and melting of metals. In all of these processes, the aim is to avoid hydrothermal waves, which are detrimental to the final product, for example, by reducing the homogeneity of crystals.

In the physics laboratory, hydrothermal waves represent an ideal experimental nonlinear waves system. They are well modeled by a complex Ginzburg–Landau equation (Garnier et al., 2003) and can be used as a robust model for the study of the transition to spatiotemporal chaos. For example, as their group velocity is finite, they are subject to the convective/absolute distinction. Modulated amplitude waves have also been reported.

NICOLAS GARNIER AND ARNAUD CHIFFAUDEL

See also **Candle; Complex Ginzburg–Landau equation; Fluid dynamics; Thermo-diffusion effects**

Further Reading

Burguete, J., et al. 2001. Buoyant-thermocapillary instabilities in extended liquid layers subjected to a horizontal temperature gradient. *Physics of Fluids*, 13: 2773–2787

Daviaud, F. & Vince, J-M. 1993. Traveling waves in a fluid layer subjected to a horizontal temperature gradient. *Physical Review* E, 48: 4432–4436

Davis, S.H. 1987. Thermocapillary Instabilities. *Annual Reviews of Fluid Mechanics*, 19: 403–435

Garnier, N. & Chiffaudel, A. 2001. Two dimensional hydrothermal waves in an extended cylindrical vessel. *European Journal of Physics* B, 19: 87–95

Garnier, N. & Normand, C. 2001. Effects of curvature on hydrothermal waves instability of radial thermocapillary flows. *Comptes Rendus de l'Academie des Sciences, Series IV*, 2(8): 1227–1233

Garnier, N., et al. 2003. Nonlinear dynamics of waves and modulated waves in 1D thermocapillary flows. I. General presentation and periodic solutions. *Physica* D, 174: 1–29

Kuhlman, D. 1999. *Thermocapillary Flows*, New York: Springer

Pearson, J.R. 1958. On convection cells induced by surface tension. *Journal of Fluid Mechanics*, 4: 489–500

Pelacho, M.A. & Burguete, J. 1999. Temperature oscillations of hydrothermal waves in thermocapillary-buoyancy convection. *Physical Review* E, 59: 835–840

Riley, R.J. & Neitzel, G.P. 1998. Instability of thermocapillary–buoyancy convection in shallow layers. Part 1. Characterization of steady and oscillatory instabilities. *Journal of Fluid Mechanics*, 359: 143–164

Schatz, M.F. & Neitzel, G.P. 2001. Experiments on thermocapillary instabilities. *Annual Reviews of Fluid Mechanics*, 33: 93–127

Smith, M.K. 1986. Instabilty mechanisms in dynamic thermocapillary liquid layers. *Physics of Fluids*, 29: 3182–3186

Smith, M.K. & Davis, S.H. 1983. Saturation of Rayleigh-Taylor instability. *Journal of Fluid Mechanics*, 132: 119–144 and 132: 145–162

Vince, J-M. & Dubois, M. 1992. Hot wire below the free surface of a liquid: Structural and dynamical properties of a secondary instability. *Europhysics Letters*, 20: 505–510

HYPERBOLIC MAPPINGS

See **Maps**

HYPERCHAOS

See **Rössler systems**

HYPERCYCLE

See **Catalytic hypercycle**

HYPERELLEPTIC FUNCTIONS

See **Elliptic functions**

HYSTERESIS

The word *hysteresis* is derived from the Greek where it meant "shortcoming." In the present physical context it describes a retardation effect when the forces acting upon a body are changed. In particular, in magnetism hysteresis represents a lagging in the values of the net magnetization in a magnetic material due to a changing magnetizing field. In general, hysteresis signifies the history of dependence of physical quantities in systems responding to changes in external conditions. The term is most commonly, but not exclusively, applied to magnetic materials; for example, there is a class of metals called shape memory alloys that can be bent or stretched plastically over large distances back and forth many times without hardening.

Consider a ferromagnetic material that is originally unmagnetized. As the external magnetic field (H) is increased, the induced magnetization (M) also increases. The induced magnetization eventually saturates. Now, if the external field is reduced, the induced magnetization also is reduced, but it does not follow the original curve. Instead, the material retains a certain permanent magnetization called the remanent magnetization M_r when $H = 0$. The remanent magnetization is the permanent magnetization that remains after the external field is removed. If the external field is reduced more, the remanent magnetization will eventually be removed. The external field applied in the opposite direction for which the remanent magnetization goes to zero is termed the coercivity H_c. The product of M_r and H_c is termed the strength of the magnet. As the external field continues to reverse, permanent magnetization of the opposite polarity is created in the magnet. A similar curve is traced for the negative direction with saturation, remanent magnetization and coercivity. The hysteresis curve then retraces the previous points as the field cycles, and the shape of the loop after the first cycle is roughly the same as after many cycles. Note that the area under the hysteresis loop corresponds to the work done on the system by an external field that reorients the magnetization in a single cycle.

When an electric field (E) is applied to a ferroelectric crystal, the domains that are favorably oriented with respect to this field grow in size at the expense of those that are misaligned. In addition, favorably oriented domains may nucleate and grow until the whole crystal becomes one domain. The relation between the resulting polarization P and E is described by a hysteresis loop in analogy to the relationship between M and H for ferromagnets.

Suppose now, in general, that the system under consideration can be described by a macroscopic order parameter η. Under the influence of an external field σ coupled linearly to η, the state of the system is determined by an equation of state that expresses a minimum condition of the associated thermodynamic potential V. Assuming the presence of at least one control parameter leads directly to the problem of catastrophes, an area investigated by René Thom. The cusp catastrophe is described by the potential

$$V(\eta) = \frac{\varepsilon}{4}\eta^4 + \frac{1}{2}a\eta^2 - \sigma\eta, \qquad (1)$$

where a is a control parameter. This potential describes second-order phase transitions both in the absence ($\sigma = 0$) and in the presence ($\sigma \neq 0$) of external fields as proposed by Lev Landau. The butterfly catastrophe, on the other hand, is described by

$$V(\eta) = \frac{1}{6}\eta^6 + \frac{a}{4}\eta^4 - \frac{b}{3}\eta^3 + \frac{c}{2}\eta^2 - \sigma\eta \qquad (2)$$

and has been used to model first-order phase transitions.

To illustrate the related phenomenon of hysteresis we first investigate the bifurcation effect by minimizing Equation (1), which yields the equation of state

$$\varepsilon\eta^3 + a\eta = \sigma. \qquad (3)$$

A transition between a single stable solution for $a > 0$ and a bistable situation for $a < 0$ takes place when $a = 0$. However, a new feature is the phenomenon of external-field-induced hysteresis and metastability. Stability corresponds to a solution for which $\partial^2 V/\partial\eta^2 > 0$ and, if more than one solution of the equation of state exists that is stable, we call the higher energy solutions metastable. Figure 1 shows the difference in the response of the order parameter to the application of an external field. In unistable situations, η as a function of σ is a smooth single-valued function. Multistability results in multi-valuedness in some ranges of the external fields. Figure 1c demonstrates the regions of multistability in the parameter space. The dividing line in the cusp catastrophe case is given by the equation

$$-4a^3 + 27\sigma^2 = 0. \qquad (4)$$

This indicates the parameter set for which a triple solution of the cubic equation in (4) is obtained. Figure 2 is an extension of these concepts to the butterfly catastrophe case that exhibits thermal hysteresis (if the control parameter a involves a temperature dependence) and a double hysteresis under the influence of an external field.

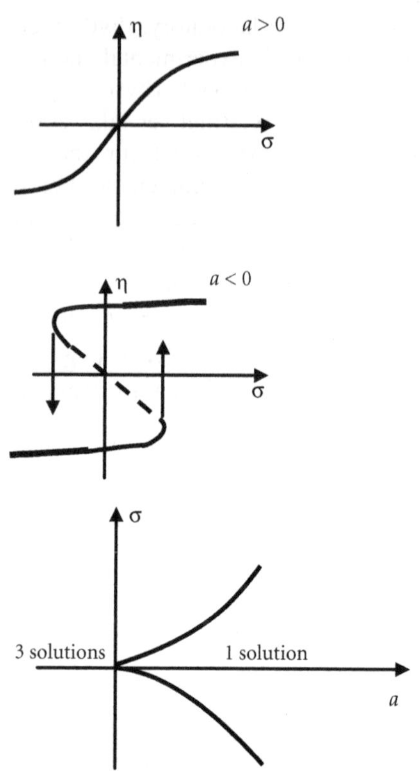

Figure 1. The order parameter's response to an externally applied field (a) in a unistable state, (b) in a bistable state (hysteresis), and (c) separation of the control parameter space into solution's multiplicity regions.

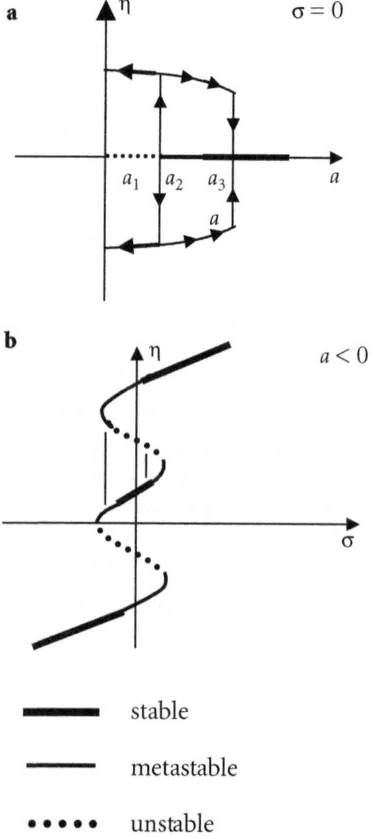

Figure 2. (a) Bifurcation plot (thermal hysteresis) and, (b) double field-induced hysteresis in the butterfly catastrophe case.

The sixth power free energy expansion for the order parameter η

$$F = F_0 + a\varepsilon\eta^2 + A_4\eta^4 + A_6\eta^6 - h\eta \qquad (5)$$

has been widely used to model first-order phase transitions provided $A_4 < 0$. The equation of state gives the order parameter in equilibrium

$$\eta_0 = \left\{\left[-A_4 - \left(A_4{}^2 - 3a\varepsilon A_6\right)^{1/2}\right]/3A_6\right\}^{1/2}. \qquad (6)$$

The transition temperature is

$$T_c^* = T_c + \frac{A_4{}^2}{4aA_6}, \qquad (7)$$

where $\varepsilon = T - T_c$, and the disordered phase ($\eta = 0$) terminates its stability at $T = T_c$ while the ordered one (η_0) does so at $T = T_0{}^*$, where

$$T_0{}^* = T_c + \frac{A_4{}^2}{3aA^6}. \qquad (8)$$

Importantly, the existence of a thermal hysteresis phenomenon is associated with first-order phase transition. Thermal hysteresis means that on cooling, the disordered phase is stable below the equilibrium transition temperature, while on heating, the ordered phase is stable above it. The associated effects are called supercooling and superheating, respectively.

JACK A. TUSZYŃSKI

See also **Catastrophe theory; Critical phenomena; Ferromagnetism and ferroelectricity; Order parameters**

Further Reading

Kittel, C. 1956. *Introduction to Solid State Physics*, New York: Wiley

Landau, L.D. & Lifshitz, E.M. 1959. *Statistical Physics*, London: Pergamon

Thompson, J.M.T. & Stewart, H.B. 1986. *Nonlinear Dynamics and Chaos*, New York: Wiley

I

IDEMPOTENT ANALYSIS

Definition and Introduction

The analyses that we all know and use (differentiation, integration, Fourier analysis, matrix analysis, etc.) are tacitly based on algebra defined by means of the conventional operations: addition, multiplication, and their inverses. In idempotent analysis, the starting point is not conventional algebra but instead the idempotent semiring.

Definition 1. *A semiring is a set $\mathcal{D}$ with two binary operations, $\oplus$ (called "sum" or "addition") and $\otimes$ (called "product" or "multiplication") and two specific elements ε and e such that*

- *$(\mathcal{D}, \oplus, \varepsilon)$ is a commutative monoid with identity element ε,*
- *$(\mathcal{D}, \otimes, e)$ is a monoid with identity element e,*
- *$a \otimes \varepsilon = \varepsilon \otimes a = \varepsilon$, for all $a \in \mathcal{D}$,*
- *$a \otimes (b \oplus c) = (a \otimes b) \oplus (a \otimes c)$, for all $a, b, c \in \mathcal{D}$.*

The semiring is called idempotent if in addition

- *$a \oplus a = a$, for all $a \in \mathcal{D}$.*

The semiring is called commutative if

- *the multiplication $\otimes$ is commutative.*

A monoid here is a set of elements with one operation. An idempotent semiring is sometimes referred to as a "dioid," which means "twice a monoid." It is the property $a \oplus a = a$ that distinguishes a dioid from conventional algebra. The most well-known example of a dioid is the so-called max-plus algebra, where $\mathcal{D} = R \cup \{-\infty\}$, the $\oplus$ operation is the maximum, with identity element $\varepsilon = -\infty$, and the $\otimes$ operation is (conventional) addition, with the identity element $e = 0$. As a numerical example, $5 \otimes (3 \oplus -2) = 8$. A noticeable difference between max-plus algebra and conventional algebra is that linear equations do not always have a solution (e.g., what is x in $5 \oplus x = 4$?). Addition is not cancellative in the sense that $a \oplus b = a \oplus c$ does not in general imply $b = c$.

The pioneering work in the max-plus algebra is Cuninghame-Green (1979), which has been extended into a system-theoretic direction in Baccelli et al. (1992). The phrase *idempotent analysis* was coined by Kolokoltsov & Maslov (1997). In Gunawardena (1998), it is shown that idempotency shows up in many different areas of mathematics; Gunawardena (1998, pp. 1–49) is an excellent survey. Applications are described in Le Boudec & Thiran (2001); and Heidergott et al. (2004).

We choose the max-plus algebra setting to introduce some analysis. Consider the recurrence relation

$$x(k+1) = A \otimes x(k), \tag{1}$$

which represents a model with state $x \in R^n$ and where A is an $n \times n$ matrix with elements a_{ij}. More explicitly, this relation is defined by its component-wise writing

$$\begin{aligned} x_i(k+1) &= \bigoplus_j (a_{ij} \otimes x_j) = (a_{i1} \otimes x_1) \\ &\quad \oplus (a_{i2} \otimes x_2) \oplus \cdots \oplus (a_{in} \otimes x_n), \\ i &= 1, 2, \ldots, n. \end{aligned}$$

The conventional way of writing this would be $x_i(k+1) = \max_j(a_{ij} + x_j)$. Such a model is sometimes referred to as a discrete event dynamic system.

If $x(0)$ is an eigenvector of A in the max-plus algebra defined as

$$A \otimes x(0) = \lambda \otimes x(0), \tag{2}$$

equivalently written as $\max_j(a_{ij} + x_j(0)) = \lambda + x_j$, with not all components $x_i(0)$ equal to ε, and where λ is called the eigenvalue, then $x_i(k+1) = \lambda + x_i(k) = x_i(0) + (k+1)\lambda$. The new state is obtained by adding λ to the previous state. In graph-theoretic terms, λ can be characterized as follows.

Theorem 1. *If the square matrix A is irreducible, equivalently, its corresponding graph is strongly connected and, there exists one and only one eigenvalue (with possibly several eigenvectors). This eigenvalue is equal to the maximum cycle mean of the graph:*

$\lambda = \max_\zeta |\zeta|_w/|\zeta|_l$, *where* ζ *ranges over the set of circuits in the graph, where* $|\zeta|_l$ *is the number of arcs of* ζ *and* $|\zeta|_w$ *its weight, that is, the addition of all* a_{ij} *which correspond to arcs of* ζ.

This theorem is referred to as the spectral theorem. A practical interpretation for this model and analysis is given below.

Examples

The first example of a dioid already encountered is $R \cup \{-\infty\}$ with $\oplus$ as max and $\otimes$ as addition. Other examples are:

- The set of $n \times n$ matrices with $A \oplus B$ and $A \otimes B$ defined as ($A = \{a_{ij}\}$, $B = \{b_{ij}\}$)

$$\{A \oplus B\}_{ij} = \max(a_{ij}, b_{ij}), \ \{A \otimes B\}_{ij} = \max_k(a_{ik} + b_{kj}).$$

- The set $R \cup \{-\infty\}$ with min as $\oplus$ and addition as $\otimes$ is isomorphic to the max-plus algebra.
- The set $R \cup \{-\infty\} \cup \{+\infty\}$ with $\oplus$ defined as max and $\otimes$ as min.
- The set $\{0, 1\}$ with $\oplus$ defined as max and $\otimes$ as min. This is the Boolean algebra.
- The set of all subsets of the R^2 plane, including $\emptyset$ and the whole R^2 itself, with $\oplus$ defined as $\cup$ and with $\oplus$ defined as the vector sum of subsets.
- Extension to polynomials or formal power series by defining (a polynomial or formal power series is given by $f(z) = \bigoplus_{k=0,1,\ldots} f_k z^k$):

$$f \oplus g: \ (f \oplus g)_k = f_k \oplus g_k; \ f \otimes g: \ (f \otimes g)_k = \max_{i+j=k} f_i \otimes g_j.$$

Relation to Nonlinear Theory and Practice

One must be careful using the word *nonlinear*. System (2) for instance is nonlinear with respect to conventional algebra, but it is considered linear in the max-plus algebra.

Relationship with Other Mathematical Disciplines

An interpretation in graph-theoretic terms is often possible. The occurrences of events in timed event graphs, the latter forming a subclass of Petri nets, can be modeled by means of an equation of the form (1). If $x \in R^n$, define $||x||$ to be its l^∞ norm, that is, $||x|| = \max_i |x_i|$. A mapping $f : R^n \to R^n$ is called non-expansive if $||f(x) - f(y)|| \leq ||x - y||$. As an example, $f(x) = A \otimes x$ is non-expansive. An important theme is the study of periodic solutions of non-expansive mappings; that is, does λ exist such that $f^k(x) = k\lambda + f(x)$? Another theme is that of the finiteness of the transient behavior (before one reaches a periodic behavior), if one starts with an arbitrary initial condition.

Some Applications

Consider the system (2). Interpret a_{ij} as the traveling time of a train on a track (if existing) between departure station j and arrival station i. Quantity $x_i(k+1)$ refers to the earliest departure time of trains at station i that depart for the $(k+1)$th time. The model states that in order to depart (for the $(k+1)$th time), the train must wait until all trains, which departed for the kth time, have arrived; $\max_j(a_{ij} + x_j)$ reflects the maximum of all arrival times. This allows passengers to change trains (one could add some minutes to all a_{ij} to account for the changeover times). If a track between two stations, say l and m, does not exist, set $a_{ml} = \varepsilon$. If one assumes that the departures take place as soon as possible subject to the restriction of passengers changing over, then the model $x(k+1) = A \otimes x(k)$, with an initial condition $x(0)$, determines all future departure times. If this $x(0)$ is an eigenvector, then one has a timetable with constant interdeparture times, this constant being λ. This constant is a lower bound for the interdeparture times of any periodic timetable obeying the changeover rule.

In the above, trains can be viewed as discrete flows on networks. If one replaces "trains" by "messages" and "all railway tracks" by "computer network" or "internet," one enters another field of application that has been described and analyzed in Le Boudec & Thiran (2001).

In discrete-time optimal control theory, or in decision processes on Markov chains, one encounters the equation

$$V(k, x) = \max_u (V(k+1, f(x, u)) + g(x, u)),$$

which is a consequence of Bellman's principle of optimality. The underlying model is $x(k+1) = f(x(k), u(k))$, and the costs during step k are $g(x(k), u(k))$. The function V is the value function. This equation, with the operations of addition and maximization can be interpreted and analyzed in the max-plus algebra sense (Akian et al., 1998). Applications in automata theory can also be found in Gunawardena (1998).

GEERT JAN OLSDER

See also **Markov partitions**

Further Reading

Akian, M., Quadrat, J.-P. & Viot, M. 1998. Duality between probability and optimization. In *Idempotency*, edited by J. Gunawardena, Cambridge and New York: Cambridge University Press

Baccelli, F., Cohen, G., Olsder, G.J. & Quadrat, J.-P. 1992. *Synchronization and Linearity*, New York: Wiley

Cuninghame-Green, R.A. 1979. *Minimax Algebra*, Berlin: Springer

Gunawardena, J. (editor). 1998. *Idempotency*, Cambridge and New York: Cambridge University Press

Heidergott, B., Olsder, G.J. & van der Woude, J.W. 2004. *Max Plus at Work*, Princeton, NJ: Princeton University Press

Kolokoltsov, V.N. & Maslov, V.P. 1997. *Idempotent Analysis and Its Applications*, Boston: Kluwer

Le Boudec, J.-Y. & Thiran, P. 2001. *Network Calculus, A Theory of Deterministic Queuing Systems for the Internet*, Berlin and New York: Springer

IKEDA MAP

See **Maps**

IMPULSE, NERVE

See **Nerve impulses**

INCOHERENT SOLITONS

Until 1996, solitons were considered to be exclusively coherent entities. However, experimental observations of self-trapping of partially spatially incoherent light (Mitchell et al., 1996) and of white light (Mitchell et al., 1997) in photorefractive media have shown that incoherent beams can also form solitons (Figure 1). These solitons are multimode or speckled beams for which the instantaneous intensity distribution varies randomly in space and/or time. They can exist only in non-instantaneous media whose response time is much longer than the random fluctuation time across the incoherent beam. Such a medium responds to the time-averaged envelope and not to the instantaneous speckles that constitute the incoherent beam. An incoherent soliton forms when the time-averaged intensity induces a multimode wave and traps itself in the waveguide by populating the guided modes in a self-consistent fashion. These random-phase and weakly correlated self-trapped entities exhibit a host of unique properties that have no analog in the coherent regime. Moreover, their existence is related to many other areas of physics, in which nonlinearities, stochastic behavior, and statistical averaging are involved. For example, incoherent modulation instability effects and incoherent pattern formation relate to many systems in nature: from clustering in a cooled atomic gas to self-supported "stripes" of electrons in semiconductors, from high-T_c superconductors to structures in fluids and gravitational effects.

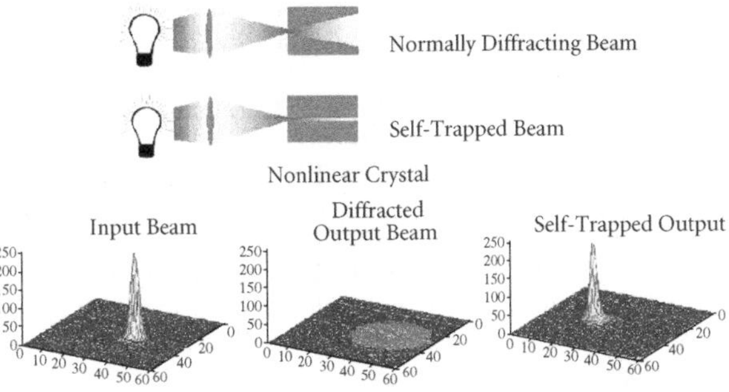

Figure 1. Self-trapping of spatially and temporally incoherent white light from an incandescent light bulb.

Numerous theoretical and experimental works followed the pioneering experiments (see Segev & Christodoulides (2001) and references within). This rapid progress in the area of incoherent solitons offers a number of interesting fundamental ideas as well as possible applications. For example, coherence engineering can be used to control modulation instabilities and to create interesting stable patterns in nonlinear materials for various applications. There is also an exciting possibility of using self-trapped light beams from incoherent sources, such as light-emitting diodes, for creating reconfigurable optical interconnects and beam steering.

Several theories have been developed to describe the propagation of partially incoherent beams and soliton formation in noninstantaneous nonlinear materials (Segev & Christodoulides, 2001). These include the coherent density theory (Christodoulides et al., 1997), the modal theory (Mitchell et al., 1997), and the mutual coherence theory (Shkunov & Anderson, 1998). Other, more approximate theories relying on ray optics (Snyder & Mitchell, 1998) and on the Wigner distribution (Hall et al., 2002) have also been suggested. The first three approaches are rather complete and equivalent to one another (Christodoulides et al., 2001). Here, we briefly overview the two most widely used theories.

Coherent Density Theory

The coherent density method is a dynamic approach suited for studying the evolution dynamics of incoherent solitons, their interactions, instabilities, and correlation statistics. In this formalism, the incoherent field is described by means of a so-called coherent density function, from which one can deduce the optical intensity distribution as well as the associated correlation statistics. Mathematically, the self-trapping process is described by an infinite set of nonlinear Schrödinger-like equations, provided that each coherent component at the input is appropriately weighted with respect to the incoherent angular power spectrum of the source (Christodoulides et al., 1997). These equations can be solved numerically by using modified beam propagation procedures.

Modal Theory

The modal theory is currently widely considered as the method of choice in terms of identifying static incoherent solitons, their range of existence, and correlation properties. The fundamental requirements of incoherent self-trapping were the starting point of this formulation. Incoherent solitons are identified by solving the underlying coupled (stationary) nonlinear Schrödinger-like equations in a self-consistent fashion (Mitchell et al., 1997). To solve these equations self-consistently, one needs to start with an arbitrary index profile, then solve for the guided modes and sum them incoherently to get the intensity profile, and finally use this intensity profile to calculate the new induced-index profile. This process should be repeated until the solutions are stationary. The stationary solutions give the incoherent soliton intensity profile as well as its modal distribution and statistical properties.

1-d and 2-d Dark Incoherent Solitons

The existence of dark incoherent solitons and requirements for their observations were predicted (Coskun et al., 1998) two years after the observation of bright incoherent solitons. Such dark incoherent solitons were experimentally observed soon thereafter (Chen et al., 1998). In that experiment, (1 + 1)-dimensional dark incoherent solitons were observed, along with the self-trapping of a (2 + 1)-d dark incoherent beam. That is, a self-trapped 2-d void nested in spatially incoherent beam was demonstrated (Chen et al., 1998). As in the coherent case, fundamental dark incoherent solitons were found to exist only in defocusing nonlinear media and to require an initial transverse phase shift at the center of the dark stripe. 2-d dark incoherent solitons were initiated from a vortex beam that was generated by replacing a step mirror with a helicoidal phase mask, which created an optical vortex of unit topological charge. For bright solitons, only guided modes (bound modes) are self-trapped, whereas dark incoherent solitons involve a belt of radiation modes (unbound states). Unlike their coherent counterparts, fundamental dark and vortex solitons were always found to be gray because of the presence of both odd and even radiation modes at the center of the notch (Christodoulides et al., 1998). Furthermore, the grayness depends on the coherence length of the carrier beam. Dark soliton splitting and "phase memory" effects were also observed theoretically and experimentally (Coskun et al., 1999).

Modulation Instability, Pattern Formation, and Soliton Clustering

Both theory and experiment have revealed that modulation instabilities can take place with incoherent light beams provided that the nonlinearity exceeds a well-defined threshold, which depends on the correlation properties of the beam (Soljacic et al., 2000). This shows that modulation instabilities (MI) can be totally suppressed by properly adjusting the correlation distance of the wave front. A natural by-product of this suppression is the observation of anti-dark incoherent solitons (a bright beam on a constant background) in self-focusing environments. Periodic trains of 1-d filaments and self-ordered 2-d lattices of light spots due to incoherent modulation instabilities were observed by carefully increasing the nonlinearity (Kip et al., 2000). Spontaneous clustering of solitons in partially coherent wave fronts initiated by random noise was also demonstrated experimentally and numerically (Chen et al., 2002). It is, in fact, an outcome of the interplay between random noise, weak correlation, and high nonlinearity. Together, this process leads to incoherent MI, formation of 2-d solitary filaments, and eventually to clustering of 2-d solitons or in other words, aggregates of fine-scale structures. This clustering phenomenon has no counterpart with solitons in coherent systems. The rapid progress in this new area of incoherent solitons brings about many interesting fundamental ideas and possible applications. We have concentrated here on the basic features, but new ideas arise very quickly in this new, dynamic area of soliton science.

TAMER COSKUN, DEMETRIOS CHRISTODOULIDES, AND MORDECHAI SEGEV

See also **Nonlinear optics; Optical fiber communications; Solitons, types of**

Further Reading

Chen, Z., Mitchell, M., Segev, M., Coskun, T.H. & Christodoulides, D.N. 1998. Self-trapping of dark incoherent light beams. *Science*, 280: 889–892

Chen, Z., Sears, S.M., Martin, H., Christodoulides D.N. & Segev, M. 2002. Clustering of solitons in weakly correlated systems. *Proceedings of the US National Academy of Science*, 99(8): 5223–5227

Christodoulides, D.N., Coskun, T.H., Mitchell, M., Chen, Z.C. & Segev, M. 1998. Theory of incoherent dark solitons. *Physical Review Letters*, 80(23): 5113–5116

Christodoulides, D.N., Coskun, T.H., Mitchell, M. & Segev, M. 1997. Theory of incoherent self-focusing in biased photorefractive media. *Physical Review* E, 59(5): R4777–R4780

Christodoulides, D.N., Eugenia, D.E., Coskun, T.H., Segev, M.& Mitchell, M. 2001. Equivalence of three approaches describing partially incoherent wave propagation in inertial nonlinear media. *Physical Review* E, 63: 035601-1–035601-4

Coskun, T.H., Christodoulides, D.N., Chen, Z. & Segev, M. 1999. Dark incoherent soliton splitting and "phase memory" effects: theory and experiment. *Optics Letters*, 23(6): 418–420

Coskun, T.H., Christodoulides, D.N., Mitchell, M., Chen, Z. & Segev, M. 1998. Dynamics of incoherent bright and dark self-trapped beams and their coherence properties in photorefractive crystals. *Optics Letters*, 23(6): 418–420

Hall, B., Lisak, M., Anderson, D., Fedele, R. & Semenov, V.E. 2002. Statistical theory for incoherent light propagation in nonlinear media. *Physical Review* E, 65: 035602-1–035602-4

Kip D., Soljacic M., Segev, M., Eugenieva, E., Christodoulides, D.N. 2000. Modulation instability and pattern formation in spatially incoherent light beams. *Science*, 290: 495–498

Mitchell, M., Chen, Z., Shih, M. & Segev, M. 1996. Self-trapping of partially spatially incoherent light. *Physical Review Letters*, 77(3): 490–493

Mitchell, M. & Segev, M. 1997. Self-trapping of incoherent white light. *Nature*, 387: 880–883

Mitchell, M., Segev, M., Coskun, T.H. & Christodoulides, D.N. 1997. Theory of self-trapped spatially incoherent light beams. *Physical Review Letters*, 79(25): 4990–4993

Segev, M. & Christodoulides, D.N. 2001. Incoherent solitons: self-trapping of weakly-correlated wave-packets. In *Spatial Solitons*, edited by S. Trillo & W.E. Torruellas, New York: Springer

Shkunov, V.V. & Anderson, D.Z. 1998. Radiation transfer model of self-trapping spatially incoherent radiation by nonlinear media. *Physical Review Letters*, 81(13): 2683–2686

Snyder, A.W. & Mitchell, D.J. 1998. Big incoherent solitons. *Physical Review Letters*, 80(7): 1422–1924

Soljacic, M., Segev, M., Coskun, T.H., Christodoulides, D.N. & Vishwanath, A. 2000. Modulation instability of incoherent beams in noninstantaneous nonlinear media. *Physical Review Letters*, 84(3): 467–470

INERTIAL MANIFOLDS

In studying the dynamics of nonlinear systems, it is sometimes observed that all trajectories rapidly converge to a subspace of the phase space, leading to a simplification of the analysis through dimensional reduction. Such subspaces are called inertial manifolds (IMs) and the underlying dynamics are called inertial. In phase spaces of infinite dimension, the question is whether a finite-dimensional dynamical system can reproduce the dynamics of an infinite-dimensional one. In finite-dimensional systems, one tries to find a system of smaller dimension.

Related concepts have been introduced in sciences. In meteorology, for example, the slow manifold introduced and studied by Lorenz and others, is expected to reproduce the slow dynamics, taking into account interaction with the rapidly changing regions of the phase space. From a somewhat different perspective, the Reynolds equations in turbulence and some large eddies simulation models are finite-dimensional systems that attempt to reproduce the dynamics of the large eddies of the fluid, taking into account the action of the small eddies.

Definition

Denote by H the phase space, a Hilbert space of finite or infinite dimension, and consider on H a semigroup of operators $\{S(t)\}_{t\geq 0}$ generating the dynamical system: $u_0 \mapsto S(t)u_0 = u(t)$.

If it exists, an inertial manifold for this system is a smooth finite-dimensional manifold $\mathcal{M}$ in H that enjoys the following properties:

(i) $S(t)\mathcal{M} \subset \mathcal{M}, \forall t > 0$ (positive invariance);
(ii) all orbits converge to $\mathcal{M}$, as $t \to \infty$, at an exponential rate;
(iii) the rate of convergence is uniform for all orbits starting in a bounded set of initial data.

If a global attractor $\mathcal{A}$ exists, then necessarily $\mathcal{A}$ is included in $\mathcal{M}$ as well as any stationary solution, periodic orbit, etc. Comparing $\mathcal{A}$ to $\mathcal{M}$ when they both exist, we see that $S(t)\mathcal{A} = \mathcal{A}$, instead of (i), and we know that orbits may converge to $\mathcal{A}$ at arbitrarily slow rates, in contrast with (ii). Finally, attractors are believed to be sometimes fractal sets, whereas smoothness is required for $\mathcal{M}$.

When an inertial manifold exists, the restriction of the semigroup to $\mathcal{M}$ is a group—due to (i)—whose dynamics are the same as that of the initial systems. In particular, their attractor is the same if $\{S(t)\}_{t\geq 0}$ has an attractor, and this reduced dynamical system is called an inertial system. When the dynamical system is defined by a differential equation, as in Equation (1), its restriction to $\mathcal{M}$ is a finite-differential system, also called an inertial system.

Examples

Inertial manifolds have been studied for dynamical systems in infinite dimensions corresponding to dissipative parabolic evolution equations, written in functional form as

$$\frac{\mathrm{d}u(t)}{\mathrm{d}t} = F(u(t)), \quad u(0) = u_0. \tag{1}$$

Here $u(t) = S(t)u_0$, where u_0 and $u(t)$ belong to H.

The space H is decomposed as the sum of a finite-dimensional space $Y = P_y H$ and of its complement $Z = P_z H$, $u = y + z$. In all examples studied, the inertial manifold is then searched for as a graph:

$$z = \Phi(y), \quad y \in Y. \tag{2}$$

Usually, Y and Z are spectral spaces for the underlying linear dissipative operator A, Y representing hence the low-frequency component of the flow and z its high-frequency component.

The existence of inertial manifolds has been proven for the Kuramoto–Sivashinsky equation, the Ginzburg–Landau equation, reaction-diffusion equations in one and two spatial dimensions, and the Navier–Stokes equations with added (hyper-)viscosity.

The initial hope (Foias et al., 1988) was to find an IM for the Navier–Stokes equation at least in space two, justifying the name coined by Foias, since the role of the inertial terms over the viscous terms would be dominant on this manifold. However, the existence of an inertial manifold for the Navier–Stokes equations has not yet been proven nor disproven.

Several proofs of existence have been derived beyond those of Foias et al. (1988); the key hypothesis and limitation in all the proofs is the so-called spectral gap condition, involving a sufficiently large gap in the spectrum of the dissipative operator A. It is believed (but not proved) that this difficulty could be overcome with IMs that are less smooth and are not graphs. Alternatively, this difficulty has been overcome by removing the condition that $\mathcal{M}$ be a manifold. The corresponding concept was introduced in Eden et al. (1994) where it was called an inertial set or an exponential attractor.

The existence of inertial sets was proven in Eden et al. (1994) for numerous dynamical systems generated by evolutionary dissipative equations, in fact, for all systems for which the existence of a compact finite-dimensional attractor has been established (Hale, 1988; Sell & You, 2002; Temam, 1997).

Despite these limitations, inertial manifolds are a natural concept, giving a precise form to the idea that infinite-dimensional dynamical systems produce a finite-dimensional dynamics. Remember that a less explicit form of this idea is the property of the attractor to be finite-dimensional.

When IMs do not exist or are not easy to construct, approximate inertial manifolds (AIMs) are sometimes used, which are finite-dimensional smooth manifolds that attract the orbits in a small neighborhood. Extensive applications of AIMs in numerical analysis have been developed (Temam, 1997; Sell & You, 2002).

ROGER TEMAM

See also **Center manifold reduction; Invariant manifolds and sets; Synergetics**

Further Reading

Eden, A., Foias, C., Nicolaenko, B. & Temam, R. 1994. *Exponential Attractors for Dissipative Evolution Equations*, Collection Recherches en Mathématiques Appliquées, Paris: Masson, Paris, and New York: Wiley

Foias, C., Sell, G.R. & Temam, R. 1988. Inertial manifolds for nonlinear evolutionary equations, *Journal of Differential Equations*, 73: 309–353

Hale, J. 1988. *Asymptotic Behavior of Dissipative Systems*, Providence, RI: American Mathematical Society

Sell, G.R. & You, Y. 2002. *Dynamics of Evolutionary Equations*, New York: Springer

Temam, R. 1988. *Infinite Dimensional Dynamical Systems in Mechanics and Physics*, New York: Springer (2nd augmented edition, 1997)

INFELD EQUATION

See **Born–Infeld equation**

INFORMATION DIMENSION

See **Dimensions**

INFORMATION THEORY

Although the word *information* is commonly used and its engineering applications influence our lives, the quantitative meaning is often unclear. Among several measures of information, the most common is called Shannon information (denoted as I_S with units of the bit), which was introduced by Claude Shannon in 1948.

As an example, suppose your friend (whose wife has been pregnant for nearly nine months) telephones with the information: It's a girl. Since the prior probability (P) of a girl being born is $\frac{1}{2}$, the Shannon information is computed as

$$I_S = \log_b\left(\frac{1}{P}\right) = \log_2\left(\frac{1}{1/2}\right) = 1\,\text{bit}, \qquad (1)$$

with the logarithmic base $b = 2$. In other words, one bit of Shannon information is gained in a message that resolves one of two equally probable events.

Strings of such events (e.g., 100110100011100) are the language of computers, and their Shannon information can be computed by thinking of a string as a set of N switches, each of which can be on (1) or off (0). Since there are 2^N possible sequences of N switches, the probability of any particular sequence is 2^{-N}, leading to a Shannon information of N bits, which is conveniently equal to the number of switches.

Shannon's information takes the particular form, Equation (1), under the assumption that the received message (an observed switch sequence) is completely reliable. A more general Shannon measure allows for generally unreliable messages; thus

$$\text{(source event } B_1 \text{ or } B_2 \text{ or } \dots B_N) \rightarrow \text{observed message } A. \qquad (2)$$

Here a particular source event $B_n \equiv B$ occurs, and the observer receives a message A, providing evidence of what was actually intended. The arrow implies transmission through a real, imperfect communication channel, in which transmission errors may occur. For example, the actual source event B might be the switch sequence given above that is transmitted to an observer. Because of transmission errors, the received message might be $A = 010110100011100$.

As when a detective observes a fingerprint, the message A implies that an event B has occurred, where each B_n is a different possible switch sequence. Shannon information was designed to answer the question: How much evidence does the observed (garbled) message A constitute for the hypothesis that the event B actually occurred?

Intuitively, the evidence provided by A will be strong if the probability of B having occurred is now greater than before receiving A. That is, if the probability $P(B|A)$ (of B in the presence of A) much exceeds $P(B)$ (the probability of B in the absence of A). This suggests that the ratio $P(B|A)/P(B)$ measures the weight of the evidence. Strong evidence would give a large ratio, weak evidence a small ratio. Because any monotonic function of this ratio works as well, taking the logarithm defines the Shannon information

$$I_S(A, B) \equiv I_S \equiv \log_b \frac{P(B|A)}{P(B)}. \qquad (3)$$

Although the choice of a logarithmic base b is at the user's discretion, probabilities in computer codes are often powers of 2 for which $b = 2$ is the most convenient choice. Then I_S is assigned the unit of a *bit*. For a fixed channel probability ($P(B|A)$), Equation (3) discloses that the information increases as $P(B)$ becomes smaller. (In editing a newspaper, for example, an article entitled "Dog Bites Man" is not unusual enough to merit publication, but "Man Bites Dog" most definitely is.)

There is a problem with this definition of information—the observer, by hypothesis, does not know which event B_n has occurred. Therefore, he or she cannot know which information quantity, $I_S(A, B_1)$ or ... or $I_S(A, B_N)$ to compute. Furthermore, the message A may be one of a set of possibilities A_m, $m = 1, \ldots, M$. All such possibilities can be taken into account by forming the grand average

$$\left\langle \log_b \frac{P(B|A)}{P(B)} \right\rangle \equiv \sum_{m=1}^{M} \sum_{n=1}^{N} P(A_m, B_n) \log_b \frac{P(B_n|A_m)}{P(B_n)} \equiv I(A, B). \quad (4)$$

Called Shannon's "mutual information," this is the quantity commonly used to evaluate the quality of communication systems. Notice that it depends upon properties of the source (the $P(B_n)$) as well as the channel (the $P(B_n|A_m)$); thus, this is a total system perspective.

If just the quality of the channel (as specified by its fixed channel probabilities $P(B_n|A_m)$) is to be evaluated, the information $I(A, B)$ is maximized through choice of a source $P(B_n)$. The resulting information, then, only depends upon the channel probabilities and is called the channel capacity since it measures the intrinsic capacity of the channel for transmitting information.

Note that Equation (4) may be evaluated in its continuous limit; thus,

$$I(X, Y) = \int\int \mathrm{d}x\,\mathrm{d}y\, p(x, y) \log_b \frac{p(y|x)}{p(y)}. \quad (5)$$

Is there an information measure that is appropriate for describing growth processes? Consider a generalization of Equation (5) to

$$I_{\mathrm{KL}}(p, r) \equiv -\int \mathrm{d}\mathbf{x}'\, p(\mathbf{x}') \log_b \frac{p(\mathbf{x}')}{r(\mathbf{x}')}, \quad (6)$$

which is called the Kullback–Leibler (K–L) information. The ratio p/r is describes a "cross entropy" between the probability law $p(\mathbf{x}')$ and the reference law $r(\mathbf{x}')$, where the vector $\mathbf{x}'$ is general. In the special case $\mathbf{x}' = (\mathbf{x}, \mathbf{y})$, $r(\mathbf{x}') = p(\mathbf{x})p(\mathbf{y})$, Equation (6) becomes the mutual information in Equation (5). Another important case of the K–L information is when $r(\mathbf{x}') = 1$, whereupon Equation (6) takes the form

$$-\int \mathrm{d}\mathbf{x}'\, p(\mathbf{x}') \log_b p(\mathbf{x}') \equiv H, \quad (7)$$

which is called the entropy.

Yet another special case of K–L information is obtained when $\mathbf{x}' \equiv x$, a single coordinate, and the reference function $r(x)$ is made to be a slightly displaced version of the probability law, $r(x) \equiv p(x+\Delta x)$, $\Delta x \rightarrow 0$. In this limit, the K–L information is found to obey

$$-\lim_{\Delta x \to 0} \frac{1}{\Delta x^2} \int \mathrm{d}x\, p(x) \log_b \frac{p(x)}{p(x+\Delta x)} = -\frac{1}{2} \int \mathrm{d}x \frac{p'^2(x)}{p(x)} \equiv -\frac{1}{2} I_{\mathrm{F}}, \quad (8)$$

where I_{F} is the Fisher information and the usual derivative notation $p'(x) \equiv \mathrm{d}p(x)/\mathrm{d}x$ is used.

A useful variant of I_{F} is obtained by using a probability amplitude $q(x)$, where $p(x) \equiv q^2(x)$. The right-hand integral of Equation (8) then gives directly

$$I_{\mathrm{F}} = 4 \int \mathrm{d}x\, q'^2(x), \quad q'(x) \equiv \mathrm{d}q(x)/\mathrm{d}x. \quad (9)$$

Thus, Fisher information measures the gradient content of the probability amplitude curve.

Because the smallest possible mean-square error one can achieve in forming an estimate of the parameter is the reciprocal of Fisher's information, this also describes our ability to know a parameter from its measurements. Assuming all knowledge is based upon measurement, Fisher information describes quite generally our ability to know.

Physicists Leon Brillouin and David Bohm long sought a measure of information that would explain the mysteries of physics in general, including those of quantum mechanics. In particular, the question of quantum entanglement has long perplexed physicists. Consider a pair of particles that are created simultaneously as reactants from a mother particle. Measurement of one member of the pair supplies important information (such as spin) about the other member, even without measuring the latter. (This is called the Einstein–Podolsky–Rosen–Bohm, or EPRB, paradox.) In what way could information about the unmeasured particle be transmitted into a measurement of the detected one? This seems to require the realities of the two particles to be somehow "entangled." Hence, an "active information" was sought that could supply the needed information. Of course, they tried the Shannon variety, but this did not work.

Surprisingly, a form of information exists for generating the probability laws of physics that is aptly called the physical information and defined as follows. Let I be the Fisher information that is acquired in an actual measurement. Depending upon the measurement scenario, there may or may not be another, unmeasured particle accompanying the measured particle. The physical information K is the difference $K \equiv I - J$ of two Fisher informations (for simplicity we now drop the subscript F), where

$$I \equiv \sum_n I_n, \quad I_n = 4 \int \mathrm{d}x\, q_n'^2(x)$$

and

$$J \equiv I/\kappa, \quad 0 \leq \kappa \leq 1. \qquad (10)$$

Notice that the information I in the measurement is actually a sum of individual Fisher terms I_n. It turns out that the additive sum also defines the maximum possible value of the Fisher information. Hence, I is called the Fisher information channel capacity (as in the Shannon case). The integral form of the second Equation (10) shows that I is also a functional of the amplitude functions $q_n(x)$. This is a scalar number whose value depends upon an integral over all values x of the functions $q_n(x)$.

By the third Equation (10), J is likewise a functional of the amplitude functions; however, its form is not fixed but dependent upon the particular measurement scenario. J must be computed in any application.

Recall that I is the information level of the particle in a measurement space. In all measurements of quantum phenomena $\kappa = 1$, so $J = I$ by the third Equation (10). Thus J represents information that arises out of entanglement with that of the measured particle. We, therefore, call J the entangled information. Mathematically, it is I as evaluated in an "entanglement space." If only the measured particle is present, entanglement space is the Fourier conjugate space to measurement space. If another, unmeasured particle is present, entanglement space is the (unused) measurement space of that particle.

In Equations (10), the form of I is a sum of squares of functions q'_n (an L^2 norm or measure). Like any squared length, such a norm is invariant to rotation, and rotations may be performed in either the space of coordinates x (which are more generally of dimension 4 or higher) or of the functions $q'_n(x)$. Because J is linear in I, it also obeys such invariance, and likewise for their difference K. Such invariance suggests a state of stationarity

$$K \equiv I - J = \text{extrem.} \qquad (11)$$

Equation (10), supplemented by Equations (11), is called the principle of extreme physical information or EPI (Frieden, 1998). Its solutions $q_n(x)$ obey corresponding Euler–Lagrange equations

$$\frac{\mathrm{d}}{\mathrm{d}x}\left(\frac{\partial k}{\partial q'_n}\right) = \frac{\partial k}{\partial q_n}, \quad k \equiv k(x),$$
$$K \equiv \int \mathrm{d}x\, k(x). \qquad (12)$$

These are wave equations whose solutions $q_n(x)$ are the amplitude functions of science. An example is the Schrodinger wave equation. Other examples are provided by growth phenomena, ranging from energy growth to crystal growth to cancer growth, with the preceding formulation leading to exponential, power law, or logarithmic growth depending on the boundary conditions and integration constants that are assumed (Gatenby & Frieden, 2002).

In summary, Shannon information and cross entropy are valuable forms of information in communication theory, and two important outgrowths of these definitions are Fisher information and physical information. Fisher information permits one to estimate the expected error in the estimate of a parameter. Physical information permits one to derive physical laws via the EPI principle.

B. Roy Frieden

See also **Algorithmic complexity; Entropy; Euler–Lagrange equations**

Further Reading

Bohm, D. & Hiley, B.J. 1993. *The Undivided Universe*, London: Routledge

Brillouin, L. 1956. *Science and Information Theory*, New York: Academic Press

Fisher, R.A. 1922. On the mathematical foundations of theoretical statistics. *Philosophical Transactions of the Royal Society of London*, 222: 309–368

Frieden, B.R. 1998. *Physics from Fisher Information*, Cambridge and New York: Cambridge University Press (see also 2nd edition, 2004)

Gatenby, R. & Frieden, B.R. 2002. Application of information theory and extreme physical information to carcinogenesis. *Cancer Research*, 62: 4675

Kullback, S. 1959. *Information Theory and Statistics*, New York: Wiley

Shannon, C.E. 1948. A mathematical theory of communication. *Bell System Technical Journal*, 27: 379–423; 623–656

INHIBITION

In what is certainly one of the best-known experiments on learning, Ivan Pavlov sounded a bell and shortly after presented food to a hungry dog. After several pairings of the bell and the food, the bell came to elicit salivation, a response that it had not elicited previously. We say that the bell become an excitatory conditioned stimulus. *Excitatory* means that it acquired the ability to excite, arouse, or elicit a new behavior. But Pavlov did not limit his work to excitatory conditioning; he performed and described (Pavlov, 1927) a variety of procedures that would endow the conditioned stimulus with exactly the opposite properties. Such conditioned stimuli acquire the ability to oppose conditioned excitation and are said to be inhibitory.

The term originates from the Latin word *inhibere* and means to restrain. *Inhibere* consists of *in* "in" and *habere* "to hold." The word *inhibition* is widely used in many disciplines, such as medicine, biochemistry, physiology, neuroscience, and psychology, and its general meaning is the arrest or restraint of a process, partial or complete. In psychoanalytic theory, it describes a conscious or unconscious restraining of an impulse or desire. In neurophysiology, to which we will return shortly in more detail, it is the process opposite to

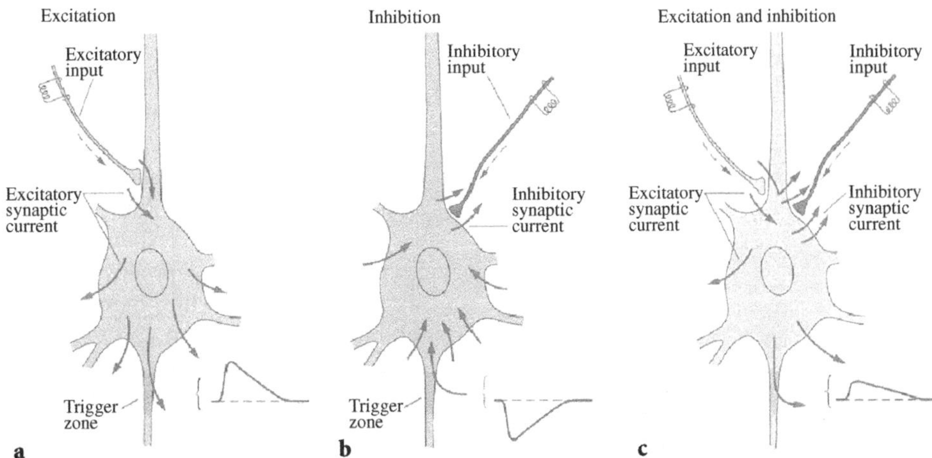

Figure 1. Excitatory and inhibitory currents have competitive effects on a single nerve cell. (a) An excitatory input in the base of a dendrite causes inward current which produces a large synaptic potential that brings the membrane potential close to its threshold. (b) An inhibitory input causes synaptic potential thus generated changes the membrane potential further below its threshold. (c) The shunting action of inhibition. When the cell receives both excitatory and inhibitory synaptic currents, the channels opened by the inhibitory pathway shunt the excitatory current, thereby reducing the excitatory synaptic potential. From Kandel et al. (2000).

excitation, or as Charles Sherrington (1933) wrote "... to refrain from an act is no less an act than to commit one, because inhibition is co-equally with excitation a nervous activity." By "nervous activity" he meant "the activity of the nervous system."

In fact, the function of all living systems and at all levels of complexity can be perceived as continuous adjustment between excitatory and inhibitory processes. The metabolic pathways are, for example, regulated by means of enzyme synthesis and inhibition. Almost all of the thousands of chemical reactions that occur in a living organism require catalysis by a specific enzyme. The character of any cell is based on its particular chemistry, which is determined by its specific enzyme composition. Genetic traits are expressed through synthesis of enzymes, which catalyze reactions that establish the phenotype. Many genetic diseases result from altered levels of enzyme production with changes in enzyme synthesis and/or inhibition. Much current drug therapy is based on the inhibition of a specific enzyme.

The nervous system continuously processes excitatory and inhibitory signals. Sherrington first pointed out that the quintessential action of the nervous system is its ability to weigh the consequences of different types of information and then decide on appropriate responses (Sherrington, 1947). He named it "the integrative action of the nervous system." The point at which one neuron communicates with another is called a synapse, and synaptic transmission is fundamental to processes such as perception, voluntary movement, or learning. Although many synaptic connections are highly specialized, all neurons make use of two basic forms of synaptic transmission: electrical and chemical.

Each neuron in the central nervous system, whether in the spinal cord or in the brain, is continuously bombarded by synaptic inputs from other neurons. Some are excitatory, others inhibitory. The effect of the synaptic potential—whether it is excitatory or inhibitory—is determined, not by the type of transmitter released from the presynaptic neuron, but by the type of ion channels gated by the transmitter in the postsynaptic cell. Most transmitters are recognized by types of receptors that mediate either excitatory or inhibitory potentials (for details see Kandel et al., 2000).

Synaptic contact can occur on the cell body, the dendrites, or the axon of a postsynaptic cell. The location of inhibitory inputs in relation to excitatory ones is critical for their functional effectiveness. Inhibitory short-circuit actions are more significant when they are initiated at the cell body near the initial axon segment. As a result, they can strongly curtail the influence of excitatory current on the membrane potential at the trigger zone, as illustrated in Figure 1c. In contrast, inhibitory action at a remote part of a dendrite is much less effective in shunting excitatory actions or in affecting the more distant trigger zones. Thus, in the brain, significant inhibitory input frequently occurs on the cell bodies of neurons.

Excitatory and inhibitory signals are integrated into a single response by the cell. In motor neurons and most interneurons, the decision to initiate an action potential is made by the initial segment of the axon (Figure 1, trigger zone). This region of cell membrane has a lower threshold for action potentials than the cell body or dendrites because it has a high density of voltage-dependent Na^+ channels. The neural integration involves the summation of synaptic potentials that spread passively to the trigger zone. It is critically affected by two passive membrane

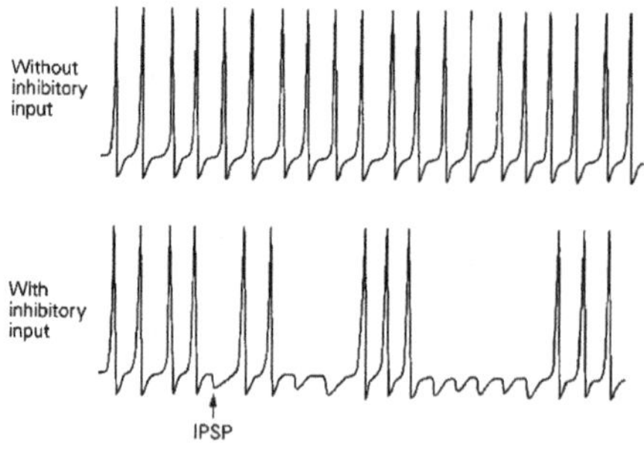

Figure 2. The *sculpturing* effect of inhibition. Without inhibitory input the neuron spontaneously fires at a fixed interval. With inhibitory input that produces an inhibitory postsynaptic potential (IPSP), some action potentials are inhibited, resulting in a distinctive pattern of impulses. From Kandel et al. (2000).

properties. First, the *time constant* affects *temporal summation*, the process by which consecutive synaptic potentials at the same site are added together in the postsynaptic cell. The larger the time constant, the greater the likelihood that two consecutive inputs from an excitatory presynaptic neuron will sum to bring the cell membrane to its threshold for an action potential. Second, the *length constant* affects *spatial summation*. Neurons with a large length constant are more likely to be brought to threshold by two different inputs arising from different sites than are neurons with a short space constant.

In addition to counteracting synaptic excitation, synaptic inhibition can exert control over spontaneously active nerve cells. Many cells in the brain are spontaneously active, as are the pacemaker cells of the heart. By suppressing the spontaneous generation of action potential in these cells, synaptic inhibition can shape the pattern of firing in a cell (Figure 2).

Many mathematical models have been introduced to describe the function of the nervous system. Their complexity ranges from that of individual nerve cells to assemblies of neurons. Hodgkin & Huxley (1952) constructed a model, known as the Hodgkin–Huxley model, to explain the electrical excitability of nerve axons in terms of discrete Na^+ and K^+ currents. The Hodgkin–Huxley model is probably the best-known and the most elegant model of a biological system, exemplifying a balance between experiment and theory that has rarely been matched. To explain how trains of impulses occur in the Hodgkin–Huxley model FitzHugh (1961) used the Hodgkin–Huxley model as one member of a large class of nonlinear systems showing excitable and oscillatory behavior. The model was further analyzed by an approximately distributed pulse transition line using analog simulations (Nagumo et al., 1962) and today is known as FitzHugh–Nagumo model.

Wilson and Cowan (1972) further increased the complexity of the system. The model they proposed was based on the properties of populations, and the single cell activity was represented not by a single spike, but rather by the spike frequency. They based their model on two main assumptions. First, the existence of spatially localized neuronal populations, based on physiological and anatomical evidence that many cells in a small volumes of cortical tissue have very nearly identical responses to identical stimuli. Second, all nervous processes of any complexity are dependent upon the interaction of excitatory and inhibitory cells. Wilson and Cowan were the first to treat inhibition as arising from exclusively inhibitory neurons. Wilson–Cowan oscillators are now widely used to describe higher functions, such as sensory information processing, learning, memory, or pattern recognition. Neural networks and chips are built for signal and image processing.

Much progress has been made in recent years toward a detailed understanding of the complex and nonlinear dynamics that underlie the Hodgkin–Huxley, FitzHugh–Nagumo, and Wilson–Cowan models. In addition, a large number of new models of synchronous oscillations have been introduced, attempting to describe features observed in the electrical activity of the brain as measured by EEG or MEG. They range from those involving details of receptors and inter-inhibitory interactions, to models depending on various nonlinear principles (see, for example, Haken, 2002).

Aneta Stefanovska

See also **FitzHugh–Nagumo equation; Flip-flop circuit; Hodgkin–Huxley equations; Integrate and fire neuron; Multiplex neuron; Myelinated nerves; Nerve impulses; Neurons**

Further Reading

FitzHugh, R. 1961. Impulses and physiological states in models of nerve membrane. *Biophysical Journal*, 1: 445–466

Haken, H. 2002. *Brain Dynamics. Synchronization and Activity Patterns in Pulse-Coupled Neural Nets with Delays and Noise*, Berlin and New York: Springer

Hodgkin, A.L. & Huxley, A.F. 1952. A quantitative description of membrane current and its application to conduction and excitation in nerve. *Journal of Physiology*, 117: 500–544

Kandel, E.R., Schwartz, J.H. & Jessell, T.M. (editors). 2000. *Principles of Neural Science*, 4nd edition, New York: McGraw-Hill

Nagumo, J.S., Arimoto, S. & Yoshizawa, S. 1962. An active pulse transmission line simulating nerve axon. *Proceedings of IRE*, 50: 2061–2071

Pavlov, I.P. 1927. *Conditioned Reflexes*, London: Oxford University Press (original Russian edition 1926)

Sherrington, C. 1933. *The Brain and Its Mechanism*, REDE Lecture, Cambridge: Cambridge University Press

Sherrington, C. 1947. *The Integrative Action of the Nervous System*, 2nd edition, New Haven: Yale University Press

Wilson, H.R. & Cowan, J.D. 1972. Excitatory and inhibitory interactions in localised populations of model neurons. *Biophysical Journal*, 12: 1–24

INSTABILITY

See **Stability**

INSTANTONS

A remarkable feature of the Yang–Mills action is that there are finite-action topological soliton solutions to the classical field equations. These solitons are known as instantons or, in early papers, as pseudoparticles. The Yang–Mills action is important in particle physics; in particular, it describes the behavior of gluons, the particles that carry the strong nuclear force. Before instanton solutions were discovered in 1975 by Belavin et al. (1975), the Yang–Mills theory of the strong force appeared to have a symmetry not found in nature. This was known as the axial U(1) problem and was solved by 't Hooft who realized that one effect of the instanton solutions was to break this unwanted symmetry. This was the first example of an extended classical solution having a physical consequence in a field theory of particle physics.

In four Euclidean dimensions, the pure SU(2) Yang–Mills action is

$$S[A] = -\frac{1}{2}\int \mathrm{d}^4x \,\mathrm{trace}\, F_{\mu\nu}F_{\mu\nu}, \tag{1}$$

where μ and ν are space indices running from 1 to 4 and repeated indices are summed. $F_{\mu\nu}$ is a field strength tensor, a skew Hermitian 2×2 matrix field in four-dimensional space, related to a gauge potential A_μ by

$$F_{\mu\nu} = \frac{\partial}{\partial x_\mu}A_\nu - \frac{\partial}{\partial x_\nu}A_\mu + [A_\mu, A_\nu]. \tag{2}$$

The action is invariant under local gauge transformations

$$A_\mu \to A'_\mu = gA_\mu g^{-1} - \frac{\partial g}{\partial x_\mu}g^{-1}, \tag{3}$$

where g is a special unitary matrix function in four dimensions. Under this transformation $F_{\mu\nu}\to F'_{\mu\nu} = gF_{\mu\nu}g^{-1}$. One key consequence of this is that $F_{\mu\nu}$ not only vanishes when $A_\mu = 0$, but also, whenever A_μ is a gauge transformation of zero. This means that the asymptotic condition required for finiteness of the action is that A_μ approaches a gauge transformation of zero:

$$A_\mu \to -\frac{\partial g}{\partial x_\mu}g^{-1} \tag{4}$$

for $r = \sqrt{x_\mu x_\mu}\to\infty$. Now, the group of special unitary 2×2 matrices is topologically a three-sphere, and so the gauge potential on the large three-sphere at infinity gives a map from that three-sphere to the three-sphere of values of g. Maps between three-sphere are classified topologically by a single integer: the winding number. This integer, therefore, classifies finite-action field configurations; in fact, it is the only gauge-invariant information determined by the asymptotic field behavior. The topologically nontrivial stationary points of the action $S[A]$ are instantons.

Since they are stationary points of the action, instantons obey the Euler–Lagrange equation for the action:

$$\frac{\partial}{\partial x_\mu}F_{\mu\nu} = 0. \tag{5}$$

This is known as the Yang–Mills equation. It can be shown that minimal action solutions obey a first-order equation called the self-dual equation:

$$F_{\mu\nu} = \tfrac{1}{2}\varepsilon_{\mu\nu\lambda\sigma}F_{\lambda\sigma}, \tag{6}$$

where $\varepsilon_{\mu\nu\lambda\sigma}$ is totally skewsymmetric in its indices with $\varepsilon_{1234} = 1$. Solutions to this equation obey the second-order Yang–Mills equation. A solution with winding number one is given by

$$A_\mu = -\frac{r^2}{r^2+R^2}\frac{\partial g}{\partial x_\mu}g^{-1}, \tag{7}$$

where R is an arbitrary scale and g has the hedgehog form

$$g = \frac{x_4}{r}\mathbf{I}_2 + \frac{i}{r}\left(x_1\sigma_1 + x_2\sigma_2 + x_3\sigma_3\right), \tag{8}$$

where $\mathbf{I}_2$ is the 2×2 identity matrix and the σ_i's are the Pauli matrices. Allowing for gauge equivalence, there is an eight-dimensional space of winding number one instantons, four of these dimensions correspond to the choice of position in space, one to the choice of scale, and three to an overall group orientation. Although these results are particular to SU(2), other groups can be studied; the main difference is in the number of overall group orientation parameters.

The self-dual equations are integrable. In fact, many of the integrable equations of mathematical physics can be derived from the self-dual equations by demanding that the solution has some translational or rotational symmetry. Because of this integrability, solutions to the self-dual equations with winding numbers greater than one may be constructed, in principle, by twistor theory and by algebraic methods (Christ et al., 1978). There is also a simple ansatz (Jackiw et al., 1977), but this does not give a general solution.

In the 1980s, the self-dual equations revolutionized the study of smooth four-dimensional manifolds: by studying the space of solutions to the self-dual equations over different manifolds, it is possible to derive invariants, known as the Donaldson invariants (Donaldson & Kronheimer, 1990). This approach has since been largely superceded by Seiberg–Witten theory.

In the path integral approach to quantum field theory, physical values are derived by certain weighted integrations over all possible field configurations. These path integrals are often intractable, and it is common to expand the integration about the stationary points of the action. This works because of the way the action appears in the integrand of the path integral. Calculations of this type are known as semi-classical calculations because they are effectively an expansion in Planck's constant, $\hbar$. This expansion must include a sum over all possible stationary points, and so, in Yang–Mills theory, it includes a sum over instanton configurations. By studying the symmetry properties of the measure in quantum chromodynamics (QCD), the Yang–Mills theory describing the strong nuclear force, it can be shown that terms in this expansion break the axial U(1) symmetry. This symmetry is unbroken if the instanton terms are omitted. The symmetry breaking allows processes that violate baryon and lepton conservation; however, the amplitudes for these effects are highly suppressed. One useful approach is to consider these processes as tunneling events between different vacua, in fact, instanton calculations in quantum field theory are very similar to Wentzel–Kramers–Brillouin (WKB) calculations in quantum mechanics.

While instantons provide a qualitative explanation for a host of phenomena in quantum chromodynamics, useful quantitative results are not available within the semi-classical approach, and even qualitatively, instanton calculations are not rigorous since they relate only to the semi-classical approximation, a truncation of the full quantum theory. The modern approach to quantum chromodynamics is lattice QCD. It is possible within lattice QCD to verify the original ideas about the role of instantons in the physics of the strong force; however, there are limits to the precision with which the lattice and semi-classical approaches can be compared (Negele, 1998).

Finite action soliton solutions in other equation systems are sometimes referred to as instantons. Examples include the finite action solutions to the forced Burgers equation, which arises in the study of turbulence, and the vortex-like solutions in the abelian Higgs model, which is related to condensed matter physics. Instantons have many similarities to the lump solitons found in certain sigma models.

CONOR HOUGHTON

See also **Burgers equation; Higgs boson; Particles and antiparticles; Quantum field theory; Solitons; Yang–Mills theory**

Further Reading

Belavin, A.A., Polyakov, A.M., Schwarz, A.S. & Tyupkin, Yu.S. 1975. Pseudoparticle solutions of the Yang–Mills equations. *Physics Letters* B, 59: 85–87

Christ, N.H., Weinberg, E.J. & Stanton, N.K. 1978. General self-dual Yang–Mills solutions. *Physical Review* D, 18: 2013–2025 (Includes a review of the complete solution to the self-dual equations originally due to Atiyah, Hitchin, Drinfeld, and Manin)

Coleman, S. 1985. *Aspects of Symmetries*, Cambridge and New York: Cambridge University Press (Chapter 7 gives a celebrated description of instantons in QCD)

Donaldson, S.K. & Kronheimer, P.B. 1990. *The Geometry of Four-manifolds*, Oxford: Clarendon Press and New York: Oxford University Press

Jackiw, R., Nohl, C. & Rebbi, C. 1977. Conformal properties of pseudoparticle configurations. *Physical Review* D, 15: 1642–1646 (A large, but incomplete, family of solutions to the self-dual equations.)

Negele, J.W. 1998. Instantons, the QCD vacuum, and hadronic physics. *Nuclear Physics* B *Proceedings Supplements*, 73: 92–104

't Hooft, G. 1976. Symmetry breaking through Bell-Jackiw anomalies. *Physical Review Letters*, 37: 8–11

Weinberg, S. 1996. *Quantum Field Theory*, vol. 2, Cambridge and New York: Cambridge University Press (Chapter 23 concerns extended field configurations in particle physics and has a very clear treatment of instantons.)

INTEGRABILITY

Central problems in the integrability theory for nonlinear ordinary differential equations (ODEs) or partial differential equations (PDEs) are knowing which systems can be solved analytically and developing appropriate solution techniques. Although in some contexts the term *integrability* is not well defined, let us begin by considering two physically motivated nonlinear ODE systems that can be analytically solved.

Nonlinear Pendulum

Here, the nonlinear system is

$$\ddot{x} = ax + bx^2 + cx^3, \quad a, b, c = \text{constants}. \qquad (1)$$

Multiplying both sides of this equation by $\dot{x}$ and integrating yields an intermediate integral $I = \dot{x}^2/2 - ax^2/2 - bx^3/3 - cx^4/4$, which is determined by the initial conditions: $x(t_0)$ and $\dot{x}(t_0)$. In terms of I, the solution $x(t)$ can be written implicitly as

$$t - t_0 = \int_{x_0}^{x} \frac{\mathrm{d}x}{\sqrt{2I + ax^2 + 2bx^3/3 + cx^4/2}}, \qquad (2)$$

where $x(t_0) = x_0$. The inversion of this formula is expressed in terms of Jacobi elliptic functions (Bryd & Friedman, 1954). Note that this system has one degree of freedom, and one intermediate integral was sufficient to obtain a solution. Because it is related to an area, the intermediate integral (I) is sometimes called a quadrature, and by extension, Equation (2) is said to be obtained by the method of quadratures.

Calogero–Moser N-Body Problem

In this example, the nonlinear system is

$$\ddot{q}_n = \sum_{m=1,m\neq n}^{N} 2g^2(q_n - q_m)^{-3}, \quad g = \text{constant}, \quad (3)$$

generalizations of which were studied extensively in the last three decades of the 20th century. Using a Lax isospectral deformation technique, the solution matrix diag $[q_1(t), \ldots, q_N(t)]$ is found to be similar to the matrix $\tilde{Q} = [\tilde{Q}_{nm}(t)]$ defined by

$$\tilde{Q}_{nm}(t) = \delta_{nm}\big[q_n(0) + \dot{q}_n(0)t\big] + \mathrm{i}(1 - \delta_{nm})g\big[q_n(0) - q_m(0)\big]^{-1}t. \quad (4)$$

Thus, the solution to the initial value (Cauchy) problem is reduced to the algebraic task of finding the N eigenvalues of the matrix $\tilde{Q}$, and the solution formula for initial data, $q_i(0)$ and $\dot{q}_i(0)$, can be constructed from linear functions, through a finite number of algebraic operations and compositions of functions (Calogero, 2001). Equations (1) and (3) are completely integrable.

In the 19th century, Evariste Galois, Niels Henrik Abel, Joseph Liouville, and Sophus Lie tried to rationalize the process of solving differential equations by quadratures, or failing that, massaging them in such a way that useful information might be extracted. Mainly based on geometry and algebra, these efforts initiated much of the mathematics—such as classifications of differential equations in term of symmetries and conservation laws—that has dominated the 20th century. Attempts by Karl Weierstrass and Henri Poincaré to create a systematic theory of integrability are based on complex function theory.

An achievement of 19th-century mathematics was the elaboration of the theory of elliptic and Abelian functions, particularly the introduction of theta functions. A solution comprising such rapidly convergent power series is quite efficient computationally; thus, a natural question for 19th-century mathematicians was this: Which differential equations admit solutions as quotients of power series that converge in large regions, independent of initial or boundary conditions? The Cauchy–Kovalevsky theorem gives local existence for such expressions. This theme traces through the work of Sophia Kovalevsky and Paul Painlevé for nonlinear ODEs and through Bernhard Riemann and Immanuel Fuchs and their successors for linear ODEs (Hermann, 1984).

There are several results about local and global solutions of differential equations that establish existence, uniqueness, smoothness, stability, approximations, and so on; yet attempts to find exact solution formulas in the 19th century were largely unsuccessful. This is because solutions to nonlinear systems may exhibit complex behaviors—such as blow-up, shocks, chaos, fractals, and bifurcation—which are obstacles to integrability. Poincaré put an end to the search for new integrable equations by showing that among all dynamical systems integrable ones are exceptional. Indeed, a small structural perturbation of an integrable system often destroys integrability.

Although Poincaré's results dampened interest in the search for new integrable equations during the first half of the 20th century, the situation changed dramatically with the discovery of the inverse scattering transform as a method of solving the initial value (Cauchy) problem for the Korteweg–de Vries (KdV) equation (Gardner et al., 1967). Rapidly extended to several other nonlinear systems of scientific interest (nonlinear Schrödinger equation, sine-Gordon equation, Toda lattice, and so on), this discovery also led to the emergence of "soliton factories" during the 1980s (Zakharov, 1991).

More generally, integrability theory aims to find global information on solutions and, if possible, to solve differential equations analytically. Integrability itself is an intrinsic characteristic of differential equations, imposing constraints on the way solutions evolve in phase space and suggesting the following working definition.

Definition A differential equation is completely integrable if all solutions to well-posed initial or boundary value problems can be presented beginning with elementary functions, using finitely many algebraic operations and compositions of functions, and evaluating limits.

Algebraic operations include inverting or diagonalizing matrices; compositions of functions include inverses; and the limits can be integrals, infinite series, or asymptotes. Thus, this definition holds for cases in which solutions can be constructed explicitly. It also generalizes the notion of integrability by quadratures (Liouville integrability) that only requires computation of intermediate integrals in addition to algebraic operations and compositions of functions.

For ODEs, the Liouville–Arnol'd theorem on finite-dimensional Hamiltonian systems gives sufficient conditions, called the Liouville conditions, for guaranteeing the integrability by quadratures (Arnol'd, 1989). For a Hamiltonian system of N degrees of freedom, these conditions are:

- the existence of N integrals of motion $\{F_1, F_2, \ldots, F_N\}$,
- that are functionally independent on the level surface $\{F_i = a_i\}$ containing the initial data, and
- that commute with each other under the associated Poisson bracket.

If the level surface $\{F_i = a_i\}$ is compact and connected, then the Liouville–Arnol'd theorem says that the underlying Hamiltonian system can be expressed as

$$\dot{I}_i = 0, \ \dot{\varphi}_i = \omega_i(I_1, \ldots, I_N), \quad (5)$$

in the action-angle coordinate system (I_i, φ_i) of a neighborhood of the level surface. Thus, the motion is conditionally periodic along an N-dimensional torus with the frequencies ω_i, and the solution to the Cauchy problem can be completely determined by the method of quadratures. The Euler top, the Lagrange top, Kowalewski's top, the Stäckel systems, and geodesic flows on a surface are Liouville-integrable systems (Perelomov, 1990).

For PDEs, there also exist some results on integrability properties from a Hamiltonian perspective, including the bi-Hamiltonian theory (Magri, 1978) from which infinitely many symmetries and conservation laws for PDEs can be deduced. Motivated by the Liouville–Arnol'd theorem, a notion of integrability for PDEs is the existence of infinitely many conservation laws. The KdV equation is integrable in this sense, and it can also be put into a Hamiltonian form in action-angle coordinates like integrable ODEs (Zakharov & Faddeev, 1971).

The existence of infinitely many conservation laws can lead to infinitely many finite-dimensional solution varieties that can be determined analytically (Fuchssteiner, 1992), but it is not clear whether these subvarieties generate the whole infinite-dimensional solution variety. Indeed, the infinitely many degrees of freedom of PDEs introduce large diversity, obscuring their solvability and integrability properties.

The notion of partial integrability arises when systems of differential equations (both ODEs and PDEs) possess more degrees of freedom than conservation laws (Conte & Boccara, 1990). Symmetry constraints developed in soliton theory (Ma & Zhou, 2001) suggest a way to show partial integrability for PDEs through relating PDEs to integrable ODEs. The method was motivated by Moser's work on Hill's equation (Moser, 1980) and nonlinearization of Lax pairs (Cao, 1990). The results generalize the theory of finite-dimensional integrable stationary equations, suggesting the possibility of establishing a Liouville–Arnol'd theorem for infinite-dimensional Hamiltonian systems.

Since the 1970s, several criteria have arisen for testing the integrability of nonlinear PDEs, and corresponding theories include infinitely many symmetries, infinitely many conservation laws, Lax structure, bi-Hamiltonian structure, the Bäcklund transform, Hirota's bilinear form, the inverse scattering transform, and the Painlevé property (Degasperis, 1998; Gu, 1995).

An ODE or a PDE is said to possess the Painlevé property, respectively, if the movable singularities of solutions of the ODE are only poles (Ablowitz & Clarkson, 1991) or if solutions of the PDE are "single-valued" in the neighborhood of noncharacteristic, movable singularity manifolds (Weiss et al., 1983), suggesting that the Painlevé property is an indication of complete integrability (Conte, 1999). One advantage of the Painlevé property is that it can be a tool for computing Lax pairs and conservation laws. Lax pairs, as we have seen in Equation (3), provide linear objects for solving Cauchy problems of nonlinear equations. A thorough understanding of the relationship between Painlevé singularities and integrability may suggest alternatives to the KAM notion of near-integrability (Zakharov, 1991).

The term S-integrability implies integrability of Cauchy problems by the inverse spectral transform technique (Calogero & Degasperis, 1982) or the inverse scattering transform (Ablowitz & Clarkson, 1991; Fokas, 1997). C-integrability, on the other hand, means that a differential equation can be transformed into a linear one by an appropriate change of variables. Examples of C-integrable equations are the Burgers equation and the Calogero–Dagasperis–Ibragimov–Shabat equation:

$$u_t = u_{xx} - 2uu_x, \qquad \text{(B)}$$

$$u_t = u_{xxx} + 3u^2u_{xx} + 9uu_x^2 + 3u^4u_x. \qquad \text{(CDIS)}$$

These two equations are, respectively, cast into the heat equation ($v_t = v_{xx}$) and the linear KdV equation ($v_t = v_{xxx}$) under appropriate dependent variable transformations. Calogero and colleagues have made a systematic study of the construction and classification of C-integrable equations, both in $1 + 1$ dimensions and $N + 1$ dimensions (Zakharov, 1991). Although the property of C-integrability often leads to an infinity of conservation laws, the Burgers equation has only one local conservation law of differential polynomial type.

WEN-XIU MA

See also **Burgers equation; Constants of motion and conservation laws; Hamiltonian systems; Hirota's method; Inverse scattering method or transform; Painlevé analysis; Solitons**

Further Reading

Ablowitz, M.J. & Clarkson, P.A. 1991. *Solitons, Nonlinear Evolution Equations and Inverse Scattering*, Cambridge and New York: Cambridge University Press

Arnol'd, V.I. 1989. *Mathematical Methods of Classical Mechanics*, 2nd edition, New York: Springer

Bryd, P.F. & Friedman, M.D. 1954. *Handbook of Elliptic Integrals*, New York: Springer

Calogero, F. 2001. *Classical Many-Body Problems Amendable to Exact Treatments*, Berlin and Heidelberg: Springer

Calogero, F. & Degasperis, A. 1982. *Spectral Transform and Solitons*, vol. I. *Tools to Solve and Investigate Nonlinear Evolution Equations*, Amsterdam and New York: North-Holland

Cao, C.W. 1990. Nonlinearization of the Lax system for AKNS hierarchy. *Science in China, Series A: Mathematics, Physics, Astronomy*, 33: 528–536

Conte, R. (editor). 1999. *The Painlevé Property, One Century Later*, New York: Springer

Conte, R. & Boccara, N. (editors). 1990. *Partially Integrable Evolution Equations in Physics*, Dordrecht: Kluwer

Degasperis, A. 1998. Resource letter Sol-1: solitons. *American Journal of Physics*, 66: 486–497

Fokas, A.S. 1997. A unified transform method for solving linear and certain nonlinear PDEs. *Proceedings of the Royal Society of London, Series A: Mathematical, Physical and Engineering Sciences*, 453: 1411–1443

Fuchssteiner, B. 1992. Hamiltonian structure and integrability. In *Nonlinear Equations in the Applied Sciences*, Boston, MA: Academic Press, pp. 211–256

Gardner, C.S., Greene, J.M., Kruskal, M.D. & Miura, R.M. 1967. Method for solving the Korteweg–de Vries equation. *Physical Review Letters*, 19: 1095–1097

Gu, C.H. (editor). 1995. *Soliton Theory and Its Applications*, Berlin and Heidelberg: Springer and Hangzhou: Zhejiang Science and Technology Publishing House

Hermann, R. 1984. *Topics in the Geometric Theory of Integrable Mechanical Systems*, Brookline, MA: Math Science Press

Ma, W.-X. & Zhou, Z. X. 2001. Binary symmetry constraints of N-wave interaction equations in $1+1$ and $2+1$ dimensions. *Journal of Mathematical Physics*, 42: 4345–4382

Magri, F. 1978. A simple model of the integrable Hamiltonian equation. *Journal of Mathematical Physics*, 19: 1156–1162

Moser, J. 1980. Various aspects of integrable Hamiltonian systems. In *Dynamical Systems*, Boston: Birkhäuser, pp. 233–289

Perelomov, A.M. 1990. *Integrable Systems of Classical Mechanics and Lie Algebras*, vol. I, Basel: Birkhäuser

Weiss, J., Tabor, M. & Carnevale, G. 1983. The Painlevé property for partial differential equations. *Journal of Mathematical Physics*, 24: 522–526

Zakharov, V. E. (editor). 1991. *What Is Integrability?* Berlin and Heidelberg: Springer

Zakharov, V.E. & Faddeev, L.D. 1971. The Korteweg–de Vries equation is a fully integrable Hamiltonian system. *Functional Analysis and Its Applications*, 5(4): 18–27

INTEGRABLE CELLULAR AUTOMATA

Integrable cellular automata (CAs) are discrete dynamical systems that possess attributes of integrability including conserved quantities, symmetries, and localized solutions. They show how to discretize a given integrable system of differential equations, maintaining the integrability property, suggesting that integrability of discrete systems and their coherent structures are inherently connected with the iterated string processing performed by automata.

The propagating solutions (or coherent objects) on certain cellular automata have been found to exhibit nondestructive collisions similar to those observed for soliton systems like the Korteweg–de Vries equation. In such cases, pulse-like disturbances propagating along a uniform nonlinear medium are represented by strings of symbols (zero-one patterns), passing through a one-way structure—the pipeline of identical automata M. The analysis of these moving objects reduces to investigating the repeated automaton action over strings, also called an iterated automaton map (IAM); $M(a^t) = a^{t+1}, t = 0, 1, \ldots$.

All known models capable of supporting such discrete localized structures are described by automata (Siwak, 2001, 2002), and a method (called ultradiscretization) leads to discrete systems in which solutions of some familiar soliton equations are preserved as soliton-like patterns (Tokihiro et al., 1999). Thus, an IAM is a fundamental discrete mechanism that supports localized soliton-like periodic structures. This is why a new term *iterons* (Siwak, 2001, 2002) has been proposed for these objects. Note also that fractals owe their existence to the iterating process of some maps.

There are two classes of iterons. The first consists of particles well known in cellular automata models (Delorme & Mazoyer, 1999) where the parallel processing of strings occurs. The second class is not widely known and consists of so-called filtrons. The filtrons are emergent coherent objects occurring in serial processing of strings (Siwak, 2001).

In parallel processing of a string a^t at a given time t all symbols a_i^{t+1} of the next string a^{t+1} are updated simultaneously, for example, by the function $a_i^{t+1} = f(a_{i-r}^t, \ldots, a_i^t, \ldots, a_{i+r}^t)$. When the same function f is used for all positions i, one has a 1-dimensional CA, with f being called its local function or rule. The arguments of f are determined by the so-called neighborhood window $N_i = (-r, \ldots, +r)$; here N_i designates r neighbors on both sides of position i.

The listing of consecutive strings a^t for $t = 0, 1, \ldots$ one under another forms what is called a space-time (ST) diagram. This diagram visualizes the evolution of a given string a^0—the dynamics of a CA system in phase space for initial global state a^0. Occasionally, some moving and periodic patterns of symbols or segments of a string are seen on ST diagrams of CA processing. These are just particles or signals of CAs (Delorme & Mazoyer, 1999).

The serial processing of strings is performed by a computational model called a finite (state) automaton. The Mealy type automaton is described by $M = (S, \Sigma, \Omega, \delta, \beta, s_0)$, where S, Σ, and Ω are nonempty finite sets of—respectively—states, inputs, and outputs; $\delta\colon S \times \Sigma \to S$ is called the next state function of M, $\beta\colon S \times \Sigma \to \Omega$ is called the output function of M, and s_0 is the initial (extinction) state of the automaton. The automaton converts sequences of symbols, preserving their length. Any input string is read sequentially from left to right, one symbol at each instant of time (pulse of clock). For all $\tau = 1, 2, \ldots$ the automaton:

(i) reads input symbol $\sigma(\tau)$,
(ii) changes its current state $s(\tau)$ onto the next one according to $\delta(s(\tau), \sigma(\tau)) = s(\tau+1)$, and
(iii) generates the symbol $\beta(s(\tau), \sigma(\tau)) = \omega(\tau)$ of the resulting string.

Thus the complete one step conversion at $a^t \to a^{t+1}$ requires a series of clock pulses τ. To consider IAMs, one has to assume the unified input-output alphabet $A = \Sigma = \Omega = 0, 1, \ldots, m$. Then the automaton's operation can be described by a

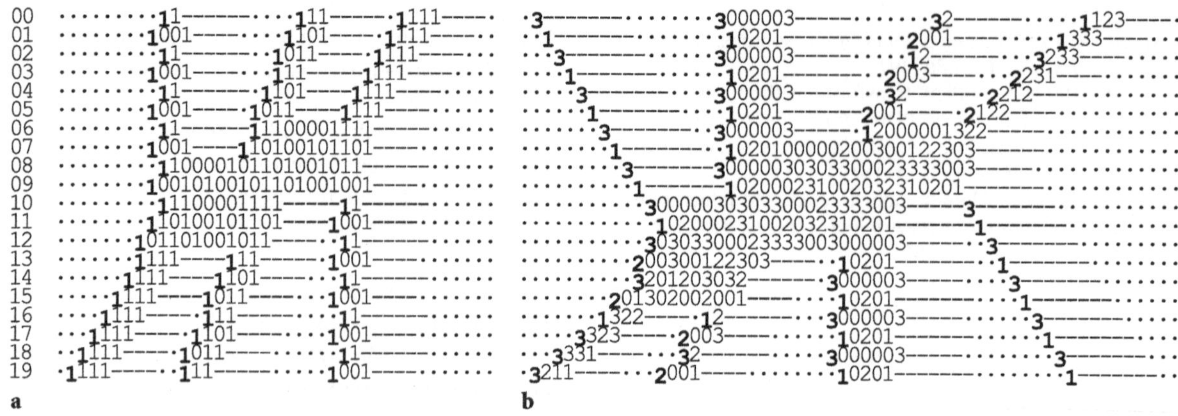

Figure 1. Nondestructive collisions of filtrons ($q=2$). (a) PST model. (b) Automaton M_1.

sequence of operations $f_s\colon A \to A$, each f_s such that $f_s(a) = \beta(s, a)$.

Coherent objects of serial string processing were not known for a long time. The first model that established a bridge between discrete systems of string processing and soliton-like phenomena was the filter CA (Park et al., 1986). The PST (Park, Steiglitz, & Thurston) model was defined for $A = \{0, 1\}$ by a special space-time window (FCA window) and a nonlinear updating Boolean function $f_{\mathrm{PST}} : A^{2r+1} \to A$. It operates by moving the window to the right along a given string a^t. At each position the function f_{PST} computes consecutive elements

$$a_i^{t+1} = f_{\mathrm{PST}}\left(a_i^t, a_{i+1}^t, \ldots, a_{i+r}^t, a_{i-r}^{t+1}, a_{i-r+1}^{t+1}, \ldots a_{i-1}^{t+1}\right)$$

of the resulting string a^{t+1} in such a way that $a_i^{t+1} = 1$ if and only if the sum of all window elements (which are arguments of f_{PST}) is even but not zero. It is assumed that the conversion $a^t \to a^{t+1}$ starts somewhere at left with the zero content of window (zero boundary conditions).

The PST model is an automaton and is classified as a digital parity infinite impulse response (IIR) filter. The dynamics of the PST model reveal discrete moving coherent objects capable of colliding in a nondestructive way. An example of such objects is shown in Figure 1(a).

Formal analysis of filter CAs has been given by Goldberg (1988). Goldberg has also given its first automaton-like description by means of a rapid updating rule.

Various generalizations of this model soon appeared. New updating functions were introduced, and the FCA window was modified (Jiang, 1992). A model with an infinite FCA window was given by Takahashi and Satsuma and called a soliton CA. Then, the working alphabet A was enlarged beyond binary (Fokas et al., 1990), and algebraic group operations were considered in computations of updating function.

Another approach was based on deriving the filter-CA-like description from the equations of motion of some systems and from so-called spectral problems (Bruschi et al., 1992).

It was observed that most of the models based on the FCA window perform very special cyclic processing. This led Siwak (2002) to a class of filter automata (FAs). FAs are described by three notions that are useful in physical interpretation of the IAM as a medium, namely:

— the activating function $\iota : A \to \{false, true\}$ such that $\iota(a) = true \leftrightarrow \delta(s_0, a) \neq s_0$;
— the cyclic sequence $(h_j, f_1, \ldots, f_k)$ of mappings f_s (the medium is excited); and
— the extinction conditions (the automaton returns to its extinction state s_0).

Note that the automaton equivalent of PST processing is activated by $a = 1$ (nonzero symbol). Its cyclic sequence of operations f_s is of the form $(N, A, \ldots, A) = NA^r$ where N denotes negate operation, and A is identity id (or accept). The extinction condition is based on a coincidence of some string's current segment $w \in \{a0^r\}$ with the cycle of activity.

The example of IAM dynamics of some cyclic automaton M_1 for $A = \{0, 1, 2, 3\}$ is shown in Figure 1(b). When the automaton M_1 is extinguished, it rewrites input zeroes; $h_0(0) = 0$, but activated by a symbol $j \neq 0$, it performs the cycles of operations $(h_0, f_1, f_2, f_3), (h_j, f_1, f_2, f_3), \ldots$ where

$$h_0 = \begin{pmatrix} 0 & 1 & 2 & 3 \\ 0 & 0 & 0 & 0 \end{pmatrix}, h_1 = \begin{pmatrix} 0 & 1 & 2 & 3 \\ 3 & 0 & 2 & 2 \end{pmatrix},$$

$$h_2 = \begin{pmatrix} 0 & 1 & 2 & 3 \\ 2 & 3 & 0 & 1 \end{pmatrix}, h_3 = \begin{pmatrix} 0 & 1 & 2 & 3 \\ 1 & 2 & 3 & 0 \end{pmatrix},$$

$$f_1 = f_3 = h_3 = \begin{pmatrix} 0 & 1 & 2 & 3 \\ 0 & 1 & 2 & 3 \end{pmatrix} = \mathrm{id},$$

$$f_2 = \begin{pmatrix} 0 & 1 & 2 & 3 \\ 0 & 1 & 3 & 2 \end{pmatrix}.$$

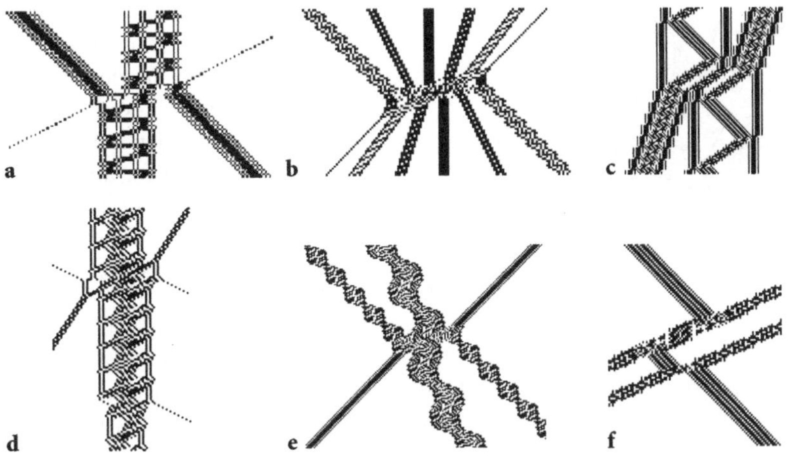

Figure 2. Exemplary images of IAM dynamics—various phenomena of colliding filtrons.

Extinction occurs when the segment $w = 0000$ is read within a cycle.

The symbols of A are shown on ST diagram in order to expose so-called M-segments within strings. Each M-segment begins with an activating symbol, which is printed in bold. The tail zeroes of an M-segment, that is all consecutive zeros immediately preceding the extinction of M are denoted by bar. All zeroes read by the automaton in its state s_0 are shown as dots, and all remaining zeroes as digits 0. With this convention one can recognize whether any two M-segments are distant, adjacent, or overlap and actually form a complex object. Also, a shift q per row to the left is sometimes applied to ST diagrams.

The filtron x of automaton M within the IAM dynamics is defined to be periodic M-segment of the string $a^0 = \ldots 0x0 \ldots$.

Among the discrete models supporting coherent objects there are also the following. For most of them the appropriate equivalent automata have been identified (Siwak, 2001, 2002):

- iterated arrays of finite state automata (Atrubin, 1965);
- filter CA, soliton CA, and other generalizations of PST model based on FCA window;
- digital recursive (IIR) filters and filter automata family (Siwak, 2001);
- classical CAs and higher order CAs (Delorme & Mazoyer, 1999);
- CAs with mixed serial and parallel scenarios of their updating schedule;
- rapid-updating rule-type procedures called also fast rules;
- integrable CAs derived from so-called spectral problems (Bruschi et al., 1992), and discrete versions of some classical soliton equations;
- box–ball systems introduced by D. Takahashi;
- crystal systems and combinatorial data processing;
- techniques of accelerating the convergence of numerical algorithms;
- affine Lie algebras.

Let us illustrate the dynamics of IAM with automata $M_{\text{BBSC}}(m, n)$, which are equivalent to the model called the box-ball system with carrier (BBSC). In BBSC each position i represents a box, and each symbol $0 < a_i \leq m$ from A is interpreted as the number of balls. A single step of the evolution of a string is performed by a carrier of capacity n that runs to the right once and moves each ball exactly once. At each box, the carrier puts as many balls into the box as possible and simultaneously takes as many balls from the box as possible.

This means that if the carrier contains $0 \leq c_j \leq n$ balls at box j, then it will take $\min(a_j, n - c_j)$ balls from this box and, simultaneously, will leave $\min(c_j, m - a_j)$ balls. The finite state automaton $M_{\text{BBSC}}(m, n)$ that performs exactly the same string processing as this BBSC is such that: the set of states $S = \{0, 1, \ldots, n\}$ and the alphabet of input/output symbols $A = \{0, 1, \ldots, m\}$ are sets of integers; $s_0 = 0$ is the initial state of $M_{\text{BBSC}}(m, n)$; and the next states of $M_{\text{BBSC}}(m, n)$ are given by $\delta(s, \sigma) = s + \min(\bar{s}, \sigma) - \min(s, \bar{\sigma})$, while the output symbols are determined by $\omega = \beta(s, \sigma) = \sigma + \min(s, \bar{\sigma}) - min(\bar{s}, \sigma)$. Here $\bar{s} = n - s$ and $\bar{\sigma} = m - \sigma$.

An example of colliding filtrons of the automaton $M_{\text{BBSC}}(16, 22)$ is given in Figure 2.

Further examples of IAM dynamics for some other automata are given in Figure 3 where among the multi-filtron nondestructive (solitonic) collisions the following phenomena can be especially seen: complex filtrons (a), (c), jumps of the filtron over a bundle (a), (b), trapped or bouncing filtrons and collision of such complexes (c), decaying object (or quasifiltron) and its earlier solitonic collision (d), vibrating filtrons (e), steady filtrons (a), (b), and attraction and repelling of filtrons (f). Also the fusion, annihilation of filtrons and

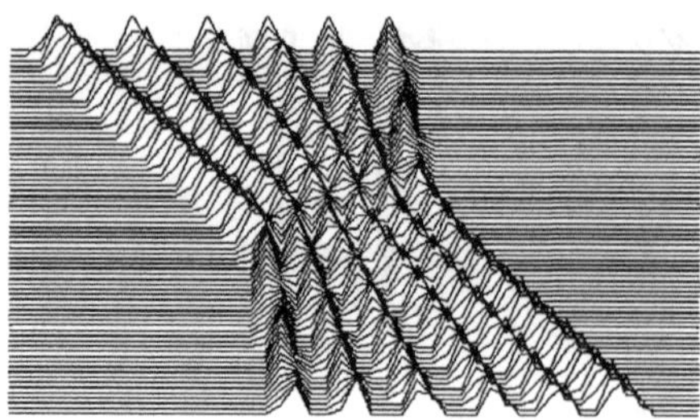

Figure 3. Six filtrons of the automaton $M_{\mathrm{BBSC}}(16, 22)$ collide solitonically (shift $q = 1$).

many other events are possible and can be seen in computer simulations—the realm of the behaviors of filtrons of automata is fascinating and amazingly rich.

The automaton approach and explanation of discrete coherent structures phenomena by IAMs leads to many applications, including computing by means of streams of filtrons (solitonic processing idea of Steiglitz), transmission and conversion of information via fiber optical lines, and signals and waves processing (e.g., by an interaction of electromagnetic waves with a shock wave background or another precisely shaped medium).

PAWEL SIWAK

See also **Cellular automata; Cellular nonlinear networks; Game of life; Integrable lattices; Solitons, types of**

Further Reading

Atrubin, A.J. 1965. An iterative one-dimensional real-time multiplier. *IEEE Transactions on Electronic Computers*, EC-14: 394–399

Adamatzky, A. (editor). 2002. *Collision-Based Computing*, London: Springer

Bruschi, M., Santini, P.M. & Ragnisco, O. 1992. Integrable cellular automata. *Physics Letters* A, 169: 151–160

Delorme, M. & Mazoyer, J. (editors). 1999. *Cellular Automata*, Dordrecht: Kluwer

Fokas, A.S., Papadopoulou, E.P. & Saridakis, Y.G. 1990. Coherent structures in cellular automata. *Physics Letters* A, 147(7): 369–379

Jiang, Z. 1992. An energy-conserved solitonic cellular automaton. *Journal of Physics* A: *Mathematical and General*, 25(11): 3369–3381

Park, J.K., Steiglitz, K. & Thurston, W.P. 1986. Soliton-like behavior in automata. *Physica* D, 19: 423–432

Siwak, P. 2001. Soliton-like dynamics of filtrons of cyclic automata. *Inverse Problems*, 17(4): 897–918

Siwak, P. 2002. Iterons of automata. In *Collision-Based Computing*, edited by A. Adamatzky, London: Springer, pp. 299–353

Tokihiro, T., Nagai, A. & Satsuma, J. 1999. Proof of solitonical nature of box and ball systems by means of inverse ultra-discretization. *Inverse Problems*, 15: 1639–1662

INTEGRABLE LATTICES

Lattices, or differential-difference equations (DDEs), are a special class of ordinary differential equations, with an infinite number of dependent variables $q_n = q_n(t)$ numbered by integer indices n, admitting a translational invariance with respect to the shift $n \to n + 1$. Due to this property, such equations are well suited for describing processes in translationally symmetric systems like crystals. By imposing suitable boundary conditions, such as periodic ones ($q_n(t) \equiv q_{n+N}(t)$ for all n), one can also model finite systems. In the notation above the dependent variable t plays the role of time; the range where the dependent variables q_n take values can be a multidimensional vector space or even a more general manifold. Sometimes it is convenient to think of n as the second, discrete independent variable, so that $q_n(t) = q(t, n\Delta x)$ is thought of as a sampling of a function $q(t, x)$ of two continuous variables (time t and space coordinate x) for the discrete sequence of values of $x = n\Delta x$. In such a context differential-difference systems appear as semi-discretizations of systems of partial differential equations (PDEs). For instance, the lattice model

$$\ddot{q}_n = \omega_0^2(q_{n+1} - 2q_n + q_{n-1}) - \sin(q_n), \qquad (1)$$

used by Yakov Frenkel and Tatiana Kontorova in their study (1939) of the relation between dislocation and deformation of crystals, can be considered as a semi-discretization of the sine-Gordon (SG) equation $q_{tt} - q_{xx} = -\sin(q)$ (in this interpretation $\omega_0 = 1/\Delta x$). Subsequent developments revealed that the discrete model (1) is non-integrable, unlike its continuous counterpart SG. Thus, in hindsight, already this early example illustrates the difficult problem of integrable discretization: how to discretize one or several independent variables in a given integrable system, maintaining the integrability property.

Another early example of a physically important system of DDEs is the model used by Enrico Fermi, John Pasta and Stan Ulam in their 1955 study of the dynamics of energy partition in crystals:

$$\begin{aligned}\ddot{q}_n &= -\omega_0^2(q_{n+1} - 2q_n + q_{n-1}) \\ &\quad -\alpha\big((q_{n+1} - q_n)^2 - (q_n - q_{n-1})^2\big). \qquad (2)\end{aligned}$$

In this model, only nearest neighbors interact with the potential depending on their relative displacement, $U(\Delta q) = \frac{\omega_0^2}{2}(\Delta q)^2 + \frac{\alpha}{3}(\Delta q)^3$. Also this model is, strictly speaking, non-integrable. However, by an interesting historical twist, the studies by Fermi, Pasta, and Ulam played the precursor role for the theory of solitons. Presumably, due to the closeness of Equation (2) to integrable models, it demonstrated a lack of thermalization (the FPU recurrence phenomenon), and in attempts to understand this, Kruskal and Zabusky studied the KdV equation and eventually discovered its solitons (*See* **Fermi–Pasta–Ulam oscillator chain**).

The story of integrable lattices began with the discovery by Toda (1989) of the lattice with an exponential interaction of nearest neighbors, which carries nowadays his name (*See* **Toda lattice**):

$$\ddot{q}_n = \mathrm{e}^{q_{n+1}-q_n} - \mathrm{e}^{q_n-q_{n-1}}. \qquad (3)$$

He found a number of exact soliton-like solutions of this systems, both in the infinite and in the periodic setting. Multisoliton solutions were constructed in 1973 by Hirota with the help of his ingenious "direct" (or bilinear) method (*See* **Hirota's method**). The integrability of this model in the sense of classical mechanics (existence of an infinite number of independent integrals of motion, which are in involution with respect to the standard Poisson bracket with canonically conjugate momenta $p_n = \dot{q}_n$), as well as in the sense of the solitons theory (Lax representation and applicability of the inverse scattering, or inverse spectral, transformation (IST)) was demonstrated in 1974 independently by Flaschka and by Manakov. Other important early work on this system includes the treatment of scattering in the finite nonperiodic Toda lattice by Moser (1975), finding the Bäcklund transformation by Wadati and Toda (1975), and the solution of the periodic lattice in terms of multidimensional theta-functions given independently by Date and Tanaka and by Krichever (1976). One has a rare possibility to read the first-hand presentation of the whole body of relevant results, including the authentic story of the original discovery, in Toda (1989).

Once the mechanism of integrability via IST for lattice systems was uncovered, new integrable DDEs began to be discovered in quick succession. The next one was the lattice

$$\dot{a}_n = a_n(a_{n+1} - a_{n-1}), \qquad (4)$$

having plenty of applications and known today under various names (Volterra, Kac–van Moerbeke, discrete KdV, etc.). Its integrability was established independently by Manakov (1974) and by Kac and van Moerbeke (1975), while the multisoliton solutions for it and for its modification,

$$\dot{a}_n = (1 + a_n^2)(a_{n+1} - a_{n-1}), \qquad (5)$$

had been found by Hirota in 1973.

An important paper by Ablowitz & Ladik (1976) contained a lattice version of the zero-curvature (Zakharov–Shabat) representation of integrable systems and integration methods based on it. This deepened understanding of the algebraic structures behind integrable lattices, and opened an opportunity to construct in a unified manner arbitrarily many examples. One of the celebrated systems introduced and studied there is an integrable spatial discretization of the nonlinear Schrödinger equation (DNLS):

$$\mathrm{i}\dot{\psi}_n = -(\psi_{n+1} - 2\psi_n + \psi_{n-1})/(\Delta x)^2 + \nu \mid \psi_n \mid^2 \times(\psi_{n+1} + \psi_{n-1}). \qquad (6)$$

A different but gauge equivalent discretization of NLS was found by Isergin and Korepin in 1981 (*See* Faddeev & Takhtajan, 1987 and **Discrete nonlinear Schrödinger equations**).

Dozens of integrable lattices have been found since then (which, of course, has led to a certain devaluation of this sort of activity). It seems, however, that only a few of them are of really fundamental importance for theory and applications. Such a "short list" depends naturally on personal tastes, but the following systems (apart from those already mentioned) should be included in any case.

A sort of a relativistic generalization of the Toda lattice is due to Ruijsenaars (1990):

$$\ddot{q}_n = (1 + \alpha\dot{q}_{n+1})(1 + \alpha\dot{q}_n)\frac{\mathrm{e}^{q_{n+1}-q_n}}{1 + \alpha^2\mathrm{e}^{q_{n+1}-q_n}} -(1 + \alpha\dot{q}_n)(1 + \alpha\dot{q}_{n-1})\frac{\mathrm{e}^{q_n-q_{n-1}}}{1 + \alpha^2\mathrm{e}^{q_n-q_{n-1}}}. \qquad (7)$$

A Toda-type lattice with nearest neighbors interactions that encounter elliptic functions but depends not only on their relative displacements was discovered by Shabat & Yamilov (1990) and independently by Krichever (2000):

$$\ddot{q}_n = (\dot{q}_n^2 - 1)\big(V(q_n, q_{n+1}) + V(q_n, q_{n-1})\big), \qquad (8)$$

where $V(q, q') = \zeta(q + q') + \zeta(q - q') - \zeta(2q)$ is an elliptic function in both arguments q, q' (here $\zeta(q)$ is the Weierstrass ζ-function).

A neat spatial discretization of the Landau–Lifshitz (LL) equation has been derived by Adler (2000):

$$\dot{s}_n = \langle s_n, Js_n\rangle\left(\frac{s_n \times s_{n+1}}{1 + \langle s_n, s_{n+1}\rangle} + \frac{s_n \times s_{n-1}}{1 + \langle s_n, s_{n-1}\rangle}\right) -2s_n \times (Js_n). \qquad (9)$$

Here $s_n \in S^2$, and J is a constant 3×3 matrix. In the isotropic case $J = I$ and this model reduces to a discrete Heisenberg magnetic discovered independently by Sklyanin and by Ishimori in 1982. Sklyanin found also a discretization of the LL equation different from (9); the relation between both models is discussed in Adler (2000).

In the rest of this article we briefly discuss some mathematical aspects of integrable lattices. We illustrate them mainly with the example of the Toda lattice (3), but almost all of this holds in one form or another for all integrable DDEs.

Lax Representations

The results by Flaschka and by Manakov of 1974 serve as a first application of IST in the differential-difference context. It is based upon the Lax representation of the Toda lattice, that is, the fact that (3) is equivalent to an operator equation

$$\dot{L} = [L, A_\pm], \qquad (10)$$

in terms of linear difference operators L and $A_\pm$ with coefficients depending on $q_n, \dot{q}_n$ and on the so-called spectral parameter λ:

$$L = \sum b_n E_{n,n} + \lambda^{-1} \sum a_n E_{n,n+1} + \lambda \sum E_{n+1,n}, \quad (11)$$

$$A_+ = \sum b_n E_{n,n} + \lambda \sum E_{n+1,n},$$
$$A_- = -\lambda^{-1} \sum a_n E_{n,n+1}, \quad (12)$$

where the Flaschka–Manakov variables a_n, b_n are expressed through q_n, $p_n = \dot{q}_n$ as

$$a_n = e^{q_{n+1}-q_n}, \quad b_n = p_n. \quad (13)$$

Here we represent difference operators as infinite matrices, with $E_{m,n}$ being the matrix with the only nonvanishing element equal to 1 in the position (m, n). In the case of the finite open-end lattice, the operators are $N \times N$ matrices; summation in the above formulas is performed for n from 1 to N, and it is assumed that $E_{N+1,N} = E_{N,N+1} = 0$. The N-periodic case differs in that it is assumed that $E_{N+1,N} = E_{1,N}$ and $E_{N,N+1} = E_{N,1}$. In the infinite and the finite open-end case, the spectral parameter λ can be removed by a conjugation with a diagonal matrix diag(λ^n); it is not so in the periodic case, which makes the latter analytically much more difficult and demanding.

An alternative to the Lax representation is the discrete zero-curvature representation

$$\dot{L}_n = M_n L_n - L_n M_{n-1}, \quad (14)$$

first developed by Ablowitz and Ladik in 1975. It uses the same second-order operators (11), (12), but in terms of the spectral parameter-dependent 2×2 matrices

$$L_n = \begin{pmatrix} b_n + \lambda & a_{n-1} \\ -1 & 0 \end{pmatrix}, \quad M_n = \begin{pmatrix} -\lambda & -a_n \\ 1 & b_n \end{pmatrix}. \quad (15)$$

Such a representation is somewhat more flexible from the technical viewpoint.

Inverse scattering

The matrix L is similar to a symmetric tri-diagonal matrix, which means that the operator L is gauge equivalent to a second-order and self-adjoint matrix. The direct and inverse spectral theory of such operators is well-developed (see Teschl, 2000), and parallel, to a large extent, to the direct and inverse spectral theory of second-order differential operators. In the rapidly decaying case ($q_n \to c$, $\dot{q}_n \to 0$ as $n \to \pm\infty$) the standard IST scheme is illustrated by the following diagram.

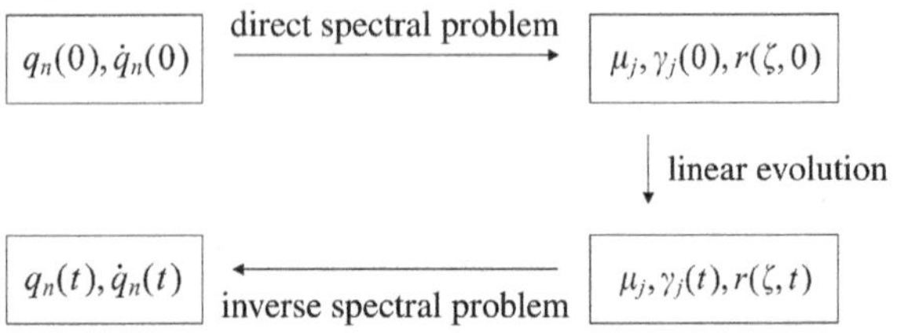

$q_n(0), \dot{q}_n(0)$	direct spectral problem →	$\mu_j, \gamma_j(0), r(\zeta, 0)$
		↓ linear evolution
$q_n(t), \dot{q}_n(t)$	← inverse spectral problem	$\mu_j, \gamma_j(t), r(\zeta, t)$

The set of spectral data allowing for a solution of the inverse problem consists of the discrete eigenvalues μ_j, the normalizing coefficients γ_j of the corresponding discrete eigenfunctions, and the reflection coefficient $r(\zeta)$ of the continuous spectrum. Equation (10) yields that the evolution of the operator L, induced by the fact that $q_n(t)$ satisfy the Toda lattice equations (3), is isospectral. More precisely, the discrete eigenvalues are constants of the motion, while the evolution of other spectral data is governed by simple linear equations:

$$\mu_j = \text{const}, \quad \gamma_j(t) = \gamma_j(0)e^{(\mu_j - \mu_j^{-1})t},$$
$$r(\zeta, t) = r(\zeta, 0)e^{(\zeta - \zeta^{-1})t}.$$

Solution of the inverse spectral problem is given in terms of the Riemann–Hilbert problem or its variants, such as the Gel'fand–Levitan equation. Multisoliton solutions correspond to reflectionless potentials, for which $r(\zeta, t) \equiv 0$.

In the periodic case, the set of the spectral data is more complicated: it includes a hyperelliptic Riemann surface $\mathcal{R}$ of genus $N-1$ determined by the eigenvalues of the periodic boundary value problem for the operator L; further, it includes $N-1$ points P_k on $\mathcal{R}$, which correspond to the eigenvalues of L with vanishing boundary conditions. Due to (10), the Riemann surface $\mathcal{R}$ itself is an integral of motion, and the evolution of points P_k is such that the image of the divisor $P_1 + \ldots + P_{N-1}$ under the Abel map moves along a straight line in the Jacobi variety of $\mathcal{R}$. Solution of the inverse spectral problem is given in terms of multidimensional theta-functions. An important technical device in this type of problems constitute Baker–Akhiezer functions, which have essential singularities at P_k and are meromorphic elsewhere on $\mathcal{R}$; their analytic properties define them uniquely.

Hierarchies

Integrable lattices (and, more generally, integrable systems) do not appear separately, but are organized in hierarchies. To describe the Toda lattice hierarchy, note that the matrices $A_\pm$ in (12) may be seen as $A_\pm = \pm\pi_\pm(L)$, where $\pi_\pm$ stands for the nonnegative (resp. negative) part of the Laurent series with respect to the spectral parameter λ. This is a particular instance of a very general construction, known as the Adler–Kostant–Symes (AKS) method, dealing with a splitting of a Lie algebra g into a direct sum of its two subspaces that are also Lie subalgebras $g_\pm$; in this general setting, $\pi_\pm : g \mapsto g_\pm$ are the corresponding projections. The higher members of the Toda lattice hierarchy (enumerated by $m \in \mathbb{N}$) are characterized by Lax equations of form (10) with the same Lax matrix L as in (11) and $A_\pm = \pm\pi_\pm(L^m)$, see also Kupershmidt (1985).

"Explicit" Solution in Terms of a Factorization Problem

A further important ingredient of the AKS method is a formula for the solution of the initial value problem for such Lax equations. It is given in terms of the factorization problem in the Lie group G corresponding to the Lie algebra g in terms of its Lie subgroups $G_\pm$ corresponding to $g_\pm$: if the elements $U_\pm(t) \in G_\pm$ solve the factorization problem $\exp(tL^m(0)) = U_+U_-$, then $L(t) = U_+^{-1}(t)L(0)U_+(t) = U_-(t)L(0)U_-^{-1}(t)$ is the solution of the mth member of the hierarchy. The factorization problem in G is, in general, a sort of Riemann–Hilbert problem, intimately related to the one participating in the inverse scattering transformation.

Darboux–Bäcklund Transformations and Discretization

The above-mentioned factorization problem in G also leads to an effective construction of Darboux–Bäcklund transformations, which are a further indispensable attribute of integrable systems. For the Toda lattice, a Bäcklund transformation $(q_n, p_n) \mapsto (\widetilde{q}_n, \widetilde{p}_n)$ with a parameter h, which serves also as an integrable discretization with the time step h, is a result of the following factorization: $I + hL = U_+U_-$, with a subsequent flipping of factors: $I + h\widetilde{L} = U_-U_+$, where the factors are given by

$$U_+ = \sum e^{\widetilde{q}_n - q_n} E_{n,n} + h\lambda \sum E_{n+1,n} \in G_+, \quad (16)$$

$$U_- = I + h\lambda^{-1} \sum e^{q_{n+1} - \widetilde{q}_n} E_{n,n+1} \in G_-. \quad (17)$$

The formulas for the Bäcklund transformation seen from these factorizations, are

$$\begin{aligned} 1 + hp_n &= e^{\widetilde{q}_n - q_n} + h^2 e^{q_n - \widetilde{q}_{n-1}}, \\ 1 + h\widetilde{p}_n &= e^{\widetilde{q}_n - q_n} + h^2 e^{q_{n+1} - \widetilde{q}_n}. \end{aligned} \quad (18)$$

This is a canonical transformation, possessing a classical generating function. This result was first obtained by Wadati and Toda in 1975, see also Suris (2003).

Hamiltonian Structure

The last immanent feature of integrable lattices (and, more generally, integrable systems) is their Hamiltonian structure. Often, they are even bi- or multi-Hamiltonian, as it is the case for the Toda lattice. In the Flaschka–Manakov variables (13) the equations of motion of the Toda lattice (3) are rewritten as

$$\dot{a}_n = a_n(b_{n+1} - b_n), \quad \dot{b}_n = a_n - a_{n-1}. \quad (19)$$

The canonical Poisson bracket for the variables q_n, p_n reads in the variables a_n, b_n as follows:

$$\{b_n, a_n\}_1 = -a_n, \quad \{a_n, b_{n+1}\}_1 = -a_n \quad (20)$$

(all other brackets of the coordinate functions vanish), and system (19) is Hamiltonian with respect to this bracket, with the Hamilton function $H_2 = \frac{1}{2}\sum b_n^2 + \sum a_n$. However, one can define also a different Poisson bracket for the variables a_n, b_n:

$$\begin{aligned} \{b_n, a_n\}_2 &= -b_n a_n, \\ \{a_n, b_{n+1}\}_2 &= -a_n b_{n+1}, \\ \{b_n, b_{n+1}\}_2 &= -a_n, \\ \{a_n, a_{n+1}\}_2 &= -a_n a_{n+1}, \end{aligned} \quad (21)$$

with the following properties: it is compatible with the first one (i.e., their linear combinations are again Poisson brackets), and system (19) is Hamiltonian with respect to this bracket, with the Hamilton function $H_1 = \sum b_n$. So, the Toda lattice in the form (19) is a bi-Hamiltonian system. This was found by M. Adler in 1979. The bi-Hamiltonian property, introduced by Magri in 1978 on the example of KdV, has been established since then as an alternative (and highly effective and informative) definition of integrability. Actually, the Toda lattice (19) is even tri-Hamiltonian, since there exists one more local Poisson bracket for the variables a_n, b_n with similar properties, discovered in Kupershmidt (1985).

Higher Dimensions

Up to now, we have considered integrable lattices with one continuous and one discrete independent variables. This clearly allows for a further generalization. Integrable systems with two continuous and one discrete independent variables are well known and widely used as the models in the field theory. For instance, the Toda field theory deals with the system

$$(q_n)_{xy} = e^{q_{n+1} - q_n} - e^{q_n - q_{n-1}}, \quad (22)$$

introduced into soliton theory by Mikhailov (1979). Later it was realized that the equivalent equation $(\log v_n)_{xy} = v_{n+1} - 2v_n + v_{n-1}$, which is obtained by setting $v_n = \exp(q_{n+1} - q_n)$, appeared already in studies by Darboux in 1888, as the equation satisfied by the so-called Laplace invariants of the chain of Laplace transformations of a given second-order hyperbolic partial differential equation.

YURI B. SURIS

See also **Bäcklund transformations; Darboux transformation; Discrete nonlinear Schrödinger equations; Hamiltonian systems; Inverse scattering method or transform; Quantum inverse scattering method; Toda lattice**

Further Reading

Ablowitz, M. & Ladik, J. 1976. Nonlinear differential-difference equations and Fourier analysis. *Journal of Mathematical Physics*, 17: 1011–1018

Adler, V.E. 2000. On discretization of the Landau–Lifshits equation. *Teoreticheskaya i Matematicheskaya Fizika*, 124: 48–61

Faddeev, L.D. & Takhtajan, L.A. 1987. *Hamiltonian Methods in the Theory of Solitons*, Berlin and New York: Springer

Flaschka, H. 1974. On the Toda lattice II. Inverse scattering solution. *Progress in Theoretical Physics*, 51: 703–716
Kac, M. & van Moerbeke, P. 1975. On an explicitly soluble system of nonlinear differential equations related to certain Toda lattices. *Advances in Mathematics*, 16: 160–169
Krichever, I. 2000. Elliptic analog of the Toda lattice. *International Mathematics Research Notices*, 8: 383–412
Kupershmidt, B.A. 1985. Discrete Lax equations and differential-difference calculus. *Asterisque*, 123
Manakov, S.V. 1974. On the complete integrability and stochastization in discrete dynamical systems. *Soviet Physics, JETP*, 40: 269–274
Moser, J. 1975. Finitely many mass points on the line under the influence of an exponential potential—an integrable system. In *Dynamic Systems Theory and Applications*, edited by J. Moser (see *Math. Reviews* MR0393426). Berlin and New York: Springer, pp. 467–497
Ruijsenaars, S.N.M. 1990. Relativistic Toda systems. *Communications in Mathematical Physics*, 133: 217–247
Suris, Yu.B. 2003. *The Problem of Integrable Discretization: Hamiltonian Approach*, Basel: Birkhäuser
Teschl, G. 2000. *Jacobi Operators and Completely Integrable Nonlinear Lattices*. Providence, RI: American Mathematical Society
Toda, M. 1989. *Theory of Nonlinear Lattices*, Berlin and New York: Springer
Wadati, M. & Toda, M. 1975. Bäcklund transformation for the exponential lattice. *Journal of the Physical Society of Japan*, 39: 1196–1203

INTEGRABLE TOPS

See **Rotating rigid bodies**

INTEGRAL EQUATIONS

See **Equations, nonlinear**

INTEGRAL TRANSFORMS

Integral transforms have their genesis in the 19th century work of Joseph Fourier and Oliver Heaviside, subsequently set into a general framework during the 20th century. The fundamental idea is to represent a function $f(x)$ in terms of a *transform* $F(p)$, using an integral transform pair,

$$F(p) = \int K(p,x) f(x)\, \mathrm{d}x,$$
$$f(x) = \int L(x,p) F(p)\, \mathrm{d}p. \qquad (1)$$

The functions $K(p,x)$ and $L(x,p)$ are *kernels*. If either of $f(x)$, $F(p)$ is discontinuous, further interpretation must be supplied at those points.

Oliver Heaviside invented his operational calculus to solve differential equations, such as those which arise in the theory of electrical transmission lines. A simple example is the diffusion problem (Whittaker, 1928),

$$\partial^2 u/\partial x^2 = k\, \partial u/\partial t, \quad x \geq 0, \quad t \geq 0,$$
$$u(0,t) = U_0. \qquad (2)$$

Heaviside treated differentiation as a symbolic operation, $p = \partial/\partial t$, to write the operational solution

$$u(x,t) = \mathrm{e}^{-x\sqrt{kp}} \cdot U_0. \qquad (3)$$

Using Euler's formula for differentiation,

$$\frac{\mathrm{d}^n(x^p)}{\mathrm{d}x^n} = \frac{p!}{(p-n)!} x^{p-n} \qquad (4)$$

with n not restricted to the integers, Heaviside was able to extract all the information he required from such an approach. However, his methods met with general disapproval, leading to lifelong difficulties for him.

Heaviside's work was formalized, using the Laplace integral, to give the Laplace transform pair:

$$F(p) = \int_0^\infty \mathrm{e}^{-px} f(x)\, \mathrm{d}x,$$
$$f(x) = \frac{1}{2\pi \mathrm{i}} \int_{c-\mathrm{i}\infty}^{c+\mathrm{i}\infty} \mathrm{e}^{+px} F(p)\, \mathrm{d}p. \qquad (5)$$

In the normal case, the transform integral converges for all p in a complex half-plane $\Re(p) > \alpha$, in which case the constant c in the inversion integral must be chosen to satisfy the restriction $c > \alpha$. Subjecting Equation (2) to the Laplace transform leads to the same formula (3) for the *transform* $U(x,p)$. Everything may be extracted from this using complex variable techniques; the complex variable p is just a placeholder for Heaviside's operational symbol $p = \partial/\partial t$.

The Laplace transform has simple properties with respect to differentiation and convolutions (Schiff, 1999). For the former, if $F(p)$ is the transform of $f(t)$, then the transform of $f'(t)$ is $pF(p) - f(0)$. Again, if $h(t)$ is the convolution of $f(t)$ with $g(t)$, defined as

$$h(t) = \int_0^t f(s) g(t-s)\, \mathrm{d}s, \qquad (6)$$

then the transform is given by simple multiplication:

$$H(p) = F(p)\, G(p). \qquad (7)$$

An important area of application is systems of ordinary differential equations with constant coefficients, such as (Barnett & Cameron, 1992)

$$\boldsymbol{x}' = A\boldsymbol{x} + B\boldsymbol{u}, \qquad \boldsymbol{y} = C\boldsymbol{x}. \qquad (8)$$

Here $\boldsymbol{x}$ is a set of n internal state variables, $\boldsymbol{u}$ a set of r inputs, $\boldsymbol{y}$ a set of s outputs; A, B, and C are matrices. Taking the Laplace transform and using elementary matrix algebra, the response of the system is given by

$$\mathbf{Y}(p) = G(p)\mathbf{U}(p),$$
$$G(p) = C(pI - A)^{-1}B. \qquad (9)$$

The transform being a product, the solution is a matrix of convolutions. However, the essential information about the system may be extracted directly, without recourse to inversion.

The Fourier transform represents functions as linear combinations of periodic functions, an idea pioneered by Fourier. A one-dimensional form is

$$F(\omega) = \int_{-\infty}^{\infty} f(t)\, \mathrm{e}^{\mathrm{i}\omega t}\, \mathrm{d}t,$$

$$f(t) = \frac{1}{2\pi \mathrm{i}} \int_{-\infty}^{\infty} F(k)\, \mathrm{e}^{-\mathrm{i}\omega t}\, \mathrm{d}\omega. \quad (10)$$

The most common uses are (i) with time as the original variable—transformation from the time domain to the frequency domain and (ii) with spatial position as the original variable—transformation from configuration space to momentum space. The Fourier transform also has simple properties with respect to differentiation and convolution (the integral in (6) now extends from $-\infty$ to $+\infty$), making it useful for investigation of partial differential equations, as well as many more complex systems.

Other important transforms are the Hankel transform, which uses Bessel functions for the kernels in pair (1), alternative forms of the Fourier transform using sine or cosine functions, and many other lesser known pairs (Davies, 2002). In fact, every Sturm–Liouville problem generates integral transforms (Antimirov et al., 1993). In addition, there are transforms that involve complex integration in an essential way, such as the Hilbert transform and Cauchy integrals, which use principal values.

The Mellin transform pair

$$F(p) = \int_{0}^{\infty} x^{p-1} f(x)\, \mathrm{d}x,$$

$$f(x) = \frac{1}{2\pi \mathrm{i}} \int_{c-\mathrm{i}\infty}^{c+\mathrm{i}\infty} x^{-p} F(p)\, \mathrm{d}p \quad (11)$$

has great utility, particularly in asymptotics (Ninham et al., 1992; Davies, 2002). For example, the Mellin transform representation of the exponential integral

$$\mathrm{Ei}(x) = \int_{x}^{\infty} \frac{\mathrm{e}^{-u}}{u}\, \mathrm{d}u, \quad (12)$$

is

$$\mathrm{Ei}(x) = \frac{1}{2\pi \mathrm{i}} \int_{c-\mathrm{i}\infty}^{c+\mathrm{i}\infty} \frac{(p-1)!x^{-p}}{p}\, \mathrm{d}p, \quad c>0, \quad (13)$$

$$= -\ln x - \gamma - \sum_{k=1}^{\infty} \frac{(-1)^k x^k}{k!\,k}, \quad (14)$$

where the infinite sum is obtained by the standard theory of residues.

The Fourier transform is particularly well suited to linear evolution problems. Typically, there is an initial configuration, $u(x,0)$, which is represented as a linear superposition of plane waves,

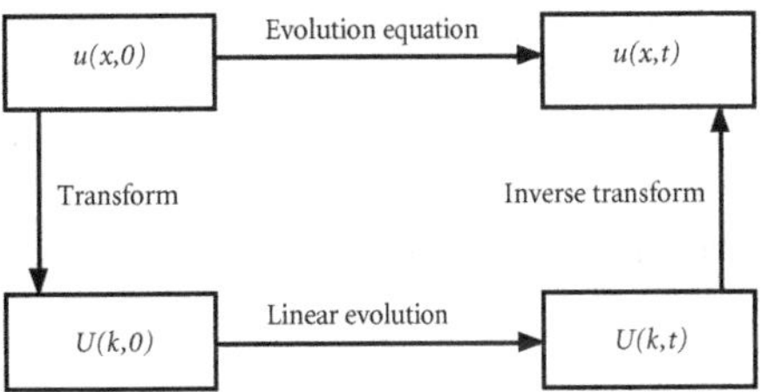

Figure 1. Solution of evolution equations using transforms.

$$u(x,0) = \frac{1}{2\pi} \int_{-\infty}^{\infty} U(k,0)\, \mathrm{e}^{-\mathrm{i}kt}\, \mathrm{d}k. \quad (15)$$

The individual waves evolve according to a simple dispersion relation, $\omega = \omega(k)$, which means that the functions $\exp[-\mathrm{i}kx + \mathrm{i}\omega(k)t]$ satisfy the partial differential equation. As a consequence, the solution at later time is

$$u(x,t) = \frac{1}{2\pi} \int_{-\infty}^{\infty} U(k,0)\, \mathrm{e}^{-\mathrm{i}kx+\mathrm{i}\omega(k)t}\, \mathrm{d}k. \quad (16)$$

The process can be visualized as in Figure 1. The top arrow indicates the required evolution of the initial state $u(x,0)$ to the later state $u(x,t)$. The circuitous route has three steps: (i) map the initial state $u(x,0)$ to some wave data $U(k,0)$, (ii) evolve the wave data under the evolution equation to the later wave data $U(k,0)\mathrm{e}^{i\omega(k)t}$, and (iii) use an inverse transform to reconstruct the new state from the new wave data.

Beginning in the 1970s, it was realized that a large number of nonlinear evolution equations may be solved using an analogous process that has come to be known as the inverse scattering method (Dodd et al., 1982). It is more complicated than the linear methods, but reduces to them in the limit of small nonlinearity. Briefly, the nonlinear equation is associated with a linear one, and there is a mechanism for mapping the data of the former onto the data of the latter. The method again proceeds according to the scheme of Figure 1. However the inverse transform, which recovers the state from the "scattering data," is not so simple, and the method derives its name from it. The first explicit use of this scheme was for the solution of the Korteweg–de Vries (KdV) equation, which is associated with a linear Schrödinger equation in which the function $u(x,t)$ plays the role of the potential. The evolution of the spectrum of the Schrödinger operator, in response to the nonlinear evolution of KdV equation, is linear; furthermore, in the limit of weak nonlinearity, the transform and its inverse reduce to the Fourier transform.

BRIAN DAVIES

See also **Ablowitz–Kaup–Newell-Segur system; Burgers equation; Gel'fand–Levitan theory; Hirota's method; Inverse scattering method or transform; Korteweg–de Vries equation; Nonlinear Schrödinger equations; Sine-Gordon equation; Solitons**

Further Reading

Antimirov, M.Ya., Kolyshkin, A.A. & Vaillancourt, R. 1993. *Applied Integral Transforms*, New York: Springer

Barnett, S. & Cameron, R.G. 1992. *Introduction to Mathematical Control Theory*, 2nd edition, Oxford: Clarendon Press

Davies, B. 2002. *Integral Transforms and Their Applications*, 3rd edition, New York: Springer

Dodd, R.K., Eilberg, J.C., Gibbon, J.D. & Morris, H.C. 1982. *Solitons and Nonlinear Wave Equations*, London and New York: Academic Press

Duffy, D.G. 1994. *Transform Methods for Solving Partial Differential Equations*, Boca Raton, FL: CRC Press

Moore, D.H. 1971. *Heaviside Operational Calculus*, New York: Elsevier

Ninham, B.W., Hughes, B.D., Frankel, N.E. & Glasser, M.L. 1992. Möbius, Mellin, and mathematical physics. *Physica* A, 186: 441–481

Schiff, J.L. 1999. *The Laplace Transform: Theory and Applications*, New York: Springer

Whittaker, E.T. 1928. Oliver Heaviside. *Bulletin of the Calcutta Mathematical Society*, 20: 199–220 (The article is reprinted in Moore, 1971.)

INTEGRALS OF MOTION

See **Constants of motion and conservation laws**

INTEGRATE AND FIRE NEURON

Classical electrophysiology using extracellular recording methods established the all-or-none nature of the nerve impulse—a stimulus has to be above a threshold θ for an action potential to be generated, and for rectangular stimuli, θ depends on stimulus duration T. A plot of θ–T, a strength-duration curve is approximated by

$$\theta = \theta_{\rm rh}/[1 - \exp(-T/\tau], \qquad (1)$$

where $\theta_{\rm rh}$ is a constant called the rheobase and τ is a time constant. Lapicque (1907) produced the strength-duration curve of Equation (1) by a model in which subthreshold inputs $x(t)$ are integrated and decay with an exponential time course

$$V(t) = \int_0^t \exp(-(t-u)/\tau)\,du, \qquad (2)$$

for $0 < V(t) < \theta$, and when $V(t)$ reaches threshold, an idealized action potential (a Dirac δ-function) is triggered and V is instantaneously reset to its resting value (say, $V = 0$). The subthreshold changes in $V(t)$ were represented as the response to an applied current of a resistance and capacitance in parallel. This separates the slow time scales of the subthreshold integrative processes from the fast time scales of the action potential mechanisms. The values of the resistance and capacitance reflect not membrane parameters, but cell parameters, and so depend via surface area on cell size. Guided by the strength-duration curve, nerve or muscle, or different diameter nerve fibers can be excited.

Lapicque's integrate-and-fire model is also known as the leaky integrator and, in the case $\tau \to \infty$, the perfect integrator. Unlike the FitzHugh–Nagumo, Hodgkin–Huxley, and Markin–Chizmadzhev models, it considers only subthreshold, integrative processes, not the processes generating the action potential, and so is more suitable for modeling neuronal information processing than action potential transmission. Originally formulated as a phenomenological model for the conditions for the initiation of an action potential, it is still widely used as a model today, but as a model for the generation of neuronal spike trains and mass activity in neuronal networks. It has been modified to include both absolute and relative refractory periods, obtained as a limiting case for the Hodgkin–Huxley equations, and extended into multiple time constant and spatially extended models (Tuckwell, 1988).

If the input $u(t)$ is a constant c, the interspike intervals t_i are a solution of

$$\theta = \int_0^{t_i} \exp(-(t-u)/\tau)\,du, \qquad (3)$$

and so the rate of firing is $1/t_i = 1/[-\tau \ln(1 - 1/c\tau)]$, showing an increased sensitivity to small amplitude inputs. Such a logarithmic relation between discharge rate and the magnitude of a constant input is observed in some primary sensory neurons, and is used as an explanation for the Weber–Fechner law of sensory psychophysics.

Gerstein & Mandelbrot (1964) introduced a random walk model for neuronal spike trains, in which the input to a perfect integrator model was a random sequence of small positive and negative steps. This was motivated by the observation that the interspike interval and scaled interval (sums of adjacent intervals) of some auditory, neurons had the same unimodal, asymmetric shape; that is, assuming the spike train was a realization of a renewal point process, the interval density function was invariant under self-convolution or has a stable, or infinitely divisible, distribution. The subthreshold response of the integrate-and-fire model to small and brief positive or negative current pulses could represent the postsynaptic response of a neuron to synaptic inputs. Superposition of a (large) number of independent point processes generates a Poisson process: with appropriate constraints on how the amplitude of the steps decreases while the rate increases, the random walk becomes a white noise, leading to diffusion models for stochastic activity of neurons. The diffusion process models for $V(t)$ are approximations of the discontinuous, random walk models and are constructed to have the same first and second infinitesimal moments. For the perfect integrator, the subthreshold $V(t)$ is a Wiener process; for the leaky integrator, an Ornstein–Uhlenbeck process. The interspike-interval probability density functions for these diffusion models are first passage time densities to an absorbing barrier, the threshold. These densities are obtained as solutions for forward or backward Chapman–Kolmogorov equations (see Holden, 1976; Tuckwell, 1988). These equations have been solved for different diffusion models, mostly numerically, but for some special cases solutions are available in closed form. The diffusion models have been extended to include time-varying threshold, temporal modulation of the rates of the excitatory and inhibitory inputs, and spatial separation filtering of the inputs. In the derivation of these diffusion models the synaptic inputs are assumed to be uncorrelated, producing a white noise input: correlation in the input, producing colored inputs, is more likely for central neurons and has profound effects on the output variability of the integrate-and-fire models (Feng, 2001).

The perfect and leaky integrators respond to a cosinusoidal input by modulated spike trains with a periodically modulated spike density. At large modulation depths both models show rectification, and the leaky integrator also shows phase-locked responses. Both these nonlinear distortions are reduced by adding noise to the input (Stein et al., 1972), and so additive noise can increase the ability of a neuron to transmit signals about narrow-band, high frequencies. This provides a functional explanation for the association between the variability in sensory mechanoreceptors and the frequency of their adequate stimulus: the discharge of auditory neurons is almost random, and velocity sensitive primary muscle spindle afferents have a higher variability than length sensitive secondary muscle spindle afferents, leading to ideas of stochastic resonance. The phase locked and aperiodic responses of the leaky integrator to a periodic input has been characterized in terms of the Arnol'd tongue structure for the dominant modes, where there are low order ratios between the spiking and forcing rhythms (Coombes & Bressloff, 1999).

The success of the integrate-and-fire models in accounting for general properties of spike trains has allowed it, with simple modifications, to be applied to model spike trains from specific neurons, see for example, Herrmann & Gerstner (2002). The simplicity of the model also allows it to be used as an element in constructing population dynamic equations for large systems of coupled neurons, as in Sirovich et al. (2000).

ARUN V. HOLDEN

See also **Excitability; FitzHugh–Nagumo equation; Hodgkin–Huxley equations; Markin–Chizmadzhev model; Nerve impulses; Neurons**

Further Reading

Coombes, S. & Bressloff, P.C. 1999. Mode locking and Arnol'd tongues in integrate-and-fire neural oscillators. *Physical Review* E, 60: 2086–2096

Feng. J. 2001. Is the integrate-and-fire model good enough? A review. *Neural Networks*, 14,955–14,975

Gerstein, G.L. & Mandelbrot. 1964. Random walk models for the spike activity of a single neuron. *Biophysical Journal*, 4: 41–68

Herrmann, A. & Gerstner, W. 2002. Noise and the PSTH response to current transients: II. Integrate-and-fire model with slow recovery and application to motoneuron data. *Journal of Computational Neuroscience*, 12: 83–95

Holden, A.V. 1976. *Models of the Stochastic Activity of Neurones*, Berlin: Springer

Lapicque, L. 1907. Recherches quantitatives sur l'excitation electrique des nerfs traite comme une polarization. *Journal de Physiologie et Pathologie Generale*, 9: 620–635

Sirovich, L., Omurtag, A. & Knight, B.W. 2000. Dynamics of neuronal populations: the equilibrium solution. *SIAM Journal on Applied Mathematics*, 6: 2009–2028

Stein, R.B., French, A.S. & Holden, A.V. 1972 The frequency response, coherence, and information capacity of two neuronal models. *Biophysical Journal*, 12: 295–322

Tuckwell, H.C. 1988. *Introduction to Theoretical Neurobiology*. vol. 1. Cambridge and New York: Cambridge University Press

INTEGRO-DIFFERENTIAL EQUATIONS

See **Equations, nonlinear**

INTERMITTENCY

Consider a dynamical system with x as the observed variable. As a function of time t, if $x(t)$ exhibits segments of relative constant values (laminar phase) interspersed by erratic bursts, we say the system dynamics is intermittent (see Figure 1). Pomeau and Manneville (1980) were the first to identify the three bifurcations in low-dimensional maps that are associated with commonly observed intermittencies. An additional mechanism for generating intermittency is related to the transverse instability of chaotic attractors confined to a manifold whose dimension is smaller than that of the full phase space, which is called on-off intermittency (Pikovsky, 1984; Fujisaka & Yamada, 1985; Platt et al., 1993).

Pomeau–Manneville type I intermittency stems from a saddle-node bifurcation. Near the bifurcating point ($x = 0$) in one-dimensional maps, the dynamics are described by $x_{n+1} = x_n + \varepsilon + x_n^2$. For positive but small ε, a narrow channel near $x = 0$ is formed between the nonlinear map function and the diagonal line $x_{n+1} = x_n$. When a trajectory enters the channel, it moves through it very slowly, giving rise to the appearance of a quiet laminar phase. After exiting the channel, the trajectory can display wild swings governed by nonlinear mechanisms, forming the bursting phase of the dynamics. As time progresses, ergodicity will bring the trajectory back to the channel again, and the process starts anew. The re-injected trajectory can land at different points inside the channel. As a result, the laminar phase has variable length, the mean of which is a statistical quantity of interest. It has been shown that the mean length of the laminar phase scales with the parameter ε as $\sim \varepsilon^{-1/2}$. As ε decreases through zero, a saddle-node bifurcation occurs, creating a stable fixed-point attractor at $-\sqrt{-\varepsilon}$ and an unstable fixed point at $\sqrt{-\varepsilon}$. Intermittency terminates at the bifurcation point $\varepsilon = 0$.

Pomeau–Manneville type II intermittency is related to a subcritical Hopf bifurcation; thus a two-dimensional map is needed for this phenomenon. In polar coordinates, the dynamics near the bifurcation point (assumed to be the origin) is $r_{n+1} = (1+\varepsilon)r_n + r_n^3$, $\theta_{n+1} = \theta_n + \alpha$, where α is a constant rotation. For $\varepsilon > 0$, the origin is an unstable spiral. Viewing the r equation as a one-dimensional map, one again finds a channel for small r. When the trajectory is randomly injected into the channel, a quiet laminar period ensues, which is followed by a burst of chaotic behavior. The mean length of the laminar phase is shown to scale with ε as $\sim \varepsilon^{-1}$. At $\varepsilon = 0$ a subcritical Hopf bifurcation occurs. For $\varepsilon < 0$, intermittency is replaced by a stable fixed-point attractor at the origin that is accompanied by an unstable invariance circle centered at the origin.

Pomeau–Manneville type III intermittency is created near an inverse period-doubling bifurcation. Near the bifurcation point ($x = 0$), the equation of motion is $x_{n+1} = -(1+\varepsilon)x_n + \alpha x_n^2 + \gamma x_n^3$. For this map, the channel dynamics is best seen by examining the second iterate of the above map that is easily obtained (after discarding higher-order terms) to be $x_{n+1} = (1+2\varepsilon)x_n + \beta x_n^3$, where $\beta = -2(\alpha^2 + \gamma)$. As ε decreases from positive values to negative values, intermittency is replaced by a stable fixed-point attractor and an unstable period-two orbit. The channel near $x = 0$ here is analogous to the channel near $r = 0$ in the type II intermittency. It is, thus, not surprising that the mean length of the laminar phase scales with the parameter ε as $\sim \varepsilon^{-1}$.

Consider an m-dimensional dynamical system that has an n-dimensional invariant manifold where $n < m$. (By invariant we mean that a trajectory initialized in the manifold stays there for all time.) Suppose that the dynamics on the manifold has a chaotic attractor that is also an attractor for the full phase space. Suppose also that, as a parameter varies across a threshold, a blowout bifurcation takes place (Ott & Sommerer, 1994) and the attractor becomes transversely unstable. If one monitors the distance between the trajectory and the bifurcating invariant manifold, this distance typically exhibits on-off intermittency (see Figure 1) immediately after the blowout bifurcation. During the laminar phase, in other words, the trajectory spends a great deal of time close to the invariant manifold, producing a nearly zero value for the distance variable. The trajectory then departs the manifold rapidly and displays erratic bursting behavior for the distance variable until the dynamics brings it back close to the invariant manifold again, and the process repeats. During the laminar phase, the distance variable can be studied via a random walk model, and the interval between successive bursts follows a distribution akin to that of the first return time (Ding & Yang, 1995).

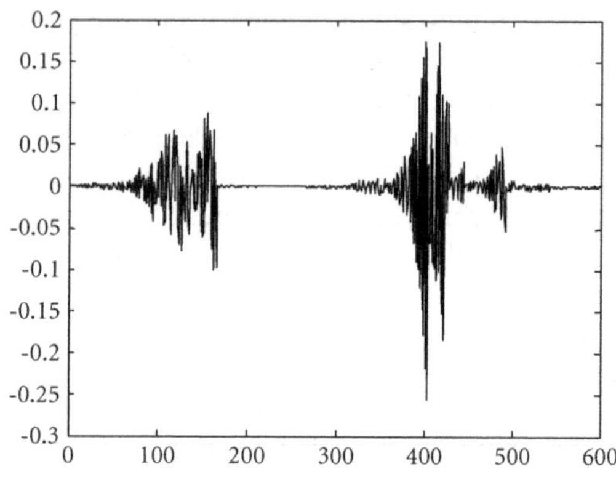

Figure 1. An intermittent function.

All four types of intermittencies have been studied in experimental systems. For example, Jeffries and Perez found Pomeau–Manneville type I intermittency in a nonlinear circuit oscillator (Jeffries & Perez, 1982). Pomeau–Manneville type II intermittency is characterized in a chemical reaction system by Herzel et al. (1991). Dubois et al. (1983) observed Pomeau–Manneville type III intermittency in Rayleigh–Bénard convection. Experimental studies of on-off intermittency are relatively recent. Hammer et al. (1994) are the first to report evidence of on-off intermittency in a nonlinear electronic circuit. On-off intermittency in a laser system was observed by Pisarchik and Pinto-Robledo in 2002.

MINGZHOU DING

See also **Bifurcations; Hénon–Heiles system; Hopf bifurcations; Phase space; Routes to chaos**

Further Reading

Berge, P., Pomeau, Y. & Vidal, C. 1986. *Order Within Chaos, Toward a Deterministic Approach to Turbulence*, New York: Wiley

Ding, M. & Yang, W. 1995. Distribution of the first return time in fractional Brownian motion and its application to the study of on–off intermittency. *Physical Review* E, 52: 207–213

Dubois, M., Rubio, M.A. & Berge, P. 1983. Experimental evidence of intermittencies associated with a subharmonic bifurcation. *Physical Review Letters*, 51: 1446–1449

Fujisaka, H. & Yamada, T. 1985. A new intermittency in coupled dynamical systems. *Progress in Theoretical Physics*, 74: 918–921

Hammer, P.W., Platt, N., Hammel, S.M., Heagy, J.F. & Lee, B.D. 1994. Experimental observation of on-off intermittency. *Physical Review Letters*, 73: 1095–1098

Herzel, H., Plath, P. & Svensson, P. 1991. Experimental evidence of homoclinic chaos and type-II intermittency during the oxidation of methanol. *Physica* D, 48: 340–352
Jeffries, C. & Perez, J. 1982. Observation of a Pomeau-Manneville intermittent route to chaos in a nonlinear oscillator. *Physical Review* A, 26: 2117–2122
Ott, E. 2002. *Chaos in Dynamical Systems*, 2nd edition, Cambridge and New York: Cambridge University Press
Ott, E. & Sommerer, J. 1994. Blowout bifurcation: the occurrence of riddled basins and on–off intermittency. *Physics Letters* A, 188: 39–47
Pikovsky, A. 1984. A new type of intermittent transition to chaos. *Journal of Physics* A, 16: L109–L112
Pisarchik, A.N. & Pinto-Robledo, V.J. 2002. Experimental observation of two-state on–off intermittency. *Physical Review* E, 027203
Platt, N., Spiegel, E.A. & Tresser, C. 1993. On–off intermittency: a mechanism for bursting. *Physical Review Letters*, 70: 279–282
Pomeau, Y. & Manneville, P. 1980. Intermittent transition to turbulence in dissipative dynamical systems. *Communications in Mathematical Physics*, 74: 189–197

INVARIANT MANIFOLDS AND SETS

An invariant set S of a dynamical system has the property that every trajectory (or orbit) that starts in S remains in S. Invariant sets are the basic building blocks of dynamical systems and almost every set that is dynamically interesting has some invariance property. Every trajectory is an invariant set, and so equilibria and periodic orbits are simple examples of such objects. Attractors are also examples of invariant sets.

An invariant set that is also a manifold (i.e., it looks like R^n locally) is called an invariant manifold. Invariant manifolds provide a natural description of the dynamics close to an equilibrium or periodic orbit (using linearization) and can also make it possible to work in lower dimensions than the phase space of the problem, since a smooth dynamical system restricted to an invariant manifold is itself a dynamical system. This reduction in dimension is at the heart of center manifold techniques and inertial manifold methods.

Invariance is a topological property and does not depend upon stability. This is often useful, and the sudden appearance or disappearance of attractors, as parameters are varied, can sometimes be understood as a change in stability of an invariant set that exists throughout the parameter region of interest. Indeed, many invariant sets, and particularly invariant manifolds, have persistence properties under perturbations of the dynamical system that make them useful when working with families of systems.

The basic definitions will be given for differential equations $\dot{\boldsymbol{x}} = \boldsymbol{f}(\boldsymbol{x})$ with $\boldsymbol{x} \in R^n$ and solutions $\boldsymbol{x}(t)$, and maps $\boldsymbol{v}_{n+1} = \boldsymbol{g}(\boldsymbol{v}_n)$, $\boldsymbol{v} \in R^n$. In both cases, the complications that arise if solutions do not exist for all time will be ignored.

For differential equations, a set S is a *forward invariant* set (resp. *backward invariant* set) if $\boldsymbol{x}(0) \in S$ implies that $\boldsymbol{x}(t) \in S$ for all $t \geq 0$ (resp. all $t \leq 0$). S is *invariant* if it is both forward invariant and backward invariant.

For maps, a set S is a *forward invariant* set (resp. *backward invariant* set) if $\boldsymbol{v} \in S$ implies that $\boldsymbol{g}(\boldsymbol{v}) \in S$ (resp. $\boldsymbol{g}^{-1}(\boldsymbol{v})$). S is *invariant* if it is both forward invariant and backward invariant.

In the remainder of this entry, some different sorts of invariant sets that arise in dynamics will be considered, illustrated by the Lorenz equations

$$\dot{x} = \sigma(y - x), \quad \dot{y} = -y + rx - xz, \quad \dot{z} = -bz + xy, \tag{1}$$

where σ, r, and b are positive constants.

Trapping Regions

A set R is a trapping region if it is forward invariant and if all solutions are eventually contained in R. Trapping regions provide the first approximation to the location of any attractors in the system.

For the Lorenz equations, (1), with $\sigma = 10$ and $b = \frac{8}{3}$ Sparrow (1982) shows that the region

$$R = \{(x, y, x) \mid rx^2 + 10y^2 + 10(z - 2r)^2 \leq 5r^2\} \tag{2}$$

is a trapping region. This can be proved using the Lyapunov function $V(x, y, z) = rx^2 + 10y^2 + 10(z - 2r)^2$.

Stable and Unstable Manifolds of Stationary Points

A stationary point of the differential equation $\dot{\boldsymbol{x}} = \boldsymbol{f}(\boldsymbol{x})$ is a solution that is constant for all time, so it must satisfy $\boldsymbol{f}(\boldsymbol{x}) = \boldsymbol{0}$. The three equations obtained by setting $\dot{x} = \dot{y} = \dot{z} = 0$ in (1) can be solved to show that the origin is always a stationary point of the Lorenz equations, and provided $r > 1$, there are two other stationary points: $(\pm\sqrt{b(r-1)}, \pm\sqrt{b(r-1)}, r-1)$. Near a stationary point, the most important terms of the differential equation are the linear terms, so it is natural to ask whether the solutions of the linearized equation at the origin

$$\dot{x} = \sigma(y - x), \quad \dot{y} = -y + rx, \quad \dot{z} = -bz \tag{3}$$

obtained by ignoring the nonlinear terms in (1) reflect properties of the full equations close to the stationary point at the origin. These results are usually stated in terms of the eigenvalues of the matrix that defines the linear equation (3):

$$M = \begin{pmatrix} -\sigma & \sigma & 0 \\ r & -1 & 0 \\ 0 & 0 & -b \end{pmatrix}. \tag{4}$$

If M has no eigenvalues with zero real parts, then the manifold (vector space) spanned by the eigenvectors with eigenvalues having negative real parts contains all the stable directions and, hence, is called the stable manifold of the linear system, (3), while the manifold spanned by the eigenvectors with eigenvalues having positive real parts contains the unstable directions. The former manifold is called the stable manifold, E^{s}, of the origin, and the latter is the unstable manifold, E^{u}. If $0 < r < 1$, then all the eigenvalues of M are negative, so the stable manifold of the linear system is the whole space, and the unstable manifold is the empty set. If $r > 1$, then M has two negative eigenvalues and one positive eigenvalue, so the unstable manifold of the linear system is a line, and the stable manifold is a plane.

This provides definitions of stable and unstable manifolds for linear systems such as (3) with no eigenvalues with zero real parts. To extend the definition to nonlinear systems, we begin by noting that in the definition used for linear systems, the stable manifold of the origin contains the solutions that tend to the origin in forward time, and the unstable manifold contains those that tend to the origin in backwards time. These properties can be used to define the stable and unstable manifolds of a stationary point $\boldsymbol{x}^*$ for a nonlinear system: the stable manifold of $\boldsymbol{x}^*$, $W^{\mathrm{s}}(\boldsymbol{x}^*)$ is the set of initial conditions for which the solution $\boldsymbol{x}(t)$ tends to $\boldsymbol{x}^*$ as $t \to \infty$, while the unstable manifold, $W^{\mathrm{u}}(\boldsymbol{x}^*)$, is the set of initial conditions for which the solution $\boldsymbol{x}(t)$ tends to $\boldsymbol{x}^*$ as $t \to -\infty$. The stable manifold theorem states that provided the matrix that defines the linearized equation near a stationary point has no eigenvalues with zero real parts, then $W^{\mathrm{s}}(\boldsymbol{x}^*)$ and $W^{\mathrm{u}}(\boldsymbol{x}^*)$ exist, are invariant sets, and are of the same dimension as the corresponding linear stable and unstable manifolds. Moreover, they are each tangential to the corresponding linear manifold at the stationary point itself. See Guckenheimer & Holmes (1983) for more details.

Calculating Stable and Unstable Manifolds

The stable manifold theorem suggests how approximations to the stable and unstable manifolds of a stationary point can be calculated close to a stationary point. First, it is tangential to the corresponding linear manifold, and second, it is invariant. This is enough to obtain a power series solution, and this will be illustrated using the unstable manifold of the origin of the Lorenz equations with $\sigma = 1$ and $r = 4$.

Step 1 (Linear approximation): The eigenvalues of M in equation (4) are 1, -3, and $-b$ if $\sigma = 1$ and $r = 4$. The unstable manifold is one-dimensional, and the unstable manifold of the linear approximation is the eigenvector corresponding to the positive eigenvalue, that is, $(1, 2, 0)$ or $y = 2x$, $z = 0$.

Step 2 (Canonical form): It is good practice to bring the linear part into canonical form at this stage, see Guckenheimer & Holmes (1983). For brevity this step will be omitted.

Step 3 (Power series): Since the unstable manifold is tangential to the linear approximation of Step 1 at the stationary point, it can be represented *locally* as a function of one of the variables. From the form of the linear approximation, either x or y can be used here. If the defining equations are smooth, then the stable and unstable manifolds are smooth, so choosing x as the independent variable we write

$$y = 2x + \alpha x^2 + O(x^3), \quad z = \beta x^2 + O(x^3), \quad (5)$$

where the linear terms are obtained from the linear approximation, α and β are constants to be determined, and $O(x^3)$ denotes the terms of order x^3 and higher that are small (locally) compared with the linear and quadratic terms. To identify the constants α and β, two different ways of calculating $\dot{y}$ and $\dot{z}$ are used and then compared in the next three steps.

Step 4 (Power series evolution): Differentiating (5) gives $\dot{y} \sim 2\dot{x} + 2\alpha x\dot{x}$ and $\dot{z} \sim 2\beta x\dot{x}$ (ignoring the higher order terms). Now, on the unstable manifold given by (5) $\dot{x} = -x + y \sim x + \alpha x^2$, so

$$\begin{aligned} \dot{y} &\sim 2(x + \alpha a x^2) + 2\alpha x^2 \sim 2x + 4\alpha x^2, \\ \dot{z} &\sim 2\beta x^2, \end{aligned} \quad (6)$$

where higher-order terms have been ignored again.

Step 5 (Differential equation evolution): From (1) and (5) with $r = 4$ and $\sigma = 1$,

$$\begin{aligned} \dot{y} &= 4x - y - xz \sim 2x - \alpha x^2, \\ \dot{z} &= -bz + xy \sim -b\beta x^2 + 2x^2. \end{aligned} \quad (7)$$

Step 6 (Equate coefficients): To determine the two constants α and β, we now simply equate coefficients of powers of x in (6) and (7). From the $\dot{y}$ equations $4\alpha = -\alpha$, so $\alpha = 0$, and from the $\dot{z}$ equations $2\beta = -b\beta + 2$, so $\beta = 2/(2 + b)$. This gives the second-order approximation to the unstable manifold:

$$y = 2x + O(x^3), \quad z = \tfrac{2}{2+b}x^2 + O(x^3), \quad (8)$$

which is valid close to the stationary point at the origin. Higher-order terms can be calculated by including more terms in the power series expansions.

Center Manifolds

If the matrix that defines the linearized differential equation at a stationary point has eigenvalues with zero real parts, then three invariant manifolds can be defined: a strong stable manifold corresponding to eigenvalues with negative real parts, a strong unstable manifold corresponding to eigenvalues with positive real parts, and a center manifold that is tangential at the stationary point to the space spanned by the eigenvectors corresponding to eigenvalues with zero real parts. The motion on the strong stable and unstable manifolds is defined by the dominant linear terms, that is, towards and away from the stationary point, respectively. The motion on the center manifold depends on the nonlinear terms of the differential equation and may be stable or unstable or neutral. Center manifolds can be approximated locally using the same ideas as the approximation of the unstable manifold above, and this leads to the technique of center manifold reduction, which is central to the development of bifurcation theory. The origin of the Lorenz equations has a one-dimensional center manifold if $r = 1$ (M in (4) has an eigenvalue of zero), which signals the bifurcation creating the nontrivial pair of stationary points if $r > 1$.

Unstable Chaotic Sets

The chaotic sets created by horseshoes are not stable, but they can be important from two points of view. First, if an unstable chaotic set exists in a system then it can manifest itself in chaotic transients before a stable attractor is reached. Second, if there are parameters in the system then the chaotic set may gain stability and so the sudden appearance of strange attractors can be explained by understanding the change of stability of a chaotic invariant set. Grebogi et al. (1982) describe some of these mechanisms as *crises*. In the Lorenz equations with $b = \frac{8}{3}$ and $\sigma = 10$, a strange invariant set (an unstable chaotic set with a fractal structure) is created by a homoclinic bifurcation as r is increased through $r \approx 13.93$, but it does not become stable until $r \approx 24.06$; see Sparrow (1982); for details.

There are many other types of invariant sets and properties of these sets that could have been described here—the selection above is only the tip of an invariant iceberg.

PAUL GLENDINNING

See also **Attractors; Bifurcations; Center manifold reduction; Chaotic dynamics; Horseshoes and hyperbolicity in dynamical systems; Inertial manifolds; Linearization; Lorenz equations; One-dimensional maps; Routes to chaos**

Further Reading

Grebogi, C., Ott, E. & Yorke, J.A. 1982. Chaotic attractors in crisis. *Physics Review Letters*, 48: 1507–1510

Guckenheimer, J. & Holmes, P. 1983. *Nonlinear Oscillations, Dynamcal Systems, and Bifurcations of Vector Fields*, New York: Springer

Sparrow, C. 1982. *The Lorenz Equations: Bifurcations, Chaos, and Strange Attractors*, New York: Springer

INVARIANT MEASURE

See **Measures**

INVERSE PROBLEMS

Scientists try to interpret the world in terms of mathematical models that depend on parameters by relating their models to the results of measurements. Thus, a model is given when a set $\mathcal{C}^0$ of parameters is mapped into a set $\mathcal{E}^0$ of experimental results. Explicit calculations of this mapping $\mathcal{M}$ are called "solving the direct problem." Providing an exact or approximate way for going back from experimental results to parameters by $\mathcal{M}$ is called "solving the inverse problem" (Sabatier, 2000).

Because the exact limits of $\mathcal{C}^0$ and $\mathcal{E}^0$ are difficult to define, they are embedded in convenient mathematical spaces, $\mathcal{C}$ and $\mathcal{E}$, usually linear normed spaces, where it is understood that $\mathcal{C}$ is the domain of $\mathcal{M}$ and $\mathcal{M}(\mathcal{C}) \subseteq \mathcal{E}$. Because of measurement errors, there may be measurement results belonging to $\mathcal{E}$ but not to $\mathcal{M}(\mathcal{C})$, and if it is so, the inverse problem is said to be overdetermined. To appraise uncertainties on measurements, the norm of $\mathcal{E}$ must be reasonably defined, corresponding to our physical knowledge of errors.

The inverse problem is said to be "well-posed" if the norm of $\mathcal{E}$ is reasonably defined, $\mathcal{C} = \mathcal{C}^0$, $\mathcal{M}(\mathcal{C}) = \mathcal{E}$, and $\mathcal{M}$ has a two-sided inverse $\mathcal{M}^{-1}$, which is continuous for the canonical distance in $\mathcal{E}$. Then given any measurement result $e \in \mathcal{E}$, $\mathcal{M}^{-1}e$ yields parameters $x \in \mathcal{C}$ such that

$$\mathcal{M}x = e. \tag{1}$$

If a small error in the experimental data e implies that the error in the parameters x is small, the solution x is stable. Unfortunately, almost all inverse problems of physics or engineering are "ill-posed" (i.e., not well-posed).

For an ill-posed problem, a regularized solution is a continuous mapping $\overline{\mathcal{M}}$ from $\mathcal{E}$ to $\mathcal{C}^0$ such that $\mathcal{M}(\overline{\mathcal{M}}e)$ is close to $e \in \mathcal{E}$. Because experimental observations define $\mathcal{C}^0$ only approximately, one often imposes additional requirements to ease the regularizing processes, for example, that $\mathcal{C}^0$ is compact (Engl et al., 2000).

Classical examples of regularized solutions for Equation (1) are obtained by minimizing over x in one of two ways: either successively as a Euclidean distance in $\mathcal{E}$ that models the misfit and a Euclidean distance in $\mathcal{C}$ that models the non-reliability or as a combination of their squares into a cost function. Typical is the Tikhonov cost

$$C(x, x_0, \lambda) = \|x - x_0\|_{\mathcal{C}}^2 + \lambda \|\mathcal{M}x - e\|_{\mathcal{E}}^2, \lambda > 0, \tag{2}$$

where two parameters (the "first guess" x_0 and the balance parameter λ) are to be chosen. (λ may be related to an evaluation of the noise/signal ratio or to the stability, whereas "physical" information on x_0 may be suggested by previous results.)

If $\mathcal{M}$ is differentiable, the cost function (2) is also differentiable, and elementary variational calculus applies. Information from statistical analyses is more conveniently represented by quadratic cost functions similar to Equation (2), but involving noncanonical distances in $\mathcal{C}$ and $\mathcal{E}$ that are constructed from covariance matrices (having many more arbitrary parameters). The minimum x_m is unique for a given quadratic cost function and Hilbert spaces $\mathcal{C}$ and $\mathcal{E}$ if $\mathcal{M}$ is linear and continuous and satisfies the requirements of a regularized solution. This minimum can be derived either by solving linear equations or by choosing among many ones an efficient iterative algorithm.

The value $C(x_m)$ is a trade-off between requirements on $\mathcal{C}$ and $\mathcal{E}$. Either it is larger than the largest number C_M allowed for this trade-off, so that there is no admissible regularized solution of this form, or it is smaller than C_M, and all values of x such that $C(x_m) \leq C(x) \leq C_M$ are also admissible solutions of Equation (2). Hence the range of this non-uniqueness can be appraised along the principal axes of the ellipsoid $C(x) = C_M$, whose center is x_m. In this linear case, the center is given by the formula $(M^*M + \lambda I)^{-1} M^*e$, where the regularising process is clearly due to adding λ^{-1} to the eigenvalues of M^*M (the "singular values of M). Extensions along the axes are trivially related to the inverse of $\mathcal{M}$ singular values, damped by means of λ.

Instabilities related to the small singular values (high spatial frequencies) make this kind of regularized solution more sensitive to accidental large deviations of data (less robust) than regularized solutions obtained from cost functions involving L^1 or L^∞ norms. However, choosing the later norms usually leads one to more complicated algorithms and to (bang–bang) solutions piecewise equal to the possible upper and lower bounds. Hence, simple and efficient analyses can be adapted to almost any linear (or linearized) inverse problem, including those important for medical tomography and most medical imaging (Bertero & Bocacci, 1998).

Among many examples of nonlinear inverse problems is the case when the a priori information on a first guess x_0 is so convincing that we seek a unique approximate solution of Equation (1) in a narrow neighborhood of x_0, allowing linearization $\mathcal{M}$. The only difference between such a linearized inverse problem and a linear one is that the conditions limiting $\mathcal{C}_0$ must involve those that ensure the relevance of our approximation.

A quite different case involves nonlinear inverse problems for which $\mathcal{M}$ is so general (with weak a priori conditions) that we cannot escape an exhaustive exploration of $\mathcal{C}^0$ for seeking the exact or approximate solutions of (1). Random walks through $\mathcal{C}^0$ can do the job, for example, Monte-Carlo methods, simulated annealing, or genetic algorithms.

Between these extreme cases are those that correspond to important physical or engineering problems, for which studies either give an exact method of solution or allow solution of the inverse problem by numerical techniques. Defining cost functionals and minimizing them by gradient iterative or other methods (Engl et al., 2000; Ghosh Roy & Couchman, 2002) is then possible, but the occurrence of several secondary minima can also make the job difficult, raising questions about their relevance as admissible solutions.

Solving an inverse problem means that we are able to derive all the admissible regularized solutions. If the non-uniqueness is due not only to small data or parameters deviations but to the structure of $\mathcal{M}$, these equivalent solutions should be classified into identified families. This is essential either to derive ways of restoring uniqueness (for instance, by making other kinds of measurements), to use the non-uniqueness (e.g., to design stealth targets), or to modify the nature of the model. In decisive as opposed to descriptive modeling, well-posed questions seek significant pieces of information instead of trying to get an image of the parameters.

A One-dimensional Example

Inverse scattering in one spatial dimension offers some completely solved examples, the simplest one being the Schrödinger problem on the line (Chadan & Sabatier, 1989; Aktosun & Klaus 2002), which may model observations of quantum, acoustic, or electromagnetic scattering. Experimental results are the scattering coefficients $R(k)$ and $T(k)$, $k \in \mathbb{R}$, related to the parameter $V(x)$ by the following formulas, where indices $\pm$ at the same level correspond :

$$\left[-\frac{\mathrm{d}^2}{\mathrm{d}x^2} + V(x)\right] f^\pm(k,x) = k^2 f^\pm(k,x), \quad (3)$$

$$\mathrm{e}^{\mp \mathrm{i}kx f^\pm}(k,x) \to 1 \quad \text{as} \quad x \to \pm\infty, \quad (4)$$

$$T(k) f^-(k,x) = f^+(-k,x) + R(k) f^+(k,x), \quad (5)$$

where $\int_{-\infty}^{+\infty} |V(x)| \left(1 + x^2\right) \mathrm{d}x < \infty$. Equation (3) shows that $\mathbb{R}$ is the continuous part of the spectrum of $-\mathrm{d}^2/\mathrm{d}x^2 + V$, which has also a finite number of eigenvalues $k = \mathrm{i}\kappa_n$, $n = 1, 2, \ldots N$, corresponding to eigenfunctions in $L^2(\mathbb{R})$, with normalizing constants

ρ_n. The set $\{R(k),\ \kappa_n,\ \rho_n\}$, called the scattering (or spectral) data, defines the following function on $\mathbb{R}^+$:

$$S(z) = \frac{1}{2\pi}\int_{-\infty}^{+\infty} \mathrm{d}k\, R(k)\mathrm{e}^{\mathrm{i}kz} + \sum_{p=1}^{N} \rho_p \mathrm{e}^{-\kappa_p z}. \quad (6)$$

Although knowledge of $R(k)$ and $T(k)$ is not sufficient to determine V, the spectral data is sufficient to determine everything $\left(V,\ f^{\pm}, T,\ \text{etc}\right)$. In fact, $\int_x^{\infty} V(t)\,\mathrm{d}t$ is equal to $2K\left(x, x^{+}\right)$, where $K(x, y)$ has been derived from S by solving the linear integral equation:

$$K(x, y) + S(x + y) + \int_x^{\infty} \mathrm{d}z K(x, z)\, S(z + y) = 0. \quad (7)$$

Only the step $K \to V$ is (weakly) instable; hence, the inverse problem ($R \to V$) is completely solved but with a structural non-uniqueness distributed into families labeled by κ_n and ρ_n.

If we let t be an evolution parameter (time) and assume that the evolution of $S(k, t)$ is described by a linear equation, the evolution of the corresponding $V(x, t)$ is generally not linear. Hence, the couple direct-inverse problem can be used to linearize special nonlinear equations. This observation provides the basis of the inverse scattering transform (IST), where the discrete eigenvalues (κ_n) correspond to solitons.

Three-dimensional Inverse Scattering

The simplest model for inverse scattering in three spatial dimensions obeys the equations:

$$(-\Delta + V)\,\Psi(\boldsymbol{k}, \boldsymbol{x}) = k^2\Psi(\boldsymbol{k}, \boldsymbol{x}) \quad (8)$$

$$\Psi(\boldsymbol{k}, \boldsymbol{x}) = \exp[\mathrm{i}\boldsymbol{k}\cdot\boldsymbol{x}] + x^{-1}\exp[\mathrm{i}kx]\, F\left(\widehat{x}, \widehat{\boldsymbol{k}}, k\right) + 0\left(x^{-1}\right), \quad (9)$$

where the bold font indicates vectors (italics their length) and the hatted letters are elements of the unit sphere.

The scattering potential V is the parameter to be determined, and F the experimental observation, which in the general case depends on 5 variables, whereas V depends only on 3. Hence, the most general inverse problem is severely ill-posed. On the other hand, spherical symmetry may reduce it to a one-dimensional problem.

If V is a piecewise continuous parameter, the general case at fixed k can be treated by standard optimization techniques. However, in most target identifications (including sonar and radar), the scatterer is not modeled by a medium function $V(\boldsymbol{x})$ but by a reflecting (or absorbing) surface. In some cases, approximate methods are relevant—ray theory, physical optics, Rytov approximation, distorted wave Born approximation (DWBA)—and they often yield linear relations between well-chosen parameters and data, for example, the (inversable) Abel integral, which relates the velocity of an earthquake signal as a function of depth to the measured travel time as a function of the ray parameter. If in mineral prospecting by seismic reflection surveys, geological information and rough identification of the signals suggest that the underground structure may be represented as a homogeneous medium with few reflectors, so-called migration methods may be useful. (Migration methods reposition returned signals that are time-distorted to their true location in time, for example, for dipping structures.)

Unfortunately, the exact relations between scatterers and data are highly nonlinear. Inversion of the DWBA fails in diffraction tomography as the structural complexity increases and so do optimization methods in seismic problems (nonconvexity of cost functions, secondary minima, etc.). Studies of stealth targets also involve nonlinear inverse problems, but of course they are not well documented in the open literature.

Full studies of the inverse problems of target identification are available in well-known works (for approximations, see Langenberg, 1987; Hopcraft & Smith, 1992; for mathematics, see Colton & Kress, 1992; for comprehensive studies, see topics 1.3 to 1.5 in Pike & Sabatier, 2002. For direct and inverse problems of elastic waves, see topics 1.7 and 1.8 in Pike & Sabatier, 2002.) Acoustic soundings used in clinical medicine (i.e., ultrasound scans) show graphs of travel times between source and reflectors (i.e., impedance discontinuities), giving evidence of imaging reflectors without need of solving inverse problems and so do more precisely devices transmitting back a time-reversed version of the recorded wave field (Fink & Prada, 2001).

A few techniques of medical imaging yield inverse problems that are either linearizable or nonlinear, according to a priori information and constraints. Among these, electrocardiography (Johnston, 2001), electroencephalography, and magnetoencephalography try to determine the electrical activity in the heart and the brain as functions of space and time. In brain imaging, the number of data points ($\sim$100) and that of unknowns ($\sim$10, 000) make the linear problem severely underdetermined, leading to arbitrary choices in the space of parameters (Bayesian approaches are commonly used) and to low resolution. From improved a priori knowledge, on the other hand, one can assume relatively few localized active groups at a given time, which are the sources of activity, but the problem becomes nonlinear (Baillet et al., 2001).

PIERRE C. SABATIER

See also **Gel'fand–Levitan theory; Integrability; Inverse scattering method or transform**

Further Reading

Aktosun, T. & Klaus, M. 2002. Inverse theory: problem on the line. In *Scattering*, edited by E.R. Pike & P.C. Sabatier, San Diego and London: Academic Press, pp. 770–785

Baillet, S., Mosher, J.C. & Leahy, R.M. 2001. Electromagnetic brain mapping. *IFEE Signal Processing Magazine*, 18(6): 14–30

Bertero, M. & Bocacci, P. 1998. *Introduction to Inverse Problems in Imaging*, Bristol and Philadelphia: Institute of Physics Publishing

Chadan, K. & Sabatier, P.C. 1989. *Inverse Problems in Quantum Scattering Theory*, 2nd edition, Berlin and New York: Springer

Colton, D. & Kress, R. 1992. *Inverse Acoustic and Electromagnetic Scattering Theory*, Berlin and New York: Springer

Engl, H.W., Hanke, M. & Neubauer, A. 2000. *Regularisation of Inverse Problems*, Dordrecht and Boston: Kluwer

Fink, M. & Prada, C. 2001. Acoustic time-reversal mirrors. *Inverse Problems*, 17(1): R1–R38

Ghosh Roy, D.N. & Couchman, L.S. 2002. *Inverse Problems and Inverse Scattering of Plane Waves*, San Diego and London: Academic Press

Hopcraft, K.I. & Smith, P.R. 1992. *An Introduction to Electromagnetic Inverse Scattering*, Boston and Dordrecht: Kluwer

Johnston, P. (editor). 2001. *Computational Inverse Problems in Electrocardiography*, Southampton: WIT Press

Langenberg, K.J. 1987. Applied inverse problems for acoustic electromagnetic and elastic wave scattering. In *Basic Methods of Tomography and Inverse Problems*, edited by P.C. Sabatier, Bristol: Adams Hilger, pp. 125–467

Pike, E.R. & Sabatier, P.C. (editors). 2002. *Scattering*, San Diego and London: Academic Press

Sabatier, P.C. 2000. Past and future of inverse problems. *Journal of Mathematical Physics*, 41(6): 4082–4124

INVERSE SCATTERING METHOD OR TRANSFORM

The story begins one August day in 1834 on the Union Canal near Edinburgh. John Scott Russell observed a mass of water, displaced by the motion of a barge, race away along the channel and disintegrate into a series of waves, the likes of which he had not seen before. Following on horseback, he noted the lead wave was about 30 ft long and about a foot high. It was a wave purely of elevation above the mean level of the canal, and it traveled at about eight miles/hr without change of shape or loss of speed for a great distance. It took another 40 years before Joseph Boussinesq and Lord Rayleigh verified that such a solitary wave was indeed a solution of the shallow water equations. Twenty years later, in 1895, Diederik Korteweg and Hendrik de Vries (KdV) wrote down the now famous, universal, and ubiquitous KdV equation

$$q_t + 6qq_x + q_{xxx} = 0 \qquad (1)$$

for the scaled elevation $q(x,t)$ of a right-going wave in a frame of reference moving with the shallow water speed $\sqrt{gh_0}$ on a layer of water of depth h_0 and its solitary wave solution

$$q(x,t) = 2\eta^2 \operatorname{sech}^2 \eta(x - x_0 - 4\eta^2 t). \qquad (2)$$

In the shallow water context, the elevation is

$$Ah_0 \operatorname{sech}^2 \left\{ \frac{\sqrt{3A}}{2h_0} \left(x - \sqrt{gh_0}\,(1 + A/2)\,t - x_o \right) \right\},$$

where g is the acceleration of gravity, A is the arbitrary dimensionless (small) height, and x_0 is the starting position of the wave.

The story continues almost 60 years later in Los Alamos. The question under address was totally different. Why do solids have finite rather than infinite heat conductivity? Peter Debye had suggested in 1916 that the anharmonicity (weak nonlinearity) of the interatomic forces hindered energy transport by coupling all the natural vibrations of the solid; thus, the relaxation time to thermal equilibrium would be a measure of conductivity. Enrico Fermi, John Pasta, and Stan Ulam (FPU) decided in the early 1950s to simulate numerically a model embracing this idea. The model consisted of 63 equal masses m attached by 64 springs with the force in each of them given by $k(\Delta + \alpha\Delta^2)$, a nonlinear modification of Hooke's law where Δ represents displacement from equilibrium. The initial energy was all in the lowest mode. As expected, very soon energy spread to the first few harmonics. But then a surprising thing occurred. The energy did not thermalize! Instead, after a while, it all recollected into the lowest mode. The pattern repeated. The negative outcome of the FPU experiment, one of the first scientific discoveries by computer experiment, was considered a great mystery (*See* **Fermi–Pasta–Ulam oscillator chain**).

Carpe Diem

The moment was seized by two young applied mathematicians, Martin Kruskal and Norman Zabusky (KZ). Because most of the energy of the FPU experiment resided in the low modes, Kruskal and Zabusky argued that they could replace the spring equations for the lateral displacements $y_n(t)$, $n = 1 \ldots 63$ by a continuum approximation $y(x = nh, t) = y_n(t)$, where h was the unstretched spring length. For disturbances traveling in one direction in a frame moving with the lattice velocity $c = \sqrt{k/m}\,h$, the suitably scaled strain $q(x,t) = y_x/6$ obeyed the KdV equation (1). They numerically simulated solutions of (1) on a periodic domain of length L beginning with $q(x,0) = (\pi a/6)\cos(2\pi x/L)$. They found that the solution evolved toward a sequence of solitary wave pulses each very close to (2) in shape that ran around the interval with the taller, faster ones periodically catching up with the slower ones, thereby producing partial, and occasionally nearly perfect, recurrences. Moreover, they made one startling observation. After a pairwise collision, the two solitary waves would emerge with their former identities (amplitude, width, and speed) intact. The only evidence that the interaction was nonlinear was a phase shift. The faster (slower) pulse seemed to advance (regress) by a fixed amount. Because the solitary waves exhibited such robust, particle-like properties, Kruskal and Zabusky called them solitons (*See* **Solitons, a brief history**).

Such behavior indicated that the KdV equation was somehow special and that its special properties were connected with hidden symmetries and conservation laws, beyond those connected with mass, momentum, and energy. A systematic search led to the Miura–Gardner transformation, $q(x,t) = w(x,t) + \mathrm{i}\varepsilon w_x(x,t) + \varepsilon^2 w^2(x,t)$, a Riccati equation mapping solutions of a modified KdV equation $w_t + 6(w + \varepsilon^2 w^2)w_x + w_{xxx} = 0$ into solutions of (1). The single conservation law $w_t + (3w^2 + 2\varepsilon^2 w^3 + w_{xx})_x = 0$ and the corresponding conserved quantity $\int w\,\mathrm{d}x$ (the integration interval is either infinite with $w(\pm\infty, t) = 0$ or finite with $w(x+L,t) = w(L,t)$) gave rise to an infinite number of conservation laws or conserved quantities for KdV, found by solving w iteratively in power of ε. Solving the Riccati equation by setting $w + 1/2\varepsilon^2 = \mathrm{i}v_x/\varepsilon v$ gives

$$v_{xx} + (q(x,t) + \lambda)v = 0, \tag{3}$$

the Schrödinger equation with energy λ and potential $-q(x,t)$. This connection of solutions of KdV with the Schrödinger operator $L(t) = -\mathrm{d}^2/\mathrm{d}x^2 - q(x,t)$ gave rise to the inverse scattering method. By this stage Kruskal had been joined by colleagues Gardner, Greene, and Miura (GGKM).

A Natural Question

In the late 1960s, GGKM began by asking: As the potential $-q(x,t)$, defined on $-\infty < x < \infty$ and decaying to zero as $x \to \pm\infty$, evolves according to (1), how does the spectrum (the set of eigenvalues) of (3) change? The character of the spectrum was already well known. It consists of a discrete number N of negative energy levels $\lambda_n < 0$, $(1 \le n \le N)$, with corresponding eigenfunctions that are square integrable and a positive continuous spectrum $\lambda > 0$. Direct substitution of $q(x,t)$ from (3) into (1) gave $\lambda_t v^2 + (uQ_x - u_x Q)_x = 0$ with $Q \equiv v_t + v_{xxx} - 3(\lambda - q)v_2$. For bound state eigenvalues, the integral of the second term vanishes giving $\lambda_{nt} \int_{-\infty}^{\infty} v_n^2\,\mathrm{d}x = 0$ so that λ_n is a motion constant. It turns out that $-2\lambda_n$ is the amplitude of the nth soliton which is contained in $q(x,t)$. The fact that it is constant is the reason for the special collision property of solitons. Because $\lambda_t = 0$, $uQ_x - u_x Q = \text{constant}$, which can easily be solved for Q to give an equation for the evolution of the eigenfunction $v(x,t)$

$$v_t = B \cdot v = -4v_{xxx} - 6qv_x + (C - 3q_x)v, \tag{4}$$

which is also equal to (remember $\lambda v = -v_{xx} - qv$)

$$v_t = B \cdot v = (q_x + C)v + (4\lambda - 2q_x)v_x. \tag{5}$$

The parameter C is a constant of integration that is used to normalize in a convenient way solutions of (3) called Jost functions. The compatibility $(v_{xx})_t = (v_t)_{xx}$ of (3) and (4) or (5) gives (1).

Peter Lax provided an elegant formalism for this compatibility. Differentiating $L(t)v = [-v_{xx} - q(x,t)v] = \lambda v$ with respect to t and using (4), one obtains $L_t = [B, L]$, which is (1) and is called the Lax equation. L and B are called the Lax pair. The Lax equation guarantees that the flow is an isospectral deformation; namely, that as the potential in (3) deforms according to (1), the spectrum remains unchanged.

The Inverse Scattering Method

The stage is now set to introduce the inverse scattering method (ISM)—or inverse scattering transform (IST)—which is done in three steps. First, one is given the potential $-q(x)$ and from this one defines solutions of (3) and calculates a set of data called the scattering data S. Second, one determines how S evolves as $q(x,t)$ evolves according to (1). Finally, one recovers the potential $-q(x)$ from the time-advanced scattering data S. This is analogous to the set of steps one uses in solving linear PDEs via the Fourier transform, and in a suitable "small" $-q(x)$ limit, S is the Fourier transform. The IST algorithm is

$$\begin{aligned} &\text{Step 1}: \; q(x,0) \to S(0). \\ &\text{Step 2}: \; S(0) \to S(t). \\ &\text{Step 3}: \; S(t) \to q(x,t). \end{aligned}$$

Interestingly, IST can be interpreted as a canonical map from the original coordinates $q(x,t)$ to suitable combinations of S that are action-angle variables.

In the first step, one defines two sets of linearly independent solutions $\{\varphi(x,k),\ \varphi(x,-k)\}$, $\{\psi(x,k),\ \psi(x,-k)\}$, $\lambda = k^2$ for real values of k (called Jost functions) by their asymptotic properties for all t as $x \to \pm\infty$,

$$\varphi(x,k) \to \mathrm{e}^{-\mathrm{i}kx}, \quad x \to -\infty; \; \psi(x,k) \to \mathrm{e}^{\mathrm{i}kx}, \quad x \to +\infty. \tag{6}$$

Because of unit (time-independent) coefficients, the choice of C in (5) for $\varphi(x,t,k)$ is $4\mathrm{i}k^3$ and the choice of C in (5) for $\psi(x,t,k)$ is $-4\mathrm{i}k^3$. Since (3) is linear, and second order, the pairs of Jost functions are related

$$\varphi(x,t,k) = a(k,t)\psi(x,t,-k) + b(k,t)\psi(x,t,k), \tag{7}$$

where $2ika(k,t)$ and $2ikb(k,t)$ are, respectively, the Wronskians $\varphi(k)\psi_x(+k)-\varphi_x(k)\psi(+k)$ and $\varphi(k)\psi_x(-k)-\varphi_x(k)\psi(-k)$. For a bound state $\lambda_n<0$ so that $k_n=\sqrt{\lambda_n}=\mathrm{i}\eta_n$. Since $\psi(x,t,-k)$ diverges as $x\to+\infty$ for $\mathrm{Im}\,k>0$, the coefficient $a(k,t)$ must vanish for a bound state so that at $\lambda=\mathrm{i}\eta_n$, $\varphi(x,t,\mathrm{i}\eta_n)=b_n(t)\psi(x,t,\mathrm{i}\eta_n)$. Dividing (7) by $a(k,t)$ and using (6), one sees that

$$\frac{\varphi}{a}\to\frac{1}{a}\,\mathrm{e}^{\mathrm{i}kx}\ \text{as}\ x\to-\infty$$
$$\to e^{\mathrm{i}kx}+\frac{b}{a}\,\mathrm{e}^{\mathrm{i}kx}\ \text{as}\ x\to+\infty. \qquad (8)$$

In wave scattering or quantum mechanical language, φ/a is a solution which represents an incoming wave of amplitude unity from $x=\infty$ and waves partially reflected and partially transmitted by the potential $-q(x)$. The coefficients $T(k)=1/a(k)$ and $R(k)=b(k)/a(k)$ are called the transmission and reflection coefficients, respectively.

It turns out that $\varphi(x,t;k)\mathrm{e}^{\mathrm{i}kx}(\varphi(x,t;-k)\mathrm{e}^{-\mathrm{i}kx})$, $\psi(x,t,k)\mathrm{e}^{-\mathrm{i}kx}(\psi(x,t,-k)\mathrm{e}^{\mathrm{i}kx})$ and $a(k,t)(a(-k,t))$, originally defined for real k, admit analytic continuation to $\mathrm{Im}\,k>0$ ($\mathrm{Im}\,k<0$). Each tends to unity as $|k|\to\infty$. The zeros of $a(k)$, $k_n=\mathrm{i}\eta_n$, $n=1,\ldots,N$ are the bound states. There is only a finite number of them. The set of data $S(t)=\{(k_n=\mathrm{i}\eta_n,b_n)_1^N;\ R(k),\ k\ \text{real}\}$ is called the scattering data. From S, $a(k)$ can be found.

In step 2, one uses (5) to calculate the time dependence of $S(t)$. Differentiate (7) with respect to t and use (5) with $C=4\mathrm{i}k^3$ for $\varphi(x,k)$ and $\psi(x,-k)$ and $C=-4\mathrm{i}k^3$ for $\psi(x,k)$ to find

$$a_t=0,\ b_t=8\mathrm{i}k^3b. \qquad (9)$$

For the bound state at $k_n=\mathrm{i}\eta_n$, differentiating $\varphi(x,k_n)=b_n\psi(x,k_n)$ gives $b_{nt}=8\eta_n^3b_n$. Therefore, the discrete eigenvalues $k_n=\mathrm{i}\eta_n$ are motion constants and $S(t)=\{(k_n=\mathrm{i}\eta_n,\ b_n(0)\exp 8\eta_n^3t)_1^N,\ R(k,0)\exp 8\mathrm{i}k^3t,\ k\ \text{real}\}$.

In step 3, one uses $S(t)$ to reconstruct $q(x,t)$ via the Riemann–Hilbert algorithm, which seeks to reconstruct a meromorphic function for $\mathrm{Im}\,k>0$ (think $\frac{\varphi(x,k)}{a(k)}\mathrm{e}^{\mathrm{i}kx}$) tending to unity as $|k|\to\infty$, $0<\arg k<\pi$ and an analytic function for $\mathrm{Im}\,k<0$ (think $\psi(x,-k)\mathrm{e}^{\mathrm{i}kx}$) tending to unity as $|k|\to\infty$, $-\pi<\arg k<0$, whose difference on the real axis $\mathrm{Im}k=0$ is given (think $R(k)\psi(x,k)\mathrm{e}^{\mathrm{i}kx}$). To achieve this, consider for $\mathrm{Im}\,k>0$,

$$I=\frac{1}{2\pi\mathrm{i}}\int_{-\infty}^{\infty}\frac{\varphi(x,k')\mathrm{e}^{\mathrm{i}k'x}}{a(k')(k'+k)}\,\mathrm{d}k' \qquad (10)$$

in two ways. First, close the contour at $|k|=\infty$ along $0<\arg k<\pi$. Second, after using (7), close the contour for that part analytic for $\mathrm{Im}\,k<0$ in the lower half-plane. Identifying the two results, one finds a linear integral equation

$$\psi(x,k)\mathrm{e}^{-\mathrm{i}kx}=1-\sum_{n=1}^{N}\frac{\gamma_k\psi(x,k_n)}{k+k_n}\,\mathrm{e}^{\mathrm{i}k_nx}+\frac{1}{2\pi\mathrm{i}}\int_{-\infty}^{\infty}R(k)\frac{\psi(x,k')\mathrm{e}^{\mathrm{i}k'x}}{k'+k}\,\mathrm{d}k', \qquad (11)$$

where $k_n=\mathrm{i}\eta_n$, $\gamma_n=b_n(a'(k_n))^{-1}$. The potential $-q(x)$ is recovered from (11) by the formula

$$q(x)=\lim_{k\to\infty}-2\mathrm{i}k\,\frac{\mathrm{d}}{\mathrm{d}x}(\psi(x,k)\mathrm{e}^{-\mathrm{i}kx}-1). \qquad (12)$$

A simple transformation converts (11) into the famous Gel'fand–Levitan equation which, since the mid-1950s, was known as a means for reconstructing potentials from scattering data.

For a special set of initial conditions $q(x,0)$ called reflectionless potentials, the reflection coefficient $R(k)$ is identically zero. In that case, the Riemann–Hilbert algorithm yields a set of linear algebraic equations found by putting $k=\mathrm{i}\eta_r, r=1,\ldots,N$ in (11). The simplest is the single soliton solution for which $R(k)=0$, $a(k)=(k-\mathrm{i}\eta)/(k+\mathrm{i}\eta)$, $b_1(0)=\exp 2\eta x_0$, $b_1(t)=\exp 2\eta\bar{x}$, $\bar{x}=x_0+4\eta^2t$, $\psi(x,k_1)=\mathrm{e}^{-\eta x}(1+\mathrm{e}^{2\eta(\bar{x}-x)})^{-1}$ and $q(x,t)=2\eta^2\mathrm{sech}^2\eta(x-\bar{x})$. Multisoliton solutions can be similarly calculated. They can also be found by a nonlinear superposition principle—the Bäcklund transformation—which is introduced below.

For $q(x,0)=Q\delta(x)$, $a(k)=(Q+2\mathrm{i}k)/2\mathrm{i}k$ and $R(k)=-Q/(Q+2\mathrm{i}k)$, and there is one bound state at $k=\mathrm{i}Q/2$. The field consists of a single soliton with $\eta=Q/2s$ and a trailing radiation component. The two components are tied together with the self-similar solution

$$q(x,t)=\frac{1}{(3t)^{2/3}}\,f\left(\frac{x}{(3t)^{1/3}}\right),$$

which we will meet again when we discuss isomonodromic deformations.

The question: "Is the complete integrability of KdV an isolated miracle or are there lots of integrable PDEs?" was quickly resolved. The pioneering GGKM work of the late 1960s was followed within a few years by the discovery of the complete integrability via ISM of the equally ubiquitous nonlinear Schrödinger (NLS) equation by Zakharov and Shabat and the characterization and integrability via ISM of families of integrable PDEs by Ablowitz, Kaup, Newell, and Segur (AKNS), (*See* **Ablowitz–Kaup–Newell–Segur system**).

More Space Dimensions

Completely integrable systems in more than one space dimension were also found. Two of the most important were the two-dimensional analogues of the KdV and NLS equations, the Kadomtsev–Petviashvili (KP) and the Benney–Roskes–Davey–Stewartson (BRDS) equations. The KP equation describes the evolution of slightly oblique waves either on shallow water or in a two-dimensional FPU lattice where the linear spring force in one direction is much weaker than it is in the other. It is given by (1) with $\int_{-\infty}^{x} q_{yy}$ added. It is also the compatibility condition of two linear operators with coefficients depending on $q(x, y, t)$ and its derivatives.

Unexpected connections of integrable PDEs with other exactly integrable models (Ising, Yang–Baxter) were found in the late 1970s and the 1980s. Most notable among them was the discovery by Sato, Miwa, and Jimbo (SMJ) that in the scaling limit, the n-point correlation function of the nearest neighbor, two-dimensional Ising model satisfies a system of very special deformation equations. These express the fact that the monodromy group of an associated linear system, which contains the correlation functions as coefficients, is preserved as the correlation functions change with their arguments. (The monodromy group of matrices encodes information about how the fundamental solution matrix of the linear system changes on taking it through a loop surrounding regular and irregular singular points.) Much of their work has deep connections with classical work on isomonodromic deformations by Ludwig Schlesinger and Paul Painlevé on systems of Fuchsian differential equations (*See* **Monodromy preserving deformations**) and by Albert Bäcklund on superposition principles for nonlinear equations (*See* **Bäcklund transformations**). Indeed, soliton theory has had a major impact in many areas of mathematical physics. It has led, for example, to the solution of the Schottky problem of algebraic geometry, to the discovery of quantum groups, to models for quantum gravity, and to useful connections with arithmetic algebraic geometry and, in particular, Jones' polynomials.

Bäcklund Transforms and Hirota's Method

To see the connection with Bäcklund transformations, it is useful to consider further the KdV family of completely integrable PDEs given by

$$q_{t_{2n+1}} = \frac{\partial}{\partial x} L^n q, \; L = -\frac{1}{4}\partial_x^2 - q - \frac{1}{2}\int_{\infty}^{x} \mathrm{d}x\, q_x, \quad n = 0, 1, \ldots, \tag{13}$$

where $t_{3/4}$ is the t of Equation (1) and $L^0 q = q$. To find them is easy. Write (5) as

$$v_t = A(\lambda; q, q_x, \ldots)v - D(\lambda; q, q_x, \ldots)v_x + Cv, \tag{14}$$

where A, D are polynomials in λ with coefficients depending on q and its x derivatives and C is the normalizing constant. The compatibility of (3) and (14), $(v_{xx})_t = (v_t)_{xx}$, tells one that $A = \frac{1}{2}D_x$ and $q_t = (-\frac{1}{2}\partial_x^3 - 2q\partial_x - q_x - 2\lambda\partial_x)D$. Solve for D as $D = -\lambda^n + D_1\lambda^{n-1} + \ldots + D_n$. Comparing powers of λ determines each D_r in terms of its predecessor as $2\partial_x D_{r+1} = (-\frac{1}{2}\partial_x^3 - 2q\partial_x - q_x)D_r$, $r = 0, \ldots, n$ with $D_0 = 1$ and D_{n+1} defined by $r = n$. The λ^0 balance gives $q_t = 2\partial_x D_{n+1} = \partial_x L^n q$.

Equations (13) are a completely integrable family of Hamiltonian commuting flows containing at level $n = 1$ the KdV equation. They all share the same constants of motion $\{H_{2r+1}\}_{-1}^{\infty}$ with $H_{-1} = \int_{-\infty}^{\infty} q\,\mathrm{d}x$, $H_1 = \int_{-\infty}^{\infty} \frac{1}{2}q^2\mathrm{d}x$, $H_3 = \frac{1}{8}\int_{-\infty}^{\infty}(q_x^2 - 2q^3)\,\mathrm{d}x$, each of which serves as the Hamiltonian for one of the flows $q_{t_{2n+1}} = \partial_x \frac{\delta H_{2n+1}}{\delta q}$ where $\frac{\delta H}{\delta q}[q]$ is the variational derivative of $H[q]$ and ∂_x is the skew symmetric symplectic operator. They are also bi-Hamiltonian in that $\partial_x \frac{\delta H_{2n+1}}{\delta q} = \left(-\frac{1}{4}\partial_x^3 - q\partial_x - \frac{1}{2}q_x\right)\frac{\delta H_{2n-1}}{\delta q}$. They commute under the Poisson bracket

$$\{H_{2m+1}, H_{2n+1}\} = \int \frac{\delta H_{2m+1}}{\delta q}\frac{\delta}{\delta n}\frac{\delta H_{2n+1}}{\delta q}\,\mathrm{d}x = 0.$$

One consequence of commutativity is that if one begins with the shape $q(x, 0, 0, \ldots)$ and evolves q with respect to t_{2n+1} for a time t_{2n+1}, $n = 0, \ldots$, in any order, one always obtains the same shape $q(x, t_1, t_3, \ldots, t_{2n+1}, \ldots)$. This means that the phase shift experienced by any two solitons is the same for each member of the family. Why? Start with the larger, faster moving solution behind the smaller, slower one (the velocity of the soliton of flow n is proportional to η^{2n}). Evolve $q(x, t_{2m+1} = 0, t_{2n+1} = 0)$ via the t_{2m+1} flow for t_{2m+1} time units sufficient for the faster soliton to overtake the slower one and then for t_{2n+1} units via the t_{2n+1} flow. Now do it in the reverse order. Because $q(x, t_{2m+1}, t_{2n+1})$ is the same no matter which order is used, the phase shift must be the same.

A second consequence of commutativity is that one can express the vector made up of the string $L^n q$ as the x derivative of the gradient of a potential which we call $2\ln\tau(x, t_1, t_3, \ldots)$, namely, $L^n q = 2\frac{\partial}{\partial x}\cdot\frac{\partial}{\partial t_{2n+1}}\ln\tau$. The function $\tau(x, t_1, t_3, \ldots)$, called the Hirota τ function, has remarkable properties. It converts each member of the family (13) into bilinear form. For $n = 1$, the KdV equation under $q = 2\frac{\partial^2}{\partial x^2}\ln\tau$ becomes (recall $q_{t_1} = q_x$ and take $4t_3 = t$)

$$D_x^2\tau\cdot\tau = \tau\tau_{xt} - \tau_x\tau_t + \tau\tau_{xxxx} - 4\tau_x\tau_{xxx} + 3\tau_{xx}^2 = 0, \tag{15}$$

where the symbol D_x acting on the ordered pair σ, τ is defined as the limit as $\varepsilon \to 0$ of $\frac{\partial}{\partial\varepsilon}\sigma(x+\varepsilon)\tau(x-\varepsilon) = \sigma_x\tau - \sigma\tau_x$. For the KP equation, one adds $\tau\tau_{yy} - \tau_y^2$ to the left-hand side of (15). From (15), one can build multisoliton solutions. The solution $\tau = 1$ corresponds to the vacuum state $q = 0$. The solution $\tau = 1 + \exp\theta(x,t)$, $\theta(x,t) = 2k(x-x_0) - 8k^3t$ corresponds to the single solution solution $q(x,t) = 2k^2\text{sech}^2 k\,(x - x_0 - 4k^2t)$. The next solution in the sequence is $\tau = 1 + \exp\theta_1(x,t) + \exp\theta_2(x,t) + \exp(\theta_1(x,t) + \theta_2(x,t) + A_{12})$, $\theta_j(x,t) = 2k_j(x - \overline{x})$, $x_j = 4k_j^2t + x_{j0}$, $j = 1, 2$, $A_{12} = 2\ln\left|\frac{k_1-k_2}{k_1+k_2}\right| < 0$, and is a superposition of two solitons with amplitudes $2k_1^2$ and $2k_2^2$, which for large negative t and large positive t are separated with the faster one (assume $k_1^2 > k_2^2$) being on the left(right) when $t = -\infty(+\infty)$. For $t \to -\infty(+\infty)$, near $x \simeq \overline{x}_2$, the center of the slower pulse, it is the second and fourth (first and third) terms that dominate τ so that for large negative (positive) t, the second soliton looks like

$$2k_2^2\text{sech}^2 k_2\left(x - \overline{x}_2 - \frac{1}{2k_2}|A_{12}|\right)$$
$$\times(2k_2^2\text{sech}^2 k_2(x - \overline{x}_2)).$$

The collision shifts the slower soliton back by $\frac{1}{2k_2}|A_{12}|$ and, by a similar argument, the faster soliton ahead by $\frac{1}{2k_1}|A_{12}|$.

What does the interaction look like? For k_1 much greater than k_2, $|A_{12}| \simeq 0$, and the faster soliton rides adiabatically over the slowly changing slower one. For k_1 greater than but close to k_2, there is an exchange of identities with the slower soliton assuming the form of the faster one as soon as the trailing edge of the former feels the leading edge of the latter.

But a century before Ryogo Hirota's work in the 1970s (*See* **Hirota's method**), the works of Bäcklund, Darboux, and Schlesinger had shown how to build, for certain classes of nonlinear equations, complicated solutions from simple ones. Applied to any member of the KdV family, the idea is this. Define

$$q = -u_x = 2\partial_x^2 \ln\tau \quad \text{and} \quad \tilde{q} = -\tilde{u}_x = 2\partial_x^2 \ln\tilde{\tau}$$

Then there are two relations, called Bäcklund transformations, which express $\tilde{u}_x + u_x$ and $\tilde{u}_t + u_t$ as functions of $\tilde{u} - u$ and a new free parameter ζ^2. Then, if q satisfies KdV, the enriched solution $\tilde{q}$ also satisfies KdV. These expressions take on a beautifully simple form that reveals the algebraic structure underlying the hidden symmetries of the KdV family when they are expressed in terms of τ functions as $\tilde{\tau} = \tau v$ where $v(x, \varsigma^2)$ solves (3) with the q corresponding to τ and $\lambda = \zeta^2$. But one can also express v as a combination of $\frac{X(\zeta)\tau}{\zeta}$ and $\frac{X(-\zeta)\tau}{\zeta}$, where the operator

$$X(\zeta) = \exp\left(\mathrm{i}\sum_{o}^{\infty}\zeta^{2k+1}t_{2k+1}\right)$$
$$\times\exp\left(\sum_{0}^{\infty}\frac{\mathrm{i}}{(2k+1)\zeta^{2k+1}}\frac{\partial}{\partial t_{2k+1}}\right),$$

so that $\tilde{\tau} = \tau_{\text{new}} = (AX(\zeta) + BX(-\zeta))\tau_{\text{old}}$.

For example, choose $\zeta = \mathrm{i}\eta$. Then if $\tau_{\text{old}} = 1$, $\tau_{\text{new}} = Ae^{\theta} + Be^{-\theta} = 2\sqrt{AB}\cosh(\theta - \theta_0)$, where $\frac{A}{B} = e^{2\theta_0}$ and $\theta = \sum_0^{\infty}(-1)^k\eta^{2k+1}t_{2k+1}$ and τ_{new} is the one soliton solution. Because of the logarithm, we can also write τ_{new} as $(1 + \beta Y(\zeta))\tau_{\text{old}}$ where $Y(\zeta)$ is called the vertex operator

$$Y(\zeta) = \exp\left(-2\mathrm{i}\sum_{0}^{\infty}\zeta^{2k+1}t_{2k+1}\right)$$
$$\times\exp\left(\sum_{k=0}^{\infty}\frac{2}{\mathrm{i}(2k+1)\zeta^{2k+1}}\frac{\partial}{\partial t_{2k+1}}\right), \quad (16)$$

which has the property that $Y(\zeta)\cdot Y(\zeta')1 = 0$ if $\zeta = \zeta'$. Thus $Y^2 = 0$ and we can replace $1 + \beta Y(\zeta)$ by $\exp\beta Y(\zeta)$. The Bäcklund transformation is therefore

$$\tau_{\text{new}} = \exp\beta Y(\zeta)\cdot\tau_{\text{old}} \quad (17)$$

namely, the action of the "group" (infinite-dimensional groups are not rigorously defined) element corresponding to the algebraic element $Y(\zeta)$, which can be expressed as a Laurent series $\sum_{-\infty}^{\infty} Y_{2k+1}\zeta^{2k+1}$. The coefficients Y_{2k+1} obey a nontrivial set of commutator relations that form an infinite dimensional, graded Lie algebra (Kac–Moody algebra), the central extension of the loop algebra of Sl (2, C). Under repeated application of (17), solutions of KdV trace out the orbit of the highest weight vector $\tau = 1$ in the basic representation of the Kac–Moody algebra. The algebra acts as an algebra of symmetries. There is also a complementary treatment where the algebra is used as the phase space on which there is defined both a natural Poisson bracket and Hamiltonian vector field.

Other Topics

Multisoliton solutions of the KdV family are a special case of finite gap solutions. The latter are defined by adding a nonlinear, constant coefficient ODE called the Lax–Novikov equation

$$\sum_{0}^{N} a_{2r+1}(L^r q)_x = \sum_{0}^{N} a_{2r+1}\, q_{t_{2r+1}} = 0 \quad (18)$$

as a constraint on (3) and (14). To illustrate the basic idea, assume traveling wave solutions

$q(x,t) = q(x - ct)$ satisfy $-cq_x + (q_{xx} + 3q^2)_x = 0$, which is (18) with $a_1 = -c$, $a_3 = 4$, and $a_{2k+1} = 0$, $n > 1$. The Lax–Novikov equation, together with (3) and (14) gives rise to a Riemann surface, the analogue of the spectrum for potentials decaying at infinity, $y^2 = \Pi_{j=1}^{2N}(\lambda - \lambda_j)$ which remains invariant on the finite gap solution family. The τ function for the general N gap solution is the Riemann theta function.

Another class of solutions, the multiphase self-similar solutions, is found by attaching a nonlinear non-autonomous ODE

$$a_0 q + \sum_0^N a_{2r+1}\, t_{2r+1} (L^r q)_x = a_0 q + \sum a_{2r+1}\, t_{2r+1}\, q_{t_{2r+1}} = 0 \quad (19)$$

as a constraint to (3) and (14). It is closely related to the string equation of modern physics, and it leads to a system of linear ODEs for

$$\lambda \frac{\partial}{\partial \lambda} \left\{ \begin{array}{c} v(x, t_{2r+1}, \lambda) \\ v_x(x, t_{2r+1}, \lambda) \end{array} \right\}$$

with coefficients depending on self-similar scalings of q and its derivatives. Solutions of the multiphase self-similar family are isomonodromic deformations of this system of ODEs in that they leave the monodromic structure invariant. At $\lambda = \infty$, an irregular singular point, the monodromic structure is expressed in terms of Stokes multipliers; at $\lambda = 0$, a regular singular point, the monodromy is that of a simple pole. Members of this family include the Painlevé equations.

For example, (19) with $a_0 = -2$, $a_1 = -1$, $a_3 = 3$, $t_3 = t$ is $-2q - xq_x + 3t(q_{xx} + 3q^2)_x = 0$ which forces $q(x,t)$ to have the form

$$\frac{1}{(3t)^{2/3}} f\left(X = \frac{x}{(3t)^{1/3}}\right),$$

which satisfies an ODE that is a close cousin of the second Painlevé equation. This solution plays a key role in joining the soliton component of the solution with the trailing wave-like radiation. Isomonodromic deformations provide the link between soliton equations and the exactly solvable models of statistical physics. For example, the two-point correlation function for the nearest-neighbor Ising model in the scaling limit satisfies the third Painlevé equation.

The existence of so many completely integrable systems of physical interest (KdV, Boussinesq, nonlinear Schrödinger, derivative nonlinear Schrödinger, massive Thirring, sine-Gordon, Maxwell–Bloch with inhomogeneous broadening, KP, BRDS, three-wave scattering, Raman scattering) leads one to ask how important to applications these integrable models are. In the set of all PDEs, they are of measure zero. On the other hand, there are two reasons for their importance. The first is that many systems such as shallow waves traveling over a channel of slowly varying depth can be treated as perturbations of PDEs integrable by ISM, and there are simple algorithms for computing how the previously constant action variables slowly evolve. The second reason is more subtle and remains to be rigorously proven. Many of the equations of physical interest arise as asymptotic approximations, as reductions of more complicated systems assuming (as one does in deriving KdV) weak nonlinearity and weak dispersion, or (as one does in deriving NLS) weak nonlinearity and strong dispersion of an almost monochromatic wavepacket. If the process of reduction preserves integrability, then as long as there is one integrable equation among the class that reduces to the universal equation of interest, the reduced equation will also be integrable.

Finally, how does one test for integrability? How does one uncover the hidden symmetries, the algebraic structure of a PDE or ODE? There is yet no foolproof method that works in all circumstances. As in the original discovery of the integrability of KdV, serendipity often plays the main role. Two of the more promising semi-algorithmic approaches are the Painlevé test (the location of any algebraic, logarithmic or essential singularity is independent of initial conditions and only the location of poles can depend on arbitrary constants of integration). and the Wahlquist–Estabrook method, which seeks to uncover the hidden symmetries by embedding the equation of interest as an integrability condition of a pair of linear systems whose algebraic structure it is the goal of the method to find. Each has its advantages and each has proven to be successful in extracting the appropriate Lax pair in several contexts. But the challenge of determining a foolproof method to determine whether a particular equation is integrable is still open.

ALAN NEWELL

See also **Ablowitz–Kaup–Newell–Segur system; Bäcklund transformations; Hirota's method; Kadomtsev–Petviashvili equation; N-soliton formulas; Painlevé analysis**

Further Reading

Ablowitz, M.J., Kaup, D.J., Newell, A.C. & Segur, H. 1974. The inverse scattering transform-Fourier analysis for nonlinear problems. *Studies in Applied Mathematics*, 53: 249–315

Ablowitz, M.J. & Segur, H. 1981. *Solitons and the Inverse Scattering Transform*, Philadelphia: SIAM

Faddeev, L.D. & Takhtajan, L.A. 1987. *Hamiltonian Methods in the theory of solitons.* Berlin: Springer

Flaschka, H. & Newell, A.C. 1980. Monodromy and spectrum preserving deformations. *Communications in Mathematical Physics*, 76: 65–116

Fokas, A.S. & Zakharov, V.E. 1993. *Important Developments in Soliton Theory*, Berlin and New York: Springer

Gardner, C.S., Greene, J.M., Kruskal, M.D. & Miura, R.M. 1974. The Korteweg de Vries equation and generalization. VI. Methods of exact solution. *Communications in Mathematical Physics*, 27: 97–133

Newell, A.C. 1985. *Solitons in Mathematics and Physics*, Philadelphia: SIAM

ION ACOUSTIC WAVES

See **Nonlinear plasma waves**

ISING MODEL

The Ising or Lenz–Ising model was introduced by Wilhelm Lenz in 1920 as a simplified model of a ferromagnet. Lenz's student Ernst Ising studied the one-dimensional case (a linear chain of coupled magnetic moments) for his doctoral dissertation at the University of Hamburg, and his results published in 1925 showed the absence of a phase transition between a ferromagnetic and paramagnetic state. In 1942, Lars Onsager applied an ingenious mathematical method to the two-dimensional model and showed analytically the existence of a phase transition in the absence of an external magnetic field. Interest in the model has increased in recent years as it became the foundation for modeling changes of state in new areas such as spin glasses and neural networks. In 1975, David Sherrington and Scott Kirkpatrick introduced long-range competing interactions in the model and showed the emergence of a new type of glassy order, while John Hopfield in 1982 used the connection between spins and neurons in extending the Ising model into a model of neuronal networks.

The original motivation for the introduction of the model came from the knowledge that certain metals, such as iron, are ferromagnetic; that is, they form macroscopic domains with nonzero magnetization when the temperature is less than a characteristic Curie temperature (T_c), but this long-range order is lost at temperatures $T > T_c$. Because magnetic properties of atoms are a result of the orbital angular momentum as well as spin of their electrons, the existence of macroscopic size domains with non-zero magnetization stems from mutual alignment of permanent atomic dipole moments induced by dipole-dipole interactions. The Lenz–Ising simplification ignores quantum mechanics and assigns to an atom located in crystal site i a magnetic moment or "spin" S_i that takes only two values, namely, $S_i = \pm 1$. The energy or Hamiltonian for the system of N spins is

$$H = -\frac{1}{2}\sum_{i,j=1}^{N} J_{ij} S_i S_j - \sum_{i=1}^{N} B_i S_i, \qquad (1)$$

where J_{ij} is the interaction between spins at sites i and j; respectively; B_i is a local magnetic field at site i; and the sums extend independently over all N sites. Any of the 2^N spin configurations can be thought of as a chemical solution of two species with N_+ and $N_- = N - N_+$ spins in up and down states, respectively, with number densities $n_\pm = N_\pm/N$; the entropy of this state is determined by the number of ways N spins are partitioned in up and down configurations, giving

$$S = -Nk_B\,(n_+ \ln n_+ + n_- \ln n_-)\,. \qquad (2)$$

Furthermore, the density of $++$, $--$, and $+-$ pairs that determine short-range order is approximated in the Bragg–Williams approximation respectively, by n_+^2, n_-^2, and $2n_+n_-$. Thus, Equation (1) becomes

$$\begin{aligned} H &= -\tfrac{1}{2} zNJ\left(n_+^2 + n_-^2 - 2n_+n_-\right) \\ &\quad -NB\,(n_+ - n_-,)\,, \end{aligned} \qquad (3)$$

where only interactions of strength J between z nearest-neighbor spins are included. Introducing the macroscopic magnetization $M = N_+ - N_- = N(n_+ - n_-)$, expressing the densities as $n_\pm = (1 \pm M/N)/2$, and substituting in the energy and entropy, respectively, one obtains

$$\begin{aligned} H &= -\frac{zJM^2}{2N} - MB \\ S &= NK\left[\ln 2 - \frac{1}{2}\left(1+\frac{M}{N}\right)\ln\left(1+\frac{M}{N}\right)\right. \\ &\quad \left. -\frac{1}{2}\left(1-\frac{M}{N}\right)\ln\left(1-\frac{M}{N}\right)\right] \end{aligned} \qquad (4)$$

The thermodynamics of the model is now completely determined from the minimization of the Helmholtz free energy $F = H - TS$ with respect to the magnetization M at constant external field B and temperature T. It yields

$$\frac{M}{N} = \tanh\beta\left(B + zJ\frac{M}{N}\right), \qquad (5)$$

where $\beta = 1/k_B T$. Equation (5) is identical to that obtained through the Weiss mean field theory predicting, after linearization and for zero external field, a Curie temperature determined by $k_B T_c = zJ$.

If the interaction J_{ij} is not fixed but taken to be random, distributed according to a Gaussian with mean J_0/N and variance J^2/N, then

$$P(J_{ij}) = \left(\frac{N}{2\pi J^2}\right)^{1/2} \exp\left[-\left(J_{ij} - \frac{J_0}{N}\right)^2 \frac{N}{2J^2}\right]. \qquad (6)$$

The resulting Ising model is referred to as the Sherrington–Kirkpatrick model, and its low-temperature phase at zero-external-field is a spin glass. This phase is not characterized by the long-range order typical in ferromagnets but by partial system freezing in time showing persistent but spatially random configurations whose thermodynamic signature is a sharp cusp in the zero field cooled susceptibility. A spin glass is typically characterized by metastability, frustration, and slow relaxation resulting from the multiple high barriers (introduced by disorder in the free energy) that force the system to remain trapped in local free energy minima for long times.

Introducing some changes in the Ising spin-glass model and with proper interpretation, one obtains a model for a network of neurons. Ising spins are now states of a neuron found in a firing ($S_i = +1$) or inhibitory state ($S_i = -1$), the interaction elements J_{ij} are synaptic strengths between neurons while the external field depends on a neuron firing threshold potential. In the learning mode, the network stores information on patterns in the synaptic strengths, while in the retrieval mode, these patterns are accessed when the network is exposed to partial information on the desired memories. If $S^{\nu} \equiv \{S_1^{\nu}, S_2^{\nu}, \ldots, S_N^{\nu}\}$ (for $\nu = 1, \ldots, p$) represent p patterns of N neurons, employment of the following version of Hebb's rule:

$$J_{ij} = \frac{1}{N} \sum_{\nu=1}^{p} S_i^{\nu} S_j^{\nu} \tag{7}$$

stores the p patterns in memory. In the Hopfield model for $p/N \leq 0.14$, these states, as well as several other spurious ones, become attractors to the evolution flow of neuron updating rules. Whenever a spin-glass phase is present, either alone or in coexistence with the memory states, information retrieval is corrupted.

While the phase diagrams of Ising spin glasses and neural networks are quite complex, the one determined analytically by Onsager for the two-dimensional Ising model with nearest-neighbor interaction J is much simpler, demonstrating the presence of a Curie temperature at $k_B T_c \approx 2.269J$ with a low-temperature ferromagnetic phase. This shows that even though quite simple, the Ising model contains the essential features of the problem that motivated its introduction, the ferromagnetic-paramagnetic transition. The absence of a transition in one dimension as found by Ising can be understood qualitatively by the following argument of Wannier. Energy equal to $2J$ can introduce a spin down defect in an ordered chain of $N + 1$ spins that are all in state up, and this can be done in N ways resulting in entropy $k_B \ln N$. The free energy for this change is $\Delta F = 2J - k_B T \ln N$ and favors always the entropic term except at $T = 0$; thus, there cannot be long-range order in one dimension at finite temperatures.

G.P. Tsironis

See also **Attractor neural network; Ferromagnetism and ferroelectricity; Frustration**

Further Reading

Baxter, R.J. 1982. *Exactly Solvable Models in Statistical Mechanics*, London: Academic Press

Binder, K. & Young, A.P. 1986. Spin glasses: experimental facts, theoretical concepts and open questions. *Reviews of Modern Physics*, 58: 801–976

Müller B., Reinhardt, J. & Strickland, M.T. 1995. *Neural Networks: An Introduction*, Berlin: Springer

Sompolinski H. 1988. Statistical mechanics of neural networks. *Physics Today*, 41 (12 Dec): 70–80

Wannier, G.H. 1966. *Statistical Physics*, New York: Wiley

ITO'S FORMULA

See **Stochastic processes**

IZERGIN–KOREPIN EQUATION

See **Discrete nonlinear Schrödinger equations**

J

JACOBI ELLIPTIC FUNCTIONS

See **Elliptic functions**

JOSEPHSON JUNCTION ARRAYS

Josephson junction arrays (JJAs) consist of islands of superconductor arranged in an ordered lattice, coupled by Josephson junctions. Large JJAs are useful model structures for studying the dynamics of coupled nonlinear oscillators, phase transitions, frustration effects, vortex dynamics, and macroscopic quantum phenomena (Newrock et al., 2000).

JJAs are generally divided into classical arrays ($E_J/E_C \gg 1$) and quantum arrays ($E_J/E_C \ll 1$), depending on the ratio of the Josephson energy, E_J, to the charging energy, E_C (Fazio & van der Zant, 2001). Classical JJAs can be divided into overdamped arrays ($\beta_C \ll 1$) and underdamped arrays ($\beta_C \gg 1$), referring to the fact that the equation of motion for a single Josephson junction is identical to a damped pendulum. The dividing line is determined by the McCumber parameter, β_C, which defines the amount of damping in terms of the junction capacitance and resistance.

Modeling of classical JJAs is essentially based on solving a system of coupled ordinary differential equations for the superconducting phase differences φ_n of Josephson junctions that constitute the array. Typically, JJAs are fabricated using photolithography and the standard fabrication processes for superconductive electronics. They are very well characterized, leading to an interesting synergy between experiments, theory, and simulations.

The properties of JJAs depend on their dimensionality and the way Josephson junctions are connected. In the series one-dimensional (1-d) JJAs shown in Figure 1a, there is no direct interaction between Josephson junctions. The junctions can be coupled, for example, via a load resistor connected between the extremities of the array. Under certain conditions (Jain et al., 1984), when biased by a bias current flowing along the array, Josephson junctions perform coherent oscillations.

Parallel 1-d arrays shown in Figure 1b have inherent mutual coupling between Josephson junctions due to screening currents and magnetic flux quantization in superconducting loops. Such an array is essentially similar to a long Josephson junction, and it is described by the discrete sine-Gordon (SG) model. Due to the similarity between the discrete SG and continuous SG models, properties of vortices in parallel 1-d JJAs are similar to those of Josephson vortices (solitons called fluxons) in long Josephson junctions (Ustinov & Parmentier, 1996; Watanabe et al., 1996).

The two-dimensional (2-d) JJA illustrated in Figure 1c has attracted a great deal of interest as it is isomorphic to a 2-d XY spin system, which is a 2-d lattice of spins free to rotate in the XY plane. In its simplest form, the Hamiltonian of an array can be written as

$$H = -\sum_{\langle n,m \rangle} E_J \cos(\varphi_m - \varphi_n - f), \qquad (1)$$

where $\langle n, m \rangle$ means summing over nearest neighbors and f is the frustration factor depending on the externally applied magnetic field. It should be noted that model (1) is valid only under the condition that both E_J and the array cell size are small enough, i.e., that inductance effects play no role. In practical JJAs, this is typically not the case and these effects significantly complicate JJA dynamics.

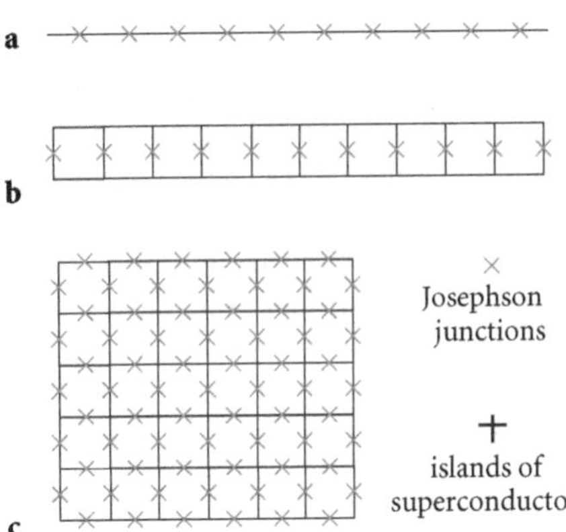

Figure 1. Schematic view of (a) 1-d series JJA; (b) 1-d parallel JJA; (c) 2-d JJA.

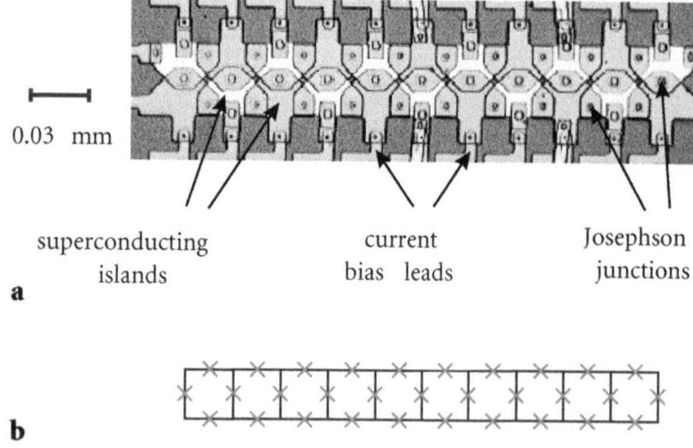

Figure 2. (a) Micro-photograph of a Josephson ladder (Binder et al., 2000). (b) The equivalent circuit of the ladder.

Starting from the mid-1990s, a special type of JJA called the Josephson ladder has received much interest. The ladder is essentially one row of cells cut from the standard 2-d JJA. Floría et al. (1996) studied a ladder driven by an ac bias current and found both oscillating and rotating localized modes. Such intrinsically localized modes, often called discrete breathers, are spatially localized time-periodic dynamical states that occur in various nonlinear lattices. The oscillating modes in Josephson arrays are difficult to detect experimentally as they are accompanied by zero average dc voltage. In contrast to oscillating discrete breathers, rotating discrete breathers induce a localized nonzero dc voltage and can be easily measured. Moreover, rotating discrete breathers can also be supported by a dc bias current. Experiments that followed the theoretical proposals successfully demonstrated dynamical localization in the form of discrete breathers in JJA (Ustinov, 2001).

There are several useful applications of JJA. At microwave frequencies, Josephson junctions can be used as sources and detectors of radiation. The Josephson relation $\Phi_0 \nu = V$ (where $\Phi_0 = h/(2e) = 2.07 \times 10^{-15}$ Vs is the magnetic flux quantum) states that at constant voltage V, the current through the junction will oscillate with a frequency ν, proportional to V. Utilizing this unique property, Josephson junction arrays have for the last two decades been studied as microwave sources (Darula et al., 1999). The advantages of using arrays as opposed to junctions are higher power outputs and better impedance matching to typical loads. Another important application of JJA, based on the same equation, is to define the voltage standard. Since frequency can be measured extremely accurately, locking Josephson junctions to a microwave source provides excellent accuracy for measuring the voltage. Using a series JJA instead of a single Josephson junction allows the upper voltage level to be brought from less than a millivolt up to the range of several volts.

Alexey V. Ustinov

See also **Breathers; Frenkel–Kontorova model; Frustration; Josephson junctions; Long Josephson junctions; Sine-Gordon equation; Superconducting quantum interference device; Superconductivity**

Further Reading

Binder, P., Abraimov, D. & Ustinov, A.V. 2000. Diversity of discrete breathers observed in a Josephson ladder. *Physical Review E*, 62: 2858–2862

Darula, M., Doderer, T. & Beuven S. 1999. Millimetre and sub-mm wavelength radiation sources based on discrete Josephson junction arrays. *Superconductor Science and Technology*, 12: R1–R25

Fazio, R. & van der Zant, H. 2001. Quantum phase transitions and vortex dynamics in superconducting networks. *Physics Reports*, 355: 235–334

Floría, L.M., Marin, P.J., Martinez, J.L., Falo, F. & Aubry, S. 1996. Intrinsic localization in the dynamics of a Josephson junction ladder. *Europhysics Letters*, 36: 539–544

Jain, A.K., Likharev, K.K., Lukens, J.E. & Sauvageau, J.E. 1984. Mutual phase-locking in Josephson junction arrays. *Physics Reports*, 109: 309–426

Newrock, R.S., Lobb, C.J., Geigenmüller, U. & Octavio, M. 2000. The two-dimensional physics of Josephson junction arrays. *Solid State Physics*, 54: 263–412

Ustinov, A.V. 2001. Experiments with discrete breathers in Josephson arrays. In *Nonlinearity and Disorder: Theory and Applications*, edited by F. Abdullaev, O. Bang & M.P. Soerensen, Dordrecht: Kluwer, pp. 183–185

Ustinov, A.V. & Parmentier, R.D. 1996. Coupled solitons in continuous and discrete Josephson transmission lines. In *Nonlinear Physics: Theory and Experiment*, edited by E. Alfinito, M. Boiti, L. Martina & F. Pempinelli, Singapore: World Scientific, pp. 582–589

Watanabe, S., van der Zant, H., Strogatz, S.H. & Orlando, T. 1996. *Physica* D, 97: 429

JOSEPHSON JUNCTIONS

Predicted theoretically by Brian Josephson in 1962, the Josephson junction (JJ) consists of two superconductors that are so weakly coupled that the Cooper pairs may quantum mechanically tunnel between the superconductors without destroying the integrity of their individual macroscopic wave functions, Ψ_1 and Ψ_2. A typical example is the trilayer tunnel junction (see Figure 1) consisting of two overlapping niobium (Nb) thin films separated by a thin (2–5 nm) insulating layer of aluminum oxide (Al_2O_3).

Extremely fast and highly nonlinear, the JJ is described by the two Josephson equations

$$I = I_c \sin \phi, \quad (1)$$

$$\frac{\partial \phi}{\partial t} = \frac{2\pi}{\Phi_0} V, \quad (2)$$

where I is the pair-current through the junction. The critical current I_c is a constant parameter given by the materials, the barrier, and the geometry of the structure. $\Phi_0 = h/2e$ is the flux quantum, $\phi = \theta_1 - \theta_2$ is the difference between the phases θ_1 and θ_2 of Ψ_1 and Ψ_2, respectively, and V is the voltage across the junction. These equations assume a constant current density over the junction area.

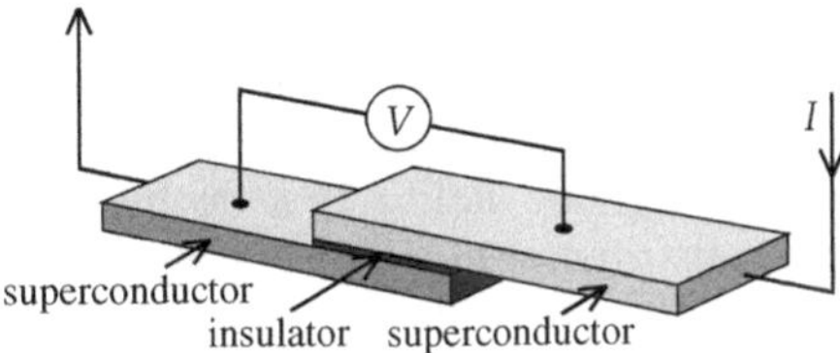

Figure 1. Typical Josephson tunnel junction consisting of two superconducting thin-films separated by a very thin insulating oxide layer.

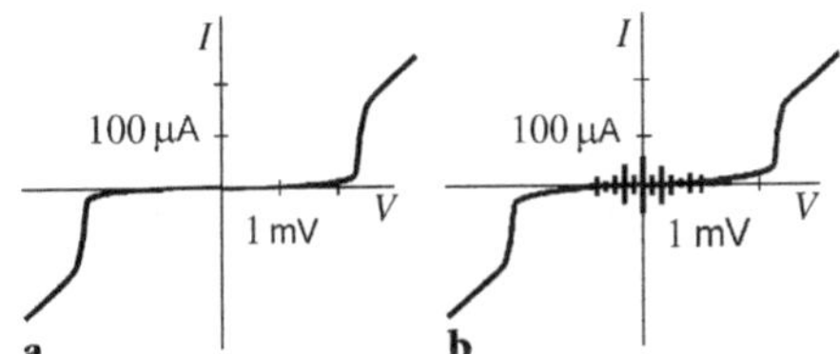

Figure 2. dc current–voltage characteristic of $Nb/Al_2O_3/Nb$ trilayer Josephson tunnel junction at 4.2 K. Note the supercurrent in the zero-voltage state and the steep gap structure near 2.7 mV. (a) No applied microwaves and (b) with applied microwaves. The small vertical zero-crossing steps are equidistantly spaced with a voltage difference of $\Delta V = hf/2e$. These nonlinear quantum phenomena allow for the practical construction of the Josephson voltage primary standard.

If one applies a dc bias current $I < I_c$ to the junction, the phase difference ϕ of the macroscopic wave functions automatically adjusts itself so that Equation (1) is satisfied; i.e., the junction remains in the zero-voltage state carrying a supercurrent up to the critical current. For $I \geq I_c$, the junction switches to the voltage state, and ϕ oscillates in time according to the second Josephson equation.

If the junction is supplied with a constant voltage, V_{DC}, the phase difference increases steadily with time, and the junction current oscillates with the frequency

$$f = \frac{1}{\Phi_0} V_{DC}; \quad (3)$$

that is, the junction functions as a voltage controlled oscillator (VCO) that may generate microwave power into the gigahertz range. (The pre-factor $1/\Phi_0 \approx 0.5$ GHz/μV.)

The capacitance between the two electrodes in Figure 1 shunts the Josephson tunneling and leads to hysteresis in the current-voltage characteristic (I–V curve) as shown in Figure 2a. Starting at zero and increasing the bias current, there is a vertical supercurrent (zero-voltage state) up to I_c where the junction switches (horizontally) to the steep so-called quasi-particle curve (near 2.7 mV), which reflects the superconducting gap of the two niobium electrodes. The quasi-particle curve is followed both when the bias current is further increased and when the current is returned to $I = 0$. The hysteretic I–V curve is point-symmetric around $(V, I) = (0, 0)$.

The I–V curve near the gap is strongly nonlinear and temperature dependent. Tunnel junctions biased close to the knee are used as low-noise bolometers to detect broad band signals in the millimeter and sub-mm range. Due to its strong nonlinearity, heterodyne receivers based on this SIS-mixer (superconductor-insulator-superconductor) can be operated near the quantum limit ($hf \approx k_B T$). The SIS-mixer may be pumped by the microwave signal emitted from a long JJ (see below). Most modern radio-telescopes employ SIS-mixers for spectral measurements in the frequency range 10–1000 GHz.

Figure 2b shows the I–V curve when a microwave signal is applied to the junction. The supercurrent is suppressed and small equidistantly spaced replica appear as vertical (Shapiro) steps with a voltage difference of $\Delta V = hf/2e$. These zero-crossing steps and the fact that voltage and frequency are related only through fundamental constants allow for the practical realization of the Josephson voltage primary standard. When pumped by a 70 GHz signal, $\Delta V \approx 140$ μV; thus a small chip with more than 20.000 dc series connected JJs can generate a reference voltage of 10 V with an accuracy of 0.1 nV. Josephson junctions are strongly nonlinear elements which when pumped by external high-frequency signals may not only generate very high-numbered higher harmonics but also via parametric processes generate both even and odd subharmonics (e.g., period-doubling bifurcations) and chaos. The stability of the Josephson voltage standard is limited by chaotic behavior.

JJs are highly sensitive to magnetic fields. The gradient of the phase difference ϕ is proportional to the magnetic field applied in the plane of the junction, and for constant current density the net critical current is zero each time ϕ has a twist of 2π, which happens when the flux Φ threading the junction is exactly one flux quantum Φ_0. Included in a superconducting ring, JJs constitute the so-called Superconducting Quantum Interference Device (SQUID), which is used, for example, for magnetoencephalographic measurements of electrical activity in the human brain.

The magnetic twist of the phase difference along the junction also leads to the so-called Fiske steps. These are nearly constant-voltage steps in the I–V curve at voltages $V_{FSn} \approx \Phi_0 c_s/2L$, where $n = 1, 2, \ldots$ and L is the junction length perpendicular to the magnetic field. Also, c_s is the Swihart velocity of electromagnetic waves propagating in the junction (about 3% of the light velocity in vacuum). The physical mechanism is as follows. The Josephson oscillations at the voltages V_{FSn} excite standing electromagnetic waves inside the junction (cavity) with wavelengths $\lambda_n = 2L/n$ and resonance frequencies $f_n = c_s/\lambda_n$. Whenever the spatial twist in $\Phi(x)$ fits to the nth mode of the standing wave, there is a strong resonant nonlinear interaction that phase-locks the Josephson oscillation at f_n, giving Fiske steps at voltages V_{FSn}.

Until now we have considered only JJs with small dimensions compared with the so-called Josephson penetration length λ_J, which is of the order of 10 µm. Longer tunnel junctions are well modeled by the perturbed sine-Gordon (SG) equation. If we consider a linear one-dimensional (1-d, x-direction) junction with dc current bias, the SG equation in normalized units reads

$$\phi_{xx} - \phi_{tt} = \sin\phi + \alpha\phi_t - \eta, \tag{4}$$

where the normalized magnetic field $\kappa_{1,2}$ (perpendicular to the x-direction) enters as the boundary condition

$$\phi_x(0,t) = \kappa_1 \text{ and } \phi_x(l,t) = \kappa_2, \tag{5}$$

specifying the magnetic field at the two ends of the junction. The normalized bias current η here is assumed to be evenly distributed along the junction (overlap geometry). In the above equations, time t is normalized to the inverse maximum plasma frequency, ω_0, length x to λ_J, currents to the maximum critical current, I_c, and magnetic fields to the critical field, H_c, needed to force the first fluxon into the junction. Magnetic fields can only enter the junction in the form of fluxons, which are individual soliton-like localized 2π phase shifts each containing one flux quantum Φ_0.

Many solutions to the nonlinear SG equation have been found by numerical integration. For zero magnetic field and low values of the damping coefficient α, fluxons oscillate resonantly back and forth inside the junction driven by the Lorentz-like force from the bias current. This gives rise to a 4π phase shift per period and leads to the so-called zero-field steps, which are nearly constant-voltage steps located at voltages $V_{ZFSn} \approx n\Phi_0 c_s L$, $n = 1, 2, ..$ in the I–V curve. Note that zero-field step voltages are twice as large as the voltages of the Fiske steps discussed above for short junctions.

For every fluxon collision at $x = l$ (where l is the normalized junction length), a small electromagnetic field is emitted and the junction may be used as a microwave oscillator when biased on either a Fiske step or a zero-field step, but applications of this resonant soliton oscillator (RSO) are rather limited. First, the resonance frequencies are fixed (given by L and c_s) and the tunability is very small (steep steps). Second, the emitted power is relatively low and has high harmonic content (fluxon collision delta-function-like in time domain). However, due to its frequency stability and rather narrow linewidth, the RSO may be used as an on-chip clock oscillator.

When a stronger external magnetic field is applied (e.g., from a dc current, I_{CL}, in an overlaying control line), the junction will contain many fluxons, and since the fluxons repel each other, they will form an equidistant chain. Under a very strong field, the fluxons are forced so close to each other that the phase gradient becomes nearly uniform and proportional to the field strength. (One may compare the phase variation to a household corkscrew where the pitch is given by the applied magnetic field.) If a dc bias current I_B (uniformly distributed along the x-direction) is also applied to the junction, the fluxon chain is forced to move with constant velocity along the junction leading to a unidirectional flux flow through the junction. In our analogy, the bias current forces the corkscrew to rotate with constant angular velocity. The total phase shift per second and, thus, the frequency of the microwave signal emitted at $x = l$ can therefore be adjusted independently by the dc bias current (rotation) and the dc magnetic field (pitch). In the I–V curve, one observes a flux flow step (FFS) where the voltage, V_{FFS}, depends on both I_B and I_{CL}. Because the pitch for a dense chain of fluxons increases proportional to the magnetic field, V_{FFS}, the oscillator frequency in this limit depends linearly on I_{CL} for fixed I_B.

This explains the dynamics of the important flux flow oscillator (FFO). Its easy tunability not only permits wide band frequency coverage but also allows for accurate phase locking of the FFO. The power emitted from the FFO depends in a complicated way on the junction parameters as well as on I_B and I_{CL}, but with appropriate microwave design the power is sufficient to pump an SIS-mixer placed on the same chip. In 2002, a fully superconducting integrated receiver (SIR) with FFO, stripline circuit, SIS-mixer, and antenna (all placed on a $5 \times 5\,\text{mm}^2$ chip) has been operated in phase-locked mode up to 712 GHz with a frequency resolution of less than 1 Hz relative to the reference oscillator used in the phase-locking loop. At 500 GHz the noise temperature was less than 100 K, just above the quantum limit. The low-noise in combination with the high-frequency resolution in the submillimeter frequency range is very promising for spectral investigations in astronomy, chemistry, and biophysics.

Numerical simulations of the SG equation have confirmed that the unidirectional flux flow mode can also be sustained for very large values of the damping parameter α. Often the damping per normalized length $\alpha l > \pi$ is used to define the flux flow range. For low values of αl the FFS consists of a distinct Fraunhöfer pattern of Fiske steps localized in the vicinity of the average FFS voltage given by I_B and I_{CL}. The correct boundary conditions, the geometry, and especially the paths along which the two bias currents are supplied to the junction are very important for the free-running linewidth of the FFO and thus for the practical implementation of the SIR.

For simplicity, we have restricted ourselves to a single Josephson junction with linear 1-d geometry. Considerable work has been devoted to 1- and 2-d arrays of both short and long junctions, and vertically stacked junctions (*See* **Josephson junction arrays**; **Long Josephson junctions**). Many structures and circuits have been fabricated in both low-T_c and high-T_c superconductors. The 1-d long Josephson junction with annular geometry is of considerable interest, not least because the cyclic boundary conditions are combined with flux quantization and flux trapping. Whole families of *single flux quantum* (SFQ) and *rapid single flux quantum* (RSFQ) electronics based on propagation of single fluxons in superconducting circuits containing Josephson junctions and SQUIDs (*See* **Superconducting quantum interference device**) have been developed and partially tested for applications. Josephson junction science and technology now is mature, and many applications are waiting to be implemented in future electronics.

JESPER MYGIND

See also **Diodes; Hysteresis; Josephson junction arrays; Long Josephson junctions; Parametric amplification; Period doubling; Sine-Gordon equation; Solitons; Superconducting quantum interference device; Superconductivity**

Further Reading

Duzer, T. Van & Turner, C.W. 1998. *Principles of Superconductive Devices and Circuits*, 2nd edition, Upper Saddle River, NJ: Prentice-Hall, pp. 158–325

Kadin, A.M. 1999. *Introduction to Superconducting Circuits*, New York: Wiley, pp. 178–340

Orlando, T.P. & Delin, K.A. 1990. *Foundations of Applied Superconductivity*, Reading, MA: Addison-Wesley, pp. 393–486

JOST FUNCTIONS

See **Inverse scattering method or transform**

JULIA SETS

See **Fractals**

JUMP PHENOMENA

Jump phenomena or surfaces of discontinuity occur when a field or gradient of a field exhibits finite jumps (discontinuities) due to the system changing nature (such as a liquid-gas interface) or the geometry of the domain changing abruptly in a given hyperplane (such as an acoustical tube with an abrupt widening). A particularly dramatic example of a hydraulic jump (see Figure 1) appeared regularly on the river Seine early in the 20th century.

In nature, systems are often governed by conservation laws of the various quantities. For example, for a fluid, we can write equations for the conservation of mass, momentum, and energy, and in an electrical system, current is conserved. There may be a hierarchy of these laws so when the system is perturbed the lower order ones are approximately maintained while the higher order ones are clearly broken.

Following the discussion of Landau (Landau & Lifschitz, 1959), one can classify discontinuities in two main groups. Let us illustrate this with the example of a continuous medium for which mass flux, momentum flux, and energy flux are continuous across a discontinuity. In the first case, mass flux is zero across the discontinuity, implying that the tangential derivative of the velocity has a jump. This is called a tangential discontinuity, a familiar example being the ocean surface separating water circulation from wind circulation. In the other case, called a shock wave (*See* **Shock Waves**), mass flow is nonzero, implying a jump in the velocity normal to the discontinuity. The existence of such discontinuous fields is strongly connected to the hyperbolic nature of the system of equations for which wave fronts can propagate along definite directions (characteristics).

In reality, the velocity normal to the interface is not discontinuous but varies on a very small scale due to viscosity and damping which are not taken into account in the original model. In particular, energy is dissipated in such an event so that the energy flux is not conserved. A way to remove this difficulty is to assume a given dissipation and modify the energy jump condition. It is remarkable that such modified jump conditions can describe accurately the state of a system where dissipation occurs even without an accurate description of the (microscopic) dissipation mechanism.

Examples

Example (i)

As an example of a normal discontinuity (shock wave), consider a one-dimensional compressible fluid flow. The conservation of mass and momentum of the fluid can be written as

$$\rho_t + (\rho v)_x = 0, \tag{1}$$

$$v_t + v v_x + \frac{c^2}{\rho} h_x = 0, \tag{2}$$

where the subscripts indicate partial derivatives, ρ is the gas density, v the velocity along the x axis, and $c(\rho)$ is the velocity of sound.

In 1860, Bernard Riemann obtained a solution of this system by assuming a single dependence of the two fields $\rho(\alpha(x,t))$ and $v(\alpha(x,t))$ on an unknown function α. He obtained the compatibility condition linking ρ and v:

$$v = \pm \int \frac{c(\rho)}{\rho} \mathrm{d}\rho, \tag{3}$$

and showed that α should satisfy the one-dimensional partial differential equation

$$\alpha_t + (v \pm c)\alpha_x = 0, \tag{4}$$

Figure 1. A tidal bore (or *mascaret*) on the Seine at Quillebeuf, France. The site is 19 km from the mouth of the river, where tides up to 10 m occur. Due to dredging in the 1960s, this striking nonlinear phenomenon has disappeared. (Courtesy of J.J. Malandain.)

whose formal solution is $\alpha = f(x - (v \pm c)t)$. This indicates that given values of v and ρ move to the right along the characteristic lines $x = (v \pm c)t$ at speeds $v \pm c$, allowing shock waves across which the velocity and density fields are discontinuous. Consequently, an initial condition with $v > c$ will only propagate to the right, whereas if $v < c$, it will propagate over the whole domain.

Consider how to derive jump conditions; from (1). Assume $c^2 = \rho g$ where g is a constant; this is not physical in the context of gases but it simplifies the derivation. The conservation of the mass flux is obtained by integrating the first equation on a small domain $-\varepsilon < x < \varepsilon$ centered on $x = 0$; thus,

$$\int_{-\varepsilon}^{\varepsilon} \rho_t \, dx + [\rho v]_{x=-\varepsilon}^{x=\varepsilon} = 0. \qquad (5)$$

Assuming that ρ_t has a finite discontinuity at $x = 0$, we obtain in the limit $\varepsilon = 0$

$$[v\rho] = 0\,. \qquad (6)$$

The momentum conservation law is obtained by examining the time evolution of the average momentum ρv:

$$(\rho v)_t + \left(\rho v^2 + \frac{g\rho^2}{2}\right)_x = 0. \qquad (7)$$

The last conservation law involves the total energy e of the fluid, which is the sum of the kinetic and potential energy

$$(\rho v^2 + g\rho^2)_t + (\rho v^3 + 2\rho^2 v g)_x = 0. \qquad (8)$$

Proceeding in a similar way as for the mass flux, we obtain jump conditions for the momentum and energy of the fluid so the entire set of jump conditions are

$$[v\rho] = 0, \left[\rho v^2 + \frac{g\rho^2}{2}\right] = 0,$$
$$[v^3\rho + 2g\rho^2 v] = 0. \qquad (9)$$

Example (ii)

The second kind of discontinuity occurs when considering compression waves in an inhomogeneous beam made of two beams of different materials soldered at some point. The equation describing the displacement u is the one-dimensional generalized wave equation

$$\rho\sigma u_{tt} - (\sigma E u_x)_x = 0, \qquad (10)$$

where σ is the beam section, ρ its density, and E the Young modulus such that $E = E_\mathrm{l}$ for $x < 0$ and $E = E_\mathrm{r}$ for $x > 0$. To obtain the jump condition at $x = 0$, one integrates equation (10) over a small domain $-\varepsilon < x < \varepsilon$ to obtain

$$\int_{-\varepsilon}^{\varepsilon} dx \rho\sigma u_{tt} - [\sigma E u_x]_{x=-\varepsilon}^{x=\varepsilon} = 0\,. \qquad (11)$$

Now consider the limit $\varepsilon = 0$, assuming the displacement u to be continuous across $x = 0$. The first term is the acceleration which should tend to zero. The second term is the stress, and relation (11) shows that it too should be continuous. At $x = 0$, we then have the following relations:

$$u_l = u_\mathrm{r}, \quad E_\mathrm{l} u_x|_\mathrm{l} = E_\mathrm{r} u_x|_\mathrm{r}, \qquad (12)$$

so the gradient u_x exhibits a finite jump.

Hydraulic Jumps

Shallow water flows for which the depth h_0 is much smaller than the length l are governed by the following equations for the water elevation h and velocity v in the direction of propagation (Landau & Lifschitz, 1959)

$$h_t + (hv)_x = 0, \qquad (13)$$
$$v_t + v v_x = -g h_x, \qquad (14)$$

which are the same as (1) if we write $h = \rho$ and assume $c^2 = \rho g = gh$.

Consider a so-called hydraulic jump solution occurring, for example, when a dam breaks. The conservation of the mass flux, momentum flux, and energy flux are given by (9). The first two conditions imply the following relations between the fields indexed l (left) or r (right) on each side of the jump.

$$h_\mathrm{l} v_\mathrm{l} = h_\mathrm{r} v_\mathrm{r} \equiv j, \;\; h_\mathrm{l} v_\mathrm{l}^2 + \frac{g h_\mathrm{l}^2}{2} = h_\mathrm{r} v_\mathrm{r}^2 + \frac{g h_\mathrm{r}^2}{2}, \qquad (15)$$

where we introduced the (continuous) mass flux j. From these we deduce the velocities $v_{\mathrm{l,r}}$:

$$v_\mathrm{l}^2 = \frac{g}{2} h_\mathrm{r}\left(1 + \frac{h_\mathrm{r}}{h_\mathrm{l}}\right), \quad v_\mathrm{r}^2 = \frac{g}{2} h_\mathrm{l}\left(1 + \frac{h_\mathrm{l}}{h_\mathrm{r}}\right). \qquad (16)$$

To gain further insight consider the energy flux across the hydraulic jump. From the original equations (13) this should be conserved; however (as we discussed above), discontinuities are in general not physical, indicating that the model should be completed by some dissipation process. Without considering the details of this, we just write that energy per unit time (power) is being taken out of the system at the jump. Integrating as above, the energy conservation equation (8), from $-\varepsilon$ to ε and passing to the limit, gives the power dissipation at the jump,

$$w_{\mathrm{lr}} \equiv \lim_{\varepsilon \to 0} \int_{-\varepsilon}^{\varepsilon} e_t \, dx = -[v^3 h + 2gh^2 v]_{\mathrm{l}}^{\mathrm{r}}. \quad (17)$$

Using (16) we obtain the final result for the power dissipation at the jump,

$$w_{\mathrm{lr}} = jg \frac{(h_{\mathrm{l}} - h_{\mathrm{r}})^3}{2h_{\mathrm{l}} h_{\mathrm{r}}}. \quad (18)$$

Assuming that energy is absorbed at the jump so $w_{\mathrm{lr}} < 0$, we get from (18) $h_{\mathrm{l}} < h_{\mathrm{r}}$, and from (16)

$$v_{\mathrm{l}}^2 = gh_{\mathrm{l}} \frac{h_{\mathrm{r}}}{h_{\mathrm{l}}} \frac{h_{\mathrm{r}} + h_{\mathrm{l}}}{2h_{\mathrm{l}}} > gh_{\mathrm{l}}$$

so that $v_{\mathrm{l}} > c_{\mathrm{l}}$, the local wave speed while $v_{\mathrm{r}} < c_{\mathrm{r}}$. These inequalities indicate the stability of the hydraulic jump.

Numerical Problems and Applications

As seen above, jump conditions come from conservation laws so an accurate numerical scheme should satisfy the conservation laws. Because the jump conditions are flux conditions, it is natural to use methods that involve integrating the operator over the spatial domain to satisfy them. One way to do this is the so-called finite volume approach, where the equation is integrated over reference volumes yielding naturally the flux conditions. Another approach is to use finite element methods, where the solution is decomposed on a basis of trial functions. The relation that gives the coefficients is obtained by integrating the equation over the whole domain.

For shock waves it is essential to have numerical schemes that do not artificially smooth the shock but respect the jump conditions. Here, special staggered finite difference schemes have been suggested by Lax, Wendroff, and Godunov.

As a final application, we present the case of a wave guide whose section changes abruptly at some point. This occurs in neurons for which varicosities are known to block the propagation of nerve impulses (Scott, 2003). To simplify the discussion, let us consider the following sine-Gordon nonlinear wave equation

$$\phi_{tt} - \phi_{xx} - \phi_{yy} + \sin(\phi) = 0, \quad (19)$$

propagating in a two-dimensional T-shaped domain as shown in Figure 2 and assuming a zero normal gradient on the boundary. This describes the electrodynamics of a Josephson junction superconducting device obtained by inserting a small oxide layer between two metallic plates (Scott, 2003).

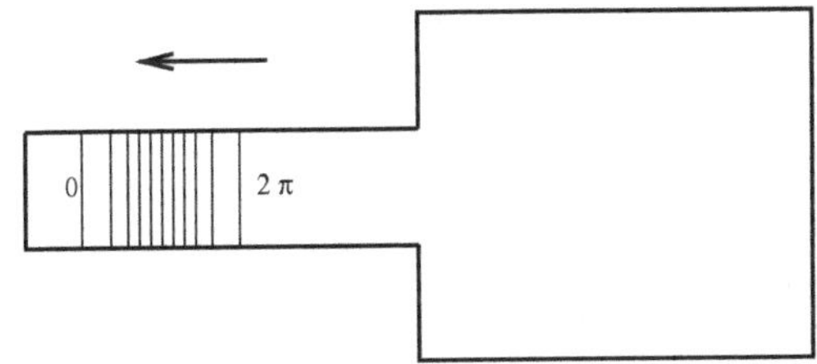

Figure 2. A T-shaped Josephson junction. The kink solution (21) is shown in the left region as contour lines. Kink propagation is easy from right to left but is inhibited in the other direction.

First note that the total energy of the system

$$E = \int dx\, dy \frac{1}{2} [\phi_t^2 + \phi_x^2 + \phi_y^2 + 2(1 - \cos(\phi))] \quad (20)$$

is a constant in time. This can be seen by multiplying (19) by ϕ_t and integrating over the domain using the boundary conditions. Consider the propagation of a front (kink)

$$\phi(x, t) = 4 \arctan \left(\exp \left(\frac{x - vt}{\sqrt{1 - v^2}} \right) \right) \quad (21)$$

which is an exact solution of the one-dimensional sine-Gordon equation for which ϕ is independent of y. Inserting (21) into (20) and calculating the integrals, it can be shown that the energy is

$$E_k = 8 \frac{w}{1 - v^2}, \quad (22)$$

where w is the width in the y direction of the domain.

Let us now consider the propagation of such a front in the domain and assume it starts at $t = 0$ from the right-hand side with a negative velocity so that it will encounter the discontinuity. First, notice that because of the boundary conditions, it travels unaffected in the straight section. When it hits the discontinuity, it will cross into the small section with no apparent change in shape apart from reflected waves (see Figure 2). On the contrary, if the kink is started from the left side whose section is narrow, it will get reflected by the discontinuity unless its velocity is quite large.

The explanation of these observations lies in the conservation of the energy of the system, assuming that most of this energy is in the front. For a kink to exist in a domain of lateral extension w, its energy should be greater than $E_k(v = 0) = 8w$. When the kink is started in the right domain, its energy $E/8 = w_{\mathrm{r}}/(1 - v^2) > w_{\mathrm{r}} > w_{\mathrm{l}}$, so we expect it to cross without trouble into the smaller section. Indeed it should gain speed, and it does. On the contrary, if the kink is started in the left domain, its speed v_{l} must be very large (so that $w_{\mathrm{l}}/(1 - v_{\mathrm{l}}^2) > w_{\mathrm{r}}$ where $8w_{\mathrm{r}}$ is the energy of a static kink in the region on the right) in order to cross into the region on the right. This provides a means to realize a rectifying gate, through which signals pass in only one direction.

JEAN GUY CAPUTO

See also **Constants of motion and conservation laws; Long Josephson junctions; Shock waves**

Further Reading

Ames, W. 1992. *Numerical Methods for Partial Differential Equations*, 3rd edition, New York: Academic Press

Caputo, J.G. & Stepanyants, Y. 2003. Bore formation, evolution and disintegration into solitons in shallow inhomogeneous channels. *Nonlinear Processes in Geophysics*, 10: 407–424

Landau, L. & Lifchitz, E. 1959. *Fluid Mechanics*, vol. 6 of *Course in Theoretical Physics*, London: Pergamon Press and Reading, MA: Addison-Wesley

LeVeque, R. 2002. *Finite Volume Methods for Hyperbolic Problems*, Cambridge and New York: Cambridge University Press

Lin, C.C., Segel, L.A. & Handelman, G.H. 1995. *Mathematics Applied to Deterministic Problems in the Natural Sciences*, 4th edition, Philadelphia: SIAM

Scott, A.C. 2003. *Nonlinear Science: Emergence and Dynamics of Coherent Structures*, 2nd edition, Oxford and New York: Oxford University Press

JUPITER'S GREAT RED SPOT

Jupiter's Great Red Spot is a large swirling cloud mass of reddish-brown appearance (see figure in color plate section). Situated in Jupiter's southern hemisphere, it straddles the south tropical zone and, to the north of this, the south equatorial belt. The Great Red Spot (GRS) is roughly elliptical in shape, with the semi-major axis zonally aligned (east-west) and with dimensions approximately 22, 000 km (twice the diameter of the Earth) by 11, 000 km. The atmospheric motions associated with the GRS are visible in the cloud layer near the tropopause. It is generally agreed to be a vortex (Mitchell et al., 1981); and Smith et al. (1979a) give an estimate of the vorticity. This vortex is anticyclonic (rotating in the opposite sense to that induced by the planetary rotation), that is, anticlockwise, but with a weakly counter-rotating, or possibly quiescent inner region. The GRS is at high pressure and low temperature relative to its surroundings. A striking feature associated with the GRS is the turbulent oscillating cloud system to the northwest.

In 1664, Robert Hooke observed a small dark spot in one of the southern belts of Jupiter, reported in Hooke (1665). This is considered by some contemporary authors to be an early manifestation of the GRS (see also Rogers, 1995). By observation of this feature, Hooke was able to make a good estimate of Jupiter's rate of rotation. Many further observations of this and other features of Jupiter were made by Hooke and by Giovanni Cassini over the next five years (e.g., Cassini, 1672), and intermittently by others over the next 50 years (see Denning, 1899). However, it was not until 1831 that Samuel H. Schwabe identified what can be clearly recognized as a feature called the Hollow, in his drawing reproduced in Rogers (1995). As the Hollow is clearly apparent above the GRS in contemporary observations, this is taken to imply the presence of the GRS. Continuous observations began in 1872 (Peek, 1958) and from these it appears that the GRS was considerably larger in the late 1800s than it is now (Beebe & Youngblood, 1979; Rogers, 1995). Very much more detailed observations became available once space probes reached Jupiter; first was Pioneer 10 in 1973, then Pioneer 11 in 1974, Voyagers 1 and 2 in 1979, Galileo in 1996, and then most recently Cassini from October 1, 2000 through until March 22, 2001. The Voyager and Galileo missions (Smith et al., 1979a, b; Vasavada et al., 1998) provided very high resolution still images, while Cassini (Porco et al., 2003) gave more dynamic information. From 1994, after corrections were made to faulty optics by the Endeavour crew, the Hubble telescope has been providing good images of Jupiter and the GRS. These, along with the Cassini sequences, clearly show vortex interactions between the GRS and smaller, intermittent vortices, thus allowing a much better understanding of the underlying dynamical processes (Morales-Juberías et al., 2002).

Attempts have been made since 1961 to model the GRS and to understand the processes giving rise to the motions observed, beginning with Hide (1961), who suggested that the GRS was the visible manifestation of a Taylor column. The quantity of relevance to these calculations is the potential vorticity q. As the atmosphere of Jupiter is stably stratified and there are significant rotational effects at the scale of the GRS, the most commonly used model is the quasi-geostrophic approximation (a simplification of the complete atmospheric dynamic equations, giving a balance of horizontal pressure gradient forces and horizontal Coriolis forces due to Jupiter's rotation). Here q is defined as

$$q = f + \nabla^2 \psi + \frac{1}{\rho_0} \frac{\partial}{\partial z} \left(\rho_0 \frac{f_0^2}{N^2} \frac{\partial \psi}{\partial z} \right), \qquad (1)$$

where $\psi(x, y, z, t)$ is the stream function for the flow and (x, y, z) are local cartesian coordinates, with x and y being the zonal and meridional directions, respectively, and z being a measure of the depth. The basic state density profile is $\rho_0(z)$ and the effects of Jupiter's rotation are introduced by the beta-plane approximation, the assumption that we can treat part of Jupiter as a plane, over which the Coriolis parameter f varies linearly: $f = f_0 + \beta y$, where f_0 is the reference value at the latitude of the GRS and β is the ambient potential vorticity gradient. The buoyancy frequency $N(z)$, the frequency with which a parcel of fluid displaced from equilibrium will oscillate, is determined by the vertical temperature variation. The potential vorticity is materially conserved, according to

$$\frac{\partial q}{\partial t} + \boldsymbol{v} \cdot \nabla q = 0, \qquad (2)$$

where the horizontal velocity $\boldsymbol{v} = (-\partial\psi/\partial y, \partial\psi/\partial x)$.

Many researchers have modeled the vortical motions associated with the GRS by assuming a thin upper layer of constant density (of order 100 km; Dowling & Ingersoll, 1989) above a much deeper lower layer of slightly higher density in which there is a uniform steady zonal flow, for example, Ingersoll and Cuong (1981), Marcus (1988), and Marcus et al. (1990). The surface of the upper layer is above the visible cloud level. These models assume barotropic motions, that is, uniform in depth with no horizontal component of vorticity, in the upper layer where isolated vortices are assumed to be generated by mechanisms such as zonal shear flow instability. Merger of vortices can lead to the formation of larger vortices, and the relevant issues are then the stability and lifetime of isolated vortices such as the GRS in the observed zonal shear. A statistical mechanics approach to such equilibria is provided in Michel and Robert (1994).

There have also been discussions on the role of vertical density stratification, allowing baroclinic instability, but generally limited to the quasi-geostrophic case of (1) and (2). For example, Achterberg and Ingersoll (1994) studied the barotropic and the first two baroclinic modes, and Cho et al. (2001) used as many as 60 vertical modes. The high horizontal resolution used in Cho et al. (2001) allowed the authors to generate features such as the counter-rotating core (Vasavada et al., 1998, Figure 7, reporting Galileo data and similar Voyager data), but as in all other studies to date, the turbulent cloud system to the northwest of the spot does not appear.

A deficiency of the quasigeostrophic model is that vortices with radius significantly greater than the Rossby radius of deformation, L_{R}, tend to decay by radiation of Rossby waves in the absence of any external forcing. (L_{R} is a measure of the scale at which Coriolis forces become significant.) This would imply a lifetime for the GRS much shorter than the observed period, which is certainly greater than 150 years. In addition, the quasigeostrophic equations have the disadvantage that neither anticyclonic nor cyclonic vortices are preferentially created or destroyed, whereas observations show that the GRS is anticyclonic, as are 90% of the other (typically much smaller) vortices observed on Jupiter (Mac Low & Ingersoll, 1986).

For two-layer models, the zonal flows in the bottom layer can be incorporated in a reduced gravity single-layer shallow water model with meridionally varying topography and a free surface, which is taken to be above the visible cloud layer (Dowling & Ingersoll, 1989). There are various levels of approximation to the shallow-water equations beyond the quasi-geostrophic approximation. The intermediate geostrophic (IG) approximation (Williams & Yamagata, 1984; Nezlin & Sutyrin, 1994) arises when there is significant nonlinearity giving rise to finite free-surface distortion, as is to be expected for vortices with a scale much greater than the Rossby radius, such as the GRS. A more complete discussion of the asymptotic treatment of the nonlinear terms, along with higher-order approximations, is given in Stegner and Zeitlin (1996).

An alternative approach to understanding the GRS as a solitary wave was first introduced by Maxworthy and Redekopp (1976a,b) and Redekopp (1977). They considered the quasigeostrophic equations (1) and (2) and were able to find elementary soliton solutions in the limit of small amplitude disturbance, which obeyed either the Korteweg–de Vries equation (KdV) or the modified KdV equation. These balance zonal shear and dispersion to give isolated disturbances that could be regarded as vortices. However, as noted by several authors (e.g., Ingersoll & Cuong, 1981), these soliton solutions of the KdV equation pass through one another unchanged in form (elastic collisions). This is contrary to the behavior of vortices generally, and more particularly the GRS, where merger with smaller vortices is observed (Morales-Juberías et al., 2002). However, Petviashvili (1980), by going beyond quasigeostrophy, was able to derive solitary axisymmetric vortex solutions, without an imposed zonal shear, for a curved shallow layer, with anticyclones living longer. Nezlin et al. (1990) refer to this anticyclonic vortex as a Rossby soliton, with weak nonlinearity balanced by Rossby wave dispersion. These solitary solutions undergo inelastic collisions and were later thought to be found experimentally in parabolic water tank experiments (Nezlin et al., 1990), although Nycander (1993) and Stegner and Zeitlin (1996) demonstrated that this weakly nonlinear balance could not in fact exist in the parabolic geometry and that stronger nonlinearity was required to explain the laboratory observations. Stegner and Zeitlin (1996) go on to show that axisymmetric soliton-like solutions of a 2-d KdV equation are found for spherical geometry, corresponding to the surface of Jupiter, using a nonlinear model. Furthermore, only an anticyclonic vortex can be supported in this case, which is in accord with the observed rotation of the GRS. However, this cannot explain the preferential formation of anticyclonic vortices smaller than the Rossby radius, where the quasigeostrophic approximation remains appropriate. Finally, as explained by Stegner and Zeitlin (1996), their 2-d KdV solutions do not apply directly to the GRS as neither the elliptical shape nor the counter-rotating inner core can be accounted for.

C. Macaskill and T.M. Schaerf

See also **Atmospheric and ocean sciences; Fluid dynamics; Korteweg–de Vries equation; Solitons**

Further Reading

Achterberg, R.K. & Ingersoll, A.P. 1994. Numerical simulation of baroclinic Jovian vortices. *Journal of the Atmospheric Sciences*, 51(4): 541–562

Beebe, R.F. & Youngblood, L.A. 1979. Pre-Voyager velocities, accelerations and shrinkage rates of Jovian cloud features. *Nature*, 280: 771–772

Cassini, G. 1672. A relation of the return of a great permanent spot in the planet Jupiter, observed by Signor Cassini, one of the Royal Parisian Academy of the Sciences. *Philosophical Transactions of the Royal Society of London*, 7: 4039–4042

Cho, J. Y-K., de la Torre Juárez, M., Ingersoll, A.P. & Dritschel, D.G. 2001. A high-resolution, three-dimensional model of Jupiter's Great Red Spot. *Journal of Geophysical Research—Planets*, 106(E3): 5099–5105

Denning, W.F. 1899. Early history of the Great Red Spot on Jupiter. *Monthly Notices of the Royal Astronomical Society*, 59: 574–584

Dowling, T.E. & Ingersoll, A.P. 1989. Jupiter's Great Red Spot as a shallow water system. *Journal of the Atmospheric Sciences*, 46(21): 3256–3278

Hide, R. 1961. Origin of Jupiter's Great Red Spot. *Nature*, 190: 895–896

Hooke, R. 1665. A spot in one of the belts of Jupiter. *Philosophical Transactions of the Royal Society of London*, 1: 3

Ingersoll, A.P & Cuong, P.G. 1981. Numerical model of long-lived Jovian vortices. *Journal of the Atmospheric Sciences*, 38(10): 2067–2076

Mac Low, M.-M. & Ingersoll, A.P. 1986. Merging of vortices in the atmosphere of Jupiter: an analysis of Voyager images. *Icarus*, 65: 353–369

Marcus, P.S. 1988. Numerical simulation of Jupiter's Great Red Spot. *Nature*, 331: 693–696

Marcus, P.S., Sommeria, J., Meyers, S.D. & Swinney, H.L. 1990. Models of the Great Red Spot. *Nature*, 343: 517–518

Maxworthy, T. & Redekopp, L.G. 1976a. A solitary wave theory of the Great Red Spot and other observed features in the Jovian atmosphere. *Icarus*, 29: 261–271

Maxworthy, T. & Redekopp, L.G. 1976b. New theory of the Great Red Spot from solitary waves in the Jovian atmosphere. *Nature*, 260: 509–511

Michel, J. & Robert, R. 1994. Statistical mechanical theory of the Great Red Spot of Jupiter. *Journal of Statistical Physics*, 77: 645–666

Mitchell, J.L., Beebe, R.F, Ingersoll, A.P. & Garneau, G.W. 1981. Flow fields within Jupiter's Great Spot and White Oval BC. *Journal of Geophysical Research—Space Physics*, 86(NA10): 8751–8757

Morales-Juberías, R., Sánchez-Lavega, A., Lecacheux, J. & Colas, F. 2002. A comparative study of Jovian anticyclone properties from a six-year (1994–2000) survey. *Icarus*, 157(1): 76–90

Nezlin, M.V., Rylov, A. Yu., Trubnikov, A.S. & Khutoreski, A.V. 1990. Cyclonic–anticyclonic asymmetry and a new soliton concept for Rossby vortices in the laboratory, oceans and the atmospheres of giant planets. *Geophysical and Astrophysical Fluid Dynamics*, 52: 211–247

Nezlin, M.V. & Sutyrin, G.G. 1994. Problems of simulation of large, long-lived vortices in the atmospheres of the giant planets (Jupiter, Saturn, Neptune). *Surveys in Geophysics*, 15: 63–99

Nycander, J. 1993. The difference between monopole vortices in planetary flows and laboratory experiments. *Journal of Fluid Mechanics*, 254: 561–577

Peek, B.M. 1958. *The Planet Jupiter*, London: Faber & Faber

Petviashvili, V.I. 1980. Red spot of Jupiter and the drift soliton in a plasma. *JETP Letters*, 32(11): 619–622

Porco, C.C., West, R.A., McEwen, A., et al. 2003. Cassini imaging of Jupiter's atmosphere, satellites and rings. *Science*, 299: 1541–1547

Rogers, J.H. 1995. *The Giant Planet Jupiter*, Cambridge and New York: University Press

Redekopp, L.G. 1977. On the theory of solitary Rossby waves. *Journal of Fluid Mechanics*, 82(4): 725–745

Smith, B.A., Soderblom, L.A., Johnson, T.V., et al. 1979a. The Jupiter system through the eyes of Voyager 1. *Science*, 204: 951–957 and 960–972

Smith, B.A., Soderblom, L.A., Beebe, R., et al. 1979b. The Galilean satellites and Jupiter: Voyager 2 imaging science results. *Science*, 206: 927–950

Stegner, A. & Zeitlin, V. 1996. Asymptotic expansions and monopolar solitary Rossby vortices in barotropic and two-layer models. *Geophysical and Astrophysical Fluid Dynamics*, 83: 159–194

Vasavada, A.R., Ingersoll, A.P., Banfield, D., et al. 1998. Galileo imaging of Jupiter's atmosphere: The Great Red Spot, equatorial region, and White Ovals. *Icarus*, 135(1): 265–275

Williams, G.P. & Yamagata, T. 1984. Geostrophic regimes, intermediate solitary vortices and Jovian eddies. *Journal of the Atmospheric Sciences*, 41(4): 453–478

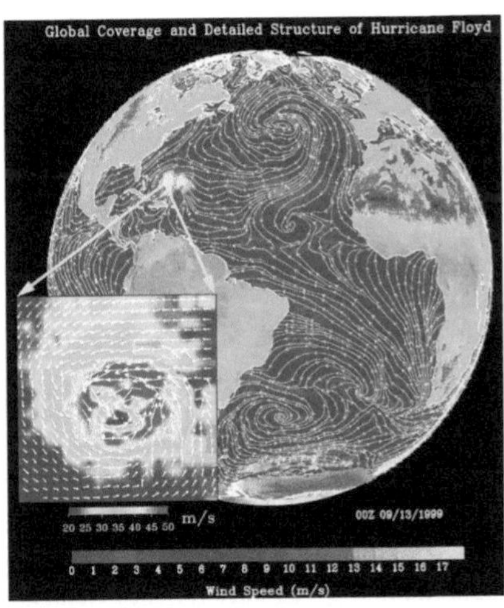

Synoptic view of surface winds over the Atlantic Ocean on September 13, 1999. The background colors indicate wind speeds and the white arrows show the direction of the wind. The high winds of Hurricane Floyd can be seen east of the Bahamas. The images are obtained from the NASA/NOAA sponsored data system Seaflux, at JPL through the courtesy of W. Timothy Liu and Wenqing Tang. (*See* **Atmospheric and ocean sciences; Hurricanes and tornadoes; Vortex dynamics in fluids**.)

Overhead view of Hurricane Andrew on 25 August 1992 at 20:20 Universal Time (UT). Different wavelengths are assigned to the red, green, and blue channels to make the hurricane appear red. Courtesy of National Oceanic and Atmospheric Administration. (*See* **Atmospheric and ocean sciences; Hurricanes and tornadoes; Vortex dynamics of fluids**.)

Aircraft wake vortex visualized by colored smoke. NASA's Wake Vortex Study at Wallops Island, 1990. Courtesy of NASA Langley Research Center. (*See* **Fluid dynamics; Vortex dynamics in fluids.**)

The planet Jupiter showing the Great Red spot, photographed by the NASA Cassini satellite on December 29, 2000. Courtesy NASA/JPL/ Space Science Institute. (*See* **Fluid dynamics; Jupiter's Great Red spot; Vortex dynamics of fluids.**)

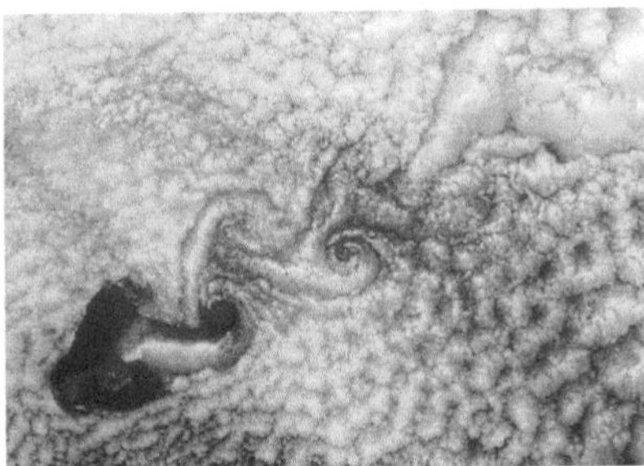

An atmospheric flow pattern known as the von Karman vortex street—a linear chain of spiral eddies. The vortex pattern is made visible by the marine stratocumulus clouds around Guadalupe Island, which is about 22 miles long and lies about 250 miles southwest of San Diego. Multi-angle Imaging Spectroradiometer (MISR) images from June 11, 2000: NASA/ GSFC/JPL, MISR Team. (*See* **Fluid dynamics; Turbulence; Vortex dynamics of fluids**.)

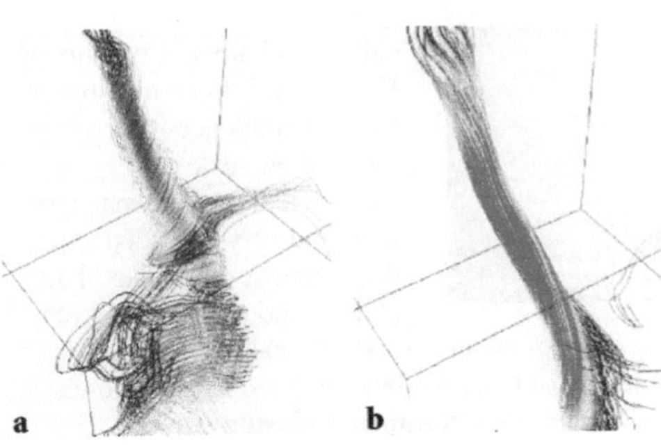

Close-up view of coherent vortex structure from direct numerical simulation of homogeneous isotropic turbulence: (a) streamlines, and (b) vortex lines—color-coded according to intensity—forming a tangle of space curves in a reference box (tropicity domain). Courtesy of Z.-S. She, E. Jackson and S.A. Orszag, Princeton University, Princeton, NJ, USA. (*See* **Structural complexity; Turbulence**.)

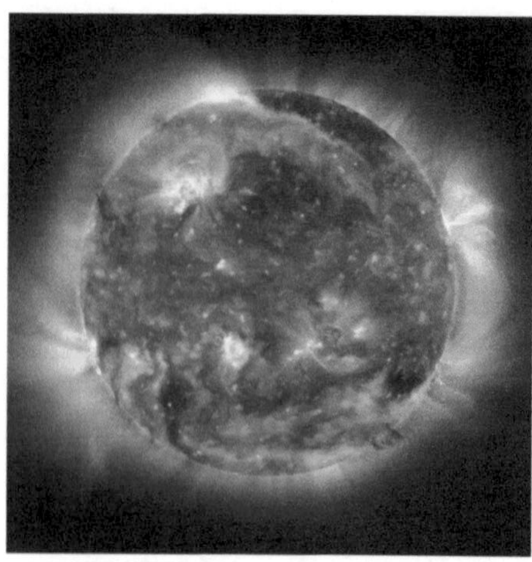

The Sun, imaged in three color-coded wavelengths (171, 195 and 284 angstrom), showing solar flares (sun spots) comprising coronal mass ejections which launch solar winds. Courtesy of SOHO/Extreme Ultraviolet Imaging Telescope (EIT) consortium. SOHO is a project of international cooperation between ESA and NASA. (*See* **Alfvén waves; Nonlinear plasma waves**.)

Pumice and ash erupting from Mount St. Helens on May 18, 1980 is a powerful example of turbulent dispersion of small particles. The plume eventually reached 12 to 15 miles (20–25 km) above sea level. Photograph by Donald A. Swanson, USGS/CVO. (*See* **Evaporation waves; Plume dynamics**.)

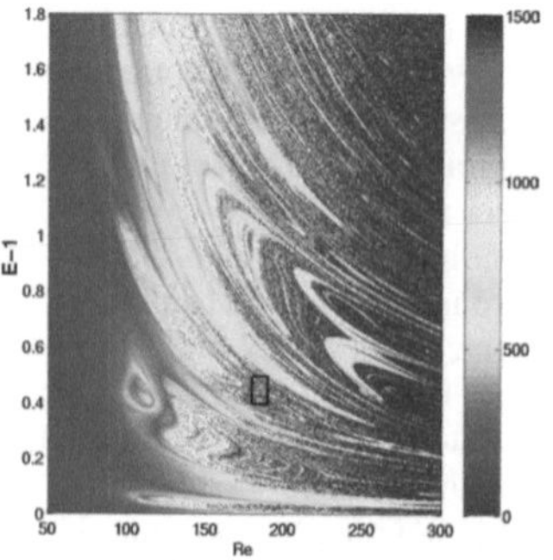

Lifetimes of perturbations in a planar shear flow for different values of initial energy (E) and Reynolds number (Re). In the transition region the lifetimes show a sensitive dependence on initial conditions and huge fluctuations. Similar behavior has been observed in plane Couette flow and in pipe flow. (*See* **Fluid dynamics; Shear flow**.)

A slice from a three-dimensional numerical simulation of turbulent flow around a square cylinder (visible at the bottom of the image). The data is visualized using the spot noise technique to show flow structure, color shows the pressure. Simulation by R. Verstappen and A. Veldman of the University of Groningen in the Netherlands, and visualization by W. de Leeuw and R. van Liere of the Center of Mathematics and Computer Science in Amsterdam. (*See* **Fluid dynamics; Turbulence**.)

Red and green colors predominate in this view of the Aurora Australis photographed from the Space Shuttle in May 1991. Long a source of myth, auroras are known from late nineteenth-century studies by Kristian Birkeland to be driven by solar wind and stabilized by Earth's magnetosphere. Photograph by the STS-39 Crew, courtesy of Earth Sciences and Image Analysis Laboratory, NASA Johnson Space Center. (*See* **Alfvén waves; Nonlinear plasma waves**.)

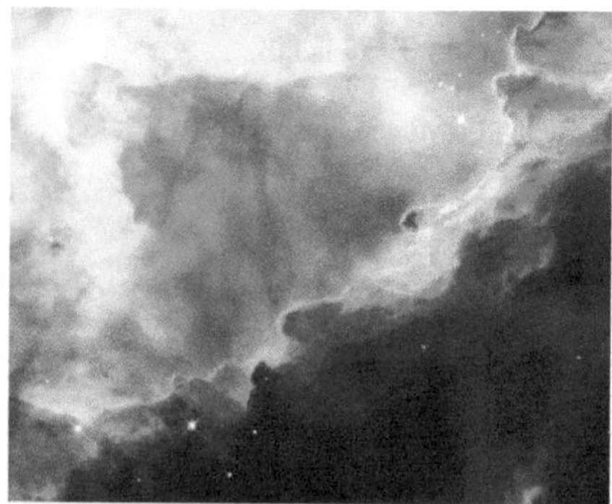

Turbulent clouds of hydrogen gas mix with small amounts of oxygen, sulfur, and other elements in this Hubble Space Telescope image of a region of M17, a nebula about 5,500 light-years away that is giving birth to new stars. Ultraviolet energy from hot, young stars in the nebula causes the gas to glow, and the intense heat and pressure cause turbulence. The different colors represent different gases. Courtesy of NASA/ESA/J. Hester (ASU). (*See* **Galaxies**.)

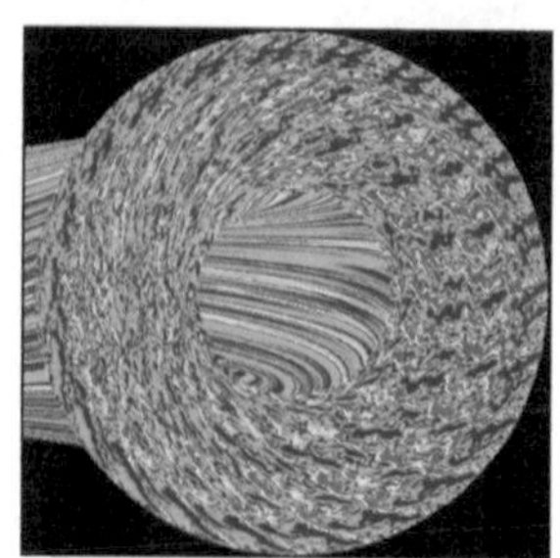

Snapshot of the results of a simulation for a toroidal magnetic confinement system called a tokamak. Contours of $\mathbf{E}\times\mathbf{B}$ reveal ballooning turbulence. Turbulence is on a relatively short scale perpendicular to the magnetic field and very extended along the magnetic field lines. The eddies tend to point preferentially in the direction of positive (bad) magnetic curvature. Photograph courtesy of Gary Kerbel, National Energy Research Supercomputer Center. (*See* **Nonlinear plasma waves**.)

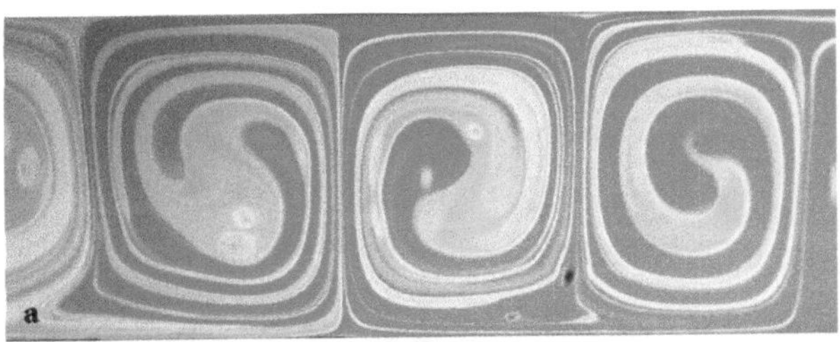

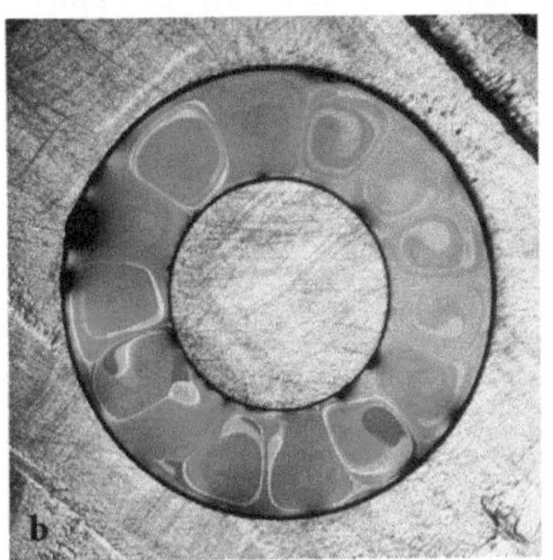

Electrically driven convection in a smectic (the most ordered phase) liquid crystal film. (a) A nonuniformly thick film is freely suspended between two wires about 1mm apart. The film itself is less than 1 micron thick, and shows bright interference colors under reflected white light, like a soap bubble. The film convects when a DC voltage of a few volts is applied to the wires. The vortex flow is visualized as it advects the thickness variations. The driving mechanism involves the field coupling to the surface charge which develops on the film's two free surfaces. This system forms a simple one-dimensional nonlinear pattern which has been quantitatively studied in films with perfectly uniform thickness, for which no color variation is visible. (b) As in the previous figure, now with annular electrodes and a radial driving force. One can also rotate the inner electrode and thus study convection superposed on a two-dimensional circular Couette flow. Images courtesy of Stephen W. Morris, University of Toronto. (*See* **Liquid crystals; Spatiotemporal chaos; Thermal convection**.)

Closed Rayleigh–Bénard convection cells in clouds over the South Atlantic Ocean. NASA/GSFC. (*See* **Atmospheric and ocean sciences; Fluid dynamics; Pattern formation; Thermal convection**.)

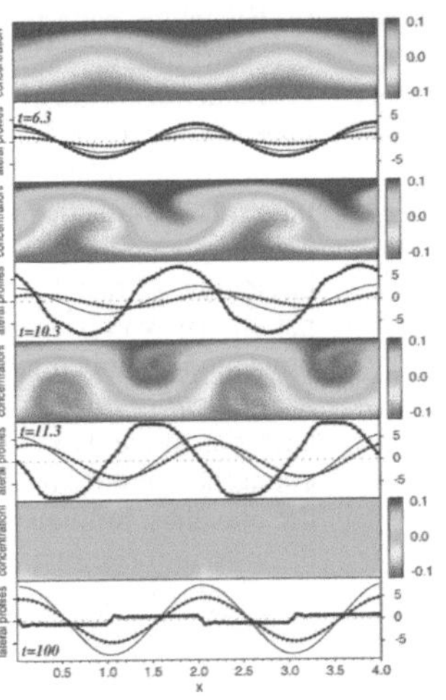

Evolution of oscillatory convection in a binary fluid layer heated from below. The concentration distribution in a vertical cross section of the fluid layer is displayed by color coded plots with blue and red denoting high and low concentration, respectively. Wave profiles at midheight are shown for the fields of vertical velocity (thin lines), temperature (lines with triangles), and concentration (lines with squares). Initially a standing wave grows ($t = 6.3$). It breaks at $t = 10.3$ and is thereby transformed into a fast traveling wave propagating to the left which then slows down. For better visibility two wavelengths are shown. (*See* **Fluid dynamics; Rayleigh–Taylor instability; Thermo-diffusion effects**.)

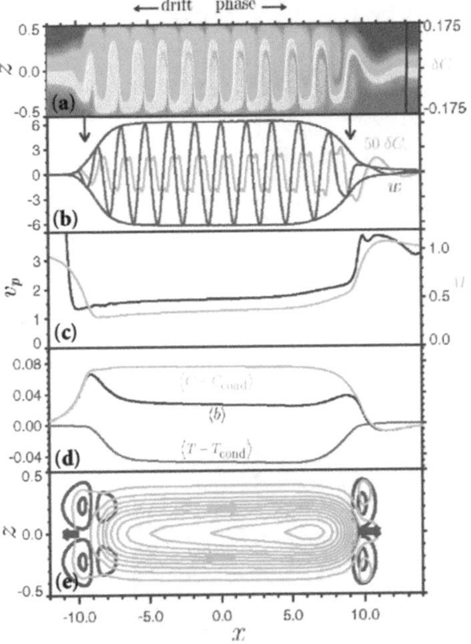

Spatially confined convection in binary fluid mixtures in the form of a localized traveling wave. (a) Concentration deviation (δC) from global mean (light green/yellow) in a vertical cross section of the layer. (b) Lateral wave profiles at mid-height, $z = 0$, of δC (green), vertical velocity w (blue), and its envelope. (c) Phase velocity v_p (black) and mixing number M (green) measuring rms deviations of concentration. (d) Time averaged deviations from the conductive state at $z = -0.25$ for concentration (green), temperature (red), and their sum ($\langle b \rangle$) measuring the convective contribution to the buoyancy. (e) Streamlines of time averaged concentration current $\langle \boldsymbol{J} \rangle$ (green) and velocity field $\langle \boldsymbol{u} \rangle$ (blue). Thick blue and green arrows indicate $\langle \boldsymbol{u} \rangle$ and transport of positive δC (alcohol surplus), respectively. Thus, in the lower half of the layer negative δC (water surplus) is transported to the right. (*See* **Fluid dynamics; Thermo-diffusion effects**.)

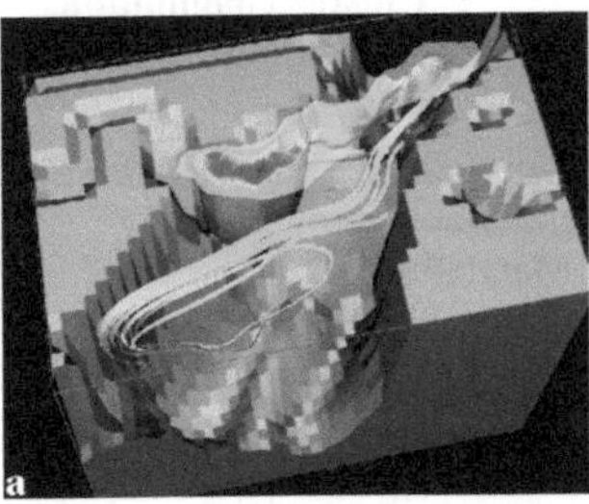

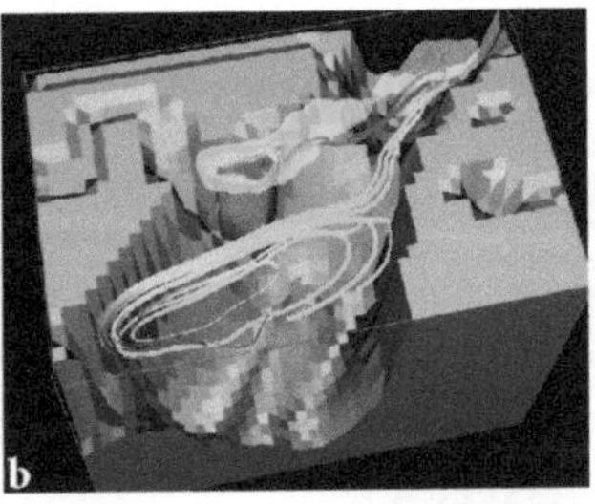

Numerical results from a coupled atmosphere-ocean model of the North Atlantic circulation including bottom topography (a) now and (b) the prediction for 100 years from now. Noticeable are the divergence of the Gulf Stream in the North Atlantic, reduction in the production of deep cold water masses near Greenland, and slowing of part of the global conveyor belt. Relevant streamlines of the general circulation are colored according to their temperature (red=warm, blue=cold). Courtesy of the Deutsches Klimarechenzentrum (Michael Böttinger, DKRZ, Hamburg, Germany). (*See* **Atmospheric and ocean sciences**.)

Snowflakes on a window pane. Ice crystals are now recognized as examples of nonlinear pattern formation. (*See* **Growth patterns; Pattern formation**.)

Self similarity of a fern frond. There is a limit to the range of scales at which the self similarity is maintained, and it occurs at only a few discrete scales. Courtesy of Paul Bourke. (*See* **Fractals; Turing patterns**.)

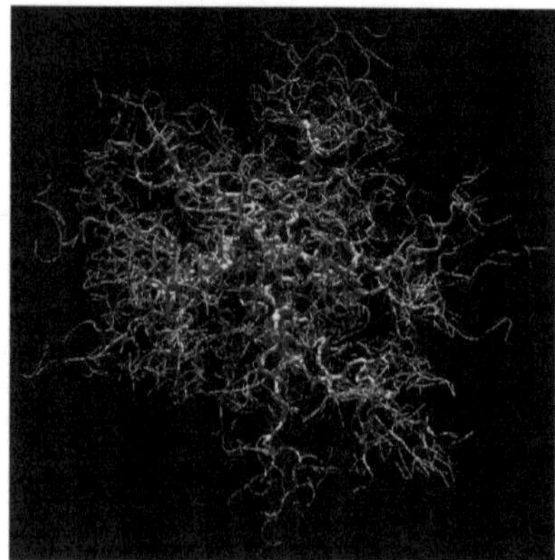

Computer simulation of growth in three dimensions by diffusion-limited aggregation. Such growth patterns are natural fractals. Courtesy of Paul Bourke. (*See* **Cluster coagulation; Fractals; Growth patterns**.)

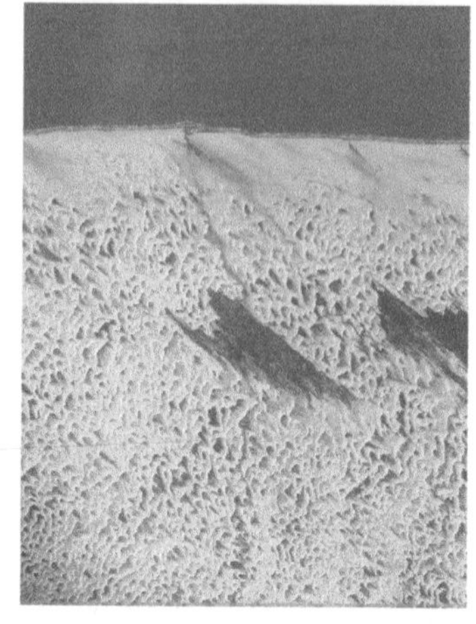

Dune fields with a regular pattern, Lençóis Maranhenses National Park, on Brazil's north coast. Image courtesy of Earth Sciences and Image Analysis Laboratory, NASA Johnson Space Center, ISS007-E-15177. (*See* **Dune Formation; Geomorphology and tectonics; Pattern formation**.)

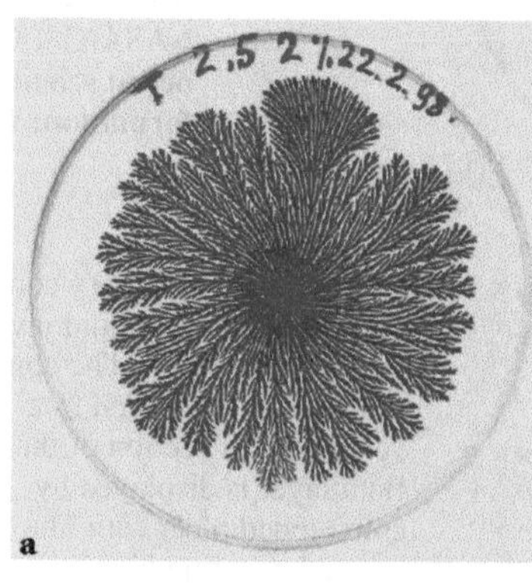

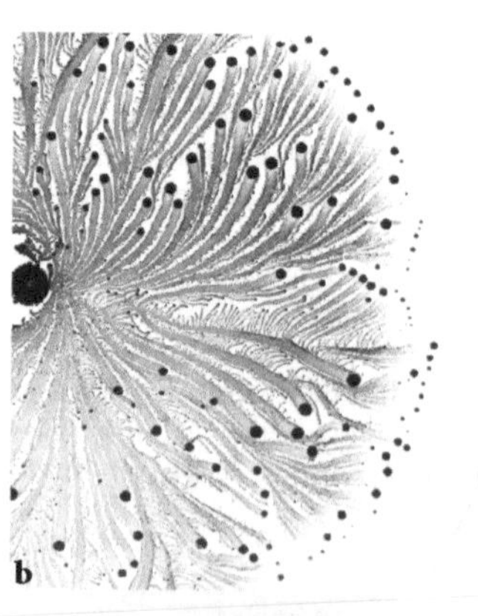

Bacterial colonies develop complex spatiotemporal patterns in response to adverse conditions such as nutrient starvation. (a) Branching pattern in a Petri dish of *Paenibacillus dendritiformis*. Growth rate is limited by diffusion of nutrients towards the colony, giving a branching or fractal pattern. (b) Vortex branching exhibited by *Paenibacillus vortex* bacteria. At the dark dot at the tip of each of the branches, millions of bacteria rotate in a highly organized vortex. Chemorepellents push the vortices outwards, allowing the colony to expand. Images courtesy of Eshel Ben Jacob, Tel Aviv University. (*See* **Cluster coagulation; Growth patterns**.)

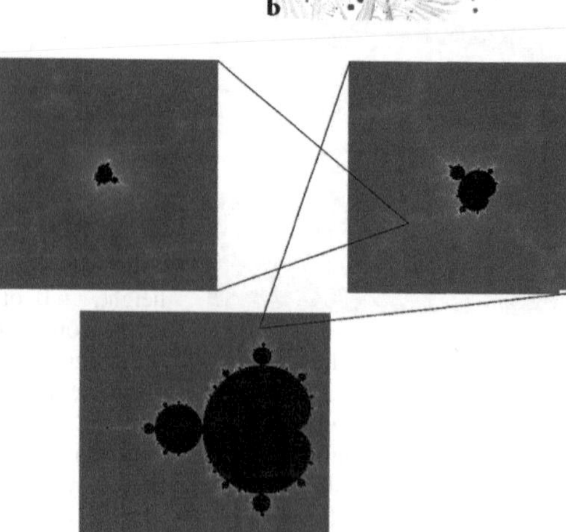

Approximate self similarity in the Mandelbrot set. The Mandelbrot set is the set of all c for which the iteration $z \to z^2 + c$, starting from $z = 0$, does not diverge to infinity. This set is named after the mathematician Benoit Mandelbrot, who discovered their fractal and self replicating structure. Three successive magnifications are shown, and at each level a structure similar, but not exactly the same, as the entire set is observed. Courtesy of Paul Bourke. (*See* **Fractals; Multifractal analysis**.)

Drainage networks have scale-invariant (fractal) symmetry. Top: Color image of elevation (white/yellow = high elevation black/red = low elevation) and drainage network (black lines) for a river basin in the Loess Plateau region of the Shanxi Province, P.R.C. River basins in homogeneous material such as loess (a silty soil deposited by wind storms) form particularly symmetric and ordered networks. Bottom: Color image of elevation for a fluvial landscape evolution simulation. Courtesy of Jon Pelletier, University of Arizona. (*See* **Geomorphology and tectonics**.)

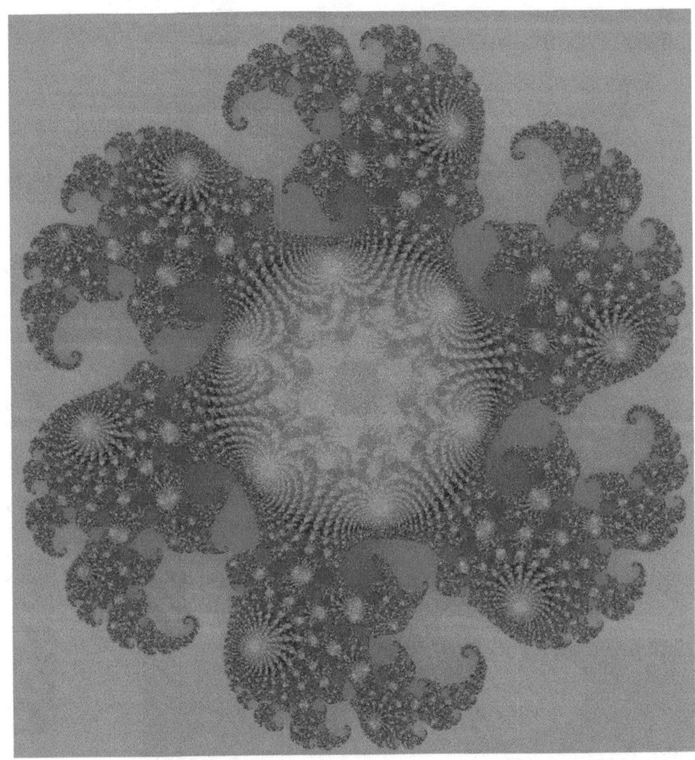

The elephants image is a Julia set with an unusual initialization scheme. The image plane is mapped to the 3rd power of the pixel coordinate. This has the effect of increasing symmetrical order. Maximim iterations/ pixel = 1024, center coordinates = $(+3.4861\times10^{-14}, -4.88498\times10^{-15})$, magnification = 0.9662319. The Julia set coordinate, p1 = (0.28767393, 0.01500042), $z = \text{pixel}^3$: $z = z^z + p1$, $(|z| \leq 4.)$ Image courtesy of Noel Giffin. (*See* **Maps in the complex plane.**)

A classic Mandelbrot image created with high iteration limits. Maximum iterations/pixel = 300,000, minimum iterations/pixel = 9587, center coordinates = (+0.37873904887513700, +0.22838619041160610), magnification = 8.279722×10^{10}. High iteration fractals can reveal striking fractal detail as the iteration count and zoom depth increase. A logarithmic coloring algorithm is used to distribute the color pallette over the large iterative range. Image courtesy of Noel Giffin. (*See* **Maps in the complex plane.**)

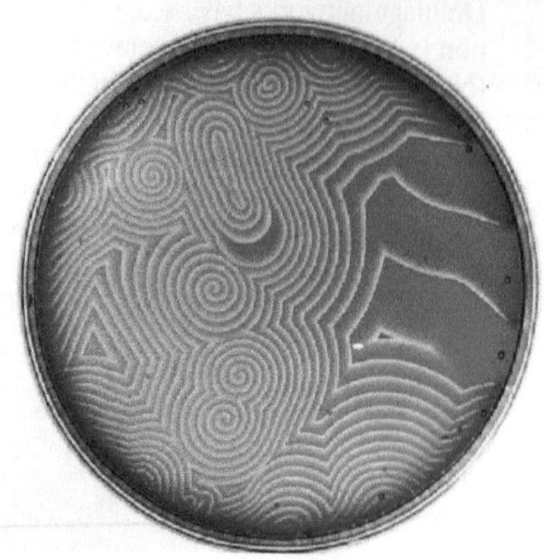

Spiral wave patterns in a BZ reaction, confined to a thin layer in a petri dish. Early on in the reaction, the fronts of the target pattern are far apart, but as the reaction proceeds they become closer and additional wave sources appear, producing a complicated pattern. Courtesy of S.K. Scott, University of Leeds. (*See* **Belousov–Zhabotinsky reaction; Chaotic dynamics; Reaction-diffusion systems**.)

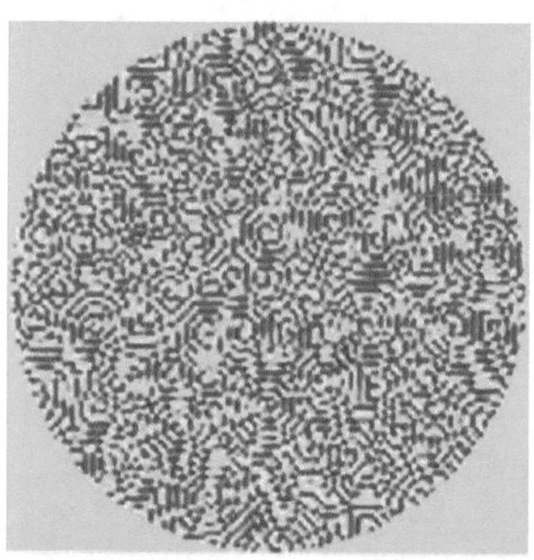

Simulation of spiral defect chaos in Rayleigh–Bénard convection. Color corresponds to temperature, increasing from violet to red. Local spirals are erratic in time—changing position and size. Courtesy of James D. Gunton, Lehigh University. (*See* **Spatiotemporal chaos; Spiral waves; Thermal convection**.)

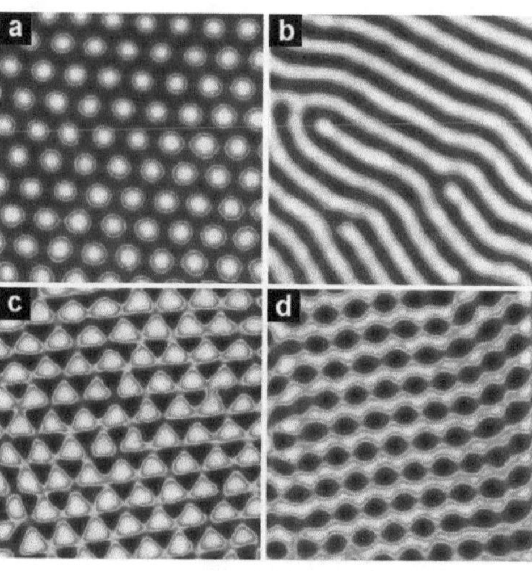

Turing structures of different symmetries obtained with the chlorite-iodide-malonic acid reaction. Dark and light regions respectively correspond to high and low iodide concentration. The wavelength (a function of kinetic parameters and diffusion coefficients) is of the order of 0.2 mm. All patterns in (a)–(d) are at the same scale: view size 1.7 mm × 1.7 mm. Courtesy P. De Kepper, CRPP. (*See* **Turing patterns**.)

A montage of planetary images taken by spacecraft managed by the Jet Propulsion Laboratory in Pasadena, CA. (Pluto is not shown.) The emergence of our planets from a cloud of gas and dust about 4.5 billion years ago was a highly nonlinear event. In his work to solve the three-body problem of planetary motion, Henri Poincaré first formulated the basic concepts of chaotic dynamics. Courtesy of NASA. (*See* ***N*-body problem; Solar system**.)

Re-creation of the famous 1834 first sighting of a soliton or solitary wave on the Union Canal near Edinburgh, 12 July 1995. Copyright Heriot-Watt University, used with permission. (*See* **Solitons, a brief history.**)

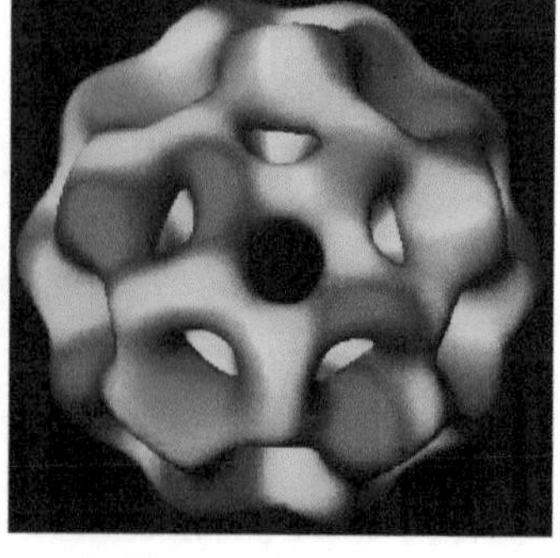

The baryon density isosurface of a 17-Skyrmion (topological charge of 17). The color represents the value of the third component of the pion field. (*See* **Skyrmions**.)

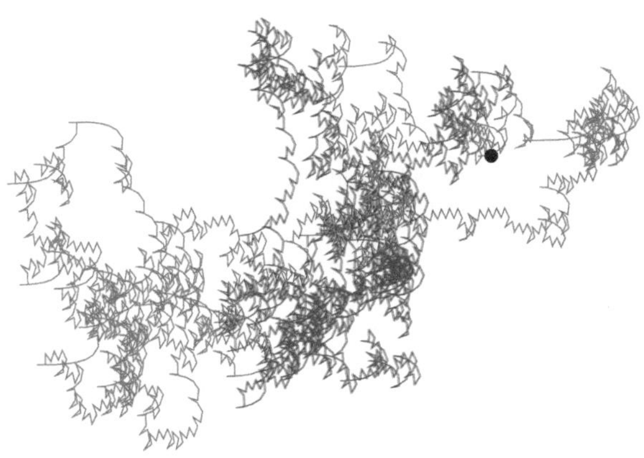

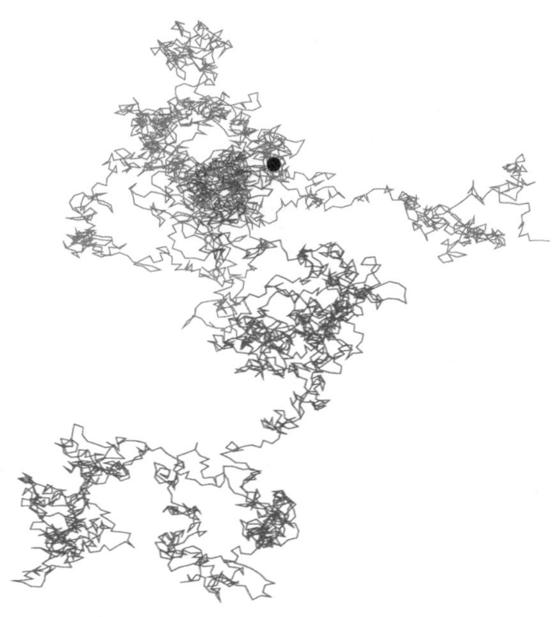

Visualization of chaos generated by the one-dimensional map $x_{n+1} = 2x_n$ (left panel), and a similar image for a truly random process (right panel). The "random walk trajectories" are plotted in the complex plane using steps of the same length but the direction is defined by x_n, i.e., $Z_{n+1} = Z_n + \exp(2\pi \mathrm{i}\, x_n)$. Initial conditions for red and blue chaotic trajectories in the equation for the one-dimensional map differ only by a distance 2×10^{-9}. (*See* **Chaotic dynamics**.)

Infrared image of a spin mode in the free-radical-frontal polymerization of an acrylate in a 1.5 cm diameter tube. Image courtesy of J. Pojman. (*See* **Polymerization**.)

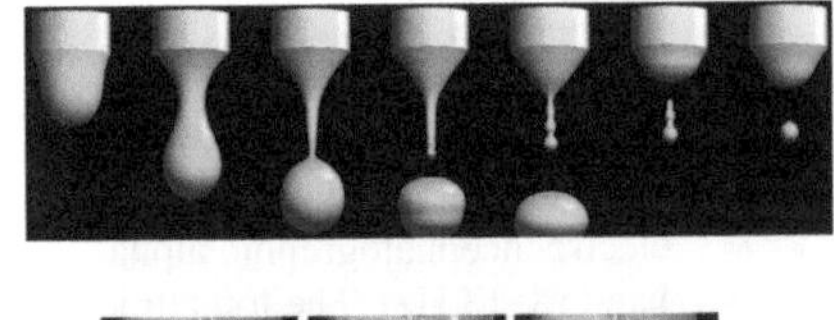

Numerical simulations of a dripping water faucet. The dripping faucet is a familiar example of chaotic dynamics. (*See* **Chaotic dynamics; Dripping faucet**.)

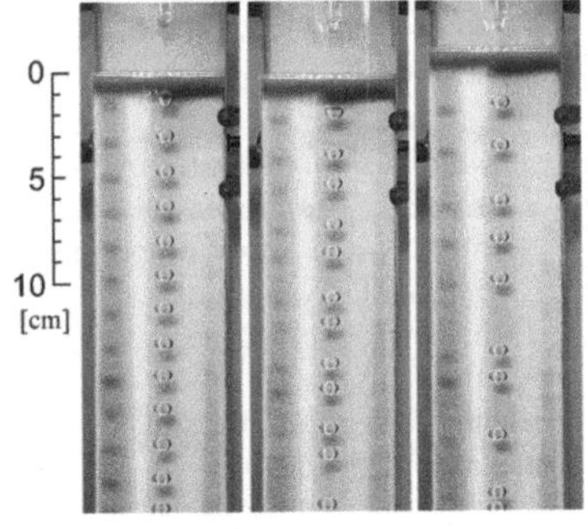

Dripping patterns of water drops falling from a nozzle (7 mm inner diameter and 10 mm outer diameter) into a transparent cylindrical container of salad oil, which sensitively depends on the flow rate. The salad oil was used only to decrease the acceleration of the drops to obtain successive pictures. Examples of period-one (left), period-two (middle) and chaos (right) are presented. (*See* **Dripping faucet**.)

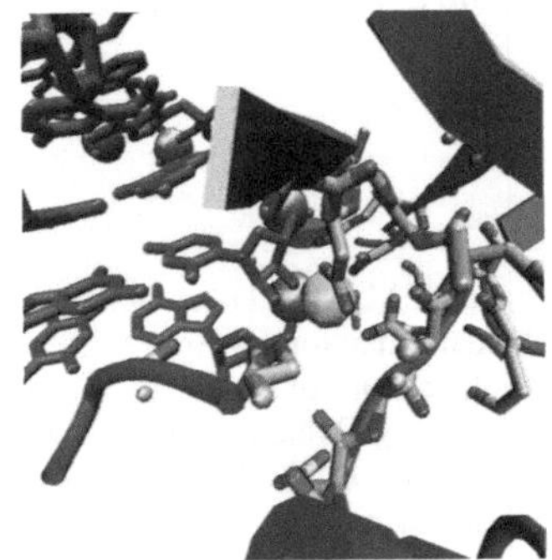

RNA polymerase active site. The DNA template and growing RNA strands are shown in purple with gold phosphorous atoms; the protein backbone is shown in blue; protein atoms are colored red for oxygen, blue-green for carbon, and blue for nitrogen. The green sphere is a magnesium ion, which coordinates both the protein and the nucleotide tri-phosphate which is being added to the growing DNA strand. Atomic coordinates are available using the protein data bank accession code 1I6H. Produced with VMD. (*See* **Protein structure**.)

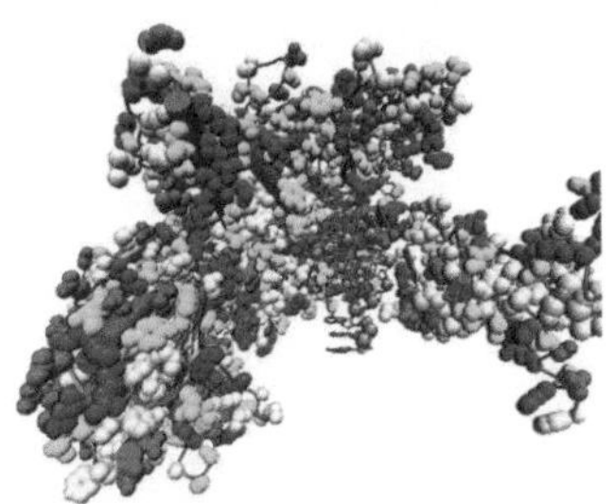

RNA polymerase catalytic subunit with DNA bound at the catalytic site. The nucleic acids are shown in purple, as in active site figure. The protein backbone is shown in red, with amino acid side chains colored according to charge and polarity; negatively charged amino acids are red, positively charged amino acids are blue, non-polar amino acids are white, and polar but uncharged amino acids are green. The patterns of charge, hydrogen bonds, and shape are recognized by other proteins in order to regulate activity of this protein. (*See* **Protein dynamics**.)

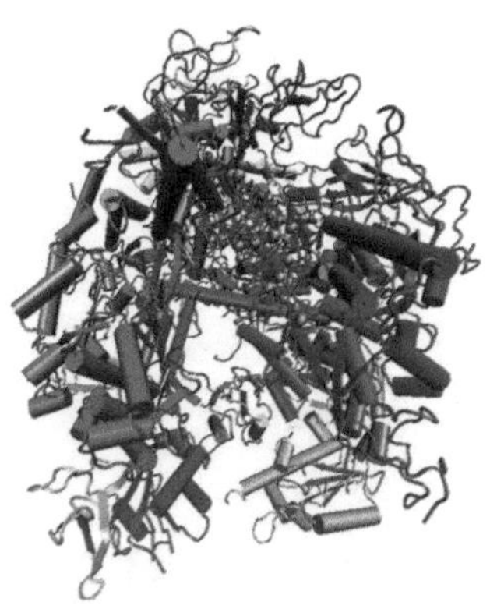

RNA polymerase transcription initiation complex, shown with several of the additional proteins necessary for it to function. Different colors represent different proteins interacting with the catalytic subunit. The interactions are quite specific. (*See* **Protein dynamics**.)

A periodically forced damped pendulum has complex behavior. These computer-generated images show initial positions that map to one of several different behaviors (one color for each). For example, orbits starting at points in the blue region would yield a different type of asymptotic motion than orbits starting in the red region. The brighter the shade of color, the longer it takes to settle into the corresponding motion. The different regions are separated by fractal basin boundaries. The pictures are made at increasing magnification levels. Courtesy, University of Maryland Chaos Group. (*See* **Attractors; Damped-driven anharmonic oscillator; Pendulum**.)

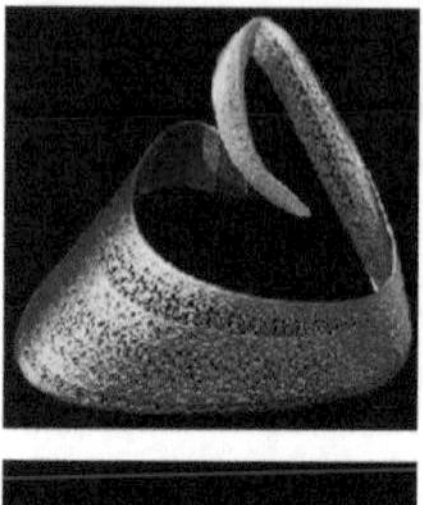

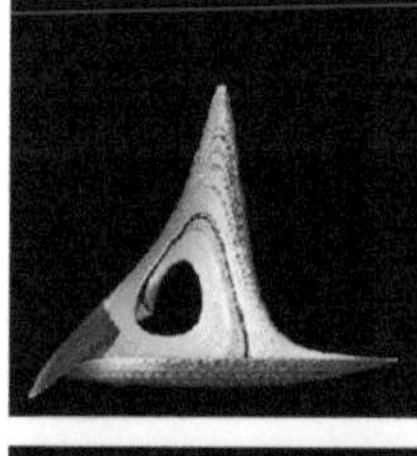
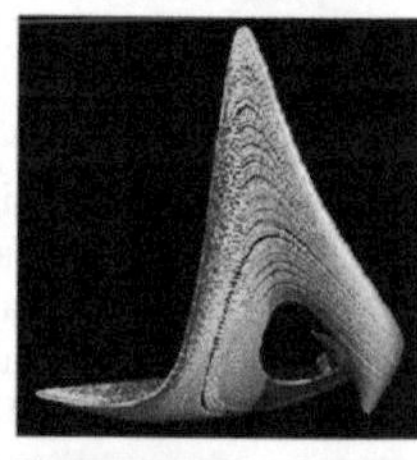
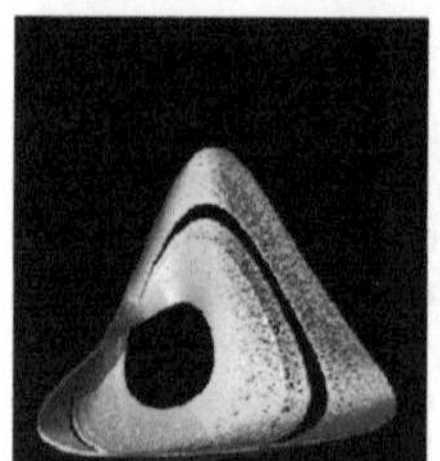

Three chaotic attractors arising from numerical solutions of a spatially homogeneous model of the mammalian electroencephalogram. Each attractor was constructed by delay-embedding a single state variable corresponding to the electroencephalographic observable. Attractors were rendered from two different viewpoints (left and right images), from 100,000 point time series of 10 seconds total duration and colored by velocity. All attractors have a dominant frequency of oscillation in the electroencephalographic alpha band (8–13 Hz). The top attractor has a Lyapunov dimension of 2.086 ± 0.003, representing an upper bound for the correlation (D_2) dimension. Graphics by Paul Bourke, Centre for Astronomy and Supercomputing, Swinburne University of Technology, Melbourne, Australia. (*See* **Electroencephalogram at mesoscopic scales**.)

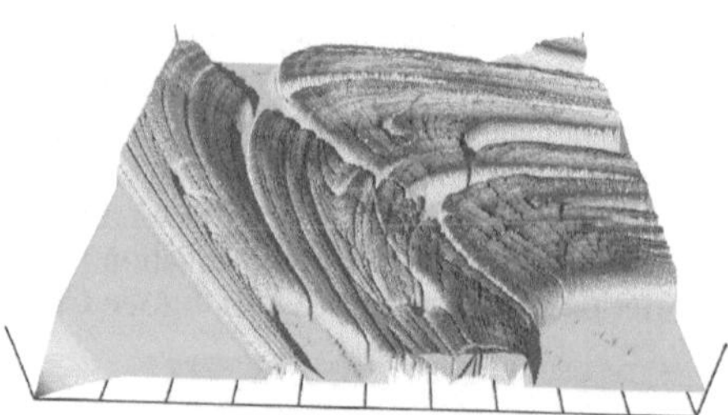

Chaos in a spatially homogeneous version of Liley et al.'s theory of the mammalian electroencephalogram. The two independent axes represent neuronal population input parameters, whereas the dependent variable (indicated by height and color) represents the largest Lyapunov exponent (LLE) of the system dynamics. Each point in this plane can be thought of as indexing a particular attractor (see previous figure). Green represents limit cycle (LLE = 0) activity, whereas points above and below this plane correspond to chaos (LLE > 0) and point attractor dynamics (LLE < 0) respectively. (*See* **Electroencephalogram at mesoscopic scales**.)

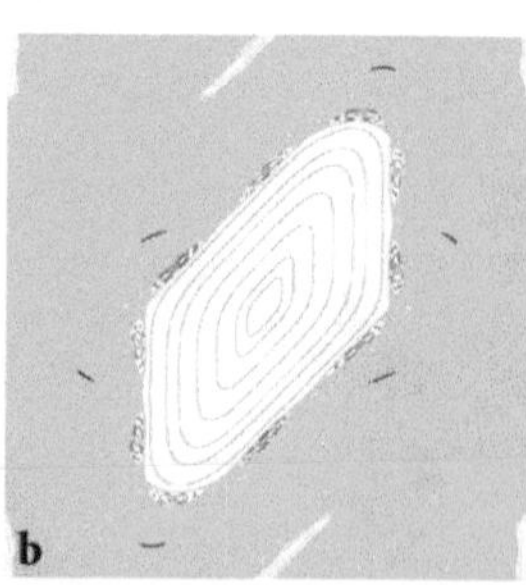

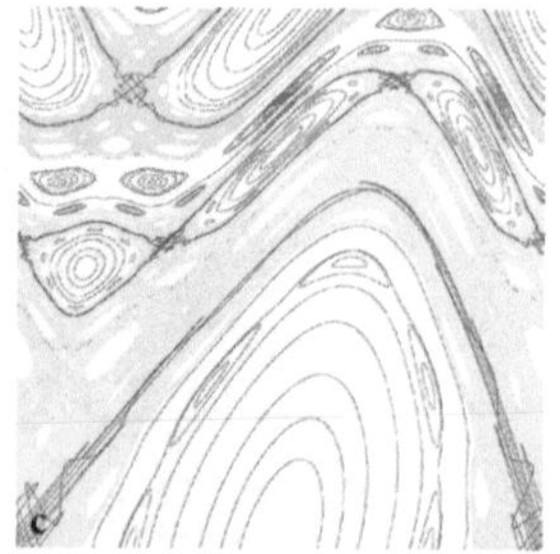

Standard Map (a): Dynamics of the standard map for $k = 0.6$;
(b). Dynamics of the standard map for $k = 2.0$; (c). Dynamics of the standard map for $k = k_{cr}(\gamma) = 0.971635406$. (*See* **Standard map.**)

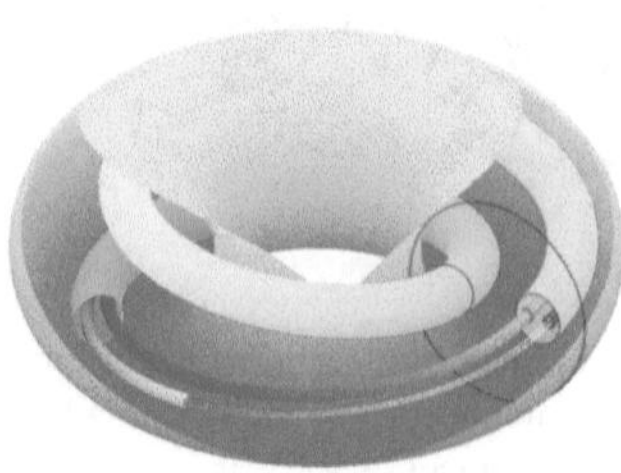

Smale's solenoid, an Axiom-A attractor. At each iteration, the fat torus is mapped into a long, thin torus which is then put back into the original torus, winding around twice. Reprinted from *A First Course in Dynamics*, Hasselblatt & Katok with the permission of Cambridge University Press. (*See* **Anosov and Axiom A systems**.)

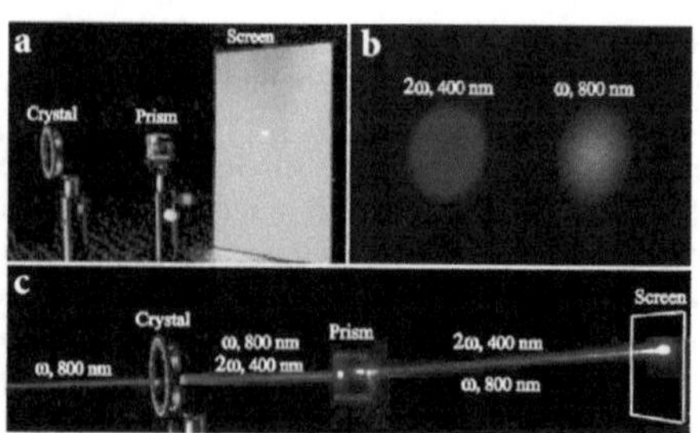

Second harmonic generation in lithium niobate: (a) Experimental setup; (b) Closeup view of the observation screen showing the fundamental and second-harmonic light; (c) Side view showing the partial conversion of the red 800-nm incident beam into the blue 400-nm beam; the separation of the colors by the prism is relatively small in the plane of this photograph. (*See* **Frequency doubling.**)

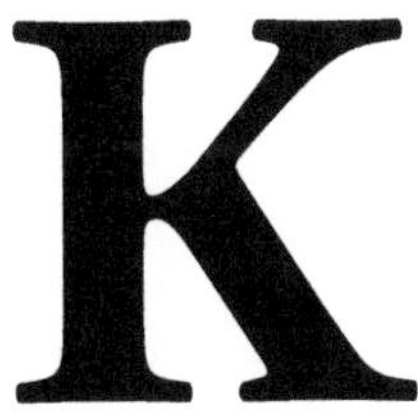

KADOMTSEV–PETVIASHVILI EQUATION

The Kadomtsev–Petviashvili (KP) equation was derived by Kadomtsev & Petviashvili (1970) to examine the stability of the one-soliton solution of the Korteweg–de Vries (KdV) equation under transverse perturbations, and it is relevant for most applications in which the KdV equation arises. After rescaling of its coefficients, the equation takes the form

$$(-4u_t + 6uu_x + u_{xxx})_x + 3\sigma^2 u_{yy} = 0, \qquad (1)$$

where indices denote differentiation, and σ is a constant parameter. If $\sigma^2 = -1$ $(+1)$, the equation is referred to as the KP1 (KP2) equation. All other real values of σ^2 can be rescaled to one of these two cases. In what follows, a reference to KP (as opposed to KP1 or KP2) implies that the result in question is independent of the sign of σ^2.

Depending on the physical context, an asymptotic derivation can result in either the KP1 or the KP2 equation. In all such derivations, the equation describes the dynamics of weakly dispersive, nonlinear waves whose wavelength is long compared with its amplitude, and whose variations in the second space dimension (rescaled y) are slow compared with their variations in the main direction of propagation (rescaled x). Two examples are as follows:

- *Surface waves in shallow water*: In this case, u is a rescaled wave amplitude with a rescaled velocity. The wavelength is long compared with the depth of the water h, which is large compared with the wave amplitude. The sign of σ^2 is determined by the magnitude of the coefficient of surface tension τ. KP1 results for large surface tension $\tau/(gh^2) > 1/3$ (thin films); otherwise, KP2 results. Here g is the acceleration of gravity. For most applications in shallow water, surface tension plays a sufficiently unimportant role, and KP2 is the relevant equation (Ablowitz &, 1979, *See* **Water waves**).
- *Magneto-elastic waves in antiferromagnetic materials*: Here u is a rescaled strain tensor with a rescaled velocity. The sign of σ^2 is determined by the difference between the linear velocities of the magnons and phonons, and the strength and direction of the external magnetic field (Turitsyn & Fal'kovich, 1986).

The KP equation has different classes of soliton solutions. The first class is a generalization of the solitons of the KdV equation. These solutions decay exponentially as $x, y \to \pm\infty$, in all but a finite number of directions along which they approach a constant value. For this reason these solutions are referred to as line solitons. By appropriately choosing their parameters, the direction of propagation of each line soliton can be chosen to be anything but the y-direction. In the simplest case, the solitons all propagate in the x-direction, adding a second dimension to the KdV solitons. Many other scenarios are possible. Two line solitons can interact with different types of interaction regions to produce two line solitons, but two line solitons can also merge to produce a single line soliton. Alternatively, a single line soliton can disintegrate in to two line solitons. The production or annihilation of a line soliton is sometimes referred to as soliton resonance. Although both KP1 and KP2 have line soliton solutions, soliton resonance occurs only for the KP2 equation. Line soliton solutions of the KP2 equation are stable, whereas line soliton solutions of the KP1 equation are unstable. More possibilities exist when more than two line solitons are involved. Two distinct line soliton interactions are illustrated in Figure 1.

Another class of soliton solutions exists for only the KP1 equation and decays algebraically in all directions as $\sqrt{x^2 + y^2} \to \infty$. These soliton solutions are referred to as lumps and are unstable. Individual lumps in multi-lump solutions do interact with each other but leave no trace of this interaction. A lump soliton is shown in Figure 2 (Ablowitz & Segur, 1981).

Another important class of solutions of the KP equation generalizes the exact periodic and quasi-periodic solutions of the KdV equation. A (quasi-) periodic solution with g phases is expressed in terms of

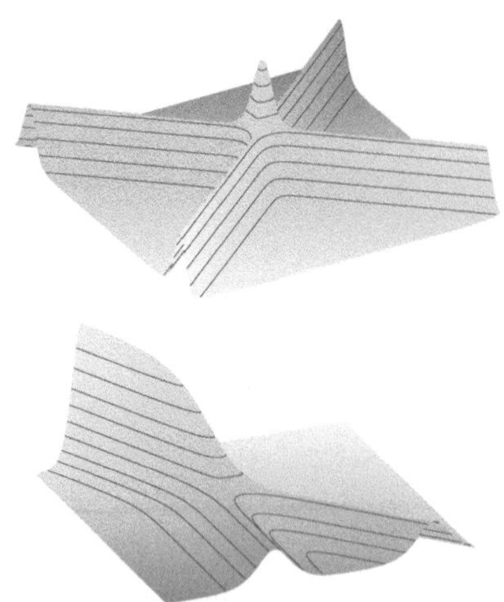

Figure 1. Two types of spatial ($t=0$) line soliton interactions for the KP2 equation: (a) Two identical line solitons with an interaction that does not change their characteristics; (b) Two line solitons merge to produce one line soliton.

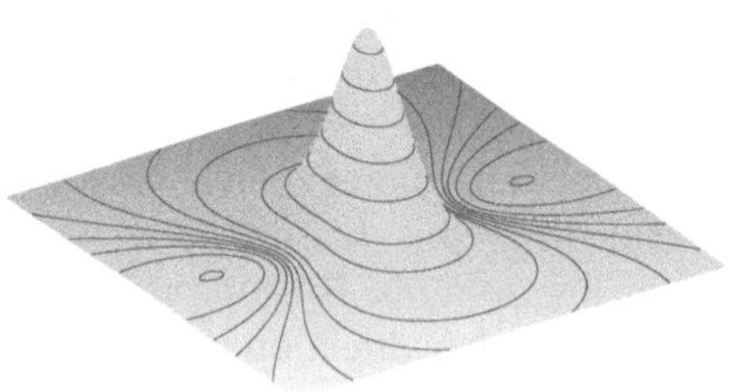

Figure 2. A lump soliton ($t=0$) solution of the KP1 equation.

the Riemann theta function $\theta(z|\boldsymbol{B})$ by

$$u(x, y, t) = c + 2\frac{\partial^2}{\partial x^2} \ln \theta(\boldsymbol{k}x + \boldsymbol{l}y + \boldsymbol{\omega}t + \boldsymbol{\phi}|\boldsymbol{B}). \quad (2)$$

Here, c is a constant, and $\boldsymbol{k}$, $\boldsymbol{l}$, $\boldsymbol{\omega}$, and $\boldsymbol{\phi}$ are g-dimensional vectors that are interpreted as wave vectors ($\boldsymbol{k}$, $\boldsymbol{l}$), frequencies ($\boldsymbol{\omega}$), and phases ($\boldsymbol{\phi}$). These parameters and the $g \times g$ Riemann matrix $\boldsymbol{B}$ are determined by a genus g compact connected Riemann surface and a set of g points on it.

For $g=1$, solution (2) generalizes the cnoidal-wave solution of the KdV equation to two spatial dimensions. For $g=2$, the solution is still periodic in space. Its basic period cell is a hexagon, which tiles the (x, y)-plane. These solutions translate along a direction in the (x, y)-plane. For $g \geq 3$, solution (2) is typically no longer periodic or translating in time. For some values of their parameters, these (quasi-)periodic solutions can be interpreted as infinite nonlinear superpositions of line solitons. Solutions with $g \leq 2$ have been compared with experiments in shallow water, with agreement being more than satisfactory (Hammack et al., 1995). A two-phase solution of the KP2 equation is shown in Figure 3. A good review of the finite-phase solutions of the KP equation is given by Dubrovin (1981).

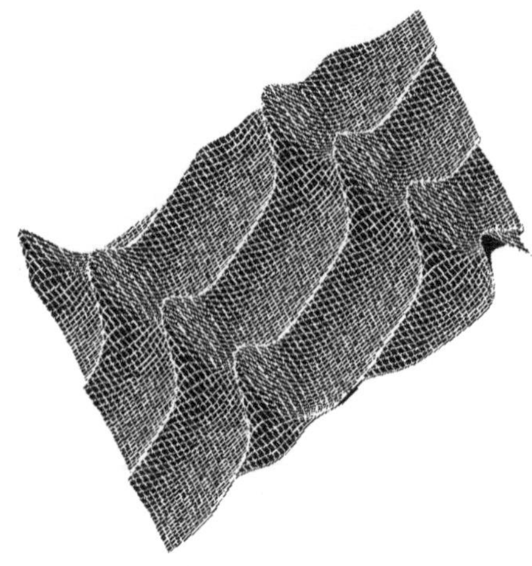

Figure 3. A two-phase periodic solution of the KP2 equation.

Unlike for the KdV equation, where only a restricted class of Riemann surfaces arises, any compact connected Riemann surface gives rise to a set of solutions of the KP equation. The reverse statement is also true: if (2) is a solution of the KP equation, then matrix $\boldsymbol{B}$ is the normalized period matrix of a genus g Riemann surface. This statement is due to Novikov. It provides a solution to the century-old Schottky problem, and its proof is by Shiota (1986).

The KP equation is the compatibility condition $\Psi_{yt} = \Psi_{ty}$ of the two linear equations

$$\begin{aligned} \sigma\Psi_y &= \Psi_{xx} + u\Psi, \\ \Psi_t &= \Psi_{xxx} + \tfrac{3}{2}u\Psi_x + \tfrac{3}{4}(u_x + w)\Psi, \end{aligned} \quad (3)$$

with $w_x = \sigma u_y$. These equations constitute the Lax pair of the KP equation. Using the inverse scattering method, it is the starting point for the solution of the initial-value problem for the KP equation on the whole (x, y)-plane with initial conditions that decay at infinity. The inclusion of line solitons is also possible (Ablowitz & Clarkson, 1991). Although the initial-value problem with periodic boundary conditions for KP2 was solved by Krichever (1989), this approach was unable to solve the same problem for the KP1 equation. In this context, solutions (2) are referred to as finite-gap solutions, as they give rise to operators (3) with spectra that have a finite number of forbidden gaps in them.

More details and different aspects of the theory of the KP equation are found in Ablowitz & Clarkson (1991), Ablowitz & Segur (1981), Dubrovin (1981), Krichever (1989), Shiota (1986), and references therein.

BERNARD DECONINCK

See also **Inverse scattering method or transform; Korteweg–de Vries equation; Multidimensional solitons; Plasma soliton experiments; Theta functions; Water waves**

Further Reading

Ablowitz, M.J. & Clarkson, P.A. 1991. *Solitons, Nonlinear Evolution Equations and Inverse Scattering*, Cambridge and New York: Cambridge University Press

Ablowitz, M.J. & Segur, H. 1979. On the evolution of packets of water waves. *Journal of Fluid Mechanics*, 92: 691–715

Ablowitz, M.J. & Segur, H. 1981. *Solitons and the Inverse Scattering Transform*, Philadelphia: SIAM

Dubrovin, B.A. 1981. Theta functions and non-linear equations. *Russian Mathematical Surveys*, 36: 11–92

Hammack, J., McCallister, D., Scheffner, N. & Segur, H. 1995. Two-dimensional periodic waves in shallow water. Part 2. Asymmetric waves. *Journal of Fluid Mechanics*, 285: 95–122

Kadomtsev, B.B. & Petviashvili, V.I. 1970. On the stability of solitary waves in weakly dispersive media. *Soviet Physics Doklady*, 15: 539–541

Krichever, I.M. 1989. Spectral theory of two-dimensional periodic operators and its applications. *Russian Mathematical Surveys*, 44: 145–225

Shiota, T. 1986. Characterization of Jacobian varieties in terms of soliton equations. *Inventiones Mathematicae*, 83: 333–382

Turitsyn, S.K. & Fal'kovich, G.E. 1986. Stability of magneto-elastic solitons and self-focusing of sound in antiferromagnets. *Soviet Physics JETP*, 62: 146–152

KELVIN–HELMHOLTZ INSTABILITY

The Kelvin–Helmholtz (KH) instability was originally investigated by Lord Kelvin (William Thomson) and Hermann von Helmholtz (Kelvin, 1871; Helmholtz, 1868; Chandrasekhar, 1981; Drazin & Reid, 1981; Gerwin, 1968) in the context of wind flowing over water. Their main motivation was to understand how wind ruffles the ocean surface to produce the ripples, eddies, and whitecaps that we see on our trips to the beach.

The KH instability has subsequently been observed in a variety of space, astrophysical, and geophysical settings involving sheared fluid and plasma flows. In addition to the original situations relating to ocean waves, relevant configurations include the interface between the solar wind and the upper magnetosphere, coronal streamers moving through the solar wind, the boundaries between adjacent sectors of the solar wind, the structure of the tails of comets, and the boundaries of the jets propagating from the core of extragalactic double radio sources into their lobes. Another interesting application of KH instability in this age of unexplained air disasters is the phenomenon known as clear air turbulence, where an aircraft flying through a clear sky is suddenly buffeted (*See* **Clear air turbulence**).

With an initial or equilibrium configuration consisting of two superposed incompressible and inviscid fluids streaming in the horizontal or z-direction at differential speeds U_1 and U_2 (see Figure 1), consider how disturbances to this configuration evolve in time. Because the velocity differential occurs at different heights (or the transition in the speed from U_1 to U_2 occurs over some finite range of x), the equilibrium configuration

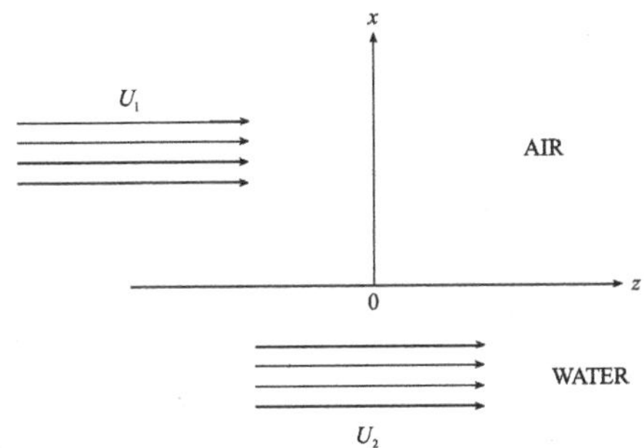

Figure 1. The basic configuration for shear flow across a vortex sheet.

has finite geometrical extent in the x-direction. Note that the inviscid assumption, which holds well for air and water as well as for very low-density interstellar media, allows an arbitrary profile for the velocity transition from U_1 to U_2 with the x transition length chosen to fit experiments. By contrast, the equilibrium is of semi-infinite extent in the horizontal or z-direction, and of infinite extent in y. Thus, one may represent field variables as Fourier series in the directions y and z which are of infinite extent, thereby assuming elementary disturbances of the form

$$\chi \sim f(x)\exp[\mathrm{i}(k_y y + k_z z - \omega t)] \tag{1}$$

for all variables χ. Here, k_y and k_z are the y and z wave numbers, and ω is the angular frequency. (Note also that only a single mode among the infinite number needed to expand the variable χ as a Fourier series has been displayed in this equation.)

Substitution of Equation (1) into the equations governing the fluid flow yields a functional relationship between the angular frequency and the wave number called the dispersion relation, which provides key information regarding the propagation of the mode of Equation (1). Importantly, there are situations where the dispersion relation yields complex angular frequencies, that is,

$$\omega = \omega_r + \mathrm{i}\omega_i\,. \tag{2}$$

The temporal exponential growth or decay of the mode is then controlled by ω_i (called the growth rate), while its oscillation frequency is governed by ω_r. Important for us are the unstable cases where $\omega_i > 0$ so the disturbance grows in time. In such cases, χ does not settle back to the original equilibrium, and one must consider the consequences of this growth.

A driving parameter for KH instability is the differential flow speed (velocity shear) of the two layers of fluid, and it suffices to consider one Fourier mode while investigating the stability of a system. The key concept here is that the wave number of the normal mode is left arbitrary, and the system is unstable if $\omega_i > 0$ for any wave number. In general, the system is stable for some range of the driving parameter and becomes unstable beyond a critical value of the parameter. The wave number that first becomes unstable as this critical parameter value is reached selects the most unstable mode, which usually guides the further development of the instability.

Let us apply these concepts by substituting Equation (1) into the dynamical equations governing the flow. In the simplest form of the analysis (linear theory), one neglects all nonlinear terms (where any of the field variables are raised to higher than linear powers, or where there are products of different field variables) to obtain a dispersion relation. This relation satisfies boundary conditions at the discontinuity corresponding to the continuity of the normal (perpendicular) speed, the normal magnetic field (if any), and the normal total (fluid plus electromagnetic) stress. The resulting dispersion relation in the incompressible fluid case is given by (Chandrasekhar, 1981)

$$\omega = k_z(\alpha_1 U_1 + \alpha_2 U_2) \pm \left[gk \left\{ (\alpha_1 - \alpha_2) + \frac{k^2 T}{g(\rho_1 + \rho_2)} \right\} - k_z^2 \alpha_1 \alpha_2 (U_1 - U_2)^2 \right]^{1/2}.$$

Here, g is the acceleration of gravity, T is the coefficient of surface tension, and

$$\alpha_1 = \frac{\rho_1}{\rho_1 + \rho_2}, \quad \alpha_2 = \frac{\rho_2}{\rho_1 + \rho_2}, \tag{3}$$

where ρ_i is the density of the i-th fluid. Analysis of Equation (3) reveals instability ($\omega_i > 0$) for (Kelvin, 1871; Chandrasekhar, 1981)

$$(U_1 - U_2)^2 > \frac{2}{\alpha_1 \alpha_2} \sqrt{\frac{Tg(\alpha_1 - \alpha_2)}{\rho_1 + \rho_2}}. \tag{4}$$

For air over sea water, this implies instability at $U_1 - U_2 > 650$ cm/s (14.8 mph) manifested as surface waves of about 1.71 cm wavelength.

The sea, however, is ruffled by winds of smaller speeds than this, and a variety of effects have been advanced to explain this discrepancy. These include continuous velocity variation rather than an abrupt jump at the surface, continuously varying density profiles, rotational effects (due to the Earth's rotation), as well as Lord Kelvin's original suggestion—the effects of viscosity (Chandrasekhar, 1981; Drazin & Reid, 1981).

In addition, compressibility and various electromagnetic fields (Choudhury, 1986; Choudhury & Lovelace, 1984; Miura, 1990; Miura & Pritchett, 1982) become important in magnetospheric, geophysical, and astrophysical settings where there are high-speed shear flows and large fields. An important effect of compressibility is to introduce a second unstable traveling-wave mode with nonzero ω_r, in addition to the standing-wave unstable mode (with zero ω_r) which exists in the incompressible limit.

There have been numerous computer simulation studies of KH instability in a variety of settings (Miura, 1995; Normal et al., 1982). The effect of anisotropic pressure has also been comprehensively considered (Brown & Choudhury, 2000).

Finally, it is important to consider nonlinear effects on the instability since the temporal growth of the linear fields in Equation (1) will eventually reach amplitudes large enough so that product terms can no longer be ignored. From physical considerations, one may also argue that exponential linear growth as predicted by Equation (1) for $\omega_i > 0$ is not tenable indefinitely because the energy is proportional to the square of the field amplitude and this energy is limited. Clearly, nonlinear effects must come into play and limit the linear growth (Debnath & Choudhury, 1997; Drazin & Reid, 1981; Weissman, 1979). Such nonlinear evolution studies reveal a variety of interesting dynamical behaviors from quasi-periodicity and chaos at one extreme (Choudhury & Brown, 2001) to self-organization and coherent structures at the other (Miura, 1999). A formulation of the problem in terms of noncanonical Hamiltonians for the fluid dynamical equations (Benjamin & Bridges, 1997) enables one to follow instability evolution well into its late stages where the interface has become appreciably folded.

S. Roy Choudhury

See also **Clear air turbulence; Coherence phenomena; Dispersion relations; Shear flow; Wave stability and instability; Water waves**

Further Reading

Batchelor, G. K., Moffatt, H. F. & Worster, M. G. (editors) 2000. *Perspectives in Fluid Dynamics*, Cambridge and New York: Cambridge University Press

Benjamin, T. B. & Bridges, T.J. 1997. Reappraisal of the Kelvin–Helmholtz problem I & II. *Journal of Fluid Mechanics*, 333: 301, 327

Brown, K.G. & Choudhury, S.R. 2000. Kelvin–Helmholtz instabilities of high-velocity magnetized shear layers with generalized polytrope laws. *Quarterly Journal of Applied Mathematics*, LVIII, 401–423

Chandrasekhar, S. 1981. *Hydrodynamic and Hydromagnetic Stability*, Oxford: Clarendon Press

Choudhury, S.R. 1986. Kelvin–Helmholtz instabilities of supersonic, magnetized shear layers. *Journal of Plasma Physics*, 35: 375–392

Choudhury, S. R. & Brown, K.G. 2001. Novel dynamics in the nonlinear evolution of the KH instability of supersonic anisotropic tangential velocity discontinuities. *Mathematics and Computers in Simulations*, 55: 377

Choudhury, S.R. & Lovelace, R.V. 1984. On the Kelvin–Helmholtz instabilities of supersonic shear layers. *Astrophysical Journal*, 283: 331–334

Debnath, L. & Choudhury, S.R. 1997. *Nonlinear Instability Analysis*, Southampton: Computational Mechanics

Drazin, P.G. & Reid, W.H. 1981. *Hydrodynamic Stability*, Cambridge and New York: Cambridge University Press

Gerwin, R.A. 1968. Stability of the interface between two fluids in relative motion. *Reviews of Modern Physics*, 40: 652–658

Helmholtz, H. von 1868. Über discontinuirliche Flüssigkeitsbewegungen. *Wissenschaftliche Abhandlungen. Monats. Konigl. Preuss. Akad. Wiss. Berlin*, 23: 215–228; as On discontinuous movements of fluids. *Philosophical Magazine*, 36: 337–36 (1868)

Kelvin, Lord (William Thomson). 1871. Hydrokinetic solutions and observations. *Philosophical Magazine*, 42(4): 362–377; with On the motion of free solids through a liquid; Influence of wind and capillarity on waves in water supposed frictionless. In *Mathematical and Physical Papers*, vol. 4: *Hydrodynamics and General Dynamics*, Cambridge: Cambridge University Press, 1910

Miura, A. 1990. Kelvin–Helmholtz instability for supersonic shear flow at the magnetospheric boundary. *Geophysical Research Letters*, 17: 749–752

Miura, A. 1995. Kelvin–Helmholtz instability at the magnetopause: computer simulations. In *Physics of the Magnetopause, Geophysical Monogr*, edited by P. Song, B.U. Sonnerup & F. Thomsen, Washington, DC: American Geophysical Union

Miura, A. 1999. Self-organization in the two-dimensional Kelvin–Helmholtz instability. *Physical Review Letters*, 83: 1586–1589

Miura, A. & Pritchett, P.L. 1982. Nonlinear stability analysis of the MHD Kelvin–Helmholtz instability in a compressible plasma. *Journal of Geophysical Research*, 87: 7431–7444

Normal, M. L., Smarr, L., Winkler, K.H. & Smith, M.D. 1982. Supersonic jets. *Astronomy and Astrophysics*, 113: 285

Weissman, M. 1979. Nonlinear wave packets in the KH instability. *Philosophical Transactions of the Royal Society, London*, 290: 639

KEPLER PROBLEM

See **Celestial mechanics**

K-EPSILON METHOD

See **Turbulence**

KERMACK–MCKENDRICK SYSTEM

See **Epidemiology**

KERR EFFECT

Discovered by John Kerr in 1875, the electro-optical effect opened new trends in the world of optics. This effect refers to the modifications that a light beam undergoes in amplitude, phase, or direction when it propagates in an optical material that responds to an external electric field. An example is the modulation of optical radiation in crystals in which two possible linearly polarized modes exist with unique directions of polarization and corresponding indices of refraction. For crystals with no inversion symmetry, $(n^2)_{ij}$ denotes the squared ray index along the directions $(i, j) = 1, 2, 3$, such as $x = 1$, $y = 2$, $z = 3$. The linear change in the coefficients $(1/n^2)_{ij}$ due to an applied low-frequency electric field $\boldsymbol{E}(E_x, E_y, E_z)$ is then defined by

$$\Delta\left(\frac{1}{n^2}\right)_{ij} = \sum_k r_{ijk} E_k \quad (k = 1, 2, 3), \qquad (1)$$

where the 6×3 matrix with elements r_{ijk} is called the electro-optical tensor. Current applications of this property are electro-optic retardation, amplitude or phase modulation, and beam deflection (Yariv, 1975).

Optical nonlinearities result from an analogous effect arising when the low-frequency electric field applied is replaced by a second optical frequency. Let us consider the electric field of an optical beam along the j-direction with frequency ω_1: $E_j^{\omega_1}(t) = \mathrm{Re}(A_j^{\omega_1} \mathrm{e}^{-\mathrm{i}\omega_1 t})$ while a second field at ω_2 is $E_k^{\omega_2}(t) = \mathrm{Re}(A_k^{\omega_2} \mathrm{e}^{-\mathrm{i}\omega_2 t})$. If the medium is nonlinear, these fields generate polarizations at frequencies $\omega_3 = n\omega_1 + m\omega_2$, where n and m are any integers. The susceptibility tensor $d_{ijk} = -\varepsilon_i \varepsilon_j r_{ijk}/2\varepsilon_0$ then enters the polarization vector as $P_i^{\omega_3} = \sum_j \sum_k d_{ijk} A_j^{\omega_1} A_k^{\omega_2}$, where ε_i and ε_j are the dielectric constants along i and j at ω_1, and ε_0 is the vacuum permittivity.

A single optical beam is able to induce nonlinear effects. The origin of the nonlinear response is then related to anharmonic motions of bound electrons under the influence of a primary optical field. As a result, the induced polarization $\boldsymbol{P}$ created from the electric dipoles is not linear in the electric field $\boldsymbol{E}$, but it satisfies the more general relation

$$\begin{aligned} \boldsymbol{P} &= \boldsymbol{P}_{\mathrm{L}} + \boldsymbol{P}_{\mathrm{NL}} \\ &= \varepsilon_0 \{ \chi^{(1)} \cdot \boldsymbol{E} + \chi^{(2)} : \boldsymbol{E}\boldsymbol{E} \\ &\quad + \chi^{(3)} \vdots \boldsymbol{E}\boldsymbol{E}\boldsymbol{E} + \cdots \}, \end{aligned} \qquad (2)$$

where $\chi^{(n)}$ $(n = 1, 2, \ldots)$ is the nth-order susceptibility tensor of rank $n+1$. The linear susceptibility $\chi^{(1)}$ represents the dominant, linear contribution to $\boldsymbol{P}$,

$$\boldsymbol{P}_{\mathrm{L}} = \varepsilon_0 \int_{-\infty}^{+\infty} \chi^{(1)}(t - t') \cdot \boldsymbol{E}(\boldsymbol{r}, t')\,\mathrm{d}t', \qquad (3)$$

including the linear refractive index n_0 and related attenuation coefficient. $\chi^{(2)}$ is the second-order susceptibility responsible, for example, for second-harmonic generation in crystals with a lack of inversion symmetry. For centro-symmetric media such as optical fibers, silica samples, and gases, this contribution is zero and the first nonlinearities are provided by the third-order susceptibility $\chi^{(3)}$ through

$$\begin{aligned} \boldsymbol{P}_{\mathrm{NL}} &= \varepsilon_0 \int\!\!\int\!\!\int_{-\infty}^{+\infty} \chi^{(3)}(t - t_1, t - t_2, t - t_3) \\ &\quad \vdots \boldsymbol{E}(\boldsymbol{r}, t_1)\boldsymbol{E}(\boldsymbol{r}, t_2)\boldsymbol{E}(\boldsymbol{r}, t_3)\,\mathrm{d}t_1\,\mathrm{d}t_2\,\mathrm{d}t_3. \end{aligned} \qquad (4)$$

This causes an intensity dependence in the total refractive index of the form

$$n = n_0 + n_2|\boldsymbol{E}|^2 \tag{5}$$

with nonlinear coefficient $n_2 = 3\chi^{(3)}_{xxxx}/8n_0$ for a linearly polarized electric field involving a single $\chi^{(3)}$ component of four-rank tensor. This intensity dependence induced in the refractive index through $\chi^{(3)}$ contributions is called the Kerr effect.

The Kerr effect affects the evolution of the electric field $\boldsymbol{E}$, which is governed by Maxwell's equations. Straightforward combination of the latter, after eliminating the magnetic and electric flux density ($\boldsymbol{D} = \varepsilon_0\boldsymbol{E} + \boldsymbol{P}$), provides the wave equation

$$\nabla \times \nabla \times \boldsymbol{E} + \frac{1}{c^2}\partial_t^2 \boldsymbol{E} = -\mu_0 \partial_t^2 \boldsymbol{P}, \tag{6}$$

where $\mu_0\varepsilon_0 = 1/c^2$ and c is the speed of light in vacuum. For treating Equation (6) conveniently, certain assumptions can be made, among which are the following:

- $|\boldsymbol{P}_{\mathrm{NL}}|$ is a small perturbation to the total induced polarization.
- Optical losses are weak, so that the imaginary part of the frequency-dependent dielectric constant $\varepsilon(\omega)$ is negligible, that is, $\varepsilon(\omega) = 1 + \tilde{\chi}^{(1)}(\omega) \simeq n^2(\omega)$.
- $\nabla \cdot \boldsymbol{D} \simeq \varepsilon\nabla \cdot \boldsymbol{E} = \boldsymbol{0}$, and thus $\nabla \times \nabla \times \boldsymbol{E} = \nabla \cdot (\nabla \boldsymbol{E}) - \nabla^2 \boldsymbol{E} \simeq -\nabla^2 \boldsymbol{E}$.
- The polarization state is preserved, so that a scalar approach is valid.
- The optical field is quasi-monochromatic; that is, its spectrum centered at $\omega_0 = 2\pi c/\lambda_0$ for the wavelength λ_0 has a spectral width $\Delta\omega = \omega - \omega_0$ satisfying $\Delta\omega/\omega_0 \ll 1$.

It is then useful to separate out the rapidly varying part of the electric field, like $\boldsymbol{E}(\boldsymbol{r}, t, z) = \hat{\boldsymbol{x}}A(\boldsymbol{r}, t, z)\, e^{-i(\omega_0 t - k_0 z)}$ for a wavefield propagating along the z axis with wave number $k_0 = k(\omega_0) = n_0\omega_0/c$. As a result, the complex envelope A is described by the nonlinear Schrödinger (NLS) model (Brabec & Krausz, 1997)

$$\mathrm{i}\partial_\xi A + \frac{1}{2k_0}T^{-1}\nabla_\perp^2 A + \hat{D}A + \frac{\omega_0 n_2}{c}T(|A|^2 A) = 0, \tag{7}$$

after passing over to the frame moving with the group-velocity, by means of the new variables $\xi = z$, $\tau = t - \beta_1 z$ where $\beta_1 \equiv \partial k/\partial\omega|_{\omega_0}$. Here, the assumptions of paraxiality ($|\partial_\xi A| \ll k_0|A|$) and slow variations in time ($|\partial_\tau A| \ll \omega_0|A|$) have been made. The transverse Laplacian $\nabla_\perp^2 = \partial_x^2 + \partial_y^2$ accounts for wave diffraction in the x–y plane [$\boldsymbol{r} = (x, y)$]. $\hat{D} \equiv \sum_{m=2}^{+\infty} \frac{1}{m!}(\partial^m k/\partial\omega^m|_{\omega_0})(\mathrm{i}\partial_\tau)^m$ is the dispersion operator, where only the leading-order contribution ($m = 2$) is considered below, which corresponds to group-velocity dispersion (GVD) with coefficient $\beta_2 \equiv \partial^2 k/\partial\omega^2|_{\omega_0}$. The operator $T = 1 + (\mathrm{i}/\omega_0)\partial_\tau$ originates from the cross derivatives $\partial^2_{\xi,\tau}$ retained in the derivation of Equation (7), and it reduces to unity, whenever $\Delta\omega/\omega_0 \ll 1$ holds. For $n_2 > 0$, the principal physical processes caused by the Kerr effect can be listed as follows:

- The Kerr effect creates spectral broadening through self-phase modulation (SPM). Solving for $\mathrm{i}\partial_\xi A = -\omega_0 n_2|A|^2 A/c$ indeed yields the exact solution $A = A_0 e^{\mathrm{i}\omega_0 n_2|A_0|^2\xi/c}$ with $A_0 = A(\xi = 0)$, which describes a self-induced phase shift experienced by the optical field during its propagation. This intensity-dependent phase shift refers to SPM, which is responsible for the spectral broadening of short pulses by virtue of the relation $\Delta\omega/\omega_0 = -\partial_\tau \arg(A)/\omega_0$. The pulse shape does not change in time, but the frequency spectrum is expanded by the nonlinearity.
- In optical fibers for which diffractive effects are negligible, pulses mainly undergo GVD and Equation (7) reduces to the standard cubic NLS equation in the limit $T \to 1$. With $\beta_2 < 0$ (so-called anomalous GVD), this equation describes the balance between 1-d dispersion and nonlinearity, which gives rise to sech-shaped, "bright" temporal solitons. With $\beta_2 > 0$ (normal GVD), "dark" solitons appear as localized dips against a uniform background. Both types of solitons have been detected experimentally (Agrawal, 1989). Solitons in dispersion-managed optical fibers, prepared with alternating spans of normal and anomalous GVD, have important applications in high-rated data communication systems over distances of many kilometers.
- For ultrashort pulses developing sharp temporal gradients, the operator T in front of the Kerr term in (7) induces shock dynamics: the field intensity without GVD and diffraction is governed by the continuity relation $\rho_\xi + 3n_2\rho\rho_\tau/c = 0$ with $\rho = |A|^2$. For localized pulses with compact distribution F, ρ has the exact implicit solution $\rho = F(\tau - 3n_2\xi\rho/c)$, which gives rise to a singular shock profile with $|A_\tau| \to +\infty$ in the trail ($\tau > 0$) of the pulse. This effect is usually called self-steepening (Anderson & Lisak, 1983). A comparable formation of a trailing shock edge, called space-time focusing, follows from the operator T^{-1} acting on diffraction (Rothenberg, 1992).
- For beams with no temporal dispersion ($T = 1$, GVD $= 0$), Equation (7) describes the self-focusing of optical wave packets (Marburger, 1975). Self-focusing arises as a shrinking of the beam in the diffraction plane, which is achieved by a transverse collapse when no saturation of the Kerr nonlinearity takes place. This process develops when the input power $P_{\mathrm{in}} = \int |A|^2\, d\boldsymbol{r}$ exceeds the critical value $P_{\mathrm{cr}} \simeq \lambda_0^2/2\pi n_0 n_2$. The beam waist then decreases

more and more, as the field amplitude $|A|$ diverges. In media such as CS_2 liquids (Campillo et al., 1973), the optical field distribution is modulationally unstable, and self-focusing often produces filamentation, along which the background beam is broken up into small-scale cells that self-focus in turn. Their growth is, however, generally arrested by higher-order nonlinearities coming from Equation (2), which can form stable spatial solitons.

- By mixing GVD and spatial diffraction when $T \simeq 1$, the Kerr nonlinearity causes defocusing in time for $\beta_2 > 0$ and temporal compression for $\beta_2 < 0$, in addition to wave focusing in the transverse direction. The interplay of these processes results in the symmetric splitting of the pulse along the time axis with normal GVD and to a 3-d spatiotemporal collapse with anomalous GVD (Bergé, 1998).
- Finally, the Kerr response of optical media favors cross-phase modulation (XPM). XPM refers to the nonlinear phase shift of an optical field induced by a copropagating beam at a different wavelength. Noting the total electric field as $\boldsymbol{E} = \hat{\boldsymbol{x}}\,\mathrm{Re}[A_1 \mathrm{e}^{-\mathrm{i}\omega_1 t} + A_2 \mathrm{e}^{-\mathrm{i}\omega_2 t}]$ for two components with different frequencies ω_1 and ω_2, the phase shift for the field at ω_1 over a length L is given by $\phi_{\mathrm{NL}} = n_2 k_0 L(|A_1|^2 + 2|A_2|^2)$. The last term refers to XPM, from which spectral broadening is larger than that for one wave alone. Coupling of several wave components belongs to the category of parametric processes, among which are the third-harmonic generation, four-wave mixing and parametric amplification (Shen, 1984; Agrawal, 1989).

Luc Bergé

See also **Development of singularities; Filamentation; Harmonic generation; Nonlinear optics**

Further Reading

Agrawal, G.P. 1989. *Nonlinear Fiber Optics*, San Diego: Academic Press, pp 14–36

Anderson, D. & Lisak, M. 1983. Nonlinear asymmetric self-phase modulation and self-steepening of pulses in long optical waveguides. *Physical Review* A, 27: 1393–1398

Bergé, L. 1998. Wave collapse in physics: Principles and applications to light and plasma waves. *Physics Reports*, 303: 259–370

Brabec, T. & Krausz, F. 1997. Nonlinear optical pulse propagation in the single-cycle regime. *Physical Review Letters*, 78: 3282–3285

Campillo, S.L., Shapiro, S.L. & Suydam, B.R. 1973. Periodic breakup of optical beams due to self-focusing. *Applied Physics Letters*, 23: 628–630

Marburger, J.H. 1975. Self-focusing: theory. *Progress in Quantum Electronics*, 4: 35–110

Rothenberg, J.E. 1992. Space-time focusing: breakdown of the slowly varying envelope approximation in the self-focusing of femtosecond pulses. *Optics Letters*, 17: 1340–1342

Shen, Y.R. 1984. *The Principles of Nonlinear Optics*, New York: Wiley, pp 242–333

Yariv, A. 1975. *Quantum Electronics*, 2nd edition, New York: Wiley, pp 407–507

KICKED ROTOR

The properties of a classical and/or quantum kicked rotor can be understood by examining how nonlinearities influence the dynamics of conservative dynamical systems. The story begins with the French mathematician Henri Poincaré and his famous study of planetary motion (Poincaré, 1892). Unlike the integrable two-body problem, a third body added to the system experiences nonperiodic motion (*See* ***N*-body problem**). Indeed, the trajectory of a light mass moving under the influence of the gravity of two heavy masses is very complex, and we know that such an orbit is fractal. This is one end of the spectrum of dynamical possibilities; on the other end are the integrable Hamiltonian equations of the Kolmogorov–Arnol'd–Moser (KAM) theory.

Thus, in one limit we have the integrable Hamiltonians that are the subject of the KAM theory, and in another we have non-integrable Hamiltonians, an example of which is the three-body problem of Poincaré. So how do we get from one side to the other using a perturbed Hamiltonian? In the unperturbed system, the various degrees of freedom are uncoupled, given by the action-angle variables, and the motion unfolds on KAM tori. Small perturbations leave the motion essentially unchanged, but even then nonlinear resonances are introduced in restricted regions of phase space. These nonlinear resonances change the dynamics of the system and deform the tori on all scales as the strength of the perturbation increases, leading to breakup of the KAM tori and chaos.

The properties of such nonlinear resonances can be examined using maps, which in a Hamiltonian system is equivalent to using a strobe lamp to register the momentum and displacement at equally spaced time intervals. One map that approximates several physical systems is the Taylor–Chirikov or standard map (Chirikov, 1979). For the nth flash we have for this map the momentum p_n and displacement x_n,

$$p_{n+1} = p_n \frac{K}{2\pi} \sin(2\pi x_n), \quad x_{n+1} = x_n + p_{n+1}. \quad (1)$$

Figure 1 shows a sequence of standard map plots for increasing values of the coupling strength K. Each of the continuous curves is the result of a mapping of a single initial condition. The orbits in the vicinity of $p = 0.5$ in Figure 1a are those of a simple pendulum, which is the "harmonic oscillator" of nonlinear dynamics. As the size of K is increased, Figure 1b shows that the nonlinear resonances generate more and more structure in the phase space as well as deforming the already existing KAM tori. The tori break up into sprays of points, beginning in the vicinity of the nonlinear resonances, whose exact locations are sensitively dependent on initial conditions. This is chaos, and as K is further increased, the chaos spreads throughout the phase space among the islands of stability, as in Figure 1c. There is no way to predict the motion of the trajectory in the chaotic sea. Further increases in the coupling parameter result in more tori disintegrating until the phase space appears to be completely dominated by chaos and all but the most stable of the KAM tori have disappeared, as in Figure 1d.

The standard map is also called a kicked rotor because it describes the strobe plot of the phase space trajectories for a rotor kicked every T units of time with a strength that depends on the angular position of the rotor. The Hamiltonian for the classical kicked rotor (CKR) can be written as (Chirikov, 1979; Reichl, 1992)

$$H = \frac{J^2}{2I} + K\cos\theta \sum_{k=-\infty}^{k=\infty} \delta(t - nT), \tag{2}$$

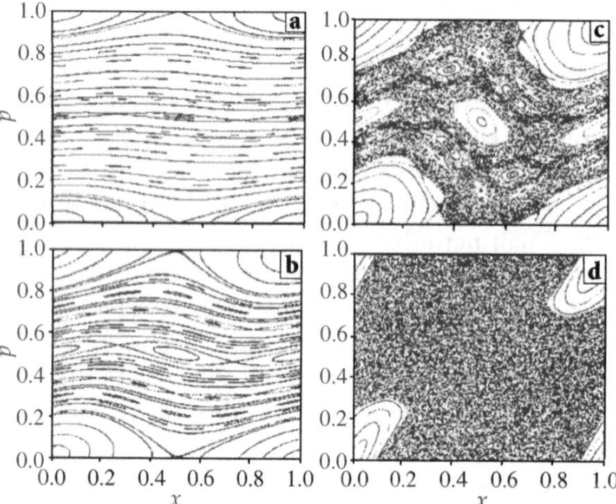

Figure 1. The phase space orbits for the kicked rotor are depicted for four values of the coupling strength: (a) $K = 0.1716354$; (b) $K = 0.4716354$; (c) $K = 1.1716354$, and (d) $K = 3.9716354$ (from Reichl, 1992).

where J is the angular momentum of the rotor, θ is its angular position, and I is the moment of inertia. Reichl (1992) reviews how to obtain the discrete map (2) by integrating the Hamiltonian equation of motion generated by (3) from one pulse to the next.

Hamiltonian (3) provides a classical description of an electron subjected to a periodically pulsed electric field. The nonlinear mechanisms are present in the motion of charged particles in accelerators and storage rings, as well as in the motion of planets and man-made celestial bodies. In the classical case, the particle dynamics in the chaotic sea are interpreted as diffusion, under which the mean-square separation of trajectories increases linearly in time with a diffusion coefficient given by $2D \approx K^2$. Thus, for the coupling strength (K) sufficiently large, the dynamical system behaves like simple Brownian motion. This has led a number of investigators to conjecture that chaos provides the dynamical foundation of thermodynamics, rather than the usual heat bath, with an infinite number of degrees of freedom (Ford & Waters, 1963; Chirikov, 1979). For example, the fluctuation-dissipation relation has been generated using nonlinear dynamical systems (Bianucci et al., 1995).

The quantum kicked rotor (QKR) employs a quantized version of the Hamiltonian, where the angular momentum J is replaced with the operator $i\hbar\,\partial/\partial\theta$. The mapping is carried out using the Floquet theory on the Schrödinger equation, because the Hamiltonian is periodic in time. There is a remarkable similarity between the QKR and the one-dimensional tight-binding model of Anderson localization in solid state physics. The dynamical evolution of the QKR can be determined analytically in terms of a Floquet map, where the Floquet momentum eigenstates are exponentially localized for irrational kicks and extended for rational kicks (Reichl, 1992). It has been established that the quantum density for the momentum states is a truncated Levy distribution in direct analogy with the classical probability density as shown in Figure 2 (Stefancich et al., 1998). The peaks on either side of the classical and quantum

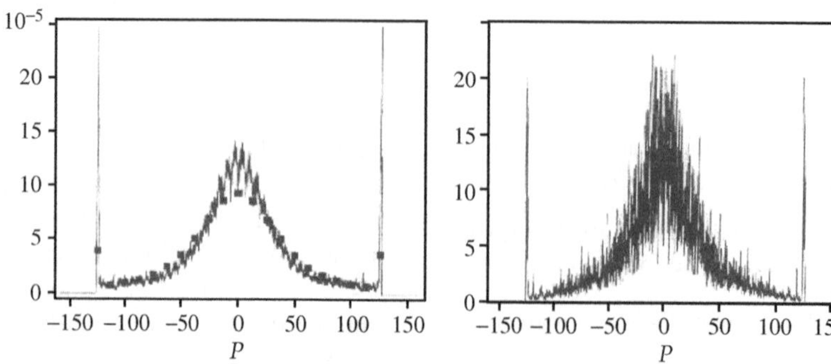

Figure 2. On the left is depicted the classical probability density for the dimensionless momentum of the CKR after 20 iterations of the map (2) and the solid curve is the theoretical prediction of the truncated Lévy distribution. On the right is the quantum probability density after 20 iterations for the QKR (Stefancich et al., 1998).

distributions correspond to the outward propagation of the initial states at maximum velocity. Finally, the CKR and QKR have been used to describe the experimental "quasifractal" structures resulting from a periodically kicked spin system in the semiclassically large-spin regions (Nakamura, 1993).

BRUCE J. WEST

See also **Chaotic dynamics; Fluctuation-dissipation theorem; Hamiltonian systems; Kolmogorov–Arnol'd–Moser theorem; Standard map**

Further Reading

Bianucci, M., Mannella, R., West, B.J. & Grigolini, P. 1995. From dynamics to thermodynamics: linear response and statistical mechanics. *Physical Review* E, 51: 3002–3022

Chirikov, B.V. 1979. A universal instability of many-dimensional oscillator systems. *Physics Reports*, 52: 263–379

Ford, J. & Waters, J. 1963. Computer studies of energy sharing and ergodicity for nonlinear oscillator systems. *J. Math. Phys.*, 4: 1293–1305.

Nakamura, K. 1993. *Quantum Chaos: A New Paradigm of Nonlinear Dynamics*, Cambridge and New York: Cambridge University Press

Ott, E. 2003. *Chaos in Dynamical Systems*, 2nd edition, Cambridge and New York: Cambridge University Press

Poincaré, H. 1892–1899. *Les Methodes nouvelles de la méchanique céleste*, Paris: Gauthier-Villars; as *New Methods of Celestial Mechanics*, 3 vols, Woodbury, NY: American Institute of Physics, 1993

Reichl, L.E. 1992. *The Transition to Chaos*, New York: Springer

Stefancich, M., Allegrini, P., Bonci, L., Grigolini, P. & West, B.J. 1998. Anomalous diffusion and ballistic peaks: a quantum perspective. *Physical Review* E, 57: 6625–6633

KINKS AND ANTIKINKS

See **Sine-Gordon equation**

KIRCHHOFF'S LAWS

Introduction

Kirchhoff's laws are conservation rules for networks; given a particular network topology, they prescribe how its currents and voltages are related. This not only facilitates understanding of the behavior of a network, but also allows an analyst to write down the equations that govern its behavior. These laws were named after Gustav Robert Kirchhoff, who in 1845 published one of the first systematic treatises on the laws of circuit analysis. Kirchhoff's current law (KCL) states that the sum of all currents into any node (or single point) of the network is zero; thus the charge flowing in and out is balanced, which makes intuitive sense. In the network of Figure 1, this means that I_1 and I_4 are equal in magnitude (KCL at node A), and that the algebraic sum of I_1, I_2, and I_3 is zero (KCL at node B).

Kirchhoff's voltage law (KVL) states that the sum of voltages around a closed loop is zero, which is expected because traversing a loop brings one back to the starting point. In Figure 1, this means that V_2 and V_3 are equal (KVL around loop D), and that the algebraic sum of V_1, V_2, and V_4 is zero (KVL around loop C). Signs and directions can be confusing in both cases, and it is important to note these correctly. A positive current entering a node is the same as a negative current leaving a node. Also, moving from the negative to positive voltage (potential) terminal of an element entails a potential rise, while moving in the opposite direction entails a drop. Passive network elements create voltage drops; active ones such as voltage sources can create voltage rises.

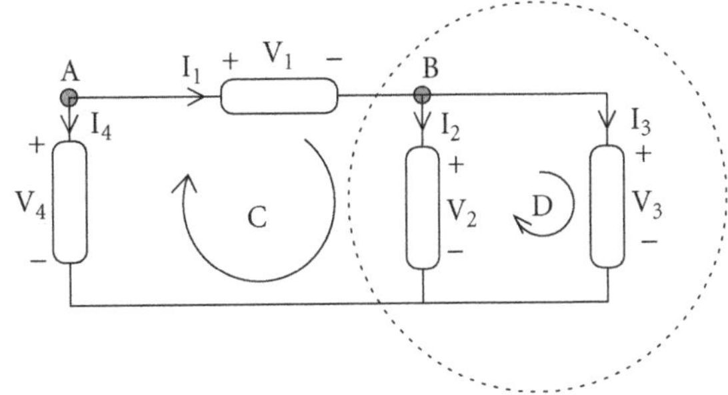

Figure 1. A simple electric network.

As with any such rules, there are some caveats. KCL actually specifies that current into a "cutset," not just a node, is zero; a cutset is any part of the network that is enclosed by a closed curve, such as the dotted circle in Figure 1. Moreover, KCL does not hold if nodes (or cutsets) can store, consume, or produce charge. KVL, on the other hand, is only valid if the voltage (or potential) between two points is independent of the path one takes between the points. If energy is lost or created as it moves around a loop, KVL does not hold.

Applications to Electronic Circuits

Kirchhoff's laws find their most common use in electronics. In Figure 2, for example, KVL around the left- and right-hand loops, respectively, yields $-V + V_1 + V_2 = 0$ and $-V_2 + V_3 = 0$, and KCL written at the point where the three resistors join yields $I_1 - I_2 - I_3 = 0$. These relationships are fairly obvious in such a simple network, assuming one gets the directions and the associated signs right. KVL and KCL become important when the network topology is complex or when one wants to automate the process in a computer algorithm.

If the constitutive relations (voltage-current characteristics) of the elements in a network are known, one can rewrite either set of equations in terms of the other state variables. Resistors, for example, obey Ohm's law ($V = IR$), so the first KVL equation can be rewritten in terms of the currents: $V = I_1R_1 + I_2R_2$.

The governing equations for a simple resistor circuit as in Figure 2 are linear and algebraic. When the

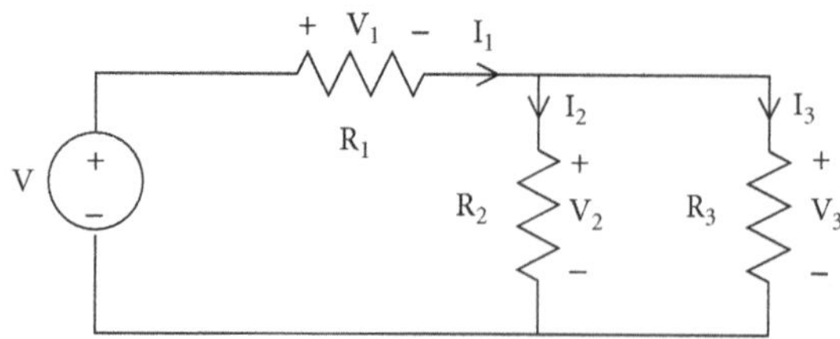

Figure 2. A resistor network.

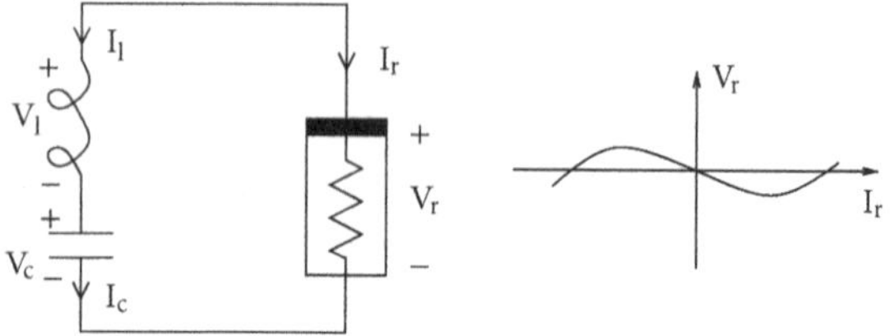

Figure 3. A simple oscillator circuit.

circuit includes inductors and capacitors, the equations become integro-differential, but Kirchhoff's laws work just as well, and they also handle nonlinear components easily. The circuit of Figure 3, for instance, is an example of the van der Pol oscillator. In this case, KCL simply states that the currents through all three elements are identical. KVL implies that $V_l + V_c - V_r = 0$. The constitutive relations for the capacitor and inductor are $I_c = C\,dV_c/dt$ and $V_l = L\,dI_l/dt$, and the resistor in this circuit is nonlinear: $V_r = -aI_r + bI_r^3$. Thus, KVL can be rewritten as

$$L\frac{dI_l}{dt} + \frac{1}{C}\int I_c dt + aI_r - bI_r^3 = 0.$$

Other Applications

These ideas generalize beyond electronic circuit analysis. Many other systems, ranging from vehicle suspensions to social groups, can be described by networks. Moreover, KVL and KCL are actually instances of a more general set of laws. In the late 1950s and early 1960s, inspired by the realization that the principles underlying KCL and Newton's third law were identical (summation of {forces, currents} at a point is zero, respectively; both are manifestations of the conservation of energy), researchers began combining multi-port methods from a number of engineering fields into a generalized engineering domain with prototypical components (Paynter, 1961). The basis of this generalized physical networks (GPN) paradigm is that the behavior of an ideal two-terminal element—the "component"—can be described by a mathematical relationship between two dependent variables: generalized flow and generalized effort, where flow × effort = power. This pair of variables is different in each domain: (flow, effort) is (current, voltage) in an electrical domain and (force, velocity) in a mechanical domain.

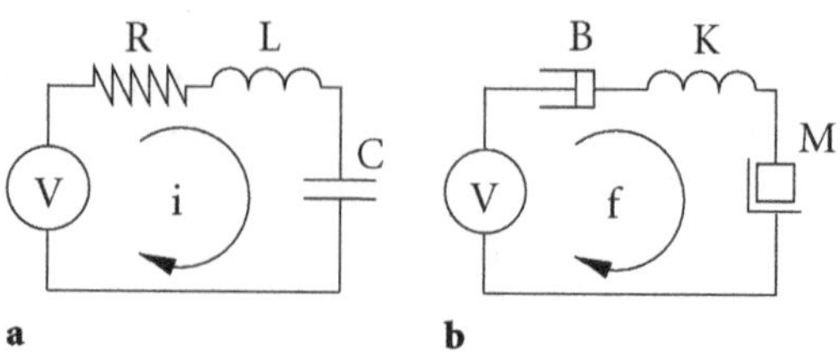

Figure 4. (a) An electrical circuit that is mathematically equivalent to (b) a mechanical circuit.

The GPN representation brings out similarities between components and properties in different domains. Electrical resistors ($v = iR$) and mechanical dampers or "dashpots" ($v = fB$) are analogous, as both dissipate energy. Both of the networks in Figure 4, for example, can be modeled by a series inertia-resistor-capacitor GPN. Thus network (a) is an electronic RLC circuit (like the van der Pol example of Figure 3), and network (b) is a mechanical mass-spring-damper system that has identical behavior. Similar analogies exist for generalized inertia, capacitance, flow, and effort source components for mechanical rotational, hydraulic, and thermal domains (Karnopp et al., 1990, Sanford, 1965). These correspondences and generalizations allow applications of KVL and KCL to mechanical structures (buildings, vehicle suspensions, aircraft, etc.), which can be modeled as interconnected networks of masses, springs, and dashpots. This approximation gives analytic insight into the vibrational modes of buildings (important for earthquake protection) and of aircraft (to keep engine frequencies from damaging wings).

ELIZABETH BRADLEY

Further Reading

Karnopp, D., Margolis, D. & Rosenberg, R. 1990. *System Dynamics: A Unified Approach*, 2nd edition, New York: Wiley

Paynter, H. 1961. *Analysis and Design of Engineering Systems*, Cambridge MA: MIT Press

Sanford, R. 1965. *Physical Networks*, Englewood Cliffs, NJ: Prentice-Hall

KLEIN–GORDON EQUATION

See **Sine-Gordon equation**

KNOT THEORY

A simple closed curve in three-dimensional space is a knot; more precisely, if M denotes a closed orientable three-manifold, then a smooth embedding of S^1 in M is called a knot in M. A link in M is a finite collection of disjoint knots, where each knot is a component of the link. Knot theory deals with the study and application of mathematical properties of knots and links in pure and applied sciences. As purely mathematical objects, knots are studied for the purpose of classifying three-dimensional surfaces according to the degree of topological complexity, regardless of their specific embedding and geometric properties (Figure 1). In this sense, knot theory is part of topology. In recent years, however, knot theory has embraced applications in dynamical systems, stimulated by the challenging difficulties associated with the study of physical knots (Kauffman, 1995). In this context knots and links are representatives of virtual and numerical objects (given by dynamical flows, phase space trajectories, and visiometric patterns), and are used to model tube-like physical systems, such as vortex filaments, magnetic loops, electric circuits, elastic cords, or even high-energy strings. For physical knots, topological issues and geometric and dynamical aspects are intimately related, influencing each other in a complex fashion. Virtual or numerical knots are studied in relation to the generating algorithms and the probability of forming knots, whereas the study of physical knots addresses questions relating topology and physics, as in the case of the topological quantum field theory (Atiyah, 1990) and topological fluid mechanics (Arnol'd & Khesin, 1998; Ricca, 2001).

Mathematical Aspects

Let us introduce some basic mathematical concepts (see, for example, Adams, 1994). A knot is said to be oriented in M, if it is a smooth embedding of an oriented curve. Two knots K and K' are said to be equivalent if there exists a smooth orientation-preserving automorphism $f: M \to M$ such that $f(K) = K'$; in particular, if the knot K is continuously deformed by f (preserving the curve orientation) to the knot K', then the two knots K and K' are said to be equivalent by ambient isotopy, and the isotopy class of K is represented by its knot type. Since knot theory deals essentially with the properties of knots and links up to isotopy, the knot parametrization, as well as any other geometric information, is irrelevant. A knot diagram of K is a plane projection with crossings marked as under or over; among the infinitely many diagrams representing the same knot K, the minimal diagram is the diagram with a minimum number of crossings. According to the type of crossing, it is customary to assign to each crossing in the knot diagram the value $\varepsilon = +1$ or $\varepsilon = -1$, as shown in Figure 2: by switching one crossing in the knot diagram from positive to negative (or the other way round), we obtain a different knot type, which is identical except for this crossing. By switching all the crossings we obtain the mirror image of the original knot. If the knot is isotopic to its mirror image, then its knot type is said to be achiral, otherwise it is chiral.

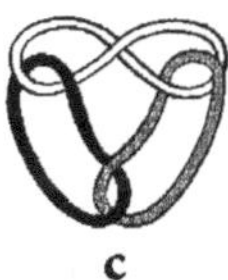

Figure 1. Three examples of knot and link types: (a) the six-crossing knot 6_3; (b) the two-component six-crossing link 6^2_3; (c) the three-component seven-crossing link 7^3_1.

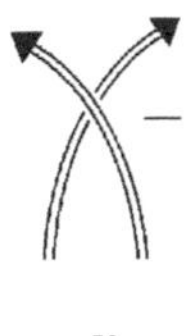

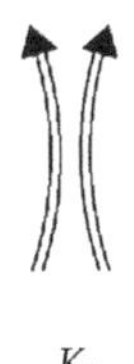

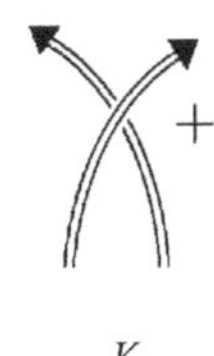

Figure 2. Standard crossing notation and algebraic sign convention for oriented strands: $\varepsilon(K_-) = -1$; $\varepsilon(K_0) = 0$; $\varepsilon(K_+) = +1$.

A knot invariant is a quantity whose value does not change when it is calculated for different isotopic knots. There are many types of invariants of knots and links, but the most common are of numerical or algebraic nature. One of the most important is the genus $g(K)$ of the knot K: recall that closed orientable surfaces are classified by genus, given by the number of handles in a handle-body decomposition. The genus $g(K)$ is defined as the minimum genus over all orientable surfaces S, which span an oriented knot K, where $\partial S = K$. One of the simplest combinatorial invariants of a knot is the minimum number of crossings of a knot K in any projection, called the crossing number $c(K)$. A fundamental invariant of links is the linking number $\mathrm{Lk}(K_1, K_2)$, that measures the topological linking between the knots K_1 and K_2; this invariant, discovered by Carl Friedrich Gauss in 1833, can be easily calculated by the crossing sign convention of Figure 2:

$$\mathrm{Lk}(K_1, K_2) = \frac{1}{2} \sum_{r \in K_1 \sqcap K_2} \varepsilon_r , \qquad (1)$$

where $\varepsilon_r = \pm 1$ and $K_1 \sqcap K_2$ denotes the total number of crossings (not necessarily minimal) between K_1 and K_2. Following the pioneering work of James W. Alexander, who used a Laurent polynomial $\Delta_K(q)$ in q to compute a polynomial invariant for the knot K by using its projection on a plane, many other polynomial invariants have been introduced; most notably the Jones polynomial $V_K(t)$ in $t^{1/2}$, defined by the following set of axioms:

(i) Let K and K' be two oriented knots (or links), which are ambient isotopic. Then

$$V_K(t) = V_{K'}(t). \qquad (2)$$

(ii) If U is the unknotted loop (that is the unknot), then

$$V_U(t) = 1. \quad (3)$$

(iii) If K_+, K_-, and K_0 are three knots (links) with diagrams that differ only as shown in the neighborhood of a single crossing site for K_+ and K_- (see Figure 2), then the polynomial satisfies the following skein relation

$$t^{-1}V_{K_+}(t) - tV_{K_-}(t) = (t^{1/2} - t^{-1/2})V_{K_0}(t)\,. \quad (4)$$

An important property of the Jones polynomial (which is not shared by previous polynomials) is that it can distinguish between a knot and its mirror image. Later work has led to other polynomial invariants, namely, the HOMFLY and Kauffman polynomials, and to a more abstract approach to algebraic invariants (Vassiliev invariants and Lie algebras). There are also invariants of different nature: among these, we mention the fundamental group $\pi_1(S^3/K)$ of the knot complement and its hyperbolic volume $v(K)$. The classification of knots and links has led to the important study of braids: these are given by a set of n interlaced strings, with ends defined on two parallel planes, placed at some distance h apart. According to specific topological characteristics, we may consider special types of knot sub-families, such as torus knots, alternating knots, two-bridge knots, tangles, and many others (see Hoste et al., 1998).

Virtual Knots

Virtual knots arise from dynamical flows, generated by the vector field of a specific ordinary differential equation (Ghrist, 1997), in connection with phase-space dynamics and statistical mechanical models (Millett & Sumners, 1994) or, as recently done, from application of ideas from the quantum field theory with an appropriate Lagrangian. This latter approach, originated in work by E. Witten in 1989, has led to the creation of a new area, called the topological quantum field theory, that has proven to be extremely fruitful in providing new results on invariants of low-dimensional manifolds. Soliton knots are given by solutions to soliton equations for one-dimensional systems: in this context there are intriguing questions relating topological invariants, integrability, and conservation laws. Virtual knots and links are generated in visiometrics by numerical simulations: in this case, smooth knots are replaced by polygonal knots, where the number of segments (or sticks) is the result of numerical discretization. Stimulating questions address the minimum number of sticks of given length for each knot type and the generation of knots and links by minimal random walks. Other questions regard charged knots: these are knots and links charged by potentials that generate self-attraction or repulsion on the knot strands (see Figure 3). Under volume-preserving diffeomorphisms, the knot is led to relax by

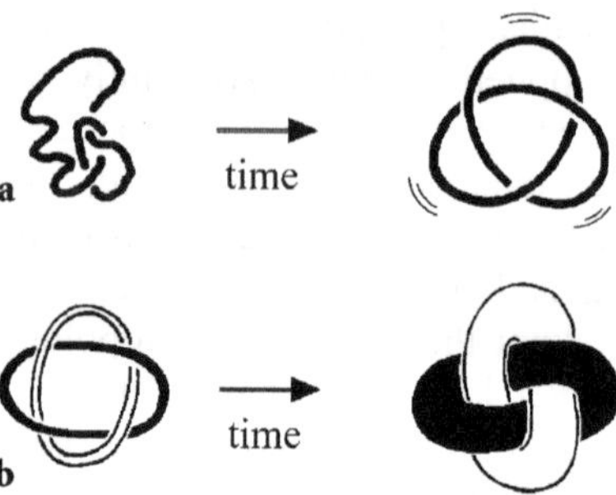

Figure 3. Examples where a topological barrier prevents further relaxation under a volume-preserving diffeomorphism: (a) an electrically charged trefoil knot is maximally extended by the Coulomb repulsion forces to its minimum energy state; (b) a magnetic link attains ground state energy by the action of the Lorentz force on the magnetic volume.

minimizing the knot energy (defined by an appropriate functional), by shrinking or extending the length as far as possible, depending on the potential, to attain an ideal shape (Stasiak et al., 1998). Questions relating to topology and geometry of ideal shapes, and uniqueness of minimum energy states, pose challenging problems at the crossroads of topology, differential geometry, functional analysis, and numerical simulation.

Physical Knots

By physical knots we mean tube models, centered around the knot K, with length $L(K)$, tubular neighborhood of radius $r(K)$, and volume $V(K)$. The tube is filled by vector field lines, whose distribution gives physical properties in terms, for example, of elasticity, vorticity, or magnetic field. A wide variety of filamentary systems present in nature at very different scales can be modeled by physical knots: from DNA molecules, polymer chains, vortex filaments, to elastic cords, strings, and magnetic flux tubes. In fluid systems, the action at a microscopic level of physical processes, such as viscosity and resistivity, may imply changes in knot topology by local recombination of the knot strands (known as knot surgery) and consequential rearrangement of energy distribution. In elastic systems, the material breaking point and internal critical twist are strongly influenced by knot strength and rope length, the latter given by the ratio L/r. All these systems are free to relax their internal energy to states of equilibrium: lower bounds on equilibrium energy for given measures of topological complexity (based, for example, on crossing number information) can be expressed by relationships of the kind

$$E_{\min} \geq h(c, \Phi, V, n), \quad (5)$$

where $E_{\min}$ is the equilibrium energy and $h(\cdot)$ gives the relationship between physical quantities—such as flux Φ, number of components n, knot volume V—and topology, given here by the crossing number c.

Understanding the interplay between topology and energy localization and redistribution can be very important in many fields of science and applications.

RENZO L. RICCA

See also **Differential geometry; Dynamical systems; Structural complexity; Topology**

Further Reading

Adams, C. 1994. *The Knot Book*, New York: W.H. Freeman
Arnol'd, V.I. & Khesin, B.A. 1998. *Topological Methods in Hydrodynamics*, Berlin and New York: Springer
Atiyah, M. 1990. *The Geometry and Physics of Knots*, Cambridge and New York: Cambridge University Press
Ghrist, R., Holmes, P. & Sullivan, M. 1997. *Knots and Links in Three-Dimensional Flows*, Berlin and New York: Springer
Hoste, J., Thistlethwaite, M. & Weeks, J. 1998. The first 1,701,936 knots. *Mathematical Intelligencer*, 20: 33–48
Kauffman, L. (editor). 1995. *Knots and Applications*, Singapore: World Scientific
Millett, K.C. & Sumners, D.W. (editors). 1994. *Random Knotting and Linking*, Singapore: World Scientific
Ricca, R.L. (editor). 2001. *An Introduction to the Geometry and Topology of Fluid Flows*, Dordrecht: Kluwer
Stasiak, A., Katritch, V. & Kauffman, L.H. (editors). 1998. *Ideal Knots*, Singapore: World Scientific

KOCH CURVE

See **Fractals**

KOLMOGOROV CASCADE

The velocity fluctuations of a high Reynolds number flow in a three-dimensional velocity field are typically dispersed over all possible wavelengths of the system, from the smallest scales, where viscosity dominates the advection and dissipates the energy of fluid motion, to the effective size of the system. This is not so bizarre: our everyday experience tells us it is so. On the corner of a city street, one might watch the fluttering and whirling of a discarded tram ticket as it is swept by an updraught, driven by localized thermal gradients from traffic or air-conditioning units; later, on the television news, one might see reports or predictions of storms on the city or district scale, and a weather map with isobars spanning whole continents. If you are a sailor you will know how to sail, or not, the multi-scaled surface of a turbulent ocean (Figure 1). The mechanism for this dispersal is vortex stretching and tilting: a conservative process whereby interactions between vorticity and velocity gradients create smaller and smaller eddies with amplified vorticity, until viscosity takes over (Tennekes & Lumley, 1972; Chorin, 1994).

An alternative, crude but picturesque, description of multi-scale turbulence was offered by the early 20th century meteorologist Lewis Fry Richardson (1922) in an evocative piece of doggerel: "big whirls have little whirls that feed on their velocity, and little whirls have lesser whirls and so on to viscosity." Richardson's often-quoted rhyme is apparently a parody of Irish satirist Jonathan Swift's verse: "So, naturalists observe, a flea—Has smaller fleas that on him prey—And these have smaller still to bite'—And so proceed *ad infinitum*."

Figure 1. Turbulent action on many different scales in a high Reynolds number flow: woodcut print by Katsushika Hokusai (1760–1849).

The statistics of the velocity fluctuation distribution in turbulent flows were quantified rather more elegantly and rigorously by the mathematician Andrei N. Kolmogorov (1941b), who derived the subsequently famous "$-5/3$ law" for the energy spectrum of the intermediate scales, or inertial scale subrange, of high Reynolds number flows which are ideally homogeneous (or statistically invariant under translation) and isotropic (or statistically invariant under rotation and reflection) in three velocity dimensions. Two thorough, but different in style and emphasis, accounts of Kolmogorov's turbulence work are Monin & Yaglom (1971) and Frisch (1995).

Kolmogorov's idea was that the velocity fluctuations in the inertial subrange are independent of initial and boundary conditions (i.e., they have no memory of the effects of anisotropic excitation at smaller wave numbers). The turbulent motions in this subrange, therefore, show universal statistics, and the flow is self-similar. From this premise Kolmogorov proposed the first hypothesis of similarity as: "For the locally isotropic turbulence the [velocity fluctuation] distributions F_n are uniquely determined by the quantities ν, the kinematic viscosity, and ε, the rate of average dispersion of energy per unit mass [energy flux]." His second hypothesis of similarity is: "For pulsations [velocity fluctuations] of intermediate orders where the length scale is large compared with the scale of the finest pulsations, whose energy is directly dispersed into heat due to viscosity, the distribution laws F_n are uniquely determined by F and do not depend on ν."

Kolmogorov derived the form of the distribution or energy spectrum, which we denote as $\mathcal{E}(k)$, where k is the wave number given by $k^2 = k_x^2 + k_y^2 + k_z^2$, over the inertial subrange simply by dimensional analysis. By the first and second hypotheses, the spectrum must be a function of the energy flux and wave number and independent of the viscosity or any other parameters:

$$\mathcal{E}(k) = f(\varepsilon, k).$$

By reference to the table below (after Vallis, 1999) we find

$$\begin{aligned}\mathcal{E}(k) &\sim \varepsilon^{2/3} g(k) \quad \text{(since } k \text{ is time-independent)} \\ &= C\varepsilon^{2/3} k^{-5/3}, \end{aligned} \qquad (1)$$

where C is a dimensionless constant which Kolmogorov (and subsequently many others, see Sreenivasan, 1995) deduced from experimental data to be of order 1.

Quantity	Dimension
Wave number	1/length
Energy per unit mass	length^2/time^2
Energy spectrum $\mathcal{E}(k)$	length^3/time^2
Energy flux ε	energy/time $\sim$ length^2/time^3

The physical picture associated with Equation (1) is that the kinetic energy of large-scale motions (whirls or eddies) is successively subdivided and redistributed among stepwise increasing wave number components (or smaller and smaller whirls and eddies), until the action of viscosity becomes competitive. Although this process has come to be known as the "Kolmogorov cascade," the cascade metaphor was not used by Kolmogorov. Its first use in this context is apparently due to Onsager (1945), who also highlights another assumption underlying the $-5/3$ law: that the modulation of a given Fourier component of the velocity field is mostly due to those others that belong to wave numbers of comparable magnitude.

So Kolmogorov's energy distribution says that ε is the only relevant parameter for turbulence in the inertial scale range. Can this really be true? Does the notorious "problem of turbulence" really boil down to such a simple relation for intermediate wavenumbers? (The fabled Problem of Turbulence was well summed up by Horace Lamb in 1932: "When I die and go to Heaven there are two matters on which I hope enlightenment. One is quantum electro-dynamics and the other is turbulence of fluids. About the former, I am really rather optimistic.") Understandably, for a turbulence result that seems so simple and universal, so flimsily derived yet so powerful, much effort has gone into verifying the wave number spectrum, Equation (1). It is difficult to create extremely high Reynolds number flows in the laboratory, but they exist naturally in the ocean. The first and still the most exciting verification of Equation (1) was carried out by Grant et al. (1962), who made a remarkable series of measurements of turbulent velocities from a ship in the Seymour Narrows, part of the Discovery Passage on the west coast of Canada, where the Reynolds number is $\sim 10^8$ (see Figure 2). A spectral exponent close to $-5/3$ has since been measured many

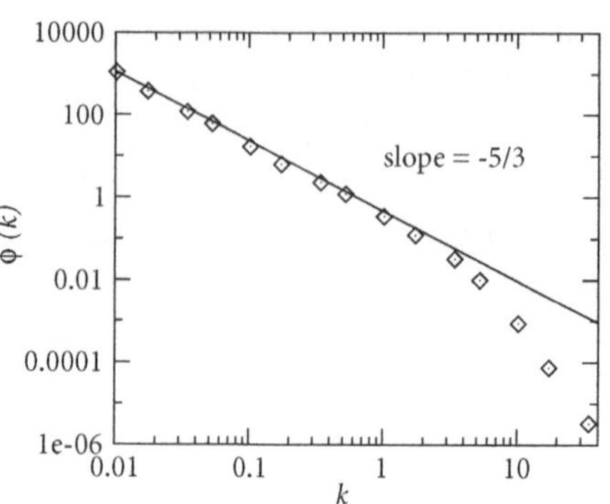

Figure 2. Data re-plotted from Grant et al. (1962), showing a Kolmogorov cascade over nearly three decades. $\phi(k)$ is the measured one-dimensional spectrum function, related to the three-dimensional spectrum function as $\mathcal{E}(k) = k^2\partial^2\phi(k)/\partial k^2 - k\partial\phi(k)/\partial k$.

times in materially different flows with high Reynolds number (e.g., Zocchi et al. (1994) in helium).

All this would seem to wrap up the problem of turbulence in the inertial scale range. Or does it? There must surely be a catch somewhere! As usual, the devil is in the details. Kolmogorov himself made a "refinement," as he delicately put it, of his hypotheses (Kolmogorov, 1962). It relates to the problem of small-scale intermittency, or the uneven distribution in space of the small scales. Clearly, intermittency is inherited from initial and boundary conditions, and the side-effects on ε, the assumed-constant rate of energy transfer, are not insignificant.

In fact, there is now quite a log of complaints about the $-5/3$ law, despite its all-pervasive influence on turbulence theoretical and experimental research:

- The hypothesis of local isotropy refers to infinite Reynolds number so is not applicable to a real fluid.
- No-one has ever extracted the $-5/3$ law from the Navier–Stokes equation or vice versa.
- Is it not a circular argument that to define an inertial subrange one has to assume a cascade process, and to postulate a cascade one has to assume that an inertial subrange exists?
- What about stochastic backscatter?
- Direct interaction between large and small scales can short-circuit the cascade.
- Katul et al. (2003) found that the effects of boundary conditions were evident in the inertial subrange of an atmospheric surface layer.
- The $-5/3$ law is demonstrably invalid in two dimensions. And so on.

What is the verdict on the Kolmogorov cascade? Chorin (1994, pp. 55–57) has a bet each way; in the light of experimental verifications of the $-5/3$ law, he considers that it may be correct *despite* flaws in the arguments supporting it. The scenario proposed as being entirely consistent with Kolmogorov's theory is that energy can and does slosh back and forth across the spectrum, once the inertial range has been set up, with the energy dissipation ε being the (presumably average) difference between energy flows in both wave number directions.

The Kolmogorov cascade is starting to sound less and less like a waterfall, which *is* one-way, and more and more like an energy exchange network.

ROWENA BALL

See also **Chaos vs. turbulence; Dimensional analysis; Navier–Stokes equation; Turbulence**

Further Reading

Chorin, A.J. 1994. *Vorticity and Turbulence*, New York: Springer

Frisch, U. 1995. *Turbulence: The Legacy of A. N. Kolmogorov*, Cambridge and New York: Cambridge University Press

Grant, H.L., Stewart, R.W. & Moilliet, A. 1962. Turbulence spectra from a tidal channel. *Journal of Fluid Mechanics*, 12: 241–268

Katul, G.G., Angelini, C., De Canditiis, D., Amato, U., Vidakovic, B. & Albertson, J.D. 2003. Are the effects of large scale flow conditions really lost through the turbulent cascade? *Geophysical Research Letters*, 30(4), 1164,doi:10.1029/2002GL015284

Kolmogorov, A.N. 1941a. Local structure of turbulence in an incompressible fluid for very large Reynolds numbers. *Comptes rendus (Doklady) de l'Academie des Sciences de l'U.R.S.S.*, 31: 301–305 Reprinted in, S.K. Friedlander & L. Topper 1961. *Turbulence: Classic Papers on Statistical Theory*, New York: Interscience Publishers

Kolmogorov, A.N. 1941b. On degeneration of isotropic turbulence in an incompressible viscous liquid. *Comptes Rendus (Doklady) de l'Academie des Sciences de l'U.R.S.S.*, 31: 538–540, *ibidem.*

Kolmogorov, A.N. 1962. A refinement of previous hypotheses concerning the local structure of turbulence in a viscous incompressible fluid at high Reynolds number. *Journal of Fluid Mechanics*, 13: 82–85

Monin, A. S. & Yaglom, A.M. 1971. *Statistical Fluid Mechanics: Mechanics of Turbulence*, vol. 2, Cambridge: MIT Press

Onsager, L. 1945. The distribution of energy in turbulence. *Physical Review*, 68: 286

Richardson, L.F. 1922. *Weather Prediction by Numerical Process*, Cambridge: Cambridge University Press

Sreenivasan, K.R. 1995. On the universality of the Kolmogorov constant. *Physics of Fluids*, 7(11), 2778–2784

Tennekes, H. & Lumley, J.A. 1972. *A First Course in Turbulence*, Cambridge: MIT Press

Vallis, G. 1999. Geostrophic turbulence: the macroturbulence of the atmosphere and ocean. Lecture notes, www.gfdl.gov/~gkv/geoturb/

Zocchi, G., Tabeling, P., Maurer, J. & Willaime, H. 1994. Measurement of the scaling of the dissipation at high Reynolds numbers. *Physical Review* E, 50(5): 3693–3700

KOLMOGOROV, PETROVSKY AND PISCOUNOFF EQUATION

See **Diffusion**

KOLMOGOROV–ARNOL'D–MOSER THEOREM

In 1954, the Russian mathematician Andrei N. Kolmogorov, already famous for his contribution to measure theory and probability theory, delivered to the International Conference of Mathematics in Amsterdam, a lecture based on a paper entitled "On the General Theory of Dynamical Systems and Classical Mechanics." In this lecture he addressed the question of how much structure remains when a "regular" (or integrable) Hamiltonian system is replaced by one that differs from the original by only a small perturbation. Kolmogorov's original paper was not completely detailed, and clarifying work was done independently in the early 1960s by Jürgen Moser and Vladimir Arnol'd (the latter had been a student of Kolmogorov). Since then, a large body of work on this subject within the mathematical theory of Hamiltonian dynamics has been assembled.

Thus, the so-called Kolmogorov–Arnol'd–Moser (KAM) theorem is not a single theorem but rather a body of work with a common theme, which can be roughly described as the persistence of quasi-periodic motions in perturbed systems. These results were a breakthrough in the understanding of the ergodic properties of dynamical systems. The original theorems dealt with Hamiltonian systems and their discrete analogues, while more recent results have been extended to volume-preserving systems, reversible systems, and dissipative systems.

The central ideas in Kolmogorov's original paper can be formulated as follows. Under some general assumptions, the phase space M^{2n} of an integrable Hamiltonian system is foliated by invariant tori: $\mathbb{T}^n$. This system will be referred to as the unperturbed system. In the neighborhood of each torus are defined action-angle variables $(I, \phi(\text{mod } 2\pi))$, such that the Hamiltonian H_0 is the function of only action variables I.

The motion on each torus is conditionally periodic with frequency vector $\omega(I) = \partial H_0/\partial I$. A torus is said to be nonresonant if all the n frequencies are rationally independent (incommensurable). The unperturbed system is called nonresonant if the frequencies are functionally independent:

$$\det\left(\frac{\partial\omega}{\partial I}\right) = \det\left(\frac{\partial^2 H_0}{\partial I^2}\right) \neq 0.$$

In a nondegenerate system, the nonresonant tori form a dense set of full measure. The resonant tori form a set of measure zero, which, however, is also dense.

The system with the Hamiltonian

$$H(I, \phi, \varepsilon) = H_0(I) + \varepsilon H_1(I, \phi, \varepsilon)$$

is called the perturbed system, where the function εH_1 is the perturbation. Note that usually the perturbed system is no longer integrable. Kolmogorov's theorem describes the fate of the nonresonant tori under perturbation.

Theorem (A.N. Kolmogorov). *If the unperturbed system is nondegenerate, then for a sufficiently small ε most nonresonant invariant tori do not vanish but are only slightly deformed, so that in the phase space of the perturbed system there are invariant tori densely filled with conditionally periodic phase curves winding around them, with the number of independent frequencies equal to the number of degrees of freedom. These invariant tori form a majority in the sense that the measure of the complement to their union is small when the perturbation is small.*

The persistence of quasi-periodic motion is consistent with numerical experiments, well-known examples being the standard map (also called the Chirikov map) and the Hénon–Heiles system.

Discussion

KAM theory has several important implications, in particular for the stability theory. Suppose that phase space is four-dimensional. A perturbed system always has one first integral, the Hamiltonian function $H(I, \phi, \varepsilon)$ itself. The energy levels $H = h$ are three-dimensional, while the invariant tori are two-dimensional. Thus, a trajectory that starts in a region between two invariant tori of the perturbed system is forever trapped in the region between the tori. This means that the values of the action variables remain forever near their initial values, which in turn implies stability.

If, however, the number n of degrees of freedom is greater than two, the n-dimensional tori do not separate the $(2n-1)$-dimensional energy level manifold into disjoint regions, and the invariant tori do not prevent a phase curve from wandering far away. There are examples of such drift, when the action variables change from their original value by a quantity of order 1, an effect known as "Arnol'd diffusion" (*See* **Arnol'd diffusion**). Although in the latter case the system is unstable, KAM theory does guarantee "metric stability," that is, stability for most initial conditions.

Examples

1. Assume the masses of the planets are sufficiently small compared with the mass of the Sun in the gravitational n-body problem. Then a large portion of the region of phase space corresponding to unperturbed motion of all planets (on identically oriented Keplerian ellipses having small eccentricities and inclinations) is filled up by conditionally periodic motions. KAM theory provides important estimates for the long-time evolution of such a system. Note that Henri Poincaré's groundbreaking work *Les méthodes nouvelles de la mécanique céleste* (1892) was motivated by questions of the stability of the solar system.
2. Consider the motion of a heavy rigid body fixed at a point. If the kinetic energy of the body is sufficiently large in comparison with its potential energy at the initial moment of time, then the length of the angular momentum vector and its inclination to the horizon remain forever near their initial values, provided that the initial values of the energy and the angular momentum differ sufficiently from values for which the body can rotate around its medium principal axis.
3. For plasma confinement in toroidal chambers, the plasma particles tend to follow magnetic field lines. The equations for these magnetic field lines can be put into Hamiltonian form. When the system is azimutally symmetric, the Hamiltonian is integrable, deviation from azimutal symmetry creates a perturbation, and KAM theory is applicable.

M.V. Deryabin and P.G. Hjorth

See also **Hénon–Heiles system; Standard map**

Further Reading

Arnol'd, V.I. 1962. The classical theory of perturbations and the problem of stability of planetary systems. *Soviet Mathematics Doklady*, 3: 1008–1012

Arnol'd, V.I. 1963. Proof of A.N. Kolmogorov's theorem on the preservation of quasiperiodic motions under small perturbations of the Hamiltonian. *Russian Mathematical Surveys*, 18(5): 9–36

Arnol'd, V.I. (editor). 1988. *Dynamical Systems III*, Berlin and New York: Springer

Kolmogorov, A.N. 1954. Théorie générale des systémes dynamiques et mécanique classique. In *Proceedings of the International Congress of Mathematicians*, vol. 1, Amsterdam: North-Holland, pp. 315–333

Moser, J.K. 1962. On invariant curves of area-preserving mappings of an annulus. *Nachrichten der Akademic der Wissenschatten in Goettingen II, Mathematisch-physikalische klasie* 1–20

KORTEWEG–DE VRIES EQUATION

Historical Introduction

The Korteweg–de Vries (KdV) equation, given here in canonical form,

$$u_t + 6uu_x + u_{xxx} = 0, \qquad (1)$$

is widely recognized as a paradigm for the description of weakly nonlinear long waves in many branches of physics and engineering. Here, $u(x, t)$ is an appropriate field variable, t is the time, and x is the space coordinate in the relevant direction. It describes how waves evolve under the competing but comparable effects of weak nonlinearity and weak dispersion. Indeed, if it is supposed that x-derivatives scale as ε where ε is the small parameter characterizing long waves (i.e., typically the ratio of a relevant background length scale to a wavelength scale), then the amplitude scales as ε^2

and the time evolution takes place on a scale of ε^{-3}. The KdV equation is characterized by its family of solitary wave solutions,

$$u = a\,\mathrm{sech}^2(\gamma(x - Vt)),\quad \text{where} \quad V = 2a = 4\gamma^2. \tag{2}$$

This solution describes a family of steady isolated wave pulses of positive polarity, characterized by the wave number γ; note that the speed V is proportional to the wave amplitude a and also to the square of the wave number γ^2.

The KdV equation (1) owes its name to the famous paper of Diederik Korteweg and Hendrik de Vries, published in 1895, in which they showed that small-amplitude long waves on the free surface of water could be described by the equation

$$\zeta_t + c\zeta_x + \frac{3c}{2h}\zeta\zeta_x + \frac{ch^2}{6}\delta\zeta_{xxx} = 0. \tag{3}$$

Here, $\zeta(x,t)$ is the elevation of the free surface relative to the undisturbed depth h, $c = (gh)^{1/2}$ is the linear long wave phase speed, and $\delta = 1 - 3B$, where $B = \sigma/gh^2$ is the Bond number measuring the effects of surface tension ($\rho\sigma$ is the coefficient of surface tension and ρ is the water density). Transformation to a reference frame moving with the speed c (i.e., (x,t) is replaced by $(x - ct, t)$, and subsequent rescaling readily establishes the equivalence of (1) and (3). Although Equation (1) now bears the name KdV, it was apparently first obtained by Joseph Boussinesq (1877) (see Miles (1980) and Pego and Weinstein (1997) for historical discussions on the KdV equation). Korteweg and de Vries found the solitary wave solutions (2), and importantly, they showed that they are the limiting members of a two-parameter family of periodic traveling-wave solutions, described by elliptic functions and commonly called cnoidal waves,

$$u = b + a\,\mathrm{cn}^2(\gamma(x - Vt)|m),\quad \text{where} \quad V = 6b + 4(2m - 1)\gamma^2,\ a = 2m\gamma^2. \tag{4}$$

Here, $\mathrm{cn}(x|m)$ is the Jacobi elliptic function of modulus m $(0 < m < 1)$. As $m \to 1$, $\mathrm{cn}(x|m) \to \mathrm{sech}(x)$, and then the cnoidal wave (4) becomes the solitary wave (2), now riding on a background level b. On the other hand, as $m \to 0$, $\mathrm{cn}(x|m) \to \cos 2x$, and so the cnoidal wave (4) collapses to a linear sinusoidal wave (note that in this limit, $a \to 0$).

This solitary wave solution found by Korteweg and de Vries had earlier been obtained directly from the governing equations (in the absence of surface tension) independently by Boussinesq (1871, 1877) and Lord Rayleigh (1876), who were motivated to explain the now very well-known observations and experiments of John Scott Russell (1844). Curiously, it was not until quite recently that it was recognized that the KdV equation is not strictly valid if surface tension is taken into account and $0 < B < \frac{1}{3}$, as then there is a resonance between the solitary wave and very short capillary waves.

After the ground-breaking work of Korteweg and de Vries, interest in solitary water waves and the KdV equation declined until the dramatic discovery of the soliton by Zabusky and Kruskal in 1965. Through numerical integrations of the KdV equation, they demonstrated that the solitary wave (2) could be generated from quite general initial conditions, and could survive intact collisions with other solitary waves, leading them to coin the term *soliton*. Their remarkable discovery, followed almost immediately by the theoretical work of Gardner et al. (1967) showing that the KdV equation was *integrable* through an inverse scattering transform, led to many other startling discoveries and marked the birth of soliton theory as we know it today (*See* **Solitons, a brief history**). The implication is that the solitary wave is the key component needed to describe the behavior of long, weakly nonlinear waves.

An alternative to the KdV equation is the Benjamin–Bona–Mahony (BBM) equation in which the linear dispersive term $c\zeta_{xxx}$ in (3) is replaced by $-\zeta_{xxt}$. It has the same asymptotic validity as the KdV equation, and since it has rather better high wave number properties, it is somewhat easier to solve numerically. However, it is not integrable and, consequently, has not attracted the same interest as the KdV equation.

Both the KdV and BBM equations are unidirectional. A two-dimensional version of the KdV equation is the KP equation (Kadomtsev & Petviashvili, 1970),

$$(u_t + 6uu_x + u_{xxx})_x \pm u_{yy} = 0. \tag{5}$$

This equation includes the effects of weak diffraction in the y-direction, in that y-derivatives scale as ε^2 whereas x-derivatives scale as ε. Like the KdV equation it is an integrable equation. When the "+"-sign holds in (5), this is the KP2 equation, and it can be shown that then the solitary wave (2) is stable to transverse disturbances. On the other hand, if the "−"-sign holds, this is the KP1 equation for which the solitary wave is unstable; instead this equation supports "lump" solitons. Both KP1 and KP2 are integrable equations. To take account of stronger transverse effects and/or to allow for bi-directional propagation in the x-direction, it is customary to replace the KdV equation with a Boussinesq system of equations; these combine the long wave approximation to the dispersion relation with the leading-order nonlinear terms and occur in several asymptotically equivalent forms.

Although the KdV equation (1) is historically associated with water waves, it in fact occurs in many other physical contexts, where it arises as an asymptotic multiscale reduction from the relevant governing equations. Typically, the outcome is

$$A_t + cA_x + \mu A A_x + \lambda A_{xxx} = 0. \quad (6)$$

Here, c is the relevant linear long wave speed for the mode whose amplitude is $A(x,t)$, while μ and λ, the coefficients of the quadratic nonlinear and linear dispersive terms, respectively, are determined from the properties of this same linear long wave mode and, like c, depend on the particular physical system being considered. Note that the linearization of (6) has the linear dispersion relation $\omega = ck - \lambda k^3$ for linear sinusoidal waves of frequency ω and wave number k; this expression is just the truncation of the full dispersion relation for the wave mode being considered, and immediately identifies the origin of the coefficient λ. Similarly, the coefficient μ can be identified with the an amplitude-dependent correction to the linear wave speed. Transformation to a reference frame moving with a speed c and subsequent rescaling shows that (6) can be transformed to the canonical form (1). Equations of the form (6) arise in the study of internal solitary waves in the atmosphere and ocean, mid-latitude and equatorial planetary waves, plasma waves, ion-acoustic waves, lattice waves, waves in elastic rods, and in many other physical contexts (see, for instance, Ablowitz & Segur, 1981; Dodd et al., 1982; Drazin & Johnson, 1989; Grimshaw, 2001).

In some physical situations, it is necessary to complement the KdV equation (6) with a higher-order cubic nonlinear term of the form $\nu A^2 A_x$. After transformation and rescaling, the amended equation (6) can be transformed to the so-called Gardner equation

$$u_t + 6uu_x + 6\delta u^2 u_x + u_{xxx} = 0. \quad (7)$$

Like the KdV equation, the Gardner equation is integrable by the inverse scattering transform. Here the coefficient δ can be either positive or negative, and the structure of the solutions depends crucially on which sign is appropriate. Again, in some physical situations, solitary waves propagate through a variable environment which means that the coefficients c, μ, and λ in (6) are functions of x, while an additional term $c(\sigma_x/2\sigma)A$ needs to be included, where $\sigma(x)$ is a magnification factor. After transforming to new variables, $\theta = (\int^x dx/c) - t, x$ with $U = \sigma^{1/2}u$, the variable-coefficient KdV equation is obtained,

$$U_x + \alpha(x)UU_\theta + \beta(x)U_{\theta\theta\theta} = 0. \quad (8)$$

Here, $\alpha = \mu/c\sigma^{1/2}$, $\beta = \lambda/c^3$. In general, this is not an integrable equation and must be solved numerically, although we shall exhibit some asymptotic solutions below. Another modification of the KdV equation occurs when it is necessary to take account of background rotation, leading to the rotation-modified KP equation (see, for instance, Grimshaw, 2001), in which a term $-f^2u$ is added to the left-hand side of Equation (5), where f is a measure of the background rotation.

Solitons

The remarkable discovery of Gardner et al. (1967) that the KdV equation was integrable through an inverse scattering transform marked the beginning of soliton theory. Their pioneering work was followed by the work of Zakharov and Shabat (1972) which showed that another well-known nonlinear wave equation, the nonlinear Schrödinger equation, was also integrable by an inverse scattering transform. Their demonstration that the integrability of the KdV equation was not an isolated result, was followed closely by analogous results for the modified KdV equation (Wadati, 1972) and the sine-Gordon equation (Ablowitz et al., 1973). In 1974, Ablowitz, Kaup, Newell, and Segur provided a generalization and unification of these results in the AKNS scheme. From this point there has been an explosive and rapid development of soliton theory in many directions (see, for instance, Ablowitz & Segur, 1981; Dodd et al., 1982; Newell, 1985; Drazin & Johnson, 1989).

For the KdV equation (1) the starting point is the Lax pair (Lax, 1968) for an auxiliary function $\phi(x,t)$,

$$L\phi \equiv -\phi_{xx} - u\phi = \lambda\phi, \quad (9)$$

$$\phi_t = B\phi \equiv (u_x + C)\phi + (4\lambda - 2u)\phi_x. \quad (10)$$

Here, $C(t)$ depends on the normalization of ϕ. The first of these equations (9), with suitable boundary conditions at infinity (see below) defines a spectral problem for ϕ in the spatial variable x with a spectral parameter λ, and with the time variable t as a parameter. The second equation (10) then describes how the spectral function ϕ evolves in time. If it is now assumed that λ is independent of time (i.e., $\lambda_t = 0$ then the KdV equation (1) is just the compatibility condition for these two equations (9, 10); that is, it emerges as a result of the condition that $(\phi_{xx})_t = (\phi_t)_{xx}$. In terms of the operators L, B defined in the Lax pair (9,10), the KdV equation can be written in the symbolic form $L_t = BL - LB$ (Lax, 1968). This form indicates the path to further generalizations, in that other nonlinear wave equations can be obtained by choosing different operators L, B. The general strategy for integration of the KdV equation now consists of three steps. Here, we will describe the process under the hypothesis that we seek solutions $u(x,t)$ of the KdV equation (1), which decay to zero sufficiently fast as $x \to \pm\infty$ and have the initial condition $u(x,0) = u_0(x)$. First, we insert the initial condition into the spectral problem (9) to obtain the scattering data (these will be defined precisely below). Then (10) is used to move the scattering data forward in time; it transpires that is a very simple process, and note in particular that the spectral parameter λ is independent of t and hence is determined by the initial condition. The third step is to invert the scattering data at time $t > 0$ and so recover $u(x,t)$; this is the most difficult step, but for the KdV equation can be reduced to solution of a linear integral equation. Thus, the three steps constitute a linear algorithm for the solution of the KdV equation, and it is in this sense that it is said that the Lax pair (9,10) constitutes integrability of the KdV equation.

The spectral problem (9) for the KdV equation consists of two parts. The *discrete* spectrum is found by seeking solutions such that $\phi \to 0$ as $x \to \pm\infty$, which requires that $\lambda < 0$. It can be shown that there then exists a *finite* set of discrete eigenvalues $\lambda = -\kappa_n^2, n = 1, 2, \ldots, N$, and corresponding real eigenfunctions ϕ_n such that

$$\phi_n \sim c_n \exp(-\kappa_n x) \quad \text{as} \quad x \to \infty. \tag{11}$$

There is a similar condition as $x \to -\infty$, namely, that $\phi_n \sim d_n \exp(\kappa_n x)$. The real constants c_n, d_n are determined once the normalization condition is satisfied, that is,

$$\int_{-\infty}^{\infty} \phi_n^2 \, \mathrm{d}x = 1. \tag{12}$$

The continuous spectrum consists of all $\lambda > 0$, and so we set $\lambda = k^2$ where k is real. Then we define the scattering problem for solutions $\phi(x; k)$ of (9) by the boundary conditions,

$$\phi \sim \exp(-\mathrm{i}kx) + R(k)\exp(\mathrm{i}kx) \quad \text{as } x \to \infty, \tag{13}$$

$$\phi \sim T(k)\exp(-\mathrm{i}kx) \quad \text{as} \quad x \to -\infty. \tag{14}$$

The scattering data then consists of the set $(\kappa_n, c_n, n = 1, 2, \ldots, N)$ together with the reflection coefficient $R(k)$. It is useful to note that $R(k)$ may be continued into the upper half of the complex k-plane, has there a set of simple poles at $k = \mathrm{i}\kappa_n$, and $R \to 1$ as $|k| \to \infty$.

The next step is to determine from (10) how the scattering data evolves in time (note that the dependence on time t has been suppressed in the preceding paragraph). First, we recall that the discrete eigenvalues κ_n are independent of t. Next, we multiply (10) by ϕ_n and integrate the result over all x; also, on using (9), it is readily found that

$$\frac{\mathrm{d}}{\mathrm{d}t}\int_{-\infty}^{\infty} \phi_n^2 \, \mathrm{d}x = C_n \int_{-\infty}^{\infty} \phi_n^2 \, \mathrm{d}x,$$

where the constant C in (10) here must be indexed with n to become C_n. But then the normalization condition (12) implies that $C_n = 0$. Now substitute (11) into (10) to show that

$$\frac{\mathrm{d}c_n}{\mathrm{d}t} = 4\kappa_n^3 c_n$$

$$\text{so that} \quad c_n(t) = c_n(0)\exp(\kappa_n^3 t). \tag{15}$$

For the continuous spectrum, the asymptotic expressions (13), (14) are substituted into (10). Now it is found that the constant $C(k) = 4\mathrm{i}k^3$, and that

$$\frac{\mathrm{d}R}{\mathrm{d}t} = 8\mathrm{i}k^3 R$$

$$\text{so that} \quad R(k; t) = R(0; k)\exp(8\mathrm{i}k^3 t). \tag{16}$$

Similarly, it can be shown that $T(k; t) = T(k; 0)$.

The final step is the inversion of the scattering data at time t to recover the potential $u(x, t)$ in 9). This is accomplished through the Marchenko integral equation for the function $K(x, y)$

$$K(x, y) + F(x + y) + \int_x^{\infty} K(x, z)\, F(y + z)\, \mathrm{d}z = 0. \tag{17}$$

Here the function $F(x)$ is known in terms of the scattering data at time t,

$$F(x) = \sum_{n=1}^{N} c_n^2(t)\exp(-\kappa_n x) + \frac{1}{2\pi}\int_{-\infty}^{\infty} R(k; t)\exp(\mathrm{i}kx)\, \mathrm{d}k. \tag{18}$$

Here the t-dependence of K, F has been suppressed as the linear integral equation (17) is solved with t fixed. Then

$$u(x, t) = 2\frac{\partial}{\partial x}\{K(x, x; t)\}, \tag{19}$$

where the t-dependence in K has been restored.

The inverse scattering transform described by (17), (18) enables one to find the solution of the KdV equation (1) for an arbitrary localized initial condition. The most important outcome is that as $t \to \infty$, the solution evolves into N rank-ordered solitons propagating to the right ($x > 0$), and some decaying radiation propagating to the left ($x < 0$),

$$u \sim \sum_{n=1}^{N} 2\kappa_n^2 \operatorname{sech}^2(\kappa_n(x - 4\kappa_n^2 t - x_n)) + \text{radiation}. \tag{20}$$

Here the N solitons are derived directly from the discrete spectrum, where each eigenvalue $-\kappa_n$ generates a soliton of amplitude $2\kappa_n^2$, while the phase shifts x_n are determined from the constants $c_n(0)$. The continuous spectrum is responsible for the decaying radiation, which decays at each fixed $x < 0$ as $t^{-1/3}$.

The important special case when the reflection coefficient $R(k) \equiv 0$ leads to the N-soliton solution, for which there is no radiation. Indeed, the N-soliton solution can be obtained as an explicit solution of the Marchenko equation (17). We illustrate the procedure for $N = 1, 2$. First, for $N = 1$, $F(x) = c^2 \exp\kappa(x - 4\kappa^2 t)$, where we have omitted the subscript $n = 1$ for simplicity. Then seek a solution of

(17) in the form $K(x, y, t) = L(x, t)\exp(-\kappa y)$, where L can be found by simple algebra. The outcome is that

$$L(x,t) = \frac{-2\kappa c(0)^2 \exp(-\kappa x + 8\kappa^2 t)}{2\kappa + c(0)^2 \exp(-2\kappa x + 8\kappa^2 t)}.$$

Finally u is found from (19),

$$u = 2\kappa^2 \mathrm{sech}^2(\kappa(x - 4\kappa^2 t) - x_1).$$

This is just the solitary wave (2) of amplitude $2\kappa^2$; the phase shift x_1 is such that $c(0)^2 = 2\kappa \exp(2\kappa x_1)$. The procedure for $N = 2$ follows a similar course. Thus, with $R \equiv 0, N = 2$ in F (18), seek a solution of the Marchenko equation (17) in the form $K(x, y, t) = L_1(x, t)\exp(-\kappa_1 y) + L_2(x, t)\exp(-\kappa_2 y)$, and again $L_{1,2}$ can be found by simple algebra. The outcome is the two-soliton solution. For instance, with $\kappa_1 = 1, \kappa_2 = 2$, this is

$$u = 12\frac{3 + 4\cosh(2x - 8t) + \cosh(4x - 64t)}{[3\cosh(x - 28t) + \cosh(3x - 36t)]^2}. \quad (21)$$

It can be readily shown that

$$u \sim 8\mathrm{sech}^2(2(x - 16t \mp x_2) + 2\mathrm{sech}^2(x - 4t \mp x_1) \quad \text{as} \quad t \to \pm\infty, \quad (22)$$

where the phase shifts $x_{1,2} = (-\frac{1}{2}, \frac{1}{4})\ln 3$. Thus, the two-soliton solution describes the elastic collision of two solitons, in which each survives the interaction intact, and the only memory of the collision is the phase shifts; note that $x_1 < 0, x_2 > 0$, so that the larger soliton is displaced forward and the smaller soliton is displaced backward. The general case of an N-soliton is analogous and is essentially a sequence of pair-wise two-soliton interactions.

The integrability of the KdV equation (1) is also characterized by the existence of an infinite set of independent conservation laws. The most transparent conservation laws are

$$\int_{-\infty}^{\infty} u \, \mathrm{d}x = \text{constant}, \quad (23)$$

$$\int_{-\infty}^{\infty} u^2 \, \mathrm{d}x = \text{constant}, \quad (24)$$

$$\int_{-\infty}^{\infty} \left(u^3 - \frac{1}{2}u_x^2\right) \mathrm{d}x = \text{constant}, \quad (25)$$

which may be associated with the conservation of mass, momentum, and energy, resepectively. Indeed (23) is obtained from the KdV equation (1) by integrating over x, while (24), (25) are obtained in an analogous manner after first multiplying (1) by u, u^2, respectively. However, it transpires that these are just the first three conservation laws in an infinite set, where each successive conservation law contains a higher power of u than the preceding one. This may be demonstrated using the inverse scattering transform (see Ablowitz & Segur, 1981; Dodd et al., 1982; Newell, 1985). However, here we use the original method based on the Miura transformation as adapted by Gardner. The Miura transformation is

$$u = -v_x - v^2, \quad (26)$$

$$v_t - 6v^2 v_x + v_{xxx} = 0. \quad (27)$$

Here (27) is the modified KdV equation. Direct substitution of (26) into the KdV equation shows that if v solves the modified KdV equation (27), then u solves the KdV equation (1). This discovery was the starting point for the discovery of the inverse scattering transform, since if one considers (26) as an equation for v and writes $v = \phi_x/\phi$, followed by a Galilean transformation for u (i.e., $u \to u - \lambda, x \to x - 6\lambda t$), one obtains the spectral problem (9). Here we follow a different route and write

$$v = \frac{1}{2\varepsilon} - \varepsilon w,$$

which (after a shift $x \to x + 3t/2\varepsilon^2$) converts the mKdV equation (27) into the Gardner equation (7) with $\delta = -\varepsilon^2$. Apart from a constant, which may be removed by a Galilean transformation, the corresponding expression for u is the Gardner transformation,

$$u = w + \varepsilon w_x - \varepsilon^2 w^2. \quad (28)$$

Thus, if w solves the Gardner equation

$$w_t + 6ww_x - 6\varepsilon^2 w^2 w_x + w_{xxx} = 0, \quad (29)$$

then u solves the KdV equation (1).

Next, we observe that the Gardner equation (29) has the conservation law

$$\int_{-\infty}^{\infty} w \, \mathrm{d}x = \text{constant}. \quad (30)$$

Since $w \to u$ as $\varepsilon \to 0$, we write the formal asymptotic expansion

$$w \sim \sum_{n=0}^{\infty} \varepsilon^n w_n.$$

It follows from (30) that then

$$\int_{-\infty}^{\infty} w_n \, \mathrm{d}x = \text{constant},$$

for each $n = 0, 1, 2, \ldots$. But substitution of this same asymptotic expansion for w into (28) generates a sequence of expressions for w_n in terms of u, of which the first few are

$$w_0 = u, \quad w_1 = -u_x, \quad w_2 = -u^2 + u_{xx}.$$

Thus, we see that $n = 0, 2$ give the conservation laws (23), (24), respectively, while $n = 1$ is an exact differential. It may now be shown that all even values of

n yield nontrivial and independent conservation laws, while all odd values of n are exact differentials.

The KdV equation belongs to a class of nonlinear wave equations, which have Lax pairs and are integrable through an inverse scattering transform. It shares with these equations several other remarkable features, such as the Hirota bilinear form, Bäcklund transformations, and the Painlevé property. Detailed descriptions of these and other properties of the KdV equation can be found in the other entries and referenced texts.

Solitary Waves in a Variable Environment

In a variable environment, the governing equation which replaces (1) is the variable-coefficient KdV equation (8). In general, this is not an integrable equation and is usually solved numerically. However, there are two distinct limiting situations in which some analytical progress can be made. First, let it be supposed that the coefficients $\alpha(x), \beta(x)$ in (8) vary rapidly with respect to the wavelength of a solitary wave, and then consider the case when these coefficients make a rapid transition from the values α_-, β_- in $x<0$ to the values α_+, β_+ in $x>0$. Then a steady solitary wave can propagate in the region $x<0$, given by

$$U = a\,\mathrm{sech}^2(\gamma(\theta - Wx)),$$
$$\text{where} \quad W = \frac{\alpha_- a}{3} = 4\beta_-\gamma^2. \qquad (31)$$

It will pass through the transition zone $x \approx 0$ essentially without change. However, on arrival into the region $x>0$, it is no longer a permissible solution of (8), which now has constant coefficients α_+, β_+. Instead, with $x=0$, expression (31) now forms an effective initial condition for the new constant-coefficient KdV equation. Using the spectral problem (9) and the inverse scattering transform, the solution in $x>0$ can now be constructed; indeed, in this case, the spectral problem (9) has an explicit solution (e.g., Drazin & Johnson, 1989). The outcome is that the initial solitary wave *fissions* into N solitons and some radiation. The number N of solitons produced is determined by the ratio of coefficients $R=\alpha_+\beta_-/\alpha_-\beta_+$. If $R>0$ (i.e., there is no change in polarity for solitary waves), then $N = 1 + [((8R+1)^{1/2} - 1)/2]$ ($[\cdots]$ denotes the integral part); as R increases from 0, a new soliton (initially of zero amplitude) is produced as R successively passes through the values $m(m+1)/2$ for $m=1,2,\ldots$. But if $R<0$ (i.e., there is a change in polarity), no solitons are produced and the solitary wave decays into radiation. For instance, for water waves, $c=(gh)^{1/2}$, $\mu = 3c/2h$, $\lambda = ch^2/6$, $\sigma = c$, and so $\alpha = 3/(2hc^{1/2})$, $\beta = h^2/(6c^2)$, where h is the water depth. It can then be shown that a solitary water wave propagating from a depth h_- to a depth h_+ will fission into N solitons where N is given as above with $R=(h_-/h_+)^{9/4}$; if $h_- > h_+$, $N \geq 2$, but if $h_- < h_+$ then $N=1$ and no further solitons are produced (Johnson, 1973).

Next, consider the opposite situation when the coefficients $\alpha(x), \beta(x)$ in (8) vary slowly with respect to the wavelength of a solitary wave. In this case a multi-scale perturbation technique (see Grimshaw, 1979, or Grimshaw & Mitsudera, 1993) can be used in which the leading term is

$$U \sim A\,\mathrm{sech}^2\gamma\left(\theta - \int_{x_0}^{x} W\,\mathrm{d}x\right), \qquad (32)$$

where

$$W = \frac{\alpha A}{3} = 4\beta\gamma^2. \qquad (33)$$

Here the wave amplitude $a(x)$ and, hence, also $W(x), \gamma(x)$ are slowly varying functions of x. Their variation is most readily determined by noting that the variable-coefficient KdV equation (8) possesses a conservation law,

$$\int_{-\infty}^{\infty} U^2\,\mathrm{d}\theta = \text{constant}, \qquad (34)$$

which expresses conservation of wave-action flux. Substitution of (32) into (34) gives

$$\frac{2A^2}{3\gamma} = \text{constant},$$
$$\text{so that} \quad A = \text{constant}\left(\frac{\beta}{\alpha}\right)^{1/3}. \qquad (35)$$

This is an explicit equation for the variation of the amplitude $A(x)$ in terms of $\alpha(x), \beta(x)$. However, the variable-coefficient KdV equation (8) also has a conservation law for mass,

$$\int_{-\infty}^{\infty} U\,\mathrm{d}\theta = \text{constant}. \qquad (36)$$

Thus, although the slowly varying solitary wave conserves wave-action flux, it cannot simultaneously conserve mass. Instead, it is accompanied by a trailing shelf of small amplitude but long length scale given by U_s, so that the conservation of mass gives

$$\int_{-\infty}^{\phi} U_\mathrm{s}\,\mathrm{d}\theta + \frac{2A}{\gamma} = \text{constant},$$

where $\phi = \int_{x_0}^{x} W\,\mathrm{d}x$ ($\theta = \phi$ gives the location of the solitary wave) and the second term is the mass of the solitary wave (32). Differentiation then yields the amplitude $U_- = U_\mathrm{s}(\theta = \phi)$ of the shelf at the rear of the solitary wave,

$$U_- = \frac{3\gamma_x}{\alpha\gamma^2}. \qquad (37)$$

This shows that if the wavelength γ^{-1} increases (decreases) as the solitary wave deforms, then the trailing shelf amplitude U_- has the opposite (same) polarity as the solitary wave. Once U_- is known, the full shelf $U_s(\theta, x)$ is found by solving (8) with the boundary condition that $U_s(\theta = \phi) = U_-$ (see El & Grimshaw, 2002, where it is shown that the trailing shelf may eventually generate secondary solitary waves).

For a solitary water wave propagating over a variable depth $h(x)$, these results show that the amplitude varies as h^{-1}, while the trailing shelf has positive (negative) polarity relative to the wave itself accordingly as $h_x < (>)\ 0$. A situation of particular interest occurs if the coefficient $\alpha(x)$ changes sign at some particular location (note that in most physical systems the coefficient β of the linear dispersive term in (8) does not vanish for any x). This commonly arises for internal solitary waves in the coastal ocean, where typically in the deeper water, $\alpha < 0$, $\beta > 0$ so that internal solitary waves propagating shorewards are waves of depression. But in shallower water, $\alpha > 0$ and so only internal solitary waves of elevation can be supported. The issue then arises as to whether an internal solitary wave of depression can be converted into one or more solitary waves of elevation as the critical point, where α changes sign, is traversed. This problem has been intensively studied (see, for instance, Grimshaw et al., 1998 and the references therein), and the solution depends on how rapidly the coefficient α changes sign. If α passes through zero rapidly compared with the local width of the solitary wave, then the solitary wave is destroyed and converted into a radiating wave train (see the discussion above in the first paragraph of this section). On the other hand, if α changes sufficiently slowly for the present theory to hold (i.e., (35) applies), we find that as $\alpha \to 0$, then $A \to 0$ in proportion to $|\alpha|^{1/3}$, while $U_- \to \infty$ as $|\alpha|^{-8/3}$. Thus, as the solitary wave amplitude decreases, the amplitude of the trailing shelf, which has the opposite polarity, grows indefinitely until a point is reached just prior to the critical point where the slowly varying solitary wave asymptotic theory fails. A combination of this trailing shelf and the distortion of the solitary wave itself then provide the appropriate "initial" condition for one or more solitary waves of the opposite polarity to emerge as the critical point is traversed. However, it is clear that in situations, as here, where $\alpha \approx 0$, it will be necessary to include a cubic nonlinear term in (8), thus converting it into a variable-coefficient Gardner equation (cf. (7)). This case has been studied by Grimshaw et al. (1999), who show that the outcome depends on the sign of the coefficient (ν) of the cubic nonlinear term at the critical point. If $\nu > 0$, so that solitary waves of either polarity can exist when $\alpha = 0$, then the solitary wave preserves its polarity (i.e., remains a wave of depression) as the critical point is traversed. On the other hand, if $\nu < 0$, so that no solitary wave can exist when $\nu = 0$, then the solitary wave of depression may be converted into one or more solitary waves of elevation.

Roger Grimshaw

See also **Inverse scattering method or transform; Kadomtsev–Petviashvili equation; Solitons; Water waves**

Further Reading

Ablowitz, M.J., Kaup, D.J., Newell, A.C. & Segur, H. 1973. Method for solving the sine-Gordon equation. *Physcs Letters*, 30: 1262–1264

Ablowitz, M.J., Kaup, D.J., Newell, A.C. & Segur, H. 1974. The inverse scattering transform–Fourier analysis for nonlinear problems. *Studies in Applied Mathematics*, 53: 249–315

Ablowitz, M.J. & Segur, H. 1981. *Solitons and the Inverse Scattering Transform*, Philadelphia: SIAM

Boussinesq, M.J. 1871. Theórie de l'intumescence liquid appellée onde solitaire ou de translation, se propageant dans un canal rectangulaire. *Comptes Rendus Acad. Sci (Paris)*, 72: 755–759

Boussinesq, M.J. 1877. Essai sur la theorie des eaux courantes, *Memoires presentees par diverse savants a l'Academie des Sciences Inst. France (Series 2)* 23: 1–680

Dodd, R.K., Eilbeck, J.C., Gibbon, J.D. & Morris, H.C. 1982. *Solitons and Nonlinear Wave Equations*, London: Academic

Drazin, P.G. & Johnson, R.S. 1989. *Solitons: An Introduction*, Cambridge and New York: Cambridge University Press

El, G.A. & Grimshaw, R. 2002. Generation of undular bores in the shelves of slowly-varying solitary waves. *Chaos*, 12: 1015–1026

Gardner, C.S., Greene, J.M., Kruskal, M.D. & Miura, R.M. 1967. Method for solving the Korteweg–de Vries equation. *Physical Review Letters*, 19: 1095–1097

Grimshaw, R. 1979. Slowly varying solitary waves. I Korteweg–de Vries equation. *Proceedings of the Royal Society*, 368A: 359–375

Grimshaw, R. 2001. Internal solitary waves. In *Environmental Stratified Flows*, edited by Boston: Kluwer, Chapter 1: 1–28

Grimshaw, R. & Mitsudera, H. 1993. Slowly-varying solitary wave solutions of the perturbed Korteweg–de Vries equation revisited. *Studies in Applied Mathematics*, 90: 75–86

Grimshaw, R., Pelinovsky, E. & Talipova, T. 1998. Solitary wave transformation due to a change in polarity. *Studies in Applied Mathematics*, 101: 357–388

Grimshaw, R., Pelinovsky, E. & Talipova, T. 1999. Solitary wave transformation in a medium with sign-variable quadratic nonlinearity and cubic nonlinearity. *Physica* D, 132: 40–62

Johnson, R.S. 1973. On the development of a solitary wave moving over an uneven bottom. *Proceedings of the Cambridge Philosophical Society*, 73: 183–203

Kadomtsev, B.B. & Petviashvili, V.I. 1970. On the stability of solitary waves in weakly dispersive media. *Soviet Physics Doklady*, 15: 539–541

Korteweg, D.J. & de Vries, H. 1895. On the change of form of long waves advancing in a rectangular canal, and on a new type of long stationary waves. *Philosophical Magazine*, 39: 422–443

Lax, P.D. 1968. Integrals of nonlinear equations of evolution and solitary waves. *Communications in Pure and Applied Mathematics*, 21: 467–490

Miles, J.W. 1980. Solitary waves. *Annual Review of Fluid Mechanics* 12: 11–43

Newell, A.C. 1985. Solitons in mathematics and physics. In *CBMS-NSF Series in Applied Mathematics*, Vol. 48, edited by Philadelphia: SIAM

Pego, R.L. & Weinstein, M.J. 1997. Convective linear stability of solitary waves for Boussinesq equations. *Studies in Applied Mathematics*, 99: 311–375

Rayleigh, Lord. 1876. On waves. *Philosophical Magazine* 1: 257–279

Russell, J.S. 1844. Report on waves, *14th Meeting of the British Association for the Advancement of Science*, pp. 311–390

Wadati, M. 1972. The exact solution of the modified Korteweg-de Vries equation. *Journal of the Physical Society of Japan*, 32: 62–69

Zabusky, N.J. & Kruskal, M.D. 1965. Interactions of solitons in a collisionless plasma and the recurrence of initial states. *Physical Review Letters*, 15: 240–243

Zakharov, V.E. & Shabat, A.B. (1972). Exact theory of two-dimensional self focussing and one dimensional self-modulation of waves in nonlinear media. *Soviet Physics JETP*, 34: 62–69

KRYLOV–BOGOLYUBOV METHOD

See **Quasilinear analyses**

KURAMOTO–SIVASHINSKY EQUATION

Derived by Yoshiki Kuramoto and Takeo Tsuzuki in the context of reaction-diffusion systems (Kuramoto & Tsuzuki, 1975, 1976), and by Gregory Sivashinsky in the study of flame front propagation (Sivashinsky, 1977), the Kuramoto–Sivashinsky equation (KS) is a prime example of a system that possesses a rich variety of spatial and temporal behaviors. The mathematical and statistical (thermodynamic) analyses and classification of both elementary solutions and observed complex behaviors have generated many advances in the understanding of the often complex patterns that arise in many experimental simulations.

The KS equation in one spatial dimension may be written

$$u_t + uu_x + u_{xx} + u_{xxxx} = 0, \qquad (1)$$

where the subscripts denote the partial derivatives w.r.t. the space variable x, and the time variable t. One important class of solutions is defined by a periodic boundary condition, say, with a $2\pi L$-periodicity. Here we may interpret such solutions as being restricted to a cell of size $2\pi L$ and the cellular state size serving as the control (or bifurcation) parameter. For small values of $L > 0$, the solutions of (1) behave periodically; then as L is increased, the system passes through a wide variety of behavior including exhibiting spatiotemporal chaos. Via a simple rescaling, the parameter dependence can be made explicit in (1), and the solutions then satisfy $u(x, t) = u(x + 2\pi, t)$. Equation (1) possesses both translation and Galilean symmetries.

Physically, the form of Equation (1) models the small perturbations from a reference Poiseuille flow of a film layer on an inclined plane (Pumir et al., 1983). The name KS often refers in the literature to the closely related equation formed by letting $u = w_x$ in (1) and integrating. Higher-dimensional equivalents of this form are found in the independent derivations of Kuramoto and Sivashinsky. The first is in the context of angular-phase turbulence for a system of reaction-diffusion equations modeling the Belouzov–Zabotinsky reaction in three spatial dimensions. Here, the solution is considered to be a small perturbation of a global periodic solution. The Sivashinsky laminar flame front derivation models small thermal diffusive instabilities, where the solution is the perturbation of an unstable planar front in the direction of propagation. Further derivations of the system can be found in the review by Nicolaenko (1986) and references therein.

The KS equation has been shown to posses an inertial manifold, that is, the infinite-dimensional solution space of the system is spanned by the solutions to a coupled system of ordinary differential equations with a low number of degrees of freedom (Foias et al., 1988). Hence, the system may be effectively studied by Fourier mode expansion, where the number of Fourier modes determining the dynamics is proportional to L.

The nontrivial behavior of the KS equation stems from the linear instability of the laminar state $u(x, t) = \text{constant}$—the evolution of the system is governed by the quadratic nonlinear coupling term uu_x and a second-order instability term u_{xx} that is balanced by the dissipation term u_{xxxx}. Long-wavelength k-numbered modes with $k < L$ are unstable. However, the linearly unstable low modes of the system are stabilized by the strong nonlinear coupling while the extremely stable high modes, with intermediate wavelengths with mode number $k \sim L$ play the important role of maintaining a "chaotic dynamical equilibrium."

For $L \geq 1$, standing and traveling waves may coexist with solutions having complicated oscillatory behavior. For example, antisymmetrically pulsating standing waves and waves that change both form and velocity periodically with time are considered in Demekhin et al. (1991). There exist windows of the parameter L in which many of these cellular states are stable. Windows of intermittency and strange (chaotic) attractors are also observed. For L not too large, these complex motions can be reached quite suddenly after extremely long transients. Figure 1 displays typical evolution once transients have died away. The KS equation also possesses persistent homoclinic and heteroclinic saddle connections which can be effectively explained via symmetry arguments (Kevrekidis et al., 1990).

The KS equation may be damped by the addition of the term νu on the left-hand side of (1), where ν parameterizes the level of damping. For zero damping the KS equation exhibits highly chaotic motions for large L. In the transition to these weakly turbulent motions, temporal intermittency is observed. Here, chaotic motions passing close to simple and weakly unstable states will feel regularizing effects for a

Figure 1. Spatiotemporally chaotic solution $u(x,t)$, computed from a 64-d Fourier truncation at $L = 10$. Coordinate x horizontal, amplitude u vertical, with plots of different times t overlapped at t progresses.

short while, for example. However, for large L and with damping, the intermittent behavior in the transition to weak turbulence also has a spatial element (Chaté & Manneville, 1986). In this scenario, a fluctuating mixture of both regular and chaotic patches with well-defined boundaries are observed in the solution surface $u(x,t)$. Choosing the parameters so that a weakly turbulent solution can be reached by spatiotemporal intermittency, one will observe chaotic domains slowly occupying the system; setting parameter values below such a threshold one will see the domains recede.

Sam Gratrix and John N. Elgin

See also **Belousov–Zhabotinsky reaction; Chaotic dynamics; Pattern formation; Turbulence**

Further Reading

Chang, H.-C. 1986. Travelling waves on fluid interfaces: normal form analysis of the Kuramoto–Sivashinsky equation. *The Physics of Fluids*, 29(10): 3142–3147

Chaté, H. & Manneville, P. 1986. Transition to turbulence via spatiotemporal intermittency. *Physical Review Letters*, 58(2): 112–115

Christiansen, F., Cvitanović, P. & Putkaradze, V. 1997. Spatiotemporal chaos in terms of unstable recurrent patterns. *Nonlinearity*, 10(1): 55–70

Demekhin, Y.A., Tokarev, G.Yu. & Shkadov, Ya.V. 1991. Hierarchy of bifurcations of space-periodic structures in a nonlinear model of active dissipative media. *Physica* D, 52 (2–3): 338–361

Elgin, J.N. & Wu, X. 1996. Stability of cellular states of the Kuramoto–Sivashinsky equation. *SIAM Journal on Applied Mathematics*, 56(6): 1621–1638

Foias, C., Nicolaenko, B., Sell, G.R. & Temam, R. 1988. Inertial manifolds for the Kuramoto–Sivashinsky equation and estimates of their dimension. *Journal de Mathématiques Pures et Appliquèes. Neuvième Série*, 67(3): 197–226

Hooper, P.A. & Grimshaw, R. 1988. Travelling wave solutions of the Kuramoto–Sivashinsky equation. *Wave Motion*, 10(5): 405–420

Kent, P. 1992. *Bifurcations of the travelling-wave solutions of the Kuramoto–Sivashinsky equation*, PhD Thesis, University of London

Kevrekidis, J.G., Nicolaenko, B. & Scovel, J.G. 1990. Back in the saddle again: A computer assisted study of the Kuramoto–Sivashinsky equation. *SIAM Journal on Applied Mathematics*, 50(3): 760–790

Kuramoto, Y. & Tsuzuki, T. 1975. On the formation of dissipative structures in reaction–diffusion systems. *Progress of Theoretical Physics*, 54(2): 687–699

Kuramoto, Y. & Tsuzuki, T. 1976. Persistent propagation of concentration waves in dissipative media far from thermal equilibrium. *Progress of Theoretical Physics*, 55(2): 356–369

Michelson, D. 1986. Steady solutions of the Kuramoto–Sivashinsky equation. *Physica* D, 19(1): 89–111

Nicolaenko, B. 1986. Some mathematical aspects of flame chaos and flame multiplicity. *Physica D*, 20(1): 109–121

Nicolaenko, B., Scheurer, B. & Temam, R. 1985. Some global dynamical properties of the Kuramoto–Sivashinsky equations: nonlinear stability and attractors. *Physica* D, 16(2): 155–183

Pumir, A., Manneville, P. & Pomeau, Y. 1983. On solitary waves running down an inclined plane. *Journal of Fluid Mechanics*, 135: 27–50

Sivashinsky, G.I. 1977. Nonlinear analysis of hydrodynamic instability in laminar flames: I. Derivation of basic equations. *Acta Astronautica*, 4(11–12): 1177–1206

L

LABORATORY MODELS OF NONLINEAR WAVES

In the decade following his 1834 discovery of the hydrodynamic solitary wave on Edinburgh's Union Canal, John Scott Russell constructed a water tank, allowing nonlinear wave phenomena to be studied in a laboratory environment (Russell, 1844). Among other results of these experiments, Russell observed, first, that the speed (v) of a solitary wave is related to its height (h) by the empirical relation $v = \sqrt{g(d+h)}$, where d is the resting depth of the water and g is the acceleration of gravity. Second, two solitary waves pass smoothly through each other without scattering. Third, two or more solitary waves can be generated from a sufficiently large "initial heap" of water. In the case sketched in Figure 1, for example, a volume (V) of water is released (by raising a sliding panel at the left-hand side of the tank) that is sufficiently large to generate two hydrodynamic solitons but not large enough to generate three of them. Also the soliton of larger amplitude is observed to have a higher velocity, leading to a separation between the two components that increases with time.

Since Russell's seminal work, hydrodynamic wave tanks have been widely used to investigate nonlinear wave propagation in a variety of settings, and several tanks suitable for undergraduate laboratories have been developed and described (Bettini et al., 1983; Olsen et al., 1984; Remoissenet, 1999). Tank experiments allow students to quantitatively investigate various properties of the Korteweg–de Vries (KdV) equation, which can be written in normalized form as

$$\frac{\partial u}{\partial t} + \frac{\partial u}{\partial x} + u\frac{\partial u}{\partial x} + \frac{\partial^3 u}{\partial x^3} = 0, \qquad (1)$$

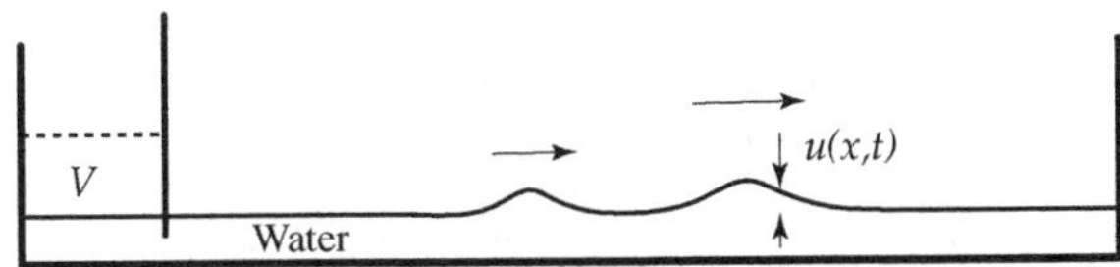

Figure 1. A sketch of John Scott Russell's wave tank, generating two solitons.

where $u(x,t)$ represents the vertical displacement of the wave from its resting level. Such experiments include quantitative comparisons of the number of solitons produced by an initial amount of water V as determined by eigenvalues of the corresponding time-independent linear Schrödinger equation through the inverse scattering transform method (Olsen et al., 1984). Similar tank experiments on deep water (where the depth is much larger than the lateral extent of the waves) allow quantitative studies of the nonlinear Schrödinger equation (Remoissenet, 1999).

In addition to wave tanks, mechanical wave models have been constructed for other nonlinear systems, including mechanical models of the normalized sine-Gordon (SG) equation

$$\frac{\partial^2 u}{\partial x^2} - \frac{\partial^2 u}{\partial t^2} = \sin u. \qquad (2)$$

As shown in the model of Figure 2, a number of pendula (dressmaker pins) are connected to a longitudinal spring (elastic band), whereupon $u(x,t)$ is the angle of rotation of the pendulum located at position x as a function of time t. The first term in Equation (2) represents the elastic restoring torque between adjacent pendula, the second term represents their angular acceleration, and the right-hand term is the angular-dependent torque of gravity. With a bit of practice, this simple model allows one to observe and demonstrate kink propagation, kink-kink collisions, breathers, and kink-antikink annihilation (Scott, 1969, 1970). The latter, in turn, is a model for electron-positron annihilation in elementary-particle physics.

More detailed mechanical models of the SG equation have been designed and constructed, which are suitable for undergraduate laboratory experiments (Scott, 1969, 1970; Remoissenet, 1999). As is evident from Figure 3, such models allow quantitative studies of the Lorentz contraction experienced by a kink as it approaches the limiting speed (Mach 1) of the system. For research purposes, Matteo Cirillo developed a mechanical model of fluxon propagation on a long Josephson junction, including an adjustable torque on the pendula (from air jets) that models the bias current acting in a

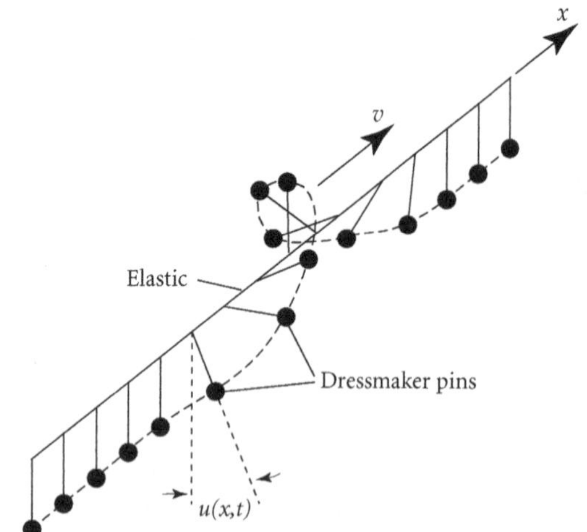

Figure 2. A kink on a simple mechanical model of the sine-Gordon equation that can be made from dressmaker pins and an elastic band.

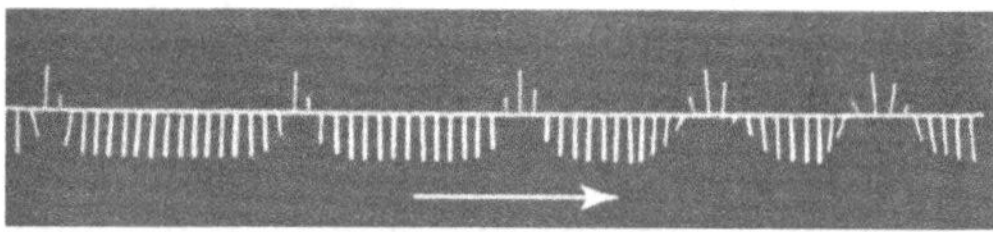

Figure 3. Strobe photos of an SG kink propagating on a mechanical model that was designed for student experiments (Scott, 1969). The kink is traveling to the right and slowing down due to friction. Spacing between the pendula is 1.59 cm and the time between successive images is 0.6 s.

typical experiment (Cirillo et al., 1981). Also, Michel Remoissenet and his colleagues have developed a model of the nonlinear Klein–Gordon equation with a double-well potential (the "phi-fourth model") to study the properties of compactons and the propagation of domain walls in ferroelectric and ferromagnetic materials (Duseul, 1998).

Complementing the family of mechanical models for nonlinear wave phenomena are nonlinear electrical transmission lines (Scott, 1970; Ostrovsky et al., 1972; Lonngren, 1978; Remoissenet, 1999). In energy-conserving versions of these models, nonlinearity is usually introduced through voltage-dependent capacitors (varactor diodes), and models of the KdV equation, Boussinesq equations, and the Toda lattice are readily constructed.

Allowing energy dissipation, it has long been known that the candle (or dynamite fuse or Japanese incense) models the leading edge of a nerve impulse as described by the Zeldovich–Frank-Kamenetsky (ZF) equation

$$\frac{\partial^2 u}{\partial x^2} - \frac{\partial u}{\partial t} = u(u-a)(u-1), \qquad (3)$$

where $u(x,t)$ represents temperature of a candle flame or transmembrane voltage of a nerve impulse. This equation can also be modeled by a nonlinear electrical transmission in which the transverse (shunt) conductive element is nonlinear with a region of negative slope (negative differential conductance) as provided by an Esaki tunnel diode or by a Giaever-type superconductive diode (Scott, 1970; Remoissenet, 1999).

In the 1920s, R.S. Lilly showed that nerve impulse propagation can be modeled by a piece of iron wire immersed in a strong nitric- or sulfuric-acid solution (Lilly, 1925). At rest, the wire is stabilized (passivated) by a thin oxide layer that prevents further oxidation. If this passivated layer is disturbed by mechanical or electrical means, however, a deoxidized region propagates along the wire, followed by re-establishment of the stabilizing oxide layer. In this manner, the iron-wire model simulates the recovery property of biological nerves that is missed by Equation (3).

A more complete electrical representation of impulse conduction along a nerve fibre was developed by Jin-ichi Nagumo and his colleagues in collaboration with Richard FitzHugh (Nagumo et al., 1962). Known as the FitzHugh–Nagumo equation, this model augments the ZF equation to allow for recovery to the initial resting state. Finally, the neuristor is an electronic device that functions like a nerve fiber and can be used as the basic element in a family of computing elements.

ALWYN SCOTT

See also **FitzHugh–Nagumo equation; Korteweg–de Vries equation; Neuristor; Sine-Gordon equation; Solitons, a brief history; Zeldovich–Frank-Kamenetsky equation**

Further Reading

Bettini, A., Minelli, T.A. & Pascoli, D. 1983. Solitons in an undergraduate laboratory. *American Journal of Physics*, 51: 977–984

Cirillo, M., Parmentier, R.D. & Savo, B. 1981. Mechanical analog studies of a perturbed sine–Gordon equation. *Physica* D, 3: 565–576

Duseul, S., Michaux, P. & Remoissenet, M. 1998. From kinks to compacton-like kinks. *Physical Review* E, 57: 2320–2326

Lilly, R.S. 1925. Factors affecting the transmission and recovery in the passive iron wire nerve model. *Journal of General Physiology*, 7: 473–507

Lonngren, K.E. 1978. Obsrvations of solitons on nonlinear dispersive transmission lines. In *Solitons in Action*, edited by K.E. Lonngren and A.C. Scott, New York: Academic Press

Nagumo, J., Arimoto, S. & Yoshizawa, S. 1962. An active pulse transmission line simulating nerve axon. *Proceedings of IRE*, 50: 2061–2071

Olsen, M., Smith, H. & Scott, A.C. 1984. Solitons in a wave tank. *American Journal of Physics*, 52: 826–830

Ostrovsky, L.A., Papko, V.V. & Pelinovsky, E.N. 1972. Solitary electromagnetic waves in nonlinear lines. *Radiophysics and Quantum Electronics*, 15: 438–446

Remoissenet, M. 1999. *Waves Called Solitons: Concepts and Experiments*, 3rd edition, Berlin and New York: Springer

Russell, J.S. 1844. Report on waves. *14th Meeting of the British Association for the Advancement of Science*, London: BAAS: pp. 311–339

Scott, A.C. 1969. A nonlinear Klein–Gordon equation. *American Journal of Physics*, 37: 52–61
Scott, A.C. 1970. *Active and Nonlinear Wave Propagation in Electronics*, New York: Wiley
Scott, A.C. 2003. *Nonlinear Science: Emergence and Dynamics of Coherent Structures*, 2nd edition, Oxford and New York: Oxford University Press

LAGRANGE–EULER EQUATIONS

See **Euler–Lagrange equations**

LAGRANGIAN DESCRIPTION

See **Fluid dynamics**

LAMB DIAGRAM

See **Bäcklund transformations**

LANDAU EQUATION

See **Equilibria**

LANDAU–LIFSHITZ EQUATION

The principal assumption of macroscopic ferromagnetism theory is that a magnetic crystal state is described unambiguously by the magnetization vector $\boldsymbol{M}$, so the dynamics and kinetics of a ferromagnet are determined by variations in its magnetization. The magnetization of a ferromagnet as a function of space coordinates and time $\boldsymbol{M}(\boldsymbol{x},t)$ is a solution of the Landau–Lifshitz (LL) equation, which was first used by Soviet scientists Lev Landau and Evgeni Lifshitz for describing the dynamics of a small velocity domain wall, the magnetic susceptibility of ferromagnets with a domain structure, and ferromagnetic resonance (Landau & Lifshitz, 1935). Later, the macroscopic theory of spin waves as small vibrations of the magnetization vector was developed on the basis of the linearized LL equation (Akhieser et al., 1968). At present, the LL equation is the theoretical foundation of phenomenological magnetization dynamics in magnetically ordered solids, including ferromagnets, antiferromagnets, and ferrites.

The LL equation in a ferromagnet has the following form:

$$\frac{\partial \boldsymbol{M}}{\partial t} = -\frac{2\mu_0}{\hbar}[\boldsymbol{M}\times\boldsymbol{H}_{\text{eff}}] - \gamma[\boldsymbol{M}\times[\boldsymbol{M}\times\boldsymbol{H}_{\text{eff}}]], \quad (1)$$

where μ_0 is the Bohr magneton and γ is the relaxation constant determining the damping motion of the vector $\boldsymbol{M}$. The effective magnetic field $\boldsymbol{H}_{\text{eff}}$ is equal to the variational derivative of the magnetic crystal energy E with respect to the vector $\boldsymbol{M}$: $\boldsymbol{H}_{\text{eff}} = -\,\delta E/\delta\boldsymbol{M}$. The energy E is assumed to be a function of $\boldsymbol{M}$, its spatial derivatives, and depends on the external magnetic field $\boldsymbol{H}$.

In the case of a many-sublattice magnet, an LL equation for the magnetization of an α-sublattice $\boldsymbol{M}^{\alpha}$ coincides with Equation (1) after the replacement $\boldsymbol{M}\to\boldsymbol{M}^{\alpha}$.

Equation (1) for ferromagnets has an integral of motion $M^2 = M_0 =$ constant and is consistent with the assumption that the $\boldsymbol{M}$-vector length in a ferromagnet is its equilibrium parameter. In the ground state, the quantity M_0 is equal to a so-called spontaneous magnetization $M_0 = 2\mu_0 S/a^3$, where S is the atomic spin and a is the interatomic separation. Conservation of the $\boldsymbol{M}$-vector length allows one to rewrite the LL equation in angular variables that are convenient for describing the magnetization dynamics in ferromagnets with axial symmetry. Assuming the external magnetic field directed along the anisotropy axis (the z-axis), let us define (Figure 1)

$$\begin{aligned} M_x + \mathrm{i}M_y &= M_0 \sin\theta \exp \mathrm{i}\psi, \\ M_z &= M_0\cos\theta. \end{aligned} \quad (2)$$

Then in a dissipativeless case ($\gamma = 0$)

$$\begin{aligned} \sin\theta\frac{\partial\theta}{\partial t} &= -\frac{2\mu_0}{\hbar M_0}\frac{\delta E}{\delta\varphi}, \\ \sin\theta\left(\frac{\partial\varphi}{\partial t} - \frac{2\mu_0 H}{\hbar}\right) &= \frac{2\mu_0}{\hbar M_0}\frac{\delta E}{\delta\theta}, \end{aligned} \quad (3)$$

where the magnetic energy E is written as a function of the angular variables.

The magnetic energy E of a ferromagnet includes two parts: the exchange energy E_{ex} and the magnetic anisotropy energy E_{a}:

$$\begin{aligned} E_{\text{ex}} &= \frac{1}{2}\alpha\int\frac{\partial\boldsymbol{M}}{\partial x_k}\frac{\partial\boldsymbol{M}}{\partial x_k}\,\mathrm{d}^3x, \\ E_a &= -\frac{1}{2}\beta_1\int M_x^2\,\mathrm{d}^3x - \frac{1}{2}\beta_3\int M_z^2\,\mathrm{d}^3x, \end{aligned} \quad (4)$$

where α is the nonuniform exchange energy constant and β_1 and β_3 are uniaxial anisotropy constants. (When all $\beta = 0$, we have an isotropic ferromagnet; when $\beta_1 = 0$, we have a uniaxial ferromagnet, the easy-axis anisotropy corresponds to $\beta = \beta_3 > 0$ and easy-plane anisotropy to $\beta < 0$; and when $\beta_1 \neq 0$ and $\beta_3 \neq 0$, we have biaxial anisotropy.)

Equations (3) for easy-axial ferromagnets can be written in the form

$$\begin{aligned} &l_0^2\Delta\theta - \left(1 + l_0^2(\nabla\varphi)^2\right)\sin\theta\cos\theta \\ &\quad + \frac{1}{\omega_0}\left(\frac{\partial\varphi}{\partial t} - \omega_H\right)\sin\theta = 0, \end{aligned} \quad (5)$$

$$l_0^2\,\mathrm{div}(\sin^2\theta\nabla\varphi) - \frac{1}{\omega_0}\frac{\partial\theta}{\partial t}\sin\theta = 0, \quad (6)$$

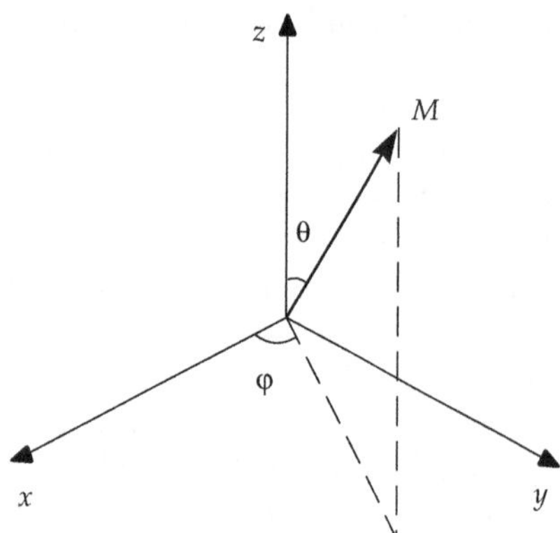

Figure 1. The angular variables for the magnetization $\boldsymbol{M}$.

where $\Delta = \nabla^2$ is the Laplacian, l_0 is the magnetic length ($l_0^2 = \alpha/\beta$), ω is the homogeneous ferromagnetic resonance frequency ($\hbar\omega_0 = 2\mu_0\beta M_0$), and $\omega_H = 2\mu_0 H/\hbar$ is the spin magnetic frequency.

Equations (5) and (6) have two additive integrals of motion, namely, the total momentum of the magnetization field $\boldsymbol{P}$ and z-component (projection) of the total magnetic moment written after normalization as a number of spin deviations N in the excited state of the magnet:

$$\boldsymbol{P} = -\frac{\hbar M_0}{2\mu_0}\int (1-\cos\theta)\nabla\varphi\, \mathrm{d}^3x,$$

$$N = \frac{M_0}{2\mu_0}\int (1-\cos\theta)\, \mathrm{d}^3x. \quad (7)$$

In a slightly excited state of ferromagnets ($\theta = \theta_0 =$ constant, $\theta_0 \ll 1$, $\psi = \omega t - \boldsymbol{kr}$) the magnetization dynamics is equivalent to a set of precessional spin waves that are solutions of linearized Equations (5), (6) with the following dispersion relation:

$$\omega(\boldsymbol{k}) = \omega_0 + \omega_H + (l_0 k)^2. \quad (8)$$

The simplest static solution of Equations (5), (6) is a domain wall (Figure 2c)

$$\sin\theta = \mathrm{sech}[(x - x_0)/l_0], \quad (9)$$

separating two semispace ferromagnet domains at $x < x_0$ and $x > x_0$. Equation (9) presents a topological soliton of the LL equation.

Interesting nontopological soliton solutions of Equations (5) and (6) are two-parametric dynamic magnetic solitons. Magnetic solitons of a general type are given by the following solutions

$$\theta = \theta(\boldsymbol{r} - \boldsymbol{V}t), \varphi = (\omega + \omega_H)t + \psi(\boldsymbol{r} - \boldsymbol{V}t), \quad (10)$$

where the function θ vanishes at infinity ($\theta = 0$ for $r = \infty$) and $\nabla\psi$ is limited for $r = \infty$, $\boldsymbol{V}$ is the translational velocity of the soliton, and $\omega + \omega_H$ is the precessional frequency of the magnetization vector in the frame of reference moving along with the soliton. Typical diagrams for the function θ in the 1-d case are presented in Figure 2. There are the following Hamilton equations for a dynamic soliton (Kosevich et al., 1977):

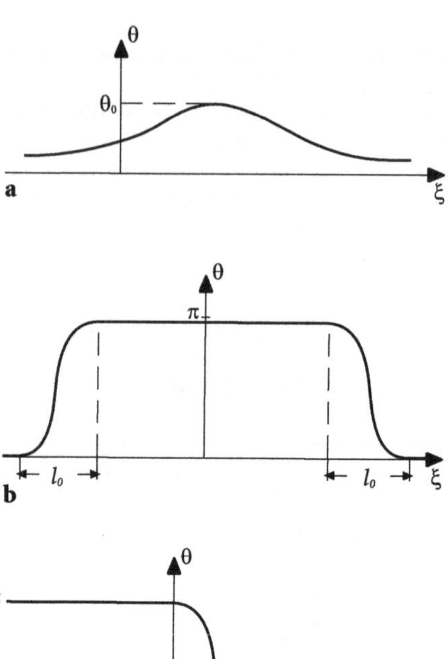

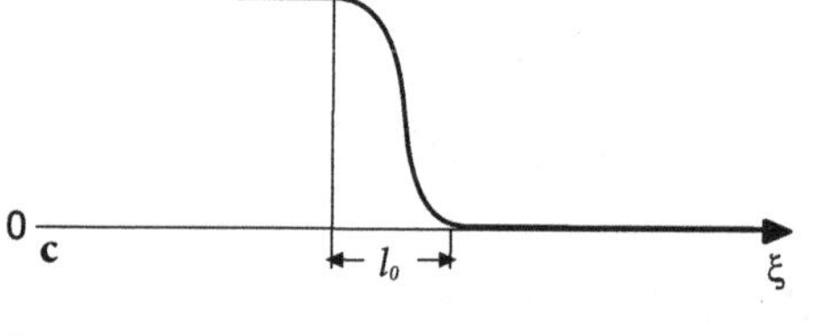

Figure 2. Magnetization distribution in (a) a low-amplitude soliton, (b) a magnetic soliton for small values V and ω, (c) a domain wall.

$$\boldsymbol{V} = \left(\frac{\partial E}{\partial \boldsymbol{P}}\right)_N, \quad \hbar\omega = \left(\frac{\partial E}{\partial N}\right)_{\boldsymbol{P}}. \quad (11)$$

The 1-dimensional version of Equations (5) and (6) is a totally integrable nonlinear equation in both isotropic (Lakshmanan, 1977; Takhtajan, 1977) and easy-axial (Borovik, 1978) cases and possesses a set of multiple-soliton solutions. The total integrability of 1-d LL equations in the case of biaxial anisotropy was proved by Sklyanin (1979).

If θ is small enough, the 1-d version of Equations (5) and (6) can be reduced to the nonlinear Schrödinger (NLS) equation for the complex function $\psi = M_x + \mathrm{i}M_y$ (Volzhan et al., 1976)

$$\frac{\mathrm{i}}{\omega_0}\frac{\partial\Psi}{\partial t} - l_0^2\frac{\partial^2\Psi}{\partial x^2} + (1+h)\Psi - \frac{1}{2M_0}|\Psi|^2\Psi = 0, \quad (12)$$

which is used for description of small amplitude long-wave excitations in one-dimensional easy-axial ferromagnets.

Another limiting case arises for a ferromagnet with the biaxial anisotropy ($\beta_1 < 0$ and $\beta_3 > 0$) under the condition $\varepsilon = -\beta_1/\beta_3 \gg 1$. Then the plane YOZ plays easy-plane, the magnetization vector lies nearly in this plane, $\chi = \pi/2 - \theta \ll 1$, and the LL equation is transformed to the following equations:

$$l_0^2\frac{\partial^2\phi}{\partial x^2} - \frac{1}{\varepsilon\omega_0}\frac{\partial^2\phi}{\partial t^2} + \sin\phi\cos\phi = 0,$$

$$\omega_0 \chi = -\frac{1}{\varepsilon}\frac{\partial \phi}{\partial t}, \tag{13}$$

where the new angle variable ϕ is introduced with the equation $M_x + \mathrm{i}M_y = M_0 \cos\chi \exp \mathrm{i}\phi$.

At present, the LL equation is widely used for models of nonlinear macroscopic dynamic phenomena such as spin waves in inhomogeneous or spatially limited magnets, magnetostatic vibrations, magnetization dynamics, interaction of magnetically ordered media with electromagnetic and elastic waves, nonlinear magnetization waves, and magnetic solitons.

ARNOLD KOSEVICH

See also **Domain walls; Ferromagnetism and ferroelectricity; Nonlinear Schrödinger equations; Sine-Gordon equation; Solitons**

Further Reading

Akhieser, A.I., Bar'yahtar V.G. & Peletminskii S.V. 1968. *Spin Waves*, Amsterdam: North-Holland

Borovik, A.E. 1978. *N*-soliton solutions of Landau-Lifshitz equation, *Pis'ma Zhurnal Experimental'noy i Teoreticheskoy Fiziki*, 28: 629 (in Russian); *JETP Letters*, 28: 581

Kosevich, A.M. 1986. Dynamical and topological solitons in ferromagnets and antiferromagnets. In *Solitons*, edited by S.E. Trullinger, V.E. Zakharov & V.L. Pokrovsky, Amsterdam: North-Holland

Kosevich, A.M., Ivanov, B.A. & Kovalev, A.S. 1977. Nonlinear localized magnetization wave of a ferromagnet as a bound-state of a large number of magnons. *Pis'ma Zhurnal Eksperimental'noy i Teoreticheskoy Fiziki* 25: 516 (in Russian); *JETP Letters*, 25: 486

Kosevich, A.M., Ivanov, B.A. & Kovalev, A.S. 1990. Magnetic solitons. *Physics Reports*, 194: 117

Lakshmanan, M. 1977. Continium spin system as an exactly solvable dynamic system. *Physics Letters* A, 61: 53

Landau, L.D. & Lifshitz, E.M. 1935. On the theory of the dispersion of magnetic permability in ferromagnet bodies. *Physikalische Zeischrift der Sowjetunion*, 8: 153

Lifshitz, E.M. & Pitaevskii, L.P. 1980. *Statistical Physics, Part 2*, Oxford: Pergamon Press

Sklyanin, E.K. 1979. On complete integrability of Landau-Lifshitz equation, *Preprit/ Academy of Sciences of USSR*, Leningrad Department Steklov Mathematical Institute, E-3

Takhtajan, L.A. 1977. Integration of the continium Heisenberg spin chain through the inverse scattering method, *Physics Letters* A, 64: 235

Volzhan, E.B., Giorgadze, N.P. & Pataraya A.D. 1976. Weakly nonlinear magnetization density waves in magnetically ordered media. *Fizika Tverdogo Tela*, 25: 516 (in Russian); *Soviet Physics, Solid State*, 18: 1487

LANGEVIN EQUATION

See **Ferromagnetism and ferroelectricity**

LANGMUIR WAVES

See **Nonlinear plasma waves**

LANGMUIR–BLODGETT FILMS

The calming effect of oil on water has been known for centuries, and one early account of the phenomenon, on clay tablets, dates from the 18th century BCE in Babylonia. In 1879, a provisional British patent was filed by John Shields, the proprietor of a Scottish linen mill, for a simple device for spreading oil from valves in undersea pipes to calm the waves at the entrances to harbors. However, the first account of an experiment to investigate this effect was probably that of Benjamin Franklin in 1773, who wrote the following in a letter to a colleague.

> At length being at Clapham, where there is on the common a large pond, which I observed one day to be very rough with the wind, I fetched out a cruet of oil, and dropped a little of it on the water, and there the oil, though not more than a teaspoonful, produced an instant calm over a space several yards square, which spread amazingly, and extended itself gradually till it reached the lee side, making all that quarter of the pond, perhaps half an acre as smooth as a looking-glass.

A quick calculation reveals that Franklin's oil film was 1–2 nm thick—about the same as the size of an oil molecule, but this implication was not realized for many years. Although Lord Rayleigh was the first to propose that such films were only one molecule in thickness, he was not able to make a direct measurement to confirm this. The simple equipment for monolayer studies, now known as a Langmuir trough, was first introduced by Agnes Pockels. In a letter to Lord Rayleigh in 1891, she described the methods that formed the foundation of monolayer research.

Irving Langmuir provided most of the early scientific evidence for the existence of monolayer films. In 1917, he published a substantial paper outlining the properties of such films on a water surface. Some years later, Katharine Blodgett, working with Langmuir at the General Electric Company Research Laboratories in Schenectady, New York, devised a method for transferring the floating monolayers onto solid surfaces. The resulting films now bear the name of these two researchers.

Until the outbreak of World War II, research into the properties of monolayer films flourished, the work being undertaken mainly by surface chemists. However, few uses were found for monolayer and multilayer structures so activity in the area declined. Interest was rekindled in the 1970s following some stimulating experiments on energy transfer in multilayer systems by Hans Kuhn working in Germany. At about this time, organic chemists became aware of the limited range of monolayer forming materials that was available. Novel electroactive compounds were synthesized (dyes, semiconductors, polymers), and a new series of investigations began. This coincided with the birth of molecular electronics, a new interdisciplinary research activity focused on the exploitation of organic materials in electronic and optoelectronic devices. Related thin film technologies, such as self-assembly and layer-by-layer electrostatic deposition, were also developed. By the end of the 20th century, a number of organic thin film deposition technologies and materials, suitable for fabricating organic molecular architectures, was available. The era of molecular nanotechnology had begun.

Materials that produce organized monomolecular layers on the surface of water invariably consist of molecules possessing both water-attracting (hydrophilic) and water-repelling (hydrophobic) chemical groups. Such organic compounds are called amphiphiles. One of the simplest materials suitable for forming such a monomolecular layer is stearic acid, $C_{17}H_{35}COOH$. The molecule consists essentially of 16 CH_2 groups forming a long hydrocarbon chain; one end of the chain terminates in a hydrophilic carboxylic acid COOH group.

Langmuir–Blodgett (LB) films are prepared by first depositing a small quantity of the amphiphilic material, dissolved in a volatile solvent such as chloroform, on the surface of carefully purified water (subphase). When the solvent has evaporated, the organic molecules may be compressed to form a floating two-dimensional solid. The hydrophilic and hydrophobic terminations of the molecules ensure that the individual molecules are aligned in the same way during this process. During compression the monolayer undergoes a number of phase transformations. The different phases are almost analogues of three-dimensional gases, liquids and solids. The phase changes may be readily identified by monitoring the surface pressure Π as a function of the area occupied by the film. This is the two-dimensional equivalent of the pressure versus volume isotherm for a gas/liquid/solid. Figure 1 shows such a plot for a hypothetical long-chain organic monolayer material (e.g., a long-chain fatty acid).

In the "gaseous" state (G in Figure 1), the molecules are far enough apart on the water surface that they exert little force on one another. As the surface area of the monolayer is reduced, the hydrocarbon chains will begin to interact. The "liquid" state that is formed is generally called the expanded monolayer phase (E). The hydrocarbon chains of the molecules in such a film are in a random, rather than a regular orientation, with their polar groups in contact with the subphase. As the molecular area is progressively reduced, condensed (C) phases may appear. There may be more than one of these, and the emergence of each condensed phase can be accompanied by constant pressure regions in the isotherm, as observed in the cases of a gas condensing to a liquid and a liquid solidifying. These regions will be associated with enthalpy changes in the monolayer. In the condensed monolayer states, the molecules are closely packed and are oriented with their hydrocarbon chains pointing away from the water surface. The area per molecule in such a state will be similar to the cross-sectional area of the hydrocarbon chain, that is, $\approx 0.19\ \text{nm}^2\ \text{molecule}^{-1}$.

The LB technique requires that the surface pressure and temperature of the floating monolayer are controlled so that the organic film is in a condensed and stable state. Figure 2 shows the commonest form of LB deposition. The substrate is hydrophilic and the first layer is transferred, like a carpet, as the substrate is raised vertically through the water. Subsequently, a monolayer is deposited on each traversal of the monolayer/air interface. As shown, these stack in a head-to-head and tail-to-tail pattern; this

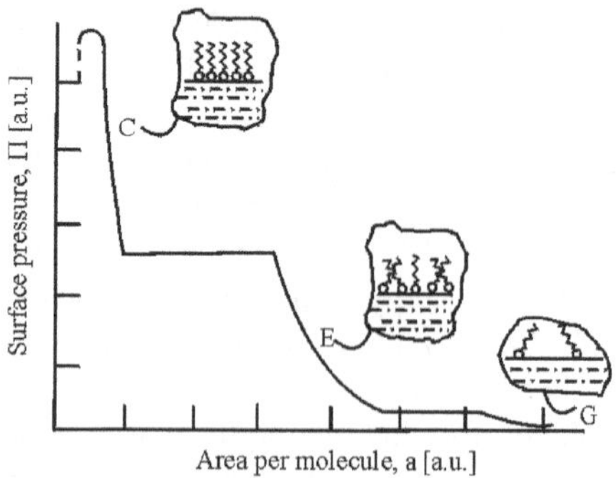

Figure 1. Surface pressure versus area per molecule for a long-chain organic compound. (The surface pressure and area are in arbitrary units (a.u.).)

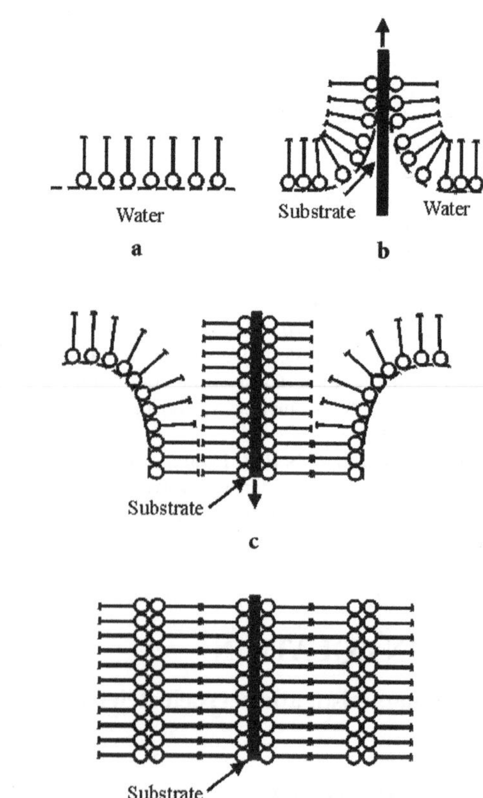

Figure 2. Y-type Langmuir–Blodgett film deposition.

deposition mode is called Y-type. Although this is the most frequently encountered situation, instances in which the monolayer is transferred to the substrate as it is being inserted into the subphase, or only as it is being removed, are often observed. These deposition modes are called X-type (monolayer transfer on the downstroke only) and Z-type (transfer on the upstroke only).

It is also possible to build up thin film architectures containing more than one type of monomolecular layer. In the simplest case, alternate-layer films may be produced by raising the substrate through a monolayer of one material (consisting of molecules of compound A, say) and then lowering the substrate through a monolayer of a second substance (compound B). An asymmetric multilayer structure consisting of ABABAB... layers is produced. This control over the molecular architecture permits the fabrication of organic superlattices with precisely defined symmetry properties. Such molecular assemblies can exhibit pyroelectric, piezoelectric, and second-order nonlinear optical phenomena.

Film transfer is characterized by measurement of the deposition ratio, τ (also called the transfer ratio). This is the decrease in the area occupied by the monolayer (held at constant pressure) on the water surface divided by the coated area of the solid substrate, that is,

$$\tau = A_L/A_S, \tag{1}$$

where A_L is the area occupied by the monolayer on the water surface and A_S is the coated area of the solid substrate. Transfer ratios significantly outside the range 0.95–1.05 suggest poor film homogeneity.

A schematic diagram of one experimental arrangement to deposit LB films is shown in Figure 3, together with a photograph of the equipment. The Langmuir trough is made from PTFE (polytetrafluoroethylene) and a working area is defined by a PTFE-coated glass fiber barrier, which can be moved using a low-geared electric motor. The barrier motor is coupled to a sensitive electronic balance, which continuously monitors, via a sensing plate (Wilhelmy plate), the surface pressure of the monolayer. Using a feedback arrangement, this pressure can be maintained at a predetermined value. The physical dimensions of the Langmuir trough arrangement are not critical (the system in the photograph is approximately 30 cm in length) and are governed by the size of the substrate used.

Many analytical techniques, such as X-ray and electron diffraction, and infrared spectroscopy may be used to study the orientation of molecules in an LB assembly. Figure 4 shows an 9 nm × 9 nm atomic force micrograph of the surface of a 12-layer fatty acid LB film. The lighter parts of the image relate to the higher part of the surface and the darker regions correspond to deeper down. Lines of individual molecules are evident at the magnification shown, confirming the highly ordered arrangement of organic molecules in the LB film.

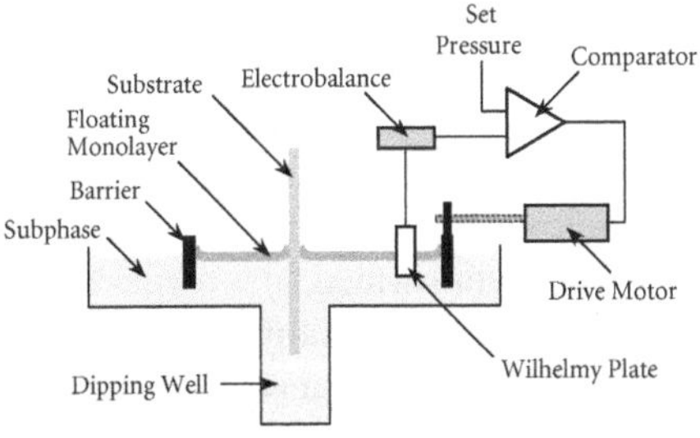

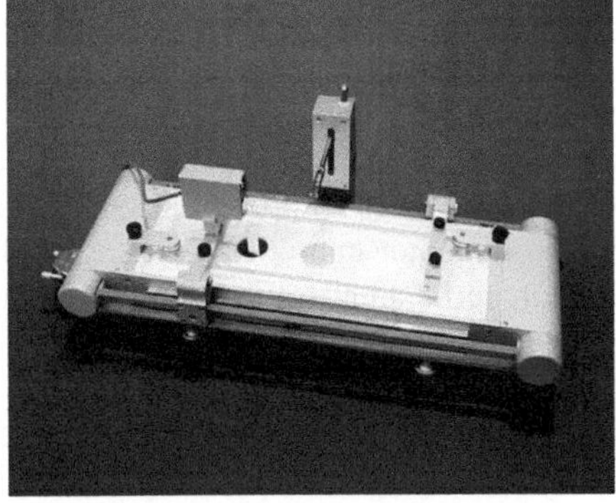

Figure 3. Langmuir–Blodgett trough (courtesy Molecular Photonics).

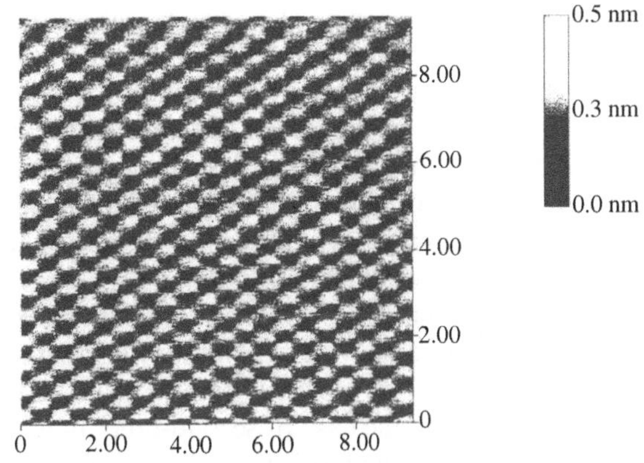

Figure 4. An atomic force micrograph of a fatty acid LB film.

The vertical dipping LB process is not the only way to transfer a floating molecular film to a solid substrate or to build-up multilayer films. Other methods are based on touching one edge of a hydrophilic substrate with the monolayer-covered subphase or lowering the substrate horizontally so that it contacts the hydrophobic ends of the floating molecules. Chemical means, for example, self-assembly of a thiol group onto a gold-coated substrate, can also be used to deposit monolayer organic films. In contrast, electrostatic layer-by-layer assembly relies on the forces between positively and negatively charged polyelectrolytes.

The earliest technical application of organic monolayer films is believed to be the Japanese printing art called *sumi-nagashi*. The dye comprising a suspension of submicron particles and protein molecules is first spread on the surface of water; the application of gelatin to the uniform layer converts the film into a patchwork of colorless and dark domains. These distinctive patterns can then be transferred by lowering a sheet of paper onto the water surface.

There are, of course, many methods available to deposit thin films of organic materials, including: thermal evaporation, sputtering, spaying, painting, dip- and spin-coating, electrodeposition, and molecular beam epitaxy. However, the LB technique is one of the few thin film technologies that actually permits the manipulation of materials at the molecular level. It should, therefore, be appropriate for exploitation by workers in nanotechnology wishing to fabricate interesting material architectures (bottom-up approach to nanotechnology) or to build up novel electronic device structures.

A range of possible applications for LB and related films is evident from the literature. Many of the ideas exploit the physical and chemical properties of the ultra-thin films to provide surface coatings with particular catalytic, adhesive, or mechanical properties (e.g., low friction). The availability of new polymeric amphiphiles has led to an interest in semi-permeable membranes. The extreme thinness of monolayer and multilayer films could provide key benefits in a variety of chemical sensing structures. For gas sensing, adequate sensitivities to some important gases and vapors have already been achieved using a variety of transduction techniques (chemiresistor, surface plasmon resonance, acousto-electric coupling, and so on).

For commercial exploitation, it is imperative to establish those areas in which LB films offer significant advantages over layers produced by other (and perhaps cheaper) means. In the case of second-order nonlinear optics, the LB method offers a means of aligning the molecules in a film of micrometer dimensions. Materials with high second-order hyperpolarizabilities (for example, leading to significant second-harmonic generation) already exist. Further work is needed on the development of practical electro-optic structures with attention to important considerations such as encapsulation and device degradation.

It is also interesting to note that a large number of biological materials form monolayers on a water surface. Chlorophyll a, the green pigment in higher plants; vitamins A, E and K; and cholesterol are all examples. Monomolecular films resemble naturally occurring biological membranes, which are based on a bilayer arrangement of long-chain phospholipid molecules. The LB technique might, therefore, be used as a means to fabricate artificial structures that emulate certain biological functions, such as photosynthesis, molecular recognition, or parallel information processing.

MICHAEL PETTY

See also **Bilayer lipid membranes; Liquid crystals; Nonlinear optics**

Further Reading

Gaines, G.L., Jr. 1966. *Insoluble Monolayers at Liquid-Gas Interfaces*, New York: Wiley-Interscience

Petty, M.C. 1996. *Langmuir–Blodgett Films: An Introduction*, Cambridge: Cambridge University Press

Roberts, G. G. (editor). 1991. *Langmuir–Blodgett Films*, New York: Plenum Press

Tredgold, R.H. 1994. *Order in Organic Films*, Cambridge and New York: Cambridge University Press

Ulman, A. 1991. *Ultrathin Organic Films*, San Diego: Academic Press

LAPLACE TRANSFORM

See **Integral transforms**

LASERS

An acronym for Light Amplification by Stimulated Emission of Radiation, a laser involves interaction of light with matter at the molecular, atomic, or nuclear levels and requires a quantum mechanical description of both light and matter. According to quantum mechanics, when an electron decays from an excited level E_+ to a lower energy level $E_- < E_+$, a photon of energy $E_+ - E_-$ and frequency $\nu = (E_+ - E_-)/h$ is emitted, where $h \doteq 6.672 \times 10^{-34}$ J s is Planck's constant.

The decay of electrons from excited states and the consequent emission of light is stimulated by the interaction of photons (of the appropriate energy) with the electrons (Einstein's stimulated emission). The emitted photons can then interact with excited electrons to produce a cascade of coherent photons, making up the familiar laser beam.

For a laser to work, three basic elements are required: (i) an energy level structure for the electrons (e.g., atomic or molecular fluids (gases or liquids), solids (crystals or glasses), or semiconducting junctures); (ii) a pumping mechanism (such as electrical current or light) to populate the upper energy level; and (iii) an optical cavity to partially confine the photons, enhancing their probability of interacting with the excited electrons.

The most common laser structures are Fabry–Perot cavities (where light is confined between two opposite mirrors) and ring lasers (in which light travels in one direction along the border of a rectangle with mirrors appropriately oriented in the corners) (Siegman, 1986) (see Figure 1). Industrial lasers are normally of Fabry–Perot type while ring lasers are often preferred in research laboratories.

The simplest laser model, describing only the first two required elements, consists of the laser rate equations, which model a laser medium in terms of the number of photons (proportional to the light

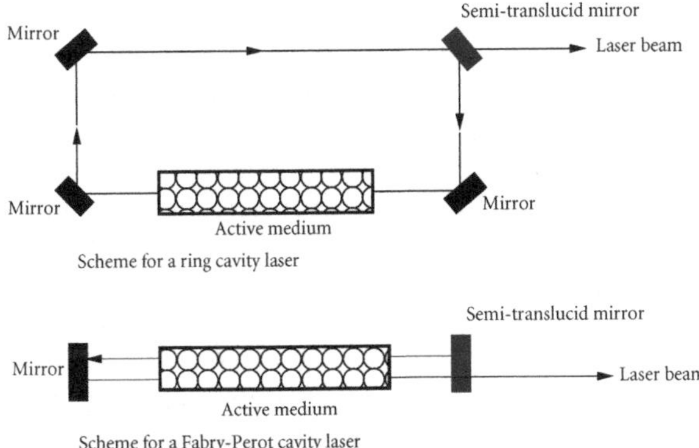

Figure 1. Ring and Fabry-Perot laser cavities.

intensity, I) and the number of excited electrons inside the laser cavity. Because the photon field can induce upward as well as downward transitions, the excited electrons are measured by the population inversion $N \equiv N_+ - N_-$, where N_+ (N_-) denotes the number of electrons in the upper (lower) level.

The photon population (I) is diminished due to the laser output as well as through other losses at a rate εI. Stimulated emission decreases the population inversion and increases the photon number at a rate αN_+, while absorption of photons produce the inverse effect at a rate αN_-. (The equivalence between emission and absorption rate per electron is a quantum mechanical result. Note that for each photon emitted, the population inversion decreases by two as N_+ decreases by one, while N_- increases by the same amount.) Excited electronic states (N_+) are also produced by the pumping mechanism at a rate $A/2$ (which may be a function of time) and diminished by nonradiative processes at a rate $\gamma(N - N_{eq})$ (where N_{eq}, a negative number, derives from the population inversion at thermal equilibrium). A simple formulation of the laser rate equations is, therefore,

$$\frac{dI}{dt} = [-\varepsilon + \alpha N] I, \tag{1}$$

$$\frac{dN}{dt} = (A + \gamma N_{eq}) - \gamma N - 2\alpha N I. \tag{2}$$

These laser rate equations have either one or two fixed points corresponding to steady operation of the laser. The laser-off state with $I = 0$ and $N = N_0 = A/\gamma + N_{eq}$ is always present. The stability of the laser-off solution depends on the sign of $\lambda = -\varepsilon + \alpha N_0$. If this quantity is positive, small disturbances such as those produced by spontaneous emission are amplified and laser action develops. If $\lambda < 0$, disturbances die out, hence $\lambda = 0$ implies a threshold value for the pumping of $A = \gamma(\varepsilon/\alpha - N_{eq})$. Steady operation of a laser above threshold is represented by the laser-on state $N = N_1 = \varepsilon/\alpha$ and $I = (\gamma/2\alpha)(N_0 - N_1)/N_1$.

These equations correspond to birth and death processes of photons and electrons that resemble those of predator and prey species in population dynamics. This analogy is somewhat tenuous, however, because quantum theory does not allow a photon or electron to be localized in time and energy simultaneously. Hence, the statistical description of a laser requires of entrained electron-photon states and the descriptions of quantum jump processes (Carmichael, 1999).

A more detailed set of equations is obtained by considering a laser operating in a single longitudinal cavity mode (i.e., having a fixed spatial pattern within the optical cavity). With a simple quantum mechanical model for electron energy levels and a classical description of the electromagnetic field (i.e., a large number of photons), the dynamics of the laser are described by Maxwell–Bloch equations, which involve complex amplitudes of the electric and polarization fields in addition to the population inversion (N) (Narducci & Abraham, 1988).

An aim of laser design has long been to achieve lasers of higher photon energies. While ultraviolet lasers were developed early, X-ray lasers have only recently been demonstrated (Rocca, 1999), and gamma-ray lasers are not yet realized.

Lasers can be classified in various ways according to their dynamics, the active media, the optical cavity, the pumping scheme, optical frequency, among others. Considering the active media, lasers can be grouped as follows.

Solid State Lasers

Solid state lasers, such as the ruby laser (the first laser, announced in 1960), which has a three-level pumping scheme for which Cr^{3+} impurities are essential, require a high-energy source for pumping and operate intermittently. In contrast, Nd-YAG (neodymium-doped yttrium aluminum garnet) and Nd-glass lasers are based in a four-level scheme, operate under moderate pump power, and display different dynamical regimes. Ruby and Nd-YAG lasers represent the two basic pumping schemes for solid state lasers.

In the three-level pumping scheme, the light emitting transition has the ground state as the final state, while in the four-level scheme the state reached by the electron after the transition is an excited state.

There are a large number of solid state lasers operating at different frequencies, each with particular features. Many of them employ a host crystal (or glass) doped with an impurity such as $Al_2O_3(Cr^{3+})$, $Y_2O_3(Eu^{3+})$, $Gd_2O_3(Nd^{3+})$, or $BaF_2(U^{3+})$.

Gas Lasers

Several active media fall in this category: helium-neon, argon, and the powerful CO_2 are among the best-known representatives. The standard method for the excitation of the molecular states is an electric discharge in the gaseous medium where collisions between exited ions provide the pumping mechanism. However, optical pumping and other pumping mechanisms have been demonstrated for some gas lasers.

Gas lasers can be operated in continuous wave and pulsed modes and are particularly important at the low-frequency (infrared) region of the electromagnetic spectrum. Other members of this class include H_2O, HCN, CH_2I, and HF lasers.

Liquid Lasers

The best-known representatives are dye lasers, where a stream of organic liquid is excited using visible or ultraviolet radiation from a flash lamp or a laser beam. The pumping mechanism involves radiative fluorescence at a lower frequency than that of the pumping source, a mechanism known as Stokes-shift.

Other lasers in this class are based on solutions of rare earth elements. In contrast with gas and solid state lasers, dye lasers usually operate in cavities with relatively high losses.

Semiconductor Lasers

Because of their small size, semiconductor lasers play an important role in electronic applications. While earlier lasers were built as semiconductor diodes using p–n junctions as in GaAs diodes (or PbS, PbSe, PbTe, SnTe, InSb, etc.) new generations of lasers, such as the Vertical Cavity Surface Emitting Laser (VCSEL) are produced modifying semiconductor surfaces and reducing the size and power requirements of the laser significantly. Semiconductor lasers are pumped by direct electrical current using the band structure of the semiconductor and electron-hole recomposition to produce the photons.

Nuclear Magnetic Resonance

The unique NMR laser relies on the (permanent) magnetic momentum of nuclei, which are pumped using a radio-frequency field. Optical pumping has also been demonstrated recently.

For gas lasers, some liquid lasers, and solid state lasers, the lasing threshold is reached only with high reflectivity mirrors minimizing the cavity losses (small ε). Such lasers operate just above threshold in a mode that is almost resonant with the frequency of the atomic transition and can be viewed as weakly nonlinear systems. Wide cavities, higher pumping levels, feedback, and other effects can induce multimode dynamics, where several empty cavity modes are required for the spatial description of the dynamics.

When laser devices are based on lossy electromagnetic cavities, as in dye and semiconductor lasers, the relation between cavity modes and monochromatic states (or solutions) is obscured. Nevertheless, these are also called multimode lasers when a description of the spatial extension is required.

Despite the diversity of active media, pumping mechanisms, and lasing frequencies, the dynamics of lasers are often rather similar, the main differences accounted for by different parameter values in Maxwell–Bloch type equations. The dynamics of most lasers are described by simple dynamical attractors (fixed points) under constant pumping. There are, however, situations in which complex behavior emerges; thus, optical reinjection of the emitted light can destabilize a laser. Such a process might occur when the laser is coupled to optic fibers in a communication system. The Lang–Kobayasi equations (Lang & Kobayashi, 1980) (a modification of the rate equations including a delayed field) describe this situation for very small amounts of reinjected signal. In the case of semiconductor lasers, even small amounts of optical feedback require a spatial description of the laser cavity (multimode operation) (Huyet et al., 1998; Duarte & Solari, 1998). Attempts to control, or mode lock, a (usually powerful) slave laser with a (usually weak) master laser also generate a rich dynamical spectrum at the unlocking transition (Zimmermann et al., 2001).

HERNÁN G. SOLARI AND MARIO NATIELLO

See also **Maxwell–Bloch equations; Nonlinear optics; Semiconductor laser**

Further Reading

Carmichael, H.J. 1999. *Statistical Methods in Quantum Optics*, vol. 1, *Master equations and Fokker–Plank equations*, Berlin and New York: Springer

Duarte, A.A. & Solari, H.G. 1998. The metamorphosis of the monochromatic spectrum in a double-cavity laser as a function of the feedback rate. *Physical Review* A, 58: 614–619

Huyet, G., Balle, S., Giudici, M., Green, C., Giacomelli, G. & Tredicce, J.R. 1998. Low frequency fluctuations and multi-mode operation of a semiconductor laser with optical feedback. *Optical Communications*, 149: 341–347

Lang, R. & Kobayashi, K. 1980. External optical feedback effects on semiconductor injection laser properties. *IEEE Journal of Quantum Electronics*, QE-16: 347

Narducci, L.M. & Abraham, N.B. 1988. *Laser Physics and Laser Instabilities*, Singapore: World Scientific

Rocca, J.J. 1999. Table-top soft X-ray lasers. *Review of Scientific Instruments*, 70: 3799

Siegman, A.E. 1986. *Lasers*, Mill Valley, CA: University Science Books

Zimmermann, M., Natiello, M. & Solari, H. 2001. Global bifurcations in a laser with inhected signal: beyond Adler's approximation. *Chaos*, 11: 500–513

LATTICE GAS METHODS

When one is interested in studying the dynamical behavior of fluid systems starting at the microscopic level, it is logical to begin with a molecular dynamics description of the interactions between the constituting particles. This is often a formidable task, as the fluid evolves into a nonlinear regime where chaos, turbulence, or reactive processes take place. But one may question whether a realistic description of the microscopic dynamics is necessary to gain insight on the underlying mechanisms of large-scale nonlinear phenomena. Around 1985, a considerable simplification was introduced (Frisch et al., 1986). These pioneering studies established (theoretically and computationally) the feasibility of simulating fluid dynamics via a microscopic approach based on a new paradigm. A virtual simplified micro-world is constructed as an automaton universe, based not on a realistic description of interacting particles but merely on the laws of symmetry and of invariance of macroscopic physics.

Suppose we implement point-like particles on a regular lattice where they move from node to node at each time step and undergo collisions when their trajectories meet at the same node. As the system evolves, we observe its collective dynamics by looking at the lattice from a distance. A remarkable fact is that—if the collisions occur according to some simple logical rules (satisfying fundamental conservations) and if the lattice has the proper symmetry—this "lattice gas automaton" (LGA) shows global behavior very similar to that of a real fluid. So we can infer that despite its simplicity at the microscopic scale, the LGA should contain the essential features that are responsible for the emergence of complex behavior and, thereby, can help us understand the basic mechanisms involved.

An LGA consists of a set of particles moving on a regular d-dimensional lattice $\mathcal{L}$ at discrete time steps, $t = n\Delta t$, with n an integer. The lattice is composed of V nodes, labeled by the d-dimensional position vectors $\boldsymbol{r} \in \mathcal{L}$. Associated to each node are b channels, labeled by indices $i, j, \ldots$, running from 1 to b. At a given time t, a channel can be either occupied by one particle or empty, so that the occupation variable $n_i(\boldsymbol{r}, t) = 1$ or 0. When channel i at node $\boldsymbol{r}$ is occupied, then the particle at the specified node $\boldsymbol{r}$ has velocity $\mathbf{c}_i$. The set of allowed velocities is such that the condition $\boldsymbol{r} + \boldsymbol{c}_i \Delta t \in \mathcal{L}$ is fulfilled. The "exclusion principle" requirement that the maximum occupation be of one particle per channel allows for a representation of the automaton configuration in terms of a set of bits $\{n_i(\boldsymbol{r}, t)\}$; $\boldsymbol{r} \in \mathcal{L}, i = \{1, b\}$. The evolution rules are thus simply logical operations over sets of bits.

The time evolution of the automaton takes place in two stages: propagation and collision. In the propagation phase, particles are moved according to their velocity vector, and in the (local) collision phase, the particles occupying a given node are redistributed among the channels associated to that node. So the microscopic evolution equation of the LGA reads

$$n_i(\boldsymbol{r} + \boldsymbol{c}_i \Delta t, t + \Delta t) = n_i(\boldsymbol{r}, t) + \Delta_i(\{n_j(\boldsymbol{r}, t)\}), \tag{1}$$

where $\Delta_i(\{n_j\})$ represents the collision term that depends on all channel occupations at node $\boldsymbol{r}$. By performing an ensemble average (denoted by angular brackets) over an arbitrary distribution of initial occupations, one obtains a hierarchy of coupled equations for the successive n-body distribution functions. This hierarchy can be truncated to yield the lattice Boltzmann equation for the single particle distribution function $f_i(\boldsymbol{r}, t) = \langle n_i(\boldsymbol{r}, t)\rangle$:

$$f_i(\boldsymbol{r} + \boldsymbol{c}_i \Delta t, t + \Delta t) - f_i(\boldsymbol{r}, t) = \Delta_i^{\mathrm{Boltz}}(\{f_j(\boldsymbol{r}, t)\}). \tag{2}$$

The left-hand side is recognized as the discrete version of the left-hand side of the classical Boltzmann equation for continuous systems, and the right-hand side denotes the collision term, where the precollisional uncorrelated state *ansatz* has been used to factorize the b-particle distribution function.

The lattice Boltzmann equation (2) is one of the most important results in LGA theory. It can be used as the starting point for the derivation (via multi-scale analysis) of the macroscopic equations describing the long wavelength behavior of the lattice gas. The LGA macroscopic equations are found to exhibit the same structure as the classical hydrodynamic equations, and under the incompressibility condition, one retrieves the Navier–Stokes equations for nonthermal fluids. Another important feature of the lattice Boltzmann equation is that it can be used as an efficient and powerful simulation algorithm. In practice, one usually prefers to use a simplified equation where the collision term is approximated by a single relaxation time process inspired by the Bhatnagar–Gross–Krook model, known in its lattice version as the LBGK equation:

$$f_i(\boldsymbol{r} + \boldsymbol{c}_i \Delta t, t + \Delta t) - f_i(\boldsymbol{r}, t) = -\frac{1}{\tau}\left[f_i(\boldsymbol{r}, t) - f_i^{\mathrm{leq}}(\boldsymbol{r}, t)\right], \tag{3}$$

where the right-hand side is proportional to the deviation from the local equilibrium distribution function.

There is a wealth of applications of the lattice gas methods that have established their validity and their usefulness. LGA simulations, based on Equation (1), are most valuable for fundamental problems in statistical mechanics, such as the study of fluctuation correlations in equilibrium and non-equilibrium systems (Rivet & Boon, 2001; Rothman & Zaleski, 1997). As an example, Figure 1 shows

Figure 1. Lattice gas simulation of the Kolmogorov flow: the tracer trajectories reflect the topology of the *ABC* flow in the regime beyond the critical Reynolds number ($\mathrm{Re} = 2.5 \times \mathrm{Re_c}$).

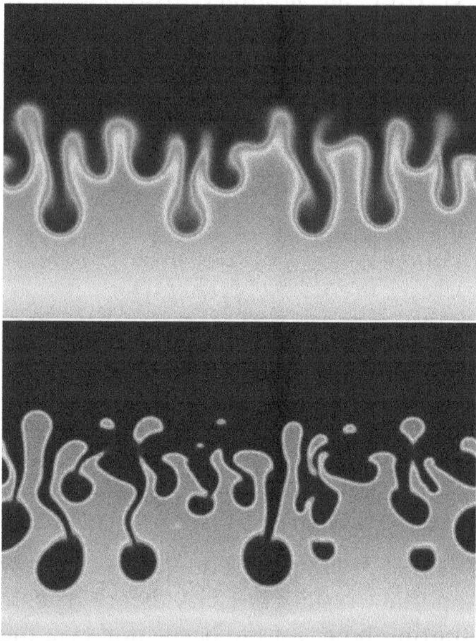

Figure 2. Lattice Boltzmann (LBGK) simulation of viscous fingering in miscible fluids (upper panel) showing the interface sharpening effect of a reactive process between the two fluids (lower panel).

the trajectories of tracer particles suspended in a Kolmogorov flow (above the critical Reynolds number) produced by a lattice gas automaton and from where turbulent diffusion was analyzed (Boon et al., 2000).

Simulations of more direct practical interest, such as, for instance, profile optimization in car design or turbulent drag problems, are most efficiently treated with the lattice Boltzmann method, in particular using the LBGK model. The example given in Figure 2 illustrates the method for the study of viscous fingering in Hele-Shaw geometry, showing the effect of reactivity between the two fluids as a determinant factor in the dynamics of the moving interface (Grosfils & Boon, 2002).

Applications of the LGA approach and of the lattice Boltzmann equation cover a wide variety of theoretical and practical problems ranging from the dynamics of thermal fluctuations and quantum lattice gas automata to multi-phase flow, complex fluids, reactive systems, and inhomogeneous turbulence.

JEAN PIERRE BOON

See also **Cellular automata; Hele-Shaw cell; Molecular dynamics; Navier–Stokes equation**

Further Reading

Boghosian, B.M., Coveney, P.V. & Emerton, A.N. 1996. A lattice gas model of microemulsions. *Proceedings of the Royal Society* A, 452: 1221–1250

Boon, J.P., Dab, D., Kapral, R. & Lawniczak, A. 1996. Lattice gas automata for reactive systems. *Physics Reports*, 273(2): 55–148

Boon, J.P., Hanon, D. & Vanden Eijnden, E. 2000. Lattice gas automaton approach to turbulent diffusion. *Chaos, Solitons and Fractals*, 11: 187–192

Coveney, P.V. & Succi, S. (editors). 2002. Discrete modeling and simulation of fluid dynamics. *Philosophical Transactions of the Royal Society*, 360: 291–573

Frisch, U., Hasslacher, B. & Pomeau, Y. 1986. Lattice gas automata for the Navier–Stokes equation. *Physical Review Letters*, 56: 1505–1508

Grosfils, P. & Boon, J.P. 2002. Viscous fingering in miscible, immiscible, and reactive fluids. *International Journal of Modern Physics* B, 17: 15–20

Meyer, D. 1996. From quantum cellular automata to quantum lattice gases. *Journal of Statistical Physics*, 85: 551–574

Rivet, J.P. & Boon, J.P. 2001. *Lattice Gas Hydrodynamics*, Cambridge and New York: Cambridge University Press

Rothman, D. & Zaleski, S. 1997. *Lattice Gas Cellular Automata*, Cambridge and New York: Cambridge University Press

Succi, S. 2001. *The Lattice Boltzmann Equation for Fluid Dynamics and Beyond*, Oxford: Clarendon Press and New York: Oxford University Press

LATTICE KINK

See **Solitons, types of**

LATTICE SOLITONS

See **Solitons, types of**

LAX OPERATORS

See **Inverse scattering method or transform**

LEADING EDGE DYNAMICS

See **Nerve impulses**

LEBESQUE MEASURE

See **Measures**

LEGENDRE TRANSFORMATION

See **Euler–Lagrange equations**

LENNARD–JONES POTENTIAL

See **Molecular dynamics**

LÉVY FLIGHTS

In 1937, the French mathematician Paul Lévy (1886–1971) introduced statistical descriptions of motion that extend beyond the more traditional Brownian motion discovered over one hundred years earlier. A diverse range of both natural and artificial phenomena is now being described in terms of Lévy statistics, from the flight of an albatross across the Antarctic skies to the trajectories followed by the abstract painter Jackson Pollock as he constructed his famous drip paintings.

In 1828, Robert Brown published his studies of the random motion of soot particles in a dish of water as they were buffeted from random directions by the thermal motion of water molecules. In 1905, Albert Einstein provided a theoretical basis for this diffusion process. A particle's Brownian motion is pictured as a sequence of jumps. For a single jump, the probability dependence on jump size x has a Gaussian distribution. A consequence of Gaussian statistics is that the size distribution for N jumps is also described by a Gaussian.

Paul Lévy generalized beyond Brownian motion by considering other distributions for which one jump and N jumps share the same mathematical form. These Lévy distributions decrease according to the power law $1/x^{1+\gamma}$ for large x values, where γ lies between 0 and 2. Compared with Gaussian distributions, Lévy distributions do not fall off as rapidly at long distances. For Brownian motion, each jump is usually small and the variance of the distribution, $\langle x^2 \rangle$, is finite. For Lévy motion, however, the small jumps are interspersed with longer jumps, or "flights," causing the variance of the distribution to diverge. As a consequence, Lévy jumps do not have a characteristic length scale.

This scale invariance is a signature of fractal patterns. Indeed, Lévy's initial question of "When does the whole look like its parts?" addresses the fractal property of self-similarity. An important parameter for assessing the scaling relationship of fractal patterns is dimension. What, then, is the dimension of the pattern traced out by Lévy motion? The short jumps making up Brownian motion build a clustered pattern that is so dense that area is a more appropriate measure than length—the pattern is actually two-dimensional. In contrast, although the short jumps of Lévy motion produce a clustering, the longer, less frequent jumps initiate new clusters. These clusters form a self-similar pattern with a dimension of less than two. Fractional dimensions are an exotic property of fractals.

Today, Lévy motion is as widely explored in nonlinear, chaotic, turbulent, and fractal systems as Brownian motion is in simpler systems. Following Mandelbrot's research in the 1970s demonstrating the prevalence of fractal patterns in nature, an increasing number of natural phenomena have been described using Lévy statistics (Mandelbrot, 1982). Lévy distributions are also having an impact on artificial systems. A recent example concerns nano-scale electronic devices in which chaotic electron trajectories produce Lévy statistics in the electrical conduction properties (Micolich et al., 2001). Other examples include diffusion in Josephson junctions (Geisel et al., 1985) and at liquid-solid interfaces (Stapf et al., 1995).

It is even possible to picture relatively simple systems in which both Brownian and Lévy motion appear and a transition between the two can be induced. Consider, for example, dropping tracer particles into a container of liquid. This, of course, is Brown's original experiment. In 1993, Harry Swinney extended this experiment by considering a rotating container of liquid shaped like a washer. As turbulence set in, vortices appeared in the liquid and Swinney's group showed that the tracer particles followed Lévy flights between the vortices with $\gamma = 1.3$ (Solomon et al., 1993).

In addition to spatial distributions, Lévy statistics can also be applied to distributions measured as a function of time. A famous example is the dripping faucet (*See* **Dripping faucet**). In 1995, Thadeu Penna's group showed that the fluctuations in the time intervals between drips follow a Lévy distribution with $\gamma = 1.66-1.85$. A significant appeal of this result lies in a comparison with earlier medical work by Ary Goldberger's group showing that fluctuations in the human heart beat follow $\gamma = 1.7$ (Goldberger, 1996). This prompted Penna to ask, "Is the heart a dripping faucet?"

Goldberger suggested that the Lévy statistics describing the human heart arise from nonlinear processes that regulate the human nervous system. He has since extended his research of the fractal dynamics of physiology to other examples of involuntary behavior. This includes studies of the human gait that show that fluctuations in stride intervals display fractal variations (Hausdorff et al., 1996). Fractal variations might therefore be a general signature of healthy human behavior, exhibited whenever conscious control is not involved.

It is interesting to consider this speculation within the context of the results of the British Antarctic Survey in 1996, which showed that albatrosses follow Lévy flights. Other species of animals, such as ants and bees, also follow Lévy flights when foraging for food. Due to the diverging variance of the flight distribution, Lévy trajectories represent an efficient way of covering large regions of space, especially when compared with Brownian motion. Significantly, these animal behavioral patterns represent yet another example of Lévy behavior generated by actions that are devoid of intellectual deliberation.

This relationship between unconscious actions and Lévy statistics has even touched on human creativity. In particular, the Surrealist art movement developed a technique called automatic painting, in which artists painted with such speed that any conscious involvement was thought to be eliminated. Jackson Pollock adopted this approach during the 1940–1950s when he dripped paint onto large horizontal canvases (see *Autumn Rhythm*, Figure 1). Remarkably, his paintings are fractal

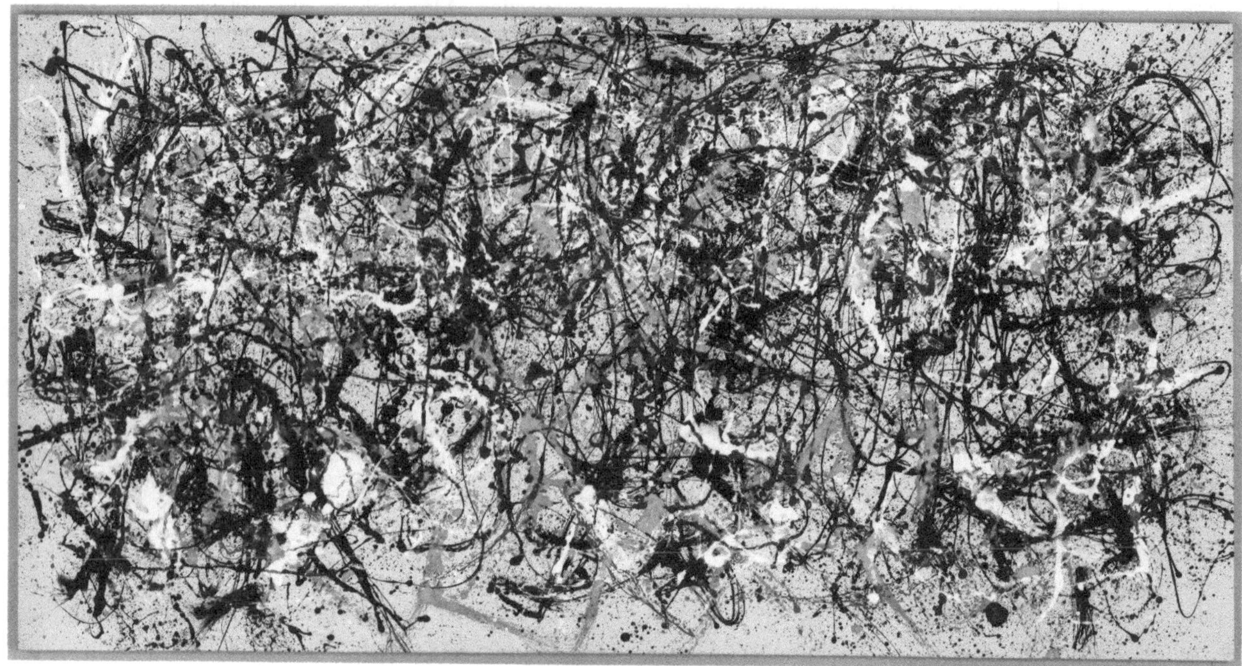

Figure 1. *Autumn Rhythm (Number 30)*, by Jackson Pollock, 1950. © 2003 The Pollock-Krasner Foundation / Artists Rights Society (ARS) New York. Courtesy The Metropolitan Museum of Art, George A. Hearn Fund, 1957. (57.92)

and his motions have been described in terms of Lévy flights (Taylor et al., 1999). This work triggered visual perception tests that identified an aesthetic preference for fractal patterns with dimensions between 1.3 and 1.5 (Taylor, 2001).

Lévy distributions represent a truly interdisciplinary phenomenon that will continue to be useful as novel artificial and natural systems are explored.

RICHARD TAYLOR

See also **Brownian motion; Dimensions; Dripping faucet; Fractals**

Further Reading

Geisel, T., Nierwetberg, J. & Zacherl, A. 1985. Accelerated diffusion in Josephson junctions and related chaotic systems. *Physical Review Letters*, 54: 616–620

Goldberger, A.L. 1996. Non-linear dynamics for clinicians: chaos theory, fractals, and complexity at the bedside. *The Lancet*, 347: 1312–1314

Haussdorff, J.M., Purdon, P.L., Peng, C.K., Ladin, Z., Wei, J.Y. & Goldberger, A.L. 1996. Fractal dynamics of human gait: stability of long-range correlations in stride interval fluctuations. *Journal of Applied Physiology*, 80: 1448–1457

Mandelbrot, B., 1982. *The Fractal Geometry of Nature*, San Francisco: Freeman

Micolich A.P., Taylor, R.P., Davies, A.G., Bird, J.P., Newbury, R., Fromhold, T.M., Ehlert, A., Linke, H., Macks, L.D., Tribe, W.R., Linfield, E.H., Ritchie, D.A., Cooper, J., Aoyagi, Y. & Wilkinson, P.B. 2001. The evolution of fractal patterns during a classical-quantum transition. *Physical Review Letters*, 87: 036802-1–036802-4

Solomon, T., Weeks, E. & Swinney, H. 1993. Observation of anomalous diffusion and Lévy flights in a two-dimensional rotating flow. *Physical Review Letters*, 71: 3975–3979

Stapf, S., Kimmich, R. & Seitter, R. 1993. Proton and deuteron field-cycling NMR relaxometry of liquids in porous glasses: evidence of Lévy-walk statistics. *Physical Review Letters*, 75: 2855–2859

Taylor, R.P., Micolich, A.P. & Jonas, D. 1999. Fractal analysis of Pollock's drip paintings. *Nature*, 399: 422 and *Physics World*, 12: 25–29

Taylor, R.P. 2001. Architect reaches for the clouds. *Nature*, 410: 18

LIE ALGEBRAS AND LIE GROUPS

Lie groups were introduced by the 19th century Norwegian mathematician Sophus Lie through his studies in geometry and integration methods for differential equations (Hawkins, 1999). Further developments by W. Killing, É. Cartan and H. Weyl established Lie's theory as a cornerstone of mathematics and its physical applications. General references include Duistermaat & Kolk (1999), Sattinger & Weaver (1986), and Varadarajan (1984).

An r parameter Lie group is defined as an r-dimensional manifold that is also a group with smooth multiplication and inversion maps. A key example is the $r = n^2$-dimensional general linear group $\mathrm{GL}(n)$ of (either real or complex) $n \times n$ nonsingular matrices, $\det A \neq 0$, under matrix multiplication. Most Lie groups can be realized as matrix groups, that is, subgroups of $\mathrm{GL}(n)$. Important examples include the

- special linear group $\mathrm{SL}(n) \subset \mathrm{GL}(n)$ with $\det A = 1$, and $r = n^2 - 1$;
- orthogonal group $\mathrm{O}(n) \subset \mathrm{GL}(n, \mathbb{R})$ with $A^{\mathrm{T}} A = I$, and $r = n(n-1)/2$;

- unitary group $\mathrm{U}(n) \subset \mathrm{GL}(n, \mathbb{C})$ with $A^\dagger A = I$, and $r = n^2$;
- symplectic group $\mathrm{Sp}(2n) \subset \mathrm{GL}(2n, \mathbb{R})$ with $A^{\mathrm{T}} J A = J = \begin{pmatrix} 0 & -I \\ I & 0 \end{pmatrix}$, and $r = n(2n+1)$.

A Lie algebra $\mathfrak{g}$ is a vector space equipped with a skew-symmetric, bilinear bracket $[\cdot, \cdot]: \mathfrak{g} \times \mathfrak{g} \to \mathfrak{g}$ that satisfies the Jacobi identity

$$[u, [v, w]] + [v, [w, u]] + [w, [u, v]] = 0.$$

The Lie algebra $\mathfrak{g}$ of left-invariant vector fields on an r-parameter Lie group G can be identified with the tangent space at the identity, and so $\dim \mathfrak{g} = r$. The Lie algebra $\mathfrak{gl}(n)$ of $\mathrm{GL}(n)$ consists of all $n \times n$ matrices under matrix commutator $[A, B] = AB - BA$. A finite-dimensional Lie algebra with basis $v_1, \ldots, v_r$ is specified by its structure constants c^i_{jk}, defined by the bracket relations $[v_j, v_k] = \sum_{i=1}^r c^i_{jk} v_i$. Each finite-dimensional Lie algebra corresponds to a unique connected and simply connected Lie group G^*; any other is obtained by quotienting by a discrete normal subgroup: $G = G^*/N$.

A subspace $\mathfrak{h} \subset \mathfrak{g}$ is a subalgebra if it is closed under the Lie bracket: $[\mathfrak{h}, \mathfrak{h}] \subset \mathfrak{h}$. Lie subalgebras are in one-to-one correspondence with connected Lie subgroups $H \subset G$. The subalgebra is an ideal if $[\mathfrak{g}, \mathfrak{h}] \subset \mathfrak{h}$. A Lie algebra is simple if it contains no nontrivial ideals, and semi-simple if it contains no nontrivial abelian (commutative) ideals. Semi-simple algebras are direct sums of simple algebras. A Lie algebra is solvable if the sequence of subalgebras $\mathfrak{g}^{(0)} = \mathfrak{g}$, $\mathfrak{g}^{(k+1)} = [\mathfrak{g}^{(k)}, \mathfrak{g}^{(k)}]$ eventually terminates with $\mathfrak{g}^{(n)} = \{0\}$. The Levi decomposition says that every Lie algebra is the semi-direct sum of a semi-simple subalgebra and its radical—the maximal solvable ideal.

The Killing–Cartan classification of complex simple Lie algebras contains four infinite families, denoted A_n, B_n, C_n, D_n, corresponding to the simple Lie groups $\mathrm{SL}(n+1), \mathrm{O}(2n+1), \mathrm{Sp}(2n+1), \mathrm{O}(2n)$. In addition, there are five exceptional simple Lie groups, G_2, F_4, E_6, E_7, E_8, of respective dimensions 14, 52, 78, 133, 248. The last plays an important role in modern theoretical physics. Extending the classification to infinite-dimensional simple Lie algebras leads to the Kac–Moody Lie algebras, of importance in integrable systems, theoretical physics, differential geometry, and topology (Kac, 1990).

Lie groups typically arise as symmetry groups of geometric objects or differential equations. A (Lie) group G acts on a manifold M, for example, Euclidean space, provided $m \mapsto g \cdot m$ for $g \in G, m \in M$, defines a sufficiently smooth invertible map that respects the group multiplication. Lie classified all transformation groups acting on one- and two-dimensional real and complex manifolds (Olver, 1995). According to Klein's Erlanger Program, geometric structure is prescribed by an underlying transformation group; thus, Euclidean, affine, conformal, projective geometries are based on the eponymous Lie groups. If G acts transitively, then $M = G/H$ is a homogeneous space, obtained by quotienting by a closed Lie subgroup. The group orbits—minimal invariant subsets—form a system of submanifolds, and the invariant functions are constant on orbits. The infinitesimal generators of the group action form a Lie algebra $\mathfrak{g}$ of vector fields tangent to the orbits whose flows generate the group action.

A linear action $\rho: G \to \mathrm{GL}(V)$ on a vector space V is known as a representation. Representation theory plays a fundamental role in quantum mechanics since linear symmetries of the Schrödinger equation induce actions on the space of solutions, which decompose into irreducible representations. The structure of atoms, nuclei, and elementary particles is governed by the representations of particular symmetry groups (Hamermesh, 1962). Important special functions, for example, Bessel and hypergeometric, arise as matrix entries of representations of particular Lie groups (Vilenkin & Klimyk, 1991). The representation theory of the orthogonal group SO(2) leads to trigonometric functions and, hence, Fourier analysis, as the simplest case of harmonic analysis on semi-simple Lie groups (Warner, 1972).

A Lie group acts on its Lie algebra $\mathfrak{g}$ by the adjoint representation and on the dual space $\mathfrak{g}^*$ by the coadjoint representation. The coadjoint orbits are symplectic submanifolds with respect to the natural Lie–Poisson structure on $\mathfrak{g}^*$, and are of importance in classifying representations (Kirillov, 1999), geometric mechanics, and geometric quantization (Woodhouse, 1992). The Euler equations of rigid body motion and of fluid mechanics are realized as the Lie–Poisson equations on, respectively, the Lie algebra of the Euclidean group and the infinite-dimensional diffeomorphism group (Marsden, 1992).

A transformation group is called a symmetry group of a system of differential equations if it maps solutions to solutions. Symmetry groups are effectively computed by solving the infinitesimal symmetry conditions, which form an overdetermined linear system of partial differential equations, usually amenable to automatic solution by computer algebra packages (Olver, 1993, 1995). Applications include integration of ordinary differential equations, determination of explicit group-invariant (similarity) solutions of partial differential equations, Noether's theorems relating symmetries of variational problems and conservation laws (Noether, 1918), bifurcation theory (Golubitsky & Schaeffer, 1985), asymptotics and blow-up (Barenblatt, 1979), and the design of geometric numerical integration schemes (Hairer et al., 2002).

Classification of differential equations and variational problems admitting a given symmetry group relies on its differential invariants. The simplest examples are the curvature and torsion of space curves, and the mean and Gaussian curvatures of surfaces under the Euclidean group acting on $\mathbb{R}^3$. Cartan's method of moving frames, and its more recent extensions to general Lie group and Lie pseudo-group actions (Olver, 2001), provides a general mechanism for construction and classification of differential invariants, with applications to differential geometry, the calculus of variations, soliton theory, computer vision, classical invariant theory, and numerical methods.

Modern developments in applications of Lie group methods have proceeded in a variety of directions. General theories of infinite-dimensional Lie groups and algebras (Kac, 1990) and Lie pseudo-groups, arising in relativity, field theory, fluid mechanics, solitons, and geometry, remain elusive. Higher order or generalized symmetries, in which the infinitesimal generators also depend upon derivative coordinates, first proposed by Noether (1918), have been used to classify integrable (soliton) systems. Recursion operators are used to generate such higher order symmetries and, via Noether's theorem, higher order conservation laws (Olver, 1993). Most recursion operators are derived from a pair of compatible Hamiltonian structures, and demonstrate the integrability of bi-Hamiltonian systems. The higher-order symmetries also appear in series expansions of Bäcklund transformations in the spectral parameter.

PETER J. OLVER

See also **Bäcklund transformations; Inverse scattering method or transform; Maps; Symmetry groups**

Further Reading

Barenblatt, G.I. 1979. *Similarity, Self-Similarity, and Intermediate Asymptotics*, New York: Consultants Bureau
Duistermaat, J.J. & Kolk, J.A.C. 1999. *Lie Groups*, Berlin and New York: Springer
Golubitsky, M. & Schaeffer, D.G. 1985. *Singularities and Groups in Bifurcation Theory*, 2 vols, New York: Springer
Hairer, E., Lubich, C. & Wanner, G. 2002. *Geometric Numerical Integration: Structure-preserving Algorithms for Ordinary Differential Equations*, Berlin and New York: Springer
Hamermesh, M. 1962. *Group Theory and Its Application to Physical Problems*, Reading, MA: Addison-Wesley, 1962
Hawkins, T. 1999. *The Emergence of the Theory of Lie Groups. An Essay on the History of Mathematics 1869–1926*, New York: Springer
Kac, V.G. 1990. *Infinite Dimensional Lie Algebras*, Cambridge and New York: Cambridge University Press
Kirillov, A.A. 1918. Merits and demerits of the orbit method, *Bulletin of the American Mathematical Society*, 36: 433–488
Marsden, J.E. 1992. *Lectures on Mechanics*, Cambridge and New York: Cambridge University Press
Noether, E. 1918. Invariante variationsprobleme. *Nachricheten von der Konig. Gesellschaft der Wissenschaften zu Gottingen, Mathematisch–Physikalische Klasse*, 235–257
Olver, P.J. 1993. *Applications of Lie Groups to Differential Equations*, 2nd edition, London and New York: Springer
Olver, P.J. 1995. *Equivalence, Invariants, and Symmetry*, Cambridge and New York: Cambridge University Press
Olver, P.J. 2001. Moving frames—in geometry, algebra, computer vision, and numerical analysis. In *Foundations of Computational Mathematics*, edited by R. DeVore, A. Iserles & E. Suli, Cambridge and New York: Cambridge University Press, pp. 267–297
Sattinger, D.H. & Weaver, O.L. 1986. *Lie Groups and Algebras with Applications to Physics, Geometry, and Mechanics*, New York: Springer
Varadarajan, V.S. 1984. *Lie Groups, Lie Algebras, and Their Representations*, New York: Springer
Vilenkin, N.J. & Klimyk, A.U. 1991. *Representation of Lie Groups and Special Functions*, Dordrecht: Kluwer
Warner, G. 1972. *Harmonic Analysis on Semi-simple Lie Groups*, Berlin and New York: Springer
Woodhouse, N.M.J. 1992. *Geometric Quantization*, 2nd edition, Oxford: Clarendon Press and New York: Oxford University Press

LIFETIME

As the word implies, *lifetime* is a measure of the time that something manages to persist and properly function. Often lifetime is a statistically defined quantity, as is evident from thinking about the failure of light bulbs. Clearly, one can exactly measure the lifetime of any particular bulb at the cost of burning it out, but it is more interesting to know the probability of failure in a given interval of time for a particular type of bulb. To obtain this information, it is necessary to measure the average burnout rate for a large number of bulbs that have been manufactured under identical conditions and use that information to characterize bulbs newly manufactured under the same conditions. A related approached is used by actuaries, who compute risks and premiums for human life insurance.

Failure of a Nuclear Power Plant

Statistical ideas are sometimes applied to more special situations, for example, a nuclear power plant, where neighbors ask: "What are the chances of it blowing up or melting down?" Engineers usually answer with soothing words. "It's very safe. This reactor can't blow up, and it has less than a fifty percent chance of melting down in the next thousand years." But the engineers have only built a few dozen plants, all of which seem to be working fine. How do they come up with such an estimate?

One procedure is to consider the various failure mechanisms (onset of corrosion, earthquakes, operators falling asleep, terrorist attacks, and so on), and assign a mean lifetime to each failure mechanism. Then the total failure probability up to time t_1 is defined as

$$P(t_1) = \int_0^{t_1} \frac{\mathrm{d}t}{\tau}, \qquad (1)$$

where τ is the mean lifetime, taking all failure mechanisms into account. If these mechanisms are independent (do not influence one another), then

$$\frac{1}{\tau} = \frac{1}{\tau_1} + \frac{1}{\tau_2} + \cdots + \frac{1}{\tau_N}, \qquad (2)$$

where $\tau_1, \tau_2, \ldots, \tau_N$ are mean lifetimes estimated for N failure mechanisms. (Note that the τ_i may be functions of time.) A key result of this simple formulation is that the total lifetime is less than the smallest constituent lifetime, making it imperative to consider all possible failure mechanisms.

Lifetime of a Quantum State

Given a quantum state $\psi(\mathbf{r})$, where $\mathbf{r}$ is a spatial coordinate, it is interesting to know the mean time (τ) before ψ makes a transition (decays) into one of N other states: $\phi_1(\mathbf{r})$, $\phi_2(\mathbf{r})$, ..., $\phi_N(\mathbf{r})$. According to "Fermi's Golden Rule", the transition probability into ϕ_i is proportional to the square of an overlap integral (Slater, 1951; Dirac, 1935)

$$|\langle\psi|\phi_i\rangle|^2 \equiv \left[\int \psi\phi^* \, \mathrm{d}\mathbf{r}\right]^2; \qquad (3)$$

thus the total lifetime is

$$\tau \propto \frac{1}{|\langle\psi|\phi_1\rangle|^2 + |\langle\psi|\phi_2\rangle|^2 + \cdots + |\langle\psi|\phi_N\rangle|^2}. \qquad (4)$$

Although this equation is simple to write, it is difficult to compute because computation requires detailed knowledge of the wave functions of all states into which ψ can decay. Just as with a nuclear power plant, the total lifetime will be less than the smallest individual decay time, but τ is easier to measure, as it is often a small faction of a second.

Soliton Lifetimes

When John Scott Russell first observed a hydrodynamic soliton on the "happiest day of [his] life" in August of 1834, he was struck by its remarkable stability, giving him time to follow it on horseback until it became "lost in the windings of the canal" (Russell, 1844). Although this and subsequent observations suggest that solitons are long-lived objects, the conclusion is somewhat misleading. It is the spatial localization of soliton energy that persists in time rather than the energy itself. Detailed perturbation calculations for the Korteweg–de Vries equation and for the nonlinear Schrödinger equation show that the times for energy decay under dissipative perturbations are less for nontopological solitons than for delocalized wave packets (Scott, 2003). Solitons of the sine-Gordon (SG) equation, on the other hand, have infinite lifetimes, as their decay is prevented by a topological constraint. Thus, the kinetic energy of a SG kink decays (the soliton slows to a stop), but its rest mass does not disappear. The property of topological stability is shared by Skyrmions, which makes these objects interesting as models for elementary particles.

In studies of nontopological solitons in molecular crystals and of biological solitons, lifetime is a central issue, as the energy must remain localized long enough to be of technical or biological interest. When such solitons are also quantum objects (as is often the case), accurate lifetime calculations require estimates of the type sketched in Equation (4). This is a daunting theoretical task for three reasons: the wave function of a quantum soliton (ψ) is quite complicated (Davydov, 1991; Scott, 2003), biological molecules are often irregular structures for which all the possible final states $\{\phi_i\}$ are not known, and the temperatures at which biological organisms exist (ca. 300 K) introduce structural uncertainties that depend irregularly on time.

Using femtosecond pump-probe spectroscopy, however, it has recently become feasible to measure lifetimes in molecular crystals of the order of a picosecond or more (Elsaesser et al., 2000). Importantly, the response of such observations is null unless nonlinear effects are present, which occurs only for localized energy; thus the measurements focus specifically on soliton-like entities. From data produced by such experiments, the relevance of various theoretical proposals for solitons in molecular crystals and biomolecules should become better understood in coming years.

ALWYN SCOTT

See also **Biomolecular solitons; Davydov soliton; Pump-probe measurements; Skyrmions**

Further Reading

Davydov, A.S. 1991. *Solitons in Molecular Systems*, 2nd edition, Dordrecht and Boston: Reidel

Dirac, P.A.M. 1935. *The Principles of Quantum Mechanics*, Oxford: Clarendon Press

Elsaesser, T., Mukamel, S., Murnanae, M.M. & Scherer N.F. (editors). 2000. *Ultrafast Phenomena XII*, Berlin: Springer

Russell, J.R. 1844. Report on waves. *14th Meeting of the British Association for the Advancement of Science*, London: BAAS: 311–339

Scott, A.C. 2003. *Nonlinear Science: Emergence and Dynamics of Coherent Structures*, 2nd edition, Oxford and New York: Oxford University Press

Slater, J.C. 1951. *Quantum Theory of Matter*, New York: McGraw-Hill

LIGHTHILL CRITERION

See **Wave stability and instability**

LINDSTEDT–POINCARÉ METHOD

See **Perturbation analysis**

LINEARIZATION

Linear systems are much easier to analyze than nonlinear systems because they satisfy the superposition principle, which can be stated as follows: if R_1 is the output of a linear system to input S_1 and R_2 is the output to input S_2, then the output to the cumulative input $aS_1 + bS_2$ is $aR_1 + bR_2$. Applied iteratively, this property allows the inputs and outputs of a linear system to be decomposed in various ways to simplify analysis. From a more general perspective, linear systems do not confuse the lines of causality between the input and output (stimulation and response) of a linear system. Fully nonlinear systems, on the other hand, do not share this useful property; thus, the causal implications of each stimulation must be individually studied.

Linearization is an attempt to carry some of the attractive properties of linear systems into the nonlinear domain. Linearization methods can be classified in two approaches: reductions of nonlinear systems to linearized approximations for small perturbations and extensions of linearized approximations to weakly nonlinear systems. Linearization methods of the first approach are used for stability analyses of special solutions of a nonlinear system, such as critical points, periodic orbits, solitary waves and periodic waves. In the neighborhood of such special solutions of nonlinear systems, linear equations are derived for small perturbations that are then studied using standard methods (Fourier or Laplace transforms, Green functions, etc.). Linearization methods of the second approach are based on specific parameters of a linearized system, such as dominant eigenvalues or resonant wave numbers. Thus, linearized systems are extended to weakly nonlinear systems, such as normal forms and amplitude equations, by means of asymptotic multiple scale expansion methods and Fredholm's alternative theorem for linear nonhomogeneous equations.

For systems of partial differential equations (PDEs), linearization methods can be illustrated with the example of a nonlinear Schrödinger (NLS) equation (Whitham, 1974):

$$\mathrm{i}u_t + u_{xx} - \left(\omega_0 + \omega_2 |u|^2\right) u = 0, \tag{1}$$

where ω_0 and ω_2 are parameters such that $\omega_2 > 0$ and $\omega_2 < 0$ occur for defocusing and focusing cases, respectively. The NLS equation (1) has a constant wave solution of the form:

$$u_0(x,t) = a\,\mathrm{e}^{-\mathrm{i}(\omega_0+\omega_2 a^2)t}, \tag{2}$$

where the wave frequency $\omega = \omega_0 + \omega_2 a^2$ depends on the constant wave amplitude a. Linearizing the NLS equation (1) at the wave solution (2), one expands $u(x,t)$ as

$$u(x,t) = a\,\mathrm{e}^{-\mathrm{i}(\omega_0+\omega_2 a^2)t}\,[1 + V(x,t) + \mathrm{i}W(x,t)], \tag{3}$$

where V and W are real. When the expansion is substituted into the NLS equation and the nonlinear terms in V and W are neglected, the linearized problem can be transformed to the linear Boussinesq equation (4) with constant coefficients:

$$W_{tt} - 2\omega_2 a^2 W_{xx} + W_{xxxx} = 0. \tag{4}$$

The Fourier spectrum of the linear Boussinesq equation consists of two counter-propagating waves:

$$W(x,t) = W_+ \mathrm{e}^{\mathrm{i}kx - \mathrm{i}\Omega(k)t} + W_- \mathrm{e}^{\mathrm{i}kx + \mathrm{i}\Omega(k)t}, \tag{5}$$

where $\Omega(k) = k\sqrt{2\omega_2 a^2 + k^2}$. When $\omega_2 > 0$ (the defocusing case), the wave spectrum is stable, that is, $\Omega(k)$ is real for any k. When $\omega_2 < 0$ (the focusing case), the wave spectrum is unstable, that is, $\Omega(k)$ is complex for an interval of k, namely for $0 < k < \sqrt{2|\omega_2|}a$. Thus, the linearization method of the first approach enables us to check stability of the constant wave solutions of the NLS equation.

The linear Boussinesq equation (4) with $\omega_2 > 0$ is a hyperbolic system. With dispersive effects neglected in a long-wave approximation, the Boussinesq equation reduces to the wave equation: $W_{tt} - c^2 W_{xx} = 0$, where $c^2 = 2\omega_2 a^2$. The wave equation has a solution in the form of two dispersionless counter-propagating waves: $W(x,t) = W_+(x - ct) + W_-(x + ct)$. Unidirectional long-wave, small-amplitude approximation can be captured by the asymptotic multi-scale expansion method in the perturbation series (Zakharov & Kuznetsov, 1986):

$$u(x,t) = a\,\mathrm{e}^{-\mathrm{i}(\omega_0+\omega_2 a^2)t} \times \left[1 + \mathrm{i}\varepsilon R(\xi,\tau) + \varepsilon^2 \left(\frac{1}{c} R_\xi - R^2\right) + \mathrm{O}(\varepsilon^3)\right], \tag{6}$$

where $\xi = \varepsilon(x - ct)$, $\tau = \varepsilon^3 t$, and ε is a small parameter. Applying the Fredholm's alternative theorem to a linear nonhomogeneous equation at order $\mathrm{O}(\varepsilon^3)$, one can derive the Korteweg–de Vries (KdV) equation for $W = R_\xi$:

$$-2cW_\tau - 6cWW_\xi + W_{\xi\xi\xi} = 0. \tag{7}$$

With quadratic nonlinear terms neglected, the KdV equation (7) reproduces the dispersion relation of the linear Boussinesq equation, which has been reduced in the unidirectional long-wave approximation. Thus, the linearized method of the second approach enables us to derive the nonlinear evolution equation for wave propagation over the background of the constant wave solution of the NLS equation (1).

Linearization methods for systems of ordinary differential equations (ODEs) are illustrated with the example of an autonomous nonlinear dynamical system (Glendinning, 1994):

$$\frac{\mathrm{d}\boldsymbol{u}}{\mathrm{d}t} = \boldsymbol{f}(\boldsymbol{u}), \tag{8}$$

where $\boldsymbol{u}=\boldsymbol{u}(t)$ is a vector observable and $\boldsymbol{f}(\boldsymbol{u})$ is a vector nonlinear function. Critical points of the dynamical system occur for $\boldsymbol{u}=\boldsymbol{u}_0$ such that $\boldsymbol{f}(\boldsymbol{u}_0)=0$. Linearizing the nonlinear system (8) at the critical point $\boldsymbol{u}_0$, one expands $\boldsymbol{u}(t)$ as

$$\boldsymbol{u}(t)=\boldsymbol{u}_0+\boldsymbol{v}(t). \tag{9}$$

Neglecting quadratic terms in $\boldsymbol{v}(t)$ reduces the linearized problem to the linear system with constant coefficients:

$$\frac{\mathrm{d}\boldsymbol{v}}{\mathrm{d}t}=\mathcal{J}\boldsymbol{v},\quad \mathcal{J}=\nabla_{\boldsymbol{u}}f\bigg|_{\boldsymbol{u}=\boldsymbol{u}_0}, \tag{10}$$

where $\mathcal{J}$ is the Jacobian matrix. Solutions of the linearized system (10) depend on eigenvalues and eigenvectors of the Jacobian matrix: $\mathcal{J}\boldsymbol{w}=\lambda\boldsymbol{w}$. If all eigenvalues are located in the left half-plane of λ, that is, $\mathrm{Re}(\lambda)<0$, then the critical point $\boldsymbol{u}_0$ is asymptotically stable and the perturbation $\boldsymbol{v}(t)$ decays to zero exponentially in t. If there exists at least one eigenvalue in the right half-plane of λ, then the critical point $\boldsymbol{u}_0$ is linearly unstable and the perturbation $\boldsymbol{v}(t)$ grows exponentially in t.

Some or all eigenvalues may be located at the imaginary axis of λ, that is, $\mathrm{Re}(\lambda)=0$. In such systems, when no other eigenvalues exist for $\mathrm{Re}(\lambda)>0$, the critical point $\boldsymbol{u}_0$ is neutrally (weakly) stable. Perturbations may however grow algebraically in t, if eigenvalues of λ are multiple with algebraic multiplicity exceeding the geometric multiplicity. Local linearization is often insufficient for full description of such weak instability and the nonlinear stability analysis is desired.

When eigenvalues cross or coalesce on the axis $\mathrm{Re}(\lambda)=0$, bifurcations may occur in the spectrum of a linearized system (Glendinning, 1994). Normal forms for bifurcations can be derived by extending the linearized system into the nonlinear domain. For example, a Hopf bifurcation occurs when two simple eigenvalues $\lambda=\Gamma(\varepsilon)+\mathrm{i}\Omega(\varepsilon)$ and $\bar{\lambda}=\Gamma(\varepsilon)-\mathrm{i}\Omega(\varepsilon)$ cross the imaginary axis at the bifurcation parameter $\varepsilon=0$, such that $\Gamma(0)=0$ and $\Omega(0)=\Omega_0$. The normal form for the Hopf bifurcation can be derived by the asymptotic multi-scale expansion method in the perturbation series:

$$\boldsymbol{u}(t)=\boldsymbol{u}_0+\varepsilon\left[A(T)\boldsymbol{w}\mathrm{e}^{\mathrm{i}\Omega_0 t}+\bar{A}(T)\bar{\boldsymbol{w}}\mathrm{e}^{-\mathrm{i}\Omega_0 t}\right]+\mathrm{O}(\varepsilon^2), \tag{11}$$

where $T=\varepsilon^2 t$ and $\boldsymbol{w}$ is the eigenvector of $\mathcal{J}\boldsymbol{w}=\lambda\boldsymbol{w}$ for $\lambda=\mathrm{i}\Omega_0$ at $\varepsilon=0$. The normal form for Hopf bifurcation follows from the Fredholm's alternative theorem at order $\mathrm{O}(\varepsilon^3)$ in the form

$$\frac{\mathrm{d}A}{\mathrm{d}T}=\left[\Gamma'(0)+\mathrm{i}\Omega'(0)\right]A-\beta(0)|A|^2A, \tag{12}$$

where $\beta(0)$ is constant. $A(T)$ bifurcates into a periodic orbit at the Hopf bifurcation when $\Gamma'(0)\mathrm{Re}(\beta(0))>0$. Thus, linearization methods allow for linear and nonlinear stability analyses, resulting in simplifications of the nonlinear dynamical system (8).

DMITRY PELINOVSKY

See also **Causality; Multiple scale analysis; Quasilinear analysis; Stability**

Further Reading

Glendinning, P. 1994. *Stability, Instability and Chaos: An Introduction to the Theory of Nonlinear Differential Equations*, Cambridge and New York: Cambridge University Press

Whitham, G.B. 1974. *Linear and Nonlinear Waves*, New York: Wiley

Zakharov, V.E. & Kuznetsov, E.A. 1986. Multi-scale expansions in the theory of systems integrable by the inverse scattering transform. *Physica* D, 18

LIOUVILLE EQUATION

See **Sine-Gordon equation**

LIOUVILLE THEOREM

See **Phase space**

LIPSCHITZ CONDITION

See **Phase space**

LIQUID CRYSTALS

Discovery and Basic Properties

Liquid crystals do not follow the conventional classification of matter into three states: gas, liquid, and solid. In solids the orientation and position of the molecular building blocks are well determined, whereas in liquids the molecules move around in a disordered mixture. In liquid crystals the molecules can change their position, but their orientation is largely determined by neighboring molecules. The liquid crystal phase was discovered by Friedrich Reinitzer in 1888, when he noticed that solid cholesterol benzoate melted at 145°C into a scattering liquid crystal phase which transformed into a clear liquid after further heating to 178°C (Priestly et al., 1976, Chapter 2).

The type of ordering is related to the shape of the molecules or mesogens building the liquid crystal (De Gennes & Prost, 1993, Chapter 1). Calamitic molecules are rod-shaped and form nematic (molecules are aligned) or smectic phases (molecules are aligned and ordered in layers) as shown in Figure 1. Discotic molecules are disc-shaped and form nematic or stacked phases (with the molecules stacked on top of each other). The ordering of the molecules is not perfect due to thermal motion. With increasing temperature, ordering decreases until the liquid crystal is transformed into an isotropic liquid. Because the phase transitions are

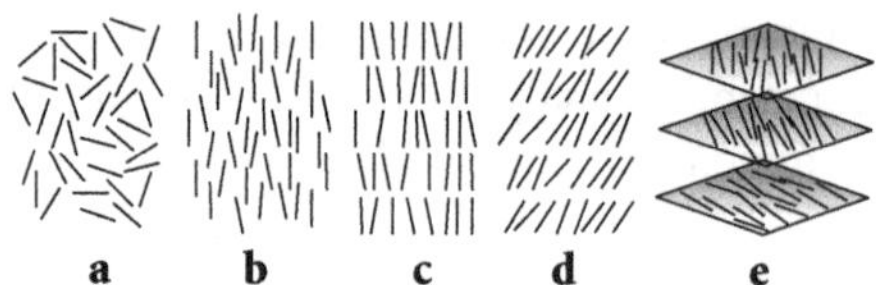

Figure 1. Different phases of liquid crystal materials. (a) isotropic, (b) nematic, (c) smectic-A, (d) smectic-C, (e) cholesteric.

related to temperature, these materials are called thermotropic liquid crystals. The organic molecules have low molecular weight (200–500) and can go through several phases from low to high temperature: solid–smectic-C–smectic-A–nematic–liquid (Figure 1). Chiral (handed) molecules do not have mirror symmetry and this may lead to a cholesteric phase in which the nematic orientation rotates slowly from one layer to the next (Figure 1e). Chiral phases have the property to reflect circularly polarized light that has the same handedness and periodicity as the molecule's structure. In lyotropic liquid crystals which have one hydrophobic and one hydrophilic part, the crystalline ordering depends on the concentration of certain molecules in the solvent. At very low concentrations individual molecules move freely in the liquid, at higher concentrations droplets, cylinders, or planes of attached molecules can be formed. Lyotropic liquid crystals are important in biological systems, in cell membranes, and dispersions (De Gennes & Prost, 1993, Chapter 1).

The nematic and smectic-A phases have uniaxial symmetry with the preferential axis parallel with the molecules. The preferential axis is characterized by a vector, the director $\bar{n}$. Other properties such as permittivity, refractive index, magnetic susceptibility, and conductivity are described by a tensor. The index of refraction of calamitic liquid crystals may, for example, vary between $n_{\perp}$=1.5 and $n_{\parallel}$=1.6. The smectic-C phase has biaxial properties. The deformation of a liquid crystal leads to elastic energy. The energy in nematic liquid crystals is due to three types of variations in the orientation: splay, twist, and bend (see De Gennes & Prost, 1993, Chapter 3; Blinov & Chigrinov, 1994, Chapter 4). The elastic energy is then expressed by the following equation:

$$F_{\text{elastic}} = \tfrac{1}{2}\left(K_{11}\,(\nabla\cdot\bar{n})^2 + K_{22}\,(\bar{n}\cdot\nabla\times\bar{n})^2 + K_{33}\,(\bar{n}\times\nabla\times\bar{n})^2\right). \quad (1)$$

Surfaces can prefer either homeotropic (perpendicular) or parallel alignment of the director. If the interface is treated by rubbing a polyimide surface layer, this dictates a preferential azimuth for the director. The minimization of the combined elastic energy and surface energy determines the orientation of the director between the surfaces. This can lead to a pure twist as shown in Figure 2.

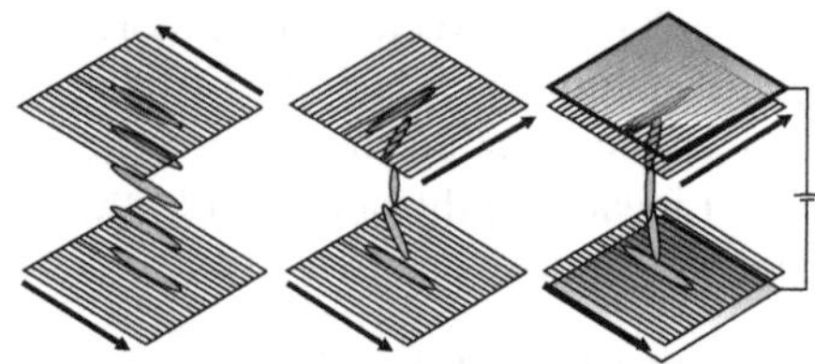

Figure 2. Nematic liquid crystal between two glass plates. The grid represents the alignment layer and the arrow the rubbing direction. Left: antiparallel rubbing with all molecules parallel; Middle: 90° twisted nematic; Right: when a voltage is applied between the electrodes, the molecules tilt.

Nonlinear Properties

In smectic liquid crystals, linear elasticity theory and linear hydrodynamics no longer hold because higher order terms come into play in the elastic energy (De Gennes & Prost, 1993, Chapter 8). The elasticity constants then depend on the deformation strength and this can lead to undulation instabilities, buckling, or focal conic structures. In a similar way, hydrodynamic nonlinearities yield frequency-dependent viscosities. If liquid crystal is illuminated with a high optical power, the physical properties may be modified. The processes leading to optical nonlinearities can be of electronic or non-electronic origin (Khoo, 1995). Electronic nonlinear effects involve processes in which the electronic wave functions of the liquid crystal molecules are perturbed by the optical field. These very fast processes are similar to electronic nonlinearities in solid crystals. These processes lead to second and third harmonic generation and optical phase conjugation (Khoo, 1995). Beside the electronic structure of the molecules, other physical properties can change because of an applied optical field, such as the molecular structure or orientation, the temperature, and the density (Elston & Sambles, 1998, Chapter 6). Slow temperature effects occur when a liquid crystal in a nematic phase is heated by an optical beam. The difference between the ordinary and extraordinary index of refraction decreases until the liquid crystal reaches the transition temperature and becomes isotropic. Another important effect is the molecular reorientation by the optical field. In the static case, the molecules will reorient due to the difference in dielectric constant ($\varepsilon_{\parallel}$ and $\varepsilon_{\perp}$); and in the optical case, due to the difference in refractive index ($n_{\parallel}$ and $n_{\perp}$).

Applications

Because of the anisotropy in the permittivity (or magnetic susceptibility), calamitic molecules with $\varepsilon_{\parallel} > \varepsilon_{\perp}$ prefer to orient their long axis parallel with an applied electric (or magnetic) field. By applying a voltage over a cell filled with liquid crystal, it is possible to make the molecules tilt out of the plane, as illustrated in Figure 2. In a uniaxial liquid crystal, linearly polarized ordinary and extra-ordinary waves propagate with refractive indices $n_\perp$ and n_e (with n_e between $n_\parallel$ and $n_\perp$) (Iizuka, 2002, Ch.4). Because of the difference in propagation speed, the polarization state of the combined wave changes continuously. In a twisted nematic display (Figure 2), the linearly polarized light roughly follows the orientation of the director (Mauguin regime) and rotates over 90° (Priestly et al., 1976, Chapter 14). If the top polarizer is also rotated over 90° compared to the bottom polarizer, all light will be transmitted. If a voltage is applied to the cell and the molecules orient perpendicularly, the linear polarization is unchanged, and no light passes through the analyser. Cholesteric liquid crystals appear in nature, for example, in the cuticle of insects. The result is a colored and polarized reflection, depending on the angle of observation.

KRISTIAAN NEYTS AND JEROEN BEECKMAN

See also **Dislocations in crystals; Harmonic generation; Nonlinear optics**

Further Reading

Blinov, L.M. & Chigrinov, V.G. 1994. *Electrooptic Effects in Liquid Crystal Materials, Partially Ordered Systems*, New York: Springer

De Gennes, P.G. & Prost, J. 1993. *The Physics of Liquid Crystals*, 2nd edition, Oxford: Clarendon Press and New York: Oxford University Press

Elston, S. & Sambles, R. 1998. *The Optics of Thermotropic Liquid Crystals*, London: Taylor & Francis

Iizuka, K. 2002. *Elements of Photonics, in Free Space and Special Media*, New York: Wiley-Interscience

Khoo, I.C. 1995. *Liquid Crystals, Physical Properties and Nonlinear Optical Phenomena*, New York: Wiley

Priestly, E.B., Wojtowicz, P.J. & Shent, P. 1976. *Introduction to Liquid Crystals*, New York and London: Plenum Press

LOCAL MODES IN MOLECULAR CRYSTALS

Energy transfer in molecular crystals occurs through exciton hopping from molecule to molecule. It is possible, however, that nonlinear localization takes place due to coupling between excitonic and vibrational degrees of freedom, transforming the exciton into a local mode. Such a mode is generated by an effective local nonlinear potential induced by the exciton-phonon coupling, and it has features similar to a polaron, a soliton, and a breather.

The experimental search for nonlinear local modes was initiated by Careri in the early 1970s in crystalline acetanilide ($CH_3CONHC_6H_5$), or ACN, which has structure similar to a polypeptide; thus, positive findings in ACN could be of biological significance. In particular, if nonlinear local modes exist and are sufficiently long-lived, they could be agents of energy transfer in biomolecules.

In ACN and in proteins, one finds hydrogen-bonded quasi-one-dimensional polypeptide chains with the following structure:

$$\cdots \mathrm{H}-\mathrm{N}-\mathrm{C}=\mathrm{O}\cdots\mathrm{H}-\mathrm{N}-\mathrm{C}=\mathrm{O}\cdots\mathrm{H}-\mathrm{N}-\mathrm{C}=\mathrm{O}\cdots. \tag{1}$$

Careri focused on the amide-I (C=O stretching) vibration and found its infrared absorption spectrum to have two peaks: a "normal" one at 1665 cm^{-1} and an "anomalous" one at 1650 cm^{-1}. The basic feature of the latter is its strong temperature dependence and its disappearance at biological temperatures, as shown in Figure 1.

In 1982, Careri and Scott hypothesized that this anomalous peak is a spectral signature of a self-trapped state similar to the Davydov soliton, while the normal peak is the delocalized free exciton of the amide-I vibration. An approximate way to represent the ACN local mode is by assuming an adiabatic separation between the faster amide-I exciton vibration and a slower local (Einstein) oscillator with which the hydrogen bond interacts. The corresponding Hamiltonian has three terms:

$$H = H_{\mathrm{ex}} + H_{\mathrm{ph}} + H_{\mathrm{int}}, \tag{2}$$

where

$$\begin{aligned} H_{\mathrm{ex}} &= \sum_n \left[\varepsilon_0 a_n^\dagger a_n - J(a_n^\dagger a_{n+1} + a_n^\dagger a_{n-1})\right], \\ H_{\mathrm{ph}} &= \frac{1}{2}\sum_n \left[\frac{{p_n}^2}{m} + m{\omega_0}^2{q_n}^2\right], \\ H_{\mathrm{int}} &= \chi \sum_n q_n a_n^\dagger a_n . \end{aligned} \tag{3}$$

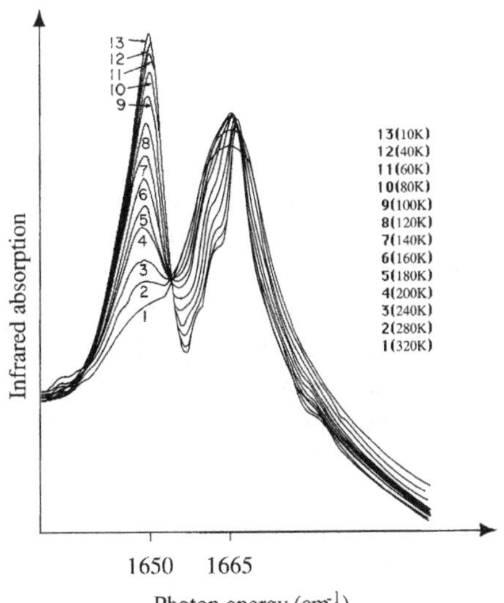

Figure 1. Infrared absorption spectra of crystalline acetanilide in the region of the amide-I mode. The free exciton peak is at 1665 cm^{-1} while the anomalous peak that is at 1650 cm^{-1} has strong temperature dependence and is associated with a self-trapped state (Careri et al., 1984).

Here $a_n^\dagger$ (a_n) creates (destroys) a quantum of amide-I excitation of energy ε_0 at the nth CO unit, J is the energy of exciton-exciton interaction between adjacent CO units, p_n and q_n are the momentum and position, respectively, of the local Einstein oscillator at the nth location representing a low frequency optical mode, and χ determines the strength of the interaction between the exciton and Einstein mode.

One way to analyze this complex problem is by treating the slow ($p_n \approx 0$) Einstein oscillators classically while evaluating the expectation value of the total Hamiltonian with respect to the exciton wave function $|\Psi\rangle = \sum_n c_n(t) a\dagger_n |0\rangle_{\text{ex}}$, where $|0\rangle_{\text{ex}}$ is the exciton system vacuum. Variational minimization of $\mathcal{H} \equiv \langle \Psi | H | \Psi \rangle$ with respect to the oscillator variable q_n leads to $q_n = -\chi/(m\omega_0{}^2)|c_n(t)|^2$, which upon substitution to $\mathcal{H}$ transforms the latter to the classical Hamiltonian

$$\mathcal{H} = \sum_n \left[\varepsilon_0 |c_n|^2 \quad -J(c*_n c_{n+1} + c*_n c_{n-1}) \right. \\ \left. - \frac{\chi^2}{2m\omega_0^2} |c_n|^4 \right]. \qquad (4)$$

Hamilton's equations then lead to a discrete self-trapping system or a discrete nonlinear Schrödinger (DNLS) equation:

$$i\hbar \dot{c}_n + J(c_{n+1} + c_{n-1}) + \frac{\chi^2}{m\omega_0{}^2} |c_n|^2 c_n = 0. \qquad (5)$$

This equation admits two types of solutions: one is an extended plane wave corresponding to the free exciton, while the other is localized and corresponds to a self-trapped state. The difference of the energies of these two solutions is

$$E_{\text{b}} = \frac{\chi^2}{2m\omega_0{}^2} - 2J, \qquad (6)$$

corresponding to the binding energy of the self-trapped state. A combination of calculations and fitting to data suggests that $J \approx 4\,\text{cm}^{-1}$, $\chi^2/m\omega_0^2 \approx 45\,\text{cm}^{-1}$, and E_{b} is equal to $1665 - 1650 = 15\,\text{cm}^{-1}$.

This picture is in accord with the presence of the two peaks in the amide-I region shown in Figure 1. Use of these parameter values in the context of a more sophisticated model gives also the correct temperature dependence of the anomalous peak. Because in the present case $J \ll \chi^2/m\omega_0^2$, the local mode can also be viewed as a DNLS discrete breather close to the anti-continuous limit (where $J \to 0$).

Although the above proposed was furnished in the 1980s, both the essence of Davydov's idea as well as its specific application to ACN were challenged during this period. The possibility that the anomalous peak is a result of Fermi resonance of the amide-I mode with an overtone of an alternative ACN vibrational mode was suggested as well as a possible structural source for the mode. These possibilities were to a large degree eliminated by the experimental work of Barthes and her collaborators. Definitive experimental information on the ACN local mode was furnished in 2002 by Edler and Hamm.

Using time-resolved pump-probe experiments with infrared beams of femtosecond duration, Edler and Hamm were able to excite the amide-I region and follow the time development of the various modes in that region. Their experiments showed a clear difference in the behavior of the normal and anomalous modes with the latter one being strongly anharmonic. Although the lifetime of the anomalous mode was only about 2 ps, return of this energy to the ground state took about 35 ps. Thus, the local mode initially relaxes into a state that is either spectroscopically dark or outside the spectral window of the probe. Interestingly, similar pump-probe experiments for the region of the NH stretching demonstrate the presence of nonlinear local mode structure in this range as well.

The initial experiments of Careri, the subsequent work of Barthes, and finally the experiments of Edler and Hamm along with Davydov's theory and its application to ACN by Scott present a convincing 30-year story that attributes the source of the anomalous ACN peak to a highly localized, relatively short-lived self-trapped state.

Another molecular crystal with different structure that is believed to form nonlinear local modes is the halide-bridged mixed-valence transition metal complex $\{[\text{Pt(en)}_2][\text{Pt(en)}_2\text{Cl}_2](\text{ClO}_4)_4\}$, where en = ethylenediamine. In PtCl, essentially one-dimensional chains of platinum alternating with chlorine atoms are formed, and an intra-molecular vibrational excitation of the $PtCl_2$ trimer caused by energy transfer between Pt^{4+} and Pt^{2+} seems to become effectively localized as a result of coupling with other crystal vibrational modes. Raman spectra show a clear red-shifted PtCl overtone spectrum induced by anharmonicity that reduces in PtBr crystals due to smaller effective anharmonicity and disappears in PtI due to absence of nonlinearity. A multiquanta generalization of the theoretical ACN approach of Equations (1)–(4) demonstrates that the overtone redshifts in the PtCl Raman spectrum can be accounted for quantitatively as well as qualitatively by a nonlinear local mode picture.

G.P. Tsironis

See also **Breathers; Davydov soliton; Discrete nonlinear Schrödinger equations; Discrete self-trapping system; Pump-probe measurements**

Further Reading

Barthes M., Nunzio, G.D. & Ribet, M. 1996. Polarons or protontransfer in chains of peptide groups? *Synthetic Metals*, 76: 337–340

Bushmann, W.E., McGrane, S.D. & Shreve, A.P., 2003. Chemical tuning of nonlinearity leading to intrinsically localized modes in halide-bridged mixed-valence platinum materials. *Journal of Physical Chemistry* A, 107(40): 8198–8207

Careri, G., Buontempo, U., Galluzzi, F., Scott, A.C., Gratton, E. & Shyamsunder, E. 1984. Spectroscopic evidence for Davydov-like solitons in acetanilide. *Physical Review* B, 30: 4689–4702

Edler, J. & Hamm, P. 2002. Self-trapping of the amide I band in a peptide model crystal. *Journal of Chemical Physics*, 117: 2415–2424

Scott, A.C. 1992. Davydov's soliton. *Physics Reports*, 217: 1–67

Swanson, B.I., Brozik, J.A., Love, S.P., Strouse, G.F., Shreve, A.P., Bishop, A.R., Zang, W.-Z. & Salkola, M.I. 1999. Observation of intrinsically localized modes in a discrete low-dimensional material. *Physical Review Letters*, 82: 3288–3291

Voulgarakis, N.K., Kalosakas, G., Bishop, A.R. & Tsironis, G.P. 2001. Multiquanta breather model for PtCl. *Physical Review* B, 64: 020301

LOCAL MODES IN MOLECULES

The local mode model was introduced in the 1970s to describe highly excited vibrational overtone states. It has been used to describe XH stretching overtone spectra where X is some heavier atom like C, N, or O. However, the genesis of the local mode model in molecules dates back to the work of J.W. Ellis (1929). Ellis measured the overtone spectrum of benzene in the liquid phase to $\Delta v_{\mathrm{CH}} = 8$ (a transition from the ground vibrational state to a state with 8 quanta in CH-stretching). He found that the spectra were pseudo-diatomic in character, in that the CH overtone frequencies could be fitted by a simple Birge–Sponer relationship that had been used to describe a diatomic anharmonic oscillator. Thus, if the observed overtone peak frequencies, $\tilde{\nu}_{v0}$, are divided by v, the number of CH-stretching quanta in the excited overtone state, and plotted versus v, a straight line is obtained. The slope yields the local mode anharmonicity, $\tilde{\omega}x$, and the intercept can be used to determine the local mode frequency, $\tilde{\omega}$.

$$\tilde{\nu}_{v0}/v = \tilde{\omega} - (v+1)\tilde{\omega}x. \tag{1}$$

In 1968, Henry and Siebrand used the localization ideas of Ellis to model the overtone spectrum of benzene on the basis of the assumption that the six local CH-stretching modes were anharmonic but only weakly coupled (Henry & Siebrand, 1968). They derived values for CH-stretching anharmonicity constants expressed in the normal mode basis and calculated the energies of all of the components of the overtone bands expressed in terms of normal modes. At $\Delta v_{\mathrm{CH}} = 6$, there are 462 such components of which 150 (the 75 doubly degenerate E_{1_u} states) are allowed. The spectra were then modeled on the basis of a weighted combination of the allowed normal modes. Subsequently, Hayward and Henry postulated that the radiation field selectively excites only the most anharmonic of the normal mode components (Hayward & Henry, 1975). In such a state, all the vibrational energy is effectively localized in one XH bond (X = C, N, O, etc.) and absorption to this pure local mode state is said to account for the overtone spectra. Some of these ideas had been published in 1936 in Germany by Mecke and his collaborators (Mecke & Ziegler, 1936). However, Mecke's work seemed to have been largely unknown and ignored. The description that emerges is that XH-stretching overtone spectra can be described by a local mode model in which the local oscillators are anharmonic but only weakly coupled. The radiation field selectively excites a state whose components have all of the vibrational energy localized in one of a set of equivalent XH bonds. Such a description predicts only a single transition for a given type of XH oscillator in agreement with experiment. The local mode model is now generally accepted, and because of the localization of energy for spectrally active states in a single chemical bond, XH-stretching overtone spectra are very sensitive to bond properties. Such spectra have been used to investigate molecular structure, molecular conformation, intermolecular forces, nonbonded intramolecular interactions, and intramolecular vibrational energy redistribution.

The Harmonically Coupled Anharmonic Oscillator Model

In a molecule like water, the two OH bonds can be treated as Morse potentials. In accord with the local mode model, the OH oscillators are anharmonic but only weakly coupled. It turns out that this off-diagonal potential and kinetic energy coupling is well approximated by the rules that govern coupling of harmonic oscillators.

The harmonically coupled anharmonic oscillator (HCAO) model was formulated independently by Mortensen et al. (1981) and by Child & Lawton (1981). Coupling is allowed only between states within a given vibrational manifold, that is, between states with the same value of the XH-stretching vibrational quantum number v. Coupling is restricted to the harmonic limit where oscillator population differs by ± 1. All of the parameters in this model are obtained either directly from the spectra or from ab initio calculations.

The model has been very successful in accounting for the observed energies of overtone peaks in XH-stretching overtone spectra and, thus, has proved useful in the assignment of these spectra. It has been generalized to more than two equivalent oscillators and to sets of non-equivalent oscillators. No identification of the wave functions is needed for the HCAO model to calculate peak energies. However, if one uses a basis set consisting of products of one-dimensional Morse oscillators, the HCAO model can be used to generate vibrational wave functions. If such wave functions are used along with ab initio calculations of dipole moment functions (expanded in local coordinates), simple but highly successful calculations of overtone intensities are possible (Kjaergaard et al., 1990).

Overtone Intensities

Typically, these calculations: (i) identify transitions to pure local mode states, whose components have all of the vibrational energy localized in one of a set of equivalent XH oscillators, as the dominant peaks for $\Delta v_{XH} \geq 3$; (ii) indicate that the intensity of these pure local mode peaks is determined primarily by second-order diagonal terms involving $\partial^2\mu/\partial R^2$ in the dipole-moment operator μ; (iii) account for the fall-off in intensity with increasing v and the intensity distribution within a given manifold; (iv) account for the intensity distribution at $\Delta v_{XH} = 2$ where coupling between the local oscillators is important; and (v) account for relative intensities for different oscillators within a molecule. For example, calculations on cyclohexane (Kjaergaard & Henry, 1992) and naphthalene (Kjaergaard & Henry, 1995) successfully predict the relative intensities of the two non-equivalent ring hydrogens. These calculations have also been applied to the determination of accurate absolute overtone intensities for a number of species that are important in atmospheric photochemistry so that effects on the absorption of solar radiation can be estimated (Fono et al., 1999; Vaida et al., 2001).

One surprising result in these intensity calculations is that electron correlation in the dipole moment function affects the calculated fundamental intensity but has no significant effect on overtone intensities. One factor in this insensitivity to correlation is likely to be the primary importance of nonlinear terms in the dipole moment operator.

The HCAO intensity model has been generalized to include torsional motion in the Hamiltonian and to include the torsional coordinate in the ab initio calculation of the dipole moment function. As a result of coupling between torsion and stretching, both in the Hamiltonian and through the dipole-moment function, a very large number of transitions carry intensity and contribute to the overall methyl spectral profile. This model has successfully accounted for the methyl spectral profiles in the overtone spectra of methyl substituted aromatic molecules (Kjaergaard et al., 2000).

One early interest in the local mode field was the possibility that these highly localized states could be pumped with high power lasers and the result would be selective bond photochemistry. To date, no practical example of such behavior has arisen. The reason probably lies in the very rapid decay of these highly vibrationally excited states on a subpicosecond timescale. The dynamics of these local mode states continues to be a very active research area.

BRYAN R. HENRY

Further Reading

Child, M.S. & Lawton, T.R. 1981. Local and normal vibrational states: a harmonically coupled anharmonic oscillator model. *Faraday Discussions of the Chemical Society*, 71: 273

Ellis, J.W. 1929. Molecular absorption spectra of liquids below 3 μ. *Transactions of the Faraday Society*, 25: 888–897

Fono, L., Donaldson, D.J., Proos, R.J. & Henry, B.R. 1999. OH overtone spectra and intensities of pernitric acid. *Chemical Physics Letters*, 311: 131–138

Hayward, R.J. & Henry, B.R. 1975. A general local-mode theory for high energy polyatomic overtone spectra and application to dichloromethane. *Journal of Molecular Spectroscopy*, 57: 221–235

Henry, B.R. & Siebrand, W. 1968. Anharmonicity in polyatomic molecules. The CH stretching overtone spectrum of benzene. *Journal of Chemical Physics*, 49: 5369–5376

Kjaergaard, H.G. & Henry, B.R. 1992. The relative intensity contributions of axial and equatorial CH bonds in the local mode overtone spectrum of cyclohexane. *Journal of Chemical Physics*, 96: 4841–4851

Kjaergaard, H.G. & Henry, B.R. 1995. CH-stretching overtone spectra and intensities of vapor phase naphthalene. *Journal of Physical Chemistry*, 99: 899–904

Kjaergaard, H.G., Rong, Z., McAlees, A.J., Howard, D.L. & Henry, B.R. 2000. Internal methyl rotation in the CH stretching overtone spectra of toluene-α-d_2, -α-d_1, and -d_0. *Journal of Physical Chemistry* A, 104: 6398–6405

Kjaergaard, H.G., Yu, H., Schattka, B.J., Henry, B.R. & Tarr, A.W. 1990. Intensities in local mode overtone spectra: propane. *Journal of Chemical Physics*, 93: 6239–6248

Mecke, R. & Ziegler, R. 1936. Das Rotationsschwingungsspektrum des Acetylens (C_2H_2). *Zeitschrift für Physik*, 101: 405–417

Mortensen, O.S., Henry, B.R. & Mohammadi, M.A. 1981. The effects of symmetry within the local mode picture: a reanalysis of the overtone spectra of the dihalomethanes. *Journal of Chemical Physics*, 75: 4800–4808

Vaida, V., Daniel, J.S., Kjaergaard, H.G., Gross, L.M. & Tuck, A.F. 2001. Atmospheric absorption of near infrared and visible solar radiation by the hydrogen bonded water dimer. *Quarterly Journal of the Royal Meteorological Society*, 127: 1627–1644

LOCALIZATION

See **Solitons, types of**

LOCKING

See **Coupled oscillators**

LOGIC, DENDRITIC

See **Multiplex neuron**

LOGISTIC EQUATION

See **Population dynamics**

LOGISTIC MAP

See **One-dimensional maps**

LONG JOSEPHSON JUNCTIONS

The long Josephson junction is an excellent example of an experimental solid-state system that in a straightforward and measurable way, supports the existence and propagation of solitons (Parmentier, 1978; Pedersen, 1986; Ustinov, 1998). In this system, the soliton is called a fluxon, because it contains one quantum of magnetic flux, $\Phi_0 = h/2e \doteq 2.064 \times 10^{-15}$ webers (h is Planck's constant and e is the electron charge), in accord with the particle nature of solitons (Scott, 2003).

A Josephson junction consists of two weakly coupled superconductors separated by a thin (typically 1–10 nm) insulating layer. An important dynamical variable is the phase difference between the macroscopic quantum mechanical phases of the two superconductors, $\phi(x, t)$. Under an appropriate time normalization, the small Josephson junction obeys approximately the same equation as the damped and driven pendulum. If the Josephson junction is long with respect to an intrinsic length scale (called the Josephson penetration depth, of order 1–1000 μm), the equation becomes the damped and driven sine-Gordon equation (often called the perturbed sine-Gordon equation) (McLaughlin & Scott, 1978):

$$-\phi_{xx} + \phi_{tt} + \alpha\phi_t + \sin\phi = i\,. \qquad (1)$$

Here, the damping parameter α (with typical experimental values in the range $0.01 < \alpha < 0.1$) is related to the Josephson junction parameters per unit area: conductance G, capacitance C, and supercurrent J; thus $\alpha = G\sqrt{\hbar/2eCJ}$. Also $i = I_{\text{bias}}/I_0$ is a bias current normalized to the maximum Josephson current, x is the direction of the long side of the junction normalized to the Josephson penetration depth, and time t is normalized to the so-called plasma frequency, which corresponds to the pendulum oscillation frequency. The plasma frequency is of order 1–500 GHz, depending on fabrication parameters (composition and thickness of insulating layer, superconducting materials, etc.).

A particularly interesting solution to Equation (1) is the sine-Gordon soliton (kink), or rather a slightly modified kink because of the damping and bias terms. Such a kink carries one quantum of magnetic flux (an antikink carries a flux quantum pointed in the opposite direction), and it has a steady state velocity determined as a balance between energy input from the bias current (i) and losses stemming from α. This solution has particle-like properties, behaving approximately as a Lorentz invariant, relativistic particle with the velocity of light in the system being the limiting velocity. This so-called Swihart velocity is determined by the junction fabrication parameters and is typically 1–5% of the velocity of light in vacuum.

The simple form of the dynamical equation shown in Equation (1) refers to the so-called overlap geometry shown in Figure 1(a), in which the externally applied

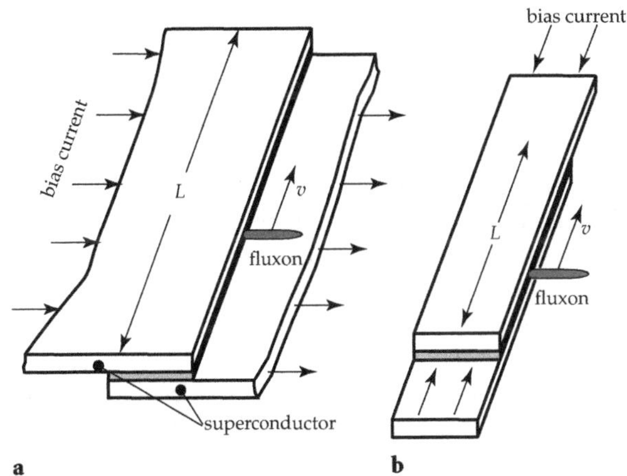

Figure 1. Rectangular-shaped long Josephson junctions (not to scale). (a) Overlap geometry. (b) Inline geometry.

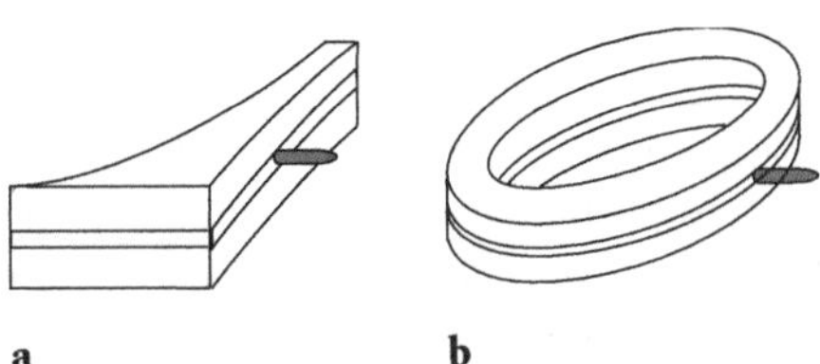

Figure 2. Other long junction geometries (bias not shown). (a) Eiffel junction. (b) Annular junction.

current is fed uniformly to the long dimension of the long Josephson junction. In addition to Equation (1), boundary constraints must be satisfied. For an overlap junction of length L, the longitudinal (x-directed) surface current (j) is zero at $x = 0$ and $x = L$ if no external magnetic field is applied.

Because the long Josephson junction is typically fabricated using photolithographic and evaporation techniques borrowed from the semiconductor chip industry, it is easy to produce different geometries. Besides the overlap junction, inline junctions, Eiffel junctions, and annular junctions are of experimental and practical interest.

For the inline junction shown in Figure 1(b), the external current is fed from the narrow ends of the junction. The right-hand side of Equation (1) becomes zero, and the bias enter through the boundary conditions at $x = 0$ and L. In the Eiffel junction shown in Figure 2(a), the width of the junction is varied in an exponentially tapered shape approximating that of the Eiffel tower. For reasons of energy conservation, the propagation becomes unidirectional, with the fluxon tending to move in the direction of decreasing junction width (Benabdallah et al., 1996, 2000). The Eiffel junction has been proposed as a dc to ac converter and a microwave oscillator.

The annular junction, shown in Figure 2(b), is like a long overlap junction where the two ends are joined. For topological reasons connected with the superconducting phase quantization, periodic boundary conditions apply; that is, $j(0,t) = j(L,t) + 2p\pi$ with $p = 0, 1, 2, \ldots$, where j is the longitudinal surface current. A relatively clean soliton (fluxon) can be studied here since there are no collisions with boundaries. The total number of fluxons minus antifluxons is a conserved number (p) and fluxons can only disappear in a fluxon-antifluxon annihilation processes, conserving the total flux in the ring. In a variant of the annular junction, periodic changes in the junction width give rise to deceleration and acceleration of the fluxons, thereby producing radiation (McLaughlin & Scott, 1978).

The long Josephson junction has a potential as a microwave oscillator in the hundreds of Gigahertz range (100–1000 GHz). This is a frequency range, interesting for applications such as fast superconducting electronics, radio astronomy, and satellite-to-satellite communication. In this frequency range, competing oscillators are often bulky, noisy, and expensive. Importantly, the line width of the fluxon oscillator with proper stabilization can become very narrow—of the order 50 kHz. Since the line width of a local oscillator sets the frequency resolution for a heterodyne receiver, this is an important property for spectroscopy and for determining the frequency of radio sources in the universe.

For radio astronomy ($\sim$100–500 GHz), the standard today is a superconducting superheterodyne receiver, for which all components (mixer, local oscillator, filters, etc.) are fabricated in superconducting electronics, and the local oscillator is typically a long Josephson junction. In practice, two different schemes for a fluxon oscillator have been used. In one scheme, the fluxon propagates back and forth within the junction. At each collision with boundary some radiation is coupled out to an external circuit. The frequency of the radiation is given by the time of flight of the fluxon in the junction.

In another mode—the so-called flux flow mode—a unidirectional stream of fluxons is created. This can be done either by the above mentioned Eiffel junction or by applying a magnetic field that breaks the symmetry, giving rise to energy absorption at one end of the junction and energy creation at the other. With proper parameter adjustment, fluxons will be continuously created at one end of the junction, propagated through the junction, and absorbed in an external load (mixer) at the other end.

The power obtainable in such schemes is at best a few microwatts at 300–500 GHz with a narrow line width (of order 100 kHz). Nonetheless, this power level is sufficient for local oscillators of superconducting receivers.

As a concluding remark, we mention that it has been known since the early days of the long Josephson junction research, that the fluxon has properties related to a relativistic particle. Recent experiments have demonstrated that it also has quantum mechanical properties (Wallraff et al., 2003).

NIELS FALSIG PEDERSEN

See also **Josephson junctions; Sine-Gordon equation; Superconductivity**

Further Reading

Benabdallah, A., Caputo, J.G. & Scott, A.C. 1996. Exponentially tapered Josephson flux flow oscillator. *Physics Review* B, 54: 16139–16146.

Benabdallah, A., Caputo, J.G. & Scott, A.C. 2000. Laminar phase flow for an exponentially tapered Josephson oscillator. *Journal Applied Physics*, 88: 16139–16146

McLaughlin, D.W. & Scott, A.C. 1978. Perturbation analysis of fluxon dynamics. *Physics Review A*, 18: 1652–1680.

Parmentier, R.D. 1978. Fluxons in long Josephson junction. In *Solitons in Action*, edited by K. Lonngren & A.C. Scott, New York: Academic Press, 173–199

Pedersen, N.F. 1986. Solitons in Josephson transmission lines. In *Solitons: Modern Problems in Condensed Matter Sciences*, edited by A.A. Maradudin & V.H. Agranovich, Amsterdam: Elsevier, 469–501

Scott, A.C. 2003. *Nonlinear Science: Emergence and Dynamics of Coherent Structures*, 2nd edition, Oxford and New York: Oxford University Press

Ustinov, A.V. 1998. Solitons in Josephson junctions. *Physica* D, 123 (1–4): 315–329

Wallraff, A., Lukashenko, A., Lisenfeld, A., Kemp, A., Fistul, M.V., Koval, Y., & Ustinov, A.V. (2003). Quantum dynamics of a single vortex. *Nature*, 425: 155–158

LONG WATER WAVE APPROXIMATIONS

See **Water waves**

LONG-WAVELENGTH APPROXIMATION

See **Continuum approximation**

LOOP SOLITONS

See **Solitons, types of**

LOOP, CLOSED CAUSAL

See **Causality**

LORENTZ GAS

The Lorentz gas is a model dynamical system generated by the motion of non-interacting point particles ("electrons") in a field of immovable (infinitely heavy) spherical particles, called scatterers. The particles move by inertia according to Newton's law and are reflected from the scatterers elastically. The scatterers may overlap and are assumed to be randomly or periodically distributed.

Hendrik Lorentz introduced this model as a simplified version of the Drude model for metallic conductance of electricity (Lorentz, 1905). Lorentz's goal was to modify the Drude model in such a way that the Boltzmann equation becomes linear. This is so for the Lorentz gas because of the lack of interactions between particles, which do not influence the positions of the scatterers. As it is sufficient to consider just one moving particle, a one particle model is now usually called the Lorentz gas.

The Lorentz gas appeared to be an inadequate model to describe a gas of electrons in metals. However, it became one of the most popular models in non-equilibrium statistical mechanics, which is used to study the most fundamental problems such as irreversibility, the existence of transport coefficients, and the derivation of kinetic and hydrodynamic equations from mechanical laws. The Lorentz gas with random distribution of scatterers also serves as a relevant model of a dilute gas in equilibrium, consisting of a mixture of two types of (spherical) particles with one much larger and heavier than the other. Almost all collisions are between smaller and larger particles, rather than among smaller particles, even though there are many more smaller particles than heavy particles (Dettmann, 2000).

The kinetic stage of the Lorentz gas evolution is described by the Boltzmann equation, which refers to dilute gases. The appropriate limit procedure was suggested by Grad (1958) and is called the Boltzmann–Grad limit. In this limit, the density of a gas (density of scatterers) tends to zero while the mean free path of the moving particle remains constant. For the Lorentz gas, the kinetic Boltzmann equation goes in the Boltzmann–Grad limit into the linear Fokker–Planck–Kolmogorov equation for the corresponding Markov process. It has been proven for a dilute random Lorentz gas in the Boltzmann–Grad limit that the Boltzmann equation holds for almost any (Poisson distributed) configuration of scatterers (Boldrighini et al., 1983). This result allows a comparison of the theory with the results of computer experiments, because it refers to an individual configuration of scatterers rather than to characteristics averaged over the ensemble of all such configurations.

Rigorous results on the hydrodynamics of the Lorentz gas are available only for periodic configurations of scatterers. This problem for a random Lorentz gas currently looks hopeless because it can be reduced to a problem of random walks in random environments that is far beyond the present abilities of this theory. Clearly, the hydrodynamic equation for the Lorentz gas must be the diffusion equation because the momentum is not preserved in this model, and the conservation of energy is just equivalent to the conservation of mass. Therefore, the diffusion coefficient is also the only transport coefficient in the Lorentz gas.

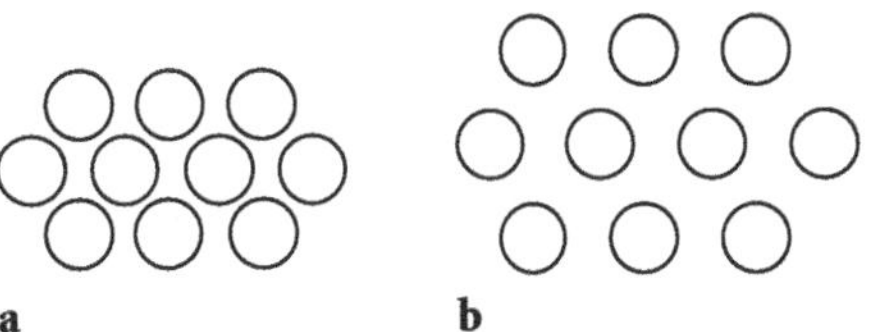

Figure 1. Periodic Lorentz gas: (a) With bounded free path. (b) With unbounded free path.

Consider a periodic Lorentz gas where a free path of the particle is uniformly bounded by some constant. The simplest such configuration of scatterers in the plane is obtained by taking the hexagonal lattice of sufficiently large scatterers (Figure 1(a)). Observe that a periodic Lorentz gas is just Sinai billiards (*See* **Billiards**) when restricted to an elementary cell of a periodic configuration of scatterers. Such a cell can always be chosen to be a rectangle $B=\{q=(q_1,q_2):0\le q_1\le b_1, 0\le q_2\le b_2\}$, where (q_1,q_2) are planar coordinates of the moving particle. Let the particle initially (at $t=0$) have a random position in B and its velocity (which in a Lorentz gas can always be taken as equal to one) be also randomly distributed in the unit circle. Suppose that this distribution μ (in the direct product Π of B and the unit circle) has a well-behaved density with respect to the volume in Π. Then $x(t)=(q(t), v(t))$ is a random vector, where $q(t)=(q_1(t), q_2(t))$ is a position of the particle at time t and $v(t)$ is its velocity. Rescale trajectories of the particle by setting $q_t(s)=\frac{1}{\sqrt{t}}q(st)$, $0\le s\le t$. The probability distribution μ on the initial conditions induces a probability distribution on μ_t on the set of all possible trajectories $q_t(s)$, $0\le s\le t$. To study a hydrodynamic behavior one should consider large times. The measures μ_t converge (weakly) as $t\to\infty$ to a Wiener measure (Bunimovich & Sinai, 1981). This result provides a rigorous derivation of (time non-invertible macroscopic) hydrodynamic diffusion equations from first principles (time invertible microscopic Newtonian equations). It shows that at large scales in space and time, trajectories of the periodic Lorentz gas look the same as trajectories of the stochastic diffusion process. Indeed, the transition probabilities of Wiener processes are the fundamental solutions of the diffusion equation. The same result holds for the periodic Lorentz gas with bounded free path in dimensions ≥ 3 (Chernov, 1994).

According to the Einstein formula,

$$D=\frac{2}{d}\int_0^\infty\int_M [v(x(0))\cdot v(x(t))\,\mathrm{d}\mu(x(0))]\,\mathrm{d}t, \quad (1)$$

where d is a dimension, M is the phase space, and $v(x(0))$ is the velocity of the particle at the moment $t=0$. The Einstein formula is the first in the infinite hierarchy of Green–Kubo formulas that relate transport coefficients with the integrals of time correlations of some phase functions. Therefore, the problem of existence of the transport coefficients is reduced to an estimate of the rate of time-correlations decay, which for the periodic Lorentz gas is fast enough (Chernov & Young, 2000).

The condition that a free pass must be bounded is important. Periodic Lorentz gases with an unbounded free path (Figure 1(b)) behave super-diffusively rather than diffusively (Bleher, 1992) and time correlations there decay only according to a power law because the particle gets trapped for a long time in the part of the phase space with arbitrarily long free paths.

Until the 1960s, it was generally believed that time correlations in many-body systems decay exponentially. This opinion was based on the analysis of completing solvable models, such as Markov processes. Therefore, the discovery of long tails of time correlations in numerical experiments came as a surprise (Alder & Wainwright, 1967). It is now believed that time correlations in a random Lorentz gas decay according to a power law. This is confirmed by numerical experiments and physical theories but not by rigorous mathematical results.

Interesting models arise by placing the Lorentz gas in an external field. A periodic Lorentz gas in a magnetic field can demonstrate ergodic as well as integrable behavior (Berglund & Kunz, 1996). The Lorentz gas in a (weak) electric field becomes a non-Hamiltonian system, with the energy of the moving particle increasing indefinitely. One can connect this system to a thermostat to keep the energy fixed and to make the dynamics time reversible, whereupon it is possible to establish rigorously Ohm's law (Chernov et al., 1993). Presently, studies of the thermostated Lorentz gas are popular both in non-equilibrium statistical mechanics and in molecular dynamics (Dettmann, 2000).

LEONID BUNIMOVICH

See also **Billiards; Deterministic walks in random environments; Drude model**

Further Reading

Alder, B.J. & Wainwright, T.E. 1967. Velocity autocorrelations for hard spheres. *Physical Review Letters*, 18: 988–990

Berglund, N. & Kunz, H. 1996. Integrability and ergodicity of classical billiards in a magnetic field. *Journal of Statistical Physics*, 83: 81–126

Bleher, P.M. 1992. Statistical properties of two-dimensional periodic Lorentz gas with infinite horizon. *Journal of Statistical Physics*, 66: 315–373

Boldrighini, C., Bunimovich, L.A. & Sinai, Ya G. 1983. On the Boltzmann equation for the Lorentz gas. *Journal of Statistical Physics*, 32: 477–501

Bunimovich, L.A. & Sinai, Ya.G. 1981. Statistical properties of Lorentz gas with periodic configuration of scatterers. *Communications in Mathematical Physics*, 78: 479–497

Chernov, N.I. 1994. Statistical properties of the periodic Lorentz gas. Multidimensional case. *Journal of Statistical Physics*, 74: 11–53

Chernov, N. & Young, L.S. 2000. Decay of correlations for Lorentz gas and hard balls. In *Hard Ball Systems and the Lorentz Gas*, edited by D. Szasz, Berlin and New York: Springer

Chernov, N.I., Eyink, G.I., Lebowitz, J.L. & Sinai, Ya.G. 1993. Steady state electrical conduction in the periodic Lorentz gas. *Communications in Mathematical Physics*, 154: 569–601

Dettmann, C.P. 2000. The Lorentz gas: a paradigm for nonequilibrium stationary states. In *Hard Ball Systems and the Lorentz Gas*, edited by D. Szasz, Berlin and New York: Springer

Grad, H. 1958. Principles of the kinetic theory of gases. In *Handbuch der Physik* 12, edited by S. Flügge, Berlin: Springer

Lorentz, H.A. 1905. The motion of electron in metallic bodies. *Proceedings of Amsterdam Akademie*, 7: 438–441

LORENTZ TRANSFORM

See **Symmetry groups**

LORENZ ATTRACTOR

See **Attractors**

LORENZ EQUATIONS

A revolutionary development in the realms of nonlinear science over the past four decades has been an understanding of how ubiquitously systems of deterministic equations exhibit complicated behavior. Although the possibility of complicated aperiodic solutions was known much earlier from the work of Henri Poincaré and George D. Birkhoff on the so-called three-body problem (the Earth, Sun, and Moon) in celestial mechanics in the late 19th and early 20th centuries, such behavior was believed to be exceptional (*See* **Poincaré theorems**).

Two crucial pieces of work in the 1960s, the first by Edward Lorenz in 1963 on a continuous model relating to weather prediction (*See* **Butterfly effect**) and the second by Stephen Smale in 1967 in the context of discrete mathematical maps, altered this perception completely, showing that complex aperiodic behavior is generic for most deterministic nonlinear systems.

The basic equations introduced by Lorenz are

$$\begin{aligned} \dot{x} &= \sigma(y - x), \\ \dot{y} &= rx - y - xz, \\ \dot{z} &= xy - bz, \end{aligned} \tag{1}$$

which were obtained by a drastic simplification of the fluid dynamical equations governing convection currents in the atmosphere. Equations (1) resulted from truncating a Fourier series description after the first few modes (Bergé et al., 1984), and the parameters σ

and r are physically important dimensionless numbers known, respectively, as the Prandtl and Rayleigh numbers in fluid mechanics.

Solutions of the Lorenz equations can be organized around a classification of their attractors (Glendinning, 1994; Strogatz, 1994; Nayfeh & Balachandran, 1995), which are regions of the space of physical variables or phase space (also called state space) to which solutions are attracted if the attractor is stable, or from whence they are repelled if the attractor is unstable.

Nonlinear systems may have several different kinds of attractors, including point attractors (also known as fixed, or critical, or equilibrium, points), isolated periodic attractors (or limit cycles), as well as more complex attractors called quasi-periodic and chaotic attractors. The nature and number of the attractors may change as the parameters (σ, r, and b) vary, leading to a qualitative change in behavior referred to as a bifurcation.

A straightforward calculation showed Lorenz that his system contracted volumes in phase space—a dissipative system. This meant that any ball or volume of initial conditions in the phase space must shrink down to zero volume at large times under evolutions governed by the Lorenz equations. In other words, any attractors of the system must have zero volume.

Also, the model has a critical point at the origin ($x=y=z=0$) and a pair of critical points $C^{\pm}$ at $x=y=\pm\sqrt{b(r-1)}$ for $r>1$. Lorenz's stability analysis showed that the first fixed point at the origin of the phase space is always stable for $r<1$ and a global attractor to which all initial points (or initial conditions) in the phase space are attracted for this parameter range. For $r>1$, he found that the first fixed point at the origin becomes linearly unstable and is longer a bona fide attractor. However, the new fixed points are stable for

$$1 < r < r_{\mathrm{H}} = \frac{\sigma(\sigma+b+3)}{\sigma-b-1}. \qquad (2)$$

In this parameter range, calculation shows that these fixed points coexist with an unstable limit cycle. At $r=r_{\mathrm{H}}$, this limit cycle is absorbed by the fixed point in a bifurcation that is known as a subcritical Hopf bifurcation. The fixed points $C^{\pm}$ go unstable in the process, being transformed into saddle points.

For $r>r_{\mathrm{H}}$, there are no simple stable point attractors, so Lorenz considered other stable limit cycles (periodic attractors) or solutions flying off to infinity. He showed that any possible limit cycles should be unstable for $r>r_{\mathrm{H}}$ and also that all trajectories must eventually enter, and be confined within, a certain large ellipsoid.

Clearly, any attractor had to be relatively complex, and the trajectories could not cross while evolving on the attractor. Together with the earlier-mentioned fact of volume contraction (which means that any attractor to which the solution trajectories or orbits are attracted must have zero volume asymptotically in time), Lorenz

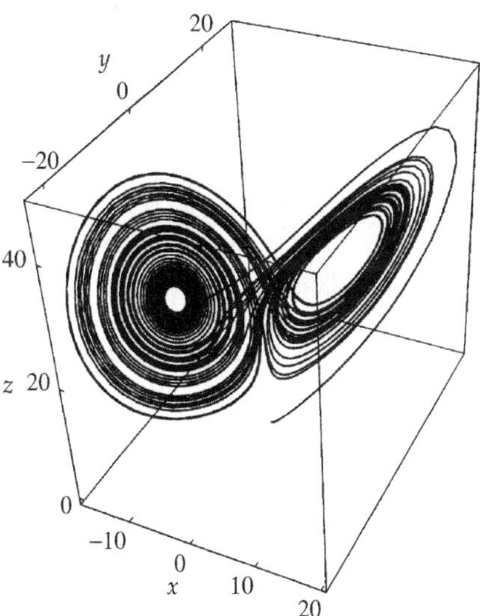

Figure 1. Lorenz attractor in (x, y, z) phase space for $\sigma=10, r=28, b=8/3$.

was left with a perplexing question: What must such an attractor be like?

He numerically studied the case $\sigma=10, r=28$, $b=\frac{8}{3}$ with the $r>r_{\mathrm{H}}=24.74$ in Equation (2). He also chose initial conditions $(x, y, z)=(0, 1, 0)$ close to the saddle point nearer the origin. The resulting solutions in (x, y, z) phase-space, and for $x(t)$ as a function of time are shown in Figures 1 and 2, respectively, where the solution trajectory settles onto a strange butterfly-shaped attractor set after an initial transient. The clearly aperiodic dynamics on this attractor is emphasized in Figure 2. In addition, the attractor appears to be infinitesimally thin and thus satisfies the requirement of zero volume. Hence, all three requirements of aperiodicity of the dynamics on the attractor, zero volume, and nonself-crossing of the orbit were reconciled on this "strange attractor"—a term subsequently coined by Ruelle and Takens. Among other observations, Lorenz noted that this strange attractor has a self-similar structure at all scales, which is now a well-known feature of chaotic attractors.

Subsequent work has revealed that trajectories on chaotic attractors exhibit sensitivity to initial conditions (SIC) where initially contiguous phase points diverge exponentially in time. Other work, some still in progress (Abraham & Shaw, 1983; Bergé et al., 1984), has elucidated the complex process leading to the structure of strange attractors and the properties of the orbits on them. In particular, there is convergence of orbits along the stable manifolds of saddle fixed points (this enables dissipation or volume contraction to be satisfied), divergence of trajectories along the corresponding unstable manifolds (this accounts for SIC), and foldings leading to bounded strange attractors.

Based on a combination of topological and numerical ideas to reconstruct chaotic attractors from

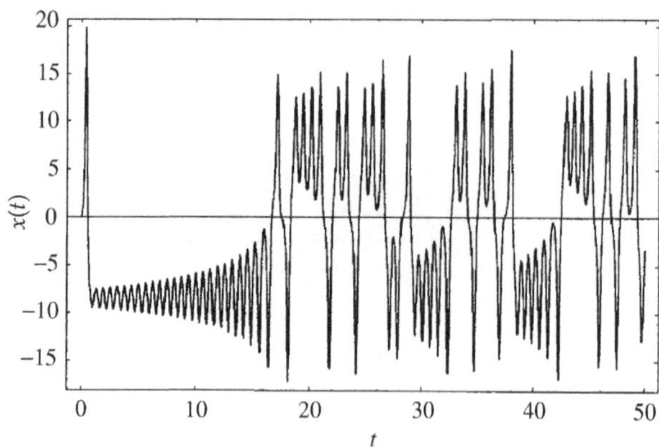

Figure 2. A solution $x(t)$ for $\sigma = 10, r = 28, b = 8/3$.

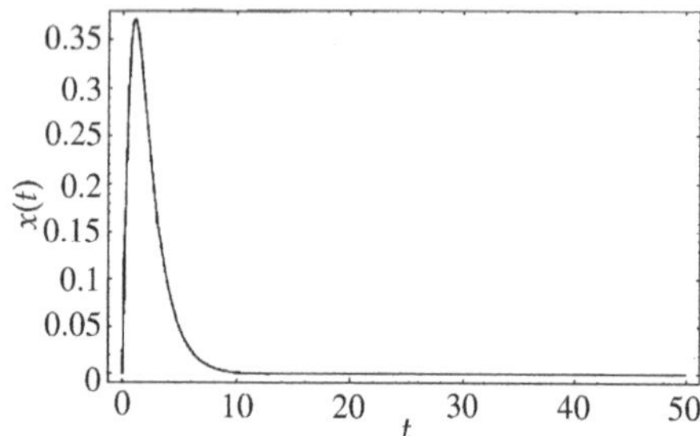

Figure 3. A solution $x(t)$ for $\sigma = 1, b = 2, r = 1/9$.

time series such as that in Figure 2, numerical procedures have been developed to compute various properties such as the fractal dimension (a measure of the global dimension of the attractor in phase space), Lyapunov exponents (measures of sensitivity to initial conditions on the attractor), power spectra and autocorrelation functions, and the so-called Kolmogorov entropy (Nayfeh & Balachandran, 1995). In particular, the chaotic attractor in Figure 1 has a dimension of about 2.05; that is, it occupies a region of the phase space that has infinite area or fills up any area (since the dimension is greater than 2), but has zero volume (since the dimension is less than 3).

The occurrence of sensitivity to initial conditions as indicated by at least one positive Lyapunov exponent or the envelope of the autocorrelation function going to zero in finite time signify the eventual loss of memory of past history (or equivalently the impossibility of prediction far into the future), which are integral features of chaotic dynamics.

Note that the Lorenz equations also exhibit completely deterministic or nonchaotic behavior for isolated parameter sets obtained via Painlevé analysis that are the so-called integrable cases (Tabor & Weiss, 1981). Figure 3 shows the completely regular dynamics for the parameter set $\sigma = 1, b = 2, r = 1/9$. Note how much more regular and orderly, the behavior of $x(t)$ is than that shown in Figure 2.

In conclusion, two features of the Lorenz model continue to make it topical. First, it remains a useful model in which to study essential basic features of chaotic systems. Second, the model has popped up in topical settings as diverse as atmospheric dynamos, lasers, and nonlinear optics, as well as the control of chaos (Ning & Haken, 1990; Pecora & Carroll, 1990; Nayfeh & Balachandran, 1995; Batchelor et al., 2000).

S. Roy Choudhury

See also **Attractors; Butterfly effect; Chaotic dynamics; Lyapunpov exponents; Phase space**

Further Reading

Abraham, R.H. & Shaw, C.D. 1983. *Dynamics: The Geometry of Behavior*, Santa Cruz: Aerial Press

Batchelor, G.K., Moffatt, H.F. & Worster, M.G. (editors). 2000. *Perspectives in Fluid Dynamics*, Cambridge and New York: Cambridge University Press

Bergé, P., Pomeau, Y. & Vidal, C. 1984. *Order Within Chaos*, New York: Wiley

Choudhury, S.R. 1992. On bifurcations and chaos in predator-prey models with delay. *Chaos, Solitons and Fractals*, 2: 393–409

Dodd, R.K., Eilbeck, J.C., Gibbon, J.D. & Morris, H.C. 1982. *Solitons and Nonlinear Wave Equations*, London: Academic Press

Glendinning, P. 1994. *Stability, Instability, and Chaos*, Cambridge and New York: Cambridge University Press

Lorenz, E.N. 1963. Deterministic nonperiodic flow. *Journal of Atmospheric Science*, 20: 130–141

Nayfeh, A.H. & Balachandran, B. 1995. *Applied Nonlinear Dynamics*, New York: Wiley

Ning, C.Z. & Haken, H, 1990. Detuned lasers and the complex Lorenz equations: Subcritical Hopf bifurcations. *Physical Review* A, 41: 3826–3837

Nusse, H. & Yorke, J.A. 1992. *Dynamics: Numerical Explorations*, New York: Springer

Pecora, L.M. & Carroll, T.L. 1990. Synchronization in chaotic systems. *Physical Review Letters*, 64: 821–824

Sparrow, C. 1982. *The Lorenz Equations*, New York: Springer

Strogatz, S.H. 1994. *Nonlinear Dynamics and Chaos*, Reading, MA: Addison-Wesley

Tabor, M. & Weiss, J. 1981. Analytic structure of the Lorenz system. *Physical Review* A, 24: 2157

LOTKA–VOLTERRA EQUATIONS

See **Population dynamics**

LYAPUNOV EXPONENTS

The sequence of powers a^k, $k = 1, 2, \ldots$, of a complex number $a \neq 0$ has one of three types of asymptotic behavior: exponential growth, exponential decay, or constant, according to the modulus $|a|$. A similar situation holds for powers A^k, $k = 1, 2, \ldots$, of an $n \times n$ matrix A. More precisely, for any vector $v \neq 0$, the following limit exists:

$$\lim_{k \to +\infty} \frac{1}{k} \log \|A^k v\| = \lambda(v). \qquad (1)$$

The possible values of $\lambda(v)$, $\lambda_1 < \cdots < \lambda_s$ are called Lyapunov exponents (LEs), and in this special case they are equal to $\log|a|$, where a is an eigenvalue of A (real or complex). Moreover, there is a strictly increasing sequence of subspaces in R^n

$$\{0\} = V_0 \subset V_1 \subset \cdots \subset V_s = R^n, \quad (2)$$

such that $\lambda(v) = \lambda_i$ for all $v \in V_i \setminus V_{i-1}$, $i = 1, 2, \ldots, s$. The difference of dimensions $m_i = \dim V_i - \dim V_{i-1}$ is called the multiplicity of the Lyapunov exponent λ_i. The rates of growth of k-dimensional volumes of k-dimensional parallelepipeds under the action of A are sums of appropriate Lyapunov exponents. For example, the exponential rate of growth of the n-dimensional volume is equal to $\log|\det A| = m_1\lambda_1 + \cdots + m_s\lambda_s$.

A similar situation occurs for products $A^k = A_k A_{k-1} \ldots A_1$ when the sequence of matrices $A_1, A_2, \ldots$ is periodic (by direct reduction to the constant case we obtain so-called Floquet multipliers, and the LEs are logarithms of their moduli).

In the general nonperiodic case, the key Oseledec Multiplicative Ergodic Theorem (OMET) states that the same scenario holds, when the matrices we multiply $(A_1, A_2, \ldots)$ are supplied by a stationary stochastic process (with matrix values). In this case, the conclusions apply with probability one.

In the context of dynamical systems, the difference equation $x_{n+1} = \Phi(x_n)$, where $\Phi : M \to M$ is a diffeomorphism of a compact smooth manifold M (the phase space). For any ergodic invariant probability measure ν, the Lyapunov exponents are the limits

$$\lim_{k\to+\infty} ||D_x\Phi^k v|| = \lambda(v), \quad (3)$$

which exist for ν almost every $x \in M$ and all nonzero tangent vectors $v \in T_xM$. Thus, Lyapunov exponents characterize the growth in linear approximation of infinitesimal perturbations of initial conditions. Positive LEs lead to instability and negative LEs to stability for perturbations in the respective directions.

Applying OMET both to the future and the past (i.e., to Φ and to Φ^{-1}), we obtain a splitting of tangent spaces

$$T_xM = W_1 \oplus W_2 \oplus \ldots \oplus W_s, \quad (4)$$

such that

$$\lim_{k\to\pm\infty} ||D_x\Phi^k v|| = \pm\lambda_i \quad (5)$$

for ν almost every $x \in M$ and all nonzero tangent vectors $v \in W_i(x)$. It is of direct interest to know if this infinitesimal behavior translates to the behavior of actual orbits of the nonlinear system. We introduce the stable $E^s(x)$ and unstable $E^u(x)$ subspaces, which are the direct sums of subspaces $W_i(x)$ with negative, respectively positive, LEs. It is the content of Pesin Theory that the subspaces E^s and E^u can be integrated almost everywhere, that is, for ν almost every point $x \in M$ there are smooth submanifolds $W^s(x)$ and $W^u(x)$, which are at every point tangent to the stable and unstable subspaces, respectively.

The stable and unstable manifolds are crucial in understanding chaotic dynamics. Their importance stems from the fact that two orbits starting on one of these submanifolds have the same asymptotic behavior either in the future or in the past.

LEs are associated with the Kolmogorov–Sinai entropy. They enter into the Ruelle Inequality, which states that for any C^1 diffeomorphism Φ and any ergodic invariant probability measure ν

$$h_\nu(\Phi) \le \sum_{\lambda_i>0} m_i\lambda_i, \quad (6)$$

where $h_\nu(\Phi)$ is the K–S entropy. This inequality turns into equality (the Pesin formula) for absolutely continuous measures ν and C^2 diffeomorphisms. It was proven by Ledrappier and Young that equality holds if and only if ν is the Sinai–Ruelle–Bowen (SRB) measure.

For Hamiltonian systems (i.e., when Φ is a symplectomorphism), the LEs come in pairs of opposite numbers, which is related to the symmetry of spectra of symplectic matrices. Furthermore the subspace V_k is skew-orthogonal to V_{s-k}, $k = 1, 2, \ldots, s$: a property shared with eigenspaces of symplectic matrices.

Although analytic formulas for LEs are almost nonexistent, they can be readily obtained by numerical methods. Analytical estimates of LEs are available where all the matrices in question are J-separated for some quadratic form (field of forms) J. A matrix A is J-separated, if $J(Av) \ge 0$ for all v such that $J(v) \ge 0$.

Maciej P. Wojtkowski

See also **Chaotic dynamics; Sinai–Ruelle–Bowen measures; Phase space**

Further Reading

Barreira, L. & Pesin, Y.B. 2002. *Lyapunov Exponents and Smooth Ergodic Theory*, Providence, RI: American Mathematical Society

Katok, A. & Hasselblatt, B. 1995. *Introduction to the Modern Theory of Dynamical Systems*, Cambridge and New York: Cambridge University Press

Wojtkowski, M.P. 2001. Monotonicity, J-algebra of Potapov and Lyapunov exponents. In *Smooth Ergodic Theory and Its Applications*, Providence, RI: American Mathematical Society, pp. 499–521

Wojtkowski, M.P. & Liverani, C. 1998. Conformally symplectic dynamics and symmetry of the Lyapunov spectrum. *Communications in Mathematical Physics*, 194: 47–60

LYAPUNOV STABILITY

See **Stability**

M

MACH STEM EFFECT

See **Shock waves**

MAGNETOHYDRODYNAMICS

Magnetohydrodynamics (MHD) is the special field of fluid dynamics that is concerned with the motions of electrically conducting fluids in the presence of magnetic fields. Flows of liquid metals or ionized gases (plasmas) are the main areas of applications of MHD.

For the theoretical description of MHD-problems, the basic equations of fluid dynamics must be supplemented by the Lorentz force $\boldsymbol{j} \times \boldsymbol{B}$ where $\boldsymbol{j}$ is the density of electric current and $\boldsymbol{B}$ is the magnetic flux density. To determine these fields Maxwell's equations are used in the magnetohydrodynamic approximation in which the displacement current is neglected. This represents a good approximation as long as the fluid velocity is small compared with the velocity of light. These equations together with Ohm's law for a moving conductor are given by

$$\nabla \cdot \boldsymbol{B} = 0, \quad \frac{\partial}{\partial t}\boldsymbol{B} = -\nabla \times \boldsymbol{E}, \quad \nabla \times \boldsymbol{B} = \mu \boldsymbol{j},$$
$$\boldsymbol{j} = \sigma(\boldsymbol{v} \times \boldsymbol{B} + \boldsymbol{E}), \tag{1}$$

where μ is the magnetic permeability of the fluid and σ is its electrical conductivity. These "pre-Maxwell" equations have the property that they are invariant with respect to a Galileo transformation, i.e., the equations remain unchanged in a new frame of reference moving with the constant velocity vector $\boldsymbol{V}$ relative to the original frame of reference. Indicating the variables of the new frame by a prime, we find

$$\boldsymbol{v}' = \boldsymbol{v} - \boldsymbol{V}, \quad \frac{\partial}{\partial t'} = \frac{\partial}{\partial t} + \boldsymbol{V} \cdot \nabla,$$
$$\boldsymbol{B}' = \boldsymbol{B}, \quad \boldsymbol{E}' = \boldsymbol{E} + \boldsymbol{V} \times \boldsymbol{B}, \quad \boldsymbol{j}' = \boldsymbol{j}. \tag{2}$$

This invariance is the basis for the combination in MHD of Equations (1) with the equations of hydrodynamics in their usual nonrelativistic form. The application of MHD to plasmas is limited to sufficiently low frequency and long wavelength phenomena. If these conditions are not satisfied, two-fluid equations may be used.

Elimination of $\boldsymbol{E}$ and $\boldsymbol{j}$ from Equation (1) yields the equation of magnetic induction

$$\nabla \times \left(\frac{1}{\sigma\mu}\nabla \times \boldsymbol{B}\right) = \frac{\partial}{\partial t}\boldsymbol{B} + \nabla \times (\boldsymbol{v} \times \boldsymbol{B}), \tag{3}$$

which for a solenoidal velocity field $\boldsymbol{v}$ and a constant magnetic diffusivity $\lambda \equiv (\sigma\mu)^{-1}$ can be further simplified,

$$\left(\frac{\partial}{\partial t} + \boldsymbol{v} \cdot \nabla\right)\boldsymbol{B} - \lambda\nabla^2\boldsymbol{B} = \boldsymbol{B} \cdot \nabla\boldsymbol{v}. \tag{4}$$

This equation has the form of a heat equation with the magnetic field line stretching term on the right-hand side acting as a heat source. This interpretation is useful in the dynamo problem of the generation of magnetic fields by fluid motions (*See* **Dynamos, homogeneous**). In order that a magnetic field $\boldsymbol{B}$ may grow, the term on the right-hand side of (3) must overcome the effect of the magnetic diffusion term on the left-hand side. Using a typical velocity U and a typical length scale d, the ratio of the two terms can be estimated by

$$Ud/\lambda \equiv \text{Rm}, \tag{5}$$

which is called the magnetic Reynolds number in analogy to the ordinary Reynolds number used in fluid dynamics. A necessary condition for dynamo action is $\text{Rm} > 1$, but this is by far not a sufficient condition (*See* **Dynamos, homogeneous**).

A typical problem of MHD is the flow in the channel in the presence of an applied homogeneous magnetic field $\boldsymbol{B}_0$. The equation for the plane parallel flow $\boldsymbol{v} = v_x(y)\boldsymbol{i}$ in the channel between two parallel plates with distance $2d$ is given by

$$\eta\frac{\mathrm{d}^2}{\mathrm{d}y^2}v_x(y) + \frac{1}{\mu}\boldsymbol{i} \cdot ((\nabla \times \boldsymbol{B}) \times \boldsymbol{B}) = -A, \tag{6}$$

where η is the dynamic viscosity, $\boldsymbol{i}$ is the unit vector in the x-direction along the channel, and A is the constant pressure gradient in the opposite direction. In accordance with the configuration of the problem, we have assumed that the steady solution depends only on the y-coordinate perpendicular to the plates. Using the

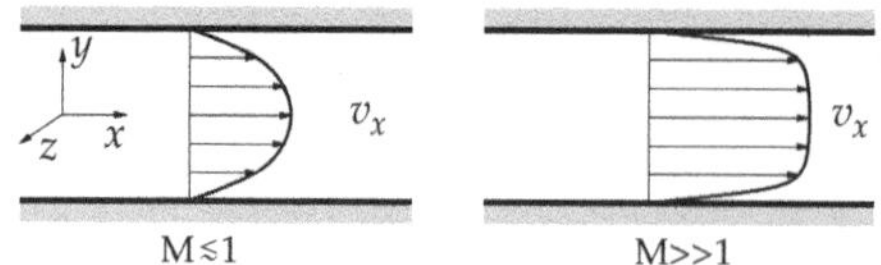

Figure 1. Profile of channel flow for small and large M.

notation $\boldsymbol{B} = \boldsymbol{B}_0 + \boldsymbol{b}$ we find after integrating Equation (4)

$$\lambda \frac{\mathrm{d}}{\mathrm{d}y} b_x(y) + B_{oy} v_x = -D, \tag{7}$$

where the constant of integration D represents the electric field directed in the z-direction; in other words, Equation (7) is the z-component of Ohm's law. Insertion of this expression into Equation (6) yields

$$\eta \frac{\mathrm{d}^2}{\mathrm{d}y^2} v_x(y) - \sigma B_{0y}^2 v_x(y) = -A + \sigma B_{oy} D \tag{8}$$

The solution of this equation is given by

$$v_x = v_0 \left(1 - \frac{\cosh(My/d)}{\cosh M}\right), \tag{9}$$

with

$$M \equiv B_{0y} d \sqrt{\sigma/\eta}$$

and

$$v_0 = (A - \sigma D B_{oy}) d^2 / \eta M^2, \tag{10}$$

where M is the Hartmann number named after the scientist who first considered this type of problem. The constant D depends on the boundary condition applied in the transverse z-direction. D can be chosen, for example, such that the average electric current in the z-direction vanishes because of the insulating side walls. It is worth noting that only the y-component of $\boldsymbol{B}_0$ enters solution (9). The flow is not affected by the x- and z-components of $\boldsymbol{B}_0$ as long as it does not depend on those coordinates. Depending on the Hartmann number M, the velocity profile $v_x(y)$ varies between the two limiting cases sketched in Figure 1. For $M \to 0$, the parabolic Poiseuille profile is recovered from expression (9), while for $M \gg 1$, boundary layers of thickness d/M develop, which are called Hartmann layers. In geophysical and astrophysical applications of MHD, the Chandrasekhar number $Q = M^2$ is often used in place of the Hartmann number.

In many applications of MHD, it is justified to neglect magnetic diffusion in first approximation. Then the relationship

$$\frac{\partial}{\partial t} \boldsymbol{B} = \nabla \times (\boldsymbol{v} \times \boldsymbol{B}) \tag{11}$$

can be obtained from Equation (3) in the limit of large magnetic Reynolds numbers. This equation is the same as that obeyed by the vorticity of an inviscid fluid (*See* **Fluid dynamics**). By analogy to the vorticity laws of Kelvin and Helmholtz it can be concluded that the field line that is attached to a fluid element at some initial time continues to be attached to that fluid element at all times; that is, the field lines are "frozen" into the fluid. This statement is known as Alfvén's theorem.

In highly electrically conducting fluids, it is appropriate to consider Equation (11) together with the Euler equation of hydrodynamics,

$$\rho \left(\frac{\partial}{\partial t} \boldsymbol{v} + \boldsymbol{v} \cdot \nabla \boldsymbol{v}\right) = -\nabla p + \mu^{-1} (\nabla \times \boldsymbol{B}) \times \boldsymbol{B}, \tag{12}$$

where the Lorentz force has been included, but viscous friction has been neglected. In the case of an incompressible fluid with $\rho =$ constant, both, $\boldsymbol{v}$ and $\boldsymbol{B}$ are solenoidal vector fields and an exact steady solution of Equations (11) and (12) is given by

$$\boldsymbol{v} = (\rho\mu)^{-1/2} \boldsymbol{B}(x, y, z)$$

$$\text{with} \qquad p = p_0 + \rho \mid \boldsymbol{v} \mid^2 /2, \tag{13}$$

where the relationship $\boldsymbol{v} \cdot \nabla \boldsymbol{v} = (\nabla \times \boldsymbol{v}) \times \boldsymbol{v} + \nabla \mid \boldsymbol{v} \mid^2 /2$ has been used. The dependence of $\boldsymbol{B}$ on the cartesian coordinates x, y, z is arbitrary except for the condition $\nabla \cdot \boldsymbol{B} = 0$. Let us write $\boldsymbol{B}$ as the sum of its average and the remainder, $\boldsymbol{B} = B_0 \boldsymbol{i} + \boldsymbol{b}$, where it has been assumed without loss of generality that the average magnetic field is parallel to the x-direction given by the unit vector $\boldsymbol{i}$. Using a Galileo transformation with the constant velocity vector $V_A \boldsymbol{i}$, we obtain a time-dependent exact solution relative to the new frame of reference,

$$\boldsymbol{v}'(x - V_{\mathrm{A}} t, y, z) = (\rho\mu)^{-1/2} \boldsymbol{b}(x - V_{\mathrm{A}} t, y, z), \tag{14}$$

where the choice $V_{\mathrm{A}} = (\rho\mu)^{-1/2} B_0$ has been made. This solution describes dispersion-free waves called Alfvén waves. The physical interpretation of these waves of arbitrary amplitude is that the field lines of the basic magnetic field are embedded in the fluid-like rubber strings and provide a restoring force whenever a fluid parcel is displaced. The phase velocity V_A is called the Alfvén velocity. In compressible electrically conducting fluids a more complex spectrum of wave phenomena is found owing to combinations of Alfvén and acoustic modes.

F.H. Busse

See also **Alfvén waves; Dynamos, homogeneous; Fluid dynamics; Nonlinear plasma waves**

Further Reading

Davidson, P.A. 2001. *An Introduction to Magnetohydrodynamics*, Cambridge and New York: Cambridge University Press

Moreau, R. 1990. *Magnetohydrodynamics*, Dordrecht: Kluwer

Roberts, P.H. 1967. *An Introduction to Magnetohydrodynamics*, New York: Elsevier and London: Longmans

MANAKOV EQUATIONS

See **Nonlinear Schrödinger equations**

MANDELBROT SETS

See **Fractals**

MANLEY-ROWE RELATIONS

In lossless linear systems, a harmonic sinusoidal input at a frequency ω remains harmonic, the system's steady-state response to a linear combination of individual harmonic inputs is a linear combination of individual responses, and an average input power is zero at each frequency. In lossless nonlinear systems, on the other hand, harmonic inputs at some frequencies also generate inputs at multiple and combination frequencies. The average input powers at each frequency may not be zero but they are predicted from the general relations found by Jack Manley and Harrison Rowe in 1956 (Manley & Rowe, 1956).

Suppose the input applied to a nonlinear system consists of two harmonic signals of frequencies ω_1 and ω_2. The nonlinear system produces an output at all frequencies $m\omega_1 + n\omega_2$, where m and n are integers. For example, $m = 2, n = 0$ and $m = 0, n = 2$ are signals at the double frequencies $2\omega_1$ and $2\omega_2$, while $m = n = 1$ a signal with the combinational frequency $\omega_1 + \omega_2$. Denote $P_{m,n}$ the total input power at frequency $m\omega_1 + n\omega_2$. In lossless systems, the total power is zero; thus

$$\sum_{m,n} P_{m,n} = 0.$$

In other words, the power $P_{m,n}$ is negative if the signal at $m\omega_1 + n\omega_2$ is generated due to nonlinearity, and the power is positive if the signal is applied from input to the system.

The Manley–Rowe relations are relations between powers $P_{m,n}$ and frequencies $m\omega_1 + n\omega_2$ (Penfield, 1960):

$$\sum_{n=-\infty}^{\infty} \sum_{m=0}^{\infty} \frac{m P_{m,n}}{m\omega_1 + n\omega_2} = 0,$$

$$\sum_{m=-\infty}^{\infty} \sum_{n=0}^{\infty} \frac{n P_{m,n}}{m\omega_1 + n\omega_2} = 0. \tag{1}$$

As an elementary example, we consider a heterodyne system, comprising an oscillator with frequency ω_1, which is mixed with the carrier incident frequency ω such that a combination frequency $\omega_2 = \omega - \omega_1$ occurs in the spectrum of the nonlinear system (Scott, 1970). We neglect signals at other frequencies and denote powers at frequencies ω_1, ω_2 and $\omega = \omega_1 + \omega_2$ as P_1, P_2, and $P = -P_1 - P_2$. The Manley–Rowe relations between the powers and frequencies of the three signals are

$$\frac{P_1}{\omega_1} = \frac{P_2}{\omega_2} = -\frac{P}{\omega}$$

Notice that these relations preserve conservation of powers $P_1 + P_2 + P = 0$ at the three resonant frequencies ω_1, ω_2, and ω. If $\omega > \omega_1, \omega_2$, the Manley–Rowe relations show that the power P_2 at the combinational frequency ω_2 is smaller than the input power P by a factor of ω_2/ω, while the power at the heterodyne frequency ω_1 is smaller than P by a factor of ω_1/ω. Therefore, when an input signal of larger frequency transforms into output signals of smaller frequencies, the powers of output signals become smaller.

As another application, we consider an electro-optical modulator that takes an input signal at frequency 1 GHz and multiplies with another signal at frequency 100 MHz to produce a modulated output signal at combinational frequency 1.1 GHz. If the power of the output signal has to be 1 mW, the Manley–Rowe relations require powers 0.0909 mW at 100 MHz and 0.9090 mW at 1 GHz. The 1 GHz signal is sometimes called the pump since it provides most of the power needed in the modulation process.

Manley–Rowe relations naturally describe constants of motion in the time evolution of the resonant nonlinear wave interactions. If three waves with frequencies ω_1, ω_2 and $\omega = \omega_1 + \omega_2$ and wavevectors $\mathbf{k}_1$, $\mathbf{k}_2$ and $\mathbf{k} = \mathbf{k}_1 + \mathbf{k}_2$ satisfy the phase matching conditions

$$\omega(\mathbf{k}_1 + \mathbf{k}_2) = \omega(\mathbf{k}_1) + \omega(\mathbf{k}_2),$$

then their interaction is resonant in time and is described by the system of three-wave interaction equations:

$$\mathrm{i}\left(\frac{\partial}{\partial t} + \mathbf{v}\cdot\nabla\right) a = \gamma a_1 a_2 \mathrm{e}^{-\mathrm{i}\Delta t},$$

$$\mathrm{i}\left(\frac{\partial}{\partial t} + \mathbf{v}_1\cdot\nabla\right) a_1 = \gamma a \bar{a}_2 \mathrm{e}^{\mathrm{i}\Delta t},$$

$$\mathrm{i}\left(\frac{\partial}{\partial t} + \mathbf{v}_2\cdot\nabla\right) a_2 = \gamma a \bar{a}_1 \mathrm{e}^{\mathrm{i}\Delta t}, \tag{2}$$

where $\Delta = \omega(\mathbf{k}) - \omega(\mathbf{k}_1) - \omega(\mathbf{k}_2)$ is the frequency detuning from exact resonance. In other words, $\mathbf{v}$, $\mathbf{v}_1$, and $\mathbf{v}_2$ are group velocities of the three waves, for example, $\mathbf{v} = \nabla\omega(\mathbf{k})$; and γ is a real-valued coupling coefficient in nonlinear systems with quadratic nonlinearities. The system of amplitude equations (2) has integrals of motions:

$$Q_1 = \int \left(|a|^2 + |a_1|^2\right) \mathrm{d}\mathbf{x} = \text{const},$$

$$Q_2 = \int \left(|a|^2 + |a_2|^2\right) \mathrm{d}\mathbf{x} = \text{const},$$

$$Q = \int \left(|a_1|^2 - |a_2|^2\right) \mathrm{d}\mathbf{x} = \text{const}. \tag{3}$$

The integrals of motions are the Manley–Rowe relations for the powers

$$P_1 = \omega_1 \int |a_1|^2 \, \mathrm{d}\mathbf{x},$$

$$P_2 = \omega_2 \int |a_2|^2 \, \mathrm{d}\mathbf{x},$$

$$P = \omega \int |a|^2 \, \mathrm{d}\mathbf{x}. \quad (4)$$

The resonant interaction of three waves results in decay of the wave with the larger frequency $\omega > \omega_1, \omega_2$ into waves of smaller frequencies ω_1; ω_2, which alternates with annihilation of the two waves of smaller frequencies ω_1, ω_2 into the wave of larger frequency ω (Gaponov-Grekhov & Rabinovich, 1992). This process has a simple quantum interpretation. Conservation of quantum momentums and energies of wave particles leads to the resonant relations:

$$\hbar\mathbf{k} = \hbar\mathbf{k}_1 + \hbar\mathbf{k}_2, \qquad \hbar\omega = \hbar\omega_1 + \hbar\omega_2,$$

where $\hbar$ is Planck's constant. If the wave of smaller frequency ω_1 is initially larger than the waves of frequencies ω and ω_2, it cannot decay into the other two waves, because there are not enough wave particles with frequencies ω_2 to merge into wave particles with frequency ω. On the other hand, if the wave of larger frequency ω is initially large, it can decay into two wave particles of smaller frequencies ω_1 and ω_2. Manley–Rowe relations follow from the conservation of number of particles in the quantum interpetation above. For instance, when a phonon with energy $\hbar\omega$ is absorbed, two phonons of energies $\hbar\omega_1$ and $\hbar\omega_2$ are emitted, such that the Manley–Rowe relations hold.

Manley–Rowe invariants play an important role in studies of properties of three-wave interactions in system (2). In particular, optical solitons are supported by dispersive and diffraction effects in nonlinear three-wave interactions. The stability of optical solitons is determined by the Vakhitov–Kolokolov criterion, which involves derivatives of the Manley–Rowe invariants (3) with respect to parameters of optical solitons (Buryak et al., 1997).

When the frequencies ω_1 and ω_2 of the two resonant waves coincide, the three-wave interactions degenerate into resonant second-harmonic generation; thus the wave $a_1 = a_2 = a_0$ with the fundamental frequency $\omega_1 = \omega_2 = \omega_0$ generates the wave a at the double frequency $\omega = 2\omega_0$. The system of equations (2) simplifies then to the form (Etrich et al., 2000)

$$\mathrm{i}\left(\frac{\partial}{\partial t} + \mathbf{v}_0 \cdot \nabla\right) a_0 = \gamma a \bar{a}_0 \mathrm{e}^{\mathrm{i}\Delta t},$$

$$\mathrm{i}\left(\frac{\partial}{\partial t} + \mathbf{v} \cdot \nabla\right) a = \gamma a_0^2 \mathrm{e}^{-\mathrm{i}\Delta t}. \quad (5)$$

System (5) for the second-harmonic generation has only one Manley–Rowe invariant:

$$Q_0 = \int \left(|a_0|^2 + |a|^2\right) \mathrm{d}\mathbf{x} = \text{const}.$$

DMITRY PELINOVSKY

See also **Frequency doubling; Harmonic generation; *N*-wave interactions**

Further Reading

Buryak, A.V., Kivshar, Yu.S. & Trillo, S. 1997. Parametric spatial solitary waves due to type II second-harmonic generation *Journal of the Optical Society of America* B, 14: 3110–3118

Etrich, C., Lederer, F., Malomed, B.A., Peschel, T. & Peschel, U. 2000. Optical solitons in media with a quadratic nonlinearity. *Progress in Optics*, 41: 483–568

Gaponov-Grekhov, A.V. & Rabinovich, M.I. 1992. *Nonlinearities in Action: Oscillations, Chaos, Order, Fractals*, Berlin and New York: Springer

Manley, J.M. & Rowe, H.E. 1956. Some general properties of nonlinear elements. *Proceedings of the Institute of Radio Engineers*, 44: 904–913

Penfield, P., Jr. 1960. *Frequency-power Formulas*, New York: Wiley

Scott, A.C. 1970. *Active and Nonlinear Wave Propagation in Electronics*, New York: Wiley

Weiss, M.T. 1957. Quantum derivation of energy relations analogous to those for nonlinear reactances. *Proceedings of the Institute of Radio Engineers*, 45: 1012–1013

MAPS

A *map* is a dynamical system with discrete time. Such dynamical systems are defined by iterating a transformation $\boldsymbol{\phi}$ of points $\mathbf{x}$

$$\mathbf{x}_{n+1} = \boldsymbol{\phi}(\mathbf{x}_n), \quad (1)$$

from a space $\mathcal{M}$ of dimension d (or a domain of this space) onto itself. $\mathbf{x}_n$ and $\mathbf{x}_{n+1}$ are thus points belonging to this so-called *phase space* $\mathcal{M}$, which can be a Euclidean space such as the space $\mathbb{R}^d$ of d-tuples of real numbers or a manifold such as a circle, a sphere, or a torus $\mathbb{T}^d$ (or a domain such as an interval or a square).

An *endomorphism* is a surjective (i.e., many-to-one) transformation $\boldsymbol{\phi}$ of the space $\mathcal{M}$ onto itself. Thus, the transformation $\boldsymbol{\phi}$ is not invertible. Examples of endomorphisms are one-dimensional maps of the interval such as the logistic map $\phi(x) = 1 - ax^2$ on $-1 \leq x \leq 1$, the Bernoulli map $\phi(x) = rx$ (modulo 1) with integer r (also called r-adic map), or the Gauss map $\phi(x) = 1/x$ (modulo 1) which generates continuous fractions. The last two maps are defined onto the unit interval $0 \leq x \leq 1$. There also exist multidimensional examples such as the exact map $\boldsymbol{\phi}(x, y) = (3x + y, x + 3y)$ (modulo 1) on the torus $\mathbb{T}^2$ (Lasota & Mackey, 1985).

An *automorphism* is a one-to-one (i.e., invertible) transformation $\boldsymbol{\phi}$ of the space $\mathcal{M}$ onto itself. Automorphisms for which the one-to-one transformation $\boldsymbol{\phi}$ is continuous on $\mathcal{M}$ are called *homeomorphisms*. We speak about C^r*-diffeomorphisms* if $\boldsymbol{\phi}$ is r-times differentiable and $\frac{\partial^r \boldsymbol{\phi}}{\partial \mathbf{x}^r}$ is continuous on $\mathcal{M}$. Examples of automorphisms are the circle maps defined with a monotonously increasing function $\phi(x) = \phi(x+1) - 1$ onto the circle; the baker map:

$$\boldsymbol{\phi}(x, y) = \begin{cases} (2x, \frac{y}{2}) & \text{if} \quad 0 \le x < 1/2, \\ \left(2x - 1, \frac{y+1}{2}\right) & \text{if} \quad 1/2 \le x < 1, \end{cases} \tag{2}$$

onto the unit square (Hopf, 1937); the cat map:

$$\boldsymbol{\phi}(x, y) = (x + y, x + 2y) \quad (\text{modulo } 1), \tag{3}$$

onto the torus $\mathbb{T}^2$ (Arnol'd & Avez, 1968); the quadratic map:

$$\boldsymbol{\phi}(x, y) = (y + 1 - ax^2, bx) \,, \tag{4}$$

onto $\mathbb{R}^2$ also called the *Hénon map* (Hénon, 1976; Gumowski & Mira, 1980), among many others.

Iterating a noninvertible transformation $\boldsymbol{\phi}$ generates a semigroup of endomorphisms $\{\boldsymbol{\phi}^n(\mathbf{x})\}_{n \in \mathbb{N}}$, where $\mathbb{N}$ is the set of nonnegative integers. Iterating an invertible transformation $\boldsymbol{\phi}$ generates a group of automorphisms $\{\boldsymbol{\phi}^n(\mathbf{x})\}_{n \in \mathbb{Z}}$, where $\mathbb{Z}$ is the set of all the integers. Such groups or semigroups are deterministic dynamical systems with discrete time, called maps.

Link Between Maps and Flows

Maps naturally arise in continuous-time dynamical systems (i.e., flows) defined with $d+1$ ordinary differential equations

$$\frac{d\mathbf{X}}{dt} = \mathbf{F}(\mathbf{X}), \tag{5}$$

by considering the successive intersections of the trajectories $\mathbf{X}(t)$ with a codimension-one Poincaré section $\sigma(\mathbf{X}) = 0$. If $\mathbf{x}$ denotes d coordinates which are intrinsic to the Poincaré section, the successive intersections $\{\mathbf{X}_n = \mathbf{X}(t_n)\}_{n \in \mathbb{Z}}$ of the trajectory correspond to a sequence of points $\{\mathbf{x}_n\}_{n \in \mathbb{Z}}$ and return times $\{t_n\}_{n \in \mathbb{Z}}$ in the Poincaré section. According to Cauchy's theorem which guarantees the unicity of the trajectory $\mathbf{X}(t)$ issued from a given initial condition $\mathbf{X}(0)$ (i.e., by the determinism of the flow), the successive points and return times are related by

$$\begin{cases} \mathbf{x}_{n+1} = \boldsymbol{\phi}(\mathbf{x}_n), \\ t_{n+1} = t_n + T(\mathbf{x}_n), \end{cases} \tag{6}$$

where $\boldsymbol{\phi}(\mathbf{x})$ is the so-called *Poincaré map* and $T(\mathbf{x})$ the return-time (or ceiling) function. The knowledge of the Poincaré map and its associated return-time function allows us to recover the flow and its properties.

Consider a similar construction for ordinary differential equations which are periodic in time

$$\frac{d\mathbf{x}}{dt} = \mathbf{F}(\mathbf{x}, t) = \mathbf{F}(\mathbf{x}, t + T), \tag{7}$$

in which case the return-time function reduces to the period T in Equation (6) and the Poincaré map becomes a *stroboscopic map*.

Examples of Poincaré maps are the Birkhoff maps in the case of billiards. Billiards are systems of particles in free flights (or more generally following Hamiltonian trajectories) interrupted by elastic collisions. The knowledge of the collisions suffices to reconstruct the full trajectories. The Birkhoff map is thus the transformation ruling the dynamics of billiards from collision to collision.

Properties

Maps can be classified according to different properties. An important question is to know if a map is locally volume-preserving or not. If the map is differentiable, the volume preservation holds if the absolute value of its Jacobian determinant is equal to unity everywhere in $\mathcal{M}$:

$$\left| \det \frac{\partial \boldsymbol{\phi}}{\partial \mathbf{x}} \right| = 1. \tag{8}$$

This is the case for the baker map (2), the cat map (3), and the quadratic map (4) if $b = \pm 1$.

Maps that contract phase space volumes on average are said to be *dissipative*. In the limit $b \to 0$, the two-dimensional automorphism (4) contracts the phase space areas so much that it becomes an endomorphism given by the one-dimensional logistic map. This explains why highly dissipative dynamical systems are often very well described in terms of endomorphisms such as the logistic map.

A map is *symplectic* if its Jacobian matrix satisfies

$$\left(\frac{\partial \boldsymbol{\phi}}{\partial \mathbf{x}}\right)^{\mathrm{T}} \cdot \boldsymbol{\Sigma} \cdot \frac{\partial \boldsymbol{\phi}}{\partial \mathbf{x}} = \boldsymbol{\Sigma}, \tag{9}$$

where ^{T} denotes the transpose and $\boldsymbol{\Sigma}$ is an antisymmetric constant matrix: $\boldsymbol{\Sigma}^{\mathrm{T}} = -\boldsymbol{\Sigma}$. Symplectic maps act onto phase spaces of even dimension. Poincaré maps of Hamiltonian systems as well as Birkhoff maps are symplectic in appropriate coordinates. Symplectic maps are volume-preserving. Area-preserving maps are symplectic, but there exist volume-preserving maps which are not symplectic in dimensions higher than two.

A map is symmetric under a group $\mathcal{G}$ of transformations $\mathbf{g} \in \mathcal{G}$ if

$$\mathbf{g} \circ \boldsymbol{\phi} = \boldsymbol{\phi} \circ \mathbf{g}. \tag{10}$$

A map is said to be *reversible* if there exists an involution, that is, a transformation $\boldsymbol{\theta}$ such that $\boldsymbol{\theta}^2 = 1$, which transforms the map into its inverse:

$$\boldsymbol{\theta} \circ \boldsymbol{\phi} \circ \boldsymbol{\theta} = \boldsymbol{\phi}^{-1}. \tag{11}$$

There exist reversible maps which are not volume-preserving (Roberts & Quispel, 1992).

Invariant Subsets

These are subsets of the phase space that are invariant under the action of the map. They include the fixed points $\boldsymbol{\phi}(\mathbf{x}_*) = \mathbf{x}_*$ (which correspond to periodic orbits of a corresponding flow) and the periodic orbits of prime period n defined as trajectories from the initial condition $\mathbf{x}_\mathrm{p}$ such that $\boldsymbol{\phi}^n(\mathbf{x}_\mathrm{p}) = \mathbf{x}_\mathrm{p}$ but $\boldsymbol{\phi}^j(\mathbf{x}_\mathrm{p}) \neq \mathbf{x}_\mathrm{p}$ for $0 < j < n$. Tori may also be invariant as in the case of KAM quasi-periodic motion.

An invariant subset $\mathcal{I}$ of the map is *attracting* if there exists an open neighborhood $\mathcal{U}$ such that $\boldsymbol{\phi}\mathcal{U} \subset \mathcal{U}$ and $\mathcal{I} = \cap_{n\in\mathbb{N}}\boldsymbol{\phi}^n\mathcal{U}$. The open set $\mathcal{B} = \cup_{n\in\mathbb{N}}\boldsymbol{\phi}^{-n}\mathcal{U}$ is called the *basin of attraction* of $\mathcal{I}$. An *attractor* is an attracting set which cannot be decomposed into smaller ones. A set which is not attracting is said to be *repelling*.

A closed invariant subset $\mathcal{I}$ is *hyperbolic* if (i) the tangent space $\mathcal{T}_\mathbf{x}\mathcal{M}$ of the phase space $\mathcal{M}$ splits into stable and unstable linear subspaces $\mathcal{E}_\mathbf{x}^{(\mathrm{s})}$ and $\mathcal{E}_\mathbf{x}^{(\mathrm{u})}$ depending continuously on $\mathbf{x} \in \mathcal{I}$,

$$\mathcal{T}_\mathbf{x}\mathcal{M} = \mathcal{E}_\mathbf{x}^{(\mathrm{s})} \oplus \mathcal{E}_\mathbf{x}^{(\mathrm{u})}; \qquad (12)$$

(ii) the linearized dynamics preserves these subspaces; and (iii) the vectors of the stable (resp. unstable) subspace are contracted (resp. expanded) by the linearized dynamics (Ott, 1993). Hyperbolicity implies sensitivity to initial conditions of exponential type, characterized by positive Lyapunov exponents. By extension, a map is said to be *hyperbolic* if its invariant subsets are hyperbolic.

For the baker map (2), the unstable linear subspace is the x-direction while the stable one is the y-direction and the unit square is hyperbolic with a positive Lyapunov exponent. Moreover, the dynamics of the baker map can be shown to be equivalent to a so-called Bernoulli shift, that is, a symbolic dynamics acting as a simple shift on all the possible infinite sequences of symbols 0 and 1, so that most of its trajectories are random. The baker map is thus an example of a hyperbolic fully chaotic map.

A diffeomorphism is said to have the *Anosov* property if its whole compact phase space $\mathcal{M}$ is hyperbolic. Examples of Anosov diffeomorphisms are the cat map (3) and its nonlinear perturbations:

$$\begin{cases} x_{n+1} = x_n + y_n + f(x_n, y_n) & \text{(modulo 1)}, \\ y_{n+1} = x_n + 2y_n + g(x_n, y_n) & \text{(modulo 1)}, \end{cases} \qquad (13)$$

with small enough periodic functions $f(x, y)$ and $g(x, y)$ defined on the torus. We notice that these nonlinear perturbations of the cat map are generally not area-preserving.

Dissipative maps may have chaotic attractors (i.e., attractors with positive Lyapunov exponents) which are not necessarily hyperbolic. This is the case for a set of $a > 0$ values of a positive Lebesgue measure in the logistic map (Jakobson, 1981), as well as in the quadratic map (4) if $b > 0$ is sufficiently small (Benedicks & Carleson, 1991).

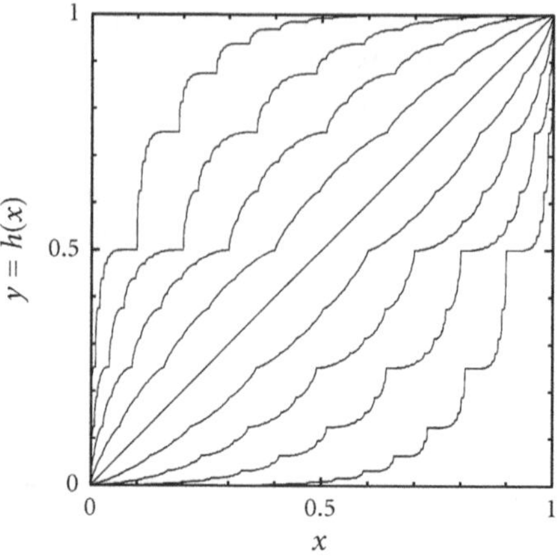

Figure 1. Homeomorphisms $y = h(x)$ transforming the dyadic map into maps (15) with $p = 0.1, 0.2, ..., 0.9$.

Maps may also have sensitivity to initial conditions of stretched-exponential type (with vanishing Lyapunov exponent) as it is the case for the intermittent maps $\phi(x) = x + ax^\zeta$ (modulo 1) with $\zeta > 2$.

Conjugacy Between Maps

In the study of maps, it is often important to modify the analytic form of the map by a change of variables $\mathbf{y} = \mathbf{h}(\mathbf{x})$. Such a conjugacy would transform map (1) into

$$\mathbf{y}_{n+1} = \boldsymbol{\psi}(\mathbf{y}_n), \quad \text{with} \quad \boldsymbol{\psi} = \mathbf{h} \circ \boldsymbol{\phi} \circ \mathbf{h}^{-1}. \qquad (14)$$

The Kolmogorov–Sinai entropy per iteration is known to remain invariant if the conjugacy $\mathbf{h}$ is a diffeomorphism, but it is only the topological entropy per iteration which is invariant if the conjugacy is a homeomorphism. For instance, the logistic map $\phi(x) = 1 - 2x^2$ is conjugated to the tent map $\psi(y) = 1 - 2|y|$ by the conjugacy $y = -1 + \frac{4}{\pi}\arcsin\sqrt{\frac{x+1}{2}}$, both maps having their Kolmogorov–Sinai entropy equal to ln 2. On the other hand, the dyadic map $\phi(x) = 2x$ (modulo 1) is conjugated to the map

$$\psi(y) = \begin{cases} \frac{y}{p} & \text{if} \quad 0 \le y \le p, \\ \frac{y-p}{1-p} & \text{if} \quad p < y \le 1, \end{cases} \qquad (15)$$

with $p \neq \frac{1}{2}$ by a homeomorphism $h(x)$ which is not differentiable (see Figure 1). These two maps have their topological entropy equal to ln 2 but different Kolmogorov–Sinai entropies.

Conjugacies are also important to transform a circle map such as $\phi(x) = x + \alpha + \varepsilon\sin(2\pi x)$ with $|\varepsilon| < \frac{1}{2\pi}$ into a pure rotation $\psi(y) = y + \omega$ of rotation number

$$\omega = \lim_{n\to\infty}\frac{1}{n}(x_n - x_0). \qquad (16)$$

According to Denjoy theory, such a conjugacy is possible if the rotation number is irrational, in which case the motion is quasi-periodic and nonchaotic. The circle map also illustrates the phenomenon of synchronization to the external frequency α, which occurs when the rotation number ω takes rational values corresponding to periodic motions. The motion may become chaotic if $|\varepsilon| > \frac{1}{2\pi}$.

Area-preserving Maps

Periodic, quasi-periodic, and chaotic motions are also the features of area-preserving maps which can be considered as Poincaré maps of Hamiltonian systems with two degrees of freedom. Area-preserving maps can be derived from a variational principle based on some Lagrangian generating function $\ell(x_{n+1}, x_n)$. The variational principle requires that the trajectories are extremals of the action

$$W = \sum_{n\in\mathbb{Z}} \ell(x_{n+1}, x_n). \tag{17}$$

The vanishing of the first variation, $\delta W = 0$, leads to the second-order recurrence equation

$$\frac{\partial \ell(x_{n+1}, x_n)}{\partial x_n} + \frac{\partial \ell(x_n, x_{n-1})}{\partial x_n} = 0. \tag{18}$$

This recurrence can be rewritten in the form of a two-dimensional map by expliciting the equations for the momenta

$$\begin{cases} p_{n+1} = \frac{\partial \ell(x_{n+1}, x_n)}{\partial x_{n+1}}, \\ p_n = -\frac{\partial \ell(x_{n+1}, x_n)}{\partial x_n}. \end{cases} \tag{19}$$

The Birkhoff map of a billiard is recovered if ℓ is the distance traveled by the particle in free flight between collisions and x_n is the arc of the perimeter at which the collision occurs. If a free particle or rotor is periodically kicked by an external driving, the Lagrangian function takes the form

$$\ell = \tfrac{1}{2}(x_{n+1} - x_n)^2 - V(x_n). \tag{20}$$

A famous map is the so-called *standard map*

$$\begin{cases} p_{n+1} = p_n + K \sin x_n, \\ x_{n+1} = x_n + p_{n+1} \quad (\text{modulo } 2\pi), \end{cases} \tag{21}$$

obtained for the kicked rotor with the potential $V(x) = K \cos x$ in Equation (20). The motivation for studying the standard map goes back to works on the origin of stochasticity in Hamiltonian systems (Chirikov, 1979; Lichtenberg & Lieberman, 1983, MacKay & Meiss, 1987).

Phase portraits of an area-preserving map typically present closed curves of KAM quasi-periodic motion, which form elliptic islands. Hierarchical structures of elliptic islands develop on smaller and smaller scales. The elliptic islands are surrounded by chaotic zones extending over finite area (see Figure 2).

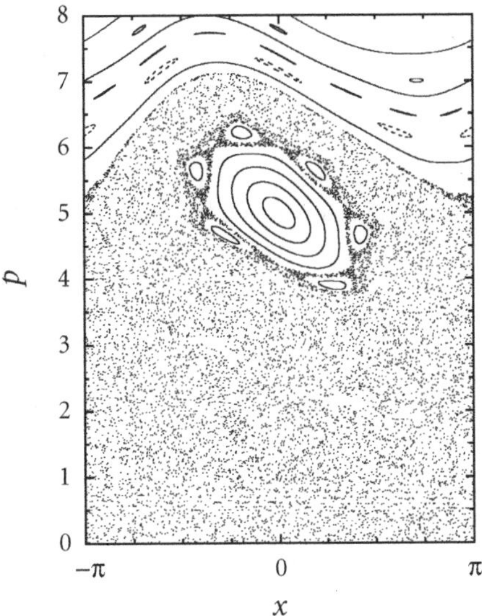

Figure 2. Phase portrait of the Fermi–Ulam area-preserving map, $p_{n+1} = |p_n + \sin x_n|$, $x_{n+1} = x_n + 2\pi M/p_{n+1}$ (modulo 2π), ruling the motion of a ball bouncing between a fixed wall and a moving wall oscillating sinusoidally in time, in the limit where the amplitude of the oscillations is much smaller than the distance between the walls. p is the velocity of the ball in units proportional to the maximum velocity of the moving wall. x is the phase of the moving wall at the time of collision. The parameter M is proportional to the ratio of the distance between the walls to the amplitude of the oscillations of the moving wall. (See Lichtenberg & Lieberman, 1983, for more details.)

Typical area-preserving maps such as the standard map (21) or the one of Figure 2 display a variety of motions that interpolate between two extremes, namely, the fully chaotic behavior of hyperbolic area-preserving maps such as the baker and cat maps and the fully regular motion of integrable maps such as the one given by the second-order recurrence:

$$x_{n+1} - 2x_n + x_{n-1} = -2\mathrm{i} \ln \frac{\kappa^2 + \mathrm{e}^{-\mathrm{i}x_n}}{\kappa^2 + \mathrm{e}^{+\mathrm{i}x_n}} \tag{22}$$

(Faddeev & Volkov, 1994).

Some area-preserving maps may have a repelling Smale horseshoe as the only invariant subset at finite distance. This is the case in the quadratic map (4) for $b = -1$ and large enough values of $a > 0$. Such horseshoes often arise in open two-degrees-of-freedom Hamiltonian systems describing the chaotic scattering of a particle in some time-periodic potential.

Complex Maps

Such maps are defined with some analytic function $\phi(z)$ of $z = x + \mathrm{i}y \in \mathbb{C}$ or some multidimensional generalizations of it. Complex maps are generally endomorphisms. An example is the complex logistic map $\phi(z) = z^2 + c$. Other examples are given by the Newton–Raphson method of finding the roots of a function $f(z)=0$:

$$z_{n+1} = z_n - \frac{f(z_n)}{f'(z_n)}. \quad (23)$$

Complex maps have invariant subsets called Julia sets which are defined as the closure of the set of repelling periodic orbits (Devaney, 1986). The motion is typically chaotic on the Julia set which is repelling and often separates the basins of attraction of the attractors. For instance, the map

$$z_{n+1} = \frac{z_n}{2} + \frac{1}{2z_n}, \quad (24)$$

derived from Equation (23) with $f(z)=z^2-1$ has the attractors $z=\pm 1$. Their respective basins of attraction $x>0$ and $x<0$ are separated by the line $x=0$ where the dynamics is ruled by Equation (24) with $z=\mathrm{i}y$. This one-dimensional map is conjugated to the dyadic map

$$\chi_{n+1} = \begin{cases} 2\chi_n - \frac{\pi}{2} & \text{if} \quad 0 \le \chi_n \le \frac{\pi}{2}, \\ 2\chi_n + \frac{\pi}{2} & \text{if} \quad -\frac{\pi}{2} < \chi_n \le 0, \end{cases} \quad (25)$$

by the transformation $y=\tan\chi$, which shows that the dynamics is chaotic on this Julia set.

However, the boundaries between the basins of attraction are typically fractal (Ott, 1993) as is the case for the Newton–Raphson map (23) with $f(z)=\mathrm{e}^z-1$ (see Figure 3).

Maps and Probability

An important issue is to understand how maps evolve probability in their phase space. The time evolution of probability densities is ruled by the so-called Frobenius–Perron equation (Lasota & Mackey, 1985). The probability density at the current point $\mathbf{x}$ comes from all the points $\mathbf{y}$ that are mapped onto $\mathbf{x}$. Since the inverse of an endomorphism is not unique, the Frobenius–Perron equation is composed of the sum

$$\rho_{n+1}(\mathbf{x}) = \sum_{\mathbf{y}:\, \boldsymbol{\phi}(\mathbf{y})=\mathbf{x}} \frac{\rho_n(\mathbf{y})}{\left| \det \frac{\partial \boldsymbol{\phi}}{\partial \mathbf{x}}(\mathbf{y}) \right|}. \quad (26)$$

For an automorphism, the sum reduces to the single term corresponding to the unique inverse. An invariant probability measure is obtained as a solution of the Frobenius–Perron equation such that $\rho_{n+1}(\mathbf{x})=\rho_n(\mathbf{x})$. The study of invariant measures is the subject of ergodic theory (Hopf, 1937; Arnol'd & Avez, 1968; Cornfeld et al., 1982). The knowledge of the ergodic invariant measure provides us with the statistics of the quantities of interest: observables, correlation functions, Lyapunov exponents, the Kolmogorov–Sinai entropy, etc. With these tools, transport properties such as normal and anomalous diffusion can also be studied in maps (Lichtenberg & Lieberman, 1983).

Some Applications

Dissipative maps are used to study chaos in hydrodynamics (Lorenz, 1963), chemical kinetics (Scott, 1991), biology (Olsen & Degn, 1985; Murray, 1993), nonlinear optics (Ikeda et al., 1980), and more. In particular, systems with time delay in some feedback can be approximated by maps, as in the nonlinear optics of a ring cavity (see Figure 4).

Dissipative maps are also used to study complex systems composed of many interacting units. The units may form a lattice or a graph and interact with each other by diffusive or global couplings. These high-dimensional maps are often called *coupled map lattices* in reference to their spatial extension.

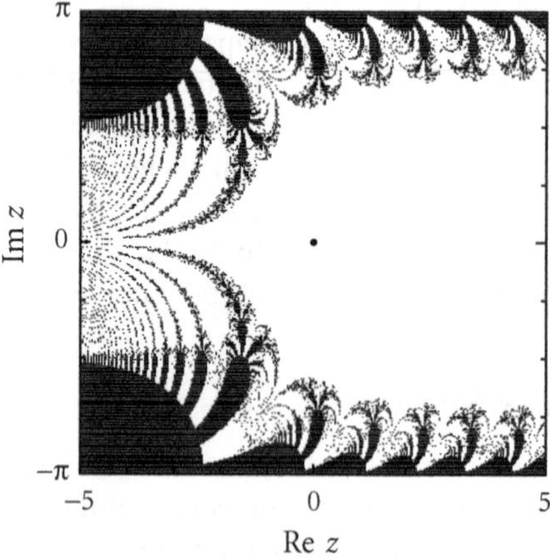

Figure 3. Complex map $z_{n+1}=z_n-1+\exp(-z_n)$: basin of attraction of the point at infinity in grey and of the attractors $z=0, \pm 2\pi\mathrm{i}, \pm 4\pi\mathrm{i}, \ldots$ in white. The dot is the attractor $z=0$.

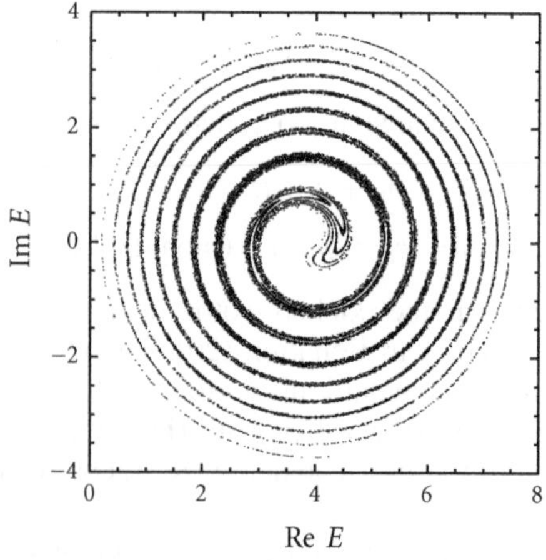

Figure 4. Chaotic attractor of the dissipative Ikeda map $E_{n+1}=a+bE_n\exp(\mathrm{i}\,|\,E_n\,|^2-\mathrm{i}c)$ ruling the complex amplitude $E_n\in\mathbb{C}$ of the electric field of light transmitted in a ring cavity containing a nonlinear dielectric medium, at each passage along the ring (Ikeda et al., 1980). The parameters take the values $a=3.9$, $b=0.5$, and $c=1$.

In addition, area-preserving and symplectic maps have become a fundamental tool to study the long-term evolution of the Solar system (Murray & Dermott, 1999).

PIERRE GASPARD

See also **Anosov and Axiom A systems; Attractors; Aubry–Mather theory; Billiards; Cat map; Chaotic dynamics; Coupled map lattice; Denjoy theory; Entropy; Ergodic theory; Hamiltonian systems; Horseshoes and hyperbolicity in dynamical systems; Kolmogorov–Arnol'd–Moser theorem; Lyapunov exponents; Maps in the complex plane; One-dimensional maps; Phase space; Symbolic dynamics**

Further Reading

Arnol'd, V.I. & Avez, A. 1968. *Ergodic Problems of Classical Mechanics*, New York: Benjamin

Benedicks, M. & Carleson, L. 1991. The dynamics of the Henon map. *Annals of Mathematics*, 133: 73–169

Chirikov, B.V. 1979. A universal instability of many-dimensional oscillator systems. *Physics Reports* 52: 263–379

Cornfeld, I.P., Fomin, S.V. & Sinai, Ya.G. 1982. *Ergodic Theory*, Berlin: Springer

Devaney, R.L. 1986. *An Introduction to Chaotic Dynamical Systems*, Menlo Park CA: Benjamin/Cummings

Faddeev, L. & Volkov, A.Yu. 1994. Hirota equation as an example of an integrable symplectic map. *Letters in Mathematical Physics*, 32: 125–135

Gumowski, I. & Mira, C. 1980. *Recurrences and Discrete Dynamical Systems*, Berlin: Springer

Hénon, M. 1976. A two-dimensional mapping with a strange attractor. *Communications in Mathematical Physics*, 50: 69–77

Hopf, E. 1937. *Ergodentheorie*, Berlin: Springer

Ikeda, K., Daido, H. & Akimoto, O. 1980. Optical turbulence: chaotic behavior of transmitted light from a ring cavity. *Physical Review Letters*, 45: 709–712

Jakobson, M.V. 1981. Absolutely continuous invariant measures for one-parameter families of one-dimensional maps. *Communications in Mathematical Physics*, 81: 39–88

Lasota, A. & Mackey, M.C. 1985. *Probabilistic Properties of Deterministic Systems*, Cambridge and New York: Cambridge University Press

Lichtenberg, A.J. & Lieberman, M.A. 1983. *Regular and Stochastic Motion*, New York: Springer

Lorenz, E.N. 1963. Deterministic nonperiodic flow. *Journal of Atmospheric Sciences*, 20: 130–141

MacKay, R.S. & Meiss, J.D. 1987. *Hamiltonian Dynamical Systems: A Reprint Selection*, Bristol: Adam Hilger

Murray, C.D. & Dermott, S.F. 1999. *Solar System Dynamics*, Cambridge and New York: Cambridge University Press

Murray, J.D. 1993. *Mathematical Biology*, Berlin and New York: Springer

Olsen, L.F. & Degn, H. 1985. Chaos in biological systems. *Quarterly Reviews of Biophysics*, 18: 165–225

Ott, E. 1993. *Chaos in Dynamical Systems*, Cambridge and New York: Cambridge University Press

Roberts, J.A.G. & Quispel, G.R.W. 1992. Chaos and time-reversal symmetry: Order and chaos in reversible dynamical systems. *Physics Reports*, 216: 63–177

Scott, S.K. 1991. *Chemical Chaos*, Oxford: Clarendon Press, and New York: Oxford University Press

MAPS IN THE COMPLEX PLANE

Iterates of rational functions in the complex plane form a class of dynamical systems that can be characterized in amazing detail and completeness. Gaston Julia (1918) and Pierre Fatou (1919/20) proved fundamental theorems on the invariant sets and the relation between critical points and attracting cycles. In some of the first numerical studies of the iterates of a quadratic polynominal, Benoit Mandelbrot discovered the set now named after him (Mandelbrot, 1980). Dennis Sullivan (1985) proved the complete classification of attractors and Adrien Douady & John Hubbard (1984) showed that the Mandelbrot set is connected. Because of their aesthetic beauty, their intricate details, and their omnipresence in iterations of complex functions, Mandelbrot and Julia sets belong to the most fascinating and most widely studied fractal objects. The book by Richter & Peitgen (1986) contains a good survey of the results and many color plates of Julia sets and Mandelbrot sets, as well as personal reflections by Mandelbrot and Douady.

In order to illustrate some of the ideas and concepts, consider the rational maps that result from Newton's method for roots of polynominals (Curry et al., 1983). Their roots can be found using Newton's method, where an initial point z_0 is iterated according to the rational map

$$z_{n+1} = g(z_n) = z_n - f(z_n)/f'(z_n)\,. \qquad (1)$$

For initial conditions sufficiently close to a root, the method converges faster than quadratically, hence its popularity in numerical mathematics. But a little numerical exploration shows that when the initial condition is further away from a root, the iterations can behave rather unpredictably.

Helpful for the investigation of the global dynamics of the iterates is a theorem due to Fatou (1919/20), according to which any attracting cycle will have a critical point in its basin of attraction. A point x_{p} on a cycle of period k returns after k iterations to its starting point, $z_{\mathrm{p}} = g^{(k)}(z_{\mathrm{p}})$. At a critical point the derivative vanishes, $g'(z_{\mathrm{c}}) = 0$. Newton's method for polynominals has $g'(z) = f(z)f''(z)/(f'(z))^2$; therefore, the critical points include the roots of the polynominal and the inflection points where $f''(z_{\mathrm{c}}) = 0$. Since the roots of the polynominals are at the same time fixed points and critical points for Newton's method, each one of them is attracting. It thus suffices to investigate the dynamics of the inflection points.

Points that do not iterate to a root or to any other attracting object form the Julia set.

In 1879, Arthur Cayley solved the simplest case, Newton's method for the quadratic polynominal $z^2 - 1 = 0$, where the iteration reads

$$z_{n+1} = \frac{z_n^2 + 1}{2z_n}\,. \qquad (2)$$

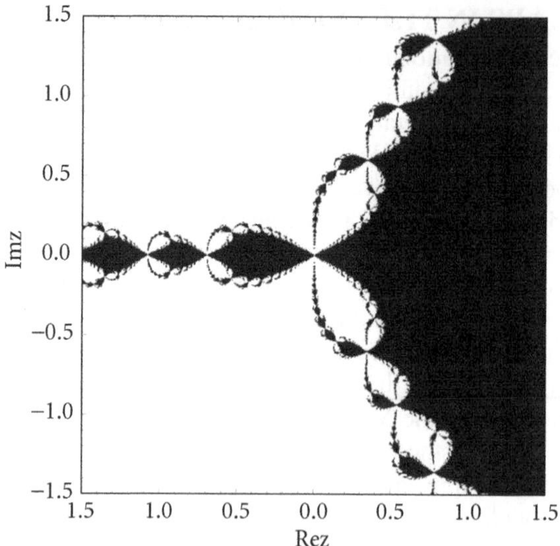

Figure 1. Initial conditions in Newton's method that converge to the root at 1 for the cubic equation $z^3 - 1$. The initial conditions for the other roots follow from symmetry by rotation through an angle of $\pm 2\pi/3$ around the origin.

There are no critical points besides the roots and hence no other attracting regions. The imaginary axis is the Julia set: it is mapped into itself and is the border between the two domains of attraction.

At the end of his paper Cayley comments that "the next succeeding case of the cubic equation appears to present considerable difficulty". Just how difficult is indicated in Figure 1 for the case $f(z) = z^3 - 1$. Evidently, not only the immediate neighborhood of 1 but also many points further away will map into the root $z_0 = 1$. The critical point $z_c = 0$ is mapped to infinity, so that there are no other attractors. The boundary of the black region is the Julia set. It has the interesting feature that in an arbitrarily small neighborhood of every point, initial conditions can be found that iterate to any one of the possible roots. That is to say, at a boundary point, the attracting regions for all roots, and not just for two roots, meet. It has self-similar and fractal features (for instance, its Hausdorff dimension is about 1.429... (Nauenberg & Schellnhuber, 1989)), but it is not of full measure.

For more general third-order polynominals, for example,

$$p_a(z) = z^3 + (a - 1)z - a\,, \tag{3}$$

it can happen that a set of initial conditions of finite measure will not converge to any one of the roots (Curry et al., 1983). Then the iterates of the critical point $z_c = 0$ remain bounded but do not approach a root. For instance, for the parameter values underlying Figure 2, the critical point maps into a period-2 cycle. For initial conditions in the set shown, Newton's method will not find a root; instead, it will converge to a cycle of period 2.

Concerning the dependence on the parameter a, one can map out regions in the parameter space where the critical point $z_c = 0$ does not iterate toward one of the roots or to infinity: for parameter a inside the black region in Figure 3, it approaches another attracting set and Newton's method fails. Besides the case of an attracting cycle of period 2 as in Figure 2, with parameters from the main cardioid of the set, one can find cycles of period 4 in the circular bud to the left, of period 8 even further to the left, and so on: the attracting periodic orbit undergoes period doubling. For parameters in other parts of the set, other sequences of bifurcations occur.

The object that appears in Figure 3 was first seen by Benoit Mandelbrot (1980) in investigations of iterates of the family of quadratic polynominals

$$z_{n+1} = z_n^2 + c\,. \tag{4}$$

This map has only one critical point, $z_c = 0$. For points outside the Mandelbrot set z_c iterates to infinity and there are no stable attracting cycles. The different compartments and regions inside the Mandelbrot set then contain parameter values where different cycles are stable, and transitions between regions correspond to various bifurcations. The cycles that are not attracting are dense in the Julia sets. For parameter values inside the Mandelbrot set, the Julia set is connected; for parameters outside, it dissolves into a Fatou dust. For $c = 0$, the Julia set is a circle of dimension 1, and for small $c \neq 0$, it becomes a fractal with dimension

$$d \approx 1 + \frac{|c|^2}{4 \ln 2} \qquad \text{for small } c\,. \tag{5}$$

For real parameters and real z_n, the map belongs to the class of unimodal maps with quadratic maxima for which the period doubling cascade with its universal scaling laws in parameter and distance between periodic points applies. These relations then translate into scalings of the diameters of the buds in the Mandelbrot set and of structures in the Julia set. A fairly complete description of the structures in both the Mandelbrot set and the associated Julia sets can be achieved using methods from conformal mappings (Douady & Hubbard, 1984/85).

The previous examples already illustrate several kinds of behaviors of iterates. They can map to a fixed point of a periodic orbit, in which case the orbit is stable; if the derivative $M_p = dg^{(k)}(x_p)/dx_p$ along the orbit vanishes, as in the case of the roots for Newton's method, the orbit is superstable. Orbits that have derivatives $|M_p| > 1$ are repelling and belong to the Julia set. To complete the classification of all possible attractors as given by Sullivan (1985), we need to add the marginal cases when the derivative is of modulus one, $M_p = \exp(2\pi\alpha)$: such orbits are called rationally or irrationally indifferent for rational and irrational α, respectively. Near irrationally indifferent

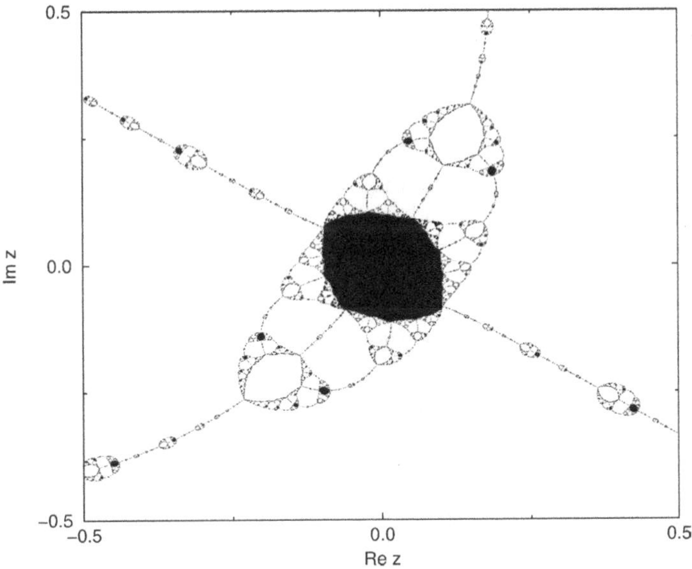

Figure 2. Initial conditions in Newton's method that do not converge to any one of the roots for the cubic polynominal (3) with parameter $a = 0.32 + 1.64\mathrm{i}$. In the big blobs initial conditions converge to a period two cycle.

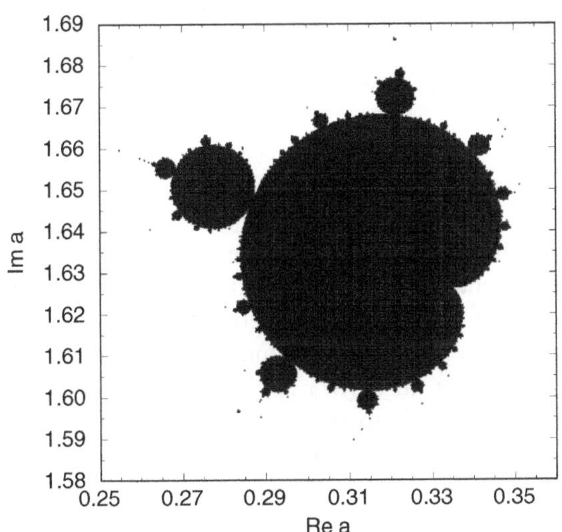

Figure 3. An example of the set of parameters a for which the iterates of the critical points for Newton's method for polynominal (3) do not approach one of the roots. In the main cardioid, the iterates approach a stable period 2 orbit, and in the buds attached to the main cardioid other orbits of higher period are stable. The small speckles outside the main object are also part of the Mandelbrot set that are connected to it by thin hairs and filaments which are not resolved in this plot.

orbits, Siegel disks, and Herman rings can appear. Sullivan's classification theorem now states that there are countably many attracting regions and that they can belong to superstable orbits, stable orbits, Siegel disks, or Herman rings. They can be identified by following iterates of critical points which will bring one to the attracting orbit or to the boundaries for the irrationally indifferent regions.

BRUNO ECKHARDT

See also **Fractals; Maps**

Further Reading

Curry, J., Garnett, L. & Sullivan, D. 1983. On the iteration of rational functions: computer experiments with Newton's method. *Communications of Mathematical Physics*, 91: 267–277

Douady, A. & Hubbard, J.H. 1984/85. Étude dynamique des polynômes complexes I et II. *Publications Mathematique D'Orsay* 84.02 (154 p.) and 85.04 (75 p.)

Fatou, P. 1919/20. Sur les Equations fonctionnelles. *Bulletin de la Société Mathematique de France*, 47: 161–271; 48: 33-94; 48: 208–314

Julia, G. 1918. Mémoire sur l'itération des fonctions rationelles. *Journal de Mathématiques Pures et Appliquées*, 4: 47–245

Mandelbrot, B.B. 1980. Fractal aspects of $z \to \lambda z(1-z)$ for complex λ and z. *Annals of the New York Academy of Sciences*, 357: 249–259

Nauenberg, M. & Schellnhuber, H.J. 1989. Analytic evaluation of the multifractal properties of a Newtonian Julia set. *Physical Review Letters*, 62: 1807–1810

Richter, P.H. & Peitgen, H.O. 1986. *The Beauty of Fractals: Images of Complex Dynamical Systems*, Berlin and New York: Springer

Sullivan, D. 1985. Quasiconformal homoemorphisms and dynamics: I. Solution of the Fatou–Julia problem on wandering domains. *Annals of Mathematics*, 122: 401–418

MARANGONI CONVECTION

See **Fluid dynamics**

MARGINAL STABILITY

See **Stability**

MARKIN–CHIZMADZHEV MODEL

The Hodgkin & Huxley (1952) equations provide a quantitatively accurate and detailed model of the currents generating the propagating nerve impulse in the squid giant axon, and as the first such model they formed a prototype for nerve excitation. The equations are nonlinear, with four dynamical variables, and are in the form of a reaction-diffusion partial differential equation. Thus, the model is analytically intractable and numerical solution—now a trivial task on a PC—in the 1960s, required mainframe facilities. Simpler models were needed, both for understanding the mechanisms of propagation, and numerical exploration of propagation; such simple models are still used to simulate propagation in the anisotropic geometry of cardiac muscle (Panfilov, 1997). One approach is the FitzHugh–Nagumo equations, in which the nonlinear current-voltage relation of excitable membranes is caricatured by a cubic function (Rinzel & Keller, 1973). Another approach is to directly specify the currents flowing during the action potential.

Starting with the nonlinear cable equation for a nonmyelinated axon with axoplasmic resistance R and membrane capacitance C (both per unit length of axon), spread of membrane potential V with distance x (cm) and time t (ms) is described by

$$C\frac{\partial V}{\partial t} = \frac{1}{R}\frac{\partial^2 V}{\partial x^2} - I_{\text{ion}}, \qquad (1)$$

where I_{ion} is a nonlinear function of V and t. Markin & Chizmadzhev (1967) assumed the following simple form for the membrane ionic current I_{ion}. This current was assumed to be switched to a constant inward current J_1 at the start of excitation and, after a time τ_1, switched to a smaller, longer constant outward current J_2 for a time period τ_2. The nonlinear diffusion equation (1) is thus replaced by a piecewise linear diffusion equation. Considering a solitary traveling-wave solution with a velocity θ ms^{-1}, moving to the coordinate system $\xi = x - \theta t$ reduces Equation (1) to an autonomous ordinary differential equation

$$\frac{\partial^2 V}{\partial \xi^2} + \theta RC\frac{\partial V}{\partial \xi} - RI(\xi) = 0. \qquad (2)$$

For the Markin–Chizmadzhev model, I_{ion} in Equation (2) is linear in the four regions: (i) $\xi > 0$, (ii) $0 > \xi > -\theta\tau_1$, (iii) $-\theta\tau_1 > \xi > -\theta(\tau_1 + \tau_2)$, and (iv) $\xi < \theta(\tau_1 + \tau_2)$, so the traveling-wave solution can be constructed analytically from four components, as shown in Figure 1.

Analytic estimates for the velocities of the two traveling wave solutions, the larger being faster and stable, were obtained, these being analogous to earlier numerical studies on propagating activity of the Hodgkin–Huxley equations. This simple Markin–Chizmadzhev model for the membrane current generator retains the ratio of inward to outward current magnitudes and

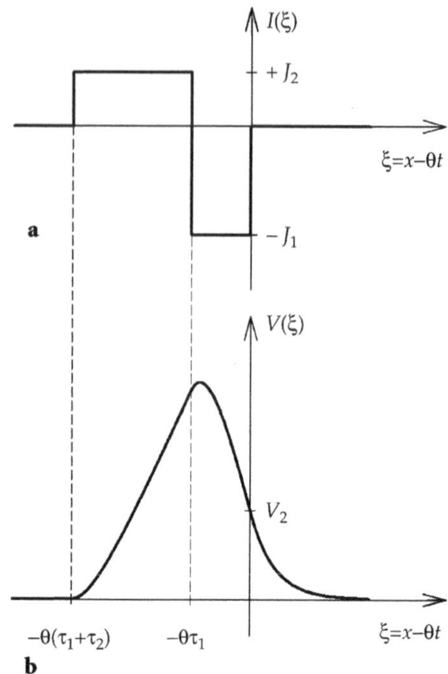

Figure 1. (a) Assumed membrane current and (b) computed potential during a solitary propagating action potential for the Markin–Chizmadzhev model for a nonmyelinated axon, with propagation velocity θ.

the relative time courses of the inward and outward membrane currents. A greater simplification is to consider the action potential as an event that is produced whenever a threshold is exceeded. Such an approach is widely used in modeling periodic and stochastic spike trains, for example, by the integrate and fire model.

Markin et al. (1987) apply the membrane current generator model to provide estimates for the effects of branching and changes in axonal diameter on propagation, and the interaction between propagating activity in axons forming nerve trunks, where extracellular conduction pathways allow the possibility of ephaptic transmission (an action potential in one fiber inducing changes in potential in neighboring fibers), and an increased synchronization of propagating activity in a bundle of nerve fibers. They also applied the model to propagation of activity in syncytia—branching networks of coupled cells. The spread of excitation in such systems depends on the relative cell sizes and connections (it is easier for a large cell to excite a smaller adjacent cell). For systems of similar cells, the syncytium merges into an excitable medium as the cell-to-cell coupling is increased. Such excitable media models are widely used to model propagation in cardiac tissue.

A behavior of excitable media in which a drifting source emits periodic waves was believed to be associated with macroscopic inhomogeneities. Markin & Chizmadzhev (1972) showed that in two homogeneous coupled one-dimensional fibers, activity propagating along one can excite activity in the other and vice-versa, setting up a reverberator that acts as a drifting source of periodic wave trains in a homogeneous system whose components all have stable equilibrium solutions. This provides a prototype, with two coupled one-dimensional fibers, for re-entry, which in two-dimensional excitable media appears as a spiral wave.

The basic phenomenology of propagation in excitable media was first studied with the Markin–Chizmadzhev model, as it allowed both piecewise linear analysis and rapid numerical solution. There is still a need for simple models for rapid simulation in three-dimensional media, but the Markin–Chizmadzhev model has been superceded by an efficient two-variable system for excitable media in general (Dowle et al., 1997), and the Fenton–Karma (1998) three current model for cardiac tissue.

ARUN V. HOLDEN

See also **FitzHugh–Nagumo equation; Hodgkin–Huxley equations; Integrate and fire neuron; Nerve impulses**

Further Reading

Dowle, M., Mantel, R.M. & Barkley, D. 1997. Fast simulations of waves in three-dimensional excitable media. *International Journal of Bifurcation and Chaos*, 7: 2529–2546

Fenton, F. & Karma, A. 1998. Vortex dynamics in three-dimesional continuous myocardium with fiber rotation: filamant instability and fibrillation. *Chaos*, 11: 20–47

Hodgkin, A.L. & Huxley, A.F. 1952. A quantitative description of membrane current and its application to conduction and excitation in nerve, *Journal of Physiology*, 117: 500–544

Markin, V.S. & Chizmadzhev, Yu. A. 1967. Spread of excitation in a model of the nerve fibre. *Biofizika*, 12: 900–907; *Biophyics GB*, 12: 1032–1040

Markin, V.S. & Chizmadzhev, Y.A. 1972. Properties of a multicomponent medium. *Journal of Theoreretical Biology*, 36: 61–80

Markin, V.V., Parushenko, V.F. & Chizmadzhev, Y.A. 1987. *Theory of Excitable Media*, New York: Wiley

Panfilov, A.V. 1997. Modelling of re-entrant patterns of excitation in an anatomical model of the heart. In *Computational Biology of the Heart*, edited by A.V. Panfilov & A.V. Holden, Chichester and New York: Wiley

Rinzel, J. & Keller, J.B. 1973. Traveling wave solutions of a nerve conduction equation. *Biophyical Journal*, 13: 1313–1337

Scott, A.C. 2002. *Neuroscience: A Mathematical Primer*, Berlin and New York: Springer

MARKOV CHAIN

See **Stochastic processes**

MARKOV PARTITIONS

To simplify analysis of a dynamical system, we often study a topologically equivalent system using symbolic dynamics, representing trajectories by infinite length sequences using a finite number of symbols. (A simple example of this idea is the writing of real numbers as sequences of digits, a finite collection of symbols.) To represent the state space of a dynamical system with a finite number of symbols, we must partition the space into a finite number of elements and assign a symbol to each one.

Definition. A topological partition of a metric space M is a finite collection $\mathcal{P}=\{P_1, P_2, ..., P_r\}$ of disjoint open sets whose closures cover M in the sense that $M=\overline{P_1} \cup \cdots \cup \overline{P_r}$ (Lind & Marcus, 1995).

In probability theory, the term Markov denotes memorylessness. In other words, the probability of each outcome conditioned on all previous history is equal to conditioning only on the current state; no previous history is necessary. The same idea has been adapted to the dynamical systems theory to denote a partitioning of the state space so that all of the past information in the symbol sequence is contained in the current symbol, giving rise to the idea of a Markov transformation.

One-dimensional Transformations

In the special, but important case that a transformation of the interval is Markov, the symbolic dynamic is simply presented as a finite directed graph. A Markov transformation in $\mathcal{R}^1$ is defined as follows (Góra & Boyarsky, 1997):

Definition. Let $I=[c,d]$ and let $\tau : I \to I$. Let $\mathcal{P}$ be a partition of I given by the points $c=c_0<c_1<\cdots<c_p=d$. For $i=1,\ldots,p$, let $I_i=(c_{i-1},c_i)$ and denote the restriction of τ to I_i by τ_i. If τ_i is a homeomorphism from I_i onto a union of intervals of $\mathcal{P}$, then τ is said to be Markov. The partition $\mathcal{P}$ is said to be a Markov partition with respect to the function τ.

There are two key elements of the Markov partition that allow the symbol dynamics to accurately represent the system. First, on each interval I_i of the partition, the map must be monotonic (a homeomorphism), which ensures that whenever the preimage of a point is inside an interval, the preimage is unique. Second, whenever the image of a partition element intersects another element, it covers that interval. Therefore, regardless of the trajectory before entering an interval I_j, the orbit may follow any allowed trajectory from I_j. (The future evolves only from the present state.)

As a one-dimensional example, consider map 1 (Figure 1a) which is a Markov map with the associated partition $\{I_1, I_2, I_3, I_4\}$. The symbol dynamics are captured by the transition graph (Figure 1b). Although map 2 (Figure 1c) is piecewise linear and is logically

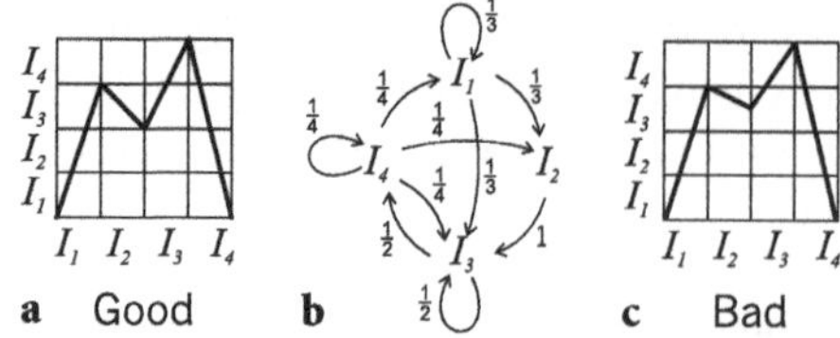

Figure 1. (a) A Markov map with partition shown. Note that on each interval, the map is a one-to-one and the image of the interval covers every interval that it intersects. (b) The transition graph for map 1. (c) The partition is not Markov ("Bad"): the image of I_2 stretches into interval I_3, but it does not completely cover that interval (and similarly for the image of I_3 with I_3).

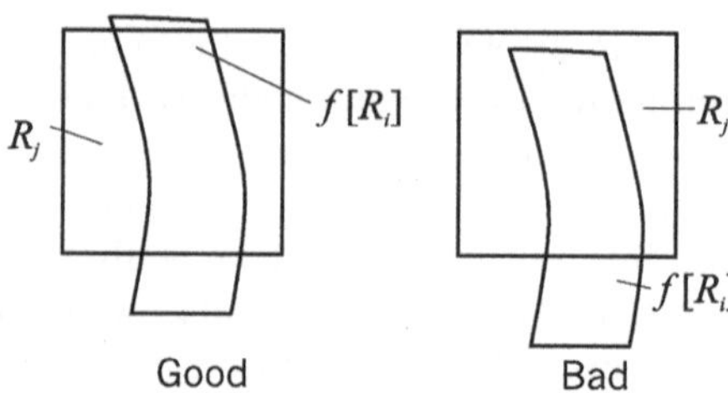

Figure 2. In the unstable (expanding) direction, the image rectangle must stretch completely across any of the partition rectangles that it intersects.

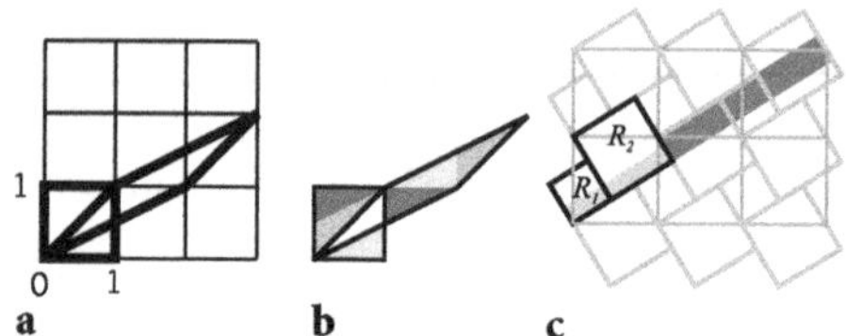

Figure 3. The cat map is a toral automorphism. (a) The operation of the linear map on the unit square. (b) Under the mod operation, the image is exactly the unit square. (c) Tessellation by rectangles R_1 and R_2 forms an infinite partition on $\mathbf{R}^2$. However, since the map is defined on the toral space $\mathbf{T}^2$, only two rectangles are required to cover the space. The filled gray boxes illustrate that R_1 and R_2 are mapped completely across a union of rectangles.

partitioned by the same intervals as map 1, the partition is not Markov because interval I_2 does not map onto (in the mathematical sense) a union of any of the intervals of the partition. However, we are not able to say that map 2 is not Markov. There may be some other partition that satisfies the Markov condition. In general, finding a Markov partition or proving that such a partition does not exist is a difficult problem.

Higher Dimensions

Any topological partitioning of the state space will create symbol dynamics for the map. In the special case where the partition is Markov, the symbol dynamics capture the essential dynamics of the original system.

Definition. Given a metric space M and a map $f: M \to M$, a Markov partition of M is a topological partition of M into rectangles $\{R_1, \ldots, R_m\}$ such that whenever $x \in R_i$ and $f(x) \in R_j$, then (Bowen, 1975; Guckenheimer & Holmes, 1983)

$$f[W^{\mathrm{u}}(x) \cap R_i] \supset W_{\mathrm{u}}[f(x)] \cap R_j$$

and

$$f[W^s(x) \cap R_i] \subset W_s[f(x)] \cap R_j. \quad (1)$$

To determine if the partition is Markov, in other words, we find the stable and unstable manifolds (W^{s} and W^{u}) of each point x and its image $f(x)$, and consider the restriction of these manifolds to the partition rectangles. Thus, whenever an image rectangle intersects a partition element, the image must stretch completely across that element in the expanding (unstable) directions, but the image must be inside that partition element in the contracting (stable) direction. (See Figure 2.)

It is important to use a "good" partition so that the resulting symbolic dynamics of orbits through the partition well represents the dynamical system. If the partition is Markov, then goodness is most easily ensured. However, a broader notion, called generating partition, may be necessary to capture the dynamics. A Markov partition is generating, but the converse is not generally true. See Bollt et al. (2001) and Rudolph (1990) for a discussion of the role of partitions in representing dynamical systems.

The cat map, defined by

$$\mathbf{x} = (\mathbf{A}\mathbf{x}) \bmod 1, \quad (2)$$

where

$$\mathbf{A} = \begin{bmatrix} 2 & 1 \\ 1 & 1 \end{bmatrix} \quad (3)$$

yields a map from the unit square onto itself. This map is said to be on the toral space $\mathbf{T}^2$ because the mod 1 operation causes the coordinate $1 + z$ to be equivalent to z. A Markov partition for this map is shown in Figure 3 (see also color plate section). The cat map is part of a larger class of functions called toral Anosov diffeomorphisms and provides a detailed description of how to construct Markov partitions for this class of maps (Robinson, 1995).

Applications

In addition to establishing the link to symbol dynamics, the Markov partition has another direct application in the one-dimensional case. In a dynamical system, we are often interested in the overall behavior of the map—the evolution of an ensemble of initial conditions. The Frobenius–Perron operator is used to describe this evolution. When the map is Markov, this operator reduces to finite-dimensional stochastic transition matrix. Following the same development as in probability theory, the stationary (invariant) density associated with these maps is described by the eigenvector for the eigenvalue $\lambda = 1$. If the system meets certain ergodic conditions, this density will describe the time average behavior of the system.

The analysis of the ensemble behavior of a dynamical system via its transition matrix is such a powerful tool that we would like to apply it to other one-dimensional systems, even when they may not be Markov. A general technique for approximating the invariant density of a map is called Ulam's method, conjectured by Ulam in 1960 and later proven by Li in 1976. The method relies upon the fact that Markov maps are dense in function space (Froyland, 2000).

ERIK M. BOLLT AND JOE D. SKUFCA

See also **Cat map; Symbolic dynamics**

Further Reading

Bollt, E., Stanford, T., Lai, Y. & Życzkowski, K. 2001. What symbol dynamics do we get with a misplaced partition? On the validity of threshold crossings analysis of chaotic time-series. *Physica* D, 154: 259–286

Bowen, R. 1975. *Equilibrium States and the Ergodic Theory of Anosov Diffeomorphisms*, Berlin and New York: Springer

Froyland, G. 2000. Extracting dynamical behavior via Markov models. In *Nonlinear Dynamics and Statistics: Proceedings, Newton Institute, Cambridge, 1998*, edited by A. Mees, Boston: Birkhäuser

Góra, P. & Boyarsky, A. 1997. *Laws of Chaos, Invariant Measures and Dynamical Systems in One Dimension*, Boston: Birkhäuser

Guckenheimer, J. & Holmes, P. 1983. *Nonlinear Oscillations, Dynamical Systems, and Bifurcations of Vector Fields*, New York: Springer

Lind, D. & Marcus, B. 1995. *An Introduction to Symbolic Dynamics and Coding*, Cambridge and New York: Cambridge University Press

Li, T.-Y. 1976. Finite approximation for the Frobenius–Perron operator. A solution to Ulam's conjecture. *Journal of Approximation Theory*, 17: 177–186

Robinson, C. 1995. *Dynamical Systems: Stability, Symbolic Dynamics, and Chaos*, Ann Arbor, MI: CRC Press

Rudolph, D.J. 1990. *Fundamentals of Measurable Dynamics, Ergodic Theory on Lebesgue Spaces*, Oxford: Clarendon Press, and New York: Oxford University Press

Ulam, S. 1960. *Problems in Modern Mathematics*, New York: Interscience Publishers

MARTINGALES

The classic setting for probability is that of independent random variables. An experiment is repeated many times, and whatever happens in one trial has no effect on what happens in future trials. However, many results of probability also hold true in a much more general setting, that of martingales. The word *martingale* is associated with the concept of a gambling scheme, but its importance in probability is quite general and goes far beyond gambling. The reason is that it is rather easy to find or construct martingales, and these give useful insights into probability problems.

In roulette, the "martingale system" is to double the bet (on black or red) after each bet, until one finally wins. (See Example 4.) The word *martingale* is also used for various arrangements of constraints: part of a horse's reins, lines controlling the jib of a yacht, and the belt at the back of a jacket. The word has been traced back to Middle French before 1600 and, in some accounts, comes from the town of Martigues, presumably a center for horses and gambling.

Definition

The mathematical definition of a martingale involves conditional expectation, and this is a nonlinear concept. If $X_1, \ldots, X_n$ are random variables, then they generate a larger set $\mathcal{F}_n$ of random variables. This set $\mathcal{F}_n$ is defined as the set of all random variables W such that there exists a function f (possibly very nonlinear) with $W = f(X_1, \ldots, X_n)$. If Z is a real random variable with well-defined expectation $E[Z]$, then the "conditional expectation" $E[Z \mid \mathcal{F}_n] = h(X_1, \ldots, X_n)$ is an element of $\mathcal{F}_n$. It is determined by the condition that for every bounded random variable W in $\mathcal{F}_n$ we have $E[E[Z \mid \mathcal{F}_n]\, W] = E[ZW]$. In other words, $E[Z \mid \mathcal{F}_n]$ is the orthogonal projection of Z onto the nonlinearly generated space $\mathcal{F}_n$.

A martingale is a fair game at each step. (There is also a concept of supermartingale, an unfavorable game, and a corresponding concept of submartingale, a favorable game. These are discussed in the references.) Thus, a martingale is a sequence of random variables S_n, one for each time step. These represent the fortune of the gambler at time n. Let $\mathcal{F}_n$ be generated by the random variables that are defined by what happens up to and including time n. Thus, in particular S_n belongs to $\mathcal{F}_n$. However, the fortune S_{n+1} at the next time step typically does not belong to S_n. Past history does not determine future performance.

The martingale condition is that the conditional expectation $E[S_{n+1} \mid \mathcal{F}_n] = S_n$. That is, given the past history up to time n, the expected change in fortune at the next step into the future is zero. It follows easily that the expected fortune $E[S_n]$ does not depend on n and is equal to $E[S_0]$.

A long-held dream of gamblers was to find a gambling scheme to play a fair game and win on the average. The situation is clarified by the following theorems. The dream can be realized, but only if one ignores constraints on time and capital.

Theorems and Examples

Some theorems involving martingales and examples are as follows.

Theorem. *A martingale that is bounded above or bounded below must converge almost surely. Thus there is a random variable S_∞ such that $S_n \to S_\infty$ as $n \to \infty$ with probability one.*

Theorem. *A martingale that is bounded above and below is fair in the limit. That is,* $E[S_\infty] = E[S_0]$.

Example 1. Symmetric random walk. A symmetric random walk is obtained by starting with zero and then making each future step equal to +1 or −1, each independently with probability $\frac{1}{2}$. The walk is the sum of the steps. This is a martingale. In fact, it is a sum of independent random variables. It is unbounded above and below, and it does not converge.

Example 2. Stop when ahead. Consider an integer b with $0 < b$. When the random walk first reaches b, make every future step equal to 0. Notice that these future steps are not independent of the past steps. This is a martingale. At each stage it is a fair game. There is no possibility of winning more than b, but one can be very far in debt. Since this martingale is bounded above, it must converge almost surely, and it can only converge to the constant value b. Thus, a gambler with unlimited credit and unlimited time can play a fair game and be almost sure to win a fixed specified amount.

Example 3. Stop when ahead or behind. Let $a < 0 < b$. When the random walk first reaches a or b, make every future step equal to 0. This is a martingale. It is bounded both above and below. This martingale converges almost surely to a random value that is either a or b. Furthermore, the expected value of the eventual winnings is zero.

Example 4. Double the bet and stop when ahead. The first step is ±1 as before. For a while, each future step is either positive or negative with equal probability, but twice the size of the previous step. After the first positive step, future steps are zero. Thus, for example, the first steps might be −1, −2, −4, −8, +16. The result is that the gambler is ahead by one. This is again a martingale. Since it is bounded above, it converges almost surely to 1. This is favorable to the gambler.

Example 5. Double the bet, start again when ahead. The strategy is the same as in the previous example. However, once the gambler is ahead by one, the game is repeated until the gambler is ahead by two. Then it is repeated again, and so on. This too is a martingale. It has the remarkable property that it diverges almost surely to $+\infty$.

The same ideas may be formulated via the concept of martingale difference. Let $X_0, X_1, X_2, \ldots, X_n, \ldots$ be a sequence of random variables. This is a stochastic process with a discrete time index. (Most of the concepts discussed below extend to continuous time stochastic processes, but this generalization is left to the references.) Let $\mathcal{F}_n$ be the set of all random variables $f(X_0, X_1, X_2, \ldots, X_n)$ that are functions of $X_0, \ldots, X_n$. A sequence of real random variables Y_n belonging to $\mathcal{F}_n$ is called a "martingale difference" if for each $n \geq 0$ the conditional expectation

$$E[Y_{n+1} \mid \mathcal{F}_n] = 0. \tag{1}$$

It is easy to see that if S_0 is in $\mathcal{F}_0$ and the Y_n form a martingale difference, then the sum $S_n = S_0 + Y_1 + Y_2 + \cdots + Y_n$ is a martingale.

The advantage of martingales is that they are easy to create. One general method is to produce martingale differences by subtracting conditional expectations. Thus, if each Z_n is in $\mathcal{F}_n$, then

$$Y_{n+1} = Z_{n+1} - E[Z_{n+1} \mid \mathcal{F}_n] \tag{2}$$

is a martingale difference. However, it is also possible to create martingales by other natural constructions, as shown by the following examples.

Example 6. A branching process. Start with W_0 individuals at stage 0. At each stage n there are W_n individuals. The jth individual in the nth generation has $X_j^{(n+1)}$ children, independently of all other individuals. This number of children is random with expectation $\mu > 0$. Thus the $n + 1$th generation has

$$W_{n+1} = X_1^{(n+1)} + \cdots + X_{W_n}^{(n+1)} \tag{3}$$

individuals. Since

$$E[W_{n+1} \mid \mathcal{F}_n] = \mu W_n, \tag{4}$$

the sequence $S_n = W_n/\mu^n$ is a martingale. In particular, the expected size of the nth generation is

$$E[W_n] = \mu^n E[W_0]. \tag{5}$$

However, this average behavior gives a rather misleading picture of the branching process. The martingale S_n is bounded below by zero, and so it converges to some random value S_∞ almost surely. When $\mu \leq 1$, then the population goes extinct, and so $S_n = W_n/\mu^n \to 0$ almost surely as $n \to \infty$. However, in the case $\mu > 1$, it may be shown that the martingale is fair in the limit. Thus $S_n = W_n/\mu^n \to S_\infty$, where $S_\infty \geq 0$ is random with expectation $E[S_\infty] = E[S_0] = E[W_0]$. If the population does not die out fairly soon, then it has exponential growth: asymptotically $W_n \sim S_\infty \mu^n$.

Example 7. Extinction of a branching process. Let $\mu > 1$ and let ρ be the probability of extinction starting with just one member of the population. Then ρ^{W_n} is a bounded martingale. It converges to 1 when the population goes extinct, and it converges to zero when the population goes to infinity. The fact that it remains fair in the limit says that the probability of extinction starting with w individuals is ρ^w. Each individual's line must die out independently.

Markov Chain Examples

Let $X_0, X_1, X_2, \ldots, X_n, \ldots$ be a Markov chain. (Continuous time Markov processes may also be treated, but that subject is also left to the references.) Let f be a real function of the state of the chain.

Then the present value $f(X_n)$ is a random variable. By the definition of a Markov chain, the expectation of a future value $f(X_{n+1})$, given the past $\mathcal{F}_n$, generated by $X_0, \ldots, X_n$, depends only on the present state X_n. That is,

$$E[f(X_{n+1}) \mid \mathcal{F}_n] = (Pf)(X_n), \tag{6}$$

where Pf is the application of the transition probability operator to the function f. (One can think of f as a column vector and P as multiplication by a square matrix, so Pf is another column vector.) Thus, in the special case when $Pf = f$, the sequence $f(X_n)$ is a martingale. It depends only on the current state. Such functions f are of particular interest when the chain has transient states that can lead to distinct recurrent classes.

Example 8. Asymmetric random walk (gambler's ruin). Let the Markov chain X_n be the random walk that starts at 0 and steps by $+1$ with probability p and steps by -1 with probability q, where $p+q=1$. To be realistic in the gambling situation, take $0<p<q<1$. It is easy to check that the modified game $S_n = (q/p)^{X_n}$ is a martingale.

Example 9. Stop when ahead. Let $0<b$. When the random walk X_n first reaches b, future steps are zero. Then $S_n = (q/p)^{X_n}$ is a bounded martingale. Therefore, it must converge almost surely and remain fair in the limit. The gambler either wins or goes further and further into debt. It follows from the fact that the modified game is fair in the limit that $1 = (q/p)^b P[X_n \to b]$. Thus, the probability of winning is $P[X_n \to b] = (p/q)^b$.

Example 10. Stop when ahead or behind. Let $a<0<b$. When the random walk X_n first reaches a or b, future steps are zero. It follows that $1 = (q/p)^a P[X_n \to a] + (q/p)^b P[X_n \to b]$. From this, it is easy to work out the probabilities of winning or losing. The probability of winning is less than in the last example, but the gambler is protected from catastrophe.

If f is a function of the state of a Markov chain, but $Pf \neq f$, then there is still an associated martingale, but it has a different character. Let

$$Y_{n+1} = f(X_{n+1}) - (Pf)(X_n). \tag{7}$$

Then Y_n forms a martingale difference sequence. Let $S_0 = f(X_0)$ and form the martingale S_n as before. Then the neighboring terms group together, and we get

$$\begin{aligned} S_n &= u(X_0) + u(X_1) + u(X_2) + \cdots + u(X_{n-1}) \\ &\quad + f(X_n), \end{aligned} \tag{8}$$

where $u = f - Pf$. The martingale is a cumulative sum over the entire history.

For this result to be useful, it is necessary to find an interesting function u for which there is a solution f of $u = f - Pf$. This often happens in the context of an irreducible Markov chain with only positive recurrent states. Let π be the invariant probability vector for the Markov chain. (One can think of π as a row vector satisfying $\pi P = \pi$.) Then a necessary condition for a solution is that $\pi u = 0$. For instance, one can take a function h and define $u = h - (\pi h)1$. If f is bounded, then $f(X_n)/n \to 0$, and in suitable circumstances the strong law of large numbers for martingales (see below) gives

$$\frac{u(X_0) + u(X_1) + u(X_2) + \cdots + u(X_{n-1})}{n} \to 0 \tag{9}$$

almost surely. In terms of the function h this says that

$$\frac{h(X_0) + h(X_1) + u(X_2) + \cdots + h(X_{n-1})}{n} \to \pi h \tag{10}$$

almost surely, where πh is the expectation computed with the invariant probability (the product of the probability row vector π with the column vector h). This is a strong law of large numbers for Markov chains. It is the idea underlying the Monte Carlo calculation of unknown invariant probabilities π.

General Results

Many classical results for sums of independent random variables have generalizations to the martingale setting. These include the strong law of large numbers and the central limit theorem.

The Kolmogorov form of the strong law of large numbers says the following. Let $Y_1, \ldots, Y_n, \ldots$ be a sequence of martingale difference random variables with finite variances $\sigma_1^2, \ldots, \sigma_n^2, \ldots$. Assume that the variances are uniformly bounded, or more generally that

$$\sum_{n=1}^{\infty} \frac{\sigma_n^2}{n^2} < \infty. \tag{11}$$

Then as $n \to \infty$ the sample means

$$\bar{Y}_n = \frac{Y_1 + \cdots + Y_n}{n} \to 0 \tag{12}$$

almost surely (that is, with probability one). This is the law of averages in a very powerful and general form.

Fluctuations about the average are described by the central limit theorem. The setting for this theorem is a sequence $Y_1, \ldots, Y_n, \ldots$ of martingale difference random variables with finite variances $\sigma_1^2, \ldots, \sigma_n^2, \ldots$. Let

$$s_n^2 = \sigma_1^2 + \cdots + \sigma_n^2 \tag{13}$$

be the variance of the sum $Y_1 + \cdots + Y_n$.

The martingale differences satisfy the conditional variance normalization condition if

$$\frac{1}{s_n^2} \sum_{k=1}^{n} E[Y_k^2 \mid \mathcal{F}_{k-1}] \to 1 \tag{14}$$

in probability as $n \to \infty$.

Fix $\varepsilon > 0$. Consider one of the random variables Y_i. Say that it is large if $|Y_i| > \varepsilon s_n$. Define the large part of Y_i to be the random variable $Y_i^{\varepsilon s_n}$ that is equal to Y_i when Y_i is large and is equal to zero otherwise. The Lindeberg condition is that the contribution of the large values to the total variance is small, in the sense that for each $\varepsilon > 0$

$$\frac{1}{s_n^2}\sum_{i=1}^{n} E[(Y_i^{\varepsilon s_n})^2] \to 0 \tag{15}$$

as $n \to \infty$.

The central limit theorem states that if $Y_1, \ldots, Y_n, \ldots$ are martingale difference random variables with finite variances that satisfy the conditional variance normalization condition and the Lindeberg condition, then the distribution of

$$Z_n = \frac{Y_1 + \cdots + Y_n}{s_n} \tag{16}$$

approaches the distribution of a standard normal random variable Z as $n \to \infty$. That is, the Gaussian distribution gives a universal description of fluctuations of martingales.

William G. Faris

See also **Random walks; Stochastic processes**

Further Reading

Doob, J.L. 1953. *Stochastic Processes*, New York: Wiley

Durrett, R. 1991. *Probability: Theory and Examples*, Belmot, CA: Duxbury Press

Hall, P. & Heyde, C.C. 1980. *Martingale Limit Theory and Its Application*, New York: Academic Press

Nelson, E. 1987. *Radically Elementary Probability Theory*, Princeton, NJ: Princeton University Press (This book presents probability through the martingale central limit theorem in the context of finite but non-standard probability spaces. It is an unusual treatment, but inspiring.)

Neveu, J. 1975. *Discrete Parameter Martingales*, Amsterdam: North-Holland

MATHIEU EQUATION

See **Surface waves**

MATTER, NONLINEAR THEORY OF

A nonlinear theory of matter was first proposed in 1912 by Gustav Mie in a prescient series of papers that aimed to derive the elementary particles of matter as localized lumps of energy in a nonlinear field (Mie, 1912). To this end, Mie suggested a nonlinear augmentation of James Maxwell's electromagnetic equations out of which the electron would arise in a natural way.

Specifically, he defined a Lagrangian density ($\mathcal{L}$) depending upon electric field intensity ($\boldsymbol{E}$) and magnetic flux density ($\boldsymbol{B}$) and the four components of the electromagnetic potential ($\boldsymbol{A}, \phi$). Requiring dependence on the parameters $\eta \equiv (\boldsymbol{B}^2 - \boldsymbol{E}^2)$ and $\chi \equiv (\phi^2 - \boldsymbol{A}^2)$ ensured relativistic invariance, and the specific choice $\mathcal{L} = \eta/2 + a\chi^3/6$ led to a static, spherically symmetric electric potential (ϕ) of the form

$$\phi(r) \approx \frac{(3r_0^2/a)^{1/4}}{\sqrt{r^2 + r_0^2}}. \tag{1}$$

Setting $4\pi(3r_0^2/a)^{1/4} = e$ (the electronic charge) yielded a spherically symmetric model for the electron with a radius of r_0 and electric potential

$$\phi(r) \to e/4\pi r$$

as $r \to \infty$. The Lorentz invariance that is built into the theory permits this solution to travel with any speed up to the limiting velocity of light with an appropriate Lorentz contraction.

Mie's approach to a nonlinear theory of matter was supported by Albert Einstein, who offered the following opinion in the mid-1930s (Einstein, 1954).

> In the foundation of any consistent field theory, the particle concept must not appear in addition to the field concept. The whole theory must be based solely on partial differential equations and their singularity-free solutions.

Empirical support for this perspective was provided in the early 1930s by Carl Anderson's discovery of positron-electron creation from cosmic radiation. In other words, massive particles were observed to emerge from and collapse back into an electromagnetic field, which is not a property of the linear Maxwell equations.

Motivated by Anderson's observation, Max Born revisited Mie's nonlinear electromagnetics. Together with Leopold Infeld, he eliminated the χ-dependence in Mie's functional formulation and chose instead the Lagrangian density (Born & Infeld, 1934)

$$\mathcal{L} = E_0^2\sqrt{1 + (\boldsymbol{B}^2 - \boldsymbol{E}^2)/E_0^2} - E_0^2, \tag{2}$$

where E_0 sets the field intensities at which nonlinearities arise. At low-field amplitudes, Equation (2) reduces to the classical Lagrangian density for Maxwell's equations: $\mathcal{L} = (\boldsymbol{B}^2 - \boldsymbol{E}^2)/2$. (Even with the currently available, high-intensity lasers, these vacuum nonlinearities would be difficult to observe, as E_0 is estimated to be about 10^{20} V/m in laboratory units.)

Among the solutions of this system, Born and Infeld found a spherically symmetric model electron with $\boldsymbol{E}$ finite everywhere, although the electric displacement ($\mathbf{D}$) exhibits a singularity at the origin. Erwin Schrödinger became interested in Born's theory as early as 1935 and continued working on it through the 1940s when—as founding director of the Dublin Institute for Advanced Studies—he attempted to move research in physics toward key areas of nonlinear science (Schrödinger, 1935).

Plane waves derived from Equation (2) obey the equation $(1 - u_t^2)u_{xx} + 2u_xu_tu_{xt} - (1 + u_x^2)u_{tt} = 0$, where u is a component of the vector potential. Called the Born–Infeld equation, this system has been studied as an interesting nonlinear wave equation (Barbashov & Chernikov, 1966).

Einstein's conviction that a consistent theory for particle physics must be based on localized solutions of nonlinear partial differential equations was shared by several of his colleagues. In addition to Mie, Born, Infeld, and Schrödinger, Werner Heisenberg (1966), Louis de Broglie (1960, 1963), and David Bohm (1957) have suggested nonlinear field theories that in their simplest representations, can be viewed as the augmentation of linear field equations by a nonlinear term of the form $|u|^2u$, as in the nonlinear Schrödinger equation. This nonlinearity conserves the integral

$$\int |u|^2 \, d\mathbf{r}, \qquad (3)$$

which can be interpreted as a mass.

The ideas of de Broglie and Bohm are related to those of the inverse scattering method (ISM). In their "theory of the double solution," the real particle is a localized solution of a nonlinear equation with the form

$$u = U e^{i\theta'}.$$

Associated with this localized nonlinear solution is the solution of a corresponding linear equation

$$\psi = \Psi e^{i\theta}$$

with

$$\theta = \theta' \qquad (4)$$

except in a small region surrounding the real particle. The function ψ is taken to be a solution of Schrödinger's quantum mechanical wave equation, and the phase condition of Equation (4) allows the particle to be guided by ψ. Similarly, in the context of the ISM, the nonlinear solution of a soliton equation is guided through space-time by the linear asymptotic solution of the associated linear operator. Although proposed almost a half century ago, the de Broglie–Bohm theory continues to offer possibilities for further studies (Holland, 1993).

During the 1960s, several investigators proposed the sine-Gordon (SG) equation as a field theory for elementary particles in one space dimension and time (Scott, 2003). This work gained momentum in the 1970s when it became known that special properties of the SG equation allow the corresponding quantum problem to be solved, showing that certain qualitative properties of the classical solution survive quantization (Dashen et al., 1974; Faddeev, 1975; Goldstone & Jakiw, 1975). In particular, the classical field energy was found to be a useful first approximation for the soliton mass, with quantum effects coming in as second-order corrections. More recently, this approach has been developed into the concept of a skyrmion, which is a generalization of the SG kink, carrying its topological stability into three space dimensions.

These examples suggest that it may be possible to develop a nonlinear theory of matter by proceeding as follows. First, guess a classical version of the correct nonlinear field. Second, solve this classical system for salient aspects of localized behavior. Third, analyze the corresponding quantum theory to obtain exact values for the mass spectrum. Finally, compare calculated values of mass with measured mass spectra. As Einstein was aware, however, this is a daunting program because there are no theoretical bounds on the range of conceivable nonlinear theories; thus, it is not surprising to find a proliferation of partially evaluated theories. In addition to the Born–Infeld, de Broglie, and skyrmion formulations, present candidates include string theory and the Yang–Mills equation. What others are out there?

ALWYN SCOTT

See also **Born–Infeld equations; Hodograph transform; Inverse scattering method or transform; Particles and antiparticles; Skyrmions; String theory; Yang–Mills theory**

Further Reading

Barbashov, B.M. & Chernikov, N.A. 1966. Solution of the two plane wave scattering problem in a nonlinear scalar field theory of the Born–Infeld type. *Soviet Physics JETP*, 23: 1025–1033

Bohm, D. 1957. *Causality and Chance in Modern Physics*, London: Routledge & Kegan Paul

Born, M. & Infeld, L. 1934. Foundations of a new field theory. *Proceedings of the Royal Society* (*London*) A, 144: 425–451

de Broglie, L. 1960. *Nonlinear Wave Mechanics*, Amsterdam: Elsevier

de Broglie, L. 1963. *Introduction to the Vigier Theory of Elementary Particles*, Amsterdam: Elsevier

Dashen, R.F., Hasslacher, B. & Neveu, A. 1974. Particle spectrum in model field theories from semiclassical functional integral techniques. *Physical Review* D, 11: 3424–3450

Einstein, A. 1954. *Ideas and Opinions*, New York: Crown

Faddeev, L.D. 1975. Hadrons from leptons? *JETP Letters*, 21: 64–65

Goldstone, J. & Jakiw, R. 1975. Quantization of nonlinear waves. *Physical Review* D, 11: 1486–1498

Heisenberg, W. 1966. *Introduction to the Unified Field Theory of Elementary Particles*, New York: Wiley

Holland, P.R. 1993. *The Quantum Theory of Motion*, Cambridge and New York: Cambridge University Press

Mie, G. 1912. Grundlagen einer Theorie der Materie. *Annalen der Physil*, 37: 511–534; 39: 1–40; 40 (1913): 1–66

Schrödinger, E. 1935. Contributions to Born's new theory of the electromagnetic field. *Proceedings of the Royal Society* (*London*) A, 150: 465–477

Scott, A.C. 2003. *Nonlinear Science: Emergence and Dynamics of Coherent Structures*, 2nd edition, Oxford and New York: Oxford University Press

MAXWELL–BLOCH EQUATIONS

The Maxwell–Bloch (MB) equations arise in nonlinear optics, where they couple "two-level atoms" to Maxwell's electromagnetic field equations to model a nonlinear dielectric. Although the two-level atom is a pseudo-atom with only two nondegenerate energy levels, it can be realized in practice by real atoms excited by light close to a suitable atomic resonance, and it can be modeled as a two-state system: $|g\rangle$ (for ground) and $|e\rangle$ (for excited). Transitions between the two states are allowed, and there is a quantum mechanical matrix element $p = q\langle g\,|r|\,e\rangle \neq 0$ between them in dipole approximation, where q is the electronic charge and qr is the dipole transition operator.

The outer products of $|e\rangle$ and $|g\rangle$ form a Lie algebra with

$$S^z = \tfrac{1}{2}\left(|e\rangle\langle e| - |g\rangle\langle g|\right),$$
$$S^+ = |e\rangle\langle g|,$$
$$S^- = |g\rangle\langle e|.$$

By choosing $\langle e \mid e\rangle = \langle g \mid g\rangle = 1$, $\langle e \mid g\rangle = \langle g \mid e\rangle = 0$ (orthonormality), one finds that the Lie algebra is

$$[S^z, S^+] = S^+, \quad [S^z, S^-] = -S^-,$$
$$[S^+, S^-] = 2S^z,$$

where [,] is the Lie bracket or "commutator" (Bullough et al., 1995b). Constructed from $|e\rangle$ and $|g\rangle$, this is a two-dimensional representation of su(2), corresponding to a spin-$\frac{1}{2}$ system in which $|e\rangle$ is spin-up and $|g\rangle$ is spin-down. In a magnetic field $\boldsymbol{B} = (B_x, B_y, B_z)$, a magnetic dipole μ has a Hamiltonian which can be taken in the form of a 2×2 Hamiltonian matrix with elements (Feynman et al., 1966, pp. 10–14)

$$H_{11} = -\mu B_z, \quad H_{12} = -\mu(B_x - \mathrm{i}B_y),$$
$$H_{21} = -\mu(B_x + \mathrm{i}B_y), \quad H_{22} = +\mu B_z.$$

There is an exact correspondence for the dipole p of the two-level atom in an electric field $\boldsymbol{E}$, and the two-level atom is a pseudo-spin-$\frac{1}{2}$ system (Bullough et al., 1995b). As any state $|\psi(t)\rangle$ at time t in the two-dimensional Hilbert space can be written $|\psi(t)\rangle = c_1(t)\,|e\rangle + c_2(t)\,|g\rangle$, the correspondence extends to the dynamics. From the Hamiltonian H_{ij}, Schrödinger's equation for the amplitudes $c_i(t)(i = 1, 2)$ becomes

$$\mathrm{i}\hbar\dot{c}_1 = -\mu[B_z c_1 + (B_x - \mathrm{i}B_y)c_2],$$
$$\mathrm{i}\hbar\dot{c}_2 = -\mu[(B_x + \mathrm{i}B_y)c_1 - B_z c_2].$$

When $B_x = B_y = 0$, $B_z =$ constant independent of t, $c_1(t), c_2(t)$ evolve as $c(0)\mathrm{e}^{\pm\mathrm{i}\omega_0 t}$, and $\omega_0 = 2\mu B_z\hbar^{-1}$ is a Larmor frequency for the spinning (precessing) magnet of moment μ. The free two-level atom with resonance frequency ω_0 thus *acts* as a true spin-$\frac{1}{2}$ in a fixed magnetic field B_z.

If we construct the Bloch vector $\boldsymbol{r}(t) = (r_1(t), r_2(t), r_3(t))$

$$r_1 = (c_1c_2^* + c_1^*c_2), \quad r_2 = -i(c_1c_2^* - c_1^*c_2),$$
$$r_3 = (|c_1|^2 - |c_2|^2)$$

($c_1^* =$ complex conjugate of c_1), the equations of motion for $c_1(t)$, $c_2(t)$ become the Bloch equation

$$\dot{\boldsymbol{r}} \equiv \mathrm{d}\boldsymbol{r}/\mathrm{d}t = \boldsymbol{\omega} \times \boldsymbol{r} \qquad (1)$$

for a spin-$\frac{1}{2}$ particle, where $\omega \equiv (-2\mu B_x\hbar^{-1}, -2\mu B_y\hbar^{-1}, -2\mu B_z\hbar^{-1})$. (For the two-level atom in a real electric field $E(t)$, $\boldsymbol{\omega} \equiv (-2pE\hbar^{-1}, 0, \omega_0)$.) The normalization $|c_1|^2 + |c_2|^2 = 1$ means $|\boldsymbol{r}|^2 = 1$, and the motion is confined to the surface of a sphere (the Bloch sphere) of unit radius: spin-up = (0,0,1), spin down = (0,0,-1) on this sphere. This description omits a Berry's phase.

Two-level atoms enter laser physics via the Jaynes–Cummings (JC) model, which couples one two-level atom to a single mode of the quantized electromagnetic field of frequency ω. The Hamiltonian is

$$H = \omega_0 S^z + \omega a^\dagger a + g(a^\dagger S^- + S^+ a),$$

where g is a (real and positive) coupling constant, S^z, $S^\pm$ satisfy the su(2) algebra given above, and a, $a^\dagger$ satisfy $[a, a^\dagger] = 1$ and the Heisenberg–Weyl algebra of standard bosons.

The number operator $N = S^z + a^\dagger a + 1/2$ commutes with H. The JC model is thus quantum integrable, there being two degrees of freedom (spin and the quantum oscillator), and there are exactly two commuting constants H and N. The model can be solved exactly in terms of 2×2 matrices (Bullough et al., 1995a) and also by the quantum inverse method in Bogoliubov et al. (1996). By coupling the JC model to a heat-bath, one obtains the master equation for a micromaser. (Such a nonlinear quantum device is in operation at the Max Planck Institute, Garching, Germany, using ^{85}Rb atoms which enter a cavity in their upper states $|e\rangle$ but may leave it in $|e\rangle$ or $|g\rangle$.)

The JC model evolves only in time and is a fundamental nonlinear quantum model. One obtains important nonlinear quantum field theories by coupling two-level atoms to Maxwell's electromagnetic field equations

$$\nabla^2\hat{\boldsymbol{E}} - c^{-2}\partial^2\hat{\boldsymbol{E}}/\partial t^2 = 4\pi n c^{-2}\,\partial^2\hat{\boldsymbol{P}}/\partial t^2. \qquad (2)$$

Here, c is the speed of light in a vacuum, n is the number of two-level atoms per unit volume, and $\hat{\boldsymbol{E}} = \hat{\boldsymbol{E}}(\boldsymbol{x}, t)$ and $n\hat{\boldsymbol{P}} = n\hat{\boldsymbol{P}}(\boldsymbol{x}, t)$ are, respectively, the electric field and dipole density operators. This operator Maxwell equation and the operator Bloch equation for the dipole density constitute the quantum operator form of the MB equations.

If the electric fields are strong enough (many photons), both $\hat{\boldsymbol{E}}$ and $\hat{\boldsymbol{P}}$ can be regarded as classical variables, while quantum mechanics still enters through the atoms. The classical electric field $\boldsymbol{E}(\boldsymbol{x}, t)$ acts on the atom at $\boldsymbol{x}$ at time t through vectors $\boldsymbol{\omega}(\boldsymbol{x}, t)$ given at points $\boldsymbol{x}$ and time t through

$$\boldsymbol{\omega}(\boldsymbol{x}, t) \equiv (-2\boldsymbol{p} \cdot \boldsymbol{E}(\boldsymbol{x}, t)\hbar^{-1}, 0, \omega_0)$$

and the Rabi frequency is $|2\boldsymbol{p} \cdot \boldsymbol{E}(\boldsymbol{x}, t)|\,\hbar^{-1}$. This form of $\boldsymbol{\omega}(\boldsymbol{x}, t)$ shows the coupling between the Maxwell equations and the Bloch equations to form the semiclassical MB equations, and the coupling is nonlinear since $\mathbf{E}(\mathbf{x}, t)$ is driven by $\boldsymbol{p}r_1(\boldsymbol{x}, t)$ for $\boldsymbol{P}(\boldsymbol{x}, t)$ in the (linear) Maxwell equation.

In one space dimension (x) one obtains the standard form of the semiclassical MB equations (Eilbeck et al., 1973). The (linear) Maxwell equations for such an $E(x, t)$ and $P(x, t) = pr_1(x, t)$ is

$$\partial^2 E/\partial x^2 - (1/c^2)\partial^2 E/\partial t^2 = (4\pi np/c^2)\partial^2 r_1/\partial t^2 . \tag{3}$$

With $\omega(x, t) \equiv (-2pE(x, t)\hbar^{-1}, 0, \omega_0)$, the Bloch equation (1) can be written out explicitly as

$$\begin{aligned} \partial r_1/\partial t &= -\omega_0 r_2, \\ \partial r_2/\partial t &= \omega_0 r_1 + 2p\hbar^{-1}Er_3, \\ \partial r_3/\partial t &= -2p\hbar^{-1}Er_2. \end{aligned} \tag{4}$$

This system of four nonlinear partial differential equations (3) with (4) is not integrable, but it becomes an integrable field theory if the Maxwell equation is replaced by the unidirectional system (valid for small densities in the one space dimension)

$$\partial E/\partial x + (1/c)\partial E/\partial t = (-2\pi np/c)\partial r_1/\partial t. \tag{5}$$

Equations (4) and (5) comprise the reduced Maxwell–Bloch (RMB) equations, which can be explicitly integrated by the AKNS inverse scattering method as in Gibbon et al. (1973).

Under a slowly varying envelope and phase approximation (SVEPA), the envelope equations become the self-induced transparency (SIT) equations

$$\begin{aligned} &\partial \mathcal{E}/\partial x + (1/c)\partial \mathcal{E}/\partial t = \alpha P, \\ &\partial P/\partial t = \mathcal{E}N + \Delta\omega' Q, \\ &\partial N/\partial t = -\mathcal{E}P, \\ &\partial Q/\partial t = -\Delta\omega' P. \end{aligned} \tag{6}$$

In these SIT equations, $\Delta\omega' \equiv \omega_0' - \omega_0$ (called inhomogeneous broadening), and ω_0' is a shifted resonance frequency for any particular atom (induced by Doppler shifts, e.g.) while P, Q, N depend on $(x, t, \Delta\omega')$ and N replaces the inversion r_3 of the atom in the Bloch vector. It is an unfortunate confusion of the literature that what we call the SIT equations are also called the Maxwell–Bloch equations. Here, we reserve MB for the Maxwell equation coupled to the three Bloch equations which are *not* envelope equations.

The SIT equations (6) must be averaged over inhomogeneous broadening $\Delta\omega'$ but remain integrable. The soliton solution is the 2π-pulse of (McCall & Hahn, 1969) and there are multi-soliton solutions. Under a sharp line resonance condition, they further reduce to the standard form of the sine-Gordon (SG) equation

$$\partial^2\phi/\partial x^2 - \partial^2\phi/\partial t^2 = \sin\phi . \tag{7}$$

Under an SVEPA, the SG equation becomes the attractive case of the nonlinear Schrödinger (NLS) equation. The RMB, SIT, SG, and attractive NLS equations form a hierarchy of integrable nonlinear field theories in which integrability is handed down by the SVEPA in the fashion described by Calogero in 1995 (see Bullough, 2001).

The two-dimensional representations of su(2) for two-level atoms extend to three-dimensional representations of su(3) for three-level atoms, and the consequent appropriately generalized SIT equations are fundamental to electromagnetically induced transparency (EIT) and the storage of quantum information (Bullough, 2001; Hau, 2001; Haroche & Raimond, 1993; Bullough & Gibbs, 2004).

ROBIN BULLOUGH

See also **Berry's phase; Lie algebras and Lie groups; Nonlinear optics; Nonlinear Schrödinger equation; Sine-Gordon equation**

Further Reading

Bogoliubov, N.M., Rybin, A.V., Bullough, R.K. & Timonen, J. 1995. Maxwell–Bloch system on a lattice. *Physical Review* A, 52: 1487–1493

Bogoliubov, N.M., Bullough, R.K. & Timonen, J. 1996. Exact solution of generalized Tavis–Cummings models in quantum optics. *Journal of Physics A: Mathematics and General*, 29: 6305–6312

Bullough, R.K. 2001. Optical solitons: twenty-seven years of the last millennium and three more years of the new? In *Mathematics and the 21st Century*, edited by A.A. Ashour & A.-S.F. Obada, Singapore: World Scientific, pp. 69–121

Bullough, R.K. & Gibbs, H.M. 2004. Information storage and retrieval by stopping pulses of light. *Journal of Modern Optics*, 51(2): 255–284

Bullough, R.K., Nayak, N. & Thompson, B.V. 1995a. Cavity quantum electrodynamics: fundamental theory of the micromaser. In *Recent Developments in Quantum Optics*, edited by R. Inguva, New York: Plenum Press

Bullough, R.K., Thompson, B.V., Nayak, N. & Boguliubov, N.M. 1995b. Microwave cavity quantum electrodynamics: I, one and many Rydberg atoms in microwave cavities; and II, fundamental theory of the micromaser. In *Studies in Classical and Quantum Nonlinear Optics*, edited by Ole Keller, New York: Nova Science Publishers

Dirac, P.A.M. 1958. *The Principles of Quantum Mechanics*, 4th edition, Oxford: Oxford University Press

Dodd, R.K., Eilbeck, J.C., Gibbon, J.D. & Morris, H.C. 1982. *Solitons and Nonlinear Wave Equations*, London: Academic Press

Eilbeck, J.C., Gibbon, J.D., Caudrey, P.J. & Bullough, R.K. 1973. Solitons in nonlinear optics I. A more accurate description of the 2π pulse in self-induced transparency. *Journal of Physics A: Mathematics and General*, 6: 1337–1347

Feynman, R.P., Leighton, R.B. & Sands, M. 1965. *The Feynman Lectures on Physics*, vol. III. *Quantum Mechanics*, Reading, MA: Addison-Wesley (second printing, 1966)

Gibbon, J.D., Caudrey, P.J., Bullough, R.K. & Eilbeck, J.C. 1973. An N-soliton solution of a nonlinear optics equation derived by a general inverse method. *Lettre al Nuovo Cimento*, 8: 775–779

Haroche, S. & Raimond, J.-M. 1993. *Scientific American*, vol. 269, April: 26–33

Hau, L.V. 2001. Frozen light. *Scientific American*, vol. 284, July: 52–59

Maimstov, A.I., Basharov, A.M., Elyatin, S.O. & Sklyarov, Yu.M. 1990. The present state of self-induced transparency theory. *Physics Reports*, 191: 1–108

McCall, S.L. & Hahn, E.L. 1969. Self-induced transparency. *Phys. Rev.*, 183(2): 457–485

MCCULLOCH–PITTS NETWORK

In 1943, Warren McCulloch and Walter Pitts published a seminal paper that described the first attempt to provide a mathematical model of a neuron (McCulloch and Pitts, 1943). This work explored the properties of networks of such mathematical neurons in relation to the working of the nervous system. The model arose from the following assumptions based on the knowledge of neurophysiology at the time

(i) The activity of the neuron is an "all-or-none" process.
(ii) A certain fixed number of synapses must be excited within the period of latent addition in order to excite a neuron at any time, and this number is independent of previous activity and position on the neuron.
(iii) The only significant delay within the nervous system is synaptic delay.
(iv) The activity of any inhibitory synapse absolutely prevents excitation of the neuron at that time.
(v) The structure of the net does not change with time.

The McCulloch–Pitts neuron is thus a binary threshold unit whose firing or activation is dependent on the values of its inputs. The neuron receives a number of excitatory (positive) or inhibitory (negative) inputs via "synaptic" connections. Excitatory synapses are equally weighted, and if the sum of these positive input signals exceeds some value, θ, the neuron fires. Otherwise, the neuron does not fire. If a signal is received on any of the negative inputs, the neuron cannot fire, regardless of the positive input values. The threshold value θ is a fixed threshold value unique to the neuron, and the model is assumed to be operating in discrete time steps.

A network of McCulloch–Pitts neurons can be assembled by connecting the outputs of neurons to the inputs of other neurons in some manner. In discussing networks of these artificial neurons, McCulloch and Pitts distinguish between nets with and without "circles." In a network without circles, it is not possible to follow a path of connections in the network from any given neuron and return to that same neuron. In other words, there are no closed causal loops. The activity in such a network will, therefore, have a feed-forward dynamic; any activity imposed on the network will propagate in a unidirectional fashion for a finite amount of time, terminating when neurons are encountered with no outgoing synaptic connections. The perceptron (Rosenblatt, 1962) and multi-layer perceptron (see e.g., Haykin, 1999) networks are examples of this kind of topology. A McCulloch–Pitts network without circles is capable of representing any statement within prepositional logic (i.e., any finite logical expression).

The dynamics of networks with circles are more complex. Activity in these networks may propagate indefinitely around the network in discrete time, so the firing activity of the network at any time may be dependent on the activity of the network at multiple stages in the past.

McCulloch and Pitts were able to show that such nets are equivalent to a Universal Turing Machine in principle. It should, however, be noted that the proofs for McCulloch–Pitts nets are existence proofs only in the sense that they do not imply an algorithm for constructing a network, with appropriate values of threshold and weight parameters, to compute a given function. Also, they do not consider computational time. Interestingly, the work of McCulloch and Pitts was also an influence in the development of the von Neumann stored program computer architecture.

Consider a McCulloch–Pitts net with circles. At any point in time, the state of the network can be defined by the current (binary) activity pattern of the neurons (i.e., which neurons are firing and which are not). The network can be thought of as a point in a configuration space of all possible activity patterns. Under some appropriately chosen scheme for updating the state of the network (synchronously or asynchronously), the activity may converge to an attractor in the configuration space, resulting in a stable state for subsequent time steps. This is the basis for attractor neural nets, as described by Hopfield in the context of simple associative memory networks (Hopfield, 1982).

MARCUS GALLAGHER

See also **Attractor neural network; Cell assemblies; Electroencephalogram at large scales; Electroencephalogram at mesoscopic scales; Neurons; Perceptron**

Further Reading

Haykin, S. 1999. *Neural Networks A Comprehensive Foundation*, 2nd edition, Upper Saddle River, NJ: Prentice-Hall

Hopfield, J.J. 1982. Neural networks and physical systems with emergent collective computational abilities. *Proceedings of National Academy of Sciences USA*, 81: 3088–3092
McCulloch, W.S. & Pitts, W. 1943. A logical calculus of ideas immanent in nervous activity. *Bulletin of Mathematical Biophysics*, 5: 115–133
Rosenblatt, F. 1962. *Principles of Neurodynamics*, New York: Spartan

MEAN FIELD THEORIES

See **Phase transitions**

MEASURES

Measures provide the means to introduce probabilistic methods into the study of a dynamical system by supplying a notion of the size or content of sets that is additive—the measure of a union of disjoint sets is the sum of the measures of these sets.

Sets correspond to events and the measure of a set to the probability of that event. The precursor to this general notion is that of the phase volume in a classical mechanical system, which is preserved under the phase flow (Liouville theorem). In other words, if a set of initial conditions is transported by the flow for a fixed time then the set of terminal conditions has the same volume as the set of initial conditions. Put differently, the phase flow consists of volume-preserving maps. In the general context of measures, this corresponds to measure-preserving transformations (see below), which are the subject of ergodic theory. For example, the Birkhoff ergodic theorem establishes a connection between the measure of a set and the proportion of time a typical orbit spends in it. The original result in ergodic theory is the Recurrence Theorem by Poincaré (1890) according to which almost every point in a volume-preserving system has to return arbitrarily close to its initial position.

The *Lebesgue measure* coincides with volume, but it is defined on a larger collection of sets. For example, the one-dimensional volume (length) of the set of rational numbers is not defined, but this set has zero Lebesgue measure. Closely related to the Lebesgue measure are *absolutely continuous* measures, that is, those defined by integrating a nonnegative density function over the set to be measured. The function can be interpreted as a probability density if the total integral is 1. (The bell curve $p(x) := \mathrm{e}^{-(x-b)^2/2a^2}/a\sqrt{2\pi}$ is a standard example.) Taking the Dirac δ-function as a (singular) "density," one obtains the *Dirac measure* defined by $\delta_p(U)=1$ if $p \in U$, $\delta_p(U)=0$ otherwise.

In general, a measure is a nonnegative additive set function. More precisely, it is a nonnegative function (with $+\infty$ an allowed value) defined on a collection of subsets (which are then said to be measurable) of the space in question, such that the union of any countable or finite union of disjoint measurable sets A_i is measurable (this is a requirement of the collection of measurable sets), and its measure is the sum of the measures of the A_i. (One cannot require an analogous property for unions of uncountably many sets because the real line is a disjoint union of points, each of which has zero Lebesgue measure.) One often is interested in finite measures, that is, those for which the measure of the whole space is finite. These can be rescaled to a *probability measure* (the whole space has measure 1). A set of measure zero is referred to as a *null set*, and a property is said to hold *almost everywhere* if it fails for only a null set of points.

If every open set is measurable and every compact set has finite measure, then the measure is said to be a *Borel measure*. This is a useful notion because a Borel measure μ is regular; that is

$$\begin{aligned}\mu(A) &= \inf\{\mu(O) : A \subset O \text{ and } O \text{ is open}\}\\ &= \sup\{\mu(B) : B \subset A \text{ and } B \text{ is compact}\}.\end{aligned}$$

Up to a scale factor, one defines the Lebesgue measure $\mu_n(A)$ of a set $A \subset \mathbb{R}^n$ by considering collections of balls whose union contains A and minimizing the sums of the volumes of these balls:

$$\mu_n(A) = \inf\left\{\sum r_i^n : A \subset \bigcup_i B(x_i, r_i)\right\}.$$

Replacing the exponent n of the radii by an arbitrary exponent α that is not necessarily related to the dimension of the ambient space, one obtains the α-dimensional *Hausdorff measure* μ_α. For example, for a smooth curve c of length $l(c)$ in the plane, we get $\mu_1(c)=l(c)$ and $\mu_2(c)=0$. Indeed, one can characterize the *Hausdorff dimension* of a set S via Hausdorff measures. It is the number α_0 such that $\mu_\alpha(S)=+\infty$ for $\alpha<\alpha_0$ and $\mu_\alpha(S)=0$ for $\alpha>\alpha_0$.

A measure is *invariant* under a map f and the map is said to be *measure-preserving*, if for any measurable set A the set $f^{-1}(A) := \{x: f(x) \in A\}$ is measurable and has the same measure as A. This preimage definition turns out to be the proper one for noninvertible maps. For example, circle rotations $z \mapsto z\mathrm{e}^{2\pi i\theta}$ preserve the Lebesgue measure because they preserve length. The doubling map $z \mapsto z^2$ of the unit circle in the complex plane preserves the Lebesgue measure because the preimage of an arc consists of two arcs of half the length. The density $1/(1+x)$ on $[0, 1]$ defines a measure μ, which is invariant under the Gauss map $G: [0, 1] \to [0, 1]$ defined by $x \mapsto \{1/x\}$ (fractional part): $a = \{1/x\}$ if and only if $1/x = a + n$ for some $n \in \mathbb{N}$, so

$$\begin{aligned}\mu(G^{-1}([a, b])) &= \sum_{n\in\mathbb{N}} \int_{1/b+n}^{1/a+n} \frac{1}{1+x}\,\mathrm{d}x\\ &= \sum_{n\in\mathbb{N}} \log\left(1 + \frac{1}{a+n}\right)\end{aligned}$$

$$
\begin{aligned}
&-\log\left(1+\frac{1}{b+n}\right)\\
&=\sum_{n\in\mathbb{N}}\log(a+n+1)-\log(a+n)\\
&\quad-\log(b+n+1)+\log(b+n)\\
&=\log(b+1)-\log(a+1)\\
&=\int_a^b\frac{1}{1+x}\,\mathrm{d}x=\mu([a,b]).
\end{aligned}
$$

The Dirac measure δ_p concentrated on a fixed point p is always invariant.

The *Birkhoff ergodic theorem* guarantees the existence of time averages of orbits. If φ is an observable (a continuous or merely measurable scalar function on phase space) and μ is an f-invariant measure, then for almost every point the time (or Birkhoff or ergodic) average

$$\lim_{n\to\infty}\sum_{i=0}^{n-1}\varphi(f^i(x))/n$$

exists. A map f is said to be *ergodic* if the phase space is indecomposable in the following sense. If A is an invariant set (i.e., $f^{-1}(A)=A$), then either A or its complement is a null set. This is the case for the Lebesgue measure and the doubling map or rotations by an irrational angle, but not for rotations by a rational angle. For ergodic systems, the time average over almost every orbit equals the space average $\int\varphi\,\mathrm{d}\mu$.

A particularly interesting measure with respect to the study of hyperbolic attractors and strange attractors is the *Sinai–Ruelle–Bowen measure*. By definition it has a positive Lyapunov exponent almost everywhere and absolutely continuous conditionals (marginals) on unstable manifolds. If such a measure is ergodic and has no zero Lyapunov exponents, then it gives a natural or physical (physically observed) measure, which is defined by the following property. While for any invariant measure μ the Birkhoff ergodic theorem implies that almost every point is μ-equidistributed, the physical measure reflects the asymptotic distribution of Lebesgue-almost every point (or at least that of a set of points of positive Lebesgue measure). This means that if one picks a point at random (with respect to Lebesgue measure), its orbit will be equidistributed uniformly with respect to the physical measure. In other words, a physical (or natural) measure represents the density of points obtained from a computed orbit. For example, the Dirac measure concentrated on an attracting fixed point is a natural measure.

BORIS HASSELBLATT

See also **Ergodic theory; Poincaré theorems; Recurrence; Sinai–Ruelle–Bowen measures**

Further Reading

Hasselblatt, B. & Katok, A. (editors). 2002: *Handbook of Dynamical Systems*, vol. 1A. Amsterdam and New York: Elsevier (See also vol. 1B, 2005)

Hasselblatt, B. & Katok, A. 2003: *Dynamics: A First Course*, Cambridge and New York: Cambridge University Press

Katok, A. & Hasselblatt, B. 1995: *Introduction to the Modern Theory of Dynamical Systems*, Cambridge and New York: Cambridge University Press [Measure theory is summarized in an Appendix chapter.]

Petersen, K. 1983: *Ergodic Theory*, Cambridge and New York: Cambridge University Press

Poincaré, H.J. 1890. Sur le probléme des trois corps et les équations de la dynamique. *Acta Mathematica*, 13: 1–270

MECHANICS OF SOLIDS

Two aspects may be discerned in the historical development of the mechanics of solids. The first of these was motivated by technical requirements, and the second was motivated by studies in theoretical physics. Sometimes these two aspects interacted, but each of them has maintained its specificity.

From the fundamental continuum hypothesis, the mechanics of solids admits an introduction of vector and tensor fields describing the strained state of a solid as a result of its deformation: the displacement vector and tensors of stress and strain. Physical features are manifested by the constitutive relations (algebraic, differential, or integro-differential), which determine the coupling between stress and strain tensors, and Newton's equations of motion can be constructed taking into account all relations mentioned above.

There are two sources of nonlinearity in the mechanics of solids. If the constitutive relations are nonlinear, one deals with an intrinsically nonlinear problem. Nonlinearity also arises from purely geometric reasons (nonlinear dependence of the strains on displacements) and is called geometric nonlinearity. Four important approximations illustrating the basic nonlinear effects in the mechanics of solids are as follows.

Prerequisites of approximations:	*Result of approximation:*
(a) Rotational and translational invariance of infinite undeformed solid	Infinite isotropic media
(b) Discrete group of symmetry	Anisotropic crystal
(c) Details of microstructure	Media with weak or strong dispersion
(d) Smallness of one or two dimensions	Bar, plate, shell

Taking account of the shear, nonlinear dynamic equations for an infinite elastic body can be considered as an extension of the hydrodynamics of a compressible non-viscous fluid. Lagrange's equations of motion in the main nonlinear approximation are written as follows (Blend, 1969):

$$\rho\frac{\partial^2 u_j}{\partial t^2}-\frac{\partial}{\partial u_j}\left(\frac{\partial U}{\partial u_{i,j}}\right)=X_{i,j},$$

$$U=\tfrac{1}{2}\lambda I_1^2+\mu I_2+T_0 S-\kappa I_1 S+\text{higher-order terms},$$

$$i,j=1,2,3, \qquad (1)$$

where ρ_0 is the initial density of the elastic media, $X_{i,j}$ are components of external loading, u_j are components of the displacement vector, U is a potential, $I_1=\varepsilon_{ii}$ and $I_2=\varepsilon_{ij}\varepsilon_{ij}$ are the first and second invariants of the strain tensor, λ and μ are Lameé coefficients characterizing the elastic properties of isotropic media, S is entropy, κ is the coefficient of heat conductivity, and T_0 is the initial temperature. Lamé coefficients can be expressed via the Young's modulus E and Poisson coefficient ν, describing longitudinal elastic resistance and transversal deformation, respectively, for a specimen subjected to uniaxial longitudinal loading.

In this case, geometric nonlinearity is caused by nonlinear dependence of ε_{ij} on derivatives of displacements, and physical nonlinearity is accounted for by high-order terms involving strains ε_{ij}. A significant consequence of nonlinearity may be the possibility of shock waves manifested by discontinuity of the first or second derivatives of displacements (strong or weak discontinuities), respectively.

The simplest and most important problem admitting discontinuous solutions in nonlinear theory is the description of plane adiabatic shock waves propagation. In this case, nonlinear partial differential equations (1) are replaced by nonlinear algebraic relations. These relations are determined by conservation laws for mechanical and thermodynamical quantities as well as by the second law of thermodynamics. The main difference from shock waves in the fluid is manifested in the presence of shear shock waves.

Taking into account microstructure and intermolecular (physical) nonlinearity in the constitutive relations, one obtains in the lowest approximation for plane waves the Korteweg–de Vries (KDV) equation (Askar, 1985)

$$\frac{\partial w}{\partial \tau}+\frac{1}{2}\alpha_1 w\frac{\partial w}{\partial \zeta}+\frac{1}{24}\frac{\partial^3 w}{\partial \zeta^3}=0,$$

where τ and ζ are dimensionless time and space coordinates, respectively, and $w(\tau,\zeta)$ is a dimensionless deformation of the media. Due to the presence of nonlinear (second) and dispersion (third) terms, this equation has a stable soliton solution (compression wave) as well as multiple soliton solutions.

A second new aspect, important for media with microstructure, is the possibility of a short wavelength continuum approximation. Such an approximation is valid for the envelopes of short waves for which the wavelengths are comparable with the characteristic length of microstructure (Manevitch, 2001). In this case, equations of motion for the plane waves can be reduced to the dimensionless nonlinear Schrödinger (NLS) equation

$$\frac{\partial \varphi_0}{\partial \tau}-\mathrm{i}\alpha|\varphi_0|^2\varphi_0-\mathrm{i}\frac{\partial^2\varphi_0}{\partial \zeta^2}=0,$$

where $\varphi_0=[\exp \mathrm{i}\tau](w-\mathrm{i}u)$, u is the displacement, w is the velocity, α is a nonlinear parameter, and τ and ζ are dimensionless time and space coordinates. An important manifestation of nonlinearity in this case is the existence of envelope solitons

$$\varphi_0(\zeta,\tau)=\sqrt{2S/\alpha}\,\mathrm{e}^{\mathrm{i}(k\zeta-\omega\tau)}\mathrm{sech}[\sqrt{S}(\zeta-\nu\tau)],$$

where $k=\nu/2$ and $\omega=\nu^2/4-S$.

Dispersion effects can be also caused by the presence of boundaries as in the case of a rod embedded in another elastic external medium. Along with physical and/or geometric nonlinearity, this may lead to formation of solitons similar to solitons in media with microstructure (Samsonov et al., 1999).

Note that the formation of shock waves in three-dimensional elastic media requires much more energy than in the common case of sound waves. Contrary to this, geometric nonlinearity in thin elastic bodies described by simplified one- and two-dimensional models provides easy manifestation of nonlinear effects due to bifurcation of equilibrium states (elastic instability). Elastic instability may lead to localization of buckling and coupling of linear modes—a sign of strong nonlinearity (Thomson & Hupt, 1973). Thin-walled structures also demonstrate numerous other effects typical of common nonlinear systems, including period doubling and internal resonance, transition to spatial, and temporal chaos. Future developments in nonlinear mechanics of solids will focus on the study of these effects as well as the propagation of nonlinear waves in anisotropic and nonhomogeneous media and on mechanics of solids at the mesoscopic level (Alexander, 1998).

LEONID MANEVITCH

See also **Korteweg–de Vries equation; Nonlinear Shrödinger equations; Shock waves; Solitons**

Further Reading

Alexander, S. 1998. Amorphous solids theory, structure, lattice dynamics and elasticity. *Physics Reports*, 296: 68–256

Askar, A. 1985. *Lattice Dynamical Foundations of Continuum Theories*, Singapore: World Scientific

Blend, D.R. 1969. *Nonlinear Dynamic Elasticity*, Waltham, MA: Blaisdell Publishing

Manevitch, L.I. 2001. Solitons in polymer physics. *Polymer Science* C, 4(2): 117–181

Samsonov, A.M. et al. 1999. Strain solitons in solids: physics, numerics and fracture. In *Dynamics of Vibro-Impact Systems*, edited by V.I. Babitsky, Berlin: Springer, pp. 215–220

Thomson, J.M. & Hupt, G.M. 1973. *A General Theory of Elastic Stability*, New York: Wiley

MEL'NIKOV METHOD

In 1963, V.K. Mel'nikov developed an analytical method for the study of time-periodic perturbations of a planar autonomous system. This method permits determination of persistence and stability of subharmonic motions and measurement of infinitesimal separation of stable and unstable manifolds of hyperbolic fixed points. The Mel'nikov method is one of the few analytical tools for establishing conditions for homoclinic chaos, and it has been generalized to many settings, including Hamiltonian systems with higher degrees of freedom and partial differential equations.

The main ideas can be introduced in the context of a periodically forced planar Hamiltonian system:

$$\dot{x} = \frac{\partial H}{\partial y} + \varepsilon f_1(x, y, t),$$
$$\dot{y} = \frac{\partial H}{\partial x} + \varepsilon f_2(x, y, t).$$

$H(x, y)$ is the Hamiltonian function of the unperturbed system ($\varepsilon = 0$), and $\mathbf{f} = (f_1, f_2)^{\mathrm{T}}$ is a time-periodic perturbation vector field. Both H and $\mathbf{f}$ are assumed to be sufficiently smooth functions of their arguments.

Assume that the unperturbed system possesses a homoclinic orbit $q_{\mathrm{h}}(t)$ to a hyperbolic saddle point p_0, as shown in Figure 1. When $\varepsilon \neq 0$, a perturbation argument guarantees persistence of a hyperbolic fixed point p_ε, and local existence of its stable and unstable manifolds, parametrized, respectively, by solutions $q_\varepsilon^{\mathrm{s}}(t) = q_{\mathrm{h}}(t - t_0) + \varepsilon q_1^{\mathrm{s}}(t) + \mathcal{O}(\varepsilon^2)$ and $q_\varepsilon^{\mathrm{u}}(t) = q_{\mathrm{h}}(t - t_0) + \varepsilon q_1^{\mathrm{u}}(t) + \mathcal{O}(\varepsilon^2)$.

As the unperturbed separatrix is expected to split under perturbation, it is convenient to introduce the distance function

$$\begin{aligned} d(t_0) &= (\partial_x H, \partial_y H)^{\mathrm{T}}|_{q_{\mathrm{h}}(0)} \cdot [q_\varepsilon^{\mathrm{u}}(t_0) - q_\varepsilon^{\mathrm{s}}(t_0)] \\ &= \varepsilon[(\partial_x H, \partial_y H)^{\mathrm{T}}|_{q_{\mathrm{h}}(0)} \wedge (q_1^{\mathrm{u}}(t_0) - q_1^{\mathrm{s}}(t_0))] \\ &\quad + \mathcal{O}(\varepsilon^2), \end{aligned}$$

which measures the displacement of the stable and unstable manifolds of p_ε along a direction normal to the unperturbed separatrix (represented by vector **N** in Figure 1). After deriving an evolution equation for the perturbation expansion of the distance function $d(t)$, we arrive at the following expression:

$$d(t_0) = \varepsilon M(t_0) + \mathcal{O}(\varepsilon^2). \qquad (1)$$

The Mel'nikov function $M(t_0)$ is defined by

$$M(t_0) = \int_{-\infty}^{+\infty} \mathbf{grad}\, H(q_{\mathrm{h}}(t - t_0)) \cdot \mathbf{f}(q_{\mathrm{h}}(t - t_0), t)\mathrm{d}t, \qquad (2)$$

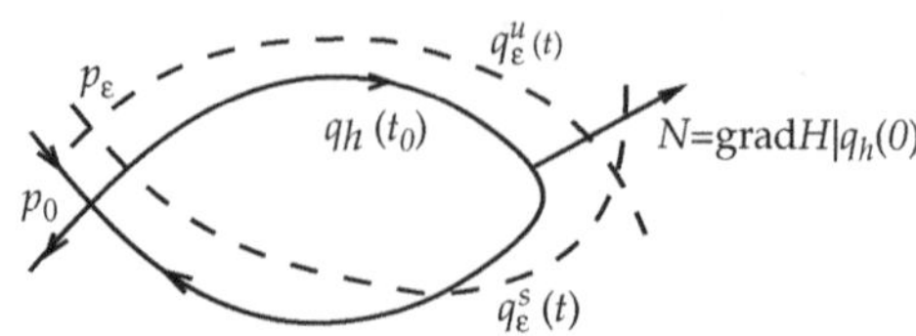

Figure 1. Unperturbed and perturbed separatrices.

where the integrand is the scalar product of vectorfields **grad** H and **f** and is evaluated along the unperturbed homoclinic orbit.

As an example, consider the following planar system of ordinary differential equations with periodic perturbation:

$$\dot{x} = y,$$
$$\dot{y} = -x + x^2 + \varepsilon \sin t.$$

The unperturbed system has Hamiltonian $H(x, y) = y^2/2 + x^2/2 - x^3/3$ and two fixed points: the origin (a center) and the point of coordinates $(1, 0)$ (a saddle point). The unperturbed homoclinic orbit (parametrizing the degenerate stable, unstable manifolds of the hyperbolic fixed point) is given by

$$q_{\mathrm{h}}(t - t_0) = \left(\frac{3}{2}\tanh^2\left(\frac{t - t_0}{2}\right) - \frac{1}{2}, \frac{3}{2}\tanh\left(\frac{t - t_0}{2}\right)\sinh^2\left(\frac{t - t_0}{2}\right)\right)^T, \qquad (3)$$

and the corresponding Mel'nikov function is computed to be

$$\begin{aligned} M(t_0) &= \frac{3}{2}\int_{-\infty}^{+\infty} \tanh(t/2)\sinh^2(t/2)\sin(t + t_0)\,\mathrm{d}t \\ &= -\frac{3\pi\cos(t_0)}{2\sinh\pi/2\cosh\pi/2}. \end{aligned} \qquad (4)$$

If, as in this example, $M(t_0)$ has an infinite sequence of nondegenerate zeros, an infinite number of transverse homoclinic orbits of the hyperbolic fixed point of the perturbed system is guaranteed to exist for all $\varepsilon \neq 0$ sufficiently small. Thus, the system exhibits homoclinic chaos.

For perturbations containing dissipative or constant forcing terms, the Mel'nikov integral may be nonvanishing. In such cases, the separatrix still splits under perturbation, but no transverse homoclinic points occur.

Generalizations of the Mel'nikov method to higher-dimensional systems have been discussed by various authors (see, for example, Yagasaki (1999), Gruendler (1992), and the monograph by Wiggins (1988, pp. 1–16)).

Autonomous Perturbations in Two Degrees of Freedom

Consider how this method applies to autonomous near-integrable Hamiltonian systems with two degrees of freedom. This case was first treated, and then generalized to integrable equations with $n+1$ degrees of freedom, by Holmes and Marsden (1982a,b). These authors assume that the unperturbed Hamiltonian has the form $H_0 = F(x, y) + G(I)$, where F is the Hamiltonian of a planar system with a homoclinic orbit to a hyperbolic fixed point, and G is the Hamiltonian of a planar system in action-angle coordinates. By restricting to the energy surface, the original equations are reduced to a single-degree-of-freedom non-autonomous system, to which the usual Mel'nikov technique is applied.

A more direct approach by Robinson (1996) assumes Hamiltonians of the general form

$$H_\varepsilon(\mathbf{x}) = H_0(\mathbf{x}) + \varepsilon H_1(\mathbf{x}), \quad \mathbf{x} \in \mathbb{R}^4. \tag{5}$$

The unperturbed system is assumed to be completely integrable, with a second constant of motion K, and to possess a hyperbolic periodic orbit γ_0. The stable and unstable manifolds of γ_0, $\mathcal{W}^{\mathrm{s}}(\gamma_0) = \mathcal{W}^{\mathrm{u}}(\gamma_0)$, are parametrized by a two-dimensional family of homoclinic orbits $q_{\mathrm{h}}(t, \mathbf{x}_0)$, where $\mathbf{x}_0$ is a point on the corresponding level set $H_0(\gamma_0) = h_0$.

Because the perturbation is Hamiltonian, the perturbed hyperbolic periodic orbit γ_ε and its stable, unstable manifolds $\mathcal{W}^{\mathrm{s,u}}(\gamma_\varepsilon)$ continue to lie within the same energy surface $H_\varepsilon^{-1}(h_0)$. Thus **grad** H_0 no longer provides a good measurement of their transversal splitting.

Letting Π_0 be a two-dimensional plane transversal to the unperturbed separatrix at $\mathbf{x_0}$ and denoting by $\zeta^{\mathrm{s,u}}(\mathbf{x}_0, h_0; \varepsilon) = \Pi_0 \cap \mathcal{W}^{\mathrm{s,u}}(\gamma_\varepsilon)$ the corresponding points on the perturbed invariant manifolds, one computes the infinitesimal displacement in terms of the second integral K:

$$\begin{aligned} M(\mathbf{x_0}, h_0) = {} & \frac{\partial}{\partial \varepsilon}[K \circ \zeta^u(\mathbf{x}_0, h_0; \varepsilon) \\ & -K \circ \zeta^s(\mathbf{x}_0, h_0; \varepsilon)]|_{\varepsilon=0}. \end{aligned} \tag{6}$$

The Mel'nikov measurement can be reduced (Robinson, 1996) to the following conditionally convergent integral:

$$\begin{aligned} & M(\mathbf{x}_0, h_0) \\ & = \lim_{j\to\infty} \int_{-T_j^*}^{T_j} \mathbf{grad}\, F \cdot \mathcal{J}\mathbf{grad}\, H_1|_{q_h(t,\mathbf{x}_0)}\, \mathrm{d}t, \end{aligned} \tag{7}$$

where $\mathcal{J}\mathbf{grad} H_1$ is the Hamiltonian vector field of the perturbation, and $-T_j^*$, T_j are chosen so that $q_h(-T_j^*, \mathbf{x}_0)$ and $q_h(T_j, \mathbf{x}_0)$ converge to the same point of the periodic orbit γ_0.

An Elastic Pendulum

Consider the application of these ideas to the following four-dimensional dynamical system modelling an elastic pendulum (Holmes and Marsden, 1982a):

$$\begin{aligned} \dot{x} &= y, \\ \dot{y} &= \sin x + \varepsilon \left(\sqrt{\frac{2I}{\omega}} \sin\theta - x \right), \\ \dot{I} &= \varepsilon \sqrt{\frac{2I}{\omega}} \left(\sqrt{\frac{2I}{\omega}} \sin\theta - x \right) \cos\theta, \\ \dot{\theta} &= \omega - \varepsilon \frac{1}{\sqrt{2I\omega}} \left(\sqrt{\frac{2I}{\omega}} \sin\theta - x \right) \sin\theta. \end{aligned}$$

The unperturbed equations are completely integrable with constants of motion $H_0(x, y, I, \theta) = \frac{1}{2}y^2 - \cos x + \omega I$ and $K(x, y, I, \theta) = I$. The energy surface $H_0(\mathbf{x}) = 1 + \omega I = h_0$ contains a hyperbolic periodic orbit $\gamma_0 = (\pi, 0, I, \omega t)^{\mathrm{T}}$ and its two-dimensional stable and unstable manifolds, parametrized by the pair of homoclinic orbits

$$\begin{aligned} q_h^{\pm}(t, \theta_0) = {} & (\pm 2 \tan^{-1}(\sinh t), \\ & \pm 2\,\mathrm{sech}\, t, \frac{h_0 - 1}{\omega}, \omega t + \theta_0)^{\mathrm{T}}. \end{aligned}$$

As **grad** $K = (0, 0, 1, 0)^{\mathrm{T}}$, the quantities $M^{\pm}(\theta_0, h_0)$ are immediately computed as the following absolutely convergent integrals ($h_0 > 1$):

$$\begin{aligned} & \frac{\sqrt{2(h_0-1)}}{\omega} \int_{-\infty}^{+\infty} \left[\frac{\sqrt{2(h_0-1)}}{\omega} \sin(\omega t + \theta_0) \right. \\ & \left. \mp 2 \tan^{-1}(\sinh t) \right] \cos(\omega t + \theta_0)\, \mathrm{d}t. \end{aligned} \tag{8}$$

The first term is odd and thus vanishes. Using integration by parts to simplify the second term yields the expression

$$\begin{aligned} M^{\pm}(\theta_0, h_0) = {} & \pm \frac{\sqrt{2(h_0-1)}}{\omega^2} \\ & \times \int_{-\infty}^{+\infty} \mathrm{sech}(t) \sin(\omega t + \theta_0)\, \mathrm{d}t, \end{aligned}$$

which can be evaluated using the method of residues as

$$M^{\pm}(\theta_0, h_0) = \pm 2\pi \frac{\sqrt{2(h_0-1)}}{\omega} \mathrm{sech}\left(\frac{\pi\omega}{2}\right) \sin(\theta_0).$$

An infinite sequence of simple zeros of $M^{\pm}(\theta_0, h_0)$ guarantees the existence of transversal homoclinic orbits on each energy surface $h_0 > 1$, for $\varepsilon \neq 0$ sufficiently small.

ANNALISA M. CALINI

See also **Chaotic dynamics; Hamiltonian systems; Horseshoes and hyperbolicity in dynamical systems; Phase space**

Further Reading

Gruendler, J. 1992. Homoclinic solutions for autonomous dynamical systems in arbitrary dimension. *SIAM Journal of Mathematical Analysis*, 23: 702–721

Guckenheimer, J. & Holmes, P. 1983. *Nonlinear Oscillations, Dynamical Systems, and Bifurcations of Vector Fields*, Berlin and New York: Springer

Holmes, P.J. & Marsden, J.E. 1982a. Mel'nikov's method and Arnol'd diffusion for perturbations of integrable Hamiltonian systems. *Journal of Mathematical Physics*, 23(4): 669–675

Holmes, P.J. & Marsden, J.E. 1982b. Horseshoes in perturbations of Hamiltonian systems with two degrees of freedom. *Communications in Mathematical Physics*, 82: 523–544

Mel'nikov, V.K. 1963. On the stability of the center for time periodic perturbations. *Transactions of Moscow Mathematical Society*, 12: 1–57

Robinson, C. 1996. Mel'nikov method for autonomous Hamiltonians. *Contemporary Mathematics*, 198: 45–53

Wiggins, S. 1988. *Global Bifurcations and Chaos: Analytical Methods*, Berlin and New York: Springer

Yagasaki, K. 1999. The method of Mel'nikov for perturbations of multi-degree-of-freedom Hamiltonian systems. *Nonlinearity*, 12(4): 799–822

METASTABILITY

See **Stability**

METEOROLOGY

See **Atmospheric and ocean sciences**

MISSING MASS (FOR KDV SOLITON)

See **Korteweg–de Vries equation**

MIURA TRANSFORMATION

See **Korteweg–de Vries equation**

MIXING

The word *mixing* is used both as a general term defining the operation of putting two or more substances together in order to achieve uniformity and as a mathematical term defining the property of random processes. Mixing as an operation is widespread as both a natural phenomenon and an industrial process. Putting milk into coffee, preparing cement, and pushing a car accelerator pedal involve mixing liquids, granular materials, and gases. On a molecular scale, it is diffusion that provides mixing. When diffusion is caused solely by the gradient of concentration $\theta(\boldsymbol{r}, t)$, it is described by the second-order partial differential equation

$$\frac{\partial\theta}{\partial t} = \operatorname{div}(\kappa\nabla\theta). \qquad (1)$$

When the diffusivity κ is constant, (1) is a linear parabolic equation which can be solved by using the Green function: $\theta(\boldsymbol{r}, t) = (4\pi\kappa t)^{-d/2} \times \int \exp[-(\boldsymbol{r} - \boldsymbol{r}')^2/4\kappa t]\theta(\boldsymbol{r}', 0)\, d\boldsymbol{r}'$.

The diffusivity of gases in gases is of order 10^{-1} cm^{2}s^{-1}, so it would take many hours for an odor to diffuse across the dinner table. Similarly, to diffuse salt to a depth of 1 km in the ocean molecular diffusion would take 10^7 years. It is the motion of fluids that provides large-scale mixing in most cases. In a moving fluid, θ satisfies the advection-diffusion equation:

$$\frac{\partial\theta}{\partial t} + (\boldsymbol{v} \cdot \nabla)\theta = \operatorname{div}(\kappa\nabla\theta). \qquad (2)$$

If the velocity gradient is λ then one can define a diffusion scale $r_{\rm d} = \sqrt{\kappa/\lambda}$ comparing advective and diffusive terms in (2). Fluid motion and molecular diffusion provide for mixing at the scales respectively larger and smaller than r_d. Inhomogeneous flow brings into contact fluid parcels with different values of θ thus producing large gradients that are then eliminated by molecular diffusivity. How fast mixing proceeds and how concentration variance decays in time depends on how inhomogeneous the flow is.

When a velocity field fluctuates, the simplest quantity (and often most important) is the concentration averaged over velocity, $\langle\theta(\boldsymbol{r}, t)\rangle$. The behavior of this quantity is determined by the properties of Lagrangian velocity $\boldsymbol{V}(t) = \boldsymbol{v}[\boldsymbol{q}(t), t]$, which is taken on the trajectory that satisfies $d\boldsymbol{q}/dt = \boldsymbol{v}[\boldsymbol{q}(t), t]$. For times longer than the Lagrangian correlation time, $\langle\theta(\boldsymbol{r}, t)\rangle$ also satisfies the diffusion equation

$$\Big[\partial_t - (\kappa\delta_{ij} + D_{ij})\nabla_i\nabla_j\Big]\langle\theta(\boldsymbol{r}, t)\rangle = 0,$$

with so-called eddy diffusivity

$$D_{ij} = \frac{1}{2}\int_0^\infty \langle V_i(0)V_j(s) + V_j(0)V_i(s)\rangle\, ds.$$

If we release a single spot of, say, a pollutant, then its average position is given by $\langle\theta(\boldsymbol{r}, t)\rangle$. On the other hand, the evolution of the spot itself depends on the spatial properties of the velocity field. In considering hydrodynamic mixing at a given scale, one usually distinguishes between two qualitatively different classes of velocity fields: spatially smooth and nonsmooth. Velocity can be considered spatially smooth on a given scale if the velocity gradient does not change much across the scale. Comparing the inertial term $(\boldsymbol{v} \cdot \nabla)\boldsymbol{v}$ with the viscous term $\nu\Delta\boldsymbol{v}$ in the Navier–Stokes equation for fluid motion, one defines the viscous scale η similarly to $r_{\rm d}$. Turbulent flows are smooth at scales smaller than η (viscous interval) and nonsmooth at larger scales (intertial interval). Fluid particles separate exponentially with time in smooth flows and according to power laws in nonsmooth flows.

Despite the fact that the fluid viscosity ν (momentum diffusivity) is caused by the same molecular motion as κ (diffusivity of a substance), their ratio varies widely depending on the type of material. That ratio is called the Schmidt number, or Prandtl number when θ is temperature. The Schmidt number is very high for viscous liquids and also for colloids and aerosols, since the diffusivity of, say, micron-size particles (e.g., cream globules in milk and smoke in the air) is six to seven orders of magnitude less than the viscosity of the ambient fluid. In those cases, $r_d \ll \eta$. At scales less than η, the flow is spatially smooth, and the velocity difference between two fluid particles can be presented as $\boldsymbol{v}(\boldsymbol{q}_1, t) - \boldsymbol{v}(\boldsymbol{q}_2, t) = \hat{\sigma}(t)\boldsymbol{R}(t)$ so that the separation $\boldsymbol{R} = \boldsymbol{q}_1 - \boldsymbol{q}_2$ obeys the ordinary differential equation

$$\dot{\boldsymbol{R}}(t) = \sigma(t)\,\boldsymbol{R}(t),$$

leading to the linear propagation $\boldsymbol{R}(t) = W(t)\,\boldsymbol{R}(0)$. The main statistical properties of $\boldsymbol{R}(t)$ can be established at the limit when t exceeds the correlation time of the strain matrix $\hat{\sigma}(t)$. The basic idea (going back to the works of Lyapunov, Furstenberg, Oseledec, and many others and developed in the theory of dynamical chaos) is to consider the positive symmetric matrix $W^{\mathrm{T}}W$ which determines R. The main result states that in almost every realization of $\hat{\sigma}(t)$, the matrix $t^{-1}\ln W^{\mathrm{T}}W$ stabilizes as $t \to \infty$. In particular, its eigenvectors tend to d fixed orthonormal eigenvectors $\boldsymbol{f}_i$. To understand that intuitively, consider some fluid volume, say, a sphere, which evolves into an elongated ellipsoid at later times. As time increases, the ellipsoid is more and more elongated, and it is less and less likely that the hierarchy of the ellipsoid axes will change. The limiting eigenvalues

$$\lambda_i = \lim_{t\to\infty} t^{-1}\ln|W\boldsymbol{f}_i| \tag{3}$$

define the so-called Lyapunov exponents. The major property of the Lyapunov exponents is that they do not depend on the starting point if the velocity field is ergodic.

Consider now a pollutant spot with size l released within a spatially smooth velocity and assume that the Peclet number l/r_d is large. The above consideration shows, in particular, that the spot will acquire an ellipsoid form. The direction that corresponds to the lowest Lyapunov exponent (necessarily negative in an incompressible flow) contracts until it reaches r_d, and further contraction is stopped by molecular diffusion. Because the exponentially growing directions continue to expand, the volume grows exponentially and the value of θ inside the spot decays exponentially in time. For an arbitrary large-scale initial distribution of θ, the concentration variance decays exponentially in a spatially smooth flow because this is how fast velocity inhomogeneity contracts θ "feeding" molecular diffusion which eventually decreases the variance. Even though it is diffusion that diminishes θ and the rate of decay is independent of κ, it is usually of order of the typical velocity gradient.

If the Schmidt number is small while the Reynolds number of the flow is large then the velocity field at scales larger than r_d cannot be considered spatially smooth. That means that the velocity difference $\delta v(r)$, measured between two points distance r apart, scales as r^a with $a < 1$ (of course, δv is random and the statement pertains to the moments). For example, for the energy cascade in incompressible fluids, a is close to $1/3$. The equation $\dot{\boldsymbol{R}} = \delta \boldsymbol{v}(\boldsymbol{R}) \propto R^a$ suggests that interparticle distance grows by a power law: $R(t) \propto t^{1/(1-a)}$. The volume of any spot also grows so that scalar variance decays by a power law: $\langle\theta^2\rangle \propto t^{d/(1-a)}$. Such estimates are supported by a rigorous theory only for a velocity field short-correlated in time. In this case, one can also show that the probability distribution $P(\theta, t)$ takes the self-similar form $t^{d/2(1-a)}Q(t^{d/2(1-a)}\theta)$ which is likely to be the case for a general scale-invariant velocity. On the contrary, $P(\theta, t)$ does not change in a self-similar way in a spatially smooth flow.

In finite vessels, the long-time properties of fluid mixing are usually determined by slowest parts, namely, the walls, where the velocity gradient may become zero, and corners with recirculating eddies.

In multiphase flows, not only mixing but also segregation can occur. The physical reason for that is a centrifugal force: when fluid streamlines are curved, heavier particles move out while lighter particles move in. It is a matter of everyday experience that air bubbles are trapped inside the sink vortex while heavy particles gather outside the vortices (which is used, in particular, for flow visualization).

Granular mixing is strikingly different from fluid mixing. In a granular flow, collisions of grains are inelastic and friction between grains makes it possible for static configurations (such as arches) to support a load and distribute stresses. As a result, granular motion has nonlocal properties, and no effective hydrodynamic description based on average over local kinetics (such as Equation (2)) is available. When a container partially filled with grains is vertically shaken with accelerations larger than the gravitational acceleration, convective rolls are observed with grains rising at the center and falling along the walls. Contrary to fluid convection, however, the grains move faster and mix better near the walls. Granular flows can also demonstrate segregation. The most celebrated example is the so-called Brazil nut effect whereby large particles (Brazil nuts) rise to the top of shaken container of mixed nuts.

The use of the term *mixing* in mathematics is based on the notion (introduced by Josiah Gibbs) that evolution is mixing when it leads asymptotically in time to some equilibrium invariant measure. Formally, one defines the evolution operator U_t acting on some phase space A and denotes the measure of any subset B as $P(B)$. The evolution is mixing if for any $B, C \in A$ one has $\lim_{t\to\infty} P(A \bigcap U_t B) = P(A)P(B)$. One can also define weak mixing property where

$\lim_{t\to\infty} t^{-1}\int P(A\bigcap U_s B)\,\mathrm{d}s = P(A)P(B)$. Mixing of a random process or dynamical system means ergodicity, that is equality between temporal and phase space average.

GREGORY FALKOVICH

See also **Chaotic advection; Diffusion; Entropy; Granular materials; Intermittency; Kolmogorov cascade; Lyapunov exponents; Turbulence**

Further Reading

Chate, H., Villermaux, E. & Chomaz, J.-M. (editors). 1999. *Mixing: Chaos and Turbulence*, New York: Kluwer

Falkovich, G., Gawędzki, K. & Vergassola, M. 2001. Particles and fields in fluid turbulence. *Reviews of Modern Physics*, 73: 913–975

MÖBIUS STRIP

See **Topology**

MODE LOCKING

See **Coupled oscillators**

MODE MIXING AND COUPLING

See **Coupled systems of partial differential equations**

MODIFIED KDV EQUATION

See **Korteweg–de Vries equation**

MODULATED WAVES

The term *modulated waves* implies waves in which some parameters vary slowly in time or in space–time compared with the main (carrier) variables. The term *modulation* comes from acoustics (music) and radio engineering, where signals of the type

$$A(t)\cos[\omega_0 t + \varphi(t)] \quad (1)$$

have long been used for transmitting and processing information. Here, the carrier frequency ω_0 is constant with respect to time t, whereas the amplitude A and phase φ are slowly varying functions of time as compared with variations of the carrier oscillations described by the function $\cos(\omega_0 t)$. If only one of the parameters A and φ is variable, the terms *amplitude modulation* or *phase modulation* are used. A more general form is $A(t)\cos[\Phi(t)]$, where Φ is the full phase, so that the frequency $\omega(t) = \mathrm{d}\Phi/\mathrm{d}t$ is slowly varying. Although this case is called frequency modulation, the terms *phase* and *frequency modulation* describe essentially the same process.

For spatially varying signals, a natural generalization of that is a modulated quasiharmonic wave

$$A(\mathbf{r},t)\cos[\omega_0 t + \varphi(\mathbf{r},t)] \text{ or } A(\mathbf{r},t)\cos[\Phi(\mathbf{r},t)], \quad (2)$$

where A, φ, and $\partial\Phi/\partial t$ are slowly varying functions of time. In the spectral representation, the first of these expressions yields a narrow frequency spectrum as compared with the carrier frequency ω_0, whereas the spectrum of the second expression can be arbitrarily wide. The cause of wave modulation can be different: modulation in initial or boundary (signaling) conditions, slow variation of the medium parameters in time and space, attenuation, and amplification.

A typical case is a one-dimensional traveling wave when functions (2) depend on time t and one spatial coordinate x, for example,

$$A(x,t)\cos[\omega_0 t - k_0 x + \varphi(x,t)], \quad (3)$$

where k_0 is the wave number. At constant A and φ, this wave is a sinusoid propagating with phase velocity $c_{\mathrm{ph}} = \omega_0/k_0$. If the envelope A and phase φ (sometimes called phase envelope, implying all slowly varying parameters of a modulated wave propagate as envelope waves) are slowly varying in space time, and the medium is linear, then up to some distance, they propagate as a wave, with the group velocity, $c_{\mathrm{gr}} = \mathrm{d}\omega_0/\mathrm{d}k_0$, where the dependence between ω_0 and k_0 (dispersion equation) is determined by the properties of the medium. Phase and group velocities are equal only in a nondispersive media, otherwise they are different. At even larger times in a dispersive medium, the wave envelopes are deformed. In particular, a finite-length impulse broadens and turns into a frequency-modulated wave in which each group propagates at its local group velocity depending on its frequency. Such dynamics can be represented by trajectories of groups on the (x,t) plane which are somewhat analogous to spatial rays in geometrical optics or geometrical acoustics and are called space-time rays.

In nonlinear dispersive media, the propagation of envelopes of a modulated dispersive wave is determined not only by its frequency but also by its amplitude (intensity). The interplay of these two factors—dispersion and nonlinearity—results in a variety of scenarios (Ostrovsky & Potapov, 1999; Scott, 1999; Whitham, 1974).

Under definite conditions, first, a non-modulated quasiharmonic wave can be unstable with respect to small modulating perturbations of its amplitude and phase (modulational or Benjamin–Feir instability). In other cases, modulation does not grow, but the envelope distorts with formation of steep (at a modulation scale) fronts (self-steepening). Finally, a steady propagation of envelopes is possible in the form of envelope solitons and envelope shocks.

The notion of modulation can be extended to significantly nonharmonic processes, such as cnoidal

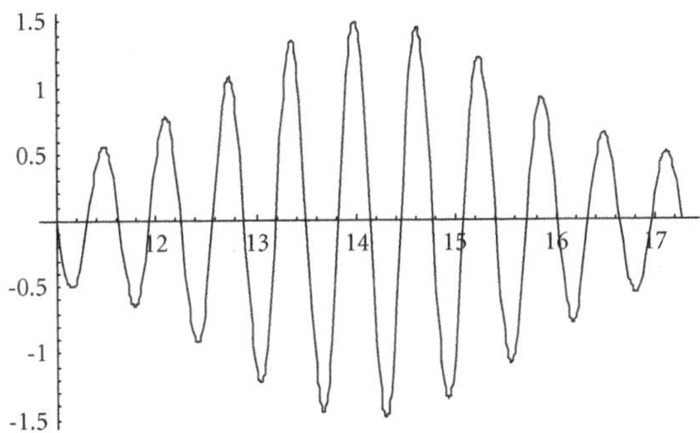

Figure 1. An example of a quasi-harmonic modulated wave.

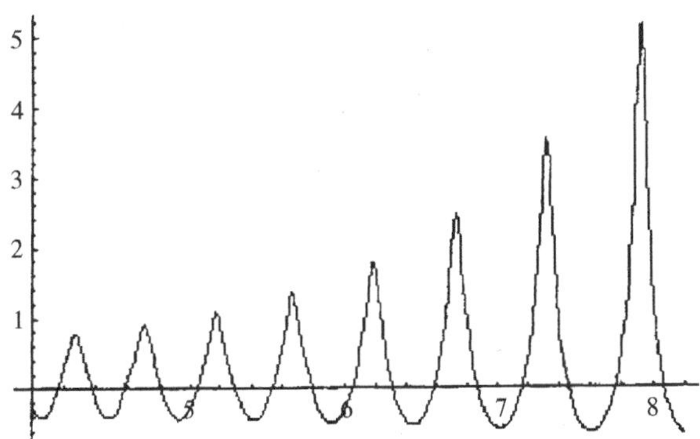

Figure 2. An example of a nonsinusoidal modulated wave.

waves which are typical elliptic function solutions of nonlinear wave equations. Also, in these solutions, slow modulation of their parameters (amplitude and period) in space and time can occur. Let the nonmodulated wave have the form $F(\omega t - kx, A_i)$, where F is a periodic function, the A_i are parameters defining the solution (such as the wave amplitude), and ω and k are effective frequency and wave number defined via the wave period T and its wavelength, Λ, as $\omega = 2\pi/T$ and $k = 2\pi/\Lambda$. Then at slow modulation of these parameters, the wave has the form

$$F[\theta(x,t), A_i(x,t)], \tag{4}$$

where F is a periodic function, A_i are slowly varying functions of x and t, and the phase θ is defined in such a way that $\omega = \partial\theta/\partial t$ and $k = -\partial\theta/\partial x$ are also slowly varying functions. Due to these slow variations, the wave profile can change continuously from an almost sinusoidal wave up to a train of pulses similar to solitary waves or solitons, as the modulus of the elliptic function increases.

Mathematical descriptions of modulated waves are based on asymptotic perturbation theory (Ostrovsky & Gorshkov, 2000) or on more heuristic descriptions. A rather general approach (Whitham's method) starts with a Lagrangian description of a wave field and employs the corresponding variational principle (Whitham, 1974; Ostrovsky & Potapov, 1999; Scott, 1999). For modulated quasi-periodic waves, substituting an expression of type (4) into the Lagrangian (L) and averaging over θ, one obtains an averaged Lagrangian $\mathcal{L}$ depending only on slowly varying parameters, A_i, ω, and k. Considering θ and A_i as new canonical variables, one then obtains from $\mathcal{L}$ variational equations describing the relation among A_i, ω, and k, and their variations in space and time.

This and other averaging methods can work for both nonharmonic nonlinear waves and for weakly nonlinear (or just linear) waves (2) or (3). For narrow-spectrum waves such as (3), a more detailed description can be developed that goes beyond the framework of space-time geometrical optics and takes into account time analogs of wave diffraction. In particular, the nonlinear Shrödinger equation has been obtained to describe such processes. This approach is useful, for example, for the study of envelope solitons having applications in nonlinear optics and for deep water waves.

Moreover, the evolution of a single soliton with slowly varying parameters (due to small losses or smooth inhomogeneities) can be represented in the form of Equation (4), where F is a localized function rather than a periodic one (Ostrovsky & Gorshkov, 2000; Ostrovsky & Potapov, 1999). This class of processes can also be referred to as modulation, and perturbation methods for slowly varying solitary waves have been constructed. Thus, such processes as damping, propagation in inhomogeneous media, and solitary-wave interactions have been analyzed. In the latter case, solitary waves can interact as particles (hence the term *soliton*) with different types of interactions, including repulsion when solitons retain their parameters after interaction and attraction when they may form an oscillating bound state or breather.

Finally, the term *modulation* can be extended to a wave that does not have a prescribed shape but a continuously but slowly distorting profile. A typical example is weakly nonlinear waves in nondispersive media—such as acoustic waves in fluids—where a cumulative steepening of a wave profile can occur up to the formation of weakly nonlinear shock waves. Provided the distortion is small at a wavelength scale, one can also refer to wave profile modulation.

LEV OSTROVSKY

See also **Averaging methods; Collective coordinates; Nonlinear acoustics; Nonlinear optics; Wave stability and instability**

Further Reading

Ostrovsky, L.A. & Gorshkov, K.A. 2000. Perturbation theories for nonlinear waves. In *Nonlinear Science at the Dawn of the 21st Century*, edited by P.L. Christiansen, M.P. Søerensen, & A.C. Scott, New York: Springer, pp. 47–65

Ostrovsky, L.A. & Potapov, A.I. 1999. *Modulated Waves: Theory and Applications*, Baltimore: John Hopkins University Press

Scott, A.C. 1999. *Nonlinear Science: Emergence and Dynamics of Coherent Structures*, Oxford and New York: Oxford University Press

Whitham, G.B. 1974. *Linear and Nonlinear Waves*, New York: Wiley

MODULATIONAL INSTABILITY

See **Wave stability and instability**

MOLECULAR CRYSTALS

See **Local modes in molecular crystals**

MOLECULAR DYNAMICS

Molecular dynamics (MD) refers to a family of computational methods aimed at simulating macroscopic behavior (thermodynamic or hydrodynamic) through the numerical integration of the classical equations of motion of a microscopic many-body system. Macroscopic properties are expressed as functions of particle coordinates and/or momenta, which are computed along a phase space trajectory generated by classical dynamics. At the core of the method is the assumption that over long times the system will reach an equilibrium state, or alternatively a steady state when simulating hydrodynamic conditions, in which temporal averages can be identified with statistical ensemble averages. This requires that mixing and Lyapunov instability be present in the phase space trajectories.

The field of MD simulations was born in the 1950s out of computational studies of the approach to equilibrium in a hard sphere system by B.J. Alder and T.E. Wainwright in 1957 (see Ciccotti et al. (1987) for a collection of seminal works illustrating the development and application of MD simulations). The first MD simulation employing continuous potentials was used to study radiation damage in a two-dimensional solid by Gibson et al. (1960). The first application of the methodology as it is used today, to an atomic fluid modeled by continuous potentials in three dimensions, was a study of structural and dynamical aspects of liquid argon by A. Rahman in 1964. Subsequently, MD simulations have been applied to the study of complex systems such as biomacromolecules and self-assembled systems, and to the elucidation of complex spatiotemporal phenomena in more simple systems. When performed under conditions corresponding to laboratory scenarios, molecular dynamics simulations can provide a detailed view of structure and dynamics at the atomic and mesoscopic levels that is not presently accessible by experimental measurements. The simulations can also be set to perform "computer experiments" that could not be carried out in the laboratory, either because they do not represent natural behavior or because the necessary controls cannot be achieved. Lastly, MD simulations can be implemented as simple test systems for condensed matter theory. In order to realize this wide spectrum of applications, several issues concerning system modeling, dynamical generation of statistical ensembles, and efficient numerical integrators must be considered.

To set the stage, consider the evolution of an isolated system of N point particles. In three dimensions, the canonical description of such system is given by a set of $6N$ first-order differential equations for the particles' positions $\boldsymbol{r}(t)=\{\boldsymbol{r}_1\ldots\boldsymbol{r}_N\}$ and momenta $\boldsymbol{p}(t)=\{\boldsymbol{p}_1\ldots\boldsymbol{p}_N\}$. The dynamics of the ith particle can be written in Hamiltonian form as

$$\dot{\boldsymbol{r}}_i = \frac{\partial\mathcal{H}}{\partial\boldsymbol{p}_i} = \frac{\boldsymbol{p}_i}{m},$$
$$\dot{\boldsymbol{p}}_i = -\frac{\partial\mathcal{H}}{\partial\boldsymbol{r}_i} = -\frac{\partial V(\boldsymbol{r})}{\partial\boldsymbol{r}_i} \qquad (1)$$

with the Hamiltonian given by

$$\mathcal{H}(\boldsymbol{r},\boldsymbol{p}) = \sum_{i=1}^{N}\frac{\boldsymbol{p}_i^2}{2m_i} + V(\boldsymbol{r}), \qquad (2)$$

where $V(\boldsymbol{r})$ is the potential. The phase space trajectory generated is constrained to a hypershell parametrized by the conserved total energy of the system $\mathcal{H}(\boldsymbol{r},\boldsymbol{p})=E$. Microcanonical ensemble averages are constructed by means of the corresponding probability density function. For any observable A, either explicitly time-dependent or independent, the microcanonical ensemble average is given by

$$\langle A\rangle = \frac{\int \mathrm{d}\mu\, A(\boldsymbol{r}(t),\boldsymbol{p}(t))\,\delta(\mathcal{H}-E)}{\int \mathrm{d}\mu\,\delta(\mathcal{H}-E)}, \qquad (3)$$

where $\mathrm{d}\mu$ is the corresponding invariant measure, in this case $\mathrm{d}\mu=\mathrm{d}\boldsymbol{p}\,\mathrm{d}\boldsymbol{r}$. The chaotic nature of the phase space trajectory, in the sense indicated previously, allows the identification of (3) with the temporal average over an equilibrated simulation of time-length T:

$$\bar{A} = \frac{1}{T}\int \mathrm{d}t A(\boldsymbol{r}(t),\boldsymbol{p}(t)) = \langle A\rangle. \qquad (4)$$

MD simulations use information (positions, velocities or momenta, and forces) at a given instant in time, t, to predict the positions and momenta at a later time, $t+\Delta t$, where Δt is the time step, usually taken to be constant throughout the simulation. Numerical solutions to the equations of motion are thus obtained by iteration of this elementary step. The most popular algorithms for propagating the equation of motion, or "integrators," are based on Taylor expansions of the positions (see Allen & Tildesley (1989) for a survey). For example, the Verlet algorithm uses the sum of third order expansions forward and backward in time of the particle positions, $\boldsymbol{r}_i$,

$$\boldsymbol{r}_i(t+\Delta t) = \boldsymbol{r}_i(t) + \Delta t \boldsymbol{v}_i(t) + \frac{\Delta t^2}{2m_i}\boldsymbol{F}_i(t) + \frac{\Delta t^3}{6}\boldsymbol{b}_i(t) + \mathcal{O}(\Delta t^4), \quad (5)$$

$$\boldsymbol{r}_i(t-\Delta t) = \boldsymbol{r}_i(t) - \Delta t \boldsymbol{v}_i(t) + \frac{\Delta t^2}{2m_i}\boldsymbol{F}_i(t) - \frac{\Delta t^3}{6}\boldsymbol{b}_i(t) + \mathcal{O}(\Delta t^4), \quad (6)$$

where $\boldsymbol{v}_i$ is the velocity, $\boldsymbol{F}_i$ is the force, and $\boldsymbol{b}_i$ is the third derivative with respect to time, to obtain an equation for the predicted position that is accurate to third order in time (because of the cancellation of the terms containing odd powers of time):

$$\boldsymbol{r}_i(t+\Delta t) = 2\boldsymbol{r}_i(t) - \boldsymbol{r}_i(t-\Delta t) + \frac{\Delta t^2}{m_i}\boldsymbol{F}_i(t) + \mathcal{O}(\Delta t^4). \quad (7)$$

An equation for the velocities is obtained by subtracting the two expansions,

$$\boldsymbol{v}_i(t) = \frac{\boldsymbol{r}_i(t+\Delta t) - \boldsymbol{r}_i(t-\Delta t)}{2\Delta t} + \mathcal{O}(\Delta t^3). \quad (8)$$

Note that, according to Equation (7), the velocities are not necessary for prediction of the positions. Because the velocities are determined at time t while the positions are determined at time $t+\Delta t$, calculation of velocity-dependent quantities (e.g., the kinetic energy) coincident with the predicted positions is awkward. The Verlet algorithm can be modified to synchronize the prediction of the positions and velocities. One popular modification, the "velocity Verlet" algorithm, is described below.

A general approach to the development of MD integrators has been formulated based on the Liouville operator formalism of Hamiltonian mechanics, in which Hamilton's equations of motion (Equations (1)) are recast as

$$\dot{\Gamma} = \mathrm{i}L\Gamma. \quad (9)$$

Here, Γ is a $6N$-dimensional phase vector of the N particle coordinates and momenta (or velocities), and $\mathrm{i}L$ is the Liouville operator

$$\mathrm{i}L = \dot{\Gamma}\cdot\nabla_\Gamma = \sum_{i=1}^{N} \boldsymbol{v}_i\cdot\nabla_{\boldsymbol{r}_i} + \sum_{i=1}^{N}\left[\frac{\boldsymbol{F}_i(\boldsymbol{r})}{m_i}\right]\cdot\nabla_{\boldsymbol{v}_i}. \quad (10)$$

The formal solution to Equation (10) is

$$\Gamma(t) = \exp(\mathrm{i}Lt)\Gamma(0). \quad (11)$$

In practice, this is not useful because the action of the propagator, $\exp(\mathrm{i}Lt)$, on the phase vector cannot be determined analytically. However, it is possible to derive accurate short-time approximations to the propagator, whose action can be evaluated analytically, by factorizing the propagator. To this end, Equation (11) is rewritten so that the trajectory from 0 to t is generated in a sequence of P discrete time steps, $\Delta t = t/P$:

$$\Gamma(t) = \prod_{j=1}^{P}\left(\prod_{s}^{n}\exp(\mathrm{i}L_s\Delta t)\right)\Gamma(0). \quad (12)$$

Then the Trotter factorization may be used to decompose the propagator (Tuckerman & Martyna, 2000):

$$\exp(\mathrm{i}L\Delta t) = \exp\left(\mathrm{i}L_1\frac{\Delta t}{2}\right)\exp(\mathrm{i}L_2\Delta t)\exp\left(\mathrm{i}L_1\frac{\Delta t}{2}\right) + \mathcal{O}(\Delta t^3) \quad (13)$$

giving an approximate, short-time evolution operator that is time-reversible and has the same accuracy as typical MD integrators. The operators $\mathrm{i}L_1$ and $\mathrm{i}L_2$ are chosen so that their action on the phase vector can be determined analytically, with the restriction that $\mathrm{i}L_1 + \mathrm{i}L_2 = \mathrm{i}L$. For example, applying (13) to the phase vector, $\Gamma = (\boldsymbol{r}, \boldsymbol{v})$ (written here and in what follows in terms of velocities), with

$$\mathrm{i}L_1 = \sum_{i=1}^{N}\left[\frac{\boldsymbol{F}_i(\boldsymbol{r})}{m_i}\right]\cdot\nabla_{\boldsymbol{v}_i}, \quad (14)$$

$$\mathrm{i}L_2 = \sum_{i=1}^{N}\boldsymbol{v}_i\cdot\nabla_{\boldsymbol{r}_i} \quad (15)$$

gives the velocity Verlet integrator:

$$\boldsymbol{r}_i(t+\Delta t) = \boldsymbol{r}_i(t) + \Delta t\boldsymbol{v}_i(t) + \frac{\Delta t^2}{2m_i}\boldsymbol{F}_i(\boldsymbol{r}(t)), \quad (16)$$

$$\boldsymbol{v}_i(t+\Delta t) = \boldsymbol{v}_i(t) + \frac{\Delta t}{2m_i}[\boldsymbol{F}_i(\boldsymbol{r}(t)) + \boldsymbol{F}_i(\boldsymbol{r}(t+\Delta t))]. \quad (17)$$

In practice, computing meaningful time averages according to (4) requires sufficient sampling of an already equilibrated trajectory. This determines the total simulation time T. Once a model system and a dynamical system (i.e., an ensemble average) have been chosen, computational efficiency will rely on the choice of the numerical integrator. The total real computational time will depend on how many steps are required to reach the total time-length T. The choice of the time step size depends on the rate of accumulation of numerical errors as evaluated by the tolerance ΔE for the conserved quantity E. In molecular systems, the time step is determined by the highest frequency vibrational mode in the system. A rule of the thumb, is that, for numerical stability, one vibrational period should be covered by roughly 20 time steps. A popular approach to increasing the time step in MD simulations of molecular systems is to use holonomic constraints to freeze the motion of the highest frequency bonds, or all of the bonds altogether. This is considered acceptable practice for most applications because the bond stretching motion is effectively decoupled from the other degrees of freedom that are typically of greater interest. To solve the equations of motion in the presence of constraints, the Lagrangian formalism of classical mechanics is used, and the constraint forces are expressed in terms of undetermined multipliers. There are a variety of algorithms for solving the resulting constraint equations and integrating the equations of motion (see Allen & Tildesley, 1989). The details depend on the choice of the integrator. The most popular are iterative methods known as SHAKE, designed to work with the Verlet integrator, and RATTLE, designed to work with the velocity Verlet integrator.

The utility of the constraint method rests on the fact that the fastest degrees of freedom place an upper bound on the time step, and the time step can therefore be increased if the fastest degrees of freedom are eliminated (constrained). A more general approach is based on the premise that the forces associated with different degrees of freedom evolve on different time scales, and the slower forces do not need to be computed as often as the faster forces. It turns out, for molecular systems, that the fastest forces, those associated with intramolecular interactions, require the least computational effort to evaluate ($\mathcal{O}(N)$), while the slowest forces, those associated with nonbonded interactions, require the most computational effort ($\mathcal{O}(N^2)$). Thus, in the interest of computational efficiency, it is clearly advantageous to calculate the rapidly evolving forces at each elementary time step, and only calculate the slowly evolving forces at a larger time interval that is several times the elementary time step. This is the essence of multiple time step MD (see Tuckerman & Martyna (2000) for a review), which is briefly summarized below.

Suppose that the total force on the particle i may be written

$$\boldsymbol{F}_i = \boldsymbol{F}_i^{\text{fast}} + \boldsymbol{F}_i^{\text{slow}}, \tag{18}$$

where $\boldsymbol{F}_i^{\text{fast}}$ and $\boldsymbol{F}_i^{\text{slow}}$ are the parts of the force that change rapidly and slowly, respectively. Now, according to Equation (13), we may write

$$\begin{aligned}\boldsymbol{\Gamma}(t+\Delta t) &= \{\exp(\mathrm{i}L_3\Delta t/2)[\exp(\mathrm{i}L_1\delta t/2)\exp(\mathrm{i}L_2\delta t) \\ &\quad \times \exp(\mathrm{i}L_1\delta t/2)]^n \exp(\mathrm{i}L_3\Delta t/2)\}\boldsymbol{x}(t) + \mathcal{O}(\Delta t^3),\end{aligned} \tag{19}$$

where $\mathrm{i}L_2$ is as before and

$$\mathrm{i}L_1 = \sum_{i=1}^{N}\left[\frac{\boldsymbol{F}_i^{\text{fast}}(\boldsymbol{r})}{m_i}\right]\cdot\nabla_{v_i}, \tag{20}$$

$$\mathrm{i}L_3 = \sum_{i=1}^{N}\left[\frac{\boldsymbol{F}_i(\boldsymbol{r}) - \boldsymbol{F}_i^{\text{fast}}(\boldsymbol{r})}{m_i}\right]\cdot\nabla_{v_i} \tag{21}$$

with $\delta t = \Delta t/n$. Thus, to evolve the system from time t to $t + \Delta t$ in one step of an MD simulation, the total force needs to be evaluated only once, and the fast component of the force n times. The small time step, δt, is determined by the time scale of the fast forces. As the computational effort required to evaluate the fast component of the force is negligible compared with that required for the slow force, this scheme results in nearly a factor of n decrease in computational effort. The resulting algorithm is straightforward to implement because each action of the evolution operators translates directly into velocity Verlet-like integration steps (Tuckerman & Martyna, 2000).

The majority of microscopic systems modeled by MD simulations are collections of atoms or molecules in a condensed phase. The usual quantum mechanical treatment of molecular systems assumes that the problem is separable between electronic and nuclear (or ion-core) degrees of freedom. Within this context, the most common approach in MD simulations is to consider the classical limit for the nuclear degrees of freedom. Thus, the phase space trajectory describes the dynamics of the atomic nuclei as classical point particles, while the electronic degrees of freedom are accounted for by the interaction potential $V(\boldsymbol{r})$. There is not a single prescription for choosing the form of the potential energy function and determining its parameters. In general, analytical forms for model systems are taken from molecular mechanics, and the specific parameters are dependent on the kind of system that is being modeled. For simple systems, the available information from zero-temperature electronic structure calculations is directly fitted to a convenient analytical function. For complex systems, electronic structure calculations for small molecular moieties are complemented with relevant experimental data pertinent to the phase being modeled. For a detailed overview of commonly used empirical potentials, see Leach (2001).

Although the number and nature of the potential energy terms varies from one application to the next, the potential for an arbitrarily complicated molecule such as a polymer is generally written as the sum of "bonded" and "nonbonded" terms. The bonded terms include energy penalties, usually represented by harmonic potentials, for deforming chemical bonds and the angles between bonds from their equilibrium values, as well as periodic potentials to describe the energy change as a function of the torsion (dihedral) angle about rotatable bonds. The bonded terms are usually taken to be diagonal, that is, there are no terms describing coupling between deformations of bonds and angles, bonds and torsions, etc. These off-diagonal terms are important for accurate reproduction of vibrational properties, as are anharmonic terms (e.g., cubic, quartic). The nonbonded terms, as the name suggests, describe the interactions between atoms in different molecules, or interactions within a molecule that are not completely accounted for by the bonded terms; that is, they are separated by more than two bonds. The nonbonded terms are typically assumed to be pairwise-additive functions of interatomic separation, r, and include van der Waals interactions and Coulomb interactions in polar or charged molecules in which the charge distribution is represented by partial charges (usually placed on the atoms). The van der Waals interactions include a term that is strongly repulsive (exponential or inverse 12th power of r), and a weakly attractive dispersion term (e.g., inverse 6th power of r, representing the induced dipole-induced dipole interaction). A popular form for the van der Waals interactions is the Lennard-Jones ("12-6") potential:

$$V(r) = 4\varepsilon\left(\left(\frac{\sigma}{r}\right)^{12} - \left(\frac{\sigma}{r}\right)^{6}\right) \tag{22}$$

Here, ε is the depth of the minimum, and σ is where the potential crosses zero and is a measure of the van der Waals diameter of the atom.

The number of bonded interactions grows in proportion to N, the number of atoms in the system, while the number of nonbonded interactions grows as N^2. Therefore, most of the computer time used for an MD simulation is spent on the evaluation of the nonbonded energies and forces. To reduce this burden, van der Waals interactions are typically ignored beyond some "cutoff" distance, usually around two or three atomic diameters. However, because the Coulomb potential is long-ranged, it is dangerous to truncate the electrostatic interactions. In condensed phase systems modeled using periodic boundary conditions (to eliminate surface effects in simulations of small systems that are meant to be in a bulk environment), the Ewald method can be used to sum all of the electrostatic interactions in an infinite, periodically replicated system. Details on the implementation of the Ewald sum may be found in Allen & Tildesley (1989).

Although MD simulations are most conveniently carried out in the microcanonical ensemble, there are disadvantages to doing so. An important aspect of performing simulation studies is to calculate experimentally measurable properties for comparison with measurements made at particular thermodynamic state points. Most experiments are carried out under the conditions of constant N and temperature, and either constant pressure or constant volume, that is, in the isobaric-isothermal or canonical statistical mechanical ensembles. In microcanonical simulations, the temperature and pressure are computed via averages of the kinetic energy and virial, respectively, and it is not known until after the simulation has been completed whether it was run at the desired state point. Moreover, in the canonical and isobaric-isothermal ensembles, the characteristic thermodynamic potentials are the Helmholtz free energy, $F(N, V, T)$, and Gibbs free energy, $G(N, P, T)$, respectively, and these free energies are easier to calculate from MD trajectories (see Frenkel & Smit (1996) for a discussion of methods for free energy calculations) than the entropy, $S(N, V, E)$, which is the characteristic thermodynamic potential of the microcanonical ensemble. In addition, constant pressure simulations are especially useful when the density of the system is not known a priori.

Several techniques for generating trajectories in ensembles other than the microcanonical ensemble have been developed (see Allen & Tildesley (1989) for an overview and Ciccotti et al. (1987) for a collection of key papers). The most powerful techniques are based on the concept of an "extended system" (ES), in which the atomic positions and momenta are supplemented by additional dynamical variables representing the coupling of the system to an external reservoir. For example, in the first application of the extended system concept to a simulation at constant pressure by H.C. Andersen in 1980, the system volume was taken to be a dynamical variable to simulate coupling to a pressure reservoir. In the Nosé implementation of constant temperature MD, a time-scaling "thermostat" variable was introduced. M. Parrinello and A. Rahman (1981) generalized Andersen's constant pressure method to allow the simulation cell to change shape as well as size by making the elements of the cell matrix dynamical variables. In these initial formulations of the ES concept, the additional dynamical variables were assigned fictitious kinetic and potential energies, and their equations of motion were derived using the Lagrangian formalism. Note that the choice of the form of the equations of motion for the additional degrees of freedom is not arbitrary, but rather should be made so that the resulting phase space distribution function is that of the desired ensemble. The Lagrangian formalism is not always the most convenient for handling extended systems. A more general approach is based on the formalism of non-Hamiltonian dynamics (Tuckerman & Martyna, 2000), which is sketched below. This approach was pioneered by W.G. Hoover, who rewrote the equations of motion for the Nosé thermostat in non-Hamiltonian form, leading to a much more straightforward implementation of constant temperature MD that was subsequently referred to as the Nosé–Hoover thermostat.

In the non-Hamiltonian dynamics approach, the equations of motion for the ES variables and a conserved Hamiltonian-like quantity, H', which includes terms for the kinetic and potential energies of the ES degrees of freedom, are chosen together such that the microcanonical distribution function generated by the extended system dynamics is formally equivalent (within a normalization constant) to the desired phase space distribution function of the atoms (e.g., canonical, isobaric-isothermal). The "non-Hamiltonian" resides in the fact that the necessary equations of motion cannot be derived from H', as in the case of Hamilton's equations (see Equations (1)). A complication that arises for non-Hamiltonian systems is that the phase space compressibility,

$$\kappa = \nabla_\Gamma \cdot \dot{\mathbf{\Gamma}}, \tag{23}$$

where Γ is now the $6N+2M$ dimensional vector of coordinates and momenta (or velocities) of the N particles and M extended system variables, does not vanish as it does for Hamiltonian systems. Consequently, a metric factor must be included in the phase space measure,

$$\mathrm{d}\mu = \sqrt{g(\mathbf{\Gamma}, t)}\mathrm{d}\mathbf{\Gamma}, \tag{24}$$

where

$$\sqrt{g(\mathbf{\Gamma}, t)} = \exp[-w(\mathbf{\Gamma}, t)], \tag{25}$$

$$\frac{\mathrm{d}w}{\mathrm{d}t} = \kappa, \tag{26}$$

so that the extended system phase space volume is conserved (i.e., Liouville's theorem is obeyed). Reversible, multiple time step integrators for extended system (Martyna et al., 1996) and non-equilibrium (Mundy et al., 2000) dynamics have been derived based on non-Hamiltonian mechanics using the Liouville operator formalism illustrated above for microcanonical dynamics.

In extended system MD simulations with temperature and pressure control, the instantaneous temperature (kinetic energy) and pressure (virial) fluctuate, and it is their average values that are equal to the values imposed by the thermal and pressure reservoirs. There are two other commonly employed approaches to temperature and pressure control (see Allen & Tildesley (1989) for a survey). In one class of methods, the instantaneous temperature and pressure are constrained to constant values. For large systems, the imposition of constraints is reasonable because fluctuations are small. By introducing Lagrange multipliers and applying Gauss's principle of least constraint, the constraints are applied in such a way so as to minimize the change in dynamics. The resulting phase space distribution functions are equal to the desired ones, multiplied by delta functions that account for the constraints on the temperature and pressure. Another technique is referred to as "coupling to an external bath," where the "bath" can be a temperature and/or pressure reservoir. This scheme drives the system to the desired temperature (pressure) at a rate determined by a preset temperature (pressure) relaxation time by scaling the velocities (volume and coordinates) at each time step. This method has the advantage that it is easily implemented, and the disadvantage that it does not generate states in any known statistical mechanical ensemble.

To date, most MD simulations have been driven by forces that were derived from assumed, empirical potentials. About two decades ago, the scope of MD simulations was greatly expanded by the introduction of a method by Car & Parrinello (1985) that, rather than relying on an empirical potential, utilizes interatomic potentials computed directly from the electronic structure which evolves simultaneously with the nuclear configuration. Thus, changes in electronic polarization and the possibility of chemical transformations are naturally included in the simulation. To this end, the wave function is expanded in a basis set, the expansion coefficients are treated as dynamical variables in an extended system Lagrangian, and the resulting equations of motion adiabatically propagate the electronic orbitals with respect to the nuclei, so that they remain on the ground state Born–Oppenheimer surface at each time step. Consequently, the need to solve the computationally burdensome electronic structure problem (self-consistent field) is avoided, and it is feasible to generate trajectories of sufficient length to study phenomena that were previously impossible to simulate. For technical details on the implementation of this so-called "ab initio MD" method, see Marx & Hütter (2000). Although the size of system (on the order of 100 atoms) and the length of trajectory (roughly tens of picoseconds) that can be simulated presently by ab initio MD are modest compared with simulations based on empirical potentials, it is clear that, as computational resources continue to evolve, applications of ab initio MD will become more and more prevalent and compelling.

DOUGLAS J. TOBIAS AND J. ALFREDO FREITES

See also **Hamiltonian systems; Local modes in molecular crystals; Lorentz gas; Newton's laws of motion; Protein dynamics**

Further Reading

Allen, M.P. & Tildesley, D.J. 1989. *Computer Simulation of Liquids*, Oxford: Oxford Science Publications

Car, R. & Parrinello, M. 1985. Unified approach for molecular dynamics and density-functional theory. *Physical Review Letters*, 55: 2471–2474

Ciccotti, G., Frenkel, D. & McDonald, I.R. (editors). 1987. *Simulation of Liquids and Solids: Molecular Dynamics and Monte Carlo Methods in Statistical Mechanics*, Amsterdam: North-Holland

Frenkel, D. & Smit, B. 1996. *Understanding Molecular Simulation: From Algorithms to Applications*, San Diego: Academic Press

Gibson, J.B., Golan, A.N., Milgram, M. & Vineyard, G.H. 1960. Dynamics of radiation damage. *Physical Review*, 120: 1229–1253

Leach, A.R. 2001. *Molecular Modelling: Principles and Applications*, 2nd edition, Harlow: Prentice-Hall

Martyna, G.J., Tuckerman, M.E., Tobias, D.J. & Klein, M.L. 1996. Explicit reversible integrators for

extended systems dynamics. *Molecular Physics*, 87: 1117–1137
Marx, D. & Hütter, J. 2000. *Ab initio* molecular dynamics: theory and implementation. In *Modern Methods and Algorithms of Quantum Chemistry*, edited by J. Grotendorst, Jülich: Forschungszentrum Jülich
Mundy, C.J., Balasubramanian, B., Bagchi, K., Tuckerman, M.E., Martyna, G.J. & Klein, M.L. 2000. Nonequilibrium molecular dynamics. *Reviews in Computational Chemistry*, 14: 291–397
Parrinello, M. & Rahman, A. 1981. Polymorphic phase transitions in single crystals: a new molecular dynamics method. *Journal of Applied Physics*, 52: 7182–7190
Tuckerman, M.E. & Martyna, G.J. 2000. Understanding modern molecular dynamics: techniques and applications. *Journal of Physical Chemistry* B, 104: 159–178

MONODROMY PRESERVING DEFORMATIONS

Suppose $z \to A(z)$ is a $p \times p$ matrix valued meromorphic function (analytic except for a finite number of poles) on the Riemann sphere (also called a "complex projective one space" and denoted as $\boldsymbol{P}^1$) with poles at the points a_ν, $\nu = 1, \ldots, n$. Suppose Ψ is a fundamental solution for the linear system,

$$\frac{\mathrm{d}\Psi}{\mathrm{d}z} = A(z)\Psi, \tag{1}$$

defined in a neighborhood of a regular point, a_0. Analytic continuation of the fundamental solution along a closed curve γ which begins and ends at a_0 produces a change,

$$\Psi \to \Psi M_\gamma,$$

where M_γ is the *monodromy* associated with the path γ and the fundamental solution Ψ. The constant matrix M_γ depends only on the homotopy type of γ, and the map $\gamma \to M_\gamma^{-1}$ is a representation of the fundamental group $\pi_1\left(\boldsymbol{P}^1\backslash\{a_\nu\}, a_0\right)$ called the monodromy representation of the differential equation.

In the 1850s, Bernhard Riemann solved the connection problem for solutions to the hypergeometric equation by exploiting the monodromy representation. The hypergeometric equation is the family of second-order equations on $\boldsymbol{P}^1$ with three regular singular points, and the connection problem is to understand how the local series solutions defined near the singularities are related to one another by analytic continuation. Riemann made some further speculations about monodromy representations, and (inspired by Riemann's work) David Hilbert asked if any representation of $\pi_1\left(\boldsymbol{P}^1\backslash\{a_\nu\}, a_0\right)$ might be realized as the monodromy representation of a suitable differential equation. This is the 21st in the list of problems presented by Hilbert at the International Congress of Mathematicians in 1900. It always has a solution for differential equations with regular singular points (Plemelj, 1963), but it does not always have a solution if the differential equation is required to have simple poles. The book by Anasov and Bolibruch (1994) has a careful discussion of the history of this problem and its solution.

Suppose that the matrix $A(z)$ has simple poles,

$$A(z) = \sum_{\nu=1}^{n} \frac{A_\nu}{z - a_\nu}, \text{ with } \sum_{\nu=1}^{n} A_\nu = 0. \tag{2}$$

The second condition guarantees that ∞ is either a simple pole or a regular point. For simplicity, we also suppose that for each ν no pair of eigenvalues for A_ν differ by an integer. In this *nonresonant* case, the differential equation (1) is holomorphically equivalent to the differential equation,

$$\frac{\mathrm{d}\Phi}{\mathrm{d}z} = \frac{A_\nu}{z - a_\nu}\Phi, \tag{3}$$

in a neighborhood of $z = a_\nu$. In 1912, Ludwig Schlesinger asked how the residue matrices A_ν should depend on the pole locations $a = (a_1, a_2, \ldots, a_n)$ in order that the monodromy representation remain unchanged. In case ∞ is a regular point, he discovered that the matrices A_ν must satisfy a nonlinear differential equation,

$$\mathrm{d}A_\mu = -\sum_{\nu \neq \mu} \frac{A_\mu A_\nu - A_\nu A_\mu}{a_\mu - a_\nu} d(a_\mu - a_\nu).$$

This equation is known as the Schlesinger equation and is an example of a *monodromy preserving deformation*.

Schlesinger's treatment of the existence question for such deformations was clarified in the work of Malgrange (1983b). Let $D_{\mu\nu} = \{a \in \boldsymbol{C}^n : a_\mu = a_\nu\}$ denote the set of points in $\boldsymbol{C}^n$ where the μ and ν poles collide—evidently such points are singular points for the Schlesinger equations. Malgrange showed that if one starts with a monodromy representation that is realized for a differential equation with simple poles at a_ν^0 and residues A_ν^0, then the deformation equation has a solution $A_\nu(a)$ which is analytic on the simply connected covering of $\boldsymbol{C}^n \backslash \cup D_{\mu\nu}$ *except possibly for poles along a hypersurface in this space*. He also showed that the hypersurface on which the solutions $A_\nu(a)$ has poles consists of those points for which a variant of the Riemann–Hilbert problem fails to have a solution. It is simplest to describe this variant in the nonresonant case (although Malgrange does not make this assumption). The point a will be a pole of $A_\nu(a)$ provided that it is not possible to find a differential equation with simple poles at a_ν for $\nu = 1, 2, \ldots, n$ which reproduces the monodromy representation of the original equation,

$$\frac{\mathrm{d}\Psi}{\mathrm{d}z} = \sum_{\nu=1}^{n} \frac{A_\nu^0}{z - a_\nu^0}\Psi,$$

and is in the local holomorphic equivalence class of

$$\frac{d}{dz} - \frac{A_\nu^0}{z - a_\nu}$$

near $z = a_\nu$. It is possible that a given monodromy representation can be realized by a differential equation with simple poles without being able to specify the local holomorphic equivalence classes (see Bolibruch, 2002).

Nonlinear differential equations in the complex plane have the property that solutions can develop singularities away from the manifest singularities of the equations. The location of singularities of this sort depends on the initial conditions for the differential equation; thus, such points are called movable singularities. Differential equations with solutions whose only movable singularities are poles are said to have the Painlevé property. Paul Painlevé and Bertrand Gambier classified the simplest such nonlinear equations and determined six new families of transcendental functions that arise in the integration of these differential equations—now referred to as Painlevé I–VI (see Ince, 1956). Schlesinger understood, and as noted above, Malgrange proved that the Schlesinger equation has the Painlevé property. In a related development, Garnier (1912) showed that the Painlevé transcendents arise in the deformation of Fuchsian equations. Later, Jimbo, & Miwa (1981) showed how to obtain all six of the Painlevé equations by suitable specialization of the Schlesinger equations for 2×2 matrix systems. Painlevé VI arises for a linear system with simple poles of the type considered by Schlesinger. The other Painlevé equations arise by successive scaling that introduces irregular singularities and requires an extension of the Schlesinger analysis that was accomplished by Miwa, Jimbo, and Ueno in Jimbo, et al. (1981). Importantly, finding such representations for Painlevé equations is analogous to finding Lax pair representations for nonlinear equations such as the KdV equations (see Kapaev, 2002; Deift et al., 1999).

Flaschka & Newell (1980) first considered a generalization of the notion of monodromy preserving deformations to linear differential equations with higher rank singularities (equations with poles of order 2 or higher are said to have irregular singularities). More subtle local invariants, called Stokes multipliers, are fixed by the deformations, and in addition to the location of the singularities, the deformation parameters include formal expansion parameters that characterize the local asymptotics of solutions near the singularities. Based on Birkhoff's generalization of the Riemann–Hilbert problem (Birkhoff, 1913), Jimbo, Miwa, and Ueno (1981) generalized Flashka and Newell's work to the consideration of nonresonant differential equations with irregular singular points. They obtained the analogue of the Schlesinger equations for generalized monodromy preserving deformations of these equations. They also introduced the notion of a τ-function for such deformations. They conjectured that the τ-function is a holomorphic function of the deformation parameters whose 0 set is the set of pole locations for the solutions to the deformation equations. Miwa (1981) later showed that the τ-function is holomorphic and the Painlevé property is satisfied for the deformation equations in the domain of convergence for his explicit solution of the Birkhoff–Riemann–Hilbert problem. Malgrange (1983b) established the full conjecture for regular singularities without restrictions.

Geometric Perspectives

Röhrl (1957) formulated and solved a version of the Riemann–Hilbert problem on Riemann surfaces. The setting for the result involves vector bundles and holomorphic connections with singularities. Deligne (1970) developed a multidimensional generalization of the Riemann–Hilbert problem which took advantage of rather abstract categorical constructions. Malgrange (1983a) introduced Stokes' sheaves at irregular singular points to describe local holomorphic equivalence classes. More recently, progress in analyzing the local structure of differential equations at irregular singular points has come by introducing sophisticated ideas from algebraic geometry (see Varadarajan, 1996). Part of the impetus for these developments is the effort to achieve clarity in a subject that is subtle enough to have inspired the misinterpretation of fundamental results and even errors in proofs by such luminaries as George D. Birkhoff (see Anasov and Bolibruch (1994) for an account).

Among these developments, there is one geometric idea introduced by Röhrl and more fully realized by Malgrange (1983c), which provides helpful insight into the nature of the Birkhoff–Riemann–Hilbert deformation problem introduced in Jimbo et al. (1981). Suppose that $a^0 = \{a_1^0, a_2^0, \ldots, a_n^0\}$ is a collection of distinct points in $\boldsymbol{P}^1$ and $A^0(z)$ is a rational matrix valued function whose principal part at $z = a_\nu$ is

$$\frac{A_{r+1}^0}{(z-a_\nu)^{r+1}} + \frac{A_r^0}{(z-a_\nu)^r} + \cdots + \frac{A_1^0}{z-a_\nu}.$$

The integer $r = r_\nu$ is called the rank of the singularity. Let

$$\nabla^0 = \frac{d}{dz} - A^0(z)$$

denote the associated *connection*. We will say that the connection ∇^0 is *nonresonant* if the leading coefficient A_{r+1}^0 has distinct eigenvalues at each of the singularities with the additional requirement that these eigenvalues do not differ by integers in case $r = 0$. Incidentally, much of the machinery that has been introduced to study irregular singular points is devoted to understanding the substantial complications that arise for *resonant* connections (Varadarajan, 1996).

It is a result of Deligne (1970) that the holomorphic equivalence class of ∇^0 on the *punctured* sphere $\boldsymbol{P}^1\backslash\{a_\nu^0\}$ is determined by its monodromy representation. On the other hand, in a neighborhood of each of the singularities, the connection ∇^0 belongs to a local holomorphic equivalence class of rank r connections. For nonresonant connections, this local equivalence class, is naturally a fiber bundle whose base consists of *diagonal* connections,

$$\frac{\mathrm{d}}{\mathrm{d}z} - \frac{H_r}{(z-a_\nu)^{r+1}} + \cdots + \frac{H_0}{z-a_\nu}, \tag{4}$$

whose leading coefficient, H_r, has distinct eigenvalues which do not differ by integers in case $r=0$. For $r=0$ the form (4) determines the local equivalence class, but for $r\geq 1$ the fibers in the bundle are essentially the Stokes multipliers, S. Each nonresonant connection ∇^0 has a *formal reduction* to a type (4) connection and the holomorphic equivalence class for ∇^0 is determined by this formal reduction together with the Stokes multipliers. The Stokes multipliers determine the relations between solutions to $\nabla^0\Psi=0$ that are sectorially asymptotic to flat sections of the connection (4) (see Sibuya, 1977; Malgrange, 1983c).

Let t denote the collection of diagonal entries for all the matrices H_j. The fiber bundle of local holomorphic equivalence classes has a natural flat connection (it is a so-called local system (Varadarajan, 1996)) and we let $t\rightarrow\sigma(t)$ denote the flat (locally constant) section of this bundle normalized to give the class of ∇^0 at the parameter value t^0 which corresponds to ∇^0. The generalized monodromy preserving deformation problem is to find a family of connections $\nabla(a,t)$ which reproduces the monodromy representation of ∇^0 in the punctured plane $\boldsymbol{P}^1\backslash\{a_\nu\}$ and whose local holomorphic equivalence class at each a_j is the equivalence class associated with the value of the flat section $\sigma(t)$ (i.e., locally constant Stokes multipliers).

The version of the Birkhoff–Riemann–Hilbert problem relevant to this deformation problem is: Given a representation of $\pi_1(\boldsymbol{P}^1\backslash a, a_0)$ and an assignment of a compatible local holomorphic equivalence class of a type r_ν connection to each of the points a_ν, find a connection on the trivial bundle over $\boldsymbol{P}^1$ which reproduces this data (compatible means that the local monodromy at a_ν should reproduce the monodromy coming from the representation). Because noncanonical choices must be made to define the Stokes multipliers, this version of the Birkhoff–Riemann–Hilbert problem is even complicated to state if the connection with local holomorphic equivalence classes is not understood. Based on ideas of Malgrange, it has been shown that the τ-function introduced by Miwa, Jimbo, and Ueno (1981) has zeros for the deformation problem described above precisely when the associated Birkhoff–Riemann–Hilbert problem fails to have a solution (Palmer, 1999). It should also be mentioned that a crucial technique used both by Jimbo et al. (1981) and by Malgrange (1983b,c) has the prolongation of the connection originally defined on $\boldsymbol{P}^1$ to an integrable connection defined on $\boldsymbol{P}^1\times\mathcal{D}$ where $\mathcal{D}$ is the space of deformation parameters. In this multidimensional setting, Malgrange adapts Röhrl's technique to prove the existence of a prolongation for the original connection on a vector bundle. The solution of the original deformation problem depends on the triviality of certain restrictions of this vector bundle.

Applications

The Riemann–Hilbert problem and the Painlevé transcendents have generated a steady stream of papers over the years, but interest in monodromy preserving deformations was largely dormant in the period from the early 1920s to the late 1970s. A rebirth of interest in the subject dates from the surprising discovery by Wu, McCoy, Tracy, and Baruch (WMTB) (1976) that the two-point scaling function for the two-dimensional Ising model can be expressed in terms of a Painlevé transcendent of type III. The two-dimensional Ising model is a statistical model of magnetism which exhibits a phase transition in the infinite volume limit. It is so far the only model with a phase transition in which the correlation functions can be studied exactly in the infinite volume limit. The scaling limit that was investigated in Wu et al. (1976) examines the asymptotics of the spin correlations at the length scale of the correlation length (which tends to ∞) as the temperature approaches the critical point. Not only is this analysis important for an understanding of critical phenomena in statistical physics, but it was understood at about the same time that the mathematics of critical phenomena in statistical physics was recognized as the same as the mathematics of renormalization for quantum field theory.

In a remarkable series of papers (Sato et al., 1978, 1979a,b,c, 1980) starting in 1978, Sato, Miwa, and Jimbo found a context for the WMTB result by showing that the n-point correlation for the Ising model could be expressed in terms of the solutions to "monodromy preserving deformations" of the Euclidean Dirac equation (with a mass term). Note that the equation which is the analogue of the Dirac equation in the earlier development is not the differential equation (1) but the Cauchy–Riemann equation $\bar{\partial}\psi=0$. The correlation functions in the Ising model were the inspiration for their introduction of the τ-function in other contexts. See Palmer & Tracy (1983), Palmer (1993), Palmer et al. (1994), and Palmer (2002) for further developments of these ideas. In Jimbo et al. (1980), Sato, Miwa, Jimbo, and Mori explained the use of their techniques to analyze the impenetrable Bose gas and incidentally remarked on a connection with an important function in random matrix theory.

The analysis of correlations in integrable models has since been developed by Its et al. (1990), who also codified some of the ideas in a theory of completely integrable integral operators (Korepin et al., 1993). The connection between the random matrix theory and Painlevé functions has been systematically worked out by Tracy & Widom (1994, 1996), who independently developed an integral equation approach to the nonlinear "deformation" equations.

Jimbo (1982) used the monodromy preserving deformation representation of Painlevé functions and the notion of the τ-function to analyze the connection problem for Painlevé functions. The problem is to understand how different families of solutions to the Painlevé equations defined by local asymptotic developments at different singular points are related to one another. Somewhat earlier the connection problem for Painlevé III was worked out by McCoy, Tracy, and Wu (1977) where the answer has consequences for the short distance behavior of the scaling function. Its and Novokshenov (1986) have developed these ideas further.

Although not quite a development in monodromy preserving deformation theory, a "steepest decent" technique based on Riemann–Hilbert ideas has emerged in work of Deift, Zhou and Its that is effective for the asymptotic analysis of solutions to nonlinear equations that arise from spectral or monodromy preserving deformations (Its, 1981; Deift et al., 1997; Bleher & Its, 1999; Deift et al., 1999).

Painlevé functions were discovered in two-dimensional quantum gravity and connections with monodromy preserving deformations are discussed in Moore (1990). Painelevé functions also arise in two-dimensional topological field theory (Dubrovin, 1999).

Further developments are discussed in Harnad & Its (2002).

JOHN PALMER

See also **Painlevé analysis; Riemann–Hilbert problem**

Further Reading

Anasov, D.V. & Bolibruch, A.A. 1994. *The Riemann–Hilbert problem*, Aspects of Mathematics E, vol. 22. Braunschweig; Wiesbaden: Vieweg

Birkhoff, G.D. 1913. The generalized Riemann problem for linear differential equations. *Proceedings of the American Academy of Arts and Sciences*, 32: 531–568

Bleher, P. & Its, A.R. 1999. Semiclassical asymptotics of orthogonal polynomials, Riemann–Hilbert problem, and universality in the matrix model. *Annals of Mathematics*, 150 (2): 185–266

Bolibruch, A. 2002. Inverse problems for linear differential equations with meromorphic coefficients. *CRM Workshop: Isomonodromic Deformations and Applications in Physics*, edited by J. Harnard & A. Its, Providence, RI: American Mathematical Society, pp. 3–25

Deift, P., Its, A., Kapaev, A. & Zhou, X. 1999. On the algebro-geometric integration of the Schlesinger equations. *Communications in Mathematical Physics*, 203: 613–633

Deift, P., Its, A. & Zhou, X. 1997. A Riemann–Hilbert approach to asymptotic problems arising in the theory of random matrix models, and also in the theory of integrable statistical mechanics. *Annals of Mathematics*, 146: 149–237

Deift, P., Kriecherbauer, T., McLaughlin, K., Venakides, S. & Zhou, X. 1999. Uniform asymptotics for polynomials orthogonal with respect to varying exponential weight and applications to universality questions in random matrix theory. *Communications in Pure and Applied Mathematics*, 52: 1335–1425

Deligne, P. 1970. *Équations différentielles à points singuliers réguliers*, Berlin and New York: Springer

Dubrovin, B. 1999. Painlevé transcendents in two-dimensional topological field theory. In *The Painlevé Property, One Hundred Years Later*, edited by R. Conte, New York: Springer, pp. 287–412

Flaschka, H. & Newell, A.C. 1980. Monodromy and spectrum preserving deformations I. *Communications in Mathematical Physics*, 76: 65–116

Garnier, R. 1912. Sur les équationes différentielles du troisième ordre dont l'intégrale générale est uniforme et sur une classe d'équations nouvelles d'ordre supérieur dont l'intégrale générale a ses points critique fixes. *Annales de l'Ecole Normale Supérieure*, (3) 29: 1–126

Harnad, J. & Its, A. 2002. *CRM Workshop: Isomonodromic Deformations and Applications in Physics*, edited by J. Harnard & A. Its, Providence, RI: American Mathematical Society

Ince, E.L. 1956. *Ordinary Differential Equations*, New York: Dover

Its, A.R. 1981. Asymptotics of solutions of the nonlinear Schrödinger equation and isomonodromic deformations of systems of linear differential equations. *Soviet Mathematics. Doklady*, 24: 454–456

Its, A.R., Izergin, A.G., Korepin, V.E. & Slavnov, N.A. 1990. Differential equations for quantum correlation functions. *International Journal of Modern Physics* B, 4: 1003–1037

Its, A.R. & Novokshenov, V.Yu. 1986. *The Isomonodromic Deformation Method in the Theory of Painlevé Equations*, Berlin: Springer

Jimbo, M. 1982. Monodromy problem and the boundary conditions for some Painlevé equations. *Publ. R.I.M.S. Kyoto University*, 18: 1137–1161

Jimbo, M. & Miwa, T. 1981. Monodromy preserving deformations of linear ordinary differential equations with rational coefficients II. *Physica* D, 2: 407–448

Jimbo, M. & Miwa, T. 1983. Monodromy preserving deformations of linear ordinary differential equations with rational coefficients III. *Physica* D, 2: 26–46

Jimbo, M., Miwa, T., Sato, M. & Mori, Y. 1980. Density matrix of an impenetrable bose gas and the fifth Painlevé transcendent. *Physica* D, 1: 80–158

Jimbo, M., Miwa, T. & Ueno, K. 1981. Monodromy preserving deformations of linear ordinary differential equations with rational coefficients I. *Physica* D, 2: 306–352

Kapaev, A.A. 2002. Lax pairs for Painlevé equations, *CRM workshop: Isomonodromic Deformations and Applications in Physics*, edited by J. Harnard & A. Its, Providence, RI: American Mathematical Society, pp. 37–47

Korepin, V.E., Bogoliubov, N.M. & Izergin, A.G. 1993. *Quantum Inverse Method and Correlation Functions*, Cambridge and New York: Cambridge University Press

Malgrange, B. 1983a. *La classification des connections irrégulieéres à une variable*, Boston: Birkhäuser, pp. 381–400
Malgrange, B. 1983b. *Sur les déformations isomonodromiques I, Singularités régulières*, Boston: Birkhäuser, pp. 401–426
Malgrange, B. 1983c. *Sur les déformations isomonodromiques II, Singularités irrégulières*, Boston: Birkhäuser, pp. 427–438
McCoy, B.M., Tracy, C.A. & Wu, T.T. 1977. Painlevé functions of the third kind. *Journal of Mathematical Physics*, 18: 1058–1092
Miwa, T. 1981. Painlevé property of monodromy preserving equations and the analyticity of τ-function. *Publ. R.I.M.S. Kyoto University*, 17: 703–721
Moore, G. 1990. Geometry of the string equations. *Communications in Mathematical Physics*, 133: 261–304
Palmer, J. 1993. Tau functions for the Dirac operator in the Euclidean plane. *Pacific Journal of Mathematics*, 160: 259–342
Palmer, J. 1999. Zeros of the Jimbo, Miwa, Ueno tau function. *Journal of Mathematical Physics*, 40: 6638–6681
Palmer, J. 2002. Short distance asymptotics of Ising correlations, *Journal of Mathematical Physics*, 43(2): 918–953
Palmer, J., Beatty, M. & Tracy, C. 1994. Tau functions for the Dirac operator on the Poincaré disk. *Communications in Mathematical Physics*, 165: 97–173
Palmer, J. & Tracy, C.A. 1983. Two dimensional Ising correlations: the SMJ analysis. *Advances in Applied Mathematics*, 4: 46–102
Plemelj, J. 1963. *Problems in the Sense of Riemann and Klein*, New York: Interscience
Röhrl, H. 1957. Das Riemann–Hilbertsche problem der theorie der linearen differentialgleichungen. *Mathematische Annalen*, 133: 1–25
Sato, M., Miwa, T. & Jimbo, M. 1978. Holonomic quantum fields I. *Publ. R.I.M.S. Kyoto University*, 14: 223–267
Sato, M., Miwa, T. & Jimbo, M. 1979a. Holonomic quantum fields II. *Publ. R.I.M.S. Kyoto University*, 15: 201–278
Sato, M., Miwa, T. & Jimbo, M. 1979b. Holonomic quantum fields III. *Publ. R.I.M.S. Kyoto University*, 15: 577–629
Sato, M., Miwa, T. & Jimbo, M. 1979c. Holonomic quantum fields IV. *Publ. R.I.M.S. Kyoto University*, 15: 871–972
Sato, M., Miwa, T. & Jimbo, M. 1980. Holonomic quantum fields V. *Publ. R.I.M.S. Kyoto University*, 16: 531–584
Schlesinger, L. 1912. Über eine klasse von differentialsystemen beliebiger ordnung mit festen kritischen punkten. *Journal für die reine und angewandte Mathematik*, 14: 95–145
Sibuya, Y. 1977. Stokes' phenomena. *Bulletin of the American Mathematical Society*, 83: 1075–1077
Tracy, C.A. & Widom, H. 1994. Fredholm determinants, differential equations and matrix models. *Communications in Mathematical Physics*, 163: 33–72
Tracy, C.A. & Widom, H. 1996. On orthogonal and symplectic matrix ensembles. *Communications in Mathematical Physics*, 177: 727–754
Varadarajan, V.S. 1996. Linear meromorphic differential equations: a modern point of view, *Bulletin of the American Mathematical Society*, 33(1): 1–41
Wu, T.T., McCoy, B., Tracy, C.A. & Barouch, E. 1976. Spin-spin correlation functions for the two dimensional Ising model: exact theory in the scaling region. *Physical Review* B, 13: 316–374

MONTE CARLO METHODS

Monte Carlo methods collectively refer to a set of computational tools based on random samples. These methods have a very long history that can be traced back to biblical times although a systematic foundation was established only in the early days of electronic computing. A group of mathematicians and physicists of the Los Alamos project during World War II, in particular, John von Neumann, Nick Metropolis, Jeffrey Kahn, Enrico Fermi, Stan Ulam, and their collaborators, developed Monte Carlo methods for estimating eigenvalues of the Schrödinger equation and for studying random neutron diffusion in fissile material used in atomic bombs. Since then, Monte Carlo methods have found applications in a vast number of scientific disciplines and are, becoming increasingly popular with the advent of cheap and powerful computers.

Starting with a space $\mathcal{X}$, let $\varphi : \mathcal{X} \to R$ be function and π be a probability density. In essence, Monte Carlo methods are concerned with

- *Integration*: Evaluate the expectation of φ under π; that is,

$$I_\pi(\varphi) = \int_{\mathcal{X}} \varphi(x)\,\pi(x)\,\mathrm{d}x. \quad (1)$$

- *Simulation*: Simulate some random samples $\{X_i\}_{i=1}^N$ distributed according to π.

Integration and simulation are essential elements of many scientific problems arising, for example, in statistical physics, statistics, engineering, and economics. The probability density π may correspond to a posterior density in the framework of Bayesian statistics or a Boltzmann–Gibbs distribution in equilibrium statistical thermodynamics.

In most applications, it is impossible to compute the expectations of interest in closed-form as the dimension of the space $\mathcal{X}$ is large; for example, $\mathcal{X} = R^{1000}$ or $\mathcal{X} = \{-1, 1\}^{5000}$. Moreover, π is often only known up to a normalizing constant, that is,

$$\pi(x) = cf(x), \quad (2)$$

where $f : \mathcal{X} \to R$ is known but c is unknown. Consequently, standard deterministic numerical integration schemes are very inefficient.

The basic idea of Monte Carlo methods is to compute the expectation(s) of interest in Equation (1) via random samples. Assume the random samples $\{X^{(i)}\}_{i=1}^N$ distributed according to π are available, then an estimate of $I_\pi(\varphi)$ can be given by the empirical average

$$\widehat{I_\pi}(\varphi) = \frac{1}{N}\sum_{i=1}^{N} \varphi\left(X^{(i)}\right).$$

In other words, π is approximated by

$$\widehat{\pi}(x) = \frac{1}{N}\sum_{i=1}^{N} \delta\left(x - X^{(i)}\right),$$

where $\delta(\cdot)$ is the Dirac delta function. This estimate has "good" theoretical properties. First, it is unbiased.

Second, if the samples $\{X^{(i)}\}_{i=1}^{N}$ are statistically independent, then the variance of this estimate is given by

$$\frac{\int [\varphi(x) - I_\pi(\varphi)]^2 \pi(x)\,dx}{N}; \tag{3}$$

that is, the convergence rate of the estimation error to zero is independent of the dimension of $\mathcal{X}$. If the numerator of (3) is "small", then only a few hundreds/thousands of samples might be necessary to achieve a good precision even if $\mathcal{X} = R^{1000}$.

Thus far, it has been shown that estimates with good theoretical properties can be easily obtained from a large set of samples distributed according to the probability density of interest, but how to generate these samples? The Monte Carlo integration problem then becomes that of simulation. Generating samples from nonstandard probability distributions/densities only known up to a proportionality factor is a central problem in Monte Carlo methods.

In 1947, von Neumann, in a letter to Ulam, described an algorithm known as the Rejection Method, to simulate samples distributed according to a "target" probability density. Given a target probability density π that is only known up to a normalizing constant, that is, π is of the form in (2) and only f is known, one chooses a candidate density g that is easy to sample from and a number M satisfying $f(x) \leq Mg(x)$ for any $x \in \mathcal{X}$. The algorithm proceeds as follows:

(a) Sample $X^* \sim g$ ("$\sim$" is a standard notation for "distributed according to") and compute

$$\alpha(X^*) = \frac{f(X^*)}{Mg(X^*)}.$$

(b) Sample a uniform random variable U on [0, 1]. If $U \leq \alpha(X^*)$, return X^* otherwise go back to (a).

It can be easily verified that the accepted sample follows the target distribution π. The problem with this method is the difficulty in finding a candidate density g satisfying $f(x) \leq Mg(x)$ that is easy to sample from. Moreover, even if such a g can be found, the acceptance probability is $(Mc)^{-1}$, which can be extremely low when $\mathcal{X}$ is a high-dimensional space. Consequently, the algorithm may take a long time to generate a single sample.

A more powerful class of algorithms are the Markov chain Monte Carlo (MCMC) techniques. Given a "target" density π, the key idea of MCMC is to build a Markov chain $\{X_i\}_{i\geq 0}$ of transition kernel $K(\cdot|\cdot)$ such that

$$\int K(x'|x)\,\pi(x)\,dx = \pi(x'), \tag{4}$$

i.e., if $X_{i-1} \sim \pi$ then any sample $X_i \sim K(\cdot|X_{i-1})$ is also distributed according to π. The target density π in (4) is known as the invariant (or stationary) distribution of the Markov chain. Under ergodicity conditions on the kernel $K(\cdot|\cdot)$, the samples $\{X_i\}_{i\geq 0}$ are distributed according to π asymptotically (as $i \to \infty$) regardless of the value of initial state X_0.

There are an infinity of kernels with invariant distribution π. However, the most frequently used are the Metropolis–Hastings kernels, first proposed by Metropolis et al. (1953), and then generalized by Hastings (1970). Green (1995) extended this to the case where $\mathcal{X}$ is a disjoint union of subspaces of different dimensions.

Given a candidate distribution $q(\cdot|\cdot)$ that is easy to sample from, the Metropolis–Hastings kernel can be constructed as follows: given X_{i-1}

(a) Sample a candidate $X^* \sim q(\cdot|X_{i-1})$ and compute

$$\alpha(X_{i-1}, X^*) = \frac{\pi(X^*)}{\pi(X_{i-1})}\frac{q(X_{i-1}|X^*)}{q(X^*|X_{i-1})}.$$

(b) Sample a uniform random variable U on [0, 1]. If $U \leq \alpha(X_{i-1}, X^*)$, then set $X_i = X^*$ otherwise set $X_i = X_{i-1}$.

It can be verified that the Markov chain $\{X_i\}_{i\geq 0}$ generated by the above kernel has invariant distribution π. Knowledge of the normalizing constant of π is not needed as it disappears in the acceptance ratio $\alpha(X_{i-1}, X^*)$. The major problem with such Markov chain techniques is that only samples from π are obtained as $i \to \infty$. Nevertheless, the Metropolis–Hastings algorithm and MCMC algorithms, in general, have been applied successfully in numerous applications. MCMC methods form the basis of global optimization algorithms such as simulated annealing (Van Laarhoven & Arts 1987). For a thorough treatment of the theory and applications of MCMC algorithms, the reader is referred to Robert & Casella, (1999).

Importance sampling (IS) is based on assigning weights (or importance) to samples generated from a candidate density q so as to approximate the target density π. Suppose that $\pi(x) = w(x)\,q(x)$, where q is a density that is easy to sample from. Then, given random samples $\{X^{(i)}\}_{i=1}^{N}$ distributed according to q, the expectation $I_\pi(\varphi)$ in Equation (1) can be estimated by the empirical average

$$\widehat{I_\pi}(\varphi) = \frac{1}{N}\sum_{i=1}^{N} \varphi\left(X^{(i)}\right) w\left(X^{(i)}\right).$$

In other words, π is approximated by

$$\widehat{\pi_1}(x) = \frac{1}{N}\sum_{i=1}^{N} w\left(X^{(i)}\right) \delta\left(x - X^{(i)}\right).$$

This strategy assumes full knowledge of π as $w\left(X^{(i)}\right) = \pi\left(X^{(i)}\right)/q\left(X^{(i)}\right)$ needs to be determined. When π is only specified upto normalizing constant, write

$$\pi(x) = \frac{w(x)\,q(x)}{\int_{\mathcal{X}} w(x)\,q(x)\,\mathrm{d}x}$$

and apply importance sampling to both numerator and denominator to yield the following approximation:

$$\widehat{\pi}_2(x) = \sum_{i=1}^{N} \frac{w\left(X^{(i)}\right)}{\sum_{j=1}^{N} w\left(X^{(j)}\right)} \delta\left(x - X^{(i)}\right).$$

An approximate sample from π can be obtained by sampling from the discrete distributions $\widehat{\pi}_1$ or $\widehat{\pi}_2$.

Applications

The range of applications of Monte Carlo methods is vast. Listed below are some of the more well-known areas.

Integral equations: IS methods have been widely used to solve linear systems and integral equations appearing in particle transport problems. The basic idea is to give a probabilistic approximation of operators of the form $(I - H)^{-1} = \sum_{i=0}^{\infty} H^i$; see Sobol (1994) for details.

Computational physics and chemistry simulation: Monte Carlo methods are used in physics and chemistry to simulate from Ising models, simulate self-avoiding random walks, and compute the free energy, entropy, and chemical potential over systems; see, for example, Frenkel & Smith (1996).

Quantum physics: To compute the dominant eigenvalue and eigenvector of a positive operator, it is possible to use a stochastic version of the power method. This is often applied to the Schrödinger equation; see Melik-Alaverdian & Nightingale (1999) for a recent review.

Statistics: Performing inference in complex statistical models invariably requires sampling from high dimensional probability distributions. See Gilks et al. (1996) for applications of MCMC and Doucet et al. (2001) for applications of IS-type methods to such problems.

Arnaud Doucet and Ba-Ngu Vo

See also **Random walks; Stochastic processes**

Further Reading

Doucet, A., De Freitas, J.F.G. & Gordon, N.J. (editors). 2001. *Sequential Monte Carlo Methods in Practice*, New York: Springer

Frenkel, D. & Smith, B. 1996. *Understanding Molecular Simulation: From Algorithms to Applications*, Boston: Academic Press

Gilks, W.R., Richardson, S. & Spiegelhalter, D.J. (editors). 1996. *Markov Chain Monte Carlo in Practice*, London: Chapman & Hall

Green, P.J. 1995. Reversible jump MCMC computation and Bayesian model determination. *Biometrika*, 82: 711–732

Hastings, W.K. 1970. Monte Carlo sampling methods using Markov chains and their applications. *Biometrika*, 57: 97–109

Melik-Alaverdian, M. & Nightingale, M.P. 1999. Quantum Monte Carlo methods in statistical mechanics. *International Journal of Modern Physics* C, 10: 1409–1418

Metropolis, N., Rosenblutt, N., Rosenblutt, A.W., Teller, M.N. & Teller, A.H. 1953. Equations of state calculations by fast computing machines. *Journal of Chemical Physics*, 21: 1087–1092

Robert, C.P. & Casella, G. 1999. *Monte Carlo Statistical Methods*, New York: Springer

Sobol, I.M. 1994. *A Primer for the Monte Carlo Method*, Boca Raton, FL: CRC Press

Van Laarhoven, P.J. & Arts, E.H.L. 1987. *Simulated Annealing: Theory and Applications*, Amsterdam: Reidel

MORPHOGENESIS, BIOLOGICAL

One of the central problems in developmental biology is to understand how patterns and structures are laid down. From the initially almost homogeneous mass of dividing cells in an embryo emerges the vast range of pattern and structure observed in animals. For example, the skeleton is laid down during chondrogenesis when chondroblast cells condense into aggregates that lead eventually to bone formation. The skin forms many specialized structures such as hair, scales, feathers, and glands. Butterfly wings exhibit spectacular colors and patterns, and many animals develop dramatic coat patterns.

Although genes play a key role, genetics say nothing about the actual mechanisms that produce pattern and structure—the process known as morphogenesis—as an organism matures from embryo to adult. Tissue movement and rearrangement are the key features of almost all morphogenetic processes and arise as the result of complex mechanical, chemical, and electrical interactions. Despite the recent vast advances in molecular biology and genetics, little is understood of how these processes conspire to produce pattern and form. There is the danger of falling into the practices of the 19th century, when biology was steeped in the mode of classification and there was a tremendous amount of list-making activity. This was recognized by D'Arcy Thompson, in his classic work first published in 1917 (see Thompson (1992) for the abridged version). He was the first to develop theories for how certain forms arose, rather than simply cataloging different forms, as was the tradition at that time.

At the heart of a number of developmental phenomena is the process of convergence-extension, in which a tissue narrows along one axis while extending along another. This process represents the integration of local cellular behavior that produces forces to change the shape of the cell population. In fact, convergence-extension is essentially responsible for the transformation of the spherical egg into the elongated, bilaterally symmetric vertebrate body axis (Keller et al., 1992).

Cell fate and position within the developing embryo can be strongly influenced by environmental factors. Therefore, to investigate the process of morphogenesis, one must really address the issue of how the embryo organizes the complex spatiotemporal sequence of signalling cues necessary to develop structure in a controlled and coordinated manner. Structure can form through tissue movement and rearrangement. Theoretical studies in this area include the early purse-string model (Odell et al., 1981) for tissue folding in which, in response to a large deformation, cells were proposed to actively contract and, in doing so, cause a large deformation in neighboring cells which, in turn, also contract, setting up a propagating contraction wave which leads to tissue folding. This model was applied to a variety of developmental problems and provided the precursor to the mechanochemical theory of developmental patterning developed by Oster, Murray, and coworkers (for review, see Murray, 2003). This approach emphasized the link between tissue mechanics and chemical regulation and has been applied widely in both developmental biology and medicine.

Discrete-cell modeling approaches have subsequently been developed in which morphogenesis is hypothesized to occur via mechanical rearrangement of neighbors in an epithelial sheet, and computational finite elements have been developed to test various theoretical explanations for morphogenesis (Weliky et al., 1991; Davidson et al., 1995).

In all these models, individual cell movements within the tissue are determined by the balance of mechanical forces acting on the cell. Such models can exhibit tissue folding, thickening, invagination, exogastrulation, and intercalation, and have been shown to capture many of the key aspects of processes such as gastrulation, neural tube formation, and ventral furrow formation in *Drosophila*. Cells can also sort out depending on their type, and this has led to the theory of differential adhesion and energy minimization (Steinberg, 1970).

Models for tissue motion are not amenable to a mathematical analysis and tend to be highly computation based. However, models for how cells differentiate can be addressed mathematically. Broadly speaking, there are two classes of such models. In one class, the chemical pre-pattern models, it is hypothesized that a chemical signal is set up in some way and cells respond to this signal by differentiating. In the other class, the cell movement models, it is hypothesized that cells respond to mechanochemical cues and form aggregates. Cells in high density aggregates are then assumed to differentiate (see Murray, 2003, for details).

The fact that such models can lead to the generation of spontaneous order was first realized by Alan Turing (1952), who showed that a system of chemicals, stable in the absence of diffusion, could be driven unstable by diffusion. He proposed that such a spatial distribution of chemicals (which he termed *morphogens*) could set up a pre-pattern to which cells could respond and differentiate accordingly. He was one of the first to postulate the existence of such chemicals, and morphogens have now been discovered. It is still not clear that morphogen patterns in biology are set up by the mechanism proposed by Turing, but Turing patterns have been found in chemistry (see Maini et al., 1997, for a review).

A variety of models based on different biology give rise to mathematical formulations in terms of coupled systems of highly nonlinear partial differential equations. The analysis of these models has, to date, yielded a number of common behaviors. This has led to the idea of using such models to determine developmental constraints. That is, independent of the underlying biology, such models predict that only certain patterns are selected at the expense of others and thus there is a limited variation. This has consequences for evolution. For example, application of mitotic inhibitors to developing limbs produces smaller limbs with reduced elements. Some of the resultant variants look very similar to the pattern of evolution in other species, suggesting that these species may be more closely related than previously thought (Oster et al., 1988). Moreover, the construction rules generated by a study of developmental constraints is another, perhaps more mechanistic, way of describing how different species are related other than the topological deformation approach of D'Arcy Thompson.

Other approaches to morphogenesis and pattern formation include cellular automata models, in which individual entities (cells, for example) behave according to a set of rules. Such models allow one to include much more biological detail and to investigate finer grain patterns than those possible in the continuum approaches discussed above (see, for example, Alt et al., 1997). However, to date they lack a detailed mathematical underpinning.

The recent spectacular advances in molecular genetics raise the issue of how we can combine the enormous amount of data now being generated at this level with the data available from the classical experiments at the cell and tissue level to provide a coherent theory for pattern formation and morphogenesis. This leads to the problem of modeling across a vast range of spatial and temporal scales. The mathematics for this has not yet been developed and is one of the challenges presently being addressed.

PHILIP K. MAINI

See also **Brusselator; Cellular automata; Pattern formation; Reaction-diffusion systems; Turing patterns**

Further Reading

Alt, W., Deutsch, A. & Dunn G. (editors). 1997. *Dynamics of Cell and Tissue Motion*, Basel: Birkhäuser

Davidson, A., Koehl, M.A.R., Keller, R. & Oster, G.F. 1995. How do sea urchins invaginate? Using biomechanics to distinguish between mechanisms of primary invagination. *Development*, 121: 2005–2018

Keller, R., Shih, J. & Domingo, C. 1992. The patterning and functioning of protrusive activity during convergence and extension of the *Xenopus* organiser. *Development Supplement*, 81–91

Maini, P.K., Painter, K.J. & Chau, H.N.P. 1997. Spatial pattern formation in chemical and biological systems. *Faraday Transactions*, 93(20): 3601–3610

Murray, J.D. 2003. *Mathematical Biology II: Spatial Models and Biomedical Applications*, 3rd edition, Berlin and New York: Springer

Odell, G.M., Oster, G., Alberch, P. & Burnside, B. 1981. The mechanical basis of morphogenesis. I. Epithelial folding and invagination. *Developmental Biology*, 85: 446–462

Oster, G.F., Shubin, N., Murray, J.D. & Alberch, P. 1988. Evolution and morphogenetic rules. The shape of the vertebrate limb in ontogeny and phylogeny. *Evolution*, 45: 862–884

Steinberg, M.S. 1970. Does differential adhesion govern self-assembly processes in histogenesis? Equilibrium configurations and the emergence of a hierarchy among populations of embryonic cells. *Journal of Experimental Zoology*, 173: 395–434

Thompson, D.W. 1992. *On Growth and Form*, Cambridge and New York: Cambridge University Press

Turing, A.M. 1952. The chemical basis of morphogenesis. *Philosophical Transactions of the Royal Society of London* B, 327: 37–72

Weliky, M., Minsuk, S., Keller, R. & Oster, G.F. 1991. Notochord morphogenesis in *Xenopus laevis*: simulation of cell behaviour underlying tissue convergence and extension. *Development*, 113: 1231–1244

MULTIDIMENSIONAL SOLITONS

Although strict analogs of the Korteweg–de Vries soliton (exponentially localized solution with a specific relation between velocity and amplitude and particular scattering properties) have not been found in the multidimensional context, solvable equations with three or more independent variables exhibit a large variety of soliton-like solutions.

As with the Kadomtsev–Petviashvili equation, wide classes of exact explicit solutions have been constructed for other $(2+1)$-dimensional nonlinear equations solvable by the inverse scattering method. We consider here two basic examples, the first being the Davey–Stewartson (DS) equation

$$\begin{aligned} \mathrm{i}q_t + \tfrac{1}{2}\left(\sigma^2 q_{xx} + q_{yy}\right) + |q|^2 q - q\phi &= 0\,, \\ \phi_{xx} - \sigma^2 \phi_{yy} &= 2\left(|q|^2\right)_{xx}, \end{aligned} \quad (1)$$

where $q(x, y, t)$ is a complex-valued function, ϕ is a real-valued function, and the parameter σ^2 takes two values $\sigma^2 = \pm 1$. The DS equation describes propagation of a two-dimensional long surface wave on water of finite depth. In the one-dimensional limit $q_y = \phi_y = 0$, it reduces to the nonlinear Schrödinger equation.

The DS equation (1) has a Lax representation with the two-dimensional Dirac operator as the Lax operator, but it has quite different properties for $\sigma^2 = 1$ (DS-I equation) and for $\sigma^2 = -1$ (DS-II equation). In both cases, there are multi-soliton solutions which do not decay in certain directions on the x, y plane. Similar to the Kadomtsev–Petviashvili equation these solutions describe elastic scattering of line solitons that decay exponentially in the direction of propagation and do not decay in the orthogonal direction. The phase shift can be explicitly calculated.

In addition, the DS equation possesses novel classes of solutions. Thus, the DS-II equation has an infinite set of nonsingular exponential-algebraic solutions, the simplest of which looks like

$$q\,(x, y, t) = \frac{2\nu \exp\left[\lambda(x+\mathrm{i}y) - \bar{\lambda}\,(x-\mathrm{i}y) - \mathrm{i}\left(\lambda^2 + \bar{\lambda}^2\right)t\right]}{|x+\mathrm{i}y+\mu - 2\mathrm{i}\lambda t|^2 + |\nu|^2}, \quad (2)$$

where λ, μ, and ν are arbitrary complex constants. It decays like $\left(x^2+y^2\right)^{-1}$ as $x, y \to \infty$.

The DS-I equation also possesses solutions for which q decays exponentially in both space dimensions. The simplest of them is of the form

$$q\,(x, y, t) = \frac{4\rho\sqrt{\lambda\mu}\exp\left[\mu\,(x+y) + \lambda\,(x-y) + \mathrm{i}\left(\mu^2+\lambda^2\right)t\right]}{\left[1+\mathrm{e}^{2\mu(x+y)}\right]\left[1+\mathrm{e}^{2\lambda(x-y)}\right] + |\rho|^2}, \quad (3)$$

where λ, μ are arbitrary real parameters and ρ is an arbitrary complex parameter. The function ϕ has the nontrivial boundary values as $x, y \to \infty$. Called dromions, such solutions exhibit not only a two-dimensional phase shift during interaction but also a change of the form. Basically, these solutions are driven by the boundary conditions on the function ϕ.

Our second example—the Ishimori equation—is of the form

$$\begin{aligned} \boldsymbol{S}_t + \boldsymbol{S} \times \left(\boldsymbol{S}_{xx} + \sigma^2 \boldsymbol{S}_{yy}\right) + \phi_x \boldsymbol{S}_y + \phi_y \boldsymbol{S}_x &= 0\,, \\ \phi_{xx} - \sigma^2 \phi_{yy} + 2\sigma^2 \boldsymbol{S} \cdot \left(\boldsymbol{S}_x \times \boldsymbol{S}_y\right) &= 0, \end{aligned} \quad (4)$$

where $\boldsymbol{S} = (S_1, S_2, S_3)$ is a unit vector $\boldsymbol{S}^2 = 1$, $\sigma^2 = \pm 1$, and ϕ is a scalar real-valued function. It represents an integrable $(2+1)$-dimensional generalization of the Heisenberg Ferromagnet model equation $\boldsymbol{S} = \boldsymbol{S} \times \boldsymbol{S}_{xx}$. An important feature of the Ishimori equation is that its solutions can be characterized by the topological invariant

$Q=(1/4\pi)\int\int dx\,dy\,\boldsymbol{S}\cdot(\boldsymbol{S}_x\times\boldsymbol{S}_y)$, which is conserved in time and takes integer values $N=0, \pm 1, \pm 2, \ldots$.

The Ishimori equation for both signs of σ^2 possesses multisoliton solutions that describe elastic scattering of line solitons. The Ishimori-I equation ($\sigma^2=1$) has a rich variety of the dromion-like solutions. An interesting feature is that the Ishimori-II equation ($\sigma^2=-1$) consists of the existence of the topologically nontrivial multi-vortex solutions. The one-vortex solution looks like

$$S_1+iS_2=\frac{2\alpha\rho e^{-i\varphi}}{|\alpha|^2+|\rho|^2}, \qquad S_3=\frac{|\alpha|^2-|\rho|^2}{|\alpha|^2+|\rho|^2}, \quad (5)$$

where $\rho e^{i\varphi}=x+iy-(x_0+iy_0)$, α is an arbitrary complex constant, and $Q=1$. An anti-vortex solution with $Q=-1$ is also given by (5) under the substitution $x+iy\to x-iy$, that is, $\varphi\to-\varphi$. General N-vortex solution has $Q=N$ and describes the time dynamics of the spin-vortices and spin-anti-vortices of the form (5) which exhibits no phase shifts. These solutions are all regular and rational functions on x and y; in other words, they are lump solutions of the Ishimori-II equation.

Both the Davey–Stewartson and the Ishimori equations also possess infinite sets of exact solutions parameterized by finite number of arbitrary functions of one variable. As for the NLS equation and the Heisenberg ferromagnet equation, the DS equation and the Ishimori equation are related by the so-called gauge transformation.

Multidimensional nonlinear equations not solvable by the inverse scattering method do not possess multisoliton-like solutions. However, some of them, such as lumps, breathers, or spherical waves, have particular solutions which are similar to one soliton solution in some respects. Such solutions play roles in various nonlinear phenomena arising in classical and quantum field theories, in nonlinear theories of matter, and even in the theory of Jupiter's Great Red Spot.

BORIS B. KONOPELCHENKO

See also **Dressing method; Inverse scattering method or transform; Kadomtsev–Petviashvili equation**

Further Reading

Ablowitz, M.J. & Clarkson, P.A. 1991. *Solitons, Nonlinear Evolution Equations and Inverse Scattering*, Cambridge and New York: Cambridge University Press

Boiti, M. et al. 1995. Multidimensional localized solitons. *Chaos, Solitons and Fractals*, 12: 2377–2417

Coleman, S. 1977. Classical lumps and their quantum descendants. In *New Phenomena in Sub-Nuclear Physics*, edited by A. Zichichi, New York: Plenum Press

Konopelchenko, B. 1993. *Solitons in Multidimensions. Inverse Spectral Transform Method*, Singapore: World Scientific

Rajaraman, R. 1982. *Solitons and Instantons*, Amsterdam: North-Holland

MULTIFRACTAL ANALYSIS

A fractal is a geometrical shape that shows self-similarity, meaning that parts appear similar to the set as a whole (*See* **Fractals**). A multifractal is a fractal with a probability measure attributed to its geometrical support set. A typical multifractal can have different fractal dimensions in different parts of its support set, depending on the multifractal measure chosen. Typical examples of multifractals are attractors of nonlinear mappings, where the relevant probability measure is given by the invariant measure of the map. Often, quite a complicated structure is observed, and the invariant density may diverge at infinitely many points in the phase space with different exponents, that is, an entire spectrum of singularities is generated.

To statistically analyze this complex behavior, it is useful to evaluate the so-called Rényi dimensions D_q (Rényi, 1970). These are generalizations of the usual box dimensions (or capacity) that contain information not only on the topological structure of the fractal but on the probability measure as well. The simplest definition of the Rényi dimensions is as follows. Cover the fractal with small d-dimensional cubes ("boxes") of volume ε^d, where ε is the side length of the box and d is an integer dimension large enough to embed the fractal. For each such box i we consider the probability p_i that is associated with it:

$$p_i=\int_{i\text{th box}}\rho(\boldsymbol{x})\,d\boldsymbol{x}. \quad (1)$$

Here $\rho(\boldsymbol{x})$ is the probability density considered. The Rényi dimensions are defined for any real parameter q as

$$D_q=\lim_{\varepsilon\to 0}\frac{1}{\log\varepsilon}\frac{1}{q-1}\log\sum_i p_i^q, \quad (2)$$

where the sum is over all i with $p_i\neq 0$. There are also more sophisticated definitions of the Rényi dimensions based on boxes of variable size (analogous to the definition of the Hausdorff dimension), see Beck & Schlögl (1993) for more details.

Generally, for a complicated multifractal (with lots of singularities of the probability density) the Rényi dimensions D_q depend on q, whereas for "simple" multifractals all D_q are the same and equal to the Hausdorff dimension. By changing the parameter q, one "scans" the structure of the multifractal. Large q values give more weight to large probabilities p_i, small q favor small probabilities p_i. Useful special cases of the Rényi dimensions are the box dimension (or capacity) D_0 (usually equal to the Hausdorff dimension, up to pathological cases), the information dimension D_1 (more precisely given by the limit $\lim_{q\to 1}D_q$), and the correlation dimension D_2. The correlation dimension can be easily extracted from experimental time series using the Grassberger–Procaccia algorithm (Grassberger & Procaccia, 1983). It is also of relevance

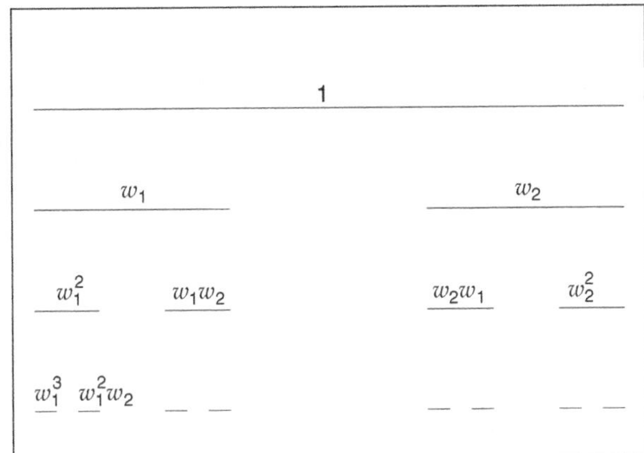

Figure 1. Construction of a classical Cantor set with a multiplicative non-uniform measure.

for the estimation of typical period lengths that are generated due to computer roundoff errors in chaotic dynamical systems (Beck, 1989). Also important are the limit dimensions $D_{\pm\infty}$ obtained for $q \to \pm\infty$. They describe the scaling behavior of the invariant density in the region of the phase space where the measure is most concentrated (D_∞) and least concentrated ($D_{-\infty}$).

From the Rényi dimensions, one can proceed to $f(\alpha)$, the spectrum of local scaling exponents α of the measure, by a Legendre transformation. The basic idea is very similar to thermodynamics. In fact, many of the ideas of the multifractal formalism are related to early work by Sinai, Ruelle, and Bowen on the so-called "thermodynamic formalism" of dynamical systems (Ruelle, 1978). For multifractals, one can regard the function $\tau(q) = (q-1)D_q$ as a kind of free energy, and q as a kind of inverse temperature (Tél, 1988; Beck & Schlögl, 1993). One then defines the variable α (the "internal energy") by $\partial\tau/\partial q =: \alpha$ and proceeds to the $f(\alpha)$ spectrum (the "entropy") by Legendre transformation:

$$f(\alpha) = q\alpha - \tau(q). \tag{3}$$

The advantage of the $f(\alpha)$ spectrum is that it has a kind of "physical meaning." Roughly speaking, it is the fractal dimension of the subset of points for which the probability density scales with a local exponent α, that is, $p_i \sim \varepsilon^{\alpha_i}$ with $\alpha_i \in [\alpha, \alpha + \mathrm{d}\alpha]$. Hence, this is a kind of statistical mechanics of local Hölder indices.

Let us consider a simple example of a multifractal, given by the classical Cantor set with a multiplicative (non-uniform) measure (Figure 1). The classical Cantor set is constructed by cutting the middle third out of the unit interval, then cutting out the middle third out of the remaining two intervals, and so on (*See* **Fractals**). In this way, at the kth construction step there are 2^k intervals of length $(\frac{1}{3})^k$. We now attribute a product measure to each of the intervals, according to the rule sketched in Figure 1, with $w_1 + w_2 = 1$ but $w_1 \neq w_2$. In the limit $k \to \infty$ we obtain a multifractal.

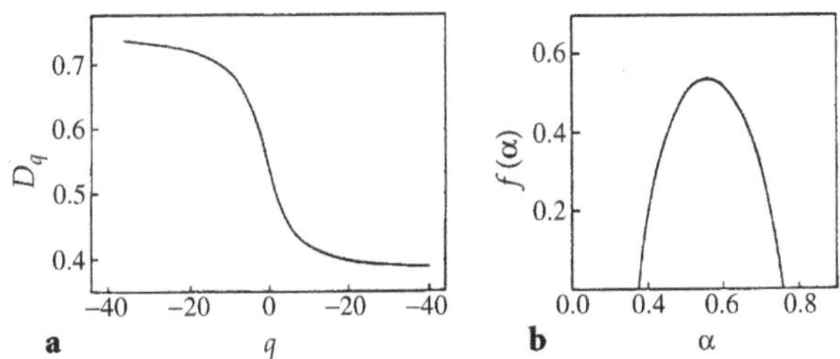

Figure 2. (a) Rényi dimensions and (b) $f(\alpha)$ spectrum of the Feigenbaum attractor.

Let us calculate the corresponding Rényi dimensions. We cover the multifractal with small intervals of size $\varepsilon_k = (\frac{1}{3})^k$. The number of boxes with probability $p_i = w_1^j w_2^{k-j}$ is

$$N_{k,j} = \left(\begin{array}{c} j \\ k \end{array}\right) = \frac{j!}{k!(k-j)!}. \tag{4}$$

Hence,

$$\sum_i p_i^q = \sum_{j=0}^{k} N_{k,j} w_1^{jq} w_2^{(k-j)q} = (w_1^q + w_2^q)^k \tag{5}$$

and for the Rényi dimensions, one obtains from definition (2)

$$D_q = \frac{\log(w_1^q + w_2^q)}{(1-q)\log 3}. \tag{6}$$

In particular, the box dimension D_0 is given by the value $D_0 = \log 2/\log 3$, independent of w_1 and w_2. The other dimensions depend on the probabilities w_j, for example, for the information dimension one obtains $D_1 = -(1/\log 3)\sum_{j=1}^{2} w_j \log w_j$ by considering the limit $q \to 1$. The Legendre transform of (6) yields the $f(\alpha)$ spectrum.

Another interesting example is the attractor of the logistic map at the critical point of period doubling accumulation. Figure 2a shows the corresponding Rényi dimensions and Figure 2b the corresponding $f(\alpha)$ spectrum. Generally, the value of α where the function $f(\alpha)$ has its maximum is equal to the Hausdorff dimension of the attractor. The value of $f(\alpha)$ where the function has slope 1 is equal to the information dimension.

In practice, one often wants to evaluate the Rényi dimensions (or the $f(\alpha)$ spectrum) for a given time series (Kantz & Schreiber, 1997) of experimental data without explicitly knowing the underlying dynamics or the invariant measure. Here, various interesting methods are known (Grassberger–Procaccia algorithm (Grassberger & Procaccia, 1983), wavelet analysis (Arneodo et al., 1995), etc.). These algorithms can be implemented without explicitly knowing the underlying dynamics. The wavelet transform of an experimental signal $s(x)$ is defined as

$$W_\Psi(x_0, a) = \frac{1}{a}\int_{-\infty}^{+\infty} \Psi^*\left(\frac{x - x_0}{a}\right) s(x)\,\mathrm{d}x, \quad (7)$$

($*$ indicates complex conjugate), where the analyzing wavelet Ψ is some localized function, often chosen to be the Nth derivative of a Gaussian function. For small a, the wavelet transform extracts local Hölder exponents from the signal s, and qth moments of W_Ψ can then be used to define suitable partition functions which yield the Rényi dimensions.

CHRISTIAN BECK

See also **Dimensions; Fractals; Free energy; Measures; Sinai–Ruelle–Bowen measures; Wavelets**

Further Reading

Arneodo, A., Bacry, E. & Muzy, J.F. 1995. The thermodynamics of fractals revisited with wavelets. *Physica* A, 213: 232–275

Beck, C. 1989. Scaling behavior of random maps. *Physics Letters* A, 136: 121–125

Beck, C. & Schlögl, F. 1993. *Thermodynamics of Chaotic Systems*, Cambridge and New York: Cambridge University Press

Grassberger, P. & Procaccia, I. 1983. Characterization of strange attractors. *Physical Review Letters*, 50: 346–349

Kantz, H. & Schreiber, T. 1997. *Nonlinear Time Series Analysis*, Cambridge and New York: Cambridge University Press

Rényi, A. 1970. *Probability Theory*, Amsterdam: North-Holland

Ruelle, D. 1978. *Thermodynamic Formalism*, Reading, MA: Addison-Wesley

Tél, T. 1988. Fractals, multifractals and thermodynamics – an introductory review. *Zeitschrift für Naturforschung* A, 43: 1154–1174

MULTIFRACTAL MEASURE

See **Dimensions**

MULTIPLE SCALE ANALYSIS

For a number of problems involving differential equations, we know methods that can provide exact solutions. However, the vast majority of modeling problems have a complexity that forbids finding an exact solution by paper and pencil. It may appear that the only way forward is to use numerical analysis. However, this is not necessarily the case since we can resort to finding good approximate solutions by various methods or strategies. In applications we usually encounter systems with dissipation and energy input, where neglecting these energy exchange terms leads to an exact solvable problem. We can then employ a strategy based on the assumption that the exact solution of the unperturbed problem is only modified slightly by adding the perturbation terms and hope that the difference between the two solutions can be estimated sufficiently accurately by assuming the perturbation is weak (Nayfeh, 1973, 1981, 2000).

Let us illustrate such a strategy by considering the damped harmonic oscillator. Denoting the displacement from equilibrium by $x(t)$ at time t, the dynamical equation reads

$$\ddot{x} + x = -\varepsilon\dot{x}. \quad (1)$$

A dot above the dependent variable $x(t)$ denotes differentiation with respect to time t. The damping parameter ε we shall vary, and accordingly it is natural to include ε in the argument list of $x = x(t;\varepsilon)$. The unperturbed oscillator corresponds to $\varepsilon = 0$, and using the initial conditions $x(0) = 1$ and $\dot{x}(0) = 0$, the solution of the unperturbed harmonic oscillator reads

$$x(t; 0) = x_0(t) = A\cos(t), \quad (2)$$

where the amplitude A equals unity. For the same initial conditions, the damped harmonic oscillator (1) with $\varepsilon \neq 0$ possesses the more complicated solution

$$x(t;\varepsilon) = \mathrm{e}^{-\varepsilon t/2}\left[\cos\left(t\sqrt{1-\varepsilon^2/4}\right) + \frac{\varepsilon/2}{\sqrt{1-\varepsilon^2/4}}\sin\left(t\sqrt{1-\varepsilon^2/4}\right)\right]. \quad (3)$$

The exact solution has features which can be used as a guide for developing a systematic way of finding an approximate solution assuming small damping, that is, $|\varepsilon| \ll 1$.

Without surprise, we recognize that the damping leads to a decreasing amplitude as time progresses but on a slow time scale of order $O(\varepsilon)$. Another result of the damping effect is an $O(\varepsilon^2)$ change of the frequency, which is $\omega = \sqrt{1-\varepsilon^2/4} \approx 1 - \varepsilon^2/8$. The damping slows down the oscillation, and accordingly the period increases, hence the frequency decreases. Suitable corrections are therefore to allow for a slow variation of the amplitude A and for an even smaller correction of the frequency. In general terms, this implies that we need to let $x(t;\varepsilon)$ depend on the slower time scales εt and $\varepsilon^2 t$. We can formalize this by introducing (Nayfeh, 1981)

$$T_0 = t, \quad T_1 = \varepsilon t, \quad T_2 = \varepsilon^2 t, \qquad x(t) = x(T_0, T_1, T_2). \quad (4)$$

Thus, the time derivatives expand as

$$\frac{\mathrm{d}}{\mathrm{d}t} = \frac{\partial}{\partial T_0} + \varepsilon\frac{\partial}{\partial T_1} + \varepsilon^2\frac{\partial}{\partial T_2} + O(\varepsilon^3), \quad (5)$$

$$\frac{d^2}{dt^2} = \frac{\partial^2}{\partial T_0^2} + 2\varepsilon\frac{\partial^2}{\partial T_0\partial T_1} + \varepsilon^2\left(\frac{\partial^2}{\partial T_1^2} + 2\frac{\partial^2}{\partial T_0\partial T_2}\right) + O(\varepsilon^3), \quad (6)$$

by invoking the chain role for partial derivatives. The idea is to introduce slowly time-varying coefficients by regarding the new scaling variables in (4) as independent variables. This is a somewhat mysterious trick, and it is not easy to understand why it works, since the scaling variables are all proportional to time, but it does. In perturbation theory, we often rely on such experience and tricks.

After having introduced the scaling into our original problem, we can add further corrections to the unperturbed solution by introducing a Taylor expansion around $\varepsilon = 0$ with respect to the perturbation parameter ε

$$x(\boldsymbol{T};\varepsilon) = x_0(\boldsymbol{T}) + x_1(\boldsymbol{T})\varepsilon + x_2(\boldsymbol{T})\varepsilon^2 + O(\varepsilon^3), \quad (7)$$

where

$$\mathbf{T} = (T_0, T_1, T_2) \quad \text{and} \quad x_n(\mathbf{T}) = \frac{1}{n!}\frac{\partial^n x(\mathbf{T};0)}{\partial\varepsilon^n}.$$

Inserting the Taylor expansion into Equation (1) results in a polynomial in ε which equals zero. The coefficient of ε^n is a differential operator working on x_n, and as all these coefficients must vanish, we have a differential equation for each x_n, which normally is simple enough that it can be solved analytically, at least for the first few equations. Inserting the Taylor expansion into Equation (1) and ordering according to powers of ε we obtain

$$x_{0,T_0T_0} + x_0 = 0, \quad (8)$$

$$x_{1,T_0T_0} + x_1 = -x_{0,T_0} - 2x_{0,T_0T_1}, \quad (9)$$

$$x_{2,T_0T_0} + x_2 = -x_{1,T_0} - 2x_{1,T_0T_1} - x_{0,T_1} - x_{0,T_1T_1} - 2x_{0,T_0T_2}. \quad (10)$$

Subscript T_0 means a partial derivative with respect to T_0 and similarly for the analogous subscripts. The solution of the first equation reads

$$x_0 = A(T_1, T_2)e^{iT_0} + B(T_1, T_2)e^{-iT_0}. \quad (11)$$

The fact that A and B depend on T_1 and T_2 expresses the slow variation of these parameters with time. The solution of the equation for x_1 can now be found by inserting x_0 from (11) into the right-hand side of Equation (9),

$$x_{1,T_0T_0} + x_1 = -i(A + 2A_{T_1})e^{iT_0} + i(B + 2B_{T_1})e^{-iT_0}. \quad (12)$$

Terms proportional to the exponentials $e^{\pm iT_0}$ are resonant forcing terms, leading to a growth of x_1 proportional to time t. When t is of order $1/\varepsilon$, the term $\varepsilon x_1(t)$ has become of order unity, and our analysis breaks down. Such terms are called secular terms. In order to get a uniform and valid expansion, we need to avoid secular terms. However, due to the dependence on T_1 of A and B, we can demand that coefficients proportional to $e^{\pm iT_0}$ vanish thereby providing simple differential equations for determining A

$$A + 2A_{T_1} = 0 \quad \Rightarrow \quad A(T_1, T_2) = A_1(T_2)e^{-T_1/2}, \quad (13)$$

and similarly for B. Equation (12) now possesses the solution $x_1 = 0$. We can proceed by solving the $O(\varepsilon^2)$ equation (10) and invoke the initial conditions $x(0) = 1$ and $\dot{x}(0) = 0$. After rather straightforward calculations, we obtain the approximate solution (Scott, 2003)

$$x(t) = e^{-\varepsilon t/2}\cos[(1 - \varepsilon^2/8)t] + \frac{\varepsilon}{2}e^{-\varepsilon t/2} \times \sin[(1 - \varepsilon^2/8)t] + O(\varepsilon^3) \quad (14)$$

in agreement with the exact solution to the desired order in ε. Carrying out the analysis to order $O(\varepsilon^n)$ requires introduction of scaled variables up to the same order for the perturbation analysis to be consistent. The above approach is called the method of multiple scales, and it has been applied to many nonlinear ordinary differential equations as well as perturbed soliton equations (McLaughlin & Scott, 1978; Scott, 2003), including derivations of nonlinear partial differential equations (Dodd, 1984). An interesting extension of the method to nonlinear difference equations is presented in Broomhead & Rowlands (1983).

MADS PETER SØRENSEN

See also **Averaging methods; Damped-driven anharmonic oscillator; Fredholm theorem; Multi-soliton perturbation theory; Perturbation theory**

Further Reading

Broomhead, D.S. & Rowlands, G. 1983. On the analytic treatment of non-integrable difference equations. *Journal of Physics* A, 16: 9–24

Dodd, R.K., Eilbeck, J.C., Gibbon, J.D. & Morris, H.C. 1984. *Solitons and Nonlinear Wave Equations*, London: Academic Press

McLaughlin, D.W. & Scott, A.C. 1978. Perturbation analysis of fluxon dynamics. *Physical Review* A, 18: 1652–1680

Nayfeh, A.H. 1973. *Perturbation Methods*, New York: Wiley-Interscience

Nayfeh, A.H. 1981. *Introduction to Perturbation Techniques*, New York: Wiley-Interscience

Nayfeh, A.H. 2000. *Nonlinear Interactions: Analytical, Computational, and Experimental Methods*, New York: Wiley-Interscience

Scott, A.C. 2003. *Nonlinear Science: Emergence and Dynamics of Coherent Structures*, 2nd edition, Oxford and New York: Oxford University Press

MULTIPLEX NEURON

Under the view that dominated neuroscience around the middle of the 20th century, neural information processing involves three components (McCulloch & Pitts, 1943). First, incoming signals from chemical synapses on the dendritic trees are gathered as a linear, weighted sum of input signals. Second, this sum is compared with a threshold level at the base (initial segment) of the outgoing axonal tree. Finally, if the threshold level is exceeded, an active (all-or-nothing) nerve impulse is launched on the axonal tree, which propagates outward without failure to all of

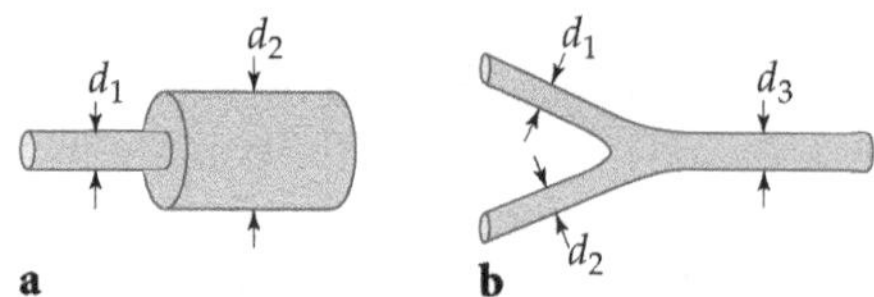

Figure 1. (a) Abrupt widening of a nerve fiber. (b) Branching region.

Membrane model	Transmission	Blockage
H–H	5.5084	5.5126
M–L	2.2521	2.2563

Table 1. Widening ratios at which isolated impulses transmit and become blocked under Hodgkin–Huxley (H–H) and Morris–Lecar (M–L) membrane models (Altenberger et al., 2001).

the distal (distant) twigs of the tree, providing inputs to other neurons or to muscle cells. A simplifying feature of this "McCulloch–Pitts model" is that the incoming dynamics are entirely linear, governed by inhomogeneous diffusion processes on the dendritic trees.

Before the end of the 1960s, there was no compelling evidence to abandon this simple view of dendritic dynamics, and two reasons for clinging to it. If all-or-nothing propagation is supposed to occur on dendrites, then the entire tree might be expected to ignite, preventing the dendrites from integrating incoming information. Additionally, the assumption of linear dendritic dynamics helps the theorist to sort out various causal influences, somewhat easing the difficulties of analyzing neural systems (Scott, 2002).

By the 1970s, three types of evidence began to indicate that dendritic and axonal dynamics are more complex. First, impulse blockage was experimentally observed in the optic nerves of cats, implying that impulses (spikes) can be extinguished at axonal branchings (Chung et al., 1970). Second, experimental studies showed that spikes do indeed propagate on the dendritic trees of some vertebrate neurons (Llinás & Nicholson, 1971). Third, numerical studies by Boris Khodorov and his colleagues in the Soviet Union on realistic models of nerve fibers confirmed that the propagation of active nerve impulses can indeed be blocked at changes in fiber geometry such as the abrupt widening and branching regions shown in Figure 1 (Khodorov, 1974).

These considerations suggested the concept of a "multiplex neuron," where the term (borrowed from communication engineering) implies the ability to handle two or more messages at the same time. As defined by Steven Waxman, a multiplex neuron has the following salient properties (Waxman, 1972).

- Impulse blockage at the branching regions of dendritic trees allows the possibility of Boolean logic, similar to the elementary operations of a digital computer. If the geometry of the branch is such that an impulse incoming on either daughter branch (one "or" the other) is able to ignite an outgoing impulse on the parent branch, this would be OR logic. If, on the other hand, coincident impulses on both incoming branches (one "and" the other) are required to ignite an outgoing impulse, it would be an example of AND logic. Thus the computational nature of a branching region may depend upon details of its geometry.
- Impulse blockage at branching regions of the axonal tree allows an impulse code transformation under which a time code on the trunk of the tree is transformed to space-time codes at the distal branches (twigs) of the tree.

Seminal Soviet studies of the propagation of impulses through the abrupt widening shown in Figure 1(a) showed that blockage of an isolated Hodgkin–Huxley impulse (Hodgkin & Huxley, 1952) (which corresponds to the dynamics of a squid giant axon) was to be expected at a widening ratio (d_2/d_1) greater than about 5.5. As indicated in Table 1, recent numerical studies by Altenberger et al. (2001) have confirmed the early Soviet results for the Hodgkin–Huxley model and extended them to the Morris–Lecar model (which represents dynamics on barnacle giant muscle fibers and more closely models the calcium dominated dynamics of typical dendritic fibers) (Morris & Lecar, 1981).

Although abrupt widenings of real nerve fibers are not observed, numerical blocking conditions in Figure 1(a) can be related to the corresponding blocking conditions in Figure 1(b) through the concept of a "geometric ratio" (GR) (Goldstein & Rall, 1974). Assuming a discontinuity with several outgoing fibers (of diameters d_{out}) and on which an impulse is incoming on a single fiber of diameter d_{in}, the GR is defined as $\sum d_{\text{out}}^{3/2}/d_{\text{in}}^{3/2}$, where the "3/2 powers" enter because branching currents divide in proportion to the characteristic admittance of the outgoing fibers (Scott, 2002).

Assuming the Morris–Lecar (M–L) model and an incoming impulse on daughter branch # 1 in the branch of Figure 1(b), the AND condition is

$$\frac{d_2^{3/2} + d_3^{3/2}}{d_1^{3/2}} > 2.2563^{3/2} = 3.389, \tag{1}$$

requiring coincident inputs on both incoming branches (# 1 and # 2) for an impulse on the outgoing branch (# 3) to ignite. If this geometric inequality is not satisfied, the branch executes OR logic, in which incoming impulse either on daughter # 1 or on daughter # 2 is able to ignite an outgoing impulse on the parent branch (# 3).

Cell type	GR range
Purkinje	2.3–3.5
Stellate	2.4–3.5
Granule	2.3–4.8
Motoneuron	2.6–3.4
Pyramidal (apical)	2.5–3.4
Pyramidal (basal)	2.4–3.1

Table 2. The GR range for some typical dendrites (Scott, 2002).

Noting the GR values for typical dendritic trees given in Table 2 and considering the several ways in which critical GRs might be lowered in real neurons (changes in ionic concentrations, variations in channel membrane density, incomplete impulse recovery from previous interactions, and so on), it seems prudent to anticipate that dendritic trees can execute logical operations on the incoming (synaptic) codes, as is assumed in the multiplex neuron model (Stuart et al., 1999).

ALWYN SCOTT

See also **McCulloch–Pitts network; Nerve impulses; Neurons**

Further Reading

Altenberger, R., Lindsay, K.A., Ogden, J.M. & Rosenberg, J.R. 2001. The interaction between membrane kinetics and membrane geometry in the transmission of action potentials in non-uniform excitable fibres: a finite element approach. *Journal of Neuroscience Methods*, 112: 101–117

Chung, S.H., Raymond, S.A. & Lettvin, J.Y. 1970. Multiple meaning in single visual units. *Brain, Behavior and Evolution*, 3: 72–101

Goldstein, S.S. & Rall, W. 1974. Changes of action potential, shape and velocity for changing core conductor geometry. *Biophysical Journal*, 14: 731–757

Hodgkin, A.L. & Huxley, A.F. 1952. A quantitative description of membrane current and its application to conduction and excitation in nerve. *Journal of Physiology*, 117: 500–544

Khodorov, B.I. 1974. *The Problem of Excitability*, New York: Plenum

Llinás, R. & Nicholson, C. 1971. Electrophysiological properties of dendrites and somata in alligator Purkinje cells. *Journal of Neurophysiology*, 34: 532–551

McCulloch, W.S. & Pitts, W.H. 1943. A logical calculus of the ideas immanent in nervous activity. *Bulletin of Mathematical Biophysics*, 5: 115–133

Morris, C. & Lecar, H. 1981. Voltage oscillations in the barnacle giant muscle fibre. *Biophysical Journal*, 35: 198–213

Scott, A.C. 2002. *Neuroscience: A Mathematical Primer*, Berlin and New York: Springer

Stuart, G., Spruston, N. & Häusser, M. (editors). 1999. *Dendrites*, Oxford and New York: Oxford University Press

Waxman, S.G. 1972. Regional differentiation of the axon, a review with special reference to the concept of the multiplex neuron. *Brain Research*, 47: 269–288

MULTIPLICATIVE NOISE

See **Stochastic processes**

MULTIPLICATIVE PROCESSES

Random processes involving accumulation and/or reduction occur widely in nature and in mathematics. At the most basic level, the laws of accumulation (and reduction) may be represented either as the result of arithmetic summation (and differencing) or as the result of multiplications. The latter such processes are the subject of this article.

Any attempt to embody all multiplicative random processes in a single mathematical definition would be futile. However, it is possible to provide large classes of examples for which rich mathematical theories and applications are possible.

The "law of proportionate effect" provides the basis for the evolution of multiplicative stochastic processes $M_0, M_1, \ldots,$ found, for example, in materials science, biology, economics, and finance, in terms of the proportionate changes $r_{n+1} = \Delta M_n/M_n \equiv (M_n + 1 - M_n)/M_n, n = 0, 1, 2, \ldots.$ In finance, M_n may be the value of a unit of stock at the nth period of time when subjected to the hypothesis that the yields $r_1, r_2, \ldots$ are stationary and independent from time period to period. In materials science, on the other hand, M_n may represent the strength of a material after n repeated independent random impacts $R_1 = 1 + r_1, R_2 = 1 + r_2, \ldots.$ Similarly, M_n may represent the size of a biological population whose subsequent growth is in (random) proportions $r_1, r_2, \ldots$ to the amount of substance available. The law of proportionate effect is a widely observed statistical law according to which the sequence of proportionate changes $r_{n+1} = \Delta M_n/M_n, n = 0, 1, 2, \ldots,$ is a sequence of independent and identically distributed (iid) random variables. Equivalently, one may iterate this law to obtain the so-called geometric random walk model:

$$M_n = M_0 \prod_{k=1}^{n} R_k, \quad k = 1, 2, \ldots, \tag{1}$$

where

$$R_k = 1 + r_k, \quad k = 1, 2, \ldots. \tag{2}$$

The special case in which R_n has two possible values, corresponding to upward and downward price movements, yields the binomial tree model for stock prices popular in the modern mathematical finance literature (Föllmer & Schied, 2002).

A more general notion of "multiplicative processes", whose origins may be traced back to early 19th century genetics, occurs by consideration of a somewhat finer scale growth rule. In particular, changes to the process occur as M_{n+1} evolves from a previous random number

M_n of constituent members, each of which individually contributes its own growth and decay amounts (Y) to the whole in successive time steps. This more general stochastic law of evolution may be precisely expressed by the following iteration scheme: If $M_n = 0$ then define $M_{n+1} = 0$, else if $M_n > 0$ define

$$M_{n+1} = \sum_{k=1}^{M_n} Y_{n+1,k} \quad n = 0, 1, \ldots, \tag{3}$$

where M_0 is a nonnegative integer valued (counting) random variable representing an initial size, and for each $n \geq 0$, the random variables $Y_{n+1,1}, Y_{n+1,2}, \ldots Y_{n+1,M_n}$ are nonnegative integer valued, representing the respective numbers of offspring of each of the M_n elements composing the nth generation.

Two special cases may be noted. In the case that the $Y_{n+1,k} \equiv R_{n+1},\ k \geq 1$, are the same for each k of the nth generation, but are independent and identically distributed (iid) from generation to generation n, Equation (3) is the law of proportionate effect. On the other hand, if for each n, the offspring sizes $Y_{n+1,1}, Y_{n+2,2}, \ldots$ are iid and independent of M_n, then Equation (3) defines the classical Bieneymé–Galton–Watson branching process (BGW) introduced in the early 19th century as a model for analyzing the survival of family names (Kendall, 1975).

A next generation of multiplicative processes arises out of these models by a still finer coding of the underlying phenomena. For simplicity, consider the BGW branching process and assume a single progenitor $M_0 = 1$. Then one may code the successive offspring as a random tree as follows: The single progenitor is coded by the empty sequence $\emptyset$. Its respective $Y_{1,1}$ offspring are coded (labeled) as $\langle 1 \rangle, \ldots, \langle Y_{1,1} \rangle$. The $Y_{2,k}$ next generation offspring of $\langle k \rangle$ can be coded as $\langle k1 \rangle, \langle k2 \rangle, \ldots, \langle kY_{2,k} \rangle$, and so on. In this way, the successive generations may be represented as a sequence of random family trees

$$\begin{aligned} \tau_0 = \{\emptyset\},\ \tau_1 &= \{\emptyset, \langle 1 \rangle, \ldots, \langle Y_{1,1} \rangle\} \\ &\equiv \tau_0 \cup \{\langle 1 \rangle, \ldots, \langle Y_{1,1} \rangle\}, \\ \tau_2 &= \tau_1 \cup \{\langle 11 \rangle, \ldots, \langle 1Y_{2,1} \rangle, \ldots, \\ &\quad \langle Y_{1,1}1 \rangle, \ldots, \langle Y_{1,1}Y_{2,Y_{1,1}} \rangle\}, \ldots. \end{aligned}$$

This, in fact, suggests an alternative description of the BGW process as a probability distribution P on a (metric) space of tree graphs. Specifically, let $\boldsymbol{T}$ be the space of (possibly infinite) labeled tree graphs rooted at $\emptyset$. As above, an element τ of $\boldsymbol{T}$ may be coded as a set of finite sequences of positive integers $\langle i_1, i_2, \ldots, i_n \rangle \in \tau$ such that:

(i) The root $\emptyset \in \tau$ is coded as the empty sequence.
(ii) If $\langle i_1, \ldots, i_k \rangle \in \tau$ then $\langle i_1, \ldots i_j \rangle \in \tau$ $\forall\, 1 \leq j \leq k$.
(iii) If $\langle i_1, i_2, \ldots, i_n \langle \in \tau$ then $\rangle i_1, \ldots i_{n-1}, j \rangle \in \tau$ $\forall\, 1 \leq j \leq i_n$.

Neighboring vertices are those of the form $\langle i_1, \ldots, i_n \rangle$ and $\langle i_1, \ldots, i_{n-1} \rangle$. The last condition (iii) specifies a left to right labeling of vertices, and (ii) connects all paths to the root via neighboring vertices without cycles. The space $\boldsymbol{T}$ of such trees may be viewed as a metric space with metric defined by $\rho(\tau, \gamma) = (\sup\{n+1 : \gamma|n = \tau|n\})^{-1}$, and $\tau|n = \{\langle i_1, \ldots, i_k \rangle \in \tau : k \leq n\}$ is the truncation of τ to the first n generations. This metric is complete and the countable dense subset $\boldsymbol{T}_0$ of finite labeled tree graphs rooted at $\emptyset$ makes $\boldsymbol{T}$ a separable metric space as well.

The BGW model may now be viewed as a probability distribution on the space $\boldsymbol{T}$, wherein the probability assigned to the ball B of radius $1/N$ centered at τ is given by

$$P(B) = \prod_{v \in \tau|N} p(\omega(\tau; v)), \tag{4}$$

where $p(j) = P(Y = j),\ j = 0, 1, 2, \ldots$ is a prescribed offspring probability distribution, and $\omega(\tau; v) = \max\{j : \langle v1 \rangle, \ldots, \langle vj \rangle \in \tau\}$ counts the number of offspring of the vertex v in the tree τ.

A number of naturally occurring geophysical and biological structures admit natural tree codings, for example, river networks, lightning discharge patterns, and arterial and neural networks in human organs. Probabilistic models correspond to probability distributions on $\boldsymbol{T}$, leading to further generalizations of multiplicatively branching models; for example, see Barndorff-Nielsen et al. (1998) and references therein.

Random multiplicative cascade models provide another class of multiplicative processes of interest for their intermittency, extreme variability, and multiscaling structure in applications to phenomena ranging from fluid turbulence and spatial oceanic rainfall distributions to internet packet data on the world wide web (Barndorff-Nielsen et al., 1998; Gilbert et al., 1999; Jouault et al., 2000). The origins of this class of models trace back to classic work of Lewis Fry Richardson (1926), Andrei N. Kolmogorov (1962), and A.M. Yaglom (1966) in statistical turbulence theory. A rich geometrical and scaling perspective was subsequently advanced by Benoit Mandelbrot, and a more complete mathematical treatment was initiated by J.-P. Kahane and Jacques Peyrière (Kahane, 1985).

In the context of turbulence, one imagines introducing energy into the fluid by a large-scale stirring motion, whereby smaller-scale eddies split off and dissipate energy in random proportions R to those available. These eddies in turn split off smaller-scale eddies, and so on. In the simplest mathematical formulation, one considers random measures $M_n(\Delta)$ on the one-dimensional unit interval $[0, 1]$ having a piecewise constant density

$\rho_n(x)$ defined by a constant value over a dyadic subinterval

$$\Delta_n(j_1,\ldots,j_n) = \left[\sum_{k=1}^{n} j_k 2^{-k}, \sum_{k=1}^{n} j_k 2^{-k} + 2^{-n}\right],$$
$$j_k \in \{0,1\},\ k \geq 1 \tag{5}$$

with constant value

$$\rho_n(x) = \prod_{k=1}^{n} R_{\langle j_1 j_2 \ldots j_k \rangle},\ \forall x \in \Delta_n(j_1, j_2, \ldots, j_k), \tag{6}$$

where the random factors R_v, $v = \langle j_1 j_2 \ldots j_k \rangle$, are iid nonnegative mean one random variables indexed by the vertices of the binary tree. The mean one condition provides conservation of the mass of M_n on average, where for any Borel subset G

$$M_n(G) = \int_G \rho_n(x)\,\mathrm{d}x. \tag{7}$$

The multiplicative cascade M_∞ is the random measure obtained by passing to the fine scale limit as $n \to \infty$.

The random cascade is defined by the branching number b ($=2$ in this exposition), and the random factors R_v, referred to as cascade generators. Kolmogorov's "log-normal hypothesis" refers to a choice of lognormal distribution for R. Tests of this hypothesis and physical arguments for alternative laws present significant challenges for modern statistical turbulence theory (e.g., She & Waymire, 1995; Jouault et al., 2000). On the mathematical side, the special choice of symmetrically distributed 0–2 valued generators yields the BGW process with binomial offspring distribution $p(j) = \binom{2}{j}\frac{1}{4}$, $j = 0, 1, 2$, for the total masses $M_n([0,1]), n = 1, 2, \ldots$. Refinements for which the branching number b may be viewed as a random parameter are given in Burd & Waymire (2000), and as a continuous parameter in Barral & Mandelbrot (2002).

Branching random walk models of the type illustrated by this next application may also be viewed within the framework of random multiplicative cascades. An explicit representation of the Fourier transform of classes of solutions to 3-d incompressible Navier–Stokes equations

$$\frac{\partial u}{\partial t} + u \cdot \nabla u = \nu \Delta u - \nabla p + g, \quad \nabla \cdot u = 0, \tag{8}$$

in the form of an expected value of a certain product of initial and forcing data evaluated at the nodes of a branching random walk was recently uncovered (LeJan & Sznitman, 1997). While probability models have long enjoyed important connections to partial differential equations, most notable being the heat equation and reaction-diffusion equations, this ranks among the most striking recent connections between multiplicative cascades and nonlinear equations of fluid motion. For a simple illustration of ideas made possible by refinements of the theory developed in Bhattacharya et al. (2003), consider the one-dimensional Burgers equation

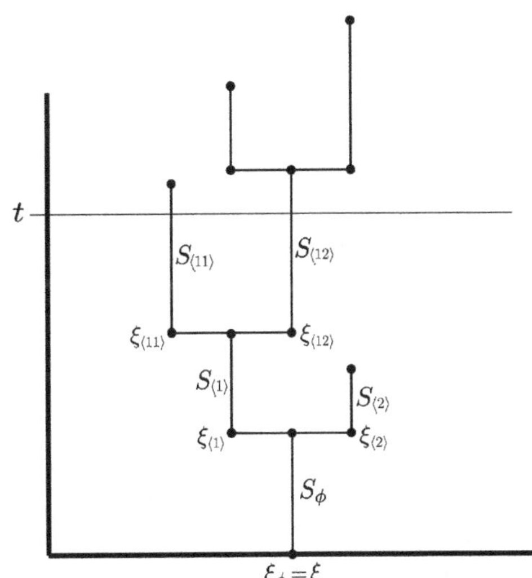

Figure 1. A sample tree graph for the Burgers equation.

$$u_t + uu_x = \nu u_{xx}, \quad u(0,x) = u_0(x). \tag{9}$$

Spatial Fourier transform will be denoted by $\hat{u}$. For simplicity of exposition, assume $\hat{u}_0(\xi) = 0$ for $\xi \leq 0$; that is, the initial data belongs to a Hardy function space. Taking spatial Fourier transforms one obtains, with the aid of an exponential integrating factor and $\xi > 0$ and writing

$$\lambda(\xi) = \nu\xi^2, \qquad m = -\frac{\mathrm{i}}{\nu}, \tag{10}$$

the result

$$\begin{aligned}\hat{u}(t,\xi) &= \mathrm{e}^{-\lambda(\xi)t}\hat{u}_0(\xi) \\ &\quad + \frac{m}{2}\int_0^t \lambda(\xi)\mathrm{e}^{-\lambda(\xi)s}\frac{1}{\xi} \\ &\quad \times \int_0^\xi \hat{u}(t-s,\eta)\hat{u}(t-s,\xi-\eta)\,\mathrm{d}\eta. \end{aligned} \tag{11}$$

Expressed in this form, Equation (11) takes on a probabilistic meaning. In particular, this is a recursive equation for the expected values of a multiplicative stochastic process initiated at $\xi_\emptyset = \xi$. The first term on the right-hand side $\mathrm{e}^{-\lambda(\xi)t}\hat{u}_0(\xi)$ is the product of the initial data $\hat{u}_0(\xi)$ times the probability $\mathrm{e}^{-\lambda(\xi)t}$ that an "exponentially distributed clock" with parameter $\lambda(\xi)$ rings after time t. The second integral term is an expected value in the complementary event of probability density $\lambda(\xi)\mathrm{e}^{-\lambda(\xi)s}$, that the clock rings at time s prior to t. Given that the clock rings at a time s prior to t, a product is formed with the factor $m(\xi)$ and a random selection of a pair of new "offspring" wave numbers (or Fourier frequencies) $\eta, \xi - \eta$ from the interval $[0, \xi]$ with (uniform) probability density $1/\xi$ to complete the recursion over the remaining time $t - s$. That is to say, the unique solution (in the appropriate function space) is furnished by the expected value

$$\hat{u}(t,\xi) = EX(t,\xi), \quad \xi > 0, \tag{12}$$

for a multiplicative cascade $X(t,\xi)$ defined by the following stochastic recursion in Fourier wave number space (see Figure 1): A particle of type $\xi_\emptyset = \xi$ waits for an exponentially distributed time $S_\emptyset$ with mean $1/\lambda(\xi) = \frac{1}{\nu\xi^2}$. If $S_\emptyset > t$ then a value $\hat{u}_0(\xi)$ is assigned to the initial vertex $\emptyset$ and the process terminates. On the other hand, if $S_\emptyset \leq t$ then an independent coin flip is made. If the outcome is a tail then the particle dies and a value 0 is assigned, but if a head occurs then the particle branches into two particles $\langle 1 \rangle$, $\langle 2 \rangle$ of respective types $\xi_{\langle 1 \rangle} = \eta$ and $\xi_{\langle 2 \rangle} = \xi - \eta$ selected according to the uniform distribution on $[0, \xi]$. Two independent exponential clocks $S_{\langle 1 \rangle}$, $S_{\langle 2 \rangle}$, with respective parameters $\lambda(\xi_{\langle 1 \rangle})$, $\lambda(\xi_{\langle 2 \rangle})$, are set, and the process is repeated independently from each of these two given types for the termination time t reduced to $t - S_\emptyset$. The multiplicative cascade $X(t,\xi)$ is, up to a (random) power of $m \equiv -\sqrt{-1}/\nu$, a product of the assigned values of $\hat{u}_0$ at the selected frequencies; for example, for the sample realization depicted in Figure 1, one has

$$X(t,\xi) = m^2 \cdot \hat{u}_0(\xi_{\langle 11 \rangle}) \cdot \hat{u}_0(\xi_{\langle 12 \rangle}) \cdot \hat{u}_0(\xi_{\langle 2 \rangle}) \cdot 0. \tag{13}$$

In particular, the premature death of $\langle 2 \rangle$ means that this particular sample realization will not contribute a positive value to the mathematical expectation in (12).

By presenting these models as a progression of modifications and extensions built on simpler structures, some sense of a mathematical theory begins to emerge, a large part of which directly rests on martingale theory. For example, for the BGW model one may observe, denoting the mean offspring number by μ, assumed positive and finite, that $M_n/\mu^n, n = 1, 2, \ldots$, is a martingale. Similarly, for each fixed Borel set G, the cascade measure $M_n(G)$, as a function of (logarithmic) scale $n = 1, 2, \ldots$, is a martingale. The latter property led Kahane to still further natural generalizations of widespread significance (Kahane, 1985). Such deep mathematical structure has made it possible to precisely analyze many of the singularities, intermittencies, and other critical phenomena associated with these models, as well as to provide precise statistical error bars required for scientifically sound empirical estimations and tests of hypothesis for multiplicative cascades. On the other hand, the overall theory is in its relative infancy and many questions of practical importance remain open (Ossiander & Waymire, 2000). Needless to say, the relationship with the incompressible Navier–Stokes equations provides a link to one of the most outstanding mathematical problems of our times.

EDWARD C. WAYMIRE

See also **Branching laws; Burgers equation; Martingales; Navier–Stokes equation**

Further Reading

Athreya, K. & Jagers, P. (editors). 1997. *Classical and Modern Branching Processes*, New York: Springer

Barndorff-Nielsen, O.E., Gupta, V.K., Perez-Abreu, V. & Waymire, E. (editors). 1998. *Stochastic Methods in Hydrology: Rain, Landforms and Floods*, Singapore: World Scientific

Barral, J. & Mandelbrot, B.B. 2002. Multifractal products of cylindrical pulses. *Probability Theory and Related Fields*, 124(3): 409–430

Bhattacharya, R., Chen, L., Dobson, S., Guenther, R., Orum, C., Ossiander, M., Thomann, E. & Waymire, E. 2003. Majorizing kernels & stochastic cascades with applications to incompressible Navier–Stokes equations. *Transactions of the American Mathematical Society*, 355: 5003–5040

Bramson, M. 1978. Maximal displacement of branching Brownian motion. *Communications in Pure and Applied Mathematics*, 31: 531–582

Burd, G. & Waymire, E. 2000. Self-similar invariance of critical binary Galton–Watson trees. *Proceedings of the American Mathematical Society*, 128: 2753–2761

Chen, L., Dobson, S., Guenther, R., Ossiander, M., Thomann, E. & Waymire, E. 2002. On Ito's complex measure condition, In *Probability, Statistics and Their Applications: Papers in Honor of Rabi Bhattacharya*, edited by K. Athreya, M. Majumdar, M. Puri & E. Waymire, Beachwood, OH: Institute of Mathematical Statistics

Föllmer, H. & Schied, A. 2002. *Stochastic Finance*, New York: Walter de Gruyter

Gilbert, A.C., Willinger, W. & Feldman, A. 1999. Scaling analysis of conservative cascades with applications to network traffic. *Institute of Electrical and Electronics Engineers Transactions on Information Theory*, 45: 971–991

Holley, R. & Waymire, E. 1992. Multifractal dimensions and scaling exponents for strongly bounded random cascades. *Annals of Applied Probability*, 2(4), 819–845

Jouault, B., Greiner, M. & Lipa, P. 2000. Fix-point multiplier distributions in discrete turbulent cascade models. *Physica* D, 136: 196–255

Kahane, J.P. 1985. Sur le chaos multiplicatif. *Annales des Sciences Mathématiques du Québec*, 9: 105–150

Kahane, J.P. & Peyrière, J. 1976. Sur certaines martingales de B. Mandelbrot. *Advances in Mathematics*, 22: 131–145

Kendall, D. 1975. Branching processes since 1873, the genealogy of genealogy: branching processes before (and after) 1873. *Bulletin of the London Mathematics Society*, 7: 385–406, 225–253

Kolmogorov, A.N. 1962. A refinement of previous hypothesis concerning the local structure of turbulence in a viscous incompressible fluid at high Reynold number. *Journal of Fluid Mechanics*, 13: 82–85

LeJan, Y. & Sznitman, A.S. 1997. Stochastic cascades and 3-dimensional Navier–Stokes equations. *Probability Theory and Related Fields*, 109: 343–366

Mandelbrot, B.B. 1974. Intermittent turbulence in self-similar cascades: divergence of high moments and dimension of the carrier. *Journal of Fluid Mechanics*, 62: 331–333

McKean, H.P. 1975. Applications of Brownian motion to the equation of Kolmogorov, Petrovskii, and Piskunov. *Communications in Pure and Applied Mathematics*, 28: 323–331

Ossiander, M. & Waymire, E. 2000. Statistical estimation theory for multiplicative cascades. *Annals of Statistics*, 28(6): 1–21

Ossiander, M. & Waymire, E. 2002. On estimation theory for multiplicative cascades. *Sankhyā: The Indian Journal of Statistics Series* A, 64(2): 323–343

Resnick, S., Samorodnitsky, G., Gilbert, A. & Willinger, G. 2003. Wavelet analysis of conservative cascades. *Bernoulli*, 9(1): 97–135

Richardson, L.F. 1926. Atmospheric diffusion shown on a distance-neighbour graph. *Proceedings of the Royal Society of London*, A110, 709–737

She, Z.S. & Waymire, E. 1995. Quantized energy cascade and logPoisson statistics in fully developed turbulence. *Physical Review Lettters*, 74(2): 262–265

Waymire, E. & Williams, S.C. 1996. A cascade decomposition theory with applications to Markov and exchangeable cascades. *Transactions of the American Mathematical Society* 348(2): 585–632

Yaglom, A.M. 1966. Effect of fluctuations in energy dissipation rate on the form of turbulence characteristics in the inertial subrange. *Dokladay Akademii Nauk SSSR*, 166: 49–52

MULTISOLITON PERTURBATION THEORY

Solitons appear as robust solutions of several important nonlinear partial differential equations and difference-differential equations, including the Korteweg–de Vries (KdV), nonlinear Schrödinger (NLS), and sine-Gordon (SG) equations and the Toda lattice (TL). It is a remarkable feature of these and other equations, integrable by means of the inverse scattering transform (IST), that one can find exact multisoliton solutions. Among other phenomena, multisoliton solutions describe collisions among several solitons and bound states of solitons (breathers).

Considering systems that conserve energy but lack integrability, multisoliton solutions, and breathers may be only approximate, their dynamics accompanied by emission of radiation. As a result, two colliding solitons may merge into a breather, and the energy of a breather gradually decreases until it completely decays.

The situation is yet more different from that in integrable models if the system is dissipative. The dissipative loss of energy may be offset by an external field (driving force). A well-known example is the damped-driven SG equation

$$\phi_{tt} - \phi_{xx} + \sin\phi = -\alpha\phi_t - \gamma, \qquad (1)$$

which models magnetic flux propagation on a long Josephson junction (LJJ) (McLaughlin & Scott, 1978). In this equation, $\partial\phi/\partial x$ is the local magnetic field, α is a coefficient of dissipation, and γ is a bias-current density, which is the driving force. In the absence of perturbations ($\alpha = \gamma = 0$), (1) is the SG equation, whose fundamental soliton solution is the kink (it represents a magnetic-flux quantum, or fluxon, in LJJ),

$$\phi_k = 4\tan^{-1}\left[\exp\left(\sigma\frac{x-\xi(t)}{\sqrt{1-c^2}}\right)\right]. \qquad (2)$$

Here, $\sigma = \pm 1, c$, and $\xi(t)$ are, respectively, the polarity, velocity, and central coordinate of the kink (or fluxon). An exact breather solution to the unperturbed SG equation is

$$\phi_{br} = 4\tan^{-1}\left[\frac{\sin(t\cos\mu)}{\cosh(x\sin\mu)}\tan\mu\right], \qquad (3)$$

where the amplitude μ takes values $0 < \mu < \pi/2$. In the limiting case $\mu \to \pi/2$, (3) becomes a solution describing collision between two kinks with opposite polarities.

In the presence of perturbations, Equation (1) still has kink-like solutions, but the kink's steady-state velocity (c_0) is no longer arbitrary. It is determined by the condition that the power input from a constant driving force (γ) is in balance with the dissipation induced by the loss term, which yields

$$c_0 = \frac{\sigma\pi\gamma}{\sqrt{(\pi\gamma)^2 + 16\alpha^2}}. \qquad (4)$$

According to Equation (4), kinks with opposite polarities move in opposite directions; hence, they may collide.

If an ac driving force is applied to the system [e.g., $\gamma = \gamma_0\cos(\omega t)$ in (1)], it may compensate the loss and support the breather whose frequency $\cos\mu$ is in resonance with the driving frequency ω. If γ in (1) is a random function of time, and α is small enough, the random force can split the breather into a free kink-antikink ($k\bar{k}$) pair.

In the general case, basic effects produced by perturbations acting on two- or multi-soliton configurations may be classified as follows. Interaction between two solitons in the presence of conservative perturbations gives rise to emission of radiation, which appears at order ε^2, where ε is the strength of the perturbation. Adiabatic effects (those that neglect radiation) may be generated by conservative perturbations in a *three-soliton* system at order ε, in the form of energy exchange between colliding solitons (this may also occur in a two-soliton system if the conservative perturbation is not spatially uniform—for instance, if it is created by a local defect). In the case of dissipative perturbations, nontrivial effects are possible at order ε for two solitons, typical examples being fusion of a $k\bar{k}$ pair into a breather or, inversely, breakup of a breather into a pair due to its collision with another kink.

If the model is a perturbed version of an integrable one, multi-soliton processes similar to those outlined above can be investigated by means of perturbation theory (PT), a powerful version of which is based on a perturbed variant of IST (Kaup & Newell, 1978). As in the integrable case, the initial configuration is mapped into scattering data of the corresponding scattering problem, but the time evolution of the scattering data is no longer trivial, discrete eigenvalues being no more time-independent. Using a perturbative expansion, it is possible to derive ordinary differential equations (ODEs) for slow variations of the scattering data, which can be solved to obtain approximate solutions of the perturbed system. A comprehensive review of the IST-based PT for solitons in nearly integrable models was given by Kivshar & Malomed (1989).

An alternative multi-soliton PT is based on using a Green function (GF) for the underlying equation linearized around the unperturbed multi-soliton solution (Keener & McLaughlin, 1977). If the zero-order approximation is integrable, this method is equivalent to the IST-based PT, as the GF can be constructed—by means of IST—around any exact multi-solution solution. Although this approach has the advantage of being directly formulated in terms of physical parameters (soliton speeds, collision delays, breather frequencies, etc.), rather than more abstract IST characteristics (bound-state eigenvalues and reflection coefficients), finding the GF can be computationally demanding. Methods based on the Bäcklund transformation may ease some of these difficulties (McLaughlin & Scott, 1978).

The GF method often works for one-soliton problems in non-integrable models, as a full set of eigenmodes can often be found for an equation linearized about a soliton even if the equation is not integrable. An example is the nonlinear Klein–Gordon equation

$$\phi_{tt} - \phi_{xx} - \phi + \phi^3 = 0, \tag{5}$$

which describes ferroelectric phase transitions, among other applications. A comprehensive account of the GF method for kinks in non-integrable Klein–Gordon equations was given by Flesch & Trullinger (1987).

Another version of PT is based on the variational approximation (VA), which only demands that the full perturbed equation be derivable from a Lagrangian (e.g., the system is conservative), and that a one-soliton solution be available in the absence of perturbations. The method applies to the description of multisoliton effects in the adiabatic approximation, representing the wave field as a linear superposition of unperturbed solitons. Inserting this approximation into the model's Lagrangian and integrating over the spatial coordinate, one arrives at an effective Lagrangian, which is a function of the solitons' parameters (amplitude, velocity, and central coordinate and phase) and their first derivatives in time. Application of the Euler–Lagrange variational procedure to the effective Lagrangian then generates a system of ODEs that govern the evolution of the solitons' parameters. Because the linear-superposition assumption underlying VA is not valid in the general case (when solitons strongly overlap in the course of the interaction), a VA is restricted to cases when the solitons interact staying far apart or when they collide with a large relative velocity.

An early review of results obtained for multisoliton interactions by means of a VA was given by Gorshkov and Ostrovsky (1981), and an up-to-date review, including the interaction problems, was given by Malomed (2002). There is also an intermediate version of PT, which is based on IST for one soliton, but treats the interaction between solitons, even in an integrable equation, as a perturbation, assuming that the solitons are far separated. By means of this technique, Karpman and Solov'ev (1981) analyzed the interaction between solitons in the NLS and SG equations. Their results for NLS solitons were later checked in a direct experiment with solitons in a nonlinear optical fiber.

The approach based on the linear-superposition approximation for widely separated solitons makes it possible to calculate an effective potential of the interaction between solitons. To this end, one should isolate a part of the model's potential energy which depends on the separation between the solitons. For instance, the potential energy corresponding to the unperturbed SG equation (1) is

$$\Pi = \int_{-\infty}^{+\infty} \left[\tfrac{1}{2}\phi_x^2 + (1 - \cos\phi)\right] \mathrm{d}x\,. \tag{6}$$

In the vicinity of the first kink, the field is approximated as $\phi_1(x-\xi_1) + \delta\phi_2(x-\xi_2)$, where $\delta\phi_2$ is a small tail of the second kink (both are taken with $c=0$). Thus, in the lowest approximation, the contribution to the interaction potential from the vicinity of the first kink [which is, say, $-\infty < x < \Xi \equiv (1/2)(\xi_1+\xi_2)$] is

$$\begin{aligned} U_1 &= \int_{-\infty}^{\Xi} \left[(\phi_1)_x (\delta\phi_2)_x + (\sin\phi_1)\,\delta\phi_2\right] \mathrm{d}x \\ &\equiv \int_{-\infty}^{\Xi} \left[-(\phi_1)_{xx} + \sin\phi_1\right] \delta\phi_2 \,\mathrm{d}x + (\phi_1)_x\, \delta\phi_2|_{-\infty}^{\Xi}, \end{aligned} \tag{7}$$

using integration by parts. A key observation, which holds in much more general situations, is that the integral term in (7) identically vanishes, as $\phi_1(x)$ is an exact stationary solution of the SG equation; hence, all the contribution comes solely from the last term in (7). This yields the interaction potential, $U = U_1 + U_2 = 32\sigma_1\sigma_2 \exp(-|\xi_1-\xi_2|)$. Noting that each kink is a quasi-particle with mass $m=8$, this potential provides for full dynamical description of two- or multi-kink systems.

The derivation of the interaction potential based on the same principles applies to several other cases, including two- and three-dimensional solitons (Malomed, 1998). In the case of NLS solitons, the interacting pair is characterized by distance and relative phase between them, a peculiarity being that an effective mass corresponding to the phase difference is *negative*, which strongly affects stability of two-soliton

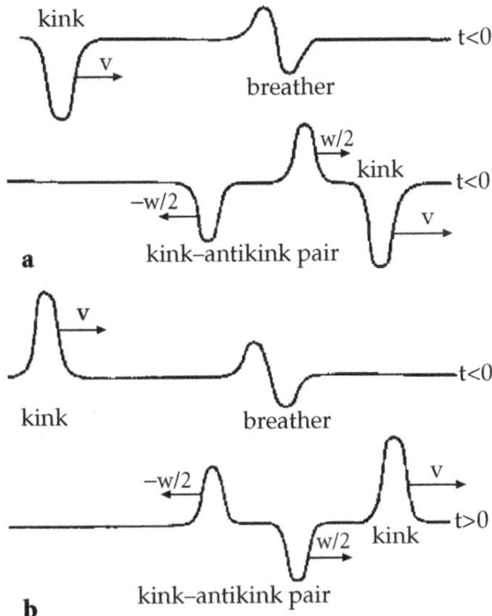

Figure 1. The breather's breakup into a kink-antikink pair as a result of the collision with another fast kink; shown is the field $\partial\phi/\partial x$. Panels (a) and (b) pertain to the polarities $\sigma = -1$ and $+1$ of the fast kink.

bound states induced by perturbations (Malomed, 2002).

The simplest nontrivial example of the two-soliton interaction is fusion of a $k\bar{k}$ pair as a result of collision. An analysis of this problem is outlined here for the case of small γ and α in (1). Because the collision-induced energy loss responsible for fusion is proportional to the dissipation (α) and can be found to be $\Delta E = 8\pi^2\alpha$, fusion occurs if the steady-state velocity given by (4) is small enough. This means that $|\gamma| \ll \alpha$ whereupon Equation (4) yields $c_0 \approx \sigma\pi\gamma\ /\ (4\alpha)$, the corresponding net kinetic energy of the pair being $E_{\rm kin} = 8c_0^2$. The collision results in pair annihilation if $E_{\rm kin} < \Delta E$, or $\gamma < \gamma_{\rm cr} \equiv 4\alpha^{3/2}$.

As shown in Figure 1, the collision of a breather with a moving kink may produce another interesting result in the same model: the breather may be split into a $k\bar{k}$ pair. The final result takes a simple form if the kink is ultrarelavitic, that is, c_0 is close to the limiting value 1: the breakup of the breather (3) into a $k\bar{k}$ pair is possible if $(\pi/2) - \mu < 2.91\sqrt{\alpha}$.

Some special types of conservative perturbations may greatly alter the dynamics. For instance, if the perturbation in the SG equation is $\varepsilon \sin(\phi\ /\ 2)$ (double SG equation), free 2π-kinks do not exist; instead, their pairs form bound states (4π-kinks), $\phi_{4\pi} = 4\tan^{-1}\left(\sqrt{|\varepsilon|\ /\ 2}\sinh x\right)$. It is possible to excite vibrations of this bound state, the eigenfrequency of which is $\sqrt{|\varepsilon|}$, as predicted by PT.

A typical example of two-soliton interactions generating radiation loss is the collision of two kinks (which may have both opposite and equal polarities) with a relative velocity c, the perturbing term in the SG equation being $\varepsilon \sin(2\phi)$. In the ultra-relativistic case, it can be shown directly that the radiation loss appears at order ε^2 and is the same for the $k\bar{k}$ and kk pairs. If the relative velocity of the collision is small, the $k\bar{k}$ pair may annihilate into a breather through the radiation loss, which takes place if $c < c_{\rm cr} \approx 2.22\,|\varepsilon|$.

Numerical simulations of $k\bar{k}$ collisions in non-integrable models, including Equation (5), reveal an interesting effect. There is a (probably infinite) set of critical relative velocities $c_{\rm cr}^{(n)}$, $n = 0, 1, 2, \ldots$, such that the annihilation takes place in intervals $0 < c < c_{\rm cr}^{(0)}$, $c_{\rm cr}^{(1)} < c < c_{\rm cr}^{(2)}$, $c_{\rm cr}^{(3)} < c < c_{\rm cr}^{(4)}$, ..., while in transmission windows between them, $c_{\rm cr}^{(0)} < c < c_{\rm cr}^{(1)}$, $c_{\rm cr}^{(2)} < c < c_{\rm cr}^{(3)}$, ..., the collision is quasi-elastic.

An explanation for this effect is that the kink has an eigenmode of its internal oscillations, and the dominant collision-induced energy loss is caused not by the emission of radiation but by excitation of the internal mode. An ODE model of a variational type makes it possible to explain the alternating annihilation and transmission windows as a result of resonant transfer of kinetic energy of the colliding kinks into their internal modes and back (Campbell et al., 1986).

BORIS A. MALOMED

See also **Collisions; Long Josephson junctions; *N*-soliton formulas; Perturbation theory; Solitons**

Further Reading

Campbell, D.K., Peyrard, M. & Sodano, P. 1986. Kink-antikink interactions in the double sine-Gordon equation. *Physica* D, 19: 165–205

Flesch, R.J. & Trullinger, S.E. 1987. Green's function for nonlinear Klein–Gordon kink perturbation theory. *Journal of Mathematical Physics*, 28: 1619–1631

Gorshkov, K.A. & Ostrovsky, L.A. 1981. Interactions of solitons in non-integrable systems: direct perturbation method and applications. *Physica* D, 3: 428–438

Karpman, V.I. & Solov'ev, V.V. 1981. A perturbational approach to the two-soliton systems. *Physica* D, 3: 487–502

Kaup, D.J. & Newell, A.C. 1978. Solitons as particles and oscillators in slowly varying media. *Proceedings of the Royal Society of London* A, 361: 413–466

Keener, J.P. & McLaughlin, D.W. 1977. Solitons under perturbations. *Physical Review* A, 16: 777–790

Kivshar, Y.S. & Malomed, B.A. 1989. Dynamics of solitons in nearly integrable systems. *Reviews of Modern Physics*, 61: 762–915

Malomed, B.A. 1998. Potential of interaction between two- and three-dimensional solitons. *Physical Review* E, 58: 7928–7933

Malomed, B.A. 2002. Variational methods in nonlinear fiber optics and related fields. *Progress in Optics*, 43: 69–191

McLaughlin, D.W. & Scott, A.C. 1978. Perturbation analysis of fluxon dynamics. *Physical Review* A, 18: 1652–1680

MULTISTABILITY

See **Stability**

MUSHROOMS

See **Fairy rings of mushrooms**

MUTUAL ENTRAINMENT

See **Coupled oscillators**

MYELINATED NERVES

While smooth nerve fibers are common to invertebrates, for example, the giant axon of the squid (Hodgkin & Huxley, 1952), myelinated nerve fibers are widely found in vertebrates. A myelinated fiber is mainly composed of a membrane—interfacing the intracellular space and the extracellular one—surrounded by a myelin sheath which acts as an insulator (Waxman et al., 1995; Koch, 1998; Scott, 2002). This sheath, formed of Schwann cells, is periodically absent in gaps called nodes of Ranvier, making a nerve fiber a discrete and periodic structure, as illustrated in Figure 1. Therefore, ion currents through the membrane can only occur at the nodes of Ranvier, which implies a "saltatory" conduction, under which the wave of activity leaps from one node to the next.

One of the main constraints of neural processing is to develop robust and high-speed traveling waves. Compared with smooth fibers, myelinated structures allow an increased velocity of nerve impulses, while decreasing the diameter of the nerve fiber. Thus, the motor nerves of vertebrates may comprise several hundred individual saltatory fibers. For instance, the information transfer rate (bits/s) in a rabbit's sciatic nerve is about 1500 times greater than in a squid axon of the same diameter. In the context of biological evolution, it is interesting to know the link between internode spacing and impulse velocity, and whether this link is optimized in terms of speed and robustness. To this end, a mathematical model can provide some answers.

A single myelinated nerve can be modeled from Kirchhoff's law analysis by the system of difference-differential equations

$$V_n - V_{n+1} = RI_n,$$
$$I_{n-1} - I_n = C\frac{\mathrm{d}V_n}{\mathrm{d}t} + I_{\mathrm{ion},n}. \qquad (1)$$

In these equations, the index n indicates successive active nodes, each of which is characterized by a transverse (inside to outside) voltage across the membrane (V_n). A second dynamic variable is the current (I_n) flowing longitudinally through the fiber from node n to node $n+1$. Also, R is the sum of the inside and outside resistances between two nodes and is inversely proportional to the internode spacing s, and C is the membrane capacitance of a node.

To study the velocity of an impulse, it is sufficient to focus on its leading edge. Thus, we assume that

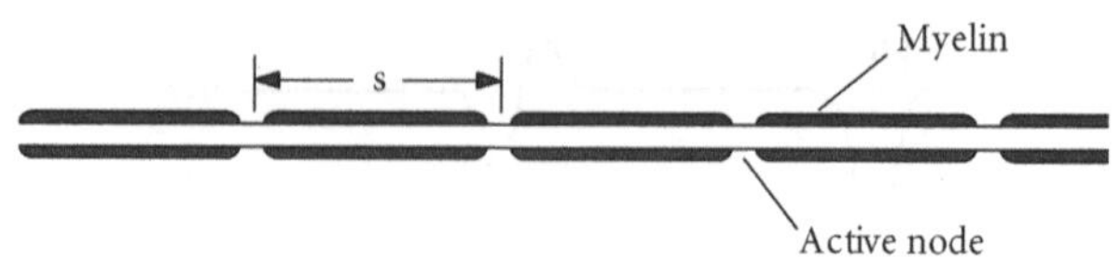

Figure 1. Sketch of a myelinated nerve fiber.

the sodium ion (Na^+) current begins without delay, and the potassium (K^+) permeability remains equal to its resting value. In this case, the expression of the ion currents developed by Alan Hodgkin and Andrew Huxley (*See* **Hodgkin–Huxley equations**) can be reduced to the simple and yet physically reasonable expression:

$$I_{\mathrm{ion},n} = I_{\mathrm{Na},n} = \left(\frac{G}{V_\mathrm{b}(V_\mathrm{b} - V_\mathrm{a})}\right) \times V_n(V_n - V_\mathrm{a})(V_n - V_\mathrm{b}), \quad (2)$$

where V_a is a threshold potential, V_b is the Nernst potential, and G is the total ionic conductance near V_b.

The parameters of this model can be related to experimental measurements on a real myelinated nerve fiber (Binczak et al., 2001). From studies on frogs' motor nerves with a diameter of 14 μm, $R = 28\,\mathrm{M\Omega}$, $C = 3.7 \pm 1\,\mathrm{pF}$, $G = 0.57\,\mu\mathrm{S}$, $V_\mathrm{a} \sim 25\,\mathrm{mV}$, and $V_\mathrm{b} = 122\,\mathrm{mV}$, while the distance between nodes equals 2 mm.

In order to study the influence of the internode spacing s on the impulse velocity, s and therefore R become variable. A discreteness parameter D can be defined by setting $R_\mathrm{f} = 28\,\mathrm{M\Omega}$, as

$$D \equiv \frac{2\,\mathrm{mm}}{s} = \frac{R_\mathrm{f}}{R}, \qquad (3)$$

so that $D = 1$ implies the discreteness of a standard frog nerve. Using dimensionless voltage variables $v_n \equiv V_n/V_\mathrm{b}$ and $a \equiv V_\mathrm{a}/V_\mathrm{b}$, the dynamic equation becomes

$$D(v_{n+1} - 2v_n + v_{n-1}) = R_\mathrm{f}C\frac{\mathrm{d}v_n}{\mathrm{d}t} + \frac{R_\mathrm{f}G}{1-a}v_n(v_n - a)(v_n - 1), \qquad (4)$$

Equation (4) has been used to compute the wave front velocity, which is plotted against D in Figure 2.

If the internodal spacing is small enough, that is, $D \gg 1$, the relative change in voltage between nodes satisfies the inequality $|(v_{n+1} - v_n)/v_n| \ll 1$ and the voltages and currents are smooth functions of distance. Letting $ns \to x$ in this continuum limit, the system can be described by the partial differential equation

$$s^2 D\frac{\partial^2 v}{\partial x^2} - R_\mathrm{f}C\frac{\partial v}{\partial t} = \frac{R_\mathrm{f}G}{1-a}v(v-a)(v-1), \qquad (5)$$

which is the Zeldovich–Frank-Kamenetsky (ZF) equation. Initially formulated as a model for flame front

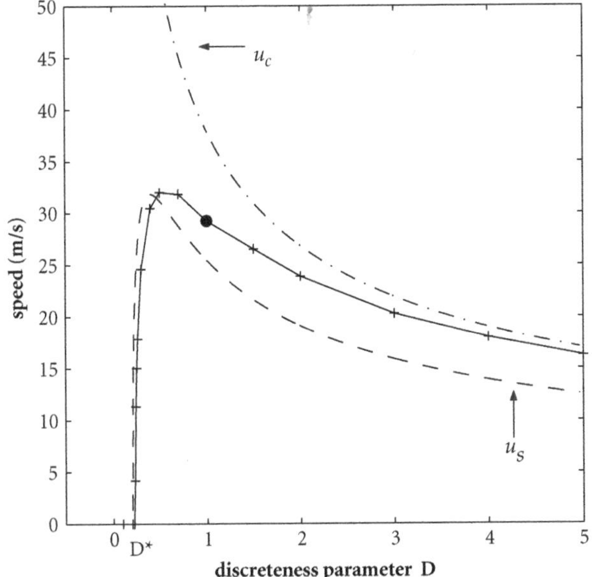

Figure 2. Leading edge impulse velocity on a myelinated axon as a function of the discreteness parameter D. The dot-dashed line indicates the continuum limit of Equation (6). The dashed line indicates the saltatory limit in Equation (8). Crosses indicates numerical results.

propagation, this equation is extensively used to study simple unmyelinated nerve models.

In the continuum limit, the wave speed u tends toward u_c (Zeldovich & Frank-Kamenetsky, 1938), where

$$u_c = s\left(\frac{1-2a}{\sqrt{2(1-a)}}\right)\sqrt{\frac{G}{RC^2}}, \tag{6}$$

corresponds to the dot-dashed line in Figure 2. As illustrated in this figure, the continuum approximation holds for $D > 5$.

Propagation failure (Keener, 1987) occurs when the distance between the active nodes becomes too large, that is, when D is above a critical value D^*. This critical value of the discreteness parameter is given to lowest order in a as (Erneux & Nicolis, 1993)

$$D^* \approx \frac{R_f G a^2}{4(1-a)}. \tag{7}$$

For D slightly larger than D^*, the impulse velocity $u \to u_s$, where

$$u_s = \frac{s}{\pi C}\sqrt{\frac{G(D-D^*)}{R_f(1-a)}}, \tag{8}$$

which corresponds to the dashed line in Figure 2.

When $D = 1$, the impulse velocity of a normal frog nerve has been calculated to be equal to 29 m/s, which matches the experimental results (Tasaki, 1982). The large dot in Figure 2 shows that the velocity is close to the maximum possible value, suggesting an optimal evolutionary design.

Failure of an impulse is expected to occur at an internode spacing of 1 cm, corresponding to $D = 0.2$, whereas the normal nerve is designed for $D = 1$, corresponding to a spacing of 2 mm. Thus, the process of evolution has provided a comfortable safety margin against accidental failure on the frog's motor nerve.

Because the impulse velocity reaches its maximum value near $D = 1$, we have an explanation for the fact that the conduction velocity of a frog's myelinated motor nerve is insensitive to the internode spacing. From an engineering perspective, operating in this region of parameter space makes the system robust. These observations are extendable to other myelinated fibers (Scott, 2002).

Finally, it should be noted that the transmission of an individual impulse expends much more energy on a smooth fiber, suggesting an additional reason that myelinated nerve fibers may have played an important role in the course of biological evolution.

STEPHANE BINCZAK

See also **Biological evolution; Nerve impulses; Zeldovich–Frank-Kamenetsky equation**

Further Reading

Binczak, S., Eilbeck, J.C. & Scott, A.C. 2001. Ephaptic coupling of myelinated nerve fiber. *Physica* D, 148: 159–174

Erneux, T. & Nicolis, G. 1993. Propagating waves in discrete bistable reaction–difusion systems. *Physica* D, 67: 237–244

Hodgkin, A.L. & Huxley, A.F. 1952. A quantitative description of membrane current and its application to conduction and excitation in nerve. *Journal of Physiology*, 117: 500–554

Keener, J.P. 1987. Propagation and its failure in coupled systems of discrete excitable cells. *SIAM Journal of Applied Mathematics*, 47: 556–572

Koch, C. 1998. *Biophysics of Computation: Information Processing in Single Neurons*, Oxford and New York: Oxford University Press

Scott, A.C. 2002. *Neuroscience: A Mathematical Primer*, Berlin and New York: Springer

Tasaki, I. 1982. *Physiology and Electrochemistry of Nerve Fibers*, New York: Academic Press

Waxman, S.G., Kocsis, J.D. & Stys, P.K. (editors). 1995. *The Axon: Structure, Function and Pathophysiology*, Oxford and New York: Oxford University Press

Zeldovich, Ya.B. & Frank-Kamenetsky, D.A. 1938. K teorii ravnomernogo rasprostranenia plameni (Toward a theory of uniformly propagating flames). *Doklady Akademii Nauk SSSR*, 19: 693–697

NATURAL MEASURE

See **Measures**

NAVIER–STOKES EQUATION

In 1822, the French engineer Claude Navier derived the Navier–Stokes equation, as an extension of Euler's equation to include viscosity. Navier was initially interested in blood flow, and he used a molecular approach to arrive at the viscous terms. Navier's equations were generalized to a compressible fluid by Poisson (1829), and one can find fully continuous derivations by De Saint-Venant (1843). A comprehensive treatment was given by George Stokes in 1845, who independently arrived at the results of Poisson and Saint-Venant using a continuous model. Stokes used the common assumption of linear relations between stress and strain rate and discovered "Stokes' law" for the terminal velocity of objects descending in fluids, which he deduced from experiments with slowing pendulums in viscous media.

The Navier–Stokes equation derives from a general equation for the conservation of momentum, balancing forces per unit volume on both sides. It is given by

$$\rho \frac{\mathrm{d}\boldsymbol{v}}{\mathrm{d}t} = \rho \boldsymbol{g} - \nabla p + \mu \left(\nabla^2 \boldsymbol{v} + \frac{1}{3} \nabla \nabla \cdot \boldsymbol{v} \right), \quad (1)$$

where $\boldsymbol{v}(\boldsymbol{r}, t)$ is the velocity vector in a cartesian coordinate system, $\rho(\boldsymbol{r}, t)$ is the density, $\boldsymbol{g}$ is the acceleration due to external forces (for example, gravitational, magnetic, electro-static forces), $p(\boldsymbol{r}, t)$ the pressure, and μ the viscosity coefficient. The material derivative is given by

$$\frac{\mathrm{d}\boldsymbol{v}}{\mathrm{d}t} = \left(\frac{\partial}{\partial t} + \boldsymbol{v} \cdot \nabla \right) \boldsymbol{v}, \quad (2)$$

where the first and second terms are the local time derivative of the quantity $\boldsymbol{v}(\boldsymbol{r}, t)$ with spatial coordinates fixed, and the convective (or advective) term accounting for the change of the quantity $\boldsymbol{v}(\boldsymbol{r}, t)$ at $\boldsymbol{r}$ due to the observer following the motion of the fluid with velocity $\boldsymbol{v}$. Both terms together are called the material derivative or convective derivative and are usually denoted by $\mathrm{d}/\mathrm{d}t$.

In an inertial coordinate system, the acceleration $\boldsymbol{g}$ originates solely from the gravitational potential of the Earth. Assuming a homogeneous mass at the center of the Earth, the acceleration $\boldsymbol{g}$ in Equation (1) is given by $\boldsymbol{g} = g\boldsymbol{k}$, where $\boldsymbol{k}$ is the local vertical direction, pointing to the center of gravity of the earth and g is the gravitational acceleration (982.1 cm s^{-2}), which is assumed constant in the first approximation. In a rotating coordinate system, the acceleration $\boldsymbol{g}$ consists of the centrifugal force and Coriolis force in addition to the gravitational acceleration, see Lamb (1906). Also, for large-scale processes, it is often necessary to account for tidal forces of the moon and sun. The term $-\nabla p$ is the force per unit volume due to a pressure gradient of the scalar pressure field $p(\boldsymbol{r}, t)$ acting on an infinitesimal volume of fluid with infinitesimal mass.

The last two terms of Equation (1)

$$\frac{\mu}{\rho} \left(\nabla^2 \boldsymbol{v} + \frac{1}{3} \nabla \nabla \cdot \boldsymbol{v} \right), \quad (3)$$

are the frictional terms which derive from a shear stress tensor for a Newtonian viscous fluid with constant viscosity coefficients. A fluid is called Newtonian if there is a linear relation between stress and rate of strain assuming isotropy, which is true for the most common conditions. Non-Newtonian fluids have more complex molecular structure or are mixtures of fluids.

The frictional term $\nabla^2 \boldsymbol{v} = \nabla \nabla \cdot \boldsymbol{v} - \nabla \times (\nabla \times \boldsymbol{v})$ contains effects due to compression and rotation. If the fluid is incompressible then the divergence terms are missing from Equations (1) and (3), since $\nabla \cdot \boldsymbol{v} = 0$. Often the coefficient of kinematic viscosity $\nu = \mu/\rho$ is used instead of the viscosity μ. The viscosity of the fluid depends on temperature, and a table for typical values of viscosity μ, kinematic viscosity ν, and density ρ is shown in Table 1 for typical temperatures. If the temperature across the medium is not uniform then a variable viscosity may have to be accounted for (see Batchelor, 1970).

T (°C)	Air ρ (g cm^{-3})	μ (g cm^{-1} s^{-1})	ν (cm^2 s^{-1})	Water ρ (g cm^{-3})	μ (g cm^{-1} s^{-1})	ν (cm^2 s^{-1})
0	1.293×10^{-3}	1.71×10^{-4}	0.132	0.999	1.787×10^{-2}	1.787×10^{-2}
10	1.247×10^{-3}	1.76×10^{-4}	0.141	0.999	1.304×10^{-2}	1.304×10^{-2}
20	1.205×10^{-3}	1.81×10^{-4}	0.150	0.998	1.002×10^{-2}	1.004×10^{-2}

Table 1. Density ρ, viscosity μ and kinematic viscosity ν of air and water.

For an ideal (perfect) fluid without internal shear stress, the momentum equation reduces to Euler's equation of motion for $\mu = 0$. Note that if the flow is irrotational ($\nabla \times \boldsymbol{v} = 0$) and incompressible ($\nabla \cdot \boldsymbol{v} = 0$), it essentially behaves as if it is inviscid ($\mu = 0$), because of the incompressibility condition $\mathrm{d}\rho/\mathrm{d}t = -\rho \nabla \cdot \boldsymbol{v} = 0$. (Note that a fluid may be incompressible yet ρ may not be constant.)

The system of Equations (1) comprises three momentum equations, together with an additional equation for mass conservation; a thermodynamic equation of state; a relation of pressure, density, and either temperature or entropy; and an energy equation for the additional thermodynamic variable (temperature or entropy). These provide six scalar equations for the determination of the six independent variables, velocity $\boldsymbol{v}$, density ρ, pressure p, and temperature or entropy as functions of space $\boldsymbol{x}$ and time t. If the fluid is incompressible, the equation of state becomes obsolete and only four equations are needed, the three momentum equations and the equation for conservation of mass.

Numerical Problems

As with most complicated nonlinear partial differential equations, the Navier–Stokes equation is solved numerically to model a specific fluid flow that is of interest experimentally and theoretically. Because there are many different ways to write the Navier–Stokes equation, the first numerical difficulty is to pick the formulation most suitable to the numerical technique that one wishes to employ and the dimensions of the model. One distinguishes between two-dimensional and three-dimensional models. For two-dimensional incompressible Navier–Stokes equation, for example, there are four different kinds of formulations: the primitive-variable (velocity and pressure), stream function-vorticity, stream function, and velocity-vorticity formulation.

It is important to distinguish between the real dissipation ($\mu \neq 0$), given by the last term of Equation (1) and numerical dissipation, which is introduced by an accumulative error due to the limited order of accuracy of any numerical model. As the numerical dissipation of a good model is often negligible, it is sometimes appropriate to introduce an artificial viscosity that acts to damp growing high wave number modes, which can lead to numerical instability.

Euler's equation conserves linear and angular momentum and energy, so it is necessary to verify that the numerical model conserves these conserved quantities and preserves symmetry. In cases where it is not possible to satisfy all these requirements, properties that are most essential to the physical problem are given priority. In order to satisfy such conserved quantities, it is important that the numerical scheme treats the nonlinear convective term

$$\boldsymbol{v} \cdot \nabla \phi \tag{4}$$

in a conservative manner, where ϕ represents the scalar quantity that is convected and could be one of the three velocity components u, v, w or density ρ in the conservation of mass. Expression (4) is known as the convection form and is nonconservative numerically, since fluxes across mesh boundaries do not cancel. The convection form thus represents a major flaw of every numerical scheme. The most commonly used corresponding conservative formulation is

$$\nabla \cdot (\boldsymbol{v}\phi), \tag{5}$$

which is also known as the divergence form. Recently it was shown that the skewsymmetric form

$$\tfrac{1}{2}\boldsymbol{v} \cdot \nabla \phi + \tfrac{1}{2}\nabla \cdot (\boldsymbol{v}\phi)$$

has advantages, because it reduces aliasing errors (see below) for yet unknown reasons and is preferable to the popular rotational form

$$(\nabla \times \boldsymbol{v}) \times \boldsymbol{v}.$$

It is important to note that although both equations (4) and (5) are formally equivalent, they are not in a corresponding numerical scheme and care must be taken to avoid this common pitfall when modeling equations that contain convective terms such as the Navier–Stokes equation or the equation for mass conservation (see Hirsch, 2000).

Nonlinear terms such as $\boldsymbol{v} \cdot \nabla \boldsymbol{v}$ also generate aliasing errors. These are high-frequency wave number modes appearing or being "aliased" as low-frequency modes that deteriorate the wave number spectrum and eventually lead to numerical blow-up.

Because a numerical model of the Navier–Stokes equation is coupled spatially and temporally, it is helpful to approximate the spatiotemporal scales of the dynamics. One should determine whether the spatial and temporal scales are equal to, smaller, or larger than each other. If, for example, the temporal scale is larger than the spatial scale, it is sufficient to use low order integration in time such as finite-difference methods, and if smaller or equal, then higher-order temporal integration such as spectral methods becomes necessary. In a temporally extensive computation, it is sometimes possible to revert to finite differences, choosing a relatively small time-step.

Especially difficult numerical problems arise when it is necessary to invert nondiagonal, nonsparse, nontrivial matrices, for example, in implicit time integrations or inverting the Laplacian. In such cases iterative techniques using preconditioning and multi-grid methods have been successfully employed. Complications arising from specifying the correct boundary conditions to use along boundaries of the physical domain should also be mentioned.

Although the Navier–Stokes equation is a very good model for a real fluid, it is important to have an understanding of the process or phenomenon that one wishes to model as well as a sound knowledge of the applicability of the numerical methods that one wishes to employ to be able to faithfully represent the physical dynamics of the fluid.

Phenomena

A salient feature of the Navier–Stokes equation is the phenomenon of turbulence. Due to viscous effects the flow of a real fluid can be observed in two very different states: a laminar state and a turbulent state.

Turbulence induces a cascade to fine scales that are eventually dissipated. It is important in many engineering applications, as a loss of energy means increased costs and can possibly cause damage to the structural body that comes in contact with the fluid or which moves through the fluid (e.g., optimization of airfoils, loss of lift force due to turbulence and boundary layer separation).

Closely related to the phenomenon of turbulence is boundary layer theory and boundary layer separation. Fluid flowing past a surface at rest relative to the fluid will experience friction at the surface of the material, and for the velocity field to be continuous, it is required that the normal as well as tangential fluid velocity vanish at the material surface ($\boldsymbol{v} = 0$). For an idealized fluid ($\mu = 0$), only the normal velocity has to vanish.

It was the problem of matching the vanishing flow at the surface boundary to some nonzero fluid flow away from the boundary that led Prandtl (1905) to introduce the concept of boundary layer, which is a small "inner" fluid layer close to the surface where viscous effects dominate the flow. This inner layer is joined to an "outer" layer, where the fluid can be considered inviscid. The concept of boundary layer has led to great progress in fluid mechanics and perturbation methods, where matched asymptotic expansions feature prominently, but a complete understanding of boundary layer separation and its connection to turbulence is still missing.

A useful concept is vorticity, which can be regarded as a beautiful generalization of fluid motion. Formally vorticity is defined as

$$\omega = \nabla \times \boldsymbol{v},$$

which simplifies to twice the angular velocity for solid body rotation. A fluid is called irrotational if its vorticity is zero and rotational if it is not. Taking the curl of Euler's equation ($\mu = 0$) for an incompressible fluid in the absence of baroclinicity (and after some vector algebra) leads to the vorticity equation

$$\frac{\mathrm{d}\omega}{\mathrm{d}t} - \omega \cdot \nabla \boldsymbol{v} = 0.$$

Vortex lines and vortex strength are conserved, vortex lines move with the fluid, and an initially irrotational flow remains irrotational. Associated to vortices and turbulence is the, still not understood, phenomenon of vortex breakdown (Benjamin, 1978).

Among wave phenomena, the most striking is definitely the solitary wave, which appears as a nonlinear wave in a fluid internally as well as at fluid surfaces or interfaces. For free-surface wave phenomena, see Johnson (1997) and the references therein, and for a colorful overview, see Lighthill (1978). A beautiful picture gallery of fluid phenomena can be found in Van Dyke (1982). Other phenomena that can only be listed here include cavitation and thermal convection (Rayleigh–Bénard convection), sub- and supersonic flow, shock waves as well as most instability mechanisms and phenomena (Rayleigh–Taylor instability, Kelvin–Helmholtz instability, barotropic and baroclinic instability).

ANDREAS A. AIGNER

See also **Alfvén waves; Boundary layers; Burgers equation; Chaos vs. turbulence; Fluid dynamics; Hydrothermal waves; Kelvin–Helmholtz instability; Korteweg–de Vries equation; Lattice gas methods; Magnetohydrodynamics; Mixing; Rayleigh–Taylor instability; Solitons; Taylor–Couette flow; Turbulence; Water waves**

Further Reading

Batchelor, G.K. 1970. *An Introduction to Fluid Dynamics*, Cambridge and New York: Cambridge University Press

Benjamin, T.B. 1978. Theory of the vortex breakdown phenomenon. *Journal of Fluid Mechanics*, 14: 593–629

Chandrasekhar, S. 1961. *Hydrodynamic and Hydromagnetic Stability*, Oxford: Clarendon Press and New York: Dover

De Saint-Venant, J.-C. 1843. Note à joindre au mémoire sur le dynamique de fluides. *Comptes Rendu de l'Academie des Sciences Paris*, 17: 1240–1243
Drazin, P.G. 1981. *Hydrodynamic Stability*, Cambridge and New York: Cambridge University Press
Hirsch, C. 2000. *Numerical Computation of Internal and External Flows*, New York: Wiley
Johnson, R.S. 1997. *A Modern Introduction to the Mathematical Theory of Water Waves*, Cambridge and New York: Cambridge University Press
Lamb, H. 1906. *Hydrodynamics*, Cambridge: Cambridge University Press; 6th edition, Cambridge and New York: Cambridge University Press, 1993
Landau, L.D. & Lifshitz, E.M. 1987. *Fluid Mechanics*, vol.6 of *Course of Theoretical Physics*, London: Pergamon Press and Reading, MA: Addison-Wesley
Lighthill, J. 1978. *Waves in Fluids*, Cambridge and New York: Cambridge University Press
Milne-Thomson, L.M. 1968. *Theoretical Hydrodynamics*, 5th edition, New York: Macmillan Press
Navier, M. 1822. Mémoire sur les lois du mouvement des fluides. *Mémoires de l'Academie Royale des Sciences*, 6: 389–440
Poisson, S.D. 1829. Mémoire sur les équations générales de l'équilibre et du nouvement des corps silides élastiques et de fluides. *Journal de l'École Polytechnique*, 13: 1–174
Prandtl, L. 1905. *Verhandlungen des dritten internationalen Mathematiker-Kongresses, Heidelberg 1904*, Leipzig: 484–491
Stokes, G.G. 1845. On the theories of the internal friction of fluids in motion, and of the equilibrium and motion of elastic solids. *Transactions, Cambridge Philosophical Society*, 8: 287–319
Streeter, V.L. & Wylie, E.B. 1985. *Fluid Mechanics*, 8th edition, New York: McGraw-Hill
Van Dyke, M. 1982. *An Album of Fluid Motion*, Stanford: Parabolic Press
Vennard, J.K. 1954. *Elementary Fluid Mechanics*, New York: Wiley

N-BODY PROBLEM

The Newtonian N-body problem of celestial mechanics is a mathematical generalization of the solar system. In the Euclidean space $\mathbb{R}^3$, consider N points of masses $m_i > 0$, $i = 1, \ldots, N$, which attract each other by a force directly proportional to the product of the masses and inversely proportional to the square of the distance. The equations of motion are given by the $6N$-dimensional system of differential equations

$$\begin{cases} \dot{q} = M^{-1}p, \\ \dot{p} = G\nabla U(q), \end{cases} \tag{1}$$

where the upper dot represents differentiation with respect to the time variable; $\boldsymbol{q} = (\boldsymbol{q}_1, \ldots, \boldsymbol{q}_N)$ is the configuration of the particle system, with $\boldsymbol{q}_i = (q_i^1, q_i^2, q_i^3)$; giving the coordinates of the point of mass m_i, $\boldsymbol{p} = M\dot{\boldsymbol{q}}$ is the momentum, where M is a $3N$-dimensional square matrix having on the diagonal the elements $m_1, m_1, m_1, \ldots, m_N, m_N, m_N$ and zeros in rest; G is the gravitational constant; and $U(\boldsymbol{q}) = \sum_{1 \le i < j \le N} \frac{m_i m_j}{|\boldsymbol{q}_i - \boldsymbol{q}_j|}$ is the potential function, $-U(\boldsymbol{q})$ representing the potential energy. Standard results of the theory of differential equations ensure the existence and uniqueness of an analytic solution for any initial value problem as long as the initial data do not belong to the collision set $\Delta = \bigcup_{1 \le i < j \le N} \Delta_{ij}$, where $\Delta_{ij} = \{\boldsymbol{q} \in \mathbb{R}^{3N} \,|\, \boldsymbol{q}_i = \boldsymbol{q}_j\}$.

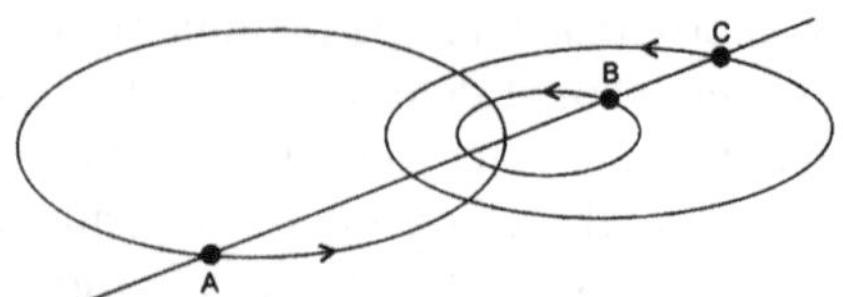

Figure 1. The Eulerian solutions of the three-body problem.

Isaac Newton formulated this problem in his master work *Principia*, but Leonhard Euler was the first to write the equations as we know them today. The case $N = 2$, also called the Kepler problem, is completely solved (see Albouy (2002) for a recent discussion of some early solutions). The relative motion of one body with respect to the other is planar and, depending on the initial conditions, can be a circle, an ellipse, a parabola, a branch of a hyperbola, or a line, in which case a collision takes place in the future or in the past. Kepler's laws (*See* **Celestial mechanics**) can be recovered from Equation (1).

For $N \ge 3$, very little is known about the N-body problem in spite of thousands of research papers written over more than three centuries. The case $N = 3$ was the most studied since many of the results obtained could be generalized to any larger N. The first attempts to understand the three-body problem were quantitative, aiming at finding explicit solutions. In 1767, Euler found the collinear periodic orbits, in which three bodies of any masses move such that they oscillate along a rotating line (Euler, 1767, Figure 1) and in 1772, Joseph-Louis Lagrange discovered some periodic solutions that lie at the vertices of a rotating equilateral triangle that shrinks and expands periodically (Lagrange, 1772, Figure 2). Those solutions led to the study of central configurations, for which $\boldsymbol{q}'' = k\boldsymbol{q}$ for some constant $k > 0$. Each central configuration provides a class of periodic orbits.

Another idea was to reduce the order of the system with the help of first integrals. Ten linearly independent integrals are known: three for the center of mass, three for the momentum, three for the angular momentum, and one for the energy (see Wintner, 1947). Together with a certain symmetry, these integrals allow the reduction of the three-body problem from dimension 18 to 7. But unfortunately the dimension of the problem cannot be further reduced. In 1887, Heinrich Bruns proved that there exist no more linearly independent integrals, algebraic with respect to $\boldsymbol{q}$, $\boldsymbol{p}$ and t (Bruns, 1887), thus showing the limitations of the quantitative methods. This led Henri Poincaré to

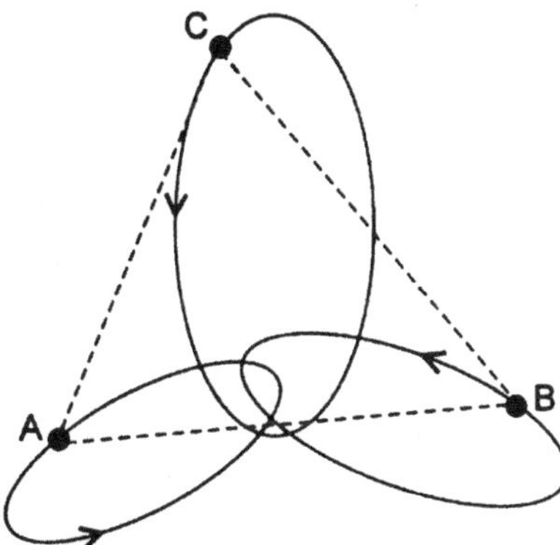

Figure 2. The Lagrangian solutions of the three-body problem.

attempt a qualitative approach. His first prolific ideas appeared in a memoir published in 1890 (Poincaré, 1890), for which he was awarded the King Oscar Prize (see Barrow-Green, 1997; Diacu & Holmes, 1996). There he laid the foundations of several branches of mathematics, including dynamical systems, chaos, KAM theory, and algebraic topology.

Poincaré aimed to understand the geometry of the phase space and the relative behavior of orbits and, thus, to answer questions regarding stability, asymptotic motion at infinity, existence of periodic orbits, etc. An important problem in this direction was that of singular solutions, that is, those that tend to the collision set Δ. It took mathematicians almost a century to prove that for $N \geq 5$, there exist singular solutions that do not end in collisions but in pseudocollisions, which are orbits that become unbounded in finite time (see Diacu & Holmes, 1996). For $N = 4$, the problem is still open.

Recently, a lot of interest has been in finding choreographies, that is, periodic orbits for which all the bodies move on the same closed curve. For more than two centuries the only known example was the class of Lagrangian solutions in the particular case that all the masses are equal and the ellipses are circles. With the help of variational methods, in 2000, a spectacular new class was proved to exist: three bodies of equal mass chase each other along a curve resembling the figure eight (Chenciner & Montgomery, 2000; Montgomery, 2001, Figure 3). There is numerical evidence that this periodic orbit is KAM stable, that is, the solutions through most initial conditions in some sufficiently small neighborhood of the orbit stay close to it for all time, while the other solutions leave the neighborhood very slowly. Unfortunately, the stability region seems to be very small. Numerical experiments suggest that the probability of finding an eight in the universe is somewhere between one per galaxy and one per universe. Hundreds of other choreographies have been numerically put into the evidence.

Still far from fully understood are the questions regarding various restricted three-body problems. In the

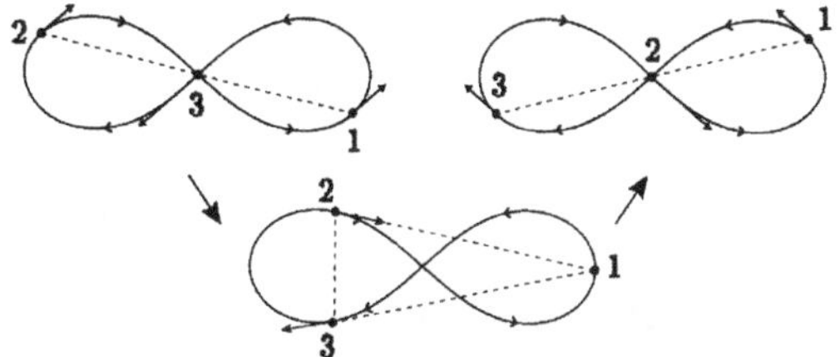

Figure 3. The figure eight solution of the 3-body problem.

elliptic one, for example, it is asked to determine the motion of one body, assumed to have zero mass, while the other two move on ellipses as in the Kepler problem with negative energy.

Numerical methods are also of help for getting insight into the problem. But due to the apparently chaotic character of the motion, they must be implemented with care. Recently, much progress has been made in successfully applying scientific computation to various aspects of the general and restricted three-body problem.

Many of the ideas of the classical N-body problem can be adapted to related problems for understanding the motion of particle systems given by other potentials, like those of Manev and Schwarzschild (also used in celestial mechanics), Coulomb (atomic and molecular theories), and Lennard-Jones (crystal formation). Based on the Coulomb potential, Niels Bohr's model of the hydrogen atom led to the development of quantum mechanics. Several other branches of science have profited from the study of the N-body problem.

FLORIN DIACU

See also **Celestial mechanics; Poincaré theorems; Solar system**

Further Reading

Albouy, A. 2002. Lectures on the two-body problem. In *Classical and Celestial Mechanics: The Recife Lectures*, edited by H. Cabral & F. Diacu, Princeton, NJ: Princeton University Press, pp. 71–135

Barrow-Green, J. 1997. *Poincaré and the Three-Body Problem*, Providence, RI: American Mathematical Society

Bruns, H. 1887. Über die Integrale des Vielkörper-Problems, *Acta Mathematica* 11: 25–96

Chenciner, A. & Montgomery, R. 2000. A remarkable periodic solution of the three-body problem in the case of equal masses. *Annals of Mathematics*, 152: 881–901

Diacu, F. & Holmes, P. 1996. *Celestial Encounters—The Origins of Chaos and Stability*, Princeton, NJ: Princeton University Press

Euler, L. 1767. De moto rectilineo trium corporum se mutuo attrahentium, *Novo Comm. Acad. Sci. Imp. Petrop.*, 11: 144–151

Lagrange, J.L. 1873. Essai sur le probl eme des trois corps. In *Ouvres de Lagrange*, vol. 6, pp. 229–324, Paris: Gauthier-Villars, 14 vols

Montgomery, R. 2001. A new solution to the three-body problem. *Notices of the American Mathematical Society* May, pp. 471–481

Poincaré, H. 1890. Sur le problème des trois corps et les équations de la dynamique. *Acta Mathematica*, 13: 1–270
Wintner, A. 1947. *The Analytical Foundations of Celestial Mechanics*, Princeton, NJ: Princeton University Press

NEEL DOMAIN WALL

See **Domain walls**

NEGATIVE RESISTANCE AND CONDUCTANCE

See **Diodes**

NEPHRON DYNAMICS

The kidneys play an important role in regulating the blood pressure and maintaining a proper environment for the cells of the body. The mammalian kidney contains a large number of functional units, the nephrons. For a human kidney the number of nephrons is of the order of 1 million, and for a rat kidney the number is approximately 30,000. The nephrons are organized in a parallel structure such that the individual nephron only processes a very small fraction of the total blood flow to the kidney. To distribute the blood that enters through the renal artery, the kidney makes use of a network of arteries and arterioles. Closest to the nephron, we have the afferent arteriole that leads the blood to the capillary network in the glomerulus where water, salts, and small molecules are filtered from the blood and into the tubular system of the nephron. On the other side of the glomerulus, the efferent arteriole leads the blood to another capillary system that receives the water and salts reabsorbed by the tubules.

Figure 1 provides a sketch of the main structural components of the nephron. Note how the terminal part of the loop of Henle passes within cellular distances of the afferent arteriole. As described below, this anatomical feature allows for a special feedback regulation.

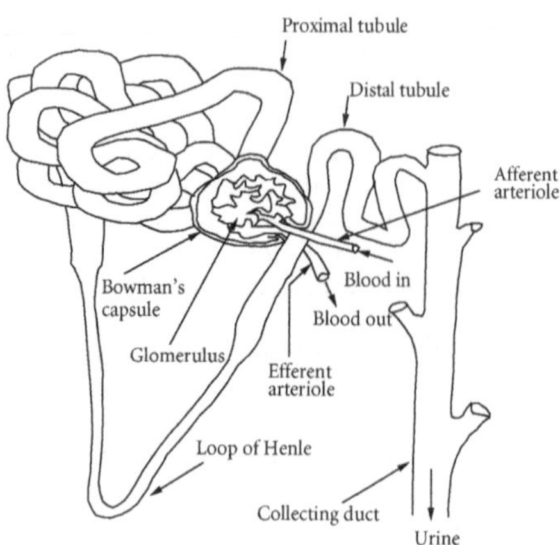

Figure 1. Sketch of the main structural components of the nephron.

In order to protect its function in the face of a varying blood pressure, the individual nephron disposes of a number of mechanisms to control the incoming blood flow. Most important is the tubuloglomerular feedback (TGF) mechanism that regulates the diameter of the afferent arteriole in response to variations in the NaCl concentration in the fluid that leaves the loop of Henle via the distal tubule. If the salt concentration in this fluid becomes too high, specialized cells (macula densa cells) near the terminal part of the loop of Henle elicit a signal that causes the smooth muscle cells at the downstream end of the afferent arteriole to contract. Hence, the incoming blood flow is reduced, and so is the glomerular filtration rate.

The TGF mechanism represents a negative feedback. However, by virtue of a delayed action associated with a finite transit time through the loop of Henle, the flow regulation tends to become unstable and produce self-sustained oscillations with a period of 30–40 s. While for rats with normal blood pressure, these oscillations have the appearance of a regular limit cycle with a sharply peaked power spectrum, highly irregular oscillations, displaying a broadband spectral distribution with strong subharmonic components are observed for spontaneously hypertensive rats. In a particular experiment, clear evidence of a period-doubling of the pressure oscillations has been found, indicating that the nephronic control system is operating close to a transition to chaos.

Figure 2 displays examples of the tubular pressure oscillations observed for normotensive rats (a) and for spontaneously hypertensive rats (b), respectively. The

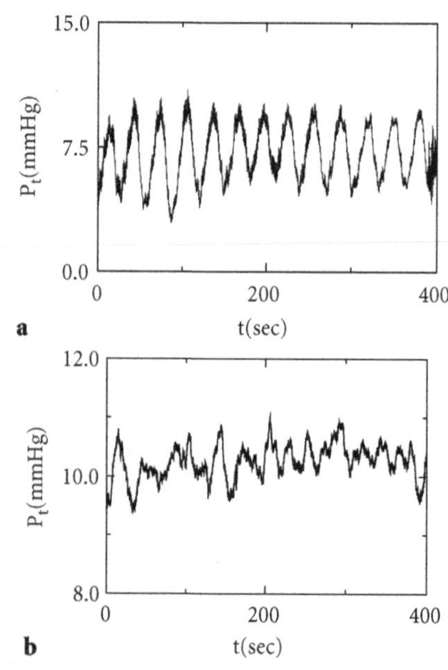

Figure 2. Regular (a) and irregular (b) tubular pressure oscillations from a normotensive and a spontaneously hypertensive rat, respectively.

pressure oscillations can be observed by means of a glass pipette inserted into the proximal tubule of a nephron at the surface of the kidney.

The processes involved in the nephron autoregulation are known in considerable detail. Besides the filtration of water and salts in the glomerulus, these processes include the passive (osmotic) and active (enzymatically controlled) processes by which water and salts are reabsorbed along the loop of Henle, the enzymatic processes through which the smooth muscle cells in the arteriolar wall are activated by the macula densa signal, and the dynamic response of the arteriolar wall to external stimulation.

The steady-state response of the TGF mechanism can be obtained from open-loop experiments in which a block of paraffin is inserted into the middle of the proximal tubule and the glomerular filtration rate is measured as a function of an externally forced flow of artificial tubular fluid into the loop of Henle. Reflecting physiological constraints on the diameter of the arteriole, this response follows an S-shaped characteristic with a maximum at low Henle flows and a lower saturation level at externally forced flows beyond 20–25 nl/min.

Together with the delay in the TGF regulation, the steepness of the response plays an essential role for the stability of the feedback system. The length of the delay can be estimated from the phase shift between the pressure oscillations in the proximal tubule and the oscillations of the NaCl concentration in the distal tubule. A typical value is 10–15 s. In addition, there is a transit time of 3–5 s for the signal from the macula densa cells to reach the smooth muscle cells in the arteriolar wall. The result is a total delay of 14–18 s. The steepness α of the steady state response curve is found to be significantly higher for spontaneously hypertensive rats than for normal rats.

By integrating the different physiological processes into a coherent, nonlinear dynamic model, it has been possible to show how these processes together produce the observed behavior, that is, the emergence of self-sustained oscillations as the slope of the response characteristic exceeds $\alpha \cong 11$ and the transition to chaos via sets of overlapping period-doubling cascades as the feedback slope exceeds $\alpha \cong 20$.

Figure 3 shows a two-dimensional bifurcation diagram for the single-nephron model. The dashed curve is a Hopf bifurcation curve. Period-doubling and saddle-node bifurcations are indicated as fully drawn and dotted curves, respectively. As before, T is the total delay in the TGF regulation, and α is the (maximal) slope of the open-loop feedback characteristic.

The single-nephron model can also be used to simulate the response to an external perturbation, for instance, the infusion of artificial tubular fluid into the loop of Henle or the administration of a drug to the rat. This last possibility is rapidly gaining in significance as the application of simulation models in the development of new drugs becomes more and more important.

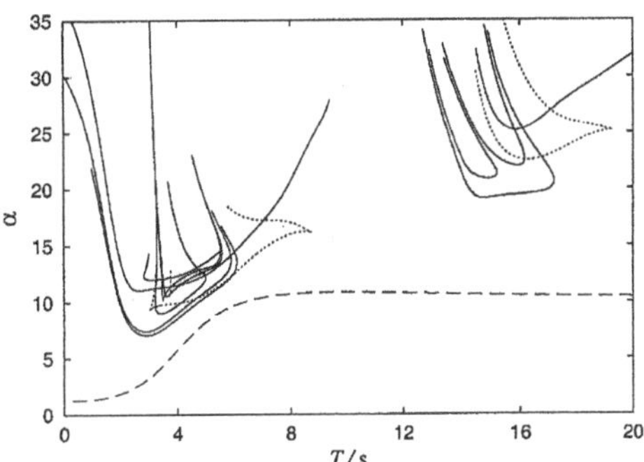

Figure 3. Two-dimensional bifurcation diagram for the single-nephron model. The diagram illustrates the complicated bifurcation structure in the region of 1:1, 1:2, and 1:3 resonances between the arteriolar dynamics and the TGF-mediated oscillations. In the physiologically interesting regime around $T = 16$ s, another set of complicated period-doubling and saddle-node bifurcations occur. Here, the nephron is operating close to a 1:4 (or 1:5) resonance.

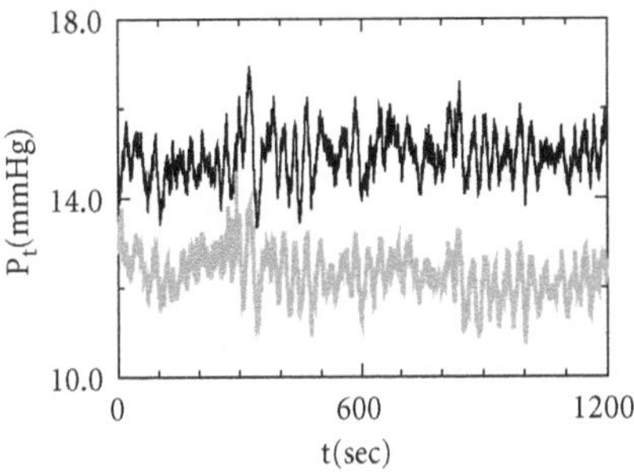

Figure 4. Example of chaotic phase synchronization for a pair of adjacent nephrons in a hypertensive rat.

A variety of cooperative phenomena that can arise from interactions among the nephrons may also be significant. The functional units are typically arranged in couples or triplets with their afferent arterioles branching off from a common interlobular artery. This structure allows neighboring nephrons to interact via signals that propagate along the arteriolar system. As experiments show, this interaction can lead to various forms of synchronization among the nephrons, including in-phase and antiphase synchronization for regularly oscillating nephrons, and chaotic phase synchronization for nephrons with irregular oscillations. By modeling these coupling phenomena in detail, one may be able to predict the typical size of the synchronization domains and to examine the role that synchronization among the nephrons plays in the overall regulation of the kidney.

Figure 4 shows an example of the tubular pressure oscillations observed for two adjacent nephrons in a hypertensive rat. The transition to synchronization can be observed as a locking of the average periods of the two signals in a 1:1 relation to one another. Alternatively, one can define and follow the temporal variation of the instantaneous phases for the two signals.

E. MOSEKILDE, N.-H. HOLSTEIN-RATHLOU, AND O. SOSNOVTSEVA

See also **Coupled oscillators**

Further Reading

Blekhman, I. 1988. *Synchronization in Science and Technology*, New York: ASME Press

Fung, Y.-C.B. 1981. *Biomechanics. Mechanical Properties of Living Tissues*, New York: Springer

Glass, L. & Makey, M.C. 1988. *From Clocks to Chaos: The Rhythms of Life*, Princeton, NJ: Princeton University Press

Keener, J. & Sneyd, J. 1998. *Mathematical Physiology, Interdisciplinary Applied Mathematics*, New York: Springer

Layton, H. & Weinstein, A. (editors). 2001. *Membrane Transport and Renal Physiology*, New York: Springer

Mosekilde, E. 1996. *Topics in Nonlinear Dynamics: Applications to Physics, Biology and Economic Systems*, Singapore: World Scientific

Mosekilde, E., Maistrenko, Yu. & Postnov, D. 2002. *Chaotic Synchronization — Applications to Living Systems*, Singapore: World Scientific

Pikovsky, A., Rosenblum, M. & Kurths, J. 2001. *Synchronization: A Universal Concept in Nonlinear Science*, Cambridge and New York: Cambridge University Press

Seldin, D.W. & Giebisch, G. (editors). 1992. *The Kidney: Physiology and Pathophysiology*, New York: Raven

NERVE IMPULSES

The scientific study of nerve impulses goes back to 1791, when Luigi Galvani reported that a frog's leg muscle twitches if the attached nerve is stimulated with a bimetallic contact (Brazier, 1961). This discovery was soon followed by Alessandro Volta's invention of the battery, which launched the science of electrophysiology and raised the question: Does animal electricity differ from chemical electricity?

Among attempts to answer this question was a key experimental study by young Hermann Helmholtz, who cleverly measured the speed of signal propagation on a frog's sciatic nerve (Helmoltz, 1850). In making this measurement, he ignored the advice of his father, a philosopher, who believed that muscular motion was identical to its motivation; thus, any time delay between thinking and doing was theoretically impossible. To the contrary, Helmholtz found a velocity of about 27 m/s, which is much less than the speed at which an electrical signals propagate along conducting wires. Although an outstanding theoretical physicist, Helmholtz was unable to understand why a nerve impulse should move so slowly. Was it the mechanical motion of some molecular substance that he had observed? Interestingly, nonlinear diffusion was suggested as an answer to this puzzle by Robert Luther at the beginning of the 20th century (Luther, 1906).

Among the wonders of electronics that appeared in the 20th century was the cathode-ray oscilloscope

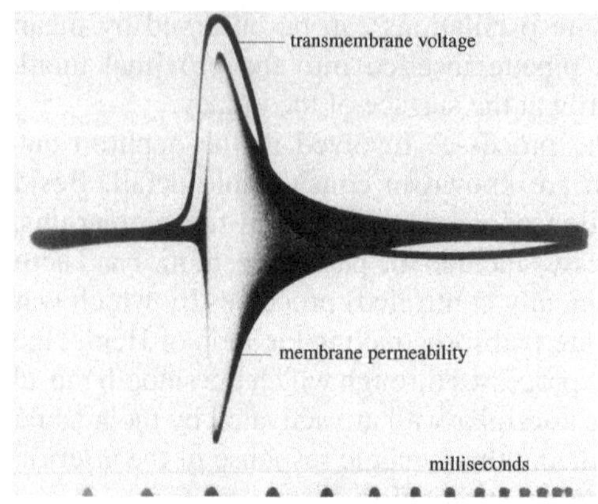

Figure 1. Time course of the transmembrane voltage and membrane permeability of an impulse on a squid nerve. (Courtesy of Kenneth Cole.)

(CRO), which Kenneth Cole used to take the first photograph of a nerve impulse on the giant axon of the squid (see Figure 1) (Cole & Curtis, 1938). Time increases to the right as indicated in milliseconds by the marks on the lower margin, showing that horizontal CRO deflections were not yet uniform in the late 1930s. The solid line is the transmembrane voltage (V), which rises rapidly from a resting level, hesitates at a peak level of about 100 mV, and then falls back more slowly. The width of the band indicates the membrane permeability (or conductivity), which evidently increases greatly during passage of the nerve impulse.

Progress toward explaining these phenomena came soon after the Second World War, taking advantage of the significant advances in electronics during those years. In 1952, Alan Hodgkin and Andrew Huxley presented a series of papers on the squid giant axon, which culminated in a formulation of nerve impulse dynamics based on an empirically determined reaction-diffusion system, from which all details of Figure 1 were computed (Hodgkin & Huxley, 1952). In this Hodgkin–Huxley (HH) model, the initial rise of transmembrane voltage is caused by an inrush of positively charged sodium ions, and the maximum voltage is attained when the inward diffusion of sodium ions is balanced by outward conduction current through the membrane. On a longer time scale, positively charged potassium ions flow out of the nerve, bringing the transmembrane voltage back to its resting value, whereupon it is ready to conduct another impulse. Thus, the means by which squid nerves carry signals are explained by the nonlinear reaction-diffusion system

$$\frac{\partial^2 V}{\partial x^2} - rc\frac{\partial V}{\partial t} = rj_{\text{ion}}, \tag{1}$$

where r is the series resistance per unit length of the axon core, c is the capacitance per unit length, and j_{ion} is a rather complicated expression for the transmembrane ionic current per unit length.

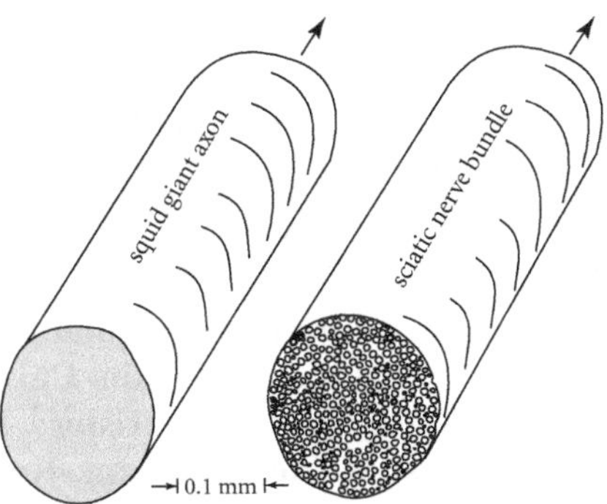

Figure 2. A squid giant axon and the sciatic nerve of a rabbit, to the same scale. (Data from Young, 1951).

In thinking about the HH formulation, it is important to be aware that a squid nerve differs from the sciatic nerve of the frog, which Galvani and Helmholtz studied. A sciatic nerve is actually a bundle of smaller fibers, whereas the squid nerve is a single "giant" axon. The squid nerve is uniform along the propagation direction; thus, the HH system is based on the partial differential equation system of Equation (1). The individual fibers of a sciatic nerve, on the other hand, are covered with an insulating layer (myelin) except at rather widely spaced active nodes (nodes of Ranvier). Thus, impulse propagation on a myelinated axon is described by a difference-differential equation, in which the wave of activity jumps from node to node (saltatory conduction). Among other differences, this means that impulse conduction velocity increases as the square root of diameter of a squid fiber, while it is roughly proportional to the first power of the diameter of a myelinated fiber. Importantly, the energy expended in transmitting a nerve impulse is much less in a myelinated fiber than in a smooth one.

These qualitative differences are emphasized in Figure 2, which compares a typical squid nerve with the sciatic nerve of a rabbit. The rabbit nerve contains about 375 small (myelinated) axons, each of which can conduct an impulse at up to 80 m/s, or about four times faster than a squid axon, leading to an increase in data transmission capacity of about three orders of magnitude. This dramatic increase in information carrying capacity is typical in vertebrate motor neurons, which are myelinated nerve bundles.

From an analytic perspective, the HH formulation is rather complicated, as it involves five dynamic variables, each of which depends on both longitudinal position and time. These are transmembrane voltage, axial current, sodium turn-on and turn-off variables, and a potassium turn-on variable. Thus, it has been of interest to consider other formulations that preserve qualitative properties of the HH system while simplifying their structure.

In the most simple approximation, the transmembrane ionic current is assumed to be a cubic nonlinear function of the transmembrane voltage; thus,

$$j_{\text{ion}} \approx \left(\frac{g}{V_{\text{th}} V_{\max}}\right) V (V - V_{\text{th}})(V - V_{\max}), \qquad (2)$$

where V_{th} is a threshold voltage and $V_{\max}$ is the amplitude of the impulse. Under this approximation, the HH dynamics reduce to the Zeldovich–Frank-Kamenetsky (ZF) equation, which describes the leading edge of an impulse but misses the return of the voltage to its resting value. From this perspective, a zero-order estimate of the impulse velocity is of order $\sqrt{g/rc^2}$. Several analytic formulas for the dependence of impulse velocity on axon parameters have been obtained under the ZF approximation (Scott, 2002).

A simple way to represent recovery was developed by Vladislav Markin and Yuri Chizmadzhev in the 1960s (Markin & Chizmadzhev, 1967). Under this MC approximation, a prescribed time course of transmembrane ionic current is assumed to be triggered if the membrane potential reaches a threshold variable. This prescribed current is a negative (inward) current (j_1) maintained for a time τ_1 followed by a positive (outward) current (j_2) maintained for a time τ_2. (The condition $j_1\tau_1 = j_2\tau_2$ then ensures that the net charge crossing the membrane during an impulse is zero.) With appropriate parameters, the MC model yields a recovering impulse having approximately the speed and threshold properties of the HH impulse.

To bring dynamics into the picture without invoking the complexities of the full HH model, Jin-Ichi Nagumo and his colleagues used a membrane model previously developed by Richard FitzHugh, in which the cubic ionic current of Equation (2) is augmented with a single dynamic recovery variable that drives the membrane voltage back to its resting level (Nagumo et al., 1962). This FitzHugh–Nagumo (FN) system was developed as an electronic equivalent of a nerve axon, which can be regarded as a *neuristor*. In the early 1970s, FN became of interest to applied mathematicians, who were beginning to study the theoretical properties of nonlinear reaction-diffusion systems, and in two or three spatial dimensions it has been used to study the emergence of spiral and scroll waves.

In addition to its inherent complexity, the HH system is driven by an initial inrush of sodium ions, whereas the exciting current in many nerve fibers, including dendrites of the human brain, are largely driven by calcium ion current. For many applications, therefore, it is currently of interest to use a version of the model developed by Catherine Morris and Harold Lecar for calcium ion induced membrane switching in the giant muscle fiber of the barnacle (Morris & Lecar, 1981). In this Morris–Lecar (ML) model, $j_{\text{ion}} \approx 2\pi a J_{\text{ml}}$, where a is the axon radius and

$$J_{\text{ml}} = G_{\text{K}} n (V - V_{\text{K}}) + G_{\text{Ca}} m (V - V_{\text{Ca}}) + G_{\text{L}}(V - V_{\text{L}})$$

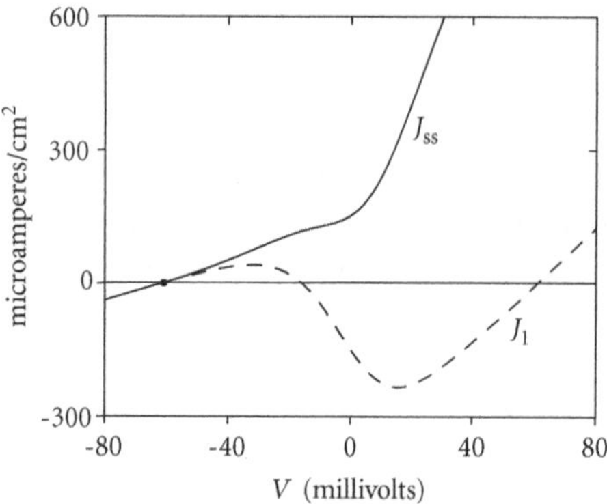

Figure 3. Plots of the initial ionic current (J_1) and the steady state current (J_{ss}) for a typical Morris–Lecar model (Fall et al., 2002; Scott, 2003).

with m and n being calcium and potassium turn-on variables, respectively. Also G_K, G_{Ca}, and G_L are membrane conductances, and V_K, V_{Ca}, and V_L are equilibrium potentials for potassium, calcium, and "leakage" ions, respectively. The turn-on variables are assumed to obey first-order dynamics as (Fall et al., 2002)

$$\frac{dm}{dt} = -[m - m_0(V)]/\tau_m(V),$$
$$\frac{dn}{dt} = -[n - n_0(V)]/\tau_n(V),$$

where

$$m_0(V) = [1 + \tanh((V - V_1)/V_2)]/2,$$
$$n_0(V) = [1 + \tanh((V - V_3)/V_4)]/2$$

and

$$\tau_m(V) = \tau_{m0}\,\mathrm{sech}[(V - V_1)/2V_2],$$
$$\tau_n(V) = \tau_{n0}\,\mathrm{sech}[(V - V_3)/2V_4].$$

At times short compared with τ_n, J_{ml} appears as the cubic function, which is plotted as a dashed line in Figure 3; thus the leading edge of an impulse will propagate as required by the ZF equation. At times long compared with τ_n, $m(t)$ remains equal to $m_0(V)$ and $n(t) \to n_0(V)$, so

$$J_{ml} \to J_{ss} = G_K n_0(V)(V - V_K) + G_{Ca} m_0(V)(V - V_{Ca}) + G_L(V - V_L).$$

This is a steady state membrane current (plotted as the solid line in Figure 3), which forces the system back to its resting state at

$$[V, m, n] = [V_R, m_0(V_R), n_0(V_R)].$$

For sufficiently large values of τ_{n0}, therefore, the ML equation supports a nerve impulse with recovery.

At the dawn of the 21st century, electrophysiology is becoming a highly sophisticated experimental science, generating data on ever more intricate neural structures; thus, it will be of interest to consider using these five approaches (HH, ZF, MC, FN, and ML) to understand the dynamics of nerve impulse propagation on axonal and dendritic branching regions of real neurons.

Alwyn Scott

See also **Candle; FitzHugh–Nagumo equation; Hodgkin–Huxley equations; Markin–Chizmadzhev model; Myelinated nerves; Neuristor; Reaction-diffusion systems; Zeldovich–Frank-Kamenetsky equation**

Further Reading

Brazier, M.A.B. 1961. *A History of Electrical Activity of the Brain*, London: Pitman

Cole, K.S. & Curtis, H.J. 1938. Electrical impedance of nerve during activity. *Nature*, 142: 209

Fall, C.P., Marland, E.S., Wagner, J.M. & Tyson, J.J. 2002. *Computational Cell Biology*, New York: Springer

Helmholtz, H. 1850. Messungen über den zeitlichen Verlauf der Zuckung animalischer Muskeln und die Fortpflanzungsgeschwindigkeit der Reizung in den Nerven. *Archiv für Anatomie und Physiologie*, 276–364

Hodgkin, A.L. & Huxley, A.F. 1952. A quantitative description of membrane current and its application to conduction and excitation in nerve. *Journal of Physiology (London)*, 117: 500–544

Luther, R. 1906. Räumliche Fortpflanzung chemischer Reaktionen. *Zeitschrift fuer Elektrochemie* 12(32): 596–600 [English translation in *Journal of Chemical Education*, 64 (1987):740–742]

Markin, V.S. & Chizmadzhev, Yu.A. 1967. On the propagation and excitation for one model of a nerve fiber. *Biophysics*, 12: 1032–1040

Morris, C. & Lecar, H. 1981. Voltage oscillations in the barnacle giant muscle. *Biophysical Journal*, 71: 193–213

Nagumo, J., Arimoto, S. & Yoshizawa, S. 1962. An active impulse transmission line simulating nerve axon. *Proceedings of the Institute of Radio Engineers*, 50: 2061–2070

Scott, A.C. 2002. *Neuroscience: A Mathematical Primer*, Berlin and New York: Springer

Scott, A.C. 2003. *Nonlinear Science: Emergence and Dynamics of Coherent Structures*, 2nd edition, Oxford and New York: Oxford University Press

Young, J.Z. 1951. *Doubt and Certainty in Science*, Oxford: Oxford University Press

NEUMANN BOUNDARY CONDITION

See **Partial differential equations, nonlinear**

NEURAL NETWORK MODELS

A neural network consists of units and connections corresponding to neurons and synapses in biological neural networks. We focus here on neural network models as a means to better understand the working principles of nervous systems, in particular the human brain. Quantitative modeling is today accepted as a very important tool in neuroscience that has potential to enable understanding of the very complex dynamical and nonlinear processes underlying the functioning of biological nervous systems. So-called artificial neural networks have been developed mainly for technical applications with little concern for biological plausibility and modeling (Haykin, 1998), and we will not consider them further here.

A computational model may help to explain experimental findings and to make new experimentally testable predictions. By now, most parts of the brain have been modeled, and many different types of models exist of a system, for example, for the hippocampus, an evolutionary old part of cortex important for memory and memory consolidation. When designing a neural network model of a particular system, the constituent neurons and their synaptic connections have to be represented with the desired biophysical detail, for example, as a set of coupled ordinary differential equations that can be solved numerically on a computer (Koch & Segev, 1998; Scott, 2002). The level of biological detail in the models studied ranges from networks of very simple threshold logic units connected by binary weights to networks comprising elaborate compartmental cell models with thousands of compartments and parameters. In addition to the signal transduction and transmission processes of the neuron such a detailed cell model may also represent intracellular processes such as biochemical second messenger cascades and calcium dynamics including, for example, diffusion.

In an accurate network model of a particular system, its constituent neuron types and proportions as well as their synaptic interactions have to be adequately represented. Available data about the system must be collected from literature and experiments and be entered as parameter values. A problem with this approach is that often some part of the information required by the model is lacking, such as the distribution of different kinds of ionic channels over the cell membrane or the details of synaptic plasticity dynamics (for example, augmentation and depression). Thus, some parameters may have to be indirectly inferred, and the models need to be tuned to fit experimental recordings, for example, of the shape of excitatory and inhibitory postsynaptic potentials or neuronal firing patterns at different levels of injected current. This introduces some uncertainty with regard to the validity of the model.

The input to the network from sources outside the model must also be represented in some way. Moreover, conditions and values measured in an *in vitro* preparation like a brain slice, a cell culture, or a piece of isolated spinal cord may differ from those of the intact in vivo system, and the latter may be unaccessible.

Nervous systems, other than the most simple ones, typically comprise a very large number of neurons. With today's computers it is feasible to simulate hundreds of thousands of compartments and millions of synapses at a reasonable level of biophysical detail. For large models the turnaround time for simulations, however, may be in the order of days, so parallel simulators are therefore useful. It is still beyond the capacity of today's supercomputers to handle full scale models of biological networks. A common practice is, therefore, to work with dramatically subsampled network models in which the number of neurons is reduced to a small fraction of those actually present in the target system. As a consequence, the number of input synaptic connections on cells in the model is also dramatically reduced. With cell models tuned to single cell data, for example, with adequate input resistances and thresholds, it will become necessary to compensate by exaggerating the synaptic conductances in order for the model to reproduce the activity seen in the real system. Having few and large synaptic interaction events in the system may, however, distort network dynamics, thus, making the model a poor quantitative representation of the actual system. Such effects should be born in mind when interpreting result from simulations using subsampled models.

Furthermore, one sometimes excludes from the model altogether neuron types known to exist in the actual biological system. For one reason or another a cell type may be considered unimportant for the questions addressed. One example is the neuroglia cells that are on average about ten times more numerous than neurons in the brain. They are thought to mainly serve the purpose of structural support and maintenance of the internal environment and are rarely included in network models. But one cannot entirely exclude that these cells in some situations subserve functions important for signal processing.

At the other extreme from networks of complex multi-compartmental model neurons are neural network models in which the biophysical detail has been reduced to a minimum. Such abstraction serves the important purpose of helping to pinpoint and elucidate the fundamental mechanisms behind the functioning of a complex system, which may enable further theoretical analysis of the phenomena under study.

A network model may be simplified in several different ways. For instance, the neurons can be modeled as point neurons (a single isopotential compartment) lacking geometric extent. Network units may use a graded output (with, for example, a sigmoid transfer function) representing an instantaneous firing frequency, or they may be spiking as real neurons. The former may, in fact, represent the average population activity in a cortical module like a minicolumn rather than an individual neuron. An integrate-and-fire model neuron is a bit more elaborate with a simple membrane dynamics and a spiking threshold (Gerstner, 1999). Whether or not temporal timing at the millisecond range in the spike trains of neurons is fundamental for the information processing in the brain or if a rate code is adequate for most purposes is presently a hotly debated issue. Yet another class of models is neural continuum field models that represent the neural structure as a continuous sheet of excitable neural tissue with a lattice connectivity (Bressloff et al., 2002).

In simplified network models, the details of synaptic transmission and plasticity may be replaced with weighted inputs and simple learning rules. Signal delays along axons are ignored, which may be reasonable at least for millimeter distances in small networks. Instead of an accurate representation of short- and long-term synaptic plasticity, one typically incorporates a correlation-based learning rule, for example, some form of Hebbian learning (Rolls & Treves, 1997). The adiabatic learning hypothesis (Caianiello, 1961) stating that synaptic weight changes occur on a much slower timescale than the neurodynamics itself, simplifies the model and network dynamics considerably. On the other hand, this assumption is now known to be invalid in, for instance, the neocortex where synaptic properties are modulated on a millisecond timescale.

A modeling strategy that has often proven fruitful is to develop a suite of models at different levels of abstraction for the same neuronal network. The most detailed model relates closely to the biological network under study, and the aim is to transform in a well-defined fashion to gradually more abstract descriptions, using the most abstract ones as the starting point for theoretical analysis.

ANDERS LANSNER

See also **Artificial intelligence; Attractor neural network; Cell assemblies; Compartmental models; Hodgkin–Huxley equations; Integrate and fire neuron; McCulloch–Pitts network; Multiplex neuron; Nerve impulses; Neurons; Perceptron**

Further Reading

Bressloff, P.C., Cowan, J.D, Golubitsky, M., Thomas, P.J. & Wiener, M. 2002. What geometric visual hallucinations tell us about the visual cortex. *Neural Computation*, 14: 473–491

Caianiello, E. 1961. Outline of a theory of thought processes and thinking machines. *Journal of Theoretical Biology*, 1: 204–235

Gerstner, W. 1999. Spiking neurons. In *Pulsed Neural Networks*, edited by W. Maass & C. Bishop, Cambridge, MA:MIT Press, pp. 3–53

Haykin, S. 1998. *Neural Networks: A Comprehensive Foundation*, Upper Saddle River, NJ: Prentice–Hall

Koch, C. & Segev, I. (editors). 1998. *Methods in Neuronal Modeling: From Ions to Networks*, Cambrigde, MA: MIT Press

Rolls, E. & Treves, A. 1997. *Neural Networks and Brain Function*, Oxford and New York: Oxford University Press

Scott, A.C. 2002. *Neuroscience. A Mathematical Primer*, Berlin and New York: Springer

NEURISTOR

Coined by Hewitt Crane in 1962, the term *neuristor* implies "the whole class of lines that exhibit attenuationless propagation with recovery" (Crane, 1962). In principal, this definition was intended to include active nerve fibers and even forest fires, but Crane's motivation was to overcome certain problems associated with the miniaturization of electronic circuits. Based upon a half-dozen reports prepared by Crane and his colleagues at the Stanford Research Institute since the late 1950s, this carefully written paper provides a window into the thinking of electrical engineers at the threshold of the integrated circuit revolution.

As the dimensions of conventional (transistor) computing circuits are greatly reduced, Crane reasoned, at least two design problems must be faced. First, the density of interconnections increases, eventually requiring "a dense set of interconnections at a point." Second, the resistance per unit length of interconnecting wires may become inconveniently large. (Crane pointed out that the resistance of a 1 μm copper wire is about a 1000 Ω/in.) Both of these problems may be mitigated by basing miniaturized computer design upon an "active wire" in which energy is stored uniformly over the system and continuously dissipated by nonlinear traveling-wave impulses, rather than at discrete and isolated amplifying elements. Some indication that neuristor design is a promising strategy is offered by the fact that it was selected by evolution for the development of our biological brains.

Being uniform in the direction of propagation, Crane's active wire supports a traveling-wave impulse in which the rate of energy release is equal to its rate of dissipation. Upon impulse passage, the line remains inactive for a certain time (the refractory period), after which it is able to transmit a new impulse. Because an impulse cannot propagate through a refractory region, two impulses will destroy one another in a head-on collision, and impulses are not reflected from the end of a neuristor line.

If two interconnection junctions are included—both of which seem feasible from an engineering perspective—a neuristor system is logically complete, meaning that it can realize all possible Boolean switching functions. These two neuristor interconnections are as follows. (a) A *T* junction is shown in Figure 1(a). An impulse entering on one of the branches proceeds outward on the other branches. Such a junction is not difficult to realize, requiring merely that the strength of the incoming impulse divided by the number of outgoing branches is above the threshold of the outgoing branches. (b) An *R* junction is shown in Figure 1(b). In an *R* junction, the refractory (or inhibitory) variable of one line is shared with an adjacent line (over the shaded area), while the excitatory variable is not. Thus, individual impulses traveling from A to B (on the upper line) can block individual impulses traveling from D to C (on the lower line) and vice versa. These two junctions can be interconnected in a variety of useful ways,

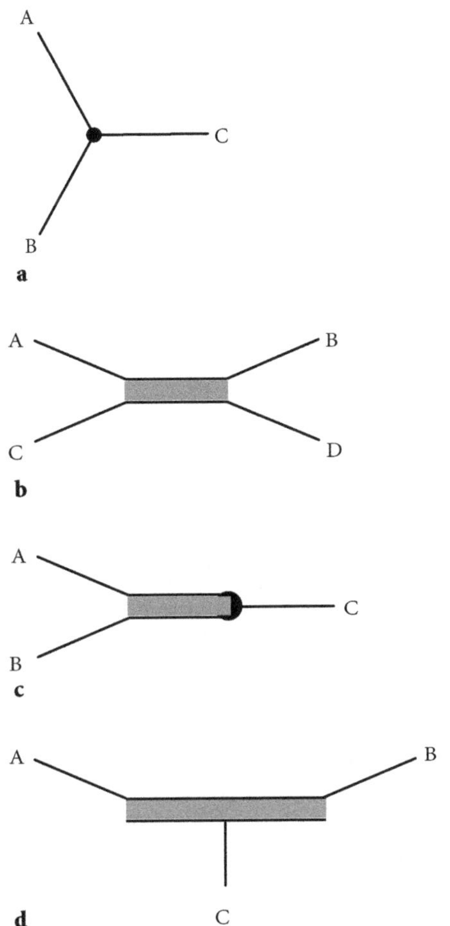

Figure 1. Some simple neuristor interconnections. (a) T junction. (b) R junction. (c) $T-R$ junction. (d) Analog of a relay or transistor.

one being the $T-R$ junction, shown in Figure 1(c). Here impulses incoming on line A will proceed through to C but will not be transmitted to B, effectively isolating the input lines (A and B) from each other.

In considering the switching possibilities of such neuristor systems, Boolean variables can be chosen in either of two ways, indicating the presence or absence of individual impulse or the presence or absence of impulse trains. From the latter perspective, the arrangement in Figure 1(d) can be viewed as equivalent to a relay or a transistor because an incoming impulse train on line C will block the transmission between A and B. As computing systems constructed from relays or transistors are known to be logically complete, it follows that neuristor systems are also logically complete.

The length of the refractory zone of an impulse is equal to the product of its velocity times the recovery interval. This is an important design parameter for a neuristor system, setting the scale for many functions. If a circular section of active line is used as a storage ring, for example, the circumference of this ring must be greater than the length of a refractory zone. Thus an insufficiently short refractory length is a major limitation on the degree to which a particular neuristor system can be miniaturized.

In his seminal paper, Crane offered several suggestions for neuristor realizations, taking advantage of the variety of interesting nonlinear diode structures that were being invented in the early 1960s—such as Esaki (tunnel) diodes and four-layer diodes. Also suggested was an ingenious combination of T and R junctions that allows two signal paths to cross without interference and without being lifted from a common substrate.

Although neuristors have been designed and fabricated in various laboratories (Beretovskii, 1963; Yoshizawa & Nagumo, 1964; Sato & Miyamoto, 1967; Parmentier, 1969; Scott, 1970; Reible & Scott, 1975; Nakajima et al., 1976), the neuristor strategy has not been important for the design of modern computing systems. Among other reasons for this failure must be counted the amazing progress in miniaturization of silicon metal-oxide-semiconductor transistors, which are now far smaller than the refractory length of any conceivable neuristor structure.

Beyond applications to electronic computing systems, however, the neuristor design concept may yet aid in understanding the behavior of neural systems, as was suggested by Crane in the early 1960s (Crane, 1964). Indeed, this possibility is more compelling today because dendritic fibers (in addition to axons) are now known to support action potentials and are, therefore, also neuristors in the original sense of the word (Stuart et al., 1999). Thus, possibilities arise for neuristor-like computations in the axonal and dendritic trees of real neurons (Scott, 2002). Neuroscientists studying the intricate networks of interwoven dendro-dendritic, currently being revealed by electron microscopy, may profit from a review of Crane's early work.

ALWYN SCOTT

See also **Multiplex neuron; Nerve impulses; Neurons**

Further Reading

Beretovskii, G.N. 1963. Study of single electric model of neuristor. *Radio Engineering and Electronic Physics*, 18: 1744–1751

Crane, H.D. 1962. Neuristor—a novel device and system concept. *Proceedings of the IRE*, 50: 2048–2060

Crane, H.D. 1964. Possibilities for signal processing in axon systems. In *Neural Theory and Modeling*, edited by R.F. Reiss. Stanford: Stanford University Press, pp. 138–153

Nakajima, K., Onodera, Y. & Ogawa, Y. 1976. Logic design of Josephson network. *Journal of Applied Physics*, 47: 1620–1627

Parmentier, R.D. 1969. Recoverable neuristor propagation on superconductive tunnel junction strip lines. *Solid-State Electronics*, 12: 287–297

Reible, S.A. & Scott, A.C. 1975. Pulse propagation on a superconductive neuristor. *Journal of Applied Physics*, 46: 4935–4945

Sato, R. & Miyamoto, H. 1967. Active transmission lines. *Electronics and Communications in Japan*, 50: 131–142
Scott, A.C. 1970. *Active and Nonlinear Wave Propagation in Electronics*, New York: Wiley
Scott, A.C. 2002. *Neuroscience: A Mathematical Primer*, Berlin and New York: Springer
Stuart, G., Spruston, N. & Häusser, M. (editors). 1999. *Dendrites*, Oxford and New York: Oxford University Press
Yoshizawa, S. & Nagumo, J. 1964. A bistable distributed line. *Proceedings of the IEEE*, 52: 308

NEURONS

Nerve cells, or neurons, are typically comprised of a cell body, dendrites, and an axon. The typical cortical neuron depicted in Figure 1 has both an apical dendrite and several basal dendrites. Dendrites and the soma are the sites at which axon terminals from other neurons make contacts and provide stimulation to a neuron. These sites of contact between neurons are termed synapses. The axon emanates from the cell body and contacts other neurons at distances varying from 1.0 mm up to almost 1 m, depending upon the the type and location of the neuron in question. The ends of axons terminate very close to the dendritic and cell body membranes of other neurons to form synpases. When the cell body of a neuron is sufficiently depolarized by its inputs, it reaches the threshold for action potential (or spike) generation, and one or more spikes are triggered. These spikes then propagate along the axon at speeds ranging from less than 1.0 ms^{-1} up to almost 100 ms^{-1}.

Neurons receive both excitatory (depolarizing) and inhibitory (hyperpolarizing) stimulation from the axon terminals that synapse onto them, with a typical cortical neuron having about 10^3–10^4 synapses. When an action potential reaches a synapse, it causes the release of neurotransmitter molecules that rapidly diffuse across the very small extracellular space to the membrane of the postsynaptic cell's dendrite or soma. There they bind to receptor molecules and cause ion channels to open resulting in either depolarization (excitatory synapse) or hyperpolarization (inhibitory synapse). These potential changes propagate down the dendrites to the cell body, where they are combined approximately linearly. If this net potential change depolarizes the cell body past a threshold, the neuron will fire one or more spikes. These spikes in turn propagate along the axon of this cell, and the process repeats itself.

Nonlinear Dynamics of Action Potential Generation

In mathematical terms, a series of action potentials is a limit cycle oscillation in the neural state space. There are several distinct dynamical patterns of action potential generation that have been observed, and these will be characterized in terms of their dynamical foundations.

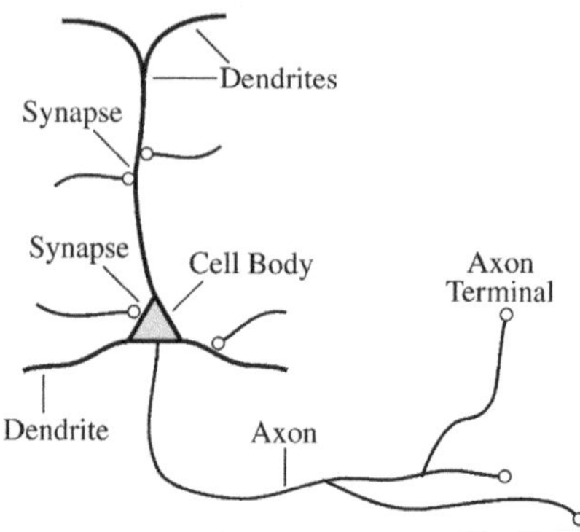

Figure 1. Schematic diagram of a typical neuron in the neocortex. The longest dendrites are between 1–3 mm in length, while the axon can be as short as a few mm or as long as 1.0 m.

As demonstrated by Hodgkin and Huxley (1952) in work that led to the Nobel Prize in 1963, the dynamics of spike generation is based upon two ionic currents: a sodium (Na^+) current that rapidly depolarizes the neuron followed by a slower potassium (K^+) current that repolarizes the neuron. Each current I_j is described by Ohm's law written as the product of a conductance (reciprocal of resistance) g_j and a voltage. For each ion, the voltage term is given by the difference between the membrane potential V and the equilibrium potential for the ion in question, E_j

$$I_j = g_j(V - E_j) \tag{1}$$

Due to the fact that the neural cell membrane functions as a capacitor, the membrane potential V is described by a differential equation of the form

$$C\frac{\mathrm{d}V}{\mathrm{d}t} = g_{\mathrm{Na}}(V)\,(V-E_{\mathrm{Na}}) + g_{\mathrm{K}}(V)\,(V-E_{\mathrm{K}}) + I_{\mathrm{ext}}, \tag{2}$$

where I_{ext} is an external stimulating current (or a synaptic input). For typical cortical neurons, $E_{\mathrm{Na}} = 55$ mV and $E_{\mathrm{K}} = -95$ mV. This equation would be linear except for one crucial observation: the two conductances $g_{\mathrm{Na}}(V)$ and $g_{\mathrm{K}}(V)$ are functions of V. This biophysical discovery by Hodgkin & Huxley (1952) means that ion conductances change with the voltage, thereby generating both positive and negative nonlinear feedback that produces the action potential.

A simple set of equations with normalized variables (Wilson, 1999a) can be used to elucidate the essential dynamics of Equation (2):

$$\frac{\mathrm{d}V}{\mathrm{d}t} = 4\left(V^2 - \frac{V}{10}\right)(V-1) + R\left(V + \frac{1}{5}\right) + I_{\mathrm{ext}},$$
$$5\frac{\mathrm{d}R}{\mathrm{d}t} = R + 3V^2 \tag{3}$$

The term $(V^2 - V/10)$ on the right in the first equation represents the V dependence of the Na^+

conductance as a quadratic function, while the variable R (Recovery variable) represents the K^+ conductance governed by the second equation. Note that the time constant for the second equation is 5 times slower than that in the first equation. Normalization has shifted the resting potential from -70 mV to zero and has set $E_{Na} = 1$ here, which corresponds to 125 mV above the resting potential. Similarly, $E_K = -\frac{1}{5}$, or 25 mV below the resting potential. Finally, the unit of time here corresponds to 0.10 ms to produce spike durations of 1.0 ms. These simplifications make the mathematical analysis significantly easier. A version of these equations with parameter values optimized to describe human neocortical excitatory and inhibitory neurons may be found elsewhere (Wilson, 1999b).

For $I_{ext} = 0$, Equation (3) has three steady states at $V = 0, R = 0$ (resting state); $V = \frac{1}{7}, R = \frac{3}{49}$; and $V = \frac{2}{5}, R = \frac{12}{25}$. In order, these are an asymptotically stable node, a saddle point, and an unstable spiral point. The spike firing threshold occurs for $I_{ext} = 0.012$, where the saddle point and node coalesce in a saddle-node bifurcation to a limit cycle. This permits spike firing at arbitrarily low rates for I_{ext} just above threshold, which is characteristic of human and mammalian cortical neurons (Wilson, 1999a). A spike train for $I_{ext} = 0.013$ is depicted in Figure 2a. Above threshold, spike rate increases monotonically as a function of I_{ext}. Neurons that begin firing at arbitrarily low rates due to a saddle-node bifurcation are known as Class I neurons.

A different form of dynamics characterizes Class II neurons. The simplest example is obtained by replacing the dR/dt equation above with

$$5\frac{dR}{dt} = R + 2V. \qquad (4)$$

For $I_{ext} = 0$, the equations now have a single steady state $V = 0$, $R = 0$. It can be shown that these neural equations undergo a subcritical Hopf bifurcation (Wilson, 1999a) to spiking at $I_{ext} = 0.062$. In this case firing begins at a relatively high spike rate, as arbitrarily low rates are precluded by the nature of the bifurcation to spiking. The original Hodgkin–Huxley (1952) equations, in fact, describe a Class II neuron (the giant axon of the squid).

Neurons in the cortex of humans and other mammals are Class I neurons. Thus, they begin firing at arbitrarily low spike rates (less than one spike per second), and this provides a greatly expanded dynamic range for encoding stimulus intensity into spike frequency. In addition, excitatory cortical neurons (but not most inhibitory neurons) typically have several additional currents that can produce even more complex spiking dynamics. For example, addition of a slow Ca^{2++} current and an even slower Ca^{2++}-driven K^+ current results in a neuron that fires bursts of spikes (Wilson, 1999a,b), as illustrated in Figure 2b.

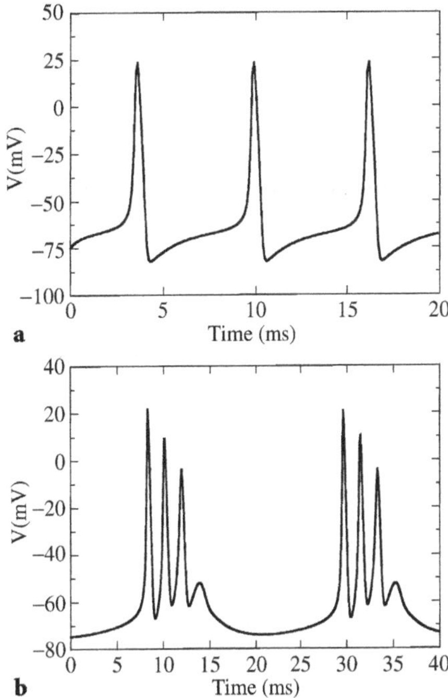

Figure 2. Spikes generated by neural equations. (a) Periodic spike train generated by Equation (3). (b) Pattern of spike bursts generated by a more complex cortical neuron with additional currents. Both graphs have been transformed back from the normalized form described in the text to reflect the mV range and ms time range actually encountered with cortical neurons.

Conclusions

Action potential dynamics are governed by two processes. Once threshold is reached, a rapid influx of Na^+ ions results in depolarization of the neuron causing the upswing of the spike. This process is described by the first term in the dV/dt equation in (3) above. Following this, the variable R increases, permitting K^+ to pass out of the neuron, thus hyperpolarizing it back toward its resting potential. The limit cycle is generated because the recovery variable R operates on a slower time scale than the very rapid Na^+ depolarization. There is also an inactivation of the Na^+ ion current that contributes to termination of the spike (Hodgkin & Huxley, 1952), but it is not essential to the dynamics of spike generation.

Virtually all brain function involves a dynamical interaction between excitatory and inhibitory neurons. For example, short-term memory circuits involve groups of neurons that are reciprocally interconnected by excitatory synapses, which enables them to continue firing following the cessation of stimulation. This ongoing neural activity makes up the short-term memory store. Inhibition is necessary both to shut off short-term memory activity and to prevent it from spreading to activate other neurons. Indeed, when there is too little inhibition in a brain area due to an imbalance or injury, the spread of excitation may generate epileptic seizures. Many other examples of excitatory and inhibitory neural circuits may be found elsewhere (Wilson, 1999a; Dayan & Abbott, 2001).

HUGH R. WILSON

See also **Hodgkin–Huxley equations; Integrate and fire neuron; Nerve impulses**

Further Reading

Dayan, P. & Abbott, L.F. 2001. *Theoretical Neuroscience*, Cambridge, MA: MIT Press

Hille, B. 1992. *Ionic Channels of Excitable Membranes*, Sunderland, MA: Sinauer

Hodgkin, A.L. & Huxley, A.F. 1952. A quantitative description of membrane current and its application to conduction and excitation in nerve. *Journal of Physiology*, 117: 500–544

Koch, C. 1999. *Biophysics of Computation: Information Processing in Single Neurons*, Oxford and New York: Oxford University Press

Wilson, H.R. 1999a. *Spikes, Decisions, and Actions: Dynamical Foundations of Neuroscience*, Oxford and New York: Oxford University Press

Wilson, H.R. 1999b. Simplified dynamics of human and mammalian neocortical neurons. *Journal of Theoretical Biology*, 200: 375–388

NEWELL–WHITEHEAD–SEGEL EQUATION

See **Complex Ginzburg–Landau equation**

NEWTON'S LAWS OF MOTION

Isaac Newton's treatise *Philosophiae Naturalis Principia Mathematica* (Mathematical Principles of Natural Philosophy), often simply referred to as the *Principia*, was completed in May 1686. Publication and printing were overseen by the astronomer Edmund Halley, and the work became available to the public in 1687. In this work Newton sets out, in effect, a template for classical mathematical physics, the dynamics of particles and rigid bodies in particular. The text starts by defining certain basic quantities such as mass, momentum, inertia, force, and acceleration. Once these have been defined and some of their basic properties described, Newton states his "axioms" or "laws of motion" thus:

> *Law I*: Every body continues in its state of rest, or of uniform motion in a right line, unless it is compelled to change that state by forces impressed upon it;
> *Law II*: The change of motion is proportional to the motive force impressed; and is made in the direction of the right line in which that force is impressed;
> *Law III*: To every action there is always opposed an equal reaction: or, the mutual actions of two bodies upon each other are always equal, and directed to contrary parts.

(We have quoted the established 1729 translation from the original Latin to English by Andrew Motte.)

In modern terms the first law, also known as the law of inertia, and already realized by Galileo, states that uniform translational motion is a "natural" state of motion for a particle or body requiring no cause or outside agent to maintain it. (This was an important departure from Aristotelian physics, which held that all motion required an explanation.) Not so with accelerated motion, according to the second law. If a change in the velocity of motion is to come about, it requires the action of a force acting on the particle or body. Finally, in the third law we have the requirement that if a body, A, is acted upon by a force due to another body, B, then B is subject to a force of the same magnitude but of opposite direction from its interaction with A.

Forces

Newtonian mechanics, then, operates with purely kinematic entities, such as position, velocity, and acceleration, and with a set of new, dynamical entities called forces. Newton's laws of motion instruct us to seek the cause of acceleration, a kinematic quantity, in the total force acting on a particle or body. Force is a dynamic quantity. The mass of the particle or body appears as the constant of proportionality in the relation between force (the cause) and acceleration (the effect). Like accelerations, forces are vectors. If two or more forces act simultaneously, they add vectorially and their resultant gives both the correct magnitude and the correct direction of the net force. Many well-known results about levers, systems with suspended masses, and various simple mechanical machines were immediately subsumed under Newtonian mechanics simply by recognizing that forces behave like vectors upon superposition.

Not only did Newton establish a framework for dynamics—and for much of physics—with his three laws, he provided also a new law of Nature for the attractive force of gravity acting between any two bodies. This development was, in some sense, extraneous to the three laws of motion but without it, the full force of Newton's dynamics might not have been appreciated and embraced. Newton's expression for the force acting on a particle, 1, of mass m_1 located at $\boldsymbol{x}_1$ due to a particle, 2, of mass m_2, located at $\boldsymbol{x}_2$, is

$$\boldsymbol{F}_{12} = Gm_1m_2\boldsymbol{e}_{12}/r^2. \qquad (1)$$

Here r is the distance between the two particles, $r = |\boldsymbol{x}_1 - \boldsymbol{x}_2|$ and $\boldsymbol{e}_{12}$ is the unit vector pointing from the location of particle "1" to the location of particle "2", that is, $\boldsymbol{e}_{12} = (\boldsymbol{x}_2 - \boldsymbol{x}_1)/r$. The coefficient G, now known as Newton's universal gravitational constant, has the approximate value $G \approx 6.672 \times 10^{-8}\,\mathrm{cm}^3\,\mathrm{g}^{-1}\,\mathrm{s}^{-2}$.

The first measurements of G were performed by Henry Cavendish in 1798 using a torsion balance. Expression (1) is fully consistent with Newton's third law, which demands that $F_{12} = -F_{21}$. Force (1) represents action-at-a-distance in the sense that it does not result from direct contact between contiguous bodies or particles. Newton was severely criticized by his contemporaries for this notion, which they viewed as fanciful and unsubstantiated.

In modern terminology, in the *Principia* Newton accomplished (among many other things) integration of the two-body problem of celestial mechanics, that is, the following system of differential equations, which result by combining Newton's law of gravitation with his second law of motion:

$$m_1\, d^2\boldsymbol{x}_1/dt^2 = Gm_1m_2(\boldsymbol{x}_2 - \boldsymbol{x}_1)/|\boldsymbol{x}_2 - \boldsymbol{x}_1|^3, \quad (2a)$$

$$m_2 d^2\boldsymbol{x}_2/dt^2 = Gm_1m_2(\boldsymbol{x}_1 - \boldsymbol{x}_2)/|\boldsymbol{x}_1 - \boldsymbol{x}_2|^3. \quad (2b)$$

Because of Newton's third law, addition of Equations (2) yields

$$m_1\, d^2\boldsymbol{x}_1/dt^2 + m_2\, d^2\boldsymbol{x}_2/dt^2 = 0, \quad (3)$$

which shows that the total momentum of the system, given by

$$\boldsymbol{P} = m_1 d\boldsymbol{x}_1/dt + m_2\, d\boldsymbol{x}_2/dt, \quad (4)$$

is a constant of the motion. Thus, the center of mass of the two particles,

$$\boldsymbol{R} = (m_1\boldsymbol{x}_1 + m_2\boldsymbol{x}_2)/(m_1 + m_2), \quad (5)$$

moves with constant velocity $\boldsymbol{P}/M$ through space, where $M = m_1 + m_2$ is the total mass of the two-body system.

On the other hand, subtracting (2a) from (2b) gives

$$m d^2\boldsymbol{r}/dt^2 = -Gm_1m_2\boldsymbol{r}/r^3, \quad (6)$$

where $\boldsymbol{r} = \boldsymbol{x}_2 - \boldsymbol{x}_1$, $r = |\boldsymbol{r}|$, $M = m_1 + m_2$ as before, and $m = m_1m_2/(m_1 + m_2)$ is called the reduced mass. Equation (5) tells us that the problem of the relative motion of the two particles is equivalent to solving for the motion of a single (fictitious) particle of mass m in the same field of force centered at the origin of coordinates. Such a fixed force field pointing toward a given point in space (here chosen as the origin of coordinates) is called a central force.

Once (6) is solved, the original positions of the two gravitationally attracting particles may be reconstructed from the formulae

$$\boldsymbol{x}_1 = \boldsymbol{R} - m_2\boldsymbol{r}/M, \quad \boldsymbol{x}_2 = \boldsymbol{R} + m_1\boldsymbol{r}/M. \quad (7)$$

The differential equation for the vector function $\boldsymbol{r}$ is nonlinear because the magnitude of the force falls off with the square of the distance. Many force laws of interest in applications have the property that they are nonlinear functions of the positions of the constituent particles of the system. The force law for small extensions of a high-quality spring, known as Hooke's law, which states that the force is directly proportional to the deviation from equilibrium, is an important and notable exception.

Newton showed in the *Principia* that the general solution of (6) is that $\boldsymbol{r}$ traces out a conic section (ellipse or hyperbola, with a parabola as the "cross-over" possibility) corresponding to bounded or unbounded relative motions of the two original particles. The bounded motions apply to planets orbiting the Sun or the Moon orbiting the Earth. The effects of bodies farther away are to be treated subsequently by adding perturbations to the above analysis. Unbounded motions are realized by comets. Newton further considered the effects of the finite extension of the attracting body, for example, Earth's finite size relative to the Moon's orbit, and the effect of the deformation of the attracting body due to tidal forces.

Hassan Aref

See also **Celestial mechanics; Determinism; *N*-body problem**

Further Reading

Chandrasekhar, S. 1995. *Newton's Principia for the Common Reader*, Oxford: Clarendon Press and New York: Oxford University Press

Newton, I. 1934. *Mathematical Principles of Natural Philosophy and System of the World*, translated by A. Motte (1729), edited by F. Cajori, 2 vols, Berkeley: University of California Press

NEWTON'S METHOD

See **Numerical methods**

NODES OF RANVIER

See **Myelinated nerves**

NOISE (WHITE, COLORED ETC.)

See **Stochastic processes**

NONATTRACTING CHAOTIC SETS

See **Chaotic dynamics; Invariant manifolds and sets**

NONAUTONOMOUS SYSTEMS

See **Phase space**

NONEQUILIBRIUM STATISTICAL MECHANICS

Nonequilibrium statistical mechanics aims to describe the behavior of large systems of particles removed from the state of thermodynamic equilibrium, in terms of the properties of the individual constituents and their interactions, as provided by the laws of classical and quantum mechanics. Such systems give rise to irreversible behavior in the form of an approach to thermodynamic equilibrium in the absence of permanent constraints (isolated systems), or to a stationary nonequilibrium state in systems under constraint (referred to as thermostatted systems).

The fundamental problem of non-equilibrium statistical mechanics is to reconcile irreversible behavior at the macroscopic level with the reversible character of the underlying microscopic laws of mechanics. For a long time it was thought that irreversibility could not be understood entirely from mechanics, an idea that led to the systematic introduction of probabilistic ideas in the traditional deterministic description. As a result, the laws of statistical mechanics have frequently been regarded as analogs of the law of large numbers and similar universal laws of probability and statistics, independent of any explicit reference to the nature of the underlying dynamics. Since the 1980s, one has witnessed a change of perspective following the realization that the trajectories of individual particles in an N-body system are typically chaotic.

Boltzmann's Kinetic Theory. Ergodic and Mixing Hypotheses

In 1872, Ludwig Boltzmann derived by what appeared to be completely mechanical arguments his famous kinetic equation for dilute gases (Brush, 1965, 1966). This equation features the time-dependent probability distribution function f, of position $\boldsymbol{r}$ and velocity $\boldsymbol{v}$, for a particle in the gas and allows one to reproduce in a very satisfactory way the transport and flow properties of the system. The kinetic theory culminates in the derivation of the H-theorem, whereby in a homogeneous system the functional $H = k \int f \ln f \, \mathrm{d}\boldsymbol{v}$ (k being the Boltzmann constant) decreases monotonously until f reaches its equilibrium form given by the Maxwell–Boltzmann distribution. The H-theorem should provide, then, a microscopic justification of the second law of thermodynamics. This claim prompted rather negative reactions, crystallized in the famous reversibility (Loschmidt's) and recurrence (Zermelo's) paradoxes. Boltzmann was unable to fully refute these objections since, as he himself recognized, his derivation makes use of a heuristic probabilistic assumption—the *Stosszahlansatz* or the assumption of molecular chaos—which allowed him to express approximately the rate at which binary collisions are taking place only in terms of the one-particle probability density f.

In trying to answer his critics Boltzmann enunciated the ergodic hypothesis, which eventually became a key concept in the entire field of statistical mechanics. Specifically, Boltzmann suggested that (a) in an isolated many-body system the overwhelming part of the phase space consists of regions where the macroscopic properties are very close to the equilibrium properties, (b) the system's trajectory will spend equal times in phase space regions of equal extent, and (c) the macroscopic properties of the system will essentially be constant throughout the allowable part of phase space and will coincide with the long-time averages of the corresponding microscopic quantities over the phase space trajectory. These statements were later completed by the equally ground-breaking discovery by Josiah W. Gibbs of the concept of mixing: the dynamical evolution of an isolated system initially occupying a limited phase space region compatible with some prescribed values of its macroscopic observables will lead it eventually to occupy uniformly (at least in a coarse-grained sense) the entire phase space available.

In their original forms, the ergodic hypothesis and the stronger mixing hypothesis had the serious drawback of relying to a considerable extent on the coarse-grained way one observes the system. Modern ergodic theory starts with the work of Henri Poincaré and George Birkhoff (Arnol'd & Avez, 1968), who were able to relate ergodicity to certain well-defined properties of the underlying deterministic evolution laws. Two examples are provided by Birkhoff's theorem on the existence of the limit $1/T \int_0^T A(t)\,\mathrm{d}t$ as $T \to \infty$ of an integrable phase space function A provided that the motion remains bounded in phase space and by Poincaré's theorem that for systems satisfying suitable resonance conditions there exist no invariants other than total energy that are analytic in some parameter. Since the 1950s, a great deal of effort has been devoted to prove whether a system described by a given Hamiltonian will or will not fulfill these properties and to what extent these properties bear a clear-cut relationship with the type of dynamics going on in phase space. An early (negative) result of considerable historical importance was obtained by Enrico Fermi, John R. Pasta, and Stanislaw Ulam (1955), who showed that in a system of coupled nonlinear oscillators energy may remain localized rather than be equipartitioned among the individual degrees of freedom. At the other extreme, one finds Yakov Sinai's result (Sinai, 1970) on the ergodic and mixing behavior of a system of hard spheres. This work signaled the beginning of a series of developments aimed at relating the foundations of non-equilibrium statistical mechanics to the instability of motion of large classes of non-integrable dynamical systems giving rise to sensitivity to initial conditions and to deterministic chaos, known to be generic since the work of Andrei Kolmogorov.

Generalized Kinetic Theories

In parallel, and largely independently of progress in ergodic theory, the need to go beyond the assumptions underlying Boltzmann's equation became a central preoccupation from the mid-1940s. Three major attempts along this line were initiated by Nikolai Bogolubov, Leon Van Hove, and Ilya Prigogine and his colleagues (see, e.g., Balescu, 1975). Their starting point was a systematic perturbation expansion of the Liouville equation and its quantum counterpart for the N-particle probability distribution or density matrix, or the hierarchy of equations for the reduced n-particle ($n=1,\ldots,N$) distribution functions obtained by integrating the Liouville equation over the $N-n$ particles, known as the BBGKY hierarchy. This procedure led as a first step to exact, formal non-Markovian equations of the form

$$\frac{\partial \rho_s}{\partial t} + K\rho_s(t) = \int_0^t \mathrm{d}\tau\, G(t-\tau)\rho_s(\tau) + D(t), \quad (1)$$

where ρ_s is a reduced distribution, K stands for the contribution of the mean field, G is a memory kernel, and $D(t)$ depends on the initial correlations in the subspace of phase space complementary to the one of ρ_s. Closed kinetic equations for reduced probability densities could then be obtained under certain assumptions, linked to first principles in a more clear-cut way than Boltzmann's *Stosszahlansatz*: Bogolubov's ansatz of higher-order distributions becoming functionals of the one-particle one, Prigogine's ansatz on initial correlations, van Hove's random phase approximation. But even when these conditions are satisfied, it has so far proved impossible to establish a general H-theorem for the corresponding equations. Still, generalized kinetic equations have been at the foundation of spectacular progress in such fields as the study of dense fluids and plasmas and the microscopic theory of transport coefficients in the linear range of irreversible phenomena close to equilibrium.

Microscopic Chaos and Nonequilibrium Statistic Mechanics

Generalized kinetic theories rest on approximations whose validity is difficult to assess. Furthermore, there is no explicit link between the structure of the kinetic equations and the nature of the microscopic dynamics in phase space. In view of the fundamental importance and the ubiquity of irreversibility in the natural world, it would certainly be desirable to arrive at a description free of both these limitations. This has been achieved since the 1980s by the cross-fertilization between dynamical systems, nonequilibrium statistical mechanics, and microscopic simulation techniques.

Mapping to a Markov Process

A first series of attempts pertains to the class of strongly unstable systems known as Kolmogorov flows, in which each phase space point lies at the intersection of stable and unstable manifolds. It takes advantage of the existence in such systems of special phase space partitions—the Markov partitions—whose boundaries remain invariant under the dynamics, each partition cell being mapped at successive time steps into a union of partition cells. If the operator projecting the full phase space dynamics on such a partition commutes with the Liouville operator, then the Liouville equation can be mapped into an exact Markovian equation exhibiting an H-theorem. This provides a rigorous microscopic basis of coarse-graining (Penrose, 1970; Nicolis et al., 1991). The mapping is, however, not one-to-one as the projection operator is not invertible reflecting the loss of information associated with this process. Ilya Prigogine, Baidyanath Misra, and Maurice Courbage (Prigogine, 1980) succeeded in constructing a non-unitary, invertible, transformation operator Λ, which transforms the unitary evolution operator U_t (essentially the exponential of the Liouville operator) into a Markov semigroup W_t

$$W_t = \Lambda U_t \Lambda^{-1} \quad (2)$$

The idea of non-unitary transformations has also been extended to more general classes of systems, such as non-integrable systems generating Poincaré resonances.

Escape Rate Formalism and Deterministic Thermostats

A second line of approach aims to express transport coefficients and other macroscopic level properties including entropy production in terms of the quantifiers of the microscopic chaos prevailing at the level of individual particle trajectories (in a classical setting). Two different methodologies have been developed.

The escape rate formalism (Gaspard, 1998; Dorfman, 1999). In this formalism, transport in Lorentz gas type systems is linked to the escape from a fractal repellor formed by trajectories for which a certain microscopic quantity associated with the macroscopic flux of interest remains confined in phase space. The corresponding transport coefficients are then expressed in terms of the Lyapunov exponents of the repellor and its Kolmogorov–Sinai entropy or its fractal dimension. No boundary conditions need to be imposed. Furthermore, the hydrodynamic modes associated with the process are computed as generalized eigenmodes of the Liouvillian and turn out to be fractal distributions.

Deterministic thermostats. The question here is how to express in purely mechanical terms a constraint maintaining a system away from equilibrium. One of the most popular ways to achieve this is to augment (in a classical system) the Hamiltonian dynamics by dynamical friction terms preserving the time-reversal symmetry (Hoover, 1999). In this way, one obtains a dissipative dynamical system where the rate of phase space volume contraction and the entropy production are shown to be related to the sum of the Lyapunov exponents. The non-equilibrium steady states generated in this formalism are fractal attractors, contrary to the equilibrium states that extend over the entire phase space available. Furthermore, they are believed to satisfy an interesting fluctuation theorem (Gallavotti &

Cohen, 1995) expressing the probability of fluctuations associated with particles moving in a direction opposite to the applied constraint.

Statistical Mechanics of Dynamical Systems

Yakov Sinai, David Ruelle, and Rufus Bowen (see, for example, Ruelle, 1999) have shown that the long-time behavior of a large class of chaotic dynamical systems is determined by an invariant probability measure characterized by a variational principle, thereby establishing a link with ergodic theory and equilibrium statistical mechanics. A major difference is that contrary to equilibrium measures, Sinai–Ruelle–Bowen (SRB) measures are smooth only along the expanding directions while they possess a fractal structure along the contracting ones. This provides a rationale for characterizing the fractal character of the non-equilibrium steady states generated by deterministic thermostats. Furthermore, it has been the starting point of a series of developments aimed at characterizing low-order deterministic dynamical systems showing complex behavior through probability densities and the spectral properties of their evolution operators, such as the Liouville or the Frobenius–Perron operators (Lasota & Mackey, 1985). In a different vein, the transport induced by the overlap of resonances arising, in particular, around a separatrix has been investigated (Lichtenberg & Lieberman, 1983; Balescu, 1997). These approaches have provided insights to questions motivated by non-equilibrium statistical mechanics that had remained unsolved for a long time, owing to the formidable difficulties arising from the presence of a large number of interacting particles.

Non-equilibrium Statistical Mechanics as a Tool to Understand Cooperative Behavior in Complex Systems

Non-equilibrium statistical mechanics has been a source of inspiration providing interesting ways to analyze complex nonlinear systems arising in chemistry, fluid mechanics, biology, or even finance and sociology. One of the earliest applications was a mesoscopic approach describing the dynamics of fluctuations of the macroscopic observables of such systems around a reference state, using master equation or Langevin–Fokker–Planck equation descriptions (Nicolis & Prigogine, 1977; Haken, 1977). It led to a theory of nonequilibrium phase transitions showing how macroscopic level bifurcation phenomena are reflected at the microscopic level through the anomalous behavior of the non-equilibrium fluctuations. When applied to nanoscale systems, this approach provides interesting insights on energy transduction by non-equilibrium devices such as biological macromolecules, referred to as molecular motors (Astumian, 1997).

Finally, the formalism of non-equilibrium statistical mechanics is well suited to studying the cooperative behavior of interacting multi-agent systems involved in, for instance, food webs and other networks of connected elements, finance, or social phenomena. In this setting, the relevance of power laws describing the statistical behavior of certain systems has attracted considerable interest, the idea being that such laws reflect well the ability of complex systems to show adaptive behavior and to optimize information flows (Albert & Barabasi, 2002). The dynamical origin of these laws and the role of non-equilibrium constraints remain major open questions.

G. NICOLIS

See also **Chaotic dynamics; Emergence; Ergodic theory; Fermi–Pasta–Ulam oscillator chain; Markov partitions; Recurrence; Sinai–Ruelle–Bowen measures; Synergetics**

Further Reading

Albert, R. & Barabasi, A.S. 2002. Statistical mechanics of complex networks. *Reviews of Modern Physics*, 74: 47–97
Arnol'd, V. & Avez, A. 1968. *Ergodic Problems of Classical Mechanics*, New York: Benjamin
Astumian, R.D. 1997. Thermodynamics and kinetics of a Brownian motor. *Science*, 276: 917–922
Balescu, R. 1975. *Equilibrium and Nonequilibrium Statistical Mechanics*, New York: Wiley
Balescu, R. 1997. *Statistical Dynamics*, London: Imperial College Press
Brush, S. 1965, 1966. *Kinetic Theory*, vols. I and II. London: Pergamon Press
Dorfman, J.R. 1999. *An Introduction to Chaos in Nonequilibrium Statistical Mechanics*, Cambridge and New York: Cambridge University Press
Fermi, E., Pasta, J.R. & Ulam S. 1955. *Studies of nonlinear problems*. Los Alamos Scientific Laboratory, Report NoLA-1940
Gallavotti, G. & Cohen, E.G.D. 1995. Dynamical ensembles in nonequilibrium statistical mechanics. *Physics Review Letters*, 74: 2694–2697
Gaspard, P. 1998. *Chaos, Scattering and Statistical Mechanics*, Cambridge and New York: Cambridge University Press
Haken, H. 1977. *Synergetics*, Berlin: Springer
Hoover, W.G. 1999. *Time Reversibility, Computer Simulation, and Chaos*, Singapore: World Scientific
Lasota, A. & Mackey, M. 1985. *Probabilistic Properties of Deterministic Systems*, Cambridge and New York: Cambridge University Press
Lichtenberg, A.J. & Lieberman, M.A. 1983. *Regular and Stochastic Motion*, Berlin: Springer
Nicolis, G., Martinez, S. & Tirapegui, E. 1991. Finite coarse-graining and Chapman–Kolmogorov equation in conservative dynamical systems. *Chaos, Solitons and Fractals*, 1: 25–37
Nicolis, G. & Prigogine, I. 1977. *Self-organization in Non-equilibrium Systems*, New York: Wiley
Penrose, O. 1970. *Foundations of Statistical Mechanics*, Oxford: Pergamon
Prigogine, I. 1980. *From Being to Becoming*, San Fransisco: Freeman
Ruelle, D. 1999. Smooth dynamics and new theoretical ideas in nonequilibrium statistical mechanics. *Journal of Statistical Physics*, 95: 393–468

Sinai, Ya. 1970. Dynamical systems with elastic reflections. *Russian Mathematical Surveys*, 25: 137–189

NONERGODICITY

See **Ergodic theory**

NONINTEGRABLE LATTICES

See **Integrable lattices**

NONLINEAR ACOUSTICS

As an area of physics and mechanics concerned with sound waves of finite amplitude, theoretical nonlinear acoustics stems from advances in classical fluid mechanics made in the 19th century. It was developed originally as a weakly nonlinear limit of gas dynamics and included the study of simple waves, weak shock waves, and their interactions. Nonlinear acoustics as a separate branch of science was formed mainly during and after World War II when, on the one hand, military applications of underwater acoustics became important, and on the other, more powerful sources of sound were developed.

The nature of finite-amplitude acoustic wave propagation in fluids depends on the value of the acoustic Reynolds number $\mathrm{Re} = \beta u_0 \lambda / \nu$, where u_0 is a characteristic amplitude of the particle velocity, λ is the wavelength, ν the kinematic viscosity (or a combination of viscosity and thermal conductivity), and β is the coefficient of nonlinearity (1.2 for air, 3.5 for water). For $\mathrm{Re} \leq 1$, the wave attenuates before significant nonlinear distortion accumulates. For $\mathrm{Re} \gg 1$, higher harmonics are generated, and the wave profile may undergo radical changes.

The physical process underlying waveform distortion for a plane (one-dimensional) traveling wave is the simple-wave characteristic relation $\mathrm{d}x/\mathrm{d}t = c_0 + \beta u$ for the propagation speed of individual points on the waveform, where c_0 is the small-signal (infinitesimal-amplitude) sound speed. Positive values of the particle velocity u (or sound pressure) in the waveform advance on the zero crossings propagating at speed c_0, and negative values recede. For $\mathrm{Re} \gg 1$, this process leads to waveform steepening up to the "breaking" point, beyond which this propagation law predicts a multivalued waveform (see Figure 1). Whereas water waves break, sound waves cannot, and instead an abrupt jump in the wave amplitude, referred to as a shock front, is formed. Shocks substantially increase attenuation of the wave, and their thickness is proportional to a combination of viscosity and thermal conductivity coefficients.

The process just discussed is well described for a plane traveling wave by the Burgers equation

$$\frac{\partial u}{\partial t} + (c_0 + \beta u)\frac{\partial u}{\partial x} = \delta \frac{\partial^2 u}{\partial x^2}, \qquad (1)$$

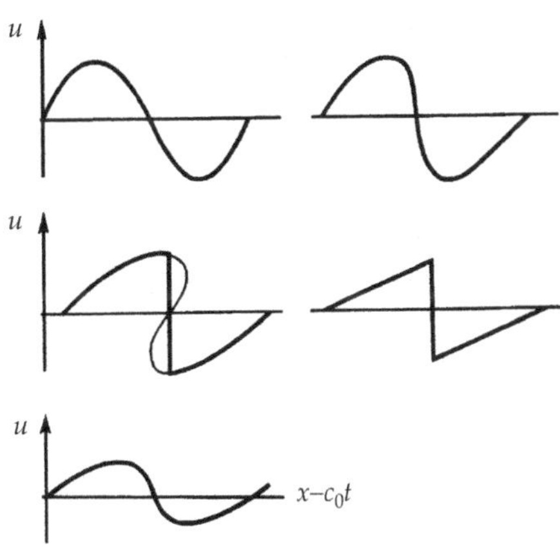

Figure 1. Schematic of the evolution of an initially sinusoidal nonlinear acoustic wave. Each plot corresponds to the one-period waveform as a function of coordinate at the successive time moments.

where δ is a dissipation coefficient that accounts for viscosity and thermal conductivity and equals $2\nu/3$ for shear viscosity alone. The left-hand side accounts for the finite-amplitude propagation speed given above. The Reynolds number, given above, characterizes the ratio of the nonlinear term to the dissipation term. Equation (1) is expressed in a form convenient for initial value problems, when an initial spatial waveform is prescribed. An alternative form of Equation (1), with the roles of x and t reversed, is used in signaling problems, in which a time waveform is prescribed on a boundary. Similar processes occur in spherical waves, such as those at sufficient distances from explosions, where the nonlinearity is weak.

An important role in the development of modern nonlinear acoustics has been played by the parametric array conceived by Westervelt in the USA and Zverev and Kalachev in Russia in the early 1960s. In the parametric array, two acoustic beams close in frequency interact nonlinearly and create a virtual antenna that radiates a secondary, low-frequency beam. Notwithstanding their low efficiency, these antennas prove useful in applications (such as sea bottom profiling) because of their high directivity and absence of side lobes.

Describing the parametric array and nonlinear sound beams, in general, is a complicated problem that requires taking into account the effect of diffraction. The combined effects of nonlinearity, attenuation, and diffraction on a sound beam may be modeled with an augmented form of Equation (1) called the Khokhlov–Zabolotskaya–Kuznetsov (KZK) equation. This equation contains an additional term that accounts for the diffraction of narrow beams. It is widely used for calculations of finite-amplitude sonars, such as the parametric array, and focused beams of high-intensity ultrasound used in imaging and medicine. Further augmentation of Equation (1) permits investigation of nonlinear propagation in inhomogeneous media and leads to a theoretical formalism called "nonlinear geometrical acoustics," which is used to study the evolution of ray patterns.

Acoustic fields in fluids can also generate time-averaged effects such as acoustic streaming (steady flow) and radiation pressure. Acoustic streaming provides a basis for some microfluidic pumps, and in other applications, it influences heat transfer. Radiation pressures generated by standing waves are used to levitate and manipulate particles and bubbles, especially in microgravity environments. Accurate prediction of these time-averaged effects requires special care when defining the problem under consideration.

Nonlinear acoustics of solids includes a variety of interactions between different wave types, such as longitudinal, shear, and surface waves in isotropic solids. Additional types of wave interactions occur in anisotropic media (crystals). The different propagation speeds of the various waves lead to selection rules for particular wave interactions, usually connected with the direction of propagation. Surface (interface) waves exhibit nonlinear propagation effects that differ fundamentally from those of bulk longitudinal and shear waves. Their penetration depth away from an interface is proportional to wavelength, and this frequency-dependent property gives rise to nonlocal nonlinearity that does not occur in bulk waves. With nonlocal nonlinearity, the nonlinear perturbation of the wave velocity at any given point on a waveform is influenced by the entire wave field.

Along with "classical" homogeneous media in which the nonlinearity is due to anharmonicity of atomic forces, there exists an important class of "structurally" nonlinear media in which the nonlinearity is due to the presence of soft inclusions in a harder main body. Examples are small gas bubbles in liquids, pores in rubber-like materials, and grain contacts, microcracks, and dislocations in solids. Acoustic nonlinearities in such media can be several orders stronger than in homogeneous media. Moreover, the nonlinear equation of state (stress-strain relation) in such materials as metals and rock can predict irreversible behavior (hysteresis), which results in some unusual effects; for example, the third-harmonic amplitude in a strain wave can be proportional to the square of the primary-wave amplitude, rather than to its cube.

From a physical viewpoint, classical nonlinear acoustics deals mainly with non- or weakly-dispersive media, in which cumulative waveform steepening and harmonic generation occurs until shocks are formed. Dispersion in acoustics can be introduced by waveguide walls (geometrical dispersion) or by inhomogeneities such as grains and bubbles. As a result, many effects known in nonlinear optics and plasma physics have been realized in acoustics, such as self-focusing, phase conjugation (wave front reversal), and parametric amplification. Acoustic solitons, in which certain effects of nonlinearity and dispersion offset each another, are also possible. Even an acoustical analog of the maser, in which randomly phased oscillators (such as resonant bubbles) self-synchronize and radiate coherently, is possible.

Applications and manifestations of nonlinear acoustics are numerous. Besides the aforementioned processes, acoustic nonlinearities are important in explosion waves, sonic booms, thermoacoustic engines (in which sound waves serve as heat pumps), therapeutic and diagnostic medical ultrasound, materials characterization, and nondestructive testing. Shock waves also form inside collapsing cavitation bubbles, and they are thought to influence the resulting flashes of light referred to as sonoluminescence.

LEV OSTROVSKY AND MARK HAMILTON

See also **Burgers equation; Shock waves; Surface waves**

Further Reading

Beyer, R.T. 1997. *Nonlinear Acoustics*, 2nd edition, New York: Acoustical Society of America

Hamilton, M.F. & Blackstock, D.T. (editors). 1998. *Nonlinear Acoustics*, San Diego: Academic Press

Naugolnykh, K.A. & Ostrovsky, L.A. 1998. *Nonlinear Wave Processes in Acoustics*, Cambridge and New York: Cambridge University Press

Rudenko, O.V. & Soluyan, S.I. 1977. *Theoretical Foundations of Nonlinear Acoustics*, New York: Plenum Press

NONLINEAR ELECTRONICS

In the early part of the 20th century, vacuum tube devices dominated electronics circuitry. However with the advent of junction transistors in 1951 and integrated chips in 1957, semiconductor devices largely replaced vacuum tubes by the 1970s. Present-day electronic circuits make extensive use of active elements/devices that may behave either linearly or nonlinearly, depending upon their applications, operating currents, and voltages (as well as various physical effects including thermal effects, dielectric breakdown, and magnetic saturation). Typical modern nonlinear devices include semiconductor diodes, bipolar junction transistors (BJTs), and field effect transistors (FETs). For analysis and design, the nonlinear devices are often replaced by equivalent basic nonlinear circuit elements such as two-terminal, multiterminal, and multiport resistors, capacitors, and inductors. Thus, a nonlinear electronic circuit is an interconnection of the various circuit elements involving at least one nonlinear element.

The major nonlinear two-terminal circuit elements are typically characterized by nonlinear functional representations in contrast to their linear counterparts: (i) nonlinear resistor: nonlinear voltage $v(t)$ vs.

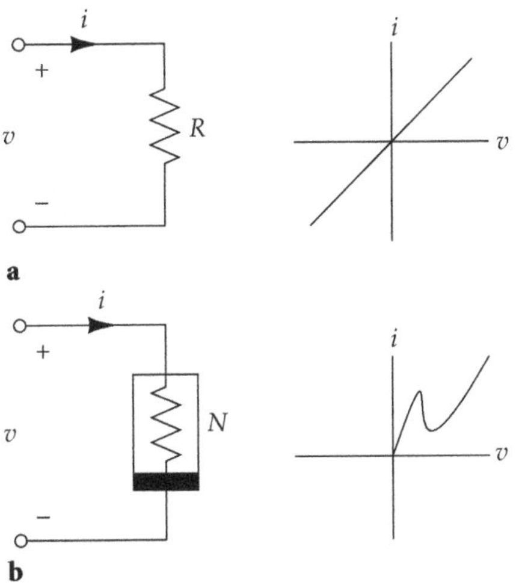

Figure 1. (a) Linear resistor and its v–i characteristic (b) Nonlinear resistor and its v–i characteristic.

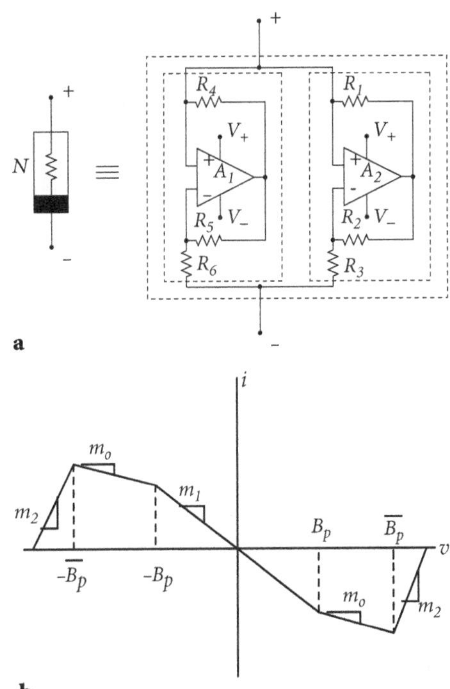

Figure 2. (a) Chua's diode and (b) its characteristic.

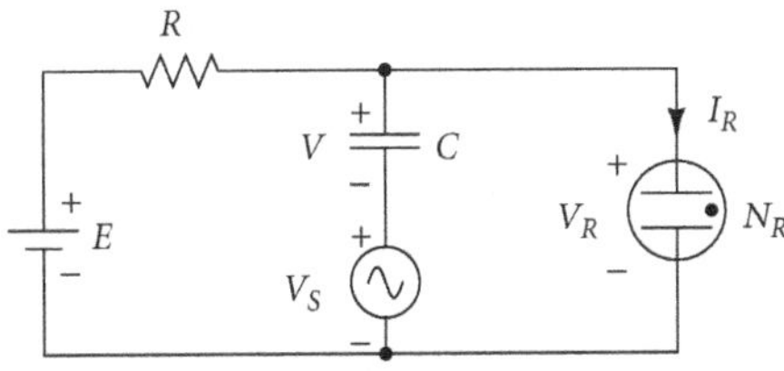

Figure 3. Van der Pol's original oscillator circuit.

current $i(t)$ characteristic curve, (ii) nonlinear capacitor: nonlinear charge $q(t)$ vs. voltage $v(t)$ curve, and (iii) nonlinear inductor: nonlinear $i(t)$ vs. magnetic flux Φ(t) curve (for an illustrative example, see Figure 1). More general circuit elements, such as three terminal transistors and multipliers, and multiterminal elements, such as operational amplifiers (opamps), are also frequently used (Chua et al., 1987; Schubert & Kim, 1996). An innovative nonlinear element is the piecewise-linear device, namely Chua's diode synthesized using opamps/diodes and linear elements (possessing a five-segment v–i characteristic curve, including three negative resistance pieces, see Figure 2), which plays a crucial role in understanding various nonlinear dynamical phenomena (Lakshmanan & Murali, 1996; Lakshmanan & Rajasekar, 2003). The state equations for the currents and voltages underlying a given nonlinear electronic circuit, which are deduced using Kirchhoff's laws for electrical circuits, turn out to be a system of coupled nonlinear differential equations. In particular, when a given circuit includes at least two energy storage elements such as capacitors and inductors and a nonlinear element (which can typically function as an amplifier), it behaves as an oscillator for appropriate feedback and circuit parameters. The underlying system of nonlinear differential equations can then be equivalently considered as a typical nonlinear oscillator dynamical system of dissipative type.

Historically, the relaxation oscillator investigated by the Dutch engineer Balthasar der Pol in his seminal paper, "Frequency Demultiplication" (van der Pol & van der Mark , 1927), may be considered as the earliest example of a nonlinear electronic circuit, exhibiting many of the characteristic features underlying bifurcations and chaos, though not fully understood at that time. The circuit typically consists of a high voltage dc source E attached via a large series resistance R to a neon bulb N_R (or a triode valve) and a capacitor C, which are connected in parallel (see Figure 3) along with an external periodic signal V_s. As the capacitance C is increased smoothly, the current exhibits "discrete jumps from one whole-submultiple of the driving frequency to the next." For a critical value of the amplitude of the driving signal, this pattern of mode-lockings has a self-similar fractal structure consisting of an infinite number of steps. In modern jargon, this is just the devil's staircase (Kennedy & Chua, 1986). Van der Pol also noted that "often an irregular noise is heard in the telephone receiver [monitoring the signal in some way] before the frequency jumps to the next level value." Now we know that this "noise" indeed corresponds to chaotic signals. Typically, the circuit is represented by the second-order non-autonomous nonlinear differential equation

$$\frac{d^2x}{dt^2} + \varepsilon\left(x^2-1\right)\frac{dx}{dt} + x = f\sin\omega t, \tag{1}$$

where ε is the damping coefficient and f and ω represent the strength and frequency, respectively, of the periodic external forcing.

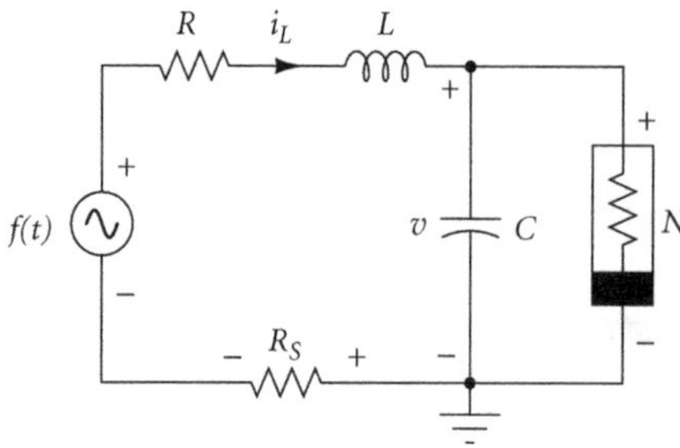

Figure 4. MLC Circuit: N is Chua's diode and $f(t)$ is the periodic signal.

Similar circuits are useful for modeling physical and biological systems; examples include flow of current across Josephson junctions or nerve impulse propagation in neuronal fiber in the form of Hodgkin–Huxley or FitzHugh–Nagumo equations (Scott, 2003). From the point of view of nonlinear dynamics, nonlinear electronic circuits arise either in analog simulation of typical nonlinear oscillators such as the Duffing oscillator or Lorenz system or as new dynamical systems per se constructed using typical nonlinear devices or ingeniously designed elements such as Chua's diode to act as black boxes to understand the various novel nonlinear phenomena. Examples of the latter include Chua's circuit/oscillator and Murali–Lakshmanan–Chua (MLC) circuit (Figure 4). In either case, these circuits are easy to build, analyse, and model, and help to scan the control parameter space quickly, thereby complimenting numerical and analytical studies of nonlinear phenomena.

Just like standard nonlinear dissipative dynamical systems, nonlinear electronic circuits typically exhibit various dynamical phenomena including bifurcations and chaos (Lakshmanan & Murali, 1996; Lakshmanan & Rajasekar, 2003; Thompson & Stewart, 1986).

- *Multistable states:* These correspond to point attractors (nodes and spiral points) as exhibited, for example, by any flip-flop or Schmitt trigger.
- *Periodic oscillations:* Autonomous nonlinear circuits can typically exhibit sinusoidal time response for weak nonlinearity (e.g., LC oscillators) and nearly square wave response (relaxation oscillators) for strong nonlinearity (e.g., square wave generators, Schmitt triggers). On the other hand, nonautonomous circuits often exhibit nonlinear resonant (harmonic, subharmonic, superharmonic, and dissipative) oscillations for weak nonlinearity (e.g., ferroresonant circuit with ac voltage source, hysteresis circuits, power amplifiers).
- *Bifurcations and chaos:* Almost the entire spectrum of bifurcations and chaos phenomena encountered in typical chaotic nonlinear dynamical systems is exhibited by numerous simple nonlinear electronic circuits/oscillators (e.g., Duffing oscillator, van der Pol oscillator, Chua's circuit, MLC circuit, RL diode circuit, Colpitts oscillator, Buck converters in power electronics).
- *Synchronization and controlling:* Nonlinear electronic circuits are versatile models to study and understand the phenomenon of synchronization in all its manifestations (such as identical, phase, lag, and generalized synchronizations) both with periodic and chaotic oscillations between two or a chain of oscillators. Similarly, to study various controlling techniques of periodic as well as chaotic oscillations, nonlinear circuits play a crucial role.
- *Secure communications and cryptography*: Because of the inherent advantage of miniaturization, nonlinear electronic circuits are natural choices for chaos application studies in secure communications, cryptography, and signal processing.

Nonlinear electronic circuits have become indispensable tools to understand a multitude of nonlinear dynamical phenomena, including chaos. In turn, nonlinear electronics has itself been enriched greatly by the advances in understanding various complex nonlinear phenomena, leading to the potential applications mentioned above.

MUTHUSAMY LAKSHMANAN

See also **Chaotic dynamics; Chua's circuit; Diodes; Duffing equation; Van der Pol equation**

Further Reading

Chua, L.O., Desoer, C.A. & Kuh, E.S. 1987. *Linear and Nonlinear Circuits*, New York: McGraw-Hill

Kennedy, M.P. & Chua, L.O. 1986. Van der Pol and chaos. *IEEE Transactions on Circuits and Systems*, 33: 974–980

Lakshmanan, M. & Murali, K. 1996. *Chaos in Nonlinear Oscillators: Controlling and Synchronization*, Singapore: World Scientific

Lakshmanan, M. & Rajasekar, S. 2003. *Nonlinear Dynamics: Integrability, Chaos and Patterns*, Berlin: Springer

Schubert, T.S. & Kim, E.M. 1996. *Active and Nonlinear Electronics*, New York: Wiley

Scott, A.C. 2003. *Nonlinear Science: Emergence and Dynamics of Coherent Structures*, 2nd edition, Oxford and New York: Oxford University Press

Thompson, J.M.T. & Stewart, H.B. 1986. *Nonlinear Dynamics and Chaos*, New York: Wiley

van der Pol, B. & van der Mark, J. 1927. Frequency demultiplication. *Nature*, 120: 363–364

NONLINEAR OPTICS

Maxwell's theory of electrical and magnetic fields and his idea that light is an electromagnetic wave were among the great milestones of scientific thought, unifying our understanding of a large and diversified range of physical phenomena. Indeed, by the late 19th century, the success of the classical electromagnetic theory of light led some to believe that there were few fundamental discoveries to be made.

Nonlinear properties of Maxwell's constitutive relations $\boldsymbol{B}=\mu(\boldsymbol{H})\boldsymbol{H}, \boldsymbol{D}=\varepsilon(\boldsymbol{E})\boldsymbol{E}$ had been recognized from the beginning. For example, the nonlinear permeability of ferromagnetic media was of prime concern in the design of electrical machinery in the 19th century. Nonlinear properties in the optical region had to await the discovery of the ruby laser in 1960 by Theodore Maiman (Maiman, 1960). The defining experiment by Peter Franken and coworkers in 1961 (Franken et al., 1961) detected ultraviolet light (l = 3470Å) at twice the frequency of a ruby laser (l = 6940Å) when this beam traversed a quartz crystal. (There is a humorous story about this discovery: the detected radiation was so weak that it appeared as a very faint spot on the photograph submitted to *Physical Review Letters*. The typesetting staff assumed that it was a glitch and erased it in the published document!) This experiment spurred a flurry of activity and many nonlinear optical phenomena were rapidly discovered. A series of monographs by Nicolaas Bloembergen in the 1960s (see Bloembergen, 1996a,b) contain an excellent historical overview of the explosive growth of activity in this field in the early days (see also Marburger, 1977; Shen, 1977).

For the most part, the study of nonlinear optical phenomena assumes a quantum material system coupled to a classical electromagnetic field. Within this approximation, noise-driven phenomena can be adequately described by a Langevin-type driving force. A proper description of quantum effects such as spontaneous emission or scattering requires that the electromagnetic field itself be quantized. Most nonlinear optical materials are nonmagnetic, so one sets the magnetic permeability $\mu(\boldsymbol{H})=\mu_0$, a constant. The procedure then is to expand the dielectric permeability $\varepsilon(\boldsymbol{E})$ as a formal power series in the electric field. This is most commonly done by defining an induced polarization field in terms of the electric displacement vector $\boldsymbol{D}=\varepsilon_0\boldsymbol{E}+\boldsymbol{P}$, where $\boldsymbol{P}=\boldsymbol{P}_{\mathrm{L}}+\boldsymbol{P}_{\mathrm{NL}}$ (linear and nonlinear contributions). The first term on the right-hand side of the expression for the electric displacement vector $\boldsymbol{D}$ is the vacuum contribution. The polarization $\boldsymbol{P}$ describes the coupling of the light to induced dipoles in the material, and this physical quantity is expanded in series in powers of the electric field, $\boldsymbol{E}$, as follows.

$$
\begin{aligned}
\frac{1}{\varepsilon_0}\boldsymbol{P}= & \int \overline{\overline{\chi}}_1(\tau)\,\boldsymbol{E}(t-\tau)\,\mathrm{d}\tau \\
& +\int\int \overline{\overline{\chi}}_2(\tau_1,\tau_2)\,\boldsymbol{E}(t-\tau_1) \\
& \times \boldsymbol{E}(t-\tau_2)\,\mathrm{d}\tau_1\,\mathrm{d}\tau_2 \\
& +\int\int\int \overline{\overline{\chi}}_3(\tau_1,\tau_2,\tau_3)\,\boldsymbol{E}(t-\tau_1) \\
& \times \boldsymbol{E}(t-\tau_2)\,\boldsymbol{E}(t-\tau_3)\,\mathrm{d}\tau_1\,\mathrm{d}\tau_2\,\mathrm{d}\tau_3+\cdots \\
= & \frac{1}{2\pi}\int \widetilde{\overline{\overline{\chi}}}_1(\omega)\,\widetilde{\boldsymbol{E}}(\omega)\,\mathrm{e}^{-\mathrm{i}\omega t}\,\mathrm{d}\omega \\
& +\frac{1}{(2\pi)^2}\int\int \widetilde{\overline{\overline{\chi}}}_2(\omega_1,\omega_2) \\
& \times \widetilde{\boldsymbol{E}}(\omega_1)\,\widetilde{\boldsymbol{E}}(\omega_2)\,\mathrm{e}^{-\mathrm{i}(\omega_1+\omega_2)t}\mathrm{d}\omega_1\mathrm{d}\omega_2 \\
& +\frac{1}{(2\pi)^3}\int\int\int \widetilde{\overline{\overline{\chi}}}_3(\omega_1,\omega_2,\omega_3)\,\widetilde{\boldsymbol{E}}(\omega_1)\,\widetilde{\boldsymbol{E}}(\omega_2) \\
& \times \widetilde{\boldsymbol{E}}(\omega_3)\,\mathrm{e}^{-\mathrm{i}(\omega_1+\omega_2+\omega_3)t}\,\mathrm{d}\omega_1\,\mathrm{d}\omega_2\,\mathrm{d}\omega_3+\cdots
\end{aligned} \tag{1}
$$

Here, the dielectric susceptibility terms $\overline{\overline{\chi}}_j$, $j=1,2,3$ are second, third, and fourth rank tensors, respectively. The latter are causal functions and tensor product notation is assumed. The second equality is written in terms of Fourier transform variables. Although, in principle, the dielectric susceptibilities at each order could be obtained from fully ab initio quantum mechanical calculations on the relevant materials, in practice, such calculations are too complex and one usually has to rely on experimentally measured data. Higher order processes than those displayed explicitly in this equation are generally unimportant because the leading order accessible nonlinear effect dominates. (Exceptions occur when nonlinear saturation becomes important.) The above induced polarization term acts as a source for the electromagnetic field in Maxwell's vector wave equation,

$$
\frac{\partial^2\boldsymbol{E}}{\partial t^2}-\frac{1}{c^2}\frac{\partial^2\boldsymbol{E}}{\partial z^2}-\nabla\cdot(\nabla\cdot\boldsymbol{E})=\frac{1}{\mu_0}\frac{\partial^2\boldsymbol{P}}{\partial t^2}. \tag{2}
$$

The leading order term $\overline{\overline{\chi}}_1$ above is a second rank tensor that describes all linear optical interactions involving propagating optical fields of any arbitrary polarization. For virtually all nonlinear optical interactions, it is appropriate to expand the electric field vector as a linear combination of products of envelope functions and optical carrier waves

$$
\boldsymbol{E}(\boldsymbol{r},t)=\sum_j \widehat{e}_j A_j(\boldsymbol{r},t)\exp\left(\mathrm{i}\left(\pm\boldsymbol{k}_j\cdot\boldsymbol{r}-\omega_j t\right)\right) + \mathrm{c.c.} \tag{3}
$$

Here A_j are slowly varying envelope functions, and $\widehat{e}_j$ is a unit vector indicating the direction of polarization. Each wave vector or frequency pair $(\boldsymbol{k}_j,\omega_j)$ satisfies a linear dispersion relation $D(\boldsymbol{k}_j,\omega_j)=0$. A goal in nonlinear optics is to write down evolution equations for the complex slowly varying envelope functions A_j. The structure of these equations, which describe how almost monochromatic, weakly nonlinear, dispersive wave trains interact, is universal and so many of the properties of lightwaves can be inferred from corresponding situations in other fields.

The leading nonlinear behavior in noncentrosymmetric crystals is due to the second order term $\overline{\overline{\chi}}_2$, a third rank tensor (with 9 components). This term is responsible for second harmonic generation as first observed by Franken et al. (1961). Significant quadratic nonlinear interactions between wave packets take place when triads of wave vectors and frequencies $\boldsymbol{k}_j, \omega_j$ obey

$$\boldsymbol{k}_1 + \boldsymbol{k}_2 + \boldsymbol{k}_3 = 0, \quad \omega_1 \pm \omega_2 \pm \omega_3 = 0, \tag{4}$$

where each $\boldsymbol{k}_j, \omega_j$ pair satisfies its own dispersion relation. Physical effects arising from this term include second harmonic generation ($2\omega_1 \rightarrow \omega_3$), dc rectification ($\omega_1 - \omega_1 \rightarrow 0$), and frequency up- and down-conversion ($\omega_1 \pm \omega_2 \rightarrow \omega_3$). The first two nonlinear interactions are termed degenerate because $\omega_1 = \omega_2$ and two photons ($2\hbar\omega_1$) in the incident laser beam are needed to create one second harmonic photon. For these three-wave interaction processes to be efficient both energy (i.e., $\hbar\omega_1 \pm \hbar\omega_2 = \hbar\omega_3$ for up-, down-conversion, $2\hbar\omega_1 = \hbar\omega_3$ for second harmonic generation) and momentum ($\hbar\boldsymbol{k}_1(\omega) \pm \hbar\boldsymbol{k}_2(\omega) = \hbar\boldsymbol{k}_3(\omega)$, for up-, down-conversion, $2\hbar\boldsymbol{k}_1(\omega) = \hbar\boldsymbol{k}_3(\omega)$ for harmonic generation) must be nearly conserved. (Here $\hbar\boldsymbol{k}$ is the photon momentum.) Three-wave interactions are generally difficult to observe in isotropic media as material dispersion precludes being able to satisfy the momentum conservation relation. A symmetry argument also shows that χ^2-processes are forbidden in materials possessing a center of symmetry. Noncentrosymmetric uniaxial and biaxial crystals (those in which the refractive index is different in different directions in the crystal) are employed to achieve efficient phase matching. This is because the magnitude of the optical wave vector is related to the material refractive index through the relation $k = n(\omega)\omega/c$, where the dispersion of the refractive index $n(\omega)$ is explicitly displayed. The complex envelopes A_1, A_2, and A_3 of the three wave packets obey the universal three-wave interaction equations of universal type

$$\frac{\partial A_j}{\partial t} + \boldsymbol{c}_j \cdot \nabla A_j = \theta_j A_l^* A_m^* \tag{5}$$

with j, l, m cycled over 1,2,3, where $\boldsymbol{c}_j$ is the linear group velocity of the jth wave packet and θ_j is a coupling coefficient proportional to $\overline{\overline{\chi_2}}$.

Four-wave χ_3-processes are the leading order nonlinear optical processes in a medium possessing a center of symmetry. Here significant exchange of energy only takes place between resonant quartets $(\boldsymbol{k}_j, \omega_j)_{j=1}^4$, each obeying its dispersion relation. Contrary to the triad resonance condition, energy conservation and phase matching can always be satisfied with several trivial choices. The nonlinear dielectric susceptibility is now a fourth rank tensor although many of its components are either identical or zero, depending on crystal symmetry properties. Degenerate four-wave interactions include self-phase modulation (or Kerr effect, $\omega_1 + \omega_1 - \omega_1 \rightarrow \omega_1$) and third harmonic generation ($\omega_1 + \omega_1 + \omega_1 \rightarrow 3\omega_1$) (Terhune et al., 1962). Common nondegenerate third-order processes include four-wave-mixing (FWM) interactions.

Four-wave-mixing interactions can be used to create a "phase-conjugated wave." A nonlinear interference grating is created within the nonlinear material through the interaction of two strong pump waves satisfying $\boldsymbol{k}_1 + \boldsymbol{k}_2 = 0$ and $\omega_1 = \omega_2 = \omega$. A weak third wave scatters off this grating and generates a time-reversed phase-conjugated replica of itself. The incident and scattered waves satisfy $\boldsymbol{k}_3 + \boldsymbol{k}_4 = 0$ and $\omega_4 = \omega_3 = \omega$. FWM interactions can also lead to degradation in wavelength-division-multiplexed (WDM) long-haul fiber transmission systems as a result of energy transfer between equally spaced channels. Another important nondegenerate interaction involves the phenomenon of nonlinear birefringence where $\boldsymbol{k}_4, \omega_4 = -\boldsymbol{k}_2, -\omega_2$, $\boldsymbol{k}_3, \omega_3 = -\boldsymbol{k}_1, -\omega_1$ and the complex slowly varying functions A_1, A_2 represent envelope wave packets of different polarization direction. Coupled nonlinear Schrödinger (NLS) equations for the envelope fields $A_1(\boldsymbol{r}, t)$ and $A_2(\boldsymbol{r}, t)$ can be derived that describe nonlinear self- (i.e., $|A_1|^2 A_1$ in the A_1 equation and $|A_2|^2 A_2$ in the A_2 equation) and cross-phase modulation (i.e., $|A_2|^2 A_1$ in the A_1 equation and $|A_1|^2 A_2$ in the A_2 equation) for co- and counter-propagating laser pulses. The relative weighting of self- and cross-phase modulation nonlinearities depends on the material properties (Marburger, 1977) and plays an important role in polarization switching of soliton pulses in optical fibers. One can write down more general universal four-wave interaction equations along the lines described above for the three-wave interaction case.

A well-known degenerate χ_3-process is the optical Kerr effect. This self-interaction term induces a nonlinear phase change proportional to the intensity $|A|^2$ of the propagating light pulse. This phase change, called self-phase modulation, is an accumulative propagation effect that is central to applications that exploit the optical Kerr effect. A canonical form of the D-dimensional NLS equation, incorporating the Kerr effect, can be written as follows:

$$\mathrm{i}\left(\frac{\partial A}{\partial t} + \sum_{j=1}^{D}\left(\frac{\partial \omega}{\partial k_j}\right)\frac{\partial A}{\partial x_j}\right) - \sum_{j,l=1}^{D}\left(\frac{\partial^2 \omega}{\partial k_j k_l}\right)\frac{\partial^2 A}{\partial x_j x_l} + \left(\frac{\partial \omega}{\partial |A|^2}\right)|A|^2 A = 0. \tag{6}$$

The first two terms combined describe the advection of a pulse (wave packet) with group velocity $v_g = \omega\,(k)$.

The second and third terms represent the balance between dispersion and weak nonlinearity. Higher order corrections terms can be derived as an asymptotic expansion in a small parameter as described in Newell & Moloney (1992). Dispersion in nonlinear optics is manifested in two physically distinct ways: diffraction describing spatial spreading of a tightly collimated beam or pulse in a direction transverse to the direction of propagation, and dispersive spreading in time (group velocity dispersion $\omega''(k)$) of an ultrashort laser pulse. The latter mechanism plays an increasingly influential role as the propagating laser pulse gets shorter and shorter.

In 1-d, the NLSE is integrable and the Kerr nonlinearity can balance group velocity dispersion to form temporal solitons in optical fibers. Both bright and dark temporal soliton pulses have been observed experimentally in fibers. Spatial soliton beams can be created in two-dimensional planar waveguides where there is strong confinement of the optical field in a single spatial dimension transverse to the propagation direction of the light beam. Spatial confinement in two (fibers) or one (planar waveguides) dimension is required to form stable soliton pulses or beams. In unconfined transparent bulk geometries, the optical Kerr effect is responsible for critical self-focusing, leading to extremely intense focused light that causes catastrophic damage in transparent glasses. Critical self-focusing was first observed experimentally by Chiao and coworkers in 1964 (Chiao et al., 1964).

There are other categories of nonlinear optical effects that can be described in terms of envelope functions and that depend sensitively on the finite response of some material oscillation. Stimulated scattering is a case in point. These processes are essentially three-wave interactions where one of the optical waves is replaced by a material oscillation. In transparent glasses for example, there are two characteristic modes of oscillation of the lattice as shown in Figure 1. The lower frequency acoustic phonon mode corresponds to an oscillation of all atoms in unison along a fixed direction in the lattice. The higher frequency optical phonon mode involves neighboring atoms in the lattice oscillating out of phase. The dispersion relation for lattice vibrations, therefore, contains an acoustic and optical branch, both of much lower frequency than the optical frequency.

Figure 1 shows a plot of the optical phonon (top curve) and acoustic phonon (bottom curve) dispersion for a diatomic crystal lattice within the first Brillouin zone. On the scale of this graph, the photon branch (light cone) is essentially vertical, making the acoustic and optical phonon dispersion essentially flat. An intense pump laser pulse at frequency ω_p entering a transparent glass can scatter off either a fluctuating low-frequency acoustic phonon (Stimulated Brillouin Scattering, SBS) or a fluctuating optical phonon (Stimulated

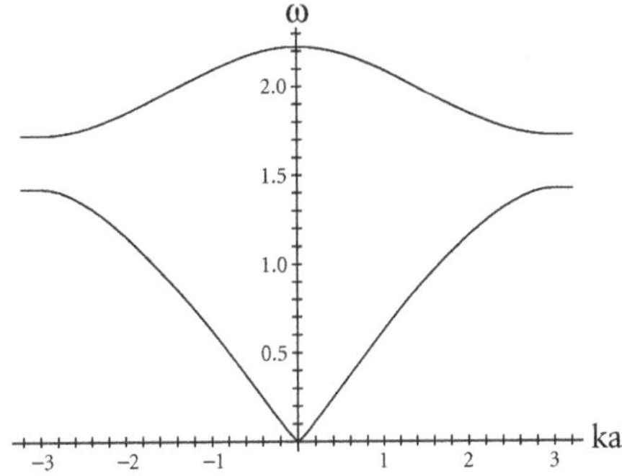

Figure 1. Acoustic (bottom)and optical (top) phonon branches of the dispersion curve for a diatomic lattice. The abscissa is scaled to the lattice constant "a."

Raman Scattering, SRS) at finite temperature. Energy conservation ensures that the scattered optical signal has a frequency $\omega_s = \omega_p - \omega_v$, where ω_v is the natural acoustic or optical phonon frequency. Technically, SBS and SRS processes look like a three-wave interaction where phase matching is easy as the dispersion of the material oscillation is locally flat on an optical frequency scale. In fact, it can be shown that these processes are essentially four-wave interactions, as the oscillator model, representing the material oscillation, is driven by the square of the optical field. Typically, SBS is generated in the backward direction and SRS can to be generated in the forward and/or backward direction relative to the propagating pump laser pulse. SRS processes in liquids and gases can generate a cascade of frequency down-shifted (Stokes) and frequency up-shifted (anti-Stokes) waves. Although both SBS and SRS processes are essentially phase matched for all directions, they can only exist by feeding off the intense pump pulse. There also exists a degenerate stimulated scattering process called Stimulated Rayleigh Scattering—this is essentially a Kerr-like process with finite memory. All of these stimulated scattering phenomena can be observed in bulk transparent materials (Marburger, 1977; Shen, 1977) and optical fibers (Agrawal, 1989). They can be stimulated from noise or seeded at the optical signal wavelength by injecting a weak laser field.

Another important class of nonlinear optical phenomena involves direct resonant coupling to some form of material oscillation. The simplest physical manifestation of this category is the two-level atom. Absorbing/amplifying media that give rise to the phenomenon of self-induced transparency (SIT) and lasers belong to this category. The dielectric susceptibility is now complex with the real part corresponding to a refractive index change induced by the optical field and the imaginary part corresponding to absorption or amplification of light. The real and imaginary parts of the dielectric susceptibility are related through a Hilbert transform—the Kramers–Kronig relation. Energy is directly transferred between optical fields and material oscillations. Self-induced transparency (SIT) solitons propagate when this coherent population or energy exchange can occur between the lower (L) and upper (U) level of a two-level atom on a timescale short relative to any irreversible energy decay processes such as spontaneous emission from the excited state (U). This population cycling between two levels occurs at a frequency $\omega_R = |\mu_{12}A|/\hbar$, where μ_{12} is the dipole matrix between the two levels and $\hbar$ is Planck's constant. The frequency ω_R is called the Rabi frequency, from the field of magnetic resonance spectroscopy. Figure 2a depicts this simple two-level atom scheme. For longer optical pulses, nonlinear saturation can occur in the presence of intense propagating optical fields. Population inversion, required for lasing action, cannot be achieved in a two-level atom (*See* **Lasers**). Optical amplification of light requires that a net inversion be achieved between the upper (U) and lower (L) atomic or molecular level involved in the light amplification process. Such inversion can be achieved by pumping atoms or molecules to energetically higher levels with a subsequent rapid nonradiative energy decay down to the upper excited level (U). Classical laser action can be described in terms of three-level and four-level quantum models. Figure 2b and c depict two classical lasing level schemes where the ground level population is pumped ($E = \hbar\omega_P$) into some upper excited state followed by a rapid (nonradiative) population decay (dashed arrows) to the upper excited lasing level (U). The original ruby laser, discovered by Maiman in 1961, is an example of a three-level (actually three band) system schematically shown in Figure 2b. Inversion is more difficult to achieve in a three-level laser as the upper lasing level must have a net population in excess of that in the ground level. This requires pumping of more than half of the atoms to the upper lasing level. In the four-level laser, the lower lasing state is not the ground state and its initial population can be zero or very small due to finite temperature populations. Efficient lasing is achieved by rapidly removing the populations from the lower lasing level (L).

Such energy level schemes are generally gross simplifications of a real laser although they provide useful models of solid state and gas lasers (Siegman, 1986). Fiber lasers with erbium and/or ytterbium-doped cores belong to the category of solid state lasers. Semiconductor lasers, on the other hand, are extremely complex many-body systems involving many bands rather than levels and the above descriptions prove inadequate.

Key current developments in the field of nonlinear optics have benefited from technical developments that have enabled experimentalists to produce ultrashort laser pulses and invent novel materials processing methodologies. We sketch some of the key areas of research where such effects have been exploited in recent years.

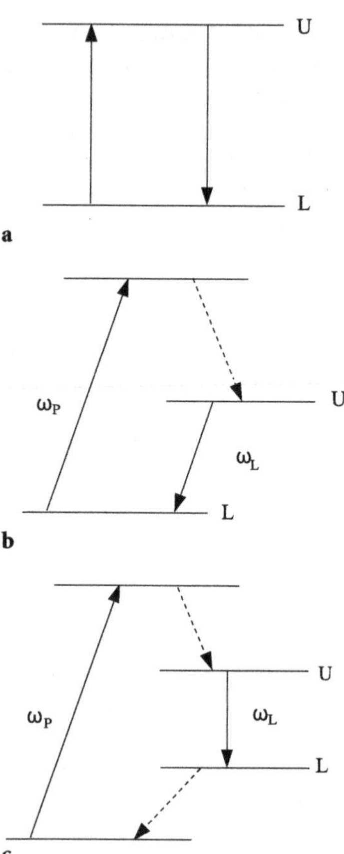

Figure 2. (a). Two-level energy level scheme. (b). Three-level lasing level scheme. (c). Four-level lasing level scheme.

Nonlinear Optical Data Processing

A cumulative intensity-induced phase shift due to the Kerr nonlinearity, of the order of p, provides one of the most useful all-optical data processing capabilities. High-speed optical sampling, switching, amplification and storage applications abound, based on this nonlinear mechanism. Optical switching and data sampling can be implemented in directional couplers and Mach–Zehnder interferometers in both waveguide and fiber arrangements. Optical amplification and storage of pulse trains representing digital information can be achieved by placing a nonlinear transparent Kerr medium in a ring or Fabry–Perot cavity arrangement. So-called optical bistable elements require an injected holding beam to maintain the cavity at some threshold level whereby an incremental input can lead to strong amplification or storage of the system in one of two stable accessible states. The latter situation is depicted in Figure 3. Here the system has a memory (hysteresis) whereby a lower (bit 0) and upper (bit 1) state can be switched by some external perturbation. Adjusting some cavity parameter can also lead to a single-valued but rapidly rising output transmission of the system. This is useful for amplifying injected pulse streams.

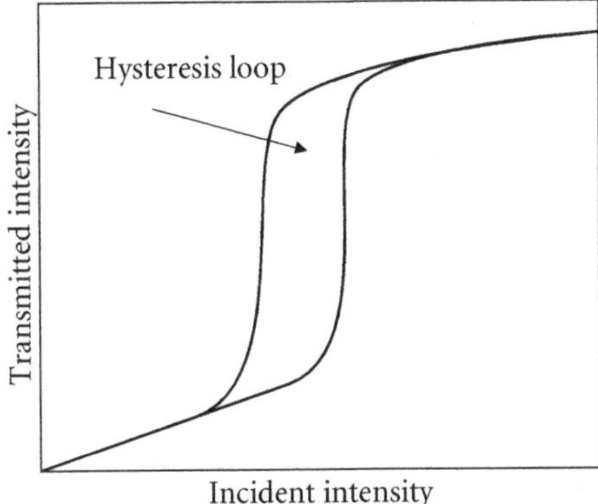

Figure 3. Hysteresis loop for a bistable optical cavity.

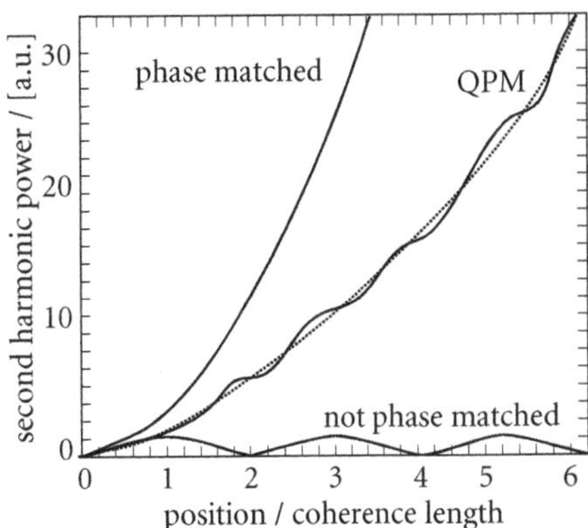

Figure 4. Comparison of energy transfer from the fundamental to the second harmonic for phase-matched, detuned and quasi-phase-matched conditions.

Such passive optical bistable cavities have also been shown to exhibit deterministic chaotic dynamics when the finite time delay of the optical signal circulating in the optical resonator has been taken into account. When transverse degrees of freedom of the external pump laser beam are included, so-called transverse spatial solitons have been observed. Although technology applications have been envisioned for both of these dynamical phenomena, there are no specific implementations to date.

Chaotic Synchronization and Communication with Lasers

Semiconductor lasers are highly susceptible to exhibiting chaotic dynamics even in the presence of extremely weak (10^{-4}) feedback from an external reflecting surface. This tendency to readily exhibit deterministic chaos means that semiconductor laser sources require greater than 40 dB optical isolation, making them the most expensive component in an optical fiber telecommunications network. Research in recent years has focused on taking advantage of this chaos by demonstrating that chaotic transmitter and receiver lasers can be synchronized and messages encoded and decoded at the transmitter and receiver end, respectively. Mathematically, these chaotic lasers are described at the simplest level by delay differential equations.

Quasi-phase Matching in Isotropic Media

Quasi-phase matching (QPM) is a means of beating the usual triad resonance condition needed to efficiently transfer energy between waves in second harmonic generation. Figure 4 shows the ideal phase-matched energy transfer to second harmonic as a function of propagation distance in a uniaxial crystal. When detuned from the phase-matching condition, energy transfer occurs less efficiently and there is a periodic transfer back and forth between the fundamental at frequency and the second harmonic signal at frequency 2ω. In this case, one observes the oscillatory nonphase-matched behavior depicted in the picture. This is typically the situation encountered in isotropic materials.

Quasi-phase matching entails taking a crystal with a nonlinear coefficient and periodically reverting the sign of the dipoles in such a way that the phase of the wave is reversed every coherence length along the crystal. The latter is proportional to the refractive index mismatch at the fundamental and second harmonic frequencies. A practical implementation of quasi-phase-matching is through the process of periodic polling whereby a ferroelectric material such as lithium niobate, for example, can effectively have its dipoles reversed periodically with patterned electrodes. QPM soliton propagation is an intense area of modern research and has the advantage that much larger nonlinear optical couplings can be achieved via the χ_2 nonlinearity.

Optical Breakdown in Intense Laser fields

Optical breakdown due to critical self-focusing of continuous wave laser beams or long pulses has been observed since the 1960s. The breakdown phenomenon itself is very complex and can involve nonlinear coupling to thermal, acoustic, vibrational, rotational, and electronic degrees of freedom. Thermally induced lensing, leading to critical self-focusing, is dominant for long microsecond-duration laser pulses. Nanosecond-duration laser pulses couple strongly to hypersonic acoustic waves and critical self-focusing is due to the physical mechanism of electrostriction. Picosecond-duration laser pulses typically generate electron/ion plasmas through the process of avalanche photo-ionization, whereas femtosecond-duration laser pulses breakdown materials predominantly via the mechanism of multiphoton ionization. The latter process is essentially instantaneous and the nonlinear ionization process is proportional to $|A|^{2N}$, where the minimum energy required to excite the bound electron to the ionization continuum is $E = N\hbar\omega$.

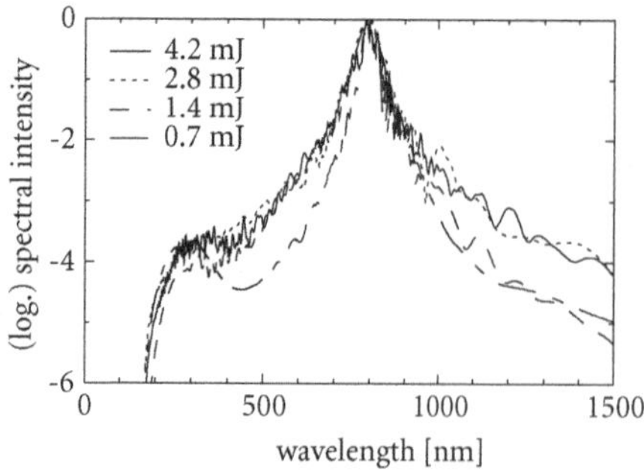

Figure 5. Calculated white-light supercontinuum for a high-power femtosecond laser pulse propagating in air.

Stimulated scattering processes usually accompany optical breakdown processes making it extremely difficult to unravel the relevant physics for longer duration laser pulses. Plasma, generated via avalanche ionization, acts as a shield, preventing the laser pulse from propagating further in the material. Shock waves generated in the nonlinear focal region can cause mechanical rupture for longer duration pulses.

Extreme Nonlinear Optics

The development of high-power femtosecond laser sources has led to some novel nonlinear interactions that exploit the optical Kerr effect. Single cycle pulses have been generated and very recent experimental evidence has been provided for the generation of attosecond pulses. These optical pulses are of such short duration that they interact with the atom or molecule on a timescale comparable to that taken by the electron to orbit the nucleus! The advantage of these laser sources is that many of the detrimental physical effects accompanying longer duration pulses are not operative in the femtosecond regime. Therefore, it is possible to achieve large local field intensities while maintaining low energies in the pulse. Recently, high-power femtosecond laser pulses, focused down to scales on the order of the wavelength of light, have been used to locally change the material properties of glass and write complicated integrated optics waveguiding structures. The actual mechanism for the transformation of the glass to a state with a higher refractive index is not yet fully understood.

High-power femtosecond laser pulses can propagate over anomalously long distances in air, achieving peak field intensities on the order of 10^{13}–10^{14} W/cm^2. These huge intensities break down the air via multiphoton ionization in the vicinity of the nonlinear focus. Plasma, generated within the 100 μm diameter focal spot, can act as a defocusing lens, causing the trailing edge of the laser pulse to defocus and refocus multiple times as it propagates through the atmosphere. Accompanying the extreme self-focusing is white-light supercontinuum generation. The relatively narrow spectral peak of the initial pulse blows out asymmetrically and encompasses wavelengths ranging from the infrared to the ultraviolet. Figure 5 shows the change in the calculated white-light supercontinuum spectrum as a function of increasing pulse energy for a 100 fs-duration laser pulse propagating in air. The experimentally measured back-reflected white-light supercontinuum from a 2.6 TW 100 fs-duration laser pulse launched vertically into the atmosphere (Rairoux et al., 2000) exhibits the same qualitative features. The initial pulse spectrum is localized around the narrow peak at 800 nm. The mechanism, by which this high-power pulse sustains itself as a waveguide, is through a combination of recurring transverse modulational instabilities that create multiple intense chaotic self-focusing light strings and accompanying plasma filament generation that acts to limit the strong focusing effect (Mlejnek et al., 1999).

Another important application of high-power femtosecond laser pulses is in the generation of highly transient X-ray pulsed sources through higher harmonic cascades accompanying extreme pulse focusing. Hundreds of harmonics have been observed experimentally (Bartels et al., 2000), and novel algorithms are being developed to reshape the initial pulse so as to be able to efficiently extract a particular higher harmonic. When the pulse peak intensities exceed around 10^{17} W/cm^2, one enters the nonlinear regime where relativistic effects become important.

The extreme conditions encountered during critical focusing of high-power femtosecond-duration laser pulses have recently spurred theorists to question whether envelope equation models such as the NLS equation, even with higher order corrections included, can adequately describe pulse propagation under such extremes. As NLS describes quasi-monochromatic laser pulses, it is implicitly assumed in its derivation that the generated spectral bandwidth satisfies the inequality, which is the underlying optical carrier wave frequency. Experimental measurements of white-light generation in condensed media show that the spectral extent of the generated supercontinuum can exceed the initial pulse spectral bandwidth by a factor of 5 or more. The picture above for air shows that the supercontinuum extends well beyond the optical carrier frequency. Pulse propagation models that explicitly include the optical carrier wave and consequently allow for optical carrier shocks are currently under active development.

Nonlinear Optics in Photonic Bragg Structures

Periodic modulation of a material refractive index or gain offers a novel means of beating the intrinsic dispersion of available nonlinear optical materials. This idea goes back to the early days of nonlinear optics where 1-dimensional periodic index modulation allowed one to manage or engineer the dispersion of a particular material. This is the photonic analog of the lattice periodicity leading to so-called Bloch states describing electron confinement in a semiconductor material. Just as the optical properties of semiconductor materials can be engineered by modifying the underlying lattice structure through the introduction of quantum wells or dots, so too can the optical confinement of photons be controlled by introducing a refractive index or gain periodicity in a transparent glass. Distributed feedback semiconductor lasers, based on this distributed feedback principal, are the workhorses of modern fiber-based telecommunications systems. 1-d Bragg or gap solitons have been experimentally generated in periodically modulated Kerr glass waveguides—these localized pulses have the property that they can be dramatically slowed down or even be nonpropagating within the nonlinear periodic structure.

2-d and 3-d photonic Bragg and crystal fiber structures have the special property that photonic bandgaps can be introduced in a structure consisting of a periodic lattice of holes or vertical columns. In principle, such structures can guide light around right angle bends in direct contrast to waveguide splitters, trap light in defects, and provide so-called omni-reflecting properties, that is, guide light at all angles and wavelengths. These properties are in marked contrast to conventional index-guided optical waveguides and, when combined with optical nonlinearity, offer many new potential technology applications (Mingaleev & Kivshar, 2002).

Electromagnetically Induced Transparency

Electromagnetically induced transparency (EIT) is a quantum interference effect that acts to reduce the usual absorption of light experienced when its frequency is tuned to the resonance frequency of the sample through which the light is propagating. The transparency is created by a second light (electromagnetic) source tuned to another resonance of the sample. Suppressing absorption through EIT leads to several other effects, including the focusing of one laser beam by another and the production of inversionless laser sources.

JEROME V. MOLONEY

See also **Filamentation; Kerr effect; Lasers; Nonlinear Schrödinger equations; Optical fiber communications; Photonic crystals; Pump-probe measurements; Rayleigh and Raman scattering and IR absorption**

Further Reading

Agrawal, G.P. 1989. *Nonlinear Fiber Optics*, San Diego: Academic Press

Bartels, R., Backus, S., Zeek, E., Misoguti, L., Vdovin, G., Christov, I.P., Murnane, M.M. & Kapteyn, H.C. 2000. Shaped-pulse optimization of coherent emission of high-harmonic soft X-rays. *Nature*, 406: 164

Bloembergen, N. 1996a. *Encounters in Nonlinear Optics: Selected Papers of Nicolaas Bloembergen*, Singapore: World Scientific

Bloembergen, N. 1996b. *Nonlinear Optics*, Singapore: World Scientific

Boyd, R.W. 1992. *Nonlinear Optics*, Boston and San Diego: Academic Press

Chiao, R.Y., Garmire, E. & Townes, C.H. 1964. Self-trapping of optical beams. *Physical Review Letters*, 13: 479–482

Franken, P.A., Hill, A.E., Peters, C.W. & Weinreich, G. 1961. Generation of optical harmonics. *Physical Review Letters*, 7: 118–120

Maiman, T.H. 1960. Stimulated optical radiation in ruby lasers. *Nature*, 187: 493

Marburger, J. 1977. Self focusing: theory. *Progress in Quantum Electronics*, 4: 35–110

Mingaleev, S. & Kivshar, Y. 2002. Nonlinear photonic crystals: toward all-optical technologies. *Optics and Photonics News*, 13: 48–51

Mlejnek, M., Kolesik, M., Moloney, J.V. & Wright, E.M. 1999. Optically turbulent femtosecond light guide in air. *Physical Review Letters*, 83: 2938–2941

Newell, A. & Moloney, J.V. 1992. *Nonlinear Optics*, Redwood City, CA: Addison-Wesley

Rairoux, P., Schillinger, H., Niedermeier, S., Rodriguez, M., Ronneberger, F., Sauerbrey, R., Stein, B., Waite, D., Wedekind, C., Wille, H. & Wöste, L. 2000. Teramobile: a nonlinear terawatt femtosecond lidar. *Applied Physics* B, 71: 573

Shen, Y.R. 1977. Self focusing: experimental. *Progress in Quantum Electronics*, 4: 1–34

Siegman, A.E. 1986. *Lasers*. Mill Valley, CA: University Science Books

Terhune, R.W., Maker, P.D. & Savage, C.M. 1962. Optical harmonic generation in calcite. *Physical Review Letters*, 8: 404–406

NONLINEAR PLASMA WAVES

In studying nonlinear plasma waves, one should recognize that there are several analytical methods, including single particle, fluid, and kinetic. Using the proper model is important, as different models can result in different nonlinear phenomena (Davidson, 1972; Horton & Ichikawa, 1996).

Because plasmas are composed of charged particles and are often immersed in external electric and magnetic fields, it is critical to determine when the self-consistent fields (fields produced by the charged particles directly) must be considered in the presence of externally applied fields. Typically, the self-consistent electric fields need to be considered, but we can often neglect self-consistent magnetic fields, because the forces produced by the self-consistent magnetic fields are usually much lower than those produced by external magnetic fields.

Transverse Plasma Waves

As an example of the single-particle approach, consider transverse electromagnetic (TEM) waves in a plasma. We begin with Maxwell's equations combining them to produce the wave equation below:

$$\nabla \times \nabla \times \boldsymbol{E} = -\mu\sigma \frac{\partial \boldsymbol{E}}{\partial t} - \mu\varepsilon \frac{\partial^2 \boldsymbol{E}}{\partial t^2}, \quad (1)$$

Where σ, ε, and μ are the conductivity, electric permittivity, and magnetic permeability of the plasma, respectively. We shall assume that all of the plasma effects can be collected in the permittivity and that the conductivity is part of the permittivity, so we can set $\sigma = 0$ in Equation (1). If we now assume cartesian coordinates and that we have a transverse electromagnetic wave propagating in the z-direction with the propagation behaving like

$$\exp(\boldsymbol{k} \cdot \boldsymbol{r} - \omega t), \quad (2)$$

then we may substitute this notation into Equation (1) and represent Equation (1) as the matrix shown below:

$$\begin{pmatrix} -k_z^2 + \omega^2\mu\varepsilon & 0 & 0 \\ 0 & -k_z^2 + \omega^2\mu\varepsilon & 0 \\ 0 & 0 & -k_z^2 + \omega^2\mu\varepsilon \end{pmatrix} \times \begin{pmatrix} E_{0x} \\ E_{0y} \\ E_{0z} \end{pmatrix} = 0. \quad (3)$$

There are three possible eigenmodes (two of which are identical) for this equation. The first two are propagating disturbances, while the third mode is a time-invariant nonpropagating mode. In a plasma, this mode represents the electrostatic oscillations generated when the electrons are displaced with respect to the ions and allowed to move under the influence of the self-consistent (charge separation) electrostatic field. Vibrations of gelatin also exhibit this same property. If we assume that E_{0z} is zero, then the remaining two modes look just as they would in an ordinary dielectric. However, the permittivity of a plasma based on single-particle motions with stationary ions and no external d.c. magnetic field is expressed as

$$\varepsilon = \varepsilon_0 \left(1 - \frac{\omega_{\mathrm{p}}^2}{\omega\,(\omega + \mathrm{i}\nu_{\mathrm{m}})}\right), \quad (4)$$

where ν_{m} is the electron momentum transfer collision frequency and ω_{p} is the electron plasma frequency defined as

$$\omega_{\mathrm{p}}^2 = \frac{ne^2}{m\varepsilon_0}. \quad (5)$$

In Equation (5), n is the electron density, e is the charge of an electron, m is its mass, and ε_0 is the permittivity of free space. Since Equation (4) now has both real and imaginary parts, an effective conductivity of the plasma can now be obtained. Figure 1 shows a plot of the real part of ω vs. wave number k for two conditions: zero collision frequency and nonzero collision frequency. Because the relationship between k and ω is not a constant as frequency changes, we see that a TEM wave in a plasma exhibits dispersion. Examination of the real part of the propagation vector also shows that the wave amplitude will decrease as long as the collision frequency is not zero.

Nonlinear Ion-acoustic Waves

In this example, we consider the representation of a plasma as a fluid. Under these conditions, it is possible to excite longitudinal acoustic-type oscillations. We assume that the plasma mass is concentrated in its ions but that any electric fields that are generated by charge separation appear because of the higher mobility of the electrons. In this representation, we may write the equation of motion of a fluid element as follows (Chen, 1984):

$$Mn\left[\frac{\partial \boldsymbol{v}_{\mathrm{i}}}{\partial t} + (\boldsymbol{v}_{\mathrm{i}} \cdot \nabla)\boldsymbol{v}_{\mathrm{i}}\right] = en\boldsymbol{E} - \nabla p, \quad (6)$$

where $\boldsymbol{v}_{\mathrm{i}}$ refers to the ion species fluid velocity, p is the pressure, M is the ion mass, and n is the plasma density. Neglecting the electron mass (for simplicity), linearizing, and assuming propagating waves as previously, $\boldsymbol{E}$ is the negative gradient of a scalar potential ϕ. Using the equation of state gives

$$Mn(\mathrm{i}\omega + \nu_{\mathrm{m}})v = -en_0\mathrm{i}k\phi - \gamma_i kT_i kn_1, \quad (7)$$

where n_0 is the steady-state density and n_1 is the perturbed density. Assuming that

$$e\phi/kT_{\mathrm{e}} = n_1/n_0, \quad (8)$$

where T_{e} is the electron temperature, we obtain

$$(\omega + \mathrm{i}\nu_{\mathrm{m}})v = kV_{\mathrm{s}}\frac{n_1}{n_0}. \quad (9)$$

In Equation (9) V_{s} is the sound speed $(kT_{\mathrm{e}}/M)^{1/2}$. Applying a similar analysis to the continuity equation and substituting Equation (9) we obtain

$$\begin{aligned} -\mathrm{i}\omega n_1 + \mathrm{i}kn_0 v &= -\mathrm{i}\omega n_1 + \mathrm{i}k\frac{n_0}{(\omega + \mathrm{i}\nu_{\mathrm{m}})}[kV_{\mathrm{s}}^2]\frac{n_1}{n_0} \\ &= 0 \end{aligned} \quad (10)$$

or

$$(\omega^2 + \mathrm{i}\nu_{\mathrm{m}}\omega - k^2V_{\mathrm{s}}^2) = 0. \quad (11)$$

Longitudinal Plasma Oscillations (Langmuir Waves)

Now consider the kinetic representation of a plasma, in the case where the electric field is parallel to the direction of propagation. Assuming propagation in the z-direction, an electric field in the z-direction can be developed from an electrostatic potential of the form

$$\phi = \phi_0 \exp[\mathrm{i}(kz - \omega t)]. \quad (12)$$

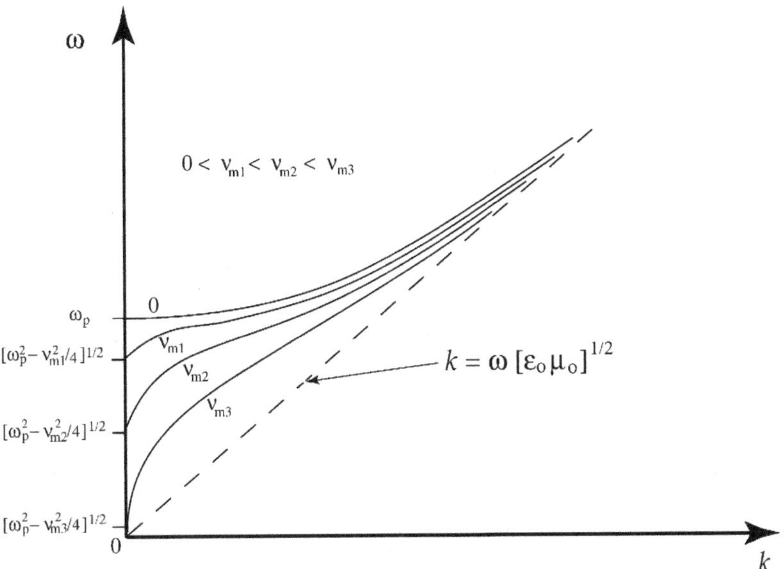

Figure 1. The real part of ω versus wave number k for increasing values of collision frequency ν_{m}.

If there are only electrostatic forces, the velocity distribution f of the electrons is governed by the Vlasov equation:

$$\frac{\partial f}{\partial t} + \boldsymbol{v} \cdot \nabla f + \frac{e}{m} \boldsymbol{E} \cdot \frac{\partial f}{\partial \boldsymbol{v}} = 0. \tag{13}$$

Writing a perturbation for f as $f = f_0 + f_1$ in which f_0 is only a function of velocity and f_1 can be a function of both position and velocity, we may write the perturbed part of the Vlasov equation as

$$\frac{\partial f_1}{\partial t} + \boldsymbol{v} \cdot \nabla f_1 + \frac{e}{m} \boldsymbol{E} \cdot \frac{\partial f_0}{\partial \boldsymbol{v}} = 0. \tag{14}$$

Writing Poisson's equation with this ordering results in

$$\nabla \cdot \boldsymbol{E} = \frac{n_0 e}{\varepsilon_0} \int f_1 \, \mathrm{d}\boldsymbol{v}. \tag{15}$$

Solving Equations (14) and (15) simultaneously yields the dispersion relation for these longitudinal plasma oscillations, often called Langmuir waves, as

$$1 = \frac{n_0 \varepsilon_0 e^2}{m} \frac{\boldsymbol{k}}{k^2} \cdot \int \frac{\partial f_0 / \partial \boldsymbol{v}}{(\boldsymbol{k} \cdot \boldsymbol{v} - \omega)} \mathrm{d}\boldsymbol{v}. \tag{16}$$

Note that the form of f_0 affects the dispersion relation. Also, the denominator in the integral of Equation (16) can have a singularity at which nonlinear effects can occur. It has been shown, for example, that the velocity distribution can draw energy from the wave or vice versa depending upon its slope. As a result, solitary waves can appear, expressing a balance between the driving and dissipative conditions.

In addition, Zakharov (1972) showed that Langmuir waves can evolve nonlinearly so that the electric field that results from the charge separation can become very large in a very small region. Under these "collapse" conditions, the field does not become infinitely large, but is limited by wave-particle interactions and often bursts of high-energy electrons are observed.

Plasmons

A plasmon is the resulting electrostatic oscillation described above when the velocity distribution can be neglected. For example, if it is assumed the ions in a plasma are stationary and the electrons are each uniformly displaced from their equilibrium position, then a Coulomb restoring force will be produced by the excess and deficiency of electrons at each end of the plasma, causing the electrons to oscillate about the ions. The frequency of oscillation is the electron plasma frequency:

$$\omega_{\mathrm{p}} = \left(\frac{ne^2}{m\varepsilon_0}\right)^{1/2}. \tag{17}$$

If the ion motion is considered, an ion plasma frequency also appears in the calculations. Plasmons, especially in solid state materials, can interact nonlinearly with electromagnetic fields and other radiation.

Plasma Turbulence

Turbulence appears when modeling for the three approaches (single particle, kinetic, and fluid) does not satisfactorily describe the experimental observations. For example, in hot plasmas, such as those appearing in stars and fusion reactors, fluctuations in the plasma produce enhanced effects on mass, momentum and heat transport that must be studied in the context of turbulence.

If we could follow the motion of each and every particle in a plasma and its interactions with other particles and external electric and magnetic fields, we would be able to understand plasma turbulence. As that is not possible, alternative approaches must be undertaken.

Turbulence is said to appear in a plasma when the oscillations and fluctuations exhibit a continuous spectrum of frequencies. As a result of this state of turbulence, interactions between species in the plasma and the fluctuations can occur. For example, electrons, in addition to interacting with charged particles, can also interact with electric fields that are generated by the fluctuations. In addition, turbulent motion in a plasma has a very interesting effect: it can generate strong magnetic fields, producing a dynamo. This has been proposed as the mechanism for the generation of earth's magnetic field.

Numerical simulation has been shown to be a very powerful tool. In this case, the simulation involves a number of assumptions in order to make the problem solvable. The key method for these simulations involves following the trajectory of many simulation particles, as they interact with each other, that represent the plasma under various assumptions which can be applied to particular problems. By use of large supercomputers, it is possible to simulate astronomical and fusion plasma phenomena and examine the structure of the turbulence. A color figure in the color plate section shows a cross section of the results of such a simulation for a toroidal magnetic confinement system called a Tokamak. Calculations of this scale of detail could not be done without numerical simulations.

Whistlers and Helical Waves

Whistler waves have been detected in the ionosphere and have been attributed to the change of phase and group velocities of a right circularly polarized plasma wave that travels along the earth's magnetic field. The dispersion relation for this mode is

$$k^2 = \omega^2 \mu_0 \varepsilon_0 \left[1 - \frac{\omega_{\mathrm{pi}}^2 + \omega_{\mathrm{pe}}^2}{(\omega + \omega_{\mathrm{ci}})(\omega - \omega_{\mathrm{ce}})} \right]. \tag{18}$$

In the ionosphere, the frequency of the waves is lower than the electron cyclotron frequency. Thus, the phase velocity (ω/k), which can be found from Equation (18) increases with frequency, and the group velocity also increases. When a lightning discharge takes place, an electromagnetic disturbance that has a large continuous spectrum of components is produced. The disturbance will travel to the ionosphere and then propagate along the Earth's magnetic field lines. However, the lower frequency components of the disturbance will take longer to arrive at a receiving station than the higher frequency components; thus, the received signal appears as a signal that decreases in frequency with time.

In examining the behavior of the electric field vector for a right circularly polarized wave, we note that as the wave propagates along an external magnetic field line, the direction of the electric field rotates around the direction of propagation which is, in this case, the direction of the magnetic field. As a result, this or any other mode can be represented as a helical wave in which the rotation of the end of the electric field vector traces out a helical path as it propagates.

MHD Kink Waves in Toroidally Confined Fusion Plasmas

The sine-Gordon equation has been shown to describe the trajectory of the "slinky" mode in a reversed-field-pinch magnetic fusion experiment (Ebraheem et al., 2002). A reversed-field-pinch experiment produces a magnetic field in a torus in which the portion of the field that is in the toroidal direction (the long way around) is reversed in its outer regions from that closer to the minor axis of the torus. Experimentally, it has been seen that during the operation of the system, magnetohydrodynamic (MHD) fluctuations in the center of the plasma can coalesce into a rotating "hot spot" that travels around the torus. Under some circumstances, the rotating mode may lock to other disturbances or to particular locations on the vacuum chamber torus.

As the mode may be envisaged as a helical solitary wave, it lends itself to being represented by the sine-Gordon equation. The mechanical transmission line analogy (Scott, 1971) can be applied here, if it is assumed that the MHD mode is a solenoid that wraps helically around the torus. The gravitational restoring torque can be represented as the gradient of the magnetic energy of the magnetic moment of the mode in the externally generated magnetic fields of the torus due to toroidal effects (the magnetic field on the inside of the torus is stronger than at the outside of the torus). The spring torque due to the adjacent pendula can be modeled by recognizing that the current flowing around the turns of the solenoid tends to make the solenoid align itself along its axis, so any disturbance from the alignment results in an equivalent spring torque.

Parametric Decay Instability

Consider two oscillators that have different resonant frequencies ω_1 and ω_2 that are nonlinearly coupled to each other through a third oscillator of resonant frequency ω_0. We assume that the oscillators are normal modes of various types in the plasma such as described above. If each of these waves is considered from the point of view of quantum theory, interactions are governed by conservation of energy:

$$\hbar\omega = \hbar\omega_1 + \hbar\omega_2 \tag{19}$$

and conservation of momentum

$$k_0 = k_1 + k_2, \tag{20}$$

where k represents the wave number for each mode. The zero subscript refers to the "pump" mode and the other two modes are referred to as the "daughter" modes (Rost et al., 2002).

Assume that two of the oscillators obey the following equation of motion where each is driven by a time-dependent force that is proportional to the amplitude of a pump oscillator E_0:

$$\frac{d^2x_1}{dt^2} + \omega_1^2 x_1 = c_1 x_2 E_0, \tag{21}$$

$$\frac{d^2x_2}{dt^2} + \omega_2^2 x_2 = c_1 x_1 E_0, \tag{22}$$

where c_1 and c_2 are nonlinear coupling constants. If E_0 is made to vary in time at frequency ω_0, it can be seen that both modes can be driven by a frequency not equal to their natural frequency and, in addition, each oscillator's motion is coupled to the other oscillator's equation of motion.

J. LEON SHOHET

See also **Alfvén waves; Magnetohydrodynamics; Plasma soliton experiments; Turbulence**

Further Reading

Chen, F.F. 1984. *Plasma Physics and Controlled Fusion*, 2nd edition, New York: Plenum Press.

Davidson, R.C. 1972. *Methods in Nonlinear Plasma Theory*, New York: Academic Press

Ebraheem, H.K., Shohet, J.L. & Scott, A.C. 2002. Mode-locking in reversed-field pinch experiments. *Physical Review Letters*, 88: 235003

Horton, C.W. & Ichikawa, Y.H. 1996. *Chaos and Structures in Nonlinear Plasmas*, River Edge, NJ: World Scientific

Rost, J.C., Porkolab, M. & Boivin, R.L. 2002. Edge ion heating and parametric decay during injection of ion cyclotron resonance frequency power on the Alcator C-Mod tokamak. *Physics of Plasmas*, 9: 1262–1270

Scott, A.C. 1971. *Active and Nonlinear Wave Propagation in Electronics*, New York: Wiley

Zakharov, V.E. 1972. Collapse of Langmuir waves. *Soviet Physics JETP*, 35: 908–912

NONLINEAR RESONANCE

See **Damped-driven anharmonic oscillator**

NONLINEAR SCHRÖDINGER EQUATIONS

The basic nonlinear Schrödinger (NLS) equations are

$$\mathrm{i}u_t \pm \tfrac{1}{2}u_{xx} + |u|^2 u = 0, \tag{1}$$

where $u(t, x)$ is a complex function, the real variables t and x are frequently (but not always, see below) time and space coordinates, and the subscripts indicate partial derivatives. The upper and lower signs in front of the second term in Equation (1) correspond, respectively, to self-focusing and self-defocusing NLS equations, to be denoted below as NLS(+) and NLS(−).

NLS equations attracted much attention after Zakharov & Shabat (1972) demonstrated that Equations (1) are integrable by means of the inverse scattering transform (IST) (Zakharov et al., 1984). Under this formulation, Equation (1) is viewed as a compatibility condition for a system of two auxiliary linear equations for a two-component function $\psi^{(1,2)}(x, t)$. The first of these equations is

$$\begin{pmatrix} -\mathrm{i}\partial_x & u^*(x) \\ -u(x) & \mathrm{i}\partial_x \end{pmatrix} \begin{pmatrix} \psi^{(1)} \\ \psi^{(2)} \end{pmatrix} = \lambda \begin{pmatrix} \psi^{(1)} \\ \psi^{(2)} \end{pmatrix}, \tag{2}$$

where λ is a spectral parameter and the asterisk stands for the complex conjugation. The first step in the application of IST is to solve the direct scattering problem for Equations (2) (also called Zakharov–Shabat (ZS) equations) with a given initial configuration $u_0(x)$ at $t = 0$, that is, to map $u_0(x)$ into a set of scattering data. In the general case, the set contains continuous and discrete components, which correspond, respectively, to real and complex λ. The temporal evolution of scattering data, generated by the temporal evolution of $u(x, t)$ under Equation (1), turns out to be trivial, so that, once the scattering data is known at $t = 0$, it is also known for any $t > 0$. Finally, at given t, one must recover the field $u(t, x)$ from the known scattering data, that is, to solve the inverse scattering problem, which is based on the Gel'fand–Levitan–Marchenko integral equation.

The simplest solution of NLS(+) represents a soliton (exponentially localized pulse)

$$\begin{aligned} u_{\mathrm{sol}} &= \eta \operatorname{sech}\left[\eta\left(x - ct - x_0\right)\right] \\ &\quad \times \exp\left(\mathrm{i}cx - \mathrm{i}\omega_{\mathrm{sol}}t + \mathrm{i}\phi_0\right), \end{aligned} \tag{3}$$

where $\omega_{\mathrm{sol}} = \left(c^2 - \eta^2\right)/2$, $\operatorname{sech} z \equiv 2\left(\mathrm{e}^z + \mathrm{e}^{-z}\right)^{-1}$, η and c are arbitrary amplitude and velocity of the soliton, and x_0 and ϕ_0 are the coordinate of the soliton's center and its phase at $t = 0$.

Note that frequencies of the zero-velocity ($c = 0$) solitons are negative ($\omega_{\mathrm{sol}} = -\eta^2/2 < 0$), while the frequencies of radiation modes, that is, solutions to the linearized version of NLS(+), $u_{\mathrm{rad}} = u_0 \exp\left(\mathrm{i}kx - \mathrm{i}\omega_{\mathrm{rad}}t\right)$, are positive ($\omega_{\mathrm{rad}} = k^2/2 > 0$, where k and u_0 are arbitrary wave number and amplitude). Thus, solitons exist in a part of the frequency space where radiation modes are absent, exemplifying a general principle: solitons cannot share frequencies with radiation, as they would otherwise be losing energy through the emission of linear waves.

In terms of the IST, the soliton corresponds to a case when the continuous component of the scattering data is absent, while the discrete component consists of a single eigenvalue $\lambda = (-c + \mathrm{i}\eta)/2$. All the solutions corresponding to an arbitrary set of complex eigenvalues can be found in an explicit form, describing collisions between solitons unless their velocities coincide. The solitons reappear after the collision with the same amplitudes and velocities, that they had prior to the collision, the only change being shifts of the constants x_0 and ϕ_0. If velocities coincide, the solution describes a bound state of solitons, which is dynamically unstable because its binding energy is zero. Thus, the simple soliton given by Equation (3) is the single stable solution to NLS(+) with a permanent wave shape.

Besides the solitons, Equations (1) have spatially uniform solutions in the form of continuous waves (CWs), $u_{\mathrm{CW}}(t) = \eta \exp\left(\pm \mathrm{i}\eta^2 t\right)$, with an arbitrary real amplitude η. An important issue is stability of the CW solutions against small perturbations. To investigate the stability, one can represent a solution as $u(t,x) = a(x,t)\exp\left(\mathrm{i}\phi(t,x)\right)$, where $a(t,x)$ and $\phi(t,x)$ are real amplitude and phase. Rewriting Equation (1) as a system of real equations for a and ϕ leads to a perturbed CW with $a(t,x) = \eta + a_1(t,x)$, and $\phi(t,x) = \pm\, \eta^2 t + \phi_1(t,x)$, with infinitesimal perturbations a_1 and ϕ_1. Linearizing the equations with respect to a_1 and ϕ_1 yields a solution in the form

$$a_1(x,t) = a_1^{(0)} \exp(\sigma t)\cos(px),\ \phi_1(x,t) = \phi_1^{(0)} \exp(\sigma t)\cos(px), \tag{4}$$

where $a_1^{(0)}$ and $\phi_1^{(0)}$ are infinitesimal amplitudes of the perturbation, p is its arbitrary wave number ($-\infty < p < +\infty$), and σ is an instability growth rate. The linearized equations imply the following relation between σ and p:

$$\sigma^2(p) = \pm p^2\left(\eta^2 \mp p^2/4\right). \tag{5}$$

The necessary and sufficient condition for the stability of CW against small perturbations is $\mathrm{Re}\,\sigma(p) \leq 0$, which must hold for all p. As is seen from Equation (5), all the CW solutions to NLS(−) are stable, but in NLS(+) they all are unstable for p^2 in the range $0 < p^2 < 4\eta^2$. It follows from Equation (4) that growing perturbations break the uniformity of the CW, initiating spatial modulation. This is called modulational instability and also Benjamin–Feir instability, after the authors who discovered it in the context of water waves.

Although NLS(−) does not support solitary-wave solutions, it has a stable solution in the form of a dark soliton (DS), which is a region of low amplitude moving on top of a CW. The DS solution is

$$u_{\mathrm{DS}} = \eta \exp\left(\mathrm{i}\eta^2 t\right) \times \left\{(\cos\theta)\tanh\left[\eta(\cos\theta)(x-ct)\right] + \mathrm{i}\sin\theta\right\}, \tag{6}$$

where η is the amplitude of the CW background, and θ, which takes values between 0 and $\pi/2$, determines the minimum value of the field at the center of DS and its velocity (Kivshar & Luther-Davies, 1998). The solution of Equation (3) is sometimes called a bright soliton to stress its difference from the DS.

A fundamental property of NLS equations which is not related to the integrability and is therefore more generic, being shared by a broad class of equations, is a possibility to represent the equations in the Lagrangian and Hamiltonian forms. The general Lagrangian representation is $\delta S/\delta u^*(t,x) = 0$, where $S\{u(t,x), u^*(t,x)\}$ is an *action functional*, which must be real, and $\delta/\delta u^*$ stands for the variational (Fréchet) derivative of the functional (u and u^* are treated as independent functional arguments). The action functional that generates Equation (1) is

$$S = \int \mathrm{d}t \int_{-\infty}^{+\infty} \mathrm{d}x \left[\mathrm{i}\left(u^* u_t - u u_t^*\right) \mp |u_x|^2 + |u|^4\right]. \tag{7}$$

The Hamiltonian representation of the NLS equations is

$$\mathrm{i}\frac{\partial u}{\partial t} = \frac{\delta H\{u, u^*\}}{\delta u^*}, \tag{8}$$

where the Hamiltonian (H) is $\int \mathrm{d}x \left(\pm |u_x|^2/2 - |u|^4/2\right)$. The representation in the form of Equation (8) is very general, applying to a large class of equations for complex fields.

For real H, it follows from Equation (8) that the Hamiltonian is conserved. From the perspective of the Noether theorem, conservation of H is a consequence of the fact that Equation (1) is invariant with respect to an arbitrary time shift, $t \to t + t_0$. The invariances with respect to an arbitrary phase shift, $u(t,x) \to u(t,x)\exp(\mathrm{i}\phi_0)$, and to an arbitrary coordinate shift, $x \to x + x_0$, generate the conservation of the number N (alias "number of quanta") and momentum P, which are given by universal expressions, irrespective of the particular form of the equation:

$$N = \int_{-\infty}^{+\infty} \mathrm{d}x\ |u(x)|^2,$$

$$P = \frac{1}{2}\mathrm{i}\int_{-\infty}^{+\infty} \mathrm{d}x \left(u u_x^* - u^* u_x\right). \tag{9}$$

If the equation is integrable by means of IST, it possesses, in addition to the dynamical invariants N,

P, and H, an infinite series of higher-order conserved quantities (Zakharov et al., 1984).

Despite the exact integrability of Equation (1) by means of IST, the solution provided by this technique may be too cumbersome; therefore, a more tractable approximate method for the construction of solutions is often desirable. To this end, a variational approximation (VA), whose applicability relies solely on the Lagrangian representation of the equation(s), has been developed for solitons of NLS(+) and for several other equations, including those described below. This method approximates the full PDE by a low-order system of ODEs, which are derived from the same action functional (Malomed, 2002).

NLS Models for Wave Envelopes in Weakly Nonlinear Dispersive Media

Equations (1) also appear as universal equations governing the evolution of slowly varying packets of quasi-monochromatic waves in weakly nonlinear media featuring dispersion (dependence of the wave's group velocity c_{gr} on its frequency ω).

To illustrate this point, one can take the sine-Gordon (SG) equation, $\psi_{tt} - \psi_{xx} + \sin\psi = 0$, for real $\psi(t, x)$. Assuming a small-amplitude wave with a slowly varying complex envelope function $u(t, x)$ carried by a high-frequency wave $\exp(-it)$, the solution is $\psi(t, x) = u(t, x)\exp(-it) + u^*(t, x)\exp(it)$, where the slow variation implies that $|u_t|^2 \ll |u|^2$. Substituting into SG, expanding $\sin\psi$ up to the cubic term, dropping the higher-order small terms u_{tt} and u^*_{tt}, and separately collecting terms in front of two independent rapidly varying functions $\exp(\pm it)$ yields two equations: NLS(+) for $u(t, x)$, and its complex conjugate for $u^*(t, x)$.

Thus, NLS equations can be derived as equations governing the evolution of weakly nonlinear wave packets in various media. An important example is light propagation in nonlinear optical fibers and planar waveguides. In the former case, NLS takes the normalized form $\mathrm{i}u_z \pm (1/2)u_{\tau\tau} + |u|^2 u = 0$, where $u(\tau, z)$ is the local amplitude of the electromagnetic wave, which is a function of the propagation distance z and $\tau \equiv t - z/c_{gr}$ (c_{gr} is the group velocity of light in the fiber). In the context of Equation (1), z plays the role of the evolutionary variable, and τ plays the role of the coordinate. The cubic term in the equation accounts for the Kerr effect (a nonlinear correction to the refractive index), while the upper and lower signs in front of $u_{\tau\tau}$ correspond to the anomalous and normal chromatic dispersion in the fiber, $\mathrm{d}c_{gr}/\mathrm{d}\omega > 0$ and $\mathrm{d}c_{gr}/\mathrm{d}\omega < 0$, respectively ($\omega$ is the frequency of the electromagnetic wave). Only anomalous dispersion gives rise to bright optical solitons (also called temporal solitons) as they are localized in τ (Agrawal, 1995). Temporal solitons are important in fiber-optic telecommunications.

In planar waveguides, spatial bright solitons are possible in the form of self-confined light beams. In this case, the propagation coordinate z again replaces t in Equation (1), while the transverse coordinate x appears the same way as in Equation (1), the term u_{xx} accounting for diffraction of light in the waveguide. The corresponding NLS equation is always self-focusing.

Other significant applications in which NLS equations arise are small-amplitude gravity waves on the surface of deep inviscid water and Langmuir waves in hot plasmas (collective oscillations of electrons relative to ions). However, these media are not really one-dimensional, which makes the corresponding NLS solitons unstable against transverse perturbations.

In plasma, NLS(+) approximates a more fundamental model—the Zakharov system

$$\begin{aligned} \mathrm{i}u_t + \tfrac{1}{2}u_{xx} + nu &= 0, \\ n_{tt} - n_{xx} &= -\left(|u|^2\right)_{xx}, \end{aligned} \tag{10}$$

where $n(t, x)$ is a perturbation of the ion density, and $u(t, x)$ is the local amplitude of the Langmuir waves. Although Equations (10) are not integrable, they generate stable solitons and reduce to NLS(+) in the adiabatic limit, when the term n_{tt} may be neglected.

Zakharov's system is a universal model describing weakly nonlinear waves in dispersive media, where a dominating process is the interaction between high- and low-frequency waves (e.g., the Langmuir and ion-acoustic waves, respectively, in plasmas). Another example is the interaction of high-frequency (near infra-red) molecular vibrations and low-frequency (far infra-red) longitudinal deformations in a model of long biological molecules, which gives rise to the Davydov soliton.

Generalized Forms of the Nonlinear Schrödinger Equations

The NLS equation can be viewed as a paradigmatic representative of a large class of nonlinear PDEs, some of which are also integrable. Examples are the derivative NLS equation, $\mathrm{i}u_t + u_{xx} + \mathrm{i}\left(|u|^{2u}\right)_x = 0$, whose integrability was discovered by Kaup and Newell (this equation applies to Alfvén waves in magnetized plasmas), the Hirota equation, $\mathrm{i}u_t + u_{xx} + \mathrm{i}\,(u_{xxx} + 6\,|u|^2\,u_x)$, and the Sasa–Satsuma equation, $\mathrm{i}u_t + u_{xx} + \mathrm{i}\left[u_{xxx} + 6\,|u|^2\,u_x + 3\left(|u|^2\right)_x u\right] = 0$.

In applications to fiber optics, an important generalization considers two polarizations of light (or two different wavelengths in one fiber), leading to a system of coupled NLS equations

$$
\begin{aligned}
&\mathrm{i}u_z \pm \tfrac{1}{2}v_{\tau\tau} + \left(|u|^2 + \beta\,|v|^2\right)u = 0, \\
&\mathrm{i}v_z \pm \tfrac{1}{2}v_{\tau\tau} + \left(|v|^2 + \beta\,|u|^2\right)u = 0.
\end{aligned} \tag{11}
$$

In the case of linear polarizations, $\beta=\frac{2}{3}$, and in the case of circular polarizations or different wavelengths, $\beta=2$ (Agrawal, 1995). As demonstrated by Manakov, Equations (11) are integrable if $\beta=1$ (Manakov, 1974).

Generalized NLS equations are not integrable in other cases of interest, including the multidimensional NLS equations

$$
\mathrm{i}u_t \pm \tfrac{1}{2}\nabla^2 u + |u|^2\,u = 0, \tag{12}
$$

where ∇^2 is the two- or three-dimensional (2-d or 3-d) Laplacian. Equation (12) describes light propagation in bulk media, and Langmuir waves in plasmas. Both the 2-d and 3-d solitons are unstable because NLS(+) gives rise to collapse, that is, formation of a singularity in finite time (Bergé, 1998) (nevertheless, the 2-d soliton, also called the Townes soliton, plays an important role, as it determines the critical power necessary for the collapse in the 2-d case).

The 2-d NLS(−) gives rise to vortex solutions, with important implications in nonlinear optics, which take the form

$$
u = \eta\exp\left(\mathrm{i}\eta^2 t + \mathrm{i}S\chi\right)U\left(\sqrt{2}\eta r\right) \tag{13}
$$

with arbitrary η and $U(\rho=\infty)=1$. Here, χ is the angular coordinate in the 2-d space, and $S\geq 1$ is an integer vorticity. Only the vortex with $S=1$ is stable, being a 2-d counterpart of the DS solution (6) with $\theta=0$.

Multidimensional bright solitons (also called light bullets in 3-d optical models) may be stable if the cubic term $|u|^2\,u$ in NLS(+) is replaced by a saturable one, $|u|^2\left(1+|u|^2/u_0^2\right)^{-1}$ ($u_0^2=\text{const}$), or by a combination of self-focusing cubic and self-defocusing quintic terms, $\left(|u|^2-|u|^4\right)u$. In the latter case, bright 2-d solitons with the embedded vorticity $S\geq 1$ and 3-d solitons with the vorticity $S=1$ may also be stable.

Similar to Equation (12) is the Gross–Pitaevskii (GP) equation for the single-atom wave function $\psi(t,\boldsymbol{r})$, on which the mean-field description of Bose–Einstein condensates is based (Dalfovo et al., 1999):

$$
\mathrm{i}\hbar\frac{\partial\psi}{\partial t} = -\frac{\hbar^2}{2m}\nabla^2\psi + U(\boldsymbol{r})\psi + g\,|\psi|^2\,\psi. \tag{14}
$$

Here $\hbar$ is Planck's constant, m is the atom mass, $U(\boldsymbol{r})$ is a trapping potential, and g is proportional to the scattering length of collisions between atoms, which may be both positive and negative.

The simplest approach to GP is to seek a stationary solution, $\psi(t,\boldsymbol{r}) = \exp(-\mathrm{i}\mu t/\hbar)\,\Psi(\boldsymbol{r})$ (μ is called the chemical potential), and apply the Thomas–Fermi approximation to $\Psi(\boldsymbol{r})$, which amounts to dropping the Laplacian in Equation (14). This yields $|\Psi(\boldsymbol{r})|^2 = g^{-1}[\mu - U(\boldsymbol{r})]$ if it is positive, or $\Psi=0$ otherwise.

Nonlinear optics motivate examples of coupled NLS equations that are different from Equations (11). Propagation of light in an optical fiber carrying a Bragg grating (a lattice of defects with the period equal to half the wavelength of light, which gives rise to mutual resonant conversion of right- and left-traveling waves) is described by equations for the amplitudes $u(t,x)$ and $v(t,x)$ of the counterpropagating waves,

$$
\begin{aligned}
&\mathrm{i}u_t + \mathrm{i}u_x + \left(|u|^2 + 2\,|v|^2\right)u + v = 0, \\
&\mathrm{i}v_t - \mathrm{i}v_x + \left(|v|^2 + 2\,|u|^2\right)v + u = 0
\end{aligned} \tag{15}
$$

with the linear coupling accounting for the resonant Bragg scattering. The spectrum of the linearized version of Equations (15) is $\omega^2=k^2+1$, hence it has a radiation-free gap, $-1<\omega<+1$, where solitons may be expected. A family of gap solitons is generated by Equations (15) that can be found exactly although Equations (15) are not integrable and not all members of this family are stable (de Sterke & Sipe, 1994; Malomed, 2002).

Another NLS system describes a basic process in nonlinear optics—second harmonic generation (SHG). If the nonlinearity is quadratic, the system is

$$
\begin{aligned}
&\mathrm{i}u_z + \tfrac{1}{2}u_{xx} + u^*v = 0, \\
&2\mathrm{i}v_z + \tfrac{1}{2}v_{xx} + \tfrac{1}{2}v^2 + qv = 0,
\end{aligned} \tag{16}
$$

where $u(z,x)$ and $v(z,x)$ are amplitudes of the fundamental-frequency and second harmonic waves, and real q is a wave number mismatch between the harmonics. Equations (16) are written for the realistic case of spatial fields in a planar waveguide. They give rise to a family of solitons in the form $u=\exp(\mathrm{i}kz)\,U(x)$, $v=\exp(2\mathrm{i}kz)\,V(x)$, with exponentially localized $U(x)$ and $V(x)$. Except for the case $k=q/3$, the solitons are not available in an analytical form, but most of them are stable (Etrich et al., 2000). Multidimensional generalizations of Equations (16) give rise to stable solitons too, as collapse does not take place in the SHG model.

Ginzburg–Landau (GL) equations, which are formally similar to NLS equations, provide general models of nonlinear dissipative media. The simplest GL equation is Equation (1) with all the coefficients made complex. In general, however, dynamics generated by GL equations is altogether different from that governed by NLS equations (Aranson & Kramer, 2002).

Boris Malomed

See also **Complex Ginzburg–Landau equation; Discrete nonlinear Schrödinger equations; Equations, nonlinear; Inverse scattering method or transform; Nonlinear optics**

Further Reading

Agrawal, G.P. 1995. *Nonlinear Fiber Optics*, San Diego: Academic Press

Aranson, I.S. & Kramer, L. 2002. The world of the complex Ginzburg–Landau equation. *Reviews of Modern Physics*, 74: 99–143

Bergé, L. 1998. Wave collapse in physics: principles and applications to light and plasma waves. *Physics Reports*, 303: 259–370

Dalfovo, F., Giorgini, S., Pitaevskii, L.P. & Stringari, S. 1999. Theory of Bose-Einstein condensation in trapped gases. *Reviews of Modern Physics*, 71: 463–512

Etrich, C., Lederer, F., Malomed, B.A, Peschel, T. & Peschel, U. 2000. Optical solitons in media with a quadratic nonlinearity. *Progress in Optics*, 41: 483–568

Kivshar, Y.S. & Luther-Davies, B. 1998. Dark optical solitons: physics and applications. *Physics Reports*, 298: 81–197

Malomed, B.A. 2002. Variational methods in nonlinear fiber optics and related fields. *Progress in Optics*, 43: 69–191

Manakov, S.V. 1974. On the theory of two-dimensional stationary self-focusing of electromagnetic waves. *Soviet Physics JETP*, 38: 248–253

de Sterke, C.M. & Sipe, J.E. 1994. Gap solitons. *Progress in Optics*, 33: 203–260

Zakharov, V.E., Manakov, S.V., Novikov, S.P. & Pitaevskii, L.P. 1984. *Solitons: The Method of Inverse Scattering Transform*, New York: Consultants Bureau

Zakharov, V.E. & Shabat, A.B. 1972. Exact theory of two-dimensional self-focusing and one-dimensional self-modulation of waves in nonlinear media. *Soviet Physics, JETP*, 34: 62–69

NONLINEAR SIGNAL PROCESSING

The field of nonlinear signal processing is an active area of research stimulated by a combination of new applications, advances in technology, and fundamental advances in mathematical theory. The breadth of applications has extended far beyond the original emphasis on communications to disciplines such as brain science, financial analysis, image and voice recognition, cryptography, virtual reality, digital photography and printing technology, and neural computing. Significant accelerations in computing speeds and the increasingly widespread availability of computer clusters have made more complex algorithms computationally feasible and motivated the investigation of a much larger range of algorithms. Included in these new algorithms are techniques extracted, for example, from chaos theory, partial differential equations, computational geometry, support vector machines, numerical analysis on manifolds, and optimization theory. Fundamental advances in nonlinear signal processing have centered around an increased mathematical understanding of the notion of information content and representation.

Nonlinear signal processing began with the realization that all components of a signal are not equal. For example, nonlinear filters have been used for compressing signals associated with telephone voice communications where the fidelity required depends on the frequency of the signal. In general, it is possible to retain fewer bits of high frequency data than lower frequency data and still maintain voice clarity. Additionally, median and min/max filters and histogram equalization represent traditional tools in nonlinear signal (and image) processing. In mathematical terms, the motivation for nonlinear signal processing techniques arises from the geometric structure, or organization, of the data. In some idealized sense, we can view data as either being flat, as possessing some degree of curvature, or having a fractal nature. The nature of this structure in the data dictates the best tool to be used for its analysis. For example, *signal dimension* is an important feature of a signal. Based on the geometry of the signal, one may employ, for example, basis dimension, topological dimension, or Hausdorff dimension. There are now a number of methods that have been proposed to detect nonlinear features in data including poly-spectral analysis, surrogate analysis, and more; see Barnett et al. (1997) for detailed comparisons.

As a consequence of the inherently complicated geometry, modeling nonlinear signals is significantly more challenging than their linear counterparts. For example, an n-dimensional linear signal may be written $f(x(t)) = Ax(t)$ where the model consists of the $m \times n$ matrix of parameters. If there are more points than unknowns to construct the model, then the approach of least squares may be used for solving over-determined systems. This theory, including the effect of perturbations on the data, is well understood. For nonlinear signals with n-dimensional domains, one encounters the well-known "curse of dimensionality" that suggests that the amount of data required to construct such nonlinear mappings is exponentially dependent on the dimension of the domain. However, the recognition that this problem is overcome by judiciously placing basis functions in the domain of the data has led to a large literature in the theory and applications of modeling nonlinear signals in high dimension. Radial basis functions (RBFs) are a popular choice and have the form

$$f(x) = w_0 + \sum_{k=1}^{N} w_k \phi(\|x - c_k\|),$$

where the vectors c_k are the special sites of the basis functions, x the domain value and the w_i the model parameters. The basis functions may be local, for example, $\phi(r) = \exp(-r^2)$ or global, for example, $\phi(r) = r^3$. Artificial neural networks (ANNs) have also been proposed for such nonlinear approximation tasks. These networks provide a flexible, or adaptive, tool to *learn* a function. Generally, a special type of ANN is employed, that is, a feed-forward sigmoidal network. Mathematically, this may be expressed as

$$y = \sum_i w_i^{(2)} \sigma \left(\sum_j w_{ij}^{(1)} x_j - \theta_i \right),$$

where σ is a monotonic nonlinear squashing function and $\{w_i^{(2)}, w_{ij}^{(1)}, \theta_i\}$ the model parameters. There exist universal approximation theorems that state under which conditions both the RBF and ANN approximations can produce models to essentially arbitrary accuracy. Of course, practical issues such as computation expense and parsimony of the model must be considered in the implementation of such methods. Further details on these issues may be found in Haykin (1994) and Kirby (2001).

Dynamical systems theory has provided a new paradigm for the analysis of nonlinear signals. Fuelled by observation that complex or chaotic signals can arise from simple nonlinear deterministic systems, a suite of new tools has been developed for their empirical detection, characterization, and design. Briefly, a discrete dynamical system may be viewed as the iteration of a mapping, that is

$$x^{(n+1)} = f(x^{(n)}). \tag{17}$$

Often the trajectory $\{x^{(0)}, x^{(1)}, x^{(2)}, \ldots\}$ traces out a complicated path. Here $x^{(n)}$ is viewed as a function of n. However, a *delay embedding* that plots a point as a function of the previous one or more points may produce a picture that reveals the structure of the simple underlying determinisitic system. For example, the well-known logistic map has the form

$$x^{(n+1)} = \alpha(1 - x^{(n)})x^{(n)}.$$

It possesses a complicated, apparently random trajectory with broad Fourier spectrum for $\alpha > 4$. However, clearly the locus of points $(x^{(n)}, x^{(n+1)})$ is a parabola. Thus, one is motivated to look for structure in a sequence of observations of a signal by constructing the delay embeddings $(x^{(n)}, x^{(n-1)})$ (one delay) or $(x^{(n)}, x^{(n-1)}, x^{(n-2)})$ (two delays) and so on. In practice, it is useful to explore delays of multiples of an integer, that is, $N, 2N, 3N, \ldots$.

This heuristic approach gains some credibility with the support of the Takens embedding theorem; see for example, Abarbanel (1996) and references therein. To get a flavor of the power of this theorem, consider a multidimensional discrete dynamical system, that is, the point $\mathbf{x}^{(n)}$ being iterated is a d-tuple. Takens's theorem basically says that if we are able to observe any one component of this system, for example $\{x_1^{(n)}\}$ then the delay embedding of this variable produces a geometrical structure that is in some sense equivalent to the original system. Furthermore, if only a function h of the first component is observed $h(x_1^{(n)})$, then we can still replicate the structure of the original system with the delay embedding approach. The number of delays required to achieve this embedding is prescribed by Takens, for example if the dynamical system resides on an m-dimensional manifold, then the number of delays need only be $2m + 1$ (Abarbanel, 1996).

The dynamical system described by Equation (17) or the delay mapping

$$x^{(n+1)} = f(x^{(n)}, x^{(n-1)}, \ldots),$$

where f is unknown but observed data are available may also be modeled using the RBF or ANN approaches. Note that the domain of this mapping of delay coordinates may be of high enough dimension to preclude the use of other standard methods. A first look at this procedure in the context of nonlinear signal modeling is described in Lapedes & Farber (1987).

Yet another promising direction of the dynamical systems approach is fractal image compression. Here a picture may be stored as an *iterated function system*, that is two or more systems of the form of Equation (17) where f is now an affine transformation. An image can be produced by iterating this system and only a small number of parameters need be stored. See Barnsley & Hurd (1993) for many interesting examples as well as mathematical details.

The richness of the nonlinear signal processing field is in part due to technology. Optical signal processing is based on the application of the characteristics of light to process and transmit information. For example, in optical image processing, holograms are used to store data while lenses are used to compute two-dimensional Fourier transforms and may also perform edge detection and filtering tasks (Poon & Banerjee, 2001). Most recently, the field of fiber optics has arisen from the observation that light emanating from a laser and passing through a tiny pure glass pipe can transmit more information than electrical signals transmitted along a wire. The ability to send and amplify multiple signals simultaneously along an optic fiber has give rise to data transmission rates of 10 Gbps with higher rates in view. This technique is referred to as dense wavelength division multiplexing (DWDM) (Ramaswami & Sivarajan, 2001). New efforts to model and exploit such technologies suggest that the field of nonlinear signal processing is at the dawn of a new era. For example, the emerging area of photonics addresses how light can be used to represent and process signals and the new discipline of biophotonics concerns how light may be used to discover the mechanisms by which the basic building blocks of living matter function. Signal processing problems such as these may prove to have significant impact on the human condition including the battle against diseases such as cancer.

MICHAEL J. KIRBY

See also **Attractor neural network; Embedding methods; McCulloch–Pitts network; One-dimensional maps; Standard map**

Further Reading

Abarbanel, H.D.I. 1996. *Analysis of Observed Chaotic Data*, Berlin and New York: Springer

Barnett, W., Gallant, A.R., Hinich, M.J., Jungeilges, J.A., Kaplan, D.T. & Jensen, M.J. 1997. A single-blind controlled competition among tests for nonlinearity and chaos. *Journal of Econometrics*, 82: 157–192

Barnsley, M.F. & Hurd, L.P. 1993. *Fractal Image Compression*, Wellesley, MA: AK Peters

Haykin, S. 1994. *Neural Networks*, New York: Macmillan

Kirby, M. 2001. *Geometric Data Analysis: An Empirical Approach to Dimensionality Reduction and the Study of Patterns*, New York: Wiley

Lapedes, A. & Farber, R. 1987. *Nonlinear signal processing using neural neworks: prediction and system modelling*, Techinical Report LA-UR-87-2662, Los Alamos National Laboratory

Poon, T.-C. & Banerjee, P.P. 2001. *Contemporary Optical Image Processing with Matlab*, New York: Elsevier

Ramaswami, R. & Sivarajan, K.N. 2001. *Optical Networks: A Practical Perspective*, 2nd edition, San Francisco: Morgan Kaufmanm

NONLINEAR SUPERPOSITION

See ***N*-solition formulas**

NONLINEAR TOYS

Toys are ubiquitous and some are unique to particular cultures. Many action toys are intriguing because they perform in unexpected ways as a result of underlying nonlinear principles, yielding both amusement and surprises for the unwary. Here we give several examples of toys and devices that operate on linear and nonlinear physical principles.

Frog on a Swing

Several toys are based on the mechanism of *parametric excitation*. Although primarily a linear mechanism, an increase in amplitude in parametric excitation leads to the nonlinear range of amplitude-dependent phenomena. One can demonstrate parametric excitation by taking a string with a weight attached to its end and make it into a pendulum by hanging it over the forefinger (see Figure 1(a)). Now, force the weight into a small amplitude oscillation, and then move the weight up and down with the same phase as the oscillation by pulling on the string periodically. The result is an increase in the amplitude of the oscillation of the pendulum. The phase of the oscillation will vary slowly as the amplitude increases and the inherently nonlinear nature of the pendulum becomes important. The basic linearized equation for parametric excitation of the nonlinear pendulum is the Mathieu equation

$$\frac{d^2\theta}{dt^2} + \left(\omega_0^2 - \alpha \sin 2\omega_0 t\right)\theta = 0$$

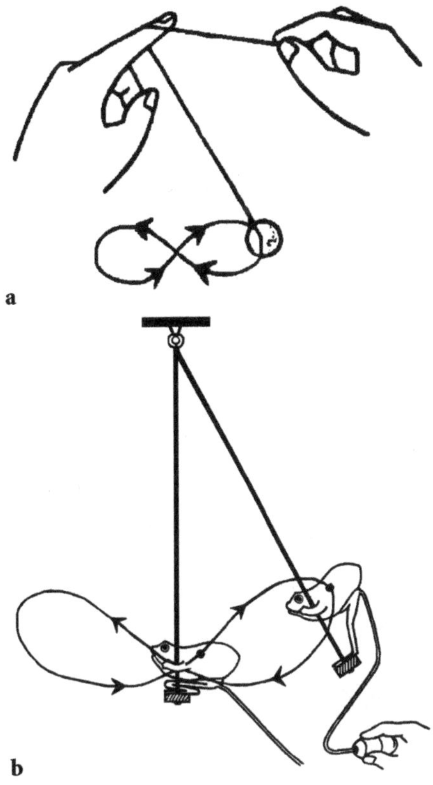

Figure 1. Parametric excitation demonstrated by (a) a pendulum over a finger and (b) a frog on a swing.

where θ is the angle of the pendulum from the vertical, ω_0 is the fundamental small amplitude frequency, and α is the amplitude of the periodically changing pendulum length.

This principle of parametric excitation is active in the toy called the "frog on a swing" (see Figure 1(b)). The frog in this toy is made of rubber and sits on a swing (pendulum). It can be inflated periodically with air using a tube and bulb. This causes the frog to stand up and sit down repeatedly and, hence, moves the center of mass of the frog up and down, as illustrated in Figure 1(b). With proper timing in inflating the bulb at the correct phase during each swing, the amplitude of the swing with the oscillating frog on it gradually increases. The same principle applies to the playground swing where children can increase the amplitude of their swing by "pumping" the swing in a standing position (Wirkus et al., 1998).

Handstand Pendulum

A related, but completely different, phenomenon occurs in the "handstand pendulum." The normal pendulum hangs stably downwards from its fulcrum as a result of gravity. However, applying a high-frequency forced oscillation to the fulcrum can cause the pendulum to be inverted, so that it now "hangs" stably pointing upwards

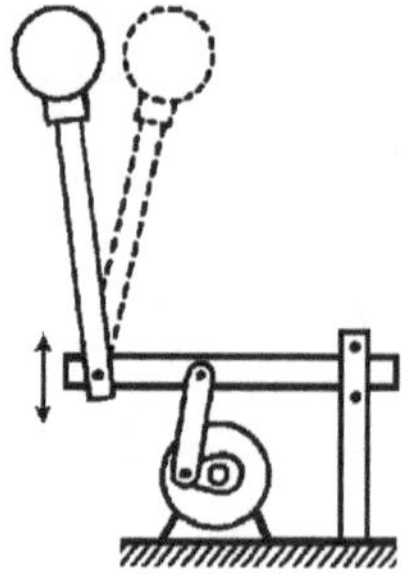

Figure 2. The handstand pendulum.

Figure 3. Bamboo deer alarm.

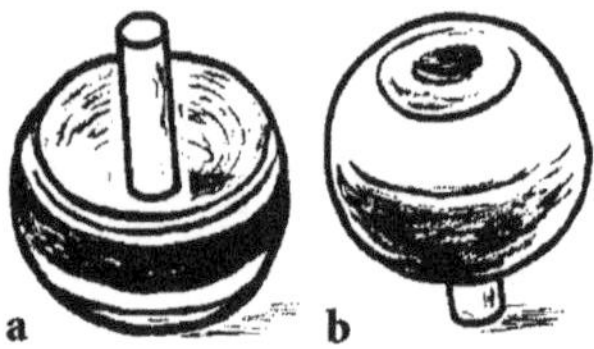

Figure 4. A tippe top in its stable resting position and stable rotating position.

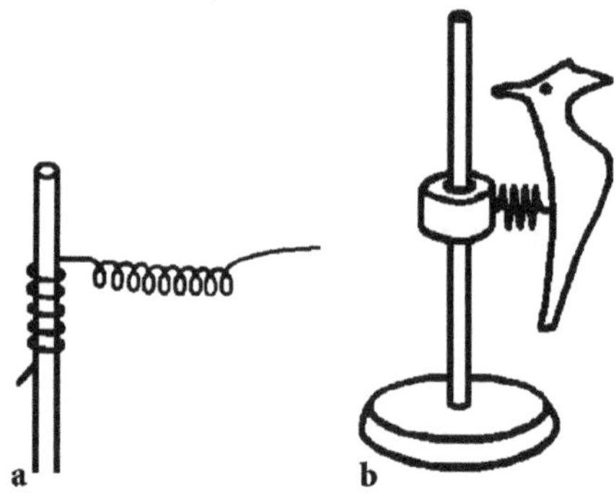

Figure 5. The pecking woodpecker with a detail of the spring in (a).

(Figure 2). Again the governing linearized equation is the Mathieu equation, but with the signs changed:

$$\frac{\mathrm{d}^2\theta}{\mathrm{d}t^2} + \left(-\frac{g}{\ell} + 2\alpha \sin 2\omega t\right)\theta = 0$$

where g is the acceleration due to gravity and ℓ is the length of the pendulum arm.

Shishi-odoshi

In some Japanese gardens, there is a device that was designed originally to scare deer and other animals away from rice fields. Now, its purpose is either to scare away the deer that may intrude into a garden to eat the plants or to serve as an aesthetic fixture in the garden for its novelty and silence breaking "clack." This device, while not strictly a toy, is based on nonlinear principles and is called a *shishi-odoshi* (deer-scarer) or *shika-oi*. The *shishi-odoshi* is made from a thick bamboo culm with a length of three internal cells, that is, four nodes (the raised rings at regular intervals along the bamboo). At one end, the last node has been cut off and the end shaped to receive water (see Figure 3). From this open end, the internal membrane of the first node also has been removed so there is a cylindrical cup that is two cells long. A pivot rod is inserted into the bamboo between the two inner nodes so that when the tube is empty, the closed end of the bamboo rests on a stone. However, the position of the pivot is chosen so that as a steady stream of water flows into the open end of the tube, the center of mass of the bamboo and water shifts across the pivot and the open end of the tube will dip down, thus pouring out the water. The empty tube now returns quickly to its stable position on the stone with a loud "clack," which is the noise that is supposed to scare the deer.

The oscillatory motion of the bamboo tube caused by the steady flow of water is called a *relaxation oscillation* and can be regarded as an example of a self-induced oscillation. There are many toys that use this principle to change their orientation periodically, namely, water filling up a reservoir, and then the water being discharged as a result of the change of orientation of the toy.

Tippe Top

A fascinating toy is the tippe top, which does not act like a conventional top. The tippe top consists of a little more than a hemisphere with a short cylindrical stem (see Figure 4(a)). It has a low center of gravity, so when placed on a surface, it will simply sit like an ordinary top on its hemispheric side. However, if the stem is held between a finger and the thumb and given a spin on the hemispheric end, then unlike a conventional top, the tippe top will readily turn over and continue to spin on the stem (see Figure 4(b)). As it turns over, the behavior of the tippe top is unusual in two respects: the center of gravity is raised and the spinning direction with respect to fixed body coordinates is reversed. When the top is spun, the low center of gravity is centrifugally moved away from the vertical spin axis. Before and after the top turns over, the angular momentum of the top about the vertical axis is the dominant momentum component. During the inversion process, the center of gravity is raised and the increase in potential energy reduces the rotational kinetic energy. The torque needed to execute

Figure 6. Two other toys that use self-induced oscillations.

this inversion comes from the sliding frictional forces between the top's hemispherical surface and the surface on which it is rotating. The mechanics of this top are described by nonlinear equations of motion that are solvable only in special cases (Cohen, 1977; Or, 1994; Gray & Nickel, 2001).

Pecking Woodpecker

A cute toy providing entertainment for young children is the pecking woodpecker going down a pole. To make this toy, coil a soft wire loosely around a smooth pole to make a soft spring. Now leave several turns of the coil on the pole and the remaining coils extended outwards from the pole (see Figure 5a). The weight of the protruding part of the spring will tilt the spring coiled around the pole so that the static friction is sufficient to keep it fixed on the pole. If one pushes the extended coil downwards and releases it, the extended coil vibrates up and down and the spring around the pole descends in a staccato fashion. The reason for the stuttering movement down the pole is that the coil around the pole oscillates periodically between being stopped by static friction and essentially free falling down the pole when the coil is aligned with the pole. Now attach a small wooden bird to the end of the extended wire, and the bird will peck the pole with its beak as it descends the pole (see Figure 5b).

There are many toys that move by the principle of self-induced excitation (see Figure 6). The basic equations for such oscillations are those of Duffing and van der Pol.

MORIKAZU TODA AND ROBERT M. MIURA

See also **Duffing equation; Equations, nonlinear; Laboratory models of nonlinear waves; Parametric amplification; Pendulum; Relaxation oscillators; Van der Pol equation**

Further Reading

Cohen, R.J. 1977. The tippe top revisited. *American Journal of Physics*, 45: 12–17

Gray, C.G. & Nickel, B.G. 2001. Constants of the motion for nonslipping tippe tops and other tops with round pegs. *American Journal of Physics*, 68: 821–828

Or, A.C. 1994. The dynamics of a tippe top. *SIAM Journal of Applied Maths*, 54: 597–609

Toda, M. 1979. *Toy Seminar.* Tokyo: Nihon-Hyoron-Sha (in Japanese)

Toda, M. 1982. *Toy Seminar (Continued).* Tokyo: Nihon-Hyoron-Sha (In Japanese)

Toda, M. 1983. *Mobile Toys.* Tokyo: Nihonkeizai-Shinbun-Sha (In Japanese)

Toda, M. 1993. *Science of Toys.* (6 vol.), Tokyo: Nihon-Hyoron-Sha, 1995 (In Japanese)

Wirkus, S., Rand, R. & Ruina, A. 1998. How to pump a swing. *The College Mathematics Journal*, 29: 266–275

NONLINEAR TRANSPARENCY

See **Nonlinear optics**

NONLINEARITY, DEFINITION OF

If a system's response to an applied force is directly proportional to the magnitude of that force, the system is said to be linear, otherwise the system is nonlinear. Nonlinear dynamics is concerned with systems whose time evolution equations are nonlinear.

As an example of a linear system, consider the one-dimensional motion of a point mass connected to a spring of force constant k, subject to a force

$$F_x(x) = -kx$$

in the x-direction. If k is changed, then the frequency and amplitude of resulting oscillations will change, but the qualitative nature of the behavior (simple harmonic oscillation in this example) remains the same. In fact, by appropriately rescaling the length and time axes, we can make the behavior for any value of k look just like that for some other value of k. For nonlinear systems, on the other hand, a small change in a parameter can lead to sudden and dramatic changes in both the qualitative and quantitative behavior of the system.

Although almost all natural systems are nonlinear, many of them respond in an approximately linear way provided that the amplitude of the force is small enough. This is one of the reasons why traditional linear approximations are so popular in science. Another reason is that the analytical solution of nonlinear equations is often difficult. With the development and availability of computers, however, numerical solutions of nonlinear equations became relatively easy in the 1970s. This focused attention on important nonlinear phenomena such as deterministic chaos, although the idea of dynamic chaos was first glimpsed by Henri Poincaré (Poincare, 1908). Nonlinearity and the phenomena that result from it have a strong bearing on the concept of determinism.

Given a set of variables (position, velocity, acceleration, pressure, the number of species in a chemical reaction, etc.) or a function of two or more independent variables that describe the state of a system, the subsequent time evolution can be represented as causal relationship between its present state and its next state in the future. The existence of causality in such relationships is suggested by all our experience of experimenting with Nature, and it corresponds to the deterministic perception of the evolution of dynamical systems.

Consider, for example, the oscillatory motion of a damped pendulum in the phase space of two variables, the angle θ and the velocity v

$$\dot{\theta} = v,$$

$$\dot{v} = -\gamma v - \omega_0^2 \sin(\theta),$$

where γ and ω_0^2 are parameters that describe damping rate and frequency of small oscillations. These time evolution equations can be written in operator form as

$$L\theta = \frac{\mathrm{d}^2\theta}{\mathrm{d}t^2} + \gamma\frac{\mathrm{d}\theta}{\mathrm{d}t} + \omega_0^2 \sin(\theta) = 0. \qquad (1)$$

Formally, the dynamics is linear if the causal relation between the present state and the next state is linear, otherwise it is nonlinear. In the above example, the dynamics will be linear if we replace the term $\beta \sin(\theta)$ with its linear approximation $\beta\theta$ near the asymptotically stable equilibrium state $\theta = 0$, $v = 0$. The corresponding linear operator in this case satisfies the property that

$$L^{(\mathrm{lin})}(ax + by) = aL^{(\mathrm{lin})}x + bL^{(\mathrm{lin})}y,$$

where a and b are real numbers and x and y are differentiable functions. The property of linear superposition plays a fundamental role in the analysis of linear equations of motion, and represents one of the basic principles of quantum mechanics (Landau & Lifshitz, 1977). If the property of linear superposition does not hold the dynamics is nonlinear.

On physical grounds, dynamical systems are in general nonlinear. Indeed, if we consider small deviations of a pendulum from its state of unstable equilibrium $\theta = \pi$, $v = 0$ and replace the nonlinear term by its linear approximation $\omega_0^2 \sin(\pi + \delta\theta) \approx -\omega_0^2 \delta\theta$, we find unbounded growth of $\delta\theta$ in time. However, the experimentally observed motion of the pendulum is bounded; that is, the character of the motion changes when the deviation from the point of unstable equilibrium is large. Thus, nonlinearity corresponds to the situation that arises when the properties of the system depend directly on its state. The latter property, on one hand, makes a nonlinear system difficult to analyze but, on the other hand, gives rise to a rich diversity of nontrivial phenomena. We now consider briefly two consequences of nonlinearity.

First, the spectrum of oscillations of a nonlinear system is a complex function of its state (see, e.g., Sagdeev et al., 1988). For example, for weakly nonlinear oscillations of the pendulum near the state of stable equilibrium, we find (Hayashi, 1964), neglecting damping, that the frequency ω varies with the amplitude of the oscillations A as

$$\omega \approx \omega_0(1 - 3/64A^3).$$

The dependence of the frequency of oscillations on the state of the system is a key feature of many phenomena, including harmonic generation, synchronization, and the formation of solitons (*See* **Harmonic generation; Solitons; Synchronization**).

Second, a prominent effect of nonlinearity is the onset of deterministic chaos. Linear dynamical systems can have only three types of invariant sets: fixed points, cycles, and quasi-periodic orbits. The occurrence of a chaotic orbit is possible only in nonlinear systems and can be viewed in simple terms as arising from the interplay between the instability and nonlinearity. The instability is responsible for the exponential divergence of two nearby trajectories, whereas the nonlinearity bounds trajectories within a finite volume of the phase space of the system. The combination of these two mechanisms gives rise to a high sensitivity of the system to the initial conditions (*See* **Butterfly effect**). Consequently, the onset of deterministic chaos is restricted to nonlinear systems and involves repeated stretching and folding. The simplest example of a nonlinear system that demonstrates chaotic behavior through this process is a Bernoulli shift: $x_{n+1} = 2x_n \pmod 1$. An important consequence of chaos is that the predictability of even simple nonlinear systems is limited, so that ergodic theory has to be used to describe their statistical properties.

From these two brief examples it is already clear that in nonlinear science, the whole is more than the sum of its parts.

The complexity and diversity of the possible types of behavior in nonlinear systems have been widely explored in physics and chemistry, including analyses of solitons, nonlinear localization, pattern formation, and formulation of the general principles of self-organization.

The formalism of nonlinear systems provides a framework for understanding and modeling of the hierarchy of complexity in the life sciences. However, there are many difficulties to be encountered on the way. In particular, Hermann Haken (in Tschacher & Dauwalder, 1999) notes:

> "While in physics or chemistry it is not too difficult to define the microscopic and macroscopic levels, with respect to the brain we must ask what adequate intermediate levels we have to choose between microscopic and macroscopic."

An example of a practical approach to the solution of this problem arose during the investigation of neurons (*See* **Neurons**). Although the ionic current through the axon membrane is successfully described by the Hodgkin–Huxley equations, heart fibrillation is more conveniently modeled by the continuous FitzHugh–Nagumo equation, which is a simplified version of Hodgkin–Huxley system. On passing to models of neural networks, the hierarchy of connections between neurons structured in time turn out to be of prime importance, but this raises new questions for nonlinear science related to the appearance of closed causal loops. Continuing this hierarchy, one may expect that the application to cognitive science of complexity theory derived from nonlinear analysis will have a major impact on our understanding of reasoning, thinking, behavior, and psychology generally (Tschacher & Dauwalder, 1999).

DMITRY G. LUCHINSKY AND ANETA STEFANOVSKA

See also **Causality; Chaotic dynamics; Determinism; Equations, nonlinear; Linearization; Pendulum; Van der Pol equation**

Further Reading

Hayashi, C. 1964. *Non-linear Oscillations in Physical Systems*, New York: McGraw-Hill

Landau, L.D. & Lifshitz, E. 1977. *Quantum Mechanics: Non-Relativistic Theory*, Oxford: Pergamon (original Russian edition 1947)

Poincaré, H. 1908. *Science et méthode*, Paris: Flammarion

Sagdeev, R.Z., Usikov, D.A. & Zaslavskii, G.M. 1988. *Nonlinear Physics: From the Pendulum to Turbulence and Chaos*, Chur: Harwood Academic

Tschacher, W. & Dauwalder, J.-P. (editors). 1999. *Dynamics, Synergetics, Autonomous Agents: Nonlinear Systems Approaches to Cognitive Psychology and Cognitive Science*, Singapore: World Scientific

NON-NEWTONIAN FLUIDS

See **Fluid dynamics**

NON-TOPOLOGICAL SOLITON

See **Solitons, types of**

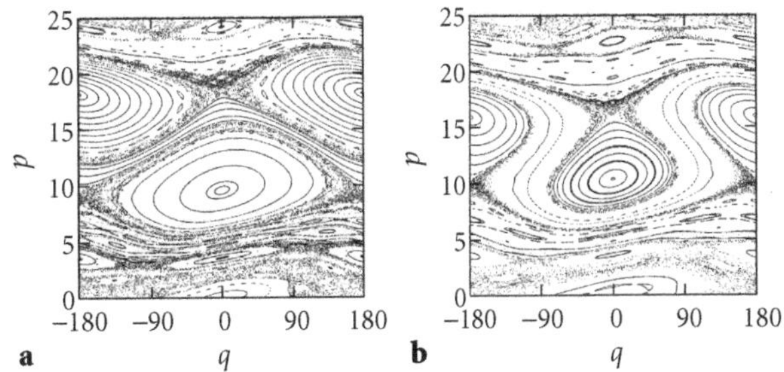

Figure 1. Standard nontwist map for $K = 1.5$ and (a) $\alpha = 0.036$, (b) $\alpha = 0.038$.

NONTWIST MAPS

Two-dimensional area-preserving mappings have become a standard tool for analyzing two-degrees-of-freedom autonomous and one-and-one-half dimensional time-periodic Hamiltonian systems (Lichtenberg & Lieberman, 1990). Prominent among such mappings are the radial twist maps

$$T : \begin{cases} p' = p - DU(q), \\ q' = q + f(p'), \end{cases} \tag{1}$$

where q and p are conjugate variables, $U(q)$ and $f(p)$ are smooth functions, and D is a derivative operator. For example, with $U = -K\cos q$ and $f(p) = p$, we recover the standard map, which has become a paradigm for chaotic Hamiltonian dynamics.

Usually $f(p)$ is monotonic, in which case the twist $\tau = \mathrm{d}f/\mathrm{d}p$ does not change sign. The twist condition, $\tau \neq 0$, is crucial for the validity of such linchpins of stability theory as the Moser twist theorem and the Kolmogorov–Arnol'd–Moser (KAM) theorem Arnol'd. Nevertheless, more and more instances have been found where the twist condition is violated, for example, plasma wave heating, zonal flows in planetary magnetospheres, beam dynamics in particle accelerators, and relativistic charged particle motion. The proliferation of such nontwist maps has challenged theorists to find alternatives to standard tools of nonlinear dynamics.

The Standard Nontwist Map

Many of the essential characteristics of nontwist maps are captured in the standard nontwist map (SNTM) (Howard and Hohs, 1984)

$$\begin{aligned} p' &= p - K\sin q, \\ q' &= q + p' - \alpha p'^2, \end{aligned} \tag{2}$$

where K and α are positive constants. Thus, the twist vanishes along the line $p = 1/2\alpha$. An equivalent version is discussed in del Castillo-Negrete et al. (1997). Generalizations employing higher order polynomial forms for $f(p)$ are discussed in Howard & Humphreys

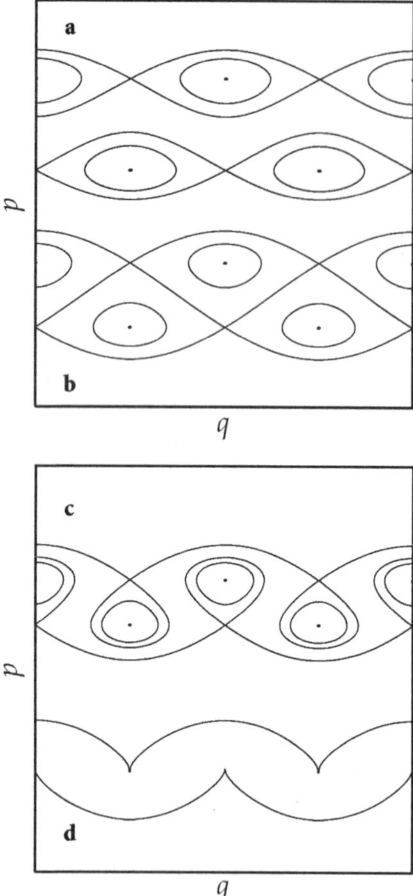

Figure 2. Reconnection scenario for standard nontwist map.

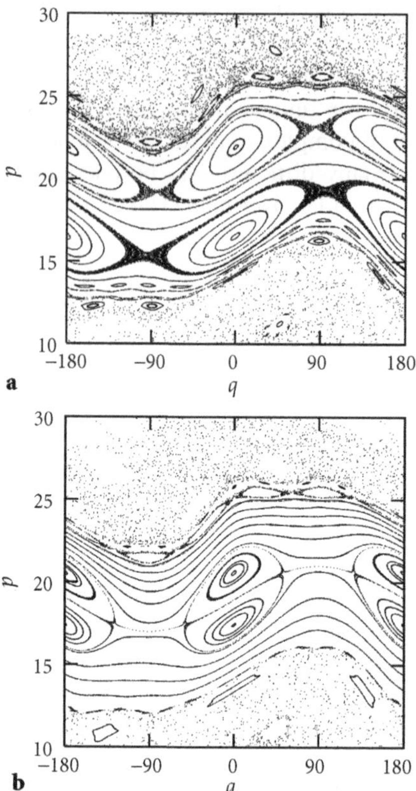

Figure 3. Vortex pair formation in the standard nontwist map for $K = 4.0$ and (a) $\alpha = 0.0260$, (b) $\alpha = 0.02634$.

(1995). Fixed points are located at $q \in (0, \pi)$ and

$$p_n^{\pm} = \frac{1 \pm \sqrt{1 - 8\pi n\alpha}}{2\alpha}. \tag{3}$$

For positive n both roots are positive real for $0 \leq 8\pi n\alpha \leq 1$, merging in a saddle-node bifurcation when $\alpha^* = (8\pi n)^{-1}$ at $p^* = 4\pi n$. Figure 1 shows the SNTM for $K = 1.5$ and two nearby values of α. Notice that, in contrast to the standard map, stable and unstable fixed points are staggered in phase rather than vertically aligned. This leads to a new kind of global bifurcation, called *reconnection* of KAM curves (by analogy with magnetic reconnection), which occurs when

$$K(\alpha) \approx \frac{1}{2} \int_{p^-(\alpha)}^{p^+(\alpha)} [f(\xi; \alpha) - 2\pi n]\, d\xi$$
$$= \frac{(1 - 8\pi n\alpha)^{3/2}}{12\alpha^2}. \tag{4}$$

The entire reconnection scenario is depicted schematically in Figure 2. We see that for a given configuration of X- and O-points, there are two topologically distinct arrangements of separatrices. Thus, in going from Figure 2(a) to (c), the upper separatrix is diverted downward, looping around the lower elliptic point. In between (b) the two separatrices merge, a process called *braiding*. Finally, Figure 2(d) shows the characteristic cusps resulting from the merging of X- and O-points via a saddle-node bifurcation. Naturally, the structure of the full mapping (Figure 1) is much more complicated than this approximate depiction. Of particular interest is the band of meandering curves seen in Figure 1(b), which constitute a new kind of barrier to chaotic transport. More complex sequences occur for more complex $f(p)$.

Higher-order fixed points also undergo reconnection. For example, the period-two islands near $p = 6\pi$ exhibit a second kind of reconnection, *vortex pair* formation, as shown in Figure 3. Here the elliptic and hyperbolic fixed points are vertically aligned, in contrast to the staggered pairs of period-one islands seen in Figure 1. An exhaustive classification of possible reconnection scenarios has not yet been accomplished.

As in the case of monotone twist maps, the location of higher order fixed points is facilitated by writing the map as a composition of involutions, $T = I_2 I_1$ ($I_1^2 = I_2^2 = I$). Fixed points of T then lie on the intersections of the invariant curves of I_1 and I_2. For the general twist map (1) we have

$$I_1 : \begin{cases} p' = p - DU, \\ q' = -q, \end{cases} \tag{5}$$

$$I_2 : \begin{cases} p' = p, \\ q' = -q + f(p). \end{cases} \quad (6)$$

Mappings with this decomposition are called *reversible*; that is, there exists a symmetry S such that $T^{-1} = STS$. Explicitly, $S = I_2$, so that $T^{-1} = I_1 I_2$.

Normal Forms

In investigating the structure of a 2-dimensional symplectic map near a stable fixed point (Howard & MacKay, 1987), it is useful to employ Birkhoff normal forms. Thus, the mapping T written in action-angle variables J and θ takes the simple form (Meyer & Hall, 1992)

$$\begin{aligned} J' &= J \\ \theta' &= \theta + 2\pi\Omega(J). \end{aligned} \quad (7)$$

Here the rotation number is given by

$$\Omega(J) = \omega + \tau_0 J + \tfrac{1}{2}\tau_1 J^2 + \cdots \quad (8)$$

and the twist by

$$\tau = \frac{\partial \Omega}{\partial J} = \tau_0 + \tau_1 J + \cdots, \quad (9)$$

where τ_0 and τ_1 are the zero and first-order twists. In this way it has been shown (Moeckel, 1990; Dullin et al., 1999) that a twistless torus is generated for any area-preserving map near a tripling bifurcation, where $\omega = \frac{1}{3}$.

Transition to Global Chaos

In many physical applications of nontwist maps, the transition to global chaos is of paramount importance, as space-spanning KAM curves (invariant rotational circles) act as barriers to particle or energy transport (Meiss, 1992; Lichtenberg & Lieberman, 1990). The breakdown of such barriers is usually studied via some kind of renormalization theory, such as that devised by Greene et al. (1981) for area-preserving maps satisfying the twist condition. Greene's method is based on the conjecture that the last KAM curve to break up with increasing perturbation strength has rotation number equal to the inverse of the golden mean, that is, $\omega_c = 1/\gamma = (\sqrt{5} - 1)/2$. Using the involution decomposition to locate a sequence of periodic orbits whose rotation number $\omega \to 1/\gamma$, he found the critical value $K_c = 0.9716\ldots$. Recently, del Castillo-Negrete et al. (1997) have succeeded in modifying Greene's semi-analytic method to obtain an accurate estimate for the breakup of the $\omega = 1/\gamma$ invariant circle for the SNTM. This calculation is complicated by the facts that (i) periodic orbits come in pairs for nontwist maps and may even fail to exist for certain parameters (K, α for the SNTM), and (ii) the sequence of periodic orbits actually has six convergent subsequences. Preliminary work has also been carried out on developing a suitable renormalization treatment for nontwist maps.

Extension to Higher Dimension

It is of interest to ask whether nontwist maps exist in higher dimension and what sort of reconnection processes might occur. This may be done in a natural way by letting $p, q \in \mathbb{R}^n$. The Birkhoff normal form then depends on the frequency map, $\Omega(J) = DS(J)$, where S is a scalar function of J. The twist, in turn, becomes the $n \times n$ Jacobian matrix $\tau(J) = D\Omega(J) = D^2 S(J)$, and the frequency map suffers a singularity whenever $\det \tau$ passes through zero. Dullin & Meiss (2003) have determined the behavior of four-dimensional symplectic maps near such a singularity. Again, a twistless torus appears near a 1 : 3 resonance but with interesting topological complications. In four dimensions the frequency map can have a fold or a cusp singularity—in higher dimensions other catastrophes can occur.

JAMES E. HOWARD

See also **Hamiltonian systems; Maps; Standard map; Symplectic maps**

Further Reading

Arnol'd, V.I. 1990. *Mathematical Methods of Classical Mechanics*, 2nd edition, New York: Springer

del Castillo-Negrete, D., Greene, J.M. & Morrison, P.J. 1997. Renormalization and transition to chaos in area preserving nontwist maps. *Physica* D, 100: 311–329

Dullin, H.R. & Meiss, J.D. 2003. Twist singularities for symplectic maps. *Chaos*, 13(1): 1–16

Dullin, H.R., Meiss, J.D. & Sterling, D. 1999. Generic twistless bifurcations. *Nonlinearity*, 13: 203–224

Greene, J., MacKay, R.S., Vivaldi, F. & Feigenbaum, M.J. 1981. Universal behavior in families of area-preserving maps. *Physica* 3D, 13: 468–486

Howard, J.E. & Hohs, S.M. 1984. Stochasticity and reconnection in Hamiltonian systems. *Physical Review* A, 29: 203–224

Howard, J.E. & Humphreys, J. 1995. Nonmonotonic twist maps. *Physics Letters* A, 29: 256–276

Howard, J.E. & MacKay, R.S. 1987. Spectral stability of symplectic maps. *Journal of Mathematical Physics*, 28: 1036–1051

Lichtenberg, A.J. & Lieberman, M.A. 1990. *Regular and Chaotic Dynamics*, 2nd edition, New York: Springer

Meiss, J.D. 1992. Symplectic maps, variational principles, and transport. *Reviews of Modern Physics*, 64: 795–848

Meyer, K.R. & Hall, G.R. 1992. *Introduction to Hamiltonian Dynamical Systems and the N-Body Problem*, New York: Springer

Moeckel, R. 1990. Generic bifurcations of the twist coefficient. *Ergodic Theory of Dynamical Systems*, 10: 185–195

NORMAL FORMS THEORY

Many nonlinear systems can be modeled by ordinary, differential, or difference equations, and a central problem of dynamical systems theory is to obtain information on the long-time behavior of typical solutions. Because it is not possible, in general, to solve these equations explicitly, we identify particular solutions (such as equilibrium solutions or periodic solutions) and try to infer global information from the behavior of nearby solutions.

As an example, consider an equilibrium state such as the downward position of a pendulum. Looking at small perturbations of this position to identify nearby periodic orbits (corresponding to small oscillations of the pendulum) shows that the downward pendulum is indeed stable. Analysis of the inverted pendulum reveals that it is unstable, in the sense that nearby solutions do not stay close to the equilibrium position.

Mathematically, an equilibrium is a fixed point of a dynamical system, and the stability analysis is carried out by linearizing the system, that is, by replacing (close to the fixed point) the nonlinear equations by linear equations for the perturbations. The resulting linear system can be solved exactly, and the analysis of these solutions may give information about the behavior of the solutions of the nonlinear system around the fixed point. This method is extremely powerful when it works, and is the basis of much dynamical system analysis. In some cases, however, the behavior of the linear system may be entirely different from that of the nonlinear system, and no information can be obtained from the linear analysis. This implies that there is crucial information contained in the nonlinear terms.

Normal forms theory is a general method designed to extract this information. The basic idea is to compute a local change of variables to transform a nonlinear system into a simpler nonlinear system that contains only the essential nonlinear terms—those that cannot be neglected without drastically changing the nature of the system. Hopefully, the new system is either linear or can be solved explicitly.

Consider the normal form analysis of the zero fixed point of systems of two differential equations of the form

$$\dot{x}_1 = a_{11}x_1 + a_{12}x_2 + f_1(x_1, x_2),$$
$$\dot{x}_2 = a_{21}x_1 + a_{22}x_2 + f_2(x_1, x_2), \qquad (1)$$

where the dots denote time derivatives and f_1 and f_2 are nonlinear analytic functions whose Taylor expansion about the origin contains no linear term. Linear analysis of the fixed point at the origin is carried out by neglecting the nonlinear terms and considering the linear system $\dot{x}_1 = a_{11}x_1 + a_{12}x_2$, $\dot{x}_2 = a_{21}x_1 + a_{22}x_2$. The dynamics of this system is governed by the eigenvalues λ_1, λ_2 of the matrix (a_{ij}), which is assumed to be diagonalizable. If the eigenvalues have nonvanishing real parts, then the dynamics of the original nonlinear system is qualitatively equivalent to the dynamics of the linear system and the fixed point is stable if both real parts are negative and unstable as soon as one of the real parts is negative. However, if the eigenvalues are imaginary, the behavior of the nonlinear system cannot be inferred from analysis of the linear system, and the information on the stability of the origin is contained in the nonlinear terms. For example, the origin of the system $\dot{x}_1 = x_2$, $\dot{x}_2 = -x_1 + \alpha x_1^2 x_2$ is either stable or unstable depending on the sign of α.

The first step in the normal form analysis of system (1) is to introduce a linear change of variables so it reads

$$\dot{y}_1 = \lambda_1 y_1 + g_1(y_1, y_2),$$
$$\dot{y}_2 = \lambda_2 y_2 + g_2(y_1, y_2). \qquad (2)$$

Second, look for a near-identity change of variables in the form of power series $y_1 = z_1 + P_1(z_1, z_2)$, $y_2 = z_2 + P_2(z_1, z_2)$ that simplifies system (2). To this end, expand g_1, g_2 in power series and choose the coefficients of the series P_1, P_2 so that in the z_1, z_2 variables, the system $\dot{z}_1 = \lambda_1 z_1 + h_1(z_1, z_2)$, $\dot{z}_2 = \lambda_2 z_2 + h_2(z_1, z_2)$ becomes simpler than the original system. Optimally, $h_1 = h_2 = 0$, which would provide an exact linearization of the original system. In general, however, some nonlinear terms remain after transformations.

It turns out that the ability to exactly linearize the original system is intimately connected to the eigenvalues λ_1, λ_2. If either $(n_1 - 1)\lambda_1 + n_2\lambda_2 = 0$ or $n_1\lambda_1 + (n_2 - 1)\lambda_2 = 0$ for some positive integers n_1, n_2, the eigenvalues are said to be resonant (or in resonance), and in general one of the functions (h_1 or h_2) contains resonant terms of the form $z_1^{n_1} z_2^{n_2}$. In the absence of resonance, therefore, no nonlinear terms remain and exact linearization is achieved.

If the eigenvalues are purely imaginary, z_2 is the complex conjugate of $z_1 = z$, and there are infinitely many resonance conditions. The stability of the fixed point is then decided by the first non-vanishing coefficient of h_1. As an example, consider again the system $\dot{x}_1 = x_2$, $\dot{x}_2 = -x_1 + \alpha x_1^2 x_2$. After the linear change of variable $x_1 = y_1 + y_2$, $x_2 = \mathrm{i}(y_1 - y_2)$, we have

$$\dot{y}_1 = \mathrm{i}y_1 + (\alpha/2)\left(y_1^2 y_2 - y_1 y_2^2 - y_2^3 + y_1^3\right),$$
$$\dot{y}_2 = -\mathrm{i}y_2 - (\alpha/2)\left(y_1^3 - y_1^2 y_2 + y_1 y_2^2 + y_2^3\right). \qquad (3)$$

The normal form transformation to third order reads:

$$y_1 = z_1 + (\mathrm{i}\alpha/8)\left(2z_1^3 + 2z_1 z_2^2 - z_2^3\right) + \text{h.o.t.},$$
$$y_2 = z_2 + (\mathrm{i}\alpha/8)\left(z_1^3 - 2z_1^2 z_2 - 2z_2^3\right) + \text{h.o.t.}, \qquad (4)$$

where h.o.t. denotes higher-order terms of degree 5 and higher. Finally, the normal form becomes

$$\dot{z}_1 = \mathrm{i}z_1 + (\alpha/2)z_1 z_2^2 + \text{h.o.t.}, \quad z_2 = z_1^*,$$
$$\dot{z}_2 = -\mathrm{i}z_2 + (\alpha/2)z_2 z_1^2 + \text{h.o.t.}. \qquad (5)$$

From Equations (5), $\dot{\rho} = (\alpha/2)\rho^3 + \text{h.o.t.}$ where $\rho^2 = z_1 z_2$, which implies that the origin of the initial system is stable if $\alpha \leq 0$ and unstable otherwise.

If all coefficients of the normal forms vanish identically, then the fixed point is surrounded by an open set of periodic orbits (a nonlinear center). The downward position of the frictionless pendulum is an example of a center. Note, however, that the convergence of the power series P_1, P_2 is not guaranteed, in general, and further conditions on the eigenvalues and the transformation must be satisfied.

For a general N-dimensional system of differential equations, we can compute the linear eigenvalues $\lambda_1, \ldots, \lambda_N$. If the real part of one of these eigenvalues vanishes, there is no guarantee that the dynamics of the linear system is equivalent to the dynamics of the nonlinear system close to the fixed point. Again, one can find explicit near-identity power series change of variables that simplify the original system. If $n_1\lambda_1 + n_2\lambda_2 + \cdots + n_N\lambda_N = \lambda_i$ for any i between 1 and N and for positive integers $n_1, n_2, \ldots, n_N$, then the eigenvalues are in resonance, and the ith new equation will contain some resonant nonlinear terms. This resonance relation is one of the most fundamental relations in nonlinear dynamics. It determines whether linear modes (determined by the eigenvectors of eigenvalues λ) are coupled by the nonlinear terms. It is the same resonance relation that appears in different guises in the analysis of forced linear systems and in the resonance between frequencies in Hamiltonian and celestial mechanics. Essentially, in the absence of resonances, the system evolves following the linear modes and no interaction is possible. When resonances occur, the linear modes interact through the nonlinear terms and create complex dynamics.

Normal forms theory provide a systematic way to include the effect of the nonlinear terms. At the practical level, the method as presented tends to be rather tedious because the number of monomials that have to be taken into account grows rapidly with the dimension of the system and the degree of the normalizing transformation. There are several equivalent alternatives to the computation of normal forms, including the method of "amplitude equation" for ordinary and differential equations and the Birkhoff–Gustavson transformation for Hamiltonian systems. Nevertheless, at the conceptual level, normal forms theory is a central tool to understand the rich dynamics of nonlinear systems and this general framework can be used to study and give a rigorous foundation to many other problems beside stability.

A particularly important use of normal form theory is the theory of bifurcations. In order to identify generic bifurcations, one considers the parameters of the system $\mu_1, \ldots, \mu_M$ as additional variables satisfying trivial differential equations $\dot{\mu}_i = 0,\ i = 1, \ldots, M$ and studies the normal forms of this extended system of dimension $N + M$, revealing the nature of the bifurcation. Other applications of normal forms include the analysis of chaotic systems in systems of differential equations, the formation of patterns for partial differential equations, exponentially small effects in the splitting of separatrices, and the Painlevé theory of integrable systems.

ALAIN GORIELY

See also **Bifurcations; Damped-driven anharmonic oscillator; Painlevé analysis**

Further Reading

Arnol'd, V.I. 1973. *Ordinary Differential Equations*, Cambridge, MA: MIT Press (originally published in Russian, 1971); 3rd edition, Berlin and New York: Springer, 1992

Guckenheimer, J. & Holmes, P. 1983. *Nonlinear Oscillations, Dynamical Systems and Bifurcations of Vector Fields*, New York: Springer

N-SOLITON FORMULAS

A one-soliton solution of an evolutionary nonlinear partial differential equation in one spatial dimension $u_t = F(u, u_x, u_{xx}, \ldots)$ is a solitary wave (pulse) of finite energy and momentum. As an example, the Korteweg–de Vries (KdV) one-soliton is the exact solution

$$u(x,t) = \frac{-2\kappa_1^2}{\cosh^2\left(\kappa_1(x-\xi_1) - 4\kappa_1^3 t\right)}, \qquad (1)$$

to the KdV equation

$$u_t = -u_{xxx} + 6uu_x. \qquad (2)$$

Note that Equation (1) has the shape of a negative well, moving at a speed that is proportional to the depth.

N-soliton solutions are characterized by the fact that, for $t \to -\infty$ and $t \to +\infty$, the shape of the solution is composed by a quasi-exact superposition of N non-interacting pulses of the form (1). This means that the shape of the solution for $t \to -\infty$ is a sequence of N wells, ordered according to their depth, the deepest being the leftmost. For $t \to +\infty$, the shape is also a sequence of wells, with now the deepest being the rightmost, each of them retaining the original shape. The solution pictorially describes a family of N individual pulses traveling at their own characteristic speeds, the faster ones catching up with the slower, and, after an interaction, reemerging with their individual shapes. The overall effect of the nonlinear interaction is a phase shift between the pulses (that is, a difference in the relative distances of the dips of the wells). It can be proved that, in the collision process of N solitons, the total phase shifts are the sum of phase shifts of two-soliton processes.

The presence of N-soliton solutions, for arbitrary N, is often regarded as the hallmark for integrability of a nonlinear PDE. Three main methods have been devised for constructing and studying N-soliton solutions:

- The inverse scattering method (ISM), based on the Gel'fand–Levitan–Marchenko (GLM) equations
- Hirota's formalism
- Bäcklund transformations

Here we present formulas making explicit reference to the ISM. For direct ways of finding N-soliton solutions, we refer to the entries on Hirota's method and on the Darboux and Bäcklund transformations.

In the ISM, the existence of N-soliton solutions is associated with the following mathematical feature.

A nonlinear integrable evolution equation (say, for one dependent variable u defined on the real line), is presented as a Lax (isospectral) equation for a linear operator $L(u)$,

$$\frac{\mathrm{d}}{\mathrm{d}t}L(u) = [B(u), L(u)], \tag{3}$$

where the bracket indicates a commutator ($[B, L] \equiv BL - LB$). The spectrum of $L(u)$, as an operator in $L^2(R)$, is the union of the point spectrum $\lambda_1, \lambda_2, \ldots$ and the continuous spectrum, consisting of the intervals $I_1 \cup I_2 \ldots$. An N-soliton solution of (3) is a solution associated with an initial data $u(x, 0)$ whose spectrum contains only a finite number N of points. Such initial data are often called reflectionless potentials.

Given N-soliton initial data, the direct spectral problem for $L_0 = L(u(x, 0))$, gives (in addition to the spectrum) normalization coefficients β_i of the eigenfunctions corresponding to the eigenvalues λ_i, $i = 1, \ldots, N$. The evolution equations of the parameters $\beta_i = \beta_i(t)$ are linear equations with coefficients that depend on the constants λ_i. Accordingly, an N-soliton solution describes $2N$-dimensional families of solutions parametrized by the point spectrum of the linear operator and by the normalization coefficients β_i.

Two Soliton Solutions for Korteweg–de Vries and Sine-Gordon Equations

For the KdV equation (2) let $\lambda_1 = -\kappa_1^2$ and $\lambda_2 = -\kappa_2^2 |\kappa_1| > |\kappa_2|$ be the points in the spectrum of the Lax operator $L = -\mathrm{d}^2/\mathrm{d}x^2 + u$, and let ξ_i be quantities associated with the normalization coefficients $\beta_i(0)$ of the eigenvector of L. In other words, κ_i, ξ_i is the scattering data of the reflectionless potential $u(x, 0)$. A two-soliton solution for KdV is

$$u(x, t) = \frac{2\kappa_1^2 (\kappa_1 - \kappa_2)(\kappa_1 + \kappa_2)\left(\cosh^2(\phi_2) + (\kappa_2/\kappa_1)^2 \cosh^2(\phi_1) - 1\right)}{(\kappa_2 \cosh(\phi_1)\cosh(\phi_2) - \kappa_1 \sinh(\phi_1)\sinh(\phi_2))^2},$$

where $\phi_i = \kappa_i (x - \xi_i) - 4\kappa_i{}^3 t$.

This is an instance of the phenomenon described above. Indeed, for $t \to -\infty$ the solution describes two pulses traveling in the positive x-direction, with the faster one (that is, the one pertaining to the data κ_1, ξ_1) being the leftmost one.

The solution describes the process whereby the fast soliton catches up to the slow one, and finally (for $t \to +\infty$), the two emerge. The shape of the individual pulses remains unchanged, the effect of the nonlinear interaction being a phase shift. The fast soliton experiences a positive phase shift of

$$\Delta\phi_1 = \frac{1}{\kappa_1} \log \left| \frac{\kappa_1 + \kappa_2}{\kappa_1 - \kappa_2} \right|,$$

while the slow soliton loses the same phase (see Figure 1).

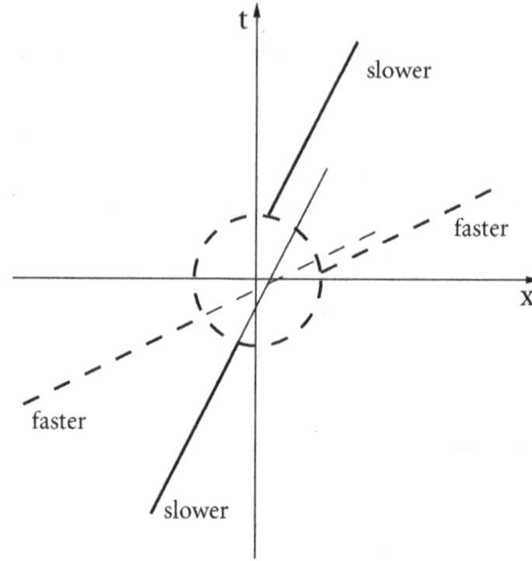

Figure 1. Phase shifts in two soliton collision.

The two-soliton solution of the sine-Gordon (SG) equation

$$\frac{\partial^2 u}{\partial^2 t} - \frac{\partial^2 u}{\partial^2 x} + \sin u = 0,$$

in terms of the points λ_1, λ_2 in the spectrum of the corresponding Lax operator and the constants c_1, c_2 is given by

$$u(x, t) = 4\operatorname{arctg} \frac{\frac{c_1}{2\lambda_1} \exp\left(-\frac{x - v_1 t}{\sqrt{1 - v_1^2}}\right) + \frac{c_2}{2\lambda_2} \exp\left(-\frac{x - v_2 t}{\sqrt{1 - v_2^2}}\right)}{1 - \frac{c_1 c_2}{4\lambda_1\lambda_2} \frac{(\lambda_1 - \lambda_2)^2}{(\lambda_1 + \lambda_2)^2} \exp\left(-\frac{x - v_1 t}{\sqrt{1 - v_1^2}} - \frac{x - v_2 t}{\sqrt{1 - v_2^2}}\right)},$$

where $v_i = (4\lambda_i^2 - 1)/(4\lambda_i^2 + 1)$, $i = 1, 2$. The fast soliton undergoes a positive phase shift given by

$$\Delta\phi_1 = \sqrt{1 - v_2^2} \log \frac{(\lambda_1 + \lambda_2)^2}{(\lambda_1 - \lambda_2)^2},$$

while the slow soliton undergoes a negative phase shift:

$$\Delta\phi_2 = -\sqrt{1 - v_1^2} \log \frac{(\lambda_1 + \lambda_2)^2}{(\lambda_1 - \lambda_2)^2}.$$

General Formulas

General formulas for N-soliton solutions of soliton equations are usually expressed in terms of determinants. For certain well-known soliton equations, one has the following results.

(1) Korteweg–de Vries (KdV) equation:

$$u_N(x, t) = -2\frac{\mathrm{d}}{\mathrm{d}x^2} \log \det(A),$$

where A is a $N \times N$ matrix with elements

$$A_{ij} = \delta_{ij} + \frac{\beta_i(t)}{\kappa_i + \kappa_j} \exp(-(\kappa_i + \kappa_j)x),$$

where $\beta_i(t) = \beta_i(0)\exp(8\kappa_i^3 t)$.

(2) Sine-Gordon (SG) equation:

$$u_N(x,t) = 2\mathrm{i}\log\frac{\det(A_+)}{\det A_-},$$

where

$$[A_\pm]_{ij} = \delta_{ij} \pm \frac{c_j}{\lambda_i + \lambda_j}\exp(\gamma_j(x,t)),$$

with $\gamma_j(x,t) = -i(\lambda_j + \frac{1}{4\lambda_j})t + i(\lambda_j - \frac{1}{4\lambda_j})x$.

(3) Nonlinear Schrödinger (NLS) equation:

$$\mathrm{i}\frac{\partial\psi}{\partial t} + \frac{\partial^2\psi}{\partial^2 x} - 2g|\psi|^2\psi, \quad g = \pm 1.$$

The N-soliton solutions are given by

$$\psi(x,t) = \frac{\mathrm{i}}{\sqrt{g}}\frac{\det M_1(x,t)}{\det M(x,t)},$$

where

$$M_{ij}(x,t) = \frac{1 + \bar{\gamma}_i(x,t)\gamma_j(x,t)}{\bar{\lambda}_i - \lambda_j}$$

and the $N+1 \times N+1$ matrix $M_1(x,t)$ is described as follows:

$$M_1 = \left[\begin{array}{ccc|c} & M & & G \\ \hline 1 & \ldots & 1 & 0 \end{array}\right]$$

where G is the vector $(\gamma_1(x,t), \ldots, \gamma_N(x,t))$, with $\gamma_i(x,t) = \gamma_i^{(0)}\exp(\mathrm{i}\lambda_i x - \mathrm{i}\lambda_i^2 t)$.

(4) Infinite Toda Lattice (ITL): The ITL equations are compactly written in the Lax form:

$$\frac{\mathrm{d}L}{\mathrm{d}t} = [B, L],$$

where L is the infinite matrix with b_i on the principal diagonal and a_i on the first lower and upper diagonal, and B is the antisymmetric matrix having a_i on the first lower diagonal; the coefficients $\{a_i, b_i\}$ are related to the canonical variables $\{Q_i, P_i\}$ by the transformation

$$a_i = \tfrac{1}{2}\exp(-(Q_{i+1} - Q_i)/2), \quad b_i = \tfrac{1}{2}P_i.$$

N-soliton solutions are given by the formula:

$$\exp(-(Q_i - Q_{i-1})) = \frac{\det B_i \det B_{i-2}}{(\det B_{i-1})^2},$$

$$P_i = \text{const.} + \frac{\mathrm{d}}{\mathrm{d}t}\log\frac{\det B_{i-1}}{\det B_i}.$$

In this formula the B_i's are $N \times N$ matrices with elements given by

$$[B_i]_{j,k} = \delta_{j,k} + \gamma_j(t)\gamma_k(t)\frac{(\lambda_j\lambda_k)^{i+1}}{1 - \lambda_j\lambda_k}, \quad j,k = 1,\ldots,N.$$

The λ_j are the N elements of the point spectrum of L, and the γ_j are normalization constants; the time dependence of the solution is given by

$$\gamma_j(t) = \gamma_j(0)\exp(\pm\sinh(\sigma_j)t), \quad \lambda_j = \pm\exp(-\sigma_j).$$

Further Comments

The formulas above can be interpreted as nonlinear superposition formulas for soliton solutions. Indeed, N-soliton solutions are constructed via the definition of suitable matrices determined by the spectral data associated with individual solitons. More precisely, the notion of nonlinear superposition principle can be better understood in the framework of the theory of Darboux–Bäcklund transformations.

In general, N-soliton solutions can be obtained as suitable limits of periodic solutions of the same system as the period tends to infinity. Periodic finite gap solutions are obtained by means of theta functions defined over Riemann surfaces. N-soliton solutions are obtained by letting suitable homology cycles of such Riemann surfaces shrink to a (degenerate) surface (with nodes and cusps) of genus equal to zero.

GREGORIO FALQUI AND TAMARA GRAVA

See also **Bäcklund transformations; Darboux transformation; Hirota's method; Inverse scattering method or transform; Theta functions**

Further Reading

Ablowitz, M.J. & Segur, H. 1981. *Solitons and the Inverse Scattering Transform*, Philadelphia: SIAM

Calogero, F. & Degasperis, A. 1982. *Spectral Transforms and Solitons: I*, Amsterdam: North-Holland

Drazin, P.G. & Johnson, R.S. 1989. *Solitons: An Introduction*, Cambridge and New York: Cambridge University Press

Faddeev, L.D. & Takhtajan, L.A. 1987. *Hamiltonian Methods in the Theory of Solitons*, Berlin: Springer

Miura, R.M. (editor). 1976. *Bäcklund Transformations*, Berlin: Springer

Newell, A.C. 1985. *Solitons in Mathematics and Physics*, Philadelphia: SIAM

Novikov, S.P., Manakov, S.V., Pitaevskii, L.P. & Zakharov, V.E. 1984. *Theory of Solitons*, New York: Consultants Bureau

Toda, M. 1988. *Theory of Nonlinear Lattices*, 2nd edition, Berlin: Springer

N-TORI

See **Quasiperiodicity**

NUMERICAL METHODS

Numerical methods are used for approximations of "exact" mathematical solutions, which either do not exist or are very complicated. With fast computer processors and advanced mathematical software, many problems become more attractive for numerical rather than for analytical solutions.

Computational Errors

Numerical computations are always uncertain because the results are only defined within the accuracy of a numerical error. Two main reasons account for errors in numerical approximations: finite precision in computer representation of numbers, and truncation of exact mathematical formulas in numerical algorithms.

Finite precision defines the tolerance interval, within which any further improvement in a numerical solution is impossible. For instance, the precision accuracy of MATLAB on the Windows platform is order of 10^{-16}. Therefore, it is meaningless to initialize irrational numbers (such as π or e) beyond the 16th digit after the period while carrying computations in MATLAB.

Round-off errors can be obstacles for accurate numerical approximations. Numerical differentiation algorithms are typically ill-posed since the round-off error increases with smaller step size of numerical discretization. Catastrophic cancellations can occur due to round-off errors, as in the example below when two large nearly identical numbers are subtracted:

$$f(x) = \sqrt{x+1} - \sqrt{x}.$$

If $x = 1.000000000000000 \times 10^{10}$, then $\sqrt{x+1} = 1.000000000050000 \times 10^5$, $\sqrt{x} = 1.000000000000000 \times 10^5$ and

$$f(x) = 4.999994416721165 \times 10^{-6}.$$

However, if the relative round-off error is order of 10^{-10}, the result is identical to zero. If the precision accuracy cannot be extended, a modification of the numerical procedure is required for a more accurate answer, as in the equivalent representation:

$$f(x) = \frac{1}{\sqrt{x+1} + \sqrt{x}}.$$

Within the same precision accuracy, the result is now $f(x) \approx 4.9999999999 \times 10^{-6}$.

Truncation errors occur due to chopping of an infinite series into a finite number of terms. For example, the Taylor series for analytical functions can be truncated with the Taylor polynomials, as in the example:

$$e^{x^2} = 1 + x^2 + \frac{1}{2!}x^4 + \frac{1}{3!}x^6 + E(x),$$

where $E(x)$ is the truncation error. The integral of e^{x^2} on $x \in [0, 1]$ is given in terms of a special function, called the error function. Equivalently, this integral can be approximated with the Taylor polynomial as

$$\int_0^1 e^{x^2}\, dx = \left(x + \frac{1}{3}x^3 + \frac{1}{10}x^5 + \frac{1}{42}x^7 \right)\Bigg|_{x=0}^{x=1} + E_{tr}$$
$$= 1.457142857142857 + E_{tr},$$

where the truncation error is $E_{tr} \approx 5.508914448636659 \times 10^{-3}$.

Numerical algorithms are classified by the rate of convergence and by their numerical stability (i.e., will a small error decay or grow through the successive iterations). Convergence and stability of numerical algorithms are studied with analysis of the truncation error. If the truncation error depends on the step size h of the finite numerical discretization and reduces as h^n, then the numerical algorithm is said to converge to an exact solution as the nth-order algorithm. As an example, we consider a numerical computation of the integral, given by irrational number e:

$$e = \int_0^1 f(x)\, dx, \qquad f(x) = 1 + e^x.$$

Using the discretization of the unit interval with N equal subintervals of the step size $h = 1/N$, we can use the composite trapezoidal rule to approximate the integral as

$$e = \frac{h}{2}\left(f(0) + 2\sum_{n=1}^{N-1} f(hn) + f(1) \right) + E(h),$$

where the theoretical truncation error is $E(h) = \alpha_2 h^2$ and α_2 is constant, such that $\lim_{h\to 0} \alpha_2 \neq 0$. The composite trapezoidal rule converges to the integral as the second-order method. Indeed, computing the algorithm for $N = 100$ and 200, we have

$$N = 100: \quad e \approx 2.718296147450418,$$
$$E(0.01) = 1.431899137260828 \times 10^{-5}$$

and

$$N = 200: \quad e \approx 2.718285408211362,$$
$$E(0.005) = 3.579752316795748 \times 10^{-6},$$

and, therefore,

$$\frac{E(0.01)}{E(0.005)} = 3.999995001169598 \approx 2^2 = 4,$$

as predicted by the theoretical formula. An improved numerical algorithm for integration is the composite Simpson's rule, defined as

$$e = \frac{h}{3}\left(f(0) + 4\sum_{n=1}^{N/2} f(h(2n-1)) + 2\sum_{n=1}^{N/2-1} f(2hn) + f(1) \right) + E(h),$$

where the theoretical truncation error is $E(h) = \alpha_4 h^4$ and N is even. The composite Simpson's rule converges to the integral as the fourth-order method. Indeed, computing the algorithm for $N = 100$ and 200, we have

$$N = 100: \quad e \approx 2.718281828554504,$$
$$E(0.01) = 9.545830792490051 \times 10^{-11}$$

and

$$N = 200: \quad e \approx 2.718281828465013,$$
$$E(0.005) = 5.967226712755291 \times 10^{-12}$$

and, therefore,

$$\frac{E(0.01)}{E(0.005)} = 15.99709756642108 \approx 2^4 = 16,$$

as predicted by the theoretical formula.

Iteration Methods

Solutions of algebraic, differential, partial differential, and integral equations can be approximated with iteration methods. Numerical errors in iteration methods may grow during iterations. When this happens, numerical approximations cannot give the exact solution because instabilities lead to huge numerical errors. If the numerical iterations are unstable, it does not matter how accurately the truncation error converges to zero with smaller discretization step size. Propagation and growth of numerical errors can be illustrated with the linear iteration map:

$$x_{n+1} = q x_n, \quad |q| \geq 1.$$

The exact solution of the linear iteration map is $x_n = q^n x_0$, where x_0 is the starting value. If two iteration sequences are obtained from almost identical values $x_0^{(1)}$ and $x_0^{(2)}$, so the initial distance $e_0 = |x_0^{(1)} - x_0^{(2)}|$ is small, then the distance grows with larger n as $e_n = |q|^n e_0$. Two iteration sequences diverge from each other, even if the small error e_0 was generated by the round-off error!

Numerical instabilities and divergences often occur in solutions of algebraic equations, which represent the simplest numerical problems. A scalar algebraic equation $f(x) = 0$ can be reformulated as a root finding algorithm:

$$x_*: \quad f(x_*) = 0.$$

If we plot $f(x)$ as a function of x, the root x_* can be immediately found from the graph $y = f(x)$, if it exists. Algorithmically, we can try to approximate the root x_* from the iteration method:

$$x_{n+1} = x_n + f(x_n).$$

If the limit $x_\infty = \lim_{n\to\infty} x_n$ exists, then the root x_* is a fixed point of the iteration method such that $x_\infty = x_*$. If the limit x_∞ does not exist but x_* is known to exist, the numerical method fails and iterations $\{x_n\}_{n=1}^{\infty}$ diverge due to numerical instabilities. Analysis of convergence of the iteration method can be performed with the linearization of the iteration method. Let $e_n = x_n - x_*$ be the small distance of x_n from the fixed point x_* such that

$$\begin{aligned} e_{n+1} &= x_n + f(x_n) - x_* = e_n + f(x_* + e_n) - f(x_*) \\ &= q e_n + \alpha_n e_n^2, \quad q = 1 + f'(x_*), \end{aligned}$$

where α_n is constant, such that $\lim_{n\to\infty} \alpha_n$ exists. If $|q| \geq 1$, the error e_n grows with larger n and iterations $\{x_n\}_{n=1}^{\infty}$ and $\{e_n\}_{n=1}^{\infty}$ diverge. In this case, the fixed point x_* cannot be found from the iteration method. For example, when $f(x) = 3x - 6$, the fixed point $x_* = 2$ cannot be found iteratively, since $q = 4 > 1$.

An improved iteration algorithm is Newton's method which approximates a new point x_{n+1} from the root of the tangent line to the graph $y = f(x)$ at the previous point x_n. Thus

$$0 = f(x_n) + f'(x_n)(x_{n+1} - x_n) + \mathrm{O}(x_{n+1} - x_n)^2,$$

such that

$$x_{n+1} = x_n - \frac{f(x_n)}{f'(x_n)}.$$

Newton's method always converges to a fixed point x_*, if the fixed point x_* exists. The rate of convergence depends on whether the root $x = x_*$ is simple or multiple. For a single root when $f'(x_*) \neq 0$, the rate of convergence is quadratic in terms of the error e_n, since

$$q = \lim_{x\to x_*} \frac{f(x)f''(x)}{(f'(x))^2} = 0.$$

For a multiple root when $f(x) \sim (x - x_*)^m$ and $m > 1$, the rate of convergence is linear in terms of e_n, since $q = (m-1)/m$. When the function $f(x)$ is linear as in the example $f(x) = 3x - 6$, the convergence of the Newton–Raphson method occurs in a single iteration, no matter what x_0 is

$$x_1 = x_0 - \frac{3x_0 - 6}{3} = 2 = x_*.$$

Differential Equations

Numerical algorithms are particularly attractive for solutions of differential equations, which are widely used in all areas of physics and engineering. It is especially important because many applied differential equations cannot be solved in terms of analytical functions. The simplest differential equation is given by a scalar first-order quasilinear equation:

$$\frac{\mathrm{d}y}{\mathrm{d}t} = f(t, y),$$

starting with the initial condition: $y(0) = y_0$. Since $y(0) = y_0$ and $y'(0) = f(0, y_0)$ are known, it makes sense to step from $t = 0$ along the tangent line to the curve $y = y(t)$ at the point $(0, y_0)$. Using a number of steps from $t = 0$ to T with small step size h, we define a numerical approximation of $y(t_n)$ as y_n, such that $y(t_n) = y_n + e_n$, where e_n is the numerical error, accumulated after n steps. If all steps are taken alone the tangent line segments to the points (t_n, y_n), we have the Euler method (also known as the slope approximation):

$$y_{n+1} = y_n + hf(t_n, y_n).$$

When the time $t = T$ is reached in N steps, such that $T = Nh$, the global error is defined as $E_T = |y(T) - y_N|$. The global error for the Euler method is theoretically given by $E_T = \alpha_1 h$, such that the Euler method is the least accurate, first-order method. Figure 1 shows that oscillations of the undamped pendulum in the nonlinear second-order equation:

$$y'' + \sin y = 0$$

are destroyed in Euler's method. The inaccurate Euler's method introduces an effective damping due to numerical discretization.

Numerical methods for differential equations also produce numerical instabilities. For example, the linear first-order equation for exponential decay:

$$\frac{dy}{dt} = -\lambda y, \quad \lambda > 0,$$

has a simple solution: $y(t) = y_0 e^{-\lambda t}$. Euler's method applied to this equation is equivalent to the linear iteration map:

$$y_{n+1} = (1 - \lambda h) y_n,$$

which diverges for $h > 2/\lambda$, when $|1 - \lambda h| > 1$. When $\lambda \to 0$, the differential equation for the exponential decay becomes stiff for numerical solution, since the step size h for numerical discretization must be very small to preserve stability of numerical approximation.

Advanced algorithms for numerical solutions of differential equations are developed with numerical integration methods. For example, integrating the first-order quasi-linear equation with the trapezoidal rule on the interval $t \in [t_n, t_{n+1}]$, we derive the implicit iteration method:

$$y_{n+1} = y_n + \frac{h}{2}\left[f(t_n, y_n) + f(t_{n+1}, y_{n+1})\right].$$

The global; truncation method of the method is $E_T = \alpha_2 h^2$, that is, it is the second-order method. The method is also implicit since the unknown values of y_{n+1} appear at the left and right sides of the equation. The implicit methods can be solved with iterations at each n, by means of root finding

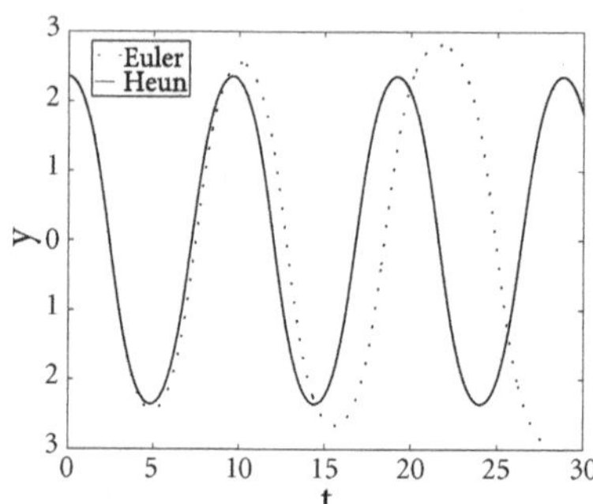

Figure 1. Oscillations of the undamped pendulum in Euler's and Heun's numerical methods.

algorithms. However, this modification makes the algorithm complicated and time-consuming. Predictor-corrector methods are used instead, either with single-step (Runge–Kutta) methods or with multi-step (Adams–Bashforth–Moulton) methods. For example, the second-order Runge–Kutta method (also known as the Heun's or improved Euler's method) is based on the trapezoidal rule in the form

$$p_{n+1} = y_n + hf(t_n, y_n),$$
$$y_{n+1} = y_n + \frac{h}{2}\left[f(t_n, y_n) + f(t_{n+1}, p_{n+1})\right],$$

where p_{n+1} is the predicted value at $t = t_{n+1}$ by using the slope approximation and y_{n+1} is the corrected value at $t = t_{n+1}$ by using the trapezoidal rule. The predictor–corrector method above has the same global truncation error $E_T = \tilde{\alpha}_2 h^2$ with a different numerical constant $\tilde{\alpha}_2$. Figure 1 shows that oscillations of the undamped pendulum are well preserved in the Heun's method based on predictions and corrections.

The most popular and accurate method is the fourth-order Runge–Kutta method, which takes four computations of the function $f(t, y)$ for a single step from $t = t_n$ to $t = t_{n+1}$ and has the global truncation error $E_T = \alpha_4 h^4$.

Higher-order predictor-corrector methods are still affected by numerical instabilities, especially in the case of stiff differential equations. More reliable methods for stiff problems are based on implicit integration formulas. Implicit methods can be rewritten as explicit methods if the function $f(t, y)$ is linear in y. For example, the equation for the exponential decay can be integrated with the implicit second-order method based on the trapezoidal rule, as follows:

$$y_{n+1} = \frac{2 - h\lambda}{2 + h\lambda} y_n.$$

Since $|(2 - h\lambda)/(2 + h\lambda)| \le 1$ for any h and $\lambda > 0$, the implicit method is unconditionally stable, no matter how large the step size h is chosen. The accuracy of the method is still the same as in the second-order Heun's method, so the step size h must not be too large, to preserve accuracy of numerical approximation.

Partial Differential Equations

Partial differential equations are defined in two- and higher-dimensional domains. They represent most complicated and time-consuming problems for numerical methods. Initial values for partial differential equations at time $t=0$ are supplemented by boundary values at the boundaries of a physical domain. A simple example of a linear partial differential equation is given by the heat equation, derived for a one-dimensional rod of finite length L:

$$\frac{\partial u}{\partial t} = \frac{\partial^2 u}{\partial x^2} + f(x,t), \quad 0 < x < L, \quad t > 0,$$

such that

$$u(0,t) = u_0(t), \quad u(L,t) = u_L(t), \quad u(x,0) = \phi(x).$$

The heat equation describes temperature $u(x,t)$ as function of time $t \geq 0$ and the point $0 \leq x \leq L$ in the rod. The heat sources $f(x,t)$, temperature values at the end points $u_0(t)$ and $u_L(t)$ and initial temperature $\phi(x)$ are all given as physical parameters of the problem. In the finite-difference methods, numerical approximations of $u(x,t)$ are defined at the equally spaced numerical grid with points at $x_n = nh$, $n = 0, 1, \ldots, N$, where h is the step size, such that $L = Nh$. Numerical approximations of $u(x,t)$ are also evaluated with equal time step τ at points $t_k = k\tau$, $k = 0, 1, \ldots$. Denoting $u_{n,k}$ as numerical approximation of $u(x_n, t_k)$, we approximate the second x-derivative of $u(x,t)$ by the central difference, derived from Taylor series expansions:

$$\begin{aligned} u_{n+1,k} &= u_{n,k} + hu_x(x_n,t_k) + \frac{1}{2}h^2 u_{xx}(x_n,t_k) \\ &\quad + \frac{1}{3!}h^3 u_{xxx}(x_n,t_k) + \mathrm{O}(h^4), \\ u_{n-1,k} &= u_{n,k} - hu_x(x_n,t_k) + \frac{1}{2}h^2 u_{xx}(x_n,t_k) \\ &\quad - \frac{1}{3!}h^3 u_{xxx}(x_n,t_k) + \mathrm{O}(h^4), \end{aligned}$$

such that

$$\frac{\partial^2 u}{\partial x^2}(x_n,t_k) = \frac{u_{n+1,k} - 2u_{n,k} + u_{n-1,k}}{h^2} + \mathrm{O}(h^2).$$

Using the slope approximation, we perform the time step from $u_{n,k}$ to $u_{n,k+1}$ according to the explicit finite-difference scheme:

$$u_{n,k+1} = (1-2r)u_{n,k} + r(u_{n+1,k} + u_{n-1,k}) + \tau f_{n,k},$$

where $r = \tau/h^2$, $n = 1, \ldots, N-1$, and $k = 0, 1, \ldots$. All boundary and initial values $u_{0,k}$, $u_{N,k}$ and $u_{n,0}$ are incorporated in computations of the explicit method for $n = 1, N-1$, and $k = 0$. Numerical approximations of $u_{n,k}$ are only computed at internal points of the grid. The total error of the explicit method is a composition of the truncation error of order $\mathrm{O}(h^2)$ for the central difference approximation and the truncation error of order $\mathrm{O}(\tau)$ for the Euler's method. The explicit method is least accurate with respect to time step size τ. It is also an unstable method for $r > 0.5$ when $\tau > h^2/2$. Figure 2 shows numerical solution of the heat equation with the explicit method for $f(x,t) = 0$, $u_0 = u_L = 0$, $\phi = \sin(\pi x)$, $L = 2$, $h = 0.1$, and $r = 0.55$. Development of instabilities of the explicit method destroys validity of the numerical approximations.

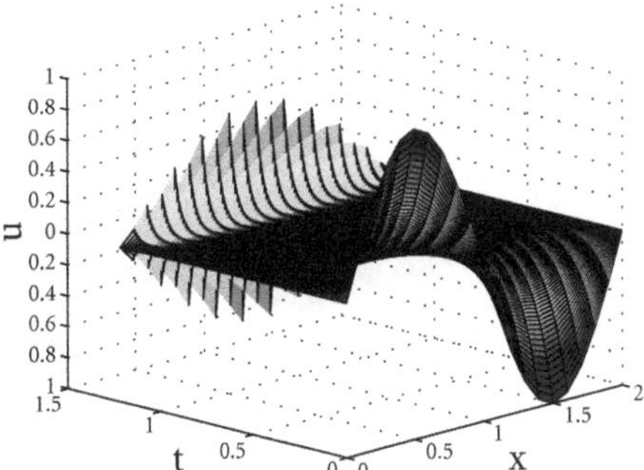

Figure 2. Solutions of the heat equation with the explicit method.

Implicit methods are more reliable in numerical computations. The implicit method, which is based on the trapezoidal rule of integration, is referred to as the Crank–Nicholson method. The method results in a linear system of equations:

$$\begin{aligned} A(r)\mathbf{u}_{k+1} &= A(-r)\mathbf{u}_k + \frac{\tau}{2}(\mathbf{f}_k + \mathbf{f}_{k+1}) \\ &\quad + \frac{r}{2}(\mathbf{b}_k + \mathbf{b}_{k+1}), \qquad k \geq 0, \end{aligned}$$

where

$$\mathbf{u}_k = \begin{bmatrix} u_{1,k} \\ u_{2,k} \\ \vdots \\ u_{N-2,k} \\ u_{N-1,k} \end{bmatrix}, \quad \mathbf{f}_k = \begin{bmatrix} f_{1,k} \\ f_{2,k} \\ \vdots \\ f_{N-2,k} \\ f_{N-1,k} \end{bmatrix},$$

$$\mathbf{b}_k = \begin{bmatrix} u_{0,k} \\ 0 \\ \vdots \\ 0 \\ u_{N,k} \end{bmatrix}$$

and

$$A(r) = \begin{bmatrix} 1+r & -r/2 & 0 & \cdots & 0 \\ -r/2 & 1+r & -r/2 & \cdots & 0 \\ 0 & -r/2 & 1+r & \cdots & 0 \\ \vdots & \vdots & \vdots & \vdots & \vdots \\ 0 & 0 & 0 & \cdots & 1+r \end{bmatrix}.$$

Although solving of the linear system at each time step is a time-consuming operation, the Crank–Nicholson

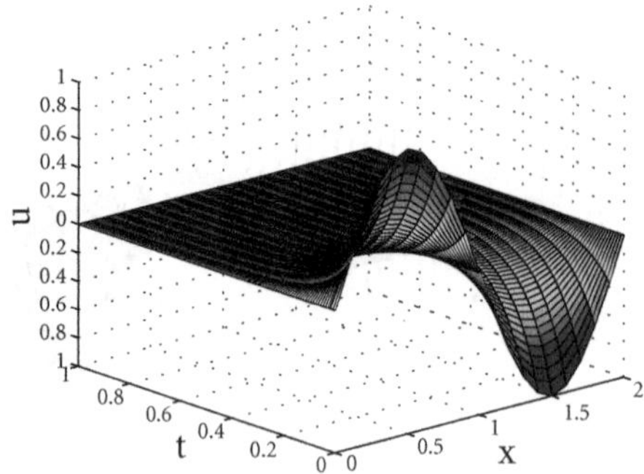

Figure 3. Solutions of the heat equation with the Crank–Nicholson method.

method is more useful compared with the explicit method. In particular, the method has a symmetric numerical error $O(\tau^2 + h^2)$ and unconditional numerical stability for any $\tau > 0$. Figure 3 shows numerical solution of the heat equation with the Crank–Nicholson method for $h = 0.1$ and $r = 1$ when $\tau = h^2$. The numerical solution is stable in time evolution.

Finite-difference methods are used successfully in numerical solution of other boundary-value problems for partial differential equations. Explicit methods have a very straightforward form. Implicit methods are unconditionally stable and result in linear systems for linear differential equations. Implicit methods are usually replaced by the semi-implicit algorithms for nonlinear differential equations. As a drawback, finite-difference methods have low accuracy. More advanced (and also more involved) shooting, finite-element, and spectral methods are used for improved solutions of boundary-value problems.

Shooting methods are based on transformations of the boundary-value problems into the initial-value problems, which are iterated with the root finding algorithms. For simplicity, we illustrate with ordinary differential equations, although multi-dimensional shooting methods are also applied to partial differential equations. We consider a numerical approximation of radially symmetrical bound states of the two-dimensional nonlinear Schrödinger equations, which satisfy the boundary-value problem:

$$\frac{d^2\Phi}{dr^2} + \frac{1}{r}\frac{d\Phi}{dr} - \Phi + 2\Phi^3 = 0, \quad 0 < r < R,$$

such that

$$\Phi'(0) = 0, \quad \Phi(R) = 0,$$

where R is large length of the approximation interval. Using a shooting numerical method, we consider a solution of the initial-value problem for the same equation $\Phi = \Phi_m(r)$, such that

$$\Phi_m(0) = m, \quad \Phi'_m(0) = 0,$$

where the value of m is unknown. Solving the differential equation on $r \in [0, R]$ with an appropriate

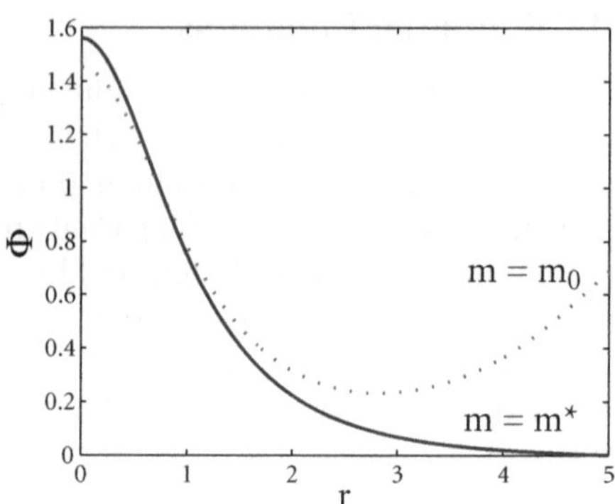

Figure 4. Bound state of the nonlinear Schrödinger equation with the numerical shooting method: starting (dotted) and final (solid) approximations.

numerical method, such as the fourth-order Runge–Kutta method, we define the function of parameter m as

$$E(m) = \Phi_m(R).$$

The function $E(m)$ has a zero at $m = m_*$ if $\Phi_{m_*}(r)$ is the bound state of the nonlinear Schrödinger equation, such that $\Phi_{m_*}(R) = 0$. Starting with any value m_0, we can approach to the zero of $E(m)$ with a root finding algorithm, such as the Newton's method:

$$m_{k+1} = m_k - \frac{E(m_k)}{E'(m_k)},$$

where the value of

$$E'(m_k) = \left.\frac{\partial \Phi_m(R)}{\partial m}\right|_{m=m_k}$$

can be found from a linearization problem or from a secant method. The starting approximation $\Phi_{m_0}(r)$ and the final solution $\Phi_{m_*}(r)$ are shown on Figure 4 by dotted and solid lines, respectively. The value of $E(m_k)$ converges to zero, such that the sequence $\Phi_{m_k}(r)$ converges to the bound state $\Phi(r)$ of the nonlinear Schrödinger equation.

Finite-element methods approximate solutions of the boundary-value problems from a variational problem, which involves minimization of the energy functionals. The bound state $\Phi(r)$ of the nonlinear Schrödinger equation coincides with a minimizer of the energy functional:

$$H(u) = \frac{1}{2}\int_0^R \left[\left(\frac{du}{dr}\right)^2 - u^4\right] r\, dr,$$

such that the variation of $H(u)$ vanishes if and only if $u = \Phi(r)$:

$$\left.\frac{\delta H}{\delta u}\right|_{u=\Phi(r)} = -\frac{d}{dr}\left[r\frac{d\Phi}{dr}\right] - 2r\Phi^3 = 0.$$

Dividing the interval $r \in [0, R]$ into n subintervals $[0, r_1] \cup [r_1, r_2] \cup \cdots \cup [r_{N-1}, r_N]$, where $r_n = hn$ and $R = hN$, we approximate $\Phi(r)$ by a linear combination of the finite elements $u_n(r)$:

$$u(r) = \sum_{n=0}^{N} \Phi_n u_n(r).$$

The finite elements $u_n(r)$ satisfy the constraints:

$$u_n(r_n) = 1, \quad \text{and} \quad u_n(r_m) = 0, \; m \neq n,$$

such that Φ_n are unknown values for the finite-element approximation of $u(r_n)$. The simplest finite elements (also known as the simplex elements) are given by the piecewise linear interpolation:

$$u_n(r) = \begin{cases} 0, & r \leq r_{n-1}, \\ \frac{r - r_{n-1}}{h}, & r_{n-1} \leq r \leq r_n, \\ \frac{r_{n+1} - r}{h}, & r_n \leq r \leq r_{n+1}, \\ 0, & r \geq r_{n+1}. \end{cases}$$

When $u(r)$ is substituted into the energy functional $H(u)$ and integrated numerically over $0 \leq r \leq R$, the function $H(u) = H(\Phi_0, \Phi_1, \ldots, \Phi_{N-1})$ with $\Phi_N = 0$ can be minimized with respect to parameters $\Phi_0, \Phi_1, \ldots, \Phi_{N-1}$. The minimization leads to a system of nonlinear algebraic equations:

$$\frac{\partial H(u)}{\partial \Phi_n} \equiv h_n(\Phi_0, \Phi_1, \ldots, \Phi_{N-1}) = 0, \quad n = 0, 1, \ldots, N-1.$$

This nonlinear system can be solved with a matrix root finding Newton's method, which completes the finite-element approximation of the bound states $\Phi(r)$. When the differential equations are linear and the exact integrations are replaced with the trapezoidal rule, the finite-element method with simplex elements recovers the finite-difference approximation. However, the finite-element method is more general and suitable for accurate numerical approximations. For example, the finite-element method can be used together with the Simpson's rule of numerical integration.

Spectral methods are based on Fourier series solutions of differential equations in finite intervals, subject to appropriate boundary conditions. Solutions of the homogeneous heat equation $u_t = u_{xx}$ on $x \in [0, L]$ with zero boundary conditions at the ends: $u(0, t) = u(L, t) = 0$ are given by the Fourier sine-series:

$$u(x, t) = \sum_{m=1}^{\infty} B_m(t) \sin \frac{\pi m x}{L},$$

where the set of Fourier amplitudes $B_m(t), m \geq 1$ solves the initial-value problems:

$$\frac{\mathrm{d}B_m}{\mathrm{d}t} = -\frac{\pi^2 m^2}{L^2} B_m, \quad m \geq 1.$$

The initial values of $B_m(0)$ are defined from $u(x, 0)$ by the Fourier integrals:

$$B_m(0) = \frac{2}{L} \int_0^L u(x, 0) \sin \frac{\pi m x}{L} \, \mathrm{d}x.$$

Truncating the Fourier series with the Galerkin method and defining the Fourier approximation at grid points $x_n = nh, n = 1, \ldots, N-1$, we replace the exact Fourier series solution with the discrete Fourier transform:

$$u(x_n, t) = \sum_{m=1}^{N-1} B_m(t) \sin \frac{\pi m n}{N},$$

where $B_m(t)$ solves the same initial-value problem for $m = 1, \ldots, N-1$ but $B_m(0)$ is defined by the inverse discrete Fourier transform:

$$B_m(0) = \frac{2}{N} \sum_{n=1}^{N-1} u(x_n, 0) \sin \frac{\pi m n}{N}.$$

We notice that the inverse discrete Fourier transform follows from Fourier integrals by means of the numerical trapezoidal rule. Solving systems of initial-value problems with Runge–Kutta methods, we define solutions of the heat equation at time steps $t_k = k\tau$, $k = 0, 1, \ldots$.

Spectral methods give more accurate solutions of the heat equation, compared with the finite-difference methods. First, all initial-value problems for Fourier coefficients $B_m(t)$ are uncoupled. Second, the spectral methods have superior accuracy, because the truncation error of the Galerkin approximation decays exponentially with larger number N of the Fourier amplitudes. Third, spectral methods can be applied to nonlinear differential equations. Performing time steps with the Runge–Kutta methods, discrete Fourier transform and its inverse can be employed for computations of the nonlinear terms. As a result, the spectral method is effectively uncoupled for Fourier amplitudes $B_n(t)$ as an explicit single-step method. Split-step spectral methods are especially important for nonlinear evolution equations, such as the nonlinear Schrödinger equations.

As computational power increases, the role of numerical methods will be increasingly important in coming years for efficient solutions of many problems in physical and engineering sciences. Presently, computations of three-dimensional equations for water waves, turbulence, and astrophysical problems can be run on workstations in reasonable time. Modern supercomputing resources based on multiprocessor clusters are designed to help scientists in their theoretical studies of nature. It is common nowadays that many research groups in physics and engineering replace analysis of a problem by numerical methods.

DMITRY PELINOVSKY

See also **Averaging methods; Compartmental models; Extremum principles; Lattice gas methods; Stability**

Further Reading

Akai, T.J. 1994. *Applied Numerical Methods for Engineers*, New York: Wiley

Chilling, R.J. & Harris, S.L. 2000. *Applied Numerical Methods for Engineers using MATLAB and C*, Pacific Grove, CA: Brooks/Cole
Fausett, L.V. 2003. *Numerical Methods: Algorithms and Applications*, Upper Saddle River, NJ: Prentice-Hall
Gerald, C.F. & Wheatley, P.O. 1999. *Applied Numerical Analysis*, 6th edition, Reading, MA: Addison-Wesley
Rao, S.S. 2002. *Applied Numerical Methods for Engineers and Scientists*, Upper Saddle River, NJ: Prentice-Hall

N-WAVE INTERACTIONS

Linear dispersive waves propagate according to a dispersion relation

$$\omega = \omega(\boldsymbol{k}),$$

which relates the frequency ω to the wave vector $\boldsymbol{k}$. In linear systems, waves of different wave vectors and frequencies do not interact due to the superposition principle. In nonlinear systems, the lowest-order nonlinear effect (expanding in the wave amplitudes) is the interaction between N waves of different wave vectors $\boldsymbol{k}_j$ and frequencies ω_j, which satisfy the spatial and temporal resonance conditions (Zaslavsky & Sagdeev, 1988)

$$\sum_{j=1}^{N} \boldsymbol{k}_j = \boldsymbol{0}, \quad \sum_{j=1}^{N} \omega(\boldsymbol{k}_j) = 0.$$

The simplest three-wave resonant interaction occurs if there are nontrivial solutions with

$$\boldsymbol{k} = \boldsymbol{k}_1 + \boldsymbol{k}_2, \quad \omega(\boldsymbol{k}) = \omega(\boldsymbol{k}_1) + \omega(\boldsymbol{k}_2). \qquad (1)$$

Resonant wave interactions are governed by the dispersion relation. Some dispersion relations do not exhibit nontrivial solutions of the resonance conditions (1) for any $\boldsymbol{k}_1$ and $\boldsymbol{k}_2$. Others may exhibit resonant three-wave interactions, for example the deep-water gravity-capillary waves satisfying the dispersion relation $\omega^2(\mathbf{k}) = g|\mathbf{k}| + T|\boldsymbol{k}|^3$, where g is the acceleration of gravity and T is the surface tension coefficient. When $T \neq 0$, the resonant configurations of the three waves occur already in one spatial dimension, as shown on a graphical solution on Figure 1. The solid curve on this figure shows the dispersion relation $\omega = \omega(k)$, while the dashed curve shows the same dispersion relation shifted relatively to the point (k_1, ω_1), such that $\omega - \omega_1 = \omega(k - k_1)$ and $\omega_1 = \omega(k_1)$. The intersection of the two curves defines a solution of the resonance equations (1) at the point $k = k_1 + k_2$, where $\omega(k) - \omega(k_1) = \omega(k_2)$.

When $T = 0$, the resonant three-wave configurations do not occur in space of any dimensions, but deep-water gravity waves satisfying the dispersion relation $\omega^2 = g|\boldsymbol{k}|$ exhibit the resonant four-wave interactions of the following type:

$$\boldsymbol{k}_1 + \boldsymbol{k}_2 = \boldsymbol{k}_3 + \boldsymbol{k}_4,$$
$$\omega(\boldsymbol{k}_1) + \omega(\boldsymbol{k}_2) = \omega(\boldsymbol{k}_3) + \omega(\boldsymbol{k}_4).$$

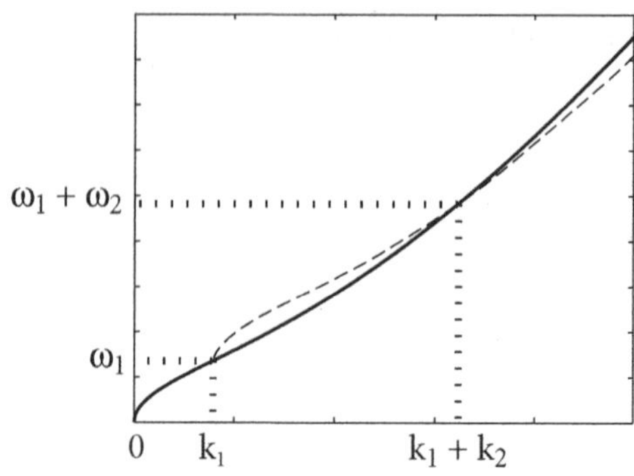

Figure 1. Graphical solution of Equations (1) for three resonant waves.

Given a dispersion relation $\omega(\boldsymbol{k})$ with one or more branches, one needs to compute all lowest-order resonances for $N = 3, 4$, and so on, in order to describe nonlinear dynamics of resonant wave interactions with the normal form analysis (Craig, 1996).

Time evolution of N resonant waves can be studied with an asymptotic multiscale expansion method. In this method, a solution of the system $u = u(\boldsymbol{x}, t)$ is assumed to be close to a linear superposition of N resonant waves with slowly varying amplitudes, where $\boldsymbol{x} = (x, y, z)$. For instance, resonant interaction of three waves is described by the following expansion (Gaponov-Grekhov & Rabinovich, 1992):

$$\begin{aligned} u(\boldsymbol{x}, t) = \varepsilon \Big[& a(\varepsilon \boldsymbol{x}, \varepsilon t) e^{i(\boldsymbol{k}\boldsymbol{x} - \omega(\boldsymbol{k})t)} \\ & + a_1(\varepsilon \boldsymbol{x}, \varepsilon t) e^{i(\boldsymbol{k}_1\boldsymbol{x} - \omega(\boldsymbol{k}_1)t)} \\ & + a_2(\varepsilon \boldsymbol{x}, \varepsilon t) e^{i(\boldsymbol{k}_2\boldsymbol{x} - \omega(\boldsymbol{k}_2)t)} \Big], \end{aligned}$$

where ε is a small parameter and ε^2 terms are dropped. The evolution equations for the wave amplitudes take the form

$$\begin{aligned} i\left(\frac{\partial}{\partial t} + \boldsymbol{v} \cdot \nabla\right) a &= \gamma a_1 a_2 e^{-i\Delta t}, \\ i\left(\frac{\partial}{\partial t} + \boldsymbol{v}_1 \cdot \nabla\right) a_1 &= \gamma_1 a \bar{a}_2 e^{i\Delta t}, \qquad (2) \\ i\left(\frac{\partial}{\partial t} + \boldsymbol{v}_2 \cdot \nabla\right) a_2 &= \gamma_2 a \bar{a}_1 e^{i\Delta t}. \end{aligned}$$

Here $\Delta = \omega(\boldsymbol{k}) - \omega(\boldsymbol{k}_1) - \omega(\boldsymbol{k}_2)$ is the frequency detuning from the exact resonance; $\boldsymbol{v}$, $\boldsymbol{v}_1$, and $\boldsymbol{v}_2$ are group velocities of the three waves, for example $\boldsymbol{v} = \nabla\omega(\boldsymbol{k})$; and γ, γ_1, and γ_2 are coupling coefficients in systems with quadratic nonlinearities. In nonlinear dispersive systems with conserved energy, parameters γ, γ_1, and γ_2 are real.

Typical dynamics of the resonant three-wave interaction can be studied from system (2) for space-

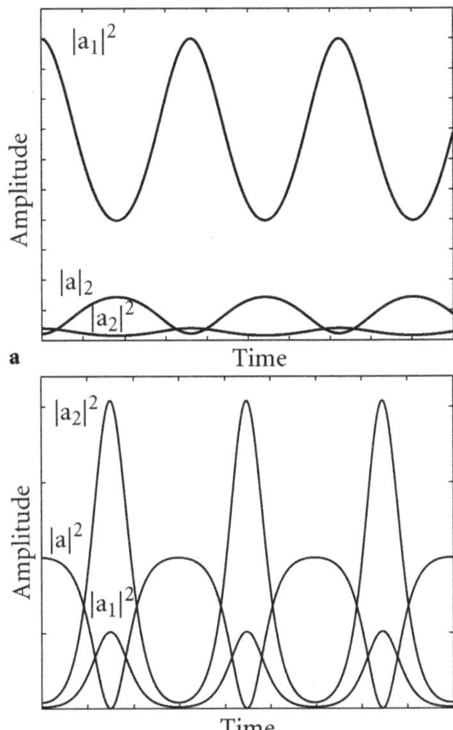

Figure 2. Two typical time evolutions of the three-wave resonant interaction: (a) the high-frequency wave a could not be generated from large low-frequency wave a_1 and small low-frequency wave a_2; (b) the large high-frequency wave a decays into pair of low-frequency waves a_1 and a_2.

independent waves $a(t)$, $a_1(t)$, and $a_2(t)$ in the case of exact resonance, when $\Delta = 0$. Without loss of generality, the real coupling coefficients can be normalized to be equal and positive: $\gamma = \gamma_1 = \gamma_2 > 0$. After these simplifications, system (2) has the following (Manley–Rowe) constants of motion:

$$|a|^2 + |a_1|^2 = C_1, \quad |a|^2 + |a_2|^2 = C_2,$$
$$|a_1|^2 - |a_2|^2 = C. \tag{3}$$

The time evolution of wave amplitudes $a(t)$, $a_1(t)$, and $a_2(t)$ displays two typical scenarios shown in Figure 2(a,b). If the wave with low frequency ($\omega_1, \omega_2 < \omega$) is initially pumped into the system (e.g., either $|a_1|^2 \gg |a|^2, |a_2|^2$ or $|a_2|^2 \gg |a|^2, |a_1|^2$ at $t = 0$), then the other two resonant waves remain small throughout the whole time evolution (see Figure 2(a)). Indeed, it is seen from Equations (3) that when $|a|^2$ grows, $|a_1|^2$ and $|a_2|^2$ decay. If either $|a_1|^2$ or $|a_2|^2$ were initially small, their decay (and the growth of $|a|^2$) is limited by a small value of C_1 and C_2.

If the wave of high frequency ω is initially pumped, however, the waves of low frequencies ω_1 and ω_2 may grow simultaneously. In this case, the resonant three-wave interactions display an exchange of energy between the high-frequency wave and the two low-frequency waves (see Figure 2(b)). This process is referred to as the "splitting" or "decaying" instability of the high-frequency wave. As seen from system (2) for small a_1 and a_2, the amplitude $a(t)$ does not change in time in the first-order approximation, while amplitudes $a_1(t)$ and $a_2(t)$ grow exponentially as $\sim e^{|\gamma a|t}$. The high-frequency wave $|a|^2$ decays into the two low-frequency waves $|a_1|^2$ and $|a_2|^2$, but the two waves merge back to the high-frequency wave later in the time evolution (see Figure 2(b)).

Resonant nonlinear three-wave interactions are known in many physical situations, such as in water waves, nonlinear optics, acoustics, and plasma physics. In nonlinear optics, resonant three-wave interactions are used in parametric amplifiers and oscillators, $\chi^{(2)}$ optical materials, self-induced transparency, and stimulated Raman scattering (Kaup et al., 1977). Four-wave mixing occur in optical communications and leads to growth of ghost pulses in an optical signal sequence. In water waves, resonant wave interactions explain effects of weak turbulence of gravity and gravity-capillary waves, as well as Rossby waves in the atmosphere (Zakharov, 1998). In plasma physics, nonlinear dynamics of high-temperature plasmas in a magnetic fields involve wave-particle and wave-wave interactions. Resonant three-wave interactions occur in ionospheric propagation, plasma heating with high-power electromagnetic sources, microwave sources, and laser sources (Kaup et al., 1977).

Dmitry Pelinovsky

See also **Dispersion relations; Frequency doubling; Manley–Rowe relations**

Further Reading

Craig, W. 1996. Birkhoff normal forms for water waves. *Contemporary Mathematics*, 200: 57–74

Gaponov-Grekhov, A.V. & Rabinovich, M.I. 1992. *Nonlinearities in Action. Oscillations, Chaos, Order, Fractals*, Berlin: Springer

Kaup, D., Reiman, A. & Bers, A. 1977. Space–time evolution of nonlinear three-wave interactions. *Reviews of Modern Physics*, 51: 915

Zakharov, V.E. (editor). 1998. *Nonlinear Waves and Weak Turbulence*, Providence, RI: American Mathematical Society

Zaslavsky, G.M. & Sagdeev, R.Z. 1988. *Introduction to Nonlinear Physics: From the Pendulum to Turbulence and Chaos*, Moscow: Nauka (in Russian)

ONE-DIMENSIONAL MAPS

The term *one-dimensional map* usually indicates a dynamical system with discrete time, generated by some map f of a one-dimensional (1-d) space (the real line or an interval of it, a circle, or a graph) onto itself or, equivalently, the iterations of this map. To specify a 1-d space, the terms *interval map*, *circle map*, or *graph map* are used. The most famous examples of a 1-d maps are the so-called logistic map $x \mapsto \lambda x(1-x)$ of the interval [0, 1] and the rotation $x \mapsto x+\alpha$ of the circle. The theory of 1-d maps can also be considered as the qualitative theory of the difference equations of the form $x_{n+1}=f(x_n)$, $n=0, 1, 2\ldots$.

Being part of the general theory of dynamical systems, the theory of 1-d maps includes studies relating to topological dynamics, symbolic dynamics, combinatorial dynamics, differential (or smooth) dynamics, and ergodic theory. The theory of 1-d maps is well developed and contains many deep and beautiful results. Even the simplest non-invertible (for example, quadratic) 1-d maps possess orbits with intricate dynamics, including those behaving like random processes. Thus, this theory offers comparatively simple tools for the understanding of general laws that govern the progression of real dynamic processes from regular to chaotic behaviors.

Many mathematical models, including those arising in population biology, can be reduced to investigations of 1-d maps. For organisms with non-overlapping generations, the population growth can be modeled with the difference equation $x_{n+1}=f(x_n)$, where x_n is the population density of the nth generation. When the density of the population is small and there are plenty of resources, the density x_n increases and follows approximately the linear law $x_{n+1}=\lambda\, x_n$, where λ is the reproduction coefficient. Because resources are bounded, density cannot increase forever; moreover, if the density x_n is very large, that is, close to 1, then x_{n+1}—the density of the next generation—must be close to 0. The logistic function $f(x)=\lambda x(1-x)$ with $0<\lambda\leq 4$ is a canonical example.

Interval Maps

Let $f: J\to J$ be a map of the interval J onto itself and f^n be its iterations, defined by $f^n(x)=f(f^{n-1}(x))$, $n=1,2,\ldots$, $f^0(x)=x$. The orbit of a point $x_0\in J$ under the map f is the sequence of points $x_n=f^n(x_0)$, $n=0, 1, 2, \ldots$.

A point $\beta\in J$ is a periodic point if $f^m(\beta)=\beta$ for some natural m; the smallest m with this property is the *period* of β; if $m=1$, β is called a fixed point. The points $f^n(\beta)$, $n=0,1,\ldots,m-1$ form a cycle of period m. To characterize the asymptotic behavior of an orbit, the set of its limiting points is used; it is called the ω-limit set of the orbit of x and denoted by $\omega(x)$. A cycle B is attracting if there exists a neighborhood U of B such that for any $x\in U$, $f^n(x)\in U$ for all $n>0$ and $\omega(x)=B$. The basin of an attracting cycle B is the set $\{x\in J:\omega(x)=B\}$; it consists of a finite or countable number of (open) intervals. A cycle B is repelling if there exists a neighborhood U of B such that for any point $x\in U\setminus B$, there exists n such that $f^n(x)\notin U$. If f is differentiable and has a cycle $B=\{\beta_1,\ldots,\beta_m\}$, the quantity $\mu(B)=f'(\beta_1)\cdot\cdots\cdot f'(\beta_m)$ is the multiplier of B. B is attracting if $|\mu(B)|<1$, and B is repelling if $|\mu(B)|>1$. The smallest closed set that contains ω-limit sets of orbits for almost all points of J (with respect to Lebesgue measure), is called the (global) attractor of f and denoted by **A** below.

If f is a monotonic (and hence, invertible) continuous function, its dynamics are very simple. If f is monotonically increasing, the ω-limit set of each orbit consists of only a fixed point. If f is monotonically decreasing, the ω-limit set of each orbit is either a fixed point or a cycle of period 2.

The orbits of nonmonotonic (non-invertible) maps can be very complicated. Maps with a single extremum point are called unimodal. Many properties of continuous interval maps can be demonstrated with the simplest representative of unimodal maps—the family of logistic maps $f: x\mapsto\lambda x(1-x)$.

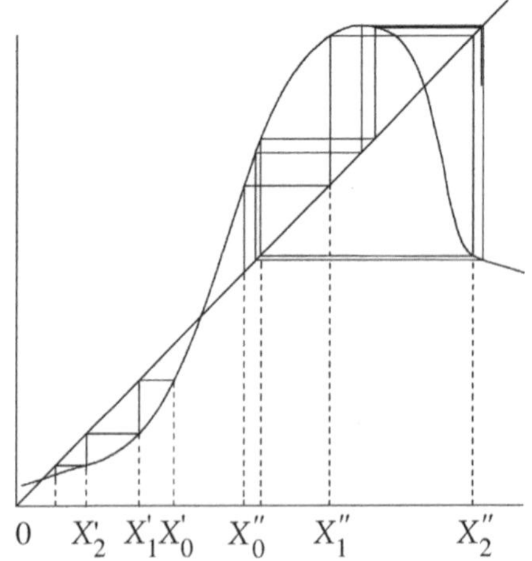

Figure 1. This Königs–Lamerey diagram gives a graphic representation of the orbit of x_0. If the graphs of two functions $y = f(x)$ and $y = x$ in the plane (x, y) are drawn, then the broken line consisting of vertical and horizontal segments that connect the points $(f^n(x_0), f^n(x_0))$ and $(f^n(x_0), f^{n+1}(x_0))$, $n = 0, 1, 2 \ldots$, demonstrates the movement along the orbit. The orbit of x'_0 tends to a fixed point, and the orbit of x''_0 tends to a cycle of period 3.

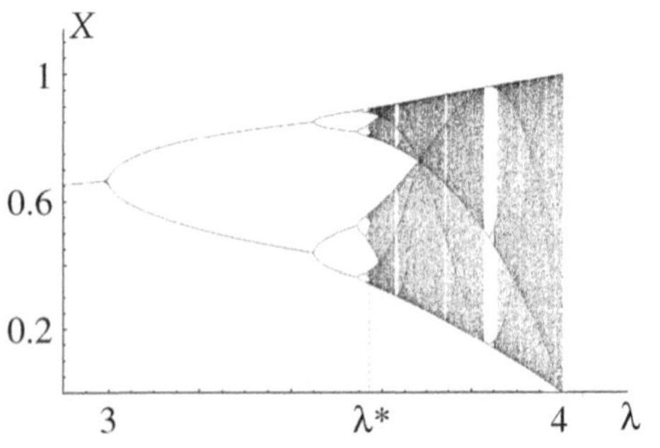

Figure 2. Bifurcation diagram of the attractors for the family of logistic maps.

Logistic Maps

The bifurcation diagram in Figure 2 shows what attractors the logistic map has when the parameter λ is changed.

With increasing λ up to $\lambda^* \approx 3.5699$, the consecutive doubling of period for attracting cycles holds and cycles of periods 2^n, $n = 0, 1, 2, \ldots$, appear: when $\lambda > 3$, the attracting fixed point $\beta_1 = 1 - 1/\lambda$ becomes repelling and a new attracting cycle of period 2 arises; for $\lambda > 1 + \sqrt{6} = 3.449\ldots$, this cycle becomes repelling and generates an attracting cycle of period 4, and so on. The period doubling for attracting cycles occurs with the speed

$$\frac{\lambda_{2^{n+1}} - \lambda_{2^n}}{\lambda_{2^{n+2}} - \lambda_{2^{n+1}}} \to 4.6692\ldots$$

(Feigenbaum–Coullet–Tresser constant), where λ_{2^n} is the parameter value when the cycle of period 2^n arises. The basin of each such attracting cycle is the interval $J = [0, 1]$ except for a countable set of repelling periodic points and their preimages.

If $\lambda = \lambda^*$, f has cycles of periods 2^n, $n = 0, 1, 2, \ldots$, only, and these cycles are repelling. The set K of limiting points for these cycles is an invariant Cantor set. The orbit of each point from K is everywhere dense on K. K is the attractor: the ω-limit set of each point from J, different from a periodic point or its preimages, coincides with K. The restriction of f^2 on the interval $[f^{-1}(\beta_1), \beta_1]$ is topologically conjugated to f on J, and hence, f is infinitely renormalizable.

For $\lambda = 3.83$, f has an attracting cycle of period 3, which is the attractor, and repelling cycles of all periods. The basin of the attracting cycle is J except for a Cantor set. This Cantor set is the closure of the set consisting of points of repelling cycles and their preimages.

If $\lambda = 4$, f is topologically expanding: for any interval $\widehat{J} \subset J$, there exists a number $m = m(\widehat{J})$ such that $f^m(\widehat{J}) = J$. Therefore, f has the property of sensitive dependence on initial conditions (sometimes called the "butterfly effect"). J is an attractor; periodic points lie everywhere dense on J. The map is topologically conjugate to the piecewise linear map

$$g : x \mapsto g(x) = \begin{cases} 2x, & 0 \le x \le 1/2, \\ 2(1-x), & 1/2 < x \le 1 \end{cases}$$

with the help of $h(x) = (2/\pi) \arcsin \sqrt{x}$. Because the Lebesgue measure is invariant under g, f has the invariant measure $\nu(\mathrm{d}x) = \mathrm{d}h(x) = \frac{1}{\pi}\mathrm{d}x / \sqrt{x(1-x)}$. This means, in particular, that for almost every orbit of f and for any $a_1, a_2 \in [0, 1]$, the frequency with which the orbit visits the interval (a_1, a_2) to $\int_{a_1}^{a_2} \nu(\mathrm{d}x)$.

The general characteristics of the logistic maps family are formulated as follows: for each $\lambda \in [0, 4]$, the attractor $\mathbf{A}$ of f is a cycle, or a Cantor set (as in the case $\lambda = \lambda^*$), or a cycle of intervals, that is, several closed non-intersecting intervals J_n, $n = 1, \ldots, m$, such that $f(J_n) = J_{(n+1) \bmod m}$ and $\cup_{n=1}^{m} J_n$ contain an everywhere dense orbit (in case $\lambda = 4$, $m = 1$).

The parameter values for which $\mathbf{A}$ is a cycle form an open and dense (on [0, 4]) set, and from the topological standpoint, a map with an attracting cycle is typical for the logistic family. The set of λ for which $\mathbf{A}$ is a Cantor set (and the map is infinitely renormalizable) has Lebesgue measure zero. The set of λ for which $\mathbf{A}$ is a cycle of intervals has positive Lebesgue measure. Moreover, for almost all value of these λ, $\mathbf{A}$ is the support of an ergodic absolutely continuous invariant measure.

The complication of dynamics in the logistic family stems, first of all, from the appearance of new attracting cycles resulting from the period-doubling bifurcation or the tangent bifurcation. With increasing λ, the multiplier of arising attracting cycle decreases from $+1$ to -1, and thereafter, this cycle becomes repelling and does not disappear in the sequel. If λ_m denotes the infinum (greatest lower bound) of the parameter values for which f has an attracting cycle of period m, then $\lambda_{m'} < \lambda_m$ for $m' \prec m$, where "$\prec$" is the following ordering of the natural numbers:

$$1 \prec 2 \prec 2^2 \prec \cdots \prec 2^n \prec \cdots \prec 7 \cdot 2^n \prec 5 \cdot 2^n \prec 3 \cdot 2^n \prec \ldots \prec 9 \prec 7 \prec 5 \prec 3,$$

called "Sharkovsky's ordering." For $\lambda = \lambda_{2^n}, n > 0$, the cycle of period 2^n arises as a result of a period-doubling bifurcation; for $\lambda = \lambda_m,\ m \neq 2^n, n = 1, 2, \ldots,$ the cycle of period m arises as a result of a tangent bifurcation.

If f has an attracting cycle of period m, then with increasing λ, the sequence of period-doubling bifurcations takes place until cycles of periods $2^n m,\ n = 1, 2, \ldots,$ appear. For the limiting parameter value for this sequence of bifurcation values, the attractor is a Cantor set (as in the case $\lambda = \lambda^*$).

Smooth Maps

Analytic unimodal maps (briefly, AU-maps) and C^3-smooth unimodal maps with negative Schwarzian

$$Sf = f'''/f' - \tfrac{3}{2}\left(f''/f'\right)^2 \qquad (1)$$

(briefly, SU-maps) are the most investigated classes of smooth maps. Any logistic map has negative Schwarzian.

As for logistic maps, the attractor **A** of each AU-map or SU-map includes (in general, not coincides with) one of the previous set: an (attracting) cycle, a Cantor set, or a cycle of intervals and always contains the ω-limit set of the extremum point. For an SU-map, **A** can have, in addition, only one attracting fixed point, and for an AU-map, **A** can have, in addition, any finite number of attracting cycles. If **A** is a cycle or a cycle of intervals, outside of **A**, f has only a finite number of isolated repelling cycles and isolated repelling invariant Cantor sets.

The period-doubling and tangent bifurcations are typical for smooth maps. The birth order of attractive cycles for the logistic family is kept, for example, for families of convex SU-maps of the form $\lambda f(x), \lambda > 0$.

For almost all members of a typical family of SU-maps, the attractor is either a cycle or a cycle of intervals; the cycle of intervals is the support of an ergodic absolutely continuous invariant measure; and there is a closed invariant interval $\widehat{J} \subset J$ containing the extremum point and $m > 0$ such that the restriction of f^m on $\widehat{J}$ is topologically conjugated to some logistic map.

For a typical family of AU-maps, there exists an open and dense set of parameters for which the map attractor consists of a finite number of cycles, and the union of basins of these cycles has full Lebesgue measure.

A smooth (at least C^2) unimodal map f is structurally stable if an attractor consists of cycles; the multiplier of each cycle is not equal to ± 1; the extremum point c is nondegenerate (i.e., $f''(c) \neq 0$); and $f^n(c)$ is not a periodic point for any $n = 0, 1, 2 \ldots$. The class of structurally stable maps is open in the C^2 topology and dense in any smooth topology.

Some properties of unimodal maps (possibly somewhat modified) hold for maps with more than two branches of monotonicity. For example, the number of attracting cycles for any C^3-smooth map with negative Schwarzian is not greater than the number of extremum points plus two.

Continuous Maps

Due to the natural ordering of points on the real line and the continuity of f, the values of f on a finite set of points provides rich information about the behavior of orbits. So, if there exist points $\beta_1 < \beta_2 < \beta_3$ such that $f(\beta_1) = \beta_1,\ f(\beta_2) = \beta_3$, and $f(\beta_3) \leq \beta_1$, then f has cycles with any period.

Each finite set $B = \{\beta_1 < \beta_2 < \cdots < \beta_m\}$ such that $f(B) \subseteq B$ generates a permutation π on the set $\{1, 2, \ldots, m\} : \pi(i) = j$, if $f(\beta_i) = \beta_j,\ i = 1, \ldots, m$. If B is a cycle, such a permutation is cyclic. For $J_i = [\beta_i, \beta_{i+1}],\ i = 1, 2, \ldots m-1,\quad f(J_i) \supseteq J_{k(i)} \cup \cdots \cup J_{K(i)}$, where $k(i) = \min\{\pi(i), \pi(i+1)\},\ K(i) = \max\{\pi(i), \pi(i+1)\} - 1$. This allows the construction of a Markov matrix of admissible transitions with elements $a_{ij} : a_{ij} = 0$, if $f(J_i) \not\supset J_j$, and $a_{ij} = 1$, if $f(J_i) \supset J_j$, and the corresponding Markov graph. An analysis of loops of Markov graph allows us to obtain the Sharkovsky theorem on the cycles coexistence: if a continuous interval map has a cycle of period m, then it has a cycle of any period m' such that $m' \prec m$.

Research on the coexistence of combinatorial objects, such as cycles, permutations, cyclic permutations, and graphs, is the subject of combinatorial dynamics. An important part of symbolic dynamics for 1-d maps is the kneading theory.

There are different criteria for the chaotic behavior of orbits. In particular, the following properties are equivalent: (i) the *topological entropy* of f is positive; (ii) there is a cycle with period $\neq 2^n,\ n \geq 0$; (iii) there is a *homoclinic orbit*; (iv) there is an ω-limit set containing a cycle, but different from it; and (v) there exist m and a closed invariant set $M \subset J$ such that the restriction of f^m on M is topologically conjugate to the *shift* on the *space of one-side sequences of two symbols*. Any of these involves the following: there is a continuum of many orbits such that for any two orbits $\{x'_n\}$ and $\{x''_n\}$, $\liminf_{n\to\infty} |x'_n - x''_n| = 0,\ \limsup_{n\to\infty} |x'_n - x''_n| > 0$.

If f has a cycle of intervals $\{J_1, J_2, \ldots, J_m\}$, then f on $\widehat{J} = \cup_{i=1}^m J_i$ is topologically expanding and has the property of the sensitive dependence on initial conditions, and the set of periodic points is everywhere dense on $\widehat{J}$.

The typical behavior of continuous maps differs from that of smooth maps. In particular, in the space of continuous maps with C^0-topology, the set of maps that have cycles of all periods contains an open and dense subset. Moreover, almost every continuous (but nowhere differentiable!) interval map has infinitely many minimal Cantor sets that attract almost all orbits.

Discontinuous interval maps constitute another important class of 1-d maps, especially for applications including expanding interval maps and so-called interval exchange maps.

Circle Maps

Many of the properties of interval maps mentioned above are true for circle maps $f : S \to S$. Along with this, there are classes of circle maps with specific properties, for example, circle homeomorphisms semi-conjugated to rotation of S with rotation number α (corresponding to the discontinuous interval map: $\widehat{f} : x \mapsto x + \alpha \bmod 1$). If α is a rational number, the attractor **A** consists of cycles; if α is irrational, **A** is a minimal set and coincides with S or a Cantor set. Bifurcations in a family of circle homeomorphisms are described by Farey's sequences.

Circle maps, semi-conjugated to the discontinuous interval map $\widehat{f} : x \mapsto k\,x \bmod 1$ with an integer $|k| > 1$, are called circle maps of degree k and represent a further important class of circle maps, including expanding circle maps with $|f'(x)| > 1$ at $x \in S$. Any expanding map is topologically conjugate to the shift on the space of one-sided sequences of finite number of symbols; it is structurally stable and possesses an absolutely continuous invariant measure.

A.N. Sharkovsky and V.V. Fedorenko

See also **Anosov and Axiom-A systems; Bifurcations; Butterfly effect; Chaotic dynamics; Denjoy theory; Horseshoes and hyperbolicity in dynamical systems; Maps in the complex plane; Markov partitions; Measures; Mixing; Period doubling; Population dynamics; Routes to chaos; Sinai–Ruelle–Bowen measures; Symbolic dynamics**

Further Reading

Alseda, L., Llibre, J. & Misiurewicz, M. 2000. *Combinatorial Dynamics and Entropy in Dimension One*, 2nd edition, River Edge, NJ: World Scientific

Argonsky, S.J., Bruckner, A.M. & Laczkovich, M. 1989. Dynamics of typical continuous functions. *Journal of the London Mathematical Society*, 40: 227–243

Avila, A., Lyubich, M. & de Melo, W. 2003. Regular or stochastic dynamics in real analytic families of unimodal maps. *Inventiones Mathematicae*, 3(154): 451–550

Block, L. & Coppel, W.A. 1992. *Dynamics in One Dimension*, Berlin: Springer

Collet, P. & Eckmann, J.-P. 1980. *Iterated Maps on the Interval as Dynamical Systems*, Boston: Birkhäuser

de Melo, W. & van Strien, S.J. 1993. *One-Dimensional Dynamics*, Berlin: Springer

Devaney, R.L. 1992. *A First Course in Chaotic Dynamical Systems: Theory and Experiment*, Reading, MA: Addison-Wesley

Hao, Bai-Lin. 1989. *Elementary Symbolic Dynamics and Chaos in Dissipative Systems*, Singapore: World Scientific

Katok, A. & Hasselblatt, B. 1995. *Introduction to the Modern Theory of Dynamical Systems*, Cambridge and New York: Cambridge University Press

Kozlovski, O.S. 2003. Axiom A maps are dense in the space of unimodal maps in the C^k topology. *Annals of Mathematics*, 1 (157): 1–43

Milnor, J. & Thurston, W. 1988. On iterated maps of the interval. In *Dynamical Systems: Proceedings University of Maryland, 1986–87*, Berlin and New York: Springer, 425–563

Peitgen, H.-O., Jürgens, H. & Saupe, D. 1993. *Chaos and Fractals: New Frontiers of Science*, Berlin and New York: Springer

Sharkovsky, A.N., Kolyada, S.F., Sivak, A.G. & Fedorenko, V.V. 1997. *Dynamics of One-Dimensional Maps*, Boston and Dordrecht: Kluwer

OPEN SETS

See **Topology**

OPTICAL FIBER COMMUNICATIONS

The existence of temporal solitons in optical fibers and their use for optical communications were suggested in 1973 (Hasegawa & Tappert, 1973), and by 1980, such solitons had been observed experimentally (Mollenauer et al., 1980; Hasegawa & Kodama, 1995). Since then, rapid progress has converted temporal solitons into a practical candidate for designing modern communication systems based on optical-fiber technology (Agrawal, 2002; Kivshar & Agrawal, 2003).

Similar to other types of solitons, those in optical fibers emerge from a balance between the group-velocity dispersion (GVD) and self-phase modulation (SPM) induced by the Kerr nonlinearity. The GVD broadens optical pulses during their propagation inside an optical fiber, except when the pulse is initially chirped (compressed with linear frequency modulation) in the right way. More specifically, a chirped pulse can be compressed during the early stage of propagation whenever the GVD parameter β_2 and the chirp parameter C happen to have opposite signs such that $\beta_2 C$ is negative. The optical pulse then propagates undistorted in the form of an optical soliton.

The nonlinear Schrödinger (NLS) equation governing pulse propagation inside optical fibers can be written in the following form:

$$\mathrm{i}\frac{\partial u}{\partial z} + \frac{1}{2}\frac{\partial^2 u}{\partial \tau^2} + |u|^2 u = 0\,, \tag{1}$$

where τ is a measure of time from the pulse center and is normalized to the input pulse width T_0, and L_D is the dispersion length. Noting that $z = Z/L_D$, the soliton period Z_0 is defined as

$$Z_0 = \frac{\pi}{2} L_D = \frac{\pi}{2}\frac{T_0^2}{|\beta_2|}. \tag{2}$$

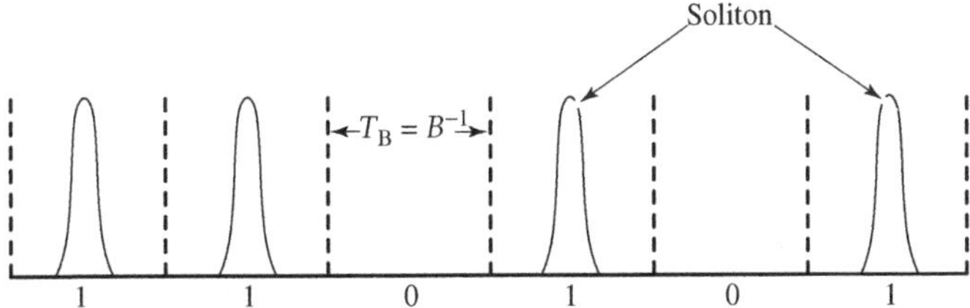

Figure 1. Soliton bit stream in the return-to-zero format where each soliton occupies a fraction of the bit slot representing 1 in a bit stream.

Temporal solitons are attractive for optical communications because they are able to maintain their width even in the presence of fiber dispersion, but their use requires substantial changes in the fiber system design, compared with conventional (nonsoliton) systems.

The basic idea is to use a soliton in each bit slot representing 1 in a bit stream. Figure 1 shows schematically a soliton bit stream in the return-to-zero (RZ) format. Typically, the spacing between two solitons exceeds a few times their full-width at half-maximum (FWHM), and individual solitons are well isolated. This requirement relates the soliton width T_0 to the bit rate B as $B = 1/T_B = 1/(2q_0T_0)$, where T_B is the duration of the bit slot and $2q_0 = T_B/T_0$ is the separation between neighboring solitons in normalized units.

The relatively large spacing necessary to avoid soliton interaction, limits the bit rate of soliton communication systems. The spacing can be reduced by up to a factor of two using unequal amplitudes for the neighboring solitons, so this scheme is feasible in practice and can be useful for increasing the system capacity.

Temporal solitons use the nonlinear phenomenon of SPM to maintain their width even in the presence of fiber dispersion. However, this property holds only if fiber losses are negligible. Optical amplifiers can be used for compensating fiber losses. Two approaches used for the management of losses through amplification of solitons are the lumped- and distributed-amplification techniques.

In the lumped-amplification scheme, optical amplifiers are placed periodically along the fiber link such that fiber losses between two amplifiers are exactly compensated by the amplifier gain. An important design parameter is the spacing L_A between amplifiers, which should be as large as possible to minimize the overall cost. For nonsoliton systems, L_A is typically 80–100 km. For soliton systems, L_A is restricted to smaller values because of the soliton nature of signal propagation.

The physical reason behind smaller values of L_A is that optical amplifiers boost soliton energy to the input level over a length of a few meters without allowing for gradual recovery of the fundamental soliton. The amplified soliton adjusts its width dynamically in the fiber section following the amplifier, but it also sheds a part of its energy as dispersive waves during this adjustment phase. The dispersive part can accumulate to significant levels over a large number of amplification stages and must be avoided. One way to reduce the dispersive part is to reduce the amplifier spacing L_A such that the soliton is not perturbed much over this short length. Numerical simulations show (Hasegawa & Kodama, 1995) that this is the case when L_A is a small fraction of the dispersion length ($L_A \ll L_D$). The dispersion length L_D depends on both the pulse width T_0 and the GVD parameter β_2 and can vary from 10 to 1000 km, depending on their values.

The condition $L_A \ll L_D$, imposed on loss-managed solitons when lumped amplifiers are used, becomes increasingly difficult to satisfy in practice as bit rates exceed 10 GB/s. This condition can be relaxed considerably when distributed amplification is used. The distributed-amplification scheme is inherently superior to lumped amplification because its use provides a nearly lossless fiber by compensating losses locally at every point along the fiber link. In fact, this scheme was used as early as 1985 using the distributed gain provided by Raman amplification (Agrawal, 2002) when the fiber carrying the signal was pumped at a wavelength of about 1.46 μm using a color-center laser (Hasegawa & Kodama, 1995). Alternatively, the transmission fiber can be doped lightly with erbium ions and pumped periodically to provide distributed gain. Several experiments have demonstrated that solitons can be propagated in such active fibers over relatively long distances.

Early soliton experiments on loss compensation used the Raman-amplification scheme. The situation changed with the application of erbium-doped fiber amplifiers from around 1989 for loss-managed soliton systems. In a typical experiment (Agrawal, 2002), 2.5 GB/s solitons were transmitted over 12,000 km by using a 75 km fiber loop containing three amplifiers, spaced 25 km apart (Agrawal, 2002). The bit rate–distance product of $BL = 30$ (TB/s)km is limited mainly by the timing jitter induced by amplifiers.

Dispersion management, which consists of a periodic change of the dispersion β_2 along the fiber length, is employed commonly for modern wavelength-division multiplexed systems. The use of dispersion management forces each soliton to propagate in the normal-dispersion regime of a fiber during each map period. If the map period is a fraction of the nonlinear length, the nonlinear effects are relatively small, and the pulse evolves in a linear fashion over one map period. On a longer length scale, solitons can still form if the nonlinear effects are balanced by the average dispersion. As a result, solitons can survive in an average sense, even though not only the peak power but also the width and shape of such solitons oscillate periodically.

Since 1996, a large number of experiments have shown the benefits of using dispersion-managed solitons for optical communication systems (Agrawal, 2002). Presently, 10 GB/s dispersion-managed solitons can be transmitted over 16 Mm of standard fiber when soliton interactions are minimized by choosing the location of amplifiers appropriately. An important application of dispersion management consists of upgrading the existing terrestrial networks designed with standard fibers and operating in the linear regime.

YURI KIVSHAR

See also **Dispersion management; Kerr effect; Nonlinear optics; Nonlinear Schrödinger equations**

Further Reading

Agrawal, G.P. 2002. *Fiber-Optic Communication Systems*, New York: Wiley

Hasegawa, A. & Kodama, Y. 1995. *Solitons in Optical Communications*, Oxford: Clarendon Press

Hasegawa, A. & Tappert, F. 1973. Transmission of stationary nonlinear optical pulses in dispersive dielectric fibers. *Applied Physics Letters*, 23: 142–144

Kivshar, Yu.S. & Agrawal, G.P. 2003. *Optical Solitons: From Fibers to Photonic Crystals*, San Diego: Academic Press

Mollenauer, L.F., Stolen, R.H. & Gordon, J.P. 1980. Experimental observation of picosecond pulse narrowing and solitons in optical fibers. *Physical Review Letters*, 45: 1095–1098

OPTICAL NONLINEARITIES

See **Nonlinear optics**

ORBIT

See **Phase space**

ORDER FROM CHAOS

At first glance, the words *chaos* and *order* seem to be contradictory. Indeed they are contradictory in terms of common definitions of the words, but not in terms of their mathematical usages. The definition of *chaos* in the Oxford American Dictionary (1986) is "great disorder or confusion." In contrast, the mathematical notion of *chaos* explains that data that seems to be random may have been generated by a deterministic and ordered process. In this sense, the field might have profitably been called "order theory" instead of "chaos theory." There are two common usages of the phrase "order from chaos." The first refers to seeking deterministic "chaotic" models to explain complex phenomenon, thus "order *in* chaos." The second refers to the fact that some chaotic dynamical systems can also exhibit regular islands of simplicity within their phase space, or "order *from* chaos." We will discuss each in turn.

Order in Chaos

The fact that seemingly simple and deterministic evolution rules can give rise to extremely complicated motion dates to Henri Poincaré (1892), in the setting of celestial mechanics. Perhaps the most famous example of unpredictable behavior is due to Edward Lorenz (1963), who discovered that even the most simplified models of the weather could produce data for which it is impossible to make long-term forecasts. Lorenz identified the "butterfly effect," or "sensitive dependence to initial conditions," whereby even small measurement errors quickly grow to swamp the signal. In 1975, Tien Yien Li and James A. Yorke published a paper entitled, "Period Three Implies Chaos," (Li & Yorke, 1975). The most famous impact of this paper was that it coined the word *chaos*, the field of study describing the nonlinear effect in which even the simple systems of Poincaré, Lorenz, and others can display sensitive dependence in a bounded domain. The chaos theory has attracted a great deal of popular attention, partly, in fact, due to its sexy title. Part of the appeal is also due to the scientific tradition since the time of Newton of searching for mechanistic explanations of reality.

The concept of phase space is useful for mapping the behavior of a dynamical system. Phase space could be described as the set of all possible relevant variables, which creates a closed description of the time evolution of the system (*See* **Phase space**). For example, for a simple pendulum, we need to specify all possible angular position and angular momentum states to uniquely define all solutions; the phase space for the periodic oscillation is a closed curve, a circle. For an electronic tank oscillator (LRC) circuit, we need to specify capacitor charge and current through the inductor. The dimension of the system is the dimension of the phase space. A billiard table's worth of balls, for example, requires many variables for a complete description—two variables for position and two for momentum for each of the balls.

As a highly interdisciplinary field, chaos theory has been successful in finding deterministic chaos, or order, in what was once thought to be mere noise, or at

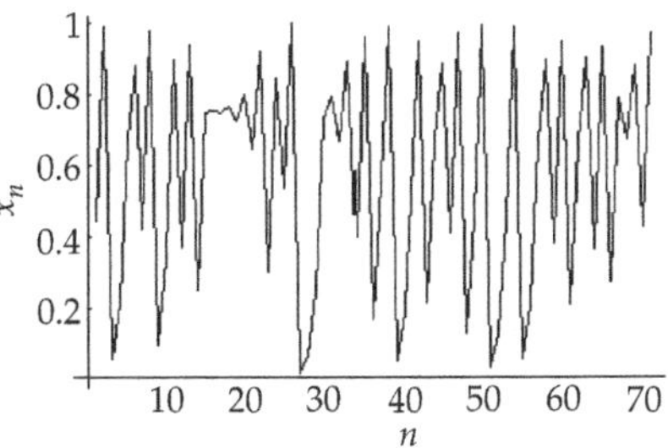

Figure 1. This seemingly stochastic data was actually generated by a "simple," low-dimensional, and deterministic process. This time-series plot of the state x_n versus time n was actually generated by the dynamical system $x_{n+1} = ax_n(1-x_n)$ called the logistic map. ($a = 4.0$, $x_0 = 0.44$).

least an extremely high-dimensional effect. There are numerous explicit examples of chaos in many areas, including biology, electrical engineering, chemistry, celestial mechanics, and brain and heart physiology. Thus, the old idea that extremely complicated data must always be due to extremely complicated effects is false (see Figure 1). One popular misuse of these ideas is a belief that all complicated data must have underlying order in chaos. This can be phrased as "Is what seems complicated always really simple?" The answer, of course, is no; it remains true that noise and other high-dimensional effects can also be responsible for complexity. What is true is that what seems complicated may *sometimes* be simple, in the sense of having a low-dimensional chaotic model. To cite one popular question, the stock market prices unarguably constitute extremely complicated data, but is there underlying order in chaos here, or is the explanation due to intractably high-dimensionality?

Order from Chaos

Some chaotic systems have an occasional tendency to exhibit simple behavior. This is often described as regular islands in an otherwise chaotic phase space. The classic mathematical example of regular islands arises in Hamiltonian systems, which can exhibit a phase space of solutions in which chaotic solutions are both intermingled and bounded by nested islands around islands around islands, and so on, of KAM-like tori (circle-like integrable solutions) (Arrowsmith & Place, 1990; Meiss, 1992).

Put differently, "order from chaos" and "regular islands" are terms commonly used to refer to the presence of an "attractor" (an imaginary point in the phase space about which the trajectories appear to orbit) in a system that one thinks should behave in a complicated manner. For example, consider a thought experiment of a game of pool in which we assume no friction so that the balls never stop moving (a two-dimensional Lorentz gas). Following the path of one specific ball, say the seven-ball, is an extremely complicated problem displaying sensitive dependence, since a small change in the velocity or position of the ball affects the next collision with the wall or with the next ball, an effect that multiplies upon subsequent collisions. However, there are several attractors near which the ball's motion becomes quite simple—the pocket holes. Once within the rim of a pocket (its basin of attraction), falling in becomes inevitable, as it would require a relatively large energy perturbation to prevent it from falling in. In this sense, we can say that the regions of phase space, corresponding to being stuck in the billiards pocket, are regular islands in a chaotic sea.

As analogies, such descriptions of islands in chaotic seas have been extended by some to explain the emergence of a coherent phenomenon from complicated processes. There is perhaps no more intriguing question than the origins of life. Proponents of emergence theories suggest that life in the original Hadean seas (the "primordial" seas on early Earth) gave rise to life through a capture process such as the billiards game (Waldrop, 1992; Kauffman, 1995). While randomly chosen initial conditions in the billiards game may each individually be unlikely to lead to capturing a ball in a pocket (and certainly any analogous attractors in chemical processes must be exceedingly unlikely), the way to win an unlikely bet is to play very quickly, over and over. This may have been the process that led to complex organic molecules, according to proponents of the emergence theory.

Emergence and order from chaos have also been used to describe processes whose attractors are surprisingly complicated themselves, as sets, but arise from surprisingly simple rules. Michael Barnsley has called the following "The Chaos Game" (Barnsley, 1993). First label the vertices of a triangle, 1, 2, 3. Using a random number generator (such as a six-sided die), assign probabilities to each triangle vertex, say die sides 1–2 to vertex 1, die sides 3–4 to vertex 2, and die sides 5–6 to vertex 3. Roll the die to randomly select one of the vertices, and record its planar coordinates (x, y). Then roll the die again to randomly select another vertex, and record the point halfway between the resulting new vertex and the current (x, y) as the new (x, y). Repeat indefinitely. For each newly defined (x, y), we record a dot for a pictorial record. Most people guess that the result will uniformly fill the triangle with a smattering of dots, but the mathematical fact is surprising. The result is an extremely intricate structure, a fractal called the Sierpinski gasket (see Figure 2). Because this simple rule gives rise to such a complicated structure, the argument goes, perhaps many of the other intricacies we see around us might have emerged from other simple rules.

Finally, we mention the meaning of "order from chaos," as developed from theories of the Nobel prize-winning physicist Ilya Prigogine, whose work on dissipative structures of systems held from thermal

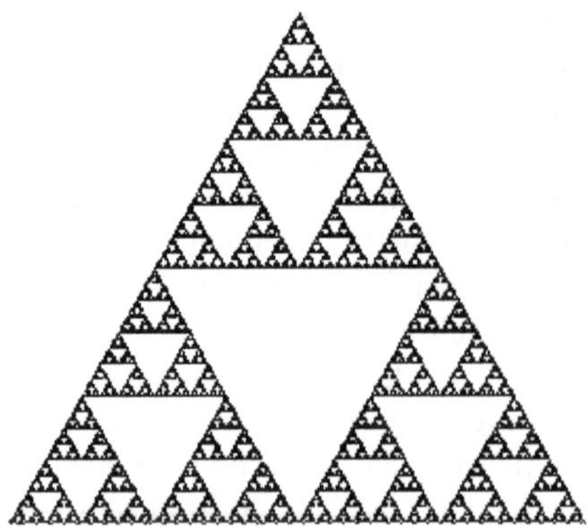

Figure 2. From such a simple algorithm as "The Chaos Game" (Barnsley, 1993) emerges the extremely intricate fractal shown, called the Sierpinski gasket.

equilibrium has been used to study self-organizing systems (Prigogine, 1984). Prigogine has defined complexity as the ability to switch between different modes of behavior as the environmental conditions vary. Thus, he has described a phenomenon in which far-from-equilibrium systems transition "from being to becoming," which some describe as order from chaos.

ERIK M. BOLLT

See also **Chaotic dynamics; Emergence; Kalmogorov–Arnol'd–Moser theorem; Phase space**

Further Reading

Arrowsmith, D.K. & Place, C.M. 1990. *An Introduction to Dynamical Systems*, Cambridge and New York: Cambridge University Press

Barnsley, M. 1993. *Fractals Everywhere*, 2nd edition, San Francisco: Morgan Kaufmann

Kauffman, S. 1995. *At Home in the Universe, the Search for the Laws of Self-Organization and Complexity*, Oxford and New York: Oxford University Press

Lorenz, E.N. 1963. Deterministic nonperiodic flow. *Journal of Atmospheric Science*, 20: 130–141

Li, T.Y. & Yorke, J.A. 1975. Period three implies chaos. *American Mathematics Monthly*, 82: 985–992

Meiss, J.D. 1992. Symplectic maps, variational principles, and transport. *Reviews of Modern Physics*, 64: 795

Poincaré, H. 1892–99. *Les méthodes nouvelles de la mécanique céleste*, Paris: Gauthier-Villars, 3 vols; as *New Methods of Celestial Mechanics*, New York: American Institute of Physics, 1993

Prigogine, S. 1984. *Order out of Chaos: Man's New Dialogue with Nature,* New York: Bantam Books

Waldrop, M. 1992. *Complexity: The Emerging Science at the Edge of Chaos and Order*, New York: Touchstone Press

ORDER PARAMETERS

Although the concept of an order parameter was introduced by Lev Landau in the 1930s, it does not yet have a precise definition. Broadly, an order parameter is a thermodynamic quantity that is invariant with respect

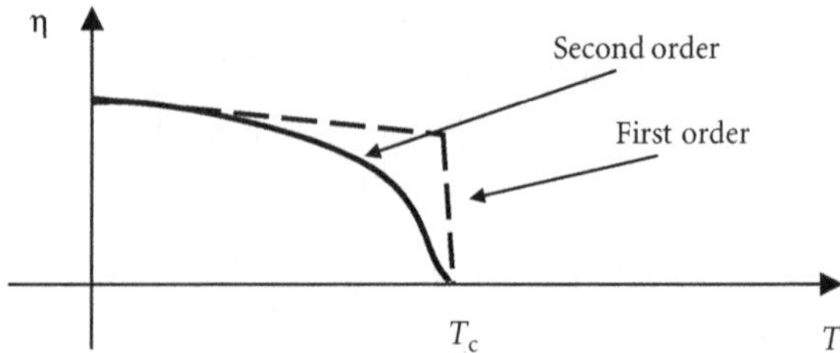

Figure 1. A typical plot of the scalar order parameter as a function of temperature.

to the symmetry group of the low-temperature phase and zero above a critical point or transition temperature, T_C. It is a measure of the amount of order that is built up in the system in the neighborhood of the critical point. In general, an order parameter has both an amplitude and a phase. To find the equation of state, a minimization procedure is followed for an appropriate thermodynamic potential.

From the original application to second-order phase transitions where it changes continuously from T_c to lower temperatures, the idea of an order parameter has been extended to first-order transitions involving an abrupt change at T_c (see Figure 1). The concept of an order parameter has been generalized from a globally defined scalar to a complex time- and space-dependent function. Note that first-order phase transitions are associated with discontinuities of the order parameter and thermal hysteresis effects. Second-order phase transitions have a continuous order parameter and show field-induced hysteresis. Diverse applications of the order parameter concept to both equilibrium and non-equilibrium critical phenomena are listed in Table 1.

Crystal formation is demonstrated by the existence of a regular diffraction pattern associated with the Fourier components of the mass density distribution $\rho(\mathbf{r})$

$$\rho(\boldsymbol{r}) = \sum_{\boldsymbol{g}} \rho(\boldsymbol{g})\mathrm{e}^{\mathrm{i}\boldsymbol{g}\cdot\boldsymbol{r}}\,, \qquad (1)$$

where the $\boldsymbol{g}$ are vectors in the reciprocal space. The set of numbers $\rho(\boldsymbol{g})$ can be used as order parameters characterizing the low-temperature (crystal) phase. The nonzero coefficients $\rho(\boldsymbol{g})$ in Equation (1) define a multi-component order parameter.

The order parameter field for a magnet is defined at each position $\boldsymbol{x}$ by a direction of the local magnetization $\boldsymbol{M}(\boldsymbol{x})$ whose length is fixed. By becoming a magnet, this material has broken the rotational symmetry and its order parameter field defines the broken symmetry directions chosen in the material.

A number of metals, alloys, and ceramics below their critical temperature T_c exhibit an ordered state in the conduction electron degrees of freedom manifested by zero resistance. The order parameter for superconductors is the wavefunction of the Cooper pair

Phenomenon	Disordered phase	Ordered phase	Order parameter
Equilibrium			
Condensation	Gas	Liquid	Density difference $\rho_L - \rho_G$
Spontaneous magnetization	Paramagnet	Ferromagnet	Net magnetization M
Antiferromagnetism	Paramagnet	Antiferromagnet	Staggered magnetization $M_1 - M_2$
Superconductivity	Conductor	Superconductor	Cooper pair, wave function η
Alloy ordering	Disordered mixture	Sublattice ordered alloy	Sublattice concentration
Ferroelectricity	Paraelectric	Ferroelectric	Polarization
Superfluidity	Fluid	Superfluid	Condensate wavefunction
Nonequilibrium			
Laser action	Lamp (incoherent)	Laser (coherent)	Electric field intensity
Super-radiant source	Noncoherent polarization	Coherent polarization	Atomic polarization
Fluid convection	Turbulent flow	Bénard cells	Amplitude of mode

Table 1. Examples of order parameters (OPs).

condensate $\eta(\boldsymbol{r})$, and it exhibits a Hopf bifurcation at $T = T_c$.

The superfluid properties in ^{4}He and ^{3}He are manifested by the absence of viscosity. The ^{4}He atoms are bosons that below a transition temperature T_λ, undergo the so-called Bose condensation into a $k = 0$ mode. The associated order parameter is the condensate's quantum wavefunction. Because ^{3}He atoms are fermions, below T_λ they form Cooper pairs.

Liquid crystals are anisotropic fluids composed of strongly elongated molecules. The nematic phase is characterized by the existence of a direction to which most of the molecules are parallel, so the order parameter is a second rank tensor describing correlations along that direction. Numerous other examples of critical phenomena, such as binary fluids, the metal-insulator transition, polymer transitions and spin- and charge-density waves, have their own order parameters.

Landau deduced that second-order phase transitions are associated with symmetry breaking and can be qualitatively described by an order parameter η. Assuming that the free energy F depends on V, T, and η, the equilibrium conditions are:

$$\left.\frac{\partial F(T, V, \eta)}{\partial \eta}\right|_{\eta=\eta_0} = 0$$

and

$$\left.\frac{\partial^2 F(T, V, \eta)}{\partial \eta^2}\right|_{\eta=\eta_0} > 0, \qquad (2)$$

where η_0 is the equilibrium value of η.

The universality hypothesis states that any two physical systems with the same spatial dimensionality, d, and the same number of order parameter components, n, belong to the same universality class having identical critical exponents (see Table 2).

Universality class		System	Order parameter (OP)
$d = 2$,	$n = 1$	Absorbed films	Surface density
$d = 2$,	$n = 2$	Superfluid ^{4}He film	Superfluid wave function
$d = 3$,	$n = 1$	Uniaxial ferromagnets	Magnetization
$d = 3$,	$n = 1$	Fluids	Density difference
$d = 3$,	$n = 1$	Mixtures, alloys	Concentration difference
$d = 3$,	$n = 2$	Planar ferromagnets	Magnetization
$d = 3$,	$n = 2$	Superfluids	Superfluid wave function
$d = 3$,	$n = 3$	Isotropic ferromagnets	Magnetization

Table 2. Examples of universality classes.

Order parameters accompany broken symmetry phenomena where the new ground state of the system does not possess the full symmetry of the Hamiltonian. A classic example is the ferromagnetic-to-paramagnetic phase transition at T_c where the full rotational symmetry of the paramagnetic phase is broken by the axiality of the ground ferromagnetic state below T_c. When a symmetry that is broken is continuous, a vibrational mode appears whose frequency vanishes at long wavelengths. (Quanta of such modes are called "Goldstone bosons.") Examples include ferromagnetic domain walls and acoustic soft modes in structural phase transitions.

There exist several different types of broken symmetries: (a) translational (crystal formation, structural transitions); (b) gauge (superfluidity, superconductivity); (c) time reversal (ferromagnets); (d) local rotational (liquid crystals); (e) rotational (some structural phase transitions); and (f) space inversion (ferroelectricity). Gauge symmetry is a universal property of Hamiltonians whenever the total number of particles is conserved or a generalized charge-like conserved quantity exists. Then, the order parameter η is complex, and its local density $\rho = \eta^* \eta(r)$ is such that a phase shift $\eta \to \eta e^{i\omega}$ leaves the Hamiltonian invariant.

Defects in the order parameter space can be topological (i.e., kinks, also referred to as domain walls) and nontopological (i.e., solitons, also called nucleation centers). They are obtained as solutions to the equations of motion for the order parameter field. Also, point defects, line defects, vortices, dislocations, vacancies, and interstitials with attendant singularities are seen experimentally in critical systems (systems close to a phase transition).

Finally, note that Haken's separation of modes in synergetic systems into masters (order parameters) and slaves has been influenced by Landau's theory of phase transitions.

JACK A. TUSZYŃSKI

See also **Bose–Einstein condensation; Critical phenomena; Domain walls; Ferromagnetism and ferroelectricity; Hysteresis; Liquid crystals; Phase transitions; Solitons; Synergetics**

Further Reading

Anderson, P.W. 1984. *Basic Notions of Condensed Matter Physics*, Menlo Park, CA: Benjamin/Cummings
Haken, H. 1980. *Synergetics*, Berlin and New York: Springer
Landau, L.D. & Lifshitz, E.M. 1959. *Statistical Physics*, London: Pergamon
Ma, S.-K. 1976. *Modern Theory of Critical Phenomena*, Reading, MA: Benjamin
Reichl, L.E. 1979. *A Modern Course in Statistical Physics*, Austin, TX: University of Texas Press
White, R.H. & Geballe, T. 1979. *Long Range Order in Solids*, New York: Academic Press
Yeomans, J.M. 1992. *Statistical Mechanics of Phase Transitions*, Oxford: Clarendon Press and New York: Oxford University Press

ORDINARY DIFFERENTIAL EQUATIONS, NONLINEAR

An ordinary differential equation (ODE) is an equation

$$\Delta(t, f, f', f'', ..., f^{(N)}) = 0 \tag{1}$$

(with $f^{(j)} = \mathrm{d}^j f/\mathrm{d}t^j$), relating a function $f(t)$ to its derivatives. The order of an ODE is the size of the highest derivative that appears, so Equation (1) is Nth order. An ODE is *linear* if it can be written as a linear combination of f and its derivatives, that is,

$$\begin{aligned}\Delta \equiv a_N(t) f^{(N)} + a_{N-1}(t) f^{(N-1)} + \cdots \\ + a_1(t) f' + a_0(t) f + b(t) = 0,\end{aligned} \tag{2}$$

including the possible addition of an inhomogeneous term $b(t)$. All other ODEs, which are not of the form (2), are referred to as *nonlinear*.

More generally, one can consider systems of ODEs relating M functions $f_0, f_1, \ldots, f_{M-1}$ and their derivatives $f_k^{(j)}$ of different orders. In fact, if a system of ODEs can be solved for the highest derivatives appearing (which generally requires the implicit function theorem), then it can always be converted to a system of first-order equations,

$$\begin{aligned} f_0' &= F_0(t, f_0, f_1, \ldots, f_{M-1}), \\ f_1' &= F_1(t, f_0, f_1, \ldots, f_{M-1}), \\ &\vdots \\ f_{M-1}' &= F_{M-1}(t, f_0, f_1, \ldots, f_{M-1}). \end{aligned} \tag{3}$$

For example, suppose Equation (1) can be solved explicitly for the Nth derivative, as $f^{(N)} = F(t, f, f', f'', \ldots, f^{(N-1)})$. In that case, the single ODE (1) may be rewritten as the first-order system

$$\begin{aligned} f_0' &= f_1, \\ f_1' &= f_2, \\ &\vdots \\ f_{N-2}' &= f_{N-1}, \\ f_{N-1}' &= F(t, f_0, f_1, f_2, \ldots, f_{N-1}). \end{aligned}$$

In most applications of nonlinear ODEs, such as in the physical sciences or biology, they appear as coupled systems of either first or second order. For example, Newton's equations in mechanics (Arnol'd, 1989) are of the form

$$\boldsymbol{q}'' = \boldsymbol{F}(t, \boldsymbol{q}, \boldsymbol{q}'),$$

relating the acceleration $\boldsymbol{q}''$ of a particle (of unit mass) to the force $\boldsymbol{F}$ acting on it, which is a function of its position $\boldsymbol{q}$ and velocity $\boldsymbol{q}'$, and possibly the time t. Mechanical systems are often derived from an action functional $S = \int_{t_0}^{t_1} L(t, \boldsymbol{q}, \boldsymbol{q}', \ldots, \boldsymbol{q}^{(j)})\,\mathrm{d}t$. In particular, if $\boldsymbol{q}'$ is the highest derivative appearing in the Lagrangian, so that $L = L(t, \boldsymbol{q}, \boldsymbol{q}')$, then the corresponding Euler–Lagrange equations form a second-order system of ODEs, viz.

$$\frac{\partial L}{\partial \boldsymbol{q}} - \frac{\mathrm{d}}{\mathrm{d}t}\left(\frac{\partial L}{\partial \boldsymbol{q}'}\right) = 0.$$

In Hamiltonian mechanics, on the other hand, the equations of motion are given as a first-order system of ODEs,

$$\frac{\mathrm{d}\boldsymbol{x}}{\mathrm{d}t} = \boldsymbol{J}(\boldsymbol{x}) \nabla H(t, \boldsymbol{x}), \tag{4}$$

describing the evolution of a point $\boldsymbol{x}$ in the phase space with Poisson tensor $\boldsymbol{J}$ and Hamiltonian H. First-order equations are commonly used both in population dynamics, for example, the Verhulst model

$$\frac{\mathrm{d}P}{\mathrm{d}t} = rP(1 - P/K) \quad (5)$$

(with r, K constants), and in reaction kinetics applying the Law of Mass Action (Murray, 1989). The ODEs describing a system of N species or N chemical reagents typically take the form of N coupled first-order ODEs, but usually these are *not* Hamiltonian.

Given an ODE such as (1), the main task is to determine the nature of the solutions, namely, those functions $f(t)$ for which all the derivatives $f^{(j)}$ exist for $j = 1, \ldots, N$ and are related by Equation (1). More specifically, one may wish to solve an initial value problem, where the values $f(t_0), f'(t_0), \ldots, f^{(N-1)}(t_0)$ of the function f and its first $N-1$ derivatives are specified at some initial time $t = t_0$, and f is to be determined at subsequent times $t > t_0$. Alternatively, one might pose a boundary value problem where the values $f(t_0)$ and $f(t_1)$ (and maybe some of the derivatives) are given at the endpoints of the interval $[t_0, t_1]$, and $f(t)$ is to be found for $t_0 < t < t_1$.

For an Nth-order linear homogeneous ODE, of the form (2) with $b \equiv 0$, the general solution is just a linear combination of N-independent solutions s_j, that is, $f(t) = \sum_{j=1}^{N} A_j s_j(t)$ with N arbitrary constants $A_1, \ldots, A_N$. Ideally, one would like to express the general solution of an Nth-order nonlinear ODE as a function of N arbitrary integration constants, but in general this is not possible. However, for systems (3) where the F_j on the right-hand sides are suitably regular functions of all their arguments in the neighborhood of the initial data, the local existence of a solution to the initial value problem near $t = t_0$ can be proved by the Cauchy–Lipschitz method (Ince, 1926). In fact, whenever the F_j are analytic functions, then the existence of a local solution to the initial value problem is guaranteed in some circle around $t = t_0$ in the complex t plane (Hille, 1976). This means that, at least locally, the solution $f(t)$ can be considered as a function of the initial data $f(t_0), f'(t_0), \ldots, f^{(N-1)}(t_0)$. Only for ODEs of Painlevé type can this local solution be extended globally to a single-valued, meromorphic function of t (Hinkkanen & Laine, 1999). In contrast, the solutions of other ODEs can display chaotic behavior, with coalescing branch points in the complex plane (Sachdev, 1991).

For first-order equations ($N = 1$), there are various classes of ODEs that are amenable to exact integration methods. The simplest example is the class of *separable* equations, which are directly solvable by a quadrature:

$$\frac{\mathrm{d}f}{\mathrm{d}t} = T(t)/F(f),$$

whence

$$\int F(f)\,\mathrm{d}f = \int T(t)\,\mathrm{d}t + \text{constant}.$$

The Verhulst model (5) is a particularly simple example for we have

$$\int \frac{\mathrm{d}P}{P(1 - P/K)} = \log\left(\frac{P}{K - P}\right) = rt + c$$
$$\Rightarrow P(t) = \frac{K}{1 + \mathrm{e}^{-rt-c}} = \frac{KP_0}{P_0 + (K - P_0)\mathrm{e}^{-rt}},$$

where the constant $P_0 = K/(1 + e^{-c}) = P(0)$ is the initial population size. Other special classes of first-order equations include the Bernoulli equations $f' = P(t)f + Q(t)f^n$ and Riccati equations $f' = A(t)f^2 + B(t)f + C(t)$; both of these types can be converted to linear equations by a substitution (Ince, 1926).

Among higher-order ODEs, exactly solvable equations are rare, but nevertheless some exact solution methods exist. In particular, if there are first integrals (constants of motion) or symmetries for a system, then it is possible to reduce the order. For autonomous Hamiltonian systems of order $N = 2n$, Liouville's theorem on integrable systems states that if there are n independent first integrals, in involution with respect to the nondegenerate Poisson bracket defined by $\boldsymbol{J}$, then the ODEs (4) can be integrated by quadratures (Arnol'd, 1989). More generally, an ODE of order N with a first integral can be reduced to an equation of order $N-1$, while if it has a Hamiltonian or Lagrangian structure invariant under a symmetry of the system, then (using Noether's theorem) the order can be reduced to $N-2$. The approach to solving ODEs using their symmetries is originally due to Lie; for a modern treatment, see Olver (1993).

As an illustration of some of these ideas, consider the second-order ODE

$$f'' = \left(kt - \frac{(n-1)}{f}\right)(f')^2 + \frac{(n-1)}{t}f' \quad (6)$$

(k, n constants), which is a radial symmetry reduction of an n-dimensional partial differential equation appearing in differential geometry (Abreu, 1998). This can be derived from the action

$$S = \int_{t_0}^{t_1} L(t, f, f')\,\mathrm{d}t,$$
$$L = t^{n-1}\log(f^{n-1}f') - kt^n f. \quad (7)$$

By a Legendre transformation, setting $q = f$ and $p = \partial L/\partial f'$, (6) can also be converted to the first-order Hamiltonian form

$$\frac{\mathrm{d}}{\mathrm{d}t}\begin{pmatrix} q \\ p \end{pmatrix} = \begin{pmatrix} 0 & 1 \\ -1 & 0 \end{pmatrix}\begin{pmatrix} \partial_q H \\ \partial_p H \end{pmatrix}$$

with $H(t, q, p) = t^{n-1}\log(q^{n-1}/p) - kt^n q$. Because this is a non-autonomous system, with t appearing explicitly, the Hamiltonian H is *not* a constant of motion. However, (6) is invariant under the one-parameter group of scaling symmetries $r \to \mu r$, $f \to \mu^{-1} f$, so by introducing the new scale-invariant independent variable $y = ft$, and the dependent variables $v = v(y) = -\log t$, $w = \mathrm{d}v/\mathrm{d}y$, it reduces to a first-order equation for $w(y)$:

$$\frac{\mathrm{d}w}{\mathrm{d}y} = y(2n - ky)w^3 + (3n - 2ky)w^2 + ((n-1)/y - k)w.$$

Unfortunately, it is not possible to reduce this to a quadrature, since the action (7) is *not* invariant under scaling, unless $k = 0$ when the general solution to (6) is $f(t) = (At^n + B)^{1/n}$ (with A, B arbitrary constants).

Other important methods for ODEs include the Painlevé analysis of movable singularities (Kruskal & Clarkson, 1992), and asymptotic expansions around regular or irregular fixed singular points (Wasow, 1965; Tovbis, 1994).

ANDREW HONE

See also **Chaotic dynamics; Constants of motion and conservation laws; Euler–Lagrange equations; Extremum principles; Hamiltonian systems; Integrability; Painlevé analysis; Partial differential equations, nonlinear; Riccati equations**

Further Reading

Abreu, M. 1998. Kähler geometry of toric varieties and extremal metrics. *International Journal of Mathematics*, 9: 641–651

Arnol'd, V.I. 1989. *Mathematical Methods of Classical Mechanics*, 2nd edition, New York: Springer

Hille, E. 1976. *Ordinary Differential Equations in the Complex Domain*, New York: Wiley

Hinkkanen, A. & Laine, I. 1999. Solutions of the first and second Painlevé equations are meromorphic. *Journal d'Analyse Mathémathique*, 79: 345–377

Ince, E.L. 1926. *Ordinary Differential Equations*, London: Longmans Green; reprinted New York: Dover, 1956

Kruskal, M.D. & Clarkson, P.A. 1992. The Painlevé–Kowalevski and Poly–Painlevé tests for integrability. *Studies in Applied Mathematics*, 86: 87–165

Murray, J.D. 1989. *Mathematical Biology*, Berlin and New York: Springer

Olver, P.J. 1993. *Applications of Lie Groups to Differential Equations*, 2nd edition, Berlin and New York: Springer

Sachdev, P.L. 1991. *Nonlinear Ordinary Differential Equations and Their Applications*, New York: Marcel Dekker

Tovbis, A. 1994. Nonlinear ordinary differential equations resolvable with respect to an irregular singular point. *Journal of Differential Equations*, 109: 201–221

Wasow, W. 1965. *Asymptotic Expansions for Ordinary Differential Equations*, New York: Wiley

ORGANIZING CENTERS

See **Spiral waves**

OSCILLATOR, CLASSICAL NONLINEAR

See **Damped-driven anharmonic oscillator**

OVERTONES

When a tonal sound, such as a note played on a flute, a human vowel sound, or a bell, is analyzed in the frequency domain by applying a Fourier transform to the acoustic pressure waveform, the spectrum consists of a large number of sharp lines. The component of lowest frequency is termed the "fundamental", and the others are "upper partials" or "overtones." If the frequencies of the overtones are all exact integer multiples of the frequency of the fundamental, then they are termed "harmonics." The partial with frequency $f_n = nf_1$, where f_1 is the frequency of the fundamental, is the nth harmonic, so that the fundamental is the first harmonic.

A one-dimensional simple harmonic oscillator, or linear oscillator, in which the restoring force is proportional to displacement y from the equilibrium position, obeys the equation

$$m\frac{\mathrm{d}^2 y}{\mathrm{d}t^2} = -ky, \tag{1}$$

where m is the mass of the moving particle and k is the restoring force constant. A damping term can also be included, but this need not concern us here. Such an oscillator vibrates with a single frequency $f_1 = (1/2\pi)(k/m)^{1/2}$ that is independent of oscillation amplitude. It is useful to think of this oscillator in terms of its potential energy function, which is quadratic as shown in Figure 1(a). In real oscillators, the restoring force is not linear for large displacements, but nonlinear so that

$$m\frac{\mathrm{d}^2 y}{\mathrm{d}t^2} = -ky(1 + \alpha_1 y + \alpha_2 y^2 + \cdots), \tag{2}$$

where the α_n are constants. The energy curve then has a distorted parabolic form such as that shown as an example in Figure 1(b). In the absence of damping, the total energy must remain constant so that the magnitude of the velocity is a simple function of the displacement, and the motion repeats cyclically. This means that the spectrum of such a nonlinear oscillator consists of exact phase-locked harmonics of the fundamental frequency, though this fundamental frequency depends upon the amplitude of the motion. For a reason that is derived from molecular physics, as discussed below, such a nonlinear oscillator is often confusingly called an "anharmonic oscillator."

A thin taut string of length L ideally obeys an equation of the form

$$m\frac{\partial^2 y}{\partial t^2} = T\frac{\partial^2 y}{\partial x^2}, \tag{3}$$

where x measures length along the string, m is the mass per unit length, and T is the string tension.

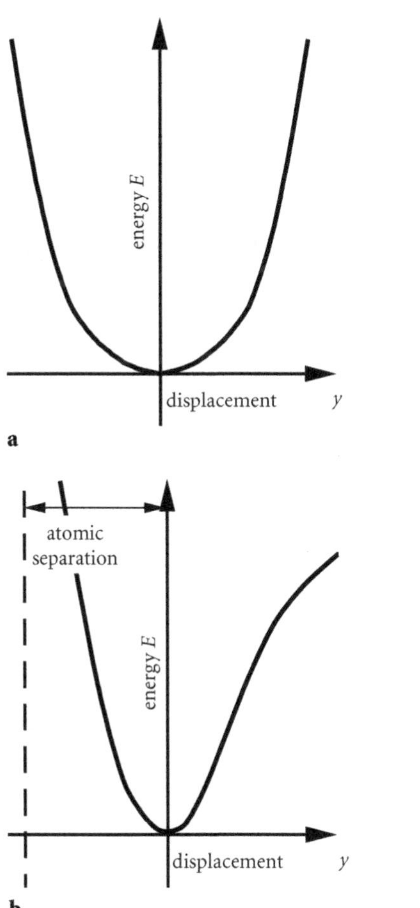

Figure 1. (a) Potential energy curve for a simple harmonic oscillator. (b) Potential energy curve for a typical nonlinear oscillator such as a diatomic molecule.

If its ends are rigidly fixed, then the mode frequencies are exact harmonics of the fundamental so that $f_n = (n/2L)(T/m)^{1/2}$. It is thus a multimode harmonic oscillator. The nonlinear frictional action of the bow on a violin reinforces the harmonicity of the modes and locks them into rigid phase relationship (Fletcher, 1999). Something very similar happens with wind instruments, which also have precisely harmonic spectra.

A thin stiff bar, on the other hand, obeys an equation of the form

$$m\frac{\partial^2 y}{\partial t^2} = K\frac{\partial^4 y}{\partial x^4}\,, \tag{4}$$

where K is the elastic stiffness. If the ends are free or rigidly clamped, then the mode frequencies are approximately $f_n \approx \frac{4}{9}(n+\frac{1}{2})^2 f_1$, and the overtones are very far from being harmonically related. Such an oscillator might be termed "inharmonic." The modes of a three-dimensional object such as a bell are even more complex (Fletcher & Rossing, 1998).

While sustained-tone musical instruments depend upon the nonlinearity of the active generator for their operation (bow, reed, or lip air-flow), the linear resonator (string or air column) determines the oscillation frequency, so that the pitch is nearly independent of loudness, and only the relative amplitudes of the harmonics change (Fletcher, 1999). Some Chinese opera gongs, however, make a virtue of nonlinearity so that, after an impulsive excitation, the pitch either rises or falls dramatically as the vibration dies away (Fletcher, 1985). The frequencies and relative intensities of upper partials determine the tone quality of a musical sound and have dictated the development of musical scales and harmonies (Sethares, 1998).

The human auditory system itself has some nonlinear aspects (Zwicker & Fastl, 1999), and, as in any forced nonlinear oscillator, these lead to the generation of harmonics ("harmonic distortion") and of multiple sum and difference tones ("intermodulation distortion"). In the ear these are chiefly apparent in the generation of the difference tone $|f_1 - f_2|$ when loud tones of frequencies f_1 and f_2 are heard simultaneously.

Optical absorption and emission spectra have many similarities to acoustic phenomena (Herzberg, 1950; Harmony, 1989). Diatomic molecules, for example, have interatomic potentials of the form shown in Figure 1(b) and thus constitute nonlinear oscillators. If the interatomic potentials were simply parabolic, as in Figure 1(a), then the quantum energy levels would have the form $E_n = (n+\frac{1}{2})h\nu$, where h is Planck's constant and ν is the classical vibration frequency labeled f_1 above. The wave functions describing the atomic vibration would then be either symmetric or antisymmetric, and the selection rule would dictate that n could change only by ± 1. There would thus be only a single absorption band consisting of the vibrational transition $0 \to 1$ surrounded by the allowed rotational transition lines.

For a more realistic model of the interatomic potential, as in Figure 1(b), the energy levels can be written as

$$\begin{aligned} E_n &= (n+\tfrac{1}{2})h\nu[1 + \beta_1(n+\tfrac{1}{2}) \\ &\quad + \beta_2(n+\tfrac{1}{2})^2 + \cdots]\,, \end{aligned} \tag{5}$$

where β_n are usually called the "coefficients of anharmonicity" and β_1 is always negative in practice. The asymmetry of the potential also relaxes the selection rule so that in addition to the strong allowed absorption transition $0 \to 1$, there are much weaker transitions from $n = 0$ to higher levels. The absorption bands associated with these transitions have frequencies that are in approximate, but not exact, harmonic relationship to the fundamental, and are called "overtone bands."

Although the quantum treatment of a nonlinear oscillator may seem to conflict with the classical treatment, and the term *anharmonic* certainly suggests this, there is not really any disagreement. The infrared spectrum is derived from transitions between two levels of different energies, and therefore different classical amplitudes, and the classical frequency depends upon amplitude, as for the Chinese opera gong.

NEVILLE FLETCHER

See also **Damped-driven anharmonic oscillator; Harmonic generation; Molecular dynamics; Nonlinearity, definition of; Ordinary differential equations, nonlinear; Partial differential equations, nonlinear; Spectral analysis**

Further Reading

Fletcher, N.H. 1985. Nonlinear frequency shifts in quasi-spherical shells: pitch glide in Chinese gongs. *Journal of the Acoustical Society of America*, 78: 2069–2073

Fletcher, N.H. 1999. The nonlinear physics of musical instruments. *Reports on Progress in Physics*, 62: 721–764

Fletcher, N.H. & Rossing, T.D. 1998. *The Physics of Musical Instruments*, 2nd edition, New York: Springer

Harmony, M.D. 1989. Molecular spectroscopy and structure. *A Physicist's Desk Reference*, edited by H.L. Anderson, New York: American Institute of Physics, pp. 238–249

Herzberg, G. 1950. *Molecular Spectra and Molecular Structure*, 2nd edition, New York: Van Nostrand

Sethares, W.A. 1998. *Tuning, Timbre, Spectrum, Scale*, London and New York: Springer

Zwicker, E. & Fastl, H. 1999. *Psychoacoustics: Facts and Models*, 2nd edition, Berlin and New York: Springer

P

PAINLEVÉ ANALYSIS

Most problems of physics, chemistry, biology, engineering, and practically every applied science are nonlinear, in the sense that effects are not related to their causes by simple proportionality. Further, in most cases, these problems are described by differential equations, which provide the rules by which the main observables (such as displacements, velocities, fields, chemical concentrations, or populations) vary with respect to changes in the continuous independent variables (space and/or time). If the independent variables are discrete, the governing rules become difference equations.

Unfortunately, there is no general theory for integrating differential equations globally, that is, for finding their solutions analytically, over the full domain of the independent variables and for arbitrary initial and boundary data. That is why scientists often concentrate their analysis in regimes where the system behaves linearly, for example, small vibrations of lattices and solid structures, Fourier's law of heat conduction, Ohm's and Kirchhoff's laws in electrical circuits, and Laplace's and Maxwell's equations in electromagnetism.

However, even in such problems, the theory of linear differential equations often meets with serious difficulties, when the solutions possess singularities, that is, space and/or time values at which these solutions or their derivatives become infinite. In linear problems, these singularities appear explicitly in the equations, and if they are of special form, the general solution can still be found analytically near the singularity, for example, in terms of convergent series expansions. The theory of Lazarus Fuchs and Ferdinand Frobenius was especially developed for this purpose in the 1860s and led to the remarkable discovery of special functions, associated with such classic second-order ordinary differential equations (ODEs), as those of Bessel, Hermite, Legendre, or the hypergeometric equation.

By contrast, nonlinear ODEs possess singular points that are not evident by inspection of the equation itself. For example, the famous Riccati equation

$$\mathrm{d}x/\mathrm{d}t = a + bx + x^2 \tag{1}$$

(a, b constant parameters) known since the 1700s (*See* **Riccati equations**), has solutions which behave as

$$x(t) = \frac{1}{t-c} + a_0 + a_1(t-c) + a_2(t-c)^2 + \cdots, \tag{2}$$

where c is an arbitrary constant. Thus, this type of singularity, at $t=c$, is called movable, since its location depends on the choice of initial conditions. Furthermore, after the pioneering work of Augustin Cauchy in the 1830s showed the importance of complex variables, a new branch of mathematics developed in which ODEs are studied in the complex domain, with the independent variable taking values $t = t_{\mathrm{R}} + \mathrm{i}t_{\mathrm{I}}$, where $t_{\mathrm{R}}, t_{\mathrm{I}} \in \Re$ are, respectively, the real and imaginary parts of t and $\mathrm{i} = \sqrt{-1}$. As is well known, in this t-plane, the singularity at $t=c$ in (2) is called a pole and limits the convergence region of the Taylor expansion of the function $x(t)$ about any nearby point $t=t_0$ (where $x(t)$ is analytic), to $| t-t_0 | < | c-t_0 |$.

Still, when moving around a circle centered at the pole $t=c$ (and enclosing only this singularity), $x(t)$ always returns to the same values, implying that it is a single-valued function in that region. This is not what happens if $t=c$ is a branch point of $x(t)$ of the type

$$x(t) = (t-c)^{\alpha} + \cdots \quad \text{as } t \to c \tag{3}$$

with α not an integer. For example, if α is a rational number, $\alpha = p\,/\,q$, it takes q turns around $t=c$ for $x(t)$ to return to its starting value, demonstrating that it is a multi-valued function, describing a finitely sheeted Riemann surface, while $t=c$ is called an algebraic singularity. If, on the other hand, α in (3) is irrational or complex, $t=c$ is called a transcendental branch point and $x(t)$ describes an infinitely sheeted Riemann surface. Finally, if an expansion of $x(t)$ about $t=c$ contains logarithmic terms of the form $\log(t-c)$, it is again an infinitely multi-valued function and $t=c$ is called a logarithmic branch point.

Paul Painlevé, the great French mathematician and statesman (he was Prime Minister of France in 1917 and 1925), pursuing the idea that the problems with the simplest singularities would be the easiest to solve, set out to determine which first-order ODEs of the form

$$\mathrm{d}x/\mathrm{d}t = f(t, x) \tag{4}$$

with f rational in x and analytic in t, are free from movable branch points; that is, their only movable singularities are poles. His remarkable discovery was that there is only one such equation: the Riccati equation (1), with the coefficients of $1, x$, and x^2 being arbitrary analytic functions of t. Then, in a remarkable series of papers in the late 1890s and early 1900s, Painlevé and his coworkers (notably Bertrand Gambier) studied the same question on all second-order ODEs

$$\mathrm{d}^2x/\mathrm{d}t^2 = f(t, x, \mathrm{d}x/\mathrm{d}t) \tag{5}$$

with f rational in $\mathrm{d}x/\mathrm{d}t$, algebraic in x, and analytic in t. After painstaking analysis, they showed that there are 50 such equations, 44 of which can be explicitly integrated and solved in terms of known functions. For the remaining six equations (called Painlevé I–VI), such reduction was not possible, and new functions had to be introduced called the Painlevé transcendents. The first two of these equations are

$$\begin{aligned} &\text{Painlevé I}: \quad \mathrm{d}^2x/\mathrm{d}t^2 = 6x^2 + t, \\ &\text{Painlevé II}: \quad \mathrm{d}^2x/\mathrm{d}t^2 = 2x^3 + tx + a \end{aligned} \tag{6}$$

with Painlevé III–VI being increasingly more complicated to write down. The interested reader will find them all discussed in the excellent books by Ince (1956) and Davis (1962), where Painlevé's approach to their discovery is also described.

It is important to recall, however, that independent of all this progress, the idea of using singularity analysis to solve systems of ODEs by requiring that their solutions have only poles was also used by the Russian mathematician Sophia Kovalevsky in 1888 to find one more solvable case of the rotating rigid body that bears her name (Kovalevsky's top). Still, the requirement that all solutions of a system of ODEs possess only poles as movable singularities in the complex t-plane has come to be referred to date as the Painlevé property.

Surprisingly enough, the Painlevé theory was not widely appreciated at first and remained largely unknown until it was revived, many decades later, in connection with exactly solvable partial differential equations (PDEs). More specifically, after the discovery that a number of PDEs are integrable by the inverse scattering transform (IST), Ablowitz and Segur observed in the mid-1970s that all ODE reductions of these PDEs had the Painlevé property and led, in fact, to some of the Painlevé I–VI equations, (6) (see Ablowitz & Segur, 1981, as well as Ablowitz & Clarkson, 1991).

Thus, the famous Ablowitz, Ramani, and Segur conjecture was formulated as follows: If a PDE is solvable by IST, all its ODE reductions obey the Painlevé property. Clearly, this conjecture provides a useful criterion to test the integrability of a given PDE by studying its ODE reductions—a much easier task than showing whether or not it is solvable by IST.

This discovery had enormous consequences, as it attracted the attention of many researchers working in the rapidly expanding field of nonlinear dynamics. The fact that most nonlinear dynamical systems of ODEs possess "irregular," "unpredictable," or chaotic solutions evidently suggests that they cannot be integrated and solved in terms of known functions, whose behavior is perfectly regular and predictable. It would thus be extremely interesting to find, in models of physical, chemical, or biological dynamics, integrable cases whose solutions would be clearly distinguished from those displaying chaotic behavior.

Such new, integrable systems could now be identified by means of the Painlevé property. Perhaps one of the most important applications occurred in Hamiltonian systems, like the famous Hénon–Heiles problem

$$H = \frac{1}{2}(\dot{x}^2 + \dot{y}^2 + Ax^2 + By^2) + x^2y + \frac{C}{3}y^3 \tag{7}$$

of two degrees of freedom (dots denote differentiation with respect to t). For $A = B = 1$ and $C = -1$, this problem had been extensively studied since the mid-1960s, as an example of a system whose chaotic regions grow dramatically as the total energy $H = E$ increases from 0 to $\frac{1}{6}$. Using the Painlevé analysis, it was first shown by Bountis et al. (1982) that (7) is completely integrable in exactly three cases:

(i) $C = 1$ and $A = B$ (known to be separable in the variables $s = x + y, d = x - y$).
(ii) $C = 6$ and any A and B.
(iii) $C = 16$ and $B = 16A$.

In each of the new cases (2) and (3), the second integral—besides Hamiltonian (7)—was also provided and their complete integrability in the sense of Liouville–Arnol'd was established.

Soon, many novel and highly nontrivial examples of Hamiltonian systems with N degrees of freedom were identified by requiring that their solutions possess the Painlevé property. Rigorous connections between integrability and the Painlevé property were also developed, mainly by Adler and van Moerbeke in the early 1980s, who used algebraic geometry to establish that if a Liouville–Arnol'd integrable Hamiltonian system has rational integrals which continue to describe tori in the complex domain, then its solutions have only movable poles (the converse, though not completely proved, is also believed to hold). A wealth of examples of integrable non-Hamiltonian systems were also discovered, related to models of physical, chemical, or biological interest. Finally, the Painlevé analysis was extended and applied by Weiss, Tabor, and Carnevale to test integrability directly on a PDE, without reference to its ODE reductions (see Ramani et al., 1989).

But the importance of Painlevé analysis does not end here. After the early observation that any deviation from the Painlevé conditions generically introduces logarithmic terms, researchers began to look for connections between the violation of the Painlevé property and non-integrability, in the sense of non-existence of analytic, single-valued integrals such as (7). In that regard, the work of the Russian mathematicians V.V. Kozlov and S.L. Ziglin in the late 1970s turned out to be extremely important (see Kozlov, 1983). They showed, using Mel'nikov's theory, that one of the most fundamental causes of chaotic behavior, that is, the transverse intersection of stable and unstable manifolds of saddle fixed points on a Poincaré map, implies the presence of infinitely multi-valued solutions and non-existence of a second integral in a two-degree-of-freedom Hamiltonian systems.

Kozlov's and Ziglin's results inspired a series of interesting papers by H. Yoshida in the 1980s, where he used them to study the variational equations of simple periodic solutions in a large class of N-degree-of-freedom Hamiltonians. Yoshida was able to prove that when these variational equations do not satisfy certain conditions imposed by the existence of a full set of global single-valued integrals, such a set of constants cannot exist, thus demonstrating non-integrability for vast parameter ranges in many Hamiltonians of physical interest (see Ramani et al., 1989).

Finally, what about systems of ODEs (or PDEs) whose solutions possess only algebraic singularities, that is, points $t = c$, near which solutions have asymptotic expansions of the form

$$x(t) = (t - c)^{p/q} + \sum_{n=0}^{\infty} a_n (t - c)^{n/q} \qquad (8)$$

without any other singularities present? In the 1980s, it was thought that this so-called weak Painlevé property, under some conditions on p / q, could also be related to complete integrability, in the Liouville–Arnol'd sense for N-degree-of-freedom Hamiltonian systems. In the 1990s, however, it was found that the situation is considerably more subtle.

First, it was shown by Goriely that many weak Painlevé examples could be transformed to systems having the usual Painlevé property after some rather general changes of coordinates (see Goriely, 2001). Then it was observed that there were weak Painlevé systems with chaotic solutions, which are evidently not integrable. Thus, the concept of globally finitely sheeted solutions (FSS) was introduced to distinguish integrable systems with algebraic singularities (8) from non-integrable ones, whose solutions are only locally finitely sheeted (see Bountis, 1992, 1995).

This is done as follows: every time a system of weak Painlevé ODEs, integrated numerically around arbitrarily large contours, showed evidence of "lattice" singularity patterns in the complex t-plane and FSS, it has been shown to be transformable to one that is completely integrable and possesses the usual Painlevé property. On the other hand, many such systems were also found which showed "dense" singularity patterns and evidence of globally infinitely sheeted solutions (ISS) for large enough contours. Although not solvable, such systems can still be integrable if they are hyperelliptically separable, that is, if they can be described by N holomorphic differentials on the Jacobian of a hyperelliptic curve of genus $g < N$ (see Abenda & Fedorov, 2000; Abenda et al., 2001).

Finally, in the 1990s, the connection between singularities and integrability was extended to discrete dynamical systems, described by difference equations. Here, since the methods of complex analysis for ODEs no longer apply, a novel criterion was introduced (see Grammaticos et al., 1991). It was based on the observation that if a difference equation such as

$$x_{n+1} = F(x_n, x_{n-1}), \quad n = 0, 1, 2 \ldots \qquad (9)$$

with F rational, possesses an integral of the form

$$I(x_n, x_{n-1}) = \text{const.}, \quad n = 0, 1, 2 \ldots \qquad (10)$$

yielding a family of curves in the x_n, x_{n+1} plane and precluding the presence of chaos, then x_n is infinite at some $n = m$ and becomes finite again at some later $n = m' > m$. This so-called singularity confinement criterion, whose success to date remains a mystery, has yielded a wealth of integrable discrete systems and has enabled many researchers to develop Painlevé difference equations, analogous to the famous Painlevé ODEs Painlevé I–VI (see Conte, 1999).

TASSOS BOUNTIS

See also **Hénon–Heiles system; Inverse scattering method or transform; Mel'nikov method; Riccati equations**

Further Reading

Abenda, S. & Fedorov, Y. 2000. On the weak Kowalevski–Painlevé property for hyperelliptically separable systems. *Acta Applicandae Mathematicae*, 60: 137

Abenda, S., Marinakis, V. & Bountis, T. 2001. On the connection between hyperelliptic separability and Painlevé integrability. *Journal of Physics* A, 34: 3521

Ablowitz, M.J. & Clarkson, P. 1991. *Solitons, Nonlinear Evolution Equations and Inverse Scattering*, Cambridge and New York: Cambridge University Press

Ablowitz, M.J. & Segur, H. 1981. *Solitons and the Inverse Scattering Transform*, Philadelphia: SIAM

Adler, M. & van Moerbeke, P. 1988. *Algebraically Completely Integrable Systems*, Boston: Academic Press,

Bountis, T. 1992. What can complex time tell us about real time dynamics? *International Journal of Bifurcation and Chaos*, 2(2): 217

Bountis, T. 1995. Investigating non-integrability and chaos in complex time. *Physica* D, 86 (1995): 256
Bountis, T., Segur H. & Vivaldi, F. 1982. Integrable Hamiltonian systems and the Painlevé property. *Physical Review* A, 25: 1257
Conte, R. (editor). 1999. *The Painlevé Property: One Century Later*, New York: Springer
Davis, H.T. 1962. *Introduction to Nonlinear Differential and Integral Equations*, London: Dover
Goriely, A. 2001. *Integrability and Non-integrability of Dynamical Systems*, Singapore: World Scientific
Grammaticos, B., Ramani, A. & Papageorgiou, V. 1991. Do integrable mappings have the Painlevé property? *Physical Review Letters*, 67(14): 1825–1828
Ince, E.L. 1956. *Ordinary Differential Equations*, London: Dover
Kozlov, V.V. 1983. Integrability and non-integrability in Hamiltonian mechanics. *Russian Mathematical Surveys* (Mathematical Review C) 38: 1
Ramani, A., Grammaticos, B. & Bountis, T. 1989. The Painlevé property and singularity analysis of integrable and non-integrable systems. *Physics Reports*, 180 (3): 160.
Tabor, M. 1989. *Chaos and Integrability in Nonlinear Dynamics*, New York: Wiley Interscience

PARAMETRIC AMPLIFICATION

Parametric generation and amplification of oscillations are based on the phenomenon of parametric resonance, using the work done by an external force under periodic variation of the parameters of an oscillatory system.

A simple mechanical system in which parametric resonance may occur is a pendulum, with the length of the cord (l) changing with time. If l is decreased when the pendulum is in the lower position and increased in its upper position, the average work done by an external force over one period is positive and oscillations of the pendulum build up. This phenomenon is the underlying principle of setting a swing into motion. When one stands on a swing, it is easier to rock it by squatting at the top position of the swing and standing up at the bottom position, thus shifting the center of mass of the person on the swing twice in one period.

An electric analog of a pendulum of variable length is an oscillatory circuit of variable capacitance. The capacitance C can be changed, for example, by mechanically moving the capacitor plates together or apart. For the amplitude of oscillations to build up, energy should be fed to the circuit by performing work against the electrostatic field forces of the capacitor. This means that the plates should be moved apart when the charge q on the capacitor is maximal and moved together when the charge on the capacitor returns to zero (Figure 1).

Such a periodic change in capacitance (pumping-up process) results in growing energy of the oscillations and appearance of instability in the system. The build-up of energy occurs at a pump frequency satisfying the relation

$$\omega_p = 2\omega_0/n, \tag{1}$$

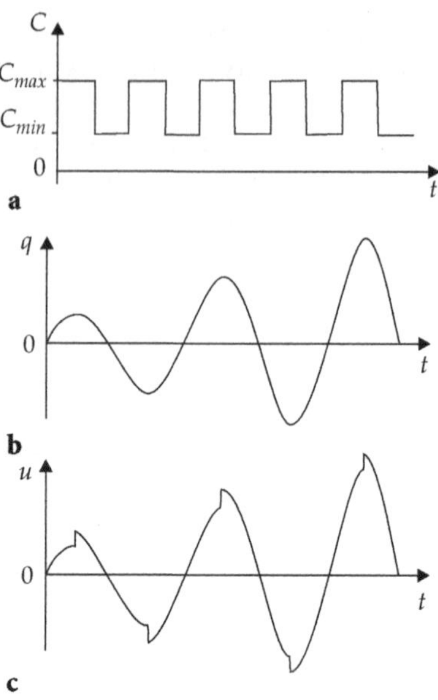

Figure 1. Periodic variations of capacitance C of capacitor (a), capacitor charge q (b), and voltage u (c) at parametric resonance.

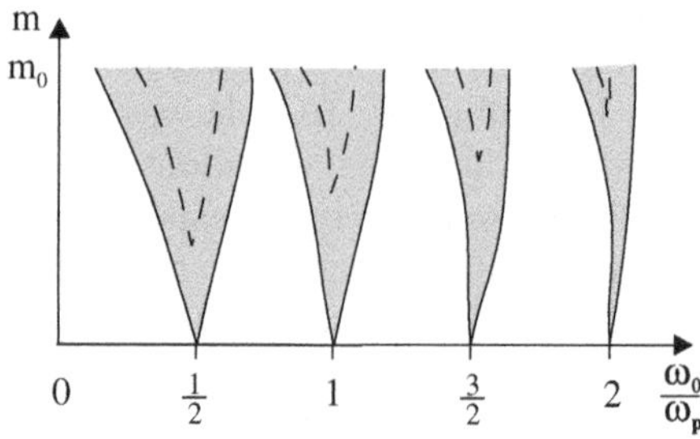

Figure 2. Mathieu instability zones within which parametric resonance may occur (solid line—without damping; dashed line—with damping; ω_0 natural frequency of the oscillatory circuit; ω_p pump frequency.

where n is an integer, ω_0 is the natural frequency of circuit oscillations, and ω_p is the pump frequency. Build-up is most effective for $n = 1$, that is, when the period of capacitance variation is half the period of natural oscillations of the circuit. Interestingly, energy build-up (parametric resonance) is possible not only for the values of ω_p satisfying Equation (1) but also at some deviations from these values within Mathieu instability zones (see Figure 2). The greater the width of Mathieu zones, the greater the extent of parameter variation $m = (C_{max} - C_{min}) / (C_{max} + C_{min})$.

In oscillatory systems with several degrees of freedom (e.g., in a system with two coupled circuits, coupled pendulums, and so on), normal modes with different natural frequencies ω_1 and ω_2 may exist. As parameters of such a system change, oscillations may build up (parametric resonance) not only at a frequency satisfying condition (1) for an arbitrary natural frequency, but also for their linear combination, for instance, when a parameter changes with sum frequency

$$\omega_p = \omega_1 + \omega_2. \quad (2)$$

Resonance coupling is also possible at $\omega_p = \omega_1 - \omega_2$. In this case, however, no build-up of oscillations occurs. Instead, energy is periodically pumped between the modes ω_1 and ω_2. For such an oscillatory system, the pump power P_p input to the system at frequency ω_p and the powers P_1 and P_2 consumed at frequencies ω_1 and ω_2 are proportional to the corresponding frequencies (the Manley–Rowe relation):

$$\frac{P_p}{\omega_p} = \frac{P_1}{\omega_1} = \frac{P_2}{\omega_2}. \quad (3)$$

Parametric resonance may also provide conditions for excitation of normal modes in oscillatory media possessing an infinite number of degrees of freedom. In 1831, Michael Faraday had observed that standing waves are excited at frequency $\omega_{p/2}$ in the liquid in a vessel vibrating in the vertical direction at frequency ω_p. The classic experiment staged by Franz Melde in 1859 revealed excitation of transverse oscillations (standing waves) in a string, one end of which was fastened to the prong of a tuning fork. When the tuning fork vibrates and the string is tight, the string performs transverse oscillations with a frequency equal to half the frequency of the tuning fork.

A distinguishing feature of parametric resonance in systems with distributed parameters is that the build up of oscillations depends strongly on the relationship between the variation in space of the system parameters and the spatial structure of the forcing oscillations (waves). For example, if the pump that changes the parameters of the medium is a traveling wave with frequency ω_p and wave vector $\boldsymbol{k}_p$, then natural waves with frequencies ω_1 and ω_2 and wave vectors $\boldsymbol{k}_1$ and $\boldsymbol{k}_2$ are excited, provided the conditions of parametric resonance are fulfilled both in space and time; thus

$$\omega_p = \omega_1 + \omega_2, \quad \boldsymbol{k}_p = \boldsymbol{k}_1 + \boldsymbol{k}_2. \quad (4)$$

The potentialities of using parametric resonance for creating a parametric generator and amplifier were studied by Leonid I. Mandelshtam and Nikolai D. Papaleksi (1931–1933). They constructed capacitive and inductive parametric machines that transformed mechanical energy into electrical energy as a result of variation of the magnitudes of capacitance and inductance during shaft rotation. Parametric generators and amplifiers found practical application in the 1950s when semiconductor parametric diodes appeared, the capacitance of which depends on applied voltage, and when variable capacitors using the properties of ferroelectrics and variable inductors using the properties of ferrites and superconductors were developed. Low-noise variable capacitance parametric amplifiers based on high-frequency parametric diodes (varactors) are used in high-frequency receiving microwave devices employed in radar and radio astronomy, among other applications. Parametric generators are capable of generating oscillations whose phase can take one of two values differing by π. This feature of parametric generators was noted by Eiichi Goto who proposed in 1954 to use such generators, which he referred to as "parametrons," as logic switching elements in computers. Parametric generators are widely employed in optics as tunable coherent generators.

Note finally that parametric action on a nonlinear oscillatory system (for example, a nonlinear pendulum with variable-length cord or a nonlinear circuit with periodically varying capacitance and inductance) may also lead to dynamic oscillations.

VLADIMIR SHALFEEV

See also **Diodes; Dynamical systems; Harmonic generation; Manley–Rowe relations; Nonlinear toys**

Further Reading

Leven, R.W. & Koch, B.P. 1981. Chaotic behavior of a parametrically excited damped pendulum. *Physics Letters*, 86(2): 71–74

Rabinovich, M.I. & Trubetskov, D.I. 1989. *Oscillations and Waves in Linear and Nonlinear Systems*, Dordrecht and Boston: Kluver

Scott, A. 1970. *Active and Nonlinear Wave Propagation in Electronics*, New York: Wiley

PARAMETRIC DECAY INSTABILITY

See **Nonlinear plasma waves**

PARTIAL DIFFERENTIAL EQUATIONS, NONLINEAR

A partial differential equation (PDE) is a functional equation of the form

$$F\left(x_1, x_2, \ldots, x_n, z_1, \ldots, z_m, \frac{\partial z_\mu}{\partial x_\nu}, \ldots, \frac{\partial^2 z_\mu}{\partial x_1^2}, \frac{\partial^2 z_\mu}{\partial x_1 \partial x_2}, \ldots\right) = 0 \quad (1)$$

with m unknown functions $z_1, z_2, \ldots, z_m$ with n independent variables $x_1, x_2, \ldots, x_n$ $(n > 1)$ and at least one of the partial derivatives of z_μ $(\mu = 1, 2, \ldots, m)$ with respect to x_ν $(\nu = 1, 2, \ldots, n)$. In general, there can be a system of coupled PDEs.

For a single equation, the order of the PDE is p if the highest derivative has order p, with an analogous definition for a system of equations. A PDE becomes an ordinary differential equation if the number of independent variables n is one. It is linear if the equation is linear with respect to z and its partial derivatives, meaning it is of the first degree in the unknown function as well as its derivatives. It is called quasilinear if it is linear with respect to the highest-order partial derivatives which have coefficients that depend only on lower-order partial derivatives and the independent variables. In addition, a PDE is called semilinear if the coefficients of the highest-order derivatives depend on the independent variables only. A PDE is called nonlinear if it does not fit into any of the above categories.

Initial value problems (IVPs) for evolution equations arise where one of the independent variables x_i of Equation (1) represents time t and an initial condition at time $t=0$ is given. Additionally, there are boundary value problems (BVPs), when spatial variables are involved and boundary conditions are specified along the n-dimensional boundary of the domain (where n is the number of independent spatial variables). There can also be mixed initial value-boundary value problems (IVBVPs), where both types of conditions are given.

For a single PDE, the number of initial conditions is generally one less than the highest temporal derivative, while the number of spatial boundary conditions is half the order of the highest spatial derivatives in the equation. For general nonlinear equations, there are no precise definitions.

Second-order semilinear PDEs have the general form

$$a\frac{\partial^2 z}{\partial x_1^2} + 2b\frac{\partial z}{\partial x_1 \partial x_2} + c\frac{\partial^2 z}{\partial x_2^2} + \text{l.o.t.} = 0,$$

where a, b, c are given functions of x_1 and x_2 or constants and l.o.t. stands for lower order terms. The coefficient matrix A of the corresponding second-order system has the determinant

$$\det A = \begin{vmatrix} a & b \\ b & c \end{vmatrix}. \tag{2}$$

A second-order semilinear PDE is said to be of elliptic type if $\det A\ (=ac-b^2)>0$, of hyperbolic type if $\det A<0$, and of parabolic type if $\det A=0$ for all points x_1, x_2 of the domain D. The determinant of A is positive if both eigenvalues of the matrix A are positive or both negative. The determinant is negative if the eigenvalues are of opposite sign and zero if one of them is zero. The normal forms of the associated differential operators L are given by

$$\det A > 0: \quad \frac{\partial^2}{\partial x_1^2} + \frac{\partial^2}{\partial x_2^2} \quad \text{(elliptic)},$$

$$\det A < 0: \quad \frac{\partial^2}{\partial x_1^2} - \frac{\partial^2}{\partial x_2^2} \quad \text{(hyperbolic)},$$

and

$$\det A = 0: \quad \frac{\partial^2}{\partial x_1^2} \quad \text{(parabolic)}.$$

Boundary value conditions may be "Dirichlet" (where the value of the solution is given along the boundary ∂D of the domain D), "Neumann" (where the normal derivative $\partial z / \partial n$ is specified along the boundary ∂D), and "Robin" (where both the value and the normal derivative of the solution $\partial z / \partial n + az$ satisfy an equation along the boundary of the domain ∂D).

The archetypical example for an elliptic PDE is the Laplace equation

$$\frac{\partial^2 z}{\partial x^2} + \frac{\partial^2 z}{\partial y^2} = 0;$$

for a hyperbolic PDE, it is the wave equation

$$\frac{\partial^2 z}{\partial t^2} - c^2\frac{\partial^2 z}{\partial x^2} = 0;$$

and for a parabolic PDE, it is the diffusion equation

$$\frac{\partial z}{\partial t} - \nu\frac{\partial^2 z}{\partial x^2} = 0.$$

The coefficient matrix A (2) can depend on first-order derivatives if the PDE is second-order quasilinear and the classes are analogous. Quasilinear and strictly linear second-order PDEs are classified according to these three classes. For a given nonlinear PDE of higher order or more than two variables, it is not useful to classify according to these three classes, instead, one classifies it according to the physical process it describes. In general, boundary value problems are often associated with elliptic equations and initial value problems with hyperbolic and parabolic equations.

Nonlinear PDEs describe many important processes in nature because often an approximate linear PDE is not sufficient to describe the process. Examples are the dynamics of fluids, gases, and elastic media; the equations from general relativity and quantum electrodynamics; and population dynamics in biology and chemistry.

The two main characteristics of such processes are the nonlinear interactions, due to the nonlinearities of the PDE and the possibility of self-organization into coherent structures.

There are three types of processes that one can distinguish. Reversible processes such as waves; irreversible; processes such as reaction, diffusion and heat flow; and stationary processes. A process is irreversible/reversible if it is impossible/possible to reverse the process in time and stationary processes are independent of time. In general, reversible processes are described by hyperbolic equations, irreversible processes by parabolic equations, and stationary processes by elliptic equations.

Most physical systems comprise a combination of all three types of processes, but often, some processes are dominant and by using a scaling argument, it is possible to neglect less dominant processes and simplify the PDE to the degree that it describes the process adequately but is still easy to work with.

The nature of irreversible processes such as diffusion, is that they tend to smooth the solution in time, whereas for reversible processes singularities can develop, for example, shock waves in gas dynamics. For stationary processes, the solutions depend on the smoothness of the boundary conditions and external forces.

Solutions to nonlinear PDEs are often obtainable only by numerical integration. In integrable cases, on the other hand, analytic solutions can be found using advanced techniques such as the inverse scattering method, the associated Lax theory, and the Bäcklund transformation. There are other techniques that are helpful although not exact, such as Floquet and Mel'nikov theory.

In general, for nonlinear PDEs common tools such as the linear superposition principle are not valid anymore, although it is sometimes possible to find a nonlinear superposition principle. Methods such as the Fourier and Laplace transforms and Green's function, in general, are not applicable to nonlinear PDEs. Alternatively, one can attempt classical solution methods such as the separation of variables or transformation of variables to reduce the nonlinear PDE to a system of equations that is solvable analytically.

ANDREAS A. AIGNER

See also **Burgers equation; Korteweg–de Vries equation; Nonlinear Schrödinger equations; Separation of variables; Sine-Gordon equation**

Further Reading

Courant, R. & Hilbert, D. 1966. *Methods of Mathematical Physics II*, New York: Wiley

Debnath, L. 1997. *Nonlinear Partial Differential Equations*, Basel: Birkhäuser

Garabedian, P. 1964. *Partial Differential Equations*, New York: Wiley

Ockendon, J., Howison, S., Lacey, A. & Movchan, A. 1999. *Applied Partial Differential Equations*, Oxford and New York: Oxford University Press

PARTICLE ACCELERATORS

Particle accelerators have been used for much of the 20th century to investigate the properties of matter. The first accelerators of charged particles used steady (d.c.) electric fields, such as those obtained from a Van de Graaff generator. These devices could accelerate particles to above an energy of a million electron volts (MeV), which was sufficient to probe atomic nuclei. Applications needing higher energies continually emerged, leading to new types of devices, particularly ones having the charged particles move in a circular orbit. In that configuration, with the additional concept that electric fields and particles can be synchronized, such that on each crossing of a gap, particles encounter a field that is accelerating, the basic accelerator configuration was created. Such devices went through a series of developments from a fixed frequency cyclotron to a frequency modulated cyclotron and finally the synchrotron. In the first two devices, the equilibrium orbit spirals out as the energy increases, while in the last device, the orbit is held fixed, and the bending magnetic field is increased.

To keep the particles on a prescribed essentially circular orbit over many traversals of the device, the orbit must be stable to transverse perturbations. Furthermore, to achieve continuous acceleration, the particle motion must be stable about the equilibrium accelerating phase of the electric field. Because the particles are kept close to an equilibrium orbit, the transverse stability can be analyzed from linear stability analysis, using a first-order expansion of the fields about the equilibrium. When the gradients of the magnetic guide field are chosen to give stable oscillations around the equilibrium orbit, they are called betatron oscillations, after the betatron, an early circular induction accelerator for which the oscillations were first analyzed (Kerst & Serber, 1941). For particles traveling in a straight line, the forces resulting from magnetic gradients cannot focus in both transverse directions simultaneously. This is not true for circular orbits, as the gradient in the centrifugal force supplies an extra focusing force to allow focusing in both directions. The resulting homogenous linear equations of motion for the simplest, assumed azimuthally symmetric, device are

$$\frac{d^2z}{d\theta^2} + nz = 0 \tag{1}$$

and

$$\frac{d^2x}{d\theta^2} + (1-n)x = 0, \tag{2}$$

where z is the vertical position and x the horizontal deviation from the equilibrium orbit r_0, and θ is the azimuthal angle around the machine. Here, n is the field index

$$n = -\frac{r_0}{B_0}\frac{\partial B_z}{\partial r} \tag{3}$$

with B_0 the equilibrium magnetic field. From (1) and (2) the stability condition is $0 < n < 1$. The azimuthally symmetric focusing of this type is now called weak focusing because the limitations on stability make the relative frequencies or tunes $\nu_z \equiv \omega_z / \omega_0 = n^{1/2}$ and $\nu_x \equiv \omega_x / \omega_0 = (1-n)^{1/2}$ both less than 1. The weak restoring forces required large vacuum chambers, limiting the size and field strength of the devices and therefore the energy to which particles could be accelerated.

The breakthrough in developing a synchrotron in which particles could be closely contained around their equilibrium orbit was the understanding that much stronger focusing could be achieved by alternating a section which strongly focused vertically while defocusing radially with a section that defocused vertically and focused radially. The net result, from simple lens theory, is that the pair can be strongly focusing. This led to the invention of the alternate gradient (AG) synchrotron (Courant et al., 1952). The simplest analysis of the linearized equations of motion is of the Hill equation

$$\frac{\mathrm{d}^2 y}{\mathrm{d}\theta^2} + K(\theta)y = 0 \tag{4}$$

with $K(\theta + \theta_L) = K(\theta)$, where θ_L is the repeating angular period. The transformation for a period can be analyzed for stability from Floquet theory by setting

$$y(\theta + \theta_L) = y(\theta)\mathrm{e}^{\pm \mathrm{i}\sigma} \tag{5}$$

and solving the transfer matrix, over the repeating period θ_L, for stability. This was done in the original paper, for a simple transfer matrix, leading to a stability diagram. However, even the early AG synchrotrons were more complicated, with bending magnets, focusing quadruple magnets, and straight sections in which resonators supplied the acceleration fields. For various reasons, described in part below, modern machines with very large radii, operating at many billions of electron volts (BeV or GeV), are much more complicated, leading to stability calculations with a very large number of elements.

However, even at the simplest level, calculating the linear stability of orbits is not sufficient. Nonlinear terms, although small, can have very important consequences. The motions given by (1) and (2), or more generally by (4), in each of the transverse directions are independent. The nonlinear terms couple the transverse dimensions and lead to a wide variety of resonances that can degrade the beam cross section. These difference resonances of various strengths must be avoided when finding a proper operating point in the $\nu_x - \nu_z$ parameter space. Additionally, magnet imperfections lead to entirely new classes of resonances. The effect of the increasing energy, coupled with the space charge of the beam of ions or electrons, leads to a movement of the operating point through the parameter space, such that all resonances cannot be avoided.

In addition to the primarily linear transverse oscillations, there are fundamentally nonlinear oscillations associated with the accelerating fields, called synchronous oscillations. For circular devices, these oscillations couple longitudinal and transverse motions. The frequencies of these electric-field-driven oscillations are generally much lower than the transverse betatron oscillations so that it is usually sufficient to consider their radial motion as adiabatic when studying the free oscillations, that is, using averaging methods. Depending on how relativistic the particles are, the phase motion may be either primarily longitudinal or radial. To simplify the treatment, we consider the synchronous motion in a linear accelerator such that transverse motions are unimportant to lowest order. The fields in both linear accelerators and synchrotrons are generated by high-power electromagnetic sources which operate in structures, for example, loaded waveguides in linear accelerators and resonant cavities in synchrotrons, that have a primary traveling-wave field synchronous with the particles. Nonsynchronous harmonic fields also exist, but are usually not important.

The equations of motion along the axis of a linear accelerator are the force equation

$$\frac{\mathrm{d}p}{\mathrm{d}t} = eE\sin\phi \tag{6}$$

and the equation for the change in phase

$$\frac{\mathrm{d}\phi}{\mathrm{d}t} = \omega\left(1 - \frac{v}{v_0}\right), \tag{7}$$

where v_0 is the wave velocity, ω the r.f. frequency, and E the applied field. The phase stable particle has $v = v_0$, and for stable acceleration, v_0 must be determined from (6), where p and v are relativistically related. The motion of particles around the stable phase is given by trajectories in the $p - \phi$ phase plane with the stable fixed point at $\phi = \phi_\mathrm{s}$, satisfying $\mathrm{d}\phi\,/\,\mathrm{d}t = 0$, and an unstable fixed point on a separatrix orbit satisfying, $\phi_\mathrm{u} = -\phi_\mathrm{s}$. The phase diagram indicates that particles injected between ϕ_u and some limiting ϕ on the separatrix trajectory with velocity v_0 oscillate nonlinearly about ϕ_s, while those outside the phase interval are not stably captured. For a general account of the basic ideas as described above, and also the many types of accelerators and variants on the concepts, the reader is referred to basic accelerator texts. Two such texts, which include many useful original references are Livingood (1961) and Kolomensky & Lebedev (1966). The early dates of publication indicate the maturity of this field.

As AG synchrotrons increased in size and energy, it became increasingly important to match the emittance phase space of the incoming particles to the acceptance phase space of the accelerator. This was to ensure the highest beam brightness, particularly as the long acceleration cycle led to low duty ratios. Phase space matching techniques were developed and exploited at the CERN high-energy physics center (Hereward et al., 1956). The transverse acceptance of a simplified AG synchrotron is illustrated in Figure 1.

The need for higher beam brightness led to the invention of the fixed field alternating gradient (FFAG) synchrotron which injected a number of pulses whose phase spaces were spatially separated by the energy increase. This phase space beam-stacking in the FFAG synchrotron further enhanced the usefulness of employing phase space matching concepts (Symon &

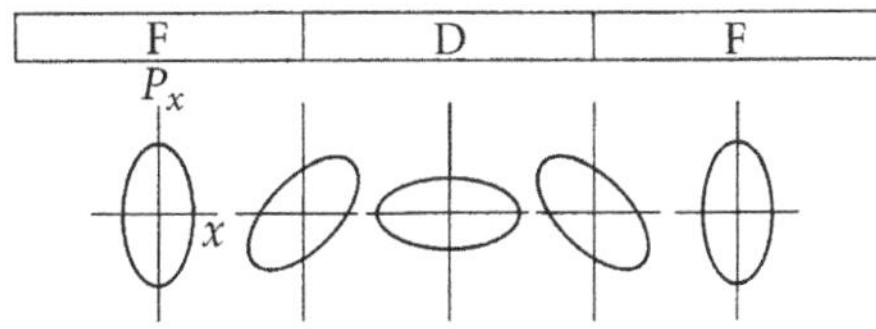

Figure 1. Phase-space transformations in an AG synchrotron.

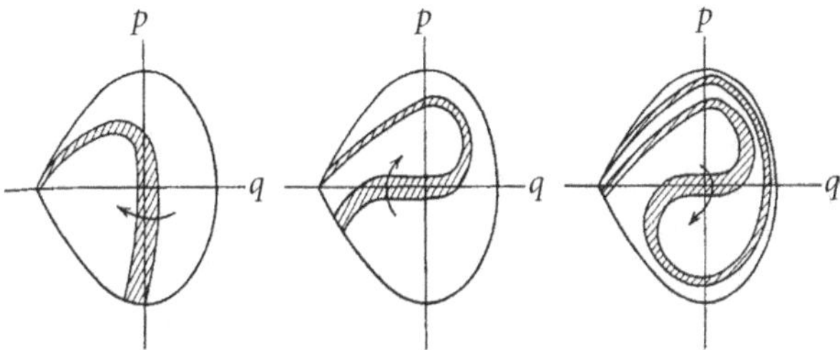

Figure 2. Filamentation of phase space due to nonlinear oscillations.

Sessler, 1956). Phase space matching was also used to increase the efficiency of linear accelerators, and of microwave devices (*See* **Electron beam microwave devices**). The effect of a mismatch of emittance and acceptance is illustrated in Figure 2 in which a beam with small momentum spread (cross-hatched region) is injected into the acceptance phase space of a synchronous oscillation in a linear accelerator or synchrotron, leading to an increase of the effective phase space area by filamentation (Lichtenberg, 1969).

The development of storage rings greatly increased the importance of keeping the phase space of the injected particles to a minimum. Conceived as a device to obtain high beam brightness, it rapidly became essential for operation of proton accelerators at energies many times the proton rest energy for which the energy in the center of mass of collisions with stationary matter was greatly reduced due to momentum conservation. This led to intersecting storage rings of particles traveling in opposite directions and a very severe requirement of high beam brightness (O'Neil, 1956). Storage rings, which store particles over many machine transits, have required the control of beam spreading. In electron storage rings, the synchrotron radiation which set a limit to electron energy is also effective in reducing the beam phase space (Robinson, 1958). For protons, which are little affected by radiation, stochastic cooling was invented to perform a similar function, a requirement that is essential for proton–antiproton colliders (Van der Meer, 1972).

The development of higher energy accelerators and higher intensity beams has led to new methods of analysis and various problems to be solved. The largest AG synchrotrons include many correcting magnets as well as the bending magnets, quadrupoles, and straight sections. Orbit calculations have required symplectic integrators to minimize numerical errors, which have involved the development of a Lie algebraic method (Dragt et al., 1988). Various resonances, coupled with machine errors, have led to exploration of resistive and Arnol'd diffusion explanations for beam spreading. Intense beams in electron linear accelerators have produced a pulse-shortening phenomenon that was analyzed as a nonlinear excitation of an electromagnetic mode in the accelerating structure, and methods of suppressing it were devised (Chao, 1993). Space-charge effects that become negligible at highly relativistic energies are again introduced by the beam-beam interaction in intersecting storage ring beams. Analysis of this subtle phenomenon within the environment of the very complicated beam dynamics has led to new exploration of simple mapping models (*See* **Fermi acceleration and Fermi map**), together with the use of averaging methods to smooth out various more rapid phenomena that do not play an essential role. In linear devices, to go beyond the 30 GeV two-mile long accelerator at SLAC (Stanford), which is field-strength limited, new ideas of structureless acceleration of ions in the wakefields produced by electron beams in plasmas and of electron acceleration, using the wakefield of an intense laser in a plasma, are being actively explored (see Esarey et al. (1996) and other articles in the same issue). The great challenges that face accelerator design above 300 GeV and intersecting beam interactions in storage rings keep a sense of excitement in a mature field. Much of the modern research and new ideas can be reviewed, with references, in Chao & Tigner (1999).

At the same time that the interaction energy has been increased to probe deeper into the fundamental properties of matter, new uses have been found for low-energy accelerators. On the atomic scale, materials are being probed with ultraviolet light and X-rays, using the natural radiation of electrons in circular devices or enhanced radiation employing undulators in combination with accelerators as advanced light sources. These latter devices typically have short period spatially oscillating magnetic fields with the accompanying electron oscillations relativistically doppler shifted to high frequencies (*See* **Electron beam microwave devices**). At still lower energies, accelerators are being used in various medical applications. Some of the technology-oriented work is treated by Scharf (1989).

ALLAN J. LICHTENBERG

See also **Arnol'd diffusion; Averaging methods; Electron beam microwave devices; Fermi acceleration and Fermi map; Hamiltonian systems**

Further Reading

Chao, A. 1993. *Physics of Collective Beam Instabilities in High Energy Accelerators*, New York: Wiley

Chao, A.W. & Tigner, M. (editors). 1999. *Handbook of Accelerator Physics and Engineering*, Singapore: World Scientific

Courant, E.D., Livingston, M.S. & Snyder, H.S. 1952. The strong-focusing synchrotron—a new high energy accelerator. *Physical Review*, 88: 1190–1196

Dragt, A.J., Neri, F. & Rangaranjan, G. 1988. Lie algebraic treatment of linear and nonlinear beam dynamics. *Annual Review of Nuclear and Particle Science*, 38: 455–496

Esarey, E., Sprangle, P., Krall, J. & Ting, A. 1996. Overview of plasma-based accelerator concepts. *IEEE Transactions on Plasma Science*, 24: 252–288

Hereward, H.G., Johnson, K. & Lapostolle, P. 1956. Problems of injection. *CERN Symposium*, 179–191

Kerst, D.W. & Serber, R. 1941. Electronic orbits in the induction accelerator. *Physical Review*, 60: 53–58

Kolomensky, A.A. & Lebedev, A.N. 1966. *Theory of Cyclic Accelerators*, Amsterdam: North-Holland

Lichtenberg, A.J. 1969. *Phase Space Dynamics of Particles*, New York: Wiley

Livingood, J.J. 1961. *Principles of Cyclic Particle Accelerators*, Princeton: Van Nostrand

O'Neil, G.K. 1956. Storage ring synchrotron: device for high energy physics research. *Physical Review*, 102: 1418–1419

Robinson, K.W. 1958. Radiation effects in circular electron accelerators. *Physical Review*, 111: 373–380

Scharf, W. 1989. *Particle Accelerator Applications in Technology and Research*, Taunton, England: Research Studies Press Ltd.

Symon, K.R. & Sessler, A.M. 1956. Methods of radiofrequency acceleration in fixed field accelerators with applications. *CERN Symposium*, 44–58

Van der Meer, S. 1972. Stochastic damping, of betatron oscillations in the ISR. *CERN Report ISR–PO*, pp. 72–81

PARTICLES AND ANTIPARTICLES

For a good understanding of the physical concept of *antiparticle* and the closely related concept of *charge*, it is important to appreciate how these notions emerged. Therefore, we begin with a brief sketch of the associated history.

In the early days of quantum mechanics, the material world was thought to be built from three elementary particles, namely, the electron, the proton, and the neutron. The idea that there might be associated particles with the same mass and opposite charge (now called antiparticles) first arose from the characteristics of the one-particle Dirac equation. Writing it in Hamiltonian form, the resulting Dirac Hamiltonian has spectrum $(-\infty, -m] \cup [m, \infty)$, where m is the particle mass. The negative part of the spectrum was already considered unphysical by Paul Dirac himself: no such negative energies had been observed, and their presence would give rise, for example, to instability of the electron.

This physical inadequacy of the "first-quantized" description quickly led to the introduction of "second quantization," as expressed in quantum field theory. In the quantum field theoretic version of the Dirac theory, the problem of unphysical negative energies is cured by a prescription that goes back to Dirac's hole theory.

Specifically, Dirac postulated that all of the negative energy states of his equation are filled by a sea of unobservable particles. In his picture, annihilating such a negative energy particle with a given charge yields a hole in the sea, which should manifest itself as a new type of positive energy particle with the same mass, but opposite charge. This intuitive idea led Dirac to the prediction that a charged particle should have an oppositely charged partner (its antiparticle).

Ever since, this prediction has been confirmed by experiment, not only for all electrically charged particles, such as the electron and proton, but also for most of the electrically neutral particles, such as the neutron. In the latter case, one still speaks of the particle having a charge, whereas the remaining electrically neutral particles are identical to their antiparticles.

As already mentioned, in the second-quantized Dirac theory no negative energies occur. In the Dirac quantum field, the creation/annihilation operators of negative energy states are replaced by annihilation/creation operators of positive energy holes. This hole theory substitution therefore leads to a physical arena with an arbitrary number of particles and antiparticles with the same positive mass, now called Fock space. (A mathematically precise account can be found in the monograph by Thaller, 1992.)

As it soon turned out, the number of particles and antiparticles in a high-energy collision is not conserved, a phenomenon that can be naturally accommodated in the Fock spaces associated with interacting relativistic quantum field theories. The quantum field theory model that is the most comprehensive description of real-world elementary particle phenomena arose in the early 1970s. During the last few decades, this so-called Standard Model has been abundantly confirmed by experiment.

In spite of these successes, the problem of obtaining nonperturbative insights into the Standard Model remains daunting. This is an important reason why its classical version and various related, but far simpler, classical nonlinear field theories have been, and still are, widely studied. It is a striking and relatively recent finding that within this classical framework, there exist localized, smooth, finite-energy solutions with characteristics that are very reminiscent of particles. The most conspicuous examples in this respect are the soliton field theories, where there exist such particle-like solutions for any given particle number and where the particle numbers and their velocities are preserved in a scattering process.

To be sure, the latter soliton equations arose independently of particle physics. They involve a lower space-time dimension (mostly two), and they have applications in a great many areas that are far removed from high-energy physics. (An early survey that is still one of the best and most comprehensive can be found in Scott et al., 1973.)

Returning briefly to the latter area, there are also various equations, typically within a gauge-theoretic context, where particle-like solutions (instantons, monopoles, vortices, and so on) have been found. A closely related field is classical gravity, where various previously known solutions (such as the Schwarzschild and Kerr black holes) came to be viewed as particle-like solutions, an interpretation strengthened by the occurrence of "many-particle" generalizations.

It should be pointed out that within this nonlinear classical context, there are no explicit many-particle solutions where "creation" and "annihilation" occur. Indeed, it is not even clear how this would manifest itself on the classical level (as compared with the quantum level, where this is a clear-cut matter).

Focusing once again on the notions of charge and particles vs. antiparticles, it is an even more striking fact that these concepts are naturally present within some of the above-mentioned nonlinear field theory models with particle-like solutions. A prime example for gauge theories is given by instanton solutions, which are accompanied by anti-instanton solutions. Roughly speaking, these are distinguished by opposite generalized winding numbers, viewed as charges of a topological nature. More precisely, these localized solutions minimize the energy in certain homotopy classes. This means they are stable under small perturbations.

Turning to soliton field theories, it should be mentioned at the outset that for most soliton equations (for instance, for their most well-known representative, the KdV equation), there exists only one type of soliton, hence no notion of charge. The sine-Gordon field theory is a paradigm for theories where more than one type of soliton occurs. We proceed by using it to exemplify the notions of particle, antiparticle, and charge at the classical level.

The sine-Gordon equation

$$\phi_{xx} - \phi_{tt} = \sin\phi \tag{1}$$

can be obtained as the Euler–Lagrange equation associated with the Lagrangian

$$\mathcal{L} = \tfrac{1}{2}(\phi_t^2 - \phi_x^2) - (1 - \cos\phi). \tag{2}$$

The corresponding energy functional reads

$$E = \int_R \left[\frac{1}{2}(\phi_t^2 + \phi_x^2) + (1 - \cos\phi)\right] \mathrm{d}x. \tag{3}$$

Obviously, the constant solutions

$$\phi_k = 2\pi k, \quad k \in Z, \tag{4}$$

yield $E = 0$. These are the so-called vacuum solutions.

There exist, however, two distinct classes of nonconstant, time-independent, finite-energy solutions, namely,

$$\phi_\pm = 4\arctan(\exp(\pm(x - x_0))), \quad x_0 \in R. \tag{5}$$

They connect the two vacuum solutions ϕ_0 and ϕ_1 for $x \to \pm\infty$. The functions $\exp(\mathrm{i}\phi_\pm(x)) : R \to S^1$ have winding numbers ± 1 and may be viewed as generators of the homotopy group $\pi_1(S^1) = Z$. They are interpreted as one-particle and one-antiparticle solutions (soliton and antisoliton) of the sine-Gordon equation, having charges ± 1. (By Lorentz invariance, they can be boosted to constant velocity v, with $|v|$ smaller than the speed of light, which is 1 for the units chosen in (1).)

The analogy with electrical charge is strengthened by the existence of soliton–antisoliton bound states, the so-called breathers. More generally, there exist solutions with N_+ solitons, N_- antisolitons, and N_0 breathers, where N_+, N_-, N_0 are arbitrary integers. These particle numbers and the velocities are conserved in the collision, the nonlinear interaction showing up only in factorized position shifts.

From the viewpoint of elementary particle physics, these findings are regarded as stepping stones for a better understanding of the associated quantum field theory. In particular, the existence of particle-like solutions stabilized by charges of a topological nature is believed to signal the existence of a corresponding stable quantum particle. A lucid survey in which this scenario is expounded is Coleman (1977).

Returning to the sine-Gordon example (also discussed in Coleman, 1977), the above scenario has not only been confirmed, but considerably enlarged: at the quantum level, the sine-Gordon theory yields a model of interacting solitons and antisolitons, whose scattering preserves particle numbers and other characteristics (such as the set of velocities), yielding an explicitly known factorized S-matrix. Thus, the classical picture is essentially preserved under quantization (cf. the review by Zamolodchikov & Zamolodchikov (1979)). It should be stressed that this absence of particle creation and annihilation is highly nongeneric for relativistic quantum field theories; it occurs for only completely integrable soliton type field theories in two space-time dimensions.

The phenomenon of oppositely charged particles and antiparticles with an attractive interaction between them has also come up in a setting quite different from the above field-theoretic one. Consider the classical Hamiltonian

$$H_N = \frac{1}{2}\sum_{j=1}^{N} p_j^2 + g^2 \sum_{1 \le j < k \le N} 1/\sinh^2(x_j - x_k) \tag{6}$$

on the phase space

$$\Omega = \{(x, p) \in R^{2N} \mid x_N < \cdots < x_1\}. \tag{7}$$

By contrast to the above partial differential equations, most of which are wave equations at face value, the Hamiltonian equations resulting from (6) have, from the outset, a point particle interpretation. To be specific, they describe N nonrelativistic particles on the line with a repulsive pair interaction.

The special character of this Hamiltonian is already manifest from the scattering to which it leads: the particle velocities are conserved and the position shifts are factorized, just as for solitons. Indeed, Hamiltonian (6) is one version of the completely integrable nonrelativistic Calogero–Moser models. (These are surveyed at the classical and quantum levels in Olshanetsky & Perelomov (1981) and Olshanetsky & Perelomov (1983), respectively.) As pointed out first by Calogero, the substitution $x_k \to x_k + i\pi$, $k = 1, \ldots, N_-$, $N_- \in \{1, \ldots, N-1\}$ has the effect of turning the repulsive $1/\sinh^2$ interaction between particles $1, \ldots, N_-$ and particles $N_- + 1, \ldots, N$ into an attractive $-1/\cosh^2$ interaction. Thus, one obtains an integrable system that can be viewed as describing N_- antiparticles and $N_+ = N - N_-$ particles in interaction.

The nonrelativistic Calogero–Moser systems have been generalized to a relativistic setting, both on the classical and on the quantum levels. Again, there are versions describing particles and antiparticles, with repulsion between like charges and attraction between opposite charges. These versions have been studied at the classical level in Ruijsenaars (1994), where the intimate connection to the sine-Gordon solutions with arbitrary numbers of solitons, antisolitons, and breathers is also detailed. More generally, Ruijsenaars (2001) is a recent survey of this relation between the solitons and antisolitons of the sine-Gordon field theory and the point particles and antiparticles of certain relativistic Calogero–Moser systems, covering both classical and quantum aspects.

Simon Ruijsenaars

See also **Higgs boson; Instantons; Quantum field theory; Skyrmions; Solitons, types of; Yang–Mills theory**

Further Reading

Coleman, S. 1977. Classical lumps and their quantum descendants. In *New Phenomena in Subnuclear Physics, Proceedings Erice 1975*, edited by A. Zichichi, New York: Plenum, pp. 297–421

Olshanetsky, M.A. & Perelomov, A.M. 1981. Classical integrable systems related to Lie algebras. *Physics Reports*, 71: 313–400

Olshanetsky, M.A. & Perelomov, A.M. 1983. Quantum integrable systems related to Lie algebras. *Physics Reports*, 94: 313–404

Ruijsenaars, S.N.M. 1994. Action-angle maps and scattering theory for some finite-dimensional integrable systems. II. Solitons, antisolitons, and their bound states. *Publications RIMS Kyoto University*, 30: 865–1008

Ruijsenaars, S.N.M. 2001. Sine-Gordon solitons vs. relativistic Calogero-Moser particles. In *Integrable Structures of Exactly Solvable Two-dimensional Models of Quantum Field Theory*, edited by S. Pakuliak & G. von Gehlen Delft: Kluwer, pp. 273–292

Scott, A.C., Chu, F.Y.F. & McLaughlin, D.W. 1973. The soliton: a new concept in applied science. *Proceedings of the IEEE*, 61: 1443–1483

Thaller, B. 1992. *The Dirac Equation*, Berlin: Springer

Zamolodchikov, A.B. & Zamolodchikov, Al.B. 1979. Factorized S-matrices in two dimensions as the exact solutions of certain relativistic quantum field theory models. *Annals of Physics (NY)*, 120: 253–291

PATTERN FORMATION

Beautiful patterns and complex structures spontaneously emerge in a large variety of natural systems. Examples can be found on all length scales, ranging in size from Jupiter's red spot, to water waves, a zebra's stripes, mussel shell patterns, snowflake structures, and down to the nanoscale patterning of the scaffolding that gives stability to individual cells.

How do patterns form? The basic laws of physics should account for all forces driving pattern formation, but even with knowledge of all equations governing the pattern-forming process, it remains difficult to compute what patterns will emerge. A closer look at how the snowflake pattern develops illustrates the key challenges:

(i) *Patterns form under nonequilibrium conditions.* In order for an initial seed crystal to form, the air generally has to reach temperatures below the dew and freezing point. Once the seed crystal forms, surrounding air is filled with excess water not in thermodynamic equilibrium. Therefore, the crystal grows under non-equilibrium conditions, which means that the shape of the whole crystal may not obey equilibrium thermodynamics principles such as minimization of free energy. Instead of a spherical shape with minimal surface free energy, complex snowflake shapes emerge. This observation is quite general: patterns form while systems are driven out of equilibrium.

(ii) *Patterns can emerge from the competition of processes on various length and/or timescales.* What physical processes govern the pattern-formation process? In the snowflake example, the crystal grows so fast that heat and impurities, rejected from the growing crystal, build up ahead of the crystal-gas interface. They interfere with the growth of the initial spherical crystal and lead to the development of many fine fins. The length scale of the fins is determined by a competition between diffusion, which would lead to many fine fins (for which the diffusion distance becomes shortest) and surface tension, which inhibits very curved interfaces. In general, a variety of physical processes can drive pattern formation including diffusion, chemical reactions, and fluid flow. Patterns emerge from the competition of processes on different length scales and timescales.

(iii) *Patterns are history-dependent.* As a non-equilibrium process, pattern formation is usually history dependent and sensitive to the initial conditions, thus it is difficult to predict patterns accurately. Small differences in temperature or water concentration in the atmosphere are amplified in snowflakes, so each snowflake looks distinct, reflecting the atmospheric conditions along the path it traveled. The theory of pattern formation provides insights into how pattern-forming systems can be analyzed and what one may learn about the underlying physical processes from observations of patterns.

General features of the pattern on the other hand, are often related to underlying symmetries in the system and can be predicted. For example, snowflakes often have six spikes, reflecting the symmetry of the anisotropy in the surface tension of ice.

However, each pattern-forming system may be driven by a unique combination of forces: similarities in the patterns may thus only reflect similar symmetries of the underlying equations, timescales, or length scales, not similarities in the underlying physics.

Driven by new mathematical tools, digital imaging, and powerful image analysis tools, numerous experimental and theoretical investigations of pattern-forming systems have been carried out over the past two decades. Cross & Hohenberg (1993) provide an excellent technical introduction into the mathematical concepts of pattern formation. Ball (1999) introduces the field of pattern formation in a non-mathematical way suitable for a broad audience. An update on the field can be found in Gollub & Langer (1999). Below, we briefly describe the basic principles and mathematical tools.

Model Systems of Pattern Formation

Pattern formation implies a change in symmetry. Spontaneous symmetry breaking is particularly evident in fluids, which are structureless in equilibrium but exhibit a surprising variety of patterns under non-equilibrium conditions. Fluid systems have yielded two model systems for detailed studies of pattern formation: surface waves and convection rolls.

Surface waves develop due to either wind or gravitational forces or, for example, when a stone is thrown into the fluid. Under vertical shaking, fluids can exhibit very ordered patterns, such as hexagonal structures as shown in Figure 1. Michael Faraday was the first to analyze ordered surface wave patterns more than a century ago (Faraday, 1831).

Spontaneous symmetry breaking into a patterned state can also be observed when a fluid is driven out of equilibrium through a thermal gradient. If a fluid is heated from below and cooled from above, the fluid near the bottom expands and rises to the top, while fluid near the top cools, contracts, and due to the increased density, drops back to the bottom. This leads to large-scale convection rolls, the so-called Rayleigh–Bénard convection, where streams of rising and falling fluid arrange into complex patterns, similar to the patterns found in surface waves shown in Figure 1.

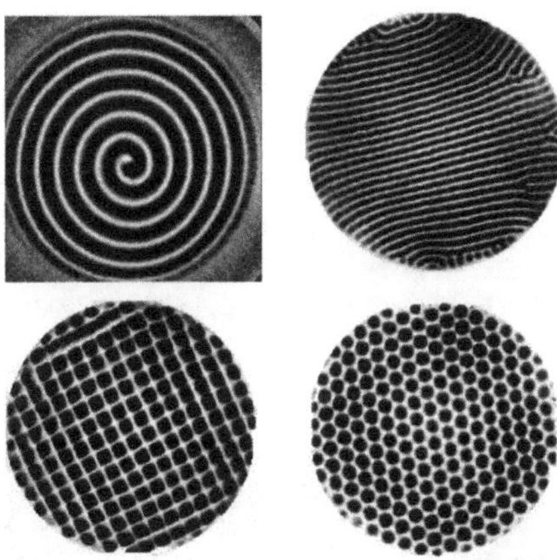

Figure 1. Surface wave patterns in a vertically vibrated fluid. Spirals, stripes, squares, and hexagonal patterns are characteristic of surface waves ("fluid" consists of 170 micron bronze spheres, a few layers deep. Courtesy of D. Goldman, P. Umbanhowar, B. Lewis, and H. Swinney, UT Austin).

While diffusion alone tends to create uniform states, Alan Turing suggested that diffusion coupled with chemical reactions may lead to spatial patterns in chemical composition. He speculated that such mechanisms would be sufficient to explain natural patterns such as zebra stripes or the regular shape and arrangement of leaves. An early experimental realization was the Belousov–Zhabotinsky reaction, which involves several intermediate chemical compounds. The original experiments were carried out in a well-stirred container and generated temporal patterns in which the concentration of intermediate components oscillated. Later experiments showed spatial patterns such as stripes and spirals.

Granular systems such as sand or coffee powder also spontaneously form patterns. Under vertical vibration, they form patterns resembling surface waves (see Figure 1). In addition, localized, stationary waves (so-called oscillons) are observed near the onset of surface waves (Umbanhowar et al., 1996). Unlike localized waves in fluids, oscillons do not spread out or dissipate over time. Mixtures of granular matter are also prone to pattern-forming instabilities. For example, when sand and sugar are mixed in a container and poured into a pile, the mixture spontaneously separates into bands under some conditions. Unlike the fleeting structure of surface waves or convection rolls in fluids, patterns in

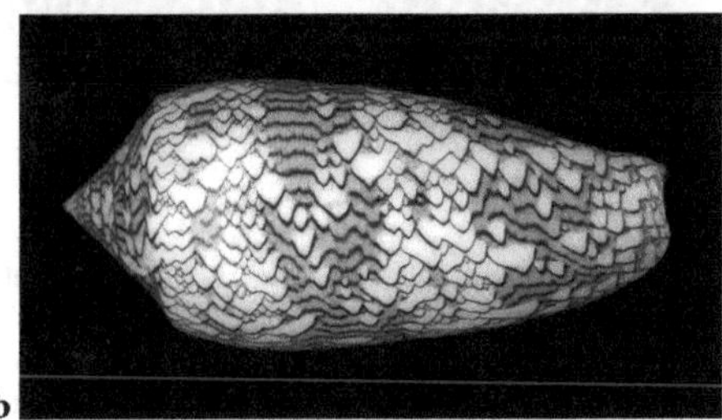

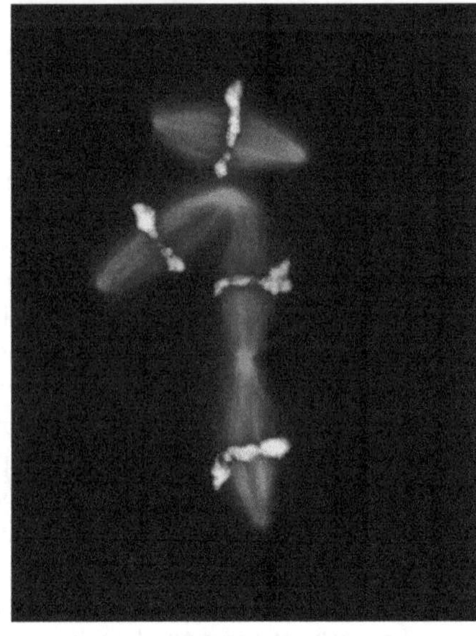

Figure 2. Pattern formation in nature on many scales: (a) Crystal growth pattern (rime on a windshield). (b) Mussel shell pattern (Conus textile, Courtesy of P. Egerton) (c) Scaffolding of a cell during cell division (Spindle assembly in *Xenopus* extracts. Microtubules (long) and DNA (rings), Courtesy of R. Ohi and T. Mitchison, Harvard Medical School, Boston).

granular flows can create lasting impressions such as bands of sugar and sand in a mixture, or sand ripples and sand dunes in wind-driven flows.

In solids, snowflakes are a well-known pattern (see Figure 2a) which we have discussed above. Similar patterns can be found in alloys and polymer blends, though they are hidden within the material micro- or nanostructure. Patterns in solids are technologically important, because they can determine material strength and flexibility.

Biological systems offer a rich variety of patterns. On large scales, patterns such as zebra stripes and mussel shell patterns (Gierer & Meinhard, 1972) (see Figure 2b) help animals blend in or send signals of intimidation or attraction. On a small scale, the microscopic mechanical structure within cells contains ordered arrangements of microtubules and other filaments, which assist, for example, in cell division (see Figure 2c).

Stability and Bifurcations

The first prerequisite for the emergence of patterns from a spatially uniform state is a symmetry-breaking instability. A linear stability analysis indicates whether a given set of equations is prone to a symmetry-breaking instability. For example, we start with a set of functions $\boldsymbol{u}(t)$:

$$\frac{\mathrm{d}\boldsymbol{u}}{\mathrm{d}t} = f(\boldsymbol{u}). \tag{1}$$

To assess the stability of a state $\boldsymbol{u}_0$, we linearize around that state $\boldsymbol{u} = \boldsymbol{u}_0 + \boldsymbol{\delta}$. This yields an equation for the perturbation $\boldsymbol{\delta}$:

$$\frac{\mathrm{d}\boldsymbol{\delta}}{\mathrm{d}t} = \left.\frac{\mathrm{d}f(\boldsymbol{u})}{\mathrm{d}u}\right|_{\boldsymbol{u}_0} \boldsymbol{\delta}. \tag{2}$$

The state $\boldsymbol{u}_0$ is linearly stable against perturbations if $\boldsymbol{\delta}$ shrinks for any shape of the initial perturbation and unstable if it grows. To test for all possible perturbations, $\boldsymbol{\delta}$ can be decomposed into Fourier modes $\delta = \delta_q \exp(\mathrm{i}\boldsymbol{q}x)$. This indicates the range of wave numbers ($\boldsymbol{q}$) that are unstable.

Generally, pattern-forming systems become unstable when they are driven further from equilibrium. The transition point toward instability is called marginal stability. In the marginally stable state, perturbations of one, or a few, wave numbers neither grow nor decay; all other perturbations decay.

When parameters are changed further, the system will become unstable for a range of wave numbers (sometimes including $\boldsymbol{q} = 0$). In the example of crystal growth, a planar crystal growth front is stable for small $\boldsymbol{q}$ (due to diffusion) and for large $\boldsymbol{q}$ (due to surface tension), but unstable for a range of intermediate $\boldsymbol{q}$.

Once the perturbations grow, the linear approximation fails since δ will diverge, and other approaches are needed to predict patterns, but the range of unstable wave number is often a rough indicator of the characteristic length scales of the final patterns.

Model Equations

Beyond linear instability, modeling of pattern-forming systems becomes interesting: rather than going back to the full equations for the pattern-forming system, which are often too complex to be simply studied numerically or analytically, model partial differential equations have been developed to describe the main qualitative features of patterns, and to highlight the universal features of many pattern-forming systems. They are discussed in great detail, for example, in the excellent review by Cross & Hohenberg (1993). Here we list the main equations.

The (complex) Ginzburg–Landau equation is an amplitude equation with an expansion in amplitudes and gradient terms. The equation applied, for example, to a two-dimensional plane can yield similar patterns as the vibrated fluid shown in Figure 1: squares, hexagons, and stripes. The equation provides insight into which gradient or higher-order terms must dominate in specific patterns. Long wavelength instabilities such as the Eckhaus instability are common and limit the range of stable patterns to a smaller range of wave numbers than the range of linearly stable wave numbers.

Other equations include the Swift–Hohenberg equation and the Kuramoto–Sivashinsky equation, which have been solved in various geometries and for different boundary conditions and parameter ranges. They provide a rich set of patterns and dynamical instabilities.

Experimentally and in simulations, the best model system has proven to be surface waves and Rayleigh–Bénard convection. The stability phase diagram of the convecting fluid—the range of stable wave numbers as a function of a measure of fluid convection such as the Rayleigh number—exhibits a region of stability, the so-called Busse balloon, bounded by various instabilities such as the Eckhaus instability, a zig-zag instability, or a cross roll instability. The range of stable patterns and the instability boundaries have been verified through extensive experiments.

Slightly different are equations that describe patterns in excitable media, which are often used to mimic biological processes. Reaction-diffusion equations yield wave fronts or spiral waves. Again, basic symmetries of the underlying processes dictate what patterns can occur, and the stability boundaries are crucial.

Finally, it should be noted that real patterns have boundaries and defects, which can strongly affect pattern selection and stability. Modeling defects and boundaries appropriately can be difficult, since it may only introduce small perturbations that change the system over long timescales (which are often beyond the reach of simulations). Recently, controlled experiments where defects are implanted deliberately into patterns have begun to allow quantitative tests of theories, but this work is only beginning.

Challenges for the future include application of these insights to real pattern forming systems, for example, to suppress or exploit crystal growth instabilities in the semiconductor industry. General theoretical descriptions of pattern formation in networks such as neural networks remains one of the open challenges.

WOLFGANG LOSERT

See also **Belousov–Zhabotinsky reaction; Dune formation; Emergence; Growth patterns; Kuramoto–Sivashinsky equation; Morphogenesis, biological; Rayleigh–Taylor instability; Reaction-diffusion systems; Turing patterns**

Further Reading

Ball, P. 1999. *The Self-made Tapestry: Pattern Formation in Nature*, Oxford and New York: Oxford University Press

Cross, M.C. & Hohenberg, P.C. 1993. Pattern formation out of equilibrium. *Reviews of Modern Physics*, 65: 851–1112

Faraday, M. 1831. On a peculiar class of acoustical figures; and on certain forms assumed by groups of particles upon vibrating elastic surfaces. *Philosophical Transactions of the Royal Society of London*, 52: 299–340

Gierer, A. & Meinhard, H. 1972. A theory of biological pattern formation. *Kybernetik*, 12: 30–39

Gollub, J.P. & Langer, J.S. 1999. Pattern formation in nonequilibrium physics. *Reviews of Modern Physics*, 71: S396–S403.

Turing, A. 1952. The chemical basis of morphogenesis. *Philosophical Transactions of the Royal Society of London B*, 237: 37–72

Umbanhowar, P.B., Melo, F. & Swinney, H.L. 1996. Localized excitations in a vertically vibrated granular layer. *Nature*, 382: 793–796

PATTERN SELECTION

See **Synergetics**

PEIERLS BARRIER

As a localized entity such as an electron, hole, polaron, soliton, or dislocation moves through a discrete lattice, it often experiences a potential barrier between one unit cell and the next. Called a Peierls barrier after the famed German/British physicist Rudolph Peierls, such potential barriers plays a key role in lattice dynamics.

More precisely, a Peierls barrier characterizes the dependence of the energy, $E(\bar{n})$, of a stationary localized entity (LE) on its position, $\bar{n}$, in a periodic lattice. The Peierls barrier, is the function $E(\bar{n}) - E_0$, where $E_0 = \min E(\bar{n})$ is the ground state energy of the LE. Because of lattice periodicity the barrier must be a periodic function with a period equal to the lattice constant a. Minima of the Peierls barrier correspond to stable positions and maxima to unstable positions. The barrier height $\Delta E = \max[E(\bar{n}) - E_0]$ is often called the pinning energy. To propagate along the chain, the LE must be activated with energy exceeding ΔE. For smaller energies, the LE vibrates near minima of the Peierls barrier, and for larger energies it can overcome the Peierls barrier and move through the lattice. The presence of a Peierls barrier also leads to the appearance of new eigenvalues corresponding to periodic oscillations of an LE near a minimum of the Peierls barrier.

The notion of the Peierls barrier goes back to the first quantitative explanation of the mechanism of plastic flow via the motion of dislocations in a lattice (Peierls, 1940; Naborro, 1967). From the perspective of the continuum theory of elasticity, a dislocation is an LE that involves a discontinuity in the displacement field, leading to singularities in strains and stresses. Peierls and Naborro showed how to overcome these artifacts of continuum idealizations by distributing the sharp discontinuity over a finite region, as in the one-dimensional nonlinear continuum Frenkel–Kontorova model of dislocation.

The dynamic influence of a Peierls barrier is greatest when the size of the LE is about equal to the lattice periodicity. Failure to include it in theoretical formulations can lead to erroneous calculations. When modeling the dynamics of protonic solitons in chains of hydrogen bonds, for example, inclusion of the Peierls barrier is necessary to find the temperature dependence of soliton mobility (Savin & Zolotaryuk, 1991).

Let us consider in more detail an autolocalized state of a quantum particle in a lattice (or molecular chain). In a simplified model, the dynamics of such a quasiparticle is described by the system of equations

$$i\hbar\dot{\phi}_n = \chi u_n \phi_n - J(\phi_{n-1} + \phi_{n+1}), \tag{1}$$
$$m\ddot{u}_n = -K u_n - \chi |\phi_n|^2, \tag{2}$$

where $-\infty < n < +\infty$ is an integer, $\{\phi_n\}$ is a complex-valued wave function normalized by the condition

$$\sum_n |\phi_n|^2 = 1, \tag{3}$$

where u_n is the intramolecular displacement of the lattice, and χ, J, m, and K are parameters of the exciton–phonon interaction: energy of resonance or exchange interaction between neighboring lattice sites, reduced mass of the molecules, and stiffness of intermolecular interaction. Neglecting the inertia of intramolecular displacements implies that $u_n = -\chi |\phi_n|^2 / K$, whereupon Equations (1) and (2) reduce to a discrete nonlinear Schrödinger equation (DNLS), which can be written as

$$-\mathrm{i}\frac{\mathrm{d}\phi_n}{\mathrm{d}\tau} = \phi_{n-1} + \phi_{n+1} + g\phi_n |\phi_n|^2, \tag{4}$$

where $\tau = Jt/\hbar$ and $g = \chi^2 / KJ$ is a dimensionless parameter of nonlinearity.

In the continuum limit, Equation (4) reduces to the nonlinear Schrödinger (NLS) equation, which is integrable. In this limit, the Peierls barrier does not play any role because an envelope soliton can propagate freely along the chain. For the discrete model, which is not integrable, the situation is quite different.

A stationary autolocalized state of the quasiparticle can be found numerically by minimizing

$$H = -\sum_n \varphi_n \varphi_{n+1} + \tfrac{1}{4} g \varphi_n^4,$$

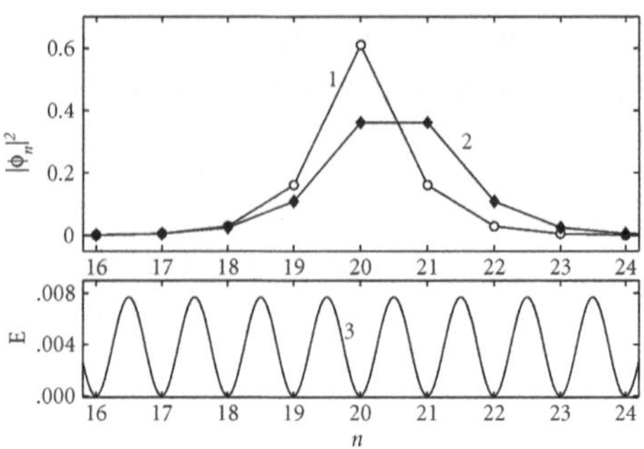

Figure 1. Stationary autolocalized state of the quantum quasiparticle with integer (curve 1) and half-integer (curve 2) centers of symmetry. Curve 3 (below) shows the Peierls barrier, i.e., the dependence of the energy of the stationary state E on the position of the center n. Nonlinear parameter $g = 3$.

subject to the constraint of Equation (3), where φ_n is the real-valued amplitude of quasiparticle wave function on nth site. If $\{\varphi_n\}_{n=-\infty}^{+\infty}$ is the solution of the minimization problem, the energy and position of the LE are $E = H$ and

$$\bar{n} = \sum_n n\varphi_n^2.$$

respectively.

Stationary states of quasiparticle have two types of symmetry. First is a solution with integer center of symmetry $\varphi_{n_0+n} \equiv \varphi_{n_0-n}$. In other words, the center of the LE is in one of the lattice sites. Second is a solution with half-integer center of symmetry, where the center of excitation is midway between two lattice sites ($\varphi_{n_0+n} \equiv \varphi_{n_0+1-n}$, $\bar{n} = n_0 + \frac{1}{2}$). The ground state is of the first type and corresponds to a minimum of the Peierls barrier. The second type is unstable and corresponds to a maximum of the Peierls barrier.

Numerical solution of the minimization problem allows one to find the energy profile of a Peierls barrier, as plotted in Figure 1. This profile has a sine-like character with minima at integer points and maxima in half-integer points. The height of the Peierls barrier depends on the value of nonlinear parameter g. If the nonlinearity parameter $g \to 0$, the size of the LE grows without bound, and the pinning energy ΔE goes monotonically to zero. One can calculate Peierls barriers analytically using the method of variational functions (Kuprievich, 1985).

Note that a specific discretization of the continuum NLS equation (Ablowitz & Ladik, 1976) leads to a discrete integrable system. In this case, the ground state energy does not depend on the position of the LE and the Peierls barrier is absent.

LEONID MANEVITCH AND ALEXANDER SAVIN

See also **Bjerrum defects; Discrete nonlinear Schrödinger equations; Dislocations in crystals; Frenkel–Kontorova model; Polarons**

Further Reading

Ablowitz, M.J. & Ladik, J.F. 1976. A nonlinear difference scheme and inverse scattering. *Studies in Applied Mathenatics*, 55(3): 213–229; Nonlinear differential-difference equations and Fourier analysis. *Journal of Mathematical Physics*, 17(6): 1011–1018

Kuprievich, V.A. 1985. Variational study of stationary autolocalized states of molecular chains in discrete model. *Theoretica Materialia Fizika (Soviet Theoretical and Mathematical Physics)*, 64(2): 269–276

Naborro, F.R.N. 1967. *Theory of Crystal Dislocations*, Oxford: Clarendon Press

Peierls, R.E. 1940. *Proceedings of the Physical Society of London*, 52: 34

Savin, A.V. & Zolotaryuk, A.V. 1991. Dynamics of ionic defects and lattice solitons in a thermalized hydrogen-bonded chain. *Physical Review* A, 44: 8167–8183

PENDULUM

With a history covering many centuries, the pendulum has been of primary importance in the development of physics. It is well known that in the 17th century, Dutch experimental physicist Christiaan Huygens discovered the synchronization phenomenon while studying pendulum clocks (Huygens, 1673). Using a pendulum as an example, Huygens formulated the conservation law for mechanical energy. In the 18th century, the French scientist Jean Charles Borda suggested using a pendulum for measuring the acceleration of gravity (Borda, 1792), and Wilhelm Bessel used a pendulum for an accurate verification of the equality of inertial and gravitational masses. There are many other applications of pendula for physical measurements; for example, a pendulum with a randomly vibrated suspension axis has turned out to be a good model for control of turbulence in subsonic submerged jets (Landa, 1996).

Two models of a pendulum are well known—mathematical and physical. The mathematical pendulum is defined as a ball suspended by a weightless thread in vacuum (Figure 1a). It is described by the equation

$$\ddot{\varphi} + \omega_0^2 \sin\varphi = 0, \tag{1}$$

where φ is the pendulum angle of deviation, $\omega_0 = \sqrt{g/l}$ is the angular frequency of small oscillations, g is the acceleration due to gravity, l is the thread length, and dots indicate derivative with respect to time. The physical pendulum can be defined as a body moving about a fixed horizontal axis O (Figure 1b).

Taking account of air friction, the motion equation for such a body is

$$I\ddot{\varphi} + h\dot{\varphi} + mga\sin\varphi = 0, \tag{2}$$

where I is the body's moment of inertia, h is the friction factor, m is the body's mass, and a is the distance between the axis and the body's center of gravity.

Equation (1) belongs to the class of nonlinear conservative (Hamiltonian) systems, whereas Equation (2) is a dissipative system. Solutions of these equations

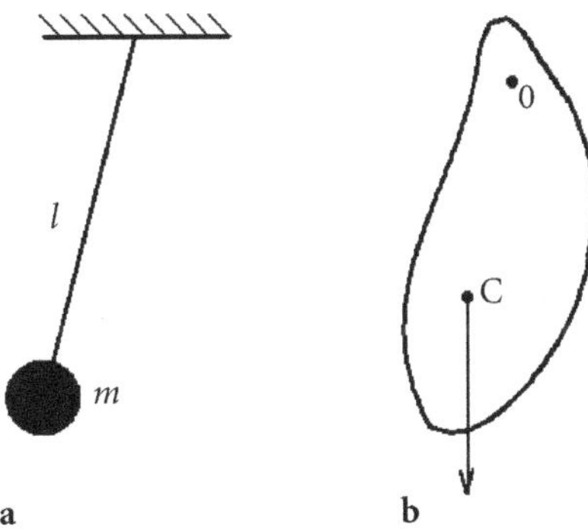

Figure 1. Schematic image of (a) mathematical pendulum and (b) physical pendulum.

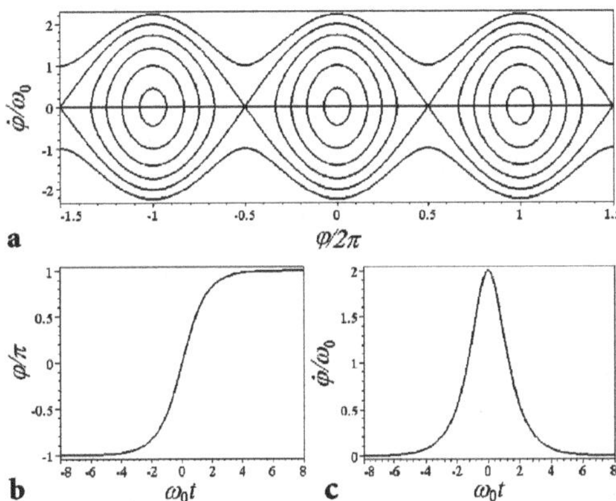

Figure 2. (a) Phase portrait of pendulum oscillations in the phase plane; (b,c) the time dependencies of the pendulum phase coordinates φ and $\dot{\varphi}$ at the motion of a representative point along a separatrix.

can be conveniently analyzed with the phase plane φ, $\dot{\varphi}$. We illustrate this for Equation (1).

Using the energy conservation law, we obtain an equation for a trajectory in the phase plane φ, $\dot{\varphi}$ in the form

$$\frac{ml^2\dot{\varphi}^2}{2} + mgl(1-\cos\varphi) = E, \tag{3}$$

where E is the total pendulum energy. For $E < 2mgl$, Equation (3) describes closed trajectories corresponding to pendulum oscillations about the stable equilibrium position ($\varphi = 0$), and for $E > 2mgl$, nonclosed trajectories corresponding to the rotation of the pendulum (Figure 2a). These two kinds of trajectories are separated by peculiar trajectories passing through singular saddle points ($\dot{\varphi} = 0$, $\varphi = \pm\pi$, $\pm 3\pi$, ...). Such trajectories are said to be *separatrices*. Because values of φ differing from each other by 2π are physically indistinguishable, the phase plane shown in Figure 2a can be rolled into a cylinder.

In general, solutions of Equations (1) and (3) are Jacobi elliptic functions, but more simple solutions can be obtained in the case of small oscillations ($E \ll mgl$) and in the case when $E = 2mgl$ (this value of E

corresponds to the motion of a representative point along a separatrix). In the latter case, Equation (3) is split into two equations:

$$\dot{\varphi}_{\pm}(t) = \pm 2\omega_0 \cos\frac{\varphi}{2}. \qquad (4)$$

By integrating (4) we find

$$\varphi_{\pm}(t) = \pm\left(4\arctan e^{\omega_0 t} - \pi\right). \qquad (5)$$

The time dependencies of $\varphi = \varphi_+$ and $\dot{\varphi} / \omega_0 = \dot{\varphi}_+ / \omega_0$ are shown in Figure 2b and c. These solutions play an important part in soliton theory.

In the case of sufficiently small oscillations, when $\sin\varphi \approx \varphi - \varphi^3 / 6$, Equation (1) reduces to the Duffing equation

$$\ddot{\varphi} + \omega_0^2\left(1 - \frac{\varphi^2}{6}\right)\varphi = 0. \qquad (6)$$

The general solution of Equation (6) is

$$\varphi = A\,\mathrm{sn}(\Omega t, k), \qquad (7)$$

where sn is a Jacobi elliptic function of modulus $k = \sqrt{1/12}\,A / \Omega$, and $\Omega = \omega_0\sqrt{1 - A^2/12}$. Solution (7) is valid for $A \le \sqrt{6}$, when $k \le 1$. This constraint is certainly fulfilled because Equation (6) follows from (1) only for $A < 1$. Solution (7) describes periodic oscillations of period $4\mathrm{K}(k) / \Omega$, where $\mathrm{K}(k)$ is the full elliptic integral of the first kind. It follows from this that the period of oscillations increases with increasing amplitude. Such a property inherent in nonlinear systems is called anisochronism. It should be noted that Galileo was the first to discover the isochronism of small pendulum oscillations.

If a pendulum consisting of an iron ball suspended by a thread of length l is placed between the opposite poles of a magnet, its behavior essentially changes (Landa, 1996, 2001). Approximating the magnetic force acting on the ball by $F(\varphi) = ml(a_1\varphi - b_1\varphi^3)$ and restricting ourselves to small oscillations of the ball, then the pendulum angular deviation φ obeys the following approximate equation:

$$\ddot{\varphi} - (a - b\varphi^2)\varphi = 0, \qquad (8)$$

where $a = a_1 - \omega_0^2$, $b = b_1 - \omega_0^2/6$, $\omega_0^2 = g/l$. In the case that $a_1 > \omega_0^2$ $(a > 0)$, the equilibrium position $x = 0$ becomes aperiodically unstable (the corresponding singular point $\varphi = 0$, $\dot{\varphi} = 0$ in the phase plane φ, $\dot{\varphi}$ becomes of saddle type). If, in addition to this, the inequality $b_1 > \omega_0^2/6$ holds, two stable equilibrium positions with coordinates $\varphi_{1,2} = \pm\sqrt{a / b}$ appear. These equilibrium positions correspond to singular points of center type. But if $b_1 < \omega_0^2/6$ and $a > 0$, then the ball adheres to one of the magnet poles. Equation (8) describes a so-called two-well oscillator, which is the subject of recent widespread interest in connection with stochastic and vibrational resonances (Landa, 2001).

If a pendulum is suspended from a uniformly rotating shaft, it can execute self-oscillations. Such a pendulum was discovered by William Froude and mentioned in the famous treatise by Lord Rayleigh (Rayleigh, 1877). Rayleigh showed that oscillations of such a pendulum are approximately described by an equation which came to be known as the Rayleigh equation. A controlled Froude pendulum was suggested by Neimark to be a model of stochastic oscillations (Neimark & Landa, 1992).

POLINA LANDA

See also **Damped-driven anharmonic oscillator; Duffing equation; Elliptic functions; Hamiltonian systems; Solitons**

Further Reading

Borda, J. 1792. *Mémoires sur la mesure du pendule (Mesure de la méridienne)*, Paris

Huygens, C. 1673. *Christiani Hvgenii Zvlichemii Horologivm oscillatorivm, sive, De motv pendvlorvm ad horologia aptato demonstrationes geometricæ*. Paris: Muguet; as *Christiaan Huygens' The Pendulum Clock, or, Geometrical Demonstrations Concerning the Motion of Pendula as Applied to Clocks*, translated with notes by R.J. Blackwell, Ames: Iowa State University Press, 1986

Landa, P.S. 1996. *Nonlinear Oscillations and Waves in Dynamical Systems*, Dordrecht and Boston: Kluwer

Landa, P.S. 2001. *Regular and Chaotic Oscillations*, Berlin and New York: Springer

Neimark, Yu.I. & Landa, P.S. 1992 *Stochastic and Chaotic Oscillations*, Dordrecht and Boston: Kluwer (original Russian edition 1987)

Rayleigh, Lord (Strutt, J.W.) 1877–78. *The Theory of Sound*, London: Macmillan; reprinted New York: Dover, 1945

PERCEPTRON

The perceptron is one of the earliest computational models inspired by the human brain. Developed by Frank Rosenblatt from 1959–1962, the perceptron model is a parallel, distributed processing network intended to provide a means of modeling the capabilities and properties of the brain at a very simplified level (Rosenblatt, 1962).

Artificial neurons as mathematical processing elements of neurological (or artificial neural) networks were described earlier by Warren McCulloch & Walter Pitts (1943). Also earlier was the work of Donald Hebb, which presented a principal of learning or self-organization in neural networks (Hebb, 1949). The perceptron brought these concepts together by providing an algorithm for adjusting the parameters or weights of the network to learn to perform mappings or predicates. This model was implemented in analog electrical hardware by Rosenblatt and colleagues as the Mark I Perceptron and as such was the first neurocomputer to perform useful functions (Block, 1962). Accepting that biological brains possess "natural" intelligence, developing

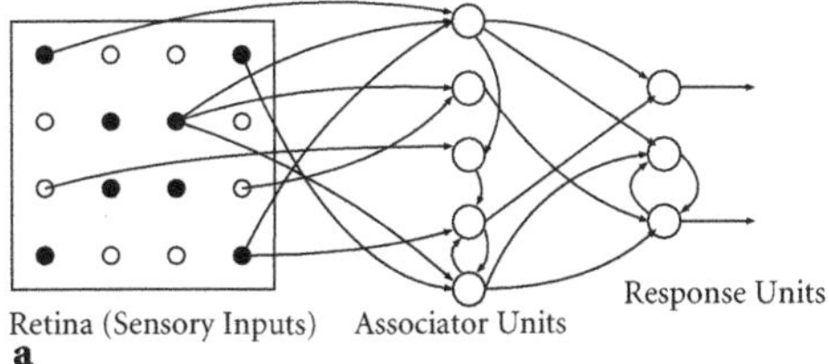

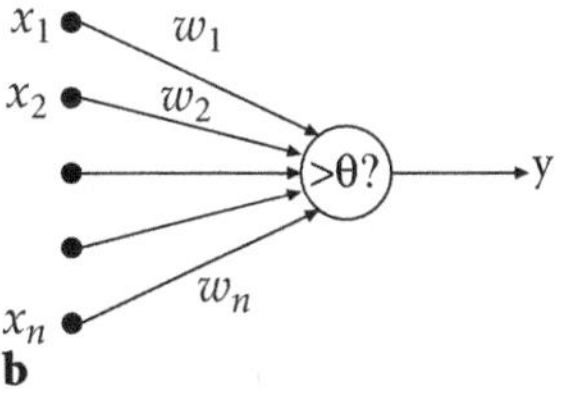

Figure 1. (a) General perceptron model. (b) Single perceptron unit.

computational models inspired by the brain represents one approach toward artificial intelligence.

The general perceptron model consists of a retina, associator units, and response units (see Figure 1a). The retina provides sensory inputs connected (many-to-many and at random) to a layer of associator units. The associator outputs are connected to response units. Connectivity can be sparse, and both associator and response units can be connected to other units in the same layer. An associator unit becomes active if its total input exceeds some threshold value and propagates a signal to the response units. This model appears to have the essential features to begin studying brain function (learning, pattern recognition, memory, generalization) in terms of brain structure.

A perceptron unit computes a threshold function of the values presented via its input connections

$$y = \begin{cases} 1 & \text{if } \sum_{i=1}^{n} w_i x_i > \theta, \\ 0 & \text{otherwise,} \end{cases} \tag{1}$$

where x_i is the value of the ith input signal. Each input connection is weighted by a real-valued weight w_i, and θ determines the threshold value. Geometrically, the perceptron unit implements a hyperplane in n-dimensional input space, where input vectors lying on one side of the line will result in an output of 1, and input vectors on the other side of the line will result in an output of 0. The position of the hyperplane is determined by the values of the weights and thresholds (zero threshold implies a hyperplane passing through the origin).

The simple perceptron unit provides a useful, simplified realization of the above model which can be implemented and trained to perform useful functions (Figure 1b). A set of inputs $X = \{x_1, \ldots, x_n\}$ are provided to a layer of associator units that compute a fixed value $\varphi(X)$. These values are then combined at the response unit to produce a binary output $\Psi(X)$. The response unit is adaptive in that there is an adjustable weight parameter α_i associated with each of its input connections. The output of the response unit Ψ is calculated as

$$\Psi = 1 \text{ if } \sum_{i=1}^{n} \alpha_i \varphi_i(X) > \theta, \tag{2}$$

$$\Psi = 0 \qquad \text{otherwise,} \tag{3}$$

where θ is a threshold value. The simple perceptron can compute any linear threshold function of the $\varphi_i(X)$ given appropriate values of the weights α_i and θ.

Given a set of data points in n-dimensional input space, labeled with desired (0 or 1) output values, Rosenblatt provided a rule for iteratively learning (modifying) the perceptron weights and thresholds, such that the outputs of the perceptron would match the desired output values. He also proved the convergence of the learning rule in a finite number of steps, provided that the perceptron has the capacity (ability) to learn an adequate linear threshold function. That is, if the n-dimensional (0,1)-labeled data is separable by an $(n-1)$-dimensional hyperplane in the input space, the perceptron can learn or be trained to find such a hyperplane.

An adaptive threshold value can be conveniently implemented by replacing θ with an additional input x_0, that always takes the value 1, and an adaptive weight w_0. Given input values and the desired (correct) value d for the output, the output of the model is computed from the inputs and the weights are adapted as follows:

$$\alpha_i = \begin{cases} \alpha_i & \text{if } d = y(X), \\ \alpha_i + x_i & \text{if } d = 1 \text{ and } y(X) = 0, \\ \alpha_i - x_i & \text{if } d = 0 \text{ and } y(X) = n1. \end{cases} \tag{4}$$

If the threshold function is changed to a continuous, differentiable function, it becomes possible to formulate a cost function for learning that is a differentiable function of the weight values. For example, a sigmoidal function is commonly used to approximate the threshold function. The delta rule can then be used to train the unit via an iterative gradient descent (Widrow & Hoff, 1960). Widrow and Hoff developed the delta (or Least Mean Squared) rule in the context of ADALINEs (ADAptive LInear NEurons), which are equivalent to McCulloch–Pitts neuron models with the bias weight (Widrow & Lehr, 1990). A quadratic error function resulting from the sum of squared error terms is used, making it possible to guarantee convergence of the delta rule under appropriate assumptions. In contrast, the perceptron learning rule does not converge if the data is nonlinearly separable.

In general, a perceptron may consist of a layer of k of the above perceptron units, each of which is fully connected to same set of inputs but has its own output. Both Rosenblatt's perceptron learning rule and the delta rule can be straightforwardly extended to train such a network.

The inability of single-layer perceptron to implement nonlinearly separable mappings and the lack of a procedure for training a multilayer perceptron model led to a lull in progress in the field for many years (Minsky & Papert, 1990). Recently, the back-propagation algorithm led to a resurgence of interest in the multilayer perceptrons and artificial neural network models. Such models have been very successfully applied to a wide range of data-driven learning tasks such as pattern recognition, classification, regression, and prediction (Haykin, 1999).

MARCUS GALLAGHER

See also **Artificial intelligence; Attractor neural network; Cell assemblies; Cellular automata; Cellular nonlinear networks; McCulloch–Pitts network; Neural network models**

Further Reading

Block, H.D. 1962. The perceptron: a model for brain functioning. I. *Reviews of Modern Physics*, 34(1): 123–135

Haykin, S. 1999. *Neural Networks A Comprehensive Foundation*, 2nd edition, Upper Saddle River, NJ: Prentice-Hall

Hebb, D.O. 1949. *The Organization of Behavior*, New York: Wiley

McCulloch, W.S. & Pitts, W. 1943. A logical calculus of ideas immanent in nervous activity. *Bulletin of Mathematical Biophysics*, 5: 115–133

Minsky, M. & Papert, S.A. 1990. *Perceptrons: An Introduction to Computational Geometry*, Cambridge, MA: MIT Press

Rosenblatt, F. 1962. *Principles of Neurodynamics*, New York: Spartan

Widrow, B. & Hoff, M.E., Jr. 1960. Adaptive switching circuits. In *1960 IRE WESCON Convention Record*, Part 4, New York: IRE, pp. 96–104

Widrow, B. & Lehr, M.A. 1990. 30 years of adaptive neural networks: perceptron, madaline, and backpropagation. *Proceedings of the IEEE*, 78: 1415–1442

PERCOLATION THEORY

For random site percolation, each site of a large lattice is randomly occupied with probability p and empty with probability $1-p$. A cluster is a set of neighboring occupied sites. A spanning cluster extends from one end of the sample to the opposite end. Percolation theory (PT) deals with the number and properties of spanning and nonspanning clusters as a function of p and of the cluster size s: the number of occupied sites in the cluster. Alternative variants are bond percolation where all sites are occupied but bonds between neighbor sites exist only with probability p. A cluster then is a set of sites connected directly or indirectly by existing bonds. One may also remove the lattice structure and place possibly overlapping circles on a piece of paper (or spheres in three dimensions). Instead of random variables, one can also study correlated percolation, such as Ising models where up spins are identified as occupied and down spins as empty sites. The following results are expected to be also valid for these variants as long as the correlations have a finite spatial extent (i.e., not at the Curie point of Ising models) (Stauffer & Aharony, 1994).

In 1941, Paul Flory published the first percolation theory (ignoring cyclic links and excluded volume effects). Physicists call this the Bethe lattice or mean field approximation, and it becomes exact for very high dimensions but is inaccurate in its critical behavior for the two or three dimensions we are mostly interested in. Percolation theory was continued by Stockmayer (1943), applied to immunology at the beginning of the 1950s, and a few years later, put onto square lattices by Broadbent & Hammersley (1957). The first computer simulations were published around 1960, and in 2001, these reached 4 million times 4 million sites (Tiggemann, 2001) and even larger lattices in Maloney & Pruessner (2003), using the Hoshen–Kopelman algorithm (Stauffer & Aharony, 1994). Exact polynomials for the number $n_s(p)$ of clusters of s sites were found for small and intermediate s, and exact results for many thresholds p_c and critical exponents are known in two dimensions.

For small p, we have only small clusters, and for p near unity, nearly the whole system forms a huge spanning cluster coexisting with rare small clusters. Thus, there is a sharp critical threshold p_c, where with increasing p for the first time, a spanning cluster appears in a very large lattice. For random site percolation on the triangular lattice with six neighbors or for random bond percolation on the square lattice, we have $p_c=\frac{1}{2}$, while for the square site problem, numerical estimates give $p_c \simeq 0.5927462\ldots$ and $p_c \simeq 0.31161\ldots$ for simple-cubic random site percolation. (Dimensions much higher than two and three were recently investigated by Grassberger (2003). Random percolation in one dimension has $p_c=1$.) Only in infinite lattices are these thresholds defined as infinitely sharp.

Quantities of interest besides n_s are the probability $R(p)$ to have a spanning cluster, the fraction P_∞ of sites in the infinite (spanning) cluster, the mean cluster size

$$S=\frac{\sum_s n_s s^2}{\sum_s n_s} \quad \text{or} \quad \sum_s n_s s^2, \tag{1}$$

the average cluster radius R_s, and the characteristic length ξ defined best through $\xi^2=\sum_s R_s^2 n_s s^2 / \sum_s n_s s^2$ among other properties. (For $p>p_c$ in these sums, the spanning cluster is omitted.) Figure 1 shows S versus p, for random site percolation on nearest-neighbor simple-cubic lattices. From scaling laws or renormalization group arguments, the critical phenomena for large clusters and small $p-p_c$ are

$$\begin{aligned} n_s &\propto s^{-\tau} f[(p-p_c)s^{\sigma}], \\ P_\infty &\propto (p-p_c)^{\beta}, \\ S &\propto |p-p_c|^{-\gamma}, \\ \xi &\propto |p-p_c|^{-\nu}, \\ R_s(p=p_c) &\propto s^{\sigma\nu} \end{aligned} \tag{2}$$

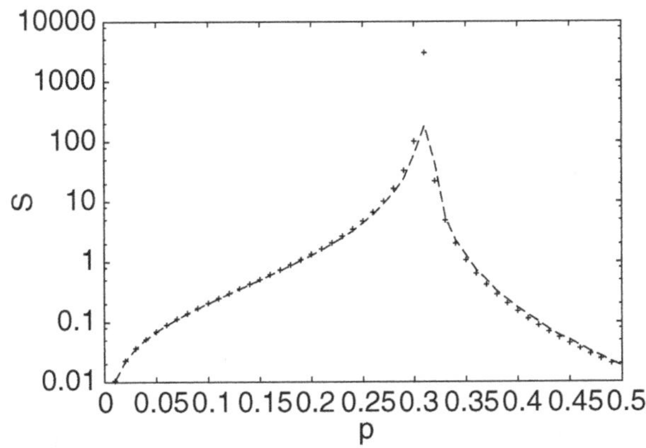

Figure 1. Mean cluster size from Equation (1) for seven nearest-neighbor simple-cubic lattices lattices $1000 \times 1000 \times 1000$ (+) compared with $100 \times 100 \times 100$ (line).

with the critical exponents given by Greek letters.

The first of these equations (Stauffer & Aharony, 1994) leads to Fisher's scaling laws $\beta = (\tau - 2)/\sigma$, $\gamma = (3 - \tau)/\sigma$, while additional hyperscaling assumptions valid for dimensionality d below six give $d\nu = \gamma + 2\beta = (\tau - 1)/\sigma$. The fractal dimensionality D of clusters at the percolation threshold, defined (Bunde & Havlin, 1996) through $s \propto R_s^D$, is then

$$D = 1/(\sigma\nu) = d - \beta/\nu = d/(\tau - 1). \quad (3)$$

Because an infinite cluster first appears for increasing p at $p = p_c$, it is plausible that for p slightly below p_c, we see finite but very large clusters; thus, the mean cluster size S diverges for $p \to p_c$ in an infinite system. In a finite system, S remains finite but the position of its maximum moves towards p_c and its height towards infinity, for system size $\to \infty$ (see Figure 1). The relation of its critical exponent γ with σ and τ can be easily derived from the above assumption for n_s by approximating the sum for S through an integral over s.

Flory's approximate solution is valid for the critical exponents only above six dimensions: $\beta = \gamma = 2\nu = 2\sigma = 1$, $\tau = \frac{5}{2}$, $D = 4$; the above scaling function f is then a Gaussian. For the more relevant case of $d = 2$, we have $\beta = 5/36$, $\gamma = 43/18$, while in three dimensions $\beta \simeq 0.42$, $\gamma \simeq 1.79$ with the other exponents following from the above scaling laws. The scaling function f is no longer a Gaussian but has a maximum below p_c; the cluster numbers n_s away from p_c decay as $\log(n_s) \propto -s$ above p_c and as $\propto -s^{1-1/d}$ below p_c. For $p < p_c$, the fractal dimension in three dimensions is $D = 2$, one of the rare cases where an exact solution was found in three but not in two dimensions.

Besides computer simulation, small-cell renormalization group methods also give reasonable approximations for thresholds and critical exponents. One divides the infinite lattice into cells of linear dimension b such that the cell centers themselves again form a lattice of the same type as the original lattice. A cell is occupied if the original occupied sites form a cluster spanning this cell. In this way, the original occupation probability p is renormalized into the cell occupation probability $p' = R(p)$. One may iterate this renormalization by computing $p'' = R(p')$ and then $R(p'')$ and so on. The fixed point p^* of this transformation obeys $p^* = R(p^*)$ and approaches, for large b, the true percolation threshold p_c. At this fixed point, the derivative $\mathrm{d}R(p)/\mathrm{d}p$ is related to the critical exponent ν. However, in contrast to many renormalization group publications, $R(p_c)$ is not equal to p_c (Ziff, 1992).

Some mathematical theorists in the 1980s claimed that the infinite cluster is unique, but this is not true at $p = p_c$ where several spanning clusters can coexist (Aizenman, 1997). Even above p_c in very elongated rectangles, one may find several clusters spanning in the short direction.

How do these theoretical and numerical results compare with applications and reality? Because the percolation thresholds p_c depend on the lattice structure, one should not expect them to agree with off-lattice experiments, but the exponents are more "universal" and should agree when comparing reality with PT. There are many applications (Sahimi, 1994) including flow in porous media, conductor-insulator mixtures, elasticity of composite materials, and breakdown of the internet. Sputtering metal drops onto a plane offers a particularly accurate realization of percolation.

Flory invented the percolation theory to describe the sol-to-gel transition of branched polymers, known from boiling your breakfast egg. But only half a century later was it experimentally confirmed that the gelation exponents are indeed those of three-dimensional lattice percolation theory.

Another application of percolation clusters is to speed up computer simulations of Ising models (ferromagnets) near their Curie point, using the clusters of correlated bond percolation formed by up spins connected by bonds with probability $p = 1 - \exp(-2J/k_B T)$. Flipping whole clusters instead of single spins equilibrates the system much faster. (In this Fortuin–Kasteleyn–Coniglio–Klein–Swendsen–Wang theory (Stauffer & Aharony, 1994), T is the absolute temperature, k_B the Boltzmann constant, and J the interaction energy such that $2J$ is the energy to change an isolated pair of parallel spins into antiparallel.) For example, right at the Curie point in a two-dimensional $L \times L$ lattice, the relaxation time no longer increases as $L^{2.17}$ but only as $\log(L)$ if these percolation clusters, instead of single spins, are flipped.

The above application stems from the relation between percolation clusters and the q-state Potts model ($q = 2$ is the Ising model again). The partition function of this Potts model is $Z = \langle q^N \rangle$, where N is the total number of bond percolation clusters at the above $p = 1 - \exp(-2J/k_{\rm B}T)$ and $\langle \cdots \rangle$ denotes the average over many realizations. Taking the limit $q \to 1$, we get $\ln(Z)/\ln(q) \to \langle N \rangle = \sum_s n_s$, recovering random percolation which is thus a $q \to 1$ Potts model.

Percolation is also a convenient teaching example. Filling a small triangular lattice at $p = p_{\rm c} = \frac{1}{2}$ by throwing coins reveals in the classroom fractals, statistical errors, and (in the resulting number of isolated sites) strong systematic errors from the lattice boundaries.

DIETRICH STAUFFER

See also **Cluster coagulation; Polymerization; Sandpile model**

Further Reading

Aizenman, M. 1997. On the number of incipient spanning clusters. *Nuclear Physics* B, 485: 551–582

Broadbent, S.R. & Hammersley, J.M. 1957. Percolation processes. *Proceedings of the Cambridge Philosophical Society*, 53: 629–645

Bunde, A. & Havlin, S. 1996. *Fractals and Disordered Systems*, Berlin: Springer

Flory, P.J. 1941. Thermodynamics of high polymer solutions. *Journal of the American Chemical Society*, 63: 3083–3100

Grassberger, P. 2003. Critical percolation in high dimensions. *Physical Review* E, 67: 036101

Maloney, N.R. & Pruessner, G. 2003. Asynchronously parallelized percolation on distributed machines. *Physical Review* E, 67: 037701

Sahimi, M. 1994. *Applications of Percolation Theory*, London: Taylor & Francis

Stauffer, D. & Aharony, A. 1994. *Introduction to Percolation Theory*, London: Taylor & Francis

Stockmayer, W.H. 1943. Theory of molecular size distribution and gel formation in branched-chain polymers. *Journal of Chemical Physics*, 11: 45–55

Tiggemann, D. 2001. Simulation of percolation on massively-parallel computers. *International Journal of Modern Physics* C, 12: 871–878

Ziff, R.M. 1992. Spanning probability in 2D percolation. *Physical Review Letters*, 69: 2670–2673

PERIOD DOUBLING

Like many terms used in the nonlinear sciences, *period doubling* has more than one meaning. Well known is the response of a system at half the driving frequency, due to nonlinear coupling. Probably the earliest observation of period doubling was by Michael Faraday, in his investigations of shallow water waves (Faraday, 1831, arts. 98–101). The phenomenon was also investigated by Lord Rayleigh (Rayleigh, 1887). He wrote of an example:

> in which a fine string is maintained in transverse vibration by connecting one of its extremities with the vibrating prong of a massive tuning-fork, *the direction of motion of the point of attachment being parallel to the length of the string* the string may settle down into a state of permanent and vigorous vibration *whose period is double that of the point of attachment.*

A more everyday example is the fact that a child may set a swing into transverse motion by standing on the seat and moving up and down at twice the natural frequency. Such phenomena are in the province of *harmonic generation* and *parametric amplification* and are not treated in this entry.

Period doubling (as discussed in this entry) is the most common of several *routes to chaos* for a nonlinear dynamical system.

The dynamics of natural processes, and the nonlinear equations used to model them, depend on externally set conditions, such as environmental or physical factors. These take fixed values over the development of the system in any particular instance, but vary from instance to instance. Expressed as numerical quantities, such factors are called parameters, and their role is vital. Often, increasing a parameter increases the nonlinearity. The simplest example is the logistic model (May, 1976)

$$x_{n+1} = r x_n (1 - x_n), \qquad (1)$$

a discrete system with state variable x_n and parameter r; n is the generation. (If x represents population, then r characterizes the underlying growth rate.)

For period doubling, it is sufficient to vary a single parameter. Over some range, the system may have a periodic attractor, that is, a periodic orbit that attracts neighboring orbits. Typically, the stability (attraction) decreases as the parameter increases, changing to instability (repulsion) at a critical value. This is a bifurcation, or change of structure, of the orbit. In period doubling, this change is accompanied by the "birth" of a new attracting period-doubled orbit, in which the system alternates between two states. In biological systems, these are known as alternans (Glass & Mackey, 1988). In the period-doubling route to chaos, each new periodic attractor loses its stability with increasing parameter value, whereupon the next period-doubled attractor is born. If the original attractor was a fixed point, this generates orbits of period 1, 2, 2^2, 2^3, ..., called the main period-doubling cascade.

A surprising *universality* was discovered by Mitchell Feigenbaum. In part, it relates to the sequence of parameter values at which successive period doubling occurs. Label the first by r_1, the second by r_2, and so on, and let the cascade end at the value r_∞. The differences $(r_\infty - r_n)$ decrease to zero; the surprising fact is that the ratios $(r_\infty - r_n)/(r_\infty - r_{n+1})$ converge to a universal constant δ, which takes the same value $\delta \approx 4.669202\ldots$ for all maps with a quadratic maximum (Feigenbaum, 1978). There is a second universal constant, $\alpha \approx 2.502908\ldots$, measuring the relative spatial scale of the orbits, which also becomes increasingly fine.

	Logistic		Hénon	
n	r_n	$\left(\frac{r_{n+1}-r_n}{r_{n+2}-r_{n+1}}\right)$	a_n	$\left(\frac{a_{n+1}-a_n}{a_{n+2}-a_{n+1}}\right)$
1	3.0000000000		0.3675000000	
2	3.4494897428		0.9125000000	
3	3.5440903596	4.751446	1.0258554050	4.807887
4	3.5644072661	4.656251	1.0511256620	4.485724
5	3.5687594195	4.668242	1.0565637582	4.646894
6	3.5696916098	4.668739	1.0577308396	4.659569

Table 1. Estimates of the Feigenbaum constant δ, for the logistic and Hénon maps.

Experiment		Observed number of period doublings	estimated value of δ	estimated value of α
Hydrodynamic:	helium	4	3.5 ± 1.5	
	mercury	4	4.4 ± 0.1	
Electronic:	diode	5	4.3 ± 0.1	2.4 ± 0.1
	Josephson	4	4.5 ± 0.3	2.7 ± 0.2
Laser:	feedback	3	4.3 ± 0.3	
Acoustic:	helium	3	4.8 ± 0.6	

Table 2. Selected experimental data on period doubling.

Numerical data for two main period-doubling sequences is given in Table 1. Columns 2–3 display data for the logistic map,

$$x_{n+1} = 1 - ax_n^2 + y_n, \qquad y_{n+1} = bx_n. \tag{2}$$

columns 4–5 for its two-dimensional cousin, the Hénon map. For Table 1, the parameter b has been kept constant. Experimental data has also been obtained in quite a few systems (Cvitanović, 1989); a selection is shown in Table 2. To appreciate the experimental difficulty involved, remember that for each successive period doubling, the significance of errors increases five-fold while the complexity of the dynamics doubles.

The mechanism for period doubling is already implicit in the fact that it is also known as a "flip bifurcation." Stability of a fixed point x^* of a smooth one-dimensional map f is the simplest case for theory. Consider a nearby point x that maps to x'. Linear approximation gives

$$(x' - x^*) \approx f'(x^*) \cdot (x - x^*). \tag{3}$$

This shows (and exact analysis confirms) that the fixed point is stable if $-1 < f'(x^*) < 1$ and unstable if $f'(x^*) < -1$ or $f'(x^*) > 1$. Successive iterates flip from side to side for negative $f'(x^*)$, so period doubling occurs at $f'(x^*) = -1$. For higher-dimensional systems, stability is determined by the eigenvalues of a matrix of partial derivatives; the occurrence of an

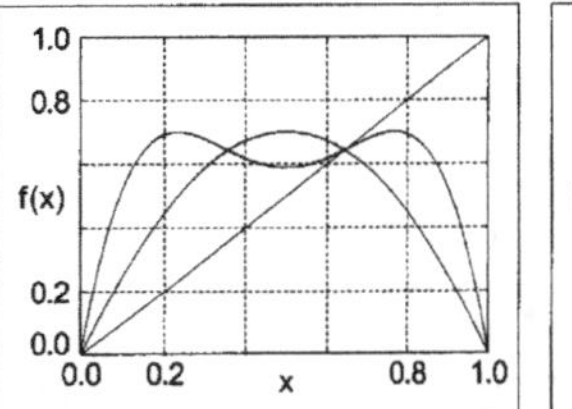

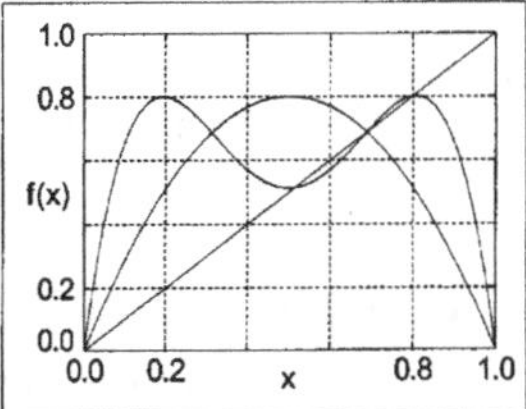

Figure 1. Mechanism: f and f_2 for $r = 2.8$ (left) and $r = 3.2$ (right).

eigenvalue $\lambda = -1$ leads to period doubling. Because $x^{**}_{n+2} = x^{**}_n$ for the new orbit, period doubling is connected with double iteration, controlled by the second composition map $f_2(x) = f(f(x))$. Elementary calculus shows that if f satisfies the two conditions

$$f(x^*) = x^*, \qquad f'(x^*) = -1, \tag{4}$$

then f_2 satisfies three conditions:

$$f_2(x^*) = x^*, \quad f_2'(x^*) = +1, \quad f_2''(x^*) = 0. \tag{5}$$

As a result, $x - f_2(x)$ must be approximated by a cubic near the bifurcation. One of its zeros is the (now unstable) fixed point of f, the other two constitute the period-doubled orbit, because they are not fixed points of f.

In the case of the logistic map, the first period doubling occurs at $r = 3$. Graphs of f and f_2 show the mechanism (Figure 1). As the cascade proceeds, the pictures and the analysis repeat but with increasing complication. The common thread is that, at each step, conditions (4) for a composition f_n imply conditions (5) for f_{2n}.

BRIAN DAVIES

See also **Attractors; Bifurcations; Dripping faucet; Harmonic generation; Hénon map; Maps; One-dimensional maps; Parametric amplification; Routes to chaos; Stability; Universality**

Further Reading

Cvitanović, P. 1989. *Universality in Chaos*, 2nd edition, Bristol: Adam Hilger

Davies, B. 1999. *Exploring Chaos: Theory and Experiment*, Reading, MA: Perseus Books

Faraday, M. 1831. On a peculiar class of acoustical figures; and on certain forms assumed by groups of particles upon vibrating elastic surfaces. *Philosophical Transactions of the Royal Society of London*, 121: 299–340

Feigenbaum, M.J. 1978. Quantitative universality for a class of nonlinear transformations. *Journal of Statistical Physics*, 19: 25–52

Glass, L. & Mackey, M.C. 1988. *From Chaos to Clocks: The Rhythms of Life*, Princeton, NJ: Princeton University Press

May, R.M. 1976. Simple mathematical models with very complicated dynamics. *Nature*, 261: 459–469

Rayleigh, Lord. 1887. Maintenance of vibrations by forces of double frequency, and propagation of waves through a medium with a periodic structure. *Philosophical Magazine*, 24: 145–159

Schuster, H.G. 1988. *Deterministic Chaos: An Introduction*, 2nd edition, Weinheim: VCH

PERIODIC BURSTING

Bursting systems show intervals of repetitive activity separated by intervals of relative quiescence, and have at least two different time scales; fast for the short period with spike-like oscillations that comprise the burst and slow for the longer period of the burst itself. Laboratory systems that exhibit bursting include the Belousov–Zhabotinsky reaction and neuronal systems. A simple conceptual model would be the bursting that is generated by switching between the active and inactive states, suggesting the coexistence of stable periodic and quiescent states. Bursting can be irregular, in which the periods between bursts are effectively random, or it can be strictly periodic. Some experimental bursting systems exhibit switching between different patterned periodic behaviors, suggesting two or more coexisting periodic states, and models have been shown to exhibit bi- or multi-rhythmicity. The exotic periodic dynamics of bursting systems—with patterning, multistability, and multi-rhythmicity provide a demonstrator for bifurcation and continuation algorithm packages.

A simple two-variable excitable or oscillatory system such as the FitzHugh–Nagumo equations can be driven to bursting by a slow periodic forcing term, so bursting can be viewed as the activity of a fast oscillatory system being modulated by a slower oscillatory system. For an autonomous bursting system, the fast and slow systems are coupled, and the burst is generated by the evolution of the slow system sweeping the dynamics of the fast subsystem between steady state and oscillatory dynamics (Chay & Rinzel, 1985). On the timescale of the fast system, the slow variable acts as a control parameter, so the fast variable approximates its attractors — its stable equilibrium during the period of quiescence and its oscillations during activity.

Bursting in neurons and other electrically excitable secretory cells may be part of their normal pattern generation behavior, as in the generation of the respiratory rhythm, or a sign of pathology, as in the abnormal bursting charges during epileptic fit. Synchronization of bursting activity between widely separate cells in the cortex excited by different features of the same visual input has been observed and proposed to underlay visual binding. Neural central pattern generators are networks of nerve cells, that generate periodic bursts as a result of their interconnections, but some single isolated cells can generate bursts. Both network and isolated cell bursting can be modeled by systems of differential equations. Specific models for isolated cells—the beta-pancreatic cell or the parabolic burster cell R15 of Aplysia—are ordinary differential systems, analogous to the Hodgkin–Huxley equations for the action potential, based on experimental measurements of currents and concentrations (Rinzel & Lee, 1986). They reproduce the observed bursting behaviors and can be reduced to simpler systems in which the dynamical mechanisms producing bursting are obtained by bifurcation analysis of these models.

In these models, the spiking is produced by oscillations of a fast subsystem that is modulated by the slower system. Bifurcation and topological analysis of the models has led to a classification of the types of bursting (Bertram et al., 1995). In type I, the fast subsystem is bistable, and the burst begins at a saddle node and ends at a homoclinic bifurcation. During the burst, the period between spikes increases illustrated by a simple three-variable polynomial model that retains the qualitative features of neuronal bursting, the Hindmarsh–Rose equations (1984) in Figure 1. The stable (thick solid line) and unstable (dotted line) Z-shaped steady state curves exchange stability at a Hopf bifurcation and a saddle-node bifurcation, and the periodic solutions that emerge at the Hopf bifurcation end at a homoclinic point. The xy subsystem has a bistable range, with stable steady state and limit cycle. During the burst, z is approximately constrained to this bistable range, each spike increments z, and the period between spikes increases as the homoclinic point is approached. Between bursts, z decreases slowly close to the lower, stable limb of the steady-state curve, as x increases until the orbit is re-injected into the spiking region.

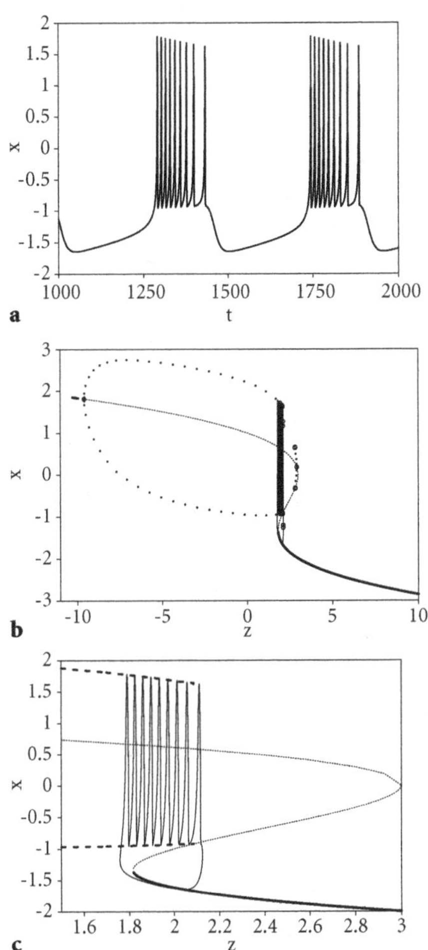

Figure 1. Example of type I burst generation in the three-variable (x, y, z) Hindmarsh & Rose (1984) model for neuronal oscillations, bursting and chaos. (a) Periodic bursting in the fast x variable representing membrane potential; (b) and (c) with z scale magnified: Bifurcation diagram for the fast $x-y$ subsystem with the slow variable z treated as a bifurcation parameter, and the periodic bursting of (a) is overlaid in the $x-z$ plane.

In type II bursting, two slow variables drive the fast system across a homoclinic bifurcation, so the spike period within the burst decreases and then increases parabolically, as the slow variables drive the system toward, and away from, the homoclinic point. This is illustrated by the parabolic burster R15 of Aplysia. The fast system has only a stable periodic limit cycle during the burst.

In type III bursting, as in type I, the fast system is bistable, but the burst ends at a saddle node rather than a homoclinic bifurcation, and the spike period need not increase during the burst.

Examples of all three types are seen in different bursting cells, and many bursting cells show more exotic, and chaotic, behaviors. An alternative approach is to reduce the models to interval maps, and to classify the bifurcations of these maps—see Fan & Holden (1993).

ARUN V. HOLDEN

See also **FitzHugh–Nagumo equation; Hodgkin–Huxley equations; Nerve impulses**

Further Reading

Bertram, R., Butte, M.J., Kiemel, T. & Sherman, A. 1995. Toplogical and phenomenological classification of bursting oscillations. *Bulletin of Mathematical Biology*, 57: 413–439

Chay, T.R. & Rinzel, J. 1985. Bursting, beating and chaos in an excitable embrane model. *Biophysical Journal*, 47: 357–366

Fan, Y-S. & Holden, A.V. 1993. Bifurcations, burstings, chaos and crises in the Rose–Hindmarsh model for neuronal activity. *Chaos Solitons and Fractals*, 3: 439–449

Hindmarsh, J.L. & Rose, R.M. 1984. A model for neuronal bursting using three coupled first order differential equations. *Proceedings of the Royal Society of London* B, 221: 87–112

Rinzel, J. & Lee, Y.S. 1986. On different mechanisms for membrane potential bursting. In *Nonlinear Oscillations in Biology and Chemistry*, edited by H.G. Othmer, New York: Springer, pp. 19–33

PERIODIC ORBIT THEORY

Many characteristic quantities for chaotic systems, such as Lyapunov exponents, correlation functions, or fractal dimensions, are defined as averages over the invariant measure (defined as a probability distribution on the phase space). For most chaotic systems, this measure is difficult to represent analytically, so it has to be approximated experimentally or numerically by averages over long trajectories. The periodic orbit theory rests on the idea that periodic orbits provide a skeleton that allows approximation of the invariant measure and that, therefore, ergodic averages can be determined from suitably weighted averages over periodic orbits. It combines mathematically established analytical expressions with well-controlled numerical input for highly accurate quantitative information about chaotic systems. It provides access to Lyapunov exponents, fractal dimensions, escape rates, diffusion constants, resonances, or even correlations between eigenvalues of quantum systems (Eckmann & Ruelle, 1985; Artuso et al., 1990; Chaos Focus Issue, 1992).

In order to motivate the basic expressions and to determine the weight of periodic orbits, consider the evolution of a density $\rho_n(\boldsymbol{x})$ at time n under the action of a discrete dynamical system, $\boldsymbol{x}_{n+1} = \boldsymbol{F}(\boldsymbol{x}_n)$ (Cvitanović & Eckhardt, 1991). After one time step, the density is given by

$$\rho_{n+1}(\boldsymbol{x}) = \int \mathrm{d}\boldsymbol{y}\, U(\boldsymbol{x}, \boldsymbol{y})\rho_n(\boldsymbol{y}) \qquad (1)$$

with the evolution kernel $U(\boldsymbol{x}, \boldsymbol{y})$. In the deterministic map, the points $\boldsymbol{y}$ at time n and $\boldsymbol{x}$ at $n+1$ are connected by a single trajectory, so that U is a δ-distribution, $U = \delta(\boldsymbol{x} - \boldsymbol{F}(\boldsymbol{y}))$. In the presence of noise, the evolution kernel will be more regular and some of the mathematical subtleties that appear when dealing with singular operators will disappear. Noise regularization can also be used to define the "natural" invariant measure for a dynamical system, the Sinai–Ruelle–Bowen measure, as the measure obtained in the limit of vanishing noise (Eckmann & Ruelle, 1985). For an attractor, this invariant measure is the eigenvector to the eigenvalue 1 of the evolution operator.

Periodic orbits appear when traces of U are calculated: the trace requires that $\boldsymbol{x}$ and $\boldsymbol{y}$ coincide, and this singles out a periodic orbit. With p the label for the different periodic orbits, n_p their primitive period (the number of iterations need to first return to the starting point), and $J_p = \mathrm{d}\boldsymbol{x}(n_p)\,/\,\mathrm{d}\boldsymbol{x}(0)$ the Jacobian after one period, the trace for m iterations becomes

$$\begin{aligned} \operatorname{tr} U^m &= \int \mathrm{d}\boldsymbol{x}\, U^m(\boldsymbol{x}, \boldsymbol{x}) \\ &= \sum_p \sum_{r=1}^{\infty} \frac{n_p}{|\det(1 - J_p^r)|} \delta(m - r n_p); \end{aligned} \quad (2)$$

powers of U are defined as multiple mappings so that U^m is the evolution operator after m time steps. The first sum extends over the primitive, nonrepeated orbits p and the second over their multiple traversals r.

The Fredholm determinant for the evolution operator U can be related to traces of the evolution operator via

$$\det(1 - zU) = \exp(\operatorname{tr} \ln(1 - zU)) \quad (3)$$

$$= \exp\left(-\sum_{n=1}^{\infty} \frac{z^n}{n} \operatorname{tr} U^n\right) = \sum_{l=0}^{\infty} c_l z^l. \quad (4)$$

The right-hand side is then an expansion in terms of periodic orbits, first through the traces in the exponent and then through their contributions to the coefficients c_l in the final power series. The left-hand side can formally be expanded in terms of eigenvalues λ_ν of the operator, $\det(1 - zF) = \prod_\nu (1 - z\lambda_\nu)$. It vanishes whenever z equals the inverse of an eigenvalue.

The formal manipulations that lead to (4) can be made precise in the context of nice hyperbolic systems in which periodic orbits are dense ("Axiom A"). Simple maps and flows on surfaces of negative curvature belong to this class. Then the theory can be developed as discussed in Ruelle (1989) and Pollicott (2002). In particular, the eigenvalues with sufficiently small damping can be given an interpretation as experimentally accessible long-lived Ruelle–Pollicott resonances.

For the calculation of averages, the relation between the leading eigenvalue and the invariant density is important. If the average of a given phase space observable A is to be calculated, one can extend the transfer operator U to include the observable, $\tilde{U} = U \exp(-qA)$, and extract the phase space average from the variation of the leading, zero $z_0(q)$ of the Fredholm determinant with q,

$$\langle A \rangle = \left. \frac{\mathrm{d}}{\mathrm{d}q} \ln z_0(q) \right|_{q=0}. \quad (5)$$

The contribution to the trace is a term $\exp(q \sum_P A(\boldsymbol{x}_P))$ with the sum extending over all points P of the orbit p: it reflects the sampling of phase space along the orbit.

The dependence of the Fredholm determinant on periodic orbits and their properties can be spelled out completely, so that it can be written as a product over contributions from periodic orbits. For the case of a one-dimensional map, we find

$$\begin{aligned} \det(1 - zU) &= \prod_p \prod_{j=0}^{\infty} \left(1 - \frac{z^{n_p}}{|J_p| J_p^j}\right) \\ &= \prod_j \zeta_j^{-1}(z). \end{aligned} \quad (6)$$

The ζ_j are called dynamical zeta functions in analogy to the Riemann zeta function $\zeta_{\mathrm{R}}(s)$, which by the Euler product representation can be expressed as a product over prime numbers,

$$\zeta_{\mathrm{R}}^{-1}(s) = \left(\sum_{n=1}^{\infty} n^{-s}\right)^{-1} = \prod_p (1 - p^{-s}). \quad (7)$$

The analytic properties of the Fredholm determinant determine the decay of the coefficients c_l in the power series expansion in (4): if the Fredholm determinant is analytic then the coefficients fall off faster than exponentially (Rugh, 1992). This is particularly useful in numerical calculations where highly accurate and quickly converging results can be obtained using the shortest orbits.

For continuous flows, the discrete period z^{n_p} is replaced by the continuous equivalent $\exp sT_p$, and the contributions of observables becomes an integral along the orbit.

When the classical evolution operator is replaced by the quantum one, a relation between the quantum eigenvalues and classical periodic orbits emerges (Gutzwiller, 1990). For instance, the semiclassical periodic orbit expression for the density of states in a chaotic system is

$$\begin{aligned} \rho(E) &= \sum_i \delta(E - E_i) \\ &\sim \mathrm{Re} \frac{1}{\pi\hbar} \sum_p \sum_{r=1}^{\infty} \frac{T_p \exp(-\mathrm{i}\mu_p \pi r/2)}{\sqrt{|\det(1 - J_p^r)|}} \mathrm{e}^{\mathrm{i} r S_p(E)/\hbar}, \end{aligned} \quad (8)$$

where the symbol $\sim$ indicates the omission of a few regularizing terms, and where S_p and μ_p are the classical action and the Maslov index, respectively. This Gutzwiller trace formula is at the heart of many developments in the field of quantum chaos.

Zeta functions for the counting of periodic orbits already appear in Smale (1967). The mathematical theory was developed prominently by David Ruelle, Rufus Bowen, and Yasha Sinai (see Eckmann & Ruelle, 1985; Ruelle, 1989; Pollicott, 2002). Practical aspects for calculating with periodic orbits are discussed by Artuso et al. (1990), by Gaspard (1998), and by many contributions to the Chaos Focus Issue (1992). Applications of this formalism to simple maps and the period-doubling attractor are given by Artuso et al. (1990), to diffusion in Lorentz gases in Chaos Focus Issue (1992), and to the dimension and resonances to the Lorenz model in Eckhardt & Ott (1994). The relation to quantum mechanics is discussed in Gaspard (1998), Gutzwiller (1990), Eckhardt (1993), and the Chaos Focus Issue (1992). As demonstrated by Flepp et al. (1991), a comparison between periodic orbits extracted from experimental data and from a numerical model can be used to fine tune and improve the model.

BRUNO ECKHARDT

See also **Phase space; Quantum chaos**

Further Reading

Artuso, R., Aurell, E. & Cvitanović, P. 1990. Recycling of strange sets: I. Cycle expansion; II. Applications. *Nonlinearity*, 3: 325–360 and 361–386

Chaos Focus Issue on Periodic Orbit Theory. 1992. *Chaos*, 2: 1–158

Cvitanović, P. & Eckhardt, B. 1991. Periodic orbit expansions for classical smooth flows. *Journal of Physics* A, 24: L237–L241

Eckhardt, B. 1993. Periodic orbit theory. *Proceedings of the International School of Physics "Enrico Fermi," Course CXIX, Quantum Chaos*, edited by G. Casati, I. Guarneri & U. Smilansky, Amsterdam: North-Holland, pp. 77–118

Eckhardt, B. & Ott, G. 1994. Periodic orbit analysis of the Lorenz attractor. *Zeitschrift für Physik* B, 94: 259–266

Eckmann, P. & Ruelle, D. 1985. Ergodic theory of chaotic systems. *Reviews of Modern Physics*, 57: 617–656 and 1115 (Addendum)

Flepp, L., Holzner, T., Brun, E., Finardi, M. & Badii, R. 1991. Model identification by periodic-orbit analysis for NMR-laser chaos. *Physical Review Letters*, 67: 2244–2247.

Gaspard, P. 1998. *Chaos, Scattering and Statistical Mechanics*, Cambridge and New York: Cambridge University Press

Gutzwiller, M.C. 1990. *Chaos in Classical and Quantum Mechanics*, New York: Springer

Pollicott, M. 2002. Periodic orbits and zeta functions. In *Handbook of Dynamical Systems*, vol 1A, edited by B. Hasselblatt & A. Katok, Amsterdam: Elsevier, pp. 409–452

Ruelle, D. 1989. *Elements of Differentiable Dynamical Systems and Bifurcation Theory*, New York: Academic Press

Rugh, H.H. 1992. The correlation spectrum for hyperbolic analytic maps. *Nonlinearity*, 5: 1237–1263

Smale, S. 1967. Differentiable dynamical systems. *Bulletin of the American Mathematical Society*, 73: 747–817

PERIODIC SPECTRAL THEORY

The direct problem of periodic spectral theory is that of constructing the spectral data of certain linear operators with periodic coefficients, that is, the determination of the spectrum of this operator and of the associated eigenfunctions. The inverse problem of periodic spectral theory is the problem of the reconstruction of such an operator (and thus its coefficients) from given spectral data. Although these questions can be asked for linear partial differential operators, this article focuses on linear ordinary differential operators.

The history of periodic spectral theory starts with the investigations of Sturm and Liouville on the eigenvalues of certain differential equations of second-order with given boundary conditions, now referred to as Sturm–Liouville theory. In 1836–1837, Charles François Sturm and Joseph Liouville examined independently different aspects of this problem, such as the asymptotics of eigenvalues, different comparison theorems on the solutions of similar equations with different coefficients, and theorems on the zeros of eigenfunctions. For the class of equations Sturm and Liouville considered, these results imply the existence of an infinite sequence of real, increasing eigenvalues, and orthogonality of eigenfunctions corresponding to different eigenvalues. Although their investigations did not as such deal with periodic spectral theory, many of their results carry over to this area.

Consider an ordinary differential operator of order n,

$$L = q_n(x)\frac{\mathrm{d}^n}{\mathrm{d}x^n} + q_{n-2}(x)\frac{\mathrm{d}^{n-2}}{\mathrm{d}x^{n-2}} + \cdots + q_1(x)\frac{\mathrm{d}}{\mathrm{d}x} + q_0(x), \tag{1}$$

where the coefficients $q_j(x)$, $j=0,\ldots,n$ are periodic functions of x, sharing a common period: $q_j(x+T)=q_j(x)$, $j=0,\ldots,n$, and $q_{n-1}(x)=0$. They are referred to as potentials. Using this operator L, define the differential equation

$$L\psi = \lambda\psi. \tag{2}$$

The direct periodic spectral problem is the problem of (i) determining the set of all $\lambda \in \mathbb{C}$ for which this differential equation has at least one bounded solution, and (ii) for each such λ, determining all bounded solutions. There are many technical issues to be dealt with: which function space do the potentials belong to? Which function space does ψ belong to? These issues will be ignored here. Sometimes one restricts attention to periodic solutions ψ: $\psi(x+T)=\psi(x)$, or antiperiodic solutions $\psi(x+T)=-\psi(x)$. These and other choices lead to spectra that are subsets of the spectrum as obtained without making these choices.

One approach to solve the direct spectral problem is Floquet theory (Amann, 1990). Rewrite Equation (2) as a first-order linear system:

$$\frac{\mathrm{d}\boldsymbol{\psi}}{\mathrm{d}x} = \boldsymbol{X}(x, \lambda)\boldsymbol{\psi}, \quad \boldsymbol{X}(x+T, \lambda) = \boldsymbol{X}(x, \lambda) \qquad (3)$$

with $\boldsymbol{\psi}_1 = \psi(x)$. Note that from $q_{n-1}(x) = 0$, it follows that $\mathrm{tr}\, \boldsymbol{X}(x, \lambda) = 0$. Thus, the flow determined by (3) is volume preserving. Define the monodromy matrix of this system as $\boldsymbol{M}(x_0, \lambda) = \boldsymbol{\Psi}(x_0 + T, x_0, \lambda)$, where $\boldsymbol{\Psi}(x, x_0, \lambda)$ is a fundamental matrix of (3) such that $\boldsymbol{\Psi}(x_0, x_0, \lambda)$ is the identity matrix. Thus, $\boldsymbol{M}(x_0, \lambda)$ is the operator of translating x by T: $\boldsymbol{M}(x_0, \lambda)\boldsymbol{\psi}(x) = \boldsymbol{\psi}(x + T)$. Note that $\boldsymbol{\psi}(x)$ also depends on x_0 and λ, but this dependence is suppressed here. This operation commutes with $\mathrm{d}/\mathrm{d}x$, since $\boldsymbol{X}(x, \lambda)$ is periodic in x with period T. Thus, (3) has a set of solutions $\boldsymbol{\phi}(x)$ which are also eigenvectors of $\boldsymbol{M}(x_0, \lambda)$. These solutions are known as Bloch functions or Floquet functions. If the eigenvalue of $\boldsymbol{M}(x_0, \lambda)$ for any Bloch function has magnitude greater than one, then this Bloch function is unbounded as $x \to +\infty$ or $x \to -\infty$. Thus, the spectrum of (3) is the set of all λ such that at least one eigenvalue of $\boldsymbol{M}(x_0, \lambda)$ has magnitude one. This spectrum is independent of the choice of x_0 due to the requirement that the flow is volume preserving; that is, $\mathrm{tr}\, \boldsymbol{X}(x, \lambda) = 0$, or $q_{n-1}(x) = 0$.

An important class of periodic spectral problems is that of self-adjoint operators. These are operators whose spectrum is contained on the real line. For self-adjoint operators, the spectrum of the periodic spectral problem consists of the union of a (possibly infinite) sequence of intervals.

For the sake of explicitness, the remainder of this article will discuss the equation (Magnus & Winkler, 1979)

$$-\psi'' + q(x)\psi = \lambda\psi, \quad q(x+T) = q(x). \qquad (4)$$

Note that many of the results stated below are true for general classes of periodic spectral problems. Depending on the literature source, (4) goes by the name of Hill's equation or (after rescaling) the time-independent Schrödinger equation. This equation is self-adjoint. Its spectrum is bounded from below. It is a collection of intervals such that the length of the separating gaps between intervals $\to 0$ as $\lambda \to \infty$. Using Floquet theory, the condition for λ to be in the spectrum is found to be $|\mathrm{tr}\boldsymbol{M}(x_0, \lambda)| \le 2$. The endpoints of the intervals are given by $|\mathrm{tr}\boldsymbol{M}(x_0, \lambda)| = 2$. Because (4) is a second-order equation, there are two linearly independent Bloch functions.

The time-independent Schrödinger equation (4) of course plays a fundamental role in quantum mechanics. In this case, $q(x)$ is the potential of the system, and λ plays the role of energy. The context of solid state physics is especially relevant here, because the potential $q(x)$ is periodic. The intervals constituting the spectrum are known as allowed (energy) bands and the gaps between them as forbidden (energy) bands.

The inverse periodic spectral problem for (4) is that of the reconstruction of $q(x)$, given a collection of spectral data. Various choices are possible for the collection of spectral data. In general, the inverse problem does not have a unique solution, using the knowledge of one spectrum. This can be resolved by also providing the eigenfunction. However, now the collection of spectral data is unnecessarily large. It is sufficient to provide two spectra (corresponding to different boundary conditions). Together with the known analyticity properties of the eigenfunction $\psi(x)$, this determines the potential $q(x)$. This is similar to the inverse scattering method where the knowledge of the scattering data and the analyticity properties of the eigenfunction determine the potential (decaying as $|x| \to \infty$) uniquely. A major difference is that in the inverse scattering method, the starting point is the asymptotic behavior as $x \to \pm\infty$. This behavior is simple, because it is governed by a differential equation with constant coefficients. This is one reason why the inverse scattering method is as efficient as it is. In the periodic problem, the role of $x \to \pm\infty$ is taken over by $x = x_0$, but there is no simple asymptotic behavior, resulting in a theory which is more technical and less explicit (Dubrovin et al., 1976; Novikov et al., 1984).

This lack of explicitness for solving the inverse periodic spectral problem is to some extent resolved by the consideration of so-called finite-gap potentials. These are potentials for which the number of intervals constituting the spectrum, and thus, the number of gaps separating these intervals, is finite. The simplest nontrivial example is that of the Lamé equation

$$-\psi'' + n(n+1)\wp(x - x_c)\psi = \lambda\psi. \qquad (5)$$

Here $\wp(x - x_c)$ is the Weierstrass elliptic function, x_c is a fixed complex number, and n is a positive integer. In this case, the number of gaps separating the intervals in the spectrum is n.

This classical example is a special case of a more recent theory of finite-gap potentials, whose development started with the works of Novikov (1974) and Lax (1975). They show that the stationary solutions of the nth member of the KdV hierarchy are n-gap potentials of (4). Here the KdV hierarchy is the collection of equations of the form $u_t = \partial_x(\delta H_n / \delta u)$, where H_n is any conserved quantity of the KdV equation, $\delta H_n / \delta u$ denotes the variational derivative of H_n with respect to u (*See* **Poisson brackets**), and indices denote differentiation. It was soon thereafter shown that all finite-gap potentials of (4) were of this nature. For example, the Lamé potential with $n = 1$ is a stationary solution of the KdV equation. This gives a nonspectral characterization of the finite-gap potentials.

To solve the direct spectral problem with an n-gap potential $q_n(x)$, one first considers the direct periodic spectral problem, as stated above. It is solved using

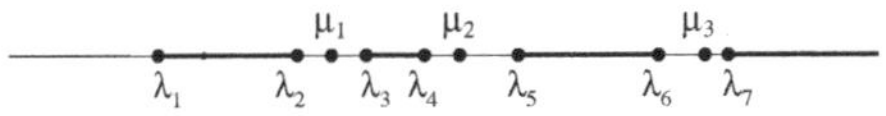

Figure 1. The main spectrum for a three-gap potential (thick solid line) and the auxiliary spectrum μ_1, μ_2 and μ_3.

Floquet theory. The outcome is the main spectrum, consisting of n finite intervals and one infinite interval. The endpoints of these intervals are labeled λ_1, λ_2, etc., in increasing order, as shown in Figure 1. At the endpoints, only one of the eigenfunctions is bounded. These eigenfunctions are periodic with period T or $2T$. There are an infinite number of isolated eigenvalues inside the interval of infinite length for which there are two bounded, periodic eigenfunctions. Next, one considers the Dirichlet problem

$$\begin{cases} -\psi'' + q_n(x)\psi = \lambda\psi, \\ \psi(x_0) = 0, \quad \psi(x_0 + T) = 0. \end{cases} \tag{6}$$

The spectrum of this problem is referred to as the auxiliary spectrum. It is discrete, and its points $\mu_k(x_0)$, $k = 1, 2, \ldots$ depend on x_0. All but n of its points lie inside the infinite interval of the main spectrum. Each remaining point lies in a different gap of the main spectrum. This is illustrated in Figure 1. The information contained in the main and auxiliary spectra determines the eigenfunction $\psi(x)$: it is a meromorphic function in the finite λ plane with zeros at $\lambda = \mu_k(x_0)$ and poles at $\lambda = \mu_k(x) = \mu_k(x_0)|_{x_0=x}$ (Dubrovin et al., 1976; Novikov et al., 1984).

Using the main and auxiliary spectra, the inverse periodic spectral problem is solved by (Novikov et al., 1984)

$$q_n(x) = \sum_{j=1}^{2n+1} \lambda_j - 2\sum_{j=1}^{n} \mu_j(x). \tag{7}$$

This is the first of the so-called trace formulae. Other trace formulae give relationships between the potential $q_n(x)$ and its derivatives and the main and auxiliary spectra.

The proposed solution of the inverse periodic spectral problem for finite-gap potentials is not effective. It requires the solution of the direct spectral problem for all x_0 in a period of the potential in order to obtain the auxiliary spectrum as a function of x_0. It is possible to avoid this by determining $\mu_k(x_0), k = 1, \ldots, n$ as a solution of a set of differential equations (Novikov et al., 1984)

$$\frac{d\mu_j}{dx_0} = \frac{\pm 2i\sqrt{\prod_{k=1}^{2n+1}(\mu_j - \lambda_k)}}{\prod_{k\neq j}^{n}(\mu_j - \mu_k)}, \quad j = 1, \ldots, n. \tag{8}$$

The choice of sign gives the direction in which $\mu_j(x_0)$ is going in between its two endpoints.

Another approach, which solves system (8), is the use of abelian functions and Riemann surfaces (Dubrovin, 1981; Belokolos, et al., 1994). An abelian function of n variables is a $2n$-periodic function. As such, abelian functions generalize elliptic functions to more than one variable. All abelian functions are expressible as ratios of homogeneous polynomials of Riemann's theta function. All finite-gap potentials of (4) are abelian functions. For example, the Lamé potentials in (5) are elliptic functions, which are special cases of abelian functions.

In the context of this method, the Bloch function $\phi(x) = \boldsymbol{\phi}_1(x)$ is often referred to as the Baker–Akhiezer function. One of the major results of the theory is the realization that the two Bloch or Baker–Akhiezer functions, regarded as a function of arbitrary complex λ, are distinct branches of one single-valued Baker–Akhiezer function, defined on a two-sheeted Riemann surface covering the complex λ plane (Krichever, 1989). This Riemann surface is already apparent in (8). It is

$$\eta^2 = \prod_{k=1}^{2n+1}(\lambda - \lambda_k), \tag{9}$$

which defines η as a double-valued function of λ. This surface has genus n. It is obtained from Figure 1 by choosing the intervals of the main spectrum as branch cuts and gluing the two resulting sheets together along these cuts. This Riemann surface defines a theta function $\theta(z_1, \ldots, z_n|\boldsymbol{B})$ through its normalized period (or Riemann) matrix $\boldsymbol{B}$. In terms of this theta function

$$q_n(x) = c - 2\frac{d^2}{dx^2}\ln\theta(k_1x + \varphi_1, \ldots, k_nx + \varphi_n|\boldsymbol{B}). \tag{10}$$

The wave numbers $k_1, \ldots, k_n$ are determined as integrals of certain differentials on the Riemann surface (9) with a pole singularity at $\lambda = \infty$. The phase constants $\varphi_1, \ldots, \varphi_n$ are determined by the Riemann constants on (9) and the Abel transform. The constant c is determined by a differential on (9) with a double pole singularity at $\lambda = \infty$. Thus, (10) gives an explicit form for all finite-gap potentials of (4), providing a complete solution of the inverse periodic spectral problem.

Some remarks:

(i) The emphasis on finite-gap potentials is justified in the sense that any T-periodic function is approximated arbitrarily well by an infinite sequence of finite-gap potentials with period T and increasing number of gaps (Dubrovin, 1981).

(ii) The periodic spectral problem (4) is the first half of the Lax pair for the Korteweg–de Vries (KdV) equation. As such, it allows the solution of the KdV equation with periodic initial data. The full solution of this problem requires the solution of both the direct and the inverse periodic spectral problems. The schematics is identical to that of the inverse scattering method. First, the initial condition $u(x, t = 0) = U(x)$ is used as a potential in (4) to solve the direct periodic spectral problem. This results in the main and auxiliary spectra. The time evolution of these spectra is implied from the second half of the Lax pair: the main spectrum is independent of time, whereas the auxiliary spectrum evolves according to differential equations similar to (8). The spectral data for any time is used to solve the inverse periodic spectral problem of (4). This gives the solution $u(x, t)$ of the KdV equation such that $u(x, t)|_{t=0} = U(x)$ (Novikov et al., 1984).

(iii) The spectral theory for the time-dependent Schrödinger equation is intimately connected to the initial-value problem for the Kadomtsev–Petviashvili equation. A solution of the inverse periodic spectral problem using Riemann's theta function and Riemann surfaces also exists here. However, here there are no restrictions on the form of Riemann surfaces that appear: all compact, connected Riemann surfaces arise. Hence, the periodic spectral theory of the time-dependent Schrödinger equation has important consequences for the theory of Riemann surfaces. It has provided, for instance, a solution to the Schottky problem, which was posed in 1903 (Novikov et al., 1984; Dubrovin, 1981).

(iv)The equation

$$\frac{d^2y}{dx^2} + (a - 2k\cos(2x))y = 0 \tag{11}$$

is Mathieu's equation. It arises from the three-dimensional Helmholtz equation by separation of variables using elliptical coordinates. It is a special case of (4), using a trigonometric potential. One is only interested in period solutions, resulting in a discrete subset of the main spectrum. The periodic solutions of this equation are referred to as Mathieu functions.

BERNARD DECONINCK

See also **Inverse scattering method or transform; Kadomtsev–Petviashvili equation; Korteweg–de Vries equation; Theta functions**

Further Reading

Amann, H. 1990. *Ordinary Differential Eequations: An Introduction to Nonlinear Analysis*, Berlin and New York: de Gruyter

Belokolos, E.D., Bobenko, A.I., Enol'skii, V.Z., Its, A.R. & Matveev, V.B. 1994. *Algebro-geometric Approach to Nonlinear Integrable Equations*, Berlin and New York: Springer

Dubrovin, B.A. 1981. Theta functions and non–linear equations. *Russian Mathematical Surveys*, 36: 11–92

Dubrovin, B.A., Matveev, V.B. & Novikov, S.P. 1976. Nonlinear equations of Korteweg–de Vries type, finite-zone linear operators, and Abelian varieties. *Russian Mathematical Surveys*, 31: 59–146

Krichever, I.M. 1989. Spectral theory of two-dimensional periodic operators and its applications. *Russian Mathematical Surveys*, 44: 145–225. (The introduction is a great overview of the development of periodic spectral theory.)

Lax, P.D. 1975. Periodic solutions of the KdV equation. *Communications on Pure and Applied Mathematics*, 28: 141–188

Magnus, W. & Winkler, S. 1979. *Hill's Equation*, New York: Dover and London: Constable

Novikov, S.P. 1974. The periodic problem for the Korteweg–de Vries equation. *Functional Analysis and Its Applications*, 8: 54–66

Novikov, S.P., Manakov, S.V., Pitaevskii, L.P. & Zakharov, V.E. 1984. *Theory of Solitons: The Inverse Scattering Method*, New York: Consultants Bureau (original Russian edition 1980)

PERMANENT WAVE

See **Wave of translation**

PERTURBATION THEORY

A limited number of nonlinear partial differential equations (PDEs) can be solved analytically for arbitrary initial conditions. Specific applications often lead to nonlinear PDEs that do not fall into the category that can be solved by means of the inverse scattering theory, Bäcklund transformations, or Hirota's bilinear method (Dodd et al., 1984). Examples of such PDEs with physical relevance include soliton equations with added perturbative terms describing energy gain and losses, where the original unperturbed problem is energy-conserving and hence Hamiltonian. Other problems involve nonconservative PDEs, such as nerve fibers and other reaction-diffusion type equations in general. In fluid mechanics, boundary layers possess a special challenge with singular perturbations requiring matching solutions close to a boundary and solutions far from a boundary layer. Perturbations can be additive or parametric (multiplicative), and they can be localized, periodic, quasi-periodic, or random, depending on the nature of the problem to be solved. In the case of a PDE that cannot be solved analytically, we resort to approximate analytical solutions or direct numerical analysis. The approximate solutions can be determined by various perturbation techniques. For soliton equations one procedure is to calculate the variation of the spectral data in the inverse scattering method due to external perturbations (Karpmann, 1979; Kivshar, & Malomed, 1989). Another method introduces slow variation of the parameters entering the soliton solution. Variants of the latter idea include multiple scales, energy methods, and the Lindsted–Poincaré technique. Here we shall illustrate the multiscale method applied to a perturbed nonlinear Schrödinger (NLS) equation. The method follows the ideas in Kaup (1990) with an extension presented in Nguyen et al. (1995). A general perturbed NLS equation is

$$i\frac{\partial u}{\partial t} + \frac{\partial^2 u}{\partial x^2} + 2|u|^2u = \varepsilon R(t, u). \tag{1}$$

The complex function $u = u(x, t)$ depends on the spatial coordinate x and time t. The absolute value of the perturbation parameter ε is assumed to be much smaller than unity, and we expect that the perturbation εR on the right-hand side only slightly modifies soliton solutions of the pure NLS equation. We introduce multiple scales of the time variable according to

$$T_0 = t, \quad T_1 = \varepsilon t. \tag{2}$$

In the simplest case, we treat the single soliton solution and invoke the solution ansatz

$$u = q \exp(\mathrm{i}\xi(\Theta - \Theta_0) + \mathrm{i}(\sigma - \sigma_0)), \tag{3}$$

$$q = \eta \operatorname{sech}(\eta(\Theta - \Theta_0)),$$

$$\text{where} \quad \Theta = x - 2X(T_0, T_1). \tag{4}$$

The functions X and σ are defined by $\partial X/\partial T_0 = -2\xi(T_1)$ and $\sigma = (\eta^2(T_1) + \xi^2(T_1))T_0$. To proceed, we expand q according to $q = q_0 + \varepsilon(\phi + \mathrm{i}\psi)$ and insert into Equation (1), finding

$$Lq_0 = (\partial_{\Theta\Theta} + 2q_0^2 - \eta^2)q_0 = 0, \tag{5}$$

$$M\phi = (\partial_{\Theta\Theta} + 6q_0^2 - \eta^2)\phi = \mathrm{Re}[F], \tag{6}$$

$$L\psi = (\partial_{\Theta\Theta} + 2q_0^2 - \eta^2)\psi = \mathrm{Im}[F]. \tag{7}$$

Here, q_0 is assumed to be real, and F is a lengthy complex expression, omitted for brevity. Re$[F]$ and Im$[F]$ denote the real and imaginary parts of F, respectively. First a solution for q_0 is obtained and the next step is to solve the inhomogeneous Equations (6) and (7). In solving these, we shall invoke Fredholm's solvability condition, which states that the null spaces of the operators M and L are orthogonal to the right-hand sides Re$[F]$ and Im$[F]$, respectively. The null space of an operator L is the space spanned by the solutions of $L^\dagger z(\Theta) = 0$, where $L^\dagger$ is the adjoint operator of L. This condition guarantees a solution without secular terms and provides evolution equations for the slowly varying parameters of the form

$$\eta \frac{\partial \xi}{\partial T_1} = \int_{-\infty}^{\infty} \mathrm{Re}[R] \frac{\partial q_0}{\partial \Theta} \, \mathrm{d}\Theta, \tag{8}$$

$$\frac{\partial \eta}{\partial T_1} = \int_{-\infty}^{\infty} \mathrm{Im}[R] q_0 \, \mathrm{d}\Theta. \tag{9}$$

For a given perturbation R, we can solve these equations for the slowly varying soliton amplitude η and ξ.

In the previous discussion, we began with a soliton as the zero-order approximation and adjusted its parameters to achieve an approximate solution of the perturbed soliton equation. In nonlinear diffusion equations, solutions emerge from balancing sources of energy with diffusion and dissipative effects. The perturbation theory outlined above can also be applied to such systems. As an example let us consider the FitzHugh–Nagumo (FN) model for nerve pulse propagation (FitzHugh, 1961):

$$\frac{\partial^2 V}{\partial x^2} - \frac{\partial V}{\partial t} = F(V) + R, \tag{10}$$

$$\frac{\partial R}{\partial t} = \varepsilon V, \tag{11}$$

where $F(V)$ is the cubic-shaped function $V(V - a)(V - 1)$. $V = V(x, t)$ is the voltage across the nerve cell membrane, and R is a recovery variable. The nerve pulse solution is localized in space and propagates by a velocity v to be determined from the model equations. Instead of using a multiscale approach, we invoke the Lindsted–Poincaré technique (Nayfeh, 1973) and expand the velocity parameter according to

$$v = v_0 + \varepsilon v_1 + O(\varepsilon^2). \tag{12}$$

The reason for introducing the above expansion is that nonlinearity will alter the propagation velocity. We also expand the dependent functions V and R

$$V = V_0 + \varepsilon V_1 + O(\varepsilon^2), \tag{13}$$

$$R = R_0 + \varepsilon R_1 + O(\varepsilon^2). \tag{14}$$

It is important to note that terms increasing to infinity for large times t, that is, secular terms in V_1 and R_1, are avoided by using the expansion of the velocity in Equation (12). Introducing a traveling wave assumption $\xi = x - vt$ in the solution and substituting expressions (12)–(14) into the FN equations (10) and (11) and ordering terms in power of ε leads to

$$\frac{\mathrm{d}^2 V_0}{\mathrm{d}\xi^2} + v_0 \frac{\mathrm{d}V_0}{\mathrm{d}\xi} - [F(V_0) - R_0] = 0, \tag{15}$$

$$\frac{\mathrm{d}R_0}{\mathrm{d}\xi} = 0. \tag{16}$$

To order $O(\varepsilon)$ we have

$$LV_1 = \frac{\mathrm{d}^2 V_1}{\mathrm{d}\xi^2} + v_0 \frac{\mathrm{d}V_1}{\mathrm{d}\xi} - F'(V_0)V_1 = R_1 - v_1 \frac{\mathrm{d}V_0}{\mathrm{d}\xi}, \tag{17}$$

$$\frac{\mathrm{d}R_1}{\mathrm{d}\xi} = -\frac{V_0}{v_0}. \tag{18}$$

Here it turns out that we can identify "boundary layers" for the determination of V and R together with appropriate matching. In the limit of $\varepsilon \to 0$, the nerve pulse voltage rises sharply from zero to a plateau value of order $O(1)$. The rapid change is equivalent to a boundary layer in fluid dynamics. Similarly a sharp decrease of the pulse appears at its rear end followed by a slow recovery interval due to a fading R. Matching of the solutions can be done in each region. The correction term v_1 in the expansion of the velocity is determined from Fredholm's solvability condition of

Equation (17). That is, the null space of the operator L should be orthogonal to the right-hand side of (17). This is equivalent to the requirement for the NLS equation in the previous section. For $a < \frac{1}{2}$ the traveling wave velocity v_0 of the leading edge is $v_0 = (1-2a)/\sqrt{2}$ and the first-order correction becomes

$$v_1 = \frac{\int_{-\infty}^{\infty} \left[\int_{\xi}^{\infty} V_0(\xi')\,\mathrm{d}\xi'\right] (\mathrm{d}V_0/\mathrm{d}\xi) \mathrm{e}^{v_0\xi}\,\mathrm{d}\xi}{v_0 \int_{-\infty}^{\infty} (\mathrm{d}V_0/\mathrm{d}\xi)^2 \mathrm{e}^{v_0\xi}\,\mathrm{d}\xi}. \quad (19)$$

As we have used a traveling wave ansatz, the obtained solution for V_0 may not be stable. However, a closer analysis reveals that stable traveling wave solutions exist (Scott, 2003). Random perturbations make up a set of conceptually different problems that may be treated by methods of stochastic differential equations and statistics. For soliton equations, studies have been performed on random parameters modeling the stochastic variation of various physical properties, such as geometry or material. The influence of thermal noise and incoherence of amplitude and phase modulations of input pulses are other important examples (Abdullaev, 1994; Abdullaev et al., 2001).

MADS PETER SØRENSEN

See also **Averaging methods; FitzHugh–Nagumo equation; Fredholm theorem; Korteweg–de Vries equation; Multisoliton perturbation theory**

Further Reading

Abdullaev, F.Kh. 1994. *Theory of Solitons in Inhomogeneous Media*, Chichester: Wiley

Abdullaev, F.Kh., Bang, O. & Sørensen, M.P. 2001. *NATO Advanced Research Workshop on Nonlinearity and Disorder: Theory and Applications*, Boston and Dordrecht: Kluwer

Dodd, R.K., Eilbeck, J.C., Gibbon, J.D. & Morris, H.C. 1984. *Solitons and Nonlinear Wave Equations*, London: Academic Press

FitzHugh, R. 1961. Impulses and physiological states in theoretical models of nerve membrane. *Biophysical Journal*, 1: 445–466

Karpmann, V.I. 1979. Soliton evolution in the presence of perturbation. *Physica Scripta*, 20: 462–478

Kaup, D.J. 1990. Perturbation theory for solitons in optical fibers. *Physical Review* A, 24: 5689–5694

Kivshar, Y.S. & Malomed, B.A. 1989. Solitons in nearly integrable systems. *Reviews of Modern Physics*, 61: 763–915

Nayfeh, A.H. 1973. *Perturbation Methods*, New York: Wiley-Interscience

Nguyen, P., Skovgaard, P., Sørensen, M.P. & Christiansen, P.L. 1995. Solitons in fibre lasers and mode-locked systems. *Physical Review* A, 24: 5689–5694

Scott, A. 2003. *Nonlinear Science: Emergence and Dynamics of Coherent Structures*, 2nd edition, Oxford and New York: Oxford University Press

PHASE DYNAMICS

Phase as a physical term is most commonly used in reference to oscillatory processes. The concept of phase dynamics is particularly relevant to self-sustained or limit-cycle oscillations. The orbital shape of a limit cycle is more or less rigid against perturbations, while the remaining degree of freedom along the closed orbit (the phase) lacks such rigidity. Consequently, when a limit-cycle oscillator is slightly perturbed, the resulting nontrivial dynamics could well be described in terms of the phase alone. This is the intuition behind the phase description of a variety of oscillatory phenomena, among others biological oscillations (Winfree, 2000).

Mathematically, limit-cycle oscillations are described with a time-periodic solution of an autonomous nonlinear ordinary differential equation $\mathrm{d}X/\mathrm{d}t = F(X)$, $X \in R^n$. Let such a solution with angular frequency ω be given by a 2π-periodic function $X_0(\omega t + \phi_0)$ involving an arbitrary phase constant ϕ_0. One may also introduce a phase variable $\phi(t)$ to represent this one-parameter family of solutions as

$$X(t) = X_0(\phi(t)), \qquad \frac{\mathrm{d}\phi}{\mathrm{d}t} = \omega. \quad (1)$$

Suppose the oscillator is slightly perturbed, by which F is slightly modified. As the resulting deformation of the closed orbit would be negligibly small, the first equation in (1) still holds approximately, while the correction to the instantaneous frequency $\mathrm{d}\phi/\mathrm{d}t$, however small it may be, would give rise to a large difference in ϕ in the long run and hence is indispensable.

Further, suppose that the perturbation represents the influence from another oscillator with state vector X' coupled weakly to the first one. If these oscillators are identical in nature, one may also use the approximation $X'(t) = X_0(\phi'(t))$ for the second oscillator, so that the current state of our oscillator pair may well be specified only with their phases $\phi(t)$ and $\phi'(t)$. Therefore, the small term to be added to the phase equation in (1) should generally be given by some function $G(\phi, \phi')$ which is 2π-periodic both in ϕ and ϕ'. Due to the assumed smallness of G relative to ω, it turns out that the effect of G over one cycle of oscillation depends only on the phase difference $\phi - \phi'$. Thus, using a 2π-periodic function Γ, one may obtain

$$\frac{\mathrm{d}\phi}{\mathrm{d}t} = \omega + \Gamma(\phi - \phi') \quad (2)$$

and, if we need, a similar equation for the second oscillator also. The function Γ can be computed in principle with the knowledge of the original dynamical-system model. Note, however, that the formula for Γ requires an extension of the definition of phase slightly outside the limit-cycle orbit (Kuramoto, 1984a).

The above equation may readily be extended to larger assemblies of weakly coupled oscillators. Furthermore, the oscillators may be slightly dissimilar, with their effect appearing only through a small difference in natural frequency ω. Thus, for such an assembly of N oscillators, we generally have

$$\frac{d\phi}{dt} = \omega_i + \sum_{j=1}^{N} \Gamma_{ij}(\phi_i - \phi_j), \quad i = 1, 2, ..., N. \quad (3)$$

This is called the phase oscillator model which is extremely useful for the study of complex collective behavior of large assemblies of coupled oscillators appearing in large varieties of fields ranging from physics to brain science (Pikovsky et al., 2001).

Phase dynamics are also useful when the oscillators constitute a continuous field. As a representative class of such continua, we will focus on oscillatory reaction-diffusion systems given by

$$\frac{\partial X}{\partial t} = F(X) + \hat{D}\nabla^2 X, \quad (4)$$

where $X(r, t)$ represents a space–time-dependent composition vector of dimension n and $\hat{D}$ is a diagonal matrix of n diffusion constants. If the spatial variation of X is of long wavelength, the diffusion term can be regarded as a small perturbation driving each local oscillator, so that the previous idea of phase dynamics should work. The approximation $X(r, t) = X_0(\phi(r, t))$ would again be valid at each spatial point. Its application to the diffusion term produces terms proportional to $\nabla^2\phi$ and $(\nabla\phi)^2$. According to a systematic perturbation theory, such terms actually appear in the phase equation. Specifically, we obtain a nonlinear phase diffusion equation

$$\frac{\partial \phi}{\partial t} = \omega + \nu\nabla^2\phi + \mu(\nabla\phi)^2 \quad (5)$$

to the nontrivial lowest order approximation with respect to the smallness in the spatial gradient of ϕ, where positive ν has been assumed. If ν is negative, the uniform oscillation is unstable. For small negative ν, the idea of phase dynamics still works. Then (5) is modified to give an equation equivalent to the Kuramoto–Sivashinsky equation, whose solution describes spatiotemporal chaos called phase turbulence.

In a more general context, phase is a degree of freedom appearing when a certain continuous symmetry of the system has been broken spontaneously (Mori & Kuramoto, 1998). For instance, the phase of a limit-cycle oscillator reflects the fact that the oscillation breaks the invariance of the governing evolution equation (assumed autonomous) with respect to temporal translations $t \to t + t_0$. Such symmetry-breaking solutions, not restricted to time-periodic solutions, generally involve an arbitrary phase constant, thus forming an infinite family recovering as a whole the original symmetry. This also means that phase has no prescribed value to which it is bound to relax or, equivalently, phase represents a neutrally stable dynamical variable. When the system is weakly perturbed, such neutrality is slightly violated causing the phase to evolve slowly, possibly apart from a constant drift. In contrast, all the other degrees of freedom will damp quickly, thus following adiabatically the slow motion of the phase. This explains why the phase so often dominates the pattern dynamics in nonlinear dissipative systems.

The general argument given above implies that the applicability of phase dynamics is by no means restricted to oscillatory systems. Two important non-oscillatory applications of phase dynamics, again in reference to reaction-diffusion systems with large spatial extension, will be touched upon next.

Reaction-diffusion systems may develop a spatially periodic static pattern out of a uniform state through the Turing instability, by which the spatial translational symmetry has been broken. When the local phase of the periodic pattern is subject to a large-scale spatial modulation, phase dynamics is applicable and describes the pattern dynamics through a slow evolution of the phase (Kuramoto, 1984b). For instance, the transient dynamics of recovery of a regular pattern can be described with a simple phase-diffusion equation.

Another important class of symmetry-breaking patterns arising in reaction-diffusion systems is localized structures such as moving domain boundaries and solitary pulses. In these cases, phase corresponds to the location of a moving front. For a planar front, the phase forms a uniform field advancing at a constant speed. When the planar front is deformed slightly in large spatial scale, the phase becomes slightly non-uniform and starts to evolve slowly. It is concluded from a systematic theory of phase dynamics that when the planar front is stable, the equation for the phase takes the same form as (5) with ∇ being replaced with a gradient of reduced dimension (Kuramoto, 1984a). A planar front that is weakly unstable due to small negative ν leads to phase turbulence described essentially by the Kuramoto–Sivashinsky equation.

YOSHI KURAMOTO

See also **Coupled oscillators; Kuramoto–Sivashinsky equation; Synchronization**

Further Reading

Kuramoto, Y. 1984a. *Chemical Oscillations, Waves, and Turbulence*, Berlin: Springer

Kuramoto, Y. 1984b. Phase dynamics of weakly unstable periodic structures. *Progress of Theoretical Physics*, 71(6): 1182–1196

Mori, H. & Kuramoto, Y. 1998. *Dissipative Structures and Chaos*, Berlin: Springer

Pikovsky, A., Rosenblum, M. & Kurths, J. 2001. *Synchronization: A Universal Concept in Nonlinear Sciences*, Cambridge and New York: Cambridge University Press

Winfree, A.T. 2000. *The Geometry of Biological Time*, 2nd edition, Berlin and New York: Springer

PHASE LOCKING

See **Coupled oscillators**

PHASE MATCHING

See **Nonlinear optics**

PHASE PLANE

The term *phase plane*, naturally enough, refers to the phase space of a two-dimensional dynamical system. The term is worthy of consideration in its own right because of the various special techniques that exist for the analysis of two-dimensional continuous-time dynamical systems (i.e., flows).

As an example, consider the two-dimensional flow induced by the following differential equations:

$$\begin{aligned} \dot{x}_1 &= x_1\,(1 - x_2)\,, \\ \dot{x}_2 &= x_2\,(x_1 - r)\,. \end{aligned} \qquad (1)$$

Each of these equations specifies one component of a two-dimensional vectorfield on the plane, tangent at each point to the induced flow. Equations (1) were first employed by Alfred Lotka (1920) to describe a hypothetical bimolecular chemical reaction capable of sustained oscillations, assuming mass action kinetics. At about the same time and independently, Vito Volterra (1926) made use of the same equations to model the interaction of biological predator and prey populations. In Equations (1), x_1 denotes the density of the prey species and x_2 that of the predator. Figure 1a shows the relevant (positive) quadrant of the phase plane of Equations (1).

The most important aspects of the dynamics of smooth flows are determined by their α- and ω-limit sets. The former are those sets approached by an orbit as $t \to -\infty$; the latter, those approached as $t \to +\infty$. In smooth planar flows, these limit sets are of only three types: singular points (or equilibria), closed (or periodic) orbits, and homo- and heteroclinic chains. The last are made up of singular points together with bounded orbits, each of which has one singular point as α-limit and another (possibly identical) singular point as ω-limit. It should be noted that more complex behaviors such as quasi-periodicity and chaotic dynamics are not possible in two-dimensional flows.

Singular points are determined as the solutions of a system of two algebraic equations and classified, on the basis of linear stability analysis, into six types: stable and unstable nodes, stable and unstable foci, centers, and saddles. Useful information about the simultaneous existence of singular points and closed orbits can be obtained from consideration of the Poincaré index, an integer associated with each singular point (*See* **Winding numbers** and Andronov et al., 1966). For example, the Poincaré index can be used to show

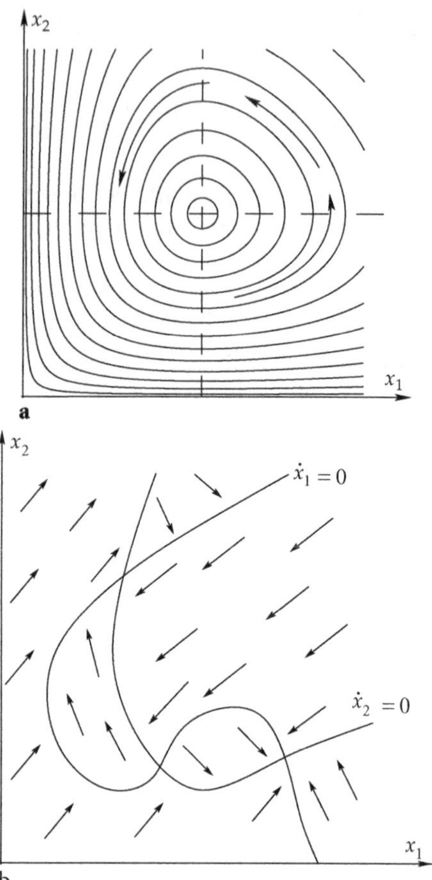

Figure 1. (a) Phase plane of the Lotka–Volterra equations (1). Representative orbits are shown as solid lines; the direction of flow is indicated by arrows. The isoclines are shown using dashed lines. The positive quadrant is filled with a family of closed orbits, corresponding to the fact that equations (1) are Hamiltonian. (b) Diagram showing a generic isocline analysis: many qualitative features of the dynamics are made plain. The isoclines show the loci where the vectorfield is horizontal or vertical. Intersections of the isoclines occur at singular points. From left to right, the three singular points are evidently of focus, saddle, and stable node types. It is not clear from the isocline analysis alone whether the focus is stable or unstable.

that every closed orbit encircles at least one singular point.

The method of isoclines is frequently invaluable in deriving qualitative information about the dynamics of a given smooth flow. An isocline is the locus where one of the vector field components vanishes. The two isoclines together split the phase plane into four divisions: within each such division each vector field component has a uniform sign. Figure 1b depicts a typical isocline analysis.

Closed orbits which attract or repel nearby orbits are called limit cycles. Systems possessing stable limit cycles are said to exhibit auto-oscillations. The method of Liénard can be used to demonstrate the existence of limit cycles in certain systems (Hartman, 2002). Sometimes, a limit cycle can also be found

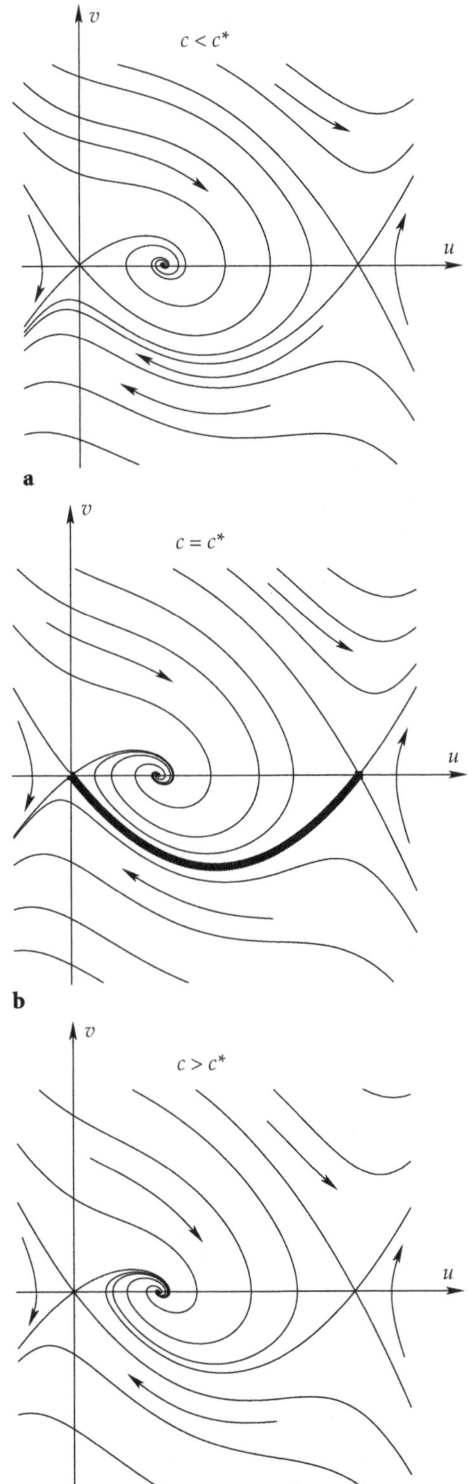

Figure 2. Determining a heteroclinic connection by shooting. Adjusting a parameter (in this case the wave speed c) causes the unstable manifold of the rightmost saddle point to move. For small positive values of c, one branch of the unstable manifold crosses the negative v-axis, never to return (a). At larger values of c, it is drawn into the stable focus point at $u = a$, $v = 0$ (c). There is precisely one intermediate value $c = c^*$ for which a branch of the unstable manifold of the rightmost saddle coincides with a branch of the stable manifold of the leftmost saddle, i.e., for which there is a heteroclinic connection (the emboldened orbit in b).

by association with a Hopf bifurcation (Arnol'd, 1973; Hale & Koçak, 1991). The Poincaré–Bendixson theorem can also be used to demonstrate the existence of limit cycles. It states that any non-empty compact α- or ω-limit set of a smooth planar flow which does not contain a singular point is a closed orbit.

Homo- and heteroclinic chains are, in general, more difficult to determine than equilibria. So-called shooting methods may sometimes be used to prove the existence of homoclinic or heteroclinic connections. For example, consider the following planar system, which arises as the traveling wave reduction of the Zeldovich–Frank-Kamenetsky equation

$$\begin{aligned} \dot{u} &= v, \\ \dot{v} &= u\,(u-a)\,(u-1) - c\,v. \end{aligned} \tag{2}$$

Here, the parameter c is a wave speed (*See* **Wave of translation**), and we assume (without loss of generality) that $0 < a < \frac{1}{2}$. It is readily seen that there are precisely three singular points (at $u = 0$, a, 1 and $v = 0$); the first and third are saddles; the intermediate is of focal type (see Figure 2). If we seek a rightward moving wave ($c > 0$), this focus is stable. For small values of c, the lower branch of the rightmost saddle's unstable manifold crosses the negative v-axis, where as for sufficiently large values of c, it crosses the positive u-axis before winding about the stable focus. It follows that there is an intermediate wave speed c^* for which a heteroclinic connection exists. It should be noted that this shooting argument depends on the fact that deleting a point from an interval in two-dimensional space topologically disconnects the interval. Shooting methods become more complicated in higher dimensions (see, e.g., Dunbar, 1984).

AARON A. KING

See also **Bifurcations; Chaotic dynamics; Chemical kinetics; Dynamical systems; Hamiltonian systems; Hopf bifurcation; Phase space; Poincaré theorems; Population dynamics; Wave of translation; Zeldovich–Frank-Kamenetsky equation**

Further Reading

Andronov, A.A., Vitt, A.A. & Khaikin, S.E. 1966. *Theory of Oscillators*, Oxford: Pergamon Press (translated from Russian)

Arnol'd, V.I. 1973. *Ordinary Differential Equations*, Cambridge, MA: MIT Press

Arnol'd, V.I. & Il'yashenko, Y.S. 1988. *Ordinary Differential Equations and Smooth Dynamical Systems. Dynamical Systems*, vol. 1, Berlin: Springer (original Russian edition 1985)

Dunbar, S.R. 1984. Traveling wave solutions of diffusive Lotka–Volterra equations: a heteroclinic connection in R^4. *Transactions of the American Mathematical Society*, 286: 557–594

Hale, J.K. & Koçak, H. 1991. *Dynamics and Bifurcations*, New York: Springer

Hartman, P. 2002. *Ordinary Differential Equations*, 2nd edition, Philadelphia: SIAM (corrected republication of edition published by Birkhäuser, Boston, 1982)
Hirsch, M.W. & Smale, S. 1974. *Differential Equations, Dynamical Systems, and Linear Algebra*, New York: Academic Press
Lotka, A.J. 1920. Undamped oscillations derived from the law of mass action. *Journal of the American Chemical Society*, 42: 1595–1598
Volterra, V. 1926. Fluctuations in abundance of a species considered mathematically. *Nature*, 118: 558–560

PHASE SPACE

The equations of motion of a mechanical system are usually of second order, and they determine the entire future from the initial *state*, which consists of both positions and velocities. The space of all states is called the *phase space*. For the mathematical pendulum (a point mass in the plane at one end of a massless rod whose other end is fixed), the configuration space is a circle, at each point of which one can choose any tangent vector as the velocity. Therefore, the phase space is a circle with a line attached to each point, that is, a cylinder. The pendulum equation for the position x and velocity v, $\ddot{x} + \sin x = 0$ converts to $(\dot{x}, \dot{v}) = (v, -\sin x)$, so the dynamics is described by integral curves of the vector field $R(x, v) = (v, -\sin x)$ on the cylinder.

That the notion of phase space is natural is also suggested by the Liouville theorem: the skew-symmetry of the Hamilton equation makes the Hamiltonian vector field divergence-free, and accordingly, a Hamiltonian flow preserves the volume on phase space.

Generally, a dynamical system consists of a phase space and a time-evolution of first order. The phase space is a set with some structure, such as differentiable (in the case of differential equations, this belongs to smooth dynamics), topological (one then speaks of topological dynamics), or measurable (this is the subject of ergodic theory, which arose from the Liouville theorem), and the time evolution is a one-parameter family of transformations that preserve this structure and that map initial states to states at another time. The time parameter may run through real numbers (continuous-time system) or integers (discrete-time system, iterations of a map and possibly its inverse). Specifically, a continuous-time system is given by a family $(f^t)_{t\in\mathbb{R}}$ of maps. If $f^t(f^s(x)) = f^{t+s}(x)$ for every s, t, x, then we say that this family is a flow. In the discrete-time case, one considers the iterates $(f^n)_{n\in\mathbb{Z}}$, where $f^0(x) = x$, $f^{n+1}(x) = f(f^n(x))$ for $n \geq 0$, and $f^n(x) = (f^{-1})^n(x)$ for $n < 0$, or if the map is not invertible, only positive iterates $(f^n)_{n\in\mathbb{N}}$. The maps $f_a(x) = ax(1-x)$ are a popular example of the latter (the so-called logistic map).

The long-term behavior of flows in the plane is well understood (*See* **Phase plane**): In the long run, any orbit either approaches fixed points or is asymptotically periodic (Poincaré–Bendixson theorem). This is ultimately due to the fact that a closed curve, such as a periodic orbit, divides the plane into separate regions. Already in three-space, one gets chaotic behavior, such as in the Lorenz attractor.

Qualitative Theory of Differential Equations and Dynamical Systems

On the phase space of a smooth continuous-time dynamical system, the time evolution is given by a first-order differential equation $\dot{x} = R(x, t)$. Suppose the right-hand side R satisfies a Lipschitz condition in x. This means that there is a constant M such that $d(R(x, t), R(x', t)) \leq Md(x, x')$ for all x, x'. The basic Picard theorem then guarantees existence and uniqueness of solutions for any initial condition. Otherwise solutions may not be unique ($\dot{x} = \sqrt[3]{x}$ has infinitely many solutions with initial value 0) or may not exist for any uniform amount of time (the solutions of $\dot{x} = x^2$ have singularities).

If solutions $x(t)$ exist for all time, as we henceforth assume, they define time-t-maps by $f^t(x(0)) = x(t)$. Each map f^t is as smooth as the right-hand side R of the differential equation (smooth dependence on initial conditions).

If R is independent of t, then the differential equation is said to be "autonomous," and R gives a vector field on the phase space that prescribes the velocity vectors of solutions. The family of time-t-maps is then a flow. The iterates of the time-1-map of a flow produce a discrete-time dynamical system whose study may yield useful information about the flow. If R depends on t, the system is said to be non-autonomous. Explicit time dependence can, for example, arise from forcing terms (forced pendulum $\ddot{x} + \sin x = \sin \omega t$) or from varying parameters (driving of a swing by parametric forcing, $\ddot{x} + \rho(t) \sin x = 0$).

An "orbit" or *trajectory* of a continuous-time system is a parametrized curve $(f^t(x))_{t\in\mathbb{R}}$. An orbit or trajectory of a map consists of the sequence of images of a point under iteration of the map: $(f^n(x))_{n\in\mathbb{Z}}$.

A singular point of a differential equation $\dot{x} = R(x, t)$ is a point x for which the right-hand side is zero for all t, that is, an equilibrium, or constant solution, or fixed point. Fixed points of a map f are those points x for which $f(x) = x$. A periodic point is a state that repeats at some positive time. For differential equations, this corresponds to solutions that are periodic functions of time; for maps, these are fixed points of an iterate. For the flow on the cylinder generated by $\ddot{x} + \sin x = 0$, all but four orbits are periodic; the point $(5 + \sqrt{5})/8$ is two-periodic for the map $4x(1-x)$.

Fixed and periodic points can be anchors for the study of the global orbit structure, and therefore, it is important to understand the behavior of nearby orbits. A fixed point is said to be attracting if orbits of nearby

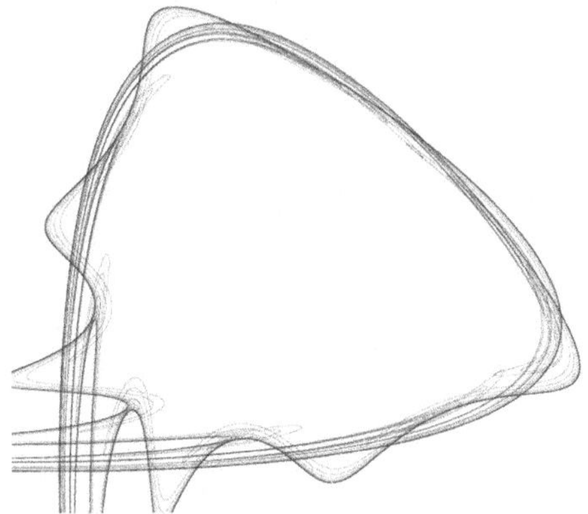

Figure 1. Stable and unstable manifold of a hyperbolic fixed point.

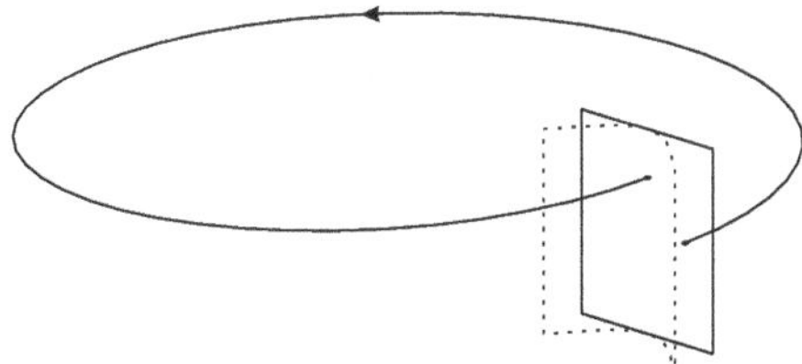

Figure 2. A section.

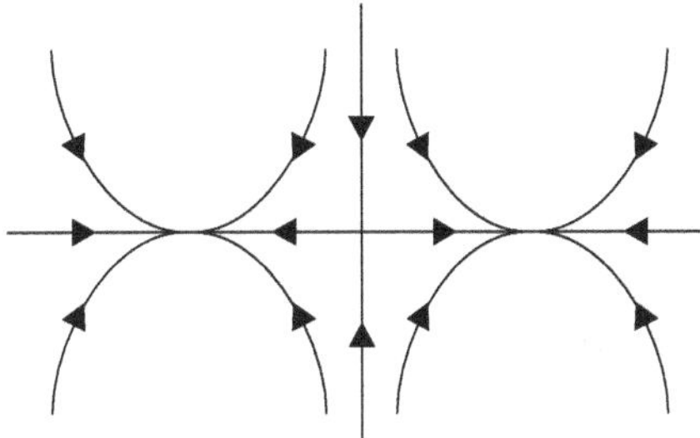

Figure 3. Two attracting points.

points stay nearby (Poisson stability) and converge to it for large positive time (asymptotic stability). (The example of a circle map like this ⟳ with a fixed point at the top illustrates that the second condition does not imply the first.) This is the case if the differential of the map (or time-1-map in the case of a flow) at that point has only eigenvalues of absolute value less than 1. If all eigenvalues have absolute value greater than 1 then the point is repelling: There is a neighborhood which every other point leaves in positive time. The map $f_2(x) = 2x(1-x)$ has 0 and $\frac{1}{2}$ as fixed points. 0 is repelling and $\frac{1}{2}$ is attracting. In fact, $\frac{1}{2}$ is superattracting: $f_2'(\frac{1}{2}) = 0$ and orbits near $\frac{1}{2}$ approach $\frac{1}{2}$ faster than exponentially.

If eigenvalues of the differential are allowed to lie both inside the unit circle and outside it but not on it, then the fixed point is said to be "hyperbolic." In this case the Hartman–Grobman theorem states that there is a continuous coordinate change that maps orbits near the fixed point to orbits of the linearized map. Moreover, tangent to the contracting and expanding subspaces of the linearization, there are the stable and unstable manifold of points positively and negatively asymptotic to the fixed point, respectively. These are smooth subspaces without self-intersections, but they may be packed into the phase space in a complicated way.

For periodic points of maps, the analysis of stability can be carried out by studying the appropriate iterate; for a flow, one likewise studies Poincaré return maps as follows. Take a small hypersurface through the periodic point transverse (for example, orthogonal) to the flow. The orbit of every point sufficiently near the periodic point returns to this surface at a time close to the period, and this defines a map from a neighborhood of the periodic point in the hypersurface into the hypersurface, with the original periodic point as a fixed point.

If, for example, a periodic orbit is an attracting fixed point for the return map, then it is a limit cycle: All nearby orbits are asymptotic to it. For the mathematical pendulum from the introduction, the circle $\{v = 0\}$ is a section of the cylinder, and the return map is defined for all of its points. It has two fixed points, and all other points are two-periodic.

A property complementary to stability of a fixed or periodic point as defined in terms of the behavior of nearby orbits (i.e., perturbations of the initial condition) is that of stability under perturbations of the dynamical system. An easy way to guarantee this is *transversality*, which is weaker than hyperbolicity: A fixed point $x = f(x)$ is said to be *transverse* if the derivative of f at x does not have 1 as an eigenvalue. (This implies that there are no other fixed points nearby.) In this case, any C^1-perturbation of f (i.e., one that changes derivatives only a little) also has a (transverse) fixed point near x. The origin is a nontransverse fixed point of $x(1-x)$, and indeed, it is absent for $x(1-x) + \varepsilon$ with $\varepsilon < 0$. The creation of two (hyperbolic) fixed points as ε changes from negative to positive is a basic local *bifurcation*. For differential equations, transversality corresponds to invertibility of the differential of the right-hand side.

An "invariant set" is a union of orbits; for example, [0, 1] is invariant under $4x(1-x)$. It is a repeller if it has a neighborhood in which only orbits of points in the invariant set stay for all positive time. It is an "attractor" if there is a neighborhood that is mapped into itself and the intersection of whose positive-time iterates is the invariant set. (Usually, one also requires that there is no proper subset with the same property. Thus, Figure 3 shows two attracting fixed points, and the interval with these as endpoints is not considered an attractor.)

The "basin of attraction" is the set of points that are asymptotic to the attractor. For example, the interval (0, 1) is the basin of attraction of $\frac{1}{2}$ for the map $2x(1-x)$.

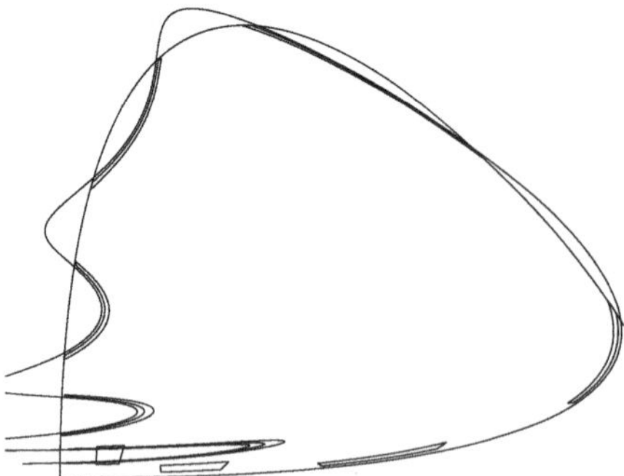

Figure 4. The Birkhoff–Smale theorem.

If two hyperbolic fixed points (saddles) in the plane are connected by a curve segment that lies in the unstable manifold of one of them and in the unstable manifold of the other, then this segment is called a "separatrix" (because it often separates two basins of attraction). More generally, the intersection of the stable manifold of one hyperbolic point with the unstable manifold of another is called a "heteroclinic intersection," and the orbit of every intersection point is called a "heteroclinic orbit." If the two fixed points coincide then one uses the terms "homoclinic intersection" and *homoclinic orbit* instead. The Birkhoff–Smale theorem asserts that if a homoclinic intersection is transverse (or if there is a pair of transverse heteroclinic intersections, that is, two hyperbolic points such that the unstable manifold of each of them intersects the stable manifold of the other point transversely), then there is a "horseshoe," that is, a rectangle that (under an iterate) gets mapped across itself in a horseshoe-like fashion as illustrated in Figure 4 and in Anosov and Axiom-A systems. This implies directly that there is an invariant Cantor set on which the dynamical system exhibits deterministic chaos.

There are several ingredients that make up chaotic behavior. One of these is recurrence. There are several recurrence properties. A point is said to be *recurrent* if it returns arbitrarily near to its initial condition. For a rigid rotation of a circle by an irrational number of degrees, all points have this property. By contrast, a point is said to be *transient* or *wandering* if it has a neighborhood all of whose images are pairwise disjoint. For the circle map ↺ with a fixed point on top, all nonfixed points are wandering. The set of nonwandering points is called the nonwandering set. Nonwandering orbits can be closed by a localized C^1 perturbation of the map (Pugh closing lemma). A dynamical system is said to be "topologically transitive" if it has a dense orbit and "minimal" if every orbit is dense. Irrational circle rotations have both properties. Minimality does not reflect chaotic behavior. Existence of a dense orbit is equivalent to the condition that for any two open sets there are arbitrarily large times at which the image of one of these sets overlaps the other. A strengthening of this property is that such overlap occurs for all sufficiently large times; this is called topological mixing and implies sensitive dependence on initial conditions. Following Devaney, one can say that a dynamical system is *chaotic* if it is topologically transitive and the set of periodic points is dense. This also implies sensitive dependence. A condition stronger than sensitive dependence is "expansivity": There is a universal positive constant by which the images of any two points, no matter how close initially, are separated at some time. The cat map and horseshoes are good examples of dynamical systems with these properties.

The Poincaré return map is not the only construction that produces a new dynamical system with a different phase space. Another straightforward one is the product of two dynamical systems. For example, the flow of rotations of the unit circle given by $x_1 = \alpha \cos t$, $x_2 = \alpha \sin t$ can be combined with a similar flow $y_1 = \omega \cos t$, $y_2 = \omega \sin t$ to a flow on the two-torus in $\mathbb{R}^4$ defined by all four equations. (Note that the plane defined by $y_1 = y_2 = 0$ is a section on which the return map is a time-$2\pi/\omega$-map of the first flow.) If one projects this to the $x_1 y_1$-plane, one gets Lissajous figures. In some applications, these readily show whether modes in weakly nonlinear oscillators are locked together.

Boris Hasselblatt

See also **Anosov and Axiom-A systems; Attractors; Cat map; Hamiltonian systems; Horseshoes and hyperbolicity in dynamical systems; Maps; Phase plane; Poincaré theorems; Population dynamics; Stability**

Further Reading

Arnol'd, V.I. 1989. *Mathematical Methods of Classical Mechanics*, 2nd edition, Berlin and New York: Springer (original Russian edition 1974)

Arnol'd, V.I. 1992. *Ordinary Differential Equations*, Berlin and New York: Springer (original Russian edition 1971, translated from 3rd edition 1984)

Fielder, B. (editor). 2002. *Handbook of Dynamical Systems*, vol. 2, Amsterdam and New York: Elsevier

Hasselblatt, B. & Katok, A. (editors). 2002. *Handbook of Dynamical Systems*, vol. IA, Amsterdam and New York: Elsevier

Hasselblatt, B. & Katok, A. 2003. *Dynamics: A First Course* (see also vol. 1B, 2005) Cambridge and New York: Cambridge University Press

Katok, A. & Hasselblatt, B. 1995. *Introduction to the Modern Theory of Dynamical Systems*, Cambridge and New York: Cambridge University Press

Katok, A., de la Llave, R., Pesin, Y. & Weiss, H. (editors). 2001. *Smooth Ergodic Theory and Its Applications (Summer Research Institute Seattle, WA, 1999), Proceedings of Symposia in Pure Mathematics*, vol. 69. Providence, RI: American Mathematical Society

Robinson, R.C. 1995. *Dynamical Systems, Stability, Symbolic Dynamics, and Chaos*, Boca Raton, FL: CRC Press
Strogatz, S. 1994. *Nonlinear Dynamics and Chaos*, Reading, MA: Addison-Wesley

PHASE-SPACE DIFFUSION AND CORRELATIONS

For problems concerned with particle acceleration and heating, which can be described in two degrees of freedom, surfaces of section in a two-dimensional phase space are usually appropriate for representing the motion (Lichtenberg & Lieberman, 1991, Chapter. 3). For periodically driven systems, a mapping representation is often convenient to describe the motion (*See* **Fermi acceleration and Fermi map; Standard map**). In the surface of section there is a characteristically divided phase space in which regular Kolmogorov–Arnol'd–Moser (KAM) curves and stochastic trajectories are intermingled. In the globally stochastic region of the phase space for a system with two degrees of freedom, in which KAM curves spanning the phase coordinate do not exist, a complete description of the motion is generally impractical. We can then seek to treat the motion in a statistical sense; that is, the evolution of certain average quantities can be determined, rather than the trajectory corresponding to a given set of initial conditions (Wang & Uhlenbeck, 1945). Such a formulation in terms of average quantities is also the basis for statistical mechanics (see Penrose, 1970, for example). The mathematical foundations for many of the results can be found in Arnol'd & Avez (1968).

In regions in which the trajectories are stochastic, nearby trajectories diverge exponentially in time. The divergence is usually measured by calculating the Lyapunov characteristic exponent σ of a trajectory $\boldsymbol{x}$ and a nearby trajectory $\boldsymbol{x}+\boldsymbol{w}$,

$$\sigma(\boldsymbol{x},\boldsymbol{w}) = \lim_{t\to\infty,\ d(0)\to 0}\left(\frac{1}{t}\right)\ln\frac{\mathrm{d}(t)}{\mathrm{d}(0)}, \tag{1}$$

where d is the distance separating the trajectories. Analytical and numerical calculations of the σ's (especially the maximum value of $\sigma=\sigma_1$ with respect to variations of $\boldsymbol{w}$) are widely used as measures of the degree of stochasticity in near-integrable systems. The commonly used numerical procedure for calculating the Lyapunov exponents was developed by Benettin et al. (1976).

In many problems, the density distribution in action space is the important quantity. Its dynamics is simplified by an average over phases to find the dynamical friction and diffusion coefficients for the action, which can then be used for determining its time evolution. If the phases are decorrelated after each mapping step, then a random phase approximation can be used which greatly simplifies the calculation. If the averaging over phase can be performed for a localized value of the action, then the resulting evolution is a Markov process, leading to the Fokker–Planck equation for the evolution of the distribution (Wang & Uhlenbeck, 1945)

$$\frac{\partial P}{\partial n} = -\frac{\partial}{\partial I}(BP) + \frac{1}{2}\frac{\partial^2}{\partial I^2}(DP), \tag{2}$$

where P is the probability distribution in the action I, B and D are the friction and diffusion coefficients, appropriately averaged over the phases, and n is a characteristic time over which the averaging can be performed. For example, using the standard map

$$\begin{aligned} I_{n+1} &= I_n + K\sin\theta_n, \\ \theta_{n+1} &= \theta_n + I_{n+1}, \end{aligned} \tag{3}$$

then one step of the mapping gives

$$\Delta I_1 = K\sin\theta_0, \tag{4}$$

and the transport coefficients, with phases randomized on each step, are

$$F_{\mathrm{QL}} = F = \frac{1}{2\pi}\int_0^{2\pi}\Delta I_1\,\mathrm{d}\theta_0 = 0, \tag{5}$$

$$D_{\mathrm{QL}} = \frac{D}{2} = \frac{1}{4\pi}\int_0^{2\pi}(\Delta I_1)^2\,\mathrm{d}\theta_0 = \frac{K^2}{4}, \tag{6}$$

where the subscripts QL refer to quasilinear (phase randomized) values with no higher-order correlations. (The factor of 2 difference between D_{QL} and D is a convention.)

Phase correlations, which always exist in a phase space with both regular and stochastic regions, complicate the calculation procedures. Close to the borders between stochastic and regular regions, the correlations become pronounced, requiring entirely different procedures for determining the diffusion. The existence of accelerator modes in the standard map also leads to nondiffusive behavior. Other phenomena of interest for diffusion calculations are the effect of noise and the effect of slow changes of system parameters. For weak correlations, for example, large K in the standard map, corrections to the single-step transport coefficients can be obtained. The corrections were obtained using Fourier techniques by Rechester et al. (1981). In the opposite limit for which a phase-spanning KAM curve (torus) has just been broken, resulting in a cantorus (the KAM curve becomes a cantor set), an approximate rate of local diffusion can be calculated. These techniques were developed to analyze the standard map, but can be used in various approximations to describe other two-degree-of-freedom systems. For a review of the various techniques and limitations see Lichtenberg & Lieberman (1991, Chapter 5).

ALLAN J. LICHTENBERG

See also **Diffusion; Fermi acceleration and Fermi map; Kolmogorov–Arnol'd–Moser theorem; Standard map**

Further Reading

Arnol'd, V.I. & Avez, A. 1968. *Ergodic Problems of Classical Mechanics*, New York: Benjamin

Benettin, G., Galgani, L. & Strelcyn, J.M. 1976. Kolmogorov entropy and numerical experiments. *Physical Review* A, 14: 2338–2345

Lichtenberg, A.J. & Lieberman, M.A. 1991. *Regular and Chaotic Dynamics*, 2nd edition, New York: Springer

Penrose, O. 1970. *Foundations of Statistical Mechanics*, Oxford: Pergamon

Rechester, A.B., Rosenbluth, M.N. & White, R.B. 1981. Fourier-space paths applied to calculation of diffusion in the Chirikov–Taylor map. *Physical Review* A, 23: 2664–2672

Wang, M.C. & Uhlenbeck, G.E. 1945. On the theory of Brownian motion II. *Reviews of Modern Physics*, 17: 323–342

PHASE SPACE RECONSTRUCTION

See **Embedding methods**

PHASE SYNCHRONISM OF CHAOS

See **Synchronization**

PHASE TRANSITIONS

A phase transition is a change in the degree or nature of order in a system and/or a change in its symmetry caused by a change in a thermodynamic variable such as temperature or pressure. When water freezes as the temperature is lowered, it undergoes a phase transition from a liquid to a solid. In the liquid state, the positions of the water molecules are disordered and the system is isotropic. In the solid state, however, the molecules are fixed at the sites of a crystal lattice, so the system has positional order and the symmetry of the ice lattice. Other examples of phase transitions include the liquid-vapor transition, in which the ordering is related to a change in density of the material; the transition from paramagnet to ferromagnet, which involves the appearance of a net magnetization as the temperature is lowered through the Curie temperature; the isotropic-nematic transition in liquid crystals, in which the rod-like liquid crystal molecules all become oriented in the same direction; and the superfluid transition in liquid helium, in which a macroscopic fraction of the helium atoms condense into a single quantum mechanical energy level. There are many others.

For any phase transition, we can define an order parameter. This is a quantity related to the degree of ordering which changes as the system goes through the transition. One measure of the order in a system is its entropy: a more ordered system has a lower entropy. While the entropy could be used as an order parameter, it is normally difficult to measure. An order parameter is usually physically measurable and is typically chosen to be zero in the less-ordered phase and nonzero in the more-ordered phase. For the liquid-vapor transition, the order parameter is proportional to the difference in density between the liquid and vapor phases. The order parameter for the ferromagnetic transition is simply the magnetization. For the superfluid transition in liquid helium, the order parameter is a complex number describing the macroscopic wave function of the superfluid.

Phase transitions can be classified as first order or second order. The terminology refers to the fact that certain derivatives of the free energy are discontinuous at a phase transition. In a first-order transition, it is the first derivatives (entropy, density, magnetization) that are discontinuous, while in a second-order transition, the discontinuity occurs in the second derivatives. A more practical classification is in terms of the behavior of the order parameter. At a first-order transition, the order parameter changes discontinuously. These transitions exhibit hysteresis and a latent heat resulting from the discontinuous change in entropy. At a second-order transition, the order parameter changes continuously; there is no hysteresis and no latent heat. In equilibrium, a thermodynamic system has a well-defined free energy and stable states correspond to absolute minima of this free energy. At a phase transition, the position of the free energy minimum changes, corresponding to a change in phase.

A phase diagram for a typical liquid-vapor system is shown in Figure 1(a). In the pressure-temperature plane, there is a line of first-order phase transitions which ends at a point at which the transition is second order. The same system is shown in the temperature-density plane in Figure 1(b). The region under the coexistence curve in Figure 1(b) is forbidden; there are no stable states in this region. When the mean properties of the system would place it inside this curve, the system phase separates into coexisting liquid and vapor phases with different densities lying on the coexistence curve itself. In the center of this region, the high-temperature phase is globally unstable, while closer to the edge, it is metastable. A supersaturated vapor would lie in this metastable region.

Points in the phase diagram at which second-order transitions occur are called critical points. At these points, the equation of state of the system takes on a special form, and the order parameter and other thermodynamic quantities show a power law dependence on the distance from the critical point. For example, if we consider a magnetic system with magnetization M, temperature T, and Curie temperature T_c, then the magnetization behaves as $M \sim [(T_c - T)/T_c]^\beta$ for $T \leq T_c$. The magnetic susceptibility $\chi = \mathrm{d}M/\mathrm{d}H$, where H is an applied magnetic field, diverges at the critical point as

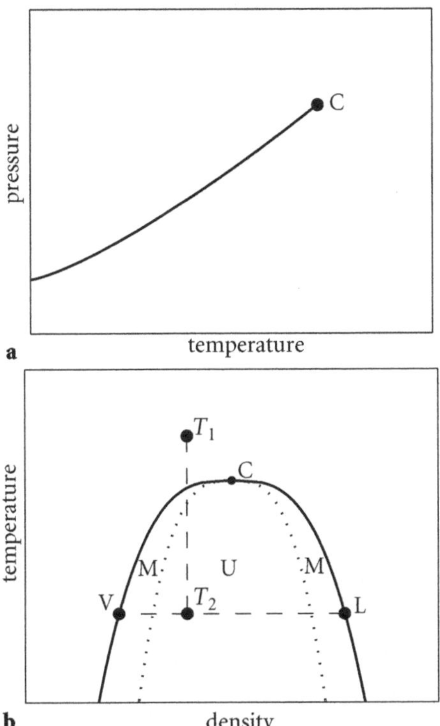

Figure 1. A typical liquid–vapor-phase diagram. (a) In the pressure–temperature plane a first-order phase transition occurs on crossing the solid line. The line of first-order transitions ends at the critical point C, at which the transition is second order. (b) The same phase diagram is shown in the temperature-density plane. The solid curve is the coexistence curve and corresponds to the solid line in (a). Outside the coexistence curve, the system exists as a single phase. In the region labeled **U** the one-phase state is unstable, and in the region labeled M it is metastable. If the system is cooled from the one-phase state at point T_1 into the unstable region at T_2, it will phase separate into a liquid component at the point L coexisting with a vapor component at V.

$\chi \sim |(T_c - T)/T_c|^{-\gamma}$. The quantities β and γ are called critical exponents. Interestingly, while T_c is different for different materials, the exponents β and γ have the same values for many different systems, including Ising ferromagnets and liquid-vapor systems, among others. This is known as "universality."

The simplest theoretical models used to describe phase transitions are called mean field theories. The van der Waals model for the liquid-gas transition and the Curie–Weiss model for the ferromagnetic transition are examples. Such theories assume that the order parameter at a particular point is determined by the average properties of the system—that is, by a mean field due to all other points of the system. Mean field theories are capable of describing first- and second-order-phase transitions qualitatively, but the values of the critical exponents are, in general, wrong. This is because mean field theories explicitly neglect the effects of spatial fluctuations in the order parameter, whereas it is these fluctuations which determine the behavior of the system near the critical point. Renormalization group theory provides a more complete description of critical point behavior.

When a system is quenched below a phase transition by suddenly changing a thermodynamic variable—for example, by suddenly dropping the temperature below the phase transition in a liquid-vapor system—it separates into two distinct phases. If the system is quenched into a region where the high-temperature phase is metastable (Figure 1), then macroscopic phase separation can only occur once large enough droplets of the unstable phase nucleate and grow. If it is quenched in the unstable region, then phase separation occurs by a process known as spinodal decomposition, whereby initially small perturbations to the local order parameter grow and coarsen with time. In a system with a conserved order parameter (e.g., a binary fluid mixture, in which the amount of each substance is fixed), the size of domains grows with time t as $L \sim t^{1/3}$, while if there is no conservation law (e.g., a ferromagnet) then $L \sim t^{1/2}$.

Phase transitions can also occur in systems out of equilibrium, for example, in the presence of a temperature gradient or a time-varying field. An important and interesting aspect of non-equilibrium systems is that the state of the system is no longer governed by the minima in the free energy, and indeed, a free energy functional may not even exist. Phase transitions can be significantly modified by non-equilibrium effects. In the case of a magnet in a time-varying field, for example, a competition between the time scale of the variations in the external field and the intrinsic relaxation time of the system itself leads to new behavior including spontaneous ordering that is not observed in the static case. Rather different phenomena, such as pattern-forming bifurcations in driven fluid systems, can also be thought of as nonequilibrium phase transitions.

JOHN R. DE BRUYN

See also **Critical phenomena; Ising model; Order parameters; Renormalization groups**

Further Reading

Chaiken, P.M. & Lubensky, T.C. 1995. *Principles of Condensed Matter Physics*, Cambridge and New York: Cambridge University Press

Chakrabarti, B.K. & Acharyya, M. 1999. Dynamic transitions and hysteresis. *Reviews of Modern Physics*, 71: 847–859

Gunton, J.D., San Miguel, M. & Sanhi, P.S. 1983. The dynamics of first order phase transitions. In *Phase Transitions and Critical Phenomena*, vol. 8, edited by C. Domb & J.L. Lebowitz, New York: Academic Press (Chapter 3). (The other volumes of this series contain review articles on many aspects of phase transitions.)

Stanley, H.E. 1971. *Introduction to Phase Transitions and Critical Phenomena*, Oxford: Oxford University Press

Tolédano, J.C. & Tolédano, P. 1987. *The Landau Theory of Phase Transitions*, Singapore: World Scientific

Uzunov, D.I. 1993. *Introduction to the Theory of Phase Transitions*, Singapore: World Scientific

PHASE TURBULENCE

See **Phase dynamics**

PHASE WINDING

See **Winding numbers**

PHI-FOUR EQUATION

See **Sine-Gordon equation**

PHOTONIC CRYSTALS

Photonic crystals are periodic dielectric (and/or metallic) structures, with periodicity comparable to a wavelength λ of interest, forming a *designable* optical medium where light propagation can exhibit unusual properties (Joannopoulos et al., 1995). The most important such property is a photonic band gap: a range of wavelengths in which there are no propagating modes in the crystal. Light in the band gap decays exponentially in the crystal, which acts like an optical "insulator." Nonlinear devices can exploit photonic crystals for enhanced phase sensitivity as well as to reduce power requirements, by means of tight spatial and long temporal confinement using the band gap and/or slow-light phenomena. Below are outlined two archetypical devices that greatly benefit from these properties of photonic crystals: Mach–Zender interferometers and bistable switches.

The simplest photonic crystals are one-dimensionally periodic multilayer films, or "Bragg mirrors," which were first studied in crystalline minerals by Lord Rayleigh (1887) (who observed that *any* periodic index variation will induce a band gap along that direction) and have since been the basis for a wide variety of applications: from reflective dielectric coatings, to distributed Bragg feedback (DFB) lasers, to fiber Bragg gratings for dispersion compensation and filters. It was not until 1987, however, that Yablonovitch (1987) and John (1987) applied the full principles of solid-state physics to electromagnetism and suggested that a three-dimensional (3-d) crystal could produce an omnidirectional gap. Since then, many photonic-crystal structures have been studied, both theoretically and experimentally, in two and three dimensions.

As a particular example, we consider a *two-dimensional* photonic crystal consisting of a square lattice of dielectric cylinders in air. This structure has a band gap for transverse-magnetic (TM) light (electric field perpendicular to the plane) and can, therefore, confine cavity and waveguide modes in point-like and linear defects of the crystal, as in Figures 1 and 2. For example, a single "defect" rod can be increased or decreased in size to trap a cavity mode within a diameter $\sim\lambda/2$, or a row of defect rods to form a waveguide. Bloch's theorem from solid-state physics (Ashcroft & Mermin, 1976) implies that modes in such a periodic waveguide propagate *without reflections*. Moreover, the waveguides form an effectively one-dimensional system, since the gap prohibits lateral scattering—this property allows photonic-crystal waveguides and cavities to be combined into complex device networks by adhering to simple design rules and symmetries. Alternatively, one can make a periodic sequence of cavities to form a coupled-cavity waveguide (CCW)—light slowly leaks from one cavity to the next, again trapping a guided mode (Yariv et al., 1999). The same principles apply in other photonic crystals, including those in three dimensions, and a direct analogue of this 2-d crystal can be made in 3-d (with an omnidirectional gap) by a stack of 2-d-like layers (Johnson & Joannopoulos, 2002).

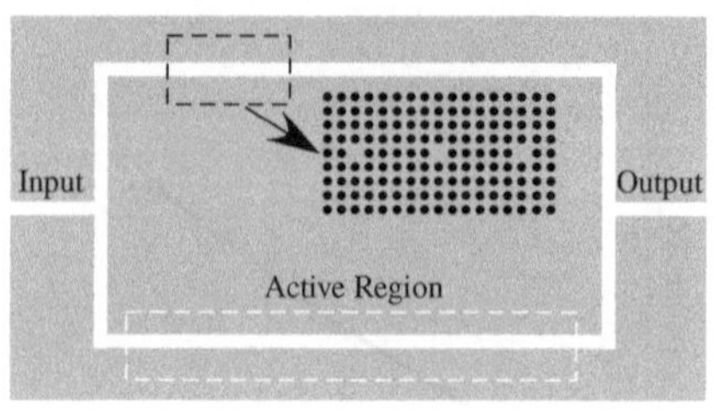

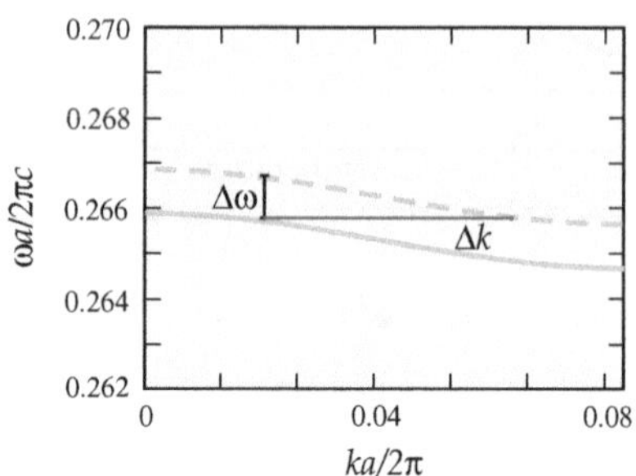

Figure 1. (Top) Schematic of a Mach–Zender interferometer: a waveguide that is split and recombined, with the relative phase modified by an active region to control the output transmission. Inset: a CCW slow-light waveguide to take advantage of photonic crystals. (Bottom) Band diagram of the CCW waveguide showing typical cosine dispersion curve. Because of the low group velocity, a slight shift of the curve (dashed) will cause a large change Δk in the wave number (horizontal axis). (The units are in terms of the distance a, the period of the square lattice.)

Mach–Zender Interferometers

The foundation of many optical device designs, from modulators to optical logic to switching, is the well-known Mach–Zender interferometer (Saleh & Teich, 1991) (Figure 1, Top).

Here, light in an incident waveguide is split into two branches and then recombined. If the two branches have the same optical path length, the recombination is in phase and light is transmitted; if their paths differ by half a period, then the recombination is out of phase and light is reflected. The relative optical path length is altered, for example, by changing the index by some Δn in the branches with a linear electro-optic (external) or Kerr (self-induced) modulation, allowing the transmission to be switched continuously from "on" to "off." By using photonic-crystal line-defect waveguides in the interferometer, one can take advantage of their low group-velocity capability to significantly lower the power and size requirements of the device by a factor of the group velocity or better (Soljačić et al., 2002b).

To attain a low group-velocity waveguide in a photonic crystal, one simple strategy is to employ a coupled-cavity waveguide (CCW) like that in Figure 1 (top). The guided band of such a waveguide has a characteristic cosine-curve dispersion relation (Figure 1, bottom) with two important properties: (i) the mid-band group velocity is low and decreases exponentially with the cavity separation, and (ii) the group-velocity dispersion (frequency derivative of the group velocity) is zero at the center of the bandwidth, minimizing signal distortion.

Because the group velocity v_g is low, when the dispersion curve $\omega(k)$ (frequency vs. wave number) is slightly shifted by altering the index n of a waveguide branch, there will be a large change in k (Figure 1, bottom), causing a corresponding large phase shift $\Delta\phi = \Delta k L$ (where L is the propagation distance). Mathematically, if the curve is shifted by $\Delta\omega \sim \Delta n\omega$, then the phase shift is $\Delta\phi \cong L\Delta\omega/v_g$, inversely proportional to group velocity $v_g = \mathrm{d}\omega/\mathrm{d}k$. This means that the device size L to achieve a fixed phase shift $\Delta\phi = \pi$ (with a given Δn) varies proportionally to v_g. Moreover, the power required to modulate the index is proportional to L, so the switching power is also proportional to v_g. Alternatively, one could keep L fixed and reduce Δn by a factor of v_g; for a linear (Pockels) modulation, this reduces the modulation power by v_g^2.

If the Mach–Zender interferometer is modulated by the optical waveguide signal itself, say for all-optical logic, then there is an analogous benefit: the light is compressed in time by a factor of v_g/c, causing a greater field $|\boldsymbol{E}|^2$ for the same input power. Combined with the above mentioned power savings from length reduction, this means that Kerr self-modulated devices have their power reduced proportionally to v_g^2.

Of course, the low group velocity comes at a price: the bandwidth of the waveguide is reduced proportional to v_g. In optical telecommunications systems, however, the required bandwidth is relatively small; for example, a 40 Gbit/s channel has a bandwidth $\Delta\omega/\omega \sim 1/3000$, meaning that the group velocity could potentially be lowered by several hundred times without limiting the bandwidth, with corresponding decreases in the device size and power relative to conventional waveguides.

Optical Bistability

Optical bistability is a dramatic nonlinear phenomenon that can be exploited to implement all-optical transistors, switches, logic gates, amplifiers, and other functions (Saleh & Teich, 1991). Bistability stems from nonlinear feedback combined with resonant transmission through a cavity, produces an output power that is a sharp nonlinear function of input power, and may even display a hysteresis loop (Figure 2, right).

In this context, the ability of photonic crystals to minimize both cavity modal volume V and lifetime Q (number of temporal periods for energy to decay by $e^{-2\pi}$) simultaneously allows them to greatly reduce the power threshold for optical bistability, in principle even down to the milliwatt level (Soljačić et al., 2002a).

An archetypical bistable device consists of an input waveguide coupled symmetrically into an output waveguide via a resonant cavity; this is shown in a photonic crystal setting of line and point defects in Figure 2 (left). In any such system, the transmission spectrum as a function of frequency will be a Lorentzian-like curve, peaked at 100% (in the absence of other loss mechanisms) at the resonant frequency (Figure 2, middle). In order to achieve bistability, one must include nonlinear feedback: the index (and thus the frequency) of the cavity depends on the field strength (e.g., via a Kerr nonlinearity $\Delta n \sim |\boldsymbol{E}|^2$). Furthermore, one must operate at a frequency ω_0 that lies *below* the linear resonant frequency ω_{res}. This combination, for continuous-wave (CW) sources, results in the bistable power-response curve shown in Figure 2 (right), which here includes a hysteresis. There are two stable system states for input powers in the bistable region (between the dashed vertical lines), and which one is realized depends upon whether one started from low or high power. (The middle, dashed, branch of the "S" curve is unstable.)

Intuitively, as the input power grows, the increasing index due to the nonlinearity will lower the resonant frequency through ω_0, as depicted by the dashed line in Figure 2 (middle), causing a rise and fall in transmission. This simple picture, however, is modified by feedback: as one moves into the resonance, coupling to the cavity is enhanced (positive feedback), creating a sharper "on" transition; and as one moves out of the resonance, the coupling is reduced (negative feedback), causing a delayed "off" transition.

The power threshold for the onset of bistability depends upon the power required to shift the cavity index sufficiently, which in turn depends upon the nonlinearity of the materials and the field strength $|\boldsymbol{E}|^2$ inside the cavity for a given input power. This field strength is inversely proportional to the modal volume V and is proportional to the lifetime Q (over which time the field builds up in the cavity). On the other hand, the required index shift of the cavity is proportional to the frequency width $1/Q$ of the Lorentzian transmission spectrum. Ultimately, therefore, the threshold power is proportional to V/Q^2; these simple arguments are confirmed by a more detailed analytical

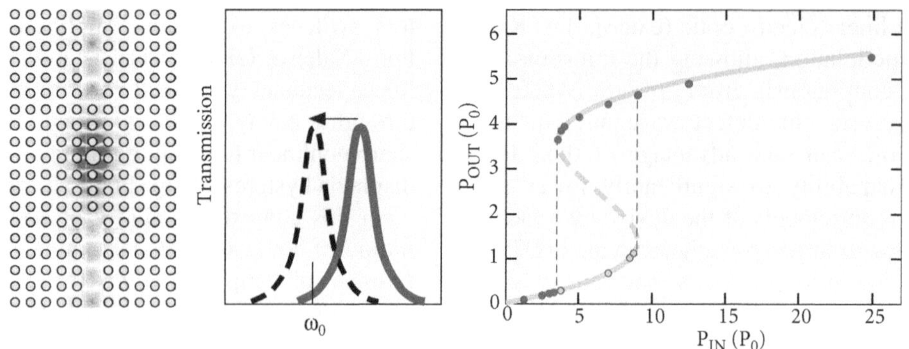

Figure 2. (Left) 100% (peak) resonant transmission from an input to an output waveguide through a cavity, formed by line and point defects, respectively, in a photonic crystal. Shaded regions indicate alternating positive/negative fields. (Middle) Lorentzian transmission spectrum (solid curve) of the linear resonant system. Increasing the power will nonlinearly shift the resonance curve (dashed) towards the operating frequency ω_0. (Right) Bistable transmission curve of output vs. input power resulting from the resonant transmission plus nonlinear feedback. The curve is an analytical theory and the dots are numerical calculations, with the vertical dashed lines indicating the region of hysteresis—for open dots the power was increased from a low value, while for closed dots the power was decreased from a high value.

theory that accurately predicts the bistability curve from the cavity characteristics (Soljačić et al., 2002a). Unlike traditional cavities such as ring resonators, photonic crystals impose no tradeoff between V and Q—the lifetime Q can be increased arbitrarily (up to the required signal bandwidth) while V is maintained near its minimum of $\sim(\lambda/2n)^3$, where n is the index of refraction. Indeed, in the example system of Figure 2, assuming reasonable material parameters and a $Q = 4000$ determined by the telecommunications bandwidth, one obtains a theoretical operating power of only a few milliwatts.

We conclude by presenting an analytical formula, derived from coupled-mode theory and perturbation theory, for the CW bistability relation in Figure 2 (right). The input/output power relation is given by

$$\frac{P_{\text{out}}}{P_{\text{in}}} = \frac{1}{1 + \left(P_{\text{out}}/P_0 - \delta\right)^2},$$

where δ is the frequency-detuning parameter $\delta \equiv 2Q(\omega_0 - \omega_{\text{res}})/\omega_{\text{res}}$ and P_0 is the characteristic power for bistability:

$$P_0 \equiv \frac{1}{\kappa Q^2(\omega_{\text{res}}/c)^{d-1}\max(n_2)}.$$

Here, d is the dimensionality of the system, n_2 is the Kerr coefficient (index change per unit intensity of light), and κ ($\sim 1/V$) is a dimensionless, scale-invariant "nonlinear feedback parameter" quantifying the concentration of the cavity field $\boldsymbol{E}$ in the nonlinear material.

Steven G. Johnson, Marin Soljačić, and J.D. Joannopoulos

See also **Hysteresis; Nonlinear optics**

Further Reading

Ashcroft, N.W. & Mermin, N.D. 1976. *Solid State Physics*, Philadelphia: Holt Saunders

Joannopoulos, J.D. Meade, R.D. & Winn, J.N. 1995. *Photonic Crystals: Molding the Flow of Light*, Princeton, NJ: Princeton University Press

John, S. 1987. Strong localization of photons in certain disordered dielectric superlattices. *Physical Review Letters*, 58: 2486–2489

Johnson, S.G. & Joannopoulos, J.D. 2002. *Photonic Crystals: The Road from Theory to Practice*, Boston: Kluwer

Rayleigh, Lord (Strutt, W.J.). 1887. On the maintenance of vibrations by forces of double frequency, and on the propagation of waves through a medium endowed with a periodic structure. *Philosophical Magazine*, 24: 145–159

Saleh, B.E.A. & Teich, M.C. 1991. *Fundamentals of Photonics*, New York: Wiley

Soljačić, M., Ibanescu, M., Johnson, S.G., Fink, Y. & Joannopoulos, J.D. 2002a. Optimal bistable switching in non-linear photonic crystals. *Physics Review E Rapid Communincations*, 66: 055601(R)

Soljačić, M., Johnson, S.G., Shanhui Fan, Ibanescu, M., Ippen, E. & Joannopoulos, J.D. 2002b. Photonic-crystal slow-light enhancement of non-linear phase sensitivity. *Journal of the Optical Society of America B*, 19: 2052–2059

Yablonovitch, E. 1987. Inhibited spontaneous emission in solid-state physics and electronics. *Physical Review Letters*, 58: 2059–2062

Yariv, A., Xu, Y., Lee, R.K, & Scherer, A. 1999. Coupled-resonator optical waveguide: a proposal and analysis. *Optics Letters*, 24: 711–713

PIECEWISE LINEARITY

See **Ratchets**

PINNING TO LATTICE

See **Peierls barrier**

PITCHFORK BIFURCATION

See **Bifurcations**

PLASMA SOLITON EXPERIMENTS

Comprising a very large number of charged particles within a confined volume, a plasma is a convenient laboratory facility in which both linear and nonlinear phenomena can be experimentally investigated. The overall plasma is electrically neutral in that the density of positive particles is equal to the density of negative particles. If the negative particles are electrons, this is called a two-component or normal plasma. If a certain fraction ε of the electrons is replaced with negative ions, this is called a three-component or a negative ion plasma.

Washimi and Taniuti (1966) were the first to demonstrate that the evolution of perturbations of the charge density in a normal plasma can be described by the nonlinear fluid equations for the ions. This description includes a Boltzmann assumption for the electrons and Poisson's equation to reflect the local charge nonneutrality in a density perturbation that can be described with the Korteweg–de Vries (KdV) equation,

$$\frac{\partial \psi}{\partial t} + \psi^{\nu} \frac{\partial \psi}{\partial x} + \beta \frac{\partial^3 \psi}{\partial x^3} = 0. \qquad (1)$$

Here the dependent variable ψ represents the perturbations in the ion density, ion velocity, or the electric potential; β is a constant; and the parameter $\nu = 1$.

In 1984, Watanabe showed that perturbations in a negative ion plasma can also be described by the same equation if the fraction ε has certain values. In particular, if this parameter is very large, then the negative ions predominate and rarefactive solitons evolve from a negative ion density perturbation. Because the mass of the negative ions could be comparable with the mass of the positive ions, he found that the parameter ε had a critical value ε_c at which the derivation led to a modified Korteweg–de Vries (mKdV) equation with the parameter $\nu = 2$. Both of these equations describe solitons that propagate in one direction.

The first extension to include effects of higher dimensions was performed in 1970 by Kadomtsev and Petviashvili, who included weak effects in a direction that was perpendicular to the dominant direction of propagation of the ion acoustic soliton. This equation is now called the Kadomtsev–Petviashvili (KP) equation for $\nu = 1$ in a normal plasma and the modified Kadomtsev–Petviashvili (mKP) equation for $\nu = 2$ in a negative ion plasma; thus

$$\frac{\partial}{\partial x}\left(\frac{\partial \psi}{\partial t} + \psi^{\nu} \frac{\partial \psi}{\partial x} + \beta \frac{\partial^3 \psi}{\partial x^3}\right) + \frac{\partial^2 \psi}{\partial y^2} = 0. \qquad (2)$$

Both of these equations have certain predictions that have been experimentally verified in a plasma.

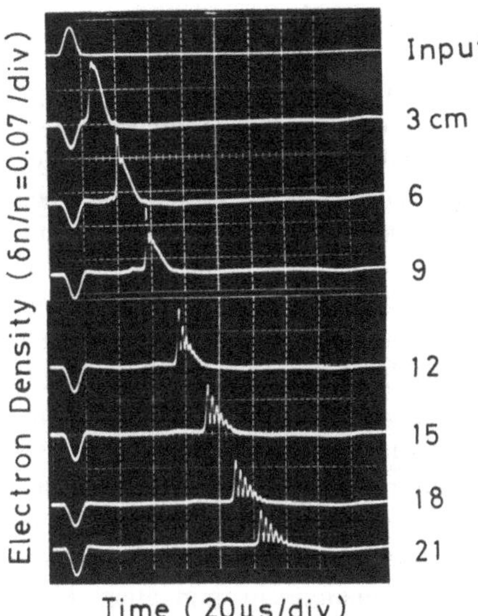

Figure 1. Evolution of a positive ion perturbation in a normal two component plasma. The pictures are taken at increasing distances from the source and illustrate the evolution of a burst of ions into a number of KdV solitons.

Laboratory experiments require the creation of a large volume of uniform collisionless plasma in which localized density perturbations are launched and movable probes monitor the evolution of the perturbation. Collisions between the particles are reduced with the evacuation of the chamber to a low pressure. Typically, a gas such as argon is inserted into the chamber and ionized to create a plasma with a volume of approximately 1 m^3. The density perturbations are created by applying a voltage signal to a fine-mesh grid or by introducing a charge density perturbation from one plasma into a second plasma. This is called a double-plasma (DP) machine and plasma solitons were first observed in a DP machine (Ikezi et al., 1970).

The spatial and temporal evolution of a compressive density perturbation in a normal plasma is illustrated in Figure 1. As this perturbation moves in the plasma, a number of solitary waves emerge for which the following KdV soliton properties have been observed: the product of the soliton amplitude (ψ) times the square of its width (W) is constant, and the soliton velocity $c_s = [1 + \psi/3]c$ where c is the linear ion acoustic velocity. In addition, the nondestructive collision of two solitons with different amplitudes was verified in the initial experiment performed in a normal plasma that existed in the DP machine.

By replacing a certain fraction of the free electrons in the normal plasma with negative ions to which these free electrons become attached, it is possible to realize a negative ion plasma. As a gas such as sulfur hexafluoride has a large attachment coefficient, a negative ion plasma can be created that consists of positive argon ions and negative fluorine ions whose masses are comparable. The parameter ε can be altered to have values of $\varepsilon < \varepsilon_c$, $\varepsilon = \varepsilon_c$, or $\varepsilon > \varepsilon_c$. In the first case, the positive ions are compressed and the normal KdV soliton is excited. KdV solitons are also excited in the third case due to compression of the negative ions. In the second case, mKdV solitons have been excited and their properties have been verified. In particular, it was observed that the product of ψW^2 is constant. The first experimental detection of solitons in a negative ion plasma was in a DP machine (Ludwig et al., 1984; Nakamura & Tsukabayashi, 1984), and both solitons were later detected in a negative ion plasma using the grid excitation mechanism (Cooney et al., 1991).

Two solitons that propagate in a normal plasma or in a negative ion plasma but in directions that are not collinear can still interact and are described by the KP equation. In particular, Miles (1977) noted that at a particular angle, the amplitude of the soliton after such an interaction would be greater than the sum of the amplitudes of the two solitons that preceded the interaction. If the amplitudes of these two initial solitons were equal, the amplitude of the new soliton would be 4 times this amplitude. The critical amplitude and the amplitude enhancement predicted from this resonant interaction was first experimentally examined in a normal plasma (Ze et al., 1979). In the negative ion plasma, an amplitude enhancement of two was anticipated and has been experimentally verified (Nakamura et al., 1999).

A laboratory plasma has been found to be a convenient venue in which several of the fundamental properties of solitons that are described with the KdV, mKdV, KP, or mKP equations can be experimentally studied and verified. A summary of other experiments has recently appeared (Lonngren, 1998).

KARL E. LONNGREN AND YOSHIHARU NAKAMURA

See also **Kadomtsev–Petviashvili equation; Korteweg–de Vries equation; Multidimensional solitons; Nonlinear plasma waves; Nonlinear Schrödinger equations**

Further Reading

Cooney, J.L., Gavin, M.T. & Lonngren, K.E. 1991. Experiments on Korteweg–de Vries solitons in a positive ion-negative ion plasma. *Physics of Fluids* B, 3: 2758–2766

Ikezi, H., Taylor, R.J. & Baker, D.R. 1970. Formation and interaction of ion acoustic solitons. *Physical Review Letters*, 25: 11–14

Kadomtsev, B.B. & Petviashvili, V.I. 1970. On the stability of solitary waves in weakly dispersing media. *Soviet Physics Doklady*, 15: 539–541

Lonngren, K.E. 1998. Ion acoustic soliton experiments in a plasma. *Optical Quantum Electronics*, 30: 615–630

Ludwig, G.O., Ferreira, J.L. & Nakamura, Y. 1984. Observation of ion acoustic rarefaction solitons in a multicomponent plasma with negative ions. *Physical Review Letters*, 52: 275–278

Miles, J.W. 1977. Resonantly interacting solitary waves. *Journal of Fluid Mechanics*, 79: 171–179

Nakamura, Y., Bailung, H. & Lonngren, K.E. 1999. Oblique collision of mKdV ion-acoustic solitons. *Physics of Plasmas*, 6: 3466–3470

Nakamura, Y. & Tsukabayashi, I. 1984. Observation of modified Korteweg–de Vries solitons in a multicomponent plasma with negative ions. *Physical Review Letters*, 52: 2356–2359

Washimi, H. & Taniuti, T. 1966. Propagation of ion acoustic solitary waves of small amplitude. *Physical Review Letters*, 17: 996–998

Watanabe, S. 1984. Ion acoustic solitons in plasma with negative ions. *Journal of the Physical Society of Japan*, 53: 950–956

Ze, F., Hershkowitz, N., Chan, C. & Lonngren, K.E. 1979. Inelastic collision of spherical ion acoustic solitons. *Physical Review Letters*, 42: 1747–1750

PLASMA TURBULENCE

See **Nonlinear plasma waves**

PLASTIC DEFORMATION

See **Frenkel–Kontorova model**

PLUME DYNAMICS

Plumes and jets are naturally and frequently occurring transport phenomena arising in a variety of settings, ranging from dry convecting atmospheric motion on hot days through explosive volcanic eruptions, for example, the 1915 eruption of Lassen Peak shown in Figure 1. The fluid dynamical purpose of a plume is dynamic equilibration of a localized unstable distortion of the fluid density, which results in vertical, coherent motion of a parcel of fluid seeking an equilibrium density. Viscosity couples and draws fluid along with the parcel on its voyage (turbulent entrainment). If the parcel is miscible with the ambient fluid, turbulent mixing will result, accelerating the equilibration process.

A further complication is that ambient fluids typically develop stable stratifications in which the fluid density is higher at the bottom. Such is the case with the Earth's atmosphere, whose density drops to zero in outer space, and the steady-state density profile is merely the thermodynamic response of an air layer under gravitational compression. Equally interesting stratification processes occur with much sharper gradients in convective boundary layers and in the thermoclines found in lakes and oceans. In these situations, stable transitions from a high-density bottom fluid layer to a low-density upper fluid layer may occur across a very sharp, nearly interfacial, layer. Typical stratifying agents include localized high temperature and/or concentration gradients (e.g., salt in the ocean). The modifications introduced by such layers can be both naturally dramatic and socio-economically challenging. The discharge of pollutants into the atmosphere, lakes, and oceans

Figure 1. May 22, 1915, eruption of Lassen Peak, taken 50 miles away in Anderson, California, by photographer, R.I. Meyers. (Thanks to Cari Kreshak and Lassen Volcanic National Park for providing the high quality image.)

is frequently accompanied by trapping phenomena directly attributed to the formation of such stable density layers (thermal inversions), in which the discharged pollutants are confined away from mixing flows and may lead to hazardous air and water quality.

Plume Mixing and Entrainment: Modifying the Large Scale Observables

A light plume of fluid in a constant density environment is expected to rise continually until the Archimedian buoyancy force is reduced through mixing of the plume with the ambient to levels at which viscous balances occur. The complete evolution requires, at minimum, the solution of the Navier–Stokes equations, with an evolving density anomaly (the plume) allowed to mix with the ambient. The mixing is turbulent, and the computational simulations of these nonlinear partial differential equations are both difficult, and necessary in making first principle predictions. Modelers have turned to alternative, somewhat ad hoc, yet nonetheless fundamental attempts to describe the evolution with fewer degrees of freedom than the complete fluid equations. As discussed below, the pioneering work of Morton et al. (1956) utilized an entrainment hypothesis with a single entrainment coefficient in an attempt to describe jet (plume) profiles by reduced, nonlinear ordinary differential equations (involving only a few degrees of freedom).

The plume dynamics in stratified fluids are dramatically different (Morton et al., 1956; Morton, 1967; Turner, 1995). In such a situation, the buoyancy of a plume of light fluid is strongly height-dependent, and an initially (low altitude) light fluid parcel may well rise to a height at which a buoyancy reversal occurs and the parcel becomes neutrally buoyant. Such a situation was originally noted by Morton et al. to cause an arrestment of vertical jets of light fluid. Their experiments and modeling for a fluid with a linear stratification (linearly decreasing density with increasing height) indeed show vertical jets arresting. The implications for functioning smokestacks, modeling of volcanic plumes (Sparks et al., 1997), and the mixing of oceanic pollutants (Fischer et al., 1979) is clear.

When the density transitions sharply between two distinct values, one finds mixing between low-density jet (or plume) fluid and the ambient fluid, which may dramatically affect the large scale observables. Experiments, performed in the UNC Applied Mathematics Fluid Lab, further exhibit the need for improved modeling that is specifically designed to better understand turbulent mixing and entrainment.

Figure 2 shows two vertical jets, fired at approximately the same volumetric flow rate (roughly 0.2 gal/min) into two identically stratified fluid tanks with a prepared sharp transition from $1.06\,\mathrm{g/cm^3}$ at the bottom to $1.015\,\mathrm{g/cm^3}$ using varying salt concentrations, with a transition of approximately 1 in, thickness, centered around the 14 in. tick on the tape. The left jet fluid is a gauge oil, with density $0.8\,\mathrm{g/cm^3}$ (lighter than all ambient tank fluid). Recall that oil and water do not mix. The right jet fluid is a dyed alcohol–water mixture, with density $0.8\,\mathrm{g/cm^3}$, also (initially) lighter than everything in the tank. In this case, the alcohol-water mixture may mix with the ambient fluid.

Observe the striking difference in large-scale observables: The nonmixing case penetrates clear to the free surface, whereas the mixing case does not penetrate, but forms, at altitudes in the vicinity of the sharp density transition, a cloud. The alcohol jet, fired in nonstratified cases of either $1.06\,\mathrm{g/cm^3}$ or $1.015\,\mathrm{g/cm^3}$ constant density tanks will reach the free surface at these flow rates and does not form a cloud, which demonstrates the powerful effect that an ambient sharp stratification can have upon plume dynamics, and the effect of the turbulent mixing and entrainment. There has been considerable work on developing plume models for studying the types of behavior shown with the alcohol jet following the original work of (Morton et al., 1956; Sparks et al., 1997; Caulfield & Woods, 1998), and some attempts have been directed at the multi-phase aspects of the oil jet example (Asaeda & Imberger, 1993; Socolofsky et al., 2001). A successful and complete modeling approach handling a full range of cases in which the mixing properties between jet fluid and ambient fluid may be continuously varied is an open challenge.

Dynamic Plumes and Solid Wall Interactions: Transient Levitation of Falling Bodies

As an extreme example in which the injected quantity cannot mix with the ambient fluid, consider recently obtained experimental results concerning the motion of falling bodies through stratified fluids (similar to

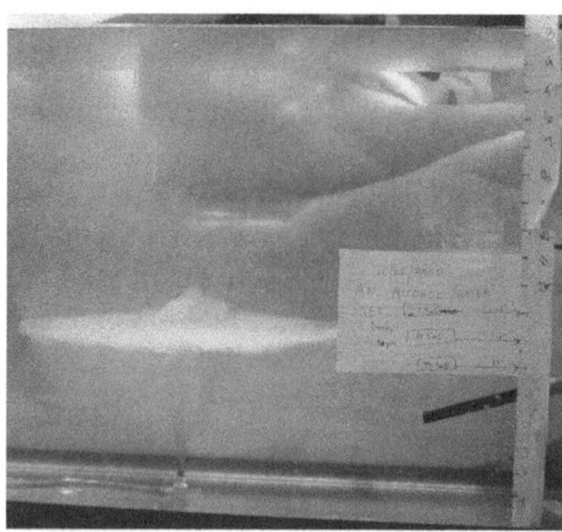
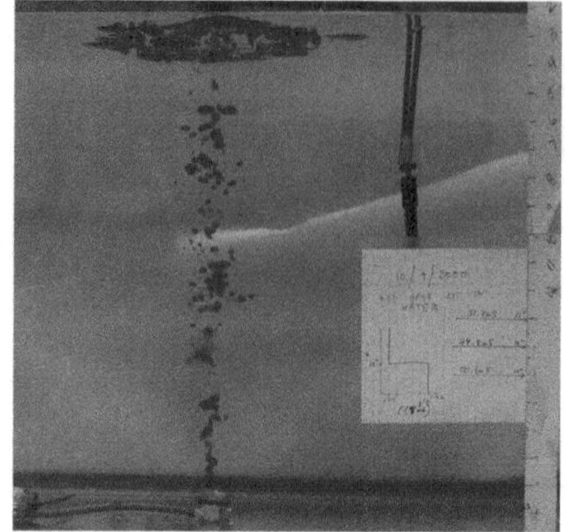

Figure 2. Vertical buoyant jets through a strong stable density step: Left is oil (0.8 g/cm^3), right is alcohol-water mixture (0.8 g/cm^3). (Thanks to former UNC undergraduates Ryan McCabe and Daniel Healion for assistance with the experiment.)

the tank setup in Figure 2) (Abaid, Adalsteinsson, Agyapong, McLaughlin, 2004). This study has focused upon the effect of self-generated plumes upon the falling body and has documented situations in which a falling body may generate a dynamic plume that through hydrodynamic coupling, may temporarily arrest the body. Of course, any body moving through a fluid experiences a hydrodynamic drag (which sets terminal velocities of falling bodies) in which the viscous boundary condition of vanishing fluid flow at the solid boundary necessarily drags a blob of ambient fluid along the moving body. In a constant density fluid, there is no potential energy cost associated with moving such a parcel of ambient fluid vertically. However, in strongly stratified fluids, a parcel of fluid moved from one altitude to another may develop a potential energy (buoyancy), as when the body falls through a sharp density transition layer. The momentum of the attached blob of fluid thrusts it into the lower (heavier) fluid, at which point the blob becomes a density anomaly and rises sharply. This motion in turn drags the falling body along with it.

Figure 3 shows three montages of a descending sphere at uniformly spaced times. The (5 mm radius) sphere in this case has a density of 1.04 g/cm^3 and is falling in a stratified tank whose top is fresh water (0.997 g/cm^3) and whose bottom is salt water (1.039 g/cm^3), again with a transition thickness of approximately 1 in. The top montage demonstrates the arrest and transient rise of the initially falling bead, and subsequent return to slow descent, each image uniformly spaced 1.5 s apart. The bead ultimately comes to rest at the tank bottom. The middle montage is the same as the top, only uniformly spaced at 0.1 s intervals. The lower montage has the same time sequence as the middle row, only focusing upon the shadow on the back of the tank, which highlights the entrained, plume-forming fluid.

The nature of this phenomenon is both nonlinear and dynamic. The nonlinear effect of such plumes upon

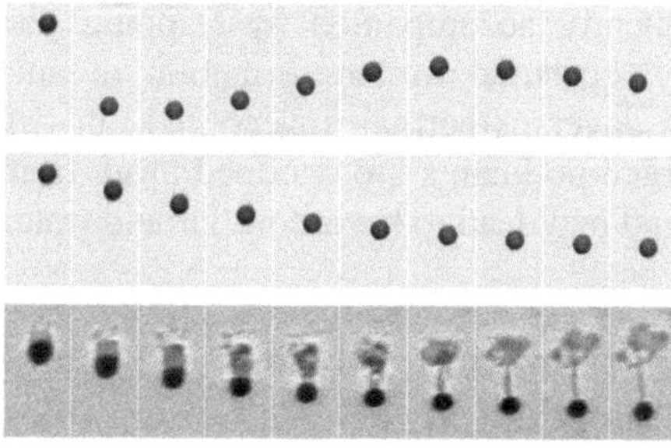

Figure 3. Top: Digital snapshots of bead position on uniform 1.5 s intervals, Middle: uniformly spaced on 0.1 s intervals, Bottom: shadowgraph depicting the dynamic plume on same time interval as middle row (Abaid, Adalsteinsson, Agyapong, McLaughlin, 2004). (Thanks to David Adalsteinsson for help with formatting the collage in his DataTank program and thanks to former UNC undergraduate Nicole Abaid for assistance with the experimental effort.)

the motion of solid bodies has been incorporated in a reduced system of ordinary differential equations in which the drag law for the falling body is modified to account for the dynamics of the plume which may modify the relative velocity of the falling sphere (Abaid, Adalsteinsson, Agyapong, McLaughlin, 2004). To describe the detailed dynamics of such transient plumes is quite difficult. Historically, there has been more success in the modeling of plume geometries under steady-state geometries. In pioneering work, Morton, Turner, and Taylor (Morton et al., 1956; Morton, 1967; Turner, 1995) were the first to model maintained plumes using an entrainment hypothesis which has become a standard in many fields (Fisher et al., 1979; Sparks et al., 1997).

Steady Plume Models in Stratified Environments

In 1956, Morton, Turner, and Taylor introduced what has become the standard maintained plume models for the shape of jet plumes (plumes emanating from a maintained source of buoyancy and momentum) (Morton et al., 1956; Morton, 1967; Turner, 1995; Fischer et al., 1979; Sparks et al., 1997; Socolofsky et al., 2001). The entrainment hypothesis assumes that the rate of inflow of diluting, ambient fluid is proportional to the vertical velocity of the jet along its centerline. Much empirical data has been collected exploring the exact dependence of the constant of proportionality upon the various physical parameters describing the jet configuration (stratified profile, jet speed, jet fluid density). A considerable effort since the original work of Morton et al. has addressed the many algebraic fits for the entrainment coefficient as a function of Richardson numbers, etc. (Fischer et al., 1979; Turner, 1995; Socolofsky et al., 2001).

Armed with this entrainment assumption, Morton et al. (1956) developed the following system of nonlinear ordinary differential equations, the solution of which yields the jet (plume) profile in steady state:

$$\frac{\mathrm{d}(b^2 w)}{\mathrm{d}z} = 2\alpha b w, \tag{1}$$

$$\frac{\mathrm{d}(b^2 w^2)}{\mathrm{d}z} = 2g\lambda^2 b^2 Q, \tag{2}$$

$$\frac{\mathrm{d}(b^2 w Q)}{\mathrm{d}z} = \left(\frac{(1+\lambda^2)b^2 w}{\lambda^2 \bar{\rho}}\right)\frac{\mathrm{d}\rho_0}{\mathrm{d}z}. \tag{3}$$

Here, the plume parameters are the center-line vertical velocity, $w(z)$, the plume radius $b(z)$, and the nondimensional plume density $Q(z)$. All are functions of the height variable z. The entrainment coefficient is α, which in neutrally stratified cases is empirically seen to be approximately 0.08 (Fischer et al., 1979; Turner, 1995). The gravitational constant is g, $\bar{\rho}$ denotes some constant reference density, and the ambient stratification is contained within the given profile $\rho_0(z)$.

This system is based on the following assumptions. First, vertical derivatives of certain horizontally averaged, low-order moments (for plume mass, momentum, and buoyancy) are simplified in terms of single point, centerline field variables. Second, radial profiles for plume vertical velocity and buoyancy are postulated in terms of "collective variables": $w(z) = \hat{w}(z) f(r/(ab(z))$ and $Q(z) = \hat{Q}(z) f(r/(cb(z))$. Through these, the integrals defining the moments may be directly calculated, leading to the closed system of differential equations given above. For both velocity and plume density, the functional forms are taken to be Gaussians, following empirical observations (Morton, 1967; Fischer et al., 1979; Turner, 1995). The ratio of the velocity length scale a, to plume length scale c is $\lambda = c/a$ is taken to be approximately 1.2, but this must certainly vary considerably upon the mixing properties of the plume fluid with the ambient. Each of these steps involves numerous approximations, an excellent list of which may be found discussed in Chapter 9 of the text by Fischer et al., along with asymptotic solutions for limiting cases (Fischer et al., 1979).

The solution of these equations gives a rough picture for plume shapes in the environment and typically shows plumes arresting at heights below their heights of static neutral buoyancy (in the absence of any mixing) in stratified environments. A more systematic mathematical reduction of this system from the complete equations, along with a numerical simulation of the complete fluid equations for multiphase fluid flow would be valuable.

RICHARD M. MCLAUGHLIN

See also **Atmospheric and ocean sciences; Mixing; Navier–Stokes equation; Turbulence; Vortex dynamics of fluids**

Further Reading

Abaid, N., Adalsteinsson, D., Agyapong, A. & McLaughlin, R.M. 2004. An internal splash: Levitation of falling spheres in stratified fluids. *Physics of Fluids*, 16(5): 1567–1580

Asaeda, T. & Imberger, J. 1993. Structure of bubble plumes in linearly stratified environments. *Journal of Fluid Mechanics*, 249: 35–57

Caulfield, C.P & Woods, A.W. 1998. Turbulent gravitational convection from a point source in a non-uniformly stratified environment. *Journal of Fluid Mechanics*, 360: 229–248

Csanady, G.T. 1973. *Turbulent Diffusion in the Environment*, Dordrecht: Reidel

Fernando, H.J.S. 1991. Turbulent mixing in stratified fluids. *Annual Reviews of Fluid Mechanics*, 23: 455–493

Fischer, H.B., List, E.J., Koh, R.C.Y., Imberger, J. & Brooks, N.H. 1979. *Mixing in Inland and Coastal Waters*, New York: Academic Press

Larsen, L.H. 1969. Oscillations of a neutrally buoyant sphere in a stratified fluid. *Deep-Sea Research*, 16: 587–603

Morton, B.R., 1967. Entrainment models for laminar jets, plumes, and wakes. *Physics of Fluids*, 10(10): 2120–2127

Morton, B.R., Taylor, G.I. & Turner, J.S. 1956. Turbulent gravitational convection from maintained and instantaneous sources. *Proceedings of the Royal Society*, A, 234: 1–23

Socolofsky, S.A., Crounse, B.C. & Adams, E.E. 2001. Multi-phase plumes in uniform, stratified, and flowing environments. In *Environmental Fluid Mechanics–Theories andApplications*, edited by H. Shen, A. Cheng, K.-H. Wang & M.H. Teng, Reston, Va: American Society of Civil Engineers, Ch. 3

Sparks, R.S.J., Burksik, M.I., Carey, S.N., Gilbert, J.S., Glaze, L.Sl, Sigurdsson, H. & Woods, A.W. 1997. *Volcanic Plumes*, New York: Wiley

Srdic-Mitrovic, A.N., Mohamed, N.A. & Fernando, H.J.S. 1999. Gravitational settling of particles through density interfaces. *Journal of Fluid Mechanics*, 381: 175–198

Torres, C.R., Hanazaki, H., Ochoa, J., Castillo, J. & Van Woert, M. 2000. Flow past a sphere moving vertically in a stratified diffusive fluid. *Journal of Fluid Mechanics*, 417: 211–236

Turner, J.S. 1995. *Buoyancy Effects in Fluids*, Cambridge and New York: Cambridge University Press

POINCARÉ INDEX

See **Winding numbers**

POINCARÉ THEOREMS

One of the greatest of all French mathematicians, Jules Henri Poincaré (1854–1912) graduated from the École Polytechnique in Paris and later studied at the École des Mines. In 1879, he became a docteur es sciences at the University of Paris, where he was appointed as a professor in 1881. Poincaré became a member of Académie des Sciences in 1887.

A mathematical genius of rare power, Poincaré's approach to science was to solve concrete problems arising from mathematics, mechanics, and physics, rather than to present his results in a "pure axiomatic form." However, a complete understanding of Poincaré's scientific works has yet to be achieved. Outside mathematics, Poincaré is also known for his works in theoretical physics including his seminal contribution in the special theory of relativity (1904–1905) and for his works on the philosophy of science.

Many of Poincaré's papers gave birth to whole new branches of mathematics, a prime example being algebraic topology, but the matter that occupied his life throughout was the geometrical approach to nonlinear differential equations—in particular, the long-time behavior of orbits of the Newtonian N-body problem in celestial mechanics.

The qualitative approach to nonlinear dynamics, introduced by Poincaré in his seminal papers "Mémoire sur les courbes définies par une équation différentielle" (1881–1886), which focuses on orbits rather than formulas was of a geometric and global nature. This is how the qualitative theory of ordinary differential equations was born.

Studying smooth vector fields on the plane, he classified their simplest equilibria (i.e., points where the given vector field vanishes): foci, nodes, saddle points, and centers. The typical smooth planar vector field has only the first three types of equilibria, but those of a more complicated nature are not excluded.

Poincaré outlined the proof that if a half trajectory γ of the planar vector field v is confined in a compact domain K in which v is free of equilibrium points, but the whole trajectory γ is not confined in K, then K contains a closed orbit of v to which γ is asymptotically attracted (the Poincaré–Bendixson theorem in its simplest form). This type of closed orbit is called a limit cycle.

To generalize, let v be a smooth vector field on a two-dimensional compact manifold M. To each isolated equilibrium p of v, Poincaré associated an integer $\mathrm{Ind}(v, p)$, called the index of v at p, which is defined as follows. Let J be a small loop surrounding the isolated equilibrium p, and let $\Delta\theta$ be the total change of the angle θ that the vector of v makes with some fixed direction when one runs counterclockwise along the loop J. This number is independent of the choice of loop. The index of p is the number $\Delta\theta/2\pi$ which is always an integer. (The index of a focus, center, or node is $+1$, and the index of a saddle point is -1.)

If v has only a finite number of equilibria $p_1, \ldots, p_r$, then Poincaré showed that

$$\sum_{i=1}^{r} \mathrm{Ind}(v, p_i) = \chi(M), \tag{1}$$

where $\chi(M)$ is the Euler–Poincaré characteristic of M and no restrictions are imposed on the nature of equilibria $p_1, \ldots, p_r$. This result was generalized later by Heinz Hopf to the case of vector fields on compact manifolds of arbitrary dimension and is now called the Poincaré–Hopf theorem.

For a sphere, we have $\chi(M) = 2$, that is, an arbitrary smooth vector field on a two-dimensional sphere must vanish in at least one point—one cannot evenly comb the hair on a sphere! Another consequence of this theorem is that, for example, on the two-dimensional sphere, one cannot have a smooth vector field having as equilibria only two saddle points or having only three centers.

Poincaré's investigations of celestial mechanics led him to the study of Hamiltonian systems with n degrees of freedom

$$\frac{\mathrm{d}q_i}{\mathrm{d}t} = \frac{\partial H}{\partial p_i}, \quad \frac{\mathrm{d}p_i}{\mathrm{d}t} = -\frac{\partial H}{\partial q_i}, \quad i = 1, \ldots, n \tag{2}$$

with an analytic Hamiltonian function $H(q, p)$, $(q, p) \in \mathbb{R}^{2n}$. His researches in this area were summarized in his epoch-making three-volume treatise "Les méthodes nouvelles de la mécanique céleste" (1892, 1893, 1899).

The study of Hamiltonian systems close to integrable ones was called the "general problem of dynamics" by Poincaré. Specifically, he studied Hamiltonian equations (2) with a perturbed Hamiltonian of the form

$$\begin{aligned} H(q, p, \varepsilon) &= H_0(p) + \varepsilon H_1(q, p) \\ &\quad + \varepsilon^2 H_2(q, p) + \cdots, \end{aligned} \tag{3}$$

where $H_1, H_2, \ldots$ are periodic functions with respect to q of the same period. The system obtained by setting $\varepsilon = 0$ is integrable, and the phase space is foliated by n-dimensional invariant tori $p =$ const. For small nonzero ε, system (2) usually becomes non-integrable. In the case of two degrees of freedom, this means that the Hamiltonian function H is the only first integral of Equations (2) that is an analytic and uniform function of variables q, p and ε.

Poincaré's participation in a mathematical competition organized in 1885 by Oskar II, King of Sweden and Norway, led him to the discovery of homoclinic orbits and related phenomena explaining how a very complicated behavior now called dynamical chaos, occurs in nonlinear dynamical systems. In his prize-winning work (1889), Poincaré discusses mainly the restricted three-body problem, where one mass is negligible compared with the other two, executing circular Keplerian motion.

Within the three-dimensional constant-energy manifold of this problem, he considered a two-dimensional surface Π transversal to most of the orbits. An orbit starting on this surface at a point x will pierce it again for the first time at some point y. The map $x \to y$ is the Poincaré return map induced on a surface of section Π.

It was in this framework that he discovered the existence of homoclinic orbits, that is, orbits which are asymptotically attracted by some periodic orbit γ when $t \to +\infty$ and $t \to -\infty$.

During his 1912 investigations on the restricted three-body problem, Poincaré conjectured that if one has a closed plane annulus Ω bordered by two concentric circles Γ_1 and Γ_2, then any area-preserving homeomorphism ϕ of Ω, such that $\phi(\Gamma_1) = \Gamma_1$, $\phi(\Gamma_2) = \Gamma_2$ and rotating these circles in opposite directions, has in Ω at least two different fixed points. This assertion, known as Poincaré's last geometric theorem, was proved in 1913 by George D. Birkhoff and is known also as the Poincaré–Birkhoff theorem.

Inspired by the analogy between a flow induced by a vector field and a flow of an incompressible fluid, Poincaré developed a theory of integral invariants, which was later refined by Elie Cartan.

Considering time t as an independent variable, we can study Hamiltonian system (2) in the extended phase space (q, p, t). A tube of trajectories is a two-dimensional cylindrical surface formed by the segments of trajectories of the vector field defined by (2) and bounded by two disjoint smooth closed curves. According to the Poincaré–Cartan theorem, in the extended phase space (q, p, t), the action integral $\oint_\gamma (p\, dq - H\, dt)$ has the same value for two different closed paths γ_1 and γ_2 encircling the same tube of trajectories and lying on it.

Poincaré also proved an important property of the long-time behavior of dynamical systems. In contemporary formulation, this proof shows that for any measure-preserving mapping of a measure space with a finite total measure, almost all trajectories starting from a given subset of positive measure eventually return to it. This is known as the Poincaré recurrence theorem, and it lies at the foundations of ergodic theory.

In his works on celestial mechanics, Poincaré provided the first formal definition of asymptotic series: divergent series giving nevertheless good numerical approximations for functions they represent.

Poincaré is the founder of the concept of normal forms in the theory of ordinary differential equations and of the contemporary bifurcation theory. His work was also at the beginning of modern variational methods in mathematics, in particular of the Morse theory which strongly links mathematical analysis to geometry and topology.

Other nonlinear problems studied by Poincaré include the problem of the existence of geodesics on convex surfaces, the problem of tides, and the stability of rotating fluid bodies.

Poincaré's impact on the theory of ordinary differential equations and dynamical systems is described in the books by Birkhoff (1927), Nemytskii & Stepanov (1960), Coddington & Levinson (1955), and Guckenheimer & Holmes (1990).

JEAN-MARIE STRELCYN AND ALEXEI TSYGVINTSEV

See also **Celestial mechanics; N-body problem; Phase plane; Phase space; Recurrence**

Further Reading

Barrow-Green, J. 1997. Poincaré and the three body problem, *History of Mathematics*, vol. 11, Providence, RI: American Mathematical Society and London: London Mathematical Society

Birkhoff, G.D. 1927. *Dynamical Systems*, Providence, RI: American Mathematical Society

Coddington, E.A. & Levinson, N. 1955. *Theory of Ordinary Differential Equations*, New York: McGraw-Hill

Guckenheimer, J. & Holmes, P. 1990. *Nonlinear Oscillations, Dynamical Systems and Bifurcations of Vector Fields*, 3rd edition, Berlin and New York: Springer

Milnor, J.W. 1965. *Topology from the Differentiable Viewpoint*, based on notes by David W. Weaver, Charlottesville, VA: University Press of Virginia; revised and reprinted, Princeton, NJ: Princeton University Press, 1997 Contains an elegant and simple proof of the Poincaré–Hopf theorem.

Nemytskii, V.V. & Stepanov, V.V. 1960. *Qualitative Theory of Differential Equations*, Princeton, NJ: Princeton University Press

Poincaré, H. 1904. L'état actuel et l'avenir de la physique mathématique. Conférence lue le 24 Septembre 1904 au Congrés d'arts et de sciences de Saint-Louis. *Bulletin des Sciences Mathématiques*, 28: 302–324; translation in *Bulletin of the American Mathematical Society*, 12 (1905–1906): 240–260, also published in Poincaré's book *La valeur de la science*, Paris: Flammarion, 1905, Chapters VII, VIII, and IX

Poincaré, H. 1905. Sur la dynamique de l'électron, *Comptes Rendus de l'Académie des Sciences*, 140: 1504–1508; also in *Œuvres*, vol. 9, pp. 489–493

Poincaré, H. 1906. Sur la dynamique de l'électron, Rendiconti del Circolo matematico di Palermo, 21: 129–176; also in *Œuvres*, vol. 9, pp. 494–550

Poincaré, H. 1916–1956. *Œuvres de Henri Poincaré*, 11 vols, Paris: Gauthier-Villars (This paper (Poincaré, 1904), which is not included in Poincaré (1916–1956), is in fact the first Poincaré's paper anticipating the special theory of relativity. The annoucement (Poincaré, 1905) and its full presentation in Poincaré (1906) is Poincaré's contribution to what is now called the special theory of relativity.)

Symposium on the Mathematical Heritage of Henri Poincaré. 1983. The Mathematical Heritage of Henri Poincaré, edited by Felix E. Browder, 2 vols., Providence, RI: American Mathematical Society

POINCARÉ–BENDIXSON THEOREM

See **Phase plane**

POISSON BRACKETS

Let M be an n-dimensional manifold (referred to as the phase space), and let f, g, and h denote analytic functions on M. A Poisson bracket of any two analytic functions on the phase space is defined as an operation which satisfies

(i) $\{\alpha f+\beta g, h\}=\alpha\{f, h\}+\beta\{g, h\}$, (linearity in the first component);
(ii) $\{f, g\}=-\{g, f\}$ (skew-symmetry);
(iii) $\{f,\{g, h\}\}+\{g,\{h, f\}\}+\{h,\{f, g\}\}=0$ (Jacobi identity);
(iv) $\{f, gh\}=g\{f, h\}+\{f, g\}h$ (Leibniz property),

where α, β are numbers. The first two properties ensure that a Poisson bracket is a bilinear operation on M. Properties (i)–(iii) imply that the analytic functions on M form a Lie algebra with respect to the Poisson bracket.

If local coordinates z_i, $i=1,\ldots,n$ are chosen on M, then the Poisson bracket has the coordinate representation

$$\{f, g\}=\sum_{j,k=1}^{n} J_{jk}(z)\frac{\partial f}{\partial z_j}\frac{\partial g}{\partial z_k}=(\nabla f)^T \boldsymbol{J}\nabla g, \quad (1)$$

where $\nabla f=(\partial f/\partial z_1,\ldots,\partial f/\partial z_n)$, and the Poisson matrix $\boldsymbol{J}(z)=(J_{jk}(z))_{j,k=1}^{n}$ is a skewsymmetric square matrix, satisfying a technical condition enforced by the Jacobi identity.

Any nonconstant function C on M that Poisson commutes with all other functions on M is called a Casimir of the Poisson bracket. From (1) it follows that the existence of a Casimir requires $\boldsymbol{J}$ to be singular, and ∇C is in the null space of $\boldsymbol{J}$. Furthermore, the number of independent Casimirs is the corank of $\boldsymbol{J}$. For a Poisson bracket with r Casimirs $C_1,\ldots,C_r$, Darboux's theorem states that it is always possible to find coordinates $(q_1,\ldots,q_N,p_1,\ldots,p_N,C_1,\ldots,C_r)$ on M such that in these coordinates

$$\boldsymbol{J}=\begin{pmatrix} \mathbf{0} & \boldsymbol{I}_N & \mathbf{0} \\ -\boldsymbol{I}_N & \mathbf{0} & \mathbf{0} \\ \mathbf{0} & \mathbf{0} & \mathbf{0} \end{pmatrix}, \quad (2)$$

where $\boldsymbol{I}_N$ is the N-dimensional identity matrix, and $\mathbf{0}$ is the zero matrix of the appropriate dimensions. In these coordinates,

$$\{f, g\}=\sum_{j=1}^{N}\left(\frac{\partial f}{\partial q_j}\frac{\partial g}{\partial p_j}-\frac{\partial f}{\partial p_j}\frac{\partial g}{\partial q_j}\right). \quad (3)$$

This representation of the Poisson bracket is called the canonical Poisson bracket, and the coordinates $(q_1,\ldots,q_n,p_1,\ldots,p_n)$ are called canonical coordinates.

The importance of Poisson brackets is derived from their relationship to Hamiltonian systems: let H be a function on M. Hamiltonian dynamics with Hamiltonian function H are defined on any function f on M by

$$\dot{f}=\{f, H\}. \quad (4)$$

Using the coordinate representation (1), the Hamiltonian dynamics for the coordinates is

$$\dot{z_j}=\{z_j, H\}=\sum_{k=1}^{n} J_{jk}\frac{\partial H}{\partial z_k}, \quad (5)$$

which reduces to the standard definition of a Hamiltonian system if canonical coordinates are used. From (4) it is clear that any function that Poisson commutes with the Hamiltonian is conserved for the Hamiltonian system defined by the Poisson bracket and the Hamiltonian H. In particular, H is conserved. Also, any Casimir is conserved. Because the conservation of the Casimirs is independent of the choice of H, they do not contain dynamical information. Rather, as is obvious from Darboux's theorem, they foliate the phase space and represent geometric restrictions on the possible motions in phase space. A Hamiltonian system can also be defined using the Hamiltonian function H and a symplectic two-form, of which $\boldsymbol{J}^{-1}$ (if it exists) is the coordinate representation (Weinstein, 1984).

As an example, consider Euler's equations of a free rigid body (Weinstein, 1984). Denote the angular momentum by (M_1, M_2, M_3) and the moments of inertia by I_1, I_2, I_3. The Poisson matrix is

$$\boldsymbol{J}=\begin{pmatrix} 0 & M_3 & -M_2 \\ -M_3 & 0 & M_1 \\ M_2 & -M_1 & 0 \end{pmatrix}. \quad (6)$$

The Hamiltonian is $H=(M_1^2/I_1+M_2^2/I_2+M_3^2/I_3)/2$. The Poisson matrix has rank 2 (except at the origin), and there is one Casimir: $C_1=M_1^2+M_2^2+M_3^2$.

The notion of Poisson brackets extends to infinite-dimensional phase spaces, so as to describe dynamics governed by evolution (partial differential) equations (Marsden & Morrison, 1984). In this case, the Poisson matrix $\boldsymbol{J}$ is replaced by a skew-adjoint differential operator B. If the evolution equation is first order in the dynamical variable t, this operator is scalar. Otherwise it is a matrix operator of the same dimension as the order of the evolution equation. Instead of functions on phase space, we consider functionals

$$F[\boldsymbol{u}]=\int f[\boldsymbol{u}]\,\mathrm{d}x. \quad (7)$$

Here $\boldsymbol{u}(x,t)$ is an infinite-dimensional coordinate on the phase space, indexed by the independent variable x. The square brackets denote that $f[\boldsymbol{u}]$ depends not only on $\boldsymbol{u}$, but possibly also on its derivatives with respect to x: $\boldsymbol{u}_x, \boldsymbol{u}_{xx}, \ldots$. The limits of integration depend on the boundary conditions imposed on the evolution equation. In the above, the variable x is assumed to be one dimensional. This is extended to higher dimensions in obvious fashion.

The Poisson bracket between any two functionals on phase space is the functional given by

$$\{F, G\}=\int \frac{\delta F}{\delta u}B\frac{\delta G}{\delta u}\,\mathrm{d}x, \quad (8)$$

where $\delta F / \delta \boldsymbol{u}$ is the variational (or Fréchet) derivative of F with respect to $\boldsymbol{u}$:

$$\frac{\delta F}{\delta \boldsymbol{u}} = \frac{\partial f}{\partial \boldsymbol{u}} - \frac{\partial}{\partial x}\frac{\partial f}{\partial \boldsymbol{u}_x} + \frac{\partial^2}{\partial x^2}\frac{\partial f}{\partial \boldsymbol{u}_{xx}} - \cdots . \quad (9)$$

The Poisson bracket defined this way satisfies properties (i), (ii), and (iv). The operator B is chosen so that the Jacobi identity (iii) is also satisfied.

A functional H on the phase space defines Hamiltonian dynamics on any functional by

$$\frac{\partial F}{\partial t} = \{F, H\} \quad \Leftrightarrow \quad \frac{\partial \boldsymbol{u}}{\partial t} = \{\boldsymbol{u}, H\} = B\frac{\delta H}{\delta \boldsymbol{u}}. \quad (10)$$

As an example, consider the Korteweg–de Vries equation $u_t = uu_x + u_{xxx}$ (Gardner, 1971; Zakharov & Faddeev, 1971). This equation with $-\infty < x < \infty$ is Hamiltonian with $B = \partial/\partial x$ and $H = \int_{-\infty}^{\infty}(u^3/6 - u_x^2/2)\,\mathrm{d}x$. Thus, the Poisson bracket is

$$\{F, G\} = \int_{-\infty}^{\infty} \frac{\delta F}{\delta u}\frac{\partial}{\partial x}\frac{\delta G}{\delta u}\,\mathrm{d}x, \quad (11)$$

and $\int_{-\infty}^{\infty} u\,\mathrm{d}x$ is its only Casimir.

The Poisson bracket formulation of a Hamiltonian system is especially significant when a quantum description of the dynamics is required. Dirac's principle of canonical quantization postulates that such a quantum description is obtained by replacing all classical quantities by their quantum mechanical operator counterparts (generalized coordinates $q \to \hat{q}$, the operation of multiplying by q, momentum $p \to -\mathrm{i}\hbar\partial/\partial q$, etc., and all Poisson brackets by commutators/$(\mathrm{i}\hbar)$). Then, in the classical limit as $\hbar \to 0$, the quantum mechanical equations reduce to classical equations, as desired by the correspondence principle.

Poisson brackets were introduced by Siméon-Denis Poisson (1809) during his investigations on perturbation theory in classical mechanics. Poisson's *Traité de mécanique* (two volumes, 1811 & 1833) were standard texts for many years.

Bernard Deconinck

See also **Constants of motion and conservation laws; Hamiltonian systems; Korteweg–de Vries equation; Lie algebras and Lie groups**

Further Reading

Gardner, C.S. 1971. The Korteweg–de Vries equation and generalizations. IV. The Korteweg–de Vries equation as a Hamiltonian system. *Journal of Mathematical Physics*, 12: 1548–1551

Marsden, J.E. & Morrison, P. 1984. Noncanonical Hamiltonian field theory and reduced MHD. In *Fluids and Plasmas: Geometry and Dynamics*, edited by J.E. Marsden, Providence, RI: American Mathematical Society

Poisson, S.D. 1809. Mémoire sur la variation des constantes arbitraires dans les questions de mécanique [Memoir on the variation of arbitrary constants in mechanics]. *Journal de l'École Polytechnique*, cahier XV: 266–298

Weinstein, A. 1984. Stability of Poisson–Hamilton equilibria. In *Fluids and Plasmas: Geometry and Dynamics*, edited by J.E. Marsden, Providence, RI: American Mathematical Society

Zakharov, V.E. & Faddeev, L.D. 1971. Korteweg–de Vries equation: a completely integrable Hamiltonian system. *Functional Analysis and Its Applications*, 5: 280-287

POLARITONS

A simple view of a dielectric material is that it is just an insulator made from atoms that have bound positive and negative charges. The ratio of an electric field measured in vacuum to that measured inside a dielectric material is called the dielectric constant. The opposite signs of the charges make the atoms electrically neutral, but these charges can be pulled apart by an applied electric field and they are displaced in opposite directions. If this happens, the material is said to be polarized. A metal, on the other hand, appears to be more complicated; it is a conductor and a solid-state plasma. Often called the fourth state of matter, a plasma is an electrically neutral assembly of separated electrons and positively charged ions. A metal is precisely like this but, unlike a gaseous plasma, it has a highly mobile "sea" of electrons carrying charges that are exactly balanced by a virtually immobile set of positive charges residing on the ionic background making up the crystal lattice. The same can be said for a heavily doped semiconductor. In spite of the plasma nature of a metal, its interaction with electromagnetic waves is well modeled by a dielectric function that turns out to be frequency-dependent (Kittel, 1995).

To understand the physical interaction between an electromagnetic wave and a dielectric, it is instructive to look at a (deceptively) simple model of how electrons behave in a metal or a heavily doped semiconductor when exposed to an electromagnetic wave. The core, realistic assumption in this model is that the electrons are highly mobile and free to move against an immobile ionic background. Suppose that the electric field carried by the electromagnetic wave is $\boldsymbol{E} = (0, 0, E)$, with a time variation $\mathrm{e}^{\mathrm{i}\omega t}$, where ω is an angular frequency. For a free electron with mass m, charge $-e$, and velocity $\boldsymbol{v}$, moving in one dimension, the equation of motion becomes

$$\mathrm{i}\omega m\boldsymbol{v} = -e\boldsymbol{E}. \quad (1)$$

If N is the number density of the electrons, this particle motion creates a current density with a magnitude $\boldsymbol{j}$,

$$\boldsymbol{j} = -Ne\boldsymbol{v} = \mathrm{i}\left(\frac{Ne^2}{\omega m}\right)\boldsymbol{E} = \mathrm{i}\omega\boldsymbol{P}, \quad (2)$$

where $\boldsymbol{P}$ is called the polarization. The latter should be thought of as being excited as a dipole moment per unit volume by the electromagnetic wave. Defining the electric displacement vector of free space as $\boldsymbol{D}$, and the magnetic field as $\boldsymbol{H}$, the appropriate Maxwell equation to incorporate (2) is

$$\begin{aligned}\operatorname{curl}\boldsymbol{E} &= \frac{\partial \boldsymbol{D}}{\partial t} + \boldsymbol{j} = \mathrm{i}\omega\varepsilon_0 \boldsymbol{E} - \mathrm{i}\frac{Ne^2}{\omega m}\boldsymbol{E} \\ &= \mathrm{i}\omega\varepsilon_0\left[1 - \frac{\omega_\mathrm{p}^2}{\omega^2}\right]\boldsymbol{E} \end{aligned} \tag{3}$$

in which t is time and ε_0 is the permittivity of free space. Equation (3) shows that a metal is neatly modeled by the frequency-dependent relative permittivity

$$\varepsilon(\omega) = 1 - \frac{\omega_\mathrm{p}^2}{\omega^2}, \tag{4}$$

where $\omega_\mathrm{p}^2 = \left(\sqrt{Ne^2/\varepsilon_0 m}\right)$ is called the plasma frequency. For this to also be a useful description of a semiconductor, or to take into account that the ionic background of a metal may contribute a frequency-independent permittivity, all that is necessary is to replace unity in Equation (4) with a quantity ε_L, which can be much greater than unity for semiconductors. This model gives good agreement with experiment, and it also serves to introduce the polariton concept. Excitations other than those of a free-electron gas are easy to capture, once this basic example is accepted. In a one-dimensional system stretching along the x-axis, the electrons move a distance $\boldsymbol{x} = e\boldsymbol{E} / (\omega^2 m)$ and the polarization is $\boldsymbol{P} = -Ne\boldsymbol{x}$.

If a plane electromagnetic wave has a wave vector $\boldsymbol{k}$ and (4) is the dielectric function, then

$$\begin{aligned}\boldsymbol{k}\times\boldsymbol{E} &= \omega\mu_0\boldsymbol{H}, \\ \boldsymbol{k}\times\boldsymbol{H} &= -\omega\varepsilon_0\varepsilon(\omega)\boldsymbol{E},\end{aligned} \tag{5}$$

where μ_0 is the magnetic permeability of free space. As a specific example, let the propagation be along the x-axis with wave vector $\boldsymbol{k} = (k, 0, 0)$ and let the electric field vector be $\boldsymbol{E} = (E_x, 0, E_z)$. The following modes are then implied (Boardman, 1982) by (5):

$$\text{Transverse}: \quad (\omega^2 - \omega_\mathrm{p}^2 - c^2k^2)E_z = 0, \tag{6}$$

$$\text{Longitudinal}: \quad (\omega^2 - \omega_\mathrm{p}^2)E_x = 0, \tag{7}$$

where $c^2 = 1/(\varepsilon_0\mu_0)$ is the velocity of light in vacuum and there are two independent solutions. The longitudinal mode is sustained when $E_x \neq 0$ and $\omega = \omega_p$. For this mode, the electrons move collectively in an oscillation that resembles the wobbling of a jelly.

At this stage, it should be pointed out that physicists are prone to call things by "ons" (Walker & Slack, 1970). All this started with the word *electron*. Unfortunately, the latter word does not, in itself, explain too much about the "on" part of the word, which is derived from the Greek word for "amber",—a fact that Benjamin Franklin would readily appreciate. Nevertheless, the electron is recognized to have a particle nature so physicists have become used to defining an "on" in association with an excited field, such as an electromagnetic wave or the jelly-like plasma oscillation. Thus, a *photon* is a particle, or quantum of energy, of light and the quantum of the wobbling electronic jelly, or plasma oscillation, is called the *plasmon*. The name *polariton* (Hopfield, 1958) is reserved for the quantum of an excitation in which electromagnetic waves cannot disentangle themselves from excitations such as plasmons.

The coupling of an electromagnetic field to the polarization excitations associated with electrons can be seen from the transverse solution (6). This shows that a finite value for E_z requires the dispersion equation to be

$$\omega^2 = \omega_\mathrm{p}^2 + c^2k^2 \tag{8}$$

Furthermore, this equation shows that the excitation is mixed because it involves not only the plasma frequency but also the velocity of light. It is for this reason that the quantum of this photon-plasmon field is called a polariton, as indicated earlier. The total energy is shared over the system because the mode has both plasmon and photon content, as can be seen by looking at the low and high wave number limits of (8). As $\boldsymbol{k} \to 0$, the frequency tends towards the plasma frequency, and the plasmon content grows at the expense of the photon contribution. The opposite is true as $\boldsymbol{k} \to \infty$, and the wave becomes a pure electromagnetic wave. The "on" labeling is not complete, however, unless the type of polariton is identified, so, in this case, they are referred to as plasmon-polaritons. This labeling is a reminder that only plasmons have been used here to introduce the concept of polaritons. The description is generic, however, and there are many types of polaritons. All that is necessary to generalize the discussion given here is to decide which polarization excitation the photons couple with and capture the material polarization properties through an effective permittivity. Well-known examples involve plasmons, excitons, and phonons.

After several decades of research, polaritons are still attracting attention (Baher & Cottam, 2003), in both guided and surface wave form, and the study of nonlinear polaritons in particular has opened new horizons (Boardman & Egan, 1985). The inclusion of nonlinearity means extending the permittivity to depend upon the amplitude of the electromagnetic wave through a functional dependence upon E, such as

$$\varepsilon(\omega, E) = \varepsilon(\omega) + \varepsilon^{\mathrm{NL}}(|E|) \tag{9}$$

Here $\varepsilon(\omega)$ is the relative permittivity for the excitation of choice, such as phonons, plasmons, or excitons. To reduce (9) to a tractable form, a Kerr-type of optical nonlinearity can be adopted, in which the total permittivity is expressed as

$$\varepsilon(\omega, E) = \varepsilon(\omega) + a\,|E|^2, \tag{10}$$

where a is a nonlinear coefficient.

This type of nonlinear process is not only popular in the literature but is genuinely applicable when the nonlinearity is well away from resonance and saturation. Immediately an interesting feature appears because the form of (10) permits an investigation of the kind of transverse electric (TE) waves that have no linear ($a \to 0$) limit. This dramatic conclusion culminates in the generation of an entirely new type of polariton. The TE waves can be guided or surface waves and for this discussion are assumed to be traveling along the x-axis. The field components are, therefore,

$$\begin{aligned} \boldsymbol{E} &= [0, E_y(k,\omega,z), 0]\mathrm{e}^{\mathrm{i}(kx-\omega t)}, \\ \boldsymbol{H} &= [H_x(k,\omega,z), 0, H_z(k,\omega,z)]\mathrm{e}^{\mathrm{i}(kx-\omega t)}. \end{aligned} \quad (11)$$

Substituting (11) into Maxwell's equations produces the following equations (Boardman et al., 1991):

$$\begin{aligned} \frac{\mathrm{d}}{\mathrm{d}z}E_y(k,\omega,z) &= -\mathrm{i}\omega\mu_0 H_x(k,\omega,z), \\ kE_y(k,\omega,z) &= \omega\mu_0 H_z(k,\omega,z), \\ \frac{\mathrm{d}}{\mathrm{d}z}H_z(k,\omega,z) &= \mathrm{i}kH_z(k,\omega,z) \\ &\quad +\mathrm{i}\omega\varepsilon_0\varepsilon(\omega, \mid E_y \mid^2)E_y(k,\omega,z). \end{aligned} \quad (12)$$

In general, E_y is complex so that $E_y = E(k,\omega,z)\, \mathrm{e}^{\mathrm{i}\phi(k,\omega,z)}$. Using $E \equiv E(k,\omega,z)$, $\phi \equiv \phi(k,\omega,z)$, and $\varepsilon \equiv \varepsilon(\omega)$ in (12) produces the following basic nonlinear equations:

$$\begin{aligned} &\frac{\mathrm{d}^2}{\mathrm{d}z^2}E - E\left(\frac{\mathrm{d}\phi}{\mathrm{d}z}\right)^2 + \left(\frac{\omega^2}{c^2}\varepsilon - k^2\right)E + \frac{\omega^2}{c^2}aE^2 = 0, \\ &\frac{\phi}{\mathrm{d}z} = \frac{K}{E^3}. \end{aligned} \quad (13)$$

The last equation expresses the conservation of energy flux along the z-axis, which is perpendicular to the propagation direction, and K is a constant of the integration process. The z-component of the time-averaged Poynting vector is

$$\langle S \rangle_z = -\frac{1}{2}\mathrm{Re}(E_y^* H_x) = \frac{c^2}{2\omega^2}\varepsilon_0 K. \quad (14)$$

Hence, $K = 0$ implies $\langle S \rangle_z = 0$. Finally,

$$\left(\frac{\mathrm{d}E}{\mathrm{d}z}\right)^2 + \left(\frac{\omega^2}{c^2}\varepsilon(\omega) - k^2\right)E + \frac{a}{2}\frac{\omega^2}{c^2}E^4 = C, \quad (15)$$

where C is another constant of integration. For a semi-infinite medium, $C = 0$, otherwise $C \neq 0$.

Consider now a $C = 0$ case describing a semi-infinite nonlinear medium interfaced to a semi-infinite linear medium. As it is well known that linear surface plasmon–polaritons are transverse magnetic (TM) polarized waves, the nonlinearity has added the extra functionality of making it possible to have TE-polarized surface plasmon-polariton waves. The field profile of these TE nonlinear waves is readily obtained from (15), since it factorizes when $C = 0$. The solution is

$$\begin{aligned} E &= \frac{c}{\omega}\sqrt{\frac{2}{a}}\sqrt{\left(k^2 - \frac{\omega^2}{c^2}\varepsilon(\omega)\right)} \\ &\quad \times \operatorname{sech}\left[\sqrt{k^2 - \frac{\omega^2}{c^2}\varepsilon(\omega)}\right](z - z_0). \end{aligned} \quad (16)$$

This is a self-focused (self-guided) beam that has a peak in the nonlinear medium below the interface. This nonlinear surface wave looks exactly like a spatial soliton. Furthermore, there is no linear limit for the nonlinear TE-polarized surface plasmon-polaritons. The frequency dependence for guided modes is more complex and can have a linear limit.

Polaritons are being studied in a wide variety important of contexts. Some examples include nonlinear exciton-polariton dynamics in the experimental study of semiconductor microcavities (Savvides et al., 2000), two-dimensional electron systems in quantizing magnetic fields (Beletskii & Bludov, 2002), and plasmon-polaritons in dielectric films (Baher & Cottam, 2003). The idea that underpins the polariton concept can be put to dramatic use in an unusual way. This is possible because to create polaritons, photons must associate themselves with the quanta or the quasiparticles of collective excitations, be they plasmons, phonons, or related excitatons. The polariton is a genuine collaborative effect between the photons and particles that are acting in a prepared organized manner, which means that the particles are not absorbing or emitting the photons. If a stream of photons (a light beam) is sent through an atomic gas (Lukin et al., 2000), for example, then provided the gas atoms are not resonant with the laser wavelength, the atoms should swarm around the photons rather like bees around a honey pot. In other words, polaritons should be formed. Furthermore, these polaritons could be organized, through an appropriate choice of laser intensity, to travel at a speed that is much less than the velocity of light. The particle swarm that has associated itself with the photon will slow down the light dramatically. This is an elegant idea and is a striking illustration of the character of the polariton. It is not surprising that the phrase "to catch a moonbeam" has been used (Sincelli, 2000) to describe the polariton-based possibility of slowing down light to walking speed.

Allan Boardman

See also **Alfvén waves; Drude model; Excitons; Nonlinear plasma waves; Plasma soliton experiments; Polarons**

Further Reading

Baher, S. & Cottam, M.G. 2003. Theory of nonlinear guided and surface plasmon-polaritons in dielectric films. *Surface Review and Letters*, 10: 13–22

Beletskii, N.N. & Bludov, Yu.V. 2002. Nonlinear surface polaritons in a two-dimensional electron system placed into a quantising magnetic field. *Telecommunications and Radioengineering*, 57: 67–78

Boardman, A.D. (editor). 1982. *Electromagnetic Surface Modes*, Chichester and New York: Wiley

Boardman, A.D. & Egan, P. 1985. s-polarised waves in a thin dielectric film asymmetrically bounded by optically nonliner media. *IEEE Quantum Electronics*, 21: 1701–1713

Boardman, A.D., Egan, P., Lederer, F., Langbein, U. & Mihalache, D. 1991. Third-order nonlinear electromagnetic TE and TM guided waves. In *Nonlinear Surface Electromagnetic Phenomena*, edited by H.-E. Ponath & G.I. Stegeman, Amsterdam and New York: North-Holland

Hopfield, J.J. 1958. Theory of the contribution of excitons to the complex dielectric constant of crystals. *Physical Review*, 112: 1555–1567

Kittel, C. 1995. *Introduction to Solid State Physics*, New York and London: Wiley

Lukin, M.D., Yelin, S. & Fleischhauer, M. 2000. Entanglement of atomic ensembles by trapping correlated photon states. *Physical Review Letters*, 84: 4232–4235

Savvides, P.G., Baumberg, J.J., Stevenson, R.M., Skolnick, M.S., Roberts, J.S. & Whittaker, D.M. 2000. Angular-asymmetric nonlinear polariton dynamics in semiconductor microcavities. *Physica Status Solidi* B, 221: 77–83

Sincelli, M. 2000. How to catch a moonbeam. *Physical Review Focus* 5, story 19

Walker, C.T. & Slack, G.A. 1970. Who named the -ON's? *American Journal of Physics*, 38: 1380–1389

POLARONS

The polaron concept goes back to 1933 when Lev Landau suggested the phenomenon of self-localization (self-trapping) of an electron (or a hole), slowly moving in a polar crystal (Landau, 1933). The electron locally polarizes and, therefore, distorts the ionic lattice in the region where it is located. This local lattice distortion, in turn, creates a potential well that traps the electron, lowering its energy. Therefore, the electron and the accompanying self-consistent polarization field can move in the crystal as a whole entity and may be considered as a quasi-particle. The term *polaron* was introduced in 1946 by Solomon Pekar, who studied the limiting (adiabatic) case of a sufficiently strong electron-lattice interaction and described the most-important properties of a stationary polaron, using a continuum model of electron motion (Pekar, 1946).

Pekar's theory of a polaron is based on the assumption that the polar coupling between electrons and optical phonons can be very large in ionic crystals. In ionic crystals, some of the ions are positively charged, while others are negatively charged, and an optical phonon has the different ions in the crystal vibrating out of phase. When the positive ions and negative ions oscillate in opposite directions (at a point $\boldsymbol{r}$ of the crystal), they create a polarization field $\boldsymbol{P}(\boldsymbol{r})$. This causes an electric field $\boldsymbol{E}(\boldsymbol{r})$, which is the source of the polar coupling. Polar coupling is only to longitudinal optical (LO) phonons, because only LO phonons set up a strong electric field when they vibrate. (This electric field is in the direction of the phonon wave vector $\boldsymbol{q}$.) Writing the expansion in normal modes:

$$\boldsymbol{E}(\boldsymbol{r}) = N^{-1/2}\sum_{\boldsymbol{q}} \boldsymbol{E}_{\boldsymbol{q}} \mathrm{e}^{\mathrm{i}\boldsymbol{q}\cdot\boldsymbol{r}}$$

$$\boldsymbol{P}(\boldsymbol{r}) = N^{-1/2}\sum_{\boldsymbol{q}} \boldsymbol{P}_{\boldsymbol{q}} \mathrm{e}^{\mathrm{i}\boldsymbol{q}\cdot\boldsymbol{r}} \qquad (1)$$

with N being the number of ions in the crystal and using the fact that there are no free charges, that is, $\nabla\cdot(\boldsymbol{E}+4\pi\boldsymbol{P})=0$, one obtains the relation

$$\boldsymbol{E}_{\boldsymbol{q}} = -4\pi\boldsymbol{P}_{\boldsymbol{q}}. \qquad (2)$$

Each $\boldsymbol{P}_{\boldsymbol{q}}$ is proportional to the (quantized) displacement field

$$\boldsymbol{P}_{\boldsymbol{q}} = U_{\boldsymbol{q}} e \left(\frac{\hbar}{2M\omega_{\boldsymbol{q}}}\right)^{1/2} \mathrm{i}\frac{\boldsymbol{q}}{|\boldsymbol{q}|}\left(a_{\boldsymbol{q}}^{\dagger}+a_{-\boldsymbol{q}}\right), \qquad (3)$$

where the coefficient $U_{\boldsymbol{q}}$ is to be determined, e is the electron charge, M the ion mass, $\omega_{\boldsymbol{q}}$ is the frequency of the $\boldsymbol{q}$th mode, and $a_{\boldsymbol{q}}$ ($a_{\boldsymbol{q}}^{\dagger}$) is the annihilation (creation) operator of the $\boldsymbol{q}$th mode.

On the other hand, the electric field can be represented as a gradient of a potential field:

$$\boldsymbol{E}(\boldsymbol{r}) = -\nabla\phi(\boldsymbol{r}) = -\mathrm{i}\sum_{\boldsymbol{q}} \boldsymbol{q}\mathrm{e}^{\mathrm{i}\boldsymbol{q}\cdot\boldsymbol{r}}\phi_{\boldsymbol{q}}\,. \qquad (4)$$

Using Equations (1)–(4), the potential field for the electron is

$$\phi(\boldsymbol{r}) = 4\pi e\sum_{\boldsymbol{q}} U_{\boldsymbol{q}}\mathrm{e}^{\mathrm{i}\boldsymbol{q}\cdot\boldsymbol{r}}\left(\frac{\hbar}{2NM\omega_{\boldsymbol{q}}}\right)^{1/2} \times\frac{1}{|\boldsymbol{q}|}\left(a_{\boldsymbol{q}}^{\dagger}+a_{-\boldsymbol{q}}\right). \qquad (5)$$

Because each normal mode $\boldsymbol{q}$ assumes a new equilibrium configuration and oscillates in the vicinity of these new equilibria, the coupling constant $U_{\boldsymbol{q}}$ can be evaluated from the potential energy between two fixed electrons. For a dispersionless polar crystal, when the frequency $\omega_{\boldsymbol{q}}$ is nearly a constant ($\omega_{\boldsymbol{q}} \doteq \omega_0$), Pekar has derived the value

$$U_0^2 = \frac{\rho\omega_0^2}{4\pi}\left(\frac{1}{\varepsilon_\infty}-\frac{1}{\varepsilon_0}\right), \qquad (6)$$

where ρ is the mass density of the crystal (so $\rho V = NM$, with V being the crystal volume). The dielectric constants ε_0 and ε_∞ are both measurable: ε_0 is measured by putting the crystal between the parallel plates of a capacitor at low frequency and ε_∞ is the square of the refractive index.

Thus, it is possible to write the total Hamiltonian of a single electron that interacts with the phonons in the form

$$H = \frac{\boldsymbol{p}^2}{2m} + \omega_0\sum_{\boldsymbol{q}} a_{\boldsymbol{q}}^{\dagger}a_{\boldsymbol{q}} + V^{-1/2}\sum_{\boldsymbol{q}}\frac{\chi_0}{|\boldsymbol{q}|}\mathrm{e}^{\mathrm{i}\boldsymbol{q}\cdot\boldsymbol{r}}\left(a_{\boldsymbol{q}}^{\dagger}+a_{-\boldsymbol{q}}\right). \qquad (7)$$

Here the unperturbed electron is taken to have free-particle motion with a momentum $\boldsymbol{p}$ and an effective mass m. Since there is only one electron in the problem, the results are independent of the statistics of the particle. The same results are obtained for any fermion or boson in the crystal, such as holes, excitons, or other particles, as long as they are free to move. Owing to Equations (5) and (6), the electron-phonon coupling constant χ_0 in Hamiltonian (7) is given by

$$\chi_0^2 = 4\pi\alpha\hbar(2m)^{-1/2}(\hbar\omega_0)^{3/2} \tag{8}$$

with the dimensionless polaron constant

$$\alpha = \frac{e^2}{\hbar}\left(\frac{m}{2\hbar\omega_0}\right)^{1/2}\left(\frac{1}{\varepsilon_\infty} - \frac{1}{\varepsilon_0}\right). \tag{9}$$

In Pekar's analysis of this model, the method of calculation is basically a variational procedure using a Gaussian wave function. The total wave function is assumed to be the product of electron and phonon coordinates (adiabatic approximation):

$$\Phi(\boldsymbol{r}; \boldsymbol{Q_q}) = \varphi(\boldsymbol{r})\psi_n(\boldsymbol{Q_q} + \delta\boldsymbol{Q_q}),$$
$$\varphi(\boldsymbol{r}) = \frac{\beta^{3/2}}{\pi^{3/4}}\exp\left(-\frac{\beta^2\boldsymbol{r}^2}{2}\right), \tag{10}$$

where β is a variational parameter to be determined. The phonon wave functions ψ_n are the wave functions for harmonic oscillators centered about new equilibrium displacements, which also need to be determined. Having carried out the variational procedure, one obtains the minimum value of β and the total energy $E = E(\beta)$ of the coupled electron–phonon system:

$$\beta_0 = \frac{2\alpha}{3}\sqrt{\frac{m\omega_0}{\pi}},$$
$$E(\beta_0) = -\frac{\alpha^2\omega_0}{3\pi} \simeq -0.106\alpha^2\omega_0. \tag{11}$$

This result shows the existence of a dynamically stable, self-localized state of an electron interacting with optical phonons in ionic crystals and having a finite binding energy, under the condition that the size of the self-trapped state is large compared with the lattice spacing constant (so the lattice discreteness becomes irrelevant in the theory).

Pekar's continuum polaron was the first model of a self-trapped state admitting a self-consistent theoretical treatment. The importance of this problem was soon understood by many theoretical and mathematical physicists working in solid state physics and quantum field theory, including N.N. Bogolyubov, S.V. Tyablikov, T.D. Lee, H. Fröhlich, R.P. Feynman, E.I. Rashba, T. Holstein, Y. Toyozawa, I.G. Lang, and Yu.A. Firsov. As a result, several pioneering works were published in the 1950s, favoring use of the methods of quantum field theory in solid state physics. Thus, the "polaron" concept has lost its original meaning as an electron interacting with long-wave-length polarized optical phonons and has been applied to a significantly wider class of self-trapped states, resulting in a large body of work that includes exciton-phonon interactions (Rashba, 1957) and even applications in biology (Scott, 1992), with very recent important contributions to the classical polaron theory (Romero et al., 1999). In this context, the following comments are relevant:

(i) The strong coupling limit in Pekar's theory is calculated in the adiabatic approximation, when the electron has a sufficient binding energy and its oscillatory motion in the potential well is much faster than the vibrational frequency of the phonons. Thus, the phonons do not have time to adjust to the individual oscillations of the electron. Instead, they adjust to the average motion of the electron. Quite a different picture applies to the weak coupling limit. In this case, the phonon energy is larger than that of the electron, so that the phonons (or ion polarization) follow the electron during its motion. The weak coupling limit was studied by Fröhlich within his polaron model, in which Hamiltonian (7) was generalized to (Fröhlich, 1954)

$$H = \sum_{\boldsymbol{k}} \frac{\boldsymbol{k}^2}{2m} c_{\boldsymbol{k}}^\dagger c_{\boldsymbol{k}} + \omega_0 \sum_{\boldsymbol{q}} a_{\boldsymbol{q}}^\dagger a_{\boldsymbol{q}}$$
$$+ V^{-1/2} \sum_{\boldsymbol{q},\boldsymbol{k}} \frac{\chi_0}{|\boldsymbol{q}|} c_{\boldsymbol{k}+\boldsymbol{q}}^\dagger c_{\boldsymbol{k}} \left(a_{\boldsymbol{q}}^\dagger + a_{\boldsymbol{q}}\right), \tag{12}$$

where $c_{\boldsymbol{k}}$ ($c_{\boldsymbol{k}}^\dagger$) is the annihilation (creation) operator of the electron with the wave number $\boldsymbol{k}$. However, the most sophisticated study of the "large" (or "continuum") polaron, using the Fröhlich Hamiltonian (12), is due to Feynman (1955) with the path-integral method, substantially extended in the past decade. For an extensive review of the large polaron studies in both the strong and weak coupling limits, see Mitra et al. (1987).

(ii) When the electron-phonon coupling constant is large, all the states in the Brillouin zone are involved in the formation of the polaron wave function. In this case, the polaron radius becomes comparatible with the lattice constant and the continuum approximation is no longer valid. For a "small" polaron, the theory should recognize the periodicity of the crystal and thereby assume that the motion of the electron, or another particle, is no longer translationally continuous. Therefore, one assumes that the electron may occupy an orbital state $\phi(\boldsymbol{r} - \boldsymbol{R}_j)$ centered on an atomic site $\boldsymbol{R}_j$. The orbital states are identical on each site, so that there is periodicity. The electron may move from site to site as in the tight-binding model. This motion may be caused by the overlap of the orbitals on adjacent sites. Thus, the following Hamiltonian is suitable for the small polaron theory:

$$H = J \sum_{j,l} C^{\dagger}_{j+l} C_j + \sum_{q} \omega_q a^{\dagger}_q a_q + \sum_{j,q} \chi_q e^{i\boldsymbol{q}\cdot\boldsymbol{R}_j} C^{\dagger}_j C_j \left(a^{\dagger}_q + a_{-q} \right), \quad (13)$$

where the factor J describes the overlap of the orbitals and C_j ($C^{\dagger}_j$) is the annihilation (creation) operator of the electron on site $\boldsymbol{R}_j$. Basic features of the small polaron were well recognized a long time ago by Tyablikov, Holstein, Lang, and Firsov, among others, and are described in several review papers and textbooks (Appel, 1968; Böttger & Bryksin, 1985; Mahan, 1990; Alexandrov & Mott, 1995).

(iii) Another interesting case is a problem of the lattice polaron with a long-range Fröhlich interaction. This polaron has a small (atomic) size of wave function but a large radius of lattice deformation. As shown recently (Alexandrov & Kornilovitch, 1999), this "small Fröhlich polaron" can propagate in a narrow band with the effective mass much smaller than that of the Holstein small polaron of the same binding energy. Alexandrov and Kornilovitch argue that small as well as large Fröhlich (bi)polarons are relevant quasiparticles in the cuprates, describing holes in the CuO_2 plane coupled with the lattice distortion created by a long-range interaction.

(iv) The polaron concept appeared so ubiquitous that it was used even in biology to describe a possible mechanism of the localization and transport of vibrational energy in molecular chains, particularly in proteins. As was suggested by Davydov, the energy of the C=O stretching (amide-I) vibrations, being localized on a peptide group of the α-helix, distorts the structure of the helix in this place. In its turn, a (contractive) distortion of the helical chain creates a potential well that traps the amide-I oscillation energy preventing its dispersion. This mobile self-localized (or self-trapped) state that can travel along the chain, carrying the energy released under the adenosine triphosphate (ATP) hydrolysis, is called a "Davydov soliton" (Scott, 1992).

Alexander V. Zolotaryuk

See also **Davydov soliton; Discrete self-trapping system; Local modes in molecular crystals**

Further Reading

Alexandrov, A.S. & Kornilovitch, P.E. 1999. Mobile small polaron. *Physical Review Letters*, 82: 807–810

Alexandrov, A.S. & Mott, N.F. 1995. *Polarons and Bipolarons*, Singapore: World Scientific

Appel, J. 1968. In *Solid State Physics*, edited by F. Seitz, D. Turnbull & H. Ehrenreich, New York: Academic Press

Böttger, H. & Bryksin, V.V. 1985. *Hopping Conduction in Solids*, Deerfield Beach, FL: VCH

Feynman, R.P. 1955. Slow electrons in a polar crystal. *Physical Review*, 97: 660–665

Fröhlich, H. 1954. Electrons in lattice fields. *Advances in Physics*, 3: 325

Landau, L.D. 1933. Über die Bewegung der Elektronen in Kristallgitter. *Physik Z. Sowjetunion*, 3: 664–665

Mahan, G.D. 1990. *Many-Particle Physics*, New York and London: Plenum Press

Mitra, T.K., Chatterjee, A. & Mukhopadhyay, S. 1987. Polarons. *Physics Reports*, 153: 91–207

Pekar, S.I. 1946. Autolocalization of an electron in a dielectric inertially polarized medium. *Soviet Physics, JETP*, 16: 335–340

Rashba, E.I. 1957. Theory of the strong interaction of electron excitations with lattice vibrations in molecular crystals. I & II. *Soviet Physics, Optika i Spektroskopiya (Optics and Spectroscopy)*, 2: 75–98

Romero, A.H., Brown, D.W. & Lindenberg, K. 1999. Polaron effective mass, band distortion, and self-trapping in the Holstein molecular-crystal model. *Physical Review* B, 59: 13728–13740; The self-trapping line of the Holstein molecular crystal model in one dimension. *Physical Review*, 60: 4618–4623

Scott, A.C. 1992. Davydov's soliton. *Physics Reports*, 217: 1–67

POLYMERIZATION

Polymers are long molecules made from one or more types of repeating units, called monomers. The simplest synthetic polymer consists of hundreds to even millions of monomers that are connected end to end in a linear chain. Polymers are essential to our modern world and our very life. Biochemical systems involve polymers because all enzymes are proteins, a type of polymer composed of about 20 different amino acids. DNA and RNA are also polymers but formed from four different bases (Nelson & Cox, 2000). Goldbeter (1995) discusses well the type of nonlinear phenomena that can occur. We focus only on synthetic polymers here.

A distribution of chain lengths always exists in a synthetic system. This molecular weight distribution can be quite broad, often spanning several orders of magnitude. However, biopolymers consist of a single molecular weight.

Linear polymers are often thermoplastic, meaning that they can flow at some temperature, which depends on the molecular weight. Common examples are poly(styrene) and poly(methyl methacrylate), whose trade name is Plexiglass.

Polymers need not be simple chains but can be branched or networked. Linear polymers are analogous to strings of spaghetti. Crosslinked polymers can be gels that swell in a solvent or thermosets, which form rigid three-dimensional networks. Home epoxy glue is an example of a crosslinked polymer. Figure 1 illustrates these variations.

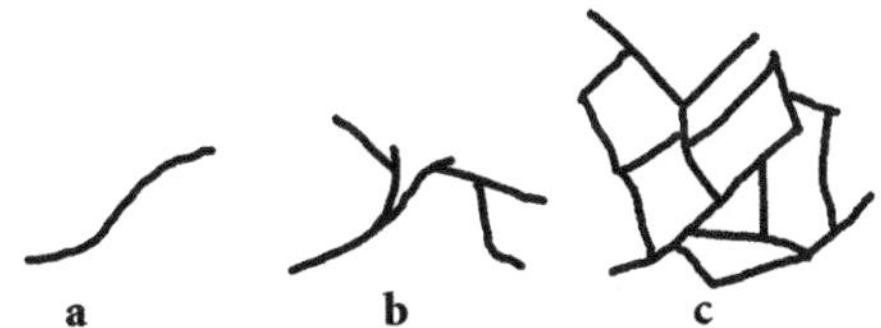

Figure 1. Illustration of three types of polymers: (a) linear, (b) branched, (c) crosslinked.

Polymer Kinetics and Mechanisms of Feedback

The mechanism of polymerization usually falls into one of two categories: chain growth or step growth. In chain-growth polymerization, a monomer is converted into a reactive species by an initiating species, either an ion or a free radical (a molecule containing an unpaired electron), that rapidly reacts with other monomers while always maintaining the reactive character. The reactive chain can be broken through some termination process. In most cases, even though relatively little of the monomer has reacted, high molecular weight molecules are produced in a few seconds. They usually do not participate further in the reaction. Plexiglass is produced via free-radical chain-growth polymerization.

For step-growth polymers, the monomers can usually react from both ends, and thus, all the monomers and polymers can react with each at any time from both ends. The molecular weight increases continuously throughout the reaction as monomers react to form dimers, and then dimers can react with each other or monomers, etc.

The physical properties of the polymer medium change dramatically during reaction. For example, the viscosity almost always increases orders of magnitude as the polymer concentration and/or molecular weight increases. This can be understood if you consider stirring a pot of penne versus a pot of spaghetti: the longer pasta pieces intertwine and resist flowing. Such physical changes will often affect the kinetic parameters of the reaction and the transport coefficients of the medium.

Synthetic polymer systems can exhibit feedback through several mechanisms. The simplest is thermal autocatalysis, which occurs in any exothermic reaction. The reaction raises the temperature of the system, which increases the rate of reaction through the Arrhenius dependence of the rate constants. Most chain-growth polymerizations, including free-radical polymerizations, are highly exothermic. Some step-growth polymerizations are highly exothermic (epoxies), but some are not (polyesters).

Free-radical polymerizations of certain monomers exhibit autoacceleration at high conversion via an additional mechanism, the isothermal "gel effect" or "Trommsdorff effect" (Odian, 1991). These reactions occur by the creation of a radical that attacks an unsaturated monomer, converting it to a radical, which can bond to another monomer, propagating the chain. The chain growth terminates when two radical chains ends encounter each other, forming a stable chemical bond. As the polymerization proceeds, the viscosity increases. The diffusion-limited termination reactions are thereby slowed down, leading to an increase in the overall polymerization rate. The increase in the polymerization rate induced by the increase in viscosity builds a positive feedback loop into the polymerizing system.

Some polymer hydrogels exhibit "phase transitions" as the pH and/or temperature are varied. Disposable diapers contain hydrogels that absorb great amounts of water because the polymers contain charged groups that bind the water. These charged groups can be affected by acid or base and are also temperature sensitive. The gel can swell significantly as the conditions are changed and can also exhibit hysteresis (Addad, 1996). Most polymers are immiscible, meaning that they will not dissolve with each other but will separate into different phases like oil and water. Introducing chemical reactions to an initially miscible polymer mixture often leads to phase separation. Autocatalytic behavior driven by chemical reactions and concentration fluctuations in miscible polymer mixtures can occur in photo-crosslinked polymer mixtures (Pojman & Tran-Cong-Miyata, 2003, Chapter 22). Concentration fluctuations increase as the reaction proceeds, leading to the reaction of photoreactive groups attached to one of the polymer components. This leads to an increase in the reaction yield that, in turn, accelerates the concentration fluctuations.

Finally, the polymer melts and the solutions are usually non-Newtonian fluids (Gupta, 2000), which means that there is no linear relationship between the stress applied to the fluid and strain that is measured. They often exhibit shear thinning, which means that the viscosity decreases as the shear is increased, but can also exhibit shear thickening. These properties make polymer additives important in the food and cosmetic industry.

Applications

There are two main scenarios by which nonlinear phenomena of possible utility arise with polymers. The first approach is to use a nonlinear system to drive the nonlinear behavior of a polymer or polymerization. Acrylonitrile will polymerize periodically when added to the Belousov–Zhabotinsky (BZ) reaction (Washington et al., 1999).

Yoshida et al. created a self-oscillating gel by coupling a pH oscillating reaction with a polymeric gel that expands and contracts with changes in pH (Yoshida et al., 1995). They have also used a gel in which the ruthenium catalyst of the BZ reaction is chemically incorporated into polymer, causing local swelling and deswelling as chemical waves pass through the gel (Pojman & Tran-Cong-Miyata, 2003, Chapter 3).

The second approach is to use the inherent nonlinearities in a polymeric system. Polymerizations in a continuously stirred flow tank reactor (CSTR) can show nonlinear phenomena through the interaction of thermal feedback and the gel effect. Teymour and Ray showed in laboratory-scale CSTR experiments on vinyl acetate polymerization that oscillations with periods of 200 min could occur (Teymour & Ray, 1992). The gel effect causes the rate of polymerization to increase, but the increase in temperature lowers the viscosity, returning the polymerization to its lower rate.

Frontal Polymerization

One of the most promising applications of nonlinear dynamics to polymer science is the phenomenon of frontal polymerization. Frontal polymerization is a process of converting monomers into a polymer via a localized reaction zone that propagates through the monomer, much like a "liquid flame."

Thermal frontal polymerization involves the coupling of thermal diffusion and Arrhenius reaction kinetics of an exothermic polymerization (Pojman et al., 1996). Front temperatures are typically 200°C, and front velocities are from 1–20 cm/min. Because of the large composition and thermal gradients created in the front, convection can be a significant issue (Pojman et al., 1996). Simple convection can occur for thermosets in ascending fronts, but for thermoplastics, the Rayleigh–Taylor instability is a significant issue unless a filler is added to increase the viscosity.

Such systems exhibit many modes of nonplanar front propagation. Particularly fascinating are the spin modes in which the front propagates as a helix, and hot spots can be seen on the surface with an infrared camera (Figure 2). Even more complicated modes can be observed depending on the chemical composition.

Figure 2. Infrared image of a spin mode in the free-radical-frontal polymerization of an acrylate in a 1.5 cm diameter tube. (Image courtesy of J. Pojman.)

Isothermal frontal polymerization (IFP), also called interfacial gel polymerization, is a slow process in which polymerization occurs at a constant temperature and a localized reaction zone propagates because of the gel effect (Epstein & Pojman, 1998, Chapter 11). Using IFP, one can control the gradient of an added material to generate materials for optical applications.

JOHN A. POJMAN

See also **Belousov–Zhabotinsky reaction; DNA premelting; DNA solitons; Fluid dynamics; Morphogenesis, biological; Protein structure; Rayleigh–Taylor instability; Rheology**

Further Reading

Addad, J.P.C. 1996. *Physical Properties of Polymer Gels*, Chichester: Wiley

Epstein, I.R. & Pojman, J.A. 1998. *An Introduction to Nonlinear Chemical Dynamics: Oscillations, Waves, Patterns and Chaos*, Oxford and New York: Oxford University Press

Goldbeter, A. 1995. *Biochemical Oscillations and Cellular Rhythms*, Cambridge and New York: Cambridge University Press

Gupta, R.K. 2000. *Polymer and Composite Rheology*, New York: Marcel Dekker

Nelson, D.L. & Cox, M.M. 2000. *Lehninger Principles of Biochemistry*, 3rd edition, New York: Worth

Odian, G. 1991. *Principles of Polymerization*, New York: Wiley

Pojman, J.A., Ilyashenko, V.M. & Khan, A.M. 1996. Free-radical frontal polymerization: Self-propagating thermal reaction waves. *Journal of the Chemical Society Faraday Transactions*, 92: 2825–2837

Pojman, J.A. & Tran-Cong-Miyata, Q. (editors). 2003. *Nonlinear Dynamics in Polymeric Systems*, Washington, DC: American Chemical Society

Teymour, F. & Ray, W.H. 1992. The dynamic behavior of continuous polymerization reactors — V. Experimental investigation of limit-cycle behavior for vinyl acetate polymerization. *Chemical Engineering Science*, 47: 4121–4132

Washington, R.P., Misra, G.P., West, W.W. & Pojman, J.A. 1999. Polymerization coupled to oscillating reactions: I. A mechanistic investigation and numerical simulation of acrylonitrile polymerization in the Belousov–Zhabotinsky reaction in a batch reactor. *Journal of the American Chemical Society*, 121: 7373–7380

Yoshida, R., Ichijo, H., Hakuta, T. & Yamaguchi, T. 1995. Self-oscillating swelling and deswelling of polymer gels. *Macromolecular Rapid Communications*, 16: 305–310

POPULATION DYNAMICS

Biological populations have a propensity for exponential growth. This Malthusian principle is expressed by the linear, ordinary differential equation $dn/dt = rn$, $r > 0$, where $n = n(t)$ is a measure of population size or density (e.g., individual numbers, biomass, or dry weight) as a function of time t. Although populations can, under appropriate conditions, exhibit exponential growth over a finite period of time, no population can grow indefinitely according to this law. The field of nonlinear population dynamics involves the study of how population numbers are regulated by biological and environmental effects. The equation

$$\frac{dn}{dt} = rn(1 - n/K) \qquad (1)$$

is an example of a modification of the exponential growth equation. It is based on the assumption that per capita growth is a decreasing, linear function of population size. In the 1920s, many researchers, including Alfred J. Lotka, Lowell J. Reed, and, most notably, Raymond Pearl, considered Equation (1) as a fundamental law of population growth. It was later discovered that a Belgian mathematician, Pierre-François Verhulst, had used the equation in his studies of population growth in the 1840s. Equation (1) is called the Pearl–Reed equation, the Pearl–Verhulst equation, or more commonly, the "logistic equation" (a name adapted from Verhulst's name *logistique* for the equation).

Equation (1) implies all population densities monotonically approach the equilibrium K, called the carrying capacity. Before leveling off at this carrying capacity, low population numbers initially grow exponentially at the inherent growth rate r, and the resulting S-shaped trajectory is called a logistic growth curve. In their studies of human population dynamics, Pearl and his colleagues fit logistic curves to census numbers from many countries and states. Although today considered rather simplistic as a model of population growth, the logistic equation nonetheless encapsulates the notion of density self-regulation in population dynamics and has inspired many basic notions in ecology (e.g., habitat carrying capacity, the classification of species as "r or K selectors").

The logistic equation is a first-order differential equation of the Kolmogorov type

$$\frac{dn}{dt} = nf(n) \qquad (2)$$

with per capita growth rate $f(n) = r(1 - n/K)$. Many theories of population growth use the Kolmogorov equations with other specialized forms for $f(n)$. One notable class of models arises from a modification of the negative feedback density regulation assumption of the logistic Equation (1) to include a positive feedback at low population numbers. This notion was put forth by W.C. Allee in order to account for increased growth rates associated with increased population size experienced by some (many biologists would say most) populations when at low numbers (although a negative feedback still comes into play at high numbers). An example is the equation

$$\frac{dn}{dt} = rn\left(1 - \frac{n}{K}\right)(n - \alpha K), \quad \alpha < 1.$$

The "Allee effect" entails a threshold αK such that populations go extinct unless their initial state $n(0)$ exceeds αK, in which case they equilibrate at the carrying capacity K.

Ordinary differential equation models based on equations of form (2) neglect important factors affecting the dynamics of real biological populations. Examples include temporal inhomogeneity (nonconstant vital rates and environment fluctuations), spatial inhomogeneity (diffusion and dispersal), and population inhomogeneity (differences among individuals). The inclusion of temporal inhomogeneities in Equation (2) produces a non-autonomous differential equation with $f = f(t, n)$. For example, a periodic carrying capacity K in the logistic Equation (1) reflects a seasonal habitat.

The movement of populations in a spatially extended habitat can be modeled by adding a diffusion or dispersal term to a growth model. For example, Fisher's equation

$$\frac{dn}{dt} = c\Delta n + rn(1 - n/K)$$

has been used to study the spread of populations by means of traveling waves. Other types of diffusion operators, including integral operators described by dispersal kernels, are often considered more appropriate for the movement of biological populations (Murray, 2003).

Differences in chronological age, body size, and other physiological characteristics of individual organisms can significantly affect birth and death rates and consequently, the entire population's dynamics. The McKendrick equation

$$\frac{\partial \rho}{\partial t} + \frac{\partial \rho}{\partial a} = -\delta \rho$$

describes the death processes of a population in terms of an age-specific density $\rho = \rho(t, a)$ and death rate δ. This equation is accompanied by a renewal or birth equation

$$\rho(t, 0) = \int \beta \rho \, da.$$

If the age-specific birth and death rates β and δ are independent of population density ρ, then these equations are linear and imply (under suitable conditions) the Fundamental Theorem of Demography. This theorem states that regardless of whether the population grows or decays exponentially, the normalized density approaches the positive, normalized eigensolution associated with the dominant eigenvalue. If β and/or δ are dependent on ρ, then the McKendrick model is nonlinear. This model can be used to study the effects that various life-cycle characteristics have on population level dynamics, effects such as maturation delays, gestation periods, and nonreproductive quiescent stages (Webb, 1985; Metz & Diekmann, 1986; Cushing, 1998).

Populations are often modeled using discrete time maps in which population numbers are predicted from one census time to the next. The discrete time exponential growth is described by the linear map (or difference equation) $x_{t+1} = \lambda x_t$, $\lambda > 1$. An analog of logistic equation for density regulated growth is the Beverton–Holt equation

$$x_{t+1} = \lambda x_t \frac{1}{1 + (\lambda - 1)\, x_t / K},$$

whose solutions monotonically approach K as $t \to +\infty$. Other types of density-dependent, population growth models can lead to non-equilibrium asymptotic states. For example, in the Ricker model

$$x_{t+1} = \lambda x_t \exp(-c x_t),$$

larger population numbers, of sufficient magnitude, produce smaller population numbers at the next census. This type of density-dependent "depensation" produces a period-doubling, bifurcation route to chaos as the inherent growth rate λ increases. In the 1970s, Robert May stimulated the interest in complex and chaotic dynamics that flowered during the last decades of the 20th century by his studies of the Ricker model and similar one-dimensional maps as models of population growth (May, 1974).

There is a long tradition of using discrete time models to model structured populations (Caswell, 2001). The Leslie matrix model

$$\boldsymbol{x}_{t+1} = L \boldsymbol{x}_t \tag{3}$$

is an example. The components of the vector $\boldsymbol{x}_t$ are the number of individuals, at time t, in a finite collection of age classes. The nonnegative "projection" matrix L contains birth and survival rates per unit time. In other models, the structuring classes are based on other categories (such as body size and life cycle stage). If L is a constant in time, matrix model (3) implies (under suitable mathematical conditions) the Fundamental Theorem of Demography. If the entries of L depend on $\boldsymbol{x}_t$, then the model is nonlinear. Nonlinear matrix models can be used to study the effects of class specific density effects on birth and death rates. For example, a model of a population with three life cycle stages and Ricker-type negative feedback nonlinearities has been used in conjunction experimental studies to document the occurrence of a bifurcation route to chaos in a laboratory population of insects (Cushing et al., 2002).

Most populations interact with other populations in ways that affect each other's vital rates. A natural extension of the logistic equation to the interaction of two species' assumes per capita growth rates are linear expressions of species densities, an assumption that results in the Volterra–Lotka system

$$\frac{dn_1}{dt} = n_1 (r_1 + c_{11} n_1 + c_{12} n_2),$$
$$\frac{dn_2}{dt} = n_2 (r_2 + c_{21} n_1 + c_{22} n_2)$$

of differential equations. A fundamental classification of two species' interactions can be based on the signs of the six coefficients in these equations. Two classical examples are

$$\frac{dn_1}{dt} = r n_1 (1 - n_1/K) - c_1 n_1 n_2,$$
$$\frac{dn_2}{dt} = -d n_2 + c_2 n_1 n_2 \tag{4}$$

and

$$\frac{dn_1}{dt} = r_1 n_1 (1 - n_1/K_1) - c_1 n_1 n_2,$$
$$\frac{dn_2}{dt} = r_2 n_2 (1 - n_2/K_2) - c_2 n_1 n_2. \tag{5}$$

The Lotka–Volterra predator-prey model (4) describes a logistically growing prey population n_1 that is preyed upon by a predator n_2 that dies out exponentially in the absence of n_1. In this model, the per predator uptake rate of prey is proportional to the amount of prey present, a mass-action type interaction that gives rise to the quadratic nonlinearity in (4). If $K < d/c_2$, this predator-prey model predicts predator extinction; if $K > d/c_2$, the model predicts coexistence (in terms of an asymptotically stable equilibrium state). The mass-action assumption is often replaced by one based on a predation rate that saturates (or even decreases) as the amount of prey increases. For example, the MacArthur–Rosenzweig predator-prey model,

$$\frac{dn_1}{dt} = r n_1 (1 - n_1/K) - c_1 \frac{n_1}{a + n_1} n_2,$$
$$\frac{dn_2}{dt} = -d n_2 + \frac{1}{\alpha} c_1 \frac{n_1}{a + n_1} n_2,$$

assumes a per predator prey uptake rate $c_1 n_1 / (a + n_1)$ of the so-called Holling II (or Monod, or Michaelis–Menten) type. This predator-prey model predicts a destabilization of the coexistence equilibrium, accompanied by a Hopf bifurcation to a stable limit cycle, for K large. Thus, an enrichment of the habitat results, paradoxically, in a destabilization of the predator-prey interaction and, for this reason, this phenomenon is called the "paradox of enrichment."

The Lotka–Volterra competition model (5) describes two logistically growing populations that adversely affect each other's growth rate through an interference competitive interaction measured by the quadratic mass–action terms. This model predicts a limited number of competitive outcomes. If the interspecific competition is weak compared with the intraspecific competition ($c_1 c_2 < r_1 r_2 / (K_1 K_2)$), then the species coexist in an asymptotically stable equilibrium state. If interspecific competition is too strong ($c_1 c_2 > r_1 r_2 / (K_1 K_2)$), then one species becomes extinct and the other survives (in an asymptotically stable equilibrium state). The Lotka–Volterra competition model gave rise to the many fundamental ecological notions, including competitive exclusion, ecological niche (there can be no more species than there are limiting resources), and limiting similarity of competitors. Although not universal for theoretical competition models, these concepts are well supported by a large number of other competition models. For example, the exploitative competition model (the chemostat model)

$$\frac{dn_1}{dt} = -dn_1 + \frac{1}{\alpha_1}c_1\frac{R}{a_1+R}n_1,$$
$$\frac{dn_2}{dt} = -dn_2 + \frac{1}{\alpha_2}c_2\frac{R}{a_2+R}n_2,$$
$$\frac{dR}{dt} = d\,(R_0 - R) - c_1\frac{R}{a_1+R}n_1 - c_2\frac{R}{a_2+R}n_2$$

describes the competition of two species for a (prey) resource R. This model also predicts the survival of only one species (Smith & Waltman, 1995).

Large-dimensional ecosystems can be modeled by coupling together any number of single species such as those described above (May, 2001). Such multispecies models can include temporal, spatial, and/or demographic inhomogeneities. For example, the dispersal and diffusion of many interacting species can be modeled by systems of reaction-diffusion equations, such as Fisher's equation, coupled together so as to describe predator-prey or competition interactions. Similarly, coupled systems of McKendrick equations or Leslie matrix models describe interactions among structured species.

J.M. CUSHING

See also **Biological evolution; Brusselator; Epidemiology**

Further Reading

Caswell, H. 2001. *Matrix Population Models: Construction, Analysis and Interpretation*, 2nd edition, Sunderland, MA: Sinauer Associates

Cushing, J.M. 1998. *An Introduction to Structured Population Dynamics*, Philadelphia: Society for Industrial and Applied Mathematics

Cushing, J.M., Costantino, R.F., Dennis, B., Desharnais, R.A. & Henson, S.M. 2002. *Chaos in Ecology: Experimental Nonlinear Dynamics*, New York: Academic Press

Kingsland, S.E. 1995. *Modeling Nature: Episodes in the History of Population Ecology*, Chicago: University of Chicago Press

May, R.M. 1974. Biological populations with nonoverlapping generations: stable points, stable cycles and chaos. *Science*, 186: 645–647

May, R.M. 2001. *Stability and Complexity in Model Ecosystems*, Landmarks in Biology, Princeton: Princeton University Press

Metz, J.A.J. & Diekmann, O. 1986. *The Dynamics of Physiologically Structured Populations*, Berlin: Springer

Murray, J.D. 2003. *Mathematical Biology*, 3rd edition, Berlin and New York: Springer

Smith, H.L. & Waltman, P. 1995. *The Theory of the Chemostat: Dynamics of Microbial Competition*, Cambridge and New York: Cambridge University Press

Webb, G.F. 1985. *Theory of Nonlinear Age-Dependent Population Dynamics*, New York: Marcel Dekker

POWER BALANCE

As the dynamics of any real system includes dissipation, steady dynamic regimes are observed if energy loss is permanently compensated by energy input from some external or intrinsic source, which is a condition of power balance (PB). In the flame of a candle, for example, power input is supplied by the burning wax and dissipation by the emission of heat and light. Starting from the seminal theoretical work by Yakov Zeldovich & David Frank-Kamenetsky (1938), the modern combustion theory has developed a detailed mathematical analysis of the flame propagation in various combustible media, based on the PB concept (Williams, 1964).

The PB analysis in its general form underlies a variety of dynamical phenomena in physics, chemistry, biophysics, and engineering. A simple model is the van der Pol oscillator, which is described by the equation

$$\ddot{\xi} + \omega^2\xi = \alpha\dot{\xi} - \beta\xi^2\dot{\xi}. \qquad (1)$$

Here, $\xi(t)$ is a dynamical variable, ω is the eigenfrequency of linear oscillations in the system, and the small positive coefficients α and β account for the linear intrinsic gain and nonlinear loss, respectively. As the energy (E) of this system is $(\dot{\xi}^2 + \omega^2\xi^2)/2$, Equation (1) implies

$$\mathrm{d}E/\mathrm{d}t = (\alpha - \beta\xi^2)\dot{\xi}^2. \qquad (2)$$

Assuming that $\xi = A\cos\omega t$, averaging Equation (2) over a cycle, and assuming that $\langle \mathrm{d}E/\mathrm{d}t\rangle = 0$, one finds that PB is established at the amplitude $A = 2\sqrt{\alpha/\beta}$.

Another generic case is the balance between intrinsic loss and energy supplied from an external source. An example is furnished by an equation for an ac-driven damped pendulum with sinusoidal nonlinearity,

$$\ddot{\xi} + \sin\xi = -\alpha\dot{\xi} + \gamma\cos(\omega_0 t), \qquad (3)$$

where $\alpha > 0$ is a friction coefficient and γ and ω_0 are the amplitude and frequency of the drive. (An important realization of this equation is a Josephson junction. In this case, ξ is the phase difference of the superconducting wave function across the junction, α accounts for ohmic loss, and the ac drive is induced by bias current applied to the junction (Barone & Paternó, 1982).) The corresponding PB equation takes the form (cf. Equation (2))

$$\mathrm{d}E/\mathrm{d}t = -\alpha\dot{\xi}^2 + \gamma\dot{\xi}\sin(\omega_0 t). \qquad (4)$$

In the lowest-order approximation, $\alpha = \gamma = 0$, two different types of exact solution are known: oscillating and rotating ones with zero and finite average velocity $\langle\dot{\xi}\rangle$, respectively. Substitution of the rotating solution into Equation (4) (where α and γ are treated as small parameters) and averaging over the period shows that PB is established for the solution whose frequency is locked to the driving frequency ω_0, so that $\omega = \omega_0/n$, $n = 1, 2, \cdots$. In other words, the PB condition selects the solution's frequency. Further, it follows from Equation (4) that the PB condition cannot be met unless the drive's strength exceeds a finite threshold value, γ_{thr}.

At $\gamma > \gamma_{\text{thr}}$, the PB between the ac drive and loss selects a constant value of the phase shift ϕ_0 between the ac drive and the phase of the oscillating part of the solution. Indeed, averaging Equation (4) and demanding $\langle \mathrm{d}E/\mathrm{d}t \rangle = 0$ yield a general result, $\cos\phi_0 = \gamma_{\text{thr}}/\gamma$; that is, there are two power-balanced solutions, $\phi_0 = \pm\cos^{-1}(\gamma_{\text{thr}}/\gamma)$. Further analysis demonstrates that one solution is stable and the other one is unstable.

A similar mechanism underlies propagation of a magnetic-flux quantum (fluxon, or topological soliton) in a long weakly damped ac-driven Josephson junction with periodic spatial modulation, which is described by the following sine-Gordon (SG) equation:

$$\xi_{tt} - \xi_{xx} + [1 + \varepsilon \sin(2\pi x/\lambda)] \sin\xi = -\alpha\xi_t + \gamma\cos(\omega_0 t). \tag{5}$$

Here x is the coordinate along the Josephson junction, the subscripts stand for partial derivatives, and ε and λ are the amplitude and period of the spatial modulation. In this case, the average velocity $\langle v \rangle$ of the ac-driven fluxon is determined by the condition of locking the frequency of the periodic passage of the spatially periodic relief by the fluxon to the ac-drive's frequency, which yields a spectrum of PB velocities: $\langle v \rangle = \lambda\omega_0/(2\pi n), n = \pm 1, \pm 2, \ldots$. Note that the sign of the velocity is determined by an initial push setting the fluxon in motion.

The latter phenomenon was observed in a direct experiment, in the form of an *inverse Josephson effect*, that is, dc voltage induced by ac bias current (Ustinov & Malomed, 2001). The periodic spatial modulation is necessary for the effect, as it opens a way to establish the PB between the ac drive and dissipation. Nevertheless, in an annular (ring-shaped) Josephson junction of a long but finite length, the same effect is possible without any spatial modulation, due to the interaction of the fluxon with its own tail (Goldobin et al., 2002). In that case, the average velocity is not "quantized," as above, but may take any value. A noteworthy feature, specific to the ring system, is the dependence of the threshold value γ_{thr} on the ring's length L (see Figure 1). (If the driving signal is slightly nonmonochromatic, progressive motion is still possible over a long time, but it is eventually destroyed by decoherence accumulating due to small fluctuations of the driving frequency (Filatrella et al., 2002.) PB for breathers in long lossy Josephson junctions is also possible (Lomdahl & Samuelsen, 1986).

Another class of PB problems can be formulated in terms of the complex Ginzburg–Landau equation, which may be regarded as a perturbed version of the

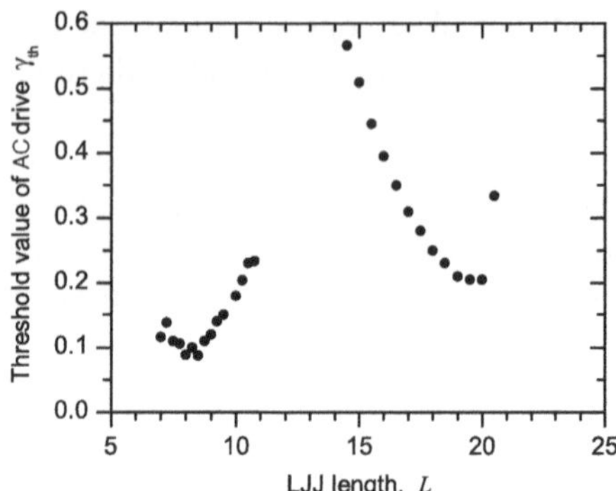

Figure 1. The minimum (threshold) value of the ac-drive's strength γ, in the case $\varepsilon = 0, \omega = 1.12$, and $\alpha = 0.1$ [see Equation (5)], which is necessary to support progressive motion (in either direction) of a fluxon in a long Josephson ring, vs. the length (L) of the ring. In regions without data, the ac-driven motion of the fluxon is impossible; in particular, the effect vanishes if L exceeds $L_{\max} \approx 21$.

nonlinear Schrödinger (NLS) equation:

$$\mathrm{i}u_t + \tfrac{1}{2}u_{xx} + |u|^2 u = -\mathrm{i}\alpha_0 u + \mathrm{i}\alpha_1 u_{xx} + \mathrm{i}\alpha_2 |u|^2 u - \mathrm{i}\alpha_3 |u|^4 u, \tag{6}$$

where all the coefficients $\alpha_{0,1,2,3}$ are assumed to be positive and small. In this model, linear and quintic nonlinear losses are accounted for by α_0, α_1, and α_3, respectively, while α_2 is a cubic-gain coefficient. Equation (6) describes transmission of soliton signals in nonlinear fiber-optic telecommunication links with loss, gain, and filtering (the last being represented by the term α_2 term) (Iannone et al., 1998) and subcritical pulses observed in binary-fluid thermal convection in narrow channels (Kolodner, 1991).

The energy of the optical field in the fiber is $E = \int_{-\infty}^{+\infty} |u(x)|^2\,\mathrm{d}x$, and the loss and gain terms on the right-hand side of Equation (6) give rise to the corresponding PB equation,

$$\mathrm{d}E/\mathrm{d}t = 2\left[-\alpha_0 E + \int_{-\infty}^{+\infty} \left(-\alpha_1 |u_x|^2 - \alpha_3 |u|^6 + \alpha_2 |u|^4\right) \mathrm{d}x\right]. \tag{7}$$

Approximating solutions by the NLS soliton, $u = \eta\,\mathrm{sech}(\eta x)\exp(\mathrm{i}\phi)$, where η is an amplitude, substituting this in Equation (7) and equating $\mathrm{d}E/\mathrm{d}t$ to zero lead to a PB condition in the form

$$\eta\left[8\alpha_3\eta^4 - 5(2\alpha_2 - \alpha_1)\eta^2 + 15\alpha_0\right] = 0. \tag{8}$$

The trivial solution, $\eta = 0$, is stable. The remaining factor in Equation (8) yields two nontrivial solutions if threshold conditions are satisfied: $\alpha_2 > \alpha_1/2$ and $5(2\alpha_2 - \alpha_1)^2 > 96\alpha_0\alpha_3$. The solution with a larger value of η gives a stable SP, and the one with smaller η is unstable.

Another manifestation of PB occurs in reaction-diffusion systems, such as the FitzHugh–Nagumo model (Cross & Hohenberg, 1993). Besides chemical systems—the Belousov–Zhabotinsky (BZ) reaction, heterogeneous catalysis, and so on—models belonging to this class find numerous other applications, including neural networks, cardiac tissue in biophysics, electron-hole plasmas in semiconductors, and gas-discharge plasmas. In this case, the PB takes place between energy release due to the chemical reaction, which is described by local terms of the model, and loss due to diffusion. As a result, various patterns may be supported as stable dynamical equilibria, including standing or traveling waves, fronts (shock waves) and periodic wave trains in one dimension, spiral vortices and localized spots in the two-dimensional case, and vortex rings in three dimensions, among others. An interesting experimental finding is the observation of two drastically different regimes of propagation of a BZ chemical wave in aqueous solution, which resemble deflagration and detonation waves in gas dynamics: a slow wave, which is driven by diffusion of chemical reactants, and a fast "big wave" (Inomoto et al., 1997), which is coupled to surface deformation and flow in the solution. The two modes of the chemical-wave propagation realize the PB in different forms, the choice between them being determined by the initial perturbation.

BORIS MALOMED

See also **Candle; Complex Ginzburg–Landau equation; Damped-driven anharmonic oscillator; FitzHugh–Nagumo equation; Flame front; Long Josephson junctions; Solitons; Zeldovich–Frank-Kamenetsky equation**

Further Reading

Barone, A. & Paternó, G. 1982. *Physics and Applications of the Josephson Effect*, New York: Wiley

Cross, M.C. & Hohenberg, P.C. 1993. Pattern-formation outside of equilibrium. *Reviews of Modern Physics*, 65: 851–1112

Filatrella, G., Malomed, B.A. & Pagano, S. 2002. Noise-induced dephasing of an ac-driven Josephson junction. *Physical Review* E, 65: 051116

Goldobin, E., Malomed, B.A. & Ustinov, A.V. 2002. Progressive motion of an ac-driven kink in an annular damped system. *Physical Review* E, 65: 056613

Iannone, E., Matera, F., Mecozzi, A. & Settembre, M. 1998. *Nonlinear Optical Communication Networks*, New York: Wiley

Inomoto, O., Kai, S., Ariyoshi, T. & Inanaga, S. 1997. Hydrodynamical effects of chemical waves in quasi-two-dimensional solution in Belousov–Zhabotinsky reaction. *International Journal of Bifurcation and Chaos*, 7: 989–996

Kolodner, P. 1991. Drifting pulses of traveling-wave convection. *Physical Review Letters*, 66: 1165–1168

Lomdahl, P.S. & Samuelsen, M.R. 1986. Persistent breather excitations in an ac-driven sine-Gordon system with loss. *Physical Review* A, 34: 664–667

Ustinov, A.V. & Malomed, B.A. 2001. Observation of progressive motion of ac-driven solitons. *Physical Review* B, 64: 020302(R)

Williams, F.A. 1964. *Combustion Theory*, Reading, MA: Addison-Wesley

Zeldozich, Ya.B. & Frank-Kamenetsky, D.A. 1938. K teorii ravnomernogo rasprostrane-niya plameni [On the theory of uniform propagation of frame]. Doklady Akademi Nauk SSSR, 19(10: 693–697.

POWER SPECTRA

See **Spectral analysis**

PRANDTL NUMBER

See **Fluid dynamics**

PREDATOR-PREY SYSTEMS

See **Population dynamics**

PREDICTABILITY OF FORECASTING

See **Forecasting**

PROTEIN DYNAMICS

Proteins carry out structural, catalytic, and molecular recognition roles that require a wide variety of motions over a broad range of timescales, from 10^{-15} to 10^3 s, as sketched in Figure 1. Motions can be as mundane as coupled vibrations or as sophisticated as gene transcription by an RNA polymerase protein. The broad range of timescales is a principal difficulty in understanding protein dynamics. We will focus here on atomic motions and conformational changes of single proteins.

Although there are a wide variety of proteins displaying many remarkable dynamical phenomena, photoactive proteins play a special role in experimental studies of protein dynamics. This is because experiments can be initiated with a femtosecond laser pulse at any temperature from absolute zero to the unfolding temperature of the protein, allowing the wide variety of motions to be observed. Proteins which have been extensively studied this way are myoglobin (Austin et al., 1975), the photosynthetic reaction center, rhodopsin, photoactive yellow protein, and cytochrome coxidase.

An essential result of such studies is that there are three fundamentally different processes involved in modeling atomic motions of proteins. This distinction can be seen in the temperature-dependent rates of various processes observed in myoglobin, sketched in Figure 2 and described in Fenimore et al. (2002). These processes are labeled as fast fluctuations, bond formation, and conformational motions.

Fast fluctuations occur on the 100 picosecond timescale over the temperature range from 200 to 300 K, but their amplitude (or number) diminishes by a factor of ten over this range. Because these motions

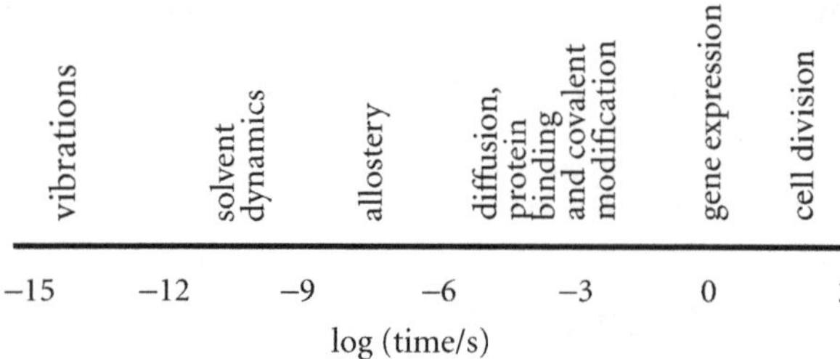

Figure 1. Types of protein motions which occur on timescales from femtoseconds to hours.

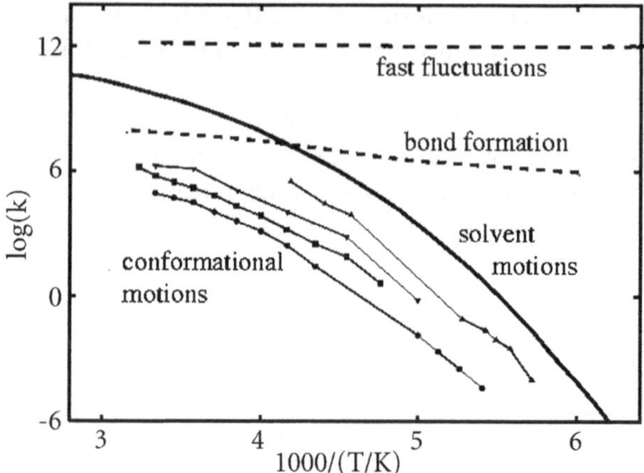

Figure 2. Arrhenius plot, showing the temperature-dependent rates of three types of experimentally measured protein motions, compared with the rate of solvent motions. (Taken from Fenimore et al., 2002, and references therein.)

occur in the absence of bulk solvent motions, they must involve relatively local rearrangements between the various possible meta-stable configurations possible in a protein. The network properties of water at the surface of the protein and the voids created by thermal expansion play an important role in these types of motion; the motions disappear when a protein sample contains less than about 30% water.

It is possible to capture the essence of these motions with a relatively simple potential energy function and Newton's law of motion, $F = ma$, $V=$ bond stretch + bond angle + dihedral twist + atom-centered point charge electrostatics + Lennard–Jones.

The bond stretch and bond angle terms are simple harmonic potentials which allow the protein to vibrate. The dihedral twist term is a cosine function which allows transitions between rotamer states of single and double bonds. Dihedral twist parameters of the backbone will also influence the relative free energy of alpha helixes and beta sheets, two of the common secondary structure motifs described in the (*See* **Protein structure**). Much of the subtlety in this potential energy function lies in the electrostatics and Lennard–Jones terms. Both are pairwise additive interactions between atoms that are not bonded to each other or through another atom. The electrostatic interaction strength falls off as $1/d$, while the the Lennard–Jones term is of the form $A/d^{12} - B/d^6$, where d is the distance between atoms.

When the solvent is treated explicitly, parameters can be developed which reproduce important dynamic and thermodynamic properties that contribute to protein function. These include solvent properties, such as dielectric response time and amplitude, self-diffusion coefficients, and heat of vaporization. For amino acids and other small molecules, heats of solvation and heats of transfer from octanol to water are accurately reproduced. Heats of solvation and some aspects of the hydration shells of ions are also correctly modeled.

When this potential energy function is applied to protein folding and dynamics, it produces a prediction for the ensemble of protein conformations, as well as rates of water penetration, motions of loops, folding temperature, and the ensemble of unfolded states. In practice, it is difficult or impossible to obtain the necessary sampling to unambiguously quantify these properties. The reason is that a solvated protein simulation involves about 30,000 atoms; thus the integration time for the dynamics is limited to about 2 fs, while many of the motions of interest occur on the nanosecond to microsecond timescales (see Figure 2). Much effort in this field over the past two decades has centered on methods to improve the sampling through efficient implementation, parallelization, and statistical physics.

Covalent bond formation is the primary function of enzymes and involves many of the complexities of fast fluctuations combined with the need for a quantum-mechanical description of electronic polarizability, bond distortion, and bond formation. Such descriptions are available through all-electron quantum chemistry techniques, such as density functional theory, but calculations require approximately 1 cpu/day to evaluate energies and forces for 200 atoms. Calculations can be efficiently parallelized, and when more than 200 protein atoms are included, computation time scales linearly with system size. It is also possible to treat the bond-formation region quantum mechanically and the rest of the protein and solvent classically in a single calculation.

The difficulties involved in computing reaction rates include proper treatment of the vibrational dynamics of bond formation, sampling over hydration states, and sampling over the ensemble of protein conformations. One useful simplification is that many enzymes exclude water from their active site, considerably reducing the disorder.

Conformational motions of proteins can involve simple shifts of helixes, repacking of amino acid side chains, or more extensive changes between active and inactive conformations. The molecular dynamics model described above makes a prediction for conformational changes of proteins, but it is only recently that computer power has reached the long times necessary to test these predictions (Karplus & McCammon, 2002; Mayor et al., 2003). It is likely that some tuning of parameters will be necessary to predict protein properties with the same reliability with which these models currently predict small molecule properties.

Because the solvent provides a cage which prevents the protein from moving freely, the fraction of possible configurations explored increases by several orders of magnitude when the solvent is considered implicitly (Takada, 1999). While it is doubtful that a single parameter set can be developed to treat all proteins with an implicit solvent, it is clear from models of protein folding that much can be learned from implicit solvent models which are explicitly, but weakly, biased toward a known folded conformation.

BENJAMIN H. MCMAHON AND PAUL W. FENIMORE

See also **Biomolecular solitons; Local modes in molecules; Molecular dynamics; Pump-probe measurements**

Further Reading

Austin, R.H., Beeson, K.W., Eisenstein, L., Frauenfelder, H. & Gunsalus, I.C. 1975. Dynamics of ligand-binding to myoglobin. *Biochemistry*, 14: 5355–5373

Fenimore, P.W., Frauenfelder, H., McMahon, B.H. & Parak, F.G. 2002. Slaving: solvent fluctuations dominate protein dynamics and functions. *Proceedings of the National Academy of Sciences, USA*, 99: 16047–16051

Karplus, M. & McCammon, J.A. 2002. Molecular dynamics simulations of biomolecules. *Nature Structural Biology*, 9: 646–652

Mayor, U., Guydosh, N.R., Johnson, C.M., Grossmann, J.G., Sato, S., Jas, G.S., Freund, S.M.V., Alonso, D.O.V., Daggett, V. & Fersht, A.R. 2003. Complete folding pathway of a protein from nanoseconds to microseconds. *Nature*, 421: 863–867

McMahon, B.H., Muller, J.D., Wraight, C.A. & Nienhaus, G.U. 1998. Electron transfer and protein dynamics in the photosynthetic reaction center. *Biophysical Journal*, 74: 2567–2587

Takada, S. 1999. Go-*ing* for the prediction of protein folding mechanisms. *Proceedings of the National Academy of Sciences, USA*, 96: 11698–11700

PROTEIN STRUCTURE

A protein's structure, flexibility, and activity depend on the primary sequence of amino acids encoded by DNA, the presence of small-molecule ligands, the presence of protein or nucleic acid binding partners, and its history of covalent modification by other enzymes. Figure 1 (in color plate section) shows the catalytic site of one protein, RNA polymerase. This protein is responsible for transcribing an organism's DNA into messenger RNA, initiating the process of protein production, an essential and highly regulated cellular process. The complex geometry of interactions which stabilize the DNA, RNA, and nucleotide being added to the RNA are highly conserved among polymerases from all organisms, from microbes to humans. It is typical of enzymes that they are precisely (within about 0.1 Å) constructed near the catalytic site in order to exclude water and polarize and orient the reactants, yet still maintain the flexibility to allow reactants to enter and exit the active site.

One reason proteins can adopt highly rigid active sites, yet still undergo the large motions necessary for substrate binding, is that the peptide backbone is stabilized by multiple interactions in particular structural motifs, such as helixes, sheets, and particular types of hairpin turns. Thus, not only are the residues directly responsible for catalysis highly conserved among organisms, but also the arrangement of helixes, sheets, and tightly packed hydrophobic residues which determine the large-scale dynamics of the protein. This is illustrated in page 7 of the color plate section, which shows the entire catalytic domain of RNA polymerase, color-coded according to the polarity and charge of the amino acids. It is useful, looking at this figure, to reflect on the numerous cooperative interactions among the 1034 amino acids of this protein which cause it to fold up in this shape, rather than the shape of any of the other 30,000 proteins, or a glob of tangled polymer. These interactions are predominantly hydrophobicity, shape, hydrogen bond formation, and charge complementarity.

Protein function entails not only catalysis, but also the ability to be regulated by other proteins. This is done either by covalent modification, such as phosphorylation (covalent attachment of PO_3^{2-}) or acetylation (covalent attachment of C_2O), or by complex formation with other proteins. In either case, all the residues at the surface and the geometry of rigid motifs communicating the surface interactions to the active site become important determinants of protein structure. This is illustrated in the color plate section, showing the complex of twelve proteins, DNA, and RNA that is required for RNA polymerase to transcribe DNA to messenger RNA, as an early step in protein synthesis.

We have used the polymerase protein to illustrate several aspects of protein structure. There are between 10,000 and 100,000 types of protein structures used in various organisms, and a significant fraction of them have had their structure experimentally determined by diffraction of X-rays from a protein crystal. These structures are collected in a publicly accessible protein data bank, and several free software programs exist to aid in visualizing various aspects of protein structures.

The single, most important determinant of a protein's structure is the primary sequence of amino acids, encoded by an organism's DNA. Advances in microbiology have allowed automated sequencing of complete genomes of numerous organisms. While tens of thousands of protein structures are known, several million protein sequences are known, from all types of living creatures, from mammals to plants, microbes, and viruses. Proteins can be identified from their sequence of amino acids by searching for patterns of amino acids with BLAST (basic local alignment search tool), or tailoring a search algorithm based on numerous examples of a particular protein with a so-called hidden Markov model. Sequence analysis has been used to annotate both complete and partial genomes, and the results are available in several public databases.

The great excess of sequence data over structural data, combined with the insight into protein function and interactions provided by knowledge of the protein structure, provokes the question of whether it is possible to predict a protein structure from its primary sequence of amino acids. The essential reason this problem is difficult is that amino acids have an average of four dihedral angles which are equally stable in any of two or three positions (0°, $+120°$ and $-120°$), creating $3^{4\times 150}$ possible structures for each protein. The second reason is that when proteins were optimized by evolution, they exploited subtleties of interactions which are not necessarily captured by simple potential energy functions. Thus, while it has long been clear what forces stabilize folded proteins (hydrophobic interactions, hydrogen bonding, and Coulomb interactions), it has only recently been possible to relate the folding of a sequence of 20 amino acids to interaction potentials derived from small-molecule thermodynamic properties.

When it is possible, the most reliable method to predict a protein's structure is to find a protein of known structure which has greater than about 30% sequence identity. The particular conformations of most amino acids can then be inferred from the known protein structure, and resulting problems can usually be corrected by a Monte Carlo search algorithm or by hand. This process is known as homology modeling. Systematic efforts have produced enough experimentally determined structures to cover most of the observed sequences.

If a suitable template protein is not available, the protein structure must be constructed from observed structural motifs. Three of the most useful rules are: (1) hydrophobic residues tend to occupy the interior of the protein; (2) beta sheets, alpha helices, and other types of turns have characteristic amino acid content; and (3) triplets of amino acids tend to make particular configurations. Progress in this so-called ab initio folding can be evaluated by observing the result of the CASP competition, where a dozen experimentally determined structures of novel proteins are held back, while competing research groups are given several months to enter predictions of their structures.

Benjamin H. McMahon and Montiago X. LaBute

See also **Hydrogen bond; Molecular dynamics; Polymerization; Protein dynamics**

Further Reading

Alberts, B. et al. 2002. *Molecular Biology of the Cell*, 4th edition, New York: Garland Science

Branden, C. & Tooze, J. 1999. *Introduction to Protein Structure*, New York: Garland Science

Cramer, P., Bushnel, D.A. & Kornberg, R.D. 2001. Structural basis of transcription: RNA polymerase II at 2.8 Angstrom resolution. *Science*, 292: 1876

Petsko, G. A. & Ringe, D. 2003. *Protein Structure and Function*, Sunderland, MA: Sinauer Associates and Oxford: Blackwell Publishing

Valuable websites to persue:

THE PROTEIN DATA BANK: www.rcsb.org

CASP5 COMPETITION: predictioncenter.llnl.gov/casp5/Casp5.html.

BASIC LINEAR ALIGNMENT SEQUENCE TOOL (BLAST): www.ncbi.nlm.nih.gov/

MODELLER: www.salilab.org/modeller/modeller.html

VMD (A PROTEIN VISUALIZATION): www.ks.uiuc.edu/Research/vmd/

(SEARCHABLE GENE DATABASES): www.ensemble.org, www.ncbi.nlm.nih.gov/Entrez

PROTEIN FAMILY CATALOG: www.sanger.ac.uk/Software/Pfam/index.shtml.

PSEUDO-DIFFERENTIAL EQUATIONS

See **Equations, nonlinear**

PULSONS

See **Solitons, types of**

PUMP-PROBE MEASUREMENTS

Figure 1 shows a prototype pump-probe experiment in which short pump-laser pulse excites the vibrational oscillator (used here as a simple example) from the $\nu=0$ vibrational state into the $\nu=1$ vibrational state. A subsequent probe-laser pulse tests the $\nu=1$ population by probing the $\nu=1\rightarrow\nu=2$ excited state absorption, the $\nu=1\rightarrow\nu=0$ stimulated emission, and/or the $\nu=0\rightarrow\nu=1$ bleach. The time resolution of such an experiment is given by the pulse duration of the laser pulses, while the detector can be slow. With recent progress in laser technology, one can now perform pump-probe experiments in almost any spectral range, starting from far-IR (10 cm^{-1}) to soft X-ray radiation (1 keV) with a time resolution of much less than a picosecond (2000).

Pump-probe spectroscopy is a special form of third-order nonlinear spectroscopy. The theory of linear and nonlinear optical spectroscopy is generally performed in a perturbative expansion, where the electric fields of the light pulses act as weak perturbation onto the molecular Hamiltonian. The linear response, which

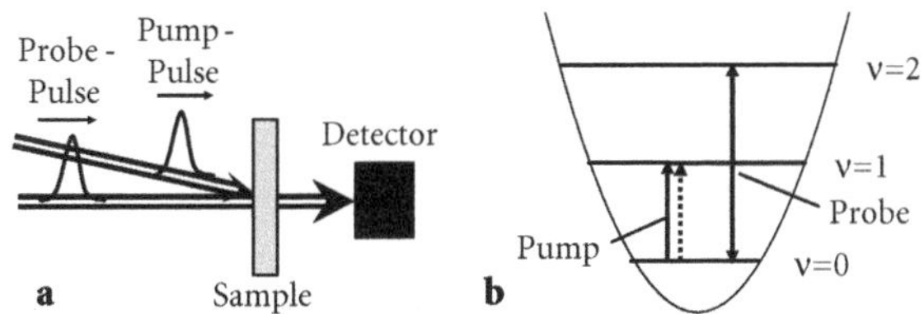

Figure 1. (a) A prototype pump-probe setup and (b) a prototype pump-probe experiment of a vibrator.

describes linear absorption spectroscopy, is given by

$$P^{(1)}(t) = \int_0^\infty \mathrm{d}t_1 R^{(1)}(t_1) \times E(t - t_1). \quad (1)$$

Expressed in simple words, the electric field E of the probe light interacts once with the sample through the first-order response function $R^{(1)}$, generating a first-order polarization $P^{(1)}$, which subsequently is measured in the detector. The first-order response function $R^{(1)}$ is a material property, which can be tested by this type of spectroscopy. For instance, if the electric field E is chosen to be a monochromatic wave, one can measure the absorption spectrum by tuning its frequency.

Third-order spectroscopy is given by

$$P^{(3)}(t) = \int_0^\infty \mathrm{d}t_3 \int_0^\infty \mathrm{d}t_2 \int_0^\infty \mathrm{d}t_1 R^{(3)}(t_1, t_2, t_3) \times E_1(t - t_3)E_2(t - t_3 - t_2)E_3(t - t_3 - t_2 - t_1). \quad (2)$$

Again expressed in simple words, the electric fields E_1, E_2, and E_3 originating from a maximum of three laser pulses (in the example of Figure1a, both E_2 and E_3 originate from one pulse, i.e., the pump pulse) interact with the sample at three different time points $t - t_3$, $t - t_3 - t_2$, and $t - t_3 - t_2 - t_1$ through the third-order response function $R^{(3)}$, generating a third-order polarization $P^{(3)}$. The third-order response function $R^{(3)}$ contains significantly more information about the molecular system than the linear response function. Both can, in principle, be calculated once all eigenstates, transition dipole moments, and relaxation pathways of a system are known.

There are different types of third-order spectroscopies, the most important of which being pump-probe and photon echo spectroscopy. Both exist in many variations. They all measure the same third-order response function $R^{(3)}$ but differ by the choice of the three field interactions E_1, E_2, and E_3 (timing, frequency, beam direction, etc.). The different methods project different aspects of the complicated third-order response function $R^{(3)}$. A comprehensive discussion of the principles of nonlinear optical spectroscopy is given by (Mukamel, 1995). In the context of nonlinear science, the following advantages of third-order spectroscopies should be noted.

- In condensed phase systems, spectroscopic transitions are often significantly broadened. The most common broadening mechanisms are (i) lifetime broadening (T_1-relaxation), (ii) homogeneous broadening (T_2-relaxation), and (iii) inhomogeneous broadening. In solution phase systems, the situation becomes even more complicated since the distinction between homogeneous and inhomogeneous broadening becomes a question of the timescale of the fluctuating forces giving rise to dephasing. As a consequence, a continuous transfer between both regimes does, in general, occur. It is a common practice to fit absorption lines to certain line profiles such as a Lorentzian profile (homogeneous and/or lifetime broadening), a Gaussian profile (inhomogeneous broadening), a Voigt profile (a mixture of both), or a Kubo profile (intermediate regime). However, the assignment to different broadening mechanisms is model-dependent and often extremely questionable. Linear absorption spectroscopy principally cannot distinguish between different broadening mechanisms.

  The information boost of third-order nonlinear spectroscopy just makes that distinction possible. For example, pump-probe spectroscopy observes T_1-relaxation directly by populating a spectroscopic state with the pump-pulse and subsequently observing its relaxation back into the ground state with the probe-pulse (see Figure 1b). Photon echo experiments and transient hole burning experiments (a special form of pump-probe experiment with a spectrally narrow pump-pulses) can distinguish between the homogeneous, inhomogeneous, and the intermediate dephasing regime.
- When considering nonlinear vibrational states, such as the Davydov soliton and/or local modes in molecular crystals, a more subtle point is the following: The nonlinear third-order response of a system of linearly coupled harmonic oscillators (i.e., a crystal described in the harmonic approximation) vanishes exactly. This is because the three contributions to the nonlinear response function (bleach, stimulated emission, and excited state absorption, see Figure 1(b) all occur at the same frequency, and the transition dipoles are such that they cancel completely. Only anharmonicity (nonlinearity) of the molecular potential surfaces gives rise to a nonzero pump-probe signal. Hence, nonlinear pump-probe spectroscopy is specifically sensitive to that part of the molecular Hamiltonian which is also responsible for collective nonlinear phenomena. A linear absorption spectrum is, of course, nonzero also in the harmonic limit.

Using these properties, one can test predictions from soliton and self-trapping theories in a much more direct way than linear absorption spectroscopy can do. For example, Davydov has speculated about vibrational solitons in protein secondary structures, with the aim to explain the capability of proteins to efficiently store small quanta of energy (Davydov, 1979). As a first experimental attempt to verify this prediction, Austin and coworkers (Xie et al., 2000) have investigated the lifetime of the amide-I band (mostly the C=O stretching mode of the protein backbone) of myoglobin, which typically relaxes on an ultrafast 1 ps timescale. Interestingly, they found a somewhat prolonged lifetime (15 ps) in the blue wing of the absorption spectrum. Pump-probe experiments on crystalline acetanilide (ACN), a simple model system used to study nonlinear collective phenomena in protein structures (Scott, 1992), have revealed useful information, including the following (Edler et al., 2002): (i) The NH free exciton self-traps on an ultrafast 400 fs timescale. (ii) The phonons that mediate self-trapping can be identified. (iii) The anharmonicity of the amide-I states is a measure of their degree of delocalization. The free exciton is delocalized at a temperature of 90 K but Anderson (disorder) localizes with increasing temperature. On the other hand, the band that has been assigned to a self-trapped state is localized at all temperatures. (iv) The lifetimes of the initially excited states are short (1–2 ps), yet the ground state recovery is somewhat longer (20–40 ps). Apparently, energy is stored in the meantime by mechanisms that still need to be explored. None of this information can be obtained from linear absorption spectroscopy.

PETER HAMM

See also **Biomolecular solitons; Davydov soliton, Local modes in molecular crystals; Nonlinear optics**

Further Reading

Davydov, A.S. 1979. Solitons in molecular systems *Physica Scripta*, 20: 387–394

Edler, J., Hamm, P. & Scott, A.C. 2002. Femtosecond study of self-trapped vibrational excitons in crystalline acetanilide. *Physical Review Letters*, 88: 067403. See also Edler, J. & Hamm, P. 2002. Self-trapping of the amide I band in a peptide model crystal. *Journal of Chemical Physics*, 117: 2415–2424

Elsaesser, T. Mukamel, S., Murnanae, M.M. & Scherer, N.F. (editors). 2000. *Ultrafast Phenomena XII*, Berlin: Springer and previous books of the same series

Mukamel, S. 1995. *Principles of Nonlinear Optical Spectroscopy*, New York: Oxford University Press

Scott, A.C. 1992. Davydov's soliton. *Physics Reports*, 217: 3–67

Xie, A., van der Meer, L., Hoff, W. & Austin, R.H. 2000. Long-lived amide I vibrational modes in myoglobin. *Physical Review Letters*, 84: 5435–5438

Q-DEFORMATION

See **Salerno equation**

QUADRATIC FAMILY

See **One-dimensional maps**

QUANTUM BILLIARDS

See **Billiards**

QUANTUM CHAOS

Classical equations of motion can have arbitrary nonlinearities and thus can show a wide variety of chaotic behavior. The Schrödinger equation for quantum systems, in contrast, is a linear partial differential equation, and the initial value problem cannot show chaotic behavior in the usual sense. However, in the transition regions between classical and quantum descriptions, especially for highly excited systems or short de Broglie wavelengths, chaos in classical dynamics will have its effects on quantum dynamics, the eigenstates, and the eigenenergies. The field of quantum chaos deals with the characteristic properties that emerge in this transitional region. Some features are typical for the time evolution of wave functions and closely related to the classical dynamics of trajectories and phase space densities. Others deal with the properties of eigenenergies and eigenstates. Both aspects can nicely be illustrated for hydrogen atoms in external fields.

A first example for the close relation between classical and quantum dynamics is provided by excitations of hydrogen in microwave fields. Exposed to a microwave field of a frequency of 9.923 GHz, a hydrogen atom starting from a state with main quantum number $n = 66$ needs to absorb about 75 photons in order to ionize. Experiments show little ionization for small intensities of the field and a rapid increase to almost complete ionization for larger fields (Figure 1a). A full quantum calculation is prohibitively complicated because each photon adds one order of perturbation theory. However, an analysis of the classical dynamics of an electron in a Coulomb field and the external microwave field, with Hamiltonian

$$H = \frac{\mathbf{p}^2}{2m} - \frac{1}{4\pi\varepsilon_0}\frac{1}{r} + Ff(t)\cos\omega t, \tag{1}$$

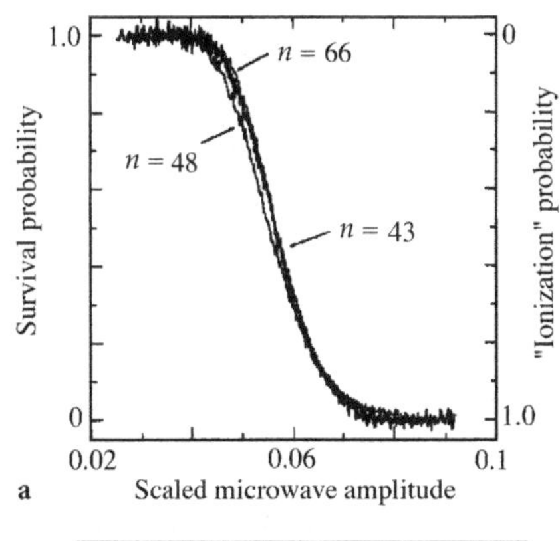

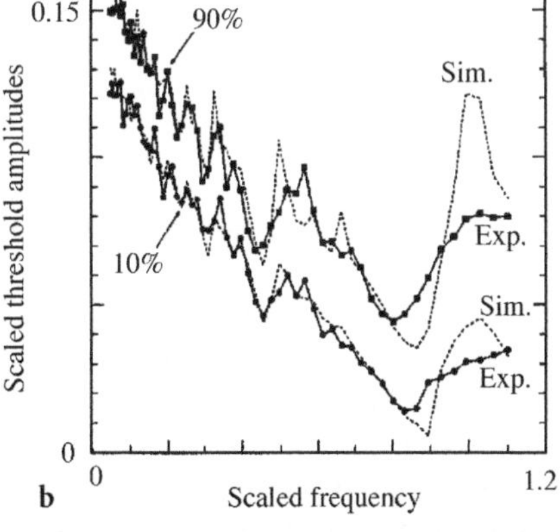

Figure 1. Microwave ionization of hydrogen atoms. The combinations $n_0\omega^3$ for the frequency and n_0^4F for the field strength are suggested by classical scaling. (Upper) Survival probability and ionization probability for different frequencies ω and different initial states n_0 that give the same scaled frequency as a function of field strengths. (Lower) Scaled threshold amplitude for the 10% and the 90% levels vs. scaled frequency. The comparison with classical simulations (dotted line) is very good and attests to the reliability of classical modelling of this process. For scaled frequencies $n_0^3\omega$ larger than 1, quantum effects become more important. (From Koch & van Leeuwen, 1995, with permission from Elsevier.)

where $f(t)$ describes the envelope of the field, reproduces the observed threshold amplitudes including many of the structures evident in Figure 1b very accurately.

The classical explanation for the fairly sharp onset is that for low field intensities, not all tori are destroyed, and the chaotic regions are separated by tori that block the transitions between them. For higher field intensities, the tori break up, the chaotic regions grow together, and electron trajectories can explore large regions of phase space, eventually escaping to infinity. This break up of tori is classically a rather dramatic event and can thus explain easily a fairly sharp onset of ionization. Many experiments since have verified the dependence on the initial state and the field intensities. Peculiar structures in the quantum ionization curves were also noted and could be explained as quantum effects due to almost degenerate quasienergy states. For the history of the problem and a careful analysis of the experiments and theories, see the review by Koch & van Leeuwen (1995).

Closely related to the microwave ionization of hydrogen is the dynamics of cold atoms in a standing light field. This problem is an experimental realization of the quantized kicked rotator, a standard model in classical and quantal chaos. Many properties of the quantized kicked rotator are described in Casati & Chirikov (1995); the first experimental realization is described in Moore et al. (1994).

For stationary problems, time can be separated and the eigenfunctions and eigenvalues of the Schrödinger equation can be analyzed. A good example is hydrogen in strong magnetic fields when the quadratic part of the vector potential cannot be neglected: the Hamiltonian then becomes

$$H = -\frac{\hbar^2}{2m}\Delta - \hbar\omega L_z - \frac{1}{4\pi\varepsilon_0}\frac{1}{r} + \frac{m\omega^2}{2}(x^2+y^2) \quad (2)$$

with the magnetic field expressed through its cyclotron frequency $\omega_c = eB/(2m)$. In order for the energy content in the cyclotron motion to be compatible to the spacing between the ground state and the first excited state, the magnetic field has to be of the order of 10^5 T, values reached and exceeded on the surface of neutron stars only. However, if the initial state is an excited state with main quantum numbers of about $n = 40$, already a few Tesla suffice to bring the system way out of the linear Zeemann regime and to scramble the spectra completely (Figure 2). Many properties of the system are described in Friedrich & Wintgen (1989), and in the contribution by Delande to Giannoni et al. (1991).

In such strong fields, there are not enough quantum numbers anymore to allow one to label the eigenstates uniquely. In a similar situation, nuclear physicists introduced statistical measures for the distribution of eigenvalues (*See* **Random matrix theory**).

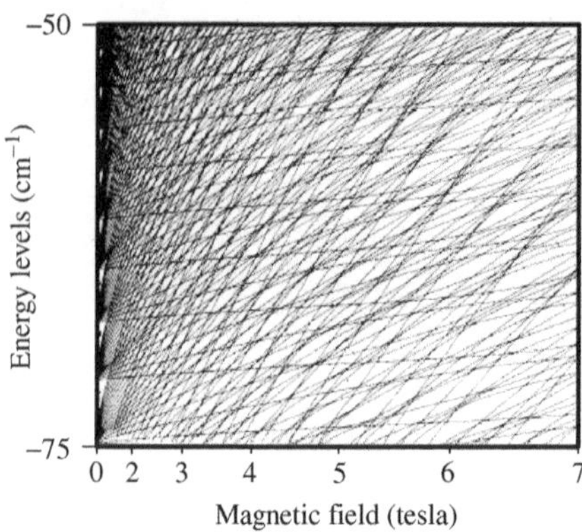

Figure 2. Energy levels for hydrogen in a magnetic field, for field strengths up to 7 T and initial quantum numbers between 38 and 45. The linear splitting within the Zeemann regime is limited to a region with very small fields. Most of the apparent crossings between eigenvalues are in fact avoided. (From Friedrich & Wintgen, 1989, with permission from Elsevier.)

Many investigations have shown that the statistics of eigenvalues for hydrogen in sufficiently strong magnetic fields, such as the level spacing distribution or two point correlation functions, are in very good agreement with random matrix theory (Friedrich & Wintgen, 1989). The same applies to many other classically chaotic systems, including vibrations of molecules, particles trapped in billiards, and model Hamiltonians with cubic and higher-order potentials, see the contribution by Bohigas to Giannoni et al. (1991) and the books by Brack & Bhaduri (1997), Haake (2001), and Stöckmann (1999).

While the distribution of eigenvalues shows statistical features, there are also deterministic ones: the Gutzwiller trace formula relates in the semiclassical limit eigenvalues E_i of the quantum system to classical periodic orbits and their properties (Gutzwiller, 1990; Chaos Focus Issue, 1992; Friedrich & Eckhardt, 1997),

$$\rho(E) = \sum_i \delta(E - E_i) \sim \mathrm{Re} \sum_p A_p \mathrm{e}^{\mathrm{i}S_p(E)/\hbar}, \quad (3)$$

$S_p = \oint \mathbf{p}\,\mathrm{d}\mathbf{q}$ is the classical action. The amplitude A_p carries information about caustics (through their phase shifts) and the instability of the classical orbits: the more unstable the smaller the weight. Because the density of states follows from a quantum amplitude (rather than a probability), the quantum amplitude is in magnitude the square root of the classical probability amplitude (*See* **Periodic orbit theory**). Expanding linearly around some reference energy E_r, the action becomes $S(E) \approx S(E_r) + T(E_r)(E - E_r)$, and a Fourier transform in E will results in peaks at the periods of periodic orbits in the system. This was noted in experiments by Karl Heinz Welge and his group and in numerical calculations by Friedrich & Wintgen (1989) (see also the comparison in Figure 3). Thus, on energy scales larger than a mean spacing

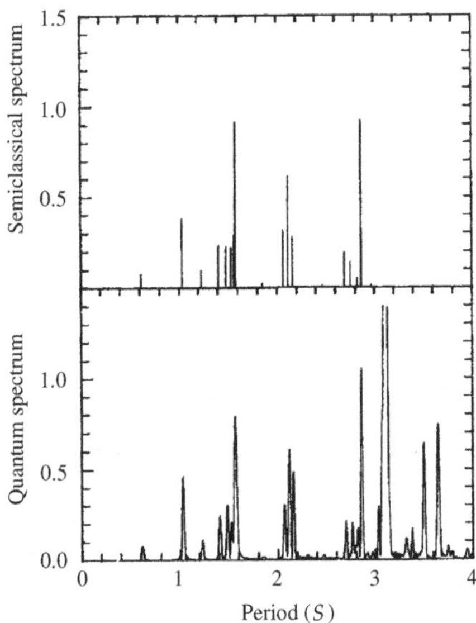

Figure 3. (Upper) The periodic orbits with periods S less than 3 and their semiclassical amplitudes and (Lower) a Fourier transform of a cross section of the quantum spectra in Figure 2.

and compatible with classical dynamics, the density of states is modulated by periodic orbits. Such periodic orbit spectroscopy can be used to extract the orbits and, thus, system specific properties.

In the presence of large-scale classical chaos and the accessibility of large parts of phase space by a single trajectory, one might expect that the wave functions are fairly uniformly spread out over the classically ergodic region. For billiards, this is the content of Shnirelman's theorem. However, as first discussed by Heller (1984) (see also Heller's contributions to Giannoni et al., 1991), periodic orbits leave their imprint on wave functions. One can find an enhanced intensity near a periodic orbit (a "scar") that can be detected in enhanced cross sections for certain processes.

The possibilities of quantum chaos are not limited to the realm of quantum phenomena proper: all that is needed is a wave theory, ideally with a classical dynamics for the propagation of wave fronts in the short wavelength limit. Examples outside the quantum world include resonances in microwave cavities, vibrations of solids, and acoustic waves in concert halls. In all cases level repulsion and other features can be found.

Many of the general properties of chaotic wave systems are discussed in the review Eckhardt (1988); in the textbooks by Brack & Bhaduri (1997), Gaspard (1998), Gutzwiller (1990), Haake (2001), and Stöckmann (1999); and in the contributions to Casati & Chirikov (1995), Giannoni et al. (1991), Chaos Focus Issue (1992), and Friedrich & Eckhardt (1997).

Bruno Eckhardt

See also **Periodic orbit theory; Random matrix theory I–IV; Regular and chaotic dynamics in atomic physics**

Further Reading

Brack, M. & Bhaduri, R.K. 1997. *Semiclassical Physics*, Reading, MA: Addison-Wesley

Casati, G. & Chirikov, B.V. 1995. *Quantum Chaos: Between Order and Disorder*, Cambridge and New York: Cambridge University Press

Cvitanoic, P. (editor). 1992. Chaos Focus Issue on Periodic Orbit Theory, *Chaos*, 2: 1–158

Eckhardt, B. 1988. Quantum mechanics of classically non-integrable systems. *Physics Reports*, 163: 205–297

Friedrich, H. and Eckhardt, B. (editors). 1997. *Classical, Semiclassical and Quantum Dynamics in Atoms*, Berlin and New York: Springer

Friedrich, H. & Wintgen, D. 1989. The hydrogen atom in a uniform magnetic field — an example of chaos. *Physics Reports*, 183: 37–79

Gaspard, P. 1998. *Chaos, Scattering and Statistical Mechanics*, Cambridge and New York: Cambridge University Press

Giannoni, M.J., Voros, A. & Zinn-Justin, J. (editors). 1991. *Chaos and Quantum Physics*. Proceedings Les Houches summer school 1989, Amsterdam and New York: North-Holland

Gutzwiller, M.C. 1990. *Chaos in Classical and Quantum Mechanics*, New York: Springer

Haake, F. 2001. *Quantum Signatures of Chaos*, Berlin: Springer

Heller, E.J. 1984. Bound-state eigenfunctions of classically chaotic Hamiltonian systems: scars of periodic orbits. *Physical Review Letters*, 53: 1515–1518

Koch, P.M. & van Leeuwen, K.A.H. 1995. The importance of resonances in microwave "ionization" of excited hydrogen atoms. *Physics Reports*, 255: 289–403

Moore, M.L., Robinson, J.C., Bharucha, C.F., Williams, P.E. & Raizen, M.G. 1994. Observation of dynamical localization in atomic momentum transfer: a new testing ground for quantum chaos. *Physical Review Letters*, 73: 2974–2977

Stöckmann, H.J. 1999. *Quantum Chaos: An Introduction*, Cambridge and New York: Cambridge University Press

QUANTUM FIELD THEORY

As our most fundamental description of physical phenomena, quantum field theory (QFT) is the natural culmination of classical mechanics, classical field theory, and quantum mechanics. To understand what QFT is and what makes it so special, it is worthwhile to briefly look at each of these structures in turn.

Classical mechanics describes the motion of objects in space and time in terms of well-defined positions and momenta that, taken together, form the phase space of the system. The location of an object in phase space can, in principle, be determined at any time, and once known suffices to predict its location at all future times given a complete enough knowledge of the forces at work. There are two main formulations of classical mechanics: the Hamiltonian formulation, which concentrates on how to find quantities at later times in terms of their known values at earlier times, and the Lagrangian one, which derives the same information from variational principles. These principles state that an object's trajectory in phase space extremizes a certain quantity called the action, which is the integral over the whole phase space history of a functional called a Lagrangian. The dimension of the phase space is usually finite, but can be enlarged beyond the usual 3 + 3 dimensions for the location and velocity of a point particle to allow for rotations of a rigid body or changes of shape of a deformable body. Relativity can be accommodated, but the ideas of quantum mechanics do not appear.

The move to an infinite number of phase space dimensions takes one from classical mechanics to classical field theory. Fluid mechanics, for example, allows for an infinite variety of shapes that a fluid can take making the phase space infinite. Sums that arose in classical mechanics are now integrals, and the general problems of analysis become much more challenging. Nevertheless, many problems are tractable, and theories based on classical fields with an infinite number of degrees of freedom have had great success. Notable among these is Maxwell's electrodynamics, which describes the electromagnetic field in terms of electric and magnetic fields that can vary in space and time and are present at all points of space and time. The phase space of classical electromagnetism is then infinite dimensional, and this accounts for much of the richness of the physics that it can describe.

Another shift away from classical mechanics is to maintain a finite number of degrees of freedom but to allow that points in phase space cannot be arbitrarily well known. The Heisenberg uncertainty principle placed constraints on the accuracy with which position and momentum could be known simultaneously, and the points of the phase space acquired a certain fuzziness. For example, in place of a point with well-defined position x and momentum p_x in the x-direction, one had something like a fuzzy disk of area roughly equal to the square of the Planck length, and shape determined by whether one tried to accurately determine x or p_x—one could only know one with a consequent sacrifice of information about the other. A system cannot now be thought of as a well-defined classical point moving around in phase space. Rather, information about the system is encoded in a function on the phase space called the wave function Ψ, and the theory allows all that can be known about the system to be recovered by applying various linear operators to Ψ.

The step to QFT involves allowing an infinite number of degrees of freedom in quantum mechanics. Thus, one might say that QFT is to quantum mechanics what classical field theory is to classical mechanics. The shift is profound and carries a number of important implications.

The first is that QFT takes as its phase space classical fields upon which an uncertainty paralleling that in quantum mechanics has been imposed. For example, the role of x could be taken by a classical field ϕ, and the role of momentum p_x by the rate of change of something (the Lagrangian) with respect to the time derivative of ϕ. This quantity is called the momentum conjugate to ϕ, and it is impossible to know both it and ϕ with arbitrary precision.

Time and space are now just dummy labels (parameters) in the theory and are integrated over, their main use being to enforce the notion of locality in a Lagrangian. This point is often misunderstood as people look for the "locations" of field quanta in physical space. In QFT, one looks at ϕ as a dynamical quantity as opposed to x. An excellent discussion of this point can be found in Schwinger (1970).

It turns out, perhaps somewhat tautologically, that one can often expand fields in Fourier series and identify each of the modes as a "particle" with momentum given by the mode numbers. It is a remarkable fact that experiments to detect fields in nature, when performed with sufficient resolution, always find them occurring in (or at least interacting as) discrete chunks or quanta. In this sense, QFT is a theory of particles, and the degree to which these quanta can be identified as isolated, discrete objects determines how well they can be thought of as particles.

As a field changes in time so too do the various Fourier modes and this is interpreted as saying that particles appear and disappear; in other words, particles can be created and destroyed in QFT. This is something foreign to both quantum and classical mechanics and was needed with the advent of high-energy physics experiments in which particles are routinely created and destroyed.

Two of many excellent texts describing how QFT is used in the calculation of physical processes are those by Peskin & Schroeder (1995) and by Weinberg (1995). The text by Itzykson and Zuber (1980) is also very good. A more philosophically inclined and less calculationally oriented reader will find much to think about in the book by Teller (1995). It is easy to suppose that QFT is intrinsically, or somehow must be, relativistic. In fact, the concepts can be very successfully applied to problems in condensed matter physics in nonrelativistic situations, and the interested reader will find the delightful introductory book by Mattuck (1992) and the more advanced book by Abrikosov et al. (1975) excellent places to start.

Several caveats are in order. First of all, in the relativistic case (the real world), these quanta do not generally admit a localization in the sense of having well-defined classical-like positions. That is, they do not behave like billiard balls. Second, if interactions between them are very strong they may not act as intuition might suggest. For example, while electrons and photons are in many ways like what one might expect particles to be, protons and neutrons seem to be comprised of particles called quarks, which cannot be removed. One says that the interactions between quarks are so strong that they are "confined." Phonons are quantized vibrations in crystals that can be detected in neutron scattering experiments and act in many ways

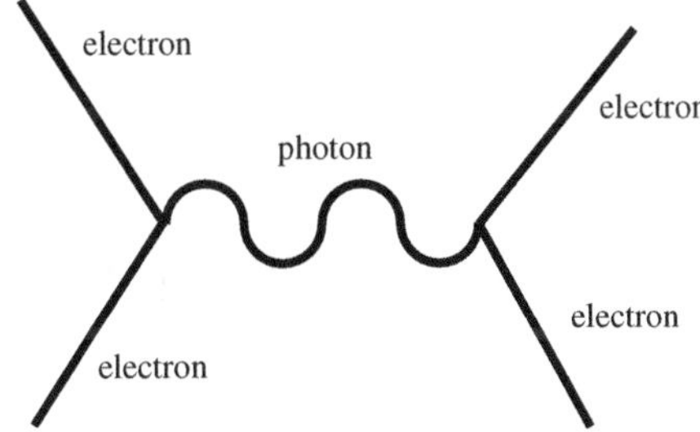

Figure 1. A Feynman diagram representing a first approximation to the scattering of one electron from another (the two straight lines) via the exchange of a photon (the wiggly line). Each diagram like this corresponds to a physically clear picture as well as to a well-defined mathematical expression contributing to the quantum mechanical amplitude for the process to take place. Only the topology of the diagram is important. Bending the lines around represents processes involving electrons and their antiparticles as different aspects of the same basic phenomenon, and connects scattering in space with scattering forward and backward in time, while not violating causality.

like particles, but they are only there insofar as there is a crystal lattice to vibrate. Any attempt to isolate a phonon by pulverizing a crystal and looking for one among the fragments is doomed to failure.

When it makes sense to think of weakly coupled particles, there is a well developed calculational framework called "perturbative quantum field theory" (pQFT), which allows one to make systematic estimates of physical quantities, often using sketches called Feynman diagrams. Figure 1 shows a Feynman diagram contributing to the scatting of one electron by another (the straight lines) via the exchange of a photon (the wiggly line).

The mathematical procedures involved are fraught with difficulties, and the series that arise are often divergent on physical grounds. In addition, the individual terms also tend to be infinite, and several procedures have been developed for handling such problems. The general approach is to first make ill-defined formal expressions finite by changing them ("regularizing" them) in a way controlled by some parameter, then comparing such regularized expressions only to one another and linking only such comparisons to observations ("renormalization"). Despite the feelings of many of the founders that there might be something deeply wrong with these procedures, it turns out that many quantities calculated in this way have given answers accurate to parts per million or better.

For strongly interacting theories the state of the art is much less well-developed. Powerful computers can be used to treat theories that are based on restricting fields to live on a grid or lattice that is meant to approximate space–time, but with a finite number of points. Analytic techniques also offer some hope, but the problems are formidable.

One of the most dramatic consequences of QFT is that the vacuum becomes a very complex object. Recalling the notion of the uncertainly principle applied to fields, one cannot set both a field and its conjugate momentum equal to zero. This, in fact is the case for each possible mode of a field and leads to an infinite energy in the vacuum due to these fluctuating, uncertain fields. As mentioned earlier, infinities like this are basically swept under the rug by insisting that only comparisons are meaningful, but in this case there are remarkable physical consequences.

For example, there are more modes that can exist for the electromagnetic field between two parallel metal plates that are far apart than for ones that are closer together. The infinite energies between the plates in these two cases can be compared and the result is finite—there is less energy between two plates that are close together than two that are far apart. This implies that two parallel metal plates, in order to reduce the energy between them (in empty space), will pull together. This is called the "Casimir effect" and has actually been observed in the laboratory, making it clear that QFT and its strange vacuum are more than theoretical fantasies.

Although it is often not made clear, one final aspect of QFT that differs from quantum mechanics is that QFT admits an infinite number of distinct vacua, all of zero energy. None of these vacua can be reached from another by a unitary transformation, making them physically distinct. This richness of "empty" space is a critical part of what makes QFT able to describe so much. To do justice to this point would require more space than is available, but the interested reader would do well to start with the book by Umezawa (1993). One consequence of this existence of unitarily inequivalent vacua is the amazing fact that the vacuum of one observer can be devoid of particles, while that of another may actually have particles present.

John David Swain

See also **Born–Infeld equations; Higgs boson; Quantum inverse scattering method; Skyrmions; String theory; Yang–Mills theory**

Further Reading

Abrikosov, A.A., Gorkov, L.P. & Dzyaloshinski, I.E. 1975. *Methods of Quantum Field Theory in Statistical Physics*, revised edition, translated and edited by Richard A. Silverman, New York: Dover

Itzykson, C. & Zuber, J.-B. 1980. *Quantum Field Theory*, New York: McGraw-Hill

Mattuck, R.D. 1992. *A Guide to Feynman Diagrams in the Many-Body Problem*, 2nd edition, New York: Dover

Peskin, M.E. & Schroeder, D.V. 1995. *An Introduction to Quantum Field Theory*, Reading, MA: Addison-Wesley

Schwinger, J. 1970. *Particles, Sources, and Fields*, Reading, MA: Addison-Wesley

Teller, P. 1995. *An Interpretative Introduction to Quantum Field Theory*, Princeton, NJ: Princeton University Press

Umezawa, H. 1993. *Advanced Field Theory*, New York: American Institute of Physics Press
Weinberg, S. 1995–2000. *The Quantum Theory of Fields*, 3 vols., Cambridge and New York: Cambridge University Press

QUANTUM INVERSE SCATTERING METHOD

The extension of the inverse scattering method to quantum theory, called the quantum inverse scattering method (QISM), was introduced at the end of the 1970s by the Leningrad (now St Petersburg) group and has been developed extensively since the 1980s. The QISM is significant in many respects. First, similar to the classical inverse scattering method, it provides a powerful method to solve the problem of quantum integrable systems. Second, it has a wide applicability to quantum particle systems, quantum field theoretic models, and quantum spin systems. Importantly, the QISM provides a unified view on exactly solvable models in physics. That is, it places completely integrable models in many-body theory and field theory and solvable models in statistical mechanics in a unified framework. Third, it has been the source of new mathematical objects and concepts such as Yangian algebra, quantum groups, q-deformation, and a revived knot theory. In these mathematical developments, the Yang–Baxter relation plays a central role.

As the simplest example, we begin with the quantum nonlinear Schrödinger (QNLS) model,

$$\mathrm{i}\phi_t + \phi_{xx} - 2\kappa\phi^\dagger\phi\phi = 0. \quad (1)$$

Here, $\phi(x,t)$ is the boson field operator, satisfying the equal-time commutation relation, $[\phi(x,t), \phi^\dagger(y,t)] = \delta(x-y)$. The Hamiltonian of the system is

$$H = \int \mathrm{d}x(\phi_x^\dagger\phi_x + \kappa\phi^\dagger\phi^\dagger\phi\phi). \quad (2)$$

The QNLS model (1) is considered to be the second-quantized theory of boson particles interacting via the pair-wise, delta-function potential. Motivated by the Zakharov–Shabat eigenvalue problem in the classical theory, we associate a linear auxiliary problem to (1),

$$\psi_{1x} + \frac{\mathrm{i}}{2}\lambda\psi_1 = \mathrm{i}\alpha\phi^\dagger\psi_2,$$

$$\psi_{2x} + \frac{\mathrm{i}}{2}\lambda\psi_2 = \mathrm{i}\varepsilon\alpha\psi_1\phi, \quad (3)$$

where $\alpha = |\kappa|^{1/2}$ and ε denotes the sign of κ. We further assume the boundary condition $\phi(x) \to 0$ as $|x| \to \infty$, which is interpreted as the weak relation (a relation for the matrix elements). The Jost function operator $\Psi(x,\lambda)$ is defined by

$$\Psi(x,\lambda) = \begin{bmatrix} 1 \\ 0 \end{bmatrix} \exp\left(-\tfrac{\mathrm{i}}{2}\lambda x\right), \qquad x \to \infty,$$

$$= \begin{bmatrix} A(\lambda)\exp\left(-\frac{\mathrm{i}}{2}\lambda x\right) \\ B(\lambda)\exp\left(\frac{\mathrm{i}}{2}\lambda x\right) \end{bmatrix}, \qquad x \to -\infty. \quad (4)$$

The coefficients $A(\lambda)$ and $B(\lambda)$, corresponding to the transmission coefficient $a(\lambda)$ and the reflection coefficient $b(\lambda)$ in the classical theory, are called scattering data operators. The auxiliary problem (3) with (4) is solved perturbatively to express $A(\lambda)$ and $B(\lambda)$ in terms of the operators ϕ and $\phi^\dagger$ (direct problem). As a result, we obtain

$$[H, A(\lambda)] = 0, \qquad [H, B(\lambda)] = -\lambda^2 B(\lambda). \quad (5)$$

The former relation proves that the QNLS model has an infinite number of conserved operators. The latter indicates that the state

$$|k_1, k_2, \ldots, k_n\rangle = B^\dagger(k_1)B^\dagger(k_2)\ldots B^\dagger(k_n)|0\rangle, \quad (6)$$

where $|0\rangle$ is the vacuum state $\phi(x)|0\rangle = 0$, is an eigenstate of the Hamiltonian. Similar to the classical theory, the quantum Gel'fand–Levitan equation can be derived (inverse problem).

The state $|k_1, k_2, \ldots, k_n\rangle$ in (6) is valid irrespective of the sign of κ and describes the continuous state (Bethe ansatz state). For the attractive case ($\kappa < 0$), there appear bound states. A bound state (n-string state) is made of the complex momenta, $k_j = P/n - \mathrm{i}(n-2j+1)\kappa/2$, $j = 1, 2, \ldots, n$ and the energy is $E = \sum_{j=1}^n k_j^2 = P^2/n - \kappa^2 n(n^2-1)/12$. This bound state corresponds to the classical bright soliton in the $n \to \infty$ limit and may be called a quantum soliton.

To make the mathematical structure clear, it is useful to consider an operator version of an auxiliary linear problem defined on a one-dimensional lattice

$$\psi_{m+1} = L_m(\lambda)\psi_m, \qquad \mathrm{d}\psi_m/\mathrm{d}t = M_m\psi_m, \quad (7)$$

where $L_m(\lambda)$ and M_m are $M \times M$ matrix operators, and λ is the spectral parameter. For a quantum integrable system, it is found that direct products of two L_n operators with different spectral parameters satisfy a similarity relation

$$R(\lambda,\mu)[L_n(\lambda) \otimes L_n(\mu)] = [L_n(\mu) \otimes L_n(\lambda)]R(\lambda,\mu). \quad (8)$$

Here, the symbol $\otimes$ denotes the direct product of matrices, and $R(\lambda,\mu)$, called R-matrix, is an $M^2 \times M^2$ c-number matrix. Relation (8) is the Yang–Baxter relation for a quantum system on a lattice. If the $L_n(\lambda)$ operators on different sites commute, (8) leads to

$$R(\lambda,\mu)[\mathcal{T}_N(\lambda) \otimes \mathcal{T}_N(\mu)] = [\mathcal{T}_N(\mu) \otimes \mathcal{T}_N(\lambda)]R(\lambda,\mu), \quad (9)$$

where $\mathcal{T}_N(\lambda) = L_N(\lambda)L_{N-1}(\lambda)\cdots L_1(\lambda)$ is called the transition matrix or monodromy matrix. From (9), the transfer matrix

$$T_N(\lambda) = \mathrm{Tr}\mathcal{T}_N(\lambda) = \sum_{i=1}^{M}[\mathcal{T}_N(\lambda)]_{ii}, \qquad (10)$$

satisfies the commutation relation

$$[T_N(\lambda), T_N(\mu)] = 0. \qquad (11)$$

Relation (11) indicates the transition matrix $T_N(\lambda)$ is a generator of conserved operators. The λ (or λ^{-1}) expansion of $T_N(\lambda)$ gives a set of conserved operators $\{I_j\}$, which are involutive, $[I_i, I_j]=0$. In addition, commutation relations among elements of the monodromy matrix offer an algebraic formulation of the Bethe ansatz method (algebraic Bethe ansatz method).

For quantum field theoretical models, the subscript N of the operator $T_N(\lambda)$ is understood as the system size that is to be sent to infinity. Then relation (11) proves the existence of an infinite number of involutive conserved operators. It is remarkable that we may consider $L_n(\lambda)$ and $R(\lambda, \mu)$ as vertices (local Boltzmann weights) of a vertex model in statistical mechanics. In this context, the Yang–Baxter relation (8) is a sufficient condition for the commutativity of the transfer matrix in statistical mechanics. Thus, quantum integrable models in $(1+1)$-dimension and solvable statistical models in 2-dimensions have a common property, a family of commuting transfer matrices.

Further, the quantum inverse scattering method and the Yang–Baxter relation have produced interesting developments in mathematics. Solutions of the Yang–Baxter relation with the difference property $R(\lambda, \mu) = R(\lambda - \mu)$ are classified into three classes, (i) elliptic, (ii) trigonometric, and (iii) rational functions. The trigonometric (rational) model is the critical case of the elliptic (trigonometric) model. Quantum groups and knot theory are due to the rich mathematical properties of the exactly solvable models at criticality. Extensive studies on the elliptic case are in progress.

MIKI WADATI

See also **Bethe ansatz; Inverse scattering method or transform; Quantum field theory; Solitons**

Further Reading

Faddeev, L.D. 1981. Quantum completely integrable models in field theory. *Soviet Science Review of Mathematics and Physics,* C1: 107–155

Korepin, V.E., Bogoliubov, N.M. & Izergin, A.G. 1993. *Quantum Inverse Scattering Method and Correlation Functions*, Cambridge and New York: Cambridge University Press

Thacker, H.B. 1981. Exact integrability in quantum field theory and statistical systems. *Reviews of Modern Physics*, 53: 253–285

Wadati, M., Deguchi, T. & Akutsu, Y. 1989. Exactly solvable models and knot theory. *Physics Reports*, 180: 247–332

Wadati, M., Kuniba, A. & Konishi, T. 1985. The quantum nonlinear Schrödinger model: Gel'fand–Levitan equation and classical soliton. *Journal of the Physical Society of Japan*, 54: 1710–1723

QUANTUM NONLINEARITY

Although quantum mechanics is a linear theory, it is often used to analyze classical problems that are nonlinear. In the course of such analyses, it is interesting to consider how the nonlinear features of the classical dynamics are represented in the context of linear quantum theory, as is expected from the correspondence principle of Niels Bohr. In this entry, we see how the correspondence principle asserts itself to describe a variety of interesting nonlinear behaviors, including anharmonic oscillation, local modes, soliton binding energy, and soliton pinning.

A Nonlinear Oscillator

Consider the nonlinear mass-spring oscillator shown in Figure 1, where the spring potential is $V(x) = Kx^2/2 - \alpha x^4/4$. Under quantum theory, the oscillator energy (E) is not a continuous variable; it takes only discrete values that are eigenvalues of the time-independent Schrödinger equation

$$\left(\frac{\hbar^2}{2M}\right)\frac{\mathrm{d}^2\psi}{\mathrm{d}x^2} + [E - V(x)]\psi = 0, \qquad (1)$$

where M is the oscillating mass and $\hbar$ is Planck's constant divided by 2π (Schiff, 1968). If α is zero, the restoring force of the spring is a linear function of its extension, and the classical oscillator of Figure 1 is linear. In this case, the energy eigenvalues are given by

$$E_n = (n + 1/2)\hbar\omega, \qquad (2)$$

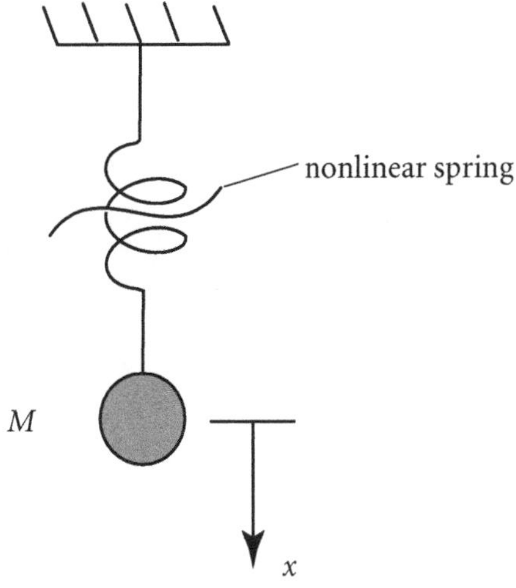

Figure 1. A nonlinear oscillator.

where $n = 0, 1, 2, \ldots$ indicates the number of quanta (called bosons) in the oscillation and $\omega = \sqrt{K/M}$ is the oscillation frequency of the classical oscillator. First derived by Erwin Schrödinger, the corresponding eigenfunctions (ψ_n) in Equation (1) are nth-order Hermite polynomials multiplied by a Gaussian factor—$\exp(-M\omega x^2/2\hbar)$—the first few of which are shown in Figure 2 (Schrödinger, 1926). A restriction of the linear ($\alpha = 0$) case is that transitions are allowed only between adjacent quantum levels. Thus, energy can only enter or leave the oscillator in quantum units of $\hbar\omega$, corresponding to Planck's radiation law.

If $\alpha \neq 0$, the classical oscillator is nonlinear, and two aspects of the quantum picture change. First, the energy eigenvalues are no longer given by the simple expression of Equation (2), and second, energetic transitions are not restricted to jumps between adjacent quantum levels. Both of these new features allow packets of energy to enter and leave the oscillator in amounts that differ from $\hbar\omega$ and, therefore, (according to Planck's law) at frequencies that differ from $\omega = \sqrt{K/M}$, in accord with observations of classical nonlinear oscillators.

In general, Equation (1) must be solved numerically to find its eigenvalues and eigenfunctions in the nonlinear case, but useful results can be obtained if the classical oscillator is considered in the rotating-wave approximation. Under this approximation, the energy eigenvalues are given by Scott (2003)

$$E_n = (\hbar\omega - \gamma/2)(n + 1/2) - \gamma n^2/2, \tag{3}$$

where $\gamma \equiv 3\alpha\hbar^2/4M^2\omega^2$, and the corresponding eigenfunctions are identical to those of the linear oscillator (as in Figure 2). Equation (3) is an example of the Birge–Sponer relation, which is widely observed for nonlinear interatomic oscillations.

To represent the quantum dynamics of an oscillator, the (linear) superposition theorem allows construction of a wave packet of the form

$$\Psi(x, t) = \sum_{n=1}^{\infty} c_n \psi_n(x) e^{-iE_n t/\hbar}, \tag{4}$$

where the c_n are complex constants that determine the initial position of the oscillator, $\Psi(x, t)$ is a solution of the linear, time-dependent Schrödinger equation, and the probability density of finding a particular position (x) is $\int |\Psi(x, t)|^2 \, dx$ (Schiff, 1968).

As the mass and spring constant become larger and the total energy of the oscillator is increased, the energy difference between adjacent levels becomes an ever smaller fraction of the total energy, and the discrete energy levels (the E_n) are better approximated as a continuous function. If the c_n in Equation (4) are chosen so that Ψ is a localized wave packet, the center of mass of this packet will follow the classical nonlinear trajectory. Thus the quantum model—which is necessary for describing interatomic oscillations—merges smoothly into the classical description that we use in our daily lives.

Local Modes in Molecules

Although empirical evidence for localization of vibrational energy in molecules has been available since the 1920s, such observations have been questioned because they seem to be at variance with the predictions of quantum theory. The benzene molecule (C_6H_6), for example, is invariant under a 60° rotation about its main axis; thus eigenfunctions of the (linear) quantum theory must be similarly invariant, but a local mode of CH-stretching vibration is not (see Figure 3).

This seeming contradiction between theory and experiment can be understood by noting that a local mode is represented not by a single eigenfunction but by a wave packet of them. For a local mode of n quanta, it turns out that the energy levels lie within a range of order (Scott, 2003)

$$\Delta E \sim \frac{n\varepsilon^n}{(n-1)!\gamma^{n-1}},$$

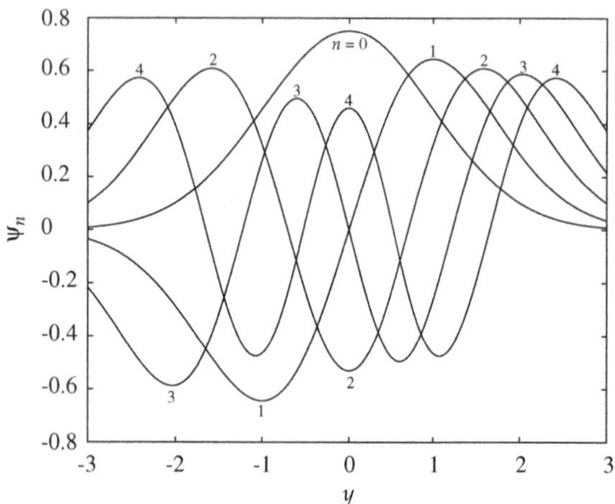

Figure 2. Harmonic oscillator wave functions, where $y = x\sqrt{M\omega/\hbar}$.

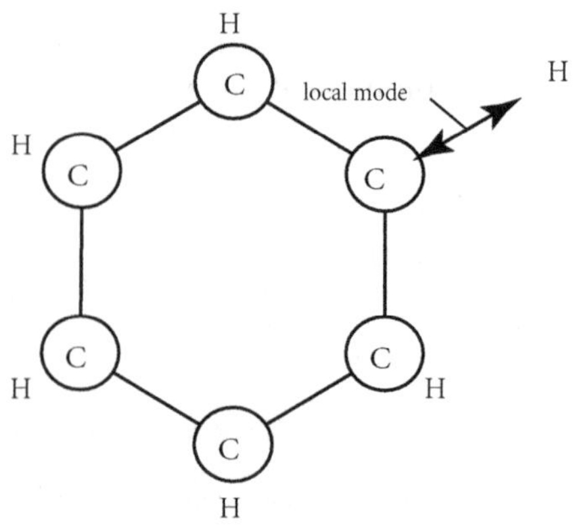

Figure 3. The planar structure of a benzene molecule, showing a local mode of the CH stretching oscillation.

where ε is the coupling energy between CH oscillations and γ is the anharmonicity as defined below Equation (3). For CH-stretching oscillations, $\varepsilon < \gamma$, so this energy range decreases very rapidly with increasing n, making the local-mode wave packet quasidegenerate at moderate quantum levels. This means that the wave packet will remain localized for times of order $\hbar/\Delta E$, which can be much longer than the time required to make an experimental observation.

Quantum Solitons

Consider next a chain (or one-dimensional lattice) of f classical nonlinear oscillators of the form

$$\left(\mathrm{i}\frac{\mathrm{d}}{\mathrm{d}t}-\omega\right)A_j+\varepsilon(A_{j+1}+A_{j-1})+\gamma|A_j|^2A_j=0, \tag{5}$$

where $A_j = A_{j+f}$ is a complex amplitude and j is an index running over periodic boundary conditions. In this formulation, ω is the oscillation frequency at one of the lattice points, ε is the interaction energy between adjacent oscillators, and γ is the anharmonicity of each oscillator in the rotating wave approximation. (With $f = 6$, Equation (5) provides a model for the benzene oscillations of Figure 3.)

This discrete nonlinear Schrödinger (DNLS) equation has solitary wave solutions that approach solitons of the continuum nonlinear Schrödinger (NLS) equation as $\gamma/\varepsilon \to 0$. (Be careful not to confuse the *linear* Schrödinger equation of quantum theory in Equation (1) with Equation (5), which is a *nonlinear, classical* equation. They are quite different.) Because Equation (5) is in the rotating wave approximation, the corresponding quantum problem can be formulated and solved (Scott et al., 1994).

For two quanta ($n = 2$) in the large f limit, energy eigenvalues are arranged as in Figure 4, where each eigenfunction changes by a factor of e^{ik} under translation by one unit of j (lattice spacing); thus k is the "crystal momentum" of an eigenstate, (Haken, 1976). This figure shows both a continuum band (the shaded area) and a soliton band given by

$$\begin{aligned} E_2(k) &= \sqrt{\gamma^2 + 16\varepsilon^2\cos^2(k/2)} \\ &= E_2(0) + k^2/2m^* + \mathrm{O}(k^4), \end{aligned} \tag{6}$$

where m^* is the "effective mass" near the band center. The soliton band is characterized by two features. First, it is displaced below the continuum band by a binding energy $E_\mathrm{b} = \sqrt{\gamma^2 + 16\varepsilon^2} - 4\varepsilon$. Second, inspection of the corresponding eigenfunctions shows that the two quanta are more likely to be on the same site for the soliton band than in the continuum band.

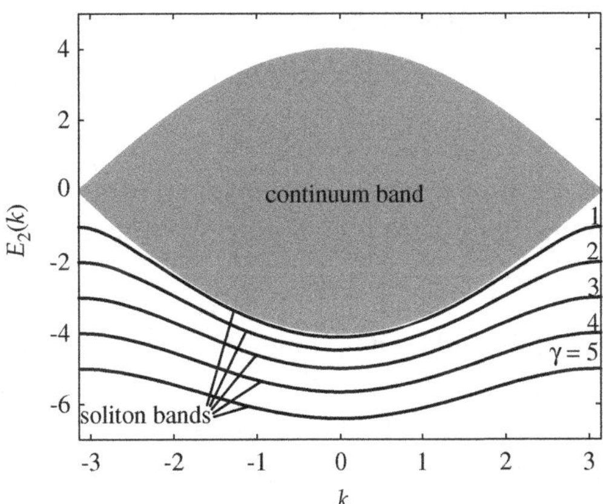

Figure 4. Energy eigenvalues of the DNLS equation at the second ($n = 2$) quantum level with $\varepsilon = 1$. There is one soliton band for each value of γ, which is plotted for several different values of γ.

For arbitrary n, $\gamma \ll \varepsilon$ and sufficiently large f, the quantum binding energy is

$$E_\mathrm{b} = \frac{\gamma^2}{48\varepsilon}n(n^2-1), \tag{7}$$

which corresponds to the binding energy of a classical NLS soliton under the identification $n = \sum |A_j|^2 \gg 1$ (Makhankov, 1990).

In the classical DNLS system with $\gamma \gg \varepsilon$, numerical studies show that the soliton becomes pinned to the lattice. Under quantum theory, this classical, nonlinear phenomenon is reflected by the fact that the effective mass,

$$m^* = \frac{(n-1)!\gamma^{n-1}}{2n\varepsilon^n},$$

becomes very large. Because the classical Ablowitz–Ladik (AL) system is completely integrable for all parameter values, its soliton is not pinned for $\gamma \gg \varepsilon$. This classical fact is reflected by an effective mass that approaches zero under the same conditions. Interestingly, the Salerno equation interpolates between the DNLS and AL limits.

Quantum Representation of a Classical Soliton

Although the quantum formulation can, in principle, be used to represent a classical soliton (on an optical fiber, for example), this is a rather complex dynamical object. The closest quantum representation of a classical soliton is provided by the coherent (or Glauber) state (Makhankov, 1990), which is constructed so that the complex wave amplitude—A_j in Equation (5)—is an eigenfunction of the bosonic lowering operator. This requires a wave packet comprising components with all values of the principal quantum number (n) (Scott, 2003). For a system with f degrees of freedom, a component at the nth level, in turn, has

$$\frac{(n+f-1)!}{(f-1)!\,n!}$$

terms, each of which is a complex construction of the eigenfunctions shown in Figure 2. (These are all of the different ways that n quanta can be placed on f freedoms.) As the number of terms at each level grows very rapidly with n, it becomes difficult to construct an exact quantum representation.

One way out of this computational dilemma is to go to the continuum approximation ($f \to \infty$) and employ the methods of quantum field theory (Lai & Haus, 1989). Another approach is to use a Hartree approximation (HA), which assumes that each boson feels the same mean-field potential as all of the other bosons (Scott, 2003). When this approximation procedure is carried through, it is found that the binding energy at the nth level is

$$E_{\mathrm{b}}^{\mathrm{HA}} = \frac{\gamma^2}{48\varepsilon} n(n-1)^2, \tag{8}$$

which is less than the exact value by the factor $(n-1)/(n+1)$.

ALWYN SCOTT

See also **Discrete nonlinear Schrödinger equations; Hartree approximation; Local modes in molecules; Quantum field theory; Quantum theory; Rotating wave approximation; Salerno equation**

Further Reading

Haken, H. 1976. *Quantum Field Theory of Solids*, Amsterdam: North-Holland

Lai, Y. & Haus, H. 1989. Quantum theory of solitons in optical fibers. I. Time-dependent Hartree approximation. II. Exact solution. *Physical Review* A, 40: 844–866

Makhankov, V.G. 1990. *Soliton Phenomenology*, Dordrecht: Amsterdam

Schiff, L.I. 1968. *Quantum Mechanics*, 3rd edition, New York: McGraw-Hill

Schrödinger, E. 1926. Quantisierung als Eigenwertproblem. *Annalen der Physik*, 79: 361–376

Scott, A.C. 2003. *Nonlinear Science: Emergence and Dynamics of Coherent Structures*, 2nd edition, Oxford and New York: Oxford University Press

Scott, A.C., Eilbeck, J.C. & Gilhøj, H. 1994. Quantum lattice solitons. *Physica* D, 78: 194–213

QUANTUM THEORY

Quantum theory was born in 1900 with an important paper by Max Planck on black-body radiation. In this paper, he introduced quanta of vibrational energy E which are proportional to their frequency ν, and his famous constant currently known to take the value (Mohr & Taylor, 2000)

$$\frac{E}{\nu} = h = 6.62606876(52)\ 10^{-34}\ \mathrm{J\,s}.$$

Ten years after its formulation, Planck's hypothesis had been extraordinarily successful, explaining the light spectrum of black-body radiation, the photoelectric effect, and the low-temperature reduction of heat capacity in solids.

Using Rutherford's planetary model of atoms, Niels Bohr showed in 1913 that Planck's hypothesis helps to understand the spectral lines of hydrogen by deriving their frequencies in terms of h, and the electron mass and charge. The success of Bohr's atomic model led to the formulation of the Bohr–Sommerfeld quantization rule for classically integrable systems possessing as many invariants of motion as degrees of freedom:

$$\oint p_j\, \mathrm{d}q_j = h\left(n_j + \frac{\mu_j}{4}\right), \qquad j = 1, 2, \ldots, f,$$

where (q_j, p_j) are the canonically conjugate position-momentum variables in terms of which the system is integrable (also said separable), n_j are integers, μ_j are the so-called Maslov indices that characterize the topology of the motion (e.g., $\mu_j = 0$ for rotation, $\mu_j = 2$ for libration), and f is the number of degrees of freedom.

In 1917, Albert Einstein pointed out that the Bohr–Sommerfeld quantization rule cannot be applied to classically non-integrable systems (such as the helium atom), and it slowly became apparent that radically new ideas were required. In 1923, Louis de Broglie suggested that massive particles should behave as waves and he completed Planck's hypothesis by his famous relation, $p\lambda = h$, between the particle's momentum p and its wavelength λ.

Finally in 1925 and 1926, Werner Heisenberg and Erwin Schrödinger established quantum mechanics in two equivalent formulations. The first represents the observable quantities as matrices and the second is based on the famous Schrödinger equation

$$\mathrm{i}\,\hbar\,\frac{\partial\psi}{\partial t} = -\frac{\hbar^2}{2m}\frac{\partial^2\psi}{\partial \boldsymbol{r}^2} + V(\boldsymbol{r}, t)\,\psi, \tag{1}$$

for the wave function $\psi(\boldsymbol{r}, t)$ of a particle of mass m moving in the energy potential $V(\boldsymbol{r}, t)$, with $\hbar = h/2\pi$ and $\mathrm{i} = \sqrt{-1}$. In 1927, Max Born proposed to interpret the square of the modulus of the wave function, $|\psi(\boldsymbol{r}, t)|^2$, as the probability density to observe the particle at position $\boldsymbol{r}$ and time t if the wave function is normalized to unity over the whole configuration space according to

$$\int |\psi(\boldsymbol{r}, t)|^2\, \mathrm{d}\boldsymbol{r} = 1. \tag{2}$$

In this formulation, physical observables are represented by linear Hermitian operators $\hat{A}$ acting on the wave function. The expectation value of an observable is given by

$$\langle A \rangle = \int \psi^*(\boldsymbol{r}, t)\ \hat{A}\psi(\boldsymbol{r}, t)\ \mathrm{d}\boldsymbol{r},$$

if the particle is in the state described by the normalized wave function ψ. The wave property prevents a quantum particle from having simultaneously well-defined position and momentum as in the classical world. This impossibility (which follows from Fourier transform theory) is expressed by Heisenberg's uncertainty relation between the uncertainties on position Δx and momentum Δp

$$\Delta x \; \Delta p \geq \hbar/2,$$

which finds its origin in the noncommutativity of position and momentum operators.

The normalization condition (2) has the pivotal role of selecting the physically acceptable wave functions among all the possible solutions of the Schrödinger equation (1). In particular, it is the normalization condition that leads to the quantization of energy into well-defined eigenvalues associated with the stationary states. Without the normalization condition, the wave function would present spatial instabilities that are not physically meaningful.

Because of their spatial extension, wave functions are allowed to penetrate into classically forbidden regions where the classical kinetic energy of the particle would be negative. In these regions, the normalization condition (2) forces the wave function to decrease exponentially and precludes its instability. A consequence is the phenomenon of quantum tunneling, which manifests itself in cold electronic emission or α radioactivity and finds technological applications in Leo Esaki's semiconductor tunneling diode, Ivar Giaever's superconducting tunneling diode, and the electron tunneling microscope.

Together with the normalization condition, the Schrödinger equation shares with nonlinear systems the general scheme:

$$\text{instability} \rightarrow \text{saturation} \rightarrow \text{structure}.$$

In quantum mechanics, the instability mechanism is spatial and the saturation is provided by the normalization condition that selects spatially stable wave functions such as the electronic orbitals of atoms, molecules, and solids. The selection of normalizable wave functions generates the molecular structures of stereochemistry.

However, Schrödinger's equation is linear and thus obeys the principle of linear superposition (Dirac, 1930). Accordingly, the linear combination

$$\psi = \sum_n c_n \psi_n$$

(with complex numbers c_n) is a physically acceptable solution of Equation (1). Following studies by Steven Weinberg and others (Weinberg, 1989), extremely stringent limits have been put on hypothetical nonlinear corrections to quantum mechanics. These limits have been obtained by searching for nonlinearly induced detuning of resonant transitions between two atomic levels, putting the following upper bounds on a hypothetical nonlinear term $\Delta E(\psi, \psi^*)\psi$ supposed to correct the right-hand side of Schrödinger Equation (1):

$$|\Delta E| < 2.4\ 10^{-20}\,\text{eV in } {}^{9}\text{Be}^{+} \text{ ion}$$
(Bollinger et al., 1989),
$$|\Delta E| < 1.6\ 10^{-20}\,\text{eV in } {}^{201}\text{Hg atom}$$
(Majumder et al., 1990).

The principle of linear superposition of quantum mechanics has thus been confirmed by these investigations.

Nevertheless, in low-temperature many-body quantum systems, effective nonlinearities may arise if some wave function describing a subset of degrees of freedom has a feedback effect onto itself. Examples of effectively nonlinear quantum equations include the Hartree–Fock equation for fermionic systems, the Ginzburg–Landau equation for superconductors, the Gross–Pitaevskii equation for Bose–Einstein condensates, and the Ginzburg–Landau equation coupled to a Chern–Simon gauge field for the fractional quantum Hall effect.

Another problem where nonlinearities are associated with Schrödinger's equation is the theory of the optimal control of quantum systems by an external electromagnetic field. The optimal external field can be obtained as the solution of coupled nonlinear Schrödinger equations, where nonlinearity arises by a feedback mechanism of the quantum system onto itself through the external field and the desired control. Such feedback mechanisms have been experimentally implemented for laser control of chemical reactions (Rice & Zhao, 2000).

Nonlinear effects also emerge out of wave mechanics in the semiclassical limit, where the motion can be described in terms of classical orbits (solutions of nonlinear Hamiltonian equations). In the 1970s, starting from Schrödinger's equation, Gutzwiller derived a semiclassical trace formula that expresses the density of energy eigenvalues in terms of periodic orbits (Gutzwiller, 1990). The periodic orbits are unstable and proliferate exponentially in chaotic systems, where semiclassical quantization can be performed, thanks to the Gutzwiller trace formula as an alternative to the Bohr–Sommerfeld quantization rule.

In summary, nonlinear effects manifest themselves in quantum systems as phenomena emerging out of the linear wave mechanics in particular limits such as the semiclassical limit or the many-body limit at low temperature.

Pierre Gaspard

See also **Bose–Einstein condensation; Hartree approximation; Nonlinear Schrödinger equations; Quantum chaos; Superconductivity; Superfluidity**

Further Reading

Bollinger, J.J., Heinzen, D.J., Itano, W.M., Gilbert, S.L., & Wineland, D.J. 1989. Test of the linearity of quantum mechanics by rf spectroscopy of the $^{9}Be^{+}$ ground state. *Physical Review Letters*, 63: 1031–1034

Dirac, P.A.M. 1930. *Principles of Quantum Mechanics*, Oxford: Clarendon Press

Gutzwiller, M.C. 1990. *Chaos in Classical and Quantum Mechanics*, New York: Springer

Majumder, P.K., Venema, B.J., Lamoreaux, S.K., Heckel, B.R. & Fortson, E.N. 1990. Test of the linearity of quantum mechanics in optically pumped ^{201}Hg. *Physical Review Letters*, 65: 2931–2934

Mohr, P.J. & Taylor, B.N. 2000. CODATA recommended values of the fundamental physical constants: 1998. *Reviews of Modern Physics*, 72: 351–495

Rice, S.A. & Zhao, M. 2000. *Optical Control of Molecular Dynamics*, New York: Wiley

Weinberg, S. 1989. Precision tests of quantum mechanics. *Physical Review Letters*, 62: 485–488; and references therein

QUASILINEAR ANALYSIS

Analytical methods in the present-day theory of nonlinear oscillations originated in the investigations by Henri Poincaré, George D. Birkhoff, and Aleksandr Lyapunov, who laid the mathematical foundations for this theory. However, it should be noted that direct application of these mathematical methods to oscillation theory as such did not occur till much later, primarily owing to the work of Alexander Andronov (Andronov et al., 1966).

An important contribution to the development of the quantitative theory of nonlinear oscillations, especially of the applied part, was made by Balthasar van der Pol (van der Pol, 1934), who studied the operation of an electronic generator and proposed his own investigative method, namely, the method of slowly time-varying amplitudes. A rigorous justification of this method was later given by Osip Mandelshtam and N. Papaleksi (Mandelshtam & Papaleksi, 1934).

Almost independently of Mandelshtam, Andronov, and other physicists, the mathematical groundwork for nonlinear oscillation theory was laid by Nikolai Krylov, Nikolai Bogolyubov, Yuri Mitropol'sky (Krylov & Bogolyubov, 1947; Bogolyubov, 1950; Bogolyubov & Mitropol'sky, 1961; Mitropol'sky, 1971), and their disciples. They worked out the most important methods for the analysis of quasilinear oscillations: the asymptotic method, the averaging method, and the method of equivalent linearization. The last can serve as a theoretical justification of the heuristic methods of harmonic balance and statistical linearization (Landa, 1980; Pervozvansky, 1962), which are well known in mechanical and electrical engineering. Indeed, in accordance with the method of equivalent linearization, for a nonlinear function $f(x)$, we substitute a linear function λx, where λ is determined from the condition of minimization of the mean-square error $\varepsilon = \overline{\left(f(x) - \lambda x\right)^2}$. Differentiating ε with respect to λ and equating the derivative to zero, we find

$$\lambda = \frac{\overline{f(x)x}}{\overline{x^2}}, \tag{1}$$

where over-line denote the averaging operation. Depending on the averaging technique, we obtain harmonic linearization (harmonic balance) or statistical linearization. The former takes place if we put $x = \cos \omega t$ and average the numerator and denominator of expression (1) over t for the period $T = 2\pi/\omega$. The latter takes place if we suppose x to be distributed according to a certain probability distribution (for example, Gaussian) and find the statistical average.

The van der Pol method, or the method of slowly time-varying amplitudes, is applicable to near-linear, near-conservative self-oscillatory systems. The method was suggested by van der Pol as applied to self-oscillatory systems with one degree of freedom. However, the method may be easily generalized to self-oscillatory systems with n degrees of freedom, which are described by equations of the form

$$\ddot{y}_k + \omega_k^2 y_k = \mu Y_k(t, y, \dot{y}) \quad (k = 1, 2, \ldots, n), \tag{2}$$

where μ is a small parameter, $\boldsymbol{y} = \{y_1, \ldots, y_n\}$, $\dot{\boldsymbol{y}} = \{\dot{y}_1, \ldots, \dot{y}_n\}$. For $\mu = 0$, Equations (2) are the equations of a linear oscillatory system with n degrees of freedom written in terms of normal coordinates.

As mentioned above, a rigorous justification of the van der Pol method was given by Mandelshtam and Papaleksi (Mandelshtam & Papaleksi, 1934). They showed that the truncated van der Pol equations can be found by averaging the initial equations over "fast time" for the period $T = 2\pi/\omega$. By doing so, Mandelshtam and Papaleksi pioneered the averaging method, the rigorous theory of which was worked out by Bogolyubov (1950) and developed by Mitropol'sky (1971). This theory concerns so-called equations in a standard form. By means of a certain change of variables, any equations describing oscillations in near-conservative systems can be reduced to equations in a standard form. More general theory is related to systems incorporating fast and slow variables (Volosov & Morgunov, 1971; Vasilyeva & Butuzov, 1990). In particular, this theory is used for analysis of relaxation oscillators (Mischenko & Rozov, 1975).

The asymptotic Krylov–Bogolyubov method is conceptually a generalization of the van der Pol method that allows us to calculate the higher approximations. The method has two modifications, depending on the form of the original equations and on the problem in question (Landa, 1980). If we are interested in the calculation of multi-frequency oscillations, we conveniently set the equations in the form of (2).

Provided we are interested in the calculation of single-frequency oscillations, we can set the original equations in the following form:

$$\dot{\boldsymbol{y}} + \mathcal{B}\boldsymbol{y} = \mu \boldsymbol{Y}(t, \boldsymbol{y}), \tag{3}$$

where $\boldsymbol{y}$ is a vector with n components, $\mathcal{B}$ is a square matrix with elements b_{jk}, μ is a small parameter, and $\boldsymbol{Y}(t, \boldsymbol{y})$ is a nonlinear vector function of time and of all components of the vector $\boldsymbol{y}$. We assume that for $\mu = 0$, Equations (3) describe a linear conservative system.

First, we consider the case when the original equations of a system are set in the form of (2). For $\mu \neq 0$, let us represent a solution of Equations (2) as a power series in μ:

$$\begin{aligned} y_k &= y_k^{(0)} + \mu u_{1k}(A, \psi, \mu t) \\ &\quad + \mu^2 u_{2k}(A, \psi, \mu t) + \cdots, \end{aligned} \tag{4}$$

where $y_k^{(0)} = A_k \cos\psi_k$, $\psi_k = \omega_k t + \varphi_k$, A_k and φ_k are slowly time-varying functions obeying the equations

$$\begin{aligned} \frac{\mathrm{d}A_k}{\mathrm{d}t} &= \mu f_{1k}(A, \varphi, \mu t) + \mu^2 f_{1k}(A, \varphi, \mu t) + \ldots, \\ \frac{\mathrm{d}\varphi_k}{\mathrm{d}t} &= \mu F_{1k}(A, \varphi, \mu t) + \mu^2 F_{1k}(A, \varphi, \mu t) + \ldots. \end{aligned} \tag{5}$$

Here $u_{1k}(A, \psi, \mu t)$, $u_{2k}(A, \psi, \mu t), \ldots$ and $f_{1k}(A, \psi, \mu t)$, $f_{2k}(A, \psi, \mu t), \ldots$, $F_{1k}(A, \psi, \mu t)$, $F_{2k}(A, \psi, \mu t), \ldots$ are unknown functions, which should be found.

Demanding the absence of resonant constituents in functions u_{1k}, we find the unknown functions f_{1k} and F_{1k}:

$$\begin{aligned} f_{1k} &= -\frac{1}{2\omega_k} X_{1k}(\mu t, A, \varphi), \\ F_{1k} &= -\frac{1}{2\omega_k A_k} Z_{1k}(\mu t, A, \varphi). \end{aligned} \tag{6}$$

Thus, we obtain the equations of the first approximation for the amplitudes and phases:

$$\begin{aligned} \frac{\mathrm{d}A_k}{\mathrm{d}t} &= -\frac{\mu}{2\omega_k} X_{1k}(\mu t, A, \varphi), \\ \frac{\mathrm{d}\varphi_k}{\mathrm{d}t} &= -\frac{\mu}{2\omega_k A_k} Z_{1k}(\mu t, A, \varphi). \end{aligned} \tag{7}$$

These equations coincide with those found by using the van der Pol method. The functions u_{1k} describe the higher harmonics and combination frequencies in the solution of the first approximation.

Using the next terms in the expansions found above, we can obtain the equations of the second and higher approximations.

Let us consider further the case when the original equations of a system are set in the form of (3) and we are interested in finding single-frequency oscillations described by

$$\boldsymbol{y} = A\left(\boldsymbol{V}\mathrm{e}^{\mathrm{i}(\omega t + \varphi)} + \text{c.c.}\right), \tag{8}$$

where ω is one of the system fundamental frequencies, $\boldsymbol{V}$ is the eigenvector of the matrix $\mathcal{B}$ corresponding to the frequency ω, and c.c. means the complex conjugate value.

Using the procedure of the method described, we obtain the equations for the unknown vector functions $\boldsymbol{k}_j$:

$$\begin{aligned} \omega\frac{\partial \boldsymbol{k}_1}{\partial\psi} + \mathcal{B}\boldsymbol{k}_1 &= -f_1\left(\boldsymbol{V}\mathrm{e}^{\mathrm{i}\psi} + \text{c.c.}\right) \\ &\quad -\mathrm{i}AF_1\left(\boldsymbol{V}\mathrm{e}^{\mathrm{i}\psi} - \text{c.c.}\right) + \boldsymbol{Y}_1, \ldots \end{aligned} \tag{9}$$

Let us further expand the vector functions $\boldsymbol{k}_1, \boldsymbol{k}_2, \ldots$, $\boldsymbol{Y}_1, \boldsymbol{Y}_2, \ldots$ into the Fourier series with slowly time-varying coefficients:

$$\begin{aligned} \boldsymbol{k}_j(A, \psi, \mu t) &= \sum_{k=-\infty}^{\infty} \boldsymbol{U}_j^{(k)}(A, \varphi, \mu t)\mathrm{e}^{\mathrm{i}k\psi}, \\ \boldsymbol{Y}_j(A, \psi, \mu t) &= \sum_{k=-\infty}^{\infty} \boldsymbol{Y}_j^{(k)}(A, \varphi, \mu t)\mathrm{e}^{\mathrm{i}k\psi}. \end{aligned} \tag{10}$$

Substituting (10) into (9) and equating the coefficients of the same harmonics, we obtain, for each j, a system of non-uniform equations for the components of the vector functions $\boldsymbol{U}_j^{(k)}$. For $k \neq 1$, the determinant of this system is nonzero, and, hence, all of $\boldsymbol{U}_j^{(k\neq 1)}$ can be determined uniquely. For $k = 1$, the system determinant is zero. In this case we should require, for all j, the fulfillment of the compatibility conditions $\mathcal{A}\boldsymbol{R}_j = 0$, where $\boldsymbol{R}_j$ is the right-hand side of the jth system and $\mathcal{A}$ is the adjoint matrix. These conditions allow us to find $f_1(A, \varphi, \mu t)$, $F_1(A, \varphi, \mu t)$, $f_2(A, \varphi, \mu t)$, $F_2(A, \varphi, \mu t), \ldots$.

Let us consider the first approximation. From (9) and (10) we obtain the following equation for the vector function $\boldsymbol{U}_1^{(k)}$:

$$(\mathrm{i}k\omega\mathcal{E} + \mathcal{B})\boldsymbol{U}_1^{(k)} = -(f_1 + \mathrm{i}AF_1)\delta_{k1}\boldsymbol{V} + \boldsymbol{Y}_1^{(k)}, \tag{11}$$

where $\mathcal{E}$ is an identity matrix and δ_{k1} is the Kronecker delta. For $k = 1$, the compatibility condition of system (11) is

$$-f_1\mathcal{A}\boldsymbol{V} - \mathrm{i}AF_1\mathcal{A}\boldsymbol{V} + \mathcal{A}\boldsymbol{Y}_1^{(1)} = 0. \tag{12}$$

Splitting the real part and the imaginary part in Equation (12) we find

$$\begin{aligned} f_1(A, \varphi, \mu t) &= \operatorname{Re}\left\{\frac{1}{\operatorname{Sp}\mathcal{A}}\sum_j \frac{A_{jj}}{V_j} Y_{1j}^{(1)}(A, \varphi, \mu t)\right\}, \\ F_1(A, \varphi, \mu t) &= \frac{1}{A}\operatorname{Im}\left\{\frac{1}{\operatorname{Sp}\mathcal{A}}\sum_j \frac{A_{jj}}{V_j} Y_{1j}^{(1)}(A, \varphi, \mu t)\right\}, \end{aligned} \tag{13}$$

where $\operatorname{Sp}\mathcal{A}$ is the spur of the matrix $\mathcal{A}$.

The second and higher approximations can be found in much the same way.

POLINA LANDA

See also **Averaging methods; Distributed oscillators; Linearization; Perturbation theory; Relaxation oscillators**

Further Reading

Andronov, A.A., Vitt, A.A. & Khaykin, S.E. 1966. *Theory of Oscillations*, Oxford and New York: Pergamon Press (original Russian edition 1959)
Bogolyubov, N.N. 1950. *Teoriya Vozmuscheniy v Nelineynoy Mechanike [Perturbation Theory in Nonlinear Mechanics]*. Sbornik Instituta Stroitel'noy Mekhaniki AN USSR, No 14, 9–34
Bogolyubov, N.N. & Mitropol'sky, Yu.A. 1961. *Asymptotic Methods in the Theory of Nonlinear Oscillations*, New York: Gordon and Breach
Krylov, N.M. & Bogolyubov, N.N. 1947. *Introduction to Nonlinear Mechanics*, Princeton, NJ: Princeton University Press (original Russian edition 1937)
Landa, P.S. 1980. *Avtokolebaniya in Systemakh s Konechnym Chislom Stepeney Svobody [Self-Oscillations in Systems with a Finite Number of Degree of Freedom]*, Moscow: Science
Mandelshtam, L.I. & Papaleksi, N.D. 1934. Obosnovanie metoda priblizhennogo resheniya differentsial'nykh uravneniy [On justification of a method of approximate solving differential equations]. *ZhETF* 4: 117–121
Mischenko, E.F. & Rozov, N.Kh. 1975. *Differentsial'nye Uravneniya s Malym Parametrom i Relaksatsionnye Kolebaniya [Differential Equations with a Small Parameter and Relaxation Oscillations]*, Moscow: Science
Mitropol'sky, Yu.A. 1971. *Metod Usredneniya v Nelineynoy Mechanike [The Averaging Method in Nonlinear Mechanics]*, Kiev: Naukova Dumka
Pervozvansky, A.A. 1962. *Sluchaynye Processy v Nelineynykh Upravlyaemykh Sistemakh [Random Processes in Nonlinear Control Systems]*, Moscow: Fizmatgiz
van der Pol, B. 1934. The nonlinear theory of electric oscillations. *Proceedings of IRE*, 22:1051–1086
Vasilyeva, A.B. & Butuzov, V.F. 1990. *Asimptoticheskie Metody v Teorii Singulyarnykh Vozmuscheniy [Asymptotic Methods in the Theory of Singular Perturbations*, Moscow: High School
Volosov, V.M. & Morgunov, B.I. 1971. *Metod Usredneniya v Teorii Nelineynykh Kolebatel'nykh Sistem [The Averaging Method in the Theory of Nonlinear Oscillatory Systems]*, Moscow: MSU Publ

QUASIPERIODICITY

The term *quasi-periodicity* implies a type of motion that is regular (nonchaotic) but consists of a combination of periodic motions with a trajectory that—after sufficient time—passes arbitrarily close to an earlier value.

For flow systems, a trajectory $x(t) \in \mathbf{R}^n$ is called k-quasi-periodic if it can be written in the form

$$x(t) = f(\omega_1 t, \ldots, \omega_k t).$$

Here f is a smooth nonlinear function of period 2π in each of its k arguments separately. The function f belongs to a class of almost periodic functions

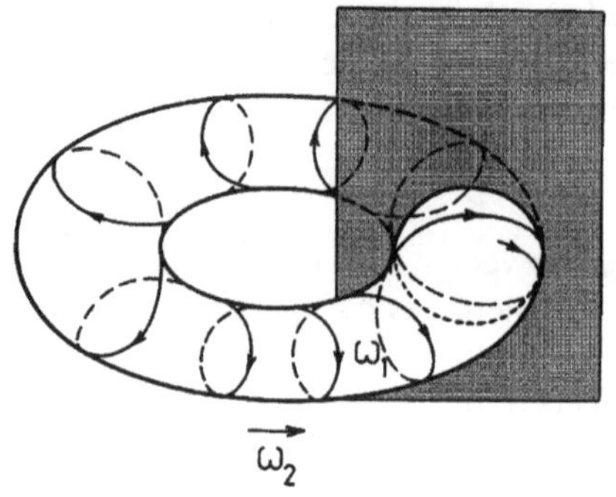

Figure 1. Schematic diagram of motion on the T^2 (two-dimensional) torus and the Poincaré cross section.

if frequencies $\omega_1, \ldots, \omega_k$ are not rationally related (are incommensurate); that is, $\sum_1^k l_i\omega_i \neq 0$ when $l_1, \ldots, l_k \in \mathbf{Z}$ and $\sum_1^k |l_i| > 0$ (Levitan & Zhikov, 1982). In other words, the ratio between two frequencies ω_i and ω_j, $i \neq j$ is irrational:

$$\frac{\omega_i}{\omega_j} \neq \frac{p}{q}, \quad p \in \mathbf{Z}, \quad q \in \mathbf{Z}. \tag{1}$$

A quasi-periodic trajectory lies on a k-dimensional torus (T^k), which can be an attractor, a repeller, or a saddle of a dynamic system. The term *quasi-periodicity* describes motion on the 2-dimensional torus (T^2) whereas *k-quasi-periodicity* is used for motion on tori of larger dimension. A two-dimensional closed surface in phase space corresponds to the two-dimensional torus T^2, and asymptotically (as $t \to \infty$), the trajectory covers its entire surface.

To describe quasi-periodic motion, it is convenient to use a Poincaré cross section (Figure 1) or a map. The T^2 torus is given by a closed invariant curve in Poincaré cross section. Maps are widely used in practice to investigate the properties of quasi-periodic motion. A typical map that serves for this purpose, as well as for investigation of the transition from quasi-periodicity to chaos, is the circle map:

$$\theta_{n+1} = \theta_n + \Omega - \frac{K}{2\pi}\sin 2\pi\theta_n, \quad \bmod 1. \tag{2}$$

Here θ is a phase angle defined in the interval [0; 1], while $\Omega \in [0; 1]$ and $K \geq 0$ are parameters of the map. Map (2) for $K < 1$ describes the dynamics of a non-autonomous self-oscillating system, for example, the van der Pol oscillator under external periodic forcing. The variable θ in this case represents the phase difference between the self-oscillations and the external force, the parameter Ω corresponds to the mismatch between the forcing frequency and the natural frequency of the oscillator, and the parameter K corresponds to the forcing amplitude.

A number of problems arise during numerical simulations and experimental investigations of attractors in quasi-periodic systems. First, it is impossible to simulate an irrational ratio in Equation (1) because each number stored in a computer is limited to a finite set of digits. Second, as a consequence of the measurement error in experimental investigations, it is, in practice, impossible to determinate reliably whether a frequency ratio is rational or irrational. Third, because the presence of noise in real systems leads to a smearing of the system's characteristics, it is possible to interpret experimental results wrongly. Thus, if a calculated invariant curve in a Poincaré cross section is smeared (has a finite width), then a noisy quasi-periodic motion can become entangled with chaos.

To obtain reliable information about the type of motion, it is necessary to use a set of characteristics such as the spectrum of Lyapunov exponents (SLE), a power spectrum, an autocorrelation function (ACF), or the smoothness of the invariant curve in the Poincaré cross section. In quasi-periodic motion, the SLE has no positive values, and the number of zeros is equal to the number of incommensurate base frequencies defined by the quasi-periodic motion. For example, the T^2 torus of a dissipative system is characterized by two Lyapunov exponents that are zero and others that are negative. The SLE of chaotic behavior includes at least one positive value. The power spectrum of quasi-periodic motion is discrete and consists of peaks at frequencies $n\omega_1 \pm m\omega_2$, with n and m taking arbitrary integer values. In reality, the peaks need not be mathematically sharp, but can be instrumentally sharp. The spectrum of a chaotic motion, on the other hand, is continuous and consists of an infinite number of base functions and their combinations. The ACF of a trajectory on a torus is an infinitely oscillating function that does not fall to zero as $t \to \infty$; in contrast, the ACF of a chaotic trajectory tends to zero as $t \to \infty$. In Poincaré cross section, a smooth curve corresponds to a torus, whereas a chaotic regime is characterized by a fractal structure.

Quasi-periodic motion can be observed in conservative and dissipative systems characterized by two or more incommensurate frequencies generated by the system and/or external sources (Glazier & Libchaber, 1988). The transition from a stationary state to quasi-periodic motion is realized as the result of at least two consecutive Hopf bifurcations, which add two incommensurate frequencies to the system, hence giving rise to a T^2 torus. Further Hopf bifurcations lead to a torus of larger dimension. Landau (1965) conjectured that turbulence arises through an infinite series of such Hopf bifurcations, increasing the degree of quasi-periodicity step-by-step up to infinity.

The possibility of such a scenario is problematic because the Kolmogorov–Arnol'd–Moser (KAM) theorem and the results of Newhouse et al. (1979) show that T^k tori can be structurally unstable when $k \geq 3$. This instability is found to increase with nonlinearity, which can lead to synchronization if an irrational frequency ratio becomes rational, or can transform a low-dimensional torus to chaos. However, there are also models (Ott, 2002) and experiments (Gollub & Benson, 1980), which prove that the tori T^k still exist for $k \geq 3$. The flow of blood through the cardiovascular system seems to be an example of dynamics characterized by a hypertorus, with $k \geq 3$ (Stefanovska & Bračič, 1999).

Several scenarios of transition from quasi-periodicity to chaos are known. The transition from T^2 to a strange chaotic attractor has been thoroughly investigated (Afraimovich & Shilnikov, 1991; Anishchenko et al., 2002). First, the torus loses smoothness, and then stable and saddle resonances (cycles corresponding to a rational frequency ratio) arise on the torus. Nonlocal bifurcations (homoclinic tangencies of manifolds) of these saddles lead to chaos later on.

A similar scenario is also observed for the T^3 torus (Anishchenko et al., 2002). The occurrence of torus doublings can precede the appearance of resonances on the torus: a finite number of bifurcations occur, after which the torus is destroyed as described above. The possible reality of an infinite series of torus doubling bifurcations (the Feigenbaum scenario of the transition to chaos) remains an open question, although renormalization group analysis (Feigenbaum et al., 1982) allows such a possibility theoretically.

Newhouse et al. (1979) have shown that small perturbations of the T^3 torus can lead to a strange chaotic attractor on the torus. Bifurcations, which lead from the torus to chaos, have not yet been studied in detail, but investigations indicate that the transition is connected with the appearance and destruction of resonances on the torus. Furthermore, the transition from T^3 to chaos has been observed only in maps, and the possibility of its realization in flow systems remains unclear.

A third scenario is the transition from quasi-periodicity to a nonstrange chaotic attractor (Arnol'd, 1983). If a chaotic attractor on T^3 covers all the surface of the torus, then the capacity of the attractor is integer and the largest Lyapunov exponent is positive.

Another way in which the transition from quasi-periodicity to chaos can occur is through the appearance of a strange nonchaotic attractor (SNA), which has a fractal structure and no positive values in the SLE (Grebogi et al., 1984). Such an evolution is observed in systems under external quasi-periodic forcing that drives the system with an irrational frequency ratio, which does not depend on parameters and properties of the system itself. The bifurcation mechanisms involved in the transition from torus to SNA, and the subsequent transition from SNA to chaos, are subjects of active investigation.

IGOR A. KHOVANOV, NATALYA A. KHOVANOVA, AND ANETA STEFANOVSKA

See also **Attractors; Bifurcations; Cat map; Kolmogorov–Arnol'd–Moser theorem; Lyapunov exponents; One-dimensional maps; Period doubling;**

Phase space; Recurrence; Routes to chaos; Synchronization

Further Reading

Afraimovich, V.S. & Shilnikov, L.P. 1991. Invariant two-dimensional tori, their breakdown and stochasticity. In *American Mathematical Society Translations*, Series 2, 149: 201–212 (original Russian edition 1983)

Anishchenko, V.S. et al. 2002. *Nonlinear Dynamics of Chaotic and Stochastic Systems: Tutorial and Modern Developments*, Berlin and New York: Springer

Arnol'd, V.I. 1983. *Geometrical methods in the Theory of Ordinary Differential Equations*, New York: Springer (original Russian edition 1978)

Feigenbaum, M.J., Kadanoff, L.P & Shenker, S.J. 1982. Quasiperiodicity in dissipative systems: a renormalization group analysis. *Physica* D, 5: 370–386

Glazier, J.A. & Libchaber A. 1988. Quasi-periodicity and dynamical systems: an experimentalist's view. *IEEE Transactions of Circuits and Systems*, 35: 790–318

Gollub, J.P. & Benson, S.M. 1980. Many routes to turbulent convection. *Journal of Fluid Mechanics*, 100: 449–470

Grebogi, C., Ott, E., Pelikan, S. & Yorke, J.A. 1984. Strange attractors that are non chaotic. *Phisica* D, 13: 261–268

Landau, L.D. 1965. On the problem turbulence. In *Collected Papers*, edited by D. ter Haar, Oxford: Pergamon (original Russian edition 1944)

Levitan, B.M. & Zhikov, V.V. 1982. *Almost Periodic Functions and Differential Equations*, Cambridge: Cambridge University Press (original Russian edition 1978)

Newhouse, S., Ruelle, D. & Takens, F. 1979. Occurence of strange Axiom A attractors near quasi-periodic flows on T^m, $m \geq 3$. *Communications in Mathematical Physics*, 64: 35–40

Ott, E. 2002. *Chaos in Dynamical Systems*, 2nd edition, Cambridge and New York: Cambridge University Press

Stefanovska, A. & Bračič, M. 1999. Physics of the human cardiovascular system. *Contemporary Physics*, 40: 31–55

R

RAMAN SOLITONS

See **Rayleigh and Raman scattering and IR absorption**

RANDOM MATRIX THEORY I: ORIGINS AND PHYSICAL APPLICATIONS

The textbook examples of quantum systems typically have a complete set of quantum numbers. For hydrogen-like atoms in a weak magnetic field, the set of radial, angular, and magnetic quantum numbers, together perhaps with the spin state, uniquely determines the energy levels of the atom. Nuclear physicists were among the first to encounter a situation where this is not possible. Scattering protons and neutrons off nuclei revealed a huge number of resonances that could no longer be completely labelled by their quantum numbers. In view of this fact and the uncertainty in the detailed interaction between the neutrons, Eugene Wigner advanced a statistical approach to the observed spectra: instead of trying to characterize each resonance individually, he proposed to look for a statistical description of their distribution and their strengths (see Wigner (1967) for an account of the early developments). Typical quantities of interest are then the mean density of resonances, probabilities for their spacings, the two- or more point correlation functions, and the distribution of transition strengths. The Hamiltonian was modeled as a random matrix, and by ingenious mathematical techniques a complete characterization of the spectral properties of certain ensembles of random matrices could be achieved. Wigner's John von Neumann lecture (Wigner, 1967), and the books by Mehta (1991) and Porter (1965) survey many of the results and methods. These sources also contain references to work in mathematics before the widespread use of random matrices in physics.

The most important ensembles are the Gaussian ones. Consider ensembles of $N \times N$ matrices H with statistically independent real or complex entries and a distribution function

$$P_N(H_{11}, H_{12}, \ldots, H_{NN}) = \prod_{i,j} p_{ij}(H_{ij}). \tag{1}$$

It is natural to demand invariance of the distribution under transformation of bases, since none of them can be singled out a priori. If the allowed transformations are orthogonal, unitary, or symplectic, the Gaussian orthogonal ensemble (GOE), Gaussian unitary ensemble (GUE), or the Gaussian symplectic ensemble (GSE) is obtained, respectively. The distribution functions are

$$P_{N,\beta} = \left(\frac{\beta}{2\pi}\right)^{N/2} \left(\frac{\beta}{\pi}\right)^{\beta N(N-1)/2} \exp\left(-\frac{\beta}{2}\mathrm{tr}\, H^2\right) \tag{2}$$

with the parameters $\beta = 1$ for GOE, $\beta = 2$ for GUE, and $\beta = 4$ for GSE.

Of the many quantities that can be studied and for which in many cases exact results can be obtained, we list only two: the level spacing distribution and the number variance (Figure 1). For others, see Brody et al. (1981), Guhr et al. (1998), Haake (2001), Mehta (1967), and Stöckmann (1999).

The level spacing distribution is the probability density to find two neighboring levels a distance s apart (in units of the mean level spacing). The expression that results from the random matrix ensembles is somewhat involved, but a very good approximation was given by Wigner:

$$P_{\mathrm{GOE}} = \frac{\pi}{2} s \mathrm{e}^{-(\pi/4)s^2}, \tag{3}$$

$$P_{\mathrm{GUE}} = \frac{32}{\pi^2} s^2 \mathrm{e}^{-(4/\pi)s^2}, \tag{4}$$

$$P_{\mathrm{GSE}} = \frac{2^{18}}{3^6 \pi^3} s^4 \mathrm{e}^{-(64/9\pi)s^2}. \tag{5}$$

The characteristic feature of these expression is that for small s the distribution increases like s^β, so that small spacings are suppressed.

The number variance measures the deviation between the true and the expected number of eigenvalues in an interval. In units of the mean spacing

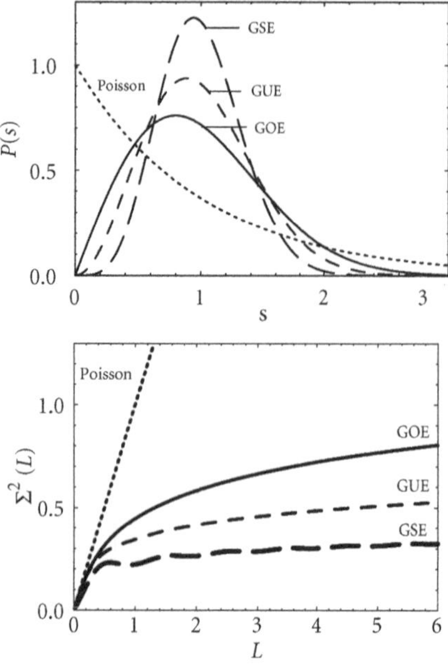

Figure 1. Level spacing distribution and number variance for the different ensembles.

the expected number equals the length L of the interval, and with $N(L)$ the true number, variance is

$$\Sigma^2(L) = \langle (N(L) - L)^2 \rangle = \langle (N(L))^2 \rangle - L^2 . \quad (6)$$

In a Poisson process for uncorrelated levels, the number variance will increase linearly, $\Sigma^2(L) = L$. In the random matrix ensembles, the increase is slower,

$$\Sigma^2(L) \sim \frac{2}{\beta\pi^2} \ln L + c_\beta \quad (7)$$

with constants c_β (for complete expressions, see, for example, Brody et al. (1981)). The slow logarithmic increase reflects the interactions between levels.

Between about 1975 and the mid-1980s, Michael Berry, Oriol Bohigas, and others applied such statistical measures to systems with a chaotic classical limit and observed level distributions and correlators in agreement with random matrix theory. They conjectured that chaotic systems will show random matrix behavior consistent with the global real symmetric, complex Hermitian, or symplectic symmetry of the Hamiltonian. The eigenvalue distribution for integrable systems, as shown earlier by Berry and Tabor, should show Poissonian statistics (Haake, 2001; Stöckmann, 1999). The argument for the two situations is roughly as follows: if the classical system is integrable, the quantum eigenstates are localized on tori and the energies can be approximated by Bohr–Sommerfeld quantization of the appropriate actions. Different tori give independent sequences of quantum eigenvalues, so that the collection of eigenvalues reflects a random appearance of eigenvalues, without correlations. As a result, the distribution of mean level spacings becomes exponential. In a chaotic system the eigenstates are spread over the connected chaotic components. They thus have the same support, but since they have to be orthogonal, they have to interact. This interaction then leads to level repulsion and the observed suppression of small spacings in $P(s)$. The conjecture is supported by semiclassical arguments for the two-point correlation function (Berry, 1985) and a large body of empirical evidence from experimental data for nuclei, atoms, and molecules and from numerical data for various systems (Stöckmann, 1999). The few exceptions (as, for instance, the arithmetical billiards on surfaces of negative curvature (Bogomolny et al., 1997)) have led to clarifications of the necessary requirements. But so far no proof has emerged, neither for integrable nor for chaotic systems.

The connection between random matrices and the chaotic behavior of the underlying wave behavior indicates why random matrix statistics should also appear in many nonquantum situations. Indeed, acoustic resonances in irregularly shaped containers, electromagnetic resonances in cavities, and mechanical vibrations of plates and other solid blocks all show statistical properties in good agreement with random matrix theory expectations when the short wavelength dynamics of wave fronts is chaotic (Stöckmann, 1999).

The statistical measures developed within random matrix theory have propagated into other fields as well, most notably into the theory of the Riemann zeta function. The Riemann zeta function as defined by

$$\zeta_R(s) = \sum_{n=1}^{\infty} n^{-s} \quad (8)$$

can be analytically continued into the complex plane. It then has a pole at $s=1$ and so-called trivial zeroes at $s = -2, -4, \ldots$. Bernhard Riemann's longstanding conjecture is that all its other zeroes lie along the line $s = \frac{1}{2} + it$. David Hilbert proposed to identify an eigenvalue problem that has these zeroes as eigenvalues. While that system is still elusive, analysis of the statistics suggests that it belongs to the universality class of the Gaussian unitary ensemble (see Berry & Keating (2000) for an account of the fascinating relations between random matrix theory, primes, and the statistics of the Riemann zeroes).

While all previous examples dealt with individual systems having a fixed Hamiltonian, one can also consider disordered systems, where as an additional statistical element, certain variations in the Hamiltonian enter. For instance, in a solid the mean density of impurities can be controlled experimentally, but the detailed positions and their effects on a specific experiment are difficult to fix. It is then possible to again set up random matrix models and to derive measurable quantities that are universal in that they depend on the global symmetry of the system only, see Beenakker (1997), Guhr et al. (1998), and Efetov (1997).

Through the connection to disordered systems, techniques from field theory, Grassmann algebras, and supersymmetry entered random matrix theory (Zirnbauer, 1996). Those tools allow analytical calculations of many quantities both in the extreme quantum limit and in the semiclassical limit. More importantly, they have indicated a link to group theory and homogeneous spaces. The classification into three universality classes (GOE, GUE, and GSE) above has to be extended to a total of 10 cases. The realizations of the other cases require additional symmetries. Three cases can be found by extending the three classical ensembles to Dirac Hamiltonians with a particle-hole symmetry. Examples for the remaining four can be found in certain normal-superconducting systems.

As suggested in the introduction to Guhr et al. (1998), the field of random matrices has opened up a new class of stochastic systems with a wide range of applications in physics and elsewhere and with intriguing mathematical connections. The unifying feature is that stochasticity combined with general symmetries leads to universal laws not based on dynamical principles. Indications are that it can also play a role in the analysis of financial data, in wireless communications, or in extracting correlations in neural signals (see, e.g., Forrester et al., 2003).

BRUNO ECKHARDT

See also **Free probability theory; Periodic orbit theory; Quantum chaos**

Further Reading

Beenakker, C.W.J. 1997. Random-matrix theory of quantum transport. *Reviews of Modern Physics*, 69: 713–808

Berry, M.V. 1985. Semiclassical theory of spectral rigidity. *Proceedings Royal Society of London* A, 400: 229–251

Berry, M.V. & Keating, J.P. 2000. The Riemann zeroes and eigenvalue asymptotics. *SIAM Review*, 41: 236–266

Bogomolny, E.B., Georgeot, B., Giannoni, M.J. & Schmit, C. 1997. Arithmetical chaos. *Physics Reports*, 291: 219–324

Brody, T.A., Floris, T., French, J.B., Mello, P.A., Pandey, A. & Wong, S.S.M. 1981. Random-matrix physics: spectrum and strength fluctuations. *Reviews of Modern Physics*, 53: 385–479

Efetov, K.B. 1997. *Supersymmetry in Disorder and Chaos*, Cambridge and New York: Cambridge University Press

Forrester, P.J., Snaith, N.C. & Verbaarschot, J.J.M. (editors). 2003. Special issue: random matrix theory. *Journal of Physics* A, 36: R1–R10 and 2859–3645

Guhr, T., Müller-Groeling, A. & Weidenmüller, H.A. 1998. Random-matrix theories in quantum physics: common concepts. *Physics Reports*, 299: 190–425

Haake, F. 2001. *Quantum Signatures of Chaos*, 2nd edition, Berlin and New York: Springer

Mehta, M.L. 1967. *Random Matrices and the Statistical Theory of Energy Levels*, New York: Academic Press

Porter, C.E. (editor). 1965. *Statistical Theory of Spectra*, New York: Academic Press

Stöckmann, H.J. 1999. *Quantum Chaos: An Introduction*, Cambridge and New York: Cambridge University Press

Wigner, E.P. 1967. Random matrices in physics. *SIAM Review*, 9: 1–23

Zirnbauer, M.R. 1996. Riemannian symmetric superspaces and their origin in random-matrix theory. *Journal of Mathematical Physics*, 37: 4986–5018

RANDOM MATRIX THEORY II: ALGEBRAIC DEVELOPMENTS

It was hypothesized by Eugene Wigner in the 1950s that the highly excited states of complex nuclei would have the same statistical properties as the eigenvalues of a large random real symmetric matrix. In pure mathematics, one finds a random matrix hypothesis in the theory of the celebrated Riemann hypothesis. Thus the Montgomery–Odlyzko law states that the statistics of the large zeros of the Riemann zeta function on the critical line (Riemann zeros) coincide with the statistics of the eigenvalues of a large complex Hermitian matrix. To test such hypotheses (for definiteness, the Montgomery–Odlyzko law), one computes a large sequence of consecutive Riemann zeros, scales the sequence so that locally the mean spacing is unity, and then empirically computes statistical quantities. A typical example of the latter is the distribution of the spacing between consecutive zeros. This must be compared against the same statistical quantity for the eigenvalues of large random complex Hermitian matrices. How then does one compute the eigenvalue spacing distribution for random matrices?

Hermitian random matrices with real, complex, and quaternion real Gaussian elements form matrix ensembles referred to as the Gaussian orthogonal ensemble (GOE) ($\beta=1$), the Gaussian unitary ensemble (GUE) ($\beta=2$), and the Gaussian symplectic ensemble (GSE) ($\beta=4$), respectively, where β is a convenient label. Consider the bulk eigenvalues of such large matrices, scaled to have unit density. Denote by $p_\beta(k;s)$ the probability density that there are exactly k eigenvalues in between two eigenvalues of spacing s, and denote by $E_\beta(k;s)$ the probability that there are exactly k eigenvalues in an interval of size s. Define the generating functions $p_\beta(s;\xi)=\sum_{k=0}^{\infty}(1-\xi)^k p_\beta(k;s)$ and $E_\beta(s;\xi)=\sum_{k=0}^{\infty}(1-\xi)^k E_\beta(k;s)$. Note from the definitions that these quantities are related by $p_\beta(s;\xi)=\frac{1}{\xi^2}\frac{\mathrm{d}^2}{\mathrm{d}s^2}E_\beta(s;\xi)$.

Gaudin observed that the determinantal form of the correlations in the case $\beta=2$ allows $E_2(s;\xi)$ to be written as a Fredholm determinant,

$$\begin{aligned} E_2(s;\xi) &= \det(1-\xi K_2) \\ &= \det(1-\xi K_2^+)\det(1-\xi K_2^-), \qquad (1) \end{aligned}$$

where K_2, $K_2^\pm$ are integral operators on $(0,s)$ with kernels

$$\frac{\sin\pi(x-y)}{\pi(x-y)},$$

and
$$\frac{1}{2}\left[\frac{\sin\pi(x-y)}{\pi(x-y)} \pm \frac{\sin\pi(x+y)}{\pi(x+y)}\right],$$
respectively. The second equality is noted in (1) because both factors therein are related to E_1. Thus, with $E_1(-1;s)=0$, define

$$E_1^{\pm}(s;\xi) = \sum_{n=0}^{\infty}(1-\xi)^n\Big(E_1(2n;s)+E_1(2n\mp 1;s)\Big).$$

Then an inter-relationship between large GOE and large GUE matrices due to Dyson implies

$$E_2(s;\xi) = E_1^{+}(s;\xi)E_1^{-}(s;\xi).$$

This factorization turns out to be the same as in (1), so one obtains Mehta's result

$$E_1^{\pm}(s;\xi) = \det(1-\xi K_2^{\pm}).$$

For the case $\beta=4$, one uses Mehta's and Dyson's inter-relationship between large GSE matrices and large GOE matrices to conclude

$$E_4(n;s) = E_1(2n;2s)+\tfrac{1}{2}\Big(E_1(2n-1;2s)+E_1(2n+1;2s)\Big).$$

A new line of study of E_β was initiated by Jimbo, Miwa, Môri, and Sato in 1980, which related the Fredholm determinant in (1) to integrable systems theory, resulting in the formula

$$E_2(s;\xi) = \exp\int_0^{\pi s}\sigma(t;\xi)\,\frac{dt}{t},$$

where σ satisfies a particular example of the σ-form of the Painlevé V equation,

$$(s\sigma'')^2 + 4(s\sigma'-\sigma)\Big(s\sigma'-\sigma+(\sigma')^2\Big) = 0,$$
$$\sigma(s;\xi) \underset{s\to 0}{\sim} -\frac{\xi s}{\pi} - \left(\frac{\xi s}{\pi}\right)^2.$$

The quantities $E_1^{\pm}$ can also be expressed in terms of Painlevé transcendents. Thus, combining results of the present author with results of Tracy and Widom, one has

$$E_1^{\pm}(s;\xi) = \exp\int_0^{(\pi s/2)^2} v(t;\xi;a)\,\frac{dt}{t}\bigg|_{a=\pm 1/2},$$

where $v(t;\xi;a)$ satisfies a particular example of the σ-form of the Painlevé III$'$ equation

$$(tv'')^2 - a^2(v')^2 - v'(4v'+1)(v-tv') = 0,$$
$$v(t;;\xi;a) \underset{t\to 0^+}{\sim} -\frac{\xi t^{a+1}}{2^{2a+2}\Gamma(a+1)\Gamma(a+2)}.$$

Another bulk spacing distribution with a Painlevé type evaluation is the nearest-neighbor spacing between

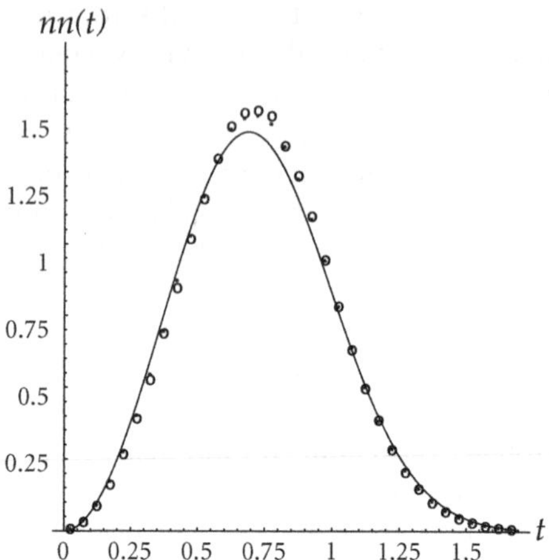

Figure 1. Comparison of $nn(t)$ for the GUE (continuous curve) and for 10^6 consecutive Riemann zeros, starting near zero number 1 (open circles), 10^6 (asterisks), and 10^{20} (filled circles).

eigenvalues (i.e., the minimum of the distances to the left neighbor and the right neighbor) for large GUE matrices, $nn(t)$ say. Forrester and Odlyzko (1996) have shown that

$$nn(t) = -\frac{y(2\pi t;a)}{t}\exp\int_0^{\pi t}\frac{y(2s;a)}{s}\,ds\bigg|_{a=1},$$

where $y(s;a)$ satisfies the differential equation

$$(sy'')^2 + 4(-a^2+sy'-y)\Big((y')^2 - \{a-(a^2-sy'+y)^{1/2}\}^2\Big) = 0$$

subject to the boundary condition

$$y(s;a) \underset{s\to 0^+}{\sim} -\frac{2(s/4)^{2a+1}}{\Gamma(1/2+a)\Gamma(3/2+a)}.$$

Comparison of a plot of this statistic, obtained from the formula in Forrester and Odlyzko (1996), against the same statistic computed, empirically for large sequences of Riemann zeros starting at three different positions along the critical line is given in Figure 1.

P.J. Forrester

See also **Random matrix theory I, III, IV**

Further Reading

Forrester, P.J. Eigenvalue probabilities and Painlevé theory, Chapter 6 of *Log-gases and random matrices*, www.ms.unimelb.edu.au/~matpjf/matpjf.html

Forrester, P.J. & Odlyzko, A.M. 1996. Gaussian unitary ensemble eigenvalues and Riemann ζ function zeros: a non-linear equation for a new statistic, *Physical Review* E, 54: R4493–R4495

Tracy, C.A. & Widom, H. 1993. Introduction to random matrices. In *Geometric and Quantum Aspects of Integrable Systems*, edited by G.F. Helminck, New York: Springer, pp. 407–424

van Moerbeke, P. 2001. Integrable lattices: random matrices and random permutations. In *Random Matrices and Their Applications*, edited by P. Bleher & A. Its, Cambridge and New York: Cambridge University Press, pp. 321–406

RANDOM MATRIX THEORY III: COMBINATORICS

As the two previous entries show, random matrix theory (RMT) is a field of research devoted to the statistical analysis of the eigenvalues of matrices selected at random from a given set of matrices (Mehta, 1991). Recently there have been surprising discoveries that connect RMT to a branch of mathematics called combinatorics.

One fundamental topic within combinatorics is the study of permutations. A permutation is a rearrangement of the elements of a set. The set is usually finite, in which case it may be taken to be the first n positive integers, and a permutation may also be represented as a one-to-one invertible mapping from $\{1, 2, \ldots, n\}$ into itself. For each positive integer n, the number of permutations of length n is $n!$. If permutations of length n are selected at random, and each permutation is equally likely, then the probability of selecting a particular permutation is $1/n!$. In thinking about permutations probabilistically, researchers have discovered new connections to random matrix theory.

Random Matrix Theory: Gaussian Unitary Ensemble

The Gaussian unitary ensemble (GUE) is a fundamental example of a RMT. The collection of matrices is all Hermitian (self-adjoint, complex) matrices of size n. The diagonal matrix entries M_{jj}, $j = 1, 2, \ldots, n$, are independent normal random variables; that is, each has probability measure $\frac{1}{\sqrt{2\pi}} \mathrm{e}^{-m^2/2}\,\mathrm{d}m$. The off-diagonal matrix entries $M_{jk} = M_{jk}^{(\mathrm{R})} + \mathrm{i}M_{jk}^{(\mathrm{I})}$, $1 \le j \ne k \le n$, are complex, and $\left\{M_{jk}^{(\mathrm{R})}, M_{jk}^{(\mathrm{I})}\right\}_{1 \le j < k \le n}$ is a collection of independent identically distributed Gaussian random variables with probability measure $\frac{1}{\sqrt{\pi}}\mathrm{e}^{-m^2}\,\mathrm{d}m$. The matrix entries below the diagonal are determined by the entries above the diagonal because the matrices satisfy the self-adjointness criterion $M^* = M$, where $(M^*)_{jk} = \overline{M_{jk}}$ and $\overline{(\cdot)}$ represents the complex conjugate of $(\cdot)$. Thus, the GUE of random matrices is merely a probability measure defined on the standard Euclidean space of dimension n^2. In standard probability terminology, the probability of finding a matrix in an "infinitesimal volume element" of matrices centered at a given matrix $M = (M_{jk})_{1 \le j,k \le n}$ is

$$\frac{1}{2^n \pi^{n^2}} \mathrm{e}^{-\frac{1}{2}\sum_{j=1}^n M_{jj}^2 - \sum_{j=1}^{n-1}\sum_{k=j+1}^n (M_{jk}^{(\mathrm{R})})^2 + (M_{jk}^{(\mathrm{I})})^2} \times \prod_1^n \mathrm{d}M_{jj} \prod_{1 \le j < k \le n} \mathrm{d}M_{jk}^{\mathrm{R}}\,\mathrm{d}M_{jk}^{\mathrm{I}}.$$

This is often more conveniently written in the form

$$\frac{1}{Z_n}\mathrm{e}^{-\frac{1}{2}\mathrm{Tr}(M^2)}\,\mathrm{d}M, \tag{1}$$

where $Z_n = 2^n \pi^{n^2}$, $\mathrm{d}M = \prod \mathrm{d}M_{jj} \prod \mathrm{d}M_{jk}^{(I)}\,\mathrm{d}M_{jk}^{(\mathrm{R})}$, and $\mathrm{Tr}(A)$ represents the trace of the matrix A, $\mathrm{Tr}(A) = \sum_{j=1}^n A_{jj}$.

This is just one example of an RMT. For example, one may replace $\mathrm{Tr}(M^2)$ in (1) by $\mathrm{Tr}(M^r)$, or $\mathrm{Tr}(V(M))$ for any reasonable function $V : \mathbb{R} \to \mathbb{R}$.

The eigenvalues $\lambda_1 \le \lambda_2 \le \ldots \le \lambda_n$ of a matrix M selected at random according to (1) are fundamental random variables, and quite a lot is known about their statistical properties. One example is the largest eigenvalue λ_N. Because of many different applications, their statistical behavior is particularly interesting when n goes to ∞. This random variable's behavior when $n \to \infty$ is as follows:

$$\lim_{n\to\infty} \mathrm{Prob}\left(\lambda_n < \sqrt{2n} + \frac{s}{\sqrt{2}n^{1/6}}\right) = F(s),$$

where

$$F(s) = \exp\left(-\int_s^\infty (x-s)^2 q(x)\,\mathrm{d}x\right) \tag{2}$$

and q is the unique solution to the Painlevé II equation

$$q'' = sq + 2q^3$$

satisfying the condition

$$q(s) \sim \mathrm{Ai}(s) \quad \text{as } s \to \infty.$$

This result is due to Craig Tracy and Harold Widom (1994), and the function F is often referred to as the Tracy–Widom distribution. Surprisingly, the Painlevé equation appears in a fundamental way in random matrix theory. It also appears, in an equally fundamental way, in combinatorics.

Increasing Subsequences of Permutations

A permutation, or an invertible mapping from the set $\{1, 2, \ldots, n\}$ to itself may be represented as a sequence of numbers. For example, the sequence 3,1,2 represents the permutation $\sigma(1) = 3$, $\sigma(2) = 1$, $\sigma(3) = 2$. An increasing subsequence of a permutation is obtained by representing a permutation as a sequence, and then selecting a subsequence in which the elements are increasing. For example, the sequence 1, 2 is an increasing subsequence of the permutation σ considered above.

Each permutation has many increasing subsequences. In combinatorics the length of the largest increasing subsequence of a permutation is an important object of study. For the permutation σ considered above, that length is 2. As another example, if π is the permutation 1,3,2,4,5, that is, $\pi(1)=1$, $\pi(2)=3$, $\pi(3)=2$, $\pi(4)=4$, $\pi(5)=5$, the length of π's longest increasing subsequence is 4. If σ is a permutation of length n, then $\ell_n(\sigma)$ denotes the length of σ's longest increasing subsequence.

If permutations of length n are selected at random, and each permutation has equal probability $1/n!$, then $\ell_n(\sigma)$ is a natural random variable arising from combinatorics. Its statistical behavior for n large has been an object of study since at least the 1960s. Interest in ℓ_n stems from basic connections between permutations and games of chance, such as "solitaire," also known as "patience sorting." In their mathematical review, Aldous & Diaconis (1999) give a precise description of such a game:

> Take a deck of cards labeled $1, 2, 3, \cdots, n$. The deck is shuffled, cards are turned up one at a time and dealt into piles on the table, according to the rule
>
> - A low card may be placed on a higher card (e.g., 2 may be placed on 7) or may be put into a new pile to the right of the existing piles.
>
> At each stage we see the top card on each pile. If the turned up card is higher than the cards showing then it *must* be put into a new pile to the right of the others. *The object of the game is to finish with as few piles as possible.* (Aldous & Diaconis, 1999, p. 413)

Suppose our deck has 5 cards, numbered 1–5, and that they are shuffled so that they are ordered as 5 3 1 4 2. During the game, the piles might evolve as follows:

		1	1	1
	3	3	3	3 2
5	5	5	5 4	5 4

and at the end of the game, we have two piles. Another (non-optimal) possibility is

			4	4 2
5	5 3	5 3 1	5 3 1	5 3 1

and we wind up with 3 piles. There is a simple algorithm (referred to as the "greedy algorithm") which always yields the minimum number of piles (see Aldous & Diaconis, 1999). The connection between the "solitaire" game described above and longest increasing subsequences is this: for any given shuffle, the minimum number of piles is exactly the longest increasing subsequence of the shuffle (when the shuffle is viewed as a permutation).

Over the past 40 years, mathematicians have studied the statistical behavior of the random variable $\ell_n(\sigma)$ when n, the size of the permutations, tends to infinity. Let $\mathbb{E}(\ell_n)$ denote the mean (or average) value of $\ell_n(\sigma)$, and let $\mathrm{Var}(\ell_n)$ denote its variance, $\mathrm{Var}(\ell_n)=\mathbb{E}(\ell_n^2) - (\mathbb{E}(\ell_n))^2$. Some natural questions are: How does $\mathbb{E}(\ell_n)$ grow with n? How fast does the variance grow? Is there a notion of a central limit theorem, or is there a limiting distribution for ℓ_n that emerges when n tends to ∞? Here are some results, both old and new, which answer these questions, and also demonstrate a striking connection between combinatorial questions and random matrix theory.

Result 1 (Logan & Shepp, 1977, see also Diaconis, 1995, Vershik & Kerov, 1997):

$$\lim_{n\to\infty} \frac{1}{\sqrt{n}}\mathbb{E}(\ell_n) = 2, \tag{3}$$

$$\lim_{n\to\infty} \frac{1}{\sqrt{n}}\mathrm{Var}(\ell_n) = 2. \tag{4}$$

Result 2 (Baik et al., 1999):

$$\lim_{n\to\infty} \mathrm{Prob}\left[\frac{\ell_n(\sigma) - 2n^{1/2}}{n^{1/6}} < x\right] = F(x), \tag{5}$$

where F is the Tracy–Widom distribution function (2). Result 2 was established in 1999. One remarkable aspect of this result is that it provides a mathematically precise connection between random permutations and random matrix theory (and Painlevé equations). There are now several proofs of this result as well as generalizations (see, e.g., Borodin et al. (2002)), and each proof contains different interconnections between areas of mathematics.

Further Developments

There have been a number of recent results that connect combinatorics to random matrix theory. Interesting generalizations of longest increasing subsequence problems have appeared. For example, Baik & Rains (2001) considered only those permutations that are their own inverses (also called involutions) and established results analogous to those mentioned above. Further generalizations to the study of "random words" have appeared (Tracy & Widom, 2001).

The Tracy–Widom distribution F has appeared in a variety of other contexts as well. For example, in the study of random domino tilings of the so-called Aztec diamond, the same distribution function appears in a natural scaling limit (Johansson, 2001). There is a notion of a "frozen region" of a domino tiling, and the boundary of this frozen region may be interpreted as a random variable, which is described, asymptotically, by the Tracy–Widom distribution function. This phenomenon is somewhat universal within random tiling problems as well—it has recently been established (Baik et al., 2004) that the same behavior appears in random rhombi tilings of hexagonal domains.

KENNETH MCLAUGHLIN

See also **Random matrix theory I, II and IV**

Further Reading

Aldous, D.J. & Diaconis, P. 1995. Hammersley's interacting particle process and longest increasing subsequences. *Probability Theory and Related Fields*, 103: 119–213

Aldous, D. & Diaconis, P. 1999. Longest increasing subsequences: from patience sorting to the Baik–Deift–Johansson theorem. *Bulletin of the American Mathematical Society (N.S.)*, 36(4): 413–432

Baik, J., Deift, P. & Johansson, K. 1999. On the distribution of the length of the longest increasing subsequence of random permutations. *Journal of the American Mathematical Society*, 12(4): 1119–1178

Baik, J., Kriecherbauer, McLaughlin, K.D.T.R. & Miller, P. 2004. Uniform asymptotics for polynomials orthogonal with respect to a general class of discrete weights and universality results for associated ensembles. *Annals of Mathematics*, submitted; Preprint CA/0310278

Baik, J. & Rains, E. 2001. The asymptotics of monotone subsequences of involutions. *Duke Mathematical Journal*, 109(2): 205–281

Borodin, A., Okounkov, A. & Olshanski, G. 2000. Asymptotics of Plancherel measures for symmetric groups. *Journal of the American Mathematical Society*, 13(3): 481–515

Johansson, K. 2001. Discrete orthogonal polynomial ensembles and the Plancherel measure. *Annals of Mathematics* (2), 153(1): 259–296

Logan, B.F. & Shepp, L.A. 1977. A variational problem for random Young tableaux. *Advances in Mathematics*, 26: 206–222

Mehta, M.L. 1991. *Random Matrices*, 2nd edition, San Diego: Academic Press

Tracy, C.A. & Widom, H. 1994. Level-spacing distributions and the Airy kernel. *Communications in Mathematical Physics*, 159: 151–174

Tracy, C.A. & Widom, H. 2001. On the distributions of the lengths of the longest monotone subsequences in random words. *Probability Theory and Related Fields*, 119: 350–380

Vershik, A.M. & Kerov, S.V. 1977. Asumptotics of the Plancherel measure of the symmetric group and the limiting form of Young tables. *Soviet Mathematics, Doklady*, 18: 527–531. Translation of *Doklady Academii Nauk, SSSR*, 233(1977): 1024–1027

RANDOM MATRIX THEORY IV: ANALYTIC METHODS

In this article, we address various analytical issues that arise in random matrix theory (RMT) using techniques from the theory of Riemann–Hilbert problems (RHPs). We refer the reader to the previous three entries for more information on the physical applications and various algebraic developments in RMT, and to the entry on Riemann–Hilbert problem for more information on RHPs. The classic text on RMT is Mehta (1991); see also Deift (2000) for a pedagogic presentation of some recent analytical results.

We discuss only the so-called unitary ensembles (UEs), that is, ensembles of $N \times N$ Hermitian matrices $M = M^* = (M_{jk} = M^{\mathrm{R}}_{jk} + \mathrm{i}M^{\mathrm{I}}_{jk})$ with probability distributions

$$P_N(M)\,\mathrm{d}M = \frac{1}{Z_N}\mathrm{e}^{-\operatorname{tr}V(M)}\,\mathrm{d}M = \frac{1}{Z_N}\mathrm{e}^{-\operatorname{tr}V(M)}\prod_{j=1}^{N}\mathrm{d}M_{jj}\prod_{j<k}\mathrm{d}M^{\mathrm{R}}_{jk}\prod_{j<k}\mathrm{d}M^{I}_{jk}, \quad (1)$$

where $V:\mathbb{B}\to\mathbb{R}$ grows sufficiently rapidly at ∞ and Z_N is a normalization constant. The name UE refers to the fact that the distribution is invariant under unitary conjugation $M \to \tilde{M} = UMU^*$, with U unitary, that is, $P(\tilde{M})\,\mathrm{d}\tilde{M} = P(M)\,\mathrm{d}M$. For information on other distributions such as orthogonal ensembles, see Tracy & Widom (1996, 1998) for some additional analytical developments. The special case in which $V(x) = x^2$ is called the Gaussian unitary ensemble (GUE), with $P_N(M)\,\mathrm{d}M = \frac{1}{Z_N}\mathrm{e}^{-\operatorname{tr}M^2}\,\mathrm{d}M$.

Under (1) the eigenvalues $\lambda_1(M) \geq \lambda_2(M) \geq \ldots \geq \lambda_N(M)$ of a matrix M in the ensemble become random variables with distribution given by the Weyl integration formula (see Mehta, 1991)

$$P_N(\lambda)\,\mathrm{d}^N\lambda = \frac{1}{Z'_N}\mathrm{e}^{-\sum_i V(\lambda_i)}\prod_{i<j}(\lambda_i-\lambda_j)^2\chi\,\mathrm{d}^N\lambda, \quad (2)$$

where χ denotes the indicator function on $\mathbb{R}^N$ for the set $\lambda_1 \geq \ldots \geq \lambda_N$. A statistic of basic importance is the *gap probability*

$$P_N(\theta; E) = \text{Prob}\ \{\ M \text{ has no eigenvalues in } (E-\theta, E+\theta)\} \quad (3)$$

for $E \in \mathbb{R}$, $\theta > 0$. Scaling the eigenvalues in the standard way around the energy E, $\lambda \mapsto \lambda' = \gamma_N(\lambda - E)$, so that the expected number of eigenvalues per unit λ'-interval is 1, one considers the limit

$$P(y; E) = \lim_{N\to\infty} P_N(y/\gamma_N; E). \quad (4)$$

In the case of GUE, a calculation of Gaudin and Mehta (Mehta, 1991) using classical estimates showed that

$$\gamma_N \sim N^{1/2} \quad (5)$$

and that

$$P(y; E=0) = \det(1 - S_y), \quad (6)$$

where S_y denotes the trace class operator with kernel

$$S_y(\xi,\eta) = \frac{\sin\pi(\xi-\eta)}{\pi(\xi-\eta)} \quad (7)$$

acting on $L^2(-y, y)$, the space of square integrable functions on the interval $(-y, y)$. The *universality conjecture* for the gap probability is the claim that under the correct scaling, (6) is true with the same right-hand side for all suitable V and E. The only thing that depends on V is the scaling parameter γ_N. Universality conjectures for other statistical quantities are discussed briefly below.

In this article we present a sketch of how to prove the universality conjecture for (6) following Deift et al. (1999a,b) in which the authors consider two kinds of potentials V

(i) $V(x) = t_{2m}x^{2m} + \ldots + t_0$, $t_{2m} > 0$,
(ii) $V(x) = NQ(x)$, where

(a) $Q(x)$ is real analytic in a neighborhood of $\mathbb{R}$
(b) $|Q(x)|/|\log x| \to +\infty$ as $|x| \to \infty$.

In order to prove (6), one must choose E in the support of the so-called equilibrium measure for V (see Step 4 Analytical considerations (ii) below).

The universality conjecture was first considered in the physics literature in Brézin & Zee (1993), in the mathematical literature in Pastur and Scherbina (1997), and in the special case $V(x) = x^4 + tx^2$, $t < 0$, in Bleher and Its (1999). The method in Bleher & Its has common elements with Deift et al. (1999a,b), but in addition Bleher & Its uses certain isomonodromy ideas.

Step 1: Representation in terms of orthogonal polynomials

Using an earlier calculation of Gaudin and (2), Gaudin and Mehta (Mehta, 1991) showed that

$$P_N(\theta, E = 0) = \det(1 - K_N), \tag{8}$$

where K_N is the finite rank operator acting on $L^2(E - \theta, E + \theta)$ with kernel

$$K_N(x, y) = \sum_{j=0}^{N-1} \phi_j(x)\phi_j(y), \tag{9}$$

where $\phi_j(x) = \mathrm{e}^{-(1/2)V(x)} p_j(x)$ and $p_j(x) = k_j x^j + \cdots$, $k_j > 0$, $j \geq 0$, are the orthonormal polynomials generated by the measure $\mathrm{e}^{-V(x)}\,\mathrm{d}x$,

$$\int_{\mathbb{R}} p_j(x) p_l(x) \mathrm{e}^{-V(x)}\,\mathrm{d}x = \delta_{jl}. \tag{10}$$

By the Christoffel–Darboux formula (Szegö, 1975), K_N can be expressed in terms of p_N and p_{N-1} and so the proof of (6) reduces to the evaluation of the asymptotics (so-called *Plancherel–Rotach* asymptotics) for $p_N(E + x/\gamma_N)$, $p_{N-1}(E + x/\gamma_N)$ as $N \to \infty$, for polynomials orthonormal with respect to a weight as in (10). For GUE, the p_N's are Hermite polynomials and such asymptotics can be found in the classical literature (Plancherel & Rotach, 1929). This is why Gaudin and Mehta were able to derive (6) for GUE, but for general weights $\mathrm{e}^{-V(x)}\,\mathrm{d}x$ the derivation of Plancherel–Rotach asymptotics is crucial.

Step 2: Introduction of Riemann–Hilbert Problem (RHP)

Write $p_j(x) = k_j \pi_j(x)$, so $\pi_j(x) = x^j + \ldots$. Following Fokas et al. (1991), the orthonormal polynomials p_j can be recovered by solving an RHP. Let the z-plane contour $\Sigma = \mathbb{R}$ be oriented from left to right so that the $(+)$ side (respectively, $(-)$ side) lies to the left (respectively, right) as one traverses Σ in the direction of the orientation. Define the jump matrix

$$v(z) = \begin{pmatrix} 1 & \mathrm{e}^{-V(z)} \\ 0 & 1 \end{pmatrix} \text{ for } z \in \Sigma,$$

and for any nonnegative integer q, let $Y = Y^{(q)} = (Y_{ij})_{1 \leq i,j \leq 2}$ be the (unique) 2×2 solution of the RHP (Σ, v).

(i) $Y(z)$ is analytic in $\mathbb{C} \setminus \Sigma$,
(ii) $Y_+(z) = Y_-(z)v(z)$, $z \in \mathbb{R}$, where $Y_\pm$ denotes the plus/minus boundary values of $Y(z)$, normalized so that
(iii) $Y(z)\begin{pmatrix} z^{-q} & 0 \\ 0 & z^q \end{pmatrix} \to \begin{pmatrix} 1 & 0 \\ 0 & 1 \end{pmatrix}$ as $z \to \infty$.

Then

$$\pi_q(z) = Y_{11}(z),\ \pi_{q-1}(z) = -Y_{21}(z)/2\pi \mathrm{i} k_{q-1}^2 \tag{11}$$

and

$$k_q = \sqrt{-(Y_{21})_1/2\pi\mathrm{i}}, \tag{12}$$

where $Y_{21}(z) = (Y_{21})_1 z^{q-1} \ldots$.

Thus, the evaluation of Plancherel–Rotach asymptotics for p_q, p_{q-1} reduces to the asymptotics of the RHP (Σ, v) as $q = N \to \infty$.

Step 3: Steepest-descent method applied to the RHP

We use the steepest-descent method for the RHPs introduced in Deift & Zhou (1993) and extended in Deift et al. (1997) to include fully nonlinear oscillations (*See* **Riemann–Hilbert problem**). We consider the case where $V(x) = NQ(x)$ with properties (a) and (b) on the left above and as in Deift et al. (1999a). The case where $V(x) = t_{2m}x^{2m} + \ldots + t_0$ can be reduced to this case with lower-order errors by scaling (see Deift et al., 1999b).

Suppose there exists a union of $n + 1$ disjoint intervals $J = \cup_{i=0}^{n} (b_i, a_{i+1})$ in $\mathbb{R}$, $b_0 < a_1 < \ldots < b_n < a_{n+1}$, a real number l, and a function $g(z)$ with the following properties:

$$g(z) \text{ is analytic in } \mathbb{C} \setminus \mathbb{R}, \tag{13}$$

$$g(z) = \log z + \mathrm{o}(1) \text{ as } z \to \infty, \tag{14}$$

$$g_+(z) + g_-(z) - V(z) - l \leq 0 \text{ for } z \in \mathbb{R} \setminus \bar{J}, \tag{15}$$

where $\bar{J}$ denotes the closure of J, and where equality holds for at most a finite number of points,

$$g_+(z) - g_-(z) \in \mathrm{i}\mathbb{R}, \tag{16}$$

for all real z and is constant in each component of $\mathbb{R}\setminus\bar{J}$,

$$g_+(z)+g_-(z)-V(z)-l=0 \text{ for } z\in\bar{J}, \quad (17)$$

$$\mathrm{i}(g'_+-g'_-)\geq 0 \text{ for } z\in J, \quad (18)$$

where equality holds for at most a finite number of points. As before, $g_\pm$ denotes the boundary values of $g(z)$.

Now define

$$M(z)\equiv \mathrm{e}^{-Nl/2\sigma_3}Y(z)\mathrm{e}^{-(g(z)-l/2)\sigma_3},$$
$$\sigma_3=\begin{pmatrix}1&0\\0&-1\end{pmatrix}. \quad (19)$$

A simple calculation using (13) and (14) shows that $M(z)$ solves a RHP normalized at infinity.

(i) $M(z)$ is analytic in $\mathbb{C}\setminus\mathbb{R}$
(ii) $M_+(z)=M_-(z)v_M(z)$, where $v_M(z)=$

$$\begin{pmatrix}\mathrm{e}^{-N(g_+(z)-g_-(z))} & \mathrm{e}^{N(g_+(z)+g_-(z)-V(z)-l)}\\ 0 & \mathrm{e}^{N(g_+(z)-g_-(z))}\end{pmatrix},$$
$z\in\mathbb{R}$,
(iii) $M(z)\to I$ as $z\to\infty$.

As we now show, the properties of $g(z)$ are chosen carefully so that the jump matrix v_M takes a specific form. By (17)

$$v_M(z)=\begin{pmatrix}\mathrm{e}^{-N(g_+(z)-g_-(z))} & 1\\ 0 & \mathrm{e}^{N(g_+(z)-g_-(z))}\end{pmatrix},$$
$$z\in J, \quad (20)$$

and by (16)

$$v_M(z)=\begin{pmatrix}\mathrm{e}^{-\mathrm{i}N\Omega_j} & \mathrm{e}^{N(g_+(z)+g_-(z)-V(z)-l)}\\ 0 & \mathrm{e}^{\mathrm{i}N\Omega_j}\end{pmatrix},$$
$$z\in(a_j,b_j),\ j=1,\ldots,n, \quad (21)$$

for some real constants Ω_j. Furthermore, if the inequality in (15) is strict then

$$v_M(z)=\begin{pmatrix}\mathrm{e}^{-\mathrm{i}N\Omega_j} & o(1)\\ 0 & \mathrm{e}^{\mathrm{i}N\Omega_j}\end{pmatrix}, z\in(a_j,b_j)$$
as $N\to\infty$. (22)

By (17), for $z\in J$,

$$G\equiv g_+-g_-=2g_+-V(z)-l=-2g_-+V(z)+l$$

and so has an analytic continuation to a suitable neighborhood U of J. If the inequality in (18) is strict, it follows by the Cauchy–Riemann conditions (make U smaller if necessary) that $\operatorname{Re}G(z)\gtrless 0$ for $z\in\{\operatorname{Im}z\gtrless 0\}\cap U$, respectively. Extend $\Sigma=\mathbb{R}$ to a new contour $\tilde{\Sigma}$ consisting of $\mathbb{R}$ together with the boundary of $n+1$ lens-shaped regions, one region surrounding each interval (b_{j-1},a_j) in J, as indicated in Figure 1.

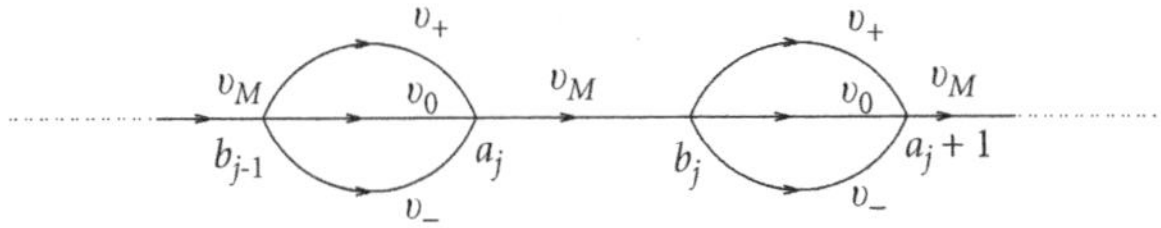

Figure 1. The Riemann–Hilbert problem $(\tilde{\Sigma},\tilde{v})$.

Note the factorization for v_M on J

$$\begin{pmatrix}\mathrm{e}^{-NG}&1\\0&\mathrm{e}^{NG}\end{pmatrix}=\begin{pmatrix}1&0\\\mathrm{e}^{NG}&1\end{pmatrix}\begin{pmatrix}0&1\\-1&0\end{pmatrix}$$
$$\begin{pmatrix}1&0\\\mathrm{e}^{-NG}&1\end{pmatrix}\equiv v_-v_0v_+. \quad (23)$$

The final algebraic step in the procedure is to define

(i) $\tilde{M}(z)\equiv M(z)$ outside the lens-shaped regions,
(ii) $\tilde{M}(z)\equiv M(z)\begin{pmatrix}1&0\\-\mathrm{e}^{-NG}&1\end{pmatrix}$ in the upper lens regions,
(iii) $\tilde{M}(z)\equiv M(z)\begin{pmatrix}1&0\\\mathrm{e}^{NG}&1\end{pmatrix}$ in the lower lens regions.

Then $\tilde{M}$ solves the normalized RHP $(\tilde{\Sigma},\tilde{v})$

(i) $\tilde{M}(z)$ is analytic in $\mathbb{C}\setminus\tilde{\Sigma}$,
(ii) $\tilde{M}_+(z)=\tilde{M}_-(z)\tilde{v}(z),\ z\in\tilde{\Sigma}$,
(iii) $\tilde{M}(z)\to I$ as $z\to\infty$,

where

(i) $\tilde{v}(z)=v_M(z)$ for $z\in\mathbb{R}\setminus\bar{J}$,
(ii) $\tilde{v}(z)=v_\pm(z)$ for $z\in\tilde{\Sigma}\cap\mathbb{C}_\pm$,
(iii) $\tilde{v}(z)=v_0(z)$ for $z\in J$

as indicated in Figure 1.

Now the situation is clear. The above computations show that as $N\to\infty$, $\tilde{v}(z)\to v^\infty(z)$, where

(i) $v^\infty(z)=\begin{pmatrix}\mathrm{e}^{-\mathrm{i}N\Omega_j}&0\\0&\mathrm{e}^{\mathrm{i}N\Omega_j}\end{pmatrix}$ for $z\in(a_j,b_j)$, $j=1,\ldots,n$,
(ii) $v^\infty(z)=I$ for $z\in(-\infty,b_0)\cup(a_0,\infty)$,
(iii) $v^\infty(z)=\begin{pmatrix}0&1\\-1&0\end{pmatrix}$ for $z\in J$,
(iv) $v^\infty(z)=I$ for z on the boundary of any of the lens-shaped regions.

Thus, we expect that as $N\to\infty$, $\tilde{M}(z)\to M^\infty(z)$, where $M^\infty(z)$ solves the RHP $(\tilde{\Sigma},v^\infty)$

(i) $M^\infty(z)$ is analytic in $\mathbb{C}\setminus\tilde{\Sigma}$,
(ii) $M^\infty_+(z)=M^\infty_-(z)v^\infty(z).\ z\in\tilde{\Sigma}$,
(iii) $M^\infty(z)\to I$ as $z\to\infty$.

It turns out that the RHP $(\tilde{\Sigma},v^\infty)$ can be solved explicitly in terms of theta functions (Deift et al., 1999a), thus yielding asymptotic formulae for $p_N(E+x/\gamma_N)$, $p_{N-1}(E+x/\gamma_N)$ as $N\to\infty$, which then leads to the proof of the universality conjecture (6) as indicated above.

Step 4: Analytical considerations

In order to make the above schema rigorous, many analytical difficulties must be addressed:

(i) One must show that J, l, and $g(z)$ with properties (13)–(18) actually exist. This can be done by considering the minimization problem

$$E_V = \inf_{\mu\in\mathcal{M}} \left(\int\int \log|s-t|^{-1}\, \mathrm{d}\mu(s)\mathrm{d}\mu(t) + \int V(t)\, \mathrm{d}\mu(t) \right), \quad (24)$$

where $\mathcal{M}$ is the space of probability measure on $\mathbb{R}$. It turns out (Saff & Totik, 1997) that the infinum is achieved at a unique measure $\mu = \mu_V$, the so-called *equilibrium measure* for V. For $V = NQ$ satisfying (i) and (ii) above, a result of Deift, Kriecherbauer and McLaughlin shows that $\mathrm{d}\mu_V(s) = \psi_V(s)\,\mathrm{d}s$ is absolutely continuous with respect to Lebesgue measure and is supported on a finite union of intervals J_V. Then quite remarkably

$$g(z) \equiv \int_{\mathbb{R}} \log(z-s)\, \mathrm{d}\mu_V(s) \quad (25)$$

has the desired properties (13)–(18) on $J = J_V$, provided l is chosen to be the Lagrange multiplier associated with the constraint $\int \mathrm{d}\mu = 1$ in the minimization problem E_V.

(ii) The scaling parameter γ_N must be determined. In order to obtain the universality limit (6) one must consider only points E in the support of μ_V, $\psi_V(E) > 0$. For such E, one takes $\gamma_N \equiv N\psi_V(E)$. If $\psi_V(E) = 0$, the limit in (4) (if it exists!) is *different* from (6). In this connection, see (26) below.

(iii) The convergence of $\tilde{v}(z)$ to $v^{\infty}(z)$ is not uniform on $\tilde{\Sigma}$ as $N \to \infty$, leading in particular to difficulties at the end points of J.

(iv) The inequalities in (15) and (18) may not be strict.

The analytical heart of the problem lies in addressing the latter three points, and we refer the reader to Deift et al. (1999a) for further details.

The asymptotic behavior of $P(y, E=0) = \det(1 - S_y)$ as $y \to \infty$ can itself be evaluated using Riemann–Hilbert techniques (see Deift et al., 1997). The work in Deift et al. (1997) follows on the earlier work of Widom (1994, 1995), who uses more classical techniques. The paper by Deift et al. (1997) is the first application of RHP techniques to random matrix theory.

Universality conjectures for other statistical quantities, such as the k-point correlation function and the nearest-neighbor spacing distribution, can also be proved using the above methods (see Deift et al., 1999a; Deift, 2000). For GUE, Tracy & Widom (1994) proved the beautiful result that if $\lambda_1(M)$ is the largest eigenvalue of a GUE matrix M, then

$$\lim_{N\to\infty} \mathrm{Prob}((\lambda_1(M) - \sqrt{2N})2^{1/2}N^{1/6} \le t) = F(t), \quad (26)$$

where $F(t)$ is the so-called *Tracy–Widom distribution*

$$F(t) = \mathrm{e}^{-\int_t^{\infty}(s-t)u^2(s)\,\mathrm{d}s},$$

where $u(s)$ is the unique solution (Hastings & McLeod, 1980) of the Painlevé II equation $u'' = 2u^3 + su$ with asymptotics $u(s) \sim \mathrm{Ai}(s)$ as $s \to \infty$. Here Ai(s) is the classical Airy function. As long as $\psi_V(s)$ behaves "generically" near the right end point of J, $\psi_V(s) \sim \sqrt{s - a_{N+1}}$, $s \sim a_{N+1}$, one can use the methods of Deift et al. (1999a) to show that the behavior of $\lambda_1(M)$ is also universal as $N \to \infty$; that is, once the correct scaling parameters at a_{N+1} are inserted, the same limit (26) is true for all UEs. As noted in the entry on Riemann–Hilbert problem, the distribution $F(t)$ arises, in particular, in the solution of Ulam's problem in Baik et al. (1999).

Percy Deift and Xin Zhou

See also **Painlevé analysis; Random matrix theory I, II, and III; Riemann–Hilbert problem**

Further Reading

Baik, J., Deift, P. & Johansson, K. 1999. On the distribution of the length of the longest increasing subsequence of random permutations. *Journal of the American Mathematical Society*, 12(4): 1119–1178

Bleher, P. & Its, A.R. 1999. Semiclassical asymptotics of orthogonal polynomials, Riemann–Hilbert problems, and universality in the matrix model. *Annals of Mathematics*, 150(2): 185–266

Brézin, E. & Zee, A. 1993. Universality of the correlations between eigenvalues of large random matrices. *Nuclear Physics* B, 402(3): 613–627

Deift, P. 2000. *Orthogonal Polynomials and Random Matrices: A Riemann–Hilbert Approach*, Providence, RI: American Mathematical Society

Deift, P., Its, A. & Zhou, X. 1997. A Riemann–Hilbert approach to asymptotic problems arising in the theory of random matrix models, and also in the theory of integrable statistical mechanics. *Annals of Mathematics*, 145: 1–88

Deift, P., Kriecherbauer, T., McLaughlin, K., Venakides, S. & Zhou, X. 1999a. Uniform asymptotics for polynomials orthogonal with respect to varying exponential weights and applications to universality questions in random matrix theory. *Communications in Mathematical Physics*, 52: 1335–1425

Deift, P., Kriecherbauer, T., McLaughlin, K., Venakides, S. & Zhou, X. 1999b. Strong asymptotics of orthogonal polynomials with respect to exponential weights. *Communications in Mathematical Physics*, 52: 1491–1552

Deift, P., Venakides, S. & Zhou, X. 1997. New results in small dispersion KdV by and extension of the steepest descent method for Riemann–Hilbert problems, *Internat. Math. Res. Notices*, No. 6, 285–299

Deift, P. & Zhou, X. 1993. A steepest descent method for oscillatory Riemann–Hilbert problems–asymptotics for the MKdV equation. *Annals of Mathematics*, 137: 295–368

Fokas, A.S., Its, A.R. & Kitaev, A.V. 1991. Discrete Painlevé equations and their appearance in quantum gravity. *Communications in Mathematical Physics*, 142(2): 313–344
Hastings, S.P. & McLeod, J.B. 1980. A boundary value problem associated with the second Painlevé transcendent and the Korteweg de Vries equation. *Archive for Rational Mechanics and Analysis*, 73: 31–51
Mehta, M.L. 1991. *Random Matrices*, 2nd edition, San Diego: Academic Press
Pastur, L. & Scherbina, M. 1997. Universality of the local eigenvalue statistics for a class of unitary random matrix ensembles. *Journal of Statistical Physics*, 86(1–2): 109–147
Plancherel, M. & Rotach, W. 1929. Sur les valeurs asymptotiques des polynomes d'Hermite $H_n(x) = (-1)^n e^{x^2/2} d^n (e^{-x^2/2})/dx^n$. *Commentarii Mathematica Helvetica*, 1: 227–254
Saff, E. & Totik, V. 1997. *Logarithmic Potentials and External Fields*, Berlin and New York: Springer
Szegö, G. 1975. *Orthogonal Polynomials*, 4th edition, Providence, RI: American Mathematical Society
Tracy, C.A. & Widom, H. 1994. Level-spacing distributions and the Airy kernel. *Communications in Mathematical Physics*, 159: 151–174
Tracy, C.A. & Widom, H. 1996. On orthogonal and symplectic matrix ensembles. *Communications in Mathematical Physics*, 177: 727–754
Tracy, C.A. & Widom, H. 1998. Correlation functions, cluster functions, and spacing distributions for random matrices. *Journal of Statistical Physics*, 92(5/6): 809–835
Widom, H. 1994. The asymptotics of a continuous analogue of orthogonal polynomials. *Journal of Approximation Theory*, 77: 51–64
Widom, H. 1995. Asymptotics for the Fredholm determinant of the sine kernel on a union of intervals. *Communications in Mathematical Physics*, 171: 159–180

RANDOM PROCESSES

See **Stochastic processes**

RANDOM WALKS

As conceptually simple stochastic processes, random walks have a plethora of practical modeling applications, leading to elegant and sometimes counter-intuitive mathematics. Consider a particle constrained to move in one dimension, moving one unit to the left or right at each unit time step, and with the direction of motion for each time step chosen at random and independently of the particle's position or history. If the particle's position after n time steps is $S(n)$, one can assert

$$S(n) = S(0) + \sum_{i=1}^{n} X_i, \tag{1}$$

where each X_i is an independent identically distributed (IID) random variable taking values $+1$ with probability (say) p and -1 with probability $1-p$. If $p = 1 - p = \frac{1}{2}$, then the simple random walk (SRW) has no "preferred direction" and is unbiased. If $p > \frac{1}{2}$, then the particle is more likely to move right than left at each time step; the SRW is biased. The traditional pedagogical example imagines a drunken student's erratic progress along a street, where bias might be caused by the street being inclined or having a pub at one end.

Basic Results

The statistics of Equation (1) are simple and intuitively reasonable; thus

$$\begin{aligned} \mathrm{E}(S(n)) &= S(0) + n(2p-1) \\ \mathrm{Var}(S(n)) &= 4np(1-p), \end{aligned} \tag{2}$$

so the SRW moves with an average speed of $2p-1$ and spreads about the mean as $\sqrt{n}$ (characteristic of diffusive processes). This spread is greatest when the SRW is unbiased.

One can readily prove that the SRW is temporally and spatially homogeneous and Markovian (i.e., it does not matter where one places the origin of time or space, and conditional on the present, the future is independent of the past). Assuming without loss of generality $S(0) = 0$,

$$\begin{aligned} P(S(n) = a) &= C^n_{(n+a)/2} p^{(1/2)(n+a)} \\ &\quad \times (1-p)^{(n-a)/2}, \end{aligned} \tag{3}$$

the particle moves $(n+a)/2$ steps right and $(n-a)/2$ steps left to give a total of n steps and a net right displacement of a, and there are $C^n_{(n+a)/2}$ equally likely paths achieving this. N.B. for a to be attainable in n steps one requires $|a| \le n$ and $(n+a)/2$ to be an integer i.e., only odd positions are attainable at odd times, hence, the mysterious factor of 2 in Equation (5).

Results exploiting symmetries, homogeneities, and combinatorics, and also applications of probability- and moment-generating functions, are summarized in Grimmett & Stirzaker (2001, Chapters 3 and 4). More interesting questions involving absorption or reflection at barriers, dealt with using linear difference equations; where the nature of the barriers, dictates the appropriate boundary conditions, are also addressed therein.

Continuous Limit: Diffusion

For an unbiased SRW, manipulation of Equation (3) reveals, for sufficiently large a and n,

$$P(S(n) = a) \approx \sqrt{\frac{2}{\pi n}} \exp\left(\frac{-a^2}{2n}\right), \tag{4}$$

resembling a normal (Gaussian) probability distribution (Murray, 2002, Chapter 9). Indeed, if one lets the discrete space and time steps have sizes Δx and Δt, respectively, and forces these to zero such that

$$\lim_{\Delta x \to 0, \Delta t \to 0} \frac{(\Delta x)^2}{2\Delta t} = D \tag{5}$$

for some finite constant D (the diffusion coefficient), then one arrives at a diffusion equation,

$$\frac{\partial P}{\partial t} = D\frac{\partial^2 P}{\partial x^2}, \tag{6}$$

where $P(x, t)$ is the probability density function for the particle's position x at time t. Solutions of Equation (6) take the form indicated by Equation (4):

$$P(x, t) = \frac{1}{\sqrt{4\pi Dt}}\exp\left(\frac{-x^2}{4Dt}\right). \tag{7}$$

An alternative derivation, using Taylor series expansions of $P(x, t)$ (Murray, 2002, Chapter 9), lacks beauty and elegance but is more readily generalizable to the nonlinear RWs discussed below (Othmer & Stevens, 1997). If the SRW is biased, then an advection term,

$$v\frac{\partial P}{\partial x}$$

(where v represents the average velocity in the preferred direction) is added to the left-hand side of (6). If one considers $P(x, t)$ to represent a density of particles (rather than a probability density for a single particle), then one can incorporate local dynamics (e.g., chemical reactions or biological reproduction) at each point; a reaction term

$$f(P, x, t)$$

appears on the right-hand side of (6). The description of SRWs using diffusion (or more generally reaction-diffusion-advection) equations allows standard partial differential equation methods to be applied, often elucidating information concerning wave-like and spatially structured solutions (Murray, 2002; Okubo & Levin, 2001).

The extension of the SRW into higher dimensions presents no great conceptual, nor mathematical, challenges. If the processes governing steps in orthogonal directions are mutually independent, then one applies the preceding results to each direction separately. More realistically, suppose that at each time step a unit distance is moved in a direction chosen randomly from the infinite number of possibilities. Again, one can show that the process is, on average, diffusive (Denny & Gaines, 2000, Chapter 6):

$$E(S(n)) = S(0) \text{ and } \mathrm{Var}(S(n)) = n\,. \tag{8}$$

Importantly, motion in more than one spatial dimension allows more realistic random walk models to be developed, some of which are mentioned below.

Random Walks in Biology

Random walks have been widely applied in the life sciences. While ostensibly describing a particle's position, the modeled quantity might equally well refer to the amount of food consumed by an animal, the frequency of an allele in a population, or the contentious notion of "fitness." Crucially, random walk models are developed using local rules of "motion" at the level of the "individual" (organism/cell/molecule, etc.) before any scaling-up to a population level description is undertaken. This encourages the modeller to adhere to realistic assumptions and provides a link between concrete experimental observation and theory.

Several accessible applications are presented in Denny & Gaines (2000, Chapters 5 and 6). For example, Equation (6) can be solved to indicate the rate at which nutrients reach an idealized organism, D being strongly dependent on the organism's environment. Such considerations help to explain relative sizes of aquatic and aerial organisms and underly the varying mechanisms of nutrient transport in plant root systems. More complicated applications deal with the distribution of receptors on cell walls, the random drift to fixation of alleles in finite populations, and the energetics of protein folding. For more mathematically demanding applications involving spatial patterns and reaction-diffusion equations, see Murray (2002) and Okubo & Levin (2001).

Nonlinear Random Walks

The basic concept of a particle "deciding" (using IID random variables) its next direction at every step may be excessively restrictive, particularly in biology. For example, a particle moving in two dimensions may be likely to choose a direction similar to that used during the previous time step or may have some fixed preferred direction toward which its trajectory is biased. The mathematics of such "correlated" RWs is discussed in Othmer et al. (1988).

Alternatively, motion may be influenced by interactions with the local environment (reinforced RWs) leading to nonlinear generalizations of the simple diffusion equation; under reasonable assumptions, Equation (6) may be more properly expressed

$$\frac{\partial P}{\partial t} = D\frac{\partial}{\partial x}\left(\frac{\partial}{\partial x}\left[\ln\frac{P}{\tau(W)}\right]\right), \tag{9}$$

where $\tau(W)$ represents a transition rate dictated by the environment W. Such systems can exhibit properties not found in diffusive SRW formulations; particles may tend to aggregate rather than disperse and may reach locally infinite or zero densities in finite time (Othmer & Stevens, 1997).

Numerical simulation of such systems can be built up from individual-level rules as in Sleeman & Wallis (2002) and Figure 1, where the authors model angiogenesis (the movement of endothelial cells to create a blood supply to an avascular tumor). The cells' motions are dictated by chemical gradients between the tumor and the blood vessel; at first the attractive signal from the tumor is weak and the motion tortuous, but

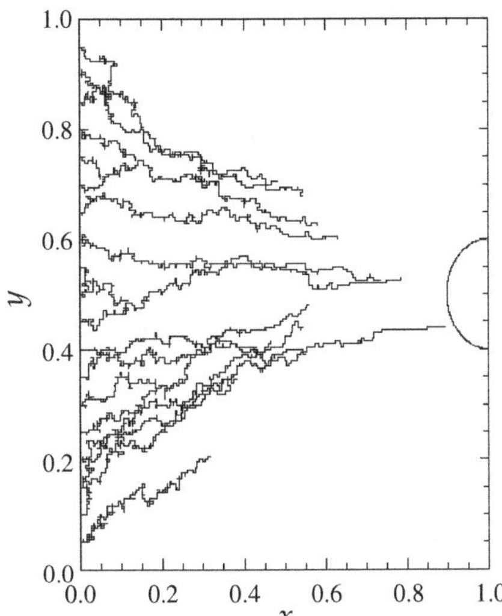

Figure 1. Tumor angiogensis modeled as a reinforced random walk. (Used with kind permission of I.P. Wallis and B.D. Sleeman, Department of Mathematics, University of Leeds, U.K.)

once the tumor's signal is sufficiently strong a more directed motion is observed.

Jon Pitchford

See also **Brownian motion; Deterministic walks in random environments; Martingales**

Further Reading

Denny, M. & Gaines, S. 2000. *Chance in Biology*, Princeton, NJ: Princeton University Press

Grimmett, G. & Stirzaker, D. 2001. *Probability and Random Processes*, Oxford and New York: Oxford University Press

Murray, J.D. 2002. *Mathematical Biology*, 3rd edition, Berlin and New York: Springer

Okubo, A. & Levin, S.A. 2001. *Diffusion and Ecological Problems*, Berlin and New York: Springer

Othmer, H.G., Dunbar, S.R. & Alt, W. 1988. Models of dispersal in biological systems, *Journal of Mathematical Biology*, 26(3): 263–298

Othmer, H.G. & Stevens, A. 1997. Aggregation, blowup and collapse: the ABC's of taxis in reinforced random walks. *SIAM Journal on Applied Mathematics*, 57(4): 1044–1081

Sleeman, B.D. & Wallis, I.P. 2002. Tumour induced angiogenesis as a reinforced random walk: modelling capillary network formation without endothelial cell proliferation. *Mathematical and Computer Modelling*, 36 (3): 339–358

RATCHETS

An essential and striking feature of living cells is their ability to generate mechanical motion and forces. Important examples are cell motility, muscle contraction, and active mass transport within cells, among other active phenomena in biology. These motions and forces are generated at the molecular level by protein molecules (called molecular motors or pumps) that are driven by chemical reactions in situations far from equilibrium (Astumian, 1997; Jülicher et al., 1997).

Two types of molecular motors are classified as linear (or translational) and rotary (or rotatory). Within the first class, motor proteins are classified into several families: myosins, kinesins, and dyneins. Each of these interacts specifically with a certain type of filament along which it is able to move in the presence of adenosine triphosphate (ATP), which is a chemical fuel. The filaments serve as guides or tracks for the motor motion. Two types of filaments play this role: microtubules and actin filaments. Both are formed by a polymerization process from identical monomers (actin and tubulin monomers, respectively), resulting in a regular and periodic one-dimensional structure. An important feature is their polarity; thus the asymmetry of the monomers that form a polar filament structure is essential for motor operation as it defines the direction of motion.

A typical representative of rotary molecular motors is ATP synthase composed of two coupled rotary engines: a membrane-embedded unit F_0 and a water-surrounded part F_1 (Oster & Wang, 2000) ATP synthase works as a reversible motor-pump machine. Thus, the proton flow through F_0 is believed to generate a mechanical torque driving the F_1 motor to synthesize ATP, and using the hydrolysis energy of ATP, the F_1 motor can drive the F_0 motor in reverse to pump protons. Direct observation of the rotation of the F_1 motor has been demonstrated in fascinating experiments by Japanese scientists (Noji et al., 1997).

Thus, molecular motors are microscopic objects that unidirectionally move along one-dimensional periodic structures. Because the motor operation is considered on the molecular level, the motor must be a Brownian object, subject to random fluctuations. A conversion (rectification) of random fluctuations $F(t)$ into useful work (i.e., a directed motion) is called a "ratchet effect." (The names *thermal ratchet*, *Brownian motor*, *Brownian rectifier*, *mechanical diode*, *stochastic ratchet*, or simply *ratchet* are also in use.) A comprehensive review of the work on ratchets and their practical applications is available in Reimann (2002).

Historically, the problem of converting Brownian motion (in general, unbiased random fluctuations) into useful work and its consistency with the second law of thermodynamics was introduced in a conference talk by Marian Smoluchowski in 1912 (Smoluchowski, 1912) and later popularized and extended by Richard Feynman (Feynman et al., 1963), using a so-called "ratchet and pawl" gadget. A simple stochastic model of this rectifying process (called a Smoluchowski–Feynman's ratchet) can be described using the overdamped Langevin equation for a Brownian particle moving in a one-dimensional viscous medium with friction coefficient η in the presence of a constant force F:

$$\eta \dot{x}(t) = -V'(x(t)) + F + \xi(t). \quad (1)$$

Here, the overdot and the prime denote differentiation with respect to the time t and the particle coordinate x, respectively. The potential $V(x)$ is a periodic function with period L, $V(x+L) = V(x)$, which has a broken parity symmetry. More precisely, this breaking of symmetry means that there is no Δx such that $V(-x) = V(x+\Delta x)$ for all x. A typical example of such a ratchet potential is a piecewise-linear (saw-tooth) function, consisting of two continuously matched linear pieces per period L, one with negative and one with positive slope, one being steeper than the other. The stochastic force $\xi(t)$ is a Gaussian white noise of zero mean, $\langle \xi(t) \rangle = 0$, satisfying the fluctuation-dissipation relation (Risken, 1989)

$$\langle \xi(t)\xi(s) \rangle = 2\eta k_B T \delta(t-s), \quad (2)$$

where k_B is Boltzmann's constant, $2\eta k_B T$ is the noise intensity or strength (T is absolute temperature), and $\delta(t)$ is Dirac's delta function.

The probability densities induced by the above equations obey a Fokker–Planck (FP) equation, which can be written in the form of a conservation law for probability, namely, a continuity equation (Risken, 1989)

$$\partial_t P(x,t) + \partial_x J(x,t) = 0. \quad (3)$$

Here the probability density $P(x,t)$ satisfies the normalization condition

$$\int_{-\infty}^{\infty} P(x,t)\,\mathrm{d}x = 1, \quad (4)$$

whereas the probability current is defined by

$$J(x,t) \doteq -\eta^{-1}[U'(x) + k_B T \partial_x] P(x,t), \quad (5)$$

with the ratchet potential $V(x)$ and the force F being incorporated into a single effective potential

$$U(x) \doteq V(x) - Fx. \quad (6)$$

The quantity of foremost interest in the context of transport in periodic systems is the particle current, defined as the time-dependent ensemble average over the velocities, $\langle \dot{x}(t) \rangle$ (where for convenience the argument t is omitted). From this definition, the general connection between the probability current and the particle current is obtained as

$$\langle \dot{x} \rangle = \int_{-\infty}^{\infty} J(x,t)\,\mathrm{d}x. \quad (7)$$

Instead of solving the FP equation (3) describing the particle transport in periodic systems but with zero boundary conditions at infinity, it is convenient to rewrite it in terms of reduced probabilities (Reimann, 2002):

$$\hat{P}(x,t) \doteq \sum_{n=-\infty}^{\infty} P(x+nL,t), \quad \hat{J}(x,t) \doteq \sum_{n=-\infty}^{\infty} J(x+nL,t), \quad (8)$$

which are evidently periodic functions with period L. Furthermore, Equations (4) and (7) are transformed to

$$\int_0^L \hat{P}(x,t)\,\mathrm{d}x = 1 \quad (9)$$

and

$$\langle \dot{x} \rangle = \int_0^L \hat{J}(x,t)\,\mathrm{d}x, \quad (10)$$

respectively. Indeed, because $V(x)$ is periodic with period L, with $P(x,t)$ being a solution of the FP equation (3), $\hat{P}(x+nL,t)$ is also a solution for any integer n. Next, since the FP equation is linear, Equations (3) and (5) are also satisfied by the reduced quantities (8). Therefore, it is sufficient to solve an FP equation with periodic boundary conditions.

If the reduced dynamics assumes a steady state $\hat{P}^{\mathrm{st}}(x)$ in Equation (3), then the reduced probability current $\hat{J}(x,t) = \hat{J}^{\mathrm{st}}$ becomes independent of x and t. Multiplying this equation by x and integrating both its sides from x to $x+L$, one finds that the particle current takes the form

$$\langle \dot{x} \rangle = L\hat{J}^{\mathrm{st}}, \quad (11)$$

as expected. Moreover, using Equations (3), (5), and (11), it can easily be checked that the steady-state solution is given by (Stratonovich, 1969; Reimann, 2002)

$$\hat{P}^{\mathrm{st}}(x) = \mathcal{N}\frac{\eta}{k_B T} \mathrm{e}^{-U(x)/k_B T} \int_x^{x+L} \mathrm{e}^{U(y)/k_B T}\,\mathrm{d}y, \quad (12)$$

$$\langle \dot{x} \rangle = L\mathcal{N}\left[1 - \mathrm{e}^{[U(L)-U(0)]/k_B T}\right], \quad (13)$$

$$\mathcal{N} \doteq \frac{k_B T}{\eta} \left[\int_0^L \mathrm{d}x \int_x^{x+L} \mathrm{e}^{[U(y)-U(x)]/k_B T}\,\mathrm{d}y\right]^{-1}. \quad (14)$$

This solution is valid for a general potential $U(x)$ provided $U'(x+L) = U'(x)$. For the specific form (6), $U(L) - U(0) = -LF$, and therefore, as expected, the sign of the particle current (13) coincides with the sign of F. The absence of an average particle current ($\langle \dot{x} \rangle = 0$) if $F = 0$, in spite of the broken spatial symmetry, agrees with the second law of thermodynamics, like in the original ratchet and pawl gadget (Feynman et al., 1963). However, if the force

F in Equation (1) depends on time, a rectified motion (ratchet effect) may occur in an asymmetric system.

For the occurrence of the ratchet effect, it is sufficient to show that $\langle \dot{x} \rangle \neq 0$ provided $\langle F(t) \rangle = 0$. If the changes of $F(t)$ are extremely slow, then at any given instant t, the particle current has practically the same value as the steady-state current (13) with a static tilt F. In this adiabatic approximation, the time t plays the role of a parameter and using result (13) for a periodic driving force $F(t+\mathcal{T}) = F(t)$, one finds the time-averaged particle current:

$$\langle \dot{x} \rangle = \frac{1}{\mathcal{T}} \int_0^{\mathcal{T}} \frac{L k_B T \left[1 - e^{-LF(t)/k_B T}\right]}{\eta \int_0^L dx \int_x^{x+L} e^{[V(y)-V(x)-F(t)(y-x)]/k_B T} dy} dt. \quad (15)$$

For general analytical conclusions, the adiabatic expression (15) is still too complicated. Only in simple special cases may one predict the direction of the current. As an example, consider the case of when the tilting force $F(t)$ takes only two possible values $\pm F_0$ with very rare deterministic or random flips for which $\langle F(t) \rangle = 0$, and $V(x)$ is a piecewise linear potential with two continuously matched linear pieces per period L, the slopes of which are different (one steeper and one flatter). Outside this fundamental cell of length L, the potential is periodically continued. In this case, the integrals in Equation (15) can be calculated analytically, yielding an explicit expression (Magnasco, 1993), which is positive if the flatter slopes are arranged in the positive direction. Intuitively, this ratchet effect can be understood as follows. For F_0 within a certain range, one of the two tilted asymmetric potentials $V(x) \mp F_0 x$ does not exhibit any extrema and, therefore, supports a permanent downhill motion (even if $T = 0$), while the other still exhibits extrema acting as motion-blocking barriers. Below this window, both $V(x) \mp F_0 x$ exhibit barriers and, thus, prohibit deterministic motion. Because the barrier induced by the steeper slope of $V(x)$ is higher than that induced by the flatter slope, a weak thermal noise induces the current in the direction of the flatter slope.

Alternatively, the ratchet effect in a spatially periodic system can be achieved under a time-dependent variation (pulsation) of the potential shape without affecting its spatial periodicity. Therefore, Equation (1) can be generalized to include both a pulsating $f(t)$ and tilting $F(t)$ forcing as follows (Jülicher et al., 1997; Reimann, 2002):

$$\eta \dot{x}(t) = -V'[x(t), f(t)] + F(t) + \xi(t), \quad (16)$$

where $V'(x, f) \doteq \partial_x V(x, f)$. Here $F(t)$ may contain a load force or torque, generally depending on time t.

Thus, two fundamental classes of ratchet models arise from Equation (16). The first comprises models with $F(t) \equiv 0$, which are called "pulsating" ratchets. Within these, the first main subclass, called "fluctuating potential" ratchets (Astumian & Bier, 1994; Prost et al., 1994), is obtained when $f(t)$ in Equation (16) is additive, i.e., $V(x, f(t)) = V(x)[1 + f(t)]$. This subclass contains as a special case the "on-off" ratchets (Ajdari & Prost, 1992) when $f(t)$ takes only two values, one of them being -1 (potential "off"). The on-off ratchets are suitable for experimental realizations and have been demonstrated (Rousselet et al., 1994) by means of colloidal polystyrene latex spheres suspended in solution and exposed to a dielectric ratchet potential, created by a series of "Christmas-tree" electrodes, which were turned on and off periodically. A similar experimental setup was used by Faucheux & Libchaber (1995) but with solutions containing *two* different species of particles at a time. As a result, it was demonstrated that they can be separated. A further experimental verification of on-off ratchets has been performed (Faucheux et al., 1995) using polystyrene spheres confined to an effective one-dimensional ratchet potential by laser-optical trapping techniques (optical tweezers). The second subclass of pulsating ratchets, called "traveling potential" ratchets, have potentials of the form $V(x, f(t)) = V(x - f(t))$.

The second class are models with $f(t) \equiv 0$, called "tilting" ratchets (Magnasco, 1993), so that $V(x, f(t)) = V(x)$ in Equation (16). When $V(x)$ is a ratchet potential, then $F(t)$ is mostly considered as a symmetric function. If $F(t)$ is a stochastic process, one speaks of a "fluctuating force" ratchet. The case of a tilting ratchet with a periodic driving $F(t)$ is of particular experimental relevance and carries the obvious name "rocking" ratchet.

Clearly, there are many possible combinations and generalizations of these classes; for example, a simultaneously pulsating and tilting ratchet or the simultaneous breaking of more than one symmetry (Reimann, 2002). For various applications in physics, chemistry, and biology, see Astumian (1997), Jülicher et al. (1997), and Reimann (2002).

ALEXANDER V. ZOLOTARYUK

See also **Brownian motion; Fokker–Planck equation; Nonequilibrium statistical mechanics**

Further Reading

Ajdari, A. & Prost, J. 1992. Mouvement induit par un potentiel périodique de basse symétrie: diélectrophorèse pulsée. *Comptes Rendus de l'Academie des Sciences, Paris, Série II*, 315: 1635–1639

Astumian, R.D. 1997. Thermodynamics and kinetics of a Brownian motor. *Science*, 276: 917–922

Astumian, R.D. & Bier, M. 1994. Fluctuation driven ratchets: molecular motors. *Physical Review Letters*, 72: 1766–1769

Faucheux, L.P., Bourdieu, L.S., Kaplan, P.D. & Libchaber, A. 1995. Optical thermal ratchet. *Physical Review Letters*, 74: 1504–1547
Faucheux, L.P. & Libchaber, A. 1995. Selection of Brownian particles. *Journal of the Chemical Society—Faraday Transactions*, 91: 3163–3170
Feynman, R.P., Leighton, R.B. & Sands, M. 1963. *The Feynman Lectures on Physics*, vol. 1, Reading, MA: Addison-Wesley (Chapter 46)
Jülicher, F., Ajdari, A. & Prost, J. 1997. Modeling molecular motors. *Reviews of Modern Physics*, 69: 1269–1281
Magnasco, M.O. 1993. Forced thermal ratchets. *Physical Review Letters*, 71: 1477–1481
Noji, H., Yasuda, R., Yoshida, M. & Kinosita, K. 1997. Direct observation of the rotation of F_1-ATPase. *Nature*, 386: 299–302
Oster, G. & Wang, H. 2000. Reverse engineering a protein: the mechanochemistry of ATP synthase. *Biochimica et Biphysica Acta*, 1458: 482–510
Prost, J., Chauwin, J.-F., Peliti, L. & Ajdari, A. 1994. Asymmetric pumping of particles. *Physical Review Letters*, 72: 2652–2675
Reimann, P. 2002. Brownian motors: noisy transport far from equilibrium. *Physics Reports*, 361: 57–265
Risken, H. 1989. *The Fokker–Planck Equation: Methods of Solution and Applications*, Berlin: Springer
Rousselet, J., Salome, L., Ajdari, A. & Prost, J. 1994. Directional motion of brownian particles induced by a periodic asymmetric potential. *Nature*, 370: 446–448
Stratonovich, R.L. 1969. *Theory of Random Noise*, London: Gordon and Breach
von Smoluchowski, M. 1912. Experimentell nachweisbare, der üblichen Thermodynamik widersprechende Molekularphänomene. *Physikalische Zeitschrift*, 13: 1069

RAYLEIGH AND RAMAN SCATTERING AND IR ABSORPTION

When an electromagnetic wave, $E = E_0 \cos 2\pi\nu t$, irradiates a molecule, a crystal, or a disordered substance, the energy may be scattered, absorbed, or transmitted. In the scattering process, the irradiated medium becomes polarized and the induced dipole moment (or electric polarization),

$$P = \alpha E + \cdots = \alpha E_0 \cos 2\pi\nu t + \ldots, \qquad (1)$$

oscillates synchronously with the electromagnetic field. The dipole radiates energy (the scattering), and α is the polarizability (usually a second-order tensor).

Irradiating with ultraviolet or visible light of frequency ν induces an electronic polarization that is modulated by the vibrational frequencies $(\nu_i, \nu_j, \nu_k, \ldots)$ of the substance. Assuming a simple diatomic molecule, vibrating with frequency ν_1, the nuclear displacement is $q = q_0 \cos 2\pi\nu_1 t$, and $\alpha = \alpha_0 + (\partial\alpha/\partial q)_0 q$, then

$$\begin{aligned} P &= \alpha_0 E_0 \cos 2\pi\nu t \\ &\quad + \frac{1}{2}\left(\frac{\partial a}{\partial q}\right)_0 q_0 E_0 \{\cos[2\pi(\nu + \nu_1)t] \\ &\quad + \cos[2\pi(\nu - \nu_1)t]\}. \end{aligned} \qquad (2)$$

The emitted intensity is proportional to the square of the polarization $I \approx P^2$.

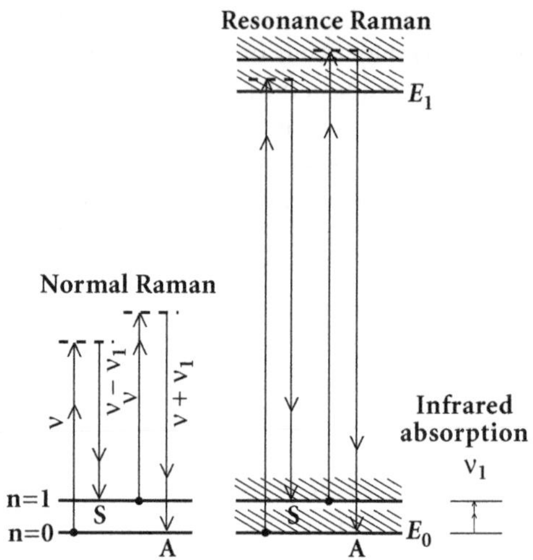

Figure 1. Level schemes for electronic and vibrational transitions involved in Raman scattering, resonance Raman scattering, and infrared absorption.

Rayleigh scattering occurs if the energy, ν, of the scattered photon is equal to that of the incident one (first term of Equation (2)). Rayleigh (1871) demonstrated that the Earth's blue sky was due to scattering of light by atmospheric particles, with a scattering intensity proportional to the inverse fourth power of the wavelength ($I \approx A/\lambda^4$). So, the blue (shorter wavelength) end of the visible spectrum is scattered more strongly, giving the sky its characteristic color.

As shown in Figure 1, the process is called Raman scattering, if the energy of the scattered photon is different from that of the incident one (the process is referred to as Brillouin scattering when the phonon emitted or absorbed is acoustic, and Raman scattering when the phonon is optical) (inelastic scattering, second term of Equation (2)). As a result, light with frequencies $(\nu \pm \nu_i)$, $(\nu \pm \nu_j)$, $(\nu \pm \nu_k) \ldots$ is emitted. In order that the total energy may be conserved, the vibrational energy of the sample is changed. If the irradiated molecule gains energy, the scattered lines in the Raman spectrum are at $(\nu - \nu_i)$ and are called "Stokes lines." If the sample loses energy, then the scattered lines are at larger frequencies than ν (at $(\nu + \nu_i)\ldots$) and are called "anti-Stokes lines."

Classical theory shows that a vibration is Raman active if the polarizability changes during the vibration, and the Stokes lines are always stronger than the anti-Stokes lines. (This latter fact is explained by quantum theory as a consequence of the Maxwell–Boltzmann distribution law.) In crystals, the photon wave vectors are small compared with the dimensions of the Brillouin zone; thus, the momentum conservation law can be obeyed only if the wave vectors of the scattered phonons are also very small (so only the zone center phonons participate in the Raman scattering). It is difficult to determine which of the phonons are Raman active, as

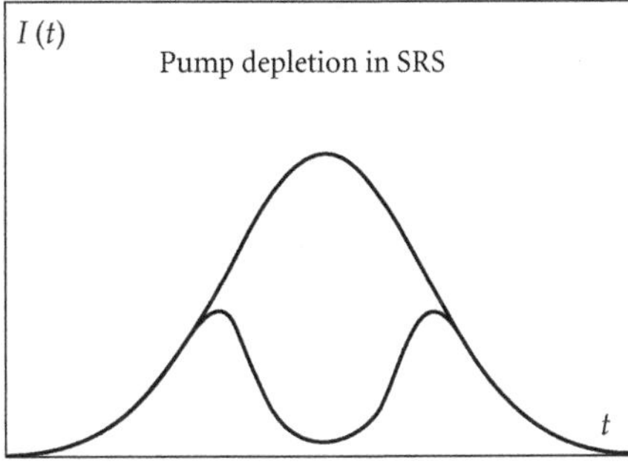

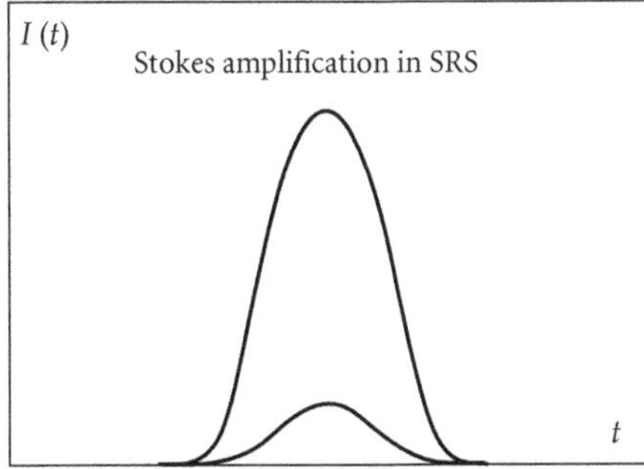

Figure 2. Stimulated Raman scattering: depletion at the pump frequency, and amplification of the Stokes beam.

changes of the components of the polarizability tensor are not obvious from inspection of the normal modes of vibration. Considering the symmetry of the molecule or the crystal and the polarization properties of the Raman lines, application of group theory gives a clear-cut solution to this problem (Poulet & Mathieu, 1970). If the polarization P contains terms of higher order (neglected in Equation (1)), these subsequent terms give hyper-Raman effects.

When the energy of the exciting line coincides with that of an allowed electronic transition of the compound, the intensity of the Raman spectra is enhanced and a series of overtones is observed—this mechanism is referred to as "resonance Raman scattering." This technique has been used in solid state and biophysics studies to identify intrinsic localized modes (Swanson et al., 1999).

Stimulated Raman scattering (SRS) is a nonlinear process in which a medium is irradiated by an intense laser beam at ν_p (pump frequency) and by a weaker beam at ν_s (Stokes frequency). The frequency spacing $\nu_p - \nu_s$ corresponds to a Raman frequency of the medium. In the simpler case, SRS amplifies the weak Stokes beam and depletes the pump beam (Raymer & Walmsley, 1991; see Figure 2).

A Raman soliton has been predicted (Chu & Scott, 1975) as a solution of the equations for transient SRS, and an experimental observation of transient SRS in H_2 gas was reported (Drühl et al., 1983). Recently, a complete interpretation of the above experiment has been given (Claude et al., 1995). The Raman soliton is still to be observed.

Picosecond time-resolved coherent anti-Stokes Raman spectroscopy (CARS) is a method for studying vibrational dephasing in the time domain. It is a special case of four-wave mixing (Levenson, 1982). Two intense simultaneous pulses at ν_p (pump) and ν_s (Stokes) specifically excite each Raman active mode at $\nu_p - \nu_s$. The third beam (probe pulse), also at the frequency ν_p, is delayed by the variable time t and stimulates coherent anti-Stokes emission. This technique is used to study the vibrational lifetime in anharmonic crystals and biomolecules to identify breathers excitations (Kosic et al., 1984).

Infrared (IR) absorption occurs when a compound is irradiated with frequencies that match its natural vibrational (or rotational) modes (Nakamoto, 1986). Absorption spectra in the infrared region originate in transitions between two vibrational levels of a sample in the electronic ground state. The absorption intensity depends on the change in dipole moment that occurs as a result of molecular vibration. If a molecule at equilibrium has a center of symmetry, the vibrations for which the center of symmetry is shared will be infrared inactive. Symmetry selection rules allow prediction of infrared activity when the associated normal coordinate belongs to the same species as one of the components of the dipole moment. When internal vibrations are coupled with phonons, nonlinear characters are observed in the infrared absorption bands: splitting, corresponding to the creation of a vibrational polaron or breather, intensity strongly dependent on the temperature, or an unusual number of overtones (Careri et al., 1984; Barthes et al., 2002). Time-resolved IR spectroscopy is of crucial importance in this field (Hamm et al., 1998; Xie et al., 2000).

MARIETTE BARTHES

See also **Nonlinear optics; Pump-probe measurements; Spectral analysis**

Further Reading

Barthes, M., Vik, A.F., Spire, A., Bordallo, H.N. & Eckert, J. 2002. Breathers or structural instability in solid L-Alanine: a new IR and inelastic neutron scattering vibrational spectroscopic study. *Journal of Physical Chemistry*, A 106(21): 5230–5241

Careri, G., Buontempo, U., Galluzzi, F., Gratton, E., Shyamsunder, E. & Scott, A.C. 1984. Spectrographic evidence for Davydov-like solitons in acetanilide. *Physical Review* B, 30: 4689–4702

Chu, F.Y.F. & Scott, A.C. 1975. Inverse scattering transform for wave-wave scattering. *Physical Review* A, 12: 2060–2064

Claude, C., Ginovart, F. & Léon, J. 1995. Nonlinear theory of transient stimulated Raman scattering and its application to long-pulse experiments. *Physical Review* A, 52: 767–782

Drühl, K., Wenzel, R. & Carlsten, J. 1983. Observation of solitons in stimulated Raman scattering. *Physical Review Letters*, 51: 1171

Hamm, P., Lim, M. & Hochstrasser, R.M. 1998. The structure of the amide I band of peptides measured by femtosecond

nonlinear IR spectroscopy. *Journal of Physical Chemistry* B, 102: 6123–6138
Kosic, T., Cline, R. & Dlott, D. 1984. Picosecond coherent Raman investigation of the relaxation of low frequency vibrational modes in amino acids and peptides. *Journal of Chemical Physics*, 81: 4932
Levenson, M. 1982. *Introduction to Nonlinear Laser Spectroscopy*, New York: Academic Press
Nakamoto, K. 1986. *Infrared and Raman Spectra of Inorganic and Coordination Compounds*, New York: Wiley
Poulet, H. & Mathieu, J.P. 1970. *Spectres de Vibration et Symétrie des Cristaux*, London: Gordon & Breach
Rayleigh, Lord (John William Strutt). 1871. On the light from the sky, its polarization and colour. *Philosophical Magazine*, 41: 107–120, 274–279
Raymer, M. & Walmsley, I.A. 1991. The quantum coherence properties of stimulated Raman scattering. *Progress in Optics*, 28: 216
Swanson, B.I., Brozik, J.A., Love, S.P., Strouse, G.F., Shreve, A.P., Bishop, A.R., Zang, W.-Z. & Salkola, M.I. 1999. Observation of intrinsically localized modes in a discrete low-dimensional material. *Physical Review Letters*, 82: 3288–3291
Xie, A.H., van der Meer, L., Hoff, W. & Austin, R.H. 2000. Long-lived amide I vibrational modes in myoglobin. *Physical Review Letters*, 84: 5435–5438

RAYLEIGH–BÉNARD CONVECTION

See **Thermal convection**

RAYLEIGH–TAYLOR INSTABILITY

The Rayleigh–Taylor instability arises when a layer of heavy fluid is accelerated into a lighter one. This occurs, for instance, if a heavy fluid is put on top of a lighter one in a uniform gravity field. Under these unstable conditions, small disturbances of the interface between the two layers grow, at first exponentially with a growth rate proportional to $\sqrt{k}$ where k is the wave number, as long as the perturbation is small compared with the wavelength. The subsequent nonlinear growth has been mostly investigated numerically or experimentally. The Rayleigh–Taylor instability was first discovered by Lord Rayleigh (John William Strutt) in 1883 for the gravitational case and then extended to the case of any accelerating fluid by Geoffrey Taylor in 1950.

One of Taylor's major observation is related to the Manhattan Project (September 1942 to January 1945), which ultimately led to the successful development of atomic bombs. Indeed, in May 1944, when he was invited to join the project, Taylor pointed out problems due to implosion instabilities (especially of the Rayleigh–Taylor type). This led to a very conservative design to minimize possible instabilities. At the same time, James Tuck brought the idea of explosive lenses for detonation wave shaping, suggesting the use of three-dimensional (3-d) lenses to create a spherical implosion and reduce instabilities.

To study an example of the Rayleigh–Taylor instability, let us look at the interface between a heavy fluid of density ρ_2 on top of a lighter fluid of density ρ_1 subject to gravity. The interface is assumed to be horizontal and both fluids are assumed to be at rest initially. For simplicity, we look at the two-dimensional case. We wish to show that a small disturbance of the interface will grow exponentially fast. Indeed, a small disturbance can engender small potential velocities $v_1 = -\nabla\phi_1$ and $v_2 = -\nabla\phi_2$ in the lower and upper fluid respectively, such that

$$\phi_2 = Ae^{-ky+nt}\cos(kx), \tag{1}$$

$$\phi_1 = -Ae^{ky+nt}\cos(kx). \tag{2}$$

Then the interface that is given by the equation $y = \eta(t, x)$ satisfies $\partial_t\eta = -\partial_y\phi_1(y=0) = -\partial_y\phi_1$ $(y=0)$. Hence,

$$\eta = \frac{Ak}{n}e^{nt}\cos(kx). \tag{3}$$

Neglecting the nonlinear term, the Euler equation reads $-\rho_2\partial_t\nabla\phi_2 + \nabla p_2 = -\rho_2 g\boldsymbol{j}$, which yields

$$p_2 = p - \rho_2 gy + n\rho_2\phi_2, \tag{4}$$

where p is the mean pressure at the interface. Similarly, the pressure in the lower fluid is given by

$$p_1 = p - \rho_1 gy + n\rho_1\phi_1. \tag{5}$$

At the interface, the pressure should be continuous, namely, $p_1 = p_2$ and, hence,

$$(\rho_2 - \rho_1)g\eta = n(\rho_2 + \rho_1)Ae^{nt}\cos(kx). \tag{6}$$

Combining (3) and (6), we get

$$n^2 = -kg\frac{(\rho_1 - \rho_2)}{(\rho_2 + \rho_1)}. \tag{7}$$

The Atwood number is defined by $At = (\rho_1 - \rho_2)/(\rho_2 + \rho_1)$. In the case $\rho_2 > \rho_1$ (heavy fluid on top of lighter fluid), $-1 < At < 0$, and

$$n = \pm\sqrt{-At\,kg}. \tag{8}$$

The unstable mode corresponds to $n = \sqrt{-At\,kg}$ and yields exponential growth as long as the nonlinear terms can be neglected, namely, as long as η is small compared with the wavelength. In experiments, surface tension and viscosity have a regularizing effect on the motion by damping the waves with large wave numbers. Indeed, for small Weber (and/or) large Reynolds numbers, the Rayleigh–Taylor instability has a short-wave character due to a balance between the exponential growth and the viscous or the surface tension effect, and hence, there exists a wave number that has a maximum growth rate. This is similar to the Rayleigh number, which represents the ratio of the destabilizing effect of buoyancy to the stabilizing effect of viscous force in the Rayleigh–Bénard instability.

Experiments and numerical simulations have given much insight into the different stages of the instability formation. However, they have yielded few quantitative results. Different numerical methods have been used such as the 3-d front tracking and the lattice Boltzmann method. In all cases, the geometric complexity of the interface is a source of difficulties for most of the algorithms. As explained in Sharp (1984), we can divide the growth of the instability into four stages. In the first, we have an exponential growth that can be quantitatively computed using the linear perturbation theory, which is no longer valid if the amplitude is about 20% of its wavelength. During the second stage, the perturbation grows nonlinearly to form bubbles of light fluid rising into the heavy one and spikes of heavy fluid falling into the light one. This stage is strongly influenced by the three-dimensional effects. Experiments and numerical simulations have shown that the bubbles rise at an approximately constant velocity during this stage. The third stage is characterized by the development of additional structures on the spikes. Moreover, the Kelvin–Helmholtz instability begins to develop due to the jump in the tangential velocity at the interface between the two fluids. The heavy fluid begins to roll up along the sides of the spikes to form "mushrooms." This phenomenon is more pronounced when the Atwood number is small and is a 3-d phenomenon. Eventually, the flow evolves into turbulent or chaotic mixing, which dominates the fourth stage.

In engineering applications, the Rayleigh–Taylor instability plays an important role in inertial confinement fusion (ICF). It is also present in the extraction of oil. Indeed, to extract oil from the Earth, water is accelerated into oil, and since oil is lighter, the interface presents the Rayleigh–Taylor instability. Some of the engineering solutions consist in adding polymers to the water. Moreover, there are many astrophysical and geophysical objects that present the Rayleigh–Taylor instability such as the supernova explosions, mantle convection, and deep convection in the ocean.

Another instability that is similar to the Rayleigh–Taylor one is the Ritchtmeyer–Meshkov instability, which arises in the case of an impulsive or instant loading as opposed to a constant acceleration (such as gravity) in the Rayleigh–Taylor case. In many applications the situation is intermediate between these two limiting cases.

NADER MASMOUDI

See also **Kelvin–Helmholtz instability; Wave stability and instability**

Further Reading

Chandrasekhar, S. 1961. *Hydrodynamic and Hydromagnetic Stability*, Oxford: Oxford University Press

Rayleigh, Lord (John William Strutt). 1883. Investigation of the character of the equilibrium of an incompressible heavy fluid of variable density. *Proceedings of the London Mathematical Society*, 14: 170–177

Sharp, D.H. 1984. An overview of Rayleigh–Taylor instability. *Physica* D, 12: 3

Taylor, G.I. 1950. The instability of liquid surfaces when accelerated in a direction perpendicular to their planes. I. *Proceedings of the Royal Society of London, Series* A, 201: 192–196

REACTION-DIFFUSION SYSTEMS

How did life appear and develop on Earth? Although there is not complete agreement, a convincing answer is that given the required conditions (which were almost surely satisfied billions of years ago), life spontaneously emerged through the endless battle of survival of the fittest from a primordial chemical soup. The theory of random catalytic networks (Kauffman, 1995) shows that autocatalytic reactions are likely in this context; thus, the theory of interactions between chemical reactions and molecular diffusion takes center stage in emergence of biological life from atoms and molecules in a system.

Today, reaction-diffusion systems have found many applications ranging from chemical and biological phenomena to medicine (physiology, diseases, etc.), genetics, physics, social science, finance, economics, weather prediction, astrophysics, and so on (Aronson & Weinberger, 1975; Grindrod, 1996; Murray, 2002; Scott, 2003). Even for phenomena that bear no initial resemblance to these processes, it is sometimes useful and productive to use the reaction-diffusion metaphor in order to gain insight into their dynamics.

An important contribution to this subject comes from the theory of pattern formation in nature. Many physical phenomena giving rise to natural patterns can be understood in terms of the interaction of a short-range self-enhancing reaction and a long-range antagonistic reaction. Take a fire, for example. It is a self-enhancing process: more heat is released as more fuel is burned. In the process oxygen and fuel which act as antagonistic factors are consumed and this may lead to a fire's extinction if fuel is not replenished. Also, the heat produced is transported from its local source through diffusion. Thus, we have the two main ingredients for pattern formation, namely, a local antagonistic interaction between two species (or reactions) coupled with a means of transport of their products. Historically, this is also one of the first examples of reaction-diffusion systems studied scientifically, for obvious reasons given the necessity for improved heating and lighting at the beginning of the industrial age. In his "Christmas Lectures" at the Royal Society in London, Michael Faraday discussed the importance of understanding the candle flame and its analogy with the process of respiration of biological organisms (Faraday, 1861). The flame of the candle is an archetypical nonlinear reaction-diffusion system that today, after decades of research, provides a basis for the science of combustion. In this article, we review the essential topics in the field of reaction-diffusion systems from a theoretical point of view and also consider some of their numerous applications.

Theoretical Aspects

A reaction-diffusion (RD) equation is typically obtained by combining Fick's law of diffusion with the chemical reaction rate law. Although the theory can be made rigorous by using the theory of stochastic differential equations leading to the Fokker–Planck equation (Øksendal, 2003), we present here a heuristic argument. If we consider for simplicity a small domain interval on the line inside which we have a concentration c of some reacting species, then the diffusive flux J_{in} of c into one side of the small region will depend on the concentration gradient, $\partial c/\partial x$, at that boundary and the diffusion coefficient, D, with $J_{\mathrm{in}} = -D(\partial c/\partial x)_{\mathrm{in}}$. The parameter $D > 0$ is called diffusivity with physical units of m^2/s. The diffusive flux out of the region at the other side J_{out} will similarly be given by $J_{\mathrm{out}} = -D(\partial c/\partial x)_{\mathrm{out}}$, where the concentration gradient is now evaluated at the other boundary. The rate at which the concentration grows due to diffusion then depends on the difference between these two fluxes—and so involves the second derivative $\partial^2 c/\partial x^2$. If we add a kinetic reaction rate term $r(c)$, then the reaction-diffusion equation, which gives the rate of change of the concentration c in time at any spatial point, has the general form

$$\frac{\partial c}{\partial t} = D\frac{\partial^2 c}{\partial x^2} + r(c). \tag{1}$$

This can be extended to any number of spatial variables to read

$$\frac{\partial c}{\partial t} = D\Delta c + Q(\boldsymbol{x}, t, c, \ldots), \tag{2}$$

where Δ denotes the n-Laplacian $(1 \le n \le 3)$ and Q accounts for other influences including sources or sink terms. Other more complicated formulations for the flux terms are possible in diffusive processes; see Okubo & Levin (2001) for an account of these ideas in biology. In many sciences, the motion of particles or living organisms is subjected to both internal and external effects often acting simultaneously. For example, in biology, bacteria are known to move randomly (akin to diffusion) but are also able to follow a chemical gradient (chemotaxis). Mathematically, this leads to a description involving reaction-diffusion-chemotaxis equations. For example, chemical attractant signaling can give rise to spiral wave pattern formation at a certain life stage in colonies of the slime mold *Dictyostelium discoideum*. Other extensions include advection, electric and/or magnetic field effects, and so on. In many of these cases, the resulting mathematical description is far more complicated than the simple RD systems above and, for many situations, their exact formulation is still an open question.

In writing an equation such as (2), we will consider that c is a vector; thus, Equation (2) will describe a system of RD equations. A simple archetypical example for a RD system is a quadratic autocatalytic reaction between two chemicals according to the rule $\mathrm{A} + \mathrm{B} \to 2\mathrm{B}$ with rate $r = kab$. Denoting by a the concentration of A and by b that of B, the two species satisfy (after suitable scaling) the equations:

$$\frac{\partial a}{\partial t} = D_1 \Delta a - ab,$$
$$\frac{\partial b}{\partial t} = D_2 \Delta b + ab. \tag{3}$$

From a mathematical point of view, systems of equations such as (3) must be well posed in order to exhibit appropriate solutions. To specify the problem, let the state variables $c(\boldsymbol{x}, t), \ldots$ represent the density or concentration of some substance at time $t \ge 0$ and position $\boldsymbol{x}$ in R^n. Then Δc denotes the Laplacian of c with respect to the space variable $\boldsymbol{x}$, and Equation (1) is an example of a parabolic equation of evolution. If (1) holds for all $\boldsymbol{x}$ in R^n, then the problem is fully specified once appropriate initial conditions

$$c(\boldsymbol{x}, 0) = c_0(\boldsymbol{x}) \tag{4}$$

are known. If (1) holds in a limited domain $\Omega \subset R^n$, then we must impose boundary conditions on c at $\partial\Omega$ (the domain boundary) compatible with the physical situation. Neumann (or no-flux), Dirichlet, or Robin conditions are usually employed in applications. Although these are linear conditions on the variable c, nonlinear conditions could also be applied. The above system of differential equations with specified initial and boundary conditions is called an initial value problem or IVP in short.

For an IVP, one naturally asks whether there are solutions. In the case of reaction-diffusion systems, there are two different aspects to consider; local existence and global existence of the solutions. By local existence we mean the existence of the solutions over a short time interval. Global existence properties are exhibited by the solutions of the IVP when they are known to exist for all positive time. In some applications global existence is precluded because the solution exhibits blow-up in finite time. This means that there is $(\boldsymbol{x}_0, t_0)$ such that $c(\boldsymbol{x}, t) \to \infty$ as $(\boldsymbol{x}, t) \to (\boldsymbol{x}_0, t_0)$. These questions are difficult to deal with in general for RD systems although there is a well-developed body of results available in the literature (Grindrod, 1996). Moreover, even if the existence of a particular type of solution is established, further important theoretical questions involve the uniqueness of the solution and the stability of this solution to small perturbations. To approach these questions, one usually transforms the given system of RD equations

$$\frac{\partial c}{\partial t} = D\Delta c + f(\boldsymbol{x}, t, c, \nabla c) \quad (5)$$

into a differential equation in an abstract Banach space (for example, $L^2[\Omega]$) as in

$$c_t + Ac = f, \quad t > 0, \quad c = c_0 \quad \text{at } t = 0. \quad (6)$$

Then, the above questions reduce the problem to studying the properties of the (linear) operator A, which are mainly resolved if one knows its spectrum. However, in practice this is a difficult question as A has an infinite-dimensional spectrum.

For many RD systems of interest, in practice, there are several rather specific and powerful analytical methods that give detailed insights into the solutions, including comparison principles, invariant regions, matched asymptotic expansions, nonlinear bifurcation, group invariant symmetries, and so on (Grindrod, 1996).

Applications of Reaction-Diffusion Systems

Historically, some of the first applications of reaction-diffusion equations were in population dynamics, combustion, and nerve impulse conduction. Thus, one of the simplest reaction-diffusion equations is the Fisher–KPP equation

$$\frac{\partial u}{\partial t} = D\Delta u + f(u), \quad (7)$$

which was proposed independently by Ronald Fisher (1937) and Andrei Kolmogorov, Ivan Petrovsky and N. Piscounoff (1937) to explain the spread of genetic influences. In those papers a quadratic function of the form

$$f(u) = ku(1-u) \quad (8)$$

was used with $k > 0$ a parameter. A year later, Yakov Zeldovich and David Frank-Kamenetsky used the same equation but with f being a cubic polynomial as a model that represented flame front propagation (Zeldovich & Frank-Kamenetsky, 1938). Due to its generic form, Equation (7) soon found new interesting applications, for example, as a model of impulse conduction along an active nerve fiber, with the solution $u(x, t)$ representing the voltage across a cell membrane.

An important class of similarity solutions in the form of traveling waves (TW) is common to both the RD system (3) (and many of the type at (5)) and scalar equation (7). A traveling wave is a solution of the form $u(x, t) = u(x - vt) = u(y)$, where $y = x - vt$ is the traveling-wave coordinate with v being the wave

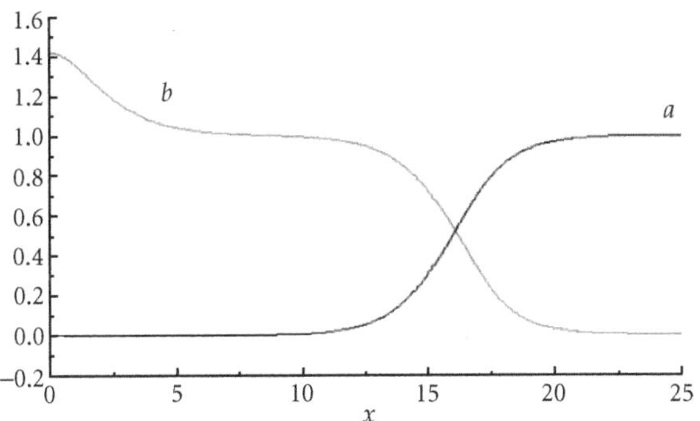

Figure 1. Typical concentration profiles for solutions a, b obtained by numerically solving the RD system (3).

speed. The problem is simplified because in the TW coordinate we have now to solve an ordinary differential equation (rather than a partial differential equation). For example, (7) generates the following TW equation to solve

$$\frac{\mathrm{d}^2 u}{\mathrm{d}t^2} + v\frac{\mathrm{d}u}{\mathrm{d}y} + f(u) = 0, \quad (9)$$

which must be supplemented with appropriate boundary conditions as $|y| \to \infty$.

Propagating waves are an important dynamical feature of many physical systems, hence, TW problems have been carefully studied. Key mathematical questions in these cases are the existence and possibly uniqueness of the waves. After the existence of a TW solution is established, it is of physical significance to study its stability under perturbations. Other questions are the influence of the initial conditions on the selection of a particular type of TW solution and the study of the shape of the TW solution. An extensive theory for TW solutions has been developed that deals with many interesting classes of RD systems (see Volpert et al., 1994; Grindrod, 1996; Scott, 2003). One particularly striking property of TW solutions for RD equations such as (3) or (7) is the existence of either a unique speed or a semi-infinite spectrum of speeds with a positive threshold boundary. The first case corresponds to a cubic nonlinearity, as in the nerve conduction application above, whereas the latter applies to a quadratic nonlinearity. These features can be contrasted with the paradox of infinite speed behavior exhibited by the classical diffusion equation. For example, with the quadratic form (8), Equation (7) has a TW solution for all $v \geq v_0 = 2\sqrt{kD}$. A similar property applies to system (3) although in practice only for a limited class of RD equations is an analytical expression known for v_0. In Figure 1, we show a typical profile of a TW solution to system (3) obtained from numerical simulations on a semi-infinite one-dimensional spatial domain via an implicit finite-difference scheme.

One of the oldest but still active fields of application is the modeling of nerve impulse conduction along a

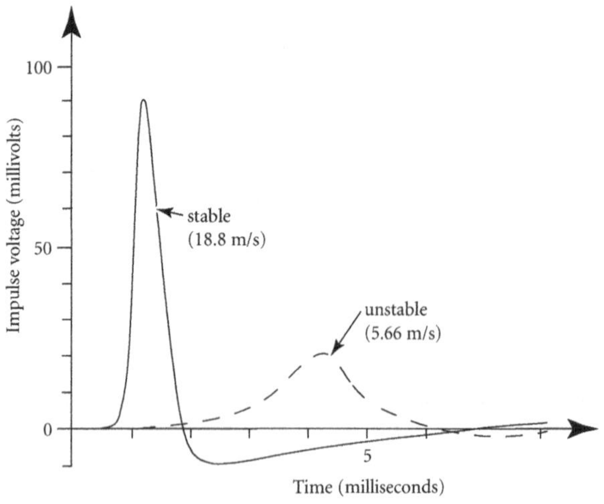

Figure 2. A full-sized action potential and an unstable threshold impulse for the HH axon: (Courtesy of A.C. Scott.)

responsive fiber (Scott, 2003). In 1952, Alan Hodgkin and Andrew Huxley formulated a detailed description of the dynamics of membrane ionic current for a dozen giant squid axons (Hodgkin & Huxley, 1952). They produced the following differential system describing the dynamics of the voltage $V(x, t)$ across the nervous cell (assumed one-dimensional):

$$rc_a\frac{\partial V}{\partial t} = \frac{\partial^2 V}{\partial x^2} + rj_i, \tag{10a}$$

where

$$j_i = 2\pi a J_i\,,\ J_i = G_{\mathrm{K}}n^4(V - V_{\mathrm{K}}) + G_{\mathrm{Na}}m^3h(V - V_{\mathrm{Na}}) + G_{\mathrm{L}}(V - V_L). \tag{10b}$$

Here, a is the radius of the fiber, r is the longitudinal resistance per unit length of the fiber, c_a is the membrane capacitance per unit length, and j_i is the total ionic current flowing across the membrane per unit length. The expressions appearing in Equation (10b) are given in terms of differential first-order rate equations describing the concentrations of potassium, sodium "turn-on" and "turn-off" variables (n, m, and h). These equations are determined from experimental data. The RD system (10a,10b) is now known as the Hodgkin–Huxley (HH) system. Using TW theory, one can show that HH is an excitable RD system for parameter ranges of interest that admit traveling pulse solutions called action potentials (see Figure 2).

From a theoretical perspective, the HH system is complicated; thus, simplified theoretical models have been proposed to capture qualitative features. One such model is the system proposed by FitzHugh and Nagumo (FitzHugh, 1961), which reads

$$\frac{\partial V}{\partial t} = \frac{\partial^2 V}{\partial x^2} + F(V) + R\,,\ \frac{\partial R}{\partial t} = \varepsilon(V - c_a - b_a R), \tag{11}$$

Figure 3. A computer generated scroll wave. (Courtesy of A. Winfree.)

where R is the slowly changing recovery variable and F has a cubic shape. The FitzHugh–Nagumo system admits periodic traveling waves and pulses both stable and unstable. In 2 or 3 spatial dimensions, target (ring) and spiral waves have been numerically produced that today are known to be characteristic types of TW solutions of an excitable system (Scott, 2003).

The most striking illustrations of target and spiral waves can be seen in excitable chemical reactions. Historically, the first such demonstration was produced in the chemical medium proposed by Belousov and Zhabotinsky, now known as the BZ reaction (Zhabotinsky, 1991). A dynamical system is called excitable when it responds in a qualitatively different way to perturbations that are below and above some threshold value, above which it typically amplifies those perturbations. After a pulse has been generated, the system takes some time (refractory state) until it can again support an impulse. In two dimensions, one can find target (ring) and spiral waves as in a typical BZ experiment (*See* photo in **Belousov–Zhabotinsky reaction** and also in color plate section).

To describe these solutions analytically, one can use a simplified model of the BZ reaction, such as the two-species Oregonator model developed by Field and Noyes (1974):

$$\begin{aligned} \varepsilon\frac{\partial a}{\partial t} &= \varepsilon^2\Delta a + a(1-a) - f\frac{a-q}{a+q}b, \\ \frac{\partial b}{\partial t} &= \varepsilon\Delta b + b - a, \end{aligned} \tag{12}$$

where ε, f, q are kinetics parameters related to the reagents rate reactions and a, b are proportional with the concentrations of $HBrO_2$ and a metal ion. Here the diffusion operator Δ denotes one-, two-, or three-dimensional diffusion. In the latter case, more complicated patterns are found such as toroidal scroll waves and multi-armed spiral waves. Figure 3 shows a computer generated scroll ring obtained in a model of the BZ reaction, which has also been observed experimentally.

To understand such solutions a geometrical theory of traveling waves propagating on arbitrary two- and three-dimensional manifolds was developed by Tyson, Keener, Grindrod, and others around the mid-1980s, (see Grindrod, 1996). The idea is to compare the problem of three-dimensional wave propagation with that for wave propagation in one dimension. Consider the equation

$$\varepsilon\frac{\partial u}{\partial t} = \varepsilon^2 \Delta u + f(u), \qquad (13)$$

which is in the Oregonator model of the BZ reaction (see Equation (12)). Suppose that u_1, u_2 are two solutions to $f(u)=0$ and that (13) has a TW solution of the form $u=u(y)$ where $y=(x-vt)/\varepsilon$, so that

$$u'' + vu' + f(u) = 0 \qquad (14)$$

satisfying $u \to u_1$ as $y \to \infty$ and $u \to u_2$ as $y \to -\infty$ for some speed $v>0$. For a TW structure to exist in three dimensions, there should be an oriented surface M moving through R^3 such that $u \sim u_1$ ahead of the surface and $u \sim u_2$ behind it. Introducing a normal stretched variable depending on a curvilinear system of coordinates, one can show that a solution $u=\phi(\xi)$ exists subject to $\phi \to u_1$ as $\xi \to \infty$ and $\phi \to u_2$ as $\xi \to -\infty$ if and only if ϕ satisfies the TW equation

$$\phi_{\xi\xi} + (N + \varepsilon K)\phi_\xi + f(\phi) = 0. \qquad (15)$$

From (14) and (15) we find that $u=\phi(\xi)$ is the inner solution of the problem when

$$v = N + \varepsilon K, \qquad (16)$$

which is called the eikonal equation (Grindrod, 1996). Here N denotes the normal velocity, and K is twice the mean curvature of M. For example for circular waves, the speed of the curved wave v should be smaller than that of the planar wave v_0. The eikonal equation gives $v=v_0-2D/r$, where r is the radius and D is the diffusion coefficient. Hence we have a condition for the initiation of circular waves that is purely geometric, namely, $r_{\text{crit}}=2D/v_0$ for the critical radius of wave initiation. When two circular waves collide, they create cusp-shaped regions with positive curvature. The theory can also be applied to armed spirals and toroidal scroll waves (Grindrod, 1996).

Spiral waves generated by reaction-diffusion processes also occur in surface reactions as tiny spiral waves, visible only through a microscope, that form out of catalytic nitric oxide reduction by hydrogen on rhodium or platinum surfaces. Such pattern formation in surface reactions can be imaged by photoemission electron microscopy (PEEM).

Flames on a spinning disk have been studied at NASA where it is shown that target and spiral waves appear in a mixture of butane in He–O_2. A simple such model is the Salnikov RD system (Scott et al., 1997):

$$a_t = \nabla^2 a + \mu - af(b), \qquad (17a)$$

$$b_t = Le\nabla^2 b + \frac{1}{\kappa}(af(b) - b). \qquad (17b)$$

Figure 4. A spiral galaxy in Centaurus. VLT, European Southern Observatory.

Here $f(b)=\exp[b/(1+\gamma b)]$ and $Le, \mu, \kappa, \gamma > 0$ are physical parameters with $\kappa \ll 1$.

Spiral forms are omnipresent throughout the visible and invisible universe in galaxies, accretion disks around black holes, coalescing interstellar clouds, and many other forms of matter and energy (see Figure 4). Lee Smolin (1996) has suggested that the formation of some spiral galaxies can be regarded as an RD process, and he produced a theory of cosmos as a self-sufficient organically developing natural system. In this scenario the galactic spirals (and everything inside them) are gargantuan relatives of the BZ waves.

Waves of spreading depression appear as electrochemical waves associated with the depolarization of the neuronal membrane of the brain as they spiral around lesions of the cortex. There is speculation that this wave activity is linked to epilepsy. Since Art Winfree's pioneering work, scroll waves have been linked quite firmly with cardiac arrhythmias (Keener & Sneyd, 2002). In particular, the onset of fibrillation seems to be connected with the development of scroll type waves (high-frequency waves of electrical activation that recirculate repeatedly, preventing normal function) in the heart muscle. In this case, the heart contractions are much smaller in amplitude than the normal coordinated contraction and also aperiodic, which has potentially fatal consequences.

In early embryo development, circular calcium waves propagate and occasionally annihilate on the surface of a fertilized egg cell as a precursor to cell division. However, in frog eggs they take the form of spiral waves whose purpose is still unclear.

The dynamics of biological populations can self-organize in propagating traveling structures akin to

RD waves (Okubo & Levin, 2001). This is a classical topic in ecology with ramifications for studies in biodiversity. Among the pioneers of the field were Alfred Lotka in 1910 (Lotka, 1925) and Vito Volterra (1926), who have shown that equations similar to those describing reacting chemicals can provide a crude description of interactions between a predator population and its multiplying prey population. Thus, Lotka–Volterra RD type interactions are the heart of modeling in much of ecological dynamics and modeling of epidemics. The spatial spread of epidemics can be usefully approached with a "prey-predator" type dynamics involving the three species: S (the susceptibles), I (infectives), and R (removed individuals) in a population (Kermack & McKendrick 1932; Murray, 2002):

$$S_t = -\beta IS, \quad I_t = \beta IS - \gamma I + D\Delta I,$$
$$R_t = \gamma I. \qquad (18)$$

Equations (18) give a simple model of the spatial spread of rabies by fox, where D is the diffusion coefficient of the infectious foxes and β represents the rate of infection per susceptible per unit density of infectives. More sophisticated versions of such models that are used today to model epidemics (both in humans and in animals) can capture other effects such as secondary outbreaks.

The above discussion has mainly concentrated on the applications of the traveling-wave solutions of the RD systems. However, a much richer solution structure exists in these systems. In fact, the spectrum of possibilities ranges from simple stationary (or time independent) solutions up to complicated spatiotemporal chaotic solutions.

Perhaps one of the most spectacular applications of RD systems is in morphogenesis. The basic question here is to understand how the complicated process of shaping and patterning in an early embryo takes place starting from a uniform structure, the initial egg. In 1952, Alan Turing proposed that this process could be accounted for in terms of the underlying chemical reactions taking place in the embryo. In his paper Turing proposed some hypothetical reactions that could generate spontaneous symmetry breaking, leading to stable spatial patterns (Turing, 1952). Today this mechanism for spatial pattern formation is called Turing instability (or diffusion-driven instability) in recognition of Turing's seminal work. A counterintuitive feature of Turing's prediction is that diffusion can act as a destabilizing force. Murray (1988, 2002) showed that activator-inhibitor RD systems can model the beautiful array of patterns seen on animal skin markings. The idea Murray presented is that at a very early stage, the embryo acquires a "pre-pattern" of chemical morphogens (in Turing's parlance) that is later read out by melanocytes (pigment producing cells). In this way, spotted or striped or a combination of both patterns can be accounted for. Furthermore, Murray showed how the geometrical aspects of the RD domain can considerably alter the final outcome of the patterns. This theory explains why spotted animals (such as cheetah, jaguars, and leopards) can have striped tails, but striped animals (such as zebras and genets) cannot have spotted tails. An equally spectacular application of Turing's theory was proposed by Meinhardt (1998) to account for many patterns seen on seashells.

The book by Murray (2002) contains further references to expanding the above RD framework on modeling crucial processes in morphogenesis, such as feather germs, the initiation of teeth primordia, cartilage and condensation in limb, and epidermis development. Most of them require expansion of the RD framework to incorporate mechanics leading to the study of mechano-chemical models.

With growing evidence of biological morphogens in living tissues, it is now clear that Turing's theory may play a crucial role in explaining biological development. Furthermore, a recent extension of the applications of RD systems in biological pattern formation has strengthened considerably the theoretical arguments in favor of the role of RD systems in biological modeling. Satnoianu et al. (2001) have shown that coupling of diffusion by advection in the presence of autocatalytic reactions can lead to an improved recipe for pattern formation, which is able to account also for biological growth. The new mechanism of patterning, now termed flow and diffusion distributed structures (or FDS), produces both stationary and traveling structures and, thus, can be viewed as an unifying theoretical construct for many of the patterning processes described above (Satnoianu, 2003). The FDS mechanism is a robust patterning process that can explain somitogenesis in vertebrates (Kaern et al., 2001).

Reaction-diffusion-advection equations have been also used in astrophysics, geochemistry, fluid dynamics, and finance, for example. Other applications of RD systems arise in chemotaxis, modeling bacterial growth patterns, the process of wound healing, or the spread of cancer cells in healthy tissue to enumerate only a few such ideas in mathematical biology.

Razvan Satnoianu

See also **Belousov–Zhabotinsky reaction; Brusselator; Cardiac arrhythmias and electrocardiogram; Chemical kinetics; Diffusion; FitzHugh–Nagumo equation; Heat conduction; Morphogenesis, biological; Pattern formation; Population dynamics; Scroll waves; Spiral waves; Turing patterns; Zeldovich–Frank-Kamenetsky equation; Vortex dynamics in excitable media**

Further Reading

Aronson, D.G. & Weinberger, H.F. 1975. Nonlinear diffusion in population genetics, combustion and nerve pulse propagation. In *Partial Differential Equations and Related Topics*, edited by J.A. Goldstein, New York: Springer

Faraday, M. 1861. *A Course of Six Lectures on the Chemical History of a Candle*. Reprinted as *Faraday's Chemical History of a Candle*, Chicago: Chicago Review Press, 1988

Field, R.J. & Noyes, R.M. 1974. Oscillations in chemical systems, IV. Limit cycle behaviour in a model of a real chemical reaction. *Journal of Chemical Physics*, 60: 1877–1884

Fisher, R.A. 1937. The wave of advance of advantageous genes. *Annals of Eugenics*, 7: 353–369

FitzHugh, R. 1961. Impulse and physiological states in theoretical models of nerve membrane. *Biophysical Journal*, 1: 445–466

Grindrod, P. 1996. *The Theory and Applications of Reaction-Diffusion Equations*, Oxford: Clarendon Press

Hodgkin, A.L. & Huxley, A.F. 1952. A quantitative description of membrane current and its application to conduction and excitation in nerve. *Journal of Physiology*, 117: 500–544

Kaern, M., Menzinger, M., Satnoianu, R.A. & Hunding, A. 2001. Chemical waves in open flows of active media: their relevance to axial segmentation in biology. *Faraday Discussions*, 120: 295–312

Kauffman, S.A. 1995. *At Home in the Universe. The Search for the Laws of Self-organization and Complexity*, New York: Oxford University Press

Keener, J. & Sneyd, J. 2002. *Mathematical Physiology*, New York: Springer

Kermack, W.O. & McKendrick, A.G. 1932. Contributions to the mathematical theory of epidemics. *Proceedings of the Royal Society, London*, A 115: 700–721

Kolmogoroff, A., Petrovsky, I. & Piscounoff, N. 1937. Étude de l'équation de la diffusion avec croissance de la quantité de matière et son application à un problème biologique. *Moscow University Bulletin Mathematics*, 1: 1–25

Lotka, A.J. 1925. *Elements of Physical Biology*, Baltimore: Williams and Wilkins

Meinhardt, H. 1998. *The Algorithmic Beauty of Seashells*, Berlin: Springer

Murray, J.D. 1988. How the leopard gets its spots. *Scientific American*, 258: 62

Murray, J.D. 2002, 2003. *Mathematical Biology*, 3rd edition, vols. 1 and 2, Berlin and New York: Springer

Øksendal, B. 2003. *Stochastic Differential Equations*, 6th edition, Berlin and New York: Springer

Okubo, A. & Levin, S.A. 2001. *Diffusion and Ecological Problems: Modern Perspectives*, Berlin and New York: Springer

Satnoianu, R.A. 2003. Coexistence of stationary and traveling waves in reaction-diffusion-advection systems. *Physical Review* E, 68: 032101

Satnoianu, R.A., Maini, P.K. & Menzinger, M. 2001. Parameter space analysis, pattern sensitivity and model comparison for Turing and stationary flow and diffusion distributed structures (FDS). *Physica* D, 160: 79–102

Scott, A. 2003. *Nonlinear Science: Emergence and Dynamics of Coherent Structure*, 2nd edition, Oxford and New York: Oxford University Press

Scott, S.K., Wang, J. & Showalter, K. 1997. Modeling studies of spiral waves and target patterns in premixed flames. *Journal Chemical Society of Faraday Transactions*, 93: 1733–1739

Smolin, L. 1996. Galactic disks as reaction-diffusion systems. eprint arXiv.org astro-ph/9612033, 22 pages

Turing, A. 1952. The chemical basis of morphogenesis. *Philosophical Transactions of the Royal Society of London* B, 327: 37–72

Volpert, A.I. & Volpert, V.A. 1994. *Travelling Wave Solutions of Parabolic Systems*. Translations of mathematical monographs, vol. 140, Providence, RI: American Mathematical Society

Volterra, V. 1926. Variations and fluctuations of a number of individuals in animal species living together. In *Animal Ecology*, New York: McGraw Hill, pp. 409-448, 1931. Translation by R.N. Chapman

Zeldovich, Ya.B. & Frank-Kamenetsky, D.A. 1938. K teorii ravnomernogo rasprostranenia plameni [toward a theory of uniformly propagating flames]. *Doklady Akademii Nauk SSSR*, 19: 693–697

Zhabotinsky, A.M. 1991. A history of chemical oscillations and waves, *Chaos*, 1: 379–386

RECURRENCE

Recurrence means repetition, and this, in turn, implies the presence of two actors: the particular event that is coming back after having occurred in the past and the law prescribing how events unfold in time.

The crossing of the vertical position by a well-serviced pendulum is a recurrent event. Here the event coincides with a particular value of the position coordinate, and the intervening law is Newton's second law in the presence of gravity. We deal here with the simplest version of recurrence, namely, strict periodicity. But one could also say that in a capitalistic economy, economic growth or slow down are recurrent events. The event is now an attribute of a whole succession of outcomes interrupted by irregular periods where this attribute is absent, and the intervening laws are no longer directly reducible to nature's dispassionate fundamental interactions but involve, rather, competing human agents each one of whom wishes to maximize his profit. In a still different vein, the need for survival has made man aware of the repeated appearance (recurrence) of weather patterns associated with different winds and/or precipitation patterns. The Greeks constructed an octagonal structure, the Tower of Winds, which can still be visited in Athens. It depicts the wind from each cardinal point and comprises bas-relief figures representing the typical type of weather associated with such a wind. The event is now a lumped, coarse-grained characterization of the state of the atmosphere that came to be known later on as *Grosswetterlagen* or "atmospheric analogs," and the laws are those of fluid mechanics and thermodynamics as applied to a rotating frame.

In the physical sciences, the description usually adopted views the system of interest as a deterministic dynamical system. At first sight this entails that an event is to be associated with the point that the system occupies in phase space at a given time. But in this view exact recurrence of a state is impossible except for the trivial case of periodic dynamics, because it would

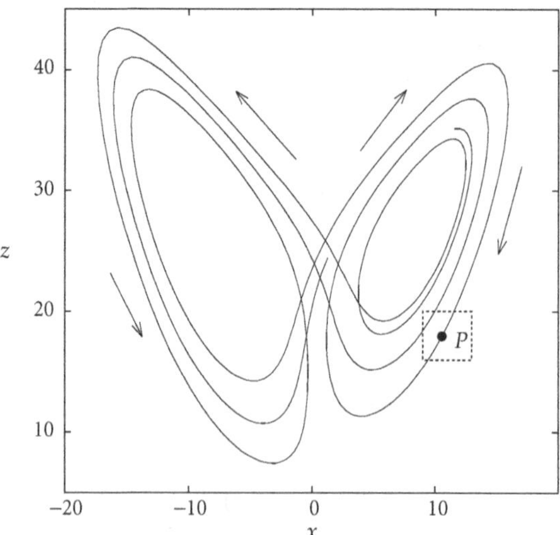

Figure 1. Illustrating recurrence on a two-dimensional projection of the Lorenz chaotic attractor. Recurrence is here reflected by the fact that the trajectory emanating from point P in the dotted square will re-enter this cell (which plays the role of a coarse-grained state) after some time.

violate the basic property of uniqueness of the solutions of the evolution equations. To cope with this limitation, one associates the concept of recurrence with repeated re-entries of the dynamical trajectory in a phase space region possessing finite volume (more technically, "measure") as indicated in Figure 1. This is, precisely, the coarse-grained state referred to earlier, as opposed to the point-like states considered in many (or even most) nonlinear science-related problems. Probabilistic description is a typical instance in which one deals essentially with coarse-grained states.

Poincaré Recurrence

In his celebrated memoir on the three-body problem Henri Poincaré, a founding father of nonlinear science, established a result that may be considered as the first quantitative and rigorous statement concerning recurrence (Poincaré, 1890). He showed that in a system of point particles under the influence of forces depending only on the spatial coordinates, a typical phase space trajectory will visit infinitely often the neighborhood, however small, of a prescribed initial state, provided that the system remains in a bounded region of phase space. The time between two such consecutive visits is referred as a "Poincaré recurrence time."

In modern terms (Kac, 1959), let A be a subset of space Ω within which the system is confined such that $\mu(A) > 0$ where μ designates the measure of A, and T_t a family of one-to-one measure preserving transformations in Ω. Then, for almost every initial point $P \in A$ (i.e., except for a set of Ps of zero μ-measure), there exists a sufficiently large t such that $T_t P \in A$. Using the machinery of the proof, one may derive an expression for the mean recurrence time,

$$\langle \theta_\tau \rangle = \tau/\mu(A), \qquad (1)$$

where τ is the time resolution between the successive observations. Actually, as observed by Marian Smoluchowski, this expression needs to be adapted, in the limit of continuous time, to

$$\langle \theta_\tau \rangle = \tau(1-\mu(A))/(\mu(A) - \mu(P \in A, T_\tau P \in A)), \qquad (2)$$

where one has introduced the measure of the intersection of A and its pre-image $T_\tau^{-1} A$.

Part of the historical role of Poincaré's theorem relates to the famous *Wiederkehreinwand*, an early objection by Ernst Zermelo challenging Ludwig Boltzmann's derivation of the second law of thermodynamics on the grounds that since states recur, the H-function (the microscopic analog of entropy) is bound to reverse its course at some stage. Boltzmann made a valiant attempt to respond, based essentially on the enormity of recurrence times in a molecular system (of the order of ten to the power of the Avogadro number). A more pertinent observation is that thermodynamic states are associated with a Gibbs ensemble rather than with single phase space points. Poincaré's theorem does not apply under these conditions; hence, Poincaré recurrences are not necessarily precluding a microscopic interpretation of irreversibility.

From Ehrenfest to Fermi–Pasta–Ulam

A first implementation of the idea that it is indeed possible to reconcile recurrence with irreversible behavior was provided by the, by now famous, "dog flea" model of Paul and Tatiana Ehrenfest. $2N$ numbered balls are distributed in boxes A and B. An integer from 1 to $2N$ is drawn at random and the ball corresponding to that number is moved from the box in which it is found to the other one. The process is then repeated to any desired number of times. One can show (Kac, 1959) that every initial state recurs with probability one, although the system possesses an H-function. Actually, this model belongs to a general class of stochastic processes known as Markov chains, whose theory was developed intensely in the first part of the 20th century (Feller, 1968, 1971). These developments allow one to put the compatibility between recurrence of states and irreversible approach to an asymptotic regime on a firm basis for this very general class of processes amenable to a Markov chain-like description. An important insight is that while recurrence is a property of trajectories (which being now stochastic and thus coarse-grained are not bound to satisfy the uniqueness theorem stated above and can therefore recur as such), the approach to an asymptotic regime is a property of probability distributions. The theory of stochastic processes has also provided a wealth of further information on recurrence independent of the problem of irreversibility, including connections with the important class of renewal processes.

Despite their power, the above results overlooked the nature of the underlying deterministic dynamics, except that the last had to be complex enough to generate somehow a stochastic game. This shortcoming became clearly recognized in the 1950s, when Poincaré's recurrence was viewed again as a problem of strictly deterministic dynamics, thanks partly to the new possibilities afforded by the availability of electronic computers. From the standpoint of nonlinear science, the most relevant among these early attempts is that by Fermi, Pasta, and Ulam (Fermi et al., 1955) who numerically examined a set of coupled differential equations modeling the motion of a linear chain of equimass particles connected by nonlinear springs

$$\begin{aligned}\frac{d^2x_j}{dt^2} &= \{1+\beta[(x_{j+1}-x_j)^2+(x_j-x_{j-1})^2\\&\quad+(x_{j+1}-x_j)(x_j-x_{j-1})]\}\\&\quad\times(x_{j+1}+x_{j-1}-2x_j),\\j &= 1,2,\ldots,N-1, x_0=x_N=0.\end{aligned} \tag{3}$$

Fermi et al. showed that with 64 particles at low energy there was no equipartition of energy among the oscillators (contrary to their expectations). Instead, a beat phenomenon was observed with ongoing recurrences of initial conditions. Recurrence seemed once again to be at odds with the tendency toward thermal equilibrium, to which is associated the stronger property of mixing and equipartition. A closer analysis shows that Fermi–Pasta–Ulam recurrence is not universal: for higher energies a phenomenon of overlap of resonances is taking place, leading to stochasticity in the sense that energy transfer between modes becomes possible. To what extent this stochasticity can invade the entire phase space as N and the energy are increased is still an unsettled question.

Modern Nonlinear Dynamics and Recurrence

The most significant advance on recurrence since the 1960s has been to dissociate the concept from the approach to equilibrium of a many-body system (and, in particular, to realize that the two concepts are not to be opposed), to extend it to cover the class of dissipative dynamical systems, and to establish a link between the deterministic and the stochastic approaches to it.

What made this advance possible is the realization, at the heart of modern nonlinear dynamics, that simple-looking, low-dimensional models may give rise to complex behavior that emulates, to a large extent, the behavior of real world, multivariate systems. Furthermore, in the presence of deterministic chaos, sensitivity to initial conditions raises the problem of long-term prediction and prompts one to adopt a probabilistic approach, perfectly compatible with (and actually induced by) the underlying dynamics, in which quantities of interest are expressed as statistical averages. In this setting, recurrence became a powerful tool providing new insights along two directions.

Recurrence as an Indicator of the Dynamical Complexity

The main point here is that dynamical systems having different ergodic properties show different recurrence patterns. An interesting example is uniform quasi-periodic motion on a two-dimensional torus or its equivalent discrete-time version given by the shift map. It can be shown (Sös, 1958; Theunissen et al., 1999) that, typically, such a system (known to be ergodic) possesses exactly three possible values of recurrence times in any prescribed phase space cell. On the other hand, in a wide class of uniformly hyperbolic systems such as one-dimensional maps, one shows that recurrence times in a sufficiently small phase space cell are generically exponentially distributed, while a sequence of successive recurrences has a Poissonian limit distribution (Collet, 1991). The situation is different in non-uniformly hyperbolic systems. In particular, in systems showing intermittent behavior the above laws are replaced, respectively, by power laws and by Lévy stable distributions (Balakrishnan et al., 1997). Work along similar lines has also led to connections between recurrence patterns and Lyapunov exponents, unstable periodic orbits, or generalized dimensions. There is, however, no firm indication about the universality of these latter results.

Recurrence as a Tool for Prediction

The question whether, and if so how frequently, the future states of a system will come close to a certain initial configuration has a direct bearing on prediction, an issue of central importance in atmospheric sciences, hydrology, and other environment-related disciplines. A dynamical approach to the recurrence of atmospheric analogs, the lumped states of the atmosphere introduced above reveals a strong dependence of recurrence times on the local properties of the attractor and a pronounced variability around their mean (Nicolis, 1998). Of crucial interest is also the recurrence of extreme events such as natural disasters, a problem approached traditionally at the level of a statistical description (Gumbel, 1958). The extension of the statistical theory of extremes and their recurrences to deterministic dynamical systems is still in its infancy.

G. Nicolis and C. Rouvas-Nicolis

See also **Dynamical systems; Fermi–Pasta–Ulam oscillator chain; Intermittancy; Nonequilibrium statistical mechanics**

Further Reading

Balakrishnan, V., Nicolis, G. & Nicolis, C. 1997. Recurrence time statistics in chaotic dynamics I. Discrete time maps. *Journal of Statistical Physics*, 86: 191–212

Collet, P. 1991. Some ergodic properties of maps of the interval, Lectures given at the CIMPA summer school *Dynamical Systems and Frustrated Systems*, edited by J.M. Gambaudo, Temuco

Feller, W. 1968, 1971. *An Introduction to Probability Theory and Its Applications*, 2 vols, 3rd edition, New York: Wiley

Fermi, E., Pasta, J.R. & Ulam, S.M. 1955. Studies of nonlinear problems. Los Alamos Scientific Laboratory Report No. LA-1940

Gumbel, E.J. 1958. *Statistics of Extremes*, New York: Columbia University Press

Kac, M. 1959. *Probability and Related Topics in Physical Sciences*, London and New York:

Nicolis, C. 1998. Atmospheric analogs and recurrence time statistics: toward a dynamical formulation. *Journal of Atmospheric Sciences*, 55: 465–475

Poincaré, H. 1890. Sur le problème des trois corps et les équations de la dynamique. *Acta Matematica*, 13: 1–270

Sös, V.T. 1958. On the distribution mod 1 of the sequence $n\alpha$. *Annales Universitatis Scientiarum Budapestinensts de Rolando Eotovos Nominatae Sectio Mathematica*, 1: 127–134

Theunissen, M., Nicolis, C. & Nicolis, G. 1999. Recurrence times in quasi-periodic motion: statistical properties, role of cell size, parameter dependence. *Journal of Statistical Physics* 94: 437–467

REGULAR AND CHAOTIC DYNAMICS IN ATOMIC PHYSICS

Many important areas of atomic physics dwell in the intriguing world of semiclassical quantum mechanics, where classical concepts such as orbits and phase space are used to calculate purely quantal quantities such as quantum numbers, wave functions, and ionization thresholds. The past two decades have witnessed renewed interest in classical interpretations of quantum phenomena (Berry & Mount, 1988; Casati et al., 1987; Blümel & Reinhardt, 1997; Jensen, 1992), a view sometimes referred to as postmodern quantum mechanics (Heller & Tomsovic, 1993). This development has led to the much debated term *quantum chaos* (*See* **Quantum chaos**). A central issue is the relevance of chaotic dynamics to the quantum world, where the sensitivity to initial conditions characteristic of classical chaos is obscured by the Heisenberg uncertainty principle. Moreover, the Schrödinger equation is linear and does not normally exhibit extreme sensitivity to initial conditions. Nevertheless, classical dynamics plays an important role in the quantum world, even in chaotic regions. Thus, the ionization of hydrogen by a microwave field is often well described in terms of chaotic diffusion of classical electron orbits, while the quantum mechanical wave function exhibits traces of classical (unstable) periodic orbits, called scarring.

When the classical dynamics are regular, semiclassical quantum numbers can be calculated according to the Einstein–Brillouin–Keller (EBK) prescription (Gutzwiller, 1990),

$$J_i = \oint p_i \mathrm{d}q_i = (n_i + \alpha_i/4)h, \tag{1}$$

where J_i is the classical action, calculated from the loop integral over coordinate q_i and canonically conjugate momentum p_i, n_i is the corresponding quantum number, h is Planck's constant, and α_i is the Maslov index, an integer that depends on the topology of the invariant torus. In this way, approximate semiclassical wave functions can also be constructed. The semiclassical description is found to work well under two conditions: (i) the quantum numbers are large, and (ii) the number of participating states is also large. Under chaotic conditions (when the classical action does not generally exist), one still has recourse to periodic orbits (Eckhardt, 1988), which can be used to organize energy levels and the form of the wave function (Gutzwiller, 1990). The principal tool under chaotic conditions is the Gutzwiller trace formula, based on the Feynman propagator.

Our unifying theme is the hydrogen atom in an electromagnetic field, for which the motion of the single electron is well described by a Hamiltonian model. We consider first two closely related integrable cases (*See* **Constants of motion and conservation laws; Integrability**), the classical problem of two fixed centers as a model for the hydrogen molecule ion H_2^+, and the H-atom in a homogeneous electrostatic field. In both cases axial symmetry implies the constancy of the azimuthal angular momentum, $p_\phi = m\rho^2\dot{\phi}$ (in cylindrical coordinates ρ, ϕ, z), allowing a reduction in dimensionality from three to two, via an effective potential formed by including the azimuthal part of the kinetic energy with the potential $U(\rho, z)$:

$$U^e(\rho, z) = U + \frac{p_\phi^2}{2m\rho^2}. \tag{2}$$

Thus, in both cases the motion is described by a two degree of freedom autonomous (time-independent) Hamiltonian.

Next we take up the non-integrable case of the H-atom in a homogeneous magnetostatic field, which has been extensively studied, both theoretically and experimentally (Born, 1960; Friedrich & Wintgen, 1989), and which is also described by a two-degree-of-freedom Hamiltonian. Then we describe fruitful experiments on microwave ionization of H-atoms, which have yielded an abundance of information on quantum physics (Blümel & Reinhardt, 1997; Koch & van Leeuwen, 1995). Various Hamiltonian models have been investigated, from one to three degrees of freedom. Here, we shall limit ourselves to a one-dimensional time-periodic model, which captures most of the relevant physics. Finally, we briefly mention other atomic systems in which a semiclassical treatment sheds light on quantal behavior.

Two Fixed Centers

The two fixed centers model (TC) consists of a test mass orbiting in the gravitational field of two massive bodies rotating about their center of mass at a fixed distance. In atomic physics the attractive force is electrostatic, furnishing a useful model for the hydrogen molecule-ion (Born, 1960; Strand & Reinhardt, 1979; Howard & Wilkerson, 1995). It also enjoys the distinction of being one of the small minority of completely integrable systems, owing to separability in confocal elliptic coordinates. With the two protons fixed at $z = \pm a$, the nonrelativistic motion of the electron is described by the Hamiltonian,

$$H = \frac{1}{2m}(p_\rho^2 + p_z^2) + U^e \tag{3}$$

with effective potential

$$U^e(\rho, z) = \frac{p_\phi^2}{2m\rho^2} - q^2\left(\frac{1}{r_1} + \frac{1}{r_2}\right), \tag{4}$$

where q is the electronic charge,

$$r_1^2 = \rho^2 + (z - a)^2, \quad r_2^2 = \rho^2 + (z + a)^2, \tag{5}$$

and $p_\rho = m\dot{\rho}$, $p_z = m\dot{z}$, and $p_\phi = m\rho^2\dot{\phi}$. All trapped orbits are confined by the two-dimensional effective potential, shown in Figure 1 for two nearby values of the control parameter $\mu = p_\phi^2/a^2$. In the first case, a single well centered on an equatorial critical point of U^e exists, which in the second case has bifurcated into a double well. Quantizable stable circular orbits exist at the elliptic critical points of U^e. Because there is no chaos in this system, it might seem straightforward to construct quantum numbers from the two conserved actions, taking care to get the correct Maslov indices. However, there is a rub: there are two disjoint classes of orbits, a situation referred to as "monodromy," resulting in separate actions, which need to be smoothly joined. This process of "uniform quantization" has been carried out by Strand & Reinhardt (1979), who obtained greatly improved values for energy levels for varying intermolecular separation $R = 2a$.

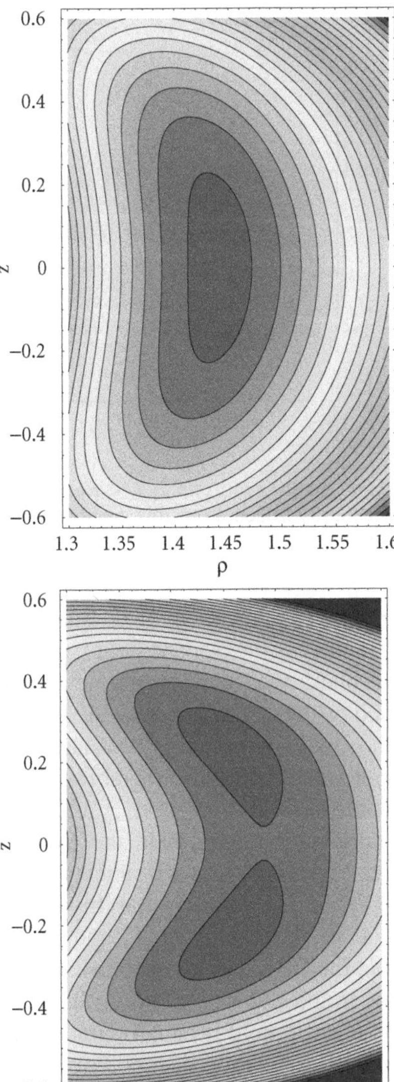

Figure 1. Contour plot of effective potential for the hydrogen molecule-ion before and after pitchfork bifurcation.

Hydrogen Atom in a Strong Electric Field

The behavior of the H-atom in a homogeneous electrostatic field is one of the oldest problems of quantum mechanics, the weak-field case giving rise to the Stark effect, traditionally treated using perturbation theory (Born, 1960). For larger fields, ionization occurs (Koch, 1978; Howard, 1995), the ionization threshold field F_c depending on the particle energy as well as a quantity called the Runge–Lenz invariant. This is an interesting problem for two reasons: (i) the required fields are too large for perturbative methods, and (ii) the motion is completely integrable, since the Hamiltonian and the Schrödinger equations both separate in parabolic coordinates. Spectral lines are then conveniently labeled by quantum numbers n, n_1, n_2, and $|m|$, where $n = n_1 + n_2 + |m| + 1$ is the principal quantum number, m is the azimuthal quantum number, and n_1 and n_2 are parabolic quantum numbers.

Field ionization may be treated either classically or via quantum mechanical tunneling (Gallas et al., 1982). With the electric field along the z-axis, the nonrelativistic motion of the electron is again described by Hamiltonian (3), with effective potential

$$U^e(\rho, z) = -\frac{q^2}{\sqrt{\rho^2 + z^2}} + \frac{p_\phi^2}{2m\rho^2} - qFz, \tag{6}$$

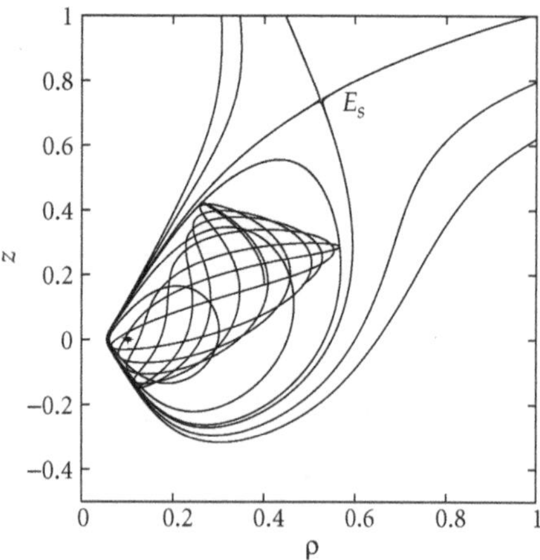

Figure 2. Electron orbit with energy $E < E_s$ trapped in effective potential for the hydrogen atom immersed in a homogeneous electric field.

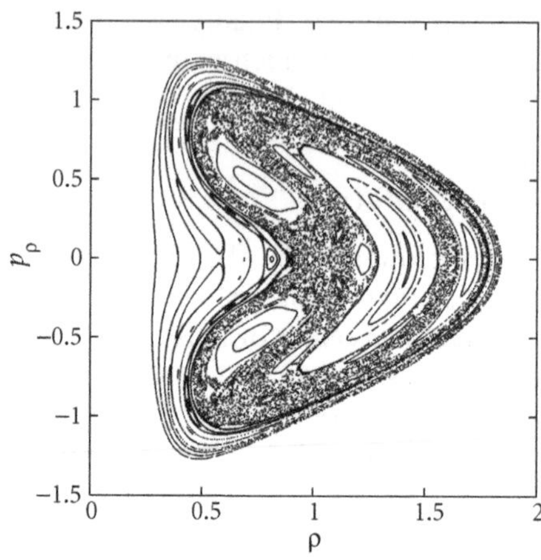

Figure 3. Poincaré section for H-atom in a uniform magnetostatic field.

where p_ϕ gives the azimuthal quantum number $m = p_\phi/\hbar$ and F is the electric field strength. Figure 2 shows an orbit trapped in the two-dimensional potential well formed by the effective potential. All orbits having $E < E_s$, the saddle point energy, are classically trapped. If the system were non-integrable, the condition $E > E_s$ would constitute a necessary and sufficient condition for ionization. However, the existence of the Runge–Lenz invariant allows a subclass of trapped orbits with $E > E_s$. Owing to separability in parabolic coordinates (ξ, η), the actions J_ξ, J_η, and $J_\phi = p_\phi$ always exist so that the EBK formula applies; there is no classical chaos in the Stark problem. In laboratory experiments the electric field is usually gradually increased from zero until ionization occurs. For $m = 0$, the saddle point criterion leads to the ionization threshold $F_c \approx 1/9n^4$, in excellent agreement with experiment.

Hydrogen Atom in a Strong Magnetic field

In contrast to the Stark problem, the electron motion in an H-atom immersed in a homogeneous magnetostatic field can be chaotic, providing a useful testing ground for ideas of quantum chaos (Friedrich & Wintgen, 1989). There are two integrable limits, (i) $B = 0$, for which the orbits are simple Kepler ellipses, and (ii) $B \to \infty$, the helical gyration of a free charged particle in a constant magnetic field. With $\boldsymbol{B} = B_0\hat{\boldsymbol{z}}$, the nonrelativistic motion of a single electron is governed by the effective potential

$$U^e(\rho, z) = \frac{p_\phi^2}{2m\rho^2} - \frac{q^2}{\sqrt{\rho^2 + z^2}} + \frac{1}{8}m\omega_c^2\rho^2 + \frac{1}{2}\omega_c p_\phi, \qquad (7)$$

where $\omega_c = qB_0/mc$ is the cyclotron frequency and now $p_\phi = m\rho^2\dot{\phi} + \frac{1}{2}\rho^2\omega_c$. The constant paramagnetic term may be removed by transforming to a rotating frame. For small magnetic field strength, perturbation theory may be used to calculate energy levels (Zeeman effect); for very large fields, the diamagnetic term proportional to ρ^2 dominates (quadratic Zeeman effect) and the spectrum resembles that of a free electron. In this case, the existence of chaotic trajectories means that a set of classical actions does not always exist, precluding EBK quantization. For small fields, however, canonical perturbation theory can be employed to calculate perturbed actions. This yields useful energy level curves $E_n = E_n(B)$, where the levels are labeled in terms of the unperturbed $(B = 0)$ states. Furthermore, at low fields an approximate constant of the motion exists, facilitating classification of the very complex spectrum and explaining avoided crossings of nearby energy level curves. This recipe fails for larger fields, where the classical motion is mainly chaotic. At very large fields, intriguing regularities persist, called quasi-Landau resonances, close to the spectrum of a free electron in a magnetostatic field. In this case, it is possible to derive a very simple formula for the ionization threshold energy.

The degree of classical chaos is conveniently revealed by the Poincaré section, which is a two-dimensional slice in the four-dimensional phase space, defined by fixing the total energy and recording unidirectional intersections with a coordinate plane (Lichtenberg & Lieberman, 1990). Figure 3 shows a section for a moderately large B-field, generated by calculating particle orbits and recording $(\rho,\ p_\rho)$ each time an orbit crosses the $z = 0$ plane with $p_z > 0$. Regular orbits appear as smooth closed curves, whose centers represent periodic orbits. A single period-one orbit appears near $\rho = 1.45$; all other fixed points are at least period-three. The scattered dots in between represent chaotic orbits (possibly just one), which ergodically fill a connected region of phase space. In this case, classical actions do not exist, and it is not possible to assign quantum numbers to the very complex spectrum.

Microwave Ionization of Rydberg Atoms

Another fruitful source of information on connections between classical dynamics and quantal behavior is the extensive experimental program on microwave ionization of highly excited Rydberg (hydrogenic) atoms (Blümel & Reinhardt, 1997; Koch & van Leeuwen, 1995). By careful control of experimental conditions, it has been found possible to produce a full range of principal quantum numbers, $n_0 = 34$ to more than 90. Ionization from such excited states can then be readily accomplished using a variety of pulsed or continuous wave sources; linearly polarized, circularly polarized, or elliptically polarized. Remarkably, classical scaling accurately describes much of the experimental results. Thus, the microwave frequency is expressed as a ratio to the Kepler frequency, which according to the Bohr–Sommerfeld model is proportional to n_0^{-3}, giving the dimensionless parameter $\hat{\omega} = n_0^3\omega$. The electric field strength may also be expressed as a ratio to the Coulomb force, giving the dimensionless parameter $\hat{F} = n_0^4 F$. Koch distinguishes several distinct frequency regimes, depending on $\hat{\omega} = n_0^3\omega$; by fixing $\omega/2\pi$ at one of several frequencies between 7 and 36 GHz and varying n_0, a range of scaled frequencies from $\hat{\omega} = 0.02$ to 2.8 can be achieved. For very low $\hat{\omega} \lesssim 0.02$ quantum tunneling dominates, significantly lowering the ionization threshold $\hat{F}_i$. At somewhat higher $\hat{\omega}$ the ionization curves (fraction ionized vs. $\hat{F}$) are primarily classical, with occasional "bumps" deriving from quantum resonances. For $0.1 \lesssim \hat{\omega} \lesssim 1.2$ classical dynamics prevails. Above this value quantum mechanisms increasingly raise the ionization thresholds, by about a factor of two at $\hat{\omega} = 2.8$. As for the Zeeman effect, scarring of the wave function occurs along classical unstable period orbits.

Theoretical and numerical analyses range from crude one-dimensional (1-d) models to elaborate 3-d Monte Carlo simulations. The simplest case to analyze theoretically is linear polarization, in which a simple one-dimensional time-periodic Hamiltonian explains much of the relevant physics,

$$H = H_0 + qFx\sin(\omega t), \tag{8}$$

where

$$H_0 = \begin{cases} \dfrac{p^2}{2m} - \dfrac{q^2}{x}, & x > 0, \\ \infty, & x \leq 0. \end{cases} \tag{9}$$

Physically this represents the limiting case of a pencil-thin orbit along the x-axis. In order to compare with experiment, we first transform to action-angle variables $J = \sqrt{a} = \sqrt{-1/2H_0}$, $\theta = u - \sin u$, where a is the semimajor axis and the eccentric anomaly u is defined

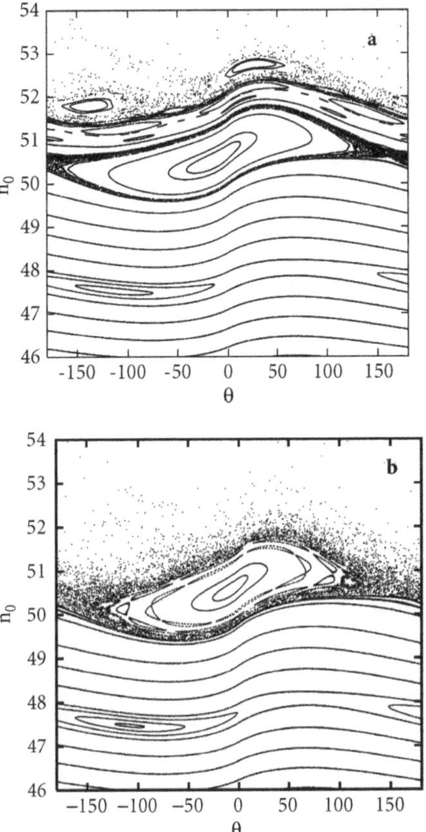

Figure 4. Poincaré section for 1-d H-atom perturbed by a time-periodic microwave field of frequency $f_{RF} = 9.923$ GHz and strength (a) $F = 62$ V/cm and (b) $F = 66.4$ V/cm.

by $x = a(1 - \cos u)$. The principal quantum number is then given by $J = n_0\hbar$. Poincaré sections can be generated by strobing the orbit at a period $\tau = 2\pi/\omega$ and recording J and θ at $t = \tau, 2\tau, \ldots$. In experiments the microwave field is gradually increased until ionization takes place. For example, for $f_{RF} = 9.923$ GHz, we obtain the two sections near the $n_0 = 51$ state shown in Figure 4 for $F = 62$ and 66.4 V/cm. At the lower field strength all orbits initialized with $n_0 < 51$ are trapped by invariant circles; at the higher field strength these curves have been destroyed, allowing most of the same orbits to escape. The experimental result for 10% ionization of the $n_0 = 51$ state is about 71 V/cm (Koch & van Leeuwen, 1995), in reasonable agreement with our numerical calculation. Other polarizations have also been investigated (Koch, 1998; Howard, 1992). The most general state is elliptic polarization, which includes circular and linear polarization as special cases. In all cases valuable insight into experiment has been obtained using classical dynamics.

Discussion

In addition to the above four examples, there are several other systems under active investigation. Combinations of fields, particularly crossed and parallel fields, $\boldsymbol{E} \times \boldsymbol{B}$ and $\boldsymbol{E} \parallel \boldsymbol{B}$, offer challenging structures that can be partially explained by semiclassical quantization (Farrelly, 1991). Three-body systems, such as the helium atom (Tanner et al., 2000) and the hydrogen negative ion H^-, long considered beyond semiclassical methods, have recently yielded important insights.

JAMES E. HOWARD

See also **Constants of motion and conservation laws; Hamiltonian systems; Quantum chaos**

Further Reading

Berry, M.V. & Mount, K.E. 1972. Semiclassical wave mechanics. *Reports on Progress in Physics*, 35: 315–800

Blümel, R. & Reinhardt, W.P. 1997. *Chaos in Atomic Physics*, Cambridge and New York: Cambridge University Press

Born, M. 1960. *The Mechanics of the Atom*, New York: Unger

Casati, B., Chirikov, B.V., Shepelyansky, D.L. & Guarneri, I. 1987. Relevance of classical chaos in quantum-mechanics—the hydrogen atom in a monochromatic field. *Physics Reports*, 154: 77–123

Eckhardt, B. 1988. Quantum mechanics of classically non-integrable systems. *Physics Reports*, 163: 205–297

Farrelly, D. 1991. Semiclassical mechanics of bounded and unbounded states of atoms and molecules, *Advances in Molecular Vibrations*, 1B, 49–79

Friedrich, H. & Wintgen, D. 1989. The hydrogen atom in a uniform magnetic field—an example of chaos. *Physics Reports*, 183: 37–79

Gallas, J.A.C., Walther, H. & Werner, E. 1982. Simple formula for the ionization rate of Rydberg states in static electric fields. *Physical Review Letters*, 49: 867–891

Gutzwiller, M.C. 1990. *Chaos in Classical and Quantum Mechanics*, New York: Springer

Heller, E.J. & Tomsovic, S. 1993. Postmodern quantum mechanics. *Physics Today*, 46: 38–46

Howard, J.E. 1992. Stochastic ionization of hydrogen atoms in a circularly polarized microwave field. *Physical Review* A, 46: 364–372

Howard, J.E. 1995. Saddle point ionization and the Runge-Lenz invariant. *Physical Review* A, 51: 3934–3946

Howard, J.E. & Wilkerson, T.W. 1995. Problem of two fixed centers and a finite dipole: a unified treatment. *Physical Review A*, 52: 4471–4492

Jensen, R.V. 1992. Quantum chaos. *Nature*, 352: 311–315

Koch, P.M. 1978. Resonant states in the nonperturbative regime: the hydrogen atom in an intense electric field, *Physical Review Letters*, 41: 99–103

Koch, P.M. 1998. Polarization dependence of microwave "ionization" of excited hydrogen atoms. *Acta Physica Polonica*, 93: 105–132

Koch, P.M. & van Leeuwen, K.A.H. 1995. The importance of resonances in microwave "ionization" of excited hydrogen atoms. *Physics Reports*, 255: 289–403

Lichtenberg, A.J. & Lieberman, M.A. 1990. *Regular and Chaotic Dynamics*, 2nd edition, New York: Springer

Strand, M.P. & Reinhardt, W.P. 1979. Semi-classical quantization of the low-lying electronic states of H_2^+. *Journal of Chemical Physics*, 70: 3812–3827

Tanner, G., Richter, K. & Rost, J. 2000. The theory of two-electron atoms: between ground state and complete fragmentation. *Reviews of Modern Physics*, 72: 497–544

RELAXATION OSCILLATORS

The first work related to relaxation oscillators was published by Balthasar van der Pol in 1926 (Van der Pol, 1926). Van der Pol, in collaboration with J. van der Mark, suggested an electrical model of the heart consisting of three relaxation generators (van der Pol & van der Mark, 1928).

An oscillator is said to be of the relaxation type if its period is inversely proportional to a relaxation time and nearly independent of other parameters. Classical examples of relaxation oscillators are a thyratron oscillator (Teodorchik, 1952) and an alternating water source known as Tantal's vessel (Panovko & Gubanova, 1964; Strelkov, 1964). Schematic images of these devices are illustrated in Figures 1a and b. The shape of oscillations of the amount of water in Tantal's vessel and of the voltage across the capacitor C is shown in Figure 1c.

An oscillator described by the Rayleigh equation (Rayleigh, 1877–78)

$$\ddot{x} - \mu(1 - \dot{x}^2)\dot{x} + x = 0 \tag{1}$$

is also of relaxation type for $\mu \gg 1$. It should be noted that the Rayleigh equation can be obtained from the van der Pol equation

$$\ddot{y} - \mu(1 - y^2)\dot{y} + y = 0 \tag{2}$$

by the substitution $y = \sqrt{3}\,\dot{x}$.

The phase portrait and the shape of oscillations for Equation (1) are illustrated in Figure 2 for $\mu = 10$. For this value of μ, the equilibrium state is an unstable node and the limit cycle has a shape close to a parallelogram. The time required for the representative point to approach the limit cycle is very short; thus the oscillations are almost discontinuous. For $\mu \ll 1$, the oscillation period is practically independent of μ and approximately equal to the period of free oscillations of the oscillatory circuit $T_0 = 2\pi$. For $\mu \gg 1$, on the other hand, the oscillation period is completely determined by the value of the parameter μ, namely $T \approx 8\mu/3\sqrt{3}$ (see below). As the relaxation time is of the order of $1/\mu$, it follows that for $\mu \gg 1$ the oscillation period is really inversely proportional to the relaxation time.

Introducing the small parameter $\varepsilon = 1/\mu$, we can rewrite Equation (1) in the form of two equations of

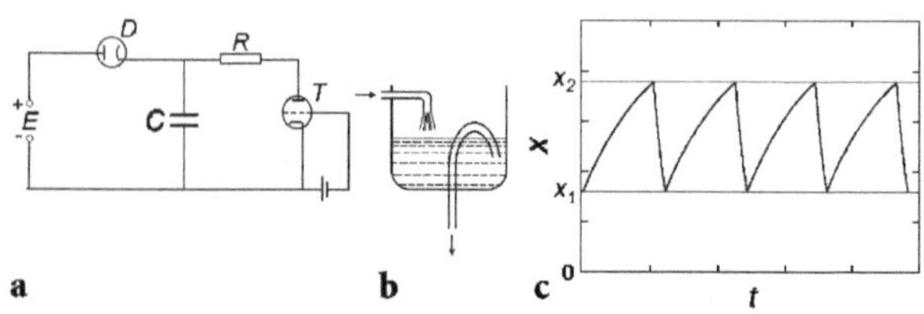

Figure 1. Schematic images (a) of a thyratron oscillator, (b) of Tantal's vessel, and (c) the shape of oscillations of the amount of water in Tantal's vessel and of the voltage across the capacitor C.

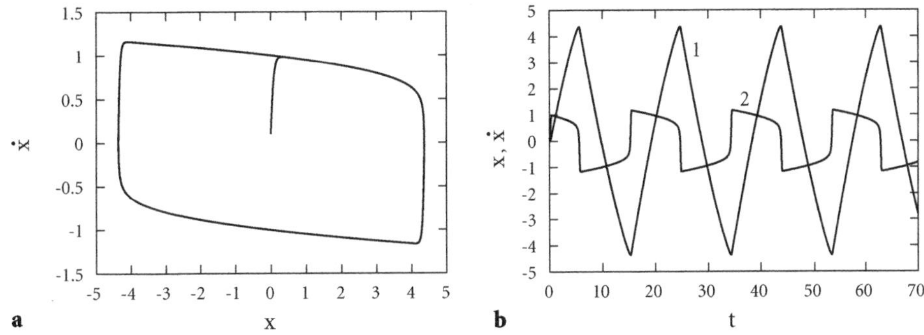

Figure 2. (a) The phase portrait for the Rayleigh equation for $\mu = 10$, and (b) the solution of this equation for $\mu = 10$: curves 1 and 2 show $x(t)$ and $\dot{x}(t)$, respectively.

first order as

$$\dot{x} = y, \tag{3}$$
$$\varepsilon \dot{y} = (1 - y^2)y - \varepsilon x. \tag{4}$$

To find an approximate solution of Equations (3) and (4) we apply the averaging method in systems incorporating fast and slow variables. It follows that the variable x is slow and the variable y is fast. Consequently, in solving Equation (4) the variable x can be considered as a constant. Stationary solutions of Equation (4) under this condition are real roots of the cubic equation

$$y^3 - y + \varepsilon x = 0. \tag{5}$$

Over the range

$$-\frac{2}{3\mu\sqrt{3}} < x < \frac{2}{3\mu\sqrt{3}} \tag{6}$$

Equation (5) has three real roots, whereas outside this range it has only one real root. For $x = \pm 2/(3\mu\sqrt{3})$ Equation (5) has two real roots, $y_1 = \pm 1/\sqrt{3}$, $y2 = \mp 2/\sqrt{3}$. Approximate solutions of Equation (5) can be found analytically under the condition $\varepsilon x \ll 1$. Expanding y sequentially about $y_{10} = 1$, $y_{20} = -1$ and $y_{30} = 0$ we find three solutions:

$$y_1 = 1 - \frac{\mu}{2}x, \quad y_2 = -1 - \frac{\mu}{2}x, \quad y_3 = \mu x. \tag{7}$$

It is seen that the approximate solutions (7) are close to the exact ones over the range (6).

Investigation of the stability of these shows that only those solutions are stable that satisfy the constraint $y_i \geq 1/\sqrt{3}$, which is fulfilled for the upper and the lower parts of the dependence $y(x)$ and is not fulfilled for its middle part.

Substituting $y_{1,2}$ from (7) into Equation (3) and integrating the latter, we find solutions of this equation under the initial condition $x(0) = x_0$:

$$x_{1,2}(t) = x_0 e^{-\varepsilon t/2} \pm \frac{2}{\varepsilon}\left(1 - e^{-\varepsilon t/2}\right). \tag{8}$$

For $\varepsilon t \ll 1$ solutions (8) can be simplified as

$$x_{1,2}(t) = (x_0 \pm t)(1 - \varepsilon t/2). \tag{9}$$

Let the initial conditions be $x_0 = -2/(3\varepsilon\sqrt{3})$, $y_0 = y_1(x_0)$. Then, as follows from (9), x will increase with an approximately constant rate, and at the instant $t = 4/(3\varepsilon\sqrt{3})$, it will attain the value $2/(3\varepsilon\sqrt{3})$. At this instant, the solution $y_1(x)$ becomes unstable because it merges with the unstable solution $y_3(x)$. As a result, y will jump to the value $y_2\left(2/(3\varepsilon\sqrt{3})\right)$. Subsequently, x will decrease to the value $-2/(3\varepsilon\sqrt{3})$ and then a jump will occur again; that is, y will take on the value $y_1\left(-2/(3\varepsilon\sqrt{3})\right)$. As the process is repeated, we obtain a limit cycle on the phase plane x, y close to the one calculated numerically (Figure 2).

If $y_0 \neq y_{1,2}(x_0)$, y takes on one of these values very rapidly, practically for a constant x. This means that the corresponding phase trajectories are nearly vertical.

An approximate expression for the oscillation period T can be found on the assumption that the jumps occur instantly and x varies with the constant velocity ± 1. Thus

$$T \approx \frac{8}{3\varepsilon\sqrt{3}} = \frac{4}{3\pi\sqrt{3}}\mu T_0, \tag{10}$$

where $T_0 = 2\pi$ is the period of natural oscillations for $\mu = 0$. Thus, in the relaxation regime, the oscillation period is approximately $\mu/4$ times the period of natural oscillations.

Importantly, one of the characteristic features of relaxation oscillators is the ease of frequency synchronization (Landa, 2001). This phenomenon was apparently first studied by Teodorchik (1943, 1945). Relaxation oscillators have long been used in radio and electrical engineering, mainly as multivibrators, but recently interest in these oscillators has quickened in connection with biological and medical applications. As early as 1953, the German physicist Karl Bonhoeffer suggested that a neuron might be modeled by equations similar to the van der Pol equation for a relaxation oscillator. The Bonhoeffer–van der Pol equations describe oscillations of the voltage across a neural membrane, including refractoriness (difficulty of refiring). These equations are

$$\dot{x} = x - x^3/3 - y + I_0, \quad \dot{y} = \varepsilon(x + a - by), \quad (11)$$

where x is the voltage across the neural membrane; y is a quantity of refractoriness; a, b, ε are membrane radius, specific resistivity of the fluid inside the membrane, and temperature factor, respectively; and I_0 is a direct component of the current across the membrane.

POLINA LANDA

See also **FitzHugh–Nagumo equation; Neurons; Nonlinear electronics; Synchronization; Van der Pol equation**

Further Reading

Bonhoeffer, K.F. 1953. Modelle der Nervenerregung. *Naturwissenschaften*, 40: 301–311

Landa, P.S. 2001. *Regular and Chaotic Oscillations*, Berlin and New York: Springer

Panovko, Ya.G. & Gubanova, I.I. 1964. *Ustoichivost' i Kolebaniya Uprugikh Sistem [Stability and Oscillations of Elastic Systems]*, Moscow: Science

Rayleigh, Lord (Strutt, J.W.) 1877–78. *The Theory of Sound*, 2 vols., London: Macmillan; reprinted New York: Dover, 1877–78

Strelkov, S.P. 1964. *Vvedenie v Teoriyu Kolebanii [Introduction to the Oscillation Theory]*, Moscow: Science

Teodorchik, K.F. 1943. On the theory of synchronization of relaxation generators. *DAN SSSR*, 40: 63–66 (in Russian)

Teodorchik, K.F. 1945. Theory of synchronization of relaxation self-oscillatory systems. *Journal of Physics*, Moscow, 9: 139–146 (in Russian)

Teodorchik, K.F. 1952. *Avtokolebatel'nye Sistemy [Self-Oscillatory Systems]*, Moscow: Gostekhizdat

van der Pol, B. 1926. On relaxation oscillation. *Philosophical Magazine*, 2: 978–992

van der Pol, B. & van der Mark, M. 1928. The heartbeat considered as a relaxation oscillation and an electrical model of the heart. *Philosophical Magazine*, 6: 763–775

RENORMALIZATION GROUPS

Systems undergoing phase transitions at a critical temperature T_c are conveniently described in terms of order parameters and critical exponents. Above T_c, for example, the specific heat is found experimentally to vary as $\varepsilon^{-\alpha}$ where $\varepsilon \equiv (T - T_c)/T_c$, and below T_c, it varies as $(-\varepsilon)^{-\alpha'}$. Thus, α and α' are called critical exponents for specific heat. Similar critical exponents are found for order parameters, isothermal susceptibility, response to an external field, the correlation length ξ, and the pair correlation function $\Gamma(r)$. These are denoted as β, γ, δ, ν, and η, respectively.

Relations among these critical exponents can be derived from scaling assumptions such as the static scaling hypothesis, which asserts that the Gibbs potential $G(T, H)$ for magnetic systems (as an example) is a generalized homogeneous function

$$G(\lambda^{a_\varepsilon}\varepsilon, \lambda^{a_{\mathrm{H}}}H) = \lambda G(\varepsilon, H). \quad (1)$$

With an arbitrary value of λ and selecting a_ε, a_{H} appropriately, relations among critical exponents are found through thermodynamic identities, for example, $M(\varepsilon, H) = -\,\mathrm{d}G(\varepsilon, H)/\mathrm{d}H$. Thus, Equation (1) leads to the following relations:

$$\beta = \frac{1 - a_{\mathrm{H}}}{a_\varepsilon}, \quad \delta = \frac{a_{\mathrm{H}}}{1 - a_{\mathrm{H}}}, \quad \gamma' = \frac{2a_{\mathrm{H}} - 1}{a_\varepsilon},$$
$$\gamma = \frac{2a_{\mathrm{H}} - 1}{a_\varepsilon}, \quad \alpha' = 2 - \frac{1}{a_\varepsilon}. \quad (2)$$

These can be recast as critical exponent equations:

$$\begin{aligned} \alpha + \beta(\delta + 1) &= 2, \\ \alpha' + 2\beta + \gamma' &= 2, \\ \gamma(\delta + 1) &= (2 - \alpha)(\delta - 1), \\ \gamma' &= \beta(\delta - 1), \\ \gamma &= \gamma', \\ \alpha &= \alpha', \end{aligned}$$

reducing the number of independent critical exponents to just four (a_ε, a_{H}, ν, and η).

The two exponents that describe spatial correlations (ν and η) also involve a scaling relationship as was shown through dynamic scaling arguments by Leo Kadanoff and Michael Fisher, among others. In the two-dimensional (2-d) Ising model above T_c, for example, spin-spin correlations show short-range order, whereas below T_c, the system exhibits long-range order. As T tends to T_c from either side, $\Gamma(r - r')$ becomes long ranged and decays slowly as $|r - r'| \gg 1$, with the correlation length diverging. This is typical of most critical systems for which long-range fluctuations accompany the onset of a second-order phase transition. However, in the 2-d XY model, the correlation function exhibits an algebraic fall-off as $\Gamma(r) \sim r^{-\eta(T)}$, demonstrating lack of long-range order in an unusual critical system—the so-called Kosterlitz–Thouless transition whose hallmark is the formation of vortex-antivortex pairs. Mermin and Wagner (1966) proved that any 2-d system with short-range interactions whose ordered phase has a continuous symmetry does *not* support long-range order. This is because at least one branch of collective excitations—called Goldstone bosons—has energy that tends to zero continuously as its wave vector (crystal momentum) vanishes.

With the exception of such unusual systems, spatial scaling applies to criticality, which according to Kadanoff is to be understood as coarse graining. This means that the essential features of the system remain unchanged as the lattice length scale is increased by a factor λ such that $1 \ll \lambda \ll \xi/a$, where ξ is the correlation length and a is the lattice spacing (see Figure 1).

If N is the number of lattice sites and d the dimensionality of the system, then $m = N/\lambda^d$ is the number of blocks obtained in the first rescaling process. Simultaneously, the lattice variables (e.g., spins) and the interaction parameters are redefined. The effective spins now represent each block and interaction parameters refer to the interacting blocks. Taking the Ising model as an example, the cell Hamiltonian is

$$H_{\text{cell}} = -J \sum_{\langle i,j \rangle}^{N} S_i S_j - h \sum_{i}^{N} S_i. \tag{3}$$

Then the block Hamiltonian becomes

$$H_{\text{block}} = -\tilde{J} \sum_{\langle \alpha,\beta \rangle}^{m} \tilde{S}_\alpha \tilde{S}_\beta - \tilde{h} \sum_{\alpha}^{m} \tilde{S}_\alpha, \tag{4}$$

where the tilde quantities refer to blocks. Crucial assumptions are that thermodynamic potentials scale

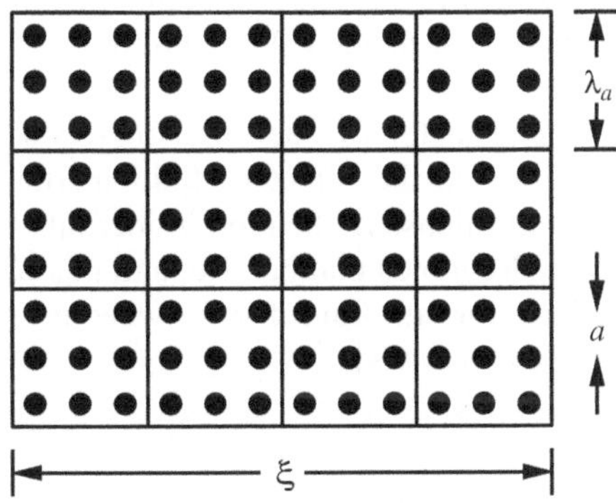

Figure 1. A schematic illustration of the Kadanoff construction.

with block size as

$$F_{\text{block}}(\tilde{\varepsilon}, \tilde{h}) = \lambda^d F_{\text{cell}}(\varepsilon, h), \tag{5}$$

where $\tilde{h} = \lambda^x h$, $\tilde{\varepsilon} = \lambda^y \varepsilon$, and F is a homogeneous function. Hence the exponents x and y are calculated as $y = a_\varepsilon d$ and $x = a_{\text{H}} d$. The Kadanoff construction applied to the pair correlation function,

$$\Gamma(i, j, \varepsilon) = \langle (S_i - \langle S \rangle)(S_j - \langle S \rangle) \rangle, \tag{6}$$

gives $y = \nu^{-1}$, and since $y = da_\varepsilon$ and $a_\varepsilon = (2 - \alpha')^{-1}$, we find that: $\alpha = \alpha'$ and $d\nu = 2 - \alpha$. Using $x = a_{\text{H}} d$ with $a_{\text{H}} = \delta/(\delta + 1)$ results in $(2 - \mu)\nu = \delta$.

Thus, static and dynamic scaling hypotheses reduce the number of independent critical exponents to just two. The renormalization group (RG) theory was developed to calculate the values of these exponents for particular models by carrying out the scaling procedure up to ξ, and because $\xi \to \infty$ as $T \to T_c$, scaling should continue "ad infinitum" as T approaches T_c.

Denoting the cell Hamiltonian by H_0 and the Hamiltonian after the nth step of rescaling as H_n, the chain of scaling transformations R is

$$\begin{aligned} R(H_0) &= H_1,\ R(H_1) = H_2, \ldots, R(H_n) \\ &= H_{n+1}, \ldots, R(H^*) = H^*, \end{aligned} \tag{7}$$

where H^* denotes a fixed point Hamiltonian characteristic of the critical state. Each step in the RG transformation chain reduces the number of degrees of freedom by λ^d. In statistical mechanics, we need to ensure that the partition function retains the same symmetry and ground-state, and hence we must re-scale the coupling constant at each step, for example, $K = J/kT$ in the Ising model. We then develop a recursion relation to compute the partition function.

Model	α	β	γ	ν	δ	η
Classical (MFT)	0 (disc.)	$\frac{1}{2}$	1	—	3	—
Spherical: $d = 3$		$\frac{1}{2}$	2	1	5	0
Spherical: $\varepsilon > 0$	$-\varepsilon/(2-\varepsilon)$	$\frac{1}{2}$	$2/(2-\varepsilon)$	$1/(2-\varepsilon)$	$1+4/(2-\varepsilon)$	0
Ising: $d = 2$ (exact)	0 (log)	$\frac{1}{8}$	$\frac{7}{4}$	1	15	$\frac{1}{4}$
Ising: $d = 3$	0.12	0.33	1.25	0.64	4.8	0.04
Heisenberg: $d = 3$	−0.12	0.36	~1.39	0.71	4.8	0.04
S^4-model: $d > 4$	$\varepsilon/2$	$\frac{1}{2}-\varepsilon/4$	1	$\frac{1}{2}$	$3+\varepsilon$	0
S^4-model: $d = 4$	0	$\frac{1}{2}$	1	$\frac{1}{2}$	3	0
S^4-model: $d < 4$	$\varepsilon/6$	$\frac{1}{2}-\varepsilon/6$	$1+\varepsilon/6$	$\frac{1}{2}+\varepsilon/12$	$3+\varepsilon$	0
S^4-model: $d = 3$	0.17	0.33	1.17	0.58	4	0
XY-model: $d = 3$	0.01	0.34	1.30	0.66	4.8	0.04

Table 1. Summary of critical exponents for key models. Note that $\varepsilon = (4 - d)$.

An iterative solution of this recursion relation for the partition function yields "roots" or fixed points that correspond to resultant critical behaviors in the model. The partition function is preserved in the RG procedure via the condition

$$Z_N(H_n) = Z_m(H_{n+1}), \tag{8}$$

where $m = N/\lambda^d$.

It is conjectured that the values of critical exponents are characteristics not of individual Hamiltonians but their sets, with numerous models leading to the same fixed point. The universality hypothesis states that any two physical systems with the same dimensionality, d, and the same number of order parameter components, n, belong to the same universality class, and each fixed point corresponds to one universality class. Table 1 is a summary of the critical exponent values obtained from RG calculations for key theoretical models.

RG ideas extend into many areas of physics, chemistry, biology, and engineering. Based on the work of Kadanoff, Kenneth Wilson proposed an algorithmic approach to the scaling problem by formulating it in reciprocal space. Although less intuitive than the real-space RG of Kadanoff, it leads to exact results for the removal of divergencies in theories of elementary particles. For this work, Wilson was awarded the 1982 Nobel Prize in Physics.

JACK A. TUSZYŃSKI

See also **Critical phenomena; Ising model; Order parameters; Phase transitions**

Further Reading

Creswick, R.J., Farach, H.A. & Poole, C.P. 1992. *Introduction to Renormalization Group Methods in Physics*, New York: Wiley

Kosterlitz, J.M. & Thouless, D.J. 1973. Ordering, metastability and phase transitions in two-dimensional systems. *Journal of Physics* C, 6: 1181–1203

Ma, S.-K. 1976. *Modern Theory of Critical Phenomena*, New York: Benjamin

Mermin, N.D. & Wagner, H. 1966. Absence of ferromagnetism or antiferromagnetism in one- or two-dimensional isotropic Heisenberg models. *Physics Review Letters*, 17: 1133–1136

Reichl, L.E. 1979. *A Modern Course in Statistical Physics*, Austin: University of Texas Press

Stanley, H.E. 1972. *Introduction to Phase Transitions and Critical Phenomena*, Oxford: Clarendon Press and New York: Oxford University Press

Wilson, K.G. 1972. Feynman-graph expansion for critical exponents. *Physics Review Letters*, 28: 548–551

Wilson, K.G. 1983. The renormalization group and critical phenomena. *Reviews of Modern Physics*, 55: 583–600

Yeomans, J.M. 1992. *Statistical Mechanics of Phase Transitions*, Oxford: Clarendon Press and New York: Oxford University Press

REYNOLDS NUMBER

See **Fluid dynamics**

RHEOLOGY

Rheology is the study of the deformation and flow of materials in general, although most real applications relate to liquids (Barnes, 2000). Deformation and flow processes are linear with respect to extent and rate of deformation only in a limited number of circumstances, and most of the liquids usually considered in practical rheology show strong nonlinear effects in typical deformation and flow situations.

Newtonian Liquids

The simplest flow behavior is that of Newtonian liquids where the viscosity, ν, is a function only of temperature and pressure, and not of the flow conditions (although the dependencies of viscosity on the pressure and temperature are themselves very nonlinear). In simple shear flow (Barnes, 2000), the behavior at any point can be adequately described by the linear equation $\sigma = \nu\dot{\gamma}$, where σ is the shear stress and $\dot{\gamma}$ is the shear rate, or velocity gradient. However, nonlinearities soon set in for the flow of low-viscosity liquids such as water, even at low flow rates. First, smooth, streamline flow is augmented by secondary flows, but these eventually give way to chaotic turbulent flow at higher flow rates. The governing factor for the onset of secondary and turbulent flows is the effect of fluid inertia, arising from the fluid's density. The ratio of inertial to viscous forces is the Reynolds number, Re, and this parameter is important for all fluid flows, whether they are Newtonian or otherwise.

Secondary flows often appear in simple flow geometries such as viscometers and are manifested as an apparent increase in viscosity, due to the extra energy dissipated in the secondary flows cells (Barnes, 2000) (see Figure 1).

The resulting increased couple in the inertial case, T_i, for a cone-and-plate instrument, compared with that for no inertia T, is given for Newtonian liquids by

$$\frac{T_i}{T} \approx 1 + \frac{6}{10^4}\left(\frac{\rho\omega\theta^2a^2}{\nu}\right)^2, \tag{1}$$

where ρ is the density of the liquid of viscosity ν and a is the cone radius whose angle in radians is θ. From this equation it is easy to see that the correction

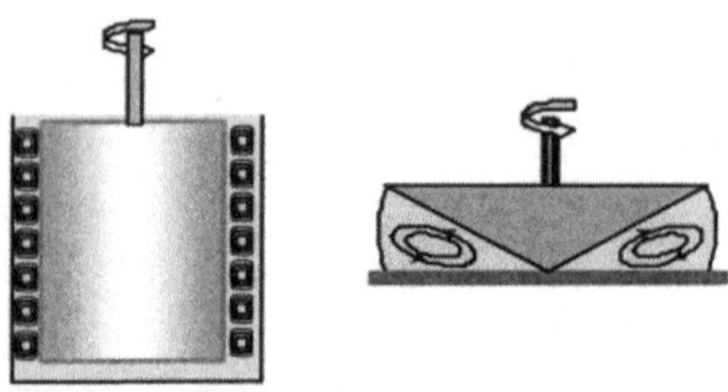

Figure 1. Inertial secondary flow patterns in concentric-cylinder and cone-and-plate geometries.

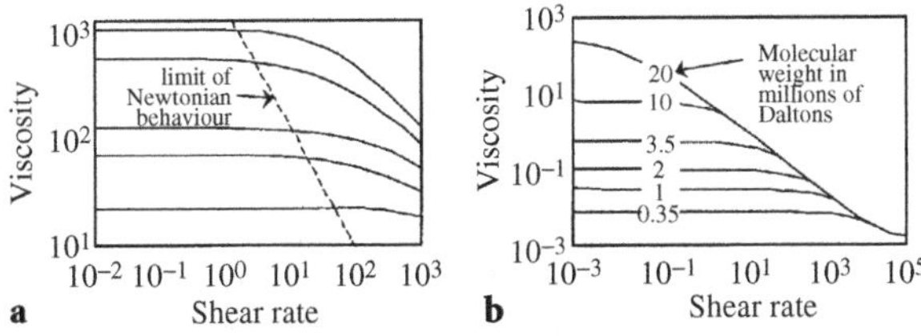

Figure 2. Viscosity (in pascal-seconds) vs. shear rate (1/seconds) for (a) a range of silicone oils, and (b) 3% by weight polystyrene in toluene.

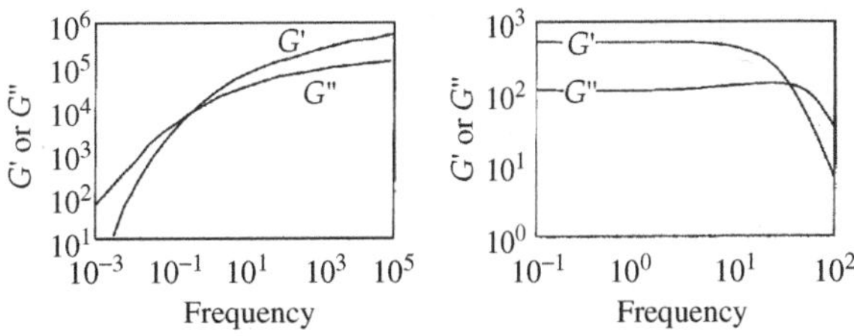

Figure 3. Storage modulus (G') and viscous behavior (G'') (in Pa) vs. Oscillatory frequency (1/s) for (a) low molecular weight polyethylene melt at 150°C, and (b) commercial ketchup.

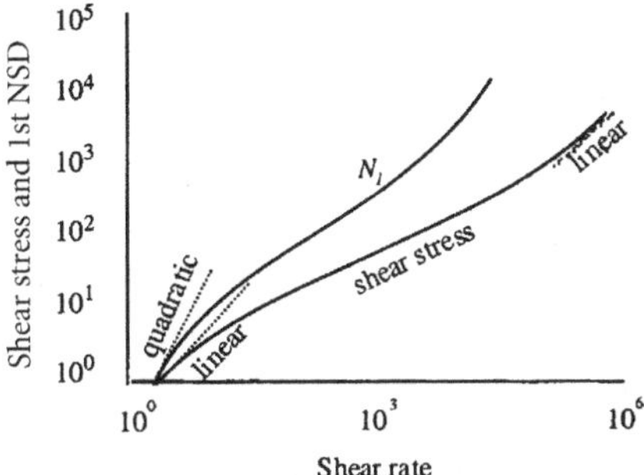

Figure 4. The shear stress (Pa) and first normal-stress difference (1st NSD) N_1(Pa) as a function of shear rate (1/s) for a 0.5% hydroxyethyl cellulose aqueous solution.

becomes important when $\rho\omega\theta^2 a^2/\nu$—which is a form of Reynolds number—exceeds 10.

Turbulent flow ensues in pipe flow when the Reynolds number given by $\rho dV/\nu$ (where V is the average velocity in a pipe of diameter d) is well above 2000, and then the pressure drop is no longer a linear function of the flow rate but approaches a quadratic dependence.

Non-Newtonian Liquids

The independence of viscosity to flow conditions is no longer valid for non-Newtonian liquids, but because their microstructure changes with the flow environment, the shear stress shows a nonlinear dependence on the shear rate. The dependence of viscosity on shear rate is such that the typical behavior is a decrease of viscosity with increase in shear rate from a value ν_0 at low shear rates, to a much lower value ν_∞ at much higher shear rates. This can often be described by the nonlinear Cross-type function

$$\frac{\nu - \nu_\infty}{\nu_0 - \nu_\infty} = \frac{1}{1 + (K\dot{\gamma})^m}, \tag{2}$$

where (written in this particular way) K has the dimensions of time and m is dimensionless. When this model is used to describe non-Newtonian liquids, the degree to which the viscosity decreases with shear rate is dictated by the value of m, with m tending to zero describing more Newtonian liquids, while the most "shear-thinning" liquids have values of m tending to unity. Figure 2 shows the departure from linearity for a number of non-Newtonian liquids.

Viscoelasticity

As well as the nonlinear behavior of viscosity, non-Newtonian liquids also display extra forces. These arise either as time or frequency effects in the linear region of behavior and are manifested, for instance, as an elastic component called the storage modulus G' in oscillatory flow as well as the viscous behavior as G''. In the simplest form of viscoelastic liquid, the parameters G' and G'' are related by a relaxation time, λ, of the liquid. These parameters are only constant over a limited region, but for deformations greater than a few percent for most non-Newtonian liquids, we enter the nonlinear region. The behavior of these G' and G'' with frequency is shown in Figure 3, along with typical departure from linearity.

For steady flows of viscoelastic liquids, the extra forces that become important are the so-called normal stresses. These arise essentially because the microstructure has a preferred configuration, which can become aligned by the flow. These normal forces manifest themselves in the linear region of slow flow as forces proportional to the square of the shear rate, but they also soon show nonlinear behavior in the region where the viscosity also becomes nonlinear. Figure 4 shows the typical behavior of both the shear stress and the first normal stress difference for a viscoelastic liquid.

Rheology of Polymeric Liquids

The microstructure that dictates the behavior of polymeric liquids is usually the state of entanglement of the polymeric chains. The chains are in incessant motion due to Brownian thermal energy, and their motions among each other give rise to viscous and elastic effects. However, as the shear rate increases, there is a gradual loss of entanglements and development of alignment, so that at high enough shear rates, linear polymer coils are transformed into aligned strings. The consequences of this change in microstructure are some very startling viscoelastic effects (Barnes, 2000).

Rheology of Suspensions

Suspension rheology depends on the relative disposition of particles relative to one another. Sometimes the particles are flocculated, and these flocs break down in size as the shear rate increases. This nonlinear phenomena can give rise to apparent yield stresses, where the viscosity is very high at low shear rates, but drops dramatically over a narrow range of stress.

Theoretical Rheology

The aim of mathematicians working in this area is to capture these complicated, and usually nonlinear, physical phenomena in constitutive equations (CEs). At their best, these describe nonlinear behaviour in viscosity—as above with the Cross-type equations—but also in the viscoelastic areas. CEs are written in the form of either differential or integral equations (see Barnes et al., 1989). A typical example of a nonlinear differential-type equation is the White–Metzner equation, where the rate-dependent viscosity $\nu_s(II_{\rm D})$ and the relaxation time $\lambda(II_{\rm D})_I I$ are written as

$$\nu_s(II_{\rm D}) = \frac{\nu_0}{[1+(K_1 II_D)^n]}$$

and

$$\lambda(II_{\rm D}) = \frac{\lambda_0}{[1+K_2 II_{\rm D}]}, \qquad (3)$$

where K_1 and K_2 are parameters with the dimensions of time, n is a numerical constant, and $II_{\rm D}$ is the second invariant of the deformation tensor (Barnes & Roberts, 1992). Barnes & Roberts (1992) have shown that this type of equation can describe many nonlinear rheological situations.

Howard A. Barnes

See also **Cluster coagulation; Fluid dynamics; Granular materials; Liquid crystals; Polymers**

Further Reading

Barnes, H.A. 2000. *A Handbook of Elementary Rheology*, Aberystwyth: The University of Wales Institute of Non-Newtonian Fluid Mechanics

Barnes, H.A., Hutton, J.F. & Walters, K. 1989. *An Introduction to Rheology*, Amsterdam: Elsevier

Barnes, H.A. & Roberts, G.P. 1992. A simple model to describe the steady state shear and extensions viscosity of polymer melts. *Journal of Non-Newtonian Fluid Mechanics*, 44: 113–126

RICCATI EQUATIONS

Count Jacopo Francesco Riccati (1676–1754) was an Italian nobleman who spent most of his life in the little town of Castelfranco, Veneto. Born in Venice and schooled at a Jesuit college in Brescia, he proceeded to the University of Padua where he read law. However, following his natural inclination and talent, he also took mathematical courses at Padua, and befriended Father Stefano degli Angeli, an astronomy lecturer. Angeli introduced him to Newton's *Principia*, which they read and studied together.

Dedicated to intellectual inquiry of every sort, Riccati is mostly remembered for his significant results in mathematics, particularly separation of variables and other solution methods for ordinary differential equations. He wrote extensively on such diverse areas as physics, architecture, and philosophy. Preferring to remain in Castelfranco, where he served as mayor for several years, he turned down numerous academic offers, including an invitation from Peter the Great to become President of the St. Petersburg Academy of Science.

Throughout his life he maintained an active correspondence with Italian scholars such as Agnesi and Rizzetti, as well as with European mathematicians such as the Bernoullis. In a letter of 1720 addressed to Rizzetti, he introduced two new differential equations, which in modern notation would be written

$$\frac{{\rm d}y}{{\rm d}x} = ay^2 + bx^m, \quad \frac{{\rm d}y}{{\rm d}x} = \alpha y^2 + \beta x + \gamma x^2, \qquad (1)$$

where a, b, m and α, β, γ are constants. His first published results on such equations appeared four years later (Riccati, 1724), and this paper is reproduced in Bittanti (1989) along with later related work of Euler and Liouville. Michieli's biographical work (Michieli, 1943) includes a detailed bibliography of Riccati's publications, letters, and manuscripts.

The scalar first order differential equation that today bears Riccati's name takes the general form

$$\frac{{\rm d}y}{{\rm d}x} = A(x)y^2 + B(x)y + C(x), \qquad (2)$$

where $A \not\equiv 0$, B, C are arbitrary functions of x. Thus, Equations (1) correspond to special cases of the Riccati equation (2) with constant A, zero B, and particular choices of C. There are two main solution methods (Ince, 1926). First, supposing that a particular solution y_1 is known, the general solution of (2) is found by the substitution

$$y = y_1 + \frac{1}{u}, \qquad (3)$$

which yields the linear equation

$$\frac{{\rm d}u}{{\rm d}x} + (2y_1 A + B)u + A = 0$$

for u. Hence u is found by quadratures, and then y is obtained (3) in terms of y_1 and u. The second method involves the substitution

$$y = -\frac{1}{A}\frac{{\rm d}}{{\rm d}x}\log\psi, \qquad (4)$$

which transforms the Riccati equation to a homogeneous second-order linear equation for ψ, namely,

$$\frac{{\rm d}^2\psi}{{\rm d}x^2} + f(x)\frac{{\rm d}\psi}{{\rm d}x} + g(x)\psi = 0 \qquad (5)$$

with

$$f = -B - \frac{\mathrm{d}}{\mathrm{d}x}\log A, \quad g = AC.$$

The scalar- second order Equation (5) can be converted to a 2×2 matrix linear system of the form

$$\frac{\mathrm{d}}{\mathrm{d}x}\begin{pmatrix}\psi_1\\ \psi_2\end{pmatrix} = \begin{pmatrix}M_{11} & M_{12}\\ M_{21} & M_{22}\end{pmatrix}\begin{pmatrix}\psi_1\\ \psi_2\end{pmatrix}. \tag{6}$$

Riccati himself considered such a system as describing the trajectory of a point (ψ_1, ψ_2) in the plane. Requesting the equation satisfied by the slope

$$y = \frac{\psi_2}{\psi_1}, \tag{7}$$

he found the answer was precisely (2) with

$$A = -M_{12}, \quad B = M_{22} - M_{11}, \quad C = M_{21}.$$

The left action of the matrix group $SL(2)$ on the linear system (6) induces a corresponding action via Möbius transformations on the projective coordinate y. This generalizes to Nth-order projective Riccati systems with nonlinear superposition principles (Anderson, 1980; Shnider & Winternitz, 1984). The latter are solved in terms of $(N+1)$th-order linear systems, with a projective action of $SL(N+1)$.

Due to this link with linear systems, (coupled) Riccati equations naturally appear in the theory of soliton-bearing integrable partial differential equations (Fordy, 1990). Within the framework of the inverse scattering method, nonlinear soliton equations (in $1+1$ dimensions) arise as the compatibility condition for a pair of matrix linear systems,

$$\Psi_x = U\Psi, \quad \Psi_t = V\Psi. \tag{8}$$

Without loss of generality the matrices U, V are taken as $sl(N+1)$ valued functions of the dependent variables and their derivatives as well as a spectral parameter, λ, say, and subscripts denote partial derivatives.

In the simplest case, $N = 1$, we consider the standard example of the Korteweg–de Vries (KdV) equation, with the first matrix in (8) given by

$$U = \begin{pmatrix}0 & 1\\ \lambda - u & 0\end{pmatrix},$$

and the spatial (x) part of pair (8) is just a 2×2 system (6). The corresponding scalar equation (5) reduces to the Schrödinger equation with potential u and spectral parameter λ; that is,

$$\psi_{xx} + u\psi = \lambda\psi,$$

and in terms of the projective variable y this yields the Riccati equation

$$y_x = -y^2 + \lambda - u. \tag{9}$$

In the special case of zero eigenvalue ($\lambda = 0$), with $y = v$ formula (9) becomes the Miura map

$$u = -v_x - v^2$$

relating a solution u of the KdV equation to a solution v of the modified KdV equation. Riccati equations are further distinguished in the theory of integrable systems by the fact that (for analytic A, B, C) they are the only first- order equations with the Painlevé property (Ince, 1926).

The fundamental problem of the calculus of variations (Gel'fand & Fomin, 1963) is to extremize a functional

$$S[q] = \int_{x_0}^{x_1} L(x, q, \dot{q})\,\mathrm{d}x, \quad \dot{q} \equiv \frac{\mathrm{d}q}{\mathrm{d}x}, \tag{10}$$

subject to $q(x_0) = q_0, q(x_1) = q_1$. Under an arbitrary small change $h(x)$ in $q(x)$,

$$q \to q + h,$$

with $h(x_0) = 0 = h(x_1)$, for a scalar function q the requirement $\delta S[h] = 0$ leads to the Euler–Lagrange equation

$$L_q - \frac{\mathrm{d}}{\mathrm{d}x}L_{\dot{q}} \equiv \frac{\partial L}{\partial q} - \frac{\mathrm{d}}{\mathrm{d}x}\frac{\partial L}{\partial \dot{q}} = 0.$$

If we further require to minimize the functional S, then a necessary condition is that the second variation should be nonnegative, that is,

$$\delta^2 S[h] = \frac{1}{2}\int_{x_0}^{x_1}\left(L_{\dot{q}\dot{q}}\dot{h}^2 + 2L_{q\dot{q}}h\dot{h} + L_{qq}h^2\right)\mathrm{d}x \geq 0.$$

After an integration by parts, this takes the form

$$\delta^2 S[h] = \int_{x_0}^{x_1}(P\dot{h}^2 + Qh^2)\,\mathrm{d}x, \tag{11}$$

where

$$P = \frac{1}{2}L_{\dot{q}\dot{q}}, \quad Q = \frac{1}{2}\left(L_{qq} - \frac{\mathrm{d}}{\mathrm{d}x}L_{q\dot{q}}\right).$$

Since h vanishes at the endpoints, there is the freedom to add any function of the form

$$\frac{\mathrm{d}}{\mathrm{d}x}(wh^2)$$

to the integrand in (11). If w is chosen to satisfy the Riccati equation

$$\frac{\mathrm{d}w}{\mathrm{d}x} = \frac{w^2}{P} - Q, \tag{12}$$

then the integrand for the second variation can be written in terms of a perfect square,

$$\delta^2 S[h] = \int_{x_0}^{x_1} P\left(\dot{h} + \frac{wh}{P}\right)^2 \mathrm{d}x.$$

Thus, to ensure nonnegativity of $\delta^2 S$, we arrive at Legendre's necessary condition

$$P(x) \geq 0, \quad x \in [x_0, x_1],$$

for a minimum of the functional $S[q]$. However, some care is needed to make this argument rigorous, since the Riccati equation (12) may not have a solution on the whole interval $[x_0, x_1]$.

In the case where q has n components, $q = (q^1, q^2, \ldots, q^n)$, the analogue of (12) is the matrix Riccati equation

$$\dot{W} = WP^{-1}W - Q,$$

where W, P, Q are symmetric matrices. Matrix Riccati equations are treated extensively in Reid (1972), and together with their discrete versions they find important applications in optimal filtering and control (Bittanti et al., 1991; Zelikin, 2000).

An n-component example is provided by the geodesics on a (pseudo-)Riemannian n-manifold with metric g_{jk} (Hughston & Tod, 1990), derived from the action

$$S[q] = \int_{\tau_0}^{\tau_1} g_{jk}(q)\dot{q}^j \dot{q}^k \, \mathrm{d}\tau$$

with affine parameter τ along the geodesics. The Euler–Lagrange equations for the geodesics, giving the trajectories of free particle motion, are

$$\frac{\mathrm{d}^2 q^j}{\mathrm{d}\tau^2} + \Gamma^j_{kl} \frac{\mathrm{d}q^k}{\mathrm{d}\tau} \frac{\mathrm{d}q^l}{\mathrm{d}\tau} = 0,$$

where Γ^j_{kl} are the Christoffel symbols. In the case of a Riemannian manifold, the metric tensor is positive definite and the Legendre condition is satisfied. The geodesics are the paths of minimum length.

Andrew Hone

See also **Euler–Lagrange equations; Extremum principles; General relativity; Integrability; Inverse scattering method or transform; Korteweg–de Vries equation; Painlevé analysis**

Further Reading

Anderson, R.L. 1980. A nonlinear superposition principle admitted by coupled Riccati equations of the projective type. *Letters in Mathematical Physics*, 4: 1–7

Bittanti, S. (editor). 1989. *Count Riccati and the Early Days of the Riccati Equation*. Bologna: Pitagora Editrice

Bittanti, S., Laub, A.J. & Willems, J.C. (editors). 1991. *The Riccati Equation*. Berlin: Springer

Fordy, A.P. (editor). 1990. *Soliton Theory: A Survey of Results*. Manchester: Manchester University Press

Gel'fand, I.M. & Fomin, S.V. 1963. *Calculus of Variations*. Englewood Cliffs, NJ: Prentice-Hall

Hughston, L.P. & Tod, K.P. 1990. *An Introduction to General Relativity*. Cambridge and New York: Cambridge University Press

Ince, E.L. 1926. *Ordinary Differential Equations*. 7th edition, Edinburgh: Oliver and Boyd and New York: Interscience, 1959

Michieli, A.A. 1943. Una famiglia di matematici e di poligrafi trivigiani: i Riccati. I. Iacopo Riccati. *Atti del Reale Istituto Veneto di Scienze, Lettere ed Arti*, 102(2): 535–587

Reid, W.T. 1972. *Riccati Differential Equations*. New York: Academic Press

Riccati, J.F. 1724. Animadversiones in aequationes differentiales secundi gradus. *Supplementa Acta Eruditorum Lipsiae*, 8(2): 66–73

Shnider, S. & Winternitz, P. 1984. Classification of nonlinear ordinary differential equations with superposition principles. *Journal of Mathematical Physics*, 25(11): 3155–3165

Zelikin, M.I. 2000. *Control Theory and Optimization I: Homogeneous Spaces and the Riccati Equation in the Calculus of Variations*. Berlin and New York: Springer

RIEMANN INVARIANTS

See **Characteristics**

RIEMANN SURFACES

See **Periodic spectral theory**

RIEMANN WAVE

See **Shock waves**

RIEMANN ZETA FUNCTION

See **Random matrix theory I: Origins and physical applications**

RIEMANN–HILBERT PROBLEM

We begin by defining what is meant by a Riemann–Hilbert problem (RHP). Let Σ be an oriented contour in $\mathbb{C}$.

By convention, if we move along the contour in the direction of the orientation, we say that the $(+/-)$ side lies to the (left/right), as indicated in Figure 1. A map v from Σ to the space of the invertible $k \times k$ matrices with complex entries is called a *jump matrix* for Σ if v and v^{-1} are bounded on Σ. We say that a

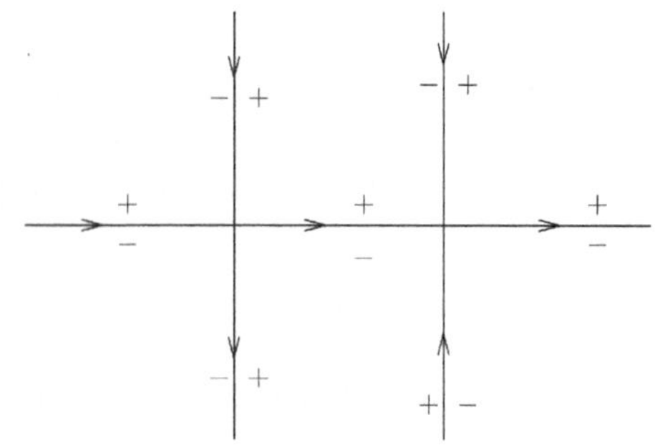

Figure 1. The contour Σ.

$j \times k$ matrix-valued function $m = m(z)$ is a *solution* of the RHP (Σ, v) if

- m is analytic in $\mathbb{C} \setminus \Sigma$,
- $m_+(z) = m_-(z)v(z)$ for $z \in \Sigma$, where $m_\pm(z) = \lim_{\substack{z' \to z \\ z' \in (\pm)\text{-side of } \Sigma}} m(z')$.

  If in addition $j = k$ and
- $m(z) \to I$ as $z \to z_0$ for some $z_0 \in \mathbb{C} \cup \{\infty\}$, we say that the RHP is *normalized* at z_0.

A discussion of technical restrictions on Σ, and the precise sense in which the boundary values, as well as the limit at z_0 in the normalized case, are attained, can be found in many texts (see, e.g., Clancey & Gohberg (1981) and the references therein). Many problems in pure and applied mathematics, and also theoretical physics, can be expressed in terms of an RHP. Riemann–Hilbert problems are closely related to the Wiener–Hopf method. We refer the reader to Clancey & Gohberg (1981) for a history of RHPs and also to the classic text by Muskhelishvili (1946). The goal of this article is to describe some recent developments.

Remark. Included in David Hilbert's famous list of 23 problems, was the following:

> Show that there always exists a linear differential equation of the Fuchsian class with given singularities and a given monodromy group.

A Fuchsian system is an nth-order, linear system of the form

$$\mathrm{d}m/\mathrm{d}z = A(z)m,$$

where A is an $n \times n$ matrix with entries that are analytic except at a finite number of simple poles. Roughly speaking, the monodromy group is the group of transformations of solutions to the above Fuchsian system under analytic continuation about these poles.

The above monodromy problem turns out to be a special case of a Riemann–Hilbert problem with jump matrices that are (essentially) piece-wise constant. In the mathematical literature, the monodromy problem is sometimes identified with the Riemann–Hilbert problem, but this is incorrect. The work of Bernhard Riemann, and later of David Hilbert, on such problems, predates Hilbert's monodromy problem and has very different origins.

Our point of departure here is the observation of Alexey Shabat in 1976 that the inverse scattering method for the Schrödinger operator can be rephrased as a RHP (see Faddeev & Takhtajan, 1987). By way of illustration, we apply Shabat's observation to a simpler case, the self-adjoint ZS–AKNS scattering problem (Zakharov & Shabat, 1971; Ablowitz et al., 1974)

$$\frac{\mathrm{d}\psi}{\mathrm{d}x} = \left(\mathrm{i}z\sigma + \begin{pmatrix} 0 & q(x) \\ \bar{q}(x) & 0 \end{pmatrix}\right)\psi, \qquad (1)$$

where $\sigma = \begin{pmatrix} 1/2 & 0 \\ 0 & -1/2 \end{pmatrix}$ and $q(x)$ is a given function on $\mathbb{R}$, which decays sufficiently fast as $|x| \to \infty$. For fixed $z \in \mathbb{C} \setminus \mathbb{R}$, one seeks so-called Beals–Coifman solutions $\psi = \psi(x, z; q)$ of (1) with the properties

$$\psi \mathrm{e}^{-\mathrm{i}xz\sigma} \to I \text{ as } x \to -\infty, \qquad (2)$$

$$\psi \mathrm{e}^{-\mathrm{i}xz\sigma} \text{ is bounded as } x \to +\infty. \qquad (3)$$

It turns out, conversely, that for fixed $x \in \mathbb{R}$, $m(z) = m(x, z; q) \equiv \psi(x, z; q)\mathrm{e}^{-\mathrm{i}xz\sigma}$ solves the RHP $(\Sigma = \mathbb{R}, v_x)$ normalized at $z_0 = \infty$, where

$$v_x(z) = \begin{pmatrix} 1 - |r(z)|^2 & r(z)\mathrm{e}^{\mathrm{i}xz} \\ -\bar{r}(z)\mathrm{e}^{-\mathrm{i}xz} & 1 \end{pmatrix}, \quad z \in \mathbb{R} \qquad (4)$$

and $r(z) = r(z; q)$ is the *reflection coefficient* for q. Define the *reflection map* $\mathcal{R}$ via $\mathcal{R}(q) \equiv r$. Given q, the direct scattering problem is given by

$$q \mapsto m(x, z; q) \mapsto v_x \mapsto r = \mathcal{R}(q).$$

On the other hand, for a given r and a fixed $x \in \mathbb{R}$, let $m = m(x, z; r)$ be the solution of the RHP $(\mathbb{R}, v_x)$ normalized at ∞. Expand $m(z)$ as $z \to \infty$,

$$m(z) = I + m_1(x; r)/z + O(z^{-2}), \qquad (5)$$

and define

$$(Q(r))(x) \equiv -\mathrm{i}(m_1(x; r))_{12}. \qquad (6)$$

Given r, the inverse scattering problem is given by

$$r \mapsto v_x \mapsto m(x, z; r) \mapsto m_1(x; r) \mapsto Q(r).$$

The basic analytical result in the subject is that $\mathcal{R}$ is bijective from a suitable space of q's onto a suitable space of r's (Beals & Coifman , 1984; Zhou, 1998) and that Q is the inverse of $\mathcal{R}$.

In terms of the classical method of scattering and inverse scattering (Zakharov & Shabat, 1971; Ablowitz et al., 1974), $r \equiv r(z; q)$ is the standard reflection coefficient. Moreover, if m solves the RHP (Σ, v_x), then the Fourier transform of the first row of $m_- \begin{pmatrix} 1 & r\mathrm{e}^{\mathrm{i}xz} \\ 0 & 1 \end{pmatrix} - I$ gives the solution of the Faddeev–Marchenko equation.

In the case of the non-self-adjoint ZS–AKNS scattering problem, the Beals–Coifman solutions $\psi(z)$ may have singularities at points $\{z_j\} \subset \mathbb{C} \setminus \mathbb{R}$, which correspond to eigenvalues with square integrable eigenfunctions. In addition, there may be singularities for $\psi_\pm(z)$ on the contour $\Sigma = \mathbb{R}$. One now associates an RHP to the scattering problem by augmenting the contour $\Sigma \to \Sigma_{\text{aug}}$ so as to enclose these singularities: information about the singularities is then contained in the augmented jump matrix v_{aug} defined on $\Sigma_{\text{aug}} \setminus \Sigma$ (see Zhou, 1989). The above ideas apply to very general first-order systems of size $N \times N$ and also to differential operators of order M. The underlying contour Σ is now more complicated and may have points of self-intersection, which leads in turn to significant new technical complications. Also, the singularities of ψ in $\mathbb{C}\setminus\Sigma$ need not correspond to eigenvalues with square integrable eigenfunctions.

We now consider the relation of RHPs to the theory of integrable systems. Again by the way of illustration, we consider the normalized RHP $(\mathbb{R}, v_x)$ associated with the self-adjoint ZS–AKNS system. If $q(t)=q(x,t)$ solves the defocusing nonlinear Schrödinger (NLS) equation

$$\begin{aligned} &\mathrm{i}q_t + q_{xx} - 2|q|^2 q = 0, \\ &q(x, t=0) = q_0(x) \to 0 \text{ as } |x| \to \infty, \end{aligned} \quad (7)$$

then by Zakharov & Shabat (1971)

$$r(t,z) = r(z)\mathrm{e}^{-\mathrm{i}tz^2}, \quad z \in \mathbb{R}. \quad (8)$$

This leads to the following solution procedure for NLS. Let $m(x, z; r(\Diamond)\mathrm{e}^{-\mathrm{i}t\Diamond^2}) = I + m_1(x; r(\Diamond)\mathrm{e}^{-\mathrm{i}t\Diamond^2})z^{-1} + O(z^{-2})$ be the solution of the normalized RHP $(\mathbb{R}, v_\theta)$, where

$$v_\theta(z) = \begin{pmatrix} 1-|r(z)|^2 & r\mathrm{e}^{\mathrm{i}\theta} \\ -\bar{r}\mathrm{e}^{-\mathrm{i}\theta} & 1 \end{pmatrix}, \quad \theta = xz - tz^2. \quad (9)$$

Then

$$q(x,t) = -\mathrm{i}(m_1(x; r(\Diamond)\mathrm{e}^{-\mathrm{i}t\Diamond^2}))_{12}. \quad (10)$$

An extraordinarily broad spectrum of problems, both dynamical and nondynamical, can be solved via a RHP as in (10) above. The list includes, in particular, the Painlevé equations (see Flaschka & Newell, 1980; Jimbo et al., 1981) where the underlying contour may be a union of intersecting lines, and also the orthogonal polynomial problem as formulated by Fokas, Its, and Kitaev, and the theory of Hankel and Toeplitz determinants (see e.g., Deift, 1999). In practice, RHPs often arise via associated *integrable operators* (Its et al., 1990; see also Deift, 1999). Let Σ be an oriented contour in $\mathbb{C}$. We say that an operator K acting in $L^p(\Sigma)$, $1<p<\infty$, is integrable if it has a kernel of the form

$$K(z,z') = (z-z')^{-1}\sum_{j=1}^{k} f_j(z)g_j(z'), \quad z, z' \in \Sigma.$$

Special examples of integrable operators began to appear in field theory and in statistical mechanics in the late 1960s and 1970s, particularly in the work of Wu, McCoy, Tracy, and Baruch, and later Sato, Miwa, and Jimbo, but integrable operators as a distinguished class were first identified by Its et al. (1990) and later Sato-Miwa-Jimbo. Some general results on integrable operators were obtained by Lev A. Sakhnovich in the late 1960s, but the full theory of integrable operators is due to Its et al. (1990).

Integrable operators have many remarkable properties, the most important being that if $(1-K)^{-1} = I + R$, then R is also an integrable operator, $R(z,z') = z-z')^{-1}\sum_{j=1}^{k} F(z)G_j(z')$, and F_i, G_j can be constructed by solving an RHP $(\Sigma, v_{f,g})$ naturally associated with f_i, g_j. It turns out that many quantities H of physical interest can be expressed in the form $H = \det(1-K)$, where K is an integrable operator. For example, let K_x be the operator acting on $L^2(0,1)$ with kernel $\frac{\sin \pi x(z-z')}{\pi(z-z')}$, $z, z' \in (0,1)$. Then $H_x = \det(1-K_x)$ is the probability that in the bulk scaling limit, a Hermitian matrix chosen from the Gaussian Unitary Ensemble has no eigenvalues in $[0,x]$. Differentiating with respect to x, one derives an expression for $\frac{\mathrm{d}}{\mathrm{d}x}\log H_x$ in terms of $R_x = (1-K_x)^{-1} - 1$, and hence we see that the analysis of H_x reduces to the analysis of an RHP. Similar examples arise in many different scientific areas, including statistical mechanics, percolation theory, combinatorics, representation theory of large groups, and approximation theory. Relatively recent bibliographies can be found in Deift et al. (1993, 1999).

Returning to the NLS equation, one expects from (9) and (10) that the leading order contribution to the solution $q(x,t)$ as $t\to\infty$ should arise from the stationary phase point $z_0 = x/2t$, $\theta'(z_0) = 0$, as in the classical steepest-descent method for the asymptotic evaluation of (scalar) integrals. In Deift & Zhou (1993), the authors introduce a general noncommutative steepest-descent method to analyze such microlocal problems in the context of RHPs. In the case of NLS, the method proceeds as follows (Deift, Its & Zhou, 1997; Deift & Zhou, 1994). Let δ solve the scalar RHP $((-\infty, z_0), 1-|r|^2)$ normalized at ∞ and set $\tilde{m} = m\delta^{-\sigma_3}$, $\sigma_3 = \begin{pmatrix} 1 & 0 \\ 0 & -1 \end{pmatrix}$. Then $\tilde{m}$ solves the normalized RHP $(\mathbb{R}, \tilde{v}_\theta)$, where $\tilde{v}_\theta$ is as in Figure 2.

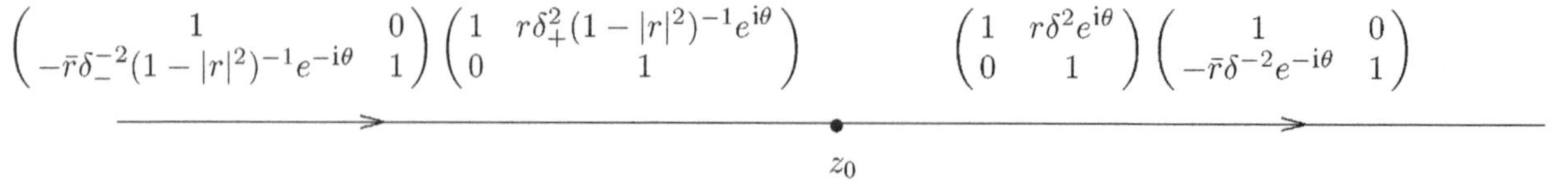

$$\begin{pmatrix} 1 & 0 \\ -\bar{r}\delta_-^{-2}(1-|r|^2)^{-1}e^{-\mathrm{i}\theta} & 1 \end{pmatrix}\begin{pmatrix} 1 & r\delta_+^2(1-|r|^2)^{-1}e^{\mathrm{i}\theta} \\ 0 & 1 \end{pmatrix} \qquad \begin{pmatrix} 1 & r\delta^2 e^{\mathrm{i}\theta} \\ 0 & 1 \end{pmatrix}\begin{pmatrix} 1 & 0 \\ -\bar{r}\delta^{-2}e^{-\mathrm{i}\theta} & 1 \end{pmatrix}$$

z_0

Figure 2. The jump matrix $\tilde{v}_\theta$.

Assuming (as we may, after suitable approximation) that the entries of $\tilde{v}_\theta$ have analytic extensions, define m_θ^d as in Figure 3. Note that, with this definition, m_θ^d solves the normalized RHP $(\Sigma_{z_0}^d, v_\theta^d)$, where v_θ^d is shown in Figure 4.

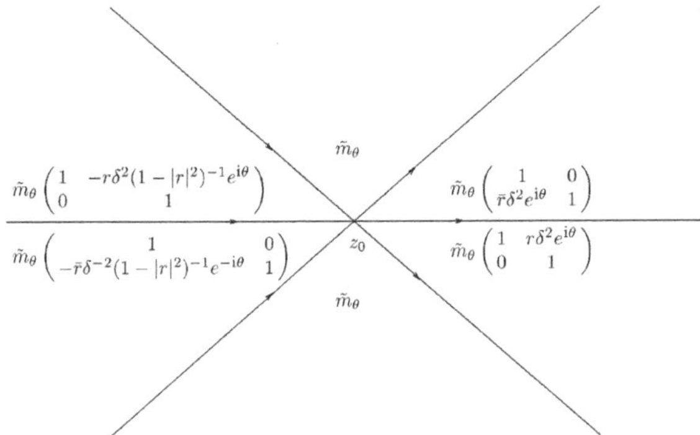

Figure 3. The solution m_θ^d.

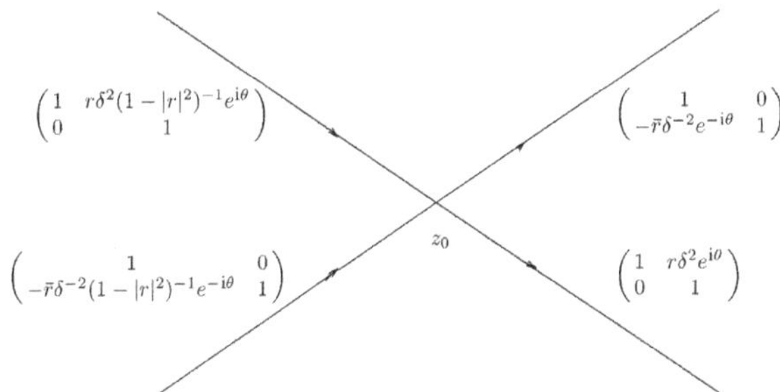

Figure 4. The jump matrix v_θ^d.

Observe that the definition of δ and the choice of the lower/upper and upper/lower factorizations of $\tilde{v}_\theta$, were made precisely to take advantage of the signature table of Re, $i\theta = -t \operatorname{Re} i (z - z_0)^2$ in Figure 5.

As $t \to \infty$, $v_\theta^d \to I$ uniformly on $\Sigma_{z_0}^d \setminus \{|z - z_0| \geq \varepsilon\}$ for any $\varepsilon > 0$, and the RHP localizes at z_0 to an RHP that can be solved explicitly. We obtain as $t \to \infty$

$$q(x,t) = t^{-1/2}\alpha(z_0)e^{i[x^2/(4t) - \nu(z_0)\log 2t]} + O(\log t/t), \tag{11}$$

where

$$\nu(z) = -\frac{1}{2\pi}\log(1 - |r(z)|^2), \quad |\alpha(z)|^2 = \nu(z)/2 \tag{12}$$

and

$$\arg\alpha(z) = \frac{1}{\pi}\int_{-\infty}^{z}\log(z-s)d\log(1-|r(s)|^2) + \frac{\pi}{4} + \arg\Gamma(i\nu(z)) + \arg r(z). \tag{13}$$

The above asymptotic form was first obtained by Zakharov and Manakov but without the error estimate. The RHP/steepest-descent method, suitably extended, has also been used by Deift and Zhou to obtain precise long-time asymptotics of solutions of perturbed NLS equations.

$\operatorname{Re} i\theta > 0$	$\operatorname{Re} i\theta < 0$
$\operatorname{Re} i\theta < 0$	z_0 $\operatorname{Re} i\theta > 0$

Figure 5. Signature of Re, $i\theta$.

In the above problem, the leading order asymptotic contribution arises from the stationary phase point, as in the classical method of stationary phase for scalar integrals. In many problems, in fact in most problems, the leading order contribution no longer arises from one (or more) critical points, but is given instead by a "critical contour." Such a contour arises, for example, in the asymptotic analysis as $x \to \infty$ of the Painlevé equation by Deift and Zhou, and also in the analysis of the collisionless shock region by Deift, Venakides and Zhou. In Deift, Venakides and Zhou (1997), in the context of the zero-dispersion limit of the Kortweg–de Vries (KdV) equation, the method in Deift & Zhou (1993) was extended significantly to include a prescription for the determination of the critical contour, making it possible in turn to describe the asymptotic development of fully nonlinear oscillations. This extension of Deift & Zhou (1993) made it possible to obtain Plancherel–Rotach-type asymptotics for a general class of orthogonal polynomials, leading in turn to a solution of the so-called universality conjecture for unitary ensembles (see Deift et al., 1999, and **Random Matrix Theory IV: Analytic methods**; see also the work of Pastur and Scherbina for a different approach, and also the work of Bleher and Its for an approach related to Deift et al. (1999), in the special case of a quartic potential). Many other long-standing problems in mathematics and in physics have been solved since the 1990s using RHP techniques and the (extended) steepest-descent method. For example, in combinatorics, Ulam's problem for increasing subsequences in random permutations was solved completely by Baik, Deift, and Johansson in terms of the Tracy–Widom distribution of random matrix theory, giving rise in turn to an explosive growth in applications linking combinatorics, statistical models, and random matrix theory. Many people are contributing to these developments, but there is as yet no adequate review. Fortunately, a comprehensive review, by T. Kriecherbauer, is due to appear in *Nonlinearity* in early 2005. Further extensions of the RHP/steepest-descent method have been given recently in the work of Kamvissis, McLaughlin and Miller on the semiclassical limit for the defocusing nonlinear Schrödinger equation.

Finally, we return to the classical Faddeev–Marchenko approach to inverse scattering theory mentioned earlier and make some comparisons with the RHP method. First, as noted above, the Faddeev–Marchenko equation arises by taking a Fourier transform of objects arising in the RHP theory. This means, in particular, that the classical method only applies in cases where the underlying contour is a group such as a line or a circle; problems such as Painlevé II, for example, cannot be handled in general by the classical method. Second, even in cases where the underlying contour is a group, such as NLS or KdV, the leading order contribution to the solution as $t \to \infty$ arises from the stationary phase points. Taking a Fourier transform "smears out" this feature and the microlocal nature of the problem is obscured.

Beals & Coifman(1984) were the first to use RHPs for the rigorous analysis of scattering and inverse scattering theory.

PERCY DEIFT AND XIN ZHOU

See also **Gel'fand–Levitan theory; Inverse scattering method or transform; Nonlinear Schrödinger equations; Random matrix theory IV: Analytic methods**

Further Reading

Ablowitz, M.J., Kaup, D.J., Newell, A.C. & Segur, H. 1974. The inverse scattering transform-Fourier analysis for nonlinear problems. *Studies in Applied Mathematics*, 53: 249–315

Beals, B. & Coifman, R. 1984. Scattering and inverse scattering for first order systems. *Communications in Pure and Applied Mathematics*, 37, 39–90

Clancey, K. & Gohberg, I.C. 1981. *Factorizations of Matrix Functions and Singular Integral Operators*, Basel and Boston: Birkhäuser

Deift, P. 1999. Integrable operators. *American Mathematical Society Translations*, 2(189): 69–84

Deift, P., Its, A. & Zhou, X. 1993. Long–time asymptotics for integrable nonlinear wave equations. In *Important Developments in Soliton Theory 1980–1990*, edited by A.S. Fokas & V.E. Zakharov, Berlin and New York: Springer, pp.181–204

Deift, P., Kriecherbauer, T., McLaughlin, K., Venakides, S. & Zhou, X. 1999. Uniform asymptotics for polynomials orthogonal with respect to varying exponential weights and applications to universality questions in random matrix theory. *Communications in Pure and Applied Mathematics*, 52: 1335–1425

Deift, P., Venakides, S. & Zhou, X. 1997. New results in small dispersion KdV by an extension of the steepest descent method for Riemann-Hilbert problems, *Internat. Math. Res. Notices*, No. 6, 285–299

Deift, P. & Zhou, X. 1993. A steepest descent method for oscillatory Riemann–Hilbert problems–asymptotics for the MKdV equation. *Annals of Mathematics*, 137: 295–368

Deift, P. & Zhou, X. 1994. *Long-term Behavior of the Non-focusing Nonlinear Schrödinger Equation—A Case Study*, Tokyo: University of Tokyo (Lectures in Mathematical Sciences, vol. 5)

Faddeev, L.D. & Takhtajan, L.A. 1987. *Hamiltonian Methods in the Theory of Solitons*, Berlin and Heidelberg: Springer

Flaschka, H. & Newell, A.C. 1980. Monodromy and spectrum preserving deformations I. *Communications in Pure and Applied Mathematics*, 76: 67–116

Its, A. 2003. The Riemann–Hilbert problem and integrable systems. *Notices of the AMS*, 50: 1389–1400

Its, A.R., Izergin, A.G., Korepin, V.E. & Slavnov, N.A. 1990. Differential equations for quantum correlation functions. *International Journal of Physics* B4: 1003–1037; The quantum correlation function as the τ function of classical differential equations. In *Important Developments in Soliton Theory*, edited by, A.S. Fokas & V. E. Zakharov, Springer, Berlin, pp. 404–417, 1993

Jimbo, M., Miwa, T. & Ueno, K. 1981. Monodromy preserving deformation of linear ordinary differential equations with rational coefficients. I. *Physica* D, 2: 306–352

Muskhelishvili, N.I. 1946. *Singular Integral Equations*, Moscow (in Russian); translation by Groningen. Leiden: P. Noordhoff, 1953

Zakharov, V.E. & Shabat, A.B. 1971. Exact theory of two-dimensional self-focusing and one-dimensional self–modulation of waves in nonlinear media. *Zhurnal Eksperimental'noi Teoreticheskoi Fiziki*, 61: 118–134 [Russian]; translated in *Soviet Physics, JETP*, 34: 62–69 (1972)

Zhou, X. 1989. Direct and inverse scattering transforms with arbitrary spectral singularities. *Communications in Pure and Applied Mathematics*, 42, 895–938

Zhou, X. 1998. The L^2-Sobolev space bijectivity of the scattering and inverse scattering transforms. *Communications in Pure and Applied Mathematics*, 51: 697–731

RING SOLITONS

See **Solitons, types of**

ROBUSTNESS

See **Stability**

ROSSBY WAVES

See **Atmospheric and ocean sciences**

RÖSSLER SYSTEMS

Rössler systems were introduced in the 1970s as prototype equations with the minimum ingredients for continuous-time chaos.

Since the Poincaré–Bendixson theorem precludes the existence of other than steady, periodic, or quasi-periodic attractors in autonomous systems defined in one- or two-dimensional manifolds such as the line, the circle, the plane, the sphere, or the torus (Hartman, 1964), the minimum dimension for chaos is three. On this basis, Otto Rössler came up with a series of prototype systems of ordinary differential equations in three-dimensional phase spaces (Rössler, 1976a,c, 1977a, 1979a). He also proposed four-dimensional systems for hyperchaos, that is, chaos with more than one positive Lyapunov exponent (Rössler, 1979a,b).

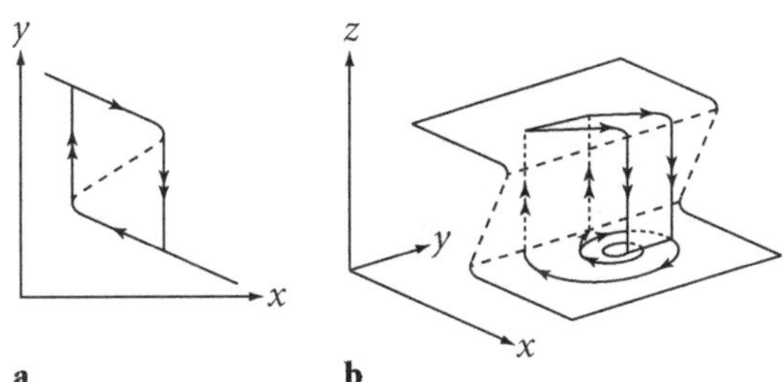

Figure 1. Illustration of the reinjection principle between the two branches of a Z-shaped slow manifold allowing (a) periodic relaxation oscillations in dimension two and (b) higher types of relaxation behavior in dimension three.

Rössler was inspired by the geometry of flows in dimension three and, in particular, by the reinjection principle, which is based on the feature of relaxation-type systems to often present a Z-shaped slow manifold in their phase space. On this manifold, the motion is slow until an edge is reached whereupon the trajectory jumps to the other branch of the manifold, allowing not only for periodic relaxation oscillations in dimension two (see Figure 1a), but also for higher types of relaxation behavior (see Figure 1b) as noted by Rössler (1979a). In dimension three, the reinjection can induce chaotic behavior if the motion is spiraling out on one branch of the slow manifold (see Figure 1b). In this way, Rössler invented a series of systems, the most famous of which is probably (Rössler 1979a):

$$\frac{\mathrm{d}x}{\mathrm{d}t} = -y - z,$$

$$\frac{\mathrm{d}y}{\mathrm{d}t} = x + ay,$$

$$\frac{\mathrm{d}z}{\mathrm{d}t} = bx - cz + xz. \quad (1)$$

This system is minimal for continuous chaos for at least three reasons: its phase space has the minimal dimension three, its nonlinearity is minimal because there is a single quadratic term, and it generates a chaotic attractor with a single lobe, in contrast to the Lorenz attractor which has two lobes. In Equation (1), (x, y, z) are the three variables that evolve in the continuous time t, and (a, b, c) are three parameters. The linear terms of the two first equations create oscillations in the variables x and y. These oscillations can be amplified if $a > 0$, which results into a spiraling-out motion. The motion in x and y is then coupled to the z variable ruled by the third equation, which contains the nonlinear term and which induces the reinjection back to the beginning of the spiraling-out motion.

System (1) possesses two steady states: one at the origin $x = y = z = 0$, around which the motion spirals out, and another one at some distance from the origin due to the quadratic nonlinearity. This system presents stationary, periodic, quasi-periodic, and chaotic attractors depending on the value of the parameters (a, b, c). These attractors are interconnected by bifurcations, in particular, a Hopf bifurcation from the stationary to periodic attractors and a period-doubling cascade from periodic to chaotic attractors. The resulting chaotic attractor has a single lobe and is referred to as spiral-type chaos, which mainly manifests itself in irregular amplitudes for the oscillations (see Figure 2a).

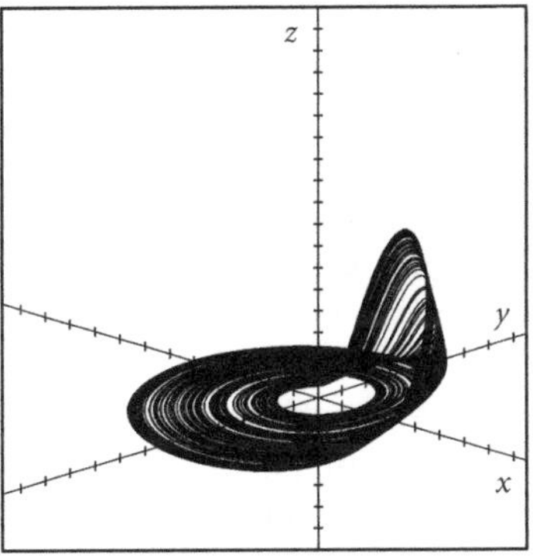

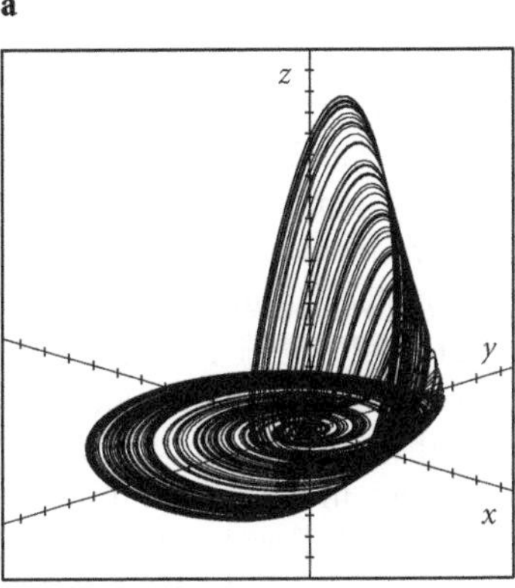

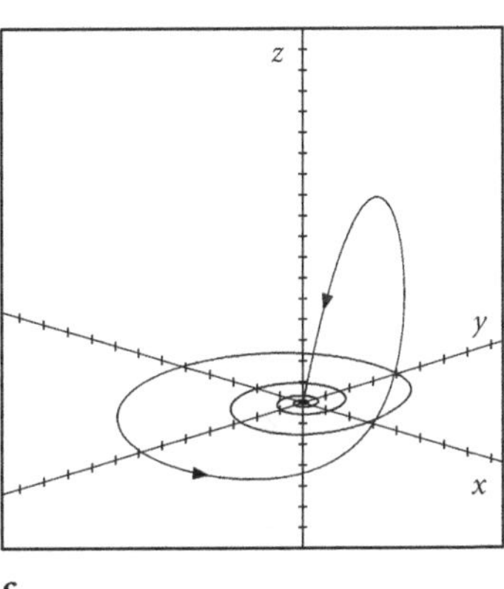

Figure 2. Phase portraits of the Rössler system (1) in the phase space of the variables (x, y, z): (a) spiral-type chaos for $a = 0.32$, $b = 0.3$, and $c = 4.5$; (b) screw-type chaos for $a = 0.38$, $b = 0.3$, and $c = 4.820$, in which case there also exists the Shil'nikov-type homoclinic orbit (c). The ticks are separated by unity. (Adapted from Gaspard & Nicolis, 1983.)

A transition occurs to a screw-type chaos in which the oscillations are irregular not only in their amplitudes but also in the reinjection times (see Figure 2b). The screw-type chaos is closely related to the presence of a Shil'nikov homoclinic orbit (see Figure 2c). This homoclinic orbit is attached to the origin $x = y = z = 0$, which is a saddle-focus with a one-dimensional stable manifold for the reinjection and a two-dimensional unstable manifold where the motion is spiraling out. The Shil'nikov criterion for chaos is that the reinjection is faster than the spiraling-out motion (Shil'nikov, 1965), and it is satisfied in the attractor of Figure 2b. As a consequence, the homoclinic system contains periodic and nonperiodic orbits belonging to multiple horseshoes that can be described in terms of symbolic dynamics. Away from homoclinicity, the system undergoes complex bifurcation cascades generating successive periodic and chaotic attractors.

Chaotic behavior and Shil'nikov homoclinic orbits in the Rössler system (1) can also be understood as originating from an oscillatory-stationary double instability taking place around the origin $x = y = z = 0$ and the parameter values $b = 1$, $-\sqrt{2} < a = c < +\sqrt{2}$.

In his work on continuous chaos, Rössler was motivated by the search for chemical chaos, that is, chaotic behavior in far-from-equilibrium chemical kinetics (Ruelle, 1973; Rössler, 1976b, 1977b; Rössler & Wegmann, 1978). With Willamowski, Rössler proposed the following chemical reaction scheme:

$$\begin{aligned} A_1 + X &\underset{k_{-1}}{\overset{k_1}{\rightleftharpoons}} 2X, \\ X + Y &\underset{k_{-2}}{\overset{k_2}{\rightleftharpoons}} 2Y, \\ A_5 + Y &\underset{k_{-3}}{\overset{k_3}{\rightleftharpoons}} A_2, \\ X + Z &\underset{k_{-4}}{\overset{k_4}{\rightleftharpoons}} A_3, \\ A_4 + Z &\underset{k_{-5}}{\overset{k_5}{\rightleftharpoons}} 2Z, \end{aligned} \tag{2}$$

which features two autocatalytic steps (reactions 1 and 5) involving the species X and Z coupled to another autocatalytic step (reaction 2) involving another species Y and two further steps (reactions 3 and 4) (Willamowski & Rössler, 1980). The concentrations of the species $A_1, \ldots, A_5$ are held fixed by large chemical reservoirs that maintain the system out of thermodynamic equilibrium. The time evolution of the concentrations (x, y, z) of the three intermediate species X, Y, and Z is ruled by a system of three coupled differential equations deriving from the mass action law of chemical kinetics. These equations have quadratic nonlinear terms because of the binary reactive steps and keep the concentrations positive as a consequence of mass action kinetics. The chemical reaction scheme (2) leads to a chaotic attractor very similar to that of the abstract system (1) (see Figure 3), and thus provides a mechanistic understanding of chemical chaos in terms of colliding and reacting particles.

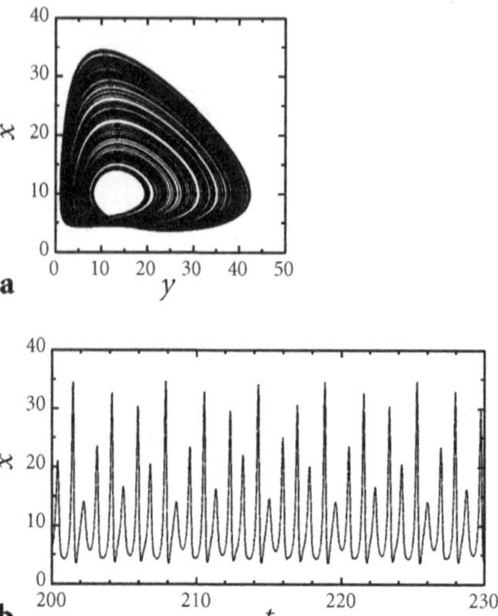

Figure 3. Chaotic time evolution of the concentrations of the Willamowski–Rössler chemical reaction scheme (2) for $k_1a_1 = 30$, $k_{-1} = 0.5$, $k_2 = 1$, $k_{-2} = 0$, $k_3a_5 = 10$, $k_{-3} = 0$, $k_4 = 1$, $k_{-4} = 0$, $k_5a_4 = 16.5$, and $k_{-5} = 0.5$: (a) phase portrait in the plane of the concentrations of species X and Y; (b) concentration of species X versus time t.

In conclusion, Rössler systems are minimal models for continuous-time chaos. The chaotic attractors of Rössler systems are prototypes for a large variety of chaotic behavior, notably, in chemical chaos (Scott, 1991).

P. GASPARD

See also **Attractors; Bifurcations; Brusselator; Chaotic dynamics; Chemical kinetics; Hopf bifurcation; Horseshoes and hyperbolicity in dynamical systems; Invariant manifolds and sets; Period doubling; Phase space; Poincaré theorems; Symbolic dynamics**

Further Reading

Gaspard, P. & Nicolis, G. 1983. What can we learn from homoclinic orbits in chaotic dynamics? *Journal of Statistical Physics*, 31: 499–518

Hartman, P. 1964. *Ordinary Differential Equations*, New York: Wiley

Rössler, O.E. 1976a. An equation for continuous chaos. *Physics Letters* A, 57: 397–398

Rössler, O.E. 1976b. Chaotic behavior in simple reaction systems. *Zeitschrift Naturforschung* A, 31: 259–264

Rössler, O.E. 1976c. Different types of chaos in two simple differential equations. *Zeitschrift für Naturforschung A* 31: 1664–1670

Rössler, O.E. 1977a. Continuous chaos. In *Synergetics: A Workshop*, edited by H. Haken, New York: Springer, pp. 184–199

Rössler, O.E. 1977b. Chaos in abstract kinetics: Two prototypes. *Bulletin of Mathematical Biology*, 39: 275–289

Rössler, O.E. 1979a. Continuous chaos—four prototype equations. *Annals of the New York Academy of Science*, 316: 376–392

Rössler, O.E. 1979b. An equation for hyperchaos. *Physics Letters* A, 71: 155–157
Rössler, O.E. & Wegmann, K. 1978. Chaos in the Zhabotinskii reaction. *Nature*, 271: 89–90
Ruelle, D. 1973. Some comments on chemical oscillators. *Transactions of the New York Academy of Science*, 35: 66–71
Scott, S.K. 1991. *Chemical Chaos*, Oxford: Clarendon Press
Shil'nikov, L.P. 1965. A case of the existence of a countable number of periodic motions. *Soviet Mathematics Doklady*, 6: 163–166
Willamowski, K.-D. & Rössler, O.E. 1980. Irregular oscillations in a realistic abstract quadratic mass action system. *Zeitschrift für Naturforschung* A, 35: 317–318

ROTATING RIGID BODIES

Rigid body motion about a fixed point is a classical mechanics problem associated with the greatest names in 18th-century mathematics. The basic equations of the motion were derived by Leonhard Euler in the 1750s and subsequently solved by him for two special cases (Euler and Lagrange tops). Development of the analytical theory of differential equations in the middle of the 19th century by Augustin Cauchy, Georg Riemann, and Karl Weierstrass inspired Sophia Kovalevsky (1889) to determine all the cases for which the integrals of rigid body dynamics were *single-valued, meromorphic* functions in the entire *complex* plane of the time variable. Pursuing this remarkable idea, she obtained a classification of all such (solvable) cases and found a new case, the Kovalevsky top, which she then integrated in quadratures. Considering time as a complex variable and imposing the above conditions upon the integrals of the equations of motion outside the real axis was a revolutionary approach to treating a mechanics problem. It appeared to be a fruitful method that led to a unified theory of (algebraically) completely integrable systems (Adler & van Moerbeke, 1989), a century after it was proposed.

The dynamics of a rigid body rotating about a fixed point can be considered in a fixed (nonmoving) frame of axes or in the moving body frame that has its origin at the fixed point and whose axes are along the principal axes of the ellipsoid of inertia. In 1750, Euler obtained the equations of motion in a fixed frame, but they appeared not to be very useful. In a series of papers during 1758–1765, he introduced the body frame, derived corresponding equations of motion, and also used Eulerian angles to relate the motion of a body frame with respect to a fixed system of coordinates.

Consider first a fixed frame $(\boldsymbol{i}', \boldsymbol{j}', \boldsymbol{k}')$. A fundamental dynamical theorem says that the time derivative of the angular momentum $\boldsymbol{J}$ of a body is equal to the moment $\boldsymbol{L}$ of the forces acting on it,

$$\frac{\mathrm{d}\boldsymbol{J}}{\mathrm{d}t} = \boldsymbol{L}. \tag{1}$$

In the body frame, $(\boldsymbol{i}, \boldsymbol{j}, \boldsymbol{k})$, the tensor of inertia is diagonal, $\boldsymbol{I} = \mathrm{diag}(A, B, C)$, and links the vector of angular velocity $\boldsymbol{\Omega} = p\,\boldsymbol{i} + q\,\boldsymbol{j} + r\,\boldsymbol{k}$ to the angular momentum $\boldsymbol{J}$ through the linear relation

$$\boldsymbol{J} = \boldsymbol{I\Omega} = Ap\,\mathbf{i} + Bq\,\boldsymbol{j} + Cr\,\boldsymbol{k}. \tag{2}$$

The time derivative of vector (2) is

$$\frac{\mathrm{d}\boldsymbol{J}}{\mathrm{d}t} = \frac{\delta\boldsymbol{J}}{\delta t} + \boldsymbol{\Omega} \times \boldsymbol{J}, \tag{3}$$

where $\frac{\delta\boldsymbol{J}}{\delta t}$ is the relative time derivative, evaluated assuming that the frame $(\boldsymbol{i}, \boldsymbol{j}, \boldsymbol{k})$ is stationary. The first three Euler equations, therefore, are

$$A\frac{\mathrm{d}p}{\mathrm{d}t} + (C-B)qr = Mg(y_0\gamma'' - z_0\gamma'),$$
$$B\frac{\mathrm{d}q}{\mathrm{d}t} + (A-C)rp = Mg(z_0\gamma - x_0\gamma''),$$
$$C\frac{\mathrm{d}r}{\mathrm{d}t} + (B-A)pq = Mg(x_0\gamma' - y_0\gamma), \tag{4}$$

where $M\boldsymbol{g} = Mg(\gamma\,\boldsymbol{i} + \gamma'\,\boldsymbol{j} + \gamma''\,\boldsymbol{k})$ is the vector of the gravitational force and $\boldsymbol{r}_0 = x_0\,\boldsymbol{i} + y_0\,\boldsymbol{j} + z_0\,\boldsymbol{k}$ is the vector originating in the fixed point and pointing at the center of mass of the body. An extra set of three equations follows from the fact that the vector $M\boldsymbol{g}$ is stationary in the fixed frame; hence, $\frac{\delta M\boldsymbol{g}}{\delta t} = -\,\boldsymbol{\Omega} \times M\boldsymbol{g}$ and

$$\frac{\mathrm{d}\gamma}{\mathrm{d}t} = r\gamma' - q\gamma'', \qquad \frac{\mathrm{d}\gamma'}{\mathrm{d}t} = p\gamma'' - r\gamma,$$
$$\frac{\mathrm{d}\gamma''}{\mathrm{d}t} = q\gamma - p\gamma'. \tag{5}$$

The Eulerian angles, namely, the angle of precession $0 \le \psi < 2\pi$, the angle of nutation $0 \le \vartheta \le \pi$, and the angle of the self rotation $0 \le \varphi < 2\pi$, define the body's position by fixing the moving frame with respect to the fixed one. We have (see Golubev, 1960)

$$\gamma = \sin\varphi\,\sin\vartheta, \qquad \gamma' = \cos\varphi\,\sin\vartheta,$$
$$\gamma'' = \cos\vartheta, \tag{6}$$

$$p = \dot\psi\sin\vartheta\,\sin\varphi + \dot\vartheta\cos\varphi,$$
$$q = \dot\psi\sin\vartheta\,\cos\varphi - \dot\vartheta\sin\varphi, \qquad r = \dot\psi\cos\vartheta + \dot\varphi \tag{7}$$

and, therefore,

$$\frac{\mathrm{d}\psi}{\mathrm{d}t} = \frac{p\gamma + q\gamma'}{\gamma^2 + \gamma'^2}. \tag{8}$$

The problem of studying the motion of a rigid body about a fixed point is solved by integration of the system of differential equations (4), (5), and (8).

Conserved Quantities

It is always useful to know a system's first integrals, for instance for controlling numerical calculations or for performing integration in quadratures. There are three important physical integrals for an arbitrary top (an arbitrary set of the six parameters $\{A, B, C, x_0, y_0, z_0\}$ in Equations (4)). They are the following.

(i) Energy integral: $E = \frac{1}{2}\left(Ap^2 + Bq^2 + Cr^2\right) - Mg(x_0\gamma + y_0\gamma' + z_0\gamma'')$.
(ii) Length of the unit vector along $\boldsymbol{Mg}$: $C_1 = \gamma^2 + \gamma'^2 + \gamma''^2 = 1$.
(iii) Projection of the angular momentum $\boldsymbol{J}$ on this vector: $C_2 = Ap\gamma + Bq\gamma' + Cr\gamma''$.

These three first integrals provide three constants of motion (since $C_1 = 1$, there are, in fact, only two arbitrary constants, the third one being the integration constant that appears when integrating (8), that is, the initial value of ψ). In addition, an autonomous system always has an extra constant corresponding to shifting of the time variable t. A remarkable feature of Equations (4) and (5) is that yet another explicit integration can be carried out using the method of Jacobi's last multiplier (Golubev, 1960), bringing in another explicit constant of integration. Therefore, for the case of a complete integration in quadratures of the equations of motion only one first integral is missing. Subsequent analysis showed that an extra algebraic integral exists only in the three special cases of Euler, Lagrange, and Kovalevsky tops. These integrable tops are given by specializing the parameters $\{A, B, C, x_0, y_0, z_0\}$.

Integrable Tops

The simpler cases of the Euler and Lagrange tops were studied, respectively, by Euler and Louis Poinsot and by Joseph-Louis Lagrange and Siméon Poisson, and later on by Gustav Jacobi, who expressed the general integrals for both systems in terms of elliptic functions of time.

Euler Top

In the Euler top $x_0 = y_0 = z_0 = 0$, and the body's center of mass is at the fixed point, so that there is no gravitational effect, meaning a free rotation of an asymmetric top. The fourth integral of motion in this case is the square of the angular momentum $\boldsymbol{J}$:

$$\boldsymbol{J}^2 = A^2p^2 + B^2q^2 + C^2r^2. \qquad (9)$$

Poinsot gave a geometric interpretation of the motion based on the intersection of two quadratics: (9) and the energy integral $2E = Ap^2 + Bq^2 + Cr^2$. The vector $\boldsymbol{J}$ is stationary in the fixed frame $(\boldsymbol{i}', \boldsymbol{j}', \boldsymbol{k}')$ and the problem is made simpler if one chooses $\boldsymbol{k}'$ along $\boldsymbol{J}$. Then

$$\cos\vartheta = \frac{Cr}{|\boldsymbol{J}|} \text{ and } \tan\varphi = \frac{Ap}{Bq}. \qquad (10)$$

As for time dependence, the variables p, q, and r can be expressed in terms of Jacobi's elliptic functions sn, cn, and dn of time and they are periodic, so are ϑ and φ obtained from (10). The dynamics of the angle of precession ψ is determined by integration of $\dot{\psi} = |\boldsymbol{J}|\frac{Ap^2+Bq^2}{A^2p^2+B^2q^2} > 0$, derived from (8). It follows that after each period, the value of ψ, in general, changes, so that the body never comes back to its initial orientation, thereby undergoing quasi-periodic motion.

Stationary rotations are those when the vector of angular velocity $\boldsymbol{\Omega}$ is constant. They correspond to uniform rotations about the principal axes of inertia. Rotation around the middle axis is unstable.

A symmetric $(A = B)$ Euler top performs regular precession. The body's symmetry axis draws a circular conic with the axis $\boldsymbol{J}$ and the angle $2\vartheta = \text{const}$. The symmetry axis goes around $\boldsymbol{J}$ with a constant angular velocity while the body rotates with a constant angular velocity around the symmetry axis.

Lagrange Top

This describes the dynamics of a symmetric body, $A = B$, with the fixed point at the symmetry axis and off the center of mass, that is, $x_0 = y_0 = 0$, $z_0 \neq 0$. The fourth constant of motion is simply r. The integration of this top can again be done in terms of elliptic functions, leading to a variety of complicated quasi-periodic motions including pseudo-regular precession, double-asymptotic, and "sleeping" tops. Simpler motions correspond to degenerations of elliptic functions into elementary functions.

The motion of the symmetry axis, described by the angles ϑ and ψ, looks like a perturbation of the corresponding uniform rotation of a symmetric Euler top in the regular precession. Illustrations of this familiar motion are available in many textbooks. Notice that now $\dot{\psi}$ can change sign, thereby giving three kinds of trajectories.

Denote $m = 1 - C/A$ and notice that $m = 0$ for a spherically symmetric body $(A = B = C)$. Consider the Lagrange top with $m \neq 0$ and the variables $(p, q, r, \gamma, \gamma', \gamma'')$ and a fully symmetric Lagrange top with $m = 0$ and the variables $(p_1, q_1, r_1, \gamma_1, \gamma_1', \gamma_1'')$. It is possible to show that the motions of these two bodies coincide when the variables are identified as follows:

$$\begin{aligned} &p_1^2 + q_1^2 = p^2 + q^2, \quad \tan^{-1}\tfrac{q_1}{p_1} = \tan^{-1}\tfrac{q}{p} + mrt, \\ &r_1 = \tfrac{C}{A} r, \\ &\varphi_1 = \varphi - mrt, \qquad \vartheta_1 = \vartheta, \\ &\psi_1 = \psi. \end{aligned} \qquad (11)$$

See Golubev (1960) for details.

Kovalevsky Top

This is the third general case of complete integrability of the Euler equations (4) and (5), which is characterized by restricting $A = B = 2C$ and $y_0 = z_0 = 0$. Hence, the Kovalevsky top is a special symmetric top whose center of mass lies in the equatorial plane of the ellipsoid of inertia ($x_0 \neq 0$). The fourth constant of motion is of order 4:

$$K = \left(p^2 - q^2 + \frac{Mgx_0}{C}\gamma\right)^2 + \left(2pq + \frac{Mgx_0}{C}\gamma'\right)^2. \tag{12}$$

Kovalevsky found this case (Kovalevsky, 1889) and integrated it in terms of hyper-elliptic functions. Later on, Kötter (1893) simplified the formulae.

There is a great difference in complexity between this top and the other two. This is why there was such a long period between the discovery and integration of Euler and Lagrange tops and that of Kovalevsky. The Paris Academy of Sciences established a special Borden Prize to promote major expansions of the theory. It was finally claimed in 1888 by Sophia Kovalevsky who made major progress toward solution of the problem. Her work required usage of theta-functions in two variables and provided an enormous boost to the creation of the modern theory of completely integrable systems. There is a large body of literature about the Kovalevsky top, of which we mention only three items. In Adler & van Moerbeke (1988), a detailed study of the algebraic geometry of the model is given. In Bobenko et al. (1989), the authors find alternative theta-function formulae by making use of the corresponding Lax matrix and the finite-gap integration technique. In Kuznetsov (2002), connections with a representation of the quadratic r-matrix algebra and with the method of separation of variables are presented.

All other general tops corresponding to other choices of the six parameters (A, B, C, x_0, y_0, z_0) are not integrable, which means that they generally exhibit a chaotic dynamics, and it is impossible to find analytic solutions valid for arbitrary initial conditions.

VADIM B. KUZNETSOV

See also **Constants of motion and conservation laws; Integrability; Newton's laws of motion; Nonlinear toys**

Further Reading

Adler, M. & van Moerbeke, P. 1988. The Kowalevski and Hénon–Heiles motions as Manakov geodesic flows on SO(4)—a two-dimensional family of Lax pairs. *Communications in Mathematical Physics*, 113: 659–700

Adler, M. & van Moerbeke, P. 1989. The complex geometry of the Kowalevski–Painlevé analysis. *Inventiones Mathematicae*, 97: 3–51

Bobenko, A.I., Reyman, A.G. & Semenov-Tian-Shansky, M.A. 1989. The Kowalevski top 99 years later: a Lax pair, generalizations and explicit solutions. *Communications in Mathematical Physics*, 122: 321–354

Golubev, V.V. 1960. *Lectures on Integration of the Equations of Motion of a Rigid Body about a Fixed Point*, Moscow, 1953. Published for the National Science Foundation by the Israel Program for Scientific Translations

Kötter, F. 1893. Sur le cas traité par Mme Kowalevski de rotation d'un corps solide autour d'un point fixe. *Acta Mathematica*, 17(1–2)

Kovalevsky, S. 1889. Sur le problème de la rotation d'un corps solide autour d'un point fixe. *Acta Mathematica*, 12: 177–232

Kuznetsov, V.B. 2002. Kowalevski top revisited. In *The Kowalevski Property*, edited by V.B. Kuznetsov, Providence, RI: American Mathematical Society

ROTATING WAVE APPROXIMATION

Energy conserving oscillators comprise pairs of energetic variables (let us call them P and Q), each being the cause of the other. In a mechanical (spring-mass) oscillator shown in Figure 1, for example, Q would be the displacement of a mass and P its momentum (mass times velocity). For a radio engineer's tank oscillator, Q is the voltage across a capacitor and P is the current through an inductor. In a laser mode, Q is electric field energy and P is magnetic field energy.

Consider the mechanical oscillator of Figure 1, assuming the spring to be linear. From Newton's second law (force equals time derivative of momentum), the dynamic equation describing this system is

$$\frac{\mathrm{d}}{\mathrm{d}t}\left(M\frac{\mathrm{d}x}{\mathrm{d}t}\right) = -Kx, \tag{1}$$

where x is the vertical displacement of the mass from its resting position and M is the oscillator mass. On the right-hand side of this equation is the vertical force acting on the mass caused by extension or compression of the spring, and on the left-hand side is the time derivative of the vertical momentum: $p \equiv M\mathrm{d}x/\mathrm{d}t$. In this case, a solution is evidently $x = a\sin(\omega_0 t)$, where a is an arbitrary amplitude and $\omega_0 = \sqrt{K/M}$ is the frequency of oscillation.

Often the force of a spring is not quite a linear function of its extension but slightly sublinear. (This

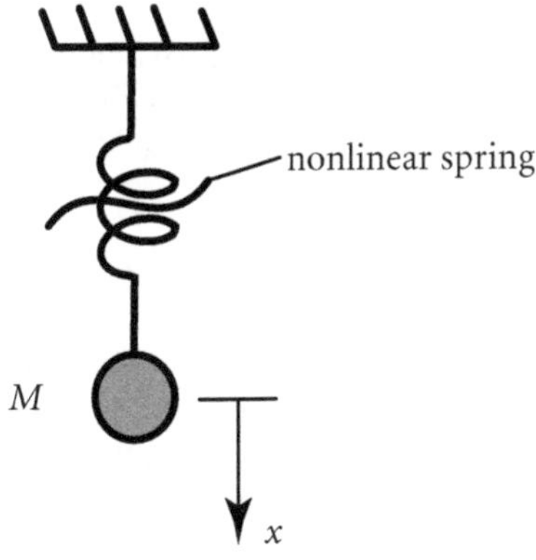

Figure 1. A spring-mass oscillator.

is the case, for example, in molecular vibrations, where the force between a pair of atoms becomes relatively weaker as the distance between them is increased because the electronic contribution to interatomic bonding is lessened.) In such cases, it is convenient to introduce anharmonicity into the formulation through the rotating wave picture (Louisell, 1960; Scott, 2003).

In the rotating wave picture of a linear oscillator, the momentum (p) and extension (x) are combined as real and imaginary parts of a single, complex amplitude

$$A \equiv \frac{(p - \mathrm{i}M\omega_0 x)}{\sqrt{2M\omega_0}}, \tag{2}$$

which obeys the first order equation

$$\mathrm{i}\frac{\mathrm{d}A}{\mathrm{d}t} = \omega_0 A \tag{3}$$

with the solution $A(t) = A(0)\mathrm{e}^{-\mathrm{i}\omega_0 t}$. Under this formulation, the total energy of the oscillator is

$$H = \omega_0 |A|^2 = \frac{1}{2}\left(\frac{p^2}{M} + Kx^2\right), \tag{4}$$

where the first term is the kinetic energy of the moving mass and the second is the potential energy stored in the spring.

If the restoring force of the spring is slightly sublinear, its potential energy may depend upon x as

$$\text{potential energy} = \tfrac{1}{2}Kx^2 - \tfrac{1}{4}\alpha x^4, \tag{5}$$

where $\alpha > 0$ is an anharmonicity parameter. With $\alpha = 0$, $x = \mathrm{i}(A - A^*)/\sqrt{2M\omega_0}$, so

$$x^4 = \frac{1}{4M^2\omega_0^2}\left[A^4 - 4A^3A^* + 6|A|^4 - 4A(A^*)^3 + (A^*)^4\right]. \tag{6}$$

Except for the middle ($6|A|^4$) term, each term in Equation (6) varies sinusoidally with time at frequencies $\pm 4\omega_0$ or $\pm 2\omega_0$. Thus, the average of the energy over a cycle of oscillation is

$$\langle H \rangle = \omega_0 |A|^2 - \frac{\gamma}{2}|A|^4, \tag{7}$$

where $\gamma = 3\alpha/4M^2\omega_0^2$.

Under the rotating wave approximation (RWA), the energy H is taken as equal to its time average $\langle H \rangle$, whereupon the complex amplitude A is governed by the first order ODE

$$\mathrm{i}\frac{\mathrm{d}A}{\mathrm{d}t} \doteq \omega_0 A - \gamma |A|^2 A. \tag{8}$$

In this expression, the symbol "$\doteq$" indicates that only terms of frequency ω_0 are included. In other words, all nonresonant terms are neglected in the RWA.

Since Equation (8) conserves $N = |A|^2$, a general solution of this equation is

$$A(t) = \sqrt{N}\mathrm{e}^{-\mathrm{i}[(\omega_0 - \gamma N)t + \varphi]}, \tag{9}$$

where φ is an arbitrary phase angle. Equation (9) gives the frequency ($\omega = \omega_0 - \gamma N$) of the nonlinear oscillation as a function of its amplitude in the RWA.

Thus, RWA accounts for all components generated by a weak nonlinearity that resonate with the fundamental oscillator frequency. (As with your radio receiver, those components not in resonance are neglected.) Because the RWA is motivated by a general oscillator formulation, it is widely employed as the initial assumption in a variety of nonlinear analyses (various versions of the nonlinear Schrödinger equation, the discrete self-trapping equation, molecular vibrations, nonlinear optics, etc.). Interestingly, odd terms in the potential function (x^3, x^5, and so on) do not generate resonant components and are, thus, neglected in the RWA.

Finally, the RWA is a convenient formulation because quantum theories are easy to construct and solve for general systems of interacting oscillators, including discrete nonlinear Schrödinger equations and the discrete self-trapping system (Louisell, 1962; Scott et al., 1994; Scott, 2003). This is so for two reasons. First, under quantum theory in the RWA, the complex amplitude (A) becomes the lowering (annihilation) operator for oscillator quanta (bosons), and its complex conjugate (A^*) becomes the raising (creation) operator. Second, the classical fact that $N = \sum |A|^2$ is conserved implies that the corresponding quantum number operator commutes with the energy operator, allowing eigenfunctions of the energy operator to be constructed as sums of linear oscillator eigenfunctions.

Alwyn Scott

See also **Damped-driven anharmonic oscillator; Discrete nonlinear Schrödinger equations; Discrete self-trapping system; Quantum nonlinearity; Salerno equation**

Further Reading

Louisell, W.H. 1960. *Coupled Mode and Parametric Electronics*, New York: Wiley

Louisell, W.H. 1962. Correspondence between Pierce's coupled mode amplitudes and quantum oscillators. *Journal of Applied Physics*, 33: 2435–2436

Scott, A.C. 2003. *Nonlinear Science: Emergence and Dynamics of Coherent Structures*, 2nd edition, Oxford and New York: Oxford University Press

Scott, A.C., Eilbeck, J.C. & Gilhøj, H. 1994. Quantum lattice solitons. *Physica* D, 78: 194–213

ROTATION-MODIFIED KORTWEG-DE VRIES EQUATION

See **Korteweg–de Vries equation**

ROUTES TO CHAOS

If a nonlinear systems has chaotic dynamics, then it is natural to ask how this complexity develops as parameters vary. For example, in the logistic map

$$x_{n+1} = r x_n (1 - x_n), \quad (1)$$

it is easy to show that if $r = \frac{1}{2}$, then there is a fixed point at $x = 0$ that attracts all solutions with initial values x_0 between 0 and 1, while if $r = 4$, the system is chaotic. How, then, does the transition to chaos occur as the parameter r varies? Indeed, is there a clean transition to chaos in any well-defined sense? The identification and description of routes to chaos has had important consequences for the interpretation of experimental and numerical observations of nonlinear systems. If an experimental system appears chaotic, it can be very difficult to determine whether the experimental data comes from a truly chaotic system, or if the results of the experiment are unreliable because there is too much external noise. Chaotic time series analysis provides one approach to this problem, but an understanding of routes to chaos provides another. In many experiments there are parameters (ambient temperature, Raleigh number, etc.) that are fixed in any realization of the experiment, but which can be changed. If recognizable routes to chaos are observed when the experiment is repeated at different values of the parameter, then there is a sense in which the presence of chaotic motion has been explained.

By the early 1980s, three "scenarios" or "routes to chaos" had been identified (e.g., Eckmann, 1981): Ruelle–Takens–Newhouse, period doubling, and intermittency (which has several variants). As we shall see, in their standard forms each of these transitions uses the term "route to chaos" in a different way, so care needs to be taken over the interpretation of experimental or numerical observations of these transitions.

Ruelle–Takens–Newhouse

In 1971, Ruelle and Takens published a mathematical paper with the provocative title "On the Nature of Turbulence." In this paper and a subsequent improvement with Newhouse (1978), they discuss the Landau scenario for the creation of turbulence by the successive addition of new frequencies to the dynamics of the fluid. They show that if the attractor of a system has three independent frequencies (four in the 1971 paper), then a small perturbation of this system has a hyperbolic strange attractor—a Plykin attractor (a solenoid in the 1971 paper). The result became known colloquially as "three frequencies implies chaos," a serious misinterpretation of the mathematical result that has been the cause of a number of misleading statements. First, the result proves the existence of chaos in systems arbitrarily close to the three frequency system in an infinite dimensional function space but gives no indication of the probability of finding chaos in any given example. Second, I know of no experimental situation where a Plykin attractor has been shown to exist even when chaotic behavior has been observed close to systems with three frequency attractors. Numerical experiments suggest that it is much more likely that the system evolves by frequency locking.

The Ruelle–Takens–Newhouse route to chaos remains somewhat of an enigma, and more work needs to be done to understand precisely how and when the strange attractors predicted by the theory come into being.

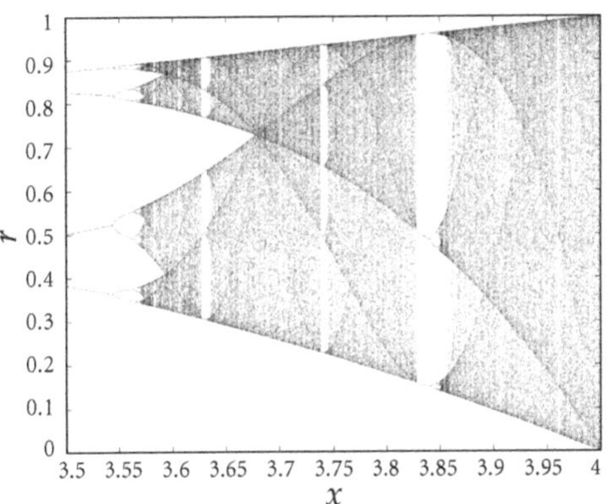

Figure 1. The attractor of the logistic equation as a function of the parameter r.

Period Doubling

Figure 1 shows the attractor of the logistic map (1) as a function of the parameter, r, for $3.5 < r < 4$. Thus, the set of points plotted on any vertical line of constant r represents the attractor of the map for that value of r, and if the set is finite, then the (numerically computed) attractor is a periodic orbit that cycles through the finite collection of points. Figure 1 suggests that for small r, the attractor is always periodic and has period 2^n, with n increasing as r increases. Beyond some critical value $r = r_c$, with $r_c \approx 3.569946$, the attractor may be more complicated. There are clearly intervals of r for which the attractor is periodic, and the attractor seems to be contained in 2^n bands that merge as r increases (the final band merging from 2 bands to one band with r just below 3.68 is particularly clear).

As r increases, the periodic orbit of period 2^n is created from the orbit of period 2^{n-1} by a period-doubling bifurcation. If this bifurcation occurs with $r = r_n$, then $r_n \to r_c$ geometrically as $n \to \infty$, with

$$\lim_{n\to\infty} \frac{r_n - r_{n-1}}{r_{n+1} - r_n} = \delta \approx 4.66920, \quad (2)$$
$$\text{i.e.,} \quad r_n \sim r_c - \kappa\delta^{-n}.$$

The really surprising feature of this period-doubling cascade, as shown in Feigenbaum (1978), is that the cascade can be observed in many maps and the accumulation rate δ of the period-doubling cascade (2)

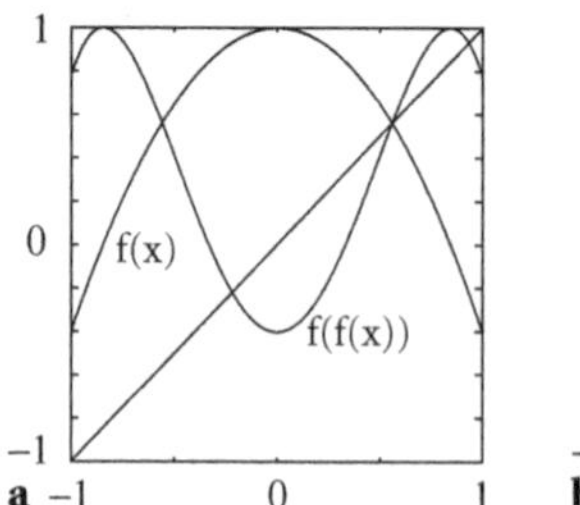

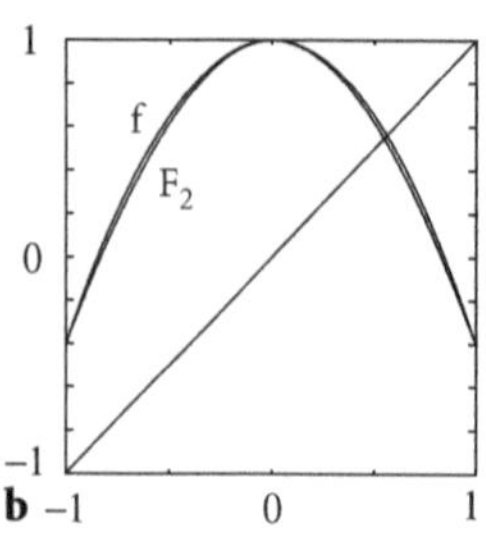

Figure 2. (a) The quadratic map $f(x)=1-mx^2$ restricted to the interval $[-1,1]$ and the second iterate, $f(f(x))$ for $m=1.40115$ which is just below the critical value m_c. (b) The original map, f, and the rescaled map, $F_2(x)=\mathcal{T}f(x)=-a^{-1}f(f(-ax))$ on $[-1,1]$ with $a=-f(1)=m-1$. Note that $a=0.40115$ is close to the universal value of $\alpha\approx 0.3995$.

is the same although the constants r_c and κ depend on the map. In fact, the universal value of δ depends on the nature of the maximum of the map: $\delta\approx 4.66920$ for maps with quadratic maximum. A complete cascade of band merging from 2^n bands to 2^{n-1} bands occurs at parameter values $\tilde{r}_n$ above r_c, and $\tilde{r}_n\to r_c$ as $n\to\infty$ at the same universal geometric rate δ.

The quantitative universality in parameter space described by the scaling δ has a counterpart in phase space. If x_n denotes the point on the periodic orbit of period 2^{n-1} that is closest to the critical point (or turning point) of the map with $r=r_n$, then

$$\lim_{n\to\infty}\frac{x_{n+1}-\frac{1}{2}}{x_n-\frac{1}{2}}=-\alpha, \tag{3}$$

where α is another universal constant, which, for maps with a quadratic turning point, takes the value $\alpha\approx 0.3995=1/2.50\ldots$.

In families of one hump (unimodal) maps this universality can be explained by a renormalization argument. Restrict attention to families of one hump maps with critical point (maximum) at $x=0$, parametrized by μ and normalized so that $f(0)c=1$. As shown in Figure 2a, for parameter values near μ_c (the accumulation of period doubling) the second iterate of the map, $f(f(x))$, restricted to an interval about the critical point is a one hump map with a minimum. So, after a rescaling (and flipping) of the coordinates, it is another one hump map with the same normalization as shown in Figure 2b. Mitchell Feigenbaum was able to show (by arguments that have been made rigorous since 1990) that the universal properties described above are due to the structure of the doubling operator $\mathcal{T}$, which is a map on one hump maps $f:[-1,1]\to[-1,1]$, with critical point at $x=0$ and $f(0)c1$, defined by

$$\mathcal{T}f(x)=-a^{-1}f(f(-ax)), \tag{4}$$

where $a=-f(1)$ so that the normalization $\mathcal{T}f(0)=1$ is preserved. This operator does the rescaling and flipping referred to above. In the appropriate universality class, for example, quadratic critical point together with some further technical conditions, there is a fixed point f_*, of $\mathcal{T}$, so $f_*=\mathcal{T}f_*$, and the universal scaling of phase space is given by $\alpha=-f_*(1)$. Furthermore, the universal accumulation rate δ of (2) is an unstable eigenvalue of the (functional) derivative of $\mathcal{T}$ at f_*.

We can now consider measures of chaos such as the topological entropy or the Lyapunov exponents of the map. If $\{f_\mu\}$ is a family of one hump maps that undergoes period doubling, then the parameter μ can be chosen so that the period-doubling cascade is for $\mu<\mu_c$. Coullet & Tresser (1980) show that the universal structure described above implies that if $H(\mu)$ is either the topological entropy (which can be thought of as the growth rate of the number of periodic orbits) or the Lyapounov exponent of f_μ with $\mu>\mu_c$, then

$$H(\mu)\sim C(\mu-\mu_c)^{(\log 2)/(\log\delta)}. \tag{5}$$

The Lyapunov exponent is a very poorly behaved function of the parameters and this scaling provides only an envelope for the graph of the exponent, but the topological entropy is continuous. Indeed, the proof of Sharkovsky's theorem (*See* **One-dimensional maps**) shows that if a continuous map of the interval has a periodic orbit which is not a power of two, then there is a horseshoe for some iterate of the map, and hence the map has positive topological entropy if $\mu>\mu_c$. The entropy is zero if $\mu<\mu_c$, so if by chaos we mean positive topological entropy, then the period-doubling route is a true route to chaos.

Intermittency

The first stable periodic orbit of each of the windows of periodic motion in $r>r_c$, which can be seen in Figure 1, is created in a saddle-node (or tangent) bifurcation. Throughout the parameter interval for which such orbits are stable, there is a repelling strange invariant set, but most solutions tend to the stable periodic orbit. Just before the creation of the stable periodic orbit, chaotic solutions spend long periods of time near the points at which the stable periodic orbit will be created (the "laminar" phase), then move away and behave erratically before returning to the laminar phase. This behavior is called intermittency by Pomeau & Manneville (1980), who were the first to describe the scaling of the time spent in the laminar phase. They looked at the average time T_A spent by solutions in the laminar phase as a function of the parameter r close to the value r_{sn}, at which the saddlenode bifurcation occurs. A simple argument based on the passage time of a trajectory of a map close to a tangency with the diagonal (the condition for the saddle-node bifurcation) establishes that the average time in the laminar phase diverges as a power law:

$$T_A\sim|r-r_{sn}|^{-1/2}. \tag{6}$$

Other types of intermittency (involving period-doubling bifurcations, etc.) can be analyzed using the same ideas. Note that a strange invariant set exists throughout the parameter regions being considered here, so in this case the term "route to chaos" refers to the stability of the chaotic invariant set, not the creation of a chaotic set. Moreover, in any open neighborhood of r_{sn}, there are parameters for which the map has other stable periodic orbits, so the full description of parameters with stable chaotic motion is much more complicated than the description above suggests.

Other Routes to Chaos

Since the pioneering work of the late 1970s, a number of other routes to chaos have been identified. New routes to chaos are still being identified, and the list provided here is by no means complete. Arnéodo et al. (1981) show that there can be cascades of homoclinic bifurcations to chaos via a mechanism closely related to period doubling. This gives the less standard convergence rates involving nonquadratic turning points immediate relevance. The bifurcation that creates the strange invariant set of the Lorenz model is another type of homoclinic bifurcation, and this strange invariant set becomes stable by a "crisis" in which the strange invariant set collides with a pair of unstable periodic orbits. Ott (2002) contains a good account of such transitions. More complicated transitions involving maps of the circle are detailed in MacKay & Tresser (1986), and Newhouse et al. (1983) give another transition.

PAUL GLENDINNING

See also **Attractors; Bifurcations; Chaotic dynamics; Intermittency; Lorenz equations; One-dimensional maps; Period doubling; Time series analysis**

Further Reading

Arnéodo, A., Coullet, P. & Tresser, C. 1981. A possible new mechanism for the onset of turbulence. *Physics Letters* A, 81: 197–201

Coullet, P. & Tresser, C. 1980. Critical transition to stochasticity for some dynamical systems. *Journal de Physique Lettres*, 41: L255–L258

Eckmann, J.-P. 1981. Roads to turbulence in dissipative dynamical systems, *Reviews of Modern Physics*, 53: 643–654

Feigenbaum, M.J. 1978. Quantitative universality for a class of nonlinear transformations. *Journal of Statistical Physics*, 19: 25–52

MacKay, R.S. & Tresser, C. 1986. Transition to topological chaos for circle maps. *Physica* D, 19: 206–237

Newhouse, S., Palis, J. & Takens, F. 1983. Bifurcations and stability of families of diffeomorphisms. *Publications Mathématiques de l'IHES*, 57: 5–72

Newhouse, S., Ruelle, D. & Takens, F. 1978. Occurrence of strange Axiom A attractors near quasi-periodic flows on $\mathbf{T}^m$, $m \geq 3$. *Communications in Mathematical Physics*, 64: 35–40

Ott, E. 2002. *Chaos in Dynamical Systems*, 2nd edition, Cambridge and New York: Cambridge University Press

Pomeau, Y. & Manneville, P. 1980. Intermittent transition to turbulence in dissipative dynamical systems. *Communications in Mathematical Physics*, 74: 189–197

Ruelle, D. & Takens, F. 1971. On the nature of turbulence. *Communications in Mathematical Physics*, 20: 167–192

RUELLE–TAKENS–NEWHOUSE

See **Routes to chaos**

RUNGE–KUTTA METHOD

See **Numerical methods**

S

SADDLE POINT

See **Phase space**

SAFFMAN–TAYLOR PROBLEM

See **Hele-Shaw cell**

SALERNO EQUATION

The Salerno equation is a q-deformed lattice model that includes, as particular cases, two known discrete versions of the continuous nonlinear Schrödinger equation (NLS): the non-integrable discrete NLS equation (DNLS) with on-site nonlinearity and the integrable Ablowitz–Ladik (AL) equation with intra-site nonlinearity. Here by q-deformed, we mean the existence, both in the Poisson bracket (commutator in the quantum case) and in the Hamiltonian, of a free parameter q that allows "tuning" the nonlinearity (interaction) of the lattice model. From a physical point of view this equation represents a generalization of the tight-binding Schrödinger model

$$\mathrm{i}\frac{\mathrm{d}A_j(t)}{\mathrm{d}t} + \Omega_j A_j(t) + J_j\ (A_{j+1}(t) + A_{j-1}(t)) = 0, \qquad (1)$$

for the propagation of a molecular excitation in a crystal. Here A_j denotes the quasi-classical complex mode amplitude of a particular molecular vibration, Ω_j is the on-site frequency of this vibration, and J_j is the next-neighbor resonance interaction energy. Equation (1) was considered by Richard Feynman as a starting point for an alternate formulation of quantum mechanics in terms of coupled probability amplitudes (Feynman, 1965) (also it is equivalent to the Schrödinger equation for the description of the wave function of an electron in a perfect crystal in the tight-binding approximation). By assuming the coupling of the mode amplitudes to low-frequency phonons (lattice distortions), one obtains, in the adiabatic and small field amplitude approximation, a dependence of the local energy and intra-site interaction upon A_j of the type

$$\begin{aligned} \Omega_j &\to \Omega_{0j} + \Omega_{1j}|A_j|^2, \\ J_j &\to J_{0j} + J_{1j}|A_j|^2. \end{aligned} \qquad (2)$$

The Salerno equation follows by substituting the above relations into Equation (1) and redefining parameters as

$$\varepsilon = \frac{\Omega_1}{J_0},\ \eta = \frac{\Omega_1}{J_1},\ \omega_j = \frac{2J_0 + \Omega_0}{J_0},\ \gamma = \frac{2J_1 + \Omega_1}{J_0},$$

thus giving

$$\begin{aligned} &\mathrm{i}\frac{\mathrm{d}A_j(t)}{\mathrm{d}t} - (2 - \omega_j - \varepsilon|A_j|^2)A_j \\ &+(1 + \frac{\varepsilon}{\eta}|A_j|^2)(A_{j+1} + A_{j-1}) = 0, \end{aligned} \qquad (3)$$

where time has been rescaled by a factor J_0 and equal local energies and intra-site resonance interactions were assumed (note that ω_j can be eliminated from Equation (3) by a rescale of time, so that in the following, we will put $\omega_j = 0$). With the above parametrization the following relationship between ε and η was introduced:

$$q \equiv \frac{\varepsilon}{\eta} = \frac{\gamma - \varepsilon}{2}. \qquad (4)$$

This allows interpolatation between the DNLS and the AL lattice as discussed below. To this end, it is worth noting that Equation (3) is a Hamiltonian system with respect to the following q-deformed Poisson bracket:

$$\begin{aligned} \{f, g\}_q = \sum &\left(\frac{\partial f}{\partial A_j}\frac{\partial g}{\partial A_j^*} - \frac{\partial g}{\partial A_j}\frac{\partial f}{\partial A_j^*} \right) \\ &\times (1 + q\,|A_j|^2), \end{aligned} \qquad (5)$$

and with Hamiltonian

$$\begin{aligned} H = -\sum_j &\Bigg\{ A_j^*(A_{j+1} + A_{j-1}) + \eta|A_j|^2 \\ &- \frac{(2+\eta)}{\log(1+\frac{\varepsilon}{\eta})}\ \log(1 + \frac{\varepsilon}{\eta}|A_j|^2) \Bigg\}. \end{aligned} \qquad (6)$$

It is easy to check that Equation (3) is obtained from Equations (5) and (6), as $\mathrm{i}(\mathrm{d}A_j/\mathrm{d}t) = \{A_j, H\}_q$. Two values of the deformation parameter are of particular interest.

Case $q = 0$. This corresponds to the limit $\varepsilon \to \gamma$ for which the Poisson bracket acquires canonical form and Hamiltonian (6) becomes

$$-\sum_j \left\{A_j^*(A_{j+1} + A_{j-1}) - 2|A_j|^2 - \frac{\gamma}{2}|A_j|^2)\right\}.$$

Equation (3) then reduces to the DNLS equation

$$\mathrm{i}\frac{\mathrm{d}A_j(t)}{\mathrm{d}t} + A_{j+1} - 2A_j + A_{j-1} + \gamma A_j|A_j|^2 = 0, \quad (7)$$

which conserves the number $N = \sum_j |A_j|^2$ and allows solitary wave propagation (Scott, 2003). Although it is non-integrable, the DNLS equation is linked to many physical problems, from propagation of self-trapped modes in biomolecules and in arrays of nonlinear optical fibers, to the tight-binding description of Bose–Einstein condensates in optical lattices.

Case $q = \frac{\gamma}{2}$. This corresponds to the limit $\varepsilon \to 0$ for which Equation (3) reduces to the Ablowitz–Ladik equation

$$\mathrm{i}\frac{\mathrm{d}A_j(t)}{\mathrm{d}t} + A_{j+1} - 2A_j + A_{j-1} + \frac{\gamma}{2} \times (A_{j+1} + A_{j-1})|A_j|^2 = 0, \quad (8)$$

which is known to have exact soliton solutions, being integrable by the Inverse Scattering Method (Scott, 2003).

The property of Salerno's equation to incorporate the above discrete versions of the NLS equation makes it an ideal general model to investigate the interplay between on-site and intra-site nonlinearity, discreteness and continuum, integrability and non-integrability. Studies performed along these lines during the past decade have shown the existence of a rich and wide range of behaviors, ranging from the existence of states localized on few lattice sites (intrinsic localized modes or discrete breathers), to the possibility of shock wave formation (Cai et al., 1994; Kivshar & Salerno, 1994; Konotop & Salerno, 1997; Mackay & Sepulchre, 2002). The existence of an integrable limit of the model also allowed clarification of the relation between intrinsic localized modes and exact discrete solitons of the AL lattice.

Signatures of these classical properties exist in the corresponding quantum model whose Hamiltonian is defined from Equation (6) as

$$\hat{H} = -\sum_j \left\{\hat{b}_j^\dagger(\hat{b}_{j+1} + \hat{b}_{j-1}) + \eta\, \hat{b}_j^\dagger \hat{b}_j - \frac{2+\eta}{\log(1+\frac{\varepsilon}{\eta})} \log[1 + \frac{\varepsilon}{\eta}\, \hat{b}_j^\dagger \hat{b}_j]\right\}, \quad (9)$$

where complex mode amplitudes A_j, A_j^* were replaced by creation and annihilation operators $\hat{b}_j, \hat{b}_j^\dagger$. Equation (5) implies the following deformed Heisenberg algebra:

$$[\hat{b}_i, \hat{b}_j^\dagger] = \delta_{i,j}(1 + q\, \hat{b}_i^\dagger \hat{b}_i),$$
$$[\hat{b}_i, \hat{b}_j] = [\hat{b}_i^\dagger, \hat{b}_j^\dagger] = 0, \quad (10)$$

with the same q as in Equation (4). Note that the on-site algebra in Equation (10) is the same as the algebra of a q-oscillator (MacFarlane, 1989), thus providing an example of occurrence of a quantum deformation algebra in a physical model. An explicit representation of the q-algebra associated with the Salerno equation can be constructed as (Salerno, 1992; Bogoliubov & Bullough, 1992)

$$b^\dagger\, |n\rangle = \sqrt{[n+1]_q}\, |n+1\rangle,$$
$$\hat{b}\, |n\rangle = \sqrt{[n]_q}\, |n-1\rangle \quad (11)$$

with $[n]_q = ((1+q)^n - 1)/q$. From this equation, it follows that the number operator is given by

$$\hat{N} = \sum_j \frac{\log(1 + \frac{\varepsilon}{\eta}\hat{b}_j^\dagger \hat{b}_j)}{\log(1 + \frac{\varepsilon}{\eta})}. \quad (12)$$

The conservation of $[\hat{N}, \hat{H}] = 0$, and the translational invariance of the system, $[\hat{T}_j, \hat{H}] = 0$ (T_j are translation operators by j sites along the lattice), allows one to block diagonalize the Salerno Hamiltonian into subspaces with fixed $\hat{N}$ eigenvalues and with fixed crystal momentum k. As for the classical model, the two limiting cases $\varepsilon = \gamma$ and $\varepsilon = 0$ correspond to the quantum DNLS lattice and to the quantum Ablowitz–Ladik system, respectively, the latter being integrable by means of the algebraic Bethe ansatz (Scott, 2003). By tuning the deformation parameter, one can get interesting physical behaviors. Thus, for example, for $q = -2$, the first commutator in Equation (10) becomes an anticommutator, so that the model describes a system of bosons with hard-core interactions (no more than one boson on a site). In this case, it was proved that in the mean field approximation (unconstrained hopping) and in the thermodynamic limit, the case $\varepsilon = 0$ displays a Bose–Einstein condensation (Salerno, 1994).

Exact diagonalizations of Equation (9) have been performed mainly for finite size and for finite number of quanta. Figure 1a depicts a typical band structure of Salerno's model in the reduced zone scheme for a chain of 15 sites, $N = 5$, $\gamma = 10$, and $\varepsilon = 5$. For

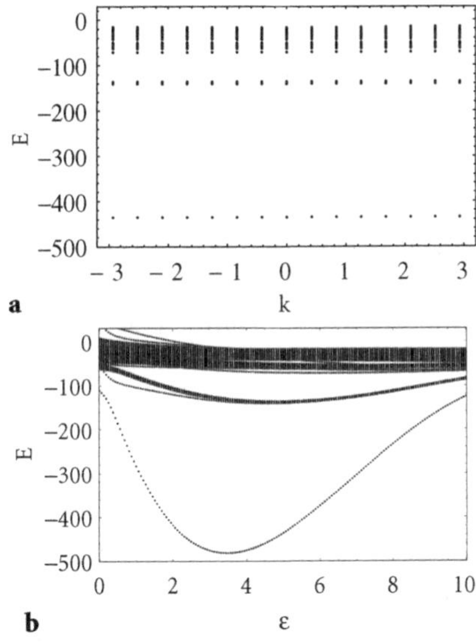

Figure 1. (a) Band structure of Salerno's model in the reduced zone scheme for a chain of 15 sites with $N = 5$, $\gamma = 10$, and $\varepsilon = 5$. (b) Energy of the states with crystal momentum $k = 0$, corresponding to the translational symmetry of the ground state, as a function of ε. Other parameters are fixed as (a).

these parameter values, the lower band corresponds to a bound state in which the five quanta are, with high probability, all on the same site. Figure 1b shows the energy of the states with crystal momentum $k = 0$ (corresponding to the translational symmetry of the ground state) as a function of ε. Note that the minimum ground state energy is achieved for an ε value in between the AL ($\varepsilon = 0$) and the DNLA case ($\varepsilon = 10$).

The presence of an integrable limit of the Salerno model gives the possibility of exploring complicated properties of this quantum many-body system starting from the exact knowledge of its integrable limit. An interesting open problem in this direction is the characterization of the quantum analogue of a discrete breather of the corresponding classical model (discrete quantum breather). Work in this direction is presently under investigation.

Mario Salerno

See also **Discrete nonlinear Schrödinger equations; Quantum nonlinearity**

Further Reading

Bogoliubov, N.M. & Bullough, R.K. 1992. A q-deformed completely integrable Bose gas model. *Journal of Physics* A, 25: 4057–4071

Cai, C., Bishop, A.R. & Grønbech-Jensen, N. 1994. *Physical Review Letters*, 72: 591

Feynman, R.P. 1965. *Lectures on Physics*, vol. III, Reading, MA: Addison-Wesley

Kivshar, Yu.S. & Salerno, M. 1994. *Physical Review* E, 49: 3543

Konotop, V.V. & Salerno, M. 1997. *Physical Review* E, 56: 3611

MacKay, R.S. & Sepulchre, J.A. 2002. *Journal of Physics* A, 35: 3985

MacFarlane, A.J. 1989. On q-analogues of the quantum harmonic oscillator and the quantum group $SU(2)_q$. *Journal of Physics* A, 22: 4581

Salerno, M. 1992. Quantum deformations of the discrete nonlinear Schrödinger equation. *Physical Review* A, 46: 6856

Salerno, M. 1994. Bose Einstein condensation in a system of q-bosons. *Physical Review* E, 50: 4528

Scott, A.C. 2003. *Nonlinear Science: Emergence and Dynamics of Coherent Structures*, 2nd edition, Oxford and New York: Oxford University Press, and references therein

SANDPILE MODEL

The concept of self-organized criticality (SOC) was introduced by Bak et al. (1987) using the example of a sandpile. If a sandpile is formed on a horizontal circular base with any arbitrary initial distribution of sand grains, a sandpile of fixed conical shape (steady state) is formed by slowly adding sand grains one after another (external drive). In the steady state, the surface of the sandpile makes on average a constant angle with the horizontal plane, known as the angle of repose. The addition of each sand grain results in some activity on the surface of the pile: an avalanche of sand mass follows, which propagates on the surface of the sandpile. In the stationary regime, avalanches are of many different sizes, and Bak et al. (1987) argued that they would have a power law distribution. If one starts with an initial uncritical state, initially most of the avalanches are small, but the range of sizes of avalanches grows with time. After a long time, the system arrives at a critical state, in which the avalanches extend over all length and time scales (Bak, 1996; Jensen, 1998; Dhar, 1999; Sornette, 2004).

Laboratory experiments on sandpiles, however, have not in general shown evidence of criticality in sandpiles due to the effects of inertia and dilatation (moving grains require more space) (Nagel, 1992), except for small avalanches (Held et al., 1990) or with elongated rice grains (Malte-Sørensen et al., 1999) where these effects are minimized. Small avalanches have small velocities (and thus negligible kinetic energy), and they activate only the first surface layer of the pile. Elongated rice grains slip at or near the surface as a result of their anisotropy (thus minimizing dilatational effects), and they also build up scaffold-like structures, which enhance the threshold nature of the dynamics.

On the theoretical front, a large number of discrete and continuous sandpile models have been studied. Among them, the so-called abelian sandpile model is the simplest and most popular (Dhar, 1999). Other variants include Zhang's model, which has modified rules for sandpile evolution (Zhang, 1989), a model for abelian distributed processors and other stochastic rule models (Dhar, 1999), and the Eulerian Walkers model (Priezzhev et al., 1996).

In the abelian sandpile model, each lattice site is characterized by its height h. Starting from an arbitrary initial distribution of heights, grains are added one at a time at randomly selected sites n: $h_n \to h_n + 1$. The sand column at any arbitrary site i becomes unstable when h_i exceeds a threshold value h_c and topples to reduce its height to $h_i \to h_i - 2d$, where d is the space dimension of the lattice. The $2d$ grains lost from the site i are redistributed on the $2d$ neighboring sites $\{j\}$, which gain a unit sand grain each: $h_j \to h_j + 1$. This toppling may make some of the neighboring sites unstable. Consequently, these sites will topple themselves, possibly making further neighbors unstable. In this way, a cascade of topplings propagates, which finally terminates when all sites in the system become stable. When this avalanche has stopped, the next grain is added on a site chosen randomly. This condition is equivalent to assuming that the rate of adding sand is much slower than the natural rate of relaxation of the system. The large separation of the driving and of the relaxation time scales is usually considered to be a defining characteristic of SOC. Finally, the system must be open to the outside; that is, it must dissipate energy or matter, for instance. An outcoming flux of grains must balance the incoming flux of grains, for a stationary state to occur. Usually, the outcoming flux occurs on the boundary of the system: even if the number of grains is conserved inside the box, it loses some grains at the boundaries. Even in a very large box, the effects of the dissipating boundaries are essential: increasing the box size will have the effect of lengthening the transient regime over which the SOC establishes itself; the SOC state is built from the long-range correlations that establish a delicate balance between internal avalanches and avalanches that touch the boundaries (Middleton & Tang, 1995).

The simplicity of the abelian model is that the final stable height configuration of the system is independent of the sequence in which sand grains are added to the system to reach this stable configuration (hence the name "abelian" referring to the mathematical property of commutativity). On a stable configuration C, if two grains are added, first at i and then at j, the resulting stable configuration C' is exactly the same as in the case where the grains were added first at j and then at i. In other sandpile models, where the stability of a sand column depends on the local slope or the local curvature, the dynamics is not abelian, since toppling of one unstable site may convert another unstable site to a stable site. Many such rules have been studied in the literature (Manna, 1991; Kadanoff et al., 1989).

An avalanche is a cascade of topplings of a number of sites created by the addition of a sand grain. The strength of an avalanche can be quantified in several ways:

- size (s): the total number of topplings in the avalanche,
- area (a): the number of distinct sites that toppled,
- lifetime (t): the duration of the avalanche, and
- radius (r): the maximum distance of a toppled site from the origin.

These four different quantities are not independent and are related to each other by scaling laws. Between any two such measures x, y belonging to the set $\{s, a, t, r\}$, one can define a mutual dependence by the scaling of the expectation of one quantity y as a function of the other x:

$$\langle y \rangle \sim x^{\gamma_{xy}}, \tag{1}$$

where γ_{xy} is called a critical exponent, index, or dimension. This equation quantifies a nonlinear generalized proportionality between the two observables x and y ($\ln\langle y \rangle$ is proportional to $\ln x$). The exponents γ_{xy} are related to one another, for example,

$$\gamma_{ts} = \gamma_{tr}\gamma_{rs}. \tag{2}$$

For the abelian sandpile model, it can be shown that the avalanche clusters cannot have any holes and in addition that $\gamma_{rs} = 2$ in two dimensions, that is,

$$\langle s \rangle \sim r^2. \tag{3}$$

In words, the size (number of toppling grains) of an avalanche is proportional to its surface. It has also been shown that $\gamma_{rt} = \frac{5}{4}$:

$$\langle t \rangle \sim r^{\frac{5}{4}}; \tag{4}$$

that is, the average duration $\langle t \rangle$ of an avalanche grows with its typical radius r faster than linearly. However, averages reflect only a part of the rich behavior of sandpile models. A significant information is provided by the full distribution function $P(x)$ for any measure $x \in \{s, a, t, r\}$. Associated with the above scaling laws (1) and (2), one often finds the finite size scaling form for $P(x)$:

$$P(x) \sim x^{-\tau_x} f_x\left(\frac{x}{L^{\sigma_x}}\right). \tag{5}$$

The exponent σ_x determines the variation of the cut-off of the tail of the distribution of the quantity x with the system size L. As long as $x < L^{\sigma_x}$, expression (5) describes a power law distribution of x, reflecting a self-similar structure of the set of avalanches. When x becomes comparable with L^{σ_x}, the function f_x ensures a fast fall-off of $P(x)$ describing the impact of the finite size L of the system on the statistics of the fluctuations of x. Scaling relations like $\gamma_{xy} = (\tau_x - 1)/(\tau_y - 1)$ connect any two measures. Scaling assumptions (5) for the avalanche sizes have not been demonstrated and may be open to doubt (Kadanoff et al., 1989). This seems to be due to the effect of rare large avalanches dissipating at the border, which strongly influence the statistics.

Many different sandpile models have been studied. However, the precise classification of various models into different universality classes in terms of their critical exponents is not yet available. Exact values of all the critical exponents of the most widely studied abelian model are still not known in two dimensions. Some effort has also been made toward the analytical calculation of avalanche size exponents (Ktitarev & Priezzhev, 1998). Blanchard et al. (2000) have developed a dynamical system theory for a certain class of SOC models (like Zhang's model, 1989), for which the whole SOC dynamics can either be described in terms of iterated function systems, or as a piecewise hyperbolic dynamical system of skew-product type where one coordinate encodes the sequence of activations. The product involves activation (corresponding to a kind of Bernouilli shift map) and relaxation (leading to contractive maps).

In summary, the sandpile model of Bak et al. (1987) and its many extensions have helped found the new concept of self-organized criticality, which is now a useful item in the toolbox and set of concepts used to study complex systems involving triggered activities.

Didier Sornette

See also **Avalanches; Critical phenomena; Granular materials; Fractals; Nonequilibrium statistical mechanics**

Further Reading

Bak, P. 1996. *How Nature Works: The Science of Self-organized Criticality*, Berlin and New York: Springer

Bak, P., Tang, C. & Weisenfeld, K. 1987. Self-organized criticality: an explanation of $1/f$ noise. *Physical Review* A, 38: 364–374

Blanchard, P., Cessac, B. & Kruger, T. 2000. What can one learn about self-organized criticality from dynamical systems theory? *Journal of Statistical Physics*, 98: 375–404

Dhar, D. 1999. The abelian sandpile and related models. *Physica* A, 263: 4–25

Held, G.A., Solina, D.H., Keane, D.T., Haag, W.J., Horn, P.M. & Grinstein, G. 1990. Experimental study of critical-mass fluctuations in an evolving sandpile. *Physical Review Letters*, 65: 1120–1123

Jensen, H.J. 1998. *Self-Organized Criticality*, Cambridge and New York: Cambridge University Press

Kadanoff, L.P., Nagel, S.R., Wu, L. & Zhou, S. 1989. Scaling and universality in avalanches. *Physical Review* A, 39: 6524–6537

Ktitarev, D.V. & Priezzhev, V.B. 1998. Expansion and contraction of avalanches in the two-dimensional abelian sandpile. *Physical Review* E, 58: 2883–2888

Malte-Sørensen, A., Feder, J., Christensen, K., Frette, V., Josang, T. & Meakin, P. 1999. Surface fluctuations and correlations in a pile of rice. *Physical Review Letters*, 83: 764–767

Manna, S.S. 1991. Critical exponents of the sandpile models in two dimensions. *Physica* A, 179: 249–268

Middleton, A.A. & Tang, C. 1995. Self-organized criticality in non-conserved systems. *Physical Review Letters*, 74: 742–745

Nagel, S.R. 1992. Instabilities in a sandpile. *Review of Modern Physics*, 64: 321–325

Priezzhev, V.B., Dhar, A., Krishnamurthy, S. & Dhar, D. 1996. Eulerian Walkers as a model of self-organized criticality. *Physical Review Letters*, 77: 5079–5082

Sornette, D. 2004. *Critical Phenomena in Natural Sciences*, 2nd edition, Berlin and New York: Springer

Zhang, Y.-C. 1989. Scaling theory of self-organized criticality. *Physical Review Letters*, 63: 470–473

SCATTERING OPERATORS

See **Inverse scattering method or transform**

SCHEIBE AGGREGATES

Scheibe aggregates (also called J-aggregates) are a type of Langmuir–Blodgett thin films that were independently discovered in 1936 by Günther Scheibe (Scheibe, 1936) in Germany and E.E. Jelly (Jelly, 1936) in England—hence their names. They comprise compact and regular arrangements of dye molecules composed of chromophores and fatty acids that can be designed to pre-selected specifications. These molecular monolayers exhibit very efficient light capture followed by energy transfer to acceptor molecules present in the film at very low acceptor-to-donor ratios (as low as 1:10,000), giving the efficiency of energy transfer to acceptors as high as 50% (Moebius, 1989).

In the late 1960s and in the 1970s, Kuhn, Moebius and co-workers (Czikkely et al., 1969) studied the effects of irradiation with ultraviolet or visible light and measured the strongly quenched donor fluorescence in these aggregates. An acceptor fluorescence line appeared whose amplitude was almost equal to that of the primary donor spectral line, but its peak was slightly redshifted, which was interpreted as evidence that the aggregate acts as a cooperative molecular array. After absorbing a photon, the energy moves in the form of an exciton traveling laterally over distances of up to 100 nm to a particular acceptor dye in the vicinity. The efficiency of this energy transfer mechanism is strongly linked with the rigidity of the aggregate and its regular order. Surprisingly, the capture of energy by an acceptor molecule improves with the ambient temperature. Optimal efficiency is achieved at fairly low acceptor-to-donor ratios as mentioned above.

Numerous technological applications of these molecular systems have been developed, including photographic and photo-detection processes as well as solar energy cell components (Inoue, 1985).

Several theoretical models of Scheibe aggregates have been developed over the years. The Frenkel exciton picture with the inclusion of diagonal disorder in molecular chains with nearest-neighbor interactions was adopted to describe the pseudo-isocyanine aggregate (Knapp, 1984) and was later extended to account for off-diagonal disorder. Subsequently, linear models of exciton propagation in the presence of shallow impurity potentials at acceptor sites showed (Bartnik et al., 1992) close quantitative agreement with experiment. These impurity potentials were usually assumed to be of a square-well type, and their depth and spatial extent were parameters subject to optimization schemes. Interestingly, shallow impurity levels gave better capture properties than deep ones.

There have also been a number of theoretical studies advocating a role played by nonlinearity in the propagation of excitonic modes in Scheibe aggregates. The main reason for this claim is due to exciton-phonon coupling whose strength for mero-cyanine was experimentally determined as approximately $\chi = 29$ meV (Inoue, 1985) and theoretically estimated as $\chi = 26$ meV (Spano et al., 1990).

In general, three types of excitonic models can be applied: (a) a nearly free delocalized case, (b) a small polaron limit, and (c) the nonlinear self-trapped exciton state. The applicability criterion for these models involves a phase diagram in terms of two characteristic parameters: $g = \pi\chi^2/\omega J$ and $\gamma = h\omega/2\pi J$, where h is Planck's constant, ω is the phonon frequency, and J is the exciton hopping constant. With the phonon energy approximately 30 meV, J in the range between 50 and 150 meV, and the exciton-phonon coupling constant ranging from 25 to 100 meV, the outcome of this parameter determination is yet inconclusive (Tuszyński et al., 1999).

It is important to stress that these systems should be modeled as two-dimensional structures. Continuous two-dimensional models with nonlinear terms have been developed that treat phonons classically and effectively eliminate them via an adiabatic approximation (Huth et al., 1989), leading to a 2-d radially symmetric cubic nonlinear Schrödinger equation, where the energy transfer takes place through soliton-like ring waves with a characteristic collapse time signifying the exciton lifetime. More recent models involving nonlinear equations of the cubic Schrödinger type also account for the presence of Gaussian impurity potentials (Christiansen et al., 1998).

Finally, it is necessary to include radiative losses in modeling the propagation of an exciton domain. Formally, this can be accomplished by adding an imaginary term to the exciton-phonon part of the Hamiltonian that describes the corresponding loss of energy of excitons as they collide with phonons in the thin-film lattice. Consequently, it has been shown that the rate of excitonic energy loss by a coherent exciton domain covering an area composed of N molecules is proportional to its size, N, and can be expressed as: $k_{\text{rad}} = N$ nanoseconds^{-1}. It has also been demonstrated that the characteristic time for radiative losses is approximately proportional to the absolute temperature T divided by 3000 K, and the result is expressed in nanoseconds (Moebius & Kuhn, 1988). In other words, the radiative decay time in nanoseconds is given by $\tau_{\text{rad}} = (T/3000)$ K. For example, at room temperature, a coherent exciton domain composed of 100 lattice sites has a decay time of 0.1 ns = 100 ps compared with a flight time of only 2 ps, which means that radiative losses will not destroy the coherence of the exciton domain unless it is very large (approximately 10,000 lattice sites). A comprehensive review of the key processes taking place in exciton energy transfer in Scheibe aggregates (time of flight, radiative losses, exciton-phonon interaction, and diffusion on the 2-d lattice) can be found in Tuszyński et al. (1999).

JACK A. TUSZYŃSKI

See also **Excitons; Langmuir–Blodgett films; Nonlinear Schrödinger equations**

Further Reading

Bartnik, E.A., Blinowska, K.J. & Tuszyński, J.A. 1992. Stability of quantum capture in Langmuir–Blodgett monolayers against positional disorder. *Physics Letters* A, 169: 46–50

Christiansen, P.L., Gaididei, Yu.B., Johansson, M., Rasmussen, K.O., Mezentsev, V.K. & Rasmussen, J.J. 1998. Solitary excitations in discrete two-dimensional nonlinear Schrödinger models with dispersive dipole-dipole interactions. *Physical Review* B, 57: 11–303

Czikkely, V., Dreizler, G., Försterling, H., Kuhn, H., Sondermann, J., Tillmann, P. & Wiegand, J. 1969. Lichtabsorption von Farbstoff-Molekülpaaren in Sandwichsystemen aus monomolekularen Schichten. *Zeitschrift für Naturforschung*, 24A: 1823

Czikkely, V., Försterling, H. & Kuhn, H. 1970. Extended dipole model for aggregates of dye molecules. *Chemical Physics Letters*, 6: 207

Huth, G.C., Gutmann, F. & Vitiello, G. 1989. Ring solitonic vibrations in Schiebe aggregates. *Physics Letters* A, 140: 339

Inoue, T. 1985. Optical absorption and luminescence in Langmuir films of merocyanine dye. *Thin Solid Films*, 132: 21

Jelly, E.E. 1936. Spectral absorption and fluorescence of dyes in the molecular state. *Nature*, 138: 1009–1010

Knapp, E.W. 1984. Lineshapes of molecular aggregates—exchange narrowing and intersite correlation. *Chemical Physics*, 85: 73

Kuhn, H. 1979. Synthetic molecular organizates. *Journal of Photochemistry*, 10: 111

Moebius, D. 1989. Monolayer assemblies. *Berichte der Bunsen-Gesellschaft—Physical Chemistry*, 82: 848

Moebius, D. & Kuhn, H. 1988. Energy transfer in monolayers with cyanine dye Scheibe aggregates. *Journal of Applied Physics*, 64: 5138–5141

Scheibe, G. 1936. Variability of the absorption spectra of some sensitizing dyes and its cause. *Angewandte Chemie*, 49: 567

Spano, F.C., Kuklinski, J.R. & Mukamel, S. 1990. Temperature dependent superradiant decay of excitons in small aggregates. *Physical Review Letters*, 65: 211

Tuszyński, J.A., Joergensen, M.F. & Moebius, D. 1999. Mechanisms of exciton energy transfer in Scheibe aggregates. *Physical Review* E, 59: 4374

SCHLESINGER EQUATIONS

See **Monodromy preserving deformations**

SCHOTTKY DIODE

See **Diodes**

SCHRÖDINGER EQUATION, LINEAR

See **Quantum nonlinearity**

SCROLL WAVES

Introduction

Scroll waves, which are observed in excitable media, can be imagined as a continuation of two-dimensional spiral waves to three dimensions. Spiral and scroll waves distribute frequency behavior to the two- and three-dimensional space, respectively, with the exception of a small subset of that space. For spiral waves this subset comprises the vicinity of a point and is called a core, whereas in three dimensions it is formed by the vicinity of a curve and is called a filament. In both cases, this exceptional subset acts as a source of waves that organizes the exhibited periodic or quasi-periodic frequency behavior; therefore, it is often referred to as an organizing center.

The intersection of the three-dimensional excitable medium containing a scroll wave with a plane that is locally perpendicular to the filament corresponds to a sheet in which a spiral wave rotates, showing the close relationship between these two types of excitation waves. But because curves (the filaments) can have complex geometrical and topological properties, the behavior of scroll waves turns out to be much more complicated than that of spirals.

Early work on rotating excitation waves was performed on physical rings of heart tissue (quasi-one-dimensional systems), where an electrical impulse can propagate around a hole without attenuation for days. The importance of this observation lies in the complex anatomy of the heart: in particular, the orifices from arteries make the surface of the heart locally similar to physical rings and, therefore, allow for the so called circus-movement reentry, that is, electrical excitation waves circulating around inhomogeneities. Modeling of this phenomenon was undertaken first by Norbert Wiener and Arturo Rosenblueth in 1946. Periodic excitation was found to be also possible in media without holes, that is, in quasi-two-dimensional media. The electrical excitation wave (see Figure 1) then assumes the shape of a spiral (Tyson & Keener, 1988). The importance of the third dimension was recognized when performing experiments with the excitable Belousov–Zhabotinsky reaction system (*See* **Belousov–Zhabotinsky reaction**), since distortions and other unexpected behavior of spiral waves in presumed two-dimensional shallow layers could be readily explained when assuming a non-uniform excitability along the small but nonnegligible vertical direction (see references in Winfree, 1987).

For the theoretical treatment of scroll waves there are several models that are most commonly used.

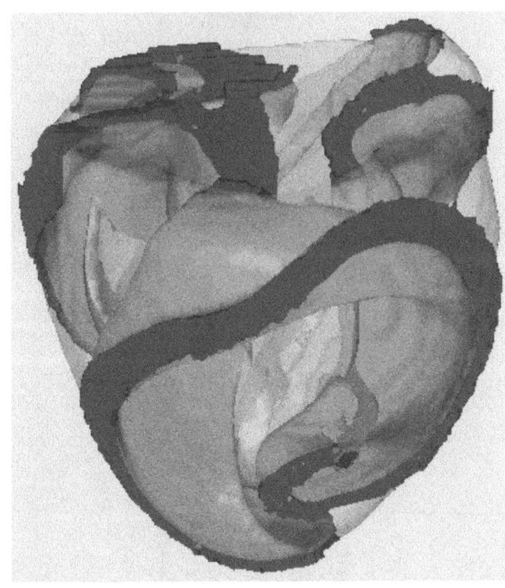

Figure 1. A scroll wave on the surface of a heart. (Image from Panfilov & Pertsov, 2001.)

Since scroll waves are structures appearing in excitable media, it is not surprising that they have been investigated by using three-dimensional FitzHugh–Nagumo equations, that is, a two-variable model designed for the simulation of neural action potentials (*See* **FitzHugh–Nagumo equation**). It was found that scroll waves always undergo twist (see below) when entering inhomogeneous media (Mikhailov et al., 1985).

Another model for the investigation of scroll waves is the Oregonator model derived from the reaction kinetics of the Belousov–Zhabotinsky reaction. The temporal development of a scroll ring, a structure with a closed filament not touching the boundary (see Figure 2, and also Welsh et al., 1983), was investigated by Winfree and Jahnke in 1989. It was found that a scroll ring decreases its size in the course of time and eventually vanishes, showing that scroll waves are topologically distinct from spirals.

A now very popular model is the so-called Barkley model, well known for its computational efficiency. Originally introduced for the investigation of the meander instability of spiral waves (Barkley, 1990), it is also used for investigating scroll waves. With this model it was possible to classify the instabilities of scroll waves in isotropic excitable media (Henry & Hakim, 2002).

While these three models give a description of the full three-dimensional concentration distribution, a fourth one deals with the geometry of an iso-concentration level. Its formulation is based on the eikonal equation, which expresses the relationship between the curvature and the normal velocity of a surface defined by such a level. Thus, the surface is described in specific coordinate systems that reflect the topological situation of the filaments involved in the

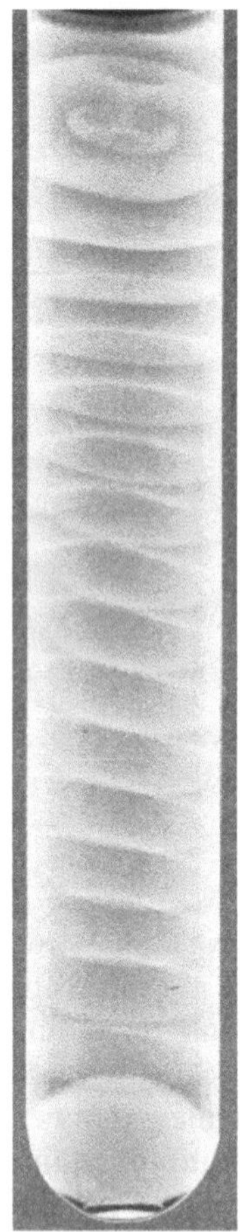

Figure 2. A scroll ring in a test tube, inner diameter: 10 mm. (Image from Winfree, 1987.)

Figure 3. An example of an impossible surface of a wave structure. (Image from Winfree, 1987.)

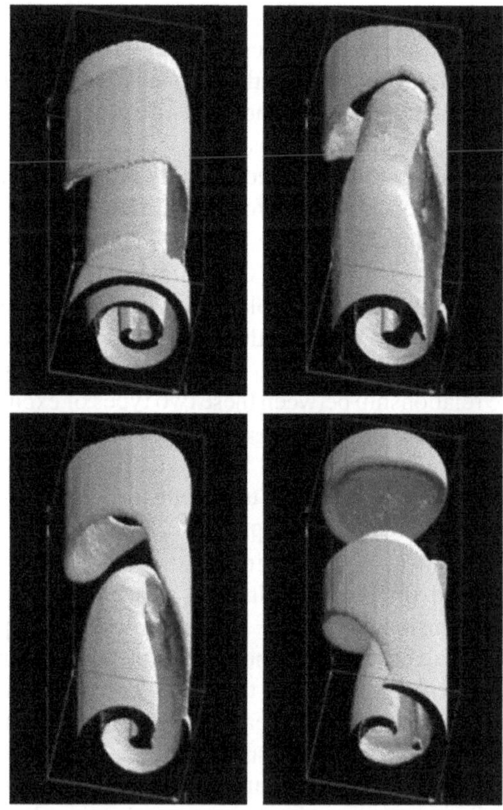

Figure 4. Decomposition of a scroll wave due to a gradient of excitability. (Image from Strob et al., 2003.)

wave structure. Investigating this type of model allows one to estimate the stability of complex, linked wave structures (McDermott et al., 2002).

Complexity of Scroll Waves

The complexity of a scroll wave becomes apparent when one imagines such a wave structure extending into a nonhomogenous excitable medium, that is, when in each slice along the filament the rotation period of the spirals is intrinsically different. Without a coupling mechanism, the phase of the spirals in these adjacent slices would evolve independently, and the differences between the phases would diverge. This is not possible, and the wave structure circumvents this inconvenience by tilting the wave fronts emanating from the organizing center. In terms of the rotation phases of the spirals, this corresponds to a phase shift along the filament, which is called twist.

While the twist takes arbitrary values in the case of scroll waves (organized by filaments reaching from one boundary of the medium to another), it has to fulfill quantization conditions for scroll rings. The first limitation is that the twist of a scroll ring along one turn along the filament has to be a multiple of 2π. A further limitation for the possible structures arises from the constraint that the surface must not have any intersections (i.e., it has to be a Seifert-surface, see Figure 3 for an impossible surface). This, for example, excludes single scroll rings of twist 2π. Instead, these always must appear as linked pairs, although in the limiting case one of them may be infinitely large or may reach from one boundary to the other (Winfree, 1987). More elaborate work on the topology and geometry of filaments has been summarized by Tyson and Strogatz (1991).

Although the computational investigation of scroll waves started in the 1980s, about ten years after their discovery, rigorous experimental research on three-dimensional wave structures did not begin before the 1990s. First measuring techniques were restricted to simple projections from one, two, or three pairwise orthogonal directions.

Three-dimensional, fully resolved observations of scroll waves and rings have been performed by optical tomography since about 1995. This technique allows one to record and evaluate time and space resolved data sequences. For instance, the decomposition of a scroll wave due to a gradient of excitability was observed in satisfactory detail (see Figure 4). Thus, experimental and theoretical tools are now available to investigate the complex interaction of the organizing centers of scroll waves.

ULRICH STORB AND STEFAN C. MÜLLER

See also **Belousov–Zhabotinsky reaction; Cardiac arrhythmias and electrocardiogram; Excitability; Geometrical optics, nonlinear; Reaction-diffusion systems; Spiral waves; Vortex dynamics in excitable media**

Further Reading

Barkley, D. 1990. Spiral-wave dynamics in a simple model of excitable media: the transition from simple to compound rotation. *Physical Review* A, 42(4): 2489–2492

Henry, H. & Hakim, V. 2002. Scroll waves in isotropic excitable media: linear instabilities, bifurcations and restabilized states. *Physical Review* E, 65: 046235-1–046235-21

McDermott, S., Mulholland, A.J. & Gomatam, J. 2002. Knotted reaction–diffusion waves. *Proceedings of the Royal Society of London Series A—Mathematical Physical and Engineering Sciences*, 458: 2947–2966

Mikhailov, A.S., Panfilov, A.V. & Rudenko, A.N. 1985. Twisted scroll waves in active three-dimensional media. *Physics Letters* A, 109(5): 246–250

Panfilov, A.V. & Pertsov, A. 2001. Ventricular fibrillation: evolution of the multiple-wavelet hypothesis. *Proceedings of the Royal Society of London Series* A, 359: 1315–1325

Storb, U., Rodrigues Neto, C., Bär, M. & Müller, S.C. 2003. A tomographic study of desynchronization and complex dynamics of scroll waves in an excitable chemical reaction with a gradient. *Physical Chemistry Chemical Physics*, 5: 2344–2353

Tyson, J.J. & Keener, J.P. 1988. Singular perturbation theory of travelling waves in excitable media. *Physica* D, 32: 327–361

Tyson, J.J. & Strogatz, S.H. 1991. The differential geometry of scroll waves. *International Journal of Bifurcation and Chaos*, 1: 723–744

Welsh, B.J., Gomatam, J. & Burgess, A.E. 1983. Three-dimensional chemical waves in the Belousov–Zhabotinskii reaction. *Nature*, 304: 611–614

Wiener, N. & Rosenblueth, A. 1946. The mathematical formulation of the problem of conduction of impulses in a network of connected excitable elements. *Specifically in Cardiac Muscle Archivos del Instituto de Cardiologica de Mexico*, 16: 205–265

Winfree, A.T. 1987. *When Time Breaks Down*, Princeton, NJ: Princeton University Press

Winfree, A.T. & Jahnke, W. 1989. Three-dimensional scroll ring dynamics in the Belousov-Zhabotinsky reagent and in the 2-variable Oregonator model. *Journal of Physical Chemistry*, 93: 2823–2832

SECOND HARMONIC GENERATION

See **Harmonic generation**

SELF-ORGANIZATION

See **Synergetics**

SELF-SIMILARITY

See **Fractals**

SEMICLASSICAL APPROXIMATIONS

See **Quantum theory**

SEMICONDUCTOR LASER

The semiconductor laser is today the most important and widespread type of laser, being a central component in many common household appliances (CD and DVD players) as well as in major industrial areas, such as measurement and sensing, materials manufacturing, and medical surgery. Not least, the semiconductor laser has enabled the rapid evolution of the Internet by providing a means for efficient and cheap conversion of digital electrical signals into optical signals, which can be transmitted at very high data rates and over very long distances in hair-thin optical fibers.

The success of the semiconductor laser to a large degree relies on its many similarities with electronic semiconductor devices such as transistors and diodes. The laser is manufactured by standard semiconductor crystal growth and processing techniques, allowing for small and cheap devices. Furthermore, the semiconductor laser distinguishes itself from other types of lasers by being electrically activated. Thus, the energy needed for pumping the laser to an excited state, from which the energy can be released by emission of photons, is achieved simply by putting direct electrical current through the device. Other kinds of lasers typically need some kind of optical pumping for populating the laser-active states.

Figure 1 shows a schematic of a typical semiconductor laser. The laser is a p-i-n structure; that is, p- and n-doped semiconductor materials sandwich an undoped intrinsic ("i") region. The structure acts as a standard pn-junction diode; when forward biased (with a positive voltage on the metallized p-side relative to the n-side), an electrical current flows. This leads to the build-up of significant electron and hole densities in the intrinsic region, and by recombination of these excited-state (conduction-band) electrons with ground-state (valence-band) holes, photons can be generated. An incoming photon may thus stimulate the emission of a new photon with identical properties. This process of stimulated emission provides optical gain, which is

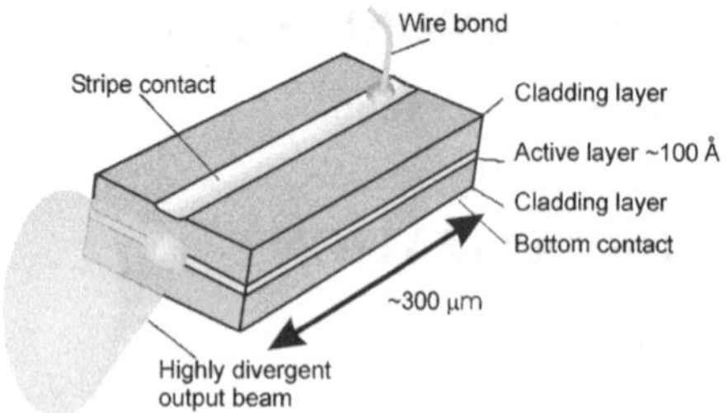

Figure 1. Schematic of a semiconductor laser. Light is confined to a waveguide of transverse dimension typically of the order of 0.2 μm × 2 μm. For a quantum well structure, stimulated emission occurs in a narrow region of width ~ 100 Å. The laser is pumped electrically by incorporating the active region in a pn-junction.

one of the two key requirements for implementing a laser.

As for any oscillator, the second requirement—besides gain—is the existence of feedback. For a laser, this is usually achieved by incorporating the gain medium in a mirror cavity. In the case of a semiconductor laser, the mirrors are particularly simple since cleaving along one of the crystal planes provides a naturally flat mirror with an intensity reflection coefficient of the order of 30%. Due to the large material gain achievable in semiconductor lasers, a laser-active region length of the order of a few hundred μm is sufficient to compensate for the corresponding 70% outcoupling loss, as well as other losses in the material, thus enabling laser oscillation.

In addition to providing feedback, the laser cavity also must confine the optical laser mode in the transverse plane. This is achieved by establishing a transverse waveguide through index-guiding. The design thus needs to ensure a larger effective index in the active region of the laser as compared with the surrounding regions. In the growth direction, index guiding is provided by use of a so-called semiconductor heterostructure. The intrinsic, "i"-region is thus composed of a material with a smaller bandgap than the surrounding materials, which leads to a larger refractive index. The incorporation of a heterojunction structure was a major achievement in the early development of semiconductor lasers and earned the inventors, Zhores Alferov and Herbert Kroemer, the Nobel Prize in Physics in 2000. Besides providing index confinement, the heterostructure also ensures efficient collection of electrons and holes in the active region.

Index guiding in the plane (lateral direction) of the semiconductor layers is more difficult and is achieved in a number of different ways, the two most important ones employing a ridge waveguide structure and a buried heterostructure. The ridge waveguide structure is obtained by processing a narrow ridge (1–2 μm wide) in the doped semiconductor material topping the active region. The material that is etched away on either side of the ridge is replaced by a material (e.g., polyimide) that is isolating and has an index of refraction less than that of the semiconductor (ca. 3.5). The ridge provides current confinement and leads to an effective refractive index that varies in the lateral direction along the active region, reaching a maximum value right below the high-index ridge, thus imposing lateral waveguiding. The refractive index contrast thus obtained is, however, modest, and ridge waveguide lasers belong to the class of weakly index-guided structures. The other class of strongly index-guided structures is exemplified by the buried heterostructure laser. By employing several growth steps, the active waveguide region can thus be surrounded, in the lateral direction, by materials with higher bandgap and lower refractive index. Furthermore, these regions, adjacent to the active region, are doped to block the current from entering.

By analogy with a standard laser cavity (etalon), a semiconductor laser using cleaved facets to define the laser cavity is denoted a Fabry–Perot laser. Due to the small difference in material gain between the longitudinal modes of such a laser cavity, the laser may oscillate in several closely spaced modes and the single-mode suppression ratio remains modest. By incorporating a grating structure in the laser, either distributed over the entire waveguide length (distributed feedback or DFB laser) or localized close to the facets (distributed Bragg reflector or DBR laser), spectral selection can be achieved, resulting in high-quality single-mode lasers. DFB lasers, in particular, have been successfully applied in optical communications systems. A more recent type of laser is the so-called vertical cavity surface emitting laser (VCSEL), where the laser end mirrors are provided by Bragg gratings parallel to the semiconductor substrate and the laser emits out of the plane of the substrate.

The choice of the materials for the active region determines the wavelength of the output optical beam. Two materials systems are particularly important: GaAlAs lasers, which cover the wavelength range of 700–900 nm, and InGaAsP lasers, which cover the wavelength range of 1000–1600 nm. The latter laser type is the most important for long-distance optical communication systems due to the low loss and/or dispersion of optical fibers in the range of 1300–1550 nm. An additional degree of freedom comes from employing a so-called quantum well structure of the active layer. Thus, by incorporating thin (about 5–10 nm) layers of semiconductor with a bandgap lower than the surrounding (barrier) material in the active region, quantum confinement effects change the allowed energy levels of electrons. This leads to a change in the effective bandgap of the laser (wavelength under lasing), as well as the number of electrons needed to reach population inversion.

Quantum well lasers have achieved low-threshold and high-power operation. Quantum dot lasers, employing three-dimensional quantum confinement of electrons, offer excellent electron control and have led to lasers with record low-threshold current density. However, semiconductor growth technology has not yet (in 2004) matured to the point of offering full control of quantum dot sizes.

The laser threshold condition and the basic features of the laser dynamics are captured by a simple set of rate equations describing the temporal evolution of the photon density, P, and the carrier density, N (Coldren & Corzine, 1995):

$$\frac{dP}{dt} = (\Gamma v_g g - \tau_p^{-1})P + \Gamma \beta_{sp} R_{sp}, \qquad (1)$$

$$\frac{dN}{dt} = \frac{I}{eV} - \frac{N}{\tau_s} - v_g g P. \qquad (2)$$

Here, v_g is the group velocity; τ_p is the photon lifetime, its inverse being the rate at which photons are lost from the laser cavity; β_{sp} is the rate of spontaneous emission; ending up in the lasing mode; I is the injected current; e is the electronic charge; V is the active region volume; and τ_s is the carrier lifetime. The confinement factor, Γ, expresses the fraction of the optical mode that overlaps with the active region; it may also be expressed as the ratio between the active region volume, V, and the effective volume of the optical mode, V_{opt}. In the form of the equations stated above, the photon density is normalized with respect to V_{opt}, whereas N is normalized with respect to V. Finally, g is the material gain. When considering a laser operating at the peak of the gain curve, it is, at least for lasers with bulk active regions, a good approximation to parameterize the gain as

$$g \cong g_N(N - N_0) \qquad (3)$$

with g_N being the differential gain, and N_0 the carrier density at transparency. For $N = N_0$, we have $g = 0$, corresponding to the case where stimulated emission and absorption exactly balance. For a further increase of the carrier density, population inversion is achieved and the material gain is positive.

The rate equations (1) and (2) are basically book-keeping equations. Equation (2) expresses the effective pumping of electrons from valence to conduction band through the applied current. Due to spontaneous emission as well as nonradiative recombination (Auger processes are particularly important in long-wavelength lasers), the excited carrier density has a lifetime τ_s of the order of a nanosecond or less. Furthermore, stimulated emission, proportional to the product of the gain and the photon density, depletes the population of excited carriers, as expressed by the last term in Equation (2). That same process leads to a generation term in the rate equation for the photon density. Also, a certain fraction, β_{sp}, which may typically be of the order of 10^{-5} of the total rate of spontaneous emission, R_{sp}, ends up in the laser mode and is accounted for by the last term in Equation (1). Finally, the drain term in Equation (1) describes all mechanisms by which photons are lost from the cavity, including output coupling at the mirror facets and internal loss (due to free-carrier absorption, waveguide scattering loss, etc.).

The steady-state solution of the rate equations, with the gain given by Equation (3), yields the light-current characteristics, laser power that is, expressed in terms of the photon density as a function of the applied current. A simple solution, yet accurate except very close to threshold, is obtained by neglecting the rate of spontaneous emission into the lasing mode, that is, the last term in Equation (1). A small-signal analysis shows that the above-threshold solution is a stable focus. The characteristic frequency is the so-called laser relaxation oscillation frequency, which is the natural frequency at which energy is exchanged between the carrier population and the photon population. The relaxation frequency is also a measure of the order of the maximum bit-rate at which a laser can be efficiently current-modulated to produce an intensity-modulated optical output signal. The square of the relaxation oscillation is approximately proportional to the laser output power, although the oscillations become strongly damped as the frequency increases. High-speed lasers have relaxation frequencies of the order of 30 GHz.

From the Kramers–Kronig relations, any change of the gain of a material (proportional to the imaginary part of the susceptibility) implies a change in the refractive index (proportional to the real part of the susceptibility). Due to the asymmetric nature of the gain spectrum of a semiconductor laser—with a transparent region below the bandgap of the material and a strongly absorbing region at large photon energies—the coupling between index and gain is particularly strong for semiconductor lasers, with profound consequences for the dynamics. This coupling is described by the so-called linewidth enhancement factor (or alpha-parameter) α. It was thus realized that the gain-index coupling gives rise to an enhancement of the linewidth of a semiconductor laser by a factor of $1 + \alpha^2$ (Henry, 1983), and also imposes a chirp on the optical signal under current modulation. These effects are not described by the rate equation (1), which only governs the magnitude squared of the electromagnetic (optical) field and is independent of the phase. Rather the dynamics need to be described by an equation for the complex electric field amplitude, E:

$$\frac{dE}{dt} = \frac{1}{2}(1 + i\alpha)\Gamma v_g g_N (N - N_0)E + F_E(t), \qquad (4)$$

where $P \propto |E|^2$. The term $F_E(t)$ is a stochastic Langevin noise term accounting for the random nature of spontaneous emission. The amplitude and phase noise properties of single-mode semiconductor lasers can be analyzed based on Equations (3) and (2).

Addition of a feedback term, proportional to $E(t-\tau)$, to the right-hand side of Equation (3) leads to the famous Lang–Kobyashi equations (Lang & Kobyashi, 1980), which govern the dynamics of a semiconductor laser coupled to an external mirror, τ being the roundtrip time in the external cavity. The semiconductor laser with feedback displays very rich dynamics, including mode-hopping, various instabilities and chaos, that to a remarkable degree are explained by Equations (4) and (2) (Mørk et al., 1992).

The rate equation model outlined above is limited to the case of single transverse mode lasers. Wide aperture lasers, in contrast, have an additional degree of freedom in the transverse direction. It has been shown that a complex semiconductor Swift–Hohenberg equation may describe the dynamics of such lasers in a single longitudinal mode mean field limit (Mercier & Moloney, 2002).

JESPER MØRK

See also **Diodes; Lasers; Nonlinear optics; Optical fiber communications**

Further Reading

Coldren, L.A. & Corzine, S.W. 1995. *Diode Lasers and Photonic Integrated Circuits*, New York: Wiley

Henry, C.H. 1983. Theory of the phase noise and power spectrum of a single mode injection laser. *IEEE Journal of Quantum Electronics*, QE-19: 1391–1397

Lang, R. & Kobyashi, K. 1980. External optical feedback effects on semiconductor injection laser properties. *IEEE Journal of Quantum Electronics*, QE-16: 347–355

Mercier, J.-F. & Moloney, J.V. 2002. Derivation of semiconductor laser mean-field and Swift-Hohenberg equations. *Physical Review* E, 66: 036221-1–036221-19

Mørk, J., Tromborg, B. & Mark, J. 1992. Chaos in semiconductor lasers with optical feedback: theory and experiment. *IEEE Journal of Quantum Electronics*, 28: 93–108

SEMICONDUCTOR OSCILLATORS

The electric transport properties of a semiconductor show up most directly in its current-voltage characteristic. It is related to a relationship between the current density j and the electric field E that is determined in a complex way by the microscopic properties of the material. Although a local, static, scalar $j(E)$ relation need not exist, it does in many cases.

Close to thermodynamic equilibrium (at sufficiently low bias voltage), the $j(E)$ relation is linear (Ohm's law), but under practical operating conditions it often becomes nonlinear and may even display a regime of negative differential conductivity (NDC), where $\sigma_{\text{diff}} = \mathrm{d}j/\mathrm{d}E < 0$. The global current-voltage characteristic $I(U)$ of a semiconductor can in principle be calculated from the local $j(E)$ relation by integrating the current density j over the cross section A of the current flow and the electric field E over the length L of the sample. Unlike the $j(E)$ relation, the $I(U)$ characteristic is not only a property of the semiconductor material, but depends also on the geometry, the boundary conditions, and the contacts of the sample. Only for the idealized case of spatially uniform states, are the $j(E)$ and the $I(U)$ characteristics identical, up to re-scaling. The $I(U)$ relation is said to display negative differential conductance if $\mathrm{d}I/\mathrm{d}U < 0$.

In the case of negative differential conductance, the current decreases with increasing voltage, and vice versa, which may lead to instability. The actual electrical response depends upon the attached circuit, which in addition to resistors may comprise capacitors and inductors. These reactive components give rise to additional degrees of freedom that are described by ordinary differential equations for I and U. If a semiconductor element with NDC is operated in such a reactive circuit, oscillatory instabilities may be induced by these reactive components. Self-sustained semiconductor oscillations, where the semiconductor itself introduces an internal unstable temporal degree of freedom, can be distinguished from those circuit-induced oscillations. The self-sustained oscillations under time-independent external bias are discussed here. Examples of internal degrees of freedom are the charge carrier density, the electron temperature, or a junction capacitance within the device.

Two important examples of NDC are described by an N-shaped or an S-shaped $j(E)$ characteristic and denoted by NNDC and SNDC, respectively (Figure 1). However, more complicated forms such as Z-shaped, loop-shaped, or disconnected characteristics are also possible. NNDC and SNDC are associated with voltage or current-controlled instabilities, respectively. In the NNDC case the current density is a single-valued function of the field, but the field is multivalued; in other words, the $E(j)$ relation has three branches in a certain range of j. The SNDC case is complementary in the sense that E and j are interchanged.

In case of NNDC, the NDC branch is often but not always (depending upon external circuit and boundary conditions) unstable against the formation of non-uniform field profiles along the charge transport direction (electric field domains). In the SNDC case, on the other hand, current filamentation generally occurs, in which the current density becomes non-uniform over the cross section of the current flow and forms a conducting channel (Ridley, 1963). These primary self-organized spatial patterns may themselves become unstable, leading to periodically or chaotically breathing, rocking, moving, or spiking filaments or domains, or even solid-state turbulence and spatiotemporal chaos (Schöll, 2001). Alternatively, the spatially uniform steady state may already become unstable with respect to uniform oscillations in a Hopf bifurcation.

Semiconductor oscillators may be classified as dominated by a bulk mechanism (drift instability,

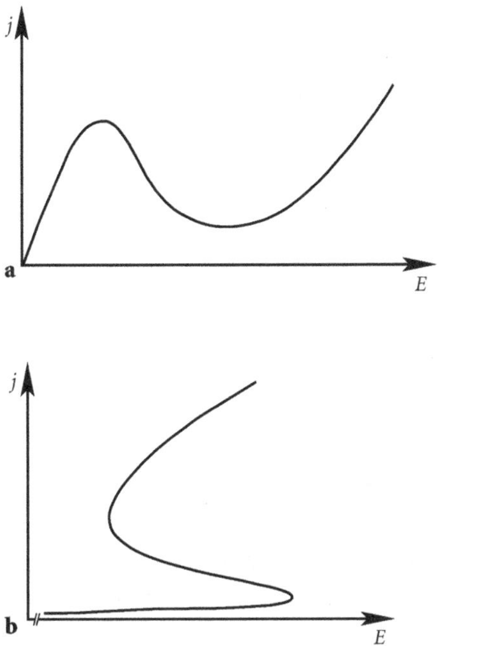

Figure 1. Current density j versus electric field E for two types of negative differential conductivity (NDC): (a) NNDC; (b): SNDC (schematic).

generation-recombination instability) or by heterojunctions and potential barriers and wells (resonant tunneling across, or thermionic emission over, barriers in nanostructures).

The first class of semiconductor oscillators includes drift instability. In the simplest extension of the Drude model, the current density is given by $j(E) = -env(E)$ where $e > 0$ is the electron charge, n is the electron density, and $v(E)$ is the field-dependent drift velocity, which may give rise to negative differential conductivity if $\mathrm{d}|v|/\mathrm{d}|E| < 0$. The best-known example is the Gunn effect, which is based upon intervalley transfer of electrons in k-space from a state of high mobility to a state of low mobility under a strong electric field in direct semiconductors like GaAs or other III–V compounds (Gunn, 1964; Ridley & Watkins, 1961). This phenomenon is used in real devices (Gunn diodes) to generate and amplify microwaves at frequencies typically beyond 1 GHz.

The class of generation-recombination (GR) instabilities is distinguished by a nonlinear dependence of the steady-state carrier concentration n upon the field E that yields a non-monotonic relation $j = en(E)\mu E$ of either NNDC or SNDC type, where μ is the mobility. This dependency is due to a redistribution of electrons between the conduction band and bound states with increasing field. The microscopic transition probabilities of the carriers between different states, and hence the GR coefficients, generally depend upon the electric field. Models of this type are relevant for a variety of materials and in various temperature ranges (Schöll, 1987) and can explain SNDC and current filamentation in the regime of low-temperature impurity breakdown and self-generated current oscillations including chaotic behavior, as observed experimentally (Peinke et al., 1992). Two important devices are also based upon GR-induced bulk negative differential conductivity, but the coupling with junction effects is essential in these cases: p-i-n diodes and impact ionization avalanche transist-time (IMPATT) diodes.

A variety of instabilities can arise due to the specific transport properties of semiconductor heterostructures. One mechanism for NNDC, which is the real space analog of the k-space intervalley transfer in the Gunn effect, uses electron transfer between a high-mobility layer and a low-mobility layer in a modulation-doped semiconductor heterostructure under a time-independent bias applied parallel to the layers (Gribnikov et al., 1995). In the NNDC regime, current oscillations of 2–200 MHz have been experimentally observed and theoretically explained.

Another class of oscillatory instabilities occurs under vertical electrical transport in layered semiconductor structures, for example, in the heterostructure hot-electron diode (HHED) or the double-barrier resonant tunneling diode (DBRT), which are associated with S-shaped and Z-shaped current-voltage characteristics, respectively (Schöll, 2001). A resonant tunneling structure is composed of alternating layers of two different semiconductor materials with different bandgaps. The energy diagram shows a modulation of the conduction band edge on a nanometer scale, forming potential barriers and quantum wells. The current density across the barrier between two wells is due to quantum mechanical tunneling and exhibits a strongly nonlinear dependency upon the electric field. It is maximum if there is maximum overlap between the occupied states in one well and the available unoccupied states in the other (the energies are in resonance). For low-fields, equivalent levels in adjacent wells are approximately in resonance. With increasing field, the energies of the two wells are shifted with respect to each other, and the available states in the collecting well are lowered with respect to the emitting well, and hence the current density drops as the overlap between the energy levels decreases, thereby displaying NDC. Upon further increase of the field, the current density rises again up to a sharp resonance peak when the ground energy level in one quantum well is aligned with the second level in the neighboring well. Thus, resonant tunneling produces NNDC.

The simplest system of this type consists of a double-barrier structure with one embedded quantum well in between, sandwiched between a highly doped emitter and collector region. However, the situation becomes more complicated if the nonlinear feedback between space charges and transport processes is taken into account. The built-up charge in the well leads to an electrostatic feedback mechanism that increases the energy of the well state supporting resonant tunneling conditions for higher applied voltages. This may result in bistability and hysteresis where a high current and a low current state coexist for the same applied voltage, and the current-voltage characteristic becomes Z-shaped. Switching between the two stable states as well as self-sustained current oscillations may occur under appropriate external circuit conditions. The bistability also provides a basis for lateral pattern formation (current filamentation) and spatiotemporal bifurcation scenarios including chaotic breathing and spiking.

Sequential resonant tunneling in a periodic structure of multiple quantum wells, a semiconductor superlattice, likewise displays NNDC (Esaki & Chang, 1974). Now, along the growth direction, the uniform field distribution may break up into a low-field domain, where the field is near the first peak of the $j(E)$ characteristic, and a high-field domain, where the field is close to the second, resonant-tunneling peak. Depending upon doping, the applied voltage, structural parameters, and the emitter contact conductivity, stationary or traveling domains occur—the latter leading to self-sustained current oscillations ranging from several hundred MHz up to 150 GHz at room temperature (Wacker, 2002).

ECKEHARD SCHÖLL

See also **Avalanche breakdown; Diodes; Drude model; Nonlinear electronics**

Further Reading

Aoki, K. 2000. *Nonlinear Dynamics and Chaos in Semiconductors*, Bristol and Philadelphia: Institute of Physics

Esaki, L. & Chang, L.L. 1974. New transport phenomenon in a semiconductor superlattice. *Physical Review Letters*, 33(8): 495–498

Gribnikov, Z.S., Hess, K. & Kosinovsky, G. 1995. Nonlocal and nonlinear transport in semiconductors: Real-space transfer effects. *Journal of Applied Physics*, 77: 1337–1373

Gunn, J.B. 1964. Instabilities of current in III–V semiconductors. *IBM Journal of Research and Development*, 8: 141–159

Peinke, J., Parisi, J., Rössler, O. & Stoop, R. 1992. *Encounter with Chaos*, Berlin and New York: Springer

Ridley, B.K. 1963. Specific negative resistance in solids. *Proceedings of the Physical Society*, 82: 954

Ridley, B.K. & Watkins, T.B. 1961. The possibility of negative resistance effects in semiconductors. *Proceedings of the Physical Society*, 78: 293–304

Schöll, E. 1987. *Nonequilibrium Phase Transitions in Semiconductors*, Berlin and New York: Springer

Schöll, E. 2001. *Nonlinear Spatio-temporal Dynamics and Chaos in Semiconductors*, Cambridge and New York: Cambridge University Press

Schöll, E., Niedernostheide, F.J., Parisi, J., Prettl, W. & Purwins, H. 1998. Formation of spatio-temporal structures in semiconductors. In *Evolution of Spontaneous Structures in Dissipative Continuous Systems*, edited by F.H. Busse & S.C. Müller, Berlin: Springer, pp. 446–494

Shaw, M.P., Mitin, V.V., Schöll, E. & Grubin, H.L. 1992. *The Physics of Instabilities in Solid State Electron Devices*, New York: Plenum Press

Wacker, A. 2002. Semiconductor superlattices: a model system for nonlinear transport. *Physics Reports*, 357: 1–111

SEMI-LINEAR PDES

See **Quasilinear analysis**

SENSITIVE DEPENDENCE ON INITIAL CONDITIONS

See **Butterfly effect**

SEPARATION OF VARIABLES

Separation of variables is the name of a general method for finding particular solutions of partial differential equations (PDEs) as a product of functions, where each factor depends on only one of the independent variables and satisfies an ordinary differential equation (ODE).

In the study of linear equations, a familiar example of the method is eigenfunction analysis. Using the superposition principle, this approach leads to expansions in products of orthogonal functions. The Fourier transform method is a special limiting case.

For a particular PDE, there may be a family of coordinate systems that admits separation of variables. The problem of finding such coordinate systems is closely connected with the group properties of differential equations. Methods from Lie group theory can be used to describe all the separable solutions of many equations from mathematical physics (Miller, 1977). In Morse and Feshbach (1953), the separable orthogonal coordinate systems for the Laplace, Helmholtz, Schrödinger, heat (diffusion), and wave equations in two and three dimensions are listed.

For example, Laplace's equation

$$u_{xx} + u_{yy} = 0 \tag{1}$$

has product solutions $u(x, y) = X(x)Y(y)$, where X (respectively, Y) is a function of only the independent variable x (respectively, y). Here X and Y satisfy the ordinary differential equations $X'' + \lambda X = 0$ and $Y'' - \lambda Y = 0$, where λ is the separation constant and prime denotes differentiation with respect to the independent variable. X and Y may thus be expressed in terms of trigonometric and exponential functions.

Separability depends on the boundary conditions. As a second example, consider the Helmholtz equation

$$u_{xx} + u_{yy} + k^2 u = 0 \tag{2}$$

in two dimensions with Dirichlet or Neumann boundary conditions ($u = 0$ or $u_n = 0$, where subscript n denotes the derivative in the direction normal to the boundary). This boundary value problem is separable in rectangular, parabolic, polar, and elliptic coordinates.

However, with the mixed (Robin) or impedance boundary condition $u_n + cu = 0$ (where the impedance c is a constant), the problem is only separable in rectangular and polar coordinates.

For nonlinear wave equations a method for separation of variables is sometimes called Lamb's method, stemming from an early analysis of the sine-Gordon (SG) equation (Lamb, 1971)

$$u_{xx} - u_{tt} - \sin u = 0. \tag{3}$$

This equation has solutions of the form $u(x,t) = 4\tan^{-1}(X(x)T(t))$, where $(X')^2 = a_1X^4 + a_2X^2 + a_3$ and $(T')^2 = -a_3T^4 + (a_2 - 1)T^2 - a_1$, and a_1, a_2, and a_3 are separation constants. In general, therefore, X and T are Jacobi elliptic functions.

In special cases simpler SG soliton solutions such as kinks, antikinks, colliding kinks and antikinks, breathers or bions, and plasma waves, are obtained. For example, choosing $a_1 = a_2 = 1/(1 - v^2)$ and $a_3 = 0$ yields colliding kinks, traveling with velocities v and $-v$, respectively, while $a_2 = -a_3 = 1/(1 - v^2)$ and $a_1 = 0$ yields colliding kinks and antikinks. Since the SG equation models long Josephson junctions, fluxons traveling on Josephson junctions of infinite length are obtained as derivatives of the kinks (Scott, 2003), and nonlinear oscillations or standing waves on junctions of finite length can also be found for boundary conditions of Dirichlet or Neumann type (Costabile et al., 1978).

Separation of variables provides solutions to other nonlinear partial differential equations such as the nonlinear Schrödinger equation (NLS)

$$\mathrm{i}u_t + u_{xx} + 2|u|^2u = 0. \tag{4}$$

Writing $u(x,t) = \phi(x,t)\mathrm{e}^{\mathrm{i}\theta(x,t)}$, the amplitude function, $\phi(x,t)$, and the phase function, $\theta(x,t)$, satisfy nonlinear coupled ordinary differential equations that permit traveling wave solutions of the form $\phi(x,t) = \phi(x - vt)$, where v is the velocity. These are the NLS envelope solitons (Scott, 2003).

Standing wave solutions to the so-called improved Boussinesq equation

$$u_{xx} - u_{tt} - (u^2)_{xx} + u_{xxtt} = O \tag{5}$$

of the form $u(x,t) = 1/2 + X(x)T(t)$, where $X(x)$ and $T(t)$ satisfy uncoupled nonlinear ordinary differential equations, have been obtained in Rosenau and Schwarzmeier (1986) and Christiansen et al. (1990).

The Hamilton–Jacobi equation is separable when Hamilton's characteristic function, W, can be written as a sum of functions where each function depends on only one of the independent variables (Goldstein, 1980). Let us look at a particle with mass m moving in central force with potential $V(r)$. Using polar coordinates (r, ϕ) in the plane of the orbit, the Hamiltonian has the form

$$H = \frac{1}{2m}\left(p_r^2 + \frac{p_\phi^2}{r^2}\right) + V(r), \tag{6}$$

and is cyclic in ϕ. Consequently, Hamilton's characteristic function

$$W = W_r(r) + W_\phi(\phi) \equiv W_r(r) + \alpha_\phi\phi, \tag{7}$$

where the momentum conjugated to r is $p_r = \partial W_r(r)/\partial r$ and the angular momentum conjugated to ϕ, $p_\phi = \partial W_\phi(\phi)/\partial\phi = \alpha_\phi$ is a constant. The Hamilton–Jacobi equation now becomes

$$\left(\frac{\partial W_r(r)}{\partial r}\right)^2 + \frac{\alpha_\phi^2}{r^2} + 2mV(r) = 2mE, \tag{8}$$

where E is the total energy of the system. Integrating (8) with respect to r, $W_r(r)$ and thus W given by (7) are obtained. The canonical equations then yield the orbitals in the form $r = r(t)$ or $r = r(\phi)$.

A new approach has been developed by Sklyanin (1995), who argues that separation of variables, understood generally enough, could be a universal tool to solve integrable models of classical and quantum mechanics. Standard construction of the action-angle variables for the poles of the Baker–Akhiezer function can be interpreted as a variant of separation of variables. The new approach has been applied to magnetic chains, the Toda lattice, the nonlinear Schrödinger equation, the sine-Gordon model, and other systems (Skyanin, 1995).

PETER L. CHRISTIANSEN

See also **Boundary value problems; Integral transforms; Long Josephson junctions**

Further Reading

Christiansen, P.L., Lomdahl, P.S. & Muto, V. 1990. On a Toda lattice model with a transversal degree of freedom. *Nonlinearity*, 4: 477–501

Costabile, G., Parmentier, R.D., Savo, B., McLaughlin, D.W. & Scott, A.C. 1978. Exact solutions of the sine-Gordon equation describing oscillations in a long (but) finite Josephson junction. *Applied Physics Letters*, 32: 587–589

Goldstein, H. 1980. *Classical Mechanics*, 2nd edition, Reading, MA: Addison-Wesley

Lamb, G.L., Jr. 1971. Analytical descriptions of ultrashort optical pulse propagation in a resonant medium. *Reviews of Modern Physics*, 43: 99–124

Miller, U. 1977. *Symmetry and Separation of Variables*, Reading, MA: Addison-Wesley

Morse, P.M. & Feshbach, H. 1953. *Methods of Theoretical Physics*, New York and London: McGraw-Hill

Rosenau, P. & Schwarzmeier, J.L. 1986. On similarity solutions of Boussinesq-type equation. *Physics Letters* A, 115: 75–77

Scott, A.C. 2003. *Nonlinear Science: Emergence and Dynamics of Coherent Structures*, 2nd edition, Oxford and New York: Oxford University Press

Sklyanin, E.K. 1995. Separation of variables: new trends. *Progress in Theoretical Physics Supplement*, 118: 35–60

SEPARATRIX

See **Phase space**

SHAPIRO STEPS

See **Josephson junction**

SHARKOVSKY'S THEOREM

See **One-dimensional Maps**

SHEAR FLOWS

Shear flows arise whenever two bodies of fluid move relative to each other, as in a jet of liquid entering a fluid at rest or the flow past a solid obstacle. When the flow speeds become too high, these flows undergo instabilities that are at the root of the vortical structures that dominate turbulent flows. For the Kelvin–Helmholtz; instability of two layers of fluid in relative motion; the instabilities of a fluid sheared between two concentric, rotating cylinders (Taylor–Couette flow); the dynamics of boundary layers; and viscous Hele-Shaw experiments, see the corresponding *Encyclopedia* entries.

Energy balance: The significance of shear as a source of energy for perturbations follows from the analysis of the energy content of a perturbation to a shear flow. Let $\boldsymbol{U}$ be a prescribed stationary flow in a domain V and let $\boldsymbol{u}$ be the perturbation added to $\boldsymbol{U}$. The Reynolds number is defined in terms of the velocity scale U of the reference flow, a length scale L, and the kinematic viscosity ν, that is, $\mathrm{Re} = UL/\nu$. For a divergence free perturbation with boundary conditions $\boldsymbol{u} = 0$ on surfaces, the energy content $E(t) = \int_V \mathrm{d}V(\boldsymbol{u}^2/2)$ satisfies the Orr–Reynolds equation

$$\frac{\mathrm{d}E}{\mathrm{d}t} = -\int_V u_i u_j \frac{\partial U_i}{\partial x_j} \mathrm{d}V - \frac{1}{\mathrm{Re}} \int_V \frac{\partial u_i}{\partial x_j} \frac{\partial u_j}{\partial x_i} \mathrm{d}V. \quad (1)$$

Since the last term on the right-hand side is negative semidefinite, all the energy input has to come from the shear $\partial U_i/\partial x_j$. If the shear is too small, then $E(t)$ will decay monotonically. The Reynolds number up to which $\dot{E} \leq 0$ for all perturbations defines the energy stability limit Re_E.

Parallel flows: A necessary condition for an instability of parallel inviscid flows was derived by Lord Rayleigh (John William Strutt) in 1880; he found that instability requires an inflection point. If $\boldsymbol{u} = U(y)\boldsymbol{e}_x$ is the unperturbed profile, then instability can occur only if there is a point y_s with $U''(y_s) = 0$. This criterion was later improved by R.Fjørtoft, who found the requirement $U''(y)(U(y) - U(y_s)) < 0$. Necessary and sufficient conditions are more complicated, as discussed in Balmforth & Morrision (1999).

For the viscous stability, it is useful to represent the perturbation with components (u, v, w) in terms of the vertical velocity component v and the vertical vorticity, $\eta = \partial_z u - \partial_x w$, and to expand in terms of Fourier modes,

$$v(x, y, z, t) = \tilde{v}(y)\mathrm{e}^{\mathrm{i}(\alpha x + \beta z - \omega t)}, \quad (2)$$

$$\eta(x, y, z, t) = \tilde{\eta}(y)\mathrm{e}^{\mathrm{i}(\alpha x + \beta z - \omega t)}. \quad (3)$$

Then the amplitudes satisfy

$$\left[(-\mathrm{i}\omega + \mathrm{i}\alpha U)(D^2 - k^2) - \mathrm{i}\alpha U'' - \frac{1}{\mathrm{Re}}(D^2 - k^2)^2\right] \times \tilde{v} = 0, \quad (4)$$

$$\left[(-\mathrm{i}\omega + \mathrm{i}\alpha U) - \frac{1}{\mathrm{Re}}(D^2 - k^2)\right] \times \tilde{\eta} = -\mathrm{i}\beta U' \tilde{v}. \quad (5)$$

The first equation is the Orr–Sommerfeld equation, the second Squire's equation. In 1933, H.B. Squire showed that if there is an instability for a mode with spanwise wave number $\beta \neq 0$, then there is another one with $\beta = 0$ that has a lower Reynolds number, where instabilities set in without spanwise modulations.

The linearized problem can show transient growth: a decaying eigenstate of the Orr–Sommerfeld equation can drive the Squire equation and cause an amplification of the normal vorticity (if $\beta \neq 0$). The most effective modes often have the form of downstream vortices, with only slight modulations in spanwise and downstream direction. They drive spanwise modulations in the downstream velocity component, so-called streaks. Fabian Waleffe (1995, 1997) pointed out that the streaks can undergo an instability themselves, forming normal vortices, which can then be fed back into downstream vortices to close the loop. This self-sustaining regeneration mechanism plays a major role in the turbulence of parallel flows. The results for the transition to turbulence in viscous planar shear flows are summarized in Table 1. The flows are shear flow between parallel plates with a linear profile in the laminar case (plane Couette), pressure-driven flow between parallel plates with a parabolic profile (plane Poiseuille), and pressure-driven flow down a pipe, also with a parabolic profile (Hagen–Poiseuille). The various Reynolds numbers are defined as follows:

- Re_E *Energy stability*: as above.
- Re_G *Global stability*: up to this Reynolds number the flow is globally stable, and any perturbation will decay, perhaps after a long transient during which the energy can grow above the initial energy.
- Re_T Reynolds number near which experiments indicate a transition to turbulence.
- Re_L Reynolds number for linear instability.

For sufficiently low Re, all flows are linearly stable. For plane Couette flow and pipe flow, we know that the flows cannot be globally stable at Reynolds numbers above the given ones because of the existence of 3-dimensional stationary states or traveling waves

Flow	Re_E	Re_G	Re_T	Re_L
Plane Couette flow	20.7	125	310	∞
Plane Poiseuille flow	49.6	≈ 1000	1000	5772
Hagen–Poiseuille flow	81.5	1250	2250	∞

Table 1. Critical Reynolds numbers for parallel shear flows.

(Nagata, 1990; Ehrenstein & Koch, 1991; Clever & Busse 1997; Faisst & Eckhardt, 2003; Wedin & Kerswell, 2004). Two of the flows do not show a linear instability, and for the third one, it appears at values far above the ones where the transition to turbulence occurs. Incidentally, plane Poiseuille flow provides an example of a flow profile that is inviscidly stable by Rayleigh's criterion but becomes unstable when viscosity is included.

In all three cases, the value for transition to turbulence is somewhat uncertain because of the absence of a sharp transition (Boberg & Brosa, 1988; Darbyshire & Mullin, 1995; Bottin et al., 1998). There is evidence that this is connected with the formation of a chaotic saddle in phase space (Schmiegel & Eckhardt, 1997). Scanning an amplitude–Reynolds number plane for perturbations in a low-dimensional model indeed shows a sensitive dependence on initial conditions and huge variations in lifetimes for neighboring trajectories (see figure in color plate section). The lifetimes of perturbations are exponentially distributed, a clear signature of a chaotic saddle.

Turbulent shear flows: The energy dissipation in a turbulent shear flow can be related to the velocity difference U and the width L of the flow as

$$\varepsilon_V = c_\varepsilon(\mathrm{Re})U^3/L \qquad (6)$$

with a dissipation factor $c_\varepsilon(\mathrm{Re})$ that depends on the Reynolds number $\mathrm{Re} = UL/\nu$. For laminar parallel flows, $c_\varepsilon \sim 1/\mathrm{Re}$. For turbulent shear flows, the variational theories of Busse (1970), Doering & Constantin (1992), Kerswell (1997), and Nicodemus et al. (1997) show that in the limit of infinite Reynolds number, $c_\varepsilon(\mathrm{Re})$ can be rigorously bounded from above. The theory bounds c_ε through a variational functional with a background profile $\phi(y)$, with the constraint that the background profile has to be energy stable. The best bound is $c_\varepsilon \leq 0.0109$. Comparing with experiment, one notes that the observed values are lower and that for smooth boundaries, they tend to decrease for increasing Reynolds number.

The presence of a large-scale shear introduces an anisotropy into the flow, which also affects the turbulent statistics. While in an isotropic turbulent flow the odd moments of the normal derivative of the downstream component vanish, this is no longer the case in a turbulent shear flow. Dimensional estimates by Lumley (1967) suggest that the anisotropy should vanish like $1/\mathrm{Re}$, but experimental and numerical evidence indicates that the decay is much slower. Current efforts aim at extracting information about the relevant process from the dynamics of passive scalars: the scalar fields develop characteristic ramps and cliffs that can be related to the asymmetry in the distributions of the gradients (Schumacher & Sreenivasan, 2003).

Other shear flows: In viscoelastic fluids, the interaction between shear and internal degrees of freedom can also give rise to much more complicated instabilities for which Squire's theorem does not hold. The reduction of turbulent drag by small additions of long flexible polymers remains a fascinating phenomenon awaiting explanation (Lumley, 1969).

In astrophysics a combination of differential motion in the plasma and the presence of a magnetic field can give rise to instabilities that are responsible for the transport of angular momentum, thus solving a long-standing puzzle about the angular momentum distribution in galaxies (Balbus & Hawley, 1998).

BRUNO ECKHARDT

See also **Boundary layers; Hele-Shaw cell; Kelvin–Helmholtz instability; Taylor–Couette flow; Turbulence**

Further Reading

Balbus, S.A. & Hawley, J.F. 1998. Instability, turbulence, and enhanced transport in accretion disks. *Reviews of Modern Physics*, 70: 1–52

Balmforth, N.J. & Morrison, P.J. 1999. A necessary and sufficient instability condition for inviscid shear flow. *Studies in Applied Mathematics*, 102: 309–344

Boberg, L. & Brosa, U. 1988. Onset of turbulence in pipe. *Zeitschrift für Naturforschung*, 43a: 697–726

Bottin, S., Daviaud, F., Manneville, P. & Dauchot, O. 1998. Discontinuous transition to spatiotemporal intermittency in plane Couette flow. *Europhysics Letters*, 43: 171–176

Busse, F.H. 1970. Bounds for turbulent shear flow. *Journal of Fluid Mechanics*, 41: 219–240

Clever, R.M. & Busse, F.H. 1997. Tertiary and quaternary solutions for plane Couette flow. *Journal of Fluid Mechanics*, 344: 137–153

Darbyshire, A.G. & Mullin, T. 1995. Transition to turbulence in constant-mass-flux pipe flow. *Journal of Fluid Mechanics*, 289: 83–114

Doering, C.R. & Constantin, P. 1992. Energy dissipation in shear driven turbulence. *Physical Review Letters*, 69: 1648–1651

Drazin, P.G. & Reid, W.H. 1981. *Hydrodynamic Stability*, 2nd edition, Cambridge and New York: University Press, 2004

Ehrenstein, U. & Koch, W. 1991. Three-dimensional wavelike equilibrium states in plane Poiseuille flow. *Journal of Fluid Mechanics*, 228: 111–148

Faisst, H. & Eckhardt, B. 2003. Travelling waves in pipe flow. *Physical Review Letters*, 91: 224502

Grossmann, S. 2000. The onset of shear flow turbulence. *Reviews of Modern Physics*, 72: 603–618

Joseph, D.D. 1976. *Stability of fluid motion*, part I and II, New York: Springer

Kerswell, R.R. 1997. Variational bounds on shear-driven turbulence and turbulent Boussinesq convection. *Physica* D, 100: 355–376
Lumley, J.L. 1967. Similarity and the turbulent energy spectrum. *Physics of Fluids*, 10: 855–858
Lumley, J.L. 1969. Drag reduction by additives. *Annual Reviews of Fluid Mechanics*, 1: 367–384
Nagata, M. 1990. Three-dimensional finite-amplitude solutions in plane Couette flow: bifurcation from infinity. *Journal of Fluid Mechanics*, 217: 519–527
Nicodemus, R., Grossmann, S. & Holthaus, M. 1997. The background flow method, part I and II. *Journal of Fluid Mechanics*, 363: 281–300 and 301–323
Rosenhead, L. (editor). 1963. *Laminar Boundary Layers*, Oxford: Oxford University Press
Schmid, P. & Henningson, D.S. 2001. *Stability and Transition in Shear Flows*, Berlin and New York: Springer
Schmiegel, A. & Eckhardt, B. 1997. Fractal stability border in plane Couette flow. *Physical Review Letters*, 79: 5250–5253
Schumacher, J. & Sreenivasan, K.R. 2003. Geometric features of the mixing of passive scalars at high Schmidt numbers. *Physical Review Letters*, 91: 174501
Waleffe, F. 1995. Transition in shear flows. Nonlinear normality versus non-normal linearity. *Physics of Fluids*, 7: 3060–3066
Waleffe, F. 1997. On a self-sustaining process in shear flows. *Physics of Fluids*, 9: 883–900
Wedin, H. & Kerswell, R.R. 2004. Exact coherent structures in pipe flow: travelling wave solutions. *Journal of Fluid Mechanics*, in press

SHOCK WAVES

In classical gas dynamics, a shock wave is a sharp, stepwise increase of density, pressure, and temperature that propagates at a supersonic speed with respect to the fluid ahead of it while remaining subsonic with respect to the fluid behind it. The fluid entropy increases after passing through the shock. Shock waves can be formed as a result of numerous processes such as supersonic motion of bodies (aircraft, meteors, bullets) in the atmosphere (see Figure 1), explosions in atmosphere and ocean, collapse of bubbles in the course of cavitation, and gas flow out of a nozzle in rockets. Also, the propagation of a nonlinear nondispersive wave (simple wave), in which each point of the profile propagates at its own velocity, generally results in the shock formation.

The changes (jumps) of different physical (thermodynamic) quantities at a shock satisfy specific relations (boundary conditions) following from the conservation of mass, momentum, and energy (mechanical plus thermodynamical). In the reference frame of the shock front, they can be presented in the form

$$[\rho v_n] = 0, \ \left[p + \rho v_n^2\right] = 0, \left[\frac{v_n^2}{2} + w\right] = 0. \quad (1)$$

Here the square brackets denote the difference of the corresponding values at the shock; v_n is fluid velocity component normal to the front; w is enthalpy; p and ρ are, respectively, pressure and density. In the reference frame in which the fluid before the shock is immovable, v_n is $-V$ where V is the shock propagation velocity.

Figure 1. Shadowgraph showing shock waves produced by a Winchester 0.308 caliber bullet traveling through air at about Mach 2.5. (With courtesy from Ruprecht Nennstiel, Wiesbaden, Germany.)

The velocity of tangential fluid motion at the shock is continuous.

Equations (1) give, in particular,

$$w_2 - w_1 = \tfrac{1}{2}(U_2 + U_1)(p_2 - p_1), \quad (2)$$

where $U = 1/\rho$ is the specific volume, and subscripts 1 and 2 refer to the gas in front of and behind the shock. The enthalpy w can be expressed in terms of p and ρ via the thermodynamic equation of state. As a result, if the gas parameters p_1 and U_1 before the shock are given, Equation (2) determines the dependence between p_2 and U_2 called the Hugoniot adiabat. It differs from the Poisson adiabat that relates p and ρ in a perfect gas in which entropy is constant.

In a stable shock wave the entropy increases due to dissipative processes occurring inside the shock front, which is actually a transient layer of a finite thickness that grows with viscosity and thermal conductivity in the medium. If the transition region remains thin compared with the outer motion scale, it can be considered a discontinuity at which the boundary conditions (1) (which do not depend on the specific dissipation mechanism) remain valid. Note that if only thermal conductivity determines the shock front width (i.e., viscosity is neglected), the shock front can contain a discontinuity (isothermal jump) inside it.

General relations concerning shock waves are simplified in the important case of a polytropic gas in which p is proportional to ρ^γ, where $\gamma = c_p/c_v$, and c_p and c_v are heat capacities at constant pressure and volume, respectively. For example, in a very strong shock when $p_2 \gg p_1$ and the Mach number, $M = V.c \gg 1$ (c is the speed of sound), the ratios of gas densities and temperatures (T) behind and before the shock are

$$\frac{\rho_1}{\rho_2} = \frac{U_2}{U_1} = \frac{\gamma - 1}{\gamma + 1}, \ \frac{T_2}{T_1} = \frac{\gamma - 1}{\gamma + 1}\frac{p_2}{p_1}. \quad (3)$$

Note that the maximal gas compression remains finite.

Among the dynamic problems associated with shock waves is that of shock reflection from a hard wall. If the incidence is normal to the wall, the pressure increases to more than twice that in an incident wave (for a linear wall the pressure is doubled). If a shock of moderate strength is incident obliquely under a sufficiently large grazing angle, the shock is reflected before reaching the wall (Mach stem effect).

Another important problem is a spherical shock from an explosion. For very strong spherical waves, a self-similar solution exists, in which the shock front radius R increases as $t^{2/5}$, whereas the pressure and particle velocity decrease as R^{-3} and $R^{-3/2}$, respectively. For a curved shock front, the solution can be found by an approximation: local shock curvature at each moment is related to the rays normal to the front, and the fluid flows along these ray tubes as in channels of variable width (Whitham, 1974).

A more general description of nonstationary flows containing shocks can be achieved for one-dimensional motions such as the waves in a tube excited by a piston moving at its end. The one-dimensional equations of motion for an ideal gas can be rewritten in terms of two Riemann invariants, $J_{\pm}$:

$$\left[\frac{\partial}{\partial t} + (v \pm c)\frac{\partial}{\partial x}\right] J_{\pm} = 0,$$

$$J_{\pm} = v \pm \int \frac{\mathrm{d}p}{\rho c} = v \pm \frac{2c}{\gamma - 1} \quad (4)$$

(the latter expressions are written for a polytropic gas). Here $c(\rho) = (\mathrm{d}p/\mathrm{d}\rho)^{1/2} = (\gamma p/\rho)^{1/2}$ is the local speed of sound. Correspondingly, there are two families of trajectories-characteristics, $\mathrm{d}x/\mathrm{d}t = v \pm c$ on the (x, t) plane, along which small perturbations that are linear sound waves propagate. In a particular case when one of the invariants is constant, there is a progressive simple (Riemann) wave that travels at a local velocity $v \pm c$. In such a wave, the variables are related by $v = \pm \int \mathrm{d}p/\rho c$. Each point of the simple wave profile propagates at its own, constant velocity, and if the velocity decreases with x, at some moment the motion becomes multivalued (the wave breaks). In gas dynamics this is physically impossible, which means the formation of a shock wave at which the energy of the continuous part of the wave dissipates.

An important particular case is that of small nonlinearity when $p(\rho) \approx p_0 + c_0^2\rho'$, where p_0 and c_0 refer to the equilibrium state (in the absence of the wave), and ρ' is density variation in the wave. This corresponds to nonlinear acoustics where the velocity of a weak shock wave is approximately the average of the linear sound speeds in front of and behind the shock. In particular, an initially sinusoidal wave transforms into a sawtooth one.

Simple waves may exist in different nondispersive, nonlinear physical systems. One example is a nonlinear surface wave in shallow water when the characteristic wavelength is much larger than the depth of the water layer, h. Such a wave propagates at the local velocity of $c(\eta) = \sqrt{g(h + \eta)}$, where g is gravitational acceleration and $\eta(x, t)$ is the local water surface displacement with respect to its unperturbed level, h. In general, this wave breaks at a finite distance so that the long-wave approximation becomes invalid in the vicinity of the breaking point. The breaking can have different outcomes. Due to dispersion, the wave can generate oscillations that eventually become solitons, or, at larger steepness, the wave crest may break, possibly forming a turbulent front (bore), somewhat similar to a shock.

Magnetohydrodynamic Shock Waves

Shock waves can form in plasma. If the plasma is sufficiently dense, it can be treated as a compressible fluid that has electrical conductivity. When such a fluid moves in a magnetic field, the motion induces electric currents. Interaction between the currents and the magnetic field can significantly affect the fluid motion. Interaction between hydrodynamic and electromagnetic phenomena is the subject of *magnetohydrodynamics*. These phenomena are important in, for example, astrophysics, where at large-scale motions, space plasma typically behaves as a conducting fluid.

In a linear approximation, magnetohydrodynamic (MHD) waves are classified as "slow" and "fast" magnetic sound in which fluid compressibility is significant, and Alfvén waves depending only on magnetic field. In these waves magnetic perturbations are polarized perpendicular to the basic constant magnetic intensity vector $\boldsymbol{H}$ and group velocity is directed along $\boldsymbol{H}$. Finite-amplitude magnetic sound is distorted with a possible formation of MHD shocks.

As in nonmagnetic gas dynamics, parameters of MHD shock waves can be determined from boundary conditions at the discontinuity, which differ from the Rankine–Hugoniot conditions in that they include the magnetic field at both sides of the shock. The corresponding generalization of the Hugoniot adiabat reads (in the CGS system)

$$(\varepsilon_2 - \varepsilon_1) + \frac{1}{2}(p_2 + p_1)(U_2 - U_1) + \frac{1}{16\pi}(U_2 - U_1)(H_{\tau 2} - H_{\tau 1})^2 = 0, \quad (5)$$

where H_{τ} is the component tangential to the shock front. In fluids with a positive thermal expansion coefficient, both pressure and density always increase at the shock, as in classical gas dynamics.

The nonshock discontinuities in MHD include tangential discontinuities, in which the vectors of fluid velocity and magnetic field are parallel to the discontinuity plane (in the comoving reference frame), and rotational (Alfvén) discontinuities in which the normal velocity component of velocity is nonzero but continuous whereas the vector $\boldsymbol{H}$ rotates around the normal with constant absolute value.

In plasma, dispersion can cause an oscillating shock structure and solitary wave formation. If plasma is rarefied so that the free path of electrons exceeds the shock front thickness, the so-called noncollisional shocks may exist.

Electromagnetic Shock Waves

Another class of shock waves occurs in media that can be macroscopically immovable but have nonlinear electromagnetic parameters; for example, the dependence between the vectors of magnetic induction and the magnetic field, $\boldsymbol{B}(\boldsymbol{H})$, or between their electric counterparts, $\boldsymbol{D}(\boldsymbol{E})$, are nonlinear. If the nonlinearity is relatively strong and dispersion effects are small, electromagnetic waves can propagate as simple waves, resulting in the formation of electromagnetic shocks.

Boundary conditions at a shock can be obtained by integrating Maxwell's equations over the shock transition layer, to give

$$[\boldsymbol{n} \times (\boldsymbol{E}_2 - \boldsymbol{E}_1)] = \frac{U_n}{c_0}(\boldsymbol{B}_2 - \boldsymbol{B}_1), \qquad (6)$$

$$[\boldsymbol{n} \times (\boldsymbol{H}_2 - \boldsymbol{H}_1)] = \frac{U_n}{c_0}(\boldsymbol{D}_2 - \boldsymbol{D}_1),$$

$$B_{n2} = B_{n1}, D_{n2} = D_{n1},$$

where $\boldsymbol{n}$ is a normal to the discontinuity surface, c_0 is the light speed in vacuum, and U_n is the normal component of shock velocity. Typically only either magnetic or electric properties of the medium are nonlinear. In the former case, an expression for the shock velocity is

$$\frac{\varepsilon U_n^2}{c^2} = \frac{H_{\tau 2} - H_{\tau 1}}{B_{\tau 2} - B_{\tau 1}}. \qquad (7)$$

Electromagnetic shock waves are observed in ferrites, ferroelectrics, and semiconductors, and they have been used for construction of powerful impulse generators. In nonlinear dispersive systems, such as nonlinear transmission lines, the shock transition can be oscillating, and as dissipation approaches zero, it becomes a solitary electromagnetic wave.

Lev Ostrovsky

See also **Alfvén waves; Characterisitics; Cherenkov radiation; Dimensional analysis; Explosions; Jump phenomena; Magnetohydrodynamics; Nonlinear acoustics**

Further Reading

Courant, R. & Friedrichs, K.O. 1948. *Supersonic Flow and Shock Waves*, reprinted New York: Springer, 1992

Gaponov, G., Ostrovsky, L. & Freidman, G. 1967. Electromagnetic shock waves (a review). *Radiophysics and Quantum Electronics*, 10: 9–10

Landau, L. & Lifshitz, E. 1960. *Electrodynamics of Continuous Media*, Oxford and New York, Pergamon Press

Landau, L. & Lifshitz, E. 1987. *Fluid Mechanics*, 2nd edition, Oxford: Pergamon Press

Whitham, G.B. 1974. *Linear and Nonlinear Waves*, New York: Wiley

SHOOTING METHOD

See **Phase plane**

SIERPINSKI GASKET

See **Order from chaos**

SIGNALING PROBLEM

See **Wave stability and instability**

SINAI BILLIARD

See **Billiards**

SINAI–RUELLE–BOWEN MEASURES

What is a natural ivariant measure? In a probabilistic approach to dynamical systems, one seeks to understand average or almost-sure behavior, particularly in the limit as time tends to infinity. In these studies, it is often fruitful to have a stationary process, or equivalently, an *invariant probability measure*. Consider a map f on a bounded domain U of a Euclidean space or a manifold with $f(U) \subset U$. (We will restrict ourselves to the discrete-time case, but the discussion is equally valid for continuous-time.) In general, f admits infinitely many mutually singular invariant probability measures, and the ergodic theory of (f, μ) depends on the choice of μ. The purpose of this entry is to discuss what constitutes a "natural" invariant measure.

Throughout this discussion, we will adopt the view that only properties that hold on positive Lebesgue measure sets are observable. A first attempt to characterize natural invariant measures is to require that they have densities. Thus for a Hamiltonian system, Liouville measure is regarded as natural. This criterion runs into difficulties with dissipative systems. Suppose that f is volume-decreasing with an attractor $\Lambda = \cap_{n \geq 0} f^n(U)$. Since there is no recurrence behavior on $U \setminus \Lambda$, by the Poincaré recurrence theorem, all the invariant probability measures are supported on Λ. Observe that with f decreasing volume, Λ must necessarily have Lebesgue measure zero. Thus f cannot have an invariant density.

For dissipative dynamical systems, then, we must relax the idea of a natural invariant measure from one that has a density to one that reflects the properties of positive Lebesgue measure sets. We discuss below three closely related sets of ideas that go in this direction. In less rigorous discussions, the first two are often confused, even though as we will see, they are different substantively and mathematically.

SRB Measures

Sinai–Ruelle–Bowen measures or *SRB measures* were first introduced for Anosov systems and Axiom A attractors; see Sinai (1972), Bowen (1975), and Ruelle (1978). The idea was later extended to the non-uniform hyperbolic setting (*See* **Horseshoes and hyperbolicity in dynamical systems**), and the name SRB measure was introduced in the review article by Eckmann-Ruelle (1985).

The idea is as follows: SRB measures live on strange attractors. The situation being chaotic, there are positive Lyapunov exponents and unstable manifolds. In the same way that the next best scenario to having a differentiable function is to have partial derivatives, if the dissipative nature of a map prevents it from having an invariant measure that is smooth, then the closest approximation would be one that is smooth in certain directions. The expanding property of a map smoothes out invariant measures along unstable manifolds.

We now give the precise definition. Consider a C^2 invertible map f and an f-invariant Borel probability measure μ. We assume that f has positive Lyapunov exponents and hence unstable manifolds μ-almost everywhere. On each k-dimensional unstable leaf γ, let m_γ denote the k-dimensional Lebesgue measure. Then we say μ is an SRB measure if the conditional measures of μ on unstable leaves have densities with respect to the measures m_γ.

The geometric definition of SRB measures given above turns out to be equivalent to the following "variational principle": SRB measures are exactly those invariant measures for which the Kolmogorov–Sinai entropy of the system is equal to the sum of its positive Lyapunov exponents. (For other invariant measures, entropy is smaller.) If there are no zero Lyapunov exponents, there is a structure theorem that says that SRB measures have at most a countable number of ergodic components, and each ergodic component is mixing up to certain permutations (*See* **Horseshoes and hyperbolicity in dynamical systems**). Relations to other notions of natural invariant measures will be mentioned as we go along.

We explained earlier that SRB measures is a notion associated with strange attractors. This does not mean that every strange attractor necessarily has an SRB measure. In practice, one often assumes that it does, but mathematically, this question is far from resolved. In fact, not many attractors have been rigorously proved to have SRB measures. The main examples are Axiom A attractors and a class of attractors with one direction of instability including certain Hénon attractors (see Young, 2002).

Physical Measures

Let f be a map and μ an invariant probability measure. We say μ is a *physical measure* if there is a positive Lebesgue measure set V in the phase space such that for every continuous observable φ, we have

$$\frac{1}{n}\sum_{i=0}^{n-1}\varphi(f^i x) \to \int \varphi \, \mathrm{d}\mu$$

for every $x \in V$. Thus, physical measures are those invariant measures that can be *observed*: Suppose we pick an initial condition $x \in V$ and plot the first n points of its trajectory. The resulting picture can be seen as that of a measure giving weight $1/n$ to each one of the points $x, f(x), f^2(x), \cdots, f^{n-1}(x)$. For n large, it is a good approximation of μ.

One of the reasons—some would say the main reason—why SRB measures are important is that all ergodic SRB measures with no zero Lyapunov exponents are physical measures.

Not all physical measures are SRB measures, however. For examples, point masses on attractive fixed points are physical measures. Unlike SRB measures, which are associated with chaotic behavior and have rich, well-defined structures, physical measures have no identifiable characteristics aside from the fact that they are observable.

Zero-noise Limits

If one subscribes to the view that the world is inherently noisy, then the following notion proposed by Kolmogorov is perhaps the most relevant notion of observability. Let f be a map on a bounded domain, and let $P^\varepsilon(\cdot\,|\,\cdot)$, $\varepsilon > 0$ be a family of Markov chains representing random perturbations of f; that is, the transition probabilities $P^\varepsilon(\cdot\,|\,x)$ have the following interpretation: instead of jumping from x to $f(x)$, $P^\varepsilon(\cdot\,|\,x)$ gives the distribution of possible images starting from x. We may think of it as the uniform distribution on an ε-disk centered at $f(x)$ or a Gaussian with ε variance. Let μ_ε be the marginal of the stationary measure of the process defined by $P^\varepsilon(\cdot\,|\,\cdot)$, and let $\varepsilon \to 0$. Limit points of μ_ε are called *zero-noise limits*.

Unlike SRB measures or physical measures, the existence of which leads to unresolved questions, systems defined on compact regions always have zero-noise limits. It is hoped that in most situations, they coincide with the other notions of natural invariant measures. This has been proved for Axiom A attractors and a handful of other examples.

Lai-Sang Young

See also **Cat map; Horseshoes and hyperbolicity in dynamical systems; Measures; Phase space**

Further Reading

Bowen, R. 1975. *Equilibrium States and the Ergodic Theory of Anosov Diffeomorphisms*, Berlin and New York: Springer

Eckmann, J.-P. & Ruelle, D. 1985. Ergodic theory of chaos and strange attractors. *Reviews of Modern Physics*, 57: 617–656
Ruelle, D. 1978. *Thermodynamic Formalism*, Reading, MA: Addison-Wesley
Sinai, Ya. G. 1972. Gibbs measures in ergodic theory. *Russian Mathematical Surveys*, 27: 21–69
Smale, S. 1967. Differentiable dynamical systems. *Bulletin of the American Mathematical Society*, 73: 747–817
Young, L.-S. 1993. Ergodic theory of differentiable dynamical systems. *Proceedings of the NATO Advanced Study Institute* held in Hillerød, June 20–July 2, 1993, edited by B. Branner & P. Hjorth, Dordrecht: Kluwer
Young, L.-S. 2002. What are SRB measures, and which dynamical systems have them? *Journal of Statistical Physics*, 108: 733–754

SINE-GORDON EQUATION

In normalized units, the classical sine-Gordon (SG) equation is the nonlinear partial differential equation

$$\frac{\partial^2\phi}{\partial x^2} - \frac{\partial^2\phi}{\partial t^2} = \sin\phi, \tag{1}$$

where $\phi(x,t)$ is a field in two dimensions—space and time. Since its introduction in 1939 by Yakov Frenkel and Tatiana Kontorova as a model for the dynamics of crystal dislocations (Frenkel & Kontorova, 1939), this equation has found a variety of applications, including Bloch wall dynamics in ferromagnetics and ferroelectrics, fluxon propagation in long Josephson (superconducting) junctions, self-induced transparency in nonlinear optics, spin waves in the A-phase of liquid ^{3}He at temperatures near to 2.6 mK, and a simple, one-dimensional model for elementary particles (Bullough, 1977; Scott, 2003). The name stems from the linear Klein–Gordon equation $\partial^2\phi\,/\,\partial x^2 - \partial^2\phi\,/\,\mathrm{d}t^2 = \phi$, where the "joke" may be due to Martin Kruskal or it may not (Bullough & Caudrey, 1980).

The SG equation can be physically modeled as a linear array of weakly coupled pendula, suggesting traveling-wave solutions of the form

$$\phi = 4\arctan\exp\left\{\pm(x-vt)\,/\sqrt{1-v^2}\right\} \tag{2}$$

with $v < 1$. These solutions are called a kink and an antikink, respectively, since at $x = -\infty, \phi \to 0$, and at $x = +\infty$, $\phi \to 2\pi$ for the $+$ sign, while at $x = -\infty, \phi \to 2\pi$ and at $x = +\infty$, $\phi \to 0$ for the $-$ sign. In the context of the physical model, ϕ twists from 0 to 2π for the kink and untwists from 2π to 0 for the antikink. Alternatively $\phi\,/\,2\pi$ has "topological charge" $+1$ for each kink and -1 for each antikink. Each of these is a one-soliton solution, but there are also N-soliton solutions for the boundary conditions $\phi \to 0$ (mod 2π) as $x \to \pm\infty$. (In "light-cone coordinates" (see below) the N-soliton solutions were given by Caudrey et al. (1973); see the detailed history in Bullough & Caudrey (1980)).

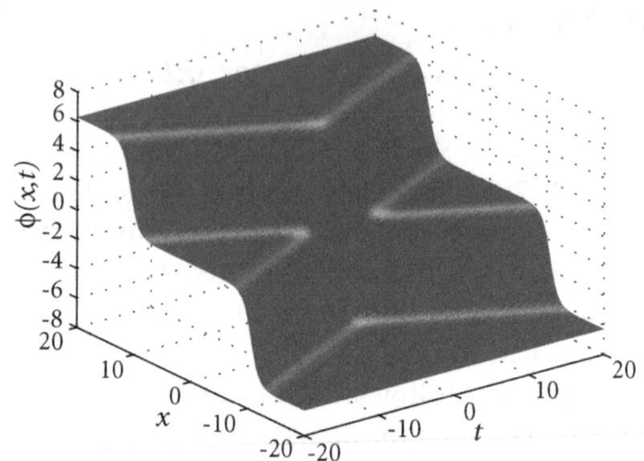

Figure 1. A kink-kink solution of the SG equation plotted from Equation (3) with $v = 0.5$.

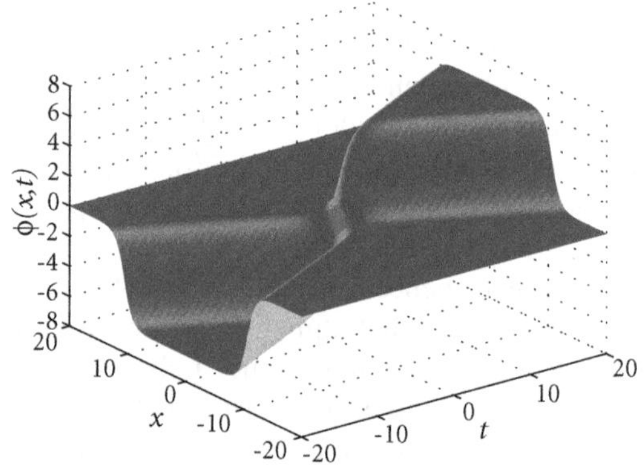

Figure 2. A kink-antikink solution of the SG equation plotted from Equation (4) with $v = 0.5$.

The simplest 2-soliton solution of SG Equation (1) is the kink-kink collision given by Perring & Skyrme (1962) as

$$\phi(x,t) = 4\arctan\left[\frac{v\sinh(x/\sqrt{1-v^2})}{\cosh\,(vt/\sqrt{1-v^2})}\right] \tag{3}$$

as shown in Figure 1 for $v = 0.5$. Similarly, Figure 2 shows a kink-antikink collision, which is plotted from

$$\phi(x,t) = 4\arctan\left[\frac{\sinh(vt/\sqrt{1-v^2})}{v\cosh\,(x/\sqrt{1-v^2})}\right] \tag{4}$$

also with $v = 0.5$.

The kink-antikink equation takes an interesting form if the velocity parameter (v) is allowed to be imaginary. For example, setting

$$v = \mathrm{i}\omega/\sqrt{1-\omega^2}, \quad \omega < 1,$$

Equation (4) becomes the stationary breather (or "bion")

$$\phi(x,t) = 4\arctan\left[\left(\frac{\sqrt{1-\omega^2}}{\omega}\right)\frac{\sin\,\omega t}{\cosh\sqrt{1-\omega^2}x}\right], \tag{5}$$

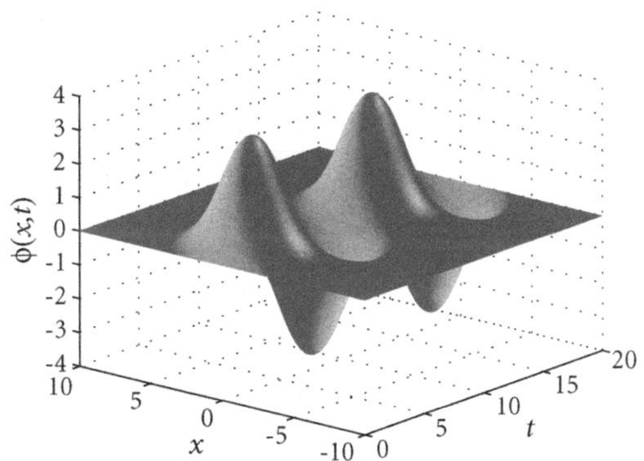

Figure 3. A stationary breather plotted from Equation (5) with $\omega = \pi/5$.

a localized, oscillating solution that is plotted in Figure 3 for $\omega = \pi/5$. Because Equation (1) is invariant under the Lorentz transformation

$$x' = (x - vt)/\sqrt{1-v^2}$$

and

$$t' = (t - xv)/\sqrt{1-v^2}, \tag{6}$$

this stationary breather can be boosted into a moving frame. Thus,

$$\phi(x,t) = 4\arctan\left\{\frac{\sqrt{1-\omega^2}}{\omega}\sin\left[\frac{\omega(t-v_{\mathrm{e}}x)}{\sqrt{1-v_{\mathrm{e}}^2}}\right]\times\operatorname{sech}\left[\frac{\sqrt{1-\omega^2}(x-v_{\mathrm{e}}t)}{\sqrt{1-v_{\mathrm{e}}^2}}\right]\right\}$$

is an exact solution of the SG equation moving with an envelope velocity (v_{e}) that is equal to the reciprocal of its carrier velocity $v_{\mathrm{c}} = 1/v_{\mathrm{e}}$.

Because $\phi(x,t)$ is defined for each value of x, the Hamiltonian description is infinite dimensional. A standard form of the Hamiltonian (or energy) is (Bullough & Timonen, 1995)

$$H = \int\left[\frac{1}{2}\Pi^2(x,t) + \frac{1}{2}(\partial\phi/\partial x)^2 + (1-\cos\phi)\right]\mathrm{d}x. \tag{7}$$

Hamilton's equations then take the form

$$\frac{\delta H}{\delta\phi} = -\dot{\Pi} \text{ and } \frac{\delta H}{\delta\Pi} = \dot{\phi}, \tag{8}$$

where the δs indicate functional (or Frechêt) derivatives. The second of Equations (8) yields $\dot{\phi} = \Pi$ and the first yields $-\ddot{\phi} = -\dot{\Pi} = -\phi_{xx} + \sin\phi$, which is the SG equation. Computing H in the rest frame, one finds rest masses equal to 8 for both kinks and antikinks, while for stationary breathers the rest masses are $16\sqrt{1-\omega^2}$. Evidently, a stationary breather is a bound pair of a kink and an antikink oscillating at frequency ω.

The SG equation is a completely integrable Hamiltonian system, which means that the number of independent constants of motion is equal to the number of degrees of freedom (Liouville's theorem). As the number of degrees of freedom is infinite (through the labels x), the theorem is not obvious, but an infinite number of commuting and independent constant action variables can be found. Equivalently, solutions of the SG equation can be obtained through a Bäcklund transformation (BT) or by using an inverse scattering method (ISM).

In BT and ISM studies, it is convenient to transform the independent variables as

$$x \to \xi = (x+t)/2 \text{ and } t \to \tau = (x-t)/2,$$

whereupon Equation (1) becomes

$$\frac{\partial^2\phi}{\partial\xi\,\partial\tau} = \sin\phi. \tag{9}$$

These new variables (ξ and τ) are sometimes called "light-cone" coordinates because they point in the directions of characteristics in the (x,t)-plane.

The SG equation first appeared in light-cone coordinates in the mathematical study of surfaces of constant negative Gaussian curvature by Albert Bäcklund (Lamb, 1976). In 1883, he discovered the first BT, which can be written as

$$\begin{aligned}\phi'_x &= \phi_x + 2k\sin(\phi+\phi')/2,\\ \phi'_t &= -\phi_t + 2k^{-1}\sin(\phi-\phi')/2\end{aligned} \tag{10}$$

for any real value of the parameter k. The integrability condition $\phi'_{xt} = \phi'_{tx}$ implies both $\phi_{xt} = \sin\phi$ and $\phi'_{xt} = \sin\phi'$; thus, Equations (10) transform a known solution ϕ of the SG in light-cone coordinates to a second solution ϕ'. As $\phi = 0$ is a solution, the single-kink solution is found to be $\phi' = 4\arctan\exp(kx - k^{-1}t)$ for this real value of k.

Bullough (1980), for example, shows how the BT Equation (10) becomes the Lax pair for an ISM analysis first written down by Ablowitz et al. (1973) (AKNS). Independently, Takhtajan & Faddeev (1974) gave the corresponding expressions for the covariant SG Equation (1). AKNS's Lax pair is

$$\frac{\partial}{\partial\xi}\begin{bmatrix}\psi_1\\ \psi_2\end{bmatrix} = \begin{bmatrix}-\mathrm{i}\lambda & -\phi_\xi/2\\ \phi_\xi/2 & \mathrm{i}\lambda\end{bmatrix}\begin{bmatrix}\psi_1\\ \psi_2\end{bmatrix},$$

where ψ_1 and ψ_2 are components of the scattering solution and λ is a complex scattering parameter together with the τ-dependence

$$\frac{\partial}{\partial\tau}\begin{bmatrix}\psi_1\\ \psi_2\end{bmatrix} = \frac{\mathrm{i}}{4\lambda}\begin{bmatrix}\cos\phi & \sin\phi\\ \sin\phi & -\cos\phi\end{bmatrix}\begin{bmatrix}\psi_1\\ \psi_2\end{bmatrix}. \tag{11}$$

It is readily checked that the cross-derivative condition

$$\frac{\partial}{\partial\tau}\frac{\partial}{\partial\xi}\begin{bmatrix}\psi_1\\ \psi_2\end{bmatrix} = \frac{\partial}{\partial\xi}\frac{\partial}{\partial\tau}\begin{bmatrix}\psi_1\\ \psi_2\end{bmatrix}$$

implies Equation (9), thus providing the basis for an ISM analysis.

Assuming that $\phi \to 0 \pmod{2\pi}$ as $\xi \to \pm\infty$, the time evolution matrix defined in Equation (11) takes the simple asymptotic form

$$\frac{\mathrm{i}}{4\lambda}\begin{bmatrix} 1 & 0 \\ 0 & -1 \end{bmatrix};$$

thus the reflection coefficient and the residues r_n of its upper-half-plane-poles (in the λ plane) evolve with time as

$$b(\lambda, \tau) = b(\lambda, 0)\mathrm{e}^{-\mathrm{i}\tau/2\lambda},$$
$$r_n(\tau) = r_n(0)\mathrm{e}^{-\mathrm{i}\tau/2\lambda_n},$$

where $b(\lambda, \tau)$ is the reflection coefficient of the solution and n is an index that runs over the number (N) of soliton components.

The evolved solution of Equation (9) turns out to be

$$\frac{\partial\phi}{\partial\xi}(\xi, \tau) = 4K(\xi, \xi; \tau), \tag{12}$$

where $K(\xi, z; \tau)$ is a solution of the (Gel'fand–Levitan–Marchenko) integral equation

$$K(\xi, z; \tau) = B^*(\xi + z; \tau) - \int_\xi^\infty \int_\xi^\infty K(\xi, y; \tau) B(y + y'; \tau) \times B^*(z + y'; \tau)\mathrm{d}y\,\mathrm{d}y' \tag{13}$$

with $z > \xi$, and

$$B(\xi + z; \tau) \equiv \frac{1}{2\pi}\int_{-\infty}^{\infty} b(\lambda, 0)\mathrm{e}^{\mathrm{i}\lambda(\xi+z)-\mathrm{i}\tau/2\lambda}\,\mathrm{d}\lambda - \mathrm{i}\sum_{n=1}^{N} r_n(0)\mathrm{e}^{\mathrm{i}\lambda_n(\xi+z)-\mathrm{i}\tau/2\lambda_n} \tag{14}$$

is determined from a scattering analysis of the initial conditions.

The simplest example of this ISM formulation is obtained by assuming that the initial potential is reflectionless $(b(\lambda, \tau) = 0)$ and has but one bound state $(N = 1)$, corresponding to a single pole at $(\lambda_1 = \mathrm{i}\kappa)$ on the imaginary axis of the upper half λ-plane with residue $r_1 = \pm 2\mathrm{i}\kappa$. Thus, from Equation (14)

$$B(\xi + z; \tau) = \pm 2\kappa\mathrm{e}^{-\kappa(\xi+z)-\tau/2\kappa},$$

so Equation (13) takes the form

$$K(\xi, z; \tau) = \pm\, 2\kappa\mathrm{e}^{-\kappa(\xi+z)-\tau/2\kappa} - 4\kappa^2\mathrm{e}^{-\tau/\kappa}\int_\xi^\infty\int_\xi^\infty K(\xi, y; \tau)\mathrm{e}^{-\kappa(y+y')} \times \mathrm{e}^{-\kappa(z+y')}\,\mathrm{d}y\,\mathrm{d}y'.$$

As $K(\xi, z; \tau) \propto \exp(-\kappa z)$, this integral equation is solved for

$$K(\xi, z; \tau) = \pm\frac{2\kappa\exp[-\kappa(\xi + z) - \tau/2\kappa]}{1 + \exp(-4\kappa\xi - \tau/\kappa)}.$$

Thus, from Equation (12)

$$\frac{\partial\phi}{\partial\xi} = \pm 4K(\xi, \xi; \tau) = \pm 4\kappa\,\mathrm{sech}(2\kappa\xi + \tau/2\kappa),$$

which integrates to

$$\phi(\xi, \tau) = 4\arctan\{\exp[\pm(2\kappa\xi + \tau/2\kappa)]\}.$$

Finally, one can transform back to the laboratory (x, t) coordinates, whereupon the corresponding solution of Equation (1) is the kink or antikink of Equation (2) with its laboratory velocity identified as

$$v = \frac{1 - 4\kappa^2}{1 + 4\kappa^2} = \frac{1 + 4\lambda_1^2}{1 - 4\lambda_1^2}. \tag{15}$$

With an appropriate choice of the constants c_1 and c_2, the kink-antikink solution of Equation (4) can be generated in a similar manner from the assumption that

$$B(\xi + z; 0) = c_1\mathrm{e}^{-\kappa_1(\xi+z)} + c_2\mathrm{e}^{-\kappa_2(\xi+z)}, \tag{16}$$

where

$$\kappa_1 = \frac{1}{2}\sqrt{\frac{1+v}{1-v}} \quad\text{and}\quad \kappa_2 = \frac{1}{2}\sqrt{\frac{1-v}{1+v}}.$$

These integrable properties of the SG equation are shared by the sinh-Gordon equation and the Liouville equation for which the sine function in Equation (1) is replaced by a hyperbolic-sine function and by an exponential function, respectively—although the Liouville equation has no inverse scattering solution. Perhaps the most striking property of SG solitons, however, is their topological stability, which is evidenced by their nonzero rest masses. This property is carried into three space dimensions by the theory of skyrmions.

Equation (1) can be embedded in a more general framework of inverse-scattering methods, with the Hamiltonian expressed in terms of finite numbers of kinks, antikinks, and breathers, in addition to a continuous spectrum of radiation (Bullough, 1980). In the mid-1970s, Roger Dashen, Brosl Hasslacher, and Andre Neveu used semiclassical quantum methods to study this system, finding the mass (or energy) spectrum of the breather of Equation (5) to be

$$M_n = 16\sin\left(\frac{n}{16}\right),$$

where n is a positive integer less than 8π (Dashen et al., 1975). They correctly conjectured this "DHN spectrum" to be exact—a result of theoretical significance because it shows that semiclassical analyses of nonlinear field theories can provide useful information about exact mass spectra.

The statistical mechanics of the SG equation is carried out in both the quantum and classical cases in Bullough & Timonen (1995), with a further appraisal of the action-angle variables in one space dimension. Bullough (1977) develops the theory of double and multiple SG equations (with $\sin\phi$ in Equation (1) replaced by $\pm[\sin\phi + (\sin\phi\,/\,2)\,/\,2]$, $\sin\phi + (\sin\phi\,/\,3)\,/\,3 + 2(\sin 2\phi\,/\,3)\,/\,3$, etc.) in connection with models for degenerate self-induced transparency of metal vapors and of spin waves of the B-phase of liquid ^{3}He near 2.6 mK.

ROBIN BULLOUGH

See also **Bäcklund transformations; Hamiltonian systems; Inverse scattering method or transform; Laboratory models of nonlinear waves; Long Josephson junctions; Maxwell–Bloch equations; Pendulum; Skyrmions**

Further Reading

Ablowitz, M.J., Kaup, D.J., Newell, A.C. & Segur, H. 1973. The initial value solution for the sine-Gordon equation. *Physical Review Letters*, 30: 1262–1264

Bullough, R.K. 1977. Solitons. In *Interaction of Radiation with Condensed Matter*, vol.1, Vienna: International Atomic Energy Agency; pp. 381–469

Bullough, R.K. 1980. Solitons: inverse scattering and its applications. Five lectures in the Proceedings of the NATO Advanced Study Institute *Bifurcation phenomena in mathematical physics and related phenomena Cargèse, Corsica, June 1977*, edited by D. Bessis and C. Bardos, Dordrecht: Reidel, pp. 295–349

Bullough, R.K. 2000. The optical solitons of QE1 are the BEC of QE14: has the quantum soliton arrived? Special Foundation Lecture: a personal view. 14th National Quantum Electronics and Photonics Conference, Owens Park, University of Manchester, 8 September 1999. *Journal of Modern Optics*, 47(11): 2029–2065 (Erratum. 2001. *Journal of Modern Optics*, 48(4): 747–748)

Bullough, R.K. & Caudrey, P.J. 1980. The soliton and its history. *Topics in Current Physics*, 17: 1–64

Bullough, R.K. & Timonen, J. 1981. Breather contributions to the dynamical form factors of the sine-Gordon systems $CsNiF_3$ and $(CH_3)_4(NiMnCl)_3$ (TMMC). *Physics Letters*, 82A(82): 182–186

Bullough, R.K. & Timonen, J.T. 1995. Quantum and classical integrable models and statistical mechanics. In *Statistical Mechanics and Field Theory*, edited by V.V. Bazhanov and C.J. Burden, Singapore: World Scientific, pp. 336–414 (Note the 'Solitons' map on pages 358–359)

Caudrey, P.J. 1989. In *Soliton Theory: A Survey of Results*, edited by A.P. Fordy, Manchester: Manchester University Press, and references therein

Caudrey, P.J., Gibbon, J.D., Eilbeck, J.C. & Bullough, R.K. 1973. Exact multi-soliton solutions of the self-induced transparency and sine-Gordon equations. *Physical Review Letters*, 30: 237–238

Dashen, R.F., Hasslacher, B. & Neveu, A. 1975. Semiclassical bound states in an asymptotically free theory. *Physical Review D*, 12: 2443–2458

Dodd, R.K. & Bullough, R.K. 1979. The generalised Marchenko equation and the canonical structure of the AKNS inverse method. *Physica Scripta*, 20: 514–530

Frenkel, J. & Kontorova, T. 1939. On the theory of plastic deformation and twinning. *Journal of Physics (USSR)* 1: 137–149

Lamb, G.L., Jr. 1976. Bäcklund transforms at the turn of the century. In *Bäcklund Transforms*, edited by R.M. Miura, New York: Springer

Perring, J.K. & Skyrme, T.R.H. 1962. A model unified field equation. *Nuclear Physics*, 31: 550–555

Scott, A.C. 2003. *Nonlinear Science: Emergence and Dynamics of Coherent Structures*, 2nd edition, Oxford and New York: Oxford University Press

Takhtajan, L.A. & Faddeev, L.D. 1974. Essentially nonlinear one-dimensional model of classical field theory. *Theoretical and Mathematical Physics*, 47: 1046–1057

SINGULAR PERTURBATION THEORY

See **Perturbation theory**

SINGULAR POINTS

See **Phase space**

SINGULARITY THEORY

Does the view from your window encompass a range of hills or mountains? If you are so blessed, you will see immediately both of the persistent singularities that can be observed generically; they are indicated on the mountains sketched in Figure 1, a section of a remote but important range called the Transversal Alps. The points marked "country" are known as fold singularities. Actually the outline of the mountains, which in reality is a mapping of a surface contour onto the retina of the eye, is composed of an infinite number of fold singularities. The two points marked "western" are known as cusp singularities. They occur where a line of folds along the edge of a mountain meets another line of folds along the gully between the mountains.

If the view from your office is more urban in character—of faces in the street, say—you will probably see many more of these two fundamental singularities. You must search a lot harder to find other types of singularities, or purposefully create one, because all other singularities dissolve under the slightest perturbation into either a fold or a cusp. This remarkable fact was first proved by Whitney (1955) whose fundamental discoveries about singularities of differentiable mappings were developed into catastrophe theory and toolkits for treating bifurcation problems with parameters, by mathematicians such as Mather (in a series of very technical papers from 1968 to 1971), Thom (1972), Martinet (1982), Arnol'd et al. (1985), and Golubitsky & Schaeffer (1985).

Singularity theory is not secret mathematicians' business though, and a more apt name for the whole business would be "theory and applications of

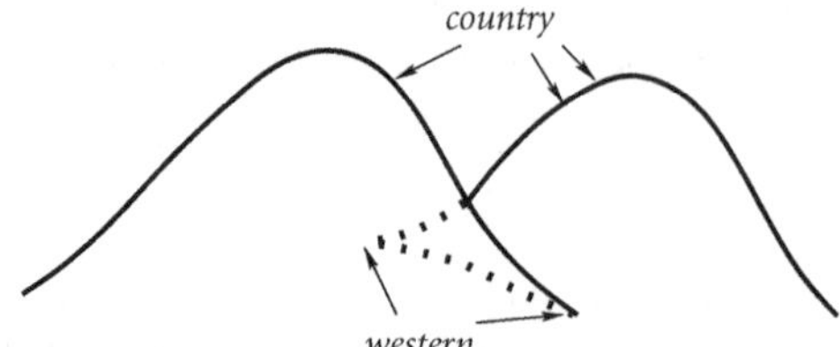

Figure 1. The Transversal Alps.

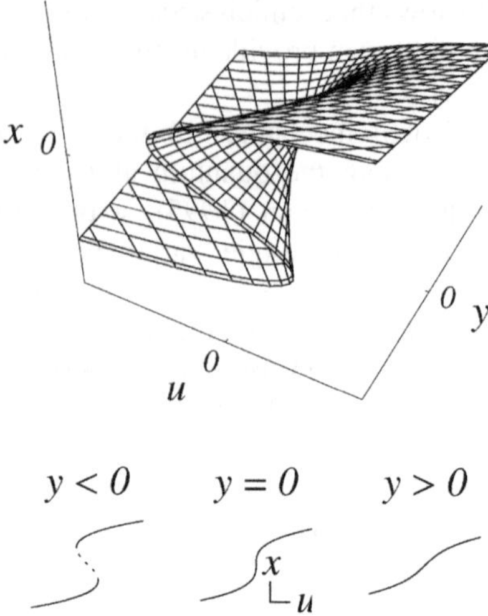

Figure 2. Slices through the cusp manifold (top) yield the three possible bifurcation diagrams (bottom).

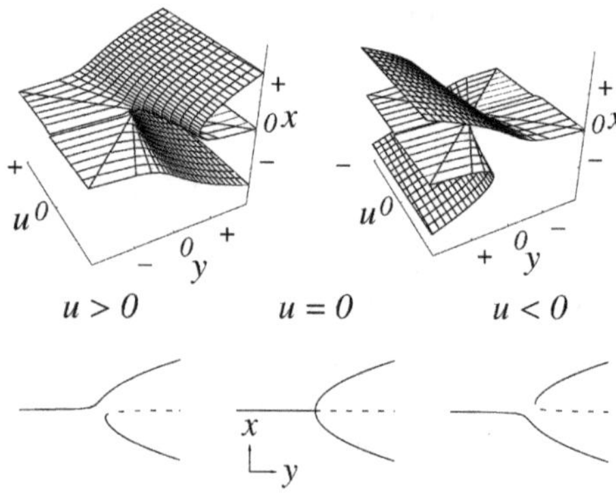

Figure 3. An orthogonal path through that cusp opens into a manifold around the pitchfork.

singularities." It is one of the more accessible entry points both to highly abstract areas of mathematics and to applied fields such as dynamical systems and bifurcations, because singularities can arise in almost any problem. The prerequisites are standard fare in first- and second-year mathematics courses: knowledge of Taylor's formula, the implicit function theorem, the theorem of existence and uniqueness, and some basic group theory, and willingness to learn some terminology.

Naturally, we should begin with a definition of singularity; (after Lu, 1976): Let f be a differentiable mapping from $\mathcal{M}$ to $\mathcal{N}$, where $\mathcal{M}$ and $\mathcal{N}$ are differentiable manifolds. A point $x_0 \in \mathcal{M}$ is a singular point of f if rank $\mathrm{d}f(x_0) < \min\{\dim \mathcal{M}, \dim \mathcal{N}\}$, where $\mathrm{d}f(x_0)$ is the Jacobian matrix of f at x_0. Otherwise x_0 is a regular point of f.

Singularity theory solves three key related problems: Given a mapping f, (i) it determines what types of singularities any good approximation $\bar{f}$ to f must have; (ii) it tells us how can we perturb f slightly to obtain a nicer and simpler, but in some sense equivalent, mapping; and (iii) it provides a taxonomy of singular objects and a binary key to identify them; hence, it is a *classification* science.

At the heart of singularity theory is a concept that is profound and yet somehow ingenuous; it is the concept of transversality. The naive (but not impercipient) version says that two curves intersect transversally if a small deformation of either one would not change the type of intersection. It is transversality that allows us to boil things down to classified normal forms.

Example 1: The mapping f from $\mathbb{R}^2$ into $\mathbb{R}^2$

$$u = x^2, \quad v = y \tag{1}$$

has a fold along $x = 0$ in the xy plane, the set of singular points is $\{(x, y) \in \mathbb{R}^2 \mid y = 2x\}$. Points along the fold remain fixed as y changes, that is, under perturbation. Equations (1) are called the *normal form* for a fold.

Example 2: The equations

$$u = xy - x^3, \quad v = y \tag{2}$$

define a mapping g from $\mathbb{R}^2$ into $\mathbb{R}^2$ for which the set of singular points is $\{(x, y) \in \mathbb{R}^2 \mid y = 3x^2\}$. The arms of the parabola are two lines of folds that come together and disappear at the cusp $(0, 0)$. Equations (2) are the normal form for the cusp, and Whitney proved that any other mapping containing a regular point satisfying the conditions

$$u_x = u_y = v_x = 0, \quad v_y = 1, u_{xx} = 0,$$
$$u_{xy} \neq 0, u_{xxx} - 3u_{xy}v_{xx} \neq 0 \tag{3}$$

can be transformed by coordinate changes into the normal form (2). In catastrophe theory this normal form becomes the *universal unfolding* $G(x, y, u)$ of the germ $g(x) = x^3$:

$$G(x, y, u) \equiv x^3 - yx + u, \tag{4}$$

where G is the gradient of a governing potential V.

One may understand the cusp by studying the surface $G(x, y, u) = 0$ shown in Figure 2. By taking slices of this surface at constant y we recover the three qualitatively different bifurcation diagrams, as shown. Now visualize a *projection* of this surface onto the (u, y) plane. We find that the two lines of folds meet at a cusp singularity, the very same that we saw in the Transversal Alps in Figure 1.

An instructive and fascinating lesson on the properties and classification of simple singularities is to parameterize the surface $G(x, y, u) = 0$ differently. In a study of singular surfaces by Ball (2001), it was observed that although the cusp is generic in the sense that all other singularities may be perturbed to either a fold or a cusp, the surface in Figure 2 is not a unique manifold of the cusp. Because all paths through the unfolding (4) are equally valid, we may choose a path in the (x, y) plane. Any such path unfurls laterally into the u-dimension to form a different surface. It is shown from two points of view in Figure 3. Constant-u slices show up the *pitchfork* singularity (center bifurcation diagram) and two of its perturbations. From this point of view Equation (4) is a *partial* unfolding of the pitchfork (but it is not a universal unfolding of the pitchfork).

In applications, the singularity theory approach has been most successful in qualitative studies of the equilibria of dynamical systems dependent on parameters. Given a dynamical system that can be reduced to a set of ordinary differential equations (by a procedure such as Lyapunov–Schmidt reduction), a general approach is to apply defining algebraic criteria systematically to the equilibria until one discovers the highest-order or most degenerate singularity, defined by its normal form. (The binary key given in Golubitsky & Schaeffer (1985, p. 201) is extremely useful for this task.) From a universal unfolding of this *organizing center*, one can "read off" all of the possible qualitatively different bifurcation behavior of the equilibria. A great many dynamical models of physical systems have been given the singularity theory treatment, often yielding results having important implications for the prediction and control of such systems. For a few but varied examples see Ball (1999) (chemical reactions), Ball et al. (2002) (plasma physics), and Broer et al. (2003) (periodic dynamics).

ROWENA BALL

See also **Bifurcations; Catastrophe theory; Development of singularities**

Further Reading

Arnol'd, V.I., Gusein-Zade, S.M. & Varchenko, A.N. 1985. *Singularities of Differentiable Maps*, vol. 1. Boston and Basel: Birkhäuser

Ball, R. 1999. The origins and limits of thermal steady state multiplicity in the continuous stirred tank reactor. *Proceedings of the Royal Society of London Series* A, 455: 141–161

Ball, R. 2001. Understanding critical behaviour through visualisation: a walk around the pitchfork. *Computer Physics Communications*, 142: 71–75

Ball, R., Dewar, R.L. & Sugama, H. 2002. Metamorphosis of plasma shear flow–turbulence dynamics through a transcritical bifurcation. *Physical Review* E, 66: 066408-1–066408-9

Broer, H.W., Golubitsky, M. & Vegter, G. 2003. The geometry of resonance tongues: a singularity theory approach. *Nonlinearity*, 16: 1511–1538

Golubitsky, M. & Schaeffer, D.G. 1985. *Singularities and Groups in Bifurcation Theory*, vol. 1, New York: Springer

Lu, Y.-C. 1976. *Singularity Theory and an Introduction to Catastrophe Theory*, Berlin and New York: Springer

Martinet, J. 1982. *Singularities of Smooth Functions and Maps*, Cambridge and New York: Cambridge University Press

Thom, R. 1972. *Stabilité Structurelle et Morphogénése: Essai d' une Théorie Générale des Modèles*, Reading, MA: W.A. Benjamin

Whitney, H. 1955. On singularities of mappings of Euclidean spaces. I. Mappings of the plane into the plane. *Annals of Mathematics*, 62: 374–410

SINK

See **Attractors**

S-INTEGRABILITY AND C-INTEGRABILITY

See **Integrability**

SKYRMIONS

As described by quantum field theories, elementary particles are associated with the quantization of small fluctuations around the vacuum and have masses that are proportional to Planck's constant, $\hbar$; thus, in the classical limit ($\hbar \to 0$), the mass goes to zero. The properties of a particle follow from the linearization of the field equations and any nonlinear terms are responsible for the interactions between particles, which are often treated perturbatively in the coupling constants. However, the 1960s saw the emergence of a new approach to quantum field theory in which the fully nonlinear classical field equations were investigated. It was found that for certain theories, the classical nonlinear field equations had static solutions that were particle-like, in the sense that they described stable, localized, finite energy field configurations. To describe elementary particles, the field theory must be Lorentz invariant, so such a static solution can simply be Lorentz boosted to describe a particle in uniform motion. These solutions, which are known as topological solitons (or sometimes just solitons for short, though they should not be confused with the solitons of integrable systems), have some novel features in comparison with elementary particles. Because they are classical solutions, then the mass of a soliton, which is identified with the energy of the static solution, is nonzero even in the limit $\hbar = 0$. Also, because the soliton is a solution of the full nonlinear field equation, it is automatically treated nonperturbatively in the coupling constants of the theory. Quantum corrections to the classical properties of a soliton can be addressed, mainly using semiclassical methods, but the important point is that these corrections are often small, so that the classical nonlinear equations largely determine the properties of a soliton. Despite the obvious differences between elementary particles and solitons, impressive research during the last decade has revealed a great deal of evidence for a remarkable connection between these two kinds of particles, with solitons in some weakly coupled theories appearing to be dual to elementary particles in strongly coupled theories (Olive, 1996). These developments have led to a renewed interest in solitons in field theory, particularly in supersymmetric theories.

As the name suggests, topological solitons arise in theories where there is a topological classification of field configurations, with the soliton being in a different topological sector to the vacuum and, hence, preventing its decay, since time evolution cannot change the conserved topological charge. This is perhaps most easily explained using a simple example. As a toy model we shall consider the static sine-Gordon theory in one space dimension, although our presentation will be slightly unusual in order to make the extension to three spatial dimensions a little more transparent. Consider a two-component unit vector $\boldsymbol{\phi}(x) = (\phi_1, \phi_2)$, so that $\boldsymbol{\phi}$ lies on the circle $\boldsymbol{\phi} \cdot \boldsymbol{\phi} = 1$. If we restrict to static fields, then the sine-Gordon model can be defined by the energy expression

$$E = \int_{-\infty}^{\infty} \left(\frac{1}{2} \frac{\mathrm{d}\boldsymbol{\phi}}{\mathrm{d}x} \cdot \frac{\mathrm{d}\boldsymbol{\phi}}{\mathrm{d}x} + 1 - \phi_2 \right) \mathrm{d}x. \qquad (1)$$

For finite energy the boundary condition is clearly that the field must take the constant vacuum value $\boldsymbol{\phi} = (0, 1)$ at spatial infinity. The fact that the field $\boldsymbol{\phi}$ takes the same value at the two points at spatial infinity implies a compactification of space from $\mathbb{R}$ to S^1, arising from the identification of the points $x = -\infty$ and $x = +\infty$. Therefore, $\boldsymbol{\phi}$ is a map $\boldsymbol{\phi} : S^1 \mapsto S^1$, where the domain is compactified space and the target circle is the set of two-component unit vectors. Such maps have an associated degree, or winding number, N, due to the homotopy group relation $\pi_1(S^1) = \mathbb{Z}$. This integer-valued winding number is simply the number of times (counted with orientation) that the field winds around the target space circle as x ranges over the circle obtained from the compactification of space. The vacuum solution, with energy $E = 0$, is given by $\boldsymbol{\phi}(x) = (0, 1)$ and clearly has $N = 0$. The soliton is the minimal energy field configuration in the $N = 1$ sector and has the explicit expression

$$\boldsymbol{\phi} = (\sin\psi, \cos\psi) \quad \text{with} \quad \psi = 4\tan^{-1} \mathrm{e}^{x-a}, \qquad (2)$$

where a is an arbitrary real constant. The soliton has energy $E = 8$ and the energy density (the integrand in Equation (1)) is a localized lump centered around the point $x = a$, which is the position of the soliton. The full time-dependent sine-Gordon theory follows from the Lorentz invariant Lagrangian associated with the energy (1). The static soliton solution (2) can be Lorentz boosted to provide a moving soliton solution (with any speed less than the speed of light) of the full second-order time-dependent field equations. The antisoliton solution, with $N = -1$, is simply obtained from the soliton solution by making the replacement $(\phi_1, \phi_2) \mapsto (-\phi_1, \phi_2)$. There are no static multisoliton solutions, that is, with $N > 1$, but time-dependent multisoltion solutions can be found in closed form and describe the elastic scattering of several individual solitons (Drazin, 1983). This remarkable fact follows from the integrability property of the time-dependent sine-Gordon equation, a feature which is not shared by more realistic theories with topological solitons in three space dimensions.

The most important aspect of the sine-Gordon model on the line is the topological classification of finite energy field configurations, which follows from the homotopy group relation $\pi_1(S^1) = \mathbb{Z}$. This is the first property that needs to be generalized when searching for a three-dimensional theory with topological solitons. If we consider a theory for which finite energy implies that the field must be constant at spatial infinity, then Euclidean space $\mathbb{R}^3$ becomes compactified to the three-sphere S^3. This is the analogue of the compactification of the line to the circle described above, and hence, we see that the simplest way to obtain a model with topological solitons is to also make the target space S^3. Then the homotopy group relation $\pi_3(S^3) = \mathbb{Z}$ ensures that there is again an integer-valued topological charge, N, which divides finite energy field configurations into distinct topological sectors. To achieve the target space S^3, the field can be taken to be a four-component unit vector, although an equivalent formulation is to take the field to be an $SU(2)$-valued matrix, since S^3 is also the manifold of the group $SU(2)$. The Skyrme model (Skyrme, 1961) is such a theory and has the static energy

$$E = \int \left\{ \mathrm{Tr}(\partial_i U \partial_i U^{-1}) - \frac{1}{8} \mathrm{Tr}\left([(\partial_i U)U^{-1}, (\partial_j U)U^{-1}]^2 \right) \right\} \mathrm{d}^3x, \qquad (3)$$

where $U(\mathbf{x}) \in SU(2)$ is the field of the model and $\partial_i = \partial/\partial x_i$ with x_i $(i = 1, 2, 3)$ is the Cartesian coordinates in Euclidean space, and we have adopted the Einstein summation convention where repeated indices are summed over. Also, the square brackets denote the commutator, $[A, B] = AB - BA$. The first term in (3), called the sigma model energy by physicists and the harmonic map energy by mathematicians, is the higher-dimensional analog of the simple gradient energy given by the first term in (1) for the toy model. In order to provide a finite nonzero scale size for the soliton, the sigma model energy needs to be balanced against an additional term with an appropriate scaling behavior under spatial dilations, as required by the Derrick–Hobart theorem. In one spatial dimension, this requires a term which contains

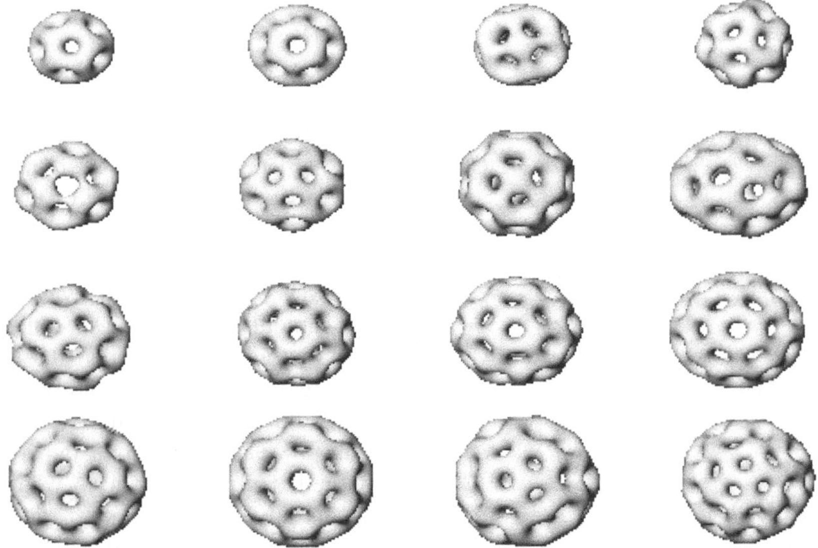

Figure 1. Skyrmions with N from 7 to 22.

no spatial derivatives of the field, such as the final term in (1), but in three spatial dimensions, the appropriate term must be at least fourth-order in the spatial derivatives. The final term in (3), known as the Skyrme term, is the unique order-four expression whose relativistic extension provides a Lagrangian in $(3+1)$-dimensions, which yields a nonlinear field equation that is only second order in time derivatives. The Skyrme model has a stable topological soliton, the solution with minimal energy in the $N=1$ sector, and this is known as a skyrmion. The solution is spherically symmetric but cannot be written in closed form and is only known numerically. Unlike the sine-Gordon toy model, the Skyrme model has static stable bound-state multi-solitons for all $N>0$. These multi-skyrmions are not spherically symmetric but often have surprising discrete symmetries, including the symmetries of the Platonic solids (Battye & Sutcliffe, 2002). Figure 1 displays skyrmions with $7 \leq N \leq 22$ by plotting surfaces around which the topological charge density is concentrated.

The Skyrme model was originally introduced (Skyrme, 1961) in the early 1960s as a model for the strong interactions of hadrons. It is a nonlinear theory of pions, with the pion particles being described in the usual quantum field theory approach by the quantization of the three degrees of freedom associated with the small fluctuations of the field around the vacuum, where U is the identity matrix, with $N=0$. Skyrme identified the conserved topological charge N with baryon number and, hence, within the nonlinear pion theory baryons appear for free as the classical soliton solutions. The Skyrme model was set aside after the advent of quantum chromodynamics (QCD), but much later it was revived by Witten (1983), who showed that it could arise from QCD as a low energy effective description in the limit in which the number of quark colors is large. Semiclassical quantization of the $N=1$ skyrmion reproduces the properties of the nucleon to within an accuracy of around 30% (Adkins et al., 1983), which is quite an achievement. There has been considerable recent progress in computing classical multi-skyrmion solutions, but it still remains to be seen whether their quantization provides a good description of nuclei.

Finally, it should be noted that in Yang–Mills–Higgs gauge theories the topological classification of field configurations arises in a slightly different way than described above for skyrmions and other modified sigma models. In gauged theories, the Higgs field is not required to be constant at infinity, and indeed, it is a nontrivial winding at infinity that provides the topological charge. In two spatial dimensions this yields vortices, and in three-dimensional space it leads to magnetic monopoles. When the Higgs field is massless, which is the situation that arises in interesting supersymmetric theories with monopoles, then the classical field equations have a particularly rich mathematical structure, and many exact results on monopole solutions are known (for a review see, e.g., Sutcliffe, 1997).

Paul Sutcliffe

See also **Derrick–Hobart theorem; Quantum field theory; Sine-Gordon equation; Solitons, types of; Yang–Mills theory**

Further Reading

Adkins, G.S., Nappi, C.R. & Witten, E. 1983. Static properties of nucleons in the Skyrme model. *Nuclear Physics* B, 228: 552–566

Battye, R.A. & Sutcliffe, P.M. 2002. Skyrmions, fullerenes and rational maps. *Reviews in Mathematical Physics*, 14: 29–85
Drazin, P.G. 1983. *Solitons*, Cambridge and New York: Cambridge University Press
Olive, D.I. 1996. Exact electromagnetic duality. *Nuclear Physics Proceedings Supplement* 45A: 88–102
Skyrme, T.H.R. 1961. A nonlinear field theory. *Proceedings of the Royal Society* A, 260: 127–138
Sutcliffe, P.M. 1997. BPS monopoles. *International Journal of Modern Physics* A, 12: 4663–4705
Witten, E. 1983. Current-algebra, baryons, and quark confinement. *Nuclear Physics* B, 223: 433–444

SLAVING PRINCIPLE

See **Synergetics**

SMALE HORSESHOE

See **Horseshoes and hyperbolicity**

SNOWFLAKES

See **Pattern formation**

SOLAR SYSTEM

The solar system shows a multitude of periodic phenomena: from the daily rotations of the Earth, to the monthly phases of the Moon, the seasonal changes during the course of a year, the phases of Venus and other planets, the regular recurrences of comets such as Halley's (every 76 years), the 25,700 year precession of the Earth's rotation axis, all the way up to changes in the parameters of the Earth's orbit, with periods of 40,000 years and more. Closer inspection reveals small modulations on top of these periods, but it is tempting to describe these by additional periods, as in the Ptolemaic theory of cycles, epicycles, and equants. However, the dynamics of the planets is governed by gravitational forces that vary as the inverse square of the distance and, hence, are strongly nonlinear. The question arises whether this nonlinearity results in some chaos.

Johannes Kepler concluded in the early 17th century that an isolated planet would move along a Kepler ellipse (such that a line joining the planet to the Sun sweeps out equal areas in equal times). For several planets their mutual perturbations leads to perihelion precessions and other variations of orbital parameters. Taking such perturbations into account was a major issue in solar mechanics in the following centuries and led to remarkable achievements. For instance, observations of the orbit of Uranus (discovered in 1781 by William Herschel) revealed significant differences to the orbit calculated in the presence of the known planets. When interpreted as due to the influence of another planet, the position of the missing planet could be predicted, and soon thereafter the efforts of Urbain V. LeVerrier, James C. Adams, and Johannes G. Galle were rewarded with the discovery of Neptune in 1846.

These results were obtained using perturbation theory. But does it actually converge? Advances in analytical mechanics in the 19th century suggested that an answer could be found and Gösta Mittag-Leffler included the issue of the stability of the solar system among the list of problems for a prize to be awarded in 1889 by King Oskar II of Sweden and Norway. Henri Poincaré was awarded the prize for his announcement that he could prove convergence, but in the course of revising his paper he noticed that there was a gap in the proof which he could not patch up. Instead, he discovered the intricate motions near a weakly perturbed hyperbolic fixed point, the so-called hyperbolic tangle, which effectively prevents quantitative continuation of trajectories that pass near a hyperbolic fixed point. Barrow-Green (1996) and Diacu & Holmes (1996) give vivid accounts of the events surrounding the prize.

A practical answer to the question of the stability of the solar system emerged in the late 20th century with the advent of powerful computers that allow integration of the equations of motions for several million years. It was then discovered that the inner planets are most susceptible to chaos, and that the Lyapunov time (inverse of the Lyapunov exponent) is about 5 million years. To illustrate the consequences of that, we quote from Laskar (1995):

> A 15 m uncertainty in the position of the Earth will grow to about 150 m in a time of 10 million years. But it will increase to 150 million km or the mean distance of Earth from the Sun within 100 million years.

As a consequence, we have difficulty predicting the Earth's orbit for times much larger than a few tens of millions of years. The consequences such an uncertainty can have are illustrated by numerical simulations for Mercury: by suitably selecting continuations, it is argued in Laskar (1995) that over a time of about 10^9 years the orbit of Mercury could change so that the planet collides with Venus and/or escapes from the solar system. It should be noted that this trajectory was especially tailored in order to show that escape is possible in principle; it does not give a clue as to how likely such an event may be.

The issue of the orbital parameters of the Earth is interesting because variations in the major rotation axis and the distance to the Sun influence our climate. In 1920, Milutin Milankovich calculated insolation data (incident solar radiation) for the Earth for long-term variations in the Earth's orbit and suggested some relationship to the appearance of past ice ages. The uncertainty in orbital parameters over periods of more than 40 million years thus indicate that reconstructing paleoclimates will be problematic (Laskar, 1999).

Besides planets there are many other objects in the solar system. Direct evidence for a chaotic trajectory has been found, for example, for the tumbling motion of Phobos and Deimos (moons of Mars) and for Saturn's moon Hyperion (Wisdom, 1987). Similarly, efforts to retrace the trajectory of Halley's comet back to its earliest recorded sighting in 163 BC gave wildly diverging results. As explained by Chirikov and Vecheslavov (1989), the orbit of Halley is chaotic with an inverse Lyapunov time of about 29 returns, so that the earliest observation is just beyond predictability. The distribution of asteroids between Mars and Jupiter shows conspicuous gaps (named after the astronomer Daniel Kirkwood) near orbits with rotation periods rationally related to the 11.9 year period of Jupiter. Such resonant interactions have strong effects on the orbits and can easily lead to collisions and escape from the resonance (Laskar, 1995; Wisdom, 1987).

The relation between the Moon, the Earth, and the Sun also holds surprises. The history of the problem, and the contributions of Babylonian, Greek, and modern astronomers to the observations and mathematical tools, is reviewed in Gutzwiller (1998). Gutzwiller also gives a quantitative example for the significance of small denominators: in order to calculate the distance between the Earth and Moon with the accuracy of 10^{-10} achievable within the Lunar Laser Ranging project, amplitudes as small as 10^{-17} have to be kept because of a significant resonance. While this does not result in positive Lyapunov exponents, it is a precursor to it, a mild form of chaos, as Gutzwiller calls it.

More surprisingly, the presence of the Moon is very important for the stability of the rotation axis of Earth: with the Moon the obliquity stays within about $\pm 1.3°$ of $23.3°$. Without it, the obliquity ends up in a resonance and can become as large as $60°$ to $90°$, with catastrophic consequences for our climate and the evolution of life (Laskar et al., 1993).

Finally, it is worthwhile to point out that the presence of hyperbolic orbits in the solar system was used in connection with the satellite GENESIS (launched in August 2001) to bring it along a stable orbit close to a Lagrange point, where it will remain for a few years to collect particles in the solar wind, before being brought back to Earth along an unstable manifold (Koon et al., 2000). Exploiting trajectories that exist in the dynamical system allows the mission to be completed with minimal requirements on fuel. This mission may thus also be considered an example of chaos control, where similar ideas are being investigated.

BRUNO ECKHARDT

See also **Celestial mechanics; Controlling chaos; Hénon–Heiles system; *N*-body problem**

Further Reading

Barrow-Green, J. 1996. *Poincaré and the Three-body Problem*, Providence, RI: American Mathematical Society

Chirikov, B.V. & Vecheslavov, V.V. 1989. Chaotic dynamics of comet Halley. *Astronomy and Astrophysics*, 221: 146–154

Diacu, F. & Holmes, P. 1996. *Celestial Encounters*, Princeton, NJ: Princeton University Press

Gutzwiller, M.C. 1998. Moon–Earth–Sun: the oldest three-body problem. *Reviews of Modern Physics*, 70: 589–639

Koon, W.S., Lo, M.W., Marsden, J.E. & Ross, S.D. 2000. Heteroclinic connections between periodic orbits and resonance transitions in celestial mechanics. *Chaos*, 10: 427–469

Laskar, J. 1995. Large scale chaos and marginal stability in the solar system. In *Constructive Methods and Results*, Proceedings of the XIth ICMP Conference, edited by D. Iagolnitzer, Boston: International Press, pp. 75–120

Laskar, J. 1999. The limits of Earth orbital calculations for geolocical time-scale use. *Philosophical Transactions Royal Society of London*, Series A, 357: 1735–1759

Laskar, J., Joutel, F. & Robutel, P. 1993. Stabilization of the Earth's obliquity by the moon. *Nature*, 361: 615–617

Wisdom, J. 1987. Chaotic motion in the solar system. *Icarus*, 72: 241–275

SOLIDIFICATION PATTERNS

See **Growth patterns**

SOLITON WAVE PACKET

See **Quantum nonlinearity**

SOLITONS

A *soliton* is a localized nonlinear wave that maintains its shape and speed as it travels, even through interaction with other waves. Another name for a localized traveling wave is a *solitary wave*. The term *soliton* was coined (by Zabusky & Kruskal, 1965) to reflect both the solitary-wave-like character and the particle-like interaction properties. Their surprising discovery has had an enormous impact on the field of nonlinear mathematics and science.

In many physical applications, however, the use of the word *soliton* has come to rely on observation of long-lived solitary waves, with little emphasis placed on interaction properties. Jupiter's Great Red Spot is often described as a soliton due to its long-lived identity (observed over hundreds of years) and the fact that it appears to maintain its identity through interaction with other disturbances. However, recent studies indicate that such Jovian vortices do not keep their identity through intrazonal interactions. In nonlinear optics, the deduction of stable, long-lived solitary waves is of great physical interest for their application as coherent light signals propagating through optical fibers, while their interaction properties are secondary. The definition of the word *soliton* in some dictionaries has come to reflect such usage by physicists, omitting the requirement of interaction. However, for mathematicians, the many surprising and miraculous discoveries of soliton theory are tied fundamentally to the special interaction properties so its definition always contains elastic interaction as an essential requirement.

The preservation of identity through interaction would not be surprising if we were describing solutions of linear nondispersive wave equations. A simple example is the linear wave equation $u_{tt} + c^2 u_{xx} = 0$, where c is a constant (assumed positive). The general solution is given by $u(x,t) = f(x - ct) + g(x + ct)$, where f and g are determined by initial conditions. The first part, $f(x - ct)$, is a wave traveling to the right with speed c, while the second part, $g(x + ct)$, is a wave traveling to the left with speed c. Two such wave profiles interact when they meet head-on but they both come out of the interaction with the same shape and speed.

However, until the discovery of solitons, it was generally believed that no such property could hold for nonlinear equations. Common understanding in mathematics and physics in the 1950s suggested that nonlinear wave solutions either break, dissipate, or thermalize, that is, distribute initial energy between different solutions over time, and, therefore, lose their identities with time. Much of this understanding was based on prototypical examples, such as the inviscid Burgers equation

$$\frac{\partial u}{\partial t} + u\frac{\partial u}{\partial x} = 0, \tag{1}$$

which describes the one-dimensional propagation of compression waves in gas or dust or automoblie traffic. For the initial condition $u(x,0) = u_0(x)$, the solution is a wave, given implicitly by $u(x,t) = u_0\big(x - t\,u(x,t)\big)$. If the initial wave profile $u_0(x)$ has a part with negative slope, the wave front steepens and eventually breaks in finite time, just like ocean waves do at a beach. (The break up time is easy to find by differentiating the solution with respect to x.)

In 1965, Zabusky and Kruskal published numerical studies of the solutions of the Korteweg–de Vries (KdV) equation

$$\frac{\partial u}{\partial t} + u\frac{\partial u}{\partial x} + \delta^2\frac{\partial^3 u}{\partial x^3} = 0, \tag{2}$$

which changed the above-described common beliefs about nonlinear waves forever (*See* **Solitons, a brief history**). If $\delta \to 0$, the limiting equation is the inviscid Burgers equation. Zabusky and Kruskal labeled Equation (2) as (1), chose $\delta = 0.022$ and the initial condition $u(x,0) = \cos(\pi x)$ and studied its periodic solutions numerically. The word *soliton* was used for the first time in their paper of 1965, in the following extract.

> (I) Initially, the first two terms of Eq. (1) dominate and the classical overtaking phenomenon occurs; that is, u steepens in regions where it has negative slope. (II) Second, after u has steepened sufficiently, the third term becomes important and serves to prevent the formation of a discontinuity. Instead, oscillations of small wavelength (of order δ) develop on the left of the front. The amplitudes of the oscillations grow, and finally *each* oscillation achieves an almost steady amplitude (that increases linearly from left to right) and has the shape of an individual solitary-wave of (1). (III) Finally, each "solitary wave pulse" or *soliton* begins to move uniformly at a rate (relative to the background value of u from which the pulse rises) that is linearly proportional to its amplitude. Thus, the solitons spread apart. Because of the periodicity, two or more solitons eventually overlap spatially and interact nonlinearly. Shortly after the interaction they reappear virtually unaffected in size or shape. In other words, solitons "pass through" one another without losing their identity.

A standard form of the KdV equation is

$$\frac{\partial u}{\partial t} + 6u\frac{\partial u}{\partial x} + \frac{\partial^3 u}{\partial x^3} = 0. \tag{3}$$

This is equivalent to Equation (2) under a scaling transformation; that is, $u(x,t) \mapsto 6\delta^{2/3}\,u\big(x\,\delta^{-2/3},\,t\big)$ maps any solution of (2) to one of (3). If we consider the initial value problem for the KdV equation on the whole real x-line and look for solutions that vanish at infinity, then solitons are characterized by the initial condition

$$u(x,0) = N(N+1)\,k^2\,\mathrm{sech}^2(k\,x), \tag{4}$$

where N is any nonnegative integer. The corresponding solutions are called N-soliton or multisoliton solutions. The case $N = 1$ gives the traveling wave solution

$$u(x,t) = 2k^2\,\mathrm{sech}^2\big(k\,(x - 4k^2 t)\big), \tag{5}$$

often called the one-soliton solution of the KdV equation. It is equivalent to the solitary wave observed by John Scott Russell in 1834 and deduced mathematically by Diederik Korteweg and Hendrik de Vries in 1895 (*See* **Solitons, a brief history**).

When $N \geq 2$ and time t is considered to be large negative (a long time into the past) or large positive (a long time into the future), the solution separates into a chain of N distinct, localized one-soliton solutions each having the form (5) for some value of k. The chain is positioned far to the left if t is negative or far to the right if t is positive. Each wave in the chain has a distinct height related to its speed. If their phases are arranged so that a taller soliton is to the left of a shorter one at some time in the past, then the taller one overtakes the shorter one, and reappears to the right with its distinctive height, shape, and speed unchanged. The only explicit sign that interaction has occurred is a phase shift, visible asymptotically as x and t become large, in each soliton. As in the one-soliton case, explicit formulae can be found for the N-soliton solution for all x and t. Moreover, the phase shifts due to pairwise interactions can be calculated exactly (*See* ***N*-soliton formulas**).

The soliton solutions of the KdV equation can be obtained explicitly by using the inverse scattering method. This method associates each solution $u(x,t)$ of Equation (3), that vanishes (faster than $1/x$) as $x \to \pm\infty$, to the stationary Schrödinger equation

$$\psi_{xx} + \big(u(x,t) + \lambda\big)\psi = 0, \tag{6}$$

where $u(x,\cdot)$ plays the role of the potential function and λ is a parameter. If we let $\lambda = \zeta^2$, we can characterize the space of solutions $\psi(x;\zeta)$ near infinity (in x) in terms of linear combinations of $\exp(\pm i\zeta x)$. (Since u vanishes there, the equation becomes simply $\psi_{xx} - \zeta^2\psi = 0$, which has such exponentials as solutions.) Quantum mechanical convention regards the solution with behavior $\exp(i\zeta x)$ at $x = +\infty$ as an *incoming wave* from $+\infty$. Part of this incoming wave reflects off the potential barrier u while part of it is transmitted through to the other side. Not all boundary conditions at infinity can be satisfied without imposing conditions on ζ. A discrete set of eigenvalues $\{\zeta_i\}$ arises when we demand that the solutions vanish at infinity. The corresponding solutions are called bound states and their amplitudes are usually normalized. The collected information about discrete eigenvalues, the normalization constants, the reflection coefficient, and the transmission coefficient is called the set of spectral or scattering data. A beautiful and fundamental result of soliton theory is that the time evolution of the spectral data can be obtained explicitly as the potential evolves according to the KdV equation. Another fundamental result is that such evolved spectral data can be inverted to give the solution $u(x,t)$ of the KdV equation at a later time.

If all waves ψ are transmitted, with none being reflected, the potential u is called a *reflectionless* potential. It is an amazing fact that the N-soliton solutions of the KdV equation are precisely the reflectionless potentials of Equation (6).

Another way to find the N-soliton solutions of the KdV is through its Bäcklund transform (*See* **Bäcklund transformations**) or equivalently the Darboux transform of the Schrödinger equation (6). The latter was introduced by Gaston Darboux in 1882, and expanded in his four-volume lecture notes on the geometry of surfaces (Darboux, 1915). The Darboux transform relates the potentials of two copies of the Schrödinger equation whose solutions are related linearly. His method shows how to change a potential so that a new discrete eigenvalue is added to those of the old potential (*See* **Darboux transformation**).

Many other nonlinear PDEs are now known to have soliton solutions. Solitons come in different shapes and many flavors (*See* **Solitons, types of**). The sine-Gordon equation

$$\frac{\partial^2 u}{\partial x^2} - \frac{\partial^2 u}{\partial t^2} = \sin u \tag{7}$$

has N-soliton solutions that asymptote to constants as $x \to \pm\infty$. One-soliton solutions can be one of two types, called a "kink" or an "anti-kink," according to whether it decays to zero to the left or the right, respectively. The explicit expression for such solutions is

$$u(x,t) = 4\arctan\left(\exp\left(\pm\frac{x - \eta t - x_0}{\sqrt{1-\eta^2}}\right)\right). \tag{8}$$

The "+" choice gives a kink and the "−" choice an antikink. Multisoliton solutions decompose into a sequence of kinks and antikinks as x and t approach $\pm\infty$.

There is another type of soliton admitted by the SG equation, which appears to make it unique within the class of nonlinear Klein–Gordon equations: $u_{xx} - u_{tt} = F(u)$. This is the "breather" solution:

$$u(x,t) = 4\arctan\left(\frac{\sqrt{1-\omega^2}}{\omega}\,\frac{\sin(\omega t)}{\cosh(\sqrt{1-\omega^2}\,(x - x_0))}\right). \tag{9}$$

This solution oscillates (or "breathes") while staying in the same location as time evolves. However, because SG is invariant under the Lorentz transform: $(x,t) \mapsto (X,T)$, $X = (x - ct)/\sqrt{1-c^2}$, $T = (t - cx)/\sqrt{1-c^2}$, where c is constant, the breather can be transformed to one that moves with speed $|c| < 1$. The existence of such spatially localized, temporally periodic solutions is rare for nonlinear wave equations.

The nonlinear Schrödinger (NLS) equation

$$i\frac{\partial u}{\partial t} + \frac{\partial^2 u}{\partial x^2} + 2|u|^2 u = 0 \tag{10}$$

provides a third example of a nonlinear PDE with N-soliton solutions. Its one-soliton solution is given by

$$u(x,t) = a\exp\left(i\eta x + (a^2 - \eta^2)t\right) \times \operatorname{sech}\left(a(x - 2\eta t - x_0)\right). \tag{11}$$

Solitons are not confined to partial differential equations. A differential-difference equation such as the Toda equation

$$\frac{\partial^2 u_n}{\partial t^2} = \exp(-(u_n - u_{n-1})) - \exp(-(u_{n+1} - u_n)), \tag{12}$$

also has N-soliton solutions for arbitrary positive integers N. See Ablowitz & Segur (1981) for inverse scattering theory extended to such discrete evolution equations.

There also exist nonlinear PDEs that have solutions that can be interpreted as 2-soliton solutions, but which do *not* have N-soliton solutions for integer $N > 2$. Hirota pointed this out after inventing a method of calculating solitons explicitly and directly without using the inverse scattering method or Bäcklund transformations. (For details, *see* Ablowitz & Segur (1981) and **Hirota's method**.) An example is the two-dimensional version of the SG equation

$$\frac{\partial^2 u}{\partial x^2} + \frac{\partial^2 u}{\partial y^2} - \frac{\partial^2 u}{\partial t^2} = \sin u\,. \tag{13}$$

Hirota's method shows that there is a traveling wave solution that can be interpreted as a two-soliton. However, no N-soliton solution with higher N appears to exist.

There are many more equations with solitary waves than those with N-soliton solutions. Traveling waves may also be known to exist without having an exact expression. An example is provided by the reduced model of nerve transmission given by the Fitzhugh–Nagumo equation (*See* **FitzHugh–Nagumo equation**). It can be shown that (for a certain range of parameter values) this equation admits traveling waves that are solitary waves. However, energy is not conserved for nerve impulses while it is for solitons.

Despite their rarity in the class of differential equations, soliton equations arise ubiquitously as models of nature. The KdV equation arises as a canonical model of water waves, under certain constraints. Consider the motion of surface waves in a fluid that is inviscid and nondispersive to leading order. Assume that the height of the waves is small compared with the depth of fluid, which in turn is much smaller than the length scale in one direction (x, say) and, moreover, that these two small ratios of scales balance. Then the resulting model is always given by the KdV equation (or its relations, such as the modified KdV equation). See Ablowitz & Segur (1981) for a detailed derivation. The soliton solutions, in the case when dimensionless surface tension is less than $\frac{1}{3}$, are the waves that travel on and raise the free surface. The general solutions are composed of N-solitons and a part called *radiation* that decays in amplitude as it moves away to infinity. When the water waves are nearly two-dimensional, the model equation becomes the Kadomtsev–Petviashvili equation

$$\frac{\partial}{\partial x}\left(\frac{\partial u}{\partial t} + 6u\frac{\partial u}{\partial x} + \frac{\partial^3 u}{\partial x^3}\right) + \frac{\partial^2 u}{\partial y^2} = 0\,, \tag{14}$$

which also possesses N-soliton solutions. Since such models always arise in an asymptotic sense (e.g., the balance of small wave height to depth of fluid), stability of solutions under perturbation is important. Solitons are stable under perturbations.

Stratified fluids are another context in which the KdV equation arises as a universal model. Consider a situation where a lower, heavier fluid rests on an impermeable solid with a lighter fluid on top that has a free surface above. A common example is the stratified ocean where colder, denser water lies underneath a warmer, lighter layer. If the sea is contained by a boundary that restricts the colder layer but there are forces present, such as tides, that force the lighter layer, interface waves appear between the two layers. Such waves are also governed by the KdV equation under appropriate assumptions. The soliton solutions are then interface waves that thicken the upper layer. The KdV and its solitons also arise in collision-free hydromagnetic waves, ion-acoustic waves, plasma physics, and lattice dynamics.

Like the KdV equation, the SG and NLS equations also arise as universal models in certain contexts. Some of the applications of the SG equation include the study of propagation of crystal defects, that of domain walls in ferromagnetic and ferroelectric materials, as a one-dimensional model for elementary particles, self-induced transparency of short optical pulses, and propagation of quantum units of magnetic flux on long Josephson (superconducting) transmission lines. See Scott (2003) for detailed descriptions. The NLS equation appears as a model of the propagation of packets of hydrodynamic waves on deep water, nonlinear pulses of light in an optical fibre, two-dimensional self-focusing of a plane wave, one-dimensional self-modulation of a monochromatic wave, propagation of a heat pulse in a solid, and Langmuir waves in plasmas.

NALINI JOSHI

See also **Bäcklund transformations; Inverse scattering method or transform; Solitons, types of; Zero-dispersion limit**

Further Reading

Ablowitz, M. & Segur, H. 1981. *Solitons and the Inverse Scattering Transform*, Philadelphia: SIAM

Darboux, G. 1915. *Leĉons sur la théorie générale des surfaces et les applications géometriques du calcul infinitésimal*, vol. 2, 2nd edition, Paris: Gauthier-Villars

Scott, A. 2003. *Nonlinear Science: Emergence and Dynamics of Coherent Structures*, 2nd edition, Oxford and New York: Oxford University Press

Zabusky, N.J. & Kruskal, M.D. 1965. Interactions of solitons in a collisionless plasma and the recurrence of initial states. *Physical Review Letters*, 15: 240–243

SOLITONS, A BRIEF HISTORY

In 1834, a young engineer named John Scott Russell was conducting experiments on the Union Canal (near Edinburgh, Scotland) to measure the relationship between the speed of a canal boat and its propelling force, with the aim of finding design parameters for conversion from horse power to steam. One August day, a rope parted in his apparatus and (Russell, 1844)

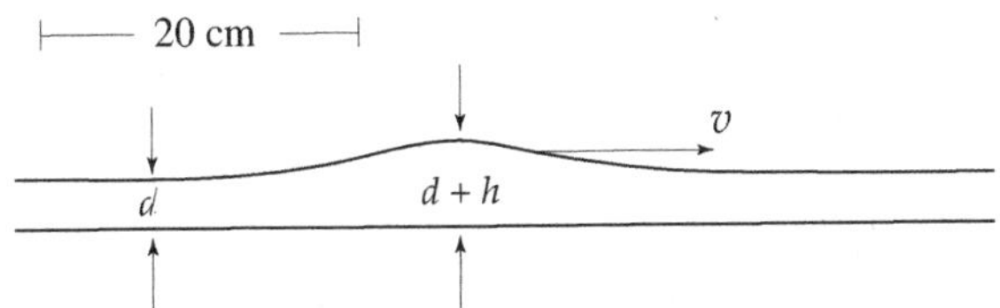

Figure 1. A hydrodynamic solitary wave (or soliton) in a tank similar to that described by John Scott Russell (1844). The wave is generated by suddenly releasing (or displacing) a mass of water at the left-hand side of the tank.

> the boat suddenly stopped—not so the mass of water in the channel which it had put in motion; it accumulated round the prow of the vessel in a state of violent agitation, then suddenly leaving it behind, rolled forward with great velocity, assuming the form of a large solitary elevation, a rounded, smooth and well-defined heap of water, which continued its course along the channel without change of form or diminution of speed.

Russell did not ignore this serendipitous phenomenon, but "followed it on horseback, and overtook it still rolling on at a rate of some eight or nine miles an hour, preserving its original figure some thirty feet long and a foot to a foot and a half in height" until the wave became lost in the windings of the channel. He continued to study the solitary wave in tanks and canals over the following decade, finding it to be an independent dynamic entity moving with constant shape and speed.

Using a wave tank, he demonstrated four facts Russell (1844). First, solitary waves have a hyperbolic secant shape. Second, a sufficiently large initial mass of water produces two or more independent solitary waves. Third, solitary waves cross each other "without change of any kind." Finally, a wave of height h and traveling in a channel of depth d has a velocity given by the expression $\sqrt{g(d+h)}$ (where g is the acceleration of gravity), implying that a large amplitude solitary wave travels faster than one of low amplitude.

Although soon confirmed by observations on the Canal de Bourgogne near Dijon, most subsequent discussions of the hydrodynamic solitary wave missed the physical significance of Russell's observations. Evidence that Russell maintained a deeper appreciation of the importance of his discovery is provided by a posthumous work where—among several provocative ideas—he correctly estimated the height of the Earth's atmosphere from the fact (well known to military engineers of the time) that "the sound of a cannon travels faster than the command to fire it" (Russell, 1885).

In 1895, Diederik Korteweg and Hendrik de Vries published a theory of shallow water waves that reduced Russell's problem to its essential features. One of their results was the nonlinear partial differential equation (PDE)

$$\frac{\partial u}{\partial t} + c\frac{\partial u}{\partial x} + \varepsilon\frac{\partial^3 u}{\partial x^3} + \gamma u\frac{\partial u}{\partial x} = 0, \tag{1}$$

which would play a key role in soliton theory (Korteweg & de Vries, 1895). In this equation, $u(x,t)$ is the wave amplitude, $c = \sqrt{gd}$ is the speed of small amplitude waves, $\varepsilon \equiv c(d^2/6 - T/2\rho g)$ is a dispersive parameter, $\gamma \equiv 3c/2d$ is a nonlinear parameter, and T and ρ are, respectively, the surface tension and the density of water. In general, Equation (1) is nonlinear with exact traveling-wave solutions

$$u(x,t) = h\,\mathrm{sech}^2[k(x - vt)], \tag{2}$$

where $k \propto \sqrt{h}$ implying that higher amplitude waves are more narrow. With this shape, the effects of dispersion balance those of nonlinearity at an adjustable value of the pulse speed; thus, the hydrodynamic solitary wave is seen to be an independent dynamic entity.

Bäcklund Transformations

Although unrecognized at the time, such an energy conserving solitary wave is related to the existence of a transform technique that was proposed by Albert Bäcklund in 1855 (Lamb, 1976). Under this Bäcklund transformation (BT), a known solution generates a new solution through an integration, after which the new solution can be used to generate yet another new solution, and so on. It is straightforward to find a BT for any linear PDE, which introduces a new eigenfunction into the total solution with each application of the transformation. Only special nonlinear PDEs have BTs, but 19th century mathematicians knew that these include

$$\frac{\partial^2 u}{\partial \xi\, \partial \tau} = \sin u, \tag{3}$$

which arose in research on the geometry of curved surfaces (Steuerwald, 1936).

In 1939, Yakov Frenkel and Tatiana Kontorova introduced a seemingly unrelated problem arising in solid state physics to model dislocation dynamics in a crystal (Frenkel & Kontorova, 1939). From this study, an equation describing dislocation motion is

$$\frac{\partial^2 u}{\partial x^2} - \frac{\partial^2 u}{\partial t^2} = \sin u, \tag{4}$$

where $u(x,t)$ is atomic displacement in the x-direction and the sine function represents periodicity of the crystal lattice. A traveling-wave solution of Equation (4), corresponding to the propagation of a dislocation, is

$$u(x,t) = 4\arctan\left[\exp\left(\frac{x - vt}{\sqrt{1 - v^2}}\right)\right], \tag{5}$$

with velocity v in the range $(-1, +1)$. Because Equation (4) is identical to Equation (3) after an independent variable transformation, exact solutions involving arbitrary numbers of dislocation components as in Equation (5) can be generated through a succession of Bäcklund transformations, but this was not known to Frenkel and Kontorova.

Numerical Discoveries of the Soliton

In the late 1940s, Enrico Fermi, John Pasta, and Stan Ulam (FPU) suggested one of the first scientific problems to be assigned to the Los Alamos MANIAC computing machine: the dynamics of energy equipartition in a slightly nonlinear crystal lattice, which is related to thermal conductivity. The system they chose was a chain of equal mass particles connected by slightly nonlinear springs, and it was expected that if all the initial energy were put into a single vibrational mode, the small nonlinearity would cause a gradual progress toward equal distribution of the energy among all modes (thermalization). But the numerical results were surprising. If all the energy is originally in the mode of lowest frequency, it returns almost entirely to that mode after a period of interaction among a few other low frequency modes. In the course of several numerical refinements, no thermalization was observed (Fermi et al., 1955).

Pursuit of an explanation for this "FPU recurrence" led Zabusky and Kruskal to approximate the nonlinear spring-mass system by the KdV equation. In 1965, they reported numerical observations that KdV solitary waves pass through each other with no change in shape or speed, and coined the term *soliton* to suggest this particle-like property (Zabusky & Kruskal, 1965).

Zabusky and Kruskal were not the first to observe nondestructive interactions of energy conserving solitary waves. Apart from Russell's tank measurements, Perring and Skyrme had studied solutions of Equation (4) comprising two solutions as in Equation (5) undergoing a collision. In 1962, they published numerical results showing perfect recovery of shapes and speeds after a collision and went on to discover an exact analytical description of this phenomenon (Perring & Skyrme, 1962).

This result would not have surprised 19th-century mathematicians; it is merely the second member of the hierarchy of solutions generated by a BT. Nor would it have been unexpected by Seeger and his colleagues, who had noted in 1953 the connections between the 19th-century work (Steuerwald, 1936) and the studies of Frenkel and Kontorova (Seeger et al., 1953). Because Perring and Skyrme were interested in Equation (4) as a nonlinear model for elementary particles of matter, however, the complete absence of scattering may have been disappointing.

Throughout the 1960s, Equation (4) arose in a variety of problems, including the propagation of ferromagnetic domain walls, self-induced transparency in nonlinear optics, and the propagation of magnetic flux quanta in long Josephson transmission lines. Eventually it became known as the "sine-Gordon" (SG) equation—a nonlinear version of the Klein–Gordon equation: $(u_{xx} - u_{tt} = u)$.

Perhaps the most important contribution made by Zabusky and Kruskal in their 1965 paper was to recognize the relation between nondestructive soliton collisions and the riddle of FPU recurrence. Viewing KdV solitons as independent and localized dynamic entities, they explained the FPU observations as follows. The initial condition generates a family of solitons with different speeds, moving apart in the x–t plane. Since the system studied was of finite length with perfect reflections at both ends, the solitons could not move infinitely far apart; instead, they eventually reassembled in the x–t plane, approximately recreating the initial condition after a surprisingly short "recurrence time."

By 1967, this insight had led Gardner, Greene, Kruskal, and Miura (GGKM) to devise a nonlinear generalization of the Fourier transform method for constructing solutions of the KdV emerging from arbitrary initial conditions (Gardner et al., 1967). Called the inverse scattering method (ISM), this approach proceeds in three steps. First, the nonlinear KdV dynamics are mapped onto an associated linear scattering problem, where each eigenvalue of the linear problem corresponds to the speed of a particular KdV soliton. Second, the time evolution of the associated linear scattering data is computed. Finally, an inverse scattering calculation determines the time evolved KdV dynamics from the evolved scattering data. Thus, the solution of a nonlinear problem is found from a series of linear computations.

Toda Lattice Solitons

Another development of the 1960s was Morikazu Toda's discovery of exact two-soliton interactions on a nonlinear spring-mass system (Toda, 1967). As in the FPU system, equal masses were assumed to be interconnected with nonlinear springs, but Toda chose the potential

$$\left(\frac{a}{b}\right)[\mathrm{e}^{-bu_j} - 1] + au_j, \tag{6}$$

where $u_j(t)$ is the longitudinal extension of the jth spring from its equilibrium value and both a and b are adjustable parameters. (In the limit $a \to \infty$ and $b \to 0$ with ab finite, this reduces to the quadratic potential of a linear spring. In the limit $a \to 0$ and $b \to \infty$ with ab finite, it describes the interaction between hard spheres.) Thus by the late 1960s, it was established that solitons were not limited to PDEs; local solutions of difference-differential equations could also exhibit the unexpected properties of unchanging shapes and speeds after collisions.

A Seminal Workshop

These events are only the salient features of a growing panorama of nonlinear wave activities that became gradually less parochial during the 1960s. Solid state physicists began to see relationships between their solitary waves (magnetic domain walls, self-shaping pulses of light, quanta of magnetic flux, polarons, etc.), and those from classical hydrodynamics and oceanography, while applied mathematicians began to suspect that the ISM (originally formulated by GGKM for the KdV equation) might be used for a broader class of nonlinear wave equations. It was amid this intellectual ferment that the first soliton research workshop was organized during the summer of 1972 (Newell, 1974). Interestingly, one of the most significant contributions to this conference came by post. From the Soviet Union arrived a paper by Vladimir Zakharov and Alexey Shabat formulating the ISM for the nonlinear PDE Zakharov & Shabat (1972)

$$\mathrm{i}\frac{\partial u}{\partial t}+\frac{\partial^2 u}{\partial x^2}+2|u|^2 u=0. \tag{7}$$

In contrast to KdV, SG, and the Toda lattice, the dependent variable in this equation is complex rather than real, so the evolutions of two quantities (magnitude and phase of u) are governed by the equation. This reflects the fact that Equation (7) is a nonlinear generalization of a linear equation $\mathrm{i}u_t+u_{xx}+u=0$, solutions of which comprise both an envelope and a carrier wave. As this linear equation is a Schrödinger equation for the quantum mechanical probability amplitude of a particle (like an electron) moving through a region of uniform potential, it is natural to call Equation (7) the nonlinear Schrödinger (NLS) equation. When the NLS equation is used to model classical wave packets in such fields as hydrodynamics, nonlinear acoustics, and plasma waves, however, its solutions are devoid of quantum character.

Upon appreciating the Zakharov and Shabat paper, participants left the 1972 workshop aware that four nonlinear equations (KdV, SG, NLS, and the Toda lattice) display solitary wave behavior with the special properties that led Zabusky and Kruskal to coin the term *soliton* (Newell, 1974). Within two years, ISM formulations had been constructed for the SG equation and also for the Toda lattice.

Since the mid-1970s, the soliton concept has become established in several areas of applied science, and dozens of nonlinear systems are now known to be integrable through the ISM. Thus, one is no longer surprised to find stable spatially localized regions of energy, balancing the opposing effects of nonlinearity and dispersion and displaying the essential properties of objects.

ALWYN SCOTT

See also **Bäcklund transformations; Fermi–Pasta–Ulam oscillator chain; Inverse scattering method or transform; Laboratory models of nonlinear waves; Solitons**

Further Reading

Fermi, E., Pasta, J.R. & Ulam, S.M. 1955. Studies of nonlinear problems, Los Alamos Scientific Laboratory Report No. LA–1940 (Reprinted in Newell, 1974.)

Frenkel, J. & Kontorova, T. 1939. On the theory of plastic deformation and twinning. *Journal Physics (USSR)* 1: 137–149

Gardner, C.S., Greene, J.M., Kruskal, M.D. & Miura, R.M. 1967. Method for solving the Korteweg–de Vries equation. *Physical Review Letters*, 19: 1095–97

Korteweg, D.J. & de Vries, H. 1895. On the change of form of long waves advancing in a rectangular canal, and on a new type of long stationary waves. *Philosophical Magazine*, 39: 422–443

Lamb, G.L., Jr. 1976. Bäcklund transforms at the turn of the century. In *Bäcklund Transforms*, edited by R.M. Miura, Berlin and New York: Springer

Newell, A.C. (editor). 1974. *Nonlinear Wave Motion*, Providence, R.I: American Mathematical Society

Perring, J.K. & Skyrme, T.R.H. 1962. A model unified field equation. *Nuclear Physics*, 31: 550–555

Russell, J.S. 1844. *Report on Waves*, 14th meeting of the British Association for the Advancement of Science, London: BAAS, 311–339

Russell, J.S. 1885. *The Wave of Translation in the Oceans of Water, Air and Ether*, London: Trübner

Scott, A.C. 2003. *Nonlinear Science: Emergence and Dynamics of Coherent Structures*, 2nd edition, Oxford and New York: Oxford University Press

Seeger, A., Donth, H. & Kochendörfer, A. 1953. Theorie der Versetzungen in eindimensionalen Atomreihen. *Zeitschrift für Physik*, 134: 173–193

Steuerwald, R. 1936. Über Enneper'sche Flächen und Bäcklund'sche Transformation. *Abhandlungen der Bayerischen Akademie der Wissenschaften* München, pp. 1–105

Toda, M. 1967. Vibration of a chain with nonlinear interactions. *Journal of the Physical Society of Japan*, 22: 431–436; Wave propagation in anharmonic lattices. *Journal of the Physical Society of Japan*, 23: 501–506

Zabusky, N.J. & Kruskal, M.D. 1965. Interactions of solitons in a collisionless plasma and the recurrence of initial states. *Physical Review Letters*, 15: 240–243

Zakharov, V.E. & Shabat, A.B. 1972. Exact theory of two-dimensional self-focusing and one-dimensional self-modulation of waves in nonlinear media. *Soviet Physics, JETP*, 34: 62–69

SOLITONS, TYPES OF

Solitons are localized solutions of nonlinear partial or difference-differential equations that preserve their integrity under collisions with other such solutions. In general, energy is conserved for soliton systems; thus, they are a special class of Hamiltonian systems. Localization may result from a dynamic balance between the effects of nonlinear and dispersion (nontopological solitons) or from a topological constraint.

Nontopological Solitons

Perhaps the most famous example among equations admitting solitary wave solution is the Korteweg–de Vries (KdV) equation

$$u_t + \alpha u u_x + u_{xxx} = 0, \quad (1)$$

which has the solitary-wave solution

$$u = a\,\mathrm{sech}^2\left\{\sqrt{a\alpha/12}[x - (a\alpha/3)t]\right\}, \quad (2)$$

where a is the wave amplitude. The quadratic nonlinear term $(\alpha u u_x)$ is balanced by the dispersive term (u_{xxx}) in the solution of Equation (2). A feature of such solitary waves of this type is that they preserve their shapes and speeds under collisions (act as particles); thus they are called "solitons."

In general, Equation (1) approximates a more detailed physical description. When the quadratic nonlinear term is small, the cubic nonlinear term may be taken into account, leading to the following equation.

$$u_t + \alpha u u_x + \beta u^2 u_x + u_{xxx} = 0, \quad (3)$$

which is known as the extended Korteweg-de Vries (eKdV) equation. With $\alpha = 0$ and $\beta = 6$, Equation (3) is known as the modified Korteweg–de Vries (mKdV) equation with negative dispersion,

$$u_t + 6u^2 u_x + u_{xxx} = 0. \quad (4)$$

Equation (4) admits solitary wave solutions tending to u_0 at infinity (Grimshaw et al., 1999),

$$u(x,t) = u_0 + \frac{2\nu^2}{u_0 + \sigma\lambda\cosh[2\nu(x - \mu t)]}, \quad (5)$$

where $\lambda > |u_0|$, $\nu = \sqrt{\lambda^2 - u_0^2}$, $\sigma = \pm 1$, $\mu = 2u_0^2 + 4\lambda^2$, and

$$u(x,t) = u_0 - \frac{4u_0}{1 + 4u_0^2(x - 6u_0^2 t)^2}. \quad (6)$$

Equation (5) reduces to a soliton solution as $u_0 \to 0$,

$$u = 2\sigma\lambda\mathrm{sech}\{2\lambda(x - 4\lambda^2 t)\}, \quad (7)$$

which is different from the KdV soliton solution (2). Equation (6) shows that the wave shape $(u - u_0)$ vanishes algebraically as $|x| \to \infty$. We call it an *algebraic soliton* or *rational soliton.*

The eKdV equation (3) has the following solitary-wave solution ((Grimshaw et al., 1999),

$$u(x,t) = \frac{(6\gamma^2/\alpha)}{1 + B\cosh\{\gamma(x - \gamma^2 t)\}}, \quad (8)$$

where $B^2 = 1 + (6\beta\gamma^2/\alpha^2)$ and γ is an arbitrary parameter, characterizing the inverse width of the solitary wave. In the case $\beta < 0$ $(0 < B < 1)$, at small wave amplitudes $(B \to 1)$, Equation (8) transforms into the KdV soliton solution. On the other hand, as the wave amplitude increases $(B \to 0)$, it approaches the critical value $a_{cr} = \alpha/|\beta|$. In the limit $(B \to 0)$, the width of the solitary wave increases to infinity and it becomes the so-called *thick soliton*. The thick soliton can be viewed as a kink-antikink combination.

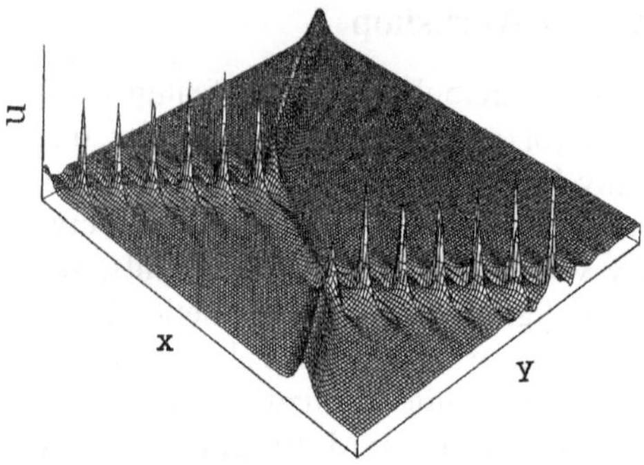

Figure 1. Interaction between line soliton and periodic soliton.

An extension of the KdV equation to motion in two space dimensions was given by Kadomtsev and Petviashvili in order to discuss the stability of the KdV soliton (line soliton in two-dimensional space) against a long transverse disturbance. This two-dimensional extension is known as the Kadomtsev–Petviashvili (KP) equation,

$$(u_t + 6uu_x + u_{xxx})_x + 3su_{yy} = 0, \quad (9)$$

which corresponds to the case of negative and positive dispersion when $s = +1$ and $s = -1$, respectively (Kadomtsev & Petviashvili, 1970). They have shown that the line soliton is stable in the case of negative dispersion and is unstable for the positive dispersion. This leads to the conjecture that a localized soliton in two-dimensional space should be formed in the positive dispersion case, since the line soliton is unstable. Such a soliton solution has been found by Manakov et al. (1977) and Ablowitz & Satsuma (1978), which is no longer exponential in character, but a rational function of space variables,

$$u = 4\frac{(L + L^*)^{-2} - \xi^2 + \eta^2}{\left[(L + L^*)^{-2} + \xi^2 + \eta^2\right]^2}, \quad (10)$$

with

$$\xi = \mathrm{Re}\{x - 2\mathrm{i}Ly - 12L^2 t\} + \xi_0,$$
$$\eta = \mathrm{Im}\{x - 2\mathrm{i}Ly - 12L^2 t\} + \eta_0,$$

where ξ_0 and η_0 are arbitrary constants and $*$ indicates the complex conjugate. This solution decays like $(x^2 + y^2)^{-1}$ as $(x^2 + y^2)^{1/2} \to \infty$; thus, it is also called an *algebraic soliton* or *rational soliton* or *lump.* Another type of localized soliton appears in the positive dispersion case as a sequence of infinite algebraic solitons, called a *periodic soliton* or *soliton chain* (Tajiri & Murakami, 1989). Figure 1 shows a typical interaction between line soliton and periodic soliton.

The Benjamin–Ono (BO) equation,

$$u_t + 4uu_x + \frac{1}{\pi} P \int_{-\infty}^{\infty} \frac{u_{\xi\xi}(\xi, t)}{\xi - x} \mathrm{d}\xi = 0, \quad (11)$$

describes a weakly nonlinear, long internal wave in a stratified fluid of great depth, where P stands for the principal value of the integration and the third term is dispersive (Ono, 1975). The BO equation has an algebraic soliton solution,

$$u = \frac{a}{a^2(x - at - x_0)^2 + 1}, \quad (12)$$

which is called a *Benjamin–Ono soliton*.

Envelope solitons

Consider next the propagation of a modulated plane wave in a nonlinear and dispersive medium where the dispersion relation is amplitude dependent, $\omega = \omega(k, |u|^2)$. Expanding around the carrier wave number k_0 and frequency ω_0, we have

$$\omega - \omega_0 = \left(\frac{\partial \omega}{\partial k}\right)_0 (k - k_0) + \frac{1}{2}\left(\frac{\partial^2 \omega}{\partial k^2}\right)_0 (k - k_0)^2 + \left(\frac{\partial \omega}{\partial |u|^2}\right)_0 |u|^2 \ldots \quad (13)$$

Replacing $\omega - \omega_0$ by $\mathrm{i}\partial/\partial t$ and $k - k_0$ by $-\mathrm{i}\partial/\partial x$ and operating on u gives

$$\mathrm{i}\left\{\frac{\partial u}{\partial t} + \left(\frac{\partial \omega}{\partial k}\right)_0 \frac{\partial u}{\partial x}\right\} + \frac{1}{2}\left(\frac{\partial^2 \omega}{\partial k^2}\right)_0 \frac{\partial^2 u}{\partial x^2} - \left(\frac{\partial \omega}{\partial |u|^2}\right)_0 |u|^2 u = 0, \quad (14)$$

which is the nonlinear Schrödinger (NLS) equation. If

$$(\partial^2 \omega/\partial k^2)_0 (\partial \omega/\partial |u|^2)_0 < 0,$$

the plane wave is unstable for the modulation, otherwise it is stable.

Under appropriate scaling and variable transformations, Equation (14) takes the standard form

$$\mathrm{i}u_t + u_{xx} + 2\varepsilon|u|^2 u = 0, \ (\varepsilon = \pm 1). \quad (15)$$

With $\varepsilon = 1$, this is called the focusing NLS (FNLS) equation and has an *envelope soliton* solution,

$$u = K\mathrm{e}^{\mathrm{i}(\tilde{k}x - \tilde{\omega}t + \theta)} \operatorname{sech}(Kx - \Omega t + \sigma), \quad (16)$$

where $-(\Omega + \mathrm{i}\tilde{\omega}) + (K + \mathrm{i}\tilde{k})^2 = 0$ and θ and σ are arbitrary constants. This is called a *bright soliton* in nonlinear optics (see Figure 2a). If the phase velocity decreases (increases) with increased amplitude, nonlinear effect results in a decrease (increase) of the wave number in the leading half of the envelope and a increase (decrease) in wave number in the trailing half. Thus, the envelope is compressed, and a soliton finds a balance between the effect of compression caused by nonlinearity and broadening caused by dispersion.

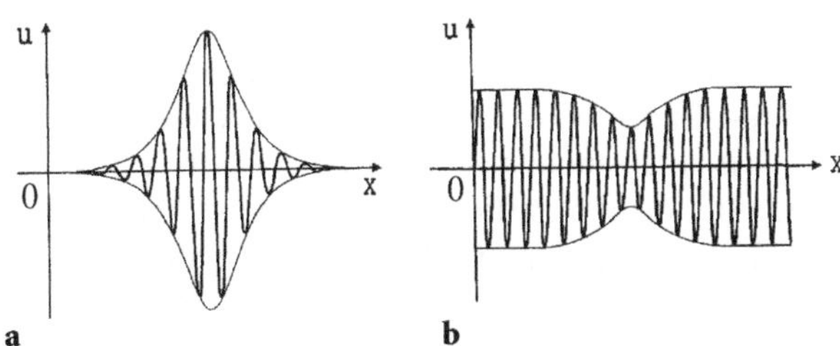

Figure 2. (a) Bright soliton and (b) dark soliton.

The defocusing NLS (DNLS) equation (with $\varepsilon = -1$) has a *dark soliton* solution (shown in Figure 2b)

$$|u|^2 = u_0^2[1 - a^2 \operatorname{sech}^2\{u_0 a(x - vt)\}], \quad (17)$$

which appears as an intensity dip in an infinitely extended constant background. The dark soliton solution with $a = 1$

$$|u|^2 = u_0^2 \tanh^2 u_0(x - vt), \quad (18)$$

is called a *black soliton* and the solution with $a < 1$ is sometimes referred to as a *gray soliton*.

Optical communication systems are degraded by the spreading out of pulses as they travel along a fiber, which is caused by dispersion. Hasagawa and Tappert proposed using the nonlinear change of dielectric constant of the fiber to compensate for this dispersive effect (Hasagawa & Tappert, 1973). The optical pulse was shown to form a bright soliton in the case of anomalous dispersion ($\partial^2\omega/\partial k^2 > 0$), a prediction that was verified by Mollenauer et al. (1980).

Writing $k = k(\omega, |u|^2)$ as the solution of the dispersion relation again leads to the NLS equation but with different variables (ω and k interchanged). The NLS equation in these variables is important in nonlinear optics.

More Space Dimensions

Spatial dark-soliton stripes are experimentally found in the transverse cross section of a continuous-wave (cw) optical beam propagating through material with a self-defocusing nonlinearity. The spatial evolution of a monochromatic transverse electric field $E(x, y, z)$ in a self-defocusing medium with Kerr nonlinearity (the intensity-dependent refractive index $n = n_0 - n_2|E|^2$) is described by the NLS equation

$$\mathrm{i}u_z + \tfrac{1}{2}\Delta_\perp u - |u|^2 u = 0, \quad (19)$$

where $\Delta_\perp = \partial_x^2 + \partial_y^2$ is the transverse diffraction operator and z is the propagation coordinate. Although dark-soliton stripes are unstable to transverse long-wavelength modulation, linear analysis predicts stability to transverse modulation having a short period.

Kivshar & Yang (1994) showed that self-defocusing nonlinear media can support a *ring soliton*, which is a dark-solitary wave with ring symmetry. If the ring radius is small enough, it is not subject to the transverse instability characteristic of dark-stripe solitons. It has also been demonstrated that in the small-amplitude limit such solitons are governed by a cylindrical KdV equation. Frantzeskakis and Malomed also showed that *anti-dark ring solitons* (humps on top of the cw background rather than dips) can exist too but only for non-Kerr saturable nonlinearities (Frantzeskakis & Malomed, 1999).

Spatial solitons are optical beams that are self-trapped in space due to a balance between the Kerr nonlinearity and diffraction, a well-studied phenomenon. A new type of spatial soliton, which was proposed by Segev et al. (1992), occurs in a photorefractive (PR) crystal biased with an external dc electric field. The presence of optical beams in a crystal leads to photoexcitation of electric charges. The space charge screens the externally applied electric field in the illuminated area of the crystal. The non-uniform screening of the electric field modifies the refractive index in such a way that the beam becomes self-trapped and propagates in a form of a *PR spatial soliton*. PR spatial solitons can occur at microwatt power levels, whereas the observation of Kerr solitons requires much higher power. Steady solitons in PR crystals are called *screening solitons*.

With the application of a slightly detuned driving field, a slightly lossy Kerr medium supports *cavity solitons*, which are then induced by writing pulses (Firth et al., 2002). This provides an optical means for writing patterns of light onto the cavity cross section and subsequently erasing them, in other words an optical memory.

It is known that the diffusion effect of photoexcited charges leads to a bending of the trajectories of PR solitons. Królikowski et al. (1996) showed by using numerical simulations that *self-bending solitons* in the presence of the diffusion processes are stable and withstand relatively large perturbations. They also showed that even a small contribution of diffusion effect leads to strong energy exchange between colliding PR solitons. In general, spatial solitons in PR media do not satisfy the mathematical definition of solitons, even when they propagate as solitary waves with an unchanging beam profile.

Self-trapping has been studied in Kerr-type, PR, quadratic, and resonant atomic nonlinear media. All of these studies have investigated self-trapping of spatially coherent light beams only. Diffraction of a spatially incoherent beam is larger than that of a coherent beam of the same width. Therefore, a spatially incoherent beam diverges much faster than a coherent beam, and self-trapping of an incoherent beam requires stronger optical nonlinearities than self-trapping a coherent beam. Recently, self-trapping of incoherent light beams has been demonstrated experimentally (Mitchell et al., 1996) and theoretically (Mitchell et al., 1997). In general, spatial *incoherent solitons* are multimode, self-trapped entities, found only in materials with non-instantaneous (internal) nonlinearity (Mitchell et al., 1997).

Another interesting phenomena is the formation of solitons by the mutual trapping of the interacting waves. Torner et al. (1996) showed that soliton-like propagation occurs in the presence of walk-off between the interacting waves. The solitons in the presence of walk-off are called *walking solitons*. Temporal walk-off is due to different group velocities of the waves forming the soliton, while spatial walk-off is due to different propagation directions of energy and phase fronts in anisotropic media.

The possibility of creating three-dimensional (3-d) localized pulses in a self-focusing medium with anomalous group-dispersion was demonstrated by Silberberg (1990), who showed that they pulses are robust in the sense that remain as separate solitary formations even after collisions. These 3-d-pulses propagating without changes in space or time are called *light bullets*.

Topological Solitons

Another mechanism leading to the emergence of solitons is a topological constraint, exemplified by the sine-Gordon (SG) equation

$$u_{tt} - u_{xx} + m^2 \sin u = 0. \tag{20}$$

The SG equation is characterized by two properties: Lorentz invariance and multiple ground states of the energy. The presence of multiple ground states aids soliton formation because the field takes different asymptotic values on both sides of the isolated wave, so the wave is unable to disperse. (It cannot decay by spreading out for the same reason a twist sealed in the boundary condition at the end of a band cannot be removed.) Thus, wave fields can incorporate *topological solitons* if there are multiple ground states with an appropriate topology. Such a soliton of the SG equation is given by

$$u = 4 \arctan \left[\exp \left(\pm m \frac{x - vt - x_0}{\sqrt{1 - v^2}} \right) \right]. \tag{21}$$

The solution represents a twist in the configuration of the field connecting one ground state $2n\pi$ to the another ground state $(2n \pm 2)\pi$. The $+$ sign and $-$ sign solutions are called a *kink* and an *antikink*, respectively, as shown in Figure 3. They are often termed *soliton* and *antisoliton* instead of kink and antikink. The energy of the SG kink solution is given by $8m/\sqrt{1 - v^2}$,

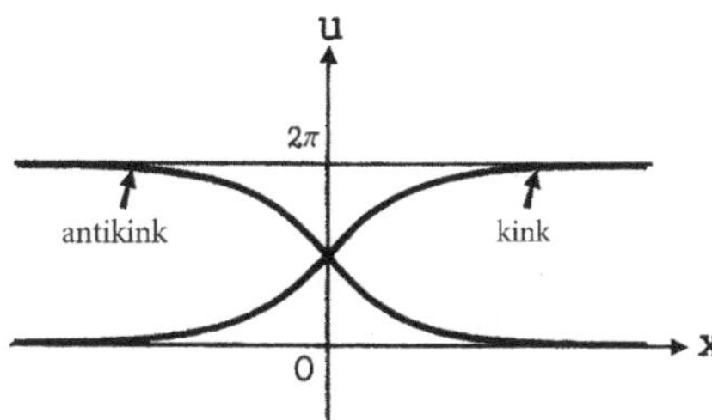

Figure 3. Kink and antikink.

which has relativistic dependence on the velocity v. There is an analogy between kink/antikink and positively/negatively charged elementary particles. The time evolution of an initial state consisting of two kinks shows that the two kinks move away from each other. On the other hand, the kink and antikink move toward each other. The twist (Q) of a kink ($+2\pi$) or antikink (-2π) is known as its *topological charge*.

Perring and Skyrme were interested in the SG equation as a model equation for physical elementary particles (Perring & Skyrme, 1962). They regarded the kink as nucleon and the linear wave (with $Q=0$) as a meson. The soliton solution of a nuclear model for which topological charge is the baryon number is called a *skyrmion*. Topological charge is a conserved quantity stemming from the geometric configuration of the field—an important feature of many field theories that admit kink solutions.

Numerous attempts were made to find particle-like solutions of relativistically invariant nonlinear field equations in higher dimensions. Under radial symmetry, ring and spherical solitons to the SG and Higgs field equations have been studied numerically by Bogolyubskii and Makhan'kov who called such waves *pulsons* due to their pulsating behavior (Bogolyubskii & Makhan'kov, 1976).

Lattice-solitons

Solitons and kinks in discrete systems are sometimes called *lattice-solitons* and *lattice-kinks*, respectively. The Toda lattice is a one-dimensional lattice of equal masses, interconnected by exponetial interaction potentials of the form (Toda, 1981)

$$\phi(r) = \frac{a}{b}\exp(-br) + ar + \text{const.}, \qquad (22)$$

where $a, b > 0$ and r is the change of a distance between adjacent masses from its equilibrium value. The equation of motion for Toda lattice is an integrable system, admitting exact N-soliton solutions, whose form for $N=1$ is given by

$$\exp(-br_n) - 1 = \sinh^2(\kappa)\text{sech}^2\left(\kappa n \pm t\sqrt{ab/m}\sinh\kappa\right), \qquad (23)$$

where κ is an amplitude parameter and m is the mass of the particles.

The theory of wave propagation in periodic structures shows the existence of forbidden frequency bands or band gaps, located around the (Bragg) reflection frequencies. A wave with overtone frequencies within the forbidden frequency bands will have little means of interaction with the periodic structure. The solitons that emerge from a balance between nonlinear self-phase modulation and dispersion associated with the periodic structure are called *gap solitons*.

Other Types of Solitons

A variable coefficient KdV equation arises in the study of a solitary water waves as it enters a region where the bottom is no longer level (Johnson, 1980). A soliton on non-uniform background generally moves with variable speed. When a soliton comes from left (or right) and goes back to left (or right) with opposite velocity—sailing back like an Australian boomerang—it is called a *boomeron*. When a soliton oscillates without ever escaping to infinity, the soliton is referred to as a *trappon* (Calogero & Degasperis, 1982).

The *Dym* equation ($r_t = r^3 r_{xxx}$) was discovered in unpublished work by Harry Dym and rediscovered in a more general form by Sabatier within the classical string problem. This equation belongs to a wide class of nonlinear equations (WKI equations) found by Wadati, Konno, and Ichikawa to be completely integrable (Wadati et al., 1979). Reciprocal links between the Dym and KdV and MKdV equations provide implicit solutions, since they include the simultaneous change of both dependent and independent variables (Kawamoto, 1985). The Dym equation has *cusp soliton* solutions. Some new approaches, which allow construction of multisoliton solutions almost explicitly have been developed for the Dym equation by (Dmitrieva, 1993). Recently, it was shown that the Dym equation on the complex plane is relevant to such physical problems as the Hele-Shaw problem and the Saffman–Taylor problem (Constantin & Kadanoff, 1991).

The nonlinear transverse oscillation of an elastic beam subject to an end-thrust is described by one of the WKI equations. If the beam is flexible enough, it deforms into a loop, with the upper half portion having negative curvature. In this case, the nonlinear oscillation can be described by an equation of the following form (Konno et al., 1981):

$$y_{xt} + \text{sgn}\left(\frac{\mathrm{d}s}{\mathrm{d}x}\right)\left(\frac{y_{xx}}{(1+y_x^2)^{3/2}}\right)_{xx} = 0, \qquad (24)$$

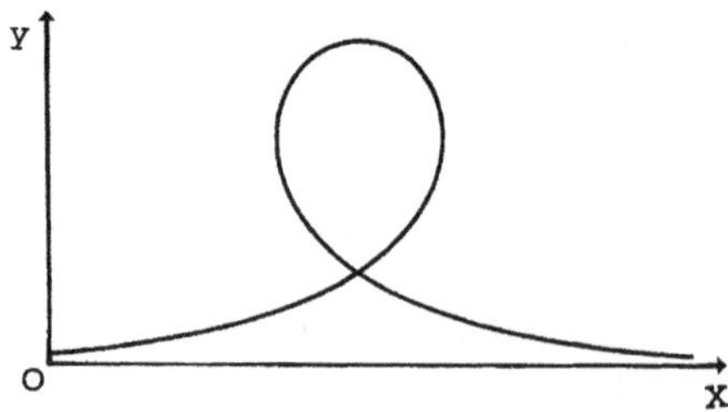

Figure 4. Loop soliton.

where s denotes arc length measured around the loop. Konno et al. showed that such a *loop soliton* (shown in Figure 4) propagating along a stretched rope can be obtained as a one soliton solution to Equation (24).

Solitary waves with finitely extended (compact) support were recently studied in various equations with nonlinear dispersion. Rosenau and Hyman showed that solitary-wave solutions may have compact support under the influence of nonlinear dispersion in various generalizations of the KdV equations with nonlinear dispersion (Rosenau & Hyman, 1993). The nonlinear dispersion is weaker for small amplitude than the linear dispersion in the KdV equation, leading to compactification. Such robust soliton-like pulses—characterized by the absence of the infinite tail—are called *compactons*.

MASAYOSHI TAJIRI

See also **Korteweg–de Vries equation; Multidimensional solitons; Nonlinear optics; Nonlinear Schrödinger equations; Sine-Gordon equation; Skyrmions; Solitons; Toda lattice**

Further Reading

Ablowitz, M.J. & Satsuma, J. 1978. Solitons and rational solutions of nonlinear evolution equations. *Journal of Mathematical Physics*, 19: 2180–2186

Bogolyubskii, I.L. 1976. Oscillating particle-like solutions of the nonlinear Klein–Gordon equation. *Soviet Physics JETP Letters*, 24: 535–538

Bogolyubskii, I.L. & Makhan'kov, V.G. 1976. Lifetime of pulsating solitons in certain classical models. *Soviet Physics JETP Letters*, 24: 12–14

Calogero, F. & Degasperis, A. 1982. *Spectral Taransform and Solitons I*, North-Holland, Amsterdam

Constantin, P. & Kadanoff, L. 1991. Dynamics of a complex interface. *Physica* D, 47: 450–460

Dmitrieva, L.A. 1993. Finite-gap solutions of the Harry Dym equation. *Physics Letters* A, 182: 65–70

Firth, W.J., Harkness, G.K., Lord, A., McSloy, J.M., Gomila, D. & Colet, P. 2002. Dynamical properties of two-dimensional Kerr cavity solitons. *Journal of Optical Society of America* B, 19(4): 747–752

Frantzeskakis, D.J. & Malomed, B.A. 1999. Multiscale expansions for a generalized cylindrical nonlinear Schrödinger equation. *Physics Letters* A, 264: 179–185

Grimshaw, R., Pelinovsky, E. & Talipova, T. 1999. Solitary wave transformation in a medium with sign-variable quadratic nonlinearity and cubic nonlinearity. *Physica* D, 132: 40–62

Hasagawa, A. & Tappert, F. 1973. Taransmission of stationary nonlinear optical pulses in dispersive dielectric fibers. I. Anomalous dispersion, II. Normal dispersion. *Applied Physics Letters*, 23: 142–144, 171–172

Johnson, R.S. 1980. Water waves and Korteweg–de Vries equations. *Journal of Fluid Mechanics*, 97: 701–719

Kadomtsev, B.B. & Petviashvili, V.I. 1970. On the stability of solitary waves in weakly dispersing media. *Soviet Physics-Doklady*, 15: 539–541

Kawamoto, S. 1985. An exact transformation from the Harry Dym equation to the modified K-dV equation. *Journal of the Physical Society of Japan*, 54: 2055–2056

Kivshar, Y.S. & Yang, X. 1994. Ring dark solitons. *Physical Review* E, 50: R40–R43

Konno, K., Ichikawa, Y.H. & Wadati, M. 1981. A loop soliton propagating along a stretched rope. *Journal of the Physical Society of Japan*, 50: 1025–1026

Królikowski, W., Akhmediev, N., Luther-Davies, B. & Gronin-Golomb, M. 1996. Self-bending photorefractive solitons. *Physical Review* E, 54: 5761–5765

Manakov, S.V., Zakharov, V.E., Bordag, L.A., Its, A.R. & Matveev, V.B. 1977. Two-dimensional solitons of the Kadomtsev–Petviashvili equation and their interaction. *Physics Letters*, 63A: 205–206

Mitchell, M., Chen, Z., Shih, M. & Segev, M. 1996. Self-trapping of partially spatially incoherent light. *Physical Review Letters*, 77: 490–493

Mitchell, M., Segev, M., Coskun, T.H. & Christodoulides, D.N. 1997. Theory of self-trapping spatially incoherent light beams. *Physical Review Letters*, 79: 4990–4993

Mollenauer, L.F., Stolen, R.H. & Gordon, J.P. 1980. Experimental observation of picosecond pulse narrowing and solitons in optical fibers. *Physical Review Letters*, 45: 1095–1098

Ono, H. 1975. Algebraic solitary waves in stratified fluids. *Journal of the Physical Society of Japan*, 39: 1082–1091

Perring, J.K. & Skyrme, T.H.R. 1962. A model unified field equation. *Nuclear Physics*, 31: 550–555

Rosenau, P. & Hyman, J. M. 1993. Compactons: solutions with finite wavelength. *Physical Review Letters*, 70: 564–567

Segev, M., Crosignani, B., Yariv, A. & Fischer, B. 1992. Spatial solitons in photorefractive media. *Physical Review Letters*, 68: 923–936

Silberberg, Y. 1990. Collapse of optical pulses. *Optics Letters*, 15: 1282–1284

Tajiri, M. & Murakami, Y. 1989. The periodic soliton resonance: solutions to the Kadomtsev–Petviashvili equation with positive dispersion. *Physics Letters* A, 143: 217–220

Toda, M. 1981. *Theory of Nonlinear Lattices*, Springer-Verlag, Berlin

Torner, L., Mazilu, D. & Mihalache, D. 1996. Walking solitons in quadratic nonlinear media. *Physical Review Letters*, 77: 2455–2458

Wadati, M., Konno, K. & Ichikawa, Y. H. 1979. New integrable nonlinear evolution equations. *Journal of the Physical Society of Japan*, 47: 1698–1700

SOLUTION TRAJECTORY

See **Phase space**

SPATIOTEMPORAL CHAOS

Spatiotemporal chaos is a dynamical regime developing in spatially distributed systems lacking long-time, large-distance coherence in spite of an organized regular behavior at the local scale. It is, so to speak, located in the middle of a triangle, the corners of which are temporal chaos, which is prevalent for a few spatially frozen degrees of freedom, spatial chaos, in disordered time-independent patterns, and turbulence, with cascading processes over a wide range of space and time scales.

Short-term local coherence is usually the result of some instability mechanism that generates dissipative structures of different kinds depending on whether or not a specific frequency ($\omega_c = 0$ or $\neq 0$) and/or a spatial periodicity ($k_c = 0$ or $\neq 0$) is introduced in the system (Cross & Hohenberg, 1993). Examples of spatiotemporal chaos may be found in every combination of these elementary cases: Rayleigh–Bénard convection produces a time-independent cellular pattern ($\omega_c = 0, k_c \neq 0$), the Belousov–Zhabotinsky (BZ) reaction-diffusion system is unstable against a homogeneous oscillatory mode ($\omega_c \neq 0, k_c = 0$), convection in binary fluid mixtures develops in the form of dissipative waves ($\omega_c \neq 0, k_c \neq 0$), and the same holds for hydrothermal waves. Transitions to spatiotemporal chaos have been observed and studied in many experimental systems, including parametrically excited surface waves, electro-hydrodynamic instabilities in nematic liquid crystals (Kramer & Pesch, 1996), and liquid films flowing down inclines (Chang, 1994). Figure 1 (left) illustrates the case of spiral defect chaos in convection.

In practice, confinement effects enter and compete with instability mechanisms in the ordering process. Their intensity can be appreciated through aspect ratios, measuring the physical size of the system in units of the instability wavelength. In small aspect-ratio experiments, confinement effects are effective in the three space directions and chaos is purely temporal. Spatiotemporal chaos develops when confinement is partially relaxed. Accordingly, phenomena developing at surfaces or in thin layers can be understood as (quasi-)two-dimensional (Gollub, 1994). In the same way, narrow channels or oriented media along a specific direction exemplify the case of (quasi-)one-dimensional systems (Daviaud, 1994).

A second important classification results from the nature of the bifurcation, either supercritical or subcritical (continuous or discontinuous), which implies either substitution or coexistence of bifurcating and bifurcated states. To a large extent, this feature dictates the type of theory most appropriate to understand the growth of spatiotemporal chaos.

Consider first the supercritical case. When confinement effects are unable to maintain order everywhere, reduced universal descriptions are generically obtained as "envelope equations" (Newell, 1974). Standing as a paradigm, the cubic complex Ginzburg–Landau (CGL3), reviewed by Aranson & Kramer (2002), spatially unfolds a local supercritical Hopf bifurcation signalling the emergence of uniform oscillations ($\omega_c \neq 0$, $k_c = 0$). In one dimension (1-d), this equation reads

$$\partial_t A = A + (1 + i\alpha)\partial_{xx} A - (1 + i\beta)|A|^2 A, \qquad (1)$$

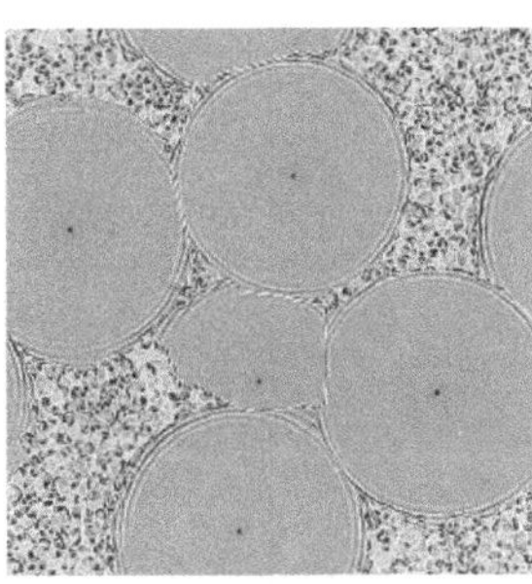

Figure 1. Left: Spiral defect chaos near threshold in Rayleigh–Bénard convection at $\mathcal{O}(1)$ Prandtl number. (Courtesy Ahlers, 1998). Right: Coexistence of spirals and defect mediated chaos in the 2-d CGL equation for $\alpha = -2$ and $\beta \simeq 0.75$ (Manneville, unpublished).

where the parameters α and β measure linear and nonlinear dispersion effects, respectively. The CGL3 equation admits trivial exact solutions in the form of plane waves $A = A_q \exp[i(qx - \omega_q t)]$, with amplitude $A_q = (1 - q^2)^{1/2}$ and angular frequency $\omega_q = \alpha q^2 + \beta(1 - q^2)$, which are stable or unstable depending on the values of α, β, and q. Other nonlinear solutions can exist; in one dimension, solitary waves called Bekki–Nozaki holes are the best known. In two dimensions (2-d) they take the form of spiral waves that are topological defects of the complex order parameter A. Figure 1 (right) illustrates the coexistence of spirals and defect-mediated chaos (see below) in the 2-d CGL equation. Figure 2 displays the different possible steady-state regimes of the 1-d CGL equation in the $(1/\beta, -\alpha)$ plane. In region I, plane waves attract most initial conditions. Phase turbulence is present in region IV slightly beyond the Benjamin–Feir instability (BF) line, as given by Newell's criterion for "$q = 0$" oscillations (Newell, 1974). In the vicinity of this line, the solution can be written as $A(x, t) = (1 + \varrho(x, t)) \exp(i\theta(x, t))$. The amplitude modulation ϱ, enslaved to the gradient of the phase perturbation θ, remains small, while θ is governed at lowest order by the Kuramoto–Sivashinsky equation (Kuramoto, 1978)

$$\partial_t \theta = D\partial_{xx}\theta - K\partial_{xxxx}\theta + g(\partial_x\theta)^2, \qquad (2)$$

where $D = 1 + \alpha\beta$ is an effective diffusion coefficient that is negative in the unstable range (Chaté & Manneville, 1994a). Deeper in the unstable domain, in region V, a "revolt" of $|A|$ ends in the formation of defects (phase singularities at zeroes of $|A|$) and amplitude turbulence or defect-mediated turbulence sets in (Coullet et al., 1989). Defects analogous to Bekki–Nozaki holes are observed to evolve in a spatiotemporal intermittent fashion in region II. As suggested by its position in the diagram, the "bi-chaos" regime in region III presents itself as a fluctuating mixture of states in regions IV and V.

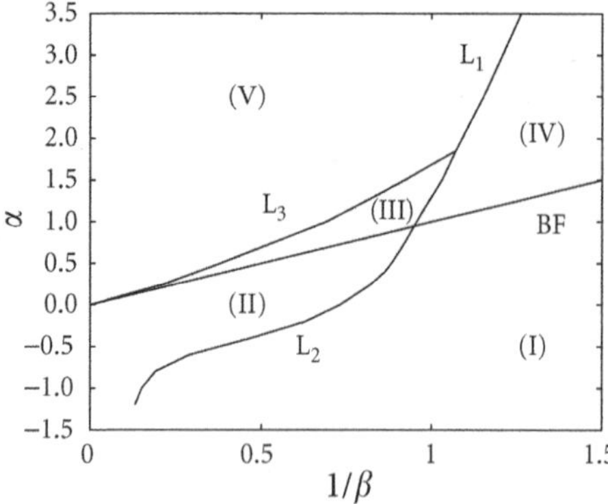

Figure 2. Bifurcation diagram of the 1-d CGL equation. (Data from Chaté, 1994.)

By contrast, subcritical instabilities are characterized by the possibility of finding the system in one of several (usually two) states at a given point in space. Short-range coherence then implies the formation of homogeneous domains of each state, separated by fronts (Pomeau, 1986). In gradient systems, these fronts move regularly so as to decrease the potential, but in nongradient systems, they may have more complicated behaviors. A particularly interesting scenario develops when one of the competing local states is a chaotic transient while the other is regular. At a given time, the whole system can be divided into so-called laminar and turbulent domains, and at a given point in space, the system is alternatively laminar or turbulent, hence the name "spatiotemporal intermittency" (STI). According to Pomeau (1986), front propagation is then akin to a time-oriented stochastic process known as "directed percolation" (DP) used to model epidemic processes (Kinzel, 1983). Directed percolation defines a critical phenomenon, with an associated universality class (a specific set of scaling exponents governing the statistical behavior of the system as a function of the distance to threshold).

The existence of separated domains and sharp fronts supports the idea of modeling extended systems in terms of identical subsystems arranged on a regular lattice, each with its own phase space, and coupled to its neighbors. In addition to space discretization, time discretization leads to the definition of coupled map lattices that have served to illustrate several transition scenarios such as cascades of spatial period doublings, defect-mediated regimes (Kaneko, 1993), or STI (Chaté & Manneville 1994b) that make explicit how local transient temporal chaos is converted into sustained spatiotemporal chaos. A last step can be taken by also discretizing the local phase space, which yields cellular automata (Wolfram, 1986). Further randomization of the dynamics then points towards an understanding of the transition to turbulence in statistical physics terms (Kinzel, 1983). Such approaches should help in understanding plane Couette flow, the simplest shear flow produced by two parallel plates sliding in opposite directions. This configuration is a particularly intriguing example of a subcritical hydrodynamic system where domains of laminar flow, known to be stable for all flow conditions, coexist in a wide range of Reynolds numbers and in continuously varying proportions with domains of small-scale turbulence.

Spatio-temporal chaos definitely relates to the process of transition to turbulence when confinement effects are weak. It appears to occupy a central position at the crossroads of nonlinear dynamics, mathematical stability theory, and statistical physics of many-body systems and non-equilibrium processes, with a wide potential for applications (Rabinovich et al., 2000).

PAUL MANNEVILLE

See also **Cellular automata; Chaos vs. turbulence; Coherence phenomena; Coupled map lattice; Development of singularities; Gradient system; Hydrothermal waves; Kuramoto–Sivashinsky equation; Modulated waves; Multiple scale analysis; Navier–Stokes equation; Nonequilibrium statistical mechanics; Order parameters; Pattern formation; Phase dynamics; Reaction-diffusion systems; Spiral waves; Surface waves; Topological defects; Turbulence**

Further Reading

Ahlers, G. 1998. Experiments on spatio-temporal chaos. *Physica* A, 249: 18–26

Aranson, I.S. and Kramer, L. 2002. The world of the complex Ginzburg–Landau equation. *Reviews of Modern Physics*, 74: 99–143

Chang, H.-C. 1994. Wave evolution on a falling film. *Annual Reviews of Fluid Mechanics*, 26: 103–136

Chaté, H. 1994. Spatiotemporal intermittency regimes of the one-dimensional complex Ginzburg–Landau equation. *Nonlinearity* 7: 185–204

Chaté, H. & Manneville, P. 1994a. Phase turbulence. In *Turbulence, A Tentative Dictionary*, edited by P. Jabeling & O. Cardoso, NewYork: Plenum press

Chaté, H. & Manneville, P. 1994b. Spatiotemporal intermittency. In *Turbulence A Tentative Dictionary*, edited by P. Jabeling & O. Cardoso, NewYork: Plenum press

Cross, M.C. & Hohenberg, P.C. 1993. Pattern formation outside equilibrium. *Reviews of Modern Physics*, 65: 851–1112

Coullet, P., Gil, L. & Lega, J. 1989. Defect-mediated turbulence. *Physical Review Letters*, 62: 1619–1622

Daviaud, F. 1994. Experiments in 1D turbulence. In *Turbulence A Tentative Dictionary*, edited by P. Jabeling & O. Cardoso, NewYork: Plenum press

Gollub, J.P. 1994. Experiments on spatiotemporal chaos (in two dimensions). In *Turbulence A Tentative Dictionary*, edited by P. Jabeling & O. Cardoso, NewYork: Plenum press

Kaneko, K. (editor). 1993. *Theory and Application of Coupled Map Lattices*. Chichester and New York: Wiley

Kinzel, W. 1983. Directed percolation. In *Percolation Structures and Processes*, edited by G. Deutscher, R. Zallen & J. Adler, Annals of the Israel Physics Society, 5: 425–445

Kramer, L. & Pesch, W. 1996. Electrohydrodynamic instabilities in nematic liquid crystals. In *Pattern Formation in Liquid*

Crystals, edited by A. Buka & K. Kramer, New York: Springer

Kuramoto, Y. 1978. Diffusion-induced chaos in reaction systems. *Suppl. Progress in Theoretical Physics*, 64: 346–367

Newell, A.C. 1974. Envelope equations *Lectures in Applied Mathematics*, 15: 157–163

Pomeau, Y. 1986. Front motion, metastability and subcritical bifurcations in hydrodynamics. *Physica* D, 23: 3–11

Rabinovich, M.I., Ezersky, A.B. & Weidman, P.D. 2000. *The Dynamics of Patterns*. Singapore: World Scientific

Tabeling, P. & Cardoso, O. (editors). 1994. *Turbulence: A Tentative Dictionary*. New York: Plenum Press

Wolfram, S. (editor). 1986. *Theory and Applications of Cellular Automata*. Singapore: World Scientific

SPECTRAL ANALYSIS

Spectral analysis is a central method in studies of linear systems (which obey the superposition principle), and spectral representations are widely used in acoustics, quantum mechanics, wave propagation, optical spectroscopy, harmonic analysis, and signal processing. If $f(x)$ is a square integrable function on real line, it can be represented in a dual (spectral) space k with the integral transforms (Hildebrand, 1976)

$$f(x) = \frac{1}{2\pi}\int_{-\infty}^{\infty} \hat{f}(k)\mathrm{e}^{\mathrm{i}kx}\mathrm{d}k \tag{1}$$

and

$$\hat{f}(k) = \int_{-\infty}^{\infty} f(x)\mathrm{e}^{-\mathrm{i}kx}\mathrm{d}x. \tag{2}$$

The Fourier transform $\hat{f}(k)$ gives the spectral density of harmonic oscillations $\mathrm{e}^{\mathrm{i}kx}$ between the wave numbers k and $k+dk$. The inverse Fourier transform $f(x)$ corresponds to a spectral decomposition of a function $f(x)$ over a continuous linear combination of the harmonic oscillations $\mathrm{e}^{\mathrm{i}kx}$.

For periodic functions with a period L, such that $f(x+L)=f(x)$, the spectral density has peaks at wave numbers $k=k_n=2\pi n/L$; that is, the spectral decomposition of a function $f(x)$ becomes a discrete sum of the harmonic oscillations $\mathrm{e}^{\mathrm{i}k_n x}$:

$$f(x) = \sum_{n=-\infty}^{\infty} \hat{f}_n \mathrm{e}^{\mathrm{i}k_n x} \tag{3}$$

and

$$\hat{f}_n = \frac{1}{2L}\int_{-L}^{L} f(x)\mathrm{e}^{-\mathrm{i}k_n x}\mathrm{d}x, \tag{4}$$

where $\hat{f}_n$ are Fourier coefficients of the complex Fourier series for $f(x)$. Thus, depending on properties of a function $f(x)$, it can be decomposed over continuous or discrete spectrum in the Fourier spectral representation.

As an example, the rectangular wave defined as

$$\Delta_\varepsilon(x) = \begin{cases} 1/2\varepsilon, & |x| < \varepsilon, \\ 0, & |x| > \varepsilon \end{cases} \tag{5}$$

can be expressed in the continuous spectral representation with the Fourier transform:

$$\hat{\Delta}_\varepsilon(k) = \frac{1}{2\varepsilon}\int_{-\varepsilon}^{\varepsilon} \mathrm{e}^{-\mathrm{i}kx}\mathrm{d}x = \frac{\sin\varepsilon k}{\varepsilon k}. \tag{6}$$

The power spectrum of a signal is defined as the squared amplitude of its Fourier spectrum: $\hat{R}(k)=|\hat{f}(k)|^2$. Equivalently, the power spectrum is the Fourier spectrum of the autocorrelation function $R(x)$ defined as

$$\begin{aligned} R(x) &= \int_{-\infty}^{\infty} f(x')f(x'-x)\,\mathrm{d}x' \\ &= \frac{1}{2\pi}\int_{-\infty}^{\infty} |\hat{f}(k)|^2\mathrm{e}^{\mathrm{i}kx}\,\mathrm{d}k \\ &= \frac{1}{2\pi}\int_{-\infty}^{\infty} \hat{R}(k)\mathrm{e}^{\mathrm{i}kx}\,\mathrm{d}k. \end{aligned} \tag{7}$$

The autocorrelation function $R(x)$ represents similarities between the function $f(x)$ and itself, and its Fourier transform can be really measured in applications.

A power spectrum measures the energy of a signal between the wave numbers k and $k+\mathrm{d}k$. The total energy of a signal is related to the power spectrum by the Parseval formula:

$$E = \int_{-\infty}^{\infty} f^2(x)\,\mathrm{d}x = \frac{1}{2\pi}\int_{-\infty}^{\infty} |\hat{f}(k)|^2\,\mathrm{d}k = R(0). \tag{8}$$

As a result, the autocorrelation function is bounded as $|R(x)| \le R(0) = E$.

Linear differential equations, especially initial-value problems for wave equations, can be easily solved with the use of spectral decompositions such as Fourier transforms. In the linear limit, for example, small-amplitude long water waves are described by the linearized KdV equation:

$$\frac{\partial u}{\partial t} + \frac{\partial^3 u}{\partial x^3} = 0, \tag{9}$$

with initial data: $u(x,0)=f(x)$. The Fourier transform method represents the solution to this problem as (Ablowitz & Fokas, 1997)

$$u(x,t) = \frac{1}{2\pi}\int_{-\infty}^{\infty} \hat{u}(k,t)\mathrm{e}^{\mathrm{i}kx}\mathrm{d}k, \tag{10}$$

where $\hat{u}(k,0)=\hat{f}(k)$. Since the Fourier transform $\hat{f}(k)$ has the property $\hat{f'}(k)=\mathrm{i}k\hat{f}(k)$, the time evolution of the spectral density $\hat{u}(k,t)$ is trivial,

$$\frac{\partial \hat{u}}{\partial t} - \mathrm{i}k^3\hat{u}(k,t) = 0, \tag{11}$$

with the solution $\hat{u}(k,t)=\hat{f}(k)\mathrm{e}^{\mathrm{i}k^3 t}$. As a result, the exact solution $u(x,t)$ of the initial-value problem for the linearized KdV equation is a spectral superposition with density $\hat{f}(k)$ of waves $\mathrm{e}^{\mathrm{i}(kx-\omega(k)t)}$, where

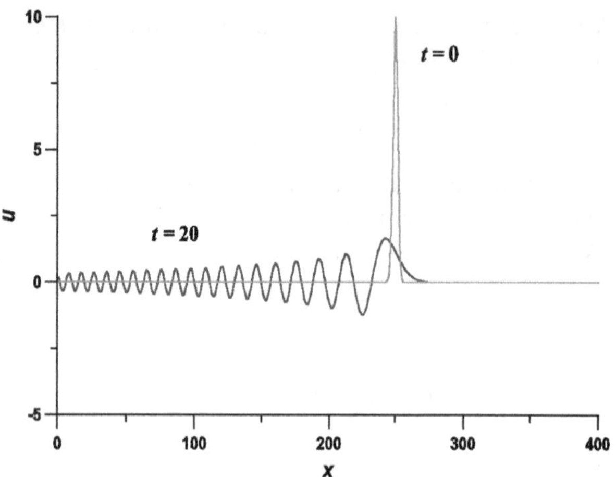

Figure 1. Exact solution $u(x,t)$ of the linear KdV equation in the Fourier spectral representation for $t>0$, compared with the initial (Gaussian) form $u(x,0)=\mathrm{e}^{-x^2}$.

$\omega(k) = -k^3$ is the dispersion relation. Figure 1 shows the exact solution $u(x,t)$ at a time $t>0$, which evolves from the initial (Gaussian) form $u(x,0)=f(x)=\mathrm{e}^{-x^2}$.

The inverse scattering transform (IST) extends spectral analysis for solutions of initial-value problems for nonlinear differential equations. For example, a solution of the nonlinear KdV equation

$$\frac{\partial u}{\partial t} + \frac{\partial^3 u}{\partial x^3} - u\frac{\partial u}{\partial x} = 0, \tag{12}$$

such that $u(x,0) = f(x)$ is related to the spectrum of the stationary Schrödinger equation on a real line of x with a t-dependent potential $u(x;t)$ (Ablowitz et al., 1974)

$$-\psi_\lambda''(x) + u(x;t)\psi_\lambda(x) = \lambda\psi_\lambda(x). \tag{13}$$

If $f(x)$ is a square integrable function on real line, the spectral data consists of the continuous spectrum at $\lambda = k^2 \geq 0$ with eigenfunctions $\psi_\lambda = \Psi(x;k)$ and a finite discrete spectrum at $\lambda = -p_n^2 < 0$ with eigenfunctions $\psi_\lambda = \Psi_n(x)$. The spectral decomposition of the potential $u(x;t)$ over eigenfunctions of the continuous and discrete spectrum takes the form (Pelinovsky & Sulem, 2000)

$$\begin{aligned} u(x;t) = {} & \frac{1}{2\pi}\int_{-\infty}^{\infty} \rho(k,t)\Psi(x;k)\,\mathrm{d}k \\ & + \sum_n \rho_n(t)\Psi_n(x), \end{aligned} \tag{14}$$

where $\rho(k,t)$ and $\rho_n(t)$ are coefficients of the spectral decomposition. The time-evolution of the spectral data has the simple solution

$$\begin{aligned} \rho(k,t) &= \rho(k,0)\mathrm{e}^{ik^3 t}, \\ \rho_n(t) &= \rho_n(0)\mathrm{e}^{p_n^3 t}. \end{aligned} \tag{15}$$

The initial values for the spectral data are found from spectral analysis of the stationary Schrödinger equation with the initial potential $u(x;0) = f(x)$.

DMITRY PELINOVSKY

See also **Generalized functions; Integral transforms; Inverse scattering method or transform; Quantum theory**

Further Reading

Ablowitz, M.J. & Fokas, A.S. 1997. *Complex Variables: Introduction and Applications*, Cambridge and New York: Cambridge University Press

Ablowitz, M.J., Kaup, D.J., Newell, A.C. & Segur, H. 1974. The inverse scattering transform-Fourier analysis for nonlinear problems. *Studies in Applied Mathematics*, 53: 249–315

Hildebrand, F.B. 1976. *Advanced Calculus for Applications*, 2nd edition, Englewood Cliffs, NJ: Prentice-Hall

Pelinovsky, D.E. & Sulem, C. 2000. Eigenfunctions and eigenvalues for a scalar Riemann–Hilbert problem associated to inverse scattering. *Communications in Mathematical Physics*, 208: 713–760

SPIN SYSTEMS

Spin systems generally refer to ordered magnetic systems. A feel for how spins behave can be obtained by observing the behavior of two or more small compasses or magnetic needles pinned close to each other but free to rotate in a plane (perpendicular to the north–south direction), so that they align in parallel corresponding to a minimum energy configuration.

Even though magnetism is one of the oldest natural phenomena known, it took many centuries to identify its origin. In 1820, the Danish physicist Hans Oersted observed that electric current flowing through a wire affects a nearby magnet, and it is now known that the motion of an electron in an orbit around the nucleus of an atom is equivalent to a small electric current loop that behaves as an atomic magnet called a magnetic dipole moment. Each electron also possesses a rotation about its own axis, which is identified as electron spin, and is equivalent to a circulating electric current with its own dipole magnetic moment.

More generally, spin angular momentum or spin is an intrinsic property, associated with quantum particles, which does not have a classical counterpart. The magnetic dipole moment due to the spinning motion with spin angular momentum operator $\boldsymbol{S}$ is given by $\boldsymbol{m}_\mathrm{S} = -(g\mu_\mathrm{B}/\hbar)\boldsymbol{S}$, where g is the gyromagnetic ratio, μ_B is the Bohr magneton and $\hbar = h/2\pi$ is Planck's constant.

Macroscopically, all substances are magnetic to some extent and every material when placed in a magnetic field acquires a magnetic moment. Magnetic materials can be classified in terms of magnetic moment in the absence of an applied field. The magnetic moment is zero for each atom in diamagnetic materials. In the case of paramagnetic materials, for each atom the moment is nonzero but still averages to zero over many atoms. In ferromagnetic materials, the moment of each atom and even the average is not zero (Mattis, 1988).

Model	Directional Coefficient
Ising/Potts	$a=b=0$
XY	$c=0$
Easy axis exchange anisotropy	$c>a=b$
Easy plane exchange anisotropy	$c<a=b$
Heisenberg	$a=b=c=1$
XYZ	$a\neq b\neq c$

Table 1. Ferromagnetic models.

Ferromagnetic materials, which are made up of domains, exhibit long-range ordering that causes the spins of the atomic ions to line up parallel to each other in a domain. The underlying interaction originates from an exchange interaction that is caused by the overlapping of electronic wave functions that has no classical analog. The exchange interaction is a quantum phenomenon of electrostatic origin first proposed by Heisenberg, and the Hamiltonian representing spin-spin exchange interaction is given by

$$H_{\rm ex} = -\sum_{\langle ij\rangle} J_{ij}\boldsymbol{S}_i\cdot\boldsymbol{S}_j. \quad (1)$$

Here $\boldsymbol{S}_i$ is the total spin of all the electrons bound to the atom or ion at the lattice site i. J_{ij} is the phenomenological exchange constant and is positive for ferromagnetic coupling. It is often sufficient to consider only interaction of nearest neighbors, represented by the bracket $\langle ij\rangle$. In the isotropic case J_{ij} is treated as a constant, J.

Different magnetic spin models can be identified by generalizing the Heisenberg Hamiltonian by introducing directional coefficients,

$$H_{\rm ex} = -\sum_{\langle ij\rangle}\left[aS_{ix}S_{jx} + bS_{iy}S_{jy} + cS_{iz}S_{jz}\right]. \quad (2)$$

The underlying models are indicated in Table 1. Among these, the Ising model is the simplest, where the interacting spins can be either parallel or antiparallel. The Potts model is a generalization of the Ising model where the number of spin states is extended to k equally spaced directions in a plane.

Other models involve additional interactions: (i) *magnetocrystalline anisotropy* is due to interaction between the crystal field and the spin-orbit coupling, either of easy axis or of easy plane. (ii) *Zeeman energy*: When a strong external magnetic field $\boldsymbol{H}$ is applied, the Hamiltonian is modified with the addition of a term $H_Z = -g\mu_{\rm B}\sum_i \boldsymbol{S}_i\cdot\boldsymbol{H}$. (iii) One can also consider interactions such as biquadratic exchange, or weak interactions such as Dzyaloshinski–Moriya type.

There are certain materials for which the exchange integrals J_{ij} become negative. Consequently, the spins align antiparallel to each other so that the net magnetization is zero. These materials are called antiferromagnets and are normally modeled to be made up of two sublattices. However, if the resulting total magnetization, which is the difference between that of the two sublattices, is nonzero, the material is a ferrimagnet.

A spin glass is another state of magnetism that is different from the long-range ordered ferro- and antiferromagnetic phases. A spin glass phase can be considered as consisting of randomly interacting spins with ferro- or antiferromagnetic coupling between them. The combination of the randomness with the competing or mixed interactions causes frustration. Consequently, the exchange integrals J_{ij} are normally distributed random variables corresponding to the so-called Ruderman, Kittel, Kasuya, and Yoshida (RRKY) interaction. The two famous models that describe spin glass are the Edwards–Anderson (EA) model and the Sherrington–Kirkpatrick (SK) mean field model.

The spin glass theory can also be considered as a statistical model of neural networks. The simple spin glass theory based on the SK model provides a qualitative description of memory. In a neural network, the minimal structure element is a neuron whose state can be described by one real variable, and a simple Ising representation $\boldsymbol{S}_i = \pm 1$ is often used. The problem of neural network systems can be formulated by writing the Hamiltonian in Ising form with the so-called Hebbian learning rule for J_{ij}, which is the Hopfield model (Dotsenko, 1994).

The Bethe ansatz is extremely useful in solving quantum spin chain models and to obtain exact results including ground state and excited states. A chain of spin one-half with spin-spin coupling has been solved using the Bethe ansatz method. However, the quantum mechanical description corresponding to the spin Hamiltonian with higher spin values is extremely complicated because it is difficult to know in which particular state one of the individual spins finds itself. Even in the ground state this is impossible. Hence in these cases, one looks for limiting procedures, either classical or semiclassical.

Owing to the presence of the exchange interaction, a small disturbance in one of the spins will be propagated in the form of spin waves. It can also be regarded as oscillations in the magnetic moment density, propagating through a magnetically ordered crystal. The spin wave energy must be equal to the excitation energy of the crystal required to cause a change in the orientation of the spin. The basic entity of the quantized form of the spin wave is called a magnon.

In semiclassical theories of magnetism it is common to approximate the unwieldy spin operators or matrices (which obey the standard spin commutation relations) by the analogous harmonic oscillator operators. Among the machineries established to handle the spin operators, namely, (i) Holstein–Primakoff (H–P) transformation, (ii) Schwinger–Boson representation, (iii) Dyson–Maleev transformation, and (iv) Jordon–

Wigner transformation. The H–P transformation is the most useful one.

Just like photons and phonons, spin waves are bosons describing collective excitations that involve all the spins to the same degree. The operators $S_n^{\pm} = S_{nx} \pm \mathrm{i}S_{ny}$ act on the ground state as field theoretical creation and annihilation operators. It is, thus, reasonable to change by means of a unitary transformation to operators a and a^+ that satisfy the Bose commutation relation $[a_i, a_j^+] = \delta_{ij}$. In this way, one can obtain the H–P transformation of spin operators, which transforms the quantum mechanical spin operators into standard boson operators. In the low-temperature limit, one can use semiclassical expansion for S_n^+ and S_n^-. The Heisenberg equation of motion for the boson operator $\mathrm{i}\hbar \mathrm{d}a_j/\mathrm{d}t = [a_j, H]$ when combined with Glauber's coherent state representation for Bose operators in the long wavelength, low temperature limit (continuum limit) leads to nonlinear evolution (partial differential) equations. Specific one-dimensional cases correspond to integrable spin models exhibiting soliton type spin excitations in addition to linear magnons (Kosevich et al., 1991).

Alternatively, in the large spin limit, the quantum description goes over to the classical vector model for spin, so that the spin operator can be replaced by an ordinary vector of length $s(s+1) > s$, which can be treated as a dynamical/canonical variable. For a ferromagnet described by Heisenberg Hamiltonian (Equation (1)) or its generalization, the classical canonical equations for the spins can be obtained in analogy with a spinless nonrelativistic particle, which is a typical nonlinear dynamical system. The resultant equation of motion $\mathrm{d}\boldsymbol{S}_n/\mathrm{d}t = \{\boldsymbol{S}_n, H\}$, where $\{\ \}$ represents spin Poisson bracket defined for functions of spins under continuum approximation, reduces to the appropriate Landau–Lifshitz (LL) equation. In real magnets, dissipative effects also play an important role, which is included in the LL equation via Gilbert damping. Thus, the LL equation with Gilbert damping (originally derived phenomenologically) is written as

$$\frac{\partial}{\partial t}\boldsymbol{S}(\boldsymbol{r}, t) = \boldsymbol{S} \times \boldsymbol{F} + \lambda \boldsymbol{S} \times (\boldsymbol{S} \times \boldsymbol{F}),$$
$$\boldsymbol{S} = (S_x, S_y, S_z), \quad \boldsymbol{S}^2 = 1, \qquad (3)$$

where $\boldsymbol{F}$ is the effective field and λ is the Gilbert damping parameter. A typical form for $\boldsymbol{F}$ is $\boldsymbol{F} = J\nabla^2\boldsymbol{S} + 2A(\boldsymbol{S} \cdot \boldsymbol{n})\boldsymbol{n} + \mu\boldsymbol{H}$, where A is the anisotropy parameter, $\boldsymbol{H}$ is the external magnetic field, and $\boldsymbol{n}$ is a unit vector.

The Landau–Lifshitz equation is a highly non-trivial vector nonlinear partial differential equation in three spatial dimensions. Several one-dimensional cases (without damping) corresponding to quasi-one-dimensional ferromagnetic spin chains (Mikesksa & Steiner, 1991; Kosevich et al., 1991) have been found to be completely integrable soliton systems. These include the isotropic case, uniaxial, and biaxial as well as circularly symmetric cases and are related to the nonlinear Schrödinger equation (Lakshmanan, 1977), (Lakshmanan et al., 1990) and its generalizations. In two dimensions, instanton and vortex-type structures have been identified. Magnetic spatiotemporal patterns have also been identified in the presence of damping and external oscillating fields.

The study of spin excitations in magnetic systems has numerous potential applications. These include (i) magnetic storage devices such as magnetic disks in computers, (ii) spintronics or spin electronics (microelectronic devices that function using the spin of the electrons), (iii) magnetic microchips (magnetic dots, magnetic waves, and magnetic computers), and (iv) magneto-optical recording. (For details, refer to the special issue on Magneto-electronics, *Physics Today*, April 1995).

MUTHUSAMY LAKSHMANAN

See also **Attractor neural network; Bethe ansatz; Ferromagnetism and ferroelectricity; Ising model; Landau–Lifshitz equation; Poisson brackets; Solitons**

Further Reading

Dotsenko, V. 1994. *An Introduction to the Theory of Spin Glasses and Neural Networks*, Singapore: World Scientific

Kosevich, A.M. Ivanov, B.A. & Kovalev, A.S. 1991. Magnetic Solitons. *Physics Reports*, 194: 117–238

Lakshmanan, M. 1977. Continuum spin systems as an exactly solvable dynamical system. *Physics Letters* A, 61: 53–54

Lakshmanan, M., Porsezian, K. & Daniel, M. 1990. On the nonlinear dynamics of the one-dimensional classical Heisenberg ferromagnetic spin chain. In *Nonlinear Evolution Equations: Integrability and Spectral Methods*, edited by A. Degasperis, A.P. Fordy & M. Lakshmanan, Manchester: Manchester University Press

Mattis, D.C. 1988. *Theory of Magnetism I: Statics and Dynamics*, Berlin and New York: Springer

Mikeska, H.J. & Steiner, M. 1991. Solitary excitations in one-dimensional magnets. *Advances in Physics*, 40: 191

SPIRAL WAVES

Rotating spiral waves occur in a wide range of chemical, biological, and physical systems that can be considered as effectively two-dimensional excitable or oscillatory media in which the local nonlinear dynamics exhibits threshold behavior, separating an excited state from a recovered state (Holden et al., 1991). In such a medium, waves of excitation spread without decrement through the stationary medium, with diffusive spread of the excitation from a region triggering excitation of neighboring regions.

A single spiral wave in a large homogeneous medium acts as an organizing center, imposing its pattern throughout the medium. It has the characteristics of

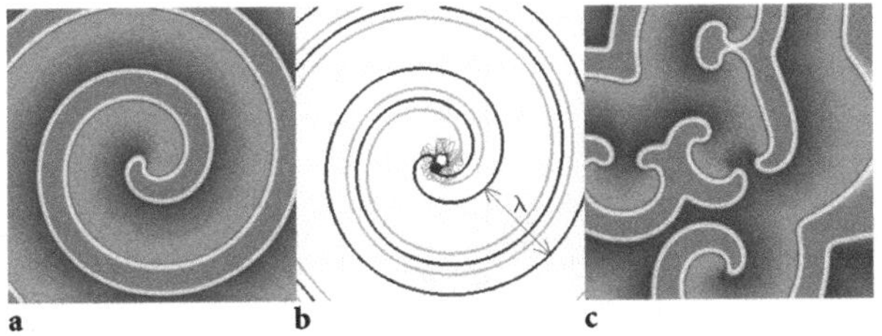

Figure 1. Spiral wave solutions of the FitzHugh–Nagumo equations for an excitable medium: (a) snapshot of solitary spiral wave; (b) intersection of excitation and recovery isolines defining the tip (solid circle), which meanders biperiodically around the central corel; and (c) evolving field of wavebreaks.

a field, in that it fills the medium, but it also has a location. The location of a spiral wave is that of its center, variously defined as the central core that is not invaded by the propagating wave, and around which the spiral rotates; or as a phase singularity; or by the tip, defined by where the wave front meets the wave back; or by the intersection of two isolines. For a rigidly rotating spiral in an isotropic medium, the tip encloses a circular core, but even in a homogeneous medium, the tip motion can be complex, meandering quasi-periodically within a bounded region, or hypermeandering in an unbounded region, and any inhomogeneities and boundaries interactions can produce drift (Barkley, 1994). Along any radius far from the core, activity appears as a periodic wave train with a temporal frequency ω (for a rigidly rotating spiral equal to the rotation frequency of the spiral) and spatial wavelength λ. These are generally single-valued functions of the medium parameters. The wave front (or any other isoline) approximates an Archimedean spiral (the involute of a circle) with a pitch λ, as illustrated in Figure 1b. The simple (one-armed) Archimedean rotating spiral in polar coordinates (θ, r) is a periodic function of phase ϕ

$$\phi = \omega t \pm \theta + a(r - r_0) \qquad (1)$$

with $a > 0$ giving the wavelength λ and pitch $dr / d\theta|_{\phi=\text{const}}$ equals to $1 / a$.

If a stable spiral wave is subject to a parametric perturbation, the form of the spiral will be restored, but its location and phase may be changed: this allows periodic perturbations, at the same frequency as the rotation frequency of the spiral, to produce a directed, resonant drift. A single spiral need not be stable—even in a parametrically homogeneous medium, dynamical heterogeneities produced by the rotating spiral wave can lead to wave breaks, resulting in pairs of new phase singularities that may develop into new spirals, collapse, or be extinguished, giving a spiral wave turbulence as illustrated in Figure 1c. In such an evolving field of wave breaks, the singularities may be considered as defects; are born and annihilate on collision, in pairs with opposite chirality; or exit the medium at boundaries.

The best studied laboratory example of spiral waves is the Belousov–Zhabotinsky reaction, a chemical excitable or oscillatory reaction originally constructed as an organic chemical analogue for oscillatory reactions in a biochemical system, in which the phase of the redox oscillation is monitored by color changes. In a shallow layer of solution in a Petri dish, spirals and interacting spirals can be initiated (Winfree, 1972), and spatiotemporal changes in activity can be monitored optically and quantified (Müller et al., 1987). Winfree's experimental and numerical modeling of simplified models and caricatures (Winfree, 1991) of two- and three-dimensional excitable systems established much of the phenomenology of spiral wave behavior.

Spiral waves occur in several biological systems. In many cell types there are intracellular waves of calcium, where calcium-induced calcium release provides the positive feedback necessary for excitability. In the slime mold, where the intercellular relaying of a chemotactic signal—cyclic AMP—between individual amoebae by cAMP-induced cAMP release provides the excitability, the social amoebae collect together in a rotating spiral wave and aggregate into a slug, which can then metamorphose into a fruiting body. An important example is from cardiac patho-physiology, where the three-dimensional analogue of a spiral, the scroll, provides the re-entrant propagation that produces high frequency arrhythmias in cardiac muscle and provides the route to ventricular fibrillation and sudden cardiac death.

In physics, spiral waves are found in surface catalysis, nonlinear optical media, and it has been suggested that spiral galaxies represent a rotating spiral wave of star formation and death.

In all these examples, the excitable medium can be modeled by a reaction-diffusion equation, in which only the excitatory variable (as in most biological examples) or both excitatory and recovery variables (as in most chemical examples) have nonzero diffusion coefficients. Spiral wave solutions of mechanistically detailed, high order models, or simplifications, can be obtained by initiation from artificial initial conditions, or by imposing a wave break or wavebreaking dynamical or parametric heterogeneity. Much of the phenomenology can be seen in the simple, two-variable FitzHugh–Nagumo system, or computationally faster simplifications (Zykov, 1987), and reproduced by eikonal equations that describe the dependence of the propagation velocity of the wave front on its curvature (Mikhailov et al., 1994).

Arun V. Holden

See also **Belousov–Zhabotinsky reaction; Excitability; FitzHugh–Nagumo equation; Geometrical optics, nonlinear; Reaction-diffusion systems; Vortex dynamics in excitable media**

Further Reading

Barkley, D. 1994. Spiral meandering. In *Chemical Waves and Patterns*, edited by R. Kapral & K. Showalter. Dordrecht: Kluwer, pp. 163–194

Holden, A.V., Markus, M. & Othmer, H.G. 1991. *Nonlinear Wave Processes in Excitable Media*, New York: Plenum Press

Mikhailov, A.S., Davydov, V.A. & Zykov, V.S. 1994. Complex dynamics of spiral waves and motion of curves. *Physica* D, 70: 1–39

Müller, S.C., Plesser, T. & Hess, B. 1987. Two-dimensional spectrophotometry of spiral wave propagation in the Belousov–Zhabotinskii reaction. *Physica* D, 24: 71–96

Winfree, A.T. 1972. Spiral waves of chemical reaction. *Science*, 175: 634–636

Winfree, A.T. 1991. Varieties of spiral wave behavior: an experimentalist's approach to the theory of excitable media. *Chaos* 1: 303–334

Zykov, V.S. 1987. *Simulation of Wave Processes in Excitable Media*, Manchester: Manchester University Press

STABILITY

The concept of stability lies at the heart of mathematical physics. Although in a perfect environment it might be thought possible to balance a pencil on its point, this is not possible in the real, noisy world—a pencil standing on its point is unstable. Meanwhile a hexagonal pencil lies easily on its side in a stable equilibrium. Intermediate is the pencil standing on its flat end, which can be achieved, albeit with difficulty, because this configuration is metastable, meaning stable against infinitesimally small disturbances, $\boldsymbol{D}_{\mathrm{S}}$, and unstable against large disturbances, $\boldsymbol{D}_{\mathrm{L}}$. Notice that under $\boldsymbol{D}_{\mathrm{S}}$, the pencil has more than one stable equilibrium (multistability). A special case is a circular pencil on its side: if it rolls a little, and it will not come back and so is marginally stable. Generalizing these experiences, a state is declared stable if it is physically observable under $\boldsymbol{D}_{\mathrm{S}}$: otherwise, it is unstable.

In 1892, Aleksandr Lyapunov formalized this intuitive idea of stability against $\boldsymbol{D}_{\mathrm{S}}$ by considering free motions of a dynamical system in phase space. Although this ignores external noise, the results agree with more realistic stochastic criteria. Consider a system of n first-order differential equations, $\dot{\boldsymbol{x}} = \boldsymbol{f}(\boldsymbol{x})$, giving a stationary vector field in the phase space spanned by the n components of vector $\boldsymbol{x}$. Trajectories fill this space with a fluid-like flow. A driven oscillator is put into this form by identifying velocity as a second phase coordinate and time as a third (with $\dot{t} = 1$).

A fixed (equilibrium) point with $\boldsymbol{f}(\boldsymbol{x}) = \boldsymbol{0}$ is asymptotically stable if all local trajectories flow into it (Figure 1). This point attractor is exemplified by the hanging state of a damped pendulum. The equilibrium of a Hamiltonian system (undamped pendulum) can be at most neutrally stable with all trajectories staying close, though not converging. A point is unstable if any single adjacent motion moves out of the neighborhood (inverted pendulum).

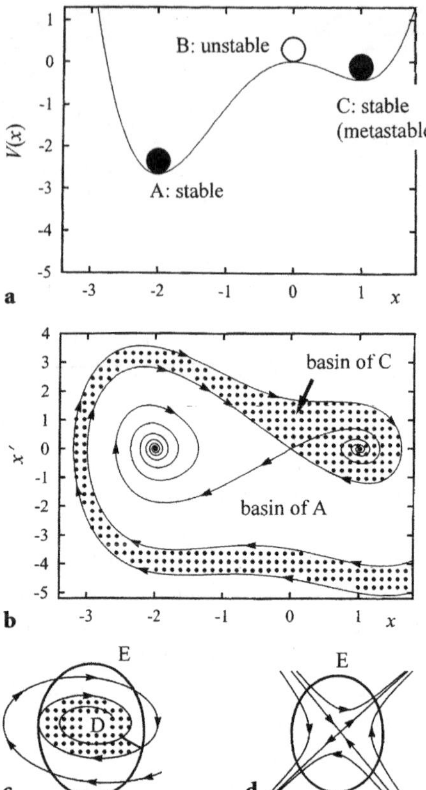

Figure 1. Potential well (a), and phase portrait (b), for $\ddot{x} + 0.4\dot{x} + \mathrm{d}V(x)/\mathrm{d}x = 0$, where $V(x) = x^4/4 + x^3/3 - x^2$, showing two asymptotically stable minima and an unstable maximum. In the enlarged focus (c), representing either minimum, we can devise a region D from which no trajectories leave the small region E, guaranteeing stability; in (d) we cannot construct a closed neighborhood around the saddle from which no trajectories leave E, proving instability.

Linear Stability

Stability analysis starts with the linearized equations describing small variations about the state. For a fixed point of a flow, stability hinges on the real parts, α_i, of the eigenvalues. Typically, the point will be hyperbolic (noncritical) with no $\alpha_i = 0$, and we then have (asymptotic) stability if and only if (*iff*) all $\alpha_i < 0$. Any one $\alpha_i > 0$ gives instability.

For a cycle we use an $(n-1)$-dimensional Poincaré mapping, $\boldsymbol{x}_{i+1} = \boldsymbol{F}(\boldsymbol{x}_i)$, where the cycle is a fixed point with $\boldsymbol{x}_{i+1} = \boldsymbol{x}_i$. Stability hinges on the moduli, ρ_i, of the eigenvalues. Stability of a (hyperbolic) periodic orbit requires all $\rho_i < 1$.

An unstable state, attracting in some directions, repelling in others, is a saddle. The instability index (degree of instability) is the number of $\alpha_i > 0$ (or $\rho_i > 1$). In nonlinear dissipative systems, saddles with index 1 play an important role organizing basin boundaries.

Nonlinear Stability

A conservative system has zero divergence in phase space (Liouville's theorem), like an incompressible fluid. With dissipation, the negative divergence ensures that an ensemble of starts will shrink onto an attracting set of zero volume. Such a post-transient can be a point, periodic, quasi-periodic, or chaotic attractor. Generically, each attractor is surrounded by its own basin of attraction. All transients from a vanishingly small neighborhood move asymptotically back to it, making it asymptotically stable.

To assess how stable an attractor is against finite disturbances, we estimate the size of its basin. A brute-force method is to make computer simulations from a grid of starts; more efficient is cell-to-cell mapping (Hsu, 1980). A refined approach is to locate a governing saddle and run simulations backwards in time to trace out the boundary formed by its stable manifold (inset). Boundaries can, however, be fractal, making their detailed location impossible (McDonald et al., 1985; Thompson & Stewart, 2002). Analytical techniques, notably Lyapunov's second (direct) method (La Salle & Lefschetz, 1961), can identify contracting (trapping) regimes of guaranteed return, but they may be excessively conservative.

Local nonlinear analysis is necessary to test the Lyapunov stability of a nonhyperbolic state at which a system is about to lose its stability. An oscillator with exclusively nonlinear negative damping is, for example, linearly (non-asymptotically) stable but nonlinearly unstable. In contrast to Lyapunov's local concept, a point is globally stable if every trajectory tends to it.

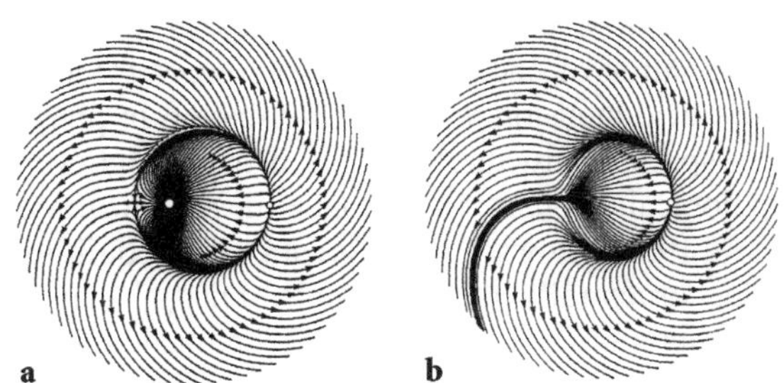

Figure 2. Two phase portraits illustrating the ABC of nonlinear dynamics. Here A stands for attractor, an example of which we see in the center of (a). B stands for basin, and we see the circular basin of attraction in (a); its boundary defined by the flow into an unstable saddle point. C stands for catastrophe, conveniently used here to mean bifurcation. Under variation of a control, (a) has changed qualitatively to become the portrait (b); the central stable node has collided with the saddle in a fold bifurcation, leaving no attractor and all trajectories flowing out to infinity. (Figure reproduced with permission of John Wiley from Thompson & Stewart 2002.)

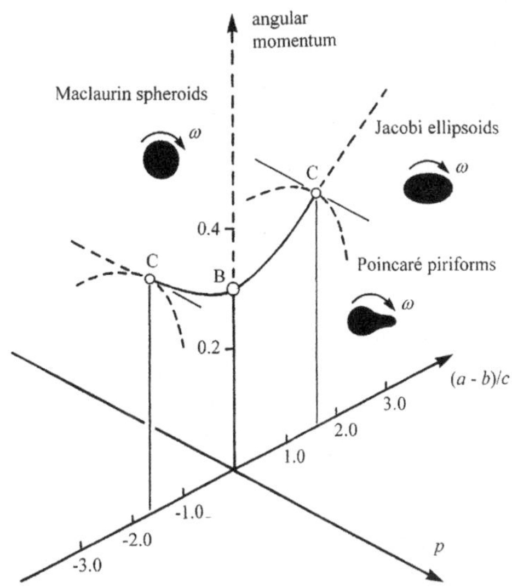

Figure 3. Stability of a rotating mass of a self-gravitating liquid is a classical problem in astrophysics, relevant to planetary evolution and the origin of satellites. It dates back to Poincaré (1899), and following the buckling studies of Euler, it gave fresh impetus to the development of stability and bifurcation theory. Stable (unstable) paths are denoted by solid (broken) lines. As the angular momentum parameter is slowly increased, the rotationally symmetric spheroids become unstable at the supercritical pitchfork at B. Emerging from B is a rising path of stable ellipsoids (with semi-axes a, b, c), which then exhibit a secondary instability at the subcritical pitchforks at C, from which unstable pear-shaped forms of "amplitude" p descend. (Figure reproduced with permission of John Wiley from Thompson, 1982.)

Loss of Stability

When linear theory indicates a loss of stability, the nonlinear system exhibits a bifurcation. Consider first an α_i passing through zero along a path. If a real eigenvalue becomes zero, we have a saddle-node bifurcation. Typically this gives a fold, at which the path folds smoothly back at a maximum of the control parameter, μ. A saddle-node coalescence at $\mu = \mu^C$ leaves no local solution; the system jumps to a distant, unrelated attractor (Figure 2). With symmetry the saddle-node manifests itself as a pitchfork (Figure 3). Two α_i of a complex conjugate pair passing through zero give the Hopf bifurcation generating stable or unstable cycles.

A mapping loses stability when a ρ_i reaches one. If a real eigenvalue equals $+1$, we have a saddle-node bifurcation, typically a cyclic fold. If it equals -1, we have a flip (period-doubling) bifurcation. If a complex conjugate pair penetrates the unit circle, we have the Neimark–Sacker (secondary Hopf) bifurcation.

Structural Stability

Because parameters of a physical system are never known precisely and may vary, a mathematical model should ideally have a phase portrait that is robust against small changes in the model itself. Small changes of parameters and functions should not qualitatively change the topology. A portrait having this robustness is structurally stable. Structurally unstable portraits encountered under variation of μ signal a bifurcation.

If we have m controls, μ_i, bifurcations observed under variation of one μ_i will lie in control space on surfaces of dimension $m-1$. A bifurcation in phase-control space is structurally stable if the whole event is topologically robust. Catastrophe theory (Poston & Stewart, 1978) shows how bifurcations of equilibria that are structurally unstable under one μ_i can be made structurally stable (unfolded) by embedding in a higher-dimensional control space. The codimension is the number of μ_i needed to achieve this. An example is the symmetry-breaking imperfection needed to unfold a pitchfork into the codimension-two cusp. The resulting saddle-node locus gives the imperfection-sensitivity of elastic stability theory (Thompson & Hunt, 1973).

MICHAEL THOMPSON

See also **Attractors; Equilibrium; Phase space; Wave stability and instability**

Further Reading

Hsu, C.S. 1980. A theory of cell-to-cell mapping for nonlinear dynamical systems. *ASME Journal of Applied Mechanics*, 47: 931–939

La Salle, J. & Lefschetz, S. 1961. *Stability by Liapunov's Direct Method with Applications*, New York: Academic Press

Lyapunov, A. 1892. *Problème général de la stabilité du mouvement*, Kharkov (in Russian); French translation in *Annals of the Faculty of Science, University of Toulouse*, vol. 9 (1907); reprinted, Princeton, NJ: Princeton, University Press (1949)

McDonald, S.W., Grebogi, C., Ott, E. & Yorke, J.A. 1985. Fractal basin boundaries. *Physica* D, 17: 125–153

Poincaré, H. 1899. *Les Méthodes nouvelles de la mécanique céleste*, vols. 1–3, Paris: Gauthier-Villars; as *New methods of celestial mechanics*, 3 vols., Woodbury, NY: American Institute of Physics, 1991

Poston, T. & Stewart, I. 1978. *Catastrophe Theory and Its Applications*, London: Pitman

Thompson, J.M.T. 1982. *Instabilities and Catastrophes in Science and Engineering*, Chichester: Wiley

Thompson, J.M.T. & Hunt, G.W. 1973. *A General Theory of Elastic Stability*, London: Wiley

Thompson, J.M.T. & Stewart, H.B. 2002. *Nonlinear Dynamics and Chaos*, 2nd edition, Chichester and New York: Wiley

STABLE AND UNSTABLE MAINFOLDS

See **Phase space**

STANDARD MAP

The standard or Taylor–Chirikov map is a family of area-preserving maps: $z' = f(z)$ where $z = (x, y)$ is the original position and $z' = (x', y')$ the new position after application of the map, which is defined by

$$\begin{aligned} x' &= x + y - \frac{k}{2\pi}\sin(2\pi x), \\ y' &= y - \frac{k}{2\pi}\sin(2\pi x). \end{aligned} \tag{1}$$

Here x is a periodic configuration variable (usually computed modulo 1), $y \in \mathbb{R}$ is the momentum variable, and the parameter k represents the strength of a nonlinear kick. This map was first proposed in 1968 by Bryan Taylor and then independently obtained by Boris Chirikov to describe the dynamics of magnetic field lines. The standard map and Hénon's area-preserving quadratic map are extensively studied paradigms for chaotic Hamiltonian dynamics.

The standard map is an "exact symplectic" map of the cylinder. Because $x'(x, y)$ is a monotone function of y for each x, it is also an example of a monotone twist map (*See* **Aubry–Mather theory**). Every twist map has a Lagrangian generating function, and the standard map is generated by $F(x, x') = \frac{1}{2}(x' - x)^2 + (k/4\pi^2)\cos(2\pi x)$, so that $y = -\partial F/\partial x$ and $y' = \partial F/\partial x'$. The map can also be obtained from a discrete Lagrangian variational principle as follows. Define the discrete action for any configuration sequence $\ldots, x_{t-1}, x_t, x_{t+1}, \ldots$ as the formal sum

$$A[\ldots, x_{t-1}, x_t, x_{t+1}, \ldots] = \sum_t F(x_t, x_{t+1}). \tag{2}$$

Then an orbit is a sequence that is a critical point of A. This gives the discrete Euler–Lagrange equation

$$x_{t+1} - 2x_t + x_{t-1} = -\frac{k}{2\pi}\sin(2\pi x_t). \tag{3}$$

This second difference equation is equivalent to (1) upon defining $y_t = x_t - x_{t-1}$.

The standard map is an exact or approximate description of many physical systems, including the "kicked rotor." Consider a rigid body with moment of inertia I that is free to rotate in a horizontal plane about its center of mass. Suppose that an impulsive torque $\Gamma(\theta) = -A\sin(\theta)$ is applied to the rotor at times nT, $n \in \mathbb{Z}$. Let (θ_j, L_j) be angular position and angular momentum at time $jT - \varepsilon$ for $\varepsilon \to 0^+$. At time T later, these become $(\theta_{j+1}, L_{j+1}) = (\theta_j + (T/I)L_{j+1}, L_j + \Gamma(\theta_j))$. Scaling variables appropriately gives (1).

The standard map also describes the relativistic cyclotron and is the equilibrium condition for a chain of masses connected by harmonic springs in a periodic potential—the Frenkel–Kontorova model introduced in 1938 (Meiss, 1992). Similar maps include Chirikov's separatrix map (valid near the separatrix of a resonance), the Kepler map (describing the motion of comets under the influence of Jupiter as well as a classical hydrogen atom in a microwave field), and the Fermi map (for a ball bouncing between oscillating walls) (Lichtenberg & Lieberman, 1992). The higher-dimensional version is the Froeshlé map (*See* **Symplectic maps**).

Symmetries

The standard map f has a number of symmetries that lead to special dynamical behavior. To see these, it is convenient to lift the map from the cylinder to the plane by extending the angle variable x to $\mathbb{R}$.

Let $T_{m,n}(x, y) = (x + m, y + n)$ be the translation by an integer vector (m, n). As f is periodic, its lift has a discrete translation symmetry $f \circ T_{m,0} = T_{m,0} \circ f$. More unusually, the standard map also has a discrete vertical translation symmetry $f \circ T_{0,n} = T_{0,n} \circ f$. Identifying orbits equivalent under these symmetries implies that standard map can be thought of as acting on the torus $\mathcal{T} = \{-\frac{1}{2} \leq x, y < \frac{1}{2}\}$.

The standard map also commutes with the reflection $S(x, y) = (-x, -y)$. This can be used to identify the lower half plane with the upper one and to restrict the map to the space $\mathcal{S} = \{(x, y) : -\frac{1}{2} \leq x < \frac{1}{2}, 0 \leq y \leq \frac{1}{2}\}$ identifying $(-\frac{1}{2}, y) \equiv (\frac{1}{2}, y)$ and each half of the upper and lower boundaries: $(x, 0) \equiv (-x, 0)$, $(x, \frac{1}{2}) \equiv (-x, \frac{1}{2})$. The map on the two sphere $\mathcal{S}$ is singular at the corners $(\pm\frac{1}{2}, 0)$ and $(\pm\frac{1}{2}, \frac{1}{2})$.

The standard map is also reversible; it is conjugate to its inverse $RfR^{-1} = f^{-1}$ (Lamb & Roberts, 1998). One reversor is $R_1(x, y) = (-x, y - (k/2\pi)\sin(2\pi x))$; this generates a family of reversors $R = f^n \circ R_1$. These reversors are involutions, $R^2 = id$, thus f can be written as the composition of two involutions $f = (f \circ R) \circ R$. Finally, the composition of a symmetry and a reversor is also a reversor, so that, for example, $R_2 = SR$ is also a reversor.

Symmetric orbits are invariant under a symmetry or a reversor. This is particularly interesting since symmetric orbits must have points on the fixed sets of the reversor, $\text{Fix}(R) = \{z : z = R(z)\}$, or on $\text{Fix}(fR)$. Because these fixed sets are curves, symmetric orbits are particularly easy to find. Rimmer showed that the bifurcations of symmetric orbits are special; for example, they undergo pitchfork bifurcations (*See* **Bifurcations**).

Dynamics

When $k = 0$, the dynamics of the standard map are integrable: the momentum y is an invariant. On each invariant circle $\mathcal{C}^0_\omega = \{(x, y) : y = \omega\}$, the angle after t iterates is given by $x_t = x_0 + \omega t \bmod 1$; thus, the dynamics are that of the constant rotation, $R_\omega(\theta) = \theta + \omega$, on the circle with rotation number ω. When ω is rational, every orbit on $\mathcal{C}^0_\omega$ is periodic; otherwise, they are quasi-periodic and densely cover the circle.

When $|k| \ll 1$, Moser's version of the KAM theorem implies that most of these invariant circles persist; that is, there is a rotational invariant circle $\mathcal{C}_\omega$ on which the dynamics is conjugate to the rotation R_ω (*See* **Hamiltonian systems**). KAM theory applies to circles with Diophantine rotation number, that is, $\omega \in \{\Omega : |n\Omega - m| > c/n^\tau; \forall m, n \in \mathbb{Z}, n \neq 0\}$ for some $\tau \geq 1$ and $c > 0$. This excludes, of course, all of the rational rotation numbers as well as intervals about each rational, but still leaves a positive measure set. While it is difficult to obtain reasonable estimates for the interval of k for which all Diophantine circles (with given c and τ) persist, in 1985 Herman showed analytically that there is at least one invariant circle when $|k| \leq 0.029$, and de la Llave & Rana (1990) used a computer-assisted proof to extend this result up to 0.91.

Some of the periodic orbits on the rational circles $\mathcal{C}^0_{m/n}$ also persist for nonzero k. Indeed, the Poincaré–Birkhoff theorem implies that there are at least two period n orbits (with positive and negative Poincaré indices). Aubry–Mather theory implies that orbits with rotation number m/n can be found variationally; one is a global minimum of the action (2), and the other is a minimax point (a saddle of A with one downward direction). For example, when $k > 0$, $(\frac{1}{2}, 0)$ is a minimizing fixed point, and $(0, 0)$ is a minimax fixed point. The reversibility of the standard map implies that there must be symmetric periodic orbits for each $\omega = m/n$ as well. Indeed, it is observed that the minimax periodic orbits always have a point on the line $\text{Fix}(R) = \{y = 0\}$, the "dominant" symmetry line.

The minimax orbits are elliptic when k is small enough. A convenient measure of stability of a period-n orbit is Greene's residue

$$R = \frac{1}{4}(2 - Tr(M)), \quad M = \prod_{t=0}^{n-1} Df(z_t).$$

An orbit is elliptic when $0 < R < 1$. For example, the fixed point $(0, 0)$ has residue $k/4$. Perturbation theory shows that the residues of the minimizing and minimax orbits are $\mathcal{O}(k^n)$.

Each nondegenerate minimum of the action (2) is a hyperbolic orbit and has unstable and stable manifolds. For each minimizing m/n orbit, these intersect and enclose the minimax orbit, forming an island chain or resonance. The intersection of the manifolds is transverse, though the angle between them is exponentially small in k (Gelfreich & Lazutkin, 2001).

A number of island chains are easily visible in computer simulations. In the color figure (*See* **Standard map**: on page 8 of color plate section), we show a number of orbits of the standard map for $k = 0.6$ on the torus $\mathcal{T}$. In the figure, each of the blue curves is formed from many iterates on a rotational invariant circle like those predicted by the KAM theorem. The green orbits are secondary and tertiary circles arising from resonances.

When stable and unstable manifolds intersect transversely, some iterate of the map has a Smale horseshoe. This implies that there is, at least, a Cantor set of chaotic orbits. Umberger & Farmer (1985) showed numerically that there is a fat fractal set on which the dynamics has a positive Lyapunov exponent. The proof of this statement is still illusive. The regions occupied by chaotic orbits appear to grow in measure as k increases. Numerically, it appears that a single initial condition densely covers each "zone of instability," a chaotic zone bounded by invariant circles. Chaotic trajectories that were only slightly visible in the previous color figure (gold orbits near the stable and unstable manifolds of the resonances) dominate the dynamics when $k = 2.0$ (*See* **Standard map**: page 8 in the color plate section). At this value of k, there are no rotational invariant circles. In the figure, the gold region is filled by a single trajectory with 1.5×10^6 iterates. It appears to densely cover most of phase space, though there are still a number of secondary and tertiary islands visible.

There are also many elliptic periodic orbits that are created for nonzero k. For example, the $(0, 0)$ fixed point undergoes a period-doubling bifurcation at $k = 4$, creating a period-two orbit. More generally, when the eigenvalues of any elliptic period-n orbit are $\lambda_{\pm} = e^{\pm 2\pi i \omega}$, then new orbits are born that encircle the original orbit and have relative rotation number ω. When $\omega = m'/n'$, these correspond to a chain of nn' islands. The color figure (*See* **Standard Map**: page 8 in the color plate section) shows red and blue island chains that encircle the fixed point with rotation number $\frac{1}{5}$; that there are two such chains is due to the reflection symmetry S. As Birkhoff realized, the newly created elliptic orbit also will undergo similar bifurcations, so that the phase space shows a structure of islands-around-islands, ad infinitum. This structure can even exhibit self-similarity (Meiss, 1992) just like the Feigenbaum period-doubling sequence for dissipative systems.

The Last Invariant Circle

In 1968, John Greene began studying the destruction of invariant circles in the standard map. He showed that sequences of periodic orbits, namely, the minimizing and minimax m/n orbits, whose rotation numbers converge on a given irrational, can be used to determine the existence of a circle with that frequency. Suppose that ω has a continued fraction expansion $[a_0, a_1, \ldots]$, $a_j \in \mathbb{Z}^+$, and let $m_j/n_j = [a_0, a_1, \ldots, a_j]$ be the jth convergent of ω. Greene conjectured that when the residues of these orbits $R_j \to 0$, as $j \to \infty$, then the invariant circle $\mathcal{C}_\omega$ exists—MacKay (1992) gives a proof of much of this.

For the standard map, it appears that each rotational invariant circle exists only up to a critical value, $k = k_{\text{cr}}(\omega)$; this graph was called the "fractal diagram" by Schmidt and Bialek in 1982. The critical k vanishes at every rational and appears to have local maxima for each *noble* irrational ω. Percival called a number noble if its continued fraction expansion has a tail that is eventually all ones. By this criterion the "most irrational" number is the golden mean $\gamma = (1 + \sqrt{5})/2 = [1, 1, 1, \ldots]$. Indeed for the standard map, Greene discovered that the invariant circles with rotation numbers $\gamma \pm m$, $m \in \mathbb{Z}$ appear to be the last circles destroyed (all such circles are destroyed simultaneously due the symmetries). Numerically it is known that the golden circle is destroyed at $k_{\text{cr}}(\gamma) \approx 0.971635406$.

This value is most efficiently computed by renormalization theory (MacKay, 1993). At the critical parameter for the destruction of a noble invariant circle, the phase space exhibits a self-similar structure. The geometric scaling of this self-similarity can be used to compute k_{cr} from the residues of the m_j/n_j orbits. This is more accurate than iteration methods—pioneered by Chirikov—which rely on finding an orbit that crosses the region containing the circle, and frequency methods—developed by Laskar—which rely on the irregularity of the numerically computed rotation number. While none of these methods proves that k_{cr} corresponds to the last invariant circle, "converse KAM theory" leads to a computer proof that there are no rotational circles for $k > \frac{63}{64}$ (MacKay & Percival, 1985). This is based on Birkhoff's theorem that every rotational invariant circle is a Lipschitz graph (Meiss, 1992). The color figure (*See* **Standard map**: page 8 in the color plate section) shows the dynamics on the sphere $\mathcal{S}$ at $k_{\text{cr}}(\gamma)$. Here the the golden circle (purple) is on the threshold of destruction. Also shown in the figure are 1.5×10^6 iterates of two chaotic trajectories (light blue and light green); the stable (blue) and unstable (red) manifolds of the $(m, n) = (0, 1)$, $(1, 2)$, and (1.3) orbits; and a number of orbits trapped in these island chains as well as the $(2, 5)$ and $(3, 8)$ chains.

Transport

Transport theory studies the motion of ensembles of trajectories from one region of phase space to another. When there are invariant circles separating the regions, then there is no transport. A Birkhoff "zone of instability" is an annular region bounded by, but otherwise not containing any, rotational invariant circles. Birkhoff showed there are orbits that traverse each zone of instability, and Mather (1991) extended this to show that there are orbits future and past asymptotic to the upper and lower bounding rotational invariant circles, respectively.

Aubry–Mather theory implies that for each irrational rotation number there is a minimizing trajectory that is dense on a circle or a cantor set. Percival proposed calling the latter sets "cantori." Thus, for $k > \frac{63}{64}$, every rotational invariant circle has become a cantorus, and vertical transport between any two momentum levels occurs. The color figure (in the color plate section) shows two such cantori (brown) with rotation numbers $(1+\gamma)/(3+4\gamma)$ and $(1+2\gamma)/(2+5\gamma)$. The rate of transport is locally governed by the flux, the area that crosses a closed loop upon iteration. The flux across a cantorus or a separatrix is given by Mather's ΔW, the difference in action between the corresponding minimax and minimizing orbits (MacKay et al., 1984). Renormalization theory shows that the flux through a noble cantorus goes to zero as $(k-k_{\mathrm{cr}})^{3.01}$; this can be very small, well beyond k_{cr}. For example in the color figure 3 (page 8 in the plate color section), the blue chaotic trajectory is bounded below by a low flux cantorus even for tens of millions of iterates. Geometrically, the flux is the area contained in a "lobe" bounded by pieces of stable and unstable manifolds. All transport occurs through lobes in two-dimensional maps (Wiggins, 1992); unfortunately, the higher-dimensional generalization is not clear.

JAMES D. MEISS

See also **Aubry–Mather theory; Cat map; Chaotic dynamics; Ergodic theory; Fermi acceleration and Fermi map; Hamiltonian systems; Hénon map; Horseshoes and hyperbolicity in dynamical systems; Lyapunov exponents; Maps; Measures; Mel'nikov method; Phase space; Symplectic maps**

Further Reading

Gelfreich, V.G. & Lazutkin, V.F. 2001. Separatrix splitting: perturbation theory and exponential smallness. *Russian Mathematical Surveys*, 56(3): 499–558

de la Llave, R. & Rana, D. 1990. Accurate strategies for small divisor problems. *Bulletin of the American Mathematical Society*, 22: 85–90

Lamb, J.W.S. & Roberts, J.A.G. 1998. Time-reversal symmetry in dynamical systems: a survey. *Physica* D, 112: 1–39

Lichtenberg, A. & Lieberman, M. 1992. *Regular and Chaotic Motion, Applied Mathematical Sciences*, vol. 38, 2nd edition, New York: Springer-Verlag

MacKay, R., Meiss, J. & Percival, I. 1984. Stochasticity and transport in Hamiltonian systems. *Physical Review Letters*, 52: 697–700

MacKay, R.S. 1992. Greene's residue criterion. *Nonlinearity*, 5: 161–187

MacKay, R.S. 1993. *Renormalisation in Area-Preserving Maps, Advanced Series in Nonlinear Dynamics*, vol. 6, Singapore: World Scientific

MacKay, R.S. & Percival, I.C. 1985. Converse KAM: theory and practice. *Communications in Mathematical Physics*, 98: 469–512

Mather, J.N. 1991. Variational construction of orbits of twist diffeomorphisms. *Journal of the American Mathematical Society*, 4(2): 207–263

Meiss, J.D. 1992. Symplectic maps, variational principles, and transport. *Reviews of Modern Physics*, 64(3): 795–848

Umberger, D.K. & Farmer, J.D. 1985. Fat fractals on the energy surface. *Physical Review Letters*, 55: 661–664

Wiggins, S. 1992. *Chaotic Transport in Dynamical Systems, Interdisciplinary Applied Mathematics*, vol. 2, New York: Springer

STATE DIAGRAMS

Consider a system comprising N elements, each of which can be set at one or the other of two values. Thinking of the elements as switches, the values (positions or settings) can be defined as one of the two Boolean variables: off (0) and on (1) (Birkhoff & MacLane, 1953). Now suppose that time progresses in discrete units and the values of the switches at time t_{n+1} depend on their values at t_n. In addition to being a rather general computer model (Fowler, 2004), such a system describes certain relay and electronic switching circuits used by electrical engineers, and it provides an approximate description of biological brains (Scott, 2002), including the "nets with circles" of Warren McCulloch and Walter Pitts (McCulloch & Pitts, 1943). If the elements are placed in a geometrical pattern (e.g., a square array), one has a cellular automaton, such as John Conway's famous "Game of life" (Gardner, 1970).

To understand how the model evolves in time, it is sometimes instructive to sketch a state diagram, which displays all states of the system and shows how these states change from one time increment to the next.

In Figure 1(a), for example, this is done for a system comprising a single switch ($N=1$), where the arrows indicate what will happen in the next increment of time. Reading from left to right, the first diagram shows a system in which the switch turns on if it is off and stays on if it is on. The second diagram shows a system that turns off if it is on and stays off if it is off. In the third diagram, the system stays off if it is off and stays on if it is on. (This corresponds to a typical wall switch.) Finally, the last diagram represents a system that turns off if it is on and turns on if it is off. (This models a "blocking oscillator" in electronic engineering.)

The number of possible state diagrams increases very rapidly with the number of elements (N) in a system. To see this, note that the number of states of the system is 2^N, and from each of these states we must choose an arrow that goes to (possibly) another state. Thus, the total number of diagrams $\mathcal{N}$ is 2^N raised to the 2^Nth power, or

$$\mathcal{N} = 2^{N2^N}. \tag{1}$$

A few calculations from this formula are given in Table 1, demonstrating a growth of $\mathcal{N}$ with N, which mathematicians term "combinatorial."

For a system comprising only six switches, the total number of state diagrams is greater than 10^{110}, which in turn equals the atomic mass of the universe (the mass of the universe measured in units of the hydrogen atom) times the age of the universe in picoseconds. In his analysis of biological organisms, the physicist Walter Elsasser has called such a finite number "immense," which implies that (although the number of them is finite) it is not—and never will be—physically possible to examine all possible diagrams (Elsasser, 1998).

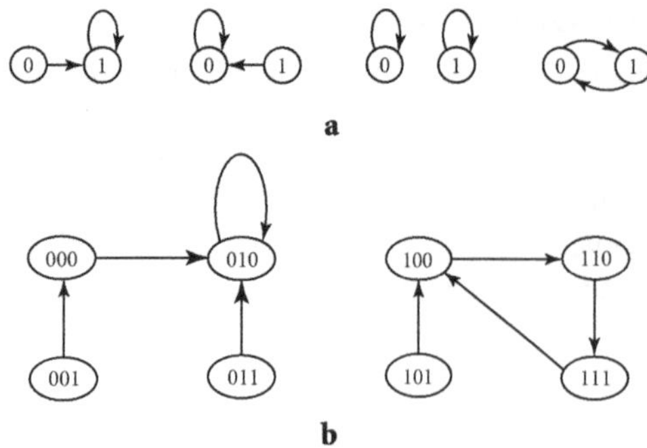

Figure 1. (a) The four Boolean state diagrams that can be constructed from a single switch. (b) One of the more than sixteen million state diagrams that can be constructed from three switches. Note that several arrows can converge on each state, but only one arrow leaves any one state.

N	$\mathcal{N} = 2^{N2^N}$
1	$2^2 = 4$
2	$4^4 = 256$
3	$8^8 \doteq 1.7 \times 10^7$
4	$16^{16} \doteq 1.8 \times 10^{19}$
5	$32^{32} \doteq 1.5 \times 10^{48}$
6	$64^{64} \doteq 3.9 \times 10^{115}$

Table 1. The number of state diagrams ($\mathcal{N}$) that can be constructed for various numbers (N) of Boolean switches.

In Figure 1(b) is shown one of the more than sixteen million state diagrams for a system of three switches. Here we find the following behaviors.

- *Transients:* $(001) \to (000) \to (010), (011) \to (010)$, and $(101) \to (100)$.
- *Periodicities:* $(010) \to (010)$ and $(100) \to (110) \to (111) \to (100)$.
- *Disjoint dynamics:* States (000), (010), (011), and (001) are dynamically disjoint from states (100), (110), (111), and (101).

The shortest transient or period is a single time unit, whereas the longest is 2^N time units, but this is unlikely for larger values of N.

Consider how the number of state diagrams depends on the way that the state of a system in the next time increment is computed. If each switch responds to the present positions of all N switches (including itself), there are

$$2^{2^N}$$

possible Boolean functions that can perform the computations. In the most general case, any one of the N switches can be assigned one of these Boolean functions. Thus, the number of possible diagrams is equal to

$$\left(2^{2^N}\right)^N = 2^{N2^N},$$

just as in Equation (1). In choosing the arrows for a particular diagram—as in Figure 1(b)—one defines a particular Boolean function.

When state diagrams are used to model biological brains, we may wish to limit the computations to threshold functions rather than Boolean functions. The total number of threshold functions of n inputs is

$$2^{kn^2},$$

where k lies between $\frac{1}{2}$ and 1 (Yajima et al., 1968). In a brain model, furthermore, it is unrealistic to assume that each neuron receives inputs from all other neurons. If each of N neurons receives inputs from n neurons, then the total number of state diagrams is

$$\mathcal{N}_{\text{th}} \geq \left(2^{n^2/2}\right)^N = 2^{n^2N/2}. \tag{2}$$

Taking $N = 10^{10}$ as a conservative estimate for the number of neurons in the human neocortex and $n = 10^4$ as an estimate for the number of synaptic inputs to each neuron, implies that the number of possible brains is about $10^{10^{17}}$.

Alwyn Scott

See also **Attractor neural network; Cellular automata; Game of life; McCulloch–Pitts network; Neural network models**

Further Reading

Birkhoff, G. & MacLane, S. 1953. *A Survey of Modern Algebra*, New York: Macmillan
Elsasser, W.M. 1998. *Reflections on a Theory of Biology: Holism in Biology*, Baltimore: Johns Hopkins University Press (first published in 1987)
Fowler, M. 2004. *UML Distilled: A Brief Guide to the Standard Object Modeling*, Boston: Addison-Wesley
Gardner, M. 1970. Mathematical games. *Scientific American*, 223 (October): 120–123
McCulloch, W.S. & Pitts, W.H. 1943. A logical calculus of the ideas immanent in nervous activity. *Bulletin of Mathematical Biology*, 5: 115–133
Scott, A.C. 2002. *Neuroscience: A Mathematical Primer*, Berlin and New York: Springer
Yajima, S., Ibaraki, T. & Kawano, I. 1968. On autonomous logic nets of threshold computers. *Transactions on IEEE Computers*, 17: 385–391

STATIONARY SOLUTIONS

See **Discrete self-trapping system; Equilibria**

STEADY STATE

See **Equipartition of energy**

STEREOSCOPIC VISION AND BINOCULAR RIVALRY

Most vertebrate predators, including humans, have their two eyes positioned in the front of the head

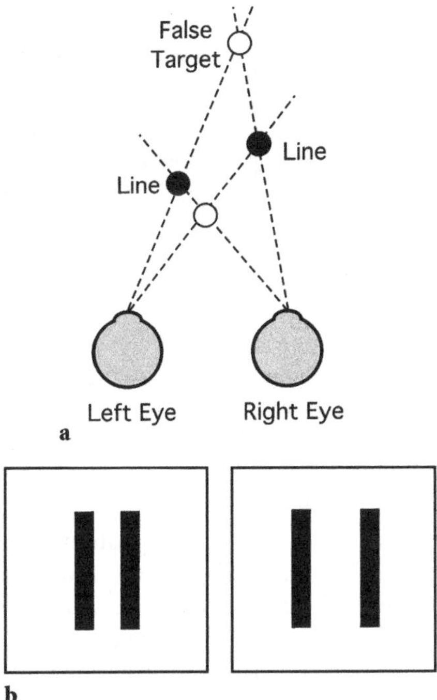

Figure 1. (a) The geometry of stereopsis. (b) The simplest two line stereogram.

looking forward. In contrast, most prey species, such as rabbits, have eyes positioned on the sides of the head. This confers on prey almost panoramic vision, thus enabling them to see predators creeping up behind them. In contrast, predators receive two slightly different views of the scene in front of them due to the horizontal offset between the two eyes (approximately 6.7 cm in humans), which confers depth information. Stereoscopic vision may be defined as the perception of depth that arises when the brain compares the images falling on the two retinas. Indeed, there are neurons in the visual cortex that receive stimulation from both eyes and consequently are sensitive to the depth of lines and contours in the world.

The geometry of stereopsis is illustrated in Figure 1a, which provides a horizontal view of two eyes viewing a scene containing two vertical rods at different depths relative to the eyes. As shown by the dashed sight lines, the retinal images of these two lines will be closer together in one eye's view than in the other. Figure 1b contains a simple two line stereogram, namely a pair of views, one for each eye, with vertical lines closer together in one view than in the other. If the viewer can either cross or diverge their eyes so that each eye fixates a different one of these views, the percept will be one in which one line lies closer to the viewer than the other line. (It should be noted that several percent of humans lack stereoscopic ability, usually due to amblyopia or "lazy eye" during early childhood.) Rather than requiring viewers to cross or diverge their eyes, laboratory research into stereopsis utilizes some version of a stereoscope, which was invented by Charles Wheatstone in 1838. A stereoscope utilizes either prisms or mirrors in front of each eye to optically diverge the views so that each eye views a different visual stimulus. Indeed, stereoscopes and stereoscopic photographs were a Victorian fascination, and many excellent stereoscopes may still be found in antique shops.

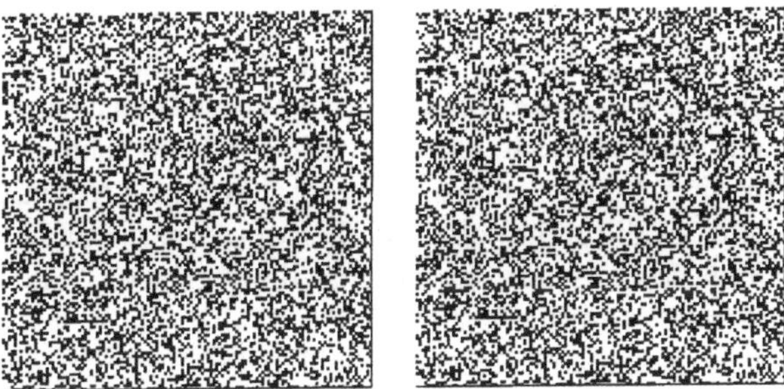

Figure 2. Random dot stereogram.

The two line stereogram in Figure 1b is the simplest possible, as two objects are the minimum for which different depths can be defined. A much more complex example is the random dot stereogram (RDS), which was developed by Bela Julesz (1971) as a tool for the study of stereopsis. To make an RDS, a computer first generates a completely random pattern of black and white dots (little squares), such as that depicted in Figure 2. This will be the stimulus for one eye. An exact copy of this is then made as the basis for the other eye's view. To generate a stimulus for depth, a rectangular (or other shaped) region of dots in the middle of this second random pattern is now shifted horizontally left or right by a distance of several dots. The space vacated by shifting this rectangular region is then filled in with random black and white dots. When each eye views one of the two random dot patterns in an RDS, the brain extracts depth information and produces a percept of a rectangular region of random dots floating in depth in front of (or behind, depending on which eye views which half of the stereogram) a random dot background. Readers who can converge or diverge their eyes while viewing Figure 2 can experience this.

Neural Basis of Stereopsis

Information originating in the two retinas remains segregated until it reaches primary visual cortex at the back of the brain. In the cortex, many individual neurons respond best when they are stimulated through both eyes at once. However, each cortical neuron receives stimulation only from a very small area on each retina, which is known as a "receptive field." If one imagines there being a coordinate system affixed to each retina with the center of the retina (the fovea) being the origin, then each cortical cell receives information from a pair of receptive fields, one from each retina, whose coordinates may be the same or different. If a cortical neuron receives stimulation from receptive fields with the same relative coordinates in each retina, it will be sensitive to lines and edges that lie in the plane on which the eyes are converged. Therefore, such a neuron will be sensitive to objects with zero depth relative to fixation. Other neurons, however, have their receptive fields at different points on the two retinas. For most of these cells, this means that the receptive field in one eye is shifted a short distance horizontally left or right relative to its receptive field in the other eye. Such a difference in receptive field coordinates on the two retinas is termed a horizontal disparity. Neurons with a particular horizontal disparity will respond to objects at a particular depth relative to the fixation plane. For example, if a cortical neurons has its receptive field in the left eye shifted left relative to that in the right eye, its disparity will correspond to lines and edges nearer than fixation. Conversely, a neuron with a left eye receptive field shifted to the right relative to the other eye will be sensitive to disparities corresponding to lines and edges farther away than the plane of fixation. Thus, different cortical neurons, in virtue of the relative placement of their two retinal receptive fields, signal the location of lines and edges in three-dimensional space, including their depth relative to the fixation plane. Although only horizontal disparities are discussed here, vertical disparities are now known to play a role in stereopsis as well, and details can be found in Howard & Rogers (2002).

The RDS demonstrates dramatically the computational problems confronted by cortical neurons in comparing the two retinal images to extract depth information. To make the point, consider a single black dot in one horizontal line of the pattern in one eye's view. The corresponding horizontal line of dots in the other eye's view will contain roughly 50% black dots, any one of which might be the appropriate match for the black dot in question. Thus, the problem has an extremely large number of potential depth solutions, only one of which is actually perceived. In other words, the existence of neurons sensitive to different disparities and, therefore, different depths only represents the initial part of the solution to the problem of extracting depth information from two retinal images.

The types of interactions among disparity-tuned cortical neurons that must be involved in the extraction of accurate depth information may be appreciated by reconsidering the simple two line stereogram in Figure 1. As shown by the dashed lines in Figure 1a, sight lines from the two eyes intersect not only at the locations of the two targets (black circles) but also at two other points in depth (open circles). These are known as "false targets," and neural interactions in the cortex are required to eliminate them. The types of interactions believed to occur include facilitation between neurons tuned to similar disparities in neighboring regions plus inhibition between neurons tuned to very different disparities at adjacent points. Examination of Figure 1a shows that if the neurons tuned to the disparities or depths represented by the black circles were mutually excitatory, whereas both of these inhibited neurons tuned to the much more extreme disparities indicated by the open circles, the neurons representing the black circles would win the competition and suppress the responses to the false targets. Many variations on this theme of facilitation between neurons tuned to similar disparities and inhibition between neurons tuned to radically different disparities have been proposed, and these have been summarized elsewhere (Blake & Wilson, 1991). Such interactions may be conceptualized as implementing an evolutionary constraint that most real world objects are bounded by surfaces that vary smoothly in depth except at their edges.

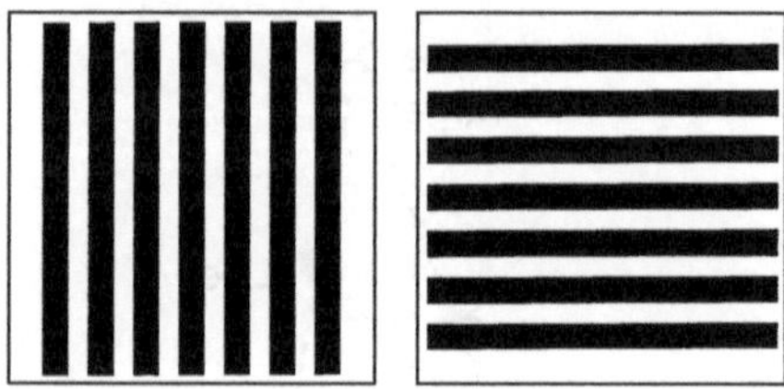

Figure 3. Stereogram illustrating binocular rivalry. Try viewing and perceiving both images simultaneously.

Binocular Rivalry

The preceding discussion assumes that the two eyes are looking ahead and viewing the same scene, albeit from slightly different viewpoints that give rise to disparities. In the laboratory, however, it is possible to generate stereograms that present radically different views to the two eyes. A simple example of this is depicted in Figure 3, where one eye is presented with horizontal lines, while the other is presented with vertical lines. Readers who can converge or diverge their eyes will discover that these lines cannot be fused to produce a depth percept. Rather, one perceives an alternation between horizontal and vertical lines such that the two are almost never seen at the same point at the same time. This perceptual alternation, which occurs at a rate of about one alternation every 2–4 s, is known as binocular rivalry. Rivalry results when the two eyes view images that could not have arisen from a real three-dimensional world in the absence of mirrors or other optical devices. It appears, therefore, that rivalry must reflect neural interactions that evolved to process depth information but that default into a perceptual oscillation when the depth task is impossible.

The neural basis of binocular rivalry requires two components. First, the different groups of rivaling neurons responding to orthogonal stimuli must be strongly and mutually inhibitory. This guarantees that only one set of these neurons can be active at any given place and time. The second ingredient is fatigue or self-adaptation. Self-adaptation, which is well documented among excitatory cortical neurons, causes neurons that are firing rapidly to slowly decrease their response rate over the course of a second or more. As they adapt, their ability to inhibit other neurons weakens. Eventually, these previously inhibited neurons overcome the inhibition and begin to respond, and they then inhibit the adapted neurons so that the percept switches. This neural mechanism has been developed into nonlinear dynamical models that accurately explain many aspects of binocular rivalry (Wilson, 2004).

A further aspect of binocular rivalry is that the reversals are not precisely periodic. Scientists have measured the duration of intervals during which the pattern presented to one eye is visible (e.g., the vertical stripes in Figure 3). The results are well described by a gamma distribution in which both short and long intervals recur in a seemingly random sequence. Recent research using neural network modeling suggests that this gamma distribution results from chaotic dynamics among cortical neurons (see Wilson, 2004).

The final and most important question concerning binocular rivalry is how the inhibitory neural interactions underlying rivalry change their role when stimuli permitting depth perception are viewed. This remains unanswered, but it is at least clear from the discussion of depth perception that inhibition is a necessary neural component. Perhaps, future research will show that the inhibition involved in elimination of false targets during depth perception is related to the inhibition causing perceptual alternations in binocular rivalry.

HUGH R. WILSON

See also **Attractor neural network; Cell assemblies; Gestalt phenomena; Neural network models; Perceptron; Synergetics**

Further Reading

Alais, D. & Blake, R. (editors). 2003. In *Binocular Rivalry and Perceptual Oscillations*, Cambridge, MA: MIT Press

Blake, R. & Wilson, H.R. 1991. Neural models of stereoscopic vision. *TINS*, 14: 445–452

Howard, I. & Rogers, B. 2002. *Seeing in Depth*, Toronto: I. Porteous

Julesz, B. 1971. *Foundations of Cyclopean Perception*, Chicago: University of Chicago Press

Wilson, H.R. 2004. Rivalry and perceptual oscillations: a dynamical synthesis. In *Binocular Rivalry and Perceptual Oscillations*, edited by D. Alais & R. Blake, Cambridge, MA: MIT Press

STIMULATED BRILLOUIN AND RAMAN SCATTERING

See **Rayleigh and Raman scattering and IR absorption**

STOCHASTIC ANALYSIS OF NEURAL SYSTEMS

Oscillatory activity abounds in the central nervous systems of animals and humans (Freeman, 1975; Steriade et al., 1990). Single neurons may act as oscillators by rhythmically generating action potentials or bursts. Central pattern generators (Cohen et al., 1988) are networks of neurons located in the spinal cord and the brain stem, which produce oscillatory activity intrinsically (Sherrington, 1947; von Holst, 1954; Cohen et al., 1988). Central pattern generators are essential for repetitive motor behavior like walking, running, flying, swimming, and chewing (Sherrington, 1947; von Holst, 1954; Cohen et al., 1988). Phase synchronization (PS) between different central pattern generators is the mechanism that controls the different types of locomotion such as walk, trot, or gallop (von Holst, 1954; Cohen et al., 1988).

In the cortex, the well-coordinated action of neural oscillations is indispensable to physiological information processing (Freeman, 1975; Steriade et al., 1990; Haken, 1996). In-phase synchronization might be one aspect of such processes (Singer & Gray, 1995).

In several neurological diseases, brain function is severely perturbed by pathologically enhanced synchronization. Resting tremor in Parkinson's disease (PD) appears to be caused by a cluster of neurons located in the thalamus and the basal ganglia, which fires synchronously at a frequency similar to that of the tremor (3–6 Hz). While under physiological conditions these neurons fire incoherently (Nini et al., 1995); in patients with PD, this cluster acts like a pacemaker and activates cortical areas, which causes the peripheral shaking.

In patients with advanced Parkinson's disease or with essential tremors that do not respond to drug therapy, depth electrodes are chronically implanted in target areas like the thalamic ventralis intermedius nucleus or the subthalamic nucleus (Benabid et al., 1991). Electrical deep brain stimulation (DBS) is performed by administering a permanent high-frequency (>100 Hz) periodic pulse train via the depth electrodes. DBS suppresses the activity of the pacemaker-like cluster, which, in turn, suppresses the peripheral tremor (Benabid et al., 1991). DBS has been developed empirically, mainly based on observations during stereotaxic neurosurgery. The advent of nonlinear dynamics and statistical physics in the field of clinically oriented neuroscience has led to an improvement of data analysis as well as a model-based development of stimulation techniques.

Stochastic Synchronization

Phase synchronization is a type of synchronization where two oscillators adjust their phases, while their amplitudes need not be correlated. Phase

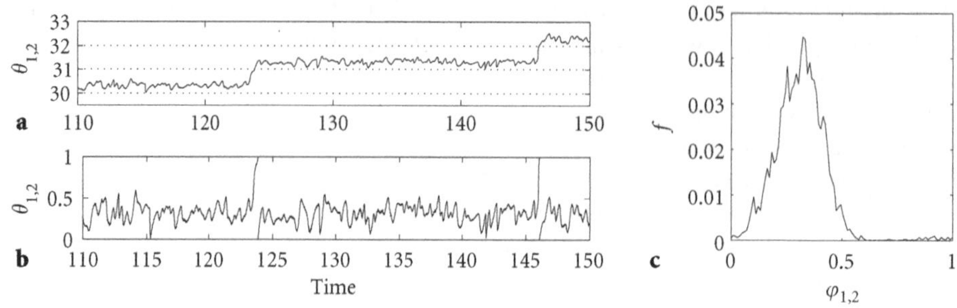

Figure 1. 1 : 2 phase difference $\theta_{1,2}$ (a) and its cyclic version $\varphi_{1,2}$ (b) from Equation (1) between the activity of the right flexor muscle of a patient with Parkinsonian resting tremor and the magnetic field measured above the corresponding left sensorimotor cortex. Corresponding distribution of $\varphi_{1,2}$ (c). (Adapted from Tass et al., 1998).

synchronization is observed in many physical, chemical, and neural systems (von Holst, 1954; Haken, 1996; Tass et al., 1998; Pikovsky et al., 2001; Tass et al., 2003). As neural systems are noisy, it is convenient to use the general concept of PS, which is valid in noisy periodic oscillators as well as in chaotic oscillators (Pikovsky et al., 2001).

To study $n:m$ phase synchronization (PS) we introduce the $n:m$ phase difference $\theta_{n,m}$ and the cyclic $n:m$ phase difference $\varphi_{n,m}$ with

$$\theta_{n,m}(t) = \frac{n\phi_1(t) - m\phi_2(t)}{2\pi},$$

$$\varphi_{n,m}(t) = \frac{n\phi_1(t) - m\phi_2(t)}{2\pi} \bmod 1, \qquad (1)$$

where ϕ_1 and ϕ_2 are the phases of the two oscillators. The time course of $\theta_{n,m}$ typically exhibits horizontal periods intersected by 2π jumps (Figure 1a), that vanish when the coupling is sufficiently strong compared to the noise. By $n:m$ PS is meant that the two oscillators maintain a stable relationship of $\varphi_{n,m}$ in a statistical sense (Figure 1b), and thus it is characterized by the appearance of one or more distinct peaks in the distribution of $\varphi_{n,m}$ (Figure 1c). Absence of $n:m$ PS corresponds to a uniform distribution of $\varphi_{n,m}$, whereas perfect $n:m$ PS is related to a Dirac-type distribution of $\varphi_{n,m}$. To quantify the strength of $n:m$ PS, the actual distribution is compared to a uniform distribution with synchronization indices that are based on the Shannon entropy or the Fourier transformation (Tass et al., 1998; Pikovsky et al., 2001). The phase of an oscillatory signal is determined with the Hilbert transform or the wavelet transform (Tass et al., 1998; Pikovsky et al., 2001). Already by the 1930s, PS was studied in coordinated movements using a similar, Poincaré-like approach, but limited phase estimation (von Holst, 1954).

In Parkinsonian patients with resting tremor, a 1 : 2 PS was found between the activity of the shaking muscles and brain areas such as the sensorimotor cortex and premotor areas (Tass et al., 1998) (Figure 1). The dominant frequency of the activity in the brain areas is approximately twice as large as that of the peripheral tremor.

Compared to PS analysis, the cross coherence method cannot distinguish between PS and amplitude modulation as well as between PS and simple linear superposition. Furthermore, cross coherence is not appropriate for the detection of $n:m$ PS with $n/m \neq 1$ (Tass et al., 1998; Pikovsky et al., 2001, 2003).

Stochastic Phase Resetting

In both single neurons and neural populations, the effect of a stimulus depends on the phase of the neural oscillation at which the stimulus is administered. To study phase-dependent effects of pulsing stimuli on neural populations and, in particular, to analyze desynchronizing effects of such stimuli, the concept of phase resetting (Winfree, 1980) has been extended to populations of non-interacting and interacting oscillators in the presence of random forces (Tass, 1999). The single neuron is modeled by a phase oscillator (Kuramoto, 1984). The dynamics of a cluster of N interacting phase oscillators subject to a stimulus S and to random forces F_j is governed by

$$\dot{\psi}_j = \omega_j + \frac{1}{N}\sum_{k=1}^{N}\Gamma(\psi_j - \psi_k) + X(t)\,S(\psi_j) + F_j(t), \qquad (2)$$

where ψ_j is the phase of the jth phase oscillator, Γ is a 2π-periodic global coupling, and S is a 2π-periodic, time-independent function modeling the stimulus, where $X(t) = 1$ during stimulation and $X(t) = 0$ without stimulation (Tass, 1999). The random forces $F_j(t)$ are modeled by Gaussian white noise fulfilling $\langle F_j(t)\rangle = 0$, $\langle F_j(t)\ F_k(t')\rangle = D\delta_{jk}\delta(t - t')$ with constant noise amplitude D.

Via the Fokker–Planck equation, one obtains an evolution equation for the average number density $n(\psi, t) = \int_0^{2\pi} \ldots \int_0^{2\pi} \mathrm{d}\psi_1 \cdots \mathrm{d}\psi_N\ \tilde{n}(\Psi;\psi) f(\Psi;t)$, where $f(\Psi, t)$ is the the probability density with $\Psi = (\psi_1, \ldots, \psi_N)$ and $\tilde{n}(\Psi;\psi) = N^{-1}\sum_{j=1}^{N}\delta(\psi - \psi_j)$. $n(\psi, t)$ tells us how many oscillators most probably have phase ψ at time t. Detailed analytical studies of that equation were performed for $X = 0$ (Kuramoto, 1984; Tass, 1999), whereas the transient stimulation-induced dynamics was analyzed numerically, mainly for the case when all oscillators have the same eigenfrequency (Tass, 1999). Since the single model neuron fires whenever its phase equals 0 (modulo 2π) the average number density of firing neurons is given by the firing density

$$p(t) = n(0, t) . \tag{3}$$

It is possible to desynchronize a synchronized cluster of oscillators with a single pulse of the right intensity and duration. To this end, however, the pulse has to hit the cluster in a vulnerable phase range, which corresponds to only a small fraction (5% or even less) of a period of the oscillation. To compound matters, different stimulation parameters have to be used to desynchronize a cluster that is not in its fully synchronized state (Tass, 1999). Hence, complex stimulation techniques have been developed that make it possible to effectively desynchronize a cluster of phase oscillators independently of the cluster's dynamical state at the beginning of the stimulation (Tass, 2002). These methods share one particular feature: the stimulus consists of two qualitatively different stimuli. The first stimulus is stronger and resets (restarts) the cluster, whereas the second, weaker stimulus is a single pulse that is administered after a constant time delay and desynchronizes by hitting the cluster in a vulnerable state. The goal of the reset is to control the dynamics of the cluster by restarting the cluster in a stereotypical way. The reset may be achieved by means of a strong single pulse, a high-frequency pulse train (HFPT) (with a pulse rate 20 times larger than the mean eigenfrequency $\bar{\omega}$) or a low-frequency pulse train (LFPT) (with a pulse rate similar to $\bar{\omega}$) (Tass, 2002). A strong pulse and an HFPT cause a hard (abrupt) reset within less than one period, whereas an LFPT causes a soft (slow) reset, where in the course of an entrainment the influence of the initial dynamic state fades away.

Since the desynchronizing effect of these stimulation techniques does not depend on the initial dynamical state of the neural population, resynchronization can effectively be blocked by administering a desynchronizing stimulus whenever the population tends to resynchronize (Figure 2a). In contrast, high-frequency stimulation not combined with a desynchronizing single pulse causes only a suppression of the firing during stimulation but no desynchronization (Figure 2b). After a simple HFPT, the extent of synchronization is particularly high, and the cluster restarts in a rebound-like manner.

Transient stimulus-induced responses of coupled neural oscillators are a key approach not only for manipulating synchronization processes, but also for studying neural information processing in basic

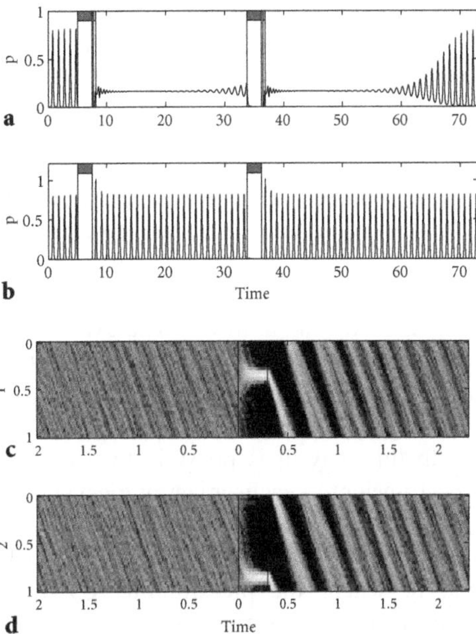

Figure 2. (a) Firing density p from Equation (3) of a neural population stimulated with two subsequent identical composite stimuli consisting of a high-frequency pulse train followed by a single pulse. (b) Two successively administered identical high-frequency pulse trains with parameters as in (a) but without following single pulses. Begin and end of a single pulse are indicated with vertical lines connected by a shaded region ((a) and (b) from Tass, 2001). CT distributions from Equation (5) of ϕ_1 (c) and ϕ_2 (d) show that the post-stimulus responses of each oscillator split into two antiphase ensembles (0 is black and maximal values are white). Begin (at $t = 0$) and end of stimulation are indicated by vertical lines. ((c) and (d) from Tass, 2003).

research as well as in diagnosis (Dawson, 1954; Steriade et al., 1990). To improve the signal-to-noise ratio, one typically averages across an ensemble of responses to reveal what is supposed to be the actual response to a stimulus (Dawson, 1954). But noise inherent in the neural dynamics makes coupled neural oscillators react to a stimulus in more than one way, which may be essential for sensory short-term adaptation (Tass, 2003). In this case, averaging causes severe artifacts.

For illustration we consider Equation (2) for two coupled oscillators ($N = 2$) undergoing an antiphase reset caused by stimuli $S_1(\psi_1) = I \cos(\psi_1)$ and $S_2(\psi_2) = I \cos(\psi_2 + \pi)$. We use the normalized cyclic phases

$$\phi_j(t) = \frac{\psi_j(t)}{2\pi} \bmod 1 \quad (j = 1, 2) \tag{4}$$

and the normalized cyclic $n : m$ phase difference $\varphi_{n,m}$ from Equation (1). To detect whether, in an ensemble of responses, there are epochs during which the phases ϕ_j and/or the phase difference $\varphi_{n,m}$ display, a stereotypical, tightly stimulus-locked time course; the stimuli S_1 and S_2 are simultaneously administered at l random times $\tau_1, \tau_2, \ldots, \tau_l$. Time-dependent cross-trial (CT) distributions of ϕ_j and $\varphi_{n,m}$ are calculated across trials for each time t relative to stimulus onset with

$$\{\phi_j(t+\tau_k)\}_{k=1,\ldots,l}, \{\varphi_{n,m}(t+\tau_k)\}_{k=1,\ldots,l}. \quad (5)$$

The time course of ϕ_j and $\varphi_{n,m}$ is perfectly stimulus-locked at time t if the corresponding CT distributions are Dirac-type distributions, whereas these distributions are uniform if ϕ_j and $\varphi_{n,m}$ are not stimulus-locked at all at time t. The extent of stimulus locking is quantified for each time t with indices characterizing the corresponding CT distributions with Shannon entropy or Fourier transformation (Tass, 1999, 2003). In this way, it is possible to detect stimulus-locked transient synchronization/desynchronization ($\varphi_{n,m}$) as well as transient cross-trial response clustering (ϕ_j). The latter is illustrated in Figures 2c and d, where the pre-stimulus CT distribution of ϕ_j is nearly uniform because of the randomized stimulus application. In all of the $l=200$ trials, the stimuli S_1 and S_2 reset the oscillators to the phases $\phi_1^{\mathrm{res}} \approx 0.36$ and $\phi_2^{\mathrm{res}} = \phi_1^{\mathrm{res}} + 0.5$, so that Dirac-like CT distributions evolve. Each oscillator reacts in two qualitatively different ways: The ensemble of post-stimulus responses starts at ϕ_j^{res} and splits into two antiphase ensembles, so that two antiphase peaks of the corresponding CT distribution emerge. The mechanism that induces cross-trial response clustering is a stochastic resonance (Tass, 2003).

Coordinated transient responses of neural oscillators such as antiphase cross-trial response clustering typically escape detection with standard methods like cross-trial averaging (Tass, 2003).

Deep Brain Stimulation Techniques

Permanent high-frequency DBS is an unphysiological type of stimulation, which in a number of patients leads to side effects like dysarthria and psychiatric sympotms. Furthermore, the therapeutic effect of DBS may vanish in the course of the treatment, possibly due to adaptation.

Demand-controlled DBS with the stimulation techniques explained in the former section has been suggested as an alternative and milder therapeutic approach (Tass, 2002). Instead of simply suppressing the neural firing of the pacemaker-like cluster, these novel stimulation techniques aim at desynchronizing the pacemaker's pathologically synchronized firing in a demand-controlled way (Tass, 2002). For this, either the depth electrode or an epicortical electrode measures the feedback signal, that is, the local field potential of the pacemaker population or the electrical activity of a cortical area that is strongly coupled to it. A desynchronizing stimulus is administered only whenever the pacemaker-like cluster becomes synchronized (Figure 2a). Repeated administration of desynchronizing stimuli may effectively maintain an uncorrelated firing. Demand-controlled DBS should, in principle, be less aggressive to the brain tissue stimulated and is currently evaluated for clinical use.

PETER A. TASS

See also **Attractor neural network; Cell assemblies; Coupled oscillators; Electroencephalogram at large scales; Electroencephalogram at mesoscopic scales; Stochastic processes; Synchronization**

Further Reading

Benabid, A.L., Pollak, P., Gervason, C., Hoffmann, D., Gao, D.M., Hommel, M., Perret, J.E. & De Rougemont, J. 1991. Long-term suppression of tremor by chronic stimulation of the ventral intermediate thalamic nucleus. *The Lancet*, 337: 403–406

Cohen, A.H., Rossignol, S. & Grillner, S. (editors). 1988. *Neural Control of Rhythmic Movements in Vertebrates*, New York: Wiley

Dawson, G.D. 1954. Summation technique for the detection of small evoked potentials. Electroencephalography and Clinical Neurophysiology, 44: 153–154.

Freeman, W.J. 1975. *Mass Action in the Nervous System*, New York: Academic Press

Haken, H. 1996. *Principles of Brain Functioning. A Synergetic Approach to Brain Activity, Behavior and Cognition*, Berlin and New York: Springer

Kuramoto, Y. 1984. *Chemical Oscillations, Waves, and Turbulence*, Berlin and New York: Springer

Nini, A., Feingold, A., Solvin, H. & Bergman, H. 1995. Neurons in the globus pallidus do not show correlated activity in the normal monkey, but phase-locked oscillations appear in the MPTP model of parkinsonism. *Journal of Neurophysiology*, 74: 1800–1805

Pikovsky, A., Rosenblum, M. & Kurths, J. 2001. *Synchronization: A Universal Concept in Nonlinear Sciences*, Cambridge and New York: Cambridge University Press

Sherrington, C. 1947. *The Integrative Action of the Nervous System*, 2nd edition, New Haven, CT: Yale University Press

Singer, W. & Gray, C.M. 1995. Visual feature integration and the temporal correlation hypothesis. *Annual Reviews of Neuroscience*, 18: 555–586

Steriade, H., Jones, E.G. & Llinás, R. 1990. *Thalamic Oscillations and Signaling*, New York: Wiley

Tass, P.A. 1999. *Phase Resetting in Medicine and Biology: Stochastic Modelling and Data Analysis*, Berlin: Springer

Tass, P.A. 2002. Effective desynchronization with bipolar double pulse stimulation. *Physical Review* E, 66: 036226

Tass, P.A. 2003. Stochastic phase resetting of two coupled phase oscillators stimulated at different times. *Physical Review* E, 67: 051902

Tass, P.A., Fieseler, T., Dammers, J., Dolan, K., Morosan, P., Majtanik, M., Boers, F., Muren, A., Zilles, K. & Fink, G.R. 2003. Synchronization tomography: a method for three-dimensional localization of phase synchronized neuronal populations in the human brain using magnetoencephalography. *Physical Review Letters*, 90: 088101

Tass, P.A., Rosenblum, M.G., Weule, J., Kurths, J., Pikovsky, A., Volkmann, J., Schnitzler, A. & Freund, H.-J. 1998. Detection of $n:m$ phase locking from noisy data: application to magnetoencephalography. *Physical Review Letters*, 81: 3291–3294

von Holst, E. 1954. Relations between the central nervous system and the peripheral organs. *British Journal of Animal Behavior*, 2: 89–94

Winfree, A.T. 1980. *The Geometry of Biological Time*, Berlin: Springer

STOCHASTIC PROCESSES

Complex processes in physics, chemistry, biology, engineering, and finance often exhibit uncertainties, fluctuations, or noises in their structures as a rule rather than as an exception. It is known that low-dimensional nonlinear dynamical systems can exhibit chaotic phenomena that appear random in nature. Nevertheless, stochastic processes involve an unmanageably large number of variables and their temporal fractal dimensions approach infinity. To interpret, analyze, model, and simulate these noises, several stochastic algorithms developed in the last century yield not only the structure or pattern of these processes, but also the statistical characteristics of the underlying uncertainties that the conventional deterministic models cannot offer.

Analysis of a stochastic process usually starts with the assumption of a Markov property. A Markov process, $X(t)$, is a stochastic process with the property that, given the value of $X(t=t_r)$, the values of $X(t=t_s)$ for $t_s > t_r$ are not influenced by the values of $X(t=t_q)$ for $t_q < t_r$. Depending on the nature of the process and the method of analysis, the random variable, $X(t)$, may assume continuous values or only discrete values. Moreover, the random variable can be analyzed with a continuous or a discrete time scale.

Algorithms have been developed for analyzing two distinct types of noise, internal and external. Internal noise is caused by the fact that the system itself consists of discrete particles and many variables associated with the particles are ignored; it is inherent in the very mechanism by which the process evolves. Small discrete systems governed by a large number of variables often exhibit notable internal fluctuations. The energy state of an electron in a molecule, the spread of an epidemic within a population, the mutation of a gene, and diffusion of a molecule into a medium cannot be predicted precisely due to their inherent complexities; thus, these processes exert internal fluctuations. The discrete state, continuous-time stochastic processes can be analyzed by the master equation in the following general form (Oppenheim et al., 1977; Gardiner, 1985; van Kampen, 2001):

$$\frac{\mathrm{d}p_n(t)}{\mathrm{d}t} = \sum_{n'} \{W_{nn'} p_{n'}(t) - W_{n'n} p_n(t)\}, \qquad (1)$$

where $p_n(t)$ denotes the probability of the system to be in state n at time t, and $W_{nn'}$ the transition probability per unit time from state n' to state n. The master equation is a gain-loss equation for the probabilities of the separate state n. The first term is the gain of state n due to transitions from other states n', and the second term is the loss due to transition from n into other states. For a system involving a set of random variables or a random vector, n is a vector and $W_{n'n}$ is a matrix.

When all elements of the transition intensity functions $W_{n'n}$ are either constant or linear functions of n, Equation (1) is considered a linear master equation; otherwise, it is considered nonlinear. Simple linear master equations such as the Poisson equation and birth and death processes can be solved by induction (see, e.g., Karlin & Taylor, 1975). For processes involving large numbers of random variables, full solutions of their joint probability distribution function are usually difficult, if not impossible, and it often suffices to determine only the first and second moments of the resultant probability distribution. These moments can be found by taking the averages of the master equations or by the method of joint probability generating function (Karlin & Taylor, 1975, 1981; Chiang, 1980). For nonlinear master equations, solutions can be approximated by resorting to the system-size expansion, a rational linearization approximation technique based on the power-series expansion (van Kampen, 1976, 2001). The system size, Ω, has been proposed as an expansion parameter because it measures the relative importance of the noises. For a linear system, fluctuations or variances are of the order of $\Omega^{1/2}$ in a collection of Ω entities. As a result, their effect on the macroscopic properties is of the order of $\Omega^{1/2}$. It is expected that the joint probability for the nonlinear system, $p_n(t)$, will have a sharp maximum around the macroscopic value with a noise of width of the order of $\Omega^{1/2}$. Thus, when the higher-order moments are ignored, the approximated joint probability becomes a time-dependent Gaussian distribution. The technique gives rise to the deterministic macroscopic equations and the equations of noises for the master equation; the latter is a set of linear Fokker–Planck equations whose characteristics and solutions will be discussed below.

For the interpretation of its relation with other stochastic algorithms, the master equation has been expressed in terms of a continuous random variable $Y(t)$ with its realization $y(t)$

$$\frac{\partial P(y,t)}{\partial t} = \int \{W(y \mid y')P(y',t) - W(y' \mid y)P(y,t)\}\,\mathrm{d}y'. \qquad (2)$$

Taylor expansion of the above master equation yields the Kramers–Moyal expansion

$$\frac{\partial P(y,t)}{\partial t} = \sum_{v=1}^{\infty} \frac{(-1)^v}{v!}\left(\frac{\partial}{\partial y}\right)^v \{a_v(y)P(y,t)\}, \qquad (3)$$

where $a_v(y)$ denotes the jump moments:

$$a_v(y) = \int_{-\infty}^{\infty} r^v W(y, r)\,\mathrm{d}r \tag{4}$$

and

$$W(y \mid y') = W(y', r), \quad r = y - y'. \tag{5}$$

A particular important class of continuous Markov processes is that for which $a_v(y) = 0$ for $v \geq 3$ in Equation (3). It is called the Fokker–Planck equation or the diffusion process (Risken, 1996):

$$\frac{\partial P(y,t)}{\partial t} = -\frac{\partial}{\partial y} a_1(y) P(y,t) + \frac{1}{2}\frac{\partial^2}{\partial y^2} a_2(y) P(y,t), \tag{6}$$

where a_1 and $a_2/2$ are called the drift and diffusion coefficients, respectively. Thus, the Fokker–Planck equation is a truncated master equation. If a_1 is a linear function of y and a_2 is constant, Equation (6) is called a linear Fokker–Planck equation, a linearity definition consistent with that of the master equation. Linear Fokker–Planck equations have been used by Rayleigh, Einstein, Smoluchowski, and Fokker for special cases. Taking the moments of Equation (6) yields the equations governing the mean trajectory and the noise around the mean. The solutions of these equations are sufficient in the construction of the macroscopic trajectory and its time-dependent Gaussian noise. An Ornstein–Uhlenbeck process is characterized by the jump coefficients, $a_1(y) = -ky$, and $a_2(y) = D$ where the constants satisfy $k > 0$ and $D \geq 0$ (Gillespie, 1992). For an Ornstein–Uhlenbeck process with the initial condition $P(y, t = 0) = \delta(y - y_0)$, where δ denotes Dirac's delta function, its evolution of the probability distribution function can be written as

$$P(y,t) = \frac{1}{[\pi \frac{D}{k}(1 - \exp[-2kt])]^{0.5}} \times \exp\left\{ -\frac{(y - y_0 \exp[-2kt])^2}{\frac{D}{k}(1 - \exp[-2kt])} \right\}. \tag{7}$$

The evolution of an Ornstein–Uhlenbeck process and its statistical features are illustrated in Figure 1.

A discrete state, discrete time stochastic process can be designated by X_i, where $X_i = X(t = t_i)$ and $i = 0, 1, 2, \ldots$. X_i is called a Markov chain if it follows the Markov property. Examples of these processes include a random walk on a line, the number of persons in a waiting line at 1 minute intervals, and the number of catalyst particles of different sizes during attrition. The propagation of a Markov chain can be expressed in the following matrix form (Taylor & Karlin, 1998):

$$p(X_n) = p(X_0) P^n, \tag{8}$$

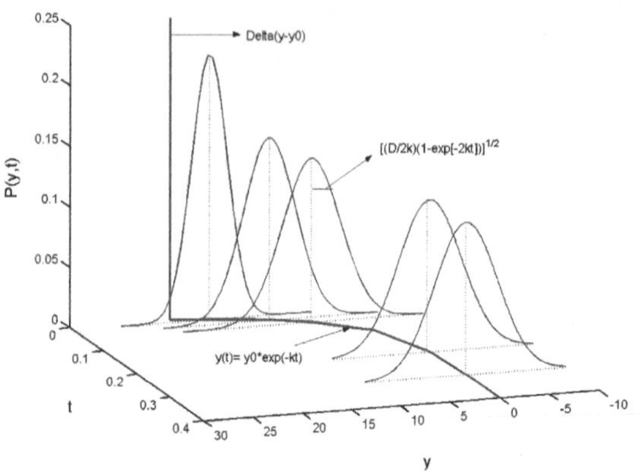

Figure 1. The evolution of an Ornstein–Uhlenbeck process, and its statistical features. The process starts with a given deterministic, initial condition, $P(y, t = 0) = \delta(y - y_0)$, and its fluctuations grow over time and eventually stabilize. The plot is drawn for $y_0 = 20$, $k = 20$, and $D = 1$. The solid line on the (y, t)-plane, $y(t) = y_0\exp(kt)$, denotes the macroscopic trajectory and its limiting distribution when $t \to \infty$ depicts the Brownian motion.

where $p(X_k)$ is a row vector representing the probability distribution at t_k and P is the one-step transition probability matrix with the following elements:

$$P_{ij} \equiv Pr\{X_{k+1} = j \mid X_k = i\}. \tag{9}$$

Equation (9) suggests that the probability distribution at a given time t_k can be estimated by successive matrix multiplication.

External noises are the fluctuations created in an otherwise deterministic system by the application of an external random force, whose stochastic properties are supposed to be known. The displacement of a mass attached to a spring with a fluctuating external force, an animal population under the influence of a fluctuating environment, the response of an electronic circuit equipped with a noise generator, the random loading of a bridge, and the yield from a chemical reactor with a fluctuating feed rate are examples of external noises. The Langevin approach is widely used for the purpose of finding the effect of noise in macroscopically known systems. The noise is introduced by adding a random term, $L(t)$, to the deterministic equation (Karlin & Taylor, 1981; van Kampen, 2001; Øksendal, 2003)

$$\frac{\mathrm{d}y}{\mathrm{d}t} = A(y) + C(y)L(t), \tag{10}$$

where $A(y)$ denotes the macroscopic characteristics of the system and $C(y)$ characterizes the interaction of the noise and the system. When $A(y)$ is a linear function of y and $C(y)$ a constant, Equation (10) is called a linear Langevin equation. When $A(y)$ is a nonlinear function of y and $C(y)$ a constant, Equation (10) is called a quasilinear Langevin equation. When $C(y)$ is a function of y, Equation (10) is called a nonlinear Langevin equation. These linearity definitions are consistent with those for the master and the Fokker–Planck equations after transformations. This consistency is important because some difficulties emerge during the integration of a Langevin equation that yields the Fokker–Planck equation (van Kampen, 2001). The treatment of the discontinuous stochastic noise, $L(t)$, is the source of the complexity. The added external noise, or the Langevin force, $L(t)$, is irregular and unpredictable, but its averaged properties over an ensemble of similar systems are simple. White noise is a commonly adopted simplification, which satisfies two averaged properties. Its ensemble has zero; that is,

$$E[L(t)] = 0. \tag{11}$$

Moreover, the individual noise takes place instantaneously with strength $\sqrt{\Gamma}$, and two successive noises arrive uncorrelated; that is,

$$E[L(t)L(t')] = \Gamma\delta(t - t'). \tag{12}$$

The spectral distribution of white noise, or the Fourier transform of Equation (12), is independent of the frequency ω. If the noise $L(t)$ is not δ correlated, the spectral density depends on the frequency and one uses the term *colored noise*.

Taking the first and second moments of Δy associated with a Langevin equation yields the jump moments of the Fokker–Planck equation. For a linear Langevin equation with $C(y) = 1$, the averaging process give rise to a linear Fokker–Planck equation (see, e.g., van Kampen, 2001)

$$\frac{\partial P(y,t)}{\partial t} = -\frac{\partial}{\partial y}A(y)P(y,t) + \frac{\Gamma}{2}\frac{\partial^2}{\partial y^2}P(y,t), \tag{13}$$

and it represents an Ornstein–Uhlenbeck process. For a quasilinear Langevin equation, the process gives rise to an expression identical to Equation (13). Nevertheless, the nonlinear nature of $A(y)$ suggests its Fokker–Planck equation is also nonlinear. Analytical solutions of the nonlinear Fokker–Planck equation are discussed by Risken (1996) and Grasman and van Herwaarden (1999).

Due to the discontinuous nature of $L(t)$, integrating $C(t)L(t)$ of a nonlinear Langevin equation in the averaging process, however, results in two different interpretations. If the averaged value of $C(y)$ during $(t, t + \Delta t)$ is used, the averaging process yields

$$\frac{\partial P(y,t)}{\partial t} = -\frac{\partial}{\partial y}[A(y)P(y,t)] + \frac{\Gamma}{2}\frac{\partial}{\partial y}\left[C(y)\frac{\partial}{\partial y}[C(y)P(y,t)]\right] \tag{14}$$

and the integration procedure is called the Stratonovich calculus. If the value of $C(y)$ before the arrival of the delta peak during $(t, t + \Delta t)$ is used, the averaging process yields

$$\frac{\partial P(y,t)}{\partial t} = -\frac{\partial}{\partial y}[A(y)P(y,t)] + \frac{\Gamma}{2}\frac{\partial^2}{\partial y^2}[C^2(y)P(y,t)] \tag{15}$$

and the integration procedure is called the Ito calculus. For practical processes experiencing external noise, the noise $L(t)$ is never infinitely sharp, and it lasts for a finite period of time; this consideration suggests that the Stratonovich interpretation is appropriate (van Kampen, 2001).

Wei-Yin Chen

See also **Averaging methods; Deterministic walks in random environments; Ergodic theory; Fokker–Planck equation; Lévy flights; Martingales; Monte Carlo methods; Nonlinearity, definition of; Random walks; Stochastic analyses of neural systems; Time series analysis**

Further Reading

Chiang, C.L. 1980. *An Introduction to Stochastic Processes and Their Applications*, Huntington, NY: Krieger

Gardiner, C.W. 1985. *Handbook of Stochastic Methods for Physics, Chemistry, and Natural Sciences*, 2nd edition, Berlin: Springer

Gillespie, D.T. 1992. *Markov Processes*, San Diego: Academic Press

Grasman, J. & van Herwaarden, O.A. 1999. *Asymptotic Methods for the Fokker–Planck Equation and the Exit Problem in Applications*, Berlin: Springer

Karlin, S. & Taylor, H.M. 1975. *A First Course in Stochastic Processes*, 2nd edition, New York: Academic Press

Karlin, S. & Taylor, H.M. 1981. *A Second Course in Stochastic Processes*, New York: Academic Press

Oppenheim, I., Shuler, K.E. & Weiss, G.H. 1977. *Stochastic Processes in Chemical Physics: The Master Equation*, Cambridge, MA: MIT Press

Øksendal, B. 2003. *Stochastic Differential Equations: An Introduction with Applications*, 6th edition, Berlin: Springer

Risken, H. 1996. *The Fokker–Planck Equation: Methods of Solution and Applications*, 2nd edition, Berlin: Springer

Taylor, H.M. & Karlin, S. 1998. *An Introduction to Stochastic Modeling*, 3rd edition, San Diego: Academic Press

van Kampen, N.G. 1976. The expansion of the master equation. *Advances in Chemical Physics*, 34: 245–309

van Kampen, N.G. 2001. *Stochastic Processes in Physics and Chemistry*, revised edition, Amsterdam: Elsevier

STOKES SCATTERING

See **Rayleigh and Raman scattering and IR absorption**

STRANGE ATTRACTORS

See **Chaotic dynamics**

STRING THEORY

String theories (Green et al., 1987; Davies & Brown, 1988; Polchinski, 1998; Gauntlett, 1998; Kaku, 1999) were developed in the 1980s and 1990s in the hope of finding a unified description of all forces in nature. The aim is to combine electroweak, strong, and gravitational interactions into one consistent theory. The basic idea is that subatomic particles are not point-like but one-dimensional extended objects: strings. These vibrate (similarly to a violin string) in various modes and can move in space-time. The various particle states that we observe in nature should then all be explained as suitable excitations of suitable strings. Interaction processes between particles are described by splitting and merging of strings. Due to the finite size of strings, many of the divergences that occur for point-like particles can be avoided. The size of the string is very small, typically of the order of a few Planck lengths (1.6×10^{-35} m), so that on large scales string-like particles look like point particles.

Different types of boundary conditions are possible; there are open strings and closed strings. The simplest example is a bosonic string. In suitable coordinates, the equation of motion is simply a wave equation of the form

$$\frac{\partial^2 X^\mu(\sigma,\tau)}{\partial\tau^2} = C^2 \frac{\partial^2 X^\mu(\sigma,\tau)}{\partial\sigma^2}. \tag{1}$$

Here X^μ is the coordinate of the string-like particle in d-dimensional space-time ($\mu = 0, 1, \ldots, d-1$), σ and τ denote an internal position and time variable parametrizing the string, and C is a constant.

The general solution can be written as a mode expansion of the form

$$X^\mu(\sigma,\tau) = x^\mu + \dot{x}^\mu\tau + \mathrm{i}\sqrt{2\alpha'} \sum_{n\neq 0} \frac{1}{n}\alpha_n^\mu \mathrm{e}^{-\mathrm{i}\frac{n\pi C\tau}{L}} \cos\frac{n\pi\sigma}{L}. \tag{2}$$

Here x^μ and $\dot{x}^\mu$ are constants independent of σ and τ that essentially represent the position and velocity components of a constantly moving point particle, whereas the sum over n describes the oscillating part of the dynamics. The string parameter $\sqrt{\alpha'}$ describes the approximate scale where string effects become relevant, and L is the length of the string. Unlike a violin string, this string is not tied down at the end points but it can move freely through space-time while it oscillates.

So far this is a classical string. First quantization is done by letting the coefficients α_n^μ of the above normal mode expansion satisfy suitable commutator relations. Second, quantization of strings, up to now, is still an active area of research with many open questions (Kaku, 1999).

The quantization of bosonic strings yields unpleasant surprises, so-called ghosts. These are quantum states with negative norm. One wants to avoid these unphysical states, which is only possible if the dimension of space-time is $d = 26$. Consequently, it has been necessary to assume that the 22 extra dimensions are "compactified," that is, curled up on small circles that are so small that we do not notice them in daily life.

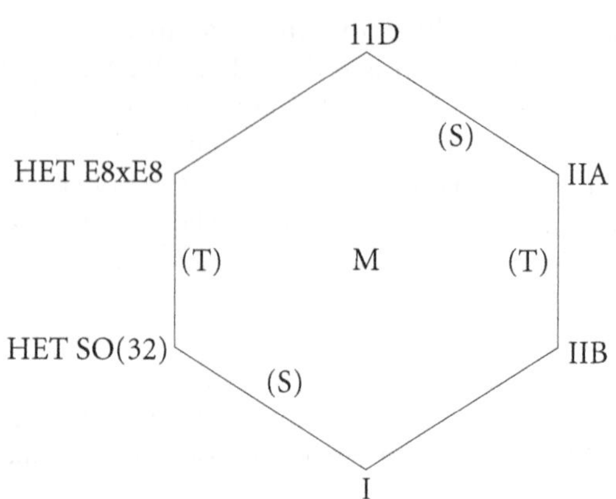

Figure 1. A schematic diagram of M-theory with its known string theory limits and the 11-dimensional theory. Some of the dualities are also shown.

But bosonic string theories have another problem. The lowest excitation mode (the ground state) is a so-called "tachyon," a state for which the mass squared is negative. To avoid this unphysical (unstable) ground state, one proceeds to supersymmetric string theories, in short "superstring theories." In these theories, for each boson there is a corresponding fermionic partner and vice versa. Technically, one introduces anti-commuting coordinates in addition to commuting ones. For supersymmetric theories, ghost states occur as well, but can now be avoided if the number of space-time dimensions is $d = 10$. So to describe low-energy physics, again six extra dimensions have to be compactified. Five different superstring theories have been found, which are denoted as type I, IIA, IIB, HET $E_8 \times E_8$, and HET SO(32). They differ in the degree of supersymmetry, the underlying gauge symmetry (if any), and whether the strings are closed or open.

Nowadays the five superstring theories are regarded as special limit cases of a more fundamental theory, called M-theory. This contains the five superstring theories and an 11-dimensional theory as special cases. These special cases are marked as the edge points of the diagram in Figure 1. Unfortunately, it is not known how M-theory should be formulated in full generality and how it should be second quantized. It is not even clear what the fundamental objects are that this theory describes. Much work needs to be done.

Some of the string theories contained in M-theory are connected by the so-called duality transformations. S-duality essentially means that a string theory with a small coupling constant α describes the same physics as another string theory with a large coupling given by $1/\alpha$. T-duality means invariance of the physics if a compactified dimension of radius R is replaced by another one of radius $1/R$.

In recent years, much research emphasis has concentrated on higher-dimensional objects contained in string and M-theory, so-called D-branes. Open strings can end and start on D-branes. Remember that in electromagnetism there are point charges, which are the sources of the electromagnetic field. D-branes are in a sense higher-dimensional generalizations of these point charges.

As already mentioned, to proceed to our four-dimensional world, six of the dimensions of superstring theory (or 7 of those of M-theory) have to be compactified. The way this should be done is unclear. One needs to compactify on special types of compact manifolds, so-called Calabi–Yau manifolds, to preserve what is generally believed to be the right amount of supersymmetry. The compactification step is clearly necessary for string theory to make contact with the real world, but almost all relevant low energy predictions depend on the numerous possibilities of how to compactify. This severely undermines the predictive power of string theories.

In principle, the ultimate theory should predict everything we want to know about particle physics, such as the values of the coupling constants of the four different forces or the masses of all fermions and bosons. In string theories, the numerical values of gauge couplings (e.g. the fine structure constant α_{el}, which is about $\frac{1}{137}$ at low energies) can be related to vacuum expectations of a scalar field contained in string theories, the dilaton field. But this vacuum expectation depends in an unknown way on second quantization effects and has not been predicted so far. As 't Hooft puts it in his book ('t Hooft, 1997), string theory, at least in its present stage, has similarities with a very uncomplete piece of furniture: "Imagine that I give you a chair while explaining that the legs are still missing, and that the seat, back, and armrest will perhaps be delivered soon; whatever I did give you, can I still call it a chair?"

While typical equations for (free) strings, such as Equation (1), are linear, a recent proposal is to amend ordinary strings (evolving in a regular way) by nonlinear versions of strings, so-called chaotic strings (Beck, 2002). These evolve in a deterministic chaotic way. In this approach, each ordinary string is shadowed by a chaotic string, which yields the "noise" for second quantization via the so-called stochastic quantization method. Mathematically, chaotic strings consist of one-dimensionally coupled Tchebyscheff maps, a very nonlinear and strongly self-interacting theory, which describes a kind of "turbulent quantum state" on a small (quantum gravity) scale. It turns out that the vacuum energy of chaotic strings is minimized for observed standard model parameters; that is, in this extended approach to second quantization, concrete predictions for vacuum expectations of dilaton-like fields and hence on masses and coupling constants can be given.

Christian Beck

See also **General relativity; Particles and antiparticles; Quantum field theory**

Further Reading

Beck, C. 2002. *Spatio-temporal Chaos and Vacuum Fluctuations of Quantized Fields*, Singapore: World Scientific

Davies, P.C.W. & Brown, J. 1988. *Superstrings: A Theory of Everything?* Cambridge and New York: Cambridge University Press

Gauntlett, J.P. 1998. M-theory: strings, duality and branes. *Contemporary Physics*, 39: 317–328

Green, M.B., Schwarz, J.H. & Witten, E. 1987. *Superstring Theory*, 2 vols, Cambridge and New York: Cambridge University Press

't Hooft, G. 1997. *In Search of the Ultimate Building Blocks*, Cambridge and New York: Cambridge University Press

Kaku, M. 1999. *Introduction to Superstrings and M-theory*, Berlin and New York: Springer

Polchinski, J. 1998. *String Theory*, 2 vols, Cambridge and New York: Cambridge University Press

STRONG COLLAPSE

See **Development of singularities**

STRUCTURAL COMPLEXITY

By *structural complexity* we mean the study of morphologically complex patterns associated with many physical entities of dynamical systems, using methods based on algebraic, geometric, and topological information (Ricca, 2001). In this respect, structural complexity offers an alternative and complementary route to study complex systems, based on morphological characteristics rather than the governing rules generating complexity. Structural complexity arises from spontaneous growth and self-organization of interacting components present in the bulk of physical systems, such as in turbulence, in biological and chemical reactions, in growth phenomena, across scales of space, time, and organizational complexity (Badii & Politi, 1997). Much of our understanding comes from observations of natural phenomena, from simulations and computational visualizations. Mathematical methods are used to extract and synthesize the relevant information, and to explore possible relationships between morphological complexity, energy localization, and other physical properties.

Dynamical systems are often characterized by the emergence and interaction of *coherent structures* (Nicolis & Prigogine, 1998; Scott, 2003). These latter are produced through filamentation mechanisms and space-time localization of characteristic physical properties. Examples of coherent filamentary structures may be as disparate as DNA macromolecules, synthetic polymer chains, nerve fibers, actin filaments, vortex filaments, magnetic flux tubes, or massive cosmic

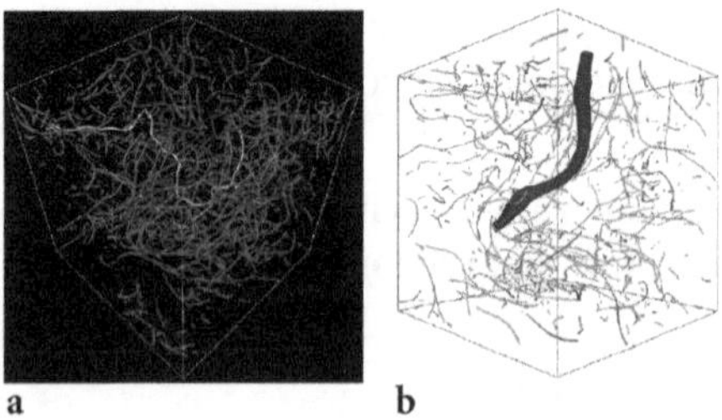

Figure 1. Computational field visualization of a pattern of vortex filaments produced by turbulent flow in a numerical box: (a) low-pressure line and (b) iso-pressure surface, obtained by numerical extraction techniques (Courtesy of S. Kida, Theoretical Division, National Center for Fusion Research, Toki, Japan).

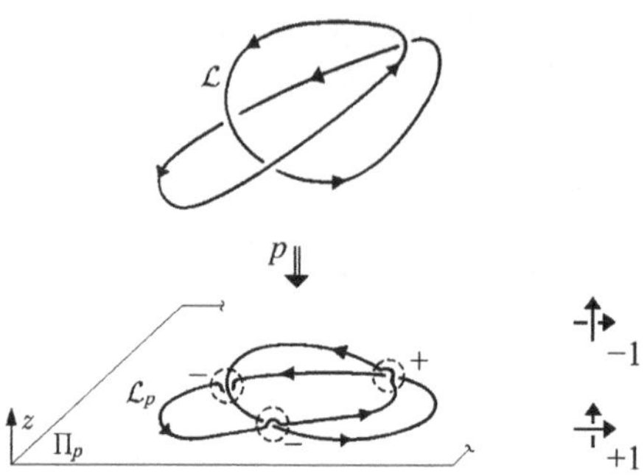

Figure 2. Example of analysis of an oriented space curve $\mathcal{L}$, based on crossing number information of the projected curve $\mathcal{L}_p$: we can calculate the total number of un-signed crossings (3, in the example shown in the figure) and, according to the sign convention shown on the right-hand side of the diagram, the algebraic sum of the $\pm$ signs ($-1-1+1=-1$, in the example shown). By averaging these two quantities over all directions p of projections, we obtain estimates of, respectively, the average crossing number and the writhing number of $\mathcal{L}$.

strings. Dislocations and discontinuities in mesoscopic and solid-state physics, fronts and cell membranes in chemistry, fluid dynamics, and biology provide other examples of complex structuring. The formation and continuous re-arrangement of coherent structures are due to local interactions and recombination, governed by specific dynamics given by nonlinear differential equations $\dot{\boldsymbol{x}} = F(\mathbf{x}, \lambda)$, with a nonlinear function F of the vector $\boldsymbol{x} = (x_1, \ldots, x_n)$, which depends on an external control parameter λ. At certain critical values, this parameter causes phase transitions and pattern changes (Levin, 2002). *Pattern recognition* techniques aim at providing mathematical methods for diagnostics of the morphological complexity of the emerging pattern, regardless of the governing fundamental physical mechanism that is responsible for complexity.

One of the great benefits of modern capability to perform complex three-dimensional simulations is the possibility to effectively analyze and visualize huge sets of three-dimensional data. Direct volume rendering, iso-surface extraction, integral convolution, and many other computational techniques allow sophisticated multi-field visualizations of complex physical patterns (Johnson et al., 2001). Progress in *visiometrics*, based on the exploitation of such techniques for an accurate identification and representation of complex three-dimensional structures, provides powerful tools and an almost inexhaustible source of information for analysis of structural complexity (see Figure 1; also see **Visiometrics**).

Physical as well as computational domains are decomposed into *tropicity* domains, defined by the characterizing properties of nulls, tubes, sheets, and blobs. These are reference sets defined by the reference physical field distribution, that allow direct measurement of the degree of *tubeness*, *sheetness*, and *bulkiness* of the coherent set by analyzing the aspect ratio of the distribution.

Algebraic, geometric, and topological measures are used to quantify structural complexity. In case of high filamentation, the original network of filaments is reduced (by appropriate threshold filtering) to a tangle $\mathcal{T}$ of space curves (representing the filament axes), hence reducing the problem to the analysis of the mutual positioning of the system of curves in space. By using methods based on crossing number information, we can evaluate the *average crossing number* $\bar{C}$ of the tangle that provides a fundamental algebraic measure of structural complexity. This quantity is defined by the formula

$$\bar{C} = \sum_{\mathcal{L}_i, \mathcal{L}_j \in \mathcal{T}} \bar{C}_{ij}, \tag{1}$$

where

$$\bar{C}_{ij} = \left\langle \sum_{r \in \mathcal{L}_i \sqcap \mathcal{L}_j} |\varepsilon_r| \right\rangle, \qquad \varepsilon_r = \pm 1 \tag{2}$$

denotes the un-signed crossings between the curve $\mathcal{L}_i$ and $\mathcal{L}_j$ of the tangle $\mathcal{T}$, and summation is made over all the crossings resulting from $\mathcal{L}_i \sqcap \mathcal{L}_j$; the average number of crossings is then evaluated by projecting the tangle onto projection planes; here the angular brackets denote averaging over all directions of projections (in the simple case of one oriented space curve, Figure 2 shows an application of this analysis). Geometric measures are based on integral measures of curvature and torsion of the reference curves and surfaces; for tangles of curves, these include *total curvature*, *average twist*, and *writhing number* which give information on the amount of coiling and entwinement of the corresponding filaments. Topological information comes from different types of measures: number of *nulls* (zero-value field singularities), *linking numbers* for open or closed curves, winding numbers and invariants associated with handle-body decomposition of reference sets are all calculable quantities. Further information may come from analysis of cell-decomposition of the network. In particular situations, when a prevailing type of geometry emerges, ad hoc analysis is used; for example, spiral spectrum analysis can be employed for scroll-wave reactions in Belousov–Zhabotinsky-type chemical systems and braid theory can be applied to study dynamics of particulates, fluid mixing, trailing vortices, astrophysical plasma loops, and tangle diagrams for polymers. Other information may come from study of dynamical systems, where we have information, for example, on Lyapunov exponents, topological entropy, multifractal properties, and topological scaling (see, for example, de Gennes, 1979; Jensen, 1998), whereas artificial intelligence provides us tools based on algorithmic complexity for neural-type networks. Measures of structural complexity are implemented computationally as time-dependent variables that change with the evolution of the physical system.

Collected information on structural complexity is analyzed against physical information. Since the evolution of any physical system is driven by variational principles that take account of energy and entropy redistributions, information on structural complexity may disclose some useful information on the physics and dynamics of the process. It is known that geometric properties strongly influence the dynamics: curvature forces (such as bending force in elastic rods or surface tension in liquid films) cause telephone cords to coil-up and foam bubbles to coalesce. We, therefore, expect relationships between geometry, dynamics, and ultimately, energy. Possible relationships between energy, entropy, and structural complexity in terms of algebraic information have, however, a more subtle rationale: network restructuring is due to local interactions and recombinations that are induced by acting potentials. These are associated with localized coherent structures, acting either on the strands of neighboring filaments (as in the case of reconnection of vortex tubes) or on the surface elements of sheet discontinuities (as in the formation of elastic films). In this context, local interaction of neighboring structures, resulting in the emergence of apparent crossings (possibly weighted by some distribution function) is a measure of the localization of internal energy. Hence, the resulting growth rate of the average crossing number provides an estimate of energy and entropy variations in the system. In absence of dissipation, topology is conserved; however, even in fully dissipative systems, topological information provides indications to detect preferred paths of energy depletion. In general, if χ denotes a measure (or a family of measures) of structural complexity, and E the energy of the system, we may expect *energy-complexity relations* of type $E = G(\chi)$, where G is a nonlinear function, where the nonlinearity is prescribed by the growth rate exponents of the physical process (see, e.g., Barenghi et al., 2001). Algebraic, geometric, and topological measures of structural complexity are, therefore, not only useful for computational implementation of diagnostics and pattern recognition, but they can also provide new tools to investigate localization and transfer of energy in complex physical systems.

RENZO L. RICCA

See also **Algorithmic complexity; Filamentation; Knot theory; Pattern formation; Topological defects; Vortex dynamics of fluids**

Further Reading

Badii, R. & Politi, A. 1997. *Complexity*, Cambridge and New York: Cambridge University Press

Barenghi, C., Ricca, R.L. & Samuels, D.C. 2001. How tangled is a tangle? *Physica* D, 157: 197–206

de Gennes, P.G. 1979. *Scaling Concepts in Polymer Physics*, Ithaca, New York: Cornell University Press

Jensen, M.H. 1998. *Self-Organized Criticality*, Cambridge and New York: Cambridge University Press

Johnson, C.R., Livnat, Y., Zhukov, L., Hart, D. & Kindlmann, G. 2001. Computational field visualization. In *Mathematics Unlimited – 2001 and Beyond*, edited by B. Engquist & W. Schmid, New York: Springer

Levin, S.A. 2002. Complex adaptive systems: exploring the known, the unknown and the unknowable. *Bulletin of the American Mathematical Society*, 40(1): 3–19

Nicolis, G. & Prigogine, I. 1998. *Exploring Complexity*, New York: W.H. Freeman

Ricca, R.L. (editor) 2001. *An Introduction to the Geometry and Topology of Fluid Flows*, Dordrecht: Kluwer

Scott, A. 2003. *Nonlinear Science: Emergence and Dynamics of Coherent Structures*, Oxford and New York: Oxford University Press

SUBHARMONIC GENERATION

See **Harmonic generation**

SUPERCONDUCTING QUANTUM INTERFERENCE DEVICE

The Superconducting QUantum Interference Device (SQUID) is the most sensitive detector for magnetic flux, Φ. It relies on the fundamental concept of flux quantization in a multiply connected superconductor, for example, a ring (torus). Flux quantization is a clear demonstration of the quantum mechanical properties of a superconductor in which the conduction electrons form (Cooper) pairs and condense in the ground state: the so-called "Bose condensation." Below the transition temperature, T_c, all the bosonic pairs can be described by a single macroscopic many-body wave function

$$\Psi(\mathbf{r}) = [n_s(\mathbf{r})]^{1/2} e^{i\theta(\mathbf{r})}, \qquad (1)$$

where n_s is the density of Cooper pairs. The phase $\theta(\mathbf{r})$ is a scalar function of the position. For small fields one may ignore the spatial dependence of n_s.

A direct consequence of the condensation is that for $T < T_c$ a superconductor exhibits zero electrical resistance and that magnetic fields are expelled from its interior. However, this diamagnetism (Meissner effect)

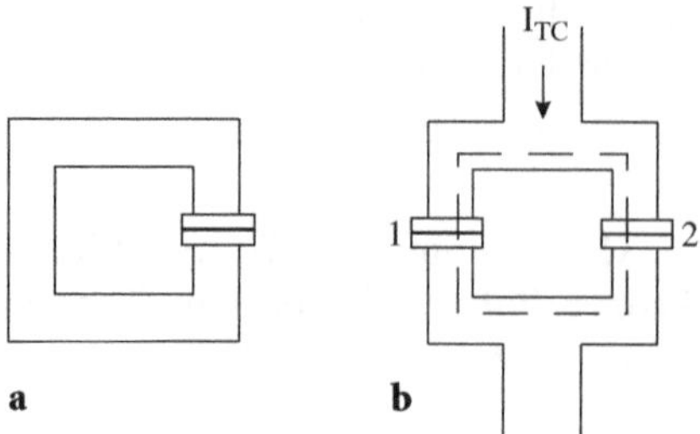

Figure 1. (a) The RF-SQUID and (b) the DC-SQUID. Dashed line is an integration contour deep inside the superconductor.

is not perfect as currents may penetrate into a thin surface layer (about 50 nm) called the London penetration depth (λ_L).

Early superconductors were metals (Hg, Al, Sn, Pb, Nb) or alloys (NbTi, Nb_3Sn) with T_c below ≈ 20 K. Since 1986, a new class of ceramic materials has been discovered with transition temperatures in the range 50–135 K, meaning that instead of requiring cooling by liquid helium (4 K), they only need cooling with liquid nitrogen (77 K) or by traditional cryocoolers. The terms low-T_c (LTS) and high-T_c (HTS) are used for these two types of superconducting materials.

The gradient of the phase θ of the macroscopic wave function describing all Cooper pairs depends on the pair-current density, $\boldsymbol{J}_s$, and the magnetic vector potential, $\boldsymbol{A}$, as

$$\hbar \nabla \theta = 2e(\Lambda \vec{\boldsymbol{J}}_s + \vec{\boldsymbol{A}}), \tag{2}$$

where the pre-factor, $\Lambda = 2m/n_s(2e)^2$, reflects the double electron charge $2e$ and mass $2m$ of the Cooper pair. If Equation (2) is integrated along a closed contour deep inside the superconducting ring where $J_s = 0$, the fact that the wave function has to be single-valued with phase (modulo 2π) leads to quantization of the magnetic flux contained in the ring so that

$$\Phi_{encl} = n\Phi_0 \,, \tag{3}$$

where $n = 0, 1, 2, 3, \ldots$ is an integer and $\Phi_0 = h/(2e) = 2.07 \times 10^{-15}$ Wb is the "flux quantum." Φ_0 is very small, corresponding to the flux generated by the Earth's magnetic field ($\approx$40 µT) in a circular loop with radius $\approx$4 µm.

A SQUID is a superconducting ring including either one or two Josephson junctions as shown in Figure 1. Named after their operational modes, these are called the RF-SQUID and the DC-SQUID. In both configurations, the response to an external magnetic flux is periodic in the flux quantum Φ_0. For simplicity, we consider only the two-junction DC-SQUID shown in Figure 1b.

Upon integrating Equation (2) along the closed contour, one may show that the difference between the phases, φ_1 and φ_2, of the macroscopic wave functions of the two Josephson junctions varies periodically with the applied flux Φ as

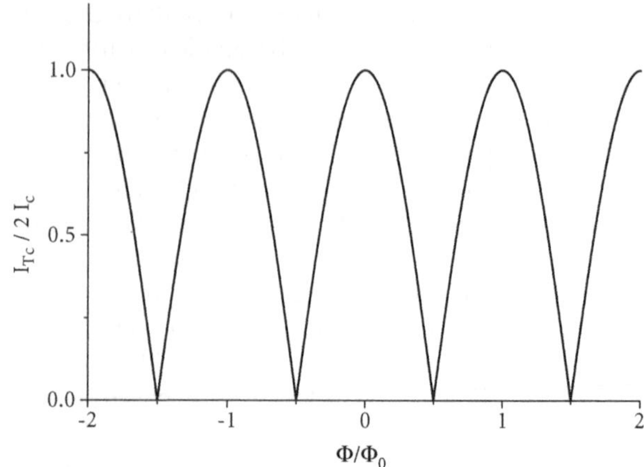

Figure 2. Magnetic interference pattern of symmetric DC-SQUID. Total critical current versus applied magnetic flux, $\beta_L = 0$.

$$\varphi_1 - \varphi_2 = 2\pi\Phi/\Phi_0, \tag{4}$$

giving a total critical (super-)current, I_{Tc}, of the SQUID

$$I_{Tc}(\Phi) = I_{c1} \sin \varphi_1 + I_{c2} \sin(\varphi_1 - 2\pi\Phi/\Phi_0) \,. \tag{5}$$

For two identical ($I_c = I_{c1} = I_{c2}$) junctions, $I_{Tc}(\Phi)$ varies from $2I_c$ to zero with a period of Φ_0 (see Figure 2). The first zero is at $\Phi = \Phi_0/2$. We say that the depth of modulation, $M = I_{Tc,max}/I_{Tc,min} = I_{Tc}/2I_c$, is 100%. M is reduced if the two junctions are different.

The variation in $I_{Tc}(\Phi)$ reflects the fact that the applied flux generates a circulating shielding supercurrent, $I_{circ}(\Phi)$ in the SQUID ring. Given the loop inductance, L, it generates a self-flux $\Phi_{self} = LI_{circ}$, which counteracts the applied flux. The effective flux in the ring is only $\Phi = \Phi_{appl} - \Phi_{self}$ leading to a reduced depth of modulation. A practical measure for a symmetric SQUID is the inductance parameter $\beta_L = \pi LI_c/\Phi_0$. For $\beta_L = 0$, $M = 100\%$ (as above) while for $\beta_L \to \infty$, $M \to 0$. Practical DC-SQUIDs are optimized with $M \approx 50\%$ corresponding to $\beta_L \approx 1$.

In order to avoid complications caused by hysteresis and flux contained in finite size junctions, one usually keeps junctions small and nonhysteretic (i.e., with relatively large damping and small capacitance). DC-SQUIDs with such junctions can be stably biased in the voltage state using a fixed DC current $I_b > I_{Tc}$. When an external flux is applied, the I-V curve is shifted following $I_{Tc}(\Phi)$, given in Equation (5), and a sensitive measurement of the oscillating voltage, $V(\Phi)|_{I_b}$, across the SQUID enables determination of Φ to a small fraction of Φ_0. Generally, the flux resolution is limited by thermal noise.

The use of a SQUID as a detector is complicated by the highly nonlinear dependence of the voltage response on the applied flux (see Figure 2). This may be overcome by operating the SQUID in the so-called "flux locked mode," where the SQUID output voltage via a feedback loop is returned as a current in a coil placed near the SQUID ring. By proper design, the coil produces a flux that exactly compensates the applied flux in the ring. The SQUID now acts as a null detector where the feedback current is a linear function of Φ over a very large range of magnetic flux.

Its impressive flux resolution does not imply that the SQUID can resolve such small magnetic fields, because the area of the SQUID ring is normally very small. An ingenious way to effectively increase the loop area without increasing L (keeping $\beta_L \leq 1$) is to use a superconducting flux transformer. This is a closed superconducting loop consisting of a series connection of a pickup coil placed in the ambient magnetic field and an *input coil* inductively coupled to the SQUID ring. When an external magnetic field is applied to the pickup coils, a shielding current is set up in the loop so that the total flux in the flux transformer is conserved (flux quantization). The current generates a flux in the input coil, which in turn couples the flux into the SQUID. To increase sensitivity, the input coil often is a multiturn thin-film coil (Washer type) placed on top of the SQUID ring. Usually, the input coil and the SQUID are enclosed in a magnetic shield to suppress external noise.

A well-designed LTS DC-SQUID magnetometer with proper pickup coil has an unsurpassed magnetic field sensitivity of $\approx 10\,\text{fT}$ (10^{-14} T), about one billionth (10^{-9}) of the Earth's magnetic field. LTS RF-SQUIDs and HTS DC-SQUIDs reach 100 fT. Most field measurements, therefore, are limited by magnetic noise from the environment and not from the magnetometer itself. Magnetic noise may be dramatically reduced by shaping the pickup coil as two matched counter-wound coils, so that a uniform magnetic field (e.g., from a distant source) generates zero net flux in the flux transformer. Effectively it measures the gradient of the magnetic field working as a first-order gradiometer with baseline equal to the distance between two counter-wound coils. Higher order gradiometers with three (or more) balanced coils have also been constructed. Modern multi-SQUID magnetometers use first-order gradiometers with advanced digital noise reduction and background field nulling.

Because of their high sensitivity, SQUIDs are now used as transducers for virtually any response that can be converted to magnetic flux, including magnetic fields, field gradients, currents, voltages, electromechanical positioning, and movements. A prominent biomedical application is three-dimensional magnetic source imaging, where an array of several hundred SQUIDs placed in a helmet fitted around the head can localize magnetic sources in the human brain to about a millimeter. This is important for precise location before or during brain surgery and for diagnosis of mental disorders. Many hospitals already use commercial SQUID instruments to record the very weak fields generated by nerve pulses in the brain (MEG, magnetoencephalogram) and heart, even the fetal heart beat (MCG, magnetocardiogram), as supplements or improved alternatives to the established electrical counterparts, EEG (electroencephalogram) and ECG (electrocardiogram) using electrodes placed on the skin.

Dedicated SQUID magnetometers are also used in geomagnetic surveying as well as for nondestructive testing (NDT), where defects in conducting samples can de detected by monitoring magnetic fields of induced eddy currents. For detection of the Earth's magnetic field, signal frequencies are typically a few millihertz, but SQUIDs may also be used for ultra-low noise microwave amplifiers as well as picosecond electronics and terahertz oscillators based on propagation at high speed of single quanta of magnetic flux in SQUID arrays.

JESPER MYGIND

See also **Josephson junctions; Superconductivity**

Further Reading

Duzer, T. Van & Turner, C.W. 1998. *Principles of Superconductive Devices and Circuits*, Upper Saddle River, NJ: Prentice-Hall, pp. 256–283

Kadin, A.M. 1999. *Introduction to Superconducting Circuits*, New York: Wiley, pp. 207–225

Orlando, T.P. & Delin, K.A. 1990. *Foundations of Applied Superconductivity*, New York: Addison-Wesley, pp. 244–251 and 410–420

SUPERCONDUCTIVITY

The phenomenon of superconductivity was discovered in 1911 by the Dutch scientist Heike Kamerlingh Onnes. In connection with work on the liquefaction of helium (normal boiling point at 4.2 K), he measured the resistance of mercury and noticed that the resistance dropped gradually with temperature as was expected. However, near 4.2 K, the resistance apparently dropped to a very small value consistent with zero. This, of course, must have been a very dramatic discovery, changing completely the general picture of the properties of solid-state matter. Today, the simple and direct consequence of zero resistance—that a large current can run indefinitely in a closed wire without energy input—intrigues many people. Even though zero resistance may be counter-intuitive, it is not forbidden by the laws of nature, and it does not imply a "perpetuum mobile." The second important property of a superconductor is that it is a perfect diamagnet. This property also has a feature that at first may seem quite unnatural. If a magnet is lowered towards a superconductor, the magnetic field of the magnet cannot penetrate into the superconductor. The magnetic field lines become compressed between the magnet and the superconductor, and eventually the magnet will levitate. This exclusion of the magnetic field is not a consequence of the

Figure 1. A levitated HTS superconductor.

induction law in connection with zero resistance of the superconductor, but an independent phenomenon. The phenomenon was discovered in 1933 by Walter Meissner and Robert Ochsenfeld long after the discovery of zero resistance. It is called the Meissner effect.

Superconductivity as discussed above occurs in many ordinary metallic elements from the periodic table. Examples are mercury, lead, tin, and niobium. The transition temperatures T_c (i.e., the highest temperature where superconductivity occur) are typically below 10 K. If the external magnetic field is too high, it will quench the superconductivity completely. This field is called the critical field H_c, and it is typically of order some hundreds of Gauss, too low for practical applications in electric machinery. The type of superconductor described above is called a type I superconductor.

Type II superconductors were discovered much later—in the 1950s. They are typically alloys of metals such as NbTi or Nb_3Sn and have transition temperatures typically in the 10–20 K range up to 23 K. They can carry much higher currents and can withstand much higher magnetic fields, thus, being much more useful for applications. A major difference from type I superconductors is the way they behave in a magnetic field. The Meissner effect as described above exists up to a so-called lower critical field H_{c_1}. At H_{c_1}, the magnetic field in the form of magnetic flux lines starts to penetrate the superconductor. As the external field is increased up to the upper critical field H_{c_2}, more and more flux lines penetrate the superconductor until at H_{c_2}, the flux lines completely fill the superconductor, which then becomes normally conducting.

In 1986, a new type of superconductor was found with a complicated structure. These superconductors are called High Temperature Superconductors (HTS, see below) and may, for lack of better understanding, be described as extreme type II. Examples are $YBa_2Cu_3O_7$ (known as YBCO) and $Bi_2Sr_2Ca_1Cu_2O_8$ (BSCCO), both of which have transition temperatures above the boiling point of liquid nitrogen at 77 K. These materials are quite useful for applications since liquid nitrogen—a cheap industrial product—may be used as a coolant. These important superconductors will be discussed separately below.

Figure 1 shows a levitated HTS superconductor. Since it is an extreme type II superconductor, magnetic flux lines penetrate the superconductor. These flux lines are "pinned" on imperfections of the crystal lattice, such as dislocations, and provide a stability not seen in type I superconductors. Thus, the HTS superconductors may even "levitate" below the magnet—hanging on the flux lines from the magnet. By symmetry a magnet can alsobe levitated over a superconductor.

Ginzburg–Landau Theory

Superconductivity is a complicated quantum mechanical phenomenon. The particles giving rise to the supercurrents may be described by a quantum mechanical wave function $\psi(r) = |\psi(r)|\exp(i\Theta)$. In a normal metal, the phases of all the electrons are different. In the superconducting state, all carriers have the same quantum mechanical phase of the wave function. This is referred to as macroscopic phase coherence. Thus, a simply connected macroscopic superconductor is characterized by a single-phase angle. The difference between a normal metal and a superconductor has some similarity to the difference between the coherent light of a laser, where all the atomic oscillators oscillate in phase and the "white" light of an ordinary bulb, where all the exited atoms are incoherent.

The interpretation of superconductivity as a macroscopic quantum phenomena and the first useful theory of superconductivity were due to Fritz and Heinz London in 1935.

A real breakthrough for applications is the still very often-used phenomenological theory published by the Russian scientists Vitaly Ginzburg and Lev Landau in 1950. Ginzburg and Landau introduced a position-dependent order parameter with a conceptual similarity to a quantum mechanical wave function. The Ginzburg–Landau theory is a masterpiece of intuition. It postulates that as the temperature is lowered and T_c is crossed from above, a (complex) order parameter $(\psi(r))$ for an ordered, superconducting phase appears—increasing from the value zero at T_c. Gibbs free energy is then expressed as a series expansion in terms of this order parameter $(\psi(r))$. Finding the equilibrium value of the Gibbs free energy leads to two coupled differential equations that connect the order parameter and the electromagnetic fields.

Variational methods are used and the Ginzburg–Landau equation is derived. It may be written

$$\alpha\psi(\boldsymbol{r}) + \beta|\psi(\boldsymbol{r})|^2\psi(\boldsymbol{r}) \\ +1/2m((h/2\pi i)\nabla - q\boldsymbol{A})2\psi(\boldsymbol{r}) = 0$$

Here α and β are coefficients to be determined by the particular system, $\boldsymbol{A}$ is the vector potential and m is a particle mass. The last term on the left side has the appearance of a kinetic energy, and we note the similarity to the Schrödinger equation (*See* **Nonlinear Schrödinger equations**).

As mentioned above, the nonlinear Ginzburg–Landau equation can be applied to other systems with an order parameter characterizing a phase transition. Examples are magnetic materials, liquid helium, and liquid crystals. Because of the series expansion, which is only good when the expansion parameter is small, the theory is formally only valid near the transition temperature.

BCS Theory

Although rather successful, the above theories are phenomenological in nature. They do not specify the nature of the physical system or why superconductivity occurs. For this, a full microscopic quantum mechanical theory must be derived. This was done by John Bardeen, Leon Cooper, and Robert Schrieffer in 1957, and their theory carries the name BCS theory. The derivation of the BCS theory is too complicated to be described here, but we note that in most practical cases a full understanding is not needed. Close to T_c the BCS theory approaches the phenomenological Ginzburg–Landau theory as it should. Central to BCS theory is the notion that there are two types of currents. Normal currents are carried by normal electrons, and supercurrents are carried by so-called "superelectrons." Just below T_c, there are only a few superelectrons and many normal electrons. At $T = 0$, all charge carriers are superelectrons. These "superelectrons" are bound pairs of normal electrons called Cooper pairs. Accordingly they have two electron masses and two electron charges. Intuitively, one would think that two electrons could not form a pair since Coulomb repulsion would rip them apart. However, the pairing mechanism has its origin in the electron-phonon interaction, which is responsible for overwhelming the Coulomb repulsion and providing the attractive potential binding the two electrons together.

The binding energy of the pair, Δ, is of order a few meV and is related to the energy scale of the phonons involved in the pairing process. Naturally in the BCS theory, there is a direct proportionality between the transition temperature T_c and the pair binding energy Δ. In fact, this way it has been suggested that T_cs cannot exceed about 30 K. Until recently the highest T_c observed experimentally was 23 K in general agreement. This is not true any more as will be described in the next section.

High T_c Superconductors

In 1986, K. Alex Muller and J. Georg Bednorz from the IBM Zurich research laboratory discovered superconductivity in a new class of materials involving typically the so-called rare earths. Within a few months after the discovery was known, researchers found a T_c of 92 K in the material $YBa_2Cu_3O_7$ and related compounds. YBCO is typical of the so-called HTS although now many other families are known, including the widely used $Bi_2Sr_2Ca_2Cu_3O_8$ compounds. To become superconducting, these materials must undergo a heat treatment at about 1000 K in an oxygen atmosphere. After this baking, the materials appear as a ceramic and are thus hard and brittle—which causes some engineering problems for making the superconductors into wires and coils. Now, many more HTSs are known and the highest T_c found so far is about 150 K. New HTS materials are found regularly, and it cannot be predicted what the limit to higher T_c will be. Maybe we will one day have room-temperature superconductors.

The theory for HTSs is not known, in spite of the efforts of many theoreticians. Obviously, the BCS theory is not directly applicable since it predicts a maximum T_c of about 30 K. An HTS typically has a layered structure with copper-oxide planes separated by isolating atoms at atomic distances, and it is speculated that the superconductivity somehow has its explanation in this structure.

Particularly following the discovery of HTS, the potential for useful applications of superconductivity is quite large. A significant benefit of the high transition temperatures is that liquid nitrogen at 77 K can be used as the coolant. Applications are both small scale (electronics, sensors, and communications) and large scale (power cables, motors, transformers, and large magnets.)

Starting with small scale, we note that Josephson junctions (*See* **Josephson junctions**) are the main component in superconducting electronics. The general advantage over semiconductors is higher speed and lower dissipation. Computers can be constructed that are much faster and much smaller than conventional computers. Sensors, in general, can be made much more sensitive with superconductor technology. This is because of the quantum mechanical nature of superconductivity and, of course, the low temperatures where thermal noise is reduced. A so-called super conducting quantum interference device (SQUID) is a commercial superconducting magnetic field sensor that is so sensitive that it can measure the magnetic field from nerve impulses in the brain. It can also be used to detect internal cracks in the aluminum-alloy plates of aircraft wings. Superconducting photon detectors can detect single photons from space. High-frequency microwave detectors can detect signals down to the quantum limit and in some spectacular cases even lower. Filters for mobile telephone ground stations can be made much sharper than conventional ones, implying more telephones, improved quality, and lower prices.

Large-scale applications include large magnets, levitated trains, power cables, motors, energy storage, and fault current limiters. Large magnets are, for example, used in the Magnetic Resonance Imaging

(MRI) systems at hospitals where a patient's whole body is introduced into the magnet. This is a very large market, and it is only feasible with superconducting magnets. A levitated train with superconducting magnets has already been built in Japan (Maglev). Superconducting (HTS) power cables have already been built in several countries. In Denmark, a 35 m section has already been installed in the Copenhagen grid. In general, the benefits of the superconducting systems is that energy savings can be obtained because there is no—or at least very low—dissipation. Thus superconducting machinery can be made much smaller than similar conventional machinery, which is important in many cases in the automotive industry. Energy can be stored in large superconducting coils or friction-free flywheels. A fault current limiter is an important protection device for the utility grid that can be made using only superconducting technology.

In the coming generation, superconductivity may play the revolutionary role that semiconductors did in the previous generation.

NIELS FALSIG PEDERSEN

See also **Bose–Einstein condensation; Josephson junctions; Nonlinear Schrödinger equations; Superconducting quantum interference device**

Further Reading

Barone, A. & Paterno, G. 1982. *Physics and Applications of the Josephson Effect*, New York: Wiley

Seeber, B. (editor). 1998. *Handbook of Applied Superconductivity*, 2 vols, Bristol and Philadelphia: Institute of Physics Publishing

Sheahen, T.B. 1994. *Introduction to High Temperature Superconductivity*, New York: Plenum Press

Tinkham, M. 1996. *Introduction to Superconductivity*, 2nd edition, New York: McGraw-Hill

SUPERFLUIDITY

Superfluidity is the state of a quantum liquid in which the liquid has the ability to flow through fine slits and capillaries without friction or production of a pressure gradient. The superfluidity of liquid helium He^4 (called "He-II") at temperatures below $T_\lambda = 2.17$ K was discovered by Peter Kapitza in 1938. The transformation of normal liquid helium into the superfluid state is a phase transition of the second order. Nevertheless, the superfluid He-II cannot be considered as a liquid without viscosity. Experiments on rotational vibrations of a disk suspended in He-II show that damping is observed at temperatures below ($T < T_\lambda$), decreasing as the temperature is lowered.

The first qualitative explanation of superfluidity phenomena was proposed by Lazlo Tisza (1938), who introduced a two-fluid model of superfluidity. Lev Landau (1941) later formulated a so-called two-fluid hydrodynamics of He-II. According to this explanation, He-II consists of two interpenetrating fluids, a normal fluid with viscosity and a superfluid fluid with zero viscosity. It is the superfluid that flows without friction through the finest of channels, while the finite viscosity of the normal component is responsible for the observed damping of the oscillating disk. Each component is assigned a density ρ_n and ρ_s, whose sum is the total density $\rho = \rho_n + \rho_s$, and each is assigned its velocity field v_n and v_s.

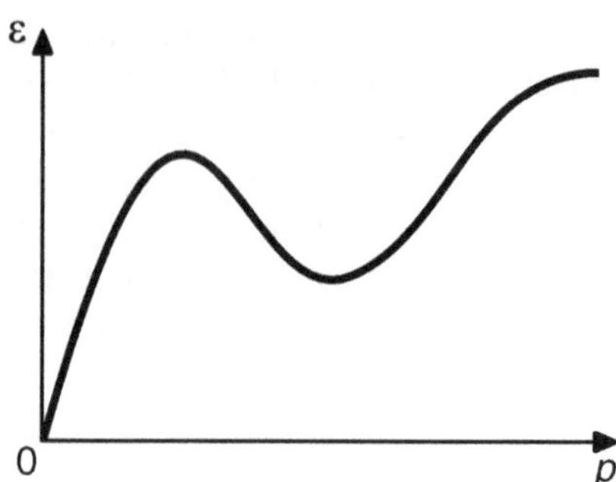

Figure 1. Dependance of the energy of elementary excitations on momentum.

Landau's theory is based on the idea of a possible description of physical properties of a weakly excited macroscopic system at low temperatures by means of elementary excitations or "quasiparticles." The elementary excitations have energy $\varepsilon(p)$ and are regarded as particles of momentum p. The normal component represents a gas of elementary quasiparticles with a dispersion curve as sketched in Figure 1. There are two types of the quasiparticles at low temperatures: sound quanta or phonons of long wavelength (with dispersion relation $\varepsilon = sp$, where s is the sound velocity) and excitations of short wavelength called "rotons" (near the minimum of the curve in Figure 1 their energy is $\varepsilon = \Delta + \frac{1}{2}(p - p_0)^2/\mu)$, where μ is the effective mass of the roton). Interactions of quasiparticles with each other and with the vessel's walls cause the viscosity of the normal component. At zero temperature, the normal density $\rho_n = 0$ and the system is in the ground state.

The remaining part of He-II is the superfluid component. It has zero viscosity and can move without friction through narrow slits and capillaries if the energy spectrum of the quasiparticles satisfies the following condition: $\varepsilon(p)/p > 0$. At $T = 0$, the superfluid density $\rho_s = \rho$. The quasiparticle concentration grows when the temperature increases and ρ_s decreases turning into zero at $T = T_\lambda$. According to the Landau theory, the liquid loses the superfluidity property if the velocity of its flux exceeds a critical velocity,

$$v_c = \min(\varepsilon(p)/p); \tag{1}$$

this condition is called the "Landau criterion." Real critical velocities are much less than (1) because the superfluidity is broken by vortices arising in the superfluid liquid at even smaller velocities. Quantum vortices have been observed during experiments on the motion of the He-II in a rotating cylindrical container.

In a quantum Bose gas and He-II at low temperatures, a large number of particles (a finite proportion of all particles) turns into one coherent state called a "condensate," which can have a stationary velocity different from zero. Such a phenomenon is called "Bose–Einstein condensation," and superfluidity is often connected with it. When the superfluidity component is a correlated many-particle quantum state; however, it may not be connected directly with Bose–Einstein condensation.

The energy spectrum of excitations in a weakly non-ideal Bose gas with repulsive interactions between particles was first derived by Nikolai Bogolubov (1947) assuming existence of the condensate

$$\varepsilon(p) = [(p^2/m)\rho_0 \nu(p) + p^4/(4m^2)]^{1/2}, \qquad (2)$$

where ρ_0 is the density of particles in the state $\boldsymbol{p}=0$, $\nu(p)$ is the Fourier component of the potential of interparticle interaction ($\nu(0) > 0$), and m is the mass of the particle. Equation (2) determines a spectrum of small vibrations of the condensate in the non-ideal Bose gas. This dispersion relation is of the phonon type at small p, where the Landau criterion is satisfied. Therefore, even the weakly non-ideal Bose gas possesses the superfluid property.

A macroscopic theory of the superfluid dynamics of the weakly non-ideal Bose gas was developed by Gross (1961) and Pitaevskii (1961). In this theory, a condensate wave function Ψ is described by the nonlinear Schrödinger equation ($\hbar = 1$):

$$\mathrm{i}\frac{\partial \Psi}{\partial t} = -\frac{1}{2m}\Delta\Psi + \nu(0)\left(|\Psi|^2 - \rho\right)\Psi. \qquad (3)$$

A solution of the linearized Equation (3) has a form of harmonic vibrations whose frequency spectrum is determined by Equation (2). Thus, Equation (3) in a total form possesses a solution in the form of the straight-line vortex

$$\Psi = \sqrt{n} f\left(\frac{r}{r_0}\right) \mathrm{e}^{\mathrm{i}\phi}, \quad r_0^2 = \frac{h^2}{2m\nu(0)n}, \qquad (4)$$

where r and ϕ are the distance from the vortex axis and the polar angle with respect to it, and the function $f(r)$ has a shape sketched in Figure 2.

Superfluidity is also possible in quantum Fermi liquids both in neutral (e.g., liquid He^3) and charged (conducting electrons in metals) liquids. Then a generation of pairs of coupled fermions is necessary (pairs of He^3 atoms or electrons with an integer spin) to create bosons. Such pairing is a result of any attraction (either van der Waals interactions between the atoms or interaction of the electrons with vibrations of the crystal lattice). Being bosons, these "Cooper pairs" can form coherent superfluid condensates. The attractive forces between He^3 atoms are very small; therefore, conditions for generation of the Cooper pairs of quasiparticles and origin of the superfluidity of He^3 appear only at superlow temperatures of the order of 10^{-6}K. Interestingly, there are two superfluid phases (A and B) of the He^3 at superlow temperatures.

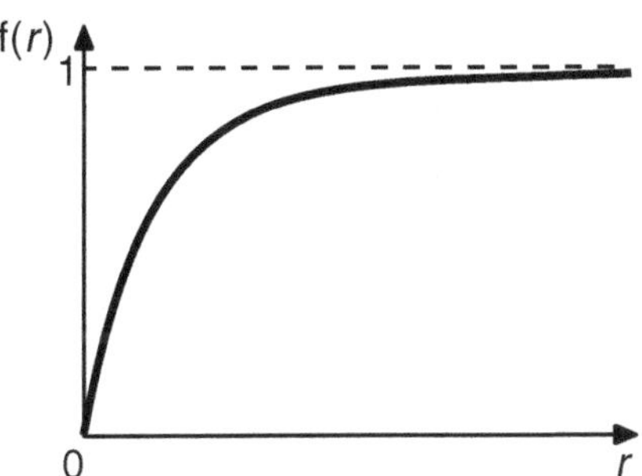

Figure 2. Distribution of the condensate near the vortex.

The emergence of electron Cooper pairs in a metal corresponds to the appearance of superconductivity. Concepts from superfluidity theory have influenced the theory of superconductivity in metals and also the development of models of the superfluidity of the atomic nucleus. As the atomic nucleus presents a Fermi system with strong interparticle interaction, both the pair correlation effects and also manifestations of some superfluid properties of Bose systems should take place.

Superfluid liquids possess several unusual properties. Besides ordinary sound (i.e., vibrations of the liquid's density), a so-called "second sound" can propagate in such a liquid. The second sound represents sound propagation in the quasiparticle gas (vibrations of the quasiparticle density and, therefore, vibrations of the temperature). Also, a thin film of superfluid He-II forms on a solid wall, rising easily up to a large height. This film can equalize liquid levels of the helium in vessels having a common wall.

ARNOLD KOSEVICH

See also **Bose–Einstein condensation; Nonlinear Schrödinger equations; Superconductivity**

Further Reading

Bogolubov, N.N. 1947. On the theory of superfluidity. *Izvestia Akademii Nauk SSSR*, Seriya Fiziko, 11: 77

Donnelly, R.J. 1967. *Experimental Superfluidity*, Chicago: University of Chicago Press

Feynman, R.P. 1957. Appliction of quantum mechanics to liquid helium. In *Progress in Low Temperature Physics*, edited by C.J. Gorter, vol. I: pp. 17–53

Gross, E.P. 1961. Structure of a quantized vortex in boson systems. *Nuovo Cimento*, 20: 454

Kapitza, P.L. 1938. Viscosity of liquid helium at temperatures below the lambda point. *Doklady Akademii Nauk SSSR*, 18: 21 (in Russian); translated in *Nature*, 141: 74

Kapitza, P.L. 1977. *Experiment. Theoriya. Practika*, Moscow: Nauka (in Russian)
Khalatnikov, I.M. 1971. *Teoriya Sverkhprovodimosti*, Moscow: Nauka (in Russian)
Landau, L.D. 1941. Theory of superfluidity of helium II. *Zhurnal Experimental'noj i Teoreticheskoj Fiziki*, 11: 592 (in Russian)
Landau, L.D. 1947. On the theory of superfluidity in helium II. *Journal of Physics of the USSR*, 11: 91
Lifshitz, E.M. & Pitajevski, L.P. 1978. *Statisticheskaya Fizika*, p.2 Moscow: Nauka (in Russian)
Pitaevskii, L.P. 1961. Vortex lines in an imperfect Bose gas. *Zhurnal Experimental'noj i Teoreticheskoj Fiziki*, 40: 646 (in Russian); translation in *Soviet Physics JETP*, 13: 451
Tisza, L. 1938. Transport phenomena in helium II. *Nature*, 141: 913
Tisza, L. 1940. On the theory of quantum liquids. Application to liquid helium. *Journal de Physic et Radium*, 1: 350 (in French)

SUPERLATTICES

Superlattices (SLs) are solid periodic structures usually made artificially and having a period much larger than the interatomic distance. The SL is called one-dimensional (1-d) if it is periodic along one axis. In such a case, the SL has the form of alternating plane-parallel layers of two or several materials differing in their physical properties (dielectric, elastic, magnetic, etc., depending on the implementation of interest) (see Figure 1). Modern techniques of crystal growth allow the creation of artificial crystals with atomic smooth boundaries between layers, which can be arranged in any order. A two-dimensional SL is periodic in two directions and homogeneous in the third, and a three-dimensional SL is periodic along three axes. There are many types of SLs, including semiconducting and superconducting SLs, dielectric or optical SLs (called photonic crystals), and elastic or acoustic SLs (called phononic crystals).

The main physical property of an SL important for applications is a band structure of the energy or frequency spectrum caused by the macroscopic periodicity analogous to the structure of electron energy or phonon frequency spectra in monocrystals.

The physical origin of the band structure can be understood by analyzing a one-dimensional SL constructed from two types of materials, where the alternating layers have thicknesses d_1 and d_2 (see Figure 1). The SL period equals $d = d_1 + d_2$, and the electromagnetic or elastic fields inside each layer are described by wave equations with phase velocities c_1 and c_2. Thus, eigenfunctions of the wave equations are characterized by a quasiwave number k, and the energy or frequency of any excitation is a periodic function of k with the period $2\pi/d$, so values of k can be restricted to the one-dimensional Brillouin zone $-\pi/d < k < \pi/d$.

If the velocities c_1 and c_2 are close, one can assume $c_1 = c_2 = c$ in the first approximation, so the dispersion relation for the wave solution is that for

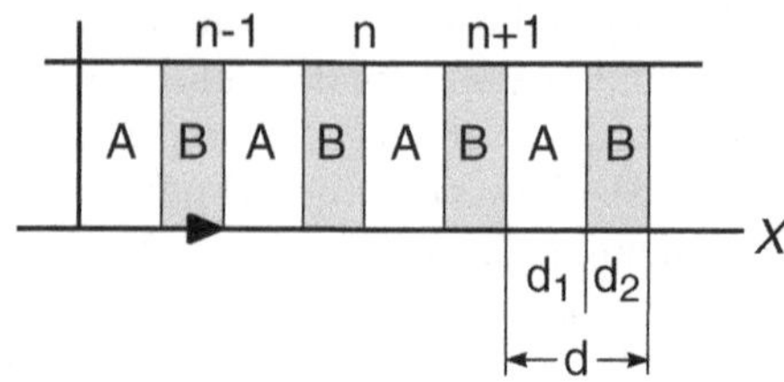

Figure 1. 1D SL made of substances of two types A and B.

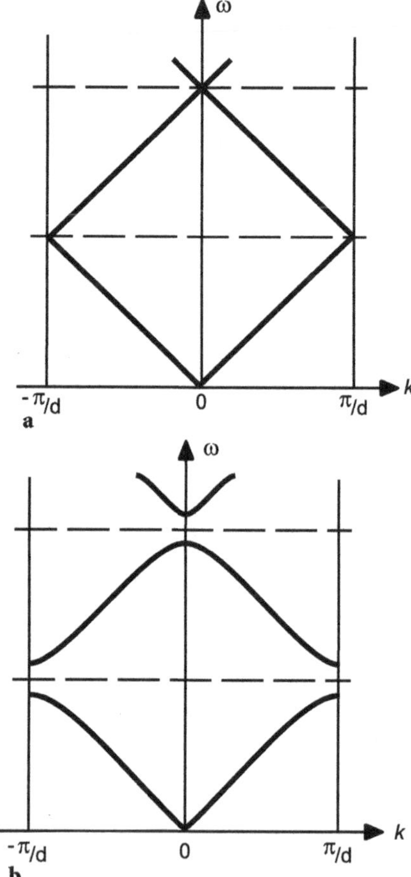

Figure 2. Dispersion relations for the SL (a) in the limiting case $c_1 = c_2$ and (b) in the case $|c_1 - c_2| \ll c_1 \approx c_2$.

a homogeneous material. In the case of an elastic medium or a dielectric transparent for electromagnetic waves, $\omega(k) = ck$ where ω is the frequency and c is either the sound or light velocity (the straight lines in Figure 2a). Because the function $\omega(k)$ repeats itself outside the Brillouin zone, the lines fold back into the zone when they reach edges. A small perturbation provided by the small difference between c_1 and c_2 takes off the degenerations at the points $k = 0, \pm\pi/d$ and as a result, a small frequency gap appears between the upper and lower branches of the lines (Figure 2b). This is called a forbidden gap as no excitation mode of the SL can exist in the gap. As the difference $c_1 - c_2$ is increased, the forbidden gap widens considerably.

Dispersion equations for the frequencies of arbitrary one-dimensional SLs made of two dielectric or elastic materials were obtained by the Soviet scientist Sergei Rytov (1955, 1956) using the standard boundary conditions at the layer boundaries

$$\cos kd = \cos k_1 d_1 \cos k_2 d_2 - \frac{1}{2}\left(\frac{k_1}{k_2} + \frac{k_2}{k_1}\right) \sin k_1 d_1 \sin k_2 d_2, \quad (1)$$

where $k_1 = \omega / c_1$ and $k_2 = \omega / c_2$. When the right-hand side of Equation (1) has values from -1 to $+1$, the equation has real roots, which give the values of ω.

If one of the alternating layers is opaque for electromagnetic waves, the corresponding wave vector is imaginary: $k_2 = \mathrm{i}\kappa_2$ and Equation (1) transforms to

$$\cos kd = \cos k_1 d_1 \cosh \kappa_2 d_2 - \frac{1}{2}\left(\frac{k_1}{\kappa_2} - \frac{\kappa_2}{k_1}\right) \sin k_1 d_1 \sinh \kappa_2 d_2. \quad (2)$$

There are two limiting cases of Equations (1) and (2) when the roots of interest can be easily obtained either graphically or analytically. The first one concerns (2) and assumes $\kappa_2 = \text{constant} \to \infty$ and $d_2 \to 0$ at $\kappa_2 d_2 \to 0$ under the condition $\kappa_2^2 d_2 = \text{constant} = q$ (Kronig & Penny, 1931). The simplified version of Equation (2) has the form

$$\cos kd = \cos kd + M \frac{\sin k_1 d}{k_1 d}, \quad M = \frac{1}{2} qd, \quad (3)$$

which can be solved graphically. In Figure 3a, $z = k_1 d$ and the roots of the equation run over values within intervals marked off on the abscissa axis. These intervals become wider and the forbidden gaps become more narrow when the frequency increases.

The second limiting case concerns Equation (1) and corresponds to the limit $d_2 \to 0$, $c_2 \to 0$ under the conditions $d_2/c_2 = 0$ and $d_2 / c_2^2 = P = \text{constant}$, leading to (Kosevich, 2001)

$$\cos kd = \cos k_1 d - Q\, k_1 d \sin k_1 d, \qquad Q = c_1^2 P/(2d). \quad (4)$$

Equation (4) can also be solved graphically (see Figure 3b), and a set of permitted intervals contracting with increasing frequency can be obtained. However, there is an analytical expression for frequencies at large numbers m of these intervals ($m^2 Q \gg 1$)

$$\omega = \frac{m\pi c}{d} + \frac{2\Omega}{m} \begin{cases} \sin^2(kd/2), & m = 2p; \\ \cos^2(kd/2), & m = 2p+1, \end{cases} \quad (5)$$

where $\Omega = c/(\pi Q d)$ and p is an integer.

The frequencies of forbidden bands correspond to a solution of the wave equation inside n-th layer of the type $\exp \pm \kappa n d$ (when $k = \mathrm{i}\kappa$) or $(-1)^n \exp(\pm \kappa n d)$

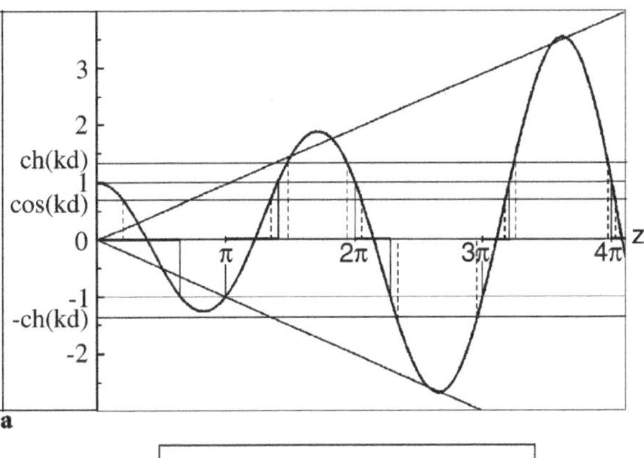

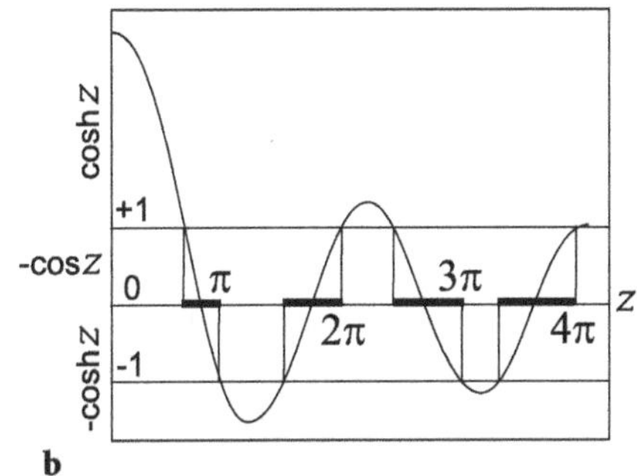

Figure 3. Graphical solutions of (a) Equation (3) and (b) Equation (4). Eigenfrequency bands are shown in heavy lines on the z-axis ($z = k_1 d$).

(when $k = \mathrm{i}\kappa + \pi$), which decrease (grow) with increasing n. The frequency dependence of the parameter κ for solutions of the first (second) type is found from Equation (1) when its right-hand side is larger then $+1$ (then $\cos kd = \cosh \kappa d$ in the left-hand side) or less than -1 (then $\cos kd = -\cosh \kappa d$). Such solutions have a physical meaning only in the x semiaxis under the condition that a solution vanishing at infinity and corresponding to certain boundary conditions at the origin is selected. The necessity of using exponentially decreasing solutions arises in describing localized states of fields under consideration in the vicinity of a local SL defect. Of course, a one-dimensional SL can have only localized states near a given plane. In certain circumstances, a localized mode can exist at the face of the SL (called a surface state).

In the general case, the spectrum of an SL is complicated and contains a system of a great number of both eigenfrequency bands and gaps corresponding to forbidden frequencies of eigenmodes. No extended states are allowed in the gap, but there exist evanescent modes, decaying exponentially with increasing the distance from the defect. In a three-dimensional SL, they are localized states.

Dependence of the physical parameters of materials (the velocities c_1 and c_2, or the parameters Q and M in the above equations) on the field strength leads to nonlinear effects in SLs. In optical SLs, a source of nonlinear effects is the dependence of the refraction coefficient on the electric field of a light wave. In the simplest case, which corresponds to the dispersion relations (5) in the linear approximation, an electromagnetic wave Ψ_n in the nth layer of the SL is governed by the discrete NLS equations for $m=2p$

$$\mathrm{i}\frac{\partial \Psi_n}{\partial t} = \frac{m\pi c}{d}\Psi_n - U_0|\Psi_n|^2\Psi_n - \frac{2\Omega}{m}(2\Psi_n - \Psi_{n+1} - \Psi_{n-1}) \qquad (6)$$

and for $m=2p+1$

$$\mathrm{i}\frac{\partial \Psi_n}{\partial t} = \frac{m\pi c}{d}\Psi_n - U_0|\Psi_n|^2\Psi_n + \frac{2\Omega}{m}(2\Psi_n + \Psi_{n+1} + \Psi_{n-1}), \qquad (7)$$

where U_0 is a parameter of the nonlinearity. Actually, Equations (6) and (7) are equations for the envelope curve of SL vibrations taken in discrete points. Such equations can be used for describing nonlinear effects and particularly dynamics of solitons in the optical SL.

ARNOLD KOSEVICH

See also **Discrete nonlinear Schrödinger equations; Nonlinear optics; Photonic crystals; Solitons**

Further Reading

Hasegawa, A. 2003. *Optical Solitons in Fibers*, 3rd edition, Berlin and New York: Springer

Jin, B.Y. & Ketterson, J.B. 1989. Artificial metallic superlattices. *Advances in Physics*, 38: 189–366

Joannopoulos, J.D., Meade, R.D. & Winn, J.N. 1995. *Photonic Crystals*, Princeton, NJ: Princeton University Press

Kivshar, Y. & Agrawal, G. 2003. *Optical Solitons: From Fibers to Photonic Crystals*, Amsterdam and New York: Academic Press

Kosevich, A.M. 2001. On a simple model of the photonic or phononic crystal. *JETP Letters*, 74: 559–563

Kronig, R. de L. & Penny, W.G. 1931. Quantum mechanics of electrons in crystal lattices. *Proceedings of the Royal Society, London, Series* A, 130: 499–513

Rytov, S.M. 1955. Electromagnetic properties of laminated media. *Zhurnal Experimental'noj i Teoreticheskoj Fiziki*, 29: 605–616; *Soviet Physics, JETP*, (1956), 2: 66

Rytov, S.M. 1956. Properties of laminated medium. *Soviet Physics-Acoustics*, 2: 68

Yeh, P. 1988. *Optical Waves in Layered Media*, New York: Wiley

SUPERSYMMETRY

See **String theory**

SURFACE WAVES

The surface of a fluid such as water lends itself most easily to the demonstration of wave phenomena. At the same time, waves on the surface of shallow water have played a very important role in the development of nonlinear science. The entries in this encyclopedia on water waves and solitons cover this aspect. Following narrower definitions, the term *surface waves* is often used for standing waves that form on the surface of a vessel filled with a liquid and moved up and down periodically in the direction normal to the liquid's surface. When adding the adjective *acoustic* (surface acoustic waves, SAW), one thinks of waves propagating on the surface of a solid elastic medium. These two types of surface waves will be discussed here. The former are called Faraday waves, and the most common species of the latter is named after Lord Rayleigh (John William Strutt).

Faraday Waves

In 1831, Michael Faraday reported on a series of experiments with various liquids that he made oscillate on glass plates and in a basin attached to a vibrating strip of wood (or lath) (Faraday, 1831). On the surface of the vibrating liquids, he observed the formation of standing wave patterns. With increasing amplitude of vibration, he found these standing waves to evolve into a regular square pattern (Figure 1). Faraday also noticed that the frequency of the surface waves was half of that of the driving force.

A proper mathematical explanation for the instability of the flat fluid surface, when being accelerated periodically, was given more than a century later, in 1954, by T.B. Benjamin and F. Ursell. They considered a sinusoidal acceleration with frequency ω. Let the initially planar surface be parallel to the x, y-plane and define the two-dimensional position vector $\boldsymbol{R}=(x, y)$. Benjamin and Ursell expanded the local excursion $\zeta(\boldsymbol{R}, t)$ of the surface from the planar initial position in real eigenfunctions S_m, $m=1, 2, \ldots,$ of the two-dimensional Laplace operator, satisfying Neumann boundary conditions at the boundary of the fluid surface with vanishing normal derivative,

$$\zeta(\boldsymbol{R}, t) = \sum_m a_m(t) S_m(\boldsymbol{R}). \qquad (1)$$

Linearizing the equations of hydrodynamics, they arrived at the Mathieu equation for the amplitudes a_m,

$$\frac{\partial^2 a_m}{\partial t^2} = [p_m + q_m \cos(\omega t)]a_m \qquad (2)$$

with coefficients q_m and p_m that depend on the corresponding eigenvalue of the Laplace operator and, hence, on the shape of the surface, on the depth of the liquid, as well as on gravity and surface tension. In addition, q_m is proportional to the amplitude of the periodic driver. The exponentially increasing solutions of the Mathieu equation that occur for certain combinations of system parameters, correspond to the instability. In particular, it becomes clear from (2) why the dominant instability is subharmonic.

The Faraday instability can thus be understood within linear theory. However, linear theory is not

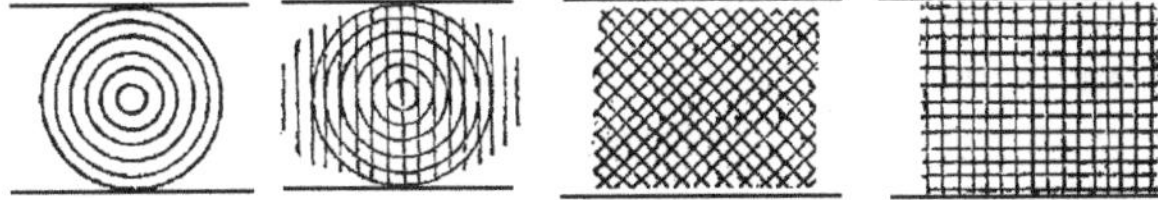

Figure 1. Sequence of surface wave patterns observed by Faraday on a quadrangular pool (Faraday, 1831).

sufficient to decide on which pattern will actually form after the instability has occurred. This can be determined by application of the modern theory of pattern formation to this damped-driven system (Miles & Henderson, 1990; Zhang & Viñals, 1997). Here, the nonlinearity comes into play. For sufficiently large spatial extension of the surface, boundary conditions at the edge of the surface become unimportant, and we may expand ζ in plane waves. Near the onset of the subharmonic instability, one may write

$$\zeta(\boldsymbol{R}, t) = \cos(\omega t/2) \sum_{j=1}^{N} \exp(\boldsymbol{k}_j \cdot \mathbf{R}) A_j(t) + \text{c.c.} \quad (3)$$

to a good approximation. (In (3) c.c. stands for the complex conjugate and a finite number N of wave vectors has been considered.) The modulus of the wave vectors $\boldsymbol{k}_j$ is fixed by the dispersion relation of linear surface waves at frequency $\omega / 2$. Viñals and his coworkers have shown that the slow variations of the amplitudes A_j are governed by a nonlinear equation of the form (Zhang & Viñals, 1997)

$$\frac{\partial A_j}{\partial \tau} = -\partial \mathcal{F} / \partial A_j^* \quad (4)$$

with the Lyapunov function

$$\mathcal{F} = \sum_{j=1}^{N} |A_j|^2 + \frac{1}{2} \sum_{j,j'=1}^{N} g(c_{jj'}) |A_j|^2 |A_{j'}|^2. \quad (5)$$

Here, τ is a stretched time, $c_{jj'}$ is the cosine of the angle between wave vectors $\boldsymbol{k}_j$ and $\boldsymbol{k}_{j'}$, and the function g depends on the system parameters. Stable stationary wave patterns correspond to minima of the Lyapunov function. For given system parameters, the pattern corresponding to the lowest value of $\mathcal{F}$ should be attained. Zhang and Viñals tested configurations with amplitudes A_j, 1, ..., M nonzero and $|A_j|$ independent of j, while the remaining $N - M$ amplitudes are zero. If the amplitudes corresponding to three or fewer different wave vectors are nonzero ($M \leq 3$), regular periodic patterns are obtained. The square pattern, already observed by Faraday, corresponds to two orthogonal wave vectors. With amplitudes of three different wave vectors nonzero, hexagonal or triangular patterns are generated. Interestingly, the theory of Zhang and Viñals also predicts patterns with more than three different wave vectors to be stable for certain system parameters. These correspond to quasi-periodic structures analogous to two-dimensional quasicrystals such as the Penrose tiling, for example. It has been a remarkable success of this nonlinear theory of pattern selection to predict the appearance of such quasi-periodic patterns of eight-fold (four wave vectors) and ten-fold (five wave vectors) symmetry, which have subsequently been found in experiments (Binks & van de Water, 1997) (see Figure 2).

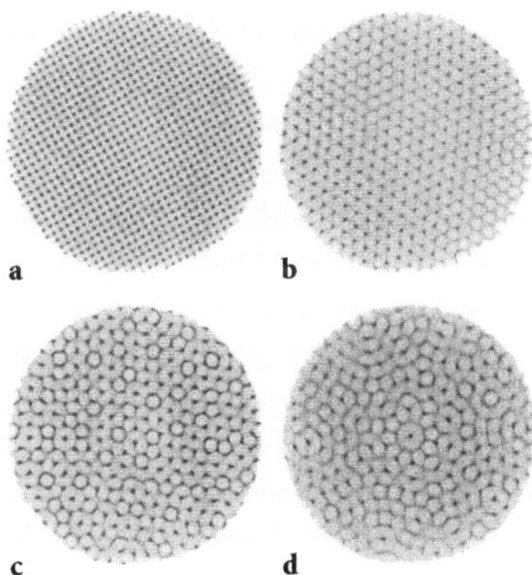

Figure 2. Faraday wave patterns of square (a), hexagonal (b), eight-fold quasi-periodic (c), and ten-fold quasi-periodic (d) symmetry observed by Binks & van de Water (1997).

In systems with length scales such that the boundary of the fluid's surface is relevant for the pattern formed, one has to consider the eigenfunctions $S_m(\boldsymbol{R})$ of the Laplace operator satisfying the correct conditions at the boundary and, hence, depending on the shape of the surface. This opens up the possibility of using Faraday waves to visualize wave-mechanical modes of stadium billiards and relating them to classical trajectories (Kudrolli et. al., 2001). At high values of the periodic acceleration of the fluid, spatiotemporal chaos has been observed.

In addition to the wave patterns arising on the surface of a moving liquid, Faraday, in his 1831 report, describes two other types of standing waves on the surfaces of liquids. The first one concerns patterns that he observed when a wave-maker was oscillating right on top of the surface or partly immersed in the liquid. Ridges are formed that are directed away from the wave-maker. The second type of standing waves occurs on a sloping beach when wind is blowing along the surface against the beach. Here, the ridges are parallel to the direction of the wind. These waves have been studied in detail in theory and by further experiments (Miles & Henderson, 1990). They are relevant in the context of oceanography.

Surface Acoustic Waves

In his landmark paper of 1885, Lord Rayleigh showed that the equations of linear elasticity theory for an isotropic homogeneous elastic half-space with a planar

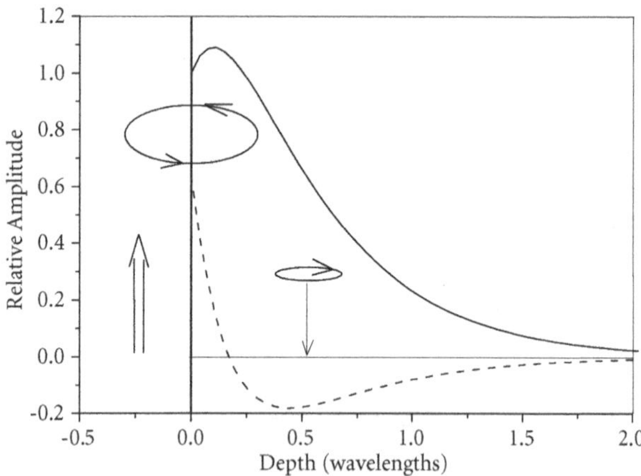

Figure 3. Displacements associated with a Rayleigh wave as function of depth. The surface wave is propagating in the direction of the double arrow. Displacement component normal (solid) and parallel (dashed) to the surface.

stress-free surface admit solutions that correspond to straight-crested waves propagating along the surface (Rayleigh, 1885). The deformations associated with these waves decay exponentially with depth. Such a Rayleigh wave is polarized in the plane spanned by the propagation direction and the surface normal (sagittal plane). It causes the mass elements of the solid to move on ellipses. This motion is retrograde near the surface and changes to prograde at greater depth (Figure 3).

Rayleigh himself thought of these waves in connection with earthquakes and their consequences, pointing out that surface wave amplitudes decrease more slowly with distance from the epicenter than those of bulk waves.

Since then, surface acoustic waves have found important applications in many branches of science and technology with wavelengths ranging from the kilometer scale in geophysics down to micrometers in ultrasonic nondestructive testing, acoustic microscopy, sensors, and micro-electronics. They are also used for basic research in solid-state physics with wavelengths down to the nanometer scale. The advent of the interdigital transducer on surfaces of piezoelectric media led to a number of signal processing devices based on SAW, especially frequency filters that are nowadays used in television sets and cell phones.

A distinctive feature of surface waves on solids as compared with ordinary liquids is the possibility of the medium to be anisotropic. For a long time it had not been known whether surface acoustic waves exist at all on surfaces of anisotropic media apart from special highly symmetric geometries. In the meantime, a mathematical theorem was established by Lothe and Barnett that guarantees their existence in most cases. However, generalized Rayleigh waves that occur in anisotropic media need no longer be polarized in the sagittal plane and may have more complicated depth profiles than the one shown in Figure 3. Their associated displacement field $\boldsymbol{u}$ is a superposition of generalized plane waves,

$$\boldsymbol{u}(x,z,t) = \sum_{j=1}^{n} \boldsymbol{W}_j \exp[\mathrm{i}k(x-vt) - |k|\alpha_j z] + \text{c.c.}, \qquad (6)$$

when the wave vector $\boldsymbol{k} = k\hat{\boldsymbol{x}}$ is along the x-direction and the elastic medium fills the half-space $z > 0$. (v is the phase velocity.) The (often complex) coefficients α_j have positive real part. In isotropic media, $n = 2$; in nonpiezoelectric media, $n \leq 3$. In very special circumstances, n may even be 1. In piezoelectric substrates, $n \leq 4$.

In anisotropic media, the group velocity of a surface acoustic wave need no longer be parallel to its wave vector. This can lead to focusing of acoustic energy propagating away from a local excitation of the surface. This effect, which is well known for acoustic bulk waves, has been demonstrated for surface waves in a comparatively simple and efficient way by Kolomenskii and Maznev. Anisotropy also gives rise to leaky waves (pseudo-surface waves) with a very high degree of localization at the surface.

When accounting for piezoelectricity and/or considering an elastic substrate in contact with other media, a whole zoo of various species of surface waves is found (Maradudin, 1985; Auld, 1990), including waves

- of shear-horizontal polarization in homogeneous piezoelectric substrates with or without metal coating (Bleustein–Gulyaev waves),
- in layered structures with shear-horizontal (Love waves) or sagittal polarization (generalized Lamb waves, Sezawa waves),
- at the interface of a solid and a liquid (Scholte waves),
- at the interface between two different solids (Stoneley waves).

The power flow associated with a Rayleigh wave is localized at the surface within a layer of approximately a wavelength. The high spatial localization favors effects generated by the elastic nonlinearity of the medium. In particular, higher harmonics are easily observed. Because neither the equations of nonlinear elasticity nor the geometry of a homogeneous half-space with a planar surface define a length scale, surface acoustic waves are nondispersive in this system. Therefore, shock formation can be observed. Theoretically, the evolution of nonlinear wave forms is studied on the basis of the following equation which follows from derivations by Lardner and Parker (Parker, 1988):

$$\mathrm{i}\frac{\partial}{\partial \tau}B(k) = k\int_0^k F(q/k)\,B(q)\,B(k-q)\,\mathrm{d}q$$
$$+2k\int_k^{\infty} (k/q)\,F^*(k/q)\,B(q)\,B^*(q-k)\,\mathrm{d}q. \qquad (7)$$

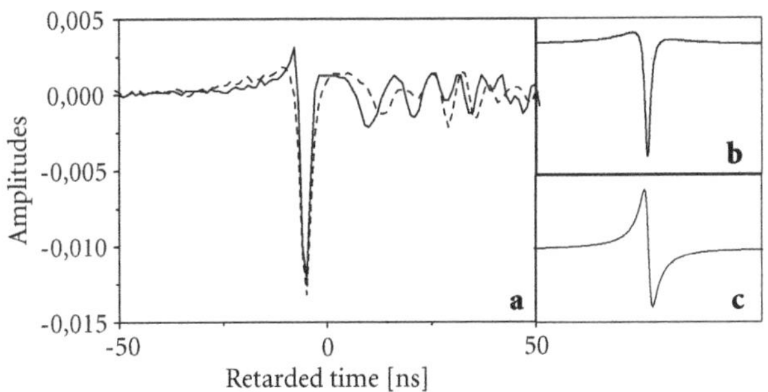

Figure 4. Solitary pulse shapes on the coated Si(111) surface (propagation along the $[\bar{1}\bar{1}2]$ direction) (a,b) and on a coated surface of isotropic quartz (c). Left: Experimental results by Lomonosov and Hess (solid) and numerical simulation (dashed). Right: Stationary pulse shapes. The local slope of the surface is shown as function of arrival time. (Compiled from data published in Lomonosov et al., 2002).

Here, τ is a stretched time and the function B is the Fourier transform of the local surface slope,

$$\frac{\partial}{\partial x} u_3(x, 0, t) = \int_0^\infty B(k, \tau) \exp[ik(x - vt)]\, dk + \text{c.c.} \quad (8)$$

The complex function F depends on the second-order and third-order elastic moduli of the medium and may often be well approximated by its value at the argument 1 / 2. When transforming (7) into real space, the nonlinearity turns out to be strongly nonlocal. This is the price one has to pay for eliminating the depth dependence of the displacement field from the evolution equation.

On the surface of homogeneous isotropic media, the velocity component of mass elements normal to the surface develops a cusp, while the component parallel to the surface steepens in a way similar to the pressure in the formation of shocks in gases and liquids. This behavior has been found in simulations starting with a sinusoidal wave form and in experiments with pulsed laser excitation. In order to generate acoustic pulses of high intensity at solid surfaces, this experimental technique has proven to be most effective (Lomonosov et al., 2001). A laser pulse focused on a line on the surface leads to rapid local heating and generates an explosive evaporation of a highly absorbing coating. Similar to the epicenter of an earthquake in the macroscopic world, the explosion leads to an acoustic pulse that is localized at the surface after having propagated some distance from the line of excitation.

Depositing on the substrate surface a thin film of a material different from that of the substrate leads to a linear dispersion term in (7). The evolution equation then has solitary wave solutions that may be regarded as solid state analogs of the solitons on shallow water. They have shapes strongly influenced by the anisotropy of the substrate, as demonstrated by the examples in Figure 4. The existence of solitary surface acoustic pulses has been verified by Lomonosov and Hess using laser excitation to generate an initial pulse that subsequently evolves into a solitary pulse and a quasilinear background (Figure 4).

ANDREAS MAYER

See also **Billiards; Nonlinear acoustics; Pattern formation; Shock waves; Solitons; Tessellation; Water waves**

Further Reading

Auld, B.A. 1990. *Acoustic Fields and Waves in Solids*. 2nd edn., Malabar, Florida: Krieger

Binks, D. & van de Water, W. 1997. Nonlinear pattern formation of Faraday waves. *Physical Review Letters*, 78: 4043–4046

Faraday, M. 1831. On a peculiar class of acoustical figures; and on certain forms assumed by groups of particles upon vibrating elastic surfaces. *Philosophical Transactions of the Royal Society of London*, 121: 299–340

Kudrolli, A., Abraham, M.C., & Gollub, J.P. 2001. Scarred patterns in surface waves. *Physical Review* E, 63: 026208/1–8

Lomonosov, A., Mayer, A.P. & Hess, P. 2001. Laser-based surface acoustic waves in materials science. In *Experimental Methods in the Physical Sciences*, vol. 3: *Modern Acoustical Techniques for the Measurement of Mechanical Properties*, edited by M. Levy, H.E. Bass, & R. Stern, San Diego: Academic, pp. 65–134

Lomonosov, A., Hess, P. & Mayer, A.P. 2002. Observation of solitary elastic surface pulses. *Physical Review Letters* 88: 076104/1–4

Maradudin, A.A. 1985. Surface acoustic waves. In *Nonequilibrium Phonon Dynamics*, edited by W.E. Bron, New York: Plenum, pp. 395–599

Miles, J. & Henderson, D. 1990. Parametrically forced surface waves. *Annual Review of Fluid Mechanics*, 22: 143–165

Parker, D.F. 1988. Waveform evolution for nonlinear surface acoustic waves. *International Journal of Engineering Science*, 26: 59–75

Rayleigh, Lord (John William Strutt). 1885. On waves propagating along the plane surface of an elastic solid. *Proceedings of the London Mathematical Society*, 17: 4–11

Zhang, W. & Viñals, J. 1997. Pattern formation in weakly damped parametric surface waves. *Journal of Fluid Mechanics*, 336: 301–330

SYMBOLIC DYNAMICS

Dynamical systems on manifolds are usually described by equations using local Euclidean variables. Symbolic dynamics has its origins in an alternative description of the orbits of a dynamical system by Jacques Hadamard at the end of the 19th century and later by Marston Morse, where it was shown that complex orbital behavior of a dynamical system could be described using sequences of symbols (Hadamard, 1898; Morse, 1921).

A simple example that shows the usefulness of this approach is the iteration $x_{n+1} = f(x_n)$ on the interval of real numbers $\mathbb{I} = [0, 1]$ defined by a piecewise linear map

$$x_{n+1} = \begin{cases} 2x_n, & x_n < 0.5, \\ 2x_n - 1, & x_n \geq 0.5. \end{cases}$$

This map can also be described by

$$x_{n+1} = 2x_n \bmod 1$$

for $x_n \in [0, 1)$ together with $f(1) = 1$. It is easy to *solve* the modular equation formally for the orbital sequence $x_n = f^n(x_0)$, $0 \leq qx_0 < 1$, as

$$x_n = 2^n x_0 \bmod 1.$$

However, the formal solution is no more illuminating than the defining iteration, because it is not at all clear how the repeated exponentiation of the solution interacts with the process of reducing mod 1. To overcome this problem and unlock the dynamical secrets held within the formula, note that any real number $x \in \mathbb{I}$ can be written in the form

$$x = \sum_{n=1}^{\infty} \frac{b_n}{2^n},$$

where $b_n = 0$ or 1. Let Σ_2 denote the set of all such binary sequences. Thus, every binary sequence $\mathbf{b} = \{b_n\}_{n=1}^{\infty} \in \Sigma_2$ represents a point $\pi(\mathbf{b}) = x \in \mathbb{I}$ by a map $\pi : \Sigma_2 \to \mathbb{I}$.

Relative to this new representation of points in $\mathbb{I}$ the map f takes the form of a shift $\sigma : \Sigma_2 \to \Sigma_2$ on binary sequences as

$$\{b_1, b_2, \ldots, b_n, \ldots\} \longmapsto \{b_2, b_3, \ldots, b_{n-1}, \ldots\},$$

that is,

$$\sigma(\{b_n\}_{n=1}^{\infty}) = \{b_{n+1}\}_{n=1}^{\infty}$$

simply because

$$2\left(\sum_{n=1}^{\infty} \frac{b_n}{2^n}\right) \bmod 1 = \sum_{n=1}^{\infty} \frac{b_{n+1}}{2^n}.$$

It is easy to check that the map π provides a *semi-conjugacy* $f\pi = \pi\sigma$ for f and σ. The map π is not a conjugacy because of the redundancy in the binary representation of the reals. This can be resolved by always choosing binary sequences representing reals in [0,1] that *do not* have infinite sequences of 1s. So, for example, 0.25 has the infinite binary representations $\{01\overline{0}\}$ and $\{00\overline{1}\}$, but we choose the former. Nevertheless, the semi-conjugacy implies that for every positive integer m, $f^m\pi = \pi\sigma^m$. In particular, if $\mathbf{b} = \{b_n\} \in \Sigma_2$ is a period-m orbit of σ, then

$$f^m(\pi(\mathbf{b})) = \pi\sigma^m(\mathbf{b}) = \pi(\mathbf{b}),$$

and so π maps the σ-periodic point $\mathbf{b}$ to an f-periodic point $\pi(\mathbf{b})$. Thus, we can obtain the periodic orbits of f by investigating the periodic orbit structure of σ. But this is straightforward because the period-m periodic points of σ in Σ_2 are precisely those binary sequences that repeat after m-digits and no fewer. Thus, period-1 points of σ are given by the repeating expansions

$$\mathbf{b} = \{\overline{0}\ldots\} = \{00000\ldots\}$$

and

$$\mathbf{b} = \{\overline{1}\ldots\} = \{11111\ldots\},$$

which both correspond by π to the fixed points $x = 0$ and $x = 1$, respectively, of the map f.

Period-2 points of σ are given by

$$\mathbf{b}_1 = \{\overline{01}\ldots\}, \qquad \mathbf{b}_2 = \{\overline{10}\ldots\}.$$

Note that $\pi(\mathbf{b}_1) = \frac{1}{3}$ and $\pi(\mathbf{b}_2) = \frac{2}{3}$ and $\left\{\frac{1}{3}, \frac{2}{3}\right\}$ is the only period-2 orbit of f. We can immediately see that periodic points of all orders can be constructed for σ, and thus also for f, in this way. Points that are on *eventually* periodic orbits of f can also be obtained by delaying the introduction of the binary recurrences in the symbol sequences. Thus, any map f defined on $\mathbb{I}$ that is described by a shift on symbol sequences has a rich periodic structure. We can extract further properties about orbits of f when we observe that π can be seen as a continuous function by taking the usual Euclidean metric on $\mathbb{I}$ and the metric

$$d(\mathbf{b}, \mathbf{b}') = \sum_{n=1}^{\infty} \frac{(b_n - b_n')}{2^n},$$

where $\mathbf{b} = \{b_n\}_{n=1}^{\infty}$ and $\mathbf{b}' = \{b_n'\}_{n=1}^{\infty}$ are elements of Σ_2. It is now possible to consider a *topology* on Σ_2.

A key feature of chaos is the requirement that the orbits of the map f should in some sense be bound together, thus, making the dynamics indecomposable. This can be achieved by finding an orbit of f that *densely* fills out the set $\mathbb{I}$. Such an orbit is given by constructing a sequence $\mathbf{b}$ obtained by listing all symbol sequences of length 1, then of length 2, and so on for all positive integers. Let $\mathbf{b}'$ be any prescribed binary sequence. Given any positive integer n, by construction $\mathbf{b}$ has within it the symbol block $\mathbf{b}'^{(n)}$ consisting of the first n symbols of $\mathbf{b}'$. Suppose this block commences at the point b_{k+1} of $\mathbf{b}$. Then $\mathbf{b}'$ and $\sigma^k(\mathbf{b})$ have the same first n entries and thus $d(\mathbf{b}', \sigma^k(\mathbf{b})) < 2^{-n}$. The positive integer n was chosen arbitrarily, and so the orbit of $\mathbf{b}$ approaches arbitrarily close to $\mathbf{b}'$. Thus, the orbit of $\mathbf{b}$ is *dense* on $\mathbb{I}$ and binds the dynamical behavior together preventing a decomposition into closed subsets with simpler dynamical behavior.

The dense orbit was, of course, produced to a special recipe. Obviously, the orbit is not unique as other dense orbits in Σ_2 can be constructed by simple permutations of the ordering of the blocks used to construct the sequence $\mathbf{b}$. What is astonishing is that "almost all" orbits are dense. This follows from a classical result of Emil Borel around 1900 (Hardy, 1983), which, when paraphrased, says that the set of real numbers in $\mathbb{I}$, which have all possible finite sequences of "0"s and "1"s in their binary expansions, called *normal* numbers, has full measure.

Thus, the doubling map can be described by a shift map σ on the space of symbolic sequences on two symbols, Σ_2. The simple ideas discussed here can be extended to provide symbolic dynamics for various types of maps. The *coding* involving two or more symbols has to be carried out so that each point x of the map domain is given by a symbol string that provides the orbit of x by merely shifting the symbols (Devaney, 2003). Note that the binary sequence gives the itinerary of an orbit where if "0" represents the interval $[0, \frac{1}{2}]$ and "1" represents $[\frac{1}{2}, 1]$, the sequence gives the sequence of intervals visited by the orbit of the map f. This approach can be applied to a wide class of one-dimensional maps (Guckenheimer, 1979; Guckenheimer et al., 1977), but it also has much greater importance in dynamical systems.

A key dynamical construction is the *Smale Horseshoe* (Smale, 1967). The simplest example is a diffeomorphism h of the square $Q \in \mathbf{R}^2$ whose image in $\mathbf{R}^2$ is folded in the shape of a horseshoe to overlay the square. The map h has an invariant Cantor set $\Lambda \subset Q$ such that $h|\Lambda$ can also be coded with *bi-infinite* binary sequences together with a shift map. This results in a complex periodic orbit structure and dense orbits within Λ, which gives some of the ingredients of chaos. The key components of the set Λ are the *homoclinic* points, that is, those points that asymptotically approach a periodic orbit of h in both forward and reverse iterations.

The Smale–Birkhoff Theorem proves that "folding maps" of this type, and the associated complex dynamical behavior, exist in the neighborhood of such homoclinic points. This is precisely stated as follows (Smale, 1967).

Theorem. Let $\mathbf{f}$ be a Kupka–Smale diffeomorphism of a compact manifold M and $x^\dagger$ be a transverse homoclinic point of a periodic point x^* of $\mathbf{f}$. Then there is a closed subset Λ containing $x^\dagger$, such that (i) Λ is a Cantor set; (ii) $\mathbf{f}^p(\Lambda) = \Lambda$ for some $p \in \mathbf{Z}^+$; and (iii) $\mathbf{f}^p|\Lambda$ is topologically conjugate to a shift on a symbol space.

Note that all possible binary sequences are allowed in the above symbol spaces and, thus, we have been considering *full* shifts. In many practical applications of data storage, there are practical restrictions on the stored patterns of sequences. The problem of transferring codes that satisfy one set of constraints into other restricted symbol sets involve *sub-shifts of finite type* and allow applications to communications coding and algebra (Coornaert & Papadopoulos, 1993; Lind & Marcus, 1996; Kitchens, 1997). Symbolic coding has been directly useful in modeling in engineering and biology, see (Daw et al., 1997) and (Voss et al., 2000).

David D. Arrowsmith

See also **Chaotic dynamics; Horseshoes and Hyperbolicity in dynamical systems; Maps; One-dimensional maps; Topology**

Further Reading

Coornaert, M. & Papadopoulos, A. 1993. *Symbolic Dynamics and Hyperbolic Groups*, Berlin & New York: Springer

Crutchfield, J.P. & Packard, N. 1982. Symbolic dynamics of one-dimensional maps: entropies, finite precision, and noise. *International Journal of Theoretical Physics*, 21: 433–466

Daw, C.S., Finney, C.E.A., Kennel, M.B. & Connolly, F.T. 1997. Cycle-by-cycle combustion variations in spark-ignited engines. *Proceedings of the Fourth Experimental Chaos Conference*, Boca Raton, FL, Singapore: World Scientific

Devaney, R.L. 2003. *An Introduction to Chaotic Dynamical Systems*, 2nd edition, Boulder, CO: Westview Press

Guckenheimer, J. 1979. Sensitive dependence to initial conditions for one-dimensional maps. *Communications in Mathematical Physics*, 70: 133–160

Guckenheimer, J., Oster, G. & Ipatchki, A. 1977. The dynamics of density dependent population models. *Journal of Mathematical Biology*, 4: 101–147

Hadamard, J. 1898. Les surfaces á courbures opposeés et leur lignes géodesiques. *Journal de Mathématiques Pures et Appliquées*, 4: 27–73

Hao, B.-L. 1986. *Elementary Symbolic Dynamics and Chaos in Dissipative Systems*, Singapore: World Scientific

Hao, B.-L. & Zheng, W.-M. 1998. *Applied Symbolic Dynamics and Chaos*, Singapore: World Scientific

Hardy, G.H. 1983. *An Introduction to the Theory of Numbers*, 5th edition, Oxford and New York: Oxford University Press (first published, 1938)

Kitchens, B.P. 1997. *Symbolic Dynamics: One-Sided, Two-Sided, and Countable State Markov Shifts*, Berlin and New York: Springer

Lind, D. & Marcus, B. 1996. *An Introduction to Symbolic Dynamics and Coding*, Cambridge and New York: Cambridge University Press

Milnor, J. & Thurston, W. 1988. On iterated maps of the interval. In *Dynamical Systems Proceedings, University of Maryland 1986–87*, edited by J.C. Alexander, Berlin and New York: Springer, pp. 425–563

Morse, M. 1921. A one-to-one representation of geodesics on a surface of negative curvature. *American Journal of Mathematics*, 43: 33–51

Smale, S. 1967. Differentiable dynamical systems. *Bulletin American Mathematical Society*, 73(6): 747–817.

Voss, A., Wessel, N., Kurths, J., Schirdewan, A., Osterziel, K.J., Malik, M. & Dietz, R. 2000. Symbolic dynamics—a powerful tool in non-invasive biomedical signal processing. In *Advances in Noninvasive Electrocardiographic Monitoring Techniques*, edited by H.H. Osterhues, V. Mombach & A.J. Moss, Dordrecht and Boston: Kluwer, pp. 429–437

SYMMETRY BREAKING

See **Bifurcations**

SYMMETRY GROUPS

Mathematically, a group is a set G together with a pair-wise "group multiplication" on elements of G that satisfies the following axioms. (i) Multiplication is associative, meaning that $xy(z) = x(yz)$ for x,

y, and z any elements of G. (ii) G contains an identity element e for which $ex = x = xe$ where x is any element of G. (iii) Each element of G has an inverse (x^{-1}), satisfying $x^{-1}x = e = xx^{-1}$ (Armstrong, 1988; Birkhoff & MacLane, 1953). Groups for which all of the multiplications commute are said to be abelian.

Easily visualized examples of finite symmetry groups are provided by operations on familiar geometrical objects, such as squares, hexagons, cubes, and tetrahedrons, which carry vertices into themselves. In general, there will be a finite set of f linearly independent functions (in the space of the geometrical object) that are carried into linear combinations of themselves by the group operations. Thus, each group element corresponds to a matrix, and the corresponding set of $f \times f$ matrices is called a representation of the group (Landau & Lifschitz, 1958).

For a square that is lying on the plane with its center at the origin, group elements are rotations about the origin (by 90°, 180°, and 270°) and reflections (in horizontal and vertical axes and in the two diagonals). Group multiplication is defined as performing two operations in order, and the identity element leaves all vertices unchanged. The reader may find it helpful to cut out a square of cardboard, number the four corners, and construct a group multiplication table for these eight elements, noting that it is not abelian.

It has long been known that symmetries of a dynamic system influence the nature of possible characteristic solutions or natural modes of behavior (*See* **Symmetry: equations vs. solutions**). To appreciate this restriction, consider a system with reflection symmetry, and note that any solution can be resolved into two components that are respectively symmetric and antisymmetric about the plane of symmetry. In a linear system, the symmetric and antisymmetric components do not interact; thus, modes must be either symmetric or antisymmetric about the plane. In a nonlinear system, however, there can also be natural modes of behavior that are neither symmetric nor antisymmetric. Such symmetry breaking is a fundamental feature of nonlinear dynamic systems, leading to the formation of solitons on optical fibers, impulses on nerve axons, and local modes in chemical molecules and molecular crystals.

Applications to Physical Chemistry

A typical chemical molecule is a simple geometric structure for which a symmetry group multiplication table has a finite number of entries (Herzberg, 1991; Wilson et al., 1980). Thus, the water molecule (H_2O) has a single reflection plane; methane (CH_4) is described by the same symmetry group as the regular tetrahedron, benzene (C_6H_6) by a planar hexagon, and so on. The physical chemist uses these finite symmetry groups to organize measurements of electronic and vibrational spectra, using the following notations: reflection group (σ), n-fold rotation symmetry (C_n), inversion symmetry (I), tetrahedral symmetry (T), and so on. A particularly lucid introduction to point symmetry notations is given by Landau & Lifschitz (1958).

Because quantum theory is linear, vibrational and electronic wave functions have the following property. If n successive applications of a symmetry operation return a molecule to its original orientation, the amplitude of the corresponding quantum wave function must change by an nth root of unity under the same operation. Under reflections, in other words, amplitudes change by factors of either $+1$ or -1 (symmetric or antisymmetric), whereas CH-stretching modes of benzene (C_6) change under rotations of 60° by factors of $\exp(\pm im\pi/3)$, where $m = 0, \pm 1, \pm 2, \pm 3, \pm 4$, or ± 5. Physical chemists label their spectral lines with notations that correspond to these factors (Landau & Lifschitz, 1958).

Although molecular vibrations can be quite nonlinear (especially those involving hydrogen atoms), this nonlinearity does not play a role in transitions from the ground state to first quantum levels of small molecules. (This is because the nonlinear operator of lowest order contains a product of two lowering operators, which annihilates first quantum states.) Thus, the excitations to the first quantum level are governed by linear symmetry considerations, and local modes in molecules are observed only for transitions to higher quantum levels.

The linear modes of a molecule with a center of inversion (I) appear either in infrared absorption or Raman scattering measurements, but not both. This "principle of mutual exclusion" helps to sort out the components of linear vibrational spectra. Local modes, which are nonlinear, can appear in both infrared and Raman measurements.

A periodic solid (or molecular crystal) with periodic boundary conditions can be viewed as a very large molecule, for which the number of elements of the corresponding symmetry group is also very large. Of particular interest are translations by lattice constants (a, b, and c) along the crystal axes, which bring the crystal back to its original configuration. The phase shift of an electronic or vibrational wave function under such a translation is $\exp(i\boldsymbol{k})$, where $\boldsymbol{k}$ is called the crystal momentum. As the number of unit cells in the model approaches infinity, the three components of $\boldsymbol{k}$ vary, respectively, from $-\pi/a$ to π/a, $-\pi/b$ to π/b, and $-\pi/c$ to π/c. In vibrational modes of such systems, nonlinearity can arise from local lattice distortion, which allows symmetry breaking (local mode formation) to be observed at the first quantum level.

Applications to Field Theories

The above definition of a group does not require the number of group elements to be finite, and many partial differential equations provide examples of infinite-order symmetry groups.

The sine-Gordon (SG) equation, for example, is invariant under the independent variable transformation $(x, t) \to (\xi, \tau)$, where $\xi = (x - vt) / \sqrt{1 - v^2}$ and $\tau = (t - vx) / \sqrt{1 - v^2}$. Taking this invariance as the property defining elements of the group, a symmetry group comprises all such transformations, parameterized by the continuous variable v with $|v| < 1$. This is a one-dimensional version of the Lorentz transformation, which is shared by Maxwell's equations. Interestingly, if $u(x)$ is a time-independent solution of SG, then $u\left[(x - vt) / \sqrt{1 - v^2}\right]$ is also a solution that demonstrates Lorentz contraction (becomes smaller as $v \to 1$).

In 1915, Emmy Noether established the following important result on the application of symmetry groups to field theories (José & Saletan, 1998).

Noether's theorem. *If a system is described by a Lagrangian that remains invariant under some continuous symmetry transformation, then there is a corresponding conservation law and constant of the motion.*

This theorem has immediate implications. As fundamental descriptions of nature are assumed to be Lagrangian and independent of time displacements (the science of today is the same as it was yesterday), the corresponding conserved quantity is energy. In other words, the law of energy conservation stems from time invariance of scientific laws. Similarly, conservation of momentum and conservation of angular momentum arise, respectively, from the assumptions that scientific laws are independent of spatial displacements and angles of rotation.

Nowadays, it is widely assumed that the fundamental fields of nature are derived from a Lorentz invariant Lagrangian density or more generally based on the Poincaré group, which incorporates independence with respect to displacements in time and space (Kim & Noz, 1986). As physicists attempt to formulate a fundamental description of nature, they use Noether's theorem to build in additional constants (charge, spin, rest mass, etc.) that have been empirically observed.

The pure soliton equations (Korteweg–de Vries, nonlinear Schrödinger, sine-Gordon, and so on) are Lagrangian systems that each have a countably infinite number of conservation laws and constants of the motion. It would be interesting to better understand the corresponding symmetries.

ALWYN SCOTT

See also **Bäcklund transformations; Dimensional analysis; Lie algebras and Lie groups; Local modes in molecules; Symmetry: equations vs. solutions**

Further Reading

Armstrong, M.A. 1988. *Groups and Symmetry*, New York: Springer

Birkhoff, G. & MacLane, S. 1953. *A Survey of Modern Algebra*, New York: Macmillan

Herzberg, G. 1991. *Molecular Spectra and Molecular Structure: Infrared and Raman of Polyatomic Molecules*, Melbourne, FL: Krieger Publishing Company

José, J.V. & Saletan, E. 1998. *Classical Dynamics: A Contemporary Approach*, Cambridge: Cambridge University Press

Kim, Y.S. & Noz, M.E. 1986. *Theory and Applications of the Poincaré Group*, Dordrecht: Reidel

Landau, L.D. & Lifshitz, E.M. 1958. *Quantum Mechanics: Non-relativistic Theory*, Reading, MA: Addison-Wesley

Weyl, H. 1952. *Symmetry*, Princeton, NJ: Princeton University Press

Wilson, E.B., Jr., Decius, J.C. & Cross, P.C. 1980. *Molecular Vibrations: The Theory of Infrared and Raman Vibrational Spectra*, New York: Dover

SYMMETRY: EQUATIONS VS. SOLUTIONS

The importance of symmetries in various fields of science is well appreciated, much beyond the scope of this entry. Here we focus on one particular aspect—how knowledge about discrete symmetry operations allows one to conclude certain properties of solutions of differential equations.

Let us start with one of the simplest examples: one oscillator that is governed by the equation

$$\frac{d^2x}{dt^2} = -\frac{dV(x)}{dx}, \tag{1}$$

where t is time, x the coordinate of the oscillator, and $V(x)$ its nonnegative potential with $V(0) = 0$. All solutions to this equation are time-periodic. If an equation is invariant under a certain symmetry operation, then the same symmetry operation applied to a solution of the equation either generates a new solution or reproduces the same old solution it is then said that the solution is also invariant.

All symmetries considered below, in fact, do not change the energy of the oscillator; thus, the solution is always invariant up to a trivial shift of the origin of time $t \to t + t_0$. One symmetry that leaves Equation (1) invariant is time reversal $t \to -t$, which implies that for each solution the origin of time can be chosen in such a way that the solution is also invariant under time reversal, that is, $x_s(t) = x_s(-t)$. For practical purposes, it means that numerical expansions of solutions into Fourier series may be restricted to cosine Fourier series. Another symmetry may hold if $V_s(x) = V_s(-x)$. Then the equation is invariant under space reflection $x \to -x$.

In order to invert the sign of a periodic function $x(t)$, we may assume either antisymmetry $x_a(t) = -x_a(-t)$ or shift symmetry $x_{sh}(t) = -x_{sh}(t + T/2)$ where T

is the period. Thus, for symmetric potentials, $x(t)$ is symmetric in time, antisymmetric in time (around a different origin), and shift symmetric. In particular, this implies that all even components of the Fourier expansion of this function vanish, including the dc component.

Consider next a more sophisticated model of a particle subject to a space-periodic force $f(x) = -\mathrm{d}V/\mathrm{d}x$ (period λ) and a time-periodic field $E(t)$ (period T) with zero mean:

$$\frac{\mathrm{d}^2x}{\mathrm{d}t^2} = f(x) + E(t). \tag{2}$$

Equation (2) corresponds to a non-integrable system with 1.5 degrees of freedom (a three-dimensional phase space). The more complicated the equations become, the more symmetries one might consider. Assume symmetry operations that change the sign of the velocity $v = \mathrm{d}x/\mathrm{d}t$. There are two symmetries: time reversal $t \to -t$ if $E(t) \equiv E_s(t)$ and a combined space reflection with time shift one $x \to -x,\ t \to t + T/2$ if $f(x) \equiv f_a(x),\ E(t) \equiv E_{sh}(t)$. For both symmetries to be valid, certain properties of the functions f and E are required. In particular, the time-periodic field $E(t)$ (e.g., an ac electric field) enters both symmetry requirements. That means that by choosing an $E(t)$ dependence that is neither symmetric nor shift symmetric, we lose both symmetries. What are the consequences?

Note first that due to the non-integrability of (2), the separatrix for $E = 0$ is replaced by a chaotic or stochastic layer in the phase space. Ergodicity inside the layer implies that the time average of a certain quantity over a trajectory, if existing, does not depend on the concrete trajectory choice. Returning to the above symmetries and choosing the time average of the velocity of the particle, we conclude that (i) the velocity average exists because the stochastic layer is bounded, and (ii) if the average velocity is nonzero for one trajectory from the stochastic layer and if any of the above symmetries apply, the velocity average will be opposite for a corresponding symmetry related trajectory from the same stochastic layer—with the only consequence that the average velocity vanishes exactly. However, choosing an $E(t)$ dependence that violates both symmetry requirements, we may expect that the average velocity of any trajectory inside the stochastic layer will become nonzero. In other words, we can predict how to generate a dc current of many non-interacting particles by a symmetry breaking choice of the field $E(t)$. This has been, indeed, predicted and demonstrated theoretically and numerically (Flach et al., 2000) and verified experimentally for directed diffusion of ultracold rubidium atoms in symmetric optical lattices (Schiavoni et al., 2003).

Next, consider applications of the above concepts to systems that are characterized by the presence of dissipation and fluctuations. Usual additive Gaussian white noise terms possess all symmetries of periodic functions; that is, if $\xi(t)$ is a realization of a stochastic process, so is $\xi(-t)$, $\xi(t+\tau)$, $-\xi(t)$, and so on. When starting with a corresponding Langevin type equation

$$\frac{\mathrm{d}^2x}{\mathrm{d}t^2} = f(x) + E(t) - \gamma\frac{\mathrm{d}x}{\mathrm{d}t} + \xi(t), \tag{3}$$

it is then possible to drop the noise term and study the symmetries of the deterministic yet dissipative equation

$$\frac{\mathrm{d}^2x}{\mathrm{d}t^2} = f(x) + E(t) - \gamma\frac{\mathrm{d}x}{\mathrm{d}t}. \tag{4}$$

While the strategy of the analysis is similar to the one outlined above, for the case of zero dissipation $\gamma = 0$, the phase space of Equation (4) is again three-dimensional, but is now composed of basins of attraction with each basin corresponding to a certain attractor. Existing symmetries of the equation imply symmetry relationships between attractors and their basins of attraction. Violation of symmetries of the equations implies desymmetrization of basins of attraction of (previously symmetry related) attractors. Thus, adding again the stochastic noise to return to the starting Equation (3), intuition suggests that the noise when leading to an average over various basins of attraction will either make certain averages vanish (in the presence of symmetries) or not vanish (in the absence of symmetries). Special care is required due to the fact that some symmetries of (4) may relate attractors with repellors. Another approach is to consider partial differential equations for the probability distributions and their symmetry properties (see Denisov et al., 2002).

The above approach has been successfully used to predict such diverse phenomena as rectification of heat currents (Flach et al., 2002) and induction of magnetizations (Flach & Ovchinnikov, 2000). Finally, note that it is easier to predict a nonzero average by breaking symmetries than to obtain an understanding for the concrete microscopic mechanisms of symmetry breaking and thus to obtain reliable estimates for the expected quantity of rectification and its dependence on essential parameters of the system (Reimann, 2002).

SERGEJ FLACH

See also **Ratchets; Symmetry groups; Stochastic processes**

Further Reading

Denisov, S., Flach, S., Ovchinnikov, A.A., Yevtushenko, O. & Zolotaryuk, Y. 2002. Broken space-time symmetries and mechanisms of rectification of ac fields by nonlinear (non)adiabatic response. *Physical Review* E, 66: 041104

Flach, S. & Ovchinnikov, A.A. 2000. Static magnetization induced by time-periodic fields with zero mean. *Physica* A, 292: 268–276

Flach, S., Yevtushenko, O. & Zolotaryuk, Y. 2000. Directed current due to broken time-space symmetry. *Physical Review Letters*, 84: 2358–2361

Flach, S., Zolotaryuk, Y., Miroshnichenko, A.E. & Fistul, M.V. 2002. Broken symmetries and directed collective energy transport. *Physical Review Letters*, 88: 184101

Reimann, P. 2002. Brownian motors: noisy transport far from equilibrium. *Physics Reports*, 361: 57–265

Schiavoni, M., Sanchez-Palencia, L., Renzoni, F. & Grynberg, G. 2003. Phase control of directed diffusion in a symmetric optical lattice. *Physical Review Letters*, 90: 094101

SYMPLECTIC MAPS

First used mathematically by Hermann Weyl, the term *symplectic* arises from a Greek word that means "twining or plaiting together." This is apt, as symplectic systems always involve a pair of n-dimensional variables, the configuration q, and momentum p, which are intertwined by the symplectic two form

$$\omega = dp \wedge dq. \tag{1}$$

This antisymmetric, bilinear form acts on a pair of tangent vectors and computes the sum of the areas of the parallelograms formed by projecting the vectors onto the planes defined by each canonical pair (q_i, p_i), $i = 1, \ldots, n$, giving

$$\omega(v, w) = \sum_{i=1}^{n} (v_{p_i} w_{q_i} - v_{q_i} w_{p_i}).$$

A diffeomorphism $f : X \to X$ on a $2n$-dimensional manifold X with coordinates $z = (q, p)$ is symplectic if it preserves the symplectic form, that is, if $f^*\omega = \omega$ (Arnol'd, 1989; McDuff & Salamon, 1995). If we write $z' = (q', p') = f(q, p)$, the symplectic condition becomes

$$Df^t J Df = J, \quad \text{where} \quad J = \begin{pmatrix} 0 & I \\ -I & 0 \end{pmatrix}. \tag{2}$$

Here $Df_{ij} = \partial f_i / \partial z_j$ is the Jacobian matrix of f, J is the Poisson matrix, and I is the $n \times n$ identity matrix. Equivalently, Stokes' theorem can be used to show that the loop action, $A[\gamma] = \oint_\gamma p dq$, is preserved by f for any contractible loop γ on X. If f preserves the loop action for all loops, even those that are not contractible, then it is *exact symplectic*.

When $n = 1$, the symplectic condition is equivalent to $\det(Df) = 1$, so that the map is area- and orientation-preserving. Examples include the much studied *standard map* and the area-preserving Hénon quadratic map $f(q, p) = (p + a - q^2, -q)$ (Meiss, 1992). When $n > 1$, the symplectic condition implies volume and orientation preservation, but as we will see, it is stronger than this. A generalization of the standard map to higher dimensions is the map

$$\begin{aligned} q' &= q + p - \nabla V(q), \\ p' &= p - \nabla V(q), \end{aligned} \tag{3}$$

where $q \in \mathbb{T}^n$ is an angle, $p \in \mathbb{R}^n$ is its conjugate momentum, and $V(q)$ is a periodic potential. This map is exact symplectic for any V. Beginning in 1972, Claude Froeschlé studied the case $n = 2$ and $V(q) = a \cos q_1 + b \cos q_2 + c \cos(q_1 + q_2)$. Similarly, the natural generalization of the Hénon map is the quadratic symplectic map whose normal form has been given by Moser (1994).

Applications

Symplectic maps arise from Hamiltonian dynamics, because these preserve the loop action. Thus, for example, the time t map of any Hamiltonian flow is symplectic, as is a Poincaré return map defined on a cross section. It is often easier to study the Poincaré map instead of the flow, because the dimension is reduced. Even though explicit construction of the map is typically impossible, approximation methods often suffice.

For example, the time T map of a periodically forced system $H(q, p, t) = H(q, p, t + T)$, such as a pendulum with an oscillating support, is symplectic (*See* **Hamiltonian systems**). An extreme example is $H = \frac{1}{2}p^2 - k\cos(q)\bar{\delta}(t)$, where $\bar{\delta}$ is the periodic Dirac delta function; the corresponding map is the standard map.

As Birkhoff showed, an ideal billiard (a free particle moving inside a rigid, convex table) is naturally written as a symplectic map. The canonical coordinates are the position and the tangential momentum at a collision point. Symplectic maps also arise naturally in systems where the forces are localized in time or space. For example, a circular particle accelerator or storage ring has a sequence of accelerating and focusing elements that can be modeled by a composition of symplectic maps, providing the damping effects of radiation can be neglected (Forest, 1998).

Area-preserving maps also arise in the study of the motion of Lagrangian tracers in incompressible fluids or of particles tightly gyrating around magnetic field lines. In particular, when one component of the field is particularly strong, such as in the plasma device called a tokamak, the transverse dynamics reduces to an area-preserving map.

Autonomous canonical transformations are symplectic maps. For example, if $F(q, q')$ is a generating function for a canonical transformation, then it generates a symplectic map. In particular, the Froeschlé map (3) is generated by $F(q, q') = \frac{1}{2}(q' - q)^2 - V(q)$.

An algorithm that respects the symplectic nature of Hamiltonian dynamics is called a symplectic integrator. A first-order symplectic algorithm with time step Δt for the Hamiltonian $H(q, p)$ is generated by $F(q, p') = qp' + \Delta t H(q, p')$, where

$dF = q'dp' + pdq$, giving the map

$$q' = q + \Delta t \frac{\partial H}{\partial p'}(q, p'), \quad p' = p - \Delta t \frac{\partial H}{\partial q}(q, p'). \quad (4)$$

Note that the map is implicit because H is evaluated at p'. However, for the case that $H = K(p) + V(q)$ this becomes a leap-frog Euler scheme, an example of a "splitting" method. Symplectic versions of many standard algorithms—such as Runge–Kutta—can be obtained (Marsden et al., 1996). While there is still some controversy on the utility of symplectic methods versus methods that, for example, conserve energy and other invariants or have variable time-stepping, they are superior for stability properties because they respect the spectral properties of the symplectic group.

The Symplectic Group

The stability of an orbit $\{...z_t, z_{t+1}, ...\}$, where $z_{t+1} = f(z_t)$, is governed by the Jacobian matrix of f evaluated along the orbit, $M = \prod_t Df(z_t)$. When f is symplectic, M obeys (2), $M^t JM = J$. The set of all such $2n \times 2n$ matrices form the symplectic group $Sp(2n)$. This group is an $n(2n+1)$-dimensional Lie group, whose Lie algebra is the set of Hamiltonian matrices—matrices of the form JS where S is symmetric. Thus, every near-identity symplectic matrix can be obtained as the exponential of a Hamiltonian matrix and corresponds to the time t-map of a linear Hamiltonian flow. There are symplectic matrices, however, that are not the exponentials of Hamiltonian matrices; for example, $-I$. As a manifold, the symplectic group has a single nontrivial loop (its fundamental group is the integers). The winding number of a loop in the symplectic group is called the Maslov index (McDuff & Salamon, 1995); it is especially important for semi-classical quantization.

If M is a symplectic matrix and λ is an eigenvalue of M with multiplicity k, then so is λ^{-1}. Moreover $\det(M) = 1$, so M is volume and orientation preserving. A consequence of this spectral theorem is that orbits of a symplectic map cannot be asymptotically stable. There are four basic stability types for symplectic maps: an eigenvalue pair (λ, λ^{-1}) is

- *hyperbolic*, if λ is real and larger than one;
- *hyperbolic with reflection*, if λ is real and less than minus one;
- *elliptic*, if $\lambda = e^{2\pi i\omega}$ has magnitude one;
- part of a *Krein quartet*, if λ is complex and has magnitude different from one, for then there is a quartet of related eigenvalues $(\lambda, \lambda^{-1}, \bar{\lambda}, \bar{\lambda}^{-1})$.

Thus, a periodic orbit can be linearly stable only when all of its eigenvalue pairs are elliptic. For this case, the linearized motion corresponds to rotation with n rotation numbers ω_i.

Symplectic Geometry

Every symplectic map is volume- and orientation-preserving, but the group Symp(X) of symplectic diffeomorphisms on X is significantly smaller than that of the volume-preserving ones. This was first shown in 1985 by Gromov in his celebrated "nonsqueezing" (or symplectic camel) theorem. Let $B(r)$ be the closed ball of radius r in $\mathbb{R}^{2n}$ and $C_1(R) = \{(q, p) : q_1^2 + p_1^2 \leq R^2\}$ be a cylinder of radius R whose circular cross section is a symplectic plane. Because the volume of C_1 is infinite, it is easy to construct a volume-preserving map that takes $B(r)$ into $C_1(R)$ regardless of their radii. What Gromov showed is that it is impossible to do this symplectically whenever $r > R$. This is one example of a *symplectic capacity*, leading to a theory of symplectic topology (McDuff & Salamon, 1995).

Another focus of this theory is to characterize the number of fixed points of a symplectic map, that is, to generalize the classical Poincaré–Birkhoff theorem for area-preserving maps on an annulus. Arnol'd conjectured in the 1960s that any Hamiltonian diffeomorphism on a compact manifold X must have at least as many fixed points as a function on X must have critical points. A Hamiltonian map is a symplectic map that can be written as a composition of maps of the form (4). Conley and Zender proved this in 1985 for the case that X is the $2n$-torus: f must have at least $2n+1$ fixed points (at least 2^{2n} if they are all nondegenerate) (Golé, 2001).

Dynamics

In general, the dynamics of a symplectic map consists of a complicated mixture of regular and chaotic motion (Meiss, 1992). Numerical studies indicate that the chaotic orbits have positive Lyapunov exponents and fill sets of positive measure that are fractal in nature. Regular orbits include periodic and quasi-periodic orbits. The latter densely cover invariant tori whose dimensions range from 1 to n. Near elliptic periodic orbits, the phase space is foliated by a positive-measure cantor set of n-dimensional invariant tori. There are chaotic regions in the *resonant* gaps between the tori, but the chaos becomes exponentially slow and exponentially small close to the periodic orbit. Some of these observations, but not all, can be proved.

The simplest case is that of an integrable symplectic map, which can be written in Birkhoff normal form: $f(\theta, J) = (\theta + \nabla S(J), J)$. Here (θ, J) are angle-action coordinates (each n-dimensional) and $\Omega = \nabla S$ is the rotation vector. Orbits for this system lie on invariant tori; thus, the structure is identical to that for integrable Hamiltonian systems.

The Birkhoff normal form is also an asymptotically valid description of the dynamics in the neighborhood of a nonresonant elliptic fixed point, one for which

$m \cdot \Omega(0) \neq n$ for any integer vector m and integer n. However, the series for the normal form is not generally convergent. Nevertheless, KAM theory implies that tori with Diophantine rotation vectors do exist near enough to the elliptic point, providing the map is more than C^3 and that the twist, $\det D\Omega(0)$, is nonzero. Each of these tori is also a Lagrangian submanifold (an n-dimensional surface on which the restriction of the symplectic form (1) vanishes). The relative measure of these tori approaches one at the fixed point.

Nevertheless, the stability of a generic, elliptic fixed point is an open question. Arnol'd showed by example in 1963 that lower-dimensional tori can have unstable manifolds that intersect the stable manifolds of nearby tori and thereby allow nearby trajectories to drift "around" the n-dimensional tori; this phenomenon is called Arnol'd diffusion (Lochak, 1993). When the map is analytic, the intersection angles become exponentially small in the neighborhood of the fixed point, and the existence of connections becomes a problem in perturbation theory *beyond all orders*.

Aubry–Mather theory gives a nonperturbative generalization of KAM theory for the case of monotone twist maps when $n = 1$. These are symplectic diffeomorphisms on the cylinder $\mathbb{S} \times \mathbb{R}$ (or on the annulus) such that $\partial q'/\partial p \geq c > 0$. For this case, Aubry–Mather theory implies that there exist orbits for all rotation numbers ω. When ω is irrational, these orbits lie on a *Lipschitz graph*, $p = P(q)$, and their iterates are ordered on the graph just as the iterates of the uniform rotation by ω. They are either dense on an invariant circle or an invariant Cantor set (called a *cantorus* when discovered by Percival). These orbits are found using a Lagrangian variational principle and turn out to be global minima of the action.

Aubry–Mather theory can be partially generalized to higher dimensions, for example, to the case of rational rotation vectors, where the orbit is periodic (Golé, 2001). Moreover, Mather (1991) has shown that action-minimizing invariant measures exist for each rotation vector, though they are not necessarily dynamically minimal. The existence of invariant cantor sets with any incommensurate rotation vector can also be proven for symplectic maps near an anti-integrable limit (MacKay & Meiss, 1992). Finally, converse KAM theory, which gives parameter domains where there are no invariant circles for the standard map, implies that, for example, the Froeschlé map has no Lagrangian invariant tori outside a closed ball in the space of its parameters (a, b, c) (MacKay et al., 1989).

James D Meiss

See also **Aubry–Mather theory; Cat map; Chaotic dynamics; Constants of motion and conservation laws; Ergodic theory; Fermi acceleration and Fermi map; Hamiltonian systems; Hénon map; Horseshoes and hyperbolicity in dynamical systems; Lyapunov exponents; Maps; Measures; Mel'nikov method; Phase space; Standard map**

Further Reading

Arnol'd, V.I. 1989. *Mathematical Methods of Classical Mechanics*, New York: Springer

Forest, E. 1998. *Beam Dynamics: A New Attitude and Framework (The Physics and Technology of Particle and Photon Beams)*, Amsterdam: Harwood Academic

Golé, C. 2001. *Symplectic Twist Maps: Global Variational Techniques*, Singapore: World Scientific

Lochak, P. 1993. Hamiltonian perturbation theory: periodic orbits, resonances and intermittancy. *Nonlinearity*, 6: 885–904

MacKay, R.S. & Meiss, J.D. 1992. Cantori for symplectic maps near the anti-integrable limit. *Nonlinearity*, 5: 149–160

MacKay, R.S., Meiss, J.D. & Stark, J. 1989. Converse KAM theory for symplectic twist maps. *Nonlinearity*, 2: 555–570

Marsden, J.E., Patrick, G.W. & Shadwick, W.F. (editiors). 1996. *Integration Algorithms and Classical Mechanics*, Providence, RI: American Mathematical Society

Mather, J.N. 1991. Action minimizing invariant measures for positive definite Lagrangian systems. *Mathematische Zeitschrift*, 207: 169–207

McDuff, D. & Salamon, D. 1995. *Introduction to Symplectic Topology*, Oxford: Clarendon Press

Meiss, J.D. 1992. Symplectic maps, variational principles, and transport. *Reviews of Modern Physics*, 64(3): 795–848

Moser, J.K. 1994. On quadratic symplectic mappings. *Mathematische Zeitschrift*, 216: 417–430

SYNAPSES

See **Neurons**

SYNCHRONIZATION

In a classical context, synchronization means adjustment of rhythms of self-sustained periodic oscillators due to their weak interaction, which can be described in terms of phase locking and frequency entrainment. The modern concept also covers such objects as rotators and chaotic systems, in which one distinguishes between different forms of synchronization, including complete, phase, and master–slave.

The history of synchronization goes back to the 17th century when the Dutch scientist Christiaan Huygens reported on his observation of synchronization of two pendulum clocks, which he had invented shortly before (Hugenii, 1673).

> ... It is quite worth noting that when we suspended two clocks so constructed from two hooks imbedded in the same wooden beam, the motions of each pendulum in opposite swings were so much in agreement that they never receded the least bit from each other and the sound of each was always heard simultaneously. Further, if this agreement was disturbed by some interference, it reestablished itself in a short time. For a long time I was amazed at this unexpected result, but after a careful examination finally found that the cause of this is due to the motion of the beam, even though this is hardly perceptible. The cause is that the oscillations of the pendula, in

proportion to their weight, communicate some motion to the clocks. This motion, impressed onto the beam, necessarily has the effect of making the pendula come to a state of exactly contrary swings if it happened that they moved otherwise at first, and from this finally the motion of the beam completely ceases. But this cause is not sufficiently powerful unless the opposite motions of the clocks are exactly equal and uniform.

Despite being among the oldest scientifically studied nonlinear effects, synchronization was understood only in the 1920s when Edward Appleton and Balthasar van der Pol theoretically and experimentally studied synchronization of triode oscillators.

The synchronization properties of periodic self-sustained oscillators are based on the existence of a special variable, phase ϕ. Mathematically, ϕ can be introduced as the variable parametrizing motion along the stable limit cycle in the state space of an autonomous continuous-time dynamical system. One can always choose phase in a way that it grows uniformly in time,

$$\frac{d\phi}{dt} = \omega_0, \qquad (1)$$

where ω_0 is the natural frequency of oscillations. The phase is neutrally stable, meaning that its perturbations neither grow nor decay. This corresponds to the invariance of solutions of autonomous dynamical systems with respect to time shifts. Thus, a small perturbation (for example, an external periodic forcing or coupling to another system) can cause large deviations of the phase—contrary to the amplitude, which is only slightly perturbed due to the transversal stability of the cycle. This property allows description of the effect of small forcing/coupling via the phase approximation.

Considering the simplest case of a limit cycle oscillator driven by a periodic force with frequency ω and amplitude ε, one can write the equation for the perturbed phase dynamics in the form

$$\frac{d\phi}{dt} = \omega_0 + \varepsilon Q(\phi, \omega t), \qquad (2)$$

where the coupling function Q is 2π-periodic in both its arguments and depends on the form of the limit cycle and the forcing. Close to the resonance $\omega \approx \omega_0$, the function Q contains fast oscillating and slow varying terms, the latter can be written as $q(\phi - \omega t)$. Upon averaging over a cycle, one obtains the following basic equation for the phase dynamics

$$\frac{d\Delta\phi}{dt} = -(\omega - \omega_0) + \varepsilon q(\Delta\phi), \qquad (3)$$

where $\Delta\phi = \phi - \omega t$ is the difference between the phases of the oscillations and of the forcing. The function q is 2π-periodic, and in the simplest case $q(\cdot) = \sin(\cdot)$ Equation (3) is called the Adler equation.

One can see that on the plane of parameters of the external forcing, there is a region

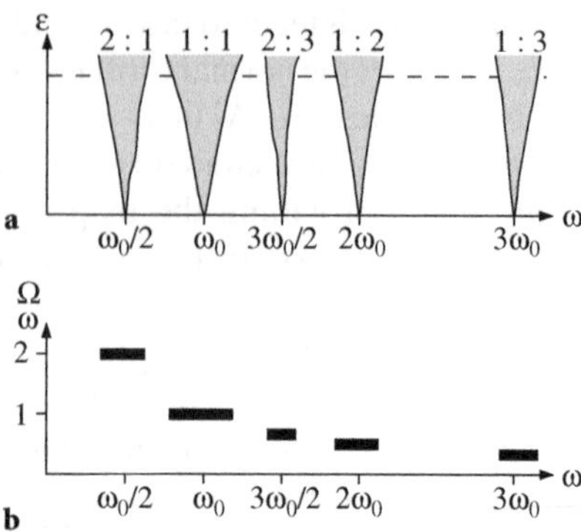

Figure 1. (a) A sketch of Arnol'd tongues. (b) Devil's staircase for a fixed amplitude of the forcing (dashed line in (a)).

($\varepsilon q_{\min} < \omega - \omega_0 < \varepsilon q_{\max}$) where Equation (3) has a stable stationary solution that exactly corresponds to phase locking (the phase ϕ just follows the phase of the forcing, so $\phi = \omega t + \text{constant}$) and frequency entrainment (the observed frequency of the oscillator $\Omega = \langle\dot{\phi}\rangle$ exactly coincides with the forcing frequency ω). Everyday examples of synchronization by external forcing are radio-controlled clocks, cardiac pacemakers, and circadian systems (the internal clocks of living objects that are synchronized to the exact 24-h periodic rhythm of sunlight).

Generally, synchronization is also observed for higher-order resonances $n\omega \approx m\omega_0$. In this case, the dynamics of the generalized phase difference $\Delta\phi = m\phi - n\omega t$ is described by an equation similar to Equation (3), namely, by $d(\Delta\phi)/dt = -(n\omega - m\omega_0) + \varepsilon\tilde{q}(\Delta\phi)$. The term *synchronous regime* then means perfect entrainment of the oscillator frequency at the rational multiple of the forcing frequency, $\Omega = n\omega/m$, as well as phase locking $m\phi = n\omega t + \text{constant}$. The overall picture can be shown on the (ω,ε) plane, where a family of triangular-shaped synchronization regions exists touching the ω-axis at the rationals of the natural frequency $m\omega_0/n$. These regions are called "Arnol'd tongues" (see Figure 1a). This picture is preserved for moderate forcing, although now the shape of the tongues generally differs from being exactly triangular.

For a fixed amplitude of the forcing ε and variable driving frequency ω, one observes different phase locking intervals where the motion is periodic, whereas in between them it is quasi-periodic. The curve Ω vs. ω, thus, consists of horizontal plateaus at all possible rational frequency ratios; this fractal curve is called a "devil's staircase" (Figure 1b).

An experimental example of such a curve is the voltage–current plot for a Josephson junction in an ac electromagnetic field, where synchronization plateaus are called Shapiro steps. As a junction can be considered as a rotator (rotations are maintained by a dc current), this example demonstrates that synchronization properties of rotators are very close to those of oscillators.

Synchronization of two coupled self-sustained oscillators can be described in a similar way. A weak interaction affects only the phases of two oscillators ϕ_1 and ϕ_2, and Equation (1) generalizes to

$$\frac{d\phi_1}{dt} = \omega_1 + \varepsilon Q_1(\phi_1, \phi_2),$$
$$\frac{d\phi_2}{dt} = \omega_2 + \varepsilon Q_2(\phi_2, \phi_1). \quad (4)$$

For the phase difference $\Delta\phi = \phi_2 - \phi_1$, one obtains after averaging an equation of the type of (3). Synchronization now means that two non-identical oscillators start to oscillate with the same frequency (or, more generally, with rationally related frequencies). This common frequency usually lies between ω_1 and ω_2. Note that locking of the phases and frequencies implies no restrictions on the amplitudes, in fact, the synchronized oscillators may have very different amplitudes and wave forms; for example, oscillations may be relaxation (integrate-and-fire) or quasiharmonic.

The mutual synchronization in a large population of oscillators (the Kuramoto transition in a population of globally coupled phase oscillators, for example) can be treated as a nonequilibrium phase transition—the mean oscillating field serving as an order parameter. Examples of synchronization in large ensembles include rhythmic applause and simultaneous flashing of fireflies, adjustment of menstrual cycles in women's dormitories, and so on. Synchronization in lattices of coupled self-sustained oscillators usually sets in via formation of clusters, that is, groups of oscillators (neighbors in a lattice) having the same frequency, and with increase of coupling the clusters grow and merge.

The concept of synchronization has been extended to include chaotic systems. One effect, called phase synchronization, is mostly close to the classical locking phenomena. Indeed, many chaotic self-sustained oscillators admit determination of the instantaneous phase and the corresponding mean frequency. Often one can find a projection of the strange attractor that looks like a smeared limit cycle, so phase is then introduced as a variable that gains 2π with each rotation. These rotations are non-uniform due to chaos, which can be modeled by an effective noise in phase dynamics. If this noise is small (i.e., the rotations are rather uniform), the mean frequency of the system can be entrained by a periodic forcing while the chaos is preserved. If two or more chaotic oscillators with different natural frequencies interact, their mean frequencies can be adjusted while the amplitudes remain chaotic and only weakly correlated.

Another type of chaotic synchronization—complete synchronization—can be observed for identical chaotic systems of any type (maps, autonomous or driven time-continuous systems). In the simplest case of two diffusively coupled in all variables systems, the dynamics is described by

$$\frac{d\boldsymbol{x}}{dt} = \boldsymbol{F}(\boldsymbol{x}) + \varepsilon(\boldsymbol{y} - \boldsymbol{x}),$$
$$\frac{d\boldsymbol{y}}{dt} = \boldsymbol{F}(\boldsymbol{y}) + \varepsilon(\boldsymbol{x} - \boldsymbol{y}), \quad (5)$$

where ε is the coupling parameter. The regime when $\boldsymbol{x}(t) = \boldsymbol{y}(t)$ for all t is called complete synchronization; because in this state the diffusive coupling vanishes, the dynamics is the same as if the systems were uncoupled. Although such symmetric solution exists for all ε, it is stable only if the coupling is sufficiently strong. To find the critical value of the coupling, one linearizes Equations (5) near the synchronized state and obtains for the mismatch $\boldsymbol{v}(t) = \boldsymbol{y}(t) - \boldsymbol{x}(t)$, the linearized system

$$\frac{d\boldsymbol{v}}{dt} = J(t)\boldsymbol{v} - 2\varepsilon\boldsymbol{v}, \quad (6)$$

where $J(t)$ is the Jacobian at the chaotic solution $\boldsymbol{x}(t)$. The ansatz $\boldsymbol{v} = e^{-2\varepsilon t}\boldsymbol{u}$ removes the last term on the right-hand side of (6), and the resulting equation coincides with the linearized equation for small perturbations of the solutions of an individual chaotic oscillator. Thus, $\boldsymbol{u}$ grows in proportion to the maximum Lyapunov exponent λ of a single system, and the critical coupling is $\varepsilon_c = \lambda / 2$. Complete synchronization occurs if $\varepsilon > \varepsilon_c$, that is, when the divergence of trajectories of interacting systems due to chaos is suppressed by the diffusive coupling. For weak coupling $\varepsilon < \varepsilon_c$, the states of two systems are different, $\boldsymbol{x}(t) \neq \boldsymbol{y}(t)$. Some other forms of synchronization in chaotic systems (generalized, master–slave) are similar to the complete one, and in all these cases synchronization appears if the coupling is strong enough.

MICHAEL ROSENBLUM AND ARKADY PIKOVSKY

See also **Commensurate-incommensurate transition; Coupled oscillators; Phase dynamics; Van der Pol equation**

Further Reading

Blekhman, I.I. 1988. *Synchronization in Science and Technology*, New York: ASME Press

Glass, L. 2001. Synchronization and rhythmic processes in physiology. *Nature*, 410: 277–284

Huygens (Hugenii), Ch. 1673. *Horologium Oscillatorium*, Paris: Apud F. Muguet; as *The Pendulum Clock*, Ames: Iowa State University Press, 1986

Kuramoto, Y. 1984. *Chemical Oscillations, Waves and Turbulence*, Berlin: Springer

Mosekilde, E., Maistrenko, Yu. & Postnov D. 2002. *Chaotic Synchronization: Applications to Living Systems*, Singapore: World Scientific

Pikovsky, A., Rosenblum, M. & Kurths, J. 2001. *Synchronization: A Universal Concept in Nonlinear Sciences*, Cambridge and New York: Cambridge University Press

SYNERGETICS

Synergetics deals with the spontaneous formation of spatiotemporal or functional structures in complex open systems by means of self-organization. The word *synergetics* is taken from Greek and means "science of cooperation." The central aim of synergetics, comprising both theoretical and experimental studies, is the search for basic principles that govern self-organization in both the physical and life sciences. To this end, a comprehensive mathematical theory has been developed that comprises both deterministic and stochastic processes.

Historical Background

Synergetics was initiated by Hermann Haken in 1969 in lectures at Stuttgart University (see also Haken & Graham, 1971). It originated from laser physics, where a pronounced transition from the disordered light of a lamp to the highly ordered light of a laser takes place. This transition can be interpreted as a nonequilibrium phase transition, on the one hand (Graham & Haken, 1968, 1970), and as a typical event of self-organization, on the other. Synergetics shows common features with and differences from a variety of interdisciplinary research fields:

1. In common with cybernetics as introduced by Norbert Wiener (1948), synergetics looks for general laws common to physical and biological systems. While cybernetics focuses on the control of a system in order to achieve its specific performance, synergetics studies the various dynamical structures a complex system can acquire.

2. Synergetics shares with general system theory as introduced by Ludwig von Bertalanffy (1968) the aim of finding general laws, in particular by seeking analogies. But while von Bertalanffy was seeking such analogies between otherwise different systems at the level of their individual elements, synergetics establishes close analogies at the level of order parameters (macroscopic field variables; see below).

3. Synergetics has used and developed methods belonging to dynamical systems theory (see, e.g., Guckenheimer & Holmes, 1983; Haken, 1983). Here in particular, emphasis is laid on the qualitative changes of the behavior of dynamical systems close to their points of instability (singular points, bifurcation points). In contrast to, for example, bifurcation theory, synergetics takes into account the pivotal role of random processes (Haken, 1983).

4. Synergetics has used and developed methods from statistical physics, such as various types of (generalized) Fokker–Planck equations, Langevin equations, and master equations (see, e.g., Stratonovich, 1963, 1967).

5. Synergetics shares with the theory of dissipative structures as introduced by Ilya Prigogine the goal of general laws (Nicolis & Prigogine, 1977). But while Prigogine's theory of dissipative structures is mainly based on thermodynamic principles, such as the entropy production principle or excess entropy production principle, synergetics is based on an approach that is close to statistical physics.

6. The concepts of synergetics have similarities with ideas developed in Gestalt theory founded by Max Wertheimer and Wolfgang Köhler (1924, 1969).

Theoretical Approach

All systems are considered subject to fixed internal or external conditions that are described by control parameters α. At particular values of α the behavior of the system may change macroscopically and qualitatively ("instability" of the previous state). As synergetics shows, close to such instability points the behavior of the system is determined by a small number of dynamical quantities, the order parameters. According to the "slaving principle" of synergetics, the order parameters determine the behavior of the individual parts of the system. In turn, the individual parts generate the order parameters by their cooperation (circular causality). Close to the instability points, nonequilibrium phase transitions occur that are characterized by symmetry breaking, critical fluctuations, and critical slowing down of the order parameters. This approach requires knowledge of the microscopic dynamics (*microscopic synergetics*). If this knowledge is absent, in *phenomenological synergetics* an order parameter dynamics is postulated.

Outline of Micoscopic Synergetics

The multi-component system is described by its state vector $\boldsymbol{q}$, whose components are labeled by the subsystem ℓ and the state or component j of each of them, $q_{\ell j}$, $\ell = 1, ..., N$; $j = 1, ..., J$. In continuous media, $\boldsymbol{q}$ becomes a function of a spatial coordinate $\boldsymbol{x}$, $\boldsymbol{q}(\boldsymbol{x})$. The state vector obeys an evolution equation

$$\mathrm{d}\boldsymbol{q}(t)/\mathrm{d}t = \boldsymbol{N}(\boldsymbol{q}(t), \nabla, \boldsymbol{x}, \alpha) + \boldsymbol{F}(\boldsymbol{q}, \boldsymbol{x}, t). \quad (1)$$

$\boldsymbol{N}$ is a nonlinear vector-valued, nonlinear function of $\boldsymbol{q}$, ∇ indicates spatial derivatives, α control parameters, and $\boldsymbol{F}$ fluctuating forces. $\boldsymbol{F}$ is mostly assumed to be δ-correlated in time ("white noise").

It is assumed that for a value $\alpha = \alpha_0$, the solution to (1) is known and the stability of $\alpha_0 : \boldsymbol{q}_0(t)$ is studied by *linear stability analysis*, putting $\boldsymbol{q}(t) = \boldsymbol{q}_0(t) + \boldsymbol{\xi}(t)$ and $\boldsymbol{F} = 0$. This leads to

$$\mathrm{d}\boldsymbol{\xi}(t)/\mathrm{d}t = L(\boldsymbol{q}_0(t), \alpha)\boldsymbol{\xi}, \quad (2)$$

where the linear operator L depends on $\boldsymbol{q}_0(t)$. The solutions to (2) are of the form

$$\boldsymbol{\xi}_k(t) = \exp(\lambda_k t)\boldsymbol{v}_k(t), \quad k = 1, ... \quad (3)$$

If λ_k has a discrete spectrum, then (Haken, 1983):

(a) q_0 is time-independent (fixed point); v_k is time-independent and contains powers of t if λ_k is degenerate;

(b) q_0 is time-periodic (limit cycle); v_k is time-periodic with the same period and contains powers of t if λ_k is degenerate and

(c) $q_0(t)$ is on torus, or arbitrary; $|v_k(t)|$ increases in time more slowly than exponential.

Close to instability points, where $\mathrm{Re}\lambda_k \geq 0$ for some k, we distinguish between unstable modes, v_u, and stable modes, v_s. In the case of (a), the wanted solution is assumed in the form

$$q(t) = q_0 + \sum_u \xi_u(t) v_u + \sum_s \xi_s(t) v_s, \qquad (4)$$

where the amplitudes ξ_u and ξ_s are still to be determined by inserting (4) into the complete Equations (1) and projecting onto the modes v_k. This yields the equations

$$\mathrm{d}\xi_u/\mathrm{d}t = \Lambda_u \xi_u + \hat{N}_u(\xi_u, \xi_s) + F_u, \qquad (5)$$

$$\mathrm{d}\xi_s/\mathrm{d}t = \Lambda_s \xi_s + \hat{N}_s(\xi_u, \xi_s) + F_s. \qquad (6)$$

In the cases, (b) and (c), the hypothesis (4) must be extended to include phase angles, ϕ. In such a case, Equations (5) and (6) must be supplemented by equations for ϕ

$$\mathrm{d}\phi/\mathrm{d}t = \hat{\mathrm{N}}_\phi(\xi_u, \xi_s, \phi) + F_\phi, \qquad (7)$$

where $\hat{N}_\phi$ is 2π-periodic in ϕ. In this case, the nonlinear functions on the right-hand side of (5) and (6) become also ϕ-dependent with periodicity 2π. The slaving principle of synergetics allows us to express the enslaved amplitudes ξ_s by means of the order parameters ξ_u, ϕ at the same time

$$\xi_s(t) = \mathbf{f}(\xi_u(t), \phi(t), t). \qquad (8)$$

The explicit time dependence stems from the action of the fluctuating forces F. By means of (8), the originally high-dimensional system (5)–(7) can be reduced to a low-dimensional system

$$\mathrm{d}\tilde{\xi}_u/\mathrm{d}t = \tilde{N}(\tilde{\xi}_u) + \tilde{F}_u, \qquad (9)$$

where $\tilde{\xi}_u$ comprises both ξ_u, ϕ. In continuous media, where

$$\xi_u(t) \rightarrow \xi_u(\mathbf{x}, t), \qquad (10)$$

the order parameter equations (9) are replaced by generalized Ginzburg–Landau equations

$$\mathrm{d}\xi_u(\mathbf{x}, t)/\mathrm{d}t = \Lambda_u(\nabla)\xi_u(x, t) + \tilde{N}(\xi_u(x, t)) + F_u, \qquad (11)$$

where the matrix of eigenvalues Λ_u depends on differential operators. In higher approximation, $\tilde{N}$ also depends on differential operators, $\tilde{N}(\xi_u, \nabla)$.

While Equations (1), (9), and (11) are of Langevin type, the stochastic problems can also be formulated by means of the Fokker–Planck equation, where the distribution function depends on the vector $[\xi, f(\xi, t)]$ and f obeys the Fokker–Planck equation

$$\partial f/\partial t = Lf,$$
$$L = -\sum_j \frac{\partial}{\partial \xi_j}(\tilde{N}_j f) + \sum_{ij} Q_{ij}\partial^2/\partial\xi_i\partial\xi_j f. \qquad (12)$$

Furthermore, another approach is based on the master equation for the distribution function P obeying

$$\mathrm{d}P(m, t)/\mathrm{d}t = \sum_{m'} w(m, m')P(m', t) - P(m, t)\sum_{m'} w(m', m), \qquad (13)$$

where $w(m, m')$ are the transition probabilities from state $m' \rightarrow m$.

Phenomenological Synergetics

In the case where microscopic dynamics of the system are not known, the analysis is based on phenomenological equations of the type (9), where the order parameters are characterized by those experimentally determined macroscopic quantities that change qualitatively when the control parameters are changed.

Applications of Concepts and Methods of Synergetics

Physics and Chemistry

Synergetics approaches are used to study stochastic properties and spatiotemporal patterns of laser light and of fluids and plasmas, current distributions in semiconductors, crystal growth, and meteorological structures, for example, baroclinic instability.

Synergetics is also used to study the formation of spatiotemporal patterns at macroscopic scales in chemical reactions, for example the Belousov–Zhabotinsky reaction.

Biology

Based on Turing's ideas of morphogenesis, synergetics calculates spatial density distributions, in particular gradients, stripes, hexagons, etc., in dependence on boundary and initial conditions. In initially undifferentiated omnipotent cells, molecules are produced as activators or inhibitors that diffuse between cells and react with each other and thus can be transformed. At places of high concentration, the activator molecules switch on genes that, eventually, lead to cell differentiation.

By means of synergetics, new kinds of analogies between evolution in biological and physical systems have been unearthed. For instance, the equations established by Manfred Eigen (1971) for prebiotic, that is, molecular evolution, turn out to be isomorphic to specific rate equations for laser light (photons), where a specific kind of photon wins the competition between different kinds.

In population dynamics, the resources, such as food, nesting places for birds, or light intensity for plants, serve as control parameters for synergetic analyses of self-organization. The numbers or densities of the individuals of species serve as order parameters. Specific examples are provided by the Verhulst equation or the predator-prey relation of the Lotka–Volterra equations. Of particular interest are dramatic changes, for instance the dying out of species under specific control parameter values. This has influences on environmental policy. If specific control parameters exceed critical values, the system's behavior can change dramatically. For instance, beyond a specific degree of evolution, the fish population of a lake may die out.

Nearly all biological systems show more or less regular rhythms—periodic oscillations or fluctuations. These can be imposed on the system from the outside, for instance, by the day/night cycle or seasons (exogen), or produced by the system itself (endogen). Endogenous rhythms that may proceed on quite different spatial and temporal scales are widely researched in synergetics. Examples are cell metabolism, circadian rhythms, brain waves in different frequency bands (see below), menstrual cycles, and cardiovascular rhythms. For instance, in the last, Stefanovska et al. (2000) were able to identify five order parameters. For certain time intervals, these order parameters can show phase and frequency couplings.

Rhythmical movements of humans and animals show well-defined patterns of coordination of the limbs, for instance, walking or running in humans or gaits of quadrupeds. Synergetics studies especially transitions between movement patterns, for instance the paradigmatic experiment by Kelso (1995). If subjects move their index fingers in parallel at a low frequency; increasing the frequency results in an abrupt involuntary transition to a new symmetric movement. The control parameter is the prescribed finger movement frequency, the order parameter is the relative phase between the index fingers. The experimentally proven properties of a nonequilibrium phase transition (critical fluctuations, critical slowing down, hysteresis) substantiate the concept of self-organization and exclude that of a fixed motor program. Numerous further coordination experiments between different limbs can be represented by the Haken–Kelso–Bunz model (Haken et al., 1985). Gaits of quadrupeds and transitions between them have been modeled in detail (Schöner et al., 1990).

In visual perception, the recognition of patterns, for example, faces, is interpreted as the action of an associative memory in accordance with usual approaches. Here incomplete data (features) with which the system is provided from the outside are complemented by means of data stored in the memory. A particular aspect of the synergetic approach is the idea that pattern recognition can be conceived as pattern formation. This is not only meant as a metaphor, but means also that specific activity patterns in the brain are established. In pattern formation, a partly ordered pattern is provided to ths system whereby several order parameters are evoked that compete with each other dynamically. The control parameters are so-called attention parameters that in cases without bias are assumed to be equal. The winning order parameter imposes a total pattern on the system according to the slaving principle. This process is the basis of the synergetic computer for pattern recognition (Haken, 1991). By means of appropriate preprocessing an invariance of the recognition process against scales, displacements and rotations, and even deformations can be achieved.

Gestalt Psychology

As is shown in Gestalt psychology (Wertheimer, Köhler), *Gestalt* is conceived as a specific organized entity to which in synergetics an order parameter with its synergetic properties (slaving principle) can be attached. The cognition or perception process of Gestalt proceeds in principle according to the synergetic process of pattern recognition. The winning order parameter generates, according to the slaving principle, an ideal percept that is the corresponding Gestalt. In ambiguous patterns, an order parameter is attached to each percept of an object. Because in ambiguous figures two or more possible interpretations are contained, several order parameters participate in the dynamics whereby the attention parameters become dynamical quantities. As already assumed by Köhler and as is shown by the synergetic equations, the corresponding attention parameter saturates; that is, it becomes zero if the corresponding object has been recognized and the other interpretation now becomes possible, where again the corresponding saturation process starts, etc. The model equations allow us also to take into account bias. (*See* the article on **Gestalt phenomena**.)

Psychology

According to the concept of synergetics, psychological behavioral patterns are generated by self-organization of neuronal activities under specific control parameter conditions and are represented by order parameters. In important special cases, the order parameter dynamic can be represented as the overdamped motion of a ball in mountainous terrain. By means of changes in the control parameters, this landscape is deformed and allows new equilibrium positions (stable behavioral patterns). This leads to new approaches to psychotherapy: destabilization of unwanted behavioral patterns by means, for example, of new external conditions or new cognitive influences and measures that support the self-organization of desired behavioral patterns. The insights of synergetics have been applied in the new field of psychosynergetics with essential contributions by Schiepek, Tschacher, Hansch, and others (Tschacher et al., 1992; Hansch, 1997; Ciompi, 1998, 1999).

Brain Theory

According to a proposal by Haken (1983), the brain of humans and animals is conceived as a synergetic, that is, a self-organizing system. This concept is supported by experiments and models on movement coordination, visual perception, and Gestalt psychology and by EEG and MEG analysis (see below). The human brain with its 10^{11} neurons (and glia cells) is a highly interconnected system with numerous feedback loops. In order to treat it as a synergetic system, control parameters and order parameters must be identified. While in synergetic systems of physics, chemistry, and partly biology the control parameters are fixed from the outside, for instance, by the experimenter, in the brain and in other biological systems, the control parameters can be fixed by the system itself. In modeling them it is assumed, however, that they are practically time-independent during the self-organization process. Such control parameters can be, among others, the synaptic strengths between neurons that can be changed by learning according to Hebb (1949), neurotransmitters, such as Dopamin, Serotonin, and drugs that block the corresponding receptors (e.g., Haloperidol, Coffein), and hormones (influencing the attention parameters). Furthermore, the control parameters may be more or less permanent external or internal stimuli.

In the frame of the given control parameters, self-organization takes place in neuronal activity whereby the activity patterns are connected with the corresponding order parameters by means of circular causality. The order parameters move for a short time in an attractor landscape whereby the attractor and also the order parameter disappear (concept of quasi-attractors). An example is the disappearance of a percept in watching ambiguous figures. The origin and disappearance of quasi-attractors and the corresponding order parameters can happen on quite different time scales, so that some of them can act as attractors practically all the time or are hard to be removed (psychotherapy in the case of behavioral disturbances). The activity patterns can be stimulated by external stimuli (exogenous activity) but can also be created spontaneously (endogenous activity), for instance in dreams, hallucinations, and, of course, thinking.

Synergetics throws a new light on the mind–body problem, for instance the percepts are conceived as order parameters, whereas the parts of a system are represented by electrochemical activities of the individual neurons. Because of circular causality, the percepts as order parameters and the neural activity (the "enslaved parts") condition each other. Beyond that, the behavior of a system can be described at the level of order parameters (information compression) or at the level of the activities of individual parts (large amount of information).

Analysis of Electroencephalograms (EEG) and Magnetoencephalograms (MEG)

Neuronal activity is accompanied by electromagnetic brain waves that cover the brain over large areas. The corresponding electric and magnetic fields are measured by the EEG and MEG, respectively. According to the ideas of synergetics, at least in situations where the macroscopic behavior changes qualitatively, the activity patterns should be connected with few order parameters. Typical experiments are the above-described finger coordination experiments by Kelso and closely related experiments, for instance, the coordination between the movement of a finger and a sequence of acoustic signals. In a typical experiment, parts of the brain or the whole brain are measured by an array of SQUIDS (superconducting quantum interference devices) that allows the determination of spatiotemporal field patterns. By means of appropriate procedures, these patterns are decomposed into fundamental patterns. As the analysis shows, two dominant basic patterns appear, whose amplitudes are the order parameters. If the coordination between finger movement and the acoustic signal changes dramatically, the dynamics of the order parameters also does so.

Sociology

Here we may distinguish between the more psychological and the more systems theoretical schools, where synergtics belongs to the second approach. We can distinguish between a qualitative and a quantitative synergetics (for a quantitative approach, see Weidlich, 2000). In the latter case, a number of sociologically relevant order parameters are identified. One example is the language of a nation. After his/her birth, a baby is exposed to the corresponding language and learns it (in technical terms of synergetics: the baby is enslaved) and then carries on this language as an adult (circular causality). These language order parameters may compete, where one wins (e.g., in the United States, the English language), they may coexist (e.g., in Switzerland), or they may cooperate (for instance, popular language and technical language). Whereas in this case the action of the slaving principle is evident, in the following examples its applicability is critically discussed by sociologists so that instead of slaving, some sociologists like to speak of binding or consensualization. Corresponding order parameters are type of state (e.g., democracy, dictatorship), public law, rituals, corporate identity, social climate in a company, and ethics. The latter example is particularly interesting, because order parameters are not prescribed from the outside or ab initio, but originate through self-organization and need not be uniquely determined.

Ethics

Conceived as order parameter means that it originates from the formation of a consensus so that the slaving principle becomes valid and also that there may exist several ethics.

Epistemology

An example for order parameters is provided by the scientific paradigms in the sense of Thomas S. Kuhn (1970), where a change of paradigms has the properties of a nonequilibrium phase transition, such as critical fluctuations and critical slowing down. Synergetics as a new scientific paradigm is evidently self-referential. It explains its own origin.

Management

The concept of self-organization is increasingly used in management theory and management praxis. Instead of fixed order structures with many hierarchical levels, now flat organizational structures with new hierarchical levels are introduced. In the latter case, a hierarchical level makes its decisions by means of its distributed intelligence. For an indirect steering of these levels by means of a higher level, specific control parameters in the sense of synergetics must be fixed, for instance, by fixing special conditions and goals. The order parameters are, for instance, the self-organized collective labor processes. In this context, the slaving principle—according to which the order parameters change slowly, whereas the enslaved parts react quickly (adaptability)—gains a new interpretation. For instance, the employees that are employed for a longer time determine the climate of labor and the style of work whereby it can also be possible that undesired cliques are established. This trend can be counteracted by job rotation.

Development of Cities

So far the development of cities was based on the concept of city planning with detailed plans for areas, but new approaches use concepts of self-organization according to synergetic principles. Instead of a detailed plan, now specific control parameters, such as a general infrastructure (streets, communication centers, and so on), are fixed. For details consider the book by Portugali (1999).

HÉRMANN HAKEN

See also **Dynamical systems; Emergence; Gestalt phenomena; Lasers; Turing patterns**

Further Reading

von Bertalanffy, L. 1968. *General System Theory*, New York: Braziller and London: Allen Lane, 1971

Ciompi, L. 1999. *Emotionale Grundlagen des Denkens*, Göttingen: Van den Hoeck; 1998. *Affektlogik*. Stuttgart: Klett-Cotta

Eigen, M. 1971. Molekulare Selbstorganisation und Evolution (Self-organization of matter and the evolution of biological macro molecules). *Naturwissenschaften*, 58: 465–523

Graham, R. & Haken, H. 1968. Quantum theory of light propagation in a fluctuating laser-active medium. *Zeitschrift für Physik*, 213: 420–450

Graham, R. & Haken, H. 1970. Laser light—first example of a second-order phase transition far away from thermal equilibrium. *Zeitschrift für Physik*, 237: 31–46

Guckenheimer, J. & Holmes, P.J. 1983. *Nonlinear Oscillations, Dynamical Systems, and Bifurcations of Vector Fields*, Berlin: Springer

Haken, H. 1983. Synopsis and introduction. In *Synergetics of the Brain*, edited by, E. Başar, H. Flohr, H. Haken & A.J. Mandell, Berlin: Springer, pp. 3–25

Haken, H. 1991. *Synergetic Computers and Cognition*, Berlin: Springer

Haken, H. 1993. *Advanced Synergetics*, 3rd edition, Berlin: Springer

Haken, H. & Graham, R. 1971. Synergetik—Die Lehre vom Zusammenwirken. *Umschau*, 6: 191

Haken, H., Kelso, J.A.S. & Bunz, H. 1985. A theoretical model of phase transitions in human hand movements. *Biological Cybernetics*, 51: 347–356

Hansch, D. 1997. *Psychosynergetik*, Wiesbaden: Westdeutscher Verlag

Hebb, D.O. 1949. *The Organization of Behavior*, New York: Wiley

Kelso, J.A.S. 1995. *Dynamic Patterns: The Self-organization of Brain and Behavior*, Boston: MIT Press

Köhler, W. 1924. *Die physischen Gestalten in Ruhe und im stationären Zustand*, Erlangen: Philosophische Akademie

Köhler, W. 1969. *The Task of Gestalt Psychology*, Princeton, NJ: Princeton University Press

Kuhn, T.S. 1970. *The Structure of Scientific Revolutions*, Chicago: University of Chicago Press

Nicolis, G. & Prigogine, I. 1977. *Self-organization in Nonequilibrium Systems*, New York: Wiley

Portugali, J. 1999. *Self-organization in the City*, Berlin: Springer

Schöner, G., Yiang, W.Y. & Kelso, J.A.S. 1990. A synergetic theory of quadrupedal gaits and gait transitions. *Journal of Theoretical Biology*, 142: 359–391

Scott, A.C. 2003. *Nonlinear Science: Emergence and Dynamics of Coherent Structures*, 2nd edition, Oxford and New York: Oxford University Press

Stefanovska, A., Haken, H., McClintock, P.V.E., Hožič, M., Bajrović, F. & Rivarič, S. 2000. Reversible transitions between synchronization states of the cardiorespiratory system. *Physical Review Letters*, 85(22): 4831–4834

Stratonovich, R.L. 1963, 1967. *Topics in the Theory of Random Noise*, vols. 1, 2, New York: Gordon and Breach

Tschacher, W., Schiepek, G. & Brunner, E.J. 1992. *Self-organization and Clinical Psychology*, Berlin: Springer

Weidlich, W. 2000. *Sociodynamics. A Systematic Approach to Mathematical Modelling in the Social Sciences*, Amsterdam: Harwood Academic Publishers

Wiener, N. 1948. *Cybernetics, or Control and Communication in the Animal and the Machine*, Cambridge, MA: MIT Press

T

TACHYONS AND SUPERLUMINAL MOTION

With a name derived from the Greek *tachys* (for swift), the tachyon is a hypothetical elementary particle that travels at speeds exceeding that of light (superluminal speed). In his 1905 paper on special relativity, Albert Einstein showed that the light speed in vacuum (c) is invariant with respect to all inertial observers, constituting a limiting value for the speed (v) of a moving mass. Consequently, tachyon research was delayed until the 1950s and 1960s, in particular till the appearance of the papers by Sudarshan and coworkers (Bilaniuk et al., 1962), and later by Recami and coworkers, among others.

If the special relativity theory is not a priori restricted to subluminal speeds, however, it seems able to accommodate tachyons. Such an extended relativity (ER) is based on the ordinary postulates of special relativity, and, therefore, does not appear to imply violations of causality (Recami, 1986). Just as photons are born, live, and die, always at the speed of light (without any need of accelerating from rest to the light speed), particles or waves may exist endowed always with speeds v larger than c.

Several areas of experimental physics suggest superluminal group velocities and energy propagation although it is not yet clear whether these phenomena imply signal and information transmissions with faster-than-light speeds (Recami, 2001).

Evanescent Waves and Tunneling Photons

In quantum mechanics, the tunneling time does not depend on the potential barrier width, implying arbitrarily large group velocities inside long barriers (Olkhovsky et al., 1992). Analogously, evanescent electromagnetic waves that travel with superluminal speeds have been predicted by extended relativity, confirmed by computer simulations based on Maxwell equations, and empirically observed (Chiao et al., 1993).

The most interesting experiment of this series was performed in 1994 by Günter Nimtz and his colleagues using two classical barriers separated by an intermediate region of width R. For nonresonant tunneling of microwaves, it was claimed that the total crossing time did not depend on R, implying the speed of transmission to be therein practically infinite. This result agrees with theoretical predictions for nonresonant tunneling through two successive opaque barriers (Olkhovsky et al., 2002) that were experimentally confirmed and have been verified by an experiment with two gratings in an optical fiber, suggesting potentially important applications.

Superluminal Localized Solutions (SLS) to the Wave Equation

Although the simplest subluminal object is a small sphere (or a point), a result of ER is that the simplest superluminal objects appear as "X-shaped" waves (double cones), rigidly moving in a homogeneous medium. Beams of this sort have been constructed (as superpositions of Bessel beams) in experiments with acoustic waves (Lu and Greenleaf, 1992), electromagnetic waves, and visible light (Saari & Reivelt, 1997). A single Bessel beam is the following solution of the

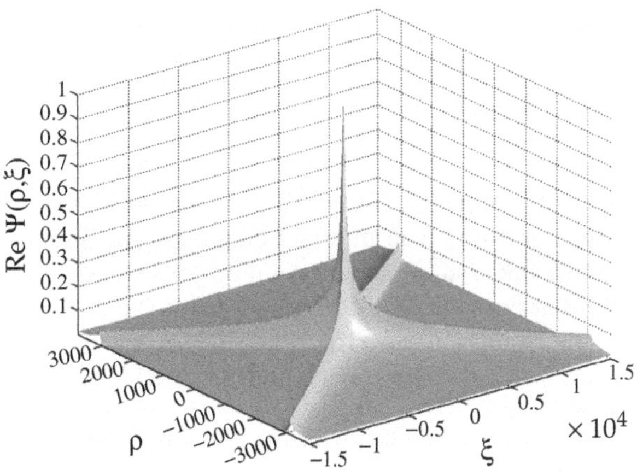

Figure 1. Illustration of the real part of a classical X-shaped wave (as a function of $\zeta \equiv z - vt$ and the radial coordinate ρ), evaluated for $v = 5c$ and $a = 5 \times 10^{-7}$ m, plotted from Equation (3).

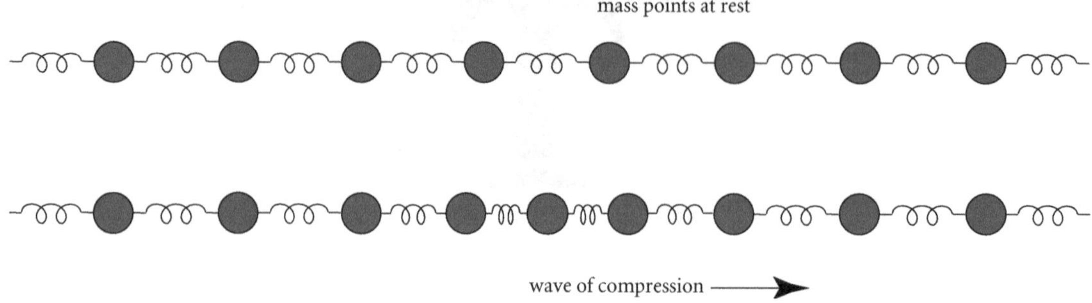

Figure 2. A spring–mass system with nonlinear springs. (Upper) Masses at rest. (Lower) A supersonic compression wave.

Maxwell equations in vacuum:

$$\psi(\rho,\zeta) = J_0[(\omega\rho/c)\sin\theta]\ \exp[\mathrm{i}(\omega\zeta/c)\cos\theta], \tag{1}$$

where J_0 is a Bessel function of the first kind of order zero, $\zeta \equiv z - vt$, $v = c/\cos\theta$, and θ is the axicone angle (angle between the direction of propagation and the propagating cone). This solution depends on z and t only through the quantity ζ; thus, it propagates without dispersion along the z-axis with speed $v > c$. By superposition of Bessel beams, for example, with variable angular frequency ω but constant θ and, therefore, constant v, one can obtain additional solutions to the Maxwell equations with any degree of transverse and longitudinal localization and centered at the desired angular frequency ω with the desired bandwidth (Zamboni-Rached et al., 2002); thus

$$\Psi(\rho,\zeta) = \int_0^\infty S(\omega) J_0[(\omega\rho/c)\sin\theta] \times \exp[\mathrm{i}(\omega\zeta/c)\cos\theta]\,\mathrm{d}\omega. \tag{2}$$

By choosing the exponential spectrum $S(\omega) = \exp[-a\omega]$, one gets the ordinary X-wave (with $v \equiv c/\cos\theta > c$)

$$\Psi(\rho,\zeta) = v\left[\sqrt{(av - \mathrm{i}\zeta)^2 + \rho^2(v^2/c^2 - 1)}\right]^{-1}, \tag{3}$$

which is a superluminal localized solution (SLS) to the wave equation. The plot of Equation (3) in Figure 1 can be understood as follows. For $\zeta < 0$, one has the familiar bow wave of a moving boat or shock wave of a supersonic aircraft. The branches for $\zeta > 0$, on the other hand, prepare the medium for the superluminal disturbance.

SLSs have also been constructed which travel without distortion along cylindrical waveguides and coaxial cables (Zamboni et al., 2002). X-shaped waves keep their localization and superluminality properties only to a certain depth of field (whose length can be determined a priori), decaying abruptly thereafter. As suggested by extended relativity, the simplest means for experimentally producing SLSs employs dynamic antennas consisting of a set of circular rings (or axicons or holographic elements). Acoustic localized supersonic beams have been used in a 3-dimensional ultrasound scanner for medical purposes: high-resolution scanning of the heart (Lu & Greenleaf, 1992).

Tachyons and SLSs in Nonlinear Media

Localized solutions for nonlinear partial differential equations that travel faster than the limiting speed for small amplitude waves are not uncommon (Conti et al., 2003). The sine-Gordon equation, for example, is Lorentz invariant and has tachyonic soliton solutions (Scott, 2003). Although dynamically unstable over long distances, these solutions may influence dynamics over shorter spans.

Another well-known example of a SLS goes back to the birth of nonlinear science, when John Scott Russell measured the speed of a hydrodynamic solitary wave in a wave tank to be $\sqrt{g(d+h)}$, where g is the acceleration of gravity, d the depth of the channel, and h the wave height. Clearly, the solitary wave speed is larger than the speed $\sqrt{gd}$ of low-amplitude waves.

Similar phenomena are observed for lattice solitary waves on spring-mass ladders (see Figure 2), which include the Fermi–Pasta–Ulam system, the Toda lattice, and various generalizations: Existence of supersonic compression-wave solutions has been mathematically proven for a general class of nonlinear intermass potentials (Friesecke & Wattis, 1994). As an example of superluminal propagation that was well known to military engineers in the early 19th century, Russell cited the fact that "the sound of a cannon travels faster than the command to fire it".

Erasmo Recami and Alwyn Scott

See also **Cherenkov radiation; Sine-Gordon equation; Skyrmions; Solitons, a brief history**

Further Reading

Bilaniuk, O.M, Deshpande, V.K. & Sudarshan, E.C.G. 1962. Meta relativity. *American Journal of Physics*, 30: 718

Chiao, R.Y., Kwiat, P.G. & Steinberg, A.M. 1993. Faster than light. *Scientific American*, 269(2): 52–60

Conti, C., Trillo, S., Di Trapani, P., Valiulis, G., Piskarskas, A., Jedrkiewicz, O. & Trull, J. 2003. Nonlinear electromagnetic X-waves. *Physical Review Letters*, 90: no.170406

Friesecke, G. & Wattis, J.A.D. 1994. Existence theorem for travelling waves on lattices. *Communications in Mathematical Physics*, 161: 391–418

Lu, J.-Y. & Greenleaf, J.F. 1992. Experimental verification of nondiffracting X-waves. *IEEE Transactions on Ultrasonic Ferroelectric Frequency Control*, 39: 441–446

Olkhovsky, V.S. et al. 1992. Recent developments in the time analysis of tunneling processes. *Physics Reports*, 214: 339–356 and references therein

Olkhovsky, V.S. et al. 2002. Superluminal tunneling through two successive barriers. *Europhysics Letters*, 57: 879–884

Recami, E. 1986. Classical tachyons and possible applications. *Rivista del Nuovo Cimento*, 9(6): 1–178 and references therein

Recami, E. 2001. Superluminal motions? A bird's-eye view of the experimental situation. *Foundations of Physics*, 31: 1119–1135

Saari, P. & Reivelt, K. 1997. Evidence of X-shaped propagation-invariant localized light waves. *Physical Review Letters*, 79: 4135–4138

Scott, A.C. 2003. *Nonlinear Science: Emergence and Dynamics of Coherent Structures.*, 2nd edition, Oxford and New York: Oxford University Press

Zamboni-Rached, M. et al. 2002. New localized superluminal solutions to the wave equations with finite total energies and arbitrary frequencies. *European Physical Journal D*, 21: 217–228

Zamboni, M. et al. 2002. Superluminal X-shaped beams propagating without distortion along a coaxial guide. *Physical Review* E, 66: no. 046617

TACOMA NARROWS BRIDGE COLLAPSE

The Tacoma Narrows suspension bridge in Washington state was opened to traffic on July 1, 1940. At the time, it was the latest in a trend in suspension bridge design toward building ever longer, lighter, and more flexible bridges. At least two other suspension bridges in America, the Golden Gate bridge in San Francisco and the Bronx–Whitestone bridge in New York, had at that time, already experienced some problems with unwanted oscillations; these bridges were stabilized by adding dampers, stiffening girders, and additional cables. The Tacoma Narrows bridge was far lighter and more flexible than its predecessors.

From its opening, the bridge experienced relatively small vertical oscillations. These were observed, and carefully recorded and the record survives in Amann et al. (1941), which is our primary source for what follows.

These oscillations were purely vertical, with no torsional component. They could be anything from no-noded (with a comparatively low frequency of about 8 cycles/min) to as many as seven-noded (with a frequency of 30 cycles/min). Amplitudes were as much as 5 f from top to bottom. Motions as large as 4 f with a frequency of 16 min^{-1} were observed

Figure 1. The Tacoma Narrows Bridge twisting, November 7, 1940. (Courtesy, Manuscripts, Special Collections, University Archives, University of Washington Libraries, UW2143.)

in winds of 3 mi/h, while at other times, the bridge remained stationary in winds of 35 mi/h. The structure had already survived winds of 48 mi/h.

These motions were not considered grounds for alarm, and it was apparently expected that with some modifications, they would eventually be eliminated. By October, hold-down cables had been installed in the side-spans, and these had effectively stopped the oscillations in these spans although, of course, they did not effect the center span. Hold-down cables for the center span had already been ordered.

On the morning of November 7, 1940, a qualitatively different phenomenon occurred. The wind was about 42 mi/h. The bridge began oscillating with somewhat more violence than usual in the vertical direction. The motion appears to have been eight- or nine-noded, with an amplitude of 4–5 f and a frequency of about 36–38 cycles/min. Although this seems rather violent, it was not initially viewed as cause for alarm. People and cars were on the bridge.

Suddenly, without warning, and virtually instantaneously, the bridge switched from the vertical motion to the famous torsional motion most often associated with the bridge (see Figure 1).

This new motion was one-noded, with an angle of rotation of nearly 45° each way, and a frequency of about 14/min. The sides of the bridge were moving with a double amplitude of 28 f, with accelerations in excess of gravity. Amazingly, this motion continued for about 45 min, after which the bridge finally began to break up. Occasionally, it would switch from one-noded to no-noded and back, but it remained primarily one-noded. The side-spans remained motionless, until the collapse of the center span, after which a torsional oscillation built up and then died down.

Thus, we are presented with at least three distinct nonlinear phenomena.

(i) The vertical oscillations: why so many and under such widely different wind conditions?
(ii) The instantaneous transition from vertical to torsional motion.
(iii) The persistence of large torsional periodic oscillation for 45 min under comparatively small aerodynamic forcing.
(iv) The peculiar shifting from one-noded torsional motion to no-noded and back.

A commission, including Othmar Amann and Theodore von Karman, tried to investigate the cause of the collapse. They made some preliminary conclusions, including "it is very improbable that resonance with alternating vortices plays an important role in the oscillations of suspension bridges." However, the main recommendation was simply to avoid designs that were light and flexible, This advice has largely been followed since that time, with the notable exception of the Millennium Bridge over the Thames in London. Moreover, other light, flexible, long-span bridges such as the Golden Gate were substantially re-engineered at great cost in the 1950s to eliminate similar unwanted oscillations.

Recent advances in nonlinear science have cast light on some of the phenomena mentioned above. The vertical oscillations may have been the result of the existence of multiple periodic solutions of a nonlinearly supported beam equation (Lazer & McKenna, 1990). They may also have been the result of some sort of negative damping induced by the aerodynamic interaction of the wind and the structure. Wind tunnel experiments on scale models (Scanlan & Tomko, 1971), predict the existence of small torsional oscillations (e.g., in torsion $0 \leq \alpha \leq \pm 3°$). However, this type of motion was never observed on the Tacoma Narrows.

The violent transition from vertical to torsional motion is now well understood. In McKenna (1999), a two-degree-of-freedom oscillator with physical constants chosen to match the bridge exhibits exactly this type of instability, as soon as the cables start alternately slackening. Also in McKenna (1999), this oscillator shows how small torsional forcing can induce large torsional periodic solutions, once one takes into account the pendulum-like nonlinearity of the torsional oscillator. The solutions are remarkably similar, in terms of amplitude and frequency, to the behavior observed on the Tacoma Narrows.

The transition from one-noded to no-noded and back has been observed in Moore (2002), where the torsional oscillations in a nonlinear beam are shown to obey a sine-Gordon equation whose long-term solutions are investigated. One can also see a numerical simulation of the bridge at www.math.lsa.umich.edu/ksmoore.

A good source of interesting behavior of early suspension bridges is Bleich et al. (1950).

JOE MCKENNA

See also **Bifurcations; Distributed oscillators; Stability**

Further Reading

Amann, O.H., von Karman, T. & Woodruff, G.B. 1941. *The Failure of the Tacoma Narrows Bridge*, Federal Works Agency Washington, D.C.

Bleich, F., McCullough, C.B., Rosecrans, R. & Vincent, G.S. 1950. *The Mathematical Theory of Suspension Bridges*, U.S. Dept. of Commerce, Bureau of Public Roads Washington, D.C.

Lazer, A.C. & McKenna, P.J. 1990. Large amplitude periodic oscillations in suspension bridges: some new connections with nonlinear analysis. *SIAM Review*, 32: 537–578

McKenna, P.J. 1999. Large torsional oscillations in suspension bridges revisited: fixing an old approximation. *American Mathematics Monthly*, 106: 1–18

Moore, K.S. 2002. Large torsional oscillations in a suspension bridge: multiple periodic solutions to a nonlinear wave equation. *SIAM Journal of Mathematical Analysis* 33(6): 1411–1429

Scanlan, R.H. & Tomko, J.J. 1971. Airfoil and bridge deck flutter derivatives. *ASCE, Journal of the Engineering Mechanics Division*, EM6: 1717–1737

TANGENT SPACE

See **Differential geometry**

TAYLOR–COUETTE FLOW

Taylor–Couette flow is generated in the gap between a pair of concentric cylinders by the rotary motion of one or both of the cylinders. If the outer cylinder alone is rotated, the torque induced by viscous drag on the inner one can be used to measure the viscosity of the fluid in the gap. This technique (studied by Arnulph Mallock in 1888 and independently by M.M. Couette in 1890) is still in use today as the basis of some commercial viscometers.

In 1916, Lord Rayleigh (John William Strutt) proposed general arguments on the stability of rotating fluids, showing that when the centrifugal force gradient decreases outwards in a rotating body of fluid, this will be an unstable situation for an inviscid fluid. In the case of Mallock's viscometer, the flow will be stable. But suppose the inner cylinder is rotated and the outer is held fixed. Here the flow will be unstable according to Rayleigh's criterion, and this question intrigued Geoffrey Taylor. In 1923, he published a brilliant theoretical and experimental investigation of a viscous version of the Rayleigh criterion, using rotating cylinders to generate what has come to be called Taylor–Couette flow. Taylor's work forms the cornerstone of much modern research on hydrodynamic stability

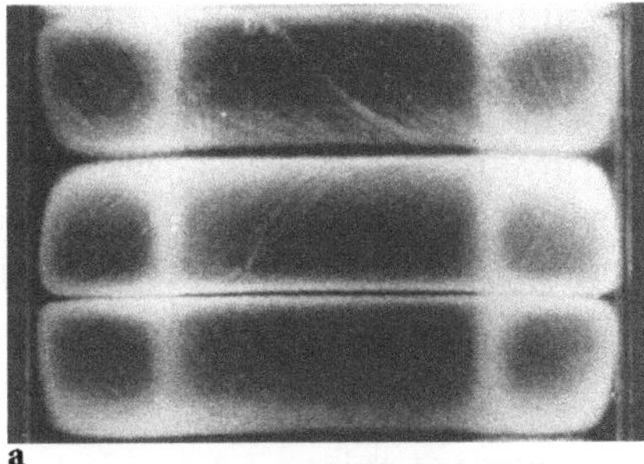
a

b

Figure 1. (a) Front view of steady Taylor–Couette cells (G. Pfister). (b) Front view of turbulent spiral flow (O. Dauchot).

and contains important ideas such as the exchange of stability between states.

In the simplest case, the outer cylinder is stationary and the inner cylinder alone rotates. When a critical Reynolds number is exceeded, the initially featureless rotary Couette flow has secondary vortices superposed on it. An example of such a Taylor vortex state is shown in Figure 1(a) where a front view of a Taylor vortex state is shown. The pattern is repeated along the length of the cylinder and has the appearance of a set of doughnuts stacked along the length the inner cylinder.

Increasing the speed of the inner cylinder leads to an instability (Davey et al., 1968) in the form of traveling waves that move at a fixed fraction of the speed of the inner cylinder (Coles, 1965). Yet, further increases in rotation rate give rise to more complicated quasi-periodic motion until low-dimensional temporal chaos ensues within each Taylor cell (Gollub & Swinney, 1975). Nowadays, it is recognized that many routes to chaos exist within this flow (Tagg, 1994), which has proved to be a rich research field for dynamical systems and bifurcation theory.

Taylor considered the general case with both cylinders rotating in co- and counter-directions, obtaining remarkable agreement between theory and experiment for the stability boundary. When the cylinders are rotated in opposite directions, new states of spiraling turbulence are sometimes found (Coles, 1965). An example of such a flow from a more recent study (Prigent et al., 2002) is shown in Figure 1(b). The general case of co- and counter-rotation produces a plethora of interesting dynamical states (Andereck et al., 1986) that can now be classified in terms of equivariant bifurcation theory (Tagg, 1994).

Coles also demonstrated dynamical non-uniqueness when the outer cylinder is stationary, finding more than 20 states at a particular Reynolds number. This multiplicity of solutions has also been observed in the steady vortex states (Burkhalter & Koschmieder, 1973; Benjamin & Mullin, 1982) where evidence was found for 42 different solutions at a single point in parameter space in the latter study.

Most of the above work is based on the notion of a theoretical model where the cylinders are considered to be infinitely long. This permits the use of periodic boundary conditions which allows analytical and numerical progress to be made in the nonlinear regime. However, the importance of end effects and their influence on global aspects of the steady flows was first recognized by Benjamin in 1978. He showed their importance in the selection process for steady flows and discovered the so called "anomalous modes" where all cells rotate in the opposite direction. These modes form an essential part of the solution set. The ideas have been developed and the role of end effects in determining the dynamics is now understood in some detail (Abshagen et al., 2001).

One clear advantage of studying fluid dynamical systems such as Taylor–Couette flow is that quantitative comparisons can be made between the results of controlled laboratory experiments and numerical calculations of the governing equations of motion—the Navier–Stokes equations (Cliffe et al., 2000). This is so not only for steady flows but also for time-dependent states, although disordered motion remains an outstanding challenge.

TOM MULLIN

See also **Bifurcations; Catastrophe theory; Chaos vs. turbulence; Hopf bifurcation; Stability**

Further Reading

Abshagen, J., Pfister, G. & Mullin, T. 2001. Gluing bifurcations in a dynamically complicated extended fluid flow. *Physical Review Letters*, 87: 4501–4505

Andereck, C.D., Liu, S.S. & Swinney, H.L. 1986. Flow regimes in a circular Couette system with independently rotating cylinders. *Journal of Fluid Mechanics*, 164: 155–183

Benjamin, T.B. & Mullin, T. 1982. Notes on the multiplicity of flows in the Taylor experiment. *Journal of Fluid Mechanics*, 121: 219–230

Benjamin, T.B. 1978. Bifurcation phenomena in steady flows of a viscous fluid. *Proceedings of the Royal Society of London*, A359, 1–43

Burkhalter, J.E. & Koschmieder, E.L. 1973. Steady supercritical Taylor vortex flow. *Journal of Fluid Mechanics*, 58: 547–560
Cliffe, K.A., Spence, A. & Tavener, S.J. 2000. The numerical analysis of bifurcation problems with application to fluid mechanics. *Acta Numerica*, 9: 39–131
Coles, D. 1965. Transition in circular Couette flow. *Journal of Fluid Mechanics*, 21: 385–425
Couette, M.M. 1890. Etudes sur le frottement des liquides. *Annales de Chimie et de Physique*, 6: 433–510
Davey, A., diPrima, R.C. & Stuart, J.T. 1968. On the instability of Taylor vortices. *Journal of Fluid Mechanics*, 31: 17–52
Gollub, J. & Swinney, H.L. 1975. Onset of turbulence in a rotating fluid. *Physical Review Letters*, 35: 927–930
Mallock, A. 1888. Determination of the viscosity of water. *Proceedings of the Royal Society of London*, A45: 126–132
Prigent, A., Gregoire, G., Chate, H., Dauchot, O. & van Saarloos, W. 2002. Large-scale finite-wavelength modulation within turbulent shear flows. *Physical Review Letters*, 89: 1501–1504
Rayleigh, Lord. 1916. On the dynamics of revolving fluids. *Proceedings of the Royal Society of London*, A93: 148–154
Tagg, R. 1994. The Couette–Taylor problem. *Nonlinear Science Today*, 4: 2–25
Taylor, G.I. 1923. Stability of a viscous liquid contained between two rotating cylinders. *Philosophical Transactions of the Royal Society of London*, 223: 289–343

TENSORS

Tensors are mathematical representations of objects that have intrinsic, geometric significances. This is a rather wider definition than many which can be found in textbooks, often referring to "sets of quantities" that "transform according to hideous formulae." The best way to understanding is likely to follow a few examples.

Readers interested in learning more will find the books of Simmon (1994) and Dodson & Poston (1991) helpful, as well as that of Bishop & Goldberg (1968). The main caveat, at the risk of repetition, is to beware of books that define tensors as some set of quantities tied to a coordinate system and then describe how they transform—tensors are there whether or not you have coordinates.

Suppose you have some apples on a table. The quantity of apples present is described completely by one number, and that number is well defined and meaningful no matter how you might orient any coordinate axes in the room. The fact that it is meaningful independent of the coordinates makes it a tensor, and the fact that it is the same no matter what coordinates might be introduced makes it a *scalar* or zeroth-rank tensor—the simplest kind of tensor.

Now, consider the weight of one of the apples. This weight is a force, equal to the mass of the apple, m, times the gravitational acceleration ($g = 9.8\,\text{m/s}^2$), and it is directed downwards. That is, its weight is mg directed downwards towards the table top. Let us call this weight $\boldsymbol{W}$, which has a clear physical meaning independent of coordinates, although it is helpful to describe its direction as "downwards." The fact that $\boldsymbol{W}$ is well defined even in the absence of coordinates makes it a tensor, and because it has a direction, it is called a *vector*, or first-rank tensor. If I introduce a set of coordinates x, y, z, I may place these axes in any way I choose. If I put the z-axis pointing downwards, the vector will have components relative to that coordinate system, with only the z component nonzero, and I might describe it as "the vector $(0, 0, mg)$." This (old) point of view is legitimate but carries with it the need to understand that the *components* are referred to as a coordinate system and as such have no intrinsic meaning.

More complicated objects can be easily imagined. For example, there could be a wind blowing across the table, and that could be described by another vector $\boldsymbol{V}$ describing its velocity. Referred to a coordinate system, it would be described by three components, and the presence of these two vectors is clearly something of intrinsic geometric significance. We can think of a product of the two vectors, meaning simply all the information needed to describe them. In terms of coordinates, if the weight is described by components w_i with i ranging from 1 to 3 and the wind velocity by v_j with j ranging from 1 to 3, we can think of a single geometric object $\boldsymbol{V} \otimes \boldsymbol{W}$ called the *tensor product* of $\boldsymbol{V}$ and $\boldsymbol{W}$ with 9 components $(\boldsymbol{V} \otimes \boldsymbol{W})^{ij} = v^i w^j$. The object $(\boldsymbol{V} \otimes \boldsymbol{W})$ is called a second-rank tensor, and this construction can be repeated to create more and more complex objects. (The reason for the upper placement of indices will be made clear shortly.)

The example given above was chosen to make it clear that tensors can turn up anytime one has vectors, but it is easy to find more physically motivated examples. Suppose one has a perfect fluid of density ρ moving with velocity $\boldsymbol{V}$ with components v^i. Density is a scalar if we allow only "proper" rotations which cannot change the signs of volumes. The flux or mass per unit volume per unit area perpendicular to direction i flows has components ρv^i. The flux of momentum in direction j is $\rho v^i v^j$. This motivates defining the second-rank "momentum tensor" tensor $\boldsymbol{T} = \rho \boldsymbol{V} \otimes \boldsymbol{V}$ with components $T^{ij} = \rho v^i v^j$.

All the tensors of a given rank have the structure of a vector space which they inherit from the operations allowed on vectors: they can be added, or multiplied by scalars. In fact, all the machinery of linear algebra carries over directly to them.

There has been an implicit indication that tensors here have something to do with symmetry or a concept of allowed transformation of the coordinates. The class of acceptable coordinate systems defines the geometry in any given situation, and we can speak then of tensors with respect to a symmetry group. For example, if we allow arbitrary rotations of a given orthogonal coordinate system, we can speak of "Cartesian tensors" or tensors under the group of orthogonal transformations in three dimensions. This means that one can pass from a representation of a tensor in terms of components with respect to one coordinate system to another by making the coordinates transform according to a matrix which represents that coordinate change.

This final step in thinking clarifies the notion of what a tensor must be. Given some space, one considers the geometry to be defined by the group of transformations which leave it invariant. For example, in flat Euclidean three-dimensional space, one might take the group of rotations, which are orthogonal matrices of determinant one, or SO(3). The vectors and tensors we have been talking about so far are then elements of spaces in which SO(3) acts linearly, i.e., they lie in representation spaces of SO(3).

Depending on the space in question and the structures imposed upon it, special tensors and operations can be defined. For example, in three-dimensional flat space, we have an operation that takes two vectors and produces from them a (coordinate-independent!) scalar called the "dot product," "inner product," or "scalar product." This is defined using the Kronecker delta δ_{ij} which is defined to have the value 1 when $i = j$ and 0 otherwise. It is a geometric object with a well-defined meaning independent of coordinates and is, thus, a tensor but of a different kind. One says that while $v^i w^j$ are components of a "contravariant tensor," δ_{ij} are the components of a "covariant tensor." The expression $\sum_i \sum_j \delta_{ij} v^i w^j$ is the scalar product of $\boldsymbol{v}$ and $\boldsymbol{w}$ and is a scalar. The terms *covariant* and *contravariant* are historical and come from the idea that for a quantity such as $\sum_i \sum_j \delta_{ij} v^i w^j$ to be invariant, the components with upper and lower indices should transform differently under a change of coordinates. The summation signs are often omitted, with a summation over repeated indices implied following the so-called "Einstein summation convention." An inner product automatically provides a dual space to that of the tensors, and one can write $v_i = \delta_{ij} v^j$ and speak of "covariant" as opposed to "contravariant" components of $\boldsymbol{v}$, and of the use of δ_{ij} to "lower indices."

The concept of a tensor is independent of any notion of a dot product, but many spaces of interest have such a product naturally present. In special relativity, for example, we consider "four-vectors" labeling differences in space and time and requiring four components. In units where the speed of light is unity, the dot product is provided by the Minkowski metric tensor η_{ab}: (a, b running from 0 to 3) where η_{ab} is 1 when $a = b = 0$, -1 when $a = b \neq 0$, and zero otherwise. This dot product is preserved by a larger group than just rotations, and this group is the Lorentz group of rotations and boosts, also called SO(3,1). Note that it is not positive-definite, so it is not an inner product in the strict sense of the word, and it defines a pseudo-Euclidean metric. In this case, we have SO(3,1) vectors as opposed to SO(3) vectors.

The most general (pseudo-)Riemannian geometry provides a dot product in terms of a metric tensor g_{ab} which is similar to η_{ab} but unrestricted other than to be nonsingular. In general relativity, we allow all invertible linear transformations in four dimensions, the symmetry group is GL(4), and we have GL(4) tensors for physical quantities.

JOHN DAVID SWAIN

See also **Einstein equations; Symmetry groups**

Further Reading

Bishop, R.L. & Goldberg, S.I. 1968. *Tensor Analysis on Manifolds*, New York: Macmillan

Dodson, C.T.J. & Poston, T. 1991. *Tensor Geometry*, 2nd edition, Berlin and New York: Springer

Simmon, J.G. 1994. *A Brief on Tensor Analysis*, 2nd edition, New York: Springer

TESSELLATION

A tessellation (or tiling) is a covering of a surface with tiles so that there are no gaps or overlaps. The tile shapes can be all different, as in the chips used to produce a complex Byzantine mosaic, or they may consist of a limited number of shapes, each congruent to one or more "prototiles" which serve as templates. Since ancient times, almost every culture has produced tessellations for utilitarian or decorative purposes—on walls, ceilings, and roofs of buildings, for pavements and plazas, and for designs to be woven, painted, printed, incised, or inlaid on every variety of surface. Tessellations adorn many churches, temples, palaces, and mosques; perhaps the most celebrated geometric tessellations are found in the Alhambra, in Granada, Spain. Tessellations also occur in nature, in the designs formed by scales or packed cells on a living surface.

Although artisans have produced tessellations for thousands of years, only recently have mathematicians undertaken a methodical study of the subject. Grünbaum and Shephard's book is the most complete reference. Intuitively, a shape "tiles" (is a prototile for a tessellation) if congruent copies of that shape can be fitted together exactly to fill a surface. Every triangle and every quadrilateral can tile the plane, and convex polygons with seven or more sides can never tile the plane. All convex hexagons that tile the plane have been determined, and 14 classes of convex pentagons that tile have been discovered, but it is not known if there are others. Tessellations by regular polygons appear frequently; those known as Archimedean tessellations are composed of regular polygons with edges matched, and have the same arrangement of tiles occurring at every vertex of the tiling (Figure 1).

An infinite variety of shapes tile the plane, many of them classified by special properties. Given an arbitrary set of shapes, many tests can attempt to answer the question: Will these tile the plane? But there is no guarantee of an answer. The question is mathematically

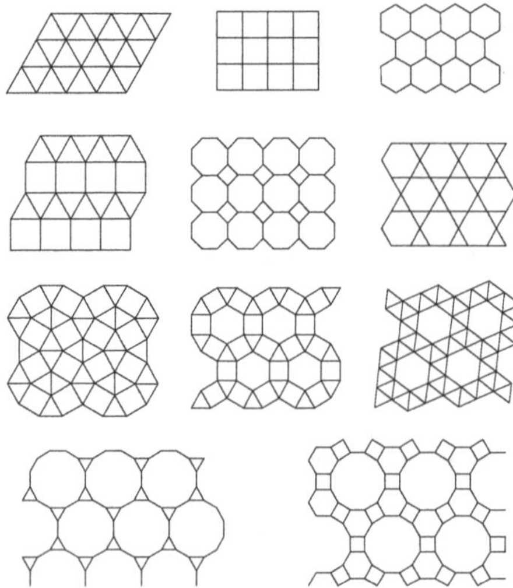

Figure 1. The 11 Archimedean tessellations. The three with only one prototile are called regular tessellations; the others are often called semiregular.

undecidable; that is, there is no algorithm that can deliver an answer of yes or no for every possible set of shapes.

M.C. Escher, a Dutch graphic artist (1898–1972), is the best-known creator of tessellations using whimsical figures as tiles (Figure 2). Inspired by the geometric tessellations in the Alhambra, he sought to answer the question: What shapes can tile the plane so that every tile is surrounded in the same way? As he discovered some answers, he developed a system of tile types that enabled him to create imaginative shapes that fit together in a prescribed manner. In his lifetime, he produced more than 150 finished tessellations; many have been used by scientists, particularly crystallographers, to illustrate their theories.

Almost all tessellations with repetition (including those in Figures 1 and 2) are periodic, that is, there is a smallest patch of the tiling that can be translated repeatedly by two independent vectors to fill out the whole tiling. It is natural to want repetition in a predictable manner to lay tile, to print, or to weave a pattern. Symmetry groups of periodic tilings are known as two-dimensional crystallographic groups, because crystals, by definition (until very recently), were defined by their periodic molecular structure. These groups consist of the translations, rotations, reflections, and glide-reflections that can act on the tiling in such a way that each tile moves to fit exactly onto another, leaving the tiling invariant. There are only 17 distinct symmetry groups for periodic tilings in the Euclidean plane; these are frequently used to classify tilings. Colored tilings (such as an extended checkerboard or Escher's tilings) can be analyzed according to color symmetries, which permute colors of tiles as well as positions of tiles in the tiling.

Nonperiodic tessellations have no translation symmetry. Although a regular tiling by squares can be made nonperiodic by shifting a few rows a small distance, the most interesting nonperiodic tessellations are produced by an aperiodic set of prototiles—every tiling formed by such a set is nonperiodic. It is not known if there is a single aperiodic tile, but there are several known aperiodic sets of two or more prototiles. The most well-known pairs were discovered by Roger Penrose in the 1970s: a kite and a dart (or a thick and a thin rhombus). In each Penrose tiling formed by these pairs (Figure 3), every patch of the tiling repeats infinitely often, but never by a translation that leaves the tiling invariant. Some properties of Penrose tilings are similar to those exhibited by unusual alloys discovered in the 1980s. These were named quasicrystals because their X-ray diffraction patterns displayed a crystal-like orderly repetition of bright spots, but also exhibited rotation symmetry forbidden in a periodic structure. Many unusual properties of Penrose and other aperiodic tilings have been discovered, and various techniques have been developed to study their structure (Senechal, 1995). Yet the study of aperiodic tilings and of quasicrystals is in its infancy, and it remains to be seen if the connections between them are more than superficial.

Figure 2. An Escher tessellation covers a column in a school in Baarn, Holland. (All M.C. Escher works © Cordon Art B.V., Baarn, The Netherlands.)

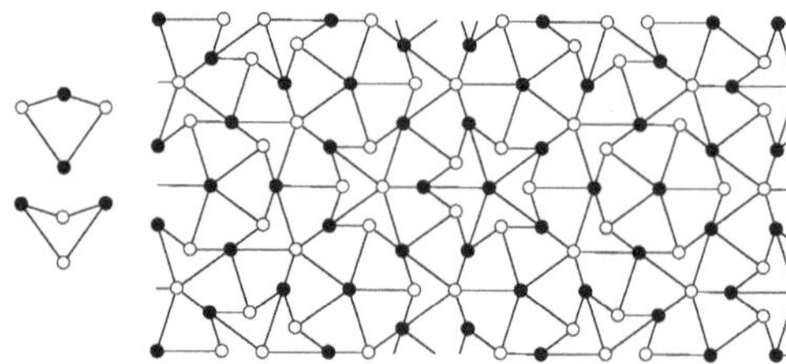

Figure 3. A Penrose tiling by kites and darts requires matching vertices of the same color.

A Voronoï (or Dirichlet) tessellation is determined by a given discrete set S of points in a surface. Each point x of S is surrounded by the region of points that are closer to x than to any others in S. The boundaries of these regions consist of those points that lie equidistant between at least two points of S. These regions with their boundaries (called Voronoï polygons or Dirichlet domains) are the tiles of the Voronoï tessellation determined by S. Such tessellations arise naturally in a wide variety of applications in physics (Wigner–Seitz regions); crystallography (Wirkungsbereich); physiology (capillary domains); urban planning (regions of service for schools or fire stations); biology (modeling cell arrangement); statistical spatial data analysis; and many other areas. Such tilings are almost always nonperiodic with many differently shaped tiles. Mathematical properties of such tilings and various algorithms to construct them are vigorous areas of research (Okabe et al., 1992).

Tessellations on surfaces such as spheres or the hyperbolic plane and tessellations of space of three or higher dimensions are equally important. Which shapes can fill space, and in what manner, is of great interest to scientists (and manufacturers). Little is known in this area; there is not even a complete list of space-filling tetrahedra. Other topics of investigation are tessellations with special tiles (e.g., polyominoes, or fractal tiles), relationships between local and global properties, classification, and construction of tessellations with special properties.

Doris Schattschneider

See also **Symmetry groups**

Further Reading

Bezdek, K. 2000. Space filling. In *Handbook of Discrete and Combinatorial Mathematics*, edited by K.H. Rosen, Boca Raton: CRC Press, pp. 824–830

Goodman, J.E. & O'Rourke, J. (editors). 2004. *Handbook of Discrete and Computational Geometry*, 2nd edition, Boca Raton: CRC Press (Relevant chapters: Tiling, D. Schattschneider & M. Senechal; Polyominoes, D. Klarner & S.W. Golomb; Voronoï diagrams and Delaunay triangulations, S. Fortune; Crystals and Quasicrystals, M. Senechal)

Gruber, P.M. & Wills, J.M. (editors). 1993. *Handbook of Convex Geometry*, Amsterdam: North-Holland (Relevant chapters: Geometric algorithms, H. Edelsbrunner, vol. A, pp. 699–735; Tilings, E. Schulte, vol. B, pp. 899–932); Geometric crystallography, P. Engel, vol. B, pp. 989–1041)

Grünbaum, B. & Shephard, G.C. 1987. *Tilings and Patterns*, New York: Freeman

Okabe, A., Boots, B. & Sugihara, K. 1992. *Spatial Tessellations: Concepts and Applications of Voronoï Diagrams*, 2nd edition, New York: Wiley (Okabe, A., Boots, B., Sugihara, K. & Chiu, S.N. 2000)

Patera, J. (editor). 1998. *Quasicrystals and Discrete Geometry*, Providence, RI: American Mathematical Society

Schattschneider, D. 2004. *M.C. Escher: Visions of Symmetry*, new edition, New York: Abrams

Schulte, E. 2002. Tilings. In *Encyclopedia of Physical Science and Technology*, 3rd edition, vol. 16, New York: Academic Press, pp. 763–782

Senechal, M. 1995. *Quasicrystals and Geometry*, Cambridge and New York: Cambridge University Press

Washburn, D.K. & Crowe, D.W. 1988. *Symmetries of Culture: Symmetry and Practice of Plane Pattern Analysis*, Seattle: University of Washington Press

THERMAL CONVECTION

Thermal convection is the transfer of heat by flow. In general, heat can be transported by convection, conduction, or radiation. In all cases, heat transfer requires the presence of a temperature gradient, and heat is transported from high to low temperature. Conduction is a diffusive process that involves the exchange of energy via collisions between molecules in a gas or a liquid, or interactions between lattice waves or electrons in a solid. Heat transport by radiation results from the emission and absorption of electromagnetic waves. Convection, finally, involves the transport of heat by the bulk physical motion of a fluid medium. The flow can arise naturally, due to thermal expansion—hot fluid is less dense than cold fluid and so tends to rise, while cold fluid is more dense and tends to fall—or it can be forced by some externally applied means.

In a layer of fluid heated from below, thermal convection is known as Rayleigh–Bénard convection, which is one of the most important systems for the study of pattern formation. This is because precise, well-controlled experiments are possible and because the equations describing the system (the Navier–Stokes equations coupled with an equation for heat transport and the appropriate boundary conditions) are well known, allowing close contact among experiment, theory, and simulations.

Figure 1(a) shows schematically an experiment for the study of Rayleigh–Bénard convection. The top plate is maintained at a temperature T_t, while the bottom plate is at a higher temperature $T_b = T_t + \Delta T$. The separation between the plates is d. For ΔT small, there is no flow and heat transport across the fluid layer is by conduction. Because of thermal expansion, however, the hotter fluid near the bottom plate is less dense than the cooler fluid above it. This is gravitationally unstable, but initiating flow costs energy due to viscous dissipation. As a result, convection does not begin until ΔT is large enough that the energy gained by starting flow offsets the cost due to dissipation. It is convenient to write ΔT in dimensionless form as the Rayleigh number, given by $\mathrm{Ra} = g\alpha d^3 \Delta T/\nu\kappa$, where g is the acceleration due to gravity, α is the thermal expansion coefficient of the fluid, ν is the fluid's viscosity, and κ is its thermal diffusivity. The onset of convection occurs when Ra reaches a critical value $\mathrm{Ra_c}$ which depends on the nature of the top and bottom boundaries. For rigid, isothermal boundaries, $\mathrm{Ra_c} = 1708$. For a rectangular cell with the same boundaries, the flow appears as a pattern of straight convection rolls oriented parallel to the short side of the cell, with a

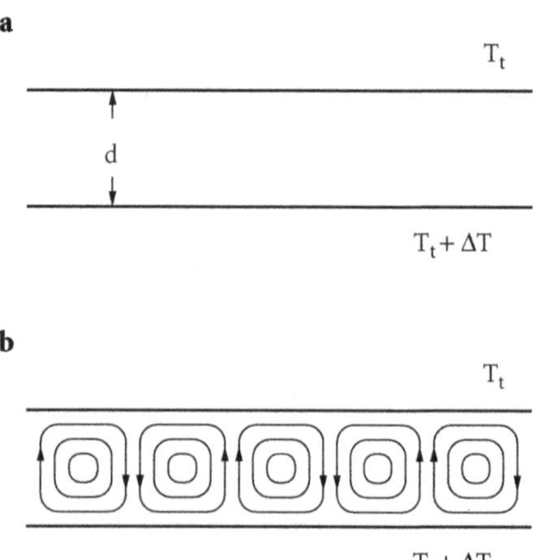

Figure 1. A schematic illustration of Rayleigh–Bénard convection. A fluid layer is bounded above and below by plates separated by a distance d, with the bottom plate at a temperature higher by ΔT than the top plate. When ΔT is below a critical value, there is no flow (a), but when ΔT is higher than the critical value, convective flow develops in the form of parallel, counter-rotating convection rolls (b).

Figure 2. Spiral defect chaos visualized in a convection experiment in a pressurized gas. Spiral defect chaos is a time-dependent steady state that appears near onset at low Prandtl number. Here dark regions indicate upflow and light regions indicate downflow. (Image courtesy of S.W. Morris.)

wavelength equal to $2.016d$. This situation is shown in Figure 1(b).

Heat transport in a convecting fluid is measured by the Nusselt number Nu, defined as $\mathrm{Nu} = \lambda_{\mathrm{eff}}/\lambda$, where λ is the thermal conductivity of the fluid and λ_{eff} is the effective thermal conductivity taking into account the heat transported by the flow. The Prandtl number, $\mathrm{Pr} = \nu/\kappa$, affects the dynamics of the flow pattern above onset.

The onset of Rayleigh–Bénard convection is an example of a bifurcation from a uniform (no-flow) state to one characterized by a spatial pattern. It can be described by the Ginzburg–Landau equation, which can be derived from the full equations describing the system. For the case of two rigid, isothermal boundaries, the system is up-down symmetric and the bifurcation is a supercritical pitchfork bifurcation. If we define $\varepsilon = (\mathrm{Ra} - \mathrm{Ra}_c)/\mathrm{Ra}_c$, then the bifurcation occurs at $\varepsilon = 0$. The Ginzburg–Landau equation predicts that close to onset, the amplitude of the convective flow field will grow as $\varepsilon^{1/2}$, while the correlation length of the pattern (the length scale over which the amplitude changes) behaves as $\varepsilon^{-1/2}$. These predictions have been confirmed experimentally.

If the up-down symmetry of the system is broken, either by making top and bottom boundaries different or due to a variation in the fluid properties with temperature across the cell (non-Boussinesq effects), then the symmetry of the bifurcation is also broken and convection appears via a subcritical bifurcation. In this case, the convection pattern at onset takes the form of an array of hexagonal cells (see photo of atmospheric hexagonal convection cells on page 3 of color plate section).

Above onset, the straight-roll pattern is stable within a range of wave numbers and Rayleigh numbers. The range of stability (known as the "Busse balloon") is limited by a variety of secondary instabilities which depend on Pr. The original straight–roll pattern becomes unstable to perturbations in phase or amplitude which lead to new flow patterns, which can have different wave numbers, orientations, and symmetries. Certain of these secondary instabilities can make the flow pattern two- or three-dimensional or time dependent.

In larger systems, when the size of the system becomes much greater than the correlation length of the pattern, the convection patterns are more complex than ideal straight rolls. The tendency of the rolls to orient perpendicular to the sidewalls results in patterns with curved rolls and a spatially varying wave number. Even close to onset, defects and localized instabilities can lead to time-dependent patterns. At low Pr, a transition to a complex, spatiotemporally chaotic state known as spiral defect chaos occurs (Figure 2). Interestingly, spiral defect chaos exists in exactly the regime where ideal straight rolls are expected to be stable, and experiments have shown that the two states can coexist. This suggests that there are two stable states in this regime, with different basins of attraction. Spiral defect chaos is obtained for most conditions, while specially chosen initial or boundary conditions are required to obtain ideal straight rolls.

At high Ra, the convective flow becomes turbulent. In this case, most of the temperature drop occurs in boundary layers near the top and bottom plates of the cell. Coherent plumes of hot and cold fluid can form at the lower and upper boundary layers, respectively, and a coherent large scale flow can exist in the convection cell. Simple theories based on the stability of the boundary layers predict that heat flow for turbulent convection should scale as $\mathrm{Nu} \sim \mathrm{Ra}^{1/3}$, while more sophisticated models give different values of the exponent. Recent research indicates that the exponent is itself a function of Ra.

There are many variants of Rayleigh–Bénard convection that have been studied in an effort to elucidate particular aspects of nonlinear dynamics. In appropriately chosen binary mixtures, one can obtain traveling-wave convection patterns. Thermal convection in anisotropic materials (e.g., liquid crystals) also leads to complex dynamics. Convection with an imposed mean flow has been used to study the transition from convective to absolute instability. Convection with rotation exhibits spatiotemporally chaotic patterns close to onset.

JOHN R. DE BRUYN

See also **Bifurcations; Pattern formation; Turbulence**

Further Reading

Bodenschatz, E., Pesch, W. & Ahlers, G. 2000. Recent developments in Rayleigh–Bénard convection. *Annual Review of Fluid Mechanics*, 32: 709–778

Busse, F. 1981. Transition to turbulence in Rayleigh–Bénard convection. In *Hydrodynamic Instabilities and the Transition to Turbulence*, edited by H.L. Swinney & J.P. Gollub, Berlin: Springer

Cross, M.C. & Hohenberg, P.C. 1993. Pattern formation outside of equilibrium. *Reviews of Modern Physics*, 65: 851–1112

de Bruyn, J.R, Bodenschatz, E., Morris, S.W., Trainoff, S.P., Hu, Y., Cannell, D.S. & Ahlers, G. 1996. Apparatus for the study of Rayleigh–Bénard convection in gases under pressure. *Review of Scientific Instruments*, 67: 2043–2067

Manneville, P. 1990. *Dissipative Structures and Weak Turbulence*, London: Academic Press

Siggia, E. 1994. High Rayleigh number convection. *Annual Review of Fluid Mechanics*, 26: 137–168

THERMO-DIFFUSION EFFECTS

In fluid mixtures, a temperature gradient can drive a concentration current or generate a concentration gradient depending on boundary conditions. This thermo-diffusion effect—nowadays referred to as the Soret effect or Ludwig–Soret effect—was first reported by Carl Ludwig in 1856 and by Charles Soret in 1879. They observed an increase (decrease) of salt concentration at the cold (hot) end of a tube filled with salty water (Ludwig, 1856; Soret, 1879). The reciprocal effect that a concentration gradient drives a heat flow or generates a temperature gradient was first reported by Louis Dufour in 1872 for gas mixtures in a porous medium. Theoretically, these and similar effects are most conveniently captured within the Onsager theory of irreversible macroscopic processes, in which generalized thermodynamic forces and resulting fluxes are linearly related to each other (de Groot & Mazur, 1962; Landau & Lifshitz, 1959).

In the last 20 years or so, it has become clear that the linear Soret effect plays a dominant role in the nonlinear behavior of convective pattern formation in binary fluid mixtures (Platten & Legros, 1984; Cross & Hohenberg, 1993; Lücke et al., 1998). Consider the typical Bénard configuration of a horizontal fluid layer of height d that is heated from below in a homogeneous gravitational field, $\boldsymbol{g} = -g\,\boldsymbol{e}_z$. Strongly heat-conducting impermeable horizontal plates impose a vertical temperature difference ($\Delta T > 0$) such that $T = T_0 \pm \Delta T/2$ at $z = \mp d/2$. T_0 is the mean temperature of the fluid layer. At small ΔT, the laterally homogeneous quiescent conductive state is stable with the linear temperature profile $T_{\text{cond}}(z) = T_0 - \Delta T z/d$. In a mixture like, for example, ethanol dissolved in water, this conductive temperature gradient generates as a consequence of the Soret effect a concentration gradient, so that

$$C_{\text{cond}}(z) = C_0 + S_T C_0(1 - C_0)\Delta T z/d. \quad (1)$$

Here, $C = \rho_1/(\rho_1 + \rho_2)$ is the mass concentration of the solute which is in our example the lighter component. C_0 is its mean, S_T the Soret coefficient, and $k_T = T_0 C_0(1 - C_0)S_T$ the thermo-diffusion ratio. The Soret coupling between temperature and concentration fields (cf. below) is most conveniently measured in terms of the separation ratio $\psi = -S_T C_0(1 - C_0)\beta/\alpha = -(k_T/T_0)\beta/\alpha$. Here, α and β are the thermal and solutal expansion coefficients of the total mass density $\rho_1 + \rho_2 = \rho = \rho_0[1 - \alpha(T - T_0) - \beta(C - C_0)]$ of the mixture for small deviations of T and C from their means. Positive S_T corresponding to negative ψ (for mixtures such as ethanol-water where α and β are positive) implies a concentration increase (of the lighter component) near the cold upper plate and a decrease near the warmer lower plate and vice versa for $S_T < 0$ ($\psi > 0$). Note that in experiments, ψ can easily be varied, say, between -0.6 and 0.25, by varying T_0 and C_0.

Convection is described by the balance equations

$$\nabla \cdot \boldsymbol{u} = 0, \quad (2a)$$

$$(\partial_t + \boldsymbol{u} \cdot \nabla)\boldsymbol{u} = \sigma\nabla^2\boldsymbol{u} + R\sigma\,(\delta T + \delta C)\,\boldsymbol{e}_z - \nabla p, \quad (2b)$$

$$(\partial_t + \boldsymbol{u} \cdot \nabla)T = \nabla^2 T, \quad (2c)$$

$$(\partial_t + \boldsymbol{u} \cdot \nabla)C = L\nabla^2 C - L\psi\nabla^2 T \quad (2d)$$

for mass (2a), momentum (2b), heat (2c), and concentration (2d) in the Oberbeck–Boussinesq approximation. δT and δC in (2b) denote deviations from the mean T_0 and C_0, respectively. Lengths are scaled with d, time with the vertical thermal diffusion time d^2/κ, and the velocity field $\boldsymbol{u} = (u, v, w)$ with κ/d, where κ is the thermal diffusivity of the mixture. Temperatures are reduced by ΔT, concentration by $\Delta T\alpha/\beta$, and pressure p by $\rho_0(\kappa/d)^2$.

The Dufour effect that provides a coupling of concentration gradients into the heat balance is discarded because it is relevant only in a few gas mixtures and possibly in liquid mixtures near the liquid–vapor critical point.

Besides the Rayleigh number $R = (\alpha g d^3/\nu\kappa)\Delta T$ measuring the thermal driving force, three additional numbers enter into the field equations: the Prandtl number $\sigma = \nu/\kappa$, which is of order 10 for ethanol–water mixtures at room temperature, the Lewis number $L = D/\kappa \simeq 0.01$, and the separation ratio ψ. Here, ν denotes the kinematic viscosity and D the concentration diffusivity.

The concentration field is responsible for the significantly larger complexity of binary mixture convection compared with pure fluids. It causes the richness of spatiotemporal properties of the convective structures, of the bifurcation behavior, and of the transient growth of convection. The Soret-generated concentration variations δC influence the buoyancy, that is, the driving force for convective flow in (2b). The flow in turn mixes by advectively redistributing concentration. This nonlinear advective mixing in developed convective flow is typically much larger than the smoothing by linear diffusion—the Péclet number measuring the strength of advective concentration transport relative to diffusion is easily of the order of a few thousand. Thus, the concentration balance is strongly nonlinear giving rise to boundary layer behavior and strongly anharmonic concentration field profiles in the horizontal direction, as in Figure 1. In contrast, the momentum and heat balances remain weakly nonlinear close to onset as in pure fluids, implying only smooth and basically harmonic variations, $\sim e^{i\boldsymbol{k}\cdot\boldsymbol{x}}$, as the critical modes (cf. Figure 1).

To summarize, the feedback interplay among (i) the Soret-generated concentration variations that are sustained against mixing and diffusion by externally imposed and internal temperature gradients, (ii) the resulting changes in the buoyancy, and (iii) the strongly nonlinear advective transport and mixing causes binary mixture convection to be rather complex not only with respect to its spatiotemporal properties but also concerning its bifurcation behavior.

Take, for example, $\psi < 0$, where the Soret-induced separation requires higher heating to destabilize the conductive state than for a pure fluid characterized by $\psi = 0$ (for a review of the multitude of convection states appearing for destabilizing positive ψ see, for example, Huke et al., 2000). Then the off-diagonal coupling between solutal buoyancy and advection of Soret-induced concentration variations described above generates oscillations—traveling waves (TWs) of horizontally propagating rolls occur via a subcritical Hopf bifurcation whenever ψ is sufficiently negative. The bifurcation properties of such oscillatory TW states are shown in Figure 2 for different ψ as a function of the reduced Rayleigh number $r = R/R_c^0$, where $R_c^0 = 1707.76$ marks the convective onset in pure fluids. With increasing flow intensity (Figure 2a), the fluid gets

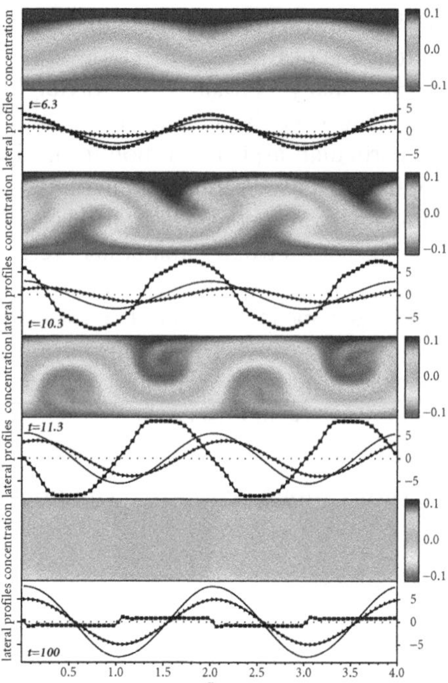

Figure 1. Evolution of convection after perturbing the quiescent conductive state. The concentration distribution in a vertical cross section of the fluid layer is displayed by color-coded plots where highest concentration was initially at the top, and lowest at the bottom. Wave profiles at midheight, $z = 0$, are shown for the fields of vertical velocity w (thin lines), $40\delta T$ (lines with triangles), and $400\delta C$ (lines with squares). The final TW propagates to the left. Parameters are $L = 0.01$, $\sigma = 10$, $\psi = -0.25$, $r = 1.42$, and wavelength $\lambda = 2$. For better visibility two wavelengths are shown. (This figure is also reproduced on page 3 of the color plate section.)

more mixed while simultaneously the TW frequency decreases (Figure 2b) as the flow intensity and the Nusselt number (Figure 2d) approach the pure fluid reference values. Here the mixing is measured by the reduced spatial variance $M = \sqrt{\langle \delta C^2 \rangle / \langle \delta C^2_{\text{cond}} \rangle}$ of the concentration.

Figure 1 shows the complex spatiotemporal concentration redistribution during the growth of oscillatory convection at slightly supercritical heating. The growth starts generically from perturbations of the conductive state that contain the two critical Hopf modes for counterpropagating TWs with roughly equal amplitudes. First, they linearly superimpose to form SW-like oscillations of growing amplitude with the large Hopf frequency. But then they compete via nonlinear advection with each other; at a critical SW amplitude, advective breaking of the concentration wave triggers a very fast flow-induced transition from SW to TW convection with anharmonic profile, large phase velocity, and large amplitude of the concentration wave. Finally, advective mixing and diffusive homogenization slow down the TW as the concentration differences between left and right turning rolls slowly decrease.

In mixtures with sufficiently negative ψ, there are also uniquely selected stable LTW states of

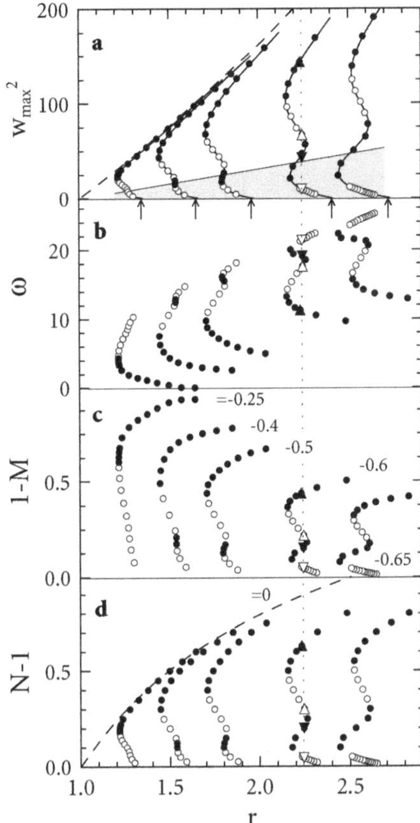

Figure 2. TW–bifurcation diagrams for different Soret coupling strength ψ: (a) squared maximal vertical flow $w_{\max}^2$, (b) frequency ω, (c) degree of mixing $1-M$, and (d) convective contribution to the Nusselt number $N-1$ vs reduced Rayleigh number r. Stable (unstable) TW states are marked by filled (open) symbols. Arrows mark Hopf thresholds $r_{\rm osc}$ for onset of TW convection. The $\psi=0$ pure fluid limit is included in (a) and (d) by the dashed line. TW states on the vertical line are discussed in more detail in Hollinger et al. (1997). Only states in the shaded region of (a) are weakly nonlinear. Parameters are $L=0.01$, $\sigma=10$, and $\lambda=2$.

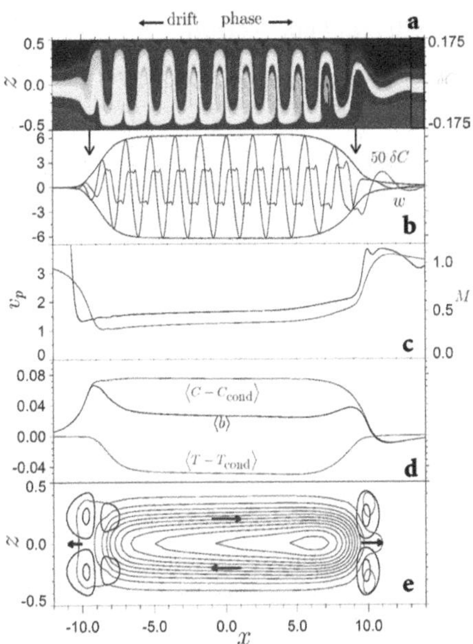

Figure 3. Broad LTW of length $l=17.4$: (a) Concentration deviation δC from global mean (pale gray) in a vertical cross section of the layer. (b) Lateral wave profiles at midheight, $z=0$, of δC (gray), vertical velocity w (black), and its envelope. At the arrows, $w_{\max}=v_{\rm p}$. (c) Mixing number M (gray) and phase velocity $v_{\rm p}$ (black). The variation of the wavelength $\lambda(x)=2\pi v_{\rm p}(x)/\omega$ is the same because the LTW frequency ω is a *global* constant. (d) Time averaged deviations from the conductive state at $z=-0.25$ for concentration (upper), temperature (lower), and their sum ($\langle b\rangle$) measuring the convective contribution to the buoyancy. (e) Streamlines of time averaged concentration current $\langle \boldsymbol{J}\rangle = \langle \boldsymbol{u}\delta C - L\nabla(\delta C-\psi\delta T)\rangle$ (gray) and velocity field $\langle \boldsymbol{u}\rangle$ (black). The latter results from $\langle b\rangle$ and documents roll shaped contributions of $\langle \boldsymbol{u}\rangle\langle \delta C\rangle$ to $\langle \boldsymbol{J}\rangle$ under the fronts and the associated $\langle \delta C\rangle$ redistribution. Thick black and gray arrows indicate $\langle \boldsymbol{u}\rangle$ and transport of positive δC (alcohol surplus), respectively. Thus, in the lower half of the layer, negative δC (water surplus) is transported to the right. Parameters are $L=0.01$, $\sigma=10$, $\psi=-0.35$, $r=1.346$. (This figure is also reproduced on page 3 of the color plate section.)

localized, that is, spatially confined, TWs. They occur at small subcritical heating where extended TWs cannot exist and where the conductive state is strongly stabilized by the Soret effect. Such a strongly nonlinear LTW (Figure 3) is robustly sustained by a complex concentration redistribution process. Therein flow-induced mixing locally reduces the Soret separation and thereby increases the buoyancy to levels that suffice to drive well-mixed fluid flow there. In Figure 3, positive (negative) δC is sucked from the top (bottom) boundary layer into right (left) turning rolls as soon as they become nonlinear under the trailing LTW front. This happens when the vertical velocity w roughly exceeds the local phase velocity v_p (left arrow in Figure 3b) so that regions with closed streamlines appear (Hollinger et al., 1997; Lücke et al., 1998). Within them "dark" ("gray") concentration is transported predominantly in the upper (lower) part of the layer to the right. Mean concentration, on the other hand, migrates mostly to the left along open streamlines that meander between the closed roll regions and that follow the global mean in Figure 3a. The *time-averaged* current of δC (gray lines in Figure 3e) reflects the mean properties of this transport. Since positive and negative (zero) δC is transported away from (towards) the left trailing front, mean concentration accumulates there and causes a strong drop of $M(x)$. In the same way, the leading front's concentration varies and with it, $M(x)$ are strongly increased even beyond the conductive state's values. Thus, unlike TWs, LTWs do not reach a balance among δC injection, advective mixing, and diffusive homogenization on a constant level of small M. Rather LTW rolls collapse under the leading front when $v_{\rm p}$ has grown up to w (right arrow in Figure 3b). Thereafter, concentration is discharged and sustains ahead of the leading front a barrier of $\langle \delta C\rangle$ that prevents the expansion of the conductive state into the LTW.

M. LÜCKE

See also **Thermal convection**

Further Reading

Cross, M.C. & Hohenberg, P.C. 1993. Pattern formation outside of equilibrium. *Reviews of Modern Physics*, 65: 851

de Groot, S.R. & Mazur, P. 1962. *Non-equilibrium Thermodynamics*, Amsterdam: North-Holland

Hollinger, St., Büchel, P. & Lücke, M. 1997. Bistability of slow and fast traveling waves in fluid mixtures. *Physical Review Letters*, 78: 235

Huke, B., Lücke, M., Büchel, P. & Jung, Ch. 2000. Stability boundaries of roll and square convection in binary fluid mixtures with positive separation ratio. *Journal of Fluid Mechanics*, 408: 121

Jung, D. & Lücke, M. 2002. Localized waves without the existence of extended waves: Oscillatory convection of binary mixtures with a strong Soret effect. *Physical Review Letters*, 89: 054502

Köhler, W. & Wiegand, S. (editors). 2002. *Thermal Nonequilibrium Phenomena in Fluid Mixtures*, Berlin and New York: Springer

Landau, L.D. & Lifshitz, E.M. 1959. *Fluid Mechanics*, Oxford: Pergamon Press and Reading, MA: Addison-Wesley (originally published in Russian)

Lücke, M., Barten, W., Büchel, P., Fütterer, C., Hollinger, St. & Jung, Ch. 1998. Pattern formation in binary fluid convection and in systems with throughflow. In *Evolution of Structures in Dissipative Continuous Systems*, edited by F.H. Busse & S.C. Müller, Berlin and New York: Springer, p. 127

Ludwig, C. 1856. Diffusion zwichen ungleich erwärmten Orten gleich zusammengesetzter Lösungen. *Sitzungsberichte der Kaiserliche. Akademie der Wissenschaften (Mathematisch-Naturwissenschaftlicheclasse), Wien*, 65: 539

Platten, J.K. & Legros, J.C. 1984. *Convection in Liquids*, Berlin: Springer

Soret, C. 1879. Sur l'état d'équilibre que prend, au point de vue de sa concentration, une dissolution saline primitivement homogène, dont deux parties sont portées à des températures différentes. *Archives des Sciences: Physiques et Naturelles, Genève*, 2: 48–61

THETA FUNCTIONS

The n-dimensional *theta function* is a function of $n+n(n+1)/2$ complex variables, $\theta(z|\tau)$. The first n coordinates form a vector in an n-dimensional complex space $z=(z_1,\dots,z_n)$ while the remaining $n(n+1)/2$ variables are entries in a symmetric n-dimensional matrix τ_{ik}; $(i,k=1,\dots,n)$. The theta function is defined by a Fourier series as

$$\theta(z|\tau)=\sum_{m\in\mathbb{Z}^n}\mathrm{e}^{\{\mathrm{i}\pi m\tau m^t+2\mathrm{i}\pi z m^t\}},$$

where the n-tuple summation runs over the whole n-dimensional set of integers $\mathbb{Z}^n$, and the imaginary part of the matrix τ is supposed to be positive definite to provide convergence of the series. The defining properties of the theta function are its periodicity and modular properties. The periodicity property is given by the relation

$$\begin{aligned}\theta(z+e_k|\tau)&=\theta(z|\tau),\quad \theta(z+e_k\tau|\tau)\\&=\mathrm{e}^{-\mathrm{i}\pi\tau_{kk}-2\mathrm{i}\pi z_k}\theta(z|\tau),\quad k=1,\dots,n,\end{aligned}$$

where only the kth component of the vector e_k is nonzero and equal to unity.

Consider the modular group Γ the group of all $2n\times 2n$ integer matrices γ such that $\gamma J\gamma^t=J$ where

$$J=\begin{pmatrix}0&-1_n\\1_n&0\end{pmatrix},$$

and 1_n is the unit n-dimensional matrix. Under the action of the modular group, the theta-function transforms as

$$\theta(z|\tau)=\frac{\varepsilon}{\sqrt{\det M(\tau)}}\exp\left\{\frac{1}{2}\sum_{i,k=1}^{n}z_iz_k\frac{\partial}{\partial\tau_{ik}}\det M(\tau)\right\}\theta\left(z'|\tau'\right),$$

where

$$\begin{aligned}M(\tau)&=c\tau+d,\, z'=(a\tau+d)^{-1}z,\\\gamma&=\begin{pmatrix}a&b\\c&d\end{pmatrix},\\\tau'&=(a\tau+b)(c\tau+d)^{-1},\end{aligned}$$

a,b,c,d are $n\times n$ matrices and $\varepsilon^8=1$.

The theta functions are introduced to construct modular functions in τ-variables of order k defined to satisfy $f(\gamma\circ\tau)=\det(c\tau+d)^kf(\tau)$ and abelian functions in z-variables. Abelian functions $F(z)$ are functions of n complex variables with $2n$ complex periods, T_i, $i=1,\dots,2n$, $F(z+T_i)=F(z)$. The advantage of using theta functions to define modular and abelian functions comes from the rapid convergence of the *theta*-series.

The most important class of abelian functions are abelian functions whose τ-variables are constructed from an algebraic curve X of genus n, given by a polynomial equation $P(\lambda,\mu)=0$. The introduction of local coordinates turns X into a one-dimensional complex analytical variety called the Riemann surface of the curve X. The Riemann surface of genus n can be topologically described as a sphere with n handles. It is always possible to draw, on such a torus, a basis of $2n$-cycles $a_1,\dots,a_n;b_1,\dots,b_n$ with intersection numbers $a_i\circ a_j=b_i\circ b_j=0$ and $a_i\circ b_j=-b_j\circ a_i=\delta_{ij}$, where $\circ$ means intersection of corresponding cycles.

Differential and integral calculus can also be developed on X. In contrast to the case of an extended complex plane, or Riemann sphere, which can be considered as a Riemann surface of genus zero, there exists n linearly independent holomorphic differentials $\mathrm{d}w_1,\dots,\mathrm{d}w_n$ which can be normalized by the conditions $\oint_{a_k}\mathrm{d}w_i=\delta_{ik}$. The period matrix τ of the curve is then given by $\tau_{ik}=\oint_{b_k}\mathrm{d}w_i$. Meromorphic differentials $\mathrm{d}\omega_k$, that is, differentials with poles of order k, can be also defined on X; usually these differentials are normalized by the condition $\oint_{a_l} d\omega_k = 0, l = 1, \ldots, n$.

The abelian function $u(\boldsymbol{t}) = u(t_1, \ldots, t_n)$ of n complex variables associated with the curve X of genus n is then defined as

$$u(t_1, \ldots, t_n) = -2\frac{\partial^2}{\partial t_1^2} \log \theta \left(\sum_{k=1}^{n} \boldsymbol{U}_k t_k + \boldsymbol{U}_0 | \tau \right) + c, \quad (1)$$

where the vector $\boldsymbol{U}_k$ is the vector of b-periods of the normalized meromorphic differential $d\omega_k$, $\boldsymbol{U}_0$ is a vector through which initial data are introduced and c is a constant.

The most developed case is the case of hyperelliptic curves, when the polynomial P is given by

$$P(\lambda, \mu) = \mu^2 - \prod_{k=1}^{2n+1} (\lambda - \lambda_k),$$

where n is the genus and the branching points λ_k, $k = 1, \ldots, 2n+1$ are supposed to be distinct. The holomorphic differentials described above are given in this case by the formula $dw_k = \lambda^{k-1} d\lambda / \mu$, $k = 1, \ldots, n$. In the case $n = 1$, the abelian function of the curve is the well-known elliptic function.

The remarkable role of θ functions in the spectral theory of the Schrödinger equation was discovered by Its & Matveev (1975). Consider the spectral problem

$$\left\{ \frac{\partial^2}{\partial x^2} - u(x) \right\} \Psi(x; \lambda, \mu) = \lambda \Psi(x; \lambda, \mu),$$

where $u(x)$ is smooth and real potential, $\Psi(x; \lambda, \mu)$ is an eigenfunction, and λ is the spectral parameter. Suppose that the spectrum consists of $n+1$ continuous segments $[\lambda_1, \lambda_2], \ldots, [\lambda_{2n+1}, \infty]$. Then the potential is given by the formula (1) with $t_1 = x$ and $t_k = \text{const}$ for $k > 1$, while the eigenfunction $\Psi(x; \lambda, \mu)$ is given by the formula

$$\Psi(x; \lambda, \mu) = C \frac{\theta\left(\int_{(\infty,\infty)}^{(\lambda,\mu)} d\boldsymbol{w} + \boldsymbol{U}_2 x + \boldsymbol{U}_0 | \tau \right)}{\theta\left(\boldsymbol{U}_2 x + \boldsymbol{U}_0 | \tau \right)} \times \exp \left\{ x \int_{(\lambda_0,\mu_0)}^{(\lambda,\mu)} d\omega_2(\lambda, \mu) \right\}, \quad (2)$$

where C is a normalizing constant, $d\omega_2(\lambda, \mu)$ is the second kind abelian differential with second-order pole at infinity and zero a_i-periods, $\boldsymbol{U}_2$ is a vector of b_i-periods of $d\omega_2(\lambda, \mu)$, and the constant vector $\boldsymbol{U}_0$ and point (λ_0, μ_0) are defined by initial conditions. The isospectral deformation of the potential $u(x)$ when the second variable $t_2 = t$ in formula (1) is switched on, while other variables $t_k, k > 2$ remain constant, turns the function $u(x, t)$ into the n-gap solution of the Korteweg–de Vries equation. Equations (1) and (2) form the foundation of the theory of finite-gap solutions of soliton equations. In the limit when the branch points collide in pairs, $\lambda_{2k-1} \to \lambda_{2k}$ $k = 1, \ldots, n$, these formulae become N-soliton formulae. Krichever (1977) generalized the whole theory to other soliton equations and nonhyperelliptic curves.

Introductions to the subject are given in Farkas & Kra (1998) for the theory of Riemann surfaces, Novikov, Chapter 11 in Zakharov et al. (1980), Dubrovin (1981). Mumford (1983, 1984) for theta functions and completely integrable equations, and Belokolos et al. (1994) for algebro-geometric methods of integration of nonlinear equations.

VICTOR ENOLSKII

See also **Elliptic functions; Inverse scattering method or transform; *N*-soliton formulas**

Further Reading

Belokolos, E.D., Bobenko, A.I., Enolskii, V.Z., Its, A.R. & Matveev, V.B. 1994. *Algebro-geometric Approach to Nonlinear Integrable Equations*, Berlin and New York: Springer

Dubrovin, B.A. 1981. Theta functions and nonlinear equations. Russian Mathematical Surveys, 36: 11–80

Farkas, H. & Kra, I. 1998. *Riemann Surfaces*, New York: Springer

Its, A.R. & Matveev, V.B. 1975. Sohrödinger operators with a finite-band spectrum and the N-soliton solutions of the Korteveg–de Vries equation. *Teoreticheskaya i Matematicheskaya Fizika*, 23: 51–68

Krichever, I.M. 1977. The method of algebraic geometry in the theory of nonlinear equations. *Russian Mathematical Surveys*, 32: 180–208

Mumford, D. 1983, 1984. *Tata Lectures on Theta*, vols. 1, 2, Boston: Birkhäuser

Zakharov, V.E., Manakov, S.V., Novikov, S.P. & Pitaevskii, L.P. 1980. *Soliton Theory: Inverse Scattering Method*, Moscow: Nauka (in Russian)

THREE-BODY PROBLEM

See ***N*-body problem**

THREE-WAVE INTERACTION

See ***N*-wave interactions**

THRESHOLD PHENOMENA

Defined in the dictionary as an "intensity below which a mental or physical stimulus cannot be perceived and can produce no response," the term *threshold* has deep roots in our language, representing a collective awareness of strong nonlinearity. Above threshold, the effect of a stimulus changes dramatically from that below, as is indicated by such common phrases as the "tipping point" and the "last straw." Examples of threshold phenomena abound in physics, engineering, nonlinear mathematics, chemistry, biology, and neuroscience.

Applications of the Threshold Concept

In nonlinear optics, the threshold for laser oscillation specifies a level of pump power below which only a small amount of incoherent light is emitted from the laser, while above threshold a brilliant, highly directed beam of output light is observed. Beginning with the wall switch, electrical engineers have devised many varieties of threshold circuits that change rapidly from one voltage level to another when an input variable exceeds a certain value; indeed, a digital computer can be viewed as a large collection of interacting threshold devices. Also the threshold logic unit (TLU), comprising a switch with input channels having adjustable weights, is a useful component of learning systems for pattern recognition (Nilsson, 1990).

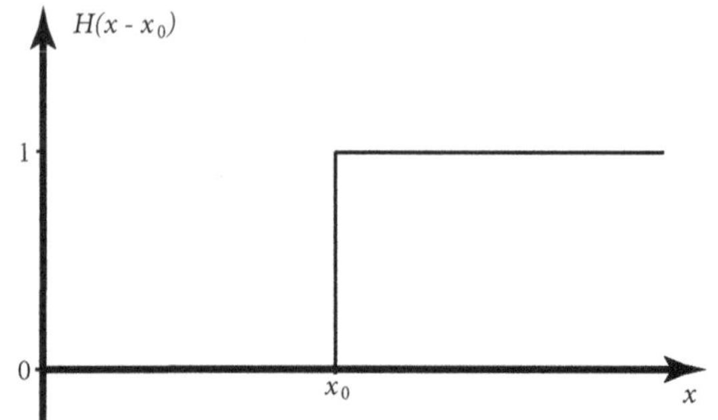

Figure 1. A Heaviside step function.

Mathematical representations of switching devices often involve the Heaviside step function,

$$H(x - x_0),$$

a generalized function that is zero for $x < x_0$ and one for $x > x_0$, where x_0 is the threshold as shown in Figure 1. In phase space models of nonlinear dynamic systems, one often finds a *separatrix*, or critical surface across which solution trajectories have very different behaviors. Noting this, mathematician cum meteorologist Edward Lorenz famously asked: "Does the flap of a butterfly's wings in Brazil set off a tornado in Texas?"—not only coining a dramatic metaphor for threshold phenomena and launching modern studies of chaos, but also leading philosophers to examine what is meant by causality. Is it only the *last* straw that should be blamed for breaking the camel's back? Or are *all* the straws that were loaded onto the beast to some degree complicit?

In the realms of physical chemistry, a reaction-diffusion (excitable) system rests quietly until stimulated above its threshold for self-supporting activity, whereupon traveling waves, spirals, and scroll waves are typically observed to emerge. In biology, examples of threshold phenomena include the germination of a seed at a critical (threshold) level of ambient moisture and the insemination of an ovum, in addition to the birth of life itself from the chemical components of the Hadean oceans some 3.5 billion years ago.

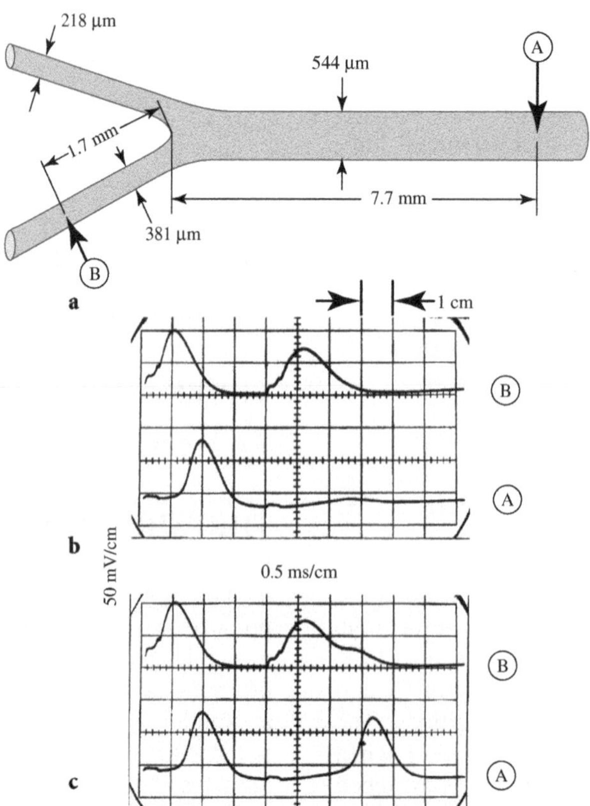

Figure 2. Threshold action in a branching region of a squid giant axon. The interpulse interval is slightly longer in (c) than in (b).

Threshold Effects in Neuroscience

As one would suppose from introspection, the field of neuroscience offers many examples of threshold phenomena which have become known since the observation of "all-or-none" response of a neuron was proposed by Edgar Adrian in 1914 (Adrian, 1914). This threshold behavior of a neuron was used by psychiatrist Warren McCulloch and mathematician Walter Pitts in 1943 to formulate the first computer model of the human brain McCulloch & Pitts (1943). Their model assumes each neuron to be represented by a single Heaviside step function, which jumps from the off (zero) to the on (one) state when a linear sum of input (dendritic) signals exceeds a threshold value. In 1958, Frank Rosenblatt introduced the "perceptron" which employs TLUs as the basic emenents of a brain model (Rosenblatt, 1958).

Recent studies of dendritic dynamics suggest far greater complexity, with the possibility of threshold effects occurring at the branching regions of incoming fibers (Stuart et al., 1999). In Figure 2 are displayed some experimental measurements of nerve impulse transmission through a branching region in the giant axon of a squid (Scott, 2002). Figure 2(a) shows the geometry of the preparation (not to scale), indicating the electrode positions where upstream (B) and downstream (A) recordings were made. The preparation was stimulated with two pulses, which are seen in the upper traces of both Figures 2(b) and 2(c). In these experiments, the spacing between the two incoming impulses was under experimental control, and below a certain threshold value of this impulse interval—that shown in Figures 2(b) and 2(c)—it is seen that the second impulse no longer makes it through. Besides interpulse interval, several other neural parameters can result in such a threshold effect, including branch geometry, ionic concentrations, fatigue, and narcotization level.

As the dendrites of many neurons are known to carry active impulses (action potentials) (Stuart et al., 1999), it appears that the branching regions of dendrites can act as threshold devices (switches), greatly increasing our estimates of the ability of a neuron to process information (Scott, 2002).

ALWYN SCOTT

See also **Butterfly effect; Excitability; Flip-flop circuit; Lasers; McCulloch–Pitts network; Multiplex neuron; Nerve impulses; Phase space; Reaction-diffusion systems**

Further Reading

Adrian, E.D. 1914. The all-or-none principle in nerve. *Journal of Physiology (London)* 47: 460–474

McCulloch, W.S. & Pitts, W.H. 1943. A logical calculus of the ideas immanent in nervous activity. *Bulletin of Mathematical Biology* 5: 115–133

Nilsson, N.J. 1990. *Learning Machines*, 2nd edition, San Mateo, CA: Morgan Kaufmann

Rosenblatt, F. 1958. A probabilistic model for information storage and organization in the brain. *Psychology Review* 65: 386–408

Scott, A.C. 2002. *Neuroscience: A Mathematical Primer*, New York: Springer-Verlag

Stuart, G., Spruston, N. & Häusser, M. 1999. *Dendrites*, Oxford: Oxford University Press

TIME-SERIES ANALYSIS

Predicting total eclipses of the moon or the sun is an art that dates back to mankind's oldest civilizations. Nonetheless, it represents a class of very modern analytic tools: the analysis of time-series data. A time-series is a set of measurements s_n, $n = 1, \ldots, T$, whose index n refers to the time t when the measurement is recorded by $n\tau = t$, τ is the sampling interval. For time-dependent phenomena, this index carries part of the information which is destroyed, for example, by a random reshuffling of the temporal order of the measurements. If one wishes to make use of this property in data analysis, particular statistical methods are required which are able to characterize temporal correlations inside the data. Typical goals in time-series analysis are predictions, data classification, signal manipulation, and system identification.

Today's technical facilities for data acquisition and data storage call for efficient time-series tools. To mention a few examples, in medicine (particularly cardiology and neurology), there is a need for automated diagnostics (i.e., data classification). In finance or weather and climate research, data-driven prediction methods are relevant, since model equations are either lacking (economy) or expensive to solve (weather). In modern telecommunications and automatic speech recognition, noise reduction is an essential issue, where time-series analysis should supply the background for signal separation.

The above examples show that unlike data of the positions of the planets, whose analysis enabled Johannes Kepler to derive the laws of planetary motion, modern time-series problems are concerned with data that have a complicated, strong aperiodic component. All data analysis methods start from some paradigm about the origin of the observed signatures and irregularities. This background is indispensable for a sound interpretation of the statistical quantities thus obtained, or for an estimate of the validity of the consequences drawn from the results. However, the large number of different approaches in time-series modeling supplies a corresponding diverse set of time-series analysis methods. We focus here on the two most general approaches, both of which represent a whole theoretical framework and not just a particular aspect.

Linear stochastic models are a well-developed class of time-series models. Being linear, their properties can be fully and rigorously derived from their model equations. However, in order to generate aperiodic time-series data, such models require stochastic inputs. Auto regressive models AR(M),

$$s_{n+1} = \sum_{k=0}^{M-1} a_k s_{n-k} + \xi_n,$$

where ξ_n is Gaussian distributed white noise ($\langle \xi_l \xi m \rangle = \delta_{l,m}$) with unit variance, represent the time discretized motion of the superposition of $M/2$ damped harmonic oscillators driven by noise, if the coefficients a_k fulfill certain stability conditions. The output is therefore characterized by $M/2$ frequencies and the corresponding damping coefficients, such that a spectral analysis is the most suitable analysis tool (*See* **Spectral analysis**). Thus, AR-models are suitable for data sets that have a few pronounced peaks in their power spectrum. If the observed power spectrum instead is broad band, another linear model, the moving average model MA(N), can be more reasonable:

$$s_{n+1} = \sum_{k=0}^{N-1} b_k \xi_{n-k},$$

where ξ_n is again white Gaussian noise. Notice that there is no feedback of the observable s, so that this model just averages over the independent noise inputs and hence creates colored noise. In the limits $N, M \to \infty$, both model classes are equivalent. For practical purposes where a small number of coefficients is desirable, a combination of both, the ARMA(M,N)-model, is often used. The coefficients of such models can be fit to observed data, for example, by solving least-squares problems (Box & Jenkins, 1976).

The well-known sunspot number time series of solar activity with its pronounced 11-year period can be well captured by an AR-model. ARMA models are also rather suited to describe many noise-dominated signals such as sound emission signals in technical environments, and they are used to model single phonemes of human speech in automatic speech recognition systems. ARMA-models are hence employed for data-driven predictions and signal classification tasks. They are often useful if the signal is either dominated by some few frequencies or when it is really noisy. However, the linearity of the model behind translates into the fact that the observables should be Gaussian random variables themselves, and that all higher-order statistics beyond the power spectrum and the auto-correlation function are fully determined by either of these two.

From many model systems and physical laboratory experiments arises a different class of sources for aperiodic time-series data: so-called chaotic dynamical systems. Aperiodicity here comes from intrinsic instabilities without random inputs, and nonlinearity destroys the superposition principle, such that a Fourier decomposition of the signal is not useful. Higher-order statistics is not trivially related to second-order statistics. Instead of a statistical characterization, one would here try to reconstruct the deterministic and dynamic origin of the signal.

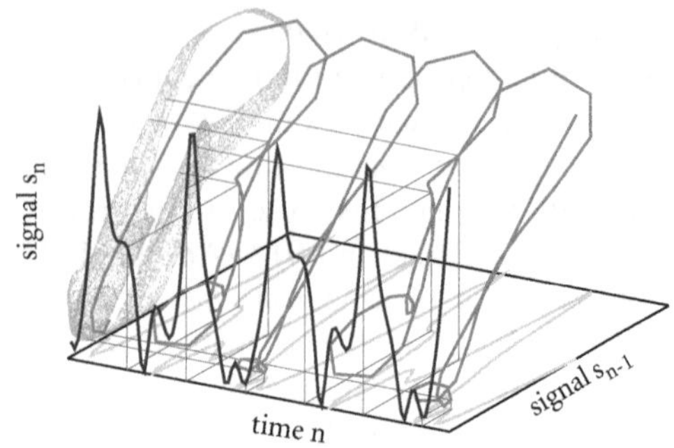

Figure 1. Sketch of the time-delay embedding procedure: The time series s_n (plotted in the frontal plane) combined with its time shifted version (in the bottom plane) forms a sequence of vectors (the curve in space), whose projection along the time axis accumulates to a set of points representing (in this particular case) a strange attractor together with an invariant measure on it. The data are voltages measured in a chaotic electric resonance circuit.

If one assumes that a given experimental observable represents the deterministic dynamics in some non-observed and even unknown phase space, the concept of phase space reconstruction by embedding is needed: As proven by Takens (1981), the set of delay vectors obtained from a scalar time series by joining successive observations, $\boldsymbol{s}_n = (s_n, s_{n-1}, s_{n-2}, \ldots, s_{n-m+1})$, is equivalent to a set of phase space vectors of the underlying dynamical system, provided that the embedding dimension m fulfills $m > 2D_f$ (see Figure 1). Here, D_f is the attractor dimension of the underlying dynamical system (*See* **Attractors; Dimensions; Fractals**). Hence, successive vectors $\boldsymbol{s}_n$ are deterministically related to each other by an unknown function $\boldsymbol{s}_{n+1} = \boldsymbol{F}(\boldsymbol{s}_n)$, which reduces to an unknown scalar function $s_{n+1} = f(\boldsymbol{s}_n)$, since the other components of $\boldsymbol{s}_{n+1}$ are just copied from $\boldsymbol{s}_n$.

In a practical analysis, one represents the time-series data in embedding spaces with increasing dimension m and searches for signatures of determinism. One model free approach for this is the estimation of the fractal dimension of the set of delay vectors in the embedding space, which should saturate at the value D_f for $m > D_f$. Of interest is predictability: one can try to extract the unknown function $f(\boldsymbol{s}_n)$ from the time series, by selecting a suitable model for f and then solving the least-squares problem $\sum(s_{n+1} - f(\boldsymbol{s}_n))^2 = \min$, where the minimization is done with respect to parameters in f. In the simplest case, f is approximated by a constant in a neighborhood of the actual observation $\boldsymbol{s}_n$. Then, the predictor is the average over the future values of all neighboring vectors

s_k of s_n, $\hat{s}_{n+1} = \langle s_{k+1} \rangle$, where $||s_k - s_n|| < \varepsilon$ with some suitable norm and some suitably small ε (also called the "Lorenz method of analogues"). Regardless of how the function f is represented, the relation $\hat{s}_{n+1} = f(s_n)$ will yield only good predictions for s_{n+1} if the embedding dimension m is large enough for the delay vectors to be equivalent to the unknown phase space vectors of the underlying dynamical system. In such an embedding space, the delay vectors form a finite sample according to the underlying invariant measure of the dynamical system, so that, in principle, all characteristics of the latter (Lyapunov exponents, entropies) can be determined (*See* **Chaotic dynamics**).

Such methods have been successfully applied to many physical laboratory experiments, but more recently successful applications to real-world data have been reported, such as noise reduction for human voice, wind speed prediction for wind farms, diagnostics on human heart rate data, epilepsy prediction, and machine wear detection in technical production systems. The extension of the theory to noise-driven nonlinear systems and to nonstationary data has recently been tackled, but highly nonrecurrent data such as economics data will probably remain untractable.

HOLGER KANTZ

See also **Attractors; Chaotic dynamics; Dimensions; Fractals; Spectral analysis**

Further Reading

Abarbanel, H.D.I. 1996. *Analysis of Observed Chaotic Data*, New York: Springer

Box, G.E.P. & Jenkins, G.M. 1976. *Time Series Analysis*, San Francisco: Holden-Day

Farmer, J.D. & Sidorowich, J.J. 1987. Predicting chaotic time series. *Physical Review Letters*, 59: 845

Kantz, H. & Schreiber, T. 1997. *Nonlinear Time Series Analysis*, Cambridge and New York: Cambridge University Press

Lorenz, E.N. 1969. Atmospheric predictability as revealed by naturally occuring analogues. *Journal of Atmospheric Sciences*, 26: 636

Takens, F. 1981. Detecting strange attractors in turbulence. In *Dynamical Systems and Turbulence*, edited by, D.A. Rand & L.S. Young Berlin and New York: Springer, p. 366

TODA LATTICE

The Hamiltonian system governing the longitudinal oscillations of a chain of unit point masses connected by identical springs is

$$\begin{aligned} \dot{q}_n &= \dot{p}_n, \\ \dot{p}_n &= V'(q_{n+1} - q_n) - V'(q_n - q_{n-1}), \end{aligned} \tag{1}$$

where $-V'$ is the restoring force of the springs and q_n is the displacement of the nth mass from equilibrium. Morikazu Toda (Toda, 1981) introduced the potential (scaled here so that two physical parameters disappear)

$$V(r) = e^{-r} + r - 1, \tag{2}$$

and discovered explicit solutions of (1). He obtained periodic traveling waves, in terms of elliptic functions, and two-soliton solutions in terms of sech^2. (The term $(r-1)$ makes $V(r) \sim r^2/2$ for small r; it cancels out in the problems considered below.)

This mass-spring chain is the Toda lattice. It is valuable as a solvable model of one-dimensional lattice dynamics; it is frequently used to illustrate fundamental constructions common to many integrable systems; and it appears, sometimes in surprising ways, in a variety of physical and mathematical problems.

The Infinite and Periodic Lattices

All the special properties of the Toda lattice flow from a Lax representation, $\dot{\mathsf{L}} = [\mathsf{B}, \mathsf{L}]$, which implies that eigenvalues and certain other spectral data of L are independent of time. For many purposes, it is convenient to write L and B in new coordinates. With the definition

$$a_n = \tfrac{1}{2} \mathrm{e}^{-(q_{n+1}-q_n)/2}, \quad b_n = \tfrac{1}{2} p_n, \tag{3}$$

the exponential nonlinearity in (2) becomes polynomial:

$$\dot{a}_n = a_n(b_n - b_{n+1}), \quad \dot{b}_n = 2\,(a_{n-1}^2 - a_n^2). \tag{4}$$

The standard Poisson bracket on q_n, p_n induces a (nonstandard) Poisson bracket on a_n, b_n, in which the only nonzero relations are

$$\{a_n, b_n\} = -\frac{1}{4} a_n, \quad \{a_n, b_{n+1}\} = \frac{1}{4} a_n. \tag{5}$$

The Lax operators are discretized Schrödinger operators with potentials a_n, b_n, acting on sequences $\mathbf{y} = (\ldots, y_{-1}, y_0, y_1, \ldots)$:

$$\begin{aligned} (\mathsf{L}\mathbf{y})_n &= a_{n-1} y_{n-1} + b_n y_n + a_n y_{n+1}, \\ (\mathsf{B}\mathbf{y})_n &= a_{n-1} y_{n-1} - a_n y_{n+1}. \end{aligned} \tag{6}$$

A (finite or infinite) tridiagonal symmetric matrix, such as L, is called a "Jacobi matrix." In terms of the shift operator $\Delta z_n = z_{n+1}$, which is a convenient abbreviation,

$$\mathsf{L} = \Delta^{-1} a + b + a\Delta, \tag{7}$$

where a and b are the diagonal matrices diag(a_n), diag(b_n). Constants of motion derived from the Jacobi operator L will be in involution, that is, they have zero Poisson bracket.

Thanks to its Lax representation, methods used in the study of other soliton equations also apply to the Toda lattice.

The Bi-infinite Lattice with Decaying Initial Condition (Toda, 1981)

According to (3), $(q_{n+1} - q_n) \to 0$ and $p_n \to 0$ translates to $a_n \to \frac{1}{2}, b_n \to 0$. The initial value problem is studied by a discrete version of the inverse scattering method. The unperturbed eigenvalue operator at $n = \pm\infty$ is $\mathsf{L}_0 = \frac{1}{2}(\Delta^{-1} + \Delta)$, which has the bounded "plane wave" eigenfunctions $y_n = z^{\pm n}$, $|z| = 1$, for $\lambda = (z + z^{-1})/2 \in [-1, 1]$. This relation between "energy" λ and "momentum" z is the analog of $\lambda = k^2$ in the Schrödinger equation. The scattering matrix is a function of momentum and is defined on $|z| = 1$. Bound states may occur for $\lambda < -1$ and $\lambda > 1$; these correspond to solitons moving to the left and to the right, respectively. In the absence of solitons (and with a certain condition on the reflection coefficient), the large-time behavior consists of a decaying wave train for $|n/t| < 1$, vanishingly small motion when $|n/t| > 1$, and a connecting regime described by a Painlevé transcendent (Kamvissis, 1993).

Constants of motion of the infinite lattice can be computed explicitly as $I_k := \mathrm{Trace}\frac{1}{k}(\mathsf{L}^k - \mathsf{L}_0^k)$ (L_0^k removes a divergent term). The Hamiltonian equations generated by the I_k are found by use of the Poisson bracket (5).

The Periodic Lattice (Toda, 1981)

The potential in the Lax operator is periodic, say, period N. As in the theory of Hill's equation, one introduces a Floquet multiplier and Floquet solutions by imposing the condition

$$y_{N+n} = \rho y_n. \tag{8}$$

L and B then reduce to matrices:

$$\mathsf{L}(\rho) = \begin{pmatrix} b_1 & a_1 & 0 & \cdots & \rho^{-1} a_N \\ a_1 & b_2 & a_2 & \cdots & 0 \\ 0 & a_2 & \ddots & \ddots & \vdots \\ \vdots & & \ddots & \ddots & a_{N-1} \\ \rho\, a_N & \cdots & \cdots & a_{N-1} & b_N \end{pmatrix}, \tag{9}$$

$$\mathsf{B}(\rho) = \begin{pmatrix} 0 & -a_1 & 0 & \cdots & \rho^{-1} a_N \\ a_1 & 0 & -a_2 & \cdots & 0 \\ 0 & a_2 & \ddots & \ddots & \vdots \\ \vdots & & \ddots & \ddots & -a_{N-1} \\ -\rho\, a_N & \cdots & \cdots & a_{N-1} & 0 \end{pmatrix}. \tag{10}$$

The parameter ρ cancels from the Lax equation. The characteristic polynomial $\det(\mathsf{L}(\rho) - \lambda\mathbb{I})$ is a rational function whose coefficients are independent of time. The function $\rho(\lambda)$ obtained by solving the characteristic equation for ρ is defined on the λ–plane with branch cuts, and determines a compact Riemann surface. For Toda's traveling wave, there are only two branch cuts; this fact is responsible for the occurrence of elliptic functions in his solution formula.

Every energy surface is compact, so that the solution curves lie on tori; abstractly, these are the (real) Jacobian varieties associated with the Riemann surface. Action variables can be written explicitly as loop integrals. The solution of the periodic Toda lattice is expressed in terms of theta functions.

The Free Toda Lattice and Lie Theory

Many features common to integrable systems are illustrated by an example due to Jürgen Moser.

The Free Toda Lattice (Toda, 1981)

Particles number 0 and $(N+1)$ are pulled to $-\infty$ and $+\infty$, respectively. (3) implies that $a_0 = a_N = 0$, and the corner entries in $\mathsf{L}(\rho)$ and $\mathsf{B}(\rho)$ disappear. The resulting matrices will be written simply as L, B.

Springs governed by potential (2) resist compression and are encouraged to expand. Therefore, there will now be no confining force on the remaining springs. The masses move apart as $t \to \pm\infty$, and behave asymptotically like free particles. As $t \to \pm\infty$, $(q_{n+1} - q_n) \to \infty$, whence $a_n \to 0$, the matrix $\mathsf{L}(t)$ becomes diagonal, and the entries $b_n = p_n/2$ approach the eigenvalues of $\mathsf{L}(0)$. Thus, the eigenvalues are the momenta of the free particles. As $t \to -\infty$, these are arranged with the slowest particle farthest to the right; after interaction and a phase shift, the particles emerge as $t \to +\infty$ with the fastest to the right. This is soliton interaction reduced to bare essentials. The diagonalization of $\mathsf{L}(t)$ as $t \to +\infty$, with the eigenvalues (momenta) ordered, is called the "sorting property," and points to a connection between isospectral flows and numerical linear algebra.

For tridiagonal L, the number of eigenvalues is exactly half the number of entries, and they suffice for integrability. Remarkably, the Lax equation $\dot{\mathsf{L}} = [\mathsf{B}, \mathsf{L}]$ (with appropriate B), is completely integrable even when L is a generic symmetric matrix. The additional constants of motion required are constructed in Deift et al. (1986). The sorting property again holds; it is related to the QR diagonalization algorithm in numerical linear algebra.

The free Toda lattice is perhaps the simplest integrable system with a clear Lie-algebraic generalization. If one sets $\mathrm{Trace}\,\mathsf{L} = 0$ in (9), that is, total momentum $= 0$, then L belongs to $\mathfrak{sl}_N$, the Lie algebra of $N \times N$ matrices with trace zero. Let $\{h_i\}$ be a basis for the diagonal matrices in $\mathfrak{sl}_N$. The matrices $e_{\pm i}$ with a single 1 in the $(i, i+1)$, resp, $(i+1, i)$ entry, and zeros elsewhere, are called "raising and lowering operators." There are analogs of $e_{\pm i}$, h_i in the class of split real semisimple Lie algebras, such as the algebra of symplectic matrices. The Lax pair, written as

$$\mathsf{L} = \sum_{i=1}^{N-1} [\beta_i h_i + a_i(e_i + e_{-i})],$$

$$\mathsf{B} = \sum_{i=1}^{N-1} a_i(-e_i + e_{-i}),$$

makes sense in those algebras. In the corresponding generalized mass–spring systems, some particles will be governed by modifications of (2) (Olshanetsky & Perelomov, 1994; Guest, 1997; Reyman & Semenov-Tyan-Shansky, 1994; Semenov-Tyan-Shansky, 1994 a,b).

The extension to Lie algebras is not just generalization for sake of generalization. It illuminates concrete matrix calculations. There is an abstractly defined Poisson bracket, the Kostant–Kirillov bracket, of which (5) is a special case. Hamiltonian equations with respect to this Poisson bracket always have the Lax form, $\dot{\mathsf{X}} = [\mathsf{Y}, \mathsf{X}]$. One of the fundamental properties of Lax equations, the representation of the solution by means of a factorization in the Lie group, is a general Lie-theoretic phenomenon that specializes to many different integrable systems and, in particular, yields involutivity of the constants of motion $\mathrm{Trace}\mathsf{L}^k$ (Guest, 1997; Semenov-Tyan-Shansky, 1994 a,b)

For the free Toda lattice, the factorization method is simple linear algebra, but it already conveys the basic idea.

Here, as well as later, it is convenient to redefine

$$a_n = \exp(q_n - q_{n+1}), b_n = -p_n. \qquad (11)$$

In terms of the shift operator (7), L and B then become

$$\mathsf{L} = \Delta^{-1}a + b + \Delta, \quad \mathsf{B} = -\Delta^{-1}a. \qquad (12)$$

($\mathsf{L} = b + \Delta$ or $\mathsf{B} = b + \Delta$ will give the same $\dot{\mathsf{L}}$.) Given the initial value $\mathsf{L}(0)$, write

$$\exp(t\mathsf{L}(0)) = n_-(t)^{-1}d(t)n_+(t) := n_-(t)^{-1}b_+(t) \qquad (13)$$

with $n_{\pm(t)}$ upper/lower triangular having 1's on the diagonal and $d(t)$ diagonal; this is just Gaussian elimination. Then

$$\mathsf{L}(t) = b(t)\mathsf{L}(0)b(t)^{-1} \text{ satisfies } \dot{\mathsf{L}}(t) = [B(t), \mathsf{L}(t)], \qquad (14)$$

where $\mathsf{B}(t) = \mathsf{L}(t)_+$ is the upper triangular part of $\mathsf{L}(t)$.

Similarly, one finds the solutions of the Hamiltonian equations generated by the constants of motion I_{k+1} (referred to as "higher Toda flows"). One introduces a separate time variable t_k for each of these systems and now factors $\exp(t_k\mathsf{L}(0)^k)$. The resulting $\mathsf{L}(t_k)$ satisfies

$$\frac{\partial \mathsf{L}}{\partial t_k} = [\mathsf{B}_k, \mathsf{L}], \quad \text{with } \mathsf{B}_k = (\mathsf{L}^k)_+. \qquad (15)$$

(L^k arises because $\nabla I_{k+1} = \mathsf{L}^k$). The simultaneous solution of (15) is a multi-parameter flow, $\mathsf{L}(t_1, \ldots, t_{N-1})$, on phase space ($t_1$ is the original t). Compatibility of these equations, that is, equality of mixed partial derivatives of L, is equivalent to the involutivity of the I_k. For an infinite lattice, there are infinitely many equations (15), and infinitely many time variables $\mathbf{t} = (t_1, t_2, \ldots)$. Many soliton hierarchies have the same general form; for the higher KdV equations, for example, $\mathsf{L} = \mathrm{d}^2/\mathrm{d}x^2 + u(x)$ and $\mathsf{B}_k = (\mathsf{L}^{(2k+1)/2})_+$.

In Gaussian elimination, as in (13), d_{nn} is known to be τ_n/τ_{n-1}, where τ_n is the upper left $n \times n$ minor determinant of $\exp[t\mathsf{L}(0)]$. Tracing through the factorization steps, one finds the Hirota formula

$$a_n(t) = a_n(0)\frac{\tau_{n+1}(t)\tau_{n-1}(t)}{\tau_n(t)^2}, \qquad (16)$$

which implies

$$a_n = \frac{\mathrm{d}^2}{\mathrm{d}t^2}\ln \tau_n, \quad b_n = \frac{\mathrm{d}}{\mathrm{d}t}\ln\frac{\tau_n}{\tau_{n-1}}.$$

These τ_n are prototypes of a fundamental object in the theory of soliton hierarchies, the "τ-function," which elsewhere occurs in far more complex settings.

The τ-functions have a representation-theoretic meaning. They are the matrix elements

$$\tau_n(t) = \langle \exp(t\mathsf{L}(0))\,\mathbf{e}_1\wedge\mathbf{e}_2\wedge\cdots\wedge\mathbf{e}_n \mid \mathbf{e}_1\wedge\mathbf{e}_2\wedge\cdots\wedge\mathbf{e}_n\rangle \qquad (17)$$

in the representation of the Lie group SL_N on the nth exterior power of $\mathbb{R}^N$ (totally antisymmetric covariant tensors of order n). This formula gives precisely the $n \times n$ minor determinant of $\exp(t(0))$. The skew-vector $\mathbf{e}_1 \wedge \ldots \wedge \mathbf{e}_n$ is the highest weight vector of the representation; it is annihilated by all raising operators. One may think of it as a vacuum vector and of (17) as a vacuum expectation value.

Loop Algebras and Affine Lie Algebras

The lower/upper factorization (13), solution (14), the τ-functions arising from the diagonal part in this factorization, and their interpretation in terms of group representations are fundamental features of integrable systems. The free Toda lattice affords a transparent illustration; more sophisticated versions of these ideas, set in other Lie algebras and groups, apply to a wide variety of equations. "Loop algebras" and "affine Lie algebras" are particularly useful in the theory of the periodic Toda lattice.

The loop algebra, $\widetilde{\mathfrak{sl}_N}$, is an infinite-dimensional Lie algebra whose elements are trace zero matrices $\mathsf{X}(\rho)$ with entries that are polynomials in ρ and ρ^{-1}. Such $\mathsf{X}(\rho)$ are called "loops," because the mapping $\mathsf{X} : \{|\rho| = 1\} \mapsto \mathfrak{sl}_N$ gives a closed curve, that is, "loop," of matrices. The Lie bracket is still the matrix commutator.

The Lax operators (9) and (10), for the periodic Toda lattice, belong to $\widetilde{\mathfrak{sl}}_N$. To solve the periodic Toda lattice by factorization, one prescribes the initial value $\mathsf{A}(\rho)$ of $\mathsf{L}(\rho)$ and seeks matrix functions $g_{\pm}(t,\rho)$ for which

$$\exp(t\mathsf{A}(\rho)) = g_{-}(t,\rho)^{-1} g_{+}(t,\rho). \tag{18}$$

These $g_{\pm}(\cdot,\rho)$ are required to be analytic inside (resp. outside) the circle $|\rho|=1$; this is the analog of lower/upper in (13). The eigenvector $v(\lambda,\rho(\lambda))$ of $\mathsf{A}(\rho)$, which is a function on the Riemann surface $\det(\mathsf{A}(\rho)-\lambda\mathbb{I})=0$, determines $g_{\pm}(t,\rho)$. The former is expressed in terms of theta functions; hence, so are $g_{\pm}$ and also $\mathsf{L}=g_{+}\mathsf{A}g_{+}^{-1}$ (Reyman & Semenov-Tyan-Shansky, 1994).

To obtain the τ-functions from a Lie algebra representation, one must introduce an extension of the loop algebra, the "affine" Lie algebra $\widehat{\mathfrak{sl}}_N$. (Affine Lie algebras give the name to affine Toda field theory, described below.)

The affine algebra has one extra element $\hbar$, which, in some respects, acts like an identity matrix. It has zero bracket with all loops; however, the ordinary matrix commutator in $\widetilde{\mathfrak{sl}}_N$ is modified so that $\hbar$ can arise as a Lie bracket of loops. This would not be possible if $\hbar$ were truly the identity, since a matrix commutator $[\mathsf{X}(\rho),\mathsf{Y}(\rho)]$ must have trace zero. The extended algebra contains loops that behave as annihilation and creation operators: they satisfy $[a_k^*,a_k]=\hbar$. Thanks to this modification, the familiar realization of the Heisenberg commutation relations by multiplication and differentiation operators becomes available in the study of representations of $\widehat{\mathfrak{sl}}_N$.

The τ-functions for the periodic Toda lattice are obtained as vacuum expectation values, analogous to (17), in certain representations of the affine Lie algebra; they turn out to be theta functions and again satisfy the Hirota equations (16). For the free Toda lattice, τ-functions also arose from the diagonal factor in $n_{-}dn_{+}$. There is a similar factorization in the affine group $\widehat{SL}_N$. This group and the diagonal factor in the lower/upper factorization are rather complicated objects. In particular, the determinant of d will be an infinite determinant.

The matrices e_0 (resp. f_0) with the corner elements ρe_{N1}, resp. $\rho^{-1}e_{1N}$ in (9), (10) have an intrinsic meaning (they are raising resp. lowering operators in $\widetilde{\mathfrak{sl}}_N$). Therefore, the periodic Toda lattice will generalize to the algebra of loops with values in a semisimple Lie algebra.

An affine Lie algebra also provides the setting for a remarkable unification of two of the most important soliton systems, the Toda lattice and the Ablowitz–Kaup–Newell–Segur (AKNS) equations. From the solution $a_n(\mathbf{t}), b_n(\mathbf{t})$ of the Lax equations (15) for the infinite lattice, one can build functions $q_n(\mathbf{t}), r_n(\mathbf{t})$, which, for every n, satisfy the general AKNS hierarchy, with t_1 playing the role of x. For example, the simultaneous solution $a_n(t_1,t_3), b_n(t_1,t_3)$ of the Toda equations and the higher Toda flow generated by Trace L^4 will be transformed into solutions $q_n(x,t_3), r_n(x,t_3)$ of the modified KdV (MKdV) equation. A kind of Bäcklund transformation (precisely, a Schlesinger transformation) connects q_n, r_n and $q_{n\pm1}, p_{n\pm1}$. As a special case, solutions of the free Toda lattice correspond to soliton-like potentials for AKNS.

This result can be obtained by operations on formal series, but the Lie-theoretic explanation is more illuminating (Bergvelt & ten Kroode, 1988). The 2×2 matrix in the AKNS scattering problem depends on a (spectral) parameter and so belongs to the loop algebra $\widetilde{\mathfrak{sl}}_2$. There is a representation of the affine extension $\widehat{\mathfrak{sl}}_2$ in which the Heisenberg subalgebra, mentioned above, has infinitely many vacuum vectors killed by the a_k^*. The vacuum expectation values are the Toda τ-functions τ_n, and the AKNS variables are determined by $q_n = -\tau_{n+1}/\tau_n$, $r_n=\tau_{n-1}/\tau_n$.

The Two-dimensional Toda Lattice

The Toda-AKNS family has a sweeping generalization related to the Kadomtsev–Petviashvili equation (KP). It consists of four Lax equations which involve two infinite families of time variables, t_k, x_k. A continuum limit of this system was encountered in the study of deformations of a two-dimensional oil–water interface (Hele-Shaw flow) and has revealed an integrable structure of conformal mappings. This is sketched in the next section.

The operator L in (12) is replaced by a formal series, and a second operator M is introduced:

$$\mathsf{L} = \Delta + u_0 + u_1\Delta^{-1} + u_2\Delta^{-2} + \cdots, \tag{19}$$
$$\mathsf{M} = \Delta^{-1} + v_0 + \Delta v_1 + \Delta v_2 + \cdots. \tag{20}$$

The u_k, v_k are infinite diagonal matrices. Two of the Lax equations are

$$\frac{\partial \mathsf{L}}{\partial t_k} = [\mathsf{B}_k,\mathsf{L}],\ \mathsf{B}_k = (\mathsf{L}^k)_+,$$
$$\frac{\partial \mathsf{L}}{\partial x_k} = [\mathsf{C}_k,\mathsf{L}],\ \mathsf{C}_k = (\mathsf{M}^k)_-. \tag{21}$$

The subscripts $\pm$ denote projections on the positive/negative powers of Δ (analog of upper/lower triangular). In the other two equations, L is replaced by M (Ueno & Takasaki, 1984).

Equations (21) specialize to the usual Toda lattice, the equations $\dot{\mathsf{L}}=[\mathsf{B},\mathsf{L}]$ for banded matrices L, and an extension of the Toda lattice in which the q_n depend on two variables. This is the "two-dimensional Toda lattice" (2DTL).

The 2DTL is obtained as follows. The compatibility conditions of the two systems in (21) are "zero curvature equations." With the abbreviations $x_1=x$, $t_1=t$, the first of these is

$$\frac{\partial \mathsf{B}}{\partial x} - \frac{\partial \mathsf{C}}{\partial t} + [\mathsf{B}, \mathsf{C}] = 0. \qquad (22)$$

Following the one-dimensional case (12), let $\mathsf{B} = -\Delta^{-1}a$ and $\mathsf{C} = b + \Delta$. a_n, b_n, still given by (11), are now functions of t and x. (22) becomes

$$(a_n)_x = a_n(b_{n+1} - b_n), \ (b_n)_t = a_n - a_{n-1}$$

or

$$(q_n)_{xt} = \exp(q_{n-1} - q_n) - \exp(q_n - q_{n+1}). \quad (23)$$

A standard change of variables converts $(q_n)_{xt}$ to the wave operator $(q_n)_{\xi\xi} - (q_n)_{\tau\tau}$.

The free and periodic boundary conditions on the 2DTL are of primary interest. The free boundary condition for 2×2 matrices B, C yields the Liouville equation

$$q_{xt} + \exp q = 0. \qquad (24)$$

Under the periodic boundary condition $q_0 = q_2, q_1 = q_3$, system (23) becomes the sinh-Gordon equation for $\Theta = q_2 - q_1$,

$$\Theta_{xt} = -4 \sinh \Theta,$$

or, if q_1, q_2 are taken to be imaginary, the sine-Gordon equation

$$\Theta_{xt} = -4 \sin \Theta.$$

Since the sine-Gordon equation is solved by the inverse scattering method, it is natural to introduce a spectral parameter ζ and an eigenvalue problem for the N-component free 2DTL:

$$\Psi_t = \mathsf{B}(\zeta)\Psi, \quad \Psi_x = \mathsf{C}(\zeta)\Psi. \qquad (25)$$

Sine-Gordon theory, in which ζ^{-1} appears, further suggests that for the free 2DTL

$$\mathsf{B}(\zeta) = -\zeta^{-1}\Delta^{-1}a, \quad \mathsf{C}(\zeta) = b + \zeta\Delta. \qquad (26)$$

Compatibility of the two equations in (25) implies (22). The free 2DTL is referred to as "Toda field theory."

Periodic boundary conditions are handled by an adaptation of (26). ζ is put in the lower left corner of $\mathsf{C}(\zeta)$ and $-\zeta^{-1}a_N$ in the upper right corner of $\mathsf{B}(\zeta)$. The inverse scattering method can be applied to (25). Indeed, under two-periodic boundary condition, (25) is precisely the AKNS system used to solve the sine-Gordon equation.

The periodic 2DTL is called "affine Toda field theory" (ATFT), because, as is the case for the one-dimensional periodic lattice, it is set in the context of loop and affine Lie algebras and groups. The structure of ATFT, however, is richer because of the new x–dependence.

In complex coordinates $t = w, x = \overline{w}$, ATFT is a system of elliptic partial differential equations which is encountered in the theory of harmonic maps. Solutions of the sinh-Gordon equation define Riemannian metrics on surfaces $z = f(x, y)$, $(x, y) \in U$, of constant mean curvature 2. The Gauss map, which sends the unit normal vector to the unit sphere S^2, is a harmonic map, meaning that it is a critical point of the "energy" $\frac{1}{2}\int_U \|\nabla\phi\|^2 \,\mathrm{d}x\,\mathrm{d}y$ of maps $\phi : U \to S^2$. (For real–valued maps, a critical ϕ is an ordinary harmonic function.) Solutions of the multi-component elliptic ATFT define harmonic maps into other symmetric spaces. A certain group of loops is the infinite–dimensional symmetry group of this 2DTL: given a solution of 2DTL and a loop, one can construct a new solution via a factorization problem (Guest, 1997; Fordy & Woods, 1994).

In wave equation form, $(q_n)_{\xi\xi} - (q_n)_{\tau\tau}$, the 2DTL equations are hyperbolic; they describe fields on two-dimensional Minkowski space. As a first step toward a quantum theory of these fields, one can ask whether they are conformally invariant. Conformal transformations scale the (indefinite) metric. In light cone coordinates $x_\pm = \xi \pm \tau$, they are given by $(x_+, x_-) \mapsto (f_+(x_+), f_-(x_-)) = (\bar{x}_+, \bar{x}_-)$. Then $(q_n)_{x_+x_-}$, the left side of (23), becomes

$$\frac{\partial x_+}{\partial \bar{x}_+}\frac{\partial x_-}{\partial \bar{x}_-}(q_n)_{\bar{x}_+\bar{x}_-}, \text{ abbreviated } \quad J \cdot (q_n)_{\bar{x}_+\bar{x}_-}. \qquad (27)$$

The 2DTL equations will be conformally invariant under a transformation of the fields, $q_n(x_\pm) \mapsto \bar{q}_n(\bar{x}_\pm)$, for which the right-hand side of (23) is also multiplied by J.

The Liouville equation (24) is conformally invariant if q transforms according to $q \mapsto \bar{q} + \ln J$. The sinh-Gordon equation, whose right side is $-2(\exp(\Theta) + \exp(-\Theta))$, admits no such transformation of Θ and is not conformally invariant. A particle described by sinh-Gordon theory has "mass": if $\sinh\Theta$ is linearized about the vaccuum state $\Theta = 0$, one obtains $\Theta_{x_+x_-} = -4\Theta$; the mass is $\sqrt{4}$. The sinh-Gordon equation is a perturbation of the conformal Liouville field equation. For example, under the change of variables $q \mapsto \Theta + \ln\varepsilon$, the equation

$$q_{x_+x_-} = -\mathrm{e}^q + \varepsilon^2\mathrm{e}^{-q},$$

becomes the sinh-Gordon equation with mass $\sqrt{2\varepsilon}$. As $\varepsilon \to 0$, which recovers the Liouville equation, the mass tends to zero. Similarly, Toda field theory is conformally invariant, while affine Toda field theory is not. Affine Toda field theory is massive in the sense described. The actual fields of interest are linear combinations of the q_n that are suitable for generalization to affine Lie algebras. This generalization is important, because the quantizations of ATFTs associated to different types of affine Lie algebras can have very different properties (Corrigan, 1999).

Continuum Limits

There are two natural continuum limits of the Toda mass–spring chain. Keeping both nonlinearity and dispersion to first order, one gets the KdV equation. The "zero dispersion limit" results in a hyperbolic system, called the "dispersionless Toda lattice."

The approximation leading to the KdV equation is taken in a right-moving coordinate system, so that the left-moving solitons of the Toda lattice disappear. The difference operator (6) becomes the Schrödinger operator, and the discrete Gel'fand–Levitan–Marchenko inverse theory limits to the inverse scattering formalism for KdV (Toda, 1981).

The zero dispersion limit retains only the quadratic nonlinearity. In the Toda equations $\dot{a}_n = a_n(b_n - b_{n+1})$, $\dot{b}_n = 2\,(a_{n-1}^2 - a_n^2)$, take $q = n\varepsilon$ and $T = t\varepsilon$. These are slow scales. With $b_n(T) \sim b(q,T)$, the difference $(b_n - b_{n+1})$ is approximately $-\varepsilon\, b_q$. The inconvenient minus signs are removed by redefining a_n, b_n. One then finds the DTL equations

$$a_T = ab_q, \quad b_T = 2(a^2)_q. \tag{28}$$

There are again Lax equations. In the Toda eigenvalue problem (6), write $\exp(\pm\varepsilon\partial/\partial n)$ for the shift by $\pm\varepsilon$, and assume a WKB ansatz for $y_n(T) \sim y(q,T)$,

$$y(q,T) = \exp[\varepsilon^{-1}S(q,T)].$$

In the WKB approximation to the Schrödinger wave function, S_q is the inverse wavelength, which is proportional to momentum, by the deBroglie relation. Set $p = S_q$. Thus, p and q are canonically conjugate variables. As $\varepsilon \to 0$, the eigenvalue problem $\mathsf{L}\mathbf{y} = \lambda\mathbf{y}$ reduces to an eikonal equation, and $\mathbf{y}_t = \mathsf{B}\mathbf{y}$ becomes the time-evolution of the momentum S_q:

$$\begin{aligned} \mathfrak{L}(p,q) :&= a\,(\mathrm{e}^p + \mathrm{e}^{-p}) + b = \lambda, \\ p_T &= \frac{\partial}{\partial q}[a(\mathrm{e}^p - \mathrm{e}^{-p})] := \frac{\partial}{\partial q}\mathfrak{B}(p,q). \end{aligned} \tag{29}$$

The commutator $[\mathsf{L},\mathsf{B}]$ is replaced by the Poisson bracket with respect to p and q, and then

$$\dot{\mathcal{L}} = \{\mathcal{B}, \mathcal{L}\}$$

yields Equations (28). Their solutions may develop shocks, but as long as they are smooth, the dispersionless limits of the Toda lattice constants of motion, I_k, are constants of motion, say H_k, for (28). For example, $\sum(2a_n^2 + b_n^2) \to H_2(a,b) = \int(2a^2 + b^2)\,\mathrm{d}q$.

For smooth solutions, the eigenvalue sorting property of the free Toda lattice remains valid. The "free" boundary conditions are $a(q,T) = 0$ for $q = 0, q = 1$. Let $a_0(q), b_0(q)$ be the initial values. As $T \to \infty$, $a(q,T) \to 0$, while $b(q,T)$ tends to a decreasing function $b^*(q)$. The conserved quantities H_k have the same values for b^* as for a_0, b_0, for example, $H_2(a_0, b_0) = H_2(0, b^*)$. In this sense, the initial "matrix" a_0, b_0 becomes diagonal, and the "eigenvalues" are sorted (Brockett & Bloch, 1990).

The two-dimensional dispersionless Toda lattice hierarchy arises in an idealized model of viscous fingering in a Hele-Shaw cell. This sketch follows Kostov et al. 2001 and references therein. Two plates confine water (zero viscosity) and oil (viscous) in the complex z-plane. The water occupies a bounded region D_+, which is surrounded by oil in the exterior domain D_-. There is a source of water at $z = 0$ and a compensating sink for the oil at $z = \infty$. The object is to find the motion of the interface $\Gamma(t)$. Under some simplifying assumptions, the time development of Γ is governed by the "Laplacean growth equation" (LGE), so called because the velocity potential satisfies $\Delta\phi = 0$.

The shape of Γ is determined by the harmonic moments

$$\begin{aligned} t_k &= -\frac{1}{k\pi}\int_{D_-} z^{-k}\,\mathrm{d}x\mathrm{d}y, \; k \geq 1, \\ t_0 &= \frac{1}{\pi} \times \text{area of } D_+. \end{aligned} \tag{30}$$

It is known that the t_k are constant under the LGE, while the area t_0 changes linearly. One, therefore, takes t_0 as time variable. Finding $\Gamma(t_0)$ amounts to finding the time-dependent conformal map $\mathfrak{L}(t_0)$ from $\{w \mid |w| > 1\}$ to $\{z \mid z \in D_-\}$. The dispersionless limit of the two-dimensional Toda hierarchy (21) enters when one allows the moments $\boldsymbol{t} = (t_1, \ldots)$ to vary and considers a family $\mathfrak{L}(t_0, \boldsymbol{t})$ of conformal maps; the LGE is then recovered as a constraint on this family.

The time-like variables t_k, x_k in (21) are taken to be the moments t_k and their conjugates $\bar{t}_k$. The area t_0 plays the role of the spatial coordinate q in (28); also set $w = \exp p$. The conformal map $\mathfrak{L}(w, \boldsymbol{t})$ and its conjugate $\bar{\mathfrak{L}}(w^{-1}, \bar{\boldsymbol{t}})$ are the dispersionless limits of L and M. The expansion of L in powers of Δ^{-1} becomes the Laurent expansion $\mathfrak{L}(w) = \text{const} \cdot w + u_0 + u_1 w^{-1} + \ldots$. The dependence of $\mathfrak{L}$, $\bar{\mathfrak{L}}$ on the deformation parameters $\boldsymbol{t}, \bar{\boldsymbol{t}}$ is given by

$$\frac{\partial\mathfrak{L}}{\partial t_k} = \{\mathfrak{B}_k, \mathfrak{L}\}, \quad \frac{\partial\mathfrak{L}}{\partial \bar{t}_k} = -\{\bar{\mathfrak{B}}_k, \mathfrak{L}\}, \tag{31}$$

plus similar equations for $\bar{\mathfrak{L}}$. As in (21), $\mathfrak{B}_k$ is $(\mathfrak{L}^k)_+$. The Poisson bracket is still $\{p,q\} = 1$ or in the new notation, $\{\ln w, t_0\} = 1$. The Laplacean growth equation can be written in the form

$$\{\mathfrak{L}(w, t_0), \bar{\mathfrak{L}}(w^{-1}, t_0)\} = 1,$$

with all $t_k, \bar{t}_k$ fixed. The constraint $\{\cdot,\cdot\} = 1$ is known as the "dispersionless string equation."

Other Topics

The topics chosen give only a hint of the importance of the Toda lattice and its generalizations. Other aspects and applications that deserve an expanded description include the following.

The Toda mass-spring chain can model dispersive lattice shocks. Two unstretched halves of the lattice move toward each other at constant speed $2c$. In the scattering problem, the boundary condition is not $b_{|n|} \to 0$, but $b_n \to \mp c$ for $n \gtrless 0$. The spectrum changes with c, and this is reflected in the shockwave behavior, which is analyzed by means of the powerful steepest descent method for Riemann–Hilbert problems (Deift et al., 1995).

The quantized Toda lattice is solvable; the eigenstates of the multi-particle Toda Hamiltonian are matrix elements of infinite-dimensional group representations. The construction generalizes the one-dimensional case, $-\mathrm{d}^2/\mathrm{d}q^2 + \exp(-2q)$, whose eigenfunctions are Whittaker functions (Semenov-Tyan-Shansky, 1994a).

Orthogonal polynomials satisfy a three-term recurrence relation such as (6). For this reason, the Toda lattice arises in random matrix theory. A probability measure

$$\mathrm{d}\mu_N(H) = Z_N^{-1} \mathrm{e}^{-N \mathrm{Trace}\, V(H)}\, \mathrm{d}H$$

on Hermitean matrices is given. It determines a family $p_j(x)$ of orthogonal monic polynomials, $\int p_i(x) p_j(x)\, \mathrm{d}\mu_N(x) = 0$ if $i \neq j$. In this basis, the shift operator $Lp(x) = xp(x)$ acting on polynomials $p(x)$ has the Jacobi form (12). If $V(x)$ depends on parameters t_k, for example, $V(x) = x^2 + t_4 x^4$, then $\mathrm{d}\mu_N$ and the p_j depend on the t_k. The change of L in the moving basis $\{p_j\}$ is described by a Lax equation (Witten, 1991).

HERMANN FLASCHKA

See also **Hele-shaw cell; Integrable lattices; Lie algebras and lie groups; Zero-dispersion limits**

Further Reading

Bergvelt, M. & ten Kroode, F. 1988. τ-functions and zero curvature equations. *Journal of Mathematical Physics*, 29: 1308–1320

Brockett, R. & Bloch, A. 1990. Sorting with the dispersionless limit of the Toda lattice. In *Proceedings of the Conference on Hamiltonian Systems, Transformation Groups and Spectral Transform Methods*, edited by J. Harnad & J. Marsden, Montréal: Publications CRM

Corrigan, E. 1999. Recent developments in affine Toda field theory. In *Particles and Fields*, edited by G.E. Semenoff & L. Vinet, New York: Springer

Deift, P., Kriecherbauer, T. & Venakides, S. 1995. Forced lattice vibrations, I, II, *Communications in Pure and Applied Mathematics*, 48: 1187–1249, 1251–1298

Deift, P., Li, L.-C., Nanda, T. & Tomei, C., 1986. The Toda lattice on a generic orbit is integrable, *Communications in Pure and Applied Mathematics*, 39: 183–232

Fordy, A. & Woods, J.C. (editors). 1994. *Harmonic Maps and Integrable Systems*. Braunschweig/Wiesbaden: Viehweg

Guest, M.A. 1997. *Harmonic Maps, Loop Groups, and Integrable Systems*, Cambridge and New York: Cambridge University Press

Kamvissis, S. 1993. On the long time behavior of the doubly infinite Toda lattice. *Communications in Mathematical Physics*, 153: 479–519

Kostov, I.K., Krichever, I., Mineev-Weinstein, M., Wiegmann, P.B. & Zabrodin, A. 2001. τ-function for analytic curves. In *Random Matrix Models and Their Applications*, edited by P. Bleher & A. Its, Mathematical Sciences Research Institute Publications, Cambridge University Press

Olshanetsky, M.A. & Perelomov, A.M. 1994. Integrable systems and finite-dimensional Lie algebras. In *Encyclopaedia of Mathematics, Dynamical Systems VII*, edited by V.I. Arnol'd & S. P. Novikov, New York: Springer (original Russian edition 1987)

Reyman, A.G. & Semenov-Tyan-Shansky, M.A. 1994. Group theoretical methods in the theory of finite dimensional integrable systems. In *Encyclopaedia of Mathematics, Dynamical Systems VII*, edited by V.I. Arnol'd & S.P. Novikov, New York: Springer (original Russian edition 1987)

Semenov-Tyan-Shansky, M.A. 1994a. Lectures on R–matrices, Poisson–Lie groups, and integrable systems. In *Lectures on Integrable Systems, In Memory of Jean-Louis Verdier*, edited by O. Babelon, P. Cartier & Y. Kosmann-Schwarzbach, Singapore: World Scientific

Semenov-Tyan-Shansky, M.A. 1994b. Quantization of open Toda lattices. In *Encyclopaedia of Mathematics, Dynamical Systems VII*, edited by V.I. Arnol'd & S.P. Novikov, New York: Springer (original Russian edition 1987)

Takasaki, K. & Takebe, T. 1995. Integrable hierarchies and dispersionless limits. *Reviews in Modern Physics*, 7: 743–808

Toda, M. 1981. *Theory of Nonlinear Lattices*, Berlin and Heidelberg: Springer (original Japanese edition 1978)

Ueno, K. & Takasaki, K. 1984. Toda lattice hierarchy. In *Group Representations and Systems of Differential Equations*, edited by K. Okamoto, Tokyo: Kinokuniya

Witten, E. 1991. Two-dimensional gravity and intersection theory on moduli space. In *Surveys in Differential Geometry*, Bethlehem, PA; Lehigh University

TOPOLOGICAL CHARGE

See **Sine-Gordon equation**

TOPOLOGICAL CONJUGACY

See **Maps**

TOPOLOGICAL DEFECTS

A *topological defect* (or *topological soliton*) represents a spatially non-uniform configuration of an order parameter field that offers topological stability and cannot be transformed into the ground state of a system under finite deformation of a field (Mermin, 1979). The structure and properties of a topological defect (TD) depend essentially on the dimensionality of a system, its symmetry, and degeneration of the ground state. Historically, vortices in liquid were the first example of a TD to be investigated (Lugt, 1996). In a two-dimensional (2-d) incompressible liquid, the equation

$\Delta\varphi = 0$ for the velocity potential φ (velocity $\boldsymbol{v} = \nabla\varphi$) has the evident solution $\varphi = \kappa\chi$ for the vortex, where κ is an arbitrary parameter and χ is the azimuthal angle in cylindrical coordinates (ρ, χ) in the x, y plane with its origin in the vortex center.

Pitaevskii vortices in superfluidity theory (Lifshitz & Pitaevskii, 1980), *magnetic vortices* in easy-plane ferro- and anti-ferromagnets (FMs and AFMS) (Mertens & Bishop, 2000), magnetic *disclinations*, and disclinations in nematic liquid crystals (de Gennes, 1974) exemplify the similar one-dimensional TD in 3-d continuous media with continuous degeneration of the ground state. The order parameters in the above examples are of the two-component type: the complex thermodynamical wave function $\psi(\boldsymbol{r}, t) = \psi_0 \exp(\mathrm{i}\phi)$ of a Bose condensate for the superfluid liquid, which, within the approach of a weakly nonideal Bose gas, satisfies the Gross–Pitaevskii equation

$$\mathrm{i}\hbar\frac{\partial\psi}{\partial t} + \frac{\hbar^2}{2m}\Delta\psi + U\{\psi - |\psi|^2\psi\} = 0, \qquad (1)$$

and two angle variables (θ, ϕ) in polar coordinates in magnetic space (associated with the hard axis z) of a magnetization vector $\boldsymbol{M} = M_0(\sin\theta\cos\phi, \sin\theta\sin\phi, \cos\theta)$, which satisfies the Landau–Lifshitz equation (similar in structure to (1)), or the director vector $\boldsymbol{L}$ in antiferromagnets and liquid crystals. These parameters are continuously degenerate in the phase of wave function ψ or in the direction of spins (or elongated molecules), defined by the angle ϕ in the easy-plane.

The solutions for the discussed TD map a plane (x, y), perpendicular to the defect line in coordinate space, onto the 2-d manifold of the order parameter (complex plane of ψ, half-sphere of radius $|\boldsymbol{M}|$, or sphere of radius $|\boldsymbol{L}|$). For example, the solution of (1) has the form $\psi = \psi_0(\rho)\exp(\mathrm{i}n\chi)$, where $n = \pm 1, \pm 2, \cdots$ and $(\pm)$ correspond to the vortex and antivortex. The density of a Bose gas tends to zero in the center of the vortex, $\psi^2(0) = 0$, and the solution has no singularity. A magnetic vortex has the same properties: $\phi = n\chi (n = \pm 1, \pm 2 \cdots)$, $\theta(\rho = 0) = 0$ or π, $\theta(\rho \to \infty) \to \pi/2$ (Kosevich et al., 1990). In AFM- and nematic disclinations $\phi = k\chi/2$, where $k = \pm 1, \pm 2 \cdots$, is called the Frank index. The mapping degree of TD is usually characterized by some integral *topological invariant* (or *topological charge*) related to the solution under consideration. The hydrodynamic vortex in an incompressible liquid is defined by its total vorticity $(1/2\pi)\oint \boldsymbol{v}\,\mathrm{d}\boldsymbol{l}$, where integration is performed over the contour enclosing the vortex center. This value coincides with κ and can be arbitrary. Other types of vortices can be characterized by the same integral $\oint \nabla\phi\,\mathrm{d}\boldsymbol{l}$. (Since a contour can be chosen with an infinite radius, these topological solitons are nonlocalized.) But in some cases, the topological charge is more conveniently defined as the 2-d integral over the coordinate plane. For example, in a ferromagnet, this invariant is

$$Q = \frac{1}{4\pi}\int m\left(\frac{\partial m}{\partial x_\alpha} \times \frac{\partial m}{\partial x_\beta}\right)\varepsilon_{\alpha\beta}\,\mathrm{d}^2x = \frac{1}{2\pi}\int \sin\theta\,\mathrm{d}\theta\,\mathrm{d}\phi, \qquad (2)$$

where $m = \boldsymbol{M}/M_0$. For a magnetic vortex $Q = pn$, where $p = m_z(\rho = 0)$, is the polarity of the vortex.

Another situation appears in systems with discrete degeneracy of the ground state. For instance, a localized magnetic 2-d topological soliton (TS) (*magnetic skyrmion*) of the type $\phi = n\chi$, $\theta = \theta(\rho)$, and $\theta(0) = 0$, $\theta(\infty) = \pi$ can exist in the easy-axis ferromagnet with two equivalent ground states $m_z = \pm 1$. At fixed n, the topological charge for this skyrmion is twice that for the vortex. Moreover, in such a system, TDs can exist in the 1-d and 3-d cases. In 3-d easy-axis FM, the TS, localized in all 3 dimensions, corresponds to the solution with the nonzero integer Hopf invariant. The 1-d TS can exist in 1-d and quasi-1-d systems or in 2-d and 3-d media as solutions depending on one spatial coordinate. Such a TD describes the *domain wall* in ferromagnets. In the framework of the Landau–Lifshitz equation, the corresponding *kink* solution is written as

$$\theta(x) = \arccos(\tanh(x/l_0)), \quad \phi = \text{const.}, \qquad (3)$$

where l_0 is the magnetic length. The domain boundary (kink) separates two half-spaces in different ground states. The topological charge, analogous to (2), can be defined in this case as follows: $Q = (1/2\pi)\int \sin\theta(x)\,\mathrm{d}\theta(x)\,\mathrm{d}\phi(x)$. If an additional anisotropy in the plane perpendicular to the easy axis is taken into account (in orthorhombic ferromagnets), more complicated TDs can exist inside the domain walls: *Bloch lines* and *Bloch points*. Similar to (3), the kink-like TD can exist in other systems with discrete degenerate ground states: kinks in antiferromagnets, kinks of incommensurate surface structures, fluxons in a long Josephson junction, phase boundaries in the problem of structural phase transitions, solitons in polyacetylene, in 1-d metals in the Peierls–Fröhlich phase, and so on. Usually these TDs are investigated within the framework of the φ^4-model or sine-Gordon equation

$$\frac{\partial^2 w}{\partial t^2} - s^2\frac{\partial^2 w}{\partial x^2} + \omega_0^2 \sin w = 0, \qquad (4)$$

where the field variable w and coefficients have different physical meanings for various systems. The topological solution to this equation is well known: $w = 4\arctan\exp((x - vt)/l\sqrt{1 - v^2/s^2})$ with $l = s/\omega_0$.

A crystal *dislocation* represents one of the most important examples of a TD in a crystal lattice. Dislocations exist due to the translational symmetry of the lattice: it transforms into itself under the translation by the interatomic distance a. (Rotational symmetry of the crystal lattice leads to the existence of another TD—crystal *disclination*.) A dislocation represents a one-dimensional TD with the following properties: the regular lattice structure is distorted only in the core of the dislocation line and the total displacement, as a closed contour goes around the dislocation line, is equal to the translation period. (In two-sublattice antiferromagnets, this translation period is twice as small as that for a magnetic lattice.) This fact leads to the appearance of the magnetic disclination and formation of the complex magnet-structural TD (Kovalev & Kosevich, 1977). In the simplest case of a screw dislocation or in a scalar model, the deformation is characterized solely by one component of displacement, u. Then, within the elasticity theory approximation, the deformation of a crystal is governed by the equation $\Delta u = 0$ with the dislocation solution $u = a\chi/2\pi$. (Here, a plays the role of the topological charge of the dislocation: $\oint du = a$.) The above solution is singular and does not describe the discrete structure of the dislocation core. It can be investigated in the simple one-dimensional model proposed by Yakov Frenkel and Tatiana Kontorova (1938) (*See also* **Frenkel–Kontorova model**). The FK model was the first 1-d model for 2-d topological defects. Within this model, the relative displacements u of the atoms from two atomic rows above and below the dislocation center are described by Equation (4), where $w = 2\pi u/a$, $\omega_0 = (2\pi/a)\sqrt{U/m}$, m is the atomic mass, s is the velocity of sound, U is the energy of the interaction between the rows, and l is the width of the dislocation core. The energy of a stationary dislocation (kink) is $E_0 = (4/\pi)\sqrt{ms^2U}$. The integral $\int \partial u/\partial x \, dx$ now plays the role of the topological charge. The FK model describes more adequately other TDs in the crystal lattice: *crowdions*, kinks of incommensurate surface structures, and dislocation kinks.

The dynamics of TDs are of great variety and depend strongly on the nature of the TD. The dynamics of dislocations, dislocation kinks, crowdions, domain walls, and phase boundaries are very simple within the framework of the Lorentz-invariant 1-d FK model: the kink can move with velocities below the velocity of sound in a system and its effective mass can be considerably smaller than the atomic mass. Actually, the dynamics of TDs in a lattice is much more complicated. Dislocations move mainly in some preferred direction (in slip planes), and crowdions propagate in closely packed atomic rows in a “relay race” manner. The discreteness of a lattice gives rise to Peierls relief for TP, and its dynamics assume diffusive features. Dislocation creeps over this relief, producing the dislocation kinks. The dynamics of vortices in media with distributed parameters has some interesting peculiarities. An isolated vortex in an infinite ideal medium cannot move: it is frozen in a liquid or superfluid flow or spin flux. The vortex can move in a bounded area or in the presence of other defects but these dynamics are non-Newtonian. Within the collective coordinates approach, the effective equation of motion for the center of a vortex $\boldsymbol{X}(t)$ is (Thiele, 1973)

$$\frac{d\boldsymbol{X}}{dt} \times \boldsymbol{G} = \boldsymbol{F}_{\mathrm{G}}, \quad (5)$$

where the force $\boldsymbol{F}_{\mathrm{G}}$ is formally equivalent to the Magnus force in fluid dynamics and the gyrocoupling vector $\boldsymbol{G}$ is parallel to the line of a vortex and depends on its topological charge.

In all the examples above, topological defects with the opposite sign of the topological charge exist (vortex-antivortex, kink-antikink, dislocations with opposite signs, etc). This implies that TDs can emerge from the ground state of a medium in pairs with zero total topological charge. At a nonzero temperature, these pairs dissociate and the finite density of TDs can be observed. Seeger & Schiller (1966) were the first to develop the thermodynamics of topological solitons in the framework of the 1-d FK model. As they showed, the equilibrium density of kinks and antikinks at a temperature T is

$$n \approx \frac{a}{l}\sqrt{\frac{2}{\pi\tau}}\exp\left(-\frac{1}{\tau}\right), \quad (6)$$

where $\tau = k_{\mathrm{B}}/E_0$, E_0 is the soliton energy, and l is the width of this kink. The situation with a TD in the 2-d case (vortices and 2-d dislocations) is essentially different. The energy of vortices and dislocations in infinite 2-d systems is infinite, but in systems of size R, it is of the order of $E_v \sim E_* \ln(R/a)$, where E_* is a characteristic energy, being specific for different TDs. The contribution from a vortex or dislocation to configuration entropy in a 2-d crystal is $\delta S = \ln(R/a)^2$, and hence, the change in free energy, if one TD is added, is $\delta F = (E_* - 2T)\ln(R/a)$. At the temperature $T_{\mathrm{c}} = E_*/2$, the value δF becomes negative and a Berezinskii–Kosterlitz–Thouless phase transition takes place: the crystal melts or magnetic ordering is broken (Berezinskii, 1971).

As indicated above, topological defects play an important role in the kinetic and thermodynamic properties of condensed matter. Dislocations and dislocation kinks cause plasticity and strengthening of a crystal, and their behavior under radiation depends to a large extent on the crowdions. Recently, the influence of dislocations on the concentration and mobility of current carriers has been widely studied experimentally in pure semiconductor and alkali-halide crystals (Osip’yan et al., 2000). Topological defects (charge-density-wave solitons) make an essential contribution to the conductivity and electrodynamics of quasi-1-d conductors such as $(CH)_x$, TaS_2, and $NbSe_2$ (Krive et al., 1986).

ALEXANDER S. KOVALEV

See also **Collective coordinates; Dislocations in crystals; Domain walls; Frenkel–Kontorova model; Landau–Lifshitz equation; Liquid crystals; Long Josephson junctions; Multidimensional solitons; Nonlinear Schrödinger equations; Sine-Gordon equation; Spin systems; Superfluidity; Topology; Vortex dynamics of fluids**

Further Reading

Berezinskii, V.L. 1971. Violation of long range order in one-dimensional and two-dimensional systems with a continuous symmetry group. I. Classical systems. *Soviet Physics-JETP*, 34: 610

de Gennes, P.G. 1974. *The Physics of Liquid Crystals*, Oxford: Clarendon Press

Frenkel, J. & Kontorova, T. 1938. On the theory of plastic deformation and twinning. *Physikalishe Zeitschrift der Sowjetunion*, 13: 1–12

Kosevich, A.M., Ivanov, B.A. & Kovalev, A.S. 1990. Magnetic solitons. *Physics Reports*, 194: 117–238

Kovalev, A.S. & Kosevich, A.M. 1977. Dislocation and domains in an antiferromagnet. *Soviet Journal of Low Temperature Physics*, 3: 125–126

Krive, I.V., Rozhavsky, A.S. & Kulik, I.O. 1986. Nonlinear conductivity mechanisms and electrodynamics of quasi-1D conductors in Peierls dielectric phase. *Soviet Journal of Low Temperature Physics*, 12: 635

Lifshitz, E.M. & Pitaevskii, L.P. 1980. *Statistical Physics*, part 2, Oxford: Pergamon Press

Lugt, H.J. 1996. *Introduction to Vortex Theory*, Potomac, MD: Vortex Flow Press

Mermin, N.D. 1979. The topological theory of defects in ordered media. *Review of Modern Physics*, 51: 591–648

Mertens, F.G. & Bishop, A.R. 2000. Dynamics of vortices in two-dimensional magnets. In *Nonlinear Science at the Dawn of the 21st Century*, Berlin and New York: Springer

Osip'yan, Yu.A. et al. 2000. *Electronic Properties of Dislocations in Semiconductors*. Moscow: Editorial USSR (in Russian)

Seeger, A. & Schiller, P. 1966. *Physical Acoustics*, New York: Academic Press

Thiele, A.A. 1973. Steady-state motion of magnetic domains. *Physical Review Letters*, 30: 239–233

TOPOLOGICAL ENTROPY

See **Entropy**

TOPOLOGICAL SOLITONS

See **Solitons, types of**

TOPOLOGY

Continuity is conventionally associated with functions defined on the real line or higher-dimensional Euclidean spaces. Topology is concerned with the abstraction of continuity to maps between more general sets. The subject is vast and its study can take many forms depending on the nature of the structures considered. They include the areas of point-set, combinatorial, algebraic, and differential topology.

A topological space has a distinguished collection of subsets known as *open sets*. An open subset U of the real line $\mathbb{R}$ is one for which every point of U is a subset of some real interval wholly contained in U. Thus, every point of U is the *interior* of U. Openness of a set U can also be expressed using the Euclidean distance, or *metric*, by saying that for every point $x \in U$, all points within a sufficiently small distance of x also lie in U. Thus, metrics can be used to create open sets. Complements of open sets are said to be *closed*. The simplest examples of open and closed sets in $\mathbb{R}$ are, respectively, the "open interval" (a, b) of all points between the numbers a and b excluding the end points, and the "closed interval" $[a, b] = (a, b) \cup \{a, b\}$. Also, sets can be neither open nor closed, for example $[a, b) = (a, b) \cup \{a\}$.

A *topology* on a set X is defined by its collection of open subsets, τ, which then makes X a *topological space* (X, τ). The collection of open subsets must include both X and the empty set ϕ, any union of elements of τ, and the intersection of any finite collection of elements of τ. The conditions for a topology can equally well be cast in terms of closed sets. Also, a set can have many different topologies.

In any sophisticated mathematical structure, there is usually a way of relating two objects. For example, if we are only considering sets, say X and Y, we consider maps $f : X \to Y$. The natural equivalence for sets X,Y would be the existence of a map $f : X \to Y$ which is both one–one and onto, that is, a bijection. When topologies are placed on X and Y, it is natural to consider maps $f : X \to Y$ that are continuous. The metric definition of continuity for maps $f : \mathbb{R} \to \mathbb{R}$, or more generally $f : \mathbb{R}^m \to \mathbb{R}^n$, can be shown to be equivalent to the topological definition: $f : X \to Y$ is continuous if V is an open subset of Y; the set $f^{-1}(V)$ is an open subset of X.

This alternative definition of continuity is the one that makes topology a key mathematical discipline of widespread importance.

The corresponding equivalence for topological spaces X and Y requires a map $h : X \to Y$, which is both (i) a bijection, and (ii) bicontinuous; that is, both h and its inverse h^{-1} are continuous. Such a map is called a *homeomorphism*. The spaces X and Y are said to be topologically equivalent or homeomorphs. Any subset S of a topological space X can be made into a topological space by declaring the intersections of open sets of X with S to be open sets of S. In fact, this collection of subsets of S forms a topology on S, called the *subspace topology*. Thus, important geometrical objects which are subsets of Euclidean spaces such as the circle, the sphere (and therefore all classical polyhedra), the torus, the pretzel, and the Klein bottle (see Figure 1) can all be seen as topological spaces when endowed with the subspace topology.

An important property of continuous functions defined on the real numbers is that a continuous function defined on a bounded closed interval attains its bounds; that is, there exist points at which the function takes its maximum and minimum value. This is not true if the "closed, bounded" condition is relaxed. For example, $f(x) = x$ is not bounded on the real line $\mathbb{R}$ which is a closed (but not bounded) set. Also, $f(x) = x$ does not attain its bounds on the bounded (but not closed) set $(0, 1)$. The bounded closed interval on $\mathbb{R}$ is called a *compact* set. Again, such a set can be defined solely in terms of open sets, and so we can define the concept of a *compact topological space* (Munkres, 2000).

Another key result in elementary analysis is that the continuous image of an interval of real numbers is also an interval. This property is often used to find roots of a continuous function $f : \mathbb{R} \to \mathbb{R}$ by finding values $a, b \in \mathbb{R}$ for which $f(a) \cdot f(b) < 0$, the so-called "intermediate value" theorem. In the generalization of this result to topological spaces, the key property of the interval is its *connectedness*. The analogous result for topological spaces is that the continuous image of a connected set is also a connected set. The characterization of connectedness in the real numbers can be described purely in terms of the properties of the open sets on the real line, that is, in terms of the Euclidean topology on $\mathbb{R}$.

Thus, both compactness and connectedness are topological properties in the sense that they can be described purely in terms of properties of the open sets of a topological space (Munkres, 2000).

In some areas of topology, the importance of the open sets is not so apparent and other features of the topological space are considered. For example, polyhedra such as the cube, tetrahedron, and dodecahedron are finite ways of building homeomorphic images of the standard sphere. We note that for all such constructs, the number of vertices (V), edges (E), and faces (F) satisfy the condition $V - E + F = 2$, the *Euler characteristic* of the sphere. The torus, when built up in terms of faces, edges, and vertices, has the property that its Euler characteristic is zero. Given that the numbers V, E, and F are conserved by homeomorphism, we see that the torus and sphere having different Euler characteristics not only makes them look "different," but ensures that they are not homeomorphic; that is, they are topologically distinct.

Note that not all spaces can be easily distinguished using topology. For example, the subsets of rational numbers, $\mathbb{Q}$, and the irrationals, I, of the real line $\mathbb{R}$, are both topologically *dense* sets in $\mathbb{R}$; that is, for both sets, the smallest closed superset is the whole interval $\mathbb{R}$. Also, both are neither open nor closed. However, note that there is no bijection between the sets $\mathbb{Q}$ and I because they have different cardinalities and, thus, cannot be homeomorphs.

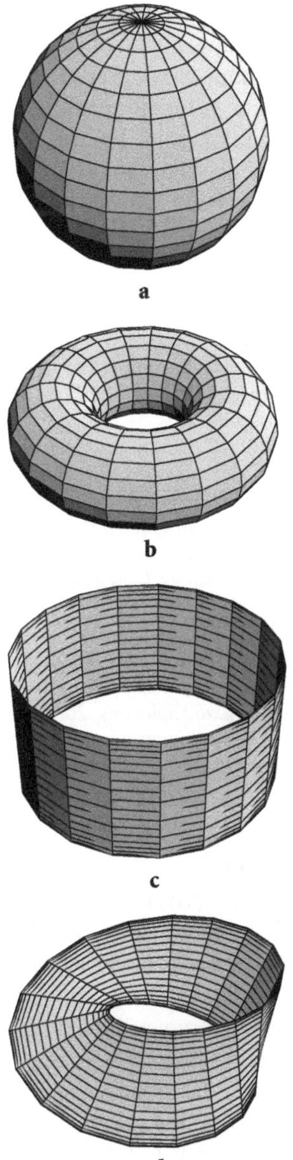

Figure 1. Four distinct topological spaces: (a) sphere; (b) torus; (c) cylinder; (d) Möbius band. See http://library.wolfram.com/graphics/

The topological spaces mentioned above, such as the torus, sphere, and pretzel, can be easily visualized within the three-dimensional Euclidean space $\mathbb{R}^3$ in which we locally live. However, it is not difficult to see how we can start to construct spaces in $\mathbb{R}^3$ which are more geometrically demanding. The simplest is the Möbius strip, M, which is derived from the rectangle by pasting together one opposite pair of edges of a band with a half-twist, see Figure 1(d). Note that the Möbius band has only one "side"; just draw a pen line along the spine of the band and the line arrives on the opposite side of the paper from its initial point. Continuing the line brings a return to the initial point of the curve. By comparison, one cannot get from one side of a cylinder to the other without passing across one of its edges. The Möbius band is topologically different from the cylinder for several reasons. For instance, the boundary of the cylinder consists of two disjoint circles whereas that of the Möbius band is a single circle.

DAVID ARROWSMITH

Further Reading

Dieudonné, J. 1985. The beginnings of topology from 1850 to 1914. *Proceedings of the Conference on Mathematical Logic 2 (Siena, 1985)*, 585–600

Dieudonné, J. 1989. *A History of Algebraic and Differential Topology, 1900–1960*, Boston: Birkhäuser

Lefschetz, S. 1970. The early development of algebraic topology. *Boletim da Sociedade Brasileira de Matematica*, 1(1): 1–48

Mendelson, B. 1990. *Introduction to Topology*, New York: Dover

Munkres, J.R. 2000. *Topology*, Upper Saddle River, NJ: Prentice-Hall

Stillwell, J.C. 1993. *Classical Topology and Combinatorial Group Theory*, 2nd edition, New York: Springer

Weil, A. 1979. Riemann, Betti and the birth of topology. *Archive for the History of Exact Sciences*, 20(2): 91–96

MATHEMATICA (http://library.wolfram.com/graphics/) Graphics of various surfaces

TRAFFIC FLOW

Most automobile drivers have encountered the widespread phenomenon of so-called "phantom traffic jams," for which there is no visible reason such as an accident or a bottleneck. Why are vehicles sometimes stopped although everyone likes to drive fast? Due to the finite adaptation time (= reaction time + acceleration time), a small disturbance in the traffic flow Q_* can cause an overreaction (overbraking) of a driver, if the safe vehicle speed $V_*(\rho)$ drops too rapidly with increasing vehicle density ρ. At high enough densities ρ, this will give rise to a chain reaction of the followers, as other vehicles will have approached before the original speed can be regained. This feedback can eventually cause the unexpected standstill of vehicles known as traffic jam.

Lighthill & Whitham (1955) described traffic flow as a function of space x and time t by means of a fluid-dynamic conservation law of vehicles, reflecting the fact that vehicles are not generated or lost in the absence of ramps, intersections, or accidents. The traffic flow $Q(x, t)$ (= vehicle density $\rho(x, t) \times$ average velocity $V(x, t)$) was specified as a function of the density $\rho(x, t)$. The corresponding "fundamental diagram" $Q_*(\rho) = \rho V_*(\rho)$ is obtained as a fit to empirical data. The conservation law $\partial\rho/\partial t + \partial Q/\partial x = 0$ leads to the nonlinear wave equation

$$\frac{\partial \rho}{\partial t} + C(\rho)\frac{\partial \rho}{\partial x} = 0, \qquad (1)$$

according to which the propagation velocity

$$C(\rho) = V_*(\rho) + \rho\frac{\mathrm{d}V_*(\rho)}{\mathrm{d}\rho} \le V_*(\rho) \qquad (2)$$

of kinematic waves depends on the vehicle density. Thus, while a density profile on a ring road keeps its amplitude, its shape is changing until shock waves (i.e., discontinuous changes in the density) have developed. The densities ρ_+ and ρ_- immediately upstream and downstream of a shock front determine its propagation speed

$$S(\rho_+, \rho_-) = \frac{Q_*(\rho_+) - Q_*(\rho_-)}{\rho_+ - \rho_-}. \qquad (3)$$

As discontinuous density changes are not fully consistent with empirical observations and a problem for efficient numerical integration, Whitham (1974) has suggested adding a diffusion term $D\,\partial^2\rho/\partial x^2$ with $D > 0$ to the right-hand side of the Lighthill–Whitham equation. For the linear velocity-density relation $V_*(\rho)$ suggested by Greenshields (1935), the resulting equation is equivalent with the Burgers equation and can be transformed into the linear heat or diffusion equation; that is, it is analytically solvable.

Experimental observations of traffic patterns show some additional features that cannot be reproduced by the above models. While traffic flow appears to be stable with respect to perturbations at small and large densities, there is a linearly unstable range at medium densities, where already small disturbances of uniform traffic flow give rise to traffic jams. Between these three density ranges, one finds meta- or multistable ranges, since there exists a density-dependent, critical amplitude $A(\rho)$, so that the resulting traffic pattern is path- or history-dependent (Kerner et al., 1994–1997). While subcritical perturbations fade away, supercritical perturbations cause a breakdown of traffic flow (nucleation effect). Consequently, traffic flows display critical points, nonequilibrium phase transitions, noise-induced transitions, and fluctuation-induced ordering phenomena.

One may view this situation as nonequilibrium analogue of the phase transitions between vapor, water, and ice. However, the breakdown and structure formation phenomena, when the "temperature" (i.e., the fluctuation strength) is increased, are sometimes counterintuitive due to the repulsive nature of vehicular interactions. From classical many-particle systems with attractive interactions, we are rather used to the idea that increasing temperature breaks up structures and destroys patterns (fluid structures are replaced by gaseous ones, not by solid ones).

The above observations in freeway traffic can be described by microscopic, mesoscopic, or macroscopic models, which are theoretically connected by means of a micro-macro link. Microscopic models are usually follow-the-leader models specifying the acceleration dv_i/dt of the single vehicles i as a function of their distance headway $d_i = x_{i-1} - x_i$, their speed v_i, and/or their relative velocity $\Delta v_i = v_i - v_{i-1}$:

$$\frac{dv_i}{dt} = f(d_i, v_i, \Delta v_i)\,. \tag{4}$$

A typical example is the non-integer car-following model:

$$\frac{dv_i(t+\Delta t)}{dt} = -\frac{\Delta v_i(t)}{T}\frac{[v_i(t+\Delta t)]^m}{[d_i(t)]^l} \tag{5}$$

with the reaction time $\Delta t \approx 1.3$ s and the parameters $T \approx \Delta t/0.55$, $m \approx 0.8$, and $l \approx 2.8$ (Gazis et al., 1961). It has a linearly unstable range for $\Delta t/T > \frac{1}{2}$. A simpler model is the optimal velocity model

$$\frac{dv_i(t)}{dt} = \frac{1}{\tau}\Big[v\big(d_i(t)\big) - v_i(t)\Big], \tag{6}$$

where $v(d_i)$ is the "optimal" velocity-distance relation and τ the adaptation time (Bando et al., 1994, 1995). This model has an unstable range for $dv(d_i)/dd_i > 1/(2\tau)$. The respective nonlinearly coupled differential equations (or stochastic differential equations, if fluctuations are taken into account) are numerically solved as in molecular dynamics. An alternative approach is rule-based cellular automata, which discretize space and time in favor of numerical efficiency: $t = i\,\Delta t, x = j\,\Delta x, d = \hat{d}\,\Delta x, v = \hat{v}\,\Delta x/\Delta t$. The Nagel–Schreckenberg model (1992), for example, can be written in the form

$$\hat{v}_{i+1} = \max\Big(0, \min(\hat{v}_{\max}, \hat{d}_i - 1, \hat{v}_i + 1) - \xi_i^{(p)}\Big), \tag{7}$$

where $\hat{v}_{\max}\Delta x/\Delta t$ is the maximum velocity and $\xi_i^{(p)}$ a Boolean random variable which is 1 with probability p and 0 otherwise. Typical parameters are $\Delta t = 1$ s, $\Delta x = 7.5$ m, $\hat{v}_{\max} = 5$, and $0.2 < p \le 0.5$.

Mesoscopic models describe the spatiotemporal change of the phase space density (= vehicle density × velocity distribution). This approach has been introduced by Prigogine et al. (1960, 1961, 1971) and is inspired by kinetic gas theory. The related equations are either of Boltzmann type (for point-like vehicles or low densities) or of Enskog type, if vehicular space requirements at moderate and high densities are taken into account (Helbing et al., 1995–1999). The equations allow a systematic derivation of a hierarchy of macroscopic equations for the vehicle density $\rho(x,t)$, the average velocity $V(x,t)$, the velocity variance $\Theta(x)$, etc. This hierarchy is usually closed after the velocity or variance equation, although the separation of time scales assumed by the underlying approximations is weak. Nevertheless, the observed traffic dynamics are rather well reproduced by the resulting coupled partial differential equations. The density equation is just the continuity equation

$$\frac{\partial \rho}{\partial t} + \frac{\partial(\rho V)}{\partial x} = \nu_+ - \nu_-\,, \tag{8}$$

where ν_+ and ν_- denote on- and off-ramp flows, respectively. The velocity equation can be cast into the form

$$\frac{\partial V}{\partial t} + V\frac{\partial V}{\partial x} = -\frac{1}{\rho}\frac{\partial P}{\partial x} + \frac{1}{\tau}(V^* - V). \tag{9}$$

In theoretically consistent macroscopic traffic models such as the gas-kinetic-based traffic model, the "traffic pressure" P and the velocity V^* are nonlocal functions of the density ρ, the average velocity V, and the variance Θ (Helbing et al., 1998, 1999). The Lighthill–Whitham model (1955) results in the unrealistic limit $\tau \to 0$ of vanishing adaptation times τ. Payne's macroscopic traffic model (1971, 1979) is obtained for $P(\rho) = [V_0 - V_*(\rho)]/(2\tau)$ and $V^* = V_*(\rho)$, where V_0 denotes the (average) desired velocity (the average velocity at very low densities). Kerner and Konhäuser's model (1993) is a variant of Kühne's model (1984) and corresponds to the specifications $P = \rho\Theta_0 - \eta_0 \partial V/\partial x$, and $V^* = V_*(\rho)$, where Θ_0 and η_0 are positive constants. The corresponding equation is a Navier–Stokes equation with a viscosity term $\eta_0 \partial^2 V/\partial x^2$ and an additional relaxation term $[V_*(\rho) - V]/\tau$ describing the delayed adaptation to the velocity–density relation $V_*(\rho)$. The condition for linear instability reads

$$\rho\left|\frac{dV_*(\rho)}{d\rho}\right| > \sqrt{\frac{dP(\rho)}{d\rho}}\,; \tag{10}$$

that is, the Payne model and the Kerner–Konhäuser model have linearly unstable ranges, if $dV_*(\rho)/d\rho$ is large, while the Lighthill–Whitham model is marginally stable. The Burgers equation, by the way, is always stable.

According to Krauß (1998), traffic models show the observed hysteretic phase transition related with metastable traffic and high flows only, if the typical maximal acceleration is not too large and the deceleration strength is moderate. In such models, the outflow Q_{out} from traffic jams is a self-organized constant of traffic flow (Kerner et al., 1994, 1996). It corresponds approximately to the intersection point of the linear jam line

$$J(\rho) = \frac{1}{T}\left(1 - \frac{\rho}{\rho_{\text{jam}}}\right) \tag{11}$$

with the free branch of the flow-density diagram, where T denotes the time headway in congested traffic and

ρ_{jam} the density inside traffic jams. The jam line corresponds to the flow-density relation for moving traffic patterns with a self-organized, stationary profile (Kerner & Konhäuser, 1994). These propagate with the velocity $C = -1/(T\rho_{jam}) \approx -15$ km/h, which is another traffic constant. (Once a traffic jam is fully developed, it moves upstream with constant velocity, as vehicles leave the downstream jam front at a constant rate, while new ones join it at the upstream front.)

With this knowledge, one can understand the various congested traffic states observed on freeway sections with bottlenecks (Helbing et al., 1998–2002). Let us assume a bottleneck due to ramp flows $\nu_+ = Q_{rmp}/(nL)$, where L is the used length of the on-ramp and n the number of freeway lanes. The corresponding bottleneck strength is, then, $\Delta Q = Q_{rmp}/n$. If Q_{up} denotes the traffic flow upstream of the bottleneck and $Q_{tot} = (Q_{up} + \Delta Q)$ is the total capacity required downstream of the ramp, we will eventually find a growing vehicle queue upstream of the bottleneck, if Q_{tot} is greater than the dynamic capacity Q_{out}. The traffic flow Q_{cong} resulting in the congested area plus the inflow or bottleneck; strength ΔQ are normally given by the outflow Q_{out}; that is,

$$Q_{cong} = Q_{out} - \Delta Q \tag{12}$$

(if vehicles cannot enter the freeway downstream of the congestion front). One can distinguish the following cases: If the density ρ_{cong} associated with the congested flow

$$Q_{cong} = Q_*(\rho_{cong}) \tag{13}$$

lies in the stable range, we find homogeneous congested traffic (HCT) such as typical traffic jams during holiday seasons or after serious accidents. For a smaller on-ramp flow or bottleneck strength ΔQ, the congested flow Q_{cong} is linearly unstable, and we either find oscillating congested traffic (OCT) or triggered stop-and-go traffic (TSG). In contrast to OCT, stop-and-go traffic is characterized by a sequence of moving jams, between which traffic flows freely. This state can either emerge from a spatial sequence of homogeneous and oscillating congested traffic (Koshi et al., 1983; called "pinch effect" by Kerner, 1998), or it can be caused by the inhomogeneity at the ramp. In the latter case, each traffic jam triggers another one by inducing a small

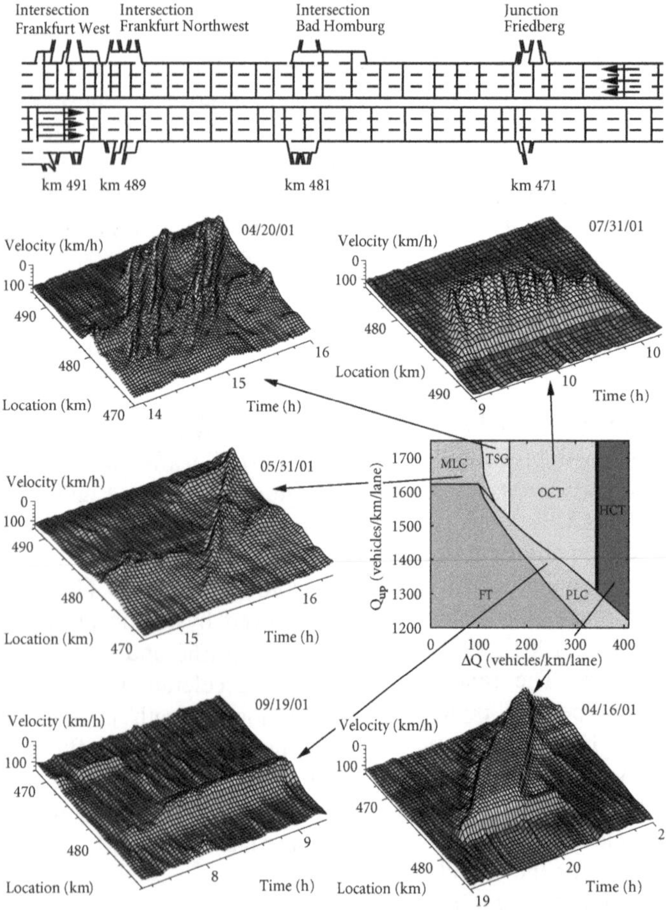

Figure 1. Numericlly determined phase diagram of traffic states in the presence of one bottleneck as a function of the upstream flow Q_{up} and the bottleneck strength ΔQ (center right), and empirical representatives for the related kinds of congested traffic (by Martin Schönhof and D. Helbing. Note that the outflow Q_{out} in the applied simulation model depends on the bottleneck strength ΔQ.

perturbation in the inhomogeneous freeway section (see Figure 1), which propagates downstream as long as it is small, but turns back when it has grown large enough (boomerang effect). This, however, requires the downstream traffic flow to be linearly unstable. If it is (meta-)stable instead (when the traffic volume Q_{tot} is further reduced), a traffic jam will usually not trigger a growing perturbation. In that case, one finds either a single moving localized cluster (MLC) or a pinned localized cluster (PLC) at the location of the ramp. The latter requires the traffic flow in the upstream section to be stable, so that no traffic jam can survive there. Finally, for sufficiently small traffic volumes Q_{tot}, we find free traffic (FT), as expected. For freeways with a single bottleneck and a large perturbation of traffic flow, these facts can be summarized by the phase diagram in Figure 1, which is universal for all microscopic and macroscopic, stochastic and deterministic traffic models with the same instability diagram (stable, metastable, and unstable density ranges). Results for more complex freeway geometries, other initial or boundary conditions, and other instability diagrams are available as well.

Current research focuses on the following open questions: Are fluctuations and psychological concepts necessary to understand the empirical observations in traffic flows? Can the large individual variation of time headways fully account for the large scattering of flow-density data in synchronized flow? (In congested traffic flow, the velocities in neighboring lanes are usually synchronized, as different speeds are balanced by lane changes.) What are the site- and country-dependent differences in traffic dynamics, and can they be adequately reflected by different model parameters for the driver-vehicle units? How can the insights regarding the laws of traffic dynamics be used for traffic optimization by variable speed limits, intelligent on-ramp controls, dynamic re-routing, and driver assistance systems? How can they be transferred to the explanation of breakdown and obstruction phenomena in socioeconomic systems?

DIRK HELBING

See also **Burgers equation; Constants of motion and conservation laws; Phase transitions; Shock waves**

Further Reading

Helbing, D. 2001. Traffic and related self-driven many-particle systems. *Reviews of Modern Physics*, 73(4): 1067–1141 [All the authors mentioned in the text above are found as references in this article.]

Lighthill, M.J. & Whitham, G.B. 1955. On kinematic waves: II. A theory of traffic on long crowded roads. *Proceedings of the Royal Society, London* A, 229: 317–345

Whitham, G.B. 1974. *Linear and Nonlinear Waves*, New York: Wiley

TRAJECTORIES

See **Phase space**

TRANSITION TO CHAOS

See **Chaotic dynamics**

TRAVELING WAVE

See **Wave of translation**

TRIAD INTERACTION

See ***N*-wave interactions**

TUNNEL DIODE ARRAYS

See **Distributed oscillators**

TURBULENCE

Turbulence is a state of a nonlinear physical system that has energy distribution over many degrees of freedom strongly deviated from equilibrium. Turbulence is irregular both in time and in space. Turbulence can be maintained by some external influence or it can decay on the way to relaxation to equilibrium. The term first appeared in fluid mechanics and was later generalized to include far-from-equilibrium states in solids and plasmas.

If an obstacle of size L is placed in a fluid of viscosity ν that is moving with velocity V, a turbulent wake emerges for sufficiently large values of the Reynolds number

$$\text{Re} \equiv VL/\nu.$$

At large Re, flow perturbations produced at scale L experience a viscous dissipation that is small compared with nonlinear effects. Nonlinearity then induces motions at smaller and smaller scales until viscous dissipation terminates the process at a scale much smaller than L, leading to a wide (so-called inertial) interval of scales where viscosity is negligible and nonlinearity plays a dominant role.

Examples of this phenomenon include waves excited on a fluid surface by wind or moving bodies and waves in plasmas and solids that are excited by external electromagnetic fields. The state of such a system is called turbulent when the wavelength of the waves excited greatly differs from the wavelength of the waves that dissipate. Nonlinear interactions excite waves in the interval of wavelengths (called the transparency window or inertial interval as in fluid turbulence) between the injection and dissipation scales.

The ensuing complicated and irregular dynamics require a statistical description based on averaging over regions of space or intervals of time. Because nonlinearity dominates in the inertial interval, it is natural to ask to what extent the statistics are universal, in the sense of being independent of the details of excitation and dissipation. The answer to this question is far from evident for nonequilibrium systems. A fundamental physical problem is to establish which statistical properties are universal in the inertial interval of scales and which are features of different turbulent systems.

Constraints on dynamics are imposed by conservation laws, and therefore, conserved quantities must play an essential role in turbulence. Although the conservation laws are broken by pumping and dissipation, these factors do not act in the inertial interval. Under incompressible turbulence, for example, the kinetic energy is pumped by external forcing and is dissipated by viscosity. As suggested by Lewis Fry Richardson in 1921, kinetic energy flows throughout the inertial interval of scales in a cascade-like process. The cascade idea explains the basic macroscopic manifestation of turbulence: the rate of dissipation of the dynamical integral of motion has a finite limit when the dissipation coefficient tends to zero. In other words, the mean rate of the viscous energy dissipation does not depend on viscosity at large Reynolds numbers. That means that symmetry of the inviscid equation (here, time-reversal invariance) is broken by the presence of the viscous term, even though the latter might have been expected to become negligible in the limit $\mathrm{Re}\to\infty$.

The cascade idea fixes only the mean flux of the respective integral of motion, requiring it to be constant across the inertial interval of scales. To describe an entire turbulence statistics, one has to solve problems on a case-by-case basis with most cases still unsolved.

Weak Wave Turbulence

From a theoretical point of view, the simplest case is the turbulence of weakly interacting waves. Examples include waves on the water surface, waves in plasma with and without a magnetic field, and spin waves in magnetics. We assume spatial homogeneity and denote by a_k the amplitude of the wave with the wave vector $\boldsymbol{k}$. When the amplitude is small, it satisfies the linear equation

$$\frac{\partial a_k}{\partial t} = -\mathrm{i}\omega_k a_k + f_k(t) - \gamma_k a_k. \tag{1}$$

Here, the dispersion law ω_k describes wave propagation, γ_k is the decrement of linear damping, and f_k describes pumping. For the linear system, a_k is different from zero only in the regions of $\boldsymbol{k}$-space where f_k is nonzero. To describe wave turbulence that involves wave numbers outside the pumping region, one must account for the interactions among different waves. Considering the wave system to be closed (no external pumping or dissipation), one can describe it as a Hamiltonian system using wave amplitudes as normal canonical variables (Zakharov et al., 1992). At small amplitudes, the Hamiltonian can be written as an expansion over a_k, where the second-order term describes noninteracting waves and high-order terms determine the interaction

$$H = \int \omega_k |a_k|^2 \,\mathrm{d}\boldsymbol{k} + \int \left(V_{123} a_1 a_2^* a_3^* + \mathrm{c.c.}\right)$$
$$\delta(\boldsymbol{k}_1 - \boldsymbol{k}_2 - \boldsymbol{k}_3)\,\mathrm{d}\boldsymbol{k}_1\,\mathrm{d}\boldsymbol{k}_2\,\mathrm{d}\boldsymbol{k}_3 + \mathrm{O}(a^4). \tag{2}$$

Here, $V_{123} = V(\boldsymbol{k}_1, \boldsymbol{k}_2, \boldsymbol{k}_3)$ is the interaction vertex, and c.c. denotes complex conjugate. In this expansion, we presume every subsequent term smaller than the previous one, in particular, $\xi_k = |V_{kkk} a_k| k^d / \omega_k \ll 1$. Wave turbulence that satisfies that condition is called weak turbulence. Also, space dimensionality d can be 1, 2, or 3.

A dynamic equation that accounts for pumping, damping, wave propagation, and interaction thus has the following form:

$$\frac{\partial a_k}{\partial t} = -\mathrm{i}\frac{\delta H}{\delta a_k^*} + f_k(t) - \gamma_k a_k. \tag{3}$$

It is likely that the statistics of the weak turbulence at $k \gg k_f$ is close to Gaussian for wide classes of pumping statistics (this has not been shown rigorously). It is definitely the case for a random force with the statistics close to Gaussian. We consider here and below a pumping by a Gaussian random force statistically isotropic and homogeneous in space and white in time. Thus,

$$\langle f_k(t) f_{k'}^*(t')\rangle = F(k)\delta(\boldsymbol{k} + \boldsymbol{k}')\delta(t - t'), \tag{4}$$

where angular brackets imply spatial averages, and $F(k)$ is assumed nonzero only around some k_f. For waves to be well defined, we assume $\gamma_k \ll \omega_k$.

Because the dynamic equation (3) contains a quadratic nonlinearity, the statistical description in terms of moments encounters the closure problem: the time derivative of the second moment is expressed via the third one, the time derivative of the third moment is expressed via the fourth one, and so on. Fortunately, weak turbulence in the inertial interval is expected to have the statistics close to Gaussian, so one can express the fourth moment as the product of two second ones. As a result, one gets a closed kinetic equation for the single-time pair correlation function $\langle a_k a_{k'}\rangle = n_k \delta(\boldsymbol{k} + \boldsymbol{k}')$ (Zakharov et al., 1992):

$$\frac{\partial n_k}{\partial t} = F_k - \gamma_k n_k + I_k^{(3)},$$
$$I_k^{(3)} = \int \left(U_{k12} - U_{1k2} - U_{2k1}\right) \mathrm{d}\boldsymbol{k}_1\,\mathrm{d}\boldsymbol{k}_2,$$
$$U_{123} = \pi\left[n_2 n_3 - n_1(n_2 + n_3)\right]|V_{123}|^2$$
$$\times\delta(\boldsymbol{k}_1 - \boldsymbol{k}_2 - \boldsymbol{k}_3)\delta(\omega_1 - \omega_2 - \omega_3). \tag{5}$$

This is called the kinetic equation for waves. The collision integral $I_k^{(3)}$ describes three-wave interactions: the first term in the integral corresponds to a decay of a given wave while the second and third terms correspond to a confluence with other waves.

One can estimate from (5) the inverse time of nonlinear interaction at a given k as $|V(k,k,k)|^2 n(k)k^d/\omega(k)$. We define k_d as the wave number where this inverse time is comparable with $\gamma(k)$ and assume nonlinearity to dominate over dissipation at $k \ll k_d$. As has been noted, wave turbulence appears when there is a wide (inertial) interval of scales where both pumping and damping are negligible, which requires $k_d \gg k_f$, the condition analogous to $\mathrm{Re} \gg 1$.

The presence of the frequency delta-function in $I_k^{(3)}$ means that wave interaction conserves the quadratic part of the energy $E = \int \omega_k n_k \, d\boldsymbol{k} = \int E_k \, dk$. For the cascade picture to be valid, the collision integral has to converge in the inertial interval which means that energy exchange is small between motions of vastly different scales, a property called interaction locality in k-space. Consider now a statistical steady state established under the action of pumping and dissipation. Let us multiply (5) by ω_k and integrate it over either interior or exterior of the ball with radius k. Taking $k_f \ll k \ll k_d$, one sees that the energy flux through any spherical surface (Ω is a solid angle),

$$P_k = \int_0^k k^{\mathrm{d}-1} \mathrm{d}k \int \mathrm{d}\Omega \, \omega_k I_k^{(3)},$$

is constant in the inertial interval and is equal to the energy production/dissipation rate:

$$P_k = \varepsilon = \int \omega_k F_k \, d\boldsymbol{k} = \int \gamma_k E_k \, dk. \qquad (6)$$

Let us assume now that the medium (characterized by ω_k and V_{123}) can be considered isotropic at the scales in the inertial interval. In addition, for scales much larger or much smaller than a typical scale in the medium (like the Debye radius in plasma or the depth of the water), the Hamiltonian coefficients are usually scale invariant: $\omega(k) = ck^\alpha$ and $|V(\boldsymbol{k}, \boldsymbol{k}_1, \boldsymbol{k}_2)|^2 = V_0^2 k^{2m} \chi(\boldsymbol{k}_1/k, \boldsymbol{k}_2/k)$ with $\chi \simeq 1$. Remember that we presumed statistically isotropic force. In this case, the pair correlation function that describes a steady cascade is also isotropic and scale invariant:

$$n_k \simeq \varepsilon^{1/2} V_0^{-1} k^{-m-d}. \qquad (7)$$

One can show that (7) reduces $I_k^{(3)}$ to zero (see Zakharov et al., 1992).

If the dispersion relation $\omega(k)$ does not allow for the resonance condition $\omega(k_1) + \omega(k_2) = \omega(|\boldsymbol{k}_1 + \boldsymbol{k}_2|)$, then the three-wave collision integral is zero and one has to account for four-wave scattering which is always resonant; that is, whatever $\omega(k)$ one can always find four wave vectors that satisfy $\omega(k_1) + \omega(k_2) = \omega(k_3) + \omega(k_4)$ and $\boldsymbol{k}_1 + \boldsymbol{k}_2 = \boldsymbol{k}_3 + \boldsymbol{k}_4$. The collision integral that describes scattering,

$$\begin{aligned} I_k^{(4)} = \frac{\pi}{2} \int & |T_{k123}|^2 \big[n_2 n_3 (n_1 + n_k) \\ & - n_1 n_k (n_2 + n_3) \big] \delta(\boldsymbol{k} + \boldsymbol{k}_1 - \boldsymbol{k}_2 - \boldsymbol{k}_3) \\ & \times \delta(\omega_k + \omega_1 - \omega_2 - \omega_2) \, d\boldsymbol{k}_1 \, d\boldsymbol{k}_2 \, d\boldsymbol{k}_3, \end{aligned} \qquad (8)$$

conserves the energy and also the wave action $N = \int n_k \, d\boldsymbol{k}$ (which can also be called the number of waves). Pumping generally provides for an input of both E and N. If there are two inertial intervals (at $k \gg k_f$ and $k \ll k_f$), then there should be two cascades. Indeed, if $\omega(k)$ grows with k, then absorbing finite amount of E at $k_d \to \infty$ corresponds to an absorption of an infinitely small N. It is thus clear that the flux of N has to go in the opposite direction, that is, to small wave numbers. A so-called inverse cascade with a constant flux of N can thus be realized at $k \ll k_f$. A sink at small k can be provided by wall friction in the container or by long waves leaving the turbulent region in open spaces (as in sea storms).

The collision integral $I_k^{(3)}$ involves products of two n_k, so that flux constancy requires $E_k \propto \varepsilon^{1/2}$ while for the four-wave case, one has $E_k \propto \varepsilon^{1/3}$. In many cases (when there is complete self-similarity), that knowledge is sufficient to obtain the scaling of E_k from a dimensional reasoning without actually calculating V and T. For example, short waves in deep water are characterized by the surface tension σ and density ρ, so the dispersion relation must be $\omega_k \sim \sqrt{\sigma k^3/\rho}$, which allows for the three-wave resonance and thus $E_k \sim \varepsilon^{1/2}(\rho\sigma)^{1/4} k^{-7/4}$. For long waves in deep water, the surface-restoring force is dominated by gravity, so that the gravitational acceleration g replaces σ as a defining parameter and $\omega_k \sim \sqrt{gk}$. Such a dispersion law does not allow for three-wave resonance, so that the dominant interaction is four-wave scattering which permits two cascades. The direct energy cascade corresponds to $E_k \sim \varepsilon^{1/3} \rho^{2/3} g^{1/2} k^{-5/2}$. The inverse cascade carries the flux of N which we denote Q; it has the dimensionality $[Q] = [\varepsilon]/[\omega_k]$ and corresponds to $E_k \sim Q^{1/3} \rho^{2/3} g^{2/3} k^{-7/3}$.

Because the statistics of weak turbulence is near Gaussian, it is completely determined by the pair correlation function, which is in turn determined by the respective flux. We thus conclude that weak turbulence is universal in the inertial interval.

Strong Wave Turbulence

One cannot treat wave turbulence as a set of weakly interacting waves when the wave amplitudes are large ($\xi_k \geq 1$) and also in the particular case of linear (acoustic) dispersion where $\omega(k)=ck$ for arbitrarily small amplitudes. Indeed, there is no dispersion of wave velocity for acoustic waves, so waves moving in the same direction interact strongly and produce shock waves when viscosity is small. Formally, there is a singularity due to the coinciding arguments of delta-functions in (5) (and in the higher terms of perturbation expansion for $\partial n_k/\partial t$), which is thus invalid at however small amplitudes. Still, some features of the statistics of acoustic turbulence can be understood even without a closed description.

Consider a one-dimensional case which pertains, for instance, to sound propagating in long pipes. Because weak shocks are stable with respect to transverse perturbations (Landau & Lifshitz, 1987), quasi-one-dimensional perturbations may propagate in two and three dimensions as well. In a reference frame that moves with the sound velocity, weakly compressible 1-d flows ($u \ll c$) are described by the Burgers equation (Landau & Lifshitz, 1987)

$$u_t + uu_x - \nu u_{xx} = 0. \tag{9}$$

The Burgers equation has a propagating shock-wave solution $u=2v\{1+\exp[v(x-vt)/\nu]\}^{-1}$ with the energy dissipation rate $\nu\int u_x^2\,\mathrm{d}x$ independent of ν. The shock width ν/v is a dissipative scale, and we consider acoustic turbulence produced by a pumping correlated on much larger scales (i.e., pumping a pipe from one end by frequencies much less than cv/ν). After some time, the system will develop shocks at random positions. Here we consider the single-time statistics of the Galilean invariant velocity difference $\delta u(x,t)=u(x,t)-u(0,t)$. The moments of δu are called structure functions $S_n(x,t)=\langle[u(x,t)-u(0,t)]^n\rangle$. Quadratic nonlinearity allows the time derivative of the second moment to be expressed via the third one:

$$\frac{\partial S_2}{\partial t} = -\frac{\partial S_3}{3\partial x} - 4\varepsilon + \nu\frac{\partial^2 S_2}{\partial x^2}. \tag{10}$$

Here $\varepsilon=\nu\langle u_x^2\rangle$ is the mean energy dissipation rate. Equation (10) describes both a free decay (then ε depends on t) and the case of a permanently acting pumping which generates turbulence statistically steady at scales less than the pumping length.

In the first case, $\partial S_2/\partial t \simeq S_2 u/L \ll \varepsilon \simeq u^3/L$ (where L is a typical distance between shocks); while in the second case, $\partial S_2/\partial t=0$ so that $S_3=12\varepsilon x+\nu\partial S_2/\partial x$. Consider now the limit $\nu\to 0$ at fixed x (and t for decaying turbulence). Shock dissipation provides for a finite limit of ε at $\nu\to 0$, then

$$S_3 = -12\varepsilon x. \tag{11}$$

This formula is a direct analog of (6). Indeed, the Fourier transform of (10) describes the energy density $E_k=\langle |u_k|^2\rangle/2$: $(\partial_t-\nu k^2)E_k=-\partial P_k/\partial k$ where the k-space flux

$$P_k = \int_0^k \mathrm{d}k' \int_{-\infty}^{\infty} \mathrm{d}x\, S_3(x)k'\sin(k'x)/24.$$

It is thus the flux constancy that fixes $S_3(x)$ which is universal (determined solely by ε) and depends neither on the initial statistics for decay nor on the pumping for steady turbulence. On the contrary, other structure functions $S_n(x)$ are not given by $(\varepsilon x)^{n/3}$. Indeed, the scaling of the structure functions can be readily understood for any dilute set of shocks (that is, when shocks do not cluster in space) which seems to be the case for both smooth initial conditions and large-scale pumping in Burgers turbulence. In this case, $S_n(x)\sim C_n|x|^n+C_n'|x|$, where the first term comes from the regular (smooth) parts of the velocity while the second comes from $O(x)$ probability to have a shock in the interval x. The scaling exponents, $\xi_n=\mathrm{d}\ln S_n/\mathrm{d}\ln x$, thus behave as follows: $\xi_n=n$ for $n\le 1$ and $\xi_n=1$ for $n>1$. That means that the probability density function (PDF) of the velocity difference in the inertial interval $P(\delta u,x)$ is not scale-invariant; that is, the function of the rescaled velocity difference $\delta u/x^a$ cannot be made scale-independent for any a. As one goes to smaller scales, the lower-order moments decrease faster than the higher-order ones, that means that the smaller the scale the more probable are large fluctuations. In other words, the level of fluctuations increases with the resolution. When the scaling exponents ξ_n do not lie on a straight line, this is called an anomalous scaling since it is related again to the symmetry (scale invariance) of the PDF broken by pumping and not restored even when $x/L\to 0$. As an alternative to the description in terms of structures (shocks), one can relate the anomalous scaling in Burgers turbulence to the additional integrals of motion. Indeed, the integrals $E_n=\int u^{2n}\,\mathrm{d}x/2$ are all conserved by the inviscid Burgers equation. Any shock dissipates the finite amount of E_n at the limit $\nu\to 0$, so that similar to (11), one denotes $\langle \dot{E}_n\rangle=\varepsilon_n$ and obtains $S_{2n+1}=-4(2n+1)\varepsilon_n x/(2n-1)$ for integer n.

Note that $S_2(x)\propto|x|$ corresponds to $E(k)\propto k^{-2}$, which is natural since every shock gives $u_k\propto 1/k$ at $k\ll v/\nu$; that is, the energy spectrum is determined by the type of structures (shocks) rather than by energy flux constancy. Similar ideas were suggested for other types of strong wave turbulence assuming them to be dominated by different structures. Weak wave turbulence, being a set of weakly interacting plane waves, can be studied uniformly for different systems (Zakharov et al., 1992). On the contrary, when nonlinearity is comparable with or exceeds dispersion, different structures appear in different systems. Identifying structures and the role they play in determining different statistical characteristics of strong wave turbulence remains to be investigated for most cases. Broadly, one distinguishes conservative structures (like solitons and vortices) from dissipative structures which usually appear as a result of finite-time singularity of the nondissipative equations (like shocks, light self-focusing, or wave collapse). For example, nonlinear wave packets are described by the nonlinear Schrödinger equation,

$$\mathrm{i}\Psi_t + \Delta\Psi + T|\Psi|^2\Psi = 0. \tag{12}$$

Weak wave turbulence is determined by $|T|^2$ and is the same for both $T<0$ (wave repulsion) and $T>0$ (wave attraction). At high levels of nonlinearity, different signs of T correspond to dramatically different physics: At $T<0$, one has a stable condensate, solitons, and vortices, while at $T>0$, instabilities dominate and wave collapse is possible at $d=2, 3$. No analytic theory is yet available for strong turbulence described by (12).

Because the parameter of nonlinearity $\xi(k)$ generally depends on k, then there may exist a weakly turbulent cascade until some k_* where $\xi(k_*)\sim 1$, and strong turbulence beyond this, wave number; thus weak and strong turbulence can coexist in the same system. Presuming that some mechanism (for instance, wave breaking) prevents the appearance of wave amplitudes that correspond to $\xi_k \gg 1$, one may hypothetize that some cases of strong turbulence correspond to the balance between dispersion and nonlinearity local in k-space so that $\xi(k)$ is constant throughout its domain in k-space. That would correspond to the spectrum $E_k \sim \omega_k^3 k^{-d}/|V_{kkk}|^2$ which is ultimately universal, that is, independent even of the flux (only the boundary k_* depends on the flux). For gravity waves, this gives $E_k = \rho g k^{-3}$, the same spectrum one obtains presuming the wave profile to have cusps (another type of dissipative structure leading to whitecaps in stormy seas—see Phillips, 1977). It is unclear if such flux-independent spectra are realized.

Incompressible Turbulence

Incompressible fluid flow is described by the Navier–Stokes equation

$$\partial_t \boldsymbol{v}(\boldsymbol{r},t) + \boldsymbol{v}(\boldsymbol{r},t)\cdot\nabla\boldsymbol{v}(\boldsymbol{r},t) - \nu\nabla^2\boldsymbol{v}(\boldsymbol{r},t)$$
$$= -\nabla p(\boldsymbol{r},t), \quad \operatorname{div}\boldsymbol{v} = 0.$$

We are again interested in the structure functions $S_n(\boldsymbol{r},t) = \langle [(\boldsymbol{v}(\boldsymbol{r},t) - \boldsymbol{v}(0,t))\cdot\boldsymbol{r}/r]^n \rangle$ and treat first the three-dimensional case. Similar to (10), one considers distance r smaller than the force correlation scale for a steady case and smaller than the size of the turbulent region for a decay case. For such r, one can derive the Karman–Howarth relation between S_2 and S_3 (see Landau & Lifshitz, 1987):

$$\frac{\partial S_2}{\partial t} = -\frac{1}{3r^4}\frac{\partial}{\partial r}(r^4 S_3) + \frac{4\varepsilon}{3} + \frac{2\nu}{r^4}\frac{\partial}{\partial r}\left(r^4\frac{\partial S_2}{\partial r}\right). \tag{13}$$

Here $\varepsilon = \nu\langle(\nabla\boldsymbol{v})^2\rangle$ is the mean energy dissipation rate. Neglecting the time derivative (which is zero in a steady state and small compared with ε for decaying turbulence), one can multiply (13) by r^4 and integrate $S_3(r) = -4\varepsilon r/5 + 6\nu\, \mathrm{d}S_2(r)/\mathrm{d}r$. Andrei Kolmogorov in 1941 considered the limit $\nu\to 0$ for fixed r and *assumed* nonzero limit for ε, which gives the so-called $\frac{4}{5}$ law (see Landau & Lifshitz, 1987; Frisch, 1995):

$$S_3 = -\tfrac{4}{5}\varepsilon r. \tag{14}$$

This relation is a direct analog of (6) and (11). It also means that the kinetic energy has a constant flux in the inertial interval of scales (the viscous scale η is defined by $\nu S_2(\eta)\simeq\varepsilon\eta^2$). Law (14) implies that the third-order moment is universal; that is, it does not depend on the details of the turbulence production but is determined solely by the mean energy dissipation rate. The rest of the structure functions have not yet been derived. Kolmogorov (and also Werner Heisenberg, Karl von Weizsacker, and Lars Onsager) presumed the pair correlation function to be determined only by ε and r which would give $S_2(r)\sim(\varepsilon r)^{2/3}$ and the energy spectrum $E_k \sim \varepsilon^{2/3}k^{-5/3}$. Experiments suggest that $\zeta_n = \mathrm{d}\ln S_n/\mathrm{d}\ln r$ lie on a smooth concave curve sketched in Figure 1. While ζ_2 is close to 2/3, it has to be a bit larger because experiments show that the slope at zero $\mathrm{d}\zeta_n/\mathrm{d}n$ is larger than $\frac{1}{3}$ while $\zeta(3)=1$ in agreement with (14). As in Burgers turbulence, the PDF of velocity differences in the inertial interval is not scale-invariant in 3-d incompressible turbulence. No one has yet found an explicit relation between the anomalous scaling for 3-d Navier–Stokes turbulence and either structures or additional integrals of motion.

While not exact, the Kolomogorov approximation $S_2(\eta)\simeq(\varepsilon\eta)^{2/3}$ can be used to estimate the viscous scale: $\eta \simeq L\mathrm{Re}^{-3/4}$. The number of degrees of freedom involved in 3-d incompressible turbulence can thus be roughly estimated as $N\sim(L/\eta)^3\sim\mathrm{Re}^{9/4}$. That means, in particular, that detailed computer simulation of water or oil pipe flows ($\mathrm{Re}\sim 10^4 - 10^7$) or turbulent clouds ($\mathrm{Re}\sim 10^6 - 10^9$) is out of question for the foreseeable future. To calculate correctly at least the large-scale part of the flow, it is desirable to have some theoretical model to parametrize the small-scale motions, the main obstacle being our lack of qualitative understanding and quantitative description of how turbulence statistics changes as one goes downscale.

Large-scale motions in a shallow fluid can be approximately considered two dimensional. When the velocities of such motions are much smaller than the velocities of the surface waves and the velocity of sound, such flows can be considered incompressible. Their description is important for understanding atmospheric and oceanic turbulence at the scales larger than atmosphere height and ocean depth.

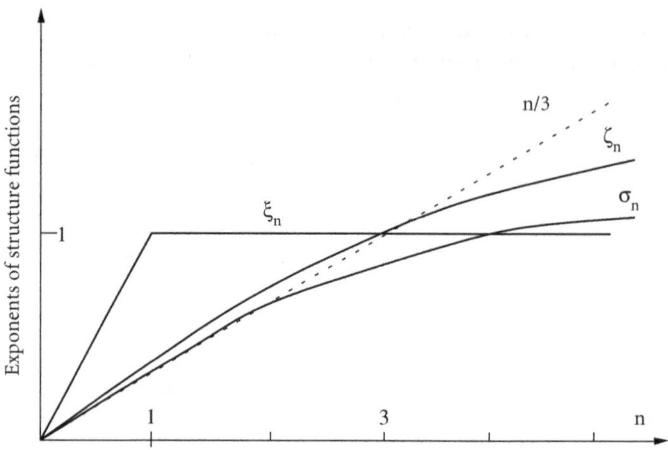

Figure 1. The scaling, exponents of the structure functions ξ_n for Burgers, ζ_n for Navier–Stokes, and σ_n for the passive scalar. The dotted straight line is the Kolmogorov hypothesis $n/3$.

Vorticity $\omega = \text{curl}\, \boldsymbol{v}$ is a scalar in a two-dimensional flow. It is advected by the velocity field and dissipated by viscosity. Taking the curl of the Navier–Stokes equation, one gets

$$\partial_t \omega + (\boldsymbol{v} \cdot \nabla)\omega = \nu \nabla^2 \omega. \tag{15}$$

Two-dimensional incompressible inviscid flow just transports vorticity from place to place and thus conserves spatial averages of any function of vorticity. In particular, we now have the second quadratic inviscid invariant (in addition to energy) which is called enstrophy: $\int \omega^2\, d\boldsymbol{r}$. Since the spectral density of the energy is $|\boldsymbol{v}_k|^2/2$ while that of the enstrophy is $|\boldsymbol{k} \times \boldsymbol{v}_k|^2$, Robert Kraichnan suggested in 1967 that the direct cascade (towards large k) is that of enstrophy while the inverse cascade is that of energy. Again, for the inverse energy cascade, there is no consistent theory except for the flux relation that can be derived similar to (14):

$$S_3(r) = 4\varepsilon r/3. \tag{16}$$

The inverse cascade is observed in the atmosphere (at scales of 30–500 km) and in laboratory experiments. Experimental data suggest that there is no anomalous scaling; thus, $S_n \propto r^{n/3}$. In particular, $S_2 \propto r^{2/3}$ which corresponds to $E_k \propto k^{-5/3}$. It is ironic that probably the most widely known statement on turbulence, the $\frac{5}{3}$ spectrum suggested by Kolmogorov for the 3-d case, is not correct in this case (even though the true scaling is close), while it is probably exact in Kraichnan's inverse 2-d cascade. Qualitatively, it is likely that the absence of anomalous scaling in the inverse cascade is associated with the growth of the typical turnover time (estimated, say, as $r/\sqrt{S_2}$) with the scale. As the inverse cascade proceeds, the fluctuations have enough time to get smoothed out as opposed to the direct cascade in three dimensions, where the turnover time decreases in the direction of the cascade.

Before discussing the direct (enstrophy) cascade, we describe a similar yet somewhat simpler problem of passive scalar turbulence, which allows one to introduce the necessary notions of Lagrangian description of the fluid flow. Consider a scalar quantity $\theta(\boldsymbol{r}, t)$ that is subject to molecular diffusion and advection by the fluid flow but has no back influence on the velocity (i.e., is passive):

$$\partial_t \theta + (\boldsymbol{v} \cdot \nabla)\theta = \kappa \nabla^2 \theta + \varphi. \tag{17}$$

Here κ is molecular diffusivity. In the same 2-d flow, ω and θ behave in the same way, but vorticity is related to velocity while the passive scalar is not. Examples of passive scalar are smoke in air, salinity in water, and temperature when one neglects thermal convection. If the source φ produces fluctuations of θ on some scale L, then the inhomogeneous velocity field stretches, contracts, and folds the field θ producing progressively smaller and smaller scales. If the rms velocity gradient is Λ, then molecular diffusion is substantial at scales less than the diffusion scale $r_d = \sqrt{\kappa/\Lambda}$. The ratio

$$\text{Pe} = L/r_d$$

is called the Péclet number. It is an analog of the Reynolds number for passive scalar turbulence. When $\text{Pe} \gg 1$, there is a long inertial interval where the flux constancy relation derived by A.M. Yaglom in 1949 holds,

$$\langle (\boldsymbol{v}_1 \cdot \nabla_1 + \boldsymbol{v}_2 \cdot \nabla_2)\theta_1\theta_2 \rangle = 2P, \tag{18}$$

where $P = \kappa \langle (\nabla\theta)^2 \rangle$ and subscripts denote the spatial points. In considering the passive scalar problem, the velocity statistics is presumed to be given. Still, the correlation function (18) mixes $\boldsymbol{v}$ and θ and does not generally allow one to make a statement on any correlation function of θ. The proper way to describe the correlation functions of the scalar at scales much larger than the diffusion scale is to employ the Lagrangian description, that is, to follow fluid trajectories. Indeed, if we neglect diffusion, then Equation (17) can be solved along the characteristics $\boldsymbol{R}(t)$ which are called Lagrangian trajectories and satisfy $\mathrm{d}\boldsymbol{R}/\mathrm{d}t = \boldsymbol{v}(\boldsymbol{R}, t)$. Presuming zero initial conditions at $t \to -\infty$, we write

$$\theta\Big(\boldsymbol{R}(t), t\Big) = \int_{-\infty}^{t} \varphi\Big(\boldsymbol{R}(t'), t'\Big)\,\mathrm{d}t'. \tag{19}$$

In that way, the correlation functions of the scalar $F_n = \langle \theta(\boldsymbol{r}_1, t) \ldots \theta(\boldsymbol{r}_n, t) \rangle$ can be obtained by integrating the correlation functions of the pumping along the trajectories that satisfy the final conditions $\boldsymbol{R}_i(t) = \boldsymbol{r}_i$.

Consider first, the case of pumping which is Gaussian, statistically homogeneous, and isotropic in space and white in time: $\langle \varphi(\boldsymbol{r}_1, t_1)\varphi(\boldsymbol{r}_2, t_2) \rangle = \Phi(|\boldsymbol{r}_1 - \boldsymbol{r}_2|)\,\delta(t_1 - t_2)$ where the function Φ is constant at $r \ll L$ and goes to zero at $r \gg L$. The pumping provides for symmetry $\theta \to -\theta$ which makes only even correlation functions F_{2n} nonzero. The pair correlation function is

$$F_2(r, t) = \int_{-\infty}^{t} \Phi\Big(R_{12}(t')\Big)\,\mathrm{d}t'. \tag{20}$$

Here $R_{12}(t') = |\boldsymbol{R}_1(t') - \boldsymbol{R}_2(t')|$ is the distance between two trajectories and $R_{12}(t) = r$. The function Φ essentially restricts the integration to the time interval when the distance $R_{12}(t') \leq L$. Simply speaking, the stationary pair correlation function of a tracer is $\Phi(0)$ (which is twice the injection rate of θ^2) times the average time $T_2(r, L)$ that two fluid particles spend within the correlation scale of the pumping. The larger r, the less time it takes for the particles to separate from r to L and the smaller is $F_2(r)$. Of course, $T_{12}(r, L)$ depends on the properties of the velocity field. A general theory is available only when the velocity field is spatially smooth at the scale of scalar pumping L. This so-called Batchelor regime happens, in particular, when the scalar cascade occurs at the scales less than the viscous scale of fluid turbulence. This requires the Schmidt number ν/κ (called the Prandtl number when θ is temperature) to be large, which is the case for very viscous liquids. In this case, one can approximate the velocity difference $\boldsymbol{v}(\boldsymbol{R}_1, t) - \boldsymbol{v}(\boldsymbol{R}_2, t) \approx \hat{\sigma}(t)\boldsymbol{R}_{12}(t)$ with the Lagrangian strain matrix $\sigma_{ij}(t) = \nabla_j v_i$. In this regime, the distance obeys the linear differential equation

$$\dot{\boldsymbol{R}}_{12}(t) = \hat{\sigma}(t)\boldsymbol{R}_{12}(t). \tag{21}$$

The theory of such equations is well developed and related to what is called Lagrangian chaos, as fluid trajectories separate exponentially as is typical for systems with dynamical chaos (see, e.g., Falkovich et al., 2001): At t much larger than the correlation time of the random process $\hat{\sigma}(t)$, all moments of R_{12} grow exponentially with time and $\langle \ln[R_{12}(t)R_{12}(0)] \rangle = \lambda t$, where λ is called a senior Lyapunov exponent of the flow (note that for the description of the scalar we need the flow taken backwards in time which is different from that taken forward because turbulence is irreversible). Dimensionally, $\lambda = \Lambda f(\mathrm{Re})$ where the limit of the function f at $\mathrm{Re} \to \infty$ is unknown. We thus obtain

$$F_2(r) = \Phi(0)\lambda^{-1}\ln(L/r) = 2P\lambda^{-1}\ln(L/r). \tag{22}$$

In a similar way, one shows that for $n \ll \ln(L/r)$, all F_n are expressed via F_2 and the structure functions $S_{2n} = \langle [\theta(\boldsymbol{r}, t) - \theta(0, t)]^{2n} \rangle \propto \ln^n(r/r_d)$ for $n \ll \ln(r/r_d)$. This can be generalized for an arbitrary statistics of pumping as long as it is finite-correlated in time (Falkovich et al., 2001).

One can use the analogy between passive scalar and vorticity in two dimensions as has been shown by Falkovich and Lebedev in 1994 following the line suggested by Kraichnan in 1967. For the enstrophy cascade, one derives the flux relation analogous to (18):

$$\langle(\boldsymbol{v}_1 \cdot \nabla_1 + \boldsymbol{v}_2 \cdot \nabla_2)\omega_1\omega_2\rangle = 2D, \tag{23}$$

where $D = \langle \nu(\nabla\omega)^2 \rangle$. The flux relation along with $\omega = \operatorname{curl} \boldsymbol{v}$ suggests the scaling $\delta v(r) \propto r$, that is, velocity being close to spatially smooth (of course, it cannot be perfectly smooth to provide for a nonzero vorticity dissipation in the inviscid limit, but the possible singularitites are indeed shown to be no stronger than logarithmic). That makes the vorticity cascade similar to the Batchelor regime of passive scalar cascade with a notable change in that the rate of stretching λ acting on a given scale is not a constant but is logarithmically growing when the scale decreases. Since λ scales as vorticity, the law of renormalization can be established from dimensional reasoning, and one gets $\langle\omega(\boldsymbol{r}, t)\omega(0, t) \rangle \sim [D\ln(L/r)]^{2/3}$ which corresponds to the energy spectrum $E_k \propto D^{2/3}k^{-3}\ln^{-1/3}(kL)$. Higher-order correlation functions of vorticity are also logarithmic, for instance, $\langle \omega^n(\boldsymbol{r}, t)\omega^n(0, t) \rangle \sim [D\ln(L/r)]^{2n/3}$. Note that both passive scalar in the Batchelor regime and vorticity cascade in two dimensions are universal, that is, determined by the single flux (P and D, respectively) despite the existence of higher-order conserved quantities. Experimental data and numeric simulations support these conclusions.

Zero Modes and Anomalous Scaling

Let us now return to the Lagrangian description and discuss it when velocity is not spatially smooth, for example, that of the energy cascades in the inertial interval. One can assume that it is Lagrangian statistics that are determined by the energy flux when the distances between fluid trajectories are in the inertial interval. That assumption leads, in particular, to the Richardson law for the asymptotic growth of the interparticle distance:

$$\langle R_{12}^2(t)\rangle \sim \varepsilon t^3, \tag{24}$$

which was first established from atmospheric observations (in 1926) and later confirmed experimentally for energy cascades both in 3-d and in 2-d. There is no consistent theoretical derivation of (24), and it is unclear whether it is exact (likely to be in 2-d) or just approximate (possible in 3-d). The semi-heuristic argument usually presented in textbooks is based on the mean-field estimate: $\dot{\boldsymbol{R}}_{12} = \delta\boldsymbol{v}(\boldsymbol{R}_{12}, t) \sim (\varepsilon R_{12})^{1/3}$, which upon integration gives $R_{12}^{2/3}(t) - R_{12}^{2/3}(0) \sim \varepsilon^{1/3} t$. For the passive scalar it gives, by virtue of (20), $F_2(r) \sim \Phi(0)\varepsilon^{-1/3}[L^{2/3} - r^{2/3}]$ which was suggested by S. Corrsin and A.M. Oboukhov. The structure function is then $S_2(r) \sim \Phi(0)\varepsilon^{-1/3} r^{2/3}$. Experiments measuring the scaling exponents $\sigma_n = \mathrm{d}\ln S_n(r)/\mathrm{d}\ln r$ generally give σ_2 close to 2/3 but higher exponents deviating from the straight line are even stronger than the exponents of the velocity in 3-d. Moreover, the scalar exponents σ_n are anomalous even when advecting velocity has a normal scaling like in the 2-d energy cascade.

To better understand the Lagrangian dynamics (and passive scalar statistics) in a spatially nonsmooth velocity, Kraichnan suggested considering the model of a velocity field as having the simplest statistical and temporal properties, namely Gaussian velocity which is white in time:

$$\langle v^i(\boldsymbol{r}, t) v^j(0, 0)\rangle = \delta(t)\Big[D_0\delta_{ij} - d_{ij}(\boldsymbol{r})\Big],$$
$$d_{ij} = D_1 r^{2-\gamma}\Big[(d+1-\gamma)\,\delta^{ij} + (\gamma-2) r^i r^j r^{-2}\Big]. \tag{25}$$

Here the exponent $\gamma \in [0, 2]$ is a measure of the velocity nonsmoothness with $\gamma = 0$ corresponding to a smooth velocity and $\gamma = 2$ corresponding to a velocity very rough in space (distributional). Richardson–Kolmogorov scaling of the energy cascade corresponds to $\gamma = 2/3$. Lagrangian flow is a Markov random process for the Kraichnan ensemble (25). Every fluid particle undergoes a Brownian random walk with the so-called eddy diffusivity D_0. The PDF for two particles to be separated by r after time t satisfies the diffusion equation (see, e.g., Falkovich et al., 2001)

$$\partial_t P(r, t) = L_2 P(r, t),$$
$$L_2 = d_{ij}(\boldsymbol{r})\nabla^i\nabla^j = D_1(d-1) r^{1-d}\partial_r r^{d+1-\gamma}\partial_r, \tag{26}$$

with the scale-dependent diffusivity $D_1(d-1)r^{2-\gamma}$. The asymptotic solution of (26) is lognormal for the Batchelor case while for $\gamma > 0$

$$P(r, t) = r^{d-1} t^{d/\gamma} \exp\Big(-\mathrm{const}\, r^\gamma/t\Big). \tag{27}$$

For $\gamma = 2/3$, it reproduces, in particular, the Richardson law. Multiparticle probability distributions also satisfy diffusion equations in the Kraichnan model as well as all the correlation functions of θ. Multiplying equation (17) by $\theta_2 \ldots \theta_{2n}$ and averaging over the Gaussian statistics of $\boldsymbol{v}$ and φ, one derives

$$\partial_t F_{2n} = L_{2n} F_{2n} + \sum_{l,m} F_{2n-2}\Phi(\boldsymbol{r}_{lm}),$$
$$L_{2n} = \sum d_{ij}(\boldsymbol{r}_{lm})\nabla_l^i\nabla_m^j. \tag{28}$$

This equation enables one, in principle, to derive inductively all steady-state F_{2n} starting from F_2. The equation $\partial_t F_2(r, t) = L_2 F_2(r, t) + \Phi(r)$ has a steady solution $F_2(r) = 2[\Phi(0)/\gamma d(d-1)D_1][dL^\gamma/(d-\gamma) - r^\gamma]$, which has the Corrsin–Oboukhov form for $\gamma = 2/3$. Further, F_4 contains the so-called forced solution having the normal scaling 2γ but also, remarkably, a zero mode Z_4 of the operator L_4: $L_4 Z_4 = 0$. Such zero modes necessarily appear (to satisfy the boundary conditions at $r \simeq L$) for all $n > 1$, and the scaling exponents of Z_{2n} are generally different from $n\gamma$ that is anomalous. In calculating the scalar structure functions, all terms cancel out except a single zero mode (called irreducible because it involves all distances between $2n$ points). Calculations of Z_n and their scaling exponents σ_n were carried out analytically at $\gamma \ll 1$, $2 - \gamma \ll 1$ and $d \gg 1$, and numerically for all γ and $d = 2, 3$ (Falkovich et al., 2001).

That gives σ_n lying on a convex curve (as in Figure 1) which saturates to a constant at large n. Such saturation (confirmed by experiments) is a signature that most singular structures in a scalar field are shocks (as in Burgers turbulence), the value σ_n at $n \to \infty$ is the fractal codimension of fronts in space. Interestingly, the Kraichnan model enables one to establish the relation between the anomalous scaling and conservation laws of a new type. Thus, the combinations of distances between points that constitute zero modes are the statistical integrals of Lagrangian evolution. To give a simple example, in a Brownian walk, the mean distance between every two particles grows with time, $\langle R_{lm}^2(t)\rangle = R_{lm}^2(0) + \kappa t$, while $\langle R_{lm}^2 - R_{pq}^2\rangle$ and $\langle 2(d+2)R_{lm}^2 R_{pq}^2 - d(R_{lm}^4 + R_{pq}^4)\rangle$ (and an infinity of similarly built harmonic polynomials) are conserved. Note that the integrals are not dynamical, they are conserved only in average. In a turbulent flow, the form of such conserved quantities is more complicated, but the essence is the same: the increase of averaged distances between fluid particles is compensated by the decrease in shape fluctuations. The existence of statistical conserved quantities breaks the scale invariance of scalar statistics in the inertial interval and explains why scalar turbulence knows more about pumping than just the value of the flux. Note that both symmetries, one broken by pumping (scale invariance) and another by damping (time reversibility) are not restored even when $r/L \to 0$ and $r_d/r \to 0$.

For the vector field (like velocity or magnetic field in magnetohydrodynamics), the Lagrangian statistical integrals of motion may involve both the coordinate of the fluid particle and the vector it carries. Such integrals of motion were built explicitly and related to the anomalous scaling for the passively advected magnetic field in the Kraichnan ensemble of velocities (Falkovich et al., 2001). Doing the same for velocity that satisfies the Navier–Stokes equation remains a task for the future.

GREGORY FALKOVICH

See also **Burgers equation; Chaos vs. turbulence; Development of singularities; Intermittency; Kolmogorov cascade; Lagrangian chaos; Magnetohydrodynamics; Mixing; Navier–Stokes equation; Nonlinear Schrödinger equations; Water waves; Wave packets, linear and nonlinear**

Further Reading

Falkovich, G., Gawędzki, K. & Vergassola, M. 2001. Particles and fields in fluid turbulence, *Reviews of Modern Physics*, 73: 913–975

Frisch, U. 1995. *Turbulence: The Legacy of A.N. Kolmogorov*, Cambridge and New York: Cambridge University Press

Landau, L. & Lifshitz, E. 1987. *Fluid Mechanics*, 2nd edition, Oxford and New York: Pergamon Press

Phillips, O. 1977. *The Dynamics of the Upper Ocean*, 2nd edition, Cambridge and New York: Cambridge University Press

Zakharov, V., L'vov, V. & Falkovich, G. 1992. *Kolmogorov Spectra of Turbulence*, Berlin and New York: Springer

TURBULENCE, IDEAL

Ideal turbulence (IT) is a mathematical phenomenon that occurs in certain infinite-dimensional deterministic dynamical systems. The attractor of an IT system lies off the phase space, and among the attractor points there are fractal or even random functions. IT is observed in various idealized models of real distributed systems (electrodynamics, acoustics, radiophysics, etc.), and it helps to understand the mathematical scenarios for features of real turbulence. Cascade processes in IT are capable of giving birth to structures of arbitrarily small scale and even causing stochastization of the systems.

A mathematically rigorous definition of ideal turbulence is based on notions of dynamical systems theory and chaos theory. Spatiotemporal chaotization in dynamical systems on spaces of smooth or piecewise smooth functions is perceived as a cascading evolution of such functions with the result that their behavior becomes more and more intricate (see Figure 1), whereupon the limiting states cannot be described with smooth functions. This implies that the attractor of the dynamical system is not contained entirely in the phase space; thus, the dynamical system needs to be extended on a wider functional space so that this new space contains whole "ω-limit" sets of all or almost all trajectories. (The ω-limit set of a trajectory is defined as the attractor of the trajectory or, more precisely, as the set of limit points of the trajectory.) The spaces of fractal and random functions are particularly appealing for use as a wider space.

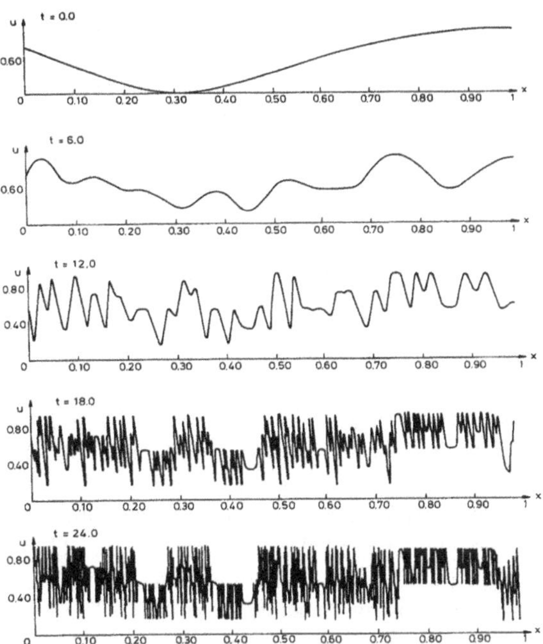

Figure 1. Start of ideal turbulence: Typical instantaneous distributions of current in a lossless transmission line described by the boundary-value problem $i_x = -Cv_t$, $v_x = -Li_t$, and $v(0,t) = 0$, $i(1,t) = G(v(1,t))$, where i and v are the current and voltage along the line, and G specifies the v-i characteristic of an Esaki (tunnel) diode fixing the boundary condition at $x = 1$.

If, with such an extension, the ω-limit set of the trajectory corresponding to some initial state contains a "point" that is a fractal function, then this initial state is said to generate IT. Similarly, if a dynamical system can be extended on a space containing both deterministic and random functions and for some initial state its associated ω-limit set contains a "point" that is a random function, then this initial state is said to generate stochastic ideal turbulence (SIT). If initial states generate IT or SIT, then IT or SIT is said to occur in the dynamical system.

For a space containing fractal functions, one may take the space of multivalued functions with the metric $\rho(\zeta_1, \zeta_2) = \mathrm{dist}_H(\mathrm{gr}\,\zeta_1, \mathrm{gr}\,\zeta_2)$, where $\mathrm{dist}_H(\cdot,\cdot)$ is the Hausdorff distance between sets and $\mathrm{gr}\,\zeta$ denotes the graph of ζ. As any function (deterministic or random) can be interpreted as the collection of all its finite-dimensional distributions, a metric for spaces containing random and deterministic functions is conveniently chosen to compare the distributions of functions (Sharkovsky & Romanenko, 1992).

This classification can be deepened. For instance, an initial state is said to generate weak ideal turbulence (WIT) if it does not generate IT but its associated

ω-limit set contains a function that is multivalued at an infinite number of points.

A simple example of a system with turbulence is the discrete dynamical system acting on the space of smooth functions $\varphi : D \to E$ according to the rule

$$S : \varphi(x) \mapsto f(\varphi(x)), \tag{1}$$

where $f : E \to E$ is a smooth function, and D and E are regions of Euclidean spaces. The trajectory through a point φ is the sequence $f^n(\varphi(x))$, $n = 0, 1, 2, \ldots$, where the superscript n denotes the nth iteration. Thus, the dynamics of the trajectory can be treated as the dynamics of a continuum of uncoupled oscillators. At every point $x \in D$, there is a "pendulum," oscillating under the law $z_n \mapsto z_{n+1} = f(z_n)$ with $z_0 = \varphi(x)$ and independently of the pendula at other points of D. The independence of the oscillators causes IT in the dynamical systems (1), and moreover, when f has the property of sensitive dependence on initial data on some open set $E' \subset E$, those φ such that $\varphi(D) \supset E'$ often generate SIT. In more general situations that occur in applied problems, the oscillation law depends on initial data φ and/or a point $x \in D$; it can also be time-dependent.

A description of long-term properties for the dynamical systems (1) is most advantageous when f is a one-dimensional map, and D and E are intervals, whereupon one has the following result:

> **Theorem.** *There occur (i) weak ideal turbulence, if f has periodic trajectories of periods $2^i, 0 \le i \le l$, with some $l > 1$, and no other periodic trajectories; (ii) ideal turbulence, if f has a periodic trajectory of period $\neq 2^i, i = 0, 1, \ldots$; and (iii) stochastic turbulence, if f possesses an ergodic smooth invariant measure.*

The map $f : z \mapsto 4z(1 - z)$, $z \in [0, 1]$, has an invariant measure with the density $p(z) = 1/\pi\sqrt{z(1 - z)}$, and almost each $\varphi : [0, 1] \to (0, 1)$ generates SIT. Its associated ω-limit set consists of a single point which is the random function with the distribution $F(x, z) = \int_0^z p(z)\,dz = (2/\pi) \arcsin \sqrt{z}$. Thus the attractor of the dynamical system (1) consists of just this one point. For $f = f_\lambda : z \mapsto \lambda z(1 - z)$, $0 < \lambda \le 4$, there exists a set of positive Lebesgue measure $\Lambda \subset (3, 4]$ such that f_λ with $\lambda \in \Lambda$ has an ergodic smooth invariant measure on $[0, 1]$; hence SIT in the dynamical systems (1) is a non-exclusive phenomenon.

Many evolutionary boundary-value problems (BVPs) for partial differential equations induce dynamical systems of shifts along solutions. It is natural to say that there arises IT or SIT in a BVP if such turbulence occurs in the corresponding dynamical system. The simplest example is the BVP

$$w_t - w_x = 0, 0 \le x \le 1, t \ge 0, \tag{2}$$

$$w(1, t) = f(w(0, t)), \tag{3}$$

Figure 2. Towards stochastic turbulence: Typical evolution of flow lines for the vector field (w^1, w^2) given by $w_t^1 = w_x^1 + w_y^1$, $w_t^2 = -w_x^2 - w_y^2$, $-\infty < x < +\infty$, $0 \le y \le 1$, and $w^1 = w^2|_{y=0}$, $w^1 = f(w^2)|_{y=1}$ with $f(z) = 1 - 2z^2$. The attractor of the BVP consists of one point—the random vector field $(\hat{w}^1, \hat{w}^2)$ whose components have the same (x, y)-independent distribution density $1/2\pi\sqrt{1 - z^2}$, $-1 < z < 1$.

where f is a smooth function from some interval E into itself. On the space of smooth functions $\varphi : [0, 1] \to E$, the BVP induces the dynamical system of shifts $S^t : \varphi(x) \mapsto w_\varphi(x, t), t \ge 0$, where $w_\varphi(x, t)$ is the solution meeting the initial condition $w(x, 0) = \varphi(x)$. For the BVP considered, the shift operator S^t is represented as

$$S^t : \varphi(x) \mapsto f^{\langle x+t \rangle}(\varphi(\{x + t\})) \tag{4}$$

with $\langle \cdot \rangle$ and $\{\cdot\}$ being the integer and fractional part of a number, respectively. Thus the dynamical system (4) is a continuous analog of the dynamical system (1), and one can formulate conditions for turbulence in the BVP according to the above theorem.

Replacing (3) with $w_t(1, t) = g(w(0, t))\, w_t(0, t))$ leads to the dynamical system (4) where f is replaced with $f_{\gamma[\varphi]} = f + \gamma[\varphi]$, f is an antiderivative of g and $\gamma[\varphi] = f(\varphi(0)) - \varphi(1)$. The type of the turbulence arising in a solution $w_\varphi(x, t)$ follows the theorem, as applied to $f_{\gamma[\varphi]}$.

BVPs for the wave equation (and related systems) provide examples of ideal turbulence. For the BVP

$$w_{tt} - w_{xx} = 0, \quad 0 \le x \le 1, \tag{5}$$

$$w(0, t) = 0, \quad w_t(1, t) = h(w_x(1, t)), \tag{6}$$

its associated shift operator S^t is expressed through a one-dimensional map $f: z_n \mapsto z_{n+1}$, defined implicitly by $z_{n+1} - z_n = h(z_n + z_{n+1})$. This allows one to find conditions for IT or SIT for particular functions h (Sharkovsky, 1994; Sharkovsky et al., 1995). When the conditions of (6) are replaced with $w(0,t) = 0$, $w_x(1,t) = h(w_x(0,t))$, there arises a two-dimensional map defined by $z_{n+1} - z_{n-1} = h(z_n)$.

There are many other one- and many-dimensional BVPs whose dynamics are described in terms of low-dimensional maps, as in the above examples. In these cases, the theory of maps suggests why and how turbulence occurs in the BVP and presents scenarios for self-structuring and self-stochastization. Of importance here are the following properties of maps: the intricate dynamical structure of the basins of attracting cycles, the local self-similarity of the set of points with unstable trajectories, and the occurrence of a smooth invariant measure. Figure 2 is an example of how processes of self-structuring lead to stochastic turbulence.

A.N. SHARKOVSKY AND E.YU. ROMANENKO

See also **Attractors; Butterfly effect; Chaotic dynamics; Dimensions; Dynamical systems; Ergodic theory; Maps; Measures; Mixing; One-dimensional maps; Phase space; Routes to chaos; Sinai–Ruelle–Bowen measures; Turbulence**

Further Reading

Romanenko, E.Yu. & Sharkovsky, A.N. 1996. From one-dimensional to infinite-dimensional dynamical systems: ideal turbulence. *Ukrainian Mathematical Journal*, 48(12): 1817–1842

Romanenko, E.Yu., Sharkovsky, A.N. & Vereikina, M.B. 1995. Self-structuring and self-similarity in boundary value problems. *International Journal of Bifurcation and Chaos*, 5(5): 145–156

Sharkovsky, A.N. 1994. Ideal turbulence in an idealized time-delayed Chua's circuit. *International Journal of Bifurcation and Chaos*, 4(2): 303–309

Sharkovsky, A.N., Deregel, Ph. & Chua, L.O. 1995. Dry turbulence and period-adding phenomena from a 1-D map. *International Journal of Bifurcation and Chaos*, 5(5): 1283–1302

Sharkovsky, A.N. & Romanenko, E.Yu. 1992. Ideal turbulence: attractors of deterministic systems may lie in the space of random fields. *International Journal of Bifurcation and Chaos*, 2(1): 31–36

TURING PATTERNS

As noted by D'Arcy Wentworth Thompson (1917) people in the early 1900s (when the study of symmetry-breaking instabilities was still in its infancy) had already considered the possibility of generating stationary regular concentration patterns through the interplay of diffusion and chemistry. At mid-century, the British mathematician and computing pioneer, Alan Turing, was the first to formulate necessary conditions for the occurrence of space symmetry breaking in the context of biological morphogenesis (Turing, 1952). Following the emergence of the self-oscillating Belousov–Zhabotinsky reaction (Epstein & Pojman, 1998) in the mid-1960s, Ilya Prigogine and coworkers revived Turing's concept, put it on sound thermodynamic and kinetic grounds, and showed that it could only be sustained in continuously fed reactors at a finite distance from equilibrium (Nicolis & Prigogine, 1977).

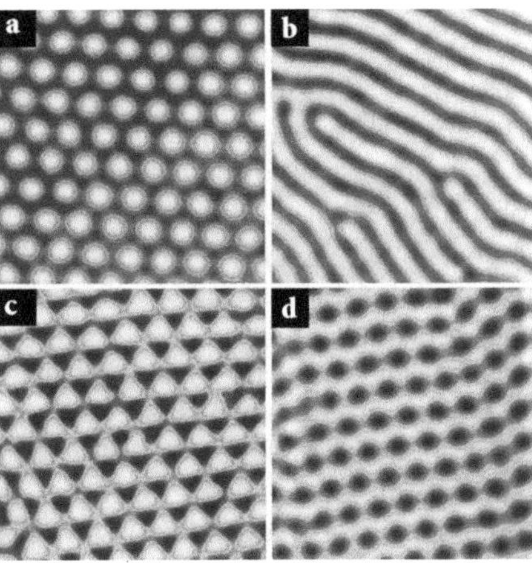

Figure 1. (a)–(d) Turing structures of different symmetries obtained with the chlorite-iodide-malonic acid reaction. Dark and light regions, respectively, correspond to high and low iodide concentration. The wavelength, a function of kinetic parameters and diffusion coefficients, is of the order of 0.2 mm. All patterns are at the same scale: view size 1.7mm × 1.7mm (Courtesy P. De Kepper, CRPP).

This work opened up a whole new field of physical chemistry. Many theoretical studies followed, and the diffusive instability that generates such dissipative structures has popped up in other domains of physics and chemistry (Ball, 1999). However, experiments in the chemical realm lagged behind, and it was only in 1989 that the first experimental evidence was obtained by De Kepper and his group (Castets et al., 1990) using the chlorite-iodide-malonic acid reactive system in so-called gel reactors. A recent detailed status of Turing patterns and other symmetry-breaking instabilities in solution chemistry is presented in Borckmans et al. (2002).

The Turing–Prigogine mechanism consists in the spontaneous instability of a homogeneous mixture of chemically reacting species, when some parameter threshold is crossed as one moves away from equilibrium conditions. It leads to stationary, space-periodic patterns for the concentrations of reactants (see Figure 1). In its minimal form, the description of all the systems that exhibit such diffusive instability can formally be cast in the common language of reaction-diffusion systems governed by the set of equations

$$\frac{\partial \boldsymbol{c}(\boldsymbol{r},t)}{\partial t} = \boldsymbol{f}(\boldsymbol{c},b) + \nabla \cdot D\nabla \boldsymbol{c}(\boldsymbol{r},t), \qquad (1)$$

where $\boldsymbol{c}(\boldsymbol{r}, t) \equiv (\ldots, c_i, \ldots)$ is the local concentration vector, $\boldsymbol{f}(\boldsymbol{c}, b)$ is a vector function representing the reaction kinetics wherein lies the source of nonlinearity, b stands for a set of control parameters, and D is the matrix of diffusive transport coefficients. Appropriate initial and boundary conditions, in relation with the experimental setup are added to complete the mathematical formulation.

To support such symmetry-breaking instability, the chemical kinetics must involve some type of positive feedback loop controlled at least by an activator species that reinforces its own changes, the latter being counterbalanced by an inhibitory process. Spatial structures can form when the inhibitory effects are transported by diffusion over a larger space range than that of the activating mechanism. An intuitive picture may be obtained when a single activator (A) and inhibitor (H) are present. A autocatalytically promotes its own production and that of H, while the latter opposes the production of A. Consider such system in a nonequilibrium homogeneous steady state (hss) and quench it beyond the instability threshold. The hss then becomes very sensitive. A slight local fluctuation of the concentration of A will increase while it also spreads to the surroundings through diffusion. It will also start producing some H that, however, will diffuse away much faster from the point where the fluctuation occurred as $D_H > D_A$. H thereby hinders the propagation of A. A localized peak of activator surrounded by a barrier of H is thus created. In extended systems, such peaks tend to emerge everywhere, randomly distributed, and their interactions lead to the periodic concentration patterns. The beauty of Turing's idea lies in the counterintuitive organization role of diffusive processes when they compete with the proper autocatalytic chemistry, while diffusion still locally strives to erase any concentration of inhomogeneity.

Theoretical work uses nonlinear kinetic models for $\boldsymbol{f}(\boldsymbol{c}, b)$ with a limited number of chemical species, typically two or three (Brusselator, Oregonator, CDIMA, etc.). These models stand as a compromise between a minimum of chemical realism and mathematical tractability. For their part, the experimental kinetic schemes usually involve many species, often not fully determined (Epstein & Pojman, 1998).

Analytical work that relies heavily on bifurcation theory (Nicolis & Prigogine, 1977; Manneville, 1990) allows one to determine, through the solution of amplitude equations, which structures of given symmetry are stable for specific conditions (pattern selection). The calculated bifurcation diagrams help to organize the results obtained by straightforward numerical integration of the reaction-diffusion equations. Both types of information may be used to interpret the experimental results. This pattern selection problem was already on Turing's mind when he stated (Turing, 1952): "Most of an organism, most of the time, is developing from one pattern into another, rather than from homogeneity into a pattern. One would like to be able to follow this more general process mathematically also."

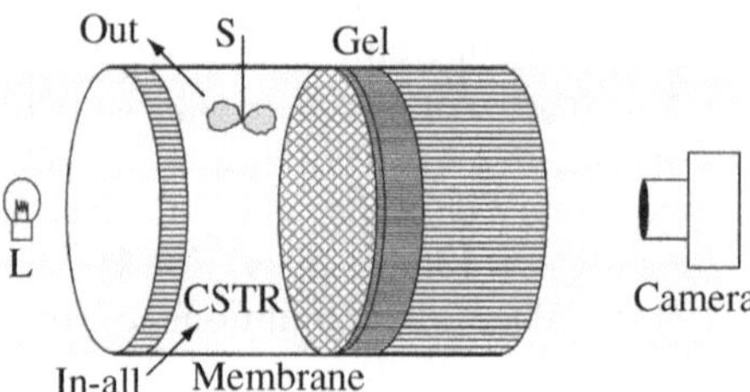

Figure 2. Schematic representation of a disc-shaped one side fed reactor (OSFR): CSTR (continuous stirred tank reactor), Membrane (mineral disc, pore size 0.02 mm) often placed to protect the gel from mechanical stress produced by the stirrer of the CSTR, Gel, In and Out (input and output ports of chemicals), L (light source), CCD camera.

The experimental work takes place in so-called open spatial reactors (Borckmans et al., 2002) which are specifically designed to control the reaction and the structures that eventually develop at a fixed distance from equilibrium and allow probing of the true asymptotic states of the reaction-diffusion systems. Experiments are now usually performed in a one-side fed reactor (OSFR) sketched in Figure 2. The core consists of a piece of soft hydrogel fed by diffusion through one of its faces with chemicals contained in a continuous stirred tank reactor (CSTR), the contents of which are continuously renewed by pumps. The other faces of the gel are pressed against impermeable transparent walls (Plexiglas). Viewing can be practiced both along the feeding axis or orthogonal to it (Ball, 1999; Ouyang & Swinney, 1991). The gel is used to avoid all perturbations induced by the hydrodynamic flows as those associated with the constant supply of fresh reactants, so that only reactive and diffusive processes compete. The necessary diffusion differential between activator and inhibitor species is obtained through the reversible binding of the activator molecules to the large molecular weight color indicator species that is included for visualization purposes. An advantage of such reactors is that they allow for direct correlations to be made between the dynamics of the CSTR, the bifurcation behaviors of which have been extensively studied in the past (Epstein & Pojman, 1998), and that of the gel.

Although scores of papers have been devoted to the application of Turing's idea to biological problems, this speculation remains to be confirmed (Epstein & Pojman, 1998; Borckmans et al., 2002).

PIERRE BORCKMANS AND GUY DEWEL

See also **Belousov–Zhabotinsky reaction; Brusselator; Morphogenesis, biological; Pattern formation; Reaction-diffusion systems**

Further Reading

Ball, P. 1999. *The Self-made Tapestry*, Oxford and New York: Oxford University Press

Borckmans, P., Dewel, G., De Wit, A., Dulos, E., Boissonade, J., Gauffre, F. & De Kepper, P. 2002. Diffusive instabilities and chemical reactions. *International Journal of Bifurcation and Chaos*, 12: 2307–2332

Castets, V., Dulos, E., Boissonade, J. & De Kepper, P. 1990. Experimental evidence of a sustained standing Turing-type nonequilibrium chemical pattern. *Physical Review Letters*, 64: 2953–2956

Cross, M.C. & Hohenberg, P.C. 1993. Pattern formation outside of equilibrium. *Reviews of Modern Physics*, 65: 851–1112

Epstein, I.R. & Pojman, J.A. 1998. *An Introduction to Nonlinear Chemical Dynamics*, Oxford and New York: Oxford University Press

Manneville, P. 1990. *Dissipative Structures and Weak Turbulence*, New York: Academic Press

Nicolis, G. & Prigogine, I. 1977. *Self-Organization in Nonequilibrium Systems*, New York: Wiley

Ouyang, Q. & Swinney, H.L. 1991. Transition from a uniform state of hexagonal and striped Turing patterns. *Nature*, 352: 610–612

Thompson, D'A.W. 1917. *On Growth and Form*, 2nd edition, Cambridge: Cambridge University Press (2nd edition, 1942); see quotes to S. Leduc

Turing, A. 1952. The chemical basis of morphogenesis *Philosophical Transactions of the Royal Society*, B 237: 37–42

Turing, A. 1992. *Morphogenesis*, Collected works of A. Turing, vol. 3, edited by P.T. Saunders, Amsterdam: Elsevier

http://www.turing.org.uk/, Alan Hodges's pages on Alan Turing

http://data.archives.ecs.soton.ac.uk/turing/, the Turing digital archive

http://www.swintons.net/jonathan/turing.htm, Alan Turing and morphogenesis

TWIST MAP

See **Nontwist maps**

TWISTOR THEORY

Introduced by Roger Penrose as a geometrical framework for the unification of quantum theory and general relativity (gravity), twistor theory brings out the complex (holomorphic) geometry that underlies real space-time. In general relativity, space-time is a four manifold with metric g. When $g = \mathrm{d}t^2 - \mathrm{d}x^2 - \mathrm{d}y^2 - \mathrm{d}z^2$, where (t, x, y, z) are coordinates on $\mathbb{R}^4$, g is said to be flat with signature (1,3) and is called Minkowski space.

The first appearance of a complex structure arises from the fact that at a given event, the celestial sphere of light rays (null directions with respect to g) naturally has the structure of the Riemann sphere, $\mathbb{CP}^1$, in such a way that Lorentz transformations (linear transformations of the tangent space preserving the metric) act on this sphere by Möbius transformations. Twistor space extends this idea to the whole of Minkowski space. Denoted $\mathbb{PT}$, the twistor space for Minkowski space is complex projective three space, $\mathbb{CP}^3$, the space of one-dimensional subspaces of $\mathbb{C}^4$; it is a three-dimensional complex manifold obtained by adding a "plane at infinity" to $\mathbb{C}^3$. Physically, points of twistor space correspond to spinning massless particles in Minkowski space. Mathematically, the correspondence can be understood as the Klein correspondence.

The Klein Correspondence

The correspondence between $\mathbb{PT}$ and Minkowski space can be extended first to complexified Minkowski space, so that the coordinates are allowed to take on values in $\mathbb{C}$, and then to its conformal compactification by including some points at infinity. It then coincides with the classical complex Klein correspondence. The Klein correspondence is the one-to-one correspondence between lines in $\mathbb{CP}^3$ and points of a four complex-dimensional quadric, $\mathbb{CM}$, in $\mathbb{CP}^5$. The four-quadric $\mathbb{CM}$ can be understood as conformally compactified complexified Minkowski space. Introducing affine coordinates (λ, z_1, z_2) on $\mathbb{PT}$, we find that a line in $\mathbb{PT}$ corresponds to a point (t, x, y, z) by

$$\begin{pmatrix} z_1 \\ z_2 \end{pmatrix} = \begin{pmatrix} t - z & x + \mathrm{i}y \\ x - \mathrm{i}y & t + z \end{pmatrix} \begin{pmatrix} 1 \\ \lambda \end{pmatrix}.$$

Alternatively, fixing (λ, z_1, z_2) in these equations gives a two-plane in complex Minkowski space corresponding to all the lines in $\mathbb{PT}$ through (λ, z_1, z_2). Such two-planes are called α-planes. They are totally null (i.e., the tangent vectors not only have zero length but are also mutually orthogonal) and also self-dual (under the differential geometer's notion of Hodge duality).

This complex correspondence can also be restricted to give correspondences for $\mathbb{R}^4$ with metrics of positive definite signature or ultra-hyperbolic (2, 2) signature.

The Penrose Transform

A basic task of twistor theory is to transform solutions to the field equations of mathematical physics into objects on twistor space. This works well for linear massless fields such as the Weyl neutrino equation, Maxwell's equations for electromagnetism, and linearized gravity. In its general form, this transform has become known as the Penrose transform. Such fields correspond to freely prescribable holomorphic functions $f(\lambda, z_1, z_2)$ (or, more precisely, analytic cohomology classes) on regions of twistor space. The field can be obtained from this function by means of a contour integral. The simplest of these integral formulae is

$$\phi(x^a) = \oint f(\lambda, t - z + \lambda(x + \mathrm{i}y), x - \mathrm{i}y + \lambda(t + z))\, \mathrm{d}\lambda, \tag{1}$$

and differentiation under the integral sign leads to the fact that ϕ satisfies the wave equation

$$\frac{\partial^2\phi}{\partial t^2} - \frac{\partial^2\phi}{\partial x^2} - \frac{\partial^2\phi}{\partial y^2} - \frac{\partial^2\phi}{\partial z^2} = 0 .$$

Equation (1) was originally discovered by Bateman (1910).

The Penrose transform has found important applications in representation theory and integral geometry. For a review, the reader is referred to Baston and Eastwood (1989), the relevant survey articles in Bailey & Baston (1990) or Chapter 1 of Mason et al. (1990).

Twistor Theory and Nonlinear Equations

The Penrose transform for the Maxwell equations and linearized gravity turns out to be linearizations of correspondences for nonlinear versions of these equations, the Einstein vacuum equations and the Yang–Mills equations, but only in the case that these fields are anti-self-dual. This is the condition that the curvature two-forms satisfy $F^* = -\mathrm{i}F$ where $*$ denotes the Hodge dual (which, up to a change of sign, has the effect of interchanging electric and magnetic fields); it is a nonlinear generalization of the right-handed circular polarization condition. In Minkowski signature, the i factor in the anti-self-duality condition implies that real fields cannot be anti-self-dual. Thus, these extensions are not sufficient to fulfill twistor theory's aim of incorporating real basic physics in Minkowski space. However, the factor of i is not present in Euclidean and ultrahyperbolic signature, so the anti-self-duality condition *is* consistent with real fields in these signatures, and this is where the main applications of these constructions have been.

The Nonlinear Graviton Construction and Its Generalizations

The first nonlinear twistor construction was due to Penrose (1976) and was inspired by Newman's construction of "heavens" from the infinities of asymptotically flat space-times in general relativity (Newman, 1976).

The nonlinear graviton construction proceeds from the definition of twistors in flat space-time as α-planes in complexified Minkowski space. It is natural to ask which complexified metrics admit a full family of α-surfaces, that is, two surfaces that are totally null and self-dual. The answer is that a full family of α-surfaces exists if and only if the conformally invariant part of the curvature tensor, the Weyl tensor, is anti-self-dual. In this case, twistor space can be defined to be the (necessarily three-dimensional) space of such α-surfaces. A remarkable fact is that the twistor space (together with its complex structure) is sufficient to determine the original space-time, and that the data defining the twistor space is effectively freely prescribable, see Penrose (1976) or Atiyah et al. (1978), for a discussion specialized to Euclidean signature.

There are now large families of extensions, generalizations and reductions of this construction. They are all based on the idea of realizing a space with a given complexified geometric structure as the parameter space of a family of holomorphically embedded submanifolds inside a twistor space. In general, the most useful of these constructions are those in which the "space-time" is obtained as the space of rational curves in a twistor space. This is because the equations that are solved on the corresponding space-time can be thought of as a completely integrable system. See Chapter 13 of Mason & Woodhouse (1996) for a more detailed discussion from this point of view.

The Anti-self-dual Yang–Mills Equations and Its Twistor Correspondence

The anti-self-dual Yang–Mills equations extend Maxwell's equations for electromagnetism in the right circularly polarized case. They are really a family of equations depending on a choice of Lie group G, usually taken to be a group of complex matrices, and Maxwell's equations arise from the case in which $G = U(1)$.

Introduce coordinates x^a, $a = 0, 1, 2, 3$, on $\mathbb{R}^4$ with metric $\mathrm{d}s^2 = \mathrm{d}x^0\,\mathrm{d}x^3 + \mathrm{d}x^1\,\mathrm{d}x^2$. The dependent variables are the components A_a of a connection $D_a = \partial_a - A_a$, where $\partial_a = \partial/\partial x^a$ and $A_a = A_a(x^b) \in$ Lie G, the Lie algebra of G. This connection defines a method of differentiating vector valued functions s in some representation of G. The freedom in changing bases for the vector bundle induce the gauge transformations $A_a \to g^{-1}A_a g - g^{-1}\partial_a g$, $g(x) \in G$ on A_a, and two connections that are related by a gauge transformation are deemed to be the same.

The self-dual Yang–Mills equations are the condition

$$[D_0, D_2] = [D_1, D_3] = [D_0, D_3] + [D_1, D_2] = 0 .$$

They are the compatibility conditions $[D_0 + \lambda D_1, D_2 + \lambda D_3] = 0$ for the linear system of equations

$$(D_0 + \lambda D_1)s = (D_2 + \lambda D_3)s = 0, \qquad (2)$$

where $\lambda \in \mathbf{C}$ and s is an n-component column vector. These last equations form a Lax pair for the system.

The Ward construction (Ward, 1977) provides a one-to-one correspondence between gauge equivalence classes of solutions of the self-dual Yang–Mills equations and holomorphic vector bundles on regions in twistor space. The key point here is that Equation (2) defines parallel propagation along α-planes. To each point Z in twistor space, we can associate the vector space E_Z of solutions to Equation (2) along the corresponding α-plane. These vector spaces vary holomorphically with Z, and that is what one means by a vector bundle $E \rightarrow \mathbb{PT}$. A remarkable fact is that the anti-self-dual Yang–Mills field can be reconstructed up to gauge from E, and E is effectively freely prescribable. See Penrose (1984, 1986); Ward & Wells (1990), or Mason & Woodhouse (1996) for a full discussion, and Atiyah (1979) for a discussion in Euclidean signature.

The Connection with Completely Integrable Systems

In effect, the twistor constructions amount to providing a geometric general local solution to the anti-self-duality equations in the sense that the twistor data is (for a local solution) freely prescribable. Thus, they demonstrate complete integrability of the anti-self-duality equations. The reconstruction of a solution on space-time from twistor data can be hard. In the anti-self-dual Yang–Mills case, it involves the solutions of a Riemann–Hilbert problem, and in the case of the anti-self-dual Einstein equations, the construction of a family of rational curves inside a complex manifold. Nevertheless, such constructions are a familiar part of the apparatus of the theory of integrable systems.

In Ward (1985), this connection with integrable systems was developed further, and the anti-self-dual Yang–Mills equations were shown to yield many important integrable systems under symmetry reduction. Ward's list has been extended and now includes many of the most famous examples of integrable systems such as the Painlevé equations, the Korteweg–de Vries equation, the nonlinear Schrödinger equation, the N-wave equations, among others (see Ablowitz & Clarkson (1992) and Mason & Woodhouse (1996) for a review). There are some notable omissions from the list such as the Kadomtsev–Petviashvili and Davey–Stewartson equations (at least if one restricts oneself to finite-dimensional gauge groups), but the list remains impressive.

One can impose symmetries on the twistor constructions for the anti-self-duality equations to obtain a reduced twistor correspondence for solutions to any of these integrable equations (see Mason & Woodhouse (1996) and Chapter 1 of Mason et al. (1995)), so there are many many twistor correspondences.

Applications

Twistor constructions have the effect of reducing problems in nonlinear differential equations to problems in complex holomorphic geometry, where there are many powerful tools. Twistor theory underlies the many appearances of algebraic geometry, loop groups, and Riemann–Hilbert problems in the theory of integrable systems, even though these structures were, for systems in one and two dimensions, usually discovered without knowledge of the twistor theory, see Mason & Woodhouse (1996). The most impressive applications of twistor theory have been in three and four dimensions, where it is difficult to imagine making such significant progress without the twistor theory.

The Ward construction was used by Atiyah, Drinfeld, Hitchin, and Manin to construct the Yang–Mills instantons on S^4; see Atiyah (1979). Its symmetry reduction was also used by both to obtain monopoles on $\mathbb{R}^3$ and to study the hyperkähler metric on their moduli spaces, see Ward & Wells (1990) and Atiyah and Hitchin (1988) (see Mason & Woodhouse (1996) for further applications).

The nonlinear graviton construction and its generalizations have been used for many constructions of Einstein manifolds and more general anti-self-dual manifolds (see Hitchin (1979) for the construction of asymptotically locally Euclidean hyperkähler spaces in four dimensions). The twistor constructions have also been an important tool in studying general properties, for a twistor construction of an anti-self-dual conformal structure on any manifold that is a connected sum of two other such manifolds (Donaldson & Friedman, 1989). Further applications and developments can be found in Mason et al. (2001).

An application that goes beyond complete integrability is the twistor framework of Merkulov for studying arbitrary geometric structures. This has led to the remarkable classification of all possible irreducible holonomies of torsion-free affine connections (Merkulov & Schwachhöfer, 1999).

It is to be hoped that these many applications will one day feed back into Penrose's original program and provide a unification between quantum theory and gravity.

LIONEL MASON

See also **Einstein equations; General relativity; Instantons; Integrability; Integral transforms; Inverse scattering method or transform; Riemann–Hilbert problem; Yang–Mills theory**

Further Reading

Ablowitz, M.J. & Clarkson, P.A. 1992. *Solitons, Nonlinear Evolution Equations and Inverse Scattering*, Cambridge and New York: Cambridge University Press

Atiyah, M.F. 1979. *Geometry of Yang-Mills Fields*, Pisa: Accademia Nazionale dei Lincei Scuola Normale Superiore

Atiyah, M.F. & Hitchin, N.J. 1988. *The Geometry and Dynamics of Monopoles*, Princeton, NJ: Princeton University Press

Atiyah, M.F., Hitchin, N.J. & Singer, I.M. 1978. Self-duality in four-dimensional Riemanninian geometry. *Proceedings of the Royal Society* A, 362: 425

Bailey, T.N. & Baston, R. (editors). 1990. *Twistors in Mathematics and Physics*, Cambridge and New York: Cambridge University Press

Baston, R.J. & Eastwood, M.G. 1989. *The Penrose Transform: Its Interaction with Representation Theory*, Oxford and New York: Oxford University Press

Bateman, H. 1910. *Partial Differential Equations of Mathematical Physics*, New York: Dover

Donaldson, S. & Friedman, R. 1989. Connected sums of self-dual manifolds and deformations of singular spaces. *Nonlinearity*, 2(2): 197–239

Hitchin, N. 1979. Polygons and gravitons. *Mathematical Proceedings of the Cambridge Philosophical Society*, 85: 456–476

Mason, L.J. & Hughston, L.P. (editors). 1990. *Further Advances in Twistor Theory, Vol. I: The Penrose Transform and Its Applications*, Harlow: Longman

Mason, L.J., Hughston, L.P. & Kobak, P.Z. 1995. *Further Advances in Twistor Theory, Vol. II: Integrable Systems, Conformal Geometry and Gravitation*, Harlow: Longman

Mason, L.J., Hughston, L.P., Kobak, P.Z. & Pulverer, K. (editors). 2001. *Further Advances in Twistor Theory, Vol. III: Curved Twistor Spaces*, Boca Raton, FL: Chapman & Hall

Mason, L.J. & Woodhouse, N.M.J. 1996. *Twistor Theory, Self-duality and Twistor Theory*, Oxford and New York: Oxford University Press

Merkulov, S., & Schwachhöfer, L. 1999. Classification of irreducible holonomies of torsion-free affine connections. *Annals of Mathematics*, (2), 150(1): 77–149

Newman, E.T. 1976. Heaven and its properties. *General Relativity and Gravitation*, 7(1): 107–111

Penrose, R. 1976. Nonlinear gravitons and curved twistor theory. *General Relativity and Gravitation*, 7: 31–52

Penrose, R. 1984, 1986. *Spinors and Space-time*, 2 vols., Cambridge and New York: Cambridge University Press

Ward, R.S. 1977. On self-dual gauge fields. *Physics Letters*, 61A: 81–82

Ward, R.S. 1985. Integrable and solvable systems and relations among them. *Philosophical Transactions of the Royal Society* A, 315: 451–457

Ward, R.S. & Wells, R.O. 1990. *Twistor Geometry and Field Theory*, Cambridge and New York: Cambridge University Press

TWO SOLITON COLLISION

See ***N*-soliton formulas**

U

UEDA EQUATION

See **Duffing equation**

UNIVERSALITY

The concept of universality serves to emphasize like behavior in seemingly unrelated systems. It became popular in the context of phase transitions when it was noted that after a suitable mapping between the thermodynamic variables, the scaling behavior of spin systems near the critical point depends on the number of spin directions, the dimensionality of the system, and the type and range of interactions only (Griffiths, 1970; Binney et al., 1992). Thus, there is no "universal" universality, but only a restricted one among systems that belong to the same universality class.

A deeper understanding of the empirically established relations emerged within the renormalization group treatment of phase transitions, which allowed identification of the parts of the interactions that are relevant for an assignment to a universality class (Wilson, 1983; Binney et al., 1992).

An elementary example of thermodynamic universality is provided by the law of corresponding states for interacting gases. They can be described over a wide range of temperature T, pressure p, and volume V by the van der Waals equation of state. For one mole of the gas it is

$$(p + a/V^2)(V - b) = RT \tag{1}$$

with R being the gas constant and a and b parameters characteristic of the gas. This equation of state has a critical point where both $\partial p/\partial V = 0$ and $\partial^2 p/\partial V^2 = 0$, given by

$$RT_c = \frac{8a}{27b}, \qquad p_c = \frac{a}{27b^2} \text{ and } V_c = 3b. \tag{2}$$

Because of the dependence on the material parameters a and b, these quantities differ from gas to gas. However, when p, T, and V are expressed in terms of the critical values, the material constants disappear. With $\tilde{p} = p/P_c$, $\tilde{V} = V/V_c$, and $\tilde{T} = T/T_c$ the reduced quantities, the van der Waals equation of state becomes

$$(\tilde{p} + 3/\tilde{V}^2)(3\tilde{V} - 1) = 8\tilde{T}. \tag{3}$$

Interestingly, this mapping between states of different gases remains useful even when the gases do not follow the predictions from van der Waals theory, for example, near the critical point and in the liquid-gas coexistence region (Stanley, 1971; Binney et al., 1992).

Universality in thermodynamic systems is often characterized quantitatively by the exponents in the power laws with which quantities, such as the specific heats, susceptibilities, and spatial or temporal correlation functions, diverge near a critical point (if the exponent vanishes, then there can be logarithmic variations or discontinuities). The ideal behavior in the thermodynamic limit is often clouded by corrections because of the finite size of the sample, and these very often show power law dependencies too. The search for universality in other systems and situations is, hence, often accompanied by investigations of scaling laws and comparisons of exponents.

The success of universality considerations in thermodynamics has triggered many investigations in nonlinear dynamical systems. Universality arguments can help to link behavior in one-dimensional maps, in chemical reactions, in electrical circuits, or in hydrodynamic systems, for instance. Here are some situations where universality and scaling appear in nonlinear systems.

Period-doubling cascade. Grossmann & Thomae (1977), Feigenbaum (1978), and Coullet & Tresser (1978) noted that the sequence of parameters λ_n in quadratic maps, like $x_{i+1} = \lambda_n(1 - 2x_i^2)$ where bifurcations from orbits of period 2^n to 2^{n+1} occur, that is, $\lambda_1 = 0.70716$, $\lambda_2 = 0.80953$, $\lambda_3 = 0.83112$, $\lambda_4 = 0.83574$, etc., form a geometric sequence, $\lambda_n = \lambda_\infty - a\delta^{-n}$ (a is a constant). The ratio

$$\delta = \frac{\lambda_n - \lambda_{n-1}}{\lambda_{n+1} - \lambda_n} \tag{4}$$

converges to a value

$$\delta = 4.669201609\ldots. \tag{5}$$

This values appears for all maps with a single quadratic maximum and negative Schwartzian derivative ($f'''/f' - (\frac{3}{2})(f''/f')^2) < 0$), as a renormalization of the map due to Cvitanovic (1984), Feigenbaum (1979), Coullet & Tresser (1978), and Lanford (1982) shows. It also appears in higher-dimensional dissipative maps, in continuous differential equations, and even in partial differential equations. Experimental evidence was first found in experiments on thermal convection (Libchaber & Maurer, 1980) and in acoustical, electrical, chemical, and many other cases since. Examples are given in Schuster (1988) and Cvitanovic (1984).

Further universal scaling laws have been established for period n-tupling, for period doubling with other forms of the maximum, for the scaling of the splitting of iterates in period doubling, for the amplitudes of higher harmonics in Fourier spectra, and for the influence of noise on period-doubling bifurcations (Cvitanovic, 1984; Schuster, 1988). Period-doubling in conservative systems has different scaling exponents (MacKay & Meiss, 1987).

Other situations with universal scaling laws arise in the case of intermittency (saddle-node bifurcations) and near the break-up of tori in conservative systems (Cvitanovic, 1984; Schuster, 1988).

Pattern formation. In many pattern forming systems, the equation for the amplitude of the pattern is of Ginzburg–Landau type, with the interactions dictated by continuous and discrete symmetries of the system (Golubitsky & Schaeffer, 1985; Golubitsky et al., 1988). A simple example for a system with a real amplitude A and invariance under $A \to -A$ is

$$\partial_t A = \varepsilon A - A^3 + \Delta A\,, \tag{6}$$

If the control parameter ε is negative, there is no pattern. For positive ε the amplitude increases like $\varepsilon^{1/2}$. This behavior is widely observed (Cross & Hohenberg, 1993).

Singularity formation. The final stages during the formation of singularities often show asymptotic scaling behavior. For instance, two gravitating bodies starting at rest will collide with distance vanishing like $t^{2/3}$ and velocities diverging like $t^{-1/3}$. The formation of the pinching off of a cylindrical column of liquid is an example of a process that is universal not only in its scaling behavior but also in the prefactors. The formation of the pinch is due to surface tension and is counteracted by viscous effects. From the relevant material parameters surface tension γ (of dimension kg/s^2), viscosity η (kg/(m s)), and density ρ (kg/m^3), one can form a scale of length $l_p = \eta/(\rho\gamma)$ and one of time $t_p = \eta^3/(\gamma^2\rho)$. The behavior of an axisymmetric column is universal in the sense that the minimal diameter $d_{\min}(t)$ and the maximal velocity $v_{\max}(t)$ at a time t before pinch off are given by (Eggers, 1993)

$$\frac{d_{\min}(t)}{l_p} = 0.0608\left(\frac{t}{t_p}\right),$$

$$\frac{v_{\max}(t)}{l_p/t_p} = 3.07\left(\frac{t}{t_p}\right)^{-1/2}. \tag{7}$$

Random matrix theories. The Hamiltonian operator that describes a given quantum system has a specific form that may be poorly known, as in the case of impurities in a solid or the interactions between nucleons in an atomic nucleus, but it certainly is not arbitrary or random. Nevertheless, for very many systems, the statistical properties of eigenvectors or the statistics of neighboring eigenvalues behave in a universal way that does not depend on the specific system anymore. When rescaled by the mean distance, neighboring energy eigenvalues of hydrogen in strong magnetic fields, electrons in quantum dots, scattering resonances in nuclei such as ^{26}Al, resonance frequencies in microwave resonators, and vibrating quartz blocks can all show the same spacing distribution (Stöckmann, 1999). The only requirement is that the systems are disordered or classically chaotic. The specific form of the distributions then depends on the global symmetry properties of the Hamiltonian, that is, on whether it is real symmetric, complex Hermitian, or symplectic (Guhr et al., 1998; Stöckmann, 1999). System-specific properties appear on much larger scales of separation between eigenvalues; they can be understood within semiclassical periodic orbit theory.

Universality in turbulence. Lewis Fry Richardson described a turbulent flow as a hierarchical arrangement of vortices of different sizes that appear and disappear in an unpredictable fashion. The famous scaling laws of Kolmogorov, Obhukov, Weizäcker, and Onsager hold that in homogeneous, isotropic turbulence, the square of the velocity difference between two points a distance r apart scales like $(\varepsilon r)^{2/3}$, where ε is now the energy dissipation (Frisch, 1995). The expectation is that this is independent of the precise mechanism of stirring, as long as it is confined to large scales and becomes exact as the Reynolds number of the flow approaches infinity. Turbulent fluctuations behind grids in wind or water tunnels or in turbulent jets support this observation.

BRUNO ECKHARDT

See also **Critical phenomena; Dimensional analysis; Fractals; Free probability theory; Kolmogorov cascade; Period doubling; Periodic orbit theory; Random matrix theories: I, II, III, IV; Renormalization groups; Turbulence**

Further Reading

Binney, J.J., Dowrick, N.J., Fisher, A.J. & Newman, M.E.J. 1992. *The Theory of Critical Phenomena: An Introduction to the Renormalization Group*, Oxford: Clarendon Press

Coullet, P. & Tresser, C. 1978. Iterations d'endomorphismes et groupe de renormalization. *Journale de Physique* 39: Colloque C5–C25

Cross, M.C. & Hohenberg, P.C. 1993. Pattern formation outside of equilibrium. *Reviews of Modern Physics*, 65: 851–1112

Cvitanovic, P. 1984. *Universality in Chaos*, Bristol: Adam Hilger

Eggers, J. 1993. Universal pinching of 3D axisymmetric free-surface flow. *Physical Review Letters*, 71: 3458–3460

Feigenbaum, M.J. 1978. Quantitative universality for a class of nonlinear transformations. *Journal of Statistical Physics*, 19: 25–52

Feigenbaum, M.J. 1979. The universal metric properties of nonlinear transformations. *Journal of Statistical Physics*, 21: 669–706

Frisch, U. 1995. *Turbulence: The Legacy of A.N. Komlogorov*, Cambridge and New York: Cambridge University Press

Griffiths, R.B. 1970. Dependence of critical indices on a parameter. *Physical Review Letters*, 24: 1479–1482

Golubitsky, M. & Schaeffer, D.G. 1985. *Singularities and Groups in Bifurcation Theory*, vol. 1, New York: Springer

Golubitsky, M., Stewart, I.N. & Schaeffer, D.G. 1988. *Singularities and Groups in Bifurcation Theory*, vol. 2, New York: Springer

Grossmann, S. & Thomae, S. 1977. Invariant distributions and stationary correlation functions of the one-dimensional discrete processes. *Zeitschrift für Naturforschung*, 32a: 1353–1363

Guckenheimer, J. & Holmes, P. 1983. *Nonlinear Oscillations, Dynamical Systems and Bifurcations of Vector Fields*, New York: Springer

Guhr, T., Müller-Groeling, A., & Weidenmüller, H.A. 1998. Random-matrix theories in quantum physics: common concepts. *Physics Reports*, 299: 190–425

Lanford, O.E. 1982. A computer assisted proof of the Feigenbaum conjectures. *Bulletin of the American Mathematical Society*, 6: 427–434

Libchaber, A. & Maurer, J. 1980. Une experience de Rayleigh-Bénard de geometrie reduite: multiplication, acchroage, et demultiplication de frequences. *Journal de Physique*, 41: Colloque C3–51

MacKay, R.S. & Meiss, J.D. (editors). 1987. *Hamiltonian Dynamical Systems: A Reprint Collection*, Bristol: Adam Hilger

Schuster, H.G. 1988. *Deterministic Chaos*, 2nd edition, Weinheim: Physik-Verlag and New York: VCH

Stanley, H.E. 1971. *Introduction to Phase Transitions and Critical Phenomena*, Oxford: Clarendon Press and New York: Oxford University Press

Stöckmann, H.J. 1999. *Quantum Chaos: An Introduction*, Cambridge and New York: Cambridge University Press

Wilson, K.G. 1983. The renormalization group and critical phenomena. *Reviews of Modern Physics*, 55: 583–600

UNSTABLE MANIFOLD

See **Phase space**

V

VAN DER POL EQUATION

In 1926, Balthasar van der Pol derived the equation (now named after him) to describe self-sustained oscillations in a triode circuit. To solve the equation, he developed a method that is based on the separation of fast and slow time dependencies and on averaging, an idea that provides the basis of various analytic approaches to nonlinear problems (van der Pol, 1926).

In dimensionless variables, the van der Pol (VDP) equation reads

$$\frac{\mathrm{d}^2x}{\mathrm{d}t^2} - \mu(1-x^2)\frac{\mathrm{d}x}{\mathrm{d}t} + x = 0. \tag{1}$$

The variable x represents the triode plate voltage in the oscillating circuit, and the frequency of the circuit is normalized to 1 by the appropriate change of the time scale. The parameter $\mu > 0$ gives the growth rate of small linear oscillatory perturbations, and the nonlinear term, approximating the nonlinear current-voltage characteristic of the triode, is normalized by scaling x. Physically, the VDP equation describes growth and saturation of oscillatory perturbations with eventual onset of periodic self-sustained oscillations. With a change of variables $x = \mathrm{d}y/\mathrm{d}t$, the VDP equation transforms to the Rayleigh equation

$$\frac{\mathrm{d}^2y}{\mathrm{d}t^2} - \mu\left[1 - \frac{1}{3}\left(\frac{\mathrm{d}y}{\mathrm{d}t}\right)^2\right]\frac{\mathrm{d}y}{\mathrm{d}t} + y = 0.$$

In the case of weak instability and nonlinearity ($\mu \ll 1$), the VDP equation can be treated analytically. Here the oscillations are nearly sinusoidal with slowly (on the time scale of order $1/\mu$) varying amplitude and phase. This key observation constitutes the essence of the method of averaging developed by van der Pol.

The first step in the solution is a transformation from $(x, \dot{x})$ to new variables: the amplitudes A, B; thus,

$$x(t) = A(t)\cos t + B(t)\sin t,$$
$$\dot{x}(t) = -A(t)\sin t + B(t)\cos t. \tag{2}$$

In these variables, Equation (1) reads

$$\frac{\mathrm{d}A}{\mathrm{d}t} = -\mu\dot{x}(1-x^2)\sin t,$$
$$\frac{\mathrm{d}B}{\mathrm{d}t} = \mu\dot{x}(1-x^2)\cos t, \tag{3}$$

where $\dot{x}, x$ on the right-hand side are to be expressed through A, B. One can see that the time variations of A, B are slow, and the major contribution to them is given by non-oscillating terms on the right-hand side of (3). Keeping only the non-oscillating terms is equivalent to averaging over the oscillation period 2π, and under the averaging procedure, the slowly varying amplitudes $A(t)$ and $B(t)$ are considered as constants. Having performed the averaging, van der Pol obtained the approximate equations in the form

$$\frac{\mathrm{d}A}{\mathrm{d}t} = \frac{\mu}{2}A\left(1 - \frac{A^2+B^2}{4}\right),$$
$$\frac{\mathrm{d}B}{\mathrm{d}t} = \frac{\mu}{2}B\left(1 - \frac{A^2+B^2}{4}\right). \tag{4}$$

This equation can be readily solved by means of a transformation to the slow amplitude and phase $A = R\cos\phi$, $B = R\sin\phi$:

$$\frac{\mathrm{d}R}{\mathrm{d}t} = \frac{\mu}{2}R\left(1 - \frac{R^2}{4}\right), \quad \frac{\mathrm{d}\phi}{\mathrm{d}t} = 0.$$

For any initial condition $R \neq 0$, the stationary amplitude $R_0 = 2$ is established as $t \to \infty$. In the original variables of Equation (1), the corresponding periodic solution reads $x(t) = 2\cos(t-\phi)$ (see Figure 1). In terms of the dynamical systems theory, the van der Pol equation (1) possesses a limit cycle; for small μ, the cycle is approximately circular with radius 2. Physically, it corresponds to periodic weakly nonlinear, self-sustained oscillations.

The solution of the VDP equation can be treated analytically also in the limiting case $\mu \gg 1$, when the equation describes relaxation oscillations. Introducing new time $\tau = t/\mu$, we can rewrite Equation (1) as a

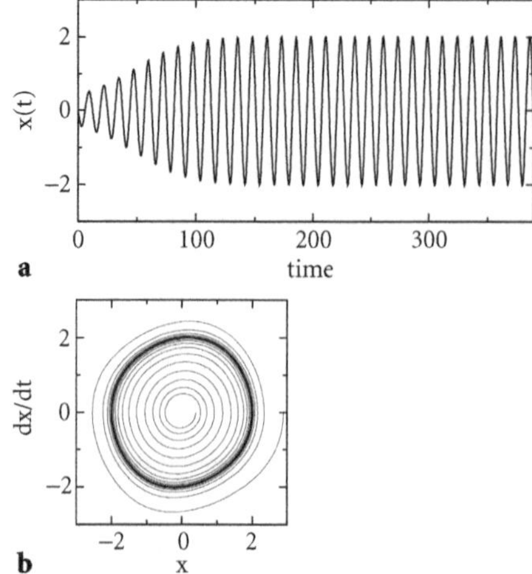

Figure 1. A solution of the van der Pol equation for $\mu = 0.1$. (a) The time dependence of x. (b) The limit cycle on the phase plane $(x, \mathrm{d}x/\mathrm{d}t)$ attracts other trajectories; its form for this small value of μ is nearly circular.

system

$$\mu^{-2}\frac{\mathrm{d}x}{\mathrm{d}\tau} = -y + x - \frac{x^3}{3}, \qquad (5)$$

$$\frac{\mathrm{d}y}{\mathrm{d}\tau} = x. \qquad (6)$$

This system has a small parameter μ^{-2} at the derivative, and it belongs to the class of singularly perturbed equations. All motions in the phase space (x, y) are divided into fast motions, when variable x jumps with the rate $\sim\mu^2$ while variable y remains nearly constant, and slow motions, when both variables x and y vary with rates of ~ 1. The slow motions are restricted to the slow manifold, where the right-hand side of Equation (5) vanishes (the curve $y = x - x^3/3$ shown as a dotted line in Figure 2b). More precisely, they are restricted to the stable branches of the curve, where $\mathrm{d}(x - x^3/3)/\mathrm{d}x < 0$. The direction of motion on these stable branches is determined by Equation (6) and is depicted in Figure 2b by arrows. When the phase point moving along a stable branch arrives at its border, it jumps to the other branch. Thus, the limit cycle consists of two pieces of slow motion connected by two pieces of fast motion (solid line in Figure 2b).

The VDP equation serves as a prototype for different dynamical models with a limit cycle. For example, an equation with a more complex nonlinearity and $\mu > 0$,

$$\frac{\mathrm{d}^2x}{\mathrm{d}t^2} - \mu(-1 + x^2 - \beta x^4)\frac{\mathrm{d}x}{\mathrm{d}t} + x = 0,$$

describes the so-called "hard excitation" of self-sustained oscillations. (This equation was derived by van der Pol and Appleton for the description of a triode generator, whose operating point is shifted from the inflection point of the current-voltage characteristics of

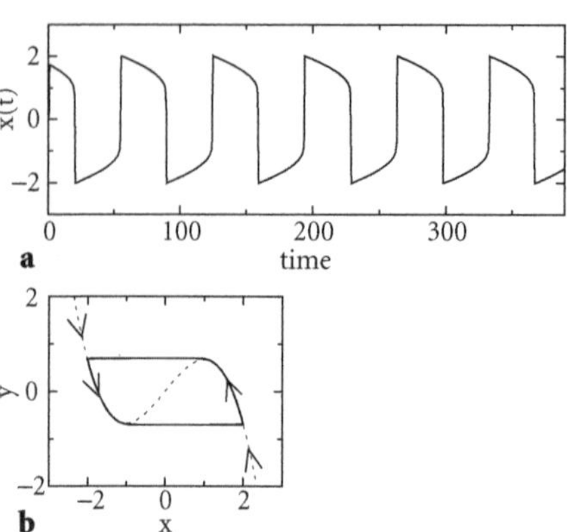

Figure 2. A solution of the van der Pol equation for $\mu = 20$. (a) The time dependence of x. (b) The limit cycle on the phase plane (x, y) consists of pieces of the slow manifold (shown with a dotted line) and fast jumps of x.

the triode.) In this case, the averaging method above described leads to the following amplitude equation:

$$\frac{\mathrm{d}R}{\mathrm{d}t} = \frac{\mu}{2}R\left(-1 + \frac{R^2}{4} - \beta\frac{R^4}{8}\right), \quad \frac{\mathrm{d}\phi}{\mathrm{d}t} = 0.$$

For $\mu > 0$ and $0 < \beta < \frac{1}{8}$, this equation possesses coexisting stable steady state $R = 0$, an unstable limit cycle at $R_{\mathrm{un}} = \beta^{-1/2}(1 - \sqrt{1 - 8\beta})^{1/2}$, and a stable limit cycle at $R_{\mathrm{st}} = \beta^{-1/2}(1 + \sqrt{1 - 8\beta})^{1/2}$. The basin of attraction of the stable limit cycle is $R > R_{\mathrm{un}}$, while the circle $R < R_{\mathrm{un}}$ is the basin of the stable fixed point. For the triode generator this means that self-sustained oscillations cannot develop from small fluctuations, but appear only if a relatively large (larger than R_{un}) perturbation is applied.

Another generalization of the van der Pol equation is often called the van der Pol–Duffing oscillator:

$$\frac{\mathrm{d}^2x}{\mathrm{d}t^2} - \mu(1 - x^2)\frac{\mathrm{d}x}{\mathrm{d}t} + x + \gamma x^3 = 0. \qquad (7)$$

This model combines the dissipative nonlinearity of the van der Pol equation (term proportional to μ) with the conservative nonlinearity of the Duffing equation (term proportional to γ). For $\mu \ll 1$ and $\gamma \ll 1$, the equations for the slowly varying amplitude and phase can be obtained by the method of averaging as

$$\frac{\mathrm{d}R}{\mathrm{d}t} = \frac{\mu}{2}R\left(1 - \frac{R^2}{4}\right), \quad \frac{\mathrm{d}\phi}{\mathrm{d}t} = \frac{3\gamma}{4}R^2.$$

The difference to the van der Pol case is in the phase dynamics. Now the oscillations are non-isochronous, because the dynamics of the slow phase depends on the amplitude. In particular, the frequency of the self-sustained oscillations differs from the frequency of linear oscillations by 3γ. In the VDP equation (1), such a frequency shift appears only in the second approximation (by taking into account the terms $\sim\mu^2$).

Another model related to Equation (1) is the Bonhoeffer–van der Pol (BVDP) oscillator, with equations

$$\frac{dx}{dt} = -y + x - \frac{x^3}{3} + I_0,$$
$$\frac{dy}{dt} = c(x + a - by).$$

Writing this system as one second-order equation, one gets a model similar to Equation (7), but with additional terms; thus,

$$\frac{d^2x}{dt^2} - (1 - bc - x^2)\frac{dx}{dt} + c(1-b)x + \frac{bc}{3}x^3 + c(a - bI_0) = 0 . \quad (8)$$

In the case $b = I_0 = 0$, the BVDP model reduces to the FitzHugh–Nagumo model of neuron spiking. Depending on the parameters, both these models demonstrate self-sustained oscillations, excitability (a stable steady state that responses to a finite perturbation by generating a spike), or bistability.

A periodically forced VDP equation,

$$\frac{d^2x}{dt^2} - \mu(1 - x^2)\frac{dx}{dt} + \omega_0^2 x = E \sin \omega t,$$

displays the phenomenon of frequency locking. For ω close to the natural frequency of the autonomous system ω_0, the forced systems starts to oscillate with the frequency of the forcing term or becomes entrained. Entrainment occurs even for relatively small E, especially when $\mu \gg 1$. For large E, the forced VDP equation can demonstrate chaotic regimes.

ARKADY PIKOVSKY AND MICHAEL ROSENBLUM

See also **Attractors; Averaging methods; Duffing equation; FitzHugh–Nagumo equation; Nonlinear electronics; Relaxation oscillators; Synchronization**

Further Reading

Andronov, A.A., Vitt, A.A. & Khaykin, S.E. 1966. *Theory of Oscillators*, Oxford and New York: Pergamon Press (original Russian edition, 1937)

Bogoliubov, N.N. & Mitropolsky, Yu.A. 1961. *Asymptotic Methods in the Theory of Nonlinear Oscillations*, New York: Gordon and Breach

van der Pol, B. 1926. On "relaxation oscillations." *Philosophical Magazine*, 2: 978–992

VECTOR FIELD

See **Phase space**

VERHULST EQUATION

See **Population dynamics**

VIRIAL THEOREM

First established by Rudolf Clausius in 1870, the virial theorem relates the average potential energy $\langle V \rangle$ to the average kinetic energy $\langle T \rangle$ of a system of particles. The particles can have arbitrary potential interactions, and the theorem holds both in classical and quantum mechanics.

For a single classical nonrelativistic particle, the theorem can be derived as a consequence of Newton's second law: $\boldsymbol{F} = m\mathrm{d}^2\boldsymbol{x}/\mathrm{d}t^2$. Thus,

$$\frac{\mathrm{d}(m\boldsymbol{x} \cdot \mathrm{d}\boldsymbol{x}/\mathrm{d}t)}{\mathrm{d}t} = \boldsymbol{x} \cdot \boldsymbol{F} + m\left(\frac{\mathrm{d}\boldsymbol{x}}{\mathrm{d}t}\right) \cdot \left(\frac{\mathrm{d}\boldsymbol{x}}{\mathrm{d}t}\right), \quad (1)$$

which in terms of the momentum $\boldsymbol{p} = m(\mathrm{d}\boldsymbol{x}/\mathrm{d}t)$ and the kinetic energy $T = m(\mathrm{d}\boldsymbol{x}/\mathrm{d}t)^2/2$ reads

$$-\boldsymbol{x} \cdot \boldsymbol{F} = -\frac{\mathrm{d}(\boldsymbol{x} \cdot \boldsymbol{p})}{\mathrm{d}t} + 2T. \quad (2)$$

In taking the average over time, the first term on the right-hand side of the equation does not contribute; thus, one obtains

$$-\langle \tfrac{1}{2}\boldsymbol{x} \cdot \boldsymbol{F} \rangle_{\mathrm{av}} = \langle T \rangle_{\mathrm{av}}, \quad (3)$$

which is the classical virial theorem. The expression on the left-hand side of this equation (called the virial by Clausius) is a measure of the net attractiveness of a system.

In the special case when the force is derivable from a potential $\boldsymbol{F} = -\nabla V$ and the potential varies as the nth power of the distance to the origin $V = C|\boldsymbol{x}|^n$, the term $\boldsymbol{x} \cdot \boldsymbol{F} = -nV$, and the virial theorem reduces to

$$n\langle V \rangle_{\mathrm{av}} = 2\langle T \rangle_{\mathrm{av}} . \quad (4)$$

The theorem can also be obtained from a variational derivation.

Applications

- For the harmonic oscillator, $V(\boldsymbol{x}) = \frac{1}{2}mw^2|\boldsymbol{x}|^2$ then $n = 2$ and $E \equiv \langle V \rangle + \langle T \rangle = 2\langle T \rangle = 2\langle V \rangle$.
- For a bouncing ball, $V(\boldsymbol{x}) = mg|\boldsymbol{x}|$ and $n = 1$, then $E = 3\langle T \rangle = 3/2\langle V \rangle$.
- If the forces vary inversely as the square of the distance (as in atomic and planetary systems), then $V(\boldsymbol{x}) = e^2/|\boldsymbol{x}|$ and $n = -1$. The average kinetic energy is then numerically equal to the total energy $E = -\langle T \rangle$ but of opposite sign.
- In astrophysics, the virial theorem is a valuable tool for studying static, non-evolving (relaxed) systems such as stars (systems of gas particles), gas clouds, star clusters, galaxies, and galaxy clusters. This theorem can be used to estimate the mass of a given system since for the gravitational case $\langle T \rangle = -\frac{1}{2}\langle V \rangle$ and

$$\frac{1}{2}M_{\mathrm{tot}}\left\langle\frac{\mathrm{d}\boldsymbol{x}}{\mathrm{d}t}\right\rangle^2 = +\frac{1}{4}\frac{GM_{\mathrm{tot}}^2}{\langle R_{\mathrm{tot}} \rangle} .$$

Thus, the virial mass of the system can be estimated as

$$M_{\text{tot}} \approx 2\frac{\langle R_{\text{tot}}\rangle \langle \mathrm{d}\boldsymbol{x}/\mathrm{d}t\rangle^2}{G}.$$

- Another application of this theorem is in the classical calculation of the state equation of gases. The system considered is that of a large number N of particles with coordinates $\boldsymbol{x}_i$ acted on by external and internal forces $\boldsymbol{F}_i = \boldsymbol{F}_{i,\text{ext}} + \boldsymbol{F}_{i,\text{int}}$ and confined within a box of volume $\mathcal{V}$. As the contribution from the external forces can be related to the pressure of the gas on the walls of the box, the virial theorem takes the form

$$P\mathcal{V} = Nk_{\mathrm{B}}T + \frac{1}{3}\left\langle \sum_i^N \boldsymbol{x}_i \cdot \boldsymbol{F}_{i,\text{int}} \right\rangle_{\text{av}},$$

  where k_{B} is the Boltzmann constant. This was the application originally considered by Clausius.
- The virial theorem can be extended to a quantum system described by the linear dimensionless Schrödinger equation

$$\mathrm{i}\psi_t = -\Delta\psi + V(\boldsymbol{x})\psi.$$

  In this case, $2T_m = nV_m$, where T_m and V_m are the kinetic and potential energy of the mth eigenstate, respectively.
- The virial theorem can also be applied to the study of the solitary waves. As an illustration, consider the one-dimensional nonlinear Schrödinger (NLS) equation

$$\mathrm{i}u_t + u_{xx} + 2(u^*u)u = 0, \qquad (5)$$

which has solitary wave solutions that preserve their shape after collision with other solitary waves. The general form of a solitary wave is

$$u(x,t) = a\exp\mathrm{i}\left[\frac{v_{\mathrm{e}}}{2}x + \left(a^2 - \left(\frac{v_{\mathrm{e}}}{2}\right)^2\right)t - \phi_0\right] \times \operatorname{sech}\left[a(x - v_e t - x_0)\right], \qquad (6)$$

where a is the wave amplitude and v_{e} the envelope velocity, while ϕ_0 and x_0 are the initial phase and position.

One of the infinite conservation laws associated to (5) is the energy

$$E = \int u_x^* u_x \,\mathrm{d}x + \int -(u^*u)^2 \,\mathrm{d}x,$$

where the first term is the kinetic energy (T) and the second term is the potential energy (V).

The NLS equation is derived from the Lagrangian density $\mathcal{L} = \frac{\mathrm{i}}{2}(uu_t^* - u_t u^*) + u_x^* u_x - (u^*u)^2$ and the corresponding action is $\mathcal{S} = \int \mathrm{d}t\,\mathrm{d}x\,\mathcal{L}$. Consider a stationary localized solution $u = \exp(\mathrm{i}\Omega t)\,u(x)$, which is the case of solution (6) with $v_{\mathrm{e}} = 0$ and $\Omega = a^2$. As the Lagrangian density is static, we can apply a dilation transformation to get the global condition $\int |u_x|^2\,\mathrm{d}x = \int (\Omega|u|^2 - |u|^4)\,\mathrm{d}x$. From Equation (5), $\int (\Omega|u|^2 + |u_x|^2 - 2|u|^4)\,\mathrm{d}x = 0$, which in turn implies that $\int |u_x|^2\,\mathrm{d}x = \frac{1}{2}\int |u|^4\,\mathrm{d}x$. Thus the total energy is $E = -T = \frac{1}{2}V$, and in the case of the stationary soliton, $E = -2a^3/3$.

Thus, the virial theorem provides nontrivial relations between quantities of physical interest, which can be used to test the accuracy of numerical simulations and to find variational solutions.

Luis Vazquez and M.P. Zorzano

See also **Damped-driven anharmonic oscillator; Nonlinear Schrödinger equations; Rotating wave approximation**

Further Reading

Goldstein, H. 1980. *Classical Mechanics*, 2nd edition, Reading, MA: Addison-Wesley

Saslaw, W.C. 1985. *Gravitational Physics of Stellar and Galactic Systems*, Cambridge and New York: Cambridge University Press

Scott, A. 2003. *Nonlinear Science: Emergence and Dynamics of Coherent Structures*, 2nd edition, Oxford and New York: Oxford University Press

VISIOMETRICS

With the advent of robust and rapid sensing devices and increasing computer speeds, memories and transfer rates, the technology for observation, measurement, and computer simulation of nonlinear processes in physical and biological systems has improved rapidly, allowing massive experimental and simulated data sets to be produced. In order to gain physical insight into the evolving phenomena and construct reduced mathematical models, the mathematical essences in this sea of data need to be determined.

This can be accomplished through the process of visiometrics, which involves visualization, projection, identification and classification, extraction, tracking, quantification, and juxtaposition of evolving amorphous coherent structures and statistical backgrounds in massive multidimensional data sets. The goal is to produce cogent images and specific, parameter-scaled (normalized) graphs for intuitive understanding and mathematical analyses (Bitz & Zabusky, 1990; Zabusky et al., 1993; Feher & Zabusky, 1996; Fernandez et al., 1996; Zabusky, 1999).

Consider the process of dealing with evolving simulation data. Most simulations use continuum partial or integro-differential equations or discrete particle descriptions and produce numerous fields, particle locations, and trajectories. (In hybrid codes both are present.) Examples of continuum fields mainly from the compressible Euler equations of fluid motion are presented here. For particle data (e.g., from Monte Carlo or plasma simulations), appropriate local averages can be used to convert to continuum quantities.

In fluid flows, we deal with scalars such as density, pressure, and temperature, vectors such as velocity $\boldsymbol{u}$ and vorticity $\boldsymbol{\omega}$, and tensors such as the rate of strain tensor $\nabla \boldsymbol{u}$. Very thin high-gradient regions such as shock waves and diffusing interfacial transition layers (ITLs) (Zabusky et al., 2003) between species with different properties may be embedded in the flows. To quantify the ITLs, we may extract a medial axis (i.e., some center line) or surface, so we can determine tangents and normals to these curves in two-dimensions (2-d) and surfaces (in 3-d), respectively. In 3-d, the curvatures of these medial surfaces may be important tensors. The visiometrics operations of visualization, projection, identification and classification, extraction, tracking, quantification, and juxtaposition are each considered below.

Visualization: Numerical fields are displayed in a variety of 2-d and 3-d images. The data are first preprocessed or filtered to make them more accessible visually. High wave number incoherent modes are removed with a filter (e.g., the wavelet process of Farge et al., 2001). The choice of appropriate color maps (palettes) (Farge, 2000) is a nontrivial operation. Farge (1992) represented the vorticity scalar with a grayscale format for lower magnitudes, a yellow intermediate contour for the region near zero, and a few graded colors for the higher magnitudes. In DAVID (see http://www.caip.rutgers.edu/~nzabusky/vizlab_cfd/david/david.html), an interactive color domain system has been developed with colors chosen, for example, with hue, saturation, and value (HSV), and with optional black/white grading superimposed in each color domain. The user sees the image automatically colored as the color map is modified. Simultaneously, at each color transition, the numerical value of the function and between neighboring color transitions, the integrated content, and the corresponding underlying pixel-area of the image are seen. (This gives the user a qualitative feeling about the magnitudes of the objects that the color renders.)

Standard visualization techniques include continuous or discrete contour maps in 2-d; volume rendering a function in 3-d as it would appear when radiating or reflecting light from various sources (Upson & Keeler, 1988); and displaying isosurfaces (contours) from connected polygons that bound regions of functions extracted by thresholding (Lorensen & Cline, 1987). There are many excellent computer vision and scientific visualization contemporary texts in the literature (both web and printed).

Projection: A general process produces abstract images and graphical information in lower dimensions, for example, the projection to 2-d from 3-d by integration with respect to a kernel. If it is a planar delta function, one extracts 2-d surfaces or curves from 3-d or 2-d spatial data, respectively. Also, integration of an appropriately weighted field variable transverse to some initial axis or flow direction (like that behind a planar shock in a shock tube) will produce a space–time diagram from a 2d+1 data set (see color plate section; Figure 2 of Hawley & Zabusky, 1989; see also Zabusky, 1999).

Identification and classification: The geometry, topology, content, moments, and distribution functions of an extracted region are examined. Are the domains in 3-d layer-like or tube-like? Are the tube-like domains right- or left- handed helices?

Extraction and setting of thresholds: Complex data sets may have a hierarchy of embedded coherent structures in a sea of incoherent very small-scale structures. Farge and colleagues (Farge et al., 2001) have applied a threshold prescription to wavelet amplitude coefficients and have developed high-compression techniques to systematically extract coherent vortex objects. Samtaney and Zabusky (2000) have examined the extraction of shocks and species ITLs in 2-d. For ITLs, one looks at absolute values of gradients and Laplacians of the density, temperature, and so on. Various heuristic and analytical methodologies have been presented by Villasenor & Vincent (1992), Melander & Hussain (1994), Jeong & Hussain (1995), Kida & Miura (1998), and Miura (2002). The results of Melander, Jeong, and Hussain are particularly noteworthy.

Tracking: Structures in space-time that have been extracted and identified are followed. One must allow for objects to collide and amalgamate, split, be created, and disappear (Samtaney et al., 1994). Post et al. (2002) and colleagues (e.g., Vrolijk et al., 2003) have carried this work further. With this information, we may be able to formulate kinematic and dynamical models.

Quantification: Graphs of projected and tracked structures are plotted and underlying physical, mathematical, and numerical parameters are varied. In the vicinity of extrema (e.g., magnitude of vorticity $\boldsymbol{\omega}$), second- and third-order moments are useful (e.g., ellipsoids in 3-d). Distribution functions of one or more variables are evaluated, and statistical characterizations of incoherent domains, after structures have been extracted, are examined.

Juxtaposition: These comparisons are of similar or different functions at the same or different times from simulations of the same or linked mathematical models in a relevant domain of parameters. Coherent structures of different functions (e.g., density or vorticity) and their quantifications with runs from different resolutions (validation) and parameters (scaling physical behavior) are compared. One often has to remove a background translation or rotation.

As nonlinear phenomena are simulated at increasingly high resolution and for longer times on adaptive meshes across parallel processors, the validity of results becomes an important issue. Numerical errors accumulate from round-off, truncation (e.g., higher-order and nonlinear dissipative and dispersive regularizations), spatially and temporally adjusting meshes, and ad-hoc filters. The visiometric process will help find these defects and provide a more rigorous basis for model building and prediction. Visiometrics will also produce new art forms (Zabusky, 2000).

N.J. ZABUSKY

See also **Contour dynamics; Fluid dynamics; Vortex dynamics of fluids; Wavelets**

Further Reading

Bitz, F. & Zabusky, N. 1990. DAVID and "Visiometrics": visualizing and quantifying evolving amorphous objects. *Computers in Physics*, 4: 603–14
Farge, M. 1992. Wavelet transforms and their application to turbulence. *Annual Reviews of Fluid Mechanics*, 24: 395–457
Farge, M. 2000. Choice of representation modes and color scales for visualization in computational fluid dynamics. In *Proceedings of the Science and Art Symposium*, edited by A. Gyr, P.D. Koumoutsakos & U. Burr, Dordrecht: Kluwer, pp. 91–100
Farge, M.G., Pellegrino, G. & Schneider, K. 2001. Coherent vortex extraction in 3D turbulent flows using orthogonal wavelets. *Physical Review Letters*, 87(5): 054501-1–054501-4
Feher, A. & Zabusky, N. 1996. An interactive imaging environment for scientific visualization and quantification. *International Journal of Imaging System Technology*, 7: 121–30
Fernandez, V.M., Silver, D. & Zabusky, N.J. 1996. Visiometrics of complex physical processes: diagnosing vortex-dominated flows. *Computers in Physics*, 10: 463–470
Hawley, J. & Zabusky, N. 1989. Vortex paradigm for shock accelerated density stratified interfaces. *Physical Review Letters*, 63: 1241–44
Jeong, J. & Hussain, F. 1995. On the identification of a vortex. *Journal of Fluid Mechanics*, 285: 69–94
Kida, S. & Miura, H. 1998. Identification and analysis of vortical structures. *European Journal of Mechanics*, B/Fluids, 17(4): 471–488
Lorensen, W.E. & Cline, H.E. 1987. Marching cubes: A high resolution 3D surface construction algorithm. *Computer Graphics*, 21(3): 163–169
Melander, M.V. & Hussain, F. 1994. Topological vortex dynamics in axisymmetric viscous flows. *Journal of Fluid Mechanics*, 260: 57–80
Miura, H. 2002. Analysis of vortex structures in compressible isotropic turbulence. *Computer and Physics Communications*, 147: 552–554
Post, F.H., Vrolijk, B., Hauser, H., Laramee, R.S. & Doleisch, H. 2002. Feature extraction and visualization of flow fields. In *Eurographics* 2002 *State-of-the-Art Reports*, edited by D. Fellner, & R. Scopigno, pp. 69–100. See also http://visualization.tudelft.nl/
Samtaney, R., Silver, D., Zabusky, N. & Cao, J. 1994. Visualizing features and tracking their evolution. *IEEE Computer*, 27(7): 20–27
Samtaney, R. & Zabusky, N.J. 2000. Visualization, feature extraction and quantification of numerical visualizations of high-gradient compressible flows. In *Flow Visualization, Techniques and Examples*, edited by A. Smits & T.T. Lim, London: Imperial College Press, pp. 317–344
Upson, C. & Keeler, M. 1988. V-BUFFER: visible volume rendering. *Computer Graphics*, 22(4): 59–64
Villasenor, J. & Vincent, A. 1992. An algorithm for space recognition and time tracking of vorticity tubes in turbulence. *CVGIP: Image Understanding*, 55: 27–35
Vrolijk, B., Reinders, F. & Post, F.H. 2003. Feature tracking with skeleton graphs. In *Data Visualization: The State of the Art*, edited by F.H. Post, G.M. Nielson & G.P. Bonneau, Boston: Kluwer, pp. 37–52
Zabusky, N.J. 1999. Vortex paradigm for accelerated inhomogeneous flows: visiometrics for the Rayleigh–Taylor and Richtmyer–Meshkov environments. *Annual Review of Fluid Mechanics*, 31: 495
Zabusky, N.J. 2000. Scientific computing visualization—a new venue in the arts. In *Proceedings of the Science and Art Symposium*, edited by A. Gyr, P.D. Koumoutsakos & U. Burr, Dordrecht: Kluwer, pp. 1–11
Zabusky, N.J., Gupta, S. & Gulak, Y. 2003. Localization and spreading of contact discontinuity layers in simulations of compressible dissipationless flows. *Journal of Computational Physics*, 188: 347—363
Zabusky, N.J., Silver, D. & Pelz, R. 1993. Vizgroup '93. Visiometrics, juxtaposition and modeling. *Physics Today*, 46(3): 24-31

VOLCANOS

See **Geomorphology and tectonics**

VOLTERRA SERIES AND OPERATORS

In addition to playing an important role in the development of theoretical biology, the Italian mathematician Vito Volterra (1860–1940) also strongly influenced the development of modern calculus. We deal here with the Volterra functional series (VFS) and associated Volterra differential operators (VDO) which provide a consistent mathematical framework for stating material properties in nonlinear wave propagation systems, for example, in nonlinear optics (Censor & Melamed, 2002; Censor, 2000; Sonnenschein & Censor, 1998). For simplicity, the present introduction is restricted to the temporal domain, adequate for time signals. Waves require a spatiotemporal domain.

Linear Systems

Modeling of physical systems requires material (or constitutive) relations. In electromagnetics, we have relations like $D=\varepsilon E$, $B=\mu H$; in acoustics, the compressibility (relation of pressure to volume) is needed; in elastodynamics, we include Hooke's law (relation of stress to strain). Here $D=\varepsilon E$ and its nonlinear extensions are treated as prototypes.

Materials are dispersive, depending (in the restricted sense of temporal dispersion discussed here) on frequency f. Consider the linear case first:

$$D(\omega) = \varepsilon(-\mathrm{i}\omega)E(\omega), \quad \omega = 2\pi f, \tag{1}$$

where ω is the angular frequency, and defining $\varepsilon(-\mathrm{i}\omega)$ instead of $\varepsilon(\omega)$ is convenient for subsequent applications.

The Fourier transform pair

$$F(t) = \frac{1}{2\pi}\int_{-\infty}^{\infty} d\omega F(\omega)e^{-i\omega t},$$
$$F(\omega) = \int_{-\infty}^{\infty} dt\, F(t)e^{i\omega t} \tag{2}$$

relates the spectral and temporal domains. We use the same symbol F, although $F(t)$ and $F(\omega)$ are different functions. Accordingly, (1) becomes a convolution integral

$$D(t) = \int_{-\infty}^{\infty} dt_1 \varepsilon(t_1)E(t-t_1), \tag{3}$$

where $D(t)$, $\varepsilon(t)$, $E(t)$, are related to $D(\omega)$, $\varepsilon(-i\omega)$, $E(\omega)$, respectively, according to (2). Note that (3) can be viewed as an integral operation, acting on $E(t)$. Also (3) is the simplest form of a VFS.

Formally we can start with (1), transform according to (2), and note that in the integral $E(-i\omega)$ (if it can be represented or approximated by a polynomial in $-i\omega$) can be considered as a polynomial operator $\varepsilon(\partial_t)$ acting on the exponential, in which every time derivative ∂_t replaces a term $-i\omega$ in $\varepsilon(-i\omega)$. Note that $\varepsilon(t)$ and $\varepsilon(\partial_t)$ are different functions, but $\varepsilon(-i\omega)$ and $\varepsilon(\partial_t)$ possess the same functional structure. Thus, instead of (3), we now have the VDO representation

$$D(t) = \varepsilon(\partial_t)E(t) = \varepsilon(\partial_\tau)E(\tau)\,|_{\tau\to t}\,. \tag{4}$$

The last expression in (4) with the instruction $\tau \to t$ is superfluous here but will be important for the nonlinear case below. The possibility of using this technique for $\varepsilon(\partial_t)$ a rational function (ratio of polynomials) is discussed elsewhere (Censor, 2001).

As a trivial example for (3) and (4), consider a harmonic signal

$$\begin{aligned} E(t) &= E_0e^{-i\omega t}, \\ D(t) &= E_0e^{-i\omega t}\int_{-\infty}^{\infty} dt_1\, \varepsilon(t_1)e^{i\omega t_1} \\ &= \varepsilon(-i\omega)E_0e^{-i\omega t} = \varepsilon(\partial_t)E_0e^{-i\omega t} \end{aligned} \tag{5}$$

clarifying the role of the VDO in (4).

Nonlinear Systems and the Volterra Series and Operators

In nonlinear systems, the material relations involve powers and products of fields. Can we simply replace (1) by a series involving powers of $E(\omega)$? A cursory analysis reveals that this leads to inconsistencies. Instead, we ask if (3) can be replaced by a "super convolution" and what form that should take. Indeed, the Volterra series provides a consistent mathematical answer to these questions. It is given by

$$\begin{aligned} D(t) &= \sum_m D^{(m)}(t), \\ D^{(m)}(t) &= \int_{-\infty}^{\infty} dt_1 \ldots \int_{-\infty}^{\infty} dt_m \\ &\quad\times \varepsilon^{(m)}(t_1,\ldots,t_m)E(t-t_1)\ldots E(t-t_m). \end{aligned} \tag{6}$$

Typically, the VFS (6) contains the products of fields expected for nonlinear systems, combined with the convolution structure (3). Various orders of nonlinear interaction are indicated by m.

Theoretically, all the orders co-exist (in practice, the series will have to be truncated within some approximation), and therefore, we cannot inject a time harmonic signal as in (5). If instead, we start with a periodic signal, $E(t) = \sum_n E_n e^{-in\omega t}$ and substitute in (6), we find

$$\begin{aligned} D^{(m)}(t) &= \sum_{n_1,\ldots,n_m} \varepsilon^{(m)}(-in_1\omega,\ldots,-in_m\omega) \\ &\quad\times E_{n_1}\ldots E_{n_m}e^{-iN\omega t} = \sum_N D_N e^{-iN\omega t}, \\ N &= n_1+\ldots+n_m, \end{aligned} \tag{7}$$

with (7) displaying the essential features of a nonlinear system, namely, the dependence on a product of amplitudes, and the creation of new frequencies as sums (including differences and harmonic multiples) of the interacting signals frequencies. In addition, (7) contains the weighting function $\varepsilon^{(m)}(-in_1\omega,\ldots,-in_m\omega)$ for each interaction mode.

The extension of (4) to the nonlinear VDO is given by

$$\begin{aligned} D^{(m)}(t) &= \varepsilon^{(m)}(\partial_{t_1},\ldots,\partial_{t_m}) \\ &\quad\times E(t_1)\ldots E(t_m)\,|_{t_1,\ldots,t_m\to t}\,. \end{aligned} \tag{8}$$

In (8), the instruction $t_1,\ldots,t_m \to t$ guarantees the separation of the differential operators, and finally renders both sides of the equation to become functions of t.

The VFS (6), including the convolution integral (3), is a global expression describing $D(t)$ as affected by integration times extending from $-\infty$ to ∞. Physically, this raises questions about causality, that is, how future times can affect past events. In the full-fledged four-dimensional generalization, causality is associated with the so-called "light cone" (Bohm, 1965). It is noted that the VDO representation (4, 8) is local, with the various time variables just serving for bookkeeping of the operators, and where this representation is justified, causality problems are not invoked.

In general, the frequency constraint of (7) is obtained from the Fourier transform of (6), having the form

$$D^{(m)}(\omega) = \frac{1}{(2\pi)^{m-1}} \int_{-\infty}^{\infty} d\omega_1 \ldots \int_{-\infty}^{\infty} d\omega_{m-1}$$
$$\times \varepsilon^{(m)}(-i\omega_1, \ldots, -i\omega_m) E(\omega_1) \ldots E(\omega_m),$$
$$\omega = \omega_1 + \ldots + \omega_m. \quad (9)$$

It is noted that in (9), we have $m-1$ integrations, one less than in (6). This tallies with the linear case where (1) and (3) involve zero and one integration, respectively. Consequently, the left- and right-hand sides of (9) are functions of ω and ω_m, respectively. The additional constraint $\omega = \omega_1 + \cdots + \omega_m$ completes the equation and renders (9) self-consistent.

Summary

The modeling of nonlinear media using the VFS and VDO provides a mathematically consistent framework which includes linear media as a limiting case. The model displays the typical ingredients of nonlinear circuits and wave systems, where the nonlinear terms are proportional to the product of the amplitudes of the interacting fields, and the newly created frequencies are sums (or differences, or harmonic multiples) of the interaction frequencies, given by $\omega = \omega_1 + \cdots + \omega_m$. In the quantum-mechanical context this is an expression of the conservation of energy. Not shown here is the associated wave propagation vector constraint $\boldsymbol{k} = \boldsymbol{k_1} + \cdots + \boldsymbol{k_m}$, which in the quantum-mechanical context expresses conservation of momentum. Schetzen (1980) is an excellent source of further reading on VFS in nonlinear systems and has many early references.

DAN CENSOR

See also **Harmonic generation; Manley–Rowe relations; Nonlinear acoustics; Nonlinear optics**

Further Reading

Bohm, D. 1965. *The Special Theory of Relativity*, New York: Benjamin

Censor, D. 2000, A quest for systematic constitutive formulations for general field and wave systems based on the Volterra differential operators. *Progress in Electromagnetics Research*, 25: 261–284

Censor, D. 2001, Constitutive relations in inhomogeneous systems and the particle–field conundrum. *Progress in Electromagnetics Research*, 30: 305–335

Censor, D. & Melamed, T. 2002. Volterra differential constitutive operators and locality considerations in electromagnetic theory. *Progress in Electromagnetic Research*, 36: 121–137

Schetzen, M. 1980. *The Volterra and Wiener Theorems of Nonlinear Systems*, Chichester and New York: Wiley

Sonnenschein, M. & Censor, D. 1998. Simulation of Hamiltonian light beam propagation in nonlinear media. *Journal of the Optical Society of America* B, 15: 1335–1345

VOLTERRA–LOTKA EQUATIONS

See **Population dynamics**

VOLUME-PRESERVING MAPS

See **Measures**

VORONOÏ DIAGRAMS

See **Tessellation**

VORTEX DYNAMICS IN EXCITABLE MEDIA

Vortices in excitable media include spiral waves in two spatial dimensions and scroll waves in three spatial dimensions. They are described by reaction-diffusion systems of equations,

$$\partial_t \boldsymbol{u} = \boldsymbol{f}(\boldsymbol{u}) + \mathbf{D}\nabla^2 \boldsymbol{u} + \varepsilon \boldsymbol{h},$$
$$\boldsymbol{u}, \boldsymbol{f}, \boldsymbol{h} \in \mathbb{R}^\ell,\ \mathbf{D} \in \mathbb{R}^{\ell \times \ell},\ \ell \geq 2, \quad (1)$$

where $\boldsymbol{u}(\boldsymbol{r}, t) = (u_1, u_2, \ldots)^{\mathrm{T}}$, is a column-vector of the reagent concentrations, $\boldsymbol{f}(\boldsymbol{u})$ is a column vector of the reaction rates, $\mathbf{D}$ is the matrix of diffusion coefficients, $\varepsilon \boldsymbol{h}(\boldsymbol{u}, \boldsymbol{r}, t)$ is some small perturbation, and $\boldsymbol{r} \in \mathbb{R}^2$ or $\mathbb{R}^3$ is the vector of coordinates on the plane or in space.

In an unbounded two-dimensional medium with $\varepsilon \boldsymbol{h} = 0$, a spiral wave solution rotating with angular velocity ω has the form

$$\boldsymbol{u} = \boldsymbol{U}(\boldsymbol{r}, t) = \boldsymbol{U}[\rho(\boldsymbol{r}), \vartheta(\boldsymbol{r}) + \omega t]$$
$$\approx \boldsymbol{P}[\rho(\boldsymbol{r}) - \frac{\lambda}{2\pi}(\vartheta(\boldsymbol{r}) + \omega t)]\Big|_{\rho \to +\infty}, \quad (2)$$

where $\rho(\boldsymbol{r})$ and $\vartheta(\boldsymbol{r})$ are the polar coordinates corresponding to the cartesian coordinates $\boldsymbol{r}$. $\boldsymbol{P}(\xi; \omega, \lambda)$ is a periodic wave solution with frequency ω and spatial period λ, so the $\rho \to +\infty$ asymptotic means that isolines are approximately Archimedian spirals with pitch λ. Solutions (2) are typically possible for isolated values of ω and corresponding λ.

Note that system (1) with $\varepsilon \boldsymbol{h} = 0$ is invariant with respect to the Euclidean group of motions of the plane $\{\boldsymbol{r}\}$. Solution (2) is a relative equilibrium, meaning that the states of the wave at all moments of time are equivalent to each other up to a Euclidean motion, namely, a rotation around the origin. If (2) is a solution, then from symmetry

$$\tilde{\boldsymbol{U}} = \boldsymbol{U}(\rho(\boldsymbol{r} - \boldsymbol{R}_\odot), \vartheta(\boldsymbol{r} - \boldsymbol{R}_\odot) + \omega t - \Phi_\odot) \quad (3)$$

is another solution for any constant displacement vector $\boldsymbol{R}_\odot = (X_\odot, Y_\odot)^{\mathrm{T}}$ and initial rotation phase $\Phi_\odot$. Thus, we have a three-dimensional manifold, parameterized by coordinates $X_\odot, Y_\odot, \Phi_\odot$, of spiral wave solutions neutrally stable with respect to each other. By "dynamics" of the vortices, we understand any deviation of the solutions from the stationary rotation (2).

Meander

A nonstationary rotation of a spiral wave accompanied by constant change of its shape is called meander. It is convenient to describe this phenomenon in terms of the spiral tip, which can be defined as an intersection of selected isolines of two components of the nonlinear field $\boldsymbol{u}$,

$$u_{j_1}(X_\bullet, Y_\bullet, t) = v_1, \quad u_{j_2}(X_\bullet, Y_\bullet, t) = v_2,$$
$$\Phi_\bullet = \arg(\partial_x + \mathrm{i}\partial_y)u_{j_3}(X_\bullet, Y_\bullet, t), \tag{4}$$

$j_1 \neq j_2$, where $X_\bullet(t)$, $Y_\bullet(t)$ are the coordinates of the tip and $\Phi_\bullet(t)$ is its orientation angle. Typically, a spiral wave in a given system develops the same kind of meander pattern $X_\bullet(t)$, $Y_\bullet(t)$ independent of the initial conditions. Changes of parameters in the same system change the meander pattern, and types of patterns can be qualitatively similar in very different excitable media models.

Possible types of meander can be classified using an orbit manifold decomposition of (1) by the Euclidean group. Evolution of the shape of the wave can be described in coordinates (ξ, η) in a moving frame of reference attached to the spiral tip,

$$\partial_t \boldsymbol{u} = \mathbf{D}(\partial_\xi^2 + \partial_\eta^2)\boldsymbol{u} + \big[C_1(t)\partial_\xi + C_2(t)\partial_\eta$$
$$+\omega(t)\left(\xi\partial_\eta - \eta\partial_\xi\right)\big]\boldsymbol{u} + \boldsymbol{f}(\boldsymbol{u})$$
$$u_{j_{1,2}}(0,0) = 0, \quad \partial_\eta u_{j_3} = 0, \tag{5}$$

and the movement of the tip is described by ordinary differential equations

$$\frac{\mathrm{d}\Phi_\bullet}{\mathrm{d}t} = \omega(t),$$
$$\frac{\mathrm{d}X_\bullet}{\mathrm{d}t} + \mathrm{i}\frac{\mathrm{d}Y_\bullet}{\mathrm{d}t} = (C_1(t) + \mathrm{i}C_2(t))\mathrm{e}^{\mathrm{i}\Phi_\bullet}. \tag{6}$$

Equations (5) define a dynamic system with the phase space $\{(\boldsymbol{u}(\xi,\eta), C_1, C_2, \omega)\}$, devoid of the Euclidean symmetry of the original system (1). Knowing the attractor in (5), one can deduce the properties of the meander patterns by integrating the ODE system (6) (see Figure 1).

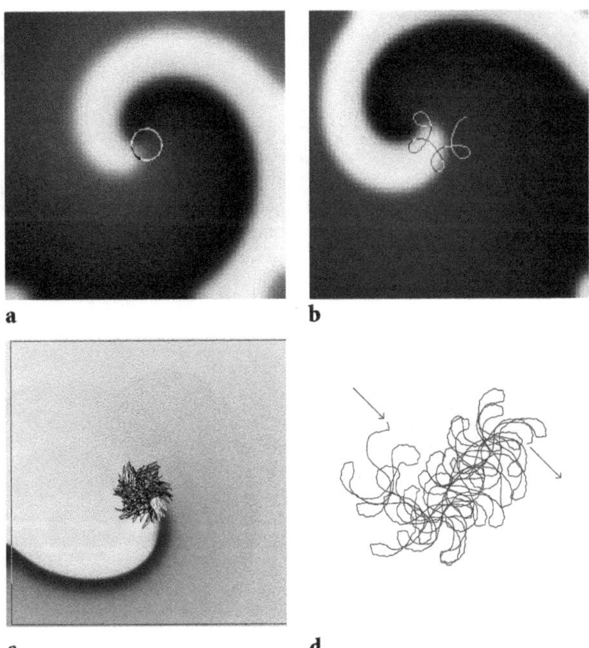

Figure 1. Typical meander patterns. Shown are snapshots of the excitation field with pieces of preceding tip paths superimposed. (a) Stationary (rigid) rotation: equilibrium in the base system (5). (b) Classical biperiodic "flower" meander: limit cycle in the base system (5). (c) Quasi-periodic hypermeander: invariant torus in the base system (5). (d) Pseudorandom walk hypermeander: chaotic attractor in the base system (5) (only the tip path shown).

Forced Drift

Another kind of deviation from (2) is drift of spirals due to perturbations $\varepsilon\boldsymbol{h} \neq 0$. As solutions of family (3) are neutrally stable with respect to each other, a small perturbation of a spiral wave caused by an $\varepsilon\boldsymbol{h}$ limited in time will die out, but it will typically result in a small change in the spiral wave coordinates $X_\odot$, $Y_\odot$ and $\Phi_\odot$.

If similar perturbations are applied repeatedly with a period equal to the period of the spiral, then small shifts of $X_\odot$ and $Y_\odot$ accumulate, which is called a resonant drift. Another type of slow drift is inhomogeneity-induced drift, occurring when $\varepsilon\boldsymbol{h}$ depends explicitly on spatial coordinates (medium properties are slightly inhomogeneous, as in Figure 2(b)). In the first order of perturbation theory, this is equivalent to a time-dependent perturbation synchronized with the spiral rotation and is therefore resonant. A third type of drift occurs if the medium is bounded, and the boundary influence on the spiral wave is not negligible. Although a boundary is not a slight perturbation, the effects of passive (nonflux) on the spiral wave can be small and similar to that of small spatial inhomogeneity (Figure 2(c)). Other kinds of perturbations breaking the Euclidean symmetry of (1) can also cause drift.

Being a first-order effect, the slow drift of a spiral due to small forces of different types obeys a superposition principle. It leads to motion equations

$$\partial_t(X_\odot + \mathrm{i}Y_\odot) = C(X_\odot, Y_\odot) + v(X_\odot, Y_\odot)\mathrm{e}^{\mathrm{i}\Theta},$$
$$\partial_t\Theta = \Omega(t) - \omega(X_\odot, Y_\odot), \tag{7}$$

where $C(X_\odot, Y_\odot)$ is the velocity of the inhomogeneity and boundary induced drift, $v(X_\odot, Y_\odot)$ is the velocity of the resonant drift, Θ is the phase difference between the spiral and the resonant forcing, $\Omega(t)$ is the

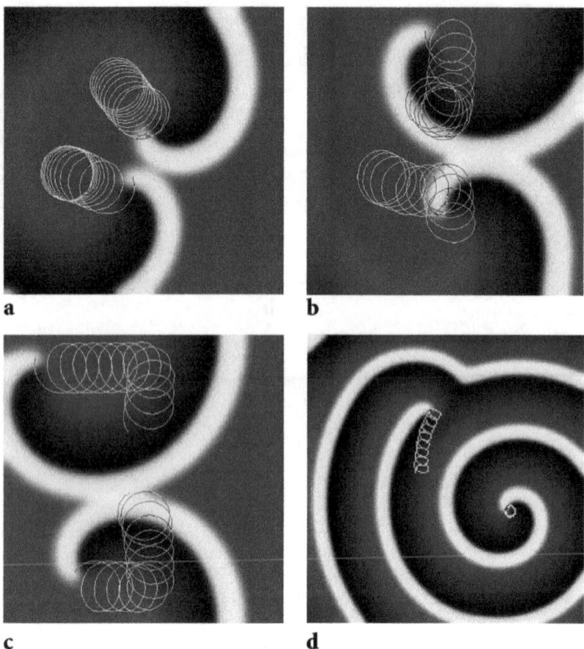

Figure 2. Different drifts of spiral waves. Shown are snapshots of the excitation field with pieces of preceding tip paths superimposed. The right half of the medium is slightly "stronger" than the left half; (a)–(c) are consecutive stages of the same numeric experiment. (a) Two close oppositely charged spirals attract with each other and form a pair drifting in SE direction. (b) The spirals have reached the inhomogeneity and are being driven apart by it. (c) The spirals have reached the medium boundary and now drift along it. (d) In a bigger medium: the right spiral has subdued the left spiral into an induced drift.

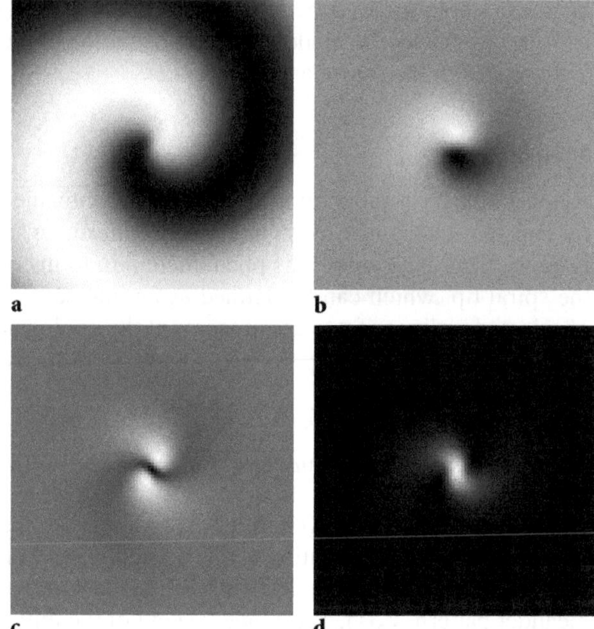

Figure 3. A spiral wave solution (a) and its temporal (b) and spatial (c,d) response functions, as density plots. Monotone gray periphery on (b–d) corresponds to zero. Thus the spiral wave is a non-local process, but its response functions are well localized.

perturbation frequency, and $\omega(X_\odot, Y_\odot)$ is the own spiral angular frequency possibly depending on the current spiral location.

These are motion equations for rigidly rotating spirals, and $X_\odot$ and $Y_\odot$ are sliding period averages of $X_\bullet$, $Y_\bullet$. Dynamics of forced meandering spirals are more complicated because of possible resonances.

Spiral Waves as Particles

Motion equations (7) are obtained by summation of the effects of elementary perturbations of different modalities localized in different sites and occurring at different moments of time, onto the spiral's location and phase. These elementary responses are described by response functions, which are critical eigenfunctions of the adjoint linearized operator. An interesting property of the response function is their localization in the vicinity of the spiral core (see Figure 3). The spiral will only drift if the perturbation is applied not too far from its core.

A paradox is thus created. Although a spiral wave appears as a significantly nonlocal process, involving in its rhythm all of the excitable medium, it behaves as a localized, particle-like object in its response to perturbations.

A spatial response function, defining the proportionality between drift velocity and inhomogeneity magnitude, typically has scalar (drift along the parameter gradient or toward the boundary) and pseudo-scalar (across the parameter gradient or along the boundary) components. The sign of the latter depends on the direction of the spiral rotation.

Bending and Twisting of Scroll Waves

As a scroll wave is a three-dimensional analogue of a spiral wave, all comments about spiral wave dynamics remain valid for the scrolls, but there are new aspects arising from the third dimension. The simplest 3-d vortex is the straight scroll wave, a spiral wave continued unchanged in the third dimension. The spiral tip, a point in the plane, becomes the edge of the scroll, a line in space, and the spiral core, a circle in the plane, becomes the scroll filament, a tube. The term *filament* also sometimes denotes the center line of the tube filament.

More interesting regimes are scrolls with bent filaments (Figure 4(a)), and with rotation phase varying along the filament (twisted scrolls, as in Figure 4(b)). Both bending and twist of scrolls are factors of their dynamics. A vortex ring will collapse or expand and, at the same time, drift along its symmetry axis, and twisted vortex will usually spread the twist evenly along its filament or, if possible, untwist.

The asymptotic motion equation for the scroll waves can be derived using response functions. If the twist is

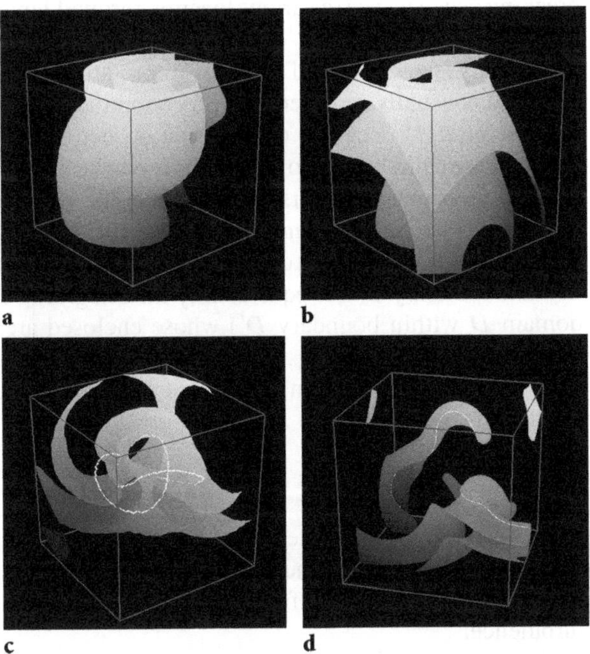

Figure 4. Scroll waves. (a) Scroll with a curved filament. (b) Twisted scroll with straight filament. (c) Scroll with a knotted filament. (d) "Turbulent regime": many scrolls developed from one via negative filament tension multiplication mechanism. On panels (c) and (d), part of the wavefronts is cut out, to make the filaments (white lines) visible.

not too strong, then the dynamics of the scroll due to bending and due to twist are decoupled. The motion equation of the filament is

$$\partial_t \boldsymbol{R}_\mathrm{f} = b_2 \partial_\mathrm{s}^2 \boldsymbol{R}_\mathrm{f} + c_3 \left[\partial_\mathrm{s} \boldsymbol{R}_\mathrm{f} \times \partial_\mathrm{s}^2 \boldsymbol{R}_\mathrm{f}\right], \tag{8}$$

where $\boldsymbol{R}_\mathrm{f} = \boldsymbol{R}_\mathrm{f}(p,t) \in \mathbb{R}^3$ is the position of the filament as function of a length parameter p and time, and ∂_s is arclength differentiation, $\partial_\mathrm{s} \equiv \left|\partial_p \boldsymbol{R}_\mathrm{f}\right|^{-1} \partial_p$. At $b_2 = 0$, Equation (8) is completely integrable; in particular, the total length of the filament is conserved. Otherwise, the total filament length decreases if $b_2 > 0$ and increases if $b_2 < 0$, and in the latter case, a straight filament is unstable.

If twist is high, it changes the filament tension and may make it negative. This causes an instability of the straight filament shape, leading to "sproing": a sudden transition from a strongly twisted scroll with straight filament to a less twisted scroll with a helical filament.

Competition and Interaction *(Divide et impera)*

Normally, two colliding excitation waves annihilate each other. Thus, if there are many periodic sources of waves (e.g., vortices), then the medium splits into domains, or regions of influence, each domain receiving waves from its source. The domains are separated by shock structures where the waves collide (see Figure 2(a–c)).

The domain boundaries work like non-flux boundaries. Thus, two spiral waves can interact with each other (cause each other's drift and frequency shift), whereas each of them actually interacts with the boundary between their domains. Such interaction between spirals may lead to the formation of linked pairs (see Figure 2(a)).

Different scroll filaments or different parts of the same filament can also interact with each other. If this interaction is repulsive, it may compensate positive tension normally causing closed filaments to contract and collapse, leading to stable "particle-like" 3-d scrolls with compact filaments (see Figure 4(c)).

Induced Drift

If colliding waves annihilate in a ratio of one-to-one, continuity of phase applies. If two vortices have different frequencies (e.g., because of a spatial inhomogeneity of the medium), then by continuity of phase the domain boundary between them moves toward the slower vortex. When it reaches its core, the slower vortex loses its identity and turns into a dislocation in the wave field emitted by the faster vortex. This dislocation, appearing as a free end of an excitation wave, periodically rejoins from one wave to another with some overall drift depending on the frequency and direction of the incident waves (see Figure 2(d)). If the incident wave packet ceases, the dislocation can develop into a vortex again.

As a dislocation is very different from a vortex, this induced drift is an example of hard, non-perturbative dynamics.

Hard Dynamics: Births, Deaths, and Multiplication of Vortices

Another kind of hard dynamics involves complete elimination of a vortex. This may happen if the wave propagation around the vortex becomes impossible, for example, if the vortex has been driven too close to a medium boundary. Alternatively, two spiral waves with opposite topological charges may annihilate if driven too close to each other. For a scroll wave, annihilation may happen to a piece of its filament, which then appears as splitting of a scroll wave into two.

Birth of a vortex may occur as a result of a temporary local block of excitation propagation. Unless this happens near the medium boundary, this means birth of a pair of oppositely rotating spirals in the plane, or a scroll with a closed filament around the perimeter of the propagation block. The block may occur as a result of external forcing or special initial conditions, or develop as a result of an instability of an existing vortex. Such instability can underlie a chain reaction of the vortex multiplication, which may lead to a turbulence of excitation vortices—a spatiotemporal chaotic state where generation of new vortices is balanced by their annihilation when they get close to each other due to overcrowding.

Several mechanisms of such instabilities have been identified, including those working in two or three dimensions, such as Eckhaus instability, zigzag/lateral instability or imposed mechanical movement of the medium, and those possible only in three dimensions. The latter include instability due to the negative "tension" b_2 of the vortex filament (see Figure 4(d)), or caused by spatially inhomogeneous anisotropy of the medium such as that observed heart ventricular muscle. Some of these types of instabilities may be responsible for the phenomenon of fibrillation of the heart (*See* **Cardiac arrhythmias**).

VADIM N. BIKTASHEV

See also **Complex Ginzburg–Landau equation; Framed space curves; Reaction-diffusion systems; Scroll waves; Spiral waves**

Further Reading

Gaponov-Grekhov, A.V., Rabinovich, M.I. & Engelbrecht, J. (editors). 1989. *Nonlinear Waves II. Dynamics and Evolution*, Berlin and New York: Springer

Holden, A.V. & El Naschie, M.S. (editors). 1985. *Nonlinear Phenomena in Excitable Physiological Systems.* Special issue: *Chaos Solitons & Fractals*, 5(3/4)

Holden, A.V., Markus, M. & Othmer, H.G. (editors). 1989. *Nonlinear Wave Processes in Excitable Media*, New York and London: Plenum Press

Kapral, R. & Showalter, K. (editors). 1995. *Chemical Waves and Patterns*, Dordrecht and Boston: Kluwer

Panfilov, A.V. & Holden, A.V. (editors). 1997. *Computational Biology of the Heart*, Chichester and New York: Wiley

Panfilov, A. & Pertsov, A. 2001. Ventricular fibrillation: evolution of the multiple wavelet hypothesis. *Philosophical Transactions of the Royal Society of London* A, 359: 1315–1325

Swinney, H.L. & Krinsky, V.I. (editors). 1991. *Waves and Patterns in Chemical and Biological Media*, Cambridge, MA: MIT Press

Winfree, A.T. (editor). 1998. *Fibrillation in normal ventricular myocardium.* Focus issue. *Chaos*, 8(1)

Winfree, A.T. 1991. Varieties of spiral wave behavior in excitable media. *Chaos*, 1(3): 303–334

Zykov, V.S. 1987. *Modelling of Wave Processes in Excitable Media*, Manchester: Manchester University Press

VORTEX DYNAMICS OF FLUIDS

The swirling and chaotic behavior of vortex-dominated fluid flows has inspired philosophers, poets, and artists from antiquity to the present, including Leonardo da Vinci, the 15th-century artist, scientist, and engineer who captured complex fluid motions as part of his applied and creative work (van Dyke, 1982; Lugt, 1983; Minahen, 1992). The traditional woodblock prints (*Ukioy-e*) of Hiroshige and Hokusai also show spiral motion produced by a vortex and "curls" on breaking ocean waves in various locations around Japan (Zabusky, 2000).

Vorticity is a measure of the "spin" or rotation of the fluid and is usually represented by a continuous vector variable $\boldsymbol{\omega}(\boldsymbol{x}, t)$. Regions of flow that do not contain vorticity are called irrotational. Vorticity can be defined in two equivalent ways. First, as the curl of the velocity $\boldsymbol{u}$, $\boldsymbol{\omega} = \nabla \times \boldsymbol{u}$. Second, and more geometrically, as a limit (when it is finite) of a closed circuit integral over a domain boundary D' (or equivalently the integral over domain D within boundary D') whose enclosed area $A \to 0$, or $\boldsymbol{\omega} = \boldsymbol{e}_{\tan}[\lim|_{A\to 0} (\Gamma/A)]$ where for finite A, Γ is the circulation or line integral of velocity around the domain,

$$\Gamma = \int_{D'} \boldsymbol{u} \cdot \mathbf{d}s = \int_{D} \boldsymbol{\omega} \cdot \mathbf{d}A \qquad (1)$$

and $\boldsymbol{e}_{\tan}$ is the tangent to the vorticity vector at the point where $A \to 0$. Note that the scalars $\frac{1}{2}\boldsymbol{\omega} \cdot \boldsymbol{\omega}$ (enstrophy) and $\boldsymbol{\omega} \cdot \boldsymbol{u}$ (helicity) arise frequently in discussions of turbulence.

Leonhard Euler derived continuum mathematical equations for an ideal (non-dissipative) fluid. In the 19th century, Claude Navier, George Stokes, and others included dissipative processes for a realistic viscous fluid (Lamb, 1932; Batchelor, 1967; Kiselev et al., 1999). These equations provide models of fluids under usual (for example nonrelativistic) conditions. Asymptotic (reduced) mathematical equations, particularly those emphasizing inviscid vortex-dominated flows (Saffman, 1992) were derived in the 19th century with works of William Thomson (Lord Kelvin) and Hermann von Helmholtz, among others (*See* **Kelvin–Helmholtz instability**). Vortex models are used today to predict the behavior of geophysical and astrophysical fluids (Nezlin, 1993), and in engineering applications (Green, 1995).

If $\boldsymbol{\omega}(\boldsymbol{x}, t)$ is modeled by one or more line filaments in three dimensions (3-d) on which the vorticity is concentrated, then the velocity produced by this vorticity can be calculated from a (Biot–Savart) line integral over the filaments. With this velocity, one can obtain the trajectories of points in the fluid merely by solving $\mathrm{d}\boldsymbol{x}/\mathrm{d}t = \boldsymbol{u}(\boldsymbol{x}, t)$. In 2-d planar flows, the vorticity vector has one component perpendicular to the plane, $(\omega = \partial_x v - \partial_y u)$, and it may be modeled by one or more points of strength Γ_i.

With these definitions, one can prove some fundamental theorems, including the following (Kiselev et al., 1999):

- *Thomson's theorem*: In an ideal barotropic moving fluid, Γ for any closed contour does not depend on time.
- *Helmholtz's theorem*: If the particles of a liquid satisfy Thomson's theorem and form a vortex filament at some moment of time, then these particles form a vortex filament at all subsequent and previous

moments of time and the vortex tube of circulation Γ will be time invariant and constant along its length.

For 2-d, the simplicity of the kinematic description for homogeneous incompressible fluids allows for simple computer models and valuable insights into the motion of ideal fluids. For example, point circulations Γ_i are usually positive and negative constants; vortex sheets along a line described by the circulation per unit length $\gamma(\boldsymbol{s}, t)$; or vorticity $\omega(\boldsymbol{x}, t)$ is constant in a domain (*See* **Contour dynamics**). (Note that for points and sheets, the equations are ill-posed mathematically, and regularization techniques must be used when computing numerically long times.) Interesting effects observed in 2-d include the merger of like-signed vortex domains if initially placed sufficiently close, binding of opposite-signed domains into "vortex projectiles" (a generic translating form whose simplest example is the dipole composed of two opposite-signed points). For vortex sheets, a linear (Kelvin–Helmholtz) stability analysis shows growth of a perturbation for all wavelengths of the disturbance. These evolve into finite-amplitude rolls with a characteristic spiral depending on their sign (see photo of an atmospheric vortex sheet on page 1 of color plate section). If one relaxes the sheet to a very thin layer then small wavelength perturbations are stabilized whereas wavelengths larger than the initial thickness remain unstable. Placing more than one harmonic perturbation on the thin layer, one observes roll-up and merger (energy cascade to large scales) as well as entrophy cascade to smaller scales; that is, the process becomes turbulent. Many details are found in the references below.

In 3-d, $\boldsymbol{\omega}$ may be concentrated on one or more filamentary lines or in finite area tubes, for example, the well-known vortex ring which translates at uniform speed and is unstable (Shariff & Leonard, 1992). If the ring is perturbed, this can lead to collapse and reconnection (Kida & Takaoka, 1994; Fernandez et al., 1995). In a dramatic experiment, Lim and Nickels (1992, 1995) fired a blue and red vortex ring toward each other in water and observed them interact and split into many blue-red smaller vortex ringlets.

The vorticity evolution equation is obtained if one applies the curl operation to the fluid momentum equation. This shows clearly many fundamental and interesting results on vorticity creation, modification, and dissipation. For the zero viscosity (ideal or Euler equation) limit, we have

$$\frac{\partial \boldsymbol{\omega}}{\partial t} + \boldsymbol{u} \cdot \nabla \boldsymbol{\omega} + \boldsymbol{\omega}(\nabla \cdot \boldsymbol{u}) = \boldsymbol{\omega} \cdot \nabla \boldsymbol{u} + \rho^{-2}(\nabla \rho \times \nabla p). \tag{2}$$

On the right-hand side the terms are: (1) vorticity "stretching" by the rate of strain tensor, which is only present in 2-d-axisymmetric and 3-d motions and is essential for turbulence seen in 3-d; and (2) baroclinic terms, which stem from the misalignment between the gradient of density and either the gravitation vector or the gradient of pressure. Both are important in incompressible inhomogeneous multi-species fluids (e.g., the Rayleigh–Taylor instability) and the latter term is important when shock waves interact with density inhomogeneities in compressible fluids (known as the Richtmyer–Meshkov (RM) instability environment). Recently, Ghoneim and colleagues (e.g., Soteriu & Ghoneim, 1995; Reinaud et al., 2000) used a point vortex model to represent inhomogeneous fluids where Γ_i and $\nabla \rho$ vary in time and Zabusky and colleagues have quantified the vortex-accelerated vorticity deposition in the compressible RM environment (Peng et al., 2003) relevant to laser fusion and supersonic combustion.

For real fluids with small dissipation or at high Reynolds number,

$$\mathrm{Re} = \frac{UL}{\nu} = \frac{\text{velocity} \times \text{length}}{\text{kinematic viscosity}}, \tag{3}$$

these insights are for short times. Vorticity is generated in a thin boundary layer in the vicinity of rigid or compliant objects in the flow—for example, cylinders (Williamson, 1996), spheres, or airfoils—or within channels because of viscosity (non-slip boundaries). Note, that the larger the Reynolds number, the more "unstable" is the fluid motion and the more likely that one finds it in a turbulent state. Cogent discussions of realistic turbulent effects including vortex phenomena are given in recent books (Pope, 2000; Tsinober, 2001).

Vortices are audible. When we hear the wind howling or the crack of a whip, we are sensing vortices in action. Aeroacoustics, the branch of fluid dynamics concerned with sound generated from vortices, is being applied to noise from jet engines, sounds in music and speech, and so on.

Currently, active fields of study include separating turbulent flows into their coherent and incoherent vortex structures (thereby simplifying prediction and control of flows) (*See* **Visiometrics**). These studies arise in geophysics (e.g., hurricanes, the Gulf stream, and Jupiter's red spot) and astrophysics (planetary nebulae, supernova, and galaxy collisions).

N.J. ZABUSKY

See also **Chaotic advection; Contour dynamics; Fluid dynamics; Turbulence; Visiometrics; Vortex dynamics in excitable media**

Further Reading

Batchelor, G. 1967. *An Introduction to Fluid Dynamics*, Cambridge: Cambridge University Press

Fernandez, V.M., Zabusky, N.J. & Gryanik, V.M. 1995. Vortex intensification and collapse of the Lissajous-elliptic ring: single and multiple-filament Biot–Savart simulations and visiometrics. *Journal of Fluid Mechanics*, 299: 289–331

Green, S.I. 1995. (editor). *Fluid Vortices*, Boston: Kluwer

Kida, S. & Takaoka, M. 1994. Vortex reconnection. *Annual Reviews of Fluid Mechanics*, 26: 169–189

Kiselev, S.P, Vorozhtsov, E.V. & Fomin, V.M. 1999. *Foundations of Fluid Mechanics with Applications*, Boston: Birkhauser

Lamb, H. 1932. *Hydrodynamics*, 6th edition, Cambridge: Cambridge University Press

Leonard, A. 1985. Computing three-dimensional incompressible flows with vortex elements. *Annual Reviews of Fluid Mechanics*, 17: 523–559

Lim, T.T. & Nickels, T.B. 1992. Instability and reconnection in the head-on collision of two vortex rings. *Nature*, 357: 225

Lim, T.T. & Nickels, T.B. 1995. Vortex rings. In *Fluid Vortices*, edited by S.I. Green, Boston: Kluwer, pp. 95–153

Lugt, H.J. 1983. *Vortex Flow in Nature and Technology*, New York: Wiley

Minahen, C. 1992. *Vortex/t: The Poetics of Turbulence*, University Park: Pennsylvania State University Press

Nezlin, M.V. 1993. *Rossby Vortices, Spiral Structures and Solitons: Astrophysics and Plasma Physics in Shallow Water Experiments*, New York: Springer

Peng, G., Zabusky, N.J. & Zhang, S. 2003. Vortex-accelerated secondary baroclinic vorticity deposition and late-intermediate times of a two-dimensional Richtmyer–Meshkov interface. *Physics of Fluids*, 15: 3730–3744

Pope, S.B. 2000. *Turbulent Flows*, Cambridge and New York: Cambridge University Press

Reinaud, J., Joly, L. & Chassaing, P. 2000. The baroclinic secondary instability of the two-dimensional shear layer. *Physics of Fluids*, 12(10): 2489–2505

Saffman, P.G. 1992. *Vortex Dynamics*, Cambridge and New York: Cambridge University Press

Shariff, K. & Leonard, A. 1992. Vortex rings. *Annual Reviews of Fluid Mechanics*, 24: 235-279

Soteriu, M.C. & Ghoneim, A. 1995. Effects of the free-stream density ratio on free and forced spatially developing shear layers. *Physics of Fluids*, 7: 2036–2051

Tsinober, A. 2001. *An Informal Introduction to Turbulence*, Boston: Kluwer

van Dyke, M. 1982. *An Album of Fluid Motion*, Stanford, CA: Parabolic Press

Williamson, C.H.K. 1996. Vortex dynamics in the cylinder wake. *Annual Reviews of Fluid Mechanics*, 28: 477–539

Zabusky, N.J. 1999 Vortex paradigm for accelerated inhomogeneous flows: visiometrics for the Rayleigh–Taylor and Richtmyer–Meshkov environments. *Annual Reviews of Fluid Mechanics*, 31: 495–536

Zabusky, N.J. 2000. Scientific computing visualization—a new venue in the arts. In *Science and Art Symposium 2000*, edited by A. Gyr, P.D. Koumoutsakos & U. Burr, Boston: Kluwer

WATER WAVES

Water waves have attracted the attention of scientists for many centuries. Although much is now understood, they are a continuing source of fascination, as many aspects of their often-complicated nonlinear behavior remain to be fully elucidated. We describe first the linear theory, much of which was developed in the 19th century, before discussing the more modern developments concerning weakly nonlinear waves. Throughout, the theory is based on the traditional assumptions that water is inviscid, incompressible with a constant density ρ, and in irrotational flow; that is, it has zero vorticity. It follows that for water waves, the governing equation is Laplace's equation, which is linear, and so all the nonlinearity in the problem resides in the free-surface boundary conditions. There is the kinematic condition that the free surface is a material surface at all times and the dynamic condition that the free surface has constant pressure at all times. The weakly nonlinear theory for water waves is described, which employs some of the basic paradigms for nonlinear waves. The special features that emerge when one considers finite-amplitude water waves are described in the review articles of Schwarz & Fenton (1982) and Dias & Kharif (1999).

Linear Waves

When the governing equations of motion are linearized about the rest state, it is customary to make a Fourier decomposition and seek solutions in which the wave elevation (deviation of the water surface from its rest position) is given by

$$\zeta = a \exp(\mathrm{i}(\boldsymbol{k} \cdot \boldsymbol{x} - \omega t)) + \text{c.c.}, \qquad (1)$$

where $\boldsymbol{x} = (x, y)$ denotes the horizontal coordinates, and t is the time variable, while $\boldsymbol{k} = (k, l)$ is the wave number, ω is the wave frequency, and a is the wave amplitude. Here, c.c. denotes the complex conjugate. This elementary disturbance represents a sinusoidal wave propagating in the direction $\boldsymbol{k}/\kappa$ where $\kappa = |\boldsymbol{k}| = \sqrt{k^2 + l^2}$, with a phase speed $c = \omega/\kappa$, a wavelength $\lambda = 2\pi/\kappa$, and a period $T = 2\pi/\omega$. The corresponding fluid velocity is $\boldsymbol{u} = (u, v)$ and is obtained from a velocity potential (so that $\boldsymbol{u} = \text{grad}\,\phi$), where

$$\phi = -\,ica \frac{\cosh \kappa(z + h)}{\sinh \kappa h} \exp(\mathrm{i}(\boldsymbol{k} \cdot \boldsymbol{x} - \omega t)) + \text{c.c.} \qquad (2)$$

Here, z is the vertical coordinate, and h is the total depth of water in the rest state (that is, the water occupies the region $-h < z < 0$). Note that the horizontal velocity $\boldsymbol{u}$ is then in phase with the surface elevation, but that the vertical velocity w is $\pi/2$ out of phase, while both velocity components decrease with depth away from the free surface.

Equations (1) and (2) provide a kinematic description of water waves, which to this point means that the conditions of incompressibility and irrotational flow have been satisfied, that the vertical velocity is zero at the bottom, and that the (linearized) kinematic boundary condition that the free surface remains a material surface for all time has been satisfied. To obtain the dynamics, these expressions are substituted into the remaining boundary condition that the free surface is one of constant pressure. This linearized formulation then yields the dispersion relation determining the wave frequency in terms of the wave number,

$$\omega^2 = (g\kappa + \sigma\kappa^3)\tanh(\kappa h), \qquad (3)$$

where $\rho\sigma$ is the coefficient of surface tension, which has a value of 74 dyn/cm at 20°C (*See* **Dispersion relations**). Detailed derivations of (3) and discussions of the consequences for the properties of water waves can be found in many classical and modern texts, see for instance, Lamb (1932), Whitham (1974), Lighthill (1978), or Mei (1983). Indeed, water waves have formed the paradigm for much of our present-day understanding of linear dispersive waves.

There are two branches of the dispersion relation (3), corresponding to waves running to either the right or the left. Note that (3) is isotropic, in that the wave frequency, and hence the phase speed, depend only on the magnitude of the wave number and not its direction. It is apparent from (3) that the effect of surface tension is significant only when $\kappa > (g/\sigma)^{1/2}$; using the above value for σ this corresponds to $\lambda < 1.73$ cm. Such waves are then usually called capillary waves, while waves with $\kappa < (g/\sigma)^{1/2}$ are called gravity waves. Another useful measure of the effect of surface tension is the Bond number $B = \sigma/gh^2$. When $B < 1/3$, the phase speed $c = \omega/\kappa$ decreases from its long wave value $c_0 = (gh)^{1/2}$, achieved at $\kappa = 0$, to a minimum value of c_m at a wave number $\kappa = \kappa_m$, and then increases without limit as κ increases to infinity. For the case when the Bond number $B > 1/3$, the phase speed increases monotonically from the long wave value c_0 as κ increases from zero. However, the critical depth below which this regime is realized occurs when the Bond number $B = 1/3$, which corresponds to $h = 0.48$ cm. In practice, this is too shallow to ignore the effects of friction and, hence, is usually not regarded as being of any practical interest.

From expressions (2) for the velocity potential and the dispersion relation (3), we see that water waves do not feel the effect of the bottom if $\kappa h \gg 1$. More precisely, if $\kappa h > 2.65$ (or $\lambda/h < 2.37$), then there is less than 1% error in supposing that $h \to \infty$ and that $\tanh(\kappa h) \to 1$. This case describes deep water waves, for which the dispersion relation (3) collapses to

$$\omega^2 = g\kappa + \sigma\kappa^3. \qquad (4)$$

In this deep water limit, the Bond number $B \to 0$ ($<1/3$), so that the dispersion relation for the phase speed $c = \omega/\kappa$ has a minimum value at $\kappa = \kappa_m$ ($= (g/\sigma)^{1/2}$ here) or $c_m = (4g\sigma)^{1/4} = 23.2$ cm/s. There are no sinusoidal waves of the form (1) for any phase speed $c < c_m$, while for any $c > c_m$ there are two classes of deep water waves, gravity waves with $\kappa < \kappa_m$ and capillary waves with $\kappa > \kappa_m$.

Long waves are characterized by the limit $\kappa h \to 0$, that is, the limit when $\lambda/h \to \infty$. In this limit, the wave frequency tends to zero, and the wave phase speed tends to $c_0 = (gh)^{1/2}$. In this case, for the reasons discussed above, it is customary to ignore the effects of surface tension, so that there is only one class of waves, namely, gravity waves, which in this limit of $\kappa h \to 0$ are often referred to as shallow water waves. The fluid velocity is then approximately horizontal and independent of the vertical coordinate z, while at $O(\kappa h)$, the vertical component of the fluid velocity is a linear function of $(z + h)$.

In this linearized system, more general solutions can be built up by Fourier superposition of the elementary solutions (1) over the wave number $\boldsymbol{k}$ in which the dispersion relation (3) is satisfied and the wave amplitude a is allowed to be a function of $\boldsymbol{k}$. From this process, it can readily be established that a localized initial state will typically evolve into wave packets, with a dominant wave number $\boldsymbol{k}$ and corresponding frequency ω given by (3), within which each wave phase propagates with the phase speed c, but whose envelope propagates with the group velocity, given by

$$\boldsymbol{c}_g = \nabla_{\boldsymbol{k}}\omega. \qquad (5)$$

The group velocity is the velocity of energy propagation, where the wave energy density is $2\rho(g + \sigma\kappa^2)|a|^2$, being composed of equal parts of potential and kinetic energy. Because water waves are isotropic, the group velocity is in the direction of the wave number $\boldsymbol{k}$, and has a magnitude

$$c_g = \partial\omega/\partial\kappa = \frac{c}{2}\left\{\frac{g + 3\sigma\kappa^2}{g + \sigma\kappa^2} + \frac{2\kappa h}{\sinh 2\kappa h}\right\}. \qquad (6)$$

Assuming that the Bond number $B < \frac{1}{3}$, it can be shown for gravity waves (defined by the wave number range $0 < \kappa < \kappa_m$) that $c_g < c$, with equality in the long wave limit, $\kappa \to 0$ when c_g and c tend to c_0, and at the value $\kappa = \kappa_m$, $c_g = c = c_m$. For capillary waves (defined by the wave number range $\kappa > \kappa_m$), on the other hand, $c_g > c$. In the absence of surface tension (i.e., $B \to 0$), $c_g < c$ for all wave numbers $\kappa > 0$, and in the deep water limit $c_g \approx c/2$.

Weakly Nonlinear Waves

The theory of linearized waves described in the previous section is valid when initial conditions are such that the waves have sufficiently small amplitudes. However, after a sufficiently long time (or if the initial conditions describe waves of moderate or large amplitudes), the effects of the nonlinear terms in the free surface boundary conditions need to be taken into account. There are three different areas where weak nonlinearity needs to be taken into account, namely, long waves, wave packets, and wave resonances.

Long Waves: Korteweg–de Vries Equation

Initially, we consider unidirectional waves propagating in the positive x-direction, so that the wave number is $\boldsymbol{k} = (k, 0)$. In the long wave limit, $kh \to 0$, the dispersion relation (3) can be approximated by

$$\omega = c_0 k - \frac{c_0 h^2}{6}\delta k^3 + \dots, \qquad \delta = 1 - 3B,$$

where we recall that B is the Bond number measuring the effect of surface tension. To leading order, the waves propagate with the linear long wave phase speed $c_0 = (gh)^{1/2}$. But, after a long time the cumulative effects of weak nonlinearity must be taken into account. When these are balanced against the leading order linear dispersive terms (the $O(k^3h^3)$ terms above), the result is the well-known Korteweg–de Vries (KdV) equation for the wave elevation

$$\zeta_t + c_0\zeta_x + \frac{3c_0}{2h}\zeta\zeta_x + \frac{c_0 h^2}{6}\delta\zeta_{xxx} = 0. \qquad (7)$$

This equation was derived by Diederik Korteweg and Hendrik de Vries in a now very famous paper published in 1895 (Korteweg & de Vries, 1895), although in fact Joseph Boussinesq had obtained it earlier in 1877 (Boussinesq, 1877; *See* **Korteweg–de Vries equation**). Relative to the leading order propagation with the speed c_0, the time evolution occurs on the nondimensional scale ε^{-3}, where the wave amplitude ζ/h scales with ε^2 and the wave dispersion kh scales with ε (that is, spatial derivatives scale with ε).

Korteweg and de Vries found a family of traveling-wave solutions, periodic waves described by elliptic functions and commonly called cnoidal waves, and the solitary wave solution

$$\zeta = a\,\mathrm{sech}^2(\gamma(x - Vt)),$$
$$\text{where} \quad V - c_0 = \frac{c_0 a}{2h} = \delta\frac{2c_0h^2}{3}\gamma^2. \quad (8)$$

The solitary wave has a free parameter, for instance, the amplitude a. When $\delta > 0$ (that is, $B < 1/3$), then the amplitude a is always positive and $V > c_0$. Further, as a increases, the wave speed V and the wave number γ also increase. Although the case when $B > \frac{1}{3}$ has little practical application, it is interesting to note that these conclusions are all reversed as then $\delta < 0$, now, $a < 0$ and $V < c_0$. This solitary wave had earlier been obtained directly from the governing equations independently by Boussinesq (1871) and Rayleigh (1876), who were motivated to explain the now very well-known observations and experiments of John Scott Russell (1844). Curiously, it was not until quite recently that it was recognized that the KdV equation is not strictly valid if surface tension is taken into account and $0 < B < \frac{1}{3}$, as there is then a resonance between the solitary wave and very short capillary waves.

After this ground-breaking work of Korteweg and de Vries, interest in solitary water waves declined until the dramatic discovery of the "soliton" by Zabusky and Kruskal in 1965 (*See* **Solitons, a brief history**). Through numerical integrations of the KdV equation, they demonstrated that the solitary wave (8) could be generated from quite general initial conditions, and could survive intact collisions with other solitary waves, leading them to coin the term *soliton*. Their remarkable discovery, followed almost immediately by the theoretical work of Gardner, Greene, Kruskal, and Miura (Gardener et al., 1967) showing that the KdV equation was integrable through an inverse scattering transform, led to many other startling discoveries and marked the birth of the soliton theory as we know it today. The implication for water waves is that the solitary wave is the key component needed to describe the behavior of long, weakly nonlinear waves.

An alternative to the KdV equation is the Benjamin–Bona–Mahony (BBM) equation in which the linear dispersive term $c_0\zeta_{xxx}$ in (7) is replaced by $-\zeta_{xxt}$. It has the same asymptotic validity as the KdV equation and, since it has rather better high wave number properties, is somewhat easier to solve numerically. However, it is not integrable and, consequently, has not attracted the same interest as the KdV equation.

Both the KdV and BBM equations are unidirectional. A two-dimensional version of the KdV equation is the KP equation (Kadomtsev & Petviashvili, 1970; *See* also **Kadomtsev–Petviashvili equation**),

$$\left(\zeta_t + c_0\zeta_x + \frac{3c_0}{2h}\zeta\zeta_x + \frac{c_0h^2}{6}\delta\zeta_{xxx}\right)_x + \frac{c_0}{2}\zeta_{yy} = 0. \quad (9)$$

This equation includes the effects of weak diffraction in the y-direction, in that y-derivatives scale as ε^2 whereas x-derivatives scale as ε. When $\delta > 0$ $(0 \le B < \frac{1}{3})$, this is the KPII equation, and it can be shown that then the solitary wave is stable to transverse disturbances. On the other hand, if $\delta < 0$ $(B > \frac{1}{3})$, this is the KPI equation for which the solitary wave (8) is unstable; instead this equation supports "lump" solitons. Both KPI and KPII are integrable equations. To take account of stronger transverse effects and/or to allow for bi-directional propagation in the x-direction, it is customary to replace the KdV equation with a Boussinesq system of equations. These combine the long wave approximation to the dispersion relation with the leading order nonlinear terms and occur in several asymptotically equivalent forms.

Wave Packets: Nonlinear Schrödinger Equation

The linear theory of water waves predicts that a localized initial state will typically evolve into wave packets, with a dominant wave number $\boldsymbol{k}$ and corresponding frequency ω given by (3), within which each wave phase propagates with the phase speed c, but whose envelope propagates with the group velocity c_g (5, 6). After a long time, the packet tends to disperse around the dominant wave number, which tendency may be opposed by cumulative nonlinear effects. In the absence of surface tension, the outcome for unidirectional waves is described by the nonlinear Schrödinger (NLS) equation, for the wave amplitude $A(x,t)$; that is, the wave elevation is $\zeta = A \exp \mathrm{i}(kx - \omega t) +$ c.c. to leading order,

$$\mathrm{i}(A_t + c_g A_x) + \tfrac{1}{2}\lambda A_{xx} + \mu|A|^2 A = 0, \quad (10)$$

and the coefficients are given by

$$\lambda = \frac{\partial^2\omega}{\partial k^2},$$
$$\mu = -\frac{\omega k^2}{16S^4}(8C^2S^2 + 9 - 2T^2) + \frac{\omega}{8C^2S^2}\frac{(2\omega C^2 + kc_g)^2}{gh - c_g^2},$$

where

$$C = \cosh(kh), \; S = \sinh(kh), \; T = \tanh(kh).$$

For water waves, the NLS equation was first derived by Zakharov (1968) for the case of deep water, and then by Hasimoto and Ono (1972) for finite depth. An analogous equation can be derived for nonzero surface tension in which the nonlinear coefficient μ will take a different value, but for reasons discussed below, it is not so useful in that case. In deep water ($kh \to \infty$) the coefficient $\mu \to -\omega k^2/2 < 0$. In general, $\mu < 0 (> 0)$ according as $kh > (<) 1.36$).

Like the KdV equation, the NLS equation is integrable with an associated inverse scattering transform, a result first shown by Zakharov and Shabat (1972). There are two cases, the so-called focusing NLS equation when $\lambda\mu > 0$ and the defocusing NLS equation when $\lambda\mu < 0$. For water waves, $\lambda = \partial c_g/\partial k < 0$, and so we have the focusing (defocusing) NLS equation according as $kh > (<) 1.36$. The focusing NLS equation has solitary wave solutions (bright solitons), given by

$$A = a \operatorname{sech}(\gamma(x - c_g t)) \exp(-i\Omega t),$$
$$\text{where } \mu a^2 = \lambda\gamma^2, \; \Omega = -\tfrac{1}{2}\mu a^2. \tag{11}$$

On the other hand, the defocusing NLS equation has no such solitary wave solutions which decay to zero at infinity. Instead, it has solitary waves riding on a nonzero background (dark solitons). A key property of the NLS equations is that plane waves are modulationally unstable (stable) in the focusing (defocusing) case. That is, the NLS equation has the exact plane wave solution,

$$A = A_0 \exp(i\mu |A|^2 t), \tag{12}$$

which is then perturbed with a small-amplitude modulation proportional to $\exp(iKx - \sigma t)$. It is readily found that the growth rate σ is given by

$$\sigma = K^2 \left(\lambda\mu A_0^2 - \frac{\lambda^2 K^2}{4} \right). \tag{13}$$

Thus, in the focusing NLS case when $\lambda\mu > 0$, there is a positive growth rate for modulation wave numbers K such that $K < 2(\lambda/\mu)^{1/2}|A_0|$. On the other hand, σ is purely imaginary for all K in the defocusing case when $\lambda\mu < 0$. The implication for water waves is that plane Stokes waves in deep water ($kh > 1.36$) are unstable. This remarkable result was first discovered by Benjamin and Feir in 1967, by a different theoretical approach, and has since been confirmed in experiments. The maximum growth rate occurs for $K = K_m = (\lambda/\mu)^{1/2}|A_0|$, and the instability is due to the generation of side bands with wave numbers $k \pm K_m$. As the instability grows, the full NLS equation (10) is needed to describe the long-time outcome of the collapse of the uniform plane wave into several soliton wave packets, each described by (11).

When the effects of modulation in the transverse y-direction are taken into account, so that the wave amplitude is now given by $A(x, y, t)$, the NLS equation is replaced by the Benney–Roskes system (Benney & Roskes, 1969), also widely known as the Davey–Stewartson equations (Davey & Stewartson, 1974),

$$i(A_t + c_g A_x) + \tfrac{1}{2}\lambda A_{xx} + \tfrac{1}{2}\delta A_{yy}$$
$$+ \mu |A|^2 A + UA = 0, \tag{14}$$
$$\alpha U_{xx} + U_{yy} + \beta(|A|^2)_{yy} = 0, \tag{15}$$

where the coefficients μ and λ are those defined in (10), while

$$\delta = \frac{c_g}{k}, \; \alpha = 1 - \frac{c_g^2}{gh},$$
$$gh\beta = \frac{\omega}{8C^2S^2}(2\omega C^2 + kc_g)^2.$$

Here, the surface tension has been set to zero. If surface tension effects are included, a similar equation holds but with different values for the coefficients (Djordjevic & Redekopp, 1977). Note that $\lambda\delta < 0$ and $\alpha > 0$, so that the equation for A is hyperbolic, but that for U it is parabolic. The variable U which appears here is a wave-induced mean flow, which tends to zero in the limit of deep-water waves, $kh \to \infty$. This system (14) again has the plane wave solution (12), whose stability can be analyzed in a manner similar to that described above in the context of the NLS equation. The outcome is that now instability can occur for all values of kh and occurs in a band in the $k - l$ plane where k and l are the modulation wave numbers. The instability is purely two-dimensional when $kh < 1.36$, and the band becomes narrower and the growth rate weaker as $kh \to 0$. For more details, see Benney and Roskes (1969) or Mei (1983).

Wave Resonant Interactions

A superposition of weakly nonlinear waves, each of which is given by (1) with a wave number $\boldsymbol{k}_n, n = 1, 2, \ldots, N$, and a corresponding frequency $\omega_n(\boldsymbol{k}), n = 1, 2, \ldots, N$, each satisfying the dispersion relation (3), interact resonantly whenever

$$\boldsymbol{k}_1 + \boldsymbol{k}_2 + \cdots + \boldsymbol{k}_N = 0,$$

and

$$\omega_1 + \omega_2 + \cdots + \omega_N = \Omega.$$

Here, Ω is a detuning term, so that exact resonance is achieved whenever $\Omega = 0$. The most prominent interactions are triad interactions; that is, $N = 3$, followed by the quartet interactions when $N = 4$, and so on. In a resonant interaction, energy is exchanged between the Fourier components in a periodic manner, assuming that dissipation is absent. For instance, the set of equations describing a triad interaction is (Craik, 1985)

$$A_{1t} + \boldsymbol{c}_{g1} \cdot \nabla A_1 = \frac{\partial D}{\partial \omega_1} \mu A_2^* A_3^* \exp(-i\Omega t), \ldots,$$

where $A_n(\boldsymbol{x}, t)$ is the amplitude of the nth mode, $D(\omega, \boldsymbol{k}) = 0$ is the dispersion relation, μ is a coefficient, and the superscript "*" denotes the complex conjugate. Remarkably, these equations are integrable for exact resonance ($\Omega = 0$) (Zakharov & Manakov, 1973).

For water waves in the absence of surface tension, the dispersion relation does not allow for triad interactions. Hence, the dominant resonance is a quartet interaction, as first shown by Phillips (1960). The four wave numbers making up the quartet are two-dimensional (i.e., the y-wave number components are generally not zero), and the allowed wave number vectors can be determined graphically from Phillips' "figure-of-eight" diagram. The interaction equations for a discrete quartet of waves is analogous to that displayed above for a triad interaction. Further, in deep water ($kh \to \infty$), Craig and Worfolk (1995) have shown that in the Birkhoff normal form for these interaction equations, the coefficients vanish for all nongeneric resonant terms and the remaining system is then integrable. However, the same is not the case for quintet interactions.

When considering a continuous spectrum of gravity waves, the resulting evolution of a spectral component is described by Zakharov's integral equation, first derived in the deep water limit by Zakharov (1968). Krasistskii (1994) later employed canonical transformations to obtain a more desirable Hamiltonian form (see also the review by Dias & Kharif, 1999). In this equation, usually truncated at the third order in wave amplitude, the evolution of the spectral component $A(\boldsymbol{k}, t)$, which is the spatial Fourier transform of $\eta(\boldsymbol{x}, t)$, is determined essentially by quartet interactions, since the nonresonant triad interactions are removed by a canonical transformation. This integral evolution equation contains much of the previous weakly nonlinear theory, in that the discrete interaction equations and the NLS equation can be derived from it. It also forms the basis for a statistical description of water waves and can then be used to describe the ocean wave spectrum.

When surface tension is taken into account, then triad interactions are possible (i.e., $N = 3$ in (??)). The most well-known example occurs when $\boldsymbol{k}_1 = \boldsymbol{k}_2 = -\boldsymbol{k}_3/2$, and is a second harmonic resonance in that then $\omega_1 = \omega_2 = -\omega_3/2$. It was first noted by Wilton (1915) and leads to a phenomenon known as Wilton's ripples. In general, the existence of triad resonances implies that capillary-gravity waves undergo wave–wave interactions on a faster time scale than pure gravity waves.

ROGER GRIMSHAW

See also **Dispersion relations; Group velocity; Korteweg–de Vries equation; Modulated waves; Nonlinear Schrödinger equations; Solitons, a brief history; Wave packets, linear and nonlinear**

Further Reading

Benjamin, T.B. & Feir, J.E. 1967. The disintegration of wave trains on deep water. *Journal of Fluid Mechanics*, 27: 417–430

Benney, D.J. & Roskes, G. 1969. Wave instabilities. *Studies in Applied Mathematics*, 48: 377–385

Boussinesq, M.J. 1871. Theórie de l'intumescence liquid appellée onde solitaire ou de translation, se propageant dans un canal rectangulaire. *Comptes Rendus de l'Académie des Sciences (Paris)*, 72: 755–759

Boussinesq, M.J. 1877. Essai sur la theorie des eaux courantes, *Memoires presentees par diverse savants a l'Academie des Sciences Inst. France (Series 2)*, 23: 1–680

Craig, W. & Worfolk, P.A. 1995. An integrable normal form for water waves in infinite depth. *Physia* D, 84: 513–531

Craik, A.D.D. 1985. *Wave Interactions and Fluid Flows*, Cambridge and New York: Cambridge University Press

Davey, A. & Stewartson, K. 1974. On three-dimensional packets of surface waves. *Proceedings of the Royal Society of London* A, 338: 101–110

Dias, F. & Kharif, C. 1999. Nonlinear gravity and capillary-gravity waves. *Annual Reviews of Fluid Mechanics*, 31: 301–346

Djordjevic, V.D. & Redekopp, L.G. 1977. On two-dimensional packets of capillary-gravity waves. *Journal of Fluid Mechanics*, 79: 703–714

Gardner, C.S., Greene, J.M., Kruskal, M.D. & Miura, R.M. 1967. Method for solving the Korteweg–de Vries equation. *Physical Review Letters*, 19: 1095–1097.

Hasimoto, H. & Ono, H. 1972. Nonlinear modulation of gravity waves. *Journal of the Physical Society of Japan*, 33: 805–811

Kadomtsev, B.B. & Petviashvili, V.I. 1970. On the stability of solitary waves in weakly dispersive media. *Soviet Physics Doklady*, 15: 539–541

Korteweg, D.J. & de Vries, H. 1895. On the change of form of long waves advancing in a rectangular canal, and on a new type of long stationary waves. *Philosophical Magazine*, 39: 422–443

Krasitskii, V.P. 1994. On reduced equations in the Hamiltonian theory of weakly nonlinear surface waves. *Journal of Fluid Mechanics*, 272: 1–30

Lamb, H. 1932. *Hydrodynamics*, Cambridge and New York: Cambridge University Press; 6th edition, Cambridge and New York, Cambridge University Press, 1993

Lighthill, M.J. 1978. *Waves in Fluids*, Cambridge and New York: Cambridge University Press

Mei, C.C. 1983. *The Applied Dynamics of Ocean Surface Waves*, New York: Wiley

Phillips, O.M. 1960. On the dynamics of unsteady gravity waves of finite amplitude. Part 1. The elementary interactions. *Journal of Fluid Mechanics*, 9: 193–217

Rayleigh, Lord (Strutt, W.J.) 1876. On waves. *Philosophical Magazine*, 1: 257–279

Russell, J.S. 1844. Report on Waves, *14th meeting of the British Association for the Advancement of Science*, London: BAAS, 311–390

Schwarz, L.W. & Fenton, J.D. 1982. Strongly nonlinear waves. *Annual Reviews of Fluid Mechanics*, 14: 39–60

Wilton, J.R. 1915. On ripples. *Philosophical magazine*, 29: 688–700

Whitham, G.B. 1974. *Linear and Nonlinear Waves*, New York: Wiley

Zabusky, N.J. & Kruskal, M.D. 1965. Interactions of solitons in a collisionless plasma and the recurrence of initial states. *Physical Review Letters*, 15: 240–243

Zakharov, V.E. 1968. Stability of periodic waves of finite amplitude on the surface of a deep fluid. *Journal of Applied Mechanics and Technical Physics*, 2: 190–194

Zakharov, V.E. & Manakov, S.V. 1973. Resonant interactions of wave packets in nonlinear media. *Soviet Physics, JETP*, 18: 243–247

Zakharov, V.E. & Shabat, A.B. 1972. Exact theory of two-dimensional self-focusing and one-dimensional self-modulation of waves in nonlinear media. *Soviet Physics, JETP*, 34: 62–69

WAVE OF TRANSLATION

Translational invariance is one of the fundamental symmetries of continuum nonlinear partial differential equations (PDEs). Such a symmetry is present when the equation of interest remains unchanged under the transformation $x \to x + \varepsilon$. Typically, partial differential equations that contain only derivatives with respect to the spatial variable x (but no explicit dependence on x) possess this symmetry.

The topic of symmetries and their role in bifurcation theory is a very large subject, a detailed discussion of which can be found, for example in Golubitsky & Schaeffer (1995) and Golubitsky (1988). Here, our scope is much narrower in giving a perspective of translational symmetry and its role in the existence, bifurcation, or absence of traveling wave solutions in some Hamiltonian and dissipative classes of PDEs.

To make our discussion of this subject more definitive, we will introduce a rather general class of dissipative or Hamiltonian PDEs of the form

$$\{u_t, u_{tt}\} = u_{xx} + f(u). \tag{1}$$

The paradigm of Equation (1) is dissipative (that is, of the reaction-diffusion type), if a one-time derivative of the field u is used in the left-hand side. On the other hand, the model is Hamiltonian if the second time derivative is used on the left-hand side. While the model is written as a single component model in a one-dimensional setting, it can be easily generalized in multiple components/dimensions. In the former case, u becomes a vector, while in the latter, the spatial second partial derivative is substituted by the Laplacian operator. We should note in passing that while here we consider the Hamiltonian and dissipative cases, there are many models which lie between the conservative and diffusion limits. Numerous examples of this type can be found in Kivshar & Malomed (1989).

It can be immediately observed that the translational symmetry mentioned above is present in the model of Equation (1). A straightforward example of its absence would be the case of a "reaction term" $f(u, x)$, i.e., one explicitly dependent on x. Let us now explore the implications of this symmetry for the static problem of Equation (1) and then for the corresponding dynamic problem.

For the static problem (i.e., for solutions $u = g(x)$ that satisfy $u_{xx} + f(u) = 0$), the presence of the invariant direction signifies that the solution can be translated along the group orbit of this invariance. Otherwise stated, there is a one parameter infinity of solutions (alternatively, a degeneracy of solutions) due to the available freedom to select solutions along the symmetry direction. In practical terms, this means that if $u = g(x)$ is a solution, then $u = g(x - x_0)$ is also a solution and arbitrary translations of the solution satisfy the original equation. Furthermore, this feature bears consequences on the linear stability around the solution. In particular, looking for the linear stability of the solution $u = g(x)$, we use the linearization $u = g(x) + \varepsilon \exp(\lambda t) w(x)$, which to $O(\varepsilon)$, yields the eigenvalue problem:

$$\{\lambda, \lambda^2\} w = w_{xx} + f'(g(x)) w. \tag{2}$$

Given that $g(x)$ satisfies the $u_{xx} + f(u) = 0$, differentiation of the latter immediately yields that (due to the absence of the explicit spatial dependence and hence due to the symmetry) $w = u_x \equiv g'(x)$ satisfies Equation (2), with $\lambda = 0$ (for the dissipative system) or $\lambda^2 = 0$ (for the Hamiltonian system). Hence, the existence of the symmetry generates a single (for the dissipative) or a pair (for the Hamiltonian) of eigenvalues at the origin of the spectral plane (i.e., with $\lambda = 0$). This is the neutral eigendirection connected with the symmetry that is often referred to as a Goldstone mode.

Furthermore, in the Hamiltonian version of the system, the presence of such eigenmodes, and of their corresponding symmetries, is intimately connected with conservation laws (through Noether's theorem; see for example the discussion in Arnol'd (1989); Sulem & Sulem (1999)). The invariance with respect to translations is directly related with the conservation of linear momentum, which in the case of Equation (1) is of the form

$$P = \int_{-\infty}^{\infty} u_t u_x \, dx. \tag{3}$$

We now turn to the dynamic consequences of the symmetry. The translational symmetry is directly related to the traveling of solutions. In the case of Hamiltonian (especially continuum) systems, there may be extra symmetries that may allow the construction of traveling solutions from stationary ones. In the case of Equation (1), such a symmetry is the Lorentz invariance [$x \to x' = \gamma(x - vt), t \to t' = \gamma (t - vx/c^2)$, $\gamma = 1/\sqrt{1 - v^2/c^2}$], which allows one to boost the solutions to any "subsonic" speed. In other cases, such as the one of the nonlinear Schrödinger equation, the corresponding symmetry, is the Galilean invariance. Hence, in Hamiltonian systems, due to the additional symmetry, traveling and standing solutions are often, in some sense, equivalent. On the other hand, for dissipative systems, such equivalence is typically absent. In the latter case, we look for traveling wave solutions in the form $u = u(\xi)$, where $\xi = x - ct$ is the traveling wave variable. This transformation leads to the so-called traveling wave frame (TWF) equation of motion (i.e., we travel together with the solution and, hence, observe it as a steady one) of the form

$$u_t = u_{\xi\xi} + cu_\xi + f(u). \tag{4}$$

By solving the steady-state problem of Equation (4) (notice: an ordinary differential equation (ODE) problem to find special solutions of the PDE), we can identify the traveling-wave solutions of Equation (1). In dissipative systems, the absence of additional symmetries typically allows for isolated solutions of the TWF ODE rather than monoparametric families of such solutions. Notice that here we have in mind fixed model parameter values (but not fixed initial condition parameters, such as, for example, energy). Furthermore, we do not discuss the mechanisms (in dissipative systems) of selection of a given speed (which is related to issues of stability). Such examples and a detailed discussion can be found in Xin (2000). A typical example is the bistable nonlinearity where $f(u)=2u(u-1)(\mu-u)$, for which a front solution of the form

$$u(x,t) = \frac{1}{2}\left[1 - \tanh\left(\frac{x - x_c(t)}{2}\right)\right], \tag{5}$$

exists, where $x_c = x_c^{\text{in}} + ct$ and x_c^{in} is the original position of the center, while the speed c is connected to the parameter μ through

$$c = 1 - 2\mu. \tag{6}$$

In view of the above comments, in energy conserving systems, typically standing and traveling solutions co-exist, while in dissipative systems one can (locally) have solutions either of the former or of the latter type. In fact, as parameters are varied in dissipative settings, traveling solutions may bifurcate from standing ones, through the so-called drift pitchfork bifurcation (Kness et al., 1992, see also Malomed & Tribelsky (1984) and Coullet & Iooss, 1990). It is worth noting that recently, a template-based technique has been proposed that dynamically factors out translational invariance (and other continuous symmetries) (Rowley & Marsden, 2000; Rowley et al., 2003).

Discrete Systems and Symmetry Breaking

An interesting variation of the above presented scenario occurs in (spatially) discrete systems. In this case, generic discretizations of the original problem of Equation (1) no longer preserve the symmetry with respect to continuum translations. Instead, only an integer shift invariance persists in this discrete limit of the equation:

$$\{\dot{u}_n, \ddot{u}_n\} = \Delta_2 u_n + f(u_n), \tag{7}$$

where $\Delta_2 u_n = (u_{n+1} + u_{n-1} - 2u_n)/h^2$ is the discrete Laplacian for a lattice of spacing h. In this case, from the informative Taylor expansion of the form

$$\Delta_2 u_n = \sum \frac{2h^{2j-2}}{(2j)!}\frac{\mathrm{d}^{2j}u}{\mathrm{d}x^{2j}}, \tag{8}$$

we deduce the following conclusions:

- discreteness is a singular higher-order derivative perturbation to the continuum limit;
- to all orders in this asymptotic expansion, the right-hand side is translationally invariant. Hence, the breaking of the translational symmetry can only occur beyond all algebraic orders and thus has to be an exponentially small effect.

A manifestation of this exponentially small symmetry breaking effect is given by the fact that there are now two (lattice shift invariant) stationary wave states in the lattice context. One of these solutions is centered on a lattice site and one is centered between two consecutive lattice sites (instead of a single translationally invariant steady state in the continuum limit). One of these solutions is stable and one is unstable. The energy difference between the two (which mirrors the amount of symmetry breaking and hence should be exponentially small, that is, $\Delta E \sim \exp(-\pi^2/h)$) is the celebrated Peierls–Nabarro barrier. For a detailed discussion of these issues, see, for example, the review by Kevrekidis et al. (2001).

From the above, we can infer that the generic effect of discreteness in breaking translational invariance is to generate an exponentially small (in the natural lattice spacing parameter) shift periodic (in fact approximately trigonometric) potential. The center of mass of the waves propagates inside the Peierls–Nabarro barrier. There is considerable interest in shearing this potential with an external (constant) field.

A constant external force F introduces a term in the potential energy $\sim Fu$, generating a washboard potential for the motion of the wave. If the external field becomes sufficiently strong, then one can infer that a saddle-node (in fact, infinite period; the so-called SNIPER) bifurcation will occur in which the stationary states (the maxima and minima of the potential, that is, the saddles and nodes) will disappear and traveling waves will arise. Scaling analysis predicts and numerical results verify (Kaldko et al., 2000; Keverkidis et al., 2001; Carpio & Bonilla, 2001) that the relevant bifurcation will yield waves of speed

$$c \sim (F - F_c)^{1/2}, \tag{9}$$

where F is the constant external field and F_c is its critical value. The energy landscapes for the cases of $F=0$, $F<F_c$, and $F>F_c$ are shown in Figure 1. This scenario happens in the dissipative case, but one can also analyze the Hamiltonian case in the same manner.

Finally, it should be noted that while the above scenario will be the generically relevant one, there

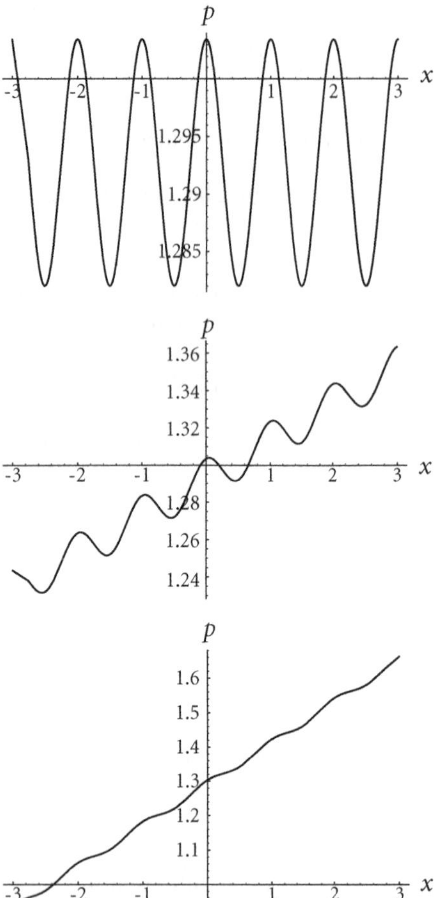

Figure 1. The (potential) energy (P) landscape for generic DDE models in the case of $F = 0$ (top panel), $F < F_c$ (middle panel) and $F > F_c$ (bottom panel), as a function of the position of the kink center.

are exceptions to the rule of absence of a continuum-like symmetry in the lattice setting. Let us consider, for example, the discretization of the Hamiltonian PDE:

$$\ddot{u}_n = \Delta_2 u_n + \frac{F(u_{n+1}) - F(u_{n-1})}{u_{n+1} - u_{n-1}}. \tag{10}$$

It can be seen that the equation of motion has a conservation law of the form

$$P = \sum \dot{u}_n (u_{n+1} - u_{n-1}) \tag{11}$$

which is the discrete analog of Equation (3). In this case, the discrete equation preserves a "ghost" of the continuum symmetry and maintains the multiplicity of Goldstone modes of the continuum problem (Kevrekidis, 2003).

In conclusion, translational invariance is a symmetry that plays a significant role in the context of both partial-differential as well as differential-difference (i.e., lattice) equations. In the former, it is typically present (unless an explicit spatial dependence occurs) and has both static as well as kinematic consequences on the solutions of the problem and their stability. In the discrete setting, the symmetry is generically absent and its absence plays a critical role in diversifying kinematic and dynamic phenomena on the lattice (see, for example, Kevrekidis, et al. (2001)) from their continuum siblings. However, it is possible to construct discretizations that respect a discrete conservation law reminiscent of the one imposed by the continuum symmetry group. In the nongeneric case, the lattice dynamics can be significantly closer to their continuum counterparts.

P.G. Kevrekidis and I.G. Kevrekidis

See also **Partial differential equations, nonlinear; Peierls barrier; Sine-Gordon equation; Zeldovich–Frank-Kamenetsky equation**

Further Reading

Arnol'd, V.I. 1989. *Mathematical Methods in Classical Mechanics*, New York: Springer

Carpio, A. & Bonilla, L.L. 2001. Wave front depinning transition in discrete one-dimensional reaction–diffusion systems. *Physical Review Letters*, 86: 6034–6037

Coullet, P. & Iooss, G. 1990. Instabilities of one-dimensional cellular patterns. *Physical Review Letters*, 64: 866–869

Dodd, R.K., Eilbeck, J.C., Gibbon, J.D. & Morris, H.C. 1982. *Solitons and Nonlinear Wave Equations*, London and New York: Academic Press

Golubitsky, M. & Schaeffer, D.G. 1985. *Singularities and Groups in Bifurcation Theory*, vol. I, New York: Springer

Golubitsky, M., Stewart, I.N. & Schaeffer, D.G. 1988. *Singularities and Groups in Bifurcation Theory*, vol. II, New York: Springer

Kevrekidis, P.G. 2003. On a class of discretizations of Hamiltonian nonlinear partial differential equations. *Physica* D, 183: 68–86

Kevrekidis, P.G., Kevrekidis, I.G. & Bishop, A.R. 2001a. Propagation failure, universal scalings and Goldstone modes. *Physics Letters* A, 279: 361–369

Kevrekidis, P.G., Rasmussen, K.Ø. & Bishop, A.R. 2001b. The discrete nonlinear Schrodinger equation: a survey of recent results. *International Journal of Modern Physics* B, 15: 2833–2900

Kivshar, Yu.S. & Malomed, B.A. 1989. Dynamics of solitons in nearly integrable systems. *Reviews of Modern Physics*, 61: 763–915

Kladko, K., Mitkov, I. & Bishop, A.R. 2000. Universal scaling of wave propagation failure in arrays of coupled nonlinear cells. *Physical Review Letters*, 84: 4505–4508

Kness, M., Tuckerman, L.S. & Barkley, D. 1992. Symmetry-breaking bifurcations in one-dimensional excitable media. *Physical Review* A, 46: 5054–5062

Malomed, B.A. & Tribelsky, M.I. 1984. Bifurcations in distributed kinetic systems with aperiodic instability. *Physica* D, 14: 67–87

Rowley, C.W. & Marsden, J.E. 2000. Reconstruction equations and the Karhunen-Loeve expansion for systems with symmetry. *Physica* D, 142: 1–19

Rowley, C.W., Kevrekidis, I.G., Marsden, J.E. & Lust, K. 2003. Reduction and reconstruction for self-similar dynamical systems. *Nonlinearity*, 16: 1257–1275

Sulem, C. & Sulem, P.L. 1999. *The Nonlinear Schrödinger Equation*, New York: Springer

Xin, J. 2000. Front propagation in heterogeneous media. *SIAM Review*, 42: 161–230

WAVE PACKETS, LINEAR AND NONLINEAR

Because linear wave systems have elementary solutions of the form $e^{i(kx-\omega t)}$, it is often convenient to write the general solution of an initial value problem as an integral of Fourier components. Thus,

$$u(x,t) = \frac{1}{2\pi}\int_{-\infty}^{\infty} F(k)e^{i(kx-\omega t)}\,dk, \qquad (1)$$

where $F(k)$ is the Fourier transform of $u(x,0)$. Initial conditions thus determine the Fourier transform, each component of which evolves independently with frequency ω related to wave number k through the dispersion relation

$$\omega = \omega(k). \qquad (2)$$

Unless $\omega = k$, different components in Equation (1) travel at different speeds (ω/k), and an initially localized wave spreads out or "disperses," hence the name.

A wave packet is a special form of Equation (1) with the largest Fourier components lying close to some wave number (k_0) and the corresponding frequency (ω_0). In other words, the initial conditions $u(x,0)$ are selected so that $F(k)$ has its maximum value at $k = k_0$, falling rapidly with increasing $|k - k_0|$. This suggests writing the dispersion relation as a power series about k_0. With the notation

$$\omega = \omega_0 + b_1(k-k_0) + b_2(k-k_0)^2 \qquad (3)$$

(which assumes that the system has no higher than second derivatives with respect to x), Equation (1) becomes

$$u(x,t) = e^{i(k_0x-\omega_0t)}\frac{1}{2\pi} \times \int_{-\infty}^{\infty} F(k)e^{i[(k-k_0)x - b_1(k-k_0)t - b_2(k-k_0)^2t]}dk, \qquad (4)$$

where the factor $e^{i(k_0x-\omega_0t)}$ is a carrier wave with a velocity $v_c = \omega_0/k_0$, shown in Figure 1(a). Riding over (or multiplying) the carrier is an envelope wave

$$\phi(x,t) = \frac{1}{2\pi}\int_{-\infty}^{\infty} F(\kappa+k_0)e^{i(\kappa x - b_1\kappa t - b_2\kappa^2 t)}\,d\kappa, \qquad (5)$$

where the variable of integration has been changed from k to $\kappa \equiv k - k_0$.

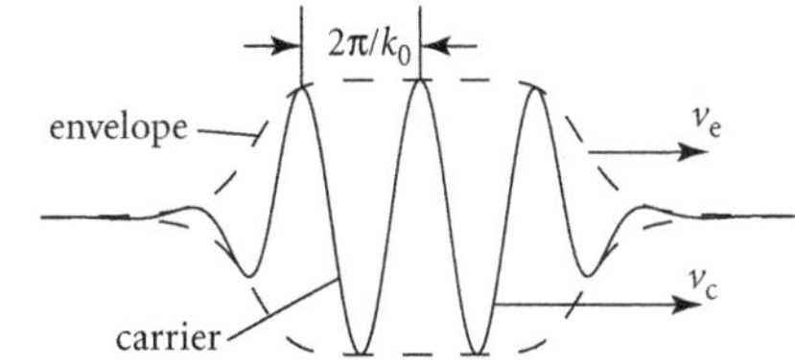

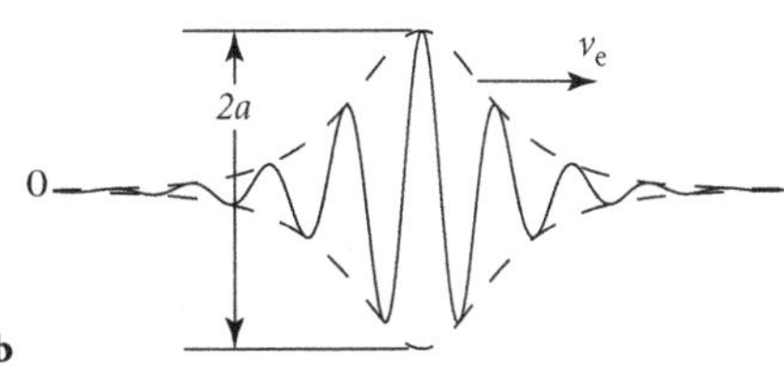

Figure 1. (a) The real part of a linear wave packet, showing the envelope (dashed lines) and the carrier (full line), as in Equation (4). (b) The real part of a soliton solution of Equation (7).

Taking the time derivative of Equation (5), one finds

$$\frac{\partial\phi}{\partial t} = \frac{1}{2\pi}\int_{-\infty}^{\infty} -i(b_1\kappa + b_2\kappa^2)F(\kappa+k_0) \times e^{i(\kappa x - b_1\kappa t - b_2\kappa^2 t)}\,d\kappa$$
$$= -b_1\frac{\partial\phi}{\partial x} + ib_2\frac{\partial^2\phi}{\partial x^2},$$

which can be written as

$$i\left(\frac{\partial\phi}{\partial t} + b_1\frac{\partial\phi}{\partial x}\right) + b_2\frac{\partial^2\phi}{\partial x^2} = 0. \qquad (6)$$

Equation (6) is a partial differential equation that governs time evolution of the envelope for a linear wave packet solution of a second-order equation. Assuming b_2 is not too large (weak dispersion), the envelope moves with the velocity $v_e \doteq b_1 = d\omega/dk|_{k=k_0}$, as in Figure 1(a).

Up to this point, the discussion has remained within the realm of linear theory, but now assume that nonlinear effects alter Equation (6) to

$$i\left(\frac{\partial\phi}{\partial t} + b_1\frac{\partial\phi}{\partial x}\right) + b_2\frac{\partial^2\phi}{\partial x^2} + \alpha|\phi|^2\phi = 0, \qquad (7)$$

with the nonlinear (amplitude-dependent) dispersion relation $\omega = b_1\kappa + b_2\kappa^2 - \alpha|\phi|^2$. Equation (7) is the nonlinear Schrödinger (NLS) equation, for which the following comments are relevant:

- If ϕ is assumed independent of x, Equation (7) has the plane wave solution

$$\phi = \phi_0 e^{i\alpha|\phi_0|^2 t}, \qquad (8)$$

which may or may not be stable.

- With $\alpha b_2 < 0$, this plane wave is stable (Whitham, 1974). If $\alpha b_2 > 0$, on the other hand, the plane wave experiences Benjamin–Feir instability, out of which emerge stable NLS solitons (Benjamin & Feir, 1967; Ostrovsky, 1967).

- The term $b_2(\partial^2\phi/\partial x^2)$ introduces wave dispersion into the problem at the lowest order of approximation. Similarly, the term $\alpha|\phi|^2\phi$ introduces nonlinearity at the lowest order of approximation. Thus, the NLS equation is generic, arising whenever one wishes to consider lowest order effects of dispersion and nonlinearity on a wave packet, including nonlinear optics (Kelley, 1965), deep water waves (Benney & Newell, 1967), and acoustics (Ostrovsky & Potapov, 1999).
- Unstable NLS wave packets decay into one or more solitons. Choosing $b_1=0$, $b_2=1$, and $\alpha=2$ in Equation (7), for example, a family of NLS solitons is (Zakharov & Shabat, 1972)

$$u(x,t) = a\exp\left[\mathrm{i}\frac{v_e}{2}x + \mathrm{i}\left(a^2 - \frac{v_e^2}{4}\right)t\right] \times \operatorname{sech}[a(x - v_e t - x_0)], \qquad (9)$$

one of which is sketched in Figure 1(b).

Beyond the superficial similarities between Figures 1(a) and (b), the differences are profound. In the linear wave packet of Figure 1(a), the shape of the envelope is determined by initial conditions and their subsequent time evolution, as in Equation (6). In the NLS soliton of Figure 1(b), on the other hand, the envelope shape is determined through a dynamic balance between the influences of dispersion and nonlinearity, as expressed by the last two terms of Equation (7).

ALWYN SCOTT

See also **Dispersion relations; Modulated waves; Nonlinear Schrödinger equations; Wave stability and instability**

Further Reading

Benjamin, T.B. & Feir, J.E. 1967. The disintegration of wave trains in deep water. *Journal of Fluid Mechanics*, 27: 417–430

Benney, D.J. & Newell, A.C. 1967. The propagation of nonlinear wave envelopes. *Journal of Mathematical Physics*, 46: 133–139

Kelley, P.L. 1965. Self-focusing of optic beams. *Physical Review Letters*, 15: 1005–1008

Ostrovsky, L.A. 1967. Propagation of wave packets and space-time self-focusing in a nonlinear medium. *Soviet Physics, JETP*, 24: 797–800

Ostrovsky, L.A. & Potapov, A.I. 1999. *Modulated Waves: Theory and Applications*, Baltimore: Johns Hopkins University Press

Whitham, G.B. 1974. *Linear and Nonlinear Waves*, New York: Wiley

Zakharov, V.E. & Shabat, A.B. 1972. Exact theory of two-dimensional self-focusing and one-dimensional self-modulation of waves in nonlinear media. *Soviet Physics, JETP*, 34: 62–69

WAVE PROPAGATION IN DISORDERED MEDIA

Although nonlinear partial differential equations with constant coefficients well describe main features of numerous nonlinear systems, understanding of many natural phenomena or experimental data requires taking into account the imperfectness of media which often have random character of two main types: time-dependent fluctuations and random spatial inhomogeneities (both may appear simultaneously). A thermal bath and quantum laser fluctuations are examples of the first type while impurities in crystal lattices, irregular variations of dielectric permittivity, and imperfectness of optical fibers are of the second type. Modeling of such phenomena requires nonlinear evolution equations with stochastic terms.

If a physical process is described by a field $u(t,\boldsymbol{r})$ that in an idealized situation is governed by a nonlinear equation $N[u]=0$, then including irregularities (fluctuations) of the medium in the consideration will result in an equation of the type $N[u]=\varepsilon(t,\boldsymbol{r})R[u]$, where $\varepsilon(t,\boldsymbol{r})$ is a random field and a (possibly nonlinear) operator $R[u]$ depends on physical nature of randomness. Such problems are difficult and still not well understood. The main difficulty arises because nonlinearity invalidates powerful methods of the linear theory (such as the Fourier transform) which allow one to relate stationary and nonstationary problems or introduce a Gaussian approach, allowing decoupling of high-order moments.

From the physical point of view, we can distinguish several important factors that must be taken into account, including generation of high harmonics resulting in nontrivial changes of the field statistics, creation of rather stable localized excitations (which have no analogy in the linear theory), and the multistability phenomenon.

Let us illustrate these issues with two examples: the nonlinear Schrödinger equation for a complex field $u(t,x)$:

$$\mathrm{i}\frac{\partial u}{\partial t} + \frac{\partial^2 u}{\partial x^2} + 2|u|^2 u = \varepsilon(t,x)R[u] \qquad (1)$$

and the nonlinear Klein–Gordon equation for a real scalar field $u(t,x)$

$$\frac{\partial^2 u}{\partial t^2} - \frac{\partial^2 u}{\partial x^2} + u - u^3 = \varepsilon(t,x)R[u]. \qquad (2)$$

These models represent some essential differences. First, at $\varepsilon=0$ Equations (1) and (2) are, respectively, integrable and nonintegrable. Second, the former model allows a solution in a form of a monochromatic wave and hence admits formulation of a stationary problem while harmonic generation is an indispensable property of the second model. Finally, their solitary wave solutions at $\varepsilon(t,x)=0$ are of different types: a dynamical soliton (in the restricted mathematical sense) in case (1) and a topological kink in case (2).

The random field $u(t,x)$, being a functional of $\varepsilon(t,x)$, is completely determined by a set of all $(n+m)$-

order moments

$$\langle u(t_1, x_1)\cdots u(t_n, x_n)\bar{u}(t_{n+1}, x_{n+1})\cdots \bar{u}(t_{n+m}, x_{n+m})\rangle, \quad (3)$$

where averaging is designated by the angular brackets and is provided over all realizations of $\varepsilon(t, x)$, when they are known. Because a Gaussian approximation is not applicable in nonlinear theory, the generic situation is that an $(n+m)$-order momentum cannot be expressed only through lower-order correlation functions. Moreover, temporal evolution of the mean field may drastically differ from the evolution of the field itself.

This is illustrated by Equation (1) with $\varepsilon(t, x)R[u] \equiv -f(t)xu(t, x)$, $f(t)$ being a Gaussian random process with

$$\langle f(t)\rangle = 0 \quad \text{and} \quad \langle f(t)f(t')\rangle = \sigma^2\delta(t-t'), \quad (4)$$

where σ is the dispersion of the fluctuations (Besieris, 1980). This equation is exactly solvable with the one-soliton solution

$$u(t, x) = \eta\frac{\exp\{\mathrm{i}\left[\mu(t)x + \eta^2 t - \beta(t)\right]\}}{\cosh\{\eta[x - 2\xi(t)]\}}, \quad (5)$$

where $\dot{\mu}(t) = f(t)$, $\mu(t) = \dot{\xi}(t)$, and $\dot{\beta} = \mu^2$ (a dot indicates a time derivative). Being interested in the evolution of the soliton intensity, which depends on $\xi(t)$ only, one can calculate the distribution

$$\begin{aligned} P(\xi, t) &= \int_{-\infty}^{\infty}\langle\delta(v - \dot{\xi}(t))\delta(\xi - \xi(t))\rangle\, \mathrm{d}v \\ &= \sqrt{\frac{3}{2\sigma^2 t^3}}\exp\left(-\frac{3\xi^2}{2\sigma^2 t^3}\right). \end{aligned} \quad (6)$$

Although the solution undergoes Brownian motion without any distortion, it follows from Equations (5) and (6) that its mean intensity is described by the Gaussian asymptotics

$$\langle u(t, x)\bar{u}(t, x)\rangle \approx \eta\sqrt{\frac{3}{2Dt^3}}\exp\left(-\frac{3x^2}{8\sigma^2 t^3}\right)$$

$$\text{at} \quad t \gg \frac{1}{(\sigma\eta)^{2/3}}.$$

In a general situation when exact solutions are not available, the problem becomes dependent on its statement and on the physical characteristics to be determined. The main statements of the problem are listed below.

Stationary Wave Scattering

This is a generalization of the problem of a wave transmission through a random slab to the nonlinear case. In the linear theory, the wave intensity decays exponentially with the slab width (Anderson localization). In the nonlinear case, one considers a monochromatic solution of Equation (1): $u(t, x) = (k/\sqrt{2})\exp(-\mathrm{i}k^2 t)\phi(x)$, where $\varepsilon(t, x) \equiv k^2\varepsilon_0(x)$, $\varepsilon_0(x)$ being a random function in the interval $0 < x < L$ and zero outside this interval, and $R[u] \equiv u(t, x)$. Then $\phi(x)$ satisfies the equation

$$\frac{\mathrm{d}^2\phi}{\mathrm{d}x^2} + k^2(1 + \varepsilon_0(x) + |\phi|^2)\phi = 0. \quad (7)$$

In order to describe an incidence of a plane wave on the random layer from the right, one imposes

$$\phi(x) = \begin{cases} a\left[\mathrm{e}^{\mathrm{i}k(L-x)} + R(L, w)\mathrm{e}^{\mathrm{i}k(x-L)}\right], & x \ge L, \\ T(L, w)\mathrm{e}^{-\mathrm{i}kx}, & x \le 0 \end{cases} \quad (8)$$

and the conditions of the continuity of the field $\phi(x)$ and of its derivative $\mathrm{d}\phi(x)/\mathrm{d}x$ at the boundaries of the layer, $x = 0$ and $x = L$ (see Figure 1).

The coefficients a, $R(L, w)$ and $T(L, w)$ are amplitudes of the incident wave, reflection and transmission coefficients. The last two quantities depend on the slab width L and on the intensity of the incident wave $w = a^2$. Using the imbedding method (Klyatskin, 1988), one can obtain a nonlinear partial differential equation for the reflection coefficient as a function of L and w, which (upon applying the method of characteristics) results in a system (Doucot & Rammal, 1987)

$$\frac{\mathrm{d}R_1}{\mathrm{d}L} = -2kR_2 - kR_2(1+R_1)(\varepsilon(L) + w|1+R|^2), \quad (9)$$

$$\begin{aligned} \frac{\mathrm{d}R_2}{\mathrm{d}L} &= 2kR_1 + \frac{k}{2}\left[(1+R_1)^2 - R_2^2\right] \\ &\quad \times(\varepsilon(L) + w|1+R|^2), \end{aligned} \quad (10)$$

where $R_1(L) = \Re(R(L, w))$, $R_2(L) = \Im(R(L, w))$,

$$w(L) = \frac{w(0)}{1 - |R(L, w)|}, \quad (11)$$

and the initial conditions are $R_1(0) = R_2(0) = 0$. Formula (11) allows us to understand two essentially different statements of the scattering problem. Indeed, (9) and (10) are dynamical equations for the characteristics that pass either through the point $(L = 0, w(0) = w_0)$, if the output intensity of the wave w_0, is given, or through the point $(L, w(L))$ if the input intensity is given (called fixed input and fixed output problems). In the former case, while increasing L, one follows the characteristic starting with its initial position. To solve a fixed output problem, one must determine a characteristic or characteristics which cross the point $(L, w(L))$ (having different starting points). As there may be more than one characteristic, multistability occurs (for another way of understanding the multistability phenomenon, see Knapp et al. (1991)).

Consider the fixed output problem (w_0 is given) in the case of weak Gaussian fluctuations, when

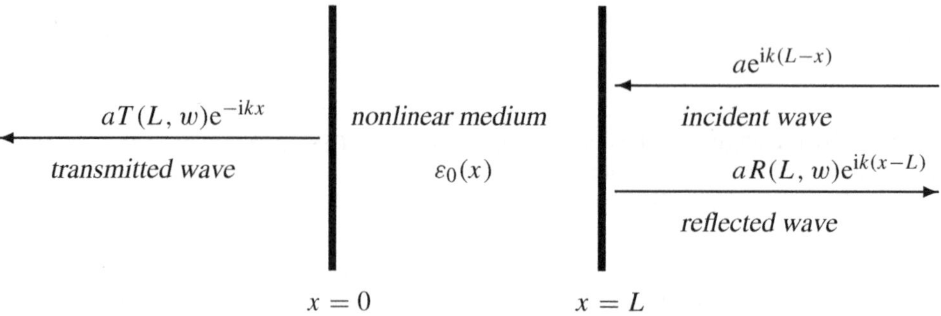

Figure 1. To the statement of the stationary scattering problem.

$\langle\varepsilon_0(x)\rangle=0$, $\langle\varepsilon_0(x)\varepsilon_0(x')\rangle=D\delta(x-x')$, and $L_{\text{loc}}\gg\lambda$, L_{nl} where $L_{\text{loc}}=1/(Dk^2)$ is the Anderson localization length of the underlying linear system, $\lambda=2\pi/k$ is the incident wavelength, and $L_{\text{nl}}=1/(w_0k)$ is the effective nonlinear length characterizing change of the reflection coefficient due to the nonlinearity. Starting with the region of a relatively strong nonlinearity, $L_{\text{nl}}\ll\lambda$, one can define a period of motion $L_{\text{p}}=\int_0^{2\pi}(\text{d}\theta/\text{d}L)^{-1}\text{d}\theta\approx L_{\text{nl}}$ (here $\theta=\arg R$) and study weak drift of trajectories of "dynamical system" (9), (10) due to the random perturbations over one period L_{p} (providing averaging over θ). The main result can be formulated as $\langle 1/T^2\rangle=O(L^2)$ and decay of the fluctuations of $\langle\ln T^2\rangle$ at $L\to\infty$ (Doucot & Rammal, 1987). Similar results, obtained for a step-like random function $\varepsilon_0(x)$ (Fröhlich et al., 1986). Thus, compared with the linear theory, the nonlinearity does not change the property of $\ln T^2$ to be a self-averaging value, but results in change of the decay law for the decay of the transmission coefficient from the exponential one to a power one. However, when the intensity of the wave becomes small enough and the effect of back-scattering becomes dominating, one can recover the exponential decay law $T\sim\exp(-L/L_{\text{loc}})$. The transition between two regions with the different laws of the decay of the reflection coefficient happens at $L_0\sim L_{\text{loc}}\ln(1/w_0)$: the law is power at $L>L_0$ and is exponential at $L<L_0$.

Interaction of a Wave Packet with a Random Layer

As this case cannot be reduced to the study of a stationary problem, the statement of the problem and physical characteristics describing wave scattering must be redefined (compared with the previous case of stationary scattering). Consider the Cauchy problem with a given initial field distribution that decays sufficiently rapidly at infinity. The main task can be formulated as a description of the evolution of the wave-packet characteristics during its propagation. Such a statement acquires a special meaning in systems that possess stable solitary wave solutions in the unperturbed ($\varepsilon=0$) limit and in which the randomness is weak enough. For relatively large temporal intervals, the nonlinear field can be represented in the form $u(t,x)\approx u_{\text{ad}}(t,x)+u_{\text{rad}}(t,x)$, where $u_{\text{ab}}(t,x)$ is a function having the same form as the unperturbed solitary wave but now with slowly varying parameters (adiabatic approximation) and $u_{\text{rad}}(t,x)$ is a small component compared with $u_{\text{ab}}(t,x)$ describing deformation of the soliton shape and radiative losses.

An advantage of the adiabatic approximation is that it reduces a stochastic partial differential equation to a system of ordinary differential equations for the soliton parameters. For Equation (1), the adiabatic approximation corresponds to the substitution (5) with time-dependent coefficients $\eta(t)$, $\xi(t)$, $\mu(t)$, and $\beta(t)$. If $R[u]\equiv u(t,x)$, the standard perturbation theory for solitons shows that (Karpman, 1979)

$$\frac{\text{d}\mu}{\text{d}t}=\eta\int_{-\infty}^{\infty}\frac{\tanh z}{\cosh^2 z}\,\varepsilon\left(t,\frac{z}{\eta}+2\xi\right)\text{d}z, \tag{12}$$

$$\frac{\text{d}\xi}{\text{d}t}=2\mu,\qquad \eta=\text{const}, \tag{13}$$

$$\frac{\text{d}\beta}{\text{d}t}=\mu^2+\eta\int_{-\infty}^{\infty}\frac{1-z\tanh z}{\cosh^2 z}\,\varepsilon\left(t,\frac{z}{\eta}+2\xi\right)\text{d}z. \tag{14}$$

If $\varepsilon(t,x)=f(t)x$, one recovers the exact solution. As the perturbation is random, this is a system of stochastic equations; it allows rather complete analysis when $\varepsilon(t,x)\equiv V_\varepsilon(x)$ is an ergodic process with a finite support localized on the interval $[0,L/\varepsilon^2]$ (Garnier, 1998). The soliton dynamics are then governed by deterministic equations for almost every realization of $V_\varepsilon(x)$. For a soliton of small amplitude (weak nonlinearity), one can define $\nu_0\ll\mu$ such that decay of the soliton amplitude follows either an exponential, for $\nu<\nu_0$, or a power, for $\nu_0\nu\ll\mu$, law. When the soliton has large amplitude and small velocity, $\mu\ll\nu$, its amplitude experiences rather weak changes, while the velocity decreases.

Solitary Wave Propagation in Media with Fluctuating Parameters

Consider $\varepsilon(t,x) \equiv f(t)$ to be a stationary Gaussian process (4). Then the adiabatic approximation allows us to obtain a Fokker–Planck equation for the distribution function of the soliton parameters, similar to Equation (6), where typical behavior is a kind of Brownian motion (Konotop & Vazquez, 1994). At large times, the adiabatic approximation fails and fluctuations of the medium may result in resonant parametrical processes. Such processes are especially interesting in systems having topological solitons which cannot be destroyed by fluctuations. If, for example, $R[u] \equiv u - u^3$ in Equation (2), long-time numerical simulations of the kink dynamics show anomalous diffusion, and the dispersion of the fluctuation of its center grows as $t^{2.087}$ while the energy of the system increases exponentially. This phenomenon is a manifestation of the stochastic parametrical resonance of linear modes.

VLADIMIR V. KONOTOP AND LUIS VÁZQUEZ

See also **Brownian motion; Characteristics; Nonlinear Schrödinger equations; Stochastic processes**

Further Reading

Besieris, I.M. 1980. Solitons in randomly inhomogeneous media. In *Nonlinear Electromagnetics*, edited by P.L.E. Uslenghi, London and New York: Academic Press, pp. 87–116

Doucot, B. & Rammal, R. 1987. On Anderson localization in nonlinear random media. *Europhysics Letters*, 3: 969–974

Fröhlich, J., Spencer, Th. & Wayne, C.E. 1986. Localization in disordered, nonlinear dynamical systems. *Journal of Statistical Physics*, 42: 247–274

Garnier, 1998. Asymptotic transmission of solitons through random media. *SIAM Journal of Applied Mathematics*, 58: 1969–1995

Karpman, V.I. 1979. Soliton evolution in the presence of perturbations. *Physica Scripta*, 20: 462–478

Klyatskin, V.I. 1988. *Imbedding Approach in the Theory of Wave Propagation*, Moscow: Nauka (in Russian)

Knapp, R., Papanicolaou, G. & White, B. 1991. Transmission of waves by a nonlinear random medium. *Journal of Statistical Physics*, 63: 567–583

Konotop, V.V. & Vázquez, L. 1994. *Nonlinear Random Waves*, Singapore: World Scientific

WAVE STABILITY AND INSTABILITY

As with other dynamic entities, a solution of a wave system is considered to be stable if it does not deviate greatly under small perturbations, and unstable if it does.

Consider first the stability of the null solution of a partial differential equation (PDE) system in an infinite and uniform medium. Small perturbations of the null solution can be taken in the form $A\exp[\mathrm{i}(kx - \omega t)]$. To satisfy the PDE, k and ω are related by a dispersion relation, which is denoted as $D(k, \omega) = 0$. The dispersion relation may be solved in the form

$$\omega = \omega_n(k), \quad (n = 1, 2, \ldots). \tag{1}$$

Thus, there may be several solutions (or modes), with different functions $\omega_n(k)$, which are referred to as different modes. In general, ω may be complex for real k, leading to real solutions of the form

$$\begin{aligned} &A\exp[\mathrm{i}(kx - \omega t)] + \text{c.c.} \\ &\quad = \{A\exp[\mathrm{i}(kx - \omega_r t)] + \text{c.c.}\}\exp(\omega_i t), \end{aligned} \tag{2}$$

where $\omega = \omega_r + \mathrm{i}\omega_i$. Therefore, if $\omega_i > 0$ for any mode, that mode will grow with time, indicating instability. If all ω_i are negative, all modes damp away, indicating stability.

Convective and Absolute Instability

In practice, an initial perturbation is often localized and can be represented as a (Fourier) wave packet of various normal modes, in which each component propagates with its own phase velocity ($v_{\mathrm{p}} = \omega/k$). The collective motion of a wave packet, on the other hand, is governed by its group velocity ($v_{\mathrm{g}} = \mathrm{d}\omega/\mathrm{d}k$). If a perturbative wave packet is localized and moving with a certain group velocity, its amplitude at a certain point will at first begin to rise and then eventually fall back to zero. This leads to two distinct concepts: convective and absolute instability, which originally arose in studies of wave instabilities in plasmas (Briggs, 1964).

A wave system is convectively unstable if the maximum amplitude of a perturbing wave packet grows without bound, but at any fixed point of the system, the disturbance eventually relaxes back to zero as the wave packet propagates away. Such behavior is useful for the design of distributed amplifiers, such as the traveling-wave tube or optical (laser) amplifiers.

If the dispersion equation contains unstable modes with zero group velocity, however, a more robust instability arises. In this case, called absolute instability, perturbations grow without bound at every point of the system. Absolutely unstable dynamics can be employed for the design of distributed oscillators (backward-wave oscillators, for example) but not distributed amplifiers.

Modulational Instability

Consider next the stability of a small amplitude component of a wave system, which is stable at infinitesimal amplitude. At finite amplitude, however, the wave is not necessarily stable against wave modulation, because the finite intensity of wave modifies the propagation properties of medium. In this case, a nonlinear dispersion relation that takes into account the finite intensity of wave may be written as $\omega = \omega(k, a)$, where a is the amplitude of wave. For a slowly modulated plane wave having a small but finite amplitude, the frequency ω may be expressed approximately as

$$\omega(k, a) = \omega_0(k) + \alpha(k)a^2 + O(a^4). \tag{3}$$

(Odd terms in a are excluded from this formulation because they would imply different values of ω when a merely changes its algebraic sign.) Modulations on a linear and weakly nonlinear wave train can be described by the equations (Whitham, 1974)

$$\frac{\partial k}{\partial t} + \frac{\partial \omega}{\partial x} \doteq 0, \tag{4}$$

$$\frac{\partial a^2}{\partial t} + \frac{\partial}{\partial x}\left(\frac{\partial \omega}{\partial k} a^2\right) \doteq 0. \tag{5}$$

Equation (4) follows from the relations $\omega = -\partial\theta/\partial t$ and $k = \partial\theta/\partial x$, where $\theta(x, t)$ is the phase. (This equation can also be viewed as a conservation law for wave crests.) As a^2 is proportional to energy density and $a^2\partial\omega/\partial k = a^2 v_g$ is proportional to power flow, Equation (5) is the law of energy conservation.

Substituting Equation (3) into these equations and assuming a to be sufficiently small leads to the following equations:

$$\frac{\partial k}{\partial t} + v_g \frac{\partial k}{\partial x} + \alpha(k_0)\frac{\partial a^2}{\partial x} \doteq 0, \tag{6}$$

$$\frac{\partial a^2}{\partial t} + v_g' a^2 \frac{\partial k}{\partial x} + v_g \frac{\partial a^2}{\partial x} \doteq 0. \tag{7}$$

Here $v_g = d\omega_0(k_0)/dk_0$, the group velocity of the linear wave with wave number k_0, and $v_g' = d^2\omega_0(k_0)/dk_0^2$. Linearizing k and a by

$$k = k_0 + \bar{k}\exp\{i(Kx - \nu t)\},$$
$$a = a_0 + \bar{a}\exp\{i(Kx - \nu t)\},$$

yields the modulational dispersion relation

$$\nu = K\left(v_g \pm \sqrt{\alpha a_0^2 v_g'}\right). \tag{8}$$

If $\alpha v_g' < 0$, ν becomes complex, which implies that the modulation becomes unstable and grows (Lighthill, 1965; Whitham, 1974). Equation (8) is called Lighthill's theorem, and the corresponding modulational instability (which has been studied analytically and experimentally in the context of deep water waves by Benjamin & Feir (1967)) is called the Benjamin–Feir (BF) instability. The nonlinear evolution of modulated envelopes is described by the nonlinear Schrödinger equation. Interestingly, the BF instability can lead to formation of stable traveling waves of modulation—envelope solitons. For further discussions, see Whitham (1974), Infeld & Rowlands (2000), Longuet-Higgins (1978), and McLean (1982).

Soliton Stability

The stability problem of traveling-wave solutions with respect to small perturbations has been studied by various methods, including the normal-mode approach (Infeld and Rowlands, 2000). Consider the solution $u(x, t)$ of a nonlinear PDE to be the stationary traveling wave $u_s(\xi)$ plus a small perturbation $p(\xi, t)$

$$u(x, t) = u_s(\xi) + p(\xi, t), \tag{9}$$

where $\xi = x - v_s t$ and v_s is a propagation velocity. Transforming the original nonlinear PDE to a coordinate system moving at v_s and linearizing, the resulting equation has parameters that are functions of ξ. This equation can then be solved for the time development of the perturbation subject to an appropriate boundary condition.

When such problems are uniform in time (as they often are), it is convenient to express $p(\xi, t)$ as product solution in the form $p(\xi, t) = f(\xi)\exp(\sigma t)$ (separation of variables). Traveling-wave solutions are linearly unstable if any product solution has $\mathrm{Re}(\sigma) > 0$ and asymptotically stable if all product solutions have $\mathrm{Re}(\sigma) < 0$. A stationary solution that is neither unstable nor asymptotically stable is said to be neutrally stable, and there is always such a case because $\sigma = 0$ corresponds to a simple displacement of the traveling wave in the direction of propagation.

As examples, consider the stability of solitons of the Korteweg–de Vries (KdV) and its two-dimensional generalization. Writing the KdV equation as $u_t + 6(u^2)_x + u_{xxx} = 0$, a solitary wave solution is given by

$$u_s = A + k^2\mathrm{sech}^2\{k(x - v_s t) + \delta\} = 0, \tag{10}$$

where $v_s = 4k^2 + 12A$ and A and k are both arbitrary. Using the method of normal modes, Jeffrey and Kakutani (1972) and also Berryman (1976) have investigated the stability of this soliton, showing that small localized perturbations do not grow without bound; thus, the KdV soliton is linearly stable. Benjamin formulated a nonlinear stability theory for the KdV soliton and also showed that the KdV soliton is stable against small but finite perturbations (Benjamin, 1972).

A soliton in a two-dimensional (2-d) space often appears as a line soliton (LS). Kadomtsev and Petviashvili studied a 2-d-generalization of the KdV equation in order to discuss the stability of the line soliton with respect to long and small transverse perturbations. They obtained the KP equation (Kadomtsev & Petviashvili, 1970)

$$(u_t + 6(u^2)_x + u_{xxx})_x + s u_{yy} = 0, (s = \pm 1), \tag{11}$$

which corresponds to the case of negative and positive dispersion when $s = +1$ and -1, respectively. The line soliton is unstable in the case of positive dispersion and is stable for negative dispersion. The KP equation with positive dispersion also has a periodic soliton (PS) solution. A spatially periodic resonance exist between the LS and PS solutions, as indicated in Figure 1. From this

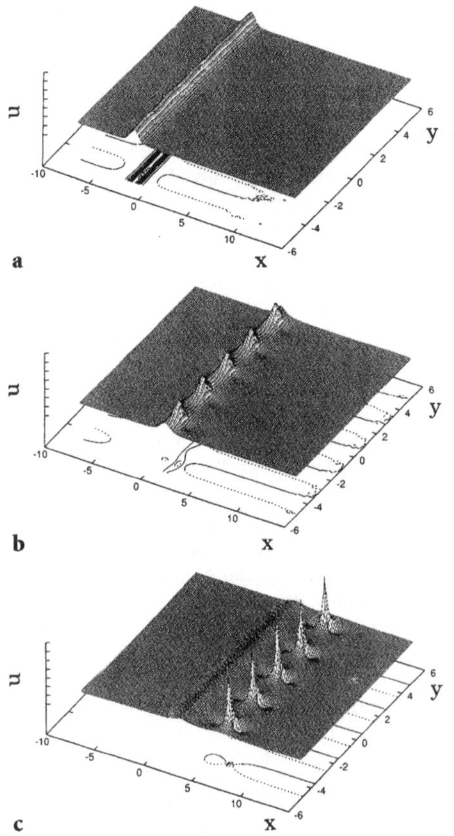

Figure 1. The sequence of snapshots of quasiresonant solution.

figure, we see how a transversely perturbed LS decays into a small LS and a PS, where the instability of the LS may be relaxed by the emission of a PS (Infeld and Rowlands, 2000, Chapter 10). Importantly, the nonlinear stage of instability of soliton is, in general, different from the conclusion of the linear stability theory. For more detailed studies of 2-d soliton stability, see Infeld & Rowlands (2000) and Zakharov et al. (1986).

Eckhaus Instability

In the neighborhood of a bifurcation (the appearance of a new solution as a parameter λ is changed), the description of a dynamical system can be greatly reduced, where the only relevant variable is a complex normalized amplitude: Z. Studies of the dissipative structures that emerge beyond the instability is often facilitated by using amplitude equations, such as the Newell–Whithead (NW) equation (Newell & Whitehead, 1969)

$$\frac{\partial Z}{\partial t} = (\lambda - \lambda_c)Z + \left(\frac{\partial}{\partial x} - \frac{i}{2k_c}\frac{\partial^2}{\partial y^2}\right)^2 Z - |Z|^2 Z. \tag{12}$$

This is the normal form of a symmetry-breaking bifurcation leading to roll or stripe patterns with wave vector parallel to the x-axis, where λ_c is the critical control parameter and k_c is the critical wave number. In the supercritical region, $(\lambda - \lambda_c) > 0$, the NW equation has the two parameter $(\delta k, \theta_0)$ family of solutions:

$$Z_k = A_k e^{i(\delta k x + \theta_0)}, \tag{13}$$

where $A_k = \sqrt{(\lambda - \lambda_c) - (\delta k)^2}$, δk is a small wave number describing the modulation of the basic dissipative structure at the critical wave number k_c, and θ_0 is an arbitrary phase. This solution expresses a stationary roll pattern with wave number $k = k_c + \delta k$.

Introducing a solution that is perturbed by small changes in amplitude and phase leads to the instability criterion (Eckhaus, 1965)

$$|\delta k| > \sqrt{\frac{\lambda - \lambda_c}{3}}. \tag{14}$$

If perturbations transverse to the basic structure are allowed, a zig-zag instability emerges (Nicolis, 1995; Mori & Kuramoto, 1998).

Nerve Impulse Stability

The stability of nerve impulses can be studied by the reduced version of the Hodgkin–Huxley system called the FitzHugh–Nagumo (FN) equation (FitzHugh, 1961; Nagumo et al., 1962):

$$\frac{\partial V}{\partial t} = \frac{\partial^2 V}{\partial x^2} - f(V) - R,$$
$$\frac{\partial R}{\partial t} = \varepsilon(V - bR),$$

where V is the nerve membrane potential, $f(V) = V(V-1)(V-a)$, and ε is a temperature parameter which controls the rate of change of the recovery variable R.

If $0 < \varepsilon < \varepsilon_c$, the FN equation has slow and fast impulse solutions with propagation speeds $c_s(\varepsilon)$ and $c_f(\varepsilon)$: $0 < c_s < c_f$. The relation between these two speeds and the temperature parameter is sketched in Figure 2, where the fast solution is stable and the slow solution is unstable (Rinzel & Keller, 1973). Thus, small positive perturbations of the slow solution will eventually grow into the fast solution, whereas small negative perturbations will cause the slow solution to collapse to zero (Scott, 2003).

At each value of ε, the FN equation has also two periodic traveling-wave solutions with the same wavelength but different propagation speeds. Viewed as a traveling wave, the fast periodic solution is stable and the slow solution is unstable, just as for a single impulse.

Suppose that a time-periodic boundary condition is imposed at $x = 0$ (where $V(0,t) = V(0, t+T)$) and periodic solutions are sought (for $x > 0$) of the form $V(x,t) = V(x+\lambda, t) = V(x, t+T)$. In this "signaling problem," an important question is whether (or not)

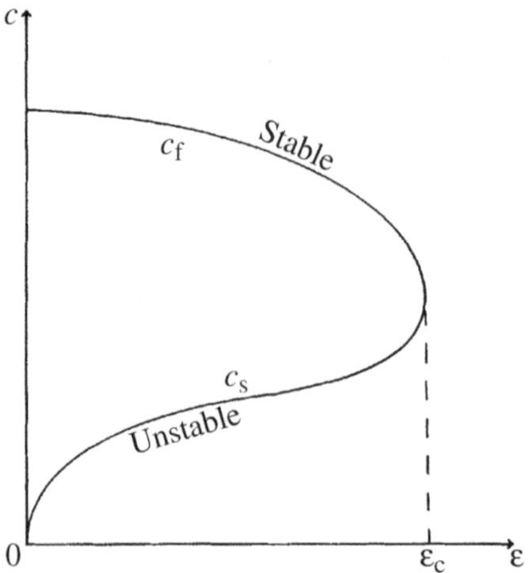

Figure 2. The typical relation between the pulse speed c and the temperature parameter ε.

small perturbations grow with x. Because $c(\lambda) = \lambda/T$, the dependence of the nonlinear frequency $(1/T)$ upon the nonlinear wave number $(1/\lambda)$ is readily calculated, and Rinzel has shown that the condition for solutions not to grow with increasing x is (Rinzel, 1975)

$$\frac{d(1/T)}{d(1/\lambda)} > 0.$$

MASAYOSHI TAJIRI

See also **Equilibrium; Kelvin–Helmholtz instability; Modulated waves; Rayleigh–Taylor instability; Stability**

Further Reading

Benjamin, T.B. 1972. The stability of solitary waves. *Proceedings of the Royal Society of London*, 328A: 153–183

Benjamin, T.B. & Feir, J.E. 1967. The disintegration of wave trains in deep water, I. *Journal of Fluid Mechanics*, 27: 417–430

Berryman, J.G. 1976. Stability of solitary waves in shallow water. *Physics of Fluids*, 19: 771–777

Briggs, R.G. 1964. *Electron-Stream Interaction with Plasmas*, Cambrige, MA: MIT Press

Eckhaus, W. 1965. *Studies in Nonlinear Stability Theory*, Berlin: Springer

FitzHugh, R. 1961. Impulses and physiological states in theoretical models of nerve membranes. *Biophysical Journal*, 1: 445–466

Infeld, E. & Rowlands, G. 2000. *Nonlinear Waves, Solitons and Chaos*, Cambridge and New York: Cambridge University Press

Jeffrey, A. & Kakutani, T. 1972. Weakly nonlinear dispersive waves: a discussion centered around the Korteweg–de Vries equation. *SIAM Review*, 14: 582–643

Kadomtsev, B.B. & Petviasvili, V.I. 1970. On the stability of solitary waves in weakly dispersive media. *Soviet Physics-Doklady*, 15: 539–541

Lighthill, J. 1965. Contributions to the theory of waves in nonlinear dispersive system. *Journal of the Institute of Mathematics and Its Applications*, 1: 269–306

Longuet-Higgins, M.S. 1978. The instabilities of gravity waves of finite amplitude. *Proceeding of Royal Society of London, I Superhamonics*, A360: 471–488; *II Subharmonics*, A360: 489–505

McLean, J.W. 1982. Instabilities of finite-amplitude water waves, *Journal of Fluid Mechamics*, 114: 315–330

Mori, H. & Kuramoto, Y. 1998. *Dissipative Structures and Chaos*, Berlin: Springer

Nagumo, J., Arimoto, S. & Yoshizawa, S. 1962. An active pulse transmission line simulating nerve axon. *Proceedings of the Institute of Radio Engineering*, 50: 2061–2070

Newell, A.C. & Whitehead, J.A. 1969. Finite bandwidth, finite amplitude convection. *Journal of Fluid Mechanics*, 38: 279–303

Nicolis, G. 1995. *Introduction to Nonlinear Science*, Cambridge and New York: Cambridge University Press

Rinzel, J. 1975. Spatial stability of traveling-wave solutions of a nerve conduction equation. *Biophysical Journal*, 15: 975–988

Rinzel, J. & Keller, J.B. 1973. Traveling-wave solutions of nerve conduction equation. *Biophysical Journal*, 13: 1313–1337

Scott, A.C. 2003. *Nonlinear Science: Emergence and Dynamics of Coherent Structures*, 2nd edition, Oxford and New York: Oxford University Press

Whitham, G.B. 1974. *Linear and Nonlinear Waves*, New York: Wiley

Zakharov, V.E., Kuznetsov, E.A. & Rubenchik, A.M. 1986. Soliton stability. In *Solitons*, edited by S.E. Trullinger, V.E. Zakharov & V.L. Pokrovsky, Amsterdam and New York: North-Holland, pp. 503–554

WAVE TANKS

See **Laboratory models of nonlinear waves**

WAVELETS

The wavelet transform is a tool that divides up data, functions, or operators into different frequency components and then studies each component with a resolution matched to its scale. The idea emerged independently in many different fields, including mathematics, quantum physics, electrical engineering, and seismology. By the end of the 1980s, French researchers Grossmann & Morlet (1984), Meyer (1990), and later Daubeschies (1992) had laid the mathematical foundations of the wavelet transform technique. Their work was motivated by the problem Morlet (a geophysicist) was facing while analyzing seismic data which comprised different features in time and frequency.

The frequency content of a signal can be obtained by taking its Fourier transform. However, in transforming to the frequency domain, time information is lost and it is impossible to tell when a particular event took place. To correct this deficiency, the Fourier transform was adapted to analyze only a small section (or window) of the signal at a time. Such a short-time Fourier transform (STFT) maps a signal into a two-dimensional function of time and frequency and provides information about both when and at what frequencies a signal event occurs. This information can only be obtained with limited precision, and that precision is determined by the size of the time window. The main drawback of STFT is that once a particular size for the time window is chosen, that window is the same for all frequencies. With Morlet's data, this approach failed either to follow the time evolution of rapid events or to estimate the frequency content in the low-frequency band.

Wavelet analysis represents the next logical step: signal cutting is performed by a window of variable length. Short windows are used at high frequencies and long windows at low frequencies. Thus, the time-frequency resolution is no longer constant but changes with frequency, allowing good time resolution for high frequencies and good frequency resolution for low frequencies to be achieved.

In Fourier analysis, the signal is decomposed into sine waves of various frequencies. Similarly, wavelet analysis is a decomposition of a signal into a set of basis functions $\Psi_{s,\tau}(t)$ called wavelets

$$\hat{f}(s,\tau) = \int f(t)\Psi^*_{s,\tau}\,dt, \tag{1}$$

where $\hat{f}(s,\tau)$ is the wavelet transform of $f(t)$ and $*$ denotes complex conjugation. The wavelets are generated from a single basic wavelet $\psi(t)$ called the mother wavelet, by scaling and translation:

$$\Psi_{s,\tau} = \sqrt{|s|}\,\psi\left(\frac{t-\tau}{s}\right). \tag{2}$$

As the scaling parameter s changes, the wavelets cover different frequency ranges: large values of s correspond to low frequencies, small values of s correspond to high frequencies. Changing the translation parameter τ allows the time localization center to be moved: each $\Psi_{s,\tau}(t)$ is localized around $t = \tau$. The factor $\sqrt{|s|}$ is for energy normalization across different scales.

An important difference between wavelet and Fourier transforms is that wavelet basis functions are not specified. The theory deals with general properties of wavelets and defines a framework within which different wavelets can be designed. Numerous families of wavelets have been proposed and proven to be useful in different applications. In order for a function to be used as the mother wavelet, it must allow for analysis and reconstruction of the signal without loss of information. This so-called admissibility condition implies that the average value of a wavelet in the time domain must be zero and it must, therefore, be oscillatory, a wave. The other important property of wavelets (the regularity condition) states that wavelets are smooth and concentrated both in time and frequency. This makes them suitable for capturing local features of a signal. Figure 1 presents an example of a wavelet, the Morlet wavelet, in time and frequency domains for two scales.

We can distinguish between the continuous and discrete wavelet transform. In a continuous transform, the parameters s and τ vary continuously. It maps a one-dimensional signal to a two-dimensional time-scale joint representation that is highly redundant. This redundancy can be either exploited or removed. To reduce the redundancy, discrete wavelets have been introduced, which can only be scaled and translated in discrete steps. For the scale, we choose integer powers of fixed dilation parameter $s_0 > 1$, $s = s_0^j$, and j is an integer. The discretization of the translation parameter τ depends on j: narrow wavelets are translated by small steps, while wider wavelets are translated by larger steps, $\tau = n\tau_0 s_0^j$, with j and n integers and $\tau_0 > 0$ is fixed. For some special choices of mother wavelets, the discrete wavelets can be made orthogonal to their own translations and dilations. In this case, they behave exactly like an orthonormal basis and redundancy is removed.

Wavelets have a bandpass-like spectrum and can be viewed as bandpass filters. Compression in time stretches the spectrum and shifts it upwards, while stretching in time compresses the bandwidth and shifts it toward zero (Figure 1). The series of dilated wavelets can be used to cover the spectrum of a signal. However, an infinite number of wavelets is needed to reach zero frequency. This problem was solved by introducing a low-pass or averaging filter with a spectrum that belongs to the scaling function $\phi(t)$. If we analyze the signal using a combination of scaling functions and wavelets, the scaling function takes care of the spectrum otherwise covered by all the wavelets up to a chosen scale, while the rest is done by wavelets. The family of scaling functions and wavelets allows for wavelet multiresolution analysis. In multiresolution analysis, the signal is split into an approximation on a coarser scale, obtained using the scaling function, and details on a current scale, obtained by wavelets. This process is repeated, giving a sequence of approximations and details removed at every scale. After N iterations, the original signal can be reconstructed by summing up the last approximations and details on all previous scales.

Interactions between the fields where wavelets were first introduced have led to many wavelet applications. Wavelets are of particular interest for the analysis of nonstationary signals with broad spectra because they permit time-frequency presentation with logarithmic resolution. Wavelet analysis is capable of revealing aspects of signals that other signal analysis techniques miss, such as trends, breakdown points, discontinuities in higher derivatives, and self-similarity. Furthermore, wavelet analysis can often compress (or reduce noise in) a signal without appreciable degradation. The above-described one-dimensional aspect can be generalized to more dimensions, for example, to handle image analysis. Wavelets are a very powerful tool for image compression, since wavelet transform clearly separates high-pass and low-pass information on a pixel-by-pixel

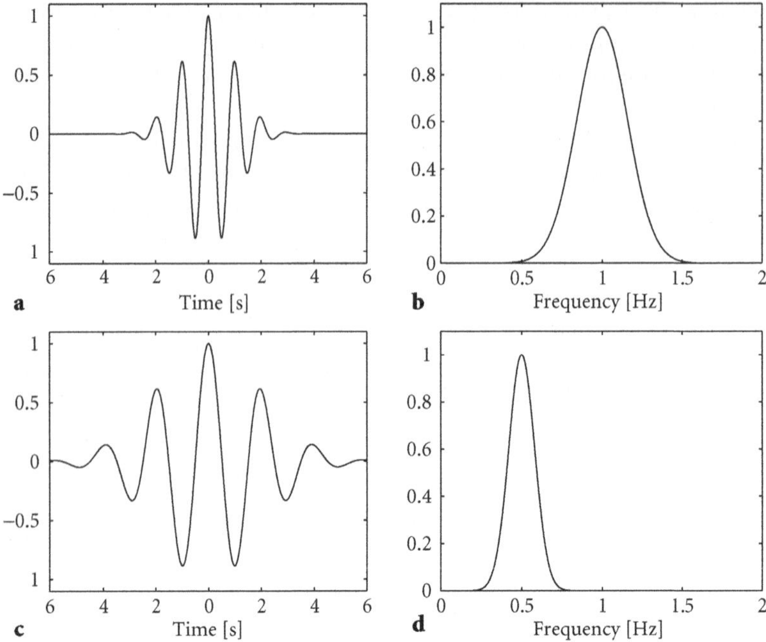

Figure 1. Morlet wavelet in time and frequency domain for scales $s = 1$ (a,b) and $s = 2$ (c,d).

basis. Among the successes of wavelet representation are the compression of digitized fingerprints by the US Federal Bureau of Investigation and image compression using the JPEG 2000 standard (the older JPEG standard uses a non-wavelet compression).

Maja Bračič Lotrič and Aneta Stefanovska

See also **Integral transforms; Nonlinear signal processing**

Further Reading

Daubeschies, I. 1992. *Ten Lectures on Wavelets*, Philadelphia, PA: Society for Industrial and Applied Mathematics

Grossmann, A. & Morlet, J. 1984. Decomposition of Hardy functions into square integrable wavelets of constant shape. *SIAM Journal of Mathematical Physics*, 27: 2437–2479

Meyer, Y. 1990. *Ondelettes et Operateurs*, 3 vols, Paris: Herrmann; vol. 1 as *Wavelets and Operators*, Cambridge and New York: Cambridge University Press, 1992 and vols. 2 and 3 as *Wavelets: Calderón-Zygmund and Multilinear Operators*, Cambridge and New York: Cambridge University Press, 1997

WAVE NUMBER SELECTION

See **Pattern formation**

WEAK COLLAPSE

See **Development of singularities**

WEAK TURBULENCE

See **Turbulence**

WEAKLY NONLINEAR ANALYSIS

See **Quasilinear analysis**

WEATHER FRONTS

See **Atmospheric and ocean sciences**

WEIERSTRASS ELLIPTIC FUNCTIONS

See **Elliptic functions**

WESTERVELT EQUATION

See **Nonlinear acoustics**

WHISTLERS

See **Nonlinear plasma waves**

WHITHAM'S METHOD

See **Modulated waves**

WIGNER STATISTICS

See **Random matrix theory I, II**

WINDING NUMBERS

If you wrap a rubber ring (rubber band) around a pencil, the intuitive idea of an integer invariant for the wrapping process arises. The number of oriented turns around the pencil is an integer, and it is independent of how tight or creased the ring is. The only way to change this integer is to wrap/unwrap one loop of the ring at the endpoints of the pencil. Using a rubber string, the number of turns could be some real number instead.

How does this fact connect with dynamics? Consider a three-dimensional (3-d) dynamical system having a torus as phase space. Let the hole in the torus represent the pencil and the rubber ring represent a closed trajectory within the torus. Small modifications to the trajectory will not alter the wrapping property.

The underlying feature in the above examples is the forbidden region given by the pencil or the hole of the torus. To further analyze its structure, decompose 3-d space ($\mathbb{R}^3$) as the product of the pencil direction times a perpendicular plane (equivalent to $\mathbb{R}^2$). The role of the pencil is to identify a special point on that plane given by the projection along the pencil direction.

Similarly, consider a periodic orbit in $\mathbb{R}^3$, as the forbidden region. Since the points in the periodic orbit are regular points in a small tube around the periodic orbit, the flow can be decomposed in a component parallel to the periodic orbit and a projected flow onto a perpendicular section. Hence, any other closed trajectory sufficiently near to the orbit will wind around it. A similar situation arises around a period-one orbit of a periodically forced flow in $\mathbb{R}^2$.

The winding number characterizes the topological properties of "the plane minus one point." Moreover, the topology of the plane minus n points gives a deeper characterization of the periodic orbit structure of 3-d dynamical systems admitting a Poincaré section.

Definition

Consider a simple continuous closed curve γ in the complex plane ($\mathbb{C}$) and a point $z_0 \in \mathbb{C}-\gamma$. The winding number n is defined as

$$n(\gamma, z_0) = \frac{1}{2\pi \mathrm{i}} \int_\gamma \frac{\mathrm{d}z}{z - z_0}. \tag{1}$$

We may regard our wrapped rubber ring as a suitable complex function γ of the unit circle, thus connecting our motivating idea with the formal definition (similarly for the second example if we project the torus along the direction perpendicular to the hole when seen as a disc). Where appropriate in the sequel, we will let $z_0 = 0$ for simplicity and recast γ as a map of the unit circle $\gamma : \mathbb{S}^1 \to \mathbb{S}^1$. It follows from Equation (1) that $n(\gamma, 0)$ is a real integer given by

$$n(\gamma, 0) = \int_{\mathbb{S}^1} \frac{\gamma'(\theta)}{\gamma(\theta)} \,\mathrm{d}\theta, \tag{2}$$

which is called the degree of γ (Rotman, 1988, p. 50).

Applications

Homotopy classes of the circle. A loop is a continuous map g of the circle to itself such that $g(0) = g(2\pi) = 2\pi$. Two loops α and β are homotopic if there exists a continuous map $H : \mathbb{S}^1 \times [0, 1] \to \mathbb{S}^1$ such that $H(\cdot, 0) = \alpha$, $H(\cdot, 1) = \beta$ and for each t, $H(\cdot, t)$ is a loop. In other words, two loops are homotopic if one can continuously deform one of them into the other, keeping it as a loop all the way throughout the deformation. The winding number classifies the homotopy classes of loops, namely, if σ is homotopic to γ, then $n(\sigma, 0) = n(\gamma, 0)$ (Rotman, 1988, p. 52).

Braids and periodic orbits. While the plane minus one point produces the winding number as a class invariant under homotopies, the homotopy classes of the plane with n special points requires a more elaborated structure which connects nicely with the dynamical properties of 3-d flows admitting a Poincaré section. In fact, periodic orbits of such flows can be regarded as imbeddings of the unit disk in phase space parametrized with time in units of $2\pi/T$ where T is the minimal period. On the Poincaré section, these special trajectories appear as an invariant set of n periodic points. The homotopy classes of loops on the plane with an invariant set of n points are classified by elements of the *Braid group on n strands* (Thurston, 1988; Hall, 1994; Natiello & Solari; 1994).

Linking number. In the same lines, given a pair of periodic orbits in phase space, we may think of the number of turns that one orbit does around the other when completing one excursion along itself. Such number is a link invariant which has a natural interpretation in terms of winding number, and it is called *linking number* (Uezu & Aizawa, 1982; Solari et al., 1996).

For 3-d flows admitting a Poincaré section, the periods of the orbits are commensurate and one may compute the average rotation per period of one orbit around another. This is called the *relative rotation rate* (Solari et al., 1996) and helps in understanding the orbit organization of such flows.

Poincaré index (PI). Consider a planar dynamical system $\dot{x} = f(x, y)$, $\dot{y} = g(x, y)$ and a simple closed counterclockwise curve C not passing through any equilibrium points. The PI k computed along C is defined as

$$\begin{aligned} k &= \frac{1}{2\pi} \int_C \mathrm{d}\left\{\arctan\left(\frac{\mathrm{d}y}{\mathrm{d}x}\right)\right\} \\ &= \frac{1}{2\pi} \int_C \frac{f\,\mathrm{d}g - g\,\mathrm{d}f}{f^2 + g^2}. \end{aligned} \tag{3}$$

(See below for a discussion of the PI in terms of complex analysis and winding number.)

In the context of planar dynamics, the PI of a node or a center is $+1$, of a hyperbolic saddle point is -1, and of a closed orbit is $+1$. Also the PI of a closed curve not containing fixed points is zero, and the PI of a closed curve equals the sum of the indices of the fixed points within (Guckenheimer & Holmes, 1983, p. 51).

Fixed point theorems. The degree of a map can be generalized to higher dimensions. In fact, this property (or the winding number when adequate) is a basic ingredient in the proof of Brouwer's fixed point theorem. An interesting discussion of this fact along with some philosophical considerations can be found in www.mathpages.com.

Complex analysis. The computation of the winding number is a standard tool in the proof of the Fundamental Theorem of Algebra.

Also, let C be a closed contour on the complex plane not passing through any singularities or zeroes of the complex function f, which is analytic inside C except at most at a finite number of poles. Then

$$\frac{1}{2\pi \mathrm{i}} \int_C \frac{f'(z)}{f(z)} \mathrm{d}z = N - P, \qquad (4)$$

where N is the number of zeroes of f and P the number of poles inside C. This is called the Principle of the Argument in standard textbooks (Wunsch, 1994, p. 458).

This result is related to Equation 2 and to the PI. Concerning Equation (2), taking the special point of the plane to be the origin (or any point inside the unit circle), n counts how many turns γ performs around this point when running along the unit circle. Assume now that f has only one zero inside C with multiplicity n and no poles. Then f restricted to C is exactly the same as γ with a suitable choice of parametrization for C and $\mathbb{S}^1$. Concerning the PI, let $z = x + \mathrm{i}y$ and $F(z) = f(x, y) + \mathrm{i}g(x, y)$, regarding the xy-plane as the complex plane. If the vector field (f, g) is continuous, F will not have poles within C and the Poincaré index reduces to the Principle of the Argument calculation for F.

Mario Natiello and Hernán Solari

See also **Conley index; Phase space; Poincaré theorems**

Further Reading

Guckenheimer, J. & Holmes, P.J. 1983. *Nonlinear Oscillators, Dynamical Systems and Bifurcations of Vector Fields*, New York and London: Springer

Hall, T. 1994. Fat one-dimensional representatives of pseudo-Anosov isotopy classes with minimal periodic orbit structure. *Nonlinearity*, 7: 367–384

Natiello, M.A. & Solari, H.G. 1994. Remarks on braid theory and the characterisation of periodic orbits. *Journal of Knot Theory and Its Ramifications*, 3: 511–539

Rotman, J.J. 1988 *An Introduction to Algebraic Topology*. New York: Springer

Solari, H.G. & R. Gilmore. 1988. Relative rotation rates for driven dynamical systems. *Physical Review* A, 37: 3096

Solari, H.G., Natiello, M.A. & Mindlin, B.G. 1996. *Nonlinear Dynamics: A Two-way Trip from Physics to Math*, Bristol: Institute of Physics Publishing

Thurston, W.P. 1988. On the geometry and dynamics of diffeomorphisms of surfaces. *Bulletin of American Mathematical Society*, 19: 417

Uezu, T. & Aizawa, Y. 1982. Topological character of a periodic solution in three-dimensional ordinary differential equation system. *Progress of Theoretical Physics*, 68: 1907

Wunsch, A.D. 1994. *Complex Variables with Applications*, Reading, MA: Addison-Wesley

www.mathpages.com see http://www.mathpages.com/home/kmath262/kmath262.htm or do a search for "Brouwer" on http://www.mathpages.com

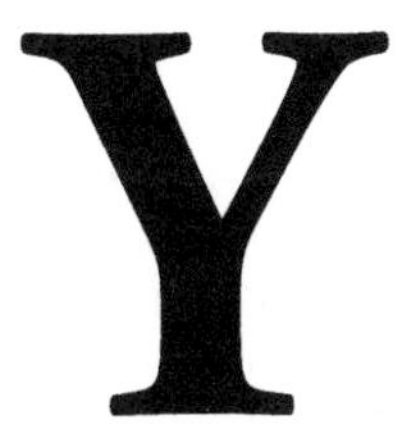

YANG–BAXTER EQUATION

See **Quantum inverse scattering method**

YANG–MILLS THEORY

Modern particle theories, such as the Standard Model, are quantum Yang–Mills theories. In a quantum field theory, space-time fields with relativistic field equations are quantized and, in many calculations, the quanta of the fields are interpreted as particles. In a Yang–Mills theory, these fields have an internal symmetry: they are acted on by space-time-dependent non-abelian group transformations in a way that leaves physical quantities, such as the action, invariant. These transformations are known as local gauge transformations, and Yang–Mills theories are also known as non-abelian gauge theories.

Yang–Mills theories, and especially quantum Yang–Mills theories, have many subtle and surprising properties and are still not fully understood, either in terms of their mathematical foundations or in terms of their physical predictions. However, the importance of Yang–Mills theory is clear; the Standard Model has produced calculations of amazing accuracy in particle physics, and in mathematics, ideas arising from Yang–Mills theory and from quantum field theory are increasingly important in geometry, algebra, and analysis.

Consider a complex doublet scalar field ϕ_a; a scalar field is one that has no Lorentz index, but, as a doublet, ϕ_a transforms under a representation of SU(2), the group represented by special unitary 2×2 matrices:

$$\phi_a(x) \to g_{ab}\phi_b(x), \tag{1}$$

where $g \in \mathrm{SU}(2)$ and the repeated index are summed over. If this is a global transformation; that is, if g is independent of x, then derivatives of ϕ_a have the same transformation property as ϕ_a itself:

$$\frac{\partial \phi_a}{\partial x_\mu} \to \frac{\partial g_{ab}\phi_b}{\partial x_\mu} = g_{ab}\frac{\partial \phi_b}{\partial x_\mu} \tag{2}$$

However, this is not true for a local, or space-time-dependent, transformations where

$$\frac{\partial \phi_a}{\partial x_\mu} \to \frac{\partial g_{ab}\phi_b}{\partial x_\mu} = g_{ab}\frac{\partial \phi_b}{\partial x_\mu} + \frac{\partial g_{ab}}{\partial x_\mu}\phi_b. \tag{3}$$

In order to construct an action that includes derivatives and that is invariant under local transformations, a new derivative is defined that transforms the same way as ϕ_a:

$$D_\mu \phi_a = \frac{\partial \phi_a}{\partial x_\mu} + (A_\mu)_{ab}\phi_b, \tag{4}$$

where A_μ is a new two-indexed space-time field, called a gauge field or gauge potential, defined to have the transformation property

$$(A_\mu)_{ab} \to g_{ac}(A_\mu)_{cd}g_{db}^{-1} - \frac{\partial g_{ac}}{\partial x_\mu}g_{cb}^{-1}. \tag{5}$$

Now, under a local transformation

$$D_\mu \phi_a \to g_{ab}D_\mu \phi_b \tag{6}$$

and so, $D_\mu \phi_a$ transforms in the same way as ϕ_a. This derivative is called a covariant derivative.

A physical theory that includes the gauge field A_μ should treat A_μ as a dynamical field, and so the action should have a kinetic term for A_μ. In other words, the action should include derivative terms for A_μ. These terms are found in the field strength

$$F_{\mu\nu} = \frac{\partial A_\nu}{\partial x^\mu} - \frac{\partial A_\mu}{\partial x^\nu} + [A_\mu, A_\nu] \tag{7}$$

that has the covariant transformation property

$$(F_{\mu\nu})_{ab} \to g_{ac}(F_{\mu\nu})_{cd}g_{db}^{-1}, \tag{8}$$

where $[A_\mu, A_\mu]$ is the normal matrix commutator. In fact, the simplest Yang–Mills theory is pure Yang–Mills theory with action

$$S[A] = -\frac{1}{2}\int \mathrm{d}^4x \,\mathrm{trace}\, F_{\mu\nu}F^{\mu\nu} \tag{9}$$

and corresponding field equation

$$\frac{\partial F_{\mu\nu}}{\partial x_\mu} = 0. \tag{10}$$

Solutions to this equation are known as instantons (*See* **Instantons**).

More generally, Yang–Mills theories contain gauge fields and matter fields like ϕ and fields with both group and Lorentz or spinor indices. Also, the group action described here can be generalized to other groups and to other representations. In the case of the Standard Model of particle physics, the gauge group is $SU(3) \times SU(2) \times U(1)$, and the group representation structure is quite intricate.

Yang–Mills theory was first discovered in the 1950s. At that time, quantum electrodynamics (QED) was known to describe electromagnetism. Quantum electrodynamics is a local gauge theory but with an abelian gauge group. It was also known that there is an approximate global non-abelian symmetry called isospin symmetry that acts on the proton and neutron fields as a doublet and on the pion fields as a triplet. This suggested that a local version of the isospin symmetry might give a quantum field theory for the strong force with the pion's fields as gauge fields (O'Raifeartaigh, 1997). This did not work because pion fields are massive whereas gauge fields are massless, and the main thrust of theoretical effort in the 1950s and 1960s was directed at other models of particle physics.

However, it is now known that the proton, neutron, and pion are not fundamental particles, but are composed of quarks and that there is, in fact, a quantum Yang–Mills theory of the strong force with quark fields and gauge particles called gluons. Furthermore, it is now known that it is possible to introduce a particle, called a Higgs boson, to break the non-abelian gauge symmetry in the physics of a symmetric action and give mass terms for gauge fields. This mechanism is part of the Weinberg–Salam model, a quantum Yang–Mills theory of the electroweak force, that is a component of the Standard Model and that includes both massive and massless gauge particles.

These theories were only discovered after several key experimental and theoretical breakthroughs in the late 1960s and early 1970s. After it became clear from collider experiments that protons have a substructure, theoretical study of the distance-dependent properties of quantum Yang–Mills theory led to the discovery that Yang–Mills fields are asymptotically free (Gross, 1999). This means that the high-energy behavior of Yang–Mills fields includes the particle-like properties seen in experiments, but the low-energy behavior may be quite different, and in fact, the quantum behavior might not be easily deduced from the classical action. Confinement and the mass gap are examples of this. The strong force is a local gauge theory with quark fields. The quark structure of particles is observed in collider experiments; but free quarks are never detected, instead, at low energies, they appear to bind together to form composite particles, such as neutrons, protons, and pions. This is called confinement. It is possible to observe this behavior in simulations of the quantum gauge theory of the strong force, but it has not been possible to prove mathematically that confinement is a consequence of the theory. The same is true of the mass gap; it is known that particles have nonzero mass, and this is observed in simulations, but there is no known way of deriving the mass gap mathematically from the original theory (Clay, 2002).

The symmetries of Yang–Mills theory can be extended to include a global symmetry between the bosonic and fermionic fields called supersymmetry. While there is no direct evidence for supersymmetry in physics, the indirect case is very persuasive, and it is commonly believed that direct evidence will be found in the future. Often, supersymmetric theories are more tractable; for example, Seiberg and Witten have found an exact formula for many quantum properties in $N = 2$ super-Yang–Mills theory (Seiberg & Witten, 1994). It is also commonly believed by theoretical physicists that the quantum Yang–Mills theories in particle physics are in fact a limit of a more fundamental string theory.

Conor Houghton

See also **Higgs boson; Instantons; Matter, nonlinear theory of; Particles and antiparticles; Quantum field theory; String theory; Tensors**

Further Reading

Clay Mathematics Institute, Millennium Prize, 2002. The Clay Institute has offered a prize for a rigorous formulation of a quantum Yang–Mills theory in which there is a mass gap. The Clay Institute web site has a description of the problem along with an essay by A. Jaffe & E. Witten http://www.claymath.org/Millennium_Prize_Problems

Davies, C. 2002. *Lattice QCD*, Bristol: Institute of Physics Publishing

Gross, D. 1999. Twenty-five years of asymptotic freedom. *Nuclear Physics Proceedings Supplements*, 74: 426–446

O'Raifeartaigh, L. 1997. *The Dawning of Gauge Theory*, Princeton, NJ: Princeton University Press

Seiberg, N. & Witten, E. 1994. Electric-magnetic duality, monopole condensation and confinement in $N = 2$ supersymmetric Yang–Mills theory. *Nuclear Physics* B, 426: 19–52, Erratum-ibid 430: 485–486 and Monopoles, duality and chiral symmetry breaking in $N = 2$ supersymmetric QCD. *Nuclear Physics* B, 431: 484–558

Weinberg, S. 1996. *Quantum Field Theory*, vol. 2, Cambridge and New York: Cambridge University Press

Z

ZAKHAROV–SHABAT EQUATION

See **Nonlinear Schrödinger equations**

ZELDOVICH–FRANK-KAMENETSKY EQUATION

In 1938, Yakov Zeldovich and David Frank-Kamenetsky published a brief theoretical paper devoted to flame propagation, presenting one of the first nonlinear traveling-wave front solutions (Zeldovich & Frank-Kamenetsky, 1938). Although both scientists later played outstanding roles in Soviet H-bomb and nuclear projects (and then both performed remarkable works in different fields of physics, including the theory of elementary particles, plasma physics, astrophysics, and cosmology), in the third decade of the last century, they were engaged with the theory of combustion and detonation and attendant problems of chemical kinetics. Their paper was intended for experts and had a rather specialized character, but one of the problems that they examined can be presented as follows.

Consider the autocatalytic production, destruction, and diffusion transfer of a substance proceeding in a homogeneous active chemical medium that occupies some region of physical space. Such processes obey two fundamental macroscopic relationships, the continuity equation

$$u_t + \operatorname{div} \boldsymbol{J} = f(u)$$

and the phenomenological Fick diffusion law

$$\boldsymbol{J} = -D \operatorname{grad} u \qquad (D = \text{const}).$$

Here, $u = u(\boldsymbol{r}, t)$ is the concentration of the substance at the point $\boldsymbol{r} = \{x, y, z\}$ and moment of time t; the literal subscripts symbolize the derivatives with respect to the corresponding variables; the symbols div and *grad* designate the spatial divergence and gradient operators; $\boldsymbol{J} = \boldsymbol{J}(\boldsymbol{r}, t)$ is the vector of diffusion flux density of the substance; the function $f(u)$ is the kinetic function of th active medium, which determines the dependence of the production/destruction rate of substance per unit volume on the concentration u; and D is diffusivity.

Generally speaking, in a medium of this type, self-sustaining nonlinear concentration waves can propagate, and their velocities will not be arbitrary but will be determined by the balance between two types of processes: the active processes of production/destruction of a substance at each local patch of medium and the passive processes of diffusion transfer between the patches. With the diffusivity taken as unity (this can always be achieved by a proper choice of units of measurements), the wave velocity will depend only on the parameters of the function $f(u)$. The problem consists in finding both the profile of a propagating wave and the wave velocity.

Zeldovich and Frank-Kamenetsky solved the last problem for the case of a one-dimensional infinite medium extending along the x-axis, whose kinetic function is described by a cubic polynomial with three zeros (see below). In this case, the diffusion flux density $\boldsymbol{J}$ has only one nonzero component, the x-component J, and the two equations written above look (at $D = 1$) like

$$u_t + J_x = f(u),$$

$$J = -u_x,$$

or, equivalently,

$$u_t = u_{xx} + f(u), \qquad (1)$$

where $f(u)$ is given as

$$f(u) = -Ku(u - b)(u - 1), \qquad (2)$$

where K is a positive constant and b is a constant $(0 < b < 1)$. The reaction-diffusion equation (1) endowed with the kinetic function (2) is referred today to as the Zeldovich–Frank-Kamenetsky (ZF) equation. As its authors have noted, the cubic polynomial structure of (2) corresponds to autocatalysis of the second order.

The active chemical medium described by the ZF equation is bistable: it has two homogeneous stable states described by two trivial solutions $u(x,t)\equiv 0$, $u(x,t)\equiv 1$ of Equation (1), which are determined by zeros $u=0,1$ of the kinetic function (2). Its third homogeneous state $u(x,t)\equiv b$, which is determined by intermediate zero b of $f(u)$ and located between 0 and 1, is unstable; it plays the role of a threshold. Testing the stability of these states proceeds as follows.

Let $u=u_*$, $u_*\in\{0,b,1\}$ be a coordinate of one of three zeros of the kinetic function, and $k_*=f_u(u_*)$, $k_*\in\{k_0,k_b,k_1\}$ be the slope of the kinetic function at this zero; note that the values of k_* satisfy the inequalities

$$k_0 = f_u(0) = -Kb < 0,$$
$$k_b = f_u(b) = Kb(1-b) > 0,$$
$$k_1 = f_u(1) = -K(1-b) < 0.$$

Next, find the solution to Equation (1) in form of $u(x,t)=u_*+w(x,t)$, where

$$w(x,t) = \omega(x,t)\exp(k_*t)$$
$$w(x,0) = \omega(x,0) = w_0(x),$$
$$|w_0(x)| \ll 1$$

is the perturbation, supposed to be small at moments of time close to $t=0$. Substituting these expressions into (1) and linearizing the kinetic function, one comes to the usual diffusion equation $\omega_t=\omega_{xx}$. The solution of the latter, defining on the infinite x-axis and satisfying the initial condition $\omega(x,0)=w_0(x)$ is well known to be given by Poisson's formula. Recopying this solution and multiplying it by the factor $\exp(k_*t)$ yields the expression

$$w(x,t) = 2^{-1}(\pi t)^{-1/2}\exp(k_*t) \times \int_{-\infty}^{+\infty} \exp\left[-\left(x-x'\right)^2/4t\right] w_0(x')\,dx',$$

which describes the time evolution of perturbation. Examination of this expression indicates that $w(x,t)$ decreases with time at the values $k_*=k_0<0$ and $k_*=k_1<0$, that is, near the states determined by zeros 0, 1 of function $f(u)$. At $k_*=k_b>0$ (near the state determined by intermediate zero b of function $f(u)$), the perturbation increases because the exponent before the integral rises with time faster than the preexponential factor $2^{-1}(\pi t)^{-1/2}$ falls.

For physical reasons, a bistable medium obeying the ZF equation must maintain the propagation of nonlinear wave fronts, which switch the medium from one of its two stable states to the other. For example, a wave of this type can be obtained numerically by setting the initial conditions $u(x,0)=1$ at $x<0$, $u(x,0)=0$ at $x>0$. In the steady-state regime, which is established after some transition time, the switching wave moves along x with constant velocity v and possesses a steady spatial profile, which is described by the traveling-wave front solution of a ZF equation of the kind

$$u = u(\xi), \quad \xi = x - vt, \tag{3}$$

$$u(-\infty) = 1, \quad u(+\infty) = 0, \tag{4}$$

$$|u(\xi)| < \infty. \tag{5}$$

Obtaining this solution is a principal goal of the ZF analysis. It includes both the problem of derivation of $u(\xi)$ and the relevant problem of determination of v. Of course, the latter problem is a central one: as a rule, it appears every time when one deals with wave propagation in nonlinear reaction-diffusion systems. Zeldovich and Frank-Kamenetsky found the expression for v in the case of (1), (2) to be

$$v = \sqrt{K/2}(1-2b) = \sqrt{2K}\,(1/2-b). \tag{6}$$

This beautiful formula connects the wave front velocity v with the position b of intermediate zero of the cubic kinetic function. In particular, it indicates that v is proportional to the deviation of b from the middle value of u, which is equal to $1/2$. If $b<1/2$ ($b>1/2$), then v is positive (negative), and the wave front, which obeys the boundary conditions (4), (5), moves in the positive (negative) direction of the x-axis. If b is exactly equal to $1/2$, then the front does not move: it is stationary.

A short derivation of (6) along with the expression for $u(\xi)$ proceeds as follows. First, substituting (3) directly into two equations written immediately before (1) (they are evidently equivalent to (1)) and allowing for the relations $\partial/\partial t=-v\,(d/d\xi)$, $\partial/\partial x=d/d\xi$, which follow from (3), one obtains

$$u_\xi = -J, \tag{7a}$$

$$J_\xi = -vJ + f(u). \tag{7b}$$

Second, dividing (7b) by (7a), one excludes the independent variable (thereupon, the evident equality $J_\xi/u_\xi = J_u$ is used) and reduces Equation (7) to the single equation

$$JJ_u = vJ - f(u). \tag{8}$$

To be integrated correctly, this differential equation must be provided with the proper boundary conditions at the points $u=0$, 1, that is, in the equilibrium states that are achieved by the traveling wave front solution at $\xi \longrightarrow \pm\infty$. To set these conditions, one should know the asymptotic behavior of diffusion flux $J=u_\xi(\xi)$ generated by the traveling wave near the states $u=0$, 1. To recognize it, one represents the unknown solution near these states in the form of $u(\xi)=u_*+w(\xi)$, $u_*\in\{0,1\}$, where $w(\xi)$ is a small perturbation necessarily satisfying the conditions $w(\pm\infty)=0$. Substituting this expression directly into (1) and linearizing the kinetic function yields

$$w_{\xi\xi} + vw_\xi - |k_*|w = 0, \quad (k_* = f_u(u_*),$$
$$k_* \in \{k_0, k_1\}, k_0 < 0, k_1 < 0),$$

where the negative parameter k_* is presented as the positive constant $|k_*|$ taken with the sign "minus." The solutions of the last linear ordinary equation, singled out by the conditions $w(\pm\infty) = 0$ look like

$$w = A_1 \exp(\lambda_1 \xi), \quad A_1 = \text{const},$$
$$\lambda_1 = -(v/2) + \sqrt{(v/2)^2 + |k_*|} > 0$$
$$(\text{at } \xi \longrightarrow -\infty),$$

$$w = A_2 \exp(\lambda_2 \xi), \quad A_2 = \text{const},$$
$$\lambda_2 = -(v/2) - \sqrt{(v/2)^2 + |k_*|} < 0$$
$$(\text{at } \xi \longrightarrow +\infty).$$

Thus, irrespective of the unknown value of velocity v, the traveling-wave front solution $u(\xi)$ approaches its limit values 0, 1 exponentially and therefore, the diffusion flux $J = -u_\xi = -w_\xi$ tends to zero near these values. Hence, the correct boundary conditions to the solution $J(u)$ of Equation (8), which are compatible with the desired traveling wave front solution (3)–(5), look like

$$J(0) = J(1) = 0. \tag{9}$$

Next, Zeldovich and Frank-Kamenetsky assumed the solution to Equation (8) to have the form of a quadratic parabola $J = -\alpha u(u-1)$ (α is a positive constant to be determined), which satisfies conditions (9) automatically. Substituting this expression into (8) and performing the cancellation yields an equation of the sort $P(\alpha, v)u + Q(\alpha, v) = 0$, where $P(\alpha, v)$ and $Q(\alpha, v)$ are set by calculations. Here, u can take any value belonging to the segment [0, 1]. Fixing the variable u on its limit value $u = 0$ yields the equation $Q(\alpha, v) = 0$; taking into account the latter and setting $u = 1$ yields the equation $P(\alpha, v) = 0$. Solving these two equations with respect to α and v leads directly to Equation (6) and to the desired expression $J = -\sqrt{K/2}u(u-1)$. Substituting the latter into (7a) and taking the integral (at the condition $u(0) = \frac{1}{2}$) yields the desired profile of the traveling-wave front:

$$u = 1/\left[1 + \exp\left(\sqrt{K/2}\xi\right)\right]$$
$$= (1/2)\left[1 - \tanh\left(\sqrt{K/8}\xi\right)\right]. \tag{10}$$

To appreciate the significance of this result in the context of current knowledge, we should stress that the problem of finding a traveling-wave solution to the parabolic reaction-diffusion equations dramatically differs from that arising in the case of nonlinear hyperbolic equations. The latter correspond to conservative physical systems and usually possess the first integrals of the kind of integrals describing energy conservation. Such equations have a Hamiltonian structure, which helps to integrate them analytically using different powerful methods (for example, by methods based on the inverse scattering problem). But the ZF equation is dissipative rather than conservative. Thus it is not Hamiltonian: it describes a gradient physical system that shows the dissipation of its free energy during its time evolution. In these circumstances, the analytical integration of reaction-diffusion equation (1) with the arbitrary kinetic function was a challenge. Zeldovich and Frank-Kamenetsky were the first to have recognized the integrable case of this equation and presented its nontrivial solution.

We should emphasize a distinction of the ZF equation from Fisher's equation, which was investigated in 1937, one year before Zeldovich and Frank-Kamenetsky's work.

The form of the Fisher equation is identical to (1), but the corresponding kinetic function is a quadratic polynomial, which possesses only two zeros. Of course, these zeros correspond to two stationary states of an active medium, but only one of them is a stable one, whereas the second is unstable. As a consequence, the Fisher equation admits not only one traveling-wave front solution, but also a continuum of such waves, and either of them is very responsive to the initial conditions. This equation is applicable only to those media in which the processes of spontaneous production of substances occur against the unstable background state.

In contrast to the Fisher equation, the ZF equation describes active media, which possess two stable states, separated by a third, unstable, state playing the role of a threshold of excitation. The natural field of application of the ZF equation covers the class of bistable active media displaying threshold properties. The linear stability analysis, which was first carried out by Zeldovich and Barenblatt in 1959, and subsequent nonlinear stability analyses performed independently by Lingren and Buratti and by Maginu show that traveling-wave fronts propagating in such media are stable (Scott, 2002).

After Zeldovich and Frank-Kamenetsky's work, decades passed before new analytical traveling-wave front solutions to the ZF-like reaction-diffusion equations appeared. They were constructed with the use of different representations of three-zero kinetic function including the piecewise linear and sinusoidal approximations (Scott, 2002, 2003). The significance of these solutions is predetermined by the fact that all of them describe various physical, chemical, and biological phenomena, which, at first glance, have no common ground. Among these are:

- Electric signals propagating along bistable transmission lines of nerve fibers and neuristors (Scott, 2002, 2003);
- Thermal waves switching boiling regimes from nucleate to film boiling near one-dimensional fuel elements (Zhukov et al., 1980);

- Waves of resistance modification in normal metals (Barelko et al., 1983) and superconductors (Gurevich & Mints, 1984), caused by thermal change;
- Gene flows and population waves in spatially distributed biological populations (Svirezev & Pasekov, 1990) and
- Nonlinear processes arising in synergetics (Loskutov & Mikhailov, 1996).

In view of these applications, the importance of Zeldovich and Frank-Kamenetsky's result is established, yet the destiny of their paper of 1938 is strange. The result obtained in it was used in the Soviet Union for processing experimental data on chemical kinetics even before World War II. After the war, when rapid development of research in the field of physiology of nervous impulses and nonlinear physical chemistry took place, Equation (6) for the velocity of a traveling-wave front became familiar to a broad audience of researchers and appeared frequently in papers and monographs. But the manuscript from which this formula was derived for the first time seemed to have been forgotten: the paper has not been cited until recently! Surprisingly, it is absent even in the two-volume edition of Zeldovich's selected works issued in Russia in 1984, in the lifetime of their author. However, as Mikhail Bulgakov has written (in his classic *Master and Margarita*): "manuscripts do not burn." One could add: even if they are devoted to the theory of combustion.

O.A. MORNEV

See also **Diffusion; Flame front; Gradient system; Nerve impulses; Reaction-diffusion systems**

Further Reading

Barelko, V.V., Beibutian, V.M., Volodin, Yu.E. & Zeldovich, Ya.B. 1983. Thermal waves and non-uniform steady states in a $Fe + H_2$ system. *Chemical Engineering Science*, 38(11): 1775–1780

Frank-Kamenetskii, D.A. 1969. *Diffusion and Heat Transfer in Chemical Kinetics*, New York: Plenum Press (original Russian editions, 1947, 1967, 1987)

Gurevich, A.V. & Mints, R.G. 1984. Localized waves in inhomogeneous media. *Soviet Physics – Uspekhi,* 27(1): 19–41

Loskutov, A.Yu. & Mikhailov, A.S. 1996. *Foundations of Synergetics*, 2nd edition, Berlin: Springer (original Russian edition 1990)

Scott, A.C. 2002. *Neuroscience: A Mathematical Primer*, Berlin and New York: Springer

Scott, A.C. 2003. *Nonlinear Science: Emergence and Dynamics of Coherent Structures*, 2nd edition, Oxford and New York: Oxford University Press

Svirezhev, Yu.M. & Pasekov, V.P. 1990. *Fundamentals of Mathematical Evolutionary Genetics*, Dordrecht: Kluwer (original Russian edition 1982)

Zeldovich, Ya.B. & Frank-Kamenetsky, D.A. 1938. K teorii ravnomernogo rasprostraneniya plameni [On the theory of uniform propagation of flame]. *Doklady Akademii Nauk SSSR*, 19(9):693–697

Zhukov, S.A., Barelko, V.V. & Merzhanov, A.G. 1980. Wave processes on heat generating surfaces in pool boiling. *International Journal of Heat Mass Transfer*, 24: 47–55

ZENER DIODE

See **Diodes**

ZERO FIELD STEPS

See **Josephson junctions**

ZERO-DISPERSION LIMITS

The Korteweg–de Vries (KdV) equation with small dispersion

$$\begin{cases} u_t - uu_x + \varepsilon^2 u_{xxx} = 0, & t, x \in R, \\ u(x, t = 0; \varepsilon) = u_0(x), \end{cases} \tag{1}$$

is a model for the formation and propagation of dispersive shock waves in one dimension. Let $u(x, t; \varepsilon)$ denote the solution of Cauchy problem (1), where the initial data $u_0(x)$ is smooth and decreases at infinity sufficiently fast. It is known that for $\varepsilon > 0$, no matter how small, the solution of (1) remains smooth for all $t > 0$. For $\varepsilon = 0$, (1) becomes the Cauchy problem for the Hopf equation

$$\begin{cases} u_t - uu_x = 0, \\ u(x, t = 0) = u_0(x). \end{cases} \tag{2}$$

The solution of the Hopf equation can be obtained by the method of characteristics. If the initial data $u_0(x)$ is somewhere increasing, the solution $u(x, t)$ of Equation (2) always has a point (x_c, t_c) of gradient catastrophe where an infinite derivative develops.

After the time of gradient catastrophe t_c, the solution $u(x, t, \varepsilon)$ of (1) develops in the neighborhood of x_c an expanding region filled with rapid modulated oscillations of wavelength of order $1/\varepsilon$. These oscillations are called dispersive shock waves.

Lax and Levermore (1998), performing the zero-dispersion asymptotics for the inverse-scattering problem for the KdV equation, showed that as ε tends to zero, $u(x, t; \varepsilon)$ tends uniformly to the smooth solution $u(x, t)$ of (2) as long as $t < t_c$. For $t > t_c$, the solution $u(x, t; \varepsilon)$ converges weakly in the oscillation region to a limit $\bar{u}(x, t)$ that is not a solution of conservation law (2).

The first example describing dispersive shock waves was proposed by Gurevich and Pitaevski (1973). Their description was rigorously proved by Venakides (1990) who derived the general form of the rapid oscillations. The oscillation zone is approximately described for a short time $t > t_c$ by a modulated periodic wave solution of the KdV equation:

$$u(x,t,\varepsilon) \simeq V(x,t) + \frac{1}{3}\alpha(x,t) + \Theta\left(\frac{x - Vt + \phi}{\varepsilon}\right), \qquad (3)$$

$$V(x,t) = \tfrac{1}{6}(u_1(x,t) + u_2(x,t) + u_3(x,t)).$$

In the above formula, the term $V(x,t) + \frac{1}{3}\alpha(x,t)$ is the weak limit $\bar{u}(x,t)$ of $u(x,t,\varepsilon)$ as $\varepsilon \to 0$, while the remaining term describes the rapid oscillations. The function Θ is 2π-periodic with zero average, and it can be expressed in terms of elliptic functions. The quantity α defined below and the phase ϕ depend on some functions $u_i(x,t)$, $i = 1, 2, 3$. The functions $u_1(x,t) > u_2(x,t) > u_3(x,t)$ solve the Whitham (1974) modulation equations

$$\partial_t u_i(x,t) - \lambda_i(u_1,u_2,u_3)\partial_x u_i(x,t) = 0, \quad i = 1,2,3, \qquad (4)$$

where

$$\lambda_i(u_1,u_2,u_3) = \frac{1}{3}(u_1+u_2+u_3) + \frac{2}{3}\frac{\prod_{j\neq i, j=1}^{3}(u_i - u_j)}{u_i + \alpha}, \qquad (5)$$

$$\alpha = -u_3 + (u_3 - u_1)\frac{E(s)}{K(s)} \qquad (6)$$

and $K(s)$ and $E(s)$ are the complete elliptic integrals of the first and second kind with modulus $s = (u_2 - u_1)/(u_3 - u_1)$.

The solution $u_1(x,t) > u_2(x,t) > u_3(x,t)$ of the Whitham equations can be plotted in the (x,u) plane as branches of a multivalued function. The solutions of the Hopf equation and the Whitham equations are connected to one another as illustrated in Figure 1(a). The function $u_2(x,t)$ can vary from $u_3(x,t)$ to $u_1(x,t)$. On the $(x, t \geq 0)$ plane, the oscillation region is bounded on one side by the curve $x^-(t)$ where $u_2(x,t) = u_1(x,t)$, and on the other side by the curve $x^+(t)$ where $u_2(x,t) = u_3(x,t)$ (see Figure 1(a)). For $x^-(t) < x < x^+(t)$, the solution of (1) for small ε is approximately given by (3) while outside the interval $[x^-(t), x^+(t)]$ is given by the solution $u(x,t)$ of the Hopf equation (2). At edge $x = x^-(t)$ of the oscillation region, the amplitude of the oscillations vanishes and (3) goes to $u(x,t,\varepsilon)|_{u_1=u_2} \simeq u_3(x,t)$. When $x = x^+(t)$, solution (3) goes to the one-soliton solution of the KdV equation. In general, the oscillation zone grows with time. For generic analytic initial data with a cubic inflection point, the growth in the (x,t) plane of the oscillation zone near the point of gradient catastrophe (x_c, t_c) is described, up to shifts and rescaling, by the semi-cubic law

$$x^{\pm}(t) = x_c \pm a^{\pm}(t - t_c)^{3/2},$$

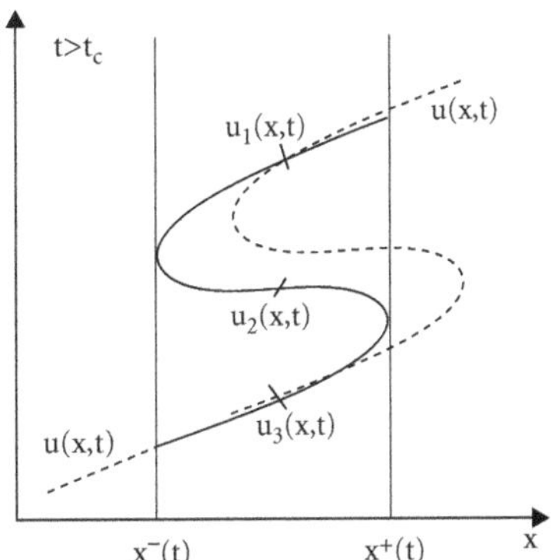

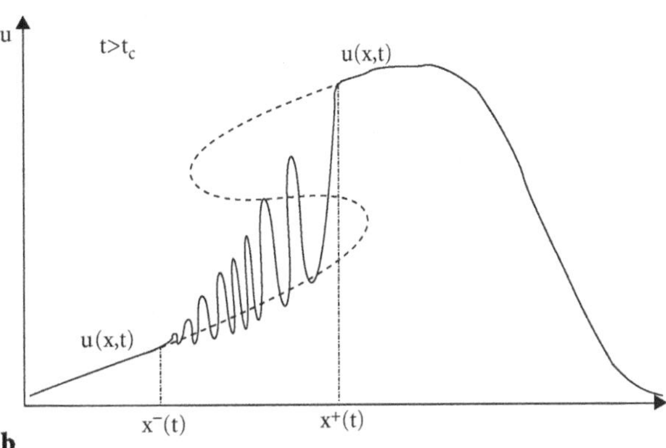

Figure 1. (a) The dashed line represents the formal solution of the Hopf equation; the continuous line represents the solution of the Whitham equations. The solution $(u_1(x,t), u_2(x,t), u_3(x,t))$ of the Whitham equations and the position of the boundaries $x^-(t)$ and $x^+(t)$ are to be determined from the conditions $u(x^-(t),t) = u_3(x^-(t),t)$, $u(x^+(t),t) = u_1(x^+(t),t)$, where $u(x,t)$ is the solution of the Hopf equation. (b) The oscillations in the region $x^-(t_1) < x < x^+(t_1)$.

where $a^{\pm}$ are two positive numbers. A completely different behavior appears in the zero-diffusion case. The simplest equation that combines nonlinearity and diffusion is the Burgers equation

$$u_t - uu_x = \varepsilon u_{xx}, \qquad (7)$$

where $\varepsilon > 0$. For smooth initial data $u_0(x)$, the Burgers equation can be integrated through the Cole–Hopf transformation to

$$u(x,t,\varepsilon) = -\frac{\int_{-\infty}^{\infty} \frac{x-\xi}{t} e^{-\frac{G(\xi)}{2\varepsilon}} d\xi}{\int_{-\infty}^{\infty} e^{-\frac{G(\xi)}{2\varepsilon}} d\xi}, \qquad (8)$$

where

$$G(\xi) = -\int_0^{\xi} u_0(\eta)\, d\eta + \frac{(x-\xi)^2}{2t}.$$

The behavior of the exact solution (8) as $\varepsilon \to 0$ can be obtained by observing that the dominant contributions to the integrals in (8) come from the neighborhood of the stationary points of G where

$$\frac{\partial G}{\partial \xi} = -u_0(\xi) - \frac{x-\xi}{t} = 0. \tag{9}$$

If (9) has only one stationary point, by the application of the steepest descent method, the asymptotic solution $u(x,t;\varepsilon)$ as $\varepsilon \to 0^+$ converges strongly to

$$u(x,t) = u_0(\xi), \quad x = \xi - u_0(\xi)t. \tag{10}$$

The above is exactly the solution of the Cauchy problem for the Hopf equation (2). The stationary point $\xi(x,t)$ of (9) becomes the characteristic variable in (10). For bump-like initial data the solution of the Hopf equation (2) has a point of gradient catastrophe (x_c, t_c). After the time $t = t_c$ of gradient catastrophe, (10) gives a multivalued solution: the characteristics of the Hopf equation begin to intersect. For a typical initial pulse, there are usually three characteristics that intersect at each point of the multivalued region; that is, (9) has three solutions $\xi_1(x,t), \xi_2(x,t), \xi_3(x,t), \xi_1 > \xi_2 > \xi_3$. The dominant behavior of the solution of the Burgers equation will be given by the following contributions:

$$u \simeq -\frac{\sum_{i=1}^{3} \frac{x-\xi_i}{t} |G''(\xi_i)|^{-\frac{1}{2}} e^{-\frac{G(\xi_i)}{2\varepsilon}}}{\sum_{i=1}^{3} |G''(\xi_i)|^{-\frac{1}{2}} e^{-\frac{G(\xi_i)}{2\varepsilon}}}. \tag{11}$$

Let us suppose that for $x_c < x < x_s$ and $t > t_c$, the function $G(\xi_1(x,t))$ is less than $G(\xi_2(x,t))$ and $G(\xi_3(x,t))$. Then the above expression for u in the limit $\varepsilon \to 0$ reads

$$u \simeq -\frac{x-\xi_1}{t}, \quad x_c < x < x_s, \tag{12}$$

while assuming that for $x > x_s$ $G(\xi_2(x,t)) < G(\xi_1(x,t)), G(\xi_3(x,t))$, we have

$$u \simeq -\frac{x-\xi_2}{t}. \tag{13}$$

In each case, (10) applies to both ξ_1 and ξ_2. Therefore, the solution of the Burgers equation converges as $\varepsilon \to 0^+$ to the solution of the Hopf equation (2) almost everywhere except at the points (x,t) where $G(\xi_i(x,t)) = G(\xi_j(x,t)),\ i \neq j,\ \ i,j = 1,2,3$. For example, in the case treated above, the change over from ξ_1 to ξ_2 occurs when $x = x_s$ where $G(\xi_1) = G(\xi_2)$. Near $x = x_s$ the solution of the Burgers equation as $\varepsilon \to 0^+$ has a transition from (12) to (13) which is called a shock wave. In other words, the solution of the Burgers equation in the zero viscosity limit is given by two different branches of the solution of the Hopf equation joined by a jump at the point x_s. The condition $G(\xi_1) = G(\xi_2)$ reads

$$-\int_0^{\xi_1} u_0(\eta)\,d\eta + \frac{(x-\xi_1)^2}{2t} = -\int_0^{\xi_2} u_0(\eta)\,d\eta + \frac{(x-\xi_2)^2}{2t}.$$

Because of (10), the above relation is equivalent to

$$\frac{1}{2}(u_0(\xi_1) + u_0(\xi_2)) = \frac{1}{\xi_1 - \xi_2}\int_{\xi_2}^{\xi_1} u_0(\eta)\,d\eta, \tag{14}$$

which describes the shock wave. Since the shock occurs at $x = x_s(t), t > t_c$, we also have

$$x_s(t) = \xi_1 - u_0(\xi_1)t, \quad x_s(t) = \xi_2 - u_0(\xi_2)t.$$

The above three equations determine the functions $x_s(t)$, $\xi_1(t)$, and $\xi_2(t)$. The values of $u(x,t)$ on the two sides of the shock are $u^-(x,t) = u_0(\xi_1(x,t))$ and $u^+(x,t) = u_0(\xi_2(x,t))$. The shock speed can be derived by taking the time derivative of the above two equations and reads

$$\frac{dx_s(t)}{dt} = -\frac{1}{2}(u_0(\xi_1) + u_0(\xi_2)).$$

Comparison of the above relation with (14) shows that the modulus of the shock speed is equal to the average value of the characteristics velocity $u_0(\eta)$ over the interval $[\xi_1, \xi_2]$.

While the zero-dispersion limits have been studied only for integrable equations such as the KdV or the nonlinear Schrödinger equation, the zero-viscosity limits have been studied for the parabolic equation of the form

$$u_t + [f(u)]_x = \varepsilon u_{xx}, \quad u \in R^n, \ (t,x) \in R \times R^m.$$

The scalar case in several spatial dimensions was investigated by Kruzhkov (1970). The two-component case in one spatial dimension has been studied by DiPerna (1983), while the n-component case in one spatial dimension has been investigated by Bressan (2002).

TAMARA GRAVA

See also **Burgers equation; Constants of motion and conservation laws; Inverse scattering method or transform; Jump phenomena; Modulated waves; Shock waves**

Further Reading

Bressan, A. 2002. Hyperbolic systems of conservation laws in one space dimension. In *Proceedings of the International Congress of Mathematicians,* Beijing, vol. I, Beijing: Higher Education Press, pp. 159–178

DiPerna, R. 1983. Convergence of approximate solutions to conservation laws. *Archive for Rational Mechanics and Analysis,* 82: 27–70

Gurevich, A.G. & Pitaevskii, L.P. 1973. Non-stationary structure of a collisionless shock waves. *JEPT Letters,* 17: 193–195

Kamvissis, S., McLaughlin, K.D.T.-R. & Miller, P.D. 2003. *Semiclassical Soliton Ensembles for the Focusing Nonlinear Schrödinger Equation,* Princeton, NJ: Princeton University Press

Kruzhkov, S. 1970. First order quasi-linear equations with several space variables. *Mathematics of the USSR Sbornik,* 10: 217–243

Lax, P.D. & Levermore, C.D. 1983. The small dispersion limit of the Korteweg de Vries equation, I, II, III. *Communications in Pure and Applied Mathematics*, 36: 253–290, 571-593, 809–830

Novikov, S., Manakov, S.V., Pitaevski, L.P. & Zakharov, V.E. 1984. *Theory of Solitons: The Inverse Scattering Method*, New York: Consultants Bureau

Venakides, S. 1990. The Korteweg–de Vries equations with small dispersion: higher order Lax–Levermore theory. *Communications in Pure and Applied Mathematics*, 43: 335–361

Whitham, G.B. 1974. *Linear and Nonlinear Waves*, New York: Wiley

ZETA FUNCTIONS

See **Randum matrix theory II**

Index

The index contains entries that refer to important concepts, scientists, equations or entities. For some very well known scientists there are cross-references to the ideas or equations associated with them. Similarly, for some very often used equations, which are referred to by the names of the scientists associated with them, there are cross-references to the scientists. The page numbers refer to the page where a mention of the relevant entry is found. Italicized page numbers refer to pages with figures that are relevant to the entry. A "*See also*" entry gives an entry related to the original one, which could also be of interest. "*see*" entries direct the reader to another entry at which the information is given.

C

D

E

F

J

K

N

S

W

Y

T - #0438 - 101024 - C8 - 279/216/59 - PB - 9781138012141 - Gloss Lamination